Acid Solutions

Prepare the following reagents by cautiously adding required amount of concentrated acid, with mixing, to designated volume of proper type of distilled water. Dilute to 1000 mL and mix thoroughly.

See Table A for preparation of HCl, H_2SO_4, and HNO_3 solutions.

Alkaline Solutions

a. Stock sodium hydroxide, NaOH, 15N (for preparing 6N, 1N, and 0.1N solutions): Cautiously dissolve 625 g solid NaOH in 800 mL distilled water to form 1 L of solution. Remove sodium carbonate precipitate by keeping solution at the boiling point for a few hours in a hot water bath or by letting particles settle for at least 48 h in an alkali-resistant container (wax-lined or polyethylene) protected from atmospheric CO_2 with a soda lime tube. Use the supernate for preparing dilute solutions listed in Table B.

Alternatively prepare dilute solutions by dissolving the weight of solid NaOH in-

TABLE B: PREPARATION OF UNIFORM SODIUM HYDROXIDE SOLUTIONS

Normality of NaOH Solution	Required Weight of NaOH to Prepare 1,000 mL of Solution	Required Volume of 15N NaOH to Prepare 1,000 mL of Solution
	g	mL
6	240	400
1	40	67
0.1	4	6.7

dicated in Table B in CO_2-free distilled water and diluting to 1000 mL.

Store NaOH solutions in polyethylene (rigid, heavy-type) bottles with polyethylene screw caps, paraffin-coated bottles with rubber or neoprene stoppers, or borosilicate-glass bottles with rubber or neoprene stoppers. Check solutions periodically. Protect them by attaching a tube of CO_2-absorbing granular material such as soda lime or a commercially available CO_2-removing agent.* Use at least 70 cm of rubber tubing to minimize vapor diffusion from bottle. Replace absorption tube before it becomes exhausted. Withdraw solution by a siphon to avoid opening bottle.

b. Ammonium hydroxide solutions: NH_4OH: Prepare 5N, 3N, and 0.2N NH_4OH solutions by diluting 333 mL, 200 mL, and 13 mL, respectively, of the concentrated reagent (sp gr 0.90, 29.0%, 15N) to 1000 mL with distilled water.

Indicator Solutions

a. Phenolphthalein indicator solution: Use either the aqueous (1) or alcoholic (2) solution.

1) Dissolve 5 g phenolphthalein disodium salt in distilled water and dilute to 1 L.

2) Dissolve 5 g phenolphthalein in 500 mL 95% ethyl or isopropyl alcohol and add 500 mL distilled water.

If necessary, add 0.02N NaOH dropwise until a faint pink color appears in solution 1) or 2).

b. Methyl orange indicator solution: Dissolve 500 mg methyl orange powder in distilled water and dilute to 1 L.

STANDARD METHODS

For the Examination of Water and Wastewater

STANDARD
METHODS

For the Examination of Water and Wastewater

SEVENTEENTH EDITION

Prepared and published jointly by:
AMERICAN PUBLIC HEALTH ASSOCIATION
AMERICAN WATER WORKS ASSOCIATION
WATER POLLUTION CONTROL FEDERATION

Joint Editorial Board

LENORE S. CLESCERI, WPCF, Chairman
ARNOLD E. GREENBERG, APHA
R. RHODES TRUSSELL, AWWA

MARY ANN H. FRANSON
  Managing Editor

Publication Office:
  American Public Health Association
  1015 Fifteenth Street NW
  Washington, DC 20005

30M9/89

The Library of Congress has cataloged this work as follows:
American Public Health Association.
    Standard methods for examination of water and wastewater.
ISBN 0-87553-161-X
ISSN 8755-3546

Printed and bound in the United States of America
  Composition and Printing: Port City Press, Baltimore, Maryland
    Set in: Times Roman

Cover Design: Donya Melanson Associates, Boston, Massachusetts

PREFACE TO THE SEVENTEENTH EDITION

The Sixteenth and Earlier Editions

The first edition of *Standard Methods* was published in 1905. Each subsequent edition presented significant improvements of methodology and enlarged its scope to include techniques suitable for examination of many types of samples encountered in the assessment and control of water quality and water pollution.

A brief history of *Standard Methods* is of interest because of its contemporary relevance. A movement for "securing the adoption of more uniform and efficient methods of water analysis" led in the 1880's to the organization of a special committee of the Chemical Section of American Association for the Advancement of Science. A report of this committee, published in 1889, was entitled: A Method, in Part, for the Sanitary Examination of Water, and for the Statement of Results, Offered for General Adoption.* Five topics were covered: (1) "free" and "albuminoid" ammonia; (2) oxygen-consuming capacity; (3) total nitrogen as nitrates and nitrites; (4) nitrogen as nitrites; and (5) statement of results.

In 1895, members of the American Public Health Association, recognizing the need for standard methods in the bacteriological examination of water, sponsored a convention of bacteriologists to discuss the problem. As a result, an APHA committee was appointed "to draw up procedures for the study of bacteria in a uniform manner and with special references to the differentiation of species." Submitted in 1897,† the procedures found wide acceptance.

In 1899, APHA appointed a Committee on Standard Methods of Water Analysis, charged with the extension of standard procedures to all methods involved in the analysis of water. The committee report, published in 1905, constituted the first edition of *Standard Methods* (then entitled *Standard Methods of Water Analysis*). Physical, chemical, microscopic, and bacteriological methods of water examination were included. In its letter of transmittal, the Committee stated:

> The methods of analysis presented in this report as "Standard Methods" are believed to represent the best current practice of American water analysts, and to be generally applicable in connection with the ordinary problems of water purification, sewage disposal and sanitary investigations. Analysts working on widely different problems manifestly cannot use methods which are identical, and special problems obviously require the methods best adapted to them; but, while recognizing these facts, it yet remains true that sound progress in analytical work will advance in proportion to the general adoption of methods which are reliable, uniform and adequate.
>
> It is said by some that standard methods within the field of applied science tend to stifle investigations and that they retard true progress. If such standards are used in the proper spirit, this ought not to be so. The Committee strongly desires that every effort shall be continued to improve the techniques of water analysis and especially to compare current methods with those herein recommended, where different, so that the results obtained may be still more accurate and reliable than they are at present.

Revised and enlarged editions were published by APHA under the title *Standard Methods of Water Analysis* in 1912 (Second Edition), 1917 (Third), 1920 (Fourth),

---

* *J. Anal. Chem.* 3:398 (1889).
† *Proc. Amer. Pub. Health Assoc.* 23:56 (1897).

and 1923 (Fifth). In 1925, the American Water Works Association joined APHA in publishing the Sixth Edition, which had the broader title, *Standard Methods of the Examination of Water and Sewage.* Joint publication was continued in the Seventh Edition, dated 1933.

In 1935, the Water Pollution Control Federation (then the Federation of Sewage Works Associations) issued a committee report, "Standard Methods of Sewage Analysis." ‡ With minor modifications, these methods were incorporated into the Eighth Edition (1936) of *Standard Methods,* which was thus the first to provide methods for the examination of "sewages, effluents, industrial wastes, grossly polluted waters, sludges, and muds." The Ninth Edition, appearing in 1946, likewise contained these methods, and in the following year the Federation became a full-fledged publishing partner. Since 1947, the work of the *Standard Methods* committees of the three associations—APHA, AWWA, and WPCF—has been coordinated by a Joint Editorial Board, on which all three are represented.

The Tenth Edition (1955) included methods specific for examination of industrial wastewaters; this was reflected by a new title: *Standard Methods for the Examination of Water, Sewage and Industrial Wastes.* To describe more accurately and concisely the contents of the Eleventh Edition (1960), the title was shortened to *Standard Methods for the Examination of Water and Wastewater.* It remained unchanged in the Twelfth Edition (1965), the Thirteenth Edition (1971), the Fourteenth Edition (1976), and the Fifteenth Edition (1981).

In the Fourteenth Edition, the separation of test methods for water from those for wastewater was discontinued. All methods for a given component or characteristic appeared under a single heading. With minor differences, the organization of the Fourteenth Edition was retained for the Fifteenth and Sixteenth (1985) Editions. Two major policy decisions of the Joint Editorial Board were implemented for the Sixteenth Edition. First, the International System of Units (SI) was adopted except where prevailing field systems or practices require English units. Second, the use of trade names or proprietary materials was eliminated insofar as possible, to avoid potential claims regarding restraint of trade or commercial favoritism.

The Seventeenth Edition

The organization of the Seventeenth Edition reflects a commitment to develop and retain a permanent numbering system. New numbers have been assigned to all sections, and numbers unused in the present edition have been reserved for future use. All part numbers have been expanded to multiples of 1000 instead of 100. The parts have retained their identity from the previous edition, with the exception of Part 6000, which now contains methods for the measurement of specific organic compounds. The more general procedures for organics are found in Part 5000.

The Seventeenth Edition has undergone a major revision in the introductory Part 1000. Sections dealing with statistical analysis, data quality, and methods

---

‡ *Sewage Works J.* 7:444 (1935).

development have been greatly expanded. The section on reagent water has been updated to include the current classification scheme for various types of reagent water. Part 1000 contains important information on the proper execution of procedures and should be studied by every user of this manual. At the beginning of each of the subsequent parts of the manual are sections discussing quality assurance and other matters of general application within the specific subject area, to minimize repetition in the succeeding text. Successful analysis rests on close adherence to the introductory recommendations and cautions. Before undertaking an analysis, read and understand the complete discussion of each procedure, including methods selection, sampling and sample storage, and interferences.

Part 2000 (physical and aggregate properties) contains a new section on dissolved gas supersaturation. It also includes analyses for certain aggregate properties (acidity, alkalinity, and hardness) that had appeared in other parts of previous editions and the tests for sludge digester gas from Part 500 of previous editions. The sections on salinity and calcium carbonate saturation have been revised to reflect the current state of the art in these measurements.

Part 3000 (metals) now includes a brief section on quality assurance and the section on sample preparation has been revised. Instrumental methods (atomic absorption spectrometric and inductively coupled plasma) that are useful for determining numerous metals are next presented; the atomic absorption methods have been revised and a new method using continuous hydride generation has been added. To the sections concerned with analysis for specific metals have been added brief sections referring the reader to instrumental methods for the determination of antimony, arsenic, bismuth, cesium, iridium, molybdenum, osmium, palladium, platinum, rhenium, rhodium, ruthenium, thallium, thorium, tin, and titanium. The section for lithium has been revised to reflect methods providing lower detection limits. New methods for selenium speciation and aluminum have been added.

Part 4000 (inorganic nonmetals) also contains a brief new quality assurance section. Analysis methods, including the ion chromatography method for measurement of a number of anions and the methods for specific anions, remain largely unchanged and have been reaffirmed by consensus ballot. New low-level amperometric titration and iodometric techniques have been added for the determination of residual chlorine, and the leuco crystal violet method has been deleted. More accurate methods have been added for chlorine dioxide and ozone. Several methods for nitrate have been deleted and a new titanium chloride method for nitrate has been added.

Part 5000 (aggregate organics) now contains only methods determining overall concentrations of groups of organic compounds. Methods for determining specific compounds (methane, pesticides and herbicides, halogenated methanes and ethanes, taste- and odor-causing compounds, and organic contaminants by gas chromatographic/mass spectrometric analysis) have been either moved to Part 6000 or superseded by new methods in that Part. Methods for DOX (formerly TOX) have been revised significantly. New methods have been added for humic substances and trihalomethane formation potential.

Part 6000 (individual organics) is new. It contains introductory information on sample collection and preservation for organics as well as operation principles and interferences for instrumental analytical methods. A method for constituent concentration by gas extraction (closed-loop stripping), included in Part 500 of the Sixteenth Edition as a method for taste- and odor-causing organics, has been revised. Each group of organic compounds is introduced by information on its presence in the environment and on selection of analytical method. Most methods in Part 6000 are completely edited and revised EPA-approved procedures that originally appeared in the supplement to the Sixteenth Edition. Several methods for the identification of specific compounds that appeared in Part 500 of the Sixteenth Edition also are included.

Part 7000 (radioactivity) contains no changes in methods. The introductory sections have been re-edited to give additional prominence to quality control.

Part 8000 (toxicity testing) also retains the methods of the previous edition. The introductory section has been revised and a short section on mutagenesis has been added. The section on zooplankton of the Sixteenth Edition no longer appears as a unit; the methods for ciliated protozoa, *Daphnia,* and *Acartia* have been placed elsewhere in the Part to reflect taxonomic relationships.

Part 9000 (microbiological examination) generally has been revised and updated. A new method for direct counts has been added and the tests for pathogenic protozoa have been separated from those for pathogenic bacteria. In testing for coliform bacteria the production of acid in a lactose-containing presumptive broth medium has been restored as a positive reaction. The rewritten section on streptococci stresses procedures for enterococci.

Part 10000 (biological examination) contains substantial revisions of methods for plankton, macrophyton, and fish. The major change in the section on plankton is the inclusion of a high-performance liquid chromatographic method for chlorophyll. Other methods have undergone general updating and revision.

Making Reagents

Following the instructions for making reagents may result in preparation of quantities larger than actually needed. In some cases these materials are toxic. To promote economy and minimize waste, the analyst should review needs and scale down solution volumes where appropriate. This conservative attitude also should extend to purchasing policies so that unused chemicals do not accumulate or need to be discarded as their shelf lives expire.

Selection and Approval of Methods

For each new edition both the technical criteria for selection of methods and the formal procedures for their approval and inclusion are reviewed critically. In regard to the approval procedures, it is considered particularly important to assure that the methods presented have been reviewed and are supported by the largest number of qualified people, so that they may represent a true consensus of expert opinion.

For the Fourteenth Edition a Joint Task Group was established for each test. This scheme has continued for each subsequent edition. Appointment of an individual to a Joint Task Group generally was based on the expressed interest or recognized expertise of the individual. The effort in every case was to assemble a group having maximum available expertise in the test methods of concern.

Each Joint Task Group was charged with reviewing the pertinent methods in the Sixteenth Edition along with other methods from the literature, recommending the methods to be included in the Seventeenth Edition, and presenting those methods in the form of a proposed section manuscript. Subsequently, each section manuscript (except for Part 1000) was ratified by vote of those members of the Standard Methods Committee who asked to review sections in that part. Every negative vote and every comment submitted in the balloting was reviewed by the Joint Editorial Board. Relevant suggestions were referred appropriately for resolution. When negative votes on the first ballot could not be resolved by the Joint Task Group or the Joint Editorial Board, the section was reballoted among all who voted affirmatively or negatively on the original ballot. Only a few issues could not be resolved in this manner and the Joint Editorial Board made the final decision.

The general and quality assurance information presented in Part 1000 was treated somewhat differently. Again, Joint Task Groups were formed, given a charge, and allowed to produce a consensus draft. This draft was reviewed by the Joint Editorial Board Liaison and subsequently by the Joint Editorial Board. The draft sections were sent to the Standard Methods Committee and comments resulting from this review were used to develop the final draft.

The methods presented here, as in previous editions, are believed to be the best available and generally accepted procedures for the analysis of water, wastewaters, and related materials. They represent the recommendations of specialists, ratified by a large number of analysts and others of more general expertise, and as such are truly consensus standards, offering a valid and recognized basis for control and evaluation.

The technical criteria for selection of methods were applied by the Joint Task Groups and by the individuals reviewing their recommendations, with the Joint Editorial Board providing only general guidelines. In addition to the classical concepts of precision, bias, and minimum detectable concentration, selection of a method also must recognize such considerations as the time required to obtain a result, needs for specialized equipment and for special training of the analyst, and other factors related to the cost of the analysis and the feasibility of its widespread use.

Status of Methods

All methods in the Seventeenth Edition are dated to assist users in determining those methods that have been changed significantly between editions. The year the section was approved by the Standard Methods Committee is indicated in a footnote at the beginning of each section. Sections or methods that appeared in the Sixteenth Edition that are unchanged, or changed only editorially in the Seventeenth Edition, show the publication date of the Sixteenth Edition, 1985. Sections or methods that were changed significantly, or that were reaffirmed by general balloting of the Standard Methods Committee, are dated 1988. If only one individual method within a section was revised, then that individual method is dated 1988, and the remaining methods retain the 1985 date.

Methods in the Seventeenth Edition are divided into fundamental classes: PROPOSED, SPECIALIZED, STANDARD, and GENERAL. Irrespective of assigned class, all methods must be approved by the Standard Methods Committee. The four classes are described below:

1. PROPOSED—A PROPOSED method must undergo development and validation that meets the requirements set forth in Section 1040A of *Standard Methods.*

2. SPECIALIZED—A procedure qualifies as a SPECIALIZED method in one of two ways: a) The procedure must undergo development and validation and collaborative testing that meet the requirements set forth in Sections 1040B and C of *Standard Methods,* respectively; or b) The procedure is the "METHOD OF CHOICE" of the members of the Standard Methods Committee actively conducting the analysis and it has appeared in TWO PREVIOUS EDITIONS of *Standard Methods.*

3. STANDARD—A procedure qualifies as a STANDARD method in one of two ways: a) The procedure must undergo development and validation and collaborative testing that meet the requirements set forth in Sections 1040B and C of *Standard Methods,* respectively, and it is "WIDELY USED" by the members of the Standard Methods Committee; or b) The procedure is "WIDELY USED" by the members of the Standard Methods Committee and it has appeared in TWO PREVIOUS EDITIONS of *Standard Methods.*

4. GENERAL—A procedure qualifies as a GENERAL method if it has appeared in TWO PREVIOUS EDITIONS of *Standard Methods.*

Assignment of a classification to a method is done by the Joint Editorial Board. When making method classifications, the Joint Editorial Board evaluates the results of the survey on method use by the Standard Methods Committee that is conducted at the time of general balloting of the method. In addition, the Joint Editorial Board considers recommendations offered by Joint Task Groups and the Part Coordinator.

Methods categorized as "PROPOSED," "SPECIALIZED," and "GENERAL" are so designated in their titles; methods with no designation are "STANDARD."

Technical progress makes advisable the establishment of a program to keep *Standard Methods* abreast of advances in research and general practice. The Joint Editorial Board has developed the following procedure for effecting interim changes in methods between editions:

1. Any method given proposed status in the current edition may be elevated by action of the Joint Editorial Board, on the basis of adequate published data supporting such a change as submitted to the Board by the appropriate Joint Task Group. Notification of such a change in status shall be accomplished by publication in the official journals of the three associations sponsoring *Standard Methods.*

2. No method may be abandoned or reduced to a lower status during the interval between editions.

3. A new method may be adopted as proposed, specialized, or standard by the Joint Editorial Board between editions, such action being based on the usual consensus procedure. Such new methods may be published in supplements to editions of *Standard Methods.*

Even more important to maintaining the current status of these standards is the intention of the sponsors and the Joint Editorial Board that subsequent editions will appear regularly at reasonably short intervals.

Reader comments and questions concerning this manual should be addressed to: Standard Methods Manager, American Water Works Association, 6666 West Quincy Avenue, Denver, CO 80235.

Acknowledgments

For the major portion of the work in preparing and revising the methods in the Sixteenth Edition, the Joint Editorial Board gives full credit to the Standard Methods Committees of the American Water Works Association and of the Water Pollution Control Federation, and to the Subcommittee on Standard Methods for the Examination of Water and Wastewater and the Committee on Laboratory Standards and Practices of the American Public Health Association. Members of these committees chair and serve as members of the Joint Task Groups. They were assisted often by advisors, not formally members of the committees, and in many cases not members of the sponsoring societies. To the advisors, special gratitude is extended in recognition of their efforts. A list of the committee members and advisors follows these pages. Robert Booth, Senior Science Advisor, U.S. Environmental Protection Agency, served as a liaison from EPA to the Joint Editorial Board; thanks are due for his interest and help.

The Joint Editorial Board expresses its appreciation to William H. McBeath, M.D., Executive Director, American Public Health Association, to John B. Mannion, Executive Director, American Water Works Association, and to Quincalee Brown, Executive Director, Water Pollution Control Federation, for their cooperation and advice in the development of this publication. Steven J. Posavec, Standard Methods Manager and Joint Editorial Board Secretary, provided a variety of important services that are vital to the preparation of a volume of this type. Jaclyn Alexander, Director of Publications, American Public Health Association, functioned as publisher, and Brigitte Coulton, also with APHA, served as production manager. Special recognition for her valuable services is due to Mary Ann H. Franson, Managing Editor, who discharged most efficiently the extensive and detailed responsibilities on which this publication depends.

Acknowledgement is made to past Executive Director Paul A. Schulte of the American Water Works Association for his extensive assistance to this and previous editions of *Standard Methods.* The late Robert A. Canham made invaluable contributions to the continuous growth and improvement of *Standard Methods* during his long tenure as Executive Director of the Water Pollution Control Federation;

his wisdom and leadership will be remembered gratefully. The Joint Editorial Board acknowledges the significant service of Frederick W. Pontius, Regulatory Engineer, American Water Works Association, who formerly was Secretary to the Joint Editorial Board, and of Adrienne Ash of the American Public Health Association, who acted as publisher during the early phases of the preparation of this edition.

Joint Editorial Board
Arnold E. Greenberg, American Public Health Association
R. Rhodes Trussell, American Water Works Association
Lenore S. Clesceri, Water Pollution Control Federation, Chairman

ix

Marvin D. Piwoni, 3030, 3110, 3111, 3112, 3113, 3114
Hugh D. Putnam, 10300
Stephen J. Randtke, 5550
Ronald L. Raschke, 10400
Donald J. Reasoner, 9211
David A. Reckhow, 5320
Donald J. Reish, 8510, 10900
Olsen J. Rogers, 4500-Br⁻

Darrell G. Rose, 3112
Robert S. Safferman, 9250
Kenneth Simon, 8010
Robert D. Swisher, 5540
Jack R. Tuschall, 3500-Li
Michael J. Vandaveer, 3030
Hugo T. Victoreen, 9215
Oleh Weres, 3500-Se
James C. Young, 5210

Standard Methods Committee and Joint Task Group Members

The following served as Standard Methods Committee Members and as Group Members of the Joint Task Group(s) developing the designated section(s).

John C. Adams, 9213
V. Dean Adams
Rose Adams-Whitehead
Cacilda Jiunko Aiba
E. Marco Aieta, 4500-ClO₂
Katherine T. Alben
Harry J. Alexander, 1060
James E. Alleman, 4500-NO₃⁻ & NO₂⁻
Herbert E. Allen
Martin J. Allen, 9212, 9215
Osman M. Aly, 5530
Gary L. Amy, 5510
Charles G. Appleby
Neal E. Armstrong
John A. Arrington, II
Robert M. Arthur
Donald B. Aulenbach, 4500-NO₃⁻ & NO₂⁻
Barry M. Austern
Warren F. Averill
Guy M. Aydlett
Robert M. Bagdigian, 1080

Rodger B. Baird, 3030, 3110, 3111, 3112, 3113, 3114
L. Malcolm Baker, 6040
Robert A. Baker
Gwendolyn L. Ball, 5710, 6040
Roy O. Ball
Edmond J. Baratta, 7010, 7020, 7110, 7500-Cs, I, Ra, Sr, ³H, & U
Michael J. Barcelona
Thomas O. Barnwell, Jr., 5210
Sylvia E. Barrett, 4500-Cl
Jerry Bashe,* 6010
Frank J. Baumann, 6210, 6220, 6230
David G. Beckwith, 1080
John W. Beland
Daniel F. Bender, 1080, 1090, 4500-Cl & ClO₂
Larry D. Benefield
Paul S. Berger, 9020, 9215
Thomas Beymer,* 6010
Robert R. Bidigare, 10200
Kenneth E. Biesinger
Star F. Birch, 1060, 1090, 9020
Stuart C. Black, 1010, 1020, 1030, 1040

---

* Individuals thus indicated are not official members of the Standard Methods Committee but contributed to the designated section(s).

H. Curt Blair, 2520, 7010, 7020, 7110, 7500-Cs, I, Ra, Sr, ³H, & U
I. Bob Blumenthal
Edwin S. Boatman*
Debra K. Bolding, 9211, 9230
Robert L. Booth†
Robert H. Bordner, 9010, 9020, 9030, 9040, 9050, 9060
Mark E. Bose, 5540
Harry Boston,‡ 10400
Robert I. Botto, 3120
Gerald R. Bouck, 2810
William H. Bouma, 1060
Theresa M. Bousquet
Russell E. Bowen
George T. Bowman, 5210
William C. Boyle
Lloyd W. Bracewell, 2530
Alvin L. Bradshaw, 2520
Brian M. Brady
Blaise J. Brazos
Geoffrey L. Brock
Joe E. Brown
Clifford Bruell
William H. Bruvold, 2160
Anthony A. Bulich
Steven M. Bupp
Gary A. Burlingame, 2160, 6040
J. Owen Callaway
Craig D. Cameron, 4500-O₃
Raul R. Cardenas, Jr.
Robert E. Carlson
Stephen R. Carpenter
Anthony R. Castorina
Paul A. Chadik, 5510
Peter M. Chapman, 8510, 10500
Eric J. Chatfield
Daniel David Chen

---

* Deceased.
† EPA Liaison to the Joint Editorial Board.
‡ Individuals thus indicated are not official members of the Standard Methods Committee but contributed to the designated section(s).

Mark G. Cherwin, 3030, 3110, 3111, 3112, 3113, 3114
Leonard L. Ciaccio
Robert R. Claeys
James A. Clark, 9221, 9225
Robert R. Clark
William H. Clement, 5530
Dennis A. Clifford
Sharon M. Cline
John Clugel
Colin E. Coggan
Robert S. Cohen, 3030, 3110, 3111, 3112, 3113, 3114
Larry D. Cole
Gary B. Collins, 10200
Michael Collins, 5510
Tom E. Collins, 6040
John Colt, 2810
Brian J. Condike, 3030, 3110, 3111, 3112, 3113, 3114
Gerald F. Connell
Wm. Bridge Cooke, 9010, 9030, 9040, 9050, 9060, 9610, 10900
Sandra D. Cooper, 10300, 10500
Robert C. Cooper, 9225, 9260, 9711
Harold S. Costa
C. Richard Cothern, 7010, 7020, 7110, 7500-Cs, I, Ra, Sr, ³H, & U
John E. Cowell
John V. Crable, 1090
Wendall H. Cross
Warren B. Crummett
William G. Crumpton, 10200
Nicholas J. Csikai
Kenneth W. Cuneo
Gregory A. Cutter, 3500-Se
Melissa S. Dale, 6040
Richard A. Danforth, 1080
Robert A. Daniels, 10600
Richard-Paul Danner
Brian G. D'Aoust, 2810
Patrick H. Davies

Charles O. Davis, III, 3500-Li
Ernst M. Davis, III, 10200
Marshall K. Davis, 3030, 3110,
 3111, 3112, 3113, 3114
Laurens M. DeBirk
Gary L. De Kock, 3120
Joseph J. Delfino
Brian A. Dempsey, 5510
W. Michael Dennis,* 10400
Steven K. Dentel
Fred L. DeRoos
David Diamond
Jack C. Dice, 2160
Denise A. Dominguez
John H. Dorsey, 10900
Ronald C. Dressman, 5320
Charles T. Driscoll, 3500-Al
Alfred P. Dufour, 9010, 9030,
 9040, 9050, 9060, 9213, 9230,
 9260, 9711
Bernard J. Dutka, 9211, 9213,
 9215
David G. Easterly, 7010, 7020,
 7110, 7500-Cs, I, Ra, Sr, ^{3}H, &
 U
Andrew D. Eaton
David L. Edelman, Jr.
James K. Edzwald, 5710
Lawrence W. Eichler, 3030, 3110,
 3111, 3112, 3113, 3114, 3500-
 Al
Gunnar Ekedahl
William M. Ellgas, 9222, 9225
G. Keith Elmund, 9020, 9211
Mohamed Elnabarawy
Alan W. Elzerman
David E. Erdmann, 4110, 4500-F$^-$
R. P. Esser, 9810
Paul D. Evans
William S. Ewell
Samuel D. Faust, 5530

Andrea F. Feagin
James C. Feeley, 9213
John M. Ferris, 9010, 9030, 9040,
 9050, 9060, 9810
V. R. Ferris, 9810
James V. Feuss, 4500-Cl
Duane H. Fickeisen, 2810
Luis Octavio Figueroa
Bradford R. Fisher
Robert P. Fisher
Marvin J. Fishman, 3030, 3110,
 3111, 3112, 3113, 3114, 4500-
 F$^-$
Arthur W. Fitchett
Ellen P. Flanagan, 9222
Charles Fogle, 2530
Verlin W. Foltz, 4110
Charles T. Ford
John A. Fournier
Steven P. Fraleigh
Martin S. Frant
Marlene Frey
Clifford Frith, 1080
John L. Fronk
Paul L. K. Fu,* 5510
Roger S. Fujioka, 9230
Leo C. Fung, 4500-O$_3$
Joseph T. Fuss, 2810
Anthony M. Gaglierd, 4500-Cl
John E. Gannon
M.E. Garza, Jr.
Anthony F. Gaudy, Jr.
Edwin E. Geldreich, 9010, 9030,
 9040, 9050, 9060, 9213, 9221,
 9222
Stephen R. Gelman, 5210
Carl J. George, 10600
Vincent A. Geraci
Michael H. Gerardi
Mriganka M. Ghosh
Sambhunath Ghosh
E. Gilbert, 4500-O$_3$
James M. Gindelberger
Thomas S. Gittelman, 6040

---

* Individuals thus indicated are not official members of
the Standard Methods Committee but contributed to the
designated section(s).

William H. Glaze
Connie Glover
C. Ellen Gonter
Reginald K.S. Goo
Arley L. Goodenkauf
Gilbert Gordon, 4500-ClO$_2$ & O$_3$
Joseph W. Gorsuch
James A. Gouck
Joseph P. Gould, 4500-Cl
W. O. K. Grabow, 9213
Nancy E. Grams, 5320
Robert L. Graves
William B. Gray
Philip M. Gschwend
Robert J. Gussman, 9020
Rufus K. Guthrie, 9213, 9225
Charles N. Haas, 4500-Cl
Earl A. Hadfield, II
Stephen W. Hager, 4500-NO$_3^-$ & NO$_2^-$
Gary E. Hahn
Bruce A. Hale, 4110, 4500-ClO$_2$
Nancy H. Hall, 9221, 9260, 9711
David J. Hansen
Robert W. Hansen
Donald W. Harper
David J. Hartman, 5710
Thomas W. Haukebo
Steven D. Hawthorne, 10500, 10900
Michael K. Hein, 10300
Charles W. Hendricks, 9225, 9260, 9711
Earl L. Henn, 3030, 3110, 3111, 3112, 3113, 3114
Edwin E. Herricks
Delbert Hicks, 10400
Anita K. Highsmith, 1080, 9010, 9030, 9040, 9050, 9060, 9213
Albert Hill,* 10400

---

* Individuals thus indicated are not official members of the Standard Methods Committee but contributed to the designated section(s).

David R. Hill, 5320
Kenneth M. Hill
Phillip A. Hill
George D.G. Hilling, 3030, 3110, 3111, 3112, 3113, 3114
George Hoag
Laura Mahoney Hodges, 1060
Jimmie A. Hodgeson
Robert C. Hoehn, 5510, 5710, 9212
Jurg Hoigne, 4500-O$_3$
G. C. Holdren
Albert C. Holler, 3030, 3110, 3111, 3112, 3113, 3114
Thomas R. Holm
Osmund Holm-Hansen
John Homa, Jr., 10600
Thomas B. Hoover, 4110
Peter C. Houle
Jack L. Hoyt, 5540
C. P. Huang
Wayne B. Huebner, 4500-Cl & ClO$_2$
Donald G. Huggins, 10900
Donald L. Hughes
R. DeLon Hull
Yung-Tse Hung
Joseph V. Hunter
Noel M. Hurley
Joseph A. Hutchinson, 7010, 7020, 7110, 7500-Cs, I, Ra, Sr, ^{3}H, & U
Cordelia J. Hwang
Veronica Y. Inouye
Kurt Irgolic
Billy G. Isom, 10500
George Izaguirre, 9250
R. Wayne Jackson, 9222
Walter Jakubowski, 9010, 9030, 9040, 9050, 9060, 9260, 9711
Carol Ruth James
Konanur G. Janardan
Harlan R. Jeche, 9020
S. Rod Jenkins

J. Charles Jennett, 3030, 3110,
3111, 3112, 3113, 3114
James N. Jensen
J. Michael Jeter
Isabel C. Johnson
J. Donald Johnson, 4500-Cl
Stephen W. Johnson
Waynon Johnson
Donald L. Johnstone, 9213, 9230
Martha N. Jones, 4500-NO$_3^-$ &
NO$_2^-$
Robert A. Jung
Swiatoslav W. Kaczmar
Sabry M. Kamhawy, 5210
Sandra A. Kaptain
Irwin J. Katz, 9250, 9260, 9711
Fred K. Kawahara, 5530
Floyd D. Kefford
Michael A. Keirn, 10300
Lawrence H. Keith
Nabih P. Kelada
William J. Kenney, 5210
Lee G. Kent, 3030, 3110, 3111,
3112, 3113, 3114
Zoltan Kerekes, 9211
Harold W. Kerster
Troy Edward King, 1090
Riley N. Kinman
Joyce S. Kippin, 9212
Norman A. Kirshen
Robert L. Klein, Jr.
Donald J. Klemm, 10500, 10900
Margaret M. Knight
Bart Koch, 5710
William F. Koch, 4500-F$^-$
Frederick C. Kopfler
Stuart W. Krasner, 2160, 6040
Richard T. Krause
Timothy I. Ladd, 9216
Lawrence E. LaFleur, 5320
Leslie E. Lancy
Dennis D. Lane
Russell W. Lane, 2330
Johan Langewis

Alexander Lapteff
Richard A. Larson, 5510
Desmond F. Lawler
James M. Lazorchak, 10500,
10900
Norman E. LeBlanc
Mark LeChevallier, 9212
G. Fred Lee
Janet M. Lee
Raymond Lee
Henry W. Leibee
Armond E. Lemke
Lawrence Y. C. Leong
Raymond D. Letterman
Philip A. Lewis
Ronald F. Lewis, 9250
Timothy E. Lewis, 3500-Al
Chun-Teh Li
Alvin Lieberman
Shundar Lin, 9212, 9221
Christopher B. Lind, 9221
Chung-King Liu
Larry B. Lobring
Stephen R. Lohman, 4500-ClO$_2$
Linda R. Lombardo, 9215
Maxine C. Long, 9225, 9260, 9711,
10200
James E. Longbottom
Karl E. Longley, 4500-Cl
Marc W. Lorenzen
Richard G. Luthy
Gerald L. Mahon, 9250
Ronald L. Malcolm, 5510
Joel Mallevialle
Bruce E. Manning
George Marinenko
Leif L. Marking
Harold G. Marshall, 10200
Theodore D. Martin, 3120
Margaret Martin-Goldberg
Maria T. Martins, 9213
John Y. Mason, 4500-ClO$_2$
Willy J. Masschelein, 4500-O$_3$
Owen B. Mathre, 3120

Howard M. May, 3500-Al
Foster L. Mayer
Lawrence M. Mayer, 5510
J. Howard McCormick
William F. McCoy, 9010, 9030,
9040, 9050, 9060, 9216
Daniel McDaniel
Gerald N. McDermott
Gordon A. McFeters, 9010, 9030,
9040, 9050, 9060, 9212, 9222
Gerald D. McKee, 5210
James J. McKeown
Gerald L. McKinney, 3120
Roy P. McKnight
Daniel A. McLean
Lilia M. McMillan, 9250
Dale D. McMurtrey
Ann Marie McNamara, 1080
Nancy E. McTigue
Jay C. Means
Robert O. Megard
Robert Meglan
Morten C. Meilgaard
Joseph L. Melnick, 9211
Lori A. Melroy, 1060
Douglas T. Merrill, 2330
Peter L. Meschi
Theodore G. Metcalf
E. J. Middlebrooks
Carl J. Miles, 5510
Arthur M. Miller
Frank J. Millero, 2520
James R. Millette
Roger A. Minear, 4500-Br$^-$, 5530
Marc W. Mittelman, 9216
Harold Moehser,* 3500-Se
James G. Moncur
William J. Montgomery, Jr.
Leown A. Moore, 5710
H. Anson Moye
Terry I. Mudder

James W. Mullins, 7010, 7020,
7110, 7500-Cs, I, Ra, Sr, ^{3}H, &
U
Andrew P. Murray, 10200
Hussein Naimie, 4500-O$_3$
Harry D. Nash, 9020, 9221
Rosemary H. Neal, 3500-Se
Stuart Neff,* 10900
Stanley A. Nichols, 10400
Teresa J. Norberg-King
James R. Nugent
Gregg Oelker, 3120
Patrick W. O'Keefe
Betty H. Olson, 9010, 9030, 9040,
9050, 9060
Cathleen S. O'Neil
Joan A. Oppenheimer
Edward D. Orme, 4110
John A. Osborne, 10400
Quentin W. Osburn, 5540
Janet G. Osteryoung
Edward B. Overton
Jeffrey L. Oxenford
Arthur T. Palin, 4500-Cl & O$_3$
Sunil P. Pande
James M. Pappenhagen
Thomas R. Parr, 9225, 9230
Patrick R. Parrish
Frances Parsons, 9010, 9030, 9040,
9050, 9060, 9225
Robert A. Paterson, 9610, 10200,
10900
Paul D. Paustian
Harry M. Pawlowski, 4500-Cl
Richard K. Peddicord
Mark E. Peden, 3500-Al
E. Michael Perdue, 5510
Robert K. Pertuit
Carol Pesch,* 8510
Deanna L. Peterson, 6040
John Peterson, 9222
William M. Peterson
Gary R. Peyton
Frederic K. Pfaender

* Individuals thus indicated are not official members of the Standard Methods Committee but contributed to the designated section(s).

John D. Pfaff, 4110
Massoud Pirbazari
Rodolfo A. Pisigan, Jr., 2330
John T. Pivinski
Marvin D. Piwoni, 3030, 3110,
  3111, 3112, 3113, 3114
Russell H. Plumb, Jr.
Robert B. Pojasek
James M. Polisini
William B. Prescott
Thomas A. Pressley
Fred T. Price
Hugh D. Putnam, 10300
James G. Quinn
Ansar A. Qureshi, 9215, 9222,
  9230
Neil M. Ram, 5710
Raman K. Raman
Steven J. Randtke, 5550
Ronald L. Raschke, 10200, 10400
Inayat A. Rashid
Judith A. Rawa
Bill T. Ray, 1090
William R. Ray
Donald J. Reasoner, 9010, 9030,
  9040, 9050, 9060, 9211, 9212
David A. Reckhow, 5320
Kurt A. Reimann, 5540
Martin Reinhard
Donald J. Reish, 8510, 10900
Vincent H. Resh, 10900
David J. Rexing
Eugene W. Rice, 9221, 9222
George G. Richards, 2330
Harry Ridgway
Robert W. Rinehart, Sr.
Michele F. Rizet, 3030, 3110,
  3111, 3112, 3113, 3114
Morris H. Roberts, Jr., 10500
Andrew Robertson, 10200

Sharon M. Roehm
Stephen C. Roesch, 1090
Gary L. Rogers
Olsen J. Rogers, 4500-Br⁻
Peter F. Rogerson
R. R. Romine
Darrell G. Rose, 3030, 3110, 3111,
  3112, 3113, 3114
Bernard Rosenberg, 9225, 9260,
  9711
Joel A. Rosenfield
William D. Rosenzweig, 9215,
  9610
John R. Rossum,* 2330
Richard B. Rubin
Peter P. Russell, 2530
Peggy A. Ryker, 9222
William A. Sack, 2520
Robert S. Safferman, 9010, 9030,
  9040, 9050, 9060, 9250
Ana M. Sancha
Ernest A. Sanchez, 7010, 7020,
  7110, 7500-Cs, I, Ra, Sr, ³H, &
  U
Rajkamal Sarin
Bernard Saunier, 4500-Cl & O₃
Larry P. Scanlan, 2330
A. David Scarchilli
David J. Schaeffer
Bernard Schmall
John A. Schmitt, 9610
Michael R. Schock, 3500-Li
Frank E. Scully, Jr.
Bette Seamonds, 1080
Alberta J. Seierstad
Kevin G. Sellner,† 10200
Roland L. Seymour, 9610
Joseph Shapiro
B. F. Shema, 9260, 9711
Joseph H. Sherrard
Marjorie G. Shovlin, 9212, 9225
Peter J. Shuba, 10300, 10500
Mark Shuman
Leonard Sideman, 1080

---

* Deceased.
† Individuals thus indicated are not official members of
the Standard Methods Committee but contributed to the
designated section(s).

Kenneth Simon, 8010
Philip C. Singer, 5710
J. Edward Singley, 2330
Robert W. Slater, Jr.
John P. Slobogin
Robert Smith
Mark D. Sobsey
William C. Sonzogni
Charles A. Sorber
R. Kent Sorrell
David T. Specht
Robert L. Spehar
Robert G. Spicher
Stuart J. Spiegel
John B. Sprague
Jack R. Stabley, Jr., 3030, 3110,
    3111, 3112, 3113, 3114
Vernon T. Stack, 5210
John F. Stafford
Alan A. Stevens
Raymond M. Stewart
Scott Stieg
I. H. Suffet, 2160
Makram T. Suidan
Bernard F. Sullivan
Richard Swartz
Robert A. Sweeney, 10200,
    10500, 10900
Robert D. Swisher, 5540
James M. Symons
John C. Synnott, 4500-NO$_3^-$ &
    NO$_2^-$
Adib F. Tabri
Yoshihiro Takahashi, 5320
Lawrence P. Talarico, 5320
Thomas A. Tamayo
Mark L. Tamplin, 9260, 9711
Theodore S. Tanaka
Donald C. Taylor
Raymond H. Taylor, 9215, 9221
Frank Thomas

Bruce M. Thomson
Earl M. Thurman
Robert V. Thurston
Edwin C. Tifft, Jr.
R. Yucel Tokuz
Michael L. Trehy
Albert R. Trussell, 6040
Leon Tsao,* 3500-Se
Jack R. Tuschall, 3500-Li, 5510
Ronald E. Twillman, 6040
Mark D. Umphres
George S. Uyesugi, 7010, 7020,
    7110, 7500-Cs, I, Ra, Sr, ^{3}H, &
    U
Michael J. Vandaveer, 3030, 3110,
    3111, 3112, 3113, 3114
Schalk W. van der Merwe
William H. van der Schalie
M. M. Varma
Roy M. Ventullo, 9216
P. Aarne Vesilind
Hugo T. Victoreen, 9010, 9030,
    9040, 9050, 9060, 9215, 9222
Craig O. Vinson
Ronald H. Voss
Frank F. Wada, 2530
Johannes E. Wajon, 6040
Gerald E. Walsh, 10200
Lawrence K. Wang
Mu Hao S. Wang
Wuncheng Wang
Timothy J. Ward
Barron L. Weand
David H. Webb, 10400
James H. Webb, 9222
Flora Mae Wellings
Oleh Weres, 3500-Se
Warren C. Westgarth, 1060, 2160,
    9020, 9211
Theodore F. Wetzler, 9221
Willis B. Wheeler
Robert E. White
James L. Whitlock
Greene R. Whitney, 2330

---

* Individuals thus indicated are not official members of the Standard Methods Committee but contributed to the designated section(s).

Donald O. Whittemore, 4500-Br⁻
Raymond C. Whittemore, 5210
George P. Whittle
Brannon H. Wilder
Robert T. Williams
Theodore J. Williams, 9020
Kenneth J. Wills, 1060
Frederic J. Winter
John A. Winter, 9020
Charles E. Woelke
Roy L. Wolfe
Richard E. Wolke, 10600
George T. F. Wong

Mark W. Wood
Ty R. Woodin
Jack Q. Word, 10900
John Wuepper, 2160
George Yamate
In Che Yang, 7010, 7020, 7110, 7500-Cs, I, Ra, Sr, ^{3}H, & U
Dennis A. Yates
Andrew Yee,* 3500-Se
Thomas L. Yohe
Roger A. Yorton
James C. Young, 5210
Michael Young, 9020, 9213
Matthew J. Zabik
Stan S. Zaworski
Carol A. Ziel, 9211, 9213
Melvin C. Zimmerman

---

* Individuals thus indicated are not official members of the Standard Methods Committee but contributed to the designated section(s).

TABLE C
INTERNATIONAL
RELATIVE
ATOMIC WEIGHTS

TABLE C. INTERNATIONAL RELATIVE ATOMIC WEIGHTS, 1987
Scaled to the relative atomic mass, $A_r(^{12}C) = 12$

The atomic weights of many elements are not invariant but depend on the origin and treatment of the material. The footnotes to this table elaborate the types of variation to be expected for individual elements. The values of $A_r(E)$ given here apply to elements as they exist naturally on earth and to certain artificial elements. When used with due regard to the footnotes they are considered reliable in the last digit to $\pm$ the figure given in parentheses.

Name	Symbol	Atomic number	Atomic weight	Footnotes	Name	Symbol	Atomic number	Atomic weight	Footnotes
Actinium*	Ac	89	227.0278	A	Fermium*	Fm	100	257.0951	A
Aluminium	Al	13	26.981539(5)		Fluorine	F	9	18.9984032(9)	
Americium*	Am	95	241.0568	A	Francium*	Fr	87	223.0197	A
Antimony	Sb	51	121.75(3)		Gadolinium	Gd	64	157.25(3)	g
Argon	Ar	18	39.948(1)	g, r	Gallium	Ga	31	69.723(1)	
Arsenic	As	33	74.92159(2)		Germanium	Ge	32	72.61(2)	
Astatine*	At	85	209.9871	A	Gold	Au	79	196.96654(3)	
Barium	Ba	56	137.327(7)	g	Hafnium	Hf	72	178.49(2)	
Berkelium*	Bk	97	247.0703	A	Helium	He	2	4.002602(2)	g, r
Beryllium	Be	4	9.012182(3)		Holmium	Ho	67	164.93032(3)	
Bismuth	Bi	83	208.98037(3)		Hydrogen	H	1	1.00794(7)	g, m, r
Boron	B	5	10.811(5)	g, m, r	Indium	In	49	114.82(1)	
Bromine	Br	35	79.904(1)		Iodine	I	53	126.90447(3)	
Cadmium	Cd	48	112.411(8)	g	Iridium	Ir	77	192.22(3)	
Calcium	Ca	20	40.078(4)	g	Iron	Fe	26	55.847(3)	g, m
Californium*	Cf	98	249.0748	A	Krypton	Kr	36	83.80(1)	g
Carbon	C	6	12.011(1)	r	Lanthanum	La	57	138.9055(2)	g
Cerium	Ce	58	140.115(4)	g	Lawrencium*	Lr	103	262.11	A
Cesium	Cs	55	132.90543(5)		Lead	Pb	82	207.2(1)	g, r
Chlorine	Cl	17	35.4527(9)		Lithium	Li	3	6.941(2)	g, m, r
Chromium	Cr	24	51.9961(6)		Lutetium	Lu	71	174.967(1)	g
Cobalt	Co	27	58.93320(1)		Magnesium	Mg	12	24.3050(6)	
Copper	Cu	29	63.546(3)	r	Manganese	Mn	25	54.93805(1)	
Curium*	Cm	96	243.0614	A	Mendelevium*	Md	101	256.094	A
Dyprosium	Dy	66	162.50(3)	g	Mercury	Hg	80	200.59(3)	
Einsteinium*	Es	99	252.083	A	Molybdenum	Mo	42	95.94(1)	g
Erbium	Er	68	167.26(3)	g	Neodymium	Nd	60	144.24(3)	g
Europium	Eu	63	151.965(9)	g	Neon	Ne	10	20.1797(6)	g, m

TABLE C. INTERNATIONAL RELATIVE ATOMIC WEIGHTS, 1987   xxi

Element	Symbol	Z	Atomic Weight	Notes
Neptunium*	Np	93	237.0482	A
Nickel	Ni	28	58.69(1)	
Niobium	Nb	41	92.90638(2)	
Nitrogen	N	7	14.00674(7)	g, r
Nobelium*	No	102	259.1009	A
Osmium	Os	76	190.2(1)	g
Oxygen	O	8	15.9994(3)	g, r
Palladium	Pd	46	106.42(1)	g
Phosphorus	P	15	30.973762(4)	
Platinum	Pt	78	195.08(3)	
Plutonium*	Pu	94	238.0496	
Polonium*	Po	84	208.9824	
Potassium	K	19	39.0983(1)	A
Praseodymium	Pr	59	140.90765(3)	A
Promethium*	Pm	61	144.9127	
Protactinium*	Pa	91	231.0359	A
Radium*	Ra	88	226.0185	A
Radon*	Rn	86	210.9906	A
Rhenium	Re	75	186.207(1)	
Rhodium	Rh	45	102.90550(3)	
Rubidium	Rb	37	85.4678(3)	g
Ruthenium	Ru	44	101.07(2)	g
Samarium	Sm	62	150.36(3)	g
Scandium	Sc	21	44.955910(9)	
Selenium	Se	34	78.96(3)	
Silicon	Si	14	28.0855(3)	r
Silver	Ag	47	107.8682(2)	g
Sodium	Na	11	22.989768(6)	
Strontium	Sr	38	87.62(1)	g, r
Sulfur	S	16	32.066(6)	r
Tantalum	Ta	73	180.9479(1)	
Technetium*	Tc	43	96.9064	A
Tellurium	Te	52	127.60(3)	g
Terbium	Tb	65	158.92534(3)	
Thallium	Tl	81	204.3833(2)	
Thorium*	Th	90	232.0381(1)	g, Z
Thulium	Tm	69	168.93421(3)	
Tin	Sn	50	118.710(7)	g
Titanium	Ti	22	47.88(3)	
Tungsten	W	74	183.85(3)	
Unnilhexium	Unh	106	263.118	A
Unnilpentium	Unp	105	262.114	A
Unnilquadium	Unq	104	261.11	A
Unnilseptium	Uns	107	262.12	A
Uranium*	U	92	238.0289(1)	g, m, Z
Vanadium	V	23	50.9415(1)	
Xenon	Xe	54	131.29(2)	g, m
Ytterbium	Yb	70	173.04(3)	
Yttrium	Y	39	88.90585(2)	
Zinc	Zn	30	65.39(2)	
Zirconium	Zr	40	91.224(2)	g

* Element has no stable nuclides.

g geological specimens are known in which the element has an isotopic composition outside the limits for normal material. The difference between the atomic weight of the element in such specimens and that given in the table may exceed considerably the implied uncertainty.

m modified isotopic compositions may be found in commercially available material because it has been subjected to an undisclosed or inadvertent isotopic separation. Substantial deviations in atomic weight of the element from that given in the table can occur.

r range in isotopic composition of normal terrestrial material prevents a more precise atomic weight being given; the tabulated $A_r(E)$ value should be applicable to any normal material.

A Radioactive element that lacks a characteristic terrestrial isotopic composition. The weight listed is that of the lowest-atomic-weight isotope.

Z An element, without stable nuclide(s), exhibiting a range of characteristic terrestrial compositions of long-lived radionuclide(s) such that a meaningful atomic weight can be given.

Source: INTERNATIONAL UNION OF PURE AND APPLIED CHEMISTRY. 1988. Atomic weights of the elements, 1987. *Pure Appl. Chem.* 60:841.

TABLE OF CONTENTS

PAGE

Part 1000 GENERAL INFORMATION

1010 INTRODUCTION. 1-1
    A. Scope and Application of Methods 1-1
    B. Statistics 1-2
    C. Glossary 1-4

1020 QUALITY ASSURANCE. 1-6
    A. Introduction 1-6
    B. Quality Control. 1-7
    C. Quality Assessment 1-12

1030 DATA QUALITY 1-14
    A. Introduction 1-14
    B. Bias 1-14
    C. Precision 1-15
    D. Total Uncertainty. 1-17
    E. Method Detection Limit 1-18
    F. Checking Correctness of Analyses 1-20

1040 METHOD DEVELOPMENT AND EVALUATION 1-22
    A. Introduction 1-22
    B. Method Validation 1-22
    C. Collaborative Testing 1-24

1050 EXPRESSION OF RESULTS. 1-27
    A. Units. 1-27
    B. Significant Figures 1-28

1060 COLLECTION AND PRESERVATION OF SAMPLES 1-30
    A. Introduction 1-30
    B. Collection of Samples 1-34
    C. Sample Preservation 1-39

1070 LABORATORY APPARATUS, REAGENTS, AND
       TECHNIQUES. 1-41
    A. Introduction 1-41
    B. Apparatus 1-41
    C. Reagents 1-42
    D. Techniques 1-45

1080 REAGENT-GRADE WATER. 1-54
    A. Introduction 1-54
    B. Methods for Preparation of Reagent-Grade Water . . 1-56
    C. Reagent Water Quality 1-57

1090 SAFETY. 1-58
    A. Introduction 1-58
    B. Safety Equipment 1-60

PAGE

    C. Laboratory Hazards.   1-62
    D. Hazard Management Practices   1-69

Part **2000** **PHYSICAL AND AGGREGATE PROPERTIES**
    2010 INTRODUCTION.   2-1
    2020 QUALITY CONTROL.   2-1
    2110 APPEARANCE.   2-2
    2120 COLOR   2-2
        A. Introduction   2-2
        B. Visual Comparison Method   2-2
        C. Spectrophotometric Method.   2-5
        D. Tristimulus Filter Method.   2-8
        E. ADMI Tristimulus Filter Method (PROPOSED). . .   2-9
    2130 TURBIDITY.   2-11
        A. Introduction   2-11
        B. Nephelometric Method   2-13
    2150 ODOR   2-16
        A. Introduction   2-16
        B. Threshold Odor Test   2-18
    2160 TASTE   2-23
        A. Introduction   2-23
        B. Flavor Threshold Test (FTT)   2-24
        C. Flavor Rating Assessment (FRA)   2-27
        D. Flavor Profile Analysis (FPA).   2-29
    2310 ACIDITY   2-30
        A. Introduction   2-30
        B. Titration Method   2-30
    2320 ALKALINITY   2-35
        A. Introduction   2-35
        B. Titration Method   2-35
    2330 CALCIUM CARBONATE SATURATION (PROPOSED) . . .   2-40
        A. Introduction   2-40
        B. Indices Indicating Tendency of a Water to
            Precipitate $CaCO_3$ or Dissolve $CaCO_3$ 2-42
        C. Indices Predicting the Quantity of $CaCO_3$ That Can
            Be Precipitated or Dissolved   2-47
        D. Diagrams and Computer Codes for $CaCO_3$ Indices. .   2-49
    2340 HARDNESS   2-52
        A. Introduction   2-52
        B. Hardness by Calculation   2-53
        C. EDTA Titrimetric Method   2-53
    2510 CONDUCTIVITY.   2-57
        A. Introduction   2-57
        B. Laboratory Method   2-59

                                                                    PAGE
2520 SALINITY.   2-61
      A. Introduction   2-61
      B. Electrical Conductivity Method   2-62
      C. Density Method   2-64
      D. Algorithm of Practical Salinity 2-65
2530 FLOATABLES   2-65
      A. Introduction   2-65
      B. Particulate Floatables (GENERAL)   2-66
      C. Trichlorotrifluoroethane-Soluble Floatable Oil and
          Grease (GENERAL)   2-69
2540 SOLIDS   2-71
      A. Introduction   2-71
      B. Total Solids Dried at 103-105°C   2-72
      C. Total Dissolved Solids Dried at 180°C   2-74
      D. Total Suspended Solids Dried at 103-105°C 2-75
      E. Fixed and Volatile Solids Ignited at 550°C 2-77
      F. Settleable Solids   2-78
      G. Total, Fixed, and Volatile Solids in Solid and Semi-
          solid Samples.   2-78
2550 TEMPERATURE   2-80
      A. Introduction   2-80
      B. Laboratory and Field Methods   2-80
2710 TESTS ON SLUDGES   2-81
      A. Introduction   2-81
      B. Oxygen-Consumption Rate   2-81
      C. Settled Sludge Volume   2-83
      D. Sludge Volume Index   2-84
      E. Zone Settling Rate   2-84
      F. Specific Gravity.   2-86
2720 SLUDGE DIGESTER GAS   2-87
      A. Introduction   2-87
      B. Volumetric Method   2-88
      C. Gas Chromatographic Method   2-91
2810 DISSOLVED GAS SUPERSATURATION.   2-94
      A. Introduction   2-94
      B. Direct-Sensing Membrane-Diffusion Method 2-94

Part 3000 DETERMINATION OF METALS
      3010 INTRODUCTION   3-1
            A. General Discussion   3-1
            B. Sampling and Sample Preservation. 3-1
            C. General Precautions.   3-3
      3020 QUALITY CONTROL.   3-3

PAGE

3030 PRELIMINARY TREATMENT OF SAMPLES   3-5
      A. Introduction   3-5
      B. Preliminary Filtration.   3-5
      C. Preliminary Treatment for Acid-Extractable Metals . 3-6
      D. Preliminary Digestion for Metals 3-6
      E. Nitric Acid Digestion. 3-8
      F. Nitric Acid-Hydrochloric Acid Digestion 3-8
      G. Nitric Acid-Sulfuric Acid Digestion 3-9
      H. Nitric Acid-Perchloric Acid Digestion 3-10
      I. Nitric Acid-Perchloric Acid-Hydrofluoric Acid
          Digestion. 3-11
      J. Dry Ashing.   3-11
3110 METALS BY ATOMIC ABSORPTION SPECTROMETRY . . .   3-12
3111 METALS BY FLAME ATOMIC ABSORPTION
      SPECTROMETRY   3-13
      A. Introduction   3-13
      B. Direct Air-Acetylene Flame Method. 3-20
      C. Extraction/Air-Acetylene Flame Method 3-23
      D. Direct Nitrous Oxide-Acetylene Flame Method . . . 3-25
      E. Extraction/Nitrous Oxide-Acetylene Flame Method . 3-27
3112 METALS BY COLD-VAPOR ATOMIC ABSORPTION
      SPECTROMETRY   3-28
      A. Introduction   3-28
      B. Cold-Vapor Atomic Absorption Spectrometric
          Method   3-29
3113 METALS BY ELECTROTHERMAL ATOMIC ABSORPTION
      SPECTROMETRY   3-32
      A. Introduction   3-32
      B. Electrothermal Atomic Absorption Spectrometric
          Method   3-36
3114 METALS BY HYDRIDE GENERATION/ATOMIC
      ABSORPTION SPECTROMETRY   3-43
      A. Introduction   3-43
      B. Manual Hydride Generation/Atomic Absorption
          Spectrometric Method 3-43
      C. Continuous Hydride Generation/Atomic Absorption
          Spectrometric Method (PROPOSED)   3-50
3120 METALS BY PLASMA EMISSION SPECTROSCOPY   3-53
      A. Introduction   3-53
      B. Inductively Coupled Plasma (ICP) Method 3-54
3500-Al ALUMINUM.   3-63
      A. Introduction   3-63
      B. Atomic Absorption Spectrometric Method 3-63

	PAGE
C. Inductively Coupled Plasma Method.	3-63
D. Eriochrome Cyanine R Method	3-64
E. Automated Pyrocatechol Violet (PCV) Method (PROPOSED)	3-68
3500-Sb ANTIMONY.	3-73
A. Introduction	3-73
B. Atomic Absorption Spectrometric Method	3-73
C. Inductively Coupled Plasma Method.	3-73
3500-As ARSENIC.	3-74
A. Introduction	3-74
B. Atomic Absorption Spectrometric Method	3-74
C. Silver Diethyldithiocarbamate Method	3-74
D. Mercuric Bromide Stain Method	3-76
E. Inductively Coupled Plasma Method.	3-78
3500-Ba BARIUM	3-78
A. Introduction	3-78
B. Atomic Absorption Spectrometric Method	3-78
C. Inductively Coupled Plasma Method.	3-79
3500-Be BERYLLIUM	3-79
A. Introduction	3-79
B. Atomic Absorption Spectrometric Method	3-79
C. Inductively Coupled Plasma Method.	3-79
D. Aluminon Method	3-80
3500-Bi BISMUTH.	3-81
A. Introduction	3-81
B. Atomic Absorption Spectrometric Method	3-82
3500-Cd CADMIUM	3-82
A. Introduction	3-82
B. Atomic Absorption Spectrometric Method	3-82
C. Inductively Coupled Plasma Method.	3-82
D. Dithizone Method	3-83
3500-Ca CALCIUM.	3-85
A. Introduction	3-85
B. Atomic Absorption Spectrometric Method	3-85
C. Inductively Coupled Plasma Method.	3-85
D. EDTA Titrimetric Method	3-85
E. Permanganate Titrimetric Method.	3-87
3500-Cs CESIUM	3-90
A. Introduction	3-90
B. Atomic Absorption Spectrometric Method	3-90
3500-Cr CHROMIUM.	3-90
A. Introduction	3-90
B. Atomic Absorption Method for Total Chromium. . .	3-91

PAGE

C. Inductively Coupled Plasma Method.   3-91
D. Colorimetric Method   3-91
3500-Co COBALT   3-94
A. Introduction   3-94
B. Atomic Absorption Spectrometric Method   3-94
C. Inductively Coupled Plasma Method.   3-94
3500-Cu COPPER   3-94
A. Introduction   3-94
B. Atomic Absorption Spectrometric Method   3-95
C. Inductively Coupled Plasma Method.   3-95
D. Neocuproine Method   3-95
E. Bathocuproine Method   3-97
3500-Au GOLD   3-99
A. Introduction   3-99
B. Atomic Absorption Spectrometric Method   3-99
3500-Ir IRIDIUM   3-100
A. Introduction 3-100
B. Atomic Absorption Spectrometric Method 3-100
3500-Fe IRON 3-100
A. Introduction 3-100
B. Atomic Absorption Spectrometric Method . . . ⁄ . . 3-102
C. Inductively Coupled Plasma Method. 3-102
D. Phenanthroline Method 3-102
3500-Pb LEAD 3-106
A. Introduction 3-106
B. Atomic Absorption Spectrometric Method 3-107
C. Inductively Coupled Plasma Method. 3-107
D. Dithizone Method 3-107
3500-Li LITHIUM 3-109
A. Introduction 3-109
B. Atomic Absorption Spectrometric Method 3-110
C. Inductively Coupled Plasma Method. 3-110
D. Flame Emission Photometric Method 3-110
3500-Mg MAGNESIUM 3-112
A. Introduction 3-112
B. Atomic Absorption Spectrometric Method 3-112
C. Inductively Coupled Plasma Method. 3-112
D. Gravimetric Method 3-113
E. Calculation Method. 3-114
3500-Mn MANGANESE 3-114
A. Introduction 3-114
B. Atomic Absorption Spectrometric Method 3-115
C. Inductively Coupled Plasma Method. 3-115
D. Persulfate Method 3-115

			PAGE
3500-Hg	MERCURY		3-118
	A.	Introduction	3-118
	B.	Cold Vapor Atomic Absorption Method	3-118
	C.	Dithizone Method	3-119
3500-Mo	MOLYBDENUM		3-120
	A.	Introduction	3-120
	B.	Atomic Absorption Spectrometric Method	3-121
	C.	Inductively Coupled Plasma Method	3-121
3500-Ni	NICKEL		3-121
	A.	Introduction	3-121
	B.	Atomic Absorption Spectrometric Method	3-121
	C.	Inductively Coupled Plasma Source Method	3-121
	D.	Heptoxime Method (GENERAL)	3-121
	E.	Dimethylglyoxime Method (GENERAL)	3-123
3500-Os	OSMIUM		3-123
	A.	Introduction	3-123
	B.	Atomic Absorption Spectrometric Method	3-123
3500-Pd	PALLADIUM		3-124
	A.	Introduction	3-124
	B.	Atomic Absorption Spectrometric Method	3-124
3500-Pt	PLATINUM		3-124
	A.	Introduction	3-124
	B.	Atomic Absorption Spectrometric Method	3-124
3500-K	POTASSIUM		3-125
	A.	Introduction	3-125
	B.	Atomic Absorption Spectrometric Method	3-125
	C.	Inductively Coupled Plasma Method	3-125
	D.	Flame Photometric Method	3-125
3500-Re	RHENIUM		3-127
	A.	Introduction	3-127
	B.	Atomic Absorption Spectrometric Method	3-127
3500-Rh	RHODIUM		3-127
	A.	Introduction	3-127
	B.	Atomic Absorption Spectrometric Method	3-127
3500-Ru	RUTHENIUM		3-128
	A.	Introduction	3-128
	B.	Atomic Absorption Spectrometric Method	3-128
3500-Se	SELENIUM		3-128
	A	Introduction	3-128
	B.	Sample Preparation	3-131
	C.	Continuous Hydride Generation/Atomic Absorption Spectrometric Method	3-134
	D.	Colorimetric Method	3-135
	E.	Fluorometric Method	3-137

PAGE

F. Determination of Volatile Selenium 3-137
G. Determination of Nonvolatile Organic Selenium
   Compounds 3-139
H. Electrothermal Atomic Absorption Spectrometric
   Method 3-141
I. Inductively Coupled Plasma Method. 3-141

3500-Ag   SILVER. 3-141
A. Introduction 3-141
B. Atomic Absorption Spectrometric Method 3-142
C. Inductively Coupled Plasma Method. 3-142
D. Dithizone Method 3-142

3500-Na   SODIUM 3-145
A. Introduction 3-145
B. Atomic Absorption Spectrometric Method 3-146
C. Inductively Coupled Plasma Method. 3-146
D. Flame Emission Photometric Method 3-146

3500-Sr   STRONTIUM 3-149
A. Introduction 3-149
B. Atomic Absorption Spectrometric Method 3-150
C. Inductively Coupled Plasma Method. 3-150
D. Flame Emission Photometric Method 3-150

3500-Tl   THALLIUM 3-152
A. Introduction 3-152
B. Atomic Absorption Spectrometric Method 3-152
C. Inductively Coupled Plasma Method. 3-152

3500-Th   THORIUM 3-153
A. Introduction 3-153
B. Atomic Absorption Spectrometric Method 3-153

3500-Sn   TIN 3-153
A. Introduction 3-153
B. Atomic Absorption Spectrometric Method 3-153

3500-Ti   TITANIUM 3-154
A. Introduction 3-154
B. Atomic Absorption Spectrometric Method 3-154

3500-V   VANADIUM. 3-154
A. Introduction 3-154
B. Atomic Absorption Spectrometric Method 3-154
C. Inductively Coupled Plasma Method. 3-154
D. Gallic Acid Method. 3-155

3500-Zn   ZINC. 3-157
A. Introduction 3-157
B. Atomic Absorption Spectrometric Method 3-157
C. Inductively Coupled Plasma Method. 3-157

			PAGE
	D.	Dithizone Method I.	3-158
	E.	Dithizone Method II	3-160
	F.	Zincon Method	3-162

Part 4000 DETERMINATION OF INORGANIC
NONMETALLIC CONSTITUTENTS

4010	INTRODUCTION	4-1
4020	QUALITY CONTROL	4-1
4110	DETERMINATION OF ANIONS BY ION CHROMATOGRAPHY	4-2
	A. Introduction	4-2
	B. Ion Chromatography with Chemical Suppression of Eluant Conductivity	4-2
4500-B	BORON	4-7
	A. Introduction	4-7
	B. Curcumin Method	4-7
	C. Carmine Method	4-10
	D. Inductively Coupled Plasma Method	4-11
4500-Br−	BROMIDE	4-11
	A. Introduction	4-11
	B. Phenol Red Colorimetric Method	4-11
	C. Ion Chromatographic Method	4-13
4500-CO$_2$	CARBON DIOXIDE	4-13
	A. Introduction	4-13
	B. Nomographic Determination of Free Carbon Dioxide and the Three Forms of Alkalinity	4-14
	C. Titrimetric Method for Free Carbon Dioxide	4-19
	D. Carbon Dioxide and Forms of Alkalinity by Calculation	4-19
4500-CN	CYANIDE	4-20
	A. Introduction	4-20
	B. Preliminary Treatment of Samples	4-25
	C. Total Cyanide after Distillation	4-28
	D. Titrimetric Method	4-30
	E. Colorimetric Method	4-31
	F. Cyanide-Selective Electrode Method	4-33
	G. Cyanides Amenable to Chlorination after Distillation	4-34
	H. Cyanides Amenable to Chlorination without Distillation (Short Cut Method)	4-36
	I. Weak and Dissociable Cyanide	4-38
	J. Cyanogen Chloride	4-39
	K. Spot Test for Sample Screening	4-40

PAGE

L. Cyanates   4-41
M. Thiocyanate   4-42

4500-Cl   CHLORINE (RESIDUAL)   4-45
A. Introduction   4-45
B. Iodometric Method I   4-48
C. Iodometric Method II.   4-51
D. Amperometric Titration Method.   4-54
E. Low-Level Amperometric Titration Method   4-57
F. DPD Ferrous Titrimetric Method   4-58
G. DPD Colorimetric Method   4-62
H. Syringaldazine (FACTS) Method   4-64
I. Iodometric Electrode Technique   4-65

4500-Cl⁻   CHLORIDE   4-67
A. Introduction   4-67
B. Argentometric Method   4-68
C. Mercuric Nitrate Method   4-69
D. Potentiometric Method   4-71
E. Automated Ferricyanide Method   4-73
F. Ion Chromatography Method   4-75

4500-ClO₂   CHLORINE DIOXIDE   4-75
A. Introduction   4-75
B. Iodometric Method   4-76
C. Amperometric Method I   4-78
D. DPD Method.   4-79
E. Amperometric Method II (PROPOSED).   4-80

4500-F⁻   FLUORIDE   4-84
A. Introduction   4-84
B. Preliminary Distillation Step   4-85
C. Ion-Selective Electrode Method   4-87
D. SPADNS Method.   4-89
E. Complexone Method   4-92

4500-H⁺   PH VALUE   4-94
A. Introduction   4-94
B. Electrometric Method.   4-95

4500-I   IODINE.   4-102
A. Introduction   4-102
B. Leuco Crystal Violet Method   4-103
C. Amperometric Titration Method.   4-105

4500-I⁻   IODIDE.   4-106
A. Introduction   4-106
B. Leuco Crystal Violet Method   4-107
C. Catalytic Reduction Method.   4-109

4500-N   NITROGEN   4-110

PAGE

4500-NH₃ NITROGEN (AMMONIA).   4-111
  A. Introduction   4-111
  B. Preliminary Distillation Step   4-115
  C. Nesslerization Method (Direct and Following
     Distillation)   4-117
  D. Phenate Method   4-120
  E. Titrimetric Method   4-121
  F. Ammonia-Selective Electrode Method   4-122
  G. Ammonia-Selective Electrode Method Using Known
     Addition   4-124
  H. Automated Phenate Method.   4-126

4500-NO₂⁻NITROGEN (NITRITE).   4-128
  A. Introduction   4-128
  B. Colorimetric Method   4-129
  C. Ion Chromatographic Method.   4-131

4500-NO₃⁻NITROGEN (NITRATE)   4-131
  A. Introduction   4-131
  B. Ultraviolet Spectrophotometric Screening Method . .   4-132
  C. Ion Chromatographic Method.   4-133
  D. Nitrate Electrode Method.   4-133
  E. Cadmium Reduction Method   4-135
  F. Automated Cadmium Reduction Method   4-137
  G. Titanous Chloride Reduction Method (PROPOSED)   4-140
  H. Automated Hydrazine Reduction Method
     (PROPOSED)   4-141

4500-N_org NITROGEN (ORGANIC)   4-143
  A. Introduction   4-143
  B. Macro-Kjeldahl Method.   4-144
  C. Semi-Micro-Kjeldahl Method   4-147

4500-O OXYGEN (DISSOLVED)   4-149
  A. Introduction   4-149
  B. Iodometric Methods   4-150
  C. Azide Modification   4-152
  D. Permanganate Modification   4-157
  E. Alum Flocculation Modification.   4-158
  F. Copper Sulfate-Sulfamic Acid Flocculation
     Modification   4-158
  G. Membrane Electrode Method   4-158

4500-O₃ OZONE (RESIDUAL) (PROPOSED)   4-162
  A. Introduction   4-162
  B. Indigo Colorimetric Method.   4-162

4500-P PHOSPHORUS   4-166
  A. Introduction   4-166

PAGE

B. Sample Preparation   4-170
C. Vanadomolybdophosphoric Acid Colorimetric
      Method   4-173
D. Stannous Chloride Method   4-175
E. Ascorbic Acid Method   4-177
F. Automated Ascorbic Acid Reduction Method   4-178

4500-Si   SILICA   4-181
A. Introduction   4-181
B. Atomic Absorption Spectrometric Method   4-182
C. Gravimetric Method   4-183
D. Molybdosilicate Method.   4-184
E. Heteropoly Blue Method   4-188
F. Automated Method for Molybdate-Reactive Silica . .   4-189
G. Inductively Coupled Plasma Method.   4-191

4500-S^{2-} SULFIDE   4-191
A. Introducton.   4-191
B. Separation of Soluble and Insoluble Sulfides   4-194
C. Sample Pretreatment to Remove Interfering
      Substances or to Concentrate the Sulfide.   4-194
D. Methylene Blue Method.   4-195
E. Iodometric Method   4-197
F. Calculation of Un-ionized Hydrogen Sulfide   4-198

4500-SO$_3^{2-}$ SULFITE   4-199
A. Introduction   4-199
B. Iodometric Method   4-200
C. Phenanthroline Method (PROPOSED).   4-201

4500-SO$_4^{2-}$ SULFATE   4-204
A. Introduction   4-204
B. Ion Chromatographic Method.   4-204
C. Gravimetric Method with Ignition of Residue   4-204
D. Gravimetric Method with Drying of Residue.   4-206
E. Turbidimetric Method   4-207
F. Automated Methylthymol Blue Method   4-208

Part   5000   DETERMINATION OF ORGANIC CONSTITUENTS
       5010   INTRODUCTION.   5-1
              A. General Discussion   5-1
              B. Sample Collection and Preservation   5-1
       5020   QUALITY CONTROL.   5-2
       5210   BIOCHEMICAL OXYGEN DEMAND   5-2
              A. Introduction   5-2
              B. 5-Day BOD Test   5-4
       5220   CHEMICAL OXYGEN DEMAND (COD)   5-10
              A. Introduction   5-10

PAGE

B. Open Reflux Method 5-12
C. Closed Reflux, Titrimetric Method 5-14
D. Closed Reflux, Colorimetric Method. 5-15

5310 TOTAL ORGANIC CARBON (TOC) 5-17
A. Introduction 5-17
B. Combustion-Infrared Method 5-18
C. Persulfate-Ultraviolet Oxidation Method 5-22
D. Wet-Oxidation Method 5-24

5320 DISSOLVED ORGANIC HALOGEN 5-26
A. Introduction 5-26
B. Adsorption-Pyrolysis-Titrimetric Method. 5-27

5510 AQUATIC HUMIC SUBSTANCES (PROPOSED) 5-37
A. Introduction 5-37
B. Diethylaminoethyl (DEAE) Method 5-38
C. XAD Method 5-40

5520 OIL AND GREASE. 5-41
A. Introduction 5-41
B. Partition-Gravimetric Method 5-43
C. Partition-Infrared Method. 5-44
D. Soxhlet Extraction Method 5-45
E. Extraction Method for Sludge Samples. 5-46
F. Hydrocarbons 5-47

5530 PHENOLS. 5-48
A. Introduction 5-48
B. Cleanup Procedure 5-50
C. Chloroform Extraction Method 5-51
D. Direct Photometric Method 5-54

5540 SURFACTANTS 5-55
A. Introduction 5-55
B. Surfactant Separation by Sublation. 5-56
C. Anionic Surfactants as MBAS. 5-59
D. Nonionic Surfactants as CTAS 5-64

5550 TANNIN AND LIGNIN. 5-67
A. Introduction 5-67
B. Colorimetric Method 5-68

5560 ORGANIC AND VOLATILE ACIDS 5-69
A. Introduction 5-69
B. Chromatographic Separation Method for Organic
Acids 5-70
C. Distillation Method. 5-72

5710 TRIHALOMETHANE FORMATION (PROPOSED) 5-73
A. Introduction 5-73
B. Trihalomethane Formation Potential (TFP) 5-75
C. Basic Trihalomethane Formation Potential (BTFP). . 5-79

PAGE

D. Ultimate Trihalomethane Formation Potential
(UTFP)   5-81

E. Simulated Distribution System Trihalomethane
Concentration (SDSTHMC).   5-82

Part 6000 AUTOMATED LABORATORY ANALYSES
6010 INTRODUCTION.   6-1
A. General Discussion   6-1
B. Sample Collection and Preservation   6-4
C. Analytical Methods.   6-6

6040 CONSTITUENT CONCENTRATION BY GAS EXTRACTION. .   6-10
A. Introduction   6-10
B. Closed-Loop Stripping, Gas-Chromatographic/Mass-
Spectrometric Analysis   6-11
C. Purge and Trap Technique   6-25

6210 VOLATILE ORGANICS   6-26
A. Introduction   6-26
B. Purge and Trap Packed-Column Gas Chromato-
graphic/Mass Spectrometric Method I.   6-27
C. Purge and Trap Packed-Column Gas Chromato-
graphic/Mass Spectrometric Method II   6-42
D. Purge and Trap Capillary-Column Gas Chromato-
graphic/Mass Spectrometric Method   6-50

6211 METHANE   6-60
A. Introduction   6-60
B. Combustible-Gas Indicator Method   6-61
C. Volumetric Method.   6-63

6220 VOLATILE AROMATIC ORGANICS   6-64
A. Introduction   6-64
B. Purge and Trap Gas Chromatographic Method I. . .   6-65
C. Purge and Trap Gas Chromatographic Method II . .   6-71
D. Purge and Trap Gas Chromatographic/Mass
Spectrometric Method   6-78
E. Liquid-Liquid Extraction Gas Chromatographic/
Mass Spectrometric Method.   6-78

6230 VOLATILE HALOCARBONS.   6-78
A. Introduction   6-78
B. Purge and Trap Packed-Column Gas Chromato-
graphic Method I.   6-79
C. Purge and Trap Packed-Column Gas Chromato-
graphic Method II   6-87
D. Purge and Trap Capillary-Column Gas Chromato-
graphic Method   6-93

   PAGE

     E. Purge and Trap Gas Chromatographic/Mass
         Spectrometric Method   6-97
6231 1,2-DIBROMOETHANE (EDB) AND 1,2-DIBROMO-3-CHLO-
     ROPROPANE (DBCP)   6-98
     A. Introduction   6-98
     B. Liquid-Liquid Extraction Gas Chromatographic
         Method   6-98
     C. Purge and Trap Gas Chromatographic/Mass
         Spectrometric Method   6-103
     D. Purge and Trap Gas Chromatographic Method . . .   6-103
6232 TRIHALOMETHANES   6-104
     A. Introduction   6-104
     B. Liquid-Liquid Extraction Gas Chromatographic
         Method   6-104
     C. Purge and Trap Gas Chromatographic/Mass
         Spectrometric Method   6-112
     D. Purge and Trap Gas Chromatographic Method . . .   6-112
6410 EXTRACTABLE BASE/NEUTRALS AND ACIDS   6-112
     A. Introduction   6-112
     B. Liquid-Liquid Extraction Gas Chromatographic/
         Mass Spectrometric Method.   6-113
6420 PHENOLS   6-137
     A. Introduction   6-137
     B. Liquid-Liquid Extraction Gas Chromatographic
         Method   6-137
     C. Liquid-Liquid Extraction Gas Chromatographic/
         Mass Spectrometric Method.   6-146
6431 POLYCHLORINATED BIPHENYLS (PCBS)   6-146
     A. Introduction   6-146
     B. Liquid-Liquid Extraction Gas Chromatographic
         Method   6-147
     C. Liquid-Liquid Extraction Gas Chromatographic/
         Mass Spectrometric Method.   6-147
6440 POLYNUCLEAR AROMATIC HYDROCARBONS   6-147
     A. Introduction   6-147
     B. Liquid-Liquid Extraction Chromatographic Method .   6-148
     C. Liquid-Liquid Extraction Gas Chromatographic/
         Mass Spectrometric Method.   6-158
6630 ORGANOCHLORINE PESTICIDES   6-158
     A. Introduction   6-158
     B. Liquid-Liquid Extraction Gas Chromatographic
         Method I.   6-158
     C. Liquid-Liquid Extraction Gas Chromatograhic
         Method II   6-170

PAGE

        D. Liquid-Liquid Extraction Gas Chromatographic/
           Mass Spectrometric Method.   6-181
  6640 CHLORINATED PHENOXY ACID HERBICIDES. 6-182
        A. Introduction 6-182
        B. Liquid-Liquid Extraction Gas Chromatographic
           Method 6-182

Part 7000 **EXAMINATION OF WATER AND WASTEWATER**
       FOR RADIOACTIVITY
  7010 INTRODUCTION.   7-1
        A. General Discussion   7-1
        B. Sample Collection and Preservation   7-3
        C. Counting Room   7-4
        D. Counting Instruments.   7-5
  7020 QUALITY ASSURANCE. 7-14
        A. Basic Quality Assurance Program 7-14
        B. Quality Assurance for Wastewater Samples 7-15
  7110 GROSS ALPHA AND GROSS BETA RADIOACTIVITY
       (TOTAL, SUSPENDED, AND DISSOLVED). 7-15
        A. Introduction 7-15
        B. Counting Method. 7-16
  7500-Cs RADIOACTIVE CESIUM 7-21
        A. Introduction 7-21
        B. Precipitation Method 7-22
  7500-I RADIOACTIVE IODINE 7-23
        A. Introduction 7-23
        B. Precipitation Method 7-24
        C. Ion-Exchange Method. 7-25
        D. Distillation Method 7-27
  7500-Ra RADIUM 7-29
        A. Introduction 7-29
        B. Precipitation Method 7-30
        C. Emanation Method 7-34
        D. Sequential Precipitation Method (PROPOSED) . . . 7-44
  7500-Sr TOTAL RADIOACTIVE STRONTIUM AND STRONTIUM
       90 7-47
        A. Introduction 7-47
        B. Precipitation Method 7-48
  7500-³H TRITIUM 7-53
        A. Introduction 7-53
        B. Liquid Scintillation Spectrometric Method 7-54
  7500-U URANIUM 7-56
        A. Introduction 7-56
        B. Radiochemical Method (PROPOSED). 7-57
        C. Fluorometric Method (PROPOSED). 7-59

PAGE

Part   8000   TOXICITY TEST METHODS FOR AQUATIC
                ORGANISMS
       8010   INTRODUCTION.   8-1
                A.   General Discussion   8-1
                B.   Terminology   8-3
                C.   Basic Requirements for Toxicity Tests   8-5
                D.   Conducting Toxicity Tests.   8-5
                E.   Preparing Organisms for Toxicity Tests   8-10
                F.   Toxicity Test Systems, Materials, and Procedures . .   8-23
                G.   Calculating, Analyzing, and Reporting Results of
                    Toxicity Tests   8-33
                H.   Interpreting and Applying Results of Toxicity Tests   8-38
       8030   MUTAGENICITY.   8-41
                A.   Introduction   8-41
                B.   *Salmonella* Test.   8-41
       8110   TEST PROCEDURES FOR ALGAE.   8-42
       8111   BIOSTIMULATION (ALGAL PRODUCTIVITY)   8-42
                A.   General Principles   8-42
                B.   Planning and Evaluating Algal Assays.   8-43
                C.   Apparatus   8-44
                D.   Sample Handling   8-45
                E.   Synthetic Algal Culture Medium   8-46
                F.   Inoculum.   8-46
                G.   Test Conditions and Procedures   8-47
                H.   Effect of Additions   8-50
                I.   Data Analysis and Interpretation   8-51
       8112   TOXICITY TESTING WITH PHYTOPLANKTON.   8-52
                A.   Introduction   8-52
                B.   Inoculum.   8-52
                C.   Test Conditions and Procedures   8-52
       8310   TOXICITY TEST PROCEDURES FOR CILIATED PROTOZOA   8-53
                A.   Introduction   8-53
                B.   Selecting and Preparing Test Organisms   8-54
                C.   Toxicity Test Procedures   8-55
                D.   Evaluating and Reporting Results   8-56
       8410   TOXICITY TEST PROCEDURES USING SCLERACTINIAN
                  CORAL.   8-56
                A.   Introduction   8-56
                B.   Selecting and Preparing Test Organisms   8-57
                C.   Toxicity Test Procedures   8-61
                D.   Evaluating and Reporting Results   8-63
       8510   TOXICITY TEST PROCEDURES FOR ANNELIDS   8-65
                A.   Introduction   8-65

PAGE

B. Selecting and Preparing Test Organisms 8-66
C. Toxicity Test Procedures 8-69
D. Data Evaluation 8-73
8610 TOXICITY TEST PROCEDURES USING MOLLUSKS. 8-73
A. Introduction 8-73
B. Selecting and Preparing Test Organisms 8-74
C. Conducting the Toxicity Tests. 8-77
D. Reporting and Analyzing Results 8-80
8710 MICROCRUSTACEANS 8-81
8711 TOXICITY TEST PROCEDURES FOR DAPHNIA. 8-81
A. Introduction 8-81
B. Selecting and Preparing Test Organisms 8-82
C. Toxicity Test Procedures 8-83
D. Evaluating and Reporting Results 8-85
8712 TOXICITY TEST PROCEDURES FOR THE CALANOID
COPEPOD, ACARTIA TONSA (DANA) 8-85
A. Introduction 8-85
B. Selecting and Preparing Test Organisms 8-85
C. Toxicity Test Procedures 8-90
D. Evaluating and Reporting Results 8-91
8720 TOXICITY TESTING PROCEDURES FOR
MACROCRUSTACEANS. 8-91
A. Introduction 8-91
B. Selecting and Preparing Test Species. 8-92
C. Conducting the Toxicity Tests. 8-104
D. Reporting Results. 8-113
8750 TOXICITY TEST PROCEDURES FOR AQUATIC INSECTS . . 8-113
A. Introduction 8-113
B. Selecting and Preparing Test Organisms 8-114
C. Toxicity Test Procedures 8-117
D. Data Evaluation 8-119
8910 TOXICITY TEST PROCEDURES FOR FISH 8-120
A. Introduction 8-120
B. Fish Selection and Preparation 8-120
C. Test Procedures. 8-133

Part 9000 MICROBIOLOGICAL EXAMINATION OF WATER
9010 INTRODUCTION. 9-1
A. General Discussion 9-1
B. U.S. EPA Regulations for Drinking Water Quality . . 9-3
9020 QUALITY ASSURANCE. 9-4
A. Introduction 9-4

			PAGE
	B.	Intralaboratory Quality Control Guidelines	9-5
	C.	Interlaboratory Quality Control	9-23
9030	LABORATORY APPARATUS		9-24
	A.	Introduction	9-24
	B.	Equipment Specifications	9-24
9040	WASHING AND STERILIZATION		9-28
9050	PREPARATION OF CULTURE MEDIA		9-29
	A.	General Procedures	9-29
	B.	Water	9-30
	C.	Media Specifications	9-31
9060	SAMPLES		9-31
	A.	Collection	9-31
	B.	Preservation and Storage	9-35
9211	RAPID DETECTION METHODS		9-36
	A.	Introduction	9-36
	B.	Seven-House Fecal Coliform Test (SPECIALIZED)	9-36
	C.	Special Techniques (SPECIALIZED)	9-37
	D.	Coliphage Detection (PROPOSED)	9-39
9212	STRESSED ORGANISMS		9-41
	A.	Introduction	9-41
	B.	Recovery Enhancement	9-43
9213	MICROBIOLOGICAL EXAMINATION OF RECREATIONAL WATERS		9-45
	A.	Introduction	9-45
	B.	Swimming Pools	9-46
	C.	Whirlpools	9-49
	D.	Natural Bathing Beaches	9-49
	E.	Membrane Filter Technique for *Pseudomonas aeruginosa*	9-52
	F.	Multiple-Tube Technique for *Pseudomonas aeruginosa*	9-53
9215	HETEROTROPHIC PLATE COUNT		9-54
	A.	Introduction	9-54
	B.	Pour Plate Method	9-58
	C.	Spread Plate Method	9-61
	D.	Membrane Filter Method	9-63
9216	DIRECT TOTAL MICROBIAL COUNT (PROPOSED)		9-64
	A.	Introduction	9-64
	B.	Epifluorescence Microscopic Method	9-65
9221	MULTIPLE-TUBE FERMENTATION TECHNIQUE FOR MEMBERS OF THE COLIFORM GROUP		9-66
	A.	Introduction	9-66
	B.	Standard Total Coliform Multiple-Tube (MPN) Fermentation Techniques	9-68

PAGE

C. Fecal Coliform MPN Procedure.   9-75
D. Estimation of Bacterial Density   9-77
E. Presence-Absence (P-A) Coliform Test.   9-80
9222 MEMBRANE FILTER TECHNIQUE FOR MEMBERS OF THE
COLIFORM GROUP   9-82
A. Introduction   9-82
B. Standard Total Coliform Membrane Filter Procedure   9-84
C. Delayed-Incubation Total Coliform Procedure   9-91
D. Fecal Coliform Membrane Filter Procedure   9-94
E. Delayed-Incubation Fecal Coliform Procedure   9-96
F. *Klebsiella* Membrane Filter Procedure   9-97
9225 DIFFERENTIATION OF THE COLIFORM BACTERIA   9-99
A. Introduction   9-99
B. Culture Purification.   9-100
C. Differentiation   9-100
D. Significance of Coliform Types   9-103
E. Media, Reagents, and Procedures   9-104
9230 FECAL STREPTOCOCCUS AND ENTEROCOCCUS GROUPS. .   9-108
A. Introduction   9-108
B. Multiple-Tube Technique   9-109
C. Membrane Filter Techniques   9-110
9240 IDENTIFICATION OF IRON AND SULFUR BACTERIA. . . .   9-114
A. Introduction   9-114
B. Iron Bacteria.   9-115
C. Sulfur Bacteria   9-118
D. Enumeration, Enrichment, and Isolation of Iron and
Sulfur Bacteria (PROPOSED).   9-121
9250 DETECTION OF ACTINOMYCETES   9-126
A. Introduction   9-126
B. Actinomycete Plate Count.   9-128
9260 DETECTION OF PATHOGENIC BACTERIA.   9-130
A. Introduction   9-130
B. General Qualitative Isolation and Identification
Procedures for *Salmonella*   9-131
C. Immunofluorescence Identification Procedure for
Salmonella   9-136
D. Quantitative *Salmonella* Procedures   9-140
E. *Shigella*   9-140
F. Pathogenic *Escherichia coli*   9-141
G. *Campylobacter jejuni*   9-143
H. *Vibrio cholerae*   9-145
I. Pathogenic Leptospires   9-147
J. Legionellaceae   9-149
K. *Yersinia enterocolitica*   9-153

                                                                    PAGE
9510 DETECTION OF ENTERIC VIRUSES.   9-155
      A. Introduction   9-155
      B. Virus Concentration from Small Sample Volumes by
          Adsorption to and Elution from Microporous
          Filters (PROPOSED).   9-159
      C. Virus Concentration from Large Sample Volumes by
          Adsorption to and Elution from Microporous
          Filters (PROPOSED).   9-163
      D. Virus Concentration by Aluminum Hydroxide
          Adsorption-Precipitation (PROPOSED)   9-170
      E. Hydroextraction-Dialysis with Polyethylene Glycol
          (PROPOSED)   9-173
      F. Recovery of Viruses from Suspended Solids in Water
          and Wastewater (PROPOSED)   9-175
      G. Assay and Identification of Viruses in Sample
          Concentrates (PROPOSED).   9-176
9610 DETECTION OF FUNGI   9-183
      A. Introduction   9-183
      B. Pour Plate Technique.   9-188
      C. Spread Plate Technique.   9-190
      D. Membrane Filter Technique.   9-191
      E. Technique for Yeasts   9-191
      F. Zoosporic Fungi   9-192
      G. Aquatic Hyphomycetes   9-194
      H. Fungi Pathogenic to Humans   9-195
9711 PATHOGENIC PROTOZOA   9-196
      A. Introduction   9-196
      B. *Giardia lamblia*.   9-197
      C. *Entamoeba histolytica*   9-203
9810 NEMATOLOGICAL EXAMINATION   9-204
      A. Introduction   9-204
      B. Technique for Nematodes.   9-206
      C. Illustrated Key to Freshwater Nematodes   9-208

Part 10000 **BIOLOGICAL EXAMINATION OF WATER**
     10010 INTRODUCTION.   10-1
     10200 PLANKTON.   10-3
            A. Introduction   10-3
            B. Sample Collection.   10-4
            C. Concentration Techniques.   10-16
            D. Preparing Slide Mounts   10-17
            E. Microscopes and Calibrations   10-19
            F. Phytoplankton Counting Techniques.   10-23

  PAGE

        G. Zooplankton Counting Techniques. 10-28
        H. Chlorophyll. 10-31
        I. Determination of Biomass (Standing Crop). 10-39
        J. Metabolic Rate Measurements. 10-42
10300 PERIPHYTON 10-48
        A. Introduction 10-48
        B. Sample Collection. 10-49
        C. Sample Analysis 10-51
        D. Productivity 10-54
        E. Interpreting and Reporting Results 10-66
10400 MACROPHYTON. 10-68
        A. Introduction 10-68
        B. Preliminary Survey 10-70
        C. Vegetation Mapping Methods 10-71
        D   Population Estimates 10-74
        E. Productivity 10-80
10500 BENTHIC MACROINVERTEBRATES 10-95
        A. Introduction 10-95
        B. Sample Collection. 10-96
        C. Sample Processing and Analysis 10-109
        D. Data Evaluation and Presentation 10-111
10600 FISH 10-113
        A. Introduction 10-113
        B. Data Acquisition 10-114
        C. Sample Preservation 10-127
        D. Analysis of Collection. 10-129
        E. Investigation of Fish Kills. 10-134
10900 IDENTIFICATION OF AQUATIC ORGANISMS. 10-136
        A. Procedure in Identification 10-137
        B. Key to Major Groups of Aquatic Organisms
             (Plates 1-38) 10-137
        C. List of Common Types of Aquatic Organisms
             (Plates 1-38), by Trophic Level 10-142
             Acknowledgements 10-143
        D. Key for Identification of Freshwater Algae Common
             in Water Supplies and Polluted Waters
             (Color Plates A-F) 10-183
        E. Recent Changes in Names of Algae 10-190
        F. Index to Illustrators 10-191
        G. Selected Taxonomic References 10-194
INDEX I-1

PAGE

FIGURES

1010:1	Normal and skewed distributions.	1-2
1020:1	Control charts for means.	1-10
1020:2	Duplicate analyses of a standard	1-11
1020:3	Range chart for variable concentrations.	1-11
1020:4	Range chart for variable ranges.	1-11
1020:5	Means control chart with out-of-control data (upper half)	1-12
1030:1	Definition of accuracy	1-14
1030:2	Detection limit relationship.	1-19
1060:1	Approximate number of samples required in estimating a mean concentration	1-36
1070:1	Ion-exchange column	1-46
2120:1	Filtration system for color determinations.	2-5
2120:2	Chromaticity diagrams.	2-7
2150:1	Odor-free-water generator	2-19
2150:2	End assembly of odor-free-water generator	2-20
2530:1	Floatables sampler with mixer	2-66
2530:2	Floatables flotation funnel and filter holder	2-67
2530:3	Floatation funnels and mixing unit	2-67
2530:4	Floatable oil tube, 1-L capacity.	2-69
2710:1	Schematic diagram of settling vessel for settled sludge volume test	2-83
2710:2	Schematic diagram of settling vessel for zone settling rate test.	2-85
2720:1	Gas collection apparatus	2-88
2810:1	Time response for the membrane-diffusion method . . .	2-96
3112:1	Schematic arrangement of equipment for measurement of mercury by cold-vapor atomic absorption technique	3-30
3114:1	Manual reaction cell for producing As and Se hydrides	3-45
3114:2	Schematic of a continuous hydride generator	3-51
3500-Al:1	Correction curves for estimation of aluminum in the presence of fluoride	3-67
3500-Al:2	Aluminum manifold for Automated System I (nonsegmented, flow-injected analyzer)	3-70
3500-Al:3	Aluminum manifold for Automated System II (segmented-flow analyzer)	3-71
3500-As:1	Arsine generator and absorber assembly	3-75
3500-As:2	Generator used with mercuric bromide stain method . .	3-77

PAGE

3500-Se:1   General scheme for speciation of selenium in water . . .   3-130

3500-Sr:1   Graphical method of computing strontium concentration   3-151

4110:1   Typical inorganic anion separation using normal-run
columns.   4-4

4110:2   Fast-run column separation   4-5

4500-CO$_2$:1   Nomograph for evaluation of hydroxide ion
concentration   4-15

4500:CO$_2$:2   Nomograph for evaluation of bicarbonate alkalinity . . .   4-16

4500:CO$_2$:3   Nomograph for evaluation of carbonate alkalinity   4-17

4500:CO$_2$:4   Nomograph for evaluation of free carbon dioxide
content   4-18

4500-CN:1   Cyanide distillation apparatus   4-29

4500-Cl$^-$:1   Example of differential titration curve (end point is 25.5
mL).   4-72

4500-Cl$^-$:2   Flow scheme for automated chloride analysis   4-74

4500-F$^-$:1   Direct distillation apparatus for fluoride   4-86

4500-F$^-$:2   Fluoride manifold   4-93

4500-H$^+$:1   Electrode potential vs. pH   4-96

4500-H$^+$:2   Typical pH electrode response as a function of
temperature   4-96

4500-NH$_3$:1   Ammonia manifold   4-127

4500-NO$_3$$^-$:1   Reduction column   4-136

4500-NO$_3$$^-$:2   Nitrate-nitrite manifold   4-138

4500-NO$_3$$^-$:3   Nitrate-nitrite manifold   4-142

4500-N$_{org}$:1   Micro-kjeldahl distillation apparatus   4-148

4500-O:1   DO and BOD sampler assembly   4-151

4500-O:2   Effect of temperature on electrode sensitivity   4-160

4500-O:3   The salting-out effect at different temperatures   4-160

4500-O:4   Typical trend of effect of stirring on electrode response   4-161

4500-P:1   Steps for analysis of phosphate fractions   4-168

4500-P:2   Phosphate manifold for automated analytical system . .   4-180

4500-Si:1   Silica manifold.   4-190

4500-S^{2-}:1   Analytical flow paths for sulfide determinations   4-193

4500-S^2-:2   Proportions of H$_2$S and HS$-$ in dissolved sulfide   4-199

4500-SO$_3$$^{2-}$:1   Apparatus for evolution of SO^2 from samples for
colorimetric analysis.   4-202

4500-SO$_4$$^{2-}$:1   Sulfate manifold   4-209

5230:1   DOX analysis system   5-30

5540:1   Sublation apparatus   5-57

		PAGE
6040:1	Schematic of closed-loop stripping apparatus	6-13
6040:2	One-liter "tall form" stripping bottle	6-13
6040:3	Gas Heater	6-14
6040:4	Extraction of filter	6-15
6040:5	Flow rate through 1.5-mg carbon filter	6-15
6040:6	Effect of filter resistance, measured as flow, on recovery of earthy-musty odorants and C1-C10 internal standard	6-16
6040:7	Mass spectra of 2-methylisoborneol under different instrumental conditions	6-21
6040:8	Mass spectrum of geosmin	6-22
6210:1	Purging device	6-29
6210:2	Trap packings and construction to include desorb capability	6-29
6210:3	Purge and trap system (purge mode)	6-31
6210:4	Purge and trap system (desorb mode)	6-31
6210:5	Gas Chromatogram of volatile organics	6-34
6210:6	Trap packings and construction to include desorb capability	6-43
6210:7	Gas chromatogram of volatile organics (wide-bore capillary column)	6-52
6210:8	Gas chromatogram of volatile organics (mega-bore capillary column)	6-54
6210:9	Chromatogram of test mixture (narrow-bore capillary column)	6-55
6211:1	Combustible gas indicator circuit and flow diagram	6-62
6220:1	Trap packings and construction to include desorb capability	6-66
6220:2	Purge and trap system (purge mode)	6-67
6220:3	Purge and trap system (dry mode)	6-68
6220:4	Purge and trap system (desorb mode)	6-68
6220:5	Gas chromatogram of purgeable aromatics	6-69
6220:6	Chromatogram of test mixture	6-72
6220:7	Chromatogram of test mixture	6-73
6230:1	Gas chromatogram of purgeable halobarbons	6-82
6230:2	Gas chromatogram of purgeable halocarbons	6-88
6230:3	Dual chromatogram of volatile organics (capillary column) with photoionization and electrolytic conductivity detectors	6-95
6231:1	Extract of reagent water with 0.114 μg/L added EDB and DBCP	6-101
6232:1	Chromatogram of finished water extract	6-107

PAGE

6232:2   Chromatogram of extract of standard.   6-108
6232:3   Chromatogram of extract of standard.   6-108
6410:1   Gas chromatogram of base/neutral fraction   6-123
6410:2   Gas chromatogram of acid fraction.   6-124
6410:3   Gas chromatogram of pesticide fraction.   6-124
6410:4   Gas chromatogram of chlordane   6-125
6410:5   Gas chromatogram of toxaphene   6-125
6410:6   Gas chromatogram of PCB-1016   6-126
6410:7   Gas chromatogram of PCB-1221   6-126
6410:8   Gas chromatogram of PCB-1232   6-127
6410:9   Gas chromatogram of PCB-1242   6-127
6410:10   Gas chromatogram of PCB-1248   6-128
6410:11   Gas chromatogram of PCB-1254   6-129
6410:12   Gas chromatogram of PCB-1260   6-130
6410:13   Tailing factor calculation.   6-131
6420:1   Gas chromatogram of phenols   6-141
6420:2   Gas chromatogram of PFB derivatives of phenols   6-143
6440:1   Liquid chromatogram of polynuclear aromatic
          hydrocarbons   6-153
6440:2   Liquid chromatogram of polynuclear aromatic
          hydrocarbons   6-154
6440:3   Gas chromatogram of polynuclear aromatic
          hydrocarbons   6-155
6630:1   Results of gas chromatogram procedure for
          organochlorine pesticide   6-161
6630:2   Results of gas chromatogram procedure for
          organochlorine pesticide   6-162
6630:3   Chromatogram of pesticide mixture.   6-163
6630:4   Chromatogram of pesticide mixture.   6-164
6630:5   Chromatogram of pesticide mixture.   6-165
6630:6   Gas chromatogram of pesticides   6-175
6630:7   Gas chromatogram of chlordane   6-175
6630:8   Gas chromatogram of toxaphene   6-176
6630:9   Gas chromatogram of PCB-1016   6-176
6630:10   Gas chromatogram of PCB-1221   6-177
6630:11   Gas chromatogram of PCB-1232   6-177
6630:12   Gas chromatogram of PCB-1242   6-178
6630:13   Gas chromatogram of PCB-1248   6-178
6630:14   Gas chromatogram of PCB-1254   6-179
6630:15   Gas chromatogram of PCB-1260   6-179
6640:1   Results of chromatographic procedure for chlorinated
          phenoxy acid herbicides   6-185
6640:2   Chromatogram of herbicide mixture   6-186

		PAGE
7010:1	Shape of counting rate—anode voltage curves.	7-7
7500-I:1	Distillation apparatus for iodine analysis	7-28
7500-Ra:1	De-emanation assembly	7-34
7500-Sr:1	Yttrium 90 vs. strontium 90 activity as a function of time.	7-51
8010:1	Holding tank design for fish and macroinvertebrates.	8-14
8010:2	Algal culture units.	8-19
8010:3	Method of lighting, free-standing frames, and placement of medium source, air pumps, and other apparatus for mass algal culture devices	8-20
8010:4	Basic components of flow-through system.	8-25
8010:5	Examples of median lethal concentration determination at two representative times by probit analysis and line of best fit	8-35
8010:6	Toxicity curve, drawn from LC50s determined in Figure 8010:5	8-36
8610:1	Diagram of constant-flow apparatus	8-79
8712:1	Algal culture system.	8-87
8712:2	Apparatus for mass copepod culture (static).	8-87
8712:3	Apparatus for mass copepod culture (flowing).	8-88
8712:4	Generation cage (after Heinle)	8-89
8720:1	Rearing and exposure beaker and automatic siphon for dungeness crab larvae	8-94
8720:2	Egg-hatching tank for lobsters	8-96
8720:3	Hughes lobster-rearing tank	8-97
9020:1	Frequency curve (positively skewed distribution)	9-21
9215:1	Preparation of dilution.	9-59
9215:2	Drying weight of 15-mL agar plates stored separately, inverted with lids on.	9-62
9215:3	Weight loss of 25-mL agar plates (100 × 15 mm) dried separately in a laminar-flow hood at room temperature (24 to 26°C), relative humidity (30 to 33%), and air velocity 0.6 m/s.	9-63
9221:1	Schematic outline of confirmed phase.	9-71
9221:2	Schematic outline of completed test for total coliform detection	9-72
9221:3	Schematic outline of optional P-A confirmations for various indicator organisms	9-83
9240:1	Filaments of Crenothrix polyspora showing variation of size and shape of cells within the sheath	9-116
9240:2	Filaments of Sphaerotilus natans showing cells within the filaments and some free "swarmer" cells	9-117

PAGE

9240:3   Laboratory culture of *Gallionella ferruginea*, showing cells, stalks excreted by cells, and branching of stalks where cells have divided   9-117

9240:4   Mixture of fragments of stalks of *Gallionella ferruginea*, and inorganic iron-manganese precipitate found in natural samples from wells.   9-118

9240:5   Single-celled iron bacterium *Siderocapsa treubii*   9-118

9240:6   Photosynthetic purple sulfur bacteria   9-119

9240:7   Colorless filamentous sulfur bacteria   9-120

9240:8   Colorless filamentous sulfur bacteria   9-120

9240:9   Colorless nonfilamentous sulfur bacteria   9-120

9250:1   Bacterial colonies   9-129

9510:1   Two-stage microporous filter adsorption-elution method for concentrating viruses from large volumes of water   9-164

9510:2   Schematic of apparatus for first-stage concentration . . .   9-165

9711:1   *Giardia* method flow chart   9-199

9711:2   *Giardia* sampling device schematic   9-200

9711:3   *Giardia* sampling device   9-200

9810:1   Life cycle of nematodes   9-205

10200:1   Structural features of common water samplers, Kemmerer and Van Dorn   10-7

10200:2   The Schindler-Patalas plankton trap   10-9

10200:3   Examples of commonly used plankton sampling nets . .   10-11

10200:4   Examples of commonly used high-speed zooplankton samplers   10-13

10200:5   Filter funnel for concentrating zooplankton   10-17

10200:6   Ocular micrometer ruling   10-21

10200:7   Calibration of Whipple Square   10-22

10200:8   Counting cell (Sedgwick-Rafter)   10-25

10200:9   A simple, efficient device for concentrating plankton . .   10-29

10200:10   The Folsom plankton splitter.   10-30

10200:11   Reverse-phase HPLC chromatogram for a fivefold dilution of EPA sample   10-37

10300:1   Periphyton sampler   10-50

10300:2   Component processes in the oxygen metabolism of a section of a hypothetical stream during the course of a cloudless day   10-58

10300:3   Gross periphytic primary production (P_G) determined by the O'Connell Thomas Chamber   10-62

10300:4   Calculation of gross primary production at a single station   10-63

10300:5   Calculation of gross periphytic primary productivity from upstream-downstream diurnal curves   10-63

PAGE

10400:1	Allen curve for a cohort of a population of aquatic macrophytes.	10-84
10500:1	Ponar grab	10-99
10500:2	Orange-peel sampler	10-99
10500:3	Peterson grab	10-100
10500:4	Van Veen grab.	10-100
10500:5	Smith-McIntyre grab.	10-101
10500:6	Shipek grab	10-101
10500:7	Ekman grab.	10-103
10500:8	Surber or square-foot sampler	10-103
10500:9	Phleger core sampler.	10-104
10500:10	KB corer	10-104
10500:11	Wilding or stovepipe sampler.	10-105
10500:12	Drift net sampler	10-105
10500:13	Hester-Dendy artificial substrate unit	10-107
10500:14	Basket sampler	10-107
10500:15	Dome sampler with serrated band and polyurethane cylinder band	10-108
10600:1	Diagram of a sunken trap net	10-118
10600:2	Bag seine in operation in a small stream	10-120
10600:3	Diagram of well-equipped electrofishing boat	10-121
10600:4	Types of tags commonly used	10-124
10600:5	Passive integrated transponder (PIT) tagging system.	10-125
10600:6	Key organs and external body parts of a soft-rayed and spiny-rayed fish	10-128
10600:7	Fish scale	10-131

TABLES

1010:I	Critical Values for 5% and 1% Tests of Discordancy for a Single Outlier in a Normal Sample	1-4
1020:I	Acceptance of Limits for Duplicate Samples and Known Additions to Water and Wastewater	1-8
1020:II	Factors for Computing Lines on Range Control Charts	1-10
1020:III	Audit of a Soil Analysis Procedure	1-13
1030:I	Precision Calculations Using Duplicates	1-15
1030:II	Precision Calculations from Known Additions	1-16
1050:I	Conversion Factors	1-27
1060:I	Summary of Special Sampling or Handling Requirements	1-37
1080:I	Reagent Water Specifications.	1-55
1080:II	Water Purification Process.	1-55

PAGE

1090:I   Threshold Limit Values for Solvents Specified in
              Standard Methods   1-64
1090:II   Threshold Limit Values for Reagents Specified in
              Standard Methods   1-65

2120:I   Selected Ordinates for Spectrophotometric Color
              Determinations   2-6
2120:II   Color Hues for Dominant Wavelength Ranges   2-8
2150:I   Threshold Odor Numbers Corresponding to Various
              Dilutions   2-21
2150:II   Dilutions for Various Odor Intensities   2-22
2160:I   Flavor Threshold Numbers Corresponding to Various
              Dilutions   2-25
2160:II   Dilutions for Determining the FTN   2-26
2320:I   Alkalinity Relationships   2-38
2330:I   Estimating Equilibrium Constants and Activity
              Coefficients   2-43
2330:II   Precalculated Values for pk and A at Selected
              Temperatures   2-44
2330:III   Graphs and Computer Software That Can Be Used to
              Calculate $CaCO_3$ Saturation Indices   2-50
2340:I   Maximum Concentrations of Interferences Permissible
              With Various Inhibitors   2-54
2510:I   Conductivity of Potassium Chloride Solutions at 25°C . . 2-60
2530:I   Coefficient of Variation and Recovery for Particulate
              Floatables Test   2-68
2710:I   Temperature Correction Factor   2-87
2810:I   Bunsen Coefficient for Oxygen in Fresh Water   2-98
2810:II   Vapor Pressure of Fresh Water   2-100

3030:I   Acids Used in Conjunction with HNO_3 for Sample
              Preparation   3-7
3111:I   Atomic Absorption Concentration Ranges with Direct
              Aspiration Atomic Absorption   3-16
3111:II   Interlaboratory Precision and Bias Data for Atomic
              Absorption Method—Direct Aspiration and Extracted
              Metals   3-18
3111:III   Single-Operator Precision and Recommended Control
              Ranges for Atomic Absorption Methods—Direct
              Aspiration and Extracted Metals   3-19
3112:I   Interlaboratory Precision and Bias of Cold-Vapor
              Atomic Absorption Spectrometric Method for
              Mercury   3-31

PAGE

3113:I   Potential Matrix Modifiers for Electrothermal Atomic Absorption Spectrometry. 3-33

3113:II   Detection Levels and Concentration Ranges for Electrothermal Atomization Atomic Absorption Spectrometry 3-35

3113:III   Interlaboratory Single-Analyst Precision Data for Electrothermal Atomization Methods. 3-40

3113:IV   Interlaboratory Overall Precision Data for Electrothermal Atomization Methods. 3-41

3113:V   Interlaboratory Relative Error Data for Electrothermal Atomization Methods 3-42

3120:I   Suggested Wavelengths, Estimated Detection Limits, Alternate Wavelengths, Calibration Concentrations, and Upper Limits 3-55

3120:II   ICP Precision and Bias Data 3-61

3500-Al:I   Single-Operator Precision and Bias for Inorganic Monomeric Aluminum Analysis 3-72

3500-Al:II   Single-Operator Precision and Bias High-Level Aluminum Determination 3-73

3500-FE:I   Selection of Light Path Length for Various Iron Concentrations 3-104

3500-V:I   Concentration at Which Various Ions Interfere in the Determination of Vanadium 3-155

4110:I   Precision and Bias Observed for Anions at Various Concentration Levels in Reagent Water. 4-6

4500-ClO₂:I   Equivalent Weights for Calculating Concentrations on the Basis of Mass 4-83

4500-F⁻:I   Concentration of Some Substances Causing 0.1-mg/L Error at 1.0 mg-F/L in Fluoride Methods 4-85

4500-H⁺:I   Preparation of pH Standard Solutions 4-98

4500-H⁺:II   Standard pH Values 4-99

4500-NH₃:I   Precision and Bias Data for Ammonia Methods. . . . 4-112

4500-NH₃:II   Precision and Bias of Ammonia-Selective Electrode . . . 4-114

4500-NH₃:III   Preparation of Permanent Color Standards for Visual Determination of Ammonia Nitrogen. 4-119

4500-NH₃:IV   Values of Q vs. ΔE (59 mv Slope) for 10% Volume Change 4-125

4500-N_org:I   Precision and Bias for Organic Nitrogen, Macro-Kjeldahl Procedure 4-147

4500-O:I   Solubility of Oxygen in Water Exposed to Water-Saturated Air at Atmosphere Pressure (101.3 kPa) . . 4-154

4500-P:I   Precision and Bias Data for Manual Phosphorus Methods 4-169

		PAGE
4500-P:II	Comparison and Bias of Ascorbic Acid Methods	4-178
4500-Si:I	Selection of Light Path Length for Various Silica Concentrations	4-185
4500-Si:II	Preparation of Permanent Color Standards for Visual Determination of Silica	4-187
4500-S^{2-}:I	Values of pK′, Logarithm of Practical Ionization Constant for Hydrogen Sulfide	4-198

5220:I	Sample and Reagent Quantities for Various Digestion Vessels	5-15
5310:I	Precision and Bias for Total Organic Carbon (TOC) by Persulfate-Ultraviolet Oxidation	5-24
5320:I	Intralaboratory, Single-Operator, Dissolved Organic Halogen (Microcolumn Procedure)—Precision and Bias Data	5-36
5540:I	Surfactant Recovery by Sublation	5-59
5710:I	Single-Operator Precision and Bias Data for TFP	5-78
5710:II	Single-Operator Precision and Bias Data For BTFP	5-80
5710:III	Single-Operator Precision of Filtered and Diluted River-Water Samples Analyzed for BTFP	5-80

6010:I	Analysis Methods for Specific Organic Compounds	6-1
6010:II	Recommended Preservation for Volatiles	6-5
6040:I	Instrumental Detection Limits for Earthy-Musty Smelling Compounds by CLSA-GC/MS	6-12
6040:II	Instrumental Detection Limits for Selected Organic Compounds by CLSA-GC/MS	6-12
6040:III	Typical Operating Conditions for GC/MS Analysis of CLSA Extracts	6-20
6040:IV	GC/MS Data for Three Internal Standards and Two Earthy-Musty Smelling Compounds	6-21
6040:V	Single-Laboratory Bias for Selected Organic Compounds Causing Taste and Odor	6-23
6040:VI	Precision Data for Selected Organic Compounds Causing Taste and Odor	6-24
6040:VII	Recovery and Precision Data for Selected Priority Pollutants	6-25
6210:I	Chromatographic Conditions and Method Detection Limits (MDL)	6-28
6210:II	BFB Key m/z Abundance Criteria	6-30
6210:III	Suggested Surrogate and Internal Standards	6-33
6210:IV	Calibration and QC Acceptance Criteria	6-36
6210:V	Characteristic Masses for Purgeable Organics	6-38

PAGE

6210:VI   Method Bias and Precision as Functions of
Concentration   6-40

6210:VII   Chromatographic Conditions for Volatile Organic
Compounds on Packed Columns   6-45

6210:VIII   Characteristic Masses (m/z) for Purgeable Organics . . .   6-47

6210:IX   Single-Laboratory Bias and Precision Data for Volatile
Organic Compounds in Reagent Water (Packed
Column)   6-49

6210:X   Chromatographic Conditions for Volatile Organic
Compounds on Wide-Bore Capillary Columns   6-51

6210:XI   Chromatographic Conditions for Volatile Organic
Compounds on Narrow-Bore Capillary Column. . . .   6-53

6210:XII   Single-Laboratory Bias and Precision Data for Volatile
Organic Compounds in Reagent Water (Wide-Bore
Capillary Column).   6-56

6210:XIII   Single-Laboratory Bias and Precision Data for Volatile
Organic Compounds in Reagent Water (Narrow-Bore
Capillary Column).   6-58

6220:I   Chromatographic Conditions and Method Detection
Limits.   6-66

6220:II   Calibration and QC Acceptance Criteria   6-70

6220:III   Method Bias and Precision as Functions of
Concentration   6-70

6220:IV   Retention Times for Aromatic Compounds   6-74

6220:V   Single-Laboratory Bias and Precision for Aromatic and
Unsaturated Compounds in Chlorinated Drinking
Water and Raw Source Water   6-76

6220:VI   Single-Analyst Precision, Overall Precision, and Bias for
Purgeable Aromatics in Drinking Water   6-77

6220:VII   Bias and Precision Data for Purgeable Aromatics from
Multilaboratory Performance Evaluation Studies . . .   6-77

6230:I   Chromatographic Conditions and Methods Detection
Limits (MDL).   6-80

6230:II   Calibration and QC Acceptance Criteria   6-83

6230:III   Method Bias and Precision as Functions of
Concentration   6-85

6230:IV   Retention Times for Organohalides   6-89

6230:V   Single-Laboratory Bias and Precision for Organohalides
in Ohio River Water and Drinking Water   6-91

6230:VI   Bias, Single-Analyst Precision, and Overall Precision for
Organohalides in Drinking Water   6-92

6230:VII   Retention Times for Volatile Organic Compounds on
Photoionization Detector (PID) and Electrolytic
Conductivity Detector (ECD)   6-93

PAGE

6230:VIII   Single-Laboratory Bias and Precision for Volatile
             Organic Compounds in Reagent Water   6-96
6231:I   Chromatographic Conditions for 1,2-Dibromoethane
             (EDB) and 1,2-Dibromo-3-Chloropropane (DPCP) . .   6-100
6231:II   Single-Laboratory Precision and Bias for EDB and
             DBCP in Tap Water.   6-102
6232:I   Retention Times for Trihalomethanes.   6-108
6232:II   Single-Laboratory Precision and Bias   6-112
6410:I   Chromatographic Conditions, Methods Detection
             Limits, and Characteristic Masses for Base/Neutral
             Extractables   6-115
6410:II   Chromatographic Conditions, Method Detection Limits,
             and Characteristic Masses for Acid Extractables . . .   6-118
6410:III   DFTPP Key Masses and Abundance Criteria   6-120
6410:IV   Suggested Internal and Surrogate Standards.   6-121
6410:V   QC Acceptance Criteria   6-132
6410:VI   Method Bias and Precision as Functions of
             Concentration   6-134
6420:I   Chromatographic Conditions and Method Detection
             Limits.   6-138
6420:II   Silica Gel Fractionation and Electron Capture Gas
             Chromatography of PFBB Derivatives   6-139
6420:III   QC Acceptance Criteria   6-144
6420:IV   Method Bias and Precision as Functions of
             Concentration   6-145
6440:I   High-Performance Liquid Chromatography Conditions
             and Method Detection Limits   6-149
6440:II   Gas Chromatographic Conditions and Retention Times   6-151
6440:III   QC Acceptance Criteria   6-156
6440:IV   Method Bias and Precision as Functions of
             Concentration   6-157
6630:I   Retention Ratios of Various Organochlorine Pesticides
             Relative to Aldrin   6-168
6630:II   Precision and Bias Data for Selected Organochlorine
             Pesticides   6-169
6630:III   Chromatographic Conditions and Method Detection
             Limits.   6-171
6630:IV   Distribution of Chlorinated Pesticides and PCBs into
             Magnesia-Silica Gel Column Fractions   6-174
6630:V   QC Acception Criteria   6-180
6630:VI   Method Precision and Bias as Functions of
             Concentration   6-181

PAGE

6640:I   Retention Times for Methyl Esters of Some Chlorinated
Phenoxy Acid Herbicides Relative to 2,4-D Methyl
Ester 6-186

6640:II   Precision of Phenoxy Acid Herbicides from Dosed
Surface Water 6-187

6640:III   Recovery of Phenoxy Acid Herbicides from Dosed
Surface Water 6-187

7010:I   Usual Distribution of Common Radioelements between
the Solid and Liquid Phases of Wastewater 7-3

7500:Ra:I   Chemical and Radiochemical Composition of Samples
Used to Determine Bias and Precision of Radium 226
Method 7-33

7500-Ra:II Factors for Decay of Radon 222, Growth of Radon 222
from Radium 226, and Correction of Radon 222
Activity for Decay During Counting 7-38

8010:I   Recommended Composition for Reconstituted Fresh
Water 8-16

8010:II   Quantities of Reagent-Grade Chemicals to be Added to
Aerated Soft Reconstituted Fresh Water for Buffering
pH 8-16

8010:III   Procedure for Preparing Reconstituted Seawater 8-17

8010:IV.A. Macronutrient Stock Solution 8-18

8010:IV.B. Micronutrient Stock Solution 8-18

8010:V   Nutrients for Algal Culture Medium in Seawater 8-19

8010:VI   Percentage of Ammonia Un-ionized in Distilled Water . 8-29

8010:VII   Experimental Data from Hypothetical Toxicity Test
Subjected to Probit Analysis 8-34

8712:I   Composition of Algal Diet and Recommended
Concentration for Adult and Naupliar Feeding and
Egg-Laying 8-88

8910:I   Recommended Prophylactic and Therapeutic for
Freshwater Fish to be Used for Experimental
Purposes 8-124

8910:II   Evansville, Indiana, Photoperiod 8-126

9020:I   Quality of Purified Water Used in Microbiology Testing 9-10

9020:II   Reagent Additions for Water Quality Test 9-11

9020:III   Time and Temperature for Autoclave Sterilization 9-16

9020:IV   Holding Times for Prepared Media 9-17

9020:V   Control Cultures for Microbiological Tests 9-18

9020:VI   Calculation of Precision Criterion 9-18

		PAGE
9020:VII	Daily Checks on Precision of Duplicate Counts	9-19
9020:VIII	Coliform Counts and Their Logarithms.	9-21
9020:IX	Comparison of Frequency of MPN Data	9-22
9020:X	Comparison of Frequency of Log MPN Data	9-22
9211:I	Special Rapid Techniques	9-38
9215:I	Effect of Temperature of Drying on Weight Loss of 15	
	ml Agar Plates Stored Separately.	9-62
9221:I	Preparation of Lauryl Tryptose Broth	9-69
9221:II	Preparation of Lactose Broth.	9-70
9221:III	MPN Index and 95% Confidence Limits for Various	
	Combination of Positive and Negative Results When	
	Five 10-mL Portions are Used	9-77
9221:IV	MPN Index and 95% Confidence Limits for Various	
	Combinations of Positive and Negative Results When	
	Ten 10-mL Portions Are Used	9-77
9221:V	MPN Index and 95% Confidence Limits for Various	
	Combinations of Positive Results When Five Tubes	
	are Used per Dilution (10 mL, 1.0 mL, 0.1 mL) . . .	9-78
9222:I	Suggested Sample Volumes for Membrane Filter Total	
	Coliform Test	9-88
9222:II	95% Confidence Limits for Membrane Filter Coliform	
	Results Using 100-mL Sample	9-91
9222:III	Suggested Sample Volumes for Membrane Filter Fecal	
	Coliform Test	9-95
9225:I	Nomenclature of Enterobacteriaceae	9-101
9225:II	Differentiation of Potential Indicator Organisms of the	
	Enterobacteriaceae.	9-102
9230:I	Selected Key Biochemical Characteristics of the	
	Streptococcus Species Within the Fecal Streptococcus	
	and Enterococcus Groups	9-114
9250:I	General Macroscopic Properties of Bacterial Colonies on	
	Solid Medium	9-130
10200:I	Characteristics of Commonly Used Plankton Nets. . . .	10-10
10200:II	Conversion Table for Membrane Filter Technique. . . .	10-27
10300:I	Sample Calculation Ledger for Computation of	
	Corrected Rate of Oxygen Change from a Single-	
	Station Diurnal Curve	10-64
10300:II	Sample Calculation Ledger for Computation of	
	Corrected Rates of Oxygen Change from the	
	Upstream-Downstream Diurnal Curve of Oxygen	
	Concentration and Temperature	10-64
10400:I	Methods Used to Determine Macrophyte Production . .	10-81

PLATES

Black and White plates of aquatic organisms

1. Blue-green algae: Coccoid (Phylum Cyanophyta) 10-145
2. Blue-green algae: Filamentous (Phylum Cyanophyta) 10-146
3. Nonmotile green algae: Coccoid (Phylum Chlorophyta) 10-147
4. Nonmotile green algae: Filamentous (Phylum Chlorophyta) 10-148
5. Diatoms: Pennate (Phylum Chrysophyta, Class
   Bacillariophyceae) 10-149
6. Diatoms: Centric (Phylum Chrysophyta, Class
   Bacillariophyceae) 10-150
7. Types of larger marine algae (green, brown, and red) 10-151
8. Higher plants: Floating plants 10-152
9. Higher plants: Submersed (all forms illustrated are
   Spermatophytes) 10-153
10. Higher plants: Emersed (all forms illustrated are
    Spermatophytes) 10-154
11. Pigmented flagellates: Single-celled (various phyla) 10-155
12. Pigmented flagellates: Colonial types (various phyla) 10-156
13. Nonpigmented flagellates (Phylum Protozoa) 10-157
14. Amoebas (Phylum Protozoa). (a) Amoeboid stages, (b)
    flagellated stages, (c) cyst stages 10-158
15. Ciliates (Phylum Protozoa) 10-159
16. Sponges (Phylum Porifera) and Bryozoans (Phylum Bryozoa) 10-160
17. Rotifers (Phylum Rotatoria) 10-161
18. Roundworms (Phylum Nemathelminthes) 10-162
19. Flatworms (Phylum Platyhelminthes) and segmented worms
    (Phylum Annelida) 10-163
20. Crustaceans (Phylum Arthropoda, Class Crustacea): Types of
    cladocerans (Order Cladocera) 10-164
21. Crustaceans (Phylum Arthropoda, Class Crustacea): Selected
    common types 10-165
22. Types of insect pupae 10-166
23. Stoneflies (Order Plecoptera) 10-167
24. Mayflies (Order Ephemeroptera) 10-168
25. Damselflies, dragonflies (Order Odonata) 10-169
26. Hellgrammite and relatives 10-170
27. Caddisflies (Order Trichoptera) 10-171
28. Two-winged flies (Order Diptera) 10-172
29. Two-winged flies (Order Diptera) 10-173
30. Beetles (Order Coleoptera) 10-174

PAGE

31. True bugs (Order Hemiptera, all adults) 10-175
32. Mollusks (Phylum Mollusca): Snails (Class Gastropoda) 10-176
33. Mollusks (Phylum Mollusca): Bivalves (Class Pelecypoda) 10-177
34. Echinoderm types (Phylum Echinodermata, all marine). 10-178
35. Miscellaneous invertebrates 10-179
36. Some types of fishes (Phylum Chordata) 10-180
37. Types of amphibians (Phylum Chordata, Class Amphibia) 10-181
38. Bacteria and Fungi 10-182

Color plates of blue-green algae (special color section) follows 10-200

A. Taste and odor algae
B. Filter clogging algae
C. Polluted water algae
D. Clean water algae
E. Plankton and other surface water algae
F. Algae growing on resrvoir walls

PART 1000

GENERAL

INTRODUCTION

1010   INTRODUCTION*

1010 A.   Scope and Application of Methods

The procedures described in these standards are intended for the examination of waters of a wide range of quality, including water suitable for domestic or industrial supplies, surface water, ground water, cooling or circulating water, boiler water, boiler feed water, treated and untreated municipal or industrial wastewater, and saline water. The unity of the fields of water supply, receiving water quality, and wastewater treatment and disposal is recognized by presenting methods of analysis for each constituent in a single section for all types of waters.

An effort has been made to present methods that apply generally. Where alternative methods are necessary for samples of different composition, the basis for selecting the most appropriate method is presented as clearly as possible. However, samples with extreme concentrations or otherwise unusual compositions may present difficulties that preclude the direct use of these methods. Hence, some modification of a procedure may be necessary in specific instances. Whenever a procedure is modified, the analyst should state plainly

the nature of modification in the report of results.

Certain procedures are intended for use with sludges and sediments. Here again, the effort has been to present methods of the widest possible application, but when chemical sludges or slurries or other samples of highly unusual composition are encountered, the methods of this manual may require modification or may be inappropriate.

Most of the methods included here have been endorsed by regulatory agencies. Procedural modification without formal approval may be unacceptable to a regulatory body.

The analysis of bulk chemicals received for water treatment is not included herein. A committee of the American Water Works Association prepares and issues standards for water treatment chemicals.

Part 1000 contains information that is common to, or useful in, laboratories desiring to produce analytical results of known quality, that is, of known accuracy and with known uncertainty in that accuracy. To accomplish this, apply the quality assurance methods described herein to the standard methods described elsewhere in this publication. Other sections of Part 1000 address laboratory equipment, laboratory safety, sampling procedures, and method development and validation, all of which provide necessary information.

---

* Part 1000 was submitted to the Standard Methods Committee on a "Review and Comment" basis and is provided for information purposes only. It is not considered a part of the procedures described in these standards.

1010 B.　Statistics

1. Normal Distribution

If a measurement is repeated many times under essentially identical conditions, the results of each measurement, x, will be distributed randomly about a mean value (arithmetic average) because of uncontrollable or experimental error. If an infinite number of such measurements were to be accumulated, the individual values would be distributed in a curve similar to those shown in Figure 1010:1. The curve at left illustrates the Gaussian or normal distribution, which is described precisely by the mean, μ, and the standard deviation, σ. The mean, or average, of the distribution is simply the sum of all values divided by the number of values so summed, i.e., $\mu = (\Sigma_i x_i)/n$. Because no measurements are repeated an infinite number of times, an *estimate* of the mean is made, using the same summation procedure but with n equal to a finite number of repeated measurements (10, or 20, or. . .). This estimate of μ is denoted by $\overline{x}$. The standard deviation of the normal distribution is defined as $\sigma = [\Sigma(x-\mu)^2/n]^{\frac{1}{2}}$. Again, the analyst can only estimate the standard deviation because the number of observations made is finite; the estimate of σ is denoted by s

and is calculated as follows:

$$s = [\Sigma(x-\overline{x})^2/n-1]^{\frac{1}{2}}$$

The standard deviation fixes the width, or spread, of the normal distribution, and also includes a fixed fraction of the values making up the curve. For example, 68.27% of the measurements lie between $\mu \pm 1\sigma$, 95.45% between $\mu \pm 2\sigma$, and 99.70% between $\mu \pm 3\sigma$. It is sufficiently accurate to state that 95% of the values are within $\pm 2\sigma$ and 99% within $\pm 3\sigma$. When values are assigned to the $\pm \sigma$ multiples, they are confidence limits. For example, 10 ± 4 indicates that the confidence limits are 6 and 14, while values from 6 to 14 represent the confidence interval.

Another useful statistic is the standard error of the mean, σ_μ, which is the standard deviation divided by the square root of the number of values, or $\sigma/\sqrt{n}$. This is an estimate of the accuracy of the mean and implies that another sample from the same population would have a mean within some multiple of this. Multiples of this statistic include the same fraction of the values as stated above for σ. In practice, a relatively small number of average values is available, so the confidence intervals of the mean are

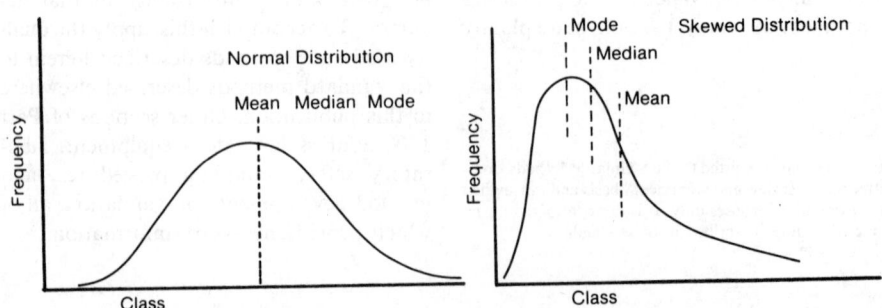

Figure 1010:1. Normal and skewed distributions.

expressed as $\bar{x} \pm ts/\sqrt{n}$ where t has the following values for 95% confidence intervals:

n	t
2	12.71
3	4.30
4	3.18
5	2.78
10	2.26
∞	1.96

The use of t compensates for the tendency of a small number of values to underestimate uncertainty. For $n > 15$, it is common to use $t = 2$ to estimate the 95% confidence interval.

Still another statistic is the relative standard deviation, also known as the coefficient of variation (CV), which commonly is expressed as a percentage. This statistic is calculated as $100\ \sigma/\mu$ and its estimate is $100\ s/\bar{x}$. This statistic normalizes the standard deviation and sometimes facilitates making direct comparisons among analyses that include a wide range of concentrations. For example, if analyses at low concentrations yield a result of 10 ± 1.5 mg/L and at high concentrations 100 ± 8 mg/L, the standard deviations do not appear comparable. However, the relative standard deviations are 100 (1.5/10) = 15% and 100 (8/100) = 8%, which indicate the smaller variability obtained by using this parameter.

2. Log-Normal Distribution

In many cases the results obtained from analysis of environmental samples will not be normally distributed, i.e., a graph of the data will be obviously skewed, as shown at right in Figure 1010:1, with the mode, median, and mean being distinctly different. To obtain a nearly normal distribution, convert the results to logarithms and then calculate $\bar{x}$ and s. The antilogarithms of these two values are estimates of the geometric mean and the geometric standard deviation, $\bar{x}_g$ and s_g.

3. Rejection of Data

Quite often in a series of measurements, one or more of the results will differ greatly from the other values. Theoretically, no result should be rejected, because it may indicate either a faulty technique that casts doubt on all results or the presence of a true variant in the distribution. In practice, reject the result of any analysis in which a known error has occurred. In environmental studies, extremely high and low concentrations of contaminants may indicate the existence of areas with problems or areas with no contamination, so they should not be rejected arbitrarily.

An objective test for outliers has been described.[1] If a set of data is ordered from low to high: $x_L, x_2 \ldots x_H$, and the average and standard deviation are calculated, then suspected high or low outliers can be tested by the following procedure. First, calculate the statistic T:

$$T = (x_H - \bar{x})/s \text{ for a high value, or}$$
$$T = (\bar{x} - x_L)/s \text{ for a low value.}$$

Second, compare the value of T with the value from Table 1010:I for either a 5% or 1% level of significance. If the calculated T is larger than the table value for the number of measurements, n, then the x_H or x_L is an outlier at that level of significance.

Further information on statistical techniques is available elsewhere.[2,3]

4. References

1. BARNETT, V. & T. LEWIS. 1984. Outliers in Statistical Data. John Wiley & Sons, New York, N.Y.
2. NATRELLA, M.G. 1966. Experimental Statistics. National Bur. Standards Handbook 91, Washington, D.C.
3. SNEDECOR, G.W. & W.G. COCHRAN. 1980. Statistical Methods. Iowa State University Press, Ames.

TABLE 1010:I. CRITICAL VALUES FOR 5% AND 1% TESTS OF DISCORDANCY FOR A SINGLE
OUTLIER IN A NORMAL SAMPLE

Number of Measurements n	Critical Value	
	5%	1%
3	1.15	1.15
4	1.46	1.49
5	1.67	1.75
6	1.82	1.94
7	1.94	2.10
8	2.03	2.22
9	2.11	2.32
10	2.18	2.41
12	2.29	2.55
14	2.37	2.66
15	2.41	2.71
16	2.44	2.75
18	2.50	2.82
20	2.56	2.88
30	2.74	3.10
40	2.87	3.24
50	2.96	3.34
60	3.03	3.41
100	3.21	3.60
120	3.27	3.66

Source: BARNETT, V. & T. LEWIS. 1984. Outliers in Statistical Data. John Wiley & Sons, New York, N.Y.

1010 C. Glossary

1. Definition of Terms

Accuracy—combination of bias and precision of an analytical procedure, which reflects the closeness of a measured value to a true value (see Figure 1030:1).

Bias—consistent deviation of measured values from the true value, caused by systematic errors in a procedure.

Calibration check standard—standard used to determine the state of calibration of an instrument between periodic recalibrations.

Confidence coefficient—the probability, %, that a measurement result will lie within the confidence interval or between the confidence limits.

Confidence interval—set of possible values within which the true value will lie with a specified level of probability.

Confidence limit—one of the boundary values defining the confidence interval.

Detection limits—Various limits in increasing order are:

Instrumental detection limit (IDL)—the constituent concentration that produces a signal greater than five times the signal/noise ratio of the instrument. This is similar, in many respects, to "critical level" and "criterion of detection." The latter limit is stated as 1.645 times the s of blank analyses.

Lower limit of detection (LLD)—the constituent concentration in reagent water that produces a signal $2(1.645)s$ above the mean of blank analyses. This sets both Type I and Type II errors at 5%. Other names for this limit are "detection limit" and "limit of detection" (LOD).

Method detection limit (MDL)—the constituent concentration that, when processed through the complete method, produces a signal with a 99% proability that it is different from the blank. For seven replicates of the sample, the mean must be $3.14s$ above the blank where s is the standard deviation of the seven replicates. The MDL will be larger than the LLD because of the few replications and the sample processing steps and may vary with constituent and matrix.

Limit of quantitation (LOQ)—the constituent concentration that produces a signal sufficiently greater than the blank that it can be detected within specified limits by good laboratories during routine operating conditions.[1] Typically it is the concentration that produces a signal $10s$ above the reagent water blank signal.

Duplicate—usually the smallest number of replicates (two) but specifically herein refers to duplicate samples, i.e., two samples taken at the same time from one location.

Internal standard—a pure compound added to a sample extract just before instrumental analysis to permit correction for inefficiencies.

Laboratory control standard—a standard, usually certified by an outside agency, used to measure the bias in a procedure. For certain constituents and matrices, use National Institute of Standards and Technology (NIST)* Standard Reference Materials when they are available.

---

* Formerly National Bureau of Standards (NBS).

Precision—measure of the degree of agreement among replicate analyses of a sample, usually expressed as the standard deviation.

Quality assessment—procedure for determining the quality of laboratory measurements by use of data from internal and external quality control measures.

Quality assurance—a definitive plan for laboratory operation that specifies the measures used to produce data of known precision and bias.

Quality control—set of measures within a sample analysis methodology to assure that the process is in control.

Random error—the deviation in any step in an analytical procedure that can be treated by standard statistical techniques.

Replicate—repeated operation occurring within an analytical procedure. Two or more analyses for the same constituent in an extract of a single sample constitute replicate extract analyses.

Surrogate standard—a pure compound added to a sample in the laboratory just before processing so that the overall efficiency of a method can be determined.

Type I error—also called alpha error, is the probability of deciding a constituent is present when it actually is absent.

Type II error—also called beta error, is the probability of not detecting a constituent when it actually is present.

2. Reference

1. U.S. ENVIRONMENTAL PROTECTION AGENCY. 1985. National Primary Drinking Water Regulations. 40 CFR Part 141; *Federal Register* 50: 46936.

1020 QUALITY ASSURANCE

1020 A. Introduction

Quality assurance (QA) is a set of operating principles that, if strictly followed during sample collection and analysis, will produce data of known and defensible quality. That is, the accuracy of the analytical result can be stated with a high level of confidence. Included in quality assurance are quality control (Section 1020B) and quality assessment (Section 1020C).

1. Quality Assurance Planning

Establish a set of operating principles that will constitute a quality-assurance program. Prepare a QA plan[1] including the following: cover sheet with plan approval signatures, staff organization and responsibilities, sample control and documentation procedures, standard operating procedure for each analytical method (SOP), analyst training requirements, equipment preventive maintenance procedures, calibration procedures, corrective actions, internal quality control activities, performance audits, data assessment procedures for bias and precision, and data reduction, validation, and reporting.

The cover sheet with approval signatures indicates that the plan has been reviewed and judged suitable, and that the organization and responsibilities section outlines the chain-of-command and assigns specific functions to each person involved.

Sample control and documentation procedures permit tracing a sample and its derivatives through all steps from collection to analysis and display of results. Documentation always is important but is especially so when chain-of-custody requirements are imposed.

A standard operating procedure for the analytical method describes the method in such detail that an experienced analyst unfamiliar with the method can obtain acceptable results. Training requirements for analysts must be specified. The number of analyses required and the uncertainty of the results will vary with the type of analysis, sample characteristics, and the experience of the analyst.

Equipment preventive maintenance procedures are required. A strict preventive maintenance program will reduce instrument malfunctions, maintain calibration, and reduce downtime.

Calibration procedures, corrective actions, internal quality control activities, performance audits, and data assessments for bias and precision are discussed in Section 1020B and C.

Data reduction, validation, and reporting are the final features of a QA program. The reading obtained from an analytical instrument must be adjusted for such factors as instrument efficiency, extraction efficiency, sample size, and background value, before it becomes a useful result. The QA plan specifies the correction factors to be applied as well as the steps to be followed in validating the result. Report results in standard units of mass, volume, or concentration. Use a prescribed method for reporting results below the method detection limit. Accompany each result or set of results by a statement of uncertainty.

2. Reference

1. STANLEY, T.W. & S.S. VERNER. 1983. Interim Guidelines and Specifications for Preparing Quality Assurance Project Plans. EPA-600/4-83-004, U.S. Environmental Protection Agency, Washington, D.C.

1020 B. Quality Control

Quality control (QC) may be either internal or external. Internal QC is the subject of this section; external QC, also known as "quality assessment," is discussed in 1020C. All analysts use some QC as an intuitive effort to produce credible results. However, a good quality control program consists of at least seven elements: certification of operator competence, recovery of known additions, analysis of externally supplied standards, analysis of reagent blanks, calibration with standards, analysis of duplicates, and maintenance of control charts. Sections 1010 and 1030 contain the necessary calculations.

1. Certification of Operator Competence

Before an analyst is permitted to do reportable work, competence in making the analysis is to be demonstrated. Requirements vary, but for most inorganic and organic chemical analyses, demonstration of acceptable single-operator precision and bias is sufficient. Make a minimum of four replicate analyses of an independently prepared check sample having a concentration between 5 and 50 times the method detection limit (MDL) for the analysis in that laboratory. General limits for acceptable work are shown in Table 1020:I; certain methods may specify more stringent limits.

2. Recovery of Known Additions

Use the recovery of known additions as part of a regular analytical protocol. Use known additions to verify the absence of matrix effects. When a new matrix type is to be analyzed, verify the amount of interference. Where duplicates are not applicable, for example, when the constituent of interest is absent, make recovery of known additions for 10% of the samples. Where

duplicates also are being analyzed, the sum of the duplicates and known additions must equal at least 10% of the number of samples. Make the known addition between 5 and 50 times the MDL or between 1 and 10 times the ambient level, whichever is greater. Do not use a known addition above the demonstrated linear range of the method; use concentrated solutions so volume change in sample is negligible. See Table 1020:I for acceptable limits.

3. Analysis of Externally Supplied Standards

As a minimum, analyze externally supplied standards whenever analysis of known additions does not result in acceptable recovery or once each day, whichever is more frequent. Use laboratory control standards with a concentration between 5 and 50 times the MDL or near sample ambient levels, whichever is greater. Where possible, use certified reference materials as laboratory control standards. National Institute of Standards and Technology (NIST)* Standard Reference Materials are preferred, if available. If internal reference materials are used, prepare them independently from the standards used for calibration. See Table 1020:I for acceptable limits for high-level duplicates.

4. Analysis of Reagent Blanks

Analyze reagent blanks whenever new reagents are used and as often as required in specific methods. Analyze a minimum of 5% of the sample load as reagent blanks; this monitors purity of reagents and the overall procedural blank. Analyze a reagent blank after any sample with a con-

---

* Formerly National Bureau of Standards (NBS).

TABLE 1020:I. ACCEPTANCE LIMITS FOR DUPLICATE SAMPLES AND KNOWN ADDITIONS TO WATER AND WASTEWATER

Analysis	Recovery of Known Additions* %	Precision of Low-Level Duplicates*† %	Precision of High-Level Duplicates*†‡ %
Metals	80–120	75–125	90–110
Volatile organics	70–130	60–140	80–120
Volatile gases	50–150	50–150	70–130
Base/neutrals	70–130	60–140	80–120
Acids	60–140	60–140	80–120
Anions	80–120	75–125	90–110
Nutrients	80–120	75–125	90–110
Other inorganics	80–120	75–125	90–110
Total organic carbon	80–120	75–125	90–110
Total organic halogens	80–120	75–125	85–115
Herbicides	40–160	60–140	80–120
Organochlorine pesticides	50–140	60–140	80–120
Captan	20–130	60–140	80–120
Endosulfans	25–140	60–140	80–120
Endrin aldehyde	25–140	60–140	80–120
Organophosphorus pesticides	50–200	60–140	80–120
Trichlorophon	20–200	60–140	80–120
Triazine pesticides	50–200	60–140	80–120
Carbamate pesticides	50–150	60–140	80–120

* Additions calculated as % of the known addition recovered, duplicates calculated as the difference as a percentage of the mean $[100(x_1 - x_2)/\bar{x}]$.
† Low-level refers to concentrations less than 20 times the MDL. High-level refers to concentrations greater than 20 times the MDL.
‡ Also acceptance limits for independent laboratory control standards and certification of operator competence.
Source: NATRELLA, M. G. 1966. Experimental Statistics. National Bur. Standards Handbook 91, Washington, D.C.

centration greater than that of the highest standard or that might result in carryover from one sample to the next.

5. Calibration with Standards

As a minimum, measure three different dilutions of the standard when an analysis is initiated. Subsequently, verify the standard curve daily by analyzing one or more standards within the linear range, as specified in the individual method. Reportable analytical results are those within the range of the standard dilutions used. Do not report values above the highest standard unless an initial demonstration of greater linear range has been made, no instrument parameters have been changed, and the value is less than 1.5 times the highest standard. The lowest reportable value is the MDL, provided that the lowest calibration standard is less than 10 times the MDL. If a blank is subtracted, report the result even if it is negative.

6. Analysis of Duplicates

When most samples have measurable levels of the constituent being determined, analysis of duplicate samples is effective for assessing precision. Analyze 5% or more of the samples in duplicate. Analyze duplicates and known additions in matrices representative of the samples analyzed in the laboratory. See Table 1020:I for acceptable limits for duplicate analyses.

7. Control Charts

Three types of control charts commonly are used in laboratories[1]: a means chart for standards—laboratory control standards (LCS) or calibration check standards (CCS); a means chart for background or reagent blank results; and a range chart for replicate analyses.

The charts are essential instruments for quality control. Each type of chart is described below.

a. Means chart: The means chart for standards is constructed from the average and standard deviation of a standard. It includes upper and lower warning levels (WL) and upper and lower control levels (CL). Common practice is to use $\pm 2s$ and $\pm 3s$ limits for the WL and CL, respectively, where s represents standard deviation. Derive these values from stated values for standard reference materials, if used for the laboratory control standard (LCS) or calibration check standard (CCS), or from replicate analyses of a CCS. The chart can be set up by using either the calculated values for mean and standard deviation or by using percentages. Percentage is necessary if the concentration varies. Construct a chart for each analytical instrument. Enter results on the chart each time the LCS or CCS is analyzed. Examples of control charts for means are given in Figure 1020:1.

b. Range chart: If the standard deviation of the method is known, use the factors from Table 1020:II to construct the central line and warning and control limits as in Figure 1020:2. Perfect agreement between duplicates results in no difference when the values are subtracted, so the base line on the chart is zero. The standard deviation is converted to the range so that the analyst need only subtract the two results to plot the value on the control chart. The mean range is computed as:

$$\overline{R} = D_2 s$$

the control limit as

$$CL = \overline{R} \pm 3s(R) = D_4 \overline{R}$$

and the warning limit as

$$WL = \overline{R} \pm 2s(R) = \overline{R} \pm 2/3\ D_4 \overline{R}$$

where:

D_2 = factor to convert s to the range (1.128 for duplicates, as given in Table 1020:II),

$s(R)$ = standard deviation of the range, and

D_4 = factor to convert mean range to $3s(R)$ (3.267 for duplicates, as given in Table 1020:II).

A range chart is rather simple when duplicate analyses of a standard are used (Figure 1020:2). For duplicate analyses of samples, the plot will appear different because of the variation in sample concentration. If a constant relative standard deviation in the concentration range of interest is assumed, the $\overline{R}$, $D_4 \overline{R}$, etc. may be computed as above for several concentrations, a smooth curve drawn through the points obtained, and so an acceptable range for duplicates for any mean concentration can be determined. Figure 1020:3 illustrates such a chart. A separate table, as suggested below the figure, will be needed to track precision over time.

More commonly, the range can be expressed as a function of the relative standard deviation (coefficient of variation). Normalize the range by dividing by the average. Determine the mean range for the pairs analyzed by

$$\overline{R} = (\Sigma\ R_i)/n$$

and the variance (square of the standard deviation) as

$$s_R{}^2 = (\Sigma R_i^2 - n\overline{R}^2)/(n-1)$$

Then draw lines on the chart at $\overline{R} + 2s_R$ and $\overline{R} + 3s_R$ and, for each duplicate analysis, calculate normalized range and enter

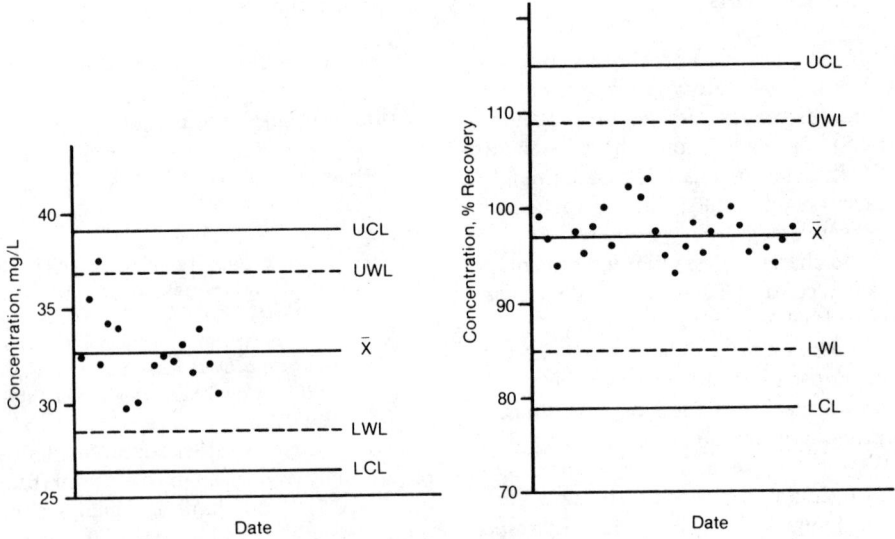

Figure 1020:1. Control charts for means.

TABLE 1020:II. FACTORS FOR COMPUTING LINES ON RANGE CONTROL CHARTS

Number of Observations n	Factor for Central Line (D_2)	Factor for Control Limits (D_4)
2	1.128	3.267
3	1.693	2.575
4	2.059	2.282
5	2.326	2.115
6	2.534	2.004

Source: ROSENSTEIN, M. & A. S. GOLDEN. 1964. Statistical Techniques for Quality Control of Environmental Radioassays. AQCS Rep. Stat-1. Public Health Serv., Winchester, Mass.

the result on the chart. Figure 1020:4 is an example of such a chart.

c. Chart analyses: If the warning limits (WL) are at the 95% level, 1 out of 20 points, on the average, would exceed that limit whereas only 1 out of 100 would exceed the control limits (CL). Take the following actions, based on these statistical parameters, which are illustrated in Figure 1020:5:

Control limit—If one measurement exceeds a CL, repeat the analysis immediately. If the repeat is within the CL, continue analyses; if it exceeds the CL, discontinue analyses and correct the problem.

Warning limit—If two out of three successive points exceed a WL, analyze another sample. If the next point is less than WL, continue analyses; if next point exceeds WL, discontinue analyses and correct the problem.

Standard deviation—If four out of five successive points exceed 1 *s*, or are in decreasing or increasing order, analyze another sample. If the next point is less than 1 *s*, or changes the order, continue analyses; otherwise, discontinue analyses and correct the problem.

Central line—If six successive samples are above the central line, analyze another

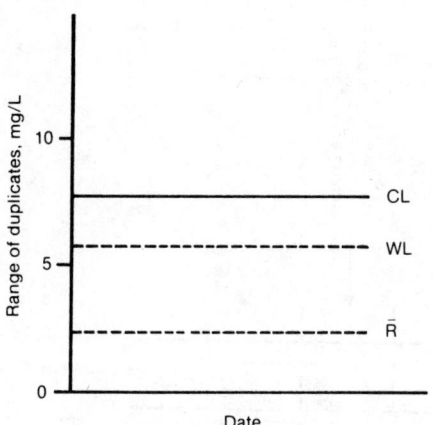

Figure 1020:2. Duplicate analyses of a standard.

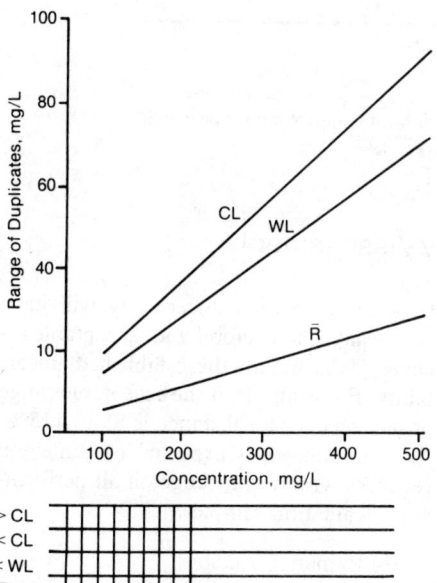

Figure 1020:3. Range chart for variable concentrations.

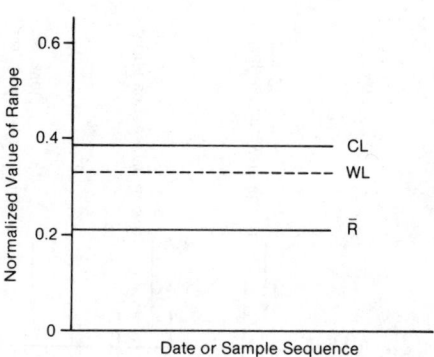

Figure 1020:4. Range chart for variable ranges.

sample. If the next point is below the central line, continue analyses; if the next point is on the same side, discontinue analyses and correct the problem.

The above considerations apply when the conditions are either above or below the central line, but not on *both* sides, e.g., four of five values must exceed either $+1\ s$ or $-1\ s$. After correcting the problem, reanalyze half the samples analyzed between the last in-control measurement and the out-of-control one.

Another important function of the control chart is assessment of improvements in method precision. In the means and range charts, if measurements never or rarely exceed the WL, recalculate the WL and CL using the 20 most recent data points. Trends in precision can be detected sooner if running averages of 20 are kept on a daily basis.

8. Reference

1. GOLDEN, A.S. 1984. Evaluation of internal control measurements in radioassay. *Health Phys.* 47:361.

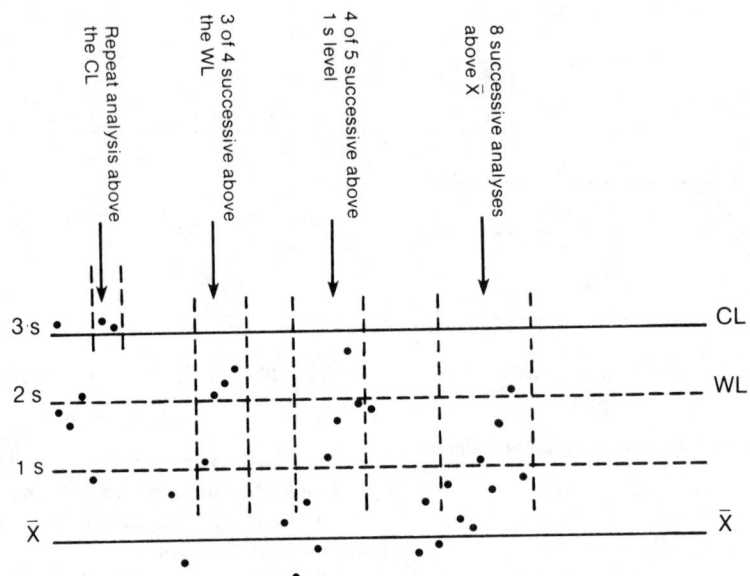

Figure 1020:5. Means control chart with out-of-control data (upper half).

1020 C.   Quality Assessment

Quality assessment is the process of using external and internal quality control measures to determine the quality of the data produced by the laboratory. It includes such items as performance evaluation samples, laboratory intercomparison samples, and performance audits as well as the internal QC described in Section 1020B. They are applied to test the recovery, bias, precision, detection limit, and adherence to standard operating procedure requirements.

1. Performance Evaluation Samples

Use samples with known amounts of the constituent of interest supplied by an outside agency or blind additions prepared independently within the laboratory to determine recovery achieved by an analyst.

In general, method uncertainty will have been established beforehand; acceptable recovery falls within the established uncertainty. For example, if the acceptable range of recovery for a substance is 85 to 115%, then the analyst is expected to achieve a recovery within that range on all performance evaluation samples.

2. Performance Audits

Make only unscheduled performance audits using a check list made to document the manner in which a sample is treated from time of receipt to final reporting of the result. The goal is to detect any deviations from the standard operating procedure so that corrective action can be taken. A recommended format with a few initial items in the check list is shown in Table 1020:III.

TABLE 1020:III. AUDIT OF A SOIL ANALYSIS PROCEDURE

Procedure	Comment	Remarks
1. Sample entered into logbook	yes	lab number assigned
2. Sample weighed	yes	dry weight
3. Drying procedure followed	no	maintenance of oven not done
4a. Balance calibrated	yes	once per year
b. Cleaned and zero adjusted	yes	weekly
5. Sample ground	yes	to pass 50 mesh
6. Ball mill cleaned	yes	should be after each sample

.
.
.

3. Laboratory Intercomparison Samples

Commercial and governmental programs supply samples containing various constituents in various matrices. A good quality assessment program requires participation in periodic laboratory intercomparison studies. Adjust frequency of participation to the quality of the results produced by the analysts being tested. For routine procedures, quarterly analyses are reasonable. Official agencies conducting such studies are listed below.

a. *Organic and inorganic compounds in water:* Samples are available from the Environmental Monitoring Systems Laboratory, U.S. EPA, 26 West St. Clair, Cincinnati, Ohio 45268. These samples can be used as performance evaluation samples or as standards for laboratory intercomparison studies.

b. *Radioisotopes in various media:* These samples are available from the Environmental Monitoring Systems Laboratory, U.S. EPA, P.O. Box 93478, Las Vegas, Nevada 89193-3478. These can be supplied for use as performance evaluation samples; the Laboratory also conducts laboratory intercomparison studies.[1] Environmental samples with background levels of radioisotopes are available from the Division of Laboratories and Research, International Atomic Energy Agency, P.O. Box 100, A-1400 Vienna, Austria.

4. Reference

1. JARVIS, A.N. & L. SIU. 1981. Environmental Radioactivity Laboratory Intercomparison Studies Program. EPA-600/4-81-004, U.S. Environmental Protection Agency, Las Vegas, Nev.

1030   DATA QUALITY

1030 A.   Introduction

1. Quality Indicators

The principal indicators of data quality are its bias (Section 1030B) and precision (Section 1030C), which, when combined, express its accuracy. The relationship among these terms is shown in Figure 1030:1. Of the four possible outcomes, only the condition of low bias and high precision is accurate.[1]

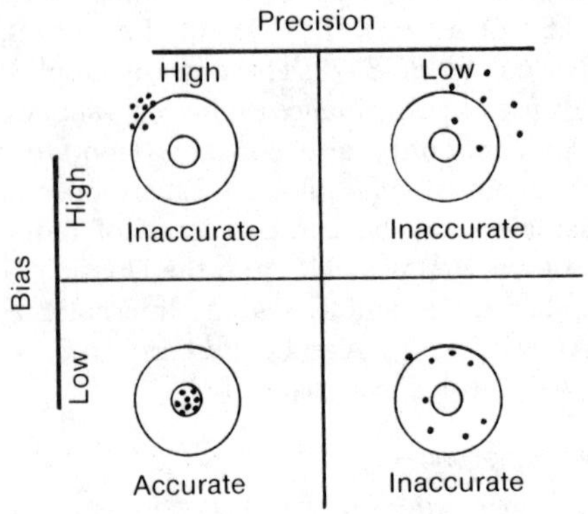

Figure 1030:1. Definition of accuracy.

Subsidiary data quality indicators are method detection limit (Section 1030E) and representativeness. Representativeness can relate both to the sample itself and to the sampled population. A method may be highly accurate, but if the results do not represent the sample or if the sample does not represent the population, the data are not useful. Representativeness of the sample is best assessed by analysis of a number of samples from the same location or from a well-mixed bulk lot. The representativeness of the method is best determined through collaborative studies using a variety of methods.

2. Reference

1. U.S. ENVIRONMENTAL PROTECTION AGENCY. 1980. Health Physics Society Committee Report HPSR-1: Upgrading Environmental Radiation Data. EPA-520/1-80-012, U.S. Environmental Protection Agency, Washington, D.C.

1030 B.   Bias

Bias is a measure of systematic error. It has two components: one due to the method, and the other to a laboratory's use of the method. The bias of a method is measured best by a laboratory intercomparison study in which the difference between the grand average and the known (or true) value is the method bias. The laboratory bias is the difference between the laboratory average recovery and the true value and is, therefore, a combination of the two biases. Assess the laboratory bias by measuring the recovery of known additions or by analyzing duplicate samples and retaining the sign of the differences when calculating the average difference. From this value, subtract the method bias from an intercomparison study to determine the bias due to the laboratory's practices as it interprets the method.

1030 C. Precision

1. Definition

Precision is a measure of the closeness with which multiple analyses of a given sample agree with each other. Assess precision by replicate analyses, by repeated analyses of a stable standard, or by analysis of known additions to samples. Precision is specified by the standard deviation of the results.[1] If overall precision of a study is desired, analyze duplicate samples. This latter precision includes the random errors involved in sampling as well as the errors in sample preparation and analysis.

When only a few replications are used, for example, duplicate sample analysis or duplicate extract analysis, the range of results, R, is nearly as efficient as the standard deviation because the two measures differ by a constant ($1.128s = R$ for duplicates, $1.693s = R$ for triplicates).

2. Computation

Table 1030:I lists the results of duplicate analyses and recovery values from repetitive analysis of a stable standard. Table 1030:II shows the precision calculation us-

TABLE 1030:I. PRECISION CALCULATIONS USING DUPLICATES

Duplicate Analyses			Standard Analyses
1st Result mg/L	2nd Result mg/L	Difference mg/L	Result mg/L
50	46	4	35.1
37	36	1	33.2
22	19	3	33.7
17	20	3	35.9
32	34	2	33.5
46	46	0	34.5
26	28	2	34.4
26	30	4	34.3
61	58	3	31.8
44	45	1	35.0
40	44	4	31.4
36	35	1	35.6
29	31	2	30.2
36	36	0	32.7
47	45	2	31.1
16	20	4	34.8
18	21	3	34.3
26	22	4	36.4
35	36	1	32.1
26	25	1	38.2
49	51	2	33.1
33	32	1	34.9
40	38	2	36.2
16	13	3	34.0
39	42	3	33.8
		$\Sigma = 56, n = 25$	$\Sigma = 850.2$

$\overline{R} = 56/25 = 2.24$

$s = 2.24/1.128 = 1.98$

$\overline{x} = 34.01$

$s = 1.83$

TABLE 1030:II. PRECISION CALCULATION FROM KNOWN ADDITIONS

(1) Value for Sample with Known Additions	(2) Sample	(3) Calculated Recovery (1)–(2)	(4) Known Additions	(5) Deviation from Expected (3)–(4)
1.91	0.68	1.23	1.30	−0.07
1.78	0.57	1.21	1.30	−0.09
1.53	0.23	1.30	1.30	0.00
1.74	0.15	1.59	1.30	0.29
2.10	0.53	1.57	1.30	0.27
1.82	0.61	1.21	1.30	−0.09
2.07	0.54	1.53	1.30	0.23
1.39	0.14	1.25	1.30	−0.05
1.16	0.20	0.96	1.30	−0.34
1.55	0.19	1.36	1.30	0.06
2.02	0.41	1.61	1.30	0.31
1.58	0.36	1.22	1.30	−0.08
13.01	11.97	1.04	1.30	−0.26
1.46	0.17	1.29	1.30	−0.01
1.63	0.31	1.32	1.30	0.02
11.95	10.98	0.97	1.30	−0.33
1.68	0.27	1.41	1.30	0.11
1.83	0.47	1.36	1.30	0.06
1.62	0.43	1.19	1.30	−0.11
5.04	3.96	1.08	1.30	−0.22
2.53	1.22	1.31	1.30	0.01
2.69	1.09	1.60	1.30	0.30
1.50	0.25	1.25	1.30	−0.05
1.77	0.51	1.26	1.30	−0.04
1.88	0.55	1.33	1.30	0.03

$$s^2 = (\Sigma \text{ deviations}^2)/(n-1)$$
$$s = \sqrt{0.8035/24} = 0.183$$

ing known additions. From duplicate analyses, the average difference, or average range, is calculated by summing all the differences (absolute values) and dividing by the number of observations: $\overline{R} = \Sigma d_i /n$. This is converted to standard deviation by dividing by 1.128. For multiple analyses of a stable standard, calculate the standard deviation as usual. The true value of the standard is irrelevant for this analysis.

In using known additions to determine precision, as in the example of Table 1030:II, subtract recovery from the known value (subtract Column 3 from Column 4) and calculate the standard deviation by the usual method as shown at the bottom of the table. If the standard deviation is a constant proportion of the amount present, then the coefficient of variation (relative standard deviation, $s/\overline{x}$) may be used instead of the standard deviation.

3. Reference

1. AMERICAN SOCIETY FOR TESTING AND MATERIALS. 1977. Standard Practice for Determination of Precision and Bias of Methods. Committee D-19 on Water, Designation D2777-77, American Soc. Testing & Materials, Philadelphia, Pa.

1030 D.   Total Uncertainty

1. Definition and Computation

This statistic is some appropriate combination of the random and systematic uncertainties in a given measurement system. The random uncertainties are assessed by calculating precision; they are ascertained statistically. The systematic uncertainties are the biases and those random uncertainties that cannot be evaluated statistically. Because the latter can be estimated only from a thorough knowledge of all steps in the measurement process, using the judgment of an experienced analyst, there is considerable vagueness in the assessment. Because the biases can be assessed, it is customary to consider only them in evaluating total uncertainty. These biases can occur at several points in the system, for example, in weighing the sample, in the result produced by the analytical instrument, in changes in quality of reagents, or in incomplete extraction. If random uncertainties have been assessed by the standard deviation (s_x) and if biases are represented by b_i, these quantities may be combined in quadratic form to calculate uncertainty[1]:

$$U_x = (s_x^2 + 1/3 \ \Sigma b_i^2)^{1/2}$$

Generally, the instrumental and extraction biases (B) are the largest and the equation simplifies to:

$$U_x = (s_x^2 + B^2)^{1/2}$$

To express the uncertainty in the form of confidence levels, assume that the bias is known with little error and add it to higher confidence levels of the variance; for example, for 95% confidence level:

$$U_{x(95\%)} = (2s_x^2 + B^2)^{1/2}$$

If the concentration of the constituent is within a narrow range, use the concentration units for B and s_x; otherwise, use the coefficient of variation for s_x and the fractional form of B.

2. Reference

1. U.S. ENVIRONMENTAL PROTECTION AGENCY. 1980. Health Physics Society Committee Report HPSR-1: Upgrading Environmental Radiation Data. EPA-520/1-80-012, U.S. Environmental Protection Agency, Washington, D.C.

1030 E.   Method Detection Limit

1. Introduction

Detection limits are controversial, principally because of inadequate definition and confusion of terms. Frequently, the instrumental detection limit is used for the method detection limit and *vice versa.* Whatever term is used, most analysts agree that the smallest amount that can be detected above the noise in a procedure and within a stated confidence limit is the detection limit. The confidence limits are set so that the probabilities of both Type I and Type II errors are acceptably small.

Current practice identifies several detection limits (see 1010C), each of which has a defined purpose. These are the instrument detection limit (IDL), the lower limit of detection (LLD), the method detection limit (MDL), and the limit of quantitation (LOQ). Occasionally the instrument detection limit is used as a guide for determining the MDL. The relationship among these limits is approximately IDL:LLD:MDL:LOQ = 1:2:4:10.

2. Determining Detection Limits

An operating analytical instrument usually produces a signal (noise) even when no sample is present or when a blank is being analyzed. Because any QA program requires frequent analysis of blanks, the mean and standard deviation become well known; the blank signal becomes very precise, i.e., the Gaussian curve of the blank distribution becomes very narrow. The IDL is the constituent concentration that produces a signal greater than three standard deviations of the mean noise level or that can be determined by injecting a standard to produce a signal that is five times the signal-to-noise ratio. The IDL is useful for estimating the constituent concentration or amount in an extract needed to produce a signal to permit calculating an estimated method detection limit.

The LLD is the amount of constituent that produces a signal sufficiently large that 99% of the trials with that amount will produce a detectable signal. Determine the LLD by multiple injections of a standard at near zero concentration (concentration no greater that five times the IDL). Determine the standard deviation by the usual method. To reduce the probability of a Type I error (false detection) to 5%, multiply s by 1.645 from a cumulative normal probability table. Also, to reduce the probability of a Type II error (false nondetection) to 5%, double this amount to 3.290. As an example, if 20 determinations of a low-level standard yielded a standard deviation of 6 μg/L, the LLD is 3.29 × 6 = 20 μg/L.[1]

The MDL differs from the LLD in that samples containing the constituent of interest are processed through the complete analytical method. The method detection limit is greater than the LLD because of extraction efficiency and extract concentration factors. The MDL can be achieved by experienced analysts operating well-calibrated instruments on a nonroutine basis. For example, to determine the MDL, add a constituent to reagent water, or to the matrix of interest, to make a concentration near the estimated MDL.[2] Analyze seven portions of this solution and calculate the standard deviation (s). From a table of the one-sided t distribution select the value of t for $7 - 1 = 6$ degrees of freedom and at the 99% level; this value is 3.14. The product 3.14 times s is the desired MDL.

Although the LOQ is useful within a

laboratory, the practical quantitation limit (PQL) has been proposed as the lowest level achievable among laboratories within specified limits during routine laboratory operations.[3] The PQL is significant because different laboratories will produce different MDLs even though using the same analytical procedures, instruments, and sample matrices. The PQL is about five times the MDL and represents a practical and routinely achievable detection limit with a relatively good certainty that any reported value is reliable.

3. Description of Limits

Figure 1030:2 illustrates the detection limits discussed above. For this figure it is assumed that the signals from an analytical instrument are distributed normally and can be represented by a normal (Gaussian) curve.[4] The curve labeled B is representative of the background or blank signal distribution. As shown, the distribution of the blank signals is nearly as broad as for the other distributions, that is $\sigma_B = \sigma_I = \sigma_L$. As blank analyses continue, this curve will become narrower because of increased degrees of freedom.

The curve labeled I represents the IDL. Its average value is located $k\sigma_B$ units distant from the blank curve, and k represents the value of t (from the one-sided t distribution) that corresponds to the confidence limit chosen to describe instrument performance. For a 95% limit and $n = 14$, $k = 1.782$ and for a 99% limit, $k = 2.68$. The overlap of the B and I curves indicates the probability of not detecting a constituent when it is present (Type II error).

The curve at the extreme right of Figure 1030:2 represents the LLD. Because only a finite number of determinations are used for calculating the IDL and LLD, the curves are broader than the blank but are similar, so it is reasonable to choose $\sigma_I = \sigma_L$. Therefore, the LLD is $k\sigma_I + k\sigma_L = 2k\sigma_L$ from the blank curve.

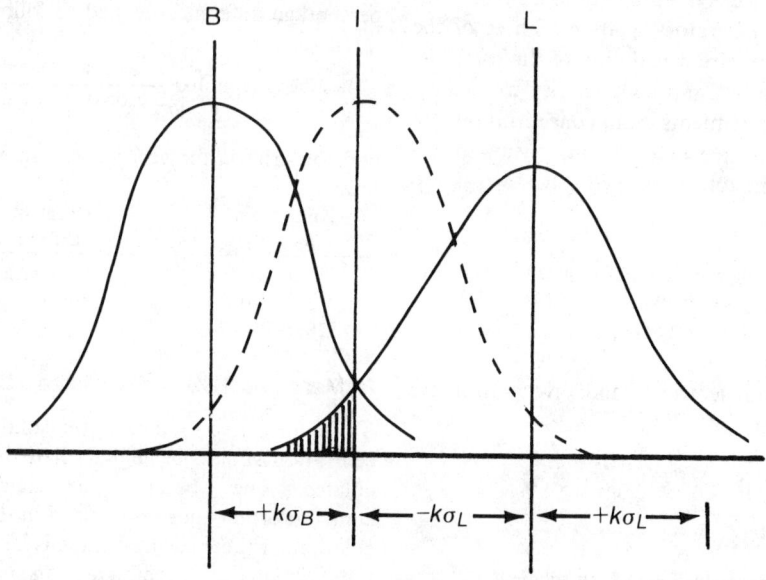

Figure 1030:2. Detection limit relationship.

4. References

1. AMERICAN SOCIETY FOR TESTING AND MA-
   TERIALS. 1983. Standard Practice for Intralaboratory Quality Control Procedures and a
   Discussion on Reporting Low-Level Data. Designation D4210-83, American Soc. Testing &
   Materials, Philadelphia, Pa.
2. GLASER, J.A., D.L. FOERST, J.D. McKEE,
   S.A. QUAVE & W.L. BUDDE. 1981. Trace analyses for wastewaters. *Environ. Sci. Technol.*
   15:1426.
3. U.S. ENVIRONMENTAL PROTECTION AGENCY.
   1985. National Primary Drinking Water Standards: Synthetic Organics, Inorganics, and Bac-
   teriologicals. 40 CFR Part 141; *Federal Register*
   50: No. 219, November 13, 1985.
4. OPPENHEIMER, J. & R. TRUSSELL. 1984. Detection limits in water quality analysis. *In* Proc.
   Water Quality Technology Conference (Denver, Colorado, December 2-5, 1984). American
   Water Works Assoc., Denver, Colo.

1030 F.   Checking Correctness of Analyses

The following procedures for checking correctness of analyses are applicable specifically to water samples for which relatively complete analyses are made.[1] These include pH, conductivity, total dissolved solids (TDS), and major anionic and cationic constituents that are indications of general water quality.

The checks described do not require additional laboratory analyses. Three of the checks require calculation of the total dissolved solids and conductivity from measured constituents. Sum concentrations (in milligrams per liter) of constituents to calculate the total dissolved solids are as follows:

$$\text{Total dissolved solids} = 0.6 \text{ (alkalinity)} + Na + K + Ca + Mg + Cl + SO_4 + SiO_3 + (NO_3\text{-}N) + F$$

Calculate electrical conductivity from the equation:

$$G = \lambda C - (k_1 \lambda + k_2)(C)^{3/2}$$

where:

G = conductivity of salt solution,
C = concentration of salt solution,
λ = equivalent conductance of salt solution at infinite dilution,

k_1, k_2 = constants for relaxation of ion cloud effect and electrophoretic effect relative to ion mobility.[1]

1. Anion-Cation Balance[2]

The anion and cation sums, when expressed as milliequivalents per liter, must balance because all potable waters are electrically neutral. The test is based on the percentage difference defined as follows:

$$\% \text{ difference} = 100 \frac{\Sigma \text{ cations} - \Sigma \text{ anions}}{\Sigma \text{ cations} + \Sigma \text{ anions}}$$

and the criteria for acceptance are as follows:

Anion Sum (meq/L)	Acceptable % Difference
0–3.0	± 0.2 meq/L
3.0–10.0	± 2%
20.0–800	± 2–5%

2. Measured TDS = Calculated TDS[2]

The measured total dissolved solids concentration should be higher than the calculated one because a significant contributor may not be included in the calculation. If the measured value is less than the calculated one, the higher ion sum and measured value are suspect; the sample should be reanalyzed. If the measured solids concentration is more than 20% higher than the calculated one, the low ion sum is suspect and selected constituents should be reanalyzed. The acceptable ratio is as follows:

$$1.0 < \frac{\text{measured TDS}}{\text{calculated TDS}} < 1.2$$

3. Measured EC = Calculated EC

If the calculated conductivity is higher than the measured value, reanalyze the higher ion sum. If the calculated EC is less than the measured one, reanalyze the lower ion sum. The acceptable ratio is as follows:

$$0.9 < \frac{\text{calculated EC}}{\text{measured EC}} < 1.1$$

4. Measured EC and Ion Sums

Both the anion and cation sums should be $\frac{1}{100}$ of the measured EC value. If either of the two sums does not meet this criterion, that sum is suspect; reanalyze the sample. The acceptable criteria are as follows:

$$100 \times \text{anion (or cation) sum, meq/L} = (0.9-1.1) \text{ EC}$$

5. Calculated TDS to EC Ratio

If the ratio of calculated TDS to conductivity falls below 0.55, the lower ion sum is suspect; reanalyze it. If the ratio is above 0.7, the higher ion sum is suspect; reanalyze it. If reanalysis causes no change in the lower ion sum, an unmeasured constituent, such as ammonia or nitrite, may be present at a significant concentration. If poorly dissociated calcium and sulfate ions are present, the TDS may be as high as 0.8 times the EC. The acceptable criterion is as follows:

$$\text{calculated TDS/conductivity} = 0.55-0.7$$

6. Measured TDS to EC Ratio

The acceptable criteria for this ratio are from 0.55 to 0.7. If the ratio of TDS to EC is outside these limits, measured TDS or measured conductivity is suspect; reanalyze.

A more complete exposition[3] of the above quality-control checks has been published.

7. References

1. ROSSUM, J.R. 1975. Checking the accuracy of water analyses through the use of conductivity. *J. Amer. Water Works Assoc.* 67:204.
2. FRIEDMAN, L.C. & D.E. ERDMANN. 1982. Quality Assurance Practices for Analyses of Water and Fluvial Sediments. Tech. Water Resources Inc., Book 5, Chapter A6. U.S. Government Printing Off., Washington, D.C.
3. OPPENHEIMER, J. & A.D. EATON. 1986. Quality control and mineral analysis. *In* Proc. Water Quality Technology Conference (Houston, Texas, December 8-11, 1985). American Water Works Assoc., Denver, Colo.

1040   METHOD DEVELOPMENT AND EVALUATION

1040 A.   Introduction

Although standard methods are available from many nationally recognized sources, there may be occasions when they cannot be used or when no standard method exists for a particular constituent.

Therefore, method development may be required. Method development is the set of experimental procedures devised for measuring a known amount of a constituent in various matrices.

1040 B.   Method Validation

Whether an entirely new method is developed by accepted research procedures or an existing method is modified to meet special requirements, validation by a three-step process is required: determination of single-operator precision and bias, analysis of independently prepared unknown samples, and determination of method ruggedness.

1. Single-Operator Characteristics

This part of the validation procedure requires determining the method detection limit (MDL) as in Section 1030; the bias of the method, i.e., the systematic error of the method; and the precision obtainable by a single operator, i.e., the random error introduced in using the method. To make these determinations, analyze at least 7 but preferably 10 or more portions of a standard at each of several concentrations in each matrix that may be used. Use one concentration at, or slightly above, the MDL and one relatively high so that the range of concentrations for which the method is applicable can be specified.

The use of several concentrations to determine bias and precision will reveal the form of the relationship between these method characteristics and the concentration of the substance. This relationship may be constant, linear, or curvilinear and is a significant characteristic of the method that should be explained clearly. Calculation of precision and bias for a single concentration in a single matrix is shown in the following table of results from eight replicate analyses of a standard with a known concentration of 1.30 mg/L.

Result mg/L	Difference (−1.30)	Squared Difference
1.23	−0.07	0.0049
1.21	−0.09	0.0081
1.30	0.0	0.0
1.59	0.29	0.0841
1.57	0.27	0.0729
1.21	−0.09	0.0081
1.53	0.23	0.0529
1.25	−0.05	0.0025
Sum	0.49	0.2335

The bias is $0.49/8 = 0.06$ mg/L and the precision is the square root of $0.2335/(8-1) = \sqrt{0.03336}$, or 0.18 mg/L (note that this is similar to the calculation for standard deviation).

2. Analysis of Unknown Samples

This step in the method validation procedure requires analysis of independently prepared standards where the value is unknown to the analyst. Analyze each unknown in replicate by following the standard operating procedure for the method. The mean amount recovered should be within three standard deviations (s) of the mean value of the standard but preferably within $2\ s$.

Obtain the unknowns from other personnel in the analyst's laboratory using either purchased analytical-grade reagents or standards available from National Institute of Standards and Technology (NIST)*, EPA, or other suitable sources. If available for the particular constituent, performance evaluation samples from EPA-Cincinnati (Section 1020) are particularly useful.

---

* Formerly National Bureau of Standards (NBS).

3. Method Ruggedness

A test of the ruggedness, i.e., stability of the result produced when steps in the method are varied, is the final validation step. It is especially important to determine this characteristic of a method if it is to be proposed as a standard or reference method. A properly conducted ruggedness test will point out those procedural steps in which rigor is critical and those in which some leeway is permissible.

The Association of Official Analytical Chemists[1] has suggested a method for this test in which eight separate analyses can be used to determine the effect of varying seven different steps in an analytical procedure. To illustrate, suppose the effect of changing the following factors is to be determined:

Factor	Nominal	Variation
Mixing time	10 min	12 min
Portion size	5 g	10 g
Acid concentration	$1M$	$1.1M$
Heat to	100°C	95°C
Hold heat for	5 min	10 min
Stirring	yes	no
pH adjust	6.0	6.5

To make the determination, denote the nominal factors by capital letters A through G and the variations by the corresponding lower-case letters. Then set up a table of the factors as follows:

Factor value	Combinations							
	1	2	3	4	5	6	7	8
A or a	A	A	A	A	a	a	a	a
B or b	B	B	b	b	B	B	b	b
C or c	C	c	C	c	C	c	C	c
D or d	D	D	d	d	d	d	D	D
E or e	E	e	E	e	e	E	e	E
F or f	F	f	f	F	F	f	f	F
G or g	G	g	g	G	G	G	G	g
Result	s	t	u	v	w	x	y	z

Source: YOUDEN, W. J. & E. H. STEINER. 1975. Statistical Manual of AOAC. Assoc. Official Analytical Chemists, Washington, D.C.

If combination 1 is analyzed, the result will be s. If combination 2 is analyzed, the result will be t, and so on until all eight combinations have been analyzed. To determine the effect of varying a factor, find the four results where the factor was nominal (all caps) and the four where it was varied (all lower case) and compare the averages of the two groups. For example, to compare the effect of changing C to c, use results (s + u + w + y)/4 and (t + v + x + z)/4. Calculate all seven pairs to get seven differences, which can then be ranked to reveal those with a significant effect on the results. If there is no outstanding difference, calculate the average and standard deviation of the eight results s through z. The standard deviation is a realistic estimate of the precision of the method. This design tests main effects, not interactions.

4. Equivalency Testing

After a new method has been validated by the procedures listed above, it may be prudent to test the method for equivalency to standard methods, unless none exist. This requires analysis of a minimum of three concentrations by the alternate and by the standard method. If the range of concentration is very broad, test more concentrations. Once an initial set of analyses (five or more) has been made at each chosen concentration, apply the following statistical steps:[2]

1. Test the distribution of data for normality and transform the data if necessary (Section 1010B).

2. Select an appropriate sample size based on an estimate of the standard deviation.[3]

3. Test the variances of the two methods using the F-ratio statistic.

4. Test the average values of the two methods using a Student-t statistic.

An explanation of each of these steps with additional techniques and examples has been published.[4] Because the number of analyses can be very large, the calculations become complex and familiarity with basic statistics is necessary. A listing of standard, reference, and equivalent methods for water analysis is available.[5]

5. References

1. YOUDEN, W.J. & E.H. STEINER. 1975. Statistical Manual of AOAC. Assoc. Official Analytical Chemists, Washington, D.C.

2. WILLIAMS, L.R. 1985. Harmonization of Biological Testing Methodology: A Performance Based Approach in Aquatic Toxicology and Hazard Assessment. 8th Symp. ASTM STP 891, R.C. Bahner & D.J. Hansen, eds. American Soc. Testing & Materials, Philadelphia, Pa.

3. NATRELLA, M.G. 1963. Experimental Statistics. National Bureau of Standards Handbook 91, Washington, D.C.

4. U.S. ENVIRONMENTAL PROTECTION AGENCY. 1983. Guidelines for Establishing Method Equivalency to Standard Methods. Rep. 600/X-83-037, Environmental Monitoring Systems Lab., Las Vegas, Nev.

5. U.S. ENVIRONMENTAL PROTECTION AGENCY. 1987. Guidelines establishing test procedures for the analysis of pollutants under the Clean Water Act. Interim final rule. 40 CFR Part 136; *Federal Register* 52:171:33542.

1040 C. Collaborative Testing

Once a new or modified method has been developed and validated it is appropriate to determine whether the method should be made a standard method. The procedure to convert a method to standard status is the collaborative test.[1] In this test, a num-

ber of laboratories use the standard operating procedure to analyze a select number of samples to determine the method's bias and precision as would occur in normal practice.

In planning for a collaborative test, consider the following factors: a precisely written standard operating procedure, the number of variables to be tested, the number of levels to be tested, and the number of replicates required. Because method precision is estimated by the standard deviation, which itself is the result of many sources of variation, the variables that affect it must be tested. These may include the laboratory, operator, apparatus, and concentration range.

1. Variables

Test at least the following variables:

Laboratory—Involve at least three different laboratories, although more are desirable to provide a better estimate of the standard deviation;

Apparatus—Because model and manufacturer differences can be sources of error, analyze at least two replicates of each concentration per laboratory;

Operators—To determine overall precision, involve at least six analysts with not more than two from each laboratory;

Levels—If the method development has indicated that the relative standard deviation is constant, test three levels covering the range of the method. If it is not constant, use more levels spread uniformly over the operating range.

If matrix effects are suspected, conduct the test in each medium for which the method was developed. If this is not feasible, use appropriate grades of reagent water as long as this is stipulated in the resulting statement of method characteristics.

2. Number of Replicates

Calculate the number of replicates after the number of variables to be tested has been determined by using the formula:

$$r > 1 + (30/P)$$

where:
r = number of replicates and
P = the product of several variables.

The minimum number of replicates is two. As an example, if three levels of a substance are to be analyzed by single operators in six laboratories on a single apparatus, then P is calculated as follows:

$$P = 3 \times 1 \times 6 \times 1 = 18$$

and the number of replicates is

$$r > 1 + (30/18) > 2.7 \text{ or } r = 3.$$

3. Illustrative Collaborative Test

Send each of five laboratories four concentrations of a compound (4.3, 11.6, 23.4, and 32.7 mg/L) with instructions to analyze in triplicate using the procedure provided. Tabulate results as shown in the table below (the results for only one concentration are shown). Because there are no obviously aberrant values (use the method in Section 1010B to reject outliers), use all the data.

Calculate the average and standard deviation for each laboratory; use all 15 results to calculate a grand average and standard deviation. The difference between the average of each laboratory and the grand average reveals any significant bias, such as that shown for Laboratories 1 and 3. The difference between the grand average and the known value is the method bias, e.g., 33.0 − 32.7 = 0.3 mg/L or 0.9%. The relative standard deviation of the grand average (1.5 mg/L) is 4.5%, which is the method precision, and the s for each laboratory is the single-operator precision.

Laboratory	Result mg/L	Experimental $x \pm s$	Deviation	
			From Known	From Grand Average
1	32.7 35.2 36.3	34.7 ± 1.8	2.0	1.7
2	32.6 33.7 33.6	33.3 ± 0.6	0.6	0.3
3	30.6 30.6 32.4	31.2 ± 1.0	−1.5	−1.8
4	32.6 32.5 33.9	33.0 ± 0.8	0.3	0
5	32.4 33.4 32.9	32.6 ± 0.8	−0.1	−0.4
$(\Sigma x)/n = 33$ $s = 1.5$			$\Sigma = 1.3$	$\Sigma = -0.2$

As noted in the table, the sum of the deviations from the known value for the laboratories was 1.3, so the average deviation (bias) was 1.3/5 = 0.26, rounded to 0.3, which is the same as the difference between the grand average and the known value.

For all four unknowns in this test, the percentage results indicated increasing bias and decreasing precision as the concentration decreased. Therefore, to describe the method in a formal statement, the precision would be given by a straight line with the formula $y = mx + b$; where y is the relative standard deviation, m is the slope of the line, x is the concentration, and b is the relative standard deviation at concentration $= 0$. The values found from the collaborative test are shown in the following table.

Known Amount mg/L	Amount Found mg/L	CV (% Standard Deviation)	Bias %
4.3	4.8	12.5	11.5
11.6	12.2	10.2	5.6
23.4	23.8	5.4	1.9
32.7	33	4.5	0.9

These results indicate that the method is acceptable. However, concentrations of less than about 10 mg/L require greater care in analysis.

4. Reference

1. YOUDEN, W.J. & E.H. STEINER. 1975. Statistical Manual of the AOAC. Assoc. Official Analytical Chemists, Washington, D.C.

1050   EXPRESSION OF RESULTS

1050 A.   Units

This text uses the International System of Units (SI) and chemical and physical results are expressed in milligrams per liter (mg/L). Record only the significant figures. If concentrations generally are less than 1 mg/L, it may be more convenient to express results in micrograms per liter (μg/L). Use μg/L when concentrations are less than 0.1 mg/L.

Express concentrations greater than 10 000 mg/L in percent, 1% being equal to 10 000 mg/L when the specific gravity is 1.00. In solid samples and liquid wastes of high specific gravity, make a correction if the results are expressed as parts per million (ppm) or percent by weight:

$$\text{ppm by weight} = \frac{mg/L}{sp\ gr}$$

$$\%\text{ by weight} = \frac{mg/L}{10\ 000 \times sp\ gr}$$

TABLE 1050:I. CONVERSION FACTORS*
(Milligrams per Liter—Milliequivalents per Liter)

Ion (Cation)	me/L = mg/L×	mg/L = me/L×	Ion (Anion)	me/L = mg/L×	mg/L = me/L×
Al^{3+}	0.111 2	8.994	BO_2^-	0.023 36	42.81
B^{3+}	0.277 5	3.603	Br^-	0.012 52	79.90
Ba^{2+}	0.014 56	68.67	Cl^-	0.028 21	35.45
Ca^{2+}	0.049 90	20.04	CO_3^{2-}	0.033 33	30.00
Cr^{3+}	0.057 70	17.33	CrO_4^{2-}	0.017 24	58.00
			F^-	0.052 64	19.00
Cu^{2+}	0.031 47	31.77	HCO_3^-	0.016 39	61.02
Fe^{2+}	0.035 81	27.92	HPO_4^{2-}	0.020 84	47.99
Fe^{3+}	0.053 72	18.62	$H_2PO_4^-$	0.010 31	96.99
H^+	0.992 2	1.008	HS^-	0.030 24	33.07
K^+	0.025 58	39.10	HSO_3^-	0.012 34	81.07
			HSO_4^-	0.010 30	97.07
Li^+	0.144 1	6.941	I^-	0.007 880	126.9
Mg^{2+}	0.082 29	12.15	NO_2^-	0.021 74	46.01
Mn^{2+}	0.036 40	27.47	NO_3^-	0.016 13	62.00
Mn^{4+}	0.072 81	13.73	OH^-	0.058 80	17.01
Na^+	0.043 50	22.99	PO_4^{3-}	0.031 59	31.66
NH_4^+	0.055 44	18.04	S^{2-}	0.062 38	16.03
Pb^{2+}	0.009 653	103.6	SiO_3^{2-}	0.026 29	38.04
Sr^{2+}	0.022 83	43.81	SO_3^{2-}	0.024 98	40.03
Zn^{2+}	0.030 59	32.69	SO_4^{2-}	0.020 82	48.03

* Factors are based on ion charge and not on redox reactions that may be possible for certain of these ions. Cations and anions are listed separately in alphabetical order.

In such cases, if the result is given as milligrams per liter, state specific gravity.

The unit equivalents per million (epm), or the identical and less ambiguous term milligram-equivalents per liter, or milliequivalents per liter (me/L), can be valuable for making water treatment calculations and checking analyses by anion-cation balance.

Table 1050:I presents factors for converting concentrations of common ions from milligrams per liter to milliequivalents per liter, and vice versa. The term milliequivalent used in this table represents 0.001 of an equivalent weight. The equivalent weight, in turn, is defined as the weight of the ion (sum of the atomic weights of the atoms making up the ion) divided by the number of charges normally associated with the particular ion. The factors for converting results from milligrams per liter to milliequivalents per liter were computed by dividing the ion charge by weight of the ion. Conversely, factors for converting results from milliequivalents per liter to milligrams per liter were calculated by dividing the weight of the ion by the ion charge.

1050 B. Significant Figures

1. Reporting Requirements

To avoid ambiguity in reporting results or in presenting directions for a procedure, it is the custom to use "significant figures." All digits in a reported result are expected to be known definitely, except for the last digit, which may be in doubt. Such a number is said to contain only significant figures. If more than a single doubtful digit is carried, the extra digit or digits are not significant. If an analytical result is reported as "75.6 mg/L," the analyst should be quite certain of the "75," but may be uncertain as to whether the ".6" should be .5 or .7, or even .4 or .8, because of unavoidable uncertainty in the analytical procedure. If the standard deviation were known from previous work to be ±2 mg/L, the analyst would have, or should have, rounded off the result to "76 mg/L" before reporting it. On the other hand, if the method were so good that a result of "75.61 mg/L" could have been conscientiously reported, then the analyst should not have rounded it off to 75.6.

Report only such figures as are justified by the accuracy of the work. Do not follow the all-too-common practice of requiring that quantities listed in a column have the same number of figures to the right of the decimal point.

2. Rounding Off

Round off by dropping digits that are not significant. If the digit 6, 7, 8, or 9 is dropped, increase preceding digit by one unit; if the digit 0, 1, 2, 3, or 4 is dropped, do not alter preceding digit. If the digit 5 is dropped, round off preceding digit to the nearest even number: thus 2.25 becomes 2.2 and 2.35 becomes 2.4.

3. Ambiguous Zeros

The digit 0 may record a measured value of zero or it may serve merely as a spacer to locate the decimal point. If the result of a sulfate determination is reported as 420 mg/L, the report recipient may be in doubt whether the zero is significant or not, because the zero cannot be deleted. If an analyst calculates a total residue of 1146 mg/L, but realizes that the 4 is somewhat doubtful and that therefore the 6 has no significance, the answer should be rounded off to 1150 mg/L and so reported but here, too, the report recipient will not know whether the zero is significant. Although the number could be expressed as a power of 10 (e.g., 11.5×10^2 or 1.15×10^3), this form is not used generally because it would not be consistent with the normal expression of results and might be confusing. In most other cases, there will be no doubt as to the sense in which the digit 0 is used. It is obvious that the zeros are significant in such numbers as 104 and 40.08. In a number written as 5.000, it is understood that all the zeros are significant, or else the number could have been rounded off to 5.00, 5.0, or 5, whichever was appropriate. Whenever the zero is ambiguous, it is advisable to accompany the result with an estimate of its uncertainty.

Sometimes, significant zeros are dropped without good cause. If a buret is read as "23.60 mL," it should be so recorded, and not as "23.6 mL." The first number indicates that the analyst took the trouble to estimate the second decimal place; "23.6 mL" would indicate a rather careless reading of the buret.

4. Standard Deviation

If a calculation yields as a result "1476 mg/L" with a standard deviation estimated as ±40 mg/L, report it as 1480 ± 40 mg/L. However, if the standard deviation is estimated as ±100 mg/L round off the answer still further and report as 1500 ±

100 mg/L. By this device, ambiguity is avoided and the report recipient can tell that the zeros are only spacers. Even if the problem of ambiguous zeros is not present, showing the standard deviation is helpful in that it provides an estimate of reliability.

5. Calculations

As a practical operating rule, round off the result of a calculation in which several numbers are multiplied or divided to as few significant figures as are present in the factor with the fewest significant figures. Suppose that the following calculations must be made to obtain the result of an analysis:

$$\frac{56 \times 0.003\ 462 \times 43.22}{1.684}$$

A ten-place calculator yields an answer of "4.975 740 998." Round off this number to "5.0" because one of the measurements that entered into the calculation, 56, has only two significant figures. It was unnecessary to measure the other three factors to four significant figures because the "56" is the "weakest link in the chain" and limits accuracy of the answer. If the other factors were measured to only three, instead of four, significant figures, the answer would not suffer and the labor might be less.

When numbers are added or subtracted, the number that has the fewest decimal places, not necessarily the fewest significant figures, puts the limit on the number of places that justifiably may be carried in the sum or difference. Thus the sum

```
     0.0072
    12.02
     4.0078
    25.9
  4886
  ─────────
  4927.9350
```

must be rounded off to "4928," no decimals, because one of the addends, 4886, has no decimal places. Notice that another addend, 25.9, has only three significant fig- ures and yet it does not set a limit to the number of significant figures in the answer.

The preceding discussion is necessarily oversimplified. The reader is referred to for more detailed mathematical texts discussion.

1060   COLLECTION AND PRESERVATION OF SAMPLES

1060 A.   Introduction

It is an old axiom that the result of any analytical determination can be no better than the sample on which it is performed. It is not practical to specify detailed procedures for the collection of all samples here because of varied purposes and analytical procedures. More detailed information appears in connection with specific methods. This section presents general considerations, applicable primarily to chemical analyses.

The objective of sampling is to collect a portion of material small enough in volume to be transported conveniently and handled in the laboratory while still accurately representing the material being sampled. This objective implies that the relative proportions or concentrations of all pertinent components will be the same in the samples as in the material being sampled, and that the sample will be handled in such a way that no significant changes in composition occur before the tests are made.

A sample may be presented to the laboratory for specific determinations with the collector taking responsibility for its validity. Often, in water and wastewater work, the laboratory conducts or prescribes the sampling program, which is determined in consultation with the user of the test results. Such consultation is essential to insure selecting samples and analytical methods that provide a true basis for an-

swering the questions that prompted the sampling.

1. General Precautions

Obtain a sample that meets the requirements of the sampling program and handle it in such a way that it does not deteriorate or become contaminated before it reaches the laboratory. Before filling, rinse sample bottle two or three times with the water being collected, unless the bottle contains a preservative or dechlorinating agent. Depending on determinations to be performed, fill container full (most organics determinations) or leave space for aeration, mixing, etc. (microbiological analyses). For samples that will be shipped, preferably leave an air space of about 1% of container capacity to allow for thermal expansion.

Special precautions are necessary for samples containing organic compounds and trace metals. Because many constituents may be present at concentrations of micrograms per liter, they may be totally or partially lost if proper sampling and preservation procedures are not followed.

Representative samples of some sources can be obtained only by making composites of samples collected over a period of time or at many different sampling points. The details of collection vary so much with local conditions that no specific recommendations would be universally applicable.

Sometimes it is more informative to analyze numerous separate samples instead of one composite so as not to obscure maxima and minima.

Sample carefully to insure that analytical results represent the actual sample composition. Important factors affecting results are the presence of suspended matter or turbidity, the method chosen for its removal, and the physical and chemical changes brought about by storage or aeration. Particular care is required when processing (grinding, blending, sieving, filtering) samples to be analyzed for trace constituents, especially metals and organic compounds. Some determinations, particularly of lead, can be invalidated by contamination from such processing. Treat each sample individually with regard to the substances to be determined, the amount and nature of turbidity present, and other conditions that may influence the results.

It is impractical to give directions covering all conditions, and the choice of technique for collecting a homogeneous sample must be left to the analyst's judgment. In general, separate any significant amount of suspended matter by decantation, centrifugation, or an appropriate filtration procedure. Often a slight turbidity can be tolerated if experience shows that it will cause no interference in gravimetric or volumetric tests and that its influence can be corrected in colorimetric tests, where it has potentially the greatest interfering effect. When relevant, state whether or not the sample has been filtered. To measure the total amount of a constituent, do not remove suspended solids but treat them appropriately.

Make a record of every sample collected and identify every bottle, preferably by attaching an appropriately inscribed tag or label. Record sufficient information to provide positive sample identification at a later date, including the name of the sample collector, the date, hour, and exact location, the water temperature, and any other data that may be needed for correlation, such as weather conditions, water level, stream flow, post-sampling handling, etc. Provide space on the label for the initials of those assuming sample custody and for the time and date of transfer. Fix sampling points by detailed description, by maps, or with the aid of stakes, buoys, or landmarks in a manner that will permit their identification by other persons without reliance on memory or personal guidance. Particularly when sample results are expected to be involved in litigation, use formal "chain-of-custody" procedures (see ¶ B.1 below), which trace sample history from collection to final reporting.

Cool hot samples collected under pressure while they are still under pressure.

Before collecting samples from distribution systems, flush lines sufficiently to insure that the sample is representative of the supply, taking into account the diameter and length of the pipe to be flushed and the velocity of flow.

Collect samples from wells only after the well has been pumped sufficiently to insure that the sample represents the groundwater source. Sometimes it will be necessary to pump at a specified rate to achieve a characteristic drawdown, if this determines the zones from which the well is supplied. Record pumping rate and drawdown.

When samples are collected from a river or stream, observed results may vary with depth, stream flow, and distance from shore and from one shore to the other. If equipment is available, take an "integrated" sample from top to bottom in the middle of the stream or from side to side at mid-depth, in such a way that the sample is integrated according to flow. If only a grab or catch sample can be collected, take it in the middle of the stream and at mid-depth.

Lakes and reservoirs are subject to considerable variations from normal causes such as seasonal stratification, rainfall, run-off, and wind. Choose location, depth, and frequency of sampling depending on local conditions and the purpose of the investigation. Avoid surface scum.

For certain constituents, sampling location is extremely important. Avoid areas of excessive turbulence because of potential loss of volatile constituents and of potential presence of toxic vapors. Avoid sampling at weirs because such locations tend to favor retrieval of lighter-than-water, immiscible compounds. Generally, collect samples beneath the surface in quiescent areas. If composite samples are required, take care that sample constituents are not lost during compositing because of improper handling of portions being pooled. For example, casual dumping together of portions rather than addition to the composite through a submerged siphon can cause unnecessary volatilization. When necessary refrigerate the composited portions to minimize volatilization[1].

Use only representative samples (or those conforming to a sampling program) for examination. The great variety of conditions under which collections must be made makes it impossible to prescribe a fixed procedure. In general, take into account the tests or analyses to be made and the purpose for which the results are needed.

2. Safety Considerations

Because sample constituents can be toxic, take adequate precautions during sampling and sample handling. Toxic substances can enter through the skin and, in the case of vapors, through the lungs. Inadvertent ingestion can occur via direct contact with foods or by adsorption of vapors onto foods. Precautions may be limited to wearing gloves or may include coveralls, aprons, or other protective apparel. Always wear eye protection. When toxic vapors might be present, sample only in well-ventilated areas or use a respirator or self-contained breathing apparatus. In a laboratory, open sample containers in a fume hood. Never have food near samples or sampling locations; always wash hands thoroughly before handling food.[1]

If flammable organic compounds may be present, take adequate precautions. Prohibit smoking near samples, sampling locations, and in the laboratory. Keep sparks, flames, and excessive heat sources away from samples and sampling locations. Avoid buildup of flammable vapors in a refrigerator storing samples because electrical arcing at contacts of the thermostat, the door-activated light switch, or other electrical components may trigger a fire or explosion. If flammable compounds are suspected or known to be present and samples are to be refrigerated, use only specially designed *explosion-proof* refrigerators.[1]

When in doubt as to the level of safety precautions needed, consult an appropriately trained industrial hygienist. Samples with radioactive contaminants require other safety considerations; consult a health physicist.

3. Types of Samples

a. Grab or catch samples: Strictly speaking, a sample collected at a particular time and place can represent only the composition of the source at that time and place. However, when a source is known to be fairly constant in composition over a considerable period of time or over substantial distances in all directions, then the sample may be said to represent a longer time period or a larger volume, or both, than the specific point at which it was collected. In such circumstances, some sources may be represented quite well by single grab samples. Examples are some water supplies, some surface waters, and rarely, some wastewater streams.

When a source is known to vary with time, grab samples collected at suitable intervals and analyzed separately can document the extent, frequency, and duration of these variations. Choose sampling intervals on the basis of the frequency with which changes may be expected, which may vary from as little as 5 min to as long as 1 h or more. Seasonal variations in natural systems may necessitate sampling over months. When the source composition varies in space rather than time, collect samples from appropriate locations.

Use great care in sampling wastewater sludges, sludge banks, and muds. No definite procedure can be given, but take every possible precaution to obtain a representative sample or one conforming to a sampling program.

b. Composite samples: In most cases, the term "composite sample" refers to a mixture of grab samples collected at the same sampling point at different times. Sometimes the term "time-composite" is used to distinguish this type of sample from others. Time-composite samples are most useful for observing average concentrations that will be used, for example, in calculating the loading or the efficiency of a wastewater treatment plant. As an alternative to the separate analysis of a large number of samples, followed by computation of average and total results, composite samples represent a substantial saving in laboratory effort and expense. For these purposes, a composite sample representing a 24-h period is considered standard for most determinations. Under certain circumstances, however, a composite sample representing one shift, or a shorter time period, or a complete cycle of a periodic operation, may be preferable. To evaluate the effects of special, variable, or irregular discharges and operations, collect composite samples representing the period during which such discharges occur.

For determining components or characteristics subject to significant and unavoidable changes on storage, do not use composite samples. Make such determinations on individual samples as soon as possible after collection and preferably at the sampling point. Analyses for all dissolved gases, residual chlorine, soluble sulfide, temperature, and pH are examples of this type of determination. Changes in such components as dissolved oxygen or carbon dioxide, pH, or temperature may produce secondary changes in certain inorganic constituents such as iron, manganese, alkalinity, or hardness. Use time-composite samples only for determining components that can be demonstrated to remain unchanged under the conditions of sample collection and preservation.

Take individual portions in a wide-mouth bottle having a diameter of at least 35 mm at the mouth and a capacity of at least 120 mL. Collect these portions every hour—in some cases every half hour or even every 5 min—and mix at the end of the sampling period or combine in a single bottle as collected. If preservatives are used, add them to the sample bottle initially so that all portions of the composite are preserved as soon as collected. Analysis of individual samples sometimes may be necessary.

It is desirable, and often essential, to combine individual samples in volumes proportional to flow. A final sample volume of 2 to 3 L is sufficient for sewage, effluents, and wastes.

Automatic sampling devices are available; however, do not use them unless the sample is preserved as described below. Clean sampling devices, including bottles, daily to eliminate biological growths and other deposits.

c. Integrated samples: For certain purposes, the information needed is provided best by analyzing mixtures of grab samples collected from different points simultaneously, or as nearly so as possible. Such mixtures sometimes are called integrated samples. An example of the need for such sampling occurs in a river or stream that varies in composition across its width and depth. To evaluate average composition or total loading, use a mixture of samples representing various points in the cross-section, in proportion to their relative flows. The need for integrated samples also may exist if combined treatment is proposed for several separate wastewater streams, the interaction of which may have a significant effect on treatability or even on composition. Mathematical prediction of the interactions may be inaccurate or impossible and testing a suitable integrated sample may provide more useful information.

Both natural and artificial lakes show variations of composition with both depth and horizontal location. However, under many conditions, neither total nor average results are especially significant; local variations are more important. In such cases,

examine samples separately rather than integrate them.

Preparation of integrated samples usually requires special equipment to collect a sample from a known depth without contaminating it with overlying water. Knowledge of the volume, movement, and composition of the various parts of the water being sampled usually is required. Therefore, collecting integrated samples is a complicated and specialized process that cannot be described in detail.

4. Reference

1. WATER POLLUTION CONTROL FEDERATION. 1986. Removal of Hazardous Wastes in Wastewater Facilities—Halogenated Organics. Manual of Practice FD-11, Water Pollution Control Fed., Alexandria, Va.

1060 B.   Collection of Samples

1. Chain-of-Custody Procedures

It is essential to ensure sample integrity from collection to data reporting. This includes the ability to trace possession and handling of the sample from the time of collection through analysis and final disposition. This is referred to as chain of custody and is important in the event of litigation involving the results. Where litigation is not involved, chain-of-custody procedures are useful for routine control of sample flow.

A sample is considered to be under a person's custody if it is in the individual's physical possession, in the individual's sight, secured in a tamper-proof way by that individual, or is secured in an area restricted to authorized personnel. The following procedures summarize the major aspects of chain of custody. More detailed discussions are available.[1,2]

a. *Sample labels:* Use labels to prevent sample misidentification. Gummed paper

labels or tags generally are adequate. Include at least the following information: sample number, name of collector, date and time of collection, and place of collection.

Affix labels to sample containers before or at the time of sampling. Fill label out with waterproof ink at time of collection.

b. *Sample seals:* Use sample seals to detect unauthorized tampering with samples up to the time of analysis. Use gummed paper seals that include, at least, the following information: sample number (identical with number on sample label), collector's name, and date and time of sampling. Plastic shrink seals also may be used.

Attach seal in such a way that it is necessary to break it to open the sample container. Affix seal to container before sample leaves custody of sampling personnel.

c. *Field log book:* Record all information pertinent to a field survey or sampling in a bound log book. As a minimum, include the following in the log book: purpose of sampling; location of sampling point; name and address of field contact; producer of material being sampled and address, if different from location; and type of sample. If sample is wastewater, identify process producing waste stream. Also provide suspected sample composition, including concentrations; number and volume of sample taken; description of sampling point and sampling method; date and time of collection; collector's sample identification number(s); sample distribution and how transported; references such as maps or photographs of the sampling site; field observations and measurements; and signatures of personnel responsible for observations. Because sampling situations vary widely no general rule can be given as to the information to be entered in the log book. It is desirable to record sufficient information so that one could reconstruct the sampling without reliance on the collector's memory. Protect the log book and keep it in a safe place.

d. Chain-of-custody record: Fill out a chain-of-custody record to accompany each sample or group of samples. The record includes the following information: sample number; signature of collector; date, time, and address of collection; sample type; signatures of persons involved in the chain of possession; and inclusive dates of possession.

e. Sample analysis request sheet: The sample analysis request sheet accompanies sample to the laboratory. The collector completes the field portion of such a form that includes most of the pertinent information noted in the log book. The laboratory portion of such a form is to be completed by laboratory personnel and includes: name of person receiving the sample, laboratory sample number, date of sample receipt, and determinations to be performed.

f. Sample delivery to laboratory: Deliver sample to laboratory as soon as practicable. Accompany sample with chain-of-custody record and a sample analysis request sheet. Deliver sample to sample custodian.

g. Receipt and logging of sample: In the laboratory, the sample custodian receives the sample and inspects its condition and seal, reconciles label information and seal against the chain-of-custody record, assigns a laboratory number, logs sample in the laboratory log book, and stores it in a secured storage room or cabinet until it is assigned to an analyst.

h. Assignment of sample for analysis: The laboratory supervisor usually assigns the sample for analysis. Once in the laboratory, the supervisor or analyst is responsible for the sample's care and custody.

2. Sampling Methods

a. Manual sampling: Manual sampling involves no equipment but may be unduly costly and time-consuming for routine or large-scale sampling programs.

b. Automatic sampling: Automatic samplers can eliminate human errors in manual sampling, can reduce labor costs, may provide the means for more frequent sampling,[3] and are used increasingly. Be sure that the automatic sampler does not contaminate the sample. For example, plastic components may be incompatible with certain organic compounds that are soluble in the plastic parts. If sample constituents are generally known, contact the manufacturer of an automatic sampler regarding potential incompatibility of plastic components. Manual sampling with a glass container and in accordance with appropriate safety procedures may be best.[3]

Program an automatic sampler in accordance with sampling needs. Carefully match pump speeds and tubing sizes to the type of sample to be taken.

3. Sample Containers

The type of sample container used is of utmost importance. Containers typically are made of plastic or glass, but one material may be preferred over the other. For example, silica and sodium may be leached from glass but not plastic, and trace levels of metals may sorb onto the walls of glass containers.[4] For samples containing organics, avoid plastic containers except those made of fluorinated polymers such as polytetrafluoroethylene (TFE).[3]

From samples containing volatile organics some compounds may dissolve into the walls of plastic containers or such compounds may even leach substances from the plastic. Container failure due to breakdown of the plastic is possible. Some organics are compatible with certain plastics (see manufacturer's literature). However, even if compatibility is assured, recognize that the walls of a plastic container can be porous to volatile organics. Glass containers generally are preferred with volatile organics.[3] Container caps, typically plastic, also can be a problem with organics. Use foil or TFE liners. Serum vials with TFE-lined rubber or plastic septa are useful.

4. Number of Samples

Given the random variations in both an analytical procedure and the occurrence of a constituent at a point of sampling, a single sample may be insufficient for a desired level of uncertainty. If an overall standard deviation is known, the required number of samples may be established by the following relationship:[4]

$$N \geq \left(\frac{ts}{U}\right)^2$$

where:

N = number of samples,
t = Student-t statistic for a given confidence level
s = overall standard deviation, and
U = acceptable level of uncertainty.

To assist in calculations, use curves such as those in Figure 1060:1. As an example, if s is 0.5 mg/L, U is $\pm$ 0.2 mg/L, and a 95% confidence level is desired, approximately 25 to 30 samples must be taken.

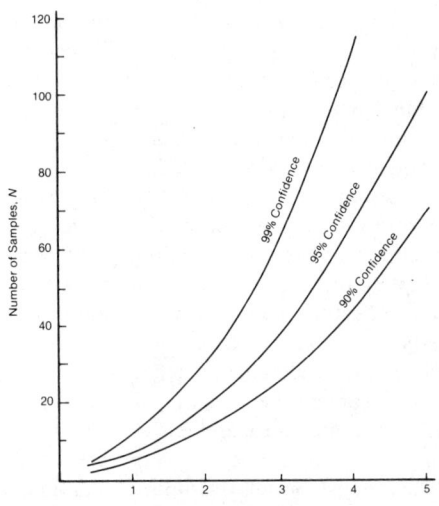

Figure 1060:1. Approximate number of samples required in estimating a mean concentration. Source: Methods for the Examination of Waters and Associated Materials: General Principles of Sampling and Accuracy of Results. 1980. Her Majesty's Stationery Off., London, England.

5. Quantity

Collect a 2-L sample for most physical and chemical analyses. For certain determinations, larger samples may be necessary. Table 1060:I shows the volumes ordinarily required for analyses.

Do not use the same sample for chemical (organic and inorganic), bacteriological, and microscopic examinations because methods of collecting and handling are different.

6. References

1. U.S. ENVIRONMENTAL PROTECTION AGENCY. 1986. Test Methods for Evaluating Solid Waste: Physical/Chemical Methods, 3rd ed. Publ. No. SW-846, Office of Solid Waste and Emergency Response, Washington, D.C.

2. U.S. ENVIRONMENTAL PROTECTION AGENCY. 1982. NEIC Policies and Procedures. EPA-330/9/78/001/-R (rev. 1982).

3. WATER POLLUTION CONTROL FEDERATION. 1986. Removal of Hazardous Wastes in Waste-

Determination	Container	Minimum Sample Size mL	Preservation	Maximum Storage Recommended/ Regulatory†
Acidity	P, G(B)	100	Refrigerate	24 h/14 d
Alkalinity	P, G	200	Refrigerate	24 h/14 d
BOD	P, G	1000	Refrigerate	6 h/48 h
Boron	P	100	None required	28 d/6 months
Bromide	P, G	—	None required	28 d/28 d
Carbon, organic, total	G	100	Analyze immediately; or refrigerate and add HCl to pH <2	7 d/28 d
Carbon dioxide	P, G	100	Analyze immediately	stat/N.S.
COD	P, G	100	Analyze as soon as possible, or add H_2SO_4 to pH <2; refrigerate	7 d/28 d
Chlorine, residual	P, G	500	Analyze immediately	0.5 h/stat
Chlorine dioxide	P, G	500	Analyze immediately	0.5 h/N.S.
Chlorophyll	P, G	500	30 d in dark	30 d/N.S.
Color	P, G	500	Refrigerate	48 h/48 h
Conductivity	P, G	500	Refrigerate	28 d/28 d
Cyanide:				
Total	P, G	500	Add NaOH to pH >12, refrigerate in dark	24 h/14 d; 24 h if sulfide present
Amenable to chlorination	P, G	500	Add 100 mg $Na_2S_2O_3$/L	stat/14 d; 24 h if sulfide present
Fluoride	P	300	None required	28 d/28 d
Hardness	P, G	100	Add HNO_3 to pH <2	6 months/6 months
Iodine	P, G	500	Analyze immediately	0.5 h/N.S.
Metals, general	P(A), G(A)	—	For dissolved metals filter immediately, add HNO_3 to pH <2	6 months/6 months
Chromium VI	P(A), G(A)	300	Refrigerate	24 h/24 h
Copper by colorimetry*				
Mercury	P(A), G(A)	500	Add HNO_3 to pH <2, 4°C, refrigerate	28 d/28 d
Nitrogen:				
Ammonia	P, G	500	Analyze as soon as possible or add H_2SO_4 to pH <2, refrigerate	7 d/28 d
Nitrate	P, G	100	Analyze as soon as possible or refrigerate	48 h/48 h (28d for chlorinated samples)
Nitrate + nitrite	P, G	200	Add H_2SO_4 to pH <2, refrigerate	none/28d
Nitrite	P, G	100	Analyze as soon as possible or refrigerate	none/48 h
Organic, Kjeldahl	P, G	500	Refrigerate; add H_2SO_4 to pH <2	7 d/28 d

TABLE 1060:I, CONT.*

Determination	Container	Minimum Sample Size mL	Preservation	Maximum Storage Recommended/Regulatory†
Odor	G	500	Analyze as soon as possible; refrigerate	6 h/N.S.
Oil and grease	G, wide-mouth calibrated	1000	Add H_2SO_4 to pH <2, refrigerate	28 d/28 d
Organic compounds:				
Pesticides	G(S), TFE-lined cap	—	Refrigerate; 1000 mg ascorbic acid/L if residual chlorine present	7 d/7 d until extraction; 40 d after extraction
Phenols	P, G	500	Refrigerate, add H_2SO_4 to pH <2	*/28 d
Purgeables by purge and trap	G, TFE-lined cap	50	Refrigerate; add HCl to pH < 2; 1000 mg ascorbic acid/L if residual chlorine present	7 d/14 d
Oxygen, dissolved:				
Electrode	G, BOD bottle	300	Analyze immediately	0.5 h/stat
Winkler			Titration may be delayed after acidification	8 h/8 h
Ozone	G	1000	Analyze immediately	0.5 h/N.S.
pH	P, G	—	Analyze immediately	2 h/stat
Phosphate	G(A)	100	For dissolved phosphate filter immediately; refrigerate	48 h/N.S.
Salinity	G, wax seal	240	Analyze immediately or use wax seal	6 months/N.S.
Silica	P	—	Refrigerate, do not freeze	28 d/28 d
Sludge digester gas	G, gas bottle	—	—	N.S.
Solids	P, G	—	Refrigerate	7 d/2–7 d; see cited reference
Sulfate	P, G	—	Refrigerate	28 d/28 d
Sulfide	P, G	100	Refrigerate; add 4 drops 2N zinc acetate/100 mL; add NaOH to pH > 9	28 d/7 d
Taste	G	500	Analyze as soon as possible; refrigerate	24 h/N.S.
Temperature	P, G	—	Analyze immediately	stat/stat
Turbidity	P, G	—	Analyze same day; store in dark up to 24 h, refrigerate	24 h/48 h

* See text for additional details. For determinations not listed, use glass or plastic containers; preferably refrigerate during storage and analyze as soon as possible. Refrigerate = storage at 4°C, in the dark. P = plastic (polyethylene or equivalent); G = glass; G(A) or P(A) = rinsed with 1 + 1 HNO_3; G(B) = glass, borosilicate; G(S) = glass, rinsed with organic solvents; stat = no storage allowed; analyze immediately.

† Environmental Protection Agency, Rules and Regulations. *Federal Register* 49; No. 209, October 26, 1984. See this citation for possible differences regarding container and preservation requirements.

water Facilities—Halogenated Organics. Manual of Practice FD-11, Water Pollution Control Fed., Alexandria, Va.

4. Methods for the Examination of Waters and Associated Materials: General Principles of Sampling and Accuracy of Results. 1980. Her Majesty's Stationery Off., London, England.

1060 C. Sample Preservation

Complete and unequivocal preservation of samples, whether domestic wastewater, industrial wastes, or natural waters, is a practical impossibility. Regardless of the sample nature, complete stability for every constituent never can be achieved. At best, preservation techniques only retard chemical and biological changes that inevitably continue after sample collection.

1. Sample Storage before Analysis

a. *Nature of sample changes:* Some determinations are more likely than others to be affected by sample storage before analysis. Certain cations are subject to loss by adsorption on, or ion exchange with, the walls of glass containers. These include aluminum, cadmium, chromium, copper, iron, lead, manganese, silver, and zinc, which are best collected in a separate clean bottle and acidified with nitric acid to a pH below 2.0 to minimize precipitation and adsorption on container walls.

Temperature changes quickly; pH may change significantly in a matter of minutes; dissolved gases (oxygen, carbon dioxide) may be lost. Determine temperature, pH, and dissolved gases in the field. With changes in the pH-alkalinity-carbon dioxide balance, calcium carbonate may precipitate and cause a decrease in the values for calcium and for total hardness.

Iron and manganese are readily soluble in their lower oxidation states but relatively insoluble in their higher oxidation states; therefore, these cations may precipitate or they may dissolve from a sediment, depending on the redox potential of the sample. Microbiological activity may be responsible for changes in the nitrate-nitrite-ammonia content, for decreases in phenol concentration and in BOD, or for reducing sulfate to sulfide. Residual chlorine is reduced to chloride. Sulfide, sulfite, ferrous iron, iodide, and cyanide may be lost through oxidation. Color, odor, and turbidity may increase, decrease, or change in quality. Sodium, silica, and boron may be leached from the glass container. Hexavalent chromium may be reduced to chromic ion.

Biological changes taking place in a sample may change the oxidation state of some constituents. Soluble constituents may be converted to organically bound materials in cell structures, or cell lysis may result in release of cellular material into solution. The well-known nitrogen and phosphorus cycles are examples of biological influences on sample composition.

Zero head-space is important in preservation of samples with volatile organics. Avoid loss of volatile materials by collecting sample in a completely filled container. Achieve this by overfilling bottle before capping or sealing. Serum vials with septum caps are particularly useful in that a sample portion for analysis can be taken through the cap by using a syringe.[1]

b. *Time interval between collection and analysis:* In general, the shorter the time that elapses between collection of a sample and its analysis, the more reliable will be the analytical results. For certain constituents and physical values, immediate analysis in the field is required. For composited samples it is common practice to use the time at the end of composite collection as the sample collection time.

It is impossible to state exactly how much elapsed time may be allowed between sample collection and analysis; this depends on the character of the sample, the analyses to be made, and the conditions of storage. Changes caused by growth of microorganisms are greatly retarded by keeping the sample in the dark and at a low temperature. When the interval between sample collection and analysis is long enough to produce changes in either the concentration or the physical state of the constituent to be measured, follow the preservation practices given in Table 1060:I. Record time elapsed between sampling and analysis, and which preservative, if any, was added.

2. Preservation Techniques

To minimize the potential for volatization or biodegradation between sampling and analysis, keep samples as cool as possible without freezing. Preferably pack samples in crushed or cubed ice or commercial ice substitutes before shipment. Avoid using dry ice because it will freeze samples and may cause glass containers to break. Dry ice also may effect a pH change in samples. Keep composite samples cool with ice or a refrigeration system set at 4°C during compositing. Analyze samples as quickly as possible on arrival at the laboratory. If immediate analysis is not possible, storage at 4°C is recommended for most samples.[1]

Use chemical preservatives only when they are shown not to interfere with the analysis being made. When they are used, add them to the sample bottle initially so that all sample portions are preserved as

soon as collected. No single method of preservation is entirely satisfactory; choose the preservative with due regard to the determinations to be made. Because a preservation method for one determination may interfere with another one, samples for multiple determinations may need to be split and preserved separately. All methods of preservation may be inadequate when applied to suspended matter. Because formaldehyde affects so many analyses, do not use it.

Methods of preservation are relatively limited and are intended generally to retard biological action, retard hydrolysis of chemical compounds and complexes, and reduce volatility of constituents.

Preservation methods are limited to pH control, chemical addition, the use of amber and opaque bottles, refrigeration, filtration, and freezing. Table 1060:I lists preservation methods by constituent.

The foregoing discussion is by no means exhaustive and comprehensive. Clearly it is impossible to prescribe absolute rules for preventing all possible changes. Additional advice will be found in the discussions under individual determinations, but to a large degree the dependability of an analytical determination rests on the experience and good judgment of the person collecting the sample.

3. Reference

1. WATER POLLUTION CONTROL FEDERATION. 1986. Removal of Hazardous Wastes in Wastewater Facilities—Halogenated Organics. Manual of Practice FD-11, Water Pollution Control Fed., Alexandria, Va.

4. Bibliography

KEITH, L.H., ed. 1988. Principles of Environmental Sampling. ACS Professional Reference Book, American Chemical Soc.

1070 LABORATORY APPARATUS, REAGENTS, AND TECHNIQUES

1070 A. Introduction

This section contains a discussion of requirements for laboratory apparatus, reagents, and techniques that are common to many of the analyses presented in this manual. Requirements that are more or less specific to the particular determination being performed are described under the methods to which they apply. In addition, observe the general considerations presented in this section. The requirements of radiological, bacteriological, biological, and bioassay methods tend to differ in many respects from those of chemical and physical tests. Special attention is directed to the descriptions of apparatus and procedures in the sections dealing with those methods.

1070 B. Apparatus

1. Containers

For general laboratory use, the most suitable material for containers is resistant borosilicate glass.* Special glassware is available with characteristics such as high resistance to alkali attack, low boron content, or exclusion of light. Choose stoppers, caps, and plugs to resist the attack of material contained in the vessel. Metal screw caps are a poor choice for samples that will cause them to corrode readily. Glass stoppers are unsatisfactory for strongly alkaline liquids because of their tendency to stick fast. Rubber stoppers are excellent for alkaline liquids but unacceptable for organic solvents, in which they swell or disintegrate, or trace metal solutions, which may be contaminated by them. Use polytetrafluoroethylene (TFE or PTFE)† for burets that contain strongly alkaline liquids. When appropriate, use other materials such as porcelain, nickel, iron, platinum, stainless steel, and high-silica glass.‡

Collect and store samples in bottles made of borosilicate glass, hard rubber, plastic, or other inert material as appropriate for specific analyses (Table 1060:I).

For relatively short storage periods, or for constituents that are not affected by storage in soft glass, such as calcium, magnesium, sulfate, chloride, and perhaps others, the 2.5-L acid-bottle "bell closure" is satisfactory. This closure holds a glass or polyethylene disk against the ground-glass surface or a polyethylene insert on the bottle lip and insures adequate protection. If part of the sample is to be analyzed later for silica, sodium, or other substances that would be affected by prolonged storage in soft glass, transfer it to a small plastic bottle, while leaving the remainder of the sample in the soft-glass bottle.

Carefully clean sample bottles before each use. Rinse glass bottles, except those to be used for chromium or manganese analyses, either with a cleaning mixture made by adding 1 L conc H_2SO_4 slowly, with stirring, to 35 mL saturated sodium dichromate solution, or with 2% $KMnO_4$ in 5% KOH solution followed by an oxalic acid solution. Commercial alternates also are available.§ Rinse with other concen-

---

*Pyrex, manufactured by Corning Glass Works; Kimax, Kimble Glass Co., Division of Owens-Illinois; or equivalent.
†Teflon or equivalent.
‡Vycor, manufactured by Corning Glass Works, or equivalent.

§Nochromix, Godax Laboratories, Inc., New York, N.Y., or equivalent.

trated acids to remove inorganic matter. Detergents are excellent cleansers for many purposes; use either detergents or conc HCl for cleaning hard-rubber and plastic bottles. After the bottles have been cleaned, rinse thoroughly with reagent-grade water.

For shipment, pack bottles in wooden, metal, plastic or heavy fiberboard cases, with a separate compartment for each bottle. Line boxes with insulating material to protect from breakage. Samples stored in plastic bottles need no protection against breakage by impact or through freezing.

2. Volumetric Glassware

Calibrate volumetric glassware or obtain a certificate of accuracy from a competent laboratory. Volumetric glassware is calibrated either "to contain" (TC) or "to deliver" (TD). Glassware designed "to deliver" will do so with accuracy only when the inner surface is so scrupulously clean that water wets it immediately and forms a uniform film upon emptying. Whenever possible, use borosilicate glassware. For accurate work use Class A volumetric glassware.

Carefully measure weights and volumes in preparing standard solutions and calibration curves. Observe similar precautions in measuring sample volumes. Use volumetric pipets or burets where the volume is designated to two decimal places (X.00 mL) in the text. Use volumetric flasks where specified and where the volume is given as 1000 mL rather than 1 L.

3. Nessler Tubes

Unless otherwise indicated use "tall"-form nessler tubes made of resistant glass and selected from uniformly drawn tubing. The glass should be clear and colorless and the tube bottoms should be plane-parallel. When the tubes are filled with liquid and viewed from the top with a light source beneath, there should be no dark spots nor any lenslike distortion of the transmitted light. The tops of the tubes should be flat, preferably fire-polished, and smooth enough to permit cover slips to be cemented on for sealing. Nessler tubes with standard-taper clear glass tops are available commercially. Graduation marks should completely encircle the tubes.

The 100-mL tubes should have a total length of approximately 375 mm. Their inside diameter should approximate 20 mm and the outside diameter 24 mm. The graduation mark should be as near as possible to 300 mm above the inside bottom. Tubes sold in sets should be of such uniformity that this distance does not vary more than 6 mm. (Sets are available commercially in which the maximum difference between tubes is not more than 2 mm.) A graduation mark at 50 mL is permissible.

The 50-mL tubes should have a total length of about 300 mm. Their inside diameter should approximate 17 mm and the outside diameter 21 mm. The graduation mark on the tube should be as near as possible to 225 mm above the inside bottom. Tubes sold in sets should be of such uniformity that this distance does not vary more than 6 mm. (Sets are available commercially in which the maximum difference between tubes is not more than 1.5 mm.) A graduation mark at 25 mL is permissible.

1070 C.   Reagents

1. Laboratory Water

See Section 1080.

2. Reagent Quality

Use only the best quality chemical reagents even though this instruction is not repeated in the description of a particular method. Order chemicals for which the American Chemical Society has published specifications in the "ACS grade." Order other chemicals as "analytical reagent grade" or "spectral grade organic solvents." Methods of checking purity of suspect reagents are found in books of reagent specifications listed in the Bibliography (¶ 4 below).

Unfortunately, many commercial dyes for which the ACS grade has not been established fail to meet exacting analytical requirements because of variations in the color response of different lots. In such cases, use dyes certified by the Biological Stain Commission.

Where neither an ACS grade nor a certified Biological Stain Commission dye is available, purify the solid dye through recrystallization.

The following standard substances, each bottle of which is accompanied by a certificate of analysis, are issued by the National Institute of Standards and Technology (NIST)*, Department of Commerce, Washington, D.C., for the purpose of standardizing analytical solutions:

Acidimetric:
    84j—Acid potassium phthalate
   350a—Benzoic acid
Oxidimetric:
    40h—Sodium oxalate
    83d—Arsenic trioxide
   136e—Potassium dichromate
Buffer:
   185f—Potassium hydrogen phthalate
  186Ic—Potassium dihydrogen phosphate
 186IIc—Disodium hydrogen phosphate
   187c—Borax
    188—Potassium hydrogen tartrate
  189a—Potassium tetroxalate
  191a—Sodium bicarbonate
  192a—Sodium carbonate

Many hundreds of other standards issued by NIST are described in Special Publication 260 (see Bibliography).

A successful dithizone test demands reagents of the highest purity. Chloroform and carbon tetrachloride are available in a grade declared to be suitable for dithizone

methods. Select reagents of this quality for the dithizone methods described in this manual.

Water-soluble sodium salts of the common indicators usually are recommended for indicator preparation because of their general availability and reasonable cost.

When alcohol or ethyl alcohol is specified for preparing such indicators as phenolphthalein, use 95% ethyl alcohol. When isopropyl alcohol is specified use a similar grade.

Certain organic reagents are somewhat unstable upon exposure to the atmosphere. If the stability of a chemical is limited or unknown, purchase small lots at frequent intervals.

Dry all anhydrous reagent chemicals required for preparation of standard calibration solutions and titrants in an oven at 105 to 110°C for at least 1 to 2 h and preferably overnight. After cooling to room temperature in an efficient desiccator, promptly weigh the proper amount for dissolution. Should a different drying temperature be necessary, this is specified for the particular chemical. For hydrated salts, substitute milder drying in an efficient desiccator for oven-drying.

3. Common Acid and Alkali Solutions

a. Concentration units used: Reagent concentrations are expressed in terms of normality, molarity, and additive volumes.

A *normal solution (N)* contains one gram equivalent weight of solute per liter of solution.

A *molar solution (M)* contains one gram molecular weight of solute per liter of solution.

In additive volumes $(a + b)$, the first number, a, refers to the volume of concentrated reagent; the second number, b, refers to the volume of distilled water required for dilution. Thus, "1 + 9 HCl" denotes that 1 volume of concentrated HCl is to be diluted with 9 volumes of distilled water.

---

*Formerly National Bureau of Standards (NBS).

To make a solution of exact normality from a chemical that cannot be measured as a primary standard, prepare a relatively concentrated stock solution and then make an exact dilution to the desired strength. Alternatively, make a solution of slightly higher concentration than that desired, standardize, and make suitable adjustments in concentration by dilution; or, use the solution as first standardized and modify the calculation factor. This last procedure is useful especially for solutions that slowly change strength and must be restandardized frequently—for example, sodium thiosulfate solution. Adjustment to exact normality specified is desirable when a laboratory makes a large number of determinations with one standard solution.

Determinations are in accord with the instructions in this manual as long as normality of a standard solution does not result in a titration volume so small as to preclude accurate measurement or so large as to cause abnormal dilution of the reaction mixture, and as long as the solution is standardized properly and the calculations are made properly.

b. Preparation and dilution of solutions: If a solution of exact normality is to be prepared by dissolving a weighed amount of a primary standard or by diluting a stronger solution, bring it up to exact volume in a volumetric flask.

Accurately prepare stock and standard solutions prescribed for colorimetric determinations in volumetric flasks. Where concentration does not need to be exact, mix the concentrated solution or solid with measured amounts of water, using graduated cylinders for these measurements. There is usually a significant change of volume when strong solutions are mixed, resulting in a total volume less than the sum of volumes used. For approximate dilutions, volume changes are negligible when concentrations of 6*N* or less are diluted.

Mix thoroughly and completely when making dilutions. One of the most common

sources of error in analyses using standard solutions diluted in volumetric flasks is failure to attain complete mixing.

c. Storage of solutions: Some standardized solutions alter slowly because of chemical or biological changes. The practical life, required frequency of standardization, or storage precautions are indicated for such standards. Others, such as dilute HCl, are nonreactive. Yet their strength, too, may change by evaporation that is not prevented by a glass stopper. Changes in temperatures cause a bottle to "breathe," and allow some evaporation.

Do not consider a standard valid for more than a year unless it is restandardized. It is valid for that length of time only if conditions minimize evaporation and it has been demonstrated previously that preservation techniques are adequate. If the bottle is opened often or it is much less than half full, significant evaporation occurs in a few months. Verify concentration of standard solutions that have been stored.

Use glass bottles of chemically resistant glass except where glass is incompatible (e.g., silica solutions). For standard solutions that do not react with rubber or neoprene, use stoppers of these materials, because they can, if properly fitted, prevent evaporation as long as the bottle is closed. Screw-cap bottles also are effective. If the cap has a gasket of reasonably resistant material, permissible usage will be about the same as for rubber stoppers.

d. Hydrochloric and sulfuric acid as alternatives: Dilute standardized H_2SO_4 and HCl are called for in various procedures. Often these solutions are interchangeable. Where one is mentioned, the other may be used if it is known to make no difference.

e. Preparation: Although instructions usually describe preparation of 1 L of solution, prepare smaller or larger volumes as needed. Instructions calling for the preparation of 100 mL usually involve either short-life reagents or those used in small amounts.

A safe general rule is to add more concentrated acid or alkali to water, with stirring, in a vessel that can withstand thermal shock, and then to dilute to final volume after cooling to room temperature.

f. Uniform reagent concentrations: An attempt has been made to establish a number of uniform common acid and base concentrations that will serve for adjustment of pH of samples before color development or final titration. The following acid concentrations are recommended for general laboratory use: the concentrated reagent of commerce, $6N$, $1N$, $0.1N$, and $0.02N$. See the inside front cover for directions for preparing these acid concentrations, as well as the required $15N$, $6N$, and $1N$ NaOH solutions, and $5N$, $3N$, and $0.2N$ NH$_4$OH solutions.

4. Bibliography

ROSIN, J. 1967. Reagent Chemicals and Standards, 5th ed. D. Van Nostrand Co., Princeton, N.J.

AMERICAN CHEMICAL SOCIETY. 1974. Reagent Chemicals—American Chemical Society Specifications, 5th ed. American Chemical Soc., Washington, D.C.

The United States Pharmacopoeia. 1975. 19th rev. U.S. Pharmacopeial Convention Inc., Rockville, Md.

NATIONAL BUREAU OF STANDARDS. 1988. Catalog of NBS Standard Reference Materials. Nat. Bur. Standards Spec. Publ. 260, 1988-89 ed.

1070 D.   Techniques

Numerous works on both general and specific analytical techniques are available (see ¶ 5, Bibliography).

1. Ion Exchange

Ion-exchange resins are useful and flexible tools. Used in atmospheric- or high-pressure columns they effect analytical separations of both inorganic and organic ions. They commonly are in the form of sodium or hydrogen counter-ions (attached to the matrix) for cation exchangers and of chloride, formate, acetate, and hydroxide counter-ions for anion exchangers. The user may substitute other counter-ions by passing regenerating solutions through a resin column as recommended by the manufacturer. The form to be used in a specific case will depend not only on the relative affinity of the resin for counter-ions and sample ions, but also on the ions that can be tolerated if the concentrated sample ions are to be eluted for analysis. Sequential elution of organics usually is done with carefully selected buffer solutions.

In water analysis, ion exchangers can be applied to: (*a*) remove interfering ions, (*b*) determine total ion content, (*c*) indicate the approximate volume of sample for certain gravimetric determinations, (*d*) concentrate trace quantities of cations, and (*e*) separate anions from cations. This manual recommends the use of ion-exchange resins for the removal of interference in the sulfate determination and for the determination of total ion content. Inasmuch as the ion-exchange process can be applied in other determinations, a brief description of typical operations follows.

a. Selection of method: The batch method of ion exchange is satisfactory for sample volumes of less than 100 mL, although it does not result in complete ion removal or exchange. The column method can be used for any sample volume. In the batch method, the resin is agitated with the sample for a given time, after which the resin is removed by filtration. The column method is more efficient in that it provides continuous contact between the sample and the resin, thereby enabling the exchange reaction to go to completion. In this modification, the solution passes slowly through the resin bed and ions are removed quantitatively from the sample. Elution of the resin permits recovery of the exchanged substances.

b. Procedure: Use resins specifically manufactured for analytical applications. Prepare the ion exchanger by rinsing the resin with several volumes of ion-free water (good-quality distilled water) to remove any fines or coloring matter and other leachable material that might interfere with subsequent colorimetric procedures.

1) *Batch method for cation removal*—Pipet a sample portion containing 0.1 to 0.2 milliequivalents (me) of cations into a 250-mL erlenmeyer flask or beaker and add enough distilled water to bring final volume to 75 mL. Add 2.0 g strongly acidic cation-exchange resin and stir at moderate speed for 15 min. Filter through a plug of glass wool placed in the neck of a 10-cm borosilicate glass funnel. When filtration is complete, wash resin with two 10-mL portions of distilled water and make up to 100 mL total volume with distilled water.

To regenerate resin, transfer spent resin from the batch procedure to a flask containing 500 mL $3N$ HNO_3. When sufficient resin has accumulated, wash into a column (Figure 1070:1) and regenerate by passing $3N$ HNO_3 through the column at a rate of 0.1 to 0.2 mL acid/mL resin/min. Use about 20 mL $3N$ HNO_3/mL resin in the column. Finally, wash resin with sufficient distilled water until the effluent pH is 5 to 7, using the same rate of flow as in the regeneration step. Remove resin from column and store under distilled water in a wide-mouth container. Should the water become colored during storage, decant and replace with fresh distilled water. Before use, filter resin through a plug of glass wool placed in the neck of a funnel, wash with distilled water, and let drain. The resin is then ready for use.

2) *Column method for cation re-*

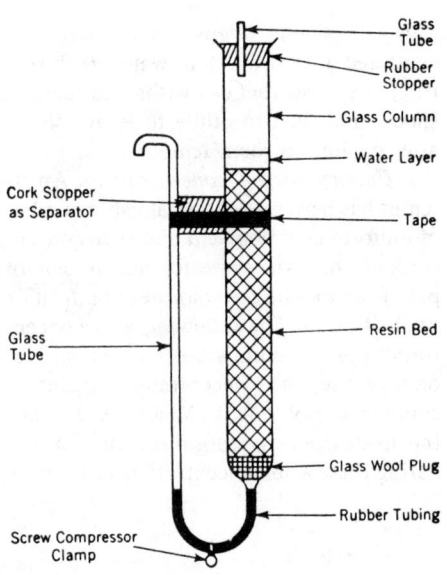

Figure 1070:1. Ion-exchange column.

moval—Prepare column as depicted in Figure 1070:1 (length of resin bed, 21.5 cm; diameter of column, 1.3 cm; representing approximately 21 mL, or 20 g resin). Other ion-exchange columns can be used equally well. One of the simplest consists of a buret containing a plug of glass wool immediately above the stopcock.

To make an effluent tube for a buret, *slowly* bend TFE tubing, 1/8 in. ID $\times$ 1/4 in. OD, into an elongated S-shape; bend to 45° angle to the straight portion, secure with rubber bands or adhesive tape, and set aside for at least 48 h to relieve stress. Repeat until desired conformation is achieved. Retain curved ends in place with plastic strips approximately 2 cm $\times$ 10 cm $\times$ 2 mm with suitably placed 3/16-in. drilled holes.

(Whatever type of column is adopted, never let liquid level in column fall below upper resin surface because trapped air causes uneven flow rates and poor efficiency of ion exchange. Adjust sample and column to the same temperature.)

Charge column by stirring resin in a

beaker with distilled water and then carefully wash the suspension through a funnel into the column already filled with distilled water. Backwash column if necessary by introducing distilled water at the bottom and passing it upward through the column until all air bubbles and channels are removed. Connect a separatory funnel to the column top or use an inverted volumetric flask to feed sample, regenerating, or rinsing solutions; make certain diameter of flask neck is large enough to permit automatic feed. More efficiently, use a small controllable peristaltic pump to apply solutions to column. Let sample flow through column at the rate of 0.2 mL solution/mL resin/min. After sample has passed through column, wash resin with distilled water until effluent pH is 5 to 7. Use a pH meter or pH test papers to determine when column has been washed free of acid. For convenience, when rinsing column or adsorbing cations from a sample of one or more liters, start this operation before the close of a workday and let exchange process proceed overnight. The column will not dry because of the curved outlet.

After distilled-water wash, elute adsorbed cations by passing 100 mL $3N$ HNO$_3$ through the column at a rate of 0.2 mL acid/mL resin /min. Because a volume of 100 mL $3N$ HNO$_3$ quantitatively removes 3 me of cations, use additional increments of 100 mL $3N$ HNO$_3$ for quantities of adsorbed cations in excess of 3 me. After elution, rinse column free of acid with enough distilled water to produce an effluent pH of 5 to 7. Wash at same flow rate for acid elution. The acid elution and washing regenerate the column for future use. The combined acid eluates contain the cations originally present in the sample.

2. Colorimetric Determinations

a. General: Many procedures depend on color determination with photometric instruments. To obtain the best possible results, understand the principles and limitations of these methods, especially because the choice of instrument and of technique is discretionary.

Photometric methods are not free from specific limitations. While an analyst will recognize that something has gone wrong on seeing an unusual color or turbidity when making a visual comparison, such a discrepancy easily may escape detection during a photometric reading, for the instrument always will yield a reading, whether meaningful or not. Check sensitivity and accuracy frequently by testing standard solutions to detect electrical, mechanical, or optical problems in the instrument and its accessories. Testing, maintaining, and repairing such instruments may call for specialized skills.

A photometer is not uniformly accurate over its entire scale. At very high absorbance the scale is crowded, so that a considerable change in relative concentration will cause only a slight change in position of the indicator dial or needle. At very low absorbance, slight differences between optical cells, the presence of condensed moisture, dust, bubbles, fingerprints, or a slight lack of reproducibility in positioning the cells can cause as great a change in readings as would a considerable change in concentration. The difficulties are minimized if readings are made to fall between 0.1 and 1.0 absorbance by diluting or concentrating the sample or varying the light path by selecting cells of appropriate size.

Some suggestions for suitable ranges and light paths are offered under individual methods in this manual, but much reliance necessarily must be placed on the knowledge and judgment of the analyst. Most photometers are capable of their best performance when readings fall in the range of approximately 1 to 0.1 absorbance with respect to a blank adjusted to read 0 absorbance. The closer the readings approach 0 or 3.0 absorbance, the less accurate they become. If it is impractical to use an optical cell with a sufficiently long light path—as in some commercial instruments—or to concentrate the sample or select a more sensitive color test, then it may be more accurate to compare very faint colors in nessler tubes than to attempt photometric readings close to 100% transmittance.

In general, the best wavelength or filter to select is that which produces the largest spread of readings between a standard and a blank. This usually corresponds to a visual color for the light beam that is complementary to that of the solution—for example, a green filter for a red solution, a violet filter for a yellow solution.

Absorptivities are useful in comparing method sensitivities and in estimating the concentration of absorbing solutions such as dithizone. The absorptivity may be computed from Beer's Law as follows:

$$a = \frac{A}{bc}$$

where:

a = absorptivity, L/(g · cm),
A = absorbance of a solution, dimensionless,
b = concentration, g/L, and
c = cell path length, cm.

Use of a photoelectric instrument makes unnecessary the preparation of a complete set of standards for each set of samples to be analyzed. However, prepare a reagent blank containing distilled water and all reagents and at least one standard in the upper end of the optimum concentration range, with every group of samples, to verify the constancy of the calibration curve. This precaution will reveal any unsuspected changes in the reagents, the instrument, or the technique. At regular intervals, or if at any time results fall under suspicion, prepare a complete set of standards—at least five or six spaced to cover the optimum concentration range—to check the calibration curve. Also valuable in this regard is the absorptivity informa-

tion given in this manual for a number of photometric methods.

Verify frequently the accuracy of the curves or permanent standards by comparing with standards prepared in the laboratory, using the same set of reagents, the same instrument, and the same procedures as those used for analyzing samples. Even if permanent calibration curves or artificial standards have been prepared accurately by the manufacturer, they may not be valid under conditions of use. Permanent standards may be subject to fading or color alteration. Their validity may depend also on certain arbitrary lighting conditions. Standards and calibration curves may be incorrect because of slight differences in reagents, instruments, or techniques between the manufacturer's and the analyst's laboratories.

Zero or null the instrument using a reagent blank or, if required, distilled water. Never zero with sample blank. Determine the absorbance of each standard; then plot absorbance against the concentrations of the standards to establish a working curve and the degree of agreement with Beer's Law. If readings are in terms of percentage transmittance, convert to absorbance before plotting or plot on semilogarithmic paper.

Should turbidity be present as an interference, several approaches are available to separate turbidity from a sample. The nature of the sample, the size of suspended particles, and the reasons for conducting the analysis will all combine to dictate the method for turbidity removal. Turbidity may be coagulated by adding zinc sulfate and an alkali, as is done in the direct nesslerization method for ammonia nitrogen. For samples of relatively coarse turbidity, centrifuging may suffice. In some instances, glass fiber filters, filter paper, or sintered-glass filters of fine porosity will serve the purpose. For very small particle sizes, membrane filters may provide the required retentiveness. Used with discretion, each of these methods will yield satisfactory results in a suitable situation. However, it must be emphasized that no single universally ideal method of turbidity removal is available. Moreover, be perpetually alert to adsorption losses possible with any flocculating or filtering procedure.

If the turbidity cannot be removed without compromising the sample, then use photometric compensation to correct for interference. Measure the sample without addition of reagents (the sample blank) against a reagent blank or a distilled water blank to determine instrument response. The response is due to sample absorbance or turbidity other than that caused by the element or compound being determined. Take into account volume changes from the omission of reagents. If the calibration curve is linear, the sample blank absorbance may be subtracted from the sample absorbance before the concentration is computed, or the concentration may be computed for the sample blank and subtracted from that of the sample. If a nonlinear calibration is used, compute the concentration of the sample blank and subtract it from the sample concentration. Subtracting the absorbance is not satisfactory in this case.

In the procedure, zero the instrument using a turbidity blank or, alternatively, determine the absorbance for this blank against a reagent blank or distilled water and then subtract it from the sample absorbance. Correct as before for volume difference caused by addition or omission of reagents in the calculation.

b. Dithizone solutions: Several colorimetric methods for metals (Cd, Pb, Hg, Ag, Zn) use dithizone (diphenylthiocarbazone) as an extractable, colored, metal-complexing agent. The methods presented later in this text have been based on three stock dithizone solutions, the preparation of which is described below. The dithizone concentration in the stock dithizone solutions is based on having a 100% pure dith-

izone reagent. Some commercial grades of dithizone are contaminated with the oxidation product diphenylthiocarbodiazone or with metals. Purify dithizone as directed below. For dithizone solutions not stronger than 0.001% (*w/v*), calculate the exact concentration by dividing the absorbance of the solution in a 1.00-cm cell at 606 nm by 40.6×10^3, the molar absorptivity.

Adjust dilutions of stock dithizone solutions to produce working dithizone solutions of the indicated strength based on the measured stock dithizone solution concentration.

1) *Stock dithizone solution I,* 100 mg dithizone/1000 mL $CHCl_3$: Dissolve 100 mg dithizone in 50 mL $CHCl_3$ in a 150-mL beaker and filter through a 7-cm-diam paper.* Receive filtrate in a 500-mL separatory funnel or in a 125-mL erlenmeyer flask under slight vacuum; use a filtering device designed to handle the $CHCl_3$ vapor. Wash beaker with two 5-mL portions $CHCl_3$, and filter. Wash the paper with three 5-mL portions $CHCl_3$, adding final portion dropwise to edge of paper. If filtrate is in flask, transfer with $CHCl_3$ to a 500-mL separatory funnel.

Add 100 mL 1 + 99 NH_4OH to separatory funnel and shake moderately for 1 min; excessive agitation produces slowly breaking emulsions. Let layers separate, swirling funnel gently to submerge $CHCl_3$ droplets held on surface of aqueous layer. Transfer $CHCl_3$ layer to 250-mL separatory funnel, retaining the orange-red aqueous layer in the 500-mL funnel. Repeat extraction, receiving $CHCl_3$ layer in another 250-mL separatory funnel and transferring aqueous layer, using 1 + 99 NH_4OH, to the 500-mL funnel holding the first extract. Repeat extraction, transferring the aqueous layer to 500-mL funnel. Discard $CHCl_3$ layer.

To combined extracts in the 500-mL separatory funnel add 1 + 1 HCl in 2-mL

---

*Whatman No. 42, or equivalent.

portions, mixing after each addition, until dithizone precipitates and solution is no longer orange-red. Extract precipitated dithizone with three 25-mL portions CHCl₃. Dilute combined extracts to 1000 mL with $CHCl_3$; 1.00 mL = 100 μg dithizone.

2) *Stock dithizone solution II*, 250 mg dithizone/250 mL $CHCl_3$: Dissolve 250 mg dithizone in 50 mL $CHCl_3$ in a 150-mL beaker and filter through a 7-cm-diam paper.† Receive filtrate in a 1000-mL separatory funnel or in a 125-mL erlenmeyer flask under slight vacuum; use a filtering device designed to handle $CHCl_3$ vapor. Wash beaker with two 5-mL portions $CHCl_3$ and filter. Wash paper with three 5-mL portions $CHCl_3$, adding the last portion dropwise to edge of paper. If filtrate is in flask, transfer with $CHCl_3$ to 1000-mL separatory funnel.

Add 200 mL 1 + 99 NH_4OH to separatory funnel and shake moderately for 1 min; excessive agitation produces slowly breaking emulsions. Let layers separate, swirling funnel gently to submerge $CHCl_3$ droplets held on surface of aqueous layer. Transfer $CHCl_3$ layer to 500-mL separatory funnel, retaining the orange-red aqueous layer in the 1000-mL funnel. Repeat extraction of $CHCl_3$ layer with 200 mL 1 + 99 NH_4OH, transferring $CHCl_3$ layer to another 500-mL separatory funnel. Transfer aqueous layer to 1000-mL funnel holding the first extract. Repeat extraction with third 200-mL portion 1 + 99 NH_4OH. Discard $CHCl_3$ layer, transfer aqueous layer to 1000-mL funnel. To the combined extracts add 1 + 1 HCl in 4-mL portions, mixing after each addition, until dithizone precipitates and solution is no longer orange-red. Extract precipitated dithizone with four 25-mL portions $CHCl_3$. Dilute combined extracts to 250 mL with $CHCl_3$; 1.00 mL = 1000 μg dithizone.

3) *Stock dithizone solution III*, 125 mg

dithizone/500 mL CCl_4: Dissolve 125 mg dithizone in 50 mL $CHCl_3$ and proceed as for dithizone solution II, but extract precipitated dithizone with 25-mL portions CCl_4. Dilute CCl_4 extracts to 500 mL; 1.00 mL = 250 μg dithizone. (CAUTION: *CCl_4 is toxic—avoid inhalation, ingestion, and contact with the skin.*)

3. Other Methods of Analysis

The use of an instrumental method of analysis not specifically described herein is permissible provided that the results so obtained are checked periodically, either against an included standard method or against a standard sample of undisputed composition. Identify any such instrumental method used in the laboratory report along with the analytical results.

a. Atomic absorption spectrometry: Atomic absorption spectrometry has been applied to the determination of metals in water without the need for prior concentration or extensive sample pretreatment. The use of organic solvents coupled with oxyacetylene, oxyhydrogen, or nitrous oxide-acetylene flames enables the determination of metals that form refractory oxides. These standards include atomic absorption methods—including certain flameless and electrothermal (heated graphite) techniques—for many metals.

b. Flame photometry: Flame photometry is used for the determination of sodium, potassium, lithium, and strontium.

c. Inductively coupled plasma (ICP) and related analytical systems: ICP is suggested for scanning samples for various metal ions, especially where the possible interferences in the colorimetric procedures are unknown. ICP has enabled the determination of multiple metals at the low microgram-per-liter level when a single 100-mL sample has been ashed with HNO_3. Polarographic techniques such as pulse polarography, differential pulse voltammetry, and differential pulse anodic stripping vol-

---

†Whatman No. 42, or equivalent.

tammetry also have gained acceptance for the determination of heavy metals, and their speciation, in water and wastewater.

A method closely allied to polarography is amperometric titration, which is suitable for determining residual chlorine, chlorine dioxide, and iodine, and in other iodometric methods.

d. Potentiometric titration: Many titrimetric methods can be performed potentiometrically, by using a millivoltmeter or pH meter with suitable electrodes.

e. Selective ion electrodes: Selective ion electrodes are available for rapid estimation of certain constituents in water. These electrodes function best in conjunction with an expanded-scale pH meter or a suitable millivoltmeter. For the most part, the electrodes operate on the ion-exchange principle. Selective ion electrodes now available are designed for measuring ammonia, cadmium, calcium, divalent copper, hardness, lead, potassium, silver, sodium, total monovalent and total divalent cations, and bromide, chloride, cyanide, fluoride, iodide, nitrate, perchlorate, and sulfide anions, among others.

These devices are subject to varying degrees of interference from other ions in the sample and many still must receive thorough study to warrant adoption as proposed and standard methods. Nonetheless, their value for monitoring activities is readily apparent. To remove all doubt of variations in reliability, check each electrode in the presence of interferences as well as the ion for which it is intended. This manual details several electrode methods.

Commercial dissolved oxygen (DO) probes vary considerably in their dependability and maintenance requirements. Despite these shortcomings, they have been applied to monitoring DO in a variety of waters and wastewaters. Most probes embody an electrolyte held in place by an oxygen-permeable membrane. The DO in solution diffuses through the membrane and electrolyte layer to react at the electrode, inducing a current proportional to the activity and hence, in solution of low or constant ionic strength, essentially proportional to concentration of DO. Satisfactory DO electrodes also are available without a membrane. In either case, keep the face of the DO sensor well agitated and provide temperature compensation to insure acceptable results.

f. Gas chromatography (GC) and gas chromatography/mass spectrometry (GC/MS): Many GC and GC/MS methods suitable for water and wastewater analysis are available. Appearing in this manual are such methods for determining chlorinated hydrocarbon pesticides, components in sludge digester gas, phenols, volatiles, PAHs, and taste- and odor-causing compounds, and generalized GC/MS analyses for volatiles and acid- and base-neutral extractables.

g. Continuous-flow analysis: Automated techniques are presented for chloride, fluoride, nitrogen (ammonia), nitrogen (nitrate), phosphate, silica, and sulfate.

h. Ion chromatography (IC): Ion chromatography is a technique for sequential determination of anions or cations using ion exchange and conductivity, amperometric, or colorimetric detectors. See Section 4110 for a generalized technique for anions.

i. Other methods of analysis: Instrumentation and new methods of analysis always are under development. The analyst will find it advantageous to keep abreast of current progress. Reviews of each branch of analytical chemistry are published regularly in *Analytical Chemistry* and the annual literature review in *Journal Water Pollution Control Federation.*

4. Interference Control

Many analytical procedures are subject to interferences from substances present in the sample. The more common and obvious interferences are known and information about them has been given in the details of individual procedures. It is inevitable that unknown or unexpected interferences may be encountered. Such occurrences are unavoidable because of the diverse nature of waters and particularly of wastewaters. Therefore, be alert to hitherto untested constituents, new treatment compounds (especially complexing agents), and new industrial wastes and their potential threat to the accuracy of chemical analyses.

Any sudden change in the apparent composition of water that has been rather constant, any abnormal color observed in a colorimetric test or during a titration, any unexpected turbidity, odor, or other laboratory finding is cause for suspicion. Such change may be due to normal variation in the relative concentrations of the usual constituents or it may be caused by the introduction of an unforeseen interfering substance.

A few substances—such as chlorine, chlorine dioxide, alum, iron salts, silicates, copper sulfate, ammonium sulfate, and polyphosphates—are so widely used that they deserve special mention as possible causes of interference. Of these, chlorine is probably the worst offender, in that it bleaches or alters the colors of many sensitive organic reagents that serve as titration indicators and as color developers for photometric methods. Among the methods that have proved effective in removing chlorine residuals are the addition of minimal amounts of sulfite, thiosulfate, or arsenite; exposure to sunlight or an artificial ultraviolet source, and prolonged storage.

Whenever interference is encountered or suspected and no specific recommendations are given for overcoming it, determine what technique, if any, will eliminate the interference without adversely affecting the analysis itself. If two or more choices of procedure are offered, often one procedure will be less affected than another by the presence of the interfering substance. If different procedures yield considerably differ-

ent results, it is likely that interference is present. Some interferences become less severe upon dilution or upon use of smaller samples; any tendency of the results to increase or decrease in a consistent manner with dilution indicates the likelihood of interference effects.

a. Types of interference: Interference may cause analytical results to be either too high or too low as a consequence of one of the following:

1) An interfering substance may react as though it were the substance sought and thus produce a high result; for example, bromide will respond to titration as though it were chloride.

2) An interfering substance may react with the substance sought and thus produce a low result.

3) An interfering substance may combine with the analytical reagent and prevent it from reacting with the substance sought; for example, chlorine will destroy many indicators and color-developing reagents.

Nearly every interference will fit one of these classes. For example, in a photometric method, turbidity may be considered as a "substance" that acts like the one being determined, that is, it reduces light transmission. Occasionally, two or more interfering substances, if present simultaneously, may interact in a nonadditive fashion, either canceling or enhancing one another's effects.

b. Counteracting interference: The best way to minimize interference is to remove the interfering substance or to render it innocuous by one of the following methods:

1) Physically remove either the substance sought or the interfering substance. For example, distill off fluoride and ammonia, leaving interferences behind. The interferences also may be absorbed on an ion-exchange resin.

2) Adjust the pH so that only the substance sought will react; for example, ad-

just the pH to 2 so that volatile acids will distill from a solution.

3) Oxidize (digest) or reduce the sample to convert the interfering substance to a harmless form. For example, reduce chlorine to chloride by adding thiosulfate; digest samples for analysis by atomic absorption spectrometry with one of a variety of digestion reagents to destroy organic matter.

4) Add a suitable agent to complex the interfering substance so that it is innocuous although still present. For example, complex iron with pyrophosphate to prevent it from interfering with the copper determination; complex copper with cyanide or sulfide to prevent interference with the titrimetric hardness determination.

5) A combination of the first four techniques may be used. For example, distill phenols from an acid solution to prevent amines from distilling; use thiosulfate in the dithizone method for zinc to prevent most of the interfering metals from passing into the CCl_4 layer.

6) Color and turbidity sometimes may be destroyed by wet or dry ashing or may be removed by using a flocculating agent. Some types of turbidity may be removed by filtration. These procedures, however, introduce the danger that the constituent that is the subject of the analysis also will be removed.

c. *Compensation for interference:* If none of these techniques is practical, several methods of compensation can be used:

1) If color or turbidity initially present interferes in a photometric determination, it may be possible to use photometric compensation. The technique is described in Section 1070D.2a.

2) Determine the concentration of interfering substances and add identical amounts to the calibration standards. This involves much labor.

3) If the interference does not continue to increase as the concentration of interfering substance increases, but tends to

level off, add a large excess of interfering substance to all samples and to all standards. This is called "swamping."

4) The presence in the chemical reagents of the substance sought may be accounted for by carrying out a blank determination.

5. Bibliography

General Analytical Techniques

FOULK, C.W., H.V. MOYER & W.M. MACNEVIN. 1952. Quantitative Chemical Analysis. McGraw-Hill Book Co., New York, N.Y.

WILLARD, H.H., N.H. FURMAN & C.E. BRICKER. 1956. Elements of Quantitative Analysis, 4th ed. D. Van Nostrand Co., Princeton, N.J.

HUGHES, J.C. 1959. Testing of Glass Volumetric Apparatus. Nat. Bur. Standards Circ. No. 602.

WILSON, C.L. & D.W. WILSON, eds. 1959, 1960, 1962. Comprehensive Analytical Chemistry, Vol. 1A, 1B, 1C. Elsevier Publishing Co., New York, N.Y.

MEITES, L., ed. 1963. Handbook of Analytical Chemistry. McGraw-Hill Book Co., New York, N.Y.

KOLTHOFF, I.M., E.J. MEEHAN, E.B. SANDELL & S. BRUCKENSTEIN. 1969. Quantitative Chemical Analysis, 4th ed. Macmillan Co., New York, N.Y.

WELCHER, F.J., ed. 1975. Standard Methods of Chemical Analysis, 6th ed. Vol. IIA, IIB, IIIA (reprint), Robert E. Krieger Publishing Co., Melbourne, Fla.

PECSOK, R. et al. 1976. Modern Methods of Chemical Analysis, 2nd ed. John Wiley & Sons, New York, N.Y.

VOGEL, A.I. 1978. Textbook of Quantitative Inorganic Analysis. Including Elementary Instrumental Analysis, 4th ed. Revised by J. Basset. Longman, New York, N.Y.

General Analytical Reviews and Bibliographies

WEIL, B.H. et al. 1948. Bibliography on Water and Sewage Analysis. State Engineering Experiment Sta., Georgia Inst. Technology, Atlanta.

Annual reviews of analytical chemistry. 1953-1987. *Anal. Chem.* 25:2; 26:2; 27:574; 28:559; 29:589; 30:553; 31:776; 32:3R; 33:3R; 34:3R; 35:3R; 36:3R; 37–59:1R.

WATER POLLUTION CONTROL FEDERATION RESEARCH COMMITTEE. 1960-1988. Annual literature review, nature and analysis of chemical species. *J. Water Pollut. Control Fed.* 32:443; 33:445; 34:419; 35:553; 36:535;

37:735; 38:869; 39:867; 40:897; 41:873;
42:863; 43:933; 44:903; 45:979; 46:1031;
47:1118; 48:998; 49:901; 50:1000; 51:1108;
52:1083; 53:620; 54:520; 55:555; 56:494;
57:438; 58:419; 59:306; 60:743.

Colorimetric Techniques

MELLON, M.G. 1947. Colorimetry and photometry in water analysis. *J. Amer. Water Works Assoc.* 39:341.

NATIONAL BUREAU OF STANDARDS. 1947. Terminology and Symbols for Use in Ultraviolet, Visible, and Infrared Absorptiometry. NBS Letter Circ. LC-857 (May 19).

GIBSON, K.S. & M. BALCOM. 1947. Transmission measurements with the Beckman quartz spectrophotometer. *J. Res. Nat. Bur. Standards* 38:601.

SNELL, F.D. & C.T. SNELL. 1948-1971. Colorimetric Methods of Analysis, 3rd ed. Vol. 1, 2, 2A, 3, 3A, 4, 4A, 4AA, 4AAA. Van Nostrand Reinhold, New York, N.Y.

MELLON, M.G., ed. 1950. Analytical Absorption Spectroscopy. John Wiley & Sons, New York, N.Y.

DISKANT, E.M. 1952. Photometric methods in water analysis. *J. Amer. Water Works Assoc.* 44:625.

SANDELL, E.B. 1978. Colorimetric Determination of Traces of Metals, 4th ed. John Wiley & Sons, New York, N.Y.

BOLTZ, D.F., ed. 1978. Colorimetric Determination of Nonmetals, 2nd ed. John Wiley & Sons, New York, N.Y.

Other Methods of Analysis

LEDERER, E. & M. LEDERER. 1957. Chromatography, 2nd ed. Elsevier Press, Houston, Tex.

LINGANE, J.J. 1958. Electroanalytical Chemistry, 2nd ed. Interscience Publishers, New York, N.Y.

SAMUELSON, O. 1963. Ion Exchangers in Analytical Chemistry. John Wiley & Sons, New York, N.Y.

HARLEY, J.H. & S.E. WIBERLEY. 1967. Instrumental Analysis, 2nd ed. John Wiley & Sons, New York, N.Y.

EWING, G.W. 1975. Instrumental Methods of Chemical Analysis, 4th ed. McGraw-Hill Book Co., New York, N.Y.

HEFTMANN, E., ed. 1975. Chromatography, 3rd ed. Reinhold Publishing Corp., New York, N.Y.

1080   REAGENT-GRADE WATER

1080 A.   Introduction

One of the most important aspects of analysis is the preparation of reagent-grade water to be used for dilution of reagents and for blank analysis. Reagent-grade water covers a range from Type I with no detectable concentration of the compound or element to be analyzed at the detection limit of the analytical method to Type III for washing and qualitative analysis (see Table 1080:I). Reagent-grade water should be free of substances that interfere with analytical methods. The quality of water required is related directly to the analysis being made. Requirements for water quality may differ for organic, inorganic, and microbiological constituents depending on the use(s) for which the water is intended.

Any method of preparation of reagent-grade water is acceptable provided that the requisite quality can be met. Improperly maintained systems may add contaminants. Reverse osmosis, distillation, and deionization in various combinations all can produce reagent-grade water when used in the proper arrangement. Ultrafiltration and/or ultraviolet treatment also may be used as part of the process. Section 1080 provides general guidelines for the preparation of reagent-grade water. Table 1080:II lists commonly available processes for water purification and major classes of contaminants removed by purification.

For details on preparing water for microbiological tests, see Section 9020B.3c.

TABLE 1080:I. REAGENT WATER SPECIFICATIONS*

Quality Parameter	Type I	Type II	Type III
Bacteria, CFU/mL	10	1000	NA
pH	NA	NA	5–8
Resistivity, megohm-cm at 25°C	> 10	> 1	0.11
Resistivity, megohm-cm at 25°C	> 10	> 1	0.1
Conductivity, μmho/cmat 25°C	< 0.1	1	10
SiO_2, mg/L	< 0.05	< 0.1	< 1
Total solids, mg/L	0.1	1	5
Total oxidizable organic carbon, mg/L	< 0.05	< 0.2	< 1

* NA = not applicable.

TABLE 1080:II. WATER PURIFICATION PROCESSES

Process	Major Classes of Contaminants*					
	Dissolved Ionized Solids	Dissolved Ionized Gases	Dissolved Organics	Particulates	Bacteria	Pyrogens
Distillation	G–E†	P	G	E	E	E
Deionization	E	E	P	P	P	P
Reverse osmosis	G‡	P	G	E	E	E
Carbon adsorption	P	P§	G–E‖	P	P	P
Filtration	P	P	P	E	E	P
Ultrafiltration	P	P	G#	E	E	G–E
Ultraviolet oxidation	P	P	G–E**	P	G††	P

Permission to use this table from C3-T2, "Preparation and Testing of Reagent Water in the Clinical Laboratory - Second Edition; Tentative Guideline," has been granted by the National Committee for Clinical Laboratory Standards. The complete current standard may be obtained from National Committee for Clinical Laboratory Standards, 771 E. Lancaster Ave., Villanova, Pa. 19085.

* E = Excellent (capable of complete or near total removal), G = Good (capable of removing large percentages), P = Poor (little or no removal).

† Resistivity of water purified by distillation is an order of magnitude less than water produced by deionization, due mainly to the presence of CO_2 and sometimes H_2S, NH_3, and other ionized gases if present in the feedwater.

‡ Resistivity of dissolved ionized solids depends on original feedwater resistivity.

§ Activated carbon removes chlorine by adsorption.

‖ When used in combination with other purification processes, special grades of activated carbon and other synthetic adsorbents exhibit excellent capabilities for removing organic contaminants. Their use, however, is targeted toward specific compounds and applications.

Ultrafilters have demonstrated usefulness in reducing specific feedwater organic contaminants based on the rated molecular weight cut-off of the membrane.

** 185 nm ultraviolet oxidation (batch process systems) is effective in removing trace organic contaminants when used as post-treatment. Feedwater makeup plays a critical role in the performance of these batch processors.

†† 254 nm UV sterilizers, while not physically removing bacteria, may have bactericidal or bacteriostatic capabilities limited by intensity, contact time, and flow rate.

1080 B. Methods for Preparation of Reagent-Grade Water

1. Distillation

Prepare laboratory-grade distilled water by distilling water from a still of all-borosilicate glass, fused quartz, tin, or titanium. To remove ammonia distill from an acid solution. Remove CO_2 by boiling the water for 15 min and cooling rapidly to room temperature; exclude atmospheric CO_2 by using a tube containing soda lime or a commercially available CO_2-removing agent.*

Boiling the water may add other impurities by leaching impurities from the container. Pretreat feedwater and provide periodic maintenance to minimize scale formation within the still. Pretreatment may be required where the feedwater contains significant concentrations of calcium, magnesium, and bicarbonate ions; it may involve demineralization via reverse osmosis or ion exchange. Resistivity of distilled water (Type II) should be > 1.0 megohm-cm at 25°C and for Type I, 10 megohm-cm. Measurements are more accurately made with in-line cells.

2. Reverse Osmosis

Reverse osmosis is a process in which water is forced under pressure through a semipermeable membrane removing a portion of dissolved constituents and suspended impurities. Product water quality depends on feedwater quality.

Select the reverse osmosis membrane module appropriate to the characteristics of the feedwater. Obtain rejection data for contaminants in the feedwater at the operating pressure to be used in preparing reagent-grade water. Set overall water production to make the most economical use of water without compromising the final quality of the permeate. Selection of spiral-wound or hollow fiber configurations depends on fouling potential of the feedwater. Regardless of configuration used, pretreatment may be required to minimize membrane fouling with colloids or particulates and to minimize introduction of chlorine, iron, and other oxidizing compounds that may degrade reverse osmosis membranes. Periodic flushing of the membrane modules is necessary.

3. Ion Exchange

Prepare deionized water by passing feedwater through a mixed-bed ion exchanger, consisting of strong anion and strong cation resins mixed together. When the system does not run continuously, recirculate product water through ion-exchange bed. Resistivity for Type I water should be 10 megohm-cm (in-line) at 25°C.

Use separate anion and cation resin beds in applications where resin regeneration is economically attractive. In such instances, position the anion exchanger downstream of the cation exchanger to remove leachates from the cation resin. Proper bed sizing is critical to the performance of the resins. In particular, set the length-to-diameter ratio of the bed in accordance with the maximum process flow rate to ensure that optimal face velocities are not exceeded and that sufficient residence time is provided.

In applications where the feedwater has significant quantities of organic matter, remove organics to minimize potential fouling of the resins. Possible pretreatments

* Ascarite, Fisher Scientific Co., or equivalent.

include prefiltration, distillation, reverse osmosis, or adsorption.

4. Adsorption

Adsorption is generally used to remove chlorine and organic impurities. It is accomplished typically with granular activated carbon. Efficiency of organics removal depends on the nature of the organic contaminants, the physical characteristics of the activated carbon, and the operating conditions. In general, organics adsorption efficiency is inversely proportional to solubility and may be inadequate for the removal of low-molecular-weight, polar compounds. Performance differences among activated carbons are attributable to the use of different raw materials and activation procedures. Select the appropriate activated carbon with regard to these differences. Even with optimum activated carbon, proper performance will not be attained unless the column is sized to give required face velocity and residence time at the maximum process flow rate.

Use of activated carbon may adversely affect resistivity. This effect may be controlled by use of reverse osmosis, mixed resins, or special adsorbents. To achieve the lowest level of organic contamination, use mixtures of polishing resins with special carbons in conjunction with additional treatment steps, such as reverse osmosis, natural carbons, ultraviolet oxidation, or ultrafiltration.

1080 C.   Reagent Water Quality

1. Quality Guidelines

Several guidelines for reagent water quality, based on contaminant levels, are available (see Table 1080:I).[1-3] Methods and uses listed below are based on National Committee for Clinical Laboratory Standards (NCCLS) guidelines.

Use Type I water in test methods requiring minimum interference and bias and maximum precision. Type II water is intended to provide the user with water in which the presence of bacteria can be tolerated. It is used for the preparation of reagents, dyes, or staining. Type III water may be used for glassware washing, preliminary rinsing of glassware, and as feedwater for production of higher-grade waters.

Type I reagent water, having a minimum resistivity of 10 megohms-cm, 25°C (in line), typically is prepared by distillation, deionization, or reverse osmosis treatment of feedwater followed by polishing with a mixed-bed deionizer and passage through a 0.2-μm-pore membrane filter. Alterna-tively treat by reverse osmosis followed by carbon adsorption and deionization. Determine quality at the time of production. Mixed bed deionizers typically add small amounts of organic matter to water, especially if the beds are fresh. Resistivity of Type I water should be > 10 megohm-cm 25°C, measured in-line. Resistivity measurements will not detect organics or nonionized contaminants, nor will they provide an accurate assessment of ionic contaminants at the microgram-per-liter level. Thus, make separate measurements of contaminants such as TOC, SiO_2, and bacterial counts.

Type II water typically is produced by distillation or deionization. Resistivity should be > 1 megohm-cm at 25°C. Observe the same precautions on measurements of other contaminants.

Type III water should have a minimum resistivity of 0.1 megohm-cm.

Other contaminants in reagent water are listed in Table 1080:I.

The pH of Type I or Type II water cannot be measured accurately without contaminating the water. Measure other constituents as required for individual tests.

Type I water cannot be stored without significant degradation; produce it continuously and use it immediately after processing. Type II water may be stored, but keep storage to a minimum and provide quality consistent with the intended use. Store only in materials that protect the water from contamination, such as TFE and glass for organics analysis or plastics for metals. Store Type III water in materials that protect the water from contamination.

2. References

1. AMERICAN SOCIETY FOR TESTING AND MATERIALS. 1987. Annual Book of ASTM Standards, Part 31, Water: Atmospheric Analysis. American Soc. Testing & Materials, Philadelphia, Pa.
2. COMMISSION ON LABORATORY INSPECTION AND ACCREDITATION. 1985. Reagent Water Specifications. College of American Pathologists, Chicago, Ill.
3. NATIONAL COMMITTEE FOR CLINICAL LABORATORY STANDARDS. 1988. Preparation and Testing of Reagent Water in the Clinical Laboratory - Second Edition; Tentative Guideline. Publ. C3-T2, National Comm. for Clinical Laboratory Standards, Villanova, Pa.

1090   SAFETY

1090 A.   Introduction

1. General Discussion

Achievement of a safe and healthy work place is the responsibility of the institution, the laboratory manager, the supervisory personnel and, finally, of the laboratory personnel themselves. All laboratory employees must make every effort to protect themselves and their fellow workers. The laboratory manager should realize that accidents have causes, and therefore can be prevented by a good safety program.[1]

Once an employee has become thoroughly familiar and competent with the new, safe techniques, the manager can be assured that the employee will propose other safety ideas.

2. Organizing for Safety

Although responsibility for establishing and enforcing a laboratory safety program ultimately rests with the laboratory director, it is desirable, except in small laboratories, to delegate responsibility to a staff member designated as the safety officer.[2] The safety officer should be part of the management team to optimize safety training, inspections, indoctrination of new employees, and acquisition and updating of safety information. The safety officer should see that personal protective equipment is available and used appropriately. The officer should make periodic inspections of emergency equipment, such as fire extinguishers, alarm systems, eye wash, and safety showers; perform periodic inspections of the laboratory to uncover overlooked hazards; observe that personnel are following safety rules; and provide reminders for individuals to be aware of safe practices.

Radiation safety is a specialized part of an overall safety program. A technically qualified individual should be designated as the Radiation Safety Officer (RSO). The RSO ensures that radioactive sources and radiation equipment are used safely and in accordance with all pertinent state and federal regulations, and that proper licensing from the appropriate state and/or federal agency is maintained.

A safety committee should be established and it must have the support of the laboratory director. The committee's recommendations must be taken seriously and acted upon promptly. Safety committees involve persons with varying expertise in making decisions relating to safety practices. For example, chemists may have insight into the handling of hazardous chemicals in a bacteriology section. This assignment can be educationally valuable. It is desirable periodically to rotate members of the safety committee so that most personnel have an opportunity to become personally involved.

Keep good records of all accidents, inspections, training, etc. Preferably use a simple standard report form. The report should contain sufficient information to enable the safety officer, supervisor, and director to determine who was involved, what happened, when and where it happened, and what injuries, if any, resulted. The Occupational Safety and Health Act (OSHA) requires that those accidents causing major disability be recorded in a log, but it is useful to record all accidents to be able to evaluate the safety program. Recording time and nature of each accident may be of further importance if disability occurs and a Worker's Compensation claim is appropriate. Maintain a file of recommendations of the safety officer or safety committee, as well as actions taken by the staff as part of the safety program.

The OSHA "Hazard Communication Standard" or "Right-to-Know Regulation"[3] specifies a method of employee notification about hazards in the workplace. Exposed laboratory personnel must be under direct supervision and regular observation of a technically qualified individual, who must have knowledge of the hazards present, their health effects, and related emergency procedures. The supervisor must educate laboratory personnel in safe work practices. Personnel have a right to know what hazardous materials are present, the specific hazards created by those materials, and the required procedures to protect against these hazards.

Training in safety techniques requires deliberate effort by management. In larger laboratories, safety seminars are important. Use educational techniques including training films, demonstrations with devices such as fire extinguishers or respirators, charts listing hazardous chemicals and proper procedures for handling them, and manuals on safety. Periodically provide refresher training. Material Safety Data Sheets (MSDSs) are provided by chemical manufacturers and are an extremely important part of the whole approach of informing personnel of hazards in their workplace.[4] These MSDSs should be made available to personnel using hazardous materials.

3. References

1. INHORN, S. L., ed. 1978. Quality Assurance Practices for Health Laboratories. American Public Health Assoc., Washington, D.C.

2. STEERE, N. V., ed. 1971. Handbook of Laboratory Safety, 2nd ed. Chemical Rubber Co., Cleveland, Ohio.

3. Hazard Communication: Final Rule. 1983. *Federal Register* 48:53280; 1985. 29 CFR Part 1910.1200.

4. Chemical Safety Data Sheets. Manufacturing Chemists' Assoc., Inc. Washington, D.C.

1090 B. Safety Equipment

1. Laboratory Equipment

a. Fire extinguishers: There are three general types of fire extinguishers:

1) Water-type extinguishers are useful for fires with ordinary combustibles, such as wood, paper, and rags.

2) Dry-chemical types are effective against most fires, but particularly those involving flammable liquids and metals and electrical fires.

3) Carbon dioxide types are useful for small fires involving flammable liquids and for limited use around electronic instrumentation and equipment.

Depending upon potential hazards, a laboratory may have more than one type in each room in an easily accessible location away from the area of greatest hazard.[1-5] Halon extinguishers are useful in special areas housing electronic or computer equipment that might be damaged by conventional extinguishers. Both Halon and carbon dioxide extinguishers create oxygen deficiencies. Use all extinguishers with care.

It is often recommended that multipurpose fire extinguishers (type ABC) be installed in the laboratory and that hoods containing organic materials be equipped with halon extinguishers. Be sure that extinguishers are properly serviced and recharged on a scheduled basis.

b. Fire blankets: Locate readily accessible fire blankets close to each laboratory.[6]

c. Safety showers: Safety showers are an integral part of the laboratory, and are to be used in accidents involving acids, caustic or other harmful liquids, clothing fires, and other emergencies. Locate showers conveniently, preferably near a doorway, keep the floor space under them uncluttered, and test regularly.[6]

d. Eye washes: Remove contact lenses if applicable. Flush eyes *immediately* and thoroughly (15 min) to prevent sight impairment or even blindness after a chemical splashes into the eye. Splashes to the face are rinsed off easily with a four-head eye wash. Some experts claim that the fountain-like stream of water from an eye wash tends to drive particulate matter (glass or metal shards, dirt, grit) into the eye rather than to wash it out. If this type of accident occurs, consult an eye expert after gentle flushing with water. Locate eye washes at or near sinks in a highly visible and central area away from electrical receptacles. Check eye washes routinely for correct function; if portable eye washes are used clean them weekly and change the water to reduce the possibility of contamination. The use of eye-wash bottles and some of their problems have been discussed elsewhere.[2]

e. Safety shields: The most generally used type of safety shield protects the worker from various forms of radiation, such as laser beams and ultraviolet emissions. Provide conventional chemical hoods with safety glass and movable doors. Consider shielding when working with glass vacuum or pressurized systems.

f. Safety containers: Safety containers are designed to minimize consequences of an accident or to prevent spread of harmful materials. Use safety containers to transport chemicals, especially concentrated acids and alkalies. For flammable solvents use safety cans approved by the Underwriters' Laboratories.

Glove boxes, laminar-flow hoods, or remotely operated confinements for radioactive material or dangerous pathogens are more complicated safety containers. Operate these under negative pressure, with primary filtration at the exit of gas or air vents leading to a major ventilation system. Periodically check filter for efficiency and change and test gloves to prevent accidental release of particulates by the pumping action created during use.

g. Storage facilities: In storing laboratory materials, it is necessary to know their properties as well as the consequences of accidents such as spilling, explosion, or fire.

As a general rule, do not store large containers of reagents in working areas, but use smaller containers holding only enough for the day's or week's work. Do not store chemicals that combine to form explosive, flammable, or dangerous compounds in a manner that would allow them to mix if an accident occurred. Store hazardous materials within a tray or catch bin area.

Store flammable solvents in properly vented cabinets approved by the National Fire Protection Association[5] or in a safety refrigerator. Use special containers for flammable solvents in quantities of more than 2 L or when the cumulative amount of flammable solvents in a room exceeds 8 L. (Flammable solvents are liquids with a flash point below 60°C and a vapor pressure exceeding 275 kPa atmospheric at 38°C.)

h. Laboratory fume hoods: Properly designed and operated fume hoods and biological safety cabinets are essential to a safe laboratory operation.[7] Use hoods to contain and vent operations with hazardous materials; do not use them primarily for storage.

Different air-flow and design classifications of fume hoods are available for different degrees of hazard. Know the air-moving capacity of the entire system; pay particular attention to the air stream exhausted to the community air. Check air flow of hoods on a scheduled basis with an anemometer. Indicate with a marking pen or tape the sash position required to obtain an air flow of 30 m/min.

i. Chemical spill kits: Provide work and storage areas with chemical spill kits obtained from commercial sources or assembled in the laboratory. Use kits of suitable size to accommodate the quantities of acids, alkalies, or solvents being used. Devise and post spill procedures before the need arises.

j. Safety wall chart: Post in central location for information on hazardous laboratory substances.

2. Personal Protective Equipment

Personal protective equipment and materials include such items as laboratory garments, gloves, shoes, hard hats, glasses, shields, and other safety items used by an individual. Their selection and use is governed by the particular tasks to be performed. Such equipment is important for protection but also serves as a constant reminder of the need for safety. If it is determined that such are needed, it is the manager's and supervisor's responsibility to insure that they are used.

a. Clothing: Personal clothing creates a barrier between the individual and the hazard. Employees using radioactive materials, suspected carcinogens, and pathogenic materials may be required to change from street to laboratory clothing when entering the work area and to change again on leaving. This not only prevents carrying hazardous materials outside the area but it also permits necessary handling and cleaning of the clothing. Disposable laboratory clothing should be fire-retardant.

b. Gloves frequently are important. Consult the material safety data sheet for a solvent or reagent to determine the type of glove to be used. Rubber gloves may be used when handling hazardous liquids, leaded gloves for handling radioactive materials, and surgical gloves for handling pathogenic material. Insulated gloves are essential for handling hot objects and extremely cold ones, but do not use asbestos gloves. White cotton gloves may be used to protect instruments.

c. Safety shoes are required in laboratories where heavy objects or equipment is to be moved. Hard hats may be required if there is overhead machinery in the laboratory.

d. Safety glasses: Even if the likelihood of accidents seems small, the consequences of an eye accident may be extremely serious. Require all laboratory personnel to wear safety glasses. Most laboratories prohibit use of contact lenses with safety glasses, but this is an area of controversy and should be evaluated in each laboratory. Safety glasses protect from splashes, flying objects, powders, or ultraviolet exposure. However, ordinary safety glasses do not provide total eye protection. If the work involves special hazard to the eyes, consider additional protection. For example, wear lenses with special filters for glass-blowing, welding, laser work, or exposure to other forms of radiation. When working in a room with ultraviolet radiation, wear safety glasses with side shields or goggles with solid side pieces. For certain activities, provide protection for exposed skin. In working with acid or caustic materials wear a face shield to protect both the eyes and the face.

e. *Respirators:* Have respirators[8] available for emergency situations in which dangerous gases, fumes, or aerosols may be formed and the room must be entered before it is thoroughly ventilated. In laboratories using toxic gases, such as boron trifluoride, chlorine, dimethylamine, ethylene oxide, fluorine, and hydrogen bromide, provide respirators, preferably Self-Contained Breathing Apparatus (SCBA), or supplied air.

3. References

1. NATIONAL INSTITUTE FOR OCCUPATIONAL SAFETY AND HEALTH. 1974. The Industrial Environment—Its Evaluation and Control, 3rd ed. U.S. Dep. Health, Education & Welfare, National Inst. Occupational Safety & Health, Rockville, Md.
2. U.S. PUBLIC HEALTH SERVICE. 1975. Lab Safety at the Center for Disease Control. U.S. Dep. Health, Education & Welfare, Publ. No. CDC 76-8118, National Center Disease Control, Atlanta, Ga.
3. THE MATHESON COMPANY. 1971. Matheson Gas Data Book, 5th ed. Matheson Co., East Rutherford, N.J.
4. MEIDL, J. H. 1970. Flammable Hazardous Materials. Glenncoe Press, Beverly Hills, Calif.
5. MCKINNON, G. P. 1976. Fire Protection Handbook, 14th ed. National Fire Protection Assoc., Boston, Mass.
6. STEERE, N.V., ed. 1971. Handbook of Laboratory Safety, 2nd ed. Chemical Rubber Co., Cleveland, Ohio.
7. AMERICAN CONFERENCE OF GOVERNMENTAL INDUSTRIAL HYGIENISTS. 1982. Industrial Ventilation, A Manual of Recommended Practices, 17th ed. American Conf. of Governmental Industrial Hygienists, Cincinnati, Ohio.
8. NATIONAL INSTITUTE FOR OCCUPATIONAL SAFETY AND HEALTH. 1976. Guide to Industrial Respiratory Protection. Publ. No. 76-189, U.S. Dep. Health, Education & Welfare, National Inst. Occupational Safety & Health, Cincinnati, Ohio.

1090 C.   Laboratory Hazards

Many laboratory hazards are fairly obvious, but identification of all likely hazards is important in developing an effective safety program. Develop a safety plan before any work is started. This is to provide all personnel working with or near hazardous materials information about them. Include Material Safety Data Sheets (MSDSs) in the safety plan and describe routine and emergency procedures; require that all personnel read and sign that they have read and understood the document.

To avoid ingestion of toxic or infectious material, prohibit eating, drinking, and smoking in all laboratory areas. Post adequate warning signs in the laboratory and insure that all containers have clear and readily recognizable labels. Avoid floor clutter so that escape routes and fire extinguishers are not blocked. Eliminate improper storage such as precarious shelving or heavy or glass objects overhead.

1. Chemical Hazards

Chemical injuries may be external or internal.[1-8] External injuries result from skin exposure to caustic or corrosive substances such as acids, bases, or reactive salts. Take care to prevent accidents, such as splashes and container spills. Coincidentally, pay attention to equipment corrosion that ultimately may lead to safety hazards from equipment failure. Internal injuries may result from toxic or corrosive effects of substances absorbed by the body.

a. Inorganic acids and bases: Many inorganic acids and bases have Occupational Safety and Health Personal Exposure Limits[9] and American Conference of Governmental Industrial Hygienists' Threshold Limit Values (TLV)[10] associated with them. These permissible exposure limits and threshold limit values indicate the maximum air concentration to which workers may be exposed. Fumes of these acids and bases are severe eye and respiratory system irritants. Liquid or solid acids and bases quickly can cause severe burns of the skin and eyes.[11] When acids are heated to increase the rate of digestion of organic materials, they pose a significantly greater hazard because fumes are produced and the hot acid reacts very quickly with the skin.

Store acids and bases separately in well-ventilated areas and away from volatile organic and oxidizable materials. Use containers (rubber or plastic buckets) to transport acids and bases. Work with strong acids and bases only in a properly functioning chemical fume hood. Slowly add acids and bases to water (with constant stirring) to avoid spattering. If skin contact is made, thoroughly flush the contaminated area with water and seek medical attention if irritation persists.[12] Do not rewear clothing until after it has been cleaned thoroughly. Leather items (e.g. belts and shoes) will retain acids even after rinsing with water and may cause severe burns if reworn. If eye contact is made, immediately flush both eyes for a full 15 min at the eye wash and seek medical attention.

Perchloric acid (see Sections 3030G and H and 4500-P.D) reacts violently or explosively on contact with organic material.[12] Do not use laboratory fume hoods used with perchloric acid for organic reagents, particularly volatile solvents. In addition to these hazards, perchloric acid produces severe burns when contact is made with the skin, eyes, or respiratory tract.[13] Preferably provide a dedicated perchloric acid hood. Follow manufacturer's instructions for proper cleaning because exhaust ducts become coated and must be washed down regularly.

The most common injuries suffered with sodium hydroxide are burns of the skin and eyes. Sodium hydroxide solutions as dilute as $2.5N$ can cause severe eye damage.[14] On dissolution, sodium hydroxide and other alkalies produce considerable heat (often sufficient to cause boiling).

b. Metals and inorganic compounds: In general, consider all laboratory chemicals hazardous and use only as prescribed. Handle chemicals in a fume hood, using eye protection and wearing a laboratory coat. In case of accidental skin contact, flush the affected area thoroughly with water. If irritation persists, medical examination is advisable. Toxicity, precautionary, and first-aid information is available from a variety of sources.[11–15]

Some hazards of specific metals and inorganic compounds are as follows: Some compounds of arsenic and nickel are highly toxic and may be carcinogenic.[14,15] Avoid inhalation, ingestion, and skin contact. Sodium azide is toxic and reacts with acid to produce still more toxic hydrazoic acid. When disposed of through a drain it may react with copper and lead plumbing to form metal azides that are extremely explosive. Azides may be destroyed by adding a concentrated solution of sodium nitrite, 1.5 g $NaNO_2$/g sodium azide.[16] The extreme toxicity of beryllium and its compounds is reflected by its low TLV of 2.0 $\mu g/m.^{3,11}$ Beryllium is a suspected carcinogen in humans.[9,15] Handle with extreme caution and only in a laboratory fume hood or glove box. Cyanides are used as reagents and may be present in samples. Hydrogen cyanide is a lethal gas. Do not acidify cyanide solutions except in a closed system or in a properly ventilated hood because hydrogen cyanide may be formed and liberated.

Mercury is unusual among the metals in that it is liquid at room temperature and has appreciable vapor pressure. One broken thermometer in a poorly ventilated room may cause the mercury TLV to be exceeded. Because of its high volatility and toxicity, handle mercury and its compounds very cautiously and provide a spill cleanup kit. Perchlorate salts are explosive when mixed with combustible material. They also are severe irritants to the skin, eyes, and respiratory system. Use care in handling and storing perchlorates. Sodium borohydride decomposes in water and liberates hydrogen, consequently creating an explosion hazard. Like many other inorganic chemicals, it is a strong skin and respiratory system irritant.

c. *Organic solvents and reagents:* Most solvents specified in this book have TLVs for workroom exposure[10] (see Table 1090:I). Many organic reagents, unlike most organic solvents, do not have TLVs (see Table 1090:II for those that do) but this does not mean that they are less hazardous. Some compounds are suspect carcinogens and should be treated with extreme caution. These compounds include both solvents and reagents such as benzene,[18] carbon tetrachloride,[11] chloroform,[11] 1,4-dioxane,[8,11,19] tetrachloroethylene,[20] and benzidine.[21] Lists of chemicals with special hazardous characteristics are available from the Occupational Safety and Health Administration and the National Institute for Occupational Safety and Health. The lists of "Regulated Carcinogens" and of "Chemicals Having Substantial Evidence of Carcinogenicity" are especially important. Determining labora-

TABLE 1090:I. THRESHOLD LIMIT VALUES*[10] FOR SOLVENTS SPECIFIED IN *Standard Methods*

Compound†	TLV ppm (v/v)
Acetic acid	10
Acetone	750
Acetonitrile (S)	40
Benzene (C)[17]	10
n-Butanol (S)	50
t-Butanol	100
Carbon disulfide (S)	10
Carbon tetrachloride (C)(S)	5
Chloroform (C)	10
Cyclohexanone (S)	25
Diethyl ether	400
1,4-Dioxane (C)(S)	25
Ethanol	1000
Ethyl acetate	400
Ethylene glycol	50
Hexane	50
Isoamyl alcohol	100
Isobutyl alcohol	50
Isopropyl alcohol	400
Isopropyl ether	250
Methanol (S)	200
2-Methoxyethanol (S)	5
Methylene chloride (C)	50
Pentane	600
Propanol (S)	200
Pyridine	5
Tetrachloroethylene	50
Toluene	100
Xylene	100

* These threshold limit values provide an indication of the maximum air concentrations to which workers may be exposed.
† (C) compound is a suspect carcinogen; (S) refers to potential contribution by skin absorption.

tory handling procedures on such authoritative lists decreases the chance for error.

Solvents used herein fall into several major categories: alcohols, chlorinated compounds, and hydrocarbons. Exposure to each of these classes of compounds can have a variety of health effects.[22–24] Alcohols, in general, are intoxicants, capable of causing irritation of the mucus membranes and drowsiness. While diols such as eth-

TABLE 1090:II. THRESHOLD LIMIT VALUES[10] FOR REAGENTS SPECIFIED IN *Standard Methods*

Compound*	TLV
2-Aminoethanol	3 ppm (v/v)
Benzidine (C)(S)[17]	—
Benzyl chloride	1 ppm (v/v)
Chlorobenzene	75 ppm (v/v)
2-Chloro-6-(trichloromethyl) pyridine	10 mg/m^3
Diethanolamine	3 ppm (v/v)
Naphthalene	10 ppm (v/v)
Oxalic acid	1 mg/m^3
Phenol	5 ppm (v/v)

* (C) compound is a suspect carcinogen; (S) refers to potential contribution by skin absorption.

ylene glycol are poisonous, triols such as glycerol are not poisonous at all. Chlorinated hydrocarbons cause narcosis and damage to the central nervous system and liver. Hydrocarbons, like the other two groups, are skin irritants and may cause dermatitis after prolonged skin exposure. Because of the volatility of these compounds, hazardous vapor concentrations can occur (fire or explosion hazard). Proper ventilation is essential.[25]

Organic reagents used in this manual fall into four major categories: acids, halogenated compounds, dyes and indicators, and pesticides. Most organic acids have irritant properties.[22,24] They are predominantly solids from which aerosols may be produced. Dyes and indicators also present an aerosol problem. Handle pesticides with caution because they are poisons[22] and avoid contact with the skin. Wear gloves and protective clothing. The chlorinated compounds present much the same hazards as the chlorinated solvents (narcosis and damage to the central nervous system and liver). Proper labeling of the compound, including a date for disposal based on the manufacturer's recommendation, permits tracking chemical usage and disposal of outdated chemicals.

2. Biological Hazards

Water laboratory safety also includes microbiological hazards.[26-30] Pathogenic microorganisms may produce human disease by accidental ingestion, inoculation or injection or other means of cutaneous penetration. Good laboratory safety techniques will control these agents.[31] Primary dangers associated with microbiological hazards involve hand-mouth contact while handling contaminated laboratory materials and aerosols created by inoculating, pipetting, centrifuging, or blending of samples or cultures.[32]

Do not mix dilutions by blowing air through a pipet into a microbiological culture. When working with grossly polluted samples, such as sewage or high-density microbial cultures, use a pipetting device attached to the mouth end of the pipet to prevent accidental ingestion. *Never pipet by mouth.* Even for drinking water samples mouth pipetting is inadvisable. Because untreated waters may contain waterborne pathogens, discard all used pipets into a jar containing disinfectant solution for decontamination before glassware washing. Do not place discarded pipets on table tops, on laboratory carts, or in sinks without adequate decontamination. Quaternary ammonium compounds that include a compatible detergent, or solutions of sodium hypochlorite are satisfactory disinfectants for pipet discard jars. Use highest concentrations recommended for these commercial products provided that this concentration does not cause a loss of markings or fogging of pipets. Replace disinfectant solution in the discard container daily. Sterilize contaminated materials (cultures, samples, used glassware, serological discards, etc.) by autoclaving before throwing them away or processing for reuse. Shattering of culture-containing tubes during centrifugation also produces microbiological aerosols.[33,34] Use leakproof blenders and keep them tightly covered during operation. Inserting a hot loop into a flask of broth culture creates a serious hazard due to aerosolized microorganisms.[35] Using electric heater incineration for sterilizing inoculating loops or needles may be desirable, but avoid possible electrical shock that could occur by touching the loop to the inside of the heater core.[36]

Good personal hygienic practices are important in the control of contact exposures. Frequently disinfect hands and working surfaces. Provide drinking water outside the laboratory, preferably from a foot-operated drinking fountain. Suggest immunization of laboratory staff against tetanus and possibly typhoid or other appropriate infectious agents, depending on nature of the work. Eliminate flies and other insects to prevent contamination of sterile equipment, media, samples, and bacterial cultures and to prevent spread of infectious organisms to personnel via this vector. Install screens in all windows and outer doors if there is no air-conditioning, and spray periodically with pesticides along toe-stripping, sink and storage cabinet areas, and utility service channels. Because laboratories also may include a chemistry section that analyzes waters for pesticides, apply these carefully within the immediate areas of microbiological activity.

3. Radiation Hazards

All persons are exposed to ionizing radiation. The average annual radiation dose to the whole body from cosmic, terrestrial, and internal sources, medical and dental X-rays, etc., is about 185 mrems/year.[37] Prevent unnecessary continuous or intermittent occupational exposure, and prevent accidents that may result in dangerous radiation exposure. In laboratories, X-rays, ultraviolet light, and radioactive material represent hazards that must be minimized. Provide a Radiation Safety Manual,[38] Handbook of Laboratory Safety,[13] or similar manual to all persons working with radioactive materials or radiation-producing machines. This manual should discuss procedure for obtaining authorization to use, order, handle, and store radionuclides. It also should include procedures for safe handling of unsealed radioactive material and procedures to follow in case of radiation accidents, for decontamination, for personnel monitoring, for laboratory monitoring, and for disposal of radioactive materials.

a. Radioactive materials: Radionuclides are used in laboratories to develop and evaluate analytical methods, to prepare counting standards, and to calibrate detectors and counting instruments (see Part 7000). Sealed sources, such as the nickel 63 detector cell used in electron capture gas chromatograph units, are common. Instruct all persons dealing with radioactive materials about associated health hazards. Establish proper procedures and incorporate radiation safety measures to minimize exposure and accident.

b. Radiation-producing machines: Minimize hazards from devices, such as X-ray diffraction apparatus or electron microscope, by adhering strictly to manufacturer's operating manual and procedures outlined in the referenced safety manuals.

c. Ultraviolet radiation (UV): UV is used frequently. With properly constructed and operated instruments, it is not a significant hazard but can be harmful when used for controlling microorganisms in laboratory rooms or for sterilizing objects. Provide proper shielding, remembering that shiny metal surfaces reflect this energy; shut off UV lamps when not in use. Post warning signs and install indicator lights to serve as a constant reminder when UV lamps are *on.* Wear safety glasses or goggles with solid side pieces whenever there is a possibility of exposure to UV radiation.[39]

4. Physical Hazards

a. Electrical: Conform electrical wiring, connections, and apparatus to the latest

National Electrical Code. Fire, explosion, power outages, and electrical shocks are all serious hazards that may result from incorrect use of electrical devices. Ground all electrical equipment or use double insulation. Use ground fault circuit breakers to the maximum extent possible. Do not locate electrical receptacles inside fume hoods, do not use equipment with frayed cords or cracked insulation nor spark-producing electrical equipment near volatile flammable solvents. Use approved safety refrigerators. Disconnect electrical equipment from the power supply before service or repair is attempted and never bypass safety interlocks. Equipment repair by employees not thoroughly acquainted with electrical principles may present particularly dangerous situations.[40]

b. Mechanical: Shield or guard drive belts, pulleys, chain drivers, rotating shafts, and other types of mechanical power transmission apparatus.[9] Laboratory equipment requiring this guarding includes vacuum pumps, mixers, blenders, and grinders. Shield portable power tools used in laboratories.[9] Guard equipment such as centrifuges, which have high-speed revolving parts, against "fly-aways." Securely fasten equipment that has a tendency to vibrate (e.g., centrifuges and air compressors) to prevent the tendency to "walk" and locate away from bottles and other items that may fall from shelves or benches because of the vibration.[13]

c. Compressed gases: Gas cylinders may explode or "rocket" if improperly handled. Leaking cylinders may present an explosion hazard if the contents are flammable, they are an obvious health hazard if the contents are toxic, and they may lead to death by suffocation if the contents are inert gases. OSHA regulations govern use and storage of compressed gases.[9] Transfer gas cylinders only by carts, hand trucks, or dollies. Secure gas cylinders properly during storage, transport, and use and leave valve safety cover on cylinders during storage and transport. Avoid the use of adapters or couplers with compressed gas. Permanently identify cylinder contents.

5. References

1. NATIONAL INSTITUTE FOR OCCUPATIONAL SAFETY AND HEALTH. 1976. Registry of Toxic Effects of Chemical Substances. U.S. Dep. Health, Education & Welfare, National Inst. Occupational Safety & Health, Rockville, Md.

2. AMERICAN CONFERENCE OF GOVERNMENTAL INDUSTRIAL HYGIENISTS. 1976. Documentation of the Threshold Limit Values for Substances in Workroom Air, 3rd ed. American Conf. of Governmental Industrial Hygienists, Cincinnati, Ohio.

3. WINDHOLZ, M. 1976. The Merck Index, 9th ed. Merck and Co., Rahway, N.J.

4. AMERICAN MUTUAL INSURANCE ALLIANCE. 1966. Handbook of Organic Solvents. Tech. Guide No. 6, American Mutual Insurance Alliance, Chicago, Ill.

5. BRAKER, W. & A. L. MOSSMAN. 1970. Effects of Exposure to Toxic Gases—First Aid and Medical Treatment. Matheson Gas Products, East Rutherford, N.J.

6. SAX, N. I. 1968. Dangerous Properties of Industrial Materials, 3rd ed. Van Nostrand Reinhold Co., New York, N.Y.

7. BROWNING, E. C. 1965. Toxicity and Metabolism of Industrial Solvents. Elsevier Scientific Publ. Co., New York, N.Y.

8. DEICHMANN, W. B. & H. W. GERARDE. 1969. Toxicology of Drugs and Chemicals. Academic Press, New York, N.Y.

9. Code of Federal Regulations. 1980. 29 Labor Part 1910. Washington, D.C.

10. AMERICAN CONFERENCE OF GOVERNMENTAL INDUSTRIAL HYGIENISTS. 1986–87. Threshold Limit Values. American Conf. Governmental Industrial Hygienists, Cincinnati, Ohio.

11. NATIONAL INSTITUTE FOR OCCUPATIONAL SAFETY AND HEALTH/OCCUPATIONAL SAFETY AND HEALTH ADMINISTRATION. 1981. Occupational Health Guidelines for Chemical Hazards. Publ. No. 81-123, U.S. Dep. Health & Human Services, U.S. Dep. Labor, Washington, D.C.

12. NATIONAL RESEARCH COUNCIL. 1981. Prudent Practices for Handling Hazardous Chem-

icals in Laboratories. National Academy Press, Washington, D.C.

13. STEERE, N.V. 1971. Handbook of Laboratory Safety, 2nd ed. Chemical Rubber Co., Cleveland, Ohio.

14. MUIR, G. C. 1977. Hazards in the Chemical Laboratory, 2nd ed. Chemical Soc., London, England.

15. CLAYTON, G. D. & F. E. CLAYTON. 1981. Patty's Industrial Hygiene and Toxicology, Vol. II A, 3rd ed. John Wiley & Sons, New York, N.Y.

16. AMERICAN PUBLIC HEALTH ASSOCIATION, AMERICAN WATER WORKS ASSOCIATION & WATER POLLUTION CONTROL FEDERATION. 1981. Standard Methods for the Examination of Water and Wastewater, 15th ed. American Public Health Assoc., Washington, D.C.

17. U.S. PUBLIC HEALTH SERVICE. 1975. Lab Safety at the Center for Disease Control. U.S. Dep. Health, Education & Welfare, Publ. No. CDC 76-8118. National Center Disease Control, Atlanta, Ga.

18. NATIONAL INSTITUTE FOR OCCUPATIONAL SAFETY AND HEALTH. 1974. Criteria for a Recommended Standard. Occupational Exposure to Benzene. Publ. No. 74-137, U.S. Dep. Health, Education & Welfare, U.S. Government Printing Off., Washington, D.C.

19. NATIONAL INSTITUTE FOR OCCUPATIONAL SAFETY AND HEALTH. 1977. Criteria for a Recommended Standard. Occupational Exposure to Dioxane. Publ. No. 77-226, U.S. Dep. Health, Education & Welfare, U.S. Government Printing Office, Washington, D.C.

20. NATIONAL INSTITUTE FOR OCCUPATIONAL SAFETY AND HEALTH. 1978. Current Intelligence Bulletin 20 Tetrachloroethylene (Perchloroethylene). Publ. No. 78-112, U.S. Dep. Health, Education & Welfare, U.S. Government Printing Off., Washington, D.C.

21. BOENIGER, M. 1980. Carcinogenicity and Metabolism of Azo Dyes, Especially Those from Benzidine. Publ. No. 80-119, U.S. Dep. Health & Human Services. U.S. Government Printing Off., Washington, D.C.

22. DOULL, J., C. D. KLASSEN & M. O. AMDUR, eds. 1980. Casarett and Doull's Toxicology, 2nd ed. Macmillan Co., New York, N.Y.

23. Toxic and Hazardous Industrial Chemicals Safety Manual for Handling and Disposal with Toxicity and Hazard Data. 1978. Interna-

tional Technical Information Inst., Tokyo, Japan.

24. PATTY, F. A. 1963. Patty's Industrial Hygiene and Toxicology, 2nd ed. John Wiley & Sons, New York, N.Y.

25. CLAYTON, G. D. & F. E. CLAYTON. 1978. Patty's Industrial Hygiene and Toxicology, 3rd ed. Vol. 1. John Wiley & Sons, New York, N.Y.

26. WEDEUN, A. G., E. HANEL, G. B. PHILLIPS & O.T. MILLER. 1956. Laboratory Design for Study of Infectious Disease. Amer. J. Pub. Health 46:1102.

27. DARLOW, H. M. 1969. Safety in the Microbiological Laboratory. In J. R. Norris & D. W. Ribbons, eds. Methods in Microbiology. Volume 1. Academic Press, New York, N.Y.

28. SCIENCE PRODUCTS DIVISION, MALLINCKRODT CHEMICAL WORKS. 1969. Laboratory Safety Handbook. Mallinckrodt Chemical Works, St. Louis, Mo.

29. SHAPTON, D. A. & R. J. BOARD. 1972. Safety in Microbiology. Soc. Applied Bacteriology, Tech. Ser. No. 6., Academic Press, New York, N.Y.

30. U.S. DEPARTMENT OF HEALTH AND HUMAN SERVICES, PUBLIC HEALTH SERVICE, CENTERS FOR DISEASE CONTROL & NATIONAL INSTITUTES OF HEALTH. 1984. Biosafety in Microbiological and Biomedical Laboratories. U.S. Government Printing Off., Washington, D.C.

31. OFFICE OF BIOSAFETY. 1974. Classification of Etiologic Agents on the Basis of Hazard, 4th ed. Center Disease Control, Atlanta, Ga.

32. PHILLIPS, G. B. 1961. Microbiological Safety in U.S. and Foreign Laboratories. Tech. Study 35, Biological Laboratories Project 4B11-05-015, U.S. Army Chemical Corps, Fort Detrick, Md.

33. REITMAN, M. & G. B. PHILLIPS. 1956. Biological hazards of common laboratory procedures. III. The centrifuge. Amer. J. Med. Technol. 22:100.

34. HALL, C. V. 1975. A biological safety centrifuge. Health Lab. Sci. 12:104.

35. PHILLIPS, G. B. & M. REITMAN. 1956. Biological hazards of common laboratory procedures. IV. The inoculating loop. Amer. J. Med. Technol. 22:16.

36. GORDON, R. C. & C. V. DAVENPORT. 1973.

Simple modification to improve usefulness of the bacti-cinerator. *Appl. Microbiol.* 26:423.

37. NATIONAL COUNCIL ON RADIATION PROTECTION AND MEASUREMENTS. 1987. Ionizing Radiation Exposure of the Population of the United States. Rep. No. 93, National Council Radiation Protection, Washington, D.C.

38. CALIFORNIA STATE DEPARTMENT OF HEALTH SERVICES. 1979. Radiation Safety Manual. Berkeley, Calif. (Similar manuals are available from most state health departments, universities, and national laboratories.)

39. WALLACE, L. A. 1981. Recent progress in developing and using personal monitors to measure human exposure to air pollutants. *Environ. International* 5:73.

40. INHORN, S.L., ed. 1978. Quality Assurance Practices for Health Laboratories. American Public Health Assoc., Washington, D.C.

6. Bibliography

NATIONAL COMMITTEE ON RADIATION PROTECTION. 1964. Safe Handling of Radioactive Materials. National Bur. Standards Handbook 92, U.S. Government Printing Off., Washington, D.C.

NATIONAL COUNCIL ON RADIATION PROTECTION AND MEASUREMENTS. 1976. Radiation Protection for Medical and Allied Health Personnel. Rep. No. 48, National Council Radiation Protection & Measurements, Washington, D.C.

WALTERS, D. B., ed. 1980. Safe Handling of Chemical Carcinogens, Mutagens, Teratogens and Highly Toxic Substances, Vols. 1 and 2. Ann Arbor Science, Ann Arbor, Mich.

FUSCALDO, A.A., B.J. ERLICK & B. HINDMAN. 1980. Laboratory Safety, Theory and Practice. Academic Press, New York, N.Y.

NATIONAL COUNCIL ON RADIATION PROTECTION AND MEASUREMENTS. 1987. Recommendations on Limits for Exposure to Ionizing Radiation. Rep. No. 91, National Council Radiation Protection & Measurements, Washington, D.C.

1090 D.   Hazard Management Practices

1. Monitoring

The establishment of policies, practices, procedures, and methods to prevent exposure of laboratory personnel to hazardous materials is only part of an effective safety program. Simultaneous establishment of a monitoring or feedback system is essential to insure that the protective features actually work.

a. Chemical monitors: Devices capable of direct measurement of concentration in a person's breathing zone have been described.[1] Use either active devices requiring a pump to pull air across the cell or passive monitors that rely on diffusion. These devices can measure inorganic as well as organic compounds with the appropriate sorbent. Particulates can be collected on a filter using an active device. A detailed discussion of the development and use of personal monitors for measuring exposure to chemical pollutants in the environment is available.[1]

b. Biohazard monitors: These are an essential part of microbiological monitoring. Include a pre-employment physical examination accompanied by hematological-serological and other tests related to exposure. Provide for subsequent annual examinations consisting of serological tests, biochemical function studies, and chest X-rays. Vaccination and other prophylactic measures are required. Archive serum specimens for future reference as necessary. Continual enforcement of basic laboratory safety rules completes the monitoring program.

c. Radiochemical monitors: These include wipes, portable survey instruments, and air samples. Multiple monitors usually are required. Measure external radiation exposure with personnel dosimeters, preferably the film dosimeter for measuring accumulated radiation over a period of time. Pocket ionization chambers, thermoluminescent dosimeters, and thimble chambers can be used to supplement the film dosimeter.

Whole body or gamma spectrometry radiation detectors determine the presence of radioactive substances in the body, but these instruments are expensive and require specialized training. Body waste also can be monitored for radionuclides.

In addition to personnel monitoring, carry out general area monitoring. Analyze all equipment and supplies that have been in contact with radioactive substances. Use dosimetric and wipe tests. Thin-windowed GM-counters are suitable for wipe samples and for monitoring skin and clothing. An alpha scintillation monitor is needed to detect alpha emitters. Steere[2] presents an excellent discussion of monitoring techniques for radioisotopes.

2. Disposal of Wastes

a. General considerations: Stringent requirements exist for the disposal of wastes involving potential criminal and civil liability on both organizations and individuals. Specifics vary by state and local area and are subject to change. Licenses and/or permits may be required; consult local and state authorities.

A plan for the safe disposal of chemical and biological substances used in the laboratory is important and should be discussed with the laboratory supervisor and, if appropriate, with the safety coordinator. Gaston[3] has discussed the care, handling, and disposal of dangerous chemicals. Install convenient collection systems, use properly labeled containers, provide fire-protected storage, and provide a special separate storage area for hazardous or highly toxic materials. Use metal safety cans for storing waste solvents and segregating incompatible materials. Consider using special containers for extremely hazardous or highly toxic wastes and special packing procedures to avoid breakage or damage to the container in transportation. Store incompatible hazardous materials separately.[4]

b. Waste disposal methods include incineration, burial, evaporation, neutralization, chemical reaction, special treatments, and use of commercial disposal specialists. Safety information including the method of choice for waste disposal of specific inorganic and organic compounds is presented in a series of Material Safety Data Sheets (MSDSs).[5]

Sometimes it may be permissible to dispose of some hazardous chemicals down the drain at certain concentrations but only with written permission of the local waste-water authority on a chemical-by-chemical basis.

1) Chemical wastes—Used solvents can be distilled and recovered for reuse. Noncombustible solvents can be evaporated if vapors do not create an environmental problem. Small quantities of flammable solvents and chemicals can be burned on the ground, in shallow metal containers, or in incinerators, provided that this does not violate air regulations. Neutralize acidic and basic materials before final disposal. Many soluble, nontoxic materials can be diluted carefully into a sewer system if it is certain that they will not harm the plumbing system or environment. Whenever possible, convert hazardous materials by chemical reaction or other processes to innocuous compounds before disposal. If this is not feasible, engage a commercial disposal specialist to dispose of hazardous or highly toxic materials off the premises. Dispose of nonreturnable gas cylinders with the help of qualified personnel only.

2) Biological wastes—Sterilize all infectious or toxic substances, and all contaminated equipment or apparatus before washing, storage, or disposal.[6] Autoclaving is the sterilization method of choice. Generally, heat in an autoclave under a pres-

sure of 103 kPa to achieve a chamber temperature of at least 121°C for a minimum of 15 min. Measure time after the material being sterilized reaches 121°C. If materials are to be autoclaved in plastic bags, add water to the contents to insure wet heat. Dry heat and chemical treatment also have been used for sterilizing nonplastic items. After sterilization, the wastes can be handled safely by conventional disposal systems. Collect contaminated combustible wastes and animal carcasses in impermeable containers for disposal by incineration.

3) Radiochemical wastes—Generalized disposal criteria for radioactive wastes have been developed by the National Committee on Radiation Protection and Measurements.[7] These recommendations have been given official status by publication in the Federal Register.[8] Two general philosophies govern the disposal of radioactive wastes: (a) dilution and dispersion to reduce the concentration of radionuclide by carrier dilution or dilution in a receiving medium and (b) concentration and confinement, usually involving reduction in waste volume with subsequent storage for decay purposes.

Airborne wastes can be treated by either method. Ventilation includes discharge from hooded operations to the atmosphere. Typical radioactive gases include iodine, krypton, and xenon. Iodine can be removed by scrubbing or by reaction with silver nitrate. Noble gases can be removed by adsorption; standard techniques can be used for particulates. Dilution methods are suitable for liquids with low activity. Intermediate levels may be treated by various physical-chemical processes to separate the waste into a nonradioactive portion that can be disposed of by dilution and a high-activity portion to be stored. Solid wastes may consist of equipment, glassware, and other materials. When possible, decontaminate these materials and reuse. Decontamination usually results in a liquid waste.

Combustible materials that cannot be decontaminated often are burned with special precautions; permits for burning may be required. Use storage for decay or permanent storage for treating radioactive wastes when alternatives are not available.

4) Other special wastes—Federal regulations concerned with polychlorinated biphenyls (PCBs), dioxins, and asbestos require special attention for wastes containing these materials.

3. References

1. Proceedings of the Symposium on the Development and Usage of Personal Monitors for Exposure and Health Effect Studies, June 1979. Publ. EPA-600/9-79-032, U.S. Environmental Protection Agency, Washington, D.C.
2. STEERE, N.V., ed. 1971. Handbook of Laboratory Safety, 2nd ed. Chemical Rubber Co., Cleveland, Ohio.
3. GASTON, P. J. 1970. The Care, Handling, and Disposal of Dangerous Chemicals. Inst. of Science Technology, Northern Publishers (Aberdeen) Ltd., Aberdeen, Scotland.
4. U.S. ENVIRONMENTAL PROTECTION AGENCY. 1980. Hazardous waste compatibility chart. Publ. No. EPA-600/2-80-076, Cincinnati, Ohio.
5. Chemical Safety Data Sheets. Manufacturing Chemists' Assoc., Inc., Washington, D.C.
6. NATIONAL INSTITUTES OF HEALTH. 1974. Biohazard Safety Guide. GPO Stock No. 1740-00383, U.S. Dep. Health, Education & Welfare, Public Health Service, National Inst. Health, Washington, D.C.
7. U.S. ENVIRONMENTAL PROTECTION AGENCY. 1985. Environmental Standards for the Management and Disposal of Spent Nuclear Fuel, High-Level and Transuranic Wastes. 40 CFR Part 191.
8. U.S. ENVIRONMENTAL PROTECTION AGENCY. 1988. Environmental Standards for the Management, Storage and Land Disposal of Low-Level Radioactive Waste. Proposed Rule. 40 CFR Part 193.

4. Bibliography

GHASSEMI, M., S. QUINLIVAN, G. GRUBER & H. CASEY. 1976. Disposing of Small Batches of Hazardous Wastes. Publ. SW 562, Office of

Solid Waste, U.S. Environmental Protection Agency, Cincinnati, Ohio.

BORDNER, R. H., J.A. WINTER & P.V. SCARPINO, eds. 1978. Microbiological Methods for Monitoring the Environment, Water and Wastes. Publ. EPA-600/8-78-017, Environmental Monitoring and Support Lab., U.S. Environ-mental Protection Agency, Cincinnati, Ohio.

NATIONAL ACADEMY OF SCIENCES, NATIONAL ACADEMY OF ENGINEERING & INSTITUTE OF MEDICINE. 1983. Prudent Practices for the Disposal of Chemicals from Laboratories. National Academy Press, Washington, D.C.

PART 2000

PHYSICAL

AND AGGREGATE

PROPERTIES

2010 INTRODUCTION

This part deals primarily with measurement of the physical properties of a sample, as distinguished from the concentrations of chemical or biological components. Many of the determinations included here, such as color, electrical conductivity, and turbidity, fit this category unequivocally. However, physical properties cannot be divorced entirely from chemical composition, and some of the techniques of this part measure aggregate properties resulting from the presence of a number of constituents. Others, for example, calcium carbonate saturation, are related to, or depend on, chemical tests. Also included here are tests for appearance, odor, and taste, which have been classified traditionally among physical properties, although the point could be argued. Finally, Section 2710, Tests on Sludges, includes certain biochemical tests. However, for convenience they are grouped with the other tests used for sludge.

With these minor exceptions, the contents of this part have been kept reasonably faithful to its name. Most of the methods included are either inherently or at least traditionally physical, as distinguished from the explicitly chemical, radiological, biological, or bacteriological methods of other parts.

2020 QUALITY CONTROL

Part 2000 contains a variety of analytical methods, many of which are not amenable to standard quality-control techniques. General information on quality control is provided in Part 1000 and specific quality-control techniques are outlined in the individual methods. The following general guidelines may be applied to many of the methods in this part:

Evaluate analyst performance for each method. Determine competence by analyses of samples containing known concentrations.

Calibrate instruments and ensure that instrument measurements do not drift.

Assess the precision of analytical procedures by analyzing at least 10% of samples in duplicate. Analyze a minimum of one duplicate with each set of samples.

Determine bias of an analytical procedure in each sample batch by analysis of blanks, known additions with a frequency of at least 5% of samples, and, if possible, an externally provided standard.

2110 APPEARANCE*

To record the general physical appearance of a sample, use any terms that briefly describe its visible characteristics. These terms may state the presence of color, turbidity, suspended solids, crustacea, larvae, worms, sediment, floating material, and similar particulate matter detectable by the unaided eye. Use numerical values when they are available, as for color, turbidity, and suspended solids.

---

*Approved by Standard Methods Committee, 1988.

2120   COLOR*

2120 A.   Introduction

Color in water may result from the presence of natural metallic ions (iron and manganese), humus and peat materials, plankton, weeds, and industrial wastes. Color is removed to make a water suitable for general and industrial applications. Colored industrial wastewaters may require color removal before discharge into watercourses.

1. Definitions

The term "color" is used here to mean true color, that is, the color of water from which turbidity has been removed. The term "apparent color" includes not only color due to substances in solution, but also that due to suspended matter. Apparent color is determined on the original sample without filtration or centrifugation. In some highly colored industrial wastewaters color is contributed principally by colloidal or suspended material. In such cases both true color and apparent color should be determined.

2. Pretreatment for Turbidity Removal

To determine color by currently accepted methods, turbidity must be removed before analysis. The optimal method for removing turbidity without removing color has not been found yet. Filtration yields results that are reproducible from day to day and among laboratories. However, some filtration procedures also may remove

---

*Approved by Standard Methods Committee, 1988.

some true color. Centrifugation avoids interaction of color with filter materials, but results vary with the sample nature and size and speed of the centrifuge. When sample dilution is necessary, whether it precedes or follows turbidity removal, it can alter the measured color if large color-bodies are present.

Acceptable pretreatment procedures are included with each method. State the pretreatment method when reporting results.

3. Selection of Method

The visual comparison method is applicable to nearly all samples of potable water. Pollution by certain industrial wastes may produce unusual colors that cannot be matched. In this case use an instrumental method. A modification of the tristimulus and the spectrophotometric methods allows calculation of a single color value representing uniform chromaticity differences even when the sample exhibits color significantly different from that of platinum cobalt standards. For comparison of color values among laboratories, calibrate the visual method by the instrumental procedures.

4. Bibliography

OPTICAL SOCIETY OF AMERICA. 1943. Committee Report. The concept of color. *J. Opt. Soc. Amer.* 33:544.
JONES, H. et al. 1952. The Science of Color. Thomas Y. Crowell Co., New York, N.Y.

2120 B.   Visual Comparison Method

1. General Discussion

a. Principle: Color is determined by visual comparison of the sample with known concentrations of colored solutions. Comparison also may be made with special, properly calibrated glass color disks. The platinum-cobalt method of measuring color is the standard method, the unit of color being that produced by 1 mg platinum/L in the form of the chloroplatinate ion. The ratio of cobalt to platinum may be varied to match the hue in special cases; the proportion given below is usually satisfactory to match the color of natural waters.

b. Application: The platinum-cobalt method is useful for measuring color of potable water and of water in which color is due to naturally occurring materials. It is not applicable to most highly colored industrial wastewaters.

c. Interference: Even a slight turbidity causes the apparent color to be noticeably higher than the true color; therefore remove turbidity before approximating true color by differential reading with different color filters[1] or by differential scattering measurements.[2] Neither technique, however, has reached the status of a standard method. Remove turbidity by centrifugation or by the filtration procedure described under Method C. Centrifuge for 1 h unless it has been demonstrated that centrifugation under other conditions accomplishes satisfactory turbidity removal.

The color value of water is extremely pH-dependent and invariably increases as the pH of the water is raised. When reporting a color value, specify the pH at which color is determined. For research purposes or when color values are to be compared among laboratories, determine the color response of a given water over a wide range of pH values.[3]

d. Field method: Because the platinum-cobalt standard method is not convenient for field use, compare water color with that of glass disks held at the end of metallic tubes containing glass comparator tubes filled with sample and colorless distilled water. Match sample color with the color of the tube of clear water plus the calibrated colored glass when viewed by looking toward a white surface. Calibrate each disk to correspond with the colors on the platinum-cobalt scale. The glass disks give results in substantial agreement with those obtained by the platinum-cobalt method and their use is recognized as a standard field procedure.

e. Nonstandard laboratory methods: Using glass disks or liquids other than water as standards for laboratory work is permissible only if these have been individually calibrated against platinum-cobalt standards. Waters of highly unusual color, such as those that may occur by mixture with certain industrial wastes, may have hues so far removed from those of the platinum-cobalt standards that comparison by the standard method is difficult or impossible. For such waters, use the methods in Sections 2120C and D. However, results so obtained are not directly comparable to those obtained with platinum-cobalt standards.

f. Sampling: Collect representative samples in clean glassware. Make the color determination within a reasonable period because biological or physical changes occurring in storage may affect color. With naturally colored waters these changes invariably lead to poor results.

2. Apparatus

a. Nessler tubes, matched, 50-mL, tall form.

b. pH meter, for determining sample pH (see Section 4500-H$^+$).

3. Preparation of Standards

a. If a reliable supply of potassium chloroplatinate cannot be purchased, use chloroplatinic acid prepared from metallic platinum. Do not use commercial chloroplatinic acid because it is very hygroscopic and may vary in platinum content. Potassium chloroplatinate is not hygroscopic.

b. Dissolve 1.246 g potassium chloroplatinate, K_2PtCl_6 (equivalent to 500 mg metallic Pt) and 1.00 g crystallized cobaltous chloride, $CoCl_2 \cdot 6H_2O$ (equivalent to about 250 mg metallic Co) in distilled water with 100 mL conc HCl and dilute to 1000 mL with distilled water. This stock standard has a color of 500 units.

c. If K_2PtCl_6 is not available, dissolve 500 mg pure metallic Pt in aqua regia with the aid of heat; remove HNO_3 by repeated evaporation with fresh portions of conc HCl. Dissolve this product, together with 1.00 g crystallized $CoCl_2 \cdot 6H_2O$, as directed above.

d. Prepare standards having colors of 5, 10, 15, 20, 25, 30, 35, 40, 45, 50, 60, and 70 by diluting 0.5, 1.0, 1.5, 2.0, 2.5, 3.0, 3.5, 4.0, 4.5, 5.0, 6.0, and 7.0 mL stock color standard with distilled water to 50 mL in nessler tubes. Protect these standards against evaporation and contamination when not in use.

4. Procedure

a. *Estimation of intact sample:* Observe sample color by filling a matched nessler tube to the 50-mL mark with sample and comparing it with standards. Look vertically downward through tubes toward a white or specular surface placed at such an angle that light is reflected upward through the columns of liquid. If turbidity is present and has not been removed, report as "apparent color." If the color exceeds 70 units, dilute sample with distilled water in known proportions until the color is within the range of the standards.

b. Measure pH of each sample.

5. Calculation

a. Calculate color units by the following equation:

$$\text{Color units} = \frac{A \times 50}{B}$$

where:

A = estimated color of a diluted sample and
B = mL sample taken for dilution.

b. Report color results in whole numbers and record as follows:

Color Units	Record to Nearest
1–50	1
51–100	5
101–250	10
251–500	20

c. Report sample pH.

6. References

1. KNIGHT, A.G. 1951. The photometric estimation of color in turbid waters. *J. Inst. Water Eng.* 5:623.
2. JULLANDER, I. & K. BRUNE. 1950. Light absorption measurements on turbid solutions. *Acta Chem. Scand.* 4:870.
3. BLACK, A.P. & R.F. CHRISTMAN. 1963. Characteristics of colored surface waters. *J. Amer. Water Works Assoc.* 55:753.

7. Bibliography

HAZEN, A. 1892. A new color standard for natural waters. *Amer. Chem. J.* 14:300.
HAZEN, A. 1896. The measurement of the colors of natural waters. *J. Amer. Chem. Soc.* 18:264.
Measurement of Color and Turbidity in Water. 1902. U.S. Geol. Surv., Div. Hydrog. Circ. 8, Washington, D.C.
RUDOLFS, W. & W.D. HANLON. 1951. Color in industrial wastes. *Sewage Ind. Wastes* 23:1125.
PALIN, A.T. 1955. Photometric determination of the colour and turbidity of water. *Water Water Eng.* 59:341.
CHRISTMAN, R.F. & M. GHASSEMI. 1966. Chemical nature of organic color in water. *J. Amer. Water Works Assoc.* 58:723.
GHASSEMI, M. & R.F. CHRISTMAN. 1968. Properties of the yellow organic acids of natural waters. *Limnol. Oceanogr.* 13:583.

2120 C.   Spectrophotometric Method

1. General Discussion

a. Principle: The color of a filtered sample is expressed in terms that describe the sensation realized when viewing the sample. The hue (red, green, yellow, etc.) is designated by the term "dominant wavelength," the degree of brightness by "luminance," and the saturation (pale, pastel, etc.) by "purity." These values are best determined from the light transmission characteristics of the filtered sample by means of a spectrophotometer.

b. Application: This method is applicable to potable and surface waters and to wastewaters, both domestic and industrial.

c. Interference: Turbidity interferes. Remove by the filtration method described below.

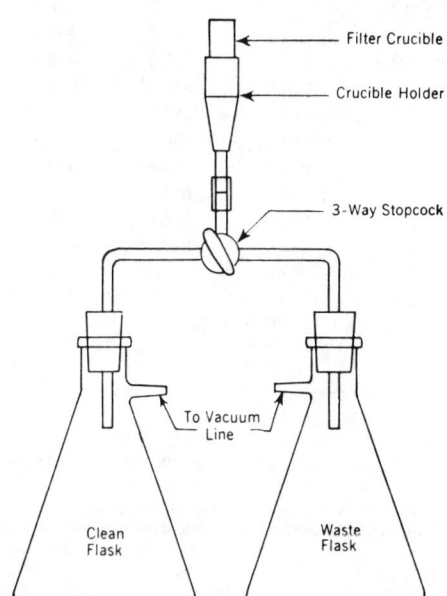

Figure 2120:1. Filtration system for color determinations.

2. Apparatus

a. Spectrophotometer, having 10-mm absorption cells, a narrow (10-nm or less) spectral band, and an effective operating range from 400 to 700 nm.

b. Filtration system, consisting of the following (see Figure 2120:1):

1) *Filtration flasks,* 250-mL, with side tubes.

2) *Walter crucible holder.*

3) *Micrometallic filter crucible,* average pore size 40 μm.

4) *Calcined filter aid.**

5) *Vacuum system.*

3. Procedure

a. Preparation of sample: Bring two 50-mL samples to room temperature. Use one sample at the original pH; adjust pH of the other to 7.6 by using sulfuric acid (H_2SO_4) and sodium hydroxide (NaOH) of such concentrations that the resulting volume change does not exceed 3%. A standard pH is necessary because of the variation of color with pH. Remove excessive quantities of suspended materials by centrifuging. Treat each sample separately, as follows:

Thoroughly mix 0.1 g filter aid in a 10-mL portion of centrifuged sample and filter to form a precoat in the filter crucible. Direct filtrate to waste flask as indicated in Figure 2120:1. Mix 40 mg filter aid in a 35-mL portion of centrifuged sample. With vacuum still on, filter through the precoat and pass filtrate to waste flask until clear; then direct clear-filtrate flow to clean flask by means of the three-way stopcock and collect 25 mL for the transmittance determination.

b. Determination of light transmission characteristics: Thoroughly clean 1-cm absorption cells with detergent and rinse with

---

*Celite No. 505, Manville Corp., or equivalent.

distilled water. Rinse twice with filtered sample, clean external surfaces with lens paper, and fill cell with filtered sample.

Determine transmittance values (in percent) at each visible wavelength value presented in Table 2120:I, using the 10 ordinates marked with an asterisk for fairly accurate work and all 30 ordinates for increased accuracy. Set instrument to read 100% transmittance on the distilled water blank and make all determinations with a narrow spectral band.

4. Calculation

a. Tabulate transmittance values corresponding to wavelengths shown in Columns X, Y, and Z in Table 2120:I. Total each transmittance column and multiply totals by the appropriate factors (for 10 or 30 ordinates) shown at the bottom of the table, to obtain tristimulus values X, Y, and Z. The tristimulus value Y is *percent luminance.*

b. Calculate the trichromatic coefficients x and y from the tristimulus values X, Y, and Z by the following equations:

$$x = \frac{X}{X + Y + Z}$$

$$y = \frac{Y}{X + Y + Z}$$

Locate point (x, y) on one of the chromaticity diagrams in Figure 2120:2 and determine the dominant wavelength (in nanometers) and the purity (in percent) directly from the diagram.

Determine hue from the dominant-wavelength value, according to the ranges in Table 2120:II.

TABLE 2120:I. SELECTED ORDINATES FOR SPECTROPHOTOMETRIC COLOR DETERMINATIONS†

Ordinate No.	X	Y	Z
	Wavelength *nm*		
1*	424.4	465.9	414.1
2*	435.5*	489.5*	422.2*
3	443.9	500.4	426.3
4	452.1	508.7	429.4
5*	461.2*	515.2*	432.0*
6	474.0	520.6	434.3
7	531.2	525.4	436.5
8*	544.3*	529.8*	438.6*
9	552.4	533.9	440.6
10	558.7	537.7	442.5
11*	564.1*	541.4*	444.4*
12	568.9	544.9	446.3
13	573.2	548.4	448.2
14*	577.4*	551.8*	450.1*
15	581.3	555.1	452.1
16	585.0	558.5	454.0
17*	588.7*	561.9*	455.9*
18	592.4	565.3	457.9
19	596.0	568.9	459.9
20*	599.6*	572.5*	462.0*
21	603.3	576.4	464.1
22	607.0	580.4	466.3
23*	610.9*	584.8*	468.7*
24	615.0	589.6	471.4
25	619.4	594.8	474.3
26*	624.2*	600.8*	477.7*
27	629.8	607.7	481.8
28	636.6	616.1	487.2
29*	645.9*	627.3*	495.2*
30	663.0	647.4	511.2

Factors When 30 Ordinates Used

0.032 69	0.033 33	0.039 38

Factors When 10 Ordinates Used

0.098 06	0.100 00	0.118 14

† Insert in each column the transmittance value (%) corresponding to the wavelength shown. Where limited accuracy is sufficient, use only the ordinates marked with an asterisk.

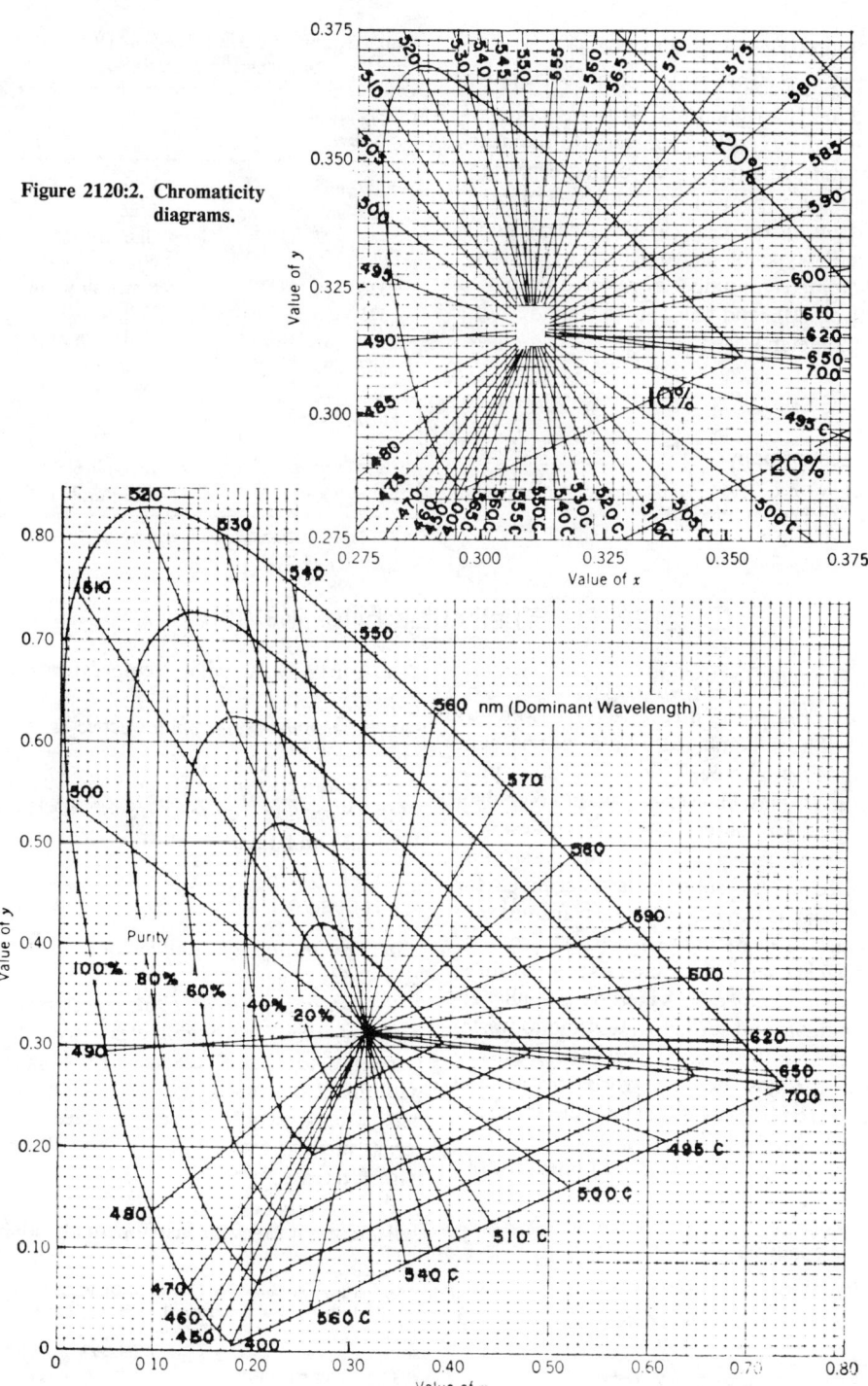

Figure 2120:2. Chromaticity diagrams.

5. Expression of Results

Express color characteristics (at pH 7.6 and at the original pH) in terms of *dominant wavelength* (nanometers, to the nearest unit), *hue* (e.g., blue, blue-green, etc.), *luminance* (percent, to the nearest tenth), and *purity* (percent, to the nearest unit). Report type of instrument (i.e., spectrophotometer), number of selected ordinates (10 or 30), and the spectral band width (nanometers) used.

6. Bibliography

HARDY, A.C. 1936. Handbook of Colorimetry. Technology Press, Boston, Mass.

TABLE 2120:II. COLOR HUES FOR DOMINANT
WAVELENGTH RANGES

Wavelength Range nm	Hue
400–465	Violet
465–482	Blue
482–497	Blue-green
497–530	Green
530–575	Greenish yellow
575–580	Yellow
580–587	Yellowish orange
587–598	Orange
598–620	Orange-red
620–700	Red
400–530c*	Blue-purple
530c–700*	Red-purple

* See Figure 2120:2 for significance of "c".

2120 D.   Tristimulus Filter Method

1. General Discussion

a. Principle: Three special tristimulus light filters, combined with a specific light source and photoelectric cell in a filter photometer, may be used to obtain color data suitable for routine control purposes.

The percentage of tristimulus light transmitted by the solution is determined for each of the three filters. The transmittance values then are converted to trichromatic coefficients and color characteristic values.

b. Application: This method is applicable to potable and surface waters and to wastewaters, both domestic and industrial. Except for most exacting work, this method gives results very similar to the more accurate Method C.

c. Interference: Turbidity must be removed.

2. Apparatus

a. Filter photometer. *

b. Filter photometer light source: Tung-

sten lamp at a color temperature of 3000°C.†

c. Filter photometer photoelectric cells, 1 cm.‡

d. Tristimulus filters. §

e. Filtration system: See Section 2120C.2*b* and Figure 2120:1.

3. Procedure

a. Preparation of sample: See Section 2120C.3*a*.

b. Determination of light transmission characteristics: Thoroughly clean (with detergent) and rinse 1-cm absorption cells with distilled water. Rinse each absorption cell twice with filtered sample, clean external surfaces with lens paper, and fill cell with filtered sample.

Place a distilled water blank in another cell and use it to set the instrument at 100%

---

*Fisher Electrophotometer or equivalent.

†General Electric lamp No. 1719 (at 6 V) or equivalent.
‡General Electric photovoltaic cell, Type PV-1, or equivalent.
§Corning CS-3-107 (No. 1), CS-4-98 (No. 2), and CS-5-70 (No. 3), or equivalent.

transmittance. Determine percentage of light transmission through sample for each of the three tristimulus light filters, with the filter photometer lamp intensity switch in a position equivalent to 4 V on the lamp.

4. Calculation

a. Determine luminance value directly as the percentage transmittance value obtained with the No. 2 tristimulus filter.

b. Calculate tristimulus values X, Y, and Z from the percentage transmittance (T_1, T_2, T_3) for filters No. 1, 2, 3, as follows:

$$X = T_3 \times 0.06 + T_1 \times 0.25$$
$$Y = T_2 \times 0.316$$
$$Z = T_3 \times 0.374$$

Calculate and determine trichromatic coefficients x and y, dominant wavelength, hue, and purity as in Section 2120C.4b above.

5. Expression of Results

Express results as prescribed in Section 2120C.5.

2120 E.   ADMI Tristimulus Filter Method (PROPOSED)

1. General Discussion

a. *Principle:* This method is an extension of Tristimulus Method 2120D. By this method a measure of the sample color, independent of hue, may be obtained. It is based on use of the Adams-Nickerson chromatic value formula[1] for calculating single number color difference values, i.e., uniform color differences. For example, if two colors, A and B, are judged visually to differ from colorless to the same degree, their ADMI color values will be the same. The modification was developed by members of the American Dye Manufacturers Institute (ADMI).[2]

b. *Application:* This method is applicable to colored waters and wastewaters having color characteristics significantly different from platinum-cobalt standards, as well as to waters and wastewaters similar in hue to the standards.

c. *Interference:* Turbidity must be removed.

2. Apparatus

a. *Filter photometer** equipped with CIE tristimulus filters (see 2120D.2d).

b. *Filter photometer light source:* Tungsten lamp at a color temperature of 3000°C (see 2120D.2b).

c. *Absorption cells and appropriate cell holders:* For color values less than 250 ADMI units, use cells with a 5.0-cm light path; for color values greater than 250, use cells with 1.0-cm light path.

d. *Filtration   system:*   See   Section

---

*Fisher Electrocolorimeter, Model 181, or equivalent.

2120C.2b and Figure 2120:1; or a centrifuge capable of achieving 1000 × g. (See Section 2120B.)

3. Procedure

a. *Instrument calibration:* Establish curves for each photometer; calibration data for one instrument cannot be applied to another one. Prepare a separate calibration curve for each absorption cell path length.

1) Prepare standards as described in 2120B.3. For a 5-cm cell length prepare standards having color values of 25, 50, 100, 200, and 250 by diluting 5.0, 10.0, 20.0, 30.0, 40.0, and 50.0 mL stock color standard with distilled water to 100 mL in volumetric flasks. For the shorter pathlength, prepare appropriate standards with higher color values.

2) Determine light transmittance (see ¶3c, below) for each standard with each filter.

3) Using the calculations described in ¶3d below, calculate the tristimulus values (X_s, Y_s, Z_s) for each standard, determine the Munsell values, and calculate the intermediate value (DE).

4) Using the DE values for each standard, calculate a calibration factor F_n for each standard from the following equation:

$$F_n = \frac{(APHA)_n (b)}{(DE)_n}$$

where:

$(APHA)_n$ = APHA color value for standard n,
$(DE)_n$ = intermediate value calculated for standard n, and
b = cell light path, cm.

Placing $(DE)_n$ on the X axis and F_n on the Y axis, plot a curve for the standard solutions. Use calibration curve to derive the F value from DE values obtained with samples.

b. *Sample preparation:* Prepare two 100-mL sample portions (one at the original

pH, one at pH 7.6) as described in Section 2120C.3a, or by centrifugation. (NOTE: Centrifugation is acceptable only if turbidity removal equivalent to filtration is achieved.)

c. *Determination of light transmission characteristics:* Thoroughly clean absorption cells with detergent and rinse with distilled water. Rinse each absorption cell twice with filtered sample. Clean external surfaces with lens paper and fill cell with sample. Determine sample light transmittance with the three filters to obtain the transmittance values: T_1 from Filter 1, T_2 from Filter 2, and T_3 from Filter 3. Standardize the instrument with each filter at 100% transmittance with distilled water.

d. *Calculation of color values:* Tristimulus values for samples are X_s, Y_s, and Z_s; for standards X_r, Y_r, and Z_r; and for distilled water X_c, Y_c, and Z_c. Munsell values for samples are V_{xs}, V_{ys}, and V_{zs}; for standards V_{xr}, V_{yr}, and V_{zr}; and for distilled water V_{xc}, Y_{yc}, and V_{zc}.

For each standard or sample calculate the tristimulus values from the following equations:

$$X = (T_3 \times 0.1899) + (T_1 \times 0.791)$$
$$Y = T_2$$
$$Z = T_3 \times 1.1835$$

Tristimulus values for the distilled water blank used to standardize the instrument are always:

$$X_c = 98.09$$
$$Y_c = 100.0$$
$$Z_c = 118.35$$

Convert the six tristimulus values $(X_s, Y_s, Z_s, X_c, Y_c, Z_c)$ to the corresponding Munsell values using published tables 2, 3, 4† or by the equation given by Bridgeman.[3]

Calculate the intermediate value of DE from the equation:

---

†Instrumental Colour Systems, Ltd., 7 Bucklebury Place, Upper Woolhampton, Berkshire RG7 5UD, England.

$$DE = \{(0.23\ \Delta V_y)^2 + [\Delta(V_x - V_y)]^2$$
$$+ [0.4\ \Delta(V_y - V_z)]^2\}^{\frac{1}{2}}$$

where:

$$V_y = V_{ys} - V_{yc}$$
$$\Delta(V_x - V_y) = (V_{xs} - V_{ys}) - (V_{xc} - V_{yc})$$
$$\Delta(V_y - V_z) = (V_{ys} - V_{zs}) - (V_{yc} - V_{zc})$$

when the sample is compared to distilled water.

With the standard calibration curve, use the DE value to determine the calibration factor F.

Calculate the final ADMI color value as follows:

$$\text{ADMI value} = \frac{(F)\,(DE)}{b}$$

where:

b = absorption cell light path, cm.

Report ADMI color values at pH 7.6 and at the original pH.

4. Alternate Method

The ADMI color value also may be determined spectrophotometrically, using a spectrophotometer with a narrow (10-nm or less) spectral band and an effective operating range of 400 to 700 nm. This method is an extension of 2120C. Tristimulus values may be calculated from transmittance measurements, preferably by using the weighted ordinate method or by the selected ordinate method. The method has been described by Allen et al.,[2] who include work sheets and worked examples.

5. References

1. McLAREN, K. 1970. The Adams-Nickerson colour-difference formula. *J. Soc. Dyers Colorists* 86:354.

2. ALLEN, W., W.B. PRESCOTT, R.E. DERBY, C.E. GARLAND, J.M. PERET & M. SALTZMAN. 1973. Determination of color of water and wastewater by means of ADMI color values. *Proc. 28th Ind. Waste Conf.*, Purdue Univ., Eng. Ext. Ser. No. 142:661.

3. BRIDGEMAN, T. 1963. Inversion of the Munsell value equation. *J. Opt. Soc. Amer.* 53:499.

6. Bibliography

JUDD, D.B. & G. WYSZECKI. 1963. Color in Business, Science, and Industry, 2nd ed. John Wiley & Sons, New York, N.Y. (See Tables A, B, and C in Appendix.)

WYSZECKI, G. & W.S. STILES. 1967. Color Science. John Wiley & Sons, New York, N.Y. (See Tables 6.4, A, B, C, pp. 462-467.)

2130 TURBIDITY*

2130 A. Introduction

1. Sources and Significance

Clarity of water is important in producing products destined for human con-

sumption and in many manufacturing uses. Beverage producers, food processors, and treatment plants drawing on a surface water supply commonly rely on coagulation, settling, and filtration to insure an acceptable product. The clarity of a natural

---

*Approved by Standard Methods Committee, 1988.

body of water is a major determinant of the condition and productivity of that system.

Turbidity in water is caused by suspended matter, such as clay, silt, finely divided organic and inorganic matter, soluble colored organic compounds, and plankton and other microscopic organisms. Turbidity is an expression of the optical property that causes light to be scattered and absorbed rather than transmitted in straight lines through the sample. Correlation of turbidity with the weight concentration of suspended matter is difficult because the size, shape, and refractive index of the particulates also affect the light-scattering properties of the suspension. Optically black particles, such as those of activated carbon, may absorb light and effectively increase turbidity measurements.

2. Selection of Method

Historically, the standard method for determination of turbidity has been based on the Jackson candle turbidimeter[1]; however, the lowest turbidity value that can be measured directly on this instrument is 25 units. Because turbidities of treated water usually fall within the range of 0 to 1 unit, indirect secondary methods also were developed to estimate turbidity. Unfortunately, no instrument could duplicate the results obtained on the Jackson candle turbidimeter for all samples. Because of fundamental differences in optical systems, the results obtained with different types of secondary instruments frequently do not check closely with one another, even though the instruments are precalibrated against the candle turbidimeter.

Most commercial turbidimeters available for measuring low turbidities give comparatively good indications of the intensity of light scattered in one particular direction, predominantly at right angles to the incident light. These nephelometers are unaffected relatively by small changes in

design parameters and therefore are specified as the standard instrument for measurement of low turbidities. Nonstandard turbidimeters, such as forward-scattering devices, are more sensitive than nephelometers to the presence of larger particles and are useful for process monitoring.

A further cause of discrepancies in turbidity analysis is the use of suspensions of different types of particulate matter for the preparation of instrumental calibration curves. Like water samples, prepared suspensions have different optical properties depending on the particle size distributions, shapes, and refractive indices. A standard reference suspension having reproducible light-scattering properties is specified for nephelometer calibration.

Because there is no direct relationship between the intensity of light scattered at a 90° angle and Jackson candle turbidity, there is no valid basis for the practice of calibrating a nephelometer in terms of candle units. To avoid misinterpretation, report the results from nephelometric measurements as nephelometric turbidity units (NTU).

Its precision, sensitivity, and applicability over a wide turbidity range make the nephelometric method preferable to visual methods. The Jackson candle method has been eliminated from this edition of *Standard Methods*.

3. Storage of Sample

Determine turbidity on the day the sample is taken. If longer storage is unavoidable, store samples in the dark for up to 24 h. Do not store for long periods because irreversible changes in turbidity may occur. Vigorously shake all samples before examination.

4. References

1. AMERICAN PUBLIC HEALTH ASSOCIATION, AMERICAN WATER WORKS ASSOCIATION &

WATER POLLUTION CONTROL FEDERATION. 1985. Standard Methods for the Examination of Water and Wastewater, 16th ed. American Public Health Assoc., Washington, D.C.

2130 B. Nephelometric Method

1. General Discussion

a. Principle: This method is based on a comparison of the intensity of light scattered by the sample under defined conditions with the intensity of light scattered by a standard reference suspension under the same conditions. The higher the intensity of scattered light, the higher the turbidity. Formazin polymer is used as the reference turbidity standard suspension. It is easy to prepare and is more reproducible in its light-scattering properties than clay or turbid natural water. The turbidity of a specified concentration of formazin suspension is defined as 40 nephelometric units. This suspension has an approximate turbidity of 40 Jackson units when measured on the candle turbidimeter; therefore, nephelometric turbidity units based on the formazin preparation will approximate units derived from the candle turbidimeter but will not be identical to them.

b. Interference: Turbidity can be determined for any water sample that is free of debris and rapidly settling coarse sediments. Dirty glassware, the presence of air bubbles, and the effects of vibrations that disturb the surface visibility of the sample will give false results. "True color," that is, water color due to dissolved substances that absorb light, causes measured turbidities to be low. This effect usually is not significant in the case of treated water.

2. Apparatus

a. Turbidimeter consisting of a nephelometer with a light source for illuminating the sample and one or more photoelectric detectors with a readout device to indicate intensity of light scattered at 90° to the path of incident light. Use a turbidimeter designed so that little stray light reaches the detector in the absence of turbidity and free from significant drift after a short warmup period. The sensitivity of the instrument should permit detecting turbidity differences of 0.02 NTU or less in waters having turbidity of less than 1 NTU with a range from 0 to 40 NTU. Several ranges are necessary to obtain both adequate coverage and sufficient sensitivity for low turbidities.

Differences in turbidimeter design will cause differences in measured values for turbidity even though the same suspension is used for calibration. To minimize such differences, observe the following design criteria:

1) Light source—Tungsten-filament lamp operated at a color temperature between 2200 and 3000°K.

2) Distance traversed by incident light and scattered light within the sample tube—Total not to exceed 10 cm.

3) Angle of light acceptance by detector—Centered at 90° to the incident light path and not to exceed ± 30° from 90°. The detector, and filter system if used, shall have a spectral peak response between 400 and 600 nm.

b. Sample tubes, clear colorless glass. Keep tubes scrupulously clean, both inside and out, and discard when they become scratched or etched. Never handle them where the light strikes them. Use tubes with sufficient extra length, or with a protective case, so that they may be handled properly. Fill tubes with samples and standards that have been agitated thoroughly and allow sufficient time for bubbles to escape.

3. Reagents

a. Turbidity-free water: Turbidity-free water is difficult to obtain. The following method is satisfactory for measuring turbidity as low as 0.02 NTU.

Pass distilled water through a membrane filter having precision-sized holes of 0.2 μm;* the usual membrane filter used for bacteriological examinations is not satisfactory. Rinse collecting flask at least twice with filtered water and discard the next 200 mL.

Some commercial bottled demineralized waters are nearly particle-free. These may be used when their turbidity is lower than can be achieved in the laboratory. Dilute samples to a turbidity not less than 1 with distilled water.

b. Stock turbidity suspension:

1) Solution I—Dissolve 1.000 g hydrazine sulfate (CAUTION: *Carcinogen; avoid inhalation, ingestion, and skin contact.*), $(NH_2)_2 \cdot H_2SO_4$, in distilled water and dilute to 100 mL in a volumetric flask.

2) Solution II—Dissolve 10.00 g hexamethylenetetramine, $(CH_2)_6N_4$, in distilled water and dilute to 100 mL in a volumetric flask.

3) In a 100-mL volumetric flask, mix 5.0 mL Solution I and 5.0 mL Solution II. Let stand 24 h at 25 ± 3°C, dilute to mark, and mix. The turbidity of this suspension is 400 NTU.

4) Prepare solutions and suspensions monthly.

c. Standard turbidity suspension: Dilute 10.00 mL stock turbidity suspension to 100 mL with turbidity-free water. Prepare daily. The turbidity of this suspension is defined as 40 NTU.

d. Alternate standards: As an alternative to preparing and diluting formazin, use commercially available standards such as styrene divinylbenzene beads† if they are

*Nuclepore Corporation, 7035 Commerce Circle, Pleasanton, Calif., or equivalent.
†AMCO-AEPA-1 Standard, Advanced Polymer Systems, 3696 C Haven Ave., Redwood City, Calif.

demonstrated to be equivalent to freshly prepared formazin.

e. Dilute turbidity standards: Dilute portions of standard turbidity suspension with turbidity-free water as required. Prepare daily.

4. Procedure

a. Turbidimeter calibration: Follow the manufacturer's operating instructions. In the absence of a precalibrated scale, prepare calibration curves for each range of the instrument. Check accuracy of any supplied calibration scales on a precalibrated instrument by using appropriate standards. Run at least one standard in each instrument range to be used. Make certain that turbidimeter gives stable readings in all sensitivity ranges used. High turbidities determined by direct measurement are likely to differ appreciably from those determined by the dilution technique, ¶ 4c.

b. Measurement of turbidities less than 40 NTU: Thoroughly shake sample. Wait until air bubbles disappear and pour sample into turbidimeter tube. When possible, pour shaken sample into turbidimeter tube and immerse it in an ultrasonic bath for 1 to 2 s, causing complete bubble release. Read turbidity directly from instrument scale or from appropriate calibration curve.

c. Measurement of turbidities above 40 NTU: Dilute sample with one or more volumes of turbidity-free water until turbidity falls between 30 and 40 NTU. Compute turbidity of original sample from turbidity of diluted sample and the dilution factor. For example, if five volumes of turbidity-free water were added to one volume of sample and the diluted sample showed a turbidity of 30 NTU, then the turbidity of the original sample was 180 NTU.

d. Calibrate continuous turbidity monitors for low turbidities by determining turbidity of the water entering or leaving them, using a laboratory-model turbidimeter. When this is not possible, use an appropriate dilute turbidity standard, ¶ 3e. For turbidities above 40 NTU use undiluted stock solution.

5. Calculation

Nephelometric turbidity units (NTU)

$$= \frac{A \times (B + C)}{C}$$

where:

A = NTU found in diluted sample,
B = volume of dilution water, mL, and
C = sample volume taken for dilution, mL.

6. Interpretation of Results

a. Report turbidity readings as follows:

Turbidity Range NTU	Report to the Nearest NTU
0–1.0	0.05
1–10	0.1
10–40	1
40–100	5
100–400	10
400–1000	50
> 1000	100

b. For comparison of water treatment efficiencies estimate turbidity more closely than is specified above. Uncertainties and discrepancies in turbidity measurements make it unlikely that two or more laboratories will duplicate results on the same sample more closely than specified.

7. Bibliography

WHIPPLE, G.C. & D.D. JACKSON. 1900. A comparative study of the methods used for the measurement of turbidity of water. *Mass. Inst. Technol. Quart.* 13:274.

AMERICAN PUBLIC HEALTH ASSOCIATION. 1901. Report of Committee on Standard Methods of Water Analysis. *Pub. Health Papers & Rep.* 27:377.

WELLS, P.V. 1922. Turbidimetry of water. *J. Amer. Water Works Assoc.* 9:488.

BAYLIS, J.R. 1926. Turbidimeter for accurate measurement of low turbidities. *Ind. Eng. Chem.* 18:311.

WELLS, P.V. 1927. The present status of turbidity measurements. *Chem. Rev.* 3:331.

BAYLIS, J.R. 1933. Turbidity determinations. *Water Works Sewage* 80:125.

ROSE, H.E. & H.B. LLOYD. 1946. On the measurement of the size characteristics of powders by photo-extinction methods. *J. Soc. Chem. Ind.* (London) 65:52 (Feb.); 65:55 (Mar.).

ROSE, H.E. & C.C.J. FRENCH. 1948. On the extinction coefficient: Particle size relationship for fine mineral powders. *J. Soc. Chem. Ind.* (London) 67:283.

GILLETT, T.R., P.F. MEADS & A.L. HOLVEN. 1949. Measuring color and turbidity of white sugar solutions. *Anal. Chem.* 21:1228.

JULLANDER, I. 1949. A simple method for the measurement of turbidity. *Acta Chem. Scand.* 3:1309.

ROSE, H.E. 1950. Powder-size measurement by a combination of the methods of nephelometry and photo-extinction. *J. Soc. Chem. Ind.* (London) 69:266.

ROSE, H.E. 1950. The design and use of photoextinction sedimentometers. *Engineering* 169:350, 405.

BRICE, B.A., M. HALWER & R. SPEISER. 1950. Photoelectric light-scattering photometer for determining high molecular weights. *J. Opt. Soc. Amer.* 40:768.

KNIGHT, A.G. 1950. The measurement of turbidity in water. *J. Inst. Water Eng.* 4:449.

HANYA, T. 1950. Study of suspended matter in water. *Bull. Chem. Soc. Jap.* 23:216.

JULLANDER, I. 1950. Turbidimetric investigations on viscose. *Svensk Papperstidn.* 22:1.

ROSE, H.E. 1951. A reproducible standard for the calibration of turbidimeters. *J. Inst. Water Eng.* 5:310.

AITKEN, R.W. & D. MERCER. 1951. Comment on "The measurement of turbidity in water." *J. Inst. Water Eng.* 5:328.

ROSE, H.E. 1951. The analysis of water by the assessment of turbidity. *J. Inst. Water Eng.* 5:521.

KNIGHT, A.G. 1951. The measurement of turbidity in water: A reply. *J. Inst. Water Eng.* 5:633.

STAATS, F.C. 1952. Measurement of color, turbidity, hardness and silica in industrial waters. Preprint 156, American Soc. Testing & Materials, Philadelphia, Pa.

PALIN, A.T. 1955. Photometric determination of the colour and turbidity of water. *Water Water Eng.* 59:341.

SLOAN, C.K. 1955. Angular dependence light scattering studies of the aging of precipitates. *J. Phys. Chem.* 59:834.

CONLEY, W.R. & R.W. PITMAN. 1957. Microphotometer turbidity analysis. *J. Amer. Water Works Assoc.* 49:63.

PACKHAM, R.F. 1962. The preparation of turbidity standards. *Proc. Soc. Water Treat. Exam.* 11:64.

BAALSRUD, K. & A. HENRIKSEN. 1964. Measurement of suspended matter in stream water. *J. Amer. Water Works Assoc.* 56:1194.

HOATHER, R.C. 1964. Comparison of different methods for measurement of turbidity. *Proc. Soc. Water Treat. Exam.* 13:89.

EDEN, G.E. 1965. The measurement of turbidity in water. A progress report on the work of the analytical panel. *Proc. Soc. Water Treat. Exam.* 14:27.

BLACK, A.P. & S.A. HANNAH. 1965. Measurement of low turbidities. *J. Amer. Water Works Assoc.* 57:901.

HANNAH, S.A., J.M. COHEN & G.G. ROBECK. 1967. Control techniques for coagulation-filtration. *J. Amer. Water Works Assoc.* 59:1149.

REBHUN, M. & H.S. SPERBER. 1967. Optical properties of diluted clay suspensions. *J. Colloid Interface Sci.* 24:131.

DANIELS, S.L. 1969. The utility of optical parameters in evaluation of processes of flocculation and sedimentation. *Chem. Eng. Progr. Symp. Ser.* No. 97, 65:171.

LIVESEY, P.J. & F.W. BILLMEYER, JR. 1969. Particle-size determination by low-angle light scattering: new instrumentation and a rapid method of interpreting data. *J. Colloid Interface Sci.* 30:447.

OSTENDORF, R.G. & J.F. BYRD. 1969. Modern monitoring of a treated industrial effluent. *J. Water Pollut. Control Fed.* 41:89.

EICHNER, D.W. & C.C. HACH. 1971. How clear is clear water? *Water Sewage Works* 118:299.

HACH, C.C. 1972. Understanding turbidity measurement. *Ind. Water Eng.* 9(2):18.

SIMMS, R.J. 1972. Industrial turbidity measurement. *ISA (Instrum. Soc. Amer.) Trans.* 11(2):146.

TALLEY, D.G., J.A. JOHNSON & J.E. PILZER. 1972. Continuous turbidity monitoring. *J. Amer. Water Works Assoc.* 64:184.

2150   ODOR*

2150 A.   Introduction

1. Discussion

Odor, like taste, depends on contact of a stimulating substance with the appropriate human receptor cell. The stimuli are chemical in nature and the term "chemical senses" often is applied to odor and taste. Water is a neutral medium, always present on or at the membranes that perceive sensory response. In its pure form, water cannot produce odor or taste sensations. No satisfactory theory of olfaction ever has been devised, although many have been formulated. Man and animals can avoid many potentially toxic foods and waters because of adverse sensory response. Without this form of primitive sensory protection many species would not have survived. Today, these same senses often provide the first warning of potential hazards in the environment.

Odor is recognized[1] as a quality factor affecting acceptability of drinking water (and foods prepared with it), tainting of fish and other aquatic organisms, and esthetics of recreational waters. Most organic and some inorganic chemicals contribute taste or odor. These chemicals may origi-

---

*Approved by Standard Methods Committee, 1985.

nate from municipal and industrial waste discharges, from natural sources such as decomposition of vegetable matter, or from associated microbial activity.

Technological expansion in varieties and quantities of waste materials, demands for water disposal of former air pollutants, and continuous population growth with consequently increased reuse of available water supplies increase the potential for impairment of sensory water quality. Domestic consumers and process industries such as food, beverage, and pharmaceutical manufacturers require water essentially free of tastes and odors.

Some substances, such as certain inorganic salts, produce taste without odor and are evaluated by taste test (Section 2160). Many other sensations ascribed to the sense of taste actually are odors, even though the sensation is not noticed until the material is taken into the mouth. Despite rapid strides in relating sensory qualities to chemical analyses,[2] most odors are too complex and are detectable at concentrations too low to permit their definition by isolating and determining the odor-producing chemicals. The ultimate odor-testing device is the human nose. Odor tests are performed to provide qualitative descriptions and approximate quantitative measurements of odor intensity. The method for intensity measurement presented here is the *threshold odor* test, based on a method of limits.[2] *Suprathreshold* methods are not included. Section 6040B provides an analytical procedure for quantifying several organic odor-producing compounds including geosmin and methylisoborneol.

Sensory tests are useful as a check on the quality of raw and finished water and for control of odor through the treatment process. They can assess the effectiveness of different treatments and provide a means of tracing the source of contamination.

2. References

1. U.S. ENVIRONMENTAL PROTECTION AGENCY. 1973. Proposed Criteria for Water Quality. Vol. 1, Washington, D.C.
2. AMERICAN SOCIETY FOR TESTING AND MATERIALS COMMITTEE E-18. 1968. STP 433, Basic principles of sensory evaluation; STP 434, Manual on sensory testing methods; STP 440, Correlation of subjective-objective methods in the study of odors and taste. ASTM, Philadelphia, Pa.

3. Bibliography

MONCRIEFF, R.W. 1946. The Chemical Senses. John Wiley & Sons, New York, N.Y.
SECHENOV, I.M. 1956 and 1958. Problem of hygenic standards for waters simultaneously polluted with harmful substances [in Russian]. *Gig. Sanit.* Nos. 10 and 8.
Taste and Odor Control in Water Purification, 2nd ed. 1959. West Virginia Pulp & Paper Co. Industrial Chemical Sales Division, New York. [Contains 1,063 classified references.]
BAKER, R.A. 1961. Problems of tastes and odors. *J. Water Pollut. Control Fed.* 33:1099.
BAKER, R.A. 1963. Odor effects of aqueous mixtures of organic chemicals. *J. Water Pollut. Control Fed.* 35:728.
ROSEN, A.A., R.T. SKEEL & M.B. ETTINGER. 1963. Relationship of river water odor to specific organic contaminants. *J. Water Pollut. Control Fed.* 35:777.
WRIGHT, R.H. 1964. The Science of Smell. Basic Books, New York, N.Y.
AMERINE, M.A., R.M. PANGBORN & E.B. ROESSLER. 1965. Principles of Sensory Evaluation of Food. Academic Press, New York, N.Y.
ROSEN, A.A. 1970. Report of research committee on tastes and odors. *J. Amer. Water Works Assoc.* 62:59.
GELDARD, F.A. 1972. The Human Senses. John Wiley & Sons, New York, N.Y.

2150 B. Threshold Odor Test

1. General Discussion

a. Principle: Determine the threshold odor by diluting a sample with odor-free water until the least definitely perceptible odor is achieved. There is no absolute threshold odor concentration, because of inherent variation in individual olfactory capability. A given person varies in sensitivity over time. Day-to-day and within-day differences occur. Furthermore, responses vary as a result of the characteristic, as well as concentration, of odorant. The number of persons selected to measure threshold odor will depend on the objective of the tests, economics, and available personnel. Larger-sized panels are needed for sensory testing when the results must represent the population as a whole or when great precision is desired. Under such circumstances, panels of not less than five persons, and preferably ten or more, are recommended.[1] Measurement of threshold levels by one person is often a necessity at water treatment plants. Interpretation of the single tester result requires knowledge of the relative acuity of that person. Some investigators have used specific odorants, such as *m*-cresol or *n*-butanol, to calibrate a tester's response.[2]

b. Application: This threshold method is applicable to samples ranging from nearly odorless natural waters to industrial wastes with threshold numbers in the thousands. There are no intrinsic difficulties with the highly odorous samples because they are reduced in concentration proportionately before being presented to the test observers.

c. Qualitative descriptions: A satisfactory system for characterizing odor has not been developed despite efforts over more than a century. Previous editions of this book contained a table of odor descriptions proposed as a guide in expressing odor quality. The reader may continue to encounter the ob-

solete standard abbreviations of that table. The 12th Edition presents an explanation of such terms.

d. Sampling and storage: Collect samples for odor testing in glass bottles with glass or TFE-lined closures. Complete tests as soon as possible after sample collection. If storage is necessary, collect at least 500 mL of sample in a bottle filled to the top; refrigerate, making sure that no extraneous odors can be drawn into the sample as it cools. Do not use plastic containers.

e. Dechlorination: Most tap waters and some wastewaters are chlorinated. Often it is desirable to determine the odor of the chlorinated sample as well as that of the same sample after dechlorination. Dechlorinate with arsenite or thiosulfate in exact stoichiometric quantity as described under Nitrogen (Ammonia), Section 4500-NH$_3$. CAUTION—*Do not use arsenic compounds as dechlorinating agents on samples to be tasted.*

f. Temperature: Threshold odor values vary with temperature. For most tap waters and raw water sources, a sample temperature of 60°C will permit detection of odors that might otherwise be missed; 60°C is the standard temperature for hot threshold tests. For some purposes—because the odor is too fleeting or there is excessive heat sensation—the hot odor test may not be applicable; where experience shows that a lower temperature is needed, use a standard test temperature of 40°C. For special purposes, other temperatures may be used. *Report temperature at which observations are made.*

2. Apparatus

To assure reliable threshold measurements, use odor-free glassware. Clean glassware shortly before use with non-odorous soap and acid cleaning solution and rinse with odor-free water. Reserve this glassware exclusively for threshold testing. Do not use rubber, cork, or plastic stoppers. Do not use narrow-mouth vessels.

a. Sample bottles, glass-stoppered or with TFE-lined closures.

b. Constant-temperature bath: A water bath or electric hot plate capable of temperature control of ± 1°C for odor tests at elevated temperatures. The bath must not contribute any odor to the odor flasks.

c. Odor flasks: Glass-stoppered, 500-mL (ST 32) erlenmeyer flasks, to hold sample dilutions during testing.

d. Pipets:

1) *Transfer and volumetric pipets or graduated cylinders:* 200-, 100-, 50-, and 25-mL.

2) *Measuring pipets:* 10-mL, graduated in tenths.

e. Thermometer: Zero to 110°C, chemical or metal-stem dial type.

3. Odor-Free Water

a. Sources: Prepare odor-free water by passing distilled, deionized, or tap water through activated carbon, insuring that the water contacts only glass or TFE. If it is impossible to make an all-glass and TFE apparatus, use the design indicated below. If product water is not odor-free, rebuild or purify the system. In all cases verify quality of product water daily.

b. Odor-free-water generator (Figures 2150:1 and 2150:2):*

1) *Borosilicate glass pipe,* 3-in. diam, 18-in. length.

2) *Asbestos inserts* (two) (CAUTION: *Carcinogen; handle with care.*), for 3-in. pipe.

3) *Flange sets* (two), for 3-in. pipe.

4) *Gaskets†* (two), 1/4-in. thickness, with 3-in. hole slotted to 3/8-in. depth to

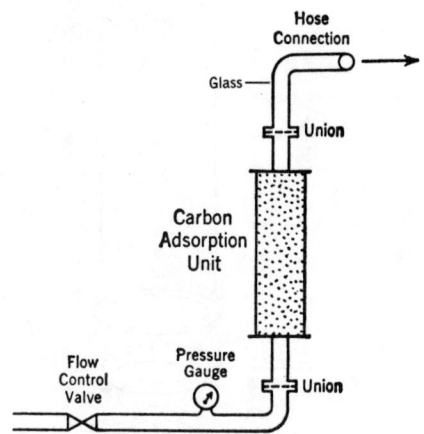

Figure 2150:1. Odor-free-water generator.

take screen. Drill three holes, 5/16-in. diam, to match flange.

5) *Stainless-steel screens* (two), 40-mesh, 3-3/4-in. diam.

6) *Brass plates* (two), 3/16-in. thickness × 6-1/4-in. diam. Tap hole in center for 3/4-in. nipple. Score a circular groove (1/16-in. depth × 1/16-in. width and 3-3/8-in. diam) into the plate to prevent leakage. Drill three holes, 5/16-in. diam, to coincide with the flange.

7) *Galvanized nipples* (two), 3/4-in. × 3-in. Thread nipple into brass plate and weld in place.

8) *Aluminum bolts and nuts* (six), 5/16-in. × 2-in., for holding assembly together.

9) *Activated carbon* of approximately 12 to 40 mesh grain size.‡

Attach end fittings of adsorption unit to the glass pipe. Draw bolts up evenly, holding brass plate to glass pipe to get a good seal on gasket. Fill unit with carbon. Tap cylinder gently but do not tamp carbon. Attach end fittings on adsorption unit and connect to water source as shown in Figure 2150:1.

Avoid organic contaminants in making pipe joints or other plumbing. Use TFE-

---

*For approximate metric dimensions in centimeters multiply dimensions in inches by 2.54.

†Neoprene, such as can be obtained from Netherland Rubber Co., Cincinnati, Ohio.

‡Nuchar WV-G, Westvaco, Covington, Va.; Filtrasorb 200, Calgon Corp., Pittsburgh, Pa.; or equivalent.

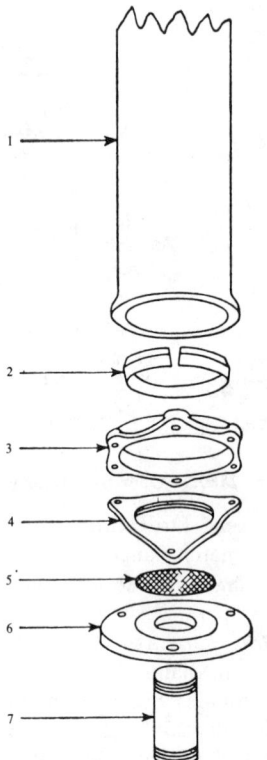

Figure 2150:2. End assembly of odor-free-water generator.

carbon indicates that a change of carbon is needed.

4. Procedure

a. Precautions: Carefully select by preliminary tests the persons to make taste or odor tests. Although extreme sensitivity is not required, exclude insensitive persons and concentrate on observers who have a sincere interest in the test. Avoid extraneous odor stimuli such as those caused by smoking and eating before the test or those contributed by scented soaps, perfumes, and shaving lotions. Insure that the tester is free from colds or allergies that affect odor response. Limit frequency of tests to a number below the fatigue level by frequent rests in an odor-free atmosphere. Keep room in which tests are conducted free from distractions, drafts, and odor.[2] If necessary, set aside a special odor-free room ventilated by air that is filtered through activated carbon and maintained at a constant comfortable temperature and humidity.[3]

For precise work use a panel of five or more testers. Do not allow persons making odor measurements to prepare samples or to know dilution concentrations being evaluated. Familiarize testers with the procedure before they participate in a panel test. Present most dilute sample first to avoid tiring the senses with the concentrated sample. Keep temperature of samples during testing within 1°C of the specified temperature.

Because many raw and waste waters are colored or have decided turbidity that may bias results, use opaque or darkly colored odor flasks, such as red actinic erlenmeyer flasks.

b. Characterization: As part of the threshold test or as a separate test, direct each observer to describe in his own words the characteristic sample odor. Compile the consensus that may appear among testers and that affords a clue to the origin of the odorous pollutant. The value of the

type tape or a paste made by mixing red lead powder and water. Clean all new fittings with kerosene and follow with a detergent wash. Rinse thoroughly with clean water.

c. Generator operation: Pass tap or distilled water through odor-free-water generator at a rate of 100 mL/min. When generator is started, flush to remove carbon fines and discard product.

Check quality of water obtained from the odor-free-water generator daily at 40 and 60°C before use. The life of the carbon will vary with the condition and amount of water filtered. Subtle odors of biological origin often are found if moist carbon filters stand idle between test periods. Detection of odor in the water coming through the characterization test increases as observers become more experienced with a particular category of odor, such as algae, chlorophenol, or mustiness.

   c. *Threshold measurement:*§ The "threshold odor number," designated by the abbreviation T.O.N., is the greatest dilution of sample with odor-free water yielding a definitely perceptible odor. Bring total volume of sample and odor-free water to 200 mL in each test. Follow dilutions and record corresponding T.O.N. presented in Table 2150:I. These numbers have been computed thus:

$$\text{T.O.N.} = \frac{A + B}{A}$$

where:

   A = mL sample and
   B = mL odor-free water.

1) Place proper volume of odor-free water in the flask first, add sample to water (avoiding contact of pipet or sample with lip or neck of flask), mix by swirling, and proceed as follows:

   Determine approximate range of the threshold number by adding 200 mL, 50 mL, 12 mL, and 2.8 mL sample to separate 500-mL glass-stoppered erlenmeyer flasks containing odor-free water to make a total volume of 200 mL. Use a separate flask containing only odor-free water as reference for comparison. Heat dilutions and reference to desired test temperature.

   2) Shake flask containing odor-free water, remove stopper, and sniff vapors. Test sample containing least amount of odor-bearing water in the same way. If

TABLE 2150:I. THRESHOLD ODOR NUMBERS CORRESPONDING TO VARIOUS DILUTIONS

Sample Volume Diluted to 200 mL mL	Threshold Odor No.	Sample Volume Diluted to 200 mL mL	Threshold Odor No.
200	1	12	17
140	1.4	8.3	24
100	2	5.7	35
70	3	4	50
50	4	2.8	70
35	6	2	100
25	8	1.4	140
17	12	1.0	200

odor can be detected in this dilution, prepare more dilute samples as described in ¶ 5) below. If odor cannot be detected in first dilution, repeat above procedure using sample containing next higher concentration of odor-bearing water, and continue this process until odor is detected clearly.

   3) Based on results obtained in the preliminary test, prepare a set of dilutions using Table 2150:II as a guide. Prepare the five dilutions shown on the appropriate line and the three next most concentrated on the next line in Table 2150:II. For example, if odor was first noted in the flask containing 50 mL sample in the preliminary test, prepare flasks containing 50, 35, 25, 17, 12, 8.3, 5.7, and 4.0 mL sample, each diluted to 200 mL with odor-free water. This array is necessary to challenge the range of sensitivities of the entire panel of testers.

   Insert two or more blanks in the series near the expected threshold, but avoid any repeated pattern. Do not let tester know which dilutions are odorous and which are blanks. Instruct tester to smell each flask in sequence, beginning with the least concentrated sample, until odor is detected with certainty.

   4) Record observations by indicating

---

§There are numerous methods of arranging and presenting samples for odor determinations. The methods offered here are practical and economical of time and personnel and generally are adequate. If extensive tests are planned and statistical analysis of data is required, become familiar with the triangle test and the methods that have been used extensively by flavor and allied industries.[4]

TABLE 2150:II. DILUTIONS FOR VARIOUS
ODOR INTENSITIES

Sample Volume in Which Odor Is First Noted mL	Volumes to Be Diluted to 200 mL mL
200	200, 140, 100, 70, 50
50	50, 35, 25, 17, 12
12	12, 8.3, 5.7, 4.0, 2.8
2.8	Intermediate dilution

whether odor is noted in each test flask. For example:

mL Sample Diluted to 200 mL	12	0	17	25	0	35	50
Response	−	−	−	+	−	+	+

5) If the sample being tested requires more dilution than is provided by Table 2150:II, prepare an intermediate dilution consisting of 20 mL sample diluted to 200 mL with odor-free water. Use this dilution for the threshold determination. Multiply T.O.N. obtained by 10 to correct for the intermediate dilution. In rare cases more than one tenfold intermediate dilution step may be required.

5. Calculation

The threshold odor number is the dilution ratio at which taste or odor is just detectable. In the example above, ¶ 4c4), the first detectable odor occurred when 25 mL sample was diluted to 200 mL. Thus the threshold is 200 divided by 25, or 8. Table 2150:I lists the threshold numbers corresponding to common dilutions.

The smallest T.O.N. that can be observed is 1, as in the case where the odor flask contains 200 mL undiluted sample. If no odor is detected at this concentration,

report "No odor observed" instead of a threshold number. (In special applications, fractional threshold numbers have been calculated.[5])

Anomalous responses sometimes occur; a low concentration may be called positive and a higher concentration in the series may be called negative. In such a case, designate the threshold as that point after which no further anomalies occur. For instance:

Increasing Concentration →

Response   −   −   +   −   +   +   +   +
                                    ↓
                            Threshold

where:
− signifies negative response and
+ signifies positive response.

Occasionally a flask contains residual odor or is contaminated inadvertently. For precise testing repeat entire threshold odor test to determine if the last flask marked "−" was actually a mislabelled blank of odor-free water or if the previous "+" was a contaminated sample.

Use appropriate statistical methods to calculate the most probable average threshold from large numbers of panel results. For most purposes, express the threshold of a group as the geometric mean of individual thresholds.

6. Interpretation of Results

A threshold number is not a precise value. In the case of the single observer it represents a judgment at the time of testing. Panel results are more meaningful because individual differences have less influence on the result. One or two observers can develop useful data if comparison with larger panels has been made to check their sensitivity. Do not make comparisons of data from time to time or place to place unless all test conditions have been standardized carefully and there is some basis for comparison of observed intensities.

7. References

1. AMERICAN SOCIETY FOR TESTING AND MATERIALS COMMITTEE E-18. 1968. STP 433, Basic principles of sensory evaluation; STP 434, Manual on sensory testing methods; STP 440, Correlation of subjective-objective methods in the study of odors and taste. ASTM, Philadelphia, Pa.
2. BAKER, R.A. 1962. Critical evaluation of olfactory measurement. *J. Water Pollut. Control Fed.* 34:582.
3. BAKER, R.A. 1963. Odor testing laboratory. *J. Water Pollut. Control Fed.* 35:1396.
4. Flavor Research and Food Acceptance. 1958. Reinhold Publishing Corp., New York, N.Y.
5. ROSEN, A.A., J.B. PETER & F.M. MIDDLETON. 1962. Odor thresholds of mixed organic chemicals. *J. Water Pollut. Control Fed.* 34:7.

8. Bibliography

HULBERT, R. & D. FEBEN. 1941. Studies on accuracy of threshold odor value. *J. Amer. Water Works Assoc.* 33:1945.

SPAULDING, C.H. 1942. Accuracy and application of threshold odor test. *J. Amer. Water Works Assoc.* 34:877.
THOMAS, H.A., JR. 1943. Calculation of threshold odor. *J. Amer. Water Works Assoc.* 35:751.
CARTWRIGHT, L.C., C.T. SNELL & P.H. KELLY. 1952. Organoleptic panel testing as a research tool. *Anal. Chem.* 24:503.
LAUGHLIN, H.F. 1954. Palatable level with the threshold odor test. *Taste Odor Control J.* 20:No. 8 (Aug.).
SHELLENBERGER, R.D. 1958. Procedures for determining threshold odor concentrations in aqueous solutions. *Taste Odor Control J.* 24: No. 5 (May).
LAUGHLIN, H.F. 1962. Influence of temperature in threshold odor evaluation. *Taste Odor Control J.* 28:No. 10 (Oct.).
The threshold odor test. 1963. *Taste Odor Control J.* 29:Nos. 6, 7, 8 (June, July, Aug.).
SUFFET, I.H. & S. SEGALL. 1971. Detecting taste and odor in drinking water. *J. Amer. Water Works Assoc.* 63:605.
STAHL, W.H., ed. 1973. Compilation of Odor and Taste Threshold Values Data. Amer. Soc. Testing & Materials Data Ser. DS 48, Philadelphia, Pa.
AMERICAN SOCIETY FOR TESTING AND MATERIALS. 1973. Annual Book of ASTM Standards. Part 23, D-1292-65, ASTM, Philadelphia, Pa.

2160   TASTE*

2160 A.   Introduction

1. General Discussion

Taste refers only to gustatory sensations called bitter, salty, sour, and sweet that result from chemical stimulation of sensory nerve endings located in the papillae of the tongue and soft palate. Flavor refers to a complex of gustatory, olfactory, and trigeminal sensations resulting from chemical stimulation of sensory nerve endings located in the tongue, nasal cavity, and oral cavity.[1] Water samples taken into the mouth for sensory analysis always produce a flavor, although taste, odor, or mouthfeel may predominate, depending on the chemical substances present. Methods for sensory analysis presented herein require that the sample be taken into the mouth, that is, be tasted, but technically the sensory analysis requires evaluation of the complex sensation called flavor. As used here, taste refers to a method of sensory

___
* Approved by Standard Methods Committee, 1988.

analysis in which samples are taken into the mouth but the resultant evaluations pertain to flavor.

Three methods have been developed for the sensory evaluation of water samples taken into the mouth: the flavor threshold test (FTT), the flavor rating assessment (FRA), and the flavor profile analysis (FPA). The FTT is the oldest. It has been used extensively and is particularly useful for determining if the overall flavor of a sample of finished water is detectably different from a defined standard.[2] The FRA is especially valuable for determining if a sample of finished water is acceptable for daily consumption,[3] and the FPA is most useful for identifying and characterizing individual flavors in a water sample.[4]

The FPA is new but not yet standardized.[4]

Make flavor tests only on samples known to be safe for ingestion. Do not use samples that may be contaminated with bacteria, viruses, parasites, or hazardous chemicals, that contain dechlorinating agents such as sodium arsenite or that are derived from an unesthetic source. Do not make flavor tests on wastewaters or similar untreated effluents. Observe all sanitary and esthetic precautions with regard to apparatus and containers contacting the sample. Properly clean and sterilize containers before using them. Conduct analyses in a laboratory free from interfering background odors and if possible provide non-odorous carbon-filtered air at constant temperature and humidity. Use the procedure described in Section 2150 with respect to taste- and odor-free water to prepare dilution water and reference samples.

2. References

1. GELDARD, F.A. 1972. The Human Senses. John Wiley & Sons, New York, N.Y.
2. BAKER, R.A. 1961. Taste and Odor in Water: A Critical Review. Manufacturing Chemists' Assoc., Washington, D.C.
3. BRUVOLD, W.H. 1968. Scales for rating the taste of water. *J. Appl. Psychol.* 52:245.
4. MALLEVIALE, J. & I.H. SUFFET, eds. 1987. The Identification and Treatment of Tastes and Odors in Drinking Water. American Water Works Association Research Foundation, Denver, Colo.

2160 B.   Flavor Threshold Test (FTT)

1. General Discussion

Use the FTT to measure detectable flavor quantitatively. More precisely, use the method to compare the sample flavor objectively with that of specified reference water used as diluent.

The flavor threshold number (FTN) is the greatest dilution of sample with reference water yielding a definitely perceptible difference. The FTN is computed as follows:

$$FTN = \frac{A + B}{A}$$

where:

A = sample volume, mL, and
B = reference water (diluent) volume, mL.

Table 2160:I gives the FTNs corresponding to various dilutions.

2. Procedure

a. Panel selection: Carefully select by preliminary trials interested persons to make flavor tests. Exclude insensitive persons and insure that the testers are free from colds or allergies. Familiarize testers with the procedure before they participate

TABLE 2160:I. FLAVOR THRESHOLD NUMBERS
CORRESPONDING TO VARIOUS DILUTIONS

Sample Volume mL	Diluent Volume mL	Flavor Threshold No. FTN
200	0	1
100	100	2
70	130	3
50	150	4
35	165	6
25	175	8
17	183	12
12	188	17
8	192	25
6	194	33
4	196	50
3	197	67
2	198	100
1	199	200

in a panel test, but do not let them prepare samples or know dilution concentrations being evaluated. For precise work use a panel of five or more testers.

b. *Taste characterization:* Have each observer describe the characteristic sample flavor of the most concentrated sample. Compile the consensus that may appear among testers. The value of characterization increases as observers become more experienced with a particular flavor category such as chlorophenolic, grassy, or musty.

c. *Preliminary test:* To determine approximate range of the FTN, add 200-, 50-, 12-, and 4-mL sample portions to volumes of reference water (see Section 2150) designated in Table 2160:I in separate 300-mL glass beakers to make a total of 200 mL in each beaker, and mix gently with clean stirrer. Use separate beaker containing only reference water for comparison. Keep sample temperature during testing within 1°C of specified temperature. Present samples to each taster in a uniform manner, with the reference water presented first, followed by the most dilute sample.

If a flavor can be detected in this dilution, prepare an intermediate sample by diluting 20 mL sample to 200 mL with reference water. Use this dilution for threshold determination and multiply FTN obtained by 10 to correct for intermediate dilution. In rare cases a higher intermediate dilution may be required.

If no flavor is detected in the most dilute sample, repeat using the next concentration. Continue this process until flavor is detected clearly.

d. *FTN determination:* Based on results obtained in the preliminary test, prepare a set of dilutions using Table 2160:II as a guide. Prepare the seven dilutions shown on the appropriate line. This array is necessary to challenge the range of sensitivities of the entire panel of testers. If the sample being tested requires more dilution than is provided by Table 2160:II, make intermediate dilutions as directed in c above.

Use a clean 50-mL beaker filled to the 25-mL level or use an ordinary restaurant-style drinking glass for each dilution and reference sample. Do not use glassware used in sensory testing for other analyses. Between tests, sanitize containers in an automatic dishwasher supplied with water at not less than 60°C.

Maintain samples at 15 ± 1°C. However, if temperature of water in the distribution system is higher than 15°C, select an appropriate temperature. Specify temperature in reporting results.

Present series of samples to each tester in order of increasing concentration. Pair each sample with a known reference. Have tester taste sample by taking into the mouth whatever volume is comfortable, moving sample throughout the mouth, holding it for several seconds, and discharging it without swallowing. Have tester compare sample with reference and record whether a flavor or aftertaste is detectable. Insert two or more reference blanks in the series near the expected threshold, but avoid any repeated pattern. Do not let tester know

TABLE 2160:II. DILUTIONS FOR DETERMINING THE FTN

Sample Volume in Which Taste Is First Noted mL	Volumes to be Diluted to 200 mL mL
200	200, 100, 70, 50, 35, 25, 17
50	50, 35, 25, 17, 12, 8, 6
12	12, 8, 6, 4, 3, 2, 1
4	Intermediate dilution

which samples have flavor and which are blanks. Instruct tester to taste each sample in sequence, beginning with the least concentrated sample, until flavor is detected with certainty.

Record observations by indicating whether flavor is noted in each test beaker. For example:

mL Sample Diluted to 200 mL	6	8	12	0	17	25	35	0	50
Response	−	−	−	−	−	+	+	−	+

where:

− signifies negative response and
+ signifies positive response.

3. Calculation

The flavor threshold number is the dilution ratio at which flavor is just detectable. In the example above, the first detectable flavor occurred when 25 mL sample was diluted to 200 mL yielding a threshold number of 8 (Table 2160:I). Reference blanks do not influence calculation of the threshold.

The smallest FTN that can be observed is 1, where the beaker contains 200 mL undiluted sample. If no flavor is detected at this concentration, report "No flavor observed" instead of a threshold number.

Anomalous responses sometimes occur; a low concentration may be called positive

and a higher concentration in the series may be called negative. In such cases, designate the threshold as that point after which no further anomalies occur. The following illustrates an approach to an anomalous series (responses to reference blanks are excluded):

Increasing Concentration →

Response: − + − + + + +
                   ↓
            Threshold

Calculate mean and standard deviation of all FTNs if the distribution is reasonably symmetrical; otherwise, express the threshold of a group as the median or geometric mean of individual thresholds.

4. Interpretation of Results

An FTN is not a precise value. In the case of the single observer it represents a judgment at the time of testing. Panel results are more meaningful because individual differences have less influence on the test result. One or two observers can develop useful data if comparison with larger panels has been made to check their sensitivity. Do not make comparisons of data from time to time or place to place unless all test conditions have been standardized carefully and there is some basis for comparison of observed FTNs.

5. Bibliography

HULBERT, R. & D. FEBEN. 1941. Studies on accuracy of threshold odor value. *J. Amer. Water Works Assoc.* 33:1945.

SPAULDING, C.H. 1942. Accuracy and application of threshold odor test. *J. Amer. Water Works Assoc.* 34:877.

THOMAS, H.A. 1943. Calculation of threshold odor. *J. Amer. Water Works Assoc.* 35:751.

COX, G.J. & J. W. NATHANS. 1952. A study of the taste of fluoridated water. *J. Amer. Water Works Assoc.* 44:940.

LAUGHLIN, H.F. 1954. Palatable level with the threshold odor test. *Taste Odor Control J.* 20:1.

COX, G.J., J.W. NATHANS & N. VONAU. 1955. Subthreshold-to-taste thresholds of sodium, potassium, calcium and magnesium ions in water. *J. Appl. Physiol.* 8:283.

LOCKHART, E.E., C.L. TUCKER & M.C. MERRITT. 1955. The effect of water impurities on the flavor of brewed coffee. *Food Res.* 20:598.

CAMPBELL, C.L., R.K. DAWES, S. DEOLALKAR & M.C. MERRITT. 1958. Effects of certain chemicals in water on the flavor of brewed coffee. *Food Res.* 23:575.

MIDDLETON, F.M., A.A. ROSEN & R.H. BRUTTSCHELL. 1958. Taste and odor research tools for water utilities. *J. Amer. Water Works Assoc.* 50:231.

SHELLENBERGER, R.D. 1958. Procedures for determining threshold odor concentrations in aqueous solutions. *Taste Odor Control J.* 24:1.

COHEN, J.N., L.J. KAMPHAKE, E.K. HARRIS & R.L. WOODWARD. 1960. Taste threshold concentrations of metals in drinking water. *J. Amer. Water Works Assoc.* 52:660.

BAKER, R.A. 1961. Problems of tastes and odors. *J. Water Pollut. Control Fed.* 33:1099.

ROSEN, A.A., J.B. PETER & F.M. MIDDLETON. 1962. Odor thresholds of mixed organic chemicals. *J. Water Pollut. Control Fed.* 34:7.

BAKER, R.A. 1962. Critical evaluation of olfactory measurement. *J. Water Pollut. Control Fed.* 34:582.

BAKER, R.A. 1963. Threshold odors of organic chemicals. *J. Amer. Water Works Assoc.* 55:913.

BAKER, R.A. 1963. Odor testing laboratory. *J. Water Pollut. Control Fed.* 35:1396.

BAKER, R.A. 1963. Odor effects of aqueous mixtures of organic chemicals. *J. Water Pollut. Control Fed.* 35:728.

ROSEN, A.A., R.T. SKEEL & M.B. ETTINGER. 1963. Relationship of river water odor to specific organic contaminants. *J. Water Pollut. Control Fed.* 35:777.

BRYAN, P.E., L.N. KUZMINSKI, F.M. SAWYER & T.H. FENG. 1973. Taste thresholds of halogens in water. *J. Amer. Water Works Assoc.* 65:363.

2160 C.   Flavor Rating Assessment (FRA)

1. General Discussion

When the purpose of the test is to estimate acceptability for daily consumption, use the flavor rating assessment described below. This procedure has been used with samples from public sources in laboratory research and consumer surveys to recommend standards governing mineral content in drinking water. Each tester is presented with a list of nine statements about the water ranging on a scale from very favorable to very unfavorable. The tester's task is to select the statement that best expresses his or her opinion. The individual rating is the scale number of the statement selected. The panel rating for a particular sample is an appropriate measure of central tendency of the scale numbers for all testers for that sample.

2. Samples

Sample finished water ready for human consumption or use experimentally treated water if the sanitary requirements given in Section 2160A.1 are met fully. Use taste- and odor-free water as described in Section 2150 and a solution of 2000 mg NaCl/L prepared with taste- and odor-free water as criterion samples.

3. Procedure

a. Panel selection and preparation: Give prospective testers thorough instructions and trial or orientation sessions followed by questions and discussion of procedures. In tasting samples, testers work alone. Select panel members on the basis of performance in these trial sessions. Do not let testers know the composition or source of specific samples.

b. Rating test: A single rating session may be used to evaluate up to 10 samples, including the criterion samples mentioned in ¶ 2 above. Allow at least 30 min rest between repeated rating sessions.

For glassware requirements, see ¶ B.2*d.*

Present samples at a temperature that the testers will find pleasant for drinking water; maintain this temperature throughout testing. A temperature of 15°C is recommended, but in any case, do not let the test temperature exceed tap water temperatures customary at the time of the test. Specify test temperature in reporting results.

Independently randomize sample order for each tester. Instruct each to complete the following steps: 1) Taste about half the sample by taking water into the mouth, holding it for several seconds, and discharging it without swallowing; 2) Form an initial judgment on the rating scale; 3) Make a second tasting in a similar manner; 4) Make a final rating and record result on an appropriate data form; 5) Rinse mouth with reference water; 6) Rest 1 min before repeating Steps 1 through 5 on next sample.

c. *Characterization:* If characterization of flavor also is required, conduct a final rating session wherein each tester is asked to describe the flavor of each sample rated (see ¶ B2b).

4. Calculation

Use the following scale for rating. Record ratings as integers ranging from one to nine, with one given the highest quality rating. Calculate mean and standard deviation of all ratings if the distribution is reasonably symmetrical, otherwise express the most typical rating of a group as the median or geometric mean of individual ratings.

Action tendency scale:

1) I would be very happy to accept this water as my everyday drinking water.

2) I would be happy to accept this water as my everyday drinking water.

3) I am sure that I could accept this water as my everyday drinking water.

4) I could accept this water as my everyday drinking water.

5) Maybe I could accept this water as my everyday drinking water.

6) I don't think I could accept this water as my everyday drinking water.

7) I could not accept this water as my everyday drinking water.

8) I could never drink this water.

9) I can't stand this water in my mouth and I could never drink it.

5. Interpretation of Results

Values representing the central tendency and dispersion of quality ratings for a laboratory panel are only estimates of these values for a defined consuming population.

6. Bibliography

BRUVOLD, W.H. & R.M. PANGBORN. 1966. Rated acceptability of mineral taste in water. *J. Appl. Psychol.* 50:22.

BRUVOLD, W.H., H.J. ONGERTH & R.C. DILLEHAY. 1967. Consumer attitudes toward mineral taste in domestic water. *J. Amer. Water Works Assoc.* 59:547.

DILLEHAY, R.C., W.H. BRUVOLD & J.P. SIEGEL. 1967. On the assessment of potability. *J. Appl. Psychol.* 51:89.

BRUVOLD, W.H. 1968. Mineral Taste in Domestic Water. Univ. California Water Resources Center, Los Angeles.

BRUVOLD, W.H. & H.J. ONGERTH. 1969. Taste quality of mineralized water. *J. Amer. Water Works Assoc.* 61:170.

BRUVOLD, W.H. & W.R. GAFFEY. 1969. Rated acceptability of mineral taste in water. II: Combinatorial effects of ions on quality and action tendency ratings. *J. Appl. Psychol.* 53:317.

DILLEHAY, R.C., W.H. BRUVOLD & J.P. SIEGEL. 1969. Attitude, object label, and stimulus factors in response to an attitude object. *J. Personal. Social Psychol.* 11:220.

BRUVOLD, W.H. 1970. Laboratory panel estimation of consumer assessments of taste and flavor. *J. Appl Psychol.* 54:326.

PANGBORN, R.M., I.M. TRABUE & R.C. BALDWIN. 1970. Sensory examination of mineralized, chlorinated waters. *J. Amer. Water Works Assoc.* 62:572.

PANGBORN, R.M., I.M. TRABUE & A.C. LITTLE. 1971. Analysis of coffee, tea and artifically flavored drinks prepared from mineralized waters. *J. Food Sci.* 36:355.

BRUVOLD, W.H. & P.C. WARD. 1971. Consumer assessment of water quality and the cost of improvements. *J. Amer. Water Works Assoc.* 63:3.

PANGBORN, R.M. & L.L. BERTOLERO. 1972. Influence of temperature on taste intensity and degree of liking of drinking water. *J. Amer. Water Works Assoc.* 64:511.

BRUVOLD, W.H. 1975. Human perception and evaluation of water quality. *Crit. Rev. Environ. Control* 5:153.

BRUVOLD, W.H. 1976. Consumer Evaluation of the Cost and Quality of Domestic Water. Univ. California Water Resources Center, Davis.

2160 D. Flavor Profile Analysis (FPA)

1. General Discussion

To identify and characterize offensive flavors that may occur against a background of documented sensory quality judged acceptable, use flavor profile analysis. This method requires careful selection and training of panelists and may be particularly useful for water utilities. The procedure has been published.[1]

2. Reference

1. MALLEVIALE, J. & I.H. SUFFET, eds. 1987. The Identification and Treatment of Tastes and Odors in Drinking Water. American Water Works Association Research Foundation, Denver, Colo.

3. Bibliography

SUFFET, I.H. & S. SEGALL. 1971. Detecting taste and odor in drinking water. *J. Amer. Water Works Assoc.* 63:605.

PERSSON, P.E. 1982. Muddy odour: a problem associated with extreme eutrophication. *Hydrobiologia* 86:161.

MEILGAARD, M.C., D.S. REID & K.A. WYBORSKI. 1982. Reference standards for beer flavor terminology system. *Amer. Soc. Brewing Chemists. J.* 40:119.

DOTY, R.L., P. SHAMAN, S.L. APPLEBAUM, R. GIBERSON, L. SIKSORSKI & L. ROSENBERG. 1984. Smell identification ability: Changes with age. *Science* 226:1441.

KRASNER, S.W., M.J. McGUIRE & V.B. FERGUSON. 1985. Tastes and odors: The flavor profile method. *J. Amer. Water Works Assoc.* 77:34.

ANSELME, C., K. N'GUYEN, A. BRUCHET & J. MALLEVIALLE. 1985. Can polyethylene pipes impart odors in drinking water? *Environ. Technol. Letters* 6:477.

AMOORE, J.E. 1986. The chemistry and physiology of odor sensitivity. *J. Amer. Water Works Assoc.* 78:70.

BARTELS, J.H.M., G.A. BURLINGAME & I.H. SUFFET. 1986. Flavor profile analysis: Taste and odor control of the future. *J. Amer. Water Works Assoc.* 78:50.

BURLINGAME, G.A., R.M. DANN & B.L. BROCK. 1986. A case study of geosmin in Philadelphia's water. *J. Amer. Water Works Assoc.* 78:56.

KRASNER, S.W. & E.G. MEANS. 1986. Returning recently covered reservoirs to service: Health and aesthetic considerations. *J. Amer. Water Works Assoc.* 78:94.

MEANS, E.G. & M.J. McGUIRE. 1986. An early warning system for taste and odor control. *J. Amer. Water Works Assoc.* 78:77.

BARTELS, J.H.M., B.M. BRADY & I.H. SUFFET. 1987. Training panelists for the flavor profile analysis method. *J. Amer. Water Works Assoc.* 78:50.

2310 ACIDITY*

2310 A. Introduction

Acidity of a water is its quantitative capacity to react with a strong base to a designated pH. The measured value may vary significantly with the end-point pH used in the determination. Acidity is a measure of an aggregate property of water and can be interpreted in terms of specific substances only when the chemical composition of the sample is known. Strong mineral acids, weak acids such as carbonic and acetic, and hydrolyzing salts such as iron or aluminum sulfates may contribute to the measured acidity according to the method of determination.

Acids contribute to corrosiveness and influence chemical reaction rates, chemical speciation, and biological processes. The measurement also reflects a change in the quality of the source water.

---

*Approved by Standard Methods Committee, 1985.

2310 B. Titration Method

1. General Discussion

a. Principle: Hydrogen ions present in a sample as a result of dissociation or hydrolysis of solutes react with additions of standard alkali. Acidity thus depends on the end-point pH or indicator used. The construction of a titration curve by recording sample pH after successive small measured additions of titrant permits identification of inflection points and buffering capacity, if any, and allows the acidity to be determined with respect to any pH of interest.

In the titration of a single acidic species, as in the standardization of reagents, the most accurate end point is obtained from the inflection point of a titration curve. The inflection point is the pH at which curvature changes from convex to concave or vice versa.

Because accurate identification of inflection points may be difficult or impossible in buffered or complex mixtures, the titration in such cases is carried to an arbitrary end-point pH based on practical considerations. For routine control titrations or rapid preliminary estimates of acidity, the color change of an indicator may be used for the end point. Samples of industrial wastes, acid mine drainage, or other solutions that contain appreciable amounts of hydrolyzable metal ions such as iron, aluminum, or manganese are treated with hydrogen peroxide to ensure oxidation of any reduced forms of polyvalent cations, and boiled to hasten hydrolysis. Acidity results may be highly variable if this procedure is not followed exactly.

b. End points: Ideally the end point of the acidity titration should correspond to the stoichiometric equivalence point for neutralization of acids present. The pH at the equivalence point will depend on the sample, the choice among multiple inflection points, and the intended use of the data.

Dissolved carbon dioxide (CO_2) usually is the major acidic component of unpolluted surface waters; handle samples from such sources carefully to minimize the loss of dissolved gases. In a sample containing only carbon dioxide-bicarbonates-carbonates, titration to pH 8.3 at 25°C corresponds to stoichiometric neutralization of carbonic acid to bicarbonate. Because the color change of phenolphthalein indicator is close to pH 8.3, this value generally is accepted as a standard end point for titration of total acidity, including CO_2 and most weak acids. Metacresol purple also has an end point at pH 8.3 and gives a sharper color change.

For more complex mixtures or buffered solutions selection of an inflection point may be subjective. Consequently, use fixed end points of pH 3.7 and pH 8.3 for standard acidity determinations via a potentiometric titration in wastewaters and natural waters where the simple carbonate equilibria discussed above cannot be assumed. Bromphenol blue has a sharp color change at its end point of 3.7. The resulting titrations are identified, traditionally, as "methyl orange acidity" (pH 3.7) and "phenolphthalein" or total acidity (pH 8.3) regardless of the actual method of measurement.

c. Interferences: Dissolved gases contributing to acidity or alkalinity, such as CO_2, hydrogen sulfide, or ammonia, may be lost or gained during sampling, storage, or titration. Minimize such effects by titrating to the end point promptly after opening sample container, avoiding vigorous shaking or mixing, protecting sample from the atmosphere during titration, and letting sample become no warmer than it was at collection.

In the potentiometric titration, oily matter, suspended solids, precipitates, or other waste matter may coat the glass electrode and cause a sluggish response. Difficulty from this source is likely to be revealed in an erratic titration curve. Do *not* remove interferences from sample because they may contribute to its acidity. Briefly pause between titrant additions to let electrode come to equilibrium or clean the electrodes occasionally.

In samples containing oxidizable or hydrolyzable ions such as ferrous or ferric iron, aluminum, and manganese, the reaction rates at room temperature may be slow enough to cause drifting end points.

Do not use indicator titrations with colored or turbid samples that may obscure the color change at the end point. Residual free available chlorine in the sample may bleach the indicator. Eliminate this source of interference by adding 1 drop of $0.1M$ sodium thiosulfate ($Na_2S_2O_3$).

d. Selection of procedure: Determine sample acidity from the volume of standard alkali required to titrate a portion to a pH of 8.3 (phenolphthalein acidity) or pH 3.7 (methyl orange acidity of wastewaters and grossly polluted waters). Titrate at room temperature using a properly calibrated pH meter, electrically operated titrator, or color indicators.

Use the hot peroxide procedure (¶ 4*a*) to pretreat samples known or suspected to contain hydrolyzable metal ions or reduced forms of polyvalent cation, such as iron pickle liquors, acid mine drainage, and other industrial wastes.

Color indicators may be used for routine and control titrations in the absence of interfering color and turbidity and for preliminary titrations to select sample size and strength of titrant (¶ 4*b*).

e. Sample size: The range of acidities found in wastewaters is so large that a single sample size and normality of base used as titrant cannot be specified. Use a sufficiently large volume of titrant (20 mL or more from a 50-mL buret) to obtain relatively good volumetric precision while keeping sample volume sufficiently small to permit sharp end points. For samples having acidities less than about 1000 mg as calcium carbonate $(CaCO_3)/L$, select a volume with less than 50 mg $CaCO_3$ equivalent acidity and titrate with 0.02N sodium hydroxide (NaOH). For acidities greater than about 1000 mg as $CaCO_3/L$, use a portion containing acidity equivalent to less than 250 mg $CaCO_3$ and titrate with 0.1N NaOH. If necessary, make a preliminary titration to determine optimum sample size and/or normality of titrant.

f. Sampling and storage: Collect samples in polyethylene or borosilicate glass bottles and store at a low temperature. Fill bottles completely and cap tightly. Because waste samples may be subject to microbial action and to loss or gain of CO_2 or other gases when exposed to air, analyze samples without delay, preferably within 1 d. If biological activity is suspected analyze within 6 h. Avoid sample agitation and prolonged exposure to air.

2. Apparatus

a. Electrometric titrator: Use any commercial pH meter or electrically operated titrator that uses a glass electrode and can be read to 0.05 pH unit. Standardize and calibrate according to the manufacturer's instructions. Pay special attention to temperature compensation and electrode care. If automatic temperature compensation is not provided, titrate at 25 ± 5°C.

b. Titration vessel: The size and form will depend on the electrodes and the sample size. Keep the free space above the sample as small as practicable, but allow room for titrant and full immersion of the indicating portions of electrodes. For conventional-sized electrodes, use a 200-mL, tall-form Berzelius beaker without a spout. Fit beaker with a stopper having three holes, to accommodate the two electrodes and the buret. With a miniature combination glass-reference electrode use a 125-mL or 250-mL erlenmeyer flask with a two-hole stopper.

c. Magnetic stirrer.

d. Pipets, volumetric.

e. Flasks, volumetric, 1000-, 200-, 100-mL.

f. Burets, borosilicate glass, 50-, 25-, 10-mL.

g. Polyolefin bottle, 1-L.

3. Reagents

a. Carbon dioxide-free water: Prepare all stock and standard solutions and dilution water for the standardization procedure with distilled or deionized water that has been freshly boiled for 15 min and cooled to room temperature. The final pH of the water should be ≥ 6.0 and its conductivity should be < 2 μmhos/cm.

b. Potassium hydrogen phthalate solution, approximately 0.05N: Crush 15 to 20 g primary standard $KHC_8H_4O_4$ to about 100 mesh and dry at 120°C for 2 h. Cool in a desiccator. Weigh 10.0 ± 0.5 g (to the nearest mg), transfer to a 1-L volumetric flask, and dilute to 1000 mL.

c. Standard sodium hydroxide titrant, 0.1N: Prepare solution approximately 0.1 N as indicated under Preparation of Desk Reagents (see inside front cover). Standardize by titrating 40.00 mL $KHC_8H_4O_4$ solution (3b), using a 25-mL buret. Titrate to the inflection point (¶ 1a), which should be close to pH 8.7. Calculate normality of NaOH:

$$\text{Normality} = \frac{A \times B}{204.2 \times C}$$

where:

A = g $KHC_8H_4O_4$ weighed into 1-L flask,
B = mL $KHC_8H_4O_4$ solution taken for titration, and
C = mL NaOH solution used.

Use the measured normality in further calculations or adjust to $0.1000N$; 1 mL = 5.00 mg $CaCO_3$.

d. Standard sodium hydroxide titrant, $0.02N$: Dilute 200 mL $0.1N$ NaOH to 1000 mL and store in a polyolefin bottle protected from atmospheric CO_2 by a soda lime tube or tight cap. Standardize against $KHC_8H_4O_4$ as directed in ¶ *3c*, using 15.00 mL $KHC_8H_4O_4$ solution and a 50-mL buret. Calculate normality as above (¶ *3c*); 1 mL = 1.00 mg $CaCO_3$.

e. Hydrogen peroxide, H_2O_2, 30%.

f. Bromphenol blue indicator solution, pH 3.7 indicator: Dissolve 100 mg bromphenol blue, sodium salt, in 100 mL water.

g. Metacresol purple indicator solution, pH 8.3 indicator: Dissolve 100 mg metacresol purple in 100 mL water.

h. Phenolphthalein indicator solution, alcoholic, pH 8.3 indicator.

i. Sodium thiosulfate, $0.1M$: Dissolve 25 g $Na_2S_2O_3 \cdot 5H_2O$ and dilute to 1000 mL with distilled water.

4. Procedure

If sample is free from hydrolyzable metal ions and reduced forms of polyvalent cations, proceed with analysis according to *b*, *c*, or *d*. If sample is known or suspected to contain such substances, pretreat according to *a*.

a. Hot peroxide treatment: Pipet a suitable sample (see ¶ *1e*) into titration flasks. Measure pH. If pH is above 4.0 add 5-mL increments of $0.02N$ sulfuric acid (H_2SO_4) (Section 2320B.3*c*) to reduce pH to 4 or less. Remove electrodes. Add 5 drops 30% H_2O_2 and boil for 2 to 5 min. Cool to room temperature and titrate with standard alkali to pH 8.3 according to the procedure of *4d*.

b. Color change: Select sample size and normality of titrant according to criteria of ¶ *1e*. Adjust sample to room temperature, if necessary, and with a pipet discharge sample into an erlenmeyer flask, while keeping pipet tip near flask bottom. If free residual chlorine is present add 0.05 mL (1 drop) $0.1M$ $Na_2S_2O_3$ solution, or destroy with ultraviolet radiation. Add 0.2 mL (5 drops) indicator solution and titrate over a white surface to a persistent color change characteristic of the equivalence point. Commercial indicator solutions or solids designated for the appropriate pH range (3.7 or 8.3) may be used. Check color at end point by adding the same concentration of indicator used with sample to a buffer solution at the designated pH.

c. Potentiometric titration curve:

1) Rinse electrodes and titration vessel with distilled water and drain. Select sample size and normality of titrant according to the criteria of ¶ *1e*. Adjust sample to room temperature, if necessary, and with a pipet discharge sample while keeping pipet tip near the titration vessel bottom.

2) Measure sample pH. Add standard alkali in increments of 0.5 mL or less, such that a change of less than 0.2 pH units occurs per increment. After each addition, mix thoroughly but gently with a magnetic stirrer. Avoid splashing. Record pH when a constant reading is obtained. Continue adding titrant and measure pH until pH 9 is reached. Construct the titration curve by plotting observed pH values versus cumulative milliliters titrant added. A smooth curve showing one or more inflections should be obtained. A ragged or erratic curve may indicate that equilibrium was not reached between successive alkali additions. Determine acidity relative to a particular pH from the curve.

d. Potentiometric titration to pH 3.7 or 8.3: Prepare sample and titration assembly as specified in ¶ *4c*1). Titrate to preselected end-point pH (¶ 1d) without recording intermediate pH values. As the end point is approached make smaller additions of alkali and be sure that pH equilibrium is reached before making the next addition.

5. Calculation

Acidity, as mg $CaCO_3/L$

$$= \frac{[(A \times B) - (C \times D)] \times 50\,000}{\text{mL sample}}$$

where:

A = mL NaOH titrant used,
B = normality of NaOH,
C = mL H_2SO_4 used (¶ 4a), and
D = normality of H_2SO_4.

Report pH of the end point used, as follows: "The acidity to pH ⎯⎯⎯ = ⎯⎯⎯ mg $CaCO_3/L$." If a negative value is obtained, determine the alkalinity according to Section 2320.

6. Precision and Bias

No general statement can be made about precision because of the great variation in sample characteristics. The precision of the titration is likely to be much greater than the uncertainties involved in sampling and sample handling before analysis.

Forty analysts in 17 laboratories analyzed synthetic water samples containing increments of bicarbonate equivalent to 20 mg $CaCO_3/L$. Titration according to the procedure of ¶ 4d gave a standard deviation of 1.8 mg $CaCO_3/L$, with negligible bias. Five laboratories analyzed two samples containing sulfuric, acetic, and formic acids and aluminum chloride by the procedures of ¶s 4b and 4d. The mean acidity of one sample (to pH 3.7) was 487 mg $CaCO_3/L$, with a standard deviation of 11 mg/L. The bromphenol blue titration of the same sample was 90 mg/L greater, with a standard deviation of 110 mg/L. The other sample had a potentiometric titration of 547 mg/ L, with a standard deviation of 54 mg/L, while the corresponding indicator result was 85 mg/L greater, with a standard deviation of 56 mg/L. The major difference between the samples was the substitution of ferric ammonium citrate, in the second sample, for part of the aluminum chloride.

7. Bibliography

WINTER, J.A. & M.R. MIDGETT. 1969. FWPCA Method Study 1. Mineral and Physical Analyses. Federal Water Pollution Control Admin., Washington, D.C.

BROWN, E., M.W. SKOUGSTAD & M.J. FISHMAN. 1970. Methods for collection and analysis of water samples for dissolved minerals and gases. Chapter A1 in Book 5, Techniques of Water-Resources Investigations of United States Geological Survey. U.S. Geological Survey, Washington, D.C.

SNOEYINK, V.L. & D. JENKINS. 1980. Water Chemistry. John Wiley & Sons, New York, N.Y.

2320 ALKALINITY*

2320 A. Introduction

1. Discussion

Alkalinity of a water is its acid-neutralizing capacity. It is the sum of all the titratable bases. The measured value may vary significantly with the end-point pH used. Alkalinity is a measure of an aggregate property of water and can be interpreted in terms of specific substances only when the chemical composition of the sample is known.

Alkalinity is significant in many uses and treatments of natural waters and wastewaters. Because the alkalinity of many surface waters is primarily a function of carbonate, bicarbonate, and hydroxide content, it is taken as an indication of the concentration of these constitutents. The

measured values also may include contributions from borates, phosphates, silicates, or other bases if these are present. Alkalinity in excess of alkaline earth metal concentrations is significant in determining the suitability of a water for irrigation. Alkalinity measurements are used in the interpretation and control of water and wastewater treatment processes. Raw domestic wastewater has an alkalinity less than, or only slightly greater than, that of the water supply. Properly operating anaerobic digesters typically have supernatant alkalinities in the range of 2000 to 4000 mg calcium carbonate ($CaCO_3$)/L.[1]

2. Reference

1. POHLAND, F.G. & D.E. BLOODGOOD. 1963. Laboratory studies on mesophilic and thermophilic anaerobic sludge digestion. *J. Water Pollut. Control Fed.* 35:11.

*Approved by Standard Methods Committee, 1985.

2320 B. Titration Method

1. General Discussion

a. Principle: Hydroxyl ions present in a sample as a result of dissociation or hydrolysis of solutes react with additions of standard acid. Alkalinity thus depends on the end-point pH used. For methods of determining inflection points from titration

curves and the rationale for titrating to fixed pH end points, see Section 2310B.1*a*.

For samples of low alkalinity (less than 20 mg $CaCO_3$/L) use an extrapolation technique based on the near proportionality of concentration of hydrogen ions to excess of titrant beyond the equivalence point. The amount of standard acid required to reduce pH exactly 0.30 pH unit is measured carefully. Because this change in pH corresponds to an exact doubling of the hydrogen ion concentration, a simple extrapolation can be made to the equivalence point.[1,2]

b. *End points:* When alkalinity is due entirely to carbonate or bicarbonate content, the pH at the equivalence point of the titration is determined by the concentration of carbon dioxide (CO_2) at that stage. CO_2 concentration depends, in turn, on the total carbonate species originally present and any losses that may have occurred during titration. The following pH values are suggested as the equivalence points for the corresponding alkalinity concentrations as milligrams $CaCO_3$ per liter. "Phenolphthalein alkalinity" is the term traditionally used for the quantity measured by titration to pH 8.3 irrespective of the colored indicator, if any, used in the determination. The sharp end-point color changes produced by metacresol purple (pH 8.3) and bromcresol green (pH 4.5) make these indicators suitable for the alkalinity titration.

	End Point pH	
	Total Alkalinity	Phenolphthalein Alkalinity
Alkalinity, mg $CaCO_3$/L:		
30	4.9	8.3
150	4.6	8.3
500	4.3	8.3
Silicates, phosphates known or suspected	4.5	8.3
Routine or automated analyses	4.5	8.3
Industrial waste or complex system	4.5	8.3

c. *Interferences:* Soaps, oily matter, suspended solids, or precipitates may coat the glass electrode and cause a sluggish response. Allow additional time between titrant additions to let electrode come to equilibrium or clean the electrodes occasionally. Do not filter, dilute, concentrate, or alter sample.

d. *Selection of procedure:* Determine sample alkalinity from volume of standard acid required to titrate a portion to a designated pH taken from ¶ 1b. Titrate at room temperature with a properly calibrated pH meter or electrically operated titrator, or use color indicators.

Report alkalinity less than 20 mg $CaCO_3$/L only if it has been determined by the low-alkalinity method of ¶ 4d.

Construct a titration curve for standardization of reagents.

Color indicators may be used for routine and control titrations in the absence of interfering color and turbidity and for preliminary titrations to select sample size and strength of titrant (see below).

e. *Sample size:* See Section 2310B.1e for selection of size sample to be titrated and normality of titrant, substituting 0.02N or 0.1N sulfuric (H_2SO_4) or hydrochloric (HCl) acid for the standard alkali of that method. For the low-alkalinity method, titrate a 200-mL sample with 0.02N H_2SO_4 from a 10-mL buret.

f. *Sampling and storage:* See Section 2310B.1f.

2. Apparatus

See Section 2310B.2.

3. Reagents

a. *Sodium carbonate solution,* approximately 0.05N: Dry 3 to 5 g primary standard Na_2CO_3 at 250°C for 4 h and cool in a desiccator. Weigh 2.5 ± 0.2 g (to the nearest mg), transfer to a 1-L volumetric flask, fill flask to the mark with distilled water, and dissolve and mix reagent. Do not keep longer than 1 week.

b. *Standard sulfuric acid or hydrochloric acid,* 0.1*N:* Prepare acid solution of approximate normality as indicated under Preparation of Desk Reagents (see inside front cover). Standardize against 40.00 mL 0.05*N* Na$_2$CO$_3$ solution, with about 60 mL water, in a beaker by titrating potentiometrically to pH of about 5. Lift out electrodes, rinse into the same beaker, and boil gently for 3 to 5 min under a watch glass cover. Cool to room temperature, rinse cover glass into beaker, and finish titrating to the pH inflection point. Calculate normality:

$$\text{Normality, } N = \frac{A \times B}{53.00 \times C}$$

where:

A = g Na$_2$CO$_3$ weighed into 1-L flask,
B = mL Na$_2$CO$_3$ solution taken for titration, and
C = mL acid used.

Use measured normality in calculations or adjust to 0.1000*N*; 1 mL 0.1000*N* solution = 5.00 mg CaCO$_3$.

c. *Standard sulfuric acid or hydrochloric acid,* 0.02*N:* Dilute 200.00 mL 0.1000*N* standard acid to 1000 mL with distilled or deionized water. Standardize by potentiometric titration of 15.00 mL 0.05*N* Na$_2$CO$_3$ according to the procedure of ¶ 3*b*; 1 mL = 1.00 mg CaCO$_3$.

d. *Bromcresol green indicator solution,* pH 4.5 indicator: Dissolve 100 mg bromcresol green, sodium salt, in 100 mL distilled water.

e. *Metacresol purple indicator solution,* pH 8.3 indicator: Dissolve 100 mg metacresol purple in 100 mL water.

f. *Phenolphthalein solution, alcoholic,* pH 8.3 indicator.

g. *Sodium thiosulfate,* 0.1*N:* See Section 2310B.3*i*.

4. Procedure

a. *Color change:* See Section 2310B.4*b*.

b. *Potentiometric titration curve:* Follow the procedure for determining acidity (Section 2310B.4*c*), substituting the appropriate normality of standard acid solution for standard NaOH, and continue titration to pH 4.5 or lower. Do not filter, dilute, concentrate, or alter the sample.

c. *Potentiometric titration to preselected pH:* Determine the appropriate end-point pH according to ¶ 1*b*. Prepare sample and titration assembly (Section 2310B.4*c*). Titrate to the end-point pH without recording intermediate pH values and without undue delay. As the end point is approached make smaller additions of acid and be sure that pH equilibrium is reached before adding more titrant.

d. *Potentiometric titration of low alkalinity:* For alkalinities less than 20 mg/L titrate 100 to 200 mL according to the procedure of ¶ 4*c*, above, using a 10-mL microburet and 0.02*N* standard acid solution. Stop the titration at a pH in the range 4.3 to 4.7 and record volume and exact pH. Carefully add additional titrant to reduce the pH exactly 0.30 pH unit and again record volume.

5. Calculations

a. *Potentiometric titration to end-point pH:*

$$\text{Alkalinity, mg CaCO}_3/\text{L} = \frac{A \times N \times 50\,000}{\text{mL sample}}$$

where:

A = mL standard acid used and
N = normality of standard acid

or

$$\text{Alkalinity, mg CaCO}_3/\text{L} = \frac{A \times t \times 1000}{\text{mL sample}}$$

where:

t = titer of standard acid, mg CaCO$_3$/mL.

Report pH of end point used as follows: "The alkalinity to pH _____ = _____ mg $CaCO_3/L$" and indicate clearly if this pH corresponds to an inflection point of the titration curve.

b. *Potentiometric titration of low alkalinity:*

Total alkalinity, mg $CaCO_3/L$

$$= \frac{(2\ B\ -\ C) \times N \times 50\ 000}{mL\ sample}$$

where:

B = mL titrant to first recorded pH,
C = total mL titrant to reach pH 0.3 unit lower, and
N = normality of acid.

c. *Calculation of alkalinity relationships:* The results obtained from the phenolphthalein and total alkalinity determinations offer a means for stoichiometric classification of the three principal forms of alkalinity present in many waters. The classification ascribes the entire alkalinity to bicarbonate, carbonate, and hydroxide, and assumes the absence of other (weak) inorganic or organic acids, such as silicic, phosphoric, and boric acids. It further presupposes the incompatibility of hydroxide and bicarbonate alkalinities. Because the calculations are made on a stoichiometric basis, ion concentrations in the strictest sense are not represented in the results, which may differ significantly from actual concentrations especially at pH > 10. According to this scheme:

1) Carbonate (CO_3^{2-}) alkalinity is present when phenolphthalein alkalinity is not zero but is less than total alkalinity.

2) Hydroxide (OH^-) alkalinity is present if phenolphthalein alkalinity is more than half the total alkalinity.

3) Bicarbonate (HCO_3^-) alkalinity is present if phenolphthalein alkalinity is less than half the total alkalinity. These relationships may be calculated by the following scheme, where P is phenolphthalein alkalinity and T is total alkalinity (¶ 1b):

Select the smaller value of P or $(T-P)$. Then, carbonate alkalinity equals twice the smaller value. When the smaller value is P, the balance $(T-2P)$ is bicarbonate. When the smaller value is $(T-P)$, the balance $(2P-T)$ is hydroxide. All results are expressed as $CaCO_3$. The mathematical conversion of the results is shown in Table 2320:I. (A modification of Table 2320:I that is more accurate when $P \simeq \frac{1}{2}T$ has been proposed.[3])

Alkalinity relationships also may be computed nomographically (see Carbon Dioxide, Section 4500-CO_2). Accurately measure pH, calculate OH^- concentration as milligrams $CaCO_3$ per liter, and calculate concentrations of CO_3^{2-} and HCO_3^- as milligrams $CaCO_3$ per liter from the OH^- concentration, and the phenolphthalein and total alkalinities by the following equations:

$$CO_3^{2-} = 2P - 2[OH^-]$$

$$HCO_3^- = T - 2P + [OH^-]$$

Similarly, if difficulty is experienced with the phenolphthalein end point, or if a check on the phenolphthalein titration is desired, calculate phenolphthalein alkalinity as $CaCO_3$ from the results of the nomographic

TABLE 2320:I. ALKALINITY RELATIONSHIPS*

Result of Titration	Hydroxide Alkalinity as CaCO₃	Carbonate Alkalinity as CaCO₃	Bicarbonate Concentration as CaCO₃
P = 0	0	0	T
P < ½T	0	2P	T − 2P
P = ½T	0	2P	0
P > ½T	2P − T	2(T − P)	0
P = T	T	0	0

*Key: P—phenolphthalein alkalinity; T—total alkalinity.

determinations of carbonate and hydroxide ion concentrations:

$$P = 1/2 [CO_3^{2-}] + [OH^-]$$

6. Precision and Bias

No general statement can be made about precision because of the great variation in sample characteristics. The precision of the titration is likely to be much greater than the uncertainties involved in sampling and sample handling before the analysis.

In the range of 10 to 500 mg/L, when the alkalinity is due entirely to carbonates or bicarbonates, a standard deviation of 1 mg $CaCO_3$/L can be achieved. Forty analysts in 17 laboratories analyzed synthetic samples containing increments of bicarbonate equivalent to 120 mg $CaCO_3$/L. The titration procedure of ¶ 4b was used, with an end point pH of 4.5. The standard deviation was 5 mg/L and the average bias (lower than the true value) was 9 mg/L.[4]

Sodium carbonate solutions equivalent to 80 and 65 mg $CaCO_3$/L were analyzed by 12 laboratories according to the procedure of ¶ 4c.[5] The standard deviations were 8 and 5 mg/L, respectively, with negligible bias.[5] Four laboratories analyzed six samples having total alkalinities of about 1000 mg $CaCO_3$/L and containing various ratios of carbonate/bicarbonate by the pro-cedures of both ¶ 4a and ¶ 4c. The pooled standard deviation was 40 mg/L, with negligible difference between the procedures.

7. References

1. LARSON, T.E. & L.M. HENLEY. 1955. Determination of low alkalinity or acidity in water. *Anal. Chem.* 27:851.
2. THOMAS, J.F.J. & J.J. LYNCH. 1960. Determination of carbonate alkalinity in natural waters. *J. Amer. Water Works Assoc.* 52:259.
3. JENKINS, S.R. & R.C. MOORE. 1977. A proposed modification to the classical method of calculating alkalinity in natural waters. *J. Amer. Water Works Assoc.* 69:56.
4. WINTER, J.A. & M.R. MIDGETT. 1969. FWPCA Method Study 1. Mineral and Physical Analyses. Federal Water Pollution Control Admin., Washington, D.C.
5. SMITH, R. 1980. Research Rep. No. 379, Council for Scientific and Industrial Research, South Africa.

8. Bibliography

AMERICAN SOCIETY FOR TESTING AND MATERIALS. 1982. Standard Methods for Acidity or Alkalinity of Water. Publ. D1067-70 (reapproved 1977), American Soc. Testing & Materials, Philadelphia, Pa.
SKOUGSTAD M.W., M.J. FISHMAN, L.C. FRIEDMAN, D.E. ERDMAN, & S.S. DUNCAN. 1979. Methods for determination of inorganic substances in water and fluvial sediments. *In* Techniques of Water-Resources Investigation of the United States Geological Survey. U.S. Geological Survey, Book 5, Chapter A1, Washington, D.C.

2330   CALCIUM CARBONATE SATURATION (PROPOSED)*

2330 A.   Introduction

1. General Discussion

Calcium carbonate ($CaCO_3$) saturation indices commonly are used to evaluate the scale-forming and scale-dissolving tendencies of water. Assessing these tendencies is useful in corrosion control programs and in preventing $CaCO_3$ scaling in piping and equipment such as industrial heat exchangers or domestic water heaters.

Waters oversaturated with respect to $CaCO_3$ tend to precipitate $CaCO_3$. Waters undersaturated with respect to $CaCO_3$ tend to dissolve $CaCO_3$. Saturated waters, i.e., waters in equilibrium with $CaCO_3$, have neither $CaCO_3$-precipitating nor $CaCO_3$-dissolving tendencies. Saturation represents the dividing line between "precipitation likely" and "precipitation not likely."

Several water quality characteristics must be measured to calculate the $CaCO_3$ saturation indices described here. Minimum requirements are total alkalinity (2320), total calcium (3500-Ca), pH (4500-H^+), and temperature (2550). The ionic strength also must be calculated or estimated from conductivity (2510) or total dissolved solids (2540C) measurements. Measure pH at the system's water temperature using a temperature-compensated pH meter. If pH is measured at a different temperature, for example in the laboratory, correct the measured pH.[1-3] In measuring

pH and alkalinity, minimize CO_2 exchange between sample and atmosphere. Ideally, seal the sample from the atmosphere during measurements[4]; at a minimum, avoid vigorous stirring of unsealed samples.

There are two general categories of $CaCO_3$ saturation indices: indices that determine whether a water has a *tendency* to precipitate $CaCO_3$ (i.e., is oversaturated) or to dissolve $CaCO_3$ (i.e., is undersaturated) and indices that estimate the *quantity* of $CaCO_3$ that can be precipitated from an oversaturated water and the amount that can be dissolved by an undersaturated water. Indices in the second category generally yield more information but are more difficult to determine.

2. Limitations

It is widely assumed that $CaCO_3$ will precipitate from oversaturated waters and that it cannot be deposited from undersaturated waters. Exceptions may occur. For example, $CaCO_3$ deposition from oversaturated waters is inhibited by the presence of phosphates (particularly polyphosphates), certain naturally occurring organics, and magnesium.[5-7] These materials can act as sequestering agents or as crystal poisons. Conversely, $CaCO_3$ deposits have been found in pipes conveying undersaturated water. This apparent contradiction is caused by high pH (relative to the bulk water pH) in the immediate vicinity of certain areas (cathodes) of corroding metal

*Approved by Standard Methods Committee, 1989.

surfaces. A locally oversaturated condition may occur even if the bulk water is undersaturated. A small, but significant, amount of $CaCO_3$ can be deposited.

The calculations referred to here, even the most sophisticated computerized calculations, do not adequately describe these exceptions. For this reason, do not consider saturation indices as absolutes. Rather, view them as guides to the behavior of $CaCO_3$ in aqueous systems and supplement them, where possible, with experimentally derived information.

Similarly, the effects predicted by the indices do not always conform to expectations. The relationship between the indices and corrosion rates is a case in point. Conceptually, piping is protected when $CaCO_3$ is precipitated on its surfaces. $CaCO_3$ is believed to inhibit corrosion by clogging reactive areas and by providing a matrix to retain corrosion products, thus further sealing the surfaces. Waters with positive indices traditionally have been assumed to be protective while waters with negative indices have been assumed to be not protective, or corrosive. The expected relationship is observed sometimes,[8,9] but not always.[10,11] Unexpected results may be due in part to limited capability to predict $CaCO_3$ behavior. Also, water characteristics not directly involved in the calculation of the indices (e.g., dissolved oxygen, buffering intensity, chloride, sulfate, and water velocity) may influence corrosion rates appreciably.[9,12–16] Thus, do not estimate corrosion rates on the basis of $CaCO_3$ indices alone.

3. References

1. MERRILL, D.T. & R.L. SANKS. 1978, 1979. Corrosion control by deposition of $CaCO_3$ films: A practical approach for plant operators. *J. Amer. Water Works Assoc.* 70:592; 70:634 & 71:12.

2. LOEWENTHAL, R.E. & G. v. R. MARAIS. 1976. Carbonate Chemistry of Aquatic Systems: Theory and Applications. Ann Arbor Science Publishers, Ann Arbor, Mich.

3. MERRILL, D.T. 1976. Chemical conditioning for water softening and corrosion control. *In* R.L. Sanks, ed. Water Treatment Plant Design. Ann Arbor Science Publishers, Ann Arbor, Mich.

4. SCHOCK, M.R., W. MUELLER & R.W. BUELOW. 1980. Laboratory techniques for measurement of pH for corrosion control studies and water not in equilibrium with the atmosphere. *J. Amer. Water Works Assoc.* 72:304.

5. PYTKOWICZ, R.M. 1965. Rates of inorganic carbon nucleation. *J. Geol.* 73:196.

6. FERGUSON, J.F. & P.L. McCARTY. 1969. Precipitation of Phosphate from Fresh Waters and Waste Waters. Tech. Rep. No. 120, Stanford Univ., Stanford, Calif.

7. MERRILL, D.T. & R.M. JORDEN. 1975. Lime-induced reactions in municipal wastewaters. *J. Water Pollut. Control Fed.* 47:2783.

8. DE MARTINI, F. 1938. Corrosion and the Langelier calcium carbonate saturation index. *J. Amer. Water Works Assoc.* 30:85.

9. LARSON, T.E. 1975. Corrosion by Domestic Waters. Bull. 59, Illinois State Water Survey.

10. JAMES M. MONTGOMERY, CONSULTING ENGINEERS, INC. 1985. Water Treatment Principles and Design. John Wiley & Sons, New York, N.Y.

11. STUMM, W. 1960. Investigations of the corrosive behavior of waters. *J. San. Eng. Div., Proc. Amer. Soc. Civil Eng.* 86:27.

12. PISIGAN, R.A., JR. & J.E. SINGLEY. 1985. Evaluation of water corrosivity using the Langelier index and relative corrosion rate models. *Materials Perform.* 24:26.

13. LANE, R.W. 1982. Control of corrosion in distribution and building water systems. *In* Proceedings AWWA Water Quality Technology Conf., Nashville, Tenn., Dec. 5–8, 1982.

14. SONTHEIMER, H., W. KOLLE & V.L. SNOEYINK. 1982. The siderite model of the formation of corrosion-resistant scales. *J. Amer. Water Works Assoc.* 73:572.

15. SCHOCK, M.R. & C.H. NEFF. 1982. Chemical aspects of internal corrosion; theory, prediction, and monitoring. *In* Proceedings AWWA Water Quality Technology Conf., Nashville, Tenn., Dec. 5–8, 1982.

16. AMERICAN WATER WORKS ASSOCIATION. 1986. Corrosion Control for Plant Operators. ISBW-O-89867-350-X. Denver, Colo.

2330 B. Indices Indicating Tendency of a Water to Precipitate CaCO₃ or Dissolve CaCO₃

1. General Discussion

Indices that indicate $CaCO_3$ precipitation or dissolution tendencies define whether a water is oversaturated, saturated, or undersaturated with respect to $CaCO_3$. The most widely used indices are the Saturation Index (SI): the Relative Saturation (RS), also known as the Driving Force Index (DFI); and the Ryznar Index (RI). The SI is by far the most commonly used and will be described here. The RS and SI are related (see Equation 6, Section 2330D). The RI[1] has been used for many years, sometimes with good results. Because it is semi-empirical it may be less reliable than the SI.

2. Saturation Index by Calculation

SI is determined from Equation 1.

$$SI = pH - pH_s \qquad (1)$$

where:

pH = measured pH and
pH_s = pH of the water if it were in equilibrium with $CaCO_3$ at the existing calcium ion $[Ca^{2+}]$ and bicarbonate ion $[HCO_3^-]$ concentrations.

A positive SI connotes a water oversaturated with respect to $CaCO_3$. A negative SI signifies an undersaturated water. An SI of zero represents a water in equilibrium with $CaCO_3$.

a. Analytical solution for pH_s: Determine pH_s as follows[2]:

$$pH_s = pK_2 - pK_s + p[Ca^{2+}] \\ + p[HCO_3^-] + 5\, pf_m \qquad (2)$$

where:

K_2 = second dissociation constant for carbonic acid, at the water temperature,

K_s = solubility product constant for $CaCO_3$ at the water temperature,
$[Ca^{2+}]$ = calcium ion concentration, g-moles/L,
$[HCO_3^-]$ = bicarbonate ion concentration, g-moles/L, and
f_m = activity coefficient for monovalent species at the specified temperature.

In Equation 2, p preceding a variable designates $-\log_{10}$ of that variable.

Calculate values of pK_2, pK_s, and pf_m required to solve Equation 2 from the equations in Table 2330:I. To save computation time, values for pK_2 and pK_s have been precalculated for selected temperatures (see Table 2330:II).

Table 2330:II gives several values for pK_s. Different isomorphs of $CaCO_3$ can form in aqueous systems, including calcite, aragonite, and vaterite. Each has somewhat different solubility properties. These differences can be accommodated when computing pH_s simply by using the pK_s for the compound most likely to form. The form of $CaCO_3$ most commonly found in fresh water is calcite. Use the pK_s for calcite unless it is clear that a different form of $CaCO_3$ controls $CaCO_3$ solubility.

Estimate calcium ion concentration from total calcium measurements with Equation 3.

$$[Ca^{2+}] = Ca_t - Ca_{ip} \qquad (3)$$

where:
Ca_t = total calcium, g-moles/L, and
Ca_{ip} = calcium associated with ion pairs such as $CaHCO_3^+$, $CaSO_4^0$, and $CaOH^+$.

Calcium associated with ion pairs is not available to form $CaCO_3$.

Estimate $[HCO_3^-]$, the bicarbonate ion concentration, from Equation 4.

$[HCO_3^-] =$

$$\frac{Alk_t - Alk_o + 10^{(pf_m - pH)} - 10^{(pH + pf_m - pK_w)}}{1 + 0.5 \times 10^{(pH - pK_2)}}$$

(4)

where:

Alk_t = total alkalinity, as determined by acid titration to the carbonic acid end point, g-equivalents/L,

K_w = dissociation constant for water, at the water temperature, and

Alk_o = alkalinity contributed by NH_3^0, $H_3SiO_4^-$, HPO_4^{2-}, $B(OH)_4^-$, CH_3COO^- (acetate), HS^-, and ion pairs such as $CaHCO_3^+$ and $MgOH^+$.

These contributions usually are small compared to the contributions of components normally considered (HCO_3^-, CO_3^{2-}, OH^-, and H^+).

Calculations can be simplified. For example, in Equation 4, terms containing exponents (e.g., $10^{(pH + pf_m - pK_w)}$) usually can be neglected for waters that are approximately neutral (pH 6.0 to 8.5) with alkalinity greater than about 50 mg/L as $CaCO_3$. The terms Ca_{ip} in Equation 3 and Alk_o in Equation 4 are difficult to calculate without computers. Therefore they usually are neglected for hand calculations. The

TABLE 2330:I. ESTIMATING EQUILIBRIUM CONSTANTS AND ACTIVITY COEFFICIENTS

Equation	Temperature Range	References
When complete mineral analysis is available:		
$I = \frac{1}{2} \sum_{i=1}^{i} [X_i] Z_i^2$	—	3
When only conductivity is available:		
$I = 1.6 \times 10^{-5}C$	—	4
When only TDS is available:		
$I = TDS/40\ 000$	—	5
$pf_m = A \dfrac{\sqrt{I}}{1 + \sqrt{I}} - 0.3I$ (valid to $I < 0.5$)	—	3
$A = 1.82 \times 10^6\ (ET)^{-1.5}$	—	3
$E = \dfrac{60\ 954}{T + 116} - 68.937$	—	6
$pK_2 = 107.8871 + 0.032\ 528\ 49T - 5151.79/T - 38.925\ 61 \log_{10}T + 563\ 713.9/T^2$	273–373	7
$pK_w = 4471/T + 0.017\ 06T - 6.0875$	280–338	8
$pK_{sc} = 171.9065 + 0.077\ 993T - 2839.319/T - 71.595 \log_{10}T$	273–363	7
$pK_{sa} = 171.9773 + 0.077\ 993T - 2903.293/T - 71.595 \log_{10}T$	273–363	7
$pK_{sv} = 172.1295 + 0.077\ 993T - 3074.688/T - 71.595 \log_{10}T$	273–363	7

* I = ionic strength
$[X_i]$ = concentration of component i, g-moles/L
Z_i = charge of species i
C = conductivity, μmhos/cm
TDS = total dissolved solids, mg/L
pY = $- \log_{10}$ of the value of any factor Y
f_m = activity coefficient for monovalent species
E = dielectric constant

T = temperature, °K (°C + 273.2)
K_2 = second dissociation constant for carbonic acid
K_w = dissociation constant for water
K_{sc} = solubility product constant for calcite
K_{sa} = solubility product constant for aragonite
K_{sv} = solubility product constant for vaterite

TABLE 2330:II. PRECALCULATED VALUES FOR pK AND A AT SELECTED TEMPERATURES

Temperature °C	pK_2	pK_s			pK_w	A
		Calcite	Aragonite	Vaterite		
5	10.55	8.39	8.24	7.77	14.73	0.494
10	10.49	8.41	8.26	7.80	14.53	0.498
15	10.43	8.43	8.28	7.84	14.34	0.502
20	10.38	8.45	8.31	7.87	14.16	0.506
25*	10.33	8.48	8.34	7.91	13.99	0.511
30	10.29	8.51	8.37	7.96	13.83	0.515
35	10.25	8.54	8.41	8.00	13.68	0.520
40	10.22	8.58	8.45	8.05	13.53	0.526
45	10.20	8.62	8.49	8.10	13.39	0.531
50	10.17	8.66	8.54	8.16	13.26	0.537
60	10.14	8.76	8.64	8.28	13.02	0.549
70	10.13	8.87	8.75	8.40	—	0.562
80	10.13	8.99	8.88	8.55	—	0.576
90	10.14	9.12	9.02	8.70	—	0.591

NOTE: All values determined from the equations of Table 2330:I.
A is used to calculate pf_m (see Table 2330:I).
*pf_m estimated from TDS values at 25°C are as follows:

TDS	pf_m
100	0.024
200	0.033
400	0.044
800	0.060
1000	0.066

simplified version of Equation 2 under such conditions is:

$$pH_s = pK_2 - pK_s + p[Ca] + p[Alk_t] + 5\, pf_m$$

1) Sample calculation—The calculation is best illustrated by working through an example. Assume that calcite controls $CaCO_3$ solubility and determine the SI for a water of the following composition:

Constituent	mg/L $\div$	$\frac{mg}{mole}$ =	g-moles/L
Calcium	152	40 000	3.80×10^{-3}
Magnesium	39	24 312	1.60×10^{-3}
Sodium	50	22 989	2.18×10^{-3}
Potassium	5	39 102	1.28×10^{-4}
Chloride	53	35 453	1.49×10^{-3}
Alkalinity (as $CaCO_3$)	130	50 000	2.60×10^{-3}*
Sulfate	430	96 060	4.48×10^{-3}
Silica (as SiO_2)	15	60 084	2.50×10^{-4}

*g-equivalents.
Water temperature = 20°C (293.2°K); pH = 9.00.

Before evaluating pf_m in Equation 2, determine the ionic strength (I) and another constant (A). Estimate ionic strength from the first equation of Table 2330:I, assuming all the alkalinity is due to bicarbonate ion. Use the alkalinity concentration (2.60×10^{-3}) and the bicarbonate charge (-1) to calculate the contribution of alkalinity to ionic strength. Assume silica is mostly H_4SiO_4. Because H_4SiO_4 has zero charge, silica does not contribute to ionic strength.

$$I = 0.5 \, [4(3.80 \times 10^{-3}) + 4(1.60 \times 10^{-3})$$
$$+ \, 2.18 \times 10^{-3} + 1.28 \times 10^{-4} + 1.49$$
$$\times \, 10^{-3} + 2.60 \times 10^{-3} + 4(4.48 \times 10^{-3})]$$
$$= 2.29 \times 10^{-2} \text{ g-moles/L}$$

In the absence of a complete water analysis, estimate ionic strength from conductivity or total dissolved solids measurements (see alternative equations, Table 2330:I).

Estimate A from the equation in Table 2330:I, after first determining the dielectric constant E from the formula in the same table. Alternatively, use precalculated values of A in Table 2330:II. In Table 2330:II, $A = 0.506$ at 20°C.

Next estimate pf_m from the equation in Table 2330:I:

$$pf_m = 0.506$$

$$\times \left[\frac{\sqrt{2.29 \times 10^{-2}}}{1 + \sqrt{2.29 \times 10^{-2}}} - 0.3 \, (2.29 \times 10^{-2}) \right]$$

$$= 0.063$$

Determine $[HCO_3^-]$ from Equation 4. Neglect Alk_o, but because the pH exceeds 8.5, calculate the other terms. From Table 2330:II, $pK_2 = 10.38$ and $pK_w = 14.16$.

$[HCO_3^-]$

$=$

$$\frac{2.60 \times 10^{-3} + 10^{(0.063 - 9.0)} - 10^{(9.0 + 0.063 - 14.16)}}{1 + 0.5 \times 10^{(9.0 - 10.38)}}$$

$= 2.54 \times 10^{-3}$ g-moles/L

Therefore $p[HCO_3^-] = 2.60$.

Determine $[Ca^{2+}]$ from Equation 3. Neglect Ca_{ip}.

$$[Ca^{2+}] = Ca_t = 3.80 \times 10^{-3} \text{ g-moles/L}$$

Therefore $p[Ca^{2+}] = 2.42$.

From Table 2330:II, pK_s for calcite is 8.45.

Determine pH_s from Equation 2:

$$pH_s = 10.38 - 8.45 + 2.42$$
$$+ \, 2.60 + 5 \, (0.063) = 7.27$$

And finally, determine SI from Equation 1:

$$SI = 9.00 - 7.27 = 1.73$$

The positive SI indicates the water is oversaturated with respect to calcite.

2) Effect of neglecting Ca_{ip} and Alk_o— If Ca_{ip} is neglected, pH_s is underestimated and SI is overestimated by an amount equal to $p(1 - Y_{Ca_{ip}})$, where $Y_{Ca_{ip}}$ is the fraction of total calcium in ion pairs. For example, if $Y_{Ca_{ip}} = 0.30$ then the estimate for SI is 0.15 units too high. Similarly, if Alk_o is neglected, SI is overestimated by an amount equal to $p(1 - Y_{Alk_o})$, where Y_{Alk_o} is the fraction of total alkalinity contributed by species other than HCO_3^-, CO_3^{2-}, OH^-, and H^+. The effects of neglecting Ca_{ip} and Alk_o are additive.

Ca_{ip} and Alk_o may be neglected if the factors $Y_{Ca_{ip}}$ and Y_{Alk_o} are small and do not interfere with interpretation of the SI. The factors are small for waters of low and neutral pH, but they increase as pH values approach and exceed 9. At high pH values, however, the SI is typically much larger than its overestimate, so neglecting Ca_{ip} and Alk_o causes no problem. To return to the example above, when calculations were done with a computerized water chemistry code (SEQUIL) (see Table 2330:III) that considers Ca_{ip} and Alk_o, the SI was 1.48, i.e., 0.25 units lower than the result obtained by hand calculations. In this instance, neglecting Ca_{ip} and Alk_o did not interfere with interpreting the result. Both calculations showed the water to be strongly oversaturated.

The potential for misinterpretation is most acute in nearly-saturated waters of high sulfate concentration. Recirculating cooling water is an example. Calcium is sequestered by the robust $CaSO_4^0$ ion pair and the SI can be overestimated by as much as 0.3 to 0.5 units, even at neutral pH. Under these conditions, the SI may be thought to be zero (neither scale-forming or corrosive) when in fact it is negative.

Resolve this problem by determining pH_s using computerized water chemistry codes that consider ion pairs and the other forms of alkalinity. Section 2330D provides information about water chemistry codes. The most accurate calculations are obtained when a complete mineral analysis is provided.

An alternative but somewhat less rigorous procedure involves direct measurement of (Ca^{2+}), the calcium ion activity, with a calcium specific-ion electrode.[9] Use Equation 5 to determine $p[Ca^{2+}]$; then use $p[Ca^{2+}]$ in Equation 2.

$$p[Ca^{2+}] = p(Ca^{2+}) - 4pf_m$$

This approach eliminates the need to determine Ca_{ip}. However, no equivalent procedure is available to bypass the determination of Alk_o.

b. *Graphical solutions for pH_s:* Caldwell-Lawrence diagrams can be used to determine pH_s.[10–12] The diagrams are particularly useful for estimating chemical dosages needed to achieve desired water conditions. Consult the references for descriptions of how to use the diagrams; see Section 2330D for additional information about the diagrams.

3. Saturation Index by Experimental Determination

a. *Saturometry:* Saturometers were developed to measure the relative saturation of seawater with respect to $CaCO_3$. A water of known calcium and pH is equilibrated with $CaCO_3$ in a sealed flask containing a pH electrode. The water temperature is controlled with a constant-temperature bath. During equilibration the pH decreases if $CaCO_3$ precipitates and increases if $CaCO_3$ dissolves. When the pH stops changing, equilibrium is said to have been achieved. The initial pH and calcium values and the final pH value are used to calculate the relative saturation (RS).[13] Equation 6, Section 2330D, may then be used to determine SI.

A major advantage of this method is that the approach to equilibrium can be tracked by measuring pH, thus minimizing uncertainty about the achievement of equilibrium. The method is most sensitive in the range of minimum buffering intensity (pH 7.5 to 8.5). The calculations do not consider ion pairs or noncarbonate alkalinity, except borate. The technique has been used for *in situ* oceanographic measurements[14] as well as in the laboratory.

The saturometry calculations discussed above use K_s of the $CaCO_3$ phase assumed to control solubility. Uncertainties occur if the identity of the controlling solid is unknown. Resolve these uncertainties by measuring K_s of the controlling solid. It is equal to the $CaCO_3$ activity product, $[Ca^{2+}] \times [CO_3^{2-}]$, at equilibrium. Calculate the latter from the equilibrium pH and initial calcium, alkalinity, and pH measurements.[15]

b. *Alkalinity difference technique:*[16] SI also can be determined by equilibrating water of known pH, calcium, and alkalinity with $CaCO_3$ in a sealed, constant-temperature system. The $CaCO_3$ activity product before equilibration is determined from initial calcium, pH, and alkalinity (or total carbonate) values. The $CaCO_3$ solubility product constant (K_s) equals the $CaCO_3$ activity product after equilibration, which is determined by using the alkalinity change that occurred during equilibration. *RS* is found by dividing the initial activity product by K_s. Calculate *SI* by using Equation 6. The advantage of this method is that it makes no assumptions about the identity of the $CaCO_3$ phase. However it is more difficult to determine when equilibrium is achieved with this method than with the saturometry method.

Whatever the method used, use the temperatures that are the same as the temperature of the water source. Alternatively, correct test results to the temperature of the water source.[16]

4. References

1. RYZNAR, J.W. 1944. A new index for determining the amount of calcium carbonate formed by water. *J. Amer. Water Works Assoc.* 36:47.

2. SNOEYINK, V.L. & D. JENKINS. 1980. Water Chemistry. John Wiley & Sons, New York, N.Y.

3. STUMM, W. & J.J. MORGAN. 1981. Aquatic Chemistry, 2nd ed. John Wiley & Sons, New York, N.Y.

4. RUSSELL, L.L. 1976. Chemical Aspects of Groundwater Recharge with Wastewaters. Ph.D. thesis, Univ. California, Berkeley.

5. LANGELIER, W.F. 1936. The analytical control of anticorrosion water treatment. *J. Amer. Water Works Assoc.* 28:1500.

6. ROSSUM, J.R. & D.T. MERRILL. 1983. An evaluation of the calcium carbonate saturation indexes. *J. Amer. Water Works Assoc.* 75:95.

7. PLUMMER, L.N. & E. BUSENBERG. 1982. The solubilities of calcite, aragonite, and vaterite in CO_2-H_2O solutions between 0 and 90 degrees C, and an evaluation of the aqueous model for the system $CaCO_3$-CO_2-H_2O. *Geochim. Cosmochim. Acta* 46:1011.

8. ELECTRIC POWER RESEARCH INSTITUTE. 1982. Design and Operating Guidelines Manual for Cooling Water Treatment: Treatment of Recirculating Cooling Water. Section 4. Process Model Documentation and User's Manual. EPRI CS-2276, Electric Power Research Inst., Palo Alto, Calif.

9. GARRELS, R.M. & C.L. CHRIST. 1965. Solutions, Minerals, and Equilibria. Harper & Row, New York, N.Y.

10. MERRILL, D.T. & R.L. SANKS. 1978, 1979. Corrosion control by deposition of $CaCO_3$ films: A practical approach for plant operators. *J. Amer. Water Works Assoc.* 70:592; 70:634 & 71:12.

11. LOEWENTHAL, R.E. & G.v.R. MARAIS. 1976. Carbonate Chemistry of Aquatic Systems: Theory and Applications. Ann Arbor Science Publishers, Ann Arbor, Mich.

12. MERRILL, D.T. 1976. Chemical conditioning for water softening and corrosion control. *In* R.L. Sanks, ed., Water Treatment Plant Design. Ann Arbor Science Publishers, Ann Arbor, Mich.

13. BEN-YAAKOV, S. & I.R. KAPLAN. 1969. Determination of carbonate saturation of seawater with a carbonate saturometer. *Limnol. Oceanogr.* 14:874.

14. BEN-YAAKOV, S. & I.R. KAPLAN. 1971. Deep-sea in situ calcium carbonate saturometry. *J. Geophys. Res.* 76:72.

15. PLATH, D.C., K.S. JOHNSON & R.M. PYTKOWICZ. 1980. The solubility of calcite—probably containing magnesium—in seawater. *Mar. Chem.* 10:9.

16. BALZAR, W. 1980. Calcium carbonate saturometry by alkalinity difference. *Oceanol. Acta.* 3:237.

2330 C. Indices Predicting the Quantity of $CaCO_3$ That Can Be Precipitated or Dissolved

The Calcium Carbonate Precipitation Potential (CCPP) predicts both tendency to precipitate or to dissolve $CaCO_3$ and quantity that may be precipitated or dissolved. The CCPP also is known by other names, e.g., Calcium Carbonate Precipitation Capacity (CCPC).

The CCPP is defined as the quantity of

CaCO₃ that theoretically can be precipitated from oversaturated waters or dissolved by undersaturated waters during equilibration.[1] The amount that actually precipitates or dissolves may be less, because equilibrium may not be achieved. The CCPP is negative for undersaturated waters, zero for saturated waters, and positive for oversaturated waters.

1. Calculating CCPP

The CCPP does not lend itself to hand calculations. Preferably calculate CCPP with computerized water chemistry models and Caldwell-Lawrence diagrams (see Section 2330D).

The most reliable calculations consider ion pairs and the contribution to alkalinity of other species besides HCO_3^-, CO_3^{2-}, OH^-, and H^+. Models that do not consider these factors overestimate the amount of $CaCO_3$ that can be precipitated and underestimate the amount of $CaCO_3$ that can be dissolved.

2. Experimental Determination of CCPP

Estimate CCPP by one of several experimental techniques.

a. Saturometry: See Section 2330B. The CCPP is determined as part of the RS calculation.

b. Alkalinity difference technique: See Section 2330B. The CCPP equals the difference between alkalinity (or calcium) values of the initial and equilibrated water, when they are expressed as $CaCO_3$.

c. Marble test: The marble test[1-5] is similar to the alkalinity difference technique. The CCPP equals the change in alkalinity (or calcium) values during equilibration, when they are expressed as $CaCO_3$.

d. Enslow test: The Enslow test[5] is a continuous version of the alkalinity difference or marble tests. Water is fed continuously to a leveling bulb or separatory funnel partly filled with $CaCO_3$. The effluent from this device is filtered through crushed mar-

ble so that the filtrate is assumed to be in equilibrium with $CaCO_3$. The CCPP equals the change in alkalinity (or calcium) values that occurs during passage through the apparatus.

e. Calcium carbonate deposition test:[6] The calcium carbonate deposition test (CCDT) is an electrochemical method that measures the electric current produced when dissolved oxygen is reduced on a rotating electrode. When an oversaturated water is placed in the apparatus, $CaCO_3$ deposits on the electrode. The deposits interfere with oxygen transfer and the current diminishes. The rate of $CaCO_3$ deposition is directly proportional to the rate at which the current declines. The CCDT and the CCPP are related, but they are not the same. The CCDT is a rate, and the CCPP is a quantity.

For realistic assessments of the CCPP (or CCDT) keep test temperature the same as the temperature of the water source. Alternatively, correct test results to the temperature of the water source.

3. References

1. MERRILL, D.T. & R.L. SANKS. 1978, 1979. Corrosion control by deposition of CaCO₃ films: A practical approach for plant operators. *J. Amer. Water Works Assoc.* 70:592; 70:634 & 71:12.

2. MERRILL, D.T. 1976. Chemical conditioning for water softening and corrosion control. *In* R.L. Sanks, ed. Water Treatment Plant Design. Ann Arbor Science Publishers, Ann Arbor, Mich.

3. DE MARTINI, F. 1938. Corrosion and the Langelier calcium carbonate saturation index. *J. Amer. Water Works Assoc.* 30:85.

4. HOOVER, C.P. 1938. Practical application of the Langelier method. *J. Amer. Water Works Assoc.* 30:1802.

5. DYE, J.F. & J.L. TUEPKER. 1971. Chemistry of the Lime-Soda Process. *In* American Water Works Association. Water Quality and Treatment. McGraw-Hill Book Co., New York, N.Y.

6. MCCLELLAND, N.I. & K.H. MANCY. 1979. CCDT bests Ryzner index as pipe CaCO₃ film predictor. *Water Sewage Works* 126:77.

2330 D.   Diagrams and Computer Codes for CaCO$_3$ Indices

1. Description

Table 2330:III lists diagrams and computer codes that can be used to determine the SI and CCPP. It also provides a brief description of their characteristics.

Many computer codes do not calculate *SI* directly, but instead calculate the relative saturation (*RS*). When *RS* data are presented, calculate the *SI* from:[1]

$$SI = \log_{10} RS$$

where:

　　RS = ratio of CaCO$_3$ activity product to CaCO$_3$ solubility product constant.

The diagrams and a few of the codes define pH$_s$ as the pH the water would exhibit if it were in equilibrium with CaCO$_3$ at the existing calcium and total alkalinity concentrations.[2] This definition of pH$_s$ differs from the definition following Equation 1 because alkalinity is used instead of bicarbonate. Within the pH range 6 to 9, alkalinity-based pH$_s$ and bicarbonate-based pH$_s$ are virtually equal, because total alkalinity is due almost entirely to bicarbonate ion. Above pH 9 they differ and Equation 6 no longer applies if *SI* is calculated with alkalinity-based pH$_s$. However, if *SI* is determined from bicarbonate-based pH$_s$, Equation 6 continues to apply.

Furthermore, calculating *SI* with alkalinity-based pH$_s$ reverses the sign of the *SI* above pH values of approximately pK$_2$, i.e., a positive, not the usual negative, *SI* connotes an undersaturated water.[3] With bicarbonate-based pH$_s$ or RS, sign reversal does not occur, thereby eliminating the confusing sign change. For these reasons, bicarbonate-based pH$_s$ or RS is preferred.

Table 2330:III lists the definition of pH$_s$ used for each code.

Some models calculate only the amount of CaCO$_3$ that can be precipitated but not the amount of CaCO$_3$ that can be dissolved. Others calculate both.

The diagrams and codes can be used to determine many more parameters than the CaCO$_3$ saturation indices. A fee may be charged for computer software or graphs. The information in Table 2330:III describes parameters each code uses to calculate *SI*. Contact the sources listed below the table for current information.

2. References

1. SNOEYINK, V.L. & D. JENKINS. 1980. Water Chemistry. John Wiley & Sons, New York, N.Y.

2. LANGELIER, W.F. 1936. The analytical control of anticorrosion water treatment. *J. Amer. Water Works Assoc.* 28:1500.

3. LOEWENTHAL, R.E. & G.v.R. MARAIS. 1976. Carbonate Chemistry of Aquatic Systems: Theory and Applications. Ann Arbor Science Publishers, Ann Arbor, Mich.

4. MERRILL, D.T. & R.L. SANKS. 1978. Corrosion Control by Deposition of CaCO$_3$ Films: A Handbook of Practical Application and Instruction. American Water Works Assoc., Denver, Colo.

5. MERRILL, D.T. 1976. Chemical conditioning for water softening and corrosion control. *In* R.L. Sanks, ed., Water Treatment Plant Design. Ann Arbor Science Publishers, Ann Arbor, Mich.

6. ELECTRIC POWER RESEARCH INSTITUTE. 1982. Design and Operating Guidelines Manual for Cooling Water Treatment: Treatment of Recirculating Cooling Water. Section 4. Process Model Documentation and User's Manual. EPRI CS-2276, Electric Power Research Inst., Palo Alto, Calif.

TABLE 2330:III. GRAPHS AND COMPUTER SOFTWARE THAT CAN BE USED TO CALCULATE $CaCO_3$ SATURATION INDICES*

Item†	$CaCO_3$ Indices		Approximate Temperature Range °C	Approximate Limit of Ionic Strength	Ion Pairs Considered?	Alk_o Considered?‡	Minimum Equipment Required
	Basis for Calculation of SI	CCPP					
1. Caldwell-Lawrence diagrams	pH_{sa}	P, D	2–25	0.030	No	No	Diagrams
2. ACAPP	RS	P, D	−10–110	6+	Yes	Yes	IBM-compatible PC, 512K bytes of RAM, MS DOS or PC DOS v.2.1 or higher
3. DRIVER	RS	P	7–65	2.5	Yes	Yes	Mainframe computer
4. INDEX C	$pH_{sa'}$ pH_{sb}	P, D	0–50	0.5	No	No	Hewlett-Packard 41C calculator, with three memory modules
5. LEQUIL	RS	No	5–90	0.5	Yes	Yes	IBM-compatible PC, 256K RAM, Lotus 1-2-3 or work-alike, PC DOS or MS DOS v.2.0 or higher
6. MINTEQA1	RS	P, D	0–100	0.5	Yes	Yes	IBM-compatible PC, 512K bytes of RAM, PC DOS v.3.0 or higher, 10 megabyte hard disk drive, math coprocessor useful but not required
7. PHREEQE Standard	RS	P, D	0–100	0.5	Yes	Yes	IBM-compatible PC, known to work with 512K RAM, PC DOS or MS DOS v.2.11 or higher. Also available for mainframe computers.
For high-salinity waters	RS	P, D	0–80	7–8	Yes	Yes	IBM-compatible PC, 640K RAM recommended, with math coprocessor, MS DOS v.3.2 or higher
8. SEQUIL	RS	P, D	7–65	2.5	Yes	Yes	IBM-compatible PC, 512K bytes of RAM, MS DOS or PC DOS v.2.1 or higher.
9. SOLMINEQ.88	RS	P, D	0–350	6	Yes	Yes	IBM-compatible PC, 640K RAM, math coprocessor, PC

TABLE 2330:III, CONT.

Item[†]	CaCO₃ Indices		Approximate Temperature Range °C	Approximate Limit of Ionic Strength	Ion Pairs Considered?	Alk₀ Considered?[‡]	Minimum Equipment Required
	Basis for Calculation of SI	CCPP					
							DOS or MS DOS v.3.0 or higher. Also available for mainframe computer.
10. WTRCHEM	pH$_{sa}$	P, D	0–100	0.5	No	No	Any PC equipped with a BASIC interpreter, 5K RAM.
11. WATEQ4F	RS	No	0–100	0.5	Yes	Yes	IBM-compatible PC, known to work with 512K RAM, PC DOS or MS DOS v.2.11 or higher.

*SI = saturation index
CCPP = CaCO₃ precipitation potential
pH$_{sa}$ = alkalinity-based pH$_s$
pH$_{sb}$ = bicarbonate-based pH$_s$
P = calculates amount of CaCO₃ theoretically precipitated

D = calculates amount of CaCO₃ theoretically dissolved
RS = relative saturation
PC = personal computer
RAM = random access memory

† 1. American Water Works Association, 6666 West Quincy Ave., Denver, Colo. 80235 provides 30.5- by 38.1-cm diagrams (order number 20204), and documentation[4] (order number 20203); Loewenthal and Marais[3] provide 10.2- by 11.4-cm diagrams, with documentation; Merrill[5] provides 10.2- by 16.5-cm diagrams, with documentation.
  2. Radian Corp., 8501 MoPac Blvd., P.O. Box 201088, Austin, Texas 78720-1088 Attn: J.G. Noblett (software and documentation).
  3. Power Computing Co., 1930 Hi Line Dr., Dallas, Texas, 74207 (software and documentation[6]).
  4. Brown and Caldwell, P.O. Box 8045, Walnut Creek, Calif. 94596-1220 Attn: D.T. Merrill (software and documentation).
  5. Illinois State Water Survey, Aquatic Chemistry Section, 2204 Griffith Dr., Champaign, Ill. 61820-7495 Attn: T.R. Holm (software and documentation).
  6. Center for Exposure Assessment Modeling, Environmental Research Laboratory, Office of Research and Development. U.S. Environmental Protection Agency, Athens, Ga. 30613 (software and documentation[7]).
  7. U.S. Geological Survey, National Center, MS 437, Reston, Va. 22902, Chief of WATSTORE Program (provides software for mainframe version of standard code); U.S. Geological Survey, Water Resources Division, MS 420, 345 Middlefield Rd., Menlo Park, Calif. 94025 Attn: J.W. Ball (provides software for personal computer version of standard code); National Water Research Institute, Canada Centre for Inland Waters, 867 Lakeshore Rd., Burlington, Ont., Canada L7R 4A6 Attn: A.S. Crowe (provides software and documentation [8,9] for personal computer versions of both standard and high-salinity codes); U.S. Geological Survey, 215 Dean McGee Ave., Oklahoma City, Okla. 73102, attention: D. L. Parkhurst (provides documentation[8,10] for mainframe and personal computer versions of standard code).
  8. Power Computing Company, 1930 Hi Line Dr., Dallas, Texas 74207 (software and documentation[11]).
  9. U.S. Geological Survey, Water Resources Division, MS 427, 345 Middlefield Rd., Menlo Park, Calif. 94025 Attn: Y.K. Kharaka (software and documentation[12]).
  10. Public Domain Software, 20760 Walsh Ave., Santa Clara, Calif. 95090 (code listing and documentation).
  11. U.S. Geological Survey, Water Resources Division, MS 420, 345 Middlefield Rd., Menlo Park, Calif. 94025 Attn: J.W. Ball (software and documentation[13]).
‡Codes differ in the species included in Alk₀.

7. BROWN, D.S. & J.D. ALLISON. 1987.
MINTEQA1 Equilibrium Metal Speciation
Model: A User's Manual. EPA-600/3-87-012.
8. PARKHURST, D.L., D.C. THORSTENSON &
L.N. PLUMMER. 1980. PHREEQE—A Computer Program for Geochemical Calculations.
USGS WRI 80-96 NTIS PB 81-167801.
9. CROWE, A.S. & F.J. LONGSTAFF. 1987. Extension of Geochemical Modeling Techniques
to Brines: Coupling of the Pitzer Equations to
PHREEQE. In Proceedings of Solving
Groundwater Problems With Models, National Water Well Assoc., Denver, Colo. Feb.
10–12, 1987.
10. FLEMING, G.W. & L.N. PLUMMER. 1983.
PHRQINPT—An Interactive Computer Program for Constructing Input Data Sets to the
Geochemical Simulation Program

PHREEQE. USGS WRI 83-4236.
11. ELECTRIC POWER RESEARCH INSTITUTE. SE-
QUIL—An Inorganic Aqueous Chemical
Equilibrium Code for Personal Computers;
Volume 1, User's Manual/Workbook, Version
1.0, GS-6234-CCML. Electric Power Re-
search Inst., Palo Alto, Calif.
12. KHARAKA, Y.K., W.D. GUNTER, P. K. AG-
GARWAL, E.H. PERKINS & J.D. DEBRAAL.
1988. Solmineq.88—A Computer Program for
Geochemical Modeling of Water–Rock Inter-
actions. USGS WRIR 88-4227, Menlo Park,
Calif.
13. BALL, J.W., D.K. NORDSTROM & D.W.
ZACHMANN. 1987. WATEQ4F—A Personal
Computer Fortran Translation of the Geo-
chemical Model WATEQ2 With Revised
Data Base. USGS OFR 87-50.

2340　HARDNESS*

2340 A.　Introduction

1. Definition

Originally, water hardness was understood to be a measure of the capacity of
water to precipitate soap. Soap is precipitated chiefly by the calcium and magnesium ions present. Other polyvalent cations
also may precipitate soap, but they often
are in complex forms, frequently with organic constituents, and their role in water
hardness may be minimal and difficult to
define. In conformity with current practice,
total hardness is defined as the sum of the
calcium and magnesium concentrations,
both expressed as calcium carbonate, in
milligrams per liter.

When hardness numerically is greater
than the sum of carbonate and bicarbonate

alkalinity, that amount of hardness equivalent to the total alkalinity is called "carbonate hardness"; the amount of hardness
in excess of this is called "noncarbonate
hardness." When the hardness numerically
is equal to or less than the sum of carbonate
and bicarbonate alkalinity, all hardness is
carbonate hardness and noncarbonate
hardness is absent. The hardness may range
from zero to hundreds of milligrams per
liter, depending on the source and treatment to which the water has been subjected.

2. Selection of Method

Two methods are presented. Method B,
hardness by calculation, is applicable to all
waters and yields the higher accuracy. If a
mineral analysis is performed, hardness by

_____
*Approved by Standard Methods Committee, 1985.

calculation can be reported. Method C, the EDTA titration method, measures the calcium and magnesium ions and may be applied with appropriate modification to any kind of water. The procedure described affords a means of rapid analysis.

3. Reporting Results

When reporting hardness, state the method used, for example, "hardness (calc.)" or "hardness (EDTA)."

2340 B.   Hardness by Calculation

1. Discussion

The preferred method for determining hardness is to compute it from the results of separate determinations of calcium and magnesium.

2. Calculation

Hardness, mg equivalent $CaCO_3/L$
$$= 2.497 \text{ [Ca, mg/L]} + 4.118 \text{ [Mg, mg/L]}$$

2340 C.   EDTA Titrimetric Method

1. General Discussion

a. *Principle:* Ethylenediaminetetraacetic acid and its sodium salts (abbreviated EDTA) form a chelated soluble complex when added to a solution of certain metal cations. If a small amount of a dye such as Eriochrome Black T or Calmagite is added to an aqueous solution containing calcium and magnesium ions at a pH of 10.0 ± 0.1, the solution becomes wine red. If EDTA is added as a titrant, the calcium and magnesium will be complexed, and when all of the magnesium and calcium has been complexed the solution turns from wine red to blue, marking the end point of the titration. Magnesium ion must be present to yield a satisfactory end point. To insure this, a small amount of complexometrically neutral magnesium salt of EDTA is added to the buffer; this automatically introduces sufficient magnesium and obviates the need for a blank correction.

The sharpness of the end point increases with increasing pH. However, the pH cannot be increased indefinitely because of the danger of precipitating calcium carbonate, $CaCO_3$, or magnesium hydroxide, $Mg(OH)_2$, and because the dye changes color at high pH values. The specified pH of 10.0 ± 0.1 is a satisfactory compromise. A limit of 5 min is set for the duration of the titration to minimize the tendency toward $CaCO_3$ precipitation.

b. *Interference:* Some metal ions interfere by causing fading or indistinct end points or by stoichiometric consumption of EDTA. Reduce this interference by adding certain inhibitors before titration. Mg-CDTA [see 2b3)], selectively complexes heavy metals, releases magnesium into the sample, and may be used as a substitute for toxic or malodorous inhibitors. It is useful only when the magnesium substituted for heavy metals does not contribute significantly to the total hardness. With heavy metal or polyphosphate concentrations below those indicated in Table 2340:I, use Inhibitor I or II. When higher concen-

TABLE 2340:I. MAXIMUM CONCENTRATIONS OF INTERFERENCES PERMISSIBLE WITH VARIOUS INHIBITORS*

Interfering Substance	Max. Interference Concentration mg/L	
	Inhibitor I	Inhibitor II
Aluminum	20	20
Barium	†	†
Cadmium	†	20
Cobalt	over 20	0.3
Copper	over 30	20
Iron	over 30	5
Lead	†	20
Manganese (Mn^{2+})	†	1
Nickel	over 20	0.3
Strontium	†	†
Zinc	†	200
Polyphosphate		10

* Based on 25-mL sample diluted to 50 mL.
† Titrates as hardness.

trations of heavy metals are present, determine calcium and magnesium by a non-EDTA method (see Sections 3500-Ca and 3500-Mg) and obtain hardness by calculation. The figures in Table 2340:I are intended as a rough guide only and are based on using a 25-mL sample diluted to 50 mL.

Suspended or colloidal organic matter also may interfere with the end point. Eliminate this interference by evaporating the sample to dryness on a steam bath and heating in a muffle furnace at 550°C until the organic matter is completely oxidized. Dissolve the residue in 20 mL 1N hydrochloric acid (HCl), neutralize to pH 7 with 1N sodium hydroxide (NaOH), and make up to 50 mL with distilled water; cool to room temperature and continue according to the general procedure.

c. Titration precautions: Conduct titrations at or near normal room temperature. The color change becomes impractically slow as the sample approaches freezing temperature. Indicator decomposition becomes a problem in hot water.

The specified pH may produce an environment conducive to CaCO$_3$ precipitation. Although the titrant slowly redissolves such precipitates, a drifting end point often yields low results. Completion of the titration within 5 min minimizes the tendency for CaCO$_3$ to precipitate. The following three methods also reduce precipitation loss:

1) Dilute sample with distilled water to reduce CaCO$_3$ concentration. This simple expedient has been incorporated in the procedure. If precipitation occurs at this dilution of 1 + 1 use modification 2) or 3). Using too small a sample contributes a systematic error due to the buret-reading error.

2) If the approximate hardness is known or is determined by a preliminary titration, add 90% or more of titrant to sample *before* adjusting pH with buffer.

3) Acidify sample and stir for 2 min to expel CO$_2$ *before* pH adjustment. Determine alkalinity to indicate amount of acid to be added.

2. Reagents

a. Buffer solution:

1) Dissolve 16.9 g ammonium chloride (NH$_4$Cl) in 143 mL conc ammonium hydroxide (NH$_4$OH). Add 1.25 g magnesium salt of EDTA (available commercially) and dilute to 250 mL with distilled water.

2) If the magnesium salt of EDTA is unavailable, dissolve 1.179 g disodium salt of ethylenediaminetetraacetic acid dihydrate (analytical reagent grade) and 780 mg magnesium sulfate (MgSO$_4$·7H$_2$O) or 644 mg magnesium chloride (MgCl$_2$·6H$_2$O) in 50 mL distilled water. Add this solution to 16.9 g NH$_4$Cl and 143 mL conc NH$_4$OH with mixing and dilute to 250 mL with distilled water. To attain the highest accuracy, adjust to exact equivalence through

appropriate addition of a small amount of EDTA or $MgSO_4$ or $MgCl_2$.

Store Solution 1) or 2) in a plastic or borosilicate glass container for no longer than 1 month. Stopper tightly to prevent loss of ammonia (NH_3) or pickup of carbon dioxide (CO_2). Dispense buffer solution by means of a bulb-operated pipet. Discard buffer when 1 or 2 mL added to the sample fails to produce a pH of 10.0 ± 0.1 at the titration end point.

3) Satisfactory alternate "odorless buffers" also are available commercially. They contain the magnesium salt of EDTA and have the advantage of being relatively odorless and more stable than the NH_4Cl-NH_4OH buffer. They usually do not provide as good an end point as NH_4Cl-NH_4OH because of slower reactions and they may be unsuitable when this method is automated. Prepare one of these buffers by mixing 55 mL conc HCl with 400 mL distilled water and then, slowly and with stirring, adding 300 mL 2-aminoethanol (free of aluminum and heavier metals). Add 5.0 g magnesium salt of EDTA and dilute to 1 L with distilled water.

b. *Complexing agents:* For most waters no complexing agent is needed. Occasionally water containing interfering ions requires adding an appropriate complexing agent to give a clear, sharp change in color at the end point. The following are satisfactory:

1) *Inhibitor I:* Adjust acid samples to pH 6 or higher with buffer or $0.1N$ NaOH. Add 250 mg sodium cyanide (NaCN) in powder form. Add sufficient buffer to adjust to pH 10.0 ± 0.1 (CAUTION: *NaCN is extremely poisonous. Take extra precautions in its use.* Flush solutions containing this inhibitor down the drain with large quantities of water after insuring that no acid is present to liberate volatile poisonous hydrogen cyanide.)

2) *Inhibitor II:* Dissolve 5.0 g sodium sulfide nonahydrate ($Na_2S \cdot 9H_2O$) or 3.7 g $Na_2S \cdot 5H_2O$ in 100 mL distilled water. Ex-

clude air with a tightly fitting rubber stopper. This inhibitor deteriorates through air oxidation. It produces a sulfide precipitate that obscures the end point when appreciable concentrations of heavy metals are present. Use 1 mL in ¶ 3b below.

3) *MgCDTA:* Magnesium salt of 1, 2-cyclohexanediaminetetraacetic acid. Add 250 mg per 100 mL sample and dissolve completely before adding buffer solution. Use this complexing agent to avoid using toxic or odorous inhibitors when interfering substances are present in concentrations that affect the end point but will not contribute significantly to the hardness value.

Commercial preparations incorporating a buffer and a complexing agent are available. Such mixtures must maintain pH 10.0 ± 0.1 during titration and give a clear, sharp end point when the sample is titrated.

c. *Indicators:* Many types of indicator solutions have been advocated and may be used if the analyst demonstrates that they yield accurate values. The prime difficulty with indicator solutions is deterioration with aging, giving indistinct end points. For example, alkaline solutions of Eriochrome Black T are sensitive to oxidants and aqueous or alcoholic solutions are unstable. In general, use the least amount of indicator providing a sharp end point. It is the analyst's responsibility to determine individually the optimal indicator concentration.

1) *Eriochrome Black T:* Sodium salt of 1-(1-hydroxy-2-naphthylazo)-5-nitro-2-naphthol-4-sulfonic acid; No. 203 in the Color Index. Dissolve 0.5 g dye in 100 g 2,2′,2″-nitrilotriethanol (also called triethanolamine) or 2-methoxymethanol (also called ethylene glycol monomethyl ether). Add 2 drops per 50 mL solution to be titrated. Adjust volume if necessary.

2) *Calmagite:* 1-(1-hydroxy-4-methyl-2-phenylazo)-2-naphthol-4-sulfonic acid. This is stable in aqueous solution and produces the same color change as Eriochrome

Black T, with a sharper end point. Dissolve 0.10 g Calmagite in 100 mL distilled water. Use 1 mL per 50 mL solution to be titrated. Adjust volume if necessary.

3) Indicators 1 and 2 can be used in dry powder form if care is taken to avoid excess indicator. Prepared dry mixtures of these indicators and an inert salt are available commercially.

If the end point color change of these indicators is not clear and sharp, it usually means that an appropriate complexing agent is required. If NaCN inhibitor does not sharpen the end point, the indicator probably is at fault.

d. Standard EDTA titrant, 0.01M: Weigh 3.723 g analytical reagent-grade disodium ethylenediaminetetraacetate dihydrate, also called (ethylenedinitrilo) tetraacetic acid disodium salt (EDTA), dissolve in distilled water, and dilute to 1000 mL. Standardize against standard calcium solution (¶ 2e) as described in ¶ 3b below.

Because the titrant extracts hardness-producing cations from soft-glass containers, store in polyethylene (preferable) or borosilicate glass bottles. Compensate for gradual deterioration by periodic restandardization and by using a suitable correction factor.

e. Standard calcium solution: Weigh 1.000 g anhydrous $CaCO_3$ powder (primary standard or special reagent low in heavy metals, alkalis, and magnesium) into a 500-mL erlenmeyer flask. Place a funnel in the flask neck and add, a little at a time, 1 + 1 HCl until all $CaCO_3$ has dissolved. Add 200 mL distilled water and boil for a few minutes to expel CO_2. Cool, add a few drops of methyl red indicator, and adjust to the intermediate orange color by adding $3N$ NH_4OH or 1 + 1 HCl, as required. Transfer quantitatively and dilute to 1000 mL with distilled water; 1 mL = 1.00 mg $CaCO_3$.

f. Sodium hydroxide, NaOH, 0.1N.

3. Procedure

a. Pretreatment of polluted water and wastewater samples: Use nitric acid-sulfuric acid or nitric acid-perchloric acid digestion (Section 3030).

b. Titration of sample: Select a sample volume that requires less than 15 mL EDTA titrant and complete titration within 5 min, measured from time of buffer addition.

Dilute 25.0 mL sample to about 50 mL with distilled water in a porcelain casserole or other suitable vessel. Add 1 to 2 mL buffer solution. Usually 1 mL will be sufficient to give a pH of 10.0 to 10.1. The absence of a sharp end-point color change in the titration usually means that an inhibitor must be added at this point (¶ 2b et seq.) or that the indicator has deteriorated.

Add 1 to 2 drops indicator solution or an appropriate amount of dry-powder indicator formulation [¶ 2c3)]. Add standard EDTA titrant slowly, with continuous stirring, until the last reddish tinge disappears. Add the last few drops at 3- to 5-s intervals. At the end point the solution normally is blue. Daylight or a daylight fluorescent lamp is recommended highly because ordinary incandescent lights tend to produce a reddish tinge in the blue at the end point.

If sufficient sample is available and interference is absent, improve accuracy by increasing sample size, as described in ¶ 3c below.

c. Low-hardness sample: For ion-exchanger effluent or other softened water and natural waters of low hardness (less than 5 mg/L), take a larger sample, 100 to 1000 mL, for titration and add proportionately larger amounts of buffer, inhibitor, and indicator. Add standard EDTA titrant slowly from a microburet and run a blank, using redistilled, distilled, or deionized water of the same volume as the sample, to which identical amounts of buffer, inhibitor, and indicator have been added.

Subtract volume of EDTA used for blank from volume of EDTA used for sample.

4. Calculation

Hardness (EDTA) as mg $CaCO_3/L$

$$= \frac{A \times B \times 1000}{mL \text{ sample}}$$

where:

A = mL titration for sample and
B = mg $CaCO_3$ equivalent to 1.00 mL EDTA titrant.

5. Precision and Bias

A synthetic sample containing 610 mg/L total hardness as $CaCO_3$ contributed by 108 mg Ca/L and 82 mg Mg/L, and the following supplementary substances: 3.1 mg K/L, 19.9 mg Na/L, 241 mg Cl^-/L, 0.25 mg NO_2^--N/L, 1.1 mg NO_3^--N/L, 259 mg SO_4^{2-}/L, and 42.5 mg total alkalinity/L (contributed by $NaHCO_3$) in distilled water was analyzed in 56 laboratories by the EDTA titrimetric method with a relative standard deviation of 2.9% and a relative error of 0.8%.

6. Bibliography

CONNORS, J.J. 1950. Advances in chemical and colorimetric methods. *J. Amer. Water Works Assoc.* 42:33.

DIEHL, H., C.A. GOETZ & C.C. HACH. 1950. The versenate titration for total hardness. *J. Amer. Water Works Assoc.* 42:40.

BETZ, J.D. & C.A. NOLL. 1950. Total hardness determination by direct colorimetric titration. *J. Amer. Water Works Assoc.* 42:49.

GOETZ, C.A., T.C. LOOMIS & H. DIEHL. 1950. Total hardness in water: The stability of standard disodium dihydrogen ethylenediaminetetraacetate solutions. *Anal. Chem.* 22:798.

DISKANT, E.M. 1952. Stable indicator solutions for complexometric determination of total hardness in water. *Anal. Chem.* 24:1856.

BARNARD, A.J., JR., W.C. BROAD & H. FLASCHKA. 1956 & 1957. The EDTA titration. *Chemist Analyst* 45:86 & 46:46.

GOETZ, C.A. & R.C. SMITH. 1959. Evaluation of various methods and reagents for total hardness and calcium hardness in water. *Iowa State J. Sci.* 34:81 (Aug. 15).

SCHWARZENBACH, G. & H. FLASCHKA. 1969. Complexometric Titrations, 2nd ed. Barnes & Noble, Inc., New York, N.Y.

2510 CONDUCTIVITY*

2510 A. Introduction

Conductivity is a numerical expression of the ability of an aqueous solution to carry an electric current. This ability depends on the presence of ions, their total concentration, mobility, valence, and relative concentrations, and on the temperature of measurement. Solutions of most inorganic acids, bases, and salts are relatively good conductors. Conversely, molecules of organic compounds that do not dissociate in aqueous solution conduct a current very poorly, if at all.

The physical measurement made in a laboratory determination of conductivity is usually of resistance, measured in ohms or megohms. The resistance of a conductor is inversely proportional to its cross-sectional area and directly proportional to its length.

---

* Approved by Standard Methods Committee, 1988.

The magnitude of the resistance measured in an aqueous solution therefore depends on the characteristics of the conductivity cell used, and is not meaningful without knowledge of these characteristics. Specific resistance is the resistance of a cube 1 cm on an edge. In aqueous solutions such a measurement is rare because of the difficulties of electrode fabrication. Practical electrodes measure a given fraction of the specific resistance, the fraction being the cell constant, C:

$$C = \frac{\text{Measured resistance, } R_m}{\text{Specific resistance, } R_s}$$

The reciprocal of resistance is conductance. It measures the ability to conduct a current and is expressed in reciprocal ohms or mhos. A more convenient unit in water analysis is micromhos. When the cell constant is known and applied, the measured conductance is converted to the specific conductance or conductivity, K_s, the reciprocal of the specific resistance:

$$K_s = \frac{1}{R_s} = \frac{C}{R_m}$$

The term "conductivity" is preferred and customarily is reported in micromhos per centimeter (μmhos/cm). In the International System of Units (SI) the reciprocal of the ohm is the siemens (S) and conductivity is reported as millisiemens per meter (mS/m); 1 mS/m = 10 μmhos/cm. To report results in SI units divide μmhos/cm by 10.

Freshly distilled water has a conductivity of 0.5 to 2 μmhos/cm, increasing after a few weeks of storage to 2 to 4 μmhos/cm. This increase is caused mainly by absorption of atmospheric carbon dioxide, and, to a lesser extent, ammonia.

The conductivity of potable waters in the United States ranges generally from 50 to 1500 μmhos/cm. The conductivity of domestic wastewaters may be near that of the local water supply, although some industrial wastes have conductivities above 10 000 μmhos/cm. Conductivity instruments are used in pipelines, channels, flowing streams, and lakes and can be incorporated in multiple-parameter monitoring stations using recorders.

Laboratory measurement of conductivity is relatively accurate but less accurate means of determining conductivity find numerous applications such as signalling exhaustion of ion-exchange resins and rapid determination of large changes in inorganic content of waters and wastewaters. Monitoring devices can give continuous, unattended records of conductivity if they are properly installed and maintained. Most problems in obtaining good records with monitoring equipment are related to electrode fouling and to inadequate sample circulation.

Laboratory conductivity measurements are used to:

a. Establish degree of mineralization to assess the effect of the total concentration of ions on chemical equilibria, physiological effect on plants or animals, corrosion rates, etc.

b. Assess degree of mineralization of distilled and deionized water.

c. Evaluate variations in dissolved mineral concentration of raw water or wastewater. Minor seasonal variations found in reservoir waters contrast sharply with the daily fluctuations in some polluted river waters. Wastewater containing significant trade wastes also may show a considerable daily variation.

d. Estimate sample size to be used for common chemical determinations and to check results of a chemical analysis.

e. Determine amount of ionic reagent needed in certain precipitation and neutralization reactions, the end point being denoted by a change in slope of the curve resulting from plotting conductivity against buret readings.

f. Estimate total dissolved solids in a

sample by multiplying conductivity (in micromhos per centimeter) by an empirical factor. This factor may vary from 0.55 to 0.9, depending on the soluble components of the water and on the temperature of measurement. Relatively high factors may be required for saline or boiler waters, whereas lower factors may apply where considerable hydroxide or free acid is present. Even though sample evaporation results in the change of bicarbonate to carbonate the empirical factor is derived for a comparatively constant water supply by dividing dissolved solids by conductivity. Approximate the milliequivalents per liter of either cations or anions in some waters by multiplying conductivity (in micromhos per centimeter) by 0.01.

Electrolytic conductivity (unlike metallic conductivity) increases with temperature at a rate of approximately 1.9%/°C. Significant errors can result from inaccurate temperature measurement. Potassium chloride (KCl) solutions have a lower temperature coefficient of conductivity than typical potable water. Sodium chloride (NaCl), on the other hand, has a temperature coefficient that closely approximates that found in most waters from wells and surface sources. Note that each ion has a different temperature coefficient; thus, for precise work, determine conductivity at 25.0°C. The significance of the temperature correction, one part in 500 for 25 ± 0.1°C, depends on available equipment and precision desired.

2510 B.   Laboratory Method

1. General Discussion

See Section 2510A.

2. Apparatus

a. *Self-contained conductivity instruments:* Use an instrument consisting of a source of alternating current, a Wheatstone bridge, a null indicator, and a conductivity cell or other instrument measuring the ratio of alternating current through the cell to voltage across it. The latter has the advantage of a linear reading of conductivity. Choose an instrument capable of measuring conductivity with an error not exceeding 1% or 1 μmho/cm, whichever is greater.

b. *Thermometer,* capable of being read to the nearest 0.1°C and covering the range 23 to 27°C. An electrical thermometer having a small thermistor sensing element is convenient because of its rapid response.

c. *Conductivity cell:*

1) Platinum-electrode   type—Conductivity cells containing platinized electrodes are available in either pipet or immersion form. Cell choice depends on expected range of conductivity and resistance range of the instrument. Experimentally check range for complete instrument assembly by comparing instrumental results with the true conductivities of the KCl solutions listed in Table 2510:I. Clean new cells with chromic-sulfuric acid cleaning mixture and platinize the electrodes before use. Subsequently, clean and replatinize them whenever the readings become erratic, when a sharp end point cannot be obtained, or when inspection shows that any platinum black has flaked off. To platinize, prepare a solution of 1 g chloroplatinic acid, $H_2PtCl_6 \cdot 6H_2O$, and 12 mg lead acetate in 100 mL distilled water. A stronger solution reduces the time required to platinize electrodes and may be used when time is a factor, e.g., when the cell constant is 1.0/cm or more. Immerse the electrodes in this solution and connect both to the neg-

TABLE 2510:I. CONDUCTIVITY OF POTASSIUM
CHLORIDE SOLUTIONS AT 25°C.*

Concentration M	Equivalent Conductivity mho/cm/equiv.	Conductivity μmhos/cm
0	149.85	
0.0001	149.43	14.94†
0.0005	147.81	73.90
0.001	146.95	147.0
0.005	143.55	717.8
0.01	141.27	1 413
0.02	138.34	2 767
0.05	133.37	6 668
0.1	128.96	12 900
0.2	124.08	24 820
0.5	117.27	58 640
1	111.87	111 900

* Data drawn from Robinson & Stokes.[1]
† Computed from equation given in Lind et al.[2]

ative terminal of a 1.5-V dry cell battery.
Connect positive side of battery to a piece
of platinum wire and dip the wire into the
solution. Use a current such that only a
small quantity of gas is evolved. Continue
electrolysis until both cell electrodes are
coated with platinum black. Save platinizing solution for subsequent use. Rinse
electrodes thoroughly and when not in use
keep immersed in distilled water.

2) Nonplatinum-electrode type—Use
conductivity cells containing electrodes
constructed from durable common metals
(stainless steel among others) for continuous monitoring and field studies. Calibrate such cells by comparing sample
conductivity with results obtained with a
laboratory instrument. Use properly designed and mated cell and instrument to
minimize errors in cell constant.

3. Reagents

a. Conductivity water: Pass distilled
water through a mixed-bed deionizer and
discard first liter. Conductivity should be
less than 1 μmho/cm.

b. Standard potassium chloride solution,

KCl, 0.0100M: Dissolve 745.6 mg anhydrous KCl in conductivity water and dilute
to 1000 mL at 25°C. This is the standard
reference solution, which at 25°C has a conductivity of 1413 μmhos/cm. It is satisfactory for most samples when the cell has
a constant between 1 and 2. For other cell
constants, use stronger or weaker KCl solutions listed in Table 2510:I. Store in a
glass-stoppered borosilicate glass bottle.

4. Procedure

a. Determination of cell constant: Rinse
conductivity cell with at least three portions of 0.01M KCl solution. Adjust temperature of a fourth portion to 25.0 ±
0.1°C. Measure resistance of this portion
and note temperature. Compute cell constant, C:

$$C = (0.001\ 413)\,(R_{KCl})\,[1 + 0.0191\,(t - 25)]$$

where:

R_{KCl} = measured resistance, ohms, and
t = observed temperature, °C.

b. Conductivity measurement: Rinse cell
with one or more portions of sample. Adjust temperature of a final portion to 25.0
± 0.1°C. Measure sample resistance or
conductivity and note temperature.

5. Calculation

The temperature coefficient of most
waters is only approximately the same as
that of standard KCl solution; the more
the temperature of measurement deviates
from 25.0°C, the greater the uncertainty in
applying the temperature correction. Report all conductivities at 25.0°C.

a. When sample resistance is measured,
conductivity at 25°C is:

$$K = \frac{(1\ 000\ 000)\,(C)}{R_m[1 + 0.0191\,(t - 25)]}$$

where:

K = conductivity, μmhos/cm,

C = cell constant, cm^{-1},
R_m = measured resistance of sample, ohms, and
t = temperature of measurement.

b. When sample conductivity is measured, conductivity at 25°C is:

$$K = \frac{(K_m)(1\ 000\ 000)(C)}{1 + 0.0191(t - 25)}$$

where:

K_m = measured conductivity, mhos at t°C, and other units are defined as above.

NOTE: If conductivity readout is in micromhos per centimeter, delete the factor 1 000 000 in the numerator.

c. For instruments giving values in SI units,

1 mS/m = 10 μmhos/cm, or conversely,
1 μmho/cm = 0.1 mS/m.

6. Precision and Bias

Three synthetic samples were tested with the following results:

Conductivity μmhos/cm	No. of Results	Relative Standard Deviation %	Relative Error %
147.0	117	8.6	9.4
303.0	120	7.8	1.9
228.0	120	8.4	3.0

7. References

1. ROBINSON, R.A. & R.H. STOKES. 1959. Electrolyte Solutions, 2nd ed. Academic Press, New York, p. 466.
2. LIND, J.E., J.J. ZWOLENIK & R.M. FUOSS. 1959. Calibration of conductance cells at 25°C with aqueous solutions of potassium chloride. J. Amer. Chem. Soc. 81:1557.

8. Bibliography

JONES, G. & B.C. BRADSHAW. 1933. The measurement of the conductance of electrolytes. V. A redetermination of the conductance of standard potassium chloride solutions in absolute units. J. Amer. Chem. Soc. 55:1780.

2520   SALINITY*

2520 A.   Introduction

1. General Discussion

Salinity is an important unitless property of industrial and natural waters. It was originally conceived as a measure of the mass of dissolved salts in a given mass of solution. The experimental determination of the salt content by drying and weighing presents some difficulties due to the loss of some components. The only reliable way to determine the true or absolute salinity of a natural water is to make a complete chemical analysis. However, this method is time-consuming and cannot yield the precision necessary for accurate work. Thus, to determine salinity, one normally uses indirect methods involving the measurement of a physical property such as conductivity, density, sound speed, or refractive index. From an empirical relationship of salinity and the physical property determined for a standard solution it is possible to calculate salinity. The resultant salinity is no more accurate than the empirical relationship. The precision of the measurement of a physical property

*Approved by Standard Methods Committee, 1988.

will determine the precision in salinity. Following are the precisions of various physical measurements and the resultant salinity presently attainable with commercial instruments:

Property	Precision of Measurement	Precision of Salinity
Conductivity	±0.0002	±0.0002
Density	±3 × 10^{-6} g/cm^3	±0.004
Sound speed	±0.02 m/s	±0.01

Although conductivity has the greatest precision, it responds only to ionic solutes. Density, although less precise, responds to all dissolved solutes.

2. Selection of Method

In the past, the salinity of seawater was determined by hydrometric and argentometric methods, both of which were included in previous editions of *Standard Methods* (see Sections 210B and C, 16th edition). In recent years the conductivity (2520B) and density (2520C) methods have been used because of their high sensitivity and precision. These two methods are recommended for precise field and laboratory work.

3. Quality Assurance

Calibrate salinometer or densimeter against standards of KCl or standard seawater. Expected precision is better than ±0.01 salinity units with careful analysis and use of bracketing standards.

2520 B. Electrical Conductivity Method

1. Determination

See Conductivity, Section 2510. Because of its high sensitivity and ease of measurement, the conductivity method is most commonly used to determine salinity.[1,2] For seawater measurements use the Practical Salinity Scale 1978.[3-5] This scale was developed relative to a KCl solution. A seawater with a conductivity at 15°C equal to that of a KCl solution containing a mass of 32.4356 g in a mass of 1 kg of solution is defined as having a practical salinity of 35. This value was determined as an average of three independent laboratory studies. The salinity dependence of the conductivity, R_t, as a function of temperature (t°C, International Practical Temperature Scale 1968) of a given sample to a standard $S = 35$ seawater is used to determine the salinity

$$S = a_0 + a_1 R_t^{1/2} + a_2 R_t + a_3 R_t^{3/2} + a_4 R_t^2 + a_5 R_t^{5/2} + \Delta S$$

where ΔS is given by

$$\Delta S = \left[\frac{t - 15}{1 + 0.0162 (t - 15)} \right] (b_0 + b_1 R_t^{1/2} + b_2 R_t + b_3 R_t^{3/2} + b_4 R_t^2 + b_5 R_t^{5/2})$$

and:

$a_0 =$ 0.0080	$b_0 =$ 0.0005
$a_1 =$ −0.1692	$b_1 =$ −0.0056
$a_2 =$ 25.3851	$b_2 =$ −0.0066
$a_3 =$ 14.0941	$b_3 =$ −0.0375
$a_4 =$ −7.0261	$b_4 =$ 0.0636
$a_5 =$ 2.7081	$b_5 =$ −0.0144

valid from $S = 2$ to 42.

To measure the conductivity, use a conductivity bridge calibrated with standard seawater* with a known conductivity relative to KCl, following manufacturer's instructions and the procedures noted in Section 2510. If the measurements are to

*Available from Standard Seawater Services, Institute of Oceanographic Services, Warmley, Godalming, Surrey, England.

be made in estuarine waters, make secondary calibrations of weight-diluted seawater of known conductivity to ensure that the bridge is measuring true conductivities.

The Practical Salinity Scale recently has been extended to low salinities[6] to give an equation that is valid from 0 to 40 salinity. The equation is:

$$S = S_{PSS} - \frac{a_0}{1 + 1.5X + X^2} - \frac{b_0 f(t)}{1 + Y^{1/2} + Y^{3/2}}$$

where:

S_{PSS} = value determined from the Practical Salinity Scale given earlier,
a_0 = 0.008,
b_0 = 0.0005,
X = $400R_t$,
Y = $100R_t$, and
$f(t)$ = $(t-15)/[1 + 0.0162 (t-15)]$

The practical salinity breaks with the old salinity-chlorinity relationship, $S = 1.806\ 55\ Cl$. Although the scale can be used for estuarine waters[7-10] and brines[11-13], there are limitations.[12,14-22]

2. References

1. LEWIS, E.L. 1978. Salinity: its definition and calculation. *J. Geophys. Res.* 83:466.

2. LEWIS, E.L. 1980. The practical salinity scale 1978 and its antecedents. *IEEE J. Oceanic Eng.* OE-5:3.

3. BRADSHAW, A.L. & K.E. SCHLEICHER. 1980. Electrical conductivity of seawater. *IEEE J. Oceanic Eng.* OE-5:50.

4. CULKIN, F. & N.D. SMITH. 1980. Determination of the concentration of potassium chloride solution having the same electrical conductivity, at 15°C and infinite frequency, as standard seawater of salinity 35.000‰ (Chlorinity 19.37394‰. *IEEE J. Oceanic Eng.* OE-5:22.

5. DAUPHINEE, T.M., J. ANCSIN, H.P. KLEIN & M.J. PHILLIPS. 1980. The effect of concentration and temperature on the conductivity ratio of potassium chloride solutions to standard seawater of salinity 35‰ (Cl.19.3740) at 15°C and 24°. *IEEE J. Oceanic Eng.* OE-5:17.

6. HILL, K.D., T.M. DAUPHINEE & D.J.

WOODS. 1986. The extension of the Practical Salinity Scale 1978 to low salinities. *IEEE J. Oceanic Eng.* OE-11:109.

7. MILLERO, F.J. 1975. The physical chemistry of estuarines. *In* T.M. Church, ed. Marine Chemistry in the Coastal Environment. American Chemical Soc. Symposium, Ser. 18.

8. MILLERO, F.J. 1978. The physical chemistry of Baltic Sea waters. *Thalassia Jugoslavica* 14:1.

9. MILLERO, F.J. 1984. The conductivity-salinity-chlorinity relationship for estuarine waters. *Limnol. Oceanogr.* 29:1318.

10. MILLERO, F.J. & K. KREMLING. 1976. The densities of Baltic Sea waters. *Deep-Sea Res.* 23:1129.

11. FERNANDEZ, F., F. VAZQUEZ & F.J. MILLERO. 1982. The density and composition of hypersaline waters of a Mexican lagoon. *Limnol. Oceanogr.* 27:315.

12. MILLERO, F.J. & P.V. CHETIRKIN. 1980. The density of Caspian Sea waters. *Deep-Sea Res.* 27:265.

13. MILLERO, F.J., A. MUCCI, J. ZULLIG & P. CHETIRKIN. 1982. The density of Red Sea brines. *Mar. Chem.* 11:477.

14. BREWER, P.G. & A. BRADSHAW. 1975. The effect of non-ideal composition of seawater on salinity and density. *J. Mar. Res.* 33:155.

15. CONNORS, D.N. & D.R. KESTER. 1974. Effect of major ion variations in the marine environment on the specific gravity-conductivity-chlorinity-salinity relationship. *Mar. Chem.* 2:301.

16. POISSON, A. 1980. The concentration of the KCl solution whose conductivity is that of standard seawater (35‰) at 15°C. *IEEE J. Oceanic Eng.* OE-5:24.

17. POISSON, A. 1980. Conductivity/salinity/temperature relationship of diluted and concentrated standard seawater. *IEEE J. Oceanic Eng.* OE-5:17.

18. MILLERO, F.J., A. GONZALEZ & G.K. WARD. 1976. The density of seawater solutions at one atmosphere as a function of temperature and salinity. *J. Mar. Res.* 34:61.

19. MILLERO, F.J., A. GONZALEZ, P.G. BREWER & A. BRADSHAW. 1976. The density of North Atlantic and North Pacific deep waters. *Earth Planet Sci. Lett.* 32:468.

20. MILLERO, F.J., D. LAWSON & A. GONZALEZ. 1976. The density of artificial river and estuary waters. *J. Geophys. Res.* 81:1177.

21. MILLERO, F.J., P. CHETIRKIN & F. CULKIN. 1977. The relative conductivity and density of standard seawaters. *Deep-Sea Res.* 24:315.

22. MILLERO, F.J., D. FORSHT, D. MEANS, J. GRIESKES & K. KENYON. 1977. The density of North Pacific ocean waters. *J. Geophys. Res.* 83:2359.

2520 C. Density Method

1. Determination

With a precise vibrating flow densimeter, it is possible to make rapid measurements of the density of natural waters. The measurements are made by passing the sample through a vibrating tube encased in a constant-temperature jacket. The solution density (ρ) is proportional to the square of the period of the vibration (τ).

$$\rho = A + B\tau^2$$

where A and B are terms determined by calibration, B being determined by calibration with a densimeter with standard seawater. The difference between the density of the sample and that of pure water is given by:

$$\rho - \rho_0 = B(\tau^2 - \tau_0^2)$$

where τ and τ_0 are, respectively, the periods of the sample and water. The system is calibrated with two solutions of known density. Follow manufacturer's recommendations for calibration. These two solutions can be nitrogen gas and water or standard seawater and water. The salinity of the sample can be determined from the 1 atm international equation of state for seawater. This equation relates ($\rho - \rho_0$) to the practical salinity (S) as a function of temperature.[1]

$$\rho \ (\text{kg/m}^3) = \rho_0 + AS + BS^{3/2} + CS^2$$

where:

$$A = 8.244\ 93 \times 10^{-1} - 4.0899 \times 10^{-3}t + 7.6438 \times 10^{-5}t^2 - 8.2467 \times 10^{-7}t^3 + 5.3875 \times 10^{-9}t^4,$$

$$B = -5.724\ 66 \times 10^{-3} + 1.0227 \times 10^{-4}t - 1.6546 \times 10^{-6}t^2,$$

$$C = 4.8314 \times 10^{-4},$$

and the density of water is given by:

$$\rho_0 = 999.842\ 594 + 6.793\ 952 \times 10^{-2}t - 9.095\ 290 \times 10^{-3}\ t^2 + 1.001\ 685 \times 10^{-4}t^3 - 1.120\ 083 \times 10^{-6}t^4 + 6.536\ 332 \times 10^{-9}t^5$$

Perform simple iteration by adjusting S until it gives the measured $\rho - \rho_0$ at a given temperature. If the measurements are made at 25°C, the salinity can be determined from the following equation:

$$S = 1.3343 \ (\rho - \rho_0) + 2.155\ 306 \times 10^{-4} (\rho - \rho_0)^2 - 1.171\ 16 \times 10^{-5} \ (\rho - \rho_0)^3$$

which has a $\tau = 0.0012$ in S. Approximate salinities also can be determined from densities or specific gravities obtained with a hydrometer at a given temperature (Section 210B, 16th edition).

2. References

1. MILLERO, F.J. & A. POISSON. 1981. International one-atmosphere equation of state of seawater. *Deep Sea Res.* 28:625.

2520 D. Algorithm of Practical Salinity

Because all practical salinity measurements are carried out in reference to the conductivity of standard seawater (corrected to $S = 35$), it is the quantity R_t that will be available for salinity calculations. R_t normally is obtained directly by laboratory salinometers, but *in situ* measurements usually produce the quantity R, the ratio of the *in situ* conductivity to the standard conductivity at $S = 35$, $t = 15°C$, $p = 0$ (where p is the pressure above one standard atmosphere and the temperature is on the 1968 International Temperature Scale). R is factored into three parts, i.e.,

$$R = R_p r_t R_t$$

where:

R_p = ratio of *in situ* conductivity to conductivity of the same sample at the same temperature, but at $p = 0$ and r_t = ratio of conductivity of reference seawater, having a practical salinity of 35, at temperature t, to its conductivity at $t = 15°C$. From R_p and r_t calculate R_t using the *in situ* results, i.e.,

$$R_t = \frac{R}{R_p r_t}$$

R_p and r_t can be expressed as functions of the numerical values of the *in situ* parameters, R, t, and p, when t is expressed in °C and p in bars (10^5 Pa), as follows:

$$R_p = 1 + \frac{p(e_1 + e_2 p + e_3 p^2)}{1 + d_1 t + d_2 t^2 + (d_3 + d_4 t)R}$$

where:

$e_1 = 2.070 \times 10^{-4}$, $d_1 = 3.426 \times 10^{-2}$,
$e_2 = -6.370 \times 10^{-8}$, $d_2 = 4.464 \times 10^{-4}$,
$e_3 = 3.989 \times 10^{-12}$, $d_3 = 4.215 \times 10^{-1}$,
and $d_4 = -3.107 \times 10^{-3}$,

and

$$r_t = c_0 + c_1 t + c_2 t^2 + c_3 t^3 + c_4 t^4$$

where:
$c_0 = 0.676\ 609\ 7$,
$c_1 = 2.005\ 64 \times 10^{-2}$,
$c_2 = 1.104\ 259 \times 10^{-4}$,
$c_3 = -6.9698 \times 10^{-7}$, and
$c_4 = 1.0031 \times 10^{-9}$.

2530 FLOATABLES*

2530 A. Introduction

One important criterion for evaluating the possible effect of waste disposal into surface waters is the amount of floatable material in the waste. Two general types of floating matter are found: particulate matter that includes "grease balls," and liquid components capable of spreading as a thin, highly visible film over large areas.

Floatable material in wastewaters is important because it accumulates on the surface, is often highly visible, is subject to wind-induced transport, may contain pathogenic bacteria and/or viruses associated with individual particles, and can significantly concentrate metals and chlorinated hydrocarbons such as pesticides and PCBs. Colloidally dispersed oil and grease behave like other dispersed organic matter and are

---

* Approved by Standard Methods Committee, 1988.

included in the material measured by the COD, BOD, and TOC tests. The floatable oil test indicates the readily separable fraction. The results are useful in designing oil and grease separators, in ascertaining the efficiency of operating separators, and in monitoring raw and treated wastewater streams. Many cities and districts have specified floatable oil and grease limits for wastewater discharged to sewers.

2530 B.   Particulate Floatables (GENERAL)

1. Discussion

a. Principle: This method depends on the gravity separation of particles having densities less than that of the surrounding water. Particles that collect on the surface and can be filtered out and dried at 103 to 105°C are defined by this test as floatable particles.

b. Application: This method is applicable to raw wastewater, treated primary and secondary effluent, and industrial wastewater. Because of the limited sensitivity, it is not applicable to tertiary effluents or receiving waters, whether freshwater or seawater.

c. Precautions: Even slight differences in sampling and handling during and after collection can give large differences in the measured amount of floatable material. Additionally, uniformity of the TFE* coating of the separation funnel is critical to obtaining reliable results. For a reproducible analysis treat all samples uniformly, preferably by mixing them in a standard manner, before flotation and use consistently prepared separation funnels as much as possible. Because the procedure relies on the difference in specific gravity between the liquid and the floating particles, temperature variations may affect the results. Conduct the test at a constant temperature the same as that of the receiving water body, and report temperature with results.

---
* Teflon or equivalent.

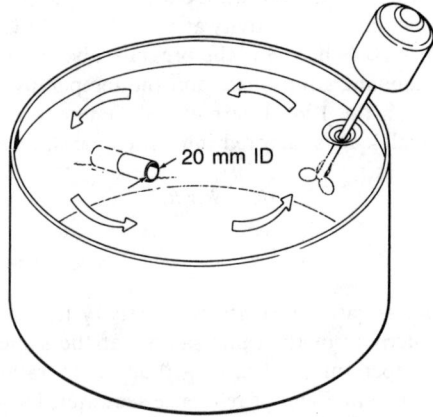

Figure 2530:1. Floatables sampler with mixer.

d. Minimum detectable concentration: The minimum reproducible detectable concentration is approximately 1 mg/L. Although the minimum levels that can be measured are below 1 mg/L, the results are not meaningful within the current established accuracy of the test.

2. Apparatus

a. Floatables sampler with mixer: Use a metal container of at least 5 L capacity equipped with a propeller mixer on a separate stand (Figure 2530:1), and with a 20-mm-ID bottom outlet cocked at an angle of 45° to the container wall in the direction of fluid movement. The 45° angle assures that even large particles will flow from the container into the flotation funnels where the sample is withdrawn. Fit exterior of

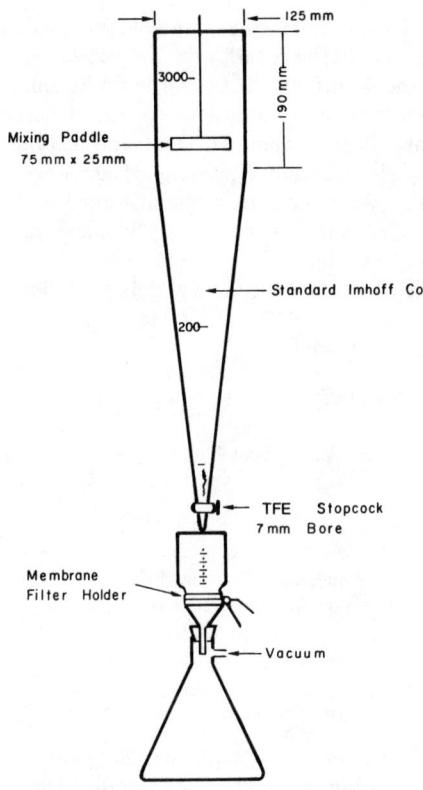

Figure 2530:2. Floatables flotation funnel and filter holder.

with TFE, again taking all possible precautions to obtain a uniform TFE coating.

d. Filters, glass fiber, fine porosity.†

e. Vacuum flask, 500 mL.

f. TFE coating: Follow instructions that accompany commercially available coating kits. Alternatively, have necessary glassware coated commercially. Uniform coatings are key to the reliability of the test results, but in practice are difficult to obtain.

3. Procedure

a. Preparation of glass fiber filters: See Section 2540D.3*a.*

b. Sample collection and treatment: Collect sample in the floatables sampler at a point of complete mixing, transport to the laboratory, and place 3.0 L in the flotation funnel within 2 h after sample collection to minimize changes in the floatable material. While the flotation funnel is being filled, mix sampler contents with a small propeller mixer. Adjust mixing speed to provide uniform distribution of floating

---

† Whatman GF/C or equivalent.

bottom outlet with a short piece of tubing and a pinch clamp to allow unrestricted flow through the outlet. Coat inside of container with TFE as uniformly as possible, using a TFE spray to prevent oil and grease from sticking to the surface.

b. Flotation funnel: Use an Imhoff cone provided with a TFE stopcock at the bottom and extended at the top to a total volume of 3.5 L (Figure 2530:2). Coat inside of flotation funnel with TFE as uniformly as possible to prevent floatable grease particles sticking to the sides. Mount flotation funnels as shown in Figure 2530:3 with a light behind the bottom of the funnels to aid in reading levels.

c. Filter holder: Coat inside of top of a standard 500-mL membrane filter holder

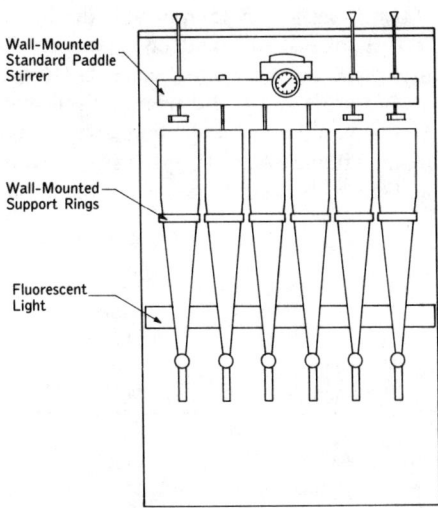

Figure 2530:3. Flotation funnels and mixing unit.

particles throughout the liquid but avoid extensive air entrapment through formation of a large vortex.

c. *Correction for density and for concentration effects:* When a receiving water has a density and ion concentration different from that of the waste, adjust sample density and ion concentration to that of the receiving water. For example, if the receiving water is ocean water, place 1.5 L sample in flotation funnel and add 1.5 L filtered seawater from the receiving area together with mixture of 39.8 g NaCl, 8.0 g $MgCl_2 \cdot 6H_2O$, and 2.3 g $CaCl_2 \cdot 2H_2O$. The final mixture contains the amount of floatables in a 1.5-L sample in a medium of approximately the same density and ion concentration as seawater.

d. *Flotation:* Mix flotation funnel contents at 40 rpm for 15 min using a paddle mixer (Figure 2530:3). Let settle for 5 min, mix at 100 rpm for 1 min, and let settle for 30 min. Discharge 2.8 L through bottom stopcock at a rate of 500 mL/min. Do not disturb the sample surface in the flotation funnel during discharge. With distilled water from a wash bottle, wash down any floatable material sticking to sides of stirring paddle and funnel. Let remaining 200 mL settle for 15 min and discharge settled solids and liquid down to the 40-mL mark on the Imhoff cone. Let settle again for 10 min and discharge until only 10 mL liquid and the floating particles remain in funnel. Add 500 mL distilled water and stir by hand to separate entrapped set-

tleable particles from the floatable particles. Let settle for 15 min, then discharge to the 40-mL mark. Let settle for 10 min, then discharge dropwise to the 10-mL mark. Filter remaining 10 mL and floating particles through a preweighed glass fiber filter. Wash sides of flotation funnel with distilled water to transfer all floatable material to filter.

e. *Weighing:* Dry and weigh glass fiber filter at 103 to 105° C for exactly 2 h (see Section 2540D.3c).

4. Calculation

$$\text{mg particulate floatables/L} = \frac{(A - B)}{C}$$

where:

  A = weight of filter + floatables, mg,
  B = weight of filter, mg, and
  C = sample volume, L. (Do not include volume used for density or concentration correction, if used.)

5. Precision and Bias

Precision varies with the concentration of suspended matter in the sample. There is no completely satisfactory procedure for determining the bias of the method for wastewater samples but approximate recovery can be determined by running a second test for floatables on all water discharged throughout the procedure, with the exception of the last 10 mL. Precision and bias are summarized in Table 2530:I. Experience with the method at one mu-

TABLE 2530:I. COEFFICIENT OF VARIATION AND RECOVERY FOR PARTICULATE FLOATABLES TEST

Type of Wastewater	Average Floatables Concentration mg/L	No. of Samples	Coefficient of Variation %	Recovery %
Raw*	49	5	5.7	96
Raw	1.0	5	20	92
Primary effluent	2.7	5	15	91

* Additional floatable material added from skimmings of a primary sedimentation basin.

nicipal treatment plant indicates that the practical lower limit of detection is approximately 1 mg/L.

6. Bibliography

HEUKELEKIAN, H. & J. BALMAT. 1956. Chemical composition of the particulate fractions of domestic sewage. *Sewage Ind. Wastes* 31:413.

ENGINEERING-SCIENCE, INC. 1965. Determination and Removal of Floatable Material from Waste Water. Rep. for U.S. Public Health Serv. contracts WPD 12-01 (R1)-63 and WPD 12-02-64, Engineering-Science, Inc., Arcadia & Oakland, Calif.

HUNTER, J.V. & H. HEUKELEKIAN. 1965. Composition of domestic sewage fractions. *J. Water Pollut. Control Fed.* 37:1142.

NUSBAUM I. & L. BURTMAN. 1965. Determination of floatable matter in waste discharges. *J. Water Pollut. Control Fed.* 37:577.

SCHERFIG, J. & H. F. LUDWIG. 1967. Determination of floatables and hexane extractables in sewage. *In* Advances in Water Pollution Research, Vol. 3, p. 217, Water Pollution Control Federation, Washington, D.C.

SELLECK, R.E., L. W. BRACEWELL & R. CARTER. 1974. The Significance and Control of Wastewater Floatables in Coastal Waters. Rep. for U.S. Environmental Protection Agency contract R-800373, SERL Rep. No. 74-1, Sanitary Engineering Research Lab., Univ. California, Berkeley.

BRACEWELL, L.W. 1976. Contribution of Wastewater Discharges to Surface Films and Other Floatables on the Ocean Surface. Thesis, Univ. California, Berkeley.

BRACEWELL, L.W., R.E. SELLECK & R. CARTER. 1980. Contribution of wastewater discharges to ocean surface particulates. *J. Water Pollut. Control Fed.* 52:2230.

2530 C.   Trichlorotrifluoroethane-Soluble Floatable Oil and Grease (GENERAL)

1. Discussion

The floatable oil and grease test does not measure a precise class of substances; rather, the results are determined by the conditions of the test. The fraction measured includes oil and grease, both floating and adhering to the sides of the test vessel. The adhering and the floating portions are of similar practical significance because it is assumed that most of the adhering portion would otherwise float under receiving water conditions. The results have been found to represent well the amount of oil removed in separators having overflow rates equivalent to test conditions.

2. Apparatus

a. Floatable oil tube (Figure 2530:4): Before use, carefully clean tube by brushing with a mild scouring power. Water must form a smooth film on inside of cleaned glass. Do not use lubricant on stopcock.

b. Conical flask, 300 mL.

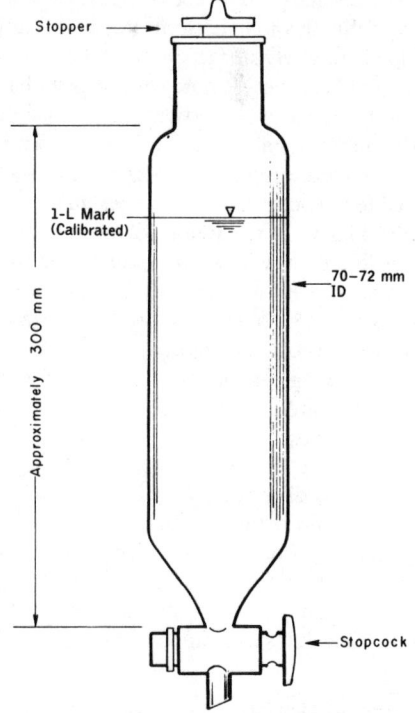

Stopper

1-L Mark (Calibrated)

70–72 mm ID

Approximately 300 mm

Stopcock

Figure 2530:4. Floatable oil tube, 1-L capacity.

3. Reagents

a. *1,1,2-trichloro-1,2,2-trifluoroethane*:* See Section 5520B.3*b*.
  b. *Hydrochloric acid,* HCl, 6*N*.
  c. *Filter paper.*†

4. Procedure

a. Sampling: Collect samples at a place where there is a strong turbulence in the water and where floating material is not trapped at the surface. Fill floatable oil tube to mark by dipping into water. *Do not use samples taken to the laboratory in a bottle, because oil and grease cannot be redispersed to their original condition.*

b. Flotation: Support tube in a vertical position. Start flotation period at sampling site immediately after filling tube. The standard flotation time is 30 min. If a different time is used, state this variation in reporting results. At end of flotation period, discharge the first 900 mL of water carefully through bottom stopcock, stopping before any surface oil or other floating material escapes. Rotate tube slightly back and forth about its vertical axis to dislodge sludge from sides, and let settle for 5 min. Completely discharge sludge that has settled to the bottom or that comes down from the sides with the liquid. Scum on top of the liquid may mix with the water as it moves down the tube. If mixing occurs, stop drawing off water before any floatables have been lost. Let settle for 5 min before withdrawing remainder of water. After removing water, return tube to laboratory to complete test.

c. Extraction: Acidify to pH 2 or lower with a few drops of 6*N* HCl, add 50 to 100 mL trichlorotrifluoroethane, and shake vigorously. Let settle and draw off solvent into a clean dry beaker. Filter solvent through a dry filter paper into a tared 300-mL conical flask, taking care not to get any

water on filter paper. Add a second 50-mL portion of trichlorotrifluoroethane and repeat extraction, settling, and filtration into the same 300-mL flask. A third extraction may be needed if the amount of floatables in sample exceeds 4 mg/L. Wash filter paper carefully with fresh solvent discharged from a wash bottle with a fine tip. Evaporate solvent from flask as described in Section 5520B.4. For each solvent batch, determine weight of residue left after evaporation from the same volume as used in the analysis.

5. Calculations

Report results as "soluble floatable oil and grease, 30 min (or other specified) settling time, mg/L."

Trichlorotrifluoroethane-soluble floatable oil and grease, 30 min settling time, mg/L

$$= \frac{(A - B) \times 1000}{\text{mL sample}}$$

where:
  A = total gain in weight of tared flask, mg, and
  B = calculated residue from solvent blank of the same volume as that used in the test, mg.

6. Precision and Bias

There is no standard against which bias of this test can be determined. Variability of replicates is influenced by sample heterogeneity. If large grease particles are present, the element of change in sampling may be a major factor. One municipal wastewater discharge and two meat-packing plant discharges, both containing noticeable particles of grease, were analyzed in triplicate. Averages for the three wastewaters were 48, 57, and 25 mg/L; standard deviations averaged 11%. An oil refinery made duplicate determinations of its sep-

---

* Freon or equivalent.
† Whatman No. 40 or equivalent.

arator effluent on 15 consecutive days, obtaining results ranging from 5.1 to 11.2 mg/L. The average difference between pairs of samples was 0.37 mg/L.

7. Bibliography

POMEROY, R.D. 1953. Floatability of oil and grease in wastewaters. *Sewage Ind. Wastes* 25:1304.

2540 SOLIDS*

2540 A. Introduction

The terms "solids," "suspended," and "dissolved," as used herein, replace the terms "residue," "nonfiltrable," and "filtrable" of editions previous to the 16th. Solids refer to matter suspended or dissolved in water or wastewater. Solids may affect water or effluent quality adversely in a number of ways. Waters with high dissolved solids generally are of inferior palatability and may induce an unfavorable physiological reaction in the transient consumer. For these reasons, a limit of 500 mg dissolved solids/L is desirable for drinking waters. Highly mineralized waters also are unsuitable for many industrial applications. Waters high in suspended solids may be esthetically unsatisfactory for such purposes as bathing. Solids analyses are important in the control of biological and physical wastewater treatment processes and for assessing compliance with regulatory agency wastewater effluent limitations.

1. Definitions

"Total solids" is the term applied to the material residue left in the vessel after evaporation of a sample and its subsequent drying in an oven at a defined temperature. Total solids includes "total suspended solids," the portion of total solids retained by a filter, and "total dissolved solids," the portion that passes through the filter.

The type of filter holder, the pore size, porosity, area, and thickness of the filter and the physical nature, particle size, and amount of material deposited on the filter are the principal factors affecting separation of suspended from dissolved solids.

"Fixed solids" is the term applied to the residue of total, suspended, or dissolved solids after ignition for a specified time at a specified temperature. The weight loss on ignition is called "volatile solids." Determinations of fixed and volatile solids do not distinguish precisely between inorganic and organic matter because the loss on ignition is not confined to organic matter. It includes losses due to decomposition or volatilization of some mineral salts. Better characterization of organic matter can be made by such tests as total organic carbon (Section 5310), BOD (Section 5210), and COD (Section 5220).

"Settleable solids" is the term applied to the material settling out of suspension within a defined period. It may include floating material, depending on the technique (2540F.3*b*).

2. Sources of Error and Variability

The temperature at which the residue is dried has an important bearing on results, because weight losses due to volatilization of organic matter, mechanically occluded water, water of crystallization, and gases from heat-induced chemical decomposi-

---

*Approved by Standard Methods Committee, 1985.

tion, as well as weight gains due to oxidation, depend on temperature and time of heating.

Residues dried at 103 to 105°C may retain not only water of crystallization but also some mechanically occluded water. Loss of CO_2 will result in conversion of bicarbonate to carbonate. Loss of organic matter by volatilization usually will be very slight. Because removal of occluded water is marginal at this temperature, attainment of constant weight may be very slow.

Residues dried at 180 ± 2°C will lose almost all mechanically occluded water. Some water of crystallization may remain, especially if sulfates are present. Organic matter may be lost by volatilization, but not completely destroyed. Loss of CO_2 results from conversion of bicarbonates to carbonates and carbonates may be decomposed partially to oxides or basic salts. Some chloride and nitrate salts may be lost. In general, evaporating and drying water samples at 180°C yields values for dissolved solids closer to those obtained through summation of individually determined mineral species than the dissolved solids values secured through drying at the lower temperature.

Results for residues high in oil or grease may be questionable because of the difficulty of drying to constant weight in a reasonable time.

Analyses performed for some special purposes may demand deviation from the stated procedures to include an unusual constituent with the measured solids. Whenever such variations of technique are introduced, record and present them with the results.

3. Sample Handling and Preservation

Use resistant-glass or plastic bottles, provided that the material in suspension does not adhere to container walls. Begin analysis as soon as possible because of the impracticality of preserving the sample. Refrigerate sample at 4°C up to analysis to minimize microbiological decomposition of solids.

4. Selection of Method

Methods B through F are suitable for the determination of solids in potable, surface, and saline waters, as well as domestic and industrial wastewaters in the range up to 20 000 mg/L.

Method G is suitable for the determination of solids in sediments, as well as solid and semisolid materials produced during water and wastewater treatment.

5. Bibliography

THERIAULT, E.J. & H.H. WAGENHALS. 1923. Studies of representative sewage plants. *Pub. Health Bull.* No. 132.

U.S. ENVIRONMENTAL PROTECTION AGENCY. 1979. Methods for Chemical Analysis of Water and Wastes. Publ. 600/4-79-020, Environmental Monitoring and Support Lab., U.S. Environmental Protection Agency, Cincinnati, Ohio.

2540 B.   Total Solids Dried at 103–105°C

1. General Discussion

a. Principle: A well-mixed sample is evaporated in a weighed dish and dried to constant weight in an oven at 103 to 105°C. The increase in weight over that of the empty dish represents the total solids. The results may not represent the weight of actual dissolved and suspended solids in wastewater samples (see above).

b. Interferences: Highly mineralized water with a significant concentration of calcium, magnesium, chloride, and/or sulfate may be hygroscopic and require prolonged drying, proper desiccation, and rapid weighing. Exclude large, floating particles or submerged agglomerates of non-homogeneous materials from the sample if it is determined that their inclusion is not desired in the final result. Disperse visible floating oil and grease with a blender before withdrawing a sample portion for analysis. Because excessive residue in the dish may form a water-trapping crust, limit sample to no more than 200 mg residue.

2. Apparatus

   a. *Evaporating dishes:* Dishes of 100-mL capacity made of one of the following materials:

   1) Porcelain, 90-mm diam.
   2) Platinum—Generally satisfactory for all purposes.
   3) High-silica glass.*

   b. *Muffle furnace* for operation at 550 ± 50°C.

   c. *Steam bath.*

   d. *Desiccator,* provided with a desiccant containing a color indicator of moisture concentration.

   e. *Drying oven,* for operation at 103 to 105°C.

   f. *Analytical balance,* capable of weighing to 0.1 mg.

3. Procedure

   a. *Preparation of evaporating dish:* If volatile solids are to be measured ignite clean evaporating dish at 550 ± 50°C for 1 h in a muffle furnace. If only total solids are to be measured, heat clean dish to 103 to

----

*Vycor, product of Corning Glass Works, Corning, N.Y., or equivalent.

105°C for 1 h. Store dish in desiccator until needed. Weigh immediately before use.

   b. *Sample analysis:* Choose a sample volume that will yield a residue between 2.5 mg and 200 mg. Transfer a measured volume of well-mixed sample to preweighed dish and evaporate to dryness on a steam bath or in a drying oven. If necessary, add successive sample portions to the same dish after evaporation. When evaporating in a drying oven, lower temperature to approximately 2°C below boiling to prevent splattering. Dry evaporated sample for at least 1 h in an oven at 103 to 105°C, cool dish in desiccator to balance temperature, and weigh. Repeat cycle of drying, cooling, desiccating, and weighing until a constant weight is obtained, or until weight loss is less than 4% of previous weight or 0.5 mg, whichever is less.

4. Calculation

$$\text{mg total solids}/\text{L} = \frac{(A - B) \times 1000}{\text{sample volume, mL}}$$

where:

   A = weight of dried residue + dish, mg, and
   B = weight of dish, mg.

5. Precision

   Single-laboratory duplicate analyses of 41 samples of water and wastewater were made with a standard deviation of differences of 6.0 mg/L.

6. Bibliography

Symons, G.E. & B. Morey. 1941. The effect of drying time on the determination of solids in sewage and sewage sludges. *Sewage Works J.* 13:936.

2540 C.   Total Dissolved Solids Dried at 180°C

1. General Discussion

a. Principle: A well-mixed sample is filtered through a standard glass fiber filter, and the filtrate is evaporated to dryness in a weighed dish and dried to constant weight at 180°C. The increase in dish weight represents the total dissolved solids.

The results may not agree with the theoretical value for solids calculated from chemical analysis of sample (see above). Approximate methods for correlating chemical analysis with dissolved solids are available.[1] The filtrate from the total suspended solids determination (Section 2540D) may be used for determination of total dissolved solids.

b. Interferences: Highly mineralized waters with a considerable calcium, magnesium, chloride, and/or sulfate content may be hygroscopic and require prolonged drying, proper desiccation, and rapid weighing. Samples high in bicarbonate require careful and possibly prolonged drying at 180°C to insure complete conversion of bicarbonate to carbonate. Because excessive residue in the dish may form a water-trapping crust, limit sample to no more than 200 mg residue.

2. Apparatus

Apparatus listed in 2540B.2*a-d* is required, and in addition:

*a. Glass-fiber filter disks** without organic binder.

b. Filtration apparatus: One of the following, suitable for filter disk selected:

1) *Membrane filter funnel.*

---

*Whatman grade 934AH; Gelman type A/E; Millipore type AP40; E-D Scientific Specialties grade 161; or equivalent. Available in diameters of 2.2 cm to 4.7 cm.

2) *Gooch crucible,* 25-mL to 40-mL capacity, with Gooch crucible adapter.

3) *Filtration apparatus* with reservoir and coarse (40- to 60-μm) fritted disk as filter support.

c. Suction flask, of sufficient capacity for sample size selected.

d. Drying oven, for operation at 180 ± 2°C.

3. Procedure

a. Preparation of glass-fiber filter disk: Insert disk with wrinkled side up into filtration apparatus. Apply vacuum and wash disk with three successive 20-mL volumes of distilled water. Continue suction to remove all traces of water. Discard washings.

b. Preparation of evaporating dish: If volatile solids are to be measured, ignite cleaned evaporating dish at 550 ± 50°C for 1 h in a muffle furnace. If only total dissolved solids are to be measured, heat clean dish to 180 ± 2°C for 1 h in an oven. Store in desiccator until needed. Weigh immediately before use.

c. Selection of filter and sample sizes: Choose sample volume to yield between 2.5 and 200 mg dried residue. If more than 10 min are required to complete filtration, increase filter size or decrease sample volume but do not produce less than 2.5 mg residue.

d. Sample analysis: Filter measured volume of well-mixed sample through glass-fiber filter, wash with three successive 10-mL volumes of distilled water, allowing complete drainage between washings, and continue suction for about 3 min after filtration is complete. Transfer filtrate to a weighed evaporating dish and evaporate to dryness on a steam bath. If filtrate volume exceeds dish capacity add successive portions to the same dish after evaporation. Dry for at least 1 h in an oven at 180 ± 2°C, cool in a desiccator to balance temperature, and weigh. Repeat drying cycle of drying, cooling, desiccating, and weighing until a constant weight is obtained or until weight loss is less than 4% of previous weight or 0.5 mg, whichever is less.

4. Calculation

$$\text{mg total dissolved solids}/\text{L} = \frac{(A - B) \times 1000}{\text{sample volume, mL}}$$

where:

A = weight of dried residue + dish, mg, and
B = weight of dish, mg.

5. Precision

Single-laboratory analyses of 77 samples of a known of 293 mg/L were made with a standard deviation of differences of 21.20 mg/L.

6. Reference

1. SOKOLOFF, V.P. 1933. Water of crystallization in total solids of water analysis. *Ind. Eng. Chem.*, Anal. Ed. 5:336.

7. Bibliography

HOWARD, C.S. 1933. Determination of total dissolved solids in water analysis. *Ind. Eng. Chem.*, Anal. Ed. 5:4.
U.S. GEOLOGICAL SURVEY. 1974. Methods for Collection and Analysis of Water Samples for Dissolved Minerals and Gases. Techniques of Water-Resources Investigations, Book 5, Chap. A1. U.S. Geological Surv., Washington, D.C.

2540 D. Total Suspended Solids Dried at 103–105°C

1. General Discussion

a. Principle: A well-mixed sample is filtered through a weighed standard glass-fiber filter and the residue retained on the filter is dried to a constant weight at 103 to 105°C. The increase in weight of the filter represents the total suspended solids. If the suspended material clogs the filter and prolongs filtration, the difference between the total solids and the total dissolved solids may provide an estimate of the total suspended solids.

b. Interferences: Exclude large floating particles or submerged agglomerates of nonhomogeneous materials from the sample if it is determined that their inclusion is not desired in the final result. Because excessive residue on the filter may form a water-entrapping crust, limit the sample size to that yielding no more than 200 mg residue. For samples high in dissolved sol-

ids thoroughly wash the filter to ensure removal of the dissolved material. Prolonged filtration times resulting from filter clogging may produce high results owing to excessive solids capture on the clogged filter.

2. Apparatus

Apparatus listed in Sections 2540B.2 and 2540C.2 is required, except for evaporating dishes, steam bath, and 180°C drying oven. In addition:

*Planchet,** aluminum or stainless steel, 65-mm diam.

3. Procedure

a. Preparation of glass-fiber filter disk: Insert disk with wrinkled side up in filtra-

---

*Available from New England Nuclear, Boston, Mass., or equivalent.

tion apparatus. Apply vacuum and wash disk with three successive 20-mL portions of distilled water. Continue suction to remove all traces of water, and discard washings. Remove filter from filtration apparatus and transfer to an aluminum or stainless steel planchet as a support. Alternatively remove crucible and filter combination if a Gooch crucible is used. Dry in an oven at 103 to 105°C for 1 h. If volatile solids are to be measured, ignite at 550 ± 50°C for 15 min in a muffle furnace. Cool in desiccator to balance temperature and weigh. Repeat cycle of drying or igniting, cooling, desiccating, and weighing until a constant weight is obtained or until weight loss is less than 0.5 mg between successive weighings. Store in desiccator until needed. Weigh immediately before use.

b. Selection of filter and sample sizes: See Section 2540C.3c. For nonhomogeneous samples such as raw wastewater, use a large filter to permit filtering a representative sample.

c. Sample analysis: Assemble filtering apparatus and filter and begin suction. Wet filter with a small volume of distilled water to seat it. Filter a measured volume of well-mixed sample through the glass fiber filter. Wash with three successive 10-mL volumes of distilled water, allowing complete drainage between washings and continue suction for about 3 min after filtration is complete. Carefully remove filter from filtration apparatus and transfer to an aluminum or stainless steel planchet as a support. Alternatively, remove the crucible and filter combination from the crucible adapter if a Gooch crucible is used. Dry for at least 1 h at 103 to 105°C in an oven, cool in a desiccator to balance temperature, and weigh. Repeat the cycle of drying, cooling, desiccating, and weighing until a constant weight is obtained or until the weight loss is less than 4% of the previous weight or 0.5 mg, whichever is less.

4. Calculation

$$\text{mg total suspended solids}/L = \frac{(A - B) \times 1000}{\text{sample volume, mL}}$$

where:

A = weight of filter + dried residue, mg, and
B = weight of filter, mg.

5. Precision

The standard deviation was 5.2 mg/L (coefficient of variation 33%) at 15 mg/L, 24 mg/L (10%) at 242 mg/L, and 13 mg/L (0.76%) at 1707 mg/L in studies by two analysts of four sets of 10 determinations each.

Single-laboratory duplicate analyses of 50 samples of water and wastewater were made with a standard deviation of differences of 2.8 mg/L.

6. Bibliography

DEGEN, J. & F.E. NUSSBERGER. 1956. Notes on the determination of suspended solids. *Sewage Ind. Wastes* 28:237.

CHANIN, G., E.H. CHOW, R.B. ALEXANDER & J. POWERS. 1958. Use of glass fiber filter medium in the suspended solids determination. *Sewage Ind. Wastes* 30:1062.

NUSBAUM, I. 1958. New method for determination of suspended solids. *Sewage Ind. Wastes* 30:1066.

SMITH, A.L. & A.E. GREENBERG. 1963. Evaluation of methods for determining suspended solids in wastewater. *J. Water Pollut. Control Fed.* 35:940.

WYCKOFF, B.M. 1964. Rapid solids determination using glass fiber filters. *Water Sewage Works* 111:277.

NATIONAL COUNCIL OF THE PAPER INDUSTRY FOR AIR AND STREAM IMPROVEMENT. 1975. A Preliminary Review of Analytical Methods for the Determination of Suspended Solids in Paper Industry Effluents for Compliance with EPA-NPDES Permit Terms. Spec. Rep. No. 75-01. National Council of the Paper Industry for Air & Stream Improvement, New York, N.Y.

NATIONAL COUNCIL OF THE PAPER INDUSTRY FOR AIR AND STREAM IMPROVEMENT. 1977.

A Study of the Effect of Alternate Procedures on Effluent Suspended Solids Measurement. Stream Improvement Tech. Bull. No. 291, National Council of the Paper Industry for Air & Stream Improvement, New York, N.Y.

TREES, C.C. 1978. Analytical analysis of the effect of dissolved solids on suspended solids determination. *J. Water Pollut. Control Fed.* 50:2370.

2540 E.   Fixed and Volatile Solids Ignited at 550°C

1. General Discussion

a. Principle: The residue from Method B, C, or D is ignited to constant weight at 550 ± 50°C. The remaining solids represent the fixed total, dissolved, or suspended solids while the weight lost on ignition is the volatile solids. The determination is useful in control of wastewater treatment plant operation because it offers a rough approximation of the amount of organic matter present in the solid fraction of wastewater, activated sludge, and industrial wastes.

b. Interferences: Negative errors in the volatile solids may be produced by loss of volatile matter during drying. Determination of low concentrations of volatile solids in the presence of high fixed solids concentrations may be subject to considerable error. In such cases, measure for suspect volatile components by another test, for example, total organic carbon (Section 5310).

2. Apparatus

See Sections 2540B.2, 2540C.2, and 2540D.2.

3. Procedure

Ignite residue produced by Method B, C, or D to constant weight in a muffle furnace at a temperature of 550 ± 50°C.

Have furnace up to temperature before inserting sample. Usually, 15 to 20 min ignition are required. Let dish or filter disk cool partially in air until most of the heat has been dissipated. Transfer to a desiccator for final cooling in a dry atmosphere. Do not overload desiccator. Weigh dish or disk as soon as it has cooled to balance temperature. Repeat cycle of igniting, cooling, desiccating, and weighing until a constant weight is obtained or until weight loss is less than 4% of previous weight.

4. Calculation

$$\text{mg volatile solids/L} = \frac{(A - B) \times 1000}{\text{sample volume, mL}}$$

$$\text{mg fixed solids/L} = \frac{(B - C) \times 1000}{\text{sample volume, mL}}$$

where:

A = weight of residue + dish before ignition, mg,
B = weight of residue + dish or filter after ignition, mg, and
C = weight of dish or filter, mg.

5. Precision

The standard deviation was 11 mg/L at 170 mg/L volatile total solids in studies by three laboratories on four samples and 10 replicates. Bias data on actual samples cannot be obtained.

2540 F. Settleable Solids

1. General Discussion

Settleable solids in surface and saline waters as well as domestic and industrial wastes may be determined and reported on either a volume (mL/L) or a weight (mg/L) basis.

2. Apparatus

The volumetric test requires only an Imhoff cone. The gravimetric test requires all the apparatus listed in Section 2540D.2 and a glass vessel with a minimum diameter of 9 cm.

3. Procedure

a. Volumetric: Fill an Imhoff cone to the 1-L mark with a well-mixed sample. Settle for 45 min, gently stir sides of cone with a rod or by spinning, settle 15 min longer, and record volume of settleable solids in the cone as milliliters per liter. If the settled matter contains pockets of liquid between large settled particles, estimate volume of these and subtract from volume of settled solids. The practical lower limit of measurement depends on sample composition and generally is in the range of 0.1 to 1.0 mL/L. Where a separation of settleable and floating materials occurs, do not estimate the floating material as settleable matter.

b. Gravimetric:
1) Determine total suspended solids of well-mixed sample (Section 2540D).
2) Pour a well-mixed sample into a glass vessel of not less than 9 cm diam using not less than 1 L and sufficient to give a depth of 20 cm. Alternatively use a glass vessel of greater diameter and a larger volume of sample. Let stand quiescent for 1 h and, without disturbing the settled or floating material, siphon 250 mL from center of container at a point halfway between the surface of the settled material and the liquid surface. Determine total suspended solids (milligrams per liter) of this supernatant liquor (Section 2540D). These are the nonsettleable solids.

4. Calculation

mg settleable solids/L

$$= \text{mg total suspended solids/L}$$
$$- \text{mg nonsettleable solids/L}$$

5. Precision and Bias

Precision and bias data are not now available.

6. Bibliography

FISCHER, A.J. & G.E. SYMONS. 1944. The determination of settleable sewage solids by weight. *Water Sewage Works* 91:37.

2540 G. Total, Fixed, and Volatile Solids in Solid and Semisolid Samples

1. General Discussion

a. Applicability: This method is applicable to the determination of total solids and its fixed and volatile fractions in such solid and semisolid samples as river and lake sediments, sludges separated from water and wastewater treatment processes, and sludge cakes from vacuum filtration, centrifugation, or other sludge dewatering processes.

b. Interferences: The determination of both total and volatile solids in these materials is subject to negative error due to loss of ammonium carbonate and volatile organic matter during drying. Although this is true also for wastewater, the effect tends to be more pronounced with sediments, and especially with sludges and sludge cakes. The mass of organic matter recovered from sludge and sediment requires a longer ignition time than that specified for wastewaters, effluents, or polluted waters. Carefully observe specified ignition time and temperature to control losses of volatile inorganic salts. Make all weighings quickly because wet samples tend to lose weight by evaporation. After drying or ignition, residues often are very hygroscopic and rapidly absorb moisture from the air.

2. Apparatus

All the apparatus listed in Section 2540B.2 is required except that a balance capable of weighing to 10 mg may be used.

3. Procedure

a. Total solids:

1) Preparation of evaporating dish—If volatile solids are to be measured, ignite a clean evaporating dish at 550 ± 50°C for 1 h in a muffle furnace. If only total solids are to be measured, heat dish at 103 to 105°C for 1 h in an oven. Cool in desiccator, weigh, and store in desiccator until ready for use.

2) Sample analysis

a) Fluid samples—If the sample contains enough moisture to flow more or less readily, stir to homogenize, place 25 to 50 g in a prepared evaporating dish, and weigh. Evaporate to dryness on a water bath, dry at 103 to 105°C for 1 h, cool to balance temperature in an individual desiccator containing fresh desiccant, and weigh.

b) Solid samples—If the sample consists of discrete pieces of solid material (dewatered sludge, for example), take cores from each piece with a No. 7 cork borer or pulverize the entire sample coarsely on a clean surface by hand, using rubber gloves. Place 25 to 50 g in a prepared evaporating dish and weigh. Place in an oven at 103 to 105°C overnight. Cool to balance temperature in an individual desiccator containing fresh desiccant and weigh.

b. Fixed and volatile solids: Transfer to a cool muffle furnace, heat furnace to 550 ± 50°C, and ignite for 1 h. (If the residue from 2) above contains large amounts of organic matter, first ignite the residue over a gas burner and under an exhaust hood in the presence of adequate air to lessen losses due to reducing conditions and to avoid odors in the laboratory.) Cool in desiccator to balance temperature and weigh.

4. Calculation

$$\% \text{ total solids} = \frac{(A - B) \times 100}{C - B}$$

$$\% \text{ volatile solids} = \frac{(A - D) \times 100}{A - B}$$

$$\% \text{ fixed solids} = \frac{(D - B) \times 100}{A - B}$$

where:

A = weight of dried residue + dish, mg,
B = weight of dish,
C = weight of wet sample + dish, mg, and
D = weight of residue + dish after ignition, mg.

5. Precision and Bias

Precision and bias data are not now available.

6. Bibliography

GOODMAN, B.L. 1964. Processing thickened sludge with chemical conditioners. Pages 78 et seq. *in* Sludge Concentration, Filtration and Incineration. Univ. Michigan Continued Education Ser. No. 113, Ann Arbor.

GRATTEAU, J.C. & R.I. DICK. 1968. Activated sludge suspended solids determinations. *Water Sewage Works* 115:468.

2550   TEMPERATURE*

2550 A.   Introduction

Temperature readings are used in the calculation of various forms of alkalinity, in studies of saturation and stability with respect to calcium carbonate, in the calculation of salinity, and in general laboratory operations. In limnological studies, water temperatures as a function of depth often are required. Elevated temperatures resulting from discharges of heated water may have significant ecological impact. Identification of source of water supply, such as deep wells, often is possible by temperature measurements alone. Industrial plants often require data on water temperature for process use or heat-transmission calculations.

___
* Approved by Standard Methods Committee, 1988.

2550 B.   Laboratory and Field Methods

1. Laboratory and Other Non-Depth Temperature Measurements

Normally, temperature measurements may be made with any good mercury-filled Celsius thermometer. As a minimum, the thermometer should have a scale marked for every 0.1°C, with markings etched on the capillary glass. The thermometer should have a minimal thermal capacity to permit rapid equilibration. Periodically check the thermometer against a precision thermometer certified by the National Institute of Standards and Technology (NIST, formerly National Bureau of Standards)* that is used with its certificate and correction chart. For field operations use a thermometer having a metal case to prevent breakage.

2. Depth Temperature Measurements

Depth temperature required for limnological studies may be measured with a reversing thermometer, thermophone, or thermistor. The thermistor is most convenient and accurate; however, higher cost

___
* Some commercial thermometers may be as much as 3°C in error.

may preclude its use. Calibrate any temperature measurement devices with a NIST-certified thermometer before field use. Make readings with the thermometer or device immersed in water long enough to permit complete equilibration. Report results to the nearest 0.1 or 1.0°C, depending on need.

The thermometer commonly used for depth measurements is of the reversing type. It often is mounted on the sample collection apparatus so that a water sample may be obtained simultaneously. Correct readings of reversing thermometers for changes due to differences between temperature at reversal and temperature at time of reading. Calculate as follows:

$$\Delta T = \left[\frac{(T' - t)(T' + V_0)}{K} \right]$$

$$\times \left[1 + \frac{(T' - t)(T' + V_0)}{K} \right] + L$$

where:

   ΔT = correction to be added algebraically to uncorrected reading,

   T' = uncorrected reading at reversal,

t = temperature at which thermometer is read,

V_0 = volume of small bulb end of capillary up to 0°C graduation,

K = constant depending on relative thermal expansion of mercury and glass (usual value of $K = 6100$), and

L = calibration correction of thermometer depending on T^1.

If series observations are made it is convenient to prepare graphs for a thermometer to obtain ΔT from any values of T^1 and t.

3. Bibliography

WARREN, H.F. & G.C. WHIPPLE. 1895. The thermophone—A new instrument for determining temperatures. *Mass. Inst. Technol. Quart.* 8:125.

SVERDRUP, H.V., M.W. JOHNSON & R.H. FLEMING. 1942. The Oceans. Prentice- Hall, Inc., Englewood Cliffs, N.J.

AMERICAN SOCIETY FOR TESTING AND MATERIALS. 1949. Standard Specifications for ASTM Thermometers. No. E1-58, ASTM, Philadelphia, Pa.

REE, W.R. 1953. Thermistors for depth thermometry. *J. Amer. Water Works Assoc.* 45:259.

2710   TESTS ON SLUDGES*

2710 A.   Introduction

This section presents a series of tests uniquely applicable to sludges or slurries.

_____

*Approved by Standard Methods Committee, 1985.

The test data are useful in designing facilities for solids separation and concentration and for assessing operational behavior, especially of the activated sludge process.

2710 B.   Oxygen-Consumption Rate

1. General Discussion

This test is used to determine the oxygen consumption rate of a sample of a biological suspension such as activated sludge. It is useful in laboratory and pilot-plant studies as well as in the operation of full-scale treatment plants. When used as a routine plant operation test, it often will indicate changes in operating conditions at an early stage. However, because test conditions are not necessarily identical to conditions at the sampling site, the observed measurement may not be identical with actual oxygen consumption rate.

2. Apparatus

a. Oxygen-consumption rate device: Either:

1) *Probe with an oxygen-sensitive electrode* (polarographic or galvanic), or

2) *Manometric or respirometric device* with appropriate readout and sample capacity of at least 300 mL. The device should have an oxygen supply capacity greater than the oxygen consumption rate of the biological suspension, or at least 150 mg/L·h.

b. Stopwatch or other suitable timing device.

c. Thermometer to read to ± 0.5°C.

3. Procedure

a. Calibration of oxygen-consumption rate device: Either:

1) Calibrate the oxygen probe and meter according to the method given in Section 4500-O.G, or

2) Calibrate the manometric or respirometric device according to manufacturer's instructions.

b. Volatile suspended solids determination: See Section 2540.

c. Preparation of sample: Adjust temperature of a suitable sample portion to that of the basin from which it was collected or to required evaluation temperature, and maintain constant during analysis. Record temperature. Increase DO concentration of sample by shaking it in a partially filled bottle or by bubbling air or oxygen through it.

d. Measurement of oxygen consumption rate:

1) Fill sample container to overflowing with an appropriate volume of a representative sample of the biological suspension to be tested.

2) If an oxygen-sensing probe is used, immediately insert it into a BOD bottle containing a magnetic stirring bar and the biological suspension. Displace enough suspension with probe to fill flared top of bottle and isolate its contents from the atmosphere. Activate probe stirring mechanism and magnetic stirrer. (NOTE: Adequate mixing is essential. For suspensions with high concentrations of suspended solids, i.e., > 5000 mg/L, more vigorous mixing than that provided by the probe stirring mechanism and magnetic stirrer may be required.) If a manometric or respirometric device is used, follow manufacturer's instructions for startup.

3) After meter reading has stabilized, record initial DO and manometric or respirometric reading, and start timing device. Record appropriate DO, manometric, or respirometric data at time intervals of less than 1 min, depending on rate of consumption. Record data over a 15-min period or until DO becomes limiting, whichever occurs first. The oxygen probe may not be accurate below 1 mg DO/L. If a manometric or respirometric device is used, refer to manufacturer's instructions for lower limiting DO value. Low DO (≤ 2 mg/L at the start of the test) may limit oxygen uptake by the biological suspension and will be indicated by a decreasing rate of oxygen consumption as the test progresses. Reject such data as being unrepresentative of suspension oxygen consumption rate and repeat test beginning with higher initial DO levels.

The results of this determination are quite sensitive to temperature variations and poor precision is obtained unless replicate determinations are made at the same temperature. When oxygen consumption is used as a plant control test, run periodic (at least monthly) replicate determinations to establish the precision of the technique. This determination also is sensitive to the time lag between sample collection and test initiation.

4. Calculations

If an oxygen probe is used, plot observed readings (DO, milligrams per liter) versus time (minutes) on arithmetic graph paper and determine the slope of the line of best fit. The slope is the oxygen consumption rate in milligrams per liter per minute.

If a manometric or respirometric device is used, refer to manufacturer's instructions for calculating the oxygen consumption rate.

Calculate specific oxygen consumption rate in milligrams per gram per hour as follows:

Specific oxygen consumption rate, (mg/g)/h

$$= \frac{\text{oxygen consumption rate, (mg/L)/min}}{\text{volatile suspended solids, g/L}} \times \frac{60 \text{ min}}{\text{h}}$$

5. Precision and Bias

Bias is not applicable. The precision for this test has not been determined.

6. Bibliography

UMBREIT, W.W., R.H. BURRIS & J.F. STAUFFER. 1964. Manometric Techniques. Burgess Publishing Co., Minneapolis, Minn.

2710 C.   Settled Sludge Volume

1. General Discussion

The settled sludge volume of a biological suspension is useful in routine monitoring of biological processes. For activated sludge plant control, a 30-min settled sludge volume or the ratio of the 15-min to the 30-min settled sludge volume has been used to determine the returned-sludge flow rate and when to waste sludge. The 30-min settled sludge volume also is used to determine sludge volume index[1] (Section 2710D).

2. Apparatus

a. Settling column: Use 1-L graduated cylinder equipped with a stirring mechanism consisting of one or more thin rods extending the length of the column and positioned within two rod diameters of the cylinder wall. Provide a stirrer able to rotate the stirring rods at no greater than 4 rpm (peripheral tip speed of approximately 1.3 cm/s). See Figure 2710:1.

b. Stopwatch.

c. Thermometer.

3. Procedure

Place 1.0 L of sample in settling column and distribute the solids by covering the top and inverting the cylinder three times. Insert stirring rods, activate stirring mechanism, and let suspension settle. Continue stirring throughout test. Maintain suspension temperature during test at that in the basin from which the sample was taken.

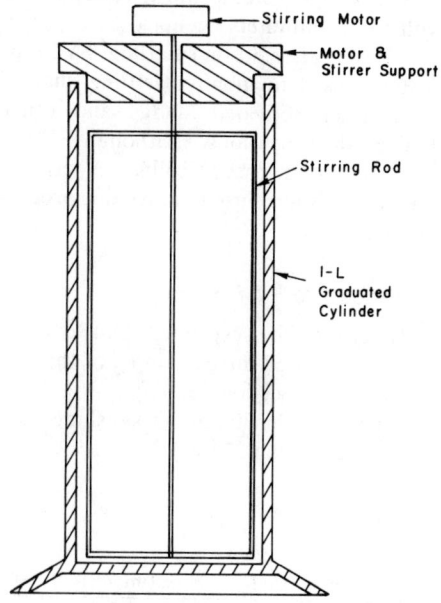

Figure 2710:1. Schematic diagram of settling vessel for settled sludge volume test.

Determine volume occupied by suspension at measured time intervals, e.g., 5, 10, 15, 20, 30, 45, and 60 min.

Report settled sludge volume of the suspension in milliliters for an indicated time interval.

Variations in suspension temperature, sampling and agitation methods, diameter of settling column, and time between sampling and start of the determination significantly affect results.

4. Precision and Bias

Bias is not applicable. The precision for this test has not been determined.

5. Reference

1. DICK, R.I. & P.A. VESILIND. 1969. The SVI—What is It? *J. Water Pollut, Control Fed.* 41:1285.

2710 D. Sludge Volume Index

1. General Discussion

The sludge volume index (SVI) is the volume in milliliters occupied by 1 g of a suspension after 30 min settling. SVI typically is used to monitor settling characteristics of activated sludge and other biological suspensions.[1] Although SVI is not supported theoretically,[2] experience has shown it to be useful in routine process control.

2. Procedure

Determine the suspended solids concentration of a well-mixed sample of the suspension (See Section 2540D).

Determine the 30 min settled sludge volume (See Section 2710C).

3. Calculations

$$SVI = \frac{\text{settled sludge volume (mL/L)} \times 1000}{\text{suspended solids (mg/L)}}$$

4. Precision and Bias

Precision is determined by the precision achieved in the suspended solids measurement, the settling characteristics of the suspension, and variables associated with the measurement of the settled sludge volume. Bias is not applicable.

5. References

1. DICK, R.I. & P.A. VESILIND. 1969. The SVI—What is it? *J. Water Pollut. Control Fed.* 41:1285.
2. FINCH, J. & H. IVES. 1950. Settleability indexes for activated sludge. *Sewage Ind. Wastes* 22:833.

6. Bibliography

DONALDSON, W. 1932. Some notes on the operation of sewage treatment works. *Sewage Works J.* 4:48.

MOHLMAN, F.W. 1934. The sludge index. *Sewage Works J.* 6:119.

RUDOLFS, W. & I.O. LACY. 1934. Settling and compacting of activated sludge. *Sewage Works J.* 6:647.

2710 E. Zone Settling Rate

1. General Discussion

At high concentrations of suspended solids, suspensions settle in the zone-settling regime. This type of settling takes place under quiescent conditions and is characterized by a distinct interface between the supernatant liquor and the sludge zone. The height of this distinct sludge interface is measured with time. Zone settling data for suspensions that undergo zone settling, e.g., activated sludge and metal hydroxide suspensions, can be used in the design, operation, and evaluation of settling basins.[1-3]

2. Apparatus

a. Settling vessel: Use a transparent cylinder at least 1 m high and 10 cm in diameter. To reduce the discrepancy between laboratory and full-scale thickener results, use larger diameters and taller cylinders.[1,3] Attach a calibrated millimeter tape to outside of cylinder. Equip cylinder with a stirring mechanism, e.g., one or more thin rods positioned within two rod diameters of the internal wall of settling vessel. Stir suspension near vessel wall over the entire depth of suspension at a peripheral speed no greater than 1 cm/s. Greater speeds may interfere with the thickening process and yield inaccurate results.[4] Provide the settling vessel with a port in the bottom plate for filling and draining. See Figure 2710:2.

b. Stopwatch.

c. Thermometer.

3. Procedure

Maintain suspension in a reservoir in a uniformly mixed condition. Adjust temperature of suspension to that of the basin from which it was collected or to required evaluation temperature. Record temperature. Remove a well-mixed sample from reservoir and measure suspended solids concentration (Section 2540D).

Activate stirring mechanism. Fill settling vessel to a fixed height by pumping suspension from reservoir or by gravity flow. Fill at a rate sufficient to maintain a uniform suspended solids concentration throughout settling vessel at end of filling. The suspension should agglomerate, i.e., form a coarse structure with visible fluid channels, within a few minutes. If suspension does not agglomerate, test is invalid and should be repeated.

Record height of solids-liquid interface at intervals of about 1 min. Collect data for sufficient time to assure that suspension is exhibiting a constant zone-settling velocity and that any initial reflocculation period, characterized by an accelerating interfacial settling velocity, has been passed.

Zone settling rate is a function of suspended solids concentration and suspen-

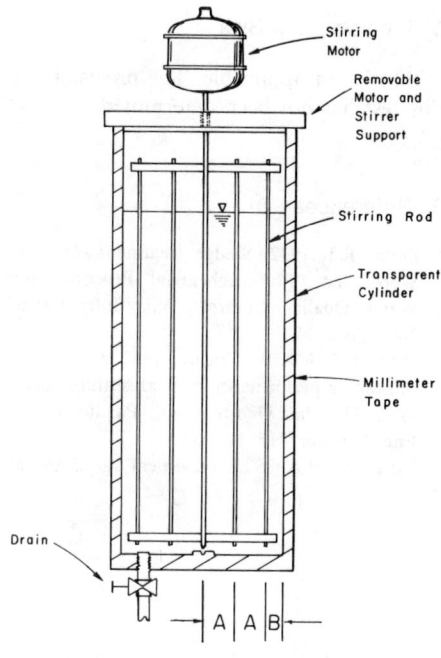

A = 10 cm minimum
B = 2 cm minimum

Figure 2710:2. Schematic diagram of settling vessel for zone settling rate test.

sion height as well as laboratory artifacts.[3] With the filling method described above and a sufficiently large cylinder, these artifacts should be minimized. However, even with careful testing suspensions often may behave erratically. Unpredictable behavior increases for sludges with high solids concentrations and poor settling characteristics, and in small cylinders.

4. Calculations

Plot interface height in centimeters vs. time in minutes.[1,3] Draw straight line through data points, ignoring initial shoulder or reflocculation period and compression shoulder. Calculate interfacial settling rate as slope of line in centimeters per minute.

5. Precision and Bias

Bias is not applicable. The precision for this test has not been determined.

6. References

1. DICK, R.I. 1972. Sludge treatment. *In* W.J. Weber, ed., Physicochemical Processes for Water Quality Control. Wiley-Interscience, New York, N.Y.
2. DICK, R.I. & K.W. YOUNG. 1972. Analysis of thickening performance of final settling tanks. *Proc. 27th Ind. Waste Conf.*, Purdue Univ., Eng. Ext. Ser. No. 141, 33.
3. VESILIND, P.A. 1975. Treatment and Disposal of Wastewater Sludges. Ann Arbor Science Publishing Co., Ann Arbor, Mich.
4. VESILIND, P.A. 1968. Discussion of Evaluation of activated sludge thickening theories. *J. San. Eng. Div., Proc. Amer. Soc. Civil Eng.* 94: SA1, 185.

7. Bibliography

DICK, R.I. & R.B. EWING. 1967. Evaluation of activated sludge thickening theories. *J. San. Eng. Div., Proc. Amer. Soc. Civil Eng.* 93:SA4, 9.
DICK, R.I. 1969. Fundamental aspects of sedimentation I & II. *Water Wastes Eng.* 3:47, 45, & 6:2.
DICK, R.I. 1970. Role of activated sludge final settling tanks. *J. San. Eng. Div., Proc. Amer. Soc. Civil Eng.* 96:SA2, 423.

2710 F.　Specific Gravity

1. General Discussion

The specific gravity of a sludge is the ratio of the masses of equal volumes of a sludge and distilled water. It is determined by comparing the mass of a known volume of a homogeneous sludge sample at a specific temperature to the mass of the same volume of distilled water at 4°C.

2. Apparatus

Container: A marked flask or bottle to hold a known sludge volume during weighing.

3. Procedure

Follow either *a* or *b*.

a. Record sample temperature, *T.* Weigh empty container and record weight, *W.* Fill empty container to mark with sample, weigh, and record weight, *S.* Fill empty container to mark with water, weigh, and record weight, *R.* Measure all masses to the nearest 10 mg.

b. If sample does not flow readily, add as much of it to container as possible without exerting pressure, record volume, weigh, and record mass, *P.* Fill container to mark with distilled water, taking care that air bubbles are not trapped in the sludge or container. Weigh and record mass, *Q.* Measure all masses to nearest 10 mg.

4. Calculation

Use *a* or *b*, matching choice of procedure above.

a. Calculate specific gravity, *SG*, from the formula

$$SG_{T/4°C} = \frac{\text{weight of sample}}{\text{weight of equal volume of water at 4°C}}$$

$$= \frac{S - W}{R - W} \times F$$

The values of the temperature correction factor F are given in Table 2710:I.

b. Calculate specific gravity, SG, from the formula

$$SG_{T/4^\circ C} = \frac{\text{weight of sample}}{\text{weight of equal volume of water at } 4^\circ C}$$

$$= \frac{(P - W)}{(R - W) - (Q - P)} \times F$$

For values of F, see Table 2710:I.

TABLE 2710:I. TEMPERATURE CORRECTION FACTOR

Temperature °C	Temperature Correction Factor
15	0.9991
20	0.9982
25	0.9975
30	0.9957
35	0.9941
40	0.9922
45	0.9903

2720   SLUDGE DIGESTER GAS*

2720 A.   Introduction

Gas produced during the anaerobic decomposition of wastes contains methane (CH_4) and carbon dioxide (CO_2) as the major components with minor quantities of hydrogen (H_2), hydrogen sulfide (H_2S), nitrogen (N_2), and oxygen (O_2). It is saturated with water vapor. Common practice is to analyze the gases produced to estimate their fuel value and to check on the treatment process. The relative proportions of CO_2, CH_4, and N_2 are normally of most concern and the easiest to determine because of the relatively high percentages of these gases.

1. Selection of Method

Two procedures are described for gas analysis, the volumetric method (B), and the gas chromatographic method (C). The volumetric analysis is suitable for the determination of CO_2, H_2, CH_4, and O_2. Nitrogen is estimated indirectly by difference. Although the method is time-consuming, the equipment is relatively simple. Because no calibration is needed before use, the procedure is particularly appropriate when analyses are conducted infrequently.

The principal advantage of gas chromatography is speed. Commercial equipment is designed specifically for ambient-temperature gas analysis and permits the routine separation and measurement of

---

*Approved by Standard Methods Committee, 1988.

CO_2, N_2, O_2, and CH_4 in less than 5 min. The requirement for a recorder, pressure-regulated bottles of carrier gas, and certified standard gas mixtures for calibration raise costs to the point where infrequent analyses by this method may be uneconomical. The advantages of this system are freedom from the cumulative errors found in sequential volumetric measurements, adaptability to other gas component analyses, adaptability to intermittent on-line sampling and analysis, and the use of samples of 1 mL or less.[1]

2. Sample Collection

When the source of gas is some distance from the apparatus used for analysis, collect samples in sealed containers and bring to the instrument. Displacement collectors are the most suitable containers. Long glass tubes with three-way glass stopcocks at each end, as indicated in Figure 2720:1, are particularly useful. These also are available with centrally located ports provided with septa for syringe transfer of samples. Connect one end of collector to gas source and vent three-way stopcock to the atmosphere. Clear line of air by passing 10 to 15 volumes of gas through vent and open stopcock to admit sample. If large quantities of gas are available, sweep air away by passing 10 to 15 volumes of gas through tube. If the gas supply is limited, fill tube with a liquid that is displaced by gas. Use either mercury or

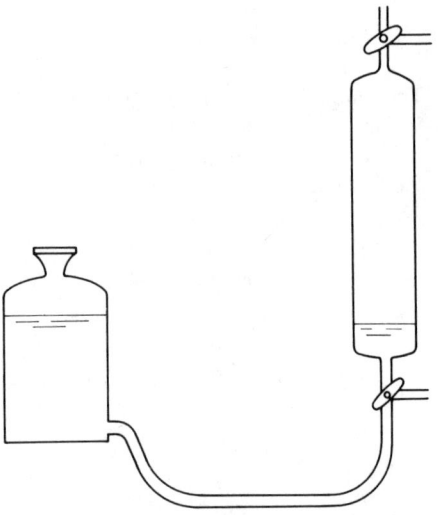

Figure 2720:1. Gas collection apparatus.

an acidified salt solution. The latter solution is easier and less expensive to use, but it dissolves gases to some extent. Therefore, fill collection tube completely with the gas and seal off from any contact with displacement fluid during temporary storage. When transferring gas to the gas-analyzing apparatus, do not transfer any fluid.

3. Reference

1. GRUNE, W.N. & C.F. CHUEH. 1962-63. Sludge gas analysis using gas chromatograph. *Water Sewage Works* 109:468; 110:43, 77, 102, 127, 171, 220, and 254.

2720 B.   Volumetric Method

1. General Discussion

a. Principle: This method may be used for the analysis either of digester gas or of methane in water (see Section 6211, Methane). A measured volume of gas is passed first through a solution of potassium hydroxide (KOH) to remove CO_2, next through a solution of alkaline pyrogallol to remove O_2, and then over heated cupric oxide, which removes H_2 by oxidation to water. After each of the above steps, the volume of gas remaining is measured; the decrease that results is a measure of the relative percentage of volume of each component in the mixture. Finally, CH_4 is determined by conversion to CO_2 and H_2O in a slow-combustion pipet or a catalytic oxidation assembly. The volume of CO_2 formed during combustion is measured to determine the fraction of methane originally present. Nitrogen is estimated by assuming that it represents the only gas remaining and equals the difference between 100% and the sum of the measured percentages of the other components.

When only CO_2 is measured, report only CO_2. No valid assumptions may be made about the remaining gases present without making a complete analysis.

Follow the equipment manufacturers' recommendations with respect to oxidation procedures.

CAUTION: *Do not attempt any slow-combustion procedure on digester gas because of the high probability of exceeding the explosive 5% by volume concentration of CH_4.*

2. Apparatus

Orsat-type gas-analysis apparatus, consisting of at least: (1) a water-jacketed gas buret with leveling bulb; (2) a CO_2-absorption pipet; (3) an O_2-absorption pipet; (4) a cupric oxide-hydrogen oxidation assembly; (5) a shielded catalytic CH_4-oxidation assembly or slow-combustion pipet assembly; and (6) a leveling bulb. With the slow-combustion pipet use a controlled source of current to heat the platinum filament electrically. Mercury is recommended as the displacement fluid; aqueous Na_2SO_4-H_2SO_4 solution also has been used successfully for sample collection. Use any

commercially available gas analyzer having these units.

3. Reagents

a. Potassium hydroxide solution: Dissolve 500 g KOH in distilled water and dilute to 1 L.

b. Alkaline pyrogallol reagent: Dissolve 30 g pyrogallol (also called pyrogallic acid) in distilled water and make up to 100 mL. Add 500 mL KOH solution.

c. Oxygen gas: Use approximately 100 mL for each gas sample analyzed.

d. Displacement liquid: Use either
1) *Mercury* or
2) *Sodium sulfate-sulfuric acid solution:* Dissolve 200 g Na_2SO_4 in 800 mL distilled water; add 30 mL conc H_2SO_4.

4. Procedure

a. Sample introduction: Transfer 5 to 10 mL gas sample into gas buret through a capillary-tube connection to the collector. Expel this sample to the atmosphere to purge the system. Transfer up to 100 mL gas sample to buret. Bring sample in buret to atmospheric or reference pressure by adjusting leveling bulb. Measure volume accurately and record as V_1.

b. Carbon dioxide absorption: Remove CO_2 from sample by passing it through the CO_2-absorption pipet charged with the KOH solution. Pass gas back and forth until sample volume remains constant. Before opening stopcocks between buret and any absorption pipet, make sure that the gas in the buret is under a slight positive pressure to prevent reagent in the pipet from contaminating stopcock or manifold. After absorption of CO_2, transfer sample to buret and measure volume. Record as V_2.

c. *Oxygen absorption:* Remove O_2 by passing sample through O_2-absorption pipet charged with alkaline pyrogallol reagent until sample volume remains constant. Measure volume and record as V_3. For digester gas samples, continue as directed in ¶ 4*d*. For CH_4 in water, store gas in CO_2 pipet and proceed to ¶ 4*e* below.

d. *Hydrogen oxidation:* Remove H_2 by passing sample through CuO assembly maintained at a temperature in the range 290 to 300°C. When a constant volume has been obtained, transfer sample back to buret, cool, and measure volume. Record as V_4.

Waste to the atmosphere all but 20 to 25 mL of remaining gas. Measure volume and record as V_5. Store temporarily in CO_2-absorption pipet.

e. *Methane oxidation:* Purge inlet connections to buret with O_2 by drawing 5 to 10 mL into buret and expelling to the atmosphere. Oxidize CH_4 either by the catalytic oxidation process for digester gas and gas phase of water samples or by the slow-combustion process for gas phase of water samples.

1) Catalytic oxidation process—For catalytic oxidation of digester gas and gas phase of water samples, transfer 65 to 70 mL O_2 to buret and record this volume as V_6. Pass O_2 into CO_2-absorption pipet so that it will mix with sample stored there. Return this mixture to buret and measure volume. Record as V_7. This volume should closely approximate V_5 plus V_6. Pass O_2-sample mixture through catalytic oxidation assembly, which should be heated in accordance with the manufacturer's directions. Keep rate of gas passage less than 30 mL/min. After first pass, transfer mixture back and forth through the assembly between buret and reservoir at a rate not faster than 60 mL/min until a constant volume is obtained. Record as V_8.

2) Slow-combustion process—For slow combustion of the gas phase of water samples, transfer 35 to 40 mL O_2 to buret and record volume as V_6. Transfer O_2 to slow-combustion pipet and then transfer sample from CO_2-absorption pipet to buret. Heat platinum coil in combustion pipet to yellow heat while controlling temperature by adjusting current. Reduce pressure of O_2 in pipet to somewhat less than atmospheric pressure by means of the leveling bulb attached to the pipet. Pass sample into slow-combustion pipet at rate of approximately 10 mL/min. After the first pass, transfer sample and O_2 mixture back and forth between pipet and buret several times at a faster rate, allowing mercury in pipet to rise to a point just below heated coil. Collect sample in combustion pipet, turn off coil, and cool pipet and sample to room temperature with a jet of compressed air. Transfer sample to buret and measure volume. Record as V_8.

f. *Measurement of carbon dioxide produced:* Determine amount of CO_2 formed in the reaction by passing sample through CO_2-absorption pipet until volume remains constant. Record volume as V_9.

Check accuracy of determination by absorbing residual O_2 from sample. After this absorption, record final volume as V_{10}.

5. Calculation

a. CH_4 and H_2 usually are the only combustible gases present in sludge digester gas. When this is the case, determine percentage by volume of each gas as follows:

$$\% \ CO_2 = \frac{(V_1 - V_2) \times 100}{V_1}$$

$$\% \ O_2 = \frac{(V_2 - V_3) \times 100}{V_1}$$

$$\% \ H_2 = \frac{(V_3 - V_4) \times 100}{V_1}$$

$$\% \ CH_4 = \frac{V_4 \times (V_8 - V_9) \times 100}{V_1 \times V_5}$$

$$\% \; N_2 = 100 - (\%CO_2 + \%O_2 + \%H_2 + \%CH_4)$$

b. Alternatively, calculate CH_4 by either of the two following equations:

$$\% \; CH_4 = \frac{V_4 \times (V_6 + V_{10} - V_9) \times 100}{2 \times V_1 \times V_5}$$

$$\% \; CH_4 = \frac{V_4 \times (V_7 - V_8) \times 100}{2 \times V_1 \times V_5}$$

Results from the calculations for CH_4 by the three equations should be in reasonable agreement. If not, repeat analysis after checking apparatus for sources of error, such as leaking stopcocks or connections. Other combustible gases, such as ethane, butane, or pentane, will cause a lack of agreement among the calculations; however, the possibility that digester gas contains a significant amount of any of these is remote.

6. Precision and Bias

A gas buret measures gas volume with a precision of 0.05 mL and a probable accuracy of 0.1 mL. With the large fractions of CO_2 and CH_4 normally present in digester gas, the overall error for their determination can be made less than $\pm 1\%$. The error in the determination of O_2 and H_2, however, can be considerable because of the small concentrations normally present. For a concentration as low as 1%, an error as large as $\pm 20\%$ can be expected. When N_2 is present in a similar low-volume percentage, the error in its determination would be even greater, because errors in each of the other determinations would be reflected in the calculation for N_2.

7. Bibliography

YANT, W.F. & L.B. BERGER. 1936. Sampling of Mine Gases and the Use of the Bureau of Mines Portable Orsat Apparatus in Their Analysis. Miner's Circ. No. 34, U.S. Bur. Mines, Washington, D.C.
MULLEN, P.W. 1955. Modern Gas Analysis. Interscience Publishers, New York, N.Y.

2720 C.   Gas Chromatographic Method

1. General Discussion

a. Principle: See Section 6630B for a discussion of gas chromatography.

b. Equipment selection: Many columns have been proposed for gas mixture analyses. Any that is capable of the desired separation is acceptable, provided that all of the exact conditions of analysis are reported with the calibration standards. The following directions are necessarily general. Follow the manufacturer's recommendations for the specific instrumentation.

2. Apparatus

a. Gas chromatograph: Use any instrument equipped with a thermal conductivity detector. With some column packings, ovens and temperature controls are necessary. Preferably use a unit with a gas sample valve.

b. Recorder: Use a 10-mV full-span strip chart recorder with the gas chromatograph. When minor components such as H_2 and H_2S are to be detected, a 1-mV full-span recorder is preferable.

c. *Column packing:** Some commercially available column packings useful for separating sludge gas components are listed below along with the routine separations possible at room temperature:[1,2]

1) Silica gel at room temperature: H_2, air ($O_2 + N_2$), CH_4, (CO_2-slow);

2) Molecular Sieve 13X: H_2, O_2, N_2, CH_4;

3) HMPA (hexamethylphosphoramide) 30% on Chromosorb P: CO_2 from (O_2, N_2, H_2, CH_4);

4) DEHS (di-2-ethylhexylsefacate) 30% on Chromosorb P: CO_2 from (O_2, N_2, H_2, CH_4).

Combinations of Columns 1 and 2, 3 and 2, or 4 and 2 when properly sized and used in the sequence: 1st column, detector, 2nd column, detector, readily will separate H_2, O_2, N_2, CH_4, and CO_2. Commercial equipment specifically designed for such operations is available.[2]

d. *Sample introduction apparatus:* An instrument equipped with gas-sampling valves is designed to permit automatic injection of a specific sample volume into the chromatograph. If such an instrument is not available, introduce samples with a 2-mL syringe fitted with a 27-gauge hypodermic needle. Reduce escape of gas by greasing plunger lightly with mineral oil or preferably by using a special gas-tight syringe.

3. Reagents

a. *Carrier gases:* Use helium for separating digester gases. If H_2 is to be determined, use argon as a carrier gas to increase the sensitivity greatly.

b. *Calibration gases:* Use samples of CH_4, CO_2, and N_2 of known purity, or mixtures of known composition, for calibration. Also use samples of O_2, H_2, and H_2S of

known purity if these gases are to be measured.

4. Procedure

a. *Preparation of gas chromatograph:* Adjust carrier gas flow rate to 60 to 80 mL/min. Turn on oven heaters, if used, and detector current and adjust to desired values. The instrument is ready for use when the recorder yields a stable base line. Silica gel and molecular sieve columns gradually lose activity because of adsorbed moisture or materials permanently adsorbed at room temperature. If insufficient separations occur, reactivate by heating or repacking.

b. *Calibration:* For accurate results, prepare a calibration curve for each gas to be measured because different gas components do not give equivalent detector responses on either a weight or a molar basis. Calibrate with synthetic mixtures or with pure gases.

1) Synthetic mixtures—Use purchased gas mixtures of known composition or prepare in the laboratory. Inject a standard volume of each mixture into the gas chromatograph and note response for each gas. Compute detector response, either as area under a peak or as height of peak, after correcting for attenuation. Read peak heights accurately and correlate with concentration of component in sample. Reproduce operating parameters exactly from one analysis to the next. If sufficient reproducibility cannot be obtained by this procedure, use peak areas for calibration. Prepare calibration curve by plotting either peak area or peak height against volume percent for each component.

2) Pure gases—Introduce pure gases into chromatograph individually with a syringe. Inject sample volumes of 0.25, 0.5, 1.0 mL, etc., and plot detector response, corrected for attenuation, against gas volume.

When the analysis system yields a linear detector response with increasing gas component concentration from zero to the

---

*Gas chromatographic methods are extremely sensitive to the materials used. Use of trade names in *Standard Methods* does not preclude the use of other existing or as-yet-undeveloped products that give *demonstrably* equivalent results.

range of interest, run standard mixtures along with samples. If the same sample size is used, calculate gas concentration by direct proportions.

c. *Sample analysis:* If samples are to be injected with a syringe, equip sample collection container with a port closed by a rubber or silicone septum. To take a sample for analysis, expel air from barrel of syringe by depressing plunger and force needle through the septum. Withdraw plunger to take gas volume desired, pull needle from collection container, and inject sample rapidly into chromatograph.

When samples are to be injected through a gas-sampling valve, connect sample collection container to inlet tube. Permit gas to flow from collection tube through the valve to purge dead air space and fill sample tube. About 15 mL normally are sufficient to clear the lines and to provide a sample of 1 to 2 mL. Transfer sample from loop into carrier gas stream by following manufacturer's instructions. Bring samples to atmospheric pressure before injection.

When calibration curves have been prepared with synthetic mixtures, use the same sample volume as that used during calibration. When calibration curves are prepared by the procedure using varying volumes of pure gases, inject any convenient gas sample volume up to about 2 mL.

5. Calculation

a. When calibration curves have been prepared with synthetic mixtures and the volume of the sample analyzed is the same as that used in calibration, read volume percent of each component directly from calibration curve after detector response for that component is computed.

b. When calibration curves are prepared with varying volumes of pure gases, calculate the percentage of each gas in the mixture as follows:

$$\text{Volume } \% = \frac{A}{B} \times 100$$

where:

A = partial volume of component (read from calibration curve) and
B = volume sample injected.

c. Where standard mixtures are run with samples and instrument response is linear from zero to the concentration range of interest:

$$\text{Volume } \% = \text{Volume } \% \text{ (std)} \times \frac{C}{D}$$

where:

C = recorder value of sample and
D = recorder value of standard.

6. Precision and Bias

Precision and bias depend on the instrument used and the techniques of operation. With proper care, a precision of 2% generally can be achieved. With digester gas the sum of the percent CH_4, CO_2, and N_2 should approximate 100%. If it does not, suspect errors in collection, handling, storage, and injection of gas, or in instrumental operation or calibration.

7. Reference

1. ANDREWS, J.F. 1968. Chromatographic analysis of gaseous products and reactants for biological processes. *Water Sewage Works* 115:54.
2. Column Systems for the Fisher Gas Partitioner. Tech. Bull. TB-154, Fisher Scientific Co., Atlanta, Ga. Catalog 77, Fisher Scientific Co.

2810 DISSOLVED GAS SUPERSATURATION*

2810 A. Introduction

Water can become supersaturated with atmospheric gases by various means, heating and air entrainment in spilled or pumped water being the most common. The primary sign of gas supersaturation is the formation of bubbles on submerged surfaces or within the vascular systems and tissues of aquatic organisms.

Gas supersaturation can limit aquatic life and interfere with water treatment processes. Levels of supersaturation lethal to aquatic organisms have been found in springs, rivers, wells, lakes, estuaries, and seawater. Gas supersaturation can be induced in pumped or processed water intended for drinking, fish hatchery supply, and laboratory bioassays. Seasonal and other temporal variations in supersaturation may occur. Because the rate of equilibration may be slow, supersaturation may persist in flowing water for days and ex-

cessive dissolved gas levels thus may persist far from the source of supersaturation.

Gas bubbles form only when the total dissolved gas pressure is greater than the sum of compensating pressures. Compensating pressures include water, barometric and, for organisms, tissue or blood pressure. The total dissolved gas pressure is equal to the sum of the partial pressures of all the dissolved gases, including water vapor. Typically, only nitrogen, oxygen, argon, carbon dioxide, and water vapor pressures need to be considered in most natural waters. Gas bubble disease, of fish or other aquatic organisms, is a result of excessive uncompensated gas pressure. A single supersaturated gas such as oxygen or nitrogen may not necessarily result in gas bubble disease because bubble formation depends largely on total dissolved gas pressure.

The degree of gas saturation should be described in terms of pressures rather than concentration or volume units.

* Approved by Standard Methods Committee, 1987.

2810 B.   Direct-Sensing Membrane-Diffusion Method

1. General Discussion

a. Principle: This method requires an instrument with a variable length of "gas permeable" tubing, connected to a pressure-measuring device. Dimethyl silicone rubber tubing often is used because it is highly permeable to dissolved gases, including water vapor. At steady state, the

gauge pressure inside the tubing is equal to the difference in gas pressure (ΔP) between the total dissolved gas pressure and the ambient barometric pressure. When the water is in equilibrium with the atmosphere, ΔP equals zero. If ΔP is greater than zero, the water is supersaturated. Conversely, ΔP is negative if the water is undersaturated.

b. Working range: The working range of this method depends on the pressure-sensing device used, but typically will range from -150 to $+600$ mm Hg. Dissolved solids in wastewater will not interfere with this method.

2. Apparatus

Several types of membrane-diffusion instruments are available commercially. Alternatively, construct a unit from commercially available parts. Several units have been described, including a direct-reading instrument using pressure transducers and a digital readout,[1] an on-line unit that can activate an alarm system,[2] and an early model of the Weiss saturometer.[3] Each of these units has specific advantages and limitations; the instrument of choice will depend on the specific application. All these instruments are portable so that data collection is completed in the field.

Test the instrument for leaks according to the manufacturer's recommendation. Even a very small leak, difficult to detect and locate, will result in useless data. Calibrate the pressure-measuring device with a mercury manometer or certified pressure gauge. If a manometer is used, include fresh mercury that flows freely in the tubing. An alternative method for directly testing membrane-diffusion instruments in a small, closed chamber where induced ΔP levels can be compared against observed ΔP levels is available.[2]

Van Slyke-Neill[4] or gas chromatography methods[1] are inappropriate for calibration but they may be used to verify results. These methods measure individual gas concentrations and require further conversion to ΔP or partial pressure and suffer from sampling and sample handling problems.[5-7]

3. Procedure

At the start of each day, test the instrument for leaks and recalibrate. At a monitoring site, completely submerge the sensing element in the water, preferably below the hydrostatic compensation depth. This is the depth where the hydrostatic and total gas pressures are equal and as a result, bubbles will not form on the tubing. Bubble formation on the silicone rubber tubing seriously reduces accuracy. Compute hydrostatic compensation depth[5] as follows:

$$Z = \frac{\Delta P}{73.42} \text{ at } 20°C \text{ and } 0 \text{ g/kg salinity}$$

and

$$Z = \frac{\Delta P}{75.40} \text{ at } 20°C \text{ and } 35 \text{ g/kg salinity}$$

where:

Z = depth, m, and
ΔP = pressure difference, mm Hg.

Dislodge formed bubbles on the tubing by gently striking the instrument or moving the instrument rapidly in the water. Movement of water across the silicone rubber tubing also facilitates establishing the equilibrium between gas pressure in the water and in the tubing.

Operate the instrument "bubble free" until a stable ΔP is observed. This may take from 5 to 30 min, depending on the ΔP, water temperature, water flow, and geometry of the system. The time response of the membrane-diffusion method is shown in Figure 2810:1 for "bubble-free" and "bubble" conditions.

If the instrument is used in heavily contaminated water containing oil or other organics, clean the silicone tubing with a mild detergent according to the manufacturer's instructions. Silicone rubber tubing has been used in uncontaminated natural water for at least 8 years without being adversely affected by attached algal growth.[2] The tubing can be damaged by abrasive grit, diatoms, biting aquatic organisms, certain organic compounds, and strong acids.[2]

Obtain barometric pressure data from a laboratory mercury barometer at least daily

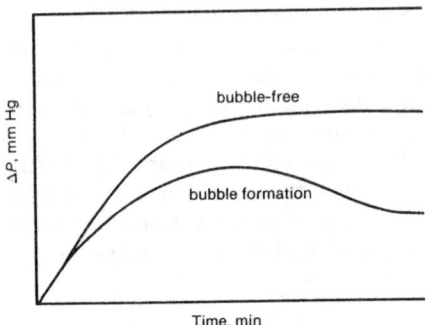

Figure 2810:1. Time response for the membrane-diffusion method.

or from a calibrated portable barometer at the sampling site. If there is more than a 50-m change in elevation between the barometer location and the sampling sites, use a portable barometer. Barometric pressures reported by weather agencies (or airports) are corrected to sea level and are unusable.

4. Calculation

a. Total gas pressure: Preferably report total gas pressure as ΔP.[2,6,8] Express pressure as millimeters of mercury.

Total gas pressure also has been reported as a percentage of local barometric pressure:

$$TGP \% = \left[\frac{P_b + \Delta P}{P_b} \right] \times 100$$

where:

P_b = true local barometric pressure, mm Hg.

The reporting of total gas pressure as a percentage is not encouraged.

b. Component gas pressures: When information on component gas supersaturation is needed, express data as partial pressures, differential pressures, or percent saturation.[5,8] This requires additional measurements of dissolved oxygen, temperature,

ture, and salinity* at the monitoring site. In a mixture of gases in a given volume, the partial pressure of a gas is the pressure that this gas would exert if it were the only gas present.

1) Oxygen partial pressure—Calculate partial pressure of oxygen as follows:

$$P_{O_2} = \frac{DO}{\beta_{O_2}} \times 0.5318$$

where:

P_{O_2} = partial pressure of oxygen,
β_{O_2} = Bunsen coefficient for oxygen (Table 2810:I) and
DO = measured concentration of oxygen, mg/L.

Bunsen coefficients for marine waters are available.[5] The factor 0.5318 equals 760/(1000 K), where K is the ratio of molecular weight to molecular volume for oxygen gas.[5]

2) Nitrogen partial pressure—Estimate the partial pressure of nitrogen by subtracting the partial pressures of oxygen and water vapor from the total gas pressure.

$$P_{N_2} = P_b + \Delta P - P_{O_2} - P_{H_2O}$$

where:

P_{H_2O} = vapor pressure of water from Table 2810:II.

This term includes a small contribution from argon and any other gases present, including carbon dioxide and methane. The partial pressure of carbon dioxide is negligible in waters of pH > 7.0.

3) Nitrogen:oxygen partial pressure ratio — The ratio of the partial pressure of nitrogen to the partial pressure of oxygen characterizes the relative contribution of the two gases to the total dissolved gas pressure. In water in equilibrium with air, this ratio is 3.77.

---
* Methods for these variables may be found in Sections 4500-O, 2550, and 2520, respectively.

c. Differential pressures: The differential pressure of a gas is the difference between the partial pressures of that gas in water and air. The oxygen differential pressure may be calculated as

$$\Delta P_{O_2} = P_{O_2} - 0.209\ 46(P_b - P_{H_2O})$$

and the nitrogen differential pressure as

$$\Delta P_{N_2} = \Delta P - \Delta P_{O_2}$$

d. Percent of nitrogen saturation: In older literature, supersaturation values have been reported as percent nitrogen saturation. This method of reporting component gases is discouraged but can be calculated as follows:

$$N_2(\%) = \left[\frac{P_{N_2}}{0.7902\ (P_b - P_{H_2O})}\right] \times 100$$

The following relationships are useful conversions:

$$TGP(\%) = 0.209\ 46\ O_2(\%) + 0.7902\ N_2(\%)$$

$$\Delta P = 0.209\ 46 \left[\frac{O_2(\%)}{100} - 1\right]\left[P_b - P_{H_2O}\right]$$

$$+ 0.7902\left[\frac{N_2(\%)}{100} - 1\right]\left[P_b - P_{H_2O}\right]$$

$$\Delta P = \frac{DO}{\beta_{O_2}}(0.5318)(1 + \%N_2/\%O_2)$$

$$- (P_b - P_{H_2O})$$

Use care with these relationships with older data because both TGP(%) and $N_2(\%)$ have been differently defined.[5]

5. Quality Control

The precision of the membrane-diffusion method depends primarily on the pressure-sensing instrument. For an experienced operator it is approximately ± 1 to 2 mm Hg with an accuracy of ± 3 to 5 mm Hg.[3,6]

Air leaks, bubble formation, incomplete equilibration, or condensation produce negative errors while direct water leaks can result in positive errors in submersible units.

6. Reporting of Results

In reporting results, include the following data:
Sensor depth, m,
Barometric pressure, mm Hg,
Water temperature, °C,
Dissolved oxygen, mm Hg or mg/L,
Salinity, g/kg, and
P, mm Hg.
If component gas information is needed add:
Partial pressure of oxygen, mm Hg,
Partial pressure of nitrogen, mm Hg, and
Oxygen:nitrogen partial pressure ratio
or
ΔP_{O_2}, mm Hg, and
ΔP_{N_2}, mm Hg.

7. Interpretation of Results

The biological effects of dissolved gas supersaturation depend on the species, age, depth in water column, length of exposure, temperature, and nitrogen:oxygen partial pressure ratio.[12] Safe limits generally are segregated into wild/natural circumstances, where behavior and hydrostatic pressure can modify the exposure by horizontal and vertical movements away from dangers, and captive environments such as aquaria, hatcheries, or laboratories, where conditions not only preclude escape but also include other significant stresses. Of these two realms, captive circumstances are more likely to cause illness or mortality from gas bubble disease and will do so sooner and at the lower ΔP levels.

In wild/natural circumstances, the limit of safe levels of gas supersaturation depends on the depth available to the species and/or species behavior, but this limit usually occurs at a ΔP between 50 and 150

TABLE 2810:I. BUNSEN COEFFICIENT FOR OXYGEN IN FRESH WATER

| | | | Bunsen Coefficient at Given Temperature (to nearest 0.1°C) L real gas/L atmosphere | | | | | | | |
Temperature °C	0.0	0.1	0.2	0.3	0.4	0.5	0.6	0.7	0.8	0.9
0	0.04914	0.04901	0.04887	0.04873	0.04860	0.04847	0.04833	0.04820	0.04807	0.04793
1	0.04780	0.04767	0.04754	0.04741	0.04728	0.04716	0.04703	0.04680	0.04678	0.04665
2	0.04653	0.04640	0.04628	0.04615	0.04603	0.04591	0.04579	0.04567	0.04555	0.04543
3	0.04531	0.04519	0.04507	0.04495	0.04484	0.04472	0.04460	0.04449	0.04437	0.04426
4	0.04414	0.04403	0.04392	0.04381	0.04369	0.04358	0.04347	0.04336	0.04325	0.04314
5	0.04303	0.04292	0.04282	0.04271	0.04260	0.04250	0.04239	0.04229	0.04218	0.04206
6	0.04197	0.04187	0.04177	0.04166	0.04156	0.04146	0.04136	0.04126	0.04116	0.04106
7	0.04096	0.04086	0.04076	0.04066	0.04056	0.04047	0.04037	0.04027	0.04018	0.04008
8	0.03999	0.03989	0.03980	0.03971	0.03961	0.03952	0.03943	0.03933	0.03924	0.03915
9	0.03906	0.03897	0.03888	0.03879	0.03870	0.03861	0.03852	0.03843	0.03835	0.03826
10	0.03817	0.03809	0.03800	0.03791	0.03783	0.03774	0.03766	0.03757	0.03749	0.03741
11	0.03732	0.03724	0.03716	0.03707	0.03699	0.03691	0.03683	0.03675	0.03667	0.03659
12	0.03651	0.03643	0.03635	0.03627	0.03619	0.03611	0.03604	0.03596	0.03588	0.03581
13	0.03573	0.03565	0.03558	0.03550	0.03543	0.03535	0.03528	0.03520	0.03513	0.03505
14	0.03498	0.03491	0.03484	0.03476	0.03469	0.03462	0.03455	0.03448	0.03441	0.03433
15	0.03426	0.03419	0.03412	0.03406	0.03399	0.03392	0.03385	0.03378	0.03371	0.03364
16	0.03358	0.03351	0.03344	0.03338	0.03331	0.03324	0.03318	0.03311	0.03305	0.03298
17	0.03292	0.03285	0.03279	0.03272	0.03266	0.03260	0.03253	0.03247	0.03241	0.03235
18	0.03228	0.03222	0.03216	0.03210	0.03204	0.03198	0.03192	0.03186	0.03180	0.03174
19	0.03168	0.03162	0.03156	0.03150	0.03144	0.03138	0.03132	0.03126	0.03121	0.03115

20	0.03109	0.03103	0.03098	0.03092	0.03086	0.03081	0.03075	0.03070	0.03064	0.03059
21	0.03053	0.03048	0.03042	0.03037	0.03031	0.03026	0.03020	0.03015	0.03010	0.03004
22	0.02999	0.02994	0.02989	0.02983	0.02978	0.02973	0.02968	0.02963	0.02958	0.02952
23	0.02947	0.02942	0.02937	0.02932	0.02927	0.02922	0.02917	0.02912	0.02907	0.02902
24	0.02897	0.02893	0.02888	0.02883	0.02878	0.02873	0.02868	0.02864	0.02859	0.02854
25	0.02850	0.02845	0.02840	0.02835	0.02831	0.02826	0.02822	0.02817	0.02812	0.02808
26	0.02803	0.02799	0.02794	0.02790	0.02785	0.02781	0.02777	0.02772	0.02768	0.02763
27	0.02759	0.02755	0.02750	0.02746	0.02742	0.02737	0.02733	0.02729	0.02725	0.02720
28	0.02716	0.02712	0.02708	0.02704	0.02700	0.02695	0.02691	0.02687	0.02683	0.02679
29	0.02675	0.02671	0.02667	0.02663	0.02659	0.02655	0.02651	0.02647	0.02643	0.02639
30	0.02635	0.02632	0.02628	0.02624	0.02620	0.02616	0.02612	0.02609	0.02605	0.02601
31	0.02597	0.02594	0.02590	0.02586	0.02582	0.02579	0.02575	0.02571	0.02568	0.02564
32	0.02561	0.02557	0.02553	0.02550	0.02546	0.02543	0.02539	0.02536	0.02532	0.02529
33	0.02525	0.02522	0.02518	0.02515	0.02511	0.02508	0.02504	0.02501	0.02498	0.02494
34	0.02491	0.02488	0.02484	0.02481	0.02478	0.02474	0.02471	0.02468	0.02465	0.02461
35	0.02458	0.02455	0.02452	0.02448	0.02445	0.02442	0.02439	0.02436	0.02433	0.02429
36	0.02426	0.02423	0.02420	0.02417	0.02414	0.02411	0.02408	0.02405	0.02402	0.02399
37	0.02396	0.02393	0.02390	0.02387	0.02384	0.02381	0.02378	0.02375	0.02372	0.02369
38	0.02366	0.02363	0.02360	0.02358	0.02355	0.02352	0.02349	0.02346	0.02343	0.02341
39	0.02338	0.02335	0.02332	0.02329	0.02327	0.02324	0.02321	0.02318	0.02316	0.02313
40	0.02310	0.02308	0.02305	0.02302	0.02300	0.02297	0.02294	0.02292	0.02289	0.02286

Based on Benson and Krause.[9,10]

TABLE 2810:II. VAPOR PRESSURE OF FRESH WATER

Temperature °C	Vapor Pressure at Given Temperature (to nearest 0.1°C) mm Hg									
	0.0	0.1	0.2	0.3	0.4	0.5	0.6	0.7	0.8	0.9
0	4.58	4.61	4.64	4.68	4.71	4.75	4.78	4.82	4.85	4.89
1	4.92	4.96	4.99	5.03	5.07	5.10	5.14	5.18	5.21	5.25
2	5.29	5.33	5.36	5.40	5.44	5.48	5.52	5.56	5.60	5.64
3	5.68	5.72	5.76	5.80	5.84	5.88	5.92	5.97	6.01	6.05
4	6.09	6.14	6.18	6.22	6.27	6.31	6.36	6.40	6.44	6.49
5	6.54	6.58	6.63	6.67	6.72	6.77	6.81	6.86	6.91	6.96
6	7.01	7.05	7.10	7.15	7.20	7.25	7.30	7.35	7.40	7.45
7	7.51	7.56	7.61	7.66	7.71	7.77	7.82	7.87	7.93	7.98
8	8.04	8.09	8.15	8.20	8.26	8.31	8.37	8.43	8.48	8.54
9	8.60	8.66	8.72	8.78	8.84	8.89	8.95	9.02	9.08	9.14
10	9.20	9.26	9.32	9.39	9.45	9.51	9.58	9.64	9.70	9.77
11	9.83	9.90	9.97	10.03	10.10	10.17	10.23	10.30	10.37	10.44
12	10.51	10.58	10.65	10.72	10.79	10.86	10.93	11.00	11.07	11.15
13	11.22	11.29	11.37	11.44	11.52	11.59	11.67	11.74	11.82	11.90
14	11.98	12.05	12.13	12.21	12.29	12.37	12.45	12.53	12.61	12.69
15	12.78	12.86	12.94	13.05	13.11	13.19	13.28	13.36	13.45	13.54
16	13.62	13.71	13.80	13.89	13.97	14.06	14.15	14.24	14.33	14.43
17	14.52	14.61	14.70	14.80	14.89	14.98	15.08	15.17	15.27	15.37
18	15.46	15.56	15.66	15.76	15.86	15.96	16.06	16.16	16.26	16.36
19	16.46	16.57	16.67	16.77	16.88	16.98	17.09	17.20	17.30	17.41

20	17.52	17.63	17.74	17.85	17.96	18.07	18.18	18.29	18.41	18.52
21	18.64	18.75	18.87	18.98	19.10	19.22	19.33	19.45	19.57	19.69
22	19.81	19.93	20.05	20.48	20.60	20.42	20.55	20.67	20.80	20.93
23	21.05	21.18	21.31	21.44	21.57	21.70	21.83	21.96	22.09	22.23
24	22.36	22.50	22.63	22.77	22.90	23.04	23.18	23.32	23.46	23.60
25	23.74	23.88	24.03	24.17	24.31	24.46	24.60	24.75	24.90	25.04
26	25.19	25.34	25.49	25.64	25.80	25.95	26.10	26.26	26.41	26.57
27	26.72	26.88	27.04	27.20	27.36	27.52	27.68	27.84	28.00	28.17
28	28.33	28.50	28.66	28.83	29.00	29.17	29.34	29.51	29.68	29.85
29	30.03	30.80	30.37	30.55	30.73	30.91	31.08	31.26	31.44	31.62
30	31.81	31.99	32.17	32.36	32.54	32.73	32.92	33.11	33.30	33.49
31	33.68	33.87	34.06	34.26	34.45	34.65	34.85	35.05	35.24	35.44
32	35.65	35.85	36.05	36.25	36.46	36.67	36.87	37.08	37.29	37.50
33	37.71	37.92	38.14	38.35	38.57	38.78	39.00	39.22	39.44	39.66
34	39.88	40.10	40.33	40.55	40.78	41.01	41.23	41.46	41.69	41.92
35	42.16	42.39	42.63	42.86	43.10	43.34	43.58	43.82	44.06	44.30
36	44.55	44.79	45.04	45.28	45.53	45.78	46.03	46.29	46.54	46.79
37	47.05	47.31	47.56	47.82	48.08	48.35	48.61	48.87	49.14	49.41
38	49.67	49.94	50.21	50.49	50.76	51.03	51.31	51.59	51.87	52.14
39	52.43	52.71	52.99	53.28	53.56	53.85	54.14	54.43	54.72	55.01
40	55.31	55.60	55.90	56.20	56.50	56.80	57.10	57.41	57.71	58.02

Based on Green and Carritt.[11]

mm Hg. Under captive conditions, the ΔP should be as close to zero as possible. For sensitive species and life stages, sublethal and lethal effects have been observed at ΔP of 10 to 50 mm Hg.[13]

8. References

1. D'AOUST, B.G., R. WHITE & H. SIEBOLD. 1975. Direct measurement of total dissolved gas partial pressure. *Undersea Biomed. Res.* 2:141.

2. BOUCK, G.R. 1982. Gasometer: an inexpensive device for continuous monitoring of dissolved gases and supersaturation. *Trans. Amer. Fish. Soc.* 111:505.

3. FICKEISEN, D.H., M.J. SCHNEIDER & J.C. MONTGOMERY. 1975. A comparative evaluation of the Weiss saturometer. *Trans. Amer. Fish. Soc.* 104:816.

4. BEININGEN, K.T. 1973. A Manual for Measuring Dissolved Oxygen and Nitrogen Gas Concentrations in Water with the Van Slyke-Neill Apparatus. Fish Commission of Oregon, Portland.

5. COLT, J. 1984. Computation of dissolved gas concentrations in water as functions of temperature, salinity, and pressure. Spec. Publ. 14, American Fisheries Soc., Bethesda, Md.

6. D'AOUST, B.G. & M.J.R. CLARK. 1980. Analysis of supersaturated air in natural waters and reservoirs. *Trans. Amer. Fish. Soc.* 109:708.

7. PIRIE, W.R. & W.A. HUBBERT. 1977. Assumptions in statistical analysis. *Trans. Amer. Fish. Soc.* 106:646.

8. COLT, J. 1983. The computation and reporting of dissolved gas levels. *Water Res.* 17:841.

9. BENSON, B.B. & D. KRAUSE. 1980. The concentration and isotopic fractionation in freshwater in equilibrium with the atmosphere. 1. Oxygen. *Limnol. Oceanogr.* 25:662.

10. BENSON, B.B. & D. KRAUSE. 1984. The concentration and isotopic fractionation of oxygen dissolved in freshwater and seawater in equilibrium with the atmosphere. *Limnol. Oceanogr.* 29:620.

11. GREEN, E.J. & D.E. CARRITT. 1967. New tables for oxygen saturation of seawater. *J. Mar. Res.* 25:140.

12. WEITKAMP, D.E. & M. KATZ. 1980. A review of dissolved gas supersaturation literature. *Trans. Amer. Fish. Soc.* 109:659.

13. COLT, J. 1986. The impact of gas supersaturation on the design and operation of aquatic culture systems. *Aquacult. Eng.* 5:49.

PART 3000

DETERMINATION OF

METALS

3010 INTRODUCTION

3010 A. General Discussion

1. Significance

The effects of metals in water and wastewater range from beneficial through troublesome to dangerously toxic. Some metals are essential, others may adversely affect water consumers, wastewater treatment systems, and receiving waters. Some metals may be either beneficial or toxic, depending on concentration.

2. Types of Methods

Metals may be determined satisfactorily by atomic absorption, inductively coupled plasma, or, with somewhat less precision and sensitivity, colorimetric methods. The absorption methods include flame and electrothermal techniques. Flame methods generally are applicable at moderate concentration levels in clean and complex-matrix systems. Electrothermal methods generally can increase sensitivity if matrix effects are not severe. Matrix modifiers may compensate for some matrix effects. Inductively coupled plasma (ICP) techniques are applicable over a broad linear range and are especially sensitive for refractory elements. In general, detection limits for ICP methods are higher than those for electrothermal methods. Colorimetric methods are applicable when interferences are known to be within the capacity of the particular method. Preliminary treatment of samples often is required. Appropriate pretreatment methods are described for each type of analysis.

3. Definition of Terms

a. Dissolved metals: Those constituents (metals) of an unacidified sample that pass through a 0.45-μm membrane filter.

b. Suspended metals: Those constituents (metals) of an unacidified sample that are retained by a 0.45-μm membrane filter.

c. Total metals: The concentration of metals determined on an unfiltered sample after vigorous digestion, or the sum of the concentrations of metals in both dissolved and suspended fractions.

d. Acid-extractable metals: The concentration of metals in solution after treatment of an unfiltered sample with hot dilute mineral acid.

To determine dissolved and suspended metals separately, filter immediately after sample collection. Do not preserve with acid until after filtration.

3010 B. Sampling and Sample Preservation

Before collecting a sample, decide what fraction is to be analyzed (dissolved, suspended, total, or acid-extractable). This decision will determine in part whether the sample is acidified with or without filtration and the type of digestion required.

Serious errors may be introduced during sampling and storage because of contamination from sampling device, failure to remove residues of previous samples from sample container, and loss of metals by adsorption on and/or precipitation in sample container caused by failure to acidify the sample properly.

1. Sample Containers

The best sample containers are made of quartz or TFE. Because these containers are expensive, the preferred sample container is made of polypropylene or linear polyethylene with a polyethylene cap. Borosilicate glass containers also may be used, but avoid soft glass containers for samples containing metals in the microgram-per-liter range. Store samples for determination of silver in light-absorbing containers. Use only containers and filters that have been acid rinsed.

2. Preservation

Preserve samples immediately after sampling by acidifying with concentrated nitric acid (HNO_3) to pH < 2. Filter samples for dissolved metals before preserving (see Section 3030). Usually 1.5 mL conc HNO_3/L sample (or 3 mL 1 + 1 HNO_3/L sample) is sufficient for short-term preservation. For samples with high buffer capacity, increase amount of acid (5 mL may be required for some alkaline or highly buffered samples). Use commercially available high-purity acid* or prepare high-purity acid by sub-boiling distillation of acid.

---

*Ultrex, J.T. Baker, or equivalent.

After acidifying sample, preferably store it in a refrigerator at approximately 4°C to prevent change in volume due to evaporation. Under these conditions, samples with metal concentrations of several milligrams per liter are stable for up to 6 months (except mercury, for which the limit is 5 weeks). For microgram-per-liter metal levels, analyze samples as soon as possible after sample collection.

Alternatively, preserve samples for mercury analysis by adding 2 mL/L 20% (w/v) $K_2Cr_2O_7$ solution (prepared in 1 + 1 HNO_3). Store in a refrigerator not contaminated with mercury. (CAUTION: Mercury concentrations may increase in samples stored in plastic bottles in mercury-contaminated laboratories.)

3. Bibliography

STRUEMPLER, A.W. 1973. Adsorption characteristics of silver, lead, calcium, zinc and nickel on borosilicate glass, polyethylene and polypropylene container surfaces. *Anal. Chem.* 45:2251.

FELDMAN, C. 1974. Preservation of dilute mercury solutions. *Anal. Chem.* 46:99.

KING, W.G., J.M. RODRIGUEZ & C.M. WAI. 1974. Losses of trace concentrations of cadmium from aqueous solution during storage in glass containers. *Anal. Chem.* 46:771.

BATLEY, G.E. & D. GARDNER. 1977. Sampling and storage of natural waters for trace metal analysis. *Water Res.* 11:745.

SUBRAMANIAN, K.S., C.L. CHAKRABARTI, J.E. SUETIAS & I.S. MAINES. 1978. Preservation of some trace metals in samples of natural waters. *Anal. Chem.* 50:444.

BERMAN, S. & P. YEATS. 1985. Sampling of seawater for trace metals. *Crit. Rev. Anal. Chem.* 16:1.

WENDLANDT, E. 1986. Sample containers and analytical accessories made of modern plastics for trace analysis. *Gewaess. Wass. Abwass.* 86:79.

3010 C. General Precautions

1. Sources of Contamination

Avoid introducing contaminating metals from containers, distilled water, or membrane filters. Some plastic caps or cap liners may introduce metal contamination; for example, zinc has been found in black bakelite-type screw caps as well as in many rubber and plastic products, and cadmium has been found in plastic pipet tips. Lead is a ubiquitous contaminant in urban air and dust.

2. Contaminant Removal

Thoroughly clean sample containers with a metal-free nonionic detergent solution, rinse with tap water, soak in acid, and then rinse with metal-free water. For quartz, TFE, or glass materials, use 1 + 1 HNO_3, 1 + 1 HCl, or aqua regia (3 parts conc HCl + 1 part conc HNO_3) for soaking. For plastic material, use 1 + 1 HNO_3 or 1 + 1 HCl. Reliable soaking conditions are 24 h at 70°C. Chromic acid or chromium-free substitutes* may be used to re-

*Nochromix, Godax Laboratories, or equivalent.

move organic deposits from containers, but rinse containers thoroughly with water to remove traces of chromium. Do not use chromic acid for plastic containers or if chromium is to be determined. Always use metal-free water in analysis and reagent preparation (see 3111B.2c). In these methods, the word "water" means metal-free water.

3. Airborne Contaminants

For analysis of microgram-per-liter concentrations of metals, airborne contaminants in the form of volatile compounds, dust, soot, and aerosols present in laboratory air may become significant. To avoid contamination use "clean laboratory" facilities such as commercially available laminar-flow clean-air benches or custom-designed work stations and analyze blanks that reflect the complete procedure.

4. Bibliography

MITCHELL, J.W. 1973. Ultrapurity in trace analysis. *Anal. Chem.* 45:492A.

GARDNER, M., D. HUNT & G. TOPPING. 1986. Analytical quality control (AQC) for monitoring trace metals in the coastal and marine environment. *Water Sci. Technol.* 18:35.

3020 QUALITY CONTROL*

Detailed recommendations and general information concerning both quality assurance and quality control (QC) are provided in Part 1000 and in individual methods. Refer to individual methods for method-

*Approved by Standard Methods Committee, 1989.

specific QC requirements. Also read Part 1000 for definitions of terms.

Always remember the overall purpose of the measurements: Use replicates to establish precision and known-additions recovery to determine bias. Use standards, control charts, blanks, calibrations, and other necessary measures. Provide adequate documentation. Quality-control measures and substantiation for operational-control determinations may differ from those for determinations made for regulatory purposes. Levels of trace metals may be orders of magnitude less than potential sources of contamination.

Before using a method, determine its detection limit (see Section 1030E). Instrumental methods generally are more sensitive than colorimetric ones. Verify that the method being used provides sufficient sensitivity for the purpose of the measurement.

Keep adequate records so that those unfamiliar with an analysis can reconstruct final results directly from raw data and have available a complete set of standard operating procedures.

To analyze a new or unfamiliar matrix, use the method of known additions to demonstrate freedom from interferences before calibrating. Verify the absence of interferences by analyzing such samples undiluted and in a 1:10 dilution; results should be comparable. Analyze a full procedural blank with each set of samples by carrying a reagent water blank through every step, including any digestion steps. Many metals are present in measurable quantities in acids and glassware and can affect results. If blank measurements result in concentrations above the method detection limit, repeat sample preparation with cleaner reagents or glassware. Verify that all glassware and materials are specially cleaned for metals analysis.

As a minimum for each analytical run have a calibration curve composed of a blank and two or more standards (depending on instrumentation), an external reference standard, a replicate, and a known addition to verify the absence of matrix interferences. Make replicate measurements on subsamples from a single sample bottle to establish method precision (as distinguished from the precision of samples). Precision and known-addition recovery should be within historical limits. Maintain control charts for more rigorous control of precision and bias.

Analyze a midpoint check standard and calibration blank at the beginning, end, and periodically (normally after each set of nine unknowns) with each group of samples to verify that the instrument calibration has not drifted. Initially, as a guideline use the following criteria: A determined value of the check standard outside 95 to 105% of the expected concentration suggests a potential problem. When a value is outside 90 to 110% of the expected concentration, the calibration usually is considered to be out of control; take corrective action. Use of calculated control limits (see Section 1020) will provide better indications of performance and is recommended for laboratories performing an analysis frequently (i.e., weekly). This may provide tighter limits.

To determine whether matrix effects exist, make known additions to samples before any digestion. Recoveries between 85 and 115% indicate that matrix effects are not significant.

Verify calibration standards and known addition solutions against an outside source. Agreement should be ± 5%.

Certain regulatory programs may require additional mandatory quality-control measures, i.e., use of a check standard at the maximum contaminant level (MCL).

3030 PRELIMINARY TREATMENT OF SAMPLES*

3030 A. Introduction

Samples containing particulates or organic material generally require pretreatment before analysis. "Total metals" includes all metals, inorganically and organically bound, both dissolved and particulate. Colorless, transparent samples (primarily drinking water) containing a turbidity of <1 NTU, no odor, and single phase may be analyzed directly by atomic absorption spectroscopy or inductively coupled plasma spectroscopy for total metals without digestion. For further verification or if changes in existing matrices are encountered, compare digested and undigested samples to ensure comparable results. On collection, acidify such samples to pH <2 with conc nitric acid (HNO$_3$) and analyze directly. Digest all other samples before determining total metals. To analyze for dissolved metals, filter sample, acidify filtrate, and analyze directly. To determine suspended metals, filter sample, digest filter and the material on it, and analyze. To determine acid-extractable metals, extract metals as indicated below and analyze extract.

This section describes general pretreatment for samples in which metals are to be determined according to Sections 3110 through 3500-Zn with several exceptions. The special digestion techniques for mercury are given in Sections 3112B.4b and c, and those for arsenic and selenium in Sections 3114 and 3500-Se.

Take care not to introduce metals into samples during preliminary treatment. During pretreatment avoid contact with rubber, metal-based paints, cigarette smoke, paper tissues, and all metal products including those made of stainless steel, galvanized metal, and brass. Conventional fume hoods can contribute significantly to sample contamination, particularly during acid digestion in open containers. Plastic pipet tips often are contaminated with copper, iron, zinc, and cadmium; before use soak in 2N HCl or HNO$_3$ for several days and rinse with deionized water. Check reagent-grade acids used for preservation, extraction, and digestion for purity. If excessive metal concentrations are found, purify the acids by distillation or use ultrapure acids. Carry blanks through all digestion and filtration steps and apply the necessary corrections to the results.

---

*Approved by Standard Methods Committee, 1989.

3030 B. Preliminary Filtration

If dissolved or suspended metals are to be determined, filter sample at time of collection using a preconditioned plastic filtering device with either vacuum or pressure, containing a filter support of plastic or TFE, through a prewashed ungridded 0.4- to 0.45-μm-pore-diam membrane filter (polycarbonate or cellulose acetate). Before use filter a blank consisting of reagent water to insure freedom from contamination. Precondition filter and filter device by rinsing with 50 mL deionized water. If the filter blank contains significant metals concentrations, soak membrane filters in approximately 0.5N HCl or 1 + 1 HNO₃ (recommended for electrothermal analysis) and rinse with water before use.

NOTE: Different filters display different sorption and filtration characteristics; for trace analysis, test filter to verify complete recovery of metals.

If filter is to be digested for suspended metals, record sample volume filtered and include a filter in determination of blank. Before filtering, centrifuge highly turbid samples in acid-washed TFE or high-density plastic tubes to reduce loading on filters. Stirred, pressure filter units foul less readily than vacuum filters; filter at a pressure of 70 to 130 kPa. After filtration acidify filtrate to pH 2 with conc HNO₃ and analyze directly. If a precipitate forms on acidification, digest acidified filtrate before analysis using Method E. Retain filter and digest it for direct determination of suspended metals.

CAUTION: *Do not use perchloric acid to digest membrane filters.*

3030 C. Preliminary Treatment for Acid-Extractable Metals

Extractable metals are lightly adsorbed on particulate material. Because some sample digestion may be unavoidable use rigidly controlled conditions to obtain meaningful and reproducible results. Maintain constant sample volume, acid volume, and contact time. Express results as extractable metals and specify extraction conditions.

At collection, acidify entire sample with 5 mL conc HNO₃/L sample. To prepare sample, mix well, transfer 100 mL to a beaker or flask, and add 5 mL 1 + 1 high-purity HCl. Heat 15 min on a steam bath. Filter through a membrane filter, adjust filtrate volume to 100 mL with water, and analyze.

3030 D. Preliminary Digestion for Metals

To reduce interference by organic matter and to convert metal associated with particulates to a form (usually the free metal) that can be determined by atomic absorption spectrometry or inductively-coupled plasma spectroscopy, use one of the digestion techniques presented below. Use the least rigorous digestion method required to provide complete and consistent recovery compatible with the analytical method and the metal being analyzed.

Nitric acid will digest most samples adequately (Section 3030E). Nitrate is an acceptable matrix for both flame and electrothermal atomic absorption. Some samples may require addition of perchloric, hydrochloric, or sulfuric acid for complete digestion. These acids may interfere in the analysis of some metals and all provide a poorer matrix for electrothermal analysis. Confirm metal recovery for each digestion and analytical procedure used. Use Table 3030:I as a guide in determining which acids (in addition to HNO_3) to use for complete digestion. As a general rule, HNO_3 alone is adequate for clean samples or easily oxidized materials; HNO_3-H_2SO_4 or HNO_3-HCl digestion is adequate for readily oxidizable organic matter; HNO_3-$HClO_4$ or HNO_3-$HClO_4$-HF digestion is necessary for difficult-to-oxidize organic matter or minerals. Dry ashing is helpful if large amounts of organic matter are present.

For determining Ag concentrations greater than 0.03 mg/L, dilute sample to contain less than 1 mg/L Ag and digest using method 3030F.3b.

Report digestion technique used.

Acid digestion techniques (Section 3030E–I) generally yield comparable precision and bias for most sample types that are totally digested by the technique. Because acids used in digestion will add metals to the samples and blanks, minimize the volume of acids used.

Suggested sample volumes are indicated below. If the recommended volume exceeds digestion vessel capacity, add sample as evaporation proceeds.

When samples are concentrated during digestion (e.g., > 100 mL sample used) determine metal recovery for each matrix digested, to verify method validity. Using larger samples will require additional acid, which also would increase the concentration of impurities.

Estimated Metal Concentration mg/L	Sample Volume* mL
<1	1000
1–10	100
10–100	10
100–1000	1

*For flame atomic absorption spectrometry. For graphite furnace, smaller volumes are appropriate.

Dry ashing (Section 3030J) yields highly variable precision and bias, depending on sample type and metal being analyzed. Dry ash only samples that have been shown to yield acceptable precision and bias.

Report results as follows:

$$\text{Metal concentration, mg/L} = A \times \frac{B}{C}$$

where:
A = concentration of metal in digested solution, mg/L,
B = final volume of digested solution, mL, and
C = sample size, mL.

Prepare solid samples or liquid sludges with high solids contents on a weight basis. Mix sample and transfer a suitable amount (typically 1 g of a sludge with 15% total solids) to a preweighed digestion vessel. Reweigh and calculate weight of sample. Proceed with one of the digestion techniques

TABLE 3030:I. ACIDS USED IN CONJUNCTION WITH HNO_3 FOR SAMPLE PREPARATION

Acid	Recommended for	May Be Helpful for	Not Recommended for
HCl	—	Sb, Ru, Sn	Ag, Th, Pb
H_2SO_4	Ti	—	Ag, Pb, Ba
$HClO_4$	—	Organic materials	—
HF	—	Siliceous materials	—

presented below. Report results on wet- or dry-weight basis as follows:

Metal concentration, mg/kg (wet-weight basis)
$$= \frac{A \times B}{\text{g sample}}$$

Metal concentration, mg/kg (dry-weight basis)
$$= \frac{A \times B}{\text{g sample}} \times \frac{100}{D}$$

where:

A = concentration of metal in digested solution, mg/L,

B = final volume of digested solution, mL, and

D = total solids, % (see Section 2540G).

Always prepare acid blanks for each type of digestion performed. Experience indicates that a blank made with the same acids and subjected to the same digestion procedure as the sample can correct for impurities present in acids, in reagent water, or on glassware.

3030 E. Nitric Acid Digestion

1. Apparatus

a. Hot plate.

b. Conical (erlenmeyer) flasks, 125-mL, or *Griffin beakers,* 150 mL, acid-washed and rinsed with water.

2. Reagents

Nitric acid, HNO_3, conc.

3. Procedure

Mix sample and transfer a suitable volume (50 to 100 mL) to a 125-mL conical flask or beaker. Add 5 mL conc HNO_3 and a few boiling chips, glass beads, or Hengar granules. Bring to a slow boil and evaporate on a hot plate to the lowest volume possible

(about 10 to 20 mL) before precipitation occurs. Continue heating and adding conc HNO_3 as necessary until digestion is complete as shown by a light-colored, clear solution. Do not let sample dry during digestion.

Wash down flask or beaker walls with water and then filter if necessary (see Section 3030B). Transfer filtrate to a 100-mL volumetric flask with two 5-mL portions of water, adding these rinsings to the volumetric flask. Cool, dilute to mark, and mix thoroughly. Take portions of this solution for required metal determinations. Alternatively, take a larger sample volume through the procedure for concentration (see 3030D).

3030 F. Nitric Acid-Hydrochloric Acid Digestion

1. Apparatus

See 3030E.1

2. Reagents

a. Nitric acid, HNO_3, conc.
b. Hydrochloric acid, 1 + 1.

3. Procedure

a. Total HNO_3/HCl: Transfer a measured volume of well-mixed, acid-preserved sample appropriate for the expected metals concentrations to a flask or beaker (see 3030E.1b). Add 3 mL conc HNO_3. Place flask or beaker on a hot plate and cautiously evaporate to less than 5 mL, making certain that sample does not boil and that no area of the bottom of the container is allowed to go dry. Cool and add 5 mL conc HNO_3. Cover container with a watch glass and return to hot plate. Increase temperature of hot plate so that a gentle reflux action occurs. Continue heating, adding additional acid as necessary, until digestion is complete (generally indicated when the digestate is light in color or does not change in appearance with continued refluxing). Evaporate to less than 5 mL and cool. Add 10 mL 1 + 1 HCl and 15 mL water per 100 mL anticipated final volume. Heat for an additional 15 min to dissolve any precipitate or residue. Cool, wash down beaker walls and watch glass with water, and filter to remove insoluble material that could clog the nebulizer. Alternatively centrifuge or let settle overnight. Adjust to a predetermined volume based on the expected metals concentrations.

b. Recoverable HNO_3/HCl: For this less rigorous digestion procedure, transfer a measured volume of well-mixed, acid-preserved sample to a flask or beaker. Add 2 mL 1 + 1 HNO_3 and 10 mL 1 + 1 HCl and heat on a steam bath or hot plate until volume has been reduced to near 25 mL, making certain sample does not boil. Cool and filter to remove insoluble material or alternatively centrifuge or let settle overnight. Adjust volume to 100 mL and mix.

3030 G.   Nitric Acid-Sulfuric Acid Digestion

1. Apparatus

See 3030E.1.

2. Reagents

a. *Methyl orange indicator solution.*
b. *Nitric acid,* HNO_3, conc.
c. *Sulfuric acid,* H_2SO_4, conc.

3. Procedure

Mix sample and pipet a suitable volume into a suitable flask or beaker (see 3030E.1b). If sample is not already acidified, acidify to methyl orange end point with conc H_2SO_4 and add 5 mL conc HNO_3 and a few boiling chips, glass beads, or Hengar granules. Bring to slow boil on hot plate and evaporate to 15 to 20 mL. Add 5 mL conc HNO_3 and 10 mL conc H_2SO_4. Evaporate on a hot plate until dense white fumes of SO_3 just appear. If solution does not clear, add 10 mL conc HNO_3 and repeat evaporation to fumes of SO_3. Heat to remove all HNO_3 before continuing treatment. All HNO_3 will be removed when the solution is clear and no brownish fumes are evident. Do not let sample dry during digestion.

Cool and dilute to about 50 mL with water. Heat to almost boiling to dissolve slowly soluble salts. Filter if necessary, then complete procedure as directed in Section 3030E.3 beginning with, "Transfer filtrate . . ."

3030 H.   Nitric Acid-Perchloric Acid Digestion

1. Apparatus

See 3030E.1. The following are also needed:
   a. *Safety shield.*
   b. *Safety goggles.*

2. Reagents

   a. *Nitric acid,* HNO_3, *conc.*
   b. *Perchloric acid,* $HClO_4$.
   c. *Methyl orange indicator aqueous solution.*
   d. *Ammonium acetate solution:* Dissolve 500 g $NH_4C_2H_3O_2$ in 600 mL water.

3. Procedure

CAUTION: *Heated mixtures of* $HClO_4$ *and organic matter may explode violently. Avoid this hazard by taking the following precautions: (a) do not add* $HClO_4$ *to a hot solution containing organic matter; (b) always pretreat samples containing organic matter with* HNO_3 *before adding* $HClO_4$; *(c) avoid repeated fuming with* $HClO_4$ *in ordinary hoods (For routine operations, use a water pump attached to a glass fume eradicator.* * *Stainless steel fume hoods with adequate water washdown facilities are available commercially and are acceptable for use with* $HClO_4$.); *and (d) never let sam-*

---

*Such as those obtainable from GFS Chemical Co., Columbus, Ohio.

ples being digested with $HClO_4$ *evaporate to dryness.*

Mix sample and transfer a suitable volume into a suitable flask or beaker (see 3030E.1b). If sample is not already acidified, acidify to methyl orange end point with conc HNO_3, add an additional 5 mL conc HNO_3 and a few boiling chips or glass beads, and evaporate on a hot plate to 15 to 20 mL. Add 10 mL each of conc HNO_3 and $HClO_4$, cooling flask or beaker between additions. Evaporate gently on a hot plate until dense white fumes of $HClO_4$ just appear. If solution is not clear, cover container with a watch glass and keep solution just boiling until it clears. If necessary, add 10 mL conc HNO_3 to complete digestion. Cool, dilute to about 50 mL with water, and boil to expel any chlorine or oxides of nitrogen. Filter, then complete procedure as directed in 3030E.3 beginning with, "Transfer filtrate . . ."

If lead is to be determined in the presence of high amounts of sulfate (e.g., determination of Pb in power plant fly ash samples), dissolve $PbSO_4$ precipitate as follows: Add 50 mL ammonium acetate solution to flask or beaker in which digestion was carried out and heat to incipient boiling. Rotate container occasionally to wet all interior surfaces and dissolve any deposited residue. Reconnect filter and slowly draw solution through it. Transfer filtrate to a 100-mL volumetric flask, cool, dilute to mark, mix thoroughly, and set aside for determination of lead.

3030 I.   Nitric Acid-Perchloric Acid-Hydrofluoric Acid Digestion

1. Apparatus

a. *Hot plate.*
b. *TFE beakers,* 250-mL, acid-washed and rinsed with water.

2. Reagents

a. *Nitric acid,* HNO_3, conc and 1 + 1.
b. *Perchloric acid,* $HClO_4$.
c. *Hydrofluoric acid,* HF, 48 to 51%.

3. Procedure

CAUTION: *See precautions for using $HClO_4$ in 3030H; handle HF with extreme care and provide adequate ventilation, especially for the heated solution. Avoid all contact with exposed skin. Provide medical attention for HF burns.*

Mix sample and transfer a suitable volume into a 250-mL TFE beaker. Add a few boiling chips and bring to a slow boil. Evaporate on a hot plate to 15 to 20 mL. Add 12 mL conc HNO_3 and evaporate to near dryness. Repeat HNO_3 addition and evaporation. Let solution cool, add 20 mL $HClO_4$ and 1 mL HF, and boil until solution is clear and white fumes of $HClO_4$ have appeared. Cool, add about 50 mL water, filter, and proceed as directed in 3030E.3 beginning with, "Transfer filtrate . . ."

3030 J.   Dry Ashing

1. Apparatus

See Sections 2540B.2 and 2540C.2.

2. Procedure

Mix sample and transfer a suitable volume into a platinum or high-silica glass* evaporating dish. Evaporate to dryness on a steam bath. Transfer dish to a muffle furnace and heat sample to a white ash. If volatile elements are to be determined, keep temperature at 400 to 450°C. If sodium only is to be determined, ash sample at a temperature up to 600°C. Dissolve ash in a minimum quantity of conc HNO_3 and warm water. Filter diluted sample and adjust to a known volume, preferably so that the final HNO_3 concentration is about 1%.

Take portions of this solution for metals determination.

---

*Vycor, a product of Corning Glass Works, Corning, N.Y., or equivalent.

3110   METALS BY ATOMIC ABSORPTION
SPECTROMETRY*

Because requirements for determining metals by atomic absorption spectrometry vary with metal and/or concentration to be determined, the method is presented as follows:

Section 3111, Metals by Flame Atomic Absorption Spectrometry, encompasses:

• Determination of antimony, bismuth, cadmium, calcium, cesium, chromium, cobalt, copper, gold, iridium, iron, lead, lithium, magnesium, manganese, nickel, palladium, platinum, potassium, rhodium, ruthenium, silver, sodium, strontium, thallium, tin, and zinc by direct aspiration into an air-acetylene flame (3111B),

• Determination of low concentrations of cadmium, chromium, cobalt, copper, iron, lead, manganese, nickel, silver, and zinc by chelation with ammonium pyrrolidine dithiocarbamate (APDC), extraction into methyl isobutyl ketone (MIBK), and

aspiration into an air-acetylene flame (3111C),

• Determination of aluminum, barium, beryllium, molybdenum, osmium, rhenium, silicon, thorium, titanium, and vanadium by direct aspiration into a nitrous oxide-acetylene flame (3111D), and

• Determination of low concentrations of aluminum and beryllium by chelation with 8-hydroxyquinoline, extraction into MIBK, and aspiration into a nitrous oxide-acetylene flame (3111E).

Section 3112 covers determination of mercury by the cold vapor technique.

Section 3113 concerns determination of micro quantities of aluminum, antimony, arsenic, barium, beryllium, cadmium, chromium, cobalt, copper, iron, lead, manganese, molybdenum, nickel, selenium, silver, and tin by electrothermal atomic absorption spectrometry.

Section 3114 covers determination of arsenic and selenium by conversion to their hydrides and aspiration into an argon-hydrogen or nitrogen-hydrogen flame.

---

* Approved by Standard Methods Committee, 1988.

3111   METALS BY FLAME ATOMIC ABSORPTION SPECTROMETRY*

3111 A.   Introduction

1. Principle

Atomic absorption spectrometry resembles emission flame photometry in that a sample is aspirated into a flame and atomized. The major difference is that in flame photometry the amount of light emitted is measured, whereas in atomic absorption spectrometry a light beam is directed through the flame, into a monochromator, and onto a detector that measures the amount of light absorbed by the atomized element in the flame. For some metals, atomic absorption exhibits superior sensitivity over flame emission. Because each metal has its own characteristic absorption wavelength, a source lamp composed of that element is used; this makes the method relatively free from spectral or radiation interferences. The amount of energy at the characteristic wavelength absorbed in the flame is proportional to the concentration of the element in the sample over a limited concentration range. Most atomic absorption instruments also are equipped for operation in an emission mode.

2. Selection of Method

See Section 3110.

---

*Approved by Standard Methods Committee, 1989.

3. Interferences

a. Chemical interference: Many metals can be determined by direct aspiration of sample into an air-acetylene flame. The most troublesome type of interference is termed "chemical" and results from the lack of absorption by atoms bound in molecular combination in the flame. This can occur when the flame is not hot enough to dissociate the molecules or when the dissociated atom is oxidized immediately to a compound that will not dissociate further at the flame temperature. Such interferences may be reduced or eliminated by adding specific elements or compounds to the sample solution. For example, the interference of phosphate in the magnesium determination can be overcome by adding lanthanum. Similarly, introduction of calcium eliminates silica interference in the determination of manganese. However, silicon and metals such as aluminum, barium, beryllium, and vanadium require the higher-temperature, nitrous oxide-acetylene flame to dissociate their molecules. The nitrous oxide-acetylene flame also can be useful in minimizing certain types of chemical interferences encountered in the air-acetylene flame. For example, the interference caused by high concentrations of phosphate in the determination of calcium in the air-acetylene flame does not occur in the nitrous oxide-acetylene flame. MIBK extractions with APDC (see

3111C) are particularly useful where a salt matrix interferes, for example, in seawater. This procedure also concentrates the sample so that the detection limits are extended.

Brines and seawater can be analyzed by direct aspiration but sample dilution is recommended. Aspiration of solutions containing high concentrations of dissolved solids often results in solids buildup on the burner head. This requires frequent shutdown of the flame and cleaning of the burner head. Preferably use background correction when analyzing waters that contain in excess of 1% solids, especially when the primary resonance line of the element of interest is below 240 nm. Make more frequent recovery checks when analyzing brines and seawaters to insure accurate results in these concentrated and complex matrices.

Barium and other metals ionize in the flame, thereby reducing the ground state (potentially absorbing) population. The addition of an excess of a cation (sodium, potassium, or lithium) having a similar or lower ionization potential will overcome this problem. The wavelength of maximum absorption for arsenic is 193.7 nm and for selenium 196.0 nm—wavelengths at which the air-acetylene flame absorbs intensely. The sensitivity for arsenic and selenium can be improved by conversion to their gaseous hydrides and analyzing them in either a nitrogen-hydrogen or an argon-hydrogen flame with a quartz tube (see Section 3114).

b. Background correction: Molecular absorption and light scattering caused by solid particles in the flame can cause erroneously high absorption values resulting in positive errors. When such phenomena occur, use background correction to obtain accurate values. Use any one of three types of background correction: continuum-source, Zeeman, or Smith-Hieftje correction.

1) Continuum-source background correction—A continuum-source background corrector utilizes either a hydrogen-filled hollow cathode lamp with a metal cathode or a deuterium arc lamp. When both the line source hollow-cathode lamp and the continuum source are placed in the same optical path and are time-shared, the broadband background from the elemental signal is subtracted electronically, and the resultant signal will be background-compensated.

Both the hydrogen-filled hollow-cathode lamp and deuterium arc lamp have lower intensities than either the line source hollow-cathode lamp or electrodeless discharge lamps. To obtain a valid correction, match the intensities of the continuum source with the line source hollow-cathode or electrodeless discharge lamp. The matching may result in lowering the intensity of the line source or increasing the slit width; these measures have the disadvantage of raising the detection limit and possibly causing nonlinearity of the calibration curve. Background correction using a continuum source corrector is susceptible to interference from other absorbing lines in the spectral bandwidth. Miscorrection occurs from significant atomic absorption of the continuum source radiation by elements other than that being determined. When a line source hollow-cathode lamp is used without background correction, the presence of an absorbing line from another element in the spectral bandwidth will not cause an interference unless it overlaps the line of interest.

Continuum-source background correction will not remove direct absorption spectral overlap, where an element other than that being determined is capable of absorbing the line radiation of the element under study.

2) Zeeman background correction—This correction is based on the principle that a magnetic field splits the spectral line into two linearly polarized light beams parallel and perpendicular to the magnetic field. One is called the pi (π) component

and the other the sigma (σ) component. These two light beams have exactly the same wavelength and differ only in the plane of polarization. The π line will be absorbed by both the atoms of the element of interest and by the background caused by broadband absorption and light scattering of the sample matrix. The σ line will be absorbed only by the background.

Zeeman background correction provides accurate background correction at much higher absorption levels than is possible with continuum source background correction systems. It also virtually eliminates the possibility of error from structured background. Because no additional light sources are required, the alignment and intensity limitations encountered using continuum sources are eliminated.

Disadvantages of the Zeeman method include reduced sensitivity for some elements, reduced linear range, and a "rollover" effect whereby the absorbance of some elements begins to decrease at high concentrations, resulting in a two-sided calibration curve.

3) Smith-Hieftje background correction—This correction is based on the principle that absorbance measured for a specific element is reduced as the current to the hollow cathode lamp is increased while absorption of nonspecific absorbing substances remains identical at all current levels. When this method is applied, the absorbance at a high-current mode is subtracted from the absorbance at a low-current mode. Under these conditions, any absorbance due to nonspecific background is subtracted out and corrected for.

Smith-Hieftje background correction provides a number of advantages over continuum-source correction. Accurate correction at higher absorbance levels is possible and error from structured background is virtually eliminated. In some cases, spectral interferences also can be eliminated. The usefulness of Smith-Hieftje background correction with electrodeless discharge lamps has not yet been established.

4. Sensitivity, Detection Limits, and Optimum Concentration Ranges

The sensitivity of flame atomic absorption spectrometry is defined as the metal concentration that produces an absorption of 1% (an absorbance of approximately 0.0044). The instrument detection limit is defined here as the concentration that produces absorption equivalent to twice the magnitude of the background fluctuation. Sensitivity and detection limits vary with the instrument, the element determined, the complexity of the matrix, and the technique selected. The optimum concentration range usually starts from the concentration of several times the sensitivity and extends to the concentration at which the calibration curve starts to flatten. To achieve best results, use concentrations of samples and standards within the optimum concentration range of the spectrometer. See Table 3111:I for indication of concentration ranges measurable with conventional atomization. In many instances the concentration range shown in Table 3111:I may be extended downward either by scale expansion or by integrating the absorption signal over a long time. The range may be extended upward by dilution, using a less sensitive wavelength, rotating the burner head, or utilizing a microprocessor to linearize the calibration curve at high concentrations.

5. Preparation of Standards

Prepare standard solutions of known metal concentrations in water with a matrix similar to the sample. Use standards that bracket expected sample concentration and are within the method's working range. Very dilute standards should be prepared daily from stock solutions in concentrations greater than 500 mg/L. Stock standard solutions can be obtained from several

TABLE 3111:I. ATOMIC ABSORPTION CONCENTRATION RANGES WITH DIRECT ASPIRATION
ATOMIC ABSORPTION

Element	Wavelength nm	Flame Gases*	Instrument Detection Limit mg/L	Sensitivity mg/L	Optimum Concentration Range mg/L
Ag	328.1	A–Ac	0.01	0.06	0.1–4
Al	309.3	N–Ac	0.1	1	5–100
Au	242.8	A–Ac	0.01	0.25	0.5–20
Ba	553.6	N–Ac	0.03	0.4	1–20
Be	234.9	N–Ac	0.005	0.03	0.05–2
Bi	223.1	A–Ac	0.06	0.4	1–50
Ca	422.7	A–Ac	0.003	0.08	0.2–20
Cd	228.8	A–Ac	0.002	0.025	0.05–2
Co	240.7	A–Ac	0.03	0.2	0.5–10
Cr	357.9	A–Ac	0.02	0.1	0.2–10
Cs	852.1	A–Ac	0.02	0.3	0.5–15
Cu	324.7	A–Ac	0.01	0.1	0.2–10
Fe	248.3	A–Ac	0.02	0.12	0.3–10
Ir	264.0	A–Ac	0.6	8	—
K	766.5	A–Ac	0.005	0.04	0.1–2
Li	670.8	A–Ac	0.002	0.04	0.1–2
Mg	285.2	A–Ac	0.0005	0.007	0.02–2
Mn	279.5	A–Ac	0.01	0.05	0.1–10
Mo	313.3	N–Ac	0.1	0.5	1–20
Na	589.0	A–Ac	0.002	0.015	0.03–1
Ni	232.0	A–Ac	0.02	0.15	0.3–10
Os	290.9	N–Ac	0.08	1	—
Pb†	283.3	A–Ac	0.05	0.5	1–20
Pt	265.9	A–Ac	0.1	2	5–75
Rh	343.5	A–Ac	0.5	0.3	—
Ru	349.9	A–Ac	0.07	0.5	—
Sb	217.6	A–Ac	0.07	0.5	1–40
Si	251.6	N–Ac	0.3	2	5–150
Sn	224.6	A–Ac	0.8	4	10–200
Sr	460.7	A–Ac	0.03	0.15	0.3–5
Ti	365.3	N–Ac	0.3	2	5–100
V	318.4	N–Ac	0.2	1.5	2–100
Zn	213.9	A–Ac	0.005	0.02	0.05–2

* A–Ac = air-acetylene; N–Ac = nitrous oxide-acetylene..
† The more sensitive 217.0 nm wavelength is recommended for instruments with background correction capabilities.
Copyright © ASTM. Reprinted with permission.

commercial sources. They also can be prepared from National Institute of Standards and Technology (NIST, formerly National Bureau of Standards) reference materials or by procedures outlined in the following sections.

For samples containing high and variable concentrations of matrix materials, make the major ions in the sample and the dilute standard similar. If the sample matrix is complex and components cannot be matched accurately with standards, use the method of standard additions, 3113B.4d2), to correct for matrix effects. If digestion is used, carry standards through the same digestion procedure used for samples.

6. Apparatus

a. *Atomic absorption spectrometer,* consisting of a light source emitting the line spectrum of an element (hollow-cathode lamp or electrodeless discharge lamp), a device for vaporizing the sample (usually a flame), a means of isolating an absorption line (monochromator or filter and adjustable slit), and a photoelectric detector with its associated electronic amplifying and measuring equipment.

b. *Burner:* The most common type of burner is a premix, which introduces the spray into a condensing chamber for removal of large droplets. The burner may be fitted with a conventional head containing a single slot; a three-slot Boling head, which may be preferred for direct aspiration with an air-acetylene flame; or a special head for use with nitrous oxide and acetylene.

c. *Readout:* Most instruments are equipped with either a digital or null meter readout mechanism. Most modern instruments are equipped with microprocessors capable of integrating absorption signals over time and linearizing the calibration curve at high concentrations.

d. *Lamps:* Use either a hollow-cathode lamp or an electrodeless discharge lamp (EDL). Use one lamp for each element being measured. Multi-element hollow-cathode lamps generally provide lower sensitivity than single-element lamps. EDLs take a longer time to warm up and stabilize.

e. *Pressure-reducing valves:* Maintain supplies of fuel and oxidant at pressures somewhat higher than the controlled operating pressure of the instrument by using suitable reducing valves. Use a separate reducing valve for each gas.

f. *Vent:* Place a vent about 15 to 30 cm above the burner to remove fumes and vapors from the flame. This precaution protects laboratory personnel from toxic vapors, protects the instrument from corrosive vapors, and prevents flame stability from being affected by room drafts. A damper or variable-speed blower is desirable for modulating air flow and preventing flame disturbance. Select blower size to provide the air flow recommended by the instrument manufacturer. In laboratory locations with heavy particulate air pollution, use clean laboratory facilities (Section 3010C).

7. Quality Assurance/Quality Control

Some data typical of the precision and bias obtainable with the methods discussed are presented in Tables 3111:II and III.

Analyze a blank between sample or standard readings to verify baseline stability. Rezero when necessary.

To one sample out of every ten (or one sample from each group of samples if less than ten are being analyzed) add a known amount of the metal of interest and reanalyze to confirm recovery. The amount of metal added should be approximately equal to the amount found. If little metal is present add an amount close to the middle of the linear range of the test. Recovery of added metal should be between 85 and 115%.

Analyze an additional standard solution after every ten samples or with each batch of samples, whichever is less, to confirm that the test is in control. Recommended concentrations of standards to be run, limits of acceptability, and reported single-operator precision data are listed in Table 3111:III.

8. References

1. AMERICAN SOCIETY FOR TESTING AND MATERIALS. 1986. Annual Book of ASTM Standards, Volume 11.01, Water and Environmental Technology. American Soc. Testing & Materials, Philadelphia, Pa.

2. U.S. DEPARTMENT HEALTH, EDUCATION AND

TABLE 3111:II. INTERLABORATORY PRECISION AND BIAS DATA FOR ATOMIC ABSORPTION
METHODS—DIRECT ASPIRATION AND EXTRACTED METALS

Metal	Conc. mg/L	SD mg/L	Relative SD %	Relative Error %	No. of Participants
Direct determination:					
Aluminum[1]	4.50	0.19	4.2	8.4	5
Barium[2]	1.00	0.089	8.9	2.7	11
Beryllium[1]	0.46	0.0213	4.6	23.0	11
Cadmium[3]	0.05	0.0108	21.6	8.2	26
Cadmium[1]	1.60	0.11	6.9	5.1	16
Calcium[1]	5.00	0.21	4.2	0.4	8
Chromium[1]	3.00	0.301	10.0	3.7	9
Cobalt[1]	4.00	0.243	6.1	0.5	14
Copper[3]	1.00	0.112	11.2	3.4	53
Copper[1]	4.00	0.331	8.3	2.8	15
Iron[1]	4.40	0.260	5.8	2.3	16
Iron[3]	0.30	0.0495	16.5	0.6	43
Lead[1]	6.00	0.28	4.7	0.2	14
Magnesium[3]	0.20	0.021	10.5	6.3	42
Magnesium[1]	1.10	0.116	10.5	10.0	8
Manganese[1]	4.05	0.317	7.8	1.3	16
Manganese[3]	0.05	0.0068	13.5	6.0	14
Nickel[1]	3.93	0.383	9.8	2.0	14
Silver[3]	0.05	0.0088	17.5	10.6	7
Silver[1]	2.00	0.07	3.5	1.0	10
Sodium[1]	2.70	0.122	4.5	4.1	12
Strontium[1]	1.00	0.05	5.0	0.2	12
Zinc[3]	0.50	0.041	8.2	0.4	48
Extracted determination:					
Aluminum[2]	300	32	10.7	0.7	15
Beryllium[2]	5	1.7	34.0	20.0	9
Cadmium[3]	50	21.9	43.8	13.3	12
Cobalt[1]	300	28.5	9.5	1.0	6
Copper[1]	100	71.7	71.7	12.0	8
Iron[1]	250	19.0	7.6	3.6	4
Manganese[1]	21.5	2.4	11.2	7.4	8
Molybdenum[1]	9.5	1.1	11.6	1.3	5
Nickel[1]	56.8	15.2	26.8	13.6	14
Lead[3]	50	11.8	23.5	19.0	8
Silver[1]	5.2	1.4	26.9	3.0	7

Source: AMERICAN SOCIETY FOR TESTING AND MATERIALS. 1986. Annual Book of ASTM Standards. Volume 11.01,
Water and Environmental Technology. American Soc. Testing & Materials, Philadelphia, Pa. Copyright ASTM.
Reprinted with permission.

WELFARE. 1970. Water Metals No. 6, Study
No. 37. U.S. Public Health Serv. Publ. No.
2029, Cincinnati, Ohio.

3. U.S. DEPARTMENT HEALTH, EDUCATION AND
WELFARE. 1968. Water Metals No. 4, Study

No. 30. U.S. Public Health Serv. Publ. No. 999-
UTH-8, Cincinnati, Ohio.

4. U.S. ENVIRONMENTAL PROTECTION AGENCY.
1983. Methods for Chemical Analysis of Water
and Wastes. Cincinnati, Ohio.

TABLE 3111:III. SINGLE-OPERATOR PRECISION AND RECOMMENDED CONTROL RANGES FOR
ATOMIC ABSORPTION METHODS—DIRECT ASPIRATION AND EXTRACTED METALS

Metal	Conc. mg/L	SD mg/L	Relative SD %	No. of Participants	QC Std. mg/L	Acceptable Range mg/L
Direct determination:						
Aluminum[1]	4.50	0.23	5.1	15	5.00	4.3–5.7
Beryllium[1]	0.46	0.012	2.6	10	0.50	0.46–0.54
Calcium[1]	5.00	0.05	1.0	8	5.00	4.8–5.2
Chromium[1]	7.00	0.69	9.9	9	5.00	3.3–6.7
Cobalt[1]	4.00	0.21	5.3	14	4.00	3.4–4.6
Copper[1]	4.00	0.115	2.9	15	4.00	3.7–4.3
Iron[1]	5.00	0.19	3.8	16	5.00	4.4–5.6
Magnesium[1]	1.00	0.009	0.9	8	1.00	0.97–1.03
Nickel[4]	5.00	0.04	0.8	—	5.00	4.9–5.1
Silver[1]	2.00	0.25	12.5	10	2.00	1.2–2.8
Sodium[4]	8.2	0.1	1.2	—	5.00	4.8–5.2
Strontium[1]	1.00	0.04	4.0	12	1.00	0.87–1.13
Potassium[4]	1.6	0.2	12.5	—	1.6	1.0–2.2
Molybdenum[4]	7.5	0.07	0.9	—	10.0	9.7–10.3
Tin[4]	20.0	0.5	2.5	—	20.0	18.5–21.5
Titanium[4]	50.0	0.4	0.8	—	50.0	48.8–51.2
Vanadium	50.0	0.2	0.4	—	50.0	49.4–50.6
Extracted determination:						
Aluminum[1]	300	12	4.0	15	300	264–336
Cobalt[1]	300	20	6.7	6	300	220–380
Copper[1]	100	21	21	8	100	22–178
Iron[1]	250	12	4.8	4	250	180–320
Manganese[1]	21.5	202	10.2	8	25	17–23
Molybdenum[1]	9.5	1.0	10.5	5	10	5.5–14.5
Nickel[1]	56.8	9.2	16.2	14	50	22–78
Silver[1]	5.2	1.2	23.1	7	5.0	0.5–9.5

Source: AMERICAN SOCIETY FOR TESTING AND MATERIALS. 1986. Annual Book of ASTM Standards. Volume 11.01, Water and Environmental Technology. American Soc. Testing & Materials, Philadelphia, Pa. Copyright ASTM. Reprinted with permission.

9. Bibliography

KAHN, H.L. 1968. Principles and Practice of Atomic Absorption. Advan. Chem. Ser. No. 73, Div. Water, Air & Waste Chemistry, American Chemical Soc., Washington, D.C.

RAMIRIZ-MUNOZ, J. 1968. Atomic Absorption Spectroscopy and Analysis by Atomic Absorption Flame Photometry. American Elsevier Publishing Co., New York, N.Y.

SLAVIN, W. 1968. Atomic Absorption Spectroscopy. John Wiley & Sons, New York, N.Y.

PAUS, P.E. 1971. The application of atomic absorption spectroscopy to the analysis of natural waters. Atomic Absorption Newsletter 10:69.

EDIGER, R.D. 1973. A review of water analysis by atomic absorption. Atomic Absorption Newsletter 12:151.

PAUS, P.E. 1973. Determination of some heavy metals in seawater by atomic absorption spectroscopy. Fresenius Zeitschr. Anal. Chem. 264:118.

BURRELL, D.C. 1975. Atomic Spectrometric Analysis of Heavy-Metal Pollutants in Water. Ann Arbor Science Publishers, Inc., Ann Arbor, Mich.

3111 B.   Direct Air-Acetylene Flame Method

1. General Discussion

This method is applicable to the determination of antimony, bismuth, cadmium, calcium, cesium, chromium, cobalt, copper, gold, iridium, iron, lead, lithium, magnesium, manganese, nickel, palladium, platinum, potassium, rhodium, ruthenium, silver, sodium, strontium, thallium, tin, and zinc.

2. Apparatus

Atomic absorption spectrometer and associated equipment: See Section 3111A.6. Use burner head recommended by the manufacturer.

3. Reagents

a. Air, cleaned and dried through a suitable filter to remove oil, water, and other foreign substances. The source may be a compressor or commercially bottled gas.

b. Acetylene, standard commercial grade. Acetone, which always is present in acetylene cylinders, can be prevented from entering and damaging the burner head by replacing a cylinder when its pressure has fallen to 689 kPa (100 psi) acetylene.

c. Metal-free water: Use metal-free water for preparing all reagents and calibration standards and as dilution water. Prepare metal-free water by deionizing tap water and/or by using one of the following processes, depending on the metal concentration in the sample: single distillation, redistillation, or sub-boiling. Always check deionized or distilled water to determine whether the element of interest is present in trace amounts. (CAUTION: *If the source water contains Hg or other volatile metals, single- or redistilled water may not be suitable for trace analysis because these metals distill over with the distilled water. In such*

cases, use sub-boiling to prepare metal-free water).

d. Calcium solution: Dissolve 630 mg calcium carbonate, $CaCO_3$, in 50 mL of 1 + 5 HCl. If necessary, boil gently to obtain complete solution. Cool and dilute to 1000 mL with water.

e. Hydrochloric acid, HCl, 1%, 10%, 20%, 1 + 5, 1 + 1, and conc.

f. Lanthanum solution: Dissolve 58.65 g lanthanum oxide, La_2O_3, in 250 mL conc HCl. Add acid slowly until the material is dissolved and dilute to 1000 mL with water.

g. Hydrogen peroxide, 30%.

h. Nitric acid, HNO_3, 2%, 1 + 1, and conc.

i. Aqua regia: Add 3 volumes conc HCl to 1 volume conc HNO_3.

j. Standard metal solutions: Prepare a series of standard metal solutions in the optimum concentration range by appropriate dilution of the following stock metal solutions with water containing 1.5 mL conc HNO_3/L. Thoroughly dry reagents before use. In general, use reagents of the highest purity. For hydrates, use fresh reagents.

1) *Antimony:* Dissolve 0.2669 g $K(SbO)C_4H_4O_6$ in water, add 10 mL 1 + 1 HCl and dilute to 1000 mL with water; 1.00 mL = 100 µg Sb.

2) *Bismuth:* Dissolve 0.100 g bismuth metal in a minimum volume of 1 + 1 HNO_3. Dilute to 1000 mL with 2% (v/v) HNO_3; 1.00 mL = 100 µg Bi.

3) *Cadmium:* Dissolve 0.100 g cadmium metal in 4 mL conc HNO_3. Add 8.0 mL conc HNO_3 and dilute to 1000 mL with water; 1.00 mL = 100 µg Cd.

4) *Calcium:* Suspend 0.2497 g $CaCO_3$ (dried at 180° for 1 h before weighing) in

water and dissolve cautiously with a minimum amount of 1 + 1 HNO_3. Add 10.0 mL conc HNO_3 and dilute to 1000 mL with water; 1.00 mL = 100 μg Ca.

5) *Cesium:* Dissolve 0.1267 g cesium chloride, CsCl, in 1000 mL water; 1.00 mL = 100 μg Cs.

6) *Chromium:* Dissolve 0.1923 g CrO_3 in water. When solution is complete, acidify with 10 mL conc HNO_3 and dilute to 1000 mL with water; 1.00 mL = 100 μg Cr.

7) *Cobalt:* Dissolve 0.1000 g cobalt metal in a minimum amount of 1 + 1 HNO_3. Add 10.0 mL 1 + 1 HCl and dilute to 1000 mL with water; 1.00 mL = 100 μg Co.

8) *Copper:* Dissolve 0.100 g copper metal in 2 mL conc HNO_3, add 10.0 mL conc HNO_3 and dilute to 1000 mL with water; 1.00 mL = 100 μg Cu.

9) *Gold:* Dissolve 0.100 g gold metal in a minimum volume of aqua regia. Evaporate to dryness, dissolve residue in 5 mL conc HCl, cool, and dilute to 1000 mL with water; 1.00 mL = 100 μg Au.

10) *Iridium:* Dissolve 0.1147 g ammonium chloroiridate, $(NH_4)_2IrCl_6$, in a minimum volume of 1% (v/v) HCl and dilute to 100 mL with 1% (v/v) HCl; 1.00 mL = 500 μg Ir.

11) *Iron:* Dissolve 0.100 g iron wire in a mixture of 10 mL 1 + 1 HCl and 3 mL conc HNO_3. Add 5 mL conc HNO_3 and dilute to 1000 mL with water; 1.00 mL = 100 μg Fe.

12) *Lead:* Dissolve 0.1598 g lead nitrate, $Pb(NO_3)_2$, in a minimum amount of 1 + 1 HNO_3, add 10 mL conc HNO_3, and dilute to 1000 mL with water; 1.00 mL = 100 μg Pb.

13) *Lithium:* Dissolve 0.5323 g lithium carbonate, Li_2CO_3, in a minimum volume of 1 + 1 HNO_3. Add 10.0 mL conc HNO_3 and dilute to 1000 mL with water; 1.00 mL = 100 μg Li.

14) *Magnesium:* Dissolve 0.1658 g MgO in a minimum amount of 1 + 1 HNO_3.

Add 10.0 mL conc HNO_3 and dilute to 1000 mL with water; 1.00 mL = 100 μg Mg.

15) *Manganese:* Dissolve 0.1000 g manganese metal in 10 mL conc HCl mixed with 1 mL conc HNO_3. Dilute to 1000 mL with water; 1.00 mL = 100 μg Mn.

16) *Nickel:* Dissolve 0.1000 g nickel metal in 10 mL hot conc HNO_3, cool, and dilute to 1000 mL with water; 1.00 mL = 100 μg Ni.

17) *Palladium:* Dissolve 0.100 g palladium wire in a minimum volume of aqua regia and evaporate just to dryness. Add 5 mL conc HCl and 25 mL water and warm until dissolution is complete. Dilute to 1000 mL with water; 1.00 mL = 100 μg Pd.

18) *Platinum:* Dissolve 0.100 g platinum metal in a minimum volume of aqua regia and evaporate just to dryness. Add 5 mL conc HCl and 0.1 g NaCl and again evaporate just to dryness. Dissolve residue in 20 mL of 1 + 1 HCl and dilute to 1000 mL with water; 1.00 mL = 100 μg Pt.

19) *Potassium:* Dissolve 0.1907 g potassium chloride, KCl, (dried at 110°C) in water and make up to 1000 mL; 1.00 mL = 100 μg K.

20) *Rhodium:* Dissolve 0.386 g ammonium hexachlororhodate, $(NH_4)_3RhCl_6 \cdot 1.5H_2O$, in a minimum volume of 10% (v/v) HCl and dilute to 1000 mL with 10% (v/v) HCl; 1.00 mL = 100 μg Rh.

21) *Ruthenium:* Dissolve 0.205 g ruthenium chloride, $RuCl_3$, in a minimum volume of 20% (v/v) HCl and dilute to 1000 mL with 20% (v/v) HCl; 1.00 mL = 100 μg Ru.

22) *Silver:* Dissolve 0.1575 g silver nitrate, $AgNO_3$, in 100 mL water, add 10 mL conc HNO_3, and make up to 1000 mL; 1.00 mL = 100 μg Ag.

23) *Sodium:* Dissolve 0.2542 g sodium chloride, NaCl, dried at 140°C, in water, add 10 mL conc HNO_3 and make up to 1000 mL; 1.00 mL = 100 μg Na.

24) *Strontium:* Suspend 0.1685 g $SrCO_3$ in water and dissolve cautiously with a

minimum amount of 1 + 1 HNO₃. Add
10.0 mL conc HNO₃ and dilute to 1000
mL with water: 1 mL = 100 µg Sr.

25) *Thallium:* Dissolve 0.1303 g thallium nitrate, TlNO₃, in water. Add 10 mL
conc HNO₃ and dilute to 1000 mL with
water; 1.00 mL = 100 µg Tl.

26) *Tin:* Dissolve 1.000 g tin metal in
100 mL conc HCl and dilute to 1000 mL
with water; 1.00 mL = 1.00 mg Sn.

27) *Zinc:* Dissolve 0.100 g zinc metal in
20 mL 1 + 1 HCl and dilute to 1000 mL
with water; 1.00 mL = 100 µg Zn.

4. Procedure

a. Sample preparation: Required sample
preparation depends on need to measure
dissolved metals only or total metals.

If dissolved metals are to be determined,
see Section 3030B for sample preparation.
If total or acid-extractable metals are to be
determined, see Sections 3030C through J.
For all samples, make certain that the concentrations of acid and matrix modifiers are
the same in both samples and standards.

When determining Ca or Mg, dilute and
mix 100 mL sample or standard with 10
mL lanthanum solution (¶ 3*f*) before atomization. When determining Fe or Mn, mix
100 mL with 25 mL of Ca solution (¶ 3*d*)
before aspirating. When determining Cr,
mix 1 mL 30% H_2O_2 with each 100 mL
before aspirating. Alternatively use proportionally smaller volumes.

b. Instrument operation: Because of differences between makes and models of
atomic absorption spectrometers, it is not
possible to formulate instructions applicable to every instrument. See manufacturer's
operating manual. In general, proceed according to the following: Install a hollow-cathode lamp for the desired metal in the
instrument and roughly set the wavelength
dial according to Table 3111:I. Set slit
width according to manufacturer's suggested setting for the element being measured. Turn on instrument, apply to the

hollow-cathode lamp the current suggested
by the manufacturer, and let instrument
warm up until energy source stabilizes,
generally about 10 to 20 min. Readjust current as necessary after warmup. Optimize
wavelength by adjusting wavelength dial
until optimum energy gain is obtained.
Align lamp in accordance with manufacturer's instructions.

Install suitable burner head and adjust
burner head position. Turn on air and adjust flow rate to that specified by manufacturer to give maximum sensitivity for
the metal being measured. Turn on acetylene, adjust flow rate to value specified,
and ignite flame. Let flame stabilize for a
few minutes. Aspirate a blank consisting of
either deionized water or an acid solution
containing the same concentration of acid
in standards and samples. Zero the instrument. Aspirate a standard solution and adjust aspiration rate of the nebulizer to
obtain maximum sensitivity. Adjust burner
both vertically and horizontally to obtain
maximum response. Aspirate blank again
and rezero the instrument. Aspirate a
standard near the middle of the linear
range. Record absorbance of this standard
when freshly prepared and with a new hollow-cathode lamp. Refer to these data on
subsequent determinations of the same element to check consistency of instrument
setup and aging of hollow-cathode lamp
and standard.

The instrument now is ready to operate.
When analyses are finished, extinguish
flame by turning off first acetylene and then
air.

c. Standardization: Select at least three
concentrations of each standard metal solution (prepared as in ¶ 3*j* above) to bracket
the expected metal concentration of a sample. Aspirate blank and zero the instrument. Then aspirate each standard in turn
into flame and record absorbance.

Prepare a calibration curve by plotting
on linear graph paper absorbance of standards versus their concentrations. For

instruments equipped with direct concentration readout, this step is unnecessary. With some instruments it may be necessary to convert percent absorption to absorbance by using a table generally provided by the manufacturer. Plot calibration curves for Ca and Mg based on original concentration of standards before dilution with lanthanum solution. Plot calibration curves for Fe and Mn based on original concentration of standards before dilution with Ca solution. Plot calibration curve for Cr based on original concentration of standard before addition of H_2O_2.

d. Analysis of samples: Rinse nebulizer by aspirating water containing 1.5 mL conc HNO_3/L. Atomize blank and zero instrument. Atomize sample and determine its absorbance.

5. Calculations

Calculate concentration of each metal ion, in micrograms per liter for trace elements, and in milligrams per liter for more common metals, by referring to the appropriate calibration curve prepared according to ¶ 4c. Alternatively, read concentration directly from the instrument readout if the instrument is so equipped. If the sample has been diluted, multiply by the appropriate dilution factor.

6. Bibliography

WILLIS, J.B. 1962. Determination of lead and other heavy metals in urine by atomic absorption spectrophotometry. *Anal. Chem.* 34:614.

Also see Section 3111A.8 and 9.

3111 C.　Extraction/Air-Acetylene Flame Method

1. General Discussion

This method is suitable for the determination of low concentrations of cadmium, chromium, cobalt, copper, iron, lead, manganese, nickel, silver, and zinc. The method consists of chelation with ammonium pyrrolidine dithiocarbamate (APDC) and extraction into methyl isobutyl ketone (MIBK), followed by aspiration into an air-acetylene flame.

2. Apparatus

a. Atomic absorption spectrometer and associated equipment: See Section 3111A.6.

b. Burner head, conventional. Consult manufacturer's operating manual for suggested burner head.

3. Reagents

　a. Air: See 3111B.3*a.*

　b. Acetylene: See 3111B.3*b.*

　c. Metal-free water: See 3111B.3*c.*

d. Methyl isobutyl ketone (MIBK), reagent grade. For trace analysis, purify MIBK by redistillation or by sub-boiling distillation.

e. Ammonium pyrrolidine dithiocarbamate (APDC) solution: Dissolve 4 g APDC in 100 mL water. If necessary, purify APDC with an equal volume of MIBK. Shake 30 s in a separatory funnel, let separate, and withdraw lower portion. Discard MIBK layer.

f. Nitric acid, HNO_3, conc, ultrapure.

g. Standard metal solutions: See 3111B.3*j.*

h. Potassium permanganate solution, $KMnO_4$, 5% aqueous.

i. Sodium sulfate, Na_2SO_4, anhydrous.

j. Water-saturated MIBK: Mix one part purified MIBK with one part water in a separatory funnel. Shake 30 s and let separate. Discard aqueous layer. Save MIBK layer.

k. Hydroxylamine hydrochloride solution, 10%.

4. Procedure

a. Instrument operation: See Section 3111B.4*b*. After final adjusting of burner position, aspirate water-saturated MIBK into flame and gradually reduce fuel flow until flame is similar to that before aspiration of solvent.

b. Standardization: Select at least three concentrations of standard metal solutions (prepared as in 3111B.3*j*) to bracket expected sample metal concentration and to be, after extraction, in the optimum concentration range of the instrument. Adjust 100 mL of each standard and 100 mL of a metal-free water blank to pH 3 by adding 1*N* HNO$_3$ or 1*N* NaOH. For individual element extraction, use the following pH ranges to obtain optimum extraction efficiency:

Element	pH Range for Optimum Extraction
Ag	2–5 (complex unstable)
Cd	1–6
Co	2–10
Cr	3–9
Cu	0.1–8
Fe	2–5
Mn	2–4 (complex unstable)
Ni	2–4
Pb	0.1–6
Zn	2–6

NOTE: For Ag and Pb extraction the optimum pH value is 2.3 ± 0.2. The Mn complex deteriorates rapidly at room temperature, resulting in decreased instrument response. Chilling the extract to 0°C may preserve the complex for a few hours. If this is not possible and Mn cannot be analyzed immediately after extraction, use another analytical procedure.

Transfer each standard solution and blank to individual 200-mL volumetric flasks, add 1 mL APDC solution, and shake to mix. Add 10 mL MIBK and shake vigorously for 30 s. (The maximum volume ratio of sample to MIBK is 40.) Let contents of each flask separate into aqueous and organic layers, then carefully add water (adjusted to the same pH at which the extraction was carried out) down the side of each flask to bring the organic layer into the neck and accessible to the aspirating tube.

Aspirate organic extracts directly into the flame (zeroing instrument on a water-saturated MIBK blank) and record absorbance.

Prepare a calibration curve by plotting on linear graph paper absorbances of extracted standards against their concentrations before extraction.

c. Analysis of samples: Prepare samples in the same manner as the standards. Rinse atomizer by aspirating water-saturated MIBK. Aspirate organic extracts treated as above directly into the flame and record absorbances.

With the above extraction procedure only hexavalent chromium is measured. To determine total chromium, oxidize trivalent chromium to hexavalent chromium by bringing sample to a boil and adding sufficient KMnO$_4$ solution dropwise to give a persistent pink color while the solution is boiled for 10 min. Destroy excess KMnO$_4$ by adding 1 to 2 drops hydroxylamine hydrochloride solution to the boiling solution, allowing 2 min for the reaction to proceed. If pink color persists, add 1 to 2 more drops hydroxylamine hydrochloride solution and wait 2 min. Heat an additional 5 min. Cool, extract with MIBK, and aspirate.

During extraction, if an emulsion forms at the water-MIBK interface, add anhydrous Na$_2$SO$_4$ to obtain a homogeneous organic phase. In that case, also add Na$_2$SO$_4$ to all standards and blanks.

To avoid problems associated with instability of extracted metal complexes, determine metals immediately after extraction.

5. Calculations

Calculate the concentration of each metal ion in micrograms per liter by referring to the appropriate calibration curve.

6. Bibliography

ALLAN, J.E. 1961. The use of organic solvents in atomic absorption spectrophotometry. *Spectrochim. Acta* 17:467.

SACHDEV, S.L. & P.W. WEST. 1970. Concentration of trace metals by solvent extraction and their determination by atomic absorption spectrophotometry. *Environ. Sci. Technol.* 4:749.

3111 D.   Direct Nitrous Oxide-Acetylene Flame Method

1. General Discussion

This method is applicable to the determination of aluminum, barium, beryllium, molybdenum, osmium, rhenium, silicon, thorium, titanium, and vanadium.

2. Apparatus

a. *Atomic absorption spectrometer and associated equipment:* See Section 3111A.6.

b. *Nitrous oxide burner head:* Use special burner head as suggested in manufacturer's manual. At roughly 20-min intervals of operation it may be necessary to dislodge the carbon crust that forms along the slit surface with a carbon rod or appropriate alternative.

c. *T-junction valve* or other switching valve for rapidly changing from nitrous oxide to air, so that flame can be turned on or off with air as oxidant to prevent flashbacks.

3. Reagents

a. *Air:* See 3111B.3a.

b. *Acetylene:* See 3111B.3b.

c. *Metal-free water:* See 3111B.3c.

d. *Hydrochloric acid,* HCl, 1N, 1+1, and conc.

e. *Nitric acid,* HNO₃, conc.

f. *Sulfuric acid,* H₂SO₄, 1%.

g. *Hydrofluoric acid,* HF, 1N.

h. *Nitrous oxide,* commercially available cylinders. Fit nitrous oxide cylinder with

a special nonfreezable regulator or wrap a heating coil around an ordinary regulator to prevent flashback at the burner caused by reduction in nitrous oxide flow through a frozen regulator. (Some atomic absorption instruments have automatic gas control systems that will shut down a nitrous oxide-acetylene flame safely in the event of a reduction in nitrous oxide flow rate.)

i. *Potassium chloride solution:* Dissolve 250 g KCl in water and dilute to 1000 mL.

j. *Aluminum nitrate solution:* Dissolve 139 g Al(NO₃)₃·9H₂O in 150 mL water. Acidify slightly with conc HNO₃ to preclude possible hydrolysis and precipitation. Warm to dissolve completely. Cool and dilute to 200 mL.

k. *Standard metal solutions:* Prepare a series of standard metal solutions in the optimum concentration ranges by appropriate dilution of the following stock metal solutions with water containing 1.5 mL conc HNO₃/L:

1) *Aluminum:* Dissolve 0.100 g aluminum metal in an acid mixture of 4 mL 1 + 1 HCl and 1 mL conc HNO₃ in a beaker. Warm gently to effect solution. Transfer to a 1-L flask, add 10 mL 1 + 1 HCl, and dilute to 1000 mL with water; 1.00 mL = 100 μg Al.

2) *Barium:* Dissolve 0.1516 g BaCl₂ (dried at 250° for 2 h), in about 10 mL water with 1 mL 1 + 1 HCl. Add 10.0 mL

1 + 1 HCl and dilute to 1000 mL with water; 1.00 mL = 100 μg Ba.

3) *Beryllium*: *Do not dry.* Dissolve 1.966 g BeSO$_4$·4H$_2$O in water, add 10.0 mL conc HNO$_3$, and dilute to 1000 mL with water; 1.00 mL = 100 μg Be.

4) *Molybdenum:* Dissolve 0.2043 g (NH$_4$)$_2$ MoO$_4$ in water and dilute to 1000 mL; 1.00 mL = 100 μg Mo.

5) *Osmium:* Obtain standard 0.1M osmium tetroxide solution* and store in glass bottle; 1.00 mL = 19.02 mg Os. Make dilutions daily as needed using 1% (v/v) H$_2$SO$_4$. CAUTION: *OsO$_4$ is extremely toxic and highly volatile.*

6) *Rhenium:* Dissolve 0.1554 g potassium perrhenate, KReO$_4$, in 200 mL water. Dilute to 1000 mL with 1% (v/v) H$_2$SO$_4$; 1.00 mL = 100 μg Re.

7) *Silica: Do not dry.* Dissolve 0.4730 g Na$_2$SiO$_3$·9H$_2$O in water. Add 10.0 mL conc HNO$_3$ and dilute to 1000 mL with water. 1.00 mL = 100 μg Si. Store in polyethylene.

8) *Thorium:* Dissolve 0.238 g thorium nitrate, Th(NO$_3$)$_4$·4H$_2$O in 1000 mL water; 1.00 mL = 100 μg Th.

9) *Titanium:* Dissolve 0.3960 g pure (99.8 or 99.9%) titanium chloride, TiCl$_4$,† in a mixture of equal volumes of 1N HCl and 1N HF. Make up to 1000 mL with this acid mixture; 1.00 mL = 100 μg Ti.

10) *Vanadium:* Dissolve 0.2297 g ammonium metavanadate, NH$_4$VO$_3$, in a minimum amount of conc HNO$_3$. Heat to dissolve. Add 10 mL conc HNO$_3$, and dilute to 1000 mL with water; 1.00 mL = 100 μg V.

4. Procedure

a. Sample preparation: See Section 3111B.4*a*.

b. Instrument operation: See Section

---

*GFS Chemical Co., P.O. Box 23214, Columbus, Ohio 43223, Cat. No. 64, or equivalent.
†Alpha Ventron, P.O. Box 299, 152 Andover St., Danvers, Mass. 01923, or equivalent.

3111B.4*b*. After adjusting wavelength, install a nitrous oxide burner head. Turn on acetylene (without igniting flame) and adjust flow rate to value specified by manufacturer for a nitrous oxide-acetylene flame. Turn off acetylene. With both air and nitrous oxide supplies turned on, set T-junction valve to nitrous oxide and adjust flow rate according to manufacturer's specifications. Turn switching valve to the air position and verify that flow rate is the same. Turn acetylene on and ignite to a bright yellow flame. With a rapid motion, turn switching valve to nitrous oxide. The flame should have a red cone above the burner. If it does not, adjust fuel flow to obtain red cone. After nitrous oxide flame has been ignited, let burner come to thermal equilibrium before beginning analysis.

Atomize water containing 1.5 mL conc HNO$_3$/L and check aspiration rate. Adjust if necessary to a rate between 3 and 5 mL/min. Atomize a standard of the desired metal with a concentration near the midpoint of the optimum concentration range and adjust burner (both horizontally and vertically) in the light path to obtain maximum response. The instrument now is ready to run standards and samples.

To extinguish flame, turn switching valve from nitrous oxide to air and turn off acetylene. This procedure eliminates the danger of flashback that may occur on direct ignition or shutdown of nitrous oxide and acetylene.

c. Standardization: Select at least three concentrations of standard metal solutions (prepared as in ¶ 3*k*) to bracket the expected metal concentration of a sample. Aspirate each in turn into the flame. Record absorbances. For Al, Ba, and Ti, add 2 mL KCl solution to 100 mL standard before aspiration. For Mo and V add 2 mL Al(NO$_3$)$_3$·9H$_2$O solution to 100 mL standard before aspiration.

Most modern instruments are equipped with microprocessors and digital readout which permit calibration in direct concentration terms. If instrument is not so equipped, prepare a calibration curve by plotting on linear graph paper absorbance of standards versus concentration. Plot calibration curves for Al, Ba, and Ti based on original concentration of standard before adding KCl solution. Plot calibration curves for Mo and V based on original concentration of standard before adding $Al(NO_3)_3$ solution.

d. Analysis of samples: Rinse atomizer by aspirating water containing 1.5 mL conc HNO_3/L and zero instrument. Atomize a sample and determine its absorbance.

When determining Al, Ba, and Ti, add 2 mL KCl solution to 100 mL sample before atomization. For Mo and V, add 2 mL $Al(NO_3)_3 \cdot 9H_2O$ solution to 100 mL sample before atomization.

5. Calculations

Calculate concentration of each metal ion in micrograms per liter by referring to the appropriate calibration curve prepared according to ¶ 4c.

Alternatively, read the concentration directly from the instrument readout if the instrument is so equipped. If sample has been diluted, multiply by the appropriate dilution factor.

6. Bibliography

WILLIS, J.B. 1965. Nitrous oxide-acetylene flame in atomic absorption spectroscopy. *Nature* 207:715.

Also see Section 3111A.8 and 9.

3111 E.   Extraction/Nitrous Oxide-Acetylene Flame Method

1. General Discussion

a. Application: This method is suitable for the determination of aluminum at concentrations less than 900 μg/L and beryllium at concentrations less than 30 μg/L. The method consists of chelation with 8-hydroxyquinoline, extraction with methyl isobutyl ketone (MIBK), and aspiration into a nitrous oxide-acetylene flame.

b. Interferences: Concentrations of Fe greater than 10 mg/L interfere by suppressing Al absorption. Iron interference can be masked by addition of hydroxylamine hydrochloride/1,10-phenanthroline. Mn concentrations up to 80 mg/L do not interfere if turbidity in the extract is allowed to settle. Mg forms an insoluble chelate with 8-hydroxyquinoline at pH 8.0 and tends to remove Al complex as a coprecipitate. However, the Mg complex forms slowly over 4 to 6 min; its interference can be avoided if the solution is extracted immediately after adding buffer.

2. Apparatus

Atomic absorption spectrometer and associated equipment: See Section 3111A.6.

3. Reagents

a. Air: See 3111B.3a.

b. Acetylene: See 3111B.3b.

c. Ammonium hydroxide, NH$_4$OH, conc.

d. Buffer: Dissolve 300 g ammonium acetate, $NH_4C_2H_3O_2$, in water, add 105 mL conc NH_4OH, and dilute to 1 L.

e. Metal-free water: See 3111B.2c.

f. Hydrochloric acid, HCl, conc.

g. 8-Hydroxyquinoline solution: Dissolve 20 g 8-hydroxyquinoline in about 200 mL water, add 60 mL glacial acetic acid, and dilute to 1 L with water.

h. Methyl isobutyl ketone: See 3111C.3d.

i. Nitric acid, HNO$_3$, conc.

j. Nitrous oxide: See 3111D.3h.

k. Standard metal solutions: Prepare a series of standard metal solutions containing 5 to 1000 μg/L by appropriate dilution of the stock metal solutions prepared according to 3111D.3k.

l. Iron masking solution: Dissolve 1.3 g hydroxylamine hydrochloride and 6.58 g 1,10-phenanthroline monohydrate in about 500 mL water and dilute to 1 L with water.

4. Procedure

a. Instrumentation operation: See Sections 3111B.4b, C.4a, and D.4b. After final adjusting of burner position, aspirate MIBK into flame and gradually reduce fuel flow until flame is similar to that before aspiration of solvent. Adjust wavelength setting according to Table 3111:I.

b. Standardization: Select at least three concentrations of standard metal solutions (prepared as in ¶ 3k) to bracket the expected metal concentration of a sample and transfer 100 mL of each (and 100 mL water blank) to four different 200-mL volumetric flasks. Add 2 mL 8-hydroxyquinoline so-lution, 2 mL masking solution (if required), and 10 mL buffer to one flask, immediately add 10 mL MIBK, and shake vigorously. The duration of shaking affects the forms of aluminum complexed. A fast, 10-s shak-ing time favors monomeric Al, whereas 5 to 10 min of shaking also will complex polymeric species. Adjustment of the 8-hydroxyquinoline to sample ratio can im-prove recoveries of extremely high or low concentrations of aluminum. Treat each blank, standard, and sample in similar fash-ion. Continue as in Section 3111C.4b.

c. Analysis of samples: Rinse atomizer by aspirating water-saturated MIBK. As-pirate extracts of samples treated as above, and record absorbances.

5. Calculations

Calculate concentration of each metal in micrograms per liter by referring to the appropriate calibration curve prepared ac-cording to ¶ 4b.

3112   METALS BY COLD-VAPOR ATOMIC ABSORPTION SPECTROMETRY*

3112 A.   Introduction

For general introductory material on atomic absorption spectrometric methods, see Section 3111A.

---

* Approved by Standard Methods Committee, 1988.

3112 B.   Cold-Vapor Atomic Absorption Spectrometric Method

1. General Discussion

This method is applicable to the determination of mercury.

2. Apparatus

a. Atomic absorption spectrometer and associated equipment: See Section 3111A.6. Instruments and accessories specifically designed for measurement of mercury by the cold vapor technique are available commercially and may be substituted.

b. Absorption cell, a glass or plastic tube approximately 2.5 cm in diameter. An 11.4-cm-long tube has been found satisfactory but a 15-cm-long tube is preferred. Grind tube ends perpendicular to the longitudinal axis and cement quartz windows in place. Attach gas inlet and outlet ports (6.4 mm diam) 1.3 cm from each end.

c. Cell support: Strap cell to the flat nitrous-oxide burner head or other suitable support and align in light beam to give maximum transmittance.

d. Air pumps: Use any peristaltic pump with electronic speed control capable of delivering 2 L air/min. Any other regulated compressed air system or air cylinder also is satisfactory.

e. Flowmeter, capable of measuring an air flow of 2 L/min.

f. Aeration tubing, a straight glass frit having a coarse porosity for use in reaction flask.

g. Reaction flask, 250-mL erlenmeyer flask or a BOD bottle, fitted with a rubber stopper to hold aeration tube.

h. Drying tube, 150-mm × 18-mm-diam, containing 20 g Mg (ClO$_4$)$_2$. A 60-W light bulb with a suitable shade may be substituted to prevent condensation of moisture inside the absorption cell. Position bulb to maintain cell temperature at 10°C above ambient.

i. Connecting tubing, glass tubing to pass mercury vapor from reaction flask to absorption cell and to interconnect all other components. Clear vinyl plastic* tubing may be substituted for glass.

3. Reagents†

a. Metal-free water: See 3111B.3c.

b. Stock mercury solution: Dissolve 1.354 g mercuric chloride, HgCl$_2$, in about 700 mL water, add 10 mL conc HNO$_3$, and dilute to 1000 mL with water; 1.00 mL = 1.00 mg Hg.

c. Standard mercury solutions: Prepare a series of standard mercury solutions containing 0 to 5 μg/L by appropriate dilution of stock mercury solution with water containing 10 mL conc HNO$_3$/L. Prepare standards daily.

d. Nitric acid, HNO$_3$, conc.

e. Potassium permanganate solution: Dissolve 50 g KMnO$_4$ in water and dilute to 1 L.

f. Potassium persulfate solution: Dissolve 50 g K$_2$S$_2$O$_8$ in water and dilute to 1 L.

g. Sodium chloride-hydroxylamine sulfate solution: Dissolve 120 g NaCl and 120 g (NH$_2$OH)$_2$·H$_2$SO$_4$ in water and dilute to 1 L. A 10% hydroxylamine hydrochloride solution may be substituted for the hydroxylamine sulfate.

h. Stannous chloride solution: Dissolve 100 g SnCl$_2$ in water containing 200 mL conc HCl and dilute to 1 L. On aging, this

---

* Tygon or equivalent.
† Use specially prepared reagents low in mercury.

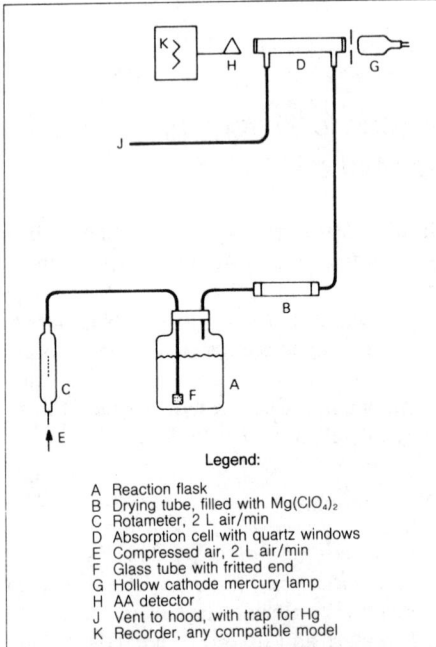

Legend:

A Reaction flask
B Drying tube, filled with Mg(ClO₄)₂
C Rotameter, 2 L air/min
D Absorption cell with quartz windows
E Compressed air, 2 L air/min
F Glass tube with fritted end
G Hollow cathode mercury lamp
H AA detector
J Vent to hood, with trap for Hg
K Recorder, any compatible model

Figure 3112:1. Schematic arrangement of equipment for measurement of mercury by cold-vapor atomic absorption technique.

solution decomposes. If a suspension forms, stir reagent continuously during use.

i. Sulfuric acid, H_2SO_4, conc.

4. Procedure

a. Instrument operation: See Section 3111B.4*b.* Install absorption cell and align in light path to give maximum transmission. Connect associated equipment to absorption cell with glass or vinyl plastic tubing as indicated in Figure 3112:1. Turn on air and adjust flow rate to 2 L/min. Allow air to flow continuously. Alternatively, follow manufacturer's directions for operation. NOTE: Fluorescent lighting may increase baseline noise.

b. Standardization: Transfer 100 mL of each of the 1.0, 2.0, and 5.0 µg/L Hg standard solutions and a blank of 100 mL water to 250-mL erlenmeyer reaction flasks. Add 5 mL conc H_2SO_4 and 2.5 mL conc HNO_3

to each flask. Add 15 mL $KMnO_4$ solution to each flask and let stand at least 15 min. Add 8 mL $K_2S_2O_8$ solution to each flask and heat for 2 h in a water bath at 95°C. Cool to room temperature.

Treating each flask individually, add enough NaCl-hydroxylamine sulfate solution to reduce excess $KMnO_4$, then add 5 mL $SnCl_2$ solution and immediately attach flask to aeration apparatus. As Hg is volatilized and carried into the absorption cell, absorbance will increase to a maximum within a few seconds. As soon as recorder returns approximately to the base line, remove stopper holding the frit from reaction flask, and replace with a flask containing water. Flush system for a few seconds and run the next standard in the same manner. Construct a standard curve by plotting peak height versus micrograms Hg.

c. Analysis of samples: Transfer 100 mL sample or portion diluted to 100 mL containing not more than 5.0 µg Hg/L to a reaction flask. Treat as in ¶ 4*b.* Seawaters, brines, and effluents high in chlorides require as much as an additional 25 mL $KMnO_4$ solution. During oxidation step, chlorides are converted to free chlorine, which absorbs at 253 nm. Remove all free chlorine before the Hg is reduced and swept into the cell by using an excess (25 mL) of hydroxylamine sulfate reagent.

Remove free chlorine by sparging sample gently with air or nitrogen after adding hydroxylamine reducing solution. Use a separate tube and frit to avoid carryover of residual stannous chloride, which could cause reduction and loss of mercury.

4. Calculation

Determine peak height of sample from recorder chart and read mercury value from standard curve prepared according to ¶ 4*b.*

TABLE 3112:I. INTERLABORATORY PRECISION AND BIAS
OF COLD-VAPOR ATOMIC ABSORPTION SPECTROMETRIC METHOD FOR MERCURY[1]

Form	Conc. $\mu g/L$	SD $\mu g/L$	Relative SD %	Relative Error %	No. of Participants
Inorganic	0.34	0.077	22.6	21.0	23
Inorganic	4.2	0.56	13.3	14.4	21
Organic	4.2	0.36	8.6	8.4	21

6. Precision and Bias

Data on interlaboratory precision and bias for this method are given in Table 3112:I.

7. Reference

1. KOPP, J.F., M.C. LONGBOTTOM & L.B. LO-BRING. 1972. "Cold vapor" method for determining mercury. *J. Amer. Water Works Assoc.* 64:20.

8. Bibliography

HATCH, W.R. & W.L. OTT. 1968. Determination of submicrogram quantities of mercury by atomic absorption spectrophotometry. *Anal. Chem.* 40:2085.

UTHE, J.F., F.A.J. ARMSTRONG & M.P. STAINTON. 1970. Mercury determination in fish samples by wet digestion and flameless atomic absorption spectrophotometry. *J. Fish. Res. Board Can.* 27:805.

FELDMAN, C. 1974. Preservation of dilute mercury solutions. *Anal. Chem.* 46:99.

BOTHNER, M.H. & D.E. ROBERTSON. 1975. Mercury contamination of sea water samples stored in polyethylene containers. *Anal. Chem.* 47:592.

HAWLEY, J.E. & J.D. INGLE, JR. 1975. Improvements in cold vapor atomic absorption determination of mercury. *Anal. Chem.* 47:719.

LO, J.M. & C.M. WAL. 1975. Mercury loss from water during storage: Mechanisms and prevention. *Anal. Chem.* 47:1869.

EL-AWADY, A.A., R.B. MILLER & M.J. CARTER. 1976. Automated method for the determination of total and inorganic mercury in water and wastewater samples. *Anal. Chem.* 48:110.

ODA, C.E. & J.D. INGLE, JR. 1981. Speciation of mercury by cold vapor atomic absorption spectrometry with selective reduction. *Anal. Chem.* 53:2305.

SUDDENDORF, R.F. 1981. Interference by selenium or tellurium in the determination of mercury by cold vapor generation atomic absorption spectrometry. *Anal. Chem.* 53:2234.

HEIDEN, R.W. & D.A. AIKENS. 1983. Humic acid as a preservative for trace mercury (II) solutions stored in polyolefin containers. *Anal. Chem.* 55:2327.

CHOU, H.N. & C.A. NALEWAY. 1984. Determination of mercury by cold vapor atomic absorption spectrometry. *Anal. Chem.* 56:1737.

3113   METALS BY ELECTROTHERMAL ATOMIC ABSORPTION SPECTROMETRY*

3113 A.   Introduction

1. Applications

Electrothermal atomic absorption permits determination of most metallic elements with sensitivities and detection limits from 20 to 1000 times better than those of conventional flame techniques without extraction or sample concentration. This increase in sensitivity results from the increased residence time of the ground-state atoms in the optical path, which is many times that of conventional flame atomic absorption. Many elements can be determined at concentrations as low as 1.0 $\mu g/L$. An additional advantage of electrothermal atomic absorption is that only a very small volume of sample is required.

Use the electrothermal technique only at concentration levels below the optimum range of direct flame atomic absorption because it is subject to more interferences than the flame procedure and requires increased analysis time. The method of standard additions often is required to insure validity of data. Because of the high sensitivity of this technique, exercise great care to avoid contamination errors.

2. Principle

Electrothermal atomic absorption spectroscopy is based on the same principle as direct flame atomization but an electrically heated atomizer or graphite furnace re-

places the standard burner head. A discrete sample volume is dispensed into the graphite sample tube (or cup). Typically, determinations are made by heating the sample in three or more stages. First, a low current heats the tube to dry the sample. The second, or charring stage, destroys organic matter and volatilizes other matrix components at an intermediate temperature. Finally, a high current heats the tube to incandescence and, in an inert atmosphere, atomizes the element being determined. Additional stages frequently are added to aid in drying and charring, and to clean and cool the tube between samples. The resultant ground-state atomic vapor absorbs monochromatic radiation from the source. A photoelectric detector measures the intensity of transmitted radiation, which is inversely proportional to the quantity of ground-state atoms in the optical path over a limited range.

3. Interference

Electrothermal atomization determinations are subject to significant interferences from molecular absorption as well as chemical and matrix effects. Molecular absorption may occur when components of the sample matrix volatilize during atomization, resulting in broadband absorption. Several background correction techniques are available commercially to compensate for this interference. A continuum source

---

*Approved by Standard Methods Committee, 1988.

such as a deuterium arc can correct for background up to absorbance levels of 0.8 or so. Zeeman effect background correctors can handle background absorbencies up to 1.5 to 2.0. The Smith-Hieftje correction technique can accommodate background absorbance levels as large as 2.5 to 3.0 (see Section 3111A.3). Use background correction when analyzing samples containing high concentrations of acid or dissolved solids and in determining elements for which an absorption line below 350 nm is used.

Matrix modification can be useful in minimizing interference. Accomplish this by adding various chemicals to the sample. Alternatively, program a modern autosampler to add matrix modifiers directly to the sample in the furnace chamber. Some matrix modifiers reduce volatility of the element being determined or increase its atomization efficiency by changing its chemical composition. This permits use of higher charring temperatures to volatilize interfering substances and increases sensitivity. Other matrix modifiers increase volatility of the matrix. Specific matrix modifiers are listed in Table 3113:I.

Temperature ramping, i.e., gradual heating, can be used to decrease background interferences and permits analysis of samples with complex matrices. Ramping permits a controlled, continuous increase of furnace temperature in any of the various

TABLE 3113:I. POTENTIAL MATRIX MODIFIERS FOR ELECTROTHERMAL
ATOMIC ABSORPTION SPECTROMETRY[1]

Element	Matrix Modifiers For Interference Removal	Matrix Modifiers as Enhancers
Al	$Mg(NO_3)_2$	$Ca(NO_3)_2$, $Ca_3(PO_4)_2$, H_2SO_4
Sb	$Ni(NO_3)_2$ & $Mg(NO_3)_2$	—
As	$Mg(NO_3)_2$, $Ni(NO_3)_2$	—
Be	$Al(NO_3)_2$, $Mg(NO_3)_2$	$Ca(NO_3)_2$
Cd	$NH_4H_2PO_4$ & $Mg(NO_3)_2$, $(NH_4)_2HPO_4$ & $Mg(NO_3)_2$, $(NH_4)_2SO_4$, HNO_3, $(NH_4)_2S_2O_8$	
Cr	$Mg(NO_3)_2$	—
Co	$Mg(NO_3)_2$, $NH_4H_2PO_4$, ascorbic acid	—
Cu	NH_4NO_3, ascorbic acid	$MgSO_4$, $LaNO_3$
Fe	NH_4NO_3	—
Pb	$NH_4H_2PO_4$, $(NH_4)_2HPO_4$, $Mg(NO_3)_2$, NH_4NO_3, ascorbic acid, oxalic acid, phosphoric acid, HNO_3, LaCl, $(NH_4)_2EDTA$	—
Mn	Ascorbic acid, $Mg(NO_3)_2$, NH_4NO_3	—
Mo	—	HNO_3
Ni	$Mg(NO_3)_2$, $NH_4H_2PO_4$	—
Se	$Ni(NO_3)_2$, $Ni(NO_3)_2$ & $Mg(NO_3)_2$, $Ni(NO_3)_2$ & $Cu(NO_3)_2$, $AgNO_3$, $(NH_4)_6Mo_7O_{24}$, $Fe(NO_3)_3$ $Fe(NO_3)_3$ & $Cu(NO_3)_2$	—
Ag	$(NH_4)_2HPO_4$, $NH_4H_2PO_4$	—
Sn	$(NH_4)_2HPO_4$ & $Mg(NO_3)_2$, $Ni(NO_3)_2$, ascorbic acid, NH_4NO_3	$Ca(NO_3)_2$

Source: SLAVIN, W. 1984. Graphite Furnace AAS—A Source Book. Perkin-Elmer Corp., Norwalk, Conn. Courtesy of the Perkin-Elmer Corporation.

steps of the temperature sequence. Use ramp drying for samples containing mixtures of solvents or for samples with a high salt content (to avoid spattering). Samples that contain a complex mixture of matrix components sometimes require ramp charring to effect controlled, complete thermal decomposition. Ramp atomization may minimize background absorption by permitting volatilization of the element being determined before the matrix. This is especially applicable in the determination of such volatile elements as cadmium and lead.

Use standard additions to compensate for matrix interferences. When making standard additions, determine whether the added species and the element being determined behave similarly under the specified conditions. [See Section 3113B.4d2)].

Chemical interaction of the graphite tube with various elements to form refractory carbides occurs at high charring and atomization temperatures. Elements that form carbides are barium, molybdenum, nickel, titanium, vanadium, and silicon. Carbide formation is characterized by broad, tailing atomization peaks and reduced sensitivity. Using pyrolytically coated tubes for these metals minimizes the problem. For the analysis of aluminum, thorium treated L'vov platforms provide sharper peaks at low concentrations and enhanced charring stability.

4. Sensitivity, Detection Limits, and Optimum Concentration Range

Estimated detection limits and optimum concentration ranges are listed in Table 3113:II. These values may vary with the chemical form of the element being determined, sample composition, or instrumental conditions.

For a given sample, increased sensitivity may be achieved by using a larger sample volume or by reducing flow rate of the purge gas or by using gas interrupt during atomization. Note, however, that these techniques also will increase the effects of any interferences present. Sensitivity can be decreased by diluting the sample, reducing sample volume, increasing purge-gas flow, or using a less sensitive wavelength. Use of argon, rather than nitrogen, as the purge gas generally improves sensitivity and reproducibility. Hydrogen mixed with the inert gas may suppress chemical interference and increase sensitivity by acting as a reducing agent, thereby aiding in producing more ground-state atoms. Using pyrolytically coated graphite tubes can increase sensitivity for the more refractory elements. The optical pyrometer/maximum power accessory available on some instruments also offers increased sensitivity with lower atomization temperatures for many elements.

Using the Stabilized Temperature Platform Furnace (STPF) technique, which is a combination of individual techniques, also offers significant interference reduction with improved sensitivity. Sensitivity changes with sample tube age. Discard graphite tubes when significant variations in sensitivity or poor reproducibility are observed. The use of high acid concentrations, brine samples, and matrix modifiers often drastically reduce tube life. Preferably use the L'vov platform in such situations.

5. Reference

1. SLAVIN, W. 1984. Graphite Furnace AAS—A Source Book. Perkin-Elmer Corp., Norwalk, Conn.

6. Bibliography

FERNANDEZ, F.J. & D.C. MANNING. 1971. Atomic absorption analyses of metal pollutants in water using a heated graphite atomizer. *Atomic Absorption Newsletter* 10:65.

SEGAR, D.A. & J.G. GONZALEZ. 1972. Evaluation of atomic absorption with a heated graphite atomizer for the direct determination of trace transition metals in sea water. *Anal. Chim. Acta* 58:7.

BARNARD, W.M. & M.J. FISHMAN. 1973. Eval-

TABLE 3113:II. DETECTION LEVELS AND CONCENTRATION RANGES FOR ELECTROTHERMAL
ATOMIZATION ATOMIC ABSORPTION SPECTROMETRY

Element	Wavelength nm	Estimated Detection Limit $\mu g/L$	Optimum Concentration Range $\mu g/L$
Al	309.3	3	20–200
Sb	217.6	3	20–300
As*	193.7	1	5–100
Ba†	553.6	2	10–200
Be	234.9	0.2	1–30
Cd	228.8	0.1	0.5–10
Cr	357.9	2	5–100
Co	240.7	1	5–100
Cu	324.7	1	5–100
Fe	248.3	1	5–100
Pb‡	283.3	1	5–100
Mn	279.5	0.2	1–30
Mo†	313.3	1	3–60
Ni†	232.0	1	5–100
Se*	196.0	2	5–100
Ag	328.1	0.2	1–25
Sn	224.6	5	20–300

* Gas interrupt utilized.
† Pyrolytic graphite tubes utilized.
‡ The more sensitive 217.0-nm wavelength is recommended for instruments with background correction capabilities.

uation of the use of heated graphite atomizer for the routine determination of trace metals in water. *Atomic Absorption Newsletter* 12:118.

KAHN, H.L. 1973. The detection of metallic elements in wastes and waters with the graphite furnace. *Int. J. Environ. Anal. Chem.* 3:121.

WALSH, P.R., J.L. FASCHING & R.A. DUCE. 1976. Matrix effects and their control during the flameless atomic absorption determination of arsenic. *Anal. Chem.* 48:1014.

HENN, E.L. 1977. Use of Molybdenum in Eliminating Matrix Interferences in Flameless Atomic Absorption. Spec. Tech. Publ. 618, American Soc. Testing & Materials, Philadelphia, Pa.

MARTIN, T.D. & J.F. KOPP. 1978. Methods for Metals in Drinking Water. U.S. Environmental Protection Agency, Environmental Monitoring and Support Lab., Cincinnati, Ohio.

HYDES, D.J. 1980. Reduction of matrix effects with a soluble organic acid in the carbon furnace atomic absorption spectrometric determination of cobalt, copper, and manganese in seawater. *Anal. Chem.* 52:289.

SOTERA, J.J. & H.L. KAHN. 1982. Background correction in AAS. *Amer. Lab.* 14:100.

SMITH, S.B. & G.M. HIEFTJE. 1983. A new background-correction method for atomic absorption spectrometry. *Appl. Spectrosc.* 37:419.

GROSSER, Z. 1985. Techniques in Graphite Furnace Atomic Absorption Spectrophotometry. Perkin-Elmer Corp., Ridgefield, Conn.

SLAVIN, W. & G.R. CARNICK. 1985. A survey of applications of the stabilized temperature platform furnace and Zeeman correction. *Atomic Spectrosc.* 6:157.

BRUEGGEMEYER, T. & F. FRICKE. 1986. Comparison of furnace & atomization behavior of aluminum from standard & thorium-treated L'vov platforms. *Anal. Chem.* 58:1143.

3113 B.   Electrothermal Atomic Absorption Spectrometric Method

1. General Discussion

This method is suitable for determination of micro quantities of aluminum, antimony, arsenic, barium, beryllium, cadmium, chromium, cobalt, copper, iron, lead, manganese, molybdenum, nickel, selenium, silver, and tin.

2. Apparatus

a. *Atomic absorption spectrometer:* See Section 3111A.6a. The instrument must have background correction capability.

b. *Source lamps:* See Section 3111A.6d.

c. *Graphite furnace:* Use an electrically heated device with electronic control circuitry designed to carry a graphite tube or cup through a heating program that provides sufficient thermal energy to atomize the elements of interest. Furnace heat controllers with only three heating steps are adequate only for fresh waters with low dissolved solids content. For salt waters, brines, and other complex matrices, use a furnace controller with up to seven individually programmed heating steps. Fit the furnace into the sample compartment of the spectrometer in place of the conventional burner assembly. Use argon as a purge gas to minimize oxidation of the furnace tube and to prevent the formation of metallic oxides. Use graphite tubes with L'vov platforms to minimize interferences and to improve sensitivity.

d. *Readout:* See Section 3111A.6c.

e. *Sample dispensers:* Use microliter pipets (5 to 100 μL) or an automatic sampling device designed for the specific instrument.

f. *Vent:* See Section 3111A.6f.

g. *Cooling water supply:* Cool with tap water flowing at 1 to 4 L/min or use a recirculating cooling device.

h. *Membrane filter apparatus:* Use an all-glass filtering device and 0.45-μm membrane filters. For trace analysis of aluminum, use device of polypropylene or TFE devices.

3. Reagents

a. *Metal-free water:* See Section 3111B.3c.

b. *Hydrochloric acid,* HCl, 1 + 1 and conc.

c. *Nitric acid,* HNO_3, 1 + 1 and conc.

d. *Matrix modifiers:*

1) *Ammonium nitrate,* 10% (w/v): Dissolve 100 g NH_4NO_3 in water. Dilute to 1000 mL with water.

2) *Ammonium phosphate,* 40%: Dissolve 40 g $(NH_4)_2HPO_4$ in water. Dilute to 100 mL with water.

3) *Calcium nitrate,* 20 000 mg Ca/L: Dissolve 11.8 g $Ca(NO_3)_2 \cdot 4H_2O$ in water. Dilute to 100 mL with water.

4) *Nickel nitrate,* 10 000 mg Ni/L: Dissolve 49.56 g $Ni(NO_3)_2 \cdot 6H_2O$ in water. Dilute to 1000 mL with water.

5) *Phosphoric acid,* 10% (v/v): Dilute 10 mL conc H_3PO_4 to 100 mL with water.

For preparation of other matrix modifiers, see references or follow manufacturers' instructions.

e. *Stock metal solutions:* Refer to Sections 3111B and 3114.

f. *Chelating resin:* 100 to 200 mesh* purified by heating at 60°C in 10N NaOH for 24 h. Cool resin and rinse 10 times each with alternating portions of 1N HCl, metal-free water, 1N NaOH, and metal-free water.

g. *Metal-free seawater (or brine):* Fill a 1.4-cm ID × 20-cm long borosilicate glass column to within 2 cm of the top with purified chelating resin. Elute resin with

---

*Chelex 100, or equivalent, available from Bio-Rad Laboratories, Richmond, Calif.

successive 50-mL portions of $1N$ HCl, metal-free water, $1N$ NaOH, and metal-free water at the rate of 5 mL/min just before use. Pass salt water or brine through the column at a rate of 5 mL/min to extract trace metals present. Discard the first 10 bed volumes (300 mL) of eluate.

4. Procedures

a. Sample pretreatment: Before analysis, pretreat all samples as indicated below. Rinse all glassware with $1 + 1$ HNO_3 and water. Carry out digestion procedures in a clean, dust-free laboratory area to avoid sample contamination. For digestion of trace aluminum, use polypropylene or TFE utensils to avoid leachable aluminum from glassware.

1) Dissolved metals—See Section 3030B. For samples requiring arsenic and/ or selenium analysis add 3 mL 30% hydrogen peroxide and an appropriate concentration of nickel before analysis. For all other metals no further pretreatment is required except for adding an optional matrix modifier (see Table 3113:I).

2) Total recoverable metals (Al, Sb, Ba, Be, Cd, Cr, Co, Cu, Fe, Pb, Mn, Mo, Ni, Ag, and Sn)—NOTE: Sb and Sn are recovered unless HCl is used in the digestion. See Section 3030D. Quantitatively transfer digested sample to a 100-mL volumetric flask, add an appropriate amount of matrix modifier (if necessary, see Table 3113:I), and dilute to volume with water.

3) Total recoverable metals (As, Se)—Transfer 100 mL of shaken sample, 1 mL conc HNO_3, and 2 mL 30% H_2O_2 to a clean, acid-washed 250-mL beaker. Heat on a hot plate without allowing solution to boil until volume has been reduced to about 50 mL. Remove from hot plate and let cool to room temperature. Add an appropriate concentration of nickel (See Table 3113:I), and dilute to volume in a 100-mL volumetric flask with water. Simultaneously prepare a digested blank by substituting

water for sample and proceed with digestion as described above.

b. Instrument operation: Mount and align furnace device according to manufacturer's instructions. Turn on instrument and strip-chart recorders. Select appropriate light source and adjust to recommended electrical setting. Select proper wavelength and set all conditions according to manufacturer's instructions, including background correction. Background correction is important when elements are determined at short wavelengths or when sample has a high level of dissolved solids. In general, background correction is usually not necessary at wavelengths longer than 350 nm. Above 350 nm deuterium arc background correction is not useful and other types must be used.

Select proper inert- or sheath-gas flow. In some cases, it is desirable to interrupt the inert-gas flow during atomization. Such interruption results in increased sensitivity by increasing residence time of the atomic vapor in the optical path. Gas interruption also increases background absorption and intensifies interference effects. Consider advantages and disadvantages of this option for each matrix when optimizing analytical conditions.

To optimize graphite furnace conditions, carefully adjust furnace temperature settings to maximize sensitivity and precision and to minimize interferences. Follow manufacturer's instructions.

Use drying temperatures slightly above the solvent boiling point to provide enough time and temperature for complete evaporation without boiling or spattering.

The charring temperature must be high enough to maximize volatilization of interfering matrix components yet too low to volatilize the element of interest. With the drying and atomization temperatures set to their optimum values, analyze a standard at a series of charring temperatures in increasing increments of 50 to 100°C. When the optimum charring temperature is exceeded, there will be a significant drop in sensitivity. Plot charring temperature versus sample absorbance: the optimum charring temperature is the highest temperature without reduced sensitivity.

Select atomization temperature by determining the temperature providing maximum sensitivity without significantly eroding precision. Optimize by a series of successive determinations at various atomization temperatures using a standard solution giving an absorbance of 0.2 to 0.5.

c. *Instrument calibration:* Prepare standards for instrument calibration by dilution of the metal stock solutions. Prepare standards fresh daily.

Prepare a blank and at least three calibration standards in the appropriate concentration range (See Table 3113:II) for correlating element concentration and instrument response. Match the matrix of the standard solutions to those of the samples as closely as possible. In most cases, this simply requires matching the acid background of the samples. For seawaters or brines, however, use the metal-free matrix (¶ 3g) as the standard solution diluent. In addition, add the same concentration of matrix modifier (if required for sample analysis) to the standard solutions.

Inject a suitable portion of each standard solution, in order of increasing concentration. Analyze each standard solution in triplicate to verify method precision.

Construct an analytical curve by plotting the average peak absorbencies or peak areas of the standard solution versus concentration on linear graph paper. Alternatively, use electronic instrument calibration if the instrument has this capability.

d. *Sample analysis:* Analyze all samples except those demonstrated to be free of matrix interferences (based on recoveries of 85%–115% for known additions) using the method of standard additions. Analyze all samples at least in duplicate or until reproducible results are obtained. A vari-

ation of $\leq 10\%$ is considered acceptable reproducibility. Average replicate values.

1) Direct determination—Inject a measured portion of pretreated sample into the graphite furnace. Use the same volume as was used to prepare the calibration curve. Dry, char, and atomize according to the preset program. Repeat until reproducible results are obtained.

Compare the average absorbance value or peak area to the calibration curve to determine concentration of the element of interest. Alternatively, read results directly if the instrument is equipped with this capability. If absorbance (or concentration) or peak area of the most concentrated sample is greater than absorbance (concentration) or peak area of the standard, dilute sample and reanalyze. If very large dilutions are required, another technique (e.g., flame AA or ICP) may be more suitable for this sample. Large dilution factors magnify small errors on final calculation. Keep acid background and concentration of matrix modifier (if required) constant. If sample is diluted with water, add acid and matrix modifier to restore the concentration of both to the original. Alternatively, dilute the sample in a blank solution of acid and matrix modifiers.

Proceed to ¶ 5a below.

2) Method of standard additions—Refer to ¶ 4c above. The method of standard additions is valid only when it falls in the linear portion of the calibration curve. Once instrument sensitivity has been optimized for the element of interest and the linear range for the element has been established, proceed with sample analyses.

Inject a measured volume of sample into furnace device. Dry, char or ash, and atomize samples according to preset program. Repeat until reproducible results are obtained. Record instrument response in absorbance or concentration as appropriate. Add a known concentration of the element of interest to a separate portion of sample so as not to change significantly the sample volume. Repeat the determination.

Add a known concentration (preferably twice that used in the first addition) to a separate sample portion. Mix well and repeat the determination.

Plot average absorbance or instrument response for the sample and the two portions with known additions on the vertical axis with the concentrations of element added on the horizontal axis of linear graph paper. Draw a straight line connecting the three points and extrapolate to zero absorbance. The intercept at the horizontal axis is the concentration of the sample. The concentration axis to the left of the origin should be a mirror image of the axis to the right.

5. Calculations

a. Direct determination:

$$\mu g \text{ metal/L} = C \times F$$

where:

C = metal concentration as read directly from the instrument or from the calibration curve, µg/L, and
F = dilution factor.

b. Method of additions:

$$\mu g \text{ metal/L} = C \times F$$

where:

C = metal concentration as read from the method of additions plot, µg/L, and
F = dilution factor.

6. Precision and Bias

Data typical of the precision and bias obtainable are presented in Tables 3113:III, IV, and V.

7. Quality Control

See Section 3020 for specific quality control procedures to be followed during analysis. Although previous indications were that very low optimum concentration ranges were attainable for most metals (see Table 3113:II), data in Table 3113:III using variations of these protocols show that this may not be so. Exercise extreme caution when applying this method to the lower concentration ranges. Verify analyst precision at the beginning of each analytical run by making triplicate analyses.

8. Reference

1. COPELAND, T.R. & J.P. MANEY. 1986. EPA Method Study 31: Trace Metals by Atomic Absorption (Furnace Techniques). EPA-600/S4-85-070, U.S. Environmental Protection Agency, Environmental Monitoring and Support Lab., Cincinnati, Ohio.

9. Bibliography

RENSHAW, G.D. 1973. The determination of barium by flameless atomic absorption spectrophotometry using a modified graphite tube atomizer. Atomic Absorption Newsletter 12:158.

YANAGISAWA, M., T. TAKEUCHI & M. SUZUKI. 1973. Flameless atomic absorption spectrometry of antimony. Anal. Chim. Acta 64:381.

RATTONETTI, A. 1974. Determination of soluble cadmium, lead, silver and indium in rainwater and stream water with the use of flameless atomic absorption. Anal. Chem. 46:739.

HENN, E.L. 1975. Determination of selenium in water and industrial effluents by flameless atomic absorption. Anal. Chem. 47:428.

MARTIN, T.D. & J.F. KOPP. 1975. Determining selenium in water, wastewater, sediment and sludge by flameless atomic absorption spectrometry. Atomic Absorption Newsletter 14:109.

MARUTA, T., K. MINEGISHI & G. SUDOH. 1976. The flameless atomic absorption spectrometric determination of aluminum with a carbon atomization system. Anal. Chim. Acta 81:313.

CRANSTON, R.E. & J.W. MURRAY. 1978. The determination of chromium species in natural waters. Anal. Chim. Acta 99:275.

HOFFMEISTER, W. 1978. Determination of iron in ultrapure water by atomic absorption spectroscopy. Z. Anal. Chem. 50:289.

LAGAS, P. 1978. Determination of beryllium, barium, vanadium and some other elements in water by atomic absorption spectrometry with electrothermal atomization. Anal. Chim. Acta 98:261.

CARRONDO, M.J.T., J.N. LESTER & R. PERRY. 1979. Electrothermal atomic absorption determination of total aluminum in waters and waste waters. Anal. Chim. Acta 111:291.

TABLE 3113:III. INTERLABORATORY SINGLE-ANALYST PRECISION DATA FOR ELECTROTHERMAL ATOMIZATION METHODS[1]

Element	Concentration μg/L	Single-Analyst Precision % RSD					
		Lab Pure	Drinking Water	Surface Water	Effluent 1	Effluent 2	Effluent 3
Al	28	66	108	70	—	—	66
	125	27	35	24	—	—	34
	11 000	11	—	—	22	—	—
	58 300	27	—	—	19	—	—
	460	9	—	—	—	30	—
	2 180	28	—	—	—	4	—
	10.5	20	13	13	13	56	18
	230	10	18	13	21	94	14
As	9.78	40	25	15	74	23	11
	227	10	6	8	11	15	6
Ba	56.5	36	21	29	59	23	27
	418	14	12	20	24	24	18
Be	0.45	18	27	15	30	2	11
	10.9	14	4	9	7	12	12
Cd	0.43	72	49	1	121	35	27
	12	11	17	22	14	11	15
Cr	9.87	24	33	10	23	15	10
	236	16	7	11	13	16	7
Co	29.7	10	17	10	19	24	12
	420	8	11	13	14	9	5
Cu	10.1	49	47	17	17	—	30
	234	8	15	6	21	—	11
	300	6	—	—	—	11	—
	1 670	11	—	—	—	6	—
Fe	26.1	144	52	153	—	—	124
	455	48	37	45	—	—	31
	1 030	17	—	—	30	—	—
	5 590	6	—	—	32	—	—
	370	14	—	—	—	19	—
	2 610	9	—	—	—	18	—
Pb	10.4	6	19	17	21	19	33
	243	17	7	17	18	12	16
Mn	0.44	187	180	—	—	—	275
	14.8	32	19	—	—	—	18
	91.0	15	—	—	48	—	—
	484.0	4	—	—	12	—	—
	111.0	12	—	—	—	21	—
	666.0	6	—	—	—	20	—
Ni	26.2	20	26	25	24	18	9
	461.0	15	11	9	8	11	4
Se	10.0	12	27	16	35	41	13
	235.0	6	6	15	6	13	14
Ag	8.48	10	—	—	15	27	16
	56.5	14	—	—	7	16	23
	0.45	27	166	48	—	—	—
	13.6	15	4	10	—	—	—

TABLE 3113:IV. INTERLABORATORY OVERALL PRECISION DATA FOR ELECTROTHERMAL ATOMIZATION METHODS[1]

Element	Concentration µg/L	Overall Precision % RSD					
		Lab Pure	Drinking Water	Surface Water	Effluent 1	Effluent 2	Effluent 3
Al	28	99	114	124	—	—	131
	125	45	47	49	—	—	40
	11 000	19	—	—	43	—	—
	58 300	31	—	—	32	—	—
	460	20	—	—	—	47	—
	2 180	30	—	—	—	15	—
	10.5	37	19	22	50	103	39
	230	26	16	16	17	180	21
As	9.78	43	26	37	72	50	39
	227	18	12	13	20	15	14
Ba	56.5	68	38	43	116	43	65
	418	35	35	28	38	48	16
Be	0.45	28	31	15	67	50	35
	10.9	33	15	26	20	9	19
Cd	0.43	73	60	5	88	43	65
	12	19	25	41	26	20	27
Cr	9.87	30	53	24	60	41	23
	236	18	14	24	20	14	20
Co	29.7	13	26	17	18	21	17
	420	21	21	17	18	13	13
Cu	10.1	58	82	31	32	—	74
	234	12	33	19	21	—	26
	300	13	—	—	—	14	—
	1 670	12	—	—	—	13	—
Fe	26.1	115	93	306	—	—	204
	455	53	46	53	—	—	44
	1 030	32	—	—	25	—	—
	5 590	10	—	—	43	—	—
	370	28	—	—	—	22	—
	2 610	13	—	—	—	22	—
Pb	10.4	27	42	31	23	28	47
	243	18	19	17	19	19	25
Mn	0.44	299	272	—	—	—	248
	14.8	52	41	—	—	—	29
	91.0	16	—	—	45	—	—
	484.0	5	—	—	17	—	—
	111.0	15	—	—	—	17	—
	666.0	8	—	—	—	24	—
Ni	26.2	35	30	49	35	37	43
	461.0	23	22	15	12	21	17
Se	10.0	17	48	32	30	44	51
	235.0	16	18	18	17	22	34
Ag	8.48	23	—	—	16	35	34
	56.5	15	—	—	24	32	28
	0.45	57	90	368	—	—	—
	13.6	19	19	59	—	—	—

TABLE 3113:V. INTERLABORATORY RELATIVE ERROR DATA FOR ELECTROTHERMAL
ATOMIZATION METHODS[1]

Element	Concentration µg/L	Relative Error %					
		Lab Pure Water	Drinking Water	Surface Water	Effluent 1	Effluent 2	Effluent 3
Al	28.0	86	150	54	—	—	126
	125.0	4	41	39	—	—	30
	11 000.0	2	—	—	14	—	—
	58 300.0	12	—	—	7	—	—
	460.0	2	—	—	—	11	—
	2 180.0	11	—	—	—	9	—
Sb	10.5	30	32	28	24	28	36
	230.0	35	14	19	13	73	39
As	9.78	36	1	22	106	13	16
	227.0	3	7	10	19	6	13
Ba	56.5	132	54	44	116	59	40
	418.0	4	0	0	13	6	60
Be	0.45	40	16	11	16	10	15
	10.9	13	2	9	7	8	8
Cd	0.43	58	45	37	66	16	19
	12.0	4	6	5	22	18	3
Cr	9.87	10	9	4	2	5	15
	236.0	11	0	9	13	5	8
Co	29.7	7	7	1	6	3	13
	420.0	12	8	8	11	5	18
Cu	10.1	16	48	2	5	—	15
	234.0	8	7	0	4	—	19
	300.0	4	—	—	—	21	—
	1 670.0	6	—	—	—	2	—
Fe	26.1	85	60	379	—	—	158
	455.0	43	22	31	—	—	18
	1 030.0	8	—	—	8	—	—
	5 590.0	2	—	—	12	—	—
	370.0	4	—	—	—	11	—
	2 610.0	35	—	—	—	2	—
Pb	10.4	16	10	17	1	34	14
	234.0	5	15	8	18	15	29
Mn	0.44	332	304	—	—	—	556
	14.8	10	1	—	—	—	36
	91.0	31	—	—	10	—	—
	484.0	42	—	—	4	—	—
	111.0	1	—	—	—	29	—
	666.0	6	—	—	—	23	—
Ni	26.2	9	16	10	7	33	54
	461.0	15	19	18	31	16	18
Se	10.0	12	9	6	36	17	37
	235.0	7	7	0	13	10	17
Ag	8.48	12	—	—	1	51	20
	56.5	16	—	—	8	51	22
	0.45	34	162	534	—	—	—
	13.6	3	12	5	—	—	—

NAKAHARA, T. & C.L. CHAKRABARTI. 1979. Direct determination of traces of molybdenum in synthetic sea water by atomic absorption spectrometry with electrothermal atomization and selective volatilization of the salt matrix. *Anal. Chim. Acta* 104:99.

TIMINAGA, M. & Y. UMEZAKI. 1979. Determination of submicrogram amounts of tin by atomic absorption spectrometry with electrothermal atomization. *Anal. Chim. Acta* 110:55.

3114   METALS BY HYDRIDE GENERATION/ATOMIC ABSORPTION SPECTROMETRY*

3114 A.   Introduction

For general introductory material on atomic absorption spectrometric methods, see Section 3111A.

Two methods are presented in this section: A manual method and a continuous-flow method especially recommended for

---

*Approved by Standard Methods Committee, 1989.

selenium. Continuous-flow automated systems are preferable to manual hydride generators because the effect of sudden hydrogen generation on light-path transparency is removed and any blank response from contamination of the HCl reagent by the elements being determined is incorporated into the background base line.

3114 B.   Manual Hydride Generation/Atomic Absorption Spectrometric Method

1. General Discussion

a. Principle: This method is applicable to the determination of arsenic and selenium by conversion to their hydrides by sodium borohydride reagent and aspiration into an atomic absorption atomizer.

Arsenous acid and selenous acid, the As(III) and Se(IV) oxidation states of arsenic and selenium, respectively, are instantaneously converted by sodium borohydride reagent in acid solution to their volatile hydrides. The hydrides are purged continuously by argon or nitrogen into an appropriate atomizer of an atomic absorption spectrometer and converted to the gas-phase atoms. The sodium borohydride reducing agent, by rapid generation

of the elemental hydrides in an appropriate reaction cell, minimizes dilution of the hydrides by the carrier gas and provides rapid, sensitive determinations of arsenic and selenium.

CAUTION: *Arsenic and selenium and their hydrides are toxic. Handle with care.*

At room temperature and solution pH values of 1 or less, arsenic acid, the As(V) oxidation state of arsenic, is reduced relatively slowly by sodium borohydride to As(III), which is then instantaneously converted to arsine. The arsine atomic absorption peaks commonly are decreased by one-fourth to one-third for As(V) when compared to As(III). Determination of total arsenic requires that all inorganic arsenic compounds be in the As(III) state. Organic and inorganic forms of arsenic are first oxidized to As(V) by acid digestion. The As(V) then is quantitatively reduced to As(III) with sodium or potassium iodide before reaction with sodium borohydride.

Selenic acid, the Se(VI) oxidation state of selenium, is not measurably reduced by sodium borohydride. To determine total selenium by atomic absorption and sodium borohydride, first reduce Se(VI) formed during the acid digestion procedure to Se(IV), being careful to prevent reoxidation by chlorine. Efficiency of reduction depends on temperature, reduction time, and HCl concentration. For 4N HCl, heat 1 h at 100°C. For 6N HCl, boiling for 10 min is sufficient.[1-3] Alternatively, autoclave samples in sealed containers at 121°C for 1 h. NOTE: Autoclaving in sealed containers may result in incomplete reduction, apparently due to the buildup of chlorine gas. To obtain equal instrument responses for reduced Se(VI) and Se (IV) solutions of equal concentrations, manipulate HCl concentration and heating time. For further details, see Section 3500-Se.

b. Equipment selection:

Certain atomic absorption atomizers and hydride reaction cells are available commercially for use with the sodium borohydride reagent. A functional system is presented in Figure 3114:1. Irrespective of the hydride reaction cell-atomizer system selected, it must meet the following quality-control considerations: (*a*) it must provide a precise and reproducible standard curve between 0 and 20 μg As or Se/L and an instrumental detection limit between 0.1 and 0.5 μg As or Se/L; (*b*) when carried through the entire procedure, oxidation state couples [As (III) - As (V) or Se (IV)-Se (VI)] must cause equal instrument response; and (*c*) sample digestion must yield 80% or greater recovery of added cacodylic acid (dimethyl arsinic acid) and 90% or greater recovery of added As(III), As(V), Se(VI), or Se(IV).

Three types of atomic absorption atomizers commonly are used in the measurement of arsenic and selenium. Most instrument manufacturers can provide a Boling-type burner for argon (or nitrogen)-air entrained-hydrogen flames. Alternatively use an externally heated quartz cell or a quartz cell with an internal fuel rich oxygen-hydrogen or air-hydrogen flame. Quartz atomization cells provide for the most sensitive arsenic and selenium hydride determinations and minimize background noise associated with the argon-air entrained-hydrogen flame.

c. Digestion techniques: Waters and wastewaters may contain varying amounts of organic arsenic compounds and inorganic compounds of As(III), As(V), Se(IV), and Se(VI). To measure total arsenic and selenium in these samples requires sample digestion to solubilize particulate forms and oxidize reduced forms of arsenic and selenium and to convert any organic compounds to inorganic ones. Organic selenium compounds rarely have been demonstrated in water. It is left to the experienced analyst's judgment whether sample digestion is required.

Two digestion procedures are provided in ¶ 4*c* below. Consider sulfuric-nitric-perchloric acid digestion or sulfuric-nitric acid digestion as providing a measure of total recoverable arsenic rather than total arsenic because they do not completely convert certain organic arsenic compounds to As(V). The sulfuric-nitric-perchloric acid digestion effectively destroys organics and most particulates in untreated wastewaters or solid samples. The potassium persulfate digestion (¶ 4*d*) is effective for converting organic arsenic and selenium compounds to As(V) and Se(VI) in potable and surface waters and in most wastewaters.[4]

The HCl-autoclave reduction of Se(VI) described above is an effective digestion procedure for total inorganic Se; however, it has not been found effective for con-

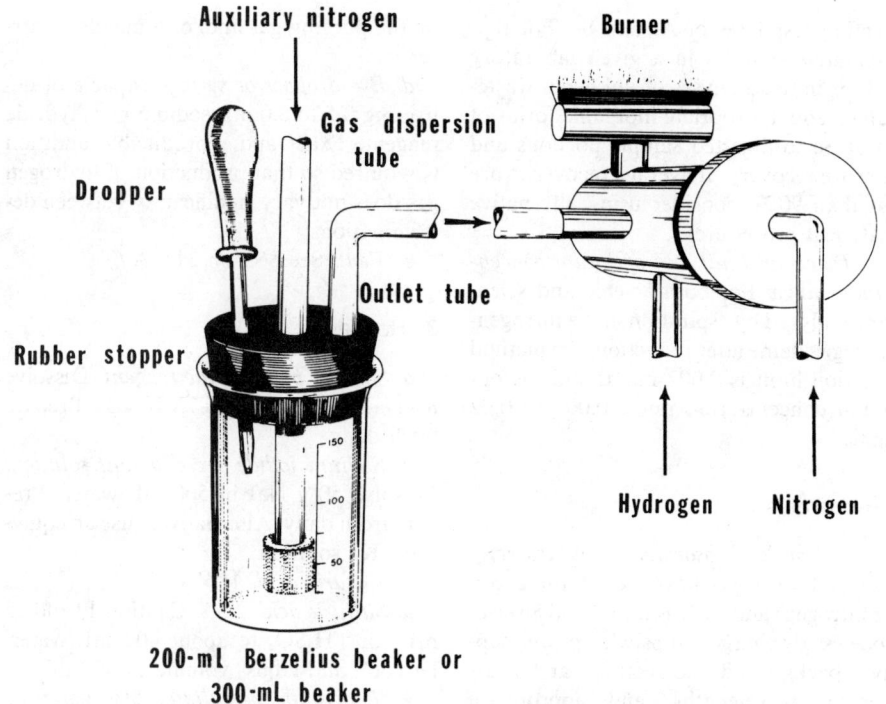

Figure 3114:1. Manual reaction cell for producing As and Se hydrides.

verting benzene substituted selenium compounds to inorganic selenium.

d. Interferences: Interferences are minimized because the As and Se hydrides are removed from the solution containing most potential interfering substances. Slight response variations occur when acid matrices are varied. Control these variations by treating standards and samples in the same manner. Low concentrations of noble metals (approximately 100 μg/L of Ag, Au, Pt, Pd, etc.), concentrations of copper, lead, and nickel at or greater than 1 mg/L, and concentrations between 0.1 and 1 mg/L of hydride-forming elements (Bi, Sb, Sn, and Te) may suppress the response of As and Se hydrides. Interference by transition metals depends strongly on HCl concentration. Interferences are less pronounced at 4 to 6N HCl than at lower concentrations.[5] The presence of As or Se

in each other's matrices can cause similar suppression. Reduced nitrogen oxides resulting from HNO_3 digestion and nitrite also can suppress instrumental response for both elements. Large concentrations of iodide interfere with the Se determination by reducing Se to its elemental form. Do not use any glassware for determining Se that has been used for iodide reduction of As(V).

To prevent chlorine gas produced in the reduction of Se(VI) to Se(IV) from reoxidizing the Se(IV), generate the hydride within a few hours of the reduction steps or purge the chlorine from the samples by sparging.[6]

Interferences depend on system design and defy quantitative description because of their synergistic effects. Certain waters and wastewaters can contain interferences in sufficient concentration to suppress absorption responses of As and Se. For representative samples in a given laboratory and for initial analyses of unknown wastewaters, add appropriate inorganic forms of As or Se to digested sample portions and measure recovery. If average recoveries are less than 90%, consider using alternative analytical procedures.

e. Detection limit and optimum concentration range: For both arsenic and selenium, analyzed by aspiration into a nitrogen-hydrogen flame after reduction, the method detection limit is 0.002 mg/L and the optimum concentration range 0.002 to 0.02 mg/L.

2. Apparatus

a. Atomic absorption spectrometer equipped with gas flow meters for argon (or nitrogen) and hydrogen, As and Se electrodeless discharge lamps with power supply, background correction at measurement wavelengths, and appropriate strip-chart recorder. A good-quality 10-mV recorder with high sensitivity and a fast response time is needed.

b. Atomizer: Use one of the following:

1) *Boling-type burner head* for argon (or nitrogen)-air entrained-hydrogen flame.

2) *Cylindrical quartz cell,* 10 to 20 cm long, electrically heated by external nichrome wire to 800 to 900°C.[7]

3) *Cylindrical quartz cell* with internal fuel rich hydrogen-oxygen (air) flame.[8]

The sensitivity of quartz cells deteriorates over several months of use. Sensitivity sometimes may be restored by treatment with 40% HF. CAUTION: *HF is extremely corrosive. Avoid all contact with exposed skin. Handle with care.*

c. Reaction cell for producing As or Se hydrides: See Figure 3114:1. A commercially available system is acceptable if it utilizes liquid sodium borohydride reagents; accepts samples digested in accordance with ¶s 4c, d, and e; accepts 4 to 6N HCl; and is efficiently and precisely stirred by the purging gas and/or a magnetic stirrer.

d. Eye dropper or syringe capable of delivering 0.5 to 3.0 mL sodium borohydride reagent. Exact and reproducible addition is required so that production of hydrogen gas does not vary significantly between determinations.

e. Vent: See Section 3111A.6f.

3. Reagents

a. Sodium borohydride reagent: Dissolve 8 g $NaBH_4$ in 200 mL 0.1N NaOH. Prepare fresh daily.

b. Sodium iodide prereductant solution: Dissolve 50 g NaI in 500 mL water. Prepare fresh daily. Alternatively use an equivalent KI solution.

c. Sulfuric acid, 18N.

d. Sulfuric acid, 2.5N: Cautiously add 35 mL conc H_2SO_4 to about 400 mL water, let cool, and adjust volume to 500 mL.

e. Potassium persulfate, 5% solution: Dissolve 25 g $K_2S_2O_8$ in water and dilute to 500 mL. Store in glass and refrigerate. Prepare weekly.

f. Nitric acid, HNO_3, conc.

g. Perchloric acid, $HClO_4$, conc.

h. Hydrochloric acid, HCl, conc.

i. Argon (or nitrogen), commercial grade.

j. Hydrogen, commercial grade.

k. Arsenic(III) solutions:

1) *Stock As(III) solution:* Dissolve 1.320 g arsenic trioxide, As_2O_3, in water containing 4 g NaOH. Dilute to 1 L; 1.00 mL = 1.00 mg As(III).

2) *Intermediate As(III) solution:* Dilute 10 mL stock As solution to 1000 mL with water containing 5 mL conc HCl; 1.00 mL = 10.0 μg As(III).

3) *Standard As(III) solution:* Dilute 10 mL intermediate As(III) solution to 1000 mL with water containing the same concentration of acid used for sample preservation (2 to 5 mL conc HNO_3); 1.00 mL = 0.100 μg As(III). Prepare diluted solutions daily.

l. Arsenic(V) solutions:

1) *Stock As(V) solution:* Dissolve 1.534 g arsenic pentoxide, As_2O_5, in distilled water containing 4 g NaOH. Dilute to 1 L; 1.00 mL = 1.00 mg As(V).

2) *Intermediate As(V) solution:* Prepare as for As(III) above; 1.00 mL = 10.0 µg As(V).

3) *Standard As(V) solution:* Prepare as for As(III) above; 1.00 mL = 0.100 µg As(V).

m. Organic arsenic solutions:

1) *Stock organic arsenic solution:* Dissolve 1.842 g dimethylarsinic acid (cacodylic acid), $(CH_3)_2AsOOH$, in water containing 4 g NaOH. Dilute to 1 L; 1.00 mL = 1.00 mg As. [NOTE: Check purity of cacodylic acid reagent against an intermediate arsenic standard (50 to 100 mg As/L) using flame atomic absorption.]

2) *Intermediate organic arsenic solution:* Prepare as for As(III) above; 1.00 mL = 10.0 µg As.

3) *Standard organic arsenic solution:* Prepare as for As(III) above; 1.00 mL = 0.100 µg As.

n. Selenium(IV) solutions:

1) *Stock Se(IV) solution:* Dissolve 2.190 g sodium selenite, Na_2SeO_3, in water containing 10 mL HCl and dilute to 1 L; 1.00 mL = 1.00 mg Se(IV).

2) *Intermediate Se(IV) solution:* Dilute 10 mL stock Se(IV) to 1000 mL with water containing 10 mL conc HCl; 1.00 mL = 10.0 µg Se(IV).

3) *Standard Se(IV) solution:* Dilute 10 mL intermediate Se(IV) solution to 1000 mL with water containing the same concentration of acid used for sample preservation (2 to 5 mL conc HNO_3). Prepare solution daily when checking the equivalency of instrument response for Se(IV) and Se(VI); 1.00 mL = 0.100 µg Se(IV).

o. Selenium(VI) solutions:

1) *Stock Se(VI) solution:* Dissolve 2.393 g sodium selenate, Na_2SeO_4, in water containing 10 mL conc HNO_3. Dilute to 1 L; 1.00 mL = 1.00 mg Se(VI).

2) *Intermediate Se(VI) solution:* Prepare as for Se(IV) above; 1.00 mL = 10.0 µg Se (VI).

3) *Standard Se(VI) solution:* Prepare as for Se(IV) above; 1.00 mL = 0.100 µg Se(VI).

4. Procedure

a. Apparatus setup: Either see Figure 3114:1 or follow manufacturer's instructions. Connect inlet of reaction cell with auxiliary purging gas controlled by flow meter. If a drying cell between the reaction cell and atomizer is necessary, use only anhydrous $CaCl_2$ but not $CaSO_4$ because it may retain SeH_2. Before using the hydride generation/analysis system, optimize operating parameters. Aspirate dilute aqueous solutions of As and Se directly into the flame to facilitate atomizer alignment. Align quartz atomizers for maximum absorbance. Aspirate a blank until memory effects are removed. Establish purging gas flow, concentration and rate of addition of sodium borohydride reagent, solution volume, and stirring rate for optimum instrument response for the chemical species to be analyzed. If a quartz atomizer is used, optimize cell temperature. If sodium borohyride reagent is added too quickly, rapid evolution of hydrogen will unbalance the system. If the volume of solution being purged is too large, the absorption signal will be decreased. Recommended wavelengths are 193.7 and 196.0 nm for As and Se, respectively.

b. Instrument calibration standards: Transfer 0.00, 1.00, 2.00, 5.00, 10.00, 15.00, and 20.00 mL standard solutions of As(III) or Se(IV) to 100-mL volumetric flasks and bring to volume with water containing the same acid concentration used for sample preservation (commonly 2 to 5 mL conc HNO_3/L). This yields blank and standard solutions of 0, 1, 2, 5, 10, 15, and 20 µg As or Se/L. Prepare fresh daily.

c. Preparation of samples and standards

for total recoverable arsenic and selenium:
Follow general procedures of Section
3030F; alternatively, add 50 mL sample,
As(III), or Se(IV) standard to 200-mL Berzelius beaker. (Alternatively, prepare
standards by adding 100 μg/L standard As
or Se solutions directly to the beaker and
dilute to 50 mL in this beaker). Add 7 mL
18N H$_2$SO$_4$ and 5 mL conc HNO$_3$. Add a
small boiling chip or glass beads if necessary. Evaporate to SO$_3$ fumes. Maintain
oxidizing conditions at all times by adding
small amounts of HNO$_3$ to prevent solution
from darkening. Maintain an excess of
HNO$_3$ until all organic matter is destroyed.
Complete digestion usually is indicated by
a light-colored solution. Cool slightly, add
25 mL water and 1 mL conc HClO$_4$ and
again evaporate to SO$_3$ fumes to expel oxides of nitrogen. CAUTION: *See Section
3030H for cautions on use of HClO$_4$.* Monitor effectiveness of digestion procedure
used by adding 5 mL of standard organic
arsenic solution or 5 mL of a standard selenium solution to a 50-mL sample and
measuring recovery, carrying standards
through entire procedure. To report total
recoverable arsenic as total arsenic, average
recoveries of cacodylic acid must exceed
80%. Alternatively, use 100-mL micro-
kjeldahl flasks for the digestion of total recoverable arsenic or selenium, thereby improving digestion effectiveness. After final
evaporation of SO$_3$ fumes, dilute to 50 mL
for arsenic measurements or to 30 mL for
selenium measurements.

*d. Preparation of samples and standards
for total arsenic and selenium:* Add 50 mL
sample or standard to a 200-mL Berzelius
beaker. Add 1 mL 2.5N H$_2$SO$_4$ and 5 mL
5% K$_2$S$_2$O$_8$. Boil gently on a pre-heated
hot plate for approximately 30 to 40 min
or until a final volume of 10 mL is reached.
Do not let sample go to dryness. Alternatively heat in an autoclave at 121°C for
1 h in capped containers. After manual
digestion, dilute to 50 mL for subsequent
arsenic measurements and to 30 mL for

selenium measurements. Monitor effectiveness of digestion by measuring recovery of
As or Se as above. If poor recovery of arsenic added as cacodylic acid is obtained,
reanalyze using double the amount of
K$_2$S$_2$O$_8$.

*e. Determination of arsenic with sodium
borohydride:* To 50 mL digested standard
or sample in a 200-mL Berzelius beaker
(see Figure 3114:1) add 5 mL conc HCl
and mix. Add 5 mL NaI prereductant solution, mix, and wait at least 30 min.
(NOTE: The NaI reagent has not been
found necessary for certain hydride reaction cell designs if a 20 to 30% loss in
instrument sensitivity is not important and
variables of solution acid conditions, temperatures, and volumes for production of
As(V) and arsine can be controlled strictly.
Such control requires an automated delivery system; see Section 3114C.)

Attach one Berzelius beaker at a time to
the rubber stopper containing the gas dispersion tube for the purging gas, the sodium borohydride reagent inlet, and the
outlet to the atomizer. Turn on strip-chart
recorder and wait until the base line is established by the purging gas and all air is
expelled from the reaction cell. Add 0.5
mL sodium borohydride reagent. After the
instrument absorbance has reached a maximum and returned to the base line, remove
beaker, rinse dispersion tube with water,
and proceed to the next sample or standard. Periodically compare standard As(III)
and As(V) curves for response consistency.
Check for presence of chemical interferences that suppress instrument response for
arsine by treating a digested sample with
10 μg/L As(III) or As(V) as appropriate.
Average recoveries should be not less than
90%.

*f. Determination of selenium with sodium
borohydride:* To 30 mL digested standard
or sample, or to 30 mL undigested standard
or sample in a 200-mL Berzelius beaker,
add 15 mL conc HCl and mix. Heat for a
predetermined period at 90 to 100°C. Al-

ternatively autoclave at 121°C in capped containers for 60 min, or heat for a predetermined time in open test tubes using a 90 to 100°C hot water bath or an aluminum block digester. Check effectiveness of the selected heating by demonstrating equal instrument responses for calibration curves prepared either from standard Se(IV) or from Se(VI) solutions. Effective heat exposure for converting Se(VI) to Se(IV), with no loss of Se(IV), ranges between 5 and 60 min when open beakers or test tubes are used. Do not digest standard Se(IV) and Se(VI) solutions used for this check of equivalency. After prereduction of Se(VI) to Se(IV), attach Berzelius beakers, one at a time, to the purge apparatus. For each, turn on the strip-chart recorder and wait until the base line is established. Add 0.50 mL sodium borohydride reagent. After the instrument absorbance has reached a maximum and returned to the base line, remove beaker, rinse dispersion tube with water, and proceed to the next sample or stand-ard. Check for presence of chemical interferences that suppress selenium hydride instrument response by treating a digested sample with 10 μg Se (IV)/L. Average recoveries should be not less than 90%.

5. Calculation

Construct a standard curve by plotting peak heights or areas of standards versus concentration of standards. Measure peak heights or areas of samples and read concentrations from curve. If sample was diluted (or concentrated) before sample digestion, apply an appropriate factor. On instruments so equipped, read concentrations directly after standard calibration.

6. Precision and Bias

Single-laboratory, single-operator data were collected for As(III) and organic arsenic by both manual and automated methods, and for the manual determination of selenium. Recovery values (%) from seven replicates are given below:

	As(III)	Org As	Se(IV)	Se(VI)
Manual with digestion	91.8	87.3	—	—
Manual without digestion	109.4	19.4	100.6	110.8
Automated with digestion	99.8	98.4	—	—
Automated without digestion	92.5	10.4	—	—

7. References

1. VIJAN, P.N. & D. LEUNG. 1980. Reduction of chemical interference and speciation studies in the hydride generation-atomic absorption method for selenium. *Anal. Chim. Acta* 120:141.

2. VOTH-BEACH, L.M. & D.E. SHRADER. 1985. Reduction of interferences in the determination of arsenic and selenium by hydride generation. *Spectroscopy* 1:60.

3. JULSHAMN, K., O. RINGDAL, K.-E. SLINNING & O.R. BRAEKKAN. 1982. Optimization of the determination of selenium in marine samples by atomic absorption spectrometry: Comparison of a flameless graphite furnace atomic absorption system with a hydride generation atomic absorption system. *Spectrochim. Acta.* 37B:473.

4. NYGAARD, D.D. & J.H. LOWRY. 1982. Sample digestion procedures for simultaneous determination of arsenic, antimony, and selenium by inductively coupled argon plasma emission spectrometry with hydride generation. *Anal. Chem.* 54:803.

5. WELZ, B. & M. MELCHER. 1984. Mechanisms of transition metal interferences in hydride generation atomic-absorption spectrometry. Part 1. Influence of cobalt, copper, iron and nickel on selenium determination. *Analyst* 109:569.

6. KRIVAN, V., K. PETRICK, B. WELZ & M. MELCHER. 1985. Radiotracer error-diagnostic investigation of selenium determination by hydride-generation atomic absorption spectrometry involving treatment with hydrogen peroxide and hydrochloric acid. *Anal. Chem.* 57:1703.

7. CHU, R.C., G.P. BARRON & P.A.W. BAUM-
GARNER. 1972. Arsenic determination at submicrogram levels by arsine evolution and flameless atomic absorption spectrophotometric
technique. *Anal. Chem.* 44:1476.
8. SIEMER, D.D., P. KOTEEL & V. JARIWALA.
1976. Optimization of arsine generation in
atomic absorption arsenic determinations.
Anal. Chem. 48:836.

8. Bibliography

FERNANDEZ, F.J. & D.C. MANNING. 1971. The
determination of arsenic at submicrogram levels by atomic absorption spectrophotometry.
Atomic Absorption Newsletter 10:86.
JARREL-ASH CORPORATION. 1971. High Sensitivity Arsenic Determination by Atomic Absorption. Jarrel-Ash Atomic Absorption
Applications Laboratory Bull. No. As-3.
MANNING, D.C. 1971. A high sensitivity arsenic-
selenium sampling system for atomic absorption spectroscopy. *Atomic Absorption Newsletter* 10:123.
BRAMAN, R.S. & C.C. FOREBACK. 1973. Meth-
ylated forms of arsenic in the environment.
Science 182:1247.
CALDWELL, J.S., R.J. LISHKA & E.F. MC-
FARREN. 1973. Evaluation of a low-cost arsenic and selenium determination of

microgram-per-liter levels. *J. Amer. Water
Works Assoc.,* 65:731.
AGGETT, J. & A.C. ASPELL. 1976. The determination of arsenic (III) and total arsenic by
atomic-absorption spectroscopy. *Analyst*
101:341.
FIORINO, J.A., J.W. JONES & S.G. CAPAR. 1976.
Sequential determination of arsenic, selenium,
antimony, and tellurium in foods via rapid
hydride evolution and atomic absorption
spectrometry. *Anal. Chem.* 48:120.
CUTTER, G.A. 1978. Species determination of selenium in natural waters. *Anal. Chem. Acta*
98:59.
GODDEN, R.G. & D.R. THOMERSON. 1980. Generation of covalent hydrides in atomic absorption spectroscopy. *Analyst* 105:1137.
BROWN, R.M., JR., R.C. FRY, J.L. MOYERS, S.J.
NORTHWAY, M.B. DENTON & G.S. WILSON.
1981. Interference by volatile nitrogen oxides
and transition-metal catalysis in the preconcentration of arsenic and selenium as hydrides. *Anal. Chem.* 53:1560.
SINEMUS, H.W., M. MELCHER & B. WELZ. 1981.
Influence of valence state on the determination of antimony, bismuth, selenium, and tellurium in lake water using the hydride AA
technique. *Atomic Spectrosc.* 2:81.
DEDINA, J. 1982. Interference of volatile hydride
forming elements in selenium determination
by atomic absorption spectrometry with hydride generation. *Anal. Chem.* 54:2097.

3114 C.   Continuous Hydride Generation/Atomic Absorption Spectrometric Method (PROPOSED)

1. General Discussion

The continuous hydride generator, introduced recently, offers the advantages of
simplicity in operation, excellent reproducibility, low detection limits, and high sample volume throughput for selenium
analysis following preparations as described in 3500-Se.B or 3114B.4c and *d*.

a. Principle: See Section 3114B.

b. Interferences: Free chlorine in hydrochloric acid is a common but difficult-to-
diagnose interference. (The amount of
chlorine varies with manufacturer and with
each lot from the same manufacturer).

Chlorine oxidizes the hydride and can contaminate the hydride generator to prevent
recoveries under any conditions. When interference is encountered, or preferably before using each new bottle of HCl,
eliminate chlorine from a 2.3-L bottle of
conc HCl by bubbling with helium (commercial grade, 100 mL/min) for 3 h.

Excess oxidant (peroxide, persulfate, or
permanganate) from the total selenium
digestion can oxidize the hydride. Follow
procedures in 3500-Se.B.2, 3, or 4 to ensure
removal of all oxidizing agents before hydride generation.

Nitrite is a common trace constituent in

natural and waste waters, and at levels as low as 10 μg/L nitrite can reduce the recovery of hydrogen selenide from Se(IV) by over 50%. Moreover, during the reduction of Se(VI) to Se(IV) by digestion with HCl (3500-Se.B.5), some nitrate is converted to nitrite, which subsequently interferes. When this interference is suspected, add sulfanilamide after sample acidification (or HCl digestion). The diazotization reaction between nitrite and sulfanilamide completely removes the interferent effect (i.e., the standard addition slope is normal).

2. Apparatus

a. Continuous hydride generator: The basic unit is composed of two parts: a precision peristaltic pump, which is used to meter and mix reagents and sample solutions, and the gas-liquid separator. At the gas-liquid separator a constant flow of argon strips out the hydrogen and metal hydride gases formed in the reaction and carries them to the heated quartz absorption cell (3114B.1*b* and 2*b*), which is supported by a metal bracket mounted on top of the regular air acetylene burner head.

The spent liquid flows out of the separator via a constant level side drain to a waste bucket. Schematics and operating parameters are shown in Figure 3114:2.

Check flow rates frequently to ensure a steady flow; an uneven flow in any tubing will cause an erratic signal. Remove tubings from pump rollers when not in use. Typical flow rates are: sample, 7 mL/min; acid, 1 mL/min; borohydride reagent, 1 mL/min. Argon flow usually is pre-fixed, typically at 90 mL/min.

b. Atomic absorption spectrometric equipment: See Section 3111A.6.

3. Reagents

a. Hydrochloric acid, HCl, 5 + 1: Handle conc HCl under a fume hood. If necessary, remove free Cl_2 by stripping conc HCl with helium as described above.

b. Borohydride reagent: Dissolve 0.6 g $NaBH_4$ and 0.5 g NaOH in 100 mL water. CAUTION: *Sodium borohydride is toxic, flammable, and corrosive.*

c. Selenium reference standard solution, 1000 mg/L: Use commercially available standard; verify that selenium is Se(IV).

d. Intermediate standard solution, 1 mg/L: Dilute 1 mL reference standard so-

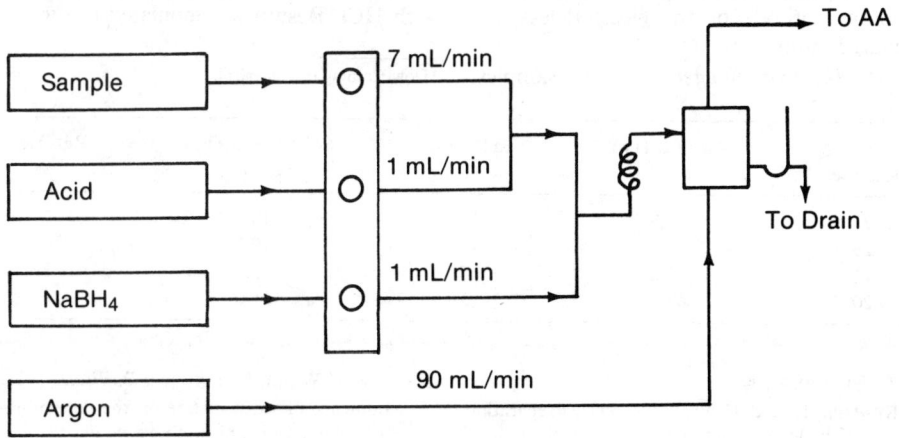

Figure 3114:2. Schematic of a continuous hydride generator.

lution to 1 L in a volumetric flask with distilled water.

e. Working standard solutions, 5, 10, 20, 30, and 40 µg/L: Dilute 0.5, 1.0, 2.0, 3.0, and 4.0 mL intermediate standard solution to 100 mL in a volumetric flask.

f. Sulfanilamide solution: Prepare a 2.5% (w/v) solution daily; add several drops conc HCl per 50 mL solution to facilitate dissolution.

4. Procedure

a. Sample preparation: See Section 3500-Se or 3114 B.4*c* and *d* for preparation steps for various Se fractions or total Se.

b. Preconditioning hydride generator: For newly installed tubing, turn on pump for at least 10 to 15 min before instrument calibration. Sample the highest standard for a few minutes to let volatile hydride react with the reactive sites in the transfer lines and on the quartz absorption cell surfaces.

c. Instrument calibration: Depending on total void volume in sample tubing, sampling time of 15 to 20 s generally is sufficient to obtain a steady signal. Between samples, submerge uptake tube in rinse water. Calibrate instrument daily after a 45-min lamp warmup time. Use either the hollow cathode or the electrodeless discharge lamp.

d. Antifoaming agents: Certain samples,

particularly wastewater samples containing a high concentration of proteinaceous substances, can cause excessive foaming that could carry the liquid directly into the heated quartz absorption cell and cause splattering of salty deposits onto the windows of the spectrometer. Add a drop of antifoaming agent* to eliminate this problem.

e. Nitrite removal: After samples have been acidified, or after acid digestion, add 0.1 mL sulfanilamide solution per 10 mL sample and let react for 2 min.

f. Analysis: Follow manufacturer's instructions for operation of analytical equipment.

5. Calculation

Construct a calibration curve based on absorbance vs. standard concentration. Apply dilution factors on diluted samples.

6. Precision and Bias

Working standards were analyzed together with batches of water samples on a routine production basis. The standards were compounded using chemically pure sodium selenite and sodium selenate. The values of Se(IV) + Se(VI) were determined by converting Se(VI) to Se(IV) by digestion with HCl. Results are tabulated below.

---

*Dow Corning or equivalent.

No. Analyses	Mean Se(IV) µg/L	Rel. Dev. %	Se(IV) + Se(VI) µg/L	Rel. Del. %
21	4.3	12	10.3	7
26	8.5	12	19.7	6
22	17.2	7	39.2	8
20	52.8	5	106.0	6

7. Bibliography

REAMER, D.C. & C. VEILOON. 1981. Preparation of biological materials for determination of selenium by hydride generation-AAS. *Anal. Chem.* 53:1192.

SINEMUS, H.W., M. MELCHER & B. WELZ. 1981. Influence of valence state on the determination of antimony, bismuth, selenium, and tellurium in lake water using the hydride AA technique. *Atomic Spectrosc.* 2:81.

RODEN, D.R. & D.E. TALLMAN. 1982. Determination of inorganic selenium species in groundwaters containing organic interferences by ion chromatography and hydride generation/atomic absorption spectrometry. *Anal. Chem.* 54:307.

CUTTER, G. 1983. Elimination of nitrite interference in the determination of selenium by hydride generation. *Anal. Chim. Acta* 149:391.

NARASAKI H. & M. IKEDA. 1984. Automated determination of arsenic and selenium by atomic absorption spectrometry with hydride generation. *Anal. Chem.* 56:2059.

WELZ, B. & M. MELCHER. 1985. Decomposition of marine biological tissues for determination of arsenic, selenium, and mercury using hy-dride-generation and cold-vapor atomic absorption spectrometries. *Anal. Chem.* 57:427.

EBDON, L. & S.T. SPARKES. 1987. Determination of arsenic and selenium in environmental samples by hydride generation-direct current plasma-atomic emission spectrometry. *Microchem. J.* 36:198.

EBDON, L. & J.R. WILKINSON. 1987. The determination of arsenic and selenium in coal by continuous flow hydride-generation atomic absorption spectroscopy and atomic fluorescence spectrometry. *Anal. Chim. Acta.* 194:177.

VOTH-BEACH, L.M. & D.E. SHRADER. 1985. Reduction of interferences in the determination of arsenic and selenium by hydride generation. *Spectroscopy* 1:60.

3120   METALS BY PLASMA EMISSION SPECTROSCOPY*

3120 A.   Introduction

1. General Discussion

Emission spectroscopy using inductively coupled plasma (ICP) was developed in the mid-1960's[1,2] as a rapid, sensitive, and convenient method for the determination of metals in water and wastewater samples.[3-6] Dissolved metals are determined in filtered and acidified samples. Total metals are determined after appropriate digestion. Care must be taken to ensure that potential interferences are dealt with, especially when dissolved solids exceed 1500 mg/L.

2. References

1. GREENFIELD, S., I.L. JONES & C.T. BERRY. 1964. High-pressure plasma-spectroscopic emission sources. *Analyst* 89: 713.
2. WENDT, R.H. & V.A. FASSEL. 1965. Induction-coupled plasma spectrometric excitation source. *Anal. Chem.* 37:920.
3. U.S. ENVIRONMENTAL PROTECTION AGENCY. 1983. Method 200.7. Inductively coupled plasma-atomic emission spectrometric method for trace element analysis of water and wastes. Methods for Chemical Analysis of Water and Wastes. EPA-600/4-79-020, revised March 1983.
4. AMERICAN SOCIETY FOR TESTING AND MATERIALS. 1987. Annual Book of ASTM Standards, Vol. 11.01. American Soc. Testing & Materials, Philadelphia, Pa.
5. FISHMAN, M.J. & W.L. BRADFORD, eds. 1982. A Supplement to Methods for the Determination of Inorganic Substances in Water and Fluvial Sediments. Rep. No. 82-272, U.S. Geological Survey, Washington, D.C.
6. GARBARINO, J.R. & H.E. TAYLOR. 1985. Trace Analysis. Recent Developments and Applications of Inductively Coupled Plasma Emission Spectroscopy to Trace Elemental Analysis of Water. Volume 4. Academic Press, New York, N.Y.

*Approved by Standard Methods Committee, 1989.

3120 B.   Inductively Coupled Plasma (ICP) Method

1. General Discussion

a. Principle: An ICP source consists of a flowing stream of argon gas ionized by an applied radio frequency field typically oscillating at 27.1 MHz. This field is inductively coupled to the ionized gas by a water-cooled coil surrounding a quartz "torch" that supports and confines the plasma. A sample aerosol is generated in an appropriate nebulizer and spray chamber and is carried into the plasma through an injector tube located within the torch. The sample aerosol is injected directly into the ICP, subjecting the constituent atoms to temperatures of about 6000 to 8000°K.[1] Because this results in almost complete dissociation of molecules, significant reduction in chemical interferences is achieved. The high temperature of the plasma excites atomic emission efficiently. Ionization of a high percentage of atoms produces ionic emission spectra. The ICP provides an optically "thin" source that is not subject to self-absorption except at very high concentrations. Thus linear dynamic ranges of four to six orders of magnitude are observed for many elements.[2]

The efficient excitation provided by the ICP results in low detection limits for many elements. This, coupled with the extended dynamic range, permits effective multielement determination of metals.[3] The light emitted from the ICP is focused onto the entrance slit of either a monochromator or a polychromator that effects dispersion. A precisely aligned exit slit is used to isolate a portion of the emission spectrum for intensity measurement using a photomultiplier tube. The monochromator uses a single exit slit/photomultiplier and may use a computer-controlled scanning mechanism to examine emission wavelengths sequentially. The polychromator uses multiple fixed exit slits and corresponding photomultiplier tubes; it simultaneously monitors all configured wavelengths using a computer-controlled readout system. The sequential approach provides greater wavelength selection while the simultaneous approach can provide greater sample throughput.

b. Applicable metals and analytical limits: Table 3120:I lists elements for which this method applies, recommended analytical wavelengths, and typical estimated instrument detection limits using conventional pneumatic nebulization. Actual working detection limits are sample-dependent. Typical upper limits for linear calibration also are included in Table 3120:I.

c. Interferences: Interferences may be categorized as follows:

1) Spectral interferences—Light emission from spectral sources other than the element of interest may contribute to apparent net signal intensity. Sources of spectral interference include direct spectral line overlaps, broadened wings of intense spectral lines, ion-atom recombination continuum emission, molecular band emission, and stray (scattered) light from the emission of elements at high concentrations.[4] Avoid line overlaps by selecting alternate analytical wavelengths. Avoid or minimize other spectral interference by judicious choice of background correction positions. A wavelength scan of the element line region is useful for detecting potential spectral interferences and for selecting positions for background correction. Make corrections for residual spectral interference using empirically determined correction factors in conjunction with the computer

TABLE 3120:I. SUGGESTED WAVELENGTHS, ESTIMATED DETECTION LIMITS, ALTERNATE
WAVELENGTHS, CALIBRATION CONCENTRATIONS, AND UPPER LIMITS

Element	Suggested Wavelength nm	Estimated Detection Limit μg/L	Alternate Wavelength* nm	Calibration Concentration mg/L	Upper Limit Concentration mg/L
Aluminum	308.22	40	237.32	10.0	100
Antimony	206.83	30	217.58	10.0	100
Arsenic	193.70	50	189.04†	10.0	100
Barium	455.40	2	493.41	1.0	50
Beryllium	313.04	0.3	234.86	1.0	10
Boron	249.77	5	249.68	1.0	50
Cadmium	226.50	4	214.44	2.0	50
Calcium	317.93	10	315.89	10.0	100
Chromium	267.72	7	206.15	5.0	50
Cobalt	228.62	7	230.79	2.0	50
Copper	324.75	6	219.96	1.0	50
Iron	259.94	7	238.20	10.0	100
Lead	220.35	40	217.00	10.0	100
Lithium	670.78	4‡	—	5.0	100
Magnesium	279.08	30	279.55	10.0	100
Manganese	257.61	2	294.92	2.0	50
Molybdenum	202.03	8	203.84	10.0	100
Nickel	231.60	15	221.65	2.0	50
Potassium	766.49	100‡	769.90	10.0	100
Selenium	196.03	75	203.99	5.0	100
Silica (SiO$_2$)	212.41	20	251.61	21.4	100
Silver	328.07	7	338.29	2.0	50
Sodium	589.00	30‡	589.59	10.0	100
Strontium	407.77	0.5	421.55	1.0	50
Thallium	190.86†	40	377.57	10.0	100
Vanadium	292.40	8	—	1.0	50
Zinc	213.86	2	206.20	5.0	100

*Other wavelengths may be substituted if they provide the needed sensitivity and are corrected for
spectral interference.
†Available with vacuum or inert gas purged optical path.
‡Sensitive to operating conditions.

software supplied by the spectrometer
manufacturer or with the calculation detailed below. The empirical correction
method cannot be used with scanning spectrometer systems if the analytical and interfering lines cannot be precisely and
reproducibly located. In addition, if using
a polychromator, verify absence of spectral
interference from an element that could
occur in a sample but for which there is
no channel in the detector array. Do this
by analyzing single-element solutions of
100 mg/L concentration and noting for
each element channel the apparent concentration from the interfering substance
that is greater than the element's instrument detection limit.

2) Nonspectral interferences

a) Physical interferences are effects associated with sample nebulization and
transport processes. Changes in the physical properties of samples, such as viscosity

and surface tension, can cause significant error. This usually occurs when samples containing more than 10% (by volume) acid or more than 1500 mg dissolved solids/L are analyzed using calibration standards containing ≤ 5% acid. Whenever a new or unusual sample matrix is encountered, use the test described in ¶ 4g. If physical interference is present, compensate for it by sample dilution, by using matrix-matched calibration standards, or by applying the method of standard addition (see ¶ 5d below).

High dissolved solids content also can contribute to instrumental drift by causing salt buildup at the tip of the nebulizer gas orifice. Using prehumidified argon for sample nebulization lessens this problem. Better control of the argon flow rate to the nebulizer using a mass flow controller improves instrument performance.

b) Chemical interferences are caused by molecular compound formation, ionization effects, and thermochemical effects associated with sample vaporization and atomization in the plasma. Normally these effects are not pronounced and can be minimized by careful selection of operating conditions (incident power, plasma observation position, etc.). Chemical interferences are highly dependent on sample matrix and element of interest. As with physical interferences, compensate for them by using matrix matched standards or by standard addition (¶ 5d). To determine the presence of chemical interference, follow instructions in ¶ 4g.

2. Apparatus

a. ICP source: The ICP source consists of a radio frequency (RF) generator capable of generating at least 1.1 KW of power, torch, tesla coil, load coil, impedance matching network, nebulizer, spray chamber, and drain. High-quality flow regulators are required for both the nebulizer argon and the plasma support gas flow. A peristaltic pump is recommended to regulate sample flow to the nebulizer. The type of nebulizer and spray chamber used may depend on the samples to be analyzed as well as on the equipment manufacturer. In general, pneumatic nebulizers of the concentric or cross-flow design are used. Viscous samples and samples containing particulates or high dissolved solids content (> 5000 mg/L) may require nebulizers of the Babington type.[5]

b. Spectrometer: The spectrometer may be of the simultaneous (polychromator) or sequential (monochromator) type with air-path, inert gas purged, or vacuum optics. A spectral bandpass of 0.05 nm or less is required. The instrument should permit examination of the spectral background surrounding the emission lines used for metals determination. It is necessary to be able to measure and correct for spectral background at one or more positions on either side of the analytical lines.

3. Reagents and Standards

Use reagents that are of ultra-high-purity grade or equivalent. Redistilled acids are acceptable. Except as noted, dry all salts at 105°C for 1 h and store in a desiccator before weighing. Use deionized water prepared by passing water through at least two stages of deionization with mixed bed cation/anion exchange resins.[6] Use deionized water for preparing all calibration standards, reagents, and for dilution.

a. Hydrochloric acid, HCl, conc and 1+1.

b. Nitric acid, HNO₃, conc.

c. Nitric acid, HNO₃, 1+1: Add 500 mL conc HNO₃ to 400 mL water and dilute to 1 L.

d. Standard stock solutions: See 3111B, 3111D, and 3114B. CAUTION: Many metal salts are extremely toxic and may be fatal if swallowed. Wash hands thoroughly after handling.

1) *Aluminum:* See 3111D.3k1).
2) *Antimony:* See 3111B.3j1).
3) *Arsenic:* See 3114B.3k1).
4) *Barium:* See 3111D.3k2).
5) *Beryllium:* See 3111D.3k3).
6) *Boron:* Do not dry but keep bottle tightly stoppered and store in a desiccator. Dissolve 0.5716 g anhydrous H_3BO_3 in water and dilute to 1000 mL; 1 mL = 100 μg B.
7) *Cadmium:* See 3111B.3j3).
8) *Calcium:* See 3111B.3j4).
9) *Chromium:* See 3111B.3j6).
10) *Cobalt:* See 3111B.3j7).
11) *Copper:* See 3111B.3j8).
12) *Iron:* See 3111B.3j11).
13) *Lead:* See 3111B.3j12).
14) *Lithium:* See 3111B.3j13).
15) *Magnesium:* See 3111B.3j14).
16) *Manganese:* See 3111B.3j15).
17) *Molybdenum:* See 3111D.3k4).
18) *Nickel:* See 3111B.3j16).
19) *Potassium:* See 3111B.3j19).
20) *Selenium:* See 3114B.3n1).
21) *Silica:* See 3111D.3k7).
22) *Silver:* See 3111B.3j22).
23) *Sodium:* See 3111B.3j23).
24) *Strontium:* See 3111B.3j24).
25) *Thallium:* See 3111B.3j25).
26) *Vanadium:* See 3111D.3k10).
27) *Zinc:* See 3111B.3j27).

e. Calibration standards: Prepare mixed calibration standards containing the concentrations shown in Table 3120:I by combining appropriate volumes of the stock solutions in 100-mL volumetric flasks. Add 2 mL 1+1 HNO_3 and 10 mL 1+1 HCl and dilute to 100 mL with water. Before preparing mixed standards, analyze each stock solution separately to determine possible spectral interference or the presence of impurities. When preparing mixed standards take care that the elements are compatible and stable. Store mixed standard solutions in an FEP fluorocarbon or unused polyethylene bottle. Verify calibration standards initially using the quality control standard; monitor weekly for sta-

bility. The following are recommended combinations using the suggested analytical lines in Table 3120:I. Alternative combinations are acceptable.

1) *Mixed standard solution I:* Manganese, beryllium, cadmium, lead, selenium, and zinc.

2) *Mixed standard solution II:* Barium, copper, iron, vanadium, and cobalt.

3) *Mixed standard solution III:* Molybdenum, silica, arsenic, strontium, and lithium.

4) *Mixed standard solution IV:* Calcium, sodium, potassium, aluminum, chromium, and nickel.

5) *Mixed standard solution V:* Antimony, boron, magnesium, silver, and thallium. If addition of silver results in an initial precipitation, add 15 mL water and warm flask until solution clears. Cool and dilute to 100 mL with water. For this acid combination limit the silver concentration to 2 mg/L. Silver under these conditions is stable in a tap water matrix for 30 d. Higher concentrations of silver require additional HCl.

f. Calibration blank: Dilute 2 mL 1+1 HNO_3 and 10 mL 1+1 HCl to 100 mL with water. Prepare a sufficient quantity to be used to flush the system between standards and samples.

g. Method blank: Carry a reagent blank through entire sample preparation procedure. Prepare method blank to contain the same acid types and concentrations as the sample solutions.

h. Instrument check standard: Prepare instrument check standards by combining compatible elements at a concentration of 2 mg/L.

i. Instrument quality control sample: Obtain a certified aqueous reference standard from an outside source and prepare according to instructions provided by the supplier. Use the same acid matrix as the calibration standards.

j. Method quality control sample: Carry the instrument quality control sample

(¶ 3*i*) through the entire sample preparation procedure.

k. Argon: Use technical or welder's grade. If gas appears to be a source of problems, use prepurified grade.

4. Procedure

a. Sample preparation: See Section 3030F.

b. Operating conditions: Because of differences among makes and models of satisfactory instruments, no detailed operating instructions can be provided. Follow manufacturer's instructions. Establish instrumental detection limit, precision, optimum background correction positions, linear dynamic range, and interferences for each analytical line. Verify that the instrument configuration and operating conditions satisfy the analytical requirements and that they can be reproduced on a day-to-day basis. An atom-to-ion emission intensity ratio [Cu(I) 324.75 nm/Mn(II) 257.61 nm] can be used to reproduce optimum conditions for multielement analysis precisely. The Cu/Mn intensity ratio may be incorporated into the calibration procedure, including specifications for sensitivity and for precision.[7] Keep daily or weekly records of the Cu and Mn intensities and/or the intensities of critical element lines. Also record settings for optical alignment of the polychromator, sample uptake rate, power readings (incident, reflected), photomultiplier tube attenuation, mass flow controller settings, and system maintenance.

c. Instrument calibration: Set up instrument as directed (¶ *b*). Warm up for 30 min. For polychromators, perform an optical alignment using the profile lamp or solution. Check alignment of plasma torch and spectrometer entrance slit, particularly if maintenance of the sample introduction system was performed. Make Cu/Mn or similar intensity ratio adjustment.

Calibrate instrument according to manufacturer's recommended procedure using calibration standards and blank. Aspirate each standard or blank for a minimum of 15 s after reaching the plasma before beginning signal integration. Rinse with calibration blank or similar solution for at least 60 s between each standard to eliminate any carryover from the previous standard. Use average intensity of multiple integrations of standards or samples to reduce random error.

Before analyzing samples, analyze instrument check standard. Concentration values obtained should not deviate from the actual values by more than ±5% (or the established control limits, whichever is lower).

d. Analysis of samples: Begin each sample run with an analysis of the calibration blank, then analyze the method blank. This permits a check of the sample preparation reagents and procedures for contamination. Analyze samples, alternating them with analyses of calibration blank. Rinse for at least 60 s with dilute acid between samples and blanks. After introducing each sample or blank let system equilibrate before starting signal integration. Examine each analysis of the calibration blank to verify that no carry-over memory effect has occurred. If carry-over is observed, repeat rinsing until proper blank values are obtained. Make appropriate dilutions and acidifications of the sample to determine concentrations beyond the linear calibration range.

e. Instrumental quality control: Analyze instrument check standard once per 10 samples to determine if significant instrument drift has occurred. If agreement is not within ± 5% of the expected values (or within the established control limits, whichever is lower), terminate analysis of samples, correct problem, and recalibrate instrument. If the intensity ratio reference is used, resetting this ratio may restore calibration without the need for reanalyzing calibration standards. Analyze instrument check standard to confirm proper recalibration. Reanalyze one or more samples analyzed just before termination of the analytical run. Results should agree to within ± 5%, otherwise all samples analyzed after the last acceptable instrument check standard analysis must be reanalyzed.

Analyze instrument quality control sample within every run. Use this analysis to verify accuracy and stability of the calibration standards. If any result is not within ± 5% of the certified value, prepare a new calibration standard and recalibrate the instrument. If this does not correct the problem, prepare a new stock solution and a new calibration standard and repeat calibration.

f. Method quality control: Analyze the method quality control sample within every run. Results should agree to within ± 5% of the certified values. Greater discrepancies may reflect losses or contamination during sample preparation.

g. Test for matrix interference: When analyzing a new or unusual sample matrix verify that neither a positive nor negative nonlinear interference effect is operative. If the element is present at a concentration above 1 mg/L, use serial dilution with calibration blank. Results from the analyses of a dilution should be within ± 5% of the original result. Alternately, or if the concentration is either below 1 mg/L or not detected, use a post-digestion addition equal to 1 mg/L. Recovery of the addition should be either between 95% and 105% or within established control limits of ± 2 standard deviations around the mean. If a matrix effect causes test results to fall outside the critical limits, complete the analysis after either diluting the sample to eliminate the matrix effect while maintaining a detectable concentration of at least twice the detection limit or applying the method of standard additions.

5. Calculations and Corrections

a. Blank correction: Subtract result of an adjacent calibration blank from each sample result to make a baseline drift correction. (Concentrations printed out should include negative and positive values to compensate for positive and negative baseline drift. Make certain that the calibration blank used for blank correction has not been contaminated by carry-over.) Use the result of the method blank analysis to correct for reagent contamination. Alternatively, intersperse method blanks with appropriate samples. Reagent blank and baseline drift correction are accomplished in one subtraction.

b. Dilution correction: If the sample was diluted or concentrated in preparation, multiply results by a dilution factor (*DF*) calculated as follows:

$$DF = \frac{\text{Final weight or volume}}{\text{Initial weight or volume}}$$

c. Correction for spectral interference: Correct for spectral interference by using computer software supplied by the instrument manufacturer or by using the manual method based on interference correction factors. Determine interference correction factors by analyzing single-element stock solutions of appropriate concentrations under conditions matching as closely as possible those used for sample analysis. Unless analysis conditions can be reproduced accurately from day to day, or for longer periods, redetermine interference correction factors found to affect the results significantly each time samples are analyzed.[7,8] Calculate interference correction factors (K_{ij}) from apparent concentrations observed in the analysis of the high-purity stock solutions:

$$K_{ij} = \frac{\text{Apparent concentration of element } i}{\text{Actual concentration of interfering element } j}$$

where the apparent concentration of element i is the difference between the observed concentration in the stock solution and the observed concentration in the blank. Correct sample concentrations observed for element i (already corrected for baseline drift), for spectral interferences from elements j, k, and l; for example:

Concentration of element i corrected for spectral interference

$$
\begin{aligned}
&\begin{array}{l}\text{Observed} \\ = \text{concentration} \\ \quad \text{of } i \end{array} - (K_{ij}) \begin{array}{l}\text{Observed} \\ \text{concentration} \\ \text{of interfering} \\ \text{element } j \end{array} \\
&- (K_{ik}) \begin{array}{l}\text{Observed} \\ \text{concentration} \\ \text{of interfering} \\ \text{element } k \end{array} - (K_{il}) \begin{array}{l}\text{Observed} \\ \text{concentration} \\ \text{of interfering} \\ \text{element } l \end{array}
\end{aligned}
$$

Interference correction factors may be negative if background correction is used for element i. A negative K_{ij} can result where an interfering line is encountered at the background correction wavelength rather than at the peak wavelength. Determine concentrations of interfering elements j, k, and l within their respective linear ranges. Mutual interferences (i interferes with j and j interferes with i) require iterative or matrix methods for calculation.

d. *Correction for nonspectral interference:* If nonspectral interference correction is necessary, use the method of standard additions. It is applicable when the chemical and physical form of the element in the standard addition is the same as in the sample, *or* the ICP converts the metal in both sample and addition to the same form; the interference effect is independent of metal concentration over the concentration range of standard additions; and the analytical calibration curve is linear over the concentration range of standard additions.

Use an addition not less than 50% nor more than 100% of the element concentration in the sample so that measurement precision will not be degraded and interferences that depend on element/interferent ratios will not cause erroneous results.

Apply the method to all elements in the sample set using background correction at carefully chosen off-line positions. Multi-element standard addition can be used if it has been determined that added elements are not interferents.

e. *Reporting data:* Report analytical data in concentration units of milligrams per liter using up to three significant figures. Report results below the determined detection limit as not detected less than the stated detection limit corrected for sample dilution.

6. Precision and Bias

As a guide to the generally expected precision and bias, see the linear regression equations in Table 3120:II.[9] Additional interlaboratory information is available.[10]

7. References

1. FAIRES, L.M., B.A. PALMER, R. ENGLEMAN, JR. & T.M. NIEMCZYK. 1984. Temperature determinations in the inductively coupled plasma using a Fourier transform spectrometer. *Spectrochim. Acta* 39B:819.
2. BARNES, R.M. 1978. Recent advances in emission spectroscopy: inductively coupled plasma discharges for spectrochemical analysis. *CRC Crit. Rev. Anal. Chem.* 7:203.
3. PARSONS, M.L., S. MAJOR & A.R. FORSTER. 1983. Trace element determination by atomic spectroscopic methods - State of the art. *Appl. Spectrosc.* 37:411.
4. LARSON, G.F., V.A. FASSEL, R. K. WINGE & R.N. KNISELEY. 1976. Ultratrace analysis by optical emission spectroscopy: The stray light problem. *Appl. Spectrosc.* 30:384.
5. GARBARINO, J.R. & H.E. TAYLOR. 1979. A Babington-type nebulizer for use in the analysis of natural water samples by inductively coupled plasma spectrometry. *Appl. Spectrosc.* 34:584.
6. AMERICAN SOCIETY FOR TESTING AND MATERIALS. 1988. Standard specification for reagent water, D1193-77 (reapproved 1983). Annual Book of ASTM Standards. American Soc. for Testing & Materials, Philadelphia, Pa.
7. BOTTO, R.I. 1984. Quality assurance in operating a multielement ICP emission spectrometer. *Spectrochim. Acta* 39B:95.

TABLE 3120:II. ICP PRECISION AND BIAS DATA

Element	Concentration Range $\mu g/L$	Total Digestion* $\mu g/L$		Recoverable Digestion* $\mu g/L$	
Aluminum	69–4792	X =	$0.9273C + 3.6$	X =	$0.9380C + 22.1$
		S =	$0.0559X + 18.6$	S =	$0.0873X + 31.7$
		SR =	$0.0507X + 3.5$	SR =	$0.0481X + 18.8$
Antimony	77–1406	X =	$0.7940C - 17.0$	X =	$0.8908C + 0.9$
		S =	$0.1556X - 0.6$	S =	$0.0982X + 8.3$
		SR =	$0.1081X + 3.9$	SR =	$0.0682X + 2.5$
Arsenic	69–1887	X =	$1.0437C - 12.2$	X =	$1.0175C + 3.9$
		S =	$0.1239X + 2.4$	S =	$0.1288X + 6.1$
		SR =	$0.0874X + 6.4$	SR =	$0.0643X + 10.3$
Barium	9–377	X =	$0.7683C + 0.47$	X =	$0.8380C + 1.68$
		S =	$0.1819X + 2.78$	S =	$0.2540X + 0.30$
		SR =	$0.1285X + 2.55$	SR =	$0.0826X + 3.54$
Beryllium	3–1906	X =	$0.9629C + 0.05$	X =	$1.0177C - 0.55$
		S =	$0.0136X + 0.95$	S =	$0.0359X + 0.90$
		SR =	$0.0203X - 0.07$	SR =	$0.0445X - 0.10$
Boron	19–5189	X =	$0.8807C + 9.0$	X =	$0.9676C + 18.7$
		S =	$0.1150X + 14.1$	S =	$0.1320X + 16.0$
		SR =	$0.0742X + 23.2$	SR =	$0.0743X + 21.1$
Cadmium	9–1943	X =	$0.9874C - 0.18$	X =	$1.0137C - 0.65$
		S =	$0.0557X + 2.02$	S =	$0.0585X + 1.15$
		SR =	$0.0300X + 0.94$	SR =	$0.0332X + 0.90$
Calcium	17–47 170	X =	$0.9182C - 2.6$	X =	$0.9658C + 0.8$
		S =	$0.1228X + 10.1$	S =	$0.0917X + 6.9$
		SR =	$0.0189X + 3.7$	SR =	$0.0327X + 10.1$
Chromium	13–1406	X =	$0.9544C + 3.1$	X =	$1.0049C - 1.2$
		S =	$0.0499X + 4.4$	S =	$0.0698X + 2.8$
		SR =	$0.0009X + 7.9$	SR =	$0.0571X + 1.0$
Cobalt	17–2340	X =	$0.9209C - 4.5$	X =	$0.9278C - 1.5$
		S =	$0.0436X + 3.8$	S =	$0.0498X + 2.6$
		SR =	$0.0428X + 0.5$	SR =	$0.0407X + 0.4$
Copper	8–1887	X =	$0.9297C - 0.30$	X =	$0.9647C - 3.64$
		S =	$0.0442X + 2.85$	S =	$0.0497X + 2.28$
		SR =	$0.0128X + 2.53$	SR =	$0.0406X + 0.96$
Iron	13–9359	X =	$0.8829C + 7.0$	X =	$0.9830C + 5.7$
		S =	$0.0683X + 11.5$	S =	$0.1024X + 13.0$
		SR =	$-0.0046X + 10.0$	SR =	$0.0790X + 11.5$
Lead	42–4717	X =	$0.9699C - 2.2$	X =	$1.0056C + 4.1$
		S =	$0.0558X + 7.0$	S =	$0.0799X + 4.6$
		SR =	$0.0353X + 3.6$	SR =	$0.0448X + 3.5$
Magnesium	34–13 868	X =	$0.9881C - 1.1$	X =	$0.9879C + 2.2$
		S =	$0.0607X + 11.6$	S =	$0.0564X + 13.2$
		SR =	$0.0298X + 0.6$	SR =	$0.0268X + 8.1$

TABLE 3120:II, CONT.

Element	Concentration Range $\mu g/L$	Total Digestion* $\mu g/L$		Recoverable Digestion* $\mu g/L$	
Manganese	4–1887	$X =$	$0.9417C + 0.13$	$X =$	$0.9725C + 0.07$
		$S =$	$0.0324X + 0.88$	$S =$	$0.0557X + 0.76$
		$SR =$	$0.0153X + 0.91$	$SR =$	$0.0400X + 0.82$
Molybdenum	17–1830	$X =$	$0.9682C + 0.1$	$X =$	$0.9707C - 2.3$
		$S =$	$0.0618X + 1.6$	$S =$	$0.0811X + 3.8$
		$SR =$	$0.0371X + 2.2$	$SR =$	$0.0529X + 2.1$
Nickel	17–47 170	$X =$	$0.9508C + 0.4$	$X =$	$0.9869C + 1.5$
		$S =$	$0.0604X + 4.4$	$S =$	$0.0526X + 5.5$
		$SR =$	$0.0425X + 3.6$	$SR =$	$0.0393X + 2.2$
Potassium	347–14 151	$X =$	$0.8669C - 36.4$	$X =$	$0.9355C - 183.1$
		$S =$	$0.0934X + 77.8$	$S =$	$0.0481X + 177.2$
		$SR =$	$-0.0099X + 144.2$	$SR =$	$0.0329X + 60.9$
Selenium	69–1415	$X =$	$0.9363C - 2.5$	$X =$	$0.9737C - 1.0$
		$S =$	$0.0855X + 17.8$	$S =$	$0.1523X + 7.8$
		$SR =$	$0.0284X + 9.3$	$SR =$	$0.0443X + 6.6$
Silicon	189–9434	$X =$	$0.5742C - 35.6$	$X =$	$0.9737C - 60.8$
		$S =$	$0.4160X + 37.8$	$S =$	$0.3288X + 46.0$
		$SR =$	$0.1987X + 8.4$	$SR =$	$0.2133X + 22.6$
Silver	8–189	$X =$	$0.4466C + 5.07$	$X =$	$0.3987C + 8.25$
		$S =$	$0.5055X - 3.05$	$S =$	$0.5478X - 3.93$
		$SR =$	$0.2086X - 1.74$	$SR =$	$0.1836X - 0.27$
Sodium	35–47 170	$X =$	$0.9581C + 39.6$	$X =$	$1.0526C + 26.7$
		$S =$	$0.2097X + 33.0$	$S =$	$0.1473X + 27.4$
		$SR =$	$0.0280X + 105.8$	$SR =$	$0.0884X + 50.5$
Thallium	79–1434	$X =$	$0.9020C - 7.3$	$X =$	$0.9238C + 5.5$
		$S =$	$0.1004X + 18.3$	$S =$	$0.2156X + 5.7$
		$SR =$	$0.0364X + 11.5$	$SR =$	$-0.0106X + 48.0$
Vanadium	13–4698	$X =$	$0.9615C - 2.0$	$X =$	$0.9551C + 0.4$
		$S =$	$0.0618X + 1.7$	$S =$	$0.0927X + 1.5$
		$SR =$	$0.0220X + 0.7$	$SR =$	$0.0472X + 0.5$
Zinc	7–7076	$X =$	$0.9356C - 0.30$	$X =$	$0.9500C + 1.22$
		$S =$	$0.0914X + 3.75$	$S =$	$0.0597X + 6.50$
		$SR =$	$-0.0130X + 10.07$	$SR =$	$0.0153X + 7.78$

*$X =$ mean recovery, $\mu g/L$,
 $C =$ true value, $\mu g/L$,
 $S =$ multi-laboratory standard deviation, $\mu g/L$,
$SR =$ single-analyst standard deviation, $\mu g/L$.

8. BOTTO, R.I. 1982. Long-term stability of spectral interference calibrations for inductively coupled plasma atomic emission spectrometry. *Anal. Chem.* 54:1654.

9. MAXFIELD, R. & B. MINDAK. 1985. EPA Method Study 27, Method 200. 7 (Trace Metals by ICP). EPA-600/S4-85/05. National Technical Information Serv., Springfield, Va.

10. GARBARINO, J.R., B.E. JONES, G. P. STEIN, W.T. BELSER & H.E. TAYLOR. 1985. Statistical evaluation of an inductively coupled plasma atomic emission spectrometric method for routine water quality testing. *Appl. Spectrosc.* 39:53.

3500-Al   ALUMINUM*

3500-Al A.   Introduction

1. Occurrence

Aluminum is the third most abundant element of the earth's crust, occurring in minerals, rocks, and clays. This wide distribution accounts for the presence of aluminum in nearly all natural water as a soluble salt, a colloid, or an insoluble compound. Soluble, colloidal, and insoluble aluminum also may appear in treated water or wastewater as a residual from coagulation with aluminum-containing material. Filtered water from a modern rapid sand filtration plant should have an aluminum concentration less than 50 $\mu g/L$.

2. Selection of Method

The atomic absorption spectrometric and inductively coupled plasma methods are free from such common interferences as fluoride and phosphate, and are preferred. The Eriochrome cyanine R colorimetric method provides a means for estimating aluminum with simpler instrumentation. The automated pyrocatechol violet method is a highly sensitive flow injection or continuous-flow analysis technique.

---

*Approved by Standard Methods Committee, 1988.

3500-Al B.   Atomic Absorption Spectrometric Method

See flame atomic absorption spectrometric method, Sections 3111D and E, and electrothermal atomic absorption spectrometric method, Section 3113.

3500-Al C.   Inductively Coupled Plasma Method

See Section 3120.

3500-Al D.   Eriochrome Cyanine R Method

1. General Discussion

a. Principle: With Eriochrome cyanine R dye, dilute aluminum solutions buffered to a pH of 6.0 produce a red to pink complex that exhibits maximum absorption at 535 nm. The intensity of the developed color is influenced by the aluminum concentration, reaction time, temperature, pH, alkalinity, and concentration of other ions in the sample. To compensate for color and turbidity, the aluminum in one portion of sample is complexed with EDTA to provide a blank. The interference of iron and manganese, two elements often found in water, is eliminated by adding ascorbic acid. The optimum aluminum range lies between 20 and 300 μg/L but can be extended upward by sample dilution.

b. Interference: Negative errors are caused by both fluoride and polyphosphates. When the fluoride concentration is constant, the percentage error decreases with increasing amounts of aluminum. Because the fluoride concentration often is known or can be determined readily, fairly accurate results can be obtained by adding the known amount of fluoride to a set of standards. A simpler correction can be determined from the family of curves in Figure 3500-Al:1. A procedure is given for the removal of complex phosphate interference. Orthophosphate in concentrations under 10 mg/L does not interfere. The interference caused by even small amounts of alkalinity is removed by acidifying the sample just beyond the neutralization point of methyl orange. Sulfate does not interfere up to a concentration of 2000 mg/L.

c. Minimum detectable concentration: The minimum aluminum concentration detectable by this method in the absence of fluorides and complex phosphates is approximately 6 μg/L.

d. Sample handling: Collect samples in clean, acid-rinsed bottles, preferably plastic, and examine them as soon as possible after collection. If only soluble aluminum is to be determined, filter a portion of sample through a 0.45-μm membrane filter; discard first 50 mL of filtrate and use succeeding filtrate for the determination. Do not use filter paper, absorbent cotton, or glass wool for filtering any solution that is to be tested for aluminum, because they will remove most of the soluble aluminum.

2. Apparatus

a. Colorimetric equipment: One of the following is required:

1) *Spectrophotometer,* for use at 535 nm, with a light path of 1 cm or longer.

2) *Filter photometer,* providing a light path of 1 cm or longer and equipped with a green filter with maximum transmittance between 525 and 535 nm.

3) *Nessler tubes,* 50-mL, tall form, matched.

b. Glassware: Treat all glassware with warm $1 + 1$ HCl and rinse with aluminum-free distilled water to avoid errors due to materials absorbed on the glass. Rinse sufficiently to remove all acid.

3. Reagents

Use reagents low in aluminum, and aluminum-free distilled water.

a. Stock aluminum solution: Use either the metal (1) or the salt (2) for preparing stock solution; 1.00 mL = 500 μg Al:

1) Dissolve 500.0 mg aluminum metal in 10 mL conc HCl by heating gently. Dilute to 1000 mL with water, or

2) Dissolve 8.791 g aluminum potassium

sulfate (also called potassium alum), $AlK(SO_4)_2 \cdot 12H_2O$, in water and dilute to 1000 mL. Correct this weight by dividing by the decimal fraction of assayed $AlK(SO_4)_2 \cdot 12H_2O$ in the reagent used.

b. Standard aluminum solution: Dilute 10.00 mL stock aluminum solution to 1000 mL with water; 1.00 mL = 5.00 μg Al. Prepare daily.

c. Sulfuric acid, H_2SO_4, 0.02N and 6N.

d. Ascorbic acid solution: Dissolve 0.1 g ascorbic acid in water and make up to 100 mL in a volumetric flask. Prepare fresh daily.

e. Buffer reagent: Dissolve 136 g sodium acetate, $NaC_2H_3O_2 \cdot 3H_2O$, in water, add 40 mL 1N acetic acid, and dilute to 1 L.

f. Stock dye solution: Use any of the following products:

1) *Solochrome cyanine R-200* or Eriochrome cyanine:†* Dissolve 100 mg in water and dilute to 100 mL in a volumetric flask. This solution should have a pH of about 2.9.

2) *Eriochrome cyanine R:‡* Dissolve 300 mg dye in about 50 mL water. Adjust pH from about 9 to about 2.9 with 1 + 1 acetic acid (approximately 3 mL will be required). Dilute with water to 100 mL.

3) *Eriochrome cyanine R:§* Dissolve 150 mg in about 50 mL water. Adjust pH from about 9 to about 2.9 with 1 + 1 acetic acid (approximately 2 mL will be required). Dilute with water to 100 mL.

Stock solutions have excellent stability and can be kept for at least a year.

g. Working dye solution: Dilute 10.0 mL of selected stock dye solution to 100 mL in a volumetric flask with water. Working solutions are stable for at least 6 months.

h. Methyl orange indicator solution, or bromcresol green indicator solution specified in the total alkalinity determination (Section 2320B.3*d*).

i. EDTA (sodium salt of ethylenediaminetetraacetic acid dihydrate), 0.01M: Dissolve 3.7 g in water, and dilute to 1 L.

j. Sodium hydroxide, NaOH, 1N and 0.1N.

4. Procedure

a. Preparation of calibration curve:

1) Prepare a series of aluminum standards from 0 to 7 μg (0 to 280 μg/L based on a 25-mL sample) by accurately measuring the calculated volumes of standard aluminum solution into 50-mL volumetric flasks or nessler tubes. Add water to a total volume of approximately 25 mL.

2) Add 1 mL 0.02N H_2SO_4 to each standard and mix. Add 1 mL ascorbic acid solution and mix. Add 10 mL buffer solution and mix. With a volumetric pipet, add 5.00 mL working dye reagent and mix. Immediately make up to 50 mL with distilled water. Mix and let stand for 5 to 10 min. The color begins to fade after 15 min.

3) Read transmittance or absorbance on a spectrophotometer, using a wavelength of 535 nm or a green filter providing maximum transmittance between 525 and 535 nm. Adjust instrument to zero absorbance with the standard containing no aluminum.

Plot concentration of Al (micrograms Al in 50 mL final volume) against absorbance.

b. Sample treatment in absence of fluoride and complex phosphates: Place 25.0 mL sample, or a portion diluted to 25 mL, in a porcelain dish or flask, add a few drops of methyl orange indicator, and titrate with 0.02N H_2SO_4 to a faint pink color. Record reading and discard sample. To two similar samples at room temperature add the same amount of 0.02N H_2SO_4 used in the titration and 1 mL in excess.

To one sample add 1 mL EDTA solution. This will serve as a blank by complexing any aluminum present and compensating for color and turbidity. To

*Arnold Hoffman & Co., Providence, R.I.
†K & K Laboratories, K & K Lab. Div., Life Sciences Group, Plainview, N.Y.
‡Pfaltz & Bauer, Inc., Stamford, Conn.
§EM Science, Gibbstown, N.J.

both samples add 1 mL ascorbic acid, 10 mL buffer reagent, and 5.00 mL working dye reagent as prescribed in ¶ *a*2) above.

Set instrument to zero absorbance or 100% transmittance using the EDTA blank. After 5 to 10 min contact time, read transmittance or absorbance and determine aluminum concentration from the calibration curve previously prepared.

c. Visual comparison: If photometric equipment is not available, prepare and treat standards and a sample, as described above, in 50-mL nessler tubes. Make up to mark with water and compare sample color with the standards after 5 to 10 min contact time. A sample treated with EDTA is not needed when nessler tubes are used. If the sample contains turbidity or color, the use of nessler tubes may result in considerable error.

d. Removal of phosphate interference: Add 1.7 mL $6N$ H_2SO_4 to 100 mL sample in a 200-mL erlenmeyer flask. Heat on a hot plate for at least 90 min, keeping solution temperature just below the boiling point. At the end of the heating period solution volume should be about 25 mL. Add water if necessary to keep it at or above that volume.

After cooling, neutralize to a pH of 4.3 to 4.5 with NaOH, using $1N$ NaOH at the start and $0.1N$ for the final fine adjustment. Monitor with a pH meter. Make up to 100 mL with water, mix, and use a 25-mL portion for the aluminum test.

Run a blank in the same manner, using 100 mL distilled water and 1.7 mL $6N$ H_2SO_4. Subtract blank reading from sample reading or use it to set instrument to zero absorbance before reading the sample.

e. Correction for samples containing fluoride: Measure sample fluoride concentration by the SPADNS or electrode method. Either:

1) Add the same amount of fluoride as in the sample to each aluminum standard, or

2) Determine fluoride correction from the set of curves in Figure 3500-Al:1.

5. Calculation

$$mg\ Al/L = \frac{\mu g\ Al\ (in\ 50\ mL\ final\ volume)}{mL\ sample}$$

6. Precision and Bias

A synthetic sample containing 520 μg Al/L and no interference in distilled water was analyzed by the Eriochrome cyanine R method in 27 laboratories. Relative standard deviation was 34.4% and relative error 1.7%.

A second synthetic sample containing 50 μg Al/L, 500 μg Ba/L, and 5 μg Be/L in distilled water was analyzed in 35 laboratories. Relative standard deviation was 38.5% and relative error 22.0%.

A third synthetic sample containing 500 μg Al/L, 50 μg Cd/L, 110 μg Cr/L, 1000 μg Cu/L, 300 μg Fe/L, 70 μg Pb/L, 50 μg Mn/L, 150 μg Ag/L, and 650 μg Zn/L in distilled water was analyzed in 26 laboratories. Relative standard deviation was 28.8% and relative error 6.2%.

A fourth synthetic sample containing 540 μg Al/L and 2.5 mg polyphosphate/L in distilled water was analyzed in 16 laboratories that hydrolyzed the sample in the prescribed manner. Relative standard deviation was 44.3% and relative error 1.3%. In 12 laboratories that applied no corrective measures, the relative standard deviation was 49.2% and the relative error 8.9%.

A fifth synthetic sample containing 480 μg Al/L and 750 μg F/L in distilled water was analyzed in 16 laboratories that relied on the curve to correct for the fluoride content. Relative standard deviation was 25.5% and relative error 2.3%. The 17 laboratories that added fluoride to the aluminum standards showed a relative standard deviation of 22.5% and a relative error of 7.1%.

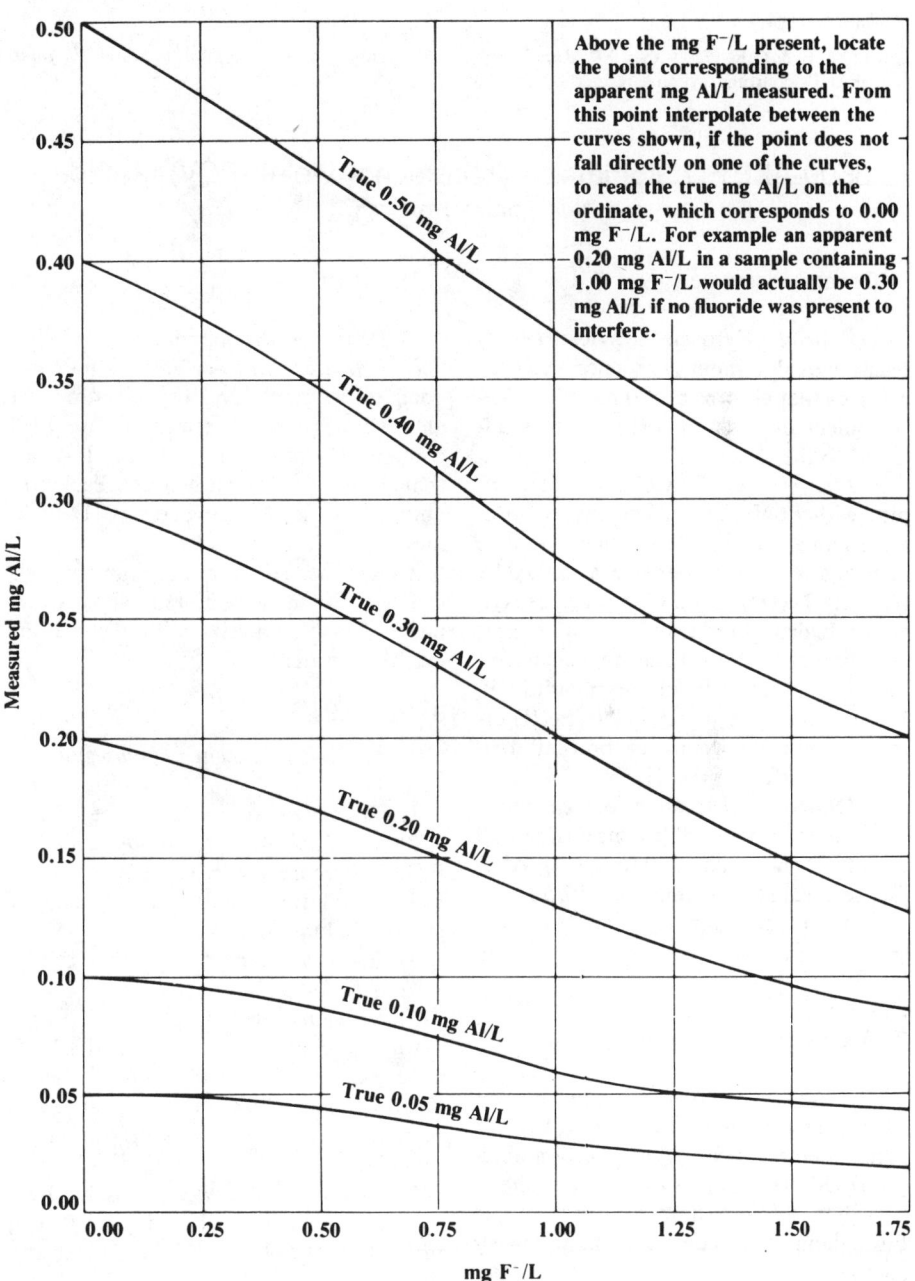

Figure 3500-Al:1. Correction curves for estimation of aluminum in the presence of fluoride.

7. Bibliography

SHULL, K.E. & G.R. GUTHAN. 1967. Rapid modified Eriochrome cyanine R method for determination of aluminum in water. *J. Amer.
Water Works Assoc.* 59:1456.

3500-Al E.   Automated Pyrocatechol Violet (PCV) Method (PROPOSED)

1. General Discussion

a. Principle: Pyrocatechol violet (PCV) reacts with aluminum to produce a colored complex that absorbs at 580 nm. Color development depends on pH, which is adjusted to the optimum, 6.1.

b. Interferences: Ferric iron will complex with PCV with an absorbance spectrum similar to Al-PCV, causing a positive interference. The interference is masked by reducing Fe(III) to Fe(II) with hydroxylamine hydrochloride and by subsequent chelation of Fe(II) with orthophenanthroline. Ferrous iron does not react with PCV. Formation of the highly stable Fe(II)-orthophenanthroline complex prevents oxidation of Fe(II) to Fe(III).

c. Minimum detectable concentration: The minimum detectable concentration of total monomeric Al is 10 and 7 μg Al/L for segmented-flow analysis (SFA) and nonsegmented, flow-injection analysis (FIA), respectively.

2. Apparatus

a. Automated analytical equipment: Use either segmented-flow analysis (SFA) or nonsegmented-flow or flow-injection analysis (FIA). The systems use the same chemistry but differ in flow rate and sample throughout. Both consist of the following components:

1) *Sampler.*
2) *Manifold.*
3) *Proportioning pump.*
4) *Colorimeter* equipped with tubular flow cell of specified length.
5) *Filters*.*
6) *Recorder.*
7) *Digital printer* (optional).

b. Cation-exchange column: If fractionation of inorganic and organic forms of aluminum is desired, preferably use TFE column 100 mm long (10 mm ID), although other sizes can be used. Pack column with cation-exchange resin† (14 to 50 mesh).

c. Acid-washed ware: Soak polypropylene or TFE utensils in 1% HNO_3 overnight. Rinse copiously with distilled or deionized water.

3. Reagents

a. Automated System I (nonsegmented FIA):

1) *Stock aluminum solution:* See ¶ D.3*a*.
2) *Standard aluminum solution:* See ¶ D.3*b*. Prepare a suitable range of standards by diluting appropriate volumes of standard aluminum solution.
3) *Hydrochloric acid*, HCl, conc.‡
4) *Ethanol*: 95%.
5) *Nitric acid*: HNO_3, conc.§
6) *Cleaning solution:* Slowly add 8.3 mL conc HCl to 500 mL water in a 1-L graduated cylinder. Add 100 mL ethanol and bring to volume with water. Prepare under fume hood.
7) *Iron masking solution:* Dissolve 7.5 g hydroxylamine hydrochloride and 0.56 g orthophenanthroline in 500 mL water and

---

* Whatman GF/C or equivalent.
† Amberlite IR-120, 1% hydrogen form, Rohm & Haas Company, or equivalent.
‡ Baker Instra-Analyzed or equivalent, d = 0.91 for conc NH_3 and d = 1.2 (38%) for conc HCl.
§ Baker Ultrex or equivalent.

dilute to 1000 mL. De-gas reagent through a 0.45-m membrane filter. Store in a polyethylene bottle. Refrigerate until use.

8) *PCV solution:* Dissolve 0.375 g 3,3′,4′-trihydroxyfuchsone-2″-sulfonic acid (PCV) in 40 mL water. Let stand for approximately 5 min with occasional swirling. Dilute to 1000 mL with water. De-gas through 0.45-μm membrane filter. Store in amber glass or polyethylene bottle. Prepare daily.

9) *Buffer solution:* Dissolve 84 g hexamethylenetetraamine in 750 mL water. Filter into a 1-L volumetric flask. Dilute to 1000 mL and transfer to a polyethylene bottle. Refrigerate until use.

b. *Automated System II (segmented-flow analyzer):*

1) *Stock aluminum solution:* See ¶ D.3*a*.

2) *Standard aluminum solution:* See ¶ *a*2), above.

3) *Cleaning solution:* See ¶ *a*6), above.

4) *Ammonia,* NH₃, conc.‡

5) *Hydrochloric acid,* HCl, conc.‡

6) *Iron masking solution:* Dissolve 17.75 g hydroxylamine hydrochloride in 200 mL water. Add and dissolve 0.176 g orthophenanthroline monohydrate. Dilute to 250 mL with water. Store refrigerated in polyethylene bottle.

7) *PCV solution:* Dissolve 0.150 g 3,3′,4′-trihydroxyfuchsone-2″-sulfonic acid (PCV) in 40 mL water. Let stand for about 5 min with occasional shaking. Dilute to 400 mL with water. Store in dark polyethylene bottle. Keep refrigerated and away from sunlight.

8) *Buffer:* Dissolve 300 g hexamethylenetetraamine in 750 mL water. Filter into a 1000-mL volumetric flask. Add 16.8 mL conc ammonia and dilute to 1000 mL with water. Use buffer solution undiluted, but verify when all reagents are running through the system that final effluent pH is between 6.0 and 6.2. If final effluent pH does not fall in this range, adjust buffer by

adding conc ammonia or 2.5*N* HCl.

9) *Working reagent HCl:* Add 51.1 mL conc HCl to 200 mL water and dilute to 250 mL final volume. Add 2.5 mL polyoxyethylene 23 lauryl ether ‖ (30% solution) to final solution. Use distilled or deionized water instead of working reagent HCl if sample does not alter effluent pH from 6.1.

4. Procedure

a. Total aluminum: Digest sample as directed in the hot nitric acid digestion procedure described in Section 3030E. Adjust sample dilution and buffer concentration to ensure that effluent pH is optimal (6.1) for color development. Remove cation-exchange column from flow path.

b. Total monomeric aluminum: Determine total monomeric aluminum without sample pretreatment. Fractionate inorganic and organic monomeric Al by measuring PCV-reactive Al before and after passage through the cation-exchange column. The difference represents the inorganic monomeric form of Al.

1) Cation-exchange column—Minimize disruption of organo-Al complexes by maintaining flow rate through column at 3.7 to 4.2 mL/min/mL resin.

2) Automated System I (nonsegmented, flow injection analysis)—Set up manifold as shown in Figure 3500-Al:2. Place cation-exchange column before sample loop to prevent dispersion of the sample bolus on the manifold. Start pumping reagents and wait to obtain a steady baseline. Never submerge effluent lines from flow cells and sample valves in the waste container because this causes back pressure and results in poor flow characteristics. Analyze sample after a steady baseline is attained.

3) Automated System II (segmented-flow analyzer)—Set up manifold as shown in Figure 3500-Al:3. The sampling rate is 30 samples/h, using 1 min for both sam-

---

‡ Baker Instra-Analyzed or equivalent, d = 0.91 for conc NH₃ and d = 1.2 (38%) for conc HCl.

‖ Brij-35, available from ICI Americas, Inc., Wilmington, Del., or Technicon Instruments Corp., Tarrytown, N.Y., or equivalent.

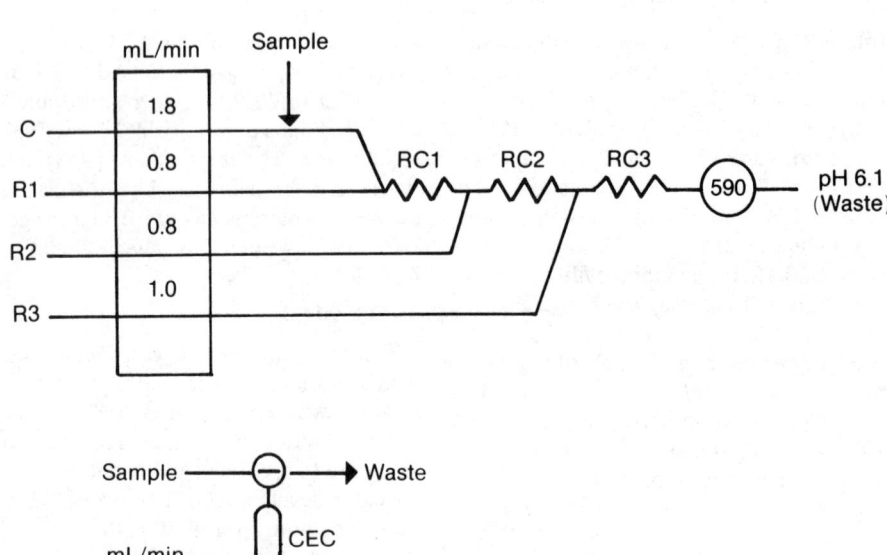

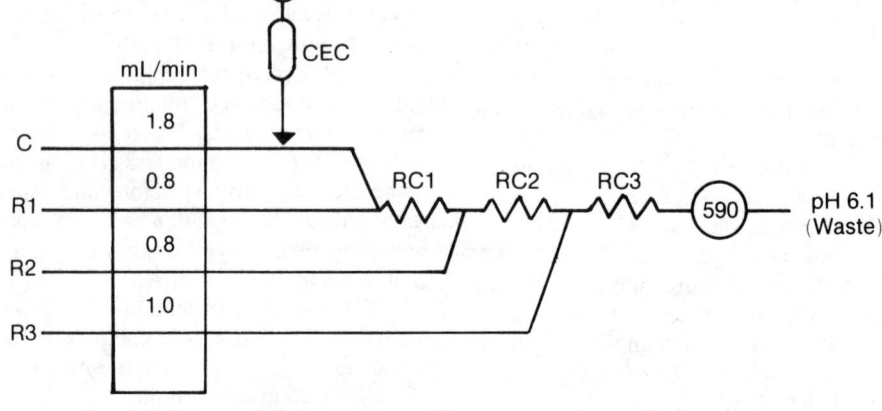

Figure 3500-Al:2. Aluminum manifold for Automated System I (nonsegmented, flow-injection analyzer).[1] (above) Channel 1—total PCV-reactive aluminum; (below) Channel 2—nonexchangeable PCV-reactive aluminum. Key: Carrier: deionized water (or 0.1N HCl); RI = masking solution: hydroxylammonium chloride; R2 = color reagent: pyrocatechol violet; R3 = buffer solution: hexamethylenetetramine; RC1 = reaction coil, 10 cm (0.5 mm ID); RC2 = reaction coil, 30 cm (0.5 mm ID); RC3 = reaction coil, 60 cm (0.5 mm ID); CEC = cation exchange column.

pling and washing. Place column in-line before air-segmentation. Because air is introduced each time the sample needle switches from wash to sample injection, place a debubbler in line before the column. Determine apparent sample dispersion caused by the column by packing column with an inert resin (e.g., polystyrene beads of comparable mesh size). Calculate dispersion factor and multiply postcolumn Al values by this factor. This correction is nec-

essary because no suitable organic Al standard is currently available.

c. Preparation of calibration curve: Prepare a series of Al standards ranging in concentration from 0 to 1000 μg/L. Measure aluminum and plot absorbance versus concentration to ascertain the linear range of the system. If linearity drops off, use two separate calibration curves for high and low concentration samples or dilute samples to within the linear range. For cali-

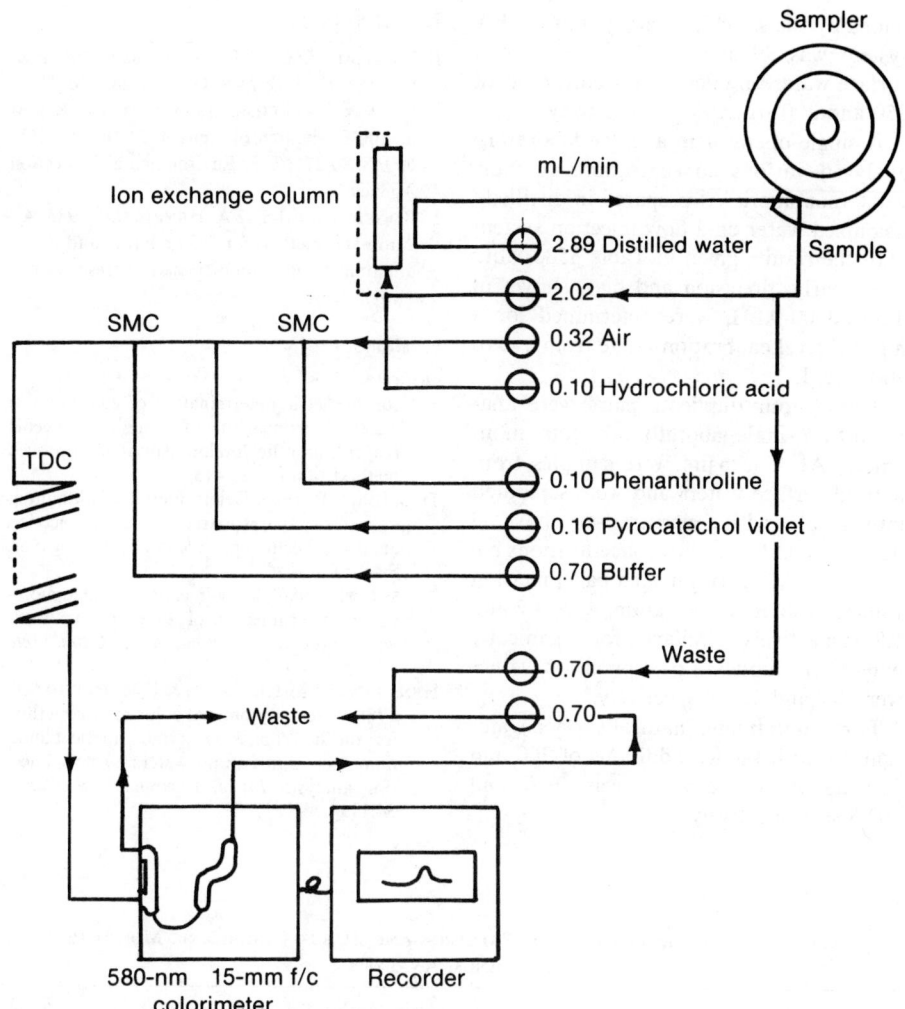

Figure 3500-Al:3. Aluminum manifold for Automated System II (segmented-flow analyzer). After Rogeberg and Henriksen, Vatten. Reprinted with permission.[2]

bration purposes disengage cation-exchange column.

5. Calculation

Determine aluminum concentration, in micrograms per liter, directly from the calibration curve using appropriate linear regression equation. Correct for dilution if necessary.

6. Precision and Bias

Precision obtained by a single laboratory was ± 3 µg Al/L for natural surface water samples ranging from 0 to 300 µg Al/L analyzed by the SFA system. Recovery obtained by the same laboratory on the SFA system was 99 and 105% for untreated surface water samples at concentrations of 150 and 200 µg Al/L, respectively.

A single operator in a single laboratory analyzed various concentrations of inorganic monomeric Al prepared in distilled/deionized water on a flow injection system with the results given in Table 3500-Al:I.

Similarly, precision and bias, shown in Table 3500-Al:II, were determined for a high Al level calibration range from 350 to 1000 µg/L:

Twenty-four duplicate pairs were analyzed by a single laboratory for total monomeric Al. The pairs were samples from natural surface waters and were separated into samples with concentrations above 85 µg Al/L and those with concentrations below 50 µg Al/L. Duplicated pair precision values, as standard deviation, were 4.2 and 1.8, respectively. Similarly, for organically bound Al, duplicate pair precision values were 4.1 and 1.1, respectively.

For two untreated natural surface water samples with known additions of 300 and 100 µg Al/L, recovery was 99.6 and 102.3%, respectively.

7. References

1. HILLMAN, D.C., T.E. LEWIS, S.H. PIA, F.X. SUAREZ, E.M. BURKE, D.E. DOBB, J.M. HENSHAW & L.J. BLUME. 1986. Summary Report on the Evaluation of Aquatic Methods. EPA-600/X-86-271, U.S. Environmental Protection Agency.
2. ROGEBERG, E.J.S. & A. HENRIKSEN. 1985. An automatic method for fractionation and determination of aluminum species in fresh-waters. *Vatten* 41:48.

8. Bibliography

DOUGAN, W.K. & A.L. WILSON. 1974. The absorptiometric determination of aluminum in water: A comparison of some chromogenic reagents and the development of an improved method. *Analyst* 99:413.
TECATOR, AB. 1985. Determination of aluminum in water and soil extracts by flow injection analysis. Tech. rep., Tecator AB, Hoganas, Sweden.
ROYSET, O. 1986. Flow-injection spectrophotometric determination of aluminum in water with pyrocatechol violet. *Anal. Chim. Acta* 185:75.
HENSHAW, J.M., T.E. LEWIS & E.M. HEITHMAR. 1988. A semi-automated colorimetric method for the determination of monomeric aluminum species in natural waters by flow injection analysis. *Int. J. Environ. Anal. Chem.* 34:119.

TABLE 3500-Al:I. SINGLE-OPERATOR PRECISION AND BIAS FOR INORGANIC MONOMERIC ALUMINUM ANALYSIS

Nominal Al Concentration µg/L	No of Determinations	Average Observed Concentration µg/L	Standard Deviation µg/L	Bias µg/L
0.0	12	4.9	3.3	4.9
10.0	13	9.2	2.5	−0.8
15.0	9	15.0	2.8	0.0
20.0	10	20.5	2.5	0.5
25.0	10	24.0	3.4	−1.0
35.0	10	34.2	2.5	−0.8
50.0	10	49.4	2.8	−0.6
75.0	10	70.0	3.1	−5.0
100.0	10	99.1	2.7	−0.9
150.0	2	150.5	4.8	0.5
350.0	5	350.8	3.8	0.8

TABLE 3500-Al:II. SINGLE-OPERATOR PRECISION AND BIAS HIGH-LEVEL ALUMINUM DETERMINATION

Nominal Al Concentration μg/L	No. of Determinations	Average Observed Concentration μg/L	Standard Deviation μg/L	Bias μg/L
350.0	5	356.9	9.2	6.9
500.0	5	494.4	11.3	−5.6
750.0	5	743.9	13.3	−6.1
1000.0	5	1004.8	16.2	4.8

3500-Sb ANTIMONY

3500-Sb A. Introduction

1. Occurrence

Antimony is present in trace amounts in natural waters (normally less than 10 μg/L) and may be present in higher concentrations in hot springs or waters draining mineralized areas. Antimony is a regulated contaminant under various federal and state programs.

2. Selection of Method

The electrothermal atomic absorption spectrometric method is the method of choice because of its sensitivity. Alternatively use the flame atomic absorption spectrometric method or the inductively coupled plasma method when high sensitivity is not required.

3500-Sb B. Atomic Absorption Spectrometric Method

See flame atomic absorption spectrometric method, Section 3111B, and electrothermal atomic absorption spectrometric method, Section 3113.

3500-Sb C. Inductively Coupled Plasma Method

See Section 3120.

3500-As   ARSENIC*

3500-As A.   Introduction

1. Occurrence and Significance

Severe poisoning can arise from the ingestion of as little as 100 mg arsenic; chronic effects can appear from its accumulation in the body at low intake levels. Carcinogenic properties also have been imputed to arsenic. The arsenic concentration of most potable waters seldom exceeds 10 μg/L, although values as high as 100 μg/L have been reported. Arsenic may occur in water as a result of mineral dissolution, industrial discharges, or the application of insecticides.

---

*Approved by Standard Methods Committee, 1988.

2. Selection of Method

The hydride atomic absorption spectrometric (AAS) method (B), which converts arsenic to its hydride and uses an argon-hydrogen flame, is the method of choice, although the direct electrothermal AAS method (B) is simpler in the demonstrated absence of interference. The silver diethyldithiocarbamate method (C) is applicable when interferences are absent. The mercuric bromide stain method (D) requires care and experience and is suitable only for qualitative or semiquantitative determinations (± 5 μg As). The inductively coupled plasma (ICP) method (E) is useful at higher concentrations (greater than 50 μg/L).

3500-As B.   Atomic Absorption Spectrometric Method

See electrothermal atomic absorption spectrometric method, Section 3113, and hydride generation atomic absorption spectrometric method, Section 3114.

3500-As C.   Silver Diethyldithiocarbamate Method

1. General Discussion

a. Principle: Inorganic arsenic is reduced to arsine, AsH$_3$, by zinc in acid solution in a Gutzeit generator. The arsine is then passed through a scrubber containing glass wool impregnated with lead acetate solution and into an absorber tube containing silver diethyldithiocarbamate dissolved in pyridine or chloroform. In the absorber, arsenic reacts with the silver salt, forming a soluble red complex suitable for photometric measurement.

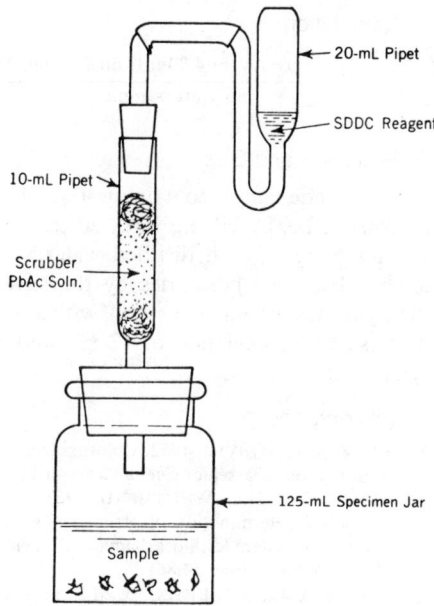

20-mL Pipet

SDDC Reagent

10-mL Pipet

Scrubber
PbAc Soln.

125-mL Specimen Jar

Sample

Figure 3500-As:1. Arsine generator and absorber assembly.

b. Interference: Although certain metals—chromium, cobalt, copper, mercury, molybdenum, nickel, platinum, and silver—interfere in the generation of arsine, the concentrations of these metals normally present in water do not interfere significantly. Antimony salts in the sample form stibine, which interferes with color development by yielding a red color with maximum absorbance at 510 nm.

c. Minimum detectable quantity: 1 μg As.

2. Apparatus

a. Arsine generator and absorption tube: See Figure 3500-As:1.*

b. Photometric equipment:

1) *Spectrophotometer,* for use at 535 nm with 1-cm cells.

2) *Filter photometer,* with green filter having a maximum transmittance in the range 530 to 540 nm, with 1-cm cells.

_____
*Fisher Scientific Co., No. 1-405 or equivalent apparatus.

3. Reagents

a. Hydrochloric acid, HCl, conc.

b. Potassium iodide solution: Dissolve 15 g KI in 100 mL distilled water. Store in a brown bottle.

c. Stannous chloride reagent: Dissolve 40 g arsenic-free $SnCl_2 \cdot 2H_2O$ in 100 mL conc HCl.

d. Lead acetate solution: Dissolve 10 g $Pb(C_2H_3O_2)_2 \cdot 3H_2O$ in 100 mL distilled water.

e. Silver diethyldithiocarbamate reagent: Prepare this reagent as described in either 1) or 2):

1) Dissolve 410 mg 1-ephedrine in 200 mL chloroform ($CHCl_3$), add 625 mg $AgSCSN(C_2H_5)_2$, and adjust volume to 250 mL with additional $CHCl_3$. Filter and store in brown bottle.

2) Dissolve 1 g $AgSCSN(C_2H_5)_2$ in 200 mL pyridine. Store in brown bottle.

f. Zinc, 20 to 30 mesh, arsenic-free.

g. Stock arsenic solution: Dissolve 1.320 g arsenic trioxide, As_2O_3, in 10 mL distilled water containing 4 g NaOH, and dilute to 1000 mL with distilled water; 1.00 mL = 1.00 mg As. (CAUTION: *Toxic—take care to avoid ingestion of arsenic solutions.*)

h. Intermediate arsenic solution: Dilute 5.00 mL stock solution to 500 mL with distilled water; 1.00 mL = 10.0 μg As.

i. Standard arsenic solution: Dilute 10.00 mL intermediate solution to 100 mL with distilled water; 1.00 mL = 1.00 μg As.

4. Procedure

For total arsenic digest sample by the procedure in ¶ D.4*a.* Report if sample has been digested or not.

a. Treatment of sample: Pipet 35.0 mL sample into a clean generator bottle. Add successively, with thorough mixing after each addition, 5 mL conc HCl, 2 mL KI solution, and 8 drops (0.40 mL) $SnCl_2$ reagent. Allow 15 min for reduction of arsenic to the trivalent state.

b. Preparation of scrubber and absorber:

Impregnate glass wool in the scrubber with lead acetate solution. Do not make too wet because water will be carried over into the reagent solution. Pipet 4.00 mL silver diethyldithiocarbamate reagent into absorber tube.

c. Arsine generation and measurement: Add 3 g zinc to generator and connect scrubber-absorber assembly immediately. Make certain that all connections are fitted tightly.

Allow 30 min for complete evolution of arsine. Warm the generator slightly to insure that all arsine is released. Pour solution from absorber directly into a 1-cm cell and measure absorbance at 535 nm, using the reagent blank as the reference.

d. Preparation of standard curve: Treat portions of standard solution containing 0, 1.0, 2.0, 5.0, and 10.0 µg As described in ¶s 4*a* through *c* above. Plot absorbance versus concentration of arsenic in the standard.

5. Calculation

$$\text{mg As/L} = \frac{\mu\text{g As (in 4.00 mL final volume)}}{\text{mL sample}}$$

6. Precision and Bias

A synthetic sample containing 40 µg As/L, 250 µg Be/L, 240 µg B/L, 20 µg Se/L, and 6 µg V/L in distilled water was analyzed in 46 laboratories by the silver diethyldithiocarbamate method, with a relative standard deviation of 13.8% and a relative error of 0%.

7. Bibliography

VASAK, V. & V. SEDIVEC. 1952. Colorimetric determination of arsenic. *Chem. Listy* 46:341.

STRATTON, G. & H.C. WHITEHEAD. 1962. Colorimetric determination of arsenic in water with silver diethyldithiocarbamate. *J. Amer. Water Works Assoc.* 54:861.

BALLINGER, D.C., R.J. LISHKA & M.E. GALES. 1962. Application of silver diethyldithiocarbamate method to determination of arsenic. *J. Amer. Water Works Assoc.* 54:1424.

3500-As D. Mercuric Bromide Stain Method

1. General Discussion

a. Principle: After sample concentration arsenic is liberated as arsine, AsH_3, by zinc in acid solution in a Gutzeit generator. The generated arsine is passed through a column containing a roll of cotton moistened with lead acetate solution. The generated arsine produces a yellow-brown stain on test paper strips impregnated with mercuric bromide. The length of the stain is roughly proportional to the amount of arsenic present.

b. Interference: Antimony (>0.10 mg) interferes by giving a similar stain.

c. Minimum detectable quantity: 1 µg As.

2. Apparatus

Arsine generator: See Figure 3500-As:2.

3. Reagents

a. Sulfuric acid, H_2SO_4, 1 + 1.

b. Nitric acid, HNO_3, conc.

c. Roll cotton: Cut a roll of dentist's cotton into 25-mm lengths.

d. Lead acetate solution: Prepare as directed in Method C, ¶ 3*d*.

e. Mercuric bromide paper: Use commercial arsenic papers cut uniformly into strips about 12 cm long and 2.5 mm wide (papers can be obtained already cut and sensitized). Soak strips for at least 1 h in filtered solution prepared by dissolving 3 to 6 g $HgBr_2$ in 100 mL 95% ethyl or isopropyl alcohol; dry by waving in air. Store in dry, dark place. For best results, make up papers just before use.

f. Potassium iodide solution: Prepare as directed in Method C, ¶ 3*b*.

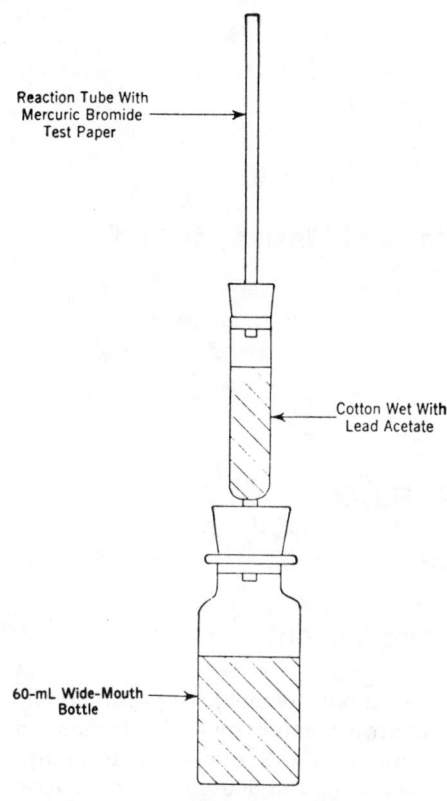

Reaction Tube With Mercuric Bromide Test Paper

Cotton Wet With Lead Acetate

60-mL Wide-Mouth Bottle

Figure 3500-As:2. Generator used with mercuric bromide stain method.

g. Stannous chloride reagent: Prepare as directed in Method C, ¶ 3c.

h. Zinc, 20 to 30 mesh, arsenic-free.

i. Standard arsenic solution: Prepare as directed in Method C, ¶ 3i.

4. Procedure

a. Concentration of sample and oxidation of organic matter: Place a suitable sample containing from 2 to 30 μg As in flask or beaker, add 7 mL 1 + 1 H_2SO_4 and 5 mL conc HNO_3. Evaporate to SO_3 fumes. Cool, add about 25 mL distilled water, and again evaporate to SO_3 fumes to expel oxides of nitrogen. Maintain an excess of HNO_3 until

the organic matter is destroyed. Do not let solution darken while organic matter is being destroyed because arsenic is likely to be reduced and lost.

Cool, add about 25 mL distilled water, and transfer to generator bottle.

b. Preparation of guard column and reaction tube: Dip one end of the 2.5-cm length of cotton into lead acetate solution and introduce into glass column. Then put the dried narrow glass tube in place and insert $HgBr_2$ test paper. Make sure paper strip is straight.

c. Treatment of sample concentrate: To the 25-mL sample concentrate in the generator, add 7 mL 1 + 1 H_2SO_4 and cool. Add 5 mL KI solution, 4 drops $SnCl_2$ reagent, and 2 to 5 g zinc. Immediately connect reaction tube to generator. Immerse apparatus to within 2.5 cm of the top of the narrow tube in a water bath kept at 20 to 25°C and allow evolution to proceed for 1.5 h. Remove strip and compute average length of stains on both sides. Using a calibration curve, the preparation of which is described below, estimate amount of arsenic present.

d. Preparation of calibration curve: Prepare a blank and standards at 3-μg intervals in the 0- to 30-μg As range with 14 mL 1 + 1 H_2SO_4 and bring total volume to 25 mL. Place in generator and treat as described for the sample concentrate. Remove strip and compute average length, in millimeters, of stains on both sides. Plot length in millimeters against micrograms arsenic and use as a standard curve.

5. Precision and Bias

A synthetic sample containing 50 μg As/ L, 400 μg Be/L, 180 μg B/L, and 50 μg Se/L in distilled water was analyzed in five laboratories by the mercuric bromide stain method with a relative standard deviation of 75.0% and a relative error of 60.0%.

6. Bibliography

FURMAN, N.H., ed. 1962. Standard Methods of Chemical Analysis, 6th ed. Vol. I. D. Van Nostrand Co., Princeton, N.J., pp. 118–124.

4500-As E.   Inductively Coupled Plasma Method

See Section 3120.

3500-Ba   BARIUM*

3500-Ba A.   Introduction

1. Occurrence and Significance

Barium stimulates the heart muscle. However, a barium dose of 550 to 600 mg is considered fatal to human beings. Afflictions arising from its consumption, inhalation, or absorption involve the heart, blood vessels, and nerves.

Despite a relative abundance in nature

(16th in order of rank), barium occurs only in trace amounts in water. The barium concentration of U.S. drinking waters ranges between 0.7 and 900 $\mu g/L$, with a mean of 49 $\mu g/L$. Higher concentrations in drinking water often signal undesirable industrial waste pollution.

2. Selection of Method

Perform analyses by the atomic absorption spectrometric method or the inductively coupled plasma method.

---

*Approved by Standard Methods Committee, 1985.

3500-Ba B.   Atomic Absorption Spectrometric Method

See flame atomic absorption spectrometric method, Section 3111D, and electrothermal atomic absorption spectrometric method, Section 3113.

3500-Ba C. Inductively Coupled Plasma Method

See Section 3120.

3500-Be   BERYLLIUM*

3500-Be A.   Introduction

1. Occurrence and Significance

Beryllium and its compounds are very poisonous and in high concentrations can cause death. Inhalation of beryllium dust can cause a serious disease called berylliosis. Beryllium disease also can take the form of dermatitis, conjunctivitis (eye disease), acute pneumonitis (lung disease), and chronic pulmonary berylliosis.

In the form of the element, compounds, or alloys, beryllium is used in atomic re-

actors, aircraft, rockets, and missile fuels. Entry into water can result from the discharges of such industries. Beryllium has been reported to occur in U.S. drinking waters in the range of 0.01 to 0.7 μg/L, with a mean of 0.013 μg/L.

2. Selection of Method

The atomic absorption spectrometric method and the inductively coupled plasma methods are the methods of choice. The colorimetric method can be used, with poorer precision and bias, if atomic absorption or ICP instrumentation is not available.

*Approved by Standard Methods Committee, 1985.

3500-Be B.   Atomic Absorption Spectrometric Method

See flame atomic absorption spectrometric method, Sections 3111D and E, and electrothermal atomic absorption spectrometric method, Section 3113.

3500-Be C.   Inductively Coupled Plasma Method

See Section 3120.

3500-Be D.   Aluminon Method

1. General Discussion

a. *Principle:* The addition of a small amount of an ethylenediaminetetraacetic acid (EDTA) complexing solution prevents interference from moderate quantities of aluminum, cobalt, copper, iron, manganese, nickel, titanium, zinc, and zirconium. An aluminon buffer reagent is added to form a beryllium lake and the color developed is measured at 515 nm.

b. *Interference:* Under the specified conditions not more than 10 mg copper can be tolerated. If more is present, increase the amount of EDTA reagent. The complexed copper absorbs slightly at 515 nm; eliminate this interference by adding an equivalent amount of copper to the standards.

c. *Minimum detectable concentration:* 5 μg/L.

d. *Molar absorptivity:* 900 L g^{-1} cm^{-1}.

2. Sample Handling

Acidify all samples at time of collection to keep metals in solution and prevent their plating out on container walls. With relatively clean waters containing no particulate matter, normally 1.5 mL conc nitric acid (HNO_3)/L is sufficient to reduce the pH to 2.0. Surface waters that may contain sediment, such as those from streams, lakes, and wastewater treatment plant effluents, require more acid. If the sample contains particulate matter and only the "dissolved" metal content is desired, filter sample through a 0.45-μm membrane filter and acidify filtrate with 1.5 mL conc HNO_3/L.

3. Apparatus

a. *Spectrophotometer,* for use at 515 nm, with a light path of 5 cm.

b. *Filter photometer,* providing a light path of 5 cm and equipped with a green filter, exhibiting a maximum transmittance near 515 nm.

4. Reagents

a. *Stock beryllium solution:* Dissolve 9.837 g beryllium sulfate tetrahydrate ($BeSO_4 \cdot 4H_2O$), purity 0.999, in 100 mL distilled water, filter if necessary, and dilute to 500 mL; 1.00 mL = 1.00 mg Be.

b. *Standard beryllium solution:* Dilute 5.00 mL stock beryllium solution with distilled water to 1000 mL in a volumetric flask; 1.00 mL = 5 μg Be.

c. *EDTA reagent:* Add 30 mL distilled water and a drop of an alcoholic solution of methyl red (50 mg/100 mL) to 2.5 g ethylenediaminetetraacetic acid. Neutralize with ammonium hydroxide (NH_4OH), cool, and dilute to 100 mL.

d. *Aluminon buffer reagent:* Transfer 500 g ammonium acetate, $NH_4C_2H_3O_2$, to 1 L distilled water in a 2-L beaker. Add 80 mL conc (glacial) acetic acid and stir until completely dissolved. Filter if necessary. Dissolve 1 g aurintricarboxylic acid triammonium salt (aluminon), in 50 mL distilled water and add to the buffer solution in the 2-L beaker. Dissolve 3.0 g benzoic acid ($C_7H_6O_2$) in 20 mL methyl alcohol and add to buffer solution while stirring. Dilute mixture to 2 L. Transfer 10 g gelatin to 250 mL distilled water in a 400-mL beaker. Place beaker in a boiling water bath and stir occasionally until gelatin has dissolved completely. Pour warm gelatin into a 1000-mL volumetric flask containing 500 mL distilled water. Cool to room temperature, dilute to mark, and mix. Transfer gelatin solution and buffer solutions to a 4-L chemical-resistant dark glass bottle. Mix and store in a cool dark place. The reagent is stable for at least a month.

5. Procedure

a. Treatment of sample: If organic matter is present and it is desired to determine total beryllium, digest sample with HNO_3 and H_2SO_4 as indicated in Section 3030G. If only dissolved beryllium is desired, filter sample through a 0.45-μm membrane filter.

b. Reaction with aluminon: Pipet 0, 0.1, 0.5, 1.00, 2.00, 3.00, and 4.00 mL standard beryllium solution to a series of 100-mL volumetric flasks. Transfer a 50-mL sample or portion containing less than 200 μg Be to a 100-mL volumetric flask. Add 2 mL EDTA reagent to each flask and dilute with distilled water to approximately 75 mL.

Add 15 mL aluminon buffer reagent, dilute to 100 mL with distilled water, and mix thoroughly. Let stand away from light for 20 min after the aluminon buffer is added. Filter if necessary. Read absorbancy of standard and unknown as compared to the blank in a spectrophotometer or filter photometer at a 515-nm wavelength using 5-cm cells. Construct a calibration curve by plotting absorbance of standards versus micrograms beryllium in 100 mL final volume. Determine amount of beryllium in sample by referring to the corresponding absorbance on the calibration curve.

6. Calculation

$$\text{mg Be/L} = \frac{\mu\text{g Be (in 100 mL final volume)}}{\text{mL sample}}$$

7. Precision and Bias

In 32 laboratories a synthetic sample of distilled water containing 250 μg Be/L, 40 μg As/L, 240 μg B/L, 20 μg Se/L, and 6 μg V/L, the beryllium was analyzed with a relative standard deviation of 7.13% and a relative error of 12%.

8. Bibliography

LUKE, C.L. & M.E. CAMPBELL. 1952. Photometric determination of beryllium in beryllium-copper alloys. *Anal. Chem.* 24:1056.

LUKE, C.L. & K.C. BROWN. 1952. Photometric determination of aluminum in manganese, bronze, zinc die casting alloys, and magnesium alloys. *Anal. Chem.* 24:1120.

3500-Bi BISMUTH

3500-Bi A. Introduction

Bismuth is extremely insoluble in natural waters and is generally present only in trace amounts (less than 10 μg/L). It may be present in higher concentration in waters draining mineralized areas.

3500-Bi B. Atomic Absorption Spectrometric Method

See flame atomic absorption spectrometric method, Section 3111B.

3500-Cd CADMIUM*

3500-Cd A. Introduction

1. Occurrence and Significance

Cadmium is highly toxic and has been implicated in some cases of poisoning through food. Minute quantities of cadmium are suspected of being responsible for adverse changes in arteries of human kidneys. Cadmium also causes generalized cancers in laboratory animals and has been linked epidemiologically with certain human cancers. A cadmium concentration of 200 $\mu g/L$ is toxic to certain fish. Cadmium may enter water as a result of industrial discharges or the deterioration of galvanized pipe.

2. Selection of Method

The electrothermal (graphite furnace) atomic absorption spectrometric method is preferred. The flame atomic absorption and inductively coupled plasma methods provide acceptable precision and bias, with higher detection limits. The dithizone method is suitable when atomic absorption spectrometric or inductively coupled plasma apparatus is unavailable and the desired precision is not as great.

---

*Approved by Standard Methods Committee, 1985.

3500-Cd B. Atomic Absorption Spectrometric Method

See flame atomic absorption spectrometric method, Sections 3111B and C, and electrothermal atomic absorption spectrometric method, Section 3113.

3500-Cd C. Inductively Coupled Plasma Method

See Section 3120.

3500-Cd D. Dithizone Method

1. General Discussion

a. Principle: Cadmium ions under suitable conditions react with dithizone to form a pink to red color that can be extracted with chloroform (CHCl$_3$). Chloroform extracts are measured photometrically and the cadmium concentration is obtained from a calibration curve prepared from a standard cadmium solution treated in the same manner as the sample.

b. Interference: Under the specified conditions, concentrations of metal ions normally found in water do not interfere. Lead up to 6 mg, zinc up to 3 mg, and copper up to 1 mg in the portion analyzed do not interfere. Ordinary room lighting (incandescent or fluorescent) does not affect the cadmium dithizonate color.

c. Minimum detectable concentration: 0.5 μg Cd in approximately 15 mL final volume with a 2-cm light path.

2. Apparatus

a. Colorimetric equipment: One of the following is required:

1) *Spectrophotometer,* for use at 518 nm with a minimum light path of 1 cm.

2) *Filter photometer,* equipped with a green filter having a maximum light transmittance near 518 nm, with a minimum light path of 1 cm.

b. Separatory funnels, 125-mL, preferably with TFE stopcocks.

c. Glassware: Clean all glassware, including sample bottles, with 1 + 1 HCl and rinse thoroughly with tap water and distilled water.

3. Reagents

a. Water, cadmium-free: Redistill distilled water in an all-glass still. Use this water to prepare all reagents and solutions.

b. Stock cadmium solution: Weigh 100.0 mg pure Cd metal and dissolve in a solution composed of 20 mL water plus 5 mL conc HCl. Use heat to assist metal dissolution. Transfer quantitatively to a 1-L volumetric flask and dilute to 1000 mL; 1.00 mL = 100 μg Cd. Store in a polyethylene container.

c. Standard cadmium solution: Pipet 10.00 mL stock cadmium solution into a 1-L volumetric flask, add 10 mL conc HCl, and dilute to 1000 mL with water. Prepare as needed and use the same day; 1.00 mL = 1.00 μg Cd.

d. Sodium potassium tartrate solution: Dissolve 250 g NaKC$_4$H$_4$O$_6$·4H$_2$O in water and make up to 1 L.

e. Sodium hydroxide-potassium cyanide solutions:

1) *Solution I:* Dissolve 400 g NaOH and 10 g KCN in water and make up to 1 L. Store in a polyethylene bottle. This solution is stable for 1 month.

2) *Solution II:* Dissolve 400 g NaOH and 0.5 g KCN in water and make up to 1 L. Store in a polyethylene bottle. This solution is stable for 1 to 2 months.

CAUTION—*Potassium cyanide is extremely poisonous. Be especially cautious when handling it. Never use mouth pipets to deliver cyanide solutions.*

f. Hydroxylamine hydrochloride solution: Dissolve 20 g NH$_2$OH·HCl in water and make up to 100 mL.

g. Stock dithizone solution I: See Section 1070D.2*b*1).

h. Working dithizone solution: Dilute stock dithizone solution I with CHCl$_3$ to produce a working solution of 10 μg/mL. Prepare daily.

i. Chloroform, ACS grade passed for "suitability for use in dithizone test." Test for a satisfactory CHCl$_3$ by adding a minute amount of dithizone to a portion of the CHCl$_3$ in a stoppered test tube so that a faint green is produced; the green color should be stable for a day.

j. Tartaric acid solution: Dissolve 20 g H$_2$C$_4$H$_4$O$_6$ in water and make up to 1 L.

Store in refrigerator and use while still cold.

k. *Hydrochloric acid,* HCl, conc.

l. *Thymol blue indicator solution:* Dissolve 0.4 g thymolsulfonephthalein sodium salt in 100 mL water.

m. *Sodium hydroxide,* NaOH, 6N.

4. Procedure

a. *Preparation of standard curve:* Prepare a blank and a series of standards from 1 to 10 μg by pipetting the appropriate amounts of standard Cd solution into separatory funnels. Dilute to 25 mL and proceed as in ¶ 4c. Plot a calibration curve.

b. *Treatment of samples:* Digest sample as directed in Section 3030. Pipet a volume of digested sample containing 1 to 10 μg Cd to a separatory funnel and dilute to 25 mL as necessary. Add 3 drops thymol blue and adjust with 6N NaOH to the first permanent yellow color, pH 2.8.

c. *Color development, extraction, and measurement:* Add reagents in the following order, mixing after each addition: 1 mL sodium potassium tartrate solution, 5 mL NaOH-KCN solution I, 1 mL NH$_2$OH·HCl solution, and 15 mL stock dithizone solution I. Stopper funnels and shake for 1 min, relieving vapor pressure in the funnels through the stopper rather than the stopcock. Drain CHCl$_3$ layer into a second funnel containing 25 mL cold tartaric acid solution. Add 10 mL CHCl$_3$ to first funnel; shake for 1 min and drain into second funnel. Do not permit aqueous layer to enter second funnel. Because time of contact of CHCl$_3$ with the strong alkali must be kept to a minimum, make the two

extractions immediately after adding dithizone (cadmium dithizonate decomposes on prolonged contact with strong alkali saturated with CHCl$_3$).

Shake second funnel for 2 min and discard CHCl$_3$ layer. Add 5 mL CHCl$_3$, shake 1 min, and discard CHCl$_3$ layer, making as close a separation as possible. In the following order, add 0.25 mL NH$_2$OH·HCl solution and 15.0 mL working dithizone solution. Add 5 mL NaOH-KCN solution II; *immediately* shake for 1 min and transfer CHCl$_3$ layer into a dry photometer tube. Read absorbance at 518 nm against the blank. Obtain Cd concentration from calibration curve.

5. Calculation

$$\text{mg Cd/L} = \frac{\mu\text{g Cd (in approx. 15 mL final volume)}}{\text{mL sample}}$$

6. Precision and Bias

A synthetic sample containing 50 μg Cd/L, 500 μg Al/L, 110 μg Cr/L, 470 μg Cu/L, 300 μg Fe/L, 70 μg Pb/L, 150 μg Ag/L, and 650 μg Zn/L was analyzed in 44 laboratories by the dithizone method with a relative standard deviation of 24.6% and a relative error of 6.0%.

7. Bibliography

SALTZMAN, B.E. 1953. Colorimetric microdetermination of cadmium with dithizone. *Anal. Chem.* 25:493.

GANOTES, J., E. LARSON & R. NAVONE. 1962. Suggested dithizone method for cadmium determination. *J. Amer. Water Works Assoc.* 54:852.

3500-Ca CALCIUM*

3500-Ca A. Introduction

1. Occurrence and Significance

The presence of calcium (fifth among the elements in order of abundance) in water supplies results from passage through or over deposits of limestone, dolomite, gypsum, and gypsiferous shale. The calcium content may range from zero to several hundred milligrams per liter, depending on the source and treatment of the water. Small concentrations of calcium carbonate combat corrosion of metal pipes by laying down a protective coating. Appreciable calcium salts, on the other hand, precipitate on heating to form harmful scale in boilers, pipes, and cooking utensils. Calcium carbonate saturation is discussed in Section 2330.

_____

*Approved by Standard Methods Committee, 1985.

Calcium contributes to the total hardness of water. Chemical softening treatment, reverse osmosis, electrodialysis, or ion exchange is used to reduce calcium and the associated hardness.

2. Selection of Method

The atomic absorption method and inductively coupled plasma method are accurate means of determining calcium. The permanganate and EDTA titration methods give good results for control and routine applications. The simplicity and rapidity of the EDTA titration procedure make it preferable to the permanganate method.

3. Storage of Samples

The customary precautions are sufficient if care is taken to redissolve any calcium carbonate that may precipitate on standing.

3500-Ca B. Atomic Absorption Spectrometric Method

See flame atomic absorption spectrometric method, Section 3111B.

3500-Ca C. Inductively Coupled Plasma Method

See Section 3120.

3500-Ca D. EDTA Titrimetric Method

1. General Discussion

  a. Principle: When EDTA (ethylenediaminetetraacetic acid or its salts) is added to water containing both calcium and magnesium, it combines first with the calcium.

Calcium can be determined directly, with EDTA, when the pH is made sufficiently high that the magnesium is largely precipitated as the hydroxide and an indicator is used that combines with calcium only. Several indicators give a color change when all of the calcium has been complexed by the EDTA at a pH of 12 to 13.

b. *Interference:* Under conditions of this test, the following concentrations of ions cause no interference with the calcium hardness determination: Cu^{2+}, 2 mg/L; Fe^{2+}, 20 mg/L; Fe^{3+}, 20 mg/L; Mn^{2+}, 10 mg/L; Zn^{2+}, 5 mg/L; Pb^{2+}, 5 mg/L; Al^{3+}, 5 mg/L; and Sn^{4+}, 5 mg/L. Orthophosphate precipitates calcium at the pH of the test. Strontium and barium give a positive interference and alkalinity in excess of 300 mg/L may cause an indistinct end point in hard waters.

2. Reagents

a. *Sodium hydroxide,* NaOH, 1N.

b. *Indicators:* Many indicators are available for the calcium titration. Some are described in the literature (see Bibliography); others are commercial preparations and also may be used. Murexide (ammonium purpurate) was the first indicator available for detecting the calcium end point, and directions for its use are presented in this procedure. Individuals who have difficulty recognizing the murexide end point may find the indicator Eriochrome Blue Black R (color index number 202) or Solochrome Dark Blue an improvement because of the color change from red to pure blue. Eriochrome Blue Black R is sodium-1-(2-hydroxy-1-naphthylazo)-2-naphthol-4-sulfonic acid. Other indicators specifically designed for use as end-point detectors in EDTA titration of calcium may be used.

1) *Murexide (ammonium purpurate) indicator:* This indicator changes from pink to purple at the end point. Prepare by dissolving 150 mg dye in 100 g absolute ethylene glycol. Water solutions of the dye are not stable for longer than 1 d. A ground mixture of dye powder and sodium chloride (NaCl) provides a stable form of the indicator. Prepare by mixing 200 mg murexide with 100 g solid NaCl and grinding

the mixture to 40 to 50 mesh. Titrate immediately after adding indicator because it is unstable under alkaline conditions. Facilitate end-point recognition by preparing a color comparison blank containing 2.0 mL NaOH solution, 0.2 g solid indicator mixture (or 1 to 2 drops if a solution is used), and sufficient standard EDTA titrant (0.05 to 0.10 mL) to produce an unchanging color.

2) *Eriochrome Blue Black R indicator:* Prepare a stable form of the indicator by grinding together in a mortar 200 mg powdered dye and 100 g solid NaCl to 40 to 50 mesh. Store in a tightly stoppered bottle. Use 0.2 g of ground mixture for the titration in the same manner as murexide indicator. During titration the color changes from red through purple to bluish purple to a pure blue with no trace of reddish or purple tint. The pH of some (not all) waters must be raised to 14 (rather than 12 to 13) by the use of 8N NaOH to get a good color change.

c. *Standard EDTA titrant,* 0.01M: Prepare standard EDTA titrant as described for the EDTA total-hardness method (Section 2340). Standard EDTA titrant, 0.0100M, is equivalent to 400.8 µg Ca/1.00 mL.

3. Procedure

a. *Pretreatment of polluted water and wastewater samples:* Follow the procedure described in Section 3030E or I.

b. *Sample preparation:* Because of the high pH used in this procedure, titrate immediately after adding alkali and indicator. Use 50.0 mL sample, or a smaller portion diluted to 50 mL so that the calcium content is about 5 to 10 mg. Analyze hard waters with alkalinity higher than 300 mg $CaCO_3$/L by taking a smaller portion and diluting to 50 mL, or by neutralizing the alkalinity with acid, boiling 1 min, and cooling before beginning the titration.

c. *Titration:* Add 2.0 mL NaOH solution

or a volume sufficient to produce a pH of 12 to 13. Stir. Add 0.1 to 0.2 g indicator mixture selected (or 1 to 2 drops if a solution is used). Add EDTA titrant slowly, with continuous stirring to the proper end point. When using murexide, check end point by adding 1 to 2 drops of titrant in excess to make certain that no further color change occurs.

4. Calculation

$$\text{mg Ca/L} = \frac{A \times B \times 400.8}{\text{mL sample}}$$

Calcium hardness as mg $CaCO_3/L$

$$= \frac{A \times B \times 1000}{\text{mL sample}}$$

where:

A = mL titrant for sample and
B = mg $CaCO_3$ equivalent to 1.00 mL EDTA titrant at the calcium indicator end point.

5. Precision and Bias

A synthetic sample containing 108 mg Ca/L, 82 mg Mg/L, 3.1 mg K/L, 19.9 mg Na/L, 241 mg Cl^-/L, 1.1 mg NO_3^--N/L, 0.25 mg NO_2^--N/L, 259 mg SO_4^{2-}/L and 42.5 mg total alkalinity/L (contributed by $NaHCO_3$) in distilled water was analyzed in 44 laboratories by the EDTA titrimetric method, with a relative standard deviation of 9.2% and a relative error of 1.9%.

6. Bibliography

DIEHL, H. & J.L. ELLINGBOE. 1956. Indicator for titration of calcium in the presence of magnesium using disodium dihydrogen ethylenediamine tetraacetate. *Anal. Chem.* 28:882.

PATTON, J. & W. REEDER. 1956. New indicator for titration of calcium with (ethylenedinitrilo) tetraacetate. *Anal. Chem.* 28:1026.

HILDEBRAND, G.P. & C.N. REILLEY. 1957. New indicator for complexometric titration of calcium in the presence of magnesium. *Anal. Chem.* 29:258.

SCHWARZENBACH, G. 1957. Complexometric Titrations. Interscience Publishers, New York, N.Y.

FURMAN, N.H. 1962. Standard Methods of Chemical Analysis, 6th ed. D. Van Nostrand Co., Inc., Princeton, N.J.

KATZ, H. & R. NAVONE. 1964. Method for simultaneous determination of calcium and magnesium. *J. Amer. Water Works Assoc.* 56:121.

3500-Ca E.   Permanganate Titrimetric Method

1. General Discussion

a. Principle: Ammonium oxalate precipitates calcium quantitatively as calcium oxalate. An excess of oxalate overcomes the adverse effects of magnesium. Optimum crystal formation and minimum occlusion are obtained only when the pH is brought slowly to the desired value. This is accomplished in two stages, with intervening digestion to promote seed crystal formation. The precipitated calcium oxalate is dissolved in acid and titrated with permanganate. The amount of permanganate required to oxidize the oxalate is proportional to the amount of calcium.

b. Interference: The sample should be free of interfering amounts of strontium, silica, aluminum, iron, manganese, phosphate, and suspended matter. Strontium may precipitate as the oxalate and cause high results. In such a case, determine strontium by flame photometry and apply the appropriate correction. Eliminate silica interference by the classical dehydration procedure. Precipitate aluminum, iron, and manganese by ammonium hydroxide after treatment with persulfate. Precipitate phosphate as the ferric salt. Remove suspended matter by centrifuging or by filtration through paper, sintered glass, or a cellulose acetate membrane (see Solids, Section 2540).

2. Apparatus

a. *Vacuum pump,* or other source of vacuum.

b. *Filter flasks.*

c. *Filter crucibles:* Medium-porosity, 30-mL crucibles are recommended. Use either glass or porcelain crucibles. Crucibles of all-porous construction are difficult to wash quantitatively.

3. Reagents

a. *Methyl red indicator solution:* Dissolve 0.1 g methyl red sodium salt and dilute to 100 mL with distilled water.

b. *Hydrochloric acid,* HCl, 1 + 1.

c. *Ammonium oxalate solution:* Dissolve 10 g $(NH_4)_2C_2O_4 \cdot H_2O$ in 250 mL distilled water. Filter if necessary.

d. *Ammonium hydroxide,* NH_4OH, 3N: Add 240 mL conc NH_4OH to about 700 mL distilled water and dilute to 1 L. Filter before use to remove suspended silica flakes.

e. *Ammonium hydroxide,* 1 + 99.

f. *Special reagents for removal of aluminum, iron, and manganese interference:*

1) *Ammonium persulfate,* solid.

2) *Ammonium chloride solution:* Dissolve 20 g NH_4Cl in 1 L distilled water. Filter if necessary.

g. *Sodium oxalate:* Use primary standard grade $Na_2C_2O_4$. Dry at 105°C overnight and store in a desiccator.

h. *Sulfuric acid,* H_2SO_4, 1 + 1.

i. *Standard potassium permanganate titrant,* 0.01M: Dissolve 1.6 g $KMnO_4$ in 1 L distilled water. Keep in a brown glass-stoppered bottle and age for at least 1 week. Carefully decant or pipet supernate without stirring up any sediment. Standardize this solution frequently by the following procedure (standard $KMnO_4$ solution, exactly 0.0100M, is equivalent to 1.002 mg Ca/1.00 mL):

Weigh to the nearest 0.1 mg several 100- to 200-mg samples of anhydrous $Na_2C_2O_4$ into 400-mL beakers. To each beaker, in turn, add 100 mL distilled water and stir to dissolve. Add 10 mL 1 + 1 H_2SO_4 and heat rapidly to 90 to 95°C. Titrate rapidly with permanganate solution to be standardized, while stirring, to a slight pink endpoint color that persists for at least 1 min. Do not let temperature fall below 85°C. If necessary, warm beaker contents during titration; 100 mg will consume about 30 mL solution. Run a blank on distilled water and H_2SO_4.

$$\text{Molarity of } KMnO_4 = \frac{\text{g } Na_2C_2O_4}{(A - B) \times 0.335\,05}$$

where:

A = mL titrant for sample and
B = mL titrant for blank.

Average the results of several titrations.

4. Procedure

a. *Pretreatment of polluted water and wastewater samples:* Follow procedure described in Section 3030E or I.

b. *Treatment of sample:* Use 200 mL sample, containing not more than 50 mg Ca (or a smaller portion diluted to 200 mL). If interfering substances are present, proceed as follows:

1) Removal of silica interference—Remove interfering amounts of silica by the gravimetric procedure described in Silica, Section 4500-Si. Discard silica precipitate and save filtrate for removing interfering amounts of combined oxides described in ¶ *c* below. If combined oxides are absent, proceed to ¶ *d*.

2) Removal of combined oxides interference—Remove interfering amounts of aluminum, iron, and manganese by concentrating filtrate from gravimetric silica determination to 120 to 150 mL. Add enough HCl so that filtrate from silica removal contains at least 10 mL conc HCl. Add 2 to 3 drops methyl red indicator and

3N NH$_4$OH until indicator color turns yellow.

Add 1 g ammonium persulfate; when boiling begins, carefully add 3N NH$_4$OH until solution becomes slightly alkaline and the steam bears a distinct but not strong odor of ammonia. Test with litmus paper. Boil for 1 to 2 min and let stand 10 min, until the hydroxides coagulate, but no longer. Filter precipitate and wash three or four times with NH$_4$Cl solution. Treat filtrate as described in the following.

c. pH adjustment of sample: To 200 mL sample, containing not more than 50 mg Ca, or to a smaller portion diluted to 200 mL, add 2 or 3 drops methyl red indicator solution. Neutralize with 1 + 1 HCl and boil for 1 min. Add 50 mL (NH$_4$)$_2$C$_2$O$_4$·H$_2$O solution and if any precipitate forms, add just enough 1 + 1 HCl to redissolve.

d. Precipitation of calcium oxalate: Keeping solution just below boiling point, add 3N NH$_4$OH dropwise from a buret, stirring constantly. Continue addition until solution is quite turbid (about 5 mL are required.) Digest for 90 min at 90°C. Filter, preferably through a filter crucible, using suction. Wash at once with 1 + 99 NH$_4$OH. Although it is not necessary to transfer all of the precipitate to the filter crucible, remove all excess (NH$_4$)$_2$C$_2$O$_4$·H$_2$O from beaker. (If magnesium is to be determined gravimetrically, set aside combined filtrate and washings for this purpose.)

e. Titration of calcium oxalate: Place filter crucible on its side in beaker and cover with distilled water. Add 10 mL 1 + 1 H$_2$SO$_4$ and, while stirring, heat rapidly to 90 to 95°C. Titrate rapidly with KMnO$_4$ titrant to a slightly pink end point that persists for at least 1 min. Do not let temperature fall below 85°C; if necessary, warm beaker contents during titration. Agitate crucible sufficiently to insure reaction of all the oxalate. Run a blank, using a clean beaker and crucible, 10 mL 1 + 1 H$_2$SO$_4$, and about the same volume of distilled water used in titrating sample.

5. Calculation

$$\text{mg Ca/L} = \frac{(A - B) \times M \times 10\,020}{\text{mL sample}}$$

where:

A = mL titrant for sample,
B = mL titrant for blank, and
M = molarity of KMnO$_4$.

6. Precision and Bias

A synthetic sample containing 108 mg Ca/L, 82 mg Mg/L, 3.1 mg K/L, 19.9 mg Na/L, 241 mg Cl$^-$/L, 1.1 mg NO$_3$$^-$-N/L, 0.25 mg NO$_2$$^-$-N/L, 259 mg SO$_4$$^{2-}$/L, and 42.5 mg total alkalinity/L (contributed by NaHCO$_3$) in distilled water was analyzed in six laboratories by the KMnO$_4$ titrimetric method, with a relative standard deviation of 3.5% and a relative error of 2.8%.

7. Bibliography

KOLTHOFF, I.M., E.J. MEEHAN, E.B. SANDELL & S. BRUCKENSTEIN. 1969. Quantitative Chemical Analysis, 4th ed. Macmillan Co., New York, N.Y.

3500-Cs   CESIUM

3500-Cs A.   Introduction

Cesium occurs at trace levels (<1 μg/ L) in most natural waters. Thermal waters may contain much higher values (up to 5 mg/L).

3500-Cs B.   Atomic Absorption Spectrometric Method

See flame atomic absorption spectrometric method, Section 3111B.

3500-Cr   CHROMIUM*

3500-Cr A.   Introduction

1. Occurrence

The hexavalent chromium concentration of U.S. drinking waters has been reported to vary between 3 and 40 μg/L with a mean of 3.2 μg/L. Chromium salts are used extensively in industrial processes and may enter a water supply through the discharge of wastes. Chromate compounds frequently are added to cooling water for corrosion control. Chromium may exist in water supplies in both the hexavalent and the trivalent state although the trivalent form rarely occurs in potable water.

2. Selection of Method

Use the colorimetric method for the determination of hexavalent chromium in a natural or treated water intended to be potable. Use the electrothermal (graphite furnace) atomic absorption spectrometric method for determination of low levels of total chromium (< 50 μg/L) in water and wastewater. Use the flame atomic absorption spectrometric method or the inductively coupled plasma method to measure concentrations up to milligram-per-liter levels.

3. Sample Handling

If only the dissolved metal content is desired, filter sample through a 0.45-μm membrane filter at the time of collection. After filtration acidify filtrate with conc nitric acid (HNO_3) to pH <2. If the total chromium content is desired, acidify unfiltered sample at time of collection with conc HNO_3 to pH <2.

---

*Approved by Standard Methods Committee, 1985.

3500-Cr B. Atomic Absorption Method for Total Chromium

See flame atomic absorption spectrometric method, Sections 3111B and C, and electrothermal atomic absorption spectrometric method, Section 3113.

3500-Cr C. Inductively Coupled Plasma Method

See Section 3120.

3500-Cr D. Colorimetric Method

1. General Discussion

a. Principle: This procedure measures only hexavalent chromium (Cr^{6+}). Therefore, to determine total chromium convert all the chromium to the hexavalent state by oxidation with potassium permanganate. The hexavalent chromium is determined colorimetrically by reaction with diphenylcarbazide in acid solution. A redviolet color of unknown composition is produced. The reaction is very sensitive, the molar absorptivity based on chromium being about 40 000 L g^{-1} cm^{-1} at 540 nm. To determine total chromium, digest the sample with a sulfuric-nitric acid mixture and then oxidize with potassium permanganate before reacting with the diphenylcarbazide.

b. Interferences: The reaction with diphenylcarbazide is nearly specific for chromium. Hexavalent molybdenum and mercury salts will react to form color with the reagent but the intensities are much lower than that for chromium at the specified pH. Concentrations as high as 200 mg Mo or Hg/L can be tolerated. Vanadium interferes strongly but concentrations up to 10 times that of chromium will not cause trouble. Potential interference from permanganate is eliminated by prior reduction with azide. Iron in concentrations greater than 1 mg/L may produce a yellow color but the ferric ion (Fe^{3+}) color is not strong and no difficulty is encountered normally if the absorbance is measured photometrically at the appropriate wavelength. Interfering amounts of molybdenum, vanadium, iron, and copper can be removed by extraction of the cupferrates of these metals into chloroform ($CHCl_3$). A procedure for this extraction is provided but do not use it unless necessary, because residual cupferron and $CHCl_3$ in the aqueous solution complicate the later oxidation. Therefore, follow the extraction by additional treatment with acid fuming to decompose these compounds.

2. Apparatus

a. Colorimetric equipment: One of the following is required:

1) *Spectrophotometer,* for use at 540 nm, with a light path of 1 cm or longer.

2) *Filter photometer,* providing a light path of 1 cm or longer and equipped with a greenish yellow filter having maximum transmittance near 540 nm.

b. Separatory funnels, 125-mL, Squibb form, with glass or TFE stopcock and stopper.

3. Reagents

Use redistilled water to prepare reagents.

a. Stock chromium solution: Dissolve 141.4 mg $K_2Cr_2O_7$ in water and dilute to 1000 mL; 1.00 mL = 50.0 μg Cr.

b. Standard chromium solution: Dilute 10.0 mL stock chromium solution to 100 mL; 1.00 mL = 5.00 μg Cr.

c. Nitric acid, HNO_3, conc.

d. Sulfuric acid, H_2SO_4, 1 + 1.

e. Methyl orange indicator solution.

f. Hydrogen peroxide, H_2O_2, 30%.

g. Redistilled water: Distilled water redistilled in all-glass apparatus.

h. Ammonium hydroxide, NH_4OH, conc.

i. Potassium permanganate solution: Dissolve 4 g $KMnO_4$ in 100 mL water.

j. Sodium azide solution: Dissolve 0.5 g NaN_3 in 100 mL water.

k. Diphenylcarbazide solution: Dissolve 250 mg 1,5-diphenylcarbazide (1,5-diphenylcarbohydrazide) in 50 mL acetone. Store in a brown bottle. Discard when solution becomes discolored.

l. Chloroform, $CHCl_3$: Avoid or redistill material that comes in containers with metal or metal-lined caps.

m. Cupferron solution: Dissolve 5 g $C_6H_5N(NO)ONH_4$ in 95 mL water.

n. Phosphoric acid, H_3PO_4, conc.

o. Sulfuric acid, H_2SO_4, 0.2N: Dilute 17 mL 6N H_2SO_4 to 500 mL with water.

4. Procedure

a. Preparation of calibration curve: To compensate for possible slight losses of chromium during digestion or other analytical operations, treat chromium standards by the same procedure as the sample. Accordingly, pipet measured volumes of standard chromium solution (5 μg/mL) ranging from 2.00 to 20.0 mL, to give standards for 10 to 100 μg Cr, into 250-mL beakers or conical flasks. Depending on pretreatment used in ¶ *b* below, proceed with subsequent treatment of standards as if they were samples, also carrying out cupferron treatment of standards if this is required for samples.

Develop color as for samples, transfer a suitable portion of each colored solution to a 1-cm absorption cell, and measure absorbance at 540 nm. As reference, use distilled water. Correct absorbance readings of standards by subtracting absorbance of a reagent blank carried through the method.

Construct a calibration curve by plotting corrected absorbance values against micrograms chromium in 102 mL final volume.

b. Treatment of sample: If sample has been filtered and acidified and only hexavalent chromium is desired, proceed to ¶ 4*e*. If total dissolved chromium is desired and there are interfering amounts of molybdenum, vanadium, copper, or iron present, proceed to ¶ 4*c*. If interferences are not present, proceed to ¶ 4*d*. If sample is unfiltered and total chromium is desired, digest with HNO_3 and H_2SO_4 as in Section 3030G. If interferences are present, proceed to ¶s 4*c*, 4*d*, and 4*e*. If there are no interferences, proceed to ¶s 4*d* and 4*e*.

c. Removal of molybdenum, vanadium, iron, and copper with cupferron: Pipet a portion of digested sample containing 10 to 100 μg Cr into a 125-mL separatory funnel. Dilute to about 40 mL with distilled water and chill in an ice bath. Add 5 mL ice-cold cupferron solution, shake well, and let stand in ice bath for 1 min. Extract in separatory funnel with three successive 5-mL portions of $CHCl_3$; shake each portion thoroughly with aqueous solution, let layers separate, and withdraw and discard

CHCl₃ extract. Transfer extracted aqueous solution to a 125-mL conical flask. Wash separatory funnel with a small amount of distilled water and add wash water to flask. Boil for about 5 min to volatilize CHCl₃ and cool. Add 5 mL HNO₃ and sufficient H₂SO₄ to have about 3 mL present. Boil samples to the appearance of SO₃ fumes. Cool slightly, carefully add 5 mL HNO₃, and again boil to fumes to complete decomposition of organic matter. Cool, wash sides of flask, and boil once more to SO₃ fumes to eliminate all HNO₃. Cool and add 25 mL water.

d. Oxidation of trivalent chromium: Pipet a portion of digested sample with or without interferences removed, and containing 10 to 100 μg Cr, into a 125-mL conical flask. Using methyl orange as indicator, add conc NH₄OH until solution is just basic to methyl orange. Add 1 + 1 H₂SO₄ dropwise until it is acidic, plus 1 mL (20 drops) in excess. Adjust volume to about 40 mL, add a boiling chip, and heat to boiling. Add 2 drops KMnO₄ solution to give a dark red color. If fading occurs, add KMnO₄ dropwise to maintain an excess of about 2 drops. Boil for 2 min longer. Add 1 mL NaN₃ solution and continue boiling gently. If red color does not fade completely after boiling for approximately 30 s, add another 1 mL NaN₃ solution. Continue boiling for 1 min after color has faded completely, then cool. Add 0.25 mL (5 drops) H₃PO₄.

e. Color development and measurement: Use 0.2N H₂SO₄ and a pH meter to adjust solution to pH 1.0 ± 0.3. Transfer solution to a 100-mL volumetric flask, dilute to 100 mL, and mix. Add 2.0 mL diphenylcarbazide solution, mix, and let stand 5 to 10 min for full color development. Transfer an appropriate portion to a 1-cm absorption cell and measure its absorbance at 540 nm. Use distilled water as reference. Correct absorbance reading of sample by subtracting absorbance of a blank carried through the method (see also note below). From the corrected absorbance, determine micrograms chromium present by reference to the calibration curve.

NOTE: If the solution is turbid after dilution to 100 mL in ¶ *e* above, take an absorbance reading before adding carbazide reagent and correct absorbance reading of final colored solution by subtracting the absorbance measured previously.

5. Calculation

$$\text{mg Cr/L} = \frac{\mu\text{g Cr (in 102 mL final volume)}}{A \times B} \times 100$$

where:

A = mL original sample, and
B = mL portion from 100 mL digested sample.

6. Precision and Bias

The dissolved (trivalent plus hexavalent) chromium was determined in 31 laboratories in a synthetic sample containing 110 μg Cr/L, 500 μg Al/L, 50 μg Cd/L, 470 μg Ca/L, 300 μg Fe/L, 70 μg Pb/L, 120 μg Mn/L, 150 μg Ag/L, and 650 μg Zn/L in distilled water. The relative standard deviation was 47.8% and the relative error 16.3%.

7. Bibliography

ROWLAND, G.P., JR. 1939. Photoelectric colorimetry—Optical study of permanganate ion and of chromium-diphenylcarbazide system. *Anal. Chem.* 11:442.

SALTZMAN, B.E. 1952. Microdetermination of chromium with diphenylcarbazide by permanganate oxidation. *Anal. Chem.* 24:1016.

URONE, P.F. 1955. Stability of colorimetric reagent for chromium, 5-diphenylcarbazide, in various solvents. *Anal. Chem.* 27:1354.

ALLEN, T.L. 1958. Microdetermination of chromium with 1,5- diphenylcarbohydrazide. *Anal. Chem.* 30:447.

SANDELL, E.B. 1959. Colorimetric Determination of Traces of Metals, 3rd ed. Interscience Publishers, New York, N.Y.

3500-Co   COBALT

3500-Co A.   Introduction

1. Occurrence

Cobalt normally occurs at levels of < 10 $\mu g/L$ in natural waters. Wastewaters may contain higher concentrations.

2. Selection of Method

Use the flame or the furnace atomic absorption spectrometric method, or the inductively coupled plasma method.

3500-Co B.   Atomic Absorption Spectrometric Method

See flame atomic absorption spectrometric method, Sections 3111B and C, and electrothermal atomic absorption spectrometric method, Section 3113.

3500-Co C.   Inductively Coupled Plasma Method

See Section 3120.

3500-Cu   COPPER*

3500-Cu A.   Introduction

1. Occurrence and Significance

Copper salts are used in water supply systems to control biological growths in reservoirs and distribution pipes and to catalyze the oxidation of manganese. Corrosion of copper-containing alloys in pipe

---

*Approved by Standard Methods Committee, 1988.

fittings may introduce measurable amounts of copper into the water in a pipe system.

Copper is essential to humans; the adult daily requirement has been estimated at 2.0 mg.

2. Selection of Method

The atomic absorption spectrometric, the inductively coupled plasma, and the neocuproine methods are recommended because of their freedom from interferences. The bathocuproine method may be used for potable waters.

3. Sampling and Storage

Copper ion tends to be adsorbed on the surface of sample containers. Therefore, analyze samples as soon as possible after collection. If storage is necessary, use 0.5 mL 1 + 1 HCl/100 mL sample to prevent this adsorption.

3500-Cu B. Atomic Absorption Spectrometric Method

See flame atomic absorption spectrometric method, Sections 3111B and C, and electrothermal atomic absorption spectrometric method, Section 3113.

3500-Cu C. Inductively Coupled Plasma Method

See Section 3120.

3500-Cu D. Neocuproine Method

1. General Discussion

a. Principle: Cuprous ion (Cu^+) in neutral or slightly acidic solution reacts with 2,9-dimethyl-1,10-phenanthroline (neocuproine) to form a complex in which 2 moles of neocuproine are bound by 1 mole of Cu^+ ion. The complex can be extracted by a number of organic solvents, including a chloroform-methanol ($CHCl_3$-CH_3OH) mixture, to give a yellow solution with a molar absorptivity of about 8000 at 457 nm. The reaction is virtually specific for copper; the color follows Beer's law up to a concentration of 0.2 mg Cu/25 mL solvent; full color development is obtained when the pH of the aqueous solution is between 3 and 9; the color is stable in $CHCl_3$-CH_3OH for several days.

The sample is treated with hydroxylamine-hydrochloride to reduce cupric ions to cuprous ions. Sodium citrate is used to complex metallic ions that might precipitate when the pH is raised. The pH is adjusted to 4 to 6 with NH_4OH, a solution of neocuproine in methanol is added, and the resultant complex is extracted into $CHCl_3$. After dilution of the $CHCl_3$ to an exact volume with CH_3OH, the absorbance of the solution is measured at 457 nm.

b. Interference: Large amounts of chromium and tin may interfere. Avoid interference from chromium by adding sulfurous acid to reduce chromate and complex chromic ion. In the presence of much tin or excessive amounts of other oxidizing ions, use up to 20 mL additional hydroxylamine-hydrochloride solution.

Cyanide, sulfide, and organic matter interfere but can be removed by a digestion procedure.

c. Minimum detectable concentration: The minimum detectable concentration, corresponding to 0.01 absorbance or 98% transmittance, is 3 μg Cu when a 1-cm cell is used and 0.6 μg Cu when a 5-cm cell is used.

2. Apparatus

a. Colorimetric equipment: One of the following is required:

1) *Spectrophotometer*, for use at 457 nm, providing a light path of 1 cm or longer.

2) *Filter photometer*, providing a light path of 1 cm or longer and equipped with a narrow-band violet filter having maximum transmittance in the range 450 to 460 nm.

b. Separatory funnels, 125-mL, Squibb form, with glass or TFE stopcock and stopper.

3. Reagents

a. Redistilled water, copper-free: Because most ordinary distilled water contains detectable amounts of copper, use redistilled water, prepared by distilling singly distilled water in a resistant-glass still, or distilled water passed through an ion-exchange unit, to prepare all reagents and dilutions.

b. Stock copper solution: To 200.0 mg polished electrolytic copper wire or foil in a 250-mL conical flask, add 10 mL water and 5 mL conc HNO_3. After the reaction has slowed, warm gently to complete dissolution of the copper and boil to expel oxides of nitrogen, using precautions to avoid loss of copper. Cool, add about 50 mL water, transfer quantitatively to a 1-L volumetric flask, and dilute to the mark with water; 1 mL = 200 μg Cu.

c. Standard copper solution: Dilute 50.00 mL stock copper solution to 500 mL with water; 1.00 mL = 20.0 μg Cu.

d. Sulfuric acid, H_2SO_4, conc.

e. Hydroxylamine-hydrochloride solution: Dissolve 50 g $NH_2OH \cdot HCl$ in 450 mL water.

f. Sodium citrate solution: Dissolve 150 g $Na_3C_6H_5O_7 \cdot 2H_2O$ in 400 mL water. Add 5 mL $NH_2OH \cdot HCl$ solution and 10 mL neocuproine reagent. Extract with 50 mL $CHCl_3$ to remove copper impurities and discard $CHCl_3$ layer.

g. Ammonium hydroxide, NH_4OH, 5N: Dilute 330 mL conc NH_4OH (28-29%) to 1000 mL with water. Store in a polyethylene bottle.

h. Congo red paper, or other pH test paper showing a color change in the pH range of 4 to 6.

i. Neocuproine reagent: Dissolve 100 mg 2,9-dimethyl-1,10-phenanthroline hemihydrate* in 100 mL methanol. This solution is stable under ordinary storage conditions for a month or more.

j. Chloroform, $CHCl_3$: Avoid or redistill material that comes in containers with metal-lined caps.

k. Methanol, CH_3OH, reagent grade.

l. Nitric acid, HNO_3, conc.

m. Hydrochloric acid, HCl, conc.

4. Procedure

a. Preparation of calibration curve: Pipet 50 mL water into a 125-mL separatory funnel for use as a reagent blank. Prepare standards by pipetting 1.00 to 10.00 mL (20.0 to 200 μg Cu) standard copper solution into a series of 125-mL separatory funnels, and dilute to 50 mL with water. Add 1 mL conc H_2SO_4 and use the extraction procedure given in ¶ 4*b* below.

Construct a calibration curve by plotting absorbance versus micrograms of copper.

To prepare a calibration curve for

---

*GFS Chemical Company, Columbus, Ohio, or equivalent.

smaller amounts of copper, dilute 10.0 mL standard copper solution to 100 mL. Carry 1.00- to 10.00-mL volumes of this diluted standard through the previously described procedure, but use 5-cm cells to measure absorbance.

b. Treatment of sample: Transfer 100 mL sample to a 250-mL beaker, add 1 mL conc H_2SO_4 and 5 mL conc HNO_3. Add a few boiling chips and cautiously evaporate to dense white SO_3 fumes on a hot plate. If solution remains colored, cool, add another 5 mL conc HNO_3, and again evaporate to dense white fumes. Repeat, if necessary, until solution becomes colorless.

Cool, add about 80 mL water, and bring to a boil. Cool and filter into a 100-mL volumetric flask. Make up to 100 mL with water using mostly beaker and filter washings.

Pipet 50.0 mL or other suitable portion containing 4 to 200 μg Cu, from the solution obtained from preliminary treatment, into a 125-mL separatory funnel. Dilute, if necessary, to 50 mL with water. Add 5 mL $NH_2OH \cdot HCl$ solution and 10 mL sodium citrate solution, and mix thoroughly. Adjust pH to approximately 4 by adding 1-mL increments of NH_4OH until Congo red paper is just definitely red (or other suitable pH test paper indicates a value between 4 and 6).

Add 10 mL neocuproine reagent and 10 mL $CHCl_3$. Stopper and shake vigorously for 30 s or more to extract the copper-neocuproine complex into the $CHCl_3$. Let mixture separate into two layers and with-draw lower $CHCl_3$ layer into a 25-mL volumetric flask, taking care not to transfer any of the aqueous layer. Repeat extraction of the water layer with an additional 10 mL $CHCl_3$ and combine extracts. Dilute combined extracts to 25 mL with CH_3OH, stopper, and mix thoroughly.

Transfer an appropriate portion of extract to a suitable absorption cell (1 cm for 40 to 200 μg Cu; 5 cm for lesser amounts) and measure absorbance at 457 nm or with a 450- to 460-nm filter. Use a sample blank prepared by carrying 50 mL water through the complete digestion and analytical procedure.

Determine micrograms copper in final solution by reference to the appropriate calibration curve.

5. Calculation

$$\text{mg Cu/L} = \frac{\mu g \text{ Cu (in 25 mL final volume)}}{\text{mL portion taken for extraction}}$$

6. Bibliography

SMITH, G.F. & W.H. McCURDY. 1952. 2,9-Dimethyl-1,10-phenanthroline: New specific in spectrophotometric determination of copper. *Anal. Chem.* 24:371.

LUKE, C.L. & M.E. CAMPBELL. 1953. Determination of impurities in germanium and silicon. *Anal. Chem.* 25:1586.

GAHLER, A.R. 1954. Colorimetric determination of copper with neocuproine. *Anal. Chem.* 26:577.

FULTON, J.W. & J. HASTINGS. 1956. Photometric determinations of copper in aluminum and lead-tin solder with neocuproine. *Anal. Chem.* 28:174.

FRANK, A.J., A.B. GOULSTON & A.A. DEACUTIS. 1957. Spectrophotometric determination of copper in titanium. *Anal. Chem.* 29:750.

3500-Cu E.   Bathocuproine Method

1. General Discussion

a. Principle: Cuprous ion forms a water-soluble orange-colored chelate with bath-ocuproine disulfonate (2,9-dimethyl-4,7-di-phenyl-1,10-phenanthrolinedisulfonic acid, disodium salt). While the color forms over the pH range 3.5 to 11.0, the recommended pH range is between 4 and 5.

The sample is buffered at a pH of about 4.3 and reduced with hydroxylamine hydrochloride. The absorbance is measured at 484 nm. The method can be applied to copper concentrations up to at least 5 mg/L with a sensitivity of 20 µg/L.

b. Interference: The following substances can be tolerated with an error of less than ±2%:

Substance	Concentration *mg/L*
Cations	
Aluminum	100
Beryllium	10
Cadmium	100
Calcium	1000
Chromium (III)	10
Cobalt (II)	5
Iron (II)	100
Iron (III)	100
Lithium	500
Magnesium	100
Manganese (II)	500
Nickel (II)	500
Sodium	1000
Strontium	200
Thorium (IV)	100
Zinc	200
Anions	
Chlorate	1000
Chloride	1000
Fluoride	500
Nitrate	200
Nitrite	200
Orthophosphate	1000
Perchlorate	1000
Sulfate	1000
Compounds	
Residual chlorine	1
Linear alkylate sulfonate (LAS)	40

Cyanide, thiocyanate, persulfate, and EDTA also can interfere.

c. Minimum detectable concentration: 20 µg/L with a 5-cm cell.

2. Apparatus

a. Colorimetric equipment: One of the following, with a light path of 1 to 5 cm (unless nessler tubes are used):

1) *Spectrophotometer,* for use at 484 nm.

2) *Filter photometer,* equipped with a blue-green filter exhibiting maximum light transmission near 484 nm.

3) *Nessler tubes,* matched, 100-mL, tall form.

b. Acid-washed glassware: Rinse all glassware with conc HCl and then with copper-free water.

3. Reagents

a. Copper-free water: See Method D, ¶ 3*a*.

b. Stock copper solution: Prepare as directed in Method D, ¶ 3*b*, but use 20.00 mg copper wire or foil; 1.00 mL = 20.00 µg Cu.

c. Standard copper solution: Dilute 250 mL stock copper solution to 1000 mL with water; 1.00 mL = 5.00 µg Cu. Prepare daily.

d. Hydrochloric acid, HCl, 1 + 1.

e. Hydroxylamine hydrochloride solution: See Method D, ¶ 3*e*.

f. Sodium citrate solution: Dissolve 300 g $Na_3C_6H_5O_7 \cdot 2H_2O$ in water and make up to 1000 mL.

g. Disodium bathocuproine disulfonate solution: Dissolve 1.000 g $C_{12}H_4N_2(CH_3)_2(C_6H_4)_2(SO_3Na)_2$ in water and make up to 1000 mL.

4. Procedure

Pipet 50.0 mL sample, or a suitable portion diluted to 50.0 mL, into a 250-mL erlenmeyer flask. In separate 250-mL erlenmeyer flasks, prepare a 50.0-mL water blank and a series of 50.0-mL copper standards containing 5.0, 10.0, 15.0, 20.0, and 25.0 µg Cu. To sample, blank, and standards add, mixing after each addition, 1.00 mL 1 + 1 HCl, 5.00 mL $NH_2OH \cdot HCl$ solution, 5.00 mL sodium citrate solution, and 5.00 mL disodium bathocuproine disulfonate solution. Transfer to cells and read sample absorbance against the blank at 484 nm. Plot absorbance against micrograms Cu in standards for the calibration curve. Estimate concentration from the calibration curve.

5. Calculation

$$\text{mg Cu/L} = \frac{\mu\text{g Cu (in 66 mL final volume)}}{\text{mL sample}}$$

6. Precision and Bias

A synthetic sample containing 1000 μg Cu/L, 500 μg Al/L, 50 μg Cd/L, 110 μg Cr/L, 300 μg Fe/L, 70 μg Pb/L, 50 μg Mn/L, 150 μg Ag/L, and 650 μg Zn/L was analyzed in 33 laboratories by the bathocuproine method, with a relative standard deviation of 4.1% and a relative error of 0.3%.

7. Bibliography

SMITH, G.F. & D.H. WILKINS. 1953. New colorimetric reagent specific for copper. *Anal. Chem.* 25:510.

BORCHARDT, L.G. & J.P. BUTLER. 1957. Determination of trace amounts of copper. *Anal. Chem.* 29:414.

ZAK, B. 1958. Simple procedure for the single sample determination of serum copper and iron. *Clinica Chim. Acta* 3:328.

BLAIR, D. & H. DIEHL. 1961. Bathophenanthrolinedisulfonic acid and bathocuproinedisulfonic acid, water soluble reagents for iron and copper. *Talanta* 7:163.

3500-Au   GOLD

3500-Au A.   Introduction

Because of its extreme insolubility, gold is present only in trace amounts in natural waters (< 1 μg/L) even in mineralized areas. Wastewaters from gold-mining operations may contain higher concentrations.

3500-Au B.   Atomic Absorption Spectrometric Method

See flame atomic absorption spectrometric method, Section 3111B.

3500-Ir IRIDIUM

3500-Ir A. Introduction

Iridium is a relatively insoluble element and normally is found at very low concentrations and associated with particulate material. Process waters may contain higher concentrations.

3500-Ir B. Atomic Absorption Spectrometric Method

See flame atomic absorption spectrometric method, Section 3111B.

3500-Fe IRON*

3500-Fe A. Introduction

1. Occurrence and Significance

In filtered samples of oxygenated surface waters iron concentrations seldom reach 1 mg/L. Some groundwaters and acid surface drainage may contain considerably more iron. Iron in water can cause staining of laundry and porcelain. A bittersweet astringent taste is detectable by some persons at levels above 1 mg/L.

Under reducing conditions, iron exists in the ferrous state. In the absence of complex-forming ions, ferric iron is not significantly soluble unless the pH is very low. On exposure to air or addition of oxidants, ferrous iron is oxidized to the ferric state and may hydrolyze to form insoluble hydrated ferric oxide.

In water samples iron may occur in true solution, in a colloidal state that may be peptized by organic matter, in inorganic or organic iron complexes, or in relatively coarse suspended particles. It may be either ferrous or ferric, suspended or dissolved.

Silt and clay in suspension may contain

acid-soluble iron. Sometimes iron oxide particles are collected with a water sample as a result of flaking of rust from pipes. Iron may come from a metal cap used to close the sample bottle.

2. Selection of Method

Sensitivity and detection limits for the atomic absorption spectrometric procedure, the inductively coupled plasma method, and the phenanthroline colorimetric procedure described here are similar and generally adequate for analysis of natural or treated waters. The complexing reagents used in the colorimetric procedures are specific for ferrous iron but the atomic absorption procedures are not. However, because of the instability of ferrous iron, which is changed easily to the ferric form in solutions in contact with air, determination of ferrous iron requires special precautions and may need to be done in the field at the time of sample collection.

The procedure for determining ferrous iron using 1,10-phenanthroline (¶ 3500-Fe.D.4c) has a somewhat limited applica-

---

*Approved by Standard Methods Committee, 1985.

bility; long storage time or exposure of samples to light must be avoided. A rigorous quantitative distinction between ferrous and ferric iron can be obtained with a special procedure using bathophenanthroline. Spectrophotometric methods using bathophenanthroline[1–6] and other organic complexing reagents such as ferrozine[7] or TPTZ[8] are capable of determining iron concentrations as low as 1 μg/L. A chemiluminescence procedure[9] is stated to have a detection limit of 5 ng/L. Additional procedures are described elsewhere.[10–13]

3. Sampling and Storage

Plan in advance the methods of collecting, storing, and pretreating samples. Clean sample container with acid and rinse with distilled water. Equipment for membrane filtration of samples in the field may be required to determine iron in solution (dissolved iron). Dissolved iron is considered to be that passing through a 0.45-μm membrane filter but colloidal iron may be included. The value of the determination depends greatly on the care taken to obtain a representative sample. Iron in well water or tap samples may vary in concentration and form with duration and degree of flushing before and during sampling. When taking a sample portion for determining iron in suspension, shake the sample bottle often and vigorously to obtain a uniform suspension of precipitated iron. Use particular care when colloidal iron adheres to the sample bottle. This problem can be acute with plastic bottles.

For a precise determination of total iron, use a separate container for sample collection. Treat with acid at the time of collection to place the iron in solution and prevent adsorption or deposition on the walls of the sample container. Take account of the added acid in measuring por-

tions for analysis. The addition of acid to the sample may eliminate the need for adding acid before digestion (¶ 3500-Fe.D.4a).

4. References

1. LEE, G.F. & W. STUMM. 1960. Determination of ferrous iron in the presence of ferric iron using bathophenanthroline. *J. Amer. Water Works Assoc.* 52:1567.
2. GHOSH, M.M., J.T. O'CONNOR & R.S. ENGELBRECHT. 1967. Bathophenanthroline method for the determination of ferrous iron. *J. Amer. Water Works Assoc.* 59:897.
3. BLAIR, D. & H. DIEHL. 1961. Bathophenanthroline-disulfonic acid and bathocuproinedisulfonic acid, water soluble reagents for iron and copper. *Talanta* 7:163.
4. SHAPIRO, J. 1966. On the measurement of ferrous iron in natural waters. *Limnol. Oceanogr.* 11:293.
5. MCMAHON, J.W. 1967. The influence of light and acid on the measurement of ferrous iron in lake water. *Limnol. Oceanogr.* 12:437.
6. MCMAHON, J.W. 1969. An acid-free bathophenanthroline method for measuring dissolved ferrous iron in lake water. *Water Res.* 3:743.
7. GIBBS, C. 1976. Characterization and application of ferrozine iron reagent as a ferrous iron indicator. *Anal. Chem.* 48:1197.
8. DOUGAN, W.K. & A. L. WILSON. 1973. Absorbtiometric determination of iron with TPTZ. *Water Treat. Exam.* 22:110.
9. SEITZ, W.R. & D.M. HERCULES. 1972. Determination of trace amounts of iron (II) using chemiluminescence analysis. *Anal. Chem.* 44:2143.
10. MOSS, M.L. & M.G. MELLON. 1942. Colorimetric determination of iron with 2,2'-bipyridine and with 2,2',2"-tripyridine. *Ind. Eng. Chem.*, Anal. Ed. 14:862.
11. WELCHER, F.J. 1947. Organic Analytical Reagents. D. Van Nostrand Co., Princeton, N.J., Vol. 3, pp. 100-104.
12. MORRIS, R.L. 1952. Determination of iron in water in the presence of heavy metals. *Anal. Chem.* 24:1376.
13. DOIG, M.T., III, & D.F. MARTIN. 1971. Effect of humic acids on iron analyses in natural water. *Water Res.* 5:689.

3500-Fe B.   Atomic Absorption Spectrometric Method

See flame atomic absorption spectrometric method, Sections 3111B and C, and electrothermal atomic absorption spectrometric method, Section 3113.

3500-Fe C.   Inductively Coupled Plasma Method

See Section 3120.

3500-Fe D.   Phenanthroline Method

1. General Discussion

a. Principle: Iron is brought into solution, reduced to the ferrous state by boiling with acid and hydroxylamine, and treated with 1,10-phenanthroline at pH 3.2 to 3.3. Three molecules of phenanthroline chelate each atom of ferrous iron to form an orange-red complex. The colored solution obeys Beer's law; its intensity is independent of pH from 3 to 9. A pH between 2.9 and 3.5 insures rapid color development in the presence of an excess of phenanthroline. Color standards are stable for at least 6 months.

b. Interference: Among the interfering substances are strong oxidizing agents, cyanide, nitrite, and phosphates (polyphosphates more so than orthophosphate), chromium, zinc in concentrations exceeding 10 times that of iron, cobalt and copper in excess of 5 mg/L, and nickel in excess of 2 mg/L . Bismuth, cadmium, mercury, molybdate, and silver precipitate phenanthroline. The initial boiling with acid converts polyphosphates to orthophosphate and removes cyanide and nitrite that otherwise would interfere. Adding excess hydroxylamine eliminates errors caused by excessive concentrations of strong oxidizing reagents. In the presence of interfering metal ions, use a larger excess of phenanthroline to replace that complexed by the interfering metals. Where excessive concentrations of interfering metal ions are present, the extraction method may be used.

If noticeable amounts of color or organic matter are present, it may be necessary to evaporate the sample, gently ash the residue, and redissolve in acid. The ashing may be carried out in silica, porcelain, or platinum crucibles that have been boiled for several hours in 1 + 1 HCl. The presence of excessive amounts of organic matter may necessitate digestion before use of the extraction procedure.

c. Minimum detectable concentration: Dissolved or total concentrations of iron as low as 10 µg/L can be determined with a spectrophotometer using cells with a 5 cm or longer light path. Carry a blank through the entire procedure to allow for correction.

2. Apparatus

a. Colorimetric equipment: One of the following is required:

1) *Spectrophotometer,* for use at 510 nm, providing a light path of 1 cm or longer.

2) *Filter photometer,* providing a light path of 1 cm or longer and equipped with a green filter having maximum transmittance near 510 nm.

3) *Nessler tubes,* matched, 100-mL, tall form.

b. Acid-washed glassware: Wash all glassware with conc hydrochloric acid (HCl) and rinse with distilled water before use to remove deposits of iron oxide.

c. Separatory funnels: 125-mL, Squibb form, with ground-glass or TFE stopcocks and stoppers.

3. Reagents

Use reagents low in iron. Use iron-free distilled water in preparing standards and reagent solutions. Store reagents in glass-stoppered bottles. The HCl and ammonium acetate solutions are stable indefinitely if tightly stoppered. The hydroxylamine, phenanthroline, and stock iron solutions are stable for several months. The standard iron solutions are not stable; prepare daily as needed by diluting the stock solution. Visual standards in nessler tubes are stable for several months if sealed and protected from light.

a. Hydrochloric acid, HCl, conc, containing less than 0.000 05% iron.

b. Hydroxylamine solution: Dissolve 10 g $NH_2OH \cdot HCl$ in 100 mL water.

c. Ammonium acetate buffer solution: Dissolve 250 g $NH_4C_2H_3O_2$ in 150 mL water. Add 700 mL conc (glacial) acetic acid. Because even a good grade of $NH_4C_2H_3O_2$ contains a significant amount of iron, prepare new reference standards with each buffer preparation.

d. Sodium acetate solution: Dissolve 200 g $NaC_2H_3O_2 \cdot 3H_2O$ in 800 mL water.

e. Phenanthroline solution: Dissolve 100 mg 1,10-phenanthroline monohydrate, $C_{12}H_8N_2 \cdot H_2O$, in 100 mL water by stirring and heating to 80°C. Do not boil. Discard the solution if it darkens. Heating is unnecessary if 2 drops conc HCl are added to the water. (NOTE: One milliliter of this reagent is sufficient for no more than 100 µg Fe.)

f. Stock iron solution: Use metal (1) or salt (2) for preparing the stock solution.

1) Use electrolytic iron wire, or "iron wire for standardizing," to prepare the so-

lution. If necessary, clean wire with fine sandpaper to remove any oxide coating and to produce a bright surface. Weigh 200.0 mg wire and place in a 1000-mL volumetric flask. Dissolve in 20 mL 6*N* sulfuric acid (H_2SO_4) and dilute to mark with water; 1.00 mL = 200 µg Fe.

2) If ferrous ammonium sulfate is preferred, slowly add 20 mL conc H_2SO_4 to 50 mL water and dissolve 1.404 g $Fe(NH_4)_2(SO_4)_2 \cdot 6H_2O$. Add 0.1*N* potassium permanganate ($KMnO_4$) dropwise until a faint pink color persists. Dilute to 1000 mL with water and mix; 1.00 mL = 200 µg Fe.

g. Standard iron solutions: Prepare daily for use.

1) Pipet 50.00 mL stock solution into a 1000-mL volumetric flask and dilute to mark with water; 1.00 mL = 10.0 µg Fe.

2) Pipet 5.00 mL stock solution into a 1000-mL volumetric flask and dilute to mark with water; 1.00 mL = 1.00 µg Fe.

h. Diisopropyl or isopropyl ether. CAUTION: *Ethers may form explosive peroxides; test before using.*

4. Procedure

a. Total iron: Mix sample thoroughly and measure 50.0 mL into a 125-mL erlenmeyer flask. If this sample volume contains more than 200 µg iron use a smaller accurately measured portion and dilute to 50.0 mL. Add 2 mL conc HCl and 1 mL $NH_2OH \cdot HCl$ solution. Add a few glass beads and heat to boiling. To insure dissolution of all the iron, continue boiling until volume is reduced to 15 to 20 mL. (If the sample is ashed, take up residue in 2 mL conc HCl and 5 mL water.) Cool to room temperature and transfer to a 50- or 100-mL volumetric flask or nessler tube. Add 10 mL $NH_4C_2H_3O_2$ buffer solution and 4 mL phenanthroline solution, and dilute to mark with water. Mix thoroughly and allow at least 10 to 15 min for maximum color development.

b. Dissolved iron: Immediately after collection filter sample through a 0.45-μm membrane filter into a vacuum flask containing 1 mL conc HCl/100 mL sample. Analyze filtrate for total dissolved iron (¶ 4a) and/or dissolved ferrous iron (¶ 4c). (This procedure also can be used in the laboratory with the understanding that normal sample exposure to air during shipment may result in precipitation of iron.)

Calculate suspended iron by subtracting dissolved from total iron.

c. Ferrous iron: Determine ferrous iron at sampling site because of the possibility of change in the ferrous-ferric ratio with time in acid solutions. To determine ferrous iron only, acidify a separate sample with 2 mL conc HCl/100 mL sample at time of collection. Fill bottle directly from sampling source and stopper. Immediately withdraw a 50-mL portion of acidified sample and add 20 mL phenanthroline solution and 10 mL $NH_4C_2H_3O_2$ solution with vigorous stirring. Dilute to 100 mL and measure color intensity within 5 to 10 min. Do not expose to sunlight. (Color development is rapid in the presence of excess phenanthroline. The phenanthroline volume given is suitable for less than 50 μg total iron; if larger amounts are present, use a correspondingly larger volume of phenanthroline or a more concentrated reagent. Excess phenanthroline is required because of the kinetics of the complexing process.)

Calculate ferric iron by subtracting ferrous from total iron.

d. Color measurement: Prepare a series of standards by accurately pipetting calculated volumes of standard iron solutions (use weaker solution to measure 1- to 10-μg portions) into 125-mL erlenmeyer flasks, diluting to 50 mL by adding measured volumes of water, and carrying out the steps in ¶ 4a.

For visual comparison, prepare a set of at least 10 standards, ranging from 1 to 100 μg Fe in the final 100-mL volume. Com-

TABLE 3500-Fe:I. SELECTION OF LIGHT PATH LENGTH FOR VARIOUS IRON CONCENTRATIONS

Fe μg		
50-mL Final Volume	100-mL Final Volume	Light Path *cm*
50–200	100–400	1
25–100	50–200	2
10–40	20–80	5
5–20	10–40	10

pare colors in 100-mL tall-form nessler tubes.

For photometric measurement, use Table 3500-Fe:I as a rough guide for selecting proper light path at 510 nm. Read standards against distilled water set at zero absorbance and plot a calibration curve, including a blank (see ¶ 3c and General Introduction).

If samples are colored or turbid, carry a second set of samples through all steps of the procedure without adding phenanthroline. Instead of distilled water, use the prepared blanks to set photometer to zero absorbance and read each developed sample with phenanthroline against the corresponding blank without phenanthroline. Translate observed photometer readings into iron values by means of the calibration curve. This procedure does *not* compensate for interfering ions.

e. Samples containing organic interferences: Digest samples containing substantial amounts of organic substances according to the directions given in Sections 3030G and H.

1) If a digested sample has been prepared according to the directions given in Section 3030G or H, pipet 10.0 mL or other suitable portion containing 20 to 500 μg Fe into a 125-mL separatory funnel. If the volume taken is less than 10 mL, add distilled water to make up to 10 mL. To the separatory funnel add 15 mL conc HCl for

a 10-mL aqueous volume; or, if the portion taken was greater than 10.0 mL, add 1.5 mL conc HCl/mL of sample. Mix, cool, and proceed with 4e3) below.

2) To prepare a sample solely for determining iron, measure a suitable volume containing 20 to 500 μg Fe and carry it through either of the digestion procedures described in Sections 3030G and H. However, use only 5 mL H_2SO_4 or $HClO_4$ and omit H_2O_2. When digestion is complete, cool, dilute with 10 mL water, heat almost to boiling to dissolve slowly soluble salts, and, if the sample is still cloudy, filter through a glass-fiber, sintered-glass, or porcelain filter, washing with 2 to 3 mL water. Quantitatively transfer filtrate or clear solution to a 25-mL volumetric flask and make up to 25 mL with water. Empty flask into a 125-mL separatory funnel, rinse with 5 mL conc HCl that is added to the funnel, and add 25 mL conc HCl measured with the same graduate or flask. Mix and cool to room temperature.

3) Extract the iron from the HCl solution in the separatory funnel by shaking for 30 s with 25 mL isopropyl ether (CAUTION). Draw off lower acid layer into a second separatory funnel. Extract acid solution again with 25 mL isopropyl ether, drain acid layer into a suitable clean vessel, and combine the two portions of isopropyl ether. Pour acid layer back into second separatory funnel and re-extract with 25 mL isopropyl ether. Withdraw and discard acid layer and add ether layer to original funnel. Persistence of a yellow color in the HCl solution after three extractions does not signify incomplete separation of iron because copper, which is not extracted, gives a similar yellow color.

Shake combined ether extracts with 25 mL water to return iron to aqueous phase and transfer lower aqueous layer to a 100-mL volumetric flask. Repeat extraction with a second 25-mL portion of water, adding this to the first aqueous extract. Discard ether layer.

4) Add 1 mL $NH_2OH \cdot HCl$ solution, 10 mL phenanthroline solution, and 10 mL $NaC_2H_3O_2$ solution. Dilute to 100 mL with water, mix thoroughly, and let stand for 10 min. Measure absorbance at 510 nm using a 5-cm absorption cell for amounts of iron less than 100 μg or 1-cm cell for quantities from 100 to 500 μg. As reference, use either distilled water or a sample blank prepared by carrying the specified quantities of acids through the entire analytical procedure. If distilled water is used as reference, correct sample absorbance by subtracting absorbance of a sample blank.

Determine micrograms of iron in the sample from the absorbance (corrected, if necessary) by reference to the calibration curve prepared by using a suitable range of iron standards containing the same amounts of phenanthroline, hydroxylamine, and sodium acetate as the sample.

5. Calculation

When the sample has been treated according to 4a, b, c, or 4e2):

$$mg\ Fe/L = \frac{\mu g\ Fe\ (in\ 100\ mL\ final\ volume)}{mL\ sample}$$

When the sample has been treated according to 4e1):

$$mg\ Fe/L = \frac{\mu g\ Fe\ (in\ 100\ mL\ final\ volume)}{mL\ sample}$$
$$\times \frac{100}{mL\ portion}$$

Report details of sample collection, storage, and pretreatment if they are pertinent to interpretation of results.

6. Precision and Bias

Precision and bias depend on the method of sample collection and storage, the method of color measurement, the iron concentration, and the presence of interfering color, turbidity, and foreign ions. In general, optimum reliability of visual comparison in nessler tubes is not better than 5% and often only 10%, whereas, under optimum conditions, photometric measurement may be reliable to 3% or 3 μg, whichever is greater. The sensitivity limit for visual observation in nessler tubes is approximately 1 μg Fe. Sample variability and instability may affect precision and bias of this determination more than will the errors of analysis itself. Serious divergences have been found in reports of different laboratories because of variations in methods of collecting and treating samples.

A synthetic sample containing 300 μg Fe/L, 500 μg Al/L, 50 μg Cd/L, 110 μg Cr/L, 470 μg Cu/L, 70 μg Pb/L, 120 μg Mn/L, 150 μg Ag/L, and 650 μg Zn/L in distilled water was analyzed in 44 laboratories by the phenanthroline method, with a relative standard deviation of 25.5% and a relative error of 13.3%.

7. Bibliography

CHRONHEIM, G. & W. WINK. 1942. Determination of divalent iron (by o-nitrosophenol). *Ind. Eng. Chem.*, Anal. Ed. 14:447.

MEHLIG, R.P. & R. H. HULETT. 1942. Spectrophotometric determination of iron with o-phenanthroline and with nitro-o-phenanthroline. *Ind. Eng. Chem.*, Anal. Ed. 14:869.

CALDWELL, D.H. & R.B. ADAMS. 1946. Colorimetric determination of iron in water with o-phenanthroline. *J. Amer. Water Works Assoc.* 38:727.

WELCHER, F.J. 1947. Organic Analytical Reagents. D. Van Nostrand Co., Princeton, N.J., Vol. 3, pp. 85-93.

KOLTHOFF, I.M., T.S. LEE & D. L. LEUSSING. 1948. Equilibrium and kinetic studies on the formation and dissociation of ferroin and ferrin. *Anal. Chem.* 20:985.

RYAN, J.A. & G. H. BOTHAM. 1949. Iron in aluminum alloys: Colorimetric determination using 1,10-phenanthroline. *Anal. Chem.* 21:1521.

REITZ, L.K., A.S. O'BRIEN & T. L. DAVIS. 1950. Evaluation of three iron methods using a factorial experiment. *Anal. Chem.* 22:1470.

SANDELL, E.B. 1959. Colorimetric Determination of Traces of Metals, 3rd ed. Interscience Publishers, New York, N.Y. Chapter 22.

SKOUGSTAD, M.W., M.J. FISHMAN, L. C. FRIEDMAN, D. E. ERDMANN & S. S. DUNCAN. 1979. Methods for Determination of Inorganic Substances in Water and Fluvial Sediment. Chapter A1 *in* Book 5, Techniques of Water Resources Investigations of the United States Geological Survey. U.S. Geological Surv., Washington, D.C.

3500-Pb   LEAD*

3500-Pb A.   Introduction

1. Occurrence

Lead is a serious cumulative body poison. Natural waters seldom contain more than 5 μg/L, although much higher values have been reported. Lead in a water supply may come from industrial, mine, and smelter discharges or from the dissolution of old lead plumbing. Tap waters that are soft, acid, and not suitably treated may contain lead resulting from an attack on lead service pipes.

2. Selection of Method

The atomic absorption spectrometric method has a relatively high detection limit in the flame mode and requires an extrac-

---

*Approved by Standard Methods Committee, 1985.

tion procedure for the low concentrations common in potable water; the electrothermal atomic absorption method is much more sensitive for low concentrations and does not require extraction. The inductively coupled plasma method has a sensitivity similar to that of the flame atomic absorption method. The dithizone method is sensitive and specific as a colorimetric procedure.

3500-Pb B. Atomic Absorption Spectrometric Method

See flame atomic absorption spectrometric method, Sections 3111B and C, and electrothermal atomic absorption spectrometric method, Section 3113.

3500-Pb C. Inductively Coupled Plasma Method

See Section 3120.

3500-Pb D. Dithizone Method

1. General Discussion

a. *Principle:* An acidified sample containing microgram quantities of lead is mixed with ammoniacal citrate-cyanide reducing solution and extracted with dithizone in chloroform ($CHCl_3$) to form a cherry-red lead dithizonate. The color of the mixed color solution is measured photometrically.[1,2] Sample volume taken for analysis may be 2 L when digestion is used.

b. *Interference:* In a weakly ammoniacal cyanide solution (pH 8.5 to 9.5) dithizone forms colored complexes with bismuth, stannous tin, and monovalent thallium. In strongly ammoniacal citrate-cyanide solution (pH 10 to 11.5) the dithizonates of these ions are unstable and are extracted only partially.[3] This method uses a high pH, mixed color, single dithizone extraction. Interference from stannous tin and monovalent thallium is reduced further when these ions are oxidized during preliminary digestion. A modification of the method allows detection and elimination of bismuth interference. Excessive quantities of bismuth, thallium, and tin may be removed.[4]

Dithizone in $CHCl_3$ absorbs at 510 nm; control its interference by using nearly equal concentrations of excess dithizone in samples, standards, and blank.

The method is without interference for the determination of 0.0 to 30.0 μg Pb in the presence of 20 μg Tl^+, 100 μg Sn^{2+}, 200 μg In^{3+}, and 1000 μg each of Ba^{2+}, Cd^{2+}, Co^{2+}, Cu^{2+}, Mg^{2+}, Mn^{2+}, Hg^{2+}, Sr^{2+}, Zn^{2+}, Al^{3+}, Sb^{3+}, As^{3+}, Cr^{3+}, Fe^{3+}, V^{3+}, PO_4^{3-}, and SO_4^{2-}. Gram quantities of alkali metals do not interfere. A modification is provided to avoid interference from excessive quantities of bismuth or tin.

c. *Preliminary sample treatment:* At time of collection acidify with conc HNO_3 to pH < 2 but avoid excess HNO_3. Add 5 mL 0.1N iodine solution to avoid losses of volatile organo-lead compounds during handling and digesting of samples. Prepare a blank of lead-free distilled water and carry through the procedure.

d. Digestion of samples: Unless digestion is shown to be unnecessary, digest all samples for dissolved or total lead as described in 3030G or H.

e. Minimum detectable concentration: 1.0 µg Pb/10 mL dithizone solution.

2. Apparatus

a. Spectrophotometer for use at 510 nm, providing a light path of 1 cm or longer.

b. pH meter.

c. Separatory funnels: 250-mL Squibb type. Clean all glassware, including sample bottles, with 1 + 1 HNO_3. Rinse thoroughly with distilled or deionized water.

d. Automatic dispensing burets: Use for all reagents to minimize indeterminate contamination errors.

3. Reagents

Prepare all reagents in lead-free distilled water.

a. Stock lead solution: Dissolve 0.1599 g lead nitrate, $Pb(NO_3)_2$ (minimum purity 99.5%), in approximately 200 mL water. Add 10 mL conc HNO_3 and dilute to 1000 mL with water. Alternatively, dissolve 0.1000 g pure Pb metal in 20 mL 1 + 1 HNO_3 and dilute to 1000 mL with water; 1.00 mL = 100 µg Pb.

b. Working lead solution: Dilute 20.0 mL stock solution to 1000 mL with water; 1 mL = 2.00 µg Pb.

c. Nitric acid, HNO_3, 1 + 4: Dilute 200 mL conc HNO_3 to 1 L with water.

d. Ammonium hydroxide, NH_4OH, 1 + 9: Dilute 10 mL conc NH_4OH to 100 mL with water.

e. Citrate-cyanide reducing solution: Dissolve 400 g dibasic ammonium citrate, $(NH_4)_2HC_6H_5O_7$, 20 g anhydrous sodium sulfite, Na_2SO_3, 10 g hydroxylamine hydrochloride, $NH_2OH \cdot HCl$, and 40 g potassium cyanide, KCN (CAUTION: *Poison*) in water and dilute to 1 L. Mix this solution with 2 L conc NH_4OH. *Do not pipet by mouth.*

f. Stock dithizone solution: See 1070D.2*b*1), stock dithizone solution I.

g. Dithizone working solution: Dilute 100 mL stock dithizone solution to 250 mL with $CHCl_3$; 1 mL = 40 µg dithizone.

h. Special dithizone solution: Dissolve 250 mg dithizone in 250 mL $CHCl_3$. This solution may be prepared without purification because all extracts using it are discarded.

i. Sodium sulfite solution: Dissolve 5 g anhydrous Na_2SO_3 in 100 mL water.

j. Iodine solution: Dissolve 40 g KI in 25 mL water, add 12.7 g resublimed iodine, and dilute to 1000 mL.

4. Procedure

a. With sample digestion: To a digested sample containing not more than 1 mL conc acid add 20 mL 1 + 4 HNO_3 and filter through lead-free filter paper* and filter funnel directly into a 250-mL separatory funnel. Rinse digestion beaker with 50 mL water and add to filter. Add 50 mL ammoniacal citrate-cyanide solution, mix, and cool to room temperature. Add 10 mL dithizone working solution, shake stoppered funnel vigorously for 30 s, and let layers separate. Insert lead-free cotton in stem of separatory funnel and draw off lower layer. Discard 1 to 2 mL $CHCl_3$ layer, then fill absorption cell. Measure absorbance of extract at 510 nm, using dithizone working solution, ¶ 3g, to zero spectrophotometer.

b. Without sample digestion: To 100 mL acidified sample (pH 2) in a 250-mL separatory funnel add 20 mL 1 + 4 HNO_3 and 50 mL citrate-cyanide reducing solution; mix. Add 10 mL dithizone working solution and proceed as in ¶ 4a.

c. Calibration curve: Plot concentration of at least five standards and a blank against absorbance. Determine concentration of

---

*Whatman No. 541 or equivalent.

lead in extract from curve. All concentrations are μg Pb/10 mL final extract.

d. *Removal of excess interferences:* The dithizonates of bismuth, tin, and thallium differ from lead dithizonate in maximum absorbance. Detect their presence by measuring sample absorbance at 510 nm and at 465 nm. Calculate corrected absorbance of sample at each wavelength by subtracting absorbance of blank at same wavelength. Calculate ratio of corrected absorbance at 510 nm to corrected absorbance at 465 nm. The ratio of corrected absorbances for lead dithizonate is 2.08 and for bismuth dithizonate is 1.07. If the ratio for the sample indicates interference, i.e., is markedly less than 2.08, proceed as follows with a new 100-mL sample: If the sample has not been digested, add 5 mL Na_2SO_3 solution to reduce iodine preservative. Adjust sample to pH 2.5 using a pH meter and $1 + 4$ HNO_3 or $1 + 9$ NH_4OH as required. Transfer sample to 250-mL separatory funnel, extract with a minimum of three 10-mL portions special dithizone solution, or until the $CHCl_3$ layer is distinctly green. Extract with 20-mL portions $CHCl_3$ to remove dithizone (absence of green). Add 20 mL $1 + 4$ HNO_3, 50 mL citrate-cyanide reducing solution, and 10 mL dithizone

working solution. Extract as in ¶ *4a* and measure absorbance.

5. Calculation

mg Pb/L

$$= \frac{\mu\text{g Pb (in 10 mL, from calibration curve)}}{\text{mL sample}}$$

6. Precision and Bias

Single-operator precision in recovering 0.0104 mg Pb/L from Mississippi River water was 6.8% relative standard deviation and -1.4% relative error. At the level of 0.026 mg Pb/L, recovery was made with 4.8% relative standard deviation and 15% relative error.

7. References

1. SNYDER, L.J. 1947. Improved dithizone method for determination of lead—mixed color method at high pH. *Anal. Chem.* 19:684.
2. SANDELL, E.B. 1959. Colorimetric Determination of Traces of Metals, 3rd ed. Interscience, New York, N.Y.
3. WICHMANN, H.J. 1939. Isolation and determination of trace metals—the dithizone system. *Ind. Eng. Chem.*, Anal. Ed. 11:66.
4. AMERICAN SOCIETY FOR TESTING AND MATERIALS. 1977. Annual Book of ASTM Standards. Part 26, Method D3112-77, American Soc. Testing & Materials, Philadelphia, Pa.

3500-Li LITHIUM*

3500-Li A. Introduction

1. Occurrence

A minor constituent of minerals, lithium

is present in fresh waters in concentrations below 0.2 mg/L. Brines and thermal waters may contain higher lithium levels. The use of lithium or its salts in dehumidifying

---

*Approved by Standard Methods Committee, 1988.

units, medicinal waters, metallurgical processes, and the manufacture of some types of glass and storage batteries may contribute to its presence in wastes. Lithium hypochlorite is available commercially as a source of chlorine and may be used in swimming pools.

2. Selection of Method

The atomic absorption spectrometric method and the inductively coupled plasma method are preferred. The flame emission photometric method is also available for laboratories not equipped to use preferred methods.

3500-Li B.   Atomic Absorption Spectrometric Method

See flame atomic absorption spectrometric method, Section 3111B.

3500-Li C.   Inductively Coupled Plasma Method

See Section 3120.

3500-Li D.   Flame Emission Photometric Method

1. General Discussion

a. *Principle:* Like the other low-atomic-weight alkali metals, sodium and potassium, lithium can be determined in trace amounts by flame photometric methods. The measurement can be made at a wavelength of 670.8 nm.

b. *Interference:* Barium, strontium, and calcium interfere in the flame photometric determination of lithium and can be removed by adding a sodium sulfate-sodium carbonate (Na_2SO_4-Na_2CO_3) solution that precipitates barium sulfate ($BaSO_4$), strontium carbonate ($SrCO_3$), and calcium carbonate ($CaCO_3$). The content of magnesium must not exceed 10 mg in the portion taken for analysis to avoid coprecipitation interference.

c. *Minimum detectable concentration:* The minimum lithium concentration detectable is approximately 0.4 µg/L for re-

agent water analyzed on an atomic absorption spectrophotometer in the emission mode.

d. *Sampling and storage:* Collect sample in a borosilicate glass or polyethylene bottle. At time of collection adjust sample to pH < 2 with nitric acid (HNO_3).

2. Apparatus

Flame photometer: An instrument demonstrated by the analyst to be suitable for the Li concentrations to be measured (currently available flame photometers usually are designed for clinical medical purposes) or an atomic absorption spectrometer operating in the emission mode using a lean air-acetylene flame.

3. Reagents

a. *Sodium sulfate and sodium carbonate reagent:* Dissolve 5 g Na_2SO_4 and 10 g

Na_2CO_3 in distilled water and dilute to 1 L.

b. Stock lithium solution: Dissolve 152.7 mg anhydrous lithium chloride, LiCl, in distilled water and dilute to 250 mL; 1.00 mL = 100 μg Li. Dry salt overnight in an oven at 105°C. Cool in a desiccator and weigh immediately after removal from desiccator.

c. Standard lithium solution: Dilute 10.00 mL stock LiCl solution to 500 mL with distilled water; 1.00 mL = 2.0 μg Li.

4. Procedure

a. Pretreatment of polluted water and wastewater samples: Use nitric acid digestion or nitric acid-perchloric acid-hydrofluoric acid digestion, as directed in Section 3030.

b. Removal of interference by barium, calcium, and strontium: Take a sample of 50.0 mL if required or less (for higher concentrations), containing not more than 10 mg Mg (to prevent competition for coprecipitation). Add 5.0 mL Na_2SO_4-Na_2CO_3 reagent. Bring to a boil to coagulate precipitate of $BaSO_4$, $SrCO_3$, $CaCO_3$, and $MgCO_3$. Continue heating to reduce volume by one third. Remove from heat and let stand for at least 30 min to complete precipitation; otherwise a feathery precipitate of $BaSO_4$ will appear after filtration. Filter,* wash with distilled water, and dilute to 50.0 mL after solution has reached room temperature for the flame photometric measurement.

c. Treatment of standard solutions: Prepare a standard of 2 μg Li/55 mL by diluting 1 mL standard lithium solution to 50.00 mL, adding 5 mL Na_2SO_4-Na_2CO_3 reagent, and mixing. Similarly, prepare dilutions of the Li standard solution to bracket sample concentration or to establish at least three points on a calibration curve of absorbance against μg Li/55 mL.

---

*Whatman No. 42 or equivalent.

d. Flame photometric measurement: Determine lithium concentration by direct intensity measurements at a wavelength of 670.8 nm. (The bracketing method can be used with some photometric instruments, while the construction of a calibration curve is necessary with others.) Run sample, distilled water, and lithium standard as nearly simultaneously as possible. For best results, average several readings on each solution.

Follow the manufacturer's instructions for instrument operation.

5. Calculation

$$\text{mg Li/L} = \frac{\mu\text{g Li/55 mL}}{\text{mL sample}} \times 0.9$$

Actual final volume is 50 mL; calibration curve assumed based on 55 mL final volume as in ¶ 4c, accounting for the 0.9 factor.

6. Quality Control

Process a QC standard through entire analytical protocol as a way of determining systematic bias. The control limits for precision of duplicate determinations at concentrations (in reagent water) of 4.0 μg/L and 10.0 μg/L were 4.09 ± 0.056 μg/L and 9.96 ± 0.094 μg/L, respectively. The single-operator RSD was 1.38% for a lithium solution containing 10 μg/L.

7. Bibliography

KUEMMEL, D.F. & H.L. KARL. 1954. Flame photometric determination of alkali and alkaline earth elements in cast iron. *Anal. Chem.* 26:386.

BRUMBAUGH, R.J. & W.E. FANUS. 1954. Determination of lithium in spodumene by flame photometry. *Anal. Chem.* 26:463.

ELLESTAD, R.B. & E.L. HORSTMAN. 1955. Flame photometric determination of lithium in silicate rocks. *Anal. Chem.* 27:1229.

WHISMAN, M. & B.H. ECCLESTON. 1955. Flame spectra of twenty metals using a recording flame spectrophotometer. *Anal. Chem.* 27:1861.

HORSTMAN, E.L. 1956. Flame photometric determination of lithium, rubidium, and cesium in
silicate rocks. *Anal. Chem.* 28:1417.

URE, A.M. & R.L. MITCHELL. 1975. Lithium,
sodium, potassium, rubidium, and cesium. *In*
J.A. Dean & T.C. Rains, eds. Flame Emission

and Atomic Absorption Spectrometry. Dekker, New York, N.Y.

PICKETT, E.E. & J.L. HAWKINS. 1981. Determination of lithium in small-animal tissues at
physiological levels by flame emission photometry. *Anal. Biochem.* 112:213.

3500-Mg   MAGNESIUM*

3500-Mg A.   Introduction

1. Occurrence

Magnesium ranks eighth among the elements in order of abundance and is a common constituent of natural water.
Important contributors to the hardness of
a water, magnesium salts break down when
heated, forming scale in boilers. Concentrations greater than 125 mg/L also can
have a cathartic and diuretic effect. Chemical softening, reverse osmosis, electrodialysis, or ion exchange reduces the
magnesium and associated hardness to acceptable levels. The magnesium concentra-

tion may vary from zero to several hundred
milligrams per liter, depending on the
source and treatment of the water.

2. Selection of Method

The four methods presented are applicable to all natural waters. Direct determinations can be made with the atomic
absorption spectrometric and inductively
coupled plasma methods. Magnesium can
be determined by the gravimetric method
only after removal of calcium salts (see Section 3500-Ca). These methods can be applied to all concentrations by selection of
suitable sample portions. Choice of method
is largely a matter of personal preference
and analyst experience.

*Approved by Standard Methods Committee, 1985.

3500-Mg B.   Atomic Absorption Spectrometric Method

See flame atomic absorption spectrometric method, Section 3111B.

3500-Mg C.   Inductively Coupled Plasma Method

See Section 3120.

3500-Mg D.   Gravimetric Method

1. General Discussion

a. *Principle:* Diammonium hydrogen phosphate quantitatively precipitates magnesium in ammoniacal solution as magnesium ammonium phosphate. The precipitate is ignited to, and weighed as, magnesium pyrophosphate. A choice is presented between: (*a*) destruction of ammonium salts and oxalate, followed by single precipitation of magnesium ammonium phosphate; and (*b*) double precipitation without pretreatment. Where time is not a factor, double precipitation is preferable because, while pretreatment is faster, it requires close attention to avoid mechanical loss.

b. *Interference:* The solution should be reasonably free from aluminum, calcium, iron, manganese, silica, strontium, and suspended matter. It should not contain more than about 3.5 g NH_4Cl.

2. Reagents

a. *Nitric acid,* HNO_3, conc.

b. *Hydrochloric acid,* HCl, conc; also 1 + 1, 1 + 9, and 1 + 99.

c. *Methyl red indicator solution:* Dissolve 100 mg methyl red sodium salt in distilled water and dilute to 100 mL.

d. *Diammonium hydrogen phosphate solution:* In distilled water, dissolve 30 g $(NH_4)_2HPO_4$ and make up to 100 mL.

e. *Ammonium hydroxide,* NH_4OH, conc; also 1 + 19.

3. Procedure

a. *By removal of oxalate and ammonium salts:* To the combined filtrate and washings from the calcium determination, containing not more than 60 mg Mg, or to a portion containing less than this amount in a 600- or 800-mL beaker, add 50 mL conc HNO_3 and evaporate carefully to dryness on a hot plate. Do not let reaction become too violent during the latter part of the evaporation; stay in constant attendance to avoid losses through spattering. Moisten residue with 2 to 3 mL conc HCl; add 20 mL distilled water, warm, filter, and wash. To the filtrate add 3 mL conc HCl, 2 to 3 drops methyl red solution, and 10 mL $(NH_4)_2HPO_4$ solution. Cool and add conc NH_4OH, drop by drop, stirring constantly, until the color changes to yellow. Stir for 5 min, add 5 mL conc NH_4OH, and stir vigorously for 10 min more. Let stand overnight and filter through filter paper.* Wash with 1 + 19 NH_4OH. Transfer to an ignited, cooled, and weighed crucible. Dry precipitate thoroughly and burn paper off *slowly,* allowing circulation of air. Heat at about 500°C until residue is white. Ignite for 30-min periods at 1100°C to constant weight.

b. *By double precipitation:* To the combined filtrate and washings from the calcium determination, containing not more than 60 mg Mg, or to a portion containing less than this amount, add 2 to 3 drops methyl red solution; adjust volume to 150 mL and acidify with 1 + 1 HCl. Add 10 mL $(NH_4)_2HPO_4$ solution. Cool. Add conc NH_4OH, drop by drop, stirring constantly, until the color changes to yellow. Stir for 5 min, add 5 mL conc NH_4OH, and stir vigorously for 10 min more. Let stand overnight and then filter through filter paper.* Wash with 1 + 19 NH_4OH. Discard filtrate and washings. Dissolve precipitate with 50 mL warm 1 + 9 HCl and wash paper well with hot 1 + 99 HCl. Add 2 to 3 drops methyl red solution, adjust volume to 100 to 150 mL, add 1 to 2 mL $(NH_4)_2HPO_4$ solution, and precipitate as before. Let stand in a cool place for at least 4 h or preferably overnight. Filter through filter paper* and wash with 1 + 19 NH_4OH. Transfer to an ignited, cooled,

---

*Carl Schleicher and Schuell Co., S & S No. 589 White Ribbon, or equivalent.

and weighed crucible. Dry precipitate thoroughly and burn paper off *slowly*, allowing circulation of air. Heat at about 500°C until residue is white. Ignite for 30-min periods at 1100°C to constant weight.

4. Calculation

$$mg\ Mg/L = \frac{mg\ Mg_2P_2O_7 \times 218.4}{mL\ sample}$$

5. Precision and Bias

A synthetic sample containing 82 mg Mg/L, 108 mg Ca/L, 3.1 mg K/L, 19.9 mg Na/L, 241 mg Cl$^-$/L, 1.1 mg NO$_3^-$-N/L, 0.250 mg NO$_2^-$-N/L, 259 mg SO$_4^{2-}$/L, and 42.5 mg total alkalinity/L (contributed by NaHCO$_3$) was analyzed in eight laboratories by the gravimetric method, with a relative standard deviation of 6.3% and a relative error of 4.9%.

6. Bibliography

EPPERSON, A.W. 1928. The pyrophosphate method for the determination of magnesium and phosphoric anhydride. *J. Amer. Chem. Soc.* 50:321.

KOLTHOFF, I.M. & E.B. SANDELL. 1952. Textbook of Quantitative Inorganic Analysis, 3rd ed. Macmillan Co., New York, N.Y., Chapter 22.

HILLEBRAND, W.F. et al. 1953. Applied Inorganic Analysis, 2nd ed. John Wiley & Sons, New York, Chapter 41 and pp. 133-134.

3500-Mg E.　Calculation Method

Magnesium may be estimated as the difference between hardness and calcium as CaCO$_3$ if interfering metals are present in noninterfering concentrations in the calcium titration (Section 3500-Ca.D) and suitable inhibitors are used in the hardness titration (Section 2340C).

mg Mg/L = [total hardness (as mg CaCO$_3$/L) − calcium hardness (as mg CaCO$_3$/L)] × 0.243

3500-Mn　MANGANESE*

3500-Mn A.　Introduction

1. Occurrence and Significance

Although manganese in groundwater generally is present in the soluble divalent ionic form because of the absence of oxygen, part or all of the manganese in a water treatment plant may be in a higher valence state. Determination of total manganese does not differentiate the various valence states. The heptavalent permanganate ion is used to oxidize manganese and/or organic matter causing taste. Excess permanganate, complexed trivalent manganese, or a suspension of quadrivalent manganese must be detected with great sensitivity to control treatment and to prevent their discharge into a distribution system. There is evidence that manganese occurs in surface waters both in suspension in the quadri-

---

*Approved by Standard Methods Committee, 1988.

valent state and in the trivalent state in a relatively stable, soluble complex. Although rarely present in excess of 1 mg/L, manganese imparts objectionable and tenacious stains to laundry and plumbing fixtures. The low manganese limits imposed on an acceptable water stem from these, rather than toxicological, considerations. Special means of removal often are necessary, such as chemical precipitation, pH adjustment, aeration, and use of special ion-exchange materials. Manganese occurs in domestic wastewater, industrial effluents, and receiving streams.

2. Selection of Method

The atomic absorption spectrometric and the inductively coupled plasma methods permit direct determination with acceptable sensitivity and are the methods of choice. Of the various colorimetric methods, the persulfate method is preferred because the use of mercuric ion can control interference from a limited chloride ion concentration.

3. Sampling and Storage

Manganese may exist in a soluble form in a neutral water when first collected, but it oxidizes to a higher oxidation state and precipitates or becomes adsorbed on the container walls. Determine manganese very soon after sample collection. When delay is unavoidable, total manganese can be determined if the sample is acidified at the time of collection with HNO_3 to pH < 2. See Section 3010B.

3500-Mn B.   Atomic Absorption Spectrometric Method

See flame atomic absorption spectrometric method, Sections 3111B and C, and electrothermal atomic absorption spectrometric method, Section 3113.

3500-Mn C.   Inductively Coupled Plasma Method

See Section 3120.

3500-Mn D.   Persulfate Method

1. General Discussion

a. Principle: Persulfate oxidation of soluble manganous compounds to form permanganate is carried out in the presence of silver nitrate. The resulting color is stable for at least 24 h if excess persulfate is present and organic matter is absent.

b. Interference: As much as 0.1 g chloride (Cl^-) in a 50-mL sample can be prevented from interfering by adding 1 g mercuric sulfate ($HgSO_4$) to form slightly dissociated complexes. Bromide and iodide still will interfere and only trace amounts may be present. The persulfate procedure can be used for potable water with trace to small amounts of organic matter if the period of heating is increased after more persulfate has been added.

For wastewaters containing organic matter, use preliminary digestion with nitric and sulfuric acids (HNO_3 and H_2SO_4) (see Section 3030G). If large amounts of Cl^- also are present, boiling with HNO_3 helps remove it. Interfering traces of Cl^- are eliminated by $HgSO_4$ in the special reagent.

Colored solutions from other inorganic ions are compensated for in the final colorimetric step.

Samples that have been exposed to air may give low results due to precipitation of manganese dioxide (MnO_2). Add 1 drop 30% hydrogen peroxide (H_2O_2) to the sample, after adding the special reagent, to redissolve precipitated manganese.

c. Minimum detectable concentration: The molar absorptivity of permanganate ion is about 2300 L g^{-1} cm^{-1}. This corresponds to a minimum detectable concentration (98% transmittance) of 210 μg Mn/L when a 1-cm cell is used or 42 μg Mn/L when a 5-cm cell is used.

2. Apparatus

Colorimetric equipment: One of the following is required:

a. Spectrophotometer, for use at 525 nm, providing a light path of 1 cm or longer.

b. Filter photometer, providing a light path of 1 cm or longer and equipped with a green filter having maximum transmittance near 525 nm.

c. Nessler tubes, matched, 100-mL, tall form.

3. Reagents

a. Special reagent: Dissolve 75 g $HgSO_4$ in 400 mL conc HNO_3 and 200 mL distilled water. Add 200 mL 85% phosphoric acid (H_3PO_4), and 35 mg silver nitrate ($AgNO_3$). Dilute the cooled solution to 1 L.

b. Ammonium persulfate, $(NH_4)_2S_2O_8$, solid.

c. Standard manganese solution: Prepare a $0.1N$ potassium permanganate ($KMnO_4$) solution by dissolving 3.2 g $KMnO_4$ in distilled water and making up to 1 L. Age for several weeks in sunlight or heat for several hours near the boiling point, then filter through a fine fritted-glass filter crucible and standardize against sodium oxalate as follows:

Weigh several 100- to 200-mg samples of $Na_2C_2O_4$ to 0.1 mg and transfer to 400-mL beakers. To each beaker, add 100 mL distilled water and stir to dissolve. Add 10 mL 1 + 1 H_2SO_4 and heat rapidly to 90 to 95°C. Titrate rapidly with the $KMnO_4$ solution to be standardized, while stirring, to a slight pink end-point color that persists for at least 1 min. Do not let temperature fall below 85°C. If necessary, warm beaker contents during titration; 100 mg $Na_2C_2O_4$ will consume about 15 mL permanganate solution. Run a blank on distilled water and H_2SO_4.

$$\text{Normality of } KMnO_4 = \frac{g\ Na_2C_2O_4}{(A - B) \times 0.067\ 01}$$

where:

A = mL titrant for sample and
B = mL titrant for blank.

Average results of several titrations. Calculate volume of this solution necessary to prepare 1 L of solution so that 1.00 mL = 50.0 μg Mn, as follows:

$$\text{mL } KMnO_4 = \frac{4.55}{\text{normality } KMnO_4}$$

To this volume add 2 to 3 mL conc H_2SO_4 and $NaHSO_3$ solution dropwise, with stirring, until the permanganate color disappears. Boil to remove excess SO_2, cool, and dilute to 1000 mL with distilled water. Dilute this solution further to measure small amounts of manganese.

d. Standard manganese solution (alternate): Dissolve 1.000 g manganese metal (99.8% min.) in 10 mL redistilled HNO_3. Dilute to 1000 mL with 1% (v/v) HCl; 1 mL = 1.000 mg Mn. Dilute 10 mL to 200 mL with distilled water; 1 mL = 0.05 mg Mn. Prepare dilute solution daily.

e. Hydrogen peroxide, H_2O_2, 30%.

f. Nitric acid, HNO_3, conc.

g. Sulfuric acid, H_2SO_4, conc.

h. Sodium nitrite solution: Dissolve 5.0 g $NaNO_2$ in 95 mL distilled water.

i. Sodium oxalate, $Na_2C_2O_4$, primary standard.

j. Sodium bisulfite: Dissolve 10 g $NaHSO_3$ in 100 mL distilled water.

4. Procedure

a. Treatment of sample: If a digested sample has been prepared according to directions for reducing organic matter and/or excessive chlorides in Section 3030G, pipet a portion containing 0.05 to 2.0 mg Mn into a 250-mL conical flask. Add distilled water, if necessary, to 90 mL and proceed as in ¶ *b*.

b. To a suitable sample portion add 5 mL special reagent and 1 drop H_2O_2. Concentrate to 90 mL by boiling or dilute to 90 mL. Add 1 g $(NH_4)_2S_2O_8$, bring to a boil, and boil for 1 min. Do not heat on a water bath. Remove from heat source, let stand 1 min, then cool under the tap. (Boiling too long results in decomposition of excess persulfate and subsequent loss of permanganate color; cooling too slowly has the same effect.) Dilute to 100 mL with distilled water free from reducing substances and mix. Prepare standards containing 0, 5.00, . . . 1500 μg Mn by treating various amounts of standard Mn solution in the same way.

c. Nessler tube comparison: Use standards prepared as in ¶ 4*b* and containing 5 to 100 μg Mn/100 mL final volume. Compare samples and standards visually.

d. Photometric determination: Use a series of standards from 0 to 1500 μg Mn/ 100 mL final volume. Make photometric measurements against a distilled water blank. The following table shows light path length appropriate for various amounts of manganese in 100 mL final volume:

Mn Range μg	Light Path cm
5–200	15
20–400	5
50–1000	2
100–1500	1

Prepare a calibration curve of manganese concentration vs. absorbance from the standards and determine Mn in the samples from the curve. If turbidity or interfering color is present, make corrections as in ¶ 4*e*.

e. Correction for turbidity or interfering color: Avoid filtration because of possible retention of some permanganate on the filter paper. If visual comparison is used, the effect of turbidity only can be estimated and no correction can be made for interfering colored ions. When photometric measurements are made, use the following "bleaching" method, which also corrects for interfering color: As soon as the photometer reading has been made, add 0.05 mL H_2O_2 solution directly to the sample in the optical cell. Mix and, as soon as the permanganate color has faded completely and no bubbles remain, read again. Deduct absorbance of bleached solution from initial absorbance to obtain absorbance due to Mn.

5. Calculation

a. When all of the original sample is taken for analysis:

$$mg\ Mn/L = \frac{\mu g\ Mn\ (in\ 100\ mL\ final\ volume)}{mL\ sample}$$

b. When a portion of the digested sample (100 mL final volume) is taken for analysis:

$$\text{mg Mn/L} = \frac{\mu\text{g Mn/100 mL}}{\text{mL sample}} \times \frac{100}{\text{mL portion}}$$

6. Precision and Bias

A synthetic sample containing 120 μg Mn/L, 500 μg Al/L, 50 μg Cd/L, 110 μg Cr/L, 470 μg Cu/L, 300 μg Fe/L, 70 μg Pb/L, 150 μg Ag/L, and 650 μg Zn/L in distilled water was analyzed in 33 laboratories by the persulfate method, with a relative standard deviation of 26.3% and a relative error of 0%.

A second synthetic sample, similar in all respects except for 50 μg Mn/L and 1000 μg Cu/L, was analyzed in 17 laboratories by the persulfate method, with a relative standard deviation of 50.3% and a relative error of 7.2%.

7. Bibliography

RICHARDS, M.D. 1930. Colorimetric determination of manganese in biological material. *Analyst* 55:554.

NYDAHL, F. 1949. Determination of manganese by the persulfate method. *Anal. Chem. Acta.* 3:144.

MILLS, S.M. 1950. Elusive manganese. *Water Sewage Works* 97:92.

SANDELL, E.B. 1959. Colorimetric Determination of Traces of Metals, 3rd ed. Interscience Publishers, New York, N.Y., Chapter 26.

3500-Hg   MERCURY*

3500-Hg A.   Introduction

1. Significance

Organic and inorganic mercury salts are very toxic and their presence in the environment, especially in water, should be monitored.

2. Selection of Method

The cold vapor atomic absorption method is the method of choice for all samples, while the dithizone method can be used for determining high levels of mercury (> 2 μg/L) in potable waters.

3. Sample Preservation

Because mercury can be lost readily from samples, preserve them by treating with HNO_3 to reduce the pH to < 2 (see Section 1060).

*Approved by Standard Methods Committee, 1985.

3500-Hg B.   Cold Vapor Atomic Absorption Method

See Section 3112.

3500-Hg C.   Dithizone Method

1. General Discussion

a. Principle: Mercury ions react with a dithizone solution in chloroform to form an orange color. The various shades of orange are measured in a spectrophotometer and unknown concentrations are calculated from a standard curve.

b. Interference: Copper, gold, palladium, divalent platinum, and silver react with dithizone in acid solution. Copper in the dithizone extract remains in the organic phase while the mercury dissolves in the aqueous phase. The other contaminants usually are not present.

The mercury dithizonate must be measured quickly because it is photosensitive.

c. Minimum detectable concentration: 1 μg Hg/10 mL final volume, corresponding to 2 μg Hg/L when a 500-mL sample is used. Acceptable precision is obtained when this concentration is exceeded.

2. Apparatus

a. Spectrophotometer, for measurements at 492 nm, providing a light path of 1 cm or longer.

b. Separatory funnels: 250- and 1000-mL, with TFE stopcocks.

c. Glassware: Clean all glassware with potassium dichromate-sulfuric acid cleaning solution.

3. Reagents

a. Mercury-free water: Use redistilled or deionized distilled water for preparing all reagents and dilutions.

b. Stock mercury solution: Dissolve 135.3 mg mercuric chloride, $HgCl_2$, in about 700 mL water, add 1.5 mL conc HNO_3, and make up to 1000 mL with water; 1.00 mL = 100 μg Hg.

c. Standard mercury solution: Dilute 10.00 mL stock solution to 1000 mL with water; 1.00 mL = 1.00 μg Hg. Prepare immediately before use.

d. Potassium permanganate solution: Dissolve 5 g $KMnO_4$ in 100 mL water.

e. Sulfuric acid, H_2SO_4, conc, low in mercury.

f. Potassium persulfate solution. Dissolve 5 g $K_2S_2O_8$ in 100 mL water.

g. Hydroxylamine hydrochloride solution: Dissolve 50 g $NH_2OH \cdot HCl$ in 100 mL water.

h. Dithizone solution: See 1070D.2*b*1). Dilute 60 mL stock dithizone solution I with $CHCl_3$ to 1000 mL; 1 mL = 6 μg dithizone.

i. Sulfuric acid, 0.25N: Dilute 250 mL 1*N* H_2SO_4 to 1 L with water.

j. Potassium bromide solution: Dissolve 40 g KBr in 100 mL water.

k. Chloroform, $CHCl_3$.

l. Phosphate-carbonate buffer solution: Dissolve 150 g $Na_2HPO_4 \cdot 12H_2O$ and 38 g anhydrous K_2CO_3 in 1 L water. Extract with 10-mL portions of dithizone until the last portion remains blue. Wash with $CHCl_3$ to remove excess dithizone.

m. Sodium sulfate, Na_2SO_4, anhydrous.

4. Procedure

a. Preparation of calibration curve: Pipet 0 (blank), 2.00, 4.00, 6.00, 8.00, and 10.00 μg mercury into separate beakers. To each beaker, add 500 mL water (or any other volume selected for sample), 1 mL $KMnO_4$ solution, and 10 mL conc H_2SO_4. Stir and bring to a boil. If necessary, add more $KMnO_4$ until a pink color persists. After boiling has ceased, cautiously add 5 mL $K_2S_2O_8$ solution and let cool for 0.5 h. Add one or more drops $NH_2OH \cdot HCl$ solution to discharge the pink color. When cool, transfer each solution to individual 1-L separatory funnels. Add about 25 mL dithizone solution. Shake separatory funnel vigorously and transfer each organic layer to a 250-mL separatory funnel. Repeat this extraction at least three times, making sure that the color in the last dithizone layer is as intense a blue as that of the original dithizone solution. Wash accumulated dithizone extracts in the 250-mL separatory funnel by shaking with 50 mL 0.25N H_2SO_4. Transfer washed dithizonate extract to another 250-mL separatory funnel. Add 50 mL 0.25N H_2SO_4 and 10 mL KBr solution and shake vigorously to transfer mercury dithizonate from organic layer to aqueous layer. Discard lower dithizone layer. Wash aqueous layer with a small volume of $CHCl_3$ and discard the $CHCl_3$. Transfer 20 mL phosphate-carbonate buffer solution to each separatory funnel and add 10 mL standard dithizone solution. Shake thoroughly, and after separation, transfer the mercury dithizone to beakers. The final dithizone extract should be slightly blue. Dry contents with anhydrous Na_2SO_4. Transfer mercury dithizonate solution to a cuvette and record absorbance at 492 nm. On linear graph paper, plot absorbance against micrograms mercury in 10 mL final volume.

b. Treatment of samples: Samples containing 1.5 mL conc HNO_3/L usually do not affect dithizone, although strong solutions of HNO_3 will oxidize it. Use a 500-mL sample to increase absorbance readings, and prepare an absorbance blank consisting of all reagents. When necessary, filter sample through glass wool into the separatory funnel after oxidation step. Complete procedure as described under ¶ 4*a* above. Read mercury content from calibration curve.

5. Calculation

$$\mu g\ Hg/L = \frac{\mu g\ Hg\ (\text{in }10\ mL\ \text{final volume})}{L\ \text{sample}}$$

6. Precision and Bias

Five portions of inorganic mercury and five portions of organic mercury as methyl mercuric chloride at concentrations of 250 $\mu g/L$ each yielded a 95% recovery. Two of the ten samples were mixed with bayou water.

7. Bibliography

SANDELL, E.B. 1959. Colorimetric Determination of Traces of Metals, 3rd ed. Interscience Publishers, New York, N.Y., pp. 637-638.
SULLIVAN, J.R. & J. J. DELFINO, 1982. The determination of mercury in fish. *J. Environ. Sci. Health.* A17:265.

3500-Mo MOLYBDENUM

3500-Mo A. Introduction

1. Occurrence

Molybdenum occurs at trace levels (< 10 $\mu g/L$) in natural waters. In waters draining mineralized areas or wastewaters from processes using molybdenum, concentrations may be much higher.

2. Selection of Method

Use the flame atomic absorption spectrometric method or the inductively coupled plasma method.

3500-Mo B.   Atomic Absorption Spectrometric Method

See flame atomic absorption spectrometric method, Section 3111D.

3500-Mo C.   Inductively Coupled Plasma Method

See Section 3120.

3500-Ni   NICKEL*

3500-Ni A.   Introduction

Selection of method: The atomic absorption spectrometric and inductively coupled plasma methods are the methods of choice for all samples. The heptoxime or dimethylglyoxime method can be used, with poorer precision and bias, if atomic absorption or ICP instrumentation is not available.

*Approved by Standard Methods Committee, 1988.

3500-Ni B.   Atomic Absorption Spectrometric Method

See flame atomic absorption spectrometric method, Sections 3111B and C, and electrothermal atomic absorption spectrometric method, Section 3113.

3500-Ni C.   Inductively Coupled Plasma Method

See Section 3120.

3500-Ni D.   Heptoxime Method (GENERAL)

1. General Discussion

Principle: After preliminary digestion with nitric acid-sulfuric acid (HNO_3-H_2SO_4) mixture, iron and copper are removed by extraction of the cupferrates with chloroform ($CHCl_3$). Nickel is separated from other ions by extraction of the nickel heptoxime complex with CHCl$_3$, re-extracted into the aqueous phase with hydrochloric acid (HCl), and determined colorimetrically in the acidic solution with heptoxime in the presence of an oxidant.

2. Apparatus

a. *Colorimetric equipment:* One of the following is required:

1) *Spectrophotometer,* for use at 445 nm, providing a light path of 1 cm or longer.

2) *Filter photometer,* providing a light path of 1 cm or longer and equipped with a violet filter with maximum transmittance near 445 nm.

b. *Separatory funnels,* 125-mL, Squibb form, with ground-glass stoppers.

3. Reagents

a. *Standard nickel sulfate solution:* Dissolve 447.9 mg NiSO$_4$·6H$_2$O in 1000 mL distilled water; 1.00 mL = 100 μg Ni.

b. *Hydrochloric acid,* HCl, 1.0N.

c. *Bromine water:* Saturate distilled water with bromine.

d. *Ammonium hydroxide,* NH$_4$OH, conc.

e. *Heptoxime reagent:* Dissolve 0.1 g 1,2-cycloheptanedionedioxime* (heptoxime) in 100 mL 95% ethyl alcohol.

f. *Ethyl alcohol,* 95%.

g. *Sodium tartrate solution:* Dissolve 10 g Na$_2$C$_4$H$_4$O$_6$·2H$_2$O in 90 mL distilled water.

h. *Methyl orange indicator solution.*

i. *Sodium hydroxide,* NaOH, 6N.

j. *Acetic acid,* conc.

k. *Cupferron solution:* Dissolve 1 g cupferron in 100 mL distilled water. Store in refrigerator or prepare fresh for each series of determinations.

l. *Chloroform,* CHCl$_3$.

m. *Hydroxylamine-hydrochloride solution:* Dissolve 10 g NH$_2$OH·HCl in 90 mL distilled water. Prepare fresh daily.

---

*Hach Chemical Company, Ames, Iowa, or equivalent.

4. Procedure

a. *Preparation of calibration curve:* Pipet portions of standard NiSO$_4$ solution into 100-mL volumetric flasks. Use a series from 50 to 250 μg Ni if 1-cm cells are used. Add 25 mL 1.0N HCl and 5 mL bromine water. Cool with cold running tap water and add 10 mL conc NH$_4$OH. Immediately add 20 mL heptoxime reagent and 20 mL ethyl alcohol. Dilute to volume with distilled water and mix.

Measure absorbance at 445 nm, 20 min after adding reagent, using a reagent blank as reference.

b. *Treatment of sample:*

1) Separation of copper and iron—Take a portion of original sample, prepared by digesting with HNO$_3$-H$_2$SO$_4$ mixture as directed in Section 3030G and containing from 50 to 250 μg Ni, place in a separatory funnel, and add 10 mL sodium tartrate solution, 2 drops (0.1 mL) methyl orange indicator, and enough 6N NaOH to make the solution basic.

Add 1 mL acetic acid and cool funnel under tap water. Add 4 mL cupferron reagent, add 10 mL CHCl$_3$, and shake. Let layers separate and if necessary add more cupferron until a white precipitate forms, indicating excess cupferron. Shake mixture again, let separate, and discard CHCl$_3$ layer. Reextract with 10 mL CHCl$_3$ and discard CHCl$_3$ layer. Add 1 mL fresh NH$_2$OH·HCl solution, mix, and let stand for 10 min.

2) Separation of nickel—Add 10 mL heptoxime reagent and extract nickel complex with at least three 10-mL portions of CHCl$_3$. Continue repetitive extractions until final CHCl$_3$ layer is colorless. Collect CHCl$_3$ layers in a separatory funnel and extract with 10 mL 1N HCl. Draw off CHCl$_3$ layer into another separatory funnel and reextract CHCl$_3$ with 10 mL 1N HCl. Combine HCl layers and determine absorbance as directed in ¶ 4a 10 min after

reagent addition. Plot absorbance against micrograms Ni in 20 mL final volume.

5. Calculation

$$mg\ Ni/L = \frac{\mu g\ Ni\ (in\ 20\ mL\ final\ volume)}{mL\ sample} \times \frac{100}{mL\ portion}$$

6. Bibliography

BUTTS, P.G., A.R. GAHLER & M.G. MELLON. 1950. Colorimetric determination of metals in sewage and industrial wastes. *Sewage Ind. Wastes* 22:1543.

FERGUSON, R.C. & C.V. BANKS. 1951. Spectro-photometric determination of nickel using 1,2-cycloheptanedionedioxime (heptoxime). *Anal. Chem.* 23:448, 1486.

SERFASS, E.J. & R.F. MURACA. 1954. Procedures for Analyzing Metal Finishing Wastes. Ohio River Valley Water Sanitation Comm., Cincinnati, Ohio.

3500-Ni E.   Dimethylglyoxime Method (GENERAL)

1. Discussion

Dimethylglyoxime may be used instead of heptoxime to develop the color with nickel. The conditions of color formation are identical, but prepare separate calibration curves. The rate of color development is slightly different for the two reagents; therefore, with dimethylglyoxime, make readings exactly 10 min after adding reagent, whereas, with heptoxime, make readings exactly 20 min after adding reagent. In both systems make measurements at 445 nm. The heptoxime system is more stable. Dimethylglyoxime cannot be substituted for heptoxime in the extraction process, Section 3500-Ni.D.4*b*2), under the conditions prescribed.

Calculate nickel concentration as in Section 3500-Ni.D.5.

2. Bibliography

AMERICAN SOCIETY FOR TESTING AND MATERIALS. 1986. Book of ASTM Standards, Part 11.01. Water. American Soc. Testing & Materials, Philadelphia, Pa.

3500-Os   OSMIUM

3500-Os A.   Introduction

Osmium is quite uncommon in natural waters and occurs only at trace levels. Process waters may contain higher concentrations.

3500-Os B.   Atomic Absorption Spectrometric Method

See flame atomic absorption spectrometric method, Section 3111D.

3500-Pd PALLADIUM

3500-Pd A. Introduction

Palladium is relatively uncommon in natural waters. Process waters may contain higher concentrations.

3500-Pd B. Atomic Absorption Spectrometric Method

See flame atomic absorption spectrometric method, Section 3111B.

3500-Pt PLATINUM

3500-Pt A. Introduction

Platinum is relatively uncommon in natural waters. Process waters may contain higher concentrations.

3500-Pt B. Atomic Absorption Spectrometric Method

See flame atomic absorption spectrometric method, Section 3111B.

3500-K POTASSIUM*

3500-K A. Introduction

1. Occurrence

Potassium ranks seventh among the elements in order of abundance, yet its concentration in most drinking waters seldom reaches 20 mg/L. However, occasional brines may contain more than 100 mg potassium/L.

2. Selection of Method

Three methods for the determination of potassium are given. They are rapid, sensitive, and accurate, and selection depends on instrument availability and analyst choice.

3. Storage of Sample

Do not store samples in soft-glass bottles because of the possibility of contamination from leaching of the glass. Use polyethylene or borosilicate glass bottles. Adjust sample to pH <2 with nitric acid. This will dissolve insoluble potassium and reduce adsorption on vessel walls.

---

*Approved by Standard Methods Committee, 1985.

3500-K B. Atomic Absorption Spectrometric Method

See flame atomic absorption spectrometric method, Section 3111B.

3500-K C. Inductively Coupled Plasma Method

See Section 3120.

3500-K D. Flame Photometric Method

1. General Discussion

a. Principle: Trace amounts of potassium can be determined in either a direct-reading or internal-standard type of flame photometer at a wavelength of 766.5 nm. Because much of the information pertaining to sodium applies equally to the potassium determination, carefully study the entire discussion dealing with the flame photometric determination of sodium (Section 3500-Na.D) before making a potassium determination.

b. Interference: Interference in the internal-standard method may occur at sodium-to-potassium ratios of 5:1 or greater. Calcium may interfere if the calcium-to-potassium ratio is 10:1 or more. Magnesium begins to interfere when the magnesium-to-potassium ratio exceeds 100:1.

c. Minimum detectable concentration: Potassium levels of approximately 0.1 mg/L can be determined.

2. Apparatus

See Section 3500-Na.D.2.

3. Reagents

To minimize potassium pickup, store all solutions in plastic bottles. Use small containers to reduce amount of dry element that may be picked up from bottle walls when the solution is poured. Shake each container thoroughly to wash accumulated salts from walls before pouring.

a. Deionized distilled water: Use this water for preparing all reagents and calibration standards, and as dilution water.

b. Stock potassium solution: Dissolve 1.907 g KCl dried at 110°C and dilute to 1000 mL with water; 1 mL = 1.00 mg K.

c. Intermediate potassium solution: Dilute 10.0 mL stock potassium solution with water to 100 mL; 1.00 mL = 0.100 mg K. Use this solution to prepare calibration curve in potassium range of 1 to 10 mg/L.

d. Standard potassium solution: Dilute 10.0 mL intermediate potassium solution with water to 100 mL; 1.00 mL = 0.010 mg K. Use this solution to prepare calibration curve in potassium range of 0.1 to 1.0 mg/L.

e. Standard lithium solution: See Section 3500-Na.D.3e.

4. Procedure

Make determination as described in Section 3500-Na.D.4, but measure emission intensity at 766.5 nm.

5. Calculation

See Section 3500-Na.D.5.

6. Precision and Bias

A synthetic sample containing 3.1 mg K/L, 108 mg Ca/L, 82 mg Mg/L, 19.9 mg Na/L, 241 mg Cl^-/L, 0.25 mg NO_2^--N/L, 1.1 mg NO_3^--N/L, 259 mg SO_4^{2-}/L, and 42.5 mg total alkalinity/L (contributed by $NaHCO_3$) was analyzed in 33 laboratories by the flame photometric method, with a relative standard deviation of 15.5% and a relative error of 2.3%.

7. Bibliography

MEHLICH, A. & R. J. MONROE. 1952. Report on potassium analyses by means of flame photometer methods. *J. Assoc. Offic. Agr. Chem.* 35:588.

Also see 3500-Na.D.7.

3500-Re RHENIUM

3500-Re A. Introduction

Rhenium is normally present in only trace levels in natural waters (probably $< 10 \ \mu g/L$).

3500-Re B. Atomic Absorption Spectrometric Method

See flame atomic absorption spectrometric method, Section 3111D.

3500-Rh RHODIUM

3500-Rh A. Introduction

Rhodium is relatively uncommon in natural waters. Process waters may contain higher concentrations.

3500-Rh B. Atomic Absorption Spectrometric Method

See flame atomic absorption spectromatic method, Section 3111B.

3500-Ru   RUTHENIUM

3500-Ru A.   Introduction

Ruthenium is relatively uncommon in natural waters. Process waters may contain higher concentrations.

3500-Ru B.   Atomic Absorption Spectrometric Method

See flame atomic absorption spectrometric method, Section 3111B.

3500-Se   SELENIUM*

3500-Se A.   Introduction

1. Occurrence

Selenium is an essential trace nutrient and selenium deficiency diseases are well known in veterinary medicine. Above trace levels, ingested selenium is toxic to animals and may be toxic to humans.

The selenium concentration of most drinking waters and natural waters is less than 10 μg/L. However, the pore water in seleniferous soils in semiarid areas may contain up to hundreds or thousands of micrograms dissolved selenium per liter. Certain plants that grow in such areas accumulate large concentrations of selenium and may poison livestock that graze on them. Water drained from such soil may cause severe environmental pollution and wildlife toxicity. Soluble selenium may be leached from coal ash and fly ash at electric power plants that burn seleniferous coal.

The inorganic fraction of dissolved selenium consists predominantly of selenium in the selenate ion (SeO_4^{2-}), designated here as Se(VI), and selenium in the selenite ion (SeO_3^{2-}), Se(IV). Selenium derived from microbial degradation of seleniferous organic matter includes selenite, selenate, and the volatile organic compounds dimethylselenide and dimethyldiselenide. Nonvolatile organic selenium compounds may be released to water by microbial processes. Selenopolysulfide ions (SSe^{2-}) may occur in the presence of hydrogen sulfide in waterlogged, anoxic soils.

2. Selection of Method

Selenium can be determined by atomic absorption (hydride generation/atomic absorption spectrometric and electrothermal atomic absorption spectrometric) and atomic emission (inductively coupled plasma) methods, and by chemical methods involving derivatization of selenite and determination of the organic derivative by colorimetry or fluorometry. The hydride generation (C) or electrothermal (H) atomization, colorimetric (D), and fluorometric (E) methods are the most sensitive currently available, and are applicable to water analysis, with the continuous hydride generation/atomic absorption spectrometric method (C) being the

*Approved by Standard Methods Committee, 1988.

method of choice. For determination of selenium at higher concentrations, the inductively coupled plasma method (I) may be used.

By using suitable preparatory steps to convert other chemical species to Se(IV), it is possible to distinguish the chemical species in the sample. In drinking water and most surface and ground waters, Se(IV), Se(VI), and particulate selenium frequently are the only significant species. However, when speciation is important, for example, when a new matrix is being analyzed, the general analytical scheme shown in Figure 3500-Se:1 may be carried out as follows: Determine volatile selenium by stripping sample with nitrogen or air and collecting selenium in alkaline hydrogen peroxide (see Method F). To obtain an estimate of selenium in suspended particles, determine total selenium, filter sample, and make a second determination of total selenium. In any case, filter sample. Occasionally, a filtered sample may have the odor of hydrogen sulfide and a yellow color; such a sample may contain selenopolysulfides, which may be estimated by comparing results of total selenium analyses before and after acidification, stripping with nitrogen, settling for 10 min, and refiltration. Determine selenite, Se(IV), by analyzing filtered water sample directly by the methods of Sections 3500-Se, C, D, or E. In principle, sample digestion with HCl will convert Se(VI) to Se(IV), and the value determined will equal the sum of the two species. In practice, samples frequently contain an unknown masking agent that produces an unduly low result. Test for this effect by analyzing samples with known additions of both species. If recovery is good, the HCl digestion followed by analyses will yield reliable results. If recovery is poor and organic selenium is to be determined subsequently, attempt to remove the interference by sample pretreatment with resin (B.1). Interference also can be eliminated by digestion with an oxidizing

agent (B.2, 3, and 4), but these procedures prevent distinguishing of Se(VI) and organic selenium and also oxidize many organic selenium compounds. To measure nonvolatile organic selenium compounds, use Method G.

The choice of digestion method for oxidizing interferences and organic selenium depends on sample matrix. The methods described in B.2, 3, and 4, in order of increasing complexity and digestive ability, use ammonium (or potassium) persulfate, hydrogen peroxide, and potassium permanganate. Ammonium persulfate digestion is adequate for most filtered ground-, drinking-, and surface water. Hydrogen peroxide digestion may be required if organic selenium compounds are present, and potassium permanganate digestion may be needed with unfiltered samples or those containing refractory organic selenium compounds. Confirm results obtained with one digestion method by using a more powerful method when characterizing a new matrix.

3. Interferences

Interferences are found in certain reagents, as well as in samples. Recognition of the presence of an interferent is critical, especially when unknown sample matrices are being analyzed. Routinely add Se(IV) and Se(VI) to test for interference. If present, characterize the interference and correct by the method of standard additions. A slope less than one indicates interference. In cases of mild interference (recoveries reduced by 25% or less), the standard additions method will largely correct determined values.

Because the hydride atomic absorption method is extremely sensitive, samples frequently need to be diluted to bring them within the linear range of the instrument. Diluting a filtered water sample frequently will eliminate sample-related interferences. Include full reagent blanks in each run to

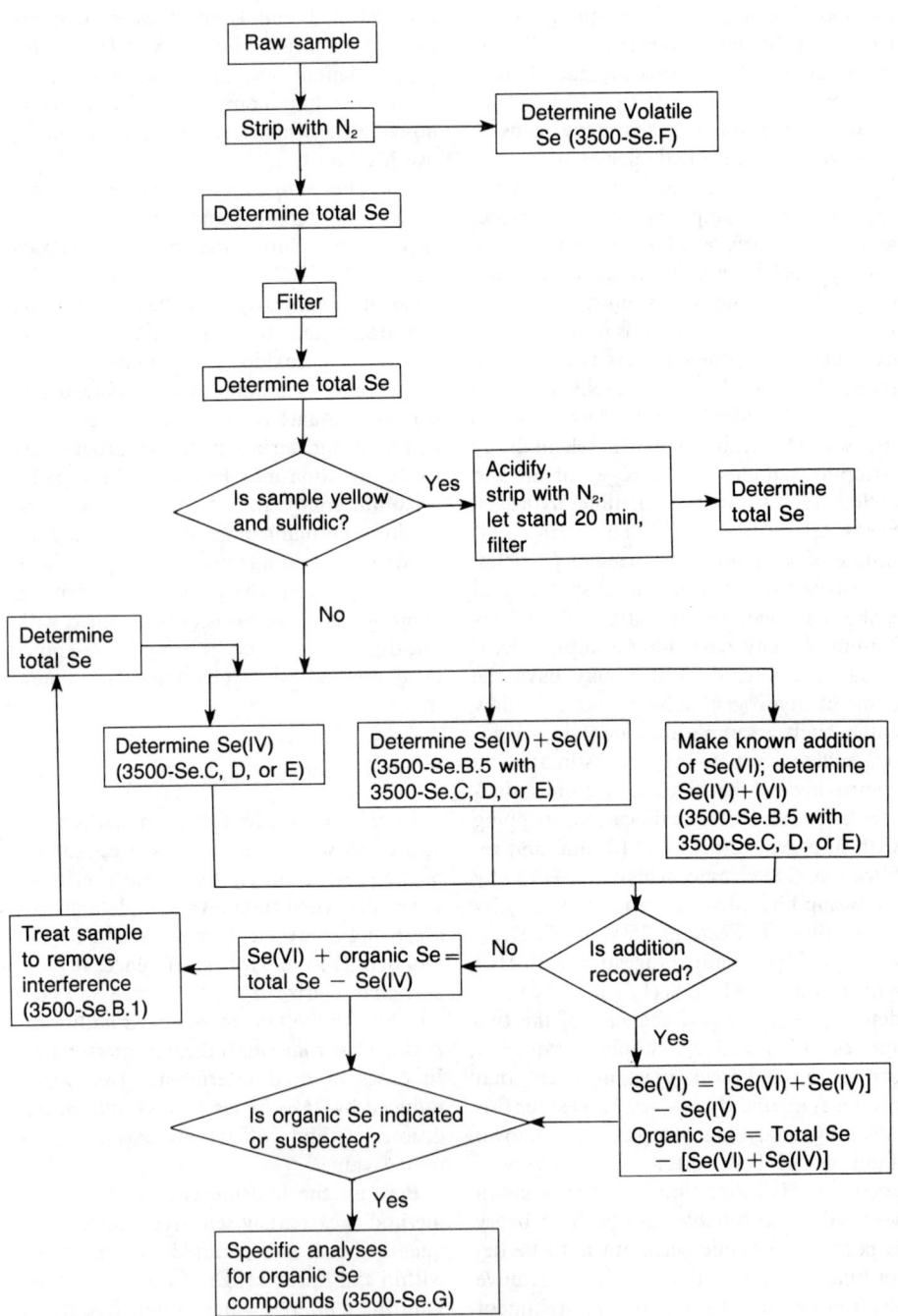

Figure 3500-Se:1. General scheme for speciation of selenium in water.

ensure absence of contamination from reagents. Hydride generator atomic absorption is susceptible to common interference problems related to nitrite in the sample or free chlorine in the reagent HCl (see Method C).

4. Bibliography

U.S. NATIONAL ACADEMY OF SCIENCES. 1976. Selenium: Medical and Biological Effects of Environmental Pollutants. National Academy of Sciences, Washington, D.C.

CUTTER, G.A. 1978. Species determination of selenium in natural waters. *Anal. Chim. Acta* 98:59.

ROBBERECHT, H. & R. VAN GRIEKEN. 1982. Selenium in environmental waters: Determination, speciation and concentration levels. *Talanta* 29:823.

KUBOTA, J. & E.E. CARY. 1982. Cobalt, molybdenum and selenium. *In* A.L. Page et al., eds.

Methods of soil analysis, Part 2, 2nd ed. *Agronomy* 9:485.

RAPTIS, S. et al. 1983. A survey of selenium in the environment and a critical review of its determination at trace levels. *Fresenius Z. Anal. Chem.* 316:105.

CUTTER, G. 1983. Elimination of nitrite interference in the determination of selenium by hydride generation. *Anal. Chim. Acta* 149:391.

CAMPBELL, A. 1984. Critical evaluation of analytical methods for the determination of trace elements in various matrices. Part 1. Determination of selenium in biological materials and water. *Anal. Chem.* 56:645.

LEMLY, A.D. 1985. Ecological basis for regulating aquatic emissions from the power industry: The case with selenium. *Regul. Toxic. Pharm.* 5:465.

OHLENDORF, H.M., D.J. HOFFMAN, M.K. SAIKI & T.W. ALDRICH. 1986. Embryonic mortality and abnormalities of aquatic birds: Apparent impacts of selenium from irrigation drainwater. *Sci. Total Environ.* 52:49.

3500-Se B. Sample Preparation

1. Removal of Organic and Iron Interference by Resin Pretreatment

Interferences are common in selenium analysis, particularly when chemical speciation is attempted. Routine pretreatment of water samples as described here is not necessary. The methods in this section should be tried when poor recovery of standard additions indicates a problem.

Many waters contain iron and/or dissolved organic matter (humic acid) in quantities sufficient to interfere. Reduction of Se(VI) to Se(IV) usually is nonquantitative. When Se(VI) standard additions show poor recovery, treat the sample before analysis. To remove dissolved organic compounds, pass an acidified sample through a resin. Because dissolved organic selenium compounds also may be removed by this treatment, also determine total selenium in the untreated water sample as well (see 2, 3, or 4 below). To remove iron use a strong

base ion exchange resin.* In this treatment the acidity and ion exchanger do not alter speciation; complete speciation of selenium is possible.

a. Apparatus:

1) *Chromatography column* for organics removal, glass, about 0.8 cm ID × 30 cm long, with fluorocarbon metering valve.

2) *Chromatography column* for ion exchange, disposable polyethylene†

3) *pH meter.*

b. Reagents:

1) *Organics-removal resin:* Thoroughly rinse 16 to 50 mesh resin‡ with deionized water and remove resin fines by decanting. Rinse three times with pH 12 solution. Store resin in pH 12 solution and refrigerate to prevent bacterial growth.

2) *Anion exchange resin:* Add 100 to 200

---

*Bio-Rad AG1-X8 or equivalent.
†Bio-Rad Econo-Columns or equivalent.
‡Amberlite XAD-8, Supelco, or equivalent.

mesh anion exchange resin§ to a beaker and thoroughly rinse with deionized water. Cover resin with 4N HCl, stir, and let settle. Decant and repeat acid rinse twice more. Store resin in 4N HCl.

3) *Hydrochloric acid,* conc: Before use, bubble helium through the acid for 3 h at rate of 100 mL/min (CAUTION: *Use a fume hood.*)

4) pH 1.6 solution: Adjust pH of deionized water to 1.6 with HCl.

5) pH 12 solution: Adjust pH of deionized water to 12 with KOH.

c. Procedure:

1) Organic removal—Place 5 cm washed resin in a 0.8-cm-ID column. Precondition column, at 1 mL/min, with 30 mL pH 12 solution and 20 mL pH 1.6 solution. Using HCl and a pH meter adjust sample to pH 1.6 to 1.8. Pass sample through preconditioned column at rate of 1 mL/min. Discard first 10 mL and use next 11 to 50 mL collected for Se(IV) determinations by Methods C, D, or E preceded, if Se(VI) also is to be determined, by preparatory step B.5. If more than 50 mL sample are needed, use another column or use a column with twice as much resin.

2) Iron removal—Place 4 cm prepared resin in a small chromatographic column (add resin to column filled with 4N HCl to avoid air bubbles). Rinse column with 10 mL 4N HCl at flow rate < 6 mL/min. Let solution drain to top of resin, but do not let the column run dry. Adjust sample to 4N HCl and pour into column. Discard first 10 mL and collect the next 11 to 100 mL for Se(IV) analysis by Methods C, D, or E preceded, if necessary, by preparatory step B.2 and if Se(VI) also is to be determined, by preparatory step B.5 below.

2. Removal of Interference by Persulfate Digestion

The combination of this procedure with step B.5 below and Method C, D, or E is,

---

§Bio-Rad AG1-X8 or equivalent.

in most cases, the preferred method for determining total selenium in filtered water. A small amount of ammonium or potassium persulfate is added to the mixture of sample and HCl to remove interference from reducing agents and to oxidize relatively labile organic selenium compounds such as selenoaminoacids and methaneseleninic acid.

If the sample contains hydrogen sulfide or a large concentration of organic matter or is otherwise suspect or to confirm method accuracy, reanalyze sample using digestion procedure 3 or 4 below.

Prepare 2% potassium or ammonium persulfate solution by dissolving 2.0 g in 100 mL deionized water (prepare weekly). Add 0.2 mL persulfate solution to the mixed sample and HCl of ¶ B.5c before heating and proceeding with pretreatment and analysis.

After completing analysis multiply concentration of selenium determined in the acidified sample by 2.04 to obtain total selenium in original sample.

3. Removal of Interference by Alkaline Hydrogen Peroxide Digestion

Occasionally, digestion with persulfate gives incomplete recovery of total selenium. In this case, digestion with hydrogen peroxide is used to remove all reducing agents that might interfere and to fully oxidize organic selenium to Se(VI). The resulting solution can be analyzed for total selenium after pretreatment according to step B.5 below.

This method is suitable for determining total selenium in unfiltered water samples, where particulate selenium is present. When working with a new matrix, confirm results obtained by reanalyzing the sample using digestion procedure 4, below.

a. Apparatus:

1) *Beakers,* 150-mL.

2) *Watch glasses.*

3) *Hot plate.*

4) *Pipetter,* 1-mL, and tips.

5) *Graduated cylinder,* 25-mL.

b. Reagents:

1) *Hydrogen peroxide,* H_2O_2, 30%. Keep refrigerated.

2) *Sodium hydroxide,* NaOH, 1*N*.

3) *Hydrochloric acid,* HCl, 1.5*N*: Dilute 125 mL conc HCl to 1 L with deionized water.

c. Procedure: Add 2 mL 30% H_2O_2 and 1 mL 1*N* NaOH to 25 mL sample in a beaker. Cover beaker to control spattering and simmer on hot plate until fine bubbles characteristic of H_2O_2 decomposition subside and are replaced by ordinary boiling. Add 1 mL 1.5*N* HCl to redissolve any precipitate that may have formed, let cool, and pour into graduated cylinder. Rinse beaker with deionized water into graduated cylinder and make volume up to 25 mL. Proceed to B.5 and chosen analytical method.

4. Removal of Interference by Permanganate Digestion

This digestion method utilizes potassium permanganate to oxidize selenium and remove interfering organic compounds. Excess $KMnO_4$ and MnO_2 are removed by reaction with hydroxylamine. HCl digestion is included here, because it is conveniently performed in the same reaction vial. Selenite may then be determined directly by Method C, D, or E.

Verify recovery for the given matrix. This method gives good recovery even with heavily contaminated water samples that contain organic selenium compounds, dissolved organic matter, and visible suspended material.

Permanganate may oxidize chloride ion to free chlorine. Part of the chlorine (which interferes with hydride analysis) is removed by reaction with hydroxylamine, but the best way to eliminate free chlorine is by prolonged heating in an open vial during the digestion step. Excess hydroxylamine

may reduce recovery by reducing selenium to Se(0).

a. Apparatus:

1) *Oven with thermostat,* for continuous operation at 110 ± 5°C.

2) *Digestion vials,* 40-mL, with fluorocarbon-lined screw caps.

3) *Metal support rack* to hold 40 digestion vials.

b. Reagents:

1) *Hydrochloric acid,* HCl, conc. [See B.1*b*3)].

2) *Hydrochloric acid,* HCl, 10*N:* Dilute 1000 mL conc HCl to 1200 mL with deionized water.

3) *Sulfuric acid,* H_2SO_4, conc and 25%. NOTE: Many brands of H_2SO_4 are contaminated with selenium. Run reagent blanks when starting a new bottle. Make 25% v/v solution by adding 250 mL conc H_2SO_4 to 500 mL deionized water, and dilute to 1 L.

4) *Potassium permanganate solution,* $KMnO_4$, 5% (w/v): Dissolve 50 g $KMnO_4$ in 1000 mL deionized water.

5) *Hydroxylamine hydrochloride solution:* Dissolve 100 g $NH_2OH \cdot HCl$ in 1000 mL deionized water.

c. Procedure: Pipe 5 mL sample into digestion vial, add 5 mL 25% H_2SO_4 and 1 mL $KMnO_4$ solution. Screw cap on and place in preheated oven at 110°C for 1 h. Remove tray with vials from the oven and cool to room temperature. Open vial, carefully add a few drops hydroxylamine hydrochloride solution, mix, and wait until sample is decolorized and residual manganese dioxide is dissolved. Avoid excess hydroxylamine solution, which can cause a low reading. Add 10 mL conc HCl to the sample and heat vial 60 min at 95°C without cap. Let cool to room temperature. Transfer sample to a 25-mL volumetric flask or graduated cylinder, rinse vial into flask, dilute to mark, and mix well. Proceed to analyze by Method C, D, or E. If Method C is used, multiply spectrometer readings by the dilution factor as follows:

Concentration, μg/L

$$= \frac{\text{final volume}}{\text{volume of sample}} \times \text{reading}$$

5. Reduction of Se(VI) to Se(IV) by Hydrochloric Acid Digestion

Se(VI) is reduced to Se(IV) by digestion with HCl. Determine Se(IV) + Se(VI) by either hydride generation atomic absorption spectrometer (Method C) or as an organic derivative (Method D or E). Test any given sample matrix to ensure recovery of added Se(VI). If recovery is poor, try to remove interference by the procedure of ¶ B.1, above, or if at least 75% recovery is achieved, use the method of standard additions. The method described here is of limited utility for direct analysis of water samples, but is useful as a step in determining total selenium in a sample where selenium has been oxidized to Se(VI).

a. Apparatus:

1) *Dispenser,* bottle type, 5-mL, suitable for dispensing concentrated HCl.

2) *Pipetters,* 0.2- and 5-mL.

3) *Screw-cap culture tubes,* borosilicate glass, 25- × 150-mm.

4) *Boiling water bath,* suitable for heating culture tubes; a 1-L beaker on a hot plate is suitable.

b. Reagents:

1) *Sodium selenate additions solution:* Dilute 1000 mg/L stock selenate solution with deionized water to prepare a solution of 1 to 10 mg/L, such that the concentration of the additions solution will be approximately 50 times greater than anticipated total selenium in the sample to be analyzed.

2) *Hydrochloric acid,* HCl, conc: See B.1*b*3).

c. Procedure: Calibrate acid dispenser using water. Preheat water bath. Pipet 5 mL filtered sample into a culture tube. Add 5 mL conc HCl. Loosely cap tube (do not tighten) and place in boiling water bath for 20 min. Let tube cool and tighten cap. Determine total Se(IV) by Method C, D, or E.

Add 0.200 mL additions solution with a microliter pipet to sample and proceed as above. Analyze a deionized water blank and a blank with the addition to ensure absence of contamination and to determine the true value of the addition.

Multiply the concentration of selenium determined in the acidified sample by 2.00 to obtain total concentration of Se(IV) + Se(VI). Multiply reading obtained for sample with addition by 2.04.

6. Bibliography

JANGHORBANI, M., B. TING, A. NAHAPETLAN & R. YOUNG. 1982. Conversion of urinary selenium to selenium IV by wet oxidation. *Anal. Chem.* 54:1188.

ADELOJU, S.B. & A.M. BOND. 1984. Critical evaluation of some wet digestion methods for the stripping voltammetric determination of selenium in biological materials. *Anal. Chem.* 56:2397.

BRIMMER, S.P., W.R. FAWCETT & K.A. KULHAVY. 1987. Quantitative reduction of selenate ion to selenite in aqueous samples. *Anal. Chem.* 59:1470.

3500-Se C. Continuous Hydride Generation/Atomic Absorption Spectrometric Method

See Section 3114C.

3500-Se D.   Colorimetric Method

1. General Discussion

a. Principle: This method is specific to determining selenite in aqueous solution. Selenite reacts with 2,3-diaminonaphthalene to produce a brightly colored and strongly fluorescent piazselenol compound, which is extracted in cyclohexane and measured colorimetrically or fluorometrically (see Method E).

The optimum pH for formation of the piazselenol complex is approximately 1.5 but should not be above 2.5 because above pH 2, the rate of formation of the colored compound is critically dependent on pH. When indicators are used to adjust pH, results frequently are erratic, but they can be improved when pH is monitored electrochemically.

b. Interference: No inorganic compounds are known to give a positive interference. Colored organic compounds extractable by cyclohexane may be encountered, but usually they are absent or can be removed by oxidizing the sample (see B.2, 3, or 4) or by treating it to remove dissolved organics (see B.1). Negative interference results from compounds that reduce the concentration of diaminonaphthalene by oxidizing it. Addition of EDTA eliminates negative interference from at least 2.5 mg Fe^{2+}.

c. Minimum detectable quantity: 10 μg Se/L.

2. Apparatus

a. Colorimetric equipment: A spectrophotometer, for use at 480 nm, providing a light path of 1 cm or longer.

b. Separatory funnel, 250-mL, preferably with a fluorocarbon stopcock.

c. Thermostatically controlled water bath (50°C) with cover.

d. pH meter.

e. Centrifuge, with rotor for 50-mL tubes (optional).

f. Centrifuge bottles, 60-mL, screw-capped, fluorocarbon.

g. Shaker, suitable for separatory funnel (optional).

3. Reagents

Use distilled and/or deionized water in preparing reagents.

a. Selenium standard reference solution: Dissolve 2.190 g sodium selenite, Na_2SeO_3, in water containing 10 mL HCl and dilute to 1 L. 1.00 mL = 1.00 mg Se(IV).

b. Working standard selenium solutions: Dilute selenium reference standard solution with water or suitable background solution to produce a series of working standards spanning the concentration range of interest.

c. Hydrochloric acid: HCl conc and 0.1N.

d. Ammonium hydroxide, NH_4OH, 50% v/v.

e. Cyclohexane, C_6H_{12}.

f. 2,3-Diaminonaphthalene (DAN) solution: Dissolve 200 mg DAN in 200 mL 0.1N HCl. Shake 5 min. Extract three times with 25-mL portions of cyclohexane retaining aqueous phase and discarding organic portions. Filter into dark container* and store in cool, dark place for no longer than 8 h. CAUTION: *Toxic, handle with extreme care.*

g. Hydroxylamine-EDTA solution (HA-EDTA): Dissolve 4.5 g Na_2EDTA in approximately 450 mL water. Add 12.5 g hydroxylamine hydrochloride ($NH_2OH \cdot HCl$) and adjust volume to 500 mL.

4. Procedure

a. Formation of piazselenol: Add 2 mL HA-EDTA solution to 10 mL sample in 60-mL centrifuge bottle (filtered if Se(IV) is to be determined; oxidized using method B.2, 3, or 4, then reduced using Method

---

*Use Whatman No. 42 filter paper, or equivalent.

B.5 for total Se). Adjust to pH 1.5 ± 0.3 with 0.1N HCl and 50% NH$_4$OH, using a pH meter. Add 5 mL DAN solution and heat in a covered water bath at 50°C for 30 min.

b. Extraction of piazselenol: Cool and add 2.0 mL cyclohexane. Cap container securely and shake vigorously for 5 min. Let solution stand for 5 min or until cyclohexane layer becomes well separated. If separation is slow, centrifuge for 5 min at 2000 rpm. Place bottle in a clamp on a ringstand at a 45° angle to the vertical. Remove aqueous phase using a disposable pipet attached to a vacuum line. Transfer organic phase to a small capped container using a clean disposable pipet, or to the spectrophotometer cuvette if absorbance is to be read immediately.

c. Determination of absorbance: Read absorbance at 480 nm using a zero standard. The piazselenol color is very stable but evaporation of the cyclohexane concentrates the color unless the container is capped. CAUTION: *Avoid inhaling cyclohexane vapors.* Beer's Law is obeyed up to 2 mg/L.

5. Calculation

Construct calibration curve using at least a three-point standard curve to bracket the expected sample concentration. Plot absorbance vs. concentration. Correct for digestion blank and any reagent blank.

6. Precision and Bias

Three standard reference materials (wheat flour, water, and a commercial standard) were used to evaluate Se recovery.[1] The wheat flour sample was digested using HNO$_3$ and HClO$_4$ to convert total selenium to Se(VI), digested with HCl to convert Se(VI) to Se(IV) and finally, the colorimetric method was used. Results were as follows:

Standard	Selenium Concentration μg Se/L	
	Expected	Recovered*
NBS, SRM 1567, wheat flour †	1097 ± 197	1113 ± 8
NBS, SRM 1543ib, water	9.7 ± 0.5	8.7 ± 0
Fisher Certified AAS Standard	1002 ± 8	1002 ± 0

*Analyses in triplicate.
†Dry weight basis.

7. Reference

1. HOLTZCLAW, K.M., R.H. NEAL, G. SPOSITO & S.J. TRAINA. 1987. A sensitive colorimetric method for the quantitation of selenite in soil solutions and natural waters. *Soil Sci. Soc. Amer. J.* 51:75.

8. Bibliography

HOSTE, J. & J. GILLIS. 1955. Spectrophotometric determination of traces of selenium with 3,3'-diaminobenzidine. *Anal. Chim. Acta.* 12:158.

CHENG, K. 1956. Determination of traces of selenium. *Anal. Chem.* 28:1738.

MAGIN, G.B. et al. 1960. Suggested modified method for colorimetric determination of selenium in natural waters. *J. Amer. Water Works Assoc.* 52:1199.

ROSSUM, J.R. & P.A. VILLARRUZ. 1962. Suggested methods for determining selenium in water. *J. Amer. Water Works Assoc.* 54:746.

OLSEN, O.E. 1973. Simplified spectrophotometric analysis of plants for selenium. *J. Assoc. Offic. Anal. Chem.* 56:1073.

3500-Se E.   Fluorometric Method

1. General Discussion

a. Principle: This procedure is comparable to the preceding one except that the piazselenol derivative is measured fluorometrically. The detection limit is 10 μg Se/L, but it may be improved to 1 μg Se/L by using a semiautomated procedure.

b. Interferences: High concentrations of nitrite may interfere with the formation of the fluorescent piazselenol complex and require removal through gentle boiling for 2 min before piazselenol formation.

2. Apparatus

In addition to the apparatus required for Method D (*c* through *f*):

a. Fluorometric equipment: A fluorometer with the capability of irradiating samples at 369 nm and measuring fluorescence at 525 nm.

b. Separatory funnels, 125-mL, with fluorocarbon stopcocks.

3. Reagents

See Method D.

4. Procedure

a. Formation of piazselenol: Add 5 mL HA-EDTA solution to 10 mL sample and adjust pH to between 1.7 and 2.0. Add 5 mL DAN solution, mix well, and heat in a 50°C water bath for 30 min.

b. Extraction of piazselenol: When cool, transfer into 125-mL separatory funnel. Add 6.0 mL cyclohexane and shake for 1 min. Discard aqueous phase and wash organic phase with 25 mL 0.1*N* HCl, shaking and discarding the aqueous phase.

c. Determination of fluorescence: Transfer organic phase to a centrifuge tube. Let solution stand for 5 min or until cyclohexane layer becomes well separated. If separation is slow, centrifuge for 5 min at 2000 rpm. Pipet a portion into fluorometer cuvette and read; zero instrument against a reagent blank.

5. Calculation

See Method D.5.

6. Precision and Bias

Precision and bias are comparable to those achieved with Method D.

7. Bibliography

OLSEN, O.E. 1969. Fluorometric analysis of selenium in plants. *J. Assoc. Offic. Anal. Chem.* 52:627.

TAKAYANAGI, K. & G. WONG. 1983. Fluorometric determination of selenium IV and total selenium in natural waters. *Anal. Chim. Acta* 148:263.

3500-Se F.   Determination of Volatile Selenium

1. General Discussion

Dimethylselenide and dimethyldiselenide are low boiling, extremely malodorous organic compounds sparingly soluble in water. They are produced by microbial processes in seleniferous soil and decaying seleniferous organic matter, and occasionally are present in natural waters. They are readily air stripped from a water sample and can be collected with high efficiency in an alkaline solution of hydrogen peroxide, which oxidizes them quantitatively to Se(VI). Total selenium is determined by digestion with HCl and analysis by the hydride atomic absorption, colorimetric, or fluorometric method.

Either nitrogen or air may be used to strip the sample. Preferably use nitrogen if air-sensitive compounds (e.g. selenopolysulfides) are suspected.

Because volatile selenium can be lost in the course of sample collection and handling, preferably air-strip the sample in the field immediately after it is collected. After boiling to decompose H_2O_2, return the alkaline peroxide solution to the laboratory for analysis.

2. Apparatus

All apparatus required for selenate reduction (B.5) and Method C, D, or E, plus:

a. *Gas washing bottles,* borosilicate glass, 250-mL, with coarse porous glass gas dispersion frit. Mark 100-mL level on side of bottle.

b. *Rotameter,* to measure 3 L/min air flow.

c. *Gas flow regulator.*

d. *Hot plate.*

e. *Graduated cylinder,* 100-mL.

f. *Beakers,* 250-mL.

g. *Rubber tubing,* to interconnect gas washing bottles and other gas equipment.

h. *Rubber gloves.*

3. Reagents

All reagents required for selenate reduction (B.5) and Method C, D, or E, plus:

a. *Hydrogen peroxide,* 30%. Refrigerate.

b. *Sodium hydroxide solution,* NaOH, 1N.

c. *Compressed air or nitrogen.*

4. Procedure

Set up air-flow train in this order:
Regulated air supply → rotameter → gas washing bottle 1 → gas washing bottle 2.

Prepare alkaline peroxide solution immediately before use by pouring 20 mL 30% H_2O_2 into a 100-mL graduated cylinder, adding 50 to 60 mL deionized water and 5 mL 1N NaOH, and making up to 100 mL. CAUTION: *Alkaline H_2O_2 is un-stable. Do not keep in glass bottle; hold at about 0 °C in oversized plastic bottle. Solution is corrosive; protect eyes and skin.* Pour into gas washing bottle 2. Pour approximately 100 mL freshly collected sample into gas washing bottle 1. Do not attempt to measure sample volume accurately before volatile selenium determination, as unnecessary handling may cause volatile selenium to be lost.

Connect and check all air lines, turn on air and adjust flow to 3 L/min. Strip for 30 min or more.

After 30 min, turn off air, disconnect gas washing bottle 2, and place it on the hot plate. Adjust heat to produce a gentle simmering of oxygen bubbles from decomposition of H_2O_2. Continue heating until the characteristic effervescence of oxygen subsides and is replaced by ordinary boiling. Remove from hot plate and let cool. Pour solution into beaker. (Volume will be very near to 100 mL, and correction will usually be unnecessary.) Analyze for total selenium using HCl digestion (B.5) and Method C, D, or E. Once boiled, this solution may be safely stored and transported in plastic bottles. Measure volume of sample in gas washing bottle 1.

5. Calculations

The concentration of volatile selenium compounds in the original water sample can be calculated as:

$$C = \frac{100}{\text{volume of original sample}} \times \text{conc of Se in solution}$$

6. Precision and Bias

Approximately 90% of dimethylselenide in samples will be recovered with 30 min air stripping. The recovery of dimethyldiselenide is not known. Loss of gases to the atmosphere during sampling and handling that precede analysis may cause a significant negative error.

3500-Se G. Determination of Nonvolatile Organic Selenium Compounds

1. General Discussion

In principle, the total amount of dissolved organic selenium plus polysulfidic selenium may be estimated by comparing "total Se", determined by oxidation and HCl digestion (B.2 or 3 and B.5, or B.4), followed by Method C, D, or E, with Se(IV) + Se(VI) determined by HCl digestion (B.5) and Method C, D, or E. In practice, this will give a meaningful estimate only if a known addition of Se(VI) is fully recovered. Even if recovery is good, this estimate may be unreliable, because it is the difference of two (frequently larger) numbers determined by slightly different methods. Comparing total Se before and after treatment with resin [B.1c1)] gives a similarly unreliable estimate of nonvolatile organic Se.

It is preferable to separate and directly determine nonvolatile organic selenium. One method involves adsorption of dissolved organic matter onto a C-18 reverse phase HPLC resin, elution with an organic solvent, and determination of selenium in this fraction. While this technique is relatively simple, it is affected by pH and small organic molecules (e.g., individual amino acids) are not retained by the resins. Adjust sample pH to 1.5 to 2.0 before using the column, but because the latter problem cannot be solved easily, the use of organic adsorbants provides only an estimate of organic selenium concentration.

Alternatively, isolate specific compounds and determine their selenium content. In some natural waters selenium may be associated with dissolved polypeptides or small proteins, and even small amounts of free selenoaminoacids may be present. Because selenoaminoacids are the most toxic form of the element, a direct determination is sometimes desirable.

To determine selenium in dissolved peptides, hydrolyze with acid and isolate the free amino acids via ligand exchange chromatography. Elute the selenoaminoacids from the column and determine selenium. Selenoaminoacids are unstable during acid hydrolysis and even using nonoxidizing methyl sulfonic acid and nitrogen-purged glass ampules, selenoaminoacid recoveries are only 50 to 80%. This method is good only for estimating protein-bound selenium. A somewhat more reliable estimate of free selenoaminoacids and selenium associated with small oligopeptides is obtained by performing a similar procedure without the hydrolysis step.

While these methods are too intricate for routine use, semiquantitative, and sensitive only to certain classes of organic compounds, at present they are the only ones available with any practical experience.

Imperfect separation of organic selenium compounds from inorganic forms of selenium may cause interference. In parallel with the actual determination, always perform the procedure using a solution compounded to resemble the actual matrix and containing a similar amount of selenium, but in the form of Se(IV) and Se(VI) to determine degree of interference.

2. Apparatus

a. *Rotary evaporator,* with temperature-control bath and 30-mL pear-shaped flasks.

b. *Glass ampule sealing apparatus,* or oxygen-gas torch.

c. *Heating block,* 100°C, or pressure cooker.

d. *Glass chromatography columns,* 15 cm long, 0.7 cm ID.*

e. *Glass syringe,* 50-mL.

---

*Bio-Rad glass Econo-Columns or equivalent.

f. Glass ampules, 10- or 20- mL. Clean by heating in a muffle furnace at 400°C for 24 h.

g. pH meter.

3. Reagents

a. Hydrochloric acid, HCl, 1*N*.

b. Methyl sulfonic acid, conc.

c. Ammonium hydroxide, NH₄OH, 1.5*N*: Dilute 100 mL conc NH₄OH to 1 L with deionized water.

d. Sodium hydroxide, NaOH pellets and 1*N* solution.

e. pH 1.6 solution: Adjust pH of deionized water to 1.6 using HCl.

f. pH 9.0 solution: Adjust pH of deionized water to 9.0 using NaOH.

g. Copper sulfate solution, CuSO₄ 1*M*: Dissolve 25 g CuSO₄·5H₂O in deionized water and dilute to 100 mL.

h. Methanol, purified.

i. C-18 cartridges†: Using a glass syringe, pass the following sequence of reagents through the cartridge to clean the resin (6 mL/min or less): 10 mL deionized water; 20 mL 1*N* HCl; 10 mL deionized water; 20 mL methanol; 10 mL deionized water; and 20 mL pH 1.6 solution. Refrigerate but do not freeze cleaned cartridges.

j. Ligand-exchange chromatographic resin, 100/200 mesh: Rinse resin‡ with deionized water to remove fines. Then rinse resin three times in the following sequence: 1*N* HCl; 1.5*N* NH₄OH; and deionized water. Store resin wet.

k. Copper-treated ligand-exchange chromatographic resin, 100/200 mesh: Rinse resin‡ with deionized water to remove fines. Then rinse three times with 1*N* HCl, followed by deionized water. Using NaOH adjust pH of supernatant above the resin to about 7. Add CuSO₄ solution to the resin and stir. After settling, decant supernatant and add more CuSO₄ solution. Decant CuSO₄ solution and rinse with deionized water until no copper is noticeable in the supernatant. Rinse resin three times with 1.5*N* NH₄OH and three times with deionized water. Store resin wet.

4. Procedures

a. Extractable organic selenium: Adjust sample (5 to 50 mL) to pH 1.5 to 2.0 using HCl and place in a clean glass syringe with an attached cleaned C-18 cartridge. Push sample through the cartridge at a rate of 6 mL/min. After removing the cartridge, draw 2 mL pH 1.6 solution into syringe as a rinse, reattach cartridge, and push the rinse through the cartridge. Repeat two additional times. The cartridge can be refrigerated for storage. To elute organic selenium, push 10 mL methanol through the cartridge at rate of 2 mL/min and collect eluate in a 30 mL pear-shaped flask. Remove methanol by rotary evaporation, with the water bath temperature less than 40°C. Use deionized water to solubilize and transfer the residue into the vessel used for total selenium digestions. Determine total dissolved selenium by digestion with persulfate (B.2) or peroxide (B.3), reduction of Se(VI) (B.5), and analysis by Method C, D, or E.

b. Hydrolysis of protein-bound selenium: Place filtered sample in a 10- or 20-mL glass ampule (depending on desired volume), and add conc methyl sulfonic acid to adjust concentration to 4*M*. Purge acidified sample with nitrogen for 10 min and seal top with a torch. Heat sealed vial at 100°C for 24 h in a heating block or pressure cooker. Transfer cooled hydrolysis solution with deionized water rinses to a 50-mL beaker and place in an ice bath. Using NaOH pellets and 1*N* NaOH, adjust to pH 9.0, taking care not to allow solution to heat to boiling.

c. Determination of selenoaminoacids: If sample is not hydrolyzed as in ¶ 4*b*, filter, and adjust pH to 9.0 using 1*N* NaOH.

Fill an empty chromatographic column with deionized water and add ammonium-

---

†Sep-Pak, Waters Associates, or equivalent.

‡Chelex 100 (ammonium form) Bio Rad, or equivalent.

form resin to a depth of 2 cm. Add copper-treated resin to form a 12-cm length of resin. (The ammonium-form resin removes any copper that bleeds from the copper-treated resin above it). Rinse with deionized water until the pH of the effluent is 9.0; maintain flow through the column by gravity. Pass sample through column, rinse sample beaker with 5 mL pH 9 solution, and place the rinse on column (after the last of the sample reaches the top of the resin). Rinse beaker twice more. Discard flow through column.

Place clean beaker under column and add 20 mL $1.5N$ NH_4OH to the column. Neutralize NH_4OH eluate with 2.5 mL conc HCl. Determine total dissolved selenium by digestion with persulfate (B.2) or peroxide (B.3), reduction of Se(VI) (B.5) and analysis by Method C, D, or E.

5. Precision and Bias

These procedures are only semiquantitative. Typical relative standard deviation is 12% for the C-18 isolation of dissolved organic selenium, and 15% for protein-bound selenium.

6. Bibliography

COOKE, T.D. & K.W. BRULAND. 1987. Aquatic chemistry of selenium: evidence of biomethylation. *Environ. Sci. Technol.* 21:1214.
CUTTER, G. 1982. Selenium in reducing waters. *Science* 217:829.

3500-Se H. Electrothermal Atomic Absorption Spectrometric Method

See Section 3113.

3500-Se I. Inductively Coupled Plasma Method

See Section 3120.

3500-Ag SILVER*

3500-Ag A. Introduction

1. Occurrence

Silver can cause argyria, a permanent, blue-gray discoloration of the skin and eyes that imparts a ghostly appearance. Concentrations in the range of 0.4 to 1 mg/L have caused pathological changes in the

---

*Approved by Standard Methods Committee, 1985.

kidneys, liver, and spleen of rats. Toxic effects on fish in fresh water have been observed at concentrations as low as 0.17 μg/L. The silver concentration of U.S. drinking waters has been reported to vary between 0 and 2 μg/L with a mean of 0.13 μg/L. Relatively small quantities of silver are bactericidal or bacteriostatic and find limited use for the disinfection of swimming pool waters.

2. Selection of Method

The atomic absorption spectrometric method and the inductively coupled plasma method are preferred. The electrothermal atomization method is the most sensitive for determining silver in natural waters. The dithizone method can be used when an atomic absorption spectrometer is unavailable.

3. Sampling and Storage

If total silver is to be determined, acidify sample with conc nitric acid (HNO_3) to pH < 2 at time of collection. If sample contains particulate matter and only the "dissolved" metal content is to be determined, filter through a 0.45-μm membrane filter at time of collection. After filtration, acidify filtrate with HNO_3 to pH < 2. Complete analysis as soon after collection as possible. Some samples may require special storage and digestion; see Section 3030D.

3500-Ag B.   Atomic Absorption Spectrometric Method

See flame atomic absorption spectrometric method, Sections 3111B and C, and electrothermal atomic absorption spectrometric method, Section 3113.

3500-Ag C.   Inductively Coupled Plasma Method

See Section 3120.

3500-Ag D.   Dithizone Method

1. General Discussion

a. Principle: Many metals can react with dithizone to produce colored coordination compounds. Under proper conditions or on removal of all interferences, the reaction can be made selective. In this mixed-color method, separation of the two colors is not attempted; either the green color of dithizone or the yellow color of silver dithizonate can be measured. In view of the sensitivity of the reaction and numerous interferences among the common metals, the method is empirical and demands careful adherence to the procedure. Final color can be evaluated visually or photometrically. The visual finish has been found as accurate as, and more efficient than, the photometric measurement because it circumvents extra handling involved with a photometer. The use of cells having a volume greater than 1 mL requires final dilution with carbon tetrachloride (CCl_4) or selection of a larger sample.

b. Interference: Ferric ion, residual chlorine, and other oxidizing agents convert dithizone to a yellow-brown color. However, extraction of silver along with other metals into a CCl_4 solution of dithizone overcomes such oxidation interference. Silver is removed selectively from other carry-over metals by using an ammonium thiocyanate solution. The extreme sensitivity of the method, as well as silver's affinity for being adsorbed, make it desirable to prepare and segregate glassware for this determination and to take extra precautions at every step. The necessity for checking and preventing contamination cannot be overemphasized. Dithizone and silver dithizonate decompose rapidly in strong light; therefore, do not leave them in the photometer light beam longer than is necessary. Avoid direct sunlight.

c. *Minimum detectable quantity:* 0.2 μg Ag in 1.0 mL final volume.

2. Apparatus

a. *Colorimetric equipment:* One of the following is required:

1) *Spectrophotometer* for measurements at either 620 nm or 462 nm, with a light path of 1 cm.

2) *Filter photometer* providing a light path of 1 cm and equipped with a red filter having maximum transmittance at or near 620 nm or a blue filter having maximum transmittance at or near 460 nm.

3) *Micro test tubes,* 10-mL capacity, 1 × 7.5-cm size.

b. *Separatory funnels,* with a capacity of 500 mL or larger, and also funnels with a capacity of 60 mL, preferably with inert TFE stopcocks.

c. *Glassware:* Treat all glassware, dishes, and crucibles with a sulfuric-chromic acid mixture and wash in 1 + 1 HNO_3 to dissolve any trace of chromium or silver adsorbed on the glassware. Thoroughly rinse with silver-free water and apply silicone coating fluid* to establish a repellent surface. Omitting these steps will result in serious errors. Oven-dry glassware and do not use acetone rinses because this solvent frequently contains interferences.

d. *High-silica-glass dishes†‡ or silica crucibles.*

---

*Desicote, manufactured by Beckman Instruments, Inc., or equivalent.
†Vycor, manufactured by Corning Glass Works, or equivalent.

3. Reagents

a. *Silver-free water:* Use silver-free redistilled or deionized distilled water for preparing all reagents and dilutions.

b. *Sulfuric acid,* H_2SO_4, 1N and conc.

c. *Carbon tetrachloride,* CCl_4: Store in a glass container and do not allow contact with any metals before use. If this reagent contains traces of an interfering metal, as evidenced by a dithizone solution that is not pure green, redistill in an all-borosilicate-glass apparatus. CAUTION: *CCl_4 is very toxic. Perform all operations with CCl_4 in a fume hood.*

d. *Working dithizone solution A:* See Section 1070D.2*b*3). Dilute 50 mL stock dithizone solution III with CCl_4 to 250 mL; 1 mL = 50 μg dithizone. Store in a brown glass bottle in the dark.

e. *Working dithizone solution B:* Dilute 2 mL working dithizone solution A with CCl_4 to 250 mL; 1 mL = 0.4 μg dithizone. Prepare daily.

f. *Ammonium thiocyanate reagent:* Dissolve 10 g NH_4CNS in water to which 5 mL conc H_2SO_4 have been added and dilute to 500 mL with water. Store in a bottle containing 25 mL dithizone solution A.

g. *Nitric acid,* HNO_3, 1N.

h. *Urea solution:* Dissolve 10 g $(NH_2)_2CO$ in water and dilute to 100 mL. Store in a bottle containing 25 mL dithizone solution A. Discard if a red film forms.

i. *Hydroxylamine sulfate solution:* Dissolve 20 g $(NH_2OH)_2 \cdot H_2SO_4$ in water and dilute to 100 mL. Store in a bottle containing 25 mL dithizone solution A.

j. *Stock silver solution:* Dissolve 157.5 mg anhydrous silver nitrate, $AgNO_3$, in water to which 14 mL conc H_2SO_4 have been added and dilute to 1000 mL; 1.00 mL = 100 μg Ag.

k. *Standard silver solution:* Immediately before use, dilute 10.00 mL stock solution to 1000 mL; 1.00 mL = 1.00 μg Ag.

4. Procedure

a. *Pretreatment of sample:* If organic matter is present and total silver is to be determined, digest sample with HNO_3 and H_2SO_4 as directed in Section 3030G. If only dissolved silver is to be determined, filter sample through a 0.45-μm membrane filter.

b. *Preliminary extraction:* To 100 mL sample in a 500-mL separatory funnel add 11 mL conc H_2SO_4. Extract silver by adding 5 mL dithizone solution A and shaking for 1 min. Collect organic phase and any scum in a 25-mL centrifuge tube. Transfer scum formed to centrifuge tube because it may contain an appreciable amount of silver. Repeat extraction twice more with 5-mL portions dithizone solution A and add extracts and scum to centrifuge tube. Reject aqueous phase. If a larger original sample is required, use two centrifuge tubes to collect dithizone extracts. For a 500-mL sample, use at least four 5-mL portions dithizone solution A. For either sample size, centrifuge, discard aqueous phase, and add 2 mL water. Recentrifuge and discard aqueous phase. Transfer CCl_4 layer to a 60-mL separatory funnel. Add 4 mL NH_4CNS reagent to centrifuge tube to collect any remaining extract, gently agitate, and transfer quantitatively to the separatory funnel. Shake for 1 min and with a suction pipet transfer as much aqueous phase as possible to a high-silica-glass dish. Repeat addition of 4 mL NH_4CNS reagent, gently agitate, and transfer aqueous phase two more times. Run off organic layer and add last few drops of aqueous phase to dish. Add 1.5 mL conc H_2SO_4 and evaporate to dryness by first evaporating to fumes by heating from above with an infrared lamp and then by heating from below with a hot plate and above with an infrared lamp. Keep heating temperature low enough to prevent bumping. Add 0.6 mL 1N HNO_3 and warm to dissolve. Add 1 mL each of urea and hydroxylamine sulfate solutions

and digest for 5 min near the boiling point, adding water dropwise to prevent caking. Let cool to room temperature. Transfer to a 10-mL micro test tube, rinsing dish twice with 2-mL portions 1N H_2SO_4.

c. *Final extraction of silver:* Add 1 mL dithizone solution B and extract silver by mixing for 2 min with the aid of a thin glass rod flattened at the bottom. If organic phase has a greenish hue, the amount of silver is less than 1.5 μg. If organic phase is clear yellow (showing that there is no excess of dithizone), add further 1-mL portions dithizone solution B and repeat extraction until a mixed color is obtained. Record total volume, B, of dithizone solution B used.

d. *Visual colorimetric estimation:* Prepare standards by placing in each of nine micro test tubes 1 mL dithizone solution B and then 0, 0.20. . . 1.60 mL standard silver solution, and 3.0, 2.8. . . 1.4 mL 1N H_2SO_4.

Extract as described in ¶ 4c above, starting with the solution of lowest concentration. Compare colors of organic phases of sample and standards in the micro test tubes.

e. *Photometric measurement:* Prepare a standard curve by adding known amounts of silver in the range 0.20 to 1.50 μg to 0.3 mL 1N HNO_3, and 1 mL each of urea and hydroxylamine sulfate solutions. Add 1 mL dithizone solution B and extract the silver, using the same method as with the samples. Measure absorbance at or near 620 nm, using special cells of 1-cm light path but of reduced width so as to contain, when full, no more than about 1 mL. Zero spectrophotometer on a cell containing dithizone solution B at an absorbance reading of 1.000 and read samples and standards against this setting. Determine absorbance of blank by carrying 100 mL (500 mL if 500-mL samples are used) silver-free water through the entire process.

Because samples and standards will give lesser absorbance readings than the dithizone solution, correct all absorbances by *adding* absorbance of the blank. Because final volumes of dithizone solution B necessary to produce a mixed color may not be equal, convert concentrations of standards to silver concentration in final dithizone extract:

$$\mu g \ Ag/mL = \frac{\mu g \ Ag \ in \ standard}{mL \ final \ extract}$$

Obtain a positive-sloping standard curve by subtracting corrected absorbances of standards from 1.000 and plotting differences against concentration of standards in micrograms Ag per milliliter.

Determine absorbance of sample and add absorbance of blank. Subtract corrected absorbance from 1.000 and from standard curve read sample concentration in micrograms Ag per milliliter final volume dithizone. Calculate sample concentration.

5. Calculation

$$mg \ Ag/L = \frac{\mu g \ Ag/mL \ (from \ standard \ curve)}{mL \ sample} \times C$$

where:

C = total volume of dithizone solution B used to extract silver for final colorimetric measurement, mL.

6. Bibliography

PIERCE, T.B. 1960. Determination of trace quantities of silver in trade effluents. *Analyst* 85:166.

WEST, F.K., P.W. WEST & F.A. IDDINGS. 1966. Adsorption of traces of silver on container surfaces. *Anal. Chem.* 38:1566.

DYCK, W. 1968. Adsorption of silver on borosilicate glass. Effect of pH and time. *Anal. Chem.* 40:454.

3500-Na   SODIUM*

3500-Na A.   Introduction

1. Occurrence

Sodium ranks sixth among the elements in order of abundance and is present in most natural waters. The levels may vary from less than 1 mg Na/L to more than 500 mg Na/L. Relatively high concentrations may be found in brines and hard waters softened by the sodium exchange process. The ratio of sodium to total cations is important in agriculture and human pathology. Soil permeability can be harmed by a high sodium ratio. Persons afflicted with certain diseases require water with low sodium concentration. A limiting concentration of 2 to 3 mg/L is recommended in feedwaters destined for high-pressure boilers. When necessary, sodium can be removed by the hydrogen-exchange process or by distillation.

2. Selection of Method

Method B uses an atomic absorption spectrometer in the flame absorption mode. Method C uses inductively coupled plasma; this method is not as sensitive as the other methods, but usually this is not an impor-

---

*Approved by Standard Methods Committee, 1985.

tant factor. Method D uses either a flame photometer or an atomic absorption spectrometer in the flame emission mode. When all of these instruments are available, the choice will depend on factors including relative quality of the instruments, precision and sensitivity required, number of samples, matrix effects, and relative ease of instrument operation. If an atomic absorption spectrometer is used, operation in the emission mode is preferred.

3. Storage of Sample

Store alkaline samples or samples containing low sodium concentrations in polyethylene bottles to eliminate the possibility of sample contamination due to leaching of the glass container.

3500-Na B.   Atomic Absorption Spectrometric Method

See flame atomic absorption spectrometric method, Section 3111B.

3500-Na C.   Inductively Coupled Plasma Method

See Section 3120.

3500-Na D.   Flame Emission Photometric Method

1. General Discussion

a. Principle: Trace amounts of sodium can be determined by flame emission photometry at a wavelength of 589 nm. The sample is sprayed into a gas flame and excitation is carried out under carefully controlled and reproducible conditions. The desired spectral line is isolated by the use of interference filters or by a suitable slit arrangement in light-dispersing devices such as prisms or gratings. The intensity of light is measured by a phototube potentiometer or other appropriate circuit. The intensity of light at 589 nm is approximately proportional to the concentration of the element. If alignment of the wavelength dial with the prism is not precise in the available photometer, the exact wavelength setting, which may be slightly more or less than 589 nm, can be determined from the maximum needle deflection and then used for the emission measurements. The calibration curve may be linear but has a tendency to level off at higher concentrations.

b. Interference: Flame photometers operating on the internal standard principle may require adding a standard lithium solution to each working standard and sample. The optimum lithium concentration may vary among individual instruments; therefore, ascertain it for the instrument used.

Minimize interference by the following:

1) Operate in the lowest practical sodium range.

2) Add radiation buffers to suppress ionization and anion interference. Among

common anions capable of causing radiation interference are Cl^-, SO_4^{2-} and HCO_3^- in relatively large amounts.

3) Introduce identical amounts of the same interfering substances present in the sample into the calibration standards.

4) Prepare a family of calibration curves embodying added concentrations of a common interferent.

5) Apply an experimentally determined correction in those instances where the sample contains a single important interference.

6) Remove interfering ions.

7) Remove burner-clogging particulate matter from the sample by filtering through a quantitative filter paper of medium retentiveness.

8) Incorporate a nonionic detergent in the standard lithium solution to assure proper aspirator function.

9) Use the standard addition technique. The standard-addition approach is described in the flame photometric method for strontium. Its use involves adding an identical portion of sample to each standard and determining the sample concentration by mathematical or graphical evaluation of the calibration data.

10) Use the internal standard technique. Potassium and calcium interfere with sodium determination by the internal-standard method if the potassium-to-sodium ratio is $\geq 5:1$ and the calcium-to-sodium ratio is $\geq 10:1$. When these ratios are exceeded, measure calcium and potassium first so that the approximate concentration of interfering ions may be added to the sodium calibration standards. Magnesium interference does not appear until the magnesium-to-sodium ratio exceeds 100, a rare occurrence.

c. *Minimum detectable concentration:* The better flame photometers can be used to determine sodium levels approximating 100 $\mu g/L$. With proper modifications in technique the range of sodium measurement can be extended to 10 $\mu g/L$ or lower.

2. Apparatus

a. *Flame photometer* (either direct-reading or internal-standard type) or atomic absorption spectrometer in the flame emission mode.

b. *Glassware:* Rinse all glassware with 1 + 15 HNO_3 followed by several portions of deionized distilled water.

3. Reagents

To minimize sodium pickup, store all solutions in plastic bottles. Use small containers to reduce the amount of dry element that may be picked up from the bottle walls when the solution is poured. Shake each container thoroughly to wash accumulated salts from walls before pouring solution.

a. *Deionized distilled water:* Use deionized distilled water to prepare all reagents and calibration standards, and as dilution water.

b. *Stock sodium solution:* Dissolve 2.542 g NaCl dried at 140°C and dilute to 1000 mL with water; 1.00 mL = 1.00 mg Na.

c. *Intermediate sodium solution:* Dilute 10.00 mL stock sodium solution with water to 100.0 mL; 1.00 mL = 100 μg Na. Use this intermediate solution to prepare calibration curve in sodium range of 1 to 10 mg/L.

d. *Standard sodium solution:* Dilute 10.00 mL intermediate sodium solution with water to 100 mL; 1.00 mL = 10.0 μg Na. Use this solution to prepare calibration curve in sodium range of 0.1 to 1.0 mg/L.

e. *Standard lithium solution:* Use either lithium chloride (1) or lithium nitrate (2) to prepare standard lithium solution containing 1.00 mg Li/1.00 mL.

1) Dry LiCl overnight in an oven at 105°C. Weigh rapidly 6.109 g, dissolve in water, and dilute to 1000 mL.

2) Dry $LiNO_3$ overnight in an oven at 105°C. Weigh rapidly 9.935 g, dissolve in water, and dilute to 1000 mL.

Prepare a new calibration curve whenever the standard lithium solution is changed. Where circumstances warrant, alternatively prepare a standard lithium solution containing 2.00 mg or even 5.00 mg Li/1.00 mL.

4. Procedure

a. *Pretreatment of polluted water and wastewater samples:* Follow the procedure described in Section 3030.

b. *Precautions:* Locate instrument in an area away from direct sunlight or constant light emitted by an overhead fixture and free of drafts, dust, and tobacco smoke. Guard against contamination from corks, filter paper, perspiration, soap, cleansers, cleaning mixtures, and inadequately rinsed apparatus.

c. *Instrument operation:* Because of differences between makes and models of instruments, it is impossible to formulate detailed operating instructions. Follow manufacturer's recommendation for selecting proper photocell and wavelength, adjusting slit width and sensitivity, appropriate fuel and air or oxygen pressures, and the steps for warm-up, correcting for interferences and flame background, rinsing of burner, igniting sample, and measuring emission intensity.

d. *Direct-intensity measurement:* Prepare a blank and sodium calibration standards in stepped amounts in any of the following applicable ranges: 0 to 1.0, 0 to 10, or 0 to 100 mg/L. Starting with the highest calibration standard and working toward the most dilute, measure emission at 589 nm. Repeat the operation with both calibration standards and samples enough times to secure a reliable average reading for each solution. Construct a calibration curve from the sodium standards. Determine sodium concentration of sample from the calibration curve. Where a large number of samples must be run routinely, the calibration curve provides sufficient accuracy. If greater precision and less bias are desired

and time is available, use the bracketing approach described in ¶ 4f below.

e. *Internal-standard measurement:* To a carefully measured volume of sample (or diluted portion), each sodium calibration standard, and a blank, add, with a volumetric pipet, an appropriate volume of standard lithium solution. Then follow all steps prescribed in ¶ 4d above for direct-intensity measurement.

f. *Bracketing approach:* From the calibration curve, select and prepare sodium standards that immediately bracket the emission intensity of the sample. Determine emission intensities of the bracketing standards (one sodium standard slightly less and the other slightly greater than the sample) and the sample as nearly simultaneously as possible. Repeat the determination on bracketing standards and sample. Calculate the sodium concentration by the equation in ¶ 5b and average the findings.

5. Calculation

a. For direct reference to the calibration curve:

$$\text{mg Na/L} = (\text{mg Na/L in portion}) \times D$$

b. For the bracketing approach:

$$\text{mg Na/L} = \left[\frac{(B - A)(s - a)}{(b - a)} + A \right] D$$

where:

B = mg Na/L in upper bracketing standard,
A = mg Na/L in lower bracketing standard,
b = emission intensity of upper bracketing standard,
a = emission intensity of lower bracketing standard,
s = emission intensity of sample, and
D = dilution ratio
$$= \frac{\text{mL sample} + \text{mL distilled water}}{\text{mL sample}}$$

6. Precision and Bias

A synthetic sample containing 19.9 mg Na/L, 108 mg Ca/L, 82 mg Mg/L, 3.1 mg K/L, 241 mg Cl^-/L, 0.25 mg NO_2^--N/L, 1.1 mg NO_3^--N/L, 259 mg SO_4^{2-}/L, and 42.5 mg total alkalinity/L was analyzed in 35 laboratories by the flame photometric method, with a relative standard deviation of 17.3% and a relative error of 4.0%.

7. Bibliography

BARNES, R.B. et al. 1945. Flame photometry: A rapid analytical method. *Ind. Eng. Chem.,* Anal. Ed. 17:605.

BERRY, J.W., D.G. CHAPPELL & R.B. BARNES. 1946. Improved method of flame photometry. *Ind. Eng. Chem.,* Anal. Ed. 18:19.

PARKS, T.D., H.O. JOHNSON & L. LYKKEN. 1948. Errors in the use of a model 18 Perkin-Elmer flame photometer for the determination of alkali metals. *Anal. Chem.* 20:822.

BILLS, C.E. et al. 1949. Reduction of error in flame photometry. *Anal. Chem.* 21:1076.

GILBERT, P.T., R.C. HAWES & A.O. BECKMAN. 1950. Beckman flame spectrophotometer. *Anal. Chem.* 22:772.

WEST, P.W., P. FOLSE & D. MONTGOMERY. 1950. Application of flame spectrophotometry to water analysis. *Anal. Chem.* 22:667.

FOX, C.L. 1951. Stable internal-standard flame photometer for potassium and sodium analyses. *Anal. Chem.* 23:137.

AMERICAN SOCIETY FOR TESTING AND MATERIALS. 1952. Symposium on flame photometry. Spec. Tech. Publ. 116, American Soc. Testing & Materials, Philadelphia, Pa.

COLLINS, C.G. & H. POLKINHORNE. 1952. An investigation of anionic interference in the determination of small quantities of potassium and sodium with a new flame photometer. *Analyst* 77:430.

WHITE, J.U. 1952. Precision of a simple flame photometer. *Anal. Chem.* 24:394.

MAVRODINEANU, R. 1956. Bibliography on analytical flame spectroscopy. *Appl. Spectrosc.* 10:51.

MELOCHE, V.W. 1956. Flame photometry. *Anal. Chem.* 28:1844.

BURRIEL-MARTI, F. & J. RAMIREZ-MUNOZ. 1957. Flame Photometry: A Manual of Methods and Applications. D. Van Nostrand Co., Princeton, N.J.

DEAN, J.A. 1960. Flame Photometry. McGraw-Hill Publishing Co., New York, N.Y.

3500-Sr STRONTIUM*

3500-Sr A. Introduction

1. Occurrence

A typical alkaline-earth element, strontium chemically resembles calcium and causes a positive error in gravimetric and titrimetric methods for the determination of calcium. Because strontium has a tendency to accumulate in bone, radioactive strontium 90, with a half-life of 28 years, presents a well-recognized peril to health. Naturally occurring strontium is not radioactive. For this reason, the determination of strontium in a water supply should be supplemented by a radiological measurement to exclude the possibility that the strontium content may originate from radioactive contamination (see Section 7500-Sr).

Although most potable supplies contain little strontium, some well waters in the midwestern United States have levels as high as 39 mg/L.

2. Selection of Method

The atomic absorption spectrometric and inductively coupled plasma methods are preferred. The flame emission photometric method also is available for those laboratories that do not have the equipment needed for one of the preferred methods.

*Approved by Standard Methods Committee, 1985.

3500-Sr B.   Atomic Absorption Spectrometric Method

See flame atomic absorption spectrometric method, Section 3111B.

3500-Sr C.   Inductively Coupled Plasma Method

See Section 3120.

3500-Sr D.   Flame Emission Photometric Method

1. General Discussion

a. Principle: The flame photometric
method makes possible the determination
of strontium in the low concentrations
prevalent in natural water supplies. The
strontium emission is measured at a wavelength of 460.7 nm. The background intensity is measured at a wavelength of 466
nm. The difference in readings obtained at
these two wavelengths measures the light
intensity emitted by strontium.

b. Interference: Emission intensity is a
linear function of strontium concentration
and concentration of other constituents.
The standard addition technique distributes the same ions throughout the standards and the sample, thereby equalizing the
radiation effect of possible interfering substances, although very low pH (<0) could
produce interferences.

c. Minimum detectable concentration:
Strontium levels of about 0.2 mg/L can be
detected by the flame photometric method
without prior sample concentration.

d. Sampling and storage: Polyethylene
bottles are preferable for sample storage,
although borosilicate glass containers also
may be used. At time of collection adjust
sample to pH <2 with nitric acid (HNO_3).

2. Apparatus

Spectrophotometer, equipped with photomultiplier tube and flame accessories; or
an atomic absorption spectrophotometer
capable of operation in flame emission
mode.

3. Reagents

a. Stock strontium solution: Dissolve
2.415 g strontium nitrate, $Sr(NO_3)_2$, dried
at 140°C, in 1000 mL 1% (v/v) HNO_3;
1.00 mL = 1.00 mg Sr.

b. Standard strontium solution: Dilute
25.00 mL stock strontium solution to 1000
mL with distilled water; 1.00 mL = 25.0
μg Sr. Use this solution for preparing Sr
standards in the 1- to 25-mg/L range.

c. Nitric acid, HNO_3, conc.

4. Procedure

*a. Pretreatment of polluted water and
wastewater samples:* Follow the procedure
described in Section 3030E or I.

b. Preparation of strontium standards:

Add 25.0 mL sample containing less than 10 mg calcium or barium and less than 1 mg strontium to 25.0 mL of each of a series of strontium standards. Use a minimum of four strontium standards from 0 mg/L to a concentration exceeding that of the sample. For most natural waters 0, 2.0, 5.0, and 10.0 mg Sr/L standards are sufficient. Brines may require a strontium series containing 0, 25, 50, and 75 mg/L. Dilute the brine sufficiently to eliminate burner splatter and clogging. Best results are obtained when the strontium concentration of the sample is less than 100 mg/L.

c. *Concentration of low-level strontium samples:* Concentrate samples containing less than 2 mg Sr/L. Add 3 to 5 drops conc HNO_3 to 250 mL sample and evaporate to about 25 mL. Cool and make up to 50.0 mL with distilled water. Proceed as in ¶ b. The HNO_3 concentration in the sample prepared for atomization can approach 0.2 mL/25 mL without producing interference.

d. *Flame photometric measurement:* Measure emission intensity of prepared samples (standards plus sample) at wavelengths of 460.7 and 466 nm. Follow manufacturer's instructions for correct instrument operation. Use a fuel-rich nitrous oxide-acetylene flame, if possible.

5. Calculation

a. Plot net intensity (reading at 460.7 nm minus reading at 466 nm) against strontium concentration added to the sample. Because the plot forms a straight calibration line that intersects the ordinate, compute sample strontium concentration from the equation:

$$\text{mg Sr/L} = \frac{A - B}{C} \times \frac{D}{E}$$

where:

A = sample emission-intensity reading at 460.7 nm,

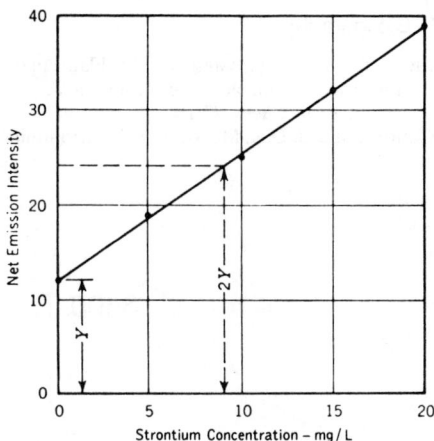

Figure 3500-Sr:1. **Graphical method of computing strontium concentration.**

B = background radiation reading at 466 nm, and
C = slope of calibration line.

Use the ratio D/E only when E mL of sample are evaporated to form a concentrate of 25.0 mL, the value for D.

b. *Graphical method:* Strontium concentration also can be evaluated by the graphical method illustrated in Figure 3500-Sr:1. Plot net intensity against strontium concentration added to sample. The calibration line in the example intersects the ordinate at 12. Thus, $Y = 12$ and $2Y = 24$. Find strontium concentration of sample by locating abscissa value of the point on the calibration line having an ordinate value of 24. In the example, the strontium concentration is 9.0 mg/L.

c. Report a strontium concentration below 10 mg/L to the nearest 0.1 mg/L and one above 10 mg/L to the nearest whole number.

6. Bias

Strontium concentrations in the range 12.0 to 16.0 mg/L can be determined with an accuracy within ±1 to 2 mg/L.

7. Bibliography

CHOW, T.J. & T.G. THOMPSON. 1955. Flame photometric determination of strontium in sea water. *Anal. Chem.* 27:18.

NICHOLS, M.S. & D.R. McNALL. 1957. Strontium content of Wisconsin municipal waters. *J. Amer. Water Works Assoc.* 49:1493.

HORR, C.A. 1959. A survey of analytical methods for the determination of strontium in natural water. U.S. Geol. Surv. Water Supply Pap. No. 1496A.

3500-Tl   THALLIUM

3500-Tl A.   Introduction

1. Occurrence and Significance

Thallium normally occurs only at trace levels (< 10 μg/L) in natural waters. Thallium is toxic and is on several regulatory lists.

2. Selection of Method

Use the flame atomic absorption spectrometric method or the inductively coupled plasma method.

3500-Tl B.   Atomic Absorption Spectrometric Method

See flame atomic absorption spectrometric method, Section 3111B.

3500-Tl C.   Inductively Coupled Plasma Method

See Section 3120.

3500-Th THORIUM

3500-Th A. Introduction

Thorium is quite unsoluble in natural waters and is normally present only at trace levels ($< 1\ \mu g/L$), except in process wastewaters.

3500-Th B. Atomic Absorption Spectrometric Method

See flame atomic absorption spectrometric method, Section 3111D.

3500-Sn TIN

3500-Sn A. Introduction

1. Occurrence

Tin normally is soluble only at trace levels in natural waters ($< 100\ \mu g/L$), except in process wastewaters or mineral waters.

2. Selection of Method

Use either the flame or the electrothermal atomic absorption spectrometric method.

3500-Sn B. Atomic Absorption Spectrometric Method

See flame atomic absorption spectrometric method, Section 3111B, and electrothermal atomic absorption spectrometric method, Section 3113.

3500-Ti   TITANIUM

3500-Ti A.   Introduction

Titanium normally is soluble only at trace levels in natural waters (< 100 $\mu g/$L), except in process wastewaters.

3500-Ti B.   Atomic Absorption Spectrometric Method

See flame atomic absorption spectrometric method, Section 3111D.

3500-V   VANADIUM*

3500-V A.   Introduction

1. Occurrence

Laboratory and epidemiological evidence suggests that vanadium may play a beneficial role in the prevention of heart disease. In New Mexico, which has a low incidence of heart disease, vanadium has been found in concentrations of 20 to 150 $\mu g/L$. In a state where incidence of heart disease is high, vanadium was not found in water supplies. However, vanadium pentoxide dust causes gastrointestinal and respiratory disturbances. The mean concentration found in U.S. drinking waters is 6 $\mu g/L$. Industrial applications of vanadium include dyeing and ceramics, ink, and catalyst manufacture. Discharges from such sources can contribute to its presence in a water supply.

2. Selection of Method

The atomic absorption spectrometric, inductively coupled plasma, and gallic acid methods are suitable for potable water samples. The atomic absorption spectrometric and inductively coupled plasma methods are preferred for polluted samples.

*Approved by Standard Methods Committee, 1985.

3500-V B.   Atomic Absorption Spectrometric Method

See flame atomic absorption spectrometric method, Section 3111D.

3500-V C.   Inductively Coupled Plasma Method

See Section 3120.

3500-V D. Gallic Acid Method

1. General Discussion

a. *Principle:* The concentration of trace amounts of vanadium in water is determined by measuring the catalytic effect it exerts on the rate of oxidation of gallic acid by persulfate in acid solution. Under the given conditions of concentrations of reactants, temperature, and reaction time, the extent of oxidation of gallic acid is proportional to the concentration of vanadium. Vanadium is determined by measuring the absorbance of the sample at 415 nm and comparing it with that of standard solutions treated identically.

b. *Interference:* The substances listed in Table 3500-V:I will interfere in the determination of vanadium if the specified concentrations are exceeded. This is not a serious problem for Cr^{6+}, Co^{2+}, Mo^{6+}, Ni^{2+}, Ag^{+}, and U^{6+} because the tolerable concentration is greater than that commonly encountered in fresh water. However, in some samples the tolerable concentration of Cu^{2+}, Fe^{2+}, and Fe^{3+} may be exceeded. Because of the high sensitivity of the method, interfering substances in concentrations only slightly above tolerance limits can be rendered harmless by dilution.

Traces of Br^{-} and I^{-} interfere seriously and dilution alone will not always reduce the concentration below tolerance limits. Mercuric ion may be added to complex these halides and minimize their interference; however, mercuric ion itself interferes if in excess. Adding 350 μg mercuric nitrate, $Hg(NO_3)_2$, per sample permits determination of vanadium in the presence of up to 100 mg Cl^{-}/L, 250 μg Br^{-}/L, and 250 μg I^{-}/L. Dilute samples containing high concentrations of these ions to concentrations below the values given above and add $Hg(NO_3)_2$.

c. *Minimum detectable concentration:* 0.025 μg V in approximately 13 mL final volume.

2. Apparatus

a. *Water bath,* capable of being operated at 25 ± 0.5°C.

b. *Colorimetric equipment:* One of the following is required:

1) *Spectrophotometer,* for measurements at 415 nm, with a light path of 1 to 5 cm.

2) *Filter photometer,* providing a light path of 1 to 5 cm and equipped with a violet filter with maximum transmittance near 415 nm.

3. Reagents

a. *Stock vanadium solution:* Dissolve 229.6 mg ammonium metavanadate, NH_4VO_3, in a volumetric flask containing approximately 800 mL distilled water and 15 mL 1 + 1 nitric acid (HNO_3). Dilute to 1000 mL; 1.00 mL = 100 μg V.

b. *Intermediate vanadium solution:* Dilute 10.00 mL stock vanadium solution

TABLE 3500-V:I. CONCENTRATION AT WHICH VARIOUS IONS INTERFERE IN THE DETERMINATION OF VANADIUM

Ion	Concentration mg/L
Cr^{6+}	1.0
Co^{2+}	1.0
Cu^{2+}	0.05
Fe^{2+}	0.3
Fe^{3+}	0.5
Mo^{6+}	0.1
Ni^{2+}	3.0
Ag^{+}	2.0
U^{6+}	3.0
Br^{-}	0.1
Cl^{-}	100.0
I^{-}	0.001

with distilled water to 1000 mL; 1.00 mL = 1.00 μg V.

c. *Standard vanadium solution:* Dilute 10.00 mL intermediate vanadium solution with distilled water to 1000 mL; 1.00 mL = 0.010 μg V.

d. *Mercuric nitrate solution:* Dissolve 350 mg $Hg(NO_3)_2 \cdot H_2O$ in 1000 mL distilled water.

e. *Ammonium persulfate-phosphoric acid reagent:* Dissolve 2.5 g $(NH_4)_2S_2O_8$ in 25 mL distilled or demineralized water. Bring just to a boil, remove from heat, and add 25 mL conc H_3PO_4. Let stand approximately 24 h before use. Discard after 48 h.

f. *Gallic acid solution:* Dissolve 2 g $H_6C_7O_5$ in 100 mL warm distilled water, heat to a temperature just below boiling, and filter through filter paper.* Prepare a fresh solution for each set of samples.

4. Procedure

a. *Preparation of standards and sample:* Prepare both blank and sufficient standards by diluting 0- to 8.0-mL portions (0 to 0.08 μg V) of standard vanadium solution to 10 mL with distilled or demineralized water. Pipet sample (10.00 mL maximum) containing less than 0.08 μg V into a suitable container and adjust volume to 10.0 mL with distilled or demineralized water. Filter colored or turbid samples. Add 1.0 mL $Hg(NO_3)_2$ solution to each blank, standard, and sample. Place containers in a water bath regulated to 25 $\pm$ 0.5°C and allow 30 to 45 min for samples to come to the bath temperature.

---

*Whatman No. 42 or equivalent.

b. *Color development and measurement:* Add 1.0 mL ammonium persulfate-phosphoric acid reagent (temperature equilibrated), swirl to mix thoroughly, and return to water bath. Add 1.0 mL gallic acid solution (temperature equilibrated), swirl to mix thoroughly, and return to water bath. Add gallic acid to successive samples at intervals of 30 s or longer to permit accurate control of reaction time. Exactly 60 min after adding gallic acid, remove sample from water bath and measure its absorbance at 415 nm, using distilled water as a reference. Subtract absorbance of blank from absorbance of each standard and sample. Construct a calibration curve by plotting absorbance values of standards versus micrograms vanadium. Determine amount of vanadium in a sample by referring to the corresponding absorbance on the calibration curve. Prepare a calibration curve with each set of samples.

5. Calculation

$$\text{mg V}/\text{L} = \frac{\mu\text{g V (in 13 mL final volume)}}{\text{mL sample}}$$

6. Precision and Bias

In a synthetic sample containing 6 μg V/L, 40 μg As/L, 250 μg Be/L, 240 μg B/L, and 20 μg Se/L in distilled water, vanadium was measured in 22 laboratories with a relative standard deviation of 20% and no relative error.

7. Bibliography

FISHMAN, M.J. & M.V. SKOUGSTAD. 1964. Catalytic determination of vanadium in water. *Anal. Chem.* 36:1643.

3500-Zn   ZINC*

3500-Zn A.   Introduction

1. Occurrence

Zinc is an essential and beneficial element in human growth. Concentrations above 5 mg/L can cause a bitter astringent taste and an opalescence in alkaline waters. The zinc concentration of U.S. drinking waters varies between 0.06 and 7.0 mg/L with a mean of 1.33 mg/L. Zinc most commonly enters the domestic water supply from deterioration of galvanized iron and dezincification of brass. In such cases lead and cadmium also may be present because they are impurities of the zinc used in galvanizing. Zinc in water also may result from industrial waste pollution.

---

*Approved by Standard Methods Committee, 1985.

2. Selection of Method

The atomic absorption spectrometric and inductively coupled plasma methods are preferred. Dithizone method I is intended for potable water whereas dithizone method II is intended for polluted water and the zincon method can be used for either.

3. Sampling and Storage

Analyze samples within 6 h after collection. Addition of HCl preserves the metallic ion content but requires that: (a) the acid be zinc-free; (b) the sample bottles be rinsed with acid before use; and (c) the samples be evaporated to dryness in silica dishes to remove excess HCl before analysis.

3500-Zn B.   Atomic Absorption Spectrometric Method

See flame atomic absorption spectrometric method, Sections 3111B and C.

3500-Zn C.   Inductively Coupled Plasma Method

See Section 3120.

3500-Zn D.   Dithizone Method I

1. General Discussion

a. Principle: Nearly 20 metals can react with dithizone to produce colored coordination compounds. These dithizonates are extractable into organic solvents such as carbon tetrachloride (CCl_4). Most interferences in the zinc-dithizone reaction can be overcome by adjusting the pH to 4.0 to 5.5 and by adding sufficient sodium thiosulfate. Zinc also forms a weak thiosulfate complex that tends to retard the slow and incomplete reaction between zinc and dithizone. For this reason, the determination is empirical and demands the use of an identical technique in standard and sample analysis. The duration and vigor of shaking, the volumes of sample, sodium thiosulfate, and dithizone, and the pH should be kept constant.

b. Interference: Interference from bismuth, cadmium, cobalt, copper, gold, lead, mercury, nickel, palladium, silver, and stannous tin in the small quantities found in potable waters is eliminated by complexing with sodium thiosulfate and by pH adjustment. Ferric iron, residual chlorine, and other oxidizing agents convert dithizone to a yellow-brown color. The zinc-dithizone reaction is extremely sensitive and unusual precautions must be taken to avoid contamination. High and erratic blanks often are traceable to glass containing zinc oxide, surface-contaminated glassware, rubber products, stopcock greases, reagent-grade chemicals, and distilled water. Because of the extreme sensitivity of the reaction, prepare and segregate glassware especially for this determination and extract reagents with dithizone solution to remove all traces of zinc and contaminating metals. Dithizone and dithizonates decompose rapidly in strong light. Perform analyses in subdued light and do not expose solutions to the light of the photometer

longer than is necessary. Avoid direct sunlight.

c. Minimum detectable quantity: 1 μg Zn.

2. Apparatus

a. Colorimetric equipment: Use one of the following:

1) *Spectrophotometer,* for use at either 535 or 620 nm, providing a light path of 1 cm or longer.

2) *Filter photometer,* providing a light path of 2 cm or longer and equipped with either a green filter having maximum transmittance near 535 nm or a red filter having maximum transmittance near 620 nm.

3) *Nessler tubes,* matched.

b. Separatory funnels, capacity 125 to 150 mL, Squibb form, preferably with inert TFE stopcocks.

c. Glassware: Rinse all glassware with 1 + 1 HNO_3 and water.

d. pH meter.

3. Reagents

a. Zinc-free water: Use redistilled or deionized distilled water for rinsing apparatus and preparing solutions and dilutions.

b. Stock zinc solution: Dissolve 100.0 mg 30-mesh zinc metal in a slight excess of 1 + 1 HCl; about 1 mL is required. Dilute to 1000 mL with water; 1.00 mL = 100 μg Zn.

c. Standard zinc solution: Dilute 10.00 mL stock zinc solution to 1000 mL with water; 1.00 mL = 1.00 μg Zn.

d. Hydrochloric acid, HCl, 0.02*N*: Dilute 1.0 mL conc HCl to 600 mL with water. If high blanks are traced to this reagent, dilute conc HCl with an equal volume of distilled water and redistill in an all-borosilicate-glass still.

e. Sodium acetate, 2*M*: Dissolve 68 g $NaC_2H_3O_2 \cdot 3H_2O$ and dilute to 250 mL with water.

f. Acetic acid, 1 + 7.

g. Acetate buffer solution: Mix equal volumes of 2*M* sodium acetate solution and 1 + 7 acetic acid solution. Extract with 10-mL portions of dithizone solution I until the last extract remains green; then extract with CCl₄ to remove excess dithizone.

h. Sodium thiosulfate solution: Dissolve 25 g $Na_2S_2O_3 \cdot 5H_2O$ in 100 mL water. Purify by dithizone extraction as in ¶ 3*g* above.

i. Stock dithizone solution (CCl₄): See Section 1070D.2*b*3). (CAUTION: *CCl₄ is toxic. Avoid inhalation, ingestion, and contact with the skin.*)

j. Dithizone solution I: Dilute 40 mL stock dithizone solution (CCl₄) to 100 mL with CCl₄. Prepare daily.

k. Dithizone solution II: Dilute 10 mL dithizone solution I to 100 mL with CCl₄. Prepare daily.

l. Carbon tetrachloride, CCl₄, ACS grade.

m. Sodium citrate solution: Dissolve 10 g $Na_3C_6H_5O_7 \cdot 2H_2O$ in 90 mL water. Purify by dithizone extraction as in ¶ 3*g* preceding. Use this reagent in final cleansing of glassware.

4. Procedure

a. Preparation of colorimetric standards: To a series of thoroughly cleansed (see ¶ 2*c* above) 125-mL Squibb separatory funnels, add 0, 1.00, 2.00, 3.00, 4.00, and 5.00 mL standard zinc solution to provide standards containing 0, 1.00, 2.00, 3.00, 4.00, and 5.00 µg Zn, respectively. Bring each volume up to 10.0 mL by adding water. To each funnel add 5.0 mL acetate buffer and 1.0 mL $Na_2S_2O_3$ solution, and mix. The pH should be between 4 and 5.5. To each funnel add 10.0 mL dithizone solution II, stopper, and shake vigorously for 4.0 min. Let layers separate, dry inside of stem below stopcock of funnel with strips of filter paper, and run lower (CCl₄) layer into a clean, *dry* absorption cell.

b. Photometric measurement: Measure either the red color of zinc dithizonate at 535 nm or the green color of unreacted dithizone at 620 nm.

Set photometer at 100% transmittance with the blank if the 535-nm wavelength is selected. If 620 nm is used, set blank at 10.0% transmittance. Plot a calibration curve. Run a new calibration curve with each set of samples.

c. Treatment of samples: If the zinc content is not within the working range, dilute sample with water or concentrate it in a silica dish. If the sample has been preserved with acid, evaporate a portion to dryness in a silica dish to remove excess acid. Do not neutralize with hydroxides because these usually contain excessive amounts of zinc. Using a pH meter, adjust sample to pH 2 to 3 with HCl. Transfer 10.0 mL to a separatory funnel. Complete analysis as in ¶ 4*a,* beginning with "To each funnel add 5.0 mL acetate buffer."

d. Visual comparison: If a photometric instrument is not available, run samples and standards at the same time and transfer to matched test tubes or nessler tubes. The range of colors obtained with various amounts of zinc are roughly:

Zinc µg	Color
0 (blank)	green
1	blue
2	blue-violet
3	violet
4	red-violet
5	red-violet

5. Calculation

$$\text{mg Zn/L} = \frac{\mu g \text{ Zn}}{\text{mL sample}}$$

6. Precision and Bias

A synthetic sample containing 650 µg Zn/L, 500 µg Al/L, 50 µg Cd/L, 110 µg Cr/L, 470 µg Cu/L, 300 µg Fe/L, 70 µg Pb/L, 120 µg Mn/L, and 150 µg Ag/L in distilled water was analyzed in 46 laboratories by the dithizone method with a relative standard deviation of 18.2% and a relative error of 25.9%.

7. Bibliography

HIBBARD, P.L. 1937. A dithizone method for measurement of small amounts of zinc. *Ind. Eng. Chem.,* Anal. Ed. 9:127.

SANDELL, E.B. 1937. Determination of copper, zinc, and lead in silicate rocks. *Ind. Eng. Chem.,* Anal. Ed. 9:464.

HIBBARD, P.L. 1938. Estimation of copper, zinc, and cobalt (with nickel) in soil extracts. *Ind. Eng. Chem.,* Anal. Ed. 10:615.

WICHMAN, H.J. 1939. Isolation and determination of traces of metals: The dithizone system. *Ind. Eng. Chem.,* Anal. Ed. 11:66.

COWLING, H. & E.J. MILLER. 1941. Determination of small amounts of zinc in plant materials: A photometric dithizone method. *Ind. Eng. Chem.,* Anal. Ed. 13:145.

ALEXANDER, O.R. & L.V. TAYLOR. 1944. Improved dithizone procedure for determination of zinc in foods. *J. Assoc. Offic. Agr. Chem.* 27:325.

SERFASS, E.J. et al. 1947. *Chem. Anal.* 35:55.

SERFASS, E.J. 1947. Research Report Serial No. 3, American Electroplaters Soc., Newark, N.J., p. 22.

SERFASS, E.J. et al. 1949. Determination of impurities in electroplating solutions. *Plating* 36:254, 818.

SNELL, F.D. & C.T. SNELL. 1949. Colorimetric Methods of Analysis, 3rd ed. D. Van Nostrand Co., Princeton, N.J., Vol. 2, pp. 1-7, 412-419.

BUTTS, P.G., A.R. GAHLER & M.G. MELLON. 1950. Colorimetric determination of metals in sewage and industrial wastes. *Sewage Ind. Wastes* 22:1543.

BARNES, H. 1951. The determination of zinc by dithizone. *Analyst* 76:220.

COOPER, S.S. & M.L. SULLIVAN. 1951. Spectrophotometric studies of dithizone and some dithizonates. *Anal. Chem.* 23:613.

SERFASS, E.J. & R.F. MURACA. 1954. Procedures for Analyzing Metal Finishing Wastes. Ohio River Valley Water Sanitation Comm., Cincinnati, Ohio.

SANDELL, E.B. 1959. Colorimetric Determination of Traces of Metals, 3rd ed. Interscience Publishers, New York, N.Y.

3500-Zn E.　Dithizone Method II

1. Principle

Zinc is separated from other metals by extraction with dithizone and is determined by measuring the color of the zinc-dithizone complex in carbon tetrachloride (CCl_4). Specificity in the separation is achieved by extracting from a nearly neutral solution containing bis(2-hydroxyethyl)dithiocarbamyl ion and cyanide ion, which prevent moderate concentrations of cadmium, copper, lead, and nickel from reacting with dithizone. If excessive amounts of these metals are present, follow the special procedure given in ¶ 4*b*2) below.

See Section 3500-Zn.D for general precautions.

2. Apparatus

a. Colorimetric equipment: One of the following is required:

1) *Spectrophotometer,* for use at 535 nm, providing a light path of 1 cm or longer.

2) *Filter photometer,* providing a light path of 1 cm or longer and equipped with a greenish yellow filter with maximum transmittance near 535 nm.

b. Separatory funnels, 125-mL, Squibb form, with ground-glass stoppers.

3. Reagents

a. Zinc-free water: See Section 3500-Zn.D.3*a*.

b. Stock zinc solution: Dissolve 1000 mg (1.000 g) zinc metal in 10 mL 1 + 1 HNO_3. Dilute and boil to expel oxides of nitrogen. Dilute to 1000 mL; 1.00 mL = 1.00 mg Zn.

c. Methyl red indicator: Dissolve 0.1 g methyl red sodium salt and dilute to 100 mL with water.

d. *Sodium citrate solution:* Dissolve 10 g Na$_3$C$_6$H$_5$O$_7 \cdot$2H$_2$O in 90 mL water. Shake with 10 mL dithizone solution to remove zinc, then filter.

e. *Ammonium hydroxide,* NH$_4$OH, conc: Place 660 mL water in a 1-L polyethylene bottle and chill by immersion in an ice bath. Pass ammonia gas from a cylinder through a glass-wool trap into chilled bottle until volume of liquid has increased to 900 mL. Alternatively, place 900 mL conc reagent-grade NH$_4$OH in a 1500-mL distillation flask and distill into a chilled 1-L polyethylene bottle initially containing 250 mL water. Continue distilling until volume of liquid in bottle has increased to 900 mL, keeping condenser tip below surface of liquid.

f. *Potassium cyanide solution:* Dissolve 5 g KCN in 95 mL water. (CAUTION: *Potassium cyanide is a deadly poison. Avoid skin contact or inhalation of vapors. Do not pipet by mouth or bring in contact with acids.)*

g. *Acetic acid,* conc.

h. *Carbon tetrachloride,* CCl$_4$: See Section 3500-Zn.D.3*l.*

i. *Bis (2-hydroxyethyl) dithiocarbamate solution:* Dissolve 4.0 g diethanolamine and 1 mL CS$_2$ in 40 mL methyl alcohol. Prepare every 3 or 4 d.

j. *Dithizone solution:* Dilute 50 mL stock dithizone solution (CCl$_4$), prepared in accordance with Section 1070D.2*b*3), to 250 mL with CCl$_4$. Prepare fresh daily.

k. *Sodium sulfide solution I:* Dissolve 3.0 g Na$_2$S$\cdot$9H$_2$O or 1.65 g Na$_2$S$\cdot$3H$_2$O in 100 mL water.

l. *Sodium sulfide solution II:* Prepare just before use by diluting 4 mL Na$_2$S solution I to 100 mL.

m. *Nitric acid,* HNO$_3$, 6*N.*

n. *Hydrogen sulfide,* H$_2$S.

4. Procedure

a. *Preparation of calibration curve:*

1) Prepare, just before use, a zinc solution containing 2.0 μg Zn/mL by diluting 5 mL standard zinc solution to 250 mL, then diluting 10 mL of the latter solution to 100 mL with water. Pipet 5.00, 10.00, 15.00, and 20.00 mL, containing 10 to 40 μg Zn, into separate 125-mL separatory funnels and adjust volume to about 20 mL. Set up another funnel containing 20 mL water as a blank.

2) Add 2 drops methyl red indicator and 2.0 mL sodium citrate solution to each funnel. If the indicator is not yellow, add conc NH$_4$OH a drop at a time until it just turns yellow. Add 1.0 mL KCN solution and acetic acid, a drop at a time, until the indicator just turns peach color.

3) Extract methyl red by shaking with 5 mL CCl$_4$. Discard yellow CCl$_4$ layer. Add 1 mL dithiocarbamate solution. Extract with 10 mL dithizone solution, shaking for 1 min.

Draw CCl$_4$ layer into another separatory funnel and repeat the extraction with successive 5-mL portions of dithizone solution until the last one shows no change from the green dithizone color. Discard aqueous layer.

4) Shake combined dithizone extracts with a 10-mL portion of Na$_2$S solution II, separate layers, and repeat the washing with further 10-mL portions of Na$_2$S solution until the unreacted dithizone solution has been removed completely, as shown by the color of the aqueous layer, which remains colorless or very pale yellow; usually three washings are sufficient.

Remove water adhering to stem of funnel with a cotton swab and drain pink CCl$_4$ solution into a dry 50-mL volumetric flask. Use a few milliliters of CCl$_4$ to rinse the last droplets from funnel and dilute to mark with CCl$_4$.

5) Determine absorbance of the zinc dithizonate solutions at 535 nm, using CCl$_4$ as a reference. Plot an absorbance-concentration curve after subtracting the blank absorbance. The calibration curve is linear if monochromatic light is used.

6) Clean separatory funnels by shaking several minutes successively with HNO_3, distilled water, and finally a mixture of 5 mL sodium citrate and 5 mL dithizone, to minimize large or erratic blanks that result from adsorption of zinc on the glass surface. If possible, reserve separatory funnels exclusively for the zinc determination.

b. Treatment of sample:

1) Digest sample as directed under Preliminary Digestion for Metals, Section 3030D. Transfer a portion containing 10 to 40 μg Zn to a clean 125-mL separatory funnel and adjust volume to about 20 mL. Determine zinc in this solution as described in ¶ *4a.*

If more than 30 mL dithizone solution is needed to extract zinc completely, the portion taken contains too much zinc or the quantity of other metals that react with dithizone exceeds the amount that can be withheld by the complexing agent. If this occurs, follow the procedure in ¶ *4b2)* below.

2) Separation of excessive amounts of cadmium, copper, and lead—When the quantity of these metals, separately or jointly, exceeds 2 mg in the portion taken, adjust volume to about 20 mL in a 100-mL beaker. Adjust acidity to 0.4 to 0.5N* by adding dilute HNO_3 or NH_4OH as necessary. Pass H_2S into cold solution for 5 min. Filter off the precipitated sulfides through a sintered-glass filter and wash precipitate with two small portions of hot water. Boil filtrate 3 to 4 min to remove H_2S, cool, transfer to a separatory funnel, and determine zinc as described in ¶ *4b1)* et seq.

5. Calculation

$$\text{mg Zn/L} = \frac{\mu\text{g Zn}}{\text{mL sample}} \times \frac{100}{\text{mL portion}}$$

6. Bibliography

See 3500-Zn.D.7.

---

*The normalities of the solutions obtained in the preliminary treatment are approximately 3N for the HNO_3-H_2SO_4 digestion and approximately 0.8N for the HNO_3-$HClO_4$ digestion.

3500-Zn F. Zincon Method

1. General Discussion

a. Principle: Zinc forms a blue complex with 2-carboxy-2'-hydroxy-5'-sulfoformazyl benzene (zincon) in a solution buffered to pH 9.0. Other heavy metals likewise form colored complexes with zincon. Cyanide is added to complex zinc and heavy metals. Cyclohexanone is added to free zinc selectively from its cyanide complex so that it can be complexed with zincon to form a blue color. Sodium ascorbate reduces manganese interference. The developed color is stable except in the presence of copper.

b. Interferences: The following ions interfere in concentrations exceeding those listed:

Ion	mg/L	Ion	mg/L
Cd^{2+}	1	Cr^{3+}	10
Al^{3+}	5	Ni^{2+}	20
Mn^{2+}	5	Cu^{2+}	30
Fe^{3+}	7	Co^{2+}	30
Fe^{2+}	9	CrO_4^{2-}	50

c. Minimum detectable concentration: 0.02 mg Zn/L.

2. Apparatus

Colorimetric equipment: One of the following is required:

a. Spectrophotometer, for measurements

at 620 nm, providing a light path of 1 cm or longer.

b. Filter photometer, providing a light path of 1 cm or longer and equipped with a red filter having maximum transmittance near 620 nm. Deviation from Beer's Law occurs when the filter band pass exceeds 20 nm.

3. Reagents

a. Zinc-free water: See Section 3500-Zn.D.3*a*.

b. Stock zinc solution: See Section 3500-Zn.E.3*b*; 1.00 mL = 1.00 mg Zn.

c. Standard zinc solution: Dilute 10.00 mL stock zinc solution to 1000 mL; 1.00 mL = 10.00 μg Zn.

d. Sodium ascorbate, fine granular powder, USP.

e. Potassium cyanide solution: Dissolve 1.00 g KCN in approximately 50 mL water and dilute to 100 mL. CAUTION: *Potassium cyanide is a deadly poison. Avoid skin contact or inhalation of vapors. Do not pipet by mouth or bring in contact with acids.*

f. Buffer solution, pH 9.0: Dissolve 8.4 g NaOH pellets in about 500 mL water. Add 31.0 g H_3BO_3 and swirl or stir to dissolve. Dilute to 1000 mL with water and mix thoroughly.

g. Zincon reagent: Dissolve 100 mg zincon (2-carboxy-2'-hydroxy-5'-sulfoformazyl benzene) in 100 mL methanol. Because zincon dissolves slowly, stir and/or let stand overnight.

h. Cyclohexanone, purified.

i. Hydrochloric acid, HCl, conc and 6*N*.

j. Sodium hydroxide, NaOH, 6*N*.

4. Procedure

a. Preparation of colorimetric standards: Add 0, 0.5, 1.0, 3.0, 5.0, 10.0, and 14.0 mL standard zinc solution to a series of clean 50-mL graduated mixing cylinders or erlenmeyer flasks. Dilute each to 20.0 mL to yield solutions containing 0, 0.25, 0.5, 1.5,

2.5, 5.0, and 7.0 mg Zn/L, respectively. Add the following to each solution in sequence, mixing thoroughly after each addition: 0.5 g sodium ascorbate, 5.0 mL buffer solution, 2.0 mL KCN solution, and 3.0 mL zincon solution. Pipet 20.0 mL of the solution into a clean 50-mL erlenmeyer flask and add 1.0 mL cyclohexanone. Swirl for 10 s and note time. Transfer portions of both solutions to clean sample cells. Use solution without cyclohexanone to zero colorimeter. Read and record absorbance for solution with cyclohexanone after 1 min. The calibration curve does not pass through zero because of the color enhancement effect of cyclohexanone on zincon.

b. Treatment of samples: To determine dissolved zinc, filter sample through a 0.45-μm membrane filter. Adjust to pH 7 with 6*N* NaOH or 6*N* HCl if necessary after filtering. For total zinc add 1 mL conc HCl to 50 mL sample and mix thoroughly. Filter and adjust to pH 7. Before analysis cool samples to less than 30°C if necessary. Analyze 20.0 mL of prepared sample as described in ¶ 4*a* above, beginning with "Add the following to each solution . . ." If the zinc concentration exceeds 7 mg/L prepare a sample dilution and analyze a 20.0-mL portion.

5. Calculation

Read zinc concentration (in milligrams per liter) directly from the calibration curve.

6. Precision and Bias

A synthetic sample containing 650 μg Zn/L, 500 μg Al/L, 50 μg Cd/L, 110 μg Cr/L, 470 μg Cu/L, 300 μg Fe/L, 70 μg Pb/L, 120 μg Mn/L, and 150 μg Ag/L in doubly demineralized water was analyzed in a single laboratory. A series of 10 replicates gave a relative standard deviation of 0.96% and a relative error of 0.15%.

A wastewater sample from an industry in Standard Industrial Classification (SIC) No. 3333, primary smelting and refining of zinc, was analyzed by 10 different persons. The mean zinc concentration was 3.36 mg Zn/L and the relative standard deviation was 1.7%. The relative error compared to results from an atomic absorption analysis of the same sample was −1.0%.

7. Bibliography

PLATTE, J.A. & V.M. MARCY. 1959. Photometric determination of zinc with zincon. *Anal. Chem.* 31:1226.

RUSH, R.M. & J.H. YOE. 1954. Colorimetric determination of zinc and copper with 2-carboxy-2′hydroxy-5′-sulfoformazyl-benzene. *Anal. Chem.* 26:1345.

MILLER, D.G. 1979. Colorimetric determination of zinc with zincon and cyclohexanone. *J. Water Pollut. Control Fed.* 51:2402.

PART 4000

DETERMINATION OF INORGANIC

NONMETALLIC CONSTITUENTS

4010 INTRODUCTION

The analytical methods included in this part make use of classical wet chemical techniques and their automated variations and such modern instrumental techniques as ion chromatography. Methods that measure various forms of chlorine, nitrogen, and phosphorus are presented. The procedures are intended for use in the assessment and control of receiving water quality, the treatment and supply of potable water, and the measurement of operation and process efficiency in wastewater treatment. The methods also are appropriate and applicable in evaluation of environmental water-quality concerns. The introduction to each procedure contains reference to special field sampling conditions, appropriate sample containers, proper procedures for sampling and storage, and the applicability of the method.

4020 QUALITY CONTROL

Detailed recommendations and general information concerning laboratory quality assurance, as well as specific details of applicable quality-control measures, are provided in Part 1000 and in individual methods. Also follow this general principle:

Keep in mind the overall purpose of the measurements. Can duplicates (replicates) adequately establish precision and can recovery of known additions determine bias? Are suitable standards, control charts, blanks, calibrations, and other necessary measures incorporated to assure the quality of data produced? Furthermore, can adequate documentation be presented to support these contentions? Obviously, the same degree of substantiation may not be necessary in all cases, e.g., requirements for an operational control determination may differ from those for data submitted for regulatory purposes. Therefore, the intended use of the data is an important factor in determining overall quality control as well as the accompanying quality assurance measures.

4110   DETERMINATION OF ANIONS BY ION
CHROMATOGRAPHY*

4110 A.   Introduction

Determination of the common anions such as bromide, chloride, fluoride, nitrate, nitrite, phosphate, and sulfate often is desirable to characterize a water and/or to assess the need for specific treatment. Although conventional colorimetric, electrometric, or titrimetric methods are available

*Approved by Standard Methods Committee, 1988.

for determining individual anions, only ion chromatography provides a single instrumental technique that may be used for their rapid, sequential measurement. Ion chromatography eliminates the need to use hazardous reagents and it effectively distinguishes among the halides (Br^-, Cl^-, and F^-) and the oxides (SO_3^{2-}-SO_4^{2-} or NO_2^--NO_3^-).

4110 B.   Ion Chromatography with Chemical Suppression of Eluant Conductivity

1. General Discussion

a. Principle: A water sample is injected into a stream of carbonate-bicarbonate eluant and passed through a series of ion exchangers. The anions of interest are separated on the basis of their relative affinities for a low capacity, strongly basic anion exchanger (guard and separator columns). The separated anions are directed onto a strongly acidic cation exchanger (packed-bed suppressor) or through a hollow fiber of cation exchanger membrane (fiber suppressor) or micromembrane suppressor bathed in continuously flowing strongly acid solution (regenerant solution). In the suppressor the separated anions are converted to their highly conductive acid forms and the carbonate-bicarbonate

eluant is converted to weakly conductive carbonic acid. The separated anions in their acid forms are measured by conductivity. They are identified on the basis of retention time as compared to standards. Quantitation is by measurement of peak area or peak height.

b. Interferences: Any substance that has a retention time coinciding with that of any anion to be determined will interfere. For example, relatively high concentrations of low-molecular-weight organic acids interfere with the determination of chloride and fluoride. A high concentration of any one ion also interferes with the resolution of others. Sample dilution overcomes many interferences. To resolve uncertainties of identification or quantitation use the method of known additions. Spurious peaks may result from contaminants in reagent water, glassware, or sample processing apparatus. Because small sample volumes are used, scrupulously avoid contamination. The method is applicable to surface, ground, and drinking water, treated mixed wastewater, and some industrial process waters, such as boiler water and cooling water, subject to the provision for filtration through a 0.2-μm membrane filter to avoid plugging the columns. Modifications such as preconcentration of samples, gradient elution, or reinjection of portions of the eluted sample may alleviate some interferences but require individual validation for precision and bias.

c. Minimum detectable concentration: The minimum detectable concentration of an anion is a function of sample size and conductivity scale used. Generally, minimum detectable concentrations are near 0.1 mg/L for Br^-, Cl^-, NO_3^-, NO_2^-, PO_4^{3-}, and SO_4^{2-} with a 100-μL sample loop and a 10-μS/cm full-scale setting on the conductivity detector. Similar values may be achieved by using a higher scale setting and an electronic integrator.

d. Limitations: This method is not recommended for the routine determination of F^- because equivalency studies have indicated positive or negative bias and poor precision in some samples. F^- can be determined accurately by ion chromatography using special techniques such as dilute eluant or gradient elution with fiber suppressor or membrane suppressors.

2. Apparatus

a. Ion chromatograph, including an injection valve, a sample loop, guard, separator, and suppressor columns or membrane suppressors, a temperature-compensated small-volume conductivity cell (6 μL or less), and a strip-chart recorder capable of full-scale response of 2 s or less. An electronic peak integrator is optional. Use an ion chromatograph capable of delivering 2 to 5 mL eluant/min at a pressure of 1400 to 6900 kPa.

b. Anion separator column, with styrene divinylbenzene-based low-capacity pellicular anion-exchange resin capable of resolving Br^-, Cl^-, NO_3^-, NO_2^-, PO_4^{3-}, and SO_4^{2-}; 4 × 250 mm.*

c. Guard column, identical to separator column except 4 × 50 mm,† to protect separator column from fouling by particulates or organics.

d. Suppressor: Use either:

1) *Suppressor column,* high-capacity cation-exchange resin capable of converting eluant and separated anions to their acid forms.‡

2) *Fiber suppressor or membrane suppressor:*§ Cation-exchange membrane capable of continuously converting eluant and separated anions to their acid forms.

3. Reagents

a. Deionized or distilled water free from interferences at the minimum detection limit of each constituent and filtered through a 0.2-μm membrane filter to avoid plugging columns.

b. Eluant solution, sodium bicarbonate-sodium carbonate, 0.003M NaHCO$_3$-0.0024M Na$_2$CO$_3$: Dissolve 1.008 g NaHCO$_3$ and 1.0176 g Na$_2$CO$_3$ in water and dilute to 4 L.

c. Regenerant solution 1, H_2SO_4, 1N: Use this regenerant when suppressor is not a continuously regenerated one.

d. Regenerant solution 2, H_2SO_4, 0.025N: Dilute 2.8 mL conc H_2SO_4 to 4 L or 100 mL regenerant solution 1 to 4 L. Use this regenerant with continuous regeneration-fiber suppressor system.

---

*Dionex P/N 030827 (normal run) or P/N 030831 (fast run), or equivalent.
†Dionex P/N 030825 (normal run) or P/N 030830 (fast run), or equivalent.
‡No longer available commercially.
§Dionex P/N 037072 (micro membrane—high capacity/low volume—suppressor), or equivalent.

e. Standard anion solutions, 1000 mg/L: Prepare a series of standard anion solutions by weighing the indicated amount of salt, dried to a constant weight at 105°C, to 1000 mL. Store in plastic bottles in a refrigerator; these solutions are stable for at least 1 month. Verify stability.

Anion ‖	Salt	Amount g/L
Cl^-	NaCl	1.6485
Br^-	NaBr	1.2876
NO_3^-	$NaNO_3$	1.3707
NO_2^-	$NaNO_2$	1.4998#
PO_4^{3-}	KH_2PO_4	1.4330
SO_4^{2-}	K_2SO_4	1.8141

f. Combined working standard solution, high range: Combine 10 mL of the Cl^-, NO_3^-, NO_2^-, and PO_4^{3-} standard anion solutions, 1 mL of the Br^-, and 100 mL of the SO_4^{2-} standard solutions, dilute to 1000 mL, and store in a plastic bottle protected from light; contains 10 mg/L each of Cl^-, NO_3^-, NO_2^-, and PO_4^{3-}, 1 mg Br^-/L, and 100 mg SO_4^{2-}/L. Prepare fresh daily.

g. Combined working standard solution, low range: Dilute 100 mL combined working standard solution, high range, to 1000 mL and store in a plastic bottle protected from light; contains 1.0 mg/L each Cl^-, NO_3^-, NO_2^-, and PO_4^{3-}, 0.1 mg Br^-/L, and 10 mg SO_4^{2-}/L. Prepare fresh daily.

h. Alternative combined working standard solutions: Prepare appropriate combinations according to anion concentration to be determined. If NO_2^- and PO_4^{3-} are not included, the combined working standard is stable for 1 month.

4. Procedure

a. System equilibration: Turn on ion chromatograph and adjust eluant flow rate to approximate the separation achieved in

‖ Expressed as compound.
\#Do not oven-dry, but dry to constant weight in a desiccator.

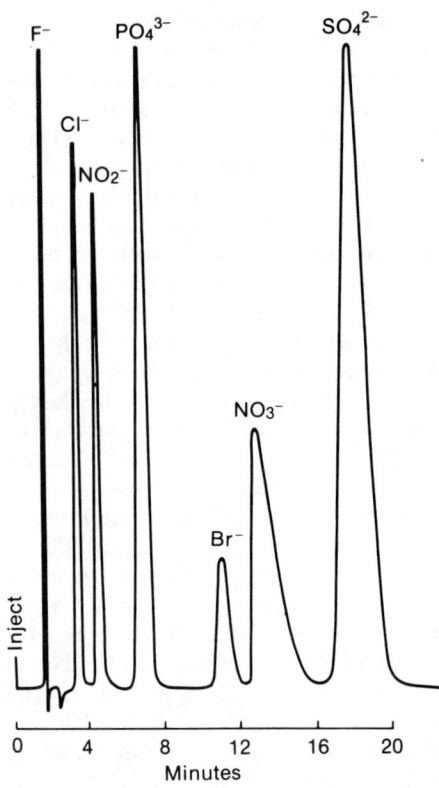

Figure 4110:1. Typical inorganic anion separation using normal-run columns. Conditions: 3mM $NaHCO_3$/2.4 mM Na_2CO_3 eluant; 2.0 mL/min flow.

Figures 4110:1 or 4110:2 (2 to 3 mL/min). Adjust detector to desired setting (usually 10 μS) and let system come to equilibrium (15 to 20 min). A stable base line indicates equilibrium conditions. Adjust detector offset to zero out eluant conductivity; with the fiber or membrane suppressor adjust the regeneration flow rate to maintain stability, usually 2.5 to 3 mL/min.

b. Calibration: Inject standards containing a single anion or a mixture and determine approximate retention times. Observed times vary with conditions but if standard eluant and anion separator column are used, retention always is in the order Cl^-, NO_2^-, HPO_4^{2-}, Br^-, NO_3^-, and SO_4^{2-}. Inject at least three different con-

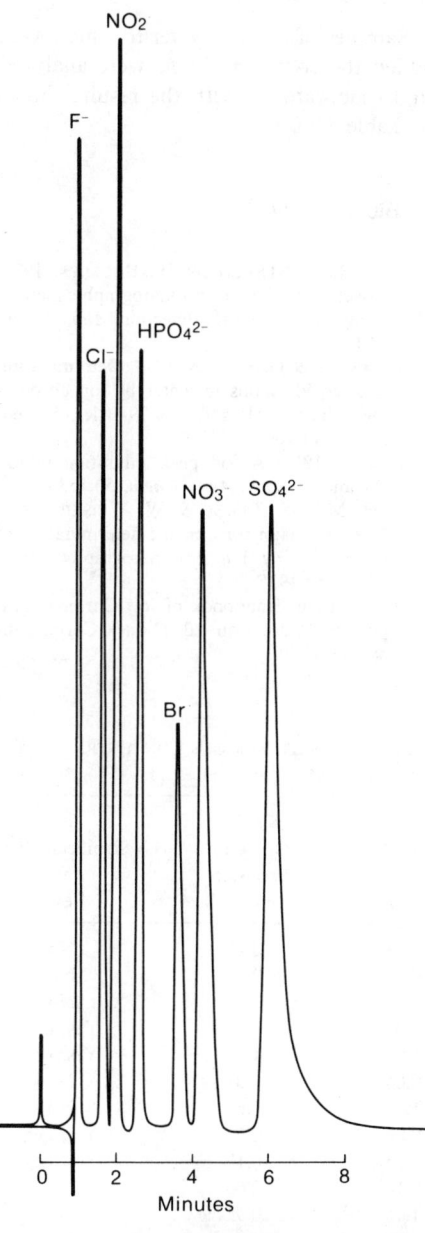

Figure 4110:2. Fast-run column separation. Conditions: 2.8 mM NaHCO$_3$/2.3 mM Na$_2$CO$_3$ eluant; 2.0 mL/min flow; 10-μL loop; 10 μS/cm full scale. Concentrations (mg/L): F$^-$, 3; Cl$^-$, 4; NO$_2^-$, 10; HPO$_4^{2-}$, 25; Br$^-$, 10; NO$_2^-$, 20; SO$_4^{2-}$, 25.

centrations for each anion to be measured and construct a calibration curve by plotting peak height or area against concentration on linear graph paper. Recalibrate whenever the detector setting is changed. With a system requiring suppressor regeneration, NO$_2^-$ interaction with the suppressor may lead to erroneous NO$_2^-$ results; make this determination only when the suppressor is at the same stage of exhaustion as during standardization or recalibrate frequently. In this type of system the water dip** may shift slightly during suppressor exhaustion and with a fast run column this may lead to slight interference for Cl$^-$. To eliminate this interference analyze standards that bracket the expected result or eliminate the water dip by diluting the sample with eluant or by adding concentrated eluant to the sample to give the same HCO$_3^-$/CO$_3^{2-}$ concentration as in the eluant. If sample adjustments are made, adjust standards and blanks identically.

If linearity is established for a given detector setting, single standard calibration is acceptable. Record peak height or area and retention time for calculation of the calibration factor, F. However, a calibration curve will result in better precision and bias.

c. Sample analysis: Remove sample particulates, if necessary, by filtering through a prewashed 0.2-μm-pore-diam membrane filter. Using a prewashed syringe of 1 to 10 mL capacity equipped with a male luer fitting inject sample or standard. Inject enough sample to flush sample loop several times: for 0.1 mL sample loop inject at least 1 mL. Switch ion chromatograph from load to inject mode and record peak heights and retention times on strip chart recorder. After the last peak (SO$_4^{2-}$) has appeared and the conductivity signal has returned to base line, another sample can be injected.

**Water dip occurs because water conductivity in sample is less than eluant conductivity (eluant is diluted by water).

d. Regeneration: For systems without fiber suppressor regenerate with $1N$ H_2SO_4 in accordance with the manufacturer's instructions when the conductivity base line exceeds 300 μS with the suppressor column on line.

5. Calculations

Calculate concentration of each anion, in milligrams per liter, by referring to the appropriate calibration curve. Alternatively, when the response is shown to be linear, use the following equation:

$$C = H \times F \times D$$

where:

C = mg anion/L,

H = peak height or area,

F = response factor = concentration of standard/height (or area) of standard, and

D = dilution factor for those samples requiring dilution.

6. Precision and Bias

Samples of reagent water to which were added the common anions were analyzed in 15 laboratories with the results shown in Table 4110:I.

7. Bibliography

SMALL, H., T. STEVENS & W. BAUMAN. 1975. Novel ion exchange chromatographic method using conductimetric detection. *Anal. Chem.* 47:1801.

FISHMAN, M. & G. S. PYEN. 1979. Determination of selected anions in water by ion chromatograph. Water Res. Invest. 79-101, U.S. Geological Survey.

JENKE, D. 1981. Anion peak migration in ion chromatography. *Anal. Chem.* 53:1536.

BYNUM, M.I., S. TYREE & W. WEISER. 1981. Effect of major ions on the determination of trace ions by ion chromatography. *Anal. Chem.* 53:1935.

WEISS, J. 1986. Handbook of Ion Chromatography. E. L. Johnson, ed. Dionex Corp., Sunnyvale, Calif.

TABLE 4110:I. PRECISION AND BIAS OBSERVED FOR ANIONS AT VARIOUS CONCENTRATION LEVELS IN REAGENT WATER*

Anion	Amount Added mg/L	Amount Found mg/L	Overall Precision mg/L	Single-Operator Precision mg/L	Significant Bias 95% Level
Cl^-	0.76	0.86	0.38	0.11	No
Cl^-	17	17.2	0.82	0.43	No
Cl^-	455	471	46	13	No
NO_2^-	0.45	0.09	0.09	0.04	Yes, neg
NO_2^-	21.8	19.4	1.9	1.3	Yes, neg
Br^-	0.25	0.25	0.04	0.02	No
Br^-	13.7	12.9	1.0	0.6	No
PO_4^{3-}	0.18	0.10	0.06	0.03	Yes, neg
PO_4^{3-}	0.49	0.34	0.15	0.17	Yes, neg
NO_3^-	0.50	0.33	0.16	0.03	No
NO_3^-	15.1	14.8	1.15	0.9	No
SO_4^{2-}	0.51	0.52	0.07	0.03	No
SO_4^{2-}	43.7	43.5	2.5	2.2	No

*Adopted, with permission, from the Annual Book of ASTM Standards. Part 31. Copyright: ASTM, 1916 Race Street, Philadelphia, Pa. 19103.

†These results were obtained with packed-column suppressors.

4500-B BORON*

4500-B A. Introduction

1. Occurrence and Significance

Although it is an element essential for plant growth, boron in excess of 2.0 mg/L in irrigation water is deleterious to certain plants and some plants may be affected adversely by concentrations as low as 1.0 mg/L (or even less in commercial greenhouses). Drinking waters rarely contain more than 1 mg B/L and generally less than 0.1 mg/L, concentrations considered innocuous for human consumption. Boron may occur naturally in some waters or may find its way into a watercourse through cleaning compounds and industrial waste effluents. Seawater contains approximately 5 mg B/L and this element is found in saline estuaries in association with other seawater salts.

The ingestion of large amounts of boron can affect the central nervous system. Protracted ingestion may result in a clinical syndrome known as borism.

2. Selection of Method

The curcumin method (B) is applicable in the 0.10- to 1.0-mg/L range, while the carmine method (C) is suitable for the determination of boron concentrations in the 1- to 10-mg/L range. The range of these methods can be extended by dilution or concentration of the sample. Boron also can be determined by the inductively coupled plasma method (D).

3. Sampling and Storage

Store samples in polyethylene bottles or alkali-resistant, boron-free glassware.

---

*Approved by Standard Methods Committee, 1988.

4500-B B. Curcumin Method

1. General Discussion

a. Principle: When a sample of water containing boron is acidified and evaporated in the presence of curcumin, a red-colored product called rosocyanine is formed. The rosocyanine is taken up in a suitable solvent and the red color is compared with standards visually or photometrically.

b. Interference: NO_3^--N concentrations above 20 mg/L interfere. Significantly high results are possible when the total of calcium and magnesium hardness exceeds 100 mg/L as calcium carbonate ($CaCO_3$). Moderate hardness levels also can cause a considerable percentage error in the low boron range. This interference springs from the insolubility of the hardness salts in 95% ethanol and consequent turbidity in the final solution. Filter the final solution or pass the original sample through a column of strongly acidic cation-exchange resin in the hydrogen form to remove interfering cations. The latter procedure permits application of the method to samples of high hardness or solids content. Phosphate does not interfere.

c. *Minimum detectable quantity:* 0.2 μg B.

2. Apparatus

a. *Colorimetric equipment:* One of the following is required:

1) *Spectrophotometer,* for use at 540 nm, with a minimum light path of 1 cm.

2) *Filter photometer,* equipped with a green filter having a maximum transmittance near 540 nm, with a minimum light path of 1 cm.

b. *Evaporating dishes,* 100- to 150-mL capacity, of high-silica glass,* platinum, or other suitable material.

c. *Water bath,* set at 55 ± 2°C.

d. *Glass-stoppered volumetric flasks,* 25- and 50-mL capacity.

e. *Ion-exchange column,* 50 cm long by 1.3 cm in diameter.

3. Reagents

Store all reagents in polyethylene or boron-free containers.

a. *Stock boron solution:* Dissolve 571.6 mg anhydrous boric acid, H_3BO_3, in distilled water and dilute to 1000 mL; 1.00 mL = 100 μg B. Because H_3BO_3 loses weight on drying at 105°C, use a reagent meeting ACS specifications and keep the bottle tightly stoppered to prevent entrance of atmospheric moisture.

b. *Standard boron solution:* Dilute 10.00 mL stock boron solution to 1000 mL with distilled water; 1.00 mL = 1.00 μg B.

c. *Curcumin reagent:* Dissolve 40 mg finely ground curcumin† and 5.0 g oxalic acid in 80 mL 95% ethyl alcohol. Add 4.2

*Vycor, manufactured by Corning Glass Works, or equivalent.
†Eastman No. 1179 or equivalent.

mL conc HCl, make up to 100 mL with ethyl alcohol in a 100-mL volumetric flask, and filter if reagent is turbid (isopropyl alcohol, 95%, may be used in place of ethyl alcohol). This reagent is stable for several days if stored in a refrigerator.

d. *Ethyl or isopropyl alcohol,* 95%.

e. *Reagents for removal of high hardness and cation interference:*

1) *Strongly acidic cation-exchange resin.*

2) *Hydrochloric acid,* HCl, 1 + 5.

4. Procedure

a. *Precautions:* Closely control such variables as volumes and concentrations of reagents, as well as time and temperature of drying. Use evaporating dishes identical in shape, size, and composition to insure equal evaporation time because increasing the time increases intensity of the resulting color.

b. *Preparation of calibration curve:* Pipet 0 (blank), 0.25, 0.50, 0.75, and 1.00 μg boron into evaporating dishes of the same type, shape, and size. Add distilled water to each standard to bring total volume to 1.0 mL. Add 4.0 mL curcumin reagent to each and swirl gently to mix contents thoroughly. Float dishes on a water bath set at 55 ± 2°C and let them remain for 80 min, which is usually sufficient for complete drying and removal of HCl. Keep drying time constant for standards and samples. After dishes cool to room temperature, add 10 mL 95% ethyl alcohol to each dish and stir gently with a polyethylene rod to insure complete dissolution of the red-colored product.

Wash contents of dish into a 25-mL volumetric flask, using 95% ethyl alcohol. Make up to mark with 95% ethyl alcohol and mix thoroughly by inverting. Read absorbance of standards and samples at a wavelength of 540 nm after setting reagent blank at zero absorbance. The calibration curve is linear from 0 to 1.00 μg boron.

Make photometric readings within 1 h of drying samples.

c. *Sample treatment:* For waters containing 0.10 to 1.00 mg B/L, use 1.00 mL sample. For waters containing more than 1.00 mg B/L, make an appropriate dilution with boron-free distilled water, so that a 1.00-mL portion contains approximately 0.50 μg boron.

Pipet 1.00 mL sample or dilution into an evaporating dish. Unless the calibration curve is being determined at the same time, prepare a blank and a standard containing 0.50 μg boron and run in conjunction with the sample. Proceed as in ¶ 4b, beginning with "Add 4.0 mL curcumin reagent. . . ." If the final solution is turbid, filter through filter paper‡ before reading absorbance. Calculate boron content from calibration curve.

d. *Visual comparison:* The photometric method may be adapted to visual estimation of low boron concentrations, from 50 to 200 μg/L, as follows: Dilute the standard boron solution 1 + 3 with distilled water; 1.00 mL = 0.20 μg B. Pipet 0, 0.05, 0.10, 0.15, and 0.20 μg boron into evaporating dishes as indicated in ¶ 4b. At the same time add an appropriate volume of sample (1.00 mL or portion diluted to 1.00 mL) to an identical evaporating dish. The total boron should be between 0.05 and 0.20 μg. Proceed as in ¶ 4b, beginning with "Add 4.0 mL curcumin reagent. . . ." Compare color of samples with standards within 1 h of drying samples.

e. *Removal of high hardness and cation interference:* Prepare an ion-exchange column of approximately 20 cm × 1.3 cm diam. Charge column with a strongly acidic cation-exchange resin. Backwash column with distilled water to remove entrained air bubbles. Keep the resin covered with liquid at all times. Pass 50 mL 1 + 5 HCl through column at a rate of 0.2 mL

acid/mL resin in column/min and wash column free of acid with distilled water.

Pipet 25 mL sample, or a smaller sample of known high boron content diluted to 25 mL, onto the resin column. Adjust rate of flow to about 2 drops/s and collect effluent in a 50-mL volumetric flask. Wash column with small portions of distilled water until flask is filled to mark. Mix and transfer 2.00 mL into evaporating dish. Add 4.0 mL curcumin reagent and complete the analysis as described in ¶ 4b preceding.

5. Calculation

Use the following equation to calculate boron concentration from absorbance readings:

$$ mg\ B/L = \frac{A_2 \times C}{A_1 \times S} $$

where:

A_1 = absorbance of standard,
A_2 = absorbance of sample,
C = μg B in standard taken, and
S = mL sample.

6. Precision and Bias

A synthetic sample containing 240 μg B/L, 40 μg As/L, 250 μg Be/L, 20 μg Se/L, and 6 μg V/L in distilled water was analyzed in 30 laboratories by the curcumin method with a relative standard deviation of 22.8% and a relative error of 0%.

7. Bibliography

SILVERMAN, L. & K. TREGO. 1953. Colorimetric microdetermination of boron by the curcumin-acetone solution method. *Anal. Chem.* 25:1264.

DIRLE, W.T., E. TRUOG & K.C. BERGER. 1954. Boron determination in soils and plants—Simplified curcumin procedure. *Anal. Chem.* 26:418.

LUKE, C.L. 1955. Determination of traces of boron in silicon, germanium, and germanium dioxide. *Anal. Chem.* 27:1150.

---

‡Whatman No. 30 or equivalent.

LISHKA, R.J. 1961. Comparison of analytical procedures for boron. *J. Amer. Water Works Assoc.* 53:1517.

BUNTON, N.G. & B.H. TAIT. 1969. Determination of boron in waters and effluents using curcumin. *J. Amer. Water Works Assoc.* 61:357.

4500-B C.   Carmine Method

1. General Discussion

a. Principle: In the presence of boron, a solution of carmine or carminic acid in concentrated sulfuric acid changes from a bright red to a bluish red or blue, depending on the concentration of boron present.

b. Interference: The ions commonly found in water and wastewater do not interfere.

c. Minimum detectable quantity: 2 μg B.

2. Apparatus

Colorimetric equipment: One of the following is required:

a. Spectrophotometer, for use at 585 nm, with a minimum light path of 1 cm.

b. Filter photometer, equipped with an orange filter having a maximum transmittance near 585 nm, with a minimum light path of 1 cm.

3. Reagents

Store all reagents in polyethylene or boron-free containers.

a. Standard boron solution: Prepare as directed in Method B, ¶ 3*b*.

b. Hydrochloric acid, HCl, conc and 1 + 11.

c. Sulfuric acid, H_2SO_4, conc.

d. Carmine reagent: Dissolve 920 mg carmine N.F. 40, or carminic acid, in 1 L conc H_2SO_4. (If unable to zero spectrophotometer, dilute carmine 1 + 1 with conc H_2SO_4 to replace above reagent.)

4. Procedure

a. Preliminary sample treatment: If sample contains less than 1 mg B/L, pipet a portion containing 2 to 20 μg B into a platinum dish, make alkaline with 1*N* NaOH plus a slight excess, and evaporate to dryness on a steam or hot water bath. If necessary, destroy any organic material by ignition at 500 to 550°C. Acidify cooled residue (ignited or not) with 2.5 mL 1 + 11 HCl and triturate with a rubber policeman to dissolve. Centrifuge if necessary to obtain a clear solution. Pipet 2.00 mL clear concentrate into a small flask or 30-mL test tube. Treat reagent blank identically.

b. Color development: Prepare a series of boron standard solutions (100, 250, 500, 750, and 1000 μg) in 100 mL with distilled water. Pipet 2.00 mL of each standard solution into a small flask or 30-mL test tube.

Treat blank and calibration standards exactly as the sample. Add 2 drops (0.1 mL) conc HCl, carefully introduce 10.0 mL conc H_2SO_4, mix, and let cool to room temperature. Add 10.0 mL carmine reagent, mix well, and after 45 to 60 min measure absorbance at 585 nm in a cell of 1-cm or longer light path, using the blank as reference.

To avoid error, make sure that no bubbles are present in the optical cell while photometric readings are being made. Bubbles may appear as a result of incomplete mixing of reagents. Because carmine reagent deteriorates, check calibration curve daily.

5. Calculation

mg B/L

$$= \frac{\mu g \ B \ (\text{in approx. } 22 \text{ mL final volume})}{\text{mL sample}}$$

6. Precision and Bias

A synthetic sample containing 180 μg B/L, 50 μg As/L, 400 μg Be/L, and 50 μg Se/L in distilled water was analyzed in nine laboratories by the carmine method with a relative standard deviation of 35.5% and a relative error of 0.6%.

7. Bibliography

HATCHER, J.T. & L.V. WILCOX. 1950. Colorimetric determination of boron using carmine. *Anal. Chem.* 22:567.

4500-B D. Inductively Coupled Plasma Method

See Section 3120.

4500-Br⁻ BROMIDE*

4500-Br⁻ A. Introduction

1. Occurrence

Bromide occurs in varying amounts in ground and surface waters in coastal areas as a result of seawater intrusion and sea-spray-affected precipitation. The bromide content of groundwaters and stream baseflows also can be affected by connate water. Industrial and oil-field brine discharges can

_____
*Approved by Standard Methods Committee, 1988.

contribute to the bromide in water sources. Under normal circumstances, the bromide content of most drinking waters is small, seldom exceeding 1 mg/L.

2. Selection of Method

Described here is a colorimetric procedure suitable for the determination of bromide in most drinking waters. Bromide also may be determined by the ion chromatography method for the determination of anions.

4500-Br⁻ B. Phenol Red Colorimetric Method

1. General Discussion

a. Principle: When a sample containing bromide ions (Br⁻) is treated with a dilute solution of Chloramine-T in the presence of phenol red, the oxidation of bromide and subsequent bromination of the phenol red occur readily. If the reaction is buffered to pH 4.5 to 4.7, the color of the brominated compound will range from reddish to violet, depending on the bromide concentration. Thus, a sharp differentiation can be made among various concentrations of bromide. The concentration of Chloramine-T and timing of the reaction before dechlorination are critical.

b. *Interference:* Most materials present in ordinary tap water do not interfere, but oxidizing and reducing agents and higher concentrations of chloride and bicarbonate can interfere. Free chlorine in samples should be destroyed as directed in Section 5210B.4e2); analyze bromide in a portion of dechlorinated sample. Addition of substantial chloride to the pH buffer solution (see ¶ 3a below) can eliminate chloride interference for waters with very low bromide/chloride ratios, such as those affected by dissolved road salt. Small amounts of dissolved iodide do not interfere, but small concentrations of ammonium ion interfere substantially. Sample dilution may reduce interferences to acceptable levels for some saline and waste waters. However, if two dilutions differing by a factor of at least five do not give comparable values, the method is inapplicable. Bromide concentration in diluted samples must be within the range of the method (0.1 to 1 mg/L).

c. *Minimum detectable concentration:* 0.1 mg Br^-/L.

2. Apparatus

a. *Colorimetric equipment:* One of the following is required:

1) *Spectrophotometer,* for use at 590 nm, providing a light path of at least 2 cm.

2) *Filter photometer,* providing a light path of at least 2 cm and equipped with an orange filter having a maximum transmittance near 590 nm.

3) *Nessler tubes,* matched, 100-mL, tall form.

b. *Acid-washed glassware:* Wash all glassware with $1 + 6$ HNO_3 and rinse with distilled water to remove all trace of adsorbed bromide.

3. Reagents

a. *Acetate buffer solution:* Dissolve 90 g NaCl and 68 g sodium acetate trihydrate, $NaC_2H_3O_2 \cdot 3H_2O$, in distilled water. Add 30 mL conc (glacial) acetic acid and make up to 1 L. The pH should be 4.6 to 4.7.

b. *Phenol red indicator solution:* Dissolve 21 mg phenolsulfonephthalein sodium salt and dilute to 100 mL with distilled water.

c. *Chloramine-T solution:* Dissolve 500 mg Chloramine-T, sodium p-toluenesulfonchloramide, and dilute to 100 mL with distilled water. Store in a dark bottle and refrigerate.

d. *Sodium thiosulfate, 2M:* Dissolve 49.6 g $Na_2S_2O_3 \cdot 5H_2O$ or 31.6 g $Na_2S_2O_3$ and dilute to 100 mL with distilled water.

e. *Stock bromide solution:* Dissolve 744.6 mg anhydrous KBr in distilled water and make up to 1000 mL; 1.00 mL = 500 µg Br^-.

f. *Standard bromide solution:* Dilute 10.00 mL stock bromide solution to 1000 mL with distilled water; 1.00 mL = 5.00 µg Br^-.

4. Procedure

a. *Preparation of bromide standards:* Prepare at least six standards, 0, 0.20, 0.40, 0.60, 0.80 and 1.00 mg Br^-/L, by diluting 0.0, 2.00, 4.00, 6.00, 8.00, and 10.00 mL standard bromide solution to 50.00 mL with distilled water. Treat standards the same as samples in ¶ 4b.

b. *Treatment of sample:* Add 2 mL buffer solution, 2 mL phenol red solution, and 0.5 mL Chloramine-T solution to 50.0 mL sample or two separate sample dilutions (see 1b above) such that the final bromide concentration is in the range of 0.1 to 1.0 mg Br^-/L. Mix thoroughly immediately after each addition. Exactly 20 min after adding Chloramine-T, dechlorinate by adding, with mixing, 0.5 mL $Na_2S_2O_3$ solution. Compare visually in nessler tubes against bromide standards prepared simultaneously, or preferably read in a photometer at 590 nm against a reagent blank. Determine the bromide values from a calibration curve of mg Br^-/L (in 55 mL final volume) against absorbance. A 2.54-cm light path yields an absorbance value of approximately 0.36 for 1 mg Br⁻/L.

5. Calculation

mg Br⁻/L = mg Br⁻/L (from calibration curve) × dilution factor (if any). Results are based on 55 mL final volume for samples and standards.

6. Bibliography

STENGER, V.A. & I.M. KOLTHOFF. 1935. Detection and colorimetric estimation of micro-quantities of bromide. *J. Amer. Chem. Soc.* 57:831.

HOUGHTON, G.U. 1946. The bromide content of underground waters. *J. Soc. Chem. Ind.* (London) 65:227.

GOLDMAN, E. & D. BYLES. 1959. Suggested revision of phenol red method for bromide. *J. Amer. Water Works Assoc.* 51:1051.

SOLLO, F.W., T.E. LARSON & F.F. McGURK. 1971. Colorimetric methods for bromine. *Environ. Sci. Technol.* 5:240.

WRIGHT, E.R., R.A. SMITH & F.G. MESSICK. 1978. *In* D.F. Boltz & J.A. Howell, eds. Colorimetric Determination of Nonmetals, 2nd ed. Wiley-Interscience, New York, N.Y.

BASEL, C.L., J.D. DEFREESE & D.O. WHITTEMORE. 1982. Interferences in automated phenol red method for determination of bromide in water. *Anal. Chem.* 54:2090.

4500-Br⁻ C.   Ion Chromatographic Method

See Section 4110.

4500-CO₂   CARBON DIOXIDE*

4500-CO₂ A.   Introduction

1. Occurrence and Significance

Surface waters normally contain less than 10 mg free carbon dioxide (CO_2) per liter while some groundwaters may easily exceed that concentration. The CO_2 content of a water may contribute significantly to corrosion. Recarbonation of a supply during the last stages of water softening is a recognized treatment process. The subject of saturation with respect to calcium carbonate is discussed in Section 2330.

2. Selection of Method

A nomographic and a titrimetric method are described for the estimation of free CO_2 in drinking water. The titration may be performed potentiometrically or with phenolphthalein indicator. Properly conducted, the more rapid, simple indicator method is satisfactory for field tests and for control and routine applications if it is understood that the method gives, at best, only an approximation.

The nomographic method (B) usually gives a closer estimation of the total free

---

*Approved by Standard Methods Committee, 1985.

CO_2 when the pH and alkalinity determinations are made immediately and correctly at the time of sampling. The pH measurement preferably should be made with an electrometric pH meter, properly calibrated with standard buffer solutions in the pH range of 7 to 8. The error resulting from inaccurate pH measurements grows with an increase in total alkalinity. For example, an inaccuracy of 0.1 in the pH determination causes a CO_2 error of 2 to 4 mg/L in the pH range of 7.0 to 7.3 and a total alkalinity of 100 mg $CaCO_3$/L. In the same pH range, the error approaches 10 to 15 mg/L when the total alkalinity is 400 mg as $CaCO_3$/L.

Under favorable conditions, agreement between the titrimetric and nomographic methods is reasonably good. When agreement is not precise and the CO_2 determination is of particular importance, state the method used.

The calculation of the total CO_2, free and combined, is given in Method D.

4500-CO_2 B.   Nomographic Determination of Free Carbon Dioxide and the Three Forms of Alkalinity*

1. General Discussion

Diagrams and nomographs enable the rapid calculation of the CO_2, bicarbonate, carbonate, and hydroxide content of natural and treated waters. These graphical presentations are based on equations relating the ionization equilibria of the carbonates and water. If pH, total alkalinity, temperature, and total mineral content are known, any or all of the alkalinity forms and CO_2 can be determined nomographically.

A set of charts, Figures 4500-CO_2:1-4,† is presented for use where their accuracy for the individual water supply is confirmed. The nomographs and the equations on which they are based are valid only when the salts of weak acids other than carbonic acid are absent or present in extremely small amounts.

Some treatment processes, such as superchlorination and coagulation, can affect significantly pH and total-alkalinity values of a poorly buffered water of low alkalinity and low total-dissolved-mineral content. In such instances the nomographs may not be applicable.

2. Precision and Bias

The precision possible with the nomographs depends on the size and range of the scales. With practice, the recommended nomographs can be read with a precision of 1%. However, the overall bias of the results depends on the bias of the analytical data applied to the nomographs and the validity of the theoretical equations and the numerical constants on which the nomographs are based. An approximate check of the bias of the calculations can be made by summing the three forms of alkalinity. Their sum should equal the total alkalinity.

3. Bibliography

MOORE, E.W. 1939. Graphic determination of carbon dioxide and the three forms of alkalinity. *J. Amer. Water Works Assoc.* 31:51.

*See also Alkalinity, Section 2320.
†Copies of the nomographs in Figures 4500-CO_2:1-4, enlarged to 2.5 times the size shown here, may be purchased as a set from The American Water Works Association, 6666 West Quincy Ave., Denver, Colorado 80235. Cat. No. 50007.

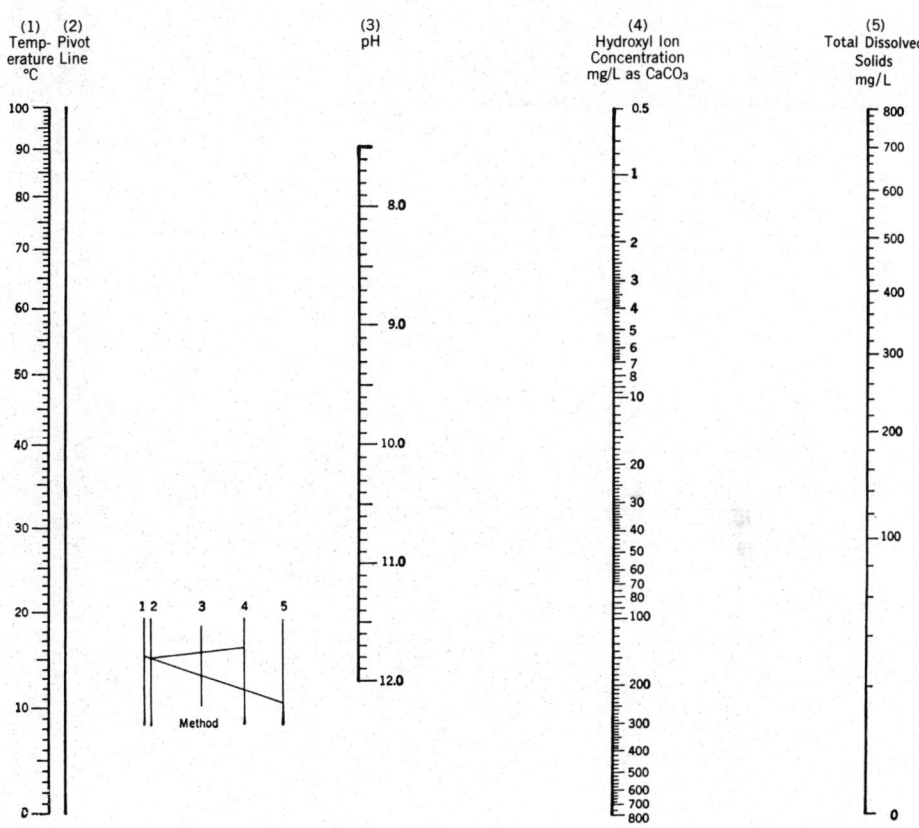

Figure 4500-CO₂:1. Nomograph for evaluation of hydroxide ion concentration. To use: Align temperature (Scale 1) and total dissolved solids (Scale 5); pivot on Line 2 to proper pH (Scale 3); read hydroxide ion concentration, as mg $CaCO_3$/L, on Scale 4. (Example: For 13°C temperature, 240 mg total dissolved solids/L, pH 9.8, the hydroxide ion concentration is found to be 1.4 mg as $CaCO_3$/L.)

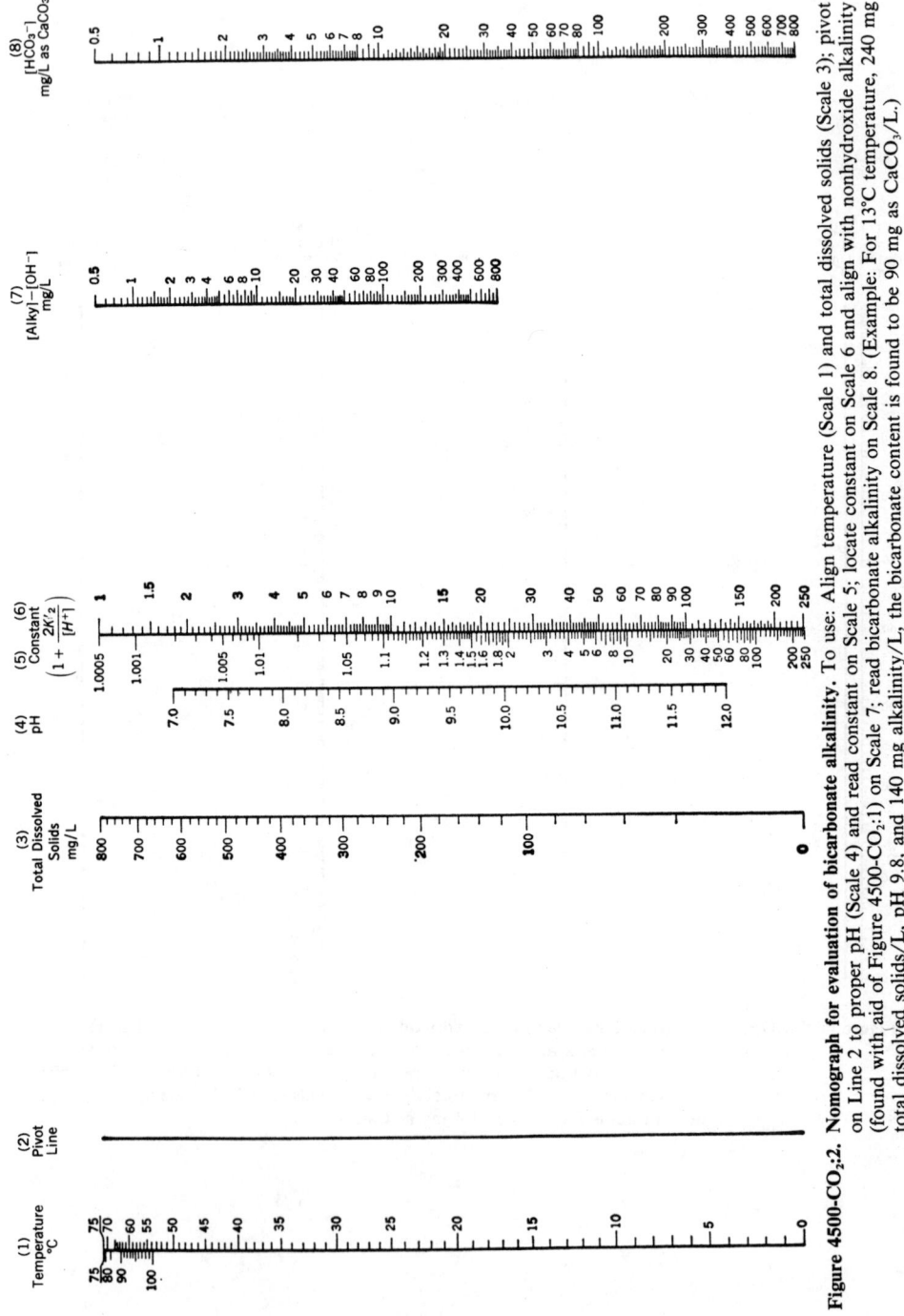

Figure 4500-CO₂:2. Nomograph for evaluation of bicarbonate alkalinity. To use: Align temperature (Scale 1) and total dissolved solids (Scale 3); pivot on Line 2 to proper pH (Scale 4) and read constant on Scale 5; locate constant on Scale 6 and align with nonhydroxide alkalinity (found with aid of Figure 4500-CO₂:1) on Scale 7; read bicarbonate alkalinity on Scale 8. (Example: For 13°C temperature, 240 mg total dissolved solids/L, pH 9.8, and 140 mg alkalinity/L, the bicarbonate content is found to be 90 mg as CaCO₃/L.)

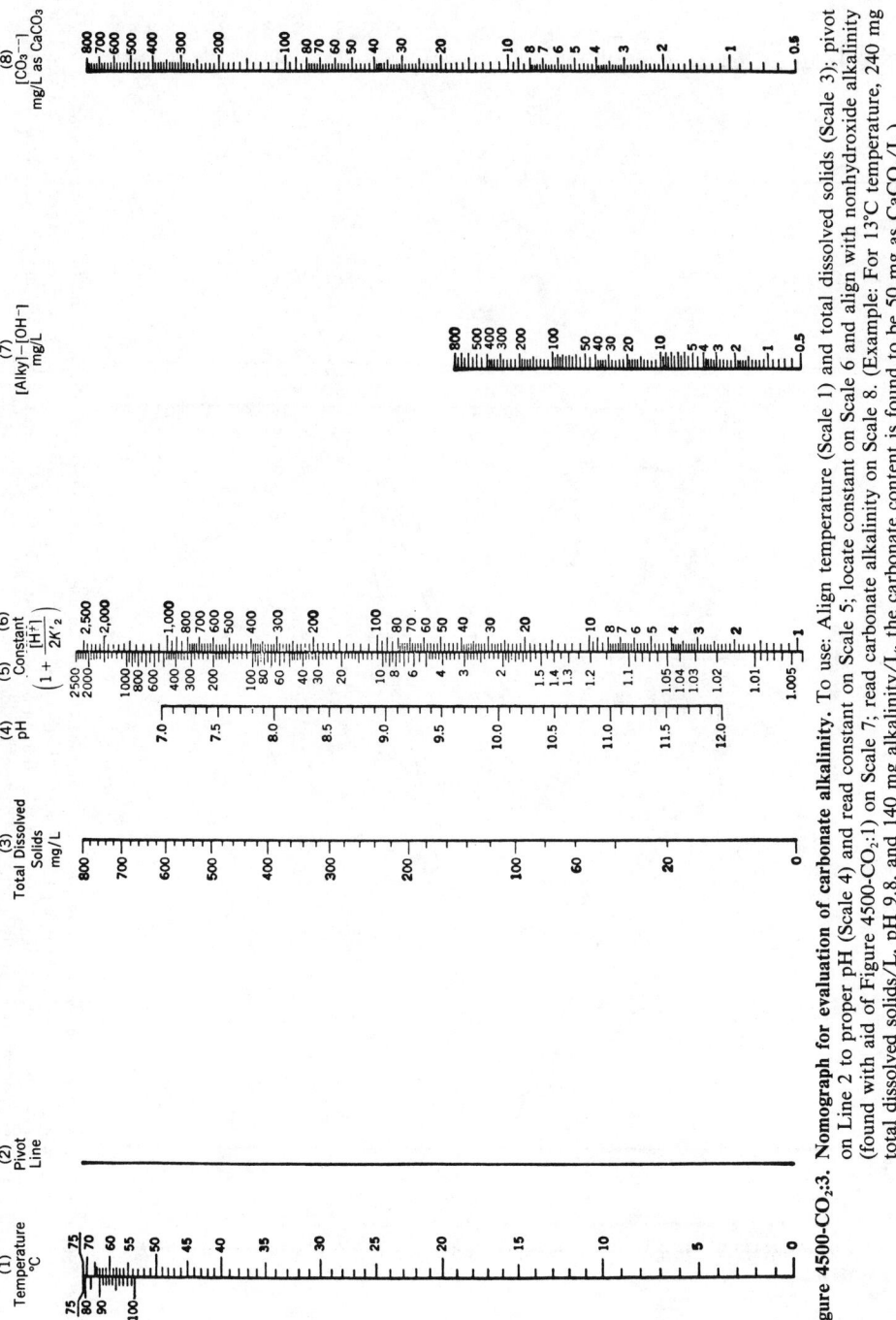

Figure 4500-CO₂:3. Nomograph for evaluation of carbonate alkalinity. To use: Align temperature (Scale 1) and total dissolved solids (Scale 3); pivot on Line 2 to proper pH (Scale 4) and read constant on Scale 5; locate constant on Scale 6 and align with nonhydroxide alkalinity (found with aid of Figure 4500-CO₂:1) on Scale 7; read carbonate alkalinity on Scale 8. (Example: For 13°C temperature, 240 mg total dissolved solids/L, pH 9.8, and 140 mg alkalinity/L, the carbonate content is found to be 50 mg as CaCO₃/L.)

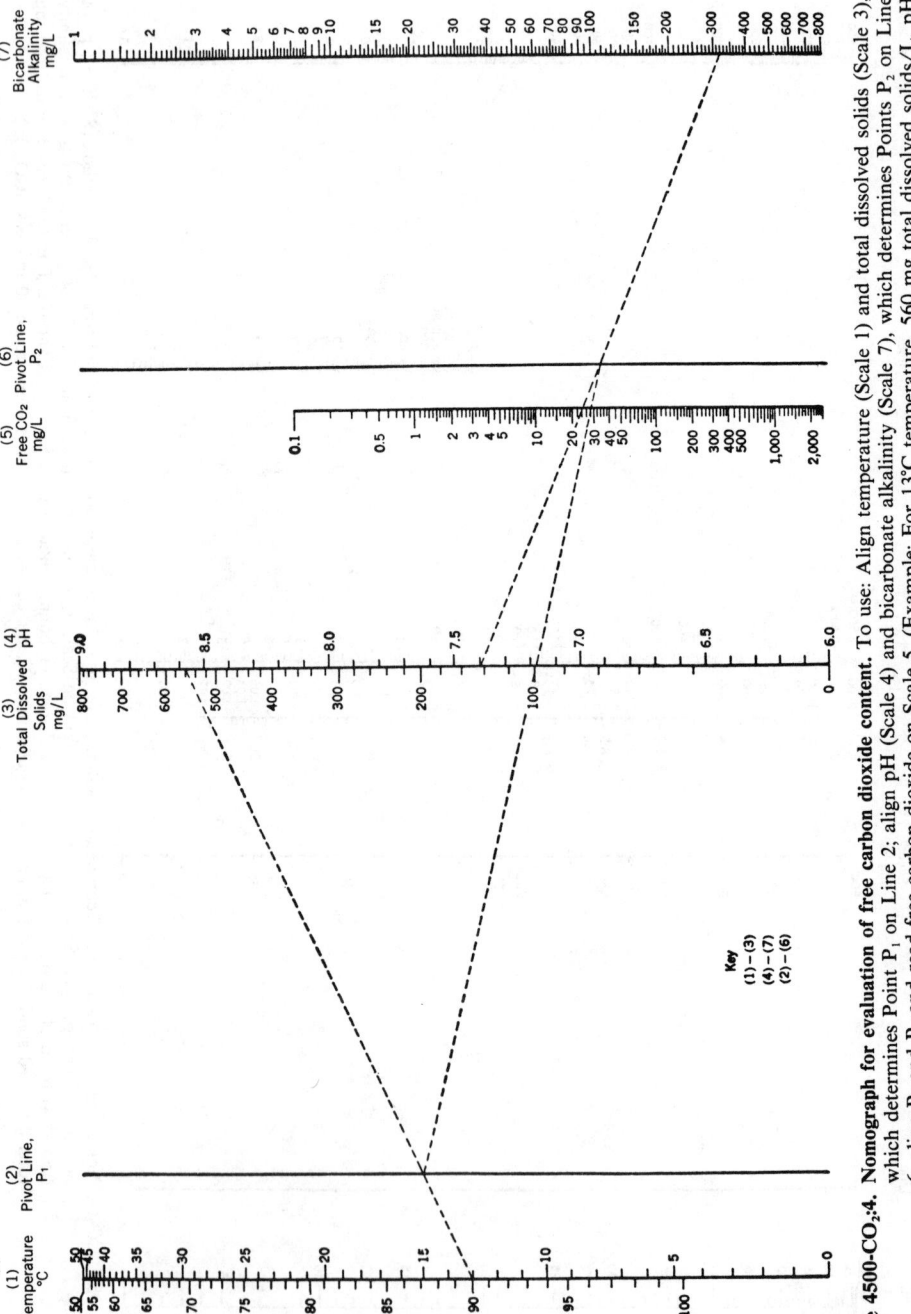

Figure 4500-CO₂:4. Nomograph for evaluation of free carbon dioxide content. To use: Align temperature (Scale 1) and total dissolved solids (Scale 3), which determines Point P_1 on Line 2; align pH (Scale 4) and bicarbonate alkalinity (Scale 7), which determines Points P_2 on Line 6; align P_1 and P_2 and read free carbon dioxide on Scale 5. (Example: For 13°C temperature, 560 mg total dissolved solids/L, pH 7.4, and 320 mg alkalinity/L, the free carbon dioxide content is found to be 28 mg/L.)

4500-CO$_2$ C.   Titrimetric Method for Free Carbon Dioxide

1. General Discussion

a. Principle: Free CO$_2$ reacts with sodium carbonate or sodium hydroxide to form sodium bicarbonate. Completion of the reaction is indicated potentiometrically or by the development of the pink color characteristic of phenolphthalein indicator at the equivalence pH of 8.3. A 0.01N sodium bicarbonate (NaHCO$_3$) solution containing the recommended volume of phenolphthalein indicator is a suitable color standard until familiarity is obtained with the color at the end point.

b. Interference: Cations and anions that quantitatively disturb the normal CO$_2$-carbonate equilibrium interfere with the determination. Metal ions that precipitate in alkaline solution, such as aluminum, chromium, copper, and iron, contribute to high results. Ferrous ion should not exceed 1.0 mg/L. Positive errors also are caused by weak bases, such as ammonia or amines, and by salts of weak acids and strong bases, such as borate, nitrite, phosphate, silicate, and sulfide. Such substances should not exceed 5% of the CO$_2$ concentration. The titrimetric method for CO$_2$ is inapplicable to samples containing acid mine wastes and effluent from acid-regenerated cation exchangers. Negative errors may be introduced by high total dissolved solids, such as those encountered in seawater, or by addition of excess indicator.

c. Sampling and storage: Even with a careful collection technique, some loss in free CO$_2$ can be expected in storage and transit. This occurs more frequently when the gas is present in large amounts. Occasionally a sample may show an increase in free CO$_2$ content on standing. Consequently, determine free CO$_2$ immediately at the point of sampling. Where a field determination is impractical, fill completely a bottle for laboratory examination. Keep the sample, until tested, at a temperature lower than that at which the water was collected. Make the laboratory examination as soon as possible to minimize the effect of CO$_2$ changes.

2. Apparatus

See Section 2310B.2.

3. Reagents

See Section 2310B.3.

4. Procedure

Follow the procedure given in Section 2310B.4b, phenolphthalein, or 2310B.4d, using end-point pH 8.3.

5. Calculation

$$\text{mg CO}_2/\text{L} = \frac{A \times N \times 44\,000}{\text{mL sample}}$$

where:

A = mL titrant and
N = normality of NaOH.

6. Precision and Bias

Precision and bias of the titrimetric method are on the order of ±10% of the known CO$_2$ concentration.

4500-CO$_2$ D.   Carbon Dioxide and Forms of Alkalinity by Calculation

1. General Discussion

When the total alkalinity of a water (Section 2320) is due almost entirely to hydroxides, carbonates, or bicarbonates, and the total dissolved solids (Section 2540) is not greater than 500 mg/L, the alkalinity forms and free CO$_2$ can be calculated from the sample pH and total alkalinity. The calculation is subject to the same limitations as the nomographic procedure given above and the additional restriction of using a single temperature, 25°C. The calculations are based on the ionization constants:

$$K_1 = \frac{[H^+][HCO_3^-]}{[H_2CO_3^*]} \qquad (K_1 = 10^{-6.36})$$

and

$$K_2 = \frac{[H^+][CO_3^{2-}]}{[HCO_3^-]} \qquad (K_2 = 10^{-10.33})$$

where:

$$[H_2CO_3^*] = [H_2CO_3] + [CO_2(aq)]$$

Activity coefficients are assumed equal to unity.

2. Calculation

Compute the forms of alkalinity and sample pH and total alkalinity using the following equations:

a. Bicarbonate alkalinity:

HCO_3^- as mg $CaCO_3/L$

$$= \frac{T - 5.0 \times 10^{(pH-10)}}{1 + 0.94 \times 10^{(pH-10)}}$$

where:

T = total alkalinity, mg $CaCO_3/L$.

b. Carbonate alkalinity:

CO_3^{2-} as mg $CaCO_3/L$
$$= 0.94 \times B \times 10^{(pH-10)}$$

where:

B = bicarbonate alkalinity, from *a*.

c. Hydroxide alkalinity:

OH^- as mg $CaCO_3/L = 5.0 \times 10^{(pH-10)}$

d. Free carbon dioxide:

mg $CO_2/L = 2.0 \times B \times 10^{(6-pH)}$

where:

B = bicarbonate alkalinity, from *a*.

e. Total carbon dioxide:

mg total $CO_2/L = A + 0.44(2B + C)$

where:

A = mg free CO_2/L,
B = bicarbonate alkalinity from *a*, and
C = carbonate alkalinity from *b*.

3. Bibliography

DYE, J.F. 1958. Correlation of the two principal methods of calculating the three kinds of alkalinity. *J. Amer. Water Works Assoc.* 50:812.

4500-CN   CYANIDE*

4500-CN A.   Introduction

1. General Discussion

"Cyanide" refers to all of the CN groups in cyanide compounds that can be determined as the cyanide ion, CN^-, by the methods used. The cyanide compounds in which cyanide can be obtained as CN^- are classed as simple and complex cyanides.

Simple cyanides are represented by the formula $A(CN)_x$, where A is an alkali (sodium, potassium, ammonium) or a metal, and x, the valence of A, is the number of CN groups. In aqueous solutions of simple alkali cyanides, the CN group is present as

CN^- and molecular HCN, the ratio depending on pH and the dissociation constant for molecular HCN ($pK_a \simeq 9.2$). In most natural waters HCN greatly predominates.[1] In solutions of simple metal cyanides, the CN group may occur also in the form of complex metal-cyanide anions of varying stability. Many simple metal cyanides are sparingly soluble or almost insoluble [CuCN, AgCN, $Zn(CN)_2$], but they form a variety of highly soluble, complex metal cyanides in the presence of alkali cyanides.

Complex cyanides have a variety of formulae, but the alkali-metallic cyanides normally can be represented by $A_yM(CN)_x$. In

---

*Approved by Standard Methods Committee, 1985.

this formula, A represents the alkali present y times, M the heavy metal (ferrous and ferric iron, cadmium, copper, nickel, silver, zinc, or others), and x the number of CN groups; x is equal to the valence of A taken y times plus that of the heavy metal. Initial dissociation of each of these soluble, alkali-metallic, complex cyanides yields an anion that is the radical $M(CN)_x^{y-}$. This may dissociate further, depending on several factors, with the liberation of CN^- and consequent formation of HCN.

The great toxicity to aquatic life of molecular HCN is well known;[2-5] it is formed in solutions of cyanide by hydrolytic reaction of CN^- with water. The toxicity of CN^- is less than that of HCN; it usually is unimportant because most of the free cyanide (CN group present as CN^- or as HCN) exists as HCN,[2,5] as the pH of most natural waters is substantially lower than the pK_a for molecular HCN. The toxicity to fish of most tested solutions of complex cyanides is attributable mainly to the HCN resulting from dissociation of the complexes.[2,4,5] Analytical distinction between HCN and other cyanide species in solutions of complex cyanides is possible.[2,5-9,10]

The degree of dissociation of the various metallocyanide complexes at equilibrium, which may not be attained for a long time, increases with decreased concentration and decreased pH, and is inversely related to their highly variable stability.[2,4,5] The zinc- and cadmium-cyanide complexes are dissociated almost totally in very dilute solutions; thus these complexes can result in acute toxicity to fish at any ordinary pH. In equally dilute solutions there is much less dissociation for the nickel-cyanide complex and the more stable cyanide complexes formed with copper (I) and silver. Acute toxicity to fish of dilute solutions containing copper-cyanide or silver-cyanide complex anions can be due mainly or entirely to the toxicity of the undissociated ions, although the complex ions are much less toxic than HCN.[2,5]

The iron-cyanide complex ions are very stable and not materially toxic; in the dark, acutely toxic levels of HCN are attained only in solutions that are not very dilute and have been aged for a long time. However, these complexes are subject to extensive and rapid photolysis, yielding toxic HCN, on exposure of dilute solutions to direct sunlight.[2,11] The photodecomposition depends on exposure to ultraviolet radiation, and therefore is slow in deep, turbid, or shaded receiving waters. Loss of HCN to the atmosphere and its bacterial and chemical destruction concurrent with its production tend to prevent increases of HCN concentrations to harmful levels. Regulatory distinction between cyanide complexed with iron and that bound in less stable complexes, as well as between the complexed cyanide and free cyanide or HCN, can, therefore, be justified.

Historically, the generally accepted physicochemical technique for industrial waste treatment of cyanide compounds is alkaline chlorination:

$$NaCN + Cl_2 \rightarrow CNCl + NaCl \qquad (1)$$

The first reaction product on chlorination is cyanogen chloride (CNCl), a highly toxic gas of limited solubility. The toxicity of CNCl may exceed that of equal concentrations of cyanide.[2,3,12] At an alkaline pH, CNCl hydrolyzes to the cyanate ion (CNO^-), which has only limited toxicity.

There is no known natural reduction reaction that may convert CNO^- to CN^-.[13] On the other hand, breakdown of toxic CNCl is pH- and time-dependent. At pH 9, with no excess chlorine present, CNCl may persist for 24 h.[14,15]

$$CNCl + 2NaOH \rightarrow NaCNO + NaCl + H_2O \qquad (2)$$

CNO^- can be oxidized further with chlorine at a nearly neutral pH to CO_2 and N_2:

$$2NaCNO + 4NaOH + 3Cl_2 \rightarrow 6NaCl$$
$$+ 2CO_2 + N_2 + 2H_2O \quad (3)$$

CNO^- also will be converted on acidification to NH_4^+:

$$2NaCNO + H_2SO_4$$
$$+ 4H_2O \rightarrow (NH_4)_2SO_4 + 2NaHCO_3 \quad (4)$$

The alkaline chlorination of cyanide compounds is relatively fast, but depends equally on the dissociation constant, which also governs toxicity. Metal cyanide complexes, such as nickel, cobalt, silver, and gold, do not dissociate readily. The chlorination reaction therefore requires more time and a significant chlorine excess.[16] Iron cyanides, because they do not dissociate to any degree, are not oxidized by chlorination. There is correlation between the refractory properties of the noted complexes, in their resistance to chlorination and lack of toxicity.

Thus, it is advantageous to differentiate between *total cyanide* and *cyanides amenable to chlorination*. When total cyanide is determined, the almost nondissociable cyanides, as well as cyanide bound in complexes that are readily dissociable and complexes of intermediate stability, are measured. Cyanide compounds that are amenable to chlorination include free cyanide as well as those complex cyanides that are potentially dissociable, almost wholly or in large degree, and therefore, potentially toxic at low concentrations, even in the dark. The chlorination test procedure is carried out under rigorous conditions appropriate for measurement of the more dissociable forms of cyanide.

Alternatively, the free and potentially dissociable cyanides also may be estimated when using the *weak acid dissociable* procedure. These methods depend on a rigorous distillation, but the solution is only slightly acidified, and elimination of iron cyanides is insured by the earlier addition of precipitation chemicals to the distillation

flask or by the avoidance of ultraviolet irradiation.

The *cyanogen chloride* procedure is common with the colorimetric test for cyanides amenable to chlorination. This test is based on the addition of chloramine-T and subsequent color complex formation with barbituric acid. Without the addition of chloramine-T, only existing CNCl is measured. CNCl is a gas that hydrolyzes to CNO^-; sample preservation is not possible. Because of this, spot testing of CNCl levels may be best. This procedure can be adapted and used when the sample is collected.

There may be analytical requirements for the determination of CNO^-, even though the reported toxicity level is low. On acidification, CNO^- decomposes to ammonia (NH_3).[3] Molecular ammonia and metal-ammonia complexes are highly toxic.[17]

Thiocyanate (SCN^-) is not very toxic to aquatic life.[2] However, upon chlorination, toxic CNCl is formed, as discussed above.[2,3,12] At least where subsequent chlorination is anticipated, the determination of SCN^- is desirable. Thiocyanate is biodegradable; ammonium is released in this reaction. Thiocyanate may be analyzed in samples properly preserved for determination of cyanide; however, thiocyanate also can be preserved in samples by acidification with H_2SO_4 to pH ≤ 2.

2. Cyanide in Solid Waste

a. Soluble cyanide: Determination of soluble cyanide requires sample leaching with distilled water until solubility equilibrium is established. One hour of stirring in distilled water should be satisfactory. Cyanide analysis is then performed on the leachate. Low cyanide concentration in the leachate may indicate presence of sparingly soluble metal cyanides. The cyanide content of the leachate is indicative of residual solubility of insoluble metal cyanides in the waste. High levels of cyanide in the leachate indicate soluble cyanide in the solid waste. When 500 mL distilled water are stirred into a 500-mg solid waste sample, the cyanide concentration (mg/L) of the leachate multiplied by 1000 will give the solubility level of the cyanide in the solid waste in milligrams per kilogram. The leachate may be analyzed for total cyanide and/or cyanide amenable to chlorination.

b. Insoluble cyanide: The insoluble cyanide of the solid waste can be determined with the total cyanide method by placing a 500-mg sample with 500 mL distilled water in the distillation flask and in general following the distillation procedure (Section 4500-CN.C). In calculating, multiply by 1000 to give the cyanide content of the solid sample in milligrams per kilogram. Insoluble iron cyanides in the solid can be leached out earlier by stirring a weighed sample for 12 to 16 h in a 10% NaOH solution. The leachate and wash waters of the solid waste will give the iron cyanide content with the distillation procedure. Prechlorination will have eliminated all cyanide amenable to chlorination. Do not expose sample to sunlight.

3. Selection of Method

a. Total cyanide after distillation: After removal of interfering substances, the metal cyanide is converted to HCN gas, which is distilled and absorbed in sodium hydroxide (NaOH) solution.[18] Because of the catalytic decomposition of cyanide in the presence of cobalt at high temperature in a strong acid solution,[19,20] cobalticyanide is not recovered completely. Indications are that cyanide complexes of the noble metals, i.e., gold, platinum, and palladium, are not recovered fully by this procedure either. Distillation also separates cyanide from other color-producing and possibly interfering organic or inorganic contaminants. Subsequent analysis is for the simple salt, sodium cyanide (NaCN). Some organic cyanide compounds, such as nitriles, are decomposed by the distillation. Aldehydes convert cyanide to nitrile. The absorption liquid is analyzed by either a titrimetric, colorimetric, or cyanide-ion-selective electrode procedure:

1) The titration method (D) is suitable for cyanide concentrations above 1 mg/L.

2) The colorimetric method (E) is suitable for cyanide concentrations to a lower limit of 20 µg/L. Analyze higher concentrations by diluting either the sample before distillation or the absorber solution before colorimetric measurement.

3) The ion-selective electrode method (F) using the cyanide ion electrode is applicable in the concentration range of 0.05 to 10 mg/L.

b. Cyanide amenable to chlorination:

1) Distillation of two samples is required, one that has been chlorinated to destroy all amenable cyanide present and the other unchlorinated. Analyze absorption liquids from both tests for total cyanide. The observed difference equals cyanides amenable to chlorination.

2) The colorimetric method, by conversion of amenable cyanide and SCN^- to CNCl and developing the color complex with barbituric acid, is used for the determination of the total of these cyanides (H). Repeating the test with the cyanide masked by the addition of formaldehyde provides a measure of the SCN^- content. When subtracted from the earlier results this provides an estimate of the amenable CN^- content. This method is useful for natural and ground waters and clean metal finishing, heat treating, and sanitary effluents.

3) The *weak acid dissociable cyanides* procedure also measures the cyanide amenable to chlorination by freeing HCN from the dissociable cyanide. After being collected in a NaOH absorption solution, HCN may be determined by one of the three finishing procedures given for the total cyanide determination.

It should be noted that although cyanide amenable to chlorination and weak acid dissociable cyanide appear to be identical, certain industrial effluents (e.g., pulp and paper, petroleum refining industry effluents) contain some poorly understood substances that may produce interference. Application of the procedure for cyanide amenable to chlorination yields negative values. For natural waters and metal-finishing effluents, the direct colorimetric determination appears to be the simplest and most economical.

c. Cyanogen chloride: The colorimetric method for measuring cyanide amenable to chlorination may be used, but omit the chloramine-T addition. The spot test also may be used.

d. Spot test for sample screening: This procedure allows a quick sample screening to establish whether more than 50 μg/L cyanide amenable to chlorination is present. The test also may be used to estimate the CNCl content at the time of sampling.

e. Cyanate: CNO$^-$ is converted to ammonium carbonate, $(NH_4)_2CO_3$, by acid hydrolysis at elevated temperature. Ammonia (NH_3) is determined before the conversion of the CNO$^-$ and again afterwards. The CNO$^-$ is estimated from the difference in NH$_3$ found in the two tests.[21-23] Measure NH$_3$ by either:

1) The selective electrode method, using the NH$_3$ gas electrode; or

2) The colorimetric method, using direct nesslerization or the phenate method for NH$_3$ (Section 4500-NH$_3$.C or D).

f. Thiocyanate: Use the colorimetric determination with ferric nitrate as a color-producing compound.

4. References

1. MILNE, D. 1950. Equilibria in dilute cyanide waste solutions. *Sewage Ind. Wastes* 23:904.
2. DOUDOROFF, P. 1976. Toxicity to fish of cyanides and related compounds. A review. EPA 600/3-76-038, U.S. Environmental Protection Agency, Duluth, Minn.
3. DOUDOROFF, P. & M. KATZ. 1950. Critical review of literature on the toxicity of industrial wastes and their components to fish. *Sewage Ind. Wastes* 22:1432.
4. DOUDOROFF, P. 1956. Some experiments on the toxicity of complex cyanides to fish. *Sewage Ind. Wastes* 28:1020.
5. DOUDOROFF, P., G. LEDUC & C.R. SCHNEIDER. 1966. Acute toxicity to fish of solutions containing complex metal cyanides, in relation to concentrations of molecular hydrocyanic acid. *Trans. Amer. Fish. Soc.* 95:6.
6. SCHNEIDER, C.R. & H. FREUND. 1962. Determination of low level hydrocyanic acid. *Anal. Chem.* 34:69.
7. CLAEYS, R. & H. FREUND. 1968. Gas chromatographic separation of HCN. *Environ. Sci. Technol.* 2:458.
8. MONTGOMERY, H.A.C., D.K. GARDINER & J.G. GREGORY. 1969. Determination of free hydrogen cyanide in river water by a solvent-extraction method. *Analyst* 94:284.
9. NELSON, K.H. & L. LYSYJ. 1971. Analysis of water for molecular hydrogen cyanide. *J. Water Pollut. Control Fed.* 43:799.
10. BRODERIUS, S.J. Determination of hydrocyanic acid and free cyanide in aqueous solution. *Anal. Chem.* 53:1472.
11. BURDICK, G.E. & M. LIPSCHUETZ. 1948. Toxicity of ferro and ferricyanide solutions to fish. *Trans. Amer. Fish. Soc.* 78:192.
12. ZILLICH, J.A. 1972. Toxicity of combined chlorine residuals to freshwater fish. *J. Water Pollut. Control Fed.* 44:212.
13. RESNICK, J.D., W. MOORE & M.E. ETTINGER. 1958. The behavior of cyanates in polluted waters. *Ind. Eng. Chem.* 50:71.
14. PETTET, A.E.J. & G.C. WARE. 1955. Disposal of cyanide wastes. *Chem. Ind.* 1955:1232.
15. BAILEY, P.L. & E. BISHOP. 1972. Hydrolysis of cyanogen chloride. *Analyst* 97:691.
16. LANCY, L. & W. ZABBAN. 1962. Analytical methods and instrumentation for determining cyanogen compounds. Spec. Tech. Publ. 337, American Soc. Testing & Materials, Philadelphia, Pa.
17. CALAMARI, D. & R. MARCHETTI. 1975. Predicted and observed acute toxicity of copper and ammonia to rainbow trout. *Progr. Water Technol.* 7, 3-4:569.
18. SERFASS, E.J. & R.B. FREEMAN. 1952. Analytical method for the determination of cyanides in plating wastes and in effluents from treatment processes. *Plating* 39:267.

19. LESCHBER, R. & H. SCHLICHTING. 1969. Über die Zersetzlichkeit Komplexer Metallcyanide bei der Cyanidbestimmung in Abwasser. Z. Anal. Chem. 245:300.

20. BASSETT, H., JR. & A.S. CORBET. 1924. The hydrolysis of potassium ferricyanide and potassium cobalticyanide by sulfuric acid. J. Chem. Soc. 125:1358.

21. DODGE, B.F. & W. ZABBAN. 1952. Analytical methods for the determination of cyanates in plating wastes. Plating 39:381.

22. GARDNER, D.C. 1956. The colorimetric determination of cyanates in effluents. Plating 43:743.

23. Procedures for Analyzing Metal Finishing Wastes. 1954. Ohio River Valley Sanitation Commission, Cincinnati, Ohio.

4500-CN B. Preliminary Treatment of Samples

CAUTION—*Use care in manipulating cyanide-containing samples because of toxicity. Process in a hood or other well-ventilated area. Avoid contact, inhalation, or ingestion.*

1. General Discussion

The nature of the preliminary treatment will vary according to the interfering substance present. Sulfides, fatty acids, and oxidizing agents are removed by special procedures. Most other interfering substances are removed by distillation. The importance of the distillation procedure cannot be overemphasized.

2. Preservation of Samples

Oxidizing agents, such as chlorine, decompose most cyanides. Test by placing a drop of sample on a strip of potassium iodide (KI)-starch paper previously moistened with acetate buffer solution, pH 4 (Section 4500-Cl.C.3e). If a bluish discoloration is noted, add 0.1 g sodium arsenite $(NaAsO_2)/L$ sample and retest. Repeat addition if necessary. Sodium thiosulfate also may be used, but avoid an excess greater than 0.1 g $Na_2S_2O_3/L$. Manganese dioxide, nitrosyl chloride, etc., if present, also may cause discoloration of the test paper. If possible, carry out this procedure before preserving sample as described below. If the following test indicates presence of sulfide, oxidizing compounds would not be expected.

Oxidized products of sulfide convert CN^- to SCN^- rapidly, especially at high pH.[1] Test for S^{2-} by placing a drop of sample on lead acetate test paper previously moistened with acetic acid buffer solution, pH 4 (Section 4500-Cl.C.3e). Darkening of the paper indicates presence of S^{2-}. Add lead acetate, or if the S^{2-} concentration is too high, add powdered lead carbonate $[Pb(CO_3)_2]$ to avoid significantly reducing pH. Repeat test until a drop of treated sample no longer darkens the acidified lead acetate test paper. Filter sample before raising pH for stabilization. When particulate, metal cyanide complexes are suspected filter solution before removing S^{2-}. Reconstitute sample by returning filtered particulates to the sample bottle after S^{2-} removal. Homogenize particulates before analyses.

Aldehydes convert cyanide to cyanohydrin. Longer contact times between cyanide and the aldehyde and the higher ratios of aldehyde to cyanide both result in increasing losses of cyanide that are not reversible during analysis. If the presence of aldehydes is suspected, stabilize with NaOH at time of collection and add 2 mL 3.5% ethylene diamine solution per 100 mL of sample.

Because most cyanides are very reactive and unstable, analyze samples as soon as possible. If sample cannot be analyzed immediately, add NaOH pellets or a strong

NaOH solution to raise sample pH to 12 to 12.5 and store in a closed, dark bottle in a cool place.

To analyze for CNCl collect a separate sample and omit NaOH addition because CNCl is converted rapidly to CNO^- at high pH. Make colorimetric estimation immediately after sampling.

3. Interferences

a. Oxidizing agents may destroy most of the cyanide during storage and manipulation. Add $NaAsO_2$ or $Na_2S_2O_3$ as directed above; avoid excess $Na_2S_2O_3$.

b. Sulfide will distill over with cyanide and, therefore, adversely affect colorimetric, titrimetric, and electrode procedures. Test for and remove S^{2-} as directed above. Treat 25 mL more than required for the distillation to provide sufficient filtrate volume.

c. Fatty acids that distill and form soaps under alkaline titration conditions make the end point almost impossible to detect. Remove fatty acids by extraction.[2] Acidify sample with acetic acid (1 + 9) to pH 6.0 to 7.0. (CAUTION—*Perform this operation in a hood as quickly as possible*). Immediately extract with iso-octane, hexane, or $CHCl_3$ (preference in order named). Use a solvent volume equal to 20% of sample volume. One extraction usually is adequate to reduce fatty acid concentration below the interference level. Avoid multiple extractions or a long contact time at low pH to minimize loss of HCN. When extraction is completed, immediately raise pH to > 12 with NaOH solution.

d. Carbonate in high concentration may affect distillation by causing excessive gasing when acid is added. The carbon dioxide (CO_2) released also may reduce significantly the NaOH content in the absorber.

When sampling effluents such as coal gasification wastes, atmospheric emission scrub waters, and other high-carbonate wastes, use hydrated lime to stabilize the sample; slowly add with stirring to raise pH to 12 to 12.5. Decant sample into sample bottle after precipitate has settled.

e. Other possible interferences include substances that might contribute color or turbidity. In most cases, distillation will remove these.

Note, however, that distillation requires using sulfuric acid with various reagents. With certain wastes, these conditions may result in reactions that otherwise would not occur in the aqueous sample. As a quality control measure, periodically conduct addition and recovery tests with industrial waste samples.

f. Aldehydes convert cyanide to cyanohydrin, which forms nitrile under the distillation conditions. Only direct titration without distillation can be used, which reveals only non-complex cyanides. Formaldehyde interference is noticeable in concentrations exceeding 0.5 mg/L. Use the following spot test to establish absence or presence of aldehydes (detection limit 0.05 mg/L):[3-5]

1) Reagents

a) *MBTH indicator solution:* Dissolve 0.05 g 3-methyl, 2-benzothiazolone hydrazone hydrochloride in 100 mL water. Filter if turbid.

b) *Ferric chloride oxidizing solution:* Dissolve 1.6 g sulfamic acid and 1 g $FeCl_3 \cdot 6H_2O$ in 100 mL water.

c) *Ethylene diamine solution, 3.5%:* Dilute 3.5 mL pharmaceutical-grade anhydrous $NH_2CH_2CH_2NH_2$ to 100 mL with water.

2) Procedure—If the sample is alkaline, add 1 + 1 H_2SO_4 to 10 mL sample to adjust pH to less than 8. Place 1 drop of sample and 1 drop distilled water for a blank in separate cavities of a white spot plate. Add 1 drop MBTH solution and then 1 drop $FeCl_3$ oxidizing solution to each spot. Allow 10 min for color development. The color change will be from a faint green-yellow to a deeper green with blue-green

to blue at higher concentrations of aldehyde. The blank should remain yellow.

To minimize aldehyde interference, add 2 mL of 3.5% ethylene diamine solution/ 100 mL sample. This quantity overcomes the interference caused by up to 50 mg/L formaldehyde.

When using a known addition in testing, 100% recovery of the CN^- is not necessarily to be expected. Recovery depends on the aldehyde excess, time of contact, and sample temperature.

g. Glucose and other sugars, especially at the pH of preservation, lead to cyanohydrin formation by reaction of cyanide with aldose.[6] Reduce cyanohydrin to cyanide with ethylene diamine (see *f* above). MBTH is not applicable.

h. Nitrite may form HCN during distillation in Methods C, G, and L, by reacting with organic compounds.[7,8] Also, NO_3^- may reduce to NO_2^-, which interferes. To avoid NO_2^- interference, add 2 g sulfamic acid to the sample before distillation. *Nitrate* also may interfere by reacting with SCN^-.[9]

i. Some sulfur compounds may decompose during distillation, releasing S, H_2S, or SO_2. Sulfur compounds may convert cyanide to thiocyanate and also may interfere with the analytical procedures for CN^-. To avoid this potential interference, add 50 mg $PbCO_3$ to the absorption solution before distillation. Filter sample before proceeding with the colorimetric or titrimetric determination.

Absorbed SO_2 forms Na_2SO_3 which consumes chloramine-T added in the colorimetric determination. The volume of chloramine-T added is sufficient to overcome 100 to 200 mg SO_3^{2-}/L. Test for presence of chloramine-T after adding it by placing a drop of sample on KI-starch test paper; add more chloramine-T if the test paper remains blank, or use Method F.

Some wastewaters, such as those from coal gasification or chemical extraction mining, contain high concentrations of sul-fites. Pretreat sample to avoid overloading the absorption solution with SO_3^{2-}. Titrate a suitable sample iodometrically (Section 4500-O) with dropwise addition of 30% H_2O_2 solution to determine volume of H_2O_2 needed for the 500 mL distillation sample. Subsequently, add H_2O_2 dropwise while stirring, but in only such volume that not more than 300 to 400 mg SO_3^{2-}/L will remain. Adding a lesser quantity than calculated is required to avoid oxidizing any CN^- that may be present.

4. References

1. LUTHY, R.G. & S. G. BRUCE, JR. 1979. Kinetics of reaction of cyanide and reduced sulfur species in aqueous solution. *Environ. Sci. Technol.* 13:1481.

2. KRUSE, J.M. & M.G. MELLON. 1951. Colorimetric determination of cyanides. *Sewage Ind. Wastes* 23:1402.

3. SAWICKI, E., T.W. STANLEY, T.R. HAUSER & W. ELBERT. 1961. The 3-methyl-2-benzothiazolone hydrazone test. Sensitive new methods for the detection, rapid estimation, and determination of aliphatic aldehydes. *Anal. Chem.* 33:93.

4. HAUSER, T.R. & R.L. CUMMINS. 1964. Increasing sensitivity of 3-methyl-2-benzothiazone hydrazone test for analysis of aliphatic aldehydes in air. *Anal. Chem.* 36:679.

5. Methods of Air Sampling and Analysis, 1st ed. 1972. Inter Society Committee, Air Pollution Control Assoc., pp. 199-204.

6. RAAF, S.F., W.G. CHARACKLIS, M.A. KESSICK & C.H. WARD. 1977. Fate of cyanide and related compounds in aerobic microbial systems. *Water Res.* 11:477.

7. RAPEAN, J.C., T. HANSON & R. A. JOHNSON. 1980. Biodegradation of cyanide-nitrate interference in the standard test for total cyanide. Proc. 35th Ind. Waste Conf., Purdue Univ., Lafayette, Ind., p. 430.

8. CASEY, J.P. 1980. Nitrosation and cyanohydrin decomposition artifacts in distillation test for cyanide. Extended Abs. American Chemical Soc., Div. Environmental Chemistry, Aug. 24-29, 1980, Las Vegas, Nev.

9. CSIKAI, N.J. & A.J. BARNARD, JR. 1983. Determination of total cyanide in thiocyanate-containing waste water. *Anal. Chem.* 55:1677.

4500-CN C. Total Cyanide after Distillation

1. General Discussion

Hydrogen cyanide (HCN) is liberated from an acidified sample by distillation and purging with air. The HCN gas is collected by passing it through an NaOH scrubbing solution. Cyanide concentration in the scrubbing solution is determined by titrimetric, colorimetric, or potentiometric procedures.

2. Apparatus

The apparatus is shown in Figure 4500-CN:1. It includes:

a. Boiling flask, 1 L, with inlet tube and provision for water-cooled condenser.

b. Gas absorber, with gas dispersion tube equipped with medium-porosity fritted outlet.

c. Heating element, adjustable.

d. Ground glass ST joints, TFE-sleeved or with an appropriate lubricant for the boiling flask and condenser. Neoprene stopper and plastic threaded joints also may be used.

3. Reagents

a. Sodium hydroxide solution: Dissolve 40 g NaOH in water and dilute to 1 L.

b. Magnesium chloride reagent: Dissolve 510 g $MgCl_2 \cdot 6H_2O$ in water and dilute to 1 L.

c. Sulfuric acid, H_2SO_4, 1 + 1.

d. Lead carbonate, $PbCO_3$, powdered.

e. Sulfamic acid, NH_2SO_3H.

4. Procedure

a. Add 500 mL sample, containing not more than 10 mg CN^-/L (diluted if necessary with distilled water) to the boiling flask. If a higher CN^- content is anticipated, use the spot test (4500-CN.K) to approximate the required dilution. Add 10 mL NaOH solution to the gas scrubber and dilute, if necessary, with distilled water to obtain an adequate liquid depth in the absorber. Do not use more than 225 mL total volume of absorber solution. When S^{2-} generation from the distilling flask is anticipated add 50 or more mg powdered $PbCO_3$ to the absorber solution to precipitate S^{2-}. Connect the train, consisting of boiling flask air inlet, flask, condenser, gas washer, suction flask trap, and aspirator. Adjust suction so that approximately 1 air bubble/s enters the boiling flask. This air rate will carry HCN gas from flask to absorber and usually will prevent a reverse flow of HCN through the air inlet. If this air rate does not prevent sample backup in the delivery tube, increase air-flow rate to 2 air bubbles/s. Observe air purge rate in the absorber where the liquid level should be raised not more than 6.5 to 10 mm. Maintain air flow throughout the reaction.

b. Add 2 g sulfamic acid through the air inlet tube and wash down with distilled water.

c. Add 50 mL 1 + 1 H_2SO_4 through the air inlet tube. Rinse tube with distilled water and let air mix flask contents for 3 min. Add 20 mL $MgCl_2$ reagent through air inlet and wash down with stream of water. A precipitate that may form redissolves on heating.

d. Heat with rapid boiling, but do not flood condenser inlet or permit vapors to rise more than halfway into condenser. Adequate refluxing is indicated by a reflux rate of 40 to 50 drops/min from the condenser lip. Reflux for at least 1 h. Discontinue heating but continue air flow. Cool for 15 min and drain gas washer contents into a separate container. Rinse connecting tube between condenser and gas washer with distilled water, add rinse water to drained liquid, and dilute to 250 mL in a volumetric flask.

e. Determine cyanide content by the titration method (D) if cyanide concentration exceeds 1 mg/L, by the colorimetric method (E) if the cyanide concentration is less, or by dilution to reduce the concentration into the applicable range (4500-

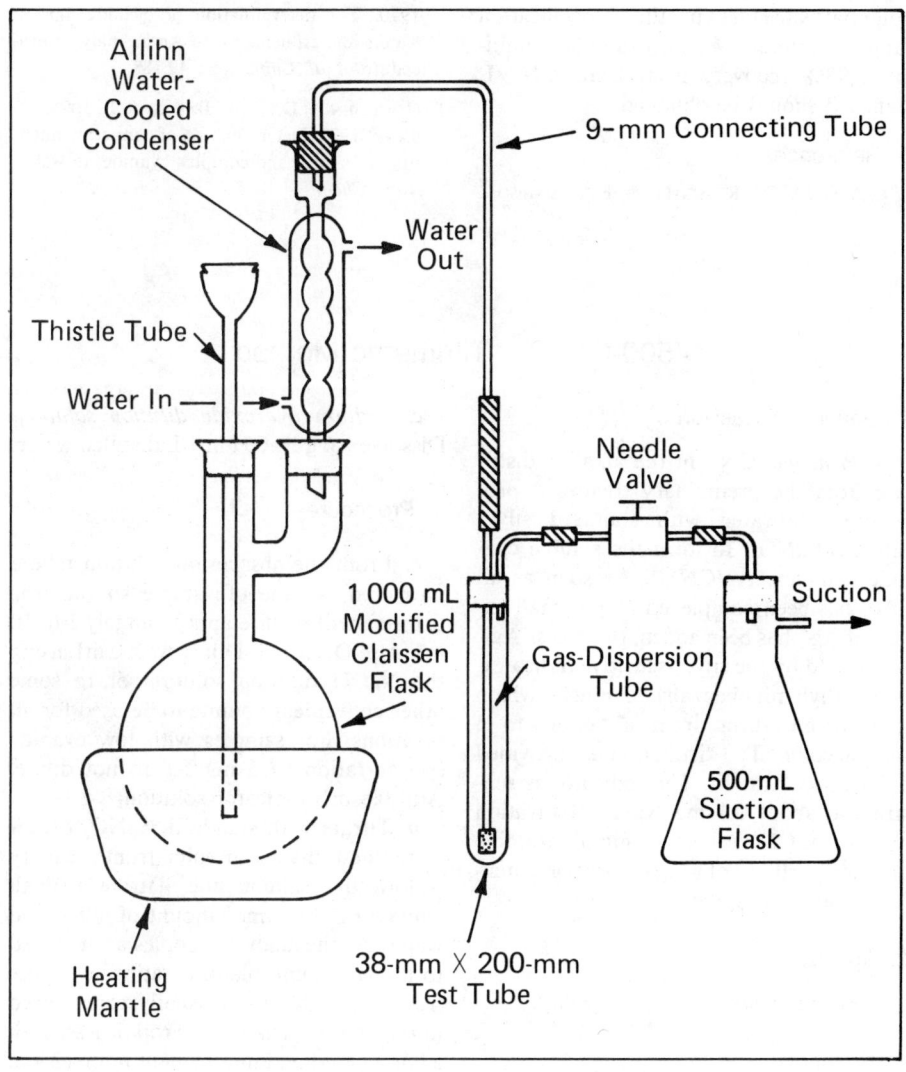

Figure 4500-CN:1. Cyanide distillation apparatus.

CN.E.4). Use titration, the electrode probe method, or the spot test to approximate CN⁻ content. Alternatively, use the cyanide-selective electrode in the concentration range 0.05 to 10 mg CN⁻/L (Method F).

f. Distillation gives quantitative recovery of even refractory cyanides such as iron complexes. To obtain complete recovery of cobalticyanide use ultraviolet radiation pretreatment.[1,2] If incomplete recovery is suspected, distill again by refilling the gas washer with a fresh charge of NaOH solution and refluxing 1 h more. The cyanide from the second reflux, if any, will indicate completeness of recovery.

g. As a quality control measure, periodically test apparatus, reagents, and other potential variables in the concentration range of interest. As an example a minimum 98% recovery from 1 mg CN^-/L standard should be obtained.

5. References

1. CASAPIERI, P., R. SCOTT & E.A. SIMPSON.

1970. The determination of cyanide ions in waters and effluents by an Auto Analyzer procedure. *Anal. Chim. Acta* 49:188.

2. GOULDEN, P.D., K.A. BADAR & P. BROOKSBANK. 1972. Determination of nanogram quantities of simple and complex cyanides in water. *Anal. Chem.* 44:1845.

4500-CN D. Titrimetric Method

1. General Discussion

a. Principle: CN^- in the alkaline distillate from the preliminary treatment procedure is titrated with standard silver nitrate ($AgNO_3$) to form the soluble cyanide complex, $Ag(CN)_2^-$. As soon as all CN^- has been complexed and a small excess of Ag^+ has been added, the excess Ag^+ is detected by the silver-sensitive indicator, *p*-dimethylaminobenzalrhodanine, which immediately turns from a yellow to a salmon color.[1] The distillation has provided a 2:1 concentration. The indicator is sensitive to about 0.1 mg Ag/L. If titration shows that CN^- is below 1 mg/L, examine another portion colorimetrically or potentiometrically.

2. Apparatus

Koch microburet, 10-mL capacity.

3. Reagents

a. Indicator solution: Dissolve 20 mg *p*-dimethylaminobenzalrhodanine in 100 mL acetone.

b. Standard silver nitrate titrant: Dissolve 3.27 g $AgNO_3$ in 1 L distilled water. Standardize against standard NaCl solution, using the argentometric method with K_2CrO_4 indicator, as directed in Chloride, Section 4500-Cl$^-$.B.

Dilute 500 mL $AgNO_3$ solution according to the titer found so that 1.00 mL is equivalent to 1.00 mg CN^-.

c. Sodium hydroxide dilution solution: Dissolve 1.6 g NaOH in 1 L distilled water.

4. Procedure

a. From the absorption solution take a measured volume of sample so that the titration will require approximately 1 to 10 mL $AgNO_3$ titrant. Dilute to 250 mL using the NaOH dilution solution or to some other convenient volume to be used for all titrations. For samples with low cyanide concentration (≤ 5 mg/L) do not dilute. Add 0.5 mL indicator solution.

b. Titrate with standard $AgNO_3$ titrant to the first change in color from a canary yellow to a salmon hue. Titrate a blank containing the same amount of alkali and water. As the analyst becomes accustomed to the end point, blank titrations decrease from the high values usually experienced in the first few trials to 1 drop or less, with a corresponding improvement in precision.

5. Calculation

$$mg\ CN^-/L = \frac{(A - B) \times 1000}{mL\ original\ sample}$$

$$\times \frac{250}{mL\ portion\ used}$$

where:

A = mL standard $AgNO_3$ for sample and
B = mL standard $AgNO_3$ for blank.

6. Precision and Bias

For samples containing more than 1 mg CN^-/L that have been distilled or for relatively clear samples without significant interference, the coefficient of variation is 2%. Extraction and removal of S^{2-} or oxidizing agents tend to increase the coefficient of variation to a degree determined by the amount of manipulation and the type of sample. The limit of sensitivity is approximately 0.1 mg CN^-/L, but at this concentration the end point is indistinct. At 0.4 mg/L the coefficient of variation is four times that at CN^- concentration levels > 1.0 mg/L.

7. Reference

1. RYAN, J.A. & G.W. CULSHAW. 1944. The use of p-dimethylaminobenzylidene rhodanine as an indicator for the volumetric determination of cyanides. *Analyst* 69:370.

4500-CN E. Colorimetric Method

1. General Discussion

a. Principle: CN^- in the alkaline distillate from preliminary treatment is converted to CNCl by reaction with chloramine-T at pH < 8 without hydrolyzing to CNO^-.[1] (CAUTION—*CNCl is a toxic gas; avoid inhalation.*) After the reaction is complete, CNCl forms a red-blue dye on addition of a pyridine-barbituric acid reagent. If the dye is kept in an aqueous solution, the absorbance is read at 578 nm. To obtain colors of comparable intensity, have the same salt content in sample and standards.

b. Interference: All known interferences are eliminated or reduced to a minimum by distillation.

2. Apparatus

Colorimetric equipment: One of the following is required:

a. Spectrophotometer, for use at 578 nm, providing a light path of 10 mm or longer.

b. Filter photometer, providing a light path of at least 10 mm and equipped with a red filter having maximum transmittance at 570 to 580 nm.

3. Reagents

a. Chloramine-T solution: Dissolve 1.0 g white, water-soluble powder in 100 mL water. Prepare weekly and store in refrigerator.

b. Stock cyanide solution: Dissolve approximately 1.6 g NaOH and 2.51 g KCN in 1 L distilled water. (CAUTION—*KCN is highly toxic; avoid contact or inhalation.*) Standardize against standard silver nitrate ($AgNO_3$) titrant as described in Section 4500-CN.D.4, using 25 mL KCN solution. Check titer weekly because the solution gradually loses strength; 1 mL = 1 mg CN^-.

c. Standard cyanide solution: Based on the concentration determined for the KCN stock solution (¶ 3b) calculate volume required (approximately 10 mL) to prepare 1 L of a 10 µg CN^-/mL solution. Dilute with the NaOH dilution solution. Dilute 10 mL of the 10 µg CN^-/mL solution to 100 mL with the NaOH dilution solution; 1.0 mL = 1.0 µg CN^-. Prepare fresh daily and keep in a glass-stoppered bottle. (CAUTION—*Toxic; take care to avoid ingestion.*)

d. Pyridine-barbituric acid reagent: Place 15 g barbituric acid in a 250-mL volumetric flask and add just enough water to wash sides of flask and wet barbituric acid. Add

75 mL pyridine and mix. Add 15 mL conc hydrochloric acid (HCl), mix, and cool to room temperature. Dilute to mark with water and mix. This reagent is stable for up to 1 month; discard if a precipitate develops.

e. Sodium dihydrogen phosphate, 1 M: Dissolve 138 g $NaH_2PO_4 \cdot H_2O$ in 1 L distilled water. Refrigerate.

f. Sodium hydroxide dilution solution: Dissolve 1.6 g NaOH in 1 L distilled water.

4. Procedure

a. Preparation of calibration curve: Prepare a blank of NaOH dilution solution. From the standard KCN solution prepare a series of standards containing from 0.2 to 6 μg CN^- in 20 mL solution using the NaOH dilution solution for all dilutions. Treat standards in accordance with ¶ *b* below. Plot absorbance of standards against CN^- concentration (micrograms).

Recheck calibration curve periodically and each time a new reagent is prepared.

On the basis of the first calibration curve, prepare additional standards containing less than 0.2 and more than 6 μg CN^- to determine the limits measurable with the photometer being used.

b. Color development: Adjust photometer to zero absorbance each time using a blank consisting of the NaOH dilution solution and all reagents. Take a portion of absorption liquid obtained in Method C, such that the CN^- concentration falls in the measurable range, and dilute to 20 mL with NaOH dilution solution. Place in a 50-mL volumetric flask. Add 4 mL phosphate buffer and mix thoroughly. Add 2.0 mL chloramine-T solution and swirl to mix. *Immediately* add 5 mL pyridine-barbituric

acid solution and swirl gently. Dilute to mark with water; mix well by inversion.

Measure absorbance with the photometer at 578 nm after 8 min but within 15 min from the time of adding the pyridine-barbituric acid reagent. Even with the specified time of 8 to 15 min there is a slight change in absorbance. To minimize this, standardize time for all readings. Using the calibration curve and the formula in ¶ 5 below, determine CN^- concentration in original sample.

5. Calculations

$$CN^-,\ mg/L = \frac{A \times B}{C \times D}$$

where:

A = μg CN^- read from calibration curve (50 mL final volume),
B = total volume of absorbing solution from the distillation, mL,
C = volume of original sample used in the distillation, mL, and
D = volume of absorbing solution used in colorimetric test, mL.

6. Precision

The analysis of a mixed cyanide solution containing sodium, zinc, copper, and silver cyanides in tap water gave a precision within the designated range as follows:

$$S_T = 0.115X + 0.031$$

where:

S_T = overall precision and
X = CN^- concentration, mg/L.

7. Reference

1. AMUS, E.& H. GARSCHAGEN. 1953. Über die Verwendung der Barbitsäure für die photometrische Bestimmund von Cyanid und Rhodanid. *Z. Anal. Chem.* 138:414.

4500-CN F. Cyanide-Selective Electrode Method

1. General Discussion

CN^- in the alkaline distillate from the preliminary treatment procedures can be determined potentiometrically by using a CN^--selective electrode in combination with a double-junction reference electrode and a pH meter having an expanded millivolt scale, or a specific ion meter. This method can be used to determine CN^- concentration in place of either the colorimetric or titrimetric procedures in the concentration range of 0.05 to 10 mg CN^-/L.[1-3] If the CN^--selective electrode method is used, the previously described titration screening step can be omitted.

2. Apparatus

a. *Expanded-scale pH meter or specific-ion meter.*

b. *Cyanide-ion-selective electrode.* *

c. *Reference electrode,* double-junction.

d. *Magnetic mixer* with TFE-coated stirring bar.

3. Reagents

a. *Stock standard cyanide solution:* See Section 4500-CN.E.3b.

b. *Sodium hydroxide diluent:* Dissolve 1.6 g NaOH in water and dilute to 1 L.

c. *Intermediate standard cyanide solution:* Dilute a calculated volume (approximately 100 mL) of stock KCN solution, based on the determined concentration, to 1000 mL with NaOH diluent. Mix thoroughly; 1 mL = 100 μg CN^-.

d. *Dilute standard cyanide solution:* Dilute 100.0 mL intermediate standard CN^- solution to 1000 mL with NaOH diluent; 1.00 mL = 10.0 μg CN^-. Prepare daily and keep in a dark, glass-stoppered bottle.

e. *Potassium nitrate solution:* Dissolve 100 g KNO_3 in water and dilute to 1 L. Adjust to pH 12 with KOH. This is the outer filling solution for the double-junction reference electrode.

4. Procedure

a. *Calibration:* Use the dilute and intermediate standard CN^- solutions and the NaOH diluent to prepare a series of three standards, 0.1, 1.0, and 10.0 mg CN^-/L. Transfer approximately 100 mL of each of these standard solutions into a 250-mL beaker prerinsed with a small portion of standard being tested. Immerse CN^- and double-junction reference electrodes. Mix well on a magnetic stirrer at 25°C, maintaining as closely as possible the same stirring rate for all solutions.

Always progress from the lowest to the highest concentration of standard because otherwise equilibrium is reached only slowly. The electrode membrane dissolves in solutions of high CN^- concentration; do not use with a concentration above 10 mg/L. After making measurements remove electrode and soak in water.

After equilibrium is reached (at least 5 min and not more than 10 min), record potential (millivolt) readings and plot CN^- concentrations versus readings on semilogarithmic graph paper. A straight line with a slope of approximately 59 mV per decade indicates that the instrument and electrodes are operating properly. Record slope of line obtained (millivolts/decade of concentration). The slope may vary somewhat from the theoretical value of 59.2 mV per decade because of manufacturing variation and reference electrode (liquid-junction) potentials. The slope should be a straight line and is the basis for calculating sample concentration.

b. *Measurement of sample:* Place 100 mL of absorption liquid obtained in Section 4500-CN.C.4d into a 250-mL beaker. When measuring low CN^- concentrations, first rinse beaker and electrodes with a small volume of sample. Immerse CN^- and dou-

---

*Orion Model 94-06A or equivalent.

ble-junction reference electrodes and mix on a magnetic stirrer at the same stirring rate used for calibration. After equilibrium is reached (at least 5 min and not more than 10 min), record values indicated on ion meter or found from graph prepared as above. Calculate concentration as directed below.

5. Calculations

$$\text{mg CN}^-/\text{L} = \frac{A \times B}{C}$$

where:
    A = mg CN⁻/L found from meter reading or graph,
    B = total volume of absorption solution after dilution, mL, and
    C = volume of original sample used in the distillation, mL.

6. Precision

The precision of the CN⁻-ion-selective electrode method using the absorption so-lution from total cyanide distillation has been found in collaborative testing to be linear within its designated range and may be expressed as follows:

Reagent water: $S_T = 0.06X + 0.0016$
Water matrix: $S_T = 0.05X - 0.0176$

where:
    S_T = overall precision and
    X = concentration, mg CN⁻/L.

The difference of overall precision from pooled single-operator precision was not statistically significant.

7. References

1. ORION RESEARCH, INC. 1975. Cyanide Ion Electrode Instruction Manual, Cambridge, Mass.

2. FRANT, M.S., J.W. ROSS & J. H. RISEMAN. 1972. An electrode indicator technique for measuring low levels of cyanide. *Anal. Chem.* 44:2227.

3. SEKERKA, J. & J.F. LECHNER. 1976. Potentiometric determination of low levels of simple and total cyanides. *Water Res.* 10:479.

4500-CN G.　Cyanides Amenable to Chlorination after Distillation

1. General Discussion

This method is applicable to the determination of cyanides amenable to chlorination, to determine the dissociable CN⁻ content of the sample.

After part of the sample is chlorinated to decompose the cyanides, both the chlorinated and the untreated sample are subjected to distillation as described in Section 4500-CN.C. The difference between the CN⁻ concentrations found in the two samples is expressed as cyanides amenable to chlorination.

The titration procedure may be used when it is known that the concentration of cyanides is more than 1 but less than 10 mg/L. With higher concentrations, use a smaller portion as described in 4500-CN.D.4a. Use a colorimetric determination when the cyanides are known to be 1 mg/L or less. The cyanide-selective electrode method is useful in the concentration range of 0.05 to 10 mg CN⁻/L. For estimation of the concentration of cyanides amenable to chlorination use the spot test procedure (4500-CN.K).

Some unidentified organic chemicals may oxidize or form breakdown products during chlorination, giving higher results for cyanide after chlorination than before chlorination. This may lead to a negative value for cyanides amenable to chlorination after distillation for wastes from, for example, the steel industry, petroleum refining, and pulp and paper processing. Where such interferences are encountered use Method 4500-CN.I for determining dissociable cyanide.

2. Apparatus

a. Distillation apparatus: See Section 4500-CN.C.2

b. Apparatus for determining cyanide by either the titrimetric method, Section 4500-CN.D.2, the colorimetric method, Section 4500-CN.E.2, or the electrode method, Section 4500-CN.F.2.

3. Reagents

a. All reagents listed in Section 4500-CN.C.3.

b. All reagents listed in Section 4500-CN.D.3, 4500-CN.E.3, or 4500-CN.F.3, depending on method of estimation.

c. Calcium hypochlorite solution: Dissolve 5 g $Ca(OCl)_2$ in 100 mL distilled water. Store in an amber-colored glass bottle in the dark. Prepare monthly.

d. Potassium iodide(KI)-starch test paper.

4. Procedure

a. Divide sample into two equal parts and chlorinate one as in ¶ *b* below. Analyze both portions for CN^-. The difference in determined concentrations is the cyanide amenable to chlorination.

b. Add $Ca(OCl)_2$ solution dropwise to sample while agitating and maintaining pH between 11 and 12 by adding NaOH solution. Test for chlorine by placing a drop of treated sample on a strip of KI-starch paper. A distinct blue color indicates sufficient chlorine (approximately 50 to 100 mg Cl_2/L. Maintain excess residual chlorine for 1 h while agitating. If necessary, add more $Ca(OCl)_2$ and/or NaOH.

c. Add approximately 500 mg/L sodium thiosulfate ($Na_2S_2O_3$) as crystals to reduce residual chlorine. Test with KI-starch paper; there should be no color change. Add approximately 0.1 g/L more $Na_2S_2O_3$ to ensure a slight excess.

d. Minimize sample exposure to ultraviolet radiation before distillation.

e. Distill both chlorinated and unchlorinated samples as in Section 4500-CN.C. Test according to Methods D, E, or F.

5. Calculation

mg CN^- amenable to chlorination/L
$$= G - H$$

where:

G = mg CN^-/L found in unchlorinated portion of sample and

H = mg CN^-/L found in chlorinated portion of sample.

For samples containing significant quantities of iron cyanides, it is possible that the second distillation will give a higher value for CN^- than the test for total cyanide, leading to a negative result. When the difference is within the precision limits of the method, report, "no determinable quantities of cyanide amenable to chlorination." If the difference is greater than the precision limit, ascertain the cause such as presence of interferences, manipulation of the procedure, etc., or use Method I.

6. Precision

Precision, with the titrimetric finish, for cyanides amenable to chlorination was determined from a mixed cyanide solution containing sodium, zinc, copper, and silver cyanides and sodium ferrocyanide. The precision of the method within its designated range may be expressed as follows:

$$S_T = 0.049X + 0.162$$

where:

S_T = overall precision and
X = CN^- concentration, mg/L.

4500-CN H. Cyanides Amenable to Chlorination without Distillation (Short-Cut Method)

1. General Discussion

This method covers the determination of HCN and of CN complexes that are amenable to chlorination and also thiocyanates (SCN^-). The procedure does not measure cyanates (CNO^-) or iron cyanide complexes, but does determine cyanogen chloride (CNCl). This test requires neither lengthy distillation nor the chlorination of one sample before distillation. The recovery of CN^- from metal cyanide complexes will be comparable to that in Methods G and I.

The cyanides are converted to CNCl by chloramine-T after the sample has been heated. In the absence of nickel, copper, silver, and gold cyanide complexes or SCN^-, the CNCl may be developed at room temperature. The pyridine-barbituric acid reagent produces a red-blue color in the sample. The color can be estimated visually against standards or photometrically at 578 nm. The limits of the determination are 0.2 to 6 µg CN^-, representing 0.01 to 0.30 mg/L in a 20-mL sample. Higher CN^- concentrations may be determined by dilution. The dissolved salt content in the standards used for the development of the calibration curve should be near the salt content of the sample, including the added NaOH and phosphate buffer. See 4500-CN.E.1a.

The test sensitivity may be extended to the 5- to 150-µg/L level if a fresh, unstabilized sample is used. In these circumstances (pH < 9), add phosphate buffer dropwise to a pH of 6.5 (pH 6.0 to 6.6) and use a 40-mL sample, minimizing dilution before color development. Add 1 g sodium chloride (NaCl) to the 40-mL sample to make up for the salt content that would have been added if sodium hydroxide (NaOH) dilution solution and the required amount of phosphate buffer had been added.

The method's usefulness is limited by thiocyanate interference. Although the procedure allows the specific determination of CN^- amenable to chlorination (see 4500-CN.H.2 and 5) by masking the CN^- content and thereby establishing a correction for the thiocyanide content, the ratio of SCN^- to CN^- should not exceed 3 to be applicable. In working with unknown samples, screen the sample for SCN^- by the spot test (4500-CN.K).

2. Interferences

a. Remove interfering agents as described in Section 4500-CN.B with the exception of NO_2^- and NO_3^- (4500-CN.B.3h).

b. The SCN^- ion reacts with chloramine-T to give a positive error equivalent to its concentration. The procedure allows the separate determination of SCN^- and subtraction of this value from the results for the total. Use the spot test (4500-CN.K) for SCN^- when its presence is suspected. If the SCN^- content is more than three times the CN^- content, use Method G or I.

c. Reducing chemical compounds, such as SO_3^{2-}, may interfere by consuming chlorine in the chloramine-T addition. A significant excess of chlorine is provided, but the procedure prescribes a test (4500-CN.H.5d) to avoid this interference.

d. Color and turbidity may interfere with the colorimetric determination. Overcome this interference by extraction with chloroform (4500-CN.B.3c) but omit reduction of the pH. Otherwise, use Method G or I.

3. Apparatus

a. Apparatus listed in 4500-CN.E.2.

b. Hot water bath.

4. Reagents

a. Reagents listed in Sections 4500-CN.B and E.3.

b. Sodium chloride, NaCl, crystals.

c. Sodium carbonate, Na_2CO_3, crystals.

d. Sulfuric acid solution, H_2SO_4, $1N$.

e. EDTA solution, 0.05M: Dissolve 18.5 g disodium salt of ethylenediamine tetraacetic acid in water and dilute to 1 L.

f. Formaldehyde solution, 10%: Dilute 10 mL formaldehyde (37% pharmaceutical grade) 1 + 9 with water.

5. Procedure

a. Calibrate as directed in Section 4500-CN.E.1*a* and 4*a*. Adjust absorbance to zero, using the NaOH dilution solution to prepare the blank. For samples with more than 3000 mg total dissolved solids/ L, prepare a calibration curve from standards and blank NaOH solutions containing 6 g NaCl/L. Samples containing total dissolved solids exceeding 10 000 mg/L require appropriate standards and a new calibration curve.

b. If it is not in the desired range, adjust pH of a small amount of sample (50 mL) to 11.4 to 11.8. If this requires acid addition, first add a small amount (0.2 to 0.4 g) Na_2CO_3 and stir to dissolve. Then add the 1N H_2SO_4 slowly while stirring. If base addition is necessary to raise the pH, use 1N NaOH. Pipet 20 mL of pH-adjusted sample into a 50-mL volumetric flask. If more than 300 mg CN⁻/L is known to be present, use a smaller sample diluted to 20 mL with NaOH dilution solution.

c. To insure uniform color development, both in calibration and testing, maintain a uniform temperature. Immerse flask in a water bath held at 27 ± 1°C for 10 min before adding reagents and hold sample in water bath until all reagents have been added. Add 4 mL phosphate buffer and swirl to mix. Add 1 drop 0.1M EDTA solution. Hold in water bath for 1 min while swirling.

d. Add 2 mL chloramine-T solution and swirl to mix. Place a drop of solution on a strip of acidified starch-iodide test paper. The test paper should show the presence of chlorine. If reducing agents in the sample consume all the chloramine-T, add once more and recheck. If the test is again negative, the method is inapplicable; use Method G or I. After 1 min, add 5 mL pyridine-barbituric acid and swirl in water bath for 1 min.

e. Remove from water bath, dilute to 50.0 mL, and allow 7 min more for color development. Cool to room temperature, if necessary, and read absorbance at 578 nm within a total of 15 min from the time the pyridine-barbituric acid solution was added.

f. Standardize instrument with an appropriate blank each time it is used. Recheck calibration curve periodically using prepared standards and each time a new reagent is prepared.

g. If the presence of SCN⁻ is suspected (4500-CN.H.2*b*), take a second, pH-adjusted sample, add 3 drops formaldehyde solution, swirl to mix, and hold for 10 min. Place in water bath for an additional 10 min. The formaldehyde addition masks up to 0.3 mg CN⁻/L thereby providing a separate determination of SCN⁻. Follow the rest of the procedure ¶s 5*b-f*.

6. Calculation

Cyanide amenable to chlorination plus thiocyanate, mg CN⁻/L = A/B

where:

A = µg CN⁻ read from calibration curve (50 mL final volume) and
B = mL sample used.

Deduct SCN⁻ value from the results found when the CN⁻ has not been masked by formaldehyde addition (total) for cyanide content.

7. Precision

Collaborative testing of a mixed cyanide solution in both reagent and matrix water, containing sodium, copper, nickel, cadmium, and zinc metal cyanide complexes and additionally thiocyanate as an interference, gave a precision of the method within the designated range that is linear with concentration and may be expressed as follows:

Reagent water: $S_T = 0.0958X + 0.006$
$S_o = 0.0681X + 0.0048$

Matrix water: $S_T = 0.0973X + 0.0085$
$S_o = 0.0201X + 0.0064$

where:
S_T = overall precision, mg/L,
S_o = pooled single-operator precision, mg/L, and
X = concentration of cyanide, mg/L.

4500-CN I.   Weak and Dissociable Cyanide

1. General Discussion

Hydrogen cyanide (HCN) is liberated from a slightly acidified (pH 4.5 to 6.0) sample under the prescribed distillation conditions. The method does not recover CN^- from tight complexes that would not be amenable to oxidation by chlorine. The acetate buffer used contains zinc salts to precipitate iron cyanide as a further assurance of the selectivity of the method. In other respects the method is similar to 4500-CN.C.

2. Apparatus

See Section 4500-CN.C.2 and Figure 4500-CN:1, and also Section 4500-CN.D.2, 4500-CN.E.2, or 4500-CN.F.2, depending on method of estimation.

3. Reagents

a. *Reagents listed in Section 4500-CN.C.3.*

b. *Reagents listed in Section 4500-CN.D.3, 4500-CN.E.3, or 4500-CN.F.3,* depending on method of estimation.

c. *Acetic acid,* 1 + 9: Mix 1 volume of glacial acetic acid with 9 volumes of water.

d. *Acetate buffer:* Dissolve 410 g sodium acetate trihydrate ($NaC_2H_3O_2 \cdot 3H_2O$) in 500 mL water. Add glacial acetic acid to

yield a solution pH of 4.5 (approximately 500 mL).

e. *Zinc acetate solution,* 100 g/L: Dissolve 100 g $Zn(C_2H_3O_2) \cdot H_2O$ in 500 mL water. Dilute to 1 L.

f. *Methyl red indicator.*

4. Procedure

Follow procedure described in 4500-CN.C.4, but with the following modifications:

a. Do not add sulfamic acid, because NO_2^- and NO_3^- do not interfere.

b. Instead of H_2SO_4 and $MgCl_2$ reagents, add 10 mL each of the acetate buffer and zinc acetate solutions through air inlet tube. Also add 2 to 3 drops methyl red indicator. Rinse air inlet tube with water and allow air to mix contents. If the solution is not pink, add acetic acid (1 + 9) dropwise through air inlet tube until a pink color persists.

c. Follow instructions beginning with 4500-CN.C.4d.

d. For determining CN^- in the absorption solution, use the preferred finish method (4500-CN.D, E, or F).

5. Precision

Collaborative testing of a mixed cyanide solution in both reagent and matrix water, using the electrode final determination, gave overall and pooled single-operator precisions that are linear within the designated range and may be expressed as follows:

Reagent water: $S_T = 0.085X + 0.0032$
$S_o = 0.023X + 0.0093$

Water matrix: $S_T = 0.068X + 0.0039$
$S_o = 0.017X + 0.0039$

where:

S_T = overall precision, mg/L,
S_o = pooled single-operator precision, mg/L, and
X = concentration, mg CN^-/L.

4500-CN J. Cyanogen Chloride

1. General Discussion

Cyanogen chloride (CNCl) is the first reaction product when cyanide compounds are chlorinated. It is a volatile gas, only slightly soluble in water, but highly toxic even in low concentrations (CAUTION: *Avoid inhalation or contact*). A mixed pyridine-barbituric acid reagent produces a red-blue color with CNCl.

Because CNCl hydrolyzes to cyanate (CNO^-) at a pH of 12 or more, collect a separate sample for CNCl analysis (See Section 4500-CN.B.2) in a closed container without sodium hydroxide (NaOH). A quick test with a spot plate or comparator as soon as the sample is collected may be the only procedure for avoiding hydrolysis of CNCl due to time lapse between sampling and analysis.

If starch-iodide (KI) test paper indicates presence of chlorine or other oxidizing agents, add sodium thiosulfate ($Na_2S_2O_3$) immediately as directed in Section 4500-CN.B.2.

2. Apparatus

See Section 4500-CN.E.2.

3. Reagents

See Sections 4500-CN.E.3 and 4500-CN.H.4.

4. Procedure

Calibrate as directed in Sections 4500-CN.E.4 and 4500-CN.H.1, second paragraph. Add a portion of unstabilized sample, diluted if necessary, to contain 0.2 to 6 μg of CN^-/40 mL to a beaker. Add phosphate buffer to a pH of 6.5 (pH 6.0 to 6.6). Record exact volume of phosphate buffer required. Prepare a second sample as before and add to a 50-mL volumetric flask. Add phosphate buffer in the previously established volume. Add 5 mL pyridine-barbituric acid solution. Dilute to mark with distilled water and mix well by inversion. Allow 8 min for color development. Measure absorbance at 578 nm within 8 to 15 min from the addition of the pyridine-barbituric acid reagent. Using the calibration curve, determine the CNCl as CN^-.

5. Calculation

$$\text{mg CNCl (as } CN^-)/L = \frac{A}{B}$$

where:

A = μg CN^- read from calibration curve (50 mL final volume) and
B = mL original unstabilized sample.

6. Precision

The instability of CNCl precludes round-robin testing procedures and a precision statement is not possible.

4500-CN K.   Spot Test for Sample Screening

1. General Discussion

The spot test procedure allows a quick screening of the sample to establish whether more than 50 μg/L of cyanide amenable to chlorination is present. The test also establishes the presence or absence of cyanogen chloride (CNCl). With practice and dilution, the test reveals the approximate concentration range of these compounds by the color development compared with that of similarly treated standards.

When chloramine-T is added to cyanides amenable to chlorination, CNCl is formed. CNCl forms a red-blue color with the mixed reagent pyridine-barbituric acid. When testing for CNCl omit the chloramine-T addition. (CAUTION: *CNCl is a toxic gas; avoid inhalation.*)

The presence of formaldehyde in excess of 0.5 mg/L interferes with the test. A spot test for the presence of aldehydes and a method for removal of this interference are given in Section 4500-CN.B.3.

Thiocyanate (SCN⁻) reacts with chloramine-T, thereby creating a positive interference. The CN⁻ can be masked with formaldehyde and the sample retested. This makes the spot test specific for SCN⁻. In this manner it is possible to determine whether the spot discoloration is due to the presence of CN⁻, SCN⁻, or both.

2. Apparatus

a. *Porcelain spot plate* with 6 to 12 cavities.

b. *Dropping pipets.*

c. *Glass stirring rods.*

3. Reagents

a. *Chloramine-T solution:* See Section 4500-CN.E.3 a.

b. *Stock cyanide solution:* See Section 4500-CN.E.3 b.

c. *Pyridine-barbituric acid reagent:* See Section 4500-CN.E.3 d.

d. *Hydrochloric acid,* HCl, 1 + 9.

e. *Phenolphthalein indicator aqueous solution.*

f. *Sodium carbonate,* Na_2CO_3, anhydrous.

g. *Formaldehyde,* 37%, pharmaceutical grade.

4. Procedure

If the solution to be tested has a pH greater than 10, neutralize a 20- to 25-mL portion. Add about 250 mg Na_2CO_3 and swirl to dissolve. Add 1 drop phenolphthalein indicator. Add 1 + 9 HCl dropwise with constant swirling until the solution becomes colorless. Place 3 drops sample and 3 drops distilled water (for blanks) in separate cavities of the spot plate. To each cavity, add 1 drop chloramine-T solution and mix with a clean stirring rod. Add 1 drop pyridine-barbituric acid solution to each cavity and again mix. After 1 min, the sample spot will turn pink to red if 50 μg/L or more of CN⁻ are present. The blank spot will be faint yellow because of the color of the reagents. Until familiarity with the spot test is gained, use, in place of the water blank, a standard solution containing 50 μg CN⁻/L for color comparison. This standard can be made by diluting the stock cyanide solution (¶ 3b).

If SCN⁻ is suspected, test a second sample pretreated as follows: Heat a 20- to 25-mL sample in a water bath at 50°C; add 0.1 mL formaldehyde and hold for 10 min. This treatment will mask up to 5 mg CN⁻/L, if present. Repeat spot testing procedure. Color development indicates presence of SCN⁻. Comparing color intensity in the two spot tests is useful in judging relative concentration of CN⁻ and SCN⁻. If deep coloration is produced, serial dilution of sample and additional testing may allow closer approximation of the concentrations.

4500-CN L. Cyanates

1. General Discussion

Cyanate (CNO^-) may be of interest in analysis of industrial waste samples because the alkaline chlorination process used for the oxidation of cyanide yields cyanate in the second reaction.

Cyanate is unstable at neutral or low pH; therefore, stabilize the sample as soon as collected by adding sodium hydroxide (NaOH) to pH > 12. Remove residual chlorine by adding sodium thiosulfate ($Na_2S_2O_3$) (see Section 4500-CN.B.2).

a. *Principle:* Cyanate hydrolyzes to ammonia when heated at low pH.

$$2NaCNO + H_2SO_4 + 4H_2O \rightarrow (NH_4)_2SO_4 + 2NaHCO_3$$

The ammonia concentration must be determined on one sample portion before acidification. The ammonia content before and after hydrolysis of cyanate may be measured by direct nesslerization (4500-NH_3.C), phenate (4500-NH_3.D), or ammonia-selective electrode (4500-NH_3.F) method.[1] The test is applicable to cyanate compounds in natural waters and industrial waste.

b. *Interferences:*

1) Organic nitrogenous compounds may hydrolyze to ammonia (NH_3) upon acidification. To minimize this interference, control acidification and heating closely.

2) Metal compounds may precipitate or form colored complexes with nessler reagent. Adding Rochelle salt or EDTA in the determination of ammonia overcomes these interferences. Metal precipitates do not interfere with the ion-selective electrode method.

3) Reduce oxidants that oxidize cyanate to carbon dioxide and nitrogen with $Na_2S_2O_3$ (see Section 4500-CN.G).

4) Industrial waste containing organic material may contain unknown interferences.

c. *Detection limit:* 1 to 2 mg CNO^-/L.

2. Apparatus

a. *Expanded-scale pH meter or selective-ion meter.*

b. *Ammonia-selective electrode.**

c. *Magnetic mixer,* with TFE-coated stirring bar.

d. *Heat barrier:* Use a 3-mm-thick insulator under beaker to insulate against heat produced by stirrer motor.

3. Reagents

a. *Stock ammonium chloride solution:* See Section 4500-NH_3.C.3d.

b. *Standard ammonium chloride solution:* From the stock NH_4Cl solution prepare standard solutions containing 1.0, 10.0, and 100.0 mg NH_3-N/L by diluting with ammonia-free water.

c. *Sodium hydroxide, 10N:* Dissolve 400 g NaOH in water and dilute to 1 L.

d. *Sulfuric acid solution,* H_2SO_4, $1 + 1$.

e. *Ammonium chloride solution:* Dissolve 5.4 g NH_4Cl in distilled water and dilute to 1 L. (Use only for soaking electrodes.)

4. Procedure

a. *Calibration:* Daily, calibrate the ammonia electrode as in 4500-NH_3.F.4b and c using standard NH_4Cl solutions.

b. *Treatment of sample:* Dilute sample, if necessary, so that the CNO^- concentration is 1 to 200 mg/L or NH_3-N is 0.5 to 100 mg/L. Take or prepare at least 200 mL. From this 200 mL, take a 100-mL portion and, following the calibration procedure, establish the potential (millivolts) developed from the sample. Check electrode reading with prepared standards and adjust instrument calibration setting daily. Record NH_3-N content of untreated sample (B).

---

*Orion Model 95-10, EIL Model 8002-2, Beckman Model 39565, or equivalent.

Acidify 100 mL of prepared sample by adding 0.5 mL 1 + 1 H_2SO_4 to a pH of 2.0 to 2.5. Heat sample to 90 to 95°C and maintain temperature for 30 min. Cool to room temperature and restore to original volume by adding ammonia-free water. Pour into a 150-mL beaker, immerse electrode, start magnetic stirrer, then add 1 mL 10N NaOH solution. With pH paper check that pH is greater than 11. If necessary, add more NaOH until pH 11 is reached.

After equilibrium has been reached (30 s) record the potential reading. Estimate NH_3-N content from calibration curve.

5. Calculations

mg NH_3-N derived from CNO^-/L = $A - B$

where:

A = mg NH_3-N/L found in the acidified and heated sample portion and

B = mg NH_3-N/L found in untreated portion.

$$\text{mg } CNO^-/L = 3.0 \times (A - B)$$

6. Precision

No data on precision of this method are available. See 4500-NH_3.F.6 for precision of ammonia-selective electrode method.

7. Reference

1. THOMAS, R.F. & R.L. BOOTH. 1973. Selective electrode determination of ammonia in water and wastes. *Environ. Sci. Technol.* 7:523.

4500-CN M. Thiocyanate

1. General Discussion

When wastewater containing thiocyanate (SCN^-) is chlorinated, highly toxic cyanogen chloride (CNCl) is formed. At an acidic pH, ferric ion (Fe^{3+}) and SCN^- form an intense red color suitable for colorimetric determination.

a. Interference:

1) Hexavalent chromium (Cr^{6+}) interferes and is removed by adding ferrous sulfate ($FeSO_4$) after adjusting to pH 1 to 2 with nitric acid (HNO_3). Raising the pH to 9 with 1N sodium hydroxide (NaOH) precipitates Fe^{3+} and Cr^{3+}, which are then filtered out.

2) Reducing agents that reduce Fe^{3+} to Fe^{2+}, thus preventing formation of ferric thiocyanate complex, are destroyed by adding a few drops of hydrogen peroxide (H_2O_2).

3) Industrial wastes may be highly colored or contain various interfering organic compounds. To eliminate these interferences,[1] use the pretreatment procedure given in ¶ 4c below. It is the analyst's re-

sponsibility to validate the method's applicability without pretreatment (¶ 4b). If in doubt, pretreat sample before proceeding with analysis (¶ 4c).

4) If sample contains cyanide amenable to chlorination and would be preserved for the cyanide determination at a high pH, sulfide could interfere by converting cyanide to SCN^-. To preserve SCN^- and CN^-, precipitate the sulfide by adding lead salts according to 4500-CN.B.2 before adding alkali; filter to remove precipitate.

5) Thiocyanate is biodegradable. Preserve samples that may contain bacteria at pH <2 by adding mineral acid and refrigerate.

b. Application: 0.1 to 2.0 mg SCN^-/L in natural or wastewaters. For higher concentrations, use a portion of diluted sample.

2. Apparatus

a. Spectrophotometer or filter photometer, for use at 460 nm, providing a light path of 5 cm.

b. *Glass adsorption column:* Use a 50-mL buret with a glass-wool plug, and pack with macroreticular resin (¶ 3f) approximately 40 cm high. For convenience, apply a powder funnel of the same diameter as the buret to the top with a short piece of plastic tubing.

3. Reagents

a. *Ferric nitrate solution:* Dissolve 404 g $Fe(NO_3)_3 \cdot 9H_2O$ in about 800 mL distilled water. Add 80 mL conc HNO_3 and dilute to 1 L.

b. *Nitric acid solution*, 0.1N: Mix 6.4 mL conc HNO_3 in about 800 mL distilled water and dilute to 1 L.

c. *Stock thiocyanate solution:* Dissolve 1.673 g potassium thiocyanate (KSCN) in distilled water and dilute to 1000 mL; 1.00 mL = 1.00 mg SCN^-.

d. *Standard thiocyanate solution:* Dilute 10 mL stock solution to 1 L with distilled water; 1.00 mL = 0.01 mg SCN^-.

e. *Sodium hydroxide solution*, 4 g/L: Dissolve 4 g NaOH in about 800 mL distilled water and dilute to 1 L.

f. *Macroreticular resin*, 18 to 50 mesh:* The resin now available is an experimental product and is not purified. Some samples have shown contamination with waxes and oil, giving poor permeability and adsorption. Purify as follows:

Place sufficient resin to fill the column or columns in a beaker and add 5 times the resin volume of acetone. Stir gently for 1 h. Pour off fines and acetone from settled resin and add 5 times the resin volume of hexane. Stir for 1 h. Pour off fines and hexane and add 5 times the resin volume of methanol. Stir for 15 min. Pour off methanol and add 3 times the resin volume of 0.1N NaOH. Stir for 15 min. Pour off NaOH solution and add 3 times the resin volume of 0.1N HNO_3. Stir for 15 min. Pour off HNO_3 solution and add 3 times

the resin volume of distilled water. Stir for 15 min. Drain excess water and use purified resin to fill the column. Store excess purified resin after covering it with distilled water. Keep in a closed jar.

g. *Methyl alcohol.*

4. Procedure

a. *Preparation of calibration curve:* Prepare a series of standards containing from 0.02 g to 0.40 mg SCN^- by pipetting measured volumes of standard KSCN solution into 200-mL volumetric flasks and diluting with distilled water. Mix well. Develop color according to ¶ b below. Plot absorbance against SCN^- concentration expressed as mg/50 mL sample. The absorbance plot should be linear.

b. *Color development:* Use a filtered sample or portion from a diluted solution so that the concentration of SCN^- is between 0.1 and 2 mg/L. Adjust pH to 2 with conc HNO_3 added dropwise. Pipet 50-mL portion to a beaker, add 2.5 mL ferric nitrate, and mix.

Fill a 5-cm absorption cell and measure absorbance against a reagent blank at 460 nm or close to the maximum absorbance found with the instrument being used. Measure absorbance of the developed color against a reagent blank within 5 min from adding the reagent. (The color develops within 30 s and fades on standing in light.)

c. *Sample pretreatment:*

1) Color and various organic compounds interfere with absorbance measurement. At pH 2, macroreticular resin removes these interfering materials by adsorption without affecting thiocyanate.

2) To prepare the adsorption column, fill it with resin, rinse with 100 mL methanol, and follow by rinses with 100 mL 0.1N NaOH, 100 mL 0.1N HNO_3, and finally with 100 mL distilled water. If previously purified resin is used, omit these preparatory steps.

3) When washing, regenerating, or pass-

---

*Amberlite® XAD-8, Rohm & Haas Company, or equivalent.

ing a sample through the column, as solution level approaches resin bed, add and drain five separate 5-mL volumes of solution or water (depending on which is used in next step) to approximate bed height. After last 5-mL volume, fill column with remaining liquid. This procedure prevents undue mixing of solutions and helps void the column of the previous solution.

4) Acidify 150 mL sample (or a dilution) to pH 2 by adding conc HNO_3 dropwise while stirring. Pass it through the column at a flow rate not to exceed 20 mL/min. If the resin becomes packed and the flow rate falls to 4 to 5 mL/min, use gentle pressure through a manually operated hand pump or squeeze bulb on the column. In this case, use a separatory funnel for the liquid reservoir instead of the powder funnel. Alternatively use a vacuum bottle as a receiver and apply gentle vacuum. Do not let liquid level drop below the adsorbent in the column.

5) When passing a sample through the column, measure 90 mL of sample in a graduated cylinder, and from this use the five 5-mL additions as directed in ¶ 3), then pour the remainder of the 90 mL into the column. Add rest of sample and collect 60 mL eluate to be tested after the first 60 mL has passed through the column.

6) Prepare a new calibration curve using standards prepared according to ¶ 4a, but acidify standards according to ¶ 4b, and pass them through the adsorption column. Develop color and measure absorbance according to ¶ 4b against a reagent blank prepared by passing acidified, distilled water through the adsorption column.

7) Pipet 50 mL from the collected eluate to a beaker, add 2.5 mL ferric nitrate solution, and mix. Measure absorbance according to ¶ 4b against a reagent blank [see ¶ 6) above].

8) From the measured absorbance value, determine thiocyanate content of the sample or dilution using the absorbance plot.

9) Each day the column is in use, test a mid-range standard to check absorption curve.

10) Regenerate column between samples by rinsing with 100 mL 0.1N NaOH; 50 mL 0.1N HNO_3; and 100 mL water. Insure that the water has rinsed empty glass section of the buret. Occasionally rinse with 100 mL methanol for complete regeneration. Adsorbed weak organic acids and thiocyanate residuals from earlier tests are eluted by the NaOH rinse. Leave the column covered with the last rinse water for storage.

5. Calculation

Calculate thiocyanate concentration as follows:

$$mg\ SCN^-/L = \frac{A \times 1000}{B}$$

where:

A = mg SCN^- from calibration curve and
B = milliliters sample used.

6. Precision

Based on the results of collaborative testing, the precision of the method, including sample pretreatment, within its designated range is linear with concentration and may be expressed as follows:

Reagent water: $S_T = 0.0930X + 0.0426$
              $S_o = 0.045X\ \ + 0.010$

Water matrix: $S_T = 0.0547X + 0.0679$
             $S_o = 0.0244X + 0.0182$

where:

S_T = overall precision, mg/L,
S_o = pooled single-operator precision, mg/L, and
X = thiocyanate concentration, mg/L.

7. Reference

1. SPENCER, R.R., J. LEENHEER, V.C. MARTI. 1980. Automated colorimetric determination of thiocyanate, thiosulfate and tetrathionate in water. 94th Annu. Meeting. Assoc. Official Agricultural Chemists, Washington, D.C. 1981.

4500-Cl CHLORINE (RESIDUAL)*

4500-Cl A. Introduction

1. Effects of Chlorination

The chlorination of water supplies and polluted waters serves primarily to destroy or deactivate disease-producing microorganisms. A secondary benefit, particularly in treating drinking water, is the overall improvement in water quality resulting from the reaction of chlorine with ammonia, iron, manganese, sulfide, and some organic substances.

Chlorination may produce adverse effects. Taste and odor characteristics of phenols and other organic compounds present in a water supply may be intensified. Potentially carcinogenic chloroorganic compounds such as chloroform may be formed. Combined chlorine formed on chlorination of ammonia- or amine-bearing waters adversely affects some aquatic life. To fulfill the primary purpose of chlorination and to minimize any adverse effects, it is essential that proper testing procedures be used with a foreknowledge of the limitations of the analytical determination.

2. Chlorine Forms and Reactions

Chlorine applied to water in its molecular or hypochlorite form initially undergoes hydrolysis to form free chlorine consisting of aqueous molecular chlorine, hypochlorous acid, and hypochlorite ion. The relative proportion of these free chlorine forms is pH- and temperature-dependent. At the pH of most waters, hypochlorous acid and hypochlorite ion will predominate.

Free chlorine reacts readily with ammonia and certain nitrogenous compounds to form combined chlorine. With ammonia, chlorine reacts to form the chloramines: monochloramine, dichloramine, and nitrogen trichloride. The presence and concentrations of these combined forms depend chiefly on pH, temperature, initial chlorine-to-nitrogen ratio, absolute chlorine demand, and reaction time. Both free and combined chlorine may be present simultaneously. Combined chlorine in water supplies may be formed in the treatment of raw waters containing ammonia or by the addition of ammonia or ammonium salts. Chlorinated wastewater effluents, as well as certain chlorinated industrial effluents, normally contain only combined chlorine. Historically, the principal analytical problem has been to distinguish between free and combined forms of chlorine.

3. Selection of Method

In two separate but related studies, samples were prepared and distributed to participating laboratories to evaluate chlorine methods. Because of poor accuracy and

---

*Approved by Standard Methods Committee, 1989.

precision and a high overall (average) total error in these studies, all orthotolidine procedures except one were dropped in the 14th edition of this work. The useful stabilized neutral orthotolidine method was deleted from the 15th edition because of the toxic nature of orthotolidine. The leuco crystal violet (LCV) procedure has been dropped from this edition because of its relative difficulty and the lack of comparative advantages.

a. *Natural and treated waters:* The iodometric methods (B and C) are suitable for measuring total chlorine concentrations greater than 1 mg/L, but the amperometric end point of Methods C and D gives greater sensitivity. All acidic iodometric methods suffer from interferences, generally in proportion to the quantity of potassium iodide (KI) and H^+ added.

The amperometric titration method (D) is a standard of comparison for the determination of free or combined chlorine. It is affected little by common oxidizing agents, temperature variations, turbidity, and color. The method is not as simple as the colorimetric methods and requires greater operator skill to obtain the best reliability. Loss of chlorine can occur because of rapid stirring in some commercial equipment. Clean and conditioned electrodes are necessary for sharp end points.

In this edition, a low-level amperometric titration procedure (E) has been added to determine total chlorine at levels below 0.2 mg/L. This method is recommended only when quantification of such low residuals is necessary. The interferences are similar to those found with the standard amperometric procedure (D). The DPD methods (Methods F and G) are operationally simpler for determining free chlorine than the amperometric titration. Procedures are given for estimating the separate mono- and dichloramine and combined fractions. High concentrations of monochloramine interfere with the free chlorine determination unless the reaction is stopped with

arsenite or thioacetamide. In addition, the DPD methods are subject to interference by oxidized forms of manganese unless compensated for by a blank.

The amperometric and DPD methods are unaffected by dichloramine concentrations in the range of 0 to 9 mg Cl as $Cl_2/$ L in the determination of free chlorine. Nitrogen trichloride, if present, may react partially as free chlorine in the amperometric, DPD, and FACTS methods. The extent of this interference in the DPD methods does not appear to be significant.

The free chlorine test, syringaldazine (FACTS, Method H) was developed specifically for free chlorine. It is unaffected by significant concentrations of monochloramine, dichloramine, nitrate, nitrite, and oxidized forms of manganese.[1]

Sample color and turbidity may interfere in all colorimetric procedures.

Organic contaminants may produce a false free chlorine reading in most colorimetric methods (see ¶ 3b below). Many strong oxidizing agents interfere in the measurement of free chlorine in all methods. Such interferences include bromine, chlorine dioxide, iodine, permanganate, hydrogen peroxide, and ozone. However, the reduced forms of these compounds—bromide, chloride, iodide, manganous ion, and oxygen, in the absence of other oxidants, do not interfere. Reducing agents such as ferrous compounds, hydrogen sulfide, and oxidizable organic matter generally do not interfere.

b. *Wastewaters:* The determination of total chlorine in samples containing organic matter presents special problems. Because of the presence of ammonia, amines, and organic compounds, particularly organic nitrogen, residual chlorine exists in a combined state. A considerable residual may exist in this form, but at the same time there may be appreciable unsatisfied chlorine demand. Addition of reagents in the determination may change these relationships so that residual chlorine is lost during the analysis. Only the DPD method for total chlorine is performed under neutral pH conditions. In wastewater, the differentiation between free chlorine and combined chlorine ordinarily is not made because wastewater chlorination seldom is carried far enough to produce free chlorine.

The determination of residual chlorine in industrial wastes is similar to that in domestic wastewater when the waste contains organic matter, but may be similar to the determination in water when the waste is low in organic matter.

None of these methods is applicable to estuarine or marine waters because the bromide is converted to bromine and bromamines, which are detected as free or total chlorine. A procedure for estimating this interference is available for the DPD method.

Although the methods given below are useful for the determination of residual chlorine in wastewaters and treated effluents, select the method in accordance with sample composition. Some industrial wastes, or mixtures of wastes with domestic wastewater, may require special precautions and modifications to obtain satisfactory results.

Determine free chlorine in wastewater by any of the methods provided that known interfering substances are absent or appropriate correction techniques are used. The amperometric method is the method of choice because it is not subject to interference from color, turbidity, iron, manganese, or nitrite nitrogen. The DPD method is subject to interference from high concentrations of monochloramine, which is avoided by adding thioacetamide immediately after reagent addition. Oxidized forms of manganese at all levels encountered in water will interfere in all methods except in the free chlorine measurement of amperometric titrations and FACTS, but a blank correction for manganese can be made in Methods F and G.

The FACTS method is unaffected by concentrations of monochloramine, dichloramine, nitrite, iron, manganese, and other interfering compounds normally found in domestic wastewaters.

For total chlorine in samples containing significant amounts of organic matter, use either the DPD methods (F and G), amperometric, or iodometric back titration method (C) to prevent contact between the full concentration of liberated iodine and the sample. With Method C, do not use the starch-iodide end point if the concentration is less than 1 mg/L. In the absence of interference, the amperometric and starch-iodide end points give concordant results. The amperometric end point is inherently more sensitive and is free of interference from color and turbidity, which can cause difficulty with the starch-iodide end point. On the other hand, certain metals, surface-active agents, and complex anions in some industrial wastes interfere in the amperometric titration and indicate the need for another method for such wastewaters. Silver in the form of soluble silver cyanide complex, in concentrations of 1.0 mg Ag/L, poisons the cell at pH 4.0 but not at 7.0. The silver ion, in the absence of the cyanide complex, gives extensive response in the current at pH 4.0 and gradually poisons the cell at all pH levels. Cuprous copper in the soluble copper cyanide ion, in concentrations of 5 mg Cu/L or less, poisons the cell at pH 4.0 and 7.0. Although iron and nitrite may interfere with this method, minimize the interference by buffering to pH 4.0 before adding KI. Oxidized forms of manganese interfere in all methods for total chlorine including amperometric titration. An unusually high content of organic matter may cause uncertainty in the end point.

Regardless of end-point detection, either phenylarsine oxide or thiosulfate may be used as the standard reducing reagent at pH 4. The former is more stable and is preferred.

The DPD titrimetric and colorimetric

methods (F and G, respectively) are applicable to determining total chlorine in polluted waters. In addition, both DPD procedures and the amperometric titration method allow for estimating monochloramine and dichloramine fractions. Because all methods for total chlorine depend on the stoichiometric production of iodine, waters containing iodine-reducing substances may not be analyzed accurately by these methods, especially where iodine remains in the solution for a significant time. This problem occurs in Methods B and D. The back titration procedure (C) and Methods F and G cause immediate reaction of the iodine generated so that it has little chance to react with other iodine-reducing substances.

In all colorimetric procedures, compensate for color and turbidity by using color and turbidity blanks.

A method (I) for total residual chlorine using a potentiometric iodide electrode is proposed. This method is suitable for analysis of chlorine residuals in natural and treated waters and wastewater effluents. No differentiation of free and combined chlorine is possible. This procedure is an adaptation of other iodometric techniques and is subject to the same inferences.

4. Sampling and Storage

Chlorine in aqueous solution is not stable, and the chlorine content of samples or solutions, particularly weak solutions, will decrease rapidly. Exposure to sunlight or other strong light or agitation will accelerate the reduction of chlorine. Therefore, start chlorine determinations immediately after sampling, avoiding excessive light and agitation. Do not store samples to be analyzed for chlorine.

5. Reference

1. COOPER, W.J., N.M. ROSCHER & R.A. SLIFER. 1982. Determining free available chlorine by DPD-colorimetric, DPD-steadifac (colorimetric) and FACTS procedures. *J. Amer. Water Works Assoc.* 74:362.

6. Bibliography

MARKS, H.C., D.B. WILLIAMS & G.U. GLASGOW. 1951. Determination of residual chlorine compounds. *J. Amer. Water Works Assoc.* 43:201.

NICOLSON, N.J. 1965. An evaluation of the methods for determining residual chlorine in water, Part 1. Free chlorine. *Analyst* 90:187.

WHITTLE, G.P. & A. LAPTEFF, JR. 1973. New analytical techniques for the study of water disinfection. *In* Chemistry of Water Supply, Treatment, and Distribution, p. 63. Ann Arbor Science Publishers, Ann Arbor, Mich.

GUTER, W.J., W.J. COOPER & C.A. SORBER. 1974. Evaluation of existing field test kits for determining free chlorine residuals in aqueous solutions. *J. Amer. Water Works Assoc.* 66:38.

4500-Cl B.   Iodometric Method I

1. General Discussion

a. Principle: Chlorine will liberate free iodine from potassium iodide (KI) solutions at pH 8 or less. The liberated iodine is titrated with a standard solution of sodium thiosulfate ($Na_2S_2O_3$) with starch as the indicator. Titrate at pH 3 to 4 because the reaction is not stoichiometric at neutral pH due to partial oxidation of thiosulfate to sulfate.

b. Interference: Oxidized forms of manganese and other oxidizing agents interfere. Reducing agents such as organic sulfides also interfere. Although the neutral titration minimizes the interfering effect of ferric and nitrite ions, the acid titration is preferred because some forms of combined chlorine do not react at pH 7. Use only acetic acid for the acid titration; sulfuric acid (H_2SO_4) will increase interferences;

never use hydrochloric acid (HCl). See Section A.3 for discussion of other interferences.

c. *Minimum detectable concentration:* The minimum detectable concentration approximates 40 μg Cl as Cl_2/L if $0.01N$ $Na_2S_2O_3$ is used with a 1000-mL sample. Concentrations below 1 mg/L cannot be determined accurately by the starch-iodide end point used in this method. Lower concentrations can be measured with the amperometric end point in Methods C and D.

2. Reagents

a. *Acetic acid,* conc (glacial).

b. *Potassium iodide,* KI, crystals.

c. *Standard sodium thiosulfate,* 0.1N: Dissolve 25 g $Na_2S_2O_3 \cdot 5H_2O$ in 1 L freshly boiled distilled water and standardize against potassium bi-iodate or potassium dichromate after at least 2 weeks storage. This initial storage is necessary to allow oxidation of any bisulfite ion present. Use boiled distilled water and add a few milliliters chloroform ($CHCl_3$) to minimize bacterial decomposition.

Standardize $0.1N$ $Na_2S_2O_3$ by one of the following:

1) Iodate method—Dissolve 3.249 g anhydrous potassium bi-iodate, $KH(IO_3)_2$, primary standard quality, or 3.567 g KIO_3 dried at 103 ± 2°C for 1 h, in distilled water and dilute to 1000 mL to yield a $0.1000N$ solution. Store in a glass-stoppered bottle.

To 80 mL distilled water, add, with constant stirring, 1 mL conc H_2SO_4, 10.00 mL $0.1000N$ $KH(IO_3)_2$, and 1 g KI. Titrate immediately with $0.1N$ $Na_2S_2O_3$ titrant until the yellow color of the liberated iodine almost is discharged. Add 1 mL starch indicator solution and continue titrating until the blue color disappears.

2) Dichromate method—Dissolve 4.904 g anhydrous potassium dichromate, $K_2Cr_2O_7$, of primary standard quality, in distilled water and dilute to 1000 mL to yield a $0.1000N$ solution. Store in a glass-stoppered bottle.

Proceed as in the iodate method, with the following exceptions: Substitute 10.00 mL $0.1000N$ $K_2Cr_2O_7$ for iodate and let reaction mixture stand 6 min in the dark before titrating with $0.1N$ $Na_2S_2O_3$ titrant.

$$\text{Normality } Na_2S_2O_3 = \frac{1}{\text{mL } Na_2S_2O_3 \text{ consumed}}$$

d. *Standard sodium thiosulfate titrant,* $0.01N$ or $0.025N$: Improve the stability of $0.01N$ or $0.025N$ $Na_2S_2O_3$ by diluting an aged $0.1N$ solution, made as directed above, with freshly boiled distilled water. Add 4 g sodium borate and 10 mg mercuric iodide/L solution. For accurate work, standardize this solution daily in accordance with the directions given above, using $0.01N$ or $0.025N$ iodate or $K_2Cr_2O_7$. Use sufficient volumes of these standard solutions so that their final dilution is not greater than 1 + 4. To speed up operations where many samples must be titrated use an automatic buret of a type in which rubber does not come in contact with the solution. Standard titrants, $0.0100N$ and $0.0250N$, are equivalent, respectively, to 354.5 μg and 886.3 μg Cl as Cl_2/1.00 mL.

e. *Starch indicator solution:* To 5 g starch (potato, arrowroot, or soluble), add a little cold water and grind in a mortar to a thin paste. Pour into 1 L of boiling distilled water, stir, and let settle overnight. Use clear supernate. Preserve with 1.25 g salicylic acid, 4 g zinc chloride, or a combination of 4 g sodium propionate and 2 g sodium azide/L starch solution. Some commercial starch substitutes are satisfactory.

f. *Standard iodine,* 0.1N: See C.3g.

g. *Dilute standard iodine,* 0.0282N: See C.3h.

3. Procedure

a. *Volume of sample:* Select a sample volume that will require no more than 20 mL $0.01N$ $Na_2S_2O_3$ and no less than 0.2 mL for the starch-iodide end point. For a chlorine range of 1 to 10 mg/L, use a 500-mL sample; above 10 mg/L, use proportionately less sample. Use smaller samples and volumes of titrant with the amperometric end point.

b. Preparation for titration: Place 5 mL acetic acid, or enough to reduce the pH to between 3.0 and 4.0, in a flask or white porcelain casserole. Add about 1 g KI estimated on a spatula. Pour sample in and mix with a stirring rod.

c. Titration: Titrate away from direct sunlight. Add $0.025N$ or $0.01N$ $Na_2S_2O_3$ from a buret until the yellow color of the liberated iodine almost is discharged. Add 1 mL starch solution and titrate until blue color is discharged.

If the titration is made with $0.025N$ $Na_2S_2O_3$ instead of $0.01N$, then, with a 1-L sample, 1 drop is equivalent to about 50 $\mu g/L$. It is not possible to discern the end point with greater accuracy.

d. Blank titration: Correct result of sample titration by determining blank contributed by oxidizing or reducing reagent impurities. The blank also compensates for the concentration of iodine bound to starch at the end point.

Take a volume of distilled water corresponding to the sample used for titration in ¶s 3*a-c*, add 5 mL acetic acid, 1 g KI, and 1 mL starch solution. Perform blank titration as in 1) or 2) below, whichever applies.

1) If a blue color develops, titrate with $0.01N$ or $0.025N$ $Na_2S_2O_3$ to disappearance of blue color and record result. *B* (see ¶ 4, below) is negative.

2) If no blue color occurs, titrate with $0.0282N$ iodine solution until a blue color appears. Back-titrate with $0.01N$ or $0.025N$ $Na_2S_2O_3$ and record the difference. *B* is positive.

Before calculating the chlorine concentration, subtract the blank titration of ¶ 1) from the sample titration; or, if necessary, add the net equivalent value of the blank titration of ¶ 2).

4. Calculation

For standardizing chlorine solution for temporary standards:

$$\text{mg Cl as } Cl_2/\text{mL} = \frac{(A \pm B) \times N \times 35.45}{\text{mL sample}}$$

For determining total available residual chlorine in a water sample:

$$\text{mg Cl as } Cl_2/\text{L} = \frac{(A \pm B) \times N \times 35\,450}{\text{mL sample}}$$

where:
  A = mL titration for sample,
  B = mL titration for blank (positive or negative), and
  N = normality of $Na_2S_2O_3$.

5. Precision and Bias

Published studies[1,2] give the results of nine methods used to analyze synthetic water samples without interferences; variations of some of the methods appear in this edition. More current data are not now available.

6. References

1. Water Chlorine (Residual) No. 1. 1969. Analytical Reference Service Rep. No. 35, U.S. Environmental Protection Agency, Cincinnati, Ohio.
2. Water Chlorine (Residual) No. 2. 1971. Analytical Reference Service Rep. No. 40, U.S. Environmental Protection Agency, Cincinnati, Ohio.

7. Bibliography

LEA, C. 1933. Chemical control of sewage chlorination. The use and value of orthotolidine test. *J. Soc. Chem. Ind.* (London) 52:245T.
AMERICAN WATER WORKS ASSOCIATION. 1943. Committee report. Control of chlorination. *J. Amer. Water Works Assoc.* 35:1315.
MARKS, H.C., R. JOINER & F.B. STRANDSKOV.

1948. Amperometric titration of residual chlorine in sewage. *Water Sewage Works* 95:175.

STRANDSKOV, F.B. , H.C. MARKS & D.H. HORCHIER. 1949. Application of a new residual chlorine method to effluent chlorination. *Sewage Works J.* 21:23.

NUSBAUM, I. & L.A. MEYERSON. 1951. Determination of chlorine demands and chlorine residuals in sewage. *Sewage Ind. Wastes* 23:968.

MARKS, H.C., & N.S. CHAMBERLIN. 1953. Determination of residual chlorine in metal finishing wastes. *Anal. Chem.* 24:1885.

4500-Cl C.   Iodometric Method II

1. General Discussion

a. Principle: In this method, used for wastewater analysis, the end-point signal is reversed because the unreacted standard reducing agent (phenylarsine oxide or thiosulfate) remaining in the sample is titrated with standard iodine or standard iodate, rather than the iodine released being titrated directly. This indirect procedure is necessary regardless of the method of end-point detection, to avoid contact between the full concentration of liberated iodine and the wastewater.

When iodate is used as a back titrant, use only phosphoric acid. Do not use acetate buffer.

b. Interference: Oxidized forms of manganese and other oxidizing agents give positive interferences. Reducing agents such as organic sulfides do not interfere as much as in Method B. Minimize iron and nitrite interference by buffering to pH 4.0 before adding potassium iodide (KI). An unusually high content of organic matter may cause some uncertainty in the end point. Whenever manganese, iron, and other interferences definitely are absent, reduce this uncertainty and improve precision by acidifying to pH 1.0. Control interference from more than 0.2 mg nitrite/L with phosphoric acid-sulfamic acid reagent. A larger fraction of organic chloramines will react at lower pH along with interfering substances. See Section A.3 for a discussion of other interferences.

2. Apparatus

For a description of the amperometric end-point detection apparatus and a discussion of its use, see D.2a.

3. Reagents

a. Standard phenylarsine oxide solution, 0.005 64N: Dissolve approximately 0.8 g phenylarsine oxide powder in 150 mL 0.3N NaOH solution. After settling, decant 110 mL into 800 mL distilled water and mix thoroughly. Bring to pH 6 to 7 with 6N HCl and dilute to 950 mL with distilled water. CAUTION: *Severe poison, cancer suspect agent.*

Standardization—Accurately measure 5 to 10 mL freshly standardized 0.0282N iodine solution into a flask and add 1 mL KI solution. Titrate with phenylarsine oxide solution, using the amperometric end point (Method D) or starch solution (see B.2e) as an indicator. Adjust to 0.005 64N and recheck against the standard iodine solution; 1.00 mL = 200 µg available chlorine. (CAUTION: *Toxic—take care to avoid ingestion.*)

b. Standard sodium thiosulfate solution, 0.1N: See B.2c.

c. Standard sodium thiosulfate solution, 0.005 64N: Prepare by diluting 0.1N $Na_2S_2O_3$. For maximum stability of the dilute solution, prepare by diluting an aged 0.1N solution with freshly boiled distilled water (to minimize bacterial action) and add 10 mg HgI_2 and 4 g $Na_4B_4O_7$/L. Standardize daily as directed in B.2c using 0.005

$64N$ $K_2Cr_2O_7$ or iodate solution. Use sufficient volume of sample so that the final dilution does not exceed $1 + 2$. Use an automatic buret of a type in which rubber does not come in contact with the solution. 1.00 mL $= 200$ μg available chlorine.

d. *Potassium iodide*, KI, crystals.

e. *Acetate buffer solution*, pH 4.0: Dissolve 146 g anhydrous $NaC_2H_3O_2$, or 243 g $NaC_2H_3O_2 \cdot 3H_2O$, in 400 mL distilled water, add 480 g conc acetic acid, and dilute to 1 L with chlorine-demand-free water.

f. *Standard arsenite solution*, $0.1N$: Accurately weigh a stoppered weighing bottle containing approximately 4.95 g arsenic trioxide, As_2O_3. Transfer without loss to a 1-L volumetric flask and again weigh bottle. Do not attempt to brush out adhering oxide. Moisten As_2O_3 with water and add 15 g NaOH and 100 mL distilled water. Swirl flask contents gently to dissolve. Dilute to 250 mL with distilled water and saturate with CO_2, thus converting all NaOH to $NaHCO_3$. Dilute to mark, stopper, and mix thoroughly. This solution will preserve its titer almost indefinitely. (CAUTION: *Severe poison. Cancer suspect agent.*)

$$\text{Normality} = \frac{\text{g As}_2\text{O}_3}{49.455}$$

g. *Standard iodine solution*, $0.1N$: Dissolve 40 g KI in 25 mL chlorine-demand-free water, add 13 g resublimed iodine, and stir until dissolved. Transfer to a 1-L volumetric flask and dilute to mark.

Standardization—Accurately measure 40 to 50 mL $0.1N$ arsenite solution into a flask and titrate with $0.1N$ iodine solution, using starch solution as indicator. To obtain accurate results, insure that the solution is saturated with CO_2 at end of titration by passing current of CO_2 through solution for a few minutes just before end point is reached, or add a few drops of HCl to liberate sufficient CO_2 to saturate solution.

tion. Alternatively standardize against $Na_2S_2O_3$; see B.2c1).

Optionally, prepare $0.1000N$ iodine solution directly as a standard solution by weighing 12.69 g primary standard resublimed iodine. Because I_2 may be volatilized and lost from both solid and solution, transfer the solid immediately to KI as specified above. Never let solution stand in open containers for extended periods.

h. *Standard iodine titrant*, $0.0282N$: Dissolve 25 g KI in a little distilled water in a 1-L volumetric flask, add correct amount of $0.1N$ iodine solution exactly standardized to yield a $0.0282N$ solution, and dilute to 1 L with chlorine-demand-free water. For accurate work, standardize daily according to directions in ¶ 3g above, using 5 to 10 mL of arsenite or $Na_2S_2O_3$ solution. Store in amber bottles or in the dark; protect solution from direct sunlight at all times and keep from all contact with rubber.

i. *Starch indicator:* See B.2e.

j. *Standard iodate titrant*, 0.005 $64N$: Dissolve 201.2 mg primary standard grade KIO_3, dried for 1 h at 103°C, or 183.3 mg primary standard anhydrous potassium biiodate in distilled water and dilute to 1 L.

k. *Phosphoric acid solution*, H_3PO_4, $1 + 9$.

l. *Phosphoric acid-sulfamic acid solution:* Dissolve 20 g NH_2SO_3H in 1 L $1 + 9$ phosphoric acid.

m. *Chlorine-demand-free water:* Prepare chlorine-demand-free water from good-quality distilled or deionized water by adding sufficient chlorine to give 5 mg/L free chlorine. After standing 2 d this solution should contain at least 2 mg/L free chlorine; if not, discard and obtain better-quality water. Remove remaining free chlorine by placing container in sunlight or irradiating with an ultraviolet lamp. After several hours take sample, add KI, and measure total chlorine with a colorimetric method using a nessler tube to increase sensitivity. Do not use before last trace of free and combined chlorine has been removed.

Distilled water commonly contains ammonia and also may contain reducing agents. Collect good-quality distilled or deionized water in a sealed container from which water can be drawn by gravity. To the air inlet of the container add an H_2SO_4 trap consisting of a large test tube half filled with $1 + 1$ H_2SO_4 connected in series with a similar but empty test tube. Fit both test tubes with stoppers and inlet tubes terminating near the bottom of the tubes and outlet tubes terminating near the top of the tubes. Connect outlet tube of trap containing H_2SO_4 to the distilled water container, connect inlet tube to outlet of empty test tube. The empty test tube will prevent discharge to the atmosphere of H_2SO_4 due to temperature-induced pressure changes. Stored in such a container, chlorine-demand-free water is stable for several weeks unless bacterial growth occurs.

4. Procedure

a. Preparation for titration:

1) Volume of sample—For chlorine concentrations of 10 mg/L or less, titrate 200 mL. For greater chlorine concentrations, use proportionately less sample and dilute to 200 mL with chlorine-demand-free water. Use a sample of such size that not more than 10 mL phenylarsine oxide solution is required.

2) Preparation for titration—Measure 5 mL 0.005 64N phenylarsine oxide or thiosulfate for chlorine concentrations from 2 to 5 mg/L, and 10 mL for concentrations of 5 to 10 mg/L, into a flask or casserole for titration with standard iodine or iodate. Start stirring. For titration by amperometry or standard iodine, also add excess KI (approximately 1 g) and 4 mL acetate buffer solution or enough to reduce the pH to between 3.5 and 4.2.

b. Titration: Use one of the following:

1) Amperometric titration—Add 0.0282N iodine titrant in small increments from a 1-mL buret or pipet. Observe meter needle response as iodine is added: the pointer remains practically stationary until the end point is approached, whereupon each iodine increment causes a temporary deflection of the microammeter, with the pointer dropping back to its original position. Stop titration at end point when a small increment of iodine titrant gives a definite pointer deflection upscale and the pointer does not return promptly to its original position. Record volume of iodine titrant used to reach end point.

2) Colorimetric (iodine) titration—Add 1 mL starch solution and titrate with 0.0282N iodine to the first appearance of blue color that persists after complete mixing.

3) Colorimetric (iodate) titration—To suitable flask or casserole add 200 mL chlorine-demand-free water and add, with agitation, the required volume of reductant, an excess of KI (approximately 0.5 g), 2 mL 10% H_3PO_4 solution, and 1 mL starch solution in the order given, and titrate immediately* with 0.005 64N iodate solution to the first appearance of a blue color that persists after complete mixing. Designate volume of iodate solution used as A. Repeat procedure, substituting 200 mL sample for the 200 mL chlorine-demand-free water. If sample is colored or turbid, titrate to the first change in color, using for comparison another portion of sample with H_3PO_4 added. Designate this volume of iodate solution as B.

5. Calculation

a. Titration with standard iodine:

$$\text{mg Cl as Cl}_2/\text{L} = \frac{(A - 5B) \times 200}{C}$$

---

*Titration may be delayed up to 10 min without appreciable error if H_3PO_4 is not added until immediately before titration.

where:

A = mL 0.005 64N reductant,
B = mL 0.0282N I$_2$, and
C = mL sample.

b. *Titration with standard iodate:*

$$\text{mg Cl as Cl}_2/\text{L} = \frac{(A - B) \times 200}{C}$$

where:

A = mL Na$_2$S$_2$O$_3$,
B = mL iodate required to titrate Na$_2$S$_2$O$_3$, and
C = mL sample.

6. Bibliography

See B.7.

4500-Cl D.   Amperometric Titration Method

1. General Discussion

Amperometric titration requires a higher degree of skill and care than the colorimetric methods. Chlorine residuals over 2 mg/L are measured best by means of smaller samples or by dilution with water that has neither residual chlorine nor a chlorine demand. The method can be used to determine total chlorine and can differentiate between free and combined chlorine. A further differentiation into monochloramine and dichloramine fractions is possible by control of KI concentration and pH.

a. *Principle:* The amperometric method is a special adaptation of the polarographic principle. Free chlorine is titrated at a pH between 6.5 and 7.5, a range in which the combined chlorine reacts slowly. The combined chlorine, in turn, is titrated in the presence of the proper amount of KI in the pH range 3.5 to 4.5. When free chlorine is determined, the pH must not be greater than 7.5 because the reaction becomes sluggish at higher pH values, nor less than 6.5 because at lower pH values some combined chlorine may react even in the absence of iodide. When combined chlorine is determined, the pH must not be less than 3.5 because of increased interferences at lower pH values, nor greater than 4.5 because the iodide reaction is not quantitative at higher pH values. The tendency of monochlora-

mine to react more readily with iodide than does dichloramine provides a means for further differentiation. The addition of a small amount of KI in the neutral pH range enables estimation of monochloramine content. Lowering the pH into the acid range and increasing the KI concentration allows the separate determination of dichloramine.

Organic chloramines can be measured as free chlorine, monochloramine, or dichloramine, depending on the activity of the chlorine in the organic compound.

Phenylarsine oxide is stable even in dilute solution and each mole reacts with two equivalents of halogen. A special amperometric cell is used to detect the end point of the residual chlorine-phenylarsine oxide titration. The cell consists of a nonpolarizable reference electrode that is immersed in a salt solution and a readily polarizable noble-metal electrode that is in contact both with the salt solution and the sample being titrated. In some applications, endpoint selectivity is improved by adding +200 mV to the platinum electrode versus silver, silver chloride. Another approach to end-point detection uses dual platinum electrodes, a mercury cell with voltage divider to impress a potential across the electrodes, and a microammeter. If there is no chlorine residual in the sample, the microammeter reading will be comparatively low because of cell polarization. The greater the residual, the greater the microammeter reading. The meter acts merely as a null-point indicator—that is, the actual meter reading is not important, but rather the relative readings as the titration proceeds. The gradual addition of phenylarsine oxide causes the cell to become more and more polarized because of the decrease in chlorine. The end point is recognized when no further decrease in meter reading can be obtained by adding more phenylarsine oxide.

b. Interference: Accurate determinations of free chlorine cannot be made in the presence of nitrogen trichloride, NCl_3, or chlorine dioxide, which titrate partly as free chlorine. When present, NCl_3 can titrate partly as free chlorine and partly as dichloramine, contributing a positive error in both fractions at a rate of approximately 0.1%/min. Some organic chloramines also can be titrated in each step. Monochloramine can intrude into the free chlorine fraction and dichloramine can interfere in the monochloramine fraction, especially at high temperatures and prolonged titration times. Free halogens other than chlorine also will titrate as free chlorine. Combined chlorine reacts with iodide ions to produce iodine. When titration for free chlorine follows a combined chlorine titration, which requires addition of KI, erroneous results may occur unless the measuring cell is rinsed thoroughly with distilled water between titrations.

Interference from copper has been noted in samples taken from copper pipe or after heavy copper sulfate treatment of reservoirs, with metallic copper plating out on the electrode. Silver ions also poison the electrode. Interference occurs in some highly colored waters and in waters containing surface-active agents. Very low temperatures slow response of measuring cell and longer time is required for the titration, but precision is not affected. A reduction in reaction rate is caused by pH

values above 7.5; overcome this by buffering all samples to pH 7.0 or less. On the other hand, some substances, such as manganese, nitrite, and iron, do not interfere. The violent stirring of some commercial titrators can lower chlorine values by volatilization. When dilution is used for samples containing high chlorine content, take care that the dilution water is free of chlorine and ammonia and possesses no chlorine demand.

See A.3 for a discussion of other interferences.

2. Apparatus

a. End-point detection apparatus, consisting of a cell unit connected to a microammeter, with necessary electrical accessories. The cell unit includes a noble-metal electrode of sufficient surface area, a salt bridge to provide an electrical connection without diffusion of electrolyte, and a reference electrode of silver-silver chloride in a saturated sodium chloride solution connected into the circuit by means of the salt bridge. Numerous commercial systems are available.

Keep platinum electrode free of deposits and foreign matter. Vigorous chemical cleaning generally is unnecessary. Occasional mechanical cleaning with a suitable abrasive usually is sufficient. Keep salt bridge in good operating condition; do not allow it to become plugged nor permit appreciable flow of electrolyte through it. Keep solution surrounding reference electrode free of contamination and maintain it at constant composition by insuring an adequate supply of undissolved salt at all times. A cell with two metal electrodes polarized by a small DC potential also may be used. (See Bibliography.)

b. Agitator, designed to give adequate agitation at the noble-metal electrode surface to insure proper sensitivity. Thoroughly clean agitator and exposed electrode system to remove all chlorine-consuming contaminants by immersing them in water containing 1 to 2 mg/L free chlorine for a few minutes. Add KI to the same water and let agitator and electrodes remain immersed for 5 min. After thorough rinsing with chlorine-demand-free water or the sample to be tested, sensitized electrodes and agitator are ready for use. Remove iodide reagent completely from cell.

c. Buret: Commercial titrators usually are equipped with suitable burets (1 mL). Manual burets are available.*

d. Glassware, exposed to water containing at least 10 mg/L chlorine for 3 h or more before use and rinsed with chlorine-demand-free water.

3. Reagents

a. Standard phenylarsine oxide titrant: See C.3*a.*

b. Phosphate buffer solution, pH 7: Dissolve 25.4 g anhydrous KH_2PO_4 and 34.1 g anhydrous Na_2HPO_4 in 800 mL distilled water. Add 2 mL sodium hypochlorite solution containing 1% chlorine and mix thoroughly. Protect from sunlight for 2 d. Determine that free chlorine still remains in the solution. Then expose to sunlight until no chlorine remains. If necessary, carry out the final dechlorination with an ultraviolet lamp. Determine that no total chlorine remains by adding KI and measuring with one of the colorimetric tests. Dilute to 1 L with distilled water and filter if any precipitate is present.

c. Potassium iodide solution: Dissolve 50 g KI and dilute to 1 L with freshly boiled and cooled distilled water. Store in the dark in a brown glass-stoppered bottle, preferably in the refrigerator. Discard when solution becomes yellow.

d. Acetate buffer solution, pH 4: See C.3*e.*

───────────────
*Kimax 17110-F, 5 mL, Kimble Products, Box 1035, Toledo, Ohio, or equivalent.

4. Procedure

a. Sample volume: Select a sample volume requiring no more than 2 mL phenylarsine oxide titrant. Thus, for chlorine concentrations of 2 mg/L or less, take a 200-mL sample; for chlorine levels in excess of 2 mg/L, use 100 mL or proportionately less.

b. Free chlorine: Unless sample pH is known to be between 6.5 and 7.5, add 1 mL pH 7 phosphate buffer solution to produce a pH of 6.5 to 7.5. Titrate with standard phenylarsine oxide titrant, observing current changes on microammeter. Add titrant in progressively smaller increments until all needle movement ceases. Make successive buret readings when needle action becomes sluggish, signaling approach of end point. Subtract last very small increment that causes no needle response because of overtitration. Alternatively, use a system involving continuous current measurements and determine end point mathematically.

Continue titrating for combined chlorine as described in ¶ 4*c* below or for the separate monochloramine and dichloramine fractions as detailed in ¶s 4*e* and 4*f.*

c. Combined chlorine: To sample remaining from free-chlorine titration add 1.00 mL KI solution and 1 mL acetate buffer solution, in that order. Titrate with phenylarsine oxide titrant to the end point, as above. Do not refill buret but simply continue titration after recording figure for free chlorine. Again subtract last increment to give amount of titrant actually used in reaction with chlorine. (If titration was continued without refilling buret, this figure represents total chlorine. Subtracting free chlorine from total gives combined chlorine.) Wash apparatus and sample cell thoroughly to remove iodide ion to avoid inaccuracies when the titrator is used subsequently for a free chlorine determination.

d. Separate samples: If desired, determine total chlorine and free chlorine on separate samples. If sample pH is between 3.5 and 9.5 and total chlorine alone is required, treat sample immediately with 1 mL KI solution followed by 1 mL acetate buffer solution, and titrate with phenylarsine oxide titrant as described in ¶ 4c preceding.

e. *Monochloramine:* After titrating for free chlorine, add 0.2 mL KI solution to same sample and, without refilling buret, continue titration with phenylarsine oxide titrant to end point. Subtract last increment to obtain net volume of titrant consumed by monochloramine.

f. *Dichloramine:* Add 1 mL acetate buffer solution and 1 mL KI solution to same sample and titrate final dichloramine fraction as described above.

5. Calculation

Convert individual titrations for free chlorine, combined chlorine, total chlorine, monochloramine, and dichloramine by the following equation:

$$\text{mg Cl as Cl}_2/\text{L} = \frac{A \times 200}{\text{mL sample}}$$

where:

A = mL phenylarsine oxide titration.

6. Precision and Bias

See B.5.

7. Bibliography

FOULK, C.W. & A.T. BAWDEN. 1926. A new type of endpoint in electrometric titration and its application to iodimetry. *J. Amer. Chem. Soc.* 48:2045.

MARKS, H.C. & J. R. GLASS. 1942. A new method of determining residual chlorine. *J. Amer. Water Works Assoc.* 34:1227.

HALLER, J.F. & S.S. LISTEK. 1948. Determination of chloride dioxide and other active chlorine compounds in water. *Anal. Chem.* 20:639.

MAHAN, W.A. 1949. Simplified amperometric titration apparatus for determining residual chlorine in water. *Water Sewage Works* 96:171.

KOLTHOFF, I.M. & J.J. LINGANE. 1952. Polarography, 2nd ed. Interscience Publishers, New York, N.Y.

MORROW, J.J. 1966. Residual chlorine determination with dual polarizable electrodes. *J. Amer. Water Works Assoc.* 58:363.

4500-Cl E. Low-Level Amperometric Titration Method

1. General Discussion

Detection and quantification of chlorine residuals below 0.2 mg/L require special modifications to the amperometric titration procedure. With these modifications chlorine concentrations at the 10-μg/L level can be measured. It is not possible to differentiate between free and combined chlorine forms. Oxidizing agents that interfere with the amperometric titration method (D) will interfere.

a. *Principle:* This method modifies D by using a more dilute titrant and a graphical procedure to determine the end point.

b. *Interference:* See D.1b.

2. Apparatus

See D.2.

3. Reagents

a. *Potassium bi-iodate,* 0.002 256N: Dissolve 0.7332 g anhydrous potassium bi-iodate, $KH(IO_3)_2$, in 500 mL chlorine-free distilled water and dilute to 1000 mL. Dilute 10.00 mL to 100.0 mL with chlorine-free distilled water. Use only freshly prepared solution for the standardization of phenylarsine oxide.

b. *Potassium iodide,* KI crystals.

c. *Low-strength phenylarsine oxide titrant,* 0.000 564N: Dilute 10.00 mL of

0.005 64N phenylarsine oxide (see C.3a) to 100.0 mL with chlorine-demand-free water (see C.3m).

Standardization—Dilute 5.00 mL 0.002 256N potassium bi-iodate to 200 mL with chlorine-free water. Add approximately 1.5 g KI and stir to dissolve. Add 1 mL acetate buffer and let stand in the dark for 6 min. Titrate using the amperometric titrator and determine the equivalence point as indicated below.

$$\text{Normality} = 0.000\ 564 \times 2.00/A$$

where A = mL phenylarsine oxide titrant required to reach the equivalence point of standard bi-iodate.

d. *Acetate buffer solution*, pH 4: See C.3e.

4. Procedure

Select a sample volume requiring no more than 2 mL phenylarsine oxide titrant. A 200-mL sample will be adequate for samples containing less than 0.2 mg total chlorine/L.

Before beginning titration, rinse buret with titrant several times. Rinse sample container with distilled water and then with sample. Add 200 mL sample to sample container and approximately 1.5 g KI. Dissolve, using a stirrer or mixer. Add 1 mL acetate buffer and place container in end-point detection apparatus. When the current signal stabilizes, record the reading. Initially adjust meter to a near full-scale deflection. Titrate by adding small,

known, volumes of titrant. After each addition, record cumulative volume added and current reading when the signal stabilizes. If meter reading falls to near or below 10% of full-scale deflection, record low reading, readjust meter to near full-scale deflection, and record difference between low amount and readjusted high deflection. Add this value to all deflection readings for subsequent titrant additions. Continue adding titrant until no further meter deflection occurs. If fewer than three titrant additions were made before meter deflection ceased, discard sample and repeat analysis using smaller titrant increments.

Determine equivalence point by plotting total meter deflection against titrant volume added. Draw straight line through the first several points in the plot and a second, horizontal straight line corresponding to the final total deflection in the meter. Read equivalence point as the volume of titrant added at the intersection of these two lines.

5. Calculation

$$\text{mg Cl as Cl}_2/\text{L} = \frac{A \times 200 \times N}{B \times 0.000\ 564}$$

where:
A = mL titrant at equivalence point,
B = sample volume, mL, and
N = phenylarsine oxide normality.

6. Bibliography

BROOKS, A. S. & G. L. SEEGERT. 1979. Low-level chlorine analysis by amperometric titration. *J. Water Pollut. Control Fed.* 51:2636.

4500-Cl F.   DPD Ferrous Titrimetric Method

1. General Discussion

a. *Principle: N,N*-diethyl-*p*-phenylenediamine (DPD) is used as an indicator in the titrimetric procedure with ferrous ammonium sulfate (FAS). Where complete differentiation of chlorine species is not required, the procedure may be simplified to give only free and combined chlorine or total chlorine.

In the absence of iodide ion, free chlorine reacts instantly with DPD indicator to produce a red color. Subsequent addition of a small amount of iodide ion acts catalytically to cause monochloramine to produce color. Addition of iodide ion to excess evokes a rapid response from dichloramine. In the presence of iodide ion, part of the nitrogen trichloride (NCl_3) is included with dichloramine and part with free chlorine. A supplementary procedure based on adding iodide ion before DPD permits estimating proportion of NCl_3 appearing with free chlorine.

Chlorine dioxide (ClO_2) appears, to the extent of one-fifth of its total chlorine content, with free chlorine. A full response from ClO_2, corresponding to its total chlorine content, may be obtained if the sample first is acidified in the presence of iodide ion and subsequently is brought back to an approximately neutral pH by adding bicarbonate ion. Bromine, bromamine, and iodine react with DPD indicator and appear with free chlorine.

Addition of glycine before determination of free chlorine converts free chlorine to unreactive forms, with only bromine and iodine residuals remaining. Subtractions of these residuals from the residual measured without glycine permits differentiation of free chlorine from bromine and iodine.

b. pH control: For accurate results careful pH control is essential. At the proper pH of 6.2 to 6.5, the red colors produced may be titrated to sharp colorless end points. *Titrate as soon as the red color is formed in each step.* Too low a pH in the first step tends to make the monochloramine show in the free-chlorine step and the dichloramine in the monochloramine step. Too high a pH causes dissolved oxygen to give a color.

c. Temperature control: In all methods for differentiating free chlorine from chloramines, higher temperatures increase the tendency for chloramines to react and lead to increased apparent free-chlorine results.

Higher temperatures also increase color fading. Complete measurements rapidly, especially at higher temperature.

d. Interference: The most significant interfering substance likely to be encountered in water is oxidized manganese. To correct for this, place 5 mL buffer solution and 0.5 mL sodium arsenite solution in the titration flask. Add 100 mL sample and mix. Add 5 mL DPD indicator solution, mix, and titrate with standard FAS titrant until red color is discharged. Subtract reading from Reading *A* obtained by the normal procedure as described in ¶ 3*a*1) of this method or from the total chlorine reading obtained in the simplified procedure given in ¶ 3*a*4). If the combined reagent in powder form (see below) is used, first add KI and arsenite to the sample and mix, then add combined buffer-indicator reagent.

As an alternative to sodium arsenite use a 0.25% solution of thioacetamide, adding 0.5 mL to 100 mL sample.

Interference by copper up to approximately 10 mg Cu/L is overcome by the EDTA incorporated in the reagents. EDTA enhances stability of DPD indicator solution by retarding deterioration due to oxidation, and in the test itself, provides suppression of dissolved oxygen errors by preventing trace metal catalysis.

Chromate in excess of 2 mg/L interferes with end-point determination. Add barium chloride to mask this interference by precipitation.

High concentrations of combined chlorine can break through into the free chlorine fraction. *If free chlorine is to be measured in the presence of more than 0.5 mg/L combined chlorine, use the thioacetamide modification.* If this modification is not used, a color-development time in excess of 1 min leads to progressively greater interference from monochloramine. Adding thioacetamide (0.5 mL 0.25% solution to 100 mL) immediately after mixing DPD reagent with sample completely stops further reaction with combined chlorine in the free chlorine measurement. Continue immediately with FAS titration to obtain free chlorine. Obtain total chlorine from the normal procedure, i.e., without thioacetamide.

Because high concentrations of iodide are used to measure combined chlorine and only traces of iodide greatly increase chloramine interference in free chlorine measurements, take care to avoid iodide contamination by rinsing between samples or using separate glassware.

See A.3 for a discussion of other interferences.

e. Minimum detectable concentration: Approximately 18 μg Cl as Cl_2/L. This detection limit is achievable under ideal conditions; normal working detection limits typically are higher.

2. Reagents

a. Phosphate buffer solution: Dissolve 24 g anhydrous Na_2HPO_4 and 46 g anhydrous KH_2PO_4 in distilled water. Combine with 100 mL distilled water in which 800 mg disodium ethylenediamine tetraacetate dihydrate (EDTA) have been dissolved. Dilute to 1 L with distilled water and add 20 mg $HgCl_2$ to prevent mold growth and interference in the free chlorine test caused by any trace amounts of iodide in the reagents. (CAUTION: *$HgCl_2$ is toxic—take care to avoid ingestion*).

b. N,N-Diethyl-p-phenylenediamine (DPD) indicator solution: Dissolve 1 g DPD oxalate,* or 1.5 g DPD sulfate pentahydrate,† or 1.1 g anhydrous DPD sulfate in chlorine-free distilled water containing 8 mL 1 + 3 H_2SO_4 and 200 mg disodium EDTA. Make up to 1 L, store in a brown glass-stoppered bottle in the dark, and discard when discolored. Periodically check solution blank for absorbance and discard

when absorbance at 515 nm exceeds 0.002/ cm. (The buffer and indicator sulfate are available commercially as a combined reagent in stable powder form.) CAUTION: *The oxalate is toxic—take care to avoid ingestion.*

c. Standard ferrous ammonium sulfate (FAS) titrant: Dissolve 1.106 g $Fe(NH_4)_2(SO_4)_2 \cdot 6H_2O$ in distilled water containing 1 mL 1 + 3 H_2SO_4 and make up to 1 L with freshly boiled and cooled distilled water. This standard may be used for 1 month, and the titer checked by potassium dichromate. For this purpose add 10 mL 1 + 5 H_2SO_4, 5 mL conc H_3PO_4, and 2 mL 0.1% barium diphenylamine sulfonate indicator to a 100-mL sample of FAS and titrate with 0.100N primary standard potassium dichromate to a violet end point that persists for 30 s. The FAS titrant is equivalent to 100 μg Cl as Cl_2/ 1.00 mL.

d. Potassium iodide, KI, crystals.

e. Potassium iodide solution: Dissolve 500 mg KI and dilute to 100 mL, using freshly boiled and cooled distilled water. Store in a brown glass-stoppered bottle, preferably in a refrigerator. Discard when solution becomes yellow.

f. Potassium dichromate solution: See B.2c2).

g. Barium diphenylaminesulfonate, 0.1%: Dissolve 0.1 g $(C_6H_5NHC_6H_4$-4-$SO_3)_2Ba$ in 100 mL distilled water.

h. Sodium arsenite solution: Dissolve 5.0 g $NaAsO_2$ in distilled water and dilute to 1 L. (CAUTION: *Toxic—take care to avoid ingestion.*)

i. Thioacetamide solution: Dissolve 250 mg CH_3CSNH_2 in 100 mL distilled water. (CAUTION: *Cancer suspect agent. Take care to avoid skin contact or ingestion.*)

j. Chlorine-demand-free water: See C.3m.

k. Glycine solution: Dissolve 20 g glycine (aminoacetic acid) in sufficient chlorine-demand-free water to bring to 100 mL total volume. Store under refrigerated conditions and discard if cloudiness develops.

---

*Eastman chemical No. 7102 or equivalent.
†Available from Gallard-Schlesinger Chemical Mfg. Corp., 584 Mineola Avenue, Carle Place, N.Y. 11514, or equivalent.

l. Barium chloride crystals, $BaCl_2 \cdot 2H_2O$.

3. Procedure

The quantities given below are suitable for concentrations of total chlorine up to 5 mg/L. If total chlorine exceeds 5 mg/L, use a smaller sample and dilute to a total volume of 100 mL. Mix usual volumes of buffer reagent and DPD indicator solution, or usual amount of DPD powder, with distilled water *before* adding sufficient sample to bring total volume to 100 mL. (If sample is added before buffer, test does not work.)

If chromate is present (> 2 mg/L) add and mix 0.2 g $BaCl_2 \cdot 2H_2O$/100 mL sample before adding other reagents. If, in addition, sulfate is > 500 mg/L, use 0.4 g $BaCl_2 \cdot 2H_2O$/100 mL sample.

a. Free chlorine or chloramine: Place 5 mL each of buffer reagent and DPD indicator solution in titration flask and mix (or use about 500 mg DPD powder). Add 100 mL sample, or diluted sample, and mix.

1) Free chlorine—Titrate rapidly with standard FAS titrant until red color is discharged (Reading *A*).

2) Monochloramine—Add one very small crystal of KI (about 0.5 mg) or 0.1 mL (2 drops) KI solution and mix. Continue titrating until red color is discharged again (Reading *B*).

3) Dichloramine—Add several crystals KI (about 1 g) and mix to dissolve. Let stand for 2 min and continue titrating until red color is discharged (Reading *C*). For dichloramine concentrations greater than 1 mg/L, let stand 2 min more if color drift-back indicates slightly incomplete reaction. When dichloramine concentrations are not expected to be high, use half the specified amount of KI.

4) Simplified procedure for free and combined chlorine or total chlorine—Omit 2) above to obtain monochloramine and dichloramine together as combined chlorine. To obtain total chlorine in one reading, add full amount of KI at the start, with the specified amounts of buffer reagent and DPD indicator, and titrate after 2 min standing.

b. Nitrogen trichloride: Place one very small crystal of KI (about 0.5 mg) or 0.1 mL KI solution in a titration flask. Add 100 mL sample and mix. Add contents to a second flask containing 5 mL each of buffer reagent and DPD indicator solution (or add about 500 mg DPD powder direct to the first flask). Titrate rapidly with standard FAS titrant until red color is discharged (Reading *N*).

c. Free chlorine in presence of bromine or iodine: Determine free chlorine as in ¶ 3*a*1). To a second 100-mL sample, add 1 mL glycine solution before adding DPD and buffer. Titrate according to ¶ 3*a*1). Subtract the second reading from the first to obtain Reading *A*.

4. Calculation

For a 100-mL sample, 1.00 mL standard FAS titrant = 1.00 mg Cl as Cl_2/L.

Reading	NCl_3 Absent	NCl_3 Present
A	Free Cl	Free Cl
B − *A*	NH_2Cl	NH_2Cl
C − *B*	$NHCl_2$	$NHCl_2 + \frac{1}{2}NCl_3$
N	—	Free Cl $+ \frac{1}{2}NCl_3$
$2(N - A)$	—	NCl_3
C − *N*	—	$NHCl_2$

In the event that monochloramine is present with NCl_3, it will be included in *N*, in which case obtain NCl_3 from $2(N-B)$.

Chlorine dioxide, if present, is included in *A* to the extent of one-fifth of its total chlorine content.

In the simplified procedure for free and combined chlorine, only *A* (free Cl) and *C* (total Cl) are required. Obtain combined chlorine from *C*−*A*.

The result obtained in the simplified total chlorine procedure corresponds to C.

5. Precision and Bias

See B.5.

6. Bibliography

PALIN, A.T. 1957. The determination of free and combined chlorine in water by the use of diethyl-p-phenylene diamine. *J. Amer. Water Works Assoc.* 49:873.

PALIN, A.T. 1960. Colorimetric determination of chlorine dioxide in water. *Water Sewage Works* 107:457.

PALIN, A.T. 1961. The determination of free residual bromine in water. *Water Sewage Works* 108:461.

NICOLSON, N.J. 1963, 1965, 1966. Determination of chlorine in water, Parts 1, 2, and 3. Water Res. Assoc. Tech. Pap. Nos. 29, 47, and 53.

PALIN, A.T. 1967. Methods for determination, in water, of free and combined available chlorine, chlorine dioxide and chlorite, bromine, iodine, and ozone using diethyl-p-phenylenediamine (DPD). *J. Inst. Water Eng.* 21:537.

PALIN, A.T. 1968. Determination of nitrogen trichloride in water. *J. Amer. Water Works Assoc.* 60:847.

PALIN, A.T. 1975. Current DPD methods for residual halogen compounds and ozone in water. *J. Amer. Water Works Assoc.* 67:32.

Methods for the Examination of Waters and Associated Materials. Chemical Disinfecting Agents in Water and Effluents, and Chlorine Demand. 1980. Her Majesty's Stationery Off., London, England.

4500-Cl G.   DPD Colorimetric Method

1. General Discussion

a. Principle: This is a colorimetric version of the DPD method and is based on the same principles. Instead of titration with standard ferrous ammonium sulfate (FAS) solution as in the titrimetric method, a colorimetric procedure is used.

b. Interference: See A.3 and F.1*d*. Compensate for color and turbidity by using sample to zero photometer. Minimize chromate interference by using the thioacetamide blank correction.

c. Minimum detectable concentration: Approximately 10 μg Cl as Cl_2/L. This detection limit is achievable under ideal conditions; normal working detection limits typically are higher.

2. Apparatus

a. Photometric equipment: One of the following is required:

1) Spectrophotometer, for use at a wavelength of 515 nm and providing a light path of 1 cm or longer.

2) Filter photometer, equipped with a filter having maximum transmission in the wavelength range of 490 to 530 nm and providing a light path of 1 cm or longer.

b. Glassware: Use separate glassware, including separate spectrophotometer cells, for free and combined (dichloramine) measurements, to avoid iodide contamination in free chlorine measurement.

3. Reagents

See F.2*a, b, c, d, e, h, i,* and *j.*

4. Procedure

a. Calibration of photometric equipment: Calibrate instrument with chlorine or potassium permanganate solutions.

1) Chlorine solutions—Prepare chlorine standards in the range of 0.05 to 4 mg/L from about 100 mg/L chlorine water standardized as follows: Place 2 mL acetic acid and 10 to 25 mL chlorine-demand-free water in a flask. Add about 1 g KI. Measure into the flask a suitable volume of chlorine solution. In choosing a convenient volume, note that 1 mL 0.025N $Na_2S_2O_3$ titrant (see B.2*d*) is equivalent to about 0.9 mg chlo-

rine. Titrate with standardized 0.025N Na$_2$S$_2$O$_3$ titrant until the yellow iodine color almost disappears. Add 1 to 2 mL starch indicator solution and continue titrating to disappearance of blue color.

Determine the blank by adding identical quantities of acid, KI, and starch indicator to a volume of chlorine-demand-free water corresponding to the sample used for titration. Perform blank titration A or B, whichever applies, according to B.3d.

$$\text{mg Cl as Cl}_2/\text{mL} = \frac{(A + B) \times N \times 35.45}{\text{mL sample}}$$

where:

N = normality of Na$_2$S$_2$O$_3$,
A = mL titrant for sample,
B = mL titrant for blank (to be added or subtracted according to required blank titration. See B.3d).

Use chlorine-demand-free water and glassware to prepare these standards. Develop color by first placing 5 mL phosphate buffer solution and 5 mL DPD indicator reagent in flask and then adding 100 mL chlorine standard with thorough mixing as described in b and c below. Fill photometer or colorimeter cell from flask and read color at 515 nm. Return cell contents to flask and titrate with standard FAS titrant as a check on chlorine concentration.

2) Potassium permanganate solutions—Prepare a stock solution containing 891 mg KMnO$_4$/1000 mL. Dilute 10.00 mL stock solution to 100 mL with distilled water in a volumetric flask. When 1 mL of this solution is diluted to 100 mL with distilled water, a chlorine equivalent of 1.00 mg/L will be produced in the DPD reaction. Prepare a series of KMnO$_4$ standards covering the chlorine equivalent range of 0.05 to 4 mg/L. Develop color by first placing 5 mL phosphate buffer and 5 mL DPD indicator reagent in flask and adding 100 mL standard with thorough mixing as described in b and c below. Fill photometer or colorimeter cell from flask and read color at 515 nm. Return cell contents to flask and titrate with FAS titrant as a check on any absorption of permanganate by distilled water.

Obtain all readings by comparison to color standards or the standard curve before use in calculation.

b. *Volume of sample:* Use a sample volume appropriate to the photometer or colorimeter. The following procedure is based on using 10-mL volumes; adjust reagent quantities proportionally for other sample volumes. Dilute sample with chlorine-demand-free water when total chlorine exceeds 4 mg/L.

c. *Free chlorine:* Place 0.5 mL each of buffer reagent and DPD indicator reagent in a test tube or photometer cell. Add 10 mL sample and mix. Read color immediately (Reading A).

d. *Monochloramine:* Continue by adding one very small crystal of KI (about 0.1 mg) and mix. If dichloramine concentration is expected to be high, instead of small crystal add 0.1 mL (2 drops) freshly prepared KI solution (0.1 g/100 mL). Read color immediately (Reading B).

e. *Dichloramine:* Continue by adding several crystals of KI (about 0.1 g) and mix to dissolve. Let stand about 2 min and read color (Reading C).

f. *Nitrogen trichloride:* Place a very small crystal of KI (about 0.1 mg) in a clean test tube or photometer cell. Add 10 mL sample and mix. To a second tube or cell add 0.5 mL each of buffer and indicator reagents; mix. Add contents to first tube or cell and mix. Read color immediately (Reading N).

g. *Chromate correction using thioacetamide:* Add 0.5 mL thioacetamide solution (F.2i) to 100 mL sample. After mixing, add buffer and DPD reagent. Read color immediately. Add several crystals of KI (about 0.1 g) and mix to dissolve. Let stand about 2 min and read color. Subtract the first reading from Reading A and the second reading from Reading C and use in calculations.

5. Calculation

Reading	NCl$_3$ Absent	NCl$_3$ Present
A	Free Cl	Free Cl
$B - A$	NH$_2$Cl	NH$_2$Cl
$C - B$	NHCl$_2$	NHCl$_2$ + $\frac{1}{2}$NCl$_3$
N	—	Free Cl + $\frac{1}{2}$NCl$_3$
$2(N - A)$	—	NCl$_3$
$C - N$	—	NHCl$_2$

In the event that monochloramine is present with NCl$_3$, it will be included in Reading N, in which case obtain NCl$_3$ from $2(N-B)$.

6. Bibliography

See F.6.

4500-Cl H.   Syringaldazine (FACTS) Method

1. General Discussion

a. Principle: The free (available) chlorine test, syringaldazine (FACTS) measures free chlorine over the range of 0.1 to 10 mg/L. A saturated solution of syringaldazine (3,5-dimethoxy-4-hydroxybenzaldazine) in 2-propanol is used. Syringaldazine is oxidized by free chlorine on a 1:1 molar basis to produce a colored product with an absorption maximum of 530 nm. The color product is only slightly soluble in water; therefore, at chlorine concentrations greater than 1 mg/L, the final reaction mixture must contain 2-propanol to prevent product precipitation and color fading.

The optimum color and solubility (minimum fading) are obtained in a solution having a pH between 6.5 and 6.8. At a pH less than 6, color development is slow and reproducibility is poor. At a pH greater than 7, the color develops rapidly but fades quickly. A buffer is required to maintain the reaction mixture pH at approximately 6.7. Take care with waters of high acidity or alkalinity to assure that the added buffer maintains the proper pH.

Temperature has a minimal effect on the color reaction. The maximum error observed at temperature extremes of 5 and 35°C is ± 10%.

b. Interferences: Interferences common to other methods for determining free chlorine do not affect the FACTS procedure. Monochloramine concentrations up to 18 mg/L, dichloramine concentrations up to 10 mg/L, and manganese concentrations (oxidized forms) up to 1 mg/L do not interfere. Trichloramine at levels above 0.6 mg/L produces an apparent free chlorine reaction. Very high concentrations of monochloramine (≥ 35 mg/L) and oxidized manganese (≥ 2.6 mg/L) produce a color with syringaldazine slowly. Ferric iron can react with syringaldazine; however, concentrations up to 10 mg/L do not interfere. Nitrite (≤ 250 mg/L), nitrate (≤ 100 mg/L), sulfate (≤ 1000 mg/L), and chloride (≤ 1000 mg/L) do not interfere. Waters with high hardness (≥ 500 mg/L) will produce a cloudy solution that can be compensated for by using a blank. Oxygen does not interfere.

Other strong oxidizing agents, such as iodine, bromine, and ozone, will produce a color.

c. Minimum detectable concentration: The FACTS procedure is sensitive to free chlorine concentrations of 0.1 mg/L or less.

2. Apparatus

Colorimetric equipment: One of the following is required:

a. *Filter photometer*, providing a light path of 1 cm for chlorine concentrations ≤ 1 mg/L or a light path from 1 to 10 mm for chlorine concentrations above 1 mg/L; also equipped with a filter having a band pass of 500 to 560 nm.

b. *Spectrophotometer*, for use at 530 nm, providing the light paths noted above.

3. Reagents

a. *Chlorine-demand-free water:* See C.3m. Use to prepare reagent solutions and sample dilutions.

b. *Syringaldazine indicator:* Dissolve 115 mg 3,5-dimethoxy-4-hydroxybenzaldazine* in 1 L 2-propanol.

c. *2-Propanol:* To aid in dissolution use ultrasonic agitation or gentle heating and stirring. Redistill reagent-grade 2-propanol to remove chlorine demand. Use a 30.5-cm Vigreux column and take the middle 75% fraction. Alternatively, chlorinate good-quality 2-propanol to maintain a free residual overnight; then expose to UV light or sunlight to dechlorinate. CAUTION: *2-Propanol is extremely flammable.*

d. *Buffer:* Dissolve 17.01 g KH_2PO_4 in 250 mL water; pH should be 4.4. Dissolve 17.75 g Na_2HPO_4 in 250 mL water; the pH should be 9.9. Mix equal volumes of these solutions to obtain FACTS buffer, pH 6.6. Verify pH with pH meter. For waters containing considerable hardness or high al-

kalinity other pH 6.6 buffers can be used, for example, 23.21 g maleic acid and 16.5 mL 50% NaOH per liter of water.

e. *Hypochlorite solution:* Dilute household hypochlorite solution, which contains about 30 000 to 50 000 mg Cl equivalent/L, to a strength between 100 and 1000 mg/L. Standardize as directed in G.4a1).

4. Procedure

a. *Calibration of photometer:* Prepare a calibration curve by making dilutions of a standardized hypochlorite solution (¶ 3e). Develop and measure colors as described in ¶ 4b, below. Check calibration regularly, especially as reagent ages.

b. *Free chlorine analysis:* Add 3 mL sample and 0.1 mL buffer to a 5-mL-capacity test tube. Add 1 mL syringaldazine indicator, cap tube, and invert twice to mix. Transfer to a photometer tube or spectrophotometer cell and measure absorbance. Compare absorbance value obtained with calibration curve and report corresponding value as milligrams free chlorine per liter.

5. Bibliography

BAUER, R. & C. RUPE. 1971. Use of syringaldazine in a photometric method for estimating "free" chlorine in water. *Anal. Chem.* 43:421.

COOPER, W.J., C.A. SORBER & E.P. MEIER. 1975. A rapid, free, available chlorine test with syringaldazine (FACTS). *J. Amer. Water Works Assoc.* 67:34.

COOPER, W.J., P.H. GIBBS, E.M. OTT & P. PATEL. 1983. Equivalency testing of procedures for measuring free available chlorine: amperometric titration, DPD, and FACTS. *J. Amer. Water Works Assoc.* 75:625.

---

*Aldrich No. 17, 753-9, Aldrich Chemical Company, Inc., Cedar Knolls, N.J. 07927, or equivalent.

4500-Cl I.   Iodometric Electrode Technique

1. General Discussion

a. *Principle:* This method involves the direct potentiometric measurement of iodine released on the addition of potassium iodide to an acidified sample. A platinum-

iodide electrode pair is used in combination with an expanded-scale pH meter.

b. *Interference:* All oxidizing agents that interfere with other iodometric procedures interfere. These include oxidized manganese and iodate, bromine, and cupric ions. Silver and mercuric ions above 10 and 20 mg/L interfere.

2. Apparatus

a. Electrodes: Use either a combination electrode consisting of a platinum electrode and an iodide ion-selective electrode or two individual electrodes. Both systems are available commercially.

b. pH/millivolt meter: Use an expanded-scale pH/millivolt meter with 0.1 mV readability or a direct-reading selective ion meter.

3. Reagents

a. pH 4 buffer solution: See C.3e.

b. Chlorine-demand-free water: See C.3m.

c. Potassium iodide solution: Dissolve 42 g KI and 0.2 g Na_2CO_3 in 500 mL chlorine-demand-free, distilled water. Store in a dark bottle.

d. Standard potassium iodate 0.002 81N: Dissolve 0.1002 g KIO_3 in chlorine-demand-free, distilled water and dilute to 1000 mL. Each 1.0 mL, when diluted to 100 mL, produces a solution equivalent to 1 mg/L as Cl_2.

4. Procedure

a. Standardization: Pipet into three 100-mL stoppered volumetric flasks 0.20, 1.00, and 5.00 mL standard iodate solution. Add to each flask, and a fourth flask to be used as a reagent blank, 1 mL each of acetate buffer solution and KI solution. Stopper, swirl to mix, and let stand 2 min before dilution. Dilute each standard to 100 mL with chlorine-demand-free distilled water. Stopper, invert flask several times to mix, and pour into separate 150-mL beakers. Stir gently without turbulence, using a magnetic stirrer, and immerse electrode(s) in the 0.2-mg/L (0.2-mL) standard. Wait for the potential to stabilize and record potential in mV. Rinse electrodes with

chlorine-demand-free water and repeat for each standard and for the reagent blank. Prepare a calibration curve by plotting, on semilogarithmic paper, potential (linear axis) against concentration. Determine apparent chlorine concentration in the reagent blank from this graph (Reading B).

b. Analysis: Select a volume of sample containing no more than 0.5 mg chlorine. Pipet 1 mL acetate buffer solution and 1 mL KI into a 100-mL glass-stoppered volumetric flask. Stopper, swirl and let stand for at least 2 min. Adjust sample pH to 4 to 5, if necessary (mid-range pH paper is adequate for pH measurement), by adding acetic acid. Add pH-adjusted sample to volumetric flask and dilute to mark. Stopper and mix by inversion several times. Let stand for 2 min. Pour into a 150-mL beaker, immerse the electrode(s), wait for the potential to stabilize, and record. If the mV reading is greater than that recorded for the 5-mg/L standard, repeat analysis with a smaller volume of sample.

5. Calculation

Determine chlorine concentration (mg/L) corresponding to the recorded mV reading from the standard curve. This is Reading A. Determine total residual chlorine from the following:

$$\text{Total residual chlorine} = A \times 100/V$$

where V = sample volume, mL. If total residual chlorine is below 0.2 mg/L, subtract apparent chlorine in reagent blank (Reading B) to obtain the true total residual chlorine value.

6. Bibliography

DIMMOCK, N.A. & D. MIDGLEY. 1981. Determination of Total Residual Chlorine in Cooling Water with the Orion 97-70 Ion Selective Electrode. Central Electricity Generating Board (U.K.) Report RD/L/2159N81.

JENKINS, R.L. & R.B. BAIRD. 1979. Determination of total chlorine residual in treated wastewaters by electrode. *Anal. Letters* 12:125.

SYNNOTT, J.C. & A.M. SMITH. 1985. Total Residual Chlorine by Ion-Selective Electrode—from Bench Top to Continuous Monitor. Paper presented at 5th International Conf. on Chemistry for Protection of the Environment, Leuven, Belgium.

4500-Cl⁻ CHLORIDE*

4500-Cl⁻ A. Introduction

1. Occurrence

Chloride, in the form of chloride (Cl^-) ion, is one of the major inorganic anions in water and wastewater. In potable water, the salty taste produced by chloride concentrations is variable and dependent on the chemical composition of water. Some waters containing 250 mg Cl^-/L may have a detectable salty taste if the cation is sodium. On the other hand, the typical salty taste may be absent in waters containing as much as 1000 mg/L when the predominant cations are calcium and magnesium.

The chloride concentration is higher in wastewater than in raw water because sodium chloride (NaCl) is a common article of diet and passes unchanged through the digestive system. Along the sea coast, chloride may be present in high concentrations because of leakage of salt water into the sewerage system. It also may be increased by industrial processes.

A high chloride content may harm metallic pipes and structures, as well as growing plants.

2. Selection of Method

Five methods are presented for the determination of chloride. Because the first two are similar in most respects, selection is largely a matter of personal preference. The argentometric method (B) is suitable for use in relatively clear waters when 0.15 to 10 mg Cl^- are present in the portion titrated. The end point of the mercuric nitrate method (C) is easier to detect. The potentiometric method (D) is suitable for colored or turbid samples in which color-indicated end points might be difficult to observe. The potentiometric method can be used without a pretreatment step for samples containing ferric ions (if not present in an amount greater than the chloride concentration), chromic, phosphate, and ferrous and other heavy-metal ions. The ferricyanide method (E) is an automated technique. Ion chromatography (F) also may be used for chloride determinations.

3. Sampling and Storage

Collect representative samples in clean, chemically resistant glass or plastic bottles. The maximum sample portion required is 100 mL. No special preservative is necessary if the sample is to be stored.

---

*Approved by Standard Methods Committee, 1985.

4500-Cl⁻ B.   Argentometric Method

1. General Discussion

a. Principle: In a neutral or slightly al-kaline solution, potassium chromate can indicate the end point of the silver nitrate titration of chloride. Silver chloride is pre-cipitated quantitatively before red silver chromate is formed.

b. Interference: Substances in amounts normally found in potable waters will not interfere. Bromide, iodide, and cyanide reg-ister as equivalent chloride concentrations. Sulfide, thiosulfate, and sulfite ions inter-fere but can be removed by treatment with hydrogen peroxide. Orthophosphate in ex-cess of 25 mg/L interferes by precipitating as silver phosphate. Iron in excess of 10 mg/L interferes by masking the end point.

2. Apparatus

a. Erlenmeyer flask, 250-mL.
b. Buret, 50-mL.

3. Reagents

a. Potassium chromate indicator solution: Dissolve 50 g K_2CrO_4 in a little distilled water. Add $AgNO_3$ solution until a definite red precipitate is formed. Let stand 12 h, filter, and dilute to 1 L with distilled water.

b. Standard silver nitrate titrant, $0.0141M$ ($0.0141N$): Dissolve 2.395 g $AgNO_3$ in distilled water and dilute to 1000 mL. Standardize against NaCl by the pro-cedure described in ¶ 4*b* below; 1.00 mL = 500 µg Cl⁻. Store in a brown bottle.

c. Standard sodium chloride, $0.0141M$ ($0.0141N$): Dissolve 824.0 mg NaCl (dried at 140°C) in distilled water and dilute to 1000 mL; 1.00 mL = 500 µg Cl⁻.

d. Special reagents for removal of inter-ference:

1) *Aluminum hydroxide suspension:* Dis-solve 125 g aluminum potassium sulfate or aluminum ammonium sulfate, $AlK(SO_4)_2 \cdot 12H_2O$ or $AlNH_4(SO_4)_2 \cdot 12H_2O$, in 1 L dis-tilled water. Warm to 60°C and add 55 mL

conc ammonium hydroxide (NH_4OH) slowly with stirring. Let stand about 1 h, transfer to a large bottle, and wash precip-itate by successive additions, with thorough mixing and decanting with distilled water, until free from chloride. When freshly pre-pared, the suspension occupies a volume of approximately 1 L.

2) *Phenolphthalein indicator solution.*
3) *Sodium hydroxide,* NaOH, $1N$.
4) *Sulfuric acid,* H_2SO_4, $1N$.
5) *Hydrogen peroxide,* H_2O_2, 30%.

4. Procedure

a. Sample preparation: Use a 100-mL sample or a suitable portion diluted to 100 mL. If the sample is highly colored, add 3 mL $Al(OH)_3$ suspension, mix, let settle, and filter.

If sulfide, sulfite, or thiosulfate is present, add 1 mL H_2O_2 and stir for 1 min.

b. Titration: Directly titrate samples in the pH range 7 to 10. Adjust sample pH to 7 to 10 with H_2SO_4 or NaOH if it is not in this range. Add 1.0 mL K_2CrO_4 indi-cator solution. Titrate with standard $AgNO_3$ titrant to a pinkish yellow end point. Be consistent in end-point recogni-tion.

Standardize $AgNO_3$ titrant and establish reagent blank value by the titration method outlined above. A blank of 0.2 to 0.3 mL is usual.

5. Calculation

$$\text{mg Cl}^-/\text{L} = \frac{(A - B) \times N \times 35\,450}{\text{mL sample}}$$

where:

A = mL titration for sample,
B = mL titration for blank, and
N = normality of $AgNO_3$.

$$\text{mg NaCl/L} = (\text{mg Cl}^-/\text{L}) \times 1.65$$

6. Precision and Bias

A synthetic sample containing 241 mg Cl^-/L, 108 mg Ca/L, 82 mg Mg/L; 3.1 mg K/L, 19.9 mg Na/L, 1.1 mg NO_3^--N/L, 0.25 mg NO_2^--N/L, 259 mg SO_4^{2-}/L, and 42.5 mg total alkalinity/L (contributed by $NaHCO_3$) in distilled water was analyzed in 41 laboratories by the argentometric method, with a relative standard deviation of 4.2% and a relative error of 1.7%.

7. Bibliography

HAZEN, A. 1889. On the determination of chlorine in water. *Amer. Chem. J.* 11:409.
KOLTHOFF, I.M. & V.A. STENGER. 1947. Volumetric Analysis, 2nd ed. Vol. 2. Interscience Publishers, New York, N.Y., pp. 242-245, 256-258.

4500-Cl⁻ C. Mercuric Nitrate Method

1. General Discussion

a. Principle: Chloride can be titrated with mercuric nitrate, $Hg(NO_3)_2$, because of the formation of soluble, slightly dissociated mercuric chloride. In the pH range 2.3 to 2.8, diphenylcarbazone indicates the titration end point by formation of a purple complex with the excess mercuric ions. Xylene cyanol FF serves as a pH indicator and end-point enhancer. Increasing the strength of the titrant and modifying the indicator mixtures extend the range of measurable chloride concentrations.

b. Interference: Bromide and iodide are titrated with $Hg(NO_3)_2$ in the same manner as chloride. Chromate, ferric, and sulfite ions interfere when present in excess of 10 mg/L.

2. Apparatus

a. Erlenmeyer flask, 250-mL.
b. Microburet, 5-mL with 0.01-mL graduation intervals.

3. Reagents

a. Standard sodium chloride, 0.0141M (0.0141N): See Method B, ¶ 3c above.
b. Nitric acid, HNO_3, 0.1N.
c. Sodium hydroxide, NaOH, 0.1N.
d. Reagents for chloride concentrations below 100 mg/L:

1) *Indicator-acidifier reagent:* The HNO_3 concentration of this reagent is an important factor in the success of the determination and can be varied as indicated in a) or b) to suit the alkalinity range of the sample. Reagent a) contains sufficient HNO_3 to neutralize a total alkalinity of 150 mg as $CaCO_3/L$ to the proper pH in a 100-mL sample. Adjust amount of HNO_3 to accommodate samples of alkalinity different from 150 mg/L.

a) Dissolve, in the order named, 250 mg s-diphenylcarbazone, 4.0 mL conc HNO_3, and 30 mg xylene cyanol FF in 100 mL 95% ethyl alcohol or isopropyl alcohol. Store in a dark bottle in a refrigerator. This reagent is not stable indefinitely. Deterioration causes a slow end point and high results.

b) Because pH control is critical, adjust pH of highly alkaline or acid samples to 2.5 ± 0.1 with 0.1N HNO_3 or NaOH, not with sodium carbonate (Na_2CO_3). Use a pH meter with a nonchloride type of reference electrode for pH adjustment. If only the usual chloride-type reference electrode is available for pH adjustment, determine amount of acid or alkali required to obtain a pH of 2.5 ± 0.1 and discard this sample portion. Treat a separate sample portion with the determined amount of acid or alkali and continue analysis. Under these circumstances, omit HNO_3 from indicator reagent.

2) *Standard mercuric nitrate titrant,* 0.007 05 M (0.0141N): Dissolve 2.3 g $Hg(NO_3)_2$ or 2.5 g $Hg(NO_3)_2 \cdot H_2O$ in 100 mL distilled water containing 0.25 mL conc HNO_3. Dilute to just under 1 L. Make a preliminary standardization by following the procedure described in ¶ 4*a*. Use replicates containing 5.00 mL standard NaCl solution and 10 mg sodium bicarbonate ($NaHCO_3$) diluted to 100 mL with distilled water. Adjust titrant to 0.0141N and make a final standardization; 1.00 mL = 500 μg Cl^-. Store away from light in a dark bottle.

e. Reagent for chloride concentrations greater than 100 mg/L:

1) *Mixed indicator reagent:* Dissolve 0.50 g diphenylcarbazone powder and 0.05 g bromphenol blue powder in 75 mL 95% ethyl or isopropyl alcohol and dilute to 100 mL with the same alcohol.

2) *Strong standard mercuric nitrate titrant,* 0.0705M (0.141N) Dissolve 25 g $Hg(NO_3)_2 \cdot H_2O$ in 900 mL distilled water containing 5.0 mL conc HNO_3. Dilute to just under 1 L and standardize by following the procedure described in ¶ 4*b*. Use replicates containing 25.00 mL standard NaCl solution and 25 mL distilled water. Adjust titrant to 0.141N and make a final standardization; 1.00 mL = 5.00 mg Cl^-.

4. Procedure

a. Titration of chloride concentrations less than 100 mg/L: Use a 100-mL sample or smaller portion so that the chloride content is less than 10 mg.

Add 1.0 mL indicator-acidifier reagent. (The color of the solution should be green-blue at this point. A light green indicates pH less than 2.0; a pure blue indicates pH more than 3.8.) For most potable waters, the pH after this addition will be 2.5 ± 0.1. For highly alkaline or acid waters, adjust pH to about 8 before adding indicator-acidifier reagent.

Titrate with 0.0141N $Hg(NO_3)_2$ titrant to a definite purple end point. The solution turns from green-blue to blue a few drops before the end point.

Determine blank by titrating 100 mL distilled water containing 10 mg $NaHCO_3$.

b. Titration of chloride concentrations greater than 100 mg/L: Use a sample portion (5 to 50 mL) requiring less than 5 mL titrant to reach the end point. Measure into a 150-mL beaker. Add approximately 0.5 mL mixed indicator reagent and mix well. The color should be purple. Add 0.1N HNO_3 dropwise until the color just turns yellow. Titrate with strong $Hg(NO_3)_2$ titrant to first permanent dark purple. Titrate a distilled water blank using the same procedure.

5. Calculation

$$\text{mg } Cl^- / L = \frac{(A - B) \times N \times 35\,450}{\text{mL sample}}$$

where:

A = mL titration for sample,
B = mL titration for blank, and
N = normality of $Hg(NO_3)_2$.

$$\text{mg } NaCl / L = (\text{mg } Cl^- / L) \times 1.65$$

6. Precision and Bias

A synthetic sample containing 241 mg Cl^-/L, 108 mg Ca/L, 82 mg Mg/L, 3.1 mg K/L, 19.9 mg Na/L, 1.1 mg NO_3^--N/L, 0.25 mg NO_2^--N/L, 259 mg SO_4^{2-}/L, and 42.5 mg total alkalinity/L (contributed by $NaHCO_3$) in distilled water was analyzed in 10 laboratories by the mercurimetric method, with a relative standard deviation of 3.3% and a relative error of 2.9%.

7. Bibliography

KOLTHOFF, I.M. & V.A. STENGER. 1947. Volumetric Analysis, 2nd ed. Vol. 2. Interscience Publishers, New York, N.Y., pp. 334-335.

DOMASK, W.C. & K.A. KOBE. 1952. Mercurimetric determination of chlorides and water-soluble chlorohydrins. *Anal. Chem.* 24:989.

GOLDMAN, E. 1959. New indicator for the mercurimetric chloride determination in potable water. *Anal. Chem.* 31:1127.

4500-Cl⁻ D.　Potentiometric Method

1. General Discussion

a. Principle: Chloride is determined by potentiometric titration with silver nitrate solution with a glass and silver-silver chloride electrode system. During titration an electronic voltmeter is used to detect the change in potential between the two electrodes. The end point of the titration is that instrument reading at which the greatest change in voltage has occurred for a small and constant increment of silver nitrate added.

b. Interference: Iodide and bromide also are titrated as chloride. Ferricyanide causes high results and must be removed. Chromate and dichromate interfere and should be reduced to the chromic state or removed. Ferric iron interferes if present in an amount substantially higher than the amount of chloride. Chromic ion, ferrous ion, and phosphate do not interfere.

Grossly contaminated samples usually require pretreatment. Where contamination is minor, some contaminants can be destroyed simply by adding nitric acid.

2. Apparatus

a. Glass and silver-silver chloride electrodes: Prepare in the laboratory or purchase a silver electrode coated with AgCl for use with specified instruments. Instructions on use and care of electrodes are supplied by the manufacturer.

b. Electronic voltmeter, to measure potential difference between electrodes: A pH meter may be converted to this use by substituting the appropriate electrode.

c. Mechanical stirrer, with plastic-coated or glass impeller.

3. Reagents

a. Standard sodium chloride solution, $0.0141M$ $(0.0141N)$: See ¶ 4500-Cl⁻.B.3c.

b. Nitric acid, HNO_3, conc.

c. Standard silver nitrate titrant, $0.0141M$ $(0.0141N)$: See ¶ 4500-Cl⁻.B.3b.

d. Pretreatment reagents:
1) *Sulfuric acid,* H_2SO_4, $1 + 1$.
2) *Hydrogen peroxide,* H_2O_2, 30%.
3) *Sodium hydroxide,* NaOH, $1N$.

4. Procedure

a. Standardization: The various instruments that can be used in this determination differ in operating details; follow the manufacturer's instructions. Make necessary mechanical adjustments. Then, after allowing sufficient time for warmup (10 min), balance internal electrical components to give an instrument setting of 0 mV or, if a pH meter is used, a pH reading of 7.0.

1) Place 10.0 mL standard NaCl solution in a 250-mL beaker, dilute to about 100 mL, and add 2.0 mL conc HNO_3. Immerse stirrer and electrodes.

2) Set instrument to desired range of millivolts or pH units. Start stirrer.

3) Add standard $AgNO_3$ titrant, recording scale reading after each addition. At the start, large increments of $AgNO_3$ may be added; then, as the end point is approached, add smaller and equal increments (0.1 or 0.2 mL) at longer intervals, so that the exact end point can be determined. Determine volume of $AgNO_3$ used at the point at which there is the greatest change in instrument reading per unit addition of $AgNO_3$.

4) Plot a differential titration curve if the exact end point cannot be determined by inspecting the data. Plot change in instrument reading for equal increments of $AgNO_3$ against volume of $AgNO_3$ added, using average of buret readings before and after each addition. The procedure is illustrated in Figure 4500-Cl⁻:1.

b. Sample analysis:
1) Pipet 100.0 mL sample, or a portion containing not more than 10 mg Cl⁻, into a 250-mL beaker. In the absence of interfering substances, proceed with ¶ 3) below.

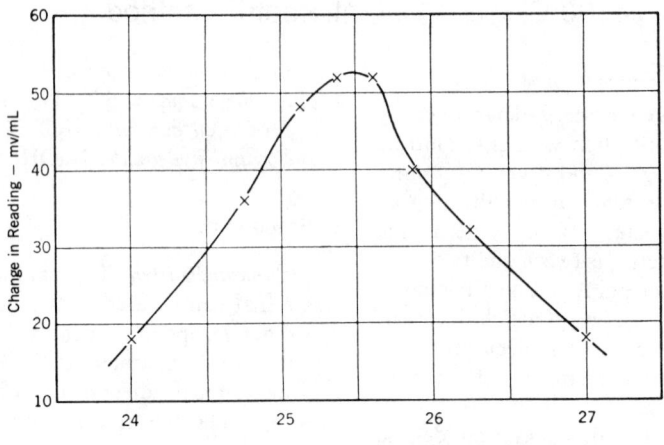

	Experimental Data Plotted Above								
Volume, mL	23.50	24.50	25.00	25.25	25.50	25.75	26.00	26.50	27.50
Change, mV/mL	18	36	48	52	52	40	32	18	

Figure 4500-Cl⁻:1. Example of differential titration curve (end point is 25.5 mL).

2) In the presence of organic compounds, sulfite, or other interferences (such as large amounts of ferric iron, cyanide, or sulfide) acidify sample with H_2SO_4, using litmus paper. Boil for 5 min to remove volatile compounds. Add more H_2SO_4, if necessary, to keep solution acidic. Add 3 mL H_2O_2 and boil for 15 min, adding chloride-free distilled water to keep the volume above 50 mL. Dilute to 100 mL, add NaOH solution dropwise until alkaline to litmus, then 10 drops in excess. Boil for 5 min, filter into a 250-mL beaker, and wash precipitate and paper several times with hot distilled water.

3) Add conc HNO_3 dropwise until acidic to litmus paper, then 2.0 mL in excess. Cool and dilute to 100 mL if necessary. Immerse stirrer and electrodes and start stirrer. Make any necessary adjustments according to the manufacturer's instructions and set selector switch to appropriate setting for measuring the difference of potential between electrodes.

4) Complete determination by titrating according to ¶ 4a4). If an end-point reading has been established from previous determinations for similar samples and conditions, use this predetermined end point. For the most accurate work, make a blank titration by carrying chloride-free distilled water through the procedure.

5. Calculation

$$\text{mg Cl}^- /\text{L} = \frac{(A - B) \times N \times 35\ 450}{\text{mL sample}}$$

where:

A = mL $AgNO_3$,
B = mL blank, and
N = normality of titrant.

6. Precision and Bias

In the absence of interfering substances, the precision and bias are estimated to be about 0.12 mg for 5 mg Cl⁻, or 2.5% of the amount present. When pretreatment is required to remove interfering substances, the precision and bias are reduced to about

0.25 mg for 5 mg Cl⁻, or 5% of amount present.

7. Bibliography

KOLTHOFF, I.M. & N.H. FURMAN. 1931. Potentiometric Titrations, 2nd ed. John Wiley & Sons, New York, N.Y.

REFFENBURG, H.B. 1935. Colorimetric determination of small quantities of chlorides in water. *Ind. Eng. Chem.*, Anal. Ed. 7:14.

CALDWELL, J.R. & H.V. MEYER. 1935. Chloride determination. *Ind. Eng. Chem.*, Anal. Ed. 7:38.

SERFASS, E.J. & R.F. MURACA. 1954. Procedures for Analyzing Metal-Finishing Wastes. Ohio River Valley Water Sanitation Commission, Cincinnati, Ohio, p. 80.

FURMAN, N.H., ed. 1962. Standard Methods of Chemical Analysis, 6th ed. D. Van Nostrand Co., Princeton, N.J., Vol. I.

WALTON, H.F. 1964. Principles and Methods of Chemical Analysis. Prentice-Hall, Inc., Englewood Cliffs, N.J.

WILLARD, H.H., L.L. MERRITT & J.A. DEAN. 1965. Instrumental Methods of Analysis, 4th ed. D. Van Nostrand Co., Princeton, N.J.

4500-Cl⁻ E.   Automated Ferricyanide Method

1. General Discussion

a. Principle: Thiocyanate ion is liberated from mercuric thiocyanate by the formation of soluble mercuric chloride. In the presence of ferric ion, free thiocyanate ion forms a highly colored ferric thiocyanate, of which the intensity is proportional to the chloride concentration.

b. Interferences: None of significance. However, use a continuous filter on turbid samples.

c. Application: The method is applicable to potable, surface, and saline waters, and domestic and industrial wastewaters. The concentration range can be varied by using the colorimeter controls.

2. Apparatus

a. Automated analytical equipment: The required continuous-flow analytical instrument* consists of the interchangeable components shown in Figure 4500-Cl⁻:2.

b. Filters, 480-nm.

3. Reagents

a. Stock mercuric thiocyanate solution: Dissolve 4.17 g Hg(SCN)₂ in about 500 mL

methanol, dilute to 1000 mL with methanol, mix, and filter through filter paper.

b. Stock ferric nitrate solution: Dissolve 202 g Fe(NO₃)₃·9H₂O in about 500 mL distilled water, then carefully add 21 mL conc HNO₃. Dilute to 1000 mL with distilled water and mix. Filter through paper and store in an amber bottle.

c. Color reagent: Add 150 mL stock Hg(SCN)₂ solution to 150 mL stock Fe(NO₃)₃ solution. Mix and dilute to 1000 mL with distilled water. Add 0.5 mL polyoxyethylene 23 lauryl ether.†

d. Stock chloride solution: Dissolve 1.6482 g NaCl, dried at 140°C, in distilled water and dilute to 1000 mL; 1.00 mL + 1.00 mg Cl⁻.

e. Standard chloride solutions: Prepare chloride standards in the desired concentration range, such as 1 to 200 mg/L, using stock chloride solution.

4. Procedure

Set up manifold as shown in Figure 4500-Cl⁻:2 and follow general procedure described by the manufacturer.

---

*Technicon™ Autoanalyzer™ II or equivalent.

†Brij 35, available from ICI Americas, Wilmington, Del., Technicon Instruments Corporation, Tarrytown, N.Y. 10591, or equivalent.

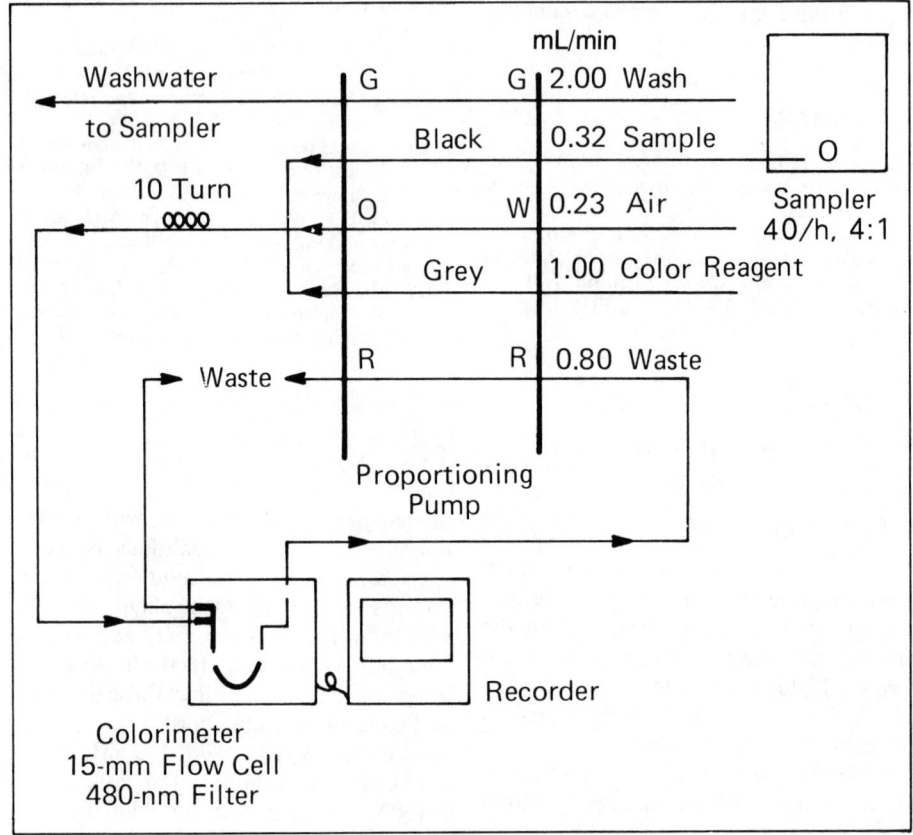

Figure 4500-Cl⁻:2. Flow scheme for automated chloride analysis.

5. Calculation

Prepare standard curves by plotting peak heights of standards processed through the manifold against chloride concentrations in standards. Compute sample chloride concentration by comparing sample peak height with standard curve.

6. Precision and Bias

With an automated system in a single laboratory six samples were analyzed in septuplicate. At a concentration ranging from about 1 to 50 mg Cl⁻/L the average standard deviation was 0.39 mg/L. The coefficient of variation was 2.2%. In two samples with added chloride, recoveries were 104% and 97%.

7. Bibliography

ZALL, D.M., D. FISHER & M.D. GARNER. 1956. Photometric determination of chlorides in water. *Anal. Chem.* 28:1665.

O'BRIEN, J.E. 1962. Automatic analysis of chlorides in sewage. *Wastes Eng.* 33:670.

4500-Cl⁻ F. Ion Chromatography Method

See Section 4110.

4500-ClO₂   CHLORINE DIOXIDE*

4500-ClO₂ A. Introduction

Because the physical and chemical properties of chlorine dioxide resemble those of chlorine in many respects, read the entire discussion of Residual Chlorine (Section 4500-Cl) before attempting a chlorine dioxide determination.

1. Occurrence and Significance

Chlorine dioxide, ClO_2, has been used widely as a bleaching agent in the paper and pulp industry. It has been applied to water supplies to combat tastes and odors due to phenolic-type wastes, actinomycetes, and algae, as well as to oxidize soluble iron and manganese to a more easily removable form. It is a disinfectant, and some results suggest that it may be stronger than free chlorine or hypochlorite.

Chlorine dioxide is a deep yellow, volatile, unpleasant-smelling gas that is toxic and under certain conditions may react explosively. It should be handled with care in a vented area. There are several methods of generating ClO_2; for laboratory purposes the acidification of a solution of sodium chlorite followed by suitable scrubbing is the most practical.

2. Selection of Method

The iodometric method (B) gives a very precise measure of total available strength of a solution in terms of its ability to lib-

erate iodine from iodide. However, ClO_2, chlorine, chlorite, and hypochlorite are not distinguished easily by this technique. It is designed primarily, and best used, for standardizing ClO_2 solutions needed for preparation of temporary standards. It often is inapplicable to industrial wastes.

The amperometric methods (C and E) are useful when a knowledge of the various chlorine fractions in a water sample is desired. They distinguish various chlorine compounds of interest with good accuracy and precision, but require specialized equipment and considerable analytical skill.

The *N,N*-diethyl-*p*-phenylenediamine (DPD) method (D) has the advantages of a relatively easy-to-perform colorimetric test with the ability to distinguish between ClO_2 and various forms of chlorine. This technique is not as accurate as the amperometric method, but should yield results adequate for most common applications.

3. Sampling and Storage

Determine ClO_2 promptly after collecting the sample. Do not expose sample to sunlight or strong artificial light and do not aerate to mix. Minimum ClO_2 losses occur when the determination is completed immediately at the site of sample collection.

---

*Approved by Standard Methods Committee, 1988.

4. Bibliography

INGOLS, R.S. & G.M. RIDENOUR. 1948. Chemical properties of chlorine dioxide in water treatment. *J. Amer. Water Works Assoc.* 40:1207.

PALIN, A.T. 1948. Chlorine dioxide in water treatment. *J. Inst. Water Eng.* 11:61.

HODGDEN, H.W. & R.S. INGOLS. 1954. Direct colorimetric method for determination of chlorine dioxide in water. *Anal. Chem.* 26:1224.

FEUSS, J.V. 1964. Problems in determination of chlorine dioxide residuals. *J. Amer. Water Works Assoc.* 56:607.

MASSCHELEIN, W. 1966. Spectrophotometric determination of chlorine dioxide with acid chrome violet K. *Anal. Chem.* 38:1839.

MASSCHELEIN, W. 1969. Les Oxydes de Chlore et le Chlorite de Sodium. Dunod, Paris, Chapter XI.

4500-ClO₂ B.   Iodometric Method

1. General Discussion

a. Principle: A pure solution of ClO_2 is prepared by slowly adding dilute H_2SO_4 to a sodium chlorite ($NaClO_2$) solution. Contaminants such as chlorine are removed by a $NaClO_2$ scrubber and passing the gas into distilled water in a steady stream of air.

ClO_2 releases free iodine from a KI solution acidified with acetic acid or H_2SO_4. The liberated iodine is titrated with a standard solution of sodium thiosulfate ($Na_2S_2O_3$), with starch as the indicator.

b. Interference: There is little interference in this method, but temperature and strong light affect solution stability. Minimize ClO_2 losses by storing stock ClO_2 solution in a dark refrigerator and by preparing and titrating dilute ClO_2 solutions for standardization purposes at the lowest practicable temperature and in subdued light.

c. Minimum detectable concentration: One drop (0.05 mL) of $0.01N$ $Na_2S_2O_3$ is equivalent to 20 μg ClO_2/L (or 40 μg/L in terms of available chlorine) when a 500-mL sample is titrated.

2. Reagents

All reagents listed for the determination of residual chlorine in Section 4500-Cl.B.2*a-g* are required. Also needed are the following:

a. Stock chlorine dioxide solution: Prepare a gas generating and absorbing system as illustrated in Figure 4500-ClO₂:1. Con-

nect aspirator flask (A), 500-mL capacity, with rubber tubing to a source of compressed air. Let air bubble through a layer of 300 mL distilled water in flask and then pass through a glass tube ending within 5 mm of the bottom of the 1-L gas-generating bottle (B). Conduct evolved gas via glass tubing through a scrubber bottle (C) containing saturated $NaClO_2$ solution or a tower packed with flaked $NaClO_2$, and finally, via glass tubing, into a 2-L borosilicate glass collecting bottle (D) where the gas is absorbed in 1500 mL distilled water. Provide an air outlet tube on collecting bottle (D) for escape of air. Select for gas generation a bottle constructed of strong borosilicate glass and having a mouth wide enough to permit insertion of three separate glass tubes: the first leading almost to the bottom for admitting air, the second reaching below the liquid surface for gradual introduction of H_2SO_4, and the third near the top for exit of evolved gas and air. Fit to second tube a graduated cylindrical separatory funnel (E) to contain H_2SO_4. Locate this system in a fume hood with an adequate shield.

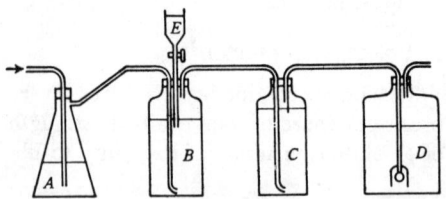

Figure 4500-ClO₂:1. Chlorine dioxide generation and absorption system.

Dissolve 10 g NaClO$_2$ in 750 mL distilled water and place in generating bottle (B). Carefully add 2 mL conc H$_2$SO$_4$ to 18 mL distilled water and mix. Transfer to funnel. Connect flask to generating bottle, generating bottle to scrubber, and the latter to collecting bottle. Pass a smooth current of air through the system, as evidenced by the bubbling rate in all bottles.

Introduce 5-mL increments of H$_2$SO$_4$ from funnel into generating bottle at 5-min intervals. Continue air flow for 30 min after last portion of acid has been added.

Store yellow stock solution in glass-stoppered dark-colored bottle in a dark refrigerator. The concentration of ClO$_2$ thus prepared varies between 250 and 600 mg/L, corresponding to approximately 500 to 1200 mg free chlorine/L.

b. *Standard chlorine dioxide solution:* Use this solution for preparing temporary ClO$_2$ standards. Dilute required volume of stock ClO$_2$ solution to desired strength with chlorine-demand-free water (see Section 4500-Cl.C.3m). Standardize solution by titrating with standard 0.01N or 0.025N Na$_2$S$_2$O$_3$ titrant in the presence of KI, acid, and starch indicator by following the procedure given in ¶ 3 below. A full or nearly full bottle of chlorine or ClO$_2$ solution retains its titer longer than a partially full one. When repeated withdrawals reduce volume to a critical level, standardize diluted solution at the beginning, midway in the series of withdrawals, and at the end of the series. Shake contents thoroughly before drawing off needed solution from middle of the glass-stoppered dark-colored bottle. Prepare this solution frequently.

3. Procedure

Select volume of sample, prepare for titration, and titrate sample and blank as described in Section 4500-Cl.B.3. The only exception is the following: *Let ClO$_2$ react in the dark with acid and KI for 5 min before starting titration.*

4. Calculations

Express ClO$_2$ concentrations in terms of ClO$_2$ or as free chlorine content. Free chlorine is defined as the total oxidizing power of ClO$_2$ measured by titrating iodine released by ClO$_2$ from an acidic solution of KI. Calculate result in terms of chlorine itself.

For standardizing ClO$_2$ solution:

$$\text{mg ClO}_2/\text{mL} = \frac{(A \pm B) \times N \times 13.49}{\text{mL sample titrated}}$$

$$\text{mg ClO}_2 \text{ as Cl}_2/\text{mL} = \frac{(A \pm B) \times N \times 35.45}{\text{mL sample titrated}}$$

For determining ClO$_2$ temporary standards:

$$\text{mg ClO}_2/\text{L} = \frac{(A \pm B) \times N \times 13\,490}{\text{mL sample}}$$

$$\text{mg ClO}_2 \text{ as Cl}_2/\text{L} = \frac{(A \pm B) \times N \times 35\,453}{\text{mL sample}}$$

where:

A = mL titration for sample,
B = mL titration for blank (positive or negative, see 4500-Cl.B.3d), and
N = normality of Na$_2$S$_2$O$_3$.

5. Bibliography

Post, M.A. & W.A. Moore. 1959. Determination of chlorine dioxide in treated surface waters. *Anal. Chem.* 31:1872.

4500-ClO₂ C. Amperometric Method I

1. General Discussion

a. Principle: The amperometric titration of ClO_2 is an extension of the amperometric method for chlorine. By performing four titrations with phenylarsine oxide, free chlorine (including hypochlorite and hypochlorous acid), chloramines, chlorite, and ClO_2 may be determined separately. The first titration step consists of conversion of ClO_2 to chlorite and chlorate through addition of sufficient NaOH to produce a pH of 12, followed by neutralization to a pH of 7 and titration of free chlorine. In the second titration KI is added to a sample that has been treated similarly with alkali and had the pH readjusted to 7; titration yields free chlorine and monochloramine. The third titration involves addition of KI and pH adjustment to 7, followed by titration of free chlorine, monochloramine, and one-fifth of the available ClO_2. In the fourth titration, addition of sufficient H_2SO_4 to lower the pH to 2 enables all available ClO_2 and chlorite, as well as the total free chlorine, to liberate an equivalent amount of iodine from the added KI and thus be titrated.

b. Interference: The interferences described in Section 4500-Cl.D.1*b* apply also to determination of ClO_2.

2. Apparatus

The apparatus required is given in Sections 4500-Cl.D.2*a* through *d*.

3. Reagents

All reagents listed for the determination of chlorine in Section 4500-Cl.D.3 are required. Also needed are the following:

a. Sodium hydroxide, NaOH, 6*N*.

b. Sulfuric acid, H_2SO_4, 6*N*, 1 + 5.

4. Procedure

Minimize effects of pH, time, and temperature of reaction by standardizing all conditions.

a. Titration of free available chlorine (hypochlorite and hypochlorous acid): Add sufficient 6*N* NaOH to raise sample pH to 12. After 10 min, add 6*N* H_2SO_4 to lower pH to 7. Titrate with standard phenylarsine oxide titrant to the amperometric end point as given in Section 4500-Cl.D. Record result as *A*.

b. Titration of free available chlorine and chloramine: Add 6*N* NaOH to raise sample pH to 12. After 10 min, add 6*N* H_2SO_4 to reduce pH to 7. Add 1 mL KI solution. Titrate with standard phenylarsine oxide titrant to the amperometric end point. Record result as *B*.

c. Titration of free available chlorine, chloramine, and one-fifth of available ClO_2: Adjust sample pH to 7 with pH 7 phosphate buffer solution. Add 1 mL KI solution. Titrate with standard phenylarsine oxide titrant to the amperometric end point. Record result as *C*.

d. Titration of free available chlorine, chloramines, ClO_2, and chlorite: Add 1 mL KI solution to sample. Add sufficient 6*N* H_2SO_4 to lower pH to 2. After 10 min, add sufficient 6*N* NaOH to raise pH to 7. Titrate with standard phenylarsine oxide titrant to the amperometric end point. Record result as *D*.

5. Calculation

Convert individual titrations (*A, B, C,* and *D*) into chlorine concentration by the following equation:

$$\text{mg Cl as } Cl_2/L = \frac{E \times 200}{\text{mL sample}}$$

where:

E = mL phenylarsine oxide titration for each individual sample *A, B, C,* or *D*.

Calculate ClO$_2$ and individual chlorine fractions as follows:

$$\text{mg ClO}_2 \text{ as ClO}_2/\text{L} = 1.9 \, (C - B)$$
$$\text{mg ClO}_2 \text{ as Cl}_2/\text{L} = 5 \, (C - B)$$
$$\text{mg free available chlorine}/\text{L} = A$$

mg chloramine/L as chlorine $= B - A$

mg chlorite/L as chlorine $= 4B - 5C + D$

6. Bibliography

HALLER, J.F. & S.S. LISTEK. 1948. Determination of chlorine dioxide and other active chlorine compounds in water. *Anal. Chem.* 20:639.

4500-ClO$_2$ D. DPD Method

1. General Discussion

a. Principle: This method is an extension of the *N,N*-diethyl-*p*-phenylenediamine (DPD) method for determining free chlorine and chloramines in water. ClO$_2$ appears in the first step of this procedure but only to the extent of one-fifth of its total available chlorine content corresponding to reduction of ClO$_2$ to chlorite ion. If the sample is then acidified in the presence of iodide the chlorite also reacts. When neutralized by subsequent addition of bicarbonate, the color thus produced corresponds to the total available chlorine content of the ClO$_2$. If chlorite is present in the sample, this will be included in the step involving acidification and neutralization. Chlorite that did not result from ClO$_2$ reduction by the procedure will cause a positive error equal to twice this chlorite concentration. In evaluating mixtures of these various chloro-compounds, it is necessary to suppress free chlorine by adding glycine before reacting the sample with DPD reagent. Differentiation is based on the fact that glycine converts free chlorine instantaneously into chloroaminoacetic acid but has no effect on ClO$_2$.

b. Interference: The interference by oxidized manganese described in Section 4500-Cl.F.1*d* applies also to ClO$_2$ determination. Manganese interference appears as an increase in the first titrations after addition of DPD, with or without KI, and irrespective of whether there has been prior addition of glycine. Titration readings must be corrected suitably. Interference by chromate in wastewaters may be corrected similarly.

Iron contributed to the sample by adding ferrous ammonium sulfate (FAS) titrant may activate chlorite so as to interfere with the first end point of the titration. Suppress this effect with additional EDTA, disodium salt.

2. Reagents

Reagents required in addition to those for the DPD free-combined chlorine method as listed in Section 4500-Cl.F.2 are as follows:

a. Glycine solution: Dissolve 10 g NH$_2$CH$_2$COOH in 100 mL distilled water.

b. Sulfuric acid solution: Dilute 5 mL conc H$_2$SO$_4$ to 100 mL with distilled water.

c. Sodium bicarbonate solution: Dissolve 27.5 g NaHCO$_3$ in 500 mL distilled water.

d. EDTA: Disodium salt of ethylenediamine tetraacetic acid, solid.

3. Procedure

For samples containing more than 5 mg/L total available chlorine follow the dilution procedure given in Section 4500-Cl.F.3.

a. Chlorine dioxide: Add 2 mL glycine solution to 100 mL sample and mix. Place 5 mL each of buffer reagent and DPD indicator solution in a separate titration flask and mix (or use about 500 mg DPD powder). Add about 200 mg EDTA, disodium salt. Then add glycine-treated sample and mix. Titrate rapidly with standard FAS titrant until red color is discharged (Reading G).

b. Free available chlorine and chloramine: Using a second 100-mL sample follow the procedures of Section 4500-Cl.F.3a adding about 200 mg EDTA, disodium salt, initially with the DPD reagents (Readings A, B, and C).

c. Total available chlorine including chlorite: After obtaining Reading C add 1 mL H_2SO_4 solution to the same sample in titration flask, mix, and let stand about 2 min. Add 5 mL $NaHCO_3$ solution, mix, and titrate (Reading D).

d. Colorimetric procedure: Instead of titration with standard FAS solution, colorimetric procedures may be used to obtain the readings at each stage. Calibrate colorimeters with standard permanganate solution as directed in Section 4500-Cl.G.4a. Use of additional EDTA, disodium salt, with the DPD reagents is not required in colorimetric procedures.

4. Calculations

For 100 mL sample, 1 mL FAS solution = 1 mg available chlorine/L.

In the absence of chlorite:

$$\text{Chlorine dioxide} = 5G \text{ (or } 1.9G \text{ expressed as } ClO_2)$$
$$\text{Free available chlorine} = A - G$$
$$\text{Monochloramine} = B - A$$
$$\text{Dichloramine} = C - B$$
$$\text{Total available chlorine} = C + 4G$$

If the step leading to Reading B is omitted, monochloramine and dichloramine are obtained together when:

$$\text{Combined available chlorine} = C - A$$

If it is desired to check for presence of chlorite in sample, obtain Reading D. Chlorite is indicated if D is greater than $C + 4G$.

In the presence of chlorite:

$$\text{Chlorine dioxide} = 5G \text{ (or } 1.9G \text{ expressed as } ClO_2)$$
$$\text{Chlorite} = D - (C + 4G)$$
$$\text{Free available chlorine} = A - G$$
$$\text{Monochloramine} = B - A$$
$$\text{Dichloramine} = C - B$$
$$\text{Total available chlorine} = D$$

If B is omitted,

$$\text{Combined available chlorine} = C - A$$

5. Bibliography

PALIN, A.T. 1960. Colorimetric determination of chlorine dioxide in water. *Water Sewage Works* 107:457.

PALIN, A.T. 1967. Methods for the determination, in water, of free and combined available chlorine, chlorine dioxide and chlorite, bromine, iodine, and ozone using diethyl-p-phenylenediamine (DPD). *J. Inst. Water Eng.* 21:537.

PALIN, A.T. 1974. Analytical control of water disinfection with special reference to differential DPD methods for chlorine, chlorine dioxide, bromine, iodine and ozone. *J. Inst. Water Eng.* 28:139.

PALIN, A.T. 1975. Current DPD methods for residual halogen compounds and ozone in water. *J. Amer. Water Works Assoc.* 67:32.

4500-ClO$_2$ E.   Amperometric Method II (PROPOSED)

1. General Discussion

a. Principle: Like Amperometric Method I (Section 4500-ClO$_2$.C), this procedure entails successive titrations of combinations of chlorine species. Subsequent calculations determine the concentration of each species. The equilibrium for reduction of the chlorine species of interest by iodide is pH-dependent.

The analysis of a sample for chlorine,

chlorine dioxide, chlorite, and chlorate requires the following steps: determination of all of the chlorine (free plus combined) and one-fifth of the chlorine dioxide at pH 7; lowering sample pH to 2 and determination of the remaining four-fifths of the ClO$_2$ and all of the chlorite (the chlorite measured in this step comes from the chlorite originally present in the sample and that formed in the first titration); preparation of a second sample by purging with nitrogen to remove ClO$_2$ and by reacting with iodide at pH 7 to remove any chlorine remaining; lowering latter sample pH to 2 and determination of all chlorite present (this chlorite only comes from the chlorite originally present in the sample); and, in a third sample, determination of all of the relevant, oxidized chlorine species—chlorine, chlorine dioxide, chlorite, and chlorate—after reduction in hydrochloric acid.[1]

This procedure can be applied to concentrated solutions (10 to 100 mg/L) or dilute solutions (0.1 to 10 mg/L) by appropriate selection of titrant concentration and sample size.

b. Interferences: At pH values above 4, significant iodate formation is possible if iodine is formed in the absence of iodide;[2] this results in a negative bias in titrating the first and second samples. Acidification of these samples causes reduction of iodate to iodine and a positive bias. To prevent formation of iodate add 1 g KI granules to stirred sample.

A positive bias results from oxidation of iodide to iodine by dissolved oxygen in strongly acidic solutions.[1] To minimize this bias, use bromide as the reducing agent in titrating the third sample (bromide is not oxidized by oxygen under these conditions). After reaction is completed, add iodide, which will be oxidized to iodine by the bromine formed from the reduction of the original chlorine species. Add iodide carefully so that bromine gas is not lost. Rapid dilution of the sample with sodium phosphate decreases sample acidity and minimizes oxidation of iodide by oxygen. The pH of the solution to be titrated should be between 1.0 and 2.0. Carry a blank through the procedure as a check on iodide oxidation.

The potential for interferences from manganese, copper, and nitrate are minimized by buffering the sample to pH ≥ 4.[3,4] For the method presented here, the low pH required for the chlorite and chlorate analyses provides conditions favorable to manganese, copper, and nitrite interferences.

2. Apparatus

a. Titrators: See Section 4500-Cl.D.2*a* through *d.* Amperometric titrators with a platinum-platinum electrode system are more stable and require less maintenance. (NOTE: Chlorine dioxide may attack adhesives used to connect the platinum plate to the electrode, resulting in poor readings.)

If a potentiometric titrator is used, provide a platinum sensing electrode and a silver chloride reference electrode for endpoint detection.

b. Glassware: Store glassware used in this method separately from other laboratory glassware and do not use for other purposes because ClO$_2$ reacts with glass to form a hydrophobic surface coating. To satisfy any ClO$_2$ demand, before first use immerse all glassware in a strong ClO$_2$ solution (200 to 500 mg/L) for 24 h and rinse only with water between uses.

c. Sampling: ClO$_2$ is volatile and will vaporize easily from aqueous solution. When sampling a liquid stream, minimize contact with air by placing a flexible sample line to reach the bottom of the sample container, letting several container volumes overflow, slowly removing sample line, and capping container with minimum headspace. Protect from sunlight. Remove sample portions with a volumetric pipet with pipet tip placed at bottom of container. Drain pipet by placing its tip below the surface of reagent or dilution water.

3. Reagents

a. Standard sodium thiosulfate, 0.100*N*: See Section 4500-Cl.B.2*c*.

b. Standard phenylarsine oxide, 0.005 64*N*: See Section 4500-Cl.C.3*a*.

c. Phosphate buffer solution, pH 7: See Section 4500-Cl.D.3*b*.

d. Potassium iodide, KI, granules.

e. Saturated sodium phosphate solution: Prepare a saturated solution of $Na_2HPO_4 \cdot 12H_2O$ with cold deionized-distilled water.

f. Potassium bromide solution, 5%: Dissolve 5 g KBr and dilute to 100 mL. Store in a brown glass-stoppered bottle. Make fresh weekly.

g. Hydrochloric acid, HCl, conc.

h. Hydrochloric acid, HCl, 2.5*N*: Cautiously add 200 mL conc HCl, with mixing, to distilled water, diluting to 1000 mL.

i. Purge gas: Use nitrogen gas for purging ClO_2 from samples. Assure that gas is free of contaminants and pass it through a 5% KI scrub solution. Discard solution at first sign of color.

4. Procedure

Use either sodium thiosulfate or phenylarsine oxide as titrant. Select concentration on basis of concentration range expected. The total mass of oxidant species should be no greater than about 15 mg. Make appropriate sample dilutions if necessary. A convenient volume for titration is 200 to 300 mL. Preferably analyze all samples and blanks in triplicate.

Minimize effects of pH, time, and temperature of reaction by standardizing all conditions.

a. Titration of residual chlorine and one-fifth of available ClO_2: Place 1 mL pH 7 phosphate buffer in beaker and add distilled-deionized dilution water if needed. Introduce sample with minimum aeration and add 1 g KI granules while stirring. Titrate to end point (see Section 4500-

Cl.D). Record reading A = mL titrant/ mL sample.

b. Titration of four-fifths of available ClO_2 and chlorite: Continuing with same sample, add 2 mL 2.5*N* HCl. Let stand in the dark for 5 min. Titrate to end point. Record reading B = mL titrant/mL sample.

c. Titration of nonvolatilized chlorine: Place 1 mL pH 7 phosphate buffer in purge vessel and add distilled-deionized dilution water if needed. Add sample and purge with nitrogen gas for 15 min. Use a gas-dispersion tube to give good gas-liquid contact. Add 1 g KI granules while stirring and titrate to end point. Record reading C = mL titrant/mL sample.

d. Titration of chlorite: Continuing with same sample, add 2 mL 2.5*N* HCl. Let stand in the dark for 5 min. Titrate to end point, and record reading D = mL titrant/ mL sample.

e. Titration of chlorine, ClO_2, chlorate, and chlorite: Add 1 mL KBr and 10 mL conc HCl to 50-mL reaction flask and mix. Carefully add 15 mL sample, with minimum aeration. Mix and stopper immediately. Let stand in the dark for 20 min. Rapidly add 1 g KI granules and shake vigorously for 5 s. Rapidly transfer to titration flask containing 25 mL saturated Na_2HPO_4 solution. Rinse reaction flask thoroughly and add rinse water to titration flask. Final titration volume should be about 200 to 300 mL. Titrate to end point.

Repeat procedure of preceding paragraph using distilled-deionized water in place of sample to determine blank value. Record reading E = (mL titrant sample − mL titrant blank)/mL sample.

NOTE: The 15-mL sample volume can be adjusted to provide an appropriate dilution, but maintain the ratio of sample to HCl.

5. Calculations

Because the combining power of the titrants is pH-dependent, all calculations are

TABLE 4500-ClO₂:I. EQUIVALENT WEIGHTS FOR CALCULATING CONCENTRATIONS ON THE BASIS OF MASS

pH	Species	Molecular Weight mg/mol	Electrons Transferred	Equivalent Weight mg/eq
7	Chlorine dioxide	67 452	1	67 452
2, 0.1	Chlorine dioxide	67 452	5	13 490
7, 2, 0.1	Chlorine	70 906	2	35 453
2, 0.1	Chlorite	67 452	4	16 863
0.1	Chlorate	83 451	6	13 909

based on the equivalents of reducing titrant required to react with equivalents of oxidant present. Use Table 4500-ClO₂:I to obtain the equivalent weights to be used in the calculations.

In the following equations, N is the normality of the titrant used in equivalents per liter and A through E are as defined previously.

Chlorite, mg ClO_2^-/L $= D \times N \times 16\ 863$
Chlorate, mg ClO_2^{2-}/L
$\qquad = [E - (A + B)] \times N \times 13\ 909$
Chlorine dioxide, mg ClO_2/L
$\qquad = (5/4) \times (B - D) \times N \times 13\ 490$
Chlorine, mg Cl_2/L
$\qquad = \{A - [(B - D)/4]\} \times N \times 35\ 453$

6. References

1. AIETA, E.M., P.V. ROBERTS & M. HERNANDEZ. 1984. Determination of chlorine dioxide, chlorine, chlorite, and chlorate in water. *J. Amer. Water Works Assoc.* 76:64.

2. WONG, G. 1982. Factors affecting the amperometric determination of trace quantities of total residual chlorine in seawater. *Environ. Sci. Technol.* 16:11.

3. WHITE, G. 1972. Handbook of Chlorination. Van Nostrand Reinhold Co., New York, N.Y.

4. JOLLEY, R. & J. CARPENTER. 1982. Aqueous Chemistry of Chlorine: Chemistry, Analysis, and Environmental Fate of Reactive Oxidant Species. ORNL/TM-788, Oak Ridge National Lab., Oak Ridge, Tenn.

7. Bibliography

AIETA, E.M. 1985. Amperometric analysis of chlorine dioxide, chlorine and chlorite in aqueous solution. Presented at American Water Works Assoc. Water Quality Technology Conf. 13, Houston, Texas.

GORDON, G. 1982. Improved methods of analysis for chlorate, chlorite, and hypochlorite ions. Presented at American Water Works Assoc. Water Quality Technology Conf., Nashville, Tenn.

TANG, T. F. & G. GORDON. 1980. Quantitative determination of chloride, chlorite, and chlorate ions in a mixture by successive potentiometric titrations. *Anal. Chem.* 52:1430.

4500-F⁻   FLUORIDE*

4500-F⁻   A.   Introduction

A fluoride concentration of approximately 1.0 mg/L in drinking water effectively reduces dental caries without harmful effects on health. Fluoride may occur naturally in water or it may be added in controlled amounts. Some fluorosis may occur when the fluoride level exceeds the recommended limits. In rare instances the naturally occurring fluoride concentration may approach 10 mg/L; such waters should be defluoridated.

Accurate determination of fluoride has increased in importance with the growth of the practice of fluoridation of water supplies as a public health measure. Maintenance of an optimal fluoride concentration is essential in maintaining effectiveness and safety of the fluoridation procedure.

1. Preliminary Treatment

Among the methods suggested for determining fluoride ion (F⁻) in water, the electrode and colorimetric methods are the most satisfactory. Because both methods are subject to errors due to interfering ions (Table 4500-F⁻:I), it may be necessary to distill the sample as directed in Section 4500-F⁻.B before making the determination. When interfering ions are not present in excess of the tolerances of the method, the fluoride determination may be made directly without distillation.

2. Selection of Method

The electrode method (Method C) is suitable for fluoride concentrations from 0.1 to more than 10 mg/L. Adding the prescribed buffer frees the electrode

method from most interferences that adversely affect the SPADNS colorimetric method and necessitate preliminary distillation. Some substances in industrial wastes, such as fluoborates, may be sufficiently concentrated to present problems in electrode measurements. Fluoride measurements can be made with a specific ion electrode and either an expanded-scale pH meter or a specific ion meter, usually without distillation, in the time necessary for electrode equilibration.

The SPADNS method (D) has an analytical range of 0 to 1.40 mg F⁻/L with virtually instantaneous color development. Color determinations are made photometrically, using either a filter photometer or a spectrophotometer. A curve developed from standards is used for determining the fluoride concentration of a sample.

Fluoride also may be determined by the automated complexone method, Method E.

Ion chromatography (Section 4110) may be an acceptable method if weaker eluants are used to separate fluoride from interfering peaks.

3. Sampling and Storage

Preferably use polyethylene bottles for collecting and storing samples for fluoride analysis. Glass bottles are satisfactory if previously they have not contained high-fluoride solutions. Always rinse bottle with a portion of sample.

Never use an excess of dechlorinating agent. Dechlorinate with sodium arsenite rather than sodium thiosulfate when using the SPADNS method because the latter may produce turbidity that causes erroneous readings.

---

*Approved by Standard Methods Committee, 1988.

TABLE 4500-F⁻:1. CONCENTRATION OF SOME SUBSTANCES CAUSING 0.1-MG/L ERROR AT
1.0 MG F/L IN FLUORIDE METHODS

Substance	Method C (Electrode)		Method D (SPADNS)	
	Conc mg/L	Type of Error*	Conc mg/L	Type of Error*
Alkalinity (CaCO$_3$)	7 000	+	5 000	−
Aluminum (Al^{3+})	3.0	−	0.1†	−
Chloride (Cl$^-$)	20 000		7 000	+
Chlorine	5 000		Remove completely with arsenite	
Color & turbidity			Remove or compensate for	
Iron	200	−	10	−
Hexametaphosphate ([NaPO$_3$]$_6$)	50 000		1.0	+
Phosphate (PO$_4^{3-}$)	50 000		16	+
Sulfate (SO$_4^{2-}$)	50 000	−	200	−

* + denotes positive error
  − denotes negative error
  Blank denotes no measurable error.
† On immediate reading. Tolerance increases with time: after 2 h, 3.0; after 4 h, 30.

4500-F⁻ B.   Preliminary Distillation Step

1. Discussion

Fluoride can be separated from other nonvolatile constituents in water by conversion to hydrofluoric or fluosilicic acid and subsequent distillation. The conversion is accomplished by using a strong, high-boiling acid. To protect against glassware etching, hydrofluoric acid is converted to fluosilicic acid by using soft glass beads. Quantitative fluoride recovery is approached by using a relatively large sample. Acid and sulfate carryover are minimized by distilling over a controlled temperature range.

Distillation will separate fluoride from most water samples. Some tightly bound fluoride, such as that in biological materials, may require digestion before distillation, but water samples seldom require such drastic treatment. Distillation produces a distillate volume equal to that of the original water sample so that usually it is not necessary to incorporate a dilution factor when expressing analytical results. The distillate will be essentially free of substances that might interfere with the fluoride determination if the apparatus used is adequate and distillation has been carried out properly. The only common volatile constituent likely to cause interference with colorimetric analysis of the distillate is chloride. When the concentration of chloride is high enough to interfere, add silver sulfate to the sulfuric acid distilling mixture to minimize the volatilization of hydrogen chloride.

Heating an acid-water mixture can be hazardous if precautions are not taken: *Mix acid and water thoroughly before heating.* Use of a quartz heating mantle and a magnetic stirrer in the distillation apparatus simplifies the mixing step.

2. Apparatus

a. Distillation apparatus consisting of a 1-L round-bottom long-neck borosilicate glass boiling flask, a connecting tube, an efficient condenser, a thermometer adapter, and a thermometer that can be read to 200°C. Use standard taper joints for all connections in the direct vapor path. Position the thermometer so that the bulb always is immersed in boiling mixture. The apparatus should be disassembled easily to permit adding sample. Substituting a thermoregulator and necessary circuitry for the thermometer is acceptable and provides some automation.

Alternative types of distillation apparatus may be used. Carefully evaluate any apparatus for fluoride recovery and sulfate carryover. The critical points are obstructions in the vapor path and trapping of liquid in the adapter and condenser. (The condenser should have a vapor path with minimum obstruction. A double-jacketed condenser, with cooling water in the outer jacket and the inner spiral tube, is ideal, but other condensers are acceptable if they have minimum obstructions. Avoid using Graham-type condensers.) Avoid using an open flame as a heat source if possible, because heat applied to the boiling flask above the liquid level causes superheating of vapor and subsequent sulfate carryover.

CAUTION: *Regardless of apparatus used, provide for thorough mixing of sample and acid; heating a non-homogenous acid-water*

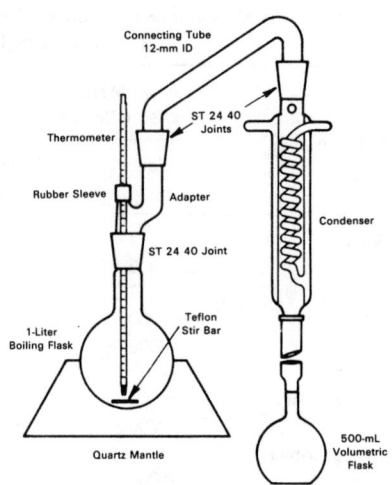

Figure 4500-F⁻:1. Direct distillation apparatus for fluoride.

mixture will result in bumping or possibly a violent explosion.

The preferred apparatus is illustrated in Figure 4500-F⁻:1.

b. Quartz hemispherical heating mantle, for full-voltage operation.

c. Magnetic stirrer, with TFE-coated stirring bar.

d. Soft glass beads.

3. Reagents

a. Sulfuric acid, H_2SO_4, conc, reagent grade.

b. Silver sulfate, Ag_2SO_4, crystals, reagent grade.

4. Procedure

a. Place 400 mL distilled water in the distilling flask and, with the magnetic stirrer operating, carefully add 200 mL conc H_2SO_4. Keep stirrer in operation throughout distillation. Add a few glass beads and connect the apparatus as shown in Figure 4500-F⁻:1, making sure all joints are tight. Begin heating and continue until flask contents reach 180°C (because of heat retention by the mantle, it is necessary to discontinue heating when the temperature reaches

178°C to prevent overheating). Discard distillate. This process removes fluoride contamination and adjusts the acid-water ratio for subsequent distillations.

b. After the acid mixture remaining in the steps outlined in ¶ 4*a*, or previous distillations, has cooled to 80°C or below, add 300 mL sample, with stirrer operating, and distill until the temperature reaches 180°C. To prevent sulfate carryover, turn off heat before 178°C. Retain the distillate for analysis.

c. Add Ag$_2$SO$_4$ to the distilling flask at the rate of 5 mg/mg Cl$^-$ when the chloride concentration is high enough to interfere (see Table 4500-F$^-$:I).

d. Use H$_2$SO$_4$ solution in the flask repeatedly until contaminants from samples accumulate to such an extent that recovery is affected or interferences appear in the distillate. Check acid suitability periodically by distilling standard fluoride samples and analyzing for both fluoride and sulfate. After distilling samples containing more than 3 mg F$^-$/L, flush still by adding 300 mL distilled water, redistill, and combine the two fluoride distillates. If necessary, repeat flushing until the fluoride content of the last distillate is at a minimum. Include additional fluoride recovered with that of the first distillation. After periods of inactivity, similarly flush still and discard distillate.

5. Interpretation of Results

The recovery of fluoride is quantitative within the accuracy of the methods used for its measurement.

6. Bibliography

BELLACK, E. 1958. Simplified fluoride distillation method. *J. Amer. Water Works Assoc.* 50:530.
BELLACK, E. 1961. Automatic fluoride distillation. *J. Amer. Water Works Assoc.* 53:98.
ZEHNPFENNIG, R.G. 1976. Letter to the editor. *Environ. Sci. Technol.* 10:1049.

4500-F$^-$ C. Ion-Selective Electrode Method

1. General Discussion

a. Principle: The fluoride electrode is an ion-selective sensor. The key element in the fluoride electrode is the laser-type doped lanthanum fluoride crystal across which a potential is established by fluoride solutions of different concentrations. The crystal contacts the sample solution at one face and an internal reference solution at the other. The cell may be represented by:

Ag|AgCl, Cl$^-$(0.3M), F$^-$(0.001M) |LaF$_3$| test solution|reference electrode

The fluoride electrode can be used with a standard calomel reference electrode and almost any modern pH meter having an expanded millivolt scale.

The fluoride electrode measures the ion activity of fluoride in solution rather than concentration. Fluoride ion activity depends on the solution total ionic strength and pH, and on fluoride complexing species. Adding an appropriate buffer provides a nearly uniform ionic strength background, adjusts pH, and breaks up complexes so that, in effect, the electrode measures concentration.

b. Interference: Table 4500-F$^-$:I lists common interferences. Fluoride forms complexes with several polyvalent cations, notably aluminum and iron. The extent to which complexation takes place depends on solution pH, relative levels of fluoride, and complexing species. However, CDTA (cyclohexylenediaminetetraacetic acid), a component of the buffer, preferentially will complex interfering cations and release free fluoride ions. Concentrations of aluminum, the most common interference, up to 3.0 mg/L can be complexed preferentially. In acid solution, F^- forms a poorly ionized $HF \cdot HF$ complex but the buffer maintains a pH above 5 to minimize hydrogen fluoride complex formation. In alkaline solution hydroxide ion also can interfere with electrode response to fluoride ion whenever the hydroxide ion concentration is greater than one-tenth the concentration of fluoride ion. At the pH maintained by the buffer, no hydroxide interference occurs.

Fluoborates are widely used in industrial processes. Dilute solutions of fluoborate or fluoboric acid hydrolyze to liberate fluoride ion but in concentrated solutions, as in electroplating wastes, hydrolysis does not occur completely. Distill such samples or measure fluoborate with a fluoborate-selective electrode. Also distill the sample if the dissolved solids concentration exceeds 10 000 mg/L.

2. Apparatus

a. Expanded-scale or digital pH meter or ion-selective meter.

b. Sleeve-type reference electrode: Do not use fiber-tip reference electrodes because they exhibit erratic behavior in very dilute solutions.

c. Fluoride electrode.

d. Magnetic stirrer, with TFE-coated stirring bar.

e. Timer.

3. Reagents

a. Stock fluoride solution: Dissolve 221.0 mg anhydrous sodium fluoride, NaF, in distilled water and dilute to 1000 mL; 1.00 mL = 100 μg F^-.

b. Standard fluoride solution: Dilute 100 mL stock fluoride solution to 1000 mL with distilled water; 1.00 mL = 10.0 μg F^-.

c. Fluoride buffer: Place approximately 500 mL distilled water in a 1-L beaker and add 57 mL glacial acetic acid, 58 g NaCl,

and 4.0 g 1,2 cyclohexylenediamine-tetraacetic acid (CDTA).* Stir to dissolve. Place beaker in a cool water bath and add slowly 6N NaOH (about 125 mL) with stirring, until pH is between 5.3 and 5.5. Transfer to a 1-L volumetric flask and add distilled water to the mark. This buffer, as well as a more concentrated version, is available commercially. In using the concentrated buffer follow the manufacturer's directions.

4. Procedure

a. Instrument calibration: No major adjustment of any instrument normally is required to use electrodes in the range of 0.2 to 2.0 mg F^-/L. For those instruments with zero at center scale adjust calibration control so that the 1.0 mg F^-/L standard reads at the center zero (100 mV) when the meter is in the expanded-scale position. This cannot be done on some meters that do not have a millivolt calibration control. To use a selective-ion meter follow the manufacturer's instructions.

b. Preparation of fluoride standards: Prepare a series of standards by diluting with distilled water 5.0, 10.0, and 20.0 mL of standard fluoride solution to 100 mL with distilled water. These standards are equivalent to 0.5, 1.0, and 2.0 mg F^-/L.

c. Treatment of standards and sample: In 100-mL beakers or other convenient containers add by volumetric pipet from 10 to 25 mL standard or sample. Bring standards and sample to same temperature, preferably room temperature. Add an equal volume of buffer. The total volume should be sufficient to immerse the electrodes and permit operation of the stirring bar.

d. Measurement with electrode: Immerse electrodes in each of the fluoride standard solutions and measure developed potential while stirring on a magnetic stirrer. Avoid stirring before immersing electrodes be-

---

*Also known as 1,2 cyclohexylenedinitrilotetraacetic acid.

cause entrapped air around the crystal can produce erroneous readings or needle fluctuations. Let electrodes remain in the solution 3 min (or until reading is constant) before taking a final millivolt reading. A layer of insulating material between stirrer and beaker minimizes solution heating. Withdraw electrodes, rinse with distilled water, and *blot dry* between readings. (CAUTION: Blotting may poison electrode if not done gently.) Repeat measurements with samples.

When using an expanded-scale pH meter or selective-ion meter, frequently recalibrate the electrode by checking potential reading of the 1.00-mg F^-/L standard and adjusting the calibration control, if necessary, until meter reads as before.

If a direct-reading instrument is not used, plot potential measurement of fluoride standards against concentration on two-cycle semilogarithmic graph paper. Plot milligrams F^- per liter on the logarithmic axis (ordinate), with the lowest concentration at the bottom of the graph. Plot millivolts on the abscissa. From the potential measurement for each sample, read the corresponding fluoride concentration from the standard curve.

The known-additions method may be substituted for the calibration method described. Follow the directions of the instrument manufacturer.

Selective-ion meters may necessitate using a slightly altered procedure, such as preparing 1.00 and 10.0 mg F^-/L standards or some other concentration. Follow the manufacturer's directions. Commercial standards, often already diluted with buffer, frequently are supplied with the meter. Verify the stated fluoride concentration of these standards by comparing them with standards prepared by the analyst.

5. Calculation

$$\text{mg } F^-/L = \frac{\mu g \ F^-}{mL \ sample}$$

6. Precision and Bias

A synthetic sample containing 0.850 mg F^-/L in distilled water was analyzed in 111 laboratories by the electrode method, with a relative standard deviation of 3.6% and a relative error of 0.7%.

A second synthetic sample containing 0.750 mg F^-/L, 2.5 mg $(NaPO_3)_6$/L, and 300 mg alkalinity/L added as $NaHCO_3$, was analyzed in 111 laboratories by the electrode method, with a relative standard deviation of 4.8% and a relative error of 0.2%.

A third synthetic sample containing 0.900 mg F^-/L, 0.500 mg Al/L, and 200 mg SO_4^{2-}/L was analyzed in 13 laboratories by the electrode method, with a relative standard deviation of 2.9% and a relative error of 4.9%.

7. Bibliography

FRANT, M.S. & J.W. ROSS, JR. 1968. Use of total ionic strength adjustment buffer for electrode determination of fluoride in water supplies. *Anal. Chem.* 40:1169.

HARWOOD, J.E. 1969. The use of an ion-selective electrode for routine analysis of water samples. *Water Res.* 3:273.

4500-F⁻ D.   SPADNS Method

1. General Discussion

a. Principle: The SPADNS colorimetric method is based on the reaction between fluoride and a zirconium-dye lake. Fluoride reacts with the dye lake, dissociating a portion of it into a colorless complex anion (ZrF_6^{2-}); and the dye. As the amount of fluoride increases, the color produced becomes progressively lighter.

The reaction rate between fluoride and zirconium ions is influenced greatly by the acidity of the reaction mixture. If the proportion of acid in the reagent is increased, the reaction can be made almost instantaneous. Under such conditions, however, the effect of various ions differs from that in the conventional alizarin methods. The selection of dye for this rapid fluoride method is governed largely by the resulting tolerance to these ions.

b. *Interference:* Table 4500-F⁻:I lists common interferences. Because these are neither linear in effect nor algebraically additive, mathematical compensation is impossible. Whenever any one substance is present in sufficient quantity to produce an error of 0.1 mg/L or whenever the total interfering effect is in doubt, distill the sample. Also distill colored or turbid samples. In some instances, sample dilution or adding appropriate amounts of interfering substances to the standards may be used to compensate for the interference effect. If alkalinity is the only significant interference, neutralize it with either hydrochloric or nitric acid. Chlorine interferes and provision for its removal is made.

Volumetric measurement of sample and reagent is extremely important to analytical accuracy. Use samples and standards at the same temperature or at least within 2°C. Maintain constant temperature throughout the color development period. Prepare different calibration curves for different temperature ranges.

2. Apparatus

Colorimetric equipment: One of the following is required:

a. *Spectrophotometer,* for use at 570 nm, providing a light path of at least 1 cm.

b. *Filter photometer,* providing a light path of at least 1 cm and equipped with a greenish yellow filter having maximum transmittance at 550 to 580 nm.

3. Reagents

a. *Standard fluoride solution:* Prepare as directed in the electrode method, Section 4500-F⁻.C.3b.

b. *SPADNS solution:* Dissolve 958 mg SPADNS, sodium 2-(parasulfophenylazo)-1,8-dihydroxy-3,6-naphthalene disulfonate, also called 4,5-dihydroxy-3-(para-sulfophenylazo)-2,7-naphthalenedisulfonic acid trisodium salt, in distilled water and dilute to 500 mL. This solution is stable for at least 1 year if protected from direct sunlight.

c. *Zirconyl-acid reagent:* Dissolve 133 mg zirconyl chloride octahydrate, $ZrOCl_2 \cdot 8H_2O$, in about 25 mL distilled water. Add 350 mL conc HCl and dilute to 500 mL with distilled water.

d. *Acid zirconyl-SPADNS reagent:* Mix equal volumes of SPADNS solution and zirconyl-acid reagent. The combined reagent is stable for at least 2 years.

e. *Reference solution:* Add 10 mL SPADNS solution to 100 mL distilled water. Dilute 7 mL conc HCl to 10 mL and add to the diluted SPADNS solution. The resulting solution, used for setting the instrument reference point (zero), is stable for at least 1 year. Alternatively, use a prepared standard of 0 mg F⁻/L as a reference.

f. *Sodium arsenite solution:* Dissolve 5.0 g $NaAsO_2$ and dilute to 1 L with distilled water. (CAUTION: *Toxic—avoid ingestion.*)

4. Procedure

a. *Preparation of standard curve:* Prepare fluoride standards in the range of 0 to 1.40 mg F⁻/L by diluting appropriate quantities of standard fluoride solution to 50 mL with distilled water. Pipet 5.00 mL each of SPADNS solution and zirconyl-acid reagent, or 10.00 mL mixed acid-zirconyl-SPADNS reagent, to each standard and mix well. Avoid contamination. Set photometer to zero absorbance with the reference solution and obtain absorbance readings of standards. Plot a curve of the milligrams fluoride-absorbance relationship. Prepare a new standard curve whenever a fresh reagent is made or a different standard temperature is desired. As an alternative to using a reference, set photometer at some convenient point (0.300 or 0.500 absorbance) with the prepared 0 mg F^-/L standard.

b. *Sample pretreatment:* If the sample contains residual chlorine, remove it by adding 1 drop (0.05 mL) $NaAsO_2$ solution/ 0.1 mg residual chlorine and mix. (Sodium arsenite concentrations of 1300 mg/L produce an error of 0.1 mg/L at 1.0 mg F^-/ L.)

c. *Color development:* Use a 50.0-mL sample or a portion diluted to 50 mL with distilled water. Adjust sample temperature to that used for the standard curve. Add 5.00 mL each of SPADNS solution and zirconyl-acid reagent, or 10.00 mL acid-zirconyl-SPADNS reagent; mix well and read absorbance, first setting the reference point of the photometer as above. If the absorbance falls beyond the range of the standard curve, repeat using a diluted sample.

5. Calculation

$$\text{mg } F^-/L = \frac{A}{\text{mL sample}} \times \frac{B}{C}$$

where:

A = μg F^- determined from plotted curve,
B = final volume of diluted sample, mL, and
C = volume of diluted sample used for color development, mL.

When the prepared 0 mg F^-/L standard is used to set the photometer, alternatively calculate fluoride concentration as follows:

$$\text{mg } F^-/L = \frac{A_0 - A_x}{A_0 - A_1}$$

where:

A_0 = absorbance of the prepared 0 mg F^-/L standard,
A_1 = absorbance of a prepared 1.0 mg F^-/L standard, and
A_x = absorbance of the prepared sample.

6. Precision and Bias

A synthetic sample containing 0.830 mg F^-/L and no interference in distilled water was analyzed in 53 laboratories by the SPADNS method, with a relative standard deviation of 8.0% and a relative error of 1.2%. After direct distillation of the sample, the relative standard deviation was 11.0% and the relative error 2.4%.

A synthetic sample containing 0.570 mg F^-/L, 10 mg Al/L, 200 mg SO_4^{2-}/L, and 300 mg total alkalinity/L was analyzed in 53 laboratories by the SPADNS method without distillation, with a relative standard deviation of 16.2% and a relative error of 7.0%. After direct distillation of the sample, the relative standard deviation was 17.2% and the relative error 5.3%.

A synthetic sample containing 0.680 mg F^-/L, 2 mg Al/L, 2.5 mg $(NaPO_3)_6$/L, 200 mg SO_4^{2-}/L, and 300 mg total alkalinity/ L was analyzed in 53 laboratories by direct distillation and SPADNS methods with a relative standard deviation of 2.8% and a relative error of 5.9%.

7. Bibliography

BELLACK, E. & P.J. SCHOUBOE. 1968. Rapid photometric determination of fluoride with SPADNS-zirconium lake. *Anal. Chem.* 30:2032.

4500-F⁻ E. Complexone Method

1. General Discussion

a. *Principle:* The sample is distilled in the automated system, and the distillate is reacted with alizarin fluorine blue-lanthanum reagent to form a blue complex that is measured colorimetrically at 620 nm.

b. *Interferences:* Interferences normally associated with the determination of fluoride are removed by distillation.

c. *Application:* This method is applicable to potable, surface, and saline waters as well as domestic and industrial wastewaters. The range of the method, which can be modified by using the adjustable colorimeter, is 0.1 to 2.0 mg F⁻/L.

2. Apparatus

The required continuous-flow analytical instrument* consists of the interchangeable components in the number and manner indicated in Figure 4500-F⁻:2.

3. Reagents

a. *Standard fluoride solution:* Prepare in appropriate concentrations from 0.10 to 2.0 mg F⁻/L using the stock fluoride solution (see Section 4500-F⁻.C.3a).

b. *Distillation reagent:* Add 50 mL conc H_2SO_4 to about 600 mL distilled water. Add 10.00 mL stock fluoride solution (see Section 4500-F⁻.C.3a; 1.00 mL = 100 μg F⁻) and dilute to 1000 mL.

c. *Acetate buffer solution:* Dissolve 60 g anhydrous sodium acetate, $NaC_2H_3O_2$, in about 600 mL distilled water. Add 100 mL conc (glacial) acetic acid and dilute to 1 L.

d. *Alizarin fluorine blue stock solution:* Add 960 mg alizarin fluorine,† $C_{14}H_7O_4 \cdot CH_2N(CH_2 \cdot COOH)_2$, to 100 mL distilled water. Add 2 mL conc NH_4OH and mix until dye is dissolved. Add 2 mL conc (glacial) acetic acid, dilute to 250 mL and store in an amber bottle in the refrigerator.

e. *Lanthanum nitrate stock solution:* Dissolve 1.08 g $La(NO_3)_3$ in about 100 mL distilled water, dilute to 250 mL, and store in refrigerator.

f. *Working color reagent:* Mix in the following order: 300 mL acetate buffer solution, 150 mL acetone, 50 mL tertiary butanol, 36 mL alizarin fluorine blue stock solution, 40 mL lanthanum nitrate stock solution, and 2 mL polyoxyethylene 23 lauryl ether.‡ Dilute to 1 L with distilled water. This reagent is stable for 2 to 4 d.

4. Procedure

No special handling or preparation of sample is required.

Set up manifold as shown in Figure 4500-F⁻:2 and follow the manufacturer's instructions.

5. Calculation

Prepare standard curves by plotting peak heights of standards processed through the manifold against constituent concentrations in standards. Compute sample concentrations by comparing sample peak height with standard curve.

6. Precision and Bias

In a single laboratory four samples of natural water containing from 0.40 to 0.82

---

*Technicon™ Autoanalyzer™ II or equivalent.
†J.T. Baker Catalog number J-112 or equivalent.

‡Brij-35, available from ICI Americas, Wilmington, Del. or Technicon Instruments Corp., Tarrytown, N.Y., or equivalent.

mg F⁻/L were analyzed in septuplicate. Average precision was ± 0.03 mg F⁻/L. To two of the samples, additions of 0.20 and 0.80 mg F⁻/L were made. Average recovery of the additions was 98%.

7. Bibliography

WEINSTEIN, L.H., R.H. MANDL, D.C. McCUNE, J.S. JACOBSON & A.E. HITCHCOCK. 1963. A semi-automated method for the determination of fluorine in air and plant tissues. *Boyce Thompson Inst.* 22:207.

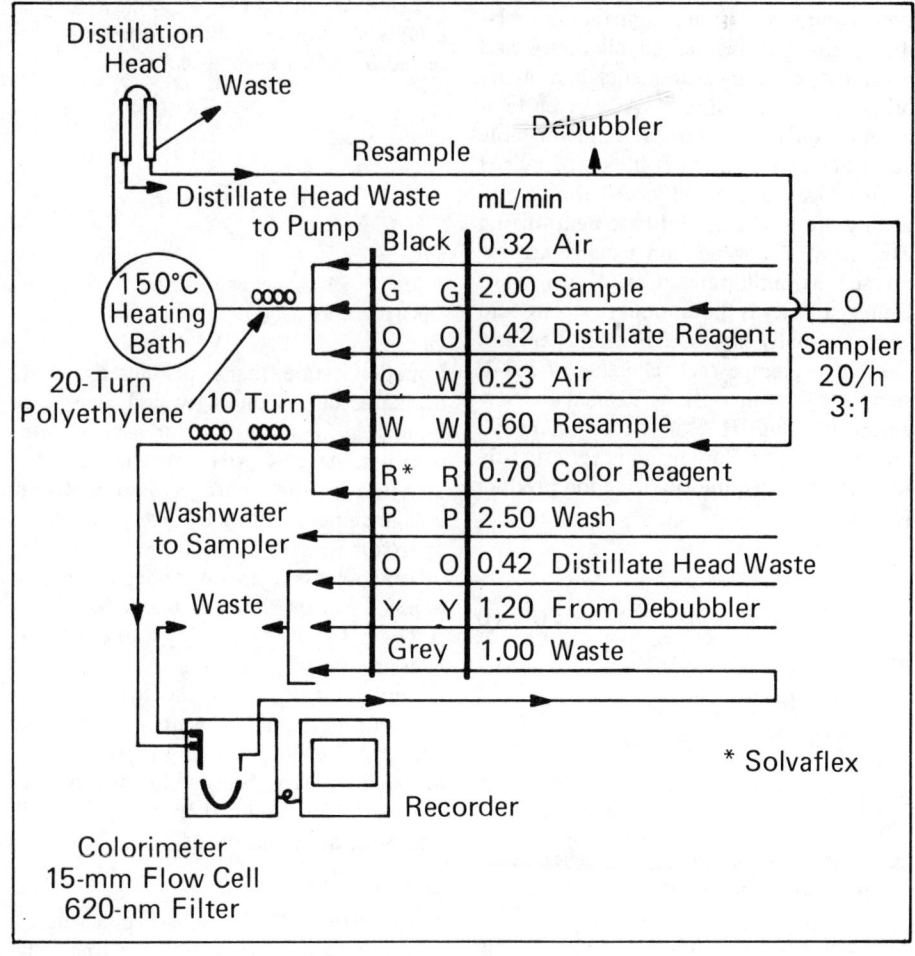

Figure 4500-F⁻:2. Fluoride manifold.

4500-H$^+$ pH VALUE*

4500-H$^+$ A.   Introduction

1. Principles

Measurement of pH is one of the most important and frequently used tests in water chemistry. Practically every phase of water supply and wastewater treatment, e.g., acid-base neutralization, water softening, precipitation, coagulation, disinfection, and corrosion control, is pH-dependent. pH is used in alkalinity and carbon dioxide measurements and many other acid-base equilibria. At a given temperature the *intensity* of the acidic or basic character of a solution is indicated by pH or hydrogen ion activity. Alkalinity and acidity are the acid- and base-neutralizing capacities of a water and usually are expressed as milligrams $CaCO_3$ per liter. Buffer capacity is the amount of strong acid or base, usually expressed in moles per liter, needed to change the pH value of a 1-L sample by 1 unit. pH as defined by Sorenson[1] is $-\log [H^+]$; it is the "intensity" factor of acidity. Pure water is very slightly ionized and at equilibrium the ion product is

$$[H^+][OH^-] = K_w$$

$$= 1.01 \times 10^{-14} \text{ at } 25°C \quad (1)$$

and

$$[H^+] = [OH^-]$$

$$= 1.005 \times 10^{-7}$$

where:

$[H^+]$ = activity of hydrogen ions, moles/L,
$[OH^-]$ = activity of hydroxyl ions, moles/L, and
K_w = ion product of water.

Because of ionic interactions in all but very dilute solutions, it is necessary to use the "activity" of an ion and not its molar

concentration. Use of the term pH assumes that the activity of the hydrogen ion, a_{H+}, is being considered. The *approximate* equivalence to molarity, $[H^+]$ can be presumed only in very dilute solutions (ionic strength < 0.1).

A logarithmic scale is convenient for expressing a wide range of ionic activities. Equation 1 in logarithmic form and corrected to reflect activity is:

$$(-\log_{10} a_{H+}) + (-\log_{10} a_{OH-}) = 14 \quad (2)$$

or

$$pH + pOH = pK_w$$

where:

pH† = $-\log_{10} a_{H+}$ and
pOH = $-\log_{10} a_{OH-}$.

Equation 2 states that as pH increases pOH decreases correspondingly and vice versa because pK_w is constant for a given temperature. At 25°C, pH 7.0 is neutral, the activities of the hydrogen and hydroxyl ions are equal, and each corresponds to an approximate activity of 10^{-7} moles/L. The neutral point is temperature-dependent and is pH 7.5 at 0°C and pH 6.5 at 60°C.

The pH value of a highly dilute solution is approximately the same as the negative common logarithm of the hydrogen ion concentration. Natural waters usually have pH values in the range of 4 to 9, and most are slightly basic because of the presence of bicarbonates and carbonates of the alkali and alkaline earth metals.

2. Reference

1. SORENSON, S. 1909. Über die Messung und die Bedeutung der Wasserstoff ionen Konzentration bei Enzymatischen Prozessen. *Biochem. Z.* 21: 131.

---

*Approved by Standard Methods Committee, 1985.

†p designates $-\log_{10}$ of a number.

4500-H⁺ B.   Electrometric Method

1. General Discussion

a. *Principle:* The basic principle of electrometric pH measurement is determination of the activity of the hydrogen ions by potentiometric measurement using a standard hydrogen electrode and a reference electrode. The hydrogen electrode consists of a platinum electrode across which hydrogen gas is bubbled at a pressure of 101 kPa. Because of difficulty in its use and the potential for poisoning the hydrogen electrode, the glass electrode commonly is used. The electromotive force (emf) produced in the glass electrode system varies linearly with pH. This linear relationship is described by plotting the measured emf against the pH of different buffers. Sample pH is determined by extrapolation.

Because single ion activities such as a_{H+} cannot be measured, pH is defined operationally on a potentiometric scale. The pH measuring instrument is calibrated potentiometrically with an indicating (glass) electrode and a reference electrode using National Institute of Standards and Technology (NIST, formerly National Bureau of Standards) buffers having assigned values so that:

$$pH_B = - \log_{10} a_{H+}$$

where:

pH_B = assigned pH of NIST buffer.

The operational pH scale is used to measure sample pH and is defined as:

$$pH_x = pH_B \pm \frac{F(E_x - E_s)}{2.303\ RT}$$

where:

pH_x = potentiometrically measured sample pH,

F = Faraday: 9.649×10^4 coulomb/mole,

E_x = sample emf, V,

E_s = buffer emf, V,

R = gas constant; 8.314 joule/(mole °K), and

T = absolute temperature, °K.

NOTE: Although the equation for pH_x appears in the literature with a plus sign, the sign of emf readings in millivolts for most pH meters manufactured in the U.S. is negative. The choice of negative sign is consistent with the IUPAC Stockholm convention concerning the sign of electrode potential.[1,2]

The activity scale gives values that are higher than those on Sorenson's scale by 0.04 units:

$$pH\ (activity) = pH\ (Sorenson) + 0.04$$

The equation for pH_x assumes that the emf of the cells containing the sample and buffer is due solely to hydrogen ion activity unaffected by sample composition. In practice, samples will have varying ionic species and ionic strengths, both affecting H^+ activity. This imposes an experimental limitation on pH measurement; thus, to obtain meaningful results, the differences between E_x and E_s should be minimal. Samples must be dilute aqueous solutions of simple solutes ($< 0.2M$). (Choose buffers to bracket the sample.) Determination of pH cannot be made accurately in nonaqueous media, suspensions, colloids, or high-ionic-strength solutions.

b. *Interferences:* The glass electrode is relatively free from interference from color, turbidity, colloidal matter, oxidants, reductants, or high salinity, except for a sodium error at pH > 10. Reduce this error by using special "low sodium error" electrodes.

pH measurements are affected by temperature in two ways: mechanical effects that are caused by changes in the properties of the electrodes and chemical effects caused by equilibrium changes. In the first instance, the Nernstian slope increases with increasing temperature and electrodes take time to achieve thermal equilibrium. This can cause long-term drift in pH. Because chemical equilibrium affects pH, standard pH buffers have a specified pH at indicated temperatures.

Always report temperature at which pH is measured.

2. Apparatus

a. pH meter consisting of potentiometer, a glass electrode, a reference electrode, and a temperature-compensating device. A circuit is completed through the potentiometer when the electrodes are immersed in the test solution. Many pH meters are capable of reading pH or millivolts and some have scale expansion that permits reading to 0.001 pH unit, but most instruments are not that precise.

For routine work use a pH meter accurate and reproducible to 0.1 pH unit with a range of 0 to 14 and equipped with a temperature-compensation adjustment.

Although manufacturers provide operating instructions, the use of different descriptive terms may be confusing. For most instruments, there are two controls: intercept (set buffer, asymmetry, standardize) and slope (temperature, offset); their functions are shown diagramatically in Figures 4500-H$^+$:1 and 2. The intercept control shifts the response curve laterally to pass through the isopotential point with no change in slope. This permits bringing the instrument on scale (0 mV) with a pH 7 buffer that has no change in potential with temperature.

The slope control rotates the emf/pH slope about the isopotential point (0 mV/pH 7). To adjust slope for temperature without disturbing the intercept, select a buffer that brackets the sample with pH 7 buffer and adjust slope control to pH of this buffer. The instrument will indicate correct millivolt change per unit pH at the test temperature.

b. Reference electrode consisting of a half cell that provides a constant electrode potential. Commonly used are calomel and silver: silver-chloride electrodes. Either is available with several types of liquid junctions.

The liquid junction of the reference electrode is critical because at this point the

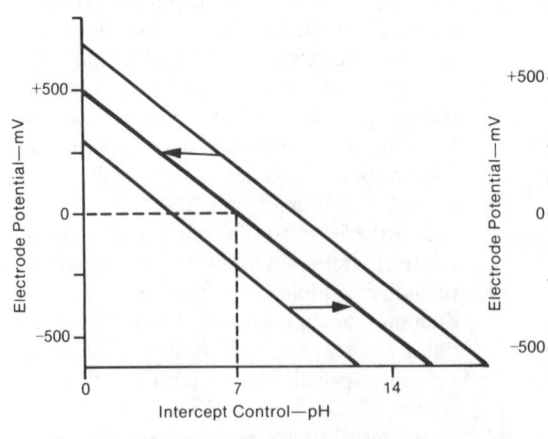

Figure 4500-H$^+$:1. Electrode potential vs. pH. Intercept control shifts response curve laterally.

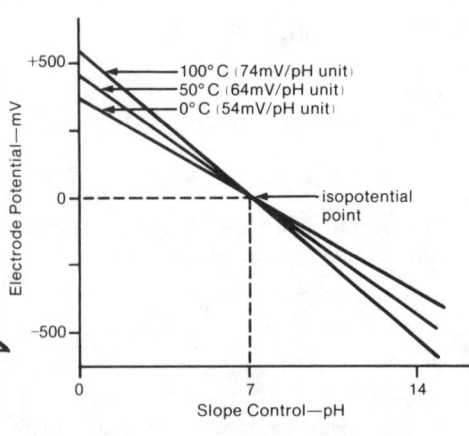

Figure 4500-H$^+$:2. Typical pH electrode response as a function of temperature.

electrode forms a salt bridge with the sample or buffer and a liquid junction potential is generated that in turn affects the potential produced by the reference electrode. Reference electrode junctions may be annular ceramic, quartz, or asbestos fiber, or the sleeve type. The quartz type is most widely used. The asbestos fiber type is not recommended for strongly basic solutions. Follow the manufacturer's recommendation on use and care of the reference electrode.

Refill nonsealed electrodes with the correct electrolyte to proper level and make sure junction is properly wetted.

c. Glass electrode: The sensor electrode is a bulb of special glass containing a fixed concentration of HCl or a buffered chloride solution in contact with an internal reference electrode. Upon immersion of a new electrode in a solution the outer bulb surface becomes hydrated and exchanges sodium ions for hydrogen ions to build up a surface layer of hydrogen ions. This, together with the repulsion of anions by fixed, negatively charged silicate sites, produces at the glass-solution interface a potential that is a function of hydrogen ion activity in solution.

Several types of glass electrodes are available. Combination electrodes incorporate the glass and reference electrodes into a single probe. Use a "low sodium error" electrode that can operate at high temperatures for measuring pH over 10 because standard glass electrodes yield erroneously low values. For measuring pH below 1 standard glass electrodes yield erroneously high values; use liquid membrane electrodes instead.

d. Beakers: Preferably use polyethylene or TFE* beakers.

e. Stirrer: Use either a magnetic, TFE-coated stirring bar or a mechanical stirrer with inert plastic-coated impeller.

---
*Teflon or equivalent.

f. Flow chamber: Use for continuous flow measurements or for poorly buffered solutions.

3. Reagents

a. General preparation: Calibrate the electrode system against standard buffer solutions of known pH. Because buffer solutions may deteriorate as a result of mold growth or contamination, prepare fresh as needed for accurate work by weighing the amounts of chemicals specified in Table 4500-H⁺:I, dissolving in distilled water at 25°C, and diluting to 1000 mL. This is particularly important for borate and carbonate buffers.

Boil and cool distilled water having a conductivity of less than 2 μmhos/cm. To 50 mL add 1 drop of saturated KCl solution suitable for reference electrode use. If the pH of this test solution is between 6.0 and 7.0, use it to prepare all standard solutions.

Dry KH_2PO_4 at 110 to 130°C for 2 h before weighing but do not heat unstable hydrated potassium tetroxalate above 60°C nor dry the other specified buffer salts.

Although ACS-grade chemicals generally are satisfactory for preparing buffer solutions, use certified materials available from the National Institute of Standards and Technology when the greatest accuracy is required. For routine analysis, use commercially available buffer tablets, powders, or solutions of tested quality. In preparing buffer solutions from solid salts, insure complete solution.

As a rule, select and prepare buffer solutions classed as primary standards in Table 4500-H⁺:I; reserve secondary standards for extreme situations encountered in wastewater measurements. Consult Table 4500-H⁺:II for accepted pH of standard buffer solutions at temperatures other than 25°C. In routine use, store buffer solutions and samples in polyethylene bottles. Replace buffer solutions every 4 weeks.

TABLE 4500-H⁺:I. PREPARATION OF pH STANDARD SOLUTIONS[3]

Standard Solution (molality)	pH at 25°C	Weight of Chemicals Needed / 1000 mL Aqueous Solution at 25°C
Primary standards:		
Potassium hydrogen tartrate (saturated at 25°C)	3.557	> 7 g KHC$_4$H$_4$O$_6$*
0.05 potassium dihydrogen citrate	3.776	11.41 g KH$_2$C$_6$H$_5$O$_7$
0.05 potassium hydrogen phthalate	4.004	10.12 g KHC$_8$H$_4$O$_4$
0.025 potassium dihydrogen phosphate + 0.025 disodium hydrogen phosphate	6.863	3.387 g KH$_2$PO$_4$ + 3.533 g Na$_2$HPO$_4$†
0.008 695 potassium dihydrogen phosphate + 0.030 43 disodium hydrogen phosphate	7.415	1.179 g KH$_2$PO$_4$ + 4.303 g Na$_2$HPO$_4$†
0.01 sodium borate decahydrate (borax)	9.183	3.80 g Na$_2$B$_4$O$_7$ · 10H$_2$O†
0.025 sodium bicarbonate + 0.025 sodium carbonate	10.014	2.092 g NaHCO$_3$ + 2.640 g Na$_2$CO$_3$
Secondary standards:		
0.05 potassium tetroxalate dihydrate	1.679	12.61 g KH$_3$C$_4$O$_8$ · 2H$_2$O
Calcium hydroxide (saturated at 25°C)	12.454	> 2 g Ca(OH)$_2$*

* Approximate solubility.
† Prepare with freshly boiled and cooled distilled water (carbon-dioxide-free).

TABLE 4500-H⁺:II. STANDARD pH VALUES[3]

Temperature °C	Primary Standards							Secondary Standards	
	Tartrate (Saturated)	Citrate (0.05 M)	Phthalate (0.05 M)	Phosphate (1:1)	Phosphate (1:3.5)	Borax (0.01 M)	Bicarbonate-Carbonate (0.025 M)	Tetroxalate (0.05 M)	Calcium Hydroxide (Saturated)
0			4.003	6.982	7.534	9.460	10.321	1.666	
5			3.998	6.949	7.501	9.392	10.248	1.668	
10			3.996	6.921	7.472	9.331	10.181	1.670	
15			3.996	6.898	7.449	9.276	10.120	1.672	
20			3.999	6.878	7.430	9.227	10.064	1.675	
25	3.557	3.776	4.004	6.863	7.415	9.183	10.014	1.679	12.454
30	3.552		4.011	6.851	7.403	9.143	9.968	1.683	
35	3.549		4.020	6.842	7.394	9.107	9.928	1.688	
37			4.024	6.839	7.392	9.093			
40	3.547		4.030	6.836	7.388	9.074	9.891	1.694	
45	3.547		4.042	6.832	7.385	9.044	9.859	1.700	
50	3.549		4.055	6.831	7.384	9.017	9.831	1.707	
55	3.554		4.070					1.715	
60	3.560		4.085					1.723	
70	3.580		4.12					1.743	
80	3.609		4.16					1.766	
90	3.650		4.19					1.792	
95	3.674		4.21					1.806	

b. Saturated potassium hydrogen tartrate solution: Shake vigorously an excess (5 to 10 g) of finely crystalline $KHC_4H_4O_6$ with 100 to 300 mL distilled water at 25°C in a glass-stoppered bottle. Separate clear solution from undissolved material by decantation or filtration. Preserve for 2 months or more by adding one thymol crystal (8 mm diam) per 200 mL solution.

c. Saturated calcium hydroxide solution: Calcine a well-washed, low-alkali grade $CaCO_3$ in a platinum dish by igniting for 1 h at 1000°C. Cool, hydrate by slowly adding distilled water with stirring, and heat to boiling. Cool, filter, and collect solid $Ca(OH)_2$ on a fritted glass filter of medium porosity. Dry at 110°C, cool, and pulverize to uniformly fine granules. Vigorously shake an excess of fine granules with distilled water in a stoppered polyethylene bottle. Let temperature come to 25°C after mixing. Filter supernatant under suction through a sintered glass filter of medium porosity and use filtrate as the buffer solution. Discard buffer solution when atmospheric CO_2 causes turbidity to appear.

d. Auxiliary solutions: $0.1N$ NaOH, $0.1N$ HCl, $5N$ HCl (dilute five volumes $6N$ HCl with one volume distilled water), and acid potassium fluoride solution (dissolve 2 g KF in 2 mL conc H_2SO_4 and dilute to 100 mL with distilled water).

4. Procedure

a. Instrument calibration: In each case follow manufacturer's instructions for pH meter and for storage and preparation of electrodes for use. Recommended solutions for short-term storage of electrodes vary with type of electrode and manufacturer, but generally have a conductivity greater than 4000 μmhos/cm. Tap water is a better substitute than distilled water, but pH 4 buffer is best for the single glass electrode and saturated KCl is preferred for a calomel and Ag/AgCl reference electrode. Saturated KCl is the preferred solution for a combination electrode. Keep electrodes wet by returning them to storage solution whenever pH meter is not in use.

Before use, remove electrodes from storage solution, rinse, blot dry with a soft tissue, place in initial buffer solution, and set the isopotential point (¶ 2a above). Select a second buffer within 2 pH units of sample pH and bring sample and buffer to same temperature, which may be the room temperature, a fixed temperature such as 25°C, or the temperature of a fresh sample. Remove electrodes from first buffer, rinse thoroughly with distilled water, blot dry, and immerse in second buffer. Record temperature of measurement and adjust temperature dial on meter so that meter indicates pH value of buffer at test temperature (this is a slope adjustment).

Use the pH value listed in the tables for the buffer used at the test temperature. Remove electrodes from second buffer, rinse thoroughly with distilled water and dry electrodes as indicated above. Immerse in a third buffer below pH 10, approximately 3 pH units different from the second; the reading should be within 0.1 unit for the pH of the third buffer. If the meter response shows a difference greater than 0.1 pH unit from expected value, look for trouble with the electrodes or potentiometer (see ¶s 5a and b below).

The purpose of standardization is to adjust the response of the glass electrode to the instrument. When only occasional pH measurements are made standardize instrument before each measurement. When frequent measurements are made and the instrument is stable, standardize less frequently. If sample pH values vary widely, standardize for each sample with a buffer having a pH within 1 to 2 pH units of the sample.

b. Sample analysis: Establish equilibrium between electrodes and sample by stirring sample to insure homogeneity; stir gently to minimize carbon dioxide entrainment. For buffered samples or those of high ionic strength, condition electrodes after cleaning by dipping them into sample for 1 min. Blot dry, immerse in a fresh portion of the same sample, and read pH.

With dilute, poorly buffered solutions, equilibrate electrodes by immersing in three or four successive portions of sample. Take a fresh sample to measure pH.

5. Trouble Shooting

a. Potentiometer: To locate trouble source disconnect electrodes and, using a short-circuit strap, connect reference electrode terminal to glass electrode terminal. Observe change in pH when instrument calibration knob is adjusted. If potentiometer is operating properly, it will respond rapidly and evenly to changes in calibration over a wide scale range. A faulty potentiometer will fail to respond, will react erratically, or will show a drift upon adjustment. Switch to the millivolt scale on which the meter should read zero. If inexperienced, do not attempt potentiometer repair other than maintenance as described in instrument manual.

b. Electrodes: If potentiometer is functioning properly, look for the instrument fault in the electrode pair. Substitute one electrode at a time and cross-check with two buffers that are about 4 pH units apart. A deviation greater than 0.1 pH unit indicates a faulty electrode. Glass electrodes fail because of scratches, deterioration, or accumulation of debris on the glass surface. Rejuvenate electrode by alternately immersing it three times each in $0.1N$ HCl and $0.1N$ NaOH. If this fails, immerse tip in KF solution for 30 s. After rejuvenation, soak in pH 7.0 buffer overnight. Rinse and store in pH 7.0 buffer. Rinse again with distilled water before use. Protein coatings can be removed by soaking glass electrodes in a 10% pepsin solution adjusted to pH 1 to 2.

To check reference electrode, oppose the emf of a questionable reference electrode against another one of the same type that is known to be good. Using an adapter, plug good reference electrode into glass electrode jack of potentiometer; then plug questioned electrode into reference electrode jack. Set meter to read millivolts and take readings with both electrodes immersed in the same electrolyte (KCl) solution and then in the same buffer solution. The millivolt readings should be 0 ± 5 mV for both solutions. If different electrodes are used, i.e., silver: silver-chloride against calomel or vice versa, the reading will be 44 ± 5 mV for a good reference electrode.

Reference electrode troubles generally are traceable to a clogged junction. Interruption of the continuous trickle of electrolyte through the junction causes increase in response time and drift in reading. Clear a clogged junction by applying suction to the tip or by boiling tip in distilled water until the electrolyte flows freely when suction is applied to tip or pressure is applied to the fill hole. Replaceable junctions are available commercially.

6. Precision and Bias

By careful use of a laboratory pH meter with good electrodes, a precision of ± 0.02 pH unit and an accuracy of ± 0.05 pH unit can be achieved. However, ± 0.1 pH unit represents the limit of accuracy under normal conditions, especially for measurement of water and poorly buffered solutions. For this reason, report pH values to the nearest 0.1 pH unit. A synthetic sample of a Clark and Lubs buffer solution of pH 7.3 was analyzed electrometrically by 30 laboratories with a standard deviation of ± 0.13 pH unit.

7. References

1. BATES, R.G. 1978. Concept and determination of pH. *In* I.M. Kolthoff & P.J. Elving, eds. Treatise on Analytical Chemistry. Part 1, Vol. 1, p. 821. Wiley-Interscience, New York, N.Y.
2. LICHT, T.S. & A.J. DE BETHUNE. 1957. Recent developments concerning the signs of electrode potentials. *J. Chem. Educ.* 34:433.
3. DURST, R.A. 1975. Standard Reference Ma-

terials: Standardization of pH Measurements. NBS Spec. Publ. 260-53, National Bur. Standards, Washington, D.C.

8. Bibliography

CLARK, W.M. 1928. The Determination of Hydrogen Ions, 3rd ed. Williams & Wilkins Co., Baltimore, Md.

DOLE, M. 1941. The Glass Electrode. John Wiley & Sons, New York, N.Y.

BATES, R.G. & S.F. ACREE. 1945. pH of aqueous mixtures of potassium dihydrogen phosphate and disodium hydrogen phosphate at 0 to 60°C. *J. Res. Nat. Bur. Standards* 34:373.

LANGELIER, W.F. 1946. Effect of temperature on the pH of natural water. *J. Amer. Water Works Assoc.* 38:179.

FELDMAN, I. 1956. Use and abuse of pH measurements. *Anal. Chem.* 28:1859.

BRITTON, H.T.S. 1956. Hydrogen Ions, 4th ed. D. Van Nostrand Co., Princeton, N.J.

KOLTHOFF, I.M. & H.A. LAITINEN. 1958. pH and Electrotitrations. John Wiley & Sons, New York, N.Y.

KOLTHOFF, I.M. & P.J. ELVING. 1959. Treatise on Analytical Chemistry. Part I, Vol. 1, Chapter 10. Wiley-Interscience, New York, N.Y.

BATES, R.G. 1962. Revised standard values for pH measurements from 0 to 95°C. *J. Res. Nat. Bur. Standards* 66A:179.

AMERICAN WATER WORKS ASSOCIATION. 1964. Simplified Procedures for Water Examination. Manual M12, American Water Works Assoc., New York, N.Y.

WINSTEAD, M. 1967. Reagent Grade Water: How, When and Why? American Soc. Medical Technologists, The Steck Company, Austin, Tex.

STAPLES, B.R. & R.G. BATES. 1969. Two new standards for the pH scale. *J. Res. Nat. Bur. Standards* 73A:37.

BATES, R.G. 1973. Determination of pH, Theory and Practice, 2nd ed. John Wiley & Sons, New York, N.Y.

4500-I   IODINE*

4500-I A.   Introduction

1. Uses and Forms

Iodine may be used to disinfect potable and swimming pool waters. For wastewaters, iodine has had limited application. Use of iodine generally is restricted to personal or remote water supplies where ease of application, storage stability, and an inertness toward organic matter are important considerations. Some swimming pool waters are treated with iodine to lessen eye burn among swimmers and to provide a stable disinfectant residual less affected by adverse environmental conditions.

Iodine is applied in the elemental form or produced in situ by the simultaneous addition of an iodide salt and a suitable oxidant. In the latter case, an excess of iodide may be maintained to serve as a reservoir for iodine production; the determination of iodide is desirable for disinfectant control (see Iodide, Section 4500-I⁻).

Because of hydrolysis, active iodine exists in the forms of elemental I_2, hypoiodous acid (HOI) or a form thereof, hypoiodite ion (OI^-), and, in the presence of excess iodide, the triiodide ion (I_3^-). Most analytical methods use the oxidizing power of all forms of active iodine for its determination and the results usually are expressed as an equivalent concentration of elemental iodine.

2. Selection of Method

For potable and swimming pool waters treated with elemental iodine, both the am-

---

*Approved by Standard Methods Committee, 1988.

perometric titration and leuco crystal violet colorimetric methods give acceptable results. However, oxidized forms of manganese interfere with the leuco crystal violet method. Where the iodide and chloride ion concentrations are above 50 mg/L and 200 mg/L, respectively, interference in color production may occur in the leuco crystal violet method and the amperometric method is preferred. However, because of the extreme sensitivity of the leuco crystal violet method, this interference may be eliminated by sample dilution to obtain halogen ion concentrations less than 50 mg/L.

For wastewaters or highly polluted waters, organic constituents normally do not interfere with either the amperometric or leuco crystal violet procedures. Determine which of the methods yields the more acceptable results, because specific substances present in these waters may interfere in one method but not in the other. Certain metallic cations such as copper and silver interfere in the amperometric titration procedure. The leuco crystal violet method is relatively free of interference from these and other cations and anions with the exceptions noted previously.

For waters containing iodine coexisting with free chlorine, combined chlorine, or other excess oxidants, of the methods described only the leuco crystal violet method can determine iodine specifically. This condition occurs in the in-situ production of iodine by the reaction of iodide and excess oxidant. Under these conditions, the amperometric method would continue to titrate the iodine produced in a cyclic reaction until exhaustion of the oxidant.

4500-I B. Leuco Crystal Violet Method

1. General Discussion

The leuco crystal violet method determines aqueous iodine present as elemental iodine and hypoiodous acid. Excess common oxidants do not interfere. While the method utilizes the sum of the oxidative power of all forms of active iodine residuals, the results are expressed as the equivalent concentration of iodine. The method also is capable of determining the sum of iodine and free iodide concentrations; the free iodide concentration can be determined by difference (see Iodide, Section 4500-I⁻).

a. Principle: Mercuric chloride added to aqueous elemental iodine solutions causes essentially complete hydrolysis of iodine and the stoichiometric production of hypoiodous acid. The compound 4,4',4"-methylidynetris (*N,N*-dimethylaniline), also known by the common name of leuco crystal violet, reacts instantaneously with the hypoiodous acid to form crystal violet dye. The maximum absorbance of the developed crystal violet dye solution is produced in the pH range of 3.5 to 4.0 and measured at a wavelength of 592 nm. The absorbance follows Beer's law over a wide range of iodine concentrations and the developed color is stable for several hours.

In the presence of certain excess oxidants such as free chlorine or chloramines, the iodine residual will exist exclusively in the form of hypoiodous acid. The leuco crystal violet is relatively insensitive to the combined forms of chlorine while any free chlorine is converted to chloramine by reaction with an ammonium salt incorporated in the test reagents. All the hypoiodous acid is determined and, when expressed as an equivalent elemental I_2 concentration, will yield a weight concentration value twice that found in an elemental I_2 solution of the same weight concentration.

b. *Interference*: Oxidized forms of manganese interfere by oxidizing the indicator to crystal violet dye and yield apparent high iodine concentrations.

Iodide and chloride ion concentrations above 50 mg/L and 200 mg/L, respectively, interfere by inhibiting full color production. Dilute the sample to eliminate this interference.

Combined chlorine residuals normally do not interfere provided that the test is completed within 5 min after adding the indicator solution. Eliminate interference from free chlorine by adding an ammonium salt buffer to form combined chlorine.

c. *Minimum detectable concentration:* 10 μg I as I_2/L.

2. Apparatus

a. *Colorimetric equipment:* One of the following is required:

1) *Filter photometer,* with a light path of 1 cm or longer, equipped with an orange filter having maximum transmittance near 592 nm.

2) *Spectrophotometer,* for use at 592 nm, with a light path of 1 cm or longer.

b. *Volumetric flasks,* 100-mL, with plastic caps or ground-glass stoppers.

c. *Glassware:* Completely remove reducing substances from glassware or plastic containers, including containers for storage of reagent solutions (see Section 4500-Cl.D.2d).

3. Reagents

a. *Iodine-demand-free water:* See Section 4500-I⁻.B.3a. Prepare all stock iodine and reagent solutions with iodine-demand-free water.

b. *Stock iodine solution:* Prepare a saturated iodine solution by dissolving 20 g elemental iodine in 300 mL water. Let stand several hours. Decant iodine solution and dilute 170 mL to 2000 mL. Standardize solution by titrating with standard sodium thiosulfate ($Na_2S_2O_3$) titrant as described in Section 4500-Cl.B.3b and c or amperometrically as in Section 4500-I.C.

Calculate iodine concentration:

$$\text{mg I as } I_2/\text{mL} = \text{normality of iodine solution} \times 126.9$$

Prepare a working solution of 10 μg I as I_2/mL by appropriate dilution of the standardized stock solution.

c. *Citric buffer solution,* pH 3.8: See Section 4500-I⁻.B.3c.

d. *Leuco crystal violet indicator:* See Section 4500-I⁻.B.3d.

e. *Sodium thiosulfate solution:* See Section 4500-I⁻.B.3f.

4. Procedure

a. *Preparation of temporary iodine standards:* For greater accuracy, standardize working solution immediately before use by the amperometric titration method (Method C). Prepare standards in the range of 0.1 to 6.0 mg I as I_2/L by adding 1 to 60 mL working solution to 100 mL glass-stoppered volumetric flasks, in increments of 1 mL or larger. Adjust these volumes if the measured iodine concentration of working solution varies by 5% or more from 10 μg I as I_2/mL.

Measure 50.0 mL of each diluted iodine working solution into a 100-mL glass-stoppered volumetric flask. Add 1.0 mL citric buffer solution, gently swirl to mix, and let stand for at least 30 s. Add 1.0 mL leuco crystal violet indicator and swirl to develop color. Dilute to 100 mL and mix.

b. *Photometric calibration:* Transfer colored temporary standards of known iodine concentrations to cells of 1-cm light path and read absorbance in a photometer or spectrophotometer at a wavelength of 592 nm against a distilled water reference. Plot absorbance values against iodine concentrations to construct a curve that follows Beer's law.

c. *Color development of iodine sample:* Measure 50.0 mL sample into a 100-mL volumetric flask and treat as described for preparation of temporary iodine standards, ¶ 4a. Match test sample visually with temporary standards or read absorbance photometrically and refer to standard calibration curve for the iodine equivalent.

d. *Samples containing > 6.0 mg I as* I_2/ *L:* Place approximately 25 mL water in a 100-mL volumetric flask. Add 1.0 mL citric buffer solution and a measured volume of 25 mL or less of sample. Mix and let stand for at least 30 s. Add 1.0 mL leuco crystal violet indicator, mix, and dilute to mark. Match visually with standards or read absorbance photometrically and compare with calibration curve from which the initial iodine is obtained by applying the dilution factor. Select one of the following sample volumes to remain within optimum iodine range:

Iodine mg/L	Sample Volume Required mL
6.0–12.0	25.0
12.0–30	10.0
30–60	5.0

e. *Samples containing both chlorine and iodine:* For samples containing free or combined chlorine and iodine, follow procedure given in ¶ 4c or d above but read absorbance within 5 min after adding leuco crystal violet indicator.

f. *Compensation for turbidity and color:* Compensate for natural color or turbidity by adding 5 mL $Na_2S_2O_3$ solution to a 50-mL sample. Add reagents to sample as described previously and use as blank to set zero absorbance on the photometer. Measure all samples in relation to this blank and, from calibration curve, determine concentrations of iodine.

5. Bibliography

BLACK, A.P. & G.P. WHITTLE. 1967. New methods for the colorimetric determination of halogen residuals. Part I. Iodine, iodide, and iodate. *J. Amer. Water Works Assoc.* 59:471.

4500-I C.   Amperometric Titration Method

1. General Discussion

The amperometric titration method for iodine is a modification of the amperometric method for residual chlorine (see Section 4500-Cl.D). Iodine residuals over 7 mg/L are best measured with smaller samples or by dilution. In most cases the titration results represent free iodine because combined iodine rarely is encountered.

a. *Principle:* The principle of the amperometric method as described for the determination of total residual chlorine is applicable to the determination of residual iodine. Iodine is determined using buffer solution, pH 4.0, and potassium iodide (KI) solution. Maintain pH at 4.0 because at pH values less than 3.5 substances such as oxidized forms of manganese interfere, while at pH values greater than 4.5, the reaction is not quantitative. Adding KI improves the sharpness of the end point.

b. *Interference:* Free chlorine and the interferences described in Section 4500-Cl.D.1b also interfere in the iodine determination.

2. Apparatus

See Section 4500-Cl.D.2a through d.

3. Reagents

With the exception of phosphate buffer solution, pH 7.0, all reagents listed for the determination of residual chlorine in Section 4500-Cl.D.3 are required. Standardized phenylarsine oxide solution (1 mL = 1 mg chlorine/L for a 200-mL sample) is equivalent to 3.58 mg I as I_2/mL for a 200-mL sample.

4. Procedure

a. Sample volume: Select a sample volume that will require no more than 2 mL phenylarsine oxide titrant. For iodine concentrations of 7 mg/L or less, take a 200-mL volume; for iodine levels above 7 mg/L, use 100 mL or proportionately less diluted to 200 mL with water.

b. Free iodine: To the sample add 1 mL KI solution and 1 mL acetate buffer, pH 4.0 solution. Titrate with phenylarsine oxide titrant to the end point described in Section 4500-Cl.D.4.

5. Calculation

Calculate the iodine concentration by the following equation:

$$\text{mg I as } I_2/L = \frac{A \times 3.58 \times 200}{\text{mL sample}}$$

where:
   A = mL phenylarsine oxide titration to the end point.

6. Bibliography

MARKS, H.C. & J.R. GLASS. 1942. A new method of determining residual chlorine. *J. Amer. Water Works Assoc.* 34:1227.

4500-I⁻   IODIDE*

4500-I⁻ A. Introduction

1. Occurrence

Only microgram-per-liter quantities of iodide (I⁻) are present in most natural waters. Higher concentrations may be found in brines, certain industrial wastes, and waters treated with iodine.

2. Selection of Method

The leuco crystal violet method (B) is applicable to iodide concentrations of 50 to 6000 μg/L and is capable of determining iodide in the presence of iodine. The catalytic reduction method (C) is applicable to iodide concentrations of 80 μg/L or less.

The choice of method depends on the sample and concentration to be determined. The high chloride concentrations of brines, seawater, and many estuarine waters will interfere with color development in the leuco crystal violet method.

*Approved by Standard Methods Committee, 1988.

4500-I⁻ B.   Leuco Crystal Violet Method

1. General Discussion

a. Principle: Iodide is selectively oxidized to iodine by the addition of potassium peroxymonosulfate, $KHSO_5$. The iodine produced reacts instantaneously with the colorless indicator reagent containing 4,4′,4″-methylidynetris (N,N-dimethylaniline), also known as leuco crystal violet, to produce the highly colored crystal violet dye. The developed color is sufficiently stable for the determination of an absorbance value and adheres to Beer's law over a wide range of iodine concentrations.

b. Interference: Chloride concentrations greater than 200 mg/L may interfere with color development. Reduce these interferences by diluting sample to contain less than 200 mg Cl^-/L.

2. Apparatus

a. Colorimetric equipment: One of the following is required:

1) *Filter photometer,* providing a light path of 1 cm or longer, equipped with an orange filter having maximum transmittance near 592 nm.

2) *Spectrophotometer,* for use at 592 nm, providing a light path of 1 cm or longer.

b. Volumetric flasks: 100-mL with plastic caps or ground-glass stoppers.

c. Glassware: Completely remove any reducing substances from all glassware or plastic containers, including containers for storing reagent solutions (see Section 4500-Cl.D.2*d*).

3. Reagents

a. Iodine-demand-free water: Prepare a 1-m ion-exchange column of 2.5 to 5 cm diam, containing strongly acid cation and strong basic anion exchange resins. If a commercial analytical-grade mixed-bed resin is used, verify that compounds that react with iodine are removed. Pass distilled water at a slow rate through the resin bed and collect in clean container that will protect the treated water from undue exposure to the atmosphere.

Prepare all stock iodide and reagent solutions with iodine-demand-free water.

b. Stock iodide solution: Dissolve 1.3081 g KI in water and dilute to 1000 mL; 1 mL = 1 mg I.

c. Citric buffer solution, pH 3.8:

1) *Citric acid:* Dissolve 192.2 g $C_6H_8O_7$ or 210.2 g $C_6H_8O_7 \cdot H_2O$ and dilute to 1 L with water.

2) *Ammonium hydroxide,* 2*N*: Add 131 mL conc NH_4OH to about 700 mL water and dilute to 1 L. Store in a polyethylene bottle.

3) *Final buffer solution:* Slowly add, with mixing, 350 mL 2*N* NH_4OH solution to 670 mL citric acid. Add 80 g ammonium dihydrogen phosphate ($NH_4H_2PO_4$) and stir to dissolve.

d. Leuco crystal violet indicator: Measure 200 mL water and 3.2 mL conc sulfuric acid (H_2SO_4) into a brown glass container of at least 1-L capacity. Introduce a magnetic stirring bar and mix at moderate speed. Add 1.5 g 4,4′,4″-methylidynetris (N,N-dimethylaniline)* and with a small amount of water wash down any reagent adhering to neck or sides of container. Mix until dissolved.

To 800 mL water, add 2.5 g mercuric chloride ($HgCl_2$) and stir to dissolve. With mixing, add $HgCl_2$ solution to leuco crystal violet solution. For maximum stability, adjust pH of final solution to 1.5 or less, adding, if necessary, conc H_2SO_4 dropwise. Store in a brown glass bottle away from direct sunlight. Discard after 6 months. Do not use a rubber stopper.

e. Potassium peroxymonosulfate solution: Obtain $KHSO_5$ as a commercial product,†

---

*Eastman chemical No. 3651 or equivalent.
†Oxone, E. I. du Pont de Nemours and Co., Inc., Wilmington, Del., or equivalent.

which is a stable powdered mixture containing 42.8% $KHSO_5$ by weight and a mixture of $KHSO_4$ and K_2SO_4. Dissolve 1.5 g powder in water and dilute to 1 L.

f. Sodium thiosulfate solution: Dissolve 5.0 g $Na_2S_2O_3 \cdot H_2O$ in water and dilute to 1 L.

4. Procedure

a. Preparation of temporary iodine standards: Add suitable portions of stock iodide solution, or of dilutions of stock iodide solution, to water to prepare a series of 0.1 to 6.0 mg I^-/L in increments of 0.1 mg/L or larger.

Measure 50.0 mL dilute KI standard solution into a 100-mL glass-stoppered volumetric flask. Add 1.0 mL citric buffer and 0.5 mL $KHSO_5$ solution. Swirl to mix and let stand approximately 1 min. Add 1.0 mL leuco crystal violet indicator, mix, and dilute to 100 mL. For best results, read absorbance as described below within 5 min after adding indicator solution.

b. Photometric calibration: Transfer colored temporary standards of known iodide concentrations to cells of 1-cm light path and read absorbance in a photometer or spectrophotometer at a wavelength of 592 nm against a water reference. Plot absorbance values against iodide concentrations to construct a curve that follows Beer's law.

c. Color development of sample: Measure a 50.0-mL sample into a 100-mL volumetric flask and treat as described for preparation of temporary iodide standards, ¶ 4a. Read absorbance photometrically and refer to standard calibration curve for iodide equivalent.

d. Samples containing > 6.0 mg I^-/L: Place approximately 25 mL water in a 100-mL volumetric flask. Add 1.0 mL citric buffer and a measured volume of 25 mL or less of sample. Add 0.5 mL $KHSO_5$ solution. Swirl to mix and let stand approximately 1 min. Add 1.0 mL leuco crystal violet indicator, mix, and dilute to 100 mL.

Read absorbance photometrically and compare with calibration curve from which the initial iodide concentration is obtained by applying the dilution factor. Select one of the following sample volumes to remain within the optimum iodide range.

Iodide mg/L	Sample Volume Required mL
6.0–12	25.0
12–30	10.0
30–60	5.0

e. Determination of iodide in the presence of iodine: On separate samples determine (1) total iodide and iodine, and (2) iodine. The iodide concentration is the difference between the iodine determined and the total iodine-iodide obtained. Determine iodine by not adding $KHSO_5$ solution in the iodide method and by comparing the absorbance value to the calibration curve developed for iodine.

f. Compensation for turbidity and color: Compensate for natural color or turbidity by adding 5 mL $Na_2S_2O_3$ solution to a 50-mL sample. Add reagents to sample as described previously and use as the blank to set zero absorbance on photometer. Measure all samples in relation to this blank and, from the calibration curve, determine concentrations of iodide or total iodine-iodide.

5. Bibliography

BLACK, A.P. & G.P. WHITTLE. 1967. New methods for the colorimetric determination of halogen residuals. Part I. Iodine, iodide, and iodate. *J. Amer. Water Works Assoc.* 59:471.

4500-I⁻ C. Catalytic Reduction Method

1. General Discussion

a. Principle: Iodide can be determined by using its ability to catalyze the reduction of ceric ions by arsenious acid. The effect is nonlinearly proportional to the amount of iodide present. The reaction is stopped after a specific time interval by the addition of ferrous ammonium sulfate. The resulting ferric ions are directly proportional to the remaining ceric ions and develop a relatively stable color complex with potassium thiocyanate.

Digestion with chromic acid and distillation are necessary to estimate the non-susceptible bound forms of iodine in addition to the usual iodide ion. Procedures for these special applications are available.[1]

b. Interferences: The formation of non-catalytic forms of iodine and the inhibitory effects of silver and mercury are reduced by adding an excess of sodium chloride (NaCl) that sensitizes the reaction.

2. Apparatus

a. Water bath, capable of temperature control to 30 ± 0.5°C.

b. Colorimetric equipment: One of the following is required:

1) *Spectrophotometer,* for use at wavelengths of 510 or 525 nm and providing a light path of 1 cm.

2) *Filter photometer,* providing a light path of 1 cm and equipped with a green filter having maximum transmittance near 525 nm.

c. Test tubes, 2 × 15 cm.

d. Stopwatch.

3. Reagents

Store all stock solutions in tightly stoppered containers in the dark. Prepare all reagent solutions in distilled water.

a. Distilled water, containing less than 0.3 μg total I/L.

b. Sodium chloride solution: Dissolve 200.0 g NaCl in water and dilute to 1 L.

Recrystallize the NaCl if an interfering amount of iodine is present, using a water-ethanol mixture.

c. Arsenious acid: Dissolve 4.946 g As_2O_3 in water, add 0.20 mL conc H_2SO_4, and dilute to 1000 mL.

d. Sulfuric acid, H_2SO_4, conc.

e. Ceric ammonium sulfate: Dissolve 13.38 g $Ce(NH_4)_4(SO_4)_4 \cdot 4H_2O$ in water, add 44 mL conc H_2SO_4, and make up to 1 L.

f. Ferrous ammonium sulfate reagent: Dissolve 1.50 g $Fe(NH_4)_2(SO_4)_2 \cdot 6H_2O$ in 100 mL distilled water containing 0.6 mL conc H_2SO_4. Prepare daily.

g. Potassium thiocyanate solution: Dissolve 4.00 g KSCN in 100 mL water.

h. Stock iodide solution: Dissolve 261.6 mg anhydrous KI in water and dilute to 1000 mL; 1.00 mL = 200 μg I⁻.

i. Intermediate iodide solution: Dilute 20.00 mL stock iodide solution to 1000 mL with water; 1.00 mL = 4.00 μg I⁻.

j. Standard iodide solution: Dilute 25.00 mL intermediate iodide solution to 1000 mL with water; 1.00 mL = 0.100 μg I⁻.

4. Procedure

a. Sample size: Add 10.00 mL sample, or a portion made up to 10.00 mL with water, to a 2-× 15-cm test tube. If possible, keep iodide content in the range 0.2 to 0.6 μg. Use thoroughly clean glassware and apparatus.

b. Color measurement: Add reagents in the following order: 1.00 mL NaCl solution, 0.50 mL As_2O_3 solution, and 0.50 mL conc H_2SO_4.

Place reaction mixture and ceric ammonium sulfate solution in 30°C water bath and let come to temperature equilibrium. Add 1.0 mL ceric ammonium sulfate solution, mix by inversion, and start stopwatch to time reaction. Use an inert clean test tube stopper when mixing. After 15 ± 0.1 min remove sample from water bath and add immediately 1.00 mL ferrous ammonium sulfate reagent with mixing, whereupon the yellow ceric ion color should disappear. Then add, with mixing, 1.00 mL KSCN solution. Replace sample in water bath. Within 1 h after adding thiocyanate read absorbance in a photometric instrument. Maintain temperature of solution and cell compartment at 30 ± 0.5°C until absorbance is determined. If several samples are run, start reactions at 1-min intervals to allow time for additions of ferrous ammonium sulfate and thiocyanate. (If temperature control of cell compartment is not possible, let final solution come to room temperature and measure absorbance with cell compartment at room temperature.)

c. *Calibration standards:* Treat standards containing 0, 0.2, 0.4, 0.6, and 0.8 μg I$^-$/10.00 mL of solution as in ¶ 4b above. Run with each set of samples to establish a calibration curve.

5. Calculation

$$\text{mg I}^-/\text{L} = \frac{\mu\text{g I (in 15 mL final volume)}}{\text{mL sample}}$$

6. Precision and Bias

Results obtained by this method are reproducible on samples of Los Angeles source waters, and have been reported to be accurate to ±0.3 μg I$^-$/L on samples of Yugoslavian water containing from 0 to 14.0 μg I$^-$/L.

7. Reference

1. Standard Methods for the Examination of Water, Sewage and Industrial Wastes, 10th ed. 1955. American Public Health Assoc., American Water Works Assoc., & Fed. of Sewage & Industrial Wastes Associations, New York, pp. 120-124.

8. Bibliography

ROGINA, B. & M. DUBRAVCIC. 1953. Microdetermination of iodides by arresting the catalytic reduction of ceric ions. *Analyst* 78:594.
DUBRAVCIC, M. 1955. Determination of iodine in natural waters (sodium chloride as a reagent in the catalytic reduction of ceric ions). *Analyst* 80:295.

4500-N   NITROGEN*

In waters and wastewaters the forms of nitrogen of greatest interest are, in order of decreasing oxidation state, nitrate, nitrite, ammonia, and organic nitrogen. All these forms of nitrogen, as well as nitrogen gas (N_2), are biochemically interconvertible and are components of the nitrogen cycle. They are of interest for many reasons.

Organic nitrogen is defined functionally as organically bound nitrogen in the trinegative oxidation state. It does not include all organic nitrogen compounds. Analytically, organic nitrogen and ammonia can be determined together and have been re-

ferred to as "kjeldahl nitrogen," a term that reflects the technique used in their determination. Organic nitrogen includes such natural materials as proteins and peptides, nucleic acids and urea, and numerous synthetic organic materials. Typical organic nitrogen concentrations vary from a few hundred micrograms per liter in some lakes to more than 20 mg/L in raw sewage.

Total oxidized nitrogen is the sum of nitrate and nitrite nitrogen. Nitrate generally occurs in trace quantities in surface water but may attain high levels in some groundwater. In excessive amounts, it contributes to the illness known as methemoglobinemia in infants. A limit of 10 mg

*Approved by Standard Methods Committee, 1988.

nitrate as nitrogen/L has been imposed on drinking water to prevent this disorder. Nitrate is found only in small amounts in fresh domestic wastewater but in the effluent of nitrifying biological treatment plants nitrate may be found in concentrations of up to 30 mg nitrate as nitrogen/L. It is an essential nutrient for many photosynthetic autotrophs and in some cases has been identified as the growth-limiting nutrient.

Nitrite is an intermediate oxidation state of nitrogen, both in the oxidation of ammonia to nitrate and in the reduction of nitrate. Such oxidation and reduction may occur in wastewater treatment plants, water distribution systems, and natural waters. Nitrite can enter a water supply system through its use as a corrosion inhibitor in industrial process water. Nitrite is the actual etiologic agent of methemoglobinemia. Nitrous acid, which is formed from nitrite in acidic solution, can react with secondary amines (RR′NH) to form nitrosamines (RR′N-NO), many of which are known to be carcinogens. The toxicologic significance of nitrosation reactions in vivo and in the natural environment is the subject of much current concern and research.

Ammonia is present naturally in surface and wastewaters. Its concentration generally is low in groundwaters because it adsorbs to soil particles and clays and is not leached readily from soils. It is produced largely by deamination of organic nitrogen-containing compounds and by hydrolysis of urea. At some water treatment plants ammonia is added to react with chlorine to form a combined chlorine residual.

In the chlorination of wastewater effluents containing ammonia, virtually no free residual chlorine is obtained until the ammonia has been oxidized. Rather, the chlorine reacts with ammonia to form mono- and dichloramines. Ammonia concentrations encountered in water vary from less than 10 μg ammonia nitrogen/L in some natural surface and groundwaters to more than 30 mg/L in some wastewaters.

In this manual, organic nitrogen is referred to as organic N, nitrate nitrogen as NO_3^--N, nitrite nitrogen as NO_2^--N, and ammonia nitrogen as NH_3-N.

4500-NH₃ NITROGEN (AMMONIA)*

4500-NH₃ A. Introduction

1. Selection of Method

The two major factors that influence selection of the method to determine ammonia are concentration and presence of interferences. In general, direct manual determination of low concentrations of ammonia is confined to drinking waters, clean surface water, and good-quality nitrified wastewater effluent. In other instances, and where interferences are present and greater precision is necessary, a preliminary distillation step (B) is required. For high ammonia concentrations a distillation and titration technique is preferred. The data presented in ¶ 4 below and Table 4500-NH₃:I should be helpful in selecting the appropriate method of analysis.

Two manual colorimetric techniques—the nesslerization (C) and phenate (D) methods—and one titration method (E) are presented. An ammonia-selective electrode method (F), which may be used either with

*Approved by Standard Methods Committee, 1985.

TABLE 4500-NH₃:I. PRECISION AND BIAS DATA FOR AMMONIA METHODS

| | | Relative Standard Deviation | | | | | Relative Error | | | | |
| | | | | Distillation Plus | | | | | Distillation Plus | | |
Number of Laboratories	Ammonia Nitrogen Concentration μg/L	Direct Nessleri-zation %	Direct Manual Phenate Method %	Nessler Method %	Manual Phenate Method %	Titri-metric Method %	Direct Nessleri-zation %	Direct Manual Phenate Method %	Nessler Method %	Manual Phenate Method %	Titri-metric Method %
20	200	38.1	—	—	—	—	0	—	—	—	—
44	200	—	—	46.3	—	—	—	—	10.0	—	—
21	200	—	—	—	—	69.8	—	—	—	—	20.0
20	800	11.2	—	—	—	—	0	—	—	—	—
42	800	—	—	21.2	—	—	—	—	8.7	—	—
20	800	—	—	—	—	28.6	—	—	—	—	5.0
21	1500	11.6	—	—	—	—	0.6	—	—	—	—
42	1500	—	—	18.0	—	—	—	—	4.0	—	—
21	1500	—	—	—	—	21.6	—	—	—	—	2.6
70	200	—	39.2	—	—	—	—	2.4	—	—	—
3	200	—	—	—	15.1	—	—	—	—	16.7	—
9	200	22.0	—	—	—	—	8.3	—	—	—	—
5	200	—	—	15.7	—	—	—	—	2.0	—	—
66	800	—	15.8	—	—	—	—	1.5	—	—	—
3	800	—	—	—	16.6	—	—	—	—	1.7	—
9	800	16.1	—	—	—	—	0.3	—	—	—	—
6	800	—	—	16.3	—	—	—	—	3.1	—	—
71	1500	—	26.0	—	—	—	—	10.0	—	—	—
3	1500	—	—	—	7.3	—	—	—	—	0.4	—
8	1500	5.3	—	—	—	—	1.2	—	—	—	—
6	1500	—	—	7.5	—	—	—	—	3.6	—	—

or without prior sample distillation, an ammonia-selective electrode method using a known addition (G), and an automated version of the phenate method (H) also are included. While the stated maximum concentration ranges for the manual methods are not rigorous limits, titration is preferred at concentrations higher than the stated maximum levels for the photometric procedure.

The nessler method is sensitive to 20 μg NH_3-N/L under optimum conditions and may be used for up to 5 mg NH_3-N/L. Turbidity, color, and substances precipitated by hydroxyl ion, such as magnesium and calcium, interfere and may be removed by preliminary distillation or, less satisfactorily, by precipitation with zinc sulfate and alkali.

The manual phenate method has a sensitivity of 10 μg NH_3-N/L and is useful for up to 500 μg NH_3-N/L. Preliminary distillation is required if the alkalinity exceeds 500 mg $CaCO_3$/L, if color or turbidity is present, or if the sample has been preserved with acid.

The distillation and titration procedure is used especially for NH_3-N concentrations greater than 5 mg/L.

Distillation into sulfuric acid (H_2SO_4) absorbent is mandatory for the phenate method when interferences are present. Boric acid must be the absorbent following distillation if the distillate is to be nesslerized or titrated.

The ammonia-selective electrode method is applicable over the range from 0.03 to 1400 mg NH_3-N/L.

2. Interferences

Glycine, urea, glutamic acid, cyanates, and acetamide hydrolyze very slowly in solution on standing but, of these, only urea and cyanates will hydrolyze on distillation at pH of 9.5. Hydrolysis amounts to about 7% at this pH for urea and about 5% for cyanates. Glycine, hydrazine, and some amines will react with nessler reagent to give the characteristic yellow color in the time required for the test. Similarly, volatile alkaline compounds such as hydrazine and amines will influence titrimetric results. Some organic compounds such as ketones, aldehydes, alcohols, and some amines may cause a yellowish or greenish off-color or a turbidity on nesslerization following distillation. Some of these, such as formaldehyde, may be eliminated by boiling off at a low pH before nesslerization. Remove residual chlorine by sample pretreatment.

3. Storage of Samples

Most reliable results are obtained on fresh samples. Destroy residual chlorine immediately after sample collection to prevent its reaction with ammonia. If prompt analysis is impossible, preserve samples with 0.8 mL conc H_2SO_4/L sample and store at 4°C. The pH of the acid-preserved samples should be between 1.5 and 2. Some wastewaters may require more conc H_2SO_4 to achieve this pH. If acid preservation is used, neutralize samples with NaOH or KOH immediately before making the determination.

4. Precision and Bias

Six synthetic samples containing ammonia and other constituents dissolved in distilled water were analyzed by five procedures. The first three samples were subjected to direct nesslerization alone, distillation followed by nesslerization, and distillation followed by titration. Samples 4 through 6 were analyzed by direct nesslerization, by distillation followed by nesslerization, by the phenate method alone, and by distillation followed by the phenate method. Results obtained by the participating laboratories are summarized in Table 4500-NH₃:I.

Sample 1 contained the following additional constituents: 10 mg Cl⁻/L, 1.0 mg

NO_3^--N/L, 1.5 mg organic N/L, 10.0 mg PO_4^{3-}/L, and 5.0 mg silica/L.

Sample 2 contained the following constituents: 200 mg Cl^-/L, 1.0 mg NO_3^--N/L, 0.8 mg organic N/L, 5.0 mg PO_4^{3-}/L, and 15.0 mg silica/L.

Sample 3 contained the following additional constituents: 400 mg Cl^-/L, 1.0 mg NO_3^--N/L, 0.2 mg organic N/L, 0.5 mg PO_4^{3-}/L, and 30.0 mg silica/L.

Sample 4 contained the following additional constituents: 400 mg Cl^-/L, 0.05 mg NO_3^--N/L, 0.23 mg organic phosphorus/L added in the form of adenylic acid, 7.00 mg orthophosphate phosphorus/L, and 3.00 mg polyphosphate phosphorus/L added as sodium hexametaphosphate.

Sample 5 contained the following additional constituents: 400 mg Cl^-/L, 5.00 mg NO_3^--N/L, 0.09 mg organic phosphorus/L added in the form of adenylic acid, 0.6 mg orthophosphate phosphorus/L, and 0.3 mg polyphosphate phosphorus/L added as sodium hexametaphosphate.

Sample 6 contained the following additional constituents: 400 mg Cl^-/L, 0.4 mg NO_3^--N/L, 0.03 mg organic phosphorus/L added in the form of adenylic acid, 0.1 mg orthophosphate phosphorus/L, and 0.08 mg polyphosphate phosphorus/L added as sodium hexametaphosphate.

For the ammonia-selective electrode in a single laboratory using surface water samples at concentrations of 1.00, 0.77, 0.19, and 0.13 mg NH_3-N/L, standard deviations were ±0.038, ±0.017, ±0.007, and ±0.003, respectively. In a single laboratory using surface water samples at concentrations of 0.10 and 0.13 NH_3-N/L, recoveries were 96% and 91%, respectively. The results of an interlaboratory study involving 12 laboratories using the ammonia-selective electrode[†] on distilled water and effluents are summarized in Table 4500-NH_3:II.

For an automated phenate system[‡] in a single laboratory using surface water samples at concentrations of 1.41, 0.77, 0.59, and 0.43 mg NH_3-N/L, the standard deviation was ±0.005, and at concentrations of 0.16 and 1.44 mg NH_3-N/L, recoveries were 107 and 99%, respectively.

[†]American Society for Testing and Materials, ASTM Method 1426-79.
[‡]AutoAnalyzer™ I, Technicon Instrument Corp., Tarrytown, N.Y. 10591.

TABLE 4500-NH_3:II. PRECISION AND BIAS OF AMMONIA-SELECTIVE ELECTRODE

Level mg/L	Matrix	Mean Recovery %	Precision	
			Overall, S_T	Operator, S_O
0.04	Distilled water	200	0.05	0.01
	Effluent water	100	0.03	0.00
0.10	Distilled water	180	0.05	0.01
	Effluent water	470	0.61	0.01
0.80	Distilled water	105	0.11	0.04
	Effluent water	105	0.30	0.06
20	Distilled water	95	2	1
	Effluent water	95	3	2
100	Distilled water	98	5	2
	Effluent water	97	—	—
750	Distilled water	97	78	12
	Effluent water	99	106	10

4500-NH₃ B.   Preliminary Distillation Step

1. General Discussion

The sample is buffered at pH 9.5 with a borate buffer to decrease hydrolysis of cyanates and organic nitrogen compounds. It is distilled into a solution of boric acid when nesslerization or titration is to be used or into H_2SO_4 when the phenate method is used. The ammonia in the distillate can be determined either colorimetrically by nesslerization or the phenate method or titrimetrically with standard H_2SO_4 and a mixed indicator or a pH meter. The choice between the colorimetric and the acidimetric method depends on the concentration of ammonia. Ammonia in the distillate also can be determined by the ammonia-selective electrode method, using $0.04N$ H_2SO_4 to trap the distillate.

2. Apparatus

a. Distillation apparatus: Arrange a borosilicate glass flask of 800- to 2000-mL capacity attached to a vertical condenser so that the outlet tip may be submerged below the surface of the receiving acid solution. Use an all-borosilicate-glass apparatus or one with condensing units constructed of block tin or aluminum tubes.

b. pH meter.

3. Reagents

a. Ammonia-free water: Prepare by ion-exchange or distillation methods:

1) Ion exchange—Prepare ammonia-free water by passing distilled water through an ion-exchange column containing a strongly acidic cation-exchange resin mixed with a strongly basic anion-exchange resin. Select resins that will remove organic compounds that interfere with the ammonia determination. Some anion-exchange resins tend to release ammonia. If this occurs, prepare ammonia-free water with a strongly acidic cation-exchange resin. Regenerate the column according to the manufacturer's instructions. Check ammonia-free water for the possibility of a high blank value.

2) Distillation—Eliminate traces of ammonia in distilled water by adding 0.1 mL conc H_2SO_4 to 1 L distilled water and redistilling. Alternatively, treat distilled water with sufficient bromine or chlorine water to produce a free halogen residual of 2 to 5 mg/L and redistill after standing at least 1 h. Discard the first 100 mL distillate. Check redistilled water for the possibility of a high blank.

It is very difficult to store ammonia-free water in the laboratory without contamination from gaseous ammonia. However, if storage is necessary, store in a tightly stoppered glass container to which is added about 10 g ion-exchange resin (preferably a strongly acidic cation-exchange resin) per liter ammonia-free water. For use, let resin settle and decant ammonia-free water. If a high blank value is produced, replace the resin or prepare fresh ammonia-free water.

Use ammonia-free distilled water for preparing all reagents, rinsing, and sample dilution.

b. Borate buffer solution: Add 88 mL $0.1N$ NaOH solution to 500 mL approximately $0.025M$ sodium tetraborate $(Na_2B_4O_7)$ solution (9.5 g $Na_2B_4O_7 \cdot$ 10 H_2O/L) and dilute to 1 L.

c. Sodium hydroxide, 6N: Dissolve 240 g NaOH in water and dilute to 1 L.

d. Dechlorinating agent: Use 1 mL of any

of the following reagents to remove 1 mg/ L residual chlorine in 500 mL sample.

1) *Sodium sulfite:* Dissolve 0.9 g Na_2SO_3 in water and dilute to 1 L. Prepare fresh daily.

2) *Sodium thiosulfate:* Dissolve 3.5 g $Na_2S_2O_3 \cdot 5H_2O$ in water and dilute to 1 L. Prepare fresh weekly.

3) *Phenylarsine oxide:* Dissolve 1.2 g C_6H_5AsO in 200 mL 0.3N NaOH solution, filter if necessary, and dilute to 1 L with water. (CAUTION: *Toxic—take care to avoid ingestion.*)

4) *Sodium arsenite:* Dissolve 0.93 g $NaAsO_2$ in water and dilute to 1 L. (CAUTION: *Toxic—take care to avoid ingestion.*) Prepare fresh weekly.

e. Neutralization agent: Prepare with ammonia-free water.

1) *Sodium hydroxide,* NaOH, 1N.

2) *Sulfuric acid,* H_2SO_4, 1N.

f. Absorbent solution, plain boric acid: Dissolve 20 g H_3BO_3 in water and dilute to 1 L.

g. Indicating boric acid solution: See Section 4500-NH_3.E.3a and b.

h. Sulfuric acid, 0.04N: Dilute 1.0 mL conc H_2SO_4 to 1 L.

4. Procedure

a. Preparation of equipment: Add 500 mL water and 20 mL borate buffer to a distillation flask and adjust pH to 9.5 with 6N NaOH solution. Add a few glass beads or boiling chips and use this mixture to steam out the distillation apparatus until distillate shows no traces of ammonia.

b. Sample preparation: Use 500 mL dechlorinated sample or a portion diluted to 500 mL with water. When NH_3-N concentration is less than 100 μg/L, use a sample volume of 1000 mL. Remove residual chlorine by adding, at the time of collection, dechlorinating agent equivalent to the chlorine residual. If necessary, neutralize

to approximately pH 7 with dilute acid or base, using a pH meter.

Add 25 mL borate buffer solution and adjust to pH 9.5 with 6N NaOH using a pH meter.

c. Distillation: To minimize contamination, leave distillation apparatus assembled after steaming out and until just before starting sample distillation. Disconnect steaming-out flask and immediately transfer sample flask to distillation apparatus. Distill at a rate of 6 to 10 mL/min with the tip of the delivery tube below the surface of acid receiving solution. Collect distillate in a 500-mL erlenmeyer flask containing 50 mL plain boric acid solution for nesslerization method. Use 50 mL indicating boric acid solution for titrimetric method. Distill ammonia into 50 mL 0.04N H_2SO_4 for the phenate method and for the ammonia-selective electrode method. Collect at least 200 mL distillate. Lower collected distillate free of contact with delivery tube and continue distillation during the last minute or two to cleanse condenser and delivery tube. Dilute to 500 mL with water.

When the phenate method is used for determining NH_3-N, neutralize distillate with 1N NaOH solution.

d. Ammonia determination: Determine ammonia by the nesslerization method (C), the phenate method (D), the titrimetric method (E) or the ammonia-selective electrode method (F).

5. Bibliography

NICHOLS, M.S. & M.E. FOOTE. 1931. Distillation of free ammonia from buffered solutions. *Ind. Eng. Chem.,* Anal. Ed. 3:311.

GRIFFIN, A.E. & N.S. CHAMBERLIN. 1941. Relation of ammonia nitrogen to breakpoint chlorination. *Amer. J. Pub. Health* 31:803.

PALIN, A.T. 1950. Symposium on the sterilization of water. Chemical aspects of chlorination. *J. Inst. Water Eng.* 4:565.

TARAS, M.J. 1953. Effect of free residual chlorination of nitrogen compounds in water. *J. Amer. Water Works Assoc.* 45:47.

4500-NH₃ C.   Nesslerization Method
(Direct and Following Distillation)

1. General Discussion

Use direct nesslerization only for purified drinking waters, natural water, and highly purified wastewater effluents, all of which should be low in color and have NH₃-N concentrations exceeding 20 μg/L. Apply the direct nesslerization method to domestic wastewaters only when errors of 1 to 2 mg/L are acceptable. Use this method only after it has been established that it yields results comparable to those obtained after distillation. Check validity of direct nesslerization measurements periodically.

Pretreatment before direct nesslerization with zinc sulfate and alkali precipitates calcium, iron, magnesium, and sulfide, which form turbidity when treated with nessler reagent. The floc also removes suspended matter and sometimes colored matter. Addition of EDTA or Rochelle salt solution inhibits precipitation of residual calcium and magnesium ions in the presence of the alkaline nessler reagent. However, use of EDTA demands an extra amount of nessler reagent to insure a sufficient nessler reagent excess for reaction with the ammonia.

The graduated yellow to brown colors produced by the nessler-ammonia reaction absorb strongly over a wide wavelength range. The yellow color characteristic of low ammonia nitrogen concentration (0.4 to 5 mg/L) can be measured with acceptable sensitivity in the wavelength region from 400 to 425 nm when a 1-cm light path is available. A light path of 5 cm extends measurements into the nitrogen concentration range of 5 to 60 μg/L. The reddish brown hues typical of ammonia nitrogen levels approaching 10 mg/L may be measured in the wavelength region of 450 to 500 nm. A judicious selection of light path and wavelength thus permits the photometric determination of ammonia nitrogen concentrations over a considerable range.

Departures from Beer's law may be evident when photometers equipped with broad-band color filters are used. For this reason, prepare the calibration curve under conditions identical with those adopted for the samples.

A carefully prepared nessler reagent may respond under optimum conditions to as little as 1 μg NH₃-N/50 mL. In direct nesslerization, this represents 20 μg/L. However, reproducibility below 100 μg/L may be erratic.

2. Apparatus

a. Colorimetric equipment: One of the following is required:

1) *Spectrophotometer,* for use at 400 to 500 nm and providing a light path of 1 cm or longer.

2) *Filter photometer,* providing a light path of 1 cm or longer and equipped with a violet filter having maximum transmittance at 400 to 425 nm. A blue filter can be used for higher NH₃-N concentrations.

3) *Nessler tubes,* matched, 50-mL, tall form.

b. pH meter, equipped with a high-pH electrode.

3. Reagents

Use ammonia-free water for preparing all reagents, rinsing, and making dilutions. All the reagents listed in Preliminary Distillation (4500-NH₃.B above), except the borate buffer and absorbent solution, are required, plus the following:

a. Zinc sulfate solution: Dissolve 100 g $ZnSO_4 \cdot 7H_2O$ and dilute to 1 L with water.

b. Stabilizer reagent: Use either EDTA or Rochelle salt to prevent calcium or magnesium precipitation in undistilled samples after addition of alkaline nessler reagent.

1) *EDTA reagent:* Dissolve 50 g disodium ethylenediamine tetraacetate dihydrate in 60 mL water containing 10 g NaOH. If necessary, apply gentle heat to complete dissolution. Cool to room temperature and dilute to 100 mL.

2) *Rochelle salt solution:* Dissolve 50 g potassium sodium tartrate tetrahydrate, $KNaC_4H_4O_6 \cdot 4H_2O$, in 100 mL water. Remove ammonia usually present in the salt by boiling off 30 mL of solution. After cooling, dilute to 100 mL.

c. Nessler reagent: Dissolve 100 g HgI_2 and 70 g KI in a small quantity of water and add this mixture slowly, with stirring, to a cool solution of 160 g NaOH dissolved in 500 mL water. Dilute to 1 L. Store in rubber-stoppered borosilicate glassware and out of sunlight to maintain reagent stability for up to a year under normal laboratory conditions. Check reagent to make sure that it yields the characteristic color with 0.1 mg NH_3-N/L within 10 min after addition and does not produce a precipitate with small amounts of ammonia within 2 h. (CAUTION: *Toxic—take care to avoid ingestion.*)

d. Stock ammonium solution: Dissolve 3.819 g anhydrous NH_4Cl, dried at 100°C, in water, and dilute to 1000 mL; 1.00 mL = 1.00 mg N = 1.22 mg NH_3.

e. Standard ammonium solution: Dilute 10.00 mL stock ammonium solution to 1000 mL with water; 1.00 mL = 10.00 μg N = 12.2 μg NH_3.

f. Permanent color solutions:

1) *Potassium chloroplatinate solution:* Dissolve 2.0 g K_2PtCl_6 in 300 to 400 mL water; add 100 mL conc HCl and dilute to 1 L.

2) *Cobaltous chloride solution:* Dissolve 12.0 g $CoCl_2 \cdot 6H_2O$ in 200 mL water. Add 100 mL conc HCl and dilute to 1 L.

4. Procedure

a. Treatment of undistilled samples: If necessary, remove residual chlorine from the freshly collected sample by adding an equivalent amount of dechlorinating agent. (Do not store chlorinated samples without prior dechlorination.) Add 1 mL $ZnSO_4$ solution to 100 mL sample and mix thoroughly. Add 0.4 to 0.5 mL 6N NaOH solution to obtain a pH of 10.5, as determined with a pH meter and a high-pH glass electrode, and mix gently. Let treated sample stand for a few minutes, whereupon a heavy flocculent precipitate should fall, leaving a clear and colorless supernate. Clarify by centrifuging or filtering. Pretest any filter paper used to be sure no ammonia is present as a contaminant. Do this by running water through the filter and testing the filtrate by nesslerization. Filter sample, discarding first 25 mL filtrate. (CAUTION: Samples containing more than about 10 mg NH_3-N/L may lose ammonia during this treatment of undistilled samples because of the high pH. Dilute such samples to the sensitive range for nesslerization before pretreatment.)

b. Color development:

1) Undistilled samples—Use 50.0 mL sample or a portion diluted to 50.0 mL with water. If the undistilled portion contains sufficient concentrations of calcium, magnesium, or other ions that produce turbidity or precipitate with nessler reagent, add 1 drop (0.05 mL) EDTA reagent or 1 to 2 drops (0.05 to 0.1 mL) Rochelle salt solution. Mix well. Add 2.0 mL nessler reagent if EDTA reagent is used or 1.0 mL nessler reagent if Rochelle salt is used.

2) Distilled samples—Neutralize the boric acid used for absorbing the ammonia distillate by adding either 2 mL nessler reagent, an excess that raises the pH to the desired high level, or, alternatively, neutralizing the boric acid with NaOH before adding 1 mL nessler reagent.

3) Mix samples by capping nessler tubes with clean rubber stoppers (washed thoroughly with water) and then inverting tubes at least six times. Keep such conditions as temperature and reaction time the same in blank, samples, and standards. Let reaction proceed for at least 10 min after

adding nessler reagent. Measure color in sample and standards. If NH₃-N is very low, use a 30-min contact time for sample, blank, and standards. Measure color either photometrically or visually as directed in ¶s c or d below.

c. *Photometric measurement:* Measure absorbance or transmittance with a spectrophotometer or filter photometer. Prepare calibration curve at the same temperature and reaction time used for samples. Measure transmittance readings against a reagent blank and run parallel checks frequently against standards in the nitrogen range of the samples. Redetermine complete calibration curve for each new batch of nessler reagent.

For distilled samples, prepare standard curve under the same conditions as the samples. Distill reagent blank and appropriate standards, each diluted to 500 mL, in the same manner as the samples. Dilute 300 mL distillate plus 50 mL boric acid absorbent to 500 mL with water and take a 50-mL portion for nesslerization.

d. *Visual comparison:* Compare colors produced in sample against those of ammonia standards. Prepare temporary or permanent standards as follows:

1) Temporary standards—Prepare a series of visual standards in nessler tubes by adding the following volumes of standard NH₄Cl solution and diluting to 50 mL with water: 0, 0.2, 0.4, 0.7, 1.0, 1.4, 1.7, 2.0, 2.5, 3.0, 3.5, 4.0, 4.5, 5.0, and 6.0 mL. Nesslerize standards and portions of distillate by adding 1.0 mL nessler reagent to each tube and mixing well.

2) Permanent standards—Measure into 50-mL nessler tubes the volumes of K₂PtCl₆ and CoCl₂ solutions indicated in Table 4500-NH₃:III, dilute to mark, and mix thoroughly. The values given in the table are *approximate;* actual equivalents of the ammonium standards will differ with the quality of nessler reagent, the kind of illumination used, and the color sensitivity of the analyst's eye. Therefore, compare color standards with nesslerized temporary

TABLE 4500-NH₃:III. PREPARATION OF PERMANENT COLOR STANDARDS FOR VISUAL DETERMINATION OF AMMONIA NITROGEN

Value in Ammonia Nitrogen μg	Approximate Volume of Platinum Solution mL*	Approximate Volume of Cobalt Solution mL*
0	1.2	0.0
2	2.8	0.0
4	4.7	0.1
7	5.9	0.2
10	7.7	0.5
14	9.9	1.1
17	11.4	1.7
20	12.7	2.2
25	15.0	3.3
30	17.3	4.5
35	19.0	5.7
40	19.7	7.1
45	19.9	8.7
50	20.0	10.4
60	20.0	15.0

*In matched 50-mL nessler tubes.

ammonia standards and modify the tints as necessary. Make such comparisons for each newly prepared nessler reagent and satisfy each analyst as to the aptness of the color match. Protect standards from dust to extend their usefulness for several months. Compare either 10 or 30 min after nesslerization, depending on reaction time used in preparing nesslerized ammonium standards against which they were matched.

5. Calculation

a. Deduct amount of NH₃-N in water used for diluting original sample before computing final nitrogen value.

b. Deduct also reagent blank for volume of borate buffer and 6N NaOH solutions used with sample.

c. Compute total NH₃-N by the following equation:

mg NH$_3$-N/L (51 mL final volume)

$$= \frac{A}{\text{mL sample}} \times \frac{B}{C}$$

where:

A = μg NH$_3$-N (51 mL final volume),
B = total volume distillate collected, mL, including acid absorbent, and
C = volume distillate taken for nesslerization, mL.

The ratio B/C applies only to distilled samples; ignore in direct nesslerization.

6. Precision and Bias

See Section 4500-NH$_3$.A.4 and Table 4500-NH$_3$:I.

7. Bibliography

JACKSON, D.D. 1900. Permanent standards for use in the analysis of water. *Mass. Inst. Technol. Quart.* 13:314.

SAWYER, C.N. 1953. pH adjustment for determination of ammonia nitrogen. *Anal. Chem.* 25:816.

JENKINS, D. 1967. The differentiation, analysis and preservation of nitrogen and phosphorus forms in natural waters. *In* Trace Inorganics in Water. American Chemical Soc., Washington, D.C.

BOLTZ, D.F. & J.A. HOWELL, EDS. 1978. Colorimetric Determination of Nonmetals. Wiley Interscience, New York, N.Y.

4500-NH$_3$ D. Phenate Method

1. General Discussion

a. Principle: An intensely blue compound, indophenol, is formed by the reaction of ammonia, hypochlorite, and phenol catalyzed by a manganous salt.

b. Interference: Alkalinity over 500 mg as CaCO$_3$/L, acidity over 100 mg as CaCO$_3$/L, and turbidity interfere. Remove these interferences by preliminary distillation.

2. Apparatus

a. Colorimetric equipment: One of the following is required:

1) *Spectrophotometer,* for use at 630 nm with a light path of approximately 1 cm.

2) *Filter photometer,* equipped with a red-orange filter having a maximum transmittance near 630 nm and providing a light path of approximately 1 cm.

b. Magnetic stirrer.

3. Reagents

a. Ammonia-free water: Prepare as directed in Section 4500-NH$_3$.B.3a. Use for making all reagents.

b. Hypochlorous acid reagent: To 40 mL water add 10 mL 5% NaOCl solution prepared from commercial bleach. Adjust pH to 6.5 to 7.0 with HCl. Prepare this unstable reagent weekly.

c. Manganous sulfate solution, 0.003M: Dissolve 50 mg MnSO$_4 \cdot$H$_2$O in 100 mL water.

d. Phenate reagent: Dissolve 2.5 g NaOH and 10 g phenol, C$_6$H$_5$OH, in 100 mL water. Because this reagent darkens on standing, prepare weekly. (CAUTION: *Handle phenol with care.*)

e. Stock ammonium solution: Dissolve 381.9 mg anhydrous NH$_4$Cl, dried at 100°C, in water, and dilute to 1000 mL; 1.00 mL = 100 μg N = 122 μg NH$_3$.

f. Standard ammonium solution: Dilute 5.00 mL stock ammonium solution to 1000 mL with water; 1.00 mL = 0.500 μg N = 0.607 μg NH$_3$.

4. Procedure

a. Treatment of sample: To a 10.0-mL sample in a 50-mL beaker, add 1 drop (0.05 mL) MnSO$_4$ solution. Place on a magnetic stirrer and add 0.5 mL hypochlorous acid reagent. Immediately add, a drop at a time, 0.6 mL phenate reagent. Add reagent without delay using a bulb pipet or a buret for convenient delivery. Mark pipet for hypochlorous acid at the 0.5-mL level and deliver the phenate reagent from a pipet or a buret that has been calibrated by counting the number of drops previously found to be equivalent to 0.6 mL. Stir vigorously during addition of reagents. Because color intensity is affected by age of reagents, carry a blank and a standard through the procedure with each batch of samples. Measure absorbance using reagent blank to zero the spectrophotometer. Color formation is complete in 10 min and is stable for at least 24 h. Although the blue color has a maximum absorbance at 630 nm, satisfactory measurements can be made in the 600- to 660-nm region.

b. Preparation of standards: Prepare a calibration curve in the NH_3-N range of 0.1 to 5 μg, treating standards exactly as the sample. Beer's Law governs.

5. Calculation

Calculate ammonia concentration as follows:

mg NH_3-N/L (11.1 mL final volume)

$$= \frac{A \times B}{C \times S} \times \frac{D}{E}$$

where:

A = absorbance of sample,
B = NH_3-N in standard, μg,
C = absorbance of standard,
S = volume of sample used, mL,
D = volume of total distillate collected, mL, including acid absorbent, neutralizing agent, and ammonia-free water added, and
E = volume of distillate used for color development, mL.

The ratio D/E applies only to distilled samples.

6. Precision and Bias

See Section 4500-NH₃.A.4 and Table 4500-NH₃:I.

7. Bibliography

ROSSUM, J.R. & P.A. VILLARRUZ. 1963. Determination of ammonia by the indophenol method. *J. Amer. Water Works Assoc.* 55:657.

WEATHERBURN, M.W. 1967. Phenolhypochlorite reaction for determination of ammonia. *Anal. Chem.* 39:971.

4500-NH₃ E.   Titrimetric Method

1. General Discussion

The titrimetric method is used only on samples that have been carried through preliminary distillation (see Section 4500-NH₃.B). The following table is useful in selecting sample volume for the distillation and titration method.

Ammonia Nitrogen in Sample mg/L	Sample Volume mL
5–10	250
10–20	100
20–50	50.0
50–100	25.0

2. Apparatus

Distillation apparatus: See Section 4500-NH₃.B.2a and b.

3. Reagents

Use ammonia-free water in making all reagents and dilutions.

a. Mixed indicator solution: Dissolve 200 mg methyl red indicator in 100 mL 95% ethyl or isopropyl alcohol. Dissolve 100 mg methylene blue in 50 mL 95% ethyl or isopropyl alcohol. Combine solutions. Prepare monthly.

b. Indicating boric acid solution: Dissolve

20 g H_3BO_3 in ammonia-free distilled water, add 10 mL mixed indicator solution, and dilute to 1 L. Prepare monthly.

c. *Standard sulfuric acid titrant, 0.02N:* Prepare and standardize as directed in Alkalinity, Section 2320B.3c. For greatest accuracy, standardize titrant against an amount of Na_2CO_3 that has been incorporated in the indicating boric acid solution to reproduce the actual conditions of sample titration; 1.00 mL = 280 μg N.

4. Procedure

a. Proceed as described in Section 4500-NH_3.B using indicating boric acid solution as absorbent for the distillate.

b. *Sludge or sediment samples:* Rapidly weigh to within ±1% an amount of wet sample, equivalent to approximately 1 g dry weight, in a weighing bottle or crucible. Wash sample into a 500-mL kjeldahl flask with water and dilute to 250 mL. Proceed as in ¶ 4a but add a piece of paraffin wax to distillation flask and collect only 100 mL distillate.

c. Titrate ammonia in distillate with standard 0.02N H_2SO_4 titrant until indicator turns a pale lavender.

d. *Blank:* Carry a blank through all steps

of the procedure and apply the necessary correction to the results.

5. Calculation

a. *Liquid samples:*

$$\text{mg NH}_3\text{-N/L} = \frac{(A - B) \times 280}{\text{mL sample}}$$

b. *Sludge or sediment samples:*

$$\text{mg NH}_3\text{-N/kg} = \frac{(A - B) \times 280}{\text{g dry wt sample}}$$

where:

A = volume of H_2SO_4 titrated for sample, mL, and

B = volume of H_2SO_4 titrated for blank, mL.

6. Precision and Bias

See Section 4500-NH_3.A.4 and Table 4500-NH_3:I.

7. Bibliography

MEEKER, E.W. & E.C. WAGNER. 1933. Titration of ammonia in the presence of boric acid. *Ind. Eng. Chem.,* Anal. Ed. 5:396.

WAGNER, E.C. 1940. Titration of ammonia in the presence of boric acid. *Ind. Eng. Chem.,* Anal. Ed. 12:711.

4500-NH_3 F.   Ammonia-Selective Electrode Method

1. General Discussion

a. *Principle:* The ammonia-selective electrode uses a hydrophobic gas-permeable membrane to separate the sample solution from an electrode internal solution of ammonium chloride. Dissolved ammonia ($NH_{3(aq)}$ and NH_4^+) is converted to $NH_{3(aq)}$ by raising pH to above 11 with a strong base. $NH_{3(aq)}$ diffuses through the membrane and changes the internal solution pH that is sensed by a pH electrode. The fixed level of chloride in the internal solution is sensed by a chloride ion-selective electrode

that serves as the reference electrode. Potentiometric measurements are made with a pH meter having an expanded millivolt scale or with a specific ion meter.

b. *Scope and application:* This method is applicable to the measurement of 0.03 to 1400 mg NH_3-N/L in potable and surface waters and domestic and industrial wastes. High concentrations of dissolved ions affect the measurement, but color and turbidity do not. Sample distillation is unnecessary. Use standard solutions and samples that have the same temperature and contain about the same total level of dissolved species. The ammonia-selective electrode responds slowly below 1 mg NH_3-N/L; hence, use longer times of electrode immersion (5 to 10 min) to obtain stable readings.

c. Interference: Amines are a positive interference. Mercury and silver interfere by complexing with ammonia.

d. Sample preservation: Do not use $HgCl_2$ as a sample preservative. Refrigerate at 4°C for samples to be analyzed within 24 h. Preserve samples high in organic and nitrogenous matter, and any other samples for a prolonged period, by lowering pH to 2 or less with conc H_2SO_4.

2. Apparatus

a. Electrometer: A pH meter with expanded millivolt scale capable of 0.1 mV resolution between −700 mV and +700 mV or a specific ion meter.

b. Ammonia-selective electrode.

c. Magnetic stirrer, thermally insulated, with TFE-coated stirring bar.

3. Reagents

a. Ammonia-free water: See Section 4500-NH_3.B.3*a.* Use for making all reagents.

b. Sodium hydroxide, 10N: Dissolve 400 g NaOH in 800 mL water. Cool and dilute to 1000 mL with water.

c. Stock ammonium chloride solution: See Section 4500-NH_3.C.3*d.*

d. Standard ammonium chloride solutions: See ¶ 4*a* below.

4. Procedure

a. Preparation of standards: Prepare a series of standard solutions covering the concentrations of 1000, 100, 10, 1, and 0.1 mg NH_3-N/L by making decimal dilutions of stock NH_4Cl solution with water.

b. Electrometer calibration: Place 100 mL of each standard solution in a 150-mL beaker. Immerse electrode in standard of lowest concentration and mix with a magnetic stirrer. Do not stir so rapidly that air bubbles are sucked into the solution because they will become trapped on the electrode membrane. Maintain the same stirring rate and a temperature of about 25°C throughout calibration and testing procedures. Add a sufficient volume of 10N NaOH solution (1 mL usually is sufficient) to raise pH above 11. Keep electrode in solution until a stable millivolt reading is obtained. CAUTION: Check electrode sensing element performance according to manufacturer's instructions to make sure that electrode is operating properly. Do not add NaOH solution before immersing electrode, because ammonia may be lost from a basic solution. Repeat procedure with remaining standards, proceeding from lowest to highest concentration. Wait for at least 5 min before recording millivolts for standards and samples containing ≤ 1 mg NH_3-N/L.

c. Preparation of standard curve: Using semilogarithmic graph paper, plot ammonia concentration in milligrams NH_3-N per liter on the log axis vs. potential in millivolts on the linear axis starting with the lowest concentration at the bottom of the scale. If the electrode is functioning properly a tenfold change of NH_3-N concentration produces a potential change of 59 mV.

d. Calibration of specific ion meter: Refer to manufacturer's instructions and proceed as in ¶s 4*a* and *b.*

e. Measurement of samples: Dilute if necessary to bring NH_3-N concentration to within calibration curve range. Place 100 mL sample in 150-mL beaker and follow procedure in ¶ 4*b* above. Record volume of 10N NaOH added in excess of 1 mL. Read NH_3-N concentration from standard curve.

5. Calculation

$$\text{mg } NH_3\text{-N/L} = A \times B \times \left[\frac{101 + C}{101}\right]$$

---

*Orion Model 95-10 or 95-12, EIL Model 8002-2, Beckman Model 39565, or equivalent.

where:

A = dilution factor,

B = concentration of NH_3-N/L, mg/L, from calibration curve, and

C = volume of added $10N$ NaOH in excess of 1 mL, mL.

6. Precision and Bias

See Section 4500-NH_3.A.4.

7. Bibliography

BANWART, W.L., J.M. BREMNER & M.A. TA-BATABAI. 1972. Determination of ammonium in soil extracts and water samples by an ammonia electrode. *Comm. Soil Sci. Plant Anal.* 3:449.

MIDGLEY, C. & K. TERRANCE. 1972. The determination of ammonia in condensed steam and boiler feed-water with a potentiometric ammonia probe. *Analyst* 97:626.

BOOTH, R.L. & R.F. THOMAS. 1973. Selective electrode determination of ammonia in water and wastes. *Environ. Sci. Technol.* 7:523.

U.S. ENVIRONMENTAL PROTECTION AGENCY. 1979. Methods for Chemical Analysis of Water and Wastes. EPA-600/4-79-020, National Environmental Research Center, Cincinnati, Ohio.

4500-NH_3 G. Ammonia-Selective Electrode Method Using Known Addition

1. General Discussion

a. Principle: When a linear relationship exists between concentration and response, known addition is convenient for measuring occasional samples because no calibration is needed. Because an accurate measurement requires that the concentration approximately double as a result of the addition, sample concentration must be known within a factor of three. Total concentration of ammonia can be measured in the absence of complexing agents down to 0.8 mg NH_3-N/L or in the presence of a large excess (50 to 100 times) of complexing agent. Known addition is a convenient check on the results of direct measurement.

b. See Section 4500-NH_3.F.1 for further discussion.

2. Apparatus

Use apparatus specified in Section 4500-NH_3.F.2.

3. Reagents

Use reagents specified in Section 4500-NH_3.F.3.

Add standard ammonium chloride solution approximately 10 times as concentrated as samples being measured.

4. Procedure

a. Dilute 1000 mg/L stock solution to make a standard solution about 10 times as concentrated as the sample concentrate.

b. Add 1 mL $10N$ NaOH to each 100 mL sample and immediately immerse electrode. When checking a direct measurement, leave electrode in 100 mL of sample solution. Use magnetic stirring throughout. Measure mV reading and record as E_1.

c. Pipet 10 mL of standard solution into sample. Thoroughly stir and immediately record new mV reading as E_2.

5. Calculation

a. $\Delta E = E_1 - E_2$.

b. From Table 4500-NH_3:IV find the concentration ratio, Q, corresponding to change in potential, ΔE. To determine original total sample concentration, multiply Q by the concentration of the added standard:

$$C_0 = Q C_s$$

TABLE 4500-NH₃:IV.* VALUES OF Q VS. ΔE (59 MV SLOPE) FOR 10% VOLUME CHANGE

ΔE	Q	ΔE	Q	ΔE	Q	ΔE	Q	ΔE	Q
5.0	0.297	9.0	0.178	16.0	0.0952	24.0	0.0556	32.0	0.0354
5.1	0.293	9.1	0.176	16.2	0.0938	24.2	0.0549	32.2	0.0351
5.2	0.288	9.2	0.174	16.4	0.0924	24.4	0.0543	32.4	0.0347
5.3	0.284	9.3	0.173	16.6	0.0910	24.6	0.0536	32.6	0.0343
5.4	0.280	9.4	0.171	16.8	0.0897	24.8	0.0530	32.8	0.0340
5.5	0.276	9.5	0.169	17.0	0.0884	25.0	0.0523	33.0	0.0335
5.6	0.272	9.6	0.167	17.2	0.0871	25.2	0.0517	33.2	0.0333
5.7	0.268	9.7	0.165	17.4	0.0858	25.4	0.0511	33.4	0.0329
5.8	0.264	9.8	0.164	17.6	0.0846	25.6	0.0505	33.6	0.0326
5.9	0.260	9.9	0.162	17.8	0.0834	25.8	0.0499	33.8	0.0336
6.0	0.257	10.0	0.160	18.0	0.0822	26.0	0.0494	34.0	0.0319
6.1	0.253	10.2	0.157	18.2	0.0811	26.2	0.0488	34.2	0.0316
6.2	0.250	10.4	0.154	18.4	0.0799	26.4	0.0482	34.4	0.0313
6.3	0.247	10.6	0.151	18.6	0.0788	26.6	0.0477	34.6	0.0310
6.4	0.243	10.8	0.148	18.8	0.0777	26.8	0.0471	34.8	0.0307
6.5	0.240	11.0	0.145	19.0	0.0767	27.0	0.0466	35.0	0.0304
6.6	0.237	11.2	0.143	19.2	0.0756	27.2	0.0461	36.0	0.0289
6.7	0.234	11.4	0.140	19.4	0.0746	27.4	0.0456	37.0	0.0275
6.8	0.231	11.6	0.137	19.6	0.0736	27.6	0.0450	38.0	0.0261
6.9	0.228	11.8	0.135	19.8	0.0726	27.8	0.0445	39.0	0.0249
7.0	0.225	12.0	0.133	20.0	0.0716	28.0	0.0440	40.0	0.0237
7.1	0.222	12.2	0.130	20.2	0.0707	28.2	0.0435	41.0	0.0226
7.2	0.219	12.4	0.128	20.4	0.0698	28.4	0.0431	42.0	0.0216
7.3	0.217	12.6	0.126	20.6	0.0689	28.6	0.0426	43.0	0.0206
7.4	0.214	12.8	0.123	20.8	0.0680	28.8	0.0421	44.0	0.0196
7.5	0.212	13.0	0.121	21.0	0.0671	29.0	0.0417	45.0	0.0187
7.6	0.209	13.2	0.119	21.2	0.0662	29.2	0.0412	46.0	0.0179
7.7	0.207	13.4	0.117	21.4	0.0654	29.4	0.0408	47.0	0.0171
7.8	0.204	13.6	0.115	21.6	0.0645	29.6	0.0403	48.0	0.0163
7.9	0.202	13.8	0.113	21.8	0.0637	29.8	0.0399	49.0	0.0156
8.0	0.199	14.0	0.112	22.0	0.0629	30.0	0.0394	50.0	0.0149
8.1	0.197	14.2	0.110	22.2	0.0621	30.2	0.0390	51.0	0.0143
8.2	0.195	14.4	0.108	22.4	0.0613	30.4	0.0386	52.0	0.0137
8.3	0.193	14.6	0.106	22.6	0.0606	30.6	0.0382	53.0	0.0131
8.4	0.190	14.8	0.105	22.8	0.0598	30.8	0.0378	54.0	0.0125
8.5	0.188	15.0	0.103	23.0	0.0591	31.0	0.0374	55.0	0.0120
8.6	0.186	15.2	0.1013	23.2	0.0584	31.2	0.0370	56.0	0.0115
8.7	0.184	15.4	0.0997	23.4	0.0576	31.4	0.0366	57.0	0.0110
8.8	0.182	15.6	0.0982	23.6	0.0569	31.6	0.0362	58.0	0.0105
8.9	0.180	15.8	0.0967	23.8	0.0563	31.8	0.0358	59.0	0.0101

*Orion Research Inc. Instruction Manual, Ammonia Electrode, Mode 95-10, Cambridge, Mass.

where:

C_0 = total sample concentration, mg/L,
Q = reading from known-addition table, and
C_s = concentration of added standard, mg/L.

c. To check a direct measurement, compare results of the two methods. If they agree within ±4%, the measurements probably are good. If the known-addition result is much larger than the direct measurement, the sample may contain complexing agents.

6. Precision and Bias

In 38 water samples analyzed by both the phenate and the known-addition ammonia-selective electrode method, the electrode method yielded a mean recovery of 102% of the values obtained by the phenate method when the NH_3-N concentrations varied between 0.30 and 0.78 mg/L. In 57 wastewater samples similarly compared, the electrode method yielded a mean recovery of 108% of the values obtained by the phenate method using distillation when the NH_3-N concentrations varied between 10.2 and 34.7 mg N/L. In 20 instances in which two to four replicates of these samples were analyzed, the mean standard deviation was 1.32 mg N/L. In three measurements at a sewer outfall, distillation did not change statistically the value obtained by the electrode method. In 12 studies using standards in the 2.5- to 30-mg N/L range, average recovery by the phenate method was 97% and by the electrode method 101%.

4500-NH$_3$ H.   Automated Phenate Method

1. General Discussion

a. *Principle:* Alkaline phenol and hypochlorite react with ammonia to form indophenol blue that is proportional to the ammonia concentration. The blue color formed is intensified with sodium nitroprusside.

b. *Interferences:* Seawater contains calcium and magnesium ions in sufficient concentrations to cause precipitation during analysis. Adding EDTA and sodium potassium tartrate reduces the problem. Eliminate any marked variation in acidity or alkalinity among samples because intensity of measured color is pH-dependent. Likewise, insure that pH of wash water and standard ammonia solutions approximates that of sample. For example, if sample has been preserved with 0.8 mL conc H_2SO_4/L, include 0.8 mL conc H_2SO_4/L in wash water and standards. Mercuric chloride used as a preservative gives a negative interference by complexing with ammonia.

Overcome this effect by adding a comparable amount of $HgCl_2$ to the ammonia standards. Remove interfering turbidity by filtration. Color in the samples that absorbs in the photometric range used for analysis interferes.

c. *Application:* Ammonia nitrogen can be determined in potable, surface, and saline waters as well as domestic and industrial wastewaters over a range of 0.02 to 2.0 mg/L when photometric measurement is made at 630 to 660 nm in a 15- or 50-mm tubular flow cell. Determine higher concentrations by diluting the sample.

2. Apparatus

Automated analytical equipment. The required continuous-flow analytical instrument* consists of the interchangeable components shown in Figure 4500-NH$_3$:1.

---

*Technicon™ Autoanalyzer™ II or equivalent.

A 50-mm flow cell and a filter for use at 630 nm may be used.

3. Reagents

a. Ammonia-free distilled water: See Section 4500-NH₃.B.3*a.* Use for preparing all reagents and dilutions.

b. Sulfuric acid, H₂SO₄, 5*N*, air scrubber solution: Carefully add 139 mL conc H₂SO₄ to approximately 500 mL water, cool to room temperature, and dilute to 1 L.

c. Sodium phenate solution: In a 1-L erlenmeyer flask, dissolve 83 g phenol in 500 mL water. In small increments and with agitation, cautiously add 32 g NaOH. Cool flask under running water and dilute to 1 L.

d. Sodium hypochlorite solution: Dilute 250 mL bleach solution containing 5.25% NaOCl to 500 mL with water.

e. EDTA reagent: Dissolve 50 g disodium ethylenediamine tetraacetate and approximately six pellets NaOH in 1 L water. For salt-water samples where EDTA reagent does not prevent precipitation of cations, use sodium potassium tartrate solution prepared as follows:

Sodium potassium tartrate solution: To 900 mL water add 100 g NaKC₄H₄O₆· 4H₂O, two pellets NaOH, and a few boiling chips, and boil gently for 45 min. Cover, cool, and dilute to 1 L. Adjust pH to 5.2 ± 0.05 with H₂SO₄. Let settle overnight in a cool place and filter to remove precipitate. Add 0.5 mL polyoxyethylene 23 lauryl ether† solution and store in stoppered bottle.

---

†Brij-35, available from ICI Americas, Wilmington, Del. or Technicon Instrument Corp., Tarrytown, N.Y. 10591.

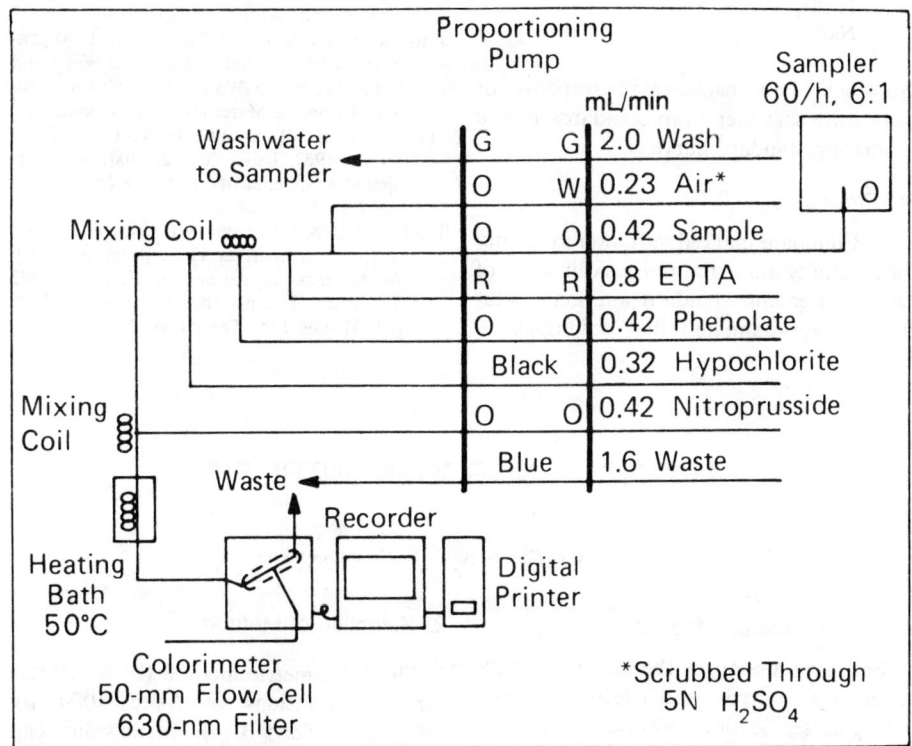

Figure 4500-NH₃:1. Ammonia manifold.

f. Sodium nitroprusside solution: Dissolve 0.5 g $Na_2(NO)Fe(CN)_5 \cdot 2H_2O$ in 1 L water.

g. Ammonia standard solutions: See Sections 4500-NH$_3$.D.3*e* and *f*. Use standard ammonia solution and water to prepare the calibration curve in the appropriate ammonia concentration range. To analyze saline waters use substitute ocean water of the following composition to prepare calibration standards:

Constituent	Concentration g/L
NaCl	24.53
MgCl$_2$	5.20
CaCl$_2$	1.16
KCl	0.70
SrCl$_2$	0.03
Na$_2$SO$_4$	4.09
NaHCO$_3$	0.20
KBr	0.10
H$_3$BO$_3$	0.03
NaF	0.003

Subtract blank background response of substitute seawater from standards before preparing standard curve.

4. Procedure

a. Eliminate marked variation in acidity or alkalinity among samples. Adjust pH of wash water and standard ammonia solutions to approximately that of sample.

b. Set up manifold and complete system as shown in Figure 4500-NH$_3$:1.

c. Obtain a stable base line with all reagents, feeding water through sample line.

d. Use a 60/h, 6:1 cam with a common wash.

5. Calculation

Prepare standard curves by plotting peak heights of standards processed through the manifold against NH$_3$-N concentrations in standards. Compute sample NH$_3$-N concentration by comparing sample peak height with standard curve.

6. Precision and Bias

See Section 4500-NH$_3$.A.4.

7. Bibliography

HILLER, A. & D. VAN SLYKE. 1933. Determination of ammonia in blood. *J. Biol. Chem.* 102:499.

FIORE, J. & J.E. O'BRIEN. 1962. Ammonia determination by automatic analysis. *Wastes Eng.* 33:352.

AMERICAN SOCIETY FOR TESTING AND MATERIALS. 1966. Manual on Industrial Water and Industrial Waste Water, 2nd ed. American Soc. Testing & Materials, Philadelphia, Pa.

O'CONNOR, B., R. DOBBS, B. VILLIERS & R. DEAN. 1967. Laboratory distillation of municipal waste effluents. *J. Water Pollut. Control Fed.* 39:25.

BOOTH, R.L. & L.B. LOBRING. 1973. Evaluation of the AutoAnalyzer II: A progress report. *In* Advances in Automated Analysis: 1972 Technicon International Congress, Vol. 8, p.7. Mediad Inc., Tarrytown, N.Y.

4500-NO$_2^-$ NITROGEN (NITRITE)*

4500-NO$_2^-$ A.   Introduction

1. Occurrence and Significance

For a discussion of the chemical characteristics, sources, and effects of nitrite nitrogen, see Section 4500-N.

*Approved by Standard Methods Committee, 1988.

2. Selection of Method

The colorimetric method (B) is suitable for concentrations of 5 to 1000 μg NO$_2^-$-N/L (See ¶ B.1*a*). Nitrite values can be obtained by the automated method given in Section 4500-NO$_3^-$.E with the Cu-Cd reduction step omitted. Additionally, nitrite nitrogen can be determined by ion chromatography (C).

4500–NO$_2^-$ B.   Colorimetric Method

1. General Discussion

a. Principle: Nitrite (NO$_2^-$) is determined through formation of a reddish purple azo dye produced at pH 2.0 to 2.5 by coupling diazotized sulfanilamide with *N*-(1-naphthyl)-ethylenediamine dihydrochloride (NED dihydrochloride). The applicable range of the method for spectrophotometric measurements is 10 to 1000 µg NO$_2^-$-N/L. Photometric measurements can be made in the range 5 to 50 µg N/L if a 5-cm light path and a green color filter are used. The color system obeys Beer's law up to 180 µg N/ L with a 1-cm light path at 543 nm. Higher NO$_2^-$ concentrations can be determined by diluting a sample.

b. Interferences: Chemical incompatibility makes it unlikely that NO$_2^-$, free chlorine, and nitrogen trichloride (NCl$_3$) will coexist. NCl$_3$ imparts a false red color when color reagent is added. The following ions interfere because of precipitation under test conditions and should be absent: Sb^{3+}, Au^{3+}, Bi^{3+}, Fe^{3+}, Pb^{2+}, Hg^{2+}, Ag$^+$, chloroplatinate (PtCl$_6^{2-}$), and metavanadate (VO$_3^{2-}$). Cupric ion may cause low results by catalyzing decomposition of the diazonium salt. Colored ions that alter the color system also should be absent. Remove suspended solids by filtration.

c. Storage of sample: Never use acid preservation for samples to be analyzed for NO$_2^-$. Make the determination promptly on fresh samples to prevent bacterial conversion of NO$_2^-$ to NO$_3^-$ or NH$_3$. For short-term preservation for 1 to 2 d, freeze at −20°C or store at 4°C.

2. Apparatus

Colorimetric equipment: One of the following is required:

a. Spectrophotometer, for use at 543 nm, providing a light path of 1 cm or longer.

b. Filter photometer, providing a light path of 1 cm or longer and equipped with a green filter having maximum transmittance near 540 nm.

3. Reagents

a. Nitrite-free water: If it is not known that the distilled or demineralized water is free from NO$_2^-$, use either of the following procedures to prepare nitrite-free water:

1) Add to 1 L distilled water one small crystal each of KMnO$_4$ and either Ba(OH)$_2$ or Ca(OH)$_2$. Redistill in an all-borosilicate-glass apparatus and discard the initial 50 mL of distillate. Collect the distillate fraction that is free of permanganate; a red color with DPD reagent (Section 4500-Cl.F.2b), indicates the presence of permanganate.

2) Add 1 mL conc H$_2$SO$_4$ and 0.2 mL MnSO$_4$ solution (36.4 g MnSO$_4$ H$_2$O/100 mL distilled water) to each 1 L distilled water, and make pink with 1 to 3 mL KMnO4 solution (400 mg KMnO$_4$/L distilled water). Redistill as described in the preceding paragraph.

Use nitrite-free water in making all reagents and dilutions.

b. Color reagent: To 800 mL water add 100 mL 85% phosphoric acid and 10 g sulfanilamide. After dissolving sulfanilamide completely, add 1 g *N*-(1-naphthyl)-

ethylenediamine dihydrochloride. Mix to dissolve, then dilute to 1 L with water. Solution is stable for about a month when stored in a dark bottle in refrigerator.

c. Sodium oxalate, 0.025M (0.05N): Dissolve 3.350 g $Na_2C_2O_4$, primary standard grade, in water and dilute to 1000 mL.

d. Ferrous ammonium sulfate, 0.05M (0.05N): Dissolve 19.607 g $Fe(NH_4)_2(SO_4)_2 \cdot 6H_2O$ plus 20 mL conc H_2SO_4 in water and dilute to 1000 mL. Standardize as in Section 5220B.3d.

e. Stock nitrite solution: Commercial reagent-grade $NaNO_2$ assays at less than 99%. Because NO_2^- is oxidized readily in the presence of moisture, use a fresh bottle of reagent for preparing the stock solution and keep bottles tightly stoppered against the free access of air when not in use. To determine $NaNO_2$ content, add a known excess of standard 0.05M (0.05N) $KMnO_4$ solution prepared (with 8.0 g $KMnO_4$) and standardized as described in 3500-Ca.C.3i, discharge permanganate color with a known quantity of standard reductant such as 0.025M $Na_2C_2O_4$ or 0.05M $Fe(NH_4)_2(SO_4)_2$, and back-titrate with standard permanganate solution.

1) Preparation of stock solution—Dissolve 1.232 g $NaNO_2$ in water and dilute to 1000 mL; 1.00 mL = 250 μg N. Preserve with 1 mL $CHCl_3$.

2) Standardization of stock nitrite solution—Pipet, in order, 50.00 mL standard 0.05M $KMnO_4$, 5 mL conc H_2SO_4, and 50.00 mL stock NO_2^- solution into a glass-stoppered flask or bottle. Submerge pipet tip well below surface of permanganate-acid solution while adding stock NO_2^- solution. Shake gently and warm to 70 to 80°C on a hot plate. Discharge permanganate color by adding sufficient 10-mL portions of standard 0.025M $Na_2C_2O_4$. Titrate excess $Na_2C_2O_4$ with 0.05M $KMnO_4$ to the faint pink end point. Carry a water blank through the entire procedure and make the necessary corrections in the final calculation as shown in the equation below.

If standard 0.05M ferrous ammonium sulfate solution is substituted for $Na_2C_2O_4$, omit heating and extend reaction period between $KMnO_4$ and Fe^{2+} to 5 min before making final $KMnO_4$ titration.

Calculate NO_2^--N content of stock solution by the following equation:

$$A = \frac{[(B \times C) - (D \times E)] \times 7}{F}$$

where:

A = mg NO_2^--N/mL in stock $NaNO_2$ solution,
B = total mL standard $KMnO_4$ used,
C = normality of standard $KMnO_4$,
D = total mL standard reductant added,
E = normality of standard reductant, and
F = mL stock $NaNO_2$ solution taken for titration.

Each 1.00 mL 0.05M $KMnO_4$ consumed by the $NaNO_2$ solution corresponds to 1750 g NO_2^--N.

f. Intermediate nitrite solution: Calculate the volume, G, of stock NO_2^- solution required for the intermediate NO_2^- solution from $G = 12.5/A$. Dilute the volume G (approximately 50 mL) to 250 mL with water; 1.00 mL = 50.0 μg N. Prepare daily.

g. Standard nitrite solution: Dilute 10.00 mL intermediate NO_2^- solution to 1000 mL with water; 1.00 mL = 0.500 μg N. Prepare daily.

4. Procedure

a. Removal of suspended solids: If sample contains suspended solids, filter through a 0.45-μm-pore-diam membrane filter.

b. Color development: If sample pH is not between 5 and 9, adjust to that range with 1N HCl or NH_4OH as required. To 50.0 mL sample, or to a portion diluted to 50.0 mL, add 2 mL color reagent and mix.

c. Photometric measurement: Between 10 min and 2 h after adding color reagent to samples and standards, measure absorbance at 543 nm. As a guide use the following light paths for the indicated NO_2^--N concentrations:

Light Path Length cm	NO_2^--N $\mu g/L$
1	2–25
5	2–6
10	<2

5. Calculation

Prepare a standard curve by plotting absorbance of standards against NO_2^--N concentration. Compute sample concentration directly from curve.

6. Precision and Bias

In a single laboratory using wastewater samples at concentrations of 0.04, 0.24,

0.55, and 1.04 mg NO^-_3 + NO_2^--N/L, the standard deviations were ±0.005, ± 0.004, ± 0.005, and ± 0.01, respectively. In a single laboratory using wastewater samples at concentrations of 0.24, 0.55, and 1.05 mg NO_3^- + NO_2^--N/L, the recoveries were 100%, 102%, and 100%, respectively.[1]

7. Reference

1. U.S. ENVIRONMENTAL PROTECTION AGENCY. 1979. Methods for Chemical Analysis of Water and Wastes. Method 353.3 U.S. Environmental Protection Agency, Washington, D.C.

8. Bibliography

BOLTZ, D.F., ed. 1958. Colorimetric Determination of Nonmetals. Interscience Publishers, New York, N.Y.

NYDAHL, F. 1976. On the optimum conditions for the reduction of nitrate by cadmium. *Talanta* 23:349.

4500-NO₂⁻ C. Ion Chromatographic Method

See Section 4110.

4500-NO₃⁻ NITROGEN (NITRATE)*

4500-NO₃⁻ A. Introduction

1. Selection of Method

Determination of nitrate (NO_3^-) is difficult because of the relatively complex procedures required, the high probability that interfering constituents will be present, and

the limited concentration ranges of the various techniques.

An ultraviolet (UV) technique (Method B) that measures the absorbance of NO_3^- at 220 nm is suitable for screening uncontaminated water (low in organic matter).

Screen a sample; if necessary, then select a method suitable for its concentration

---

* Approved by Standard Methods Committee, 1988.

range and probable interferences. Nitrate may be determined by ion chromatography (C). Applicable ranges for other methods are: nitrate electrode method (D), 0.14 to 1400 mg NO_3^--N/L; cadmium reduction method (E), 0.01 to 1.0 mg NO_3^--N/L; titanous chloride method (G), 0.01 to 10 mg NO_3^--N/L; hydrazine reduction method (H), 0.01 to 10 mg NO_3^--N/L; automated cadmium reduction method (F), 0.5 to 10 mg NO_3^--N/L. For higher NO_3^--N concentrations, dilute into the range of the selected method.

Colorimetric methods require an optically clear sample. Filter turbid sample through 0.45-μm-pore-diam membrane filter. Test filters for nitrate contamination.

2. Storage of Samples

Start NO_3^- determinations promptly after sampling. If storage is necessary, store for up to 24 h at 4°C; for longer storage, preserve with 2 mL conc H_2SO_4/L and store at 4°C. NOTE: When sample is preserved with acid, NO_3^- and NO_2^- cannot be determined as individual species.

4500-NO_3^- B.   Ultraviolet Spectrophotometric Screening Method

1. General Discussion

a. Principle: Use this technique only for screening samples that have low organic matter contents, i.e., uncontaminated natural waters and potable water supplies. The NO_3^- calibration curve follows Beer's law up to 11 mg N/L.

Measurement of UV absorption at 220 nm enables rapid determination of NO_3^-. Because dissolved organic matter also may absorb at 220 nm and NO_3^- does not absorb at 275 nm, a second measurement made at 275 nm may be used to correct the NO_3^- value. The extent of this empirical correction is related to the nature and concentration of organic matter and may vary from one water to another. Consequently, this method is not recommended if a significant correction for organic matter absorbance is required, although it may be useful in monitoring NO_3^- levels within a water body with a constant type of organic matter. Correction factors for organic matter absorbance can be established by the method of additions in combination with analysis of the original NO_3^- content by another method. Sample filtration is in-

tended to remove possible interference from suspended particles. Acidification with $1N$ HCl is designed to prevent interference from hydroxide or carbonate concentrations up to 1000 mg $CaCO_3$/L. Chloride has no effect on the determination.

b. Interference: Dissolved organic matter, surfactants, NO_2^-, and Cr^{6+} interfere. Various inorganic ions not normally found in natural water, such as chlorite and chlorate, may interfere. Inorganic substances can be compensated for by independent analysis of their concentrations and preparation of individual correction curves. For turbid samples, see ¶ A.1.

2. Apparatus

Spectrophotometer, for use at 220 nm and 275 nm with matched silica cells of 1-cm or longer light path.

3. Reagents

a. Nitrate-free water: Use redistilled or distilled, deionized water of highest purity to prepare all solutions and dilutions.

b. Stock nitrate solution: Dry potassium

nitrate (KNO₃) in an oven at 105°C for 24 h. Dissolve 0.7218 g in water and dilute to 1000 mL; 1.00 mL $= 100 \mu g$ NO₃⁻-N. Preserve with 2 mL CHCl₃/L. This solution is stable for at least 6 months.

c. *Intermediate nitrate solution:* Dilute 100 mL stock nitrate solution to 1000 mL with water; 1.00 mL $= 10.0 \mu g$ NO₃⁻-N. Preserve with 2 mL CHCl₃/L. This solution is stable for 6 months.

d. *Hydrochloric acid solution,* HCl, 1N.

4. Procedure

a. *Treatment of sample:* To 50 mL clear sample, filtered if necessary, add 1 mL HCl solution and mix thoroughly.

b. *Preparation of standard curve:* Prepare NO₃⁻ calibration standards in the range 0 to 7 mg NO₃⁻-N/L by diluting to 50 mL the following volumes of intermediate nitrate solution: 0, 1.00, 2.00, 4.00, 7.00 . . . 35.0 mL. Treat NO₃⁻ standards in same manner as samples.

c. *Spectrophotometric measurement:* Read absorbance or transmittance against redistilled water set at zero absorbance or 100% transmittance. Use a wavelength of 220 nm to obtain NO₃⁻ reading and a wavelength of 275 nm to determine interference due to dissolved organic matter.

5. Calculation

For samples and standards, subtract two times the absorbance reading at 275 nm from the reading at 220 nm to obtain absorbance due to NO₃⁻. Construct a standard curve by plotting absorbance due to NO₃⁻ against NO₃⁻-N concentration of standard. Using corrected sample absorbances, obtain sample concentrations directly from standard curve. NOTE: If correction value is more than 10% of the reading at 220 nm, do not use this method.

6. Bibliography

HOATHER, R.C. & R.F. RACKMAN. 1959. Oxidized nitrogen and sewage effluents observed by ultraviolet spectrophotometry. *Analyst* 84:549.

GOLDMAN, E. & R. JACOBS. 1961. Determination of nitrates by ultraviolet absorption. *J. Amer. Water Works Assoc.* 53:187.

ARMSTRONG, F.A.J. 1963. Determination of nitrate in water by ultraviolet spectrophotometry. *Anal. Chem.* 35:1292.

NAVONE, R. 1964. Proposed method for nitrate in potable waters. *J. Amer. Water Works Assoc.* 56:781.

4500-NO₃⁻ C. Ion Chromatographic Method

See Section 4110.

4500-NO₃⁻ D. Nitrate Electrode Method

1. General Discussion

a. *Principle:* The NO₃⁻ ion electrode is a selective sensor that develops a potential across a thin, porous, inert membrane that holds in place a water-immiscible liquid ion exchanger. The electrode responds to NO₃⁻ ion activity between about 10^{-5} and 10^{-1} M (0.14 to 1400 mg NO₃⁻-N/L). The lower limit of detection is determined by the small but finite solubility of the liquid ion exchanger.

b. *Interferences:* Chloride and bicarbonate ions interfere when their weight ratios to NO_3^--N are > 10 or > 5, respectively. Ions that are potential interferences but do not normally occur at significant levels in potable waters are NO_2^-, CN^-, S^{2-}, Br^-, I^-, ClO_3^-, and ClO_4^-. Although the electrodes function satisfactorily in buffers over the range pH 3 to 9, erratic responses have been noted where pH is not held constant. Because the electrode responds to NO_3^- activity rather than concentration, ionic strength must be constant in all samples and standards. Minimize these problems by using a buffer solution containing Ag_2SO_4 to remove Cl^-, Br^-, I^-, S^{2-}, and CN^-, sulfamic acid to remove NO_2^-, a buffer at pH 3 to eliminate HCO_3^- and to maintain a constant pH and ionic strength, and $Al_2(SO_4)_3$ to complex organic acids.

2. Apparatus

a. pH meter, expanded-scale or digital, capable of 0.1 mV resolution.

*b. Double-junction reference electrode.** Fill outer chamber with $(NH_4)_2SO_4$ solution.

c. Nitrate ion electrode:† Carefully follow manufacturer's instructions regarding care and storage.

d. Magnetic stirrer: TFE-coated stirring bar.

3. Reagents

a. Nitrate-free water: Prepare as described in ¶ B.3a. Use for all solutions and dilutions.

b. Stock nitrate solution: Prepare as described in ¶ B.3b.

c. Standard nitrate solutions: Dilute 1.0, 10, and 50 mL stock nitrate solution to 100 mL with water to obtain standard solutions of 1.0, 10, and 50 mg NO_3^--N/L, respectively.

d. Buffer solution: Dissolve 17.32 g

---

* Orion Model 90-02, or equivalent.
† Orion Model 93-07, Corning Model 476134, or equivalent.

$Al_2(SO_4)_3 \cdot 18H_2O$, 3.43 g Ag_2SO_4, 1.28 g H_3BO_3, and 2.52 g sulfamic acid (H_2NSO_3H), in about 800 mL water. Adjust to pH 3.0 by slowly adding 0.10N NaOH. Dilute to 1000 mL and store in a dark glass bottle.

e. Sodium hydroxide, NaOH, 0.1N.

f. Reference electrode filling solution: Dissolve 0.53 g $(NH_4)_2SO_4$ in water and dilute to 100 mL.

4. Procedure

a. Preparation of calibration curve: Transfer 10 mL of 1 mg NO_3^--N/L standard to a 50-mL beaker, add 10 mL buffer, and stir with a magnetic stirrer. Immerse tips of electrodes and record millivolt reading when stable (after about 1 min). Remove electrodes, rinse, and blot dry. Repeat for 10-mg NO_3^--N/L and 50-mg NO_3^--N/L standards. Plot potential measurements against NO_3^--N concentration on semilogarithmic graph paper, with NO_3^--N concentration on the logarithmic axis (abscissa) and potential (in millivolts) on the linear axis (ordinate). A straight line with a slope of $+ 57 \pm 3$ mV/decade at 25°C should result. Recalibrate electrodes several times daily by checking potential reading of the 10 mg NO_3^--N standard and adjusting the calibration control until the reading plotted on the calibration curve is displayed again.

b. Measurement of sample: Transfer 10 mL sample to a 50-mL beaker, add 10 mL buffer solution, and stir (for about 1 min) with a magnetic stirrer. Measure standards and samples at about the same temperature. Immerse electrode tips in sample and record potential reading when stable (after about 1 min). Read concentration from calibration curve.

5. Precision

Over the range of the method, precision of ± 0.4 mV, corresponding to 2.5% in concentration, is expected.

6. Bibliography

LANGMUIR, D. & R.I. JACOBSON. 1970. Specific ion electrode determination of nitrate in some fresh waters and sewage effluents. *Environ. Sci. Technol.* 4:835.

KEENEY, D.R., B.H. BYRNES & J.J. GENSON. 1970. Determination of nitrate in waters with nitrate-selective ion electrode. *Analyst* 95:383.

MILHAM, P.J., A.S. AWAD, R.E. PAUL & J.H. BULL. 1970. Analysis of plants, soils and waters for nitrate by using an ion-selective electrode. *Analyst* 95:751.

SYNNOTT, J.C., S.J. WEST & J.W. ROSS. 1984. Comparison of ion-selective electrode and gas-sensing electrode technique for measurement of nitrate in environmental samples. *In* Pawlowski et al., eds., Studies in Environmental Science, No. 23, Chemistry for Protection of the Environment, Elsevier Press, New York, N.Y.

4500-NO₃⁻ E.   Cadmium Reduction Method

1. General Discussion

a. Principle: NO_3^- is reduced almost quantitatively to nitrite (NO_2^-) in the presence of cadmium (Cd). This method uses commercially available Cd granules treated with copper sulfate ($CuSO_4$) and packed in a glass column.

The NO_2^- produced thus is determined by diazotizing with sulfanilamide and coupling with N-(1-naphthyl)-ethylenediamine dihydrochloride to form a highly colored azo dye that is measured colorimetrically. A correction may be made for any NO_2^- present in the sample by analyzing without the reduction step. The applicable range of this method is 0.01 to 1.0 mg NO_3^--N/L. The method is recommended especially for NO_3^- levels below 0.1 mg N/L where other methods lack adequate sensitivity.

b. Interferences: Suspended matter in the column will restrict sample flow. For turbid samples, see ¶ A.1. Concentrations of iron, copper, or other metals above several milligrams per liter lower reduction efficiency. Add EDTA to samples to eliminate this interference. Oil and grease will coat the Cd surface. Remove by pre-extraction with an organic solvent (see Section 5520). Residual chlorine can interfere by oxidizing the Cd column, reducing its efficiency. Check samples for residual chlorine (see DPD methods in Section 4500-Cl). Remove residual chlorine by adding sodium thiosulfate ($Na_2S_2O_3$) solution [Section 4500-NH₃.B.3*d*4)]. Sample color that absorbs at about 540 nm interferes.

2. Apparatus

a. Reduction column: Purchase or construct the column* (Figure 4500-NO₃⁻:1) from a 100-mL volumetric pipet by removing the top portion. The column also can be constructed from two pieces of tubing joined end to end: join a 10-cm length of 3-cm-ID tubing to a 25-cm length of 3.5-mm-ID tubing. Add a TFE stopcock with metering valve[1] to control flow rate.

b. Colorimetric equipment: One of the following is required:

1) *Spectrophotometer,* for use at 543 nm, providing a light path of 1 cm or longer.

2) *Filter photometer,* with light path of 1 cm or longer and equipped with a filter having maximum transmittance near 540 nm.

3. Reagents

a. Nitrate-free water: See ¶ B.3*a*. The absorbance of a reagent blank prepared with this water should not exceed 0.01. Use for all solutions and dilutions.

b. Copper-cadmium granules: Wash 25

---

* Tudor Scientific Glass Co., 555 Edgefield Road, Belvedere, S.C. 29841, Cat. TP-1730, or equivalent.

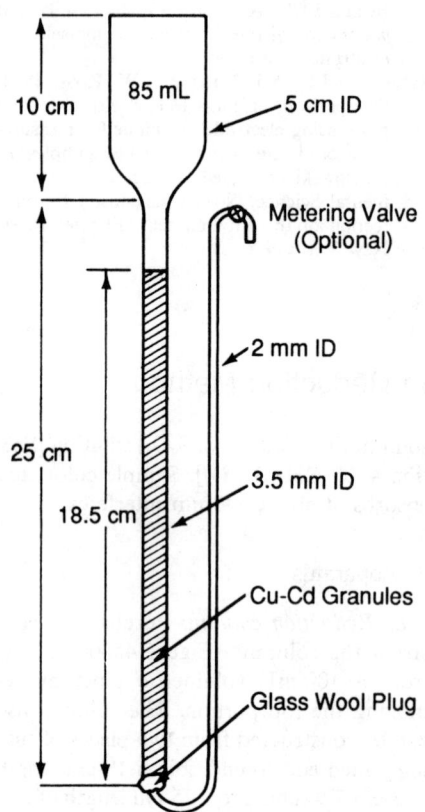

Figure 4500-NO$_3$⁻:1. Reduction column.

g new or used 40- to 60-mesh Cd granules†
with 6*N* HCl and rinse with water. Swirl
Cd with 100 mL 2% CuSO$_4$ solution for
5 min or until blue color partially fades.
Decant and repeat with fresh CuSO$_4$ until
a brown colloidal precipitate begins to develop. Gently flush with water to remove
all precipitated Cu.

c. Color reagent: Prepare as directed in
Section 4500-NO$_2$⁻B.3*b.*

d. Ammonium chloride-EDTA solution:
Dissolve 13 g NH$_4$Cl and 1.7 g disodium
ethylenediamine tetraacetate in 900 mL
water. Adjust to pH 8.5 with conc NH$_4$OH
and dilute to 1 L.

† EM Laboratories, Inc., 500 Exec. Blvd., Elmsford,
N.Y., Cat. 2001, or equivalent.

*e. Dilute ammonium chloride-EDTA so-
lution:* Dilute 300 mL NH$_4$Cl-EDTA solution to 500 mL with water.

f. Hydrochloric acid, HCl, 6*N.*

g. Copper sulfate solution, 2%: Dissolve
20 g CuSO$_4$.5H$_2$O in 500 mL water and
dilute to 1 L.

h. Stock nitrate solution: Prepare as directed in ¶ B.3*b.*

i. Intermediate nitrate solution: Prepare
as directed in ¶ B.3*c.*

j. Stock nitrite solution: See Section
4500-NO$_2$⁻.B.3*e.*

k. Intermediate nitrite solution: See Section 4500-NO$_2$⁻.B.3*f.*

l. Working nitrite solution: Dilute 50.0
mL intermediate nitrite solution to 500 mL
with nitrite-free water; 1.00 mL = 5 μg
NO$_2$⁻-N.

4. Procedure

a. Preparation of reduction column: Insert a glass wool plug into bottom of reduction column and fill with water. Add
sufficient Cu-Cd granules to produce a column 18.5 cm long. Maintain water level
above Cu-Cd granules to prevent entrapment of air. Wash column with 200 mL
dilute NH$_4$Cl-EDTA solution. Activate
column by passing through it, at 7 to 10
mL/min, at least 100 mL of a solution
composed of 25% 1.0 mg NO$_3$⁻-N/L
standard and 75% NH$_4$Cl-EDTA solution.

b. Treatment of sample:

1) Turbidity removal—For turbid samples, see ¶ A.1.

2) pH adjustment—Adjust pH to between 7 and 9, as necessary, using a pH
meter and dilute HCl or NaOH. This insures a pH of 8.5 after adding NH$_4$Cl-
EDTA solution.

3) Sample reduction—To 25.0 mL sample or a portion diluted to 25.0 mL, add
75 mL NH$_4$Cl-EDTA solution and mix.
Pour mixed sample into column and collect
at a rate of 7 to 10 mL/min. Discard first
25 mL. Collect the rest in original sample

flask. There is no need to wash columns between samples, but if columns are not to be reused for several hours or longer, pour 50 mL dilute NH$_4$Cl-EDTA solution on to the top and let it pass through the system. Store Cu-Cd column in this solution and never let it dry.

4) Color development and measurement—As soon as possible, and not more than 15 min after reduction, add 2.0 mL color reagent to 50 mL sample and mix. Between 10 min and 2 h afterward, measure absorbance at 543 nm against a distilled water-reagent blank. NOTE: If NO$_3^-$ concentration exceeds the standard curve range (about 1 mg N/L), use remainder of reduced sample to make an appropriate dilution and analyze again.

c. *Standards:* Using the intermediate NO$_3^-$- N solution, prepare standards in the range 0.05 to 1.0 mg NO$_3^-$-N/L by diluting the following volumes to 100 mL in volumetric flasks: 0.5, 1.0, 2.0, 5.0, and 10.0 mL. Carry out reduction of standards exactly as described for samples. Compare at least one NO$_2^-$ standard to a reduced NO$_3^-$ standard at the same concentration to verify reduction column efficiency. Reactivate Cu-Cd granules as described in ¶ 3*b* above when efficiency of reduction falls below about 75%.

5. Calculation

Obtain a standard curve by plotting absorbance of standards against NO$_3^-$-N concentration. Compute sample concentra-

tions directly from standard curve. Report as milligrams oxidized N per liter (the sum of NO$_3^-$-N plus NO$_2^-$-N) unless the concentration of NO$_2^-$-N is separately determined and subtracted.

6. Precision and Bias

In a single laboratory using wastewater samples at concentrations of 0.04, 0.24, 0.55, and 1.04 mg NO$_3^-$ + NO$_2^-$-N/L, the standard deviations were ±0.005, ±0.004, ±0.005, and ±0.01, respectively. In a single laboratory using wastewater with additions of 0.24, 0.55, and 1.05 mg NO$_3^-$ + NO$_2^-$-N/L, the recoveries were 100%, 102%, and 100%, respectively.[2]

7. References

1. WOOD, E.D., F.A.J. ARMSTRONG & F.A. RICHARDS. 1967. Determination of nitrate in sea water by cadmium-copper reduction to nitrite. *J. Mar. Biol. Assoc. U.K.* 47:23.
2. U.S. ENVIRONMENTAL PROTECTION AGENCY. 1979. Methods for Chemical Analysis of Water and Wastes, Method 353.3. U.S. Environmental Protection Agency, Washington, D.C.

8. Bibliography

STRICKLAND, J.D.H. & T.R. PARSONS. 1972. A Practical Handbook of Sea Water Analysis, 2nd ed. Bull. No. 167, Fisheries Research Board Canada, Ottawa, Ont.
NYDAHL, F. 1976. On the optimum conditions for the reduction of nitrate by cadmium. *Talanta* 23:349.
AMERICAN SOCIETY FOR TESTING AND MATERIALS. 1987. Annual Book of ASTM Standards, Vol. 11.01. American Soc. Testing & Materials, Philadelphia, Pa.

4500-NO$_3^-$ F.   Automated Cadmium Reduction Method

1. General Discussion

a. *Principle*: See ¶ E.1*a*.

b. *Interferences:* Sample turbidity may interfere. Remove by filtration before analysis. Sample color that absorbs in the pho-

tometric range used for analysis also will interfere.

c. *Application:* Nitrate and nitrite, singly or together in potable, surface, and saline waters and domestic and industrial wastewaters, can be determined over a range of 0.5 to 10 mg N/L.

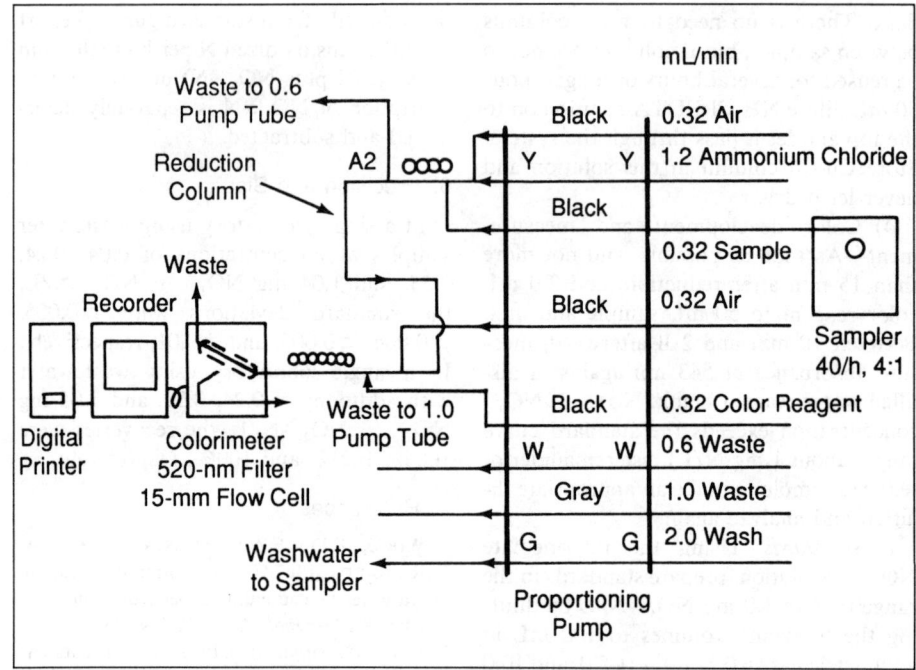

Figure 4500-NO₃⁻:2. Nitrate-nitrite manifold.

2. Apparatus

Automated analytical equipment: The required continuous-flow analytical instrument* consists of the components shown in Figure 4500-NO₃⁻:2.

3. Reagents

a. Deionized distilled water: See ¶ B.3*a*.

b. Copper sulfate solution: Dissolve 20 g CuSO₄·5H₂O in 500 mL water and dilute to 1 L.

c. Wash solution: Use water for unpreserved samples. For samples preserved with H₂SO₄, add 2 mL conc H₂SO₄/L wash water.

d. Copper-cadmium granules: See ¶ E.3*b*.

e. Hydrochloric acid, HCl, conc.

f. Ammonium hydroxide, NH₄OH, conc.

g. Color reagent: To approximately 800 mL water, add, while stirring, 100 mL conc H₃PO₄, 40 g sulfanilamide, and 2 g *N*-(1-naphthyl)-ethylenediamine dihydrochloride. Stir until dissolved and dilute to 1 L. Store in brown bottle and keep in the dark when not in use. This solution is stable for several months.

h. Ammonium chloride solution: Dissolve 85 g NH₄Cl in water and dilute to 1 L. Add 0.5 mL polyoxyethylene 23 lauryl ether.†

i. Stock nitrate solution: See ¶ B.3*b*.

---

* AutoAnalyzer™II: Technicon Instrument Corp., Tarrytown, N.Y., or equivalent.

† Brij-35, available from ICI Americas, Inc., Wilmington, Del., or Technicon Instruments Corp., Tarrytown, N.Y., or equivalent.

j. Intermediate nitrate solution: See ¶ B.3*c*.

k. Standard nitrate solutions: Using intermediate NO_3^--N solution and water, prepare standards for calibration curve in appropriate nitrate range. Compare at least one NO_2^- standard to a NO_3^- standard at the same concentration to verify column reduction efficiency. To examine saline waters prepare standard solutions with the substitute ocean water described in Section 4500-NH_3.G.3*g*.

l. Standard nitrite solution: See 4500-NH_3.3*g*.

4. Procedure

Set up manifold as shown in Figure 4500-NO_3^-:2 and follow general procedure described by the manufacturer.

If sample pH is below 5 or above 9, adjust to between 5 and 9 with either conc HCl or conc NH₄OH.

5. Calculation

Prepare standard curves by plotting peak heights of standards processed through the manifold against NO_3^--N concentration in standards. Compute sample NO_3^--N concentration by comparing sample peak height with standard curve.

6. Precision and Bias

Data obtained in three laboratories with an automated system based on identical chemical principles but having slightly different configurations‡ are given in the table

‡ Automated System I, described in the 16th edition of this publication, Section 418F.

below. Analyses were conducted on four natural water samples containing exact increments of inorganic nitrate:

Increment as NO_3^--N µg/L	Standard Deviation µg N/L	Bias %	Bias µg N/L
290	12	+5.75	+17
350	92	+18.10	+63
2310	318	+4.47	+103
2480	176	−2.69	−67

In a single laboratory using surface water samples at concentrations of 100, 200, 800, and 2100 µg N/L, the standard deviations were 0, ±40, ±50, and ±50 µg/L, respectively, and at concentrations of 200 and 2200 µg N/L, recoveries were 100 and 96%, respectively.

Precision and bias for the system described herein are believed to be comparable.

7. Bibliography

FIORE, J. & J.E. O'BRIEN. 1962. Automation in sanitary chemistry—Parts 1 and 2. Determination of nitrates and nitrites. *Wastes Eng.* 33:128 & 238.

ARMSTRONG, F.A., C.R. STEARNS & J.D. STRICKLAND. 1967. The measurement of upwelling and subsequent biological processes by means of the Technicon AutoAnalyzer and associated equipment. *Deep Sea Res.* 14:381.

U.S. ENVIRONMENTAL PROTECTION AGENCY. 1979. Methods for Chemical Analysis of Water and Wastes. U.S. Environmental Protection Agency, Washington, D.C.

AMERICAN SOCIETY FOR TESTING AND MATERIALS. 1987. Annual Book of ASTM Standards, Vol. 11.01. American Soc. Testing & Materials, Philadelphia, Pa.

4500-NO$_3^-$ G.   Titanous Chloride Reduction Method
(PROPOSED)

1. General Discussion

a. Principle: NO$_3^-$ is determined potentiometrically using an NH$_3$ gas-sensing electrode after reduction of the NO$_3^-$ to NH$_3$. Reduction is accomplished by use of titanous chloride reagent. The electrode is calibrated with NO$_3^-$ solutions and the concentrations of samples are determined in the same background solution containing sodium hydroxide and titanous chloride.

This method is applicable to the determination of NO$_3^-$ in water containing 0.1 to 20 mg NO$_3^-$-N/L. The concentration range can be extended by appropriate sample dilution.

b. Interferences: NH$_3$ and NO$_2^-$, if present, are measured with NO$_3^-$. If either concentration is significant, relative to the NO$_3^-$ concentration, measure separately and subtract.

2. Apparatus

a. pH meter, with 0.1 mV resolution, or a specific ion meter.

b. NH$_3$ gas-sensing electrode*.

c. Magnetic stirrer, with a TFE-coated stirring bar.

3. Reagents

a. Titanous chloride reagent solution, TiCl$_3$, practical grade, 20%.

b. Sodium hydroxide solution, NaOH, 10N: Dissolve 400 g reagent-grade NaOH pellets and dilute to 1000 mL with water.

c. Stock nitrate solution: See ¶ B.3*b*.

d. NH$_3$ electrode filling solution, NH$_4$Cl and NaNO$_3$: Dissolve 0.535 g NH$_4$Cl and 8.500 g NaNO$_3$ and dilute to 100 mL with distilled water. Add 0.3 mL 0.1M AgNO$_3$. Solution will be slightly cloudy.

4. Procedure

a. Preparation of calibration curve: Assemble and check NH$_3$ electrode according to manufacturer's instructions, but use the filling solution prepared in ¶ 3*d*. By serial dilution of the NO$_3^-$ stock solution, prepare NO$_3^-$ standards at 0.1, 1.0, and 10.0 mg/L. Transfer 100 mL 0.1 mg/L standard into a 150-mL beaker, add a stirring bar, and stir moderately. Add 10 mL 10N NaOH and 2 mL TiCl$_3$ reagent. Immerse electrode in stirring solution, let potential stabilize, and record the reading in millivolts. Plot as 0.1 mg NO$_3^-$-N/L and repeat for each remaining standard. Prepare a calibration curve by plotting smoothly, on semilogarithmic graph paper, the potential observed on the linear axis (ordinate) and the NO$_3^-$-N concentration on the logarithmic axis (abscissa). NOTE: Volume corrections are incorporated into calibration, so that readings can be made directly from the graph.

b. Measurement of samples: Transfer 100 mL sample to a 150-mL beaker, add a stirring bar, and stir moderately. Add 10 mL 10N NaOH and 2 mL TiCl$_3$ reagent. Immerse electrode in stirring solution, let potential stabilize, and use the voltage to read NO$_3^-$/L directly from the calibration curve. Rinse electrode with distilled water between samples.

5. Precision

Over the range of the method, precision of ± 0.6 mV or less, corresponding to about ± 3.0% in concentration, is expected.

6. Bibliography

BRAUNSTEIN, L.K. HOCHMULLER & K. SPENGLER. 1980. Combined ammonium/nitrate analysis using an ion-selective electrode. Special printing from *Vom Wasser,* 54, Verlag Chemie GmbH.

---

* Orion Model 95-12, or equivalent.

4500-NO$_3^-$ H.   Automated Hydrazine Reduction Method (PROPOSED)

1. General Discussion

a. Principle: NO_3^- is reduced to NO_2^- with hydrazine sulfate. The NO_2^- (originally present plus reduced NO_3^-) is determined by diazotization with sulfanilamide and coupling with *N*-(1-naphthyl)-ethylenediamine dihydrochloride to form a highly colored azo dye that is measured colorimetrically.

b. Interferences: Sample color that absorbs in the photometric range used will interfere. Concentrations of sulfide ion of less than 10 mg/L cause variations of NO_3^- and NO_2^- concentrations of $\pm 10\%$.

c. Application: $NO_3^- + NO_2^-$ in potable and surface water and in domestic and industrial wastes can be determined over a range of 0.01 to 10 mg N/L.

2. Apparatus

Automated analytical equipment: The required continuous-flow analytical instrument* consists of the components shown in Figure 4500-NO$_3^-$:3.

3. Reagents

a. Color developing reagent: To approximately 500 mL water add 200 mL conc phosphoric acid and 10 g sulfanilamide. After sulfanilamide is dissolved completely, add 0.8 g *N*-(1-naphthyl)-ethylenediamine dihydrochloride. Dilute to 1 L with water, store in a dark bottle, and refrigerate. Solution is stable for approximately 1 month.

b. Copper sulfate stock solution: Dissolve 2.5 g $CuSO_4 \cdot 5H_2O$ in water and dilute to 1 L.

c. Copper sulfate dilute solution: Dilute 20 mL stock solution to 2 L.

d. Sodium hydroxide stock solution, 10N: Dissolve 400 g NaOH in 750 mL water, cool, and dilute to 1 L.

e. Sodium hydroxide, 1.0N: Dilute 100 mL stock NaOH solution to 1 L.

f. Hydrazine sulfate stock solution: Dissolve 27.5 g $N_2H_4 \cdot H_2SO_4$ in 900 mL water and dilute to 1 L. This solution is stable for approximately 6 months. CAUTION: *Toxic if ingested. Mark container with appropriate warning.*

g. Hydrazine sulfate dilute solution: Dilute 22 mL stock solution to 1 L.

h. Stock nitrate solution: See ¶ B.3*b*.

i. Intermediate nitrate solution: See ¶ B.3*c*.

j. Standard nitrate solutions: Prepare NO_3^- calibration standards in the range 0 to 10 mg/L by diluting to 100 mL the following volumes of stock nitrate solution: 0, 0.5, 1.0, 2.0...10.0 mL. For standards in the range of 0.01 mg/L use intermediate nitrate solution. Compare at least one nitrite standard to a nitrate standard at the same concentration to verify the efficiency of the reduction.

k. Standard nitrite solution: See ¶s B.3*e*, *f,* and *g*.

4. Procedure

Set up manifold as shown in Figure 4500-NO$_3^-$:3 and follow general procedure described by manufacturer. Run a 2.0-mg NO_3^--N/L and a 2.0-mg NO_2^--N/L

---

* AutoAnalyzer II: Technicon Instrument Corp., Tarrytown. N.Y., or equivalent.

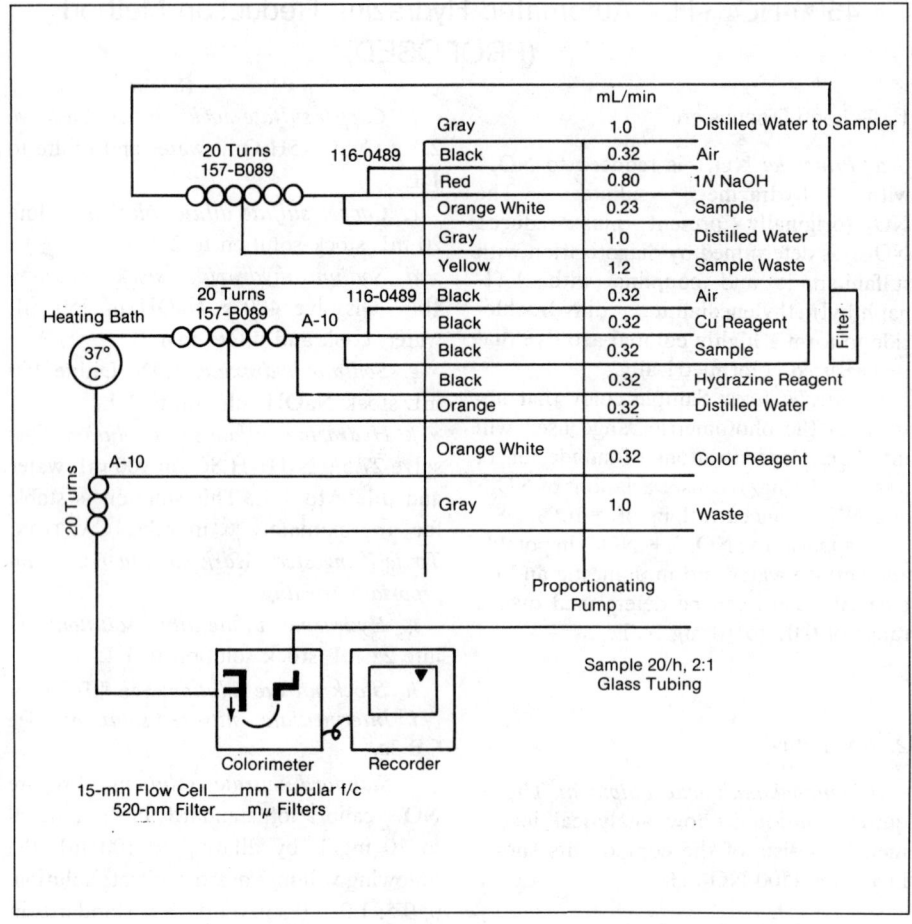

Figure 4500-NO$_3^-$:3. Nitrate-nitrite manifold.

standard through the system to check for 100% reduction of nitrate to nitrite. The two peaks should be of equal height; if not, adjust concentration of the hydrazine sulfate solution: If the NO$_3^-$ peak is lower than the NO$_2^-$ peak, increase concentration of hydrazine sulfate until they are equal; if the NO$_3^-$ peak is higher than the NO$_2^-$ reduce concentration of hydrazine sulfate. When correct concentration has

been determined, no further adjustment should be necessary.

5. Calculation

Prepare a standard curve by plotting peak heights of processed standards against known concentrations. Compute concentrations of samples by comparing peak heights with standard curve.

6. Precision and Bias

In a single laboratory using drinking water, surface water, and industrial waste at concentrations of 0.39, 1.15, 1.76, and 4.75 mg NO_3^--N/L, the standard deviations were ±0.02, ±0.01, ±0.02, and ±0.03, respectively. In a single laboratory using drinking water at concentrations of 0.75 and 2.97 the recoveries were 99% and 101%.[1]

7. Reference

1. U.S. ENVIRONMENTAL PROTECTION AGENCY. 1979. Methods for Chemical Analysis of Water and Wastes. U.S. Environmental Protection Agency, Washington, D.C.

8. Bibliography

KAMPHAKE, L., S. HANNAH & J. COHEN. 1967. Automated analysis for nitrate by hydrazine reduction. *Water Res.* 1:205.

4500-N$_{org}$ NITROGEN (ORGANIC)*

4500-N$_{org}$ A.   Introduction

1. Selection of Method

The major factor that influences the selection of a macro- or semi-micro-kjeldahl method to determine organic nitrogen is the concentration of organic nitrogen. The macro-kjeldahl method is applicable for samples containing either low or high concentrations of organic nitrogen but requires a relatively large sample volume for low concentrations. In the semi-micro-kjeldahl method, which is applicable to samples containing high concentrations of organic nitrogen, the sample volume should be chosen to contain organic plus ammonia nitrogen (kjeldahl nitrogen) in the range of 0.2 to 2 mg.

2. Storage of Samples

The most reliable results are obtained on fresh samples. If an immediate analysis is not possible, preserve samples by acidifying to pH 1.5 to 2.0 with concentrated H_2SO_4 and storing at 4°C. Do not use $HgCl_2$ be-

cause it will interfere with ammonia removal.

3. Interferences

a. Nitrate: During digestion, nitrate in excess of 10 mg/L can oxidize a portion of the ammonia released from the digested organic nitrogen, producing N_2O and resulting in a negative interference. When sufficient organic matter in a low state of oxidation is present, nitrate can be reduced to ammonia, resulting in a positive interference. The conditions under which significant interferences occur are not well defined and there is no proven way to eliminate the interference in conjunction with the methods described herein.

b. Inorganic salts and solids: The acid and salt content of the digestion reagent is intended to produce a digestion temperature of about 360 to 370°C. If the sample contains a very large quantity of salt or inorganic solids that dissolve during digestion, the temperature may rise above 400°C, at which point pyrolytic loss of nitrogen

---

*Approved by Standard Methods Committee, 1985.

begins to occur. To prevent an excessive digestion temperature, add more H_2SO_4 to maintain the acid-salt balance. Not all salts cause precisely the same temperature rise, but adding of 1 mL H_2SO_4/g salt in the sample gives reasonable results. Add the extra acid and the digestion reagent to both sample and reagent blank. Too much acid will lower the digestion temperature below 360°C and result in incomplete digestion and recovery. If necessary, add more sodium hydroxide-sodium thiosulfate before the final distillation step to neutralize the excess acid.

Large amounts of salt or solids also may cause bumping during distillation. If this occurs, add more dilution water to the samples after digestion.

c. Organic matter: During digestion, H_2SO_4 oxidizes organic matter to CO_2 and H_2O. If a large amount of organic matter is present, a large amount of acid will be consumed, the ratio of salt to acid will increase, and the digestion temperature will increase. If enough organic matter is present, the temperature will rise above 400°C, resulting in pyrolytic loss of nitrogen. To prevent this, add to the digestion flask 10 mL conc H_2SO_4/3 g COD. (For most organic substances, 3 g COD equals about 1 g organic matter). Alternately, add 50 mL more of digestion reagent/g COD. Additional sodium hydroxide-sodium thiosulfate reagent may be necessary to keep the distillation pH high. Because reagents may contain traces of ammonia, treat the reagent blank identically with the samples.

4. Use of a Catalyst

Although it generally is desirable to avoid using mercury because of its toxicity and the problems associated with disposal of residues, mercury is the catalyst of choice. Only selenium is as effective as mercury, but selenium is highly toxic and there are potential interferences associated with its use. Digestion of some samples may be complete or nearly complete without the use of a catalyst or with the use of a less toxic catalyst, such as copper. If copper is substituted for mercury, add 10 mL of a solution containing 25.115 g $CuSO_4$/L to each macro-kjeldahl digestion flask with 50 mL digestion reagent from which the HgO has been omitted. Use 2 mL of $CuSO_4$ solution for the semi-micro method. If the mercury catalyst is omitted, report this deviation and indicate, if possible, the percentage recovery relative to the results for similar samples analyzed using the mercury catalyst.

4500-N$_{org}$ B.   Macro-Kjeldahl Method

1. General Discussion

The kjeldahl method determines nitrogen in the trinegative state. It fails to account for nitrogen in the form of azide, azine, azo, hydrazone, nitrate, nitrite, nitrile, nitro, nitroso, oxime, and semi-carbazone. If ammonia nitrogen is not removed in the initial phase (¶ 4*b* below) of the procedure, the term "kjeldahl nitrogen" is applied to the result. Should kjeldahl nitrogen and ammonia nitrogen be determined individually, "organic nitrogen" can be obtained by difference.

a. Principle: In the presence of H_2SO_4, potassium sulfate (K_2SO_4), and mercuric sulfate (HgSO$_4$) catalyst, amino nitrogen of many organic materials is converted to ammonium sulfate [$(NH_4)_2SO_4$]. Free ammonia and ammonium-nitrogen also are converted to $(NH_4)_2SO_4$. During sample digestion, a mercury ammonium complex is formed and then decomposed by sodium thiosulfate ($Na_2S_2O_3$). After decomposition the ammonia is distilled from an alkaline medium and absorbed in boric or sulfuric acid. The ammonia is determined colorimetrically or by titration with a standard mineral acid.

b. Selection of ammonia measurement method: The sensitivity of colorimetric methods makes them particularly useful for determining organic nitrogen levels below 5 mg/L. The titrimetric and selective electrode methods of measuring ammonia in the distillate are suitable for determining a wide range of organic nitrogen concentrations.

2. Apparatus

a. Digestion apparatus: Kjeldahl flasks with a total capacity of 800 mL yield the best results. Digest over a heating device adjusted so that 250 mL water at an initial temperature of 25°C can be heated to a rolling boil in approximately 5 min. For testing, preheat heaters for 10 min if gas or 30 min if electric. A heating device meeting this specification should provide the temperature range of 365 to 370°C for effective digestion.

b. Distillation apparatus: See Section 4500-NH$_3$.B.2*a*.

c. Apparatus for ammonia determination: See Sections 4500-NH$_3$.C.2, D.2, E.2, or F.2.

3. Reagents

Prepare all reagents and dilutions in ammonia-free water.

All of the reagents listed for the determination of Nitrogen (Ammonia), Sections 4500-NH$_3$.C.3, D.3, E.3, or F.3, are required, plus the following:

a. Mercuric sulfate solution: Dissolve 8 g red mercuric oxide, HgO, in 100 mL 6*N* H$_2$SO$_4$.

b. Digestion reagent: Dissolve 134 g K$_2$SO$_4$ in 650 mL water and 200 mL conc H$_2$SO$_4$. Add, with stirring, 25 mL mercuric sulfate solution. Dilute the combined solution to 1 L with water. Keep at a temperature close to 20°C to prevent crystallization.

c. Sodium hydroxide-sodium thiosulfate reagent: Dissolve 500 g NaOH and 25 g Na$_2$S$_2$O$_3$·5H$_2$O in water and dilute to 1 L.

d. Borate buffer solution: See Section 4500-NH$_3$.B.3*b*.

e. Sodium hydroxide, NaOH, 6*N*.

4. Procedure

a. Selection of sample volume and sample preparation: Place a measured volume of sample in an 800-mL kjeldahl flask. Select sample size from the following tabulation:

Organic Nitrogen in Sample mg/L	Sample Size mL
0–1	500
1–10	250
10–20	100
20–50	50.0
50–100	25.0

If necessary, dilute sample to 300 mL, neutralize to pH 7, and dechlorinate as described in Section 4500-NH$_3$.B.4*b*.

b. Ammonia removal: Add 25 mL borate buffer and then 6*N* NaOH until pH 9.5 is reached. Add a few glass beads or boiling chips and boil off 300 mL. If desired, distill this fraction and determine ammonia nitrogen. Alternately, if ammonia has been determined by the distillation method, use residue in distilling flask for organic nitrogen determination.

For sludge and sediment samples, weigh wet sample in a crucible or weighing bottle, transfer contents to a kjeldahl flask, and determine kjeldahl nitrogen. Follow a similar procedure for ammonia nitrogen and organic nitrogen determined by difference. Determinations of organic and kjeldahl nitrogen on dried sludge and sediment samples are not accurate because drying results in loss of ammonium salts. Measure dry weight of sample on a separate portion.

c. Digestion: Cool and add carefully 50 mL digestion reagent (or substitute 10 mL conc H_2SO_4, 6.7 g K_2SO_4, and 1.25 mL $HgSO_4$ solution) to distillation flask. Add a few glass beads and, after mixing, heat under a hood or with suitable ejection equipment to remove acid fumes. Boil briskly until the volume is greatly reduced (to about 25 to 50 mL) and copious white fumes are observed (fumes may be dark for samples high in organic matter). Then continue to digest for an additional 30 min. As digestion continues, colored or turbid samples will turn clear or straw-colored. After digestion, let flask and contents cool, dilute to 300 mL with water, and mix. Tilt flask and carefully add 50 mL hydroxide-thiosulfate reagent to form an alkaline layer at flask bottom. Connect flask to steamed-out distillation apparatus and shake flask to insure complete mixing. A black precipitate, HgS, will form, and the pH should exceed 11.0.

d. Distillation: Distill and collect 200 mL distillate below surface of 50 mL absorbent solution. Use plain boric acid solution when ammonia is to be determined by nesslerization and use indicating boric acid for a titrimetric finish. Use 50 mL 0.04N H_2SO_4 solution for collecting distillate for manual phenate, nesslerization, or electrode methods. Extend tip of condenser well below level of absorbent solution and do not let temperature in condenser rise above 29°C. Lower collected distillate free of contact with delivery tube and continue distillation during last 1 or 2 min to cleanse condenser.

e. Final ammonia measurement: Use the nesslerization, manual phenate, titration, or ammonia-selective electrode method, Sections 4500-NH$_3$.C, D, E, and F, respectively.

f. Blank: Carry a reagent blank through all steps of the procedure and apply necessary corrections to the results.

5. Calculation

See Section 4500-NH$_3$.C.5, D.5, E.5, or F.5.

6. Precision and Bias

Three synthetic samples containing various organic nitrogen concentrations and other constituents were analyzed by three procedural modifications of the macro-kjeldahl method: kjeldahl-nessler finish, kjeldahl-titrimetric finish, and calculation of the difference between kjeldahl nitrogen and ammonia nitrogen, both determined by a nessler finish. The results obtained by participating laboratories are summarized in Table 4500-N$_{org}$:I.

No data on the precision of the macro-kjeldahl-phenate method are available.

Sample 1 contained the following additional constituents: 400 mg Cl$^-$/L, 1.50 mg NH$_3$-N/L, 1.0 mg NO$_3^-$-N/L, 0.5 mg PO$_4^{3-}$/L, and 30.0 mg SiO$_2$/L.

Sample 2 contained the following additional constituents: 200 mg Cl$^-$/L, 0.8 mg NH$_3$-N/L, 1.0 mg NO$_3^-$-N/L, 5.0 mg PO$_4^{3-}$/L, and 15.0 mg SiO$_2$/L.

Sample 3 contained the following additional constituents: 10 mg Cl$^-$/L, 0.2 mg NH$_3$-N/L, 1.0 mg NO$_3^-$-N/L, 10.0 mg PO$_4^{3-}$/L, and 5.0 mg SiO$_2$/L.

7. Bibliography

Kjeldahl, J. 1883. A new method for the determination of nitrogen in organic matter. *Z. Anal. Chem.* 22:366.

Phelps, E.B. 1905. The determination of organic nitrogen in sewage by the Kjeldahl process. *J. Infect. Dis.* (Suppl.) 1:225.

Meeker, E.W. & E.C. Wagner. 1933. Titration of ammonia in the presence of boric acid. *Ind. Eng. Chem.,* Anal. Ed. 5:396.

Wagner, E.C. 1940. Titration of ammonia in the presence of boric acid. *Ind. Eng. Chem.,* Anal. Ed. 12:771.

McKenzie, H.A. & H.S. Wallace. 1954. The

TABLE 4500-N$_{org}$:I. PRECISION AND BIAS DATA FOR ORGANIC NITROGEN, MACRO-KJELDAHL PROCEDURE

Sample	No. of Labora-tories	Organic Nitrogen Concen-tration $\mu g/L$	Relative Standard Deviation			Relative Error		
			Nessler Finish %	Titri-metric Finish %	Calculation of Total Kjeldahl N Minus NH_3-N %	Nessler Finish %	Titri-metric Finish %	Calculation of Total Kjeldahl N Minus NH_3-N %
1	26	200	94.8	—	—	55.0	—	—
	29	200	—	104.4	—	—	70.0	—
	15	200	—	—	68.8	—	—	70.0
2	26	800	52.1	—	—	12.5	—	—
	31	800	—	44.8	—	—	3.7	—
	16	800	—	—	52.6	—	—	8.7
3	26	1500	43.1	—	—	9.3	—	—
	30	1500	—	54.7	—	—	22.6	—
	16	1500	—	—	45.9	—	—	4.0

Kjeldahl determination of nitrogen: A critical study of digestion conditions. *Aust. J. Chem.* 7:55.

MORGAN, G.B., J.B. LACKEY & F.W. GILCREAS. 1957. Quantitative determination of organic nitrogen in water, sewage, and industrial wastes. *Anal. Chem.* 29:833.

BOLTZ, D.F., ed. 1978. Colorimetric Determination of Nonmetals. Interscience Publishers, New York, N.Y.

4500-N$_{org}$ C. Semi-Micro-Kjeldahl Method

1. General Discussion

See Section 4500-N$_{org}$.B.1.

2. Apparatus

a. Digestion apparatus: Use kjeldahl flasks with a capacity of 100 mL in a semi-micro-kjeldahl digestion apparatus* equipped with heating elements to accomodate kjeldahl flasks and a suction outlet to vent fumes. The heating elements should provide the temperature range of 365 to 380°C for effective digestion.

b. Distillation apparatus: Use an all-glass

unit equipped with a steam-generating vessel containing an immersion heater† (Figure 4500-N$_{org}$:1).

c. pH meter.

3. Reagents

All of the reagents listed for the determination of Nitrogen (Ammonia) (Section 4500-NH$_3$.B.3) and Nitrogen (Organic) macro-kjeldahl (Section 4500-N$_{org}$.B.3) are required. Prepare all reagents and dilutions with ammonia-free water.

---

*Rotary kjeldahl digestion unit, Kontes, Model K551100, or equivalent.

†ASTM E-147 or equivalent.

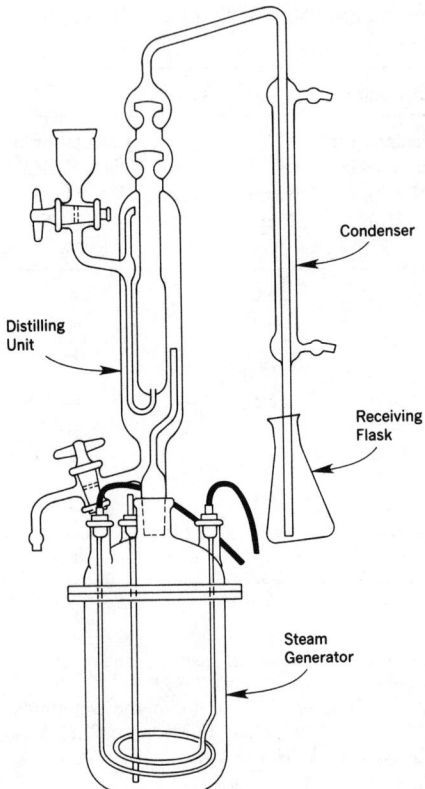

Figure 4500-N$_{org}$:1. Micro-kjeldahl distillation apparatus.

4. Procedure

a. Selection of sample volume: Determine the sample size from the following tabulation:

Organic Nitrogen in Sample mg/L	Sample Size mL
4–40	50
8–80	25
20–200	10
40–400	5

For sludge and sediment samples weigh a portion of wet sample containing between 0.2 and 2 mg organic nitrogen in a crucible

or weighing bottle. Transfer sample quantitatively to a 100-mL beaker by diluting it and rinsing the weighing dish several times with small quantities of water. Make the transfer using as small a quantity of water as possible and do not exceed a total volume of 50 mL. Measure dry weight of sample on a separate portion.

b. Ammonia removal: Pipet 50 mL sample or an appropriate volume diluted to 50 mL with water into a 100-mL beaker. Add 3 mL borate buffer and adjust to pH 9.5 with 6N NaOH, using a pH meter. Quantitatively transfer sample to a 100-mL kjeldahl flask and boil off 30 mL.

Alternatively, if ammonia removal is not required, digest samples directly as described in ¶ *c* below. Distillation following this direct digestion yields kjeldahl nitrogen concentration rather than organic nitrogen.

c. Digestion: Carefully add 10 mL digestion reagent to kjeldahl flask containing sample. Add 5 or 6 glass beads (3- to 4-mm size) to prevent bumping during digestion. Set each heating unit on the micro-kjeldahl digestion apparatus to its medium setting and heat flasks under a hood or with suitable ejection equipment to remove fumes of SO$_3$. Continue to boil briskly until solution clears (becomes colorless or a pale straw color) and copious fumes are observed. Then turn each heating unit up to its maximum setting and digest for an additional 30 min. Cool. Quantitatively transfer digested sample by diluting and rinsing several times into micro-kjeldahl distillation apparatus so that total volume in distillation apparatus does not exceed 30 mL. Add 10 mL hydroxide-thiosulfate reagent and turn on steam.

d. Distillation: Control rate of steam generation to boil contents in distillation unit so that neither escape of steam from tip of condenser nor bubbling of contents in receiving flask occurs. Distill and collect 30 to 40 mL distillate below surface of 10 mL boric acid solution contained in a 125-mL erlenmeyer flask. Use plain boric acid solution when ammonia is to be determined by nesslerization and use indicating boric acid for a titrimetric finish. Use 10 mL 0.04N H$_2$SO$_4$ solution for collecting distillate for the phenate, nessler, or electrode methods. Extend tip of condenser well below level of boric acid solution and do not let temperature in condenser rise above 29°C. Lower collected distillate free of contact with delivery tube and continue distillation during last 1 or 2 min to cleanse condenser.

e. Blank: Carry a reagent blank through all steps of procedure and apply necessary correction to results.

f. Final ammonia measurement: Determine ammonia by nesslerization, manual phenate, titration, or ammonia-selective electrode method.

5. Calculation

See Section 4500-NH$_3$.C.5, D.5, E.5, or F.5.

6. Precision and Bias

No data on the precision and bias of the semi-micro-kjeldahl method are available.

7. Bibliography

See Section 4500-N$_{org}$.B.7.

4500-O OXYGEN (DISSOLVED)*

4500-O A. Introduction

1. Significance

Dissolved oxygen (DO) levels in natural and wastewaters depend on the physical, chemical, and biochemical activities in the water body. The analysis for DO is a key test in water pollution and waste treatment process control.

2. Selection of Method

Two methods for DO analysis are described: the Winkler or iodometric method and its modifications and the electrometric method using membrane electrodes. The iodometric method[1] is a titrimetric procedure based on the oxidizing property of DO while the membrane electrode procedure is based on the rate of diffusion of molecular oxygen across a membrane.[2] The choice of procedure depends on the interferences present, the accuracy desired, and, in some cases, convenience or expedience.

3. References

1. WINKLER, L.W. 1888. The determination of dissolved oxygen in water. *Berlin. Deut. Chem. Ges.* 21:2843.
2. MANCY, K.H. & T. JAFFE. 1966. Analysis of Dissolved Oxygen in Natural and Waste Waters. Publ. No. 999-WP-37, U.S. Public Health Serv., Washington, D.C.

---

*Approved by Standard Methods Committee, 1988.

4500-O B. Iodometric Methods

1. Principle

The iodometric test is the most precise and reliable titrimetric procedure for DO analysis. It is based on the addition of divalent manganese solution, followed by strong alkali, to the sample in a glass-stoppered bottle. DO rapidly oxidizes an equivalent amount of the dispersed divalent manganous hydroxide precipitate to hydroxides of higher valency states. In the presence of iodide ions in an acidic solution, the oxidized manganese reverts to the divalent state, with the liberation of iodine equivalent to the original DO content. The iodine is then titrated with a standard solution of thiosulfate.

The titration end point can be detected visually, with a starch indicator, or electrometrically, with potentiometric or dead-stop techniques.[1] Experienced analysts can maintain a precision of ±50 µg/L with visual end-point detection and a precision of ±5 µg/L with electrometric end-point detection.[1,2]

The liberated iodine also can be determined directly by simple absorption spectrophotometers.[3] This method can be used on a routine basis to provide very accurate estimates for DO in the microgram-per-liter range provided that interfering particulate matter, color, and chemical interferences are absent.

2. Selection of Method

Before selecting a method consider the effect of interferences, particularly oxidizing or reducing materials that may be present in the sample. Certain oxidizing agents liberate iodine from iodides (positive interference) and some reducing agents reduce iodine to iodide (negative interference). Most organic matter is oxidized partially when the oxidized manganese precipitate is acidified, thus causing negative errors.

Several modifications of the iodometric method are given to minimize the effect of interfering materials.[2] Among the more commonly used procedures are the azide modification,[4] the permanganate modification,[5] the alum flocculation modification,[6] and the copper sulfate-sulfamic acid flocculation modification.[7,8] The azide modification (C) effectively removes interference caused by nitrite, which is the most common interference in biologically treated effluents and incubated BOD samples. Use the permanganate modification (D) in the presence of ferrous iron. When the sample contains 5 or more mg ferric iron salts/L, add potassium fluoride (KF) as the first reagent in the azide modification or after the permanganate treatment for ferrous iron. Alternately, eliminate Fe(III) interference by using 85 to 87% phosphoric acid (H_3PO_4) instead of sulfuric acid (H_2SO_4) for acidification. This procedure has not been tested for Fe(III) concentrations above 20 mg/L.

Use the alum flocculation modification (E) in the presence of suspended solids that cause interference and the copper sulfate-sulfamic acid flocculation modification (F) on activated-sludge mixed liquor.

3. Collection of Samples

Collect samples very carefully. Methods of sampling are highly dependent on source to be sampled and, to a certain extent, on method of analysis. Do not let sample remain in contact with air or be agitated, because either condition causes a change in its gaseous content. Samples from any depth in streams, lakes, or reservoirs, and samples of boiler water, need special precautions to eliminate changes in pressure and temperature. Procedures and equipment have been developed for sampling waters under pressure and unconfined waters (e.g., streams, rivers, and reservoirs). Sampling procedures and equipment needed are described in American Society for Testing and Materials Special Technical

Publication No. 148-1 and in U.S. Geological Survey Water Supply Paper No. 1454.

Collect surface water samples in narrow-mouth glass-stoppered BOD bottles of 300-mL capacity with tapered and pointed ground-glass stoppers and flared mouths. Avoid entraining or dissolving atmospheric oxygen. In sampling from a line under pressure, attach a glass or rubber tube to the tap and extend to bottom of bottle. Let bottle overflow two or three times its volume and replace stopper so that no air bubbles are entrained.

Suitable samplers for streams, ponds, or tanks of moderate depth are of the APHA type shown in Figure 4500-O:1. Use a Kemmerer-type sampler for samples collected from depths greater than 2 m. Bleed sample from bottom of sampler through a tube extending to bottom of a 250- to 300-mL BOD bottle. Fill bottle to overflowing (overflow for approximately 10 s), and prevent turbulence and formation of bubbles while filling. Record sample temperature to nearest degree Celsius or more precisely.

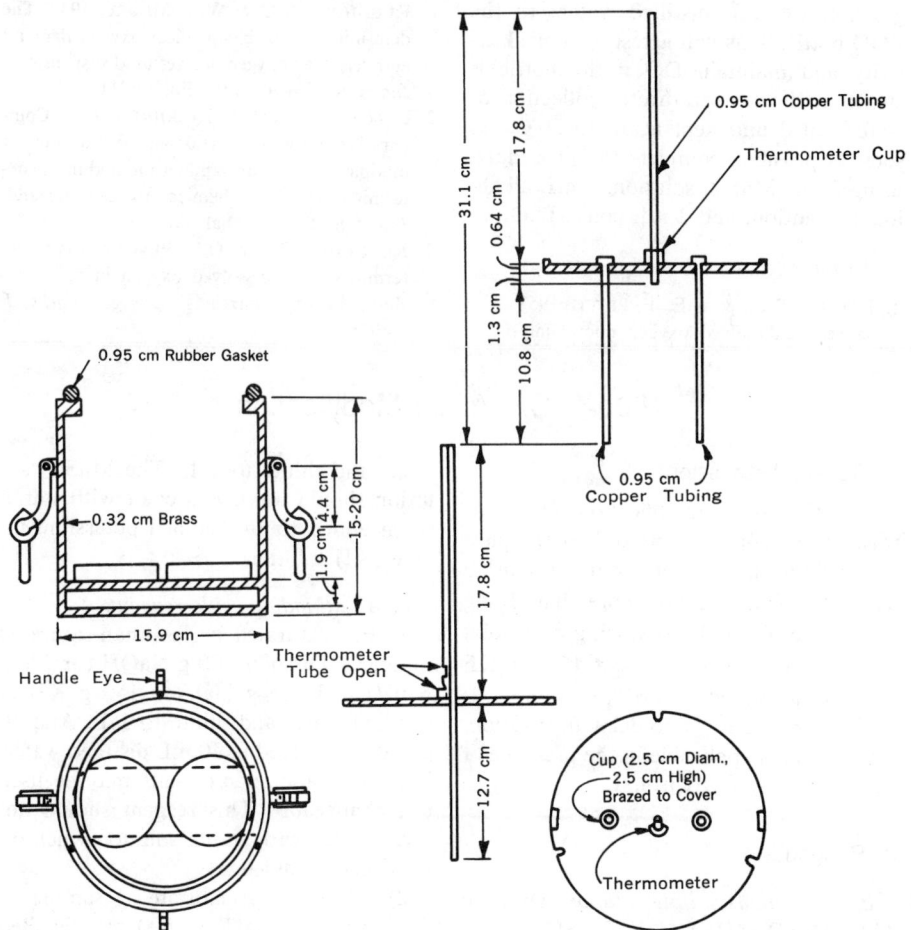

Figure 4500-O:1. DO and BOD sampler assembly.

4. Preservation of Samples

Determine DO immediately on all samples containing an appreciable oxygen or iodine demand. Samples with no iodine demand may be stored for a few hours without change after adding manganous sulfate ($MnSO_4$) solution, alkali-iodide solution, and H_2SO_4, followed by shaking in the usual way. Protect stored samples from strong sunlight and titrate as soon as possible.

For samples with an iodine demand, preserve for 4 to 8 h by adding 0.7 mL conc H_2SO_4 and 1 mL sodium azide solution (2 g NaN_3/100 mL distilled water) to the BOD bottle. This will arrest biological activity and maintain DO if the bottle is stored at the temperature of collection or water-sealed and kept at 10 to 20°C. As soon as possible, complete the procedure, using 2 mL $MnSO_4$ solution, 3 mL alkali-iodide solution, and 2 mL conc H_2SO_4.

5. References

1. POTTER, E.C. & G.E. EVERITT. 1957. Advances in dissolved oxygen microanalysis. *J. Appl. Chem.* 9:642.
2. MANCY, K.H. & T. JAFFE. 1966. Analysis of Dissolved Oxygen in Natural and Waste Waters. Publ. No. 99-WP-37, U.S. Public Health Serv., Washington, D.C.
3. OULMAN, C.S. & E.R. BAUMANN. 1956. A colorimetric method for determining dissolved oxygen. *Sewage Ind. Wastes* 28:1461.
4. ALSTERBERG, G. 1925. Methods for the determination of elementary oxygen dissolved in water in the presence of nitrite. *Biochem. Z.* 159:36.
5. RIDEAL, S. & G.G. STEWART. 1901. The determination of dissolved oxygen in waters in the presence of nitrites and of organic matter. *Analyst* 26:141.
6. RUCHHOFT, C.C. & W.A. MOORE. 1940. The determination of biochemical oxygen demand and dissolved oxygen of river mud suspensions. *Ind. Eng. Chem.*, Anal. Ed. 12:711.
7. PLACAK, O.R. & C.C. RUCHHOFT. 1941. Comparative study of the azide and Rideal-Stewart modifications of the Winkler method in the determination of biochemical oxygen demand. *Ind. Eng. Chem.*, Anal. Ed. 13:12.
8. RUCHHOFT, C.C. & O.R. PLACAK. 1942. Determination of dissolved oxygen in activated-sludge sewage mixtures. *Sewage Works J.* 14:638.

4500-O C.   Azide Modification

1. General Discussion

Use the azide modification for most wastewater, effluent, and stream samples, especially if samples contain more than 50 μg NO_2^--N/L and not more than 1 mg ferrous iron/L. Other reducing or oxidizing materials should be absent. If 1 mL KF solution is added before the sample is acidified and there is no delay in titration, the method is applicable in the presence of 100 to 200 mg ferric iron/L.

2. Reagents

a. *Manganous sulfate solution:* Dissolve 480 g $MnSO_4 \cdot 4H_2O$, 400 g $MnSO_4 \cdot 2H_2O$, or 364 g $MnSO_4 \cdot H_2O$ in distilled water, filter, and dilute to 1 L. The $MnSO_4$ solution should not give a color with starch when added to an acidified potassium iodide (KI) solution.

b. *Alkali-iodide-azide reagent:*
1) For saturated or less-than-saturated samples—Dissolve 500 g NaOH (or 700 g KOH) and 135 g NaI (or 150 g KI) in distilled water and dilute to 1 L. Add 10 g NaN_3 dissolved in 40 mL distilled water. Potassium and sodium salts may be used interchangeably. This reagent should not give a color with starch solution when diluted and acidified.

2) For supersaturated samples—Dissolve 10 g NaN_3 in 500 mL distilled water. Add 480 g sodium hydroxide

(NaOH) and 750 g sodium iodide (NaI), and stir until dissolved. There will be a white turbidity due to sodium carbonate (Na_2CO_3), but this will do no harm. CAUTION—*Do not acidify this solution because toxic hydrazoic acid fumes may be produced.*

c. Sulfuric acid, H_2SO_4, conc: One milliliter is equivalent to about 3 mL alkali-iodide-azide reagent.

d. Starch: Use either an aqueous solution or soluble starch powder mixtures.

To prepare an aqueous solution, dissolve 2 g laboratory-grade soluble starch and 0.2 g salicylic acid, as a preservative, in 100 mL hot distilled water.

e. Standard sodium thiosulfate titrant: Dissolve 6.205 g $Na_2S_2O_3 \cdot 5H_2O$ in distilled water. Add 1.5 mL 6N NaOH or 0.4 g solid NaOH and dilute to 1000 mL. Standardize with bi-iodate solution.

f. Standard potassium bi-iodate solution, 0.0021M: Dissolve 812.4 mg $KH(IO_3)_2$ in distilled water and dilute to 1000 mL.

Standardization—Dissolve approximately 2 g KI, free from iodate, in an erlenmeyer flask with 100 to 150 mL distilled water. Add 1 mL 6N H_2SO_4 or a few drops of conc H_2SO_4 and 20.00 mL standard bi-iodate solution. Dilute to 200 mL and titrate liberated iodine with thiosulfate titrant, adding starch toward end of titration, when a pale straw color is reached. When the solutions are of equal strength, 20.00 mL 0.025M $Na_2S_2O_3$ should be required. If not, adjust the $Na_2S_2O_3$ solution to 0.025M.

g. Potassium fluoride solution: Dissolve 40 g $KF \cdot 2H_2O$ in distilled water and dilute to 100 mL.

3. Procedure

a. To the sample collected in a 250- to 300-mL bottle, add 1 mL $MnSO_4$ solution, followed by 1 mL alkali-iodide-azide reagent. If pipets are dipped into sample, rinse them before returning them to reagent bottles. Alternatively, hold pipet tips just above liquid surface when adding reagents. Stopper carefully to exclude air bubbles and mix by inverting bottle a few times. When precipitate has settled sufficiently (to approximately half the bottle volume) to leave clear supernate above the manganese hydroxide floc, add 1.0 mL conc H_2SO_4. Restopper and mix by inverting several times until dissolution is complete. Titrate a volume corresponding to 200 mL original sample after correction for sample loss by displacement with reagents. Thus, for a total of 2 mL (1 mL each) of $MnSO_4$ and alkali-iodide-azide reagents in a 300-mL bottle, titrate $200 \times 300/(300 - 2) = 201$ mL.

b. Titrate with 0.025M $Na_2S_2O_3$ solution to a pale straw color. Add a few drops of starch solution and continue titration to first disappearance of blue color. If end point is overrun, back-titrate with 0.0021M bi-iodate solution added dropwise, or by adding a measured volume of treated sample. Correct for amount of bi-iodate solution or sample. Disregard subsequent recolorations due to the catalytic effect of nitrite or to traces of ferric salts that have not been complexed with fluoride.

4. Calculation

a. For titration of 200 mL sample, 1 mL 0.025M $Na_2S_2O_3$ = 1 mg DO/L.

b. To express results as percent saturation at 101.3 kPa, use the solubility data in Table 4500-O:I. Equations for correcting solubilities to barometric pressures other than mean sea level and for various chlorinities are given below the table.

5. Precision and Bias

DO can be determined with a precision, expressed as a standard deviation, of about 20 $\mu g/L$ in distilled water and about 60 $\mu g/L$ in wastewater and secondary effluents. In the presence of appreciable interference, even with proper modifications,

TABLE 4500-O:I. SOLUBILITY OF OXYGEN IN WATER EXPOSED TO WATER-SATURATED AIR AT ATMOSPHERIC PRESSURE (101.3 KPA)[1]

Temperature °C	Oxygen Solubility mg/L					
	Chlorinity: 0	5.0	10.0	15.0	20.0	25.0
0.0	14.621	13.728	12.888	12.097	11.355	10.657
1.0	14.216	13.356	12.545	11.783	11.066	10.392
2.0	13.829	13.000	12.218	11.483	10.790	10.139
3.0	13.460	12.660	11.906	11.195	10.526	9.897
4.0	13.107	12.335	11.607	10.920	10.273	9.664
5.0	12.770	12.024	11.320	10.656	10.031	9.441
6.0	12.447	11.727	11.046	10.404	9.799	9.228
7.0	12.139	11.442	10.783	10.162	9.576	9.023
8.0	11.843	11.169	10.531	9.930	9.362	8.826
9.0	11.559	10.907	10.290	9.707	9.156	8.636
10.0	11.288	10.656	10.058	9.493	8.959	8.454
11.0	11.027	10.415	9.835	9.287	8.769	8.279
12.0	10.777	10.183	9.621	9.089	8.586	8.111
13.0	10.537	9.961	9.416	8.899	8.411	7.949
14.0	10.306	9.747	9.218	8.716	8.242	7.792
15.0	10.084	9.541	9.027	8.540	8.079	7.642
16.0	9.870	9.344	8.844	8.370	7.922	7.496
17.0	9.665	9.153	8.667	8.207	7.770	7.356
18.0	9.467	8.969	8.497	8.049	7.624	7.221
19.0	9.276	8.792	8.333	7.896	7.483	7.090
20.0	9.092	8.621	8.174	7.749	7.346	6.964
21.0	8.915	8.456	8.021	7.607	7.214	6.842
22.0	8.743	8.297	7.873	7.470	7.087	6.723
23.0	8.578	8.143	7.730	7.337	6.963	6.609
24.0	8.418	7.994	7.591	7.208	6.844	6.498
25.0	8.263	7.850	7.457	7.083	6.728	6.390
26.0	8.113	7.711	7.327	6.962	6.615	6.285
27.0	7.968	7.575	7.201	6.845	6.506	6.184
28.0	7.827	7.444	7.079	6.731	6.400	6.085
29.0	7.691	7.317	6.961	6.621	6.297	5.990
30.0	7.559	7.194	6.845	6.513	6.197	5.896
31.0	7.430	7.073	6.733	6.409	6.100	5.806
32.0	7.305	6.957	6.624	6.307	6.005	5.717
33.0	7.183	6.843	6.518	6.208	5.912	5.631
34.0	7.065	6.732	6.415	6.111	5.822	5.546
35.0	6.950	6.624	6.314	6.017	5.734	5.464
36.0	6.837	6.519	6.215	5.925	5.648	5.384
37.0	6.727	6.416	6.119	5.835	5.564	5.305
38.0	6.620	6.316	6.025	5.747	5.481	5.228
39.0	6.515	6.217	5.932	5.660	5.400	5.152
40.0	6.412	6.121	5.842	5.576	5.321	5.078
41.0	6.312	6.026	5.753	5.493	5.243	5.005
42.0	6.213	5.934	5.667	5.411	5.167	4.933
43.0	6.116	5.843	5.581	5.331	5.091	4.862
44.0	6.021	5.753	5.497	5.252	5.017	4.793
45.0	5.927	5.665	5.414	5.174	4.944	4.724

TABLE 4500-O:I, CONT.

| Temperature °C | Oxygen Solubility mg/L | | | | | |
	Chlorinity: 0	5.0	10.0	15.0	20.0	25.0
46.0	5.835	5.578	5.333	5.097	4.872	4.656
47.0	5.744	5.493	5.252	5.021	4.801	4.589
48.0	5.654	5.408	5.172	4.947	4.730	4.523
49.0	5.565	5.324	5.094	4.872	4.660	4.457
50.0	5.477	5.242	5.016	4.799	4.591	4.392

NOTE:
1. The table provides three decimal places to aid interpolation. When computing saturation values to be used with measured values, such as in computing DO deficit in a receiving water, precision of measured values will control choice of decimal places to be used.
2. Equations are available to compute DO concentration in fresh water[1-3] and in seawater[1] at equilibrium with water-saturated air. Figures and tables also are available.[3]

Calculate the equilibrium oxygen concentration, C^*, from the equation:

$$\ln C^* = -139.344\ 11 + (1.575\ 701 \times 10^5/T) - (6.642\ 308 \\ \times 10^7/T^2) + (1.243\ 800 \times 10^{10}/T^3) \\ - (8.621\ 949 \times 10^{11}/T^4) - \text{Chl}\ [(3.1929) \times 10^{-2}) \\ - (1.9428 \times 10^1/T) + (3.8673 \times 10^3/T^2)]$$

where:

C^* = equilibrium oxygen concentration at 101.325 kPa, mg/L,

T = temperature (°K) = °C + 273.150, (°C is between 0.0 and 40.0 in the equation; the table is accurate up to 50.0), and

Chl = chlorinity (see definition in Note 4, below).

Example 1: At 20°C and 0.000 Chl, $\ln C^* = 2.207\ 442$ and
$\quad\quad C^* = 9.092$ mg/L;
Example 2: At 20°C and 15.000 Chl,
$\quad\quad \ln C^* = (2.207\ 442) - 15.000\ (0.010\ 657)$
$\quad\quad = 2.0476$ and $C^* = 7.749$ mg/L.
When salinity is used, replace the chlorinity term $(-\text{Chl}[...])$ by:
$-S\ (1.7674 \times 10^{-2}) - (1.0754 \times 10^1/T)$
$+ (2.1407 \times 10^3/T^2)$

where:

S = salinity (see definition in Note 4, below).

3. For nonstandard conditions of pressure:

$$C_p = C^* P \left[\frac{(1 - P_{wv}/P)\ (1 - \theta P)}{(1 - P_{wv})\ (1 - \theta)} \right]$$

where:

C_p = equilibrium oxygen concentration at nonstandard pressure, mg/L,

C^* = equilibrium oxygen concentration at standard pressure of 1 atm, mg/L.

P = nonstandard pressure, atm,

P_{wv} = partial pressure of water vapor, atm, computed from: $\ln P_{wv} = 11.8571 - (3840.70/T) - (216\ 961/T^2)$,

T = temperature, °K,

$\theta = 0.000\ 975 - (1.426 \times 10^{-5}t) + (6.436 \times 10^{-8}t^2)$, and

t = temperature, °C.

N.B.: Although not explicit in the above, the quantity in brackets in the equation for C_p has dimensions of atm^{-1} per Reference 4, so that P multiplied by this quantity is dimensionless.

Also, the equation for ln P_{wv} is strictly valid for fresh water only, but for practical purposes no error is made by neglecting the effect of salinity. An equation for P_{wv} that includes the salinity factor may be found in Reference 1.

Example 3: At 20°C, 0.000 Chl, and 0.700 atm,

$C_p = C* \ P \ (0.990\ 092) = 6.30$ mg/L.

4. Definitions:

Salinity: Although salinity has been defined traditionally as the total solids in water after all carbonates have been converted to oxides, all bromide and iodide have been replaced by chloride, and all organic matter has been oxidized (see Section 2520), the new scale used to define salinity is based on the electrical conductivity of seawater relative to a specified solution of KCl in water.[5] The scale is dimensionless and the traditional dimension of parts per thousand (i.e., g/kg of solution) no longer applies.

Chlorinity: Chlorinity is defined in relation to salinity as follows:

Salinity = 1.806 55 × chlorinity

Although chlorinity is not equivalent to chloride concentration, the factor for converting a chloride concentration in seawater to include bromide, for example, is only 1.0045 (based on the relative molecular weights and amounts of the two ions). Therefore, for practical purposes, chloride concentration (in g/kg of solution) is nearly equal to chlorinity in seawater. For wastewater, it is necessary to know the ions responsible for the solution's electrical conductivity to correct for their effect on oxygen solubility and use of the tabular value. If this is not done, the equation is inappropriate unless the relative composition of the wastewater is similar to that of seawater.

the standard deviation may be as high as 100 μg/L. Still greater errors may occur in testing waters having organic suspended solids or heavy pollution. Avoid errors due to carelessness in collecting samples, prolonging the completion of test, or selecting an unsuitable modification.

6. References

1. BENSON, B.B. & D. KRAUSE, JR. 1984. The concentration and isotopic fractionation of oxygen dissolved in freshwater and seawater in equilibrium with the atmosphere. *Limnol. Oceanogr.* 29:620.

2. BENSON, B.B. & D. KRAUSE, JR. 1980. The concentration and isotopic fractionation of gases dissolved in fresh water in equilibrium with the atmosphere: I. Oxygen. *Limnol. Oceanogr.* 25:662.

3. MORTIMER, C.H. 1981. The oxygen content of air-saturated fresh waters over ranges of temperature and atmospheric pressure of limnological interest. *Int. Assoc. Theoret. Appl. Limnol.*, Communication No. 22, Stuttgart, West Germany.

4. SULZER, F. & W.M. WESTGARTH. 1962. Continuous D. O. recording in activated sludge. *Water Sewage Works* 109: 376.

5. UNITED NATIONS EDUCATIONAL, SCIENTIFIC & CULTURAL ORGANIZATION. 1981. Background Papers and Supporting Data on the Practical Salinity Scale 1978. Tech. Paper Mar. Sci. No. 37.

4500-O D.   Permanganate Modification

1. General Discussion

Use the permanganate modification only on samples containing ferrous iron. Interference from high concentrations of ferric iron (up to several hundred milligrams per liter), as in acid mine water, may be overcome by the addition of 1 mL potassium fluoride (KF) and azide, provided that the final titration is made immediately after acidification.

This procedure is ineffective for oxidation of sulfite, thiosulfate, polythionate, or the organic matter in wastewater. The error with samples containing 0.25% by volume of digester waste from the manufacture of sulfite pulp may amount to 7 to 8 mg DO/L. With such samples, use the alkali-hypochlorite modification.[1] At best, however, the latter procedure gives low results, the deviation amounting to 1 mg/L for samples containing 0.25% digester wastes.

2. Reagents

All the reagents required for Method C, and in addition:

a. Potassium permanganate solution: Dissolve 6.3 g $KMnO_4$ in distilled water and dilute to 1 L.

b. Potassium oxalate solution: Dissolve 2 g $K_2C_2O_4 \cdot H_2O$ in 100 mL distilled water; 1 mL will reduce about 1.1 mL permanganate solution.

3. Procedure

a. To a sample collected in a 250- to 300-mL bottle add, below the surface, 0.70 mL conc H_2SO_4, 1 mL $KMnO_4$ solution, and 1 mL KF solution. Stopper and mix by inverting. Never add more than 0.7 mL conc H_2SO_4 as the first step of pretreatment. Add acid with a 1-mL pipet graduated to 0.1 mL. Add sufficient $KMnO_4$ solution to obtain a violet tinge that persists for 5 min. If the permanganate color is destroyed in a shorter time, add additional $KMnO_4$ solution, but avoid large excesses.

b. Remove permanganate color completely by adding 0.5 to 1.0 mL $K_2C_2O_4$ solution. Mix well and let stand in the dark to facilitate the reaction. Excess oxalate causes low results; add only enough $K_2C_2O_4$ to decolorize the $KMnO_4$ completely without an excess of more than 0.5 mL. Complete decolorization in 2 to 10 min. If it is impossible to decolorize the sample without adding a large excess of oxalate, the DO result will be inaccurate.

c. From this point the procedure closely parallels that in Section 4500-O.C.3. Add 1 mL $MnSO_4$ solution and 3 mL alkali-iodide-azide reagent. Stopper, mix, and let precipitate settle a short time; acidify with 2 mL conc H_2SO_4. When 0.7 mL acid, 1 mL $KMnO_4$ solution, 1 mL $K_2C_2O_4$ solution, 1 mL $MnSO_4$ solution, and 3 mL alkali-iodide-azide (or a total of 6.7 mL reagents) are used in a 300-mL bottle, take $200 \times 300/(300 - 6.7) = 205$ mL for titration.

This correction is slightly in error because the $KMnO_4$ solution is nearly saturated with DO and 1 mL would add about 0.008 mg oxygen to the DO bottle. However, because precision of the method (standard deviation, 0.06 mL thiosulfate titration, or 0.012 mg DO) is 50% greater than this error, a correction is unnecessary. When substantially more $KMnO_4$ solution is used routinely, use a solution several times more concentrated so that 1 mL will satisfy the permanganate demand.

4. Reference

1. THERIAULT, E.J. & P.D. MCNAMEE. 1932. Dissolved oxygen in the presence of organic matter, hypochlorites, and sulfite wastes. *Ind. Eng. Chem.*, Anal. Ed. 4:59.

4500-O E. Alum Flocculation Modification

1. General Discussion

Samples high in suspended solids may consume appreciable quantities of iodine in acid solution. The interference due to solids may be removed by alum flocculation.

2. Reagents

All the reagents required for the azide modification (Section 4500-O.C.2) and in addition:

a. Alum solution: Dissolve 10 g aluminum potassium sulfate, $AlK(SO_4)_2 \cdot 12H_2O$, in distilled water and dilute to 100 mL.

b. Ammonium hydroxide, NH_4OH, conc.

3. Procedure

Collect sample in a glass-stoppered bottle of 500 to 1000 mL capacity, using the same precautions as for regular DO samples. Add 10 mL alum solution and 1 to 2 mL conc NH_4OH. Stopper and invert gently for about 1 min. Let sample settle for about 10 min and siphon clear supernate into a 250- to 300-mL DO bottle until it overflows. Avoid sample aeration and keep siphon submerged at all times. Continue sample treatment as in Section 4500-O.C.3 or an appropriate modification.

4500-O F. Copper Sulfate-Sulfamic Acid Flocculation Modification

1. General Discussion

This modification is used for biological flocs such as activated sludge mixtures, which have high oxygen utilization rates.

2. Reagents

All the reagents required for the azide modification (Section 4500-O.C.2) and, in addition:

Copper sulfate-sulfamic acid inhibitor solution: Dissolve 32 g technical-grade NH_2SO_2OH without heat in 475 mL distilled water. Dissolve 50 g $CuSO_4 \cdot 5H_2O$ in

500 mL distilled water. Mix the two solutions and add 25 mL conc acetic acid.

3. Procedure

Add 10 mL $CuSO_4$-NH_2SO_2OH inhibitor to a 1-L glass-stoppered bottle. Insert bottle in a special sampler designed so that bottle fills from a tube near bottom and overflows only 25 to 50% of bottle capacity. Collect sample, stopper, and mix by inverting. Let suspended solids settle and siphon relatively clear supernatant liquor into a 250- to 300-mL DO bottle. Continue sample treatment as rapidly as possible by the azide (Section 4500-O.C.3) or other appropriate modification.

4500-O G. Membrane Electrode Method

1. General Discussion

Various modifications of the iodometric method have been developed to eliminate

or minimize effects of interferences; nevertheless, the method still is inapplicable to a variety of industrial and domestic wastewaters.[1] Moreover, the iodometric method is not suited for field testing and cannot be adapted easily for continuous monitoring or for DO determinations in situ.

Polarographic methods using the dropping mercury electrode or the rotating platinum electrode have not been reliable always for the DO analysis in domestic and industrial wastewaters because impurities in the test solution can cause electrode poisoning or other interferences.[2,3] With membrane-covered electrode systems these problems are minimized, because the sensing element is protected by an oxygen-permeable plastic membrane that serves as a diffusion barrier against impurities.[4-6] Under steady-state conditions the current is directly proportional to the DO concentration.*

Membrane electrodes of the polarographic[4] as well as the galvanic[5] type have been used for DO measurements in lakes and reservoirs,[8] for stream survey and control of industrial effluents,[9,10] for continuous monitoring of DO in activated sludge units,[11] and for estuarine and oceanographic studies.[12] Being completely submersible, membrane electrodes are suited for analysis in situ. Their portability and ease of operation and maintenance make them particularly convenient for field applications. In laboratory investigations, membrane electrodes have been used for continuous DO analysis in bacterial cultures, including the BOD test.[5,13]

Membrane electrodes provide an excellent method for DO analysis in polluted waters, highly colored waters, and strong waste effluents. They are recommended for use especially under conditions that are unfavorable for use of the iodometric method, or when that test and its modifications are subject to serious errors caused by interferences.

a. Principle: Oxygen-sensitive membrane electrodes of the polarographic or galvanic type are composed of two solid metal electrodes in contact with supporting electrolyte separated from the test solution by a selective membrane. The basic difference between the galvanic and the polarographic systems is that in the former the electrode reaction is spontaneous (similar to that in a fuel cell), while in the latter an external source of applied voltage is needed to polarize the indicator electrode. Polyethylene and fluorocarbon membranes are used commonly because they are permeable to molecular oxygen and are relatively rugged.

Membrane electrodes are commercially available in some variety. In all these instruments the "diffusion current" is linearly proportional to the concentration of molecular oxygen. The current can be converted easily to concentration units (e.g., milligrams per liter) by a number of calibration procedures.

Membrane electrodes exhibit a relatively high temperature coefficient largely due to changes in the membrane permeability.[6] The effect of temperature on the electrode sensitivity, ϕ (microamperes per milligram per liter), can be expressed by the following simplified relationship:[6]

$$\log \phi = 0.43 \, mt + b$$

where:
t = temperature, °C,
m = constant that depends on the membrane material, and
b = constant that largely depends on membrane thickness.

If values of ϕ and m are determined for one temperature (ϕ_0 and t_0), it is possible to calculate the sensitivity at any desired temperature (ϕ and t) as follows:

$$\log \phi = \log \phi_0 + 0.43 \, m \, (t - t_0)$$

---

*Fundamentally, the current is directly proportional to the activity of molecular oxygen.[7]

Nomographic charts for temperature correction can be constructed easily[7] and are available from some manufacturers. An example is shown in Figure 4500-O:2, in which, for simplicity, sensitivity is plotted versus temperature on semilogarithmic coordinates. Check one or two points frequently to confirm original calibration. If calibration changes, the new calibration should be parallel to the original, provided that the same membrane material is used.

Temperature compensation also can be made automatically by using thermistors in the electrode circuit.[4] However, thermistors may not compensate fully over a wide temperature range. For certain applications where high accuracy is required, use calibrated nomographic charts to correct for temperature effect.

To use the DO membrane electrode in estuarine waters or in wastewaters with varying ionic strength, correct for effect of salting-out on electrode sensitivity.[6,7] This effect is particularly significant for large changes in salt content. Electrode sensitivity varies with salt concentration according to the following relationship:

$$\log \phi_s = 0.43\, m_s C_s + \log \phi_0$$

where:

ϕ_s, ϕ_0 = sensitivities in salt solution and distilled water, respectively,

C_s = salt concentration (preferably ionic strength), and

m_s = constant (salting-out coefficient).

If ϕ_0 and m_s are determined, it is possible to calculate sensitivity for any value of C_s. Conductivity measurements can be used to approximate salt concentration (C_s). This is particularly applicable to estuarine waters. Figure 4500-O:3 shows calibration curves for sensitivity of varying salt solutions at different temperatures.

b. Interference: Plastic films used with membrane electrode systems are permeable to a variety of gases besides oxygen, although none is depolarized easily at the indicator electrode. Prolonged use of membrane electrodes in waters containing such gases as hydrogen sulfide (H_2S) tends to lower cell sensitivity. Eliminate this interference by frequently changing and calibrating the membrane electrode.

c. Sampling: Because membrane electrodes offer the advantage of analysis in situ they eliminate errors caused by sample handling and storage. If sampling is required, use the same precautions suggested for the iodometric method.

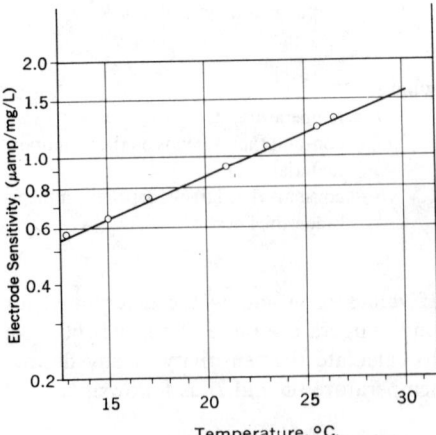

Figure 4500-O:2. Effect of temperature on electrode sensitivity.

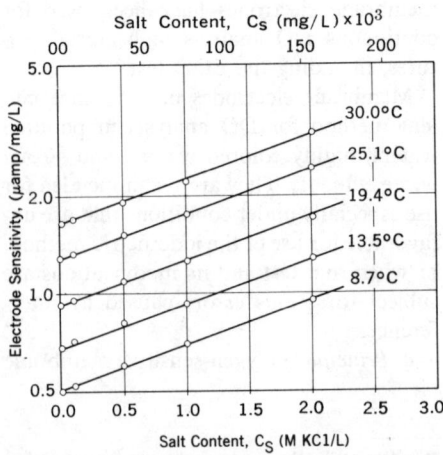

Figure 4500-O:3. The salting-out effect at different temperatures.

2. Apparatus

Oxygen-sensitive membrane electrode, polarographic or galvanic, with appropriate meter.

3. Procedure

a. Calibration: Follow manufacturer's calibration procedure exactly to obtain guaranteed precision and accuracy. Generally, calibrate membrane electrodes by reading against air or a sample of known DO concentration (determined by iodometric method) as well as in a sample with zero DO. (Add excess sodium sulfite, Na_2SO_3, and a trace of cobalt chloride, $CoCl_2$, to bring DO to zero.) Preferably calibrate with samples of water under test. Avoid an iodometric calibration where interfering substances are suspected. The following illustrate the recommended procedures:

1) Fresh water—For unpolluted samples where interfering substances are absent, calibrate in the test solution or distilled water, whichever is more convenient.

2) Salt water—Calibrate directly with samples of seawater or waters having a constant salt concentration in excess of 1000 mg/L.

3) Fresh water containing pollutants or interfering substances—Calibrate with distilled water because erroneous results occur with the sample.

4) Salt water containing pollutants or interfering substances—Calibrate with a sample of clean water containing the same salt content as the sample. Add a concentrated potassium chloride (KCl) solution (see Conductivity, Section 2510 and Table 2510:I) to distilled water to produce the same specific conductance as that in the sample. For polluted ocean waters, calibrate with a sample of unpolluted seawater.

5) Estuary water containing varying quantities of salt—Calibrate with a sample of uncontaminated seawater or distilled or tap water. Determine sample chloride or salt concentration and revise calibration to account for change of oxygen solubility in the estuary water.[7]

b. Sample measurement: Follow all precautions recommended by manufacturer to insure acceptable results. Take care in changing membrane to avoid contamination of sensing element and also trapping of minute air bubbles under the membrane, which can lead to lowered response and high residual current. Provide sufficient sample flow across membrane surface to overcome erratic response (see Figure 4500-O:4 for a typical example of the effect of stirring).

c. Validation of temperature effect: Check frequently one or two points to verify temperature correction data.

4. Precision and Bias

With most commercially available membrane electrode systems an accuracy of ± 0.1 mg DO/L and a precision of ± 0.05 mg DO/L can be obtained.

5. References

1. McKeown, J.J., L.C. Brown & G.W. Gove. 1967. Comparative studies of dissolved oxygen analysis methods. *J. Water Pollut. Control Fed.* 39:1323.

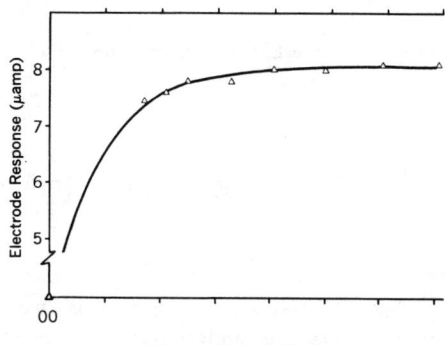

Figure 4500-O:4. Typical trend of effect of stirring on electrode response.

2. LYNN, W.R. & D.A. OKUN. 1955. Experience with solid platinum electrodes in the determination of dissolved oxygen. *Sewage Ind. Wastes* 27:4.

3. MANCY, K.H. & D.A. OKUN. 1960. Automatic recording of dissolved oxygen in aqueous systems containing surface active agents. *Anal. Chem.* 32:108.

4. CARRITT, D.E. & J.W. KANWISHER. 1959. An electrode system for measuring dissolved oxygen. *Anal. Chem.* 31:5.

5. MANCY, K.H. & W.C. WESTGARTH. 1962. A galvanic cell oxygen analyzer. *J. Water Pollut. Control Fed.* 34:1037.

6. MANCY, K.H., D.A. OKUN & C.N. REILLEY. 1962. A galvanic cell oxygen analyzer. *J. Electroanal. Chem.* 4:65.

7. MANCY, K.H. & T. JAFFE. 1966. Analysis of Dissolved Oxygen in Natural and Waste Waters. Publ. No. 999-WP-37, U.S. Public Health Serv., Washington, D.C.

8. WEISS, C.M. & R.T. OGLESBY. 1963. Instrumentation for monitoring water quality in reservoirs. American Water Works Assoc. 83rd Annual Conf., New York, N.Y.

9. CLEARY, E.J. 1962. Introducing the ORSANCO robot monitor. *Proc. Water Quality Meas. Instrum.* Publ. No. 108, U.S. Public Health Serv., Washington, D.C.

10. MACKERETH, F.J.H. 1964. An improved galvanic cell for determination of oxygen concentrations in fluids. *J. Sci. Instrum.* 41:38.

11. SULZER, F. & W.M. WESTGARTH. 1962. Continuous D.O. recording in activated sludge. *Water Sewage Works* 109:376.

12. DUXBURY, A.C. 1963. Calibration and use of a galvanic type oxygen electrode in field work. *Limnol. Oceanogr.* 8:483.

13. LIPNER, H.J., L.R. WITHERSPOON & V.C. CHAMPEAUS. 1964. Adaptation of a galvanic cell for microanalysis of oxygen. *Anal. Chem.* 36:204.

4500-O₃   OZONE (RESIDUAL)* (PROPOSED)

4500-O₃ A.   Introduction

Ozone, a potent germicide, is used also as an oxidizing agent for the destruction of

---

*Approved by Standard Methods Committee, 1988.

organic compounds producing taste and odor in drinking water, for the destruction of organic coloring matter, and for the oxidation of reduced iron or manganese salts to insoluble oxides.

4500-O₃ B.   Indigo Colorimetric Method

1. General Discussion

The indigo colorimetric method is quantitative, selective, and simple; it replaces methods based on the measurement of total oxidant. The method is applicable to lake water, river infiltrate, manganese-contain-

ing groundwaters, extremely hard groundwaters, and even biologically treated domestic wastewaters.

a. Principle: In acidic solution, ozone rapidly decolorizes indigo. The decrease in absorbance is linear with increasing concentration. The proportionality constant at

600 nm is 0.42 $\pm$ 0.01/cm/mg/L ($\Delta\epsilon$ = 20 000/$M \cdot$cm) compared to the ultraviolet absorption of pure ozone of ϵ = 2950/$M \cdot$cm at 258 nm).[1]

b. *Interferences:* Hydrogen peroxide (H_2O_2) and organic peroxides decolorize the indigo reagent very slowly. H_2O_2 does not interfere if ozone is measured in less than 6 h after adding reagents. Organic peroxides may react more rapidly. Fe(III) does not interfere. Mn(II) does not interfere but it is oxidized by ozone to forms that decolorize the reagent. Correct for this interference by making the measurement relative to a blank in which the ozone has been destroyed selectively. Without the corrective procedure, 0.1 mg/L ozonated manganese gives a response of about 0.08 mg/L apparent ozone. Chlorine also interferes but it can be masked by malonic acid. Bromine, which can be formed by oxidation of Br⁻, interferes (1 mole HOBr corresponds to 0.4 mole ozone).

c. *Minimum detectable concentration:* For the spectrophotometric procedure using thermostated cells and a high-quality photometer, the low-range procedure will measure down to 2 μg O₃/L. For the visual method the detection limit is 10 μg/L.

2. Apparatus

a. *Photometer:* Spectrophotometer or filter colorimeter for use at 600 $\pm$ 5 nm.

b. *Glass cylinders* (for visual procedure): 100-mL graduated glass cylinders, preferably with flat bottoms.

3. Reagents

a. *Indigo stock solution:* Add about 500 mL distilled water and 1 mL conc phosphoric acid to a 1-L volumetric flask. With stirring, add 770 mg potassium indigo trisulfonate, $C_{16}H_7N_2O_{11}S_3K_3$ (commercially available at about 80 to 85% purity). Fill to mark with distilled water. A 1:100 dilution exhibits an absorbance of 0.20 $\pm$ 0.010 cm at 600 nm. The stock solution is

stable for about 4 months when stored in the dark. Discard when absorbance of a 1:100 dilution falls below 0.16/cm.

b. *Indigo reagent I:* To a 1-L volumetric flask add 20 mL indigo stock solution, 10 g sodium dihydrogen phosphate (NaH_2PO_4), and 7 mL conc phosphoric acid. Dilute to mark. Prepare solution fresh when its absorbance decreases to less than 80% of its initial value, typically within a week.

c. *Indigo reagent II:* Proceed as with indigo reagent I, but add 100 mL indigo stock solution instead of 20 mL.

d. *Malonic acid reagent:* Dissolve 5 g malonic acid in water and dilute to 100 mL.

e. *Glycine reagent:* Dissolve 7 g glycine in water and dilute to 100 mL.

4. Procedure

a. *Spectrophotometric procedure:*

1) Concentration range 0.01 to 0.1 mg O₃/L—Add 10.0 mL indigo reagent I to two 100-mL volumetric flasks. Fill one flask (blank) to mark with distilled water. Fill other flask to mark with sample. Add sample so that completely decolorized zones are eliminated quickly by stirring but no ozone degassing occurs. Measure absorbance of both solutions at 600 $\pm$ 5 nm as soon as possible but at least within 4 h. Preferably use 10-cm cells. Calculate the ozone concentration from the difference between the absorbances found in sample and blank (¶ 5a below). (NOTE: A maximum delay of 4 h before spectrophotometric reading can be tolerated only for drinking water samples. For other sample types test the time drift.)

2) Range 0.05 to 0.5 mg O₃/L—Proceed as above using 10.0 mL indigo reagent II instead of reagent I. Preferably measure absorbance in 4- or 5-cm cells.

3) Concentrations greater than 0.3 mg O₃/L—Proceed using indigo reagent II, but for these higher ozone concentrations

use a correspondingly smaller sample volume. Dilute resulting mixture to 100 mL with distilled water. Use a glass pipet for dosing sample; let sample flow through an erlenmeyer flask for at least 1 min without generating bubbles. Rinse pipet with sample and add measured amount to flask while keeping the pipet tip below the surface.

4) Control of interferences—In presence of chlorine, place 1 mL malonic acid reagent in both flasks before adding sample and/or filling to mark. Measure absorbance as soon as possible, within 60 min (Br^-, Br_2, and HOBr are only partially masked by malonic acid).

In presence of manganese prepare a blank solution using sample, in which ozone is selectively destroyed by addition of glycine. Place 0.1 mL glycine reagent in 100-mL volumetric flask (blank) and 10.0 mL indigo reagent II in second flask (sample). Pipet exactly the same volume of sample into each flask. Adjust dose so that decolorization in second flask is easily visible but complete bleaching does not result (maximum 80 mL).

Insure that pH of glycine/sample mixture in blank flask (before adding indigo) is not below 6 because reaction between ozone and glycine becomes very slow at low pH. Stopper flasks and mix by carefully inverting. Add 10.0 mL indigo reagent II to blank flask only 30 to 60 s after sample addition. Fill both flasks to the mark with ozone-free water and mix thoroughly. Measure absorbance of both solutions at comparable contact times of approximately 30 to 60 min (after this time, residual manganese oxides further discolor indigo only slowly and the drift of absorbance in blank and sample become comparable). Reduced absorbance in blank flask results from manganese oxides while that in sample flask is due to ozone plus manganese oxide.

5) Calibration—Because ozone is uns, base measurements on known and t loss of absorbance of the indigo

reagent ($f = 0.42 \pm 0.01$/cm/mg O_3/L). For maximum accuracy analyze the lot of potassium indigo trisulfonate (no commercial lot has been found to deviate from $f = 0.42$) using the iodometric procedure.

When using a filter photometer, readjust the conversion factor, f, by comparing photometer sensitivity with absorbance at 600 nm by an accurate spectrophotometer.

b. Visual procedure:

1) Concentration range 0.01 to 0.1 mg O_3/L—Add 10.0 mL indigo reagent I to each of two identical 100-mL graduated glass cylinders. Fill reference cylinder (blank) to the mark with distilled water and other cylinder with sample. Add sample to cylinder so that completely decolorized zones are eliminated quickly by mixing but no degassing occurs. Pour off blank by portions until liquid height gives the same apparent color intensity as the sample when viewed from top. Record volume in blank cylinder. Color comparisons may be made up to 4 h after sample addition.

2) Concentrations greater than 0.1 mg O_3/L—Proceed as above, adding either 30 or 45 mL of sample and dilute to the mark.

3) Manganese-containing waters—The visual method is not suitable for these waters when the manganese concentration is comparable to that of the ozone because the difference measurement becomes too inaccurate.

5. Calculations

a. Spectrophotometric procedure:

$$\text{mg } O_3/L = \frac{100 \times \Delta A}{f \times b \times V}$$

where:

ΔA = difference in absorbance between sample and blank,

b = path length of cell, cm,

V = volume of sample, mL (normally 90 mL), and

f = 0.42.

The factor f is based on a sensitivity factor of 20 000/cm for the change of absorbance (600 nm) per mole of added ozone per liter. It was calibrated by iodometric titration. The UV absorbance of ozone in pure water may serve as a secondary standard: the factor $f = 0.42$ corresponds to an absorption coefficient for aqueous ozone, $\epsilon = 2950/M \cdot cm$ at 258 nm.

  b. *Visual procedure:*

$$\text{mg } O_3/L = \frac{(100 - V) \times k}{100}$$

where:

  $V = $ volume of reference solution in blank cylinder, mL, and

  $k = $ conversion factor for indigo stock solution, calibrated by a spectrophotometric analysis of ozone. The value is about 0.10 mg O_3/L if the 1:100 dilution gives an absorbance of 0.19/cm.

When adding only 45 or 30 mL of sample, the conversion factor becomes $2k$ or $3k$, respectively.

6. Precision and Bias

  a. *Spectrophotometric procedure:* In the absence of interferences, the relative error is less than 5% without special sampling setups. In laboratory testing this may be reduced to 1%.

Because this method is based on the differences in absorbance between the sample and blank (ΔA) the method is not applicable in the presence of chlorine. If the manganese content exceeds the ozone, precision is reduced. If the ratio of manganese to ozone is less than 10:1, ozone concentrations above 0.02 mg/L may be determined with a relative error of less than 20%.

  b. *Visual procedure:* Duplicate determinations gave an average deviation of 1 to 1.5% within the pair. If the manganese concentration is comparable to that of ozone, this method is not applicable.

7. Reference

1. HOIGNÉ, J. & H. BADER. 1980. Bestimmung von Ozon und Chlordioxid im Wasser mit der Indigo-Methode. *Vom Wasser* 55:261.

8. Bibliography

THÉNARD, A. & P. THÉNARD. 1872. Mémoire sur l'action comparée de l'ozone sur le sulfate d'indigo et l'acide arsénieux. *Comptes Rend. Acad. Sci.* 75:458.

BADER, H. & J. HOIGNÉ. 1981. Determination of ozone in water by the indigo method. *Water Res.* 15:449.

BADER, H. & J. HOIGNÉ. 1982. Colorimetric method for the measurement of aqueous ozone based on the decolorization of indigo derivatives. *In* W.J. Masschelein, ed. Ozonization Manual for Water and Wastewater Treatment. John Wiley & Sons, New York, N.Y.

BADER, H. & J. HOIGNÉ. 1982. Determination of ozone in water by the indigo method: A submitted standard method. *Ozone: Sci. Eng.* 4:169.

GILBERT, E. & J. HOIGNÉ. 1983. Messung von Ozon in Wasserwerken; Vergleich der DPD-und Indigo-Methode. *GWF-Wasser/Abwasser* 124:527.

HAAG, W.R. & J. HOIGNÉ. 1983. Ozonation of bromide-containing waters: kinetics of formation of hypobromous acid and bromate. *Environ. Sci. Technol.* 17:261.

STRAKA, M.R., G.E. PACEY & G. GORDON. 1984. Residual ozone determination by flow injection analysis. *Anal. Chem.* 56:1973.

STRAKA, M.R., G. GORDON & G.E. PACEY. 1985. Residual aqueous ozone determination by gas diffusion flow injection analysis. *Anal. Chem.* 57:1799.

GORDON, G. & G.E. PACEY. 1986. An introduction to the chemical reactions of ozone pertinent to its analysis. *In* R.G. Rice, L.J. Bollyky & W.J. Lacy, eds. Analytical Aspects of Ozone Treatment of Water and Wastewater—A Monograph. Lewis Publishers, Inc., Chelsea, Mich.

4500-P PHOSPHORUS*

4500-P A. Introduction

1. Occurrence

Phosphorus occurs in natural waters and in wastewaters almost solely as phosphates. These are classified as orthophosphates, condensed phosphates (pyro-, meta-, and other polyphosphates), and organically bound phosphates. They occur in solution, in particles or detritus, or in the bodies of aquatic organisms.

These forms of phosphate arise from a variety of sources. Small amounts of certain condensed phosphates are added to some water supplies during treatment. Larger quantities of the same compounds may be added when the water is used for laundering or other cleaning, because these materials are major constituents of many commercial cleaning preparations. Phosphates are used extensively in the treatment of boiler waters. Orthophosphates applied to agricultural or residential cultivated land as fertilizers are carried into surface waters with storm runoff and to a lesser extent with melting snow. Organic phosphates are formed primarily by biological processes. They are contributed to sewage by body wastes and food residues, and also may be formed from orthophosphates in biological treatment processes or by receiving water biota.

Phosphorus is essential to the growth of organisms and can be the nutrient that limits the primary productivity of a body of water. In instances where phosphate is a growth-limiting nutrient, the discharge of raw or treated wastewater, agricultural drainage, or certain industrial wastes to that water may stimulate the growth of photosynthetic aquatic micro- and macro-organisms in nuisance quantities.

Phosphates also occur in bottom sediments and in biological sludges, both as precipitated inorganic forms and incorporated into organic compounds.

2. Definition of Terms

Phosphorus analyses embody two general procedural steps: (*a*) conversion of the phosphorus form of interest to dissolved orthophosphate, and (*b*) colorimetric determination of dissolved orthophosphate. The separation of phosphorus into its various forms is defined analytically but the analytical differentiations have been selected so that they may be used for interpretive purposes.

Filtration through a 0.45-μm-pore-diam membrane filter separates dissolved from suspended forms of phosphorus. No claim is made that filtration through 0.45-μm filters is a true separation of suspended and dissolved forms of phosphorus; it is merely a convenient and replicable analytical technique designed to make a gross separation.

Membrane filtration is selected over depth filtration because of the greater likelihood of obtaining a consistent separation of particle sizes. Prefiltration through a glass fiber filter may be used to increase the filtration rate.

Phosphates that respond to colorimetric tests without preliminary hydrolysis or oxidative digestion of the sample are termed "reactive phosphorus." While reactive phosphorus is largely a measure of orthophosphate, a small fraction of any condensed phosphate present usually is hydrolyzed unavoidably in the procedure. Reactive phosphorus occurs in both dissolved and suspended forms.

Acid hydrolysis at boiling-water temperature converts dissolved and particulate condensed phosphates to dissolved ortho-

*Approved by Standard Methods Committee, 1988.

phosphate. The hydrolysis unavoidably releases some phosphate from organic compounds, but this may be reduced to a minimum by judicious selection of acid strength and hydrolysis time and temperature. The term "acid-hydrolyzable phosphorus" is preferred over "condensed phosphate" for this fraction.

The phosphate fractions that are converted to orthophosphate only by oxidation destruction of the organic matter present are considered "organic" or "organically bound" phosphorus. The severity of the oxidation required for this conversion depends on the form—and to some extent on the amount—of the organic phosphorus present. Like reactive phosphorus and acid-hydrolyzable phosphorus, organic phosphorus occurs both in the dissolved and suspended fractions.

The total phosphorus as well as the dissolved and suspended phosphorus fractions each may be divided analytically into the three chemical types that have been described: reactive, acid-hydrolyzable, and organic phosphorus. Figure 4500-P:1 shows the steps for analysis of individual phosphorus fractions. As indicated, determinations usually are conducted only on the unfiltered and filtered samples. Suspended fractions generally are determined by difference.

3. Selection of Method

a. Digestion methods: Because phosphorus may occur in combination with organic matter, a digestion method to determine total phosphorus must be able to oxidize organic matter effectively to release phosphorus as orthophosphate. Three digestion methods are given in Section 4500-P.B.3, 4, and 5. The perchloric acid method, the most drastic and time-consuming method, is recommended only for particularly difficult samples such as sediments. The nitric acid-sulfuric acid method is recommended for most samples. By far the simplest method is the persulfate oxidation tech-

nique. It is recommended that this method be checked against one or more of the more drastic digestion techniques and be adopted if identical recoveries are obtained.

After digestion, determine liberated orthophosphate by Method C, D, or E. The colorimetric method used, rather than the digestion procedure, governs in matters of interference and minimum detectable concentration.

b. Colorimetric methods: Three methods of orthophosphate determination are described. Selection depends largely on the concentration range of orthophosphate. The vanadomolybdophosphoric acid method (C) is most useful for routine analyses in the range of 1 to 20 mg P/L. The stannous chloride method (D) or the ascorbic acid method (E) is more suited for the range of 0.01 to 6 mg P/L. An extraction step is recommended for the lower levels of this range and when interferences must be overcome. An automated version of the ascorbic acid method F also is presented.

4. Precision and Bias

To aid in method selection, Table 4500-P:I presents the results of various combinations of digestions, hydrolysis, and colorimetric techniques for three synthetic samples of the following compositions:

Sample 1: 100 μg orthophosphate phosphorus (PO_4^{3-}-P)/L, 80 μg condensed phosphate phosphorus/L (sodium hexametaphosphate), 30 μg organic phosphorus/L (adenylic acid), 1.5 mg NH_3-N/L, 0.5 mg NO_3-N/L, and 400 mg Cl^-/L.

Sample 2: 600 μg PO_4^{3-}-P/L, 300 μg condensed phosphate phosphorus/L (sodium hexametaphosphate), 90 μg organic phosphorus/L (adenylic acid), 0.8 mg NH_3-N/L, 5.0 mg NO^-_3-N/L, and 400 mg Cl^-/L.

Sample 3: 7.00 mg PO_4^{3-}-P/L, 3.00 mg condensed phosphate phosphorus/L (sodium hexametaphosphate), 0.230 mg organic phosphorus/L (adenylic acid), 0.20 mg NH_3-N/L, 0.05 mg NO_3^--N/L, and 400 mg Cl^-/L.

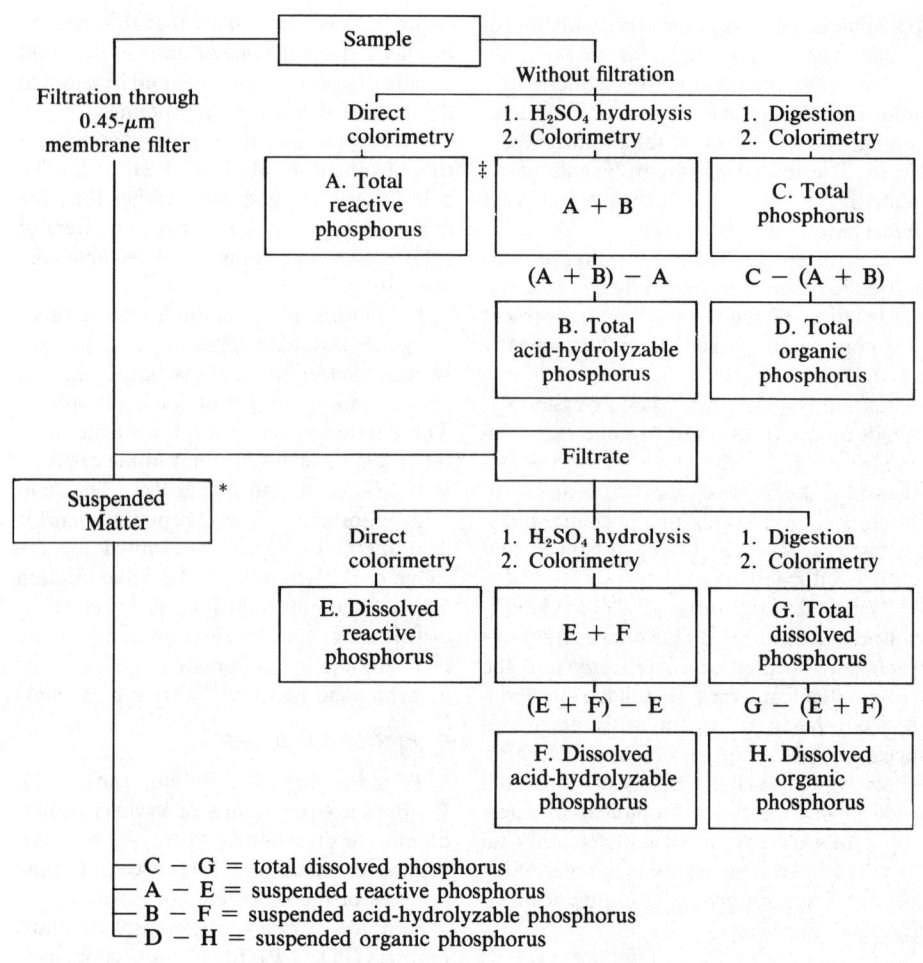

Figure 4500-P:1. Steps for analysis of phosphate fractions.

_____

*Direct determination of phosphorus on the membrane filter containing suspended matter will be required where greater precision than that obtained by difference is desired. Digest filter with HNO_3 and follow by perchloric acid. Then perform colorimetry.

†Total phosphorus measurements on highly saline samples may be difficult because of precipitation of large quantities of salt as a result of digestion techniques that drastically reduce sample volume. For total phosphorus analyses on such samples, directly determine total dissolved phosphorus and total suspended phosphorus and add the results.

‡In determination of total dissolved or total suspended reactive phosphorus, anomalous results may be obtained on samples containing large amounts of suspended sediments. Very often results depend largely on the degree of agitation and mixing to which samples are subjected during analysis because of a time-dependent desorption of orthophosphate from the suspended particles.

5. Sampling and Storage

If phosphorus forms are to be differentiated, filter sample immediately after collection. Preserve by freezing at or below −10°C. Add 40 mg $HgCl_2$/L to the samples, especially when they are to be stored for long periods. CAUTION: *HgCl₂ is a hazardous substance; take appropriate precautions in disposal.* Do not add either acid or

TABLE 4500-P:I. PRECISION AND BIAS DATA FOR MANUAL PHOSPHORUS METHODS

Method	Phosphorus Concentration			No. of Laboratories	Relative Standard Deviation %	Relative Error %
	Ortho-phosphate $\mu g/L$	Poly-phosphate $\mu g/L$	Total $\mu g/L$			
Vanadomolybdophosphoric	100			45	75.2	21.6
acid	600			43	19.6	10.8
	7000			44	8.6	5.4
Stannous chloride	100			45	25.5	28.7
	600			44	14.2	8.0
	7000			45	7.6	4.3
Ascorbic acid	100			3	9.1	10.0
	600			3	4.0	4.4
	7000			3	5.2	4.9
Acid hydrolysis +		80		37	106.8	7.4
vanadomolybdophosphoric		300		38	66.5	14.0
acid		3000		37	36.1	23.5
Acid hydrolysis + stannous		80		39	60.1	12.5
chloride		300		36	47.6	21.7
		3000		38	37.4	22.8
Persulfate +			210	32	55.8	1.6
vanadomolybdophosphoric			990	32	23.9	2.3
acid			10 230	31	6.5	0.3
Sulfuric-nitric acids +			210	23	65.6	20.9
vanadomolybdophosphoric			990	22	47.3	0.6
acid			10 230	20	7.0	0.4
Perchloric acid +			210	4	33.5	45.2
vanadomolybdophosphoric			990	5	20.3	2.6
acid			10 230	6	11.7	2.2
Persulfate + stannous			210	29	28.1	9.2
chloride			990	30	14.9	12.3
			10 230	29	11.5	4.3
Sulfuric-nitric acids +			210	20	20.8	1.2
stannous chloride			990	17	8.8	3.2
			10 230	19	7.5	0.4

$CHCl_3$ as a preservative when phosphorus forms are to be determined. If total phosphorus alone is to be determined, add 1 mL conc HCl/L or freeze without any additions.

Do not store samples containing low concentrations of phosphorus in plastic bottles unless kept in a frozen state because phosphates may be adsorbed onto the walls of plastic bottles.

Rinse all glass containers with hot dilute HCl, then rinse several times in distilled water. Never use commercial detergents containing phosphate for cleaning glassware used in phosphate analysis.

6. Bibliography

BLACK, C.A., D.D. EVANS, J.L. WHITE, L.E. ENSMINGER & F.E. CLARK, eds. 1965. Methods of Soil Analysis, Part 2. Chemical and Microbiological Properties. American Soc. Agronomy, Madison, Wisc.

JENKINS, D. 1965. A study of methods suitable for the analysis and preservation of phosphorus forms in an estuarine environment. SERL Rep. No. 65-18, Sanitary Engineering Research Lab., Univ. California, Berkeley.

LEE, G.F. 1967. Analytical chemistry of plant nutrients. *In* Proc. Int. Conf. Eutrophication, Madison, Wisc.

FITZGERALD, G.P. & S.L. FAUST. 1967. Effect of water sample preservation methods on the release of phosphorus from algae. *Limnol. Oceanogr.* 12:332.

4500-P B.   Sample Preparation

For information on selection of digestion method (¶s 3 through 5 below), see 4500-P.A.3*a*.

1. Preliminary Filtration

Filter samples for determination of dissolved reactive phosphorus, dissolved acid-hydrolyzable phosphorus, and total dissolved phosphorus through 0.45-μm membrane filters. A glass fiber filter may be used to prefilter hard-to-filter samples.

Wash membrane filters by soaking in distilled water before use because they may contribute significant amounts of phosphorus to samples containing low concentrations of phosphate. Use one of two washing techniques: (*a*) soak 50 filters in 2 L distilled water for 24 h; (*b*) soak 50 filters in 2 L distilled water for 1 h, change distilled water, and soak filters an additional 3 h. Membrane filters also may be washed by running several 100-mL portions of distilled water through them. This procedure requires more frequent determination of blank values to ensure consistency in washing and to evaluate different lots of filters.

2. Preliminary Acid Hydrolysis

The acid-hydrolyzable phosphorus content of the sample is defined operationally as the difference between reactive phosphorus as measured in the untreated sample and phosphate found after mild acid hydrolysis. Generally, it includes condensed phosphates such as pyro-, tripoly-, and higher-molecular-weight species such as hexametaphosphate. In addition, some natural waters contain organic phosphate compounds that are hydrolyzed to orthophosphate under the test conditions. Polyphosphates generally do not respond to reactive phosphorus tests but can be hydrolyzed to orthophosphate by boiling with acid.

After hydrolysis, determine reactive phosphorus by a colorimetric method (C, D, or E). Interferences, precision, bias, and sensitivity will depend on the colorimetric method used.

a. *Apparatus:*
Autoclave or pressure cooker, capable of operating at 98 to 137 kPa.

b. *Reagents:*

1) *Phenolphthalein indicator aqueous solution.*

2) *Strong acid solution:* Slowly add 300 mL conc H_2SO_4 to about 600 mL distilled water. When cool, add 4.0 mL conc HNO_3 and dilute to 1 L.

3) *Sodium hydroxide,* NaOH, 6*N*.

c. *Procedure:* To 100-mL sample or a portion diluted to 100 mL, add 0.05 mL (1 drop) phenolphthalein indicator solution. If a red color develops, add strong acid solution dropwise, to just discharge the color. Then add 1 mL more.

Boil gently for at least 90 min, adding distilled water to keep the volume between 25 and 50 mL. Alternatively, heat for 30 min in an autoclave or pressure cooker at 98 to 137 kPa. Cool, neutralize to a faint pink color with NaOH solution, and restore to the original 100-mL volume with distilled water.

Prepare a calibration curve by carrying a series of standards containing orthophosphate (see colorimetric method C, D, or E) through the hydrolysis step. Do not use orthophosphate standards without hydrolysis, because the salts added in hydrolysis cause an increase in the color intensity in some methods.

Determine reactive phosphorus content of treated portions, using Method C, D, or E. This gives the sum of polyphosphate and orthophosphate in the sample. To calculate its content of acid-hydrolyzable phosphorus, determine reactive phosphorus in a sample portion that has not been hydrolyzed, using the same colorimetric method as for treated sample, and subtract.

3. Perchloric Acid Digestion

a. *Apparatus:*

1) *Hot plate:* A 30- × 50-cm heating surface is adequate.

2) *Safety shield.*

3) *Safety goggles.*

4) *Erlenmeyer flasks,* 125-mL, acid-washed and rinsed with distilled water.

b. *Reagents:*

1) *Nitric acid,* HNO_3, conc.

2) *Perchloric acid,* $HClO_4 \cdot 2H_2O$, purchased as 70 to 72% $HClO_4$, reagent-grade.

3) *Sodium hydroxide,* NaOH, 6*N*.

4) *Methyl orange indicator solution.*

5) *Phenolphthalein indicator aqueous solution.*

c. *Procedure:* CAUTION—*Heated mixtures of $HClO_4$ and organic matter may explode violently. Avoid this hazard by taking the following precautions: (a) Do not add $HClO_4$ to a hot solution that may contain organic matter. (b) Always initiate digestion of samples containing organic matter with HNO_3. Complete digestion using the mixture of HNO_3 and $HClO_4$. (c) Do not fume with $HClO_4$ in ordinary hoods. Use hoods especially constructed for $HClO_4$ fuming or a glass fume eradicator* connected to a water pump. (d) Never let samples being digested with $HClO_4$ evaporate to dryness.*

Measure sample containing the desired amount of phosphorus (this will be determined by whether Method C, D, or E is to be used) into a 125-mL erlenmeyer flask. Acidify to methyl orange with conc HNO_3, add another 5 mL conc HNO_3, and evaporate on a steam bath or hot plate to 15 to 20 mL.

Add 10 mL each of conc HNO_3 and $HClO_4$ to the 125-mL conical flask, cooling the flask between additions. Add a few boiling chips, heat on a hot plate, and evaporate gently until dense white fumes of $HClO_4$ just appear. If solution is not clear, cover neck of flask with a watch glass and keep solution barely boiling until it clears. If necessary, add 10 mL more HNO_3 to aid oxidation.

Cool digested solution and add 1 drop aqueous phenolphthalein solution. Add 6*N*

_____

*GFS Chemical Co., Columbus, Ohio, or equivalent.

NaOH solution until the solution just turns pink. If necessary, filter neutralized solution and wash filter liberally with distilled water. Make up to 100 mL with distilled water.

Determine the PO_4^{3-}-P content of the treated sample by Method C, D, or E.

Prepare a calibration curve by carrying a series of standards containing orthophosphate (see Method C, D, or E) through digestion step. Do not use orthophosphate standards without treatment.

4. Sulfuric Acid-Nitric Acid Digestion

a. Apparatus:

1) *Digestion rack:* An electrically or gas-heated digestion rack with provision for withdrawal of fumes is recommended. Digestion racks typical of those used for micro-kjeldahl digestions are suitable.

2) *Micro-kjeldahl flasks.*

b. Reagents:

1) *Sulfuric acid,* H_2SO_4, conc.

2) *Nitric acid,* HNO_3, conc.

3) *Phenolphthalein indicator aqueous solution.*

4) *Sodium hydroxide,* NaOH, 1N.

c. Procedure: Into a micro-kjeldahl flask, measure a sample containing the desired amount of phosphorus (this is determined by the colorimetric method used). Add 1 mL conc H_2SO_4 and 5 mL conc HNO_3.

Digest to a volume of 1 mL and then continue until solution becomes colorless to remove HNO_3.

Cool and add approximately 20 mL distilled water, 0.05 mL (1 drop) phenolphthalein indicator, and as much 1N NaOH solution as required to produce a faint pink tinge. Transfer neutralized solution, filtering if necessary to remove particulate material or turbidity, into a 100-mL volumetric flask. Add filter washings to flask and adjust sample volume to 100 mL with distilled water.

Determine phosphorus by Method C, D,

or E, for which a separate calibration curve has been constructed by carrying standards through the acid digestion procedure.

5. Persulfate Digestion Method

a. Apparatus:

1) *Hot plate:* A 30- × 50-cm heating surface is adequate.

2) *Autoclave:* An autoclave or pressure cooker capable of developing 98 to 137 kPa may be used in place of a hot plate.

3) *Glass scoop,* to hold required amounts of persulfate crystals.

b. Reagents:

1) *Phenolphthalein indicator aqueous solution.*

2) *Sulfuric acid solution:* Carefully add 300 mL conc H_2SO_4 to approximately 600 mL distilled water and dilute to 1 L with distilled water.

3) *Ammonium persulfate,* $(NH_4)_2S_2O_8$, solid, or potassium persulfate, $K_2S_2O_8$, solid.

4) *Sodium hydroxide,* NaOH, 1N.

c. Procedure: Use 50 mL or a suitable portion of thoroughly mixed sample. Add 0.05 mL (1 drop) phenolphthalein indicator solution. If a red color develops, add H_2SO_4 solution dropwise to just discharge the color. Then add 1 mL H_2SO_4 solution and either 0.4 g solid $(NH_4)_2S_2O_8$ or 0.5 g solid $K_2S_2O_8$.

Boil gently on a preheated hot plate for 30 to 40 min or until a final volume of 10 mL is reached. Organophosphorus compounds such as AMP may require as much as 1.5 to 2 hr for complete digestion. Cool, dilute to 30 mL with distilled water, add 0.05 mL (1 drop) phenolphthalein indicator solution, and neutralize to a faint pink color with NaOH. Alternatively, heat for 30 min in an autoclave or pressure cooker at 98 to 137 kPa. Cool, add 0.05 mL (1 drop) phenolphthalein indicator solution, and neutralize to a faint pink color with NaOH. Make up to 100 mL with distilled water. In some samples a precipitate may form at this stage, but do not filter. For any subsequent subdividing of the sample, shake well. The precipitate (which is possibly a calcium phosphate) redissolves under the acid conditions of the colorimetric reactive phosphorus test. Determine phosphorus by Method C, D, or E, for which a separate calibration curve has been constructed by carrying standards through the persulfate digestion procedure.

6. Bibliography

LEE, G.F., N.L. CLESCERI & G.P. FITZGERALD. 1965. Studies on the analysis of phosphates in algal cultures. *J. Air Water Pollut.* 9:715.

SHANNON, J.E. & G.F. LEE. 1966. Hydrolysis of condensed phosphates in natural waters. *J. Air Water Pollut.* 10:735.

GALES, M.E., JR., E.C. JULIAN & R.C. KRONER. 1966. Method for quantitative determination of total phosphorus in water. *J. Amer. Water Works Assoc.* 58:1363.

4500-P C.   Vanadomolybdophosphoric Acid Colorimetric Method

1. General Discussion

a. Principle: In a dilute orthophosphate solution, ammonium molybdate reacts under acid conditions to form a heteropoly acid, molybdophosphoric acid. In the presence of vanadium, yellow vanadomolybdophosphoric acid is formed. The intensity of the yellow color is proportional to phosphate concentration.

b. Interference: Positive interference is caused by silica and arsenate only if the sample is heated. Negative interferences are caused by arsenate, fluoride, thorium, bismuth, sulfide, thiosulfate, thiocyanate, or excess molybdate. Blue color is caused by ferrous iron but this does not affect results if ferrous iron concentration is less than 100 mg/L. Sulfide interference may be removed by oxidation with bromine water. Ions that do not interfere in concentrations up to 1000 mg/L are Al^{3+}, Fe^{3+}, Mg^{2+}, Ca^{2+}, Ba^{2+}, Sr^{2+}, Li^+, Na^+, K^+, NH_4^+, Cd^{2+}, Mn^{2+}, Pb^{2+}, Hg^+, Hg^{2+}, Sn^{2+}, Cu^{2+}, Ni^{2+}, Ag^+, U^{4+}, Zr^{4+}, AsO_3^-, Br^-, CO_3^{2-}, ClO_4^-, CN^-, IO_3^-, SiO_4^{4-}, NO_3^-, NO_2^-, SO_4^{2-}, SO_3^{2-}, pyrophosphate, molybdate, tetraborate, selenate, benzoate, citrate, oxalate, lactate, tartrate, formate, and salicylate. If HNO_3 is used in the test, Cl^- interferes at 75 mg/L.

c. Minimum detectable concentration: The minimum detectable concentration is 200 µg P/L in 1-cm spectrophotometer cells.

2. Apparatus

a. Colorimetric equipment: One of the following is required:

1) *Spectrophotometer,* for use at 400 to 490 nm.

2) *Filter photometer,* provided with a blue or violet filter exhibiting maximum transmittance between 400 and 470 nm.

The wavelength at which color intensity is measured depends on sensitivity desired, because sensitivity varies tenfold with wavelengths 400 to 490 nm. Ferric iron causes interference at low wavelengths, particularly at 400 nm. A wavelength of 470 nm usually is used. Concentration ranges for different wavelengths are:

P Range *mg/L*	Wavelength *nm*
1.0– 5.0	400
2.0–10	420
4.0–18	470

b. Acid-washed glassware: Use acid-washed glassware for determining low concentrations of phosphorus. Phosphate contamination is common because of its absorption on glass surfaces. Avoid using commercial detergents containing phosphate. Clean all glassware with hot dilute HCl and rinse well with distilled water. Preferably, reserve the glassware only for phosphate determination, and after use, wash and keep filled with water until needed. If this is done, acid treatment is required only occasionally.

c. Filtration apparatus and filter paper. [*]

3. Reagents

a. Phenolphthalein indicator aqueous solution.

b. Hydrochloric acid, HCl, 1 + 1. H_2SO_4, $HClO_4$, or HNO_3 may be substituted for HCl. The acid concentration in the determination is not critical but a final sample concentration of $0.5N$ is recommended.

c. Activated carbon. [†] Remove fine particles by rinsing with distilled water.

d. Vanadate-molybdate reagent:

1) *Solution A:* Dissolve 25 g ammonium molybdate, $(NH_4)_6Mo_7O_{24} \cdot 4H_2O$, in 300 mL distilled water.

2) *Solution B:* Dissolve 1.25 g ammonium metavanadate, NH_4VO_3, by heating to boiling in 300 mL distilled water. Cool and add 330 mL conc HCl. Cool Solution B to room temperature, pour Solution A into Solution B, mix, and dilute to 1 L.

e. Standard phosphate solution: Dissolve in distilled water 219.5 mg anhydrous KH_2PO_4 and dilute to 1000 mL; 1.00 mL = 50.0 μg PO_4^{3-}-P.

4. Procedure

a. Sample pH adjustment: If sample pH is greater than 10, add 0.05 mL (1 drop) phenolphthalein indicator to 50.0 mL sample and discharge the red color with 1 + 1 HCl before diluting to 100 mL.

b. Color removal from sample: Remove excessive color in sample by shaking about 50 mL with 200 mg activated carbon in an erlenmeyer flask for 5 min and filter to remove carbon. Check each batch of car-

bon for phosphate because some batches produce high reagent blanks.

c. Color development in sample: Place 35 mL or less of sample, containing 0.05 to 1.0 mg P, in a 50-mL volumetric flask. Add 10 mL vanadate-molybdate reagent and dilute to the mark with distilled water. Prepare a blank in which 35 mL distilled water is substituted for the sample. After 10 min or more, measure absorbance of sample versus a blank at a wavelength of 400 to 490 nm, depending on sensitivity desired (see ¶ 2a above). The color is stable for days and its intensity is unaffected by variation in room temperature.

d. Preparation of calibration curve: Prepare a calibration curve by using suitable volumes of standard phosphate solution and proceeding as in ¶ 4c. When ferric ion is low enough not to interfere, plot a family of calibration curves of one series of standard solutions for various wavelengths. This permits a wide latitude of concentrations in one series of determinations. Analyze at least one standard with each set of samples.

5. Calculation

$$\text{mg P/L} = \frac{\text{mg P(in 50 mL final volume)} \times 1000}{\text{mL sample}}$$

6. Precision and Bias

See Table 4500-P:I.

7. Bibliography

KITSON, R.E. & M.G. MELLON. 1944. Colorimetric determination of phosphorus as molybdovanadophosphoric acid. *Ind. Eng. Chem.,* Anal. Ed. 16:379.

BOLTZ, D.F. & M.G. MELLON. 1947. Determination of phosphorus, germanium, silicon, and arsenic by the heteropoly blue method. *Ind. Eng. Chem.,* Anal. Ed. 19:873.

GREENBERG, A.E., L.W. WEINBERGER & C.N. SAWYER. 1950. Control of nitrite interference in colorimetric determination of phosphorus. *Anal. Chem.* 22:499.

---

[*] Whatman No. 42 or equivalent.
[†] Darco G60 or equivalent.

Young, R.S. & A. Golledge. 1950. Determination of hexametaphosphate in water after threshold treatment. *Ind. Chem.* 26:13.

Griswold, B.L., F.L. Humoller & A.R. McIntyre. 1951. Inorganic phosphates and phosphate esters in tissue extracts. *Anal. Chem.* 23:192.

Boltz, D.F., ed. 1958. Colorimetric Determination of Nonmetals. Interscience Publishers, New York, N.Y.

American Water Works Association. 1958. Committee report. Determination of orthophosphate, hydrolyzable phosphate, and total phosphate in surface waters. *J. Amer. Water Works Assoc.* 50:1563.

Jackson, M.L. 1958. Soil Chemical Analysis. Prentice-Hall, Englewood Cliffs, N.J.

Abbot, D.C., G.E. Emsden & J.R. Harris. 1963. A method for determining orthophosphate in water. *Analyst* 88:814.

Gottfried, P. 1964. Determination of total phosphorus in water and wastewater as molybdovanadophosphoric acid. *Limnologica* 2:407.

4500-P D.   Stannous Chloride Method

1. General Discussion

a. Principle: Molybdophosphoric acid is formed and reduced by stannous chloride to intensely colored molybdenum blue. This method is more sensitive than Method C and makes feasible measurements down to 7 μg P/L by use of increased light path length. Below 100 μg P/L an extraction step may increase reliability and lessen interference.

b. Interference: See Section 4500-P.C.1*b*.

c. Minimum detectable concentration: The minimum detectable concentration is about 3 μg P/L. The sensitivity at 0.3010 absorbance is about 10 μg P/L for an absorbance change of 0.009.

2. Apparatus

The same apparatus is required as for Method C, except that a pipetting bulb is required for the extraction step. Set spectrophotometer at 625 nm in the measurement of benzene-isobutanol extracts and at 690 nm for aqueous solutions. If the instrument is not equipped to read at 690 nm, use a wavelength of 650 nm for aqueous solutions, with somewhat reduced sensitivity and precision.

3. Reagents

a. Phenolphthalein indicator aqueous solution.

b. Strong-acid solution: Prepare as directed in Section 4500-P.B.2*b*2).

c. Ammonium molybdate reagent I: Dissolve 25 g $(NH_4)_6Mo_7O_{24} \cdot 4H_2O$ in 175 mL distilled water. Cautiously add 280 mL conc H_2SO_4 to 400 mL distilled water. Cool, add molybdate solution, and dilute to 1 L.

d. Stannous chloride reagent I: Dissolve 2.5 g fresh $SnCl_2 \cdot 2H_2O$ in 100 mL glycerol. Heat in a water bath and stir with a glass rod to hasten dissolution. This reagent is stable and requires neither preservatives nor special storage.

e. Standard phosphate solution: Prepare as directed in Section 4500-P.C.3*e*.

f. Reagents for extraction:

1) *Benzene-isobutanol solvent:* Mix equal volumes of benzene and isobutyl alcohol. (Caution—*This solvent is highly flammable.*)

2) *Ammonium molybdate reagent II:* Dissolve 40.1 g $(NH_4)_6Mo_7O_{24} \cdot 4H_2O$ in approximately 500 mL distilled water. Slowly add 396 mL ammonium molybdate reagent I. Cool and dilute to 1 L.

3) *Alcoholic sulfuric acid solution:* Cautiously add 20 mL conc H_2SO_4 to 980 mL methyl alcohol with continuous mixing.

4) *Dilute stannous chloride reagent II:* Mix 8 mL stannous chloride reagent I with

50 mL glycerol. This reagent is stable for at least 6 months.

4. Procedure

a. Preliminary sample treatment: To 100 mL sample containing not more than 200 µg P and free from color and turbidity, add 0.05 mL (1 drop) phenolphthalein indicator. If sample turns pink, add strong acid solution dropwise to discharge the color. If more than 0.25 mL (5 drops) is required, take a smaller sample and dilute to 100 mL with distilled water after first discharging the pink color with acid.

b. Color development: Add, with thorough mixing after each addition, 4.0 mL molybdate reagent I and 0.5 mL (10 drops) stannous chloride reagent I. Rate of color development and intensity of color depend on temperature of the final solution, each 1°C increase producing about 1% increase in color. Hence, hold samples, standards, and reagents within 2°C of one another and in the temperature range between 20 and 30°C.

c. Color measurement: After 10 min, but before 12 min, using the same specific interval for all determinations, measure color photometrically at 690 nm and compare with a calibration curve, using a distilled water blank. Light path lengths suitable for various concentration ranges are as follows:

Approximate P Range mg/L	Light Path cm
0.3–2	0.5
0.1–1	2
0.007–0.2	10

Always run a blank on reagents and distilled water. Because the color at first develops progressively and later fades, maintain equal timing conditions for samples and standards. Prepare at least one standard with each set of samples or once each day that tests are made. The calibration curve may deviate from a straight line at the upper concentrations of the 0.3 to 2.0-mg/L range.

d. Extraction: When increased sensitivity is desired or interferences must be overcome, extract phosphate as follows: Pipet a 40-mL sample, or one diluted to that volume, into a 125-mL separatory funnel. Add 50.0 mL benzene-isobutanol solvent and 15.0 mL molybdate reagent II. Close funnel at once and shake vigorously for exactly 15 s. If condensed phosphate is present, any delay will increase its conversion to orthophosphate. Remove stopper and withdraw 25.0 mL of separated organic layer, using a pipet with safety bulb. Transfer to a 50-mL volumetric flask, add 15 to 16 mL alcoholic H_2SO_4 solution, swirl, add 0.50 mL (10 drops) dilute stannous chloride reagent II, swirl, and dilute to the mark with alcoholic H_2SO_4. Mix thoroughly. After 10 min, but before 30 min, read against the blank at 625 nm. Prepare blank by carrying 40 mL distilled water through the same procedure used for the sample. Read phosphate concentration from a calibration curve prepared by taking known phosphate standards through the same procedure used for samples.

5. Calculation

Calculate as follows:

a. Direct procedure:

$$\text{mg P/L} = \frac{\text{mg P (in approximately 104.5 mL final volume)} \times 1000}{\text{mL sample}}$$

b. Extraction procedure:

$$\text{mg P/L} = \frac{\text{mg P (in 50 mL final volume)} \times 1000}{\text{mL sample}}$$

6. Precision and Bias

See Table 4500-P:I.

4500-P E. Ascorbic Acid Method

1. General Discussion

a. Principle: Ammonium molybdate and potassium antimonyl tartrate react in acid medium with orthophosphate to form a heteropoly acid—phosphomolybdic acid—that is reduced to intensely colored molybdenum blue by ascorbic acid.

b. Interference: Arsenates react with the molybdate reagent to produce a blue color similar to that formed with phosphate. Concentrations as low as 0.1 mg As/L interfere with the phosphate determination. Hexavalent chromium and NO_2^- interfere to give results about 3% low at concentrations of 1 mg/L and 10 to 15% low at 10 mg/L. Sulfide (Na_2S) and silicate do not interfere at concentrations of 1.0 and 10 mg/L.

c. Minimum detectable concentration: Approximately 10 μg P/L. P ranges are as follows:

Approximate P Range mg/L	Light Path cm
0.30–2.0	0.5
0.15–1.30	1.0
0.01–0.25	5.0

2. Apparatus

a. Colorimetric equipment: One of the following is required:

1) *Spectrophotometer,* with infrared phototube for use at 880 nm, providing a light path of 2.5 cm or longer.

2) *Filter photometer,* equipped with a red color filter and a light path of 0.5 cm or longer.

b. Acid-washed glassware: See Section 4500-P.C.2*b*.

3. Reagents

a. Sulfuric acid, H_2SO_4, 5*N*: Dilute 70 mL conc H_2SO_4 to 500 mL with distilled water.

b. Potassium antimonyl tartrate solution: Dissolve 1.3715 g $K(SbO)C_4H_4O_6 \cdot 1/2H_2O$ in 400 mL distilled water in a 500-mL volumetric flask and dilute to volume. Store in a glass-stoppered bottle.

c. Ammonium molybdate solution: Dissolve 20 g $(NH_4)_6Mo_7O_{24} \cdot 4H_2O$ in 500 mL distilled water. Store in a glass-stoppered bottle.

d. Ascorbic acid, 0.01*M:* Dissolve 1.76 g ascorbic acid in 100 mL distilled water. The solution is stable for about 1 week at 4°C.

e. Combined reagent: Mix the above reagents in the following proportions for 100 mL of the combined reagent: 50 mL 5*N* H_2SO_4, 5 mL potassium antimonyl tartrate solution, 15 mL ammonium molybdate solution, and 30 mL ascorbic acid solution. *Mix after addition of each reagent.* Let all reagents reach room temperature before they are mixed and mix in the order given. If turbidity forms in the combined reagent, shake and let stand for a few minutes until turbidity disappears before proceeding. The reagent is stable for 4 h.

f. Stock phosphate solution: See Section 4500-P.C.3*e*.

g. Standard phosphate solution: Dilute 50.0 mL stock phosphate solution to 1000 mL with distilled water; 1.00 mL = 2.50 μg P.

4. Procedure

a. Treatment of sample: Pipet 50.0 mL sample into a clean, dry test tube or 125-mL erlenmeyer flask. Add 0.05 mL (1 drop) phenolphthalein indicator. If a red color develops add 5*N* H_2SO_4 solution dropwise to just discharge the color. Add 8.0 mL combined reagent and mix thoroughly. After at least 10 min but no more than 30 min, measure absorbance of each sample at 880 nm, using reagent blank as the reference solution.

b. Correction for turbidity or interfering color: Natural color of water generally does not interfere at the high wavelength used.

TABLE 4500-P:II. COMPARISON OF PRECISION AND BIAS OF ASCORBIC ACID METHODS

Ascorbic Acid Method	Phosphorus Concentration, Dissolved Orthophosphate $\mu g/L$	No. of Laboratories	Relative Standard Deviation %		Relative Error %	
			Distilled Water	River Water	Distilled Water	River Water
13th Edition[1]	228	8	3.87	2.17	4.01	2.08
Current method[2]	228	8	3.03	1.75	2.38	1.39

For highly colored or turbid waters, prepare a blank by adding all reagents except ascorbic acid and antimonyl potassium tartrate to the sample. Subtract blank absorbance from absorbance of each sample.

c. *Preparation of calibration curve:* Prepare individual calibration curves from a series of six standards within the phosphate ranges indicated in ¶ 1c above. Use a distilled water blank with the combined reagent to make photometric readings for the calibration curve. Plot absorbance vs. phosphate concentration to give a straight line passing through the origin. Test at least one phosphate standard with each set of samples.

5. Calculation

$$mg\ P/L = \frac{mg\ P\ (in\ approximately\ 58\ mL\ final\ volume) \times 1000}{mL\ sample}$$

6. Precision and Bias

The precision and bias values given in Table 4500-P:I are for a single-solution procedure given in the 13th edition. The present procedure differs in reagent-to-sample ratios, no addition of solvent, and acidity conditions. It is superior in precision and bias to the previous technique in the analysis of both distilled water and river water at the 228-μg P/L level (Table 4500-P:II).

7. References

1. EDWARDS, G.P., A.H. MOLOF & R.W. SCHNEEMAN. 1965. Determination of orthophosphate in fresh and saline waters. *J. Amer. Water Works Assoc.* 57:917.
2. MURPHY, J. & J. RILEY. 1962. A modified single solution method for the determination of phosphate in natural waters. *Anal. Chim. Acta* 27:31.

8. Bibliography

SLETTEN, O. & C.M. BACH. 1961. Modified stannous chloride reagent for orthophosphate determination. *J. Amer. Water Works Assoc.* 53:1031.
STRICKLAND, J.D.H. & T.R. PARSONS. 1965. A Manual of Sea Water Analysis, 2nd ed. Fisheries Research Board of Canada, Ottawa.

4500-P F.   Automated Ascorbic Acid Reduction Method

1. General Discussion

a. *Principle:* Ammonium molybdate and potassium antimonyl tartrate react with orthophosphate in an acid medium to form an antimony-phosphomolybdate complex, which, on reduction with ascorbic acid, yields an intense blue color suitable for photometric measurement.

b. *Interferences:* As much as 50 mg

Fe^{3+}/L, 10 mg Cu/L, and 10 mg SiO_2/L can be tolerated. High silica concentrations cause positive interference.

In terms of phosphorus, the results are high by 0.005, 0.015, and 0.025 mg/L for silica concentrations of 20, 50, and 100 mg/L, respectively. Salt concentrations up to 20% (w/v) cause an error of less than 1%. Arsenate (AsO_4^{3-}) is a positive interference.

Eliminate interference from NO_2^- and S^{2-} by adding an excess of bromine water or a saturated potassium permanganate ($KMnO_4$) solution. Remove interfering turbidity by filtration before analysis. Filter samples for total or total hydrolyzable phosphorus only after digestion. Sample color that absorbs in the photometric range used for analysis also will interfere. See also Section 4500-P.E.1*b*.

c. Application: Orthophosphate can be determined in potable, surface, and saline waters as well as domestic and industrial wastewaters over a range of 0.001 to 10.0 mg P/L when photometric measurements are made at 650 to 660 or 880 nm in a 15-mm or 50-mm tubular flow cell. Determine higher concentrations by diluting sample. Although the automated test is designed for orthophosphate only, other phosphorus compounds can be converted to this reactive form by various sample pretreatments described in Section 4500-P.B.1, 2, and 5.

2. Apparatus

a. Automated analytical equipment: The required continuous-flow analytical instrument† consists of the interchangeable components shown in Figure 4500-P:2. A flow cell of 15 or 50 mm and a filter of 650 to 660 or 880 nm may be used.

b. Hot plate or autoclave.

c. Acid-washed glassware: See Section 4500-P.C.2*b*.

---

†AutoAnalyzer™ II, Technicon Instrument Co., Tarrytown, N.Y. 10591, or equivalent.

3. Reagents

a. Potassium antimonyl tartrate solution: Dissolve 0.3 g $K(SbO)C_2H_4O_6 \cdot 1/2H_2O$ in approximately 50 mL distilled water and dilute to 100 mL. Store at 4°C in a dark, glass-stoppered bottle.

b. Ammonium molybdate solution: Dissolve 4 g $(NH_4)_6Mo_7O_{24} \cdot 4H_2O$ in 100 mL distilled water. Store in a plastic bottle at 4°C.

c. Ascorbic acid solution: See Section 4500-P.E.3*d*.

d. Combined reagent: See Section 4500-P.E.3*e*.

e. Dilute sulfuric acid solution: Slowly add 140 mL conc H_2SO_4 to 600 mL distilled water. When cool, dilute to 1 L.

f. Ammonium persulfate, $(NH_4)_2S_2O_8$, crystalline.

g. Phenolphthalein indicator aqueous solution.

h. Stock phosphate solution: Dissolve 439.3 mg anhydrous KH_2PO_4, dried for 1 h at 105°C, in distilled water and dilute to 1000 mL; 1.00 mL = 100 μg P.

i. Intermediate phosphate solution: Dilute 100.0 mL stock phosphate solution to 1000 mL with distilled water; 1.00 mL = 10.0 μg P.

j. Standard phosphate solutions: Prepare a suitable series of standards by diluting appropriate volumes of intermediate phosphate solution.

4. Procedure

Set up manifold as shown in Figure 4500-P:2 and follow the general procedure described by the manufacturer.

Add 0.05 mL (1 drop) phenolphthalein indicator solution to approximately 50 mL sample. If a red color develops, add H_2SO_4 (¶ 3*e*) dropwise to just discharge the color.

5. Calculation

Prepare standard curves by plotting peak heights of standards processed through the manifold against P concentration in stand-

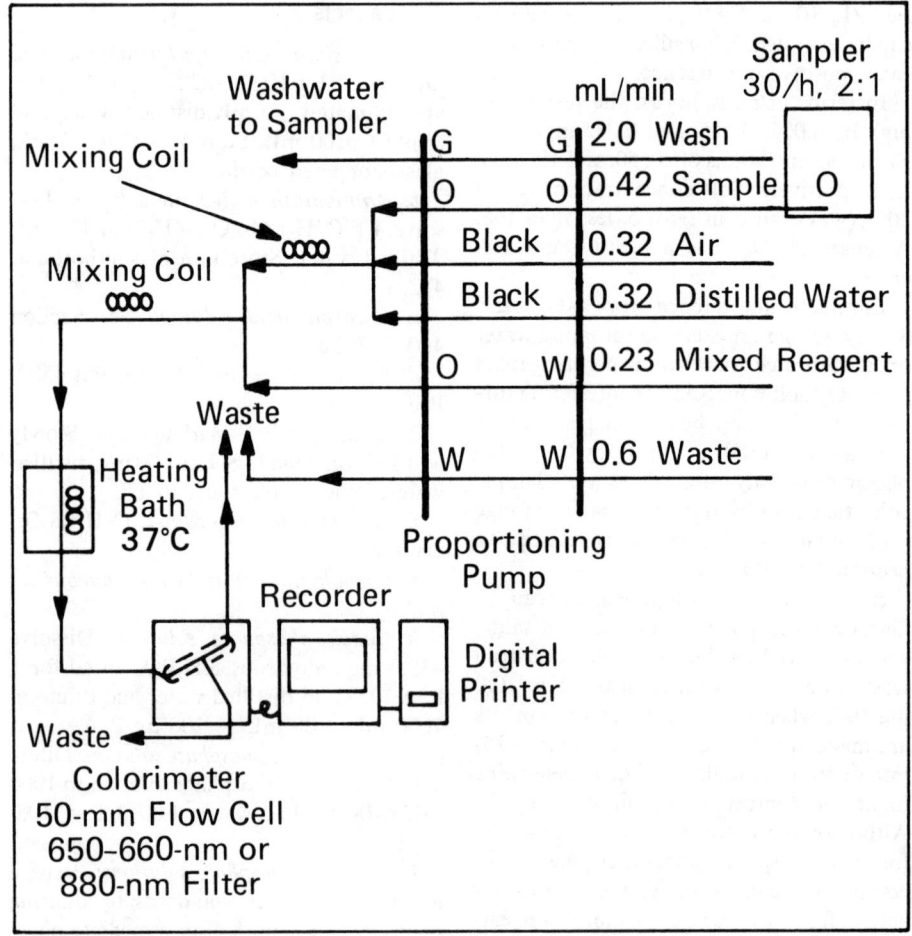

Figure 4500-P:2. Phosphate manifold for automated analytical system.

ards. Compute sample P concentration by comparing sample peak height with standard curve.

6. Precision and Bias

Six samples were analyzed in a single laboratory in septuplicate. At an average PO_4^{3-} concentration of 0.340 mg/L, the average deviation was 0.015 mg/L. The coefficient of variation was 6.2%. In two samples with added PO_4^{3-}, recoveries were 89 and 96%.

7. Bibliography

HENRIKSEN, A. 1966. An automatic method for determining orthophosphate in sewage and highly polluted waters. *Analyst* 91:652.

LOBRING, L.B. & R.L. BOOTH. 1973. Evaluation of the AutoAnalyzer II; A progress report. *In* Advances in Automated Analysis: 1972 Technicon International Congress. Vol. 8, p. 7. Mediad, Inc., Tarrytown, N.Y.

U.S. ENVIRONMENTAL PROTECTION AGENCY. 1979. Methods for Chemical Analysis of Water and Wastes. EPA-600/4-79-020, National Environmental Research Center, Cincinnati, Ohio.

U.S. ENVIRONMENTAL PROTECTION AGENCY. MDQARL Method Study 4, Automated Methods. National Environmental Research Center, Cincinnati, Ohio (in preparation).

4500-Si   SILICA*

4500-Si A.   Introduction

1. Occurrence and Significance

Silicon ranks next to oxygen in abundance in the earth's crust. It appears as the oxide (silica) in quartz and sand and is combined with metals in the form of many complex silicate minerals, particularly igneous rocks. Degradation of silica-containing rocks results in the presence of silica in natural waters as suspended particles, in a colloidal or polymeric state, and as silicic acids or silicate ions. Volcanic and geothermally heated waters often contain an abundance of silica.

A more complete discussion of the occurrence and chemistry of silica in natural waters is available.[1]

The silica (SiO_2) content of natural water most commonly is in the 1- to 30-mg/L range, although concentrations as high as 100 mg/L are not unusual and concentrations exceeding 1000 mg/L are found in some brackish waters and brines.

Silica in water is undesirable for a number of industrial uses because it forms difficult-to-remove silica and silicate scales in equipment, particularly on high-pressure steam-turbine blades. Silica is removed most often by the use of strongly basic anion-exchange resins in the deionization process, by distillation, or by reverse osmosis. Some plants use precipitation with magnesium oxide in either the hot or cold lime softening process.

2. Selection of Method

Method C determines total silica. Methods D, E, and F determine molybdate-reactive silica. As noted in Section 4500-Si.D.4, it is possible to convert other forms of silica to the molybdate-reactive form for determination by these methods. Method B, like Method C, determines more than one form of silica. It will determine all dissolved silica and some colloidally dispersed silica. The determination of silica present in micrometer and submicrometer particles will depend on the size distribution, composition, and structure of the particles; thus Method B cannot be said to determine total silica.

Use Method C to standardize sodium silicate solutions used as standards for Methods D, E, and F. Do not use Method C for water samples that contain less than 20 mg SiO_2/L. Method D is recommended for relatively pure waters containing from 0.4 to 25 mg SiO_2/L. As with most colorimetric methods, the range can be extended, if necessary, by diluting, by concentrating, or by varying the light path. Interferences due to tannin, color, and turbidity are more severe with this method than with Method E. Moreover, the yellow color produced by Method D has a limited stability and attention to timing is necessary. When applicable, however, it offers

---

*Approved by Standard Methods Committee, 1988.

greater speed and simplicity than Method E because one reagent fewer is used; one timing step is eliminated; and many natural waters can be analyzed without dilution, which is not often the case with Method E. Method E is recommended for the low range, from 0.04 to 2 mg SiO_2/L. This range also can be extended if necessary. Such extension may be desirable if interference is expected from tannin, color, or turbidity. A combination of factors renders Methods E and F less susceptible than Method D to those interferences; also, the blue color in Methods E and F is more stable than the yellow color in Method D. However, many samples will require dilution because of the high sensitivity of the method. Permanent artificial color standards are not available for the blue color developed in Method E.

The yellow color produced by Method D and the blue color produced by Methods E and F are affected by high concentrations of salts. With sea water the yellow color intensity is decreased by 20 to 25% and the blue color intensity is increased by 10 to 15%. When waters of high ionic strength are analyzed by these methods, use silica standards of approximately the same ionic strengths.[2]

Method F may be used where large numbers of samples are analyzed regularly. Method B is recommended for broad-range use. Although Method B is usable from 1 to 300 mg SiO_2/L, optimal results are obtained from about 20 to 300 mg/L. The range can be extended upward by dilution if necessary. This method is rapid and does not require any timing step.

The inductively coupled plasma method (G) also may be used in analyses for silica.

3. Sampling and Storage

Collect samples in bottles of polyethylene, other plastic, or hard rubber, especially if there will be a delay between collection and analysis. Borosilicate glass is a less desirable choice, particularly with waters of pH above 8 or with seawater, in which cases a significant amount of silica in the glass can dissolve. Freezing to preserve samples for analysis of other constituents can lower soluble silica values by as much as 20 to 40% in waters that have a pH below 6. Do not acidify samples for preservation because silica precipitates in acidic solutions.

4. References

1. HEM, J.D. 1959. Study and interpretation of the chemical characteristics of natural water. U.S. Geol. Surv. Water Supply Pap. No. 1473.
2. FANNING, K.A. & M.E.Q. PILSON. 1973. On the spectrophotometric determination of dissolved silica in natural waters. *Anal. Chem.* 45:136.

5. Bibliography

ROY, C.J. 1945. Silica in natural waters. *Amer. J. Sci.* 243:393.
VAIL, J.G. 1952. The Soluble Silicates, Their Properties and Uses. Reinhold Publishing Corp., New York, N.Y. Vol. 1, pp. 95-97, 100-161.

4500-Si B.   Atomic Absorption Spectrometric Method

See Section 3111D.

4500-Si C.   Gravimetric Method

1. General Discussion

a. Principle: Hydrochloric acid decomposes silicates and dissolved silica, forming silicic acids that are precipitated as partially dehydrated silica during evaporation and baking. Ignition completes dehydration of the silica, which is weighed and then volatilized as silicon tetrafluoride, leaving any impurities behind as nonvolatile residue. The residue is weighed and silica is determined as loss on volatilization. Perchloric acid ($HClO_4$) may be used to dehydrate the silica instead of HCl. A single fuming with $HClO_4$ will recover more silica than one with HCl, although for complete silica recovery two dehydrations with either acid are necessary. The use of $HClO_4$ lessens the tendency to spatter, yields a silica precipitate that is easier to filter, and shortens the time required for the determination.

b. Interference: Because glassware may contribute silica, avoid its use as much as possible. Use reagents and distilled or deionized water low in silica. Carry out a blank determination to correct for silica introduced by the reagents and apparatus.

2. Apparatus

a. Platinum crucibles, with covers.

b. Platinum evaporating dishes, 200-mL. In dehydration steps, acid-leached glazed porcelain evaporating dishes free from etching may be substituted for platinum, but for greatest accuracy, platinum is preferred.

3. Reagents

For maximum accuracy, set aside batches of chemicals low in silica for this method. Store all reagents in plastic containers and run blanks.

a. Hydrochloric acid, HCl, 1 + 1 and 1 + 50.

b. Sulfuric acid, H_2SO_4, 1 + 1.

c. Hydrofluoric acid, HF, 48%.

d. Perchloric acid, $HClO_4$, 72%.

4. Procedure

Before determining silica, test H_2SO_4 and HF for interfering nonvolatile matter by carrying out the procedure of ¶ 4a5) below. Use a clean empty platinum crucible. If any increase in weight is observed, make a correction in the silica determinations.

a. HCl dehydration:

1) Sample evaporation—To a clear sample containing at least 10 mg silica, add 5 mL 1 + 1 HCl. Evaporate to dryness in a 200-mL platinum or acid-leached glazed porcelain evaporating dish, in several portions if necessary, on a water bath or suspended on an asbestos ring over a hot plate. Protect against contamination by atmospheric dust. During evaporation, add a total of 15 mL 1 + 1 HCl in several portions. Dry dish and place it in a 110°C oven or over a hot plate to bake for 30 min.

2) First filtration—Add 5 mL 1 + 1 HCl, warm, and add 50 mL hot distilled water. While mixture is hot, filter through an ashless medium-texture filter paper, decanting as much liquid as possible. Wash dish and residue with hot 1 + 50 HCl and then with a minimum volume of distilled water until washings are chloride-free. Save all washings. Set aside filter paper with its residue.

3) Second filtration—Evaporate filtrate and washings from the above operation to dryness in the original platinum dish. Bake residue in a 110°C oven or over a hot plate for 30 min. Repeat steps in ¶ 2) above. Use a separate filter paper and a rubber policeman to aid in transferring residue from dish to filter. Take special care with porcelain dishes because silica adheres to the dish.

4) Ignition—Transfer the two filter papers (one if dehydrated by 4b) and residues to a covered platinum crucible, dry at 110°C, and ignite at 1200°C to constant weight. Avoid mechanical loss of residue when first charring and burning off the pa-

per by gradual heating at minimum temperature. Too rapid heating may form black silicon carbide. Cool in desiccator, weigh, and repeat ignition and weighing until constant weight is attained. Record weight of crucible and contents.

5) Volatilization with HF—Thoroughly moisten weighed residue with distilled water. Add 4 drops 1 + 1 H_2SO_4, followed by 10 mL HF, measuring the latter in a plastic graduated cylinder or pouring an estimated 10 mL directly from the reagent bottle. Slowly evaporate to dryness over an air bath or hot plate in a hood and avoid loss by splattering. Ignite crucible to constant weight at 1200°C. Record weight of crucible and contents.

b. *HClO$_4$ dehydration:* Follow procedure in ¶ 4a1) above until all but 50 mL of sample has been evaporated. Add 5 mL $HClO_4$ and evaporate until dense white fumes appear. (CAUTION: *Explosive—Place a shield between analyst and fuming dish.*)

Continue dehydration for 10 min. Cool, add 5 mL 1 + 1 HCl and 50 mL hot distilled water. Bring to a boil and filter through an ashless quantitative filter paper. Wash thoroughly ten times with hot distilled water and proceed as directed in ¶s 4a4) and 5) preceding. For many purposes, the silica precipitate often is sufficiently

pure for the purpose intended and may be weighed directly, omitting HF volatilization. Make an initial check against the longer procedure, however, to be sure that the result is within the limits of accuracy required.

5. Calculation

Subtract weight of crucible and contents after HF treatment from the corresponding weight before HF treatment. The difference, A, in milligrams is "loss on volatilization" and represents silica:

$$\text{mg SiO}_2/\text{L} = \frac{A \times 1000}{\text{mL sample}}$$

6. Precision and Bias

The accuracy is limited both by the finite solubility of silica in water under the conditions of analysis and by the analytical balance sensitivity. Under optimum conditions, an experienced analyst can obtain a precision of approximately ±0.2 mg SiO_2.

7. Bibliography

HILLEBRAND, W.F. et al. 1953. Applied Inorganic Analysis, 2nd ed. John Wiley & Sons, New York, N.Y. Chapter 43.

KOLTHOFF, I.M., E.J. MEEHAN, E.B. SANDELL & S. BRUCKENSTEIN. 1969. Quantitative Chemical Analysis, 4th ed. Macmillan Co., New York, N.Y.

4500-Si D.   Molybdosilicate Method

1. General Discussion

a. *Principle:* Ammonium molybdate at pH approximately 1.2 reacts with silica and any phosphate present to produce heteropoly acids. Oxalic acid is added to destroy the molybdophosphoric acid but not the molybdosilicic acid. Even if phosphate is known to be absent, the addition of oxalic acid is highly desirable and is a mandatory step in both this method and Method E.

The intensity of the yellow color is proportional to the concentration of "molybdate-reactive" silica. In at least one of its forms, silica does not react with molybdate even though it is capable of passing through filter paper and is not noticeably turbid. It is not known to what extent such "unreactive" silica occurs in waters. Terms such as "colloidal," "crystalloidal," and "ionic" have been used to distinguish among various forms of silica but such terminology cannot be substantiated. "Molybdate-unreactive" silica can be converted to the "molybdate-reactive" form by heating or fusing with alkali. Molybdate-reactive or unreactive does not imply reactivity, or lack of it, toward *other* reagents or processes.

b. *Interference:* Because both apparatus and reagents may contribute silica, avoid using glassware as much as possible and use reagents low in silica. Also, make a blank determination to correct for silica so introduced. In both this method and Method E, tannin, large amounts of iron, color, turbidity, sulfide, and phosphate interfere. Treatment with oxalic acid eliminates interference from phosphate and decreases interference from tannin. If necessary, use photometric compensation to cancel interference from color or turbidity.

c. *Minimum detectable concentration:* Approximately 1 mg SiO_2/L can be detected in 50-mL nessler tubes.

2. Apparatus

a. *Platinum dishes,* 100-mL.

b. *Colorimetric equipment:* One of the following is required:

1) *Spectrophotometer,* for use at 410 nm, providing a light path of 1 cm or longer. See Table 4500-Si:I for light path selection.

2) *Filter photometer,* providing a light path of 1 cm or longer and equipped with a violet filter having maximum transmittance near 410 nm.

3) *Nessler tubes,* matched, 50-mL, tall form.

3. Reagents

For best results, set aside and use batches of chemicals low in silica. Store all reagents in plastic containers to guard against high blanks.

a. *Sodium bicarbonate,* $NaHCO_3$, powder.

b. *Sulfuric acid,* H_2SO_4, 1N.

c. *Hydrochloric acid,* HCl, 1 + 1.

TABLE 4500-Si:I. SELECTION OF LIGHT PATH LENGTH FOR VARIOUS SILICA CONCENTRATIONS

Light Path cm	Method D Silica in 55 mL Final Volume μg	Method E Silica in 55 mL Final Volume μg	
		650 nm Wavelength	815 nm Wavelength
1	200–1300	40–300	20–100
2	100–700	20–150	10–50
5	40–250	7–50	4–20
10	20–130	4–30	2–10

d. *Ammonium molybdate reagent:* Dissolve 10 g $(NH_4)_6Mo_7O_{24} \cdot 4H_2O$ in distilled water, with stirring and gentle warming, and dilute to 100 mL. Filter if necessary. Adjust to pH 7 to 8 with silica-free NH_4OH or NaOH and store in a polyethylene bottle to stabilize. (If the pH is not adjusted, a precipitate gradually forms. If the solution is stored in glass, silica may leach out and cause high blanks.) If necessary, prepare silica-free NH_4OH by passing gaseous NH_3 into distilled water contained in a plastic bottle.

e. *Oxalic acid solution:* Dissolve 7.5 g $H_2C_2O_4 \cdot H_2O$ in distilled water and dilute to 100 mL.

f. *Stock silica solution:* Dissolve 4.73 g sodium metasilicate nonahydrate, $Na_2SiO_3 \cdot 9H_2O$, in distilled water and dilute to 1000 mL. Analyze 100.0-mL portions by Method C to determine concentration. Store in a tightly stoppered plastic bottle.

g. *Standard silica solution:* Dilute 10.00 mL stock solution to 1000 mL with distilled water; 1.00 mL = 10.0 μg SiO_2. Calculate exact concentration from concentration of stock silica solution. Store in a tightly stoppered plastic bottle.

h. Permanent color solutions:

1) *Potassium chromate solution:* Dissolve 630 mg K_2CrO_4 in distilled water and dilute to 1 L.

2) *Borax solution:* Dissolve 10 g sodium borate decahydrate, $Na_2B_4O_7 \cdot 10H_2O$, in distilled water and dilute to 1 L.

4. Procedure

a. Color development: To 50.0 mL sample add in rapid succession 1.0 mL 1 + 1 HCl and 2.0 mL ammonium molybdate reagent. Mix by inverting at least six times and let stand for 5 to 10 min. Add 2.0 mL oxalic acid solution and mix thoroughly. Read color after 2 min but before 15 min, measuring time from addition of oxalic acid. Because the yellow color obeys Beer's law, measure photometrically or visually.

b. To detect the presence of molybdate-unreactive silica, digest sample with $NaHCO_3$ before color development. This digestion is not necessarily sufficient to convert all molybdate-unreactive silica to the molybdate-reactive form. Complex silicates and higher silica polymers may require extended fusion with alkali at high temperatures or digestion under pressure for complete conversion. Omit digestion if all the silica is known to react with molybdate.

Prepare a clear sample by filtration if necessary. Place 50.0 mL, or a smaller portion diluted to 50 mL, in a 100-mL platinum dish. Add 200 mg silica-free $NaHCO_3$ and digest on a steam bath for 1 h. Cool and add slowly, with stirring, 2.4 mL 1N H_2SO_4. Do not interrupt analysis but proceed *at once* with remaining steps. Transfer quantitatively to a 50-mL nessler tube and make up to mark with distilled water. (Tall-form 50-mL nessler tubes are convenient for mixing even if the solution subsequently is transferred to an absorption cell for photometric measurement.)

c. Preparation of standards: If $NaHCO_3$ pretreatment is used, add to the standards (approximately 45 mL total volume) 200 mg $NaHCO_3$ and 2.4 mL 1N H_2SO_4, to compensate both for the slight amount of silica introduced by the reagents and for the effect of the salt on color intensity. Dilute to 50.0 mL.

d. Correction for color or turbidity: Prepare a special blank for every sample that needs such correction. Carry two identical portions of each such sample through the procedure, including $NaHCO_3$ treatment if this is used. To one portion add all reagents as directed in ¶ 4a preceding. To the other portion add HCl and oxalic acid but no molybdate. Adjust photometer to zero absorbance with the blank containing no molybdate before reading absorbance of molybdate-treated sample.

e. Photometric measurement: Prepare a calibration curve from a series of approximately six standards to cover the optimum ranges cited in Table 4500-Si:I. Follow directions of ¶ 4a above on suitable portions of standard silica solution diluted to 50.0 mL in nessler tubes. Set photometer at zero absorbance with distilled water and read all standards, including a reagent blank, against distilled water. Plot micrograms silica in the final (55 mL) developed solution against photometer readings. Run a reagent blank and at least one standard with each group of samples to confirm that the calibration curve previously established has not shifted.

f. Visual comparison: Make a set of permanent artificial color standards, using K_2CrO_4 and borax solutions. Mix liquid volumes specified in Table 4500-Si:II and place them in well-stoppered, appropriately labeled 50-mL nessler tubes. Verify correctness of these permanent artificial standards by comparing them visually against standards prepared by analyzing portions of the standard silica solution. Use permanent artificial color standards only for visual comparison.

TABLE 4500-Si:II. PREPARATION OF
PERMANENT COLOR STANDARDS FOR VISUAL
DETERMINATION OF SILICA

Value in Silica μg	Potassium Chromate Solution mL	Borax Solution mL	Water mL
0	0.0	25	30
100	1.0	25	29
200	2.0	25	28
400	4.0	25	26
500	5.0	25	25
750	7.5	25	22
1000	10.0	25	20

5. Calculation

$$\text{mg } SiO_2/L = \frac{\mu g \ SiO_2 \ (\text{in 55 mL final volume})}{mL \ \text{sample}}$$

Report whether NaHCO$_3$ digestion was used.

6. Precision and Bias

A synthetic sample containing 5.0 mg SiO$_2$/L, 10 mg Cl$^-$/L, 0.20 mg NH$_3$-N/L, 1.0 mg NO$_3^-$-N/L, 1.5 mg organic N/L, and 10.0 mg PO$_4^{3-}$/L in distilled water was analyzed in 19 laboratories by the molybdosilicate method with a relative standard deviation of 14.3% and a relative error of 7.8%.

Another synthetic sample containing 15.0 mg SiO$_2$/L, 200 mg Cl$^-$/L, 0.800 mg NH$_3$-N/L, 1.0 mg NO$_3^-$-N/L, 0.800 mg organic N/L, and 5.0 mg PO$_4^{3-}$/L in distilled water was analyzed in 19 laboratories by the molybdosilicate method, with a relative standard deviation of 8.4% and a relative error of 4.2%.

A third synthetic sample containing 30.0 mg SiO$_2$/L, 400 mg Cl$^-$/L, 1.50 mg NH$_3$-N/L, 1.0 mg NO$_3^-$-N/L, 0.200 mg organic N/L, and 0.500 mg PO$_4^{3-}$/L, in distilled water was analyzed in 20 laboratories by the molybdosilicate method, with a relative standard deviation of 7.7% and a relative error of 9.8%.

All results were obtained after sample digestion with NaHCO$_3$.

7. Bibliography

DIENERT, F.& F. WANDENBULCKE. 1923. On the determination of silica in waters. *Bull. Soc. Chim. France* 33:1131, *Compt. Rend.* 176:1478.

DIENERT, F. & F. WANDENBULCKE. 1924. A study of colloidal silica. *Compt. Rend.* 178:564.

SWANK, H.W. & M.G. MELLON. 1934. Colorimetric standards for silica. *Ind. Eng. Chem.*, Anal. Ed. 6:348.

TOURKY, A.R. & D.H. BANGHAM. 1936. Colloidal silica in natural waters and the "silicomolybdate" colour test. *Nature* 138:587.

BIRNBAUM, N. & G.H. WALDEN. 1938. Co-precipitation of ammonium silicomolybdate and ammonium phosphomolybdate. *J. Amer. Chem. Soc.* 60:66.

KAHLER, H.L. 1941. Determination of soluble silica in water: A photometric method. *Ind. Eng. Chem.*, Anal. Ed. 13:536.

NOLL, C.A. & J.J. MAGUIRE. 1942. Effect of container on soluble silica content of water samples. *Ind. Eng. Chem.*, Anal. Ed. 14:569.

SCHWARTZ, M.C. 1942. Photometric determination of silica in the presence of phosphates. *Ind. Eng. Chem.*, Anal. Ed. 14:893.

GUTTER, H. 1945. Influence of pH on the composition and physical aspects of the ammonium molybdates. *Compt. Rend.* 220:146.

MILTON, R.F. 1951. Formation of silicomolybdate. *Analyst* 76:431.

KILLEFFER, D.H. & A. LINZ. 1952. Molybdenum Compounds, Their Chemistry and Technology. Interscience Publishers, New York, N.Y. pp. 1-2, 42-45, 67-82, 87-92.

STRICKLAND, J.D.H. 1952. The preparation and properties of silicomolybdic acid. *J. Amer. Chem. Soc.* 74:862, 868, 872.

CHOW, D.T.W. & R.J. ROBINSON. 1953. The forms of silicate available for colorimetric determination. *Anal. Chem.* 25:646.

4500-Si E.   Heteropoly Blue Method

1. General Discussion

a. *Principle:* The principles outlined under Method D, ¶ 1a, also apply to this method. The yellow molybdosilicic acid is reduced by means of aminonaphtholsulfonic acid to heteropoly blue. The blue color is more intense than the yellow color of Method D and provides increased sensitivity.

b. *Interference:* See Section 4500-Si.D.1b.

c. *Minimum detectable concentration:* Approximately 20 μg SiO_2/L can be detected in 50-mL nessler tubes and 50 μg SiO_2/L spectrophotometrically with a 1-cm light path at 815 nm.

2. Apparatus

a. *Platinum dishes,* 100-mL.

b. *Colorimetric equipment:* One of the following is required:

1) *Spectrophotometer,* for use at approximately 815 nm. The color system also obeys Beer's law at 650 nm, with appreciably reduced sensitivity. Use light path of 1 cm or longer. See Table 4500-Si:I for light path selection.

2) *Filter photometer,* provided with a red filter exhibiting maximum transmittance in the wavelength range of 600 to 815 nm. Sensitivity improves with increasing wavelength. Use light path of 1 cm or longer.

3) *Nessler tubes,* matched, 50-mL, tall form.

3. Reagents

For best results, set aside and use batches of chemicals low in silica. Store all reagents in plastic containers to guard against high blanks. Use distilled water that does not contain detectable silica after storage in glass.

All of the reagents listed in Section 4500-Si.D.3 are required, and in addition:

Reducing agent: Dissolve 500 mg 1-amino-2-naphthol-4-sulfonic acid and 1 g

Na_2SO_3 in 50 mL distilled water, with gentle warming if necessary; add this to a solution of 30 g $NaHSO_3$ in 150 mL distilled water. Filter into a plastic bottle. Discard when solution becomes dark. Prolong reagent life by storing in a refrigerator and away from light. Do not use aminonaphtholsulfonic acid that is incompletely soluble or that produces reagents that are dark even when freshly prepared.*

4. Procedure

a. *Color development:* Proceed as in ¶ 4500-Si.D.4a up to and including the words, "Add 2.0 mL oxalic acid solution and mix thoroughly." Measuring time from the moment of adding oxalic acid, wait at least 2 min but not more than 15 min, add 2.0 mL reducing agent, and mix thoroughly. After 5 min, measure blue color photometrically or visually. If $NaHCO_3$ pretreatment is used, follow ¶ 4500-Si.D.4b.

b. *Photometric measurement:* Prepare a calibration curve from a series of approximately six standards to cover the optimum range indicated in Table 4500-Si:I. Carry out the steps described above on suitable portions of standard silica solution diluted to 50.0 mL in nessler tubes; pretreat standards if $NaHCO_3$ digestion is used (see 4500-Si.D.4b). Adjust photometer to zero absorbance with distilled water and read all standards, including a reagent blank, against distilled water. If necessary to correct for color or turbidity in a sample, see ¶ 4500-Si.D.4d. To the special blank add HCl and oxalic acid, but no molybdate or reducing agent. Plot micrograms silica in the final 55 mL developed solution against absorbance. Run a reagent blank and at least one standard with each group of samples to check the calibration curve.

c. *Visual comparison:* Prepare a series of

---

*Eastman No. 360 has been found satisfactory.

not less than 12 standards, covering the range 0 to 120 μg SiO_2, by placing the calculated volumes of standard silica solution in 50-mL nessler tubes, diluting to mark with distilled water, and developing color as described in ¶ *a* preceding.

5. Calculation

$$\text{mg } SiO_2/L = \frac{\mu\text{g } SiO_2 \text{ (in 55 mL final volume)}}{\text{mL sample}}$$

Report whether $NaHCO_3$ digestion was used.

6. Precision and Bias

A synthetic sample containing 5.0 mg SiO_2/L, 10 mg Cl^-/L, 0.200 mg NH_3-N/L, 1.0 mg NO_3^--N/L, 1.5 mg organic N/L, and 10.0 mg PO_4^{3-}/L in distilled water was analyzed in 11 laboratories by the heteropoly blue method, with a relative standard deviation of 27.2% and a relative error of 3.0%.

A second synthetic sample containing 15 mg SiO_2/L, 200 mg Cl^-/L, 0.800 mg NH_3-N/L, 1.0 mg NO_3^--N/L, 0.800 mg organic N/L, and 5.0 mg PO_4^{3-}/L in distilled water was analyzed in 11 laboratories by the heteropoly blue method, with a relative standard deviation of 18.0% and a relative error of 2.9%.

A third synthetic sample containing 30.0 mg SiO_2/L, 400 mg Cl^-/L, 1.50 mg NH_3-N/L, 1.0 mg NO_3^--N/L, 0.200 mg organic N/L, and 0.500 mg PO_4^{3-}/L in distilled water was analyzed in 10 laboratories by the heteropoly blue method with a relative standard deviation of 4.9% and a relative error of 5.1%.

All results were obtained after sample digestion with $NaHCO_3$.

7. Bibliography

BUNTING, W.E. 1944. Determination of soluble silica in very low concentrations. *Ind. Eng. Chem.*, Anal. Ed. 16:612.

STRAUB, F.G. & H. GRABOWSKI. 1944. Photometric determination of silica in condensed steam in the presence of phosphates. *Ind. Eng. Chem.*, Anal. Ed. 16:574.

BOLTZ, D.F. & M.G. MELLON. 1947. Determination of phosphorus, germanium, silicon, and arsenic by the heteropoly blue method. *Ind. Eng. Chem.*, Anal. Ed. 19:873.

MILTON, R.F. 1951. Estimation of silica in water. *J. Appl. Chem.* (London) 1: (Supplement No. 2) 126.

CARLSON, A.B. & C.V. BANKS. 1952. Spectrophotometric determination of silicon. *Anal. Chem.* 24:472.

4500-Si F.   Automated Method for Molybdate-Reactive Silica

1. General Discussion

a. *Principle:* This method is an adaptation of the heteropoly blue method (Method E) utilizing a continuous-flow analytical instrument.

b. *Interferences:* See Section 4500-Si.D.1*b*. If particulate matter is present, filter sample or use a continuous filter as an integral part of the system.

c. *Application:* This method is applicable to potable, surface, domestic, and other waters containing 0 to 20 mg SiO_2/L. The range of concentration can be broadened to 0 to 80 mg/L by substituting a 15-mm flow cell for the 50-mm flow cell shown in Figure 4500-Si:1.

2. Apparatus

Automated analytical equipment: The required continuous-flow analytical instrument* consists of the interchangeable components shown in Figure 4500-Si:1.

---

*Technicon™ Autoanalyzer™ or equivalent.

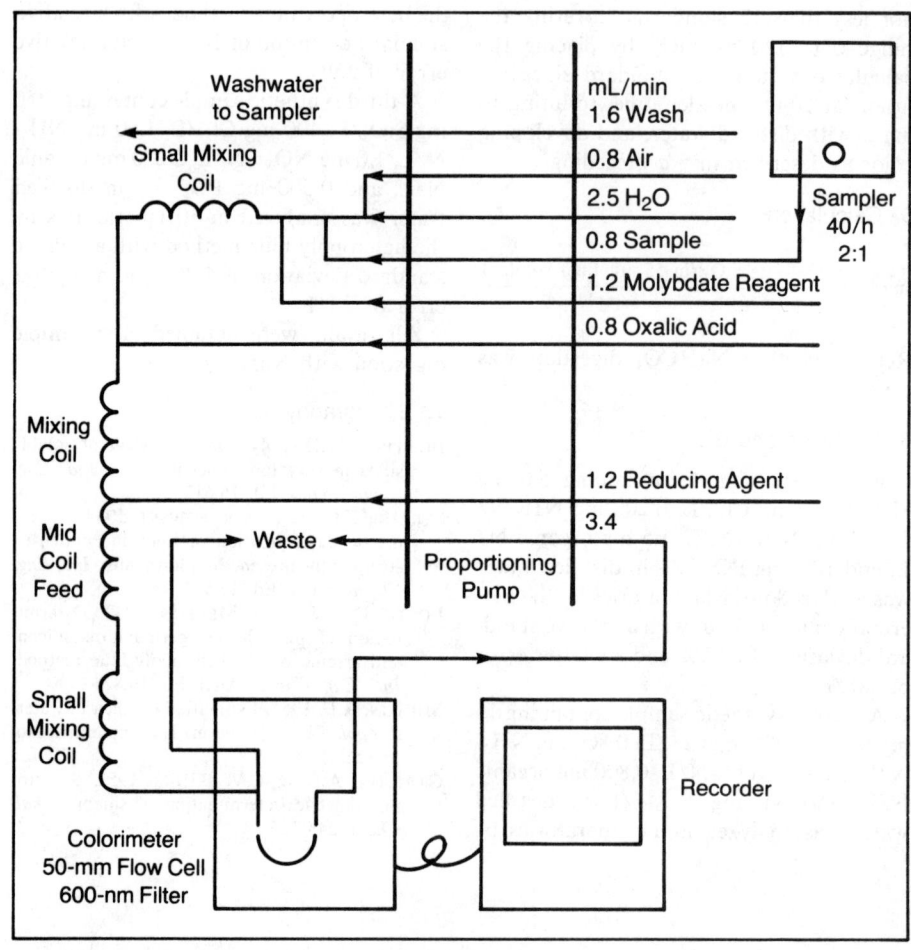

Figure 4500-Si:1. Silica manifold.

3. Reagents

a. Sulfuric acid, H_2SO_4, 0.1*N.*

b. Ammonium molybdate reagent: Dissolve 10 g $(NH_4)_6Mo_7O_{24} \cdot 4H_2O$ in 1 L 0.1*N* H_2SO_4. Filter and store in an amber plastic bottle.

c. Oxalic acid solution: Dissolve 50 g oxalic acid in 900 mL distilled water and dilute to 1 L.

d. Reducing agent: Dissolve 120 g $NaHSO_3$ and 4 g Na_2SO_3 in 800 mL warm distilled water. Add 2 g 1-amino-2-naphthol-4-sulfonic acid, mix well, and dilute to 1 L. Filter into amber plastic bottle for storage.

To prepare working reagent, dilute 100 mL to 1 L with distilled water. Make working reagent daily.

e. Standard silica solution: See 4500-Si.C.3*f.*

4. Procedure

Set up manifold as shown in Figure 4500-Si:1 and follow the general procedure described by the manufacturer. Determine absorbance at 660 nm.

5. Calculation

Prepare standard curves by plotting peak heights of standards processed through the manifold against SiO$_2$ concentration in standards. Compute sample SiO$_2$ concentration by comparing sample peak height with standard curve.

6. Precision and Bias

For 0 to 20 mg SiO$_2$/L, when a 50-mm flow cell was used at 40 samples/h, the detection limit was 0.1 mg/L, sensitivity (concentration giving 0.398 absorbance) was 7.1 mg/L, and the coefficient of variation (95% confidence level at 7.1 mg/L) was 1.6%. For 0 to 80 mg SiO$_2$/L, when a 15-mm flow cell was used at 50 samples/h, detection limit was 0.5 mg/L, sensitivity was 31 mg/L, and coefficient of variation at 31 mg/L was 1.5%.

4500-Si G. Inductively Coupled Plasma Method

See Section 3120.

4500-S^{2-} SULFIDE*

4500-S^{2-} A. Introduction

1. Occurrence and Significance

Sulfide often is present in groundwater, especially in hot springs. Its common presence in wastewaters comes partly from the decomposition of organic matter, sometimes from industrial wastes, but mostly from the bacterial reduction of sulfate. Hydrogen sulfide escaping into the air from sulfide-containing wastewater causes odor nuisances. The threshold odor concentration of H$_2$S in clean water is between 0.025 and 0.25 μg/L. H$_2$S is very toxic and has claimed the lives of numerous workers in sewers. It attacks metals directly and indirectly has caused serious corrosion of concrete sewers because it is oxidized biologically to H$_2$SO$_4$ on the pipe wall.

2. Categories of Sulfides

From an analytical standpoint, three categories of sulfide in water and wastewater are distinguished:

a. Total sulfide includes dissolved H$_2$S and HS$^-$, as well as acid-soluble metallic sulfides present in suspended matter. The S^{2-} is negligible, amounting to less than 0.5% of the dissolved sulfide at pH 12, less than 0.05% at pH 11, etc. Copper and silver sulfides are so insoluble that they do not respond in ordinary sulfide determinations; they can be ignored for practical purposes.

b. Dissolved sulfide is that remaining after suspended solids have been removed by flocculation and settling.

c. Un-ionized hydrogen sulfide may be

---

*Approved by Standard Methods Committee, 1988.

calculated from the concentration of dissolved sulfide, the sample pH, and the practical ionization constant of H_2S.

3. Sampling and Storage

Take samples with minimum aeration. To preserve a sample for a total sulfide determination put zinc acetate solution into bottle before filling it with sample. Use 4 drops of 2N zinc acetate solution per 100 mL sample. Fill bottle completely and stopper.

4. Qualitative Tests

A qualitative test for sulfide often is useful. It is advisable in the examination of industrial wastes containing interfering substances that may give a false negative result in the methylene blue procedure.

a. Antimony test: To about 200 mL sample, add 0.5 mL saturated solution of potassium antimony tartrate and 0.5 mL 6N HCl in excess of phenolphthalein alkalinity.

Yellow antimony sulfide (Sb_2S_3) is discernible at a sulfide concentration of 0.5 mg/L. Comparisons with samples of known sulfide concentration make the technique roughly quantitative. The only known interferences are metallic ions such as lead, which hold the sulfide so firmly that it does not produce Sb_2S_3, and dithionite, which decomposes in acid solution to produce sulfide.

b. Silver-silver sulfide electrode test: The potential of a silver-silver sulfide electrode assembly relative to a reference electrode varies with the activity of the sulfide ion in solution. By correcting for the ion activity coefficient and pH, this potential estimates sulfide concentration. Standardize electrode frequently against a sulfide solution of known strength. An electrode of this type can be used as an end-point indicator for titrating dissolved sulfide with a standard solution of a silver or lead salt, but slow response always is a problem.

c. Lead acetate paper and silver foil tests: Confirm odors attributed to H_2S with lead acetate paper. On exposure to the vapor of a slightly acidified sample, the paper becomes blackened by formation of PbS. A strip of silver foil is more sensitive than lead acetate paper. Clean the silver by dipping in NaCN solution and rinse. Silver is suitable particularly for long-time exposure in the vicinity of possible H_2S sources because black Ag_2S is permanent whereas PbS slowly oxidizes.

5. Selection of Quantitative Methods

Iodine reacts with sulfide in acid solution, oxidizing it to sulfur. A titration based on this reaction is an accurate method for determining sulfide at concentrations above 1 mg/L if interferences are absent and if loss of H_2S is avoided. The iodometric method (E) is useful for standardizing the methylene blue colorimetric method and is suitable for analyzing samples freshly taken from wells or springs. The method can be used for wastewater and partly oxidized water from sulfur springs if interfering substances are removed first.

The methylene blue method (D) is based on the reaction of sulfide, ferric chloride, and dimethyl-*p*-phenylenediamine to produce methylene blue. Ammonium phosphate is added after color development to remove ferric chloride color. The procedure is applicable at sulfide concentrations up to 20 mg/L.

Potentiometric methods utilizing a silver electrode may be suitable. From the potential of the electrode relative to a reference electrode an estimate can be made of the sulfide concentration, but careful attention to details of procedures and frequent standardizations are needed to secure good results. The electrode is useful particularly as an end-point indicator for titration of dissolved sulfide with silver nitrate.

Figure 4500-S^{2-}:1 shows analytical flow

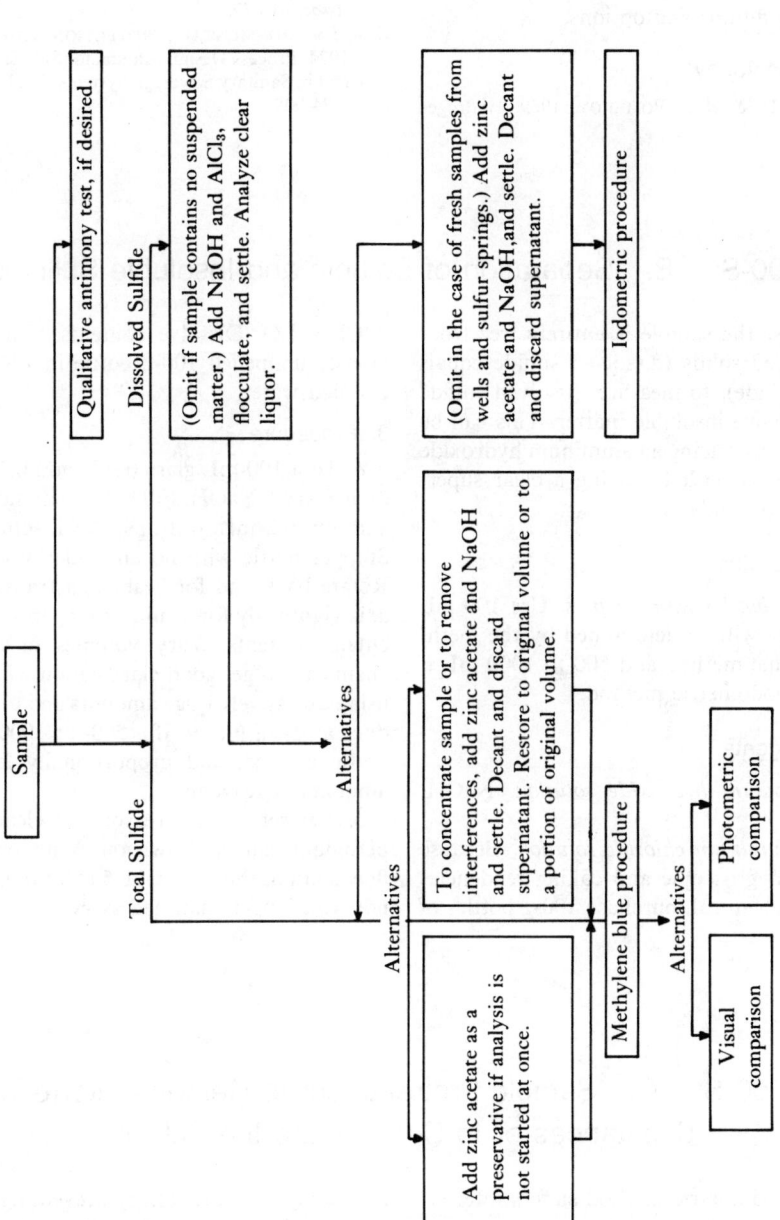

Figure 4500-S²⁻:1. Analytical flow paths for sulfide determinations.

paths for sulfide determinations under various conditions and options.

5. Bibliography

CRUSE, H. & R.D. POMEROY. 1969. Hydrogen sulfide odor threshold. *J. Amer. Water Works Assoc.* 61:677.

U.S. ENVIRONMENTAL PROTECTION AGENCY. 1974. Process Design Manual for Sulfide Control in Sanitary Sewerage Systems. Publ. 625/1-74-005.

4500-S^{2-} B.   Separation of Soluble and Insoluble Sulfides

Unless the sample is entirely free from suspended solids (dissolved sulfide equals total sulfide), to measure dissolved sulfide first remove insoluble matter. This can be done by producing an aluminum hydroxide floc that is settled, leaving a clear supernatant for analysis.

1. Apparatus

Glass bottles with stoppers: Use 100 mL if sulfide will be determined by the methylene blue method and 500 to 1000 mL if by the iodometric method.

2. Reagents

a. Sodium hydroxide solution, NaOH, 6N.

b. Aluminum chloride solution: Because of the hygroscopic and caking tendencies of this chemical, purchase 100-g bottles of AlCl$_3 \cdot$6H$_2$O. Dissolve contents of a previously unopened 100-g bottle in 144 mL distilled water.

3. Procedure

a. To a 100-mL glass bottle add 0.2 mL (4 drops) 6N NaOH. Fill bottle with sample and add 0.2 mL (4 drops) AlCl$_3$ solution. Stopper bottle with no air under stopper. Rotate back and forth about a transverse axis vigorously for 1 min or longer to flocculate contents. Vary volumes of these chemicals to get good clarification without using excessively large amounts and to produce a pH of 6 to 9. If a 500- or 1000-mL bottle is used, add proportionally larger amounts of reagents.

b. Let settle until reasonably clear supernatant can be drawn off. With proper flocculation, this may take 5 to 15 min. Do not wait longer than necessary.

4500-S^{2-} C.   Sample Pretreatment to Remove Interfering Substances or to Concentrate the Sulfide

The iodometric method suffers interference from reducing substances that react with iodine, including thiosulfate, sulfite, and various organic compounds, both solid and dissolved.

Strong reducing agents also interfere in the methylene blue test by preventing formation of the blue color. Thiosulfate at concentrations above 10 mg/L may retard color formation or completely prevent it. Sulfide itself prevents the reaction if its concentration is very high, in the range of several hundred milligrams per liter. To avoid the possibility of false negative results, use the antimony method to obtain a qualitative result in industrial wastes likely to contain sulfide but showing no color by the methylene blue method. Iodide, which is likely to be present in oil-field wastewaters, may diminish color formation if its concentration exceeds 2 mg/L. Ferrocyanide produces a blue color.

Eliminate interferences due to sulfite, thiosulfate, iodide, and many other soluble substances, but not ferrocyanide, by first precipitating ZnS, removing the supernatant, and replacing it with distilled water. Use the same procedure, even when not needed for removal of interferences, to concentrate sulfide.

1. Apparatus

Glass bottles with stoppers: See Section 4500-S²⁻.B.1.

2. Reagents

a. Zinc acetate solution: Dissolve 220 g $Zn(C_2H_3O_2)_2 \cdot 2H_2O$ in 870 mL water; this makes 1 L solution.

b. Sodium hydroxide solution, NaOH, 6N.

3. Procedure

a. Put 0.15 mL (3 drops) zinc acetate solution into a 100-mL glass bottle, fill with sample, and add 0.10 mL (2 drops) 6N NaOH solution. Stopper with no air bubbles under stopper and mix by rotating

back and forth vigorously about a transverse axis. For the iodometric procedure, use a 500-mL bottle or other convenient size, with proportionally larger volumes of reagents. Vary volume of reagents added according to sample so that the resulting precipitate is not excessively bulky and settles readily. Add enough NaOH to produce a pH above 9. Let precipitate settle for 30 min. The treated sample is relatively stable and can be held for several hours. However, if much iron is present, oxidation may be fairly rapid.

b. If the iodometric method is to be used, filter precipitate through glass fiber filter paper and continue at once with titration according to the procedure of Method E. If the methylene blue method is used, let precipitate settle for 30 min and decant as much supernatant as possible without loss of precipitate. Refill bottle with distilled water, resuspend precipitate, and withdraw a sample. If interfering substances are present in high concentration, settle, decant, and refill a second time. If sulfide concentration is known to be low, add only enough water to bring volume to one-half or one-fifth of original volume. Use this technique for analyzing samples of very low sulfide concentrations. After determining the sulfide concentration colorimetrically, multiply the result by the ratio of final to initial volume.

Cadmium salts sometimes are used instead of zinc, but CdS is more susceptible to oxidation than ZnS.

4500-S²⁻   D.   Methylene Blue Method

1. Apparatus

a. Matched test tubes, approximately 125 mm long and 15 mm OD.

b. Droppers, delivering 20 drops/mL methylene blue solution. To obtain uniform

drops hold dropper in a vertical position and let drops form slowly.

c. If photometric rather than visual color determination will be used, either:

1) *Spectrophotometer,* for use at a wavelength of 664 nm with cells providing light paths of 1 cm and 1 mm, or

2) *Filter photometer,* with a filter providing maximum transmittance near 600 nm.

2. Reagents

a. Amine-sulfuric acid stock solution: Dissolve 27 g N,N- dimethyl-*p*-phenylenediamine oxalate* in a cold mixture of 50 mL conc H_2SO_4 and 20 mL distilled water. Cool and dilute to 100 mL with distilled water. Use fresh oxalate because an old supply may be oxidized and discolored to a degree that results in interfering colors in the test. Store in a dark glass bottle. When this stock solution is diluted and used in the procedure with a sulfide-free sample, it first will be pink but then should become colorless within 3 min.

b. Amine-sulfuric acid reagent: Dilute 25 mL amine-sulfuric acid stock solution with 975 mL 1 + 1 H_2SO_4. Store in a dark glass bottle.

c. Ferric chloride solution: Dissolve 100 g $FeCl_3 \cdot 6H_2O$ in 40 mL water.

d. Sulfuric acid solution, H_2SO_4, 1 + 1.

e. Diammonium hydrogen phosphate solution: Dissolve 400 g $(NH_4)_2HPO_4$ in 800 mL distilled water.

f. Methylene blue solution I: Use USP grade dye or one certified by the Biological Stain Commission. The dye content should be reported on the label and should be 84% or more. Dissolve 1.0 g in distilled water and make up to 1 L. This solution will be approximately the correct strength, but because of variation between different lots of dye, standardize against sulfide solutions of known strength and adjust its concentration so that 0.05 mL (1 drop) = 1.0 mg sulfide/L.

Standardization—Put several grams of clean, washed crystals of $Na_2S \cdot 9H_2O$ into a small beaker. Add somewhat less than enough water to cover crystals. Stir occasionally for a few minutes, then pour so-

lution into another vessel. This solution reacts slowly with oxygen but the change is unimportant within a few hours. Make solution daily. To 1 L distilled water add 1 drop of solution and mix. Immediately determine sulfide concentration by the methylene blue procedure and by the iodometric procedure. Repeat, using more than 1 drop Na_2S solution or smaller volumes of water, until at least five tests have been made, with a range of sulfide concentrations between 1 and 8 mg/L. Calculate average percent error of the methylene blue result as compared to the iodometric result. If the average error is negative, that is, methylene blue results are lower than iodometric results, dilute methylene blue solution by the same percentage, so that a greater volume will be used in matching colors. If methylene blue results are high, increase solution strength by adding more dye.

g. Methylene blue solution II: Dilute 10.00 mL of adjusted methylene blue solution I to 100 mL.

3. Procedure

a. Color development: Transfer 7.5 mL sample to each of two matched test tubes, using a special wide-tip pipet or filling to marks on test tubes. Add to Tube A 0.5 mL amine-sulfuric acid reagent and 0.15 mL (3 drops) $FeCl_3$ solution. Mix immediately by inverting slowly, only once. (Excessive mixing causes low results by loss of H_2S as a gas before it has had time to react). To Tube B add 0.5 mL 1 + 1 H_2SO_4 and 0.15 mL (3 drops) $FeCl_3$ solution and mix. The presence of S^{2-} will be indicated by the appearance of blue color in Tube A. Color development usually is complete in about 1 min, but a longer time often is required for fading out of the initial pink color. Wait 3 to 5 min and add 1.6 mL $(NH_4)_2HPO_4$ solution to each tube. Wait 3 to 15 min and make color comparisons. If

---

*Eastman catalog No. 5672 has been found satisfactory for this purpose.

zinc acetate was used, wait at least 10 min before making a visual color comparison.

b. Color determination:

1) Visual color estimation—Add methylene blue solution I or II, depending on sulfide concentration and desired accuracy, dropwise, to the second tube, until color matches that developed in first tube. If the concentration exceeds 20 mg/L, repeat test with a portion of sample diluted to one tenth.

With methylene blue solution I, adjusted so that 0.05 mL (1 drop) = 1.0 mg S^{2-}/ L when 7.5 mL of sample are used:

mg S^{2-}/L = no. drops solution I
+ 0.1 (no. drops solution II)

2) Photometric color measurement—A cell with a light path of 1 cm is suitable for measuring sulfide concentrations from 0.1 to 2.0 mg/L. Use shorter or longer light paths for higher or lower concentrations. The upper limit of the method is 20 mg/ L. Zero instrument with a portion of treated sample from Tube B. Prepare calibration curves on basis of colorimetric tests made on Na$_2$S solutions simultaneously analyzed by the iodometric method, plotting concentration vs. absorbance. A straight-line relationship between concentration and absorbance can be assumed from 0 to 1.0 mg/L.

Read sulfide concentration from calibration curve.

4. Precision and Bias

The accuracy is about ±10%. The standard deviation has not been determined.

5. Bibliography

POMEROY, R.D. 1936. The determination of sulfides in sewage. *Sewage Works J.* 8:572.

NUSBAUM, I. 1965. Determining sulfides in water and waste water. *Water Sewage Works* 112:113.

4500-S^{2-} E.   Iodometric Method

1. Reagents

a. Hydrochloric acid, HCl, 6*N*.

b. Standard iodine solution, 0.0250*N:* Dissolve 20 to 25 g KI in a little water and add 3.2 g iodine. After iodine has dissolved, dilute to 1000 mL and standardize against 0.0250*N* Na$_2$S$_2$O$_3$, using starch solution as indicator.

c. Standard sodium thiosulfate solution, 0.0250*N:* See Section 4500-O.C.2*e.*

d. Starch solution: See Section 4500-O.C.2*d.*

2. Procedure

a. Measure from a buret into a 500-mL flask an amount of iodine solution estimated to be an excess over the amount of sulfide present. Add distilled water, if necessary, to bring volume to about 20 mL. Add 2 mL 6*N* HCl. Pipet 200 mL sample into flask, discharging sample under solution surface. If iodine color disappears, add more iodine so that color remains. Back-titrate with Na$_2$S$_2$O$_3$ solution, adding a few drops of starch solution as end point is approached, and continuing until blue color disappears.

b. If sulfide was precipitated with zinc and ZnS filtered out, return filter with precipitate to original bottle and add about 100 mL water. Add iodine solution and HCl and titrate as in ¶ 2*a* above.

3. Calculation

One milliliter 0.0250N iodine solution reacts with 0.4 mg S^{2-}:

$$\text{mg } S^{2-}/L = \frac{[(A \times B) - (C \times D)] \times 16\,000}{\text{mL sample}}$$

where:

A = mL iodine solution,
B = normality of iodine solution,
C = mL $Na_2S_2O_3$ solution, and
D = normality of $Na_2S_2O_3$ solution.

4. Precision

The precision of the end point varies with the sample. In clean waters it should be determinable within 1 drop, which is equivalent to 0.1 mg/L in a 200-mL sample.

4500-S^{2-} F. Calculation of Un-ionized Hydrogen Sulfide

1. Calculation

Hydrogen sulfide and HS^-, which together constitute dissolved sulfide, are in equilibrium with hydrogen ions:

$$H_2S \rightleftarrows H^+ + HS^-$$

The ionization constant of H_2S is used to calculate the distribution of dissolved sulfide between the two forms. The practical constant written in logarithmic form, pK', is used. The constant varies with temperature and ionic strength of the solution. The ionic strength effect can be estimated most easily from the conductivity. Because the effect of ionic strength is not large, values that are sufficiently dependable generally can be assumed if the nature of the sample is known. Table 4500-S^{2-}:I gives approximate pK' values for various temperatures and conductivities. The temperature effect is practically linear from 15°C to 35°C; interpolations or extrapolations can be used. The last line of Table 4500-S^{2-}:I corresponds approximately to seawater.

From sample pH and appropriate value of pK', calculate pH − pK'. From Figure 4500-S^{2-}:2 read proportions of dissolved sulfide present as H_2S (left-side scale of Figure 4500-S^{2-}:2). Let this proportion equal J,

TABLE 4500-S^{2-}:I. VALUES OF pK', LOGARITHM OF PRACTICAL IONIZATION CONSTANT FOR HYDROGEN SULFIDE

Conductivity at 25°C $\mu mhos/cm$	pK' at Given Temperature		
	20°C	25°C	30°C
0	—	7.03*	—
100	7.08	7.01	6.94
200	7.07	7.00	6.93
400	7.06	6.99	6.92
700	7.05	6.98	6.91
1 200	7.04	6.97	6.90
2 000	7.03	6.96	6.89
3 000	7.02	6.95	6.88
4 000	7.01	6.94	6.87
5 200	7.00	6.93	6.86
7 200	6.99	6.92	6.85
10 000	6.98	6.91	6.84
14 000	6.97	6.90	6.83
22 000	6.96	6.89	6.82
50 000	6.95	6.88	6.81

*Theoretical.

$J \times$ (dissolved sulfide)
= un-ionized H_2S expressed as S^{2-}

2. Bibliography

PLATFORD, R.F. 1965. The activity coefficient of sodium chloride in seawater. *J. Mar. Res.* 23:55.

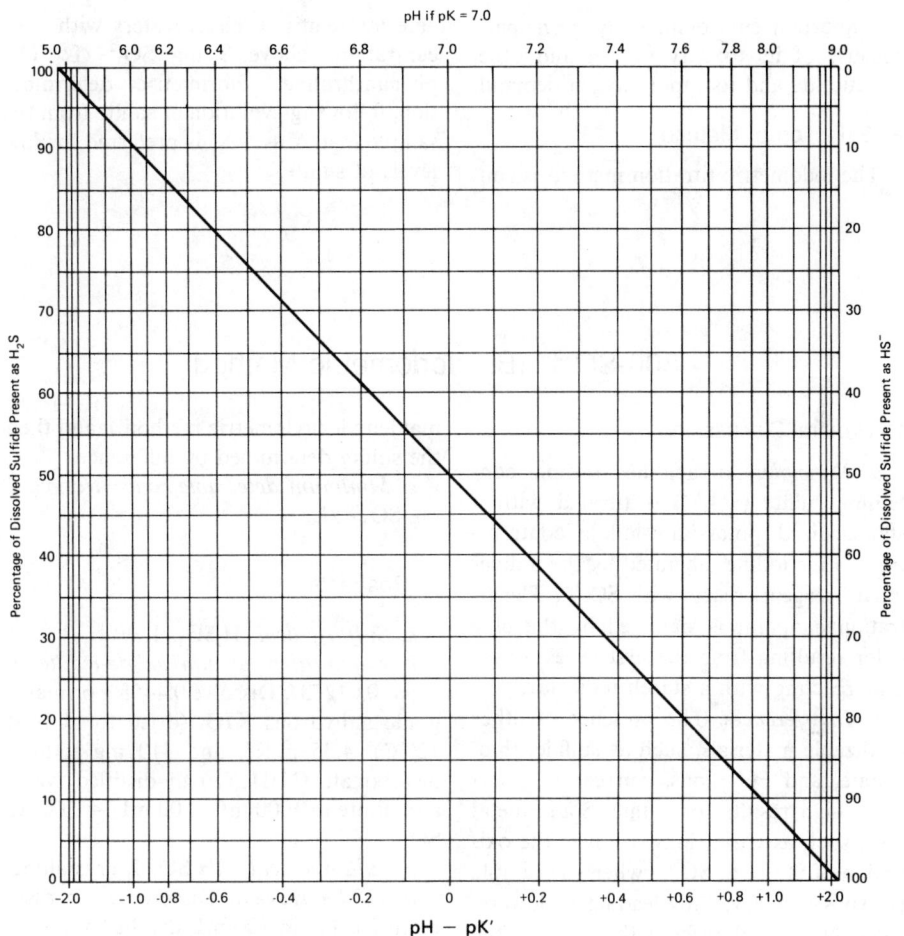

Figure 4500-S²⁻:2. Proportions of H₂S and HS⁻ in dissolved sulfide.

4500-SO₃²⁻   SULFITE*

4500-SO₃²⁻   A.   Introduction

1. Occurrence

Sulfite ions (SO_3^{2-}) may occur in boilers and boiler feedwaters treated with sulfite for dissolved oxygen control, in natural waters or wastewaters as a result of industrial pollution, and in treatment plant effluents dechlorinated with sulfur dioxide (SO_2). Excess sulfite ion in boiler waters is deleterious because it lowers the pH and promotes corrosion. Control of sulfite ion in wastewater treatment and discharge may

*Approved by Standard Methods Committee, 1985.

be important environmentally, principally because of its toxicity to fish and other aquatic life and its rapid oxygen demand.

2. Selection of Method

The iodometric titration method is suit-able for relatively clean waters with concentrations above 2 mg SO_3^{2-}/L. The phenanthroline colorimetric determination, following evolution of sulfite from the sample matrix as SO_2, is preferred for low levels of sulfite.

4500-SO_3^{2-} B. Iodometric Method

1. General Discussion

a. Principle: An acidified sample containing sulfite (SO_3^{2-}) is titrated with a standardized potassium iodide-iodate titrant. Free iodine, liberated by the iodide-iodate reagent, reacts with SO_3^{2-}. The titration endpoint is signalled by the blue color resulting from the first excess of iodine reacting with a starch indicator.

b. Interferences: The presence of other oxidizable materials, such as sulfide, thiosulfate, and Fe^{2+} ions, can cause apparently high results for sulfite. Some metal ions, such as Cu^{2+}, may catalyze the oxidation of SO_3^{2-} to SO_4^{2-} when the sample is exposed to air, thus leading to low results. NO_2^- will react with SO_3^{2-} in the acidic reaction medium and lead to low sulfite results unless sulfamic acid is added to destroy nitrite. Addition of EDTA as a complexing agent at the time of sample collection inhibits Cu^{2+} catalysis and promotes oxidation of ferrous to ferric iron before analysis. Sulfide and thiosulfate ions normally would be expected only in samples containing certain industrial discharges, but must be accounted for if present. Sulfide may be removed by adding about 0.5 g zinc acetate and analyzing the supernatant of the settled sample. However, thiosulfate may have to be determined by an independent method (e.g., the for-

maldehyde/iodometric method[1]), and then the sulfite determined by difference.

c. Minimum detectable concentration: 2 mg SO_3^{2-}/L.

2. Reagents

a. Sulfuric acid: H_2SO_4, 1 + 1.

b. Standard potassium iodide-iodate titrant, 0.0125*M:* Dissolve 0.4458 g primary-grade anhydrous KIO_3 (dried for 4 h at 120°C), 4.35 g KI, and 310 mg sodium bicarbonate ($NaHCO_3$) in distilled water and dilute to 1000 mL; 1.00 mL = 500 μg SO_3^{2-}.

c. Sulfamic acid, NH_2SO_3H, crystalline.

d. EDTA reagent: Dissolve 2.5 g disodium EDTA in 100 mL distilled water.

e. Starch indicator: To 5 g starch (potato, arrowroot, or soluble) in a mortar, add a little cold distilled water and grind to a paste. Add mixture to 1 L boiling distilled water, stir, and let settle overnight. Use clear supernatant. Preserve by adding either 1.3 g salicylic acid, 4 g $ZnCl_2$, or a combination of 4 g sodium propionate and 2 g sodium azide to 1 L starch solution.

3. Procedure

a. Sample collection: Collect a fresh sample, taking care to minimize contact with air. Fix cooled samples (< 50°C) immediately by adding 1 mL EDTA solution/100 mL sample. Cool hot samples to 50°C or below. Do not filter.

b. Titration: Add 1 mL H$_2$SO$_4$ and 0.1 g NH$_2$SO$_3$H crystals to a 250-mL erlenmeyer flask or other suitable titration vessel. Accurately measure 50 to 100 mL EDTA-stabilized sample into flask, keeping pipet tip below liquid surface. Add 1 mL starch indicator solution. Titrate immediately with standard KI-KIO$_3$ titrant, while swirling flask, until a faint permanent blue color develops. Analyze a reagent blank using distilled water instead of sample.

4. Calculation

$$\text{mg SO}_3^{2-}/\text{L} = \frac{(A - B) \times M \times 40\,000}{\text{mL sample}}$$

where:

A = mL titrant for sample,
B = mL titrant for blank, and
M = molarity of KI-KIO$_3$ titrant.

5. Precision and Bias

Three laboratories analyzed five replicate portions of a standard sulfite solution and of secondary treated wastewater effluent to which sulfite was added. The data are summarized below. Individual analyst's precision ranged from 0.7 to 3.6% standard deviation ($N = 45$).

Sample	$\overline{X}$ mg/L	Standard Deviation, σ mg/L	Relative Standard Deviation %
Standard, 6.3 mg SO$_3^{2-}$/L	4.5	0.25	5.5
Secondary effluent with 2.0 mg SO$_3^{2-}$/L	2.1	0.28	13.4
Secondary effluent with 4.0 mg SO$_3^{2-}$/L	3.6	0.17	4.8

6. Reference

1. KURTENACKER, A. 1924. The aldehyde-bisulfite reaction in mass analysis. *Z. Anal. Chem.* 64:56.

4500-SO$_3^{2-}$ C. Phenanthroline Method (PROPOSED)

1. General Discussion

a. Principle: An acidified sample is purged with nitrogen gas and the liberated SO$_2$ is trapped in an absorbing solution containing ferric ion and 1,10-phenanthroline. Ferric iron is reduced to the ferrous state by SO$_2$, producing the orange tris(1,10-phenanthroline) iron(II) complex. After excess ferric iron is removed with ammonium bifluoride, the phenanthroline complex is measured colorimetrically at 510 nm.[1]

b. Interferences: See Section 4500-SO$_3^{2-}$.B.1*b*.

c. Minimum detectable concentration: 0.01 mg SO$_3^{2-}$/L.

2. Apparatus

a. Apparatus for evolution of SO$_2$: Figure 4500-SO$_3^{2-}$:1 shows the following components:

1) *Gas flow meter,* with a capacity to measure 2 L/min of pure nitrogen gas.

2) *Gas washing bottle,* 250-mL, with coarse-porosity, 12-mm-diam fritted cylinder gas dispersion tube.

3) *Tubing connectors,* quick-disconnect, polypropylene.

4) *Tubing,* flexible PVC, for use in all connections.

5) *Nessler tube,* 100-mL.

b. Colorimetric equipment: One of the following is required:

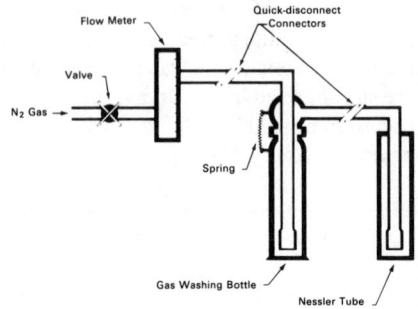

Figure 4500-SO₃²⁻:1. Apparatus for evolution of SO₂ from samples for colorimetric analysis.

1) *Spectrophotometer,* for use at 510 nm, providing a light path of 1 cm or longer.

2) *Filter photometer,* providing a light path of 1 cm or longer and equipped with a green filter having maximum transmittance near 510 nm.

3. Reagents

a. *1,10-phenanthroline solution,* 0.03*M:* Dissolve 5.95 g 1,10-phenanthroline in 100 mL 95% ethanol. Dilute to 1 L with distilled water. Discard if solution becomes colored.

b. *Ferric ammonium sulfate solution,* 0.01*M:* Dissolve 4.82 g $NH_4Fe(SO_4)_2 \cdot 12H_2O$ in 1 L distilled water to which has been added 1 mL conc H_2SO_4 to suppress ferric hydrolysis. Filter through a glass fiber filter if insoluble matter is visible. If necessary, adjust volume of acid so that a mixture of 10 parts of phenanthroline solution and one part of ferric ammonium sulfate solution will have a pH between 5 and 6.

c. *Ammonium bifluoride,* 5%: Dissolve 25 g NH_4HF_2 in 500 mL distilled water. Store in a polyethylene bottle and dispense with a plastic pipet.

d. *Potassium tetrachloromercurate, (TCM),* K_2HgCl_4, 0.04*M:* Dissolve 10.86 g $HgCl_2$, 5.96 g KCl, and 0.066 g disodium EDTA in distilled water and dilute to 1 L. Adjust pH to 5.2. This reagent normally is

stable for 6 months, but discard if a precipitate forms.[2]

e. *Dilute TCM-stabilized sulfite standard:* Dissolve 0.5 g Na_2SO_3 in 500 mL distilled water. Standardize on the day of preparation, but wait at least 30 min to allow the rate of oxidation to slow. Determine molarity by titrating with standard 0.0125*M* potassium iodide-iodate titrant using starch indicator (see Section 4500-SO_3^{2-}.B). Calculate molarity of working standard as follows:

$$\text{Molarity of } SO_3^{2-} \text{ standard} = \frac{(A - B) \times M}{\text{mL sample}}$$

where:

A = titrant for sample, mL,
B = titrant for blank, mL,
M = molarity of potassium iodide-iodate titrant.

Because stock Na_2SO_3 solution is unstable, immediately after standardization, pipet 10 mL into a 500-mL volumetric flask partially filled with TCM and dilute to mark with TCM. Calculate the concentration of this dilute sulfite solution by multiplying the stock solution concentration by 0.02. This TCM-stabilized standard is stable for 30 d if stored at 5°C. Discard as soon as any precipitate is noticed at the bottom.

f. *Standard potassium iodide-iodate titrant,* 0.0125*M:* See Section 4500-SO_3^{2-}.B.2*b*.

g. *Hydrochloric acid,* 1 + 1.

h. *Octyl alcohol,* reagent-grade.

i. *Sulfamic acid,* 10%: Dissolve 10 g NH_2SO_3H in 100 mL distilled water. This reagent can be kept for a few days if protected from air.

j. *EDTA reagent:* See Section 4500-SO_3^{2-}.B.2*d*.

4. Procedure

a. *Sample collection:* Collect a fresh sample taking care to minimize contact with air. Fix cooled samples (< 50°C) immediately by adding 1 mL EDTA solution for each 100 mL sample.

b. SO₂ evolution: Prepare the absorbing solution by adding 5 mL 1,10-phenanthroline solution, 0.5 mL ferric ammonium sulfate solution, 25 mL distilled water, and 5 drops octyl alcohol (to act as defoamer) to a 100-mL nessler tube; insert a gas dispersion tube. Add 1 mL sulfamic acid solution to the gas washing bottle and 100 mL of sample or a portion containing less than 100 μg SO_3^{2-} diluted to 100 mL. Add 10 mL 1 + 1 HCl and immediately connect the gas washing bottle to the gas train as shown in Figure 4500-SO_3^{2-}:1. Place a spring or rubber band on the gas washing bottle to keep the top securely closed during gas flow. Adjust nitrogen flow to 2.0 L/min and purge for 60 min.

c. Colorimetric measurement: After exactly 60 min, turn off nitrogen flow, disconnect nessler tube, and immediately add 1 mL ammonium bifluoride solution. Remove gas dispersion tube after rinsing it with distilled water into the tube and forcing the rinse water into the nessler tube with a rubber bulb. Dilute to 50 mL in the nessler tube and mix by rapidly moving the tube in a circular motion. Do not let rubber stoppers or PVC tubing come in contact with the absorbing solution. After at least 5 min from the time of adding ammonium bifluoride, read the absorbance versus distilled water at 510 nm using either a 5-cm cell for a range of 0 to 30 μg SO_3^{2-} per portion or a 1-cm cell for a range of 0 to 100 μg SO_3^{2-}. Avoid transferring octyl alcohol into the cell by letting it rise to the surface of the absorbing solution and transferring the clear lower solution to the cell with a pipet. Make a calibration curve by analyzing a procedure blank and at least three standards. Run at least one standard with each set of samples. For maximum accuracy hold samples and standards at the

same temperature and keep the time interval from start of purging to the addition of ammonium bifluoride constant. This is easier to do if several gas trains are used simultaneously in parallel. If ambient temperatures are subject to frequent fluctuation, a water bath may be used to control color development at a fixed temperature.

5. Calculation

mg SO_3^{2-}/L

$$= \frac{\mu g \ SO_3^{2-} \ \text{from calibration curve}}{\text{mL sample}}$$

6. Precision and Bias

Three laboratories analyzed five replicate portions of a standard sulfite solution and of secondary treated wastewater effluent to which sulfite was added. The data are summarized below. Individual analyst's precision ranged from 4.1 to 10.5% standard deviation ($N = 45$).

Sample	X̄ mg/L	Standard Deviation, σ mg/L	Relative Standard Deviation %
Standard, 4.7 mg SO_3^{2-}/L	3.7	0.78	21
Secondary effluent with 0.12 mg SO_3^{2-}/L	0.12	0.03	25
Secondary effluent with 4.0 mg SO_3^{2-}/L	3.7	0.30	8.0

7. References

1. STEPHENS, B.G. & F. LINDSTROM. 1964. Spectrophotometric determination of sulfur dioxide suitable for atmospheric analysis. *Anal. Chem.* 36:1308.
2. WEST, P. W. & G.C. GAEKE. 1956. Fixation of sulfur dioxide as sulfitomercurate and subsequent colorimetric determination. *Anal. Chem.* 28:1816.

4500-SO₄²⁻   SULFATE*

4500-SO₄²⁻ A.   Introduction

1. Occurrence

Sulfate (SO_4^{2-}) is widely distributed in nature and may be present in natural waters in concentrations ranging from a few to several thousand milligrams per liter. Mine drainage wastes may contribute large amounts of SO_4^{2-} through pyrite oxidation. Sodium and magnesium sulfate exert a cathartic action.

2. Selection of Method

The ion chromatographic method (B) is suitable for sulfate concentrations above

0.1 mg/L. The gravimetric methods (C and D) are suitable for SO_4^{2-} concentrations above 10 mg/L; use one of these methods for accurate results. The turbidimetric method (E) is applicable in the range of 1 to 40 mg SO_4^{2-}/L. The automated methylthymol blue method (F) is the procedure for analyzing large numbers of samples for sulfate alone when the equipment is available; about 30 samples can be analyzed per hour.

3. Sampling and Storage

In the presence of organic matter certain bacteria may reduce SO_4^{2-} to S^{2-}. To avoid this, store heavily polluted or contaminated samples at 4°C.

---

*Approved by Standard Methods Committee, 1985.

4500-SO₄²⁻ B.   Ion Chromatographic Method

See Section 4110.

4500-SO₄²⁻ C.   Gravimetric Method with Ignition of Residue

1. General Discussion

a. Principle: Sulfate is precipitated in a hydrochloric acid (HCl) solution as barium sulfate ($BaSO_4$) by the addition of barium chloride ($BaCl_2$).

The precipitation is carried out near the boiling temperature, and after a period of digestion the precipitate is filtered, washed with water until free of Cl^-, ignited or dried, and weighed as $BaSO_4$.

b. Interference: The gravimetric determination of SO_4^{2-} is subject to many errors, both positive and negative. In potable waters where the mineral concentration is low, these may be of minor importance.

1) Interferences leading to high results—Suspended matter, silica, $BaCl_2$ pre-

cipitant, NO_3^-, SO_3^{2-} and occluded mother liquor in the precipitate are the principal factors in positive errors. Suspended matter may be present in both the sample and the precipitating solution; soluble silicate may be rendered insoluble and SO_3^{2-} may be oxidized to SO_4^{2-} during analysis. Barium nitrate [$Ba(NO_3)_2$], $BaCl_2$, and water are occluded to some extent with the $BaSO_4$ although water is driven off if the temperature of ignition is sufficiently high.

2) Interferences leading to low results—Alkali metal sulfates frequently yield low results. This is true especially of alkali hydrogen sulfates. Occlusion of alkali sulfate with $BaSO_4$ causes substitution of an element of lower atomic weight than

barium in the precipitate. Hydrogen sulfates of alkali metals act similarly and, in addition, decompose on being heated. Heavy metals, such as chromium and iron, cause low results by interfering with the complete precipitation of SO$_4^{2-}$ and by formation of heavy metal sulfates. BaSO$_4$ has small but significant solubility, which is increased in the presence of acid. Although an acid medium is necessary to prevent precipitation of barium carbonate and phosphate, it is important to limit its concentration to minimize the solution effect.

2. Apparatus

a. *Steam bath.*

b. *Drying oven,* equipped with thermostatic control.

c. *Muffle furnace,* with temperature indicator.

d. *Desiccator.*

e. *Analytical balance,* capable of weighing to 0.1 mg.

f. *Filter:* Use one of the following:

1) *Filter paper,* acid-washed, ashless hard-finish, sufficiently retentive for fine precipitates.

2) *Membrane filter,* with a pore size of about 0.45 μm.

g. *Filtering apparatus,* appropriate to the type of filter selected. (Coat membrane filter holder with silicone fluid to prevent precipitate from adhering.)

3. Reagents

a. *Methyl red indicator solution:* Dissolve 100 mg methyl red sodium salt in distilled water and dilute to 100 mL.

b. *Hydrochloric acid,* HCl, 1 + 1.

c. *Barium chloride solution:* Dissolve 100 g BaCl$_2$·2H$_2$O in 1 L distilled water. Filter through a membrane filter or hard-finish filter paper before use; 1 mL is capable of precipitating approximately 40 mg SO$_4^{2-}$.

d. *Silver nitrate-nitric acid reagent:* Dissolve 8.5 g AgNO$_3$ and 0.5 mL conc HNO$_3$ in 500 mL distilled water.

e. *Silicone fluid.* *

4. Procedure

a. *Removal of silica:* If the silica concentration exceeds 25 mg/L, evaporate sample nearly to dryness in a platinum dish on a steam bath. Add 1 mL HCl, tilt, and rotate dish until the acid comes in complete contact with the residue. Continue evaporation to dryness. Complete drying in an oven at 180°C and if organic matter is present, char over flame of a burner. Moisten residue with 2 mL distilled water and 1 mL HCl, and evaporate to dryness on a steam bath. Add 2 mL HCl, take up soluble residue in hot water, and filter. Wash insoluble silica with several small portions of hot distilled water. Combine filtrate and washings. Discard residue.

b. *Precipitation of barium sulfate:* Adjust volume of clarified sample to contain approximately 50 mg SO$_4^{2-}$ in a 250-mL volume. Lower concentrations of SO$_4^{2-}$ may be tolerated if it is impracticable to concentrate sample to the optimum level, but in such cases limit total volume to 150 mL. Adjust pH with HCl to pH 4.5 to 5.0, using a pH meter or the orange color of methyl red indicator. Add 1 to 2 mL HCl. Heat to boiling and, while stirring gently, slowly add warm BaCl$_2$ solution until precipitation appears to be complete; then add about 2 mL in excess. If amount of precipitate is small, add a total of 5 mL BaCl$_2$ solution. Digest precipitate at 80 to 90°C, preferably overnight but for not less than 2 h.

c. *Filtration and weighing:* Mix a small amount of ashless filter paper pulp with the BaSO$_4$, quantitatively transfer to a filter, and filter at room temperature. The pulp aids filtration and reduces the tendency of the precipitate to creep. Wash precipitate with small portions of warm distilled water until washings are free of Cl$^-$ as indicated by testing with AgNO$_3$-HNO$_3$ reagent.

---

* "Desicote" (Beckman), or equivalent.

Place filter and precipitate in a weighed platinum crucible and ignite at 800°C for 1 h. *Do not let filter paper flame.* Cool in desiccator and weigh.

5. Calculation

$$\text{mg } SO_4^{2-}/L = \frac{\text{mg } BaSO_4 \times 411.6}{\text{mL sample}}$$

6. Precision and Bias

A synthetic sample containing 259 mg SO_4^{2-}/L, 108 mg Ca^{2+}/L, 82 mg Mg^{2+}/L, 3.1 mg K^+/L, 19.9 mg Na^+/L, 241 mg Cl^-/L, 0.250 mg NO_2^--N/L, 1.1 mg NO_3^--N/L, and 42.5 mg total alkalinity/L (contributed by $NaHCO_3$) was analyzed in 32 laboratories by the gravimetric method, with a relative standard deviation of 4.7% and a relative error of 1.9%.

7. Bibliography

HILLEBRAND, W.F. et al. 1953. Applied Inorganic Analysis, 2nd ed. John Wiley & Sons, New York, N.Y.

KOLTHOFF, I.M., E.J. MEEHAN, E.B. SANDELL & S. BRUCKENSTEIN. 1969. Quantitative Chemical Analysis, 4th ed. Macmillan Co., New York, N.Y.

4500-SO₄²⁻ D. Gravimetric Method with Drying of Residue

1. General Discussion

See Method C, preceding.

2. Apparatus

With the exception of the filter paper, all of the apparatus cited in Section 4500-SO_4^{2-}.C.2 is required, plus the following:

a. *Filters:* Use one of the following:

1) *Fritted-glass filter,* fine ("F") porosity, with a maximum pore size of 5 μm.

2) *Membrane filter,* with a pore size of about 0.45 μm.

b. *Vacuum oven.*

3. Reagents

All the reagents listed in Section 4500-SO_4^{2-}.C.3 are required.

4. Procedure

a. *Removal of interference:* See Section 4500-SO_4^{2-}.C.4a.

b. *Precipitation of barium sulfate:* See Section 4500-SO_4^{2-}.C.4b.

c. *Preparation of filters:*

1) Fritted glass filter—Dry to constant weight in an oven maintained at 105°C or higher, cool in desiccator, and weigh.

2) Membrane filter—Place filter on a piece of filter paper or a watch glass and dry to constant weight* in a vacuum oven at 80°C, while maintaining a vacuum of at least 85 kPa or in a conventional oven at a temperature of 103 to 105°C. Cool in desiccator and weigh membrane only.

d. *Filtration and weighing:* Filter $BaSO_4$ at room temperature. Wash precipitate with several small portions of warm distilled water until washings are free of Cl^-, as indicated by testing with $AgNO_3$-HNO_3 reagent. If a membrane filter is used add a few drops of silicone fluid to the suspension before filtering, to prevent adherence of precipitate to holder. Dry filter and precipitate by the same procedure used in preparing filter. Cool in a desiccator and weigh.

5. Calculation

$$\text{mg } SO_4^{2-}/L = \frac{\text{mg } BaSO_4 \times 411.6}{\text{mL sample}}$$

6. Bibliography

See Section 4500-SO_4^{2-}.C.7.

---

*Constant weight is defined as a change of not more than 0.5 mg in two successive operations consisting of heating, cooling in desiccator, and weighing.

4500-SO$_4^{2-}$ E. Turbidimetric Method

1. General Discussion

a. Principle: Sulfate ion (SO$_4^{2-}$) is precipitated in an acetic acid medium with barium chloride (BaCl$_2$) so as to form barium sulfate (BaSO$_4$) crystals of uniform size. Light absorbance of the BaSO$_4$ suspension is measured by a photometer and the SO$_4^{2-}$ concentration is determined by comparison of the reading with a standard curve.

b. Interference: Color or suspended matter in large amounts will interfere. Some suspended matter may be removed by filtration. If both are small in comparison with the SO$_4^{2-}$ concentration, correct for interference as indicated in ¶ 4d below. Silica in excess of 500 mg/L will interfere, and in waters containing large quantities of organic material it may not be possible to precipitate BaSO$_4$ satisfactorily.

In potable waters there are no ions other than SO$_4^{2-}$ that will form insoluble compounds with barium under strongly acid conditions. Make determination at room temperature; variation over a range of 10°C will not cause appreciable error.

c. Minimum detectable concentration: Approximately 1 mg SO$_4^{2-}$/L.

2. Apparatus

a. Magnetic stirrer: Use a constant stirring speed. It is convenient to incorporate a fixed resistance in series with the motor operating the magnetic stirrer to regulate stirring speed. Use magnets of identical shape and size. The exact speed of stirring is not critical, but keep it constant for each run of samples and standards and adjust it to prevent splashing.

b. Photometer: One of the following is required, with preference in the order given:

1) Nephelometer.

2) Spectrophotometer, for use at 420 nm, providing a light path of 2.5 to 10 cm.

3) Filter photometer, equipped with a violet filter having maximum transmittance near 420 nm and providing a light path of 2.5 to 10 cm.

c. Stopwatch or electric timer.

d. Measuring spoon, capacity 0.2 to 0.3 mL.

3. Reagents

a. Buffer solution A: Dissolve 30 g magnesium chloride, MgCl$_2 \cdot$6H$_2$O, 5 g sodium acetate, CH$_3$COONa$\cdot$3H$_2$O, 1.0 g potassium nitrate, KNO$_3$, and 20 mL acetic acid, CH$_3$COOH (99%), in 500 mL distilled water and make up to 1000 mL.

b. Buffer solution B (required when the sample SO$_4^{2-}$ concentration is less than 10 mg/L): Dissolve 30 g MgCl$_2 \cdot$6H$_2$O, 5 g CH$_3$COONa$\cdot$3H$_2$O, 1.0 g KNO$_3$, 0.111 g sodium sulfate, Na$_2$SO$_4$, and 20 mL acetic acid (99%) in 500 mL distilled water and make up to 1000 mL.

c. Barium chloride, BaCl$_2$, crystals, 20 to 30 mesh. In standardization, uniform turbidity is produced with this mesh range and the appropriate buffer.

d. Standard sulfate solution: Prepare a standard sulfate solution as described in 1) or 2) below; 1.00 mL = 100 μg SO$_4^{2-}$.

1) Dilute 10.4 mL standard 0.0200N H$_2$SO$_4$ titrant specified in Alkalinity, Section 2320B.3c, to 100 mL with distilled water.

2) Dissolve 0.1479 g anhydrous Na$_2$SO$_4$ in distilled water and dilute to 1000 mL.

4. Procedure

a. Formation of barium sulfate turbidity: Measure 100 mL sample, or a suitable portion made up to 100 mL, into a 250-mL erlenmeyer flask. Add 20 mL buffer solution and mix in stirring apparatus. While stirring, add a spoonful of $BaCl_2$ crystals and begin timing immediately. Stir for 60 ± 2 s at constant speed.

b. Measurement of barium sulfate turbidity: After stirring period has ended, pour solution into absorption cell of photometer and measure turbidity at 5 ± 0.5 min.

c. Preparation of calibration curve: Estimate SO_4^{2-} concentration in sample by comparing turbidity reading with a calibration curve prepared by carrying SO_4^{2-} standards through the entire procedure. Space standards at 5-mg/L increments in the 0- to 40-mg/L SO_4^{2-} range. Above 40 mg/L accuracy decreases and $BaSO_4$ suspensions lose stability. Check reliability of calibration curve by running a standard with every three or four samples.

d. Correction for sample color and turbidity: Correct for sample color and turbidity by running blanks to which $BaCl_2$ is not added.

5. Calculation

$$\text{mg } SO_4^{2-}/L = \frac{\text{mg } SO_4^{2-} \times 1000}{\text{mL sample}}$$

If buffer solution A was used, determine SO_4^{2-} concentration directly from the calibration curve after subtracting sample absorbance before adding $BaCl_2$. If buffer solution B was used subtract SO_4^{2-} concentration of blank from apparent SO_4^{2-} concentration as determined above; because the calibration curve is not a straight line, this is not equivalent to subtracting blank absorbance from sample absorbance.

6. Precision and Bias

With a turbidimeter,* in a single laboratory with a sample having a mean of 7.45 mg SO_4^{2-}/L, a standard deviation of 0.13 mg/L and a coefficient of variation of 1.7% were obtained. Two samples dosed with sulfate gave recoveries of 85 and 91%.

7. Bibliography

SHEEN, R.T., H.L. KAHLER & E.M. ROSS. 1935. Turbidimetric determination of sulfate in water. *Ind. Eng. Chem.,* Anal. Ed. 7:262.

THOMAS, J.F. & J.E. COTTON. 1954. A turbidimetric sulfate determination. *Water Sewage Works* 101:462.

ROSSUM, J.R. & P. VILLARRUZ. 1961. Suggested methods for turbidimetric determination of sulfate in water. *J. Amer. Water Works Assoc.* 53:873.

_____
*Hach 2100 A.

4500-SO_4^{2-} F.   Automated Methylthymol Blue Method

1. General Discussion

a. Principle: Barium sulfate is formed by the reaction of the SO_4^{2-} with barium chloride ($BaCl_2$) at a low pH. At high pH excess barium reacts with methylthymol blue to produce a blue chelate. The uncomplexed methylthymol blue is gray. The amount of gray uncomplexed methylthymol blue indicates the concentration of SO_4^{2-}.

b. Interferences: Because many cations interfere, use an ion-exchange column to remove interferences.

c. Application: This method is applicable to potable, surface, and saline waters as well as domestic and industrial wastewaters over a range from about 10 to 300 mg SO_4^{2-}/L.

2. Apparatus

a. Automated analytical equipment: The required continuous-flow analytical instrument* consists of the interchangeable components shown in Figure 4500-SO₄²⁻:1.

b. Ion-exchange column: Fill a piece of 2-mm-ID glass tubing about 20 cm long with the ion-exchange resin.† To simplify filling column put resin in distilled water and aspirate it into the tubing, which contains a glass-wool plug. After filling, plug other end of tube with glass wool. Avoid trapped air in the column.

3. Reagents

a. Barium chloride solution: Dissolve 1.526 g $BaCl_2 \cdot 2H_2O$ in 500 mL distilled water and dilute to 1 L. Store in a polyethylene bottle.

b. Methylthymol blue reagent: Dissolve 118.2 mg methylthymol blue‡ in 25 mL $BaCl_2$ solution. Add 4 mL $1N$ HCl and 71 mL distilled water and dilute to 500 mL with 95% ethanol. Store in a brown glass bottle. Prepare fresh daily.

*Technicon™ Autoanalyzer™ or equivalent.
†Ion-exchange resin Bio-Rex 70, 20-50 mesh, sodium form, available from Bio-Rad Laboratories, Richmond, Calif. 94804, or equivalent.

‡Eastman Organic Chemicals, Rochester, N.Y. 14615. No. 8068 3′,3″ bis [*N,N*-bis(carboxymethyl)-aminolmethyl] thymolsulfonphthalein pentasodium salt.

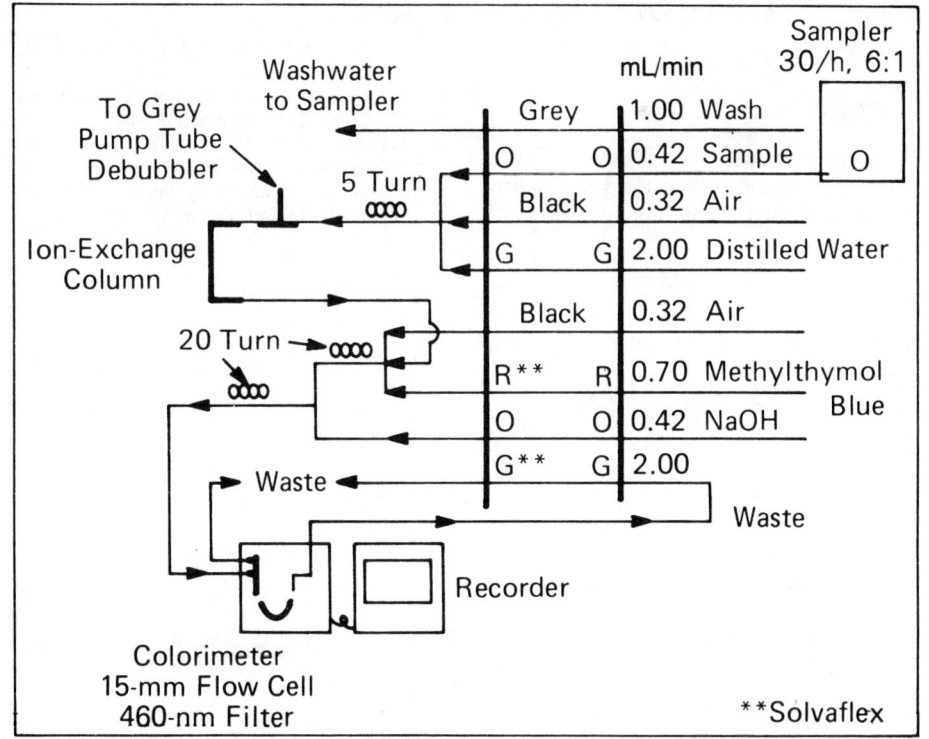

Figure 4500-SO₄²⁻:1. Sulfate manifold.

c. Buffer solution, pH 10.1: Dissolve 6.75 g NH_4Cl in 500 mL distilled water. Add 57 mL conc NH_4OH and dilute to 1 L with distilled water. Adjust pH to 10.1 and store in a polyethylene bottle. Prepare fresh monthly.

d. EDTA reagent: Dissolve 40 g tetrasodium ethylenediaminetetraacetate in 500 mL pH 10.1 buffer solution. Dilute to 1 L with pH 10.1 buffer solution and store in a polyethylene bottle.

e. Sodium hydroxide solution, 0.36*N:* Dissolve 7.2 g NaOH in 500 mL distilled water. Cool and make up to 1 L with distilled water.

f. Stock sulfate solution: Dissolve 1.479 g anhydrous Na_2SO_4 in 500 mL distilled water and dilute to 1000 mL; 1.00 mL = 1.00 mg SO_4^{2-}.

g. Standard sulfate solutions: Prepare in appropriate concentrations from 10 to 300 mg SO_4^{2-}/L, using the stock sulfate solution.

4. Procedure

Set up the manifold as shown in Figure 4500-SO_4^{2-}:1 and follow the general procedure described by the manufacturer.

After use, rinse methylthymol blue and NaOH reagent lines in water for a few minutes, rinse them in the EDTA solution for 10 min, and then rinse in distilled water.

5. Calculation

Prepare standard curves by plotting peak heights of standards processed through the manifold against SO_4^{2-} concentrations in standards. Compute sample SO_4^{2-} concentration by comparing sample peak height with standard curve.

6. Precision and Bias

With a Technicon Autoanalyzer II in a single laboratory a sample with an average concentration of about 28 mg SO_4^{2-}/L had a standard deviation of 0.68 mg/L and a coefficient of variation of 2.4%. In two samples with added SO_4^{2-}, recoveries were 91% and 100%.

7. Bibliography

LAZRUS, A.L., K.C. HILL & J. P. LODGE. 1965. A new colorimetric microdetermination of sulfate ion. *In* Automation in Analytical Chemistry. Technicon Symposium.

COLOROS, E., M. R. PANESAR & F.P. PERRY. 1976. Linearizing the calibration curve in determination of sulfate by the methylthymol blue method. *Anal. Chem.* 48:1693.

PART 5000

DETERMINATION OF

ORGANIC CONSTITUENTS

5010 INTRODUCTION

5010 A. General Discussion

Analyses for organic matter in water and wastewater can be classified into two general types of measurements: those that quantify an aggregate amount of organic matter comprising organic constituents with a common characteristic and those that quantify individual organic compounds. The latter can be found in Part 6000. The former, described here in Part 5000, have been grouped into four categories: oxygen-demanding substances, organically bound elements, classes of compounds, and formation potentials.

Methods for total organic carbon and chemical oxygen demand are used to assess the total amount of organics present. Gross fractions of the organic matter can be identified analytically, as in the measurement of BOD, which is an index of the biodegradable organics present, oil and grease, which represents material extractable from a sample by a nonpolar solvent, or total organic halide (TOX), which measures organically bound halogens. Trihalomethane formation potential is an aggregate measure of the total concentration of trihalomethanes formed upon chlorination of a water sample.

Analyses of organics are made to assess the concentration and general composition of organic matter in raw water supplies, wastewaters, treated effluents, and receiving waters; and to determine the efficiency of treatment processes.

5010 B. Sample Collection and Preservation

The sampling, field treatment, preservation, and storage of samples taken for organic matter analysis are covered in detail in the individual introductions to the methods. If possible, analyze samples immediately because preservatives often interfere with the tests. Otherwise, store at a low temperature (4°C) immediately after collection to preserve most samples. Use chemical preservatives only when they are shown not to interfere with the examinations to be made (see Section 1060). Never use preservatives for samples to be analyzed for BOD. When preservatives are used, add them to the sample bottle initially so that all portions are preserved as soon as collected. No single method of preservation is entirely satisfactory; choose the preservative with due regard to the determinations that are to be made. All methods of preservation may be inadequate when applied to samples containing significant amounts of suspended matter.

5020   QUALITY CONTROL

Part 1000 contains important information relevant to analyses included in Part 5000. Give particular attention to Sections 1020B (Quality Control), 1060 (Collection and Preservation of Samples), 1080 (Reagent-Grade Water), and 1090 (Safety), all of which are critical for many of the Part 5000 methods.

Take special precautions when analyses are performed by independent laboratories. Reliable use of independent laboratories deserves the same quality assurance procedures observed for in-house analyses: replicate samples, samples with known additions, and blanks.

Preparation of samples with known additions may not be feasible for certain analyses. In such cases, consider using a mixture, in varying ratios, of several samples. Use the reported concentrations in the samples and the proportions in which they were mixed to calculate the expected concentration in the mixture. Examine laboratory performance using externally prepared standards and check samples (see Section 1020B).

Type I reagent water (Section 1080) should give satisfactory results for most of the analyses in Part 5000, but additional purification steps may be needed for certain methods, such as total organic halide (TOX) and trihalomethane formation potential (THMFP).

5210   BIOCHEMICAL OXYGEN DEMAND*

5210 A.   Introduction

1. General Discussion

The biochemical oxygen demand (BOD) determination is an empirical test in which standardized laboratory procedures are used to determine the relative oxygen requirements of wastewaters, effluents, and polluted waters. The test has its widest application in measuring waste loadings to treatment plants and in evaluating the BOD-removal efficiency of such treatment systems. The test measures the oxygen utilized during a specified incubation period for the biochemical degradation of organic material (carbonaceous demand) and the oxygen used to oxidize inorganic material such as sulfides and ferrous iron. It also may measure the oxygen used to oxidize reduced forms of nitrogen (nitrogenous demand) unless their oxidation is prevented by an inhibitor. The seeding and dilution procedures provide an estimate of the BOD at pH 6.5 to 7.5.

Although only the 5-d BOD (BOD_5) is described here, many variations of oxygen demand measurements exist. These include using shorter and longer incubation periods, tests to determine rates of oxygen uptake, and continuous oxygen-uptake measurements by respirometric techniques. Alternative seeding, dilution, and incubation conditions can be chosen to mimic receiving-water conditions, thereby providing an estimate of the environmental effects of wastewaters and effluents.

2. Carbonaceous Versus Nitrogenous BOD

Oxidation of reduced forms of nitrogen, mediated by microorganisms, exerts nitrog-

---

*Approved by Standard Methods Committee, 1988.

enous demand. Nitrogenous demand historically has been considered an interference in the determination of BOD, as clearly evidenced by the inclusion of ammonia in the dilution water. The interference from nitrogenous demand can now be prevented by an inhibitory chemical.[1] If an inhibiting chemical is not used, the oxygen demand measured is the sum of carbonaceous and nitrogenous demands.

Measurements that include nitrogenous demand generally are not useful for assessing the oxygen demand associated with organic material. Nitrogenous demand can be estimated directly from ammonia nitrogen (Section 4500-NH$_3$); and carbonaceous demand can be estimated by subtracting the theoretical equivalent of the reduced nitrogen oxidation from uninhibited test results. However, this method is cumbersome and is subject to considerable error. Chemical inhibition of nitrogenous demand provides a more direct and more reliable measure of carbonaceous demand.

The extent of oxidation of nitrogenous compounds during the 5-d incubation period depends on the presence of microorganisms capable of carrying out this oxidation. Such organisms usually are not present in raw sewage or primary effluent in sufficient numbers to oxidize significant quantities of reduced nitrogen forms in the 5-d BOD test. Many biological treatment plant effluents contain significant numbers of nitrifying organisms. Because oxidation of nitrogenous compounds can occur in such samples, inhibition of nitrification as directed in ¶ B.4e6) is recommended for samples of secondary effluent, for samples seeded with secondary effluent, and for samples of polluted waters.

Report results as CBOD$_5$ when inhibiting the nitrogenous oxygen demand. When nitrification is not inhibited, report results as BOD$_5$.

3. Dilution Requirements

The BOD concentration in most wastewaters exceeds the concentration of dissolved oxygen (DO) available in an air-saturated sample. Therefore, it is necessary to dilute the sample before incubation to bring the oxygen demand and supply into appropriate balance. Because bacterial growth requires nutrients such as nitrogen, phosphorus, and trace metals, these are added to the dilution water, which is buffered to ensure that the pH of the incubated sample remains in a range suitable for bacterial growth. Complete stabilization of a sample may require a period of incubation too long for practical purposes; therefore, 5 d has been accepted as the standard incubation period.

If the dilution water is of poor quality, effectively, dilution water will appear as sample BOD. This effect will be amplified by the dilution factor. A positive bias will result. The method included below contains both a dilution-water check and a dilution-water blank. Seeded dilution waters are checked further for acceptable quality by measuring their consumption of oxygen from a known organic mixture, usually glucose and glutamic acid.

The source of dilution water is not restricted and may be distilled, tap, or receiving-stream water free of biodegradable organics and bioinhibitory substances such as chlorine or heavy metals. Distilled water may contain ammonia or volatile organics; deionized waters often are contaminated with soluble organics leached from the resin bed. Use of copper-lined stills or copper fittings attached to distilled water lines may produce water containing excessive amounts of copper (see Section 3500-Cu).

4. Reference

1. YOUNG, J.C. 1973. Chemical methods for nitrification control. *J. Water Pollut. Control Fed.* 45:637.

5. Bibliography

THERIAULT, E.J., P.D. MCNAMEE & C.T. BUT-
    TERFIELD. 1931. Selection of dilution water
    for use in oxygen demand tests. *Pub. Health
    Rep.* 46:1084.
LEA, W.L. & M.S. NICHOLS. 1937. Influence of
    phosphorus and nitrogen on biochemical ox-
    ygen demand. *Sewage Works J.* 9:34.

RUCHHOFT, C.C. 1941. Report on the cooperative
    study of dilution waters made for the Stand-
    ard Methods Committee of the Federation of
    Sewage Works Associations. *Sewage Works
    J.* 13:669.
MOHLMAN, F.W., E. HURWITZ, G.R. BARNETT
    & H.K. RAMER. 1950. Experience with mod-
    ified methods for BOD. *Sewage Ind. Wastes*
    22:31.

5210 B. 5-Day BOD Test

1. General Discussion

a. Principle: The method consists of filling with sample, to overflowing, an airtight
bottle of the specified size and incubating
it at the specified temperature for 5 d. Dissolved oxygen is measured initially and
after incubation, and the BOD is computed
from the difference between initial and final
DO. Because the initial DO is determined
immediately after the dilution is made, all
oxygen uptake, including that occurring
during the first 15 min, is included in the
BOD measurement.

b. Sampling and storage: Samples for
BOD analysis may degrade significantly
during storage between collection and analysis, resulting in low BOD values. Minimize reduction of BOD by analyzing
sample promptly or by cooling it to near-
freezing temperature during storage. However, even at low temperature, keep holding
time to a minimum. Warm chilled samples
to 20°C before analysis.

1) Grab samples—If analysis is begun
within 2 h of collection, cold storage is
unnecessary. If analysis is not started
within 2 h of sample collection, keep sample at or below 4°C from the time of collection. Begin analysis within 6 h of
collection; when this is not possible because
the sampling site is distant from the laboratory, store at or below 4°C and report
length and temperature of storage with the
results. In no case start analysis more than
24 h after grab sample collection. When
samples are to be used for regulatory purposes make every effort to deliver samples
for analysis within 6 h of collection.

2) Composite samples—Keep samples
at or below 4°C during compositing. Limit
compositing period to 24 h. Use the same
criteria as for storage of grab samples, starting the measurement of holding time from
end of compositing period. State storage
time and conditions as part of the results.

2. Apparatus

a. Incubation bottles, 250- to 300-mL capacity. Clean bottles with a detergent, rinse
thoroughly, and drain before use. As a precaution against drawing air into the dilution bottle during incubation, use a water-
seal. Obtain satisfactory water seals by
inverting bottles in a water bath or by adding water to the flared mouth of special
BOD bottles. Place a paper or plastic cup
or foil cap over flared mouth of bottle to
reduce evaporation of the water seal during
incubation.

b. Air incubator or water bath, thermostatically controlled at 20 ± 1°C. Exclude
all light to prevent possibility of photosynthetic production of DO.

3. Reagents

a. Phosphate buffer solution: Dissolve
8.5 g KH_2PO_4, 21.75 g K_2HPO_4, 33.4 g
$Na_2HPO_4 \cdot 7H_2O$, and 1.7 g NH_4Cl in about

about 500 mL distilled water and dilute to 1 L. The pH should be 7.2 without further adjustment. Discard reagent (or any of the following reagents) if there is any sign of biological growth in the stock bottle.

b. Magnesium sulfate solution: Dissolve 22.5 g $MgSO_4 \cdot 7H_2O$ in distilled water and dilute to 1 L.

c. Calcium chloride solution: Dissolve 27.5 g $CaCl_2$ in distilled water and dilute to 1 L.

d. Ferric chloride solution: Dissolve 0.25 g $FeCl_3 \cdot 6H_2O$ in distilled water and dilute to 1 L.

e. Acid and alkali solutions, $1N$, for neutralization of caustic or acidic waste samples.

1) Acid—Slowly and while stirring, add 28 mL conc sulfuric acid to distilled water. Dilute to 1 L.

2) Alkali—Dissolve 40 g sodium hydroxide in distilled water. Dilute to 1 L.

f. Sodium sulfite solution: Dissolve 1.575 g Na_2SO_3 in 1000 mL distilled water. This solution is not stable; prepare daily.

g. Nitrification inhibitor, 2-chloro-6-(trichloro methyl) pyridine.*

h. Glucose-glutamic acid solution: Dry reagent-grade glucose and reagent-grade glutamic acid at 103°C for 1 h. Add 150 mg glucose and 150 mg glutamic acid to distilled water and dilute to 1 L. Prepare fresh immediately before use.

i. Ammonium chloride solution: Dissolve 1.15 g NH_4Cl in about 500 mL distilled water, adjust pH to 7.2 with NaOH solution, and dilute to 1 L. Solution contains 0.3 mg N/mL.

4. Procedure

a. Preparation of dilution water: Place desired volume of water in a suitable bottle and add 1 mL each of phosphate buffer, $MgSO_4$, $CaCl_2$, and $FeCl_3$ solutions/L of water. Seed dilution water, if desired, as

_____
*Nitrification Inhibitor 2579-24 (2.2% TCMP), Hach Co., or equivalent.

described in ¶ *4d*. Test and store dilution water as described in ¶s *4b* and *c* so that water of assured quality always is on hand.

Before use bring dilution water temperature to 20°C. Saturate with DO by shaking in a partially filled bottle or by aerating with organic-free filtered air. Alternatively, store in cotton-plugged bottles long enough for water to become saturated with DO. Protect water quality by using clean glassware, tubing, and bottles.

b. Dilution water check: Use this procedure as a rough check on quality of dilution water.

If the oxygen depletion of a candidate water exceeds 0.2 mg/L obtain a satisfactory water by improving purification or from another source. Alternatively, if nitrification inhibition is used, store the dilution water, seeded as prescribed below, in a darkened room at room temperature until the oxygen uptake is sufficiently reduced to meet the dilution-water check criteria. Check quality of stored dilution water on use, but do not add seed to dilution water stored for quality improvement. Storage is not recommended when BODs are to be determined without nitrification inhibition because nitrifying organisms may develop during storage. Check stored dilution water to determine whether sufficient ammonia remains after storage. If not, add ammonium chloride solution to provide a total of 0.45 mg ammonia/L as nitrogen. If dilution water has not been stored for quality improvement, add sufficient seeding material to produce a DO uptake of 0.05 to 0.1 mg/L in 5 d at 20°C. Incubate a BOD bottle full of dilution water for 5 d at 20°C. Determine initial and final DO as in ¶s *4g* and *j*. The DO uptake in 5 d at 20°C should not be more than 0.2 mg/L and preferably not more than 0.1 mg/L.

c. Glucose-glutamic acid check: Because the BOD test is a bioassay its results can be influenced greatly by the presence of toxicants or by use of a poor seeding material. Distilled waters frequently are contaminated with copper; some sewage seeds are relatively inactive. Low results always are obtained with such seeds and waters. Periodically check dilution water quality, seed effectiveness, and analytical technique by making BOD measurements on pure organic compounds and samples with known additions. In general, for BOD determinations not requiring an adapted seed, use a mixture of 150 mg glucose/L and 150 mg glutamic acid/L as a "standard" check solution. Glucose has an exceptionally high and variable oxidation rate but when it is used with glutamic acid, the oxidation rate is stabilized and is similar to that obtained with many municipal wastes. Alternatively, if a particular wastewater contains an identifiable major constituent that contributes to the BOD, use this compound in place of the glucose-glutamic acid.

Determine the 5-d 20°C BOD of a 2% dilution of the glucose-glutamic acid standard check solution using the techniques outlined in ¶s 4d-j. Evaluate data as described in ¶ 6, Precision and Bias.

d. Seeding:

1) Seed source—It is necessary to have present a population of microorganisms capable of oxidizing the biodegradable organic matter in the sample. Domestic wastewater, unchlorinated or otherwise-undisinfected effluents from biological waste treatment plants, and surface waters receiving wastewater discharges contain satisfactory microbial populations. Some samples do not contain a sufficient microbial population (for example, some untreated industrial wastes, disinfected wastes, high-temperature wastes, or wastes with extreme pH values). For such wastes seed the dilution water by adding a population of microorganisms. The preferred seed is effluent from a biological treatment system processing the waste. Where this is not available, use supernatant from domestic wastewater after settling at room temperature for at least 1 h but no longer

than 36 h. When effluent from a biological treatment process is used, inhibition of nitrification is recommended.

Some samples may contain materials not degraded at normal rates by the microorganisms in settled domestic wastewater. Seed such samples with an adapted microbial population obtained from the undisinfected effluent of a biological process treating the waste. In the absence of such a facility, obtain seed from the receiving water below (preferably 3 to 8 km) the point of discharge. When such seed sources also are not available, develop an adapted seed in the laboratory by continuously aerating a sample of settled domestic wastewater and adding small daily increments of waste. Optionally use a soil suspension or activated sludge, or a commercial seed preparation to obtain the initial microbial population. Determine the existence of a satisfactory population by testing the performance of the seed in BOD tests on the sample. BOD values that increase with time of adaptation to a steady high value indicate successful seed adaptation.

2) Seed control—Determine BOD of the seeding material as for any other sample. This is the *seed control*. From the value of the seed control and a knowledge of the seeding material dilution (in the dilution water) determine seed DO uptake. Ideally, make dilutions of seed such that the largest quantity results in at least 50% DO depletion. A plot of DO depletion, in milligrams per liter, versus milliters seed should present a straight line for which the slope indicates DO depletion per milliliter of seed. The DO-axis intercept is oxygen depletion caused by the dilution water and should be less than 0.1 mg/L (¶ 4h). To determine a sample DO uptake subtract seed DO uptake from total DO uptake. The DO uptake of seeded dilution water should be between 0.6 and 1.0 mg/L.

Techniques for adding seeding material to dilution water are described for two sample dilution methods (¶ 4f).

e. Sample pretreatment:

1) Samples containing caustic alkalinity or acidity— Neutralize samples to pH 6.5 to 7.5 with a solution of sulfuric acid (H_2SO_4) or sodium hydroxide (NaOH) of such strength that the quantity of reagent does not dilute the sample by more than 0.5%. The pH of seeded dilution water should not be affected by the lowest sample dilution.

2) Samples containing residual chlorine compounds—If possible, avoid samples containing residual chlorine by sampling ahead of chlorination processes. If the sample has been chlorinated but no detectable chlorine residual is present, seed the dilution water. If residual chlorine is present, dechlorinate sample and seed the dilution water (¶ 4*f*). Do not test chlorinated/dechlorinated samples without seeding the dilution water. In some samples chlorine will dissipate within 1 to 2 h of standing in the light. This often occurs during sample transport and handling. For samples in which chlorine residual does not dissipate in a reasonably short time, destroy chlorine residual by adding Na_2SO_3 solution. Determine required volume of Na_2SO_3 solution on a 100- to 1000-mL portion of neutralized sample by adding 10 mL of 1 + 1 acetic acid or 1 + 50 H_2SO_4, 10 mL potassium iodide (KI) solution (10 g/100 mL), and titrating with Na_2SO_3 solution to the starch-iodine end point for residual. Add to neutralized sample the relative volume of Na_2SO_3 solution determined by the above test, mix, and after 10 to 20 min check sample for residual chlorine. (NOTE: Excess Na_2SO_3 exerts an oxygen demand and reacts slowly with certain organic chloramine compounds that may be present in chlorinated samples).

3) Samples containing other toxic substances—Certain industrial wastes, for example, plating wastes, contain toxic metals. Such samples often require special study and treatment.

4) Samples supersaturated with DO— Samples containing more than 9 mg DO/L at 20°C may be encountered in cold waters or in water where photosynthesis occurs. To prevent loss of oxygen during incubation of such samples, reduce DO to saturation at 20°C by bringing sample to about 20°C in partially filled bottle while agitating by vigorous shaking or by aerating with clean, filtered compressed air.

5) Sample temperature adjustment— Bring samples to 20 ± 1°C before making dilutions.

6) Nitrification inhibition—If nitrification inhibition is desired add 3 mg 2-chloro-6-(trichloro methyl) pyridine (TCMP) to each 300-mL bottle before capping or add sufficient amounts to the dilution water to make a final concentration of 10 mg/L. (NOTE: Pure TCMP may dissolve slowly and can float on top of the sample. Some commercial formulations dissolve more readily but are not 100% TCMP; adjust dosage accordingly.) Samples that may require nitrification inhibition include, but are not limited to, biologically treated effluents, samples seeded with biologically treated effluents, and river waters. Note the use of nitrogen inhibition in reporting results.

f. Dilution technique: Dilutions that result in a residual DO of at least 1 mg/L and a DO uptake of at least 2 mg/L after 5 d incubation produce the most reliable results. Make several dilutions of prepared sample to obtain DO uptake in this range. Experience with a particular sample will permit use of a smaller number of dilutions. A more rapid analysis, such as COD, may be correlated approximately with BOD and serve as a guide in selecting dilutions. In the absence of prior knowledge, use the following dilutions: 0.0 to 1.0% for strong industrial wastes, 1 to 5% for raw and settled wastewater, 5 to 25% for biologically treated effluent, and 25 to 100% for polluted river waters.

Prepare dilutions either in graduated cylinders and then transfer to BOD bottles or prepare directly in BOD bottles. Either dilution method can be combined with any DO measurement technique. The number of bottles to be prepared for each dilution depends on the DO technique and the number of replicates desired.

When using graduated cylinders to prepare dilutions, and when seeding is necessary, add seed either directly to dilution water or to individual cylinders before dilution. Seeding of individual cylinders avoids a declining ratio of seed to sample as increasing dilutions are made. When dilutions are prepared directly in BOD bottles and when seeding is necessary, add seed directly to dilution water or directly to the BOD bottles.

1) Dilutions prepared in graduated cylinders—If the azide modification of the titrimetric iodometric method (Section 4500-O.C) is used, carefully siphon dilution water, seeded if necessary, into a 1- to 2-L-capacity graduated cylinder. Fill cylinder half full without entraining air. Add desired quantity of carefully mixed sample and dilute to appropriate level with dilution water. Mix well with a plunger-type mixing rod; avoid entraining air. Siphon mixed dilution into two BOD bottles. Determine initial DO on one of these bottles. Stopper the second bottle tightly, water-seal, and incubate for 5 d at 20°C. If the membrane electrode method is used for DO measurement, siphon dilution mixture into one BOD bottle. Determine initial DO on this bottle and replace any displaced contents with sample dilution to fill the bottle. Stopper tightly, water-seal, and incubate for 5 d at 20°C.

2) Dilutions prepared directly in BOD bottles—Using a wide-tip volumetric pipet, add the desired sample volume to individual BOD bottles of known capacity. Add appropriate amounts of seed material to the individual BOD bottles or to the dilution water. Fill bottles with enough dilution water, seeded if necessary, so that insertion of stopper will displace all air, leaving no bubbles. For dilutions greater than 1:100 make a primary dilution in a graduated cylinder before making final dilution in the bottle. When using titrimetric iodometric methods for DO measurement, prepare two bottles at each dilution. Determine initial DO on one bottle. Stopper second bottle tightly, water-seal, and incubate for 5 d at 20°C. If the membrane electrode method is used for DO measurement, prepare only one BOD bottle for each dilution. Determine initial DO on this bottle and replace any displaced contents with dilution water to fill the bottle. Stopper tightly, water-seal, and incubate for 5 d at 20°C. Rinse DO electrode between determinations to prevent cross-contamination of samples.

g. Determination of initial DO: If the sample contains materials that react rapidly with DO, determine initial DO immediately after filling BOD bottle with diluted sample. If rapid initial DO uptake is insignificant, the time period between preparing dilution and measuring initial DO is not critical.

Use the azide modification of the iodometric method (Section 4500-O.C) or the membrane electrode method (Section 4500-O.G) to determine initial DO on all sample dilutions, dilution water blanks, and where appropriate, seed controls.

h. Dilution water blank: Use a dilution water blank as a rough check on quality of unseeded dilution water and cleanliness of incubation bottles. Together with each batch of samples incubate a bottle of unseeded dilution water. Determine initial and final DO as in ¶s 4g and j. The DO uptake should not be more than 0.2 mg/L and preferably not more than 0.1 mg/L.

i. Incubation: Incubate at 20°C ± 1°C BOD bottles containing desired dilutions, seed controls, dilution water blanks, and glucose-glutamic acid checks. Water-seal bottles as described in ¶ 4f.

j. Determination of final DO: After 5 d

incubation determine DO in sample dilutions, blanks, and checks as in ¶ 4g.

5. Calculation

When dilution water is not seeded:

$$BOD_5, \text{ mg/L} = \frac{D_1 - D_2}{P}$$

When dilution water is seeded:

$$BOD_5, \text{ mg/L} = \frac{(D_1 - D_2) - (B_1 - B_2) f}{P}$$

where:
  D_1 = DO of diluted sample immediately after preparation, mg/L,
  D_2 = DO of diluted sample after 5 d incubation at 20°C, mg/L,
  P = decimal volumetric fraction of sample used,
  B_1 = DO of seed control before incubation, mg/L (¶ 4d),
  B_2 = DO of seed control after incubation mg/L (¶ 4d), and
  f = ratio of seed in diluted sample to seed in seed control = (% seed in diluted sample)/(% seed in seed control).

If seed material is added directly to sample or to seed control bottles:

$$f = (\text{volume of seed in diluted sample})/(\text{volume of seed in seed control})$$

Report results as CBOD$_5$ if nitrification is inhibited.

If more than one sample dilution meets the criteria of a residual DO of at least 1 mg/L and a DO depletion of at least 2 mg/L and there is no evidence of toxicity at higher sample concentrations or the existence of an obvious anomaly, average results in the acceptable range.

In these calculations, do not make corrections for DO uptake by the dilution water blank during incubation. This correction is unnecessary if dilution water meets the blank criteria stipulated above. If the dilution water does not meet these criteria, proper corrections are difficult and results become questionable.

6. Precision and Bias

There is no measurement for establishing bias of the BOD procedure. The glucose-glutamic acid check prescribed in ¶ 4c is intended to be a reference point for evaluation of dilution water quality, seed effectiveness, and analytical technique. Single-laboratory tests using a 300-mg/L mixed glucose-glutamic acid solution provided the following results:[1]

Number of months: 14
Number of triplicates: 421
Average monthly recovery: 204   mg/L
Average monthly
  standard deviation: 10.4 mg/L

In a series of interlaboratory studies,[2] each involving 2 to 112 laboratories (and as many analysts and seed sources), 5-d BOD measurements were made on synthetic water samples containing a 1:1 mixture of glucose and glutamic acid in the total concentration range of 3.3 to 231 mg/L. The regression equations for mean value, $\overline{X}$, and standard deviation, S, from these studies were:

$$\overline{X} = 0.658 \text{ (added level, mg/L)} + 0.280 \text{ mg/L}$$
$$S = 0.100 \text{ (added level, mg/L)} + 0.547 \text{ mg/L}$$

For the 300-mg/L mixed primary standard, the average 5-d BOD would be 198 mg/L with a standard deviation of 30.5 mg/L.

a. Control limits: Because of many factors affecting BOD tests in multilaboratory studies and the resulting extreme variability in test results, one standard deviation, as determined by interlaboratory tests, is recommended as a control limit for individual laboratories. Alternatively, for each laboratory, establish its control limits by performing a minimum of 25 glucose-glutamic acid checks (¶ 4c) over a period of several weeks or months and calculating the mean and standard deviation. Use the mean $\pm$ 3 standard deviations as the control limit for future glucose-glutamic acid checks. Compare calculated control limits to the single-laboratory tests presented above and to interlaboratory results. If control limits are outside the range of 198 $\pm$ 30.5, re-evaluate the control limits and investigate source of the problem. If measured BOD for a glucose-glutamic acid check is outside the accepted control limit range, reject tests made with that seed and dilution water.

b. *Working range and detection limit:* The working range is equal to the difference between the maximum initial DO (7 to 9 mg L) and minimum DO residual of 1 mg/L multiplied by the dilution factor. A lower detection limit of 2 mg/L is established by the requirement for a minimum DO depletion of 2 mg/L.

7. References

1. KENNEY, W. & J. RESNICK. 1987. Personal communication with J. C. Young. Water Pollution Control Plant, Davenport, Iowa.

2. U.S. ENVIRONMENTAL PROTECTION AGENCY, OFFICE OF RESEARCH AND DEVELOPMENT. 1986. Method-by-Method Statistics from Water Pollution (WP) Laboratory Performance Evaluation Studies. Quality Assurance Branch, Environmental Monitoring and Support Lab., Cincinnati, Ohio.

8. Bibliography

SAWYER, C.N. & L. BRADNEY. 1946. Modernization of the BOD test for determining the efficiency of the sewage treatment process. *Sewage Works J.* 18:1113.

RUCHHOFT, C.C., O.R. PLACAK, J. KACHMAR & C.E. CALBERT. 1948. Variations in BOD velocity constant of sewage dilutions. *Ind. Eng. Chem.* 40:1290.

ABBOTT, W.E. 1948. The bacteriostatic effects of methylene blue on the BOD test. *Water Sewage Works* 95:424.

SAWYER, C.N., P. CALLEJAS, M. MOORE & A.Q.Y. TOM. 1950. Primary standards for BOD work. *Sewage Ind. Wastes* 22:26.

YOUNG, J.C., G.N. MCDERMOTT & D. JENKINS. 1981. Alterations in the BOD procedure for the 15th edition of Standard Methods for the Examination of Water and Wastewater. *J. Water Pollut. Control Fed.* 53:1253.

5220   CHEMICAL OXYGEN DEMAND (COD)*

5220 A.   Introduction

The chemical oxygen demand (COD) is used as a measure of the oxygen equivalent of the organic matter content of a sample that is susceptible to oxidation by a strong chemical oxidant. For samples from a specific source, COD can be related empirically to BOD, organic carbon, or organic matter. The test is useful for monitoring

*Approved by Standard Methods Committee, 1985.

and control after correlation has been established. The dichromate reflux method is preferred over procedures using other oxidants because of superior oxidizing ability, applicability to a wide variety of samples, and ease of manipulation. Oxidation of most organic compounds is 95 to 100% of the theoretical value. Pyridine and related compounds resist oxidation and volatile organic compounds are oxidized only to the extent that they remain in contact with the oxidant. Ammonia, present either in the waste or liberated from nitrogen-containing organic matter, is not oxidized in the absence of significant concentration of free chloride ions.

1. Selection of Method

The open reflux method (B) is suitable for a wide range of wastes where a large sample size is preferred. The closed reflux methods (C and D) are more economical in the use of metallic salt reagents, but require homogenization of samples containing suspended solids to obtain reproducible results. Ampules and culture tubes with premeasured reagents are available commercially. Follow instructions furnished by the manufacturer.

Determine COD values of > 50 mg $O_2/$L by using procedures 5220B.4a, C.4, or D.4. Use procedure 5220B.4b to determine, with lesser accuracy, COD values from 5 to 50 mg O_2/L.

2. Interferences and Limitations

Volatile straight-chain aliphatic compounds are not oxidized to any appreciable extent. This failure occurs partly because volatile organics are present in the vapor space and do not come in contact with the oxidizing liquid. Straight-chain aliphatic compounds are oxidized more effectively when silver sulfate (Ag_2SO_4) is added as a catalyst. However, Ag_2SO_4 reacts with chloride, bromide, and iodide to produce precipitates that are oxidized only partially. The difficulties caused by the presence of the halides can be overcome largely, though not completely, by complexing with mercuric sulfate ($HgSO_4$) before the refluxing procedure. Although 1 g $HgSO_4$ is specified for 50 mL sample, a lesser amount may be used where sample chloride concentration is known to be less than 2000 mg/L, as long as a 10:1 ratio of $HgSO_4$:Cl^- is maintained. Do not use the test for samples containing more than 2000 mg Cl^-/L. Techniques designed to measure COD in saline waters are available.[1,2]

Nitrite (NO_2^-) exerts a COD of 1.1 mg O_2/mg NO_2^--N. Because concentrations of NO_2^- in waters rarely exceed 1 or 2 mg NO_2^--N/L, the interference is considered insignificant and usually is ignored. To eliminate a significant interference due to NO_2^-, add 10 mg sulfamic acid for each mg NO_2^--N present in the sample volume used; add the same amount of sulfamic acid to the reflux vessel containing the distilled water blank.

Reduced inorganic species such as ferrous iron, sulfide, manganous manganese, etc., are oxidized quantitatively under the test conditions. For samples containing significant levels of these species, stoichiometric oxidation can be assumed from known initial concentration of the interfering species and corrections can be made to the COD value obtained.

3. Sampling and Storage

Preferably collect samples in glass bottles. Test unstable samples without delay. If delay before analysis is unavoidable, preserve sample by acidification to pH ≤ 2 using conc H_2SO_4. Blend samples containing settleable solids with a homogenizer to permit representative sampling. Make preliminary dilutions for wastes containing a high COD to reduce the error inherent in measuring small sample volumes.

4. References

1. BURNS, E.R. & C. MARSHALL. 1965. Correction for chloride interference in the chemical oxygen demand test. *J. Water Pollut. Control Fed.* 37:1716.

2. BAUMANN, F.I. 1974. Dichromate reflux chemical oxygen demand: A proposed method for chloride correction in highly saline waters. *Anal. Chem.* 46:1336.

5220 B. Open Reflux Method

1. General Discussion

a. Principle: Most types of organic matter are oxidized by a boiling mixture of chromic and sulfuric acids. A sample is refluxed in strongly acid solution with a known excess of potassium dichromate ($K_2Cr_2O_7$). After digestion, the remaining unreduced $K_2Cr_2O_7$ is titrated with ferrous ammonium sulfate to determine the amount of $K_2Cr_2O_7$ consumed and the oxidizable organic matter is calculated in terms of oxygen equivalent. Keep ratios of reagent weights, volumes, and strengths constant when sample volumes other than 50 mL are used. The standard 2-h reflux time may be reduced if it has been shown that a shorter period yields the same results.

2. Apparatus

Reflux apparatus, consisting of 500- or 250-mL erlenmeyer flasks with ground-glass 24/40 neck* and 300-mm jacket Liebig, West, or equivalent condenser† with 24/40 ground-glass joint, and a hot plate having sufficient power to produce at least 1.4 W/cm^2 of heating surface, or equivalent.

3. Reagents

a. Standard potassium dichromate solution, 0.0417M: Dissolve 12.259 g $K_2Cr_2O_7$, primary standard grade, previously dried at 103°C for 2 h, in distilled water and dilute to 1000 mL.

b. Sulfuric acid reagent: Add Ag_2SO_4, reagent or technical grade, crystals or powder, to conc H_2SO_4 at the rate of 5.5 g Ag_2SO_4/kg H_2SO_4. Let stand 1 to 2 d to dissolve Ag_2SO_4.

c. Ferroin indicator solution: Dissolve 1.485 g 1,10-phenanthroline monohydrate and 695 mg $FeSO_4 \cdot 7H_2O$ in distilled water and dilute to 100 mL. This indicator so-

lution may be purchased already prepared.‡

d. Standard ferrous ammonium sulfate (FAS) titrant, approximately 0.25M: Dissolve 98 g $Fe(NH_4)_2(SO_4)_2 \cdot 6H_2O$ in distilled water. Add 20 mL conc H_2SO_4, cool, and dilute to 1000 mL. Standardize this solution daily against standard $K_2Cr_2O_7$ solution as follows:

Dilute 10.0 mL standard $K_2Cr_2O_7$ to about 100 mL. Add 30 mL conc H_2SO_4 and cool. Titrate with FAS titrant using 0.10 to 0.15 mL (2 to 3 drops) ferroin indicator.

Molarity of FAS solution

$$= \frac{\text{Volume } 0.0417M \text{ } K_2Cr_2O_7 \text{ solution titrated, mL}}{\text{Volume FAS used in titration, mL}} \times 0.25$$

e. Mercuric sulfate, $HgSO_4$, crystals or powder.

f. Sulfamic acid: Required only if the interference of nitrites is to be eliminated (see 5220A.2 above).

g. Potassium hydrogen phthalate (KHP) standard: Lightly crush and then dry potassium hydrogen phthalate ($HOOCC_6H_4COOK$) to constant weight at 120°C. Dissolve 425 mg in distilled water and dilute to 1000 mL. KHP has a theoretical COD[1] of 1.176 mg O_2/mg and this solution has a theoretical COD of 500 µg O_2/mL. This solution is stable when refrigerated for up to 3 months in the absence of visible biological growth.

4. Procedure

a. Treatment of samples with COD of > 50 mg O_2/L: Place 50.0 mL sample (for samples with COD of > 900 mg O_2/L, use smaller sample portion diluted to 50.0 mL) in a 500-mL refluxing flask. Add 1 g $HgSO_4$, several glass beads, and very slowly add 5.0 mL sulfuric acid reagent, with mix-

---

*Corning 5000 or equivalent.
†Corning 2360, 91548, or equivalent.

‡GFS Chemical Co., Columbus, Ohio.

ing to dissolve $HgSO_4$. Cool while mixing to avoid possible loss of volatile materials. Add 25.0 mL $0.0417M$ $K_2Cr_2O_7$ solution and mix. Attach flask to condenser and turn on cooling water. Add remaining sulfuric acid reagent (70 mL) through open end of condenser. Continue swirling and mixing while adding the sulfuric acid reagent. CAUTION: *Mix reflux mixture thoroughly before applying heat to prevent local heating of flask bottom and a possible blowout of flask contents.*

Cover open end of condenser with a small beaker to prevent foreign material from entering refluxing mixture and reflux for 2 h. Cool and wash down condenser with distilled water. Disconnect reflux condenser and dilute mixture to about twice its volume with distilled water. Cool to room temperature and titrate excess $K_2Cr_2O_7$ with FAS, using 0.10 to 0.15 mL (2 to 3 drops) ferroin indicator. Although the quantity of ferroin indicator is not critical, use the same volume for all titrations. Take as the end point of the titration the first sharp color change from blue-green to reddish brown. The blue-green may reappear. In the same manner, reflux and titrate a blank containing the reagents and a volume of distilled water equal to that of sample.

b. *Alternate procedure for low-COD samples:* Follow procedure of ¶ 4a, with two exceptions: (i) use standard $0.00417M$ $K_2Cr_2O_7$, and (ii) titrate with $0.025M$ FAS. Exercise extreme care with this procedure because even a trace of organic matter on the glassware or from the atmosphere may cause gross errors. If a further increase in sensitivity is required, concentrate a larger volume of sample before digesting under reflux as follows: Add all reagents to a sample larger than 50 mL and reduce total volume to 150 mL by boiling in the refluxing flask open to the atmosphere without the condenser attached. Compute amount of $HgSO_4$ to be added (before concentration) on the basis of a weight ratio of 10:1, $HgSO_4$:Cl^-, using the amount of Cl^- present in the original volume of sample. Carry a blank reagent through the same procedure. This technique has the advantage of concentrating the sample without significant losses of easily digested volatile materials. Hard-to-digest volatile materials such as volatile acids are lost, but an improvement is gained over ordinary evaporative concentration methods.

c. *Determination of standard solution:* Evaluate the technique and quality of reagents by conducting the test on a standard potassium hydrogen phthalate solution.

5. Calculation

$$\text{COD as mg } O_2/L = \frac{(A - B) \times M \times 8000}{\text{mL sample}}$$

where:
  A = mL FAS used for blank,
  B = mL FAS used for sample, and
  M = molarity of FAS.

6. Precision and Bias

A set of synthetic samples containing potassium hydrogen phthalate and NaCl was tested by 74 laboratories. At a COD of 200 mg O_2/L in the absence of chloride, the standard deviation was ±13 mg/L (coefficient of variation, 6.5%). At COD of 160 mg O_2/L and 100 mg Cl^-/L, the standard deviation was ±14 mg/L (coefficient of variation, 10.8%).

7. Reference

1. PITWELL, L.R. 1983. Standard COD. *Chem. Brit.* 19:907.

8. Bibliography

MOORE, W.A., R.C. KRONER & C.C. RUCHHOFT. 1949. Dichromate reflux method for determination of oxygen consumed. *Anal. Chem.* 21:953.

MEDALIA, A.I. 1951. Test for traces of organic matter in water. *Anal. Chem.* 23:1318.

MOORE, W.A., F.J. LUDZACK & C.C. RUCHHOFT.

1951. Determination of oxygen-consumed values of organic wastes. *Anal. Chem.* 23:1297.

DOBBS, R.A. & R.T. WILLIAMS. 1963. Elimination of chloride interference in the chemical oxygen demand test. *Anal. Chem.* 35:1064.

5220 C. Closed Reflux, Titrimetric Method

1. General Discussion

a. Principle: See 5220B.1*a*.

b. Interferences and limitations: See 5220A.2. Volatile organic compounds are more completely oxidized in the closed system because of longer contact with the oxidant. Before each use inspect culture-tube caps for breaks in the TFE liner. Select culture-tube size for the degree of sensitivity desired. Use the 25- $\times$ 150-mm tube for samples with low COD content because a larger volume sample can be treated.

2. Apparatus

a. Digestion vessels: Preferably use borosilicate culture tubes, 16- $\times$ 100-mm, 20- $\times$ 150-mm, or 25- $\times$ 150-mm, with TFE-lined screw caps. Alternatively, use borosilicate ampules, 10-mL capacity, 19- to 20-mm diam.

b. Heating block, cast aluminum, 45 to 50 mm deep, with holes sized for close fit of culture tubes or ampules.

c. Block heater or oven, to operate at 150 $\pm$ 2°C. NOTE: Severe damage of most culture tube closures from oven digestion introduces a potential source of contamination and increases the probability of leakage. Use an oven for culture-tube digestion only when it has been determined that 2 h exposure at 150°C will not damage the caps.

d. Ampule sealer: Use only a mechanical sealer to insure strong, consistent seals.

3. Reagents

a. Standard potassium dichromate digestion solution, 0.0167*M:* Add to about 500 mL distilled water 4.913 g $K_2Cr_2O_7$, pri-mary standard grade, previously dried at 103°C for 2 h, 167 mL conc H_2SO_4, and 33.3 g $HgSO_4$. Dissolve, cool to room temperature, and dilute to 1000 mL.

b. Sulfuric acid reagent: See Section 5220B.3*b*.

c. Ferroin indicator solution: See Section 5220B.3*c*.

d. Standard ferrous ammonium sulfate titrant (FAS), approximately 0.10*M:* Dissolve 39.2 g $Fe(NH_4)_2(SO_4)_2 \cdot 6H_2O$ in distilled water. Add 20 mL conc H_2SO_4, cool, and dilute to 1000 mL. Standardize solution daily against standard $K_2Cr_2O_7$ digestion solution as follows:

Add reagents according to Table 5220:I to a culture tube containing the correct volume of distilled water substituted for sample. Cool tube to room temperature and add 0.05 to 0.10 mL (1 to 2 drops) ferroin indicator and titrate with FAS titrant.

Molarity of FAS solution

$$= \frac{\substack{\text{Volume } 0.0167M\ K_2Cr_2O_7 \\ \text{solution titrated, mL}}}{\text{Volume FAS used in titration, mL}} \times 0.10$$

e. Sulfamic acid: See Section 5220B.3*f*.

f. Potassium hydrogen phthalate standard: See Section 5220B.3*g*.

4. Procedure

Wash culture tubes and caps with 20% H_2SO_4 before first use to prevent contamination. Refer to Table 5220:I for proper sample and reagent volumes. Place sample in culture tube or ampule and add digestion solution. Carefully run sulfuric acid reagent down inside of vessel so an acid layer

TABLE 5220:I. SAMPLE AND REAGENT QUANTITIES FOR VARIOUS DIGESTION VESSELS

Digestion Vessel	Sample mL	Digestion Solution mL	Sulfuric Acid Reagent mL	Total Final Volume mL
Culture tubes:				
16 × 100 mm	2.5	1.5	3.5	7.5
20 × 150 mm	5.0	3.0	7.0	15.0
25 × 150 mm	10.0	6.0	14.0	30.0
Standard 10-mL ampules	2.5	1.5	3.5	7.5

is formed under the sample-digestion solution layer. Tightly cap tubes or seal ampules, and invert each several times to mix completely. CAUTION: *Wear face shield and protect hands from heat produced when contents of vessels are mixed. Mix thoroughly before applying heat to prevent local heating of vessel bottom and possible explosive reaction.*

Place tubes or ampules in block digester or oven preheated to 150°C and reflux for 2 h. Cool to room temperature and place vessels in test tube rack. Remove culture tube caps and add small TFE-covered magnetic stirring bar. If ampules are used, transfer contents to a larger container for titrating. Add 0.05 to 0.10 mL (1 to 2 drops) ferroin indicator and stir rapidly on magnetic stirrer while titrating with 0.10M FAS. The end point is a sharp color change from blue-green to reddish brown, although the blue-green may reappear within minutes. In the same manner reflux and titrate a blank containing the reagents and a volume of distilled water equal to that of the sample.

5. Calculation

$$\text{COD as mg } O_2/L = \frac{(A - B) \times M \times 8000}{\text{mL sample}}$$

where:
A = mL FAS used for blank,
B = mL FAS used for sample, and
M = molarity of FAS.

6. Precision and Bias

Sixty synthetic samples containing potassium hydrogen phthalate and NaCl were tested by six laboratories. At an average COD of 195 mg O_2/L in the absence of chloride, the standard deviation was ±11 mg O_2/L (coefficient of variation, 5.6%). At an average COD of 208 mg O_2/L and 100 mg Cl^-/L, the standard deviation was ±10 mg O_2/L (coefficient of variation, 4.8%).

5220 D.   Closed Reflux, Colorimetric Method

1. General Discussion

a. Principle: See Section 5220B.1*a.* Colorimetric reaction vessels are sealed glass ampules or capped culture tubes. Oxygen consumed is measured against standards at 600 nm with a spectrophotometer.

b. Interferences and limitations: See Section 5220C.1*b.*

2. Apparatus

a. See Section 5220C.2.

b. Spectrophotometer, for use at 600 nm with access opening adapter for ampule or 16-, 20-, or 25-mm tubes.

3. Reagents

a. Digestion solution: Add to about 500 mL distilled water 10.216 g $K_2Cr_2O_7$, primary standard grade, previously dried at 103°C for 2 h, 167 mL conc H_2SO_4, and 33.3 g $HgSO_4$. Dissolve, cool to room temperature, and dilute to 1000 mL.

b. Sulfuric acid reagent: See 5220B.3*b.*

c. Sulfamic acid: See Section 5220B.3*f.*

d. Potassium hydrogen phthalate standard: See Section 5220B.3*g.*

4. Procedure

a. Treatment of samples: Measure suitable volume of sample and reagents into tube or ampule as indicated in Table 5220:I. Prepare, digest, and cool samples, blank, and one or more standards as directed in Section 5220C.4.

b. Measurement of dichromate reduction: Invert cooled samples, blank, and standards several times and allow solids to settle before measuring absorbance. Dislodge solids that adhere to container wall by gentle tapping and settling. Insert unopened tube or ampule through access door into light path of spectrophotometer set at 600 nm. Read absorbance and compare to calibration curve. Use optically matched culture tubes or ampules for greater sensitivity; discard scratched or blemished glassware.

c. Preparation of calibration curve: Prepare at least five standards from potassium hydrogen phthalate solution with COD equivalents from 20 to 900 μg O_2/L. Make up to volume with distilled water; use same reagent volumes, tube, or ampule size, and digestion procedure as for samples. Prepare calibration curve for each new lot of tubes or ampules or when standards prepared in ¶ 4*a* differ by $\geq 5\%$ from calibration curve.

5. Calculation

COD as mg O_2/L

$$= \frac{\text{mg } O_2 \text{ in final volume} \times 1000}{\text{mL sample}}$$

6. Precision and Bias

Forty-eight synthetic samples containing potassium hydrogen phthalate and NaCl were tested by five laboratories. At an average COD of 193 mg O_2/L in the absence of chloride, the standard deviation was ± 17 mg O_2/L (coefficient of variation 8.7%). At an average COD of 212 mg O_2/L and 100 mg Cl^-/L, the standard deviation was ± 20 mg O_2/L (coefficient of variation, 9.6%).

7. Bibliography

JIRKA, A.M. & M.J. CARTER. 1975. Micro semi-automated analysis of surface and wastewaters for chemical oxygen demand. *Anal. Chem.* 47:1397.

HIMEBAUGH, R.R. & M.J. SMITH. 1979. Semi-micro tube method for chemical oxygen demand. *Anal. Chem.* 51:1085.

5310 TOTAL ORGANIC CARBON (TOC)*

5310 A. Introduction

1. General Discussion

The organic carbon in water and wastewater is composed of a variety of organic compounds in various oxidation states. Some of these carbon compounds can be oxidized further by biological or chemical processes, and the biochemical oxygen demand (BOD) and chemical oxygen demand (COD) may be used to characterize these fractions. The presence of organic carbon that does not respond to either the BOD or COD test makes them unsuitable for the measurement of total organic carbon. Total organic carbon (TOC) is a more convenient and direct expression of total organic content than either BOD or COD, but does not provide the same kind of information. If a repeatable empirical relationship is established between TOC and BOD or COD, then TOC can be used to estimate the accompanying BOD or COD. This relationship must be established independently for each set of matrix conditions, such as various points in a treatment process. Unlike BOD or COD, TOC is independent of the oxidation state of the organic matter and does not measure other organically bound elements, such as nitrogen and hydrogen, and inorganics that can contribute to the oxygen demand measured by BOD and COD. TOC measurement does not replace BOD and COD testing.

To determine the quantity of organically bound carbon, the organic molecules must be broken down to single carbon units and converted to a single molecular form that can be measured quantitatively. TOC methods utilize heat and oxygen, ultraviolet irradiation, chemical oxidants, or combinations of these oxidants to convert organic carbon to carbon dioxide (CO_2). The CO_2 may be measured directly by a nondispersive infrared analyzer, it may be reduced to methane and measured with a flame ionization detector, or CO_2 may be titrated chemically.

2. Fractions of Total Carbon

The methods and instruments used in measuring TOC analyze fractions of total carbon (TC) and measure TOC by two or more determinations. These fractions of total carbon are defined as: inorganic carbon (IC)—the carbonate, bicarbonate, and dissolved CO_2; total organic carbon (TOC)—all carbon atoms covalently bonded in organic molecules; dissolved organic carbon (DOC)—the fraction of TOC that passes through a 0.45-μm-pore-diam filter; nondissolved organic carbon (NDOC)—also referred to as particulate organic carbon, the fraction of TOC retained by a 0.45-μm filter; purgeable organic carbon (POC)—also referred to as volatile organic carbon, the fraction of TOC removed from an aqueous solution by gas stripping under specified conditions; and nonpurgeable organic carbon (NPOC)—the fraction of TOC not removed by gas stripping.

In most water samples, the IC fraction is many times greater than the TOC fraction. Eliminating or compensating for IC interferences requires multiple determinations to measure true TOC. IC interference can be eliminated by acidifying samples to

---

*Approved by Standard Methods Committee, 1985.

pH 2 or less to convert IC species to CO_2. Subsequently, purging the sample with a purified gas removes the CO_2 by volatilization. Sample purging also removes POC so that the organic carbon measurement made after eliminating IC interferences is actually a NPOC determination; determine POC to measure true TOC. In many surface and ground waters the POC contribution to TOC is negligible. Therefore, in practice, the NPOC determination is substituted for TOC.

Alternatively, IC interference may be compensated for by separately measuring total carbon (TC) and inorganic carbon. The difference between TC and IC is TOC.

The purgeable fraction of TOC is a function of the specific conditions and equipment employed. Sample temperature and salinity, gas-flow rate, type of gas diffuser, purging-vessel dimensions, volume purged, and purging time affect the division of TOC into purgeable and nonpurgeable fractions. When separately measuring POC and NPOC on the same sample, use identical conditions for purging during the POC measurement as in purging to prepare the NPOC portion for analysis. Consider the conditions of purging when comparing POC or NPOC data from different laboratories or different instruments.

3. Selection of Method

The combustion-infrared method (B) is suitable for samples with TOC ≥ 1 mg/L.

For lower concentrations, use the persulfate-ultraviolet oxidation method (C) or the wet-oxidation method (D). See 5310B.1, C.1, and D.1.

4. Bibliography

FORD, D.L. 1968. Total organic carbon as a wastewater parameter. *Pub. Works* 99:89.

FREDERICKS, A.D. 1968. Concentration of organic carbon in the Gulf of Mexico. Off. Naval Res. Rep. 68-27T.

WILLIAMS, P.M. 1969. The determination of dissolved organic carbon in seawater: A comparison of two methods. *Limnol. Oceanogr.* 14:297.

CROLL, B.T. 1972. Determination of organic carbon in water. *Chem. Ind.* (London), 110:386.

SHARP, J. 1973. Total organic carbon in seawater—comparison of measurements using persulfate oxidation and high temperature combustion. *Mar. Chem.* 1:211.

BORTLIJZ, J. 1976. Instrumental TOC analysis. *Vom Wasser* 46:35.

VAN STEENDEREN, R.A. 1976. Parameters which influence the organic carbon determination in water. *Water SA* 2:156.

TAKAHASHI, Y. 1979. Analysis Techniques for Organic Carbon and Organic Halogen. Proc. EPA/NATO-CCMS Conf. on Adsorption Techniques, Reston, Va.

STANDING COMMITTEE OF ANALYSIS, DEPARTMENT OF THE ENVIRONMENT, NATIONAL WATER COUNCIL. 1980. The instrumental determination of total organic carbon, total oxygen demand and related determinants, 1979. Her Majesty's Stationery Off., London.

WERSHAW, R.L., M.J. FISHMAN, R.R. GRABBE & L.E. LOWE, eds. 1983. Methods for the Determination of Organic Substances in Water and Fluvial Sediments. U.S. Geological Survey Techniques of Water-Resources Investigations, Book 5, Laboratory Analysis, Chapter A3.

5310 B.   Combustion-Infrared Method

1. General Discussion

The combustion-infrared method has been used for a wide variety of samples,

but its accuracy is dependent on particle size reduction because it uses small-orifice syringes.

a. Principle: The sample is homogenized

and diluted as necessary and a microportion is injected into a heated reaction chamber packed with an oxidative catalyst such as cobalt oxide. The water is vaporized and the organic carbon is oxidized to CO_2 and H_2O. The CO_2 from oxidation of organic and inorganic carbon is transported in the carrier-gas streams and is measured by means of a nondispersive infrared analyzer.

Because total carbon is measured, IC must be measured separately and TOC obtained by difference.

Measure IC by injecting the sample into a separate reaction chamber packed with phosphoric acid-coated quartz beads. Under the acidic conditions, all IC is converted to CO_2, which is measured. Under these conditions organic carbon is not oxidized and only IC is measured.

Alternatively, convert inorganic carbonates to CO_2 with acid and remove the CO_2 by purging before sample injection. The sample contains only the NPOC fraction of total carbon; a POC determination also is necessary to measure true TOC.

b. Interference: Removal of carbonate and bicarbonate by acidification and purging with purified gas results in the loss of volatile organic substances. The volatiles also can be lost during sample blending, particularly if the temperature is allowed to rise. Another important loss can occur if large carbon-containing particles fail to enter the needle used for injection. Filtration, although necessary to eliminate particulate organic matter when only DOC is to be determined, can result in loss or gain of DOC, depending on the physical properties of the carbon-containing compounds and the adsorption of carbonaceous material on the filter, or its desorption from it. Check filters for their contribution to DOC by analyzing a filtered blank. Note that any contact with organic material may contaminate a sample. Avoid contaminated glassware, plastic containers, and rubber tubing. Analyze treatment, system, and reagent blanks.

c. Minimum detectable concentration: 1 mg carbon/L. This can be achieved with most combustion-infrared analyzers although instrument performance varies. The minimum detectable concentration may be reduced by concentrating the sample, or by increasing the portion taken for analysis.

d. Sampling and storage: Collect and store samples in amber glass bottles with TFE-lined cap. Before use, wash bottles with acid, seal with aluminum foil, and bake at 400°C for at least 1 h. Wash TFE septa with detergent, rinse repeatedly with organic-free water, wrap in aluminum foil, and bake at 100°C for 1 h. Preferably use thick silicone rubber-backed TFE septa with open ring caps to produce a positive seal. Because the detection limit is relatively high, less rigorous cleaning may be acceptable; use bottle blanks with each set of samples. Use a Kemmerer or similar type sampler for collecting samples from a depth exceeding 2 m. Preserve samples that cannot be examined promptly by holding at 4°C with minimal exposure to light and atmosphere. Acidification with phosphoric or sulfuric acid to a pH ≤ 2 may be used only if inorganic carbon subsequently is purged. Under any conditions, minimize storage time.

2. Apparatus

*a. Total organic carbon analyzer.**

b. Syringes: 0 to 50-µL, 0 to 200-µL, 0 to 500-µL, and 0 to 1-mL capacity.†

c. Sample blender or homogenizer.

d. Magnetic stirrer and TFE-coated stirring bars.

e. Filtering apparatus and 0.45-µm-pore-diam filters.

---

*Beckman Instruments Model No. 915B or equivalent.
†Hamilton No. 705 N or 750 N: CR-700-20 or CR-700-200 with needle point style No. 3, or equivalent. Spring-loaded syringe may improve reproducibility.

3. Reagents

a. Reagent water: Prepare blanks and standard solutions from carbon-free water; preferably use carbon-filtered, redistilled water.

b. Phosphoric acid, H_3PO_4, conc. Alternatively use sulfuric acid, H_2SO_4, but not hydrochloric acid.

c. Organic carbon stock solution: Dissolve 2.1254 g anhydrous potassium biphthalate, $C_8H_5KO_4$, in carbon-free water and dilute to 1000 mL; 1.00 mL = 1.00 mg carbon. Alternatively, use any other organic-carbon-containing compound of adequate purity, stability, and water solubility. Preserve by acidifying with H_3PO_4 or H_2SO_4 to pH ≤ 2.

d. Inorganic carbon stock solution: Dissolve 4.4122 g anhydrous sodium carbonate, Na_2CO_3, in water, add 3.497 g anhydrous sodium bicarbonate, $NaHCO_3$, and dilute to 1000 mL; 1.00 mL = 1.00 mg carbon. Alternatively, use any other inorganic carbonate compound of adequate purity, stability, and water solubility. Keep tightly stoppered. Do not acidify.

e. Carrier gas: Purified oxygen or air, CO_2-free and containing less than 1 ppm hydrocarbon (as methane).

f. Purging gas: Any gas free of CO_2 and hydrocarbons.

4. Procedure

a. Instrument operation: Follow manufacturer's instructions for analyzer assembly, testing, calibration, and operation. Adjust to optimum combustion temperature (900°C) before using instrument; monitor temperature to insure stability.

b. Sample treatment: If a sample contains gross solids or insoluble matter, homogenize until satisfactory replication is obtained. Analyze a homogenizing blank consisting of reagent water carried through the homogenizing treatment.

If inorganic carbon must be removed before analysis, transfer a representative portion of 10 to 15 mL to a 30-mL beaker, add conc H_3PO_4 to reduce pH to 2 or less, and purge with gas for 10 min. Do not use plastic tubing. Inorganic carbon also may be removed by stirring the acidified sample in a beaker while directing a stream of purified gas into the beaker. Because volatile organic carbon will be lost during purging of the acidified solution, report organic carbon as total nonpurgeable organic carbon.

If the available instrument provides for a separate determination of inorganic carbon (carbonate, bicarbonate, free CO_2) and total carbon, omit decarbonation and proceed according to the manufacturer's directions to determine TOC by difference between TC and IC.

If dissolved organic carbon is to be determined, filter sample through 0.45-μm-pore-diam filter with vacuum; analyze a filtering blank.

c. Sample injection: Withdraw a portion of prepared sample using a syringe fitted with a blunt-tipped needle. Select sample volume according to manufacturer's direction. Stir samples containing particulates with a magnetic stirrer. Select needle size consistent with sample particulate size. Inject samples and standards into analyzer according to manufacturer's directions and record response. Repeat injection until consecutive peaks are obtained that are reproducible to within $\pm 10\%$.

d. Preparation of standard curve: Prepare standard organic and inorganic carbon series by diluting stock solutions to cover the expected range in samples. Inject and record peak height of these standards and a dilution water blank. Plot carbon concentration in milligrams per liter against corrected peak height in millimeters on rectangular coordinate paper. This is unnecessary for instruments provided with a digital readout of concentration. If desirable, prepare a standard curve having concentrations of 1 to 10 mg/L by making appropriate dilutions of the standards.

With most TOC analyzers, it is not possible to determine separate blanks for reagent water, reagents, and the entire system. In addition, some TOC analyzers produce a variable and erratic blank that cannot be corrected reliably. In many laboratories, reagent water is the major contributor to the blank value. Correcting only the peak heights of standards (which contain reagent water + reagents + system blank) creates a positive error, while also correcting samples (which contain only reagents and system blank contributions) for the reagent water blank creates a negative error. Minimize errors by using reagent water and reagents low in carbon.

Inject samples and procedural blanks (consisting of reagent water taken through any pre-analysis steps—values are typically higher than those for reagent water) and determine sample organic carbon concentrations by comparing corrected peak heights to the calibration curve.

5. Calculations

a. When reagent water is a major portion of the total blank:

1) Calculate corrected peak height of standards by subtracting the reagent-water blank peak height from the standard peak heights.

2) Prepare a standard curve of corrected peak height vs. TOC concentration.

3) Subtract the procedural blank from each sample peak height and compare to the standard curve to determine carbon content.

NOTE: There will be a positive error if the TOC of the reagent water is significant in comparison to the TOC of the sample. Make a special effort to obtain carbon-free reagent water.

4) Apply the appropriate dilution factor when necessary.

5) Subtract the inorganic carbon from the total carbon when TOC is determined by difference.

b. When reagent water is a minor portion of the total blank:

1) Calculate corrected peak height of standards and samples by subtracting the reagent-water blank peak height from the standard and sample peak heights.

2) Prepare a standard curve of corrected peak height vs. TOC concentration.

3) Subtract the procedural blank from each sample peak height and compare to the standard curve to determine the carbon content. Values will have a negative error equal to the blank contribution from the reagent water.

4) Apply the appropriate dilution factor when necessary.

5) Subtract the inorganic carbon from the total carbon when TOC is determined by difference.

NOTE: If the TOC analyzer design permits isolation of each of the contributions to the total blank, apply appropriate blank corrections to peak heights of standards (reagent blank, water blank, system blank) and sample (reagent blank and system blank).

6. Precision

The difficulty of sampling particulate matter on unfiltered samples limits the precision of the method to approximately 5 to 10%. On clear samples or on those that have been filtered before analysis, precision approaches 1 to 2% or ± 1 to 2 mg carbon/L, whichever is greater.

7. Bibliography

KATZ, J., S. ABRAHAN & N. BAKER. 1954. Analytical procedure using a combined combustion-diffusion vessel: Improved method for combustion of organic compounds in aqueous solution. *Anal. Chem.* 26:1503.

VAN HALL, C.E., J. SAFRANKO & V. A. STENGER. 1963. Rapid combustion method for the determination of organic substances in aqueous solutions. *Anal. Chem.* 35:315.

SCHAFFER, R.B. et al. 1965. Application of a carbon analyzer in waste treatment. *J. Water Pollut. Control Fed.* 37:1545.

VAN HALL, C.E., D. BARTH & V. A. STENGER. 1965. Elimination of carbonates from aqueous solutions prior to organic carbon determinations. *Anal. Chem.* 37:769.

BUSCH, A.W. 1966. Energy, total carbon and oxygen demand. *Water Resour. Res.* 2:59.

BLACKMORE, R.H. & D. VOSHEL. 1967. Rapid determination of total organic carbon (TOC) in sewage. *Water Sewage Works* 114:398.

WILLIAMS, R.T. 1967. Water-pollution instrumentation—Analyzer looks for organic carbon. *Instrum. Technol.* 14:63.

5310 C.   Persulfate-Ultraviolet Oxidation Method

1. General Discussion

Many instruments utilizing persulfate oxidation of organic carbon are available. They depend either on heat or ultraviolet irradiation activation of the reagents. The persulfate-ultraviolet oxidation method is a rapid, precise method for the measurement of trace levels of organic carbon in water and is of particular interest to the electronic, pharmaceutical, and steam-power-generation industries where even trace concentrations of organic compounds may degrade ion-exchange capacity, serve as a nutrient source for biological growth, or be detrimental to the process for which the water is being utilized.

a. Principle: Organic carbon is oxidized to carbon dioxide, CO_2, by persulfate in the presence of ultraviolet light. The CO_2 produced may be measured directly by a nondispersive infrared analyzer, be reduced to methane and measured by a flame ionization detector, or be chemically titrated.

Instruments are available that utilize an ultraviolet lamp submerged in a continuously gas-purged reactor that is filled with a constant-feed persulfate solution. The samples are introduced serially into the reactor by an autosampler or they are injected manually. The CO_2 produced is sparged continuously from the solution and is carried in the gas stream to an infrared analyzer that is specifically tuned to the absorptive wavelength of CO_2. The instrument's microprocessor calculates the area of the peaks produced by the analyzer, compares them to the peak area of the calibration standard stored in its memory, and prints out a calibrated organic carbon value in milligrams per liter.

b. Interferences: See Section 5310B.1. Excessive acidification of sample, producing a reduction in pH of the persulfate solution to 1 or less, can result in sluggish and incomplete oxidation of organic carbon.

The intensity of the ultraviolet light reaching the sample matrix may be reduced by highly turbid samples or with aging of the ultraviolet source, resulting in sluggish or incomplete oxidation. Large organic particles or very large or complex organic molecules such as tannins, lignins, and humic acid may be oxidized slowly because persulfate oxidation is rate-limited. Because the efficiency of conversion of organic carbon to CO_2 may be affected by many factors, check efficiency of oxidation with selected model compounds representative of the sample matrix.

Persulfate oxidation of organic molecules is slowed in samples containing significant concentrations of chloride by the preferential oxidation of chloride; at a concentration of 0.1%, chloride oxidation of organic matter may be inhibited completely. To remove this interference add mercuric nitrate to the persulfate solution.

With any organic carbon measurement, contamination during sample handling and treatment is a likely source of interference. This is especially true of trace analysis. Take extreme care in sampling, handling, and analysis of samples below 1 mg TOC/L.

c. *Minimum detectable concentration:* Concentration of 0.05 mg organic carbon/L can be measured if scrupulous attention is given to minimizing sample contamination and method background. Use the combustion-infrared method (B) for high concentrations of TOC.

d. *Sampling and storage:* See Section 5310B.1d.

2. Apparatus

a. *Total organic carbon analyzer.**

b. *Syringes:* 0 to 50-µL, 0 to 250-µL, and 0 to 1-mL capacity, fitted with a blunt-tipped needle.

3. Reagents

See Section 5310B.3.

4. Procedure

a. *Instrument operation:* Follow manufacturer's instructions for assembly, testing, calibration, and operation.

b. *Sample preparation:* If a sample contains gross particulates or insoluble matter, homogenize until a representative portion can be withdrawn through the syringe needle or autosampler tubing.

If dissolved organic carbon is to be determined, filter sample and a reagent water blank through 0.45-µm glass fiber filter with vacuum. Pretreat filter by submerging overnight in a 1:1 solution of HNO_3 and reagent water and collect in an acid-washed, baked flask. Use a clean filter and flask for each sample.

To determine NPOC, transfer 15 to 30 mL sample to a flask or test tube and acidify to a pH of 2. Purge according to manufacturer's recommendations.

c. *Sample injection:* See Section 5310B.4c.

d. *Standard curve preparation:* Prepare an organic carbon standard series over the range of organic carbon concentrations in the samples. Inject standards and blanks and record analyzer's response. Determine peak area for each standard and blank. Determinations based on peak height may be inadequate because of differences in the rate of oxidation of standards and samples. Correct peak area of standards by subtracting reagent water blank and plot organic carbon concentration in milligrams per liter against corrected peak area. For instruments providing a digital readout of concentration, this is not necessary.

Inject samples, treatment blanks, if applicable, and instrument blank. Subtract appropriate blank's peak area from each sample's peak area and determine organic carbon from the standard curve.

5. Calculation

See Section 5310B.5, but use peak area rather than peak height.

6. Precision and Bias

See Table 5310:I.

7. Bibliography

BEATTIE, J., C. BRICKER & D. GARVIN. 1961. Photolytic determination of trace amounts of organic material in water. *Anal. Chem.* 33:1890.

ARMSTRONG, F.A.J., P.M. WILLIAMS & J.D.H. STRICKLAND. 1966. Photooxidation of organic matter in sea water by ultraviolet radiation, analytical and other applications. *Nature* 211:481.

BRAMER, H.C., M.J. WALSH & S.C. CARUSO. 1966. Instrument for monitoring trace organic compounds in water. *Water Sewage Works* 113:275.

DOBBS, R.A., R. H. WISE & R. B. DEAN. 1967. Measurement of organic carbon in water using the hydrogen flame ionization detector. *Anal. Chem.* 39:1255.

MONTGOMERY, H.A.C. & N.S. THOM. 1967. The determination of low concentrations of organic carbon in water. *Analyst* 87:689.

ARMSTRONG, F.A.J. & S. TIBBITS. 1968. Photochemical combustion of organic matter in seawater for nitrogen, phosphorus and carbon

---

*Xertex-Dohrmann DC-80 or equivalent.

TABLE 5310:I. PRECISION AND BIAS FOR TOTAL ORGANIC CARBON (TOC) BY PERSULFATE-ULTRAVIOLET OXIDATION

Characteristic of Analysis	Spring Water	Spring Water +0.15 mg/L KHP*	Tap Water	Tap Water +10 mg/L KHP*	Municipal Wastewater Effluent
Concentration determined, mg/L:					
Replicate 1	0.402	0.559	2.47	11.70	5.88
Replicate 2	0.336	0.491	2.49	11.53	5.31
Replicate 3	0.340	0.505	2.47	11.70	5.21
Replicate 4	0.341	0.523	2.47	11.64	5.17
Replicate 5	0.355	0.542	2.46	11.55	5.10
Replicate 6	0.366	0.546	2.46	11.68	5.33
Replicate 7	0.361	0.548	2.42	11.55	5.35
Mean, mg/L	0.35	0.53	2.46	11.53	5.32
Standard deviation:					
mg/L	0.02	0.03	0.02	0.21	0.23
%	6	6	1	2	4
Actual value, mg/L	—	0.50	—	12.46	—
Recovery, %	—	106	—	93	—
Error, %	—	6	—	7	—

*KHP = potassium acid phthalate.

determination. *J. Mar. Biol. Assoc. U.K.* 48:143.

JONES, R.H. & A.F. DAGEFORD. 1968. Application of a high sensitivity total organic carbon analyzer. *Proc. Instr. Soc. Amer.* 6:43.

TAKAHASHI, Y., R.T. MOORE & R.J. JOYCE. 1972. Direct determination of organic carbon in water by reductive pyrolysis. *Amer. Lab.* 4:31.

COLLINS, K.J. & P.J. LEB. WILLIAMS. 1977. An automated photochemical method for the determination of dissolved organic carbon in sea and estuarine waters. *Mar. Chem.* 5:123.

GRAVELET-BLONDIN, L.R., H.R. VAN VLIET & P.A. MYNHARDT. 1980. An automated method for the determination of dissolved organic carbon in fresh water. *Water SA* 6:138.

OAKE, R.J. 1981. A Review of Photo-Oxidation for the Determination of Total Organic Carbon in Water. Water Research Center, Technical Rep. (TR 160), Medmenham, England.

VAN STEENDEREN, R.A. & J.S. LIN. 1981. Determination of dissolved organic carbon in water. *Anal. Chem.* 53:521.

5310 D. Wet-Oxidation Method

1. General Discussion

The wet-oxidation method is suitable for the analyses of water, water-suspended sediment mixtures, brines, and wastewaters containing at least 0.1 mg nonpurgeable organic carbon (NPOC)/L. The method is not suitable for the determination of volatile organic constituents.

a. Principle: The sample is acidified, purged to remove inorganic carbon, and oxidized with persulfate in an autoclave at temperatures from 116 to 130°C. The resultant carbon dioxide (CO_2) is measured by nondispersive infrared spectrometry.

b. *Interferences:* See Section 5310B.1 and C.1.

c. *Minimum detectable concentrations:* High concentrations of reducing agents may interfere. Concentration of 0.10 mg organic carbon/L can be measured if scrupulous attention is given to minimizing sample contamination and method background. Use the combustion-infrared method (B) for high concentrations of TOC.

d. *Sampling and storage:* See Section 5310B.1d.

2. Apparatus

a. *Ampules,* precombusted, 10-mL, glass.*

b. *Ampule purging and sealing unit.* *

c. *Autoclave.* *

d. *Carbon analyzer.* *

e. *Homogenizer.*†

3. Reagents

In addition to the reagents specified in Section 5310B.3a, c, e, and f, the following reagents are required:

a. *Phosphoric acid solution,* H_3PO_4, 1.2N: Add 83 mL H_3PO_4 (85%) to water and dilute to 1 L with water. Store in a tightly stoppered glass bottle.

b. *Potassium persulfate,* reagent-grade, granular. Avoid using finely divided forms.

4. Procedure

a. *Instrument operation:* Follow manufacturer's instructions for assembly, testing calibration, and operation. Add 0.2 g po-

tassium persulfate using a dipper calibrated to deliver 0.2 g and 0.5 mL 1.2N H_3PO_4 solution to precombusted ampules.

To analyze for dissolved organic carbon, pass sample and reagent water blank under vacuum through a separate 0.45-μm glass fiber filter that was pretreated by submerging overnight in a 1:1 solution of HNO_3 and reagent water and collect in acid-washed, baked flasks. Use clean filter and flask for each sample.

Homogenize sample to produce a uniform suspension. Rinse homogenizer with reagent water after each use. Pipet water sample (10.0 mL maximum) into an ampule. Adjust smaller volumes to 10 mL with reagent water. Prepare one reagent blank (10 mL reagent water plus acid and oxidant) for every 15 to 20 water samples. Prepare standards covering the range of 0.1 to 40 mg carbon/L by diluting the carbon standard solution. Immediately place filled ampules on purging and sealing unit and purge them at rate of 60 mL/min for 6 min with purified oxygen. Seal samples according to the manufacturer's instructions. Place sealed samples, blanks, and a set of standards in ampule racks in an autoclave and digest 4 h at temperature between 116 and 130°C.

Set sensitivity range of carbon analyzer by adjusting the zero and span controls in accordance with the manufacturer's instructions. Break combusted ampules in the cutter assembly of the carbon analyzer, sweep CO_2 into the infrared cell with nitrogen gas, and record area of each CO_2 peak. CAUTION: *Because combusted ampules are under positive pressure, handle with care to prevent explosion.*

5. Calculations

Prepare an analytical standard curve by plotting peak area of each standard versus concentration (mg/L) of organic carbon standards. The relationship between peak ·

_____
*Oceanographics International Corp. or equivalent.
†Willems Polytron PT-105T, Brinkman Inst., or equivalent.

area and carbon concentration is curvilinear. Define operating curves each day samples are analyzed.

Report nonpurgeable organic carbon concentration (NPOC) as follows: 0.1 mg/L to 0.9 mg/L, one significant figure; 1.0 mg/L and above, two significant figures.

Number of Replicates	Mean mg/L	Relative Standard Deviation %
9	2.2	5.9
10	5.3	2.8
10	9.9	1.1
10	38.0	3.7

6. Precision and Bias

Multiple determinations of four different concentrations of aqueous potassium acid phthalate samples at 2.00, 5.00, 10.0, and 40.0 mg carbon/L resulted in mean values of 2.2, 5.3, 9.9, and 38 mg/L and standard deviations of 0.13, 0.15, 0.11, and 1.4, respectively.

Precision also may be expressed in terms of percent relative standard deviation as follows:

7. Bibliography

WILLIAMS, P.M. 1969. The determination of dissolved organic carbon in seawater: A comparison of two methods. *Limnol. Oceanogr.* 14:297.

OCEANOGRAPHY INTERNATIONAL CORP. 1970. The total carbon system operating manual. College Station, Tex.

MACKINNON, M.D. 1981. The measurement of organic carbon in seawater. *In* E.K. Duursma & R. Dawson, eds. Marine Organic Chemistry: Evolution, Composition, Interactions and Chemistry of Organic Matter in Seawater. Elsevier Scientific Publishing Co., New York, N.Y.

5320 DISSOLVED ORGANIC HALOGEN*

5320 A. Introduction

Dissolved organic halogen (DOX) is a measurement used to estimate the total quantity of dissolved halogenated organic material in a water sample. This is similar to previous literature references to "TOX." The presence of halogenated organic molecules is indicative of synthetic chemical contamination. Halogenated compounds that contribute to a DOX result include, but are not limited to: the trihalomethanes (THMs); organic solvents such as trichloroethene, tetrachloroethene, and other halogenated alkanes and alkenes; chlorinated and brominated pesticides and herbicides; polychlorinated biphenyls (PCBs); chlorinated aromatics such as hexachloroben-

zene and 2,4-dichlorophenol; and high-molecular-weight, partially chlorinated aquatic humic substances. Compound-specific methods such as gas chromatography typically are more sensitive than DOX measurements.

The adsorption-pyrolysis-titrimetric method for DOX measures only the total molar amount of dissolved organically bound halogen retained on the carbon adsorbent; it yields no information about the structure or nature of the organic compounds to which the halogens are bound or about the individual halogens present. It is sensitive to organic chloride, bromide, and iodide, but does not detect fluorinated organics.

DOX measurement is an inexpensive

---

* Approved by Standard Methods Committee, 1988.

and useful method for screening large numbers of samples before specific (and often more complex) analyses; for extensive field surveying for pollution by certain classes of synthetic organic compounds in natural waters; for mapping the extent of organohalide contamination in groundwater; for monitoring the breakthrough of some synthetic organic compounds in water treatment processes; and for estimating the level of formation of chlorinated organic by-products after disinfection with chlorine. When used as a screening tool, a large positive (i.e., above background measurements) DOX test result indicates the need for identifying and quantifying specific substances. In saline or brackish waters the high inorganic halogen concentrations interfere. The possibility of overestimating DOX concentration because of inorganic halide interference always should be considered when interpreting results.

5320 B. Adsorption-Pyrolysis-Titrimetric Method

1. General Discussion

a. Principle: The method consists of four processes. First, dissolved organic material is separated from inorganic halides and concentrated from aqueous solution by adsorption onto activated carbon. Second, inorganic halides present on the activated carbon are removed by competitive displacement by nitrate ions. Third, the activated carbon with adsorbed organic material is introduced into a furnace that pyrolyzes organic carbon to carbon dioxide (CO_2) and the bound halogens to hydrogen halide (HX). Fourth, the HX is transported in a carrier gas stream to a microcoulometric titration cell where the amount of halide is quantified by measuring the current produced by silver-ion precipitation of the halides. The microcoulometric detector operates by maintaining a constant silver-ion concentration in a titration cell. An electric potential is applied to a solid silver electrode to produce silver ions in the cell solution. As hydrogen halide from the pyrolysis furnace enters the cell in the carrier gas, it is partitioned into the acetic acid solution where it precipitates as silver halide. The current that is produced is integrated over the period of the pyrolysis. The integrated area under the curve is proportional to the number of moles of halogen recovered. The mass concentration of organic halides is reported as an equivalent concentration of organically bound chloride in micrograms per liter.

When a sample is purged with inert gas before activated carbon adsorption, analysis of that sample determines the nonpurgeable dissolved organic halogen (NPDOX) fraction of DOX. The purgeable organic halogen concentration (POX) may be calculated by subtracting the NPDOX value from the DOX value. Alternatively, the POX fraction may be determined directly by purging the sample with carrier gas and introducing that gas stream and the volatilized organics directly into the pyrolysis furnace. From the analysis of POX and DOX, NPDOX may be determined by difference.

b. Interferences: The method is applicable only to aqueous samples free of particulate matter. Inorganic substances such as chloride, chlorite, chlorate, bromide, and iodide will adsorb on activated carbon to an extent dependent on their original concentration in the aqueous solution and the volume of sample adsorbed. Positive interference will result if inorganic halides are not removed. Treating the activated carbon with a concentrated aqueous solution of nitrate ion causes competitive desorption from the activated carbon of inorganic halide species and washes inorganic halides from other surfaces. However, if the inorganic halide concentration is greater than 20 000 times the concentration of organic halides, the DOX results may be affected significantly. In general, this procedure may not be applicable to samples with inorganic halide concentrations above 500 mg Cl⁻/L, based on carbon quality testing results. Therefore, consider both the results of mineral analysis for inorganic halides and the results of the carbon quality test (see ¶ 5, below) when interpreting results.

Halogenated organic compounds that are weakly adsorbed on activated carbon are recovered only partially. These include certain alcohols and acids (e.g., chloroethanol), and such compounds as chloroacetic acid, that can be removed from activated carbon by the nitrate ion wash. However, for most halogenated organic molecules, recovery is very good; the activated carbon adsorbable organic halide (CAOX) therefore is a good estimate of true DOX.

Failure to acidify samples with nitric acid may result in reduced adsorption efficiency for some halogenated organic compounds and may intensify the inorganic halide interference. Further, if the water contains residual chlorine, reduce it before adsorption to eliminate positive interference resulting from continued chlorination reactions with organic compounds adsorbed on the activated carbon surface or with the activated carbon surface itself. The sulfite dechlorinating agent may cause decomposition of a small fraction of the DOX; do not add in great excess.

Highly volatile components of the POX fraction may be lost during sampling, shipment, sample storage, sample handling, and sample preparation, or during sample adsorption. A laboratory quality-control program to insure sample integrity from time of sampling until analysis is vital. During sample filtration for the analysis of samples containing undissolved solids major losses of POX are to be expected. Therefore, analyze for POX before sample filtration and analyze for NPDOX after filtration; the sum of POX and NPDOX is the true DOX. In preparing samples for DOX analysis, process a blank and a standard solution to determine effect of this procedure on DOX measurement. If an insignificant loss of POX occurs during the removal of particulate matter by filtration, DOX may be measured directly.

Granular activated carbon used to concentrate organic material from the sample is a major source of variability in the analysis and it has a dramatic effect on the minimum detectable concentration. Ideally, activated carbon should have a low halide content, readily release adsorbed inorganic halides on nitrate washing, be homogeneous, and readily adsorb *all* organic halide compounds even in the presence of large excesses of other organic material. An essential element of quality control for DOX requires testing and monitoring of activated carbon (see ¶ 5 below). Nonhomogeneous activated carbon or activated carbon with a high background value affects the method reliability at low concentrations of DOX. A high and/or variable blank value raises the minimum detectable concentration. Random positive bias, in part due to the ease of carbon contamination during use, necessitates analyzing duplicates of each sample. Because activated carbon from different sources may vary widely in the ease of releasing inorganic halides, test for this quality before using activated carbon. Proper quantification also may be affected by the adsorptive capacity of the activated carbon. If excessive organic loading occurs, some DOX may break through and not be recovered. For this reason, make serial adsorptions of each sample portion and individual analyses.

c. Sampling and storage: Collect and store samples in amber glass bottles with TFE-lined caps. If amber bottles are not available, store samples in the dark. To prepare sample bottles acid wash, rinse with deionized water, seal with aluminum foil, and bake at 400°C for at least 1 h. If bottle blanks without baking show no detectable DOX, baking may be omitted. Wash septa with detergent, rinse repeatedly in organic-free, deionized water, wrap in aluminum foil, and bake for 1 h at 100°C. Preferably use thick silicone rubber-backed TFE septa and open ring caps to produce a positive seal that prevents loss of POX and contamination. Store sealed sample bottles in a clean environment until use. Completely fill sample bottles but take care not to volatilize any organic halogen compounds. Preserve samples that cannot be analyzed promptly by acidifying with conc nitric acid to pH 2. Refrigerate samples at 4°C with minimal exposure to light. Reduce any residual chlorine by adding sodium sulfite crystals (minimum: 5 mg/L). NOTE: Some organic chloramines are not completely dechlorinated by sodium sulfite, particularly at pH > 7. This may affect reported concentrations.[1] Analyze all samples within 14 d.

d. Minimum detectable concentration: For nonsaline waters free of particulate matter, 5 μg organic Cl^-/L is considered a typical minimum detection limit. The minimum detectable concentration may be influenced by the analytical repeatability, equipment used, activated carbon quality, and the operator. Determine the detection limit for each procedure, instrument, and operator.

2. Apparatus

a. Adsorption assembly, including gas-tight sample reservoir, activated carbon-packed adsorption columns, column housings, and nitrate solution reservoir. In particular, note the following:

1) *Noncombustible insulating material (microcolumn method only):* Form into plugs to hold activated carbon in columns. CAUTION: *Do not touch with fingers.*

2) *Activated carbon columns (microcolumn method only):* Pack 40 ± 5 mg activated carbon (¶ 3*k*) into dry glass tubing approximately 2 to 3 mm ID × 6 mm OD × 40 to 50 mm long. CAUTION: *Protect these columns from all sources of halogenated organic vapors.* Clean glass tubes before use with a small-diameter pipe cleaner to remove residual carbon, then soak in chromate cleaning solution for 15 min and dry at 400°C. Rinse between steps with deionized water.

b. Analyzer assembly, including carrier gas source, boat sampler, and pyrolysis furnace, that can oxidatively pyrolyze halogenated organics at a temperature of 800°C ± 25°C to produce hydrogen halides and deliver them to the titration cell with a minimum overall efficiency of 90% for 2,4,6-trichlorophenol; including a microcoulometric titration system with integrator, digital display, and chart recorder connection; including (optional) purging apparatus (see Figure 5320:1).

c. Chart recorder.

3. Reagents and Materials

Use ACS reagent-grade chemicals. Other grades may be used if it can be demonstrated that the reagent is of sufficiently high purity to permit its use without lessening accuracy of the determination.

a. Carbon dioxide, argon, or nitrogen, as recommended by the equipment manufacturer, purity 99.99%.

b. Oxygen, purity 99.99%.

c. Aqueous acetic acid, 70 to 85%, as recommended by the equipment manufacturer.

d. Sodium chloride standard, NaCl: Dissolve 0.1648 g NaCl and dilute to 100 mL with reagent water; 1 μL = 1 μg Cl^-.

e. Ammonium   chloride   standard,

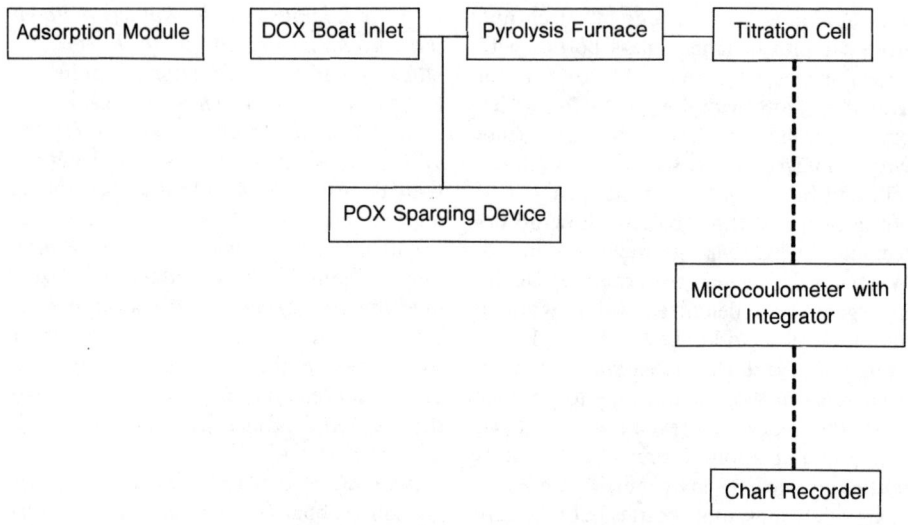

Figure 5230:1 DOX analysis system.

NH₄Cl: Dissolve 0.1509 g NH₄Cl and dilute to 100 mL with reagent water; 1 µL = 1 µg Cl⁻.

f. Trichlorophenol stock solution: Dissolve 1.856 g trichlorophenol and dilute to 100 mL with methanol; 1 µL = 10 µg Cl⁻.

g. Trichlorophenol standard solution: Make a 1:20 dilution of the trichlorophenol stock solution with methanol; 1 µL = 0.5 µg Cl⁻.

h. Chloroform standard solution, CHCl₃: Dilute 100 mg CHCl₃ to 100 mL with methanol; 1 µL = 1 µg CHCl₃.

i. Blank standard: Use reagent water. Reagent water preferably is carbon-filtered, deionized water that has been heated and purged.

j. Nitrate wash solution, 0.08N: Dilute 8.2 g KNO₃ to 1000 mL with reagent water. Adjust to pH 2 with HNO₃. 1 L = 5000 mg NO₃⁻.

k. Activated carbon, 100 to 200 mesh: Ideally use activated carbon having a very low apparent halide background that readily releases adsorbed inorganic halides on nitrate washing, and reliably adsorbs organic halides in the presence of a large excess or other organics.* See ¶ 5 below for preparation and evaluation of activated carbon. CAUTION: *Protect activated carbon from contact with halogenated organic vapors.*

l. Sodium sulfite, Na₂SO₃, crystals.

m. Nitric acid, HNO₃, conc.

n. Humic acid.†

4. Procedure

Use either the microcolumn (4*a*) or batch adsorption (4*b*) method. The microcolumn method utilizes small glass columns packed with activated carbon through which the sample is passed under positive pressure to adsorb the organic halogen compounds. The batch adsorption

---

* Westvaco or Calgon Filtrasorb 400 or equivalent.

† Aldrich Chemical Co. or equivalent.

method directly measures POX. A small quantity of activated carbon then is added to the NPDOX portion. After stirring, activated carbon is removed by filtration and analyzed.

   a. Microcolumn procedure:

   1) Apparatus setup—Adjust equipment in accordance with the manufacturer's instructions. Include several injections of NaCl solution directly into the titration cell [¶ 5cl)] as a microcoulometer/titration cell check at the start of each day.

   2) Sample pretreatment for DOX analysis—Adjust sample pH to 2 with HNO₃. If the samples contain undissolved solids, filter through a glass-fiber filter (other means of removing particulate matter may be used, if it can be demonstrated that they do not cause significant interferences). Also filter a blank and standard. Analyze these to determine the contribution of filtration to the organic halogen measurement. Vacuum filtration will cause some loss of volatile organic halogen. Analyze for POX before filtration and NPDOX after filtration, unless it is shown that POX losses during filtration are insignificant.

   3) POX determination (optional)—Select sample volume by comparing expected POX value (if known) with the optimum instrument range. Using a gastight syringe, inject sample through the septum into the purge vessel and purge as recommended by the equipment manufacturer. Carefully control gas flow rate, sample temperature, and purging time.

   4) Sample adsorption—Transfer a representative portion of sample to the cleaned sample reservoir with two activated carbon adsorption columns in series attached by the column housings to the reservoir outlet. Seal the reservoir. Adjust to produce a flow rate of about 3 mL/min. When the desired volume has been processed, stop the flow, detach the activated carbon housings and columns, and rinse the sample reservoir twice with reagent-grade water. Vary volume processed to produce optimum quan-

tities of adsorbed DOX on the columns. Suggested volumes are as follows:

Volume Processed mL	Instrument Optimum Range µg Cl⁻	Conc of DOX in Waters µg/L
100	0.5–50	5–500
50	12.5–50	250–1000
25	12.5–50	500–2000

   If possible, avoid using volumes greater than 100 mL because the maximum adsorptive capacity of the activated carbon may be exceeded, leading to adsorbate breakthrough and loss of DOX. Larger sample volumes processed lead to an increased quantity of inorganic halide accumulated on the activated carbon and may result in a positive interference. Do not use a sample less than 25 mL to minimize volumetric errors. For samples exceeding 2000 µg DOX/L dilute before adsorption. Protect columns from the atmosphere until DOX is determined.

   5) Inorganic halide removal—Attach columns through which sample has been processed in series to the nitrate wash reservoir and pass 2 to 5 mL NO₃⁻ solution through the columns at a rate of approximately 1 mL/min.

   6) DOX determination—After concentrating sample on activated carbon and removing inorganic halogens by nitrate washing, pyrolyze contents of each microcolumn and determine organic halogen content. Remove top glass microcolumn from the column housing, taking care not to contaminate the sample with inorganic halides. Using a clean ejector rod, eject the activated carbon and noncombustible insulating material plugs into the sample boat. Prepare sample boat during the preceding 4 h by heating at 400 to 800°C for at least 4 min in an oxygen-rich atmosphere (i.e., in the pyrolysis furnace). Remove residual ash. Place ejector rod on the plug of the effluent end of the carbon microcolumn

and place the influent end of the carbon microcolumn in the quartz boat first. Seal sample inlet tube and let instrument stabilize. After NO_3^- wash avoid contact with inorganic halides such as the chloride from perspiration on the hands or contaminated bench surfaces. Wash hands frequently, especially after eating, and clean work area with deionized water often.

Pyrolyze the activated carbon and determine halide content. Repeat for each microcolumn. Check for excess breakthrough (¶ 5b) and repeat analysis as necessary.

7) Replicates—When DOX determination is used strictly as a screening tool, replicates are not necessary. Otherwise, repeat steps 4, 5 and 6 using a different dilution of the same sample. Use dilutions that differ so that the concentration ratios are either less than 0.7 or greater than 1.4.

8) Blanks—Analyze one method blank [¶ 5e2)] with each set of eight samples. Analyze no fewer than two blanks each day.

9) Preparation and analysis of calibration standard—Run several calibration standards daily in accordance with ¶ 5c3) for POX analysis or ¶ 5c5) for microcolumn-adsorption DOX analysis. Accompany by a suitable blank [¶ 5e3) or ¶ 5e4)]. Be certain that analytical conditions and procedures (e.g., purging temperature) are the same for the analysis of calibration standards as for the analysis of samples.

b. Batch adsorption procedure:

1) Apparatus setup—Adjust equipment in accordance with the manufacturer's instructions.

2) Sample pretreatment—Adjust sample pH to 2 with conc HNO_3.

3) POX determination—Select sample volume by comparing expected POX value (if known) with the optimum instrument range. Using a gas-tight syringe, inject sample through the septum into the purge vessel and purge as recommended by the equipment manufacturer. To compare POX values from different laboratories

standardized purging is essential. Adjust instrument temperature, purge rate, and purge time to just produce complete purging of bromoform using a bromoform standard. Take replicate samples from different sample bottles to minimize volatilization losses.

4) NPDOX determination—Prepare carbon suspension by adding high-quality activated carbon to high-purity, deionized, granular activated carbon (GAC)-treated water to produce a uniform suspension of 10 mg carbon/mL. To an erlenmeyer flask, transfer prepurged sample of optimum size from a purging flask standardized in the same manner as the instrument's purging vessel. Add 20 mg activated carbon (2 mL carbon suspension). Using a high-speed mixer (20 000 rpm), stir for 45 min in an organohalide vapor-free environment. Filter through a membrane filter under vacuum and collect filtrate. Remove flask containing filtrate. Wash carbon cake and filter with 10 mL NO_3^- wash solution. Add portions of wash solution serially to keep activated carbon and NO_3^- solution in contact for 15 min. Using clean instruments, transfer carbon cake and membrane filter to pyrolysis unit sample boat. Let instrument stabilize, pyrolyze, and determine the halide content of the first serial filter.

Add 20 mg more activated carbon to filtrate in erlenmeyer flask. Repeat carbon mixing, filtering, and washing procedures. Pyrolyze and determine halide content of second serial filter. If the second value is greater than 10% of the total value (first plus second), perform the NPDOX determination on an additional sample portion.

5. Quality Control

a. Carbon quality: Prepare activated carbon by milling and sieving high-quality activated carbon. Use only 100- to 200-mesh carbon in the microcolumn method. During preparation, take care not to expose the activated carbon to organic vapors. Use of a clean room is helpful. Prepare only small quantities (a month's supply or less) at one time. Discard the activated carbon if its DOX background concentration has increased significantly from the time of preparation or if the background is greater than 1 μg apparent organic Cl^-/40 mg activated carbon. Uniformity of activated carbon is important; therefore, after sieving small portions, combine and mix thoroughly. Transfer representative portions to clean glass bottles with ground-glass stoppers or with rubber-backed TFE septa and open ring caps. Store bottles in a gas-purged, evacuated, sealed desiccator.

Test each newly prepared batch of activated carbon to insure adequate quality before use. Use only carbon meeting the guidelines outlined below.

1) Check carbon particle size by applying deionized water to two 40-mg activated carbon microcolumns. If flow rate is significantly less than 3 mL/min, resieve carbon to remove excess fines.

2) Analyze a pair of method blanks, ¶ 5e2). Reject carbon if the apparent organic halogen exceeds 1.2 μg/40 mg activated carbon.

If the activated carbon originated from a previously untested batch from a commercial supplier, test it for adsorption efficiency and inorganic halide rejection.

3) Adsorb replicate 100-mL portions of solutions containing 100, 500, and 1000 mg inorganic Cl^-/L deionized water. Wash with nitrate solution and analyze. The apparent organic halogen yield should not increase by more than 0.50 μg over the value determined in 2) above. A greater increase indicates significant interference at that concentration.

4) Prepare an adsorption test standard solution in deionized water to contain 10 mg organic carbon (as humic acids or equivalent)/L and an organic halide concentration of 100 μg organic chloride (from 2,4,6-trichlorophenol)/L. Prepare a blank solution containing only the 10 mg organic

carbon. Adsorb, nitrate wash, and analyze replicate 100-mL portions of these solutions. Directly inject 10 μg organic chloride (from 2,4,6-trichlorophenol) onto nitrate-washed 40-mg activated carbon microcolumns, in replicate. Analyze replicate method blanks. Recovery of the *aqueous* organic chloride should be $\geq$ 90% on the first microcolumn or filter and 5% or less on the second. Total recovery of organic chloride from water should be 95% or more (after subtraction of the organic carbon blank) of the amount applied.

b. Serial adsorption: Each aqueous standard and sample is serially adsorbed on activated carbon in both procedures given above. Of the net organic halide, 90% or more should be adsorbed on the first activated carbon portion and the remaining 10% or less on the second. If, upon separate analysis of the two serial activated carbon portions, the second shows more than 10% of the net (after subtracting the method blank), reanalyze sample. Inorganic halogen interference or organic breakthrough are the most common reasons for a high second activated carbon value. Sample dilution before adsorption may improve recovery on the first activated carbon in series, but the minimum detectable concentration will be affected.

c. Standards: The standards used in routine analysis, quality control testing, and isolating specific causes during corrective maintenance include:

1) Sodium chloride standard (¶ 3d)— Use to check functioning of the titration cell and microcoulometer by injecting directly into the acetic acid solution of the titration cell. By examining the height and shape of the peak produced on the chart recorder and from the integrated value, problems associated with the cell and coulometer may be isolated. Use this standard at startup each day and after cell cleaning throughout the day. At daily startup consecutive duplicates should be within 3% of the historical mean. Depending on sample

loading and number of analyses performed, it may be necessary to clean the titration cell several times per day. After cleaning, cell performance may be very unstable; therefore, inject a single NaCl standard before analyzing an instrument calibration standard [see ¶ 4) below]. Do *not* introduce NaCl standards into the pyrolysis furnace by application to the sample boat.

2) Ammonium chloride standard (¶ 3e)—Apply this standard to the sample boat to check for loss of halide in the pyrolysis furnace and entrance of the titration cell. When injection of a NaCl standard indicates proper titration cell and microcoulometer function but the recovery of the calibration standard is poor, suspect either poor conversion of organic chloride to hydrogen chloride or loss of hydrogen halide after conversion but before partitioning into the cell solution. To isolate the possible loss of hydrogen halides inject NH_4Cl standard directly onto the quartz sample boat. Recovery should be better than 95%, with a single peak of uniform shape produced. Use only a new quartz sample boat free of any residue; an encrusted boat dramatically reduces recovery. Use this standard for corrective maintenance problem isolation but not for routine analyses.

3) Purgeable organic halide calibration standards—For the POX analysis use aqueous chloroform solutions for instrument calibration. Also for POX analysis use an aqueous bromoform standard to insure acceptable purging conditions. Develop a standard curve over the dynamic range of the microcoulometer and check daily as in ¶ 5c5). Recovery of chloroform and bromoform should exceed 90% and 80%, respectively.

4) Instrument calibration standard—Direct injection of trichlorophenol working standard onto the nitrate-washed method blank in concentrations over the working range of the instrument determines linearity and calibration of the analyzer module. After checking for proper microcoulometer

function by injecting NaCl standard, pyrolyze duplicate instrument calibration standards and then duplicate method blanks. The net response to the calibration standards should be within 3% of the calibration curve value. If not, check for loss of halide in the pyrolysis furnace using the ammonium chloride standard [¶ 5c2)].

5) Nonvolatile organic halide calibration standards—Develop a standard curve by analyzing aqueous solutions of 2,4,6-trichlorophenol or another appropriate halogenated organic compound over the dynamic range of the microcoulometer. This dynamic range typically is from 0.5 to 50 µg chloride, but will vary between microcoulometers and titration cells. Because of the limited throughput of the DOX procedure, use single-instrument calibration standards at 50, 30, 15, 10, 5, 2.5, and 0.5 µg organic chloride to construct a standard curve after changes in an instrument's configuration, such as replacement of a titration cell or major instrument maintenance. Daily, analyze additional replicates of several nonvolatile organic halide calibration standards to insure reproducibility, linearity, and proper function of the instrumentation and procedures. Select standard concentrations in the range of samples to be analyzed that day. When sample filtration is used to remove particulate matter, also use this pretreatment with the calibration standards. If DOX recovery is less than 95%, analyze a set of instrument calibration standards [¶ 5c4)].

d. Standard addition recovery: During routine analyses, ideally make standard additions to every tenth sample. Where the compounds constituting the DOX are known, use standards of these compounds. Where the compounds constituting the DOX are wholly or partially unknown (e.g., halogenated humic acids), use standards reflecting the relative abundance of the halogens, the molecular size, and the volatility of the halogenated compounds presumed to be present. Recovery of 90% or more of the added amount indicates that the analyses are in control.

e. *Blanks:* High precision and accuracy of the background or blank value is important to the accurate measurement of DOX. Make blank measurements daily. Blanks that may be required are:

1) System blank—Analyze organic-free reagent water. The blank should have less than the minimum detectable concentration. Use this blank to insure that the equipment and procedures are not contributing to the DOX.

2) Method blank—Analyze activated carbon that has been nitrate-washed. Analyze duplicate method blanks daily before sample analysis and after each eight sample pyrolyses.

3) Standard blank—Analyze reagent water to determine the blank for standards.

4) Purgeable organic halogen blank—Analyze organic-free, pre-purged, reagent water to determine the POX blank.

6. Calculation

Calculate the net organic halide content as chloride (C_4) of each replicate of each sample and standard:

$$C_4 = \frac{C_1 - C_3 + C_2 - C_3}{V}$$

where:

C_1 = organic halide as Cl^- on the first activated carbon column or activated carbon cake, μg,

C_2 = organic halide as Cl^- on the second activated carbon column or activated carbon cake, μg,

C_3 = mean of method blanks on the same day and same instrument, μg X as Cl^-,

C_4 = uncorrected net organic halide as Cl^- of absorbed sample, μg organic halide as Cl^-/L, and

V = volume of sample absorbed, L.

If $C_2 \leq C_3$, then use:

$$C_4 = \frac{C_1 - C_3}{V}$$

If applicable, calculate net purgeable organic halide as Cl^- content (P_3):

$$P_3 = \frac{P_1 - P_3}{V}$$

where:

P_1 = sample purgeable organic halide as Cl^-, μg,

P_2 = blank purgeable organic halide as Cl^-, μg,

P_3 = uncorrected net purgeable organic halide as Cl^-, μg X as Cl^-/L, and

V = volume of sample or standard purged, L.

Report sample results and percent recovery of the corresponding calibration standards [¶ 5c3) or ¶ 5c5)]. Also report the calibration standard curve if it is significantly nonlinear.

7. Precision and Bias

Precision and bias depend on specific procedures, equipment, and analyst. Develop and routinely update precision and bias data for each procedure, each instrument configuration, and each analyst. Table 5230:I shows sample calculations of precision expressed as the standard deviation among replicates and bias in the recovery of 2,4,6-trichlorophenol.

8. Reference

1. STANBORO, W.D. & M.J. LENKEVICH. 1982. Slowly dechlorinated organic chloramines. *Science* 215:967.

9. Bibliography

KUHN, W. 1974. Thesis, Univ. Karlsruhe, West Germany.

KUHN, W., F. FUCHS & H. SONTHEIMER. 1977. Untersuchungen zur Bestimmung des organisch gebundenen Chlors mit Hilfe eines neuartigen Anreicherungsverfahrens. *Z. Wasser-Abwasser Forsch.* 10:6:162.

DRESSMAN, R.C., B.A. NAJAR & R. REDZI-

TABLE 5320:I. INTRALABORATORY, SINGLE-OPERATOR, DISSOLVED ORGANIC HALOGEN (MICROCOLUMN PROCEDURE)—PRECISION AND BIAS DATA

Characteristic of Analysis	Tap Water	Tap Water + 43.5 μg Organic Chloride	Ground Water (50:1)	Wastewater	Waste-water + 1000 μg Organic Chloride
Concentration determined, μg Cl⁻/L:					
Replicate 1	38.5	89.0	123.6	186.0	1178.0
Replicate 2	36.7	90.9	124.8	195.0	1183.0
Replicate 3	43.1	88.4	125.2	195.0	1185.5
Replicate 4	35.9	90.1	123.3	204.0	1196.5
Replicate 5	41.1	91.7	125.3	185.0	1183.0
Replicate 6	48.5	93.0	127.0	236.5	1204.0
Replicate 7	52.8	97.0	123.5	204.0	1138.0
Mean, μg Cl⁻/L	42.37	91.5	124.7	200.8	1181.1
Standard deviation:					
μg Cl⁻/L	±6.29	±3.0	±1.3	±17.47	±21.04
%	15	3	1	9	2
Value of blank + standard addition, μg Cl⁻/L	—	85.87	—	—	1200.8
Recovery, %	—	107	—	—	98
Error, %	—	7	—	—	2

KOWSKI. 1979. The analysis of organohalides (OX) in water as a group parameter. Proc. 7th Annual Water Quality Technology Conf., Philadelphia, Pa. American Water Works Assoc., Denver, Colo.

TAKAHASHI, Y., et al. 1980. Measurement of total organic halides (TOX) and purgeable organic halides (POX) in water using carbon adsorption and microcoulometric detection. Proc. Symp. on Chemistry and Chemical Analysis of Water and Waste Water Intended for Reuse, Houston, Tex. American Chemical Soc., Washington, D.C.

DRESSMAN, R. C. 1980. Total Organic Halide, Method 450.1—Interim. Drinking Water Research Div., Municipal Environmental Research Lab., U.S. Environmental Protection Agency, Cincinnati, Ohio.

JEKEL, M.R. & P.V. ROBERTS. 1980. Total organic halogen as a parameter for the characterization of reclaimed waters: measurement, occurence, formation, and removal. Environ. Sci. Technol. 14:970.

DRESSMAN, R.C. & A. STEVENS. 1983. Analysis of organohalides in water—An evaluation update. J. Amer. Water Works Assoc. 75:431.

5510   AQUATIC HUMIC SUBSTANCES (PROPOSED)*

5510 A.   Introduction

1. General Discussion

Aquatic humic substances (AHS) are heterogeneous, yellow to black, organic materials that include most of the naturally occurring dissolved organic matter in water. Aquatic humic substances have been shown to produce trihalomethanes (THMs) on chlorination and to affect the transport and fate of other organic and inorganic species through partition/adsorption, catalytic, and photolytic reactions.

Humic substances, the major fraction of soil organic matter, are mixtures; their chemical composition is poorly understood. They have been classified into three fractions based on water "solubility"†: humin is the fraction not soluble in water at any pH value; humic acid is not soluble under acidic conditions (pH < 2) but becomes soluble at higher pH; and fulvic acid is soluble at all pH conditions.

AHS have the solubility characteristics of fulvic acids but they should not be referred to as such unless they have been fractionated by precipitation at pH < 2. Avoid using the terms "humic acid" and "tannic acid" to describe AHS because they represent other classifications of natural organic materials.

The heterogeneity of AHS requires an operational definition. Isolation by the methods included herein most likely will be incomplete and compounds that are not AHS may be isolated incidentally. Users of these methods are cautioned in the interpretation of results; the bibliography suggests several sources for more information.

Measurement of AHS begins by separation of the sample into dissolved (containing AHS) and particulate organic carbon fractions. Although there is no distinct size that separates these two groups, 0.45 μm is used as the compromise between acceptable flow rate and rejection of small colloidal materials. Low-pressure liquid chromatography serves to concentrate these materials and to isolate them from interfering substances. AHS are quantified by measuring dissolved organic carbon (DOC), Method 5310.

2. Selection of Method

Concentration/isolation of AHS may be achieved by sorption on the nonpolar resin XAD-8 (Method 5510C) or by anion-exchange on diethylaminoethyl (DEAE) cellulose (Method 5510B). In a collaborative study with seven laboratories using deionized water fortified with about 10 mg AHS/L (previously isolated with XAD), the DEAE method gave better recoveries. Nevertheless, the XAD method has been used extensively; refer to the discussions of interferences and minimum detectable concentrations to assist in method selection. Both methods require further quality control development.

3. Bibliography

CHRISTMAN, R. F. & E. T. GJESSING, eds. 1983. Aquatic and Terrestrial Humic Substances. Ann Arbor Science, Ann Arbor, Mich.

MILES, C. J., J. R. TUSCHALL, JR. & P. L. BREZONIK. 1983. Isolation of aquatic humus with diethylaminoethyl cellulose. *Anal. Chem.* 55:410.

---

* Approved by Standard Methods Committee, 1988.
† "Solubility" is here used as a general description of whether or not the material can be uniformly dispersed in an aqueous phase rather than as an expression of equilibrium between a pure solute and its aqueous solution.

THURMAN, E. M. 1984. Determination of aquatic humic substances in natural waters. *In* E. L. Meyer, ed. Selected Papers in the Hydrologic Sciences. U.S. Geological Survey Water Supply Paper 2262, p. 47.

TUSCHALL, J. R., JR. & G. GEORGE. 1984. Selective Isolation of Dissolved Organic Matter from Aquatic Systems. UILUWRC-84-190, Water Resources Center, Univ. Illinois, Urbana.

AIKEN, G. R., D. M. MCKNIGHT, R. L. WERSHAW & P. L. MACCARTHY, eds. 1985. Humic Substances in Soil, Sediment, and Water. John Wiley & Sons, New York, N.Y.

5510 B.   Diethylaminoethyl (DEAE) Method

1. General Discussion

a. Principle: AHS are concentrated by column chromatography on diethylaminoethyl (DEAE) cellulose and measured as dissolved organic carbon (DOC). AHS are weak organic acids that bind to anion-exchange materials, such as DEAE cellulose, at neutral pH values. The method is based on the assumption that AHS are the major dissolved organic acids present.

b. Interferences: Any carbonaceous nonhumic materials that are concentrated and isolated by the chromatographic method will interfere (false positive response). Substances that have been shown to interfere include fatty acids, phenols, surfactants, proteinaceous materials, and DOC leached from cellulose.

c. Minimum detectable concentration: Estimated limit of detection is 1.1 mg/L using a 50-mL water sample. The detection limit can be decreased by increasing sample volume. The major limitation is blank contamination.

d. Standard substance: Eliminate documentation of false negatives by analyses of a sample of known humic concentration at regular intervals (at least once per batch of samples).

2. Apparatus

a. Membrane filtration apparatus: Use an all-glass filtering device and 0.45-μm silver membrane filters. Consult manufacturer's specifications for filter details. Do not use filters that sorb AHS or are contaminated with detergents and other organic material.

b. Glass column, approximately 1 $\times$ 20 cm with silanized glass wool.

c. Dye-impregnated paper or strips for approximate pH measurements.

d. Organic carbon analyzer capable of measuring concentrations as low as 0.1 mg/L (see Section 5310).

*e. Buchner funnel and filter paper.**

3. Reagents

a. Water, DOC-free: Preferably use activated-carbon-filtered, redistilled water.

b. DEAE cellulose, exchange capacity 0.22–1.0 meq/g.† Do not use high-exchange-capacity cellulose, which may decrease recovery of AHS. Take care not to overload low-exchange-capacity cellulose.

c. Hydrochloric acid, HCl, 0.1N: Add 8.3 mL conc HCl to 1000 mL water.

d. Hydrochloric acid, HCl, 0.5N: Add 41.5 mL conc HCl to 1000 mL water.

e. Sodium hydroxide, NaOH, 0.1N: Dissolve 4.0 g NaOH in 1000 mL water.

---

* Whatman No. 1 or equivalent.
† Whatman pre-swollen microgranular DE 52 or DE 51, or equivalent.

f. Sodium hydroxide, NaOH, 0.5*N:* Dissolve 20 g NaOH in 1000 mL water.

g. DOC standards: See Section 5310.

h. Potassium chloride, KCl, 0.01*N:* Add 0.75 g KCl to 1000 mL water.

i. Phosphoric acid, H_3PO_4, conc.

4. Procedure

a. Sample concentration and preservation: AHS are sensitive to biodegradation and photodegradation. Collect and store samples in organic-free glass containers. Filter at least duplicate portions through a 0.45-μm silver membrane filter as soon after collection as possible. Store samples in the dark at 4°C.

Use care to avoid overloading chromatographic columns and losing AHS. A rough guideline for sample volume selection is as follows:

Sample DOC *mg/L*	Sample Volume *mL*
0–2	250
2–10	50
10–50	25

b. Preparation of DEAE cellulose: Add 70 g DEAE cellulose to 1000 mL 0.5*N* HCl and stir gently for 1 h. Rinse cellulose with water in a Buchner funnel until funnel effluent pH is about 4. Resuspend DEAE in 1000 mL 0.5*N* NaOH and stir for 1 h. Rinse in a Buchner funnel with water until pH is about 6. Remove fines by suspending the treated DEAE in a 1000-mL graduated cylinder filled with water. Let mixture stand undisturbed for 1 h, then decant and discard the supernatant. Repeat removal of fines. Filter remaining DEAE using a Buchner funnel and store in a refrigerated glass container. Avoid prolonged storage, which may lead to microbial contamination.

c. Chromatography: Add 10 mL water to about 1 g DEAE to make a slurry. Carefully pipet enough into a 1- × 20-cm column fitted with a small (0.5-cm) glass-wool plug to make a 1-cm-deep column bed. Avoid getting DEAE on the sides of the column. Carefully place another 0.5-cm glass-wool plug on top of the bed. Rinse column with 50 mL 0.01*N* KCl (adjusted to pH 6 with 0.1*N* HCl or NaOH) just before sample concentration.

Adjust sample to pH 6 and pass it through the column at a flow rate of about 2 mL/min. Rinse with 5 mL water (pH 6). Elute AHS by adding about 3 mL 0.1*N* NaOH to the top of the column. Start collecting column effluent when it appears colored. (This will occur after about 1 mL has passed out of the column). Collect eluate in a graduated, conical test tube until it becomes colorless (about 2 mL). Acidify with conc H_3PO_4 to a pH of 2 or less (about 2 to 3 drops) and remove dissolved carbon dioxide (inorganic carbon) by purging with nitrogen for 10 min. Avoid exposure of alkaline samples to air (i.e. acidify immediately) to minimize contamination with CO_2. Determine volume and DOC of acidified eluate.

Process two portions of water and a second portion of sample by the same procedure. Pack a fresh column of DEAE for each sample and each control (DEAE cannot be reused).

5. Calculation

Calculate the concentration of AHS as:

$$\text{AHS, mg DOC/L} = [(A - B) \times C]/D$$

where:

A = average DOC concentration of the two sample NaOH eluates, mg C/L,

B = average DOC concentration of the two control NaOH eluates, mg C/L,

C = volume of eluate, L, and

D = volume of sample, L.

Multiplication of AHS, mg DOC/L, by 2 converts concentration to AHS, mg/L, if it is assumed that AHS contain 50% carbon. This will be the minimum concentration of AHS because recoveries are less than 100%.

6. Precision and Bias

For seven single-operator analyses, the relative standard deviation of triplicate samples (about 10 mg/L as AHS) ranged from 2.5 to 14.4% with an average of 4.9% ($n = 7$).

For seven single-operator analyses, recoveries ranged from 59.3 to 97.3% with an average of 77.4% and a relative standard deviation of 18.1%.

5510 C.   XAD Method

1. General Discussion

a. Principle: AHS are concentrated by column chromatography on XAD resin and measured as dissolved organic carbon (DOC). Acidification of AHS decreases polarity, allowing partition into the nonpolar XAD matrix. The method is based on the assumption that AHS are the major dissolved organic acids present.

b. Interferences: Any carbonaceous nonhumic materials that are concentrated and isolated by the chromatographic method will interfere. This includes fatty acids, phenols, surfactants, proteinaceous materials, and DOC leached from the resin, chromatography pump, or tubing.

c. Minimum detectable concentration: Estimated limit of detection is 1.4 mg/L using a 50-mL water sample. The detection limit can be decreased by increasing sample volume. The major limitation is blank contaminations.

2. Apparatus

See Section 5510B.2*a*, *c*, and *d*. In addition, the following are required:

a.   Glass column, 0.2 × 25 cm with silanized glass wool.

b.   Pump, with inert internal parts and tubing, capable of flow rates of 0.2 to 1.0 mL/min.*

c.   TFE tubing, 0.2 cm ID.

d. Extraction apparatus, Soxhlet.

3. Reagents

In addition to reagents *a, c, e, g,* and *i* of Section 5510B:

a. XAD resin,† approximately 250 μm size.

b. Hexane.

c. Methanol.

d. Acetonitrile.

4. Procedure

a. Sample collection and preservation: See Section 5510B.4*a*.

b. Preparation of XAD resin: Clean resin by successive washing with 0.1*N* NaOH for 5 d. Extract resin sequentially in a Soxhlet extractor with hexane, methanol, acetonitrile, and methanol, for 24 h each. Pack clean resin into a 0.2- × 25-cm glass column that has a 2-mm length of glass wool in one end. After filling, cap column with another 2-mm length of glass wool.

Wet dry column with methanol. When the air has been displaced, pump distilled water through the column until the effluent concentration of DOC decreases to 0.5 mg/L (approximately 20 bed volumes).

c. Chromatography: Preclean column with three cycles of 0.1*N* NaOH and 0.1*N* HCl just before pumping sample into col-

---

† XAD-8, Rohm and Haas, Philadelphia, Pa., or equivalent.

umn. Leave column saturated with $0.1N$ HCl. Acidify sample to pH 2.0 with concentrated HCl, and pump it onto the column at rate of 1.0 mL/min. Save column effluent for DOC analysis. Significant concentrations of DOC in the effluent can indicate that the column was overloaded and that a smaller sample volume should be used. Colored organic acids adsorb to the top of the column. Back-elute (reverse flow) the column with $0.1N$ NaOH at 0.2 mL/min and collect eluate in a graduated, conical test tube until it becomes colorless (about 2 mL). Acidify with conc H_3PO_4 to a pH of 2 or less (about 2 to 3 drops) and remove dissolved carbon dioxide (inorganic carbon) by purging with nitrogen for 10 min. Avoid exposure of alkaline samples to air (i.e. acidify immediately) to minimize contamination with CO_2. Determine volume and DOC of acidified column effluent.

After eluting and collecting AHS from the column with back-elution using $0.1N$ NaOH, continue rinsing with about 20 bed volumes of the basic solution. Rinse with water for about 20 bed volumes. Repeat the triplicate acid/base column precleaning procedure described above, then reuse the column to analyze a replicate sample. Process two portions of water by the same procedure to serve as controls.

The XAD column may be reused to analyze subsequent samples and controls if the triplicate acid/base precleaning procedure is repeated immediately before analysis of each replicate. Replace the column if recovery is poor or the resin becomes discolored.

5. Calculation

Calculate the concentration of AHS as given in 5510B.5.

6. Precision and Bias

For seven single-operator analyses, the relative standard deviation of triplicate samples (about 10 mg/L as AHS) ranged from 0.9 to 20.7% with an average of 5.4% ($n = 7$).

For seven single-operator analyses, recoveries ranged from 15.1 to 71.0% with an average of 51.6% and a relative standard deviation of 35.1%.

5520   OIL AND GREASE*

5520 A.   Introduction

In the determination of oil and grease, an absolute quantity of a specific substance is not measured. Rather, groups of substances with similar physical characteristics are determined quantitatively on the basis of their common solubility in trichlorotrifluoroethane. "Oil and grease" is any material recovered as a substance soluble in trichlorotrifluoroethane. It includes other material extracted by the solvent from an acidified sample (such as sulfur compounds, certain organic dyes, and chlorophyll) and not volatilized during the test. It is important that this limitation be understood clearly. Unlike some constituents that represent distinct chemical ele-

---

*Approved by Standard Methods Committee, 1985.

ments, ions, compounds, or groups of compounds, oils and greases are defined by the method used for their determination.

The methods presented here are suitable for biological lipids and mineral hydrocarbons. They also may be suitable for most industrial wastewaters or treated effluents containing these materials, although sample complexity may result in either low or high results because of lack of analytical specificity. The method is not applicable to measurement of low-boiling fractions that volatilize at temperatures below 70°C.

1. Significance

Certain constituents measured by the oil and grease analysis may influence wastewater treatment systems. If present in excessive amounts, they may interfere with aerobic and anaerobic biological processes and lead to decreased wastewater treatment efficiency. When discharged in wastewater or treated effluents, they may cause surface films and shoreline deposits leading to environmental degradation.

A knowledge of the quantity of oil and grease present is helpful in proper design and operation of wastewater treatment systems and also may call attention to certain treatment difficulties.

2. Selection of Method

For liquid samples, three methods are presented: the partition-gravimetric method (B), the partition-infrared method (C), and the Soxhlet method (D). Method C is designed for samples that might contain volatile hydrocarbons that otherwise would be lost in the solvent-removal operations of the gravimetric procedure. Method D is the method of choice when relatively polar, heavy petroleum fractions are present, or when the levels of nonvolatile greases may challenge the solubility limit of the solvent. For low levels of oil and grease (< 10 mg/L), Method C is the method of choice because gravimetric methods do not provide the needed precision.

Method E is a modification of the Soxhlet method and is suitable for sludges and similar materials. Method F can be used in conjunction with Methods B, C, D, or E to obtain a hydrocarbon measurement in addition to, or instead of, the oil and grease measurement. This method separates hydrocarbons from the total oil and grease on the basis of polarity.

3. Sampling and Storage

Collect a representative sample in a wide-mouth glass bottle that has been rinsed with the solvent to remove any detergent film, and acidify in the sample bottle. Collect a separate sample for an oil and grease determination and do not subdivide in the laboratory. When information is required about average grease concentration over an extended period, examine individual portions collected at prescribed time intervals to eliminate losses of grease on sampling equipment during collection of a composite sample.

In sampling sludges, take every possible precaution to obtain a representative sample. When analysis cannot be made immediately, preserve samples with 1 mL conc HCl/80 g sample. Never preserve samples with $CHCl_3$ or sodium benzoate.

5520 B.   Partition-Gravimetric Method

1. General Discussion

a. Principle: Dissolved or emulsified oil and grease is extracted from water by intimate contact with trichlorotrifluoroethane. Some extractables, especially unsaturated fats and fatty acids, oxidize readily; hence, special precautions regarding temperature and solvent vapor displacement are included to minimize this effect.

b. Interference: Trichlorotrifluoroethane has the ability to dissolve not only oil and grease but also other organic substances. No known solvent will dissolve selectively only oil and grease. Solvent removal results in the loss of short-chain hydrocarbons and simple aromatics by volatilization. Significant portions of petroleum distillates from gasoline through No. 2 fuel oil are lost in this process. In addition, heavier residuals of petroleum may contain a significant portion of materials that are not extractable with the solvent.

2. Apparatus

a. Separatory funnel, 1-L, with TFE* stopcock.
b. Distilling flask, 125-mL.
c. Water bath.
d. Filter paper, 11-cm diam.†

3. Reagents

a. Hydrochloric acid, HCl, 1 + 1.
b. Trichlorotrifluoroethane‡ (1,1,2-trichloro-1,2,2-trifluoroethane), boiling point 47°C. The solvent should leave no measurable residue on evaporation; distill if necessary. Do not use any plastic tubing to transfer solvent between containers.

c. Sodium sulfate, Na_2SO_4, anhydrous crystal.

4. Procedure

Collect about 1 L of sample and mark sample level in bottle for later determination of sample volume. Acidify to pH 2 or lower; generally, 5 mL HCl is sufficient. Transfer to a separatory funnel. Carefully rinse sample bottle with 30 mL trichlorotrifluoroethane and add solvent washings to separatory funnel. Preferably shake vigorously for 2 min. However, if it is suspected that a stable emulsion will form, shake gently for 5 to 10 min. Let layers separate. Drain solvent layer through a funnel containing solvent-moistened filter paper into a clean, tared distilling flask. If a clear solvent layer cannot be obtained, add 1 g Na_2SO_4 to the filter paper cone and slowly drain emulsified solvent onto the crystals. Add more Na_2SO_4 if necessary. Extract twice more with 30 mL solvent each but first rinse sample container with each solvent portion. Combine extracts in tared distilling flask and wash filter paper with an additional 10 to 20 mL solvent. Distill solvent from distilling flask in a water bath at 70°C. Place flask on a water bath at 70°C for 15 min and draw air through it with an applied vacuum for the final 1 min. Cool in a desiccator for 30 min and weigh.

5. Calculation

If the organic solvent is free of residue, the gain in weight of the tared distilling flask is mainly due to oil and grease. Total gain in weight, *A*, of tared flask less calculated residue, *B*, from solvent blank is the amount of oil and grease in the sample:

$$\text{mg oil and grease/L} = \frac{(A - B) \times 1000}{\text{mL sample}}$$

---

*Teflon or equivalent.
†Whatman No. 40 or equivalent.
‡Freon or equivalent.

6. Precision and Bias

Methods B, C, and D were tested by a single laboratory on a sewage sample. By this method the oil and grease concentration was 12.6 mg/L. When 1-L portions of the sewage were dosed with 14.0 mg of a mixture of No. 2 fuel oil and Wesson oil, recovery of added oils was 93% with a standard deviation of 0.9 mg.

7. Bibliography

KIRSCHMAN, H.D. & R. POMEROY. 1949. Determination of oil in oil field waste waters. *Anal. Chem.* 21:793.

5520 C.   Partition-Infrared Method

1. General Discussion

a. Principle: Although the extraction procedure for this method is identical to that of Method B, infrared detection permits the measurement of many relatively volatile hydrocarbons. Thus, the lighter petroleum distillates, with the exception of gasoline, may be measured accurately. Adequate instrumentation allows for the measurement of as little as 0.2 mg oil and grease/L.

b. Interference: Some degree of selectivity is offered by this method to overcome some of the coextracted interferences discussed in Method B. Heavier residuals of petroleum may contain a significant portion of materials insoluble in trichlorotrifluoroethane.

c. Definitions: A "known oil" is defined as a sample of oil and/or grease that represents the only material of that type used or manufactured in the processes represented by a wastewater. An "unknown oil" is defined as one for which a representative sample of the oil or grease is not available for preparation of a standard.

2. Apparatus

a. Separatory funnel, 1-L, with TFE* stopcock.

b. Infrared spectrophotometer, double beam, recording.

c. Cells, near-infrared silica.

d. Filter paper, 11-cm diam.†

3. Reagents

a. Hydrochloric acid, HCl, 1 + 1.

b. Trichlorotrifluoroethane: See Section 5520B.3*b*.

c. Sodium sulfate, Na_2SO_4, anhydrous, crystal.

d. Reference oil: Prepare a mixture, by volume, of 37.5% iso-octane, 37.5% hexadecane, and 25% benzene. Store in sealed container to prevent evaporation.

4. Procedure

Refer to Method B for sample collection, acidification, and extraction. Collect combined extracts in a 100-mL volumetric flask and adjust final volume to 100 mL with solvent.

Prepare a stock solution of known oil by rapidly transferring about 1 mL (0.5 to 1.0 g) of the oil or grease to a tared 100-mL volumetric flask. Stopper flask and weigh to nearest milligram. Add solvent to dissolve and dilute to mark. If the oil identity is unknown (¶ 1*c*) use the reference oil (¶ 3*d*) as the standard. Using volumetric techniques, prepare a series of standards over the range of interest. Select a pair of

---

*Teflon or equivalent.

†Whatman No. 40 or equivalent.

matched near-infrared silica cells. A 1-cm-path-length cell is appropriate for a working range of about 4 to 40 mg. Scan standards and samples from 3200 cm^{-1} to 2700 cm^{-1} with solvent in the reference beam and record results on absorbance paper. Measure absorbances of samples and standards by constructing a straight base line over the scan range and measuring absorbance of the peak maximum at 2930 cm^{-1} and subtracting base-line absorbance at that point. If the absorbance exceeds 0.8 for a sample, select a shorter path length or dilute as required. Use scans of standards to prepare a calibration curve.

5. Calculation

$$\text{mg oil and grease/L} = \frac{A \times 1000}{\text{mL sample}}$$

where:

A = mg of oil or grease in extract as determined from calibration curve.

6. Precision and Bias

See 5520B.6. By this method the oil and grease concentration was 17.5 mg/L. When 1-L portions of the sewage were dosed with 14.0 mg of a mixture of No. 2 fuel oil and Wesson oil, the recovery of added oils was 99% with a standard deviation of 1.4 mg.

7. Bibliography

GRUENFELD, M. 1973. Extraction of dispersed oils from water for quantitative analysis by infrared spectrophotometry. *Environ. Sci. Technol.* 7:636.

5520 D.   Soxhlet Extraction Method

1. General Discussion

a. Principle: Soluble metallic soaps are hydrolyzed by acidification. Any oils and solid or viscous grease present are separated from the liquid samples by filtration. After extraction in a Soxhlet apparatus with trichlorotrifluoroethane, the residue remaining after solvent evaporation is weighed to determine the oil and grease content. Compounds volatilized at or below 103°C will be lost when the filter is dried.

b. Interference: The method is entirely empirical and duplicate results can be obtained only by strict adherence to all details. By definition, any material recovered is oil and grease and any filtrable trichlorotrifluoroethane-soluble substances, such as elemental sulfur and certain organic dyes, will be extracted as oil and grease. The rate and time of extraction in the Soxhlet apparatus must be exactly as directed because of varying solubilities of different greases. In addition, the length of time required for drying and cooling extracted material cannot be varied. There may be a gradual increase in weight, presumably due to the absorption of oxygen, and/or a gradual loss of weight due to volatilization.

2. Apparatus

a. Extraction apparatus, Soxhlet.

b. Vacuum pump or other source of vacuum.

c. Buchner funnel, 12-cm.

d. Electric heating mantle.

e. Extraction thimble, paper.

f. Filter paper, 11-cm diam.*

g. Muslin cloth disks, 11-cm diam.

---

*Whatman No. 40 or equivalent.

3. Reagents

   a. Hydrochloric acid, HCl, 1 + 1.

   b. Trichlorotrifluoroethane: See Section 5520B.3*b.*

   c. Diatomaceous-silica filter aid suspension,† 10 g/L distilled water.

4. Procedure

Collect about 1 L of sample in a wide-mouth glass bottle and mark sample level in bottle for later determination of sample volume. Acidify to pH 2 or lower; generally, 5 mL HCl is sufficient. Prepare a filter consisting of a muslin cloth disk overlaid with filter paper. Wet paper and muslin and press down edges of paper. Using a vacuum, pass 100 mL filter aid suspension through prepared filter and wash with 1 L distilled water. Apply vacuum until no more water passes filter. Filter acidified sample. Apply vacuum until no more water passes through filter. Using forceps, transfer filter paper to a watch glass. Add material adhering to edges of muslin cloth disk. Wipe sides and bottom of collecting vessel and Buchner funnel with pieces of filter paper soaked in solvent, taking care to remove all films caused by grease and to collect all solid material. Add pieces of filter paper to filter paper on watch glass. Roll all filter paper containing sample and fit into a paper extraction thimble. Add any pieces of material remaining on watch glass. Wipe watch glass with a filter paper soaked in solvent and place in paper extraction thimble. Dry filled thimble in a

---

†Hyflo Super-Cel, Johns-Manville Corp., or equivalent.

hot-air oven at 103°C for 30 min. Fill thimble with glass wool or small glass beads. Weigh extraction flask. Extract oil and grease in a Soxhlet apparatus, using trichlorotrifluoroethane at a rate of 20 cycles/h for 4 h. Time from first cycle. Distill solvent from extraction flask in a water bath at 70°C. Place flask on a water bath at 70°C for 15 min and draw air through it using an applied vacuum for the final 1 min. Cool in a desiccator for 30 min and weigh.

5. Calculation

See Section 5520B.5.

6. Precision and Bias

See Section 5520B.6. By this method the oil and grease concentration was 14.8 mg/L. When 1-L portions of the sewage were dosed with 14.0 mg of a mixture of No. 2 fuel oil and Wesson oil, the recovery of added oils was 88% with a standard deviation of 1.1 mg.

7. Bibliography

HATFIELD, W.D. & G.E. SYMONS. 1945. The determination of grease in sewage. *Sewage Works J.* 17:16.

GILCREAS, F.W., W.W. SANDERSON & R.P. ELMER. 1953. Two new methods for the determination of grease in sewage. *Sewage Ind. Wastes* 25:1379.

ULLMANN, W.W. & W.W. SANDERSON. 1959. A further study of methods for the determination of grease in sewage. *Sewage Ind. Wastes* 31:8.

CHANIN, G., E.H. CHOW, R.B. ALEXANDER & J.F. POWERS. 1967. A safe solvent for oil and grease analyses. *J. Water Pollut. Control Fed.* 39:1892.

5520 E.   Extraction Method for Sludge Samples

1. General Discussion

   a. Principle: Drying acidified sludge by heating leads to low results. Magnesium

sulfate monohydrate is capable of combining with 75% of its own weight in water in forming $MgSO_4 \cdot 7H_2O$ and is used to dry sludge. After drying, the oil and grease

can be extracted with trichlorotrifluoroethane.

b. Interference: See 5520D.1*b.*

2. Apparatus

a. Extraction apparatus, Soxhlet.

b. Vacuum pump or other source of vacuum.

c. Extraction thimble, paper.

d. Grease-free cotton: Extract nonabsorb-
ent cotton with solvent.

3. Reagents

a. Hydrochloric acid, HCl, conc.

b. Magnesium sulfate monohydrate: Prepare $MgSO_4 \cdot H_2O$ by overnight drying of
a thin layer in an oven at 150°C.

c. Trichlorotrifluoroethane: See Section
5520B.3*b.*

4. Procedure

In a 150-mL beaker weigh a sample of
wet sludge, 20 ± 0.5 g, of which the dry-
solids content is known. Acidify to pH 2.0
(generally, 0.3 mL conc HCl is sufficient).
Add 25 g $MgSO_4 \cdot H_2O$. Stir to a smooth
paste and spread on sides of beaker to facilitate subsequent removal. Let stand until
solidified, 15 to 30 min. Remove solids and
grind in a porcelain mortar. Add the powder to a paper extraction thimble. Wipe
beaker and mortar with small pieces of filter paper moistened with solvent and add
to thimble. Fill thimble with glass wool or
small glass beads. Extract in a Soxhlet apparatus, using trichlorotrifluoroethane, at
a rate of 20 cycles/h for 4 h. If any turbidity
or suspended matter is present in the extraction flask, remove by filtering through
grease-free cotton into another weighed
flask. Rinse flask and cotton with solvent.
Distill solvent from extraction flask in
water at 70°C. Place flask on a water bath
at 70°C for 15 min and draw air through
it using an applied vacuum for the final 1
min. Cool in a desiccator for 30 min and
weigh.

5. Calculation

$$\text{Oil and grease as } \% \text{ of dry solids}$$
$$= \frac{\text{gain in weight of flask, g} \times 100}{\text{weight of wet solids, g} \times \text{dry solids fraction}}$$

6. Precision

The examination of six replicate samples
of sludge yielded a standard deviation of
4.6%.

5520 F.   Hydrocarbons

1. Significance

In the absence of specially modified industrial products, oil and grease is composed primarily of fatty matter from
animal and vegetable sources and hydrocarbons of petroleum origin. A knowledge
of the percentage of each of these constituents in the total oil and grease minimizes
the difficulty in determining the major
source of the material and simplifies the
correction of oil and grease problems in
wastewater treatment plant operation and
stream pollution abatement.

2. General Discussion

a. Principle: Silica gel has the ability to
adsorb polar materials. If a solution of hydrocarbons and fatty materials in trichlorotrifluoroethane is mixed with silica gel,
the fatty acids are selectively removed from
solution. The materials not eliminated by
silica gel adsorption are designated hydrocarbons by this test.

b. Interference: The more polar hydrocarbons, such as complex aromatic compounds and hydrocarbon derivatives of

chlorine, sulfur, and nitrogen, may be adsorbed by the silica gel. Any compounds other than hydrocarbons and fatty matter recovered by the procedures for the determination of oil and grease also interfere.

3. Reagents

a. *Trichlorotrifluoroethane:* See Section 5520B.3b.

b. *Silica gel,* 100 to 200 mesh.* Dry at 110°C for 24 h and store in a tightly sealed container.

4. Procedure

Use the oil and grease extracted by Method B, C, D, or E for this test. When only hydrocarbons are of interest, introduce this procedure in any of the previous methods before final measurement. When hydrocarbons are to be determined after total oil and grease has been measured, redissolve, if necessary, the extracted oil and grease in trichlorotrifluoroethane. To 100 mL solvent containing less than 100 mg fatty material, add 3.0 g silica gel. Stopper container and stir on a magnetic stirrer for 5 min. For infrared measurement of

hydrocarbons no further treatment is required before measurement as described in Method C. For gravimetric determinations, filter solution through filter paper and wash silica gel and filter paper with 10 mL solvent and combine with filtrate before performing the solvent stripping steps outlined in Methods B, D, or E.

5. Calculation

Calculate hydrocarbon concentration, in milligrams per liter, as in oil and grease (Method B, C, D, or E).

6. Precision and Bias

The bias of this determination cannot be measured directly in wastewaters. The following data, obtained on synthetic samples, are indicative for natural animal, vegetable, and mineral products, but cannot be applied to the specialized industrial products previously discussed.

For hydrocarbon determinations on 10 synthetic solvent extracts containing known amounts of a wide variety of petroleum products, average recovery was 97.2%. Similar synthetic extracts of Wesson oil, olive oil, Crisco, and butter gave 0.0% recovery as hydrocarbons measured by infrared analysis.

---

*Davidson Grade 923 or equivalent.

5530   PHENOLS*

5530 A.   Introduction

Phenols, defined as hydroxy derivatives of benzene and its condensed nuclei, may occur in domestic and industrial wastewaters, natural waters, and potable water

---

*Approved by Standard Methods Committee, 1988.

supplies. Chlorination of such waters may produce odorous and objectionable-tasting chlorophenols. Phenol removal processes in water treatment include superchlorination, chlorine dioxide or chloramine treatment, ozonation, and activated carbon adsorption.

1. Selection of Method

The analytical procedures offered here use the 4-aminoantipyrine colorimetric method that determines phenol, ortho- and meta-substituted phenols, and, under proper pH conditions, those para-substituted phenols in which the substitution is a carboxyl, halogen, methoxyl, or sulfonic acid group. The 4-aminoantipyrine method does not determine those para-substituted phenols where the substitution is an alkyl, aryl, nitro, benzoyl, nitroso, or aldehyde group. A typical example of these latter groups is paracresol, which may be present in certain industrial wastewaters and in polluted surface waters.

The 4-aminoantipyrine method is given in two forms: Method C, for extreme sensitivity, is adaptable for use in water samples containing less than 1 mg phenol/L. It concentrates the color in a nonaqueous solution. Method D retains the color in the aqueous solution. Because the relative amounts of various phenolic compounds in a given sample are unpredictable, it is not possible to provide a universal standard containing a mixture of phenols. For this reason, phenol (C_6H_5OH) itself has been selected as a standard for colorimetric procedures and any color produced by the reaction of other phenolic compounds is reported as phenol. Because substitution generally reduces response, this value represents the minimum concentration of phenolic compounds. A gas-liquid chromatographic procedure is included in Section 6420B and may be applied to samples or concentrates to quantify individual phenolic compounds.

2. Interferences

Interferences such as phenol-decomposing bacteria, oxidizing and reducing substances, and alkaline pH values are dealt with by acidification. Some highly contaminated wastewaters may require specialized techniques for eliminating interferences and for quantitative recovery of phenolic compounds.

Eliminate major interferences as follows (see Section 5530B for reagents):

Oxidizing agents, such as chlorine and those detected by the liberation of iodine on acidification in the presence of potassium iodide (KI)—Remove immediately after sampling by adding excess ferrous sulfate ($FeSO_4$). If oxidizing agents are not removed, the phenolic compounds will be oxidized partially.

Sulfur compounds—Remove by acidifying to pH 4.0 with H_3PO_4 and aerating briefly by stirring. This eliminates the interference of hydrogen sulfide (H_2S) and sulfur dioxide (SO_2).

Oils and tars—Make an alkaline extraction by adjusting to pH 12 to 12.5 with NaOH pellets. Extract oil and tar from aqueous solution with 50 mL chloroform ($CHCl_3$). Discard oil- or tar-containing layer. Remove excess $CHCl_3$ in aqueous layer by warming on a water bath before proceeding with the distillation step.

3. Sampling

Sample in accordance with the instructions of Section 1060.

4. Preservation and Storage of Samples

Phenols in concentrations usually encountered in wastewaters are subject to biological and chemical oxidation. Preserve and store samples at 4°C or lower unless analyzed within 4 h after collection.

Acidify with 2 mL conc H_2SO_4/L. Samples may be stored for 4 weeks at 4°C.

Analyze preserved and stored samples within 28 d after collection.

5. Bibliography

ETTINGER, M.B., S. SCHOTT & C.C. RUCHHOFT. 1943. Preservation of phenol content in polluted river water samples previous to analysis. *J. Amer. Water Works Assoc.* 35:299.

CARTER, M.J. & M.T. HUSTON. 1978. Preservation of phenolic compounds in wastewaters. *Environ. Sci. Technol.* 12:309.

NEUFELD, R.D. & S.B. POLADINO. 1985. Comparison of 4-aminoantipyrine and gas-liquid chromatography techniques for analysis of phenolic compounds. *J. Water Pollut. Control Fed.* 57:1040.

5530 B. Cleanup Procedure

1. Principle

Phenols are distilled from nonvolatile impurities. Because the volatilization of phenols is gradual, the distillate volume must ultimately equal that of the original sample.

2. Apparatus

a. Distillation apparatus, all-glass, consisting of a 1-L borosilicate glass distilling apparatus with Graham condenser.*

b. pH meter.

3. Reagents

Prepare all reagents with distilled water free of phenols and chlorine.

a. Phosphoric acid solution, H_3PO_4, 1 + 9: Dilute 10 mL 85% H_3PO_4 to 100 mL with water.

b. Methyl orange indicator solution.

c. Special reagents for turbid distillates:

1) *Sulfuric acid,* H_2SO_4, 1N.

2) *Sodium chloride,* NaCl.

3) *Chloroform,* $CHCl_3$, *or methylene chloride,* CH_2Cl_2.

4) *Sodium hydroxide,* NaOH, 2.5N: Dilute 41.7 mL 6N NaOH to 100 mL or dissolve 10 g NaOH pellets in 100 mL water.

4. Procedure

a. Measure 500 mL sample into a beaker, adjust pH to approximately 4.0 with H_3PO_4 solution using methyl orange indicator or a pH meter, and transfer to distillation apparatus. Use a 500-mL graduated cylinder as a receiver. Omit adding H_3PO_4 and adjust pH to 4.0 with 2.5N NaOH if sample was preserved as described in 5530A.4.

b. Distill 450 mL, stop distillation and, when boiling ceases, add 50 mL warm water to distilling flask. Continue distillation until a total of 500 mL has been collected.

c. One distillation should purify the sample adequately. Occasionally, however, the distillate is turbid. If so, acidify with H_3PO_4 solution and distill as described in ¶ 4*b*. If second distillate is still turbid, use extraction process described in ¶ 4*d* before distilling sample.

d. Treatment when second distillate is turbid: Extract a 500-mL portion of original sample as follows: Add 4 drops methyl orange indicator and make acidic to methyl orange with 1N H_2SO_4. Transfer to a separatory funnel and add 150 g NaCl. Shake with five successive portions of $CHCl_3$, using 40 mL in the first portion and 25 mL in each successive portion. Transfer $CHCl_3$ layer to a second separatory funnel and shake with three successive portions of 2.5N NaOH solution, using 4.0 mL in the first portion and 3.0 mL in each of the next two portions. Combine alkaline extracts, heat on a water bath until $CHCl_3$ has been removed, cool, and dilute to 500 mL with distilled water. Proceed with distillation as described in ¶s 4*a* and *b*.

NOTE: CH_2Cl_2 may be used instead of $CHCl_3$, especially if an emulsion forms when the $CHCl_3$ solution is extracted with NaOH.

*Corning No. 3360 or equivalent.

5530 C. Chloroform Extraction Method

1. General Discussion

a. *Principle:* Steam-distillable phenols react with 4-aminoantipyrine at pH 7.9 ± 0.1 in the presence of potassium ferricyanide to form a colored antipyrine dye. This dye is extracted from aqueous solution with $CHCl_3$ and the absorbance is measured at 460 nm. This method covers the phenol concentration range from 1.0 μg/L to over 250 μg/L with a sensitivity of 1 μg/L.

b. *Interference:* All interferences are eliminated or reduced to a minimum if the sample is preserved, stored, and distilled in accordance with the foregoing instructions.

c. *Minimum detectable quantity:* The minimum detectable quantity for clean samples containing no interferences is 0.5 μg phenol when a 25-mL $CHCl_3$ extraction with a 5-cm cell or a 50-mL $CHCl_3$ extraction with a 10-cm cell is used in the photometric measurement. This quantity is equivalent to 1 μg phenol/L in 500 mL distillate.

2. Apparatus

a. *Photometric equipment:* A spectrophotometer for use at 460 nm equipped with absorption cells providing light paths of 1 to 10 cm, depending on the absorbances of the colored solutions and the individual characteristics of the photometer.

b. *Filter funnels:* Buchner type with fritted disk.*

c. *Filter paper:* Alternatively use an appropriate 11-cm filter paper for filtering $CHCl_3$ extracts instead of the Buchner-type funnels and anhydrous Na_2SO_4.

d. *pH meter.*

e. *Separatory funnels,* 1000-mL, Squibb form, with ground-glass stoppers and TFE stopcocks. At least eight are required.

---

*15-mL Corning No. 36060 or equivalent.

3. Reagents

Prepare all reagents with distilled water free of phenols and chlorine.

a. *Stock phenol solution:* Dissolve 1.00 g phenol in freshly boiled and cooled distilled water and dilute to 1000 mL. CAUTION— *Toxic; handle with extreme care.* Ordinarily this direct weighing yields a standard solution; if extreme accuracy is required, standardize as follows:

1) To 100 mL water in a 500-mL glass-stoppered conical flask, add 50.0 mL stock phenol solution and 10.0 mL bromate-bromide solution. Immediately add 5 mL conc HCl and swirl gently. If brown color of free bromine does not persist, add 10.0-mL portions of bromate-bromide solution until it does. Keep flask stoppered and let stand for 10 min; then add approximately 1 g KI. Usually four 10-mL portions of bromate-bromide solution are required if the stock phenol solution contains 1000 mg phenol/L.

2) Prepare a blank in exactly the same manner, using distilled water and 10.0 mL bromate-bromide solution. Titrate blank and sample with 0.025 M sodium thiosulfate, using starch solution indicator.

3) Calculate the concentration of phenol solution as follows:

$$\text{mg phenol/L} = 7.842 \, [(A \times B) - C]$$

where:

A = mL thiosulfate for blank,
B = mL bromate-bromide solution used for sample divided by 10, and
C = mL thiosulfate used for sample.

b. *Intermediate phenol solution:* Dilute 10.0 mL stock phenol solution in freshly boiled and cooled distilled water to 1000 mL; 1 mL = 10.0 μg phenol. Prepare daily.

c. *Standard phenol solution:* Dilute 50.0 mL intermediate phenol solution to 500 mL with freshly boiled and cooled distilled water; 1 mL = 1.0 µg phenol. Prepare within 2 h of use.

d. *Bromate-bromide solution:* Dissolve 2.784 g anhydrous $KBrO_3$ in water, add 10 g KBr crystals, dissolve, and dilute to 1000 mL.

e. *Hydrochloric acid,* HCl, conc.

f. *Standard sodium thiosulfate titrant,* 0.025 *M:* See Section 4500-O.C.2e.

g. *Starch solution:* See Section 4500-O.C.2d.

h. *Ammonium hydroxide,* NH_4OH, 0.5*N:* Dilute 35 mL fresh, conc NH_4OH to 1 L with water.

i. *Phosphate buffer solution:* Dissolve 104.5 g K_2HPO_4 and 72.3 g KH_2PO_4 in water and dilute to 1 L. The pH should be 6.8.

j. *4-Aminoantipyrine solution:* Dissolve 2.0 g 4-aminoantipyrine in water and dilute to 100 mL. Prepare daily.

k. *Potassium ferricyanide solution:* Dissolve 8.0 g $K_3Fe(CN)_6$ in water and dilute to 100 mL. Filter if necessary. Store in a brown glass bottle. Prepare fresh weekly.

l. *Chloroform,* $CHCl_3$.

m. *Sodium sulfate,* anhydrous Na_2SO_4, granular.

n. *Potassium iodide,* KI, crystals.

4. Procedure

Ordinarily, use Procedure *a*; however, Procedure *b* may be used for infrequent analyses.

a. Place 500 mL distillate, or a suitable portion containing not more than 50 µg phenol, diluted to 500 mL, in a 1-L beaker. Prepare a 500-mL distilled water blank and a series of 500-mL phenol standards containing 5, 10, 20, 30, 40, and 50 µg phenol.

Treat sample, blank, and standards as follows: Add 12.0 mL 0.5*N* NH_4OH and *immediately* adjust pH to 7.9 ± 0.1 with phosphate buffer. Under some circumstances, a higher pH may be required.[†] About 10 mL phosphate buffer are required. Transfer to a 1-L separatory funnel, add 3.0 mL aminoantipyrine solution, mix well, add 3.0 mL $K_3Fe(CN)_6$ solution, mix well, and let color develop for 15 min. The solution should be clear and light yellow.

Extract immediately with $CHCl_3$ using 25 mL for 1- to 5-cm cells and 50 mL for a 10-cm cell. Shake separatory funnel at least 10 times, let $CHCl_3$ settle, shake again 10 times, and let $CHCl_3$ settle again. Filter each $CHCl_3$ extract through filter paper or fritted glass funnels containing a 5-g layer of anhydrous Na_2SO_4. Collect dried extracts in clean cells for absorbance measurements; do not add more $CHCl_3$ or wash filter papers or funnels with $CHCl_3$.

Read absorbance of sample and standards against the blank at 460 nm. Plot absorbance against micrograms phenol concentration. Construct a separate calibration curve for each photometer and check each curve periodically to insure reproducibility.

b. For infrequent analyses prepare only one standard phenol solution. Prepare 500 mL standard phenol solution of a strength approximately equal to the phenolic content of that portion of original sample used for final analysis. Also prepare a 500-mL distilled water blank.

Continue as described in ¶ *a*, above, but measure absorbances of sample and standard phenol solution against the blank at 460 nm.

5. Calculation

a. For Procedure *a:*

$$\mu g \text{ phenol/L} = \frac{A}{B} \times 1000$$

---

†For NPDES permit analyses, pH 10 ± 0.1 is required.

where:

$A = \mu g$ phenol in sample, from calibration curve, and

$B = mL$ original sample.

b. For Procedure *b,* calculate the phenol content of the original sample:

$$\mu g\ phenol/L = \frac{C \times D \times 1000}{E \times B}$$

where:

$C = \mu g$ standard phenol solution,
$D =$ absorbance reading of sample,
$E =$ absorbance of standard phenol solution, and
$B = mL$ original sample.

6. Precision and Bias

Because the "phenol" value is based on C_6H_5OH, this method yields only an approximation and represents the minimum amount of phenols present. This is true because the phenolic reactivity to 4-aminoantipyrine varies with the types of phenols present.

In a study of 40 refinery wastewaters analyzed in duplicate at concentrations from 0.02 to 6.4 mg/L the average relative standard deviation was $\pm$ 12%. Data are not available for precision at lower concentrations.

7. Bibliography

SCOTT, R.D. 1931. Application of a bromine method in the determination of phenols and cresols. *Ind. Eng. Chem.*, Anal. Ed. 3:67.

EMERSON, E., H.H. BEACHAM & L.C. BEEGLE. 1943. The condensation of aminoantipyrine. II. A new color test for phenolic compounds. *J. Org. Chem.* 8:417.

ETTINGER, M.B. & R.C. KRONER. 1949. The determination of phenolic materials in industrial wastes. *Proc. 5th Ind. Waste Conf.,* Purdue Univ., p. 345.

ETTINGER, M.B., C.C. RUCHHOFT & R.J. LISHKA. 1951. Sensitive 4-aminoantipyrine method for phenolic compounds. *Anal. Chem.* 23:1783.

DANNIS, M. 1951. Determination of phenols by the aminoantipyrine method. *Sewage Ind. Wastes* 23:1516.

MOHLER, E.F., JR. & L.N. JACOB. 1957. Determination of phenolic-type compounds in water and industrial waste waters: Comparison of analytical methods. *Anal. Chem.* 29:1369.

BURTSCHELL, R.H., A.A. ROSEN, F.M. MIDDLETON & M.B. ETTINGER. 1959. Chlorine derivatives of phenol causing taste and odor. *J. Amer. Water Works Assoc.* 51:205.

GORDON, G.E. 1960. Colorimetric determination of phenolic materials in refinery waste waters. *Anal. Chem.* 32:1325.

OCHYNSKI, F.W. 1960. The absorptiometric determination of phenol. *Analyst* 85:278.

FAUST, S.D. & O.M. ALY. 1962. The determination of 2,4-dichlorophenol in water. *J. Amer. Water Works Assoc.* 54:235.

FAUST, S.D. & E.W. MIKULEWICZ. 1967. Factors influencing the condensation of 4-aminoantipyrine with derivatives of hydroxybenzene. II. Influence of hydronium ion concentration on absorptivity. *Water Res.* 1:509.

FAUST, S.D. & P.W. ANDERSON. 1968. Factors influencing the condensation of 4-aminoantipyrine with derivations of hydroxy benzene. III. A study of phenol content in surface waters. *Water Res.* 2:515.

SMITH, L.S. 1976. Evaluation of Instrument for the (Ultraviolet) Determination of Phenol in Water. EPA-600/4-76-048, U.S. Environmental Protection Agency, Cincinnati, Ohio.

5530 D.   Direct Photometric Method

1. General Discussion

a. Principle: Steam-distillable phenolic compounds react with 4-aminoantipyrine at pH 7.9 ± 0.1 in the presence of potassium ferricyanide to form a colored antipyrine dye. This dye is kept in aqueous solution and the absorbance is measured at 500 nm.

b. Interference: Interferences are eliminated or reduced to a minimum by using the distillate from the preliminary distillation procedure.

c. Minimum detectable quantity: This method has less sensitivity than Method C. The minimum detectable quantity is 10 μg phenol when a 5-cm cell and 100 mL distillate are used.

2. Apparatus

a. Photometric equipment: Spectrophotometer equipped with absorption cells providing light paths of 1 to 5 cm for use at 500 nm.

b. pH meter.

3. Reagents

See Section 5530C.3.

4. Procedure

Place 100 mL distillate, or a portion containing not more than 0.5 mg phenol diluted to 100 mL, in a 250-mL beaker. Prepare a 100-mL distilled water blank and a series of 100-mL phenol standards containing 0.1, 0.2, 0.3, 0.4, and 0.5 mg phenol.

Treat sample, blank, and standards as follows: Add 2.5 mL 0.5N NH$_4$OH solution and immediately adjust to pH 7.9 ± 0.1 with phosphate buffer. Add 1.0 mL 4-aminoantipyrine solution, mix well, add 1.0 mL K$_3$Fe(CN)$_6$ solution, and mix well.

After 15 min, transfer to cells and read absorbance of sample and standards against the blank at 500 nm.

5. Calculation

a. Use of calibration curve: Estimate sample phenol content from photometric readings by using a calibration curve constructed as directed in Section 5530C.4*a.*

$$\text{mg phenol/L} = \frac{A}{B} \times 1000$$

where:

A = mg phenol in sample, from calibration curve, and

B = mL original sample.

b. Use of single phenol standard:

$$\text{mg phenol/L} = \frac{C \times D \times 1000}{E \times B}$$

where:

C = mg standard phenol solution,

D = absorbance of sample, and

E = absorbance of standard phenol solution.

6. Precision and Bias

Precision and bias data are not available.

5540 SURFACTANTS*

5540 A. Introduction

1. Occurrence and Significance

Surfactants enter waters and waste-waters mainly by discharge of aqueous wastes from household and industrial laundering and other cleansing operations. A surfactant combines in a single molecule a strongly hydrophobic group with a strongly hydrophilic one. Such molecules tend to congregate at the interfaces between the aqueous medium and the other phases of the system such as air, oily liquids, and particles, thus imparting properties such as foaming, emulsification, and particle suspension.

The surfactant hydrophobic group generally is a hydrocarbon radical (R) containing about 10 to 20 carbon atoms. The hydrophilic groups are of two types, those that ionize in water and those that do not. Ionic surfactants are subdivided into two categories, differentiated by the charge. An anionic surfactant ion is negatively charged, e.g., $(RSO_3)^- Na^+$, and a cationic one is positively charged, e.g., $(RMe_3N)^+ Cl^-$. Nonionizing (nonionic) surfactants commonly contain a polyoxyethylene hydrophilic group $(ROCH_2CH_2OCH_2CH_2......OCH_2CH_2OH$, often abbreviated RE_n, where n is the average number of $-OCH_2CH_2-$ units in the hydrophilic group). Hybrids of these types exist also.

In the United States, ionic surfactants amount to about two thirds of the total surfactants used and nonionics to about one third. Cationic surfactants amount to less than one tenth of the ionics and are used generally for disinfecting, fabric softening, and various cosmetic purposes rather than for their detersive properties. At current detergent and water usage levels the surfactant content of raw domestic wastewater is in the range of about 1 to 20 mg/L. Most domestic wastewater surfactants are dissolved in equilibrium with proportional amounts adsorbed on particulates. Primary sludge concentrations range from 1 to 20 mg adsorbed anionic surfactant per gram dry weight.[1] In environmental waters the surfactant concentration generally is below 0.1 mg/L except in the vicinity of an outfall or other point source of entry.[2]

2. Analytical Precautions

Because of inherent properties of surfactants, special analytical precautions are necessary. Avoid foam formation because the surfactant concentration is higher in the foam phase than in the associated bulk aqueous phase and the latter may be significantly depleted. If foam is formed, let it subside by standing, or collapse it by other appropriate means, and remix the liquid phase before sampling. Adsorption of surfactant from aqueous solutions onto the walls of containers, when concentrations below about 1 mg/L are present, may seriously deplete the bulk aqueous phase. Minimize adsorption errors, if necessary, by rinsing container with sample, and for anionic surfactants by adding alkali phosphate (e.g., $0.03N$ KH_2PO_4).[3]

3. References

1. SWISHER, R.D. 1987. Surfactant Biodegradation, 2nd ed. Marcel Dekker, New York, N.Y.

---

*Approved by Standard Methods Committee, 1988.

2. ARTHUR D. LITTLE, INC. 1977. Human Safety and Environmental Aspects of Major Surfactants. Rep. No. PB-301193, National Technical Information Serv., Springfield, Va.

3. WEBER, W.J., JR., J.C. MORRIS & W. STUMM. 1962. Determination of alkylbenzenesulfonate by ultraviolet spectrophotometry. *Anal. Chem.* 34:1844.

5540 B.   Surfactant Separation by Sublation

1. General Discussion

a. Principle: The sublation process isolates the surfactant, regardless of type, from dilute aqueous solution, and yields a dried residue relatively free of nonsurfactant substances. It is accomplished by bubbling a stream of nitrogen up through a column containing the sample and an overlying layer of ethyl acetate. The surfactant is adsorbed at the water-gas interfaces of the bubbles and is carried into the ethyl acetate layer. The bubbles escape into the atmosphere leaving behind the surfactant dissolved in ethyl acetate. The solvent is separated, dehydrated, and evaporated, leaving the surfactant as a residue suitable for analysis. This procedure is the same as that used by the Organization for Economic Co-operation and Development (OECD),[1] following the development by Wickbold. [2,3]

b. Interferences: The sublation method is specific for surfactants, because any substance preferentially adsorbed at the water-gas interface is by definition a surfactant. Although nonsurfactant substances largely are rejected in this separation process, some amounts will be carried over mechanically into the ethyl acetate.

c. Limitations: The sublation process separates only dissolved surfactants. If particulate matter is present it holds back an equilibrium amount of adsorbed surfactant. As sublation removes the initially dissolved surfactant, the particulates tend to re-equilibrate and their adsorbed surfactants redissolve. Thus, continued sublation eventually should remove substantially all adsorbed surfactant. However, if the particulates adsorb the surfactant tightly, as sewage particulates usually do, complete removal may take a very long time. The procedure given herein calls for preliminary filtration and measures only dissolved surfactant. Determine adsorbed surfactant content by analyzing particulates removed by filtration; no standard method is available now.

d. Operating conditions: Make successive 5-min sublations from 1 L of sample containing 5 g $NaHCO_3$ and 100 g NaCl. Under the conditions specified, extensive transfer of surfactant occurs in the first sublation and is substantially complete in the second.[2-4]

e. Quantitation: Quantitate the surfactant residue by the procedures in 5540C or D. Direct weighing of the residue is not useful because the weight of surfactant isolated generally is too low, less than a milligram, and varied amounts of mechanically entrained nonsurfactants may be present. The procedure is applicable to water and wastewater samples.

2. Apparatus

a. Sublator: A glass column with dimensions as shown in Figure 5540:1. For the

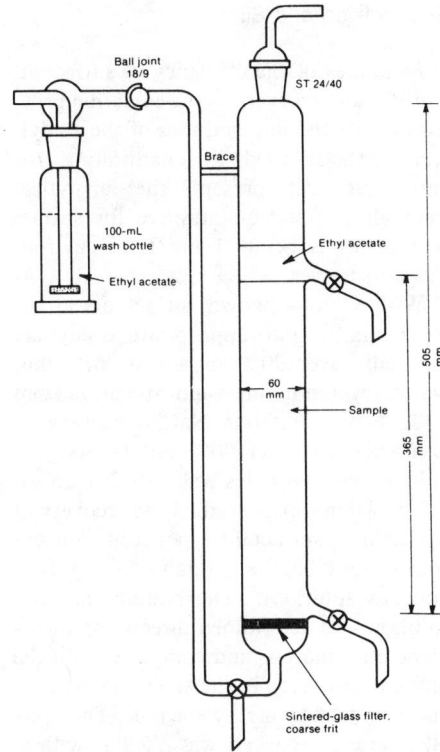

Figure 5540:1. Sublation apparatus.[1] See Section 5540B.2*a* and *b* and 4*c*. Bottom stopcock: TFE plug, 4-mm bore; side stopcocks: TFE plug, 2-mm bore.

sintered glass disk use a coarse-porosity frit (designation "c" - nominal maximum pore diam 40 to 60 μm as measured by ASTM E-128) of the same diameter as the column internal diameter. Volume between disk and upper stopcock should be approximately 1 L.

b. Gas washing bottle, as indicated in Figure 5540:1, working volume 100 mL or more.

c. Separatory funnel, working volume 250 mL, preferably with inert TFE stopcock.

d. Filtration equipment, suitable for 1-L samples, using medium-porosity qualitative-grade filter paper.

e. Gas flowmeter, for measuring flows up to 1 L/min.

3. Reagents

a. Nitrogen, standard commercial grade.

b. Ethyl acetate: CAUTION: *Ethyl acetate is flammable and its vapors can form explosive mixtures with air.*

c. Sodium bicarbonate, $NaHCO_3$.

d. Sodium chloride, NaCl.

e. Water, surfactant-free.

4. Procedure

a. Sample size: Select a sample to contain not more than 1 to 2 mg surfactant.[4] For most waters the sample volume will be about 1 L; for wastewater use a smaller volume.

b. Filtration: Filter sample through medium-porosity qualitative filter paper. Wash filter paper by discarding the first few hundred milliliters of filtrate.

c. Assembly: Refer to Figure 5540:1.

Connect nitrogen cylinder through flowmeter to inlet of gas washing bottle. Connect gas outlet at top of sublator to means for disposing ethyl acetate vapor such as by a gas scrubber or by venting to a hood or directly outdoors. In the absence of a flowmeter, ensure proper gas flow rate by measuring volume of gas leaving the sublator, with a water-displacement system.

d. Charging: Fill gas washing bottle about two-thirds full with ethyl acetate. Rinse sublation column with ethyl acetate and discard rinse. Place measured filtered sample in sublator and add 5 g $NaHCO_3$, 100 g NaCl, and sufficient water to bring the level up to or slightly above the upper stopcock (about 1 L total volume). If sample volume permits, add salts as a solution in 400 mL water or dissolve them in the sample and quantitatively transfer to the sublator. Add 100 mL ethyl acetate by running it carefully down the wall of the sublator to form a layer on top of the sample.

e. *Sublation:* Start the nitrogen flow, increasing the rate carefully to 1 L/min initially but do not exceed a rate at which the liquid phases begin vigorous intermixing at their interface. Avoid overly vigorous intermixing, which will lead to back-extraction of the surfactant into the aqueous phase and to dissolution of ethyl acetate. Continue sublation for 5 min at 1 L/min. If a lower flow rate is necessary to avoid phase intermixing, prolong sublation time proportionally. If the volume of the upper phase has decreased by more than about 20%, repeat the operation on a new sample but avoid excessive intermixing at the interface. Draw off entire ethyl acetate layer through upper stopcock into the separatory funnel; return any transferred water layer to the sublator. Filter ethyl acetate layer into a 250-mL beaker through a dry, medium-porosity, qualitative filter paper (prewashed with ethyl acetate to remove any adventitious surfactant) to remove any remaining aqueous phase.

Repeat process of preceding paragraph with a second 100-mL layer of ethyl acetate, using the same separatory funnel and filter, and finally rinse sublator wall with another 20 mL, all into the original beaker.

Evaporate ethyl acetate from the beaker on a steam bath in a hood, blowing a gentle stream of nitrogen or air over the liquid surface to speed evaporation and to minimize active boiling. Evaporate the first 100 mL during the second sublation to avoid overfilling the beaker. To avoid possible solute volatilization, discontinue heating after removing the ethyl acetate. The sublated surfactant remains in the beaker as a film of residue.

Draw off aqueous layer in the sublator and discard, using the stopcock just above the sintered disk to minimize disk fouling.

5. Precision and Bias

Estimates of the efficiency of surfactant transfer and recovery in the sublation process include the uncertainties of the analytical methods used in quantitating the surfactant. At present the analytical methods are semiquantitative for surfactant at levels below 1 mg/L in environmental samples.

With various known surfactants at 0.2 to 2 mg/L and appropriate analytical methods, over 90% of added surfactant was recovered in one 5-min sublation from 10% NaCl. Without NaCl, recovery of nonionics was over 90% but recovery of anionics and cationics was only 2 to 25%.[4]

Five laboratories studied the recovery of five anionic surfactant types from concentrations of 0.05, 0.2, 1.0, and 5.0 mg/L in aqueous solutions.[5] The amount in each solution was determined directly by methylene blue analysis and compared with the amount recovered in the sublation process, also analyzed by methylene blue. The overall average recovery was 95.9% with a standard deviation of ± 7.4 ($n = 100$). The extreme individual values for recovery were 65% and 115% and the other 98 values ranged from 75% to 109%. Recovery did not depend on surfactant concentration (average recoveries ranging from 94.7% at 5.0 mg/L to 96.8% at 1.0 mg/L) nor on the surfactant type (average recoveries ranging from 94.7% to 96.6%). Average recoveries at the five laboratories ranged from 90.0% to 98.0%.

Application of the sublation method in three laboratories to eight different samples of raw wastewater in duplicate gave the results shown in Table 5540:I. Methylene blue active substances (MBAS) recovery in double sublation averaged 87 ± 16% of that determined directly on the filtered wastewater; these results would have been influenced by any nonsurfactant MBAS that might have been present. Repeating

TABLE 5540:I. SURFACTANT RECOVERY BY SUBLATION

Variable	MBAS	CTAS
Sample volume, mL	200–300	500
Concentration without sublation, mg/L	2.2–4.7	—
Concentration found in sublate,* mg/L	1.8–4.4	0.3–0.6
Recovery in sublate, %	87 ± 16†	—
Amount in second sublate,‡ mg	0.02 ± 0.02†	0.08 ± 0.01†
Amount added, mg	0.05–0.10§	0.50–0.67‖
Recovery in sublation,# %	94 ± 17†	92 ± 6†

* Two 5-min sublations.
† Average ± SD ($n = 8$).
‡ Two more 5-min sublations.
§ Reference LAS.
‖ Linear alcohol ethoxylate $C_{12\text{-}18}E_{11}$.
Fifth and sixth 5-min sublations.

double sublation on the spent aqueous phase yielded another 0.02 mg MBAS and another 0.08 mg cobalt thiocyanate active substances (CTAS). Adding 0.05 to 0.10 mg of known linear alkylbenzene sulfonate (LAS) or 0.50 to 0.67 mg of known linear alcohol-based $C_{12-18}E_{11}$ to the same sublator contents and again running double sublation resulted in over 90% recovery of the amount added.

6. References

1. ORGANIZATION FOR ECONOMIC CO-OPERA-TION AND DEVELOPMENT ENVIRONMENTAL DIRECTORATE. 1976. Proposed Method for the Determination of the Biodegradability of Surfactants Used in Synthetic Detergents. Org. for Economic Co-Operation & Development, Paris.

2. WICKBOLD, R. 1971. Enrichment and separation of surfactants from surface waters through transport in the gas/water interface. *Tenside* 8:61.

3. WICKBOLD, R. 1972. Determination of nonionic surfactants in river- and wastewaters. *Tenside* 9:173.

4. KUNKEL, E., G. PEITSCHER & K. ESPETER. 1977. New developments in trace- and microanalysis of surfactants. *Tenside* 14:199.

5. DIVO, C., S. GAFA, T. LA NOCE, A. PARIS, C. RUFFO & M. SANNA. 1980. Use of nitrogen blowing technique in microdetermination of anionic surfactants. *Riv. Ital. Sost. Grasse* 57:329.

5540 C. Anionic Surfactants as MBAS

1. General Discussion

a. Definition and principle: Methylene blue active substances (MBAS) bring about the transfer of methylene blue, a cationic dye, from an aqueous solution into an immiscible organic liquid upon equilibration. This occurs through ion pair formation by the MBAS anion and the methylene blue cation. The intensity of the resulting blue color in the organic phase is a measure of MBAS. Anionic surfactants are among the most prominent of many substances, natural and synthetic, showing methylene blue activity. The MBAS method is useful for estimating the anionic surfactant content of waters and wastewaters, but the possible presence of other types of MBAS always must be kept in mind.

This method is relatively simple and precise. It comprises three successive extractions from acid aqueous medium containing excess methylene blue into chloroform ($CHCl_3$), followed by an aqueous backwash and measurement of the blue color in the $CHCl_3$ by spectrophotometry at 652 nm. The method is applicable at MBAS concentrations down to about 0.025 mg/L.

b. Anionic surfactant responses: Soaps do not respond in the MBAS method. Those used in or as detergents are alkali salts of C_{10-20} fatty acids $[RCO_2]^-Na^+$, and though anionic in nature they are so weakly ionized that an extractable ion pair is not formed under the conditions of the test. Nonsoap anionic surfactants commonly used in detergent formulations are strongly responsive. These include principally surfactants of the sulfonate type $[RSO_3]^-Na^+$, the sulfate ester type $[ROSO_3]^-Na^+$, and sulfated nonionics $[RE_nOSO_3]^-Na^+$. They are recovered almost completely by a single $CHCl_3$ extraction.

Linear alkylbenzene sulfonate (LAS) is the most widely used anionic surfactant and is used to standardize the MBAS method. LAS is not a single compound, but may comprise any or all of 26 isomers and homologs with structure $[R'C_6H_4SO_3]^-Na^+$, where R' is a linear secondary alkyl group ranging from 10 to 14 carbon atoms in length. The manufacturing process defines the mixture, which may be modified further by the wastewate treatment process.

Sulfonate- and sulfate-type surfactants respond together in MBAS analysis, but they can be differentiated by other means. The sulfate type decomposes upon acid hydrolysis; the resulting decrease in MBAS corresponds to the original sulfate surfactant content while the MBAS remaining corresponds to the sulfonate surfactants. Alkylbenzene sulfonate can be identified and quantified by infrared spectrometry after purification.[1] LAS can be distinguished from other alkylbenzene sulfonate surfactants by infrared methods.[2] LAS can be identified unequivocally and its detailed isomer-homolog composition determined by desulfonation-gas chromatography.[3]

c. Interferences: Positive interferences result from all other MBAS species present; if a direct determination of any individual MBAS species, such as LAS, is sought, all others interfere. Substances such as organic sulfonates, sulfates, carboxylates and phenols, and inorganic thiocyanates, cyanates, nitrates, and chlorides also may transfer more or less methylene blue into the chloroform phase. The poorer the extractability of their ion pairs, the more effective is the aqueous backwash step in removing these positive interferences; interference from chloride is eliminated almost entirely and from nitrate largely so by the backwash. Because of the varied extractability of nonsurfactant MBAS, deviations in $CHCl_3$ ratio and backwashing procedure may lead to significant differences in the total MBAS observed, although the recovery of sulfonate- and sulfate-type surfactants will be substantially complete in all cases.

Negative interferences can result from the presence of cationic surfactants and other cationic materials, such as amines, because they compete with the methylene blue in the formation of ion pairs. Particulate matter may give negative interference through adsorption of MBAS. Although some of the adsorbed MBAS may be desorbed and paired during the $CHCl_3$ extractions, recovery may be incomplete and variable.

Minimize interferences by nonsurfactant materials by sublation if necessary (Section 5540B). Other countermeasures are nonstandard. Remove interfering cationic surfactants and other cationic materials by using a cation-exchange resin under suitable conditions.[3] Handle adsorption of MBAS by particulates preferably by filtering and analyzing the insolubles. With or without filtration, adsorbed MBAS can be desorbed by acid hydrolysis; however, MBAS originating in any sulfate ester-type surfactant present is destroyed simultaneously.[1] Sulfides, often present in raw or primary treated wastewater, may react with methylene blue to form a colorless reduction product, making the analysis impossible. Eliminate this interference by prior oxidation with hydrogen peroxide.

d. Molecular weight: Test results will appear to differ if expressed in terms of weight rather than in molar quantities. Equimolar amounts of two anionic surfactants with different molecular weights should give substantially equal colors in the $CHCl_3$ layer, although the amounts by weight may differ significantly. If results are to be expressed by weight, as generally is desirable, the average molecular weight of the surfactant measured must be known or a calibration curve made with that particular compound must be used. Because such detailed information generally is lacking, report results in terms of a suitable standard calibration curve, for example "0.65 mg MBAS/L (calculated as LAS, mol wt 318)."

e. Minimum detectable quantity: About 10 μg MBAS (calculated as LAS).

f. Application: The MBAS method has been applied successfully to drinking water samples. In wastewater, industrial wastes, and sludge, numerous materials normally present can interfere seriously if direct determination of MBAS is attempted. Most nonsurfactant aqueous-phase interferences can be removed by sublation.

2. Apparatus

a. Colorimetric equipment: One of the following is required:

1) *Spectrophotometer,* for use at 652 nm, providing a light path of 1 cm or longer.

2) *Filter photometer,* providing a light path of 1 cm or longer and equipped with a red color filter exhibiting maximum transmittance near 652 nm.

b. Separatory funnels: 500-mL, preferably with inert TFE stopcocks and stoppers.

3. Reagents

a. Stock LAS solution: Weigh an amount of the reference material* equal to 1.00 g LAS on a 100% active basis. Dissolve in water and dilute to 1000 mL; 1.00 mL = 1.00 mg LAS. Store in a refrigerator to minimize biodegradation. If necessary, prepare weekly.

b. Standard LAS solution: Dilute 10.00 mL stock LAS solution to 1000 mL with water; 1.00 mL = 10.0 μg LAS. Prepare daily.

c. Phenolphthalein indicator solution, alcoholic.

d. Sodium hydroxide, NaOH, 1N.

e. Sulfuric acid, H_2SO_4, 1N and 6N.

f. Chloroform, $CHCl_3$: CAUTION: *Chloroform is toxic and a suspected carcinogen. Take appropriate precautions against inhalation and skin exposure.*

g. Methylene blue reagent: Dissolve 100 mg methylene blue† in 100 mL water. Transfer 30 mL to a 1000-mL flask. Add 500 mL water, 41 mL 6N H_2SO_4, and 50 g sodium phosphate, monobasic, monohydrate, $NaH_2PO_4 \cdot H_2O$. Shake until dissolved. Dilute to 1000 mL.

h. Wash solution: Add 41 mL 6 N H_2SO_4 to 500 mL water in a 1000-mL flask. Add 50 g $NaH_2PO_4 \cdot H_2O$ and shake until dissolved. Dilute to 1000 mL.

i. Methanol, CH_3OH. CAUTION: *Methanol vapors are flammable and toxic; take appropriate precautions.*

j. Hydrogen peroxide, H_2O_2, 30%.

k. Glass wool: Pre-extract with $CHCl_3$ to remove interferences.

l. Water, distilled or deionized, MBAS-

_____

*A suitable reference material can be obtained from U.S. Environmental Protection Agency, Environmental Monitoring and Support Laboratory, Cincinnati, Ohio 45268.
†Eastman No. P573 or equivalent.

free. Use for making all reagents and dilutions.

4. Procedure

a. Preparation of calibration curve: Prepare a series of 10 separatory funnels with 0, 1.00, 3.00, 5.00, 7.00, 9.00, 11.00, 13.00, 15.00, and 20.00 mL standard LAS solution. Add sufficient water to make the total volume 100 mL in each separatory funnel. Treat each standard as described in ¶s 4*d* and *e* following, and plot a calibration curve of absorbance vs. micrograms LAS taken, specifying the molecular weight of the LAS used.

b. Sample size: For direct analysis of waters and wastewaters, select sample volume on the basis of expected MBAS concentration:

Expected MBAS Concentration *mg/L*	Sample Taken *mL*
0.025–0.080	400
0.08 –0.40	250
0.4 –2.0	100

If expected MBAS concentration is above 2 mg/L, dilute sample containing 40 to 200 µg MBAS to 100 mL with water.

For analysis of samples purified by sublation, dissolve sublate residue (Section 5540B.4*e*) in 10 to 20 mL methanol, quantitatively transfer the entire amount (or a suitable portion if more than 200 µg MBAS is expected) to 25 to 50 mL water, evaporate without boiling until methanol is gone, adding water as necessary to avoid going to dryness, and dilute to about 100 mL with water.

c. Peroxide treatment: If necessary to avoid decolorization of methylene blue by sulfides, add a few drops of 30% H_2O_2.

d. Ion pairing and extraction:

1) Add sample to a separatory funnel. Make alkaline by dropwise addition of 1*N* NaOH, using phenolphthalein indicator.

Discharge pink color by dropwise addition of 1*N* H_2SO_4.

2) Add 10 mL $CHCl_3$ and 25 mL methylene blue reagent. Rock funnel vigorously for 30 s and let phases separate. Alternatively, place a magnetic stirring bar in the separatory funnel; lay funnel on its side on a magnetic mixer and adjust speed of stirring to produce a rocking motion. Excessive agitation may cause emulsion formation. To break persistent emulsions add a small volume of isopropyl alcohol (< 10 mL); add same volume of isopropyl alcohol to all standards. Some samples require a longer period of phase separation than others. Before draining $CHCl_3$ layer, swirl gently, then let settle.

3) Draw off $CHCl_3$ layer into a second separatory funnel. Rinse delivery tube of first separatory funnel with a small amount of $CHCl_3$. Repeat extraction two additional times, using 10 mL $CHCl_3$ each time. If blue color in water phase becomes faint or disappears, discard and repeat, using a smaller sample.

4) Combine all $CHCl_3$ extracts in the second separatory funnel. Add 50 mL wash solution and shake vigorously for 30 s. Emulsions do not form at this stage. Let settle, swirl, and draw off $CHCl_3$ layer through a funnel containing a plug of glass wool into a 100-mL volumetric flask; filtrate must be clear. Extract wash solution twice with 10 mL $CHCl_3$ each and add to flask through the glass wool. Rinse glass wool and funnel with $CHCl_3$. Collect washings in volumetric flask, dilute to mark with $CHCl_3$, and mix well.

e. Measurement: Determine absorbance at 652 nm against a blank of $CHCl_3$.

5. Calculation

From the calibration curve (¶ 4*a*) read micrograms of apparent LAS (mol wt _) corresponding to the measured absorbance.

$$\text{mg MBAS/L} = \frac{\mu\text{g apparent LAS}}{\text{mL original sample}}$$

Report as "MBAS, calculated as LAS, mol wt __."

6. Precision and Bias

A synthetic sample containing 270 μg LAS/L in distilled water was analyzed in 110 laboratories with a relative standard deviation of 14.8% and a relative error of 10.6%.

A tap water sample to which was added 480 μg LAS/L was analyzed in 110 laboratories with a relative standard deviation of 9.9% and a relative error of 1.3%.

A river water sample with 2.94 mg LAS/L added was analyzed in 110 laboratories with a relative standard deviation of 9.1% and a relative error of 1.4%.[4]

7. References

1. AMERICAN PUBLIC HEALTH ASSOCIATION, AMERICAN WATER WORKS ASSOCIATION & WATER POLLUTION CONTROL FEDERATION. 1981. Carbon adsorption-infrared method. *In* Standard Methods for the Examination of Water and Wastewater, 15th ed. American Public Health Assoc., Washington, D.C.
2. OGDEN, C.P., H.L. WEBSTER & J. HALLIDAY. 1961. Determination of biologically soft and hard alkylbenzenesulfonates in detergents and sewage. *Analyst* 86:22.
3. OSBURN, Q.W. 1986. Analytical methodology for LAS in waters and wastes. *J. Amer. Oil Chem. Soc.* 63:257.
4. LISHKA, R.J. & J.H. PARKER. 1968. Water Surfactant No. 3, Study No. 32. U.S. Public Health Serv. Publ. No. 999-UIH-11, Cincinnati, Ohio.

8. Bibliography

BARR, T., J. OLIVER & W.V. STUBBINGS. 1948. The determination of surface-active agents in solution. *J. Soc. Chem. Ind.* (London) 67:45.

EPTON, S.R. 1948. New method for the rapid titrimetric analysis of sodium alkyl sulfates and related compounds. *Trans. Faraday Soc.* 44:226.
EVANS, H.C. 1950. Determination of anionic synthetic detergents in sewage. *J. Soc. Chem. Ind.* (London) 69:Suppl. 2, S76.
DEGENS, P.N., JR., H.C. EVANS, J.D. KOMMER & P.A. WINSOR. 1953. Determination of sulfate and sulfonate anion-active detergents in sewage. *J. Appl. Chem.* (London) 3:54.
AMERICAN WATER WORKS ASSOCIATION. 1954. Task group report. Characteristics and effects of synthetic detergents. *J. Amer. Water Works Assoc.* 46:751.
EDWARDS, G.P. & M.E. GINN. 1954. Determination of synthetic detergents in sewage. *Sewage Ind. Wastes* 26:945.
LONGWELL, J. & W.D. MANIECE. 1955. Determination of anionic detergents in sewage, sewage effluents, and river water. *Analyst* 80:167.
MOORE, W.A. & R.A. KOLBESON. 1956. Determination of anionic detergents in surface waters and sewage with methyl green. *Anal. Chem.* 28:161.
ROSEN, A.A., F.M. MIDDLETON & N. TAYLOR. 1956. Identification of synthetic detergents in foams and surface waters. *J. Amer. Water Works Assoc.* 48:1321.
SALLEE, E.M., et al. 1956. Determination of trace amounts of alkyl benzenesulfonates in water. *Anal. Chem.* 28:1822.
AMERICAN WATER WORKS ASSOCIATION. 1958. Task group report. Determination of synthetic detergent content of raw water supplies. *J. Amer. Water Works Assoc.* 50:1343.
McGUIRE, O.E., F. KENT, LaR. L. MILLER & G.J. PAPENMEIER. 1962. Field test for analysis of anionic detergents in well waters. *J. Amer. Water Works Assoc.* 54:665.
ABBOTT, D.C. 1962. The determination of traces of anionic surface-active materials in water. *Analyst* 87:286.
ABBOTT, D.C. 1963. A rapid test for anionic detergents in drinking water. *Analyst* 88:240.
REID, V.W., G.F. LONGMAN & E. HEINERTH. 1967. Determination of anionic-active detergents by two-phase titration. *Tenside* 4:292.
WANG, L.K., P.J. PANZARDI, W.W. SHUSTER & D. AULENBACH. 1975. Direct two-phase titration method for analyzing anionic nonsoap surfactants in fresh and saline waters. *J. Environ. Health* 38:159.

5540 D. Nonionic Surfactants as CTAS

1. General Discussion

a. *Definition and principle:* Cobalt thiocyanate active substances (CTAS) are those that react with aqueous cobalt thiocyanate solution to give a cobalt-containing product extractable into an organic liquid in which it can be measured. Nonionic surfactants exhibit such activity, as may other natural and synthetic materials; thus, estimation of nonionic surfactants as CTAS is possible only if substantial freedom from interfering CTAS species can be assured.

The method requires sublation to remove nonsurfactant interferences and ion exchange to remove cationic and anionic surfactants, partition of CTAS into methylene chloride from excess aqueous cobalt thiocyanate by a single extraction, and measurement of CTAS in the methylene chloride by spectrophotometry at 620 nm. Lower limit of detectability is around 0.1 mg CTAS, calculated as $C_{12-18}E_{11}$. Beyond the sublation step the procedure is substantially identical to that of the Soap and Detergent Association (SDA).[1]

b. *Nonionic surfactant responses:* For pure individual molecular species the CTAS response is negligible up to about RE_5, where it increases sharply and continues to increase more gradually for longer polyether chains.[2,3] Fewer than about six oxygens in the molecule do not supply enough cumulative coordinate bond strength to hold the complex together. Commercial nonionic surfactants generally range from about RE_7 to RE_{15}; however, each such product, because of synthesis process constraints, is actually a mixture of many individual species ranging from perhaps RE_0 to RE_{2n} in a Poisson distribution averaging RE_n.

The hydrophobes used for nonionic surfactants in the U.S. household detergent industry are mainly linear primary and linear secondary alcohols with chain lengths ranging from about 12 to about 18 carbon

atoms. Nonionics used in industrial operations include some based on branched octyl- and nonylphenols. These products give strong CTAS responses that may differ from each other, on a weight basis, by as much as a factor of 2. Specifically, eight such products showed responses from 0.20 to 0.36 absorbance units/mg by the SDA procedure.[1]

As with anionic surfactants measured as MBAS, the nonionic surfactants found in water and wastewater might have CTAS responses at least as varied as their commercial precursors because the proportions of the individual molecular species will have been changed by biochemical and physicochemical removal at varied rates, and further because their original molecular structures may have been changed by biodegradation processes.

c. *Reference nonionic surfactant:* Until it is practical to determine the nature and molecular composition of an unknown mixed CTAS, and to calculate or determine the CTAS responses of its component species, exact quantitation of uncharacterized CTAS in a sample in terms of weight is not possible. Instead, express the analytical result in terms of some arbitrarily chosen reference nonionic surfactant, i.e., as the weight of the reference that gives the same amount of CTAS response. The reference is the nonionic surfactant $C_{12-18}E_{11}$, derived from a mixture of linear primary alcohols ranging from 12 to 18 carbon atoms in chain length by reaction with ethylene oxide in a molar ratio of 1:11. $C_{12-18}E_{11}$ is reasonably representative of nonionic surfactants in commercial use; its CTAS response is about 0.21 absorbance units/mg.

If the identity of the nonionic surfactant in the sample is known, use that same material in preparing the calibration curve.

d. *Interferences:* Both anionic and cationic surfactants may show positive CTAS response[1,4] but both are removed in the ion-exchange step. Sublation removes nonsurfactant interferences. Physical interferences occur if some of the CTAS is adsorbed on particulate matter. Avoid such interference by filtering out the particulates for the sublation step; this will measure only dissolved CTAS.

e. Minimum detectable quantity: About 0.1 mg CTAS, calculated as $C_{12-18}E_{11}$, which corresponds to 0.1 mg/L in a 1-L sample.

f. Application: The method is suitable for determining dissolved nonionic surfactants of the ethoxylate type in most aqueous systems.

2. Apparatus

a. Sublation apparatus: See Section 5540B.2.

b. Ion-exchange column, glass, about 1- $\times$ 30-cm. Slurry anion-exchange resin in methanol and pour into column to give a bed about 10 cm deep. Insert plug of glass wool and then add a 10-cm bed of cation-exchange resin on top in the same manner. One column may be used for treating up to six sublated samples before repacking.

c. Spectrophotometer and 2.0-cm stoppered cells, suitable for measuring absorbance at 620 nm.

d. Separatory funnels, 125-mL, preferably with TFE stopcock and stopper.

e. Extraction flasks, Soxhlet type, 150-mL.

3. Reagents

a. Sublation reagents: See Section 5540B.3.

b. Anion-exchange resin, polystyrene-quaternary ammonium-type,* 50- to 100-mesh, hydroxide form. To convert chloride form to hydroxide, elute with 20 bed volumes of $1N$ NaOH and wash with methanol until free alkali is displaced.

c. Cation-exchange resin, polystyrene-sulfonate type,† 50- to 100-mesh, hydrogen form.

d. Cobaltothiocyanate reagent: Dissolve 30 g $Co(NO_3)_2 \cdot 6H_2O$ and 200 g NH_4SCN in water and dilute to 1 L. This reagent is stable for at least 1 month at room temperature.

e. Reference nonionic surfactant, $C_{12-18}E_{11}$: Reaction product of C_{12-18} linear primary alcohol with ethylene oxide in 1:11 molar ratio.‡

f. Reference nonionic surfactant stock solution, methanolic, approximately 2 mg nonionic/mL methanol: Quantitatively transfer entire contents (approximately 1 g nonionic) from preweighed ampule into 500-mL volumetric flask, thoroughly rinse ampule with methanol, make up to volume with methanol, and reweigh dried ampule. Calculate concentration in milligrams per milliter as in ¶ 5a. Because of possible phase separation, use all material in the ampule.

g. Reference nonionic surfactant standard solution, methanolic, approximately 0.1 mg nonionic/mL methanol: Dilute 10.00 mL stock solution to 200 mL with methanol. Exact concentration is 1/20 that of the stock solution.

h. Sodium hydroxide, NaOH: $1N$.

i. Glass wool: Pre-extract with chloroform or methylene chloride.

j. Methanol, CH_3OH: CAUTION: *Methanol vapors are flammable and toxic; take appropriate precautions.*

k. Methylene chloride, CH_2Cl_2: CAUTION: *Methylene chloride vapors are toxic; take adequate precautions.*

l. Water: Use distilled or deionized, CTAS-free water for making reagents and dilutions.

4. Procedure

a. Purification by sublation: Proceed according to Section 5540B, using sample containing no more than 2 mg CTAS. (NOTE: For samples of known character

---

*Bio-Rad, AGl-X2, or equivalent.
†Bio-Rad AG 50W-X8, or equivalent.

‡This material can be obtained from U.S. Environmental Protection Agency, Environmental Monitoring and Support Laboratory, Cincinnati, Ohio 45268.

containing no interfering materials, omit this step.)

b. *Ion-exchange removal of anionic and cationic surfactants:* Dissolve sublation residue in 5 to 10 mL methanol and transfer quantitatively to ion-exchange column. Elute with methanol at 1 drop/s into a clean, dry 150-mL extraction flask until about 125 mL is collected. Evaporate methanol on a steam bath aided by a gentle stream of clean, dry nitrogen or air, taking care to avoid loss by entrainment; remove from heat as soon as the methanol is completely evaporated. (NOTE: With samples of known character containing no anionic or cationic materials, omit step *b.*)

c. *CTAS calibration curve:* Into a series of 150-mL extraction flasks containing 10 to 20 mL methanol place 0.00, 5.00, 10.00, 20.00, and 30.00 mL reference nonionic surfactant standard solution and evaporate just to dryness. Continue as in ¶s *4d* and *e*, below, and plot a calibration curve of absorbance against milligrams of reference nonionic taken, specifying its identity (e.g., $C_{12-18}E_{11}$ and lot number).

d. *Cobalt complexing and extraction:* Charge a 125-mL separatory funnel with 5 mL cobaltothiocyanate reagent. With precautions against excessive and variable evaporation of the methylene chloride, dissolve residue from ion-exchange operation, ¶ *4b*, by adding 10.00 mL methylene chloride and swirling for a few seconds. Immediately transfer by pouring into the separatory funnel. *Do not rinse flask.* (NOTE: Because of the volatility of methylene chloride, rigidly standardize these operations with respect to handling and elapsed time; alternatively, evaporate the methanol in 200-mL erlenmeyer flasks to be stoppered with glass or TFE stoppers during dissolution. Transfer as directed here is incomplete, but in this case it will not introduce error because the loss of nonionics is exactly compensated for by the diminished volume of the organic layer in the extraction.) Shake separatory funnel

vigorously for 60 s and let layers separate. Run lower layer into a 2.0-cm cell through a funnel containing a plug of pre-extracted glass wool and stopper. Be sure filtrate is absolutely clear. (NOTE: If desired, clarify by running the lower layer into a 12-mL centrifuge tube, stopper, spin at or above $1000 \times g$ for 3 min, and transfer to the cell by a Pasteur pipet; use same procedure for both calibration and samples.)

e. *Measurement:* Determine absorbance at 620 nm against a blank of methylene chloride. (NOTE: If haze develops in the cell, warm slightly with a hot air gun or heat lamp to clarify.)

5. Calculations

a. *Nonionic surfactant in reference nonionic stock solution ¶ 3f:*

mg nonionic/mL methanol

$$= \text{mg reference sample}/500 \text{ mL}$$

b. *Nonionic surfactant in sample:* From the calibration curve read milligrams of reference nonionic corresponding to the measured absorbance:

mg CTAS/L = mg apparent nonionic/L sample

Report as "CTAS, calculated as nonionic surfactant $C_{12-18}E_{11}$."

6. Precision and Bias

Twenty-four samples of 6.22% w/v solution of reference nonionic surfactant $C_{12-18}E_{11}$ were analyzed in three laboratories by CTAS alone, without sublation or ion exchange. The overall relative standard deviation was about 3%. Results of the three laboratories individually were:

Laboratory	% w/w ± SD
A	6.08 ± 0.14 ($n = 36$)
B	6.56 ± 0.17 ($n = 6$)
C	6.25 ± 0.14 ($n = 36$)
Overall	6.20 ± 0.19 ($n = 78$)

Samples of raw wastewater were freed of surfactants by four successive sublations, then 0.50 or 0.67 mg reference nonionic surfactant $C_{12-18}E_{11}$ was added and carried through the entire sequence of sublation, ion exchange, and CTAS extraction. Recoveries averaged 92% with overall standard deviation around 6%:

Laboratory	% Recovery ± SD
A	87 ± 4 ($n = 4$)
B	97 ± 1 ($n = 4$)
Overall	92 ± 6 ($n = 8$)

The above data relate to the bias and precision of the method when applied to a known nonionic surfactant. When the nature of the nonionic surfactant is unknown, there is greater uncertainty. The response of the reference $C_{12-18}E_{11}$ is about 0.21 absorbance units/mg, while that of the eight nonionic types mentioned under ¶ 1b ranged from 0.20 to 0.36, and environmental nonionics might differ still more. If the nonionic surfactant in the sample has a response of 0.42, the result calculated in terms of milligrams $C_{12-18}E_{11}$ would be double the actual milligrams of the unknown nonionic.

7. References

1. SOAP AND DETERGENT ASSOCIATION ANALYTICAL SUBCOMMITTEE. 1977. Analytical methods for nonionic surfactants in laboratory biodegradation and environmental studies. *Environ. Sci. Technol.* 11:1167.
2. CRABB, N.T. & H.E. PERSINGER. 1964. The determination of polyoxyethylene nonionic surfactants in water at the parts per million level. *J. Amer. Oil Chem. Soc.* 41:752.
3. CRABB, N.T. & H.E. PERSINGER. 1968. A determination of the apparent molar absorption coefficients of the cobalt thiocyanate complexes of nonylphenol ethylene oxide adducts. *J. Amer. Oil Chem. Soc.* 45:611.
4. GREFF, R.A., E.A. SETZKORN & W.D. LESLIE. 1965. A colorimetric method for the determination of parts per million of nonionic surfactants. *J. Amer Oil Chem. Soc.* 42:180.

8. Bibliography

TABAK, H.H. & R.L. BUNCH. 1981. Measurement of non-ionic surfactants in aqueous environments. Proc. 36th Ind. Waste Conf., p. 888. Purdue Univ., Lafayette, Ind.

5550   TANNIN AND LIGNIN*

5550 A.   Introduction

Lignin is a plant constituent that often is discharged as a waste during the manufacture of paper pulp. Another plant constituent, tannin, may enter the water supply through the process of vegetable matter degradation or through the wastes of the tanning industry. Tannin also is applied in the so-called internal treatment of boiler waters, where it reduces scale formation by causing the production of a more easily handled sludge.

* Approved by Standard Methods Committee, 1988.

5550 B.   Colorimetric Method

1. General Discussion

a. Principle: Both lignin and tannin contain aromatic hydroxyl groups that react with Folin phenol reagent (tungstophosphoric and molybdophosphoric acids) to form a blue color suitable for estimation of concentrations up to at least 9 mg/L. However, the reaction is not specific for lignin or tannin, nor for compounds containing aromatic hydroxyl groups, inasmuch as many other reducing materials, both organic and inorganic, respond similarly.

b. Applicability: This method is generally suitable for the analysis of any organic chemical that will react with Folin phenol reagent to form measurable blue color at the concentration of interest. However, many compounds are reactive (see ¶ 1c) and each yields a different molar extinction coefficient (color intensity). Hence, the analyst must demonstrate conclusively the absence of interfering substances.

c. Interferences: Any substance able to reduce Folin phenol reagent will produce a false positive response. Organic chemicals known to interfere include hydroxylated aromatics, proteins, humic substances, nucleic acid bases, fructose, and amines. Inorganic substances known to interfere include iron (II), manganese (II), nitrite, cyanide, bisulfite, sulfite, sulfide, hydrazine, and hydroxylamine hydrochloride. Both 2 mg ferrous iron/L and 125 mg sodium sulfite/L individually produce a color equivalent to 1 mg tannic acid/L.

d. Minimum detectable concentrations: Approximately 0.025 mg/L for phenol and tannic acid and 0.1 mg/L for lignin with a 1-cm-path-length spectrophotometer.

2. Apparatus

Colorimetric equipment: One of the following is required:

a. Spectrophotometer, for use at 700 nm. A light path of 1 cm or longer yields satisfactory results.

b. Filter photometer, provided with a red filter exhibiting maximum transmittance in the wavelength range of 600 to 700 nm. Sensitivity improves with increasing wavelength. A light path of 1 cm or longer yields satisfactory results.

c. Nessler tubes, matched, 100-mL, tall form, marked at 50-mL volume.

3. Reagents

a. Folin phenol reagent: Transfer 100 g sodium tungstate, $Na_2WO_4 \cdot 2H_2O$, and 25 g sodium molybdate, $Na_2MoO_4 \cdot 2H_2O$, together with 700 mL distilled water, to a 2000-mL flat-bottom boiling flask. Add 50 mL 85% H_3PO_4 and 100 mL conc HCl. Connect to a reflux condenser and boil gently for 10 h. Add 150 g Li_2SO_4, 50 mL distilled water, and a few drops of liquid bromine. Boil without condenser for 15 min to remove excess bromine. Cool to 25°C, dilute to 1 L, and filter. Store finished reagent, which should have no greenish tint, in a tightly stoppered bottle to protect against reduction by air-borne dust and organic materials.

Alternatively, purchase commercially prepared Folin phenol reagent and use before the recommended expiration date.

b. Carbonate-tartrate reagent: Dissolve 200 g Na_2CO_3 and 12 g sodium tartrate, $Na_2C_4H_4O_6 \cdot 2H_2O$, in 750 mL hot distilled water, cool to 20°C, and dilute to 1 L.

c. Stock solution: The nature of the substance present in the sample dictates the choice of chemical used to prepare the standard, because each substance produces a different color intensity. Weigh 1.000 g tannic acid, tannin, lignin, or other compound being used for boiler water treatment or known to be a contaminant of the water sample. Dissolve in distilled water and dilute to 1000 mL. If the identity of the compound in the water sample is not known, use phenol and report results as "substances reducing Folin phenol reagent" in mg phenol/L. Interpret such results with caution.

Note that tannin and lignin are not individual chemical species of known molecular weight and structure; rather, they are substances containing a spectrum of chemicals of different molecular weights. Their chemical properties depend on source and method of isolation. If a particular substance is being added to the water, use it to prepare the stock solution.

    d. Standard solution: Dilute 10.00 mL or 50.00 mL stock solution to 1000 mL with distilled water; 1.00 mL = 10.0 or 50.0 μg active ingredient.

4. Procedure

Bring 50-mL portions of clear sample and standards to a temperature above 20°C and maintain within a ± 2°C range. Add in rapid succession 1 mL Folin phenol reagent and 10 mL carbonate-tartrate reagent. Allow 30 min for color development. Compare visually against simultaneously prepared standards in matched Nessler tubes or make photometric readings against a reagent blank prepared at the same time. Use the following guide for instrumental measurement at a wavelength of 700 nm:

Tannic Acid in 61-mL Final Volume	Lignin in 61-mL Final Volume	Light Path
μg	μg	cm
50–600	100–1500	1
10–150	30–400	5

Report results in mg/L of the compound known to be present or as "substances reducing Folin phenol reagent" in mg phenol/L.

5. Precision and Bias

In a single laboratory analyzing seven replicates for phenol at 0.1 mg/L the precision was ± 7% and recovery was 107%.

6. Bibliography

FOLIN, O. & V. CIOCALTEU. 1927. On tyrosine and tryptophane determinates in proteins. *J. Biol. Chem.* 73:627.

BERK, A.A. & W.C. SCHROEDER. 1942. Determination of tannin substances in boiler water. *Ind. Eng. Chem.,* Anal. Ed. 14:456.

KLOSTER, M.B. 1973. Determination of tannin and lignin. *J. Amer. Water Works Assoc.* 66:44.

BOX, J.D. 1983. Investigation of the Folin-Ciocalteu phenol reagent for the determination of polyphenolic substances in natural waters. *Water Res.* 17:511; discussion by S.J. Randtke & R.A. Larson. 1984. *Water Res.* 18:1597.

5560 ORGANIC AND VOLATILE ACIDS*

5560 A. Introduction

The measurement of organic acids, either by adsorption and elution from a chromatographic column or by distillation, can be used as a control test for anaerobic digestion. The chromatographic separation method is presented for organic acids (B), while a method using distillation (C) is presented for volatile acids.

Volatile fatty acids are classified as water-soluble fatty acids that can be distilled at atmospheric pressure. These vol-

---

*Approved by Standard Methods Committee, 1985.

atile acids can be removed from aqueous solution by distillation, despite their high boiling points, because of co-distillation with water. This group includes water-soluble fatty acids with up to six carbon atoms.

The distillation method is empirical and gives incomplete and somewhat variable recovery. Factors such as heating rate and proportion of sample recovered as distillate affect the result, requiring the determination of a recovery factor for each apparatus and set of operating conditions. However, it is suitable for routine control purposes. Removing sludge solids from the sample reduces the possibility of hydrolysis of complex materials to volatile acids.

5560 B. Chromatographic Separation Method for Organic Acids

1. General Discussion

a. Principle: An acidified aqueous sample containing organic acids is adsorbed on a column of silicic acid and the acids are eluted with *n*-butanol in chloroform ($CHCl_3$). The eluate is collected and titrated with standard base. All short-chain (C_1 to C_6) organic acids are eluted by this solvent system and are reported collectively as total organic acids.

b. Interference: The $CHCl_3$-butanol solvent system is capable of eluting organic acids other than the volatile acids and also some synthetic detergents. Besides the so-called volatile acids, crotonic, adipic, pyruvic, phthalic, fumaric, lactic, succinic, malonic, gallic, aconitic, and oxalic acids; alkyl sulfates; and alkyl-aryl sulfonates are adsorbed by silicic acid and eluted.

c. Precautions: Basic alcohol solutions decrease in strength with time, particularly when exposed repeatedly to the atmosphere. These decreases usually are accompanied by the appearance of a white precipitate. The magnitude of such changes normally is not significant in process control if tests are made within a few days of standardization. To minimize this effect, store standard sodium hydroxide (NaOH) titrant in a tightly stoppered borosilicate glass bottle and protect from atmospheric carbon dioxide (CO_2) by attaching a tube of CO_2-absorbing material, as described in the inside front cover. For more precise analyses, standardize titrant or prepare before each analysis.

Although the procedure is adequate for routine analysis of most sludge samples, volatile-acids concentrations above 5000 mg/L may require an increased amount of organic solvent for quantitative recovery. Elute with a second portion of solvent and titrate to reveal possible incomplete recoveries.

2. Apparatus

a. Centrifuge or filtering assembly.

b. Crucibles, Gooch or medium-porosity fritted-glass, with filtering flask and vacuum source. Use crucibles of sufficient size (30 to 35 mL) to hold 12 g silicic acid.

c. Separatory funnel, 1000-mL.

3. Reagents

a. Silicic acid, specially prepared for chromatography, 50 to 200 mesh: Remove fines by slurrying in distilled water and decanting supernatant after settling for 15 min. Repeat several times. Dry washed acid in an oven at 103°C until *absolutely dry,* then store in a desiccator.

b. Chloroform-butanol reagent: Mix 300 mL reagent-grade $CHCl_3$, 100 mL *n*-butanol, and 80 mL 0.5N H_2SO_4 in a separatory funnel. Let water and organic layers separate. Drain off lower organic layer through a fluted filter paper into a dry bottle.

c. *Thymol blue indicator solution:* Dissolve 80 mg thymol blue in 100 mL absolute methanol.

d. *Phenolphthalein indicator solution:* Dissolve 80 mg phenolphthalein in 100 mL absolute methanol.

e. *Sulfuric acid,* H_2SO_4, conc.

f. *Standard sodium hydroxide,* NaOH, 0.02N: Dilute 20 mL 1.0N NaOH stock solution to 1 L with absolute methanol. Prepare stock in water and standardize in accordance with the methods outlined in Section 2310B.3d.

4. Procedure

a. *Pretreatment of sample:* Centrifuge or vacuum-filter enough sludge to obtain 10 to 15 mL clear sample in a small test tube or beaker. Add a few drops of thymol blue indicator solution, then conc H_2SO_4 dropwise, until definitely red to thymol blue (pH = 1.0 to 1.2).

b. *Column chromatography:* Place 12 g silicic acid in a Gooch or fritted-glass crucible and apply suction to pack column. Tamp column while applying suction to reduce channeling when the sample is applied. With a pipet, distribute 5.0 mL acidified sample as uniformly as possible over column surface. Apply suction momentarily to draw sample into silicic acid. Release vacuum as soon as last portion of sample has entered column. Quickly add 65 mL $CHCl_3$-butanol reagent and apply suction. Discontinue suction just before the last of reagent enters column. Use a new column for each sample.

c. *Titration:* Remove filter flask and purge eluted sample with N_2 gas or CO_2-free air immediately before titrating. (Obtain CO_2-free air by passing air through a CO_2 absorbant.*)

Titrate sample with standard 0.02N NaOH to phenolphthalein end point, using a fine-tip buret and taking care to avoid aeration. The fine-tip buret aids in improving accuracy and precision of the titration. Use N_2 gas or CO_2-free air delivered through a small glass tube to purge and mix sample and to prevent contact with atmospheric CO_2 during titration.

d. *Blank:* Carry a distilled water blank through steps ¶s 4a through 4c.

5. Calculation

Total organic acids (mg as acetic acid/L)

$$= \frac{(a - b) \times N \times 60\,000}{\text{mL sample}}$$

where:

    a = mL NaOH used for sample,
    b = mL NaOH used for blank, and
    N = normality of NaOH.

6. Precision

Average recoveries of about 95% are obtained for organic acid concentrations above 200 mg as acetic acid/L. Individual tests generally vary from the average by approximately 3%. A greater variation results when lower concentrations of organic acids are present. Titration precision expressed as the standard deviation is about ±0.1 mL (approximately ±24 mg as acetic acid/L).

7. Bibliography

MUELLER, H.F., A.M. BUSWELL & T.E. LARSON. 1956. Chromatographic determination of volatile acids. *Sewage Ind. Wastes* 28:255.

MUELLER, H.F., T.E. LARSON & M. FERRETTI. 1960. Chromatographic separation and identification of organic acids. *Anal. Chem.* 32:687.

WESTERHOLD, A.F. 1963. Organic acids in diges-

---

*Ascarite or equivalent.

ter liquor by chromatography. *J. Water Pollut. Control Fed.* 35:1431.

HATTINGH, W.H.J. & F.V. HAYWARD. 1964. An improved chromatographic method for the determination of total volatile fatty acid con-

tent in anaerobic digester liquors. *Int. J. Air Water Pollut.* 8:411.

POHLAND, F.G. & B.H. DICKSON, JR. 1964. Organic acids by column chromatography. *Water Works Wastes Eng.* 1:54.

5560 C.　Distillation Method

1. General Discussion

a. Principle: This technique recovers acids containing up to six carbon atoms. Fractional recovery of each acid increases with increasing molecular weight. Calculations and reporting are on the basis of acetic acid. The method often is applicable for control purposes. Because it is empirical, carry it out exactly as described. Because the still-heating rate, presence of sludge solids, and final distillate volume affect recovery, determine a recovery factor.

b. Interference: Hydrogen sulfide (H_2S) and CO_2 are liberated during distillation and will be titrated to give a positive error. Eliminate this error by discarding the first 15 mL of distillate and account for this in the recovery factor.

2. Apparatus

a. Centrifuge, with head to carry four 50-mL tubes or 250-mL bottles.

b. Distillation flask, 500-mL capacity.

c. Condenser, about 76 cm long.

d. Adapter tube.

e. pH meter or recording titrator: See Section 2310B.2a.

f. Distillation assembly: Use a conventional distilling apparatus. To minimize fluctuations in distillation rate, supply heat with a variable-wattage electrical heater.

3. Reagents

a. Sulfuric acid, H_2SO_4, 1 + 1.

b. Standard sodium hydroxide titrant, 0.1*N:* See Section 2310B.3c.

c. Phenolphthalein indicator solution.

d. Acetic acid stock solution, 2000 mg/L: Dilute 1.9 mL conc CH_3COOH to 1000 mL with deionized water. Standardize against 0.1*N* NaOH.

4. Procedure

a. Recovery factor: To determine the recovery factor, *f,* for a given apparatus, dilute an appropriate volume of acetic acid stock solution to 250 mL in a volumetric flask to approximate the expected sample concentration and distill as for a sample. Calculate the recovery factor

$$f = \frac{a}{b}$$

where:

a = volatile acid concentration recovered in distillate, mg/L, and

b = volatile acid concentration in standard solution used, mg/L.

b. Sample analysis: Centrifuge 200 mL sample for 5 min. Pour off and combine supernatant liquors. Place 100 mL supernatant liquor in a 500-mL distillation flask. Add 100 mL distilled water, four to five clay chips or similar material to prevent bumping, and 5 mL H_2SO_4. Mix so that acid does not remain on bottom of flask. Connect flask to a condenser and adapter tube and distill at the rate of about 5 mL/min. Discard the first 15 mL and collect exactly 150 mL distillate in a 250-mL graduated cylinder. Titrate with $0.1N$ NaOH, using phenolphthalein indicator, a pH meter, or an automatic titrator. The end points of these three methods are, respectively, the first pink coloration that persists on standing a short time, pH 8.3, and the inflection point of the titration curve (see Section 2310). Titration at 95°C produces a stable end point.

5. Calculation

mg volatile acids as acetic acid/L

$$= \frac{mL\ NaOH \times N \times 60\ 000}{mL\ sample \times f}$$

where:

N = normality of NaOH, and
f = recovery factor.

6. Bibliography

OLMSTEAD, W.H., W.M. WHITAKER & C.W. DUDEN. 1929-1930. Steam distillation of the lower volatile fatty acids from a saturated salt solution. *J. Biol. Chem.* 85:109.

OLMSTEAD, W.H., C.W. DUDEN, W.M. WHITAKER & R.F. PARKER. 1929-1930. A method for the rapid distillation of the lower volatile fatty acids from stools. *J. Biol. Chem.* 85:115.

BUSWELL, A.M. & S.L. NEAVE. 1930. Laboratory studies of sludge digestion. *Ill. State Water Surv. Bull.* 30:76.

HEUKELEKIAN, H. & A.J. KAPLOVSKY. 1949. Improved method of volatile-acid recovery from sewage sludges. *Sewage Works J.* 21:974.

KAPLOVSKY, A.J. 1951. Volatile-acid production during the digestion of seeded, unseeded, and limed fresh solids. *Sewage Ind. Wastes* 23:713.

5710   TRIHALOMETHANE FORMATION (PROPOSED)*

5710 A.   Introduction

Trihalomethanes (THMs) are produced during chlorination of water. There are many predictive models for estimating/calculating THM formation, but because eventual THM concentrations cannot be calculated precisely from conventional analyses, a method to determine the potential for forming THMs is useful in evaluating water treatment processes or water sources or for predicting THM concentrations in a distribution system.

To obtain reproducible and meaningful results, control such variables as temperature, reaction time, chlorine dose and residual, and pH. THM formation is enhanced by elevated temperatures and

pH; a longer reaction time generally increases THM formation.[1,2]

1. Definition of Terms

Trihalomethane formation potential (TFP) is the concentration of THMs formed in water buffered at pH 7.0, containing an excess of free chlorine with a chlorine residual of 1 to 5 mg/L after being held 7 d at 25°C. This test does not simulate the water treatment process but is useful in estimating THM precursors.

Basic trihalomethane formation potential (BTFP) is identical to TFP except that the sample is buffered at pH 9.2, simulating lime-softened or naturally high-pH water and accelerating THM formation. BTFP is very similar to maximum total trihalome-

*Approved by Standard Methods Committee, 1988.

thane potential,[3] the primary differences being the reagent preparation procedures and required residual chlorine concentration.

Ultimate trihalomethane formation potential (UTFP) estimates the highest THM concentrations that might be produced at the extreme conditions of a public water supply system (chlorine dose, temperature, pH, and contact time). These results may be of regulatory or monitoring value.

Maximum trihalomethane potential (MTP) is determined by taking triplicate samples from a point in the distribution system that reflects maximum residence time, storing them for 7 d at or above 25°C, and determining the total THM concentration if a detectable free chlorine residual remains in one of the samples.[4,5] MTP determination is not included in the methods of this section because of concern that a combined chlorine residual could be mistaken for a free chlorine residual.

Simulated distribution system trihalomethane concentration (SDSTHMC) is the concentration of THMs in a previously chlorinated water sample after storage that represents the time and conditions in the utility's distribution system. It includes pre-existing THMs plus those produced during storage. This method can be used in conjunction with laboratory, pilot, or full-scale studies of treatment processes to assess their effectiveness in reducing THM formation.

2. Sampling and Storage

Collect samples in 1-L glass bottles sealed with TFE-lined screw caps. One liter volume provides enough sample to deter-mine chlorine demand and duplicate THM analyses. Store unused sample at 4°C and use for repeat analysis, if necessary. For accuracy, use only fresh samples.

The procedures for TFP, BTFP, and UTFP are intended for previously unchlorinated water samples. If the sample has been chlorinated previously collect the sample with minimum turbulence and fill the sample bottle completely to avoid loss of THMs already present. Determine the instantaneous THM concentration, if desired, on another sample collected at the same time and dechlorinated immediately.

3. References

1. STEVENS, A.A. & J.M. SYMONS. 1977. Measurement of trihalomethanes and precursor concentration changes. *J. Amer. Water Works Assoc.* 69:546.

2. SYMONS, J.M., A.A. STEVENS, R.M. CLARK, E.E. GELDREICH, O.T. LOVE, JR. & J. DE-MARCO. 1981. Treatment Techniques for Controlling Trihalomethanes in Drinking Water. EPA-600/2-81-156, U.S. Environmental Protection Agency, Cincinnati, Ohio.

3. BELLAR, T.A. 1982. The Determination of the Maximum Total Trihalomethane Potential, Method 510.1. EPA-600/4-81-044, U.S. Environmental Protection Agency.

4. U.S. ENVIRONMENTAL PROTECTION AGENCY. July 1, 1980. Determination of Maximum Trihalomethane Potential (MTP). 40 CFR Part 141, Appendix C, Part III.

5. AMERICAN PUBLIC HEALTH ASSOCIATION, AMERICAN WATER WORKS ASSOCIATION & WATER POLLUTION CONTROL FEDERATION. 1981. Supplement to the Fifteenth Edition of Standard Methods for the Examination of Water and Wastewater: Selected Analytical Methods Approved and Cited by the United States Environmental Protection Agency. American Public Health Assoc., Washington, D.C.

5710 B. Trihalomethane Formation Potential (TFP)

1. General Discussion

a. Principle: Samples are buffered at pH 7.0, chlorinated with an excess of free chlorine, and stored at 25°C for 7 d to allow the reaction to approach completion. THM concentration is determined by using liquid-liquid extraction (Method 6232B) or purge and trap (Method 6230C or D).

b. Interference: If the water was exposed to free chlorine before sample collection (e.g., in a water treatment plant), a fraction of the precursor material may have been converted to THM. Take special precautions to avoid loss of volatile THMs by minimizing turbulence and filling sample bottle completely.

Interference will be caused by any organic THM-precursor materials present in the reagents or adsorbed on glassware. Heat nonvolumetric glassware to 400°C for 1 h, unless routine analysis of blanks demonstrates that this precaution is unnecessary. Reagent impurity is a difficult interference to control. It usually is traceable to reagent water containing bromide ion or organic impurities. Use high-grade reagent water as free of organic contamination and chlorine demand as possible. If anion exchange is used to remove bromide or organic ions, follow such treatment by carbon adsorption (see Section 1080).

Other interferences include volatile organic chemicals, including THMs and chlorine-demanding substances. Volatile organics may co-elute with THMs during analysis while THMs present as the result of a chemical spill, etc., will bias the results.

Nitrogenous species and other constituents may interfere in the determination of free residual chlorine concentration. Add enough free chlorine in Methods 5710B, C, and D to oxidize the chorine-demanding substances and leave a free chlorine residual of at least 1 mg/L and preferably 2 to 5 mg/L. A free chlorine residual of at least 1 mg/L is specified to reduce the likelihood that a combined residual will be mistaken for a free residual.

c. Minimum detectable quantity: The sensitivity of the method is determined by the analytical procedure used for THM.

2. Apparatus

a. Incubator: To maintain temperature of 25 ± 2°C during the reaction period.

b. Bottles, glass, with TFE-lined screw caps to contain 245 to 255 mL.

c. Vials, glass, 25-mL with TFE-lined screw caps.

d. pH meter, accurate to within ± 0.1 unit.

3. Reagents

a. Hypochlorite solution: Dilute 1 mL 5% aqueous sodium hypochlorite (NaOCl) solution to 25 mL with chlorine-demand-free water (see ¶ 3*f* below), mix well, and titrate to a starch-iodide end point using 0.100*N* sodium thiosulfate titrant (see Section 4500Cl.B.2*c*). Calculate chlorine concentration as:

Hypochlorite, mg Cl_2/mL

$$= \frac{N \times 35.45 \times \text{mL titrant}}{\text{mL hypochlorite added}}$$

where *N* is the normality of the titrant. The titration should require at least 10 mL titrant; if not, use 2 mL hypochlorite solution. Measure chlorine concentration each time a dosing solution is made; discard the hypochlorite solution if chlorine concentration is less than 20 mg Cl_2/mL.

b. Chlorine dosing solution, 5 mg Cl_2/mL: Calculate volume of hypochlorite solution required:

$$\text{mL required} = \frac{1250}{\text{hypochlorite conc (mg } Cl_2/mL)}$$

Dilute this volume in a 250-mL volumetric flask to the mark with chlorine-demand-free water. Mix and transfer to an amber bottle, seal with a TFE-lined screw cap, and refrigerate. Keep away from sunlight. Discard after 1 week.

c. *Phosphate buffer:* Dissolve 68.1 g potassium dihydrogen phosphate, KH_2PO_4, and 11.7 g sodium hydroxide, NaOH, in 1 L organic-free water. Refrigerate in airtight container. Discard after 1 week.

d. *Sodium sulfite solution:* Dissolve 10 g sodium sulfite, Na_2SO_3, in 100 mL organic-free water. Use for dechlorination: 0.1 mL will destroy about 5 mg residual chlorine. Make fresh every 2 weeks.

e. *Organic-free water:* Pass distilled or deionized water through granular-activated-carbon columns. A commercial system may be used.* Special techniques such as preoxidation, carbon adsorption (perhaps accompanied by acidification and subsequent reneutralization), or purging with an inert gas may be necessary.

f. *Chlorine-demand-free water:* Follow the procedure outlined in 4500-Cl.C.3m, starting with organic-free water. After residual chlorine has been completely destroyed, purge by passing a clean, inert gas through the water.

g. *DHBA solution:* Dissolve 0.078 g anhydrous 3,5-dihydroxybenzoic acid (DHBA) in 2 L chlorine-demand-free water. This solution is not stable; make fresh before each use.

h. *Nitric or hydrochloric acid,* HNO_3 or HCl, conc, 1.0, and 0.1N: Prepare with organic-free water.

i. *Sodium hydroxide,* NaOH, 1.0 and 0.1N: Prepare with organic-free water.

---

*Milli-Q, Millipore Corp., or equivalent.

4. Procedure

a. *Chlorine demand determination:* Determine or accurately estimate sample chlorine demand. A high chlorine dose is specified below to drive the reaction close to completion quickly, but smaller chlorine doses or shorter reaction times may be used if background knowledge of the sample is available.

Pipet 5 mL chlorine dosing solution into a 250-mL bottle, fill completely with chlorine-demand-free water, and cap with a TFE-lined screw cap. Titrate 100 mL with 0.025N sodium thiosulfate to determine the initial chlorine concentration. This should be about 100 mg Cl_2/L. Pipet 5 mL phosphate buffer and 5 mL chlorine dosing solution into a second 250-mL bottle, fill completely with sample, and seal with a TFE-lined screw cap. Store in the dark for at least 4 h at 25°C. After storage, determine chlorine residual. Calculate chlorine demand as follows:

$$CD = D_0 - R$$

where:
  CD = chlorine demand, mg Cl_2/L,
  R = chlorine residual of sample after at least 4 h storage, mg Cl_2/L, and
  D_0 = dosed chlorine concentration, mg Cl_2/L.

b. *Sample chlorination:* If sample contains more than 200 mg/L alkalinity or acidity, adjust pH to 7.0 using 0.1 or 1.0N HNO_3, HCl, or NaOH and a pH meter. With a graduated pipet, transfer appropriate volume of chlorine dosing solution, $[(CD \times 5)/D_0] + 0.1$, to a 250-mL bottle. Add 5 mL phosphate buffer solution and fill completely with sample. Immediately seal with a TFE-lined screw cap and let stand in the dark at 25°C for 7 d. Analyze a reagent blank (¶ 4d) with each batch of samples.

c. *Sample analysis:* After the 7-d reaction period, place 0.1 mL of sulfite reducing solution in a 25-mL vial and gently and completely fill vial with sample. If THMs will not be analyzed immediately, reduce pH to <2 by adding 5 drops conc acid. Seal vial with TFE-lined screw cap and analyze for THMs. Store samples at 4°C until ready for THM analysis, but warm to room temperature before analysis.

Measure free chlorine residual of the sample remaining in the 250-mL bottle using a method accurate to 0.1 mg/L and able to distinguish free and combined chlorine (see Section 4500-Cl). Adjust pH as directed before determining free chlorine.

If the sample does not contain at least 1 mg Cl_2/L, discard and treat duplicate of original sample with a higher chlorine dose or, preferably, start with a fresh sample. Obtain a free chlorine residual of at least 1 mg/L and preferably 2 to 5 mg/L as Cl_2.

d. *Reagent blank:* Add 1 mL chlorine dosing solution to 50 mL phosphate buffer, completely fill a 25-mL vial, seal with a TFE-lined screw cap, and store with samples. (NOTE: This reagent blank is for quality control and is not a true blank, because the reagent concentrations are considerably higher than they are in samples. The THM concentrations in the reagent blank will be biased high and cannot be subtracted from sample values. No further dilutions are made before the reaction because the reagent water itself would likely contribute to THM formation.) After reaction for 7 d, pipet 1 mL of sulfite reducing solution into a 250-mL bottle and add, without stirring, 5.0 mL of the reacted reagent mixture. Immediately fill bottle with organic-free water that has been purged free of THMs and seal with a TFE-lined screw cap. Analyze a portion for THMs with the same procedure used for samples. The sum of all THM compounds should be less than 5 μg $CHCl_3$/L.

The reagent blank is a rough measure of THMs contributed by reagents added to the samples but it cannot be used as a correction factor. If the reagent blank is greater than 5% of the sample value or greater than 5 μg THM/L, whichever is larger, additional treatment for reagent water is necessary. See Section 1080. It also may be necessary to obtain reagents of higher purity. Analyze a reagent blank each time samples are analyzed and each time fresh reagents are prepared.

5. Calculation

Report concentration of each of the four common THM compounds separately because it is desirable to know their relative concentrations. Larger amounts of brominated compounds, relative to $CHCl_3$, indicate a higher concentration of dissolved bromide in the water. Also report free chlorine concentration at the end of the reaction period.

TFP may be reported as a single value as micrograms per liter as $CHCl_3$, or micromoles per liter. Use of the units micrograms per liter is discouraged except for regulatory purposes, in which case it is required. Compute TFP using one of the following equations:

$$\text{TFP, } \mu g/L = A + B + C + D$$

TFP, μg/L as $CHCl_3$
$$= A + 0.728B + 0.574C + 0.472D$$

$$\text{TFP, } \mu M = \frac{\text{TFP, } \mu g/L \text{ as } CHCl_3}{119}$$

where:

$A = $ μg $CHCl_3$/L,
$B = $ μg $CHBrCl_2$/L,
$C = $ μg $CHBr_2Cl$/L, and
$D = $ μg $CHBr_3$/L

Do not make blank correction or a correction for sample dilution by addition of reagent. If conditions (pH, temperature, time, etc.) different from those stated above are used for nonstandard measurements tailored to local needs or conditions, report any such differences with the results.

6. Quality Control

a. Use dihydroxybenzoic acid solution (DHBA) as a quality-control check, especially for the presence of interfering bromides in reagents or reagent water.

Dilute 1.0 mL chlorine dosing solution to 1000 mL with chlorine-demand-free water. Pipet 5 mL phosphate buffer solution into each of two 250-mL bottles; add 1.00 mL DHBA solution to one bottle and fill both bottles completely with diluted chlorine dosing solution; and seal with TFE-lined screw caps. After holding in the dark for 7 d at 25°C, analyze as directed in ¶ 4c.

b. The THM concentration of the solution containing the added DHBA minus the THM concentration of the blank (which is a true blank) should equal 119 $\mu g/L$ THM as $CHCl_3$, with essentially no contribution from bromide-containing THMs. If there is a significant contribution from brominated THMs, 10% or more of the total THM, it may be necessary to remove bromide from the reagent water or to obtain higher-purity reagents containing less bromide. The source of bromide must be determined and the problem corrected. If the THM concentration of the water blank exceeds 20 $\mu g/L$, treat reagent water to reduce contamination.

7. Precision and Bias

The precision of this method is determined almost entirely by the analytical precision and bias of the method used for measuring THM. Method bias can be determined only for synthetic solutions (e.g., the DHBA solution), because TFP is not an intrinsic property of the sample but rather a quantity defined by this method.

Table 5710:I presents single-operator

TABLE 5710:I. SINGLE-OPERATOR PRECISION AND BIAS DATA FOR TFP*

Sample	THM $\mu g/L$				TFP $\mu g/L$ as $CHCl_3$	Recovery %
	$CHCl_3$	$CHCl_2Br$	$CHClBr_2$	$CHBr_3$		
Blank 1	0.8	—	—	—	0.8	—
Blank 2	1.9	—	—	—	1.9	—
Blank 3	0.1	0.1	—	—	0.2	—
Blank 4	0.7	—	—	—	0.7	—
Blank 5	0.5	—	—	—	0.5	—
Blank 6	0.7	—	—	—	0.7	—
Average					0.8	—
Standard deviation					± 0.6	—
DHBA 1	114.1	0.1	—	—	114.2	97.8
DHBA 2	113.2	—	—	—	113.2	96.9
DHBA 3	107.8	—	—	—	107.8	92.2
DHBA 4	108.3	—	—	—	108.3	92.7
DHBA 5	109.6	0.1	—	—	109.7	93.9
DHBA 6	111.8	0.1	—	—	111.9	95.8
DHBA 7	112.6	—	—	—	112.6	96.4
Average					111.1	95.1
Standard deviation					± 2.5	2.2

*Source: MOORE, L., Unpublished data. U.S. Environmental Protection Agency, Cincinnati, Ohio.

precision and bias data. The values were obtained by analyzing DHBA solutions and blanks. The expected value for the samples listed is 116 μg/L (as CHCl$_3$), rather than 119 μg/L, because the DHBA reagent used was only 97% pure. Percent recovery was calculated by the formula:

$$\% \text{ recovery} = \frac{\text{DHBA sample} - \text{average blank}}{116} \times 100$$

Because THM formation varies directly (though not proportionally) with both the initial dosage and the residual concentration of free chlorine, exercise caution in comparing results for samples having different residual chlorine concentrations. The recommended procedure for comparing samples is to chlorinate each sample with several different dosages of chlorine, yielding a range of free residual chlorine concentrations, and to compare values of TFP for equivalent residual chlorine concentrations.

5710 C.   Basic Trihalomethane Formation Potential (BTFP)

1. General Discussion

a. Principle: Samples are chlorinated and stored at pH 9.2.
b. Interference: See 5710B.1*b*.
c. Minimum detectable quantity: See 5710B.1*c*.

2. Apparatus

See 5710B.2.

3. Reagents

Use reagents as in 5710B.3, but substitute borate for phosphate buffer solution.

Borate buffer: Dissolve 30.9 g anhydrous boric acid, H$_3$BO$_3$, and 10.8 g sodium hydroxide, NaOH, in 1 L organic-free water. Refrigerate and prepare fresh weekly.

4. Procedure

Follow procedure described in 5710B.4, but adjust sample pH to 9.2 $\pm$ 0.2 after buffer addition.

5. Calculation

See 5710B.5, but substitute BTFP for TFP.

6. Quality Control

See 5710B.6, but use borate instead of phosphate buffer.

7. Precision and Bias

Table 5710:II lists single-operator precision and bias data for the BTFP procedure. The expected value for these samples is 46.4 μg/L (as CHCl$_3$) because 0.4 mL of DHBA solution (prepared from 97% pure DHBA) was used per bottle. Percent recovery was calculated as described in 5710B.7.

Table 5710:III lists results for single-operator precision of analyses of filtered river-water samples that had been diluted with 2 parts organic-free water to 1 part filtrate.

TABLE 5710:II. SINGLE-OPERATOR PRECISION AND BIAS DATA FOR BTFP*

Sample	THM μg/L				BTFP μg/L as CHCl₃	Recovery %
	CHCl₃	CHCl₂Br	CHClBr₂	CHBr₃		
Blank 1	3.0	0.3	—	—	3.2	—
Blank 2	1.7	0.1	—	—	1.8	—
Blank 3	1.3	0.1	—	—	1.4	—
Blank 4	1.6	0.1	—	—	1.7	—
Blank 5	2.3	0.2	—	—	2.4	—
Blank 6	2.6	0.1	—	—	2.7	—
Blank 7	2.5	0.2	—	—	2.6	—
Average					2.3	—
Standard deviation					± 0.6	—
DHBA 1	45.4	3.3	0.1	—	47.9	98.3
DHBA 2	51.0	3.9	0.1	—	53.9	111.2
DHBA 3	39.2	3.0	0.1	—	41.4	84.3
DHBA 4	48.3	3.6	0.1	—	51.0	105.0
DHBA 5	47.6	3.7	0.1	—	50.4	103.7
DHBA 6	43.4	3.2	0.1	—	45.8	93.8
DHBA 7	46.0	3.6	0.1	—	48.7	100.0
Average					48.4	99.5
Standard deviation					± 3.7	8.7

*Source: MOORE, L., Unpublished data. U.S. Environmental Protection Agency, Cincinnati, Ohio.

TABLE 5710:III. SINGLE-OPERATOR PRECISION OF FILTERED AND DILUTED RIVER-WATER SAMPLES ANALYZED FOR BTFP*

Sample	THM μg/L				BTFP μg/L as CHCl₃
	CHCl₃	CHCl₂Br	CHClBr₂	CHBr₃	
RWS 1	33.1	17.2	11.3	0.5	52.3
RWS 2	31.7	16.1	10.6	0.5	49.7
RWS 3	38.7	18.4	11.7	0.6	59.1
RWS 4	35.1	18.0	11.7	0.8	55.3
RWS 5	36.0	17.9	11.7	0.6	56.0
RWS 6	38.7	18.7	11.7	0.6	59.3
RWS 7	37.7	18.1	11.2	0.6	57.6
Average					55.6
Standard deviation					± 3.3

*Source: MOORE, L., Unpublished data. U.S. Environmental Protection Agency, Cincinnati, Ohio.

5710 D. Ultimate Trihalomethane Formation Potential (UTFP)

1. General Discussion

a. Principle: Samples are chlorinated and stored under conditions that are expected to result in formation of the greatest concentration of THMs that might reasonably be expected to occur in a particular distribution system. This involves adjusting temperature, time, chlorine concentration, and pH to maximum expected values.

b. Interference: See 5710B.1*b*.

c. Minimum detectable quality: See 5710B.1*c*.

2. Apparatus

See 5710B.2.

3. Reagents

See 5710B.3. Use borate buffer as in 5710C, or a 1:1 mixture of borate and phosphate buffers. Use of borate buffer to obtain pH values above 8 and phosphate buffer for pH values below 8 is recommended.

4. Procedure

The procedure is similar to that described in 5710B.4, except that pH, temperature, time, and chlorine dosage are altered to simulate worst-case conditions.

Add 5 mL buffer solution to 250 mL sample and adjust pH, using acid or alkali, to the highest value expected to occur in the water system. The buffer may be omitted for naturally buffered samples, but confirm that the pH remains at the desired value following chlorine addition.

Use the greater of the following two chlorine dosages: the highest total chlorine dosage to which the water reasonably can be expected to be exposed or the dosage necessary to leave the incubated sample with a free residual chlorine concentration at least as great as the highest concentration expected in the distribution system. The volume of chlorine dosing solution to be added to a 250-mL bottle is:

$$V, \text{mL} = \text{chlorine dosage} \times 0.05$$

Choose an incubation time equal to the longest time that water is expected to remain in the distribution system or 7 d. Choose a temperature equal to the greatest water temperature expected to occur anywhere in the distribution system.

Dose, incubate, and analyze samples as in 5710B.4.

5. Calculation

See 5710B.5, but substitute UTFP for TFP. Report conditions of the analysis (pH, temperature, chlorine dosage, incubation time, and residual free chlorine concentration).

6. Quality Control

See 5710B.6, but use conditions selected for UTFP determination unless those stated in 5710B.6 are more severe.

7. Precision and Bias

See 5710B.7 and 5710C.7.

5710 E.   Simulated Distribution System Trihalomethane Concentration (SDSTHMC)

1. General Discussion

a. Principle: Samples of treated water are stored under conditions representative of those occurring in the distribution system to estimate concentration of THMs reaching consumers.

b. Interference: See 5710B.1*b.*

c. Minimum detectable quantity: See 5710B.1*c.*

2. Apparatus

See 5710B.2.

3. Reagents

None required.

4. Procedure

Collect a chlorinated, treated water sample at the water treatment plant. Fill bottle completely with a minimum of turbulence to avoid loss of preformed THMs. Store sample in the dark at the same temperature as the water in the distribution system and for the longest period of time that the water is expected to remain in the distribution system (or 7 d, whichever is less). After incubation analyze for THMs as in 5710B.4. On remaining sample, determine form and concentration of residual chlorine, which should be similar to form and concentration of chlorine generally observed in the farthest reaches of the distribution system.

5. Calculation

See 5710B.5, but substitute SDSTHMC for TFP. Report pH, temperature, and incubation time used, as well as form and concentration of residual chlorine.

6. Precision and Bias

See 5710B.7. If analysis of replicate samples yields results varying by more than the limits specified for the THM method, investigate source and correct problem.

PART 6000

AUTOMATED

LABORATORY ANALYSES

6010 INTRODUCTION

6010 A. General Discussion

The methods presented in Part 6000 are intended for the determination of individual organic compounds. Methods for determination of aggregate concentrations of groups of organic compounds are presented in Part 5000.

Most of the methods presented herein are highly sophisticated instrumental methods for determining very low concentrations of the organic constituents. Stringent quality control requirements are given with each method and require careful attention.

Many compounds are determinable by two or more of the methods presented in Part 6000. Table 6010:I shows the specific analytical methods applicable to each compound. Guidance on selection of method is provided in the introduction to each section.

TABLE 6010:I. ANALYSIS METHODS FOR SPECIFIC ORGANIC COMPOUNDS*

Compound	Analysis Methods (section number)
Acenaphthene	6040B; 6410B; 6440B
Acenaphthylene	6410B; 6440B
Aldrin	6410B; 6630B,C
Anthracene	6040B; 6410B; 6440B
Benzene	6210B,C,D; 6220B,C; 6230D
Benzidine	6410B
Benzo(a)anthracene	6040B; 6410B; 6440B
Benzo(a)pyrene	6410B, 6440B
Benzo(b)fluoranthene	6410B, 6440B
Benzo(ghi)perylene	6410B, 6440B
Benzo(k)fluoranthene	6410B, 6440B
BHC(s)	6410B; 6630C
Bromobenzene	6040B; 6210C,D; 6220C; 6230C,D
Bromochloromethane	6210C,D; 6230C,D
Bromodichloromethane	6040B; 6210B,C,D; 6230B,C,D; 6232B
Bromoform	6040B; 6210B,C,D; 6230B,C,D; 6232B
Bromomethane	6210B,C,D; 6230B,C,D
Bromophenoxybenzene	6040B
Bromophenyl phenyl ether	6410B

*Compounds are listed under the names by which they are most commonly known and called in specific methods.

Compound	Analysis Methods (section number)
Butyl benzyl phthalate	6410B
Butylbenzene(s)	6210C,D; 6220C; 6230C,D
Captan	6630B
Carbon tetrachloride	6210B,C,D; 6230B,C,D
Chlordane	6410B; 6630B,C
Chlorobenzene	6040B; 6210B,C,D; 6220B,C; 6230B,C,D
Chloroethane	6210B,C,D; 6230B,C,D
Chloroethoxy methane	6040B; 6410B
Chloroethyl ether	6040B; 6410B
Chloroethylvinyl ether	6210B; 6230B
Chloroform	6210B,C,D; 6230B,C,D; 6232B
Chloroisopropyl ether	6410B
Chloromethane	6210B,C,D; 6230B,C,D
Chloromethyl benzene	6040B
Chloromethylphenol	6410B; 6420B
Chloronaphthalene(s)	6040B; 6410B
Chlorophenol(s)	6401B; 6420B
Chlorophenoxy benzene	6040B
Chlorophenyl phenyl ether	6410B
Chlorotoluene	6210C,D; 6220C; 6230C,D
Chrysene	6040B; 6410B; 6440B
2,4-D (dichlorophenoxyacetic acid)	6640B
DDD	6410B; 6630B,C
DDE	6410B; 6630B,C
DDT	6410B; 6630B,C
Dibenzo(a,h) anthracene	6410B; 6440B
Dibromochloromethane	6040B; 6210B,C,D; 6230B,C,D; 6232B
Dibromochloropropane	6210C,D; 6230D; 6231B
Dibromoethane	6040B; 6210C,D; 6230C,D; 6231B
Dibromomethane	6210C,D; 6230C,D
Dibutyl phthalate	6410B
Dichloran	6630B
Dichlorobenzene(s)	6040B; 6210B,C,D; 6220B,C; 6230B,C,D; 6410B
Dichlorobenzidine	6410B
Dichlorodifluoromethane	6210C,D; 6230B,C,D
Dichloroethane	6210B,C,D; 6230B,C,D
Dichloroethene(s)	6210B,C,D; 6230B,C,D
Dichlorophenol(s)	6410B; 6420B
Dichloropropane(s)	6210B,C,D; 6230B,C,D
Dichloropropene	6040B; 6210B,C,D; 6230B,C,D
Dieldrin	6410B; 6630B,C
Diethyl phthalate	6040B; 6410B
Dimethyl phthalate	6410B
Dimethylphenol(s)	6410B; 6420B
Dinitrophenol(s)	6410B; 6420B
Dinitrotoluene(s)	6410B
Di-*n*-octyl phthalate	6410B

TABLE 6010:I, CONT.

Compound	Analysis Methods (section number)
Diphenyl hydrazine	6040B
Endosulfan	6410B; 6630B,C
Endosulfan sulfate	6410B; 6630C
Endrin	6410B; 6630B,C
Endrin aldehyde	6410B; 6630C
Ethenyl benzene (styrene)	6040B
Ethylbenzene	6040B; 6210B,C,D; 6220B,C; 6230D
Ethylhexyl phthalate	6410B
Fluoranthene	6040B; 6410B; 6440B
Fluorene	6040B; 6410B; 6440B
Geosmin	6040B
Heptachlor	6410B; 6630B,C
Heptachlor epoxide	6410B; 6630B,C
Hexachlorobenzene	6040B; 6410B
Hexachlorobutadiene	6040B; 6210C,D; 6220C; 6230D; 6410B
Hexachlorocyclopentadiene	6410B
Hexachloroethane	6040B; 6410B
Indeno(1,2,3-cd)pyrene	6410B; 6440B
Isobutylmethoxy pyrazine	6040B
Isophorone	6410B
Isopropylbenzene	6210C,D; 6220C; 6230D
Isopropyl methoxy pyrazine	6040B
Isopropyltoluene	6210D; 6220C; 6230D
Lindane (γ-BHC)	6630B
Malathion	6630B
Methane	6211
Methoxychlor	6630B
Methyldinitrophenol(s)	6410B; 6420B
Methylene chloride	6210B,C,D; 6230B,C,D
Methylisoborneol	6040B
Methyl parathion	6630B
Mirex	6630B
Naphthalene	6040B; 6210D; 6220C; 6230D; 6410B; 6440B
Nitrobenzene	6410B
Nitrophenol(s)	6410B; 6420B
Nitrosodi-n-propylamine	6410B
Nitrosodimethylamine	6410B
Nitrosodiphenylamine	6410B
Parathion	6630B
PCB-1016, 1221, 1232, 1242, 1248, 1254, 1260	6410B, 6630C
Pentachloronitrobenzene	6630B
Pentachlorophenol	6410B; 6420B
Phenanthrene	6040B; 6410B; 6440B
Phenol	6410B; 6420B
Phenylbenzamine	6040B
Propylbenzene	6040B; 6210C,D; 6220C; 6230D
Pyrene	6040B; 6410B; 6440B

TABLE 6010:I, CONT.

Compound	Analysis Methods (section number)
Silvex (trichlorophenoxy propionic acid)	6640B
Strobane	6630B
Styrene (ethenyl benzene)	6210C,D; 6220C; 6230D
2,4,5-T (trichlorophenoxy acetic acid)	6640B
Tetrachloroethane(s)	6040B; 6210B,C,D; 6230B,C,D
Tetrachloroethene	6040B; 6210B,C,D; 6220C; 6230B,C,D
Toluene	6210B,C,D; 6220B,C; 6230D
Toxaphene	6410B; 6630B,C
Trichloanisole	6040B
Trichlorobenzene(s)	6040B; 6210D; 6220C; 6230D; 6410B
Trichloroethane(s)	6040B; 6210B,C,D; 6230B,C,D
Trichloroethene	6040B; 6210B,C,D; 6220C; 6230B,C,D
Trichlorofluoromethane	6210B,C,D; 6230B,C,D
Trichlorophenol	6410B; 6420B
Trichloropropane	6210CD; 6230C,D
Trifluralin	6630B
Trimethylbenzene(s)	6210D; 6220C; 6230D
Vinyl chloride	6210B,C,D; 6230B,C,D
Xylene(s)	6040B; 6210C,D; 6220C; 6230D

6010 B.   Sample Collection and Preservation

1. Volatile Organic Compounds

Use 25- or 40-mL vial equipped with a screw cap with a hole in the center* and TFE-faced silicone septum.† Wash vials, caps, and septa with detergent, rinse with tap and distilled water, and dry at 105°C for 1 h before use in an area free of organic vapors. NOTE—Do not heat seals for extended periods of time (> 1 h) because the silicone layer slowly degrades at 105°C. When bottles are cool, seal with TFE seals. Alternatively purchase precleaned vials free from volatile organics.

Collect all samples in duplicate and prepare replicate field reagent blanks with each sample set. A sample set is all samples collected from the same general sampling site at approximately the same time. Prepare field reagent blanks in the laboratory by filling a minimum of two sample bottles with reagent water, sealing, and shipping to the sampling site along with empty sample bottles.

Fill sample bottle just to overflowing without passing air bubbles through sample or trapping air bubbles in sealed bottle. When sampling from a water tap, open tap and flush until water temperature has stabilized (usually about 10 min). Adjust flow rate to about 500 mL/min and collect duplicate samples from flowing stream. When sampling from an open body of water, fill a 1-L, wide-mouth bottle or breaker with a representative sample and carefully fill duplicate sample bottles from the container.

Preservation of samples is highly dependent on target constituents and sample matrix. Ongoing research indicates the following areas of concern: rapid biodegradation of aromatic compounds, even at

---
* Pierce 13075 or equivalent.
† Pierce 12722 or equivalent.

low temperatures;[1] dehydrohalogenation reactions such as conversion of pentachloroethane to tetrachloroethane;[2] reactions of alkylbenzenes in chlorinated samples, even after acidification; and possible interactions among preservatives and reductants when dechlorination is used to prevent artifact formation, especially in samples potentially containing many target compounds.

There is as yet no single preservative that can be recommended. Ideally, maintain a sample at 4°C and analyze it immediately. In practice, delays between sampling and analysis often necessitate preservation. The recommended preservation technique is summarized in Table 6010:II.

1) For samples and field blanks that contain volatile constituents but do not contain residual chlorine, add HCl (4 drops 6N HCl/40 mL) to prevent biodegradation and dehydrohalogenation. CAUTION: HCl may contain traces of organic solvents. Verify freedom from contamination before using a specific lot for preservation.

2) For samples and field blanks that contain residual chlorine, also add a reducing agent. Ascorbic acid (25 mg/40 mL) appears to be optimal, but demonstrate that this reductant is appropriate for the specific sample matrix. Sodium thiosulfate (3 mg/40 mL) or sodium sulfite (3 mg/40 mL) also may be appropriate reducing agents, but when either of these is added in the presence of HCl, SO$_2$ formation may interfere with certain packed-column gas chromatographic or GC/MS techniques.

In all cases, run reagent blanks to insure absence of interferences. Add either ascorbic acid or HCl to the sample bottle immediately before shipping it to the sample site or immediately before filling sample bottle. When both preservatives are needed, add only one before filling the sample bottle, to prevent interactions between the acid and the reductant. Add the second preservative once the bottle is almost full. However, if there is evidence that interactions of acid and reducing agent will not create analytical or preservation problems, they may be added simultaneously.

Tightly seal sample bottles, TFE face down. After sampling and preservation invert several times to mix. Chill samples to 4°C immediately after collection and hold at that temperature in an atmosphere free of organic solvent vapors until analysis. Normally analyze all samples within 14 d of collection. Shorter or longer holding times may be appropriate, depending on constituents and sample matrix. Develop data to show that alternate holding times are appropriate.

2. Other Organic Compounds

See individual methods for sampling and preservation requirements.

3. References

1. BELLAR, T. & J. LICHTENBERG. 1978. Semi-automated headspace analysis of drinking waters and industrial waters for purgeable volatile organic compounds. In C. E. Van Hall, ed. Measurement of Organic Pollutants in Water and Wastewater. STP 686, American Soc. Testing & Materials. Philadelphia, Pa.

2. BELLAR, T. & J. LICHTENBERG. 1985. The Determination of Synthetic Organic Compounds in Water by Purge and Sequential Trapping Capillary Column Gas Chromatography. U.S. Environmental Protection Agency, Cincinnati, Ohio.

4. Bibliography

KEITH, L. H., ed. 1988. Principles of Environmental Testing. American Chemical Soc., Washington, D.C.

TABLE 6010:II. RECOMMENDED PRESERVATION FOR VOLATILES

Constituents	Chlorinated Matrix	Non-Chlorinated Matrix
Halocarbons	HCl + reducing agent	HCl
Aromatics	HCl + reducing agent	HCl
THMs	Reducing agent (HCl optional)*	None required
EDB/DBCP	None required	None required

* See 6232B.2.

6010 C.   Analytical Methods

1. General Discussion

The methods presented in Part 6000 for identification and quantitation of trace organic constituents in water generally involve isolation and concentration of the organics from a sample by solvent or gas extraction (see Section 6040 and individual methods), separation of the components, and identification and quantitation of the compounds with a detector.

2. Gas Chromatographic Methods

Gas chromatographic (GC) methods are highly sophisticated microanalytical procedures. They should be used only by analysts experienced in the techniques required and competent to evaluate and interpret the data.

a. Gas chromatograph:

1) Principle—In gas chromatography a mobile phase (a carrier gas) and a stationary phase (column packing or capillary column coating) are used to separate individual compounds. The carrier gas is nitrogen, argon-methane, helium, or hydrogen. For packed columns, the stationary phase is a liquid that has been coated on an inert granular solid, called the column packing, that is held in borosilicate glass tubing. The column is installed in an oven with the inlet attached to a heated injector block and the outlet attached to a detector. Precise and constant temperature control of the injector block, oven, and detector is maintained. Stationary-phase material and concentration, column length and diameter, oven temperature, carrier-gas flow, and detector type are the controlled variables.

When the sample solution is introduced into the column, the organic compounds are vaporized and moved through the column by the carrier gas. They travel through the column at different rates, depending on differences in partition coefficients between the mobile and stationary phases.

2) Interferences—Some interferences in GC analyses occur as a result of sample, solvent, or carrier gas contamination, or because large amounts of a compound may be injected into the GC and linger in the detector. Methylene chloride, chloroform, and other halocarbon and hydrocarbon solvents are ubiquitous contaminants in environmental laboratories. Make strenuous efforts to isolate the analytical system from laboratory areas where these or other solvents are in use. An important sample contaminant is sulfur, which is encountered generally only in base/neutral extracts of water, although anaerobic groundwaters and certain wastewaters and sediment/sludge extracts may contain reduced sulfur compounds, elemental sulfur, or polymeric sulfur. Eliminate this interference by adding a small amount of mercury or copper filings to precipitate the sulfur as metallic sulfide. Sources of interference originating in the chromatograph, and countermeasures, are as follows:

• Septum bleed—This occurs when compounds used to make the septum on the injection port of the GC bleed from the heated septum. These high-molecular-weight silicon compounds are distinguished readily from compounds normally encountered in environmental samples. Nevertheless, minimize septum bleed by using septum sweep, in which clean carrier gas passes over the septum to flush out the "bleed" compounds.

• Column bleed—This term refers to loss of column coating or breakdown products when the column is heated. This interference is more prevalent in packed columns, but also occurs to a much lesser extent in capillary columns. It occurs when the column temperature is high or when water or oxygen are introduced into the system. Solvent injection can damage the stationary phase by displacing it. Certain organic compounds acting as powerful solvents, acids, or bases can degrade the column coating. Injection of large amounts of certain surface-active agents may destroy GC columns.

• Ghost peaks—These peaks occur when an injected sample contains either a large amount of a given compound, or a compound that adsorbs to the column coating or injector parts (e.g., septum). When a subsequent sample is injected, peaks can appear as a result of the previous injection. Eliminate ghost peaks by injecting a more dilute sample, by producing less reactive derivatives of a compound that may interact strongly with the column material, by selecting a column coating that precludes these interactions, or by injecting solvent blanks between samples.

b. *Detectors:* Various detectors are available for use with gas chromatographic systems. See individual methods for recommendations on appropriate detectors.

1) Electrolytic conductivity detector—The electrolytic conductivity detector is a sensitive and element-specific detector that has gained considerable attention because of its applicability to the gas chromatographic analysis of environmentally significant compounds. It is utilized in the analysis of purgeable halocarbons, pesticides, herbicides, pharmaceuticals, and nitrosamines. This detector is capable of operation in each of four specific modes: halogen (X), nitrogen (N), sulfur (S), and nitrosamine (NO). Only organic compounds containing these elements will be detected.

Compounds eluting from a gas chromatographic column enter a reactor tube heated to 800°C. They are mixed with a reaction gas, hydrogen for X, N, or NO modes, and air for the S mode. The hydrogen catalytically reduces the eluants while the air oxidizes them. The gaseous products are transferred to the detector through a conditioned ion exchange resin or scrubber. In the halogen mode, only HX is detected, while NH_3 or H_2S are eliminated on the resin. In the nitrogen or nitrosamine mode, the NH_3 formed is ionized while HX and H_2S, if present, are eliminated with a KOH/quality wool scrubber. The sulfur mode produces SO_2 or SO_3, which is ionized while HX is removed with a silver wire scrubber. All other products either are not ionizable or are produced in such low yield that they are not detectable.

The electrolytic conductivity detector contains reference and analytical electrodes, a gas-liquid contactor, and a gas-liquid separator. The conductivity solvent enters the cell and flows by the reference electrode. It combines with the gaseous reaction products in the gas-liquid contactor. This heterogeneous mixture is separated into gas and liquid phases in the gas-liquid separator, with the liquid phase flowing past the analytical electrode. The electrometer monitors the difference in conductivity at the reference electrode (solvent) and the analytical electrode (solvent + carrier + reaction products).

2) Electron capture detector—The electron capture detector (ECD) usually is used for the analysis of compounds that have high electron affinities, such as chlorinated pesticides, drugs, and their metabolites. This detector is somewhat selective in its response, being highly sensitive toward molecules containing electronegative groups: halogens, peroxides, quinones, and nitro groups. It is insensitive toward functional groups, such as amines, alcohols, and hydrocarbons.

The detector is operated by passing the effluent from the gas chromatographic column over a radioactive beta particle emitter, usually nickel-63 or tritium adsorbed on platinum or titanium foil. An electron from the emitter ionizes the carrier gas, preferably nitrogen, and produces a burst of electrons. About 100 secondary electrons are produced for each initial beta particle. After further collisions, the energy of these electrons is reduced to the thermal level and they can be captured by electrophilic sample molecules.

The electron population in the ECD cell is collected periodically by applying a short voltage pulse to the cell electrodes and the resulting current is compared with a reference current. The pulse interval is adjusted automatically to keep the cell current constant, even when some of the electrons are being captured by the sample. The change in the pulse rate when a sample enters the ECD is then related to the sample concentration. The ECD offers linearity in the range of 10^4 and subpicogram detection limits for compounds with high electron affinities.

3) Flame ionization detector—The flame ionization detector (FID) is widely used because of its high sensitivity to organic carbon-containing compounds. The detector consists of a small hydrogen/air diffusion flame burning at the end of a jet. When organic compounds enter the flame from the column, electrically charged intermediates are formed. These are collected by applying a voltage across the flame. The resulting current is amplified by an electrometer and measured. The response of the detector is directly proportional to the total mass entering the detector per unit time and is independent of the concentration in the carrier gas.

The FID is perhaps the most widely used detector for gas chromatography because of several advantages: (a) it responds to virtually all organic carbon-containing compounds with high sensitivity (approximately 10^{-13}g/mL); (b) it does not respond to common carrier gas impurities such as water and carbon dioxide; (c) it has a large linear response range (approximately 10^7) and excellent baseline stability; (d) it is relatively insensitive to small column flow-rate changes during temperature programming; (e) it is highly reliable, rugged, and easy to use; and (f) it has low detector dead volume effects and fast response. Its limitations include: (a) it gives little or no response to noncombustible gases and all noble gases; and (b) it is a destructive detector that changes the physical and chemical properties of the sample irreversibly.

4) Photoionization detector—Photoionization occurs when a molecular species absorbs a photon of light energy and dissociates into a parent ion and an electron. The photoionization detector (PID) detects organic and some inorganic species in the effluent of a gas chromatograph with detection limits as low as the picogram range. The PID is equipped with a sealed ultraviolet light source that emits photons which pass through an optically transparent window (made of LiF, MgF_2, NaF, or sapphire) into an ionization chamber where photons are absorbed by the eluted species. Compounds having ionization potential less than the UV source energy are ionized. A positively biased high-voltage electrode accelerates the resulting ions to a collecting electrode and the resulting current is measured by an electrometer. This current is proportional to the concentration.

The PID has high sensitivity, low noise (approximately 10^{-14}A), and excellent linearity (10^7), is nondestructive, and can be used in series with a second detector for more selective detection. The PID can be operated as a universal detector or a selective detector by simply changing the photon energy of the ionization source. Tables of ionization potentials are used to select the appropriate UV source for a given measurement.

5) Mass spectrometer—The mass spectrometer (MS) has the ability to detect a wide variety of compounds, coupled with a capacity to deduce compound structures from fragmentation patterns. Among the different types of mass spectrometers, the quadrupole has become the most widely used in water and wastewater analysis.

The mass spectrometer detects compounds by ionizing molecules into charged species with a 70-eV beam. The ions are accelerated toward the quadrupole mass filter through a series of lenses held at 0 to 200 V. The differently sized, charged fragments are separated according to mass-to-charge ratio (related to molecular weight) by means of the quadrupole, which uses varying electric and radiofrequency (rf) fields. The quadrupole is connected to a computer, which varies these fields so that only fragments of one particular mass-to-charge ratio (± 0.5) can traverse the quadrupole at any one time. As the ions leave the quadrupole they are attracted to the electron multiplier through an electrical potential of several thousand volts. The charge fragments, in turn, are detected by the electron multiplier. Because the electric and the rf fields are cycled every few seconds, a fragmentation pattern is obtained. Each cycle is called a mass scan. Most chemicals have unique fragmentation patterns, called mass spectra. The computer contains, and can search, a library of known mass spectra to identify tentatively an unknown compound exhibiting a particular spectrum. Use authentic compounds for confirmation after tentative identifications are made.

Background mass interference can result from the ability of the mass spectrometer to detect any ions created in its ion volume (up to a specified mass). Any compounds continuously present in the source will be detected. Some mass ions always present are due to air components that leak into the system, such as oxygen (masses 16 and 32), nitrogen (masses 14 and 28), carbon dioxide (mass 44), argon (mass 40), and water (mass 18), or to helium carrier gas (masses 4 and 8), or to diffusion pump oil vapors.

3. High-Performance Liquid Chromatographic (HPLC) Methods

a. Principle: HPLC is an analytical technique in which a liquid mobile phase transports a sample through a column containing a liquid stationary phase. The interaction of the sample with the stationary phase selectively retains individual compounds and permits separation of sample components. Detection of the separated sample compounds is achieved mainly through the use of absorbance detectors for organic compounds and through conductivity and electrochemical detectors for metal and inorganic components.

b. Detectors:

1) Photodiode array detector (PDAD) —The PDAD measures the absorbance of a sample from an incident light source (UV-VIS). After passing through the sample cell, the light is directed through a holographic grating that separates the beam into its component wavelengths reflected on a linear array of photodiodes. This permits the complete absorbance spectrum to be obtained in 1 s or less and simultaneous multiwavelength analysis.

The PDAD is subject to the interference encountered with all absorbance detectors. Of special concern for HPLC is the masking of the absorbance region of the HPLC mobile phase and its additives. This may reduce the range and sensitivity of the detector to the sample components. Most interferences occur in monitoring the shorter wavelengths (200–230 nm). In this region, many organic compounds absorb light energy and can be sources of interference.

2) Post column reactor (PCR)—The PCR consists of in-line sample derivatizing/reacting equipment that permits chemical alteration of certain organic compounds. This equipment is used to enhance detection by attaching a chromophore to the compound(s) of interest. Sensitivity and selectivity of compounds that were initially undetectable are altered to make them detectable.

Interferences from this technique usually arise from the impurities in the reagents used in the reaction. When this technique is coupled with a selective detector such as fluorescence, these interferences are minimized. Generally, only compounds of the same class as the compounds of interest will cause interference.

3) Fluorescence detector—The fluorescence detector is an absorbance detector in which the sample is energized by a monochromatic light source. Compounds capable of absorbing the light energy do so and release it as fluorescence emission. Filters permit the detector to respond only to the fluorescent energy. The fluorescence detector is the most sensitive of the current HPLC detectors available and often is used in conjunction with a post column reactor.

Because of instrument sensitivity, minute quantities of contaminants can cause interferences to fluorescence detectors. Contamination can happen from glassware, mobile phase solvents, post-column reagents, etc. These sources will raise the background signal and thus narrow the range of the detector. Interference from individual compounds is minimal because of detector specificity (i.e., all interferences must fluoresce).

6040   CONSTITUENT CONCENTRATION BY GAS EXTRACTION*

6040 A.   Introduction

The ability to analyze ultratrace levels of organic pollutants in water has been limited, in part, by the concentration technique. With the development of closed-loop stripping analysis (CLSA) (Method B), organic compounds of intermediate volatility and molecular weight, i.e., from the heavier volatiles to the lighter polynuclear aromatic hydrocarbons, can be extracted from water and concentrated to allow quantitative and semiquantitative analysis (depending on the compound) at parts-per-trillion levels. This extract can be analyzed on a gas chromatograph (GC) connected to one of several detectors. A CLSA technique coupled with gas chromatographic/mass spectrometric (GC/MS) analysis for the determination of trace organic compounds is presented here. It is applicable to both treated and natural waters.

The purge and trap technique (Method C) is a valuable concentration method applicable to volatile organic compounds. The compounds are concentrated by bubbling of an inert gas through the sample followed by collection in, and desorption from, a sorbent trap. This extract may be analyzed by GC or GC/MS methods. The technique is applicable to both water and wastewater.

*Approved by Standard Methods Committee, 1988.

6040 B. Closed-Loop Stripping, Gas-Chromatographic/ Mass-Spectrometric Analysis

1. General Discussion

a. Principle: This CLSA-GC/MS procedure is suitable for the analysis of a broad spectrum of organic compounds in water. It can be used for the identification and quantification of specific compounds, such as earthy-musty-smelling compounds (e.g., 2-methylisoborneol (MIB) and geosmin)[1–3] or U.S. Environmental Protection Agency (EPA) priority pollutants[4,5].

In closed-loop stripping, volatile organic compounds of intermediate molecular weight are stripped from water by a recirculating stream of air. The organics are removed from the gas phase by an activated carbon filter. They are extracted from the filter with carbon disulfide (CS_2). A portion of the extract is injected into a capillary-column GC/MS for identification of the organic compounds by retention time and spectrum matching; quantification is done by single-ion current integration.

b. Interference: Organic compounds that are stripped during this procedure may coelute with the compounds of interest. The uniqueness of the mass spectrum of each target compound makes it possible to confirm compound identity with a high probability when coeluting components are present. Problems may arise if several isomers of a compound are present that are not resolvable chromatographically.

c. Detection limits: Trace organics can be detected at low nanogram-per-liter levels. The CLSA-GC/MS detection limits are affected by many factors; especially important are the stripping efficiency and the condition of the GC/MS. Stripping efficiencies can be improved by using an elevated stripping temperature and/or the salting-out technique.

The instrumental detection limits for five earthy-musty-smelling compounds are shown in Table 6040:I. Detection limits for the salted CLSA method are less than half those for the unsalted method for each compound. Using the elevated stripping temperature rather than the salting-out technique produces comparable recoveries[6] and similar detection limits. Detection limits for various organic compounds of interest, obtained with an elevated stripping temperature/salting-out technique, ranged from 0.1 to 100 ng/L (see Table 6040:II).[7]

2. Apparatus

Use clean glassware in sample collection and calibration standard preparation. Wash with soapy water, rinse with tap water, with demineralized water, and finally with reagent-grade acetone. Air-dry and bake at 180°C for 6 to 12 h. Do not bake sample bottle caps or volumetric ware. After drying and baking, store inverted or cover mouths with aluminum foil to prevent accumulation of dust or other contaminants.

a. Sample bottles, 1-L capacity or larger, glass, with TFE-lined screw caps.

b. CLSA apparatus, equipped with the following components (Figure 6040:1) or their equivalents[4].*

1) *Stripping bottle,* with mark at 1-L level and stainless-steel quick-connect stems (Figure 6040:2) or unpolished spherical glass joints sealed with TFE-covered silicone rubber O-rings and secured with metal clamps.†

2) *Gas heater,* with aluminum heating cylinder and soldering iron (25 W) controlled by a variable transformer (Figure 6040:3). Alternatively, use a temperature-controlled heater block to maintain temperature at the filter that is at least 10 to 20°C above temperature of thermostatic water bath.

---

*Model CLS 1, Tekmar, Cincinnati, Ohio; Brechbühler AG, 8952 Schlieren ZH, Switzerland, available from Chromapon, Whittier, Calif.; or equivalent.
†Rotulex Sovirel, Brechbühler AG or equivalent.

TABLE 6040:I. INSTRUMENTAL DETECTION LIMITS FOR EARTHY-MUSTY SMELLING COMPOUNDS BY CLSA-GC/MS

Compound	Detection Limit ng/L*	
	Unsalted Method[1]	Salting-Out Technique[3]
Geosmin	2	0.8
2-Methylisoborneol	2	0.8
2-Isopropyl-3-methoxy pyrazine	2	0.8
2-Isobutyl-3-methoxy pyrazine	2	0.8
2,3,6-Trichloranisole	5	0.8

*Stripping at 25°C.

TABLE 6040:II. INSTRUMENTAL DETECTION LIMITS FOR SELECTED ORGANIC COMPOUNDS BY CLSA-GC/MS[7]

Compound	Detection Limit ng/L*†	Compound	Detection Limit ng/L*†
1,1,1-Trichloroethane	2.0	1,2,3-Trichlorobenzene	2.0
Trichloroethene	100	bis(2-Chloro-ethoxy)methane	10
Dichlorobromomethane	5.0	Methylisoborneol (MIB)	0.5
1,3-Dichloropropene	2.0	Geosmin	0.2
1,1,2-Trichloroethane	2.0	Naphthalene	100
Chlorodibromomethane	1.0	1,1,2,3,4,4-Hexachloro-1,3-butadiene	2.0
1,2-Dibromoethane	2.0		
Tetrachloroethene	100	1-Chloronaphthalene	0.5
Chlorobenzene	10	2-Chloronaphthalene	0.5
Ethylbenzene	50	Acenaphthene	0.5
m,p-Xylene	100	Fluorene	2.0
Bromoform	1.0	Diethylphthalate	100
Ethylbenzene	5.0	1-Chloro-4-phenoxybenzene	0.5
o-Xylene	50	N-Phenylbenzamine	20
1,1,2,2-Tetrachloroethane	50	1,2-Diphenylhydrazine (as azobenzene)	1.0
Bromobenzene	0.5		
Propylbenzene	0.5	1-Bromo-4-phenoxybenzene	0.5
1-Chloro-3-methylbenzene	0.5	Hexachlorobenzene	1.0
bis (2-Chloroethyl)ether	1.0	Phenanthrene	10
o-Dichlorobenzene	0.1	Anthracene	50
m-Dichlorobenzene	10	Fluoranthene	20
p-Dichlorobenzene	10	Pyrene	20
Hexachloroethane	20	Chrysene	50
N-Nitrosodi-n-propylamine	5.0	Benzo(a)anthracene	50
1,3,5-Trichlorobenzene	0.1		
1,2,4-Trichlorobenzene	10		

*Elevated stripping temperature and salting-out both utilized.
†Instrument detection limit based on a 2:1 signal: noise ratio (where a background interference existed, the target compound was required to be at least twice background).

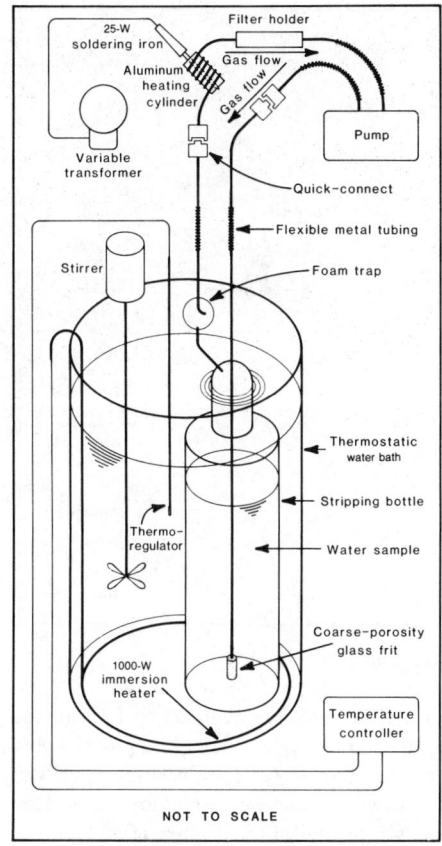

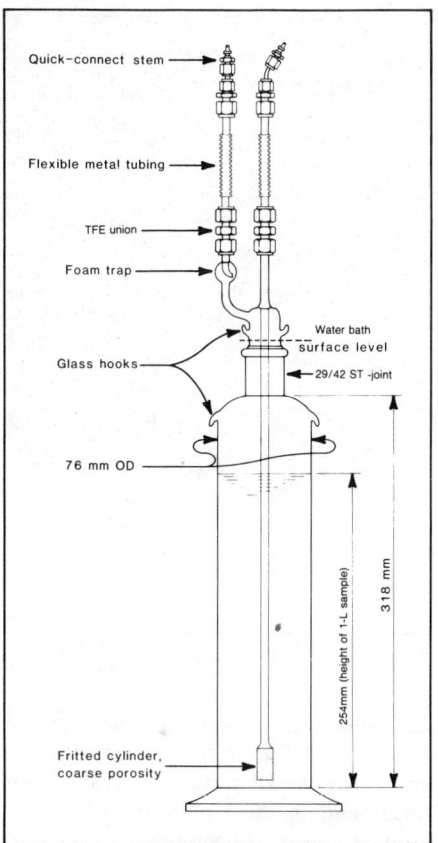

Figure 6040:1. Schematic of closed-loop stripping apparatus. Source: KRASNER, S.W., C.J. HWANG & M.J. MCGUIRE. 1983. A standard method for quantification of earthy-musty odorants in water. *Water Sci. Technol.* 15(6/7):127.

Figure 6040:2. One-liter "tall form" stripping bottle. Source: KRASNER, S.W., C.J. HWANG & M.J. MCGUIRE. 1983. A standard method for quantification of earthy-musty odorants in water. *Water Sci. Technol.* 15(6/7):127.

3) *Filter holder,* stainless steel or glass.‡ If glass is used, also use an auxiliary heating device, e.g., an infrared light, to maintain proper filter temperature.

4) *Pump,* with stainless-steel bellows,§ providing air flow in the range of 1 to 1.5 L/min.

5) *Automatic timer* (optional), connected to pump.

6) *Circuit,* with stainless-steel parts: 1/8-in.-OD tubing, 4-in. × 1/4-in.-OD flexible tubing, tube fittings, and quick-connect bodies‖ or glass joints described in ¶ 1) above. Glass sample lines can be used except where circuit enters and exits pump. Use TFE ferrules in making connections to glass and flexible metal tubing.

7) *Thermostatic water bath,* with 222-mm-OD × 457-mm chromatography jar

---

‡Brechbühler AG or equivalent.
§Metal Bellows Model MB-21, Sharon, Mass., or equivalent.

‖Swagelok fittings or equivalent.

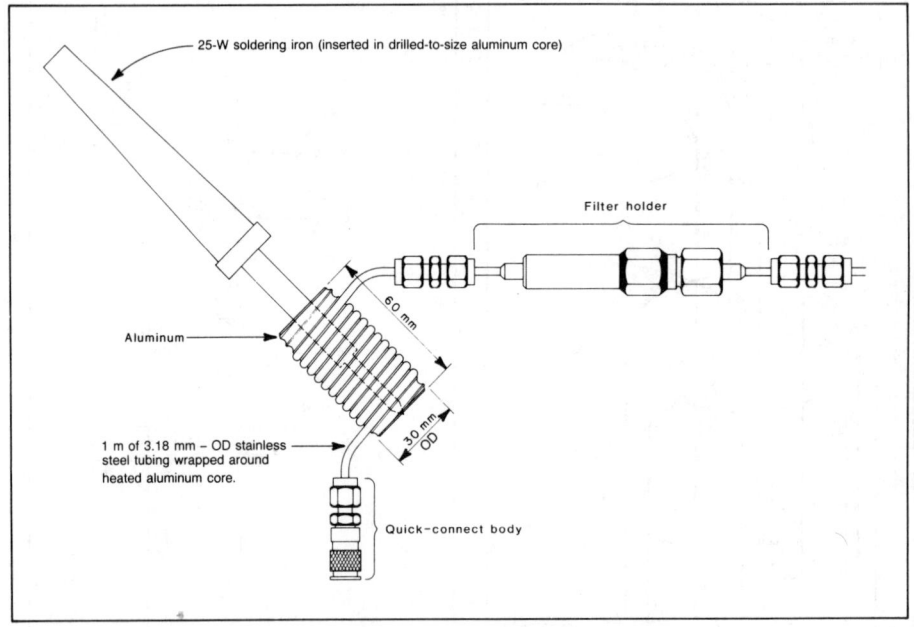

Figure 6040:3. Gas heater.

and thermoregulating system accurate to at least ±0.5°C. When the ambient temperature of the laboratory is greater than 25°C, maintain water bath at 25°C by inserting a coil of copper tubing connected to a cold water tap to recirculate cold water.

8) *Filters,* with 1.5 mg activated carbon # (Figure 6040:4). Use a set of filters matched in resistance for each group of samples and calibration standards. Determine filter resistance by measuring solvent flow rate through a solvent-wetted filter. Fill the longer glass tube above the charcoal with solvent and let solvent flow by gravity through filter disk. After several rinses with solvent to wet the filter, measure time necessary to empty the solvent (0.3-mL volume) from top of filter tube to surface of carbon. Rates for new, commercially prepared filters vary significantly, and decrease with use. Do not use filters

with flow rate less than 0.6 mL/min; for optimal recoveries, use filters with flow greater than 0.9 mL/min. Figure 6040:5 shows the reduction in air flow caused by using a "slow" filter. Figure 6040:6 shows the effect of filter resistance on recovery of earthy-musty odorants and one of the internal standards.

c. Stirrer (optional), with 5-cm-long TFE stirring bar.

d. Microsyringes, 5-, 10-, 25-, and 50-μL capacity.

e. Vials, 50-μL capacity (Figure 6040:4) with gastight stoppers or caps. Vials can be produced by a custom glass-blowing company: use a 1.6-mm-ID precision-bore capillary glass and grind to 5 mm OD, then heat-constrict to close off bore at approximately 29 mm from top. Mark vial at 20-μL level with glass scribe. Prepare gastight cap by purchasing 1-cm-OD TFE cylindrical bar stock, cutting it into 1-cm-long pieces, and drilling out a hole that is 5 mm ID × 7 mm deep. In practice, make the hole slightly smaller than the 5-mm OD of

---

#Brechbühler AG, Chromapon, Inc., or equivalent.

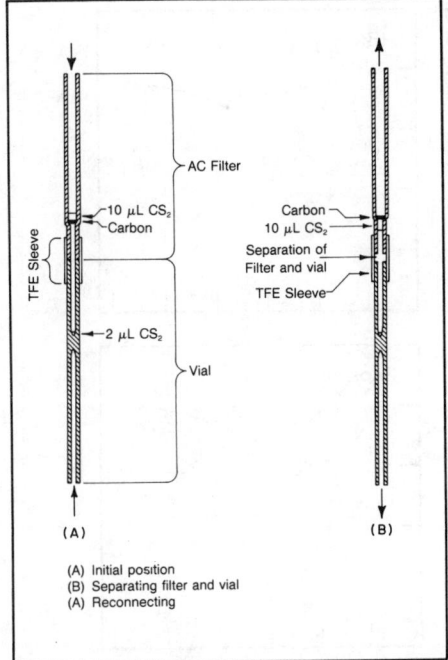

(A)

(B)

(A) Initial position
(B) Separating filter and vial
(A) Reconnecting

Figure 6040:4. Extraction of filter. Source: KRAS-NER, S.W., C.J. HWANG & M.J. MCGUIRE. 1981. Development of a closed-loop stripping technique for the analysis of taste- and odor-causing substances in drinking water. *In* L.H. Keith, ed. Advances in the Identification and Analysis of Organic Pollutants in Water, Vol. 2. Ann Arbor Science Publishers, Ann Arbor, Mich.

the vial to get a gastight seal. Alternatively, use a solid TFE tapered stopper. If cap or stopper does not form gastight closure, wrap vial with TFE tape. For permanent storage or shipment of extract, transfer to 100-μL vial with TFE septum.**

f. TFE sleeve, 5-mm-ID TFE flexible tubing approximately 19 mm long. If a 5-mm-OD vial is not prepared as described above, then connect filter and vial with a piece of heat-shrink TFE tubing that is custom-shrunk to the dimensions of the filter and vial.

** Pierce Chemical Company #13100 or equivalent.

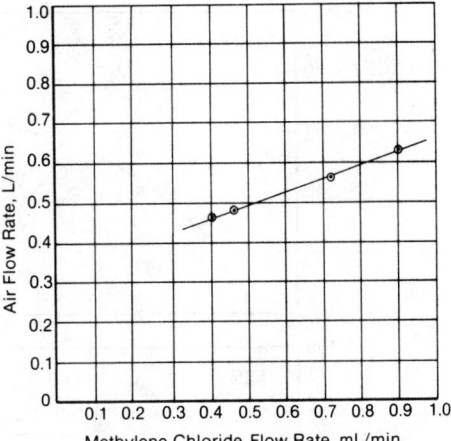

Figure 6040:5. Flow rate through 1.5-mg carbon filter. Air flow rate with no filter is 0.86 L/min.

g. Gas chromatograph (GC)/mass spectrometer (MS)/data system, equipped with:

1) *Capillary injector,* Grob-designed split-splitless injector or equivalent with 2.5-mm-ID glass insert or nonvaporizing, septumless, cold on-column injector.

2) *Capillary column,* 30-m or 60-m $\times$ 0.25-mm-ID DB-1 or DB-5 fused silica or other capillary column capable of producing adequate and reproducible resolution. Use a 0.32-mm-ID column for on-column injection when a stainless-steel needle is used. Use a 0.25-mm-ID column for on-column work with a fused-silica needle.

3) *Microsyringes,* 5- and 10-μL capacity, with 75-mm-long needles. Use 0.23-mm-OD stainless-steel or 0.17-mm-OD fused-silica needle for on-column injection.

4) *Mass spectrometer analyzer:* See Section 6210B.3 for suggested specifications.

5) *Data system,* with software capable of performing reverse-library searches (optional).

3. Reagents

Use reagent-grade solvents or better and obtain purest standards available.

a. Carbon disulfide, CS$_2$: Use only after gas chromatographic verification of purity

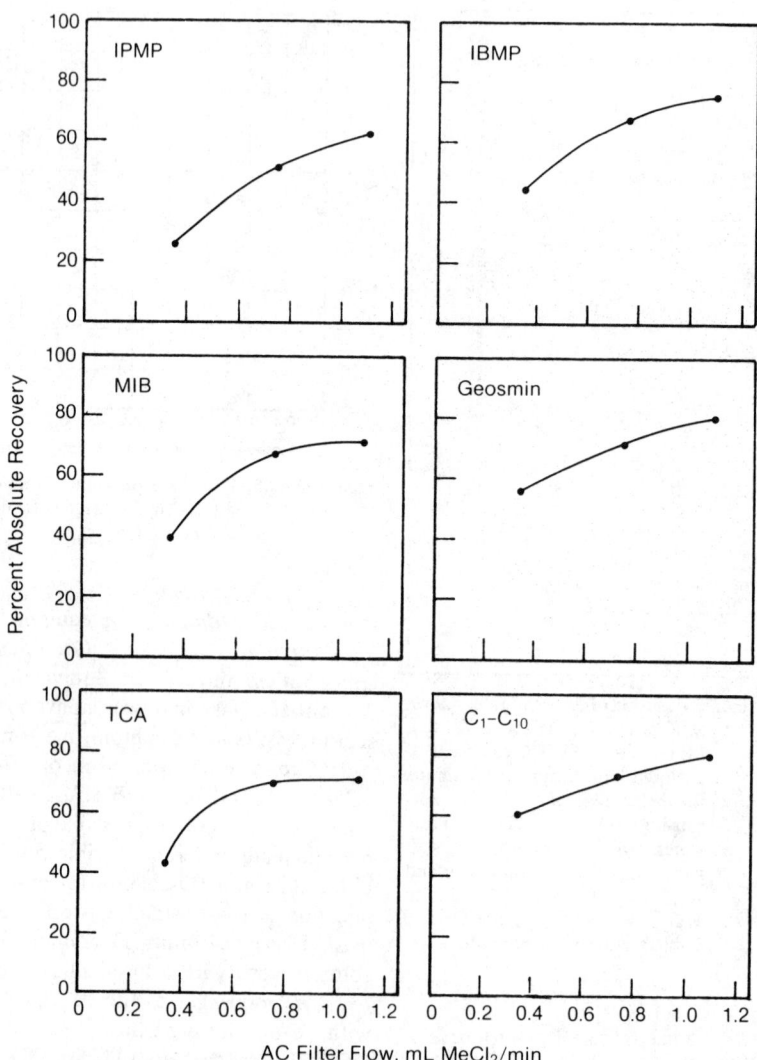

Figure 6040:6. Effect of filter resistance, measured as flow, on recovery of earthy-musty odorants and C_1-C_{10} internal standard. Reprinted with permission from: Hwang, C.J., S.W. Krasner, M.J. McGuire, M.S. Moylan & M.S. Dale. 1984. Determination of subnanogram per liter levels of earthy-musty odorants in water by the salted closed-loop stripping method. *Environ. Sci. Technol.* 18:535. Copyright 1984, American Chemical Society.

to ensure that the solvent does not contain components that coelute with the compounds of interest. CAUTION: *Use proper safety procedures; explosive, toxic, and occasionally allergenic.*

   b. Acetone.

   c. Methylene chloride.

   d. Carrier gas: Helium gas, ultrapurified grade, moisture- and oxygen-free.

   e. Internal standards:

   1) *1-Chlorooctane.*

   2) *1-Chlorodecane.*

3) *1-Chlorododecane.*

4) *1-Chlorohexane, 1-chlorohexadecane,* and *1-chlorooctadecane* can be added for broad-spectrum analysis.

f. Stock internal standard solutions: Dissolve 1 mL of each internal standard†† in acetone and dilute to 25 mL with acetone in a volumetric flask; 1 μL = 35 μg.

g. Combination internal standard solution: Combine 7.2 μL of each stock solution and dilute to 25 mL with acetone; 1 μL = 10 ng each.

h. Reference standards: Compounds of interest may be available commercially.‡‡

i. Stock reference solutions: Dissolve 20 mg of each target compound in acetone and dilute to 10 mL with acetone in a volumetric flask; 1 μL = 2 μg.

j. Combination reference standards solution: Combine 10 μL of each stock reference solution and dilute to 5 mL with acetone; 1 μL = 4 ng each.

k. Organic-free water: Prepare water by treating with activated carbon, mixed-bed deionization, and filtration through a membrane filter. Irradiate under ultraviolet light for 1 h (optional) and prestrip in the CLSA apparatus for 1 h, using a clean activated carbon filter or, alternatively, prestrip large quantities of water with nitrogen just before use. Store in a closed bottle tightly capped with TFE-lined screw cap, under nitrogen (optional), in a refrigerator and away from solvent contamination for not longer than 1 week.

l. Sodium sulfate, Na_2SO_4 (optional), granular, anhydrous. Bake at 625°C for 2 h before use; store at room temperature in desiccator.

---

†† 1-Chlorohexadecane and 1-chlorooctadecane solidify upon refrigeration. Warm before removing a portion.
‡‡ Geosmin and 2-methylisoborneol are available from Wako Chemicals, 12300 Ford Rd., Suite 130, Dallas, Texas 75234. NOTE: This synthetic geosmin includes nonodorous isomers that preclude its use in quantitative sensory analysis; however, its GC/MS characteristics (i.e., retention time and spectrum) are the same as those of natural geosmin.

4. Procedure

a. Sample collection and storage: Rinse sample bottle with sample, fill to overflowing, flush out air bubbles, and cap tightly. Collect duplicate samples and in the field keep in an insulated container stocked with ice. In the laboratory store at 4°C, but analyze as soon as possible, preferably within 3 d. For holding longer than 3 d, add 40 mg $HgCl_2$/L to inhibit biological activity. Adding a dechlorinating agent is optional, because disinfection byproducts may be affected. CAUTION: *$HgCl_2$ - containing samples must be disposed of as hazardous waste. See Section 1090 for precautions.*

b. Treatment of samples:

1) Stripping—Rinse stripping bottle with sample and fill to the 1-L mark, wetting the glass joint with sample. Fill stripping bottle slowly, with minimal aeration, to prevent loss of volatile compounds. Add 10 μL combination internal standard solution with the syringe needle tip immersed. Stopper tightly and attach springs. Place in thermostatic water bath at 25°C with glass joint below water level and connect bottle to the circuit. Operate gas heater at 45 to 50°C. Put an "auxiliary" carbon filter in the holder and prestrip for 10 s to flush air contaminants from system. Exchange auxiliary filter for a clean one and strip for 2 h. Auxiliary filter may be reused many times before cleaning. If stripping bottle has a smaller height-to diameter ratio than shown in Figure 6040:2, more than 2 h may be required for stripping. Optionally use an automatic timer to terminate each stripping run.

If sample contains a large amount of algae or turbidity or foaming agents, use only 900 mL sample and 9 μL combination internal standard solution. Because this additional headspace can result in different stripping efficiencies, comparably analyze a calibration standard.

Immediately after use clean stripping bottle by rinsing twice with demineralized water and once with organic-free water. For particularly adherent impurities, clean with acetone and methylene chloride and bake at 180°C for at least 2 h. Whenever sample carryover is observed, clean the circuit and pump as follows: Connect fittings to the quick-connect bodies on both ends of the circuit to open system. Turn on pump and flush with approximately 100 mL each organic-free water, acetone, and methanol. After the last rinse, dry with a heat gun until there is no residual methanol. If there is a noticeable drop in pump performance, clean valve assembly with distilled trichlorotrifluoroethane or replace.

2) Alternate stripping techniques—To improve stripping recovery, use either modification: a combination of a) and b) below can reduce stripping time. Optimize combination, depending on compounds to be analyzed.

a) Elevated stripping temperature—Increase temperature of thermostatic water bath to 45°C to increase recovery of many organic compounds.[5,6] Raise temperature of gas heater to at least 55°C (for a 45°C stripping temperature) to avoid condensation of water vapor on the activated carbon filter. Further increases in stripping temperature reduce recovery.[5]

b) Salting-out technique—Raising the ionic strength of the sample with Na_2SO_4 before stripping increases the stripping rate of many organic compounds.[3] Bring sample to room temperature before analysis by immersing it in a water bath at 25°C for approximately 15 min. Transfer 800 mL to the stripping bottle and add stirring bar. Using a glass funnel and with the stirring bar (at intermediate setting) in motion, add 72 g Na_2SO_4. Remove funnel and replace with a glass stopper. Continue stirring until Na_2SO_4 has dissolved (not more than 1 min). Remove stopper and stirring bar, then add remaining 100 mL of sample,§§

§§A total sample volume of 900 mL is preferred to minimize foaming-over due to salt addition.

rinsing and wetting inside neck of bottle. Add 9 μL combination internal standard solution and strip at 25°C as described in standard stripping procedure above. Strip for 1 h and 30 min. Alternatively, combine salt and sample by pouring salt directly into 900 mL sample, stopper tightly, shake vigorously, let stand for several minutes, and add internal standards.

3) Extracting the filter—Remove activated carbon filter from holder. In a fume hood, extract with CS_2 as indicated in Figure 6040:4. Keep solvents well within the hood to avoid inhalation by analyst or contamination of stripping apparatus. Add 2 μL CS_2 to a clean vial and connect filter and vial with a TFE sleeve so as to leave no dead space between glass parts. Place 10 μL CS_2 above the carbon, taking care not to touch carbon with the needle. Warm vial with hands and alternately pull and push the CS_2 through the carbon 10 times by separating and reconnecting the filter and vial while still within the TFE sleeve. With filter and vial tightly butted, cool vial with ice or liquid nitrogen, taking care not to freeze CS_2. This cooling draws the CS_2 below the carbon. Tap the filter/vial assembly gently on a hard surface to complete transfer of CS_2 to bottom of vial.

Repeat filter extraction with a 10-μL and a 5-μL portion of CS_2 to yield approximately 20 μL of combined extract. Separate vial from filter and stopper. Label and store at −20°C until analysis.

Filter may be extracted while maintaining tight seal between filter and vial during the procedure. Using ice chips, cool closed volume in the vial; solvent accumulates on lower side of the filter disk. Push solvent back to upper side by warming the closed vial between two fingers. Repeat and then extract with more solvent as above.

Because methylene chloride is easier to obtain in higher purity than CS_2, and is less hazardous, it can be used as the extraction solvent. Changing solvent may reduce extraction efficiency for some compounds.[8]

Clean filter as soon as possible after use.

Fill glass tube with CS_2 and let solvent flow through filter disk. Repeat three times with acetone and three times with methylene chloride. If solvent flow is slow (lower than 0.6 mL/min) because of salt deposits, pull $1N$ HNO_3 through filter, using a vacuum connection. After washing with acid, rinse with distilled water and acetone and continue with cleaning as above. After final rinse, remove residual solvent by connecting filter to a vacuum for approximately 5 min. Before reusing filter, repeat cleaning procedure. Clean auxiliary filter after 40 uses or 2 weeks, whichever comes first.

If the salting-out technique is used, carried over Na_2SO_4 ultimately will clog the filter. To avoid, routinely use a water rinse to remove deposited salts. For initial cleaning of filter only, after use, add solvents as follows: acetone, organic-free water, acetone, CS_2, acetone (three times), and methylene chloride (three times). Between samples repeat cleaning just before filter use beginning with CS_2 wash.

Rinse vials seven times with CS_2 and bake at 180°C overnight. Rinse several more times with CS_2 before using. Clean stoppers with methylene chloride, immersing at least overnight. Rinse TFE sleeve with methylene chloride and acetone and store in acetone until ready to use.

c. Gas chromatography/mass spectrometry:

1) "Hot-needle" injection technique—To reduce discrimination against higher-boiling compounds by distillation from the needle, use a hot-needle injection technique when the injector is a hot vaporizing type. (Do not use the following procedure for cold on-column injection.) Wet syringe needle and barrel with solvent and expel as much as possible. Pull syringe plunger back, leaving an air gap. Pull up approximately 1.5 μL sample and pull sample totally into syringe barrel. Close the split on the GC injector, wait 10 s, insert syringe needle into the injector, and let needle warm up for 1 to 2 s (analyst should op-timize time). Rapidly push plunger to bottom of syringe barrel to inject sample. Remove syringe and rinse well with solvent. Open split valve approximately 30 s after the injection.

2) On-column injection technique—To more fully reduce discrimination against higher-boiling compounds, use an on-column injector. A cold on-column injector also can be used to avoid decomposition of thermally labile compounds, e.g., dimethyl polysulfides.[9] Determine thermally labile compounds quantitatively by using a cold on-column injector or an inactive, vaporizing injector.

With an on-column injector, increase sensitivity by injecting large sample volumes (up to 8 μL). To prevent problems from a heavy condensation of solvent with such large-volume injections, use a 2-m retention gap (an empty, deactivated piece of fused silica tubing connected to the head of the column with a zero-dead-volume connector).[10] To preclude backpressure from large-volume injections, inject slowly at about 1 μL/5 s. Keep initial column temperature at 10°C above boiling point of solvent for a full solvent effect and to produce sharp peaks (low peak widths).[10] Because the entire injection is deposited directly into the head of the column, the column can become active after as few as 50 to 80 injections. Check activity by injecting a polarity test mixture at least weekly. Breaking approximately 30 cm off the head of the column can restore inertness.

3) Operating conditions for GC/MS—After initial installation of the capillary column, condition it according to the manufacturer's instruction. Daily, make a conditioning run with a CS_2 injection before injecting any samples (optional). Typical instrument conditions are given in Table 6040:III.

d. Calibration standard: The method is semiquantitative for a large number of compounds, but has been shown to be

TABLE 6040:III. TYPICAL OPERATING
CONDITIONS FOR GC/MS ANALYSIS OF CLSA
EXTRACTS[6]

Variable	Description or Value
Column	30- or 60-m × 0.25-mm-ID DB-1 or DB-5 fused silica capillary column*
Column temperature program	35°C, 16 min; 35 to 260°C @ 4°C/min; 260°C, 18 min
Carrier gas	Helium
Carrier gas flow rate	1 mL/min
Sample size	About 1.5 μL (splitless injection)
Injector temperature	250°C
Transfer line temperature	250°C
Separator temperature	265°C
Ionizer temperature	280°C
Source pressure	About 7 × 10^{-5} Pa
Electron energy	70 eV
Mass range scanned	41 to 453 amu
Scan time	1 s

*J&W Scientific, Inc., or equivalent.

Reprinted from *Identification and Treatment of Tastes and Odors in Drinking Water*,[6] by permission. Copyright 1987, American Water Works Association.

quantitative for many of the compounds listed in this section. Prepare a 20-ng/L target-compound calibration standard by dosing 1 L organic-free water in the stripping bottle with 10 μL combination internal standard solution plus 5 μL combination reference standards solution. (Internal standards concentration is 100 ng/L each.) If the salting-out technique is used, add 72 g Na_2SO_4 to a total volume of 900 mL organic-free water before dosing with 9 μL combination internal standard solution plus 4.5 μL combination reference standards solution. Analyze as directed

above. Inject the calibration standard extract at least weekly, preferably daily, to determine GC/MS response factors and verify spectra.

Verify working linear range by analyzing standards and representative samples with added organics at different concentrations.

e. Blanks: Run a procedural blank daily to assess contamination from reagents, apparatus, and other sources. Run a blank immediately after analyzing any very high-level sample or after installing new parts in the system. Analyze organic-free water with internal standards under the same conditions as samples.

5. Calculations

a. Identification: Identify a compound by matching both retention time and spectra of sample and standard. If they are available, use both a reverse-search computer program with a target-compound library and a forward-search program with the National Institute of Standards and Technology (formerly National Bureau of Standards) library for tentative identification of other compounds present.

1) Retention times—Use each internal standard to calculate relative retention times for all the compounds in the same part of the chromatogram (Table 6040:IV). For compounds eluting on the solvent tail, use an early-eluting internal standard (e.g., 1-chlorohexane). Sample retention times should match predicted retention times within ± 15 s.

$$\text{Predicted } T_{z,x} = \frac{T_{z,s}}{T_{I,s}} \times T_{I,x}$$

where:

T = retention time,
z = target compound,
I = internal standard,
x = sample analysis, and
s = calibration standard analysis.

2) Spectra—Peaks of at least three characteristic ions should all maximize at the

TABLE 6040:IV. GC/MS DATA FOR THREE INTERNAL STANDARDS AND TWO EARTHY-MUSTY-SMELLING COMPOUNDS

Compound	Retention Time* min	Quantification Mass amu	Characteristic Ions (with relative intensities)
1-Chlorooctane	30.8	91	43 (100), 91 (86, 93 (27)
2-Methylisoborneol	36.4	95, 107†	95 (100), 107 (26), 135 (9)
1-Chlorodecane	39.8	91	43 (100), 91 (87), 93 (28)
Geosmin	45.1	112	112 (100), 111 (28), 125 (18)
1-Chlorododecane	47.2	85	43 (100), 91 (61), 93 (19), 85 (12)

*See Table 6040:III for GC conditions. Data accumulated using 30-m DB-5 capillary column.

†Quantify using two different masses and obtain an average value.

same retention time and have standard intensity ratios (spectra) within ±20% of those of the calibration-standard compounds. Characteristic ions and their typical relative intensities for three of the internal standards and two earthy-musty-smelling compounds are given in Table 6040:IV. Preferably, use reference spectra of 10 to 14 key masses. Determine reference spectra by analysis of standards; verify these frequently. The spectra of MIB are particularly dependent on instrument condition; both 107 and 95 amu have been reported as base peaks (Figure 6040:7). Figure 6040:8 shows the mass spectrum for geosmin.

b. *Quantitation:* Determine concentrations by comparison of peak areas of specific quantitation ions. A quantitation ion should be relatively intense in the mass spectrum, yet be free from interference problems caused by closely eluting compounds (see Table 6040:IV). Calculate a response factor for each compound from CLSA of a calibration standard as follows:

$$R_z = \frac{A_z \times C_I}{C_z \times A_I}$$

where:

R_z = response factor for target compound Z,
I = internal standard, 1-chlorodecane,
C = concentration, ng/L, and
A = peak area in consistent units.

Compound concentration in the sample (x) is:

$$C_{z,x}, \text{ng/L} = \frac{A_{z,x} \times C_{I,x}}{R_z \times A_{I,x}}$$

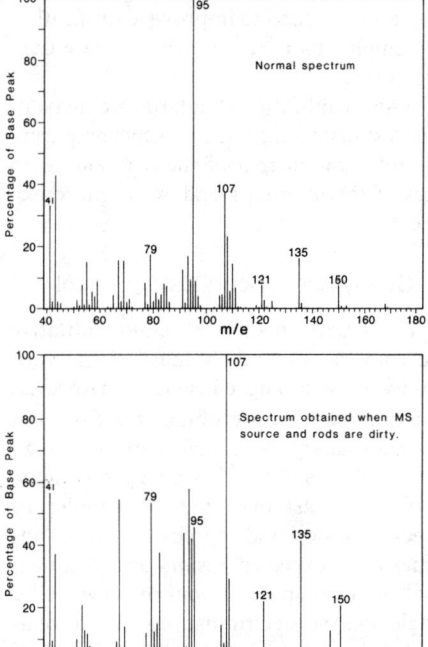

Figure 6040:7. Mass spectra of 2-methylisoborneol under different instrumental conditions.

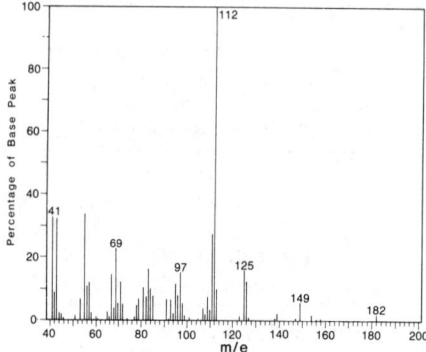

Figure 6040:8. Mass spectrum of geosmin.

Use the internal standard 1-chlorodec-ane for determining response factors. Use the other internal standards as a check on the system; calculated values should be to ± 20%. Computerized reverse-search spectral matching and automatic quantitation are recommended to improve identification in complex matrices and to facilitate data processing.

Where calibration standards are unavailable, estimate concentrations by comparing the total ion current of the compounds to that of the internal standard 1-chlorodec-ane.

6. Quality Assurance/Quality Control

The CLSA method is semiquantitative for some compounds because of the variability of stripping efficiencies. However, quantitative data are obtainable for compounds that are reproducibly stripped (e.g., MIB and geosmin).[1-3, 6] Analyze a replicate sample at least once per 10 samples to check precision and reproducibility. If unusual or unexpected results are obtained, analyze a replicate to confirm. Typically, single-analyst determinations for a relatively simple matrix have a coefficient of variation less than or equal to 10%. Otherwise, precision is usually within 20%. For compounds that are poorly stripped, a higher coefficient of variation may be obtained.

Analyze a sample with a known addition at least once per 10 samples to check accuracy and recovery. If matrix problems exist, this will confirm the accuracy of results. Adjust these recoveries against the calibration standards results. Even when absolute recoveries are less than 50%, standard adjusted recoveries, which correct for stripping efficiencies, are usually between 80 and 120%.

7. Precision and Bias

Precision and bias data are given in Tables 6040:V and VI for the analysis of earthy-musty-smelling compounds. Table 6040:VII shows recovery and precision data for selected pollutants.

8. References

1. KRASNER, S.W., C.J. HWANG & M.J. McGUIRE. 1981. Development of a closed-loop stripping technique for the analysis of taste- and odor-causing substances in drinking water. *In* L.H. Keith, ed. Advances in the Identification and Analysis of Organic Pollutants in Water, Vol. 2. Ann Arbor Science Publishers, Ann Arbor, Mich.

2. KRASNER, S.W., C.J. HWANG & M.J. McGUIRE. 1983. A standard method for quantification of earthy-musty odorants in water. *Water Sci. Technol.* 15(6/7):127.

3. HWANG, C.J., S.W. KRASNER, M.J. McGUIRE, M.S. MOYLAN & M.S. DALE. 1984. Determination of subnanogram per liter levels of earthy-musty odorants in water by the salted closed-loop stripping method. *Environ. Sci Technol.* 18:535.

4. COLEMAN, W.E., J.W. MUNCH, R.W. SLATER, R.G. MELTON & F.C. KOPFLER. 1983. Optimization of purging efficiency and quantification of organic contaminants from water using a 1-L closed-loop-stripping apparatus and computerized capillary column GC/MS. *Environ. Sci. Technol.* 17:571.

5. THOMASON, M.M. & W. BERTSCH. 1983. Evaluation of sampling methods for the determination of trace organics in water. *J. Chromatogr.* 279:383.

6. MALLEVIALLE, J. & I.H. SUFFET, eds. 1987. Identification and Treatment of Tastes and Odors in Drinking Water. American Water

TABLE 6040:V. SINGLE-LABORATORY BIAS FOR SELECTED ORGANIC COMPOUNDS CAUSING TASTE AND ODOR

Stripping Technique Compound	Dose Level ng/L	Number of Samples*	Mean Recovery† %	Standard Deviation %
Unsalted method‡				
2-Isobutyl-3-methoxy pyrazine	4–20	23	89	21
2-Isobutyl-3-methoxy pyrazine	4–20	22	101	28
2-Methylisoborneol	4–120	30	101	21
2,3,6-Trichloroanisole	4–20	22	88	27
Geosmin	4–120	28	109	20
Salting-out method‡				
2-Isobutyl-3-methoxy pyrazine	4–20	44	120	24
2-Isobutyl-3-methoxy pyrazine	4–20	48	106	18
2-Methylisoborneol	4–20	48	106	15
2,3,6-Trichloroanisole	4–20	45	99	22
Geosmin	4–20	48	105	15

*Finished and natural surface waters.
†Standard-adjusted recovery.
‡Stripping at 25°C.
Reprinted with permission from *Environmental Science and Technology,* 18:535.[3] Copyright 1984. American Chemical Society.

Works Association & American Water Works Association Research Foundation, Denver, Colo.

7. Report of Analysis for Semivolatile Organics by Closed-Loop Stripping GC/MS. 1982. Environmental Research Lab., James M. Montgomery, Consulting Engineers, Inc., Pasadena, Calif.

8. GROB, K. & F. ZÜRCHER. 1976. Stripping of trace organic substances from water. Equipment and procedure. *J. Chromatogr.* 117:285.

9. WAJON, J.E., R. ALEXANDER & R.I. KAGI. 1985. Determination of trace levels of dimethyl polysulphides by capillary gas chromatography. *J. Chromatogr.* 319:187.

10. GROB, K. 1978. On-column injection onto capillary columns. Part 2: Study of sampling conditions; practical recommendations. *J.*

High Resolution Chromatogr. Chromatogr. Commun. 1:263.

9. Bibliography

BUDDE, W.L. & J.W. EICHELBERGER. 1979. Organic Analysis Using Gas Chromatography-Mass Spectrometry. Ann Arbor Science Publishers, Ann Arbor, Mich.

GROB, K. & K. GROB, JR. 1978. Splitless injection and the solvent effect. *J. High Resolution Chromatogr. Chromatogr. Commun.* 1:57.

GROB, K. & G. GROB. 1979. Practical gas chromatography—a systematic approach. *J. High Resolution Chromatogr. Chromatogr. Commun.* 2:109.

REINHARD, M., J.E. SCHREINER, T. EVERHART & J. GRAYDON. 1987. Specific analysis of trace organics in water using gas chromatography and mass spectroscopy. *J. Environ. Pathol., Toxicol. & Oncol.* 7:417.

TABLE 6040:VI. PRECISION DATA FOR SELECTED ORGANIC COMPOUNDS CAUSING TASTE AND ODOR*

Compound	Multiple Laboratories†				Single Laboratory‡		
	Dose Level ng/L	Mean ng/L	Standard Deviation ng/L	Coefficient of Variation %	Mean ng/L	Standard Deviation ng/L	Coefficient of Variation %
Sample A§							
2-Isopropyl-3-methoxy pyrazine	5.9	5.6	1.6	28	6.6	0.6	9
2-Isobutyl-3-methoxy pyrazine	3.0	3.0	0.7	24	3.2	0.3	11
2-Methylisoborneol	4.3	4.8	1.0	20	5.1	0.1	2
2,3,6-Trichloroanisole	8.7	7.3	3.1	43	9.3	2.0	22
Geosmin	2.9	3.3	0.9	27	3.6	0.5	14
Sample B§							
2-Isopropyl-3-methoxy pyrazine	25	22.3	8.0	36	27.5	2.8	10
2-Isobutyl-3-methoxy pyrazine	15	14.3	4.2	30	17.1	2.5	14
2-Methylisoborneol	20	18.3	6.2	34	22.6	2.0	9
2,3,6-Trichloroanisole	35	32.0	9.7	30	37.2	3.3	9
Geosmin	16	15.9	5.9	37	19.9	2.8	14

* Stripping at 25°C, unsalted method.
† Five analysts at three laboratories.
‡ Three analysts at one laboratory.
§ Organic-free water dosed with taste and odor compounds.

TABLE 6040:VII. RECOVERY AND PRECISION DATA FOR SELECTED PRIORITY POLLUTANTS*

Compound	Amount ng	Mean Recovered Amount† ng	Range	Recovery Efficiency %	RSD %
Thiophene	25.4	8.8	7.2 – 10.9	34.6	15.9
Dibromochloromethane	29.4	16.6	13.3 – 21.0	56.5	12.6
Styrene	21.8	17.4	15.9 – 19.6	79.8	6.9
Isopropylbenzene	24.2	25.9	23.8 – 28.7	107.0	7.7
2-Chlorotoluene	25.9	23.3	21.6 – 26.9	90.0	7.7
bis(2-Chloroethyl)ether	24.4	2.8	2.6 – 3.5	11.6	10.7
α-Methylstyrene	21.6	19.4	18.1 – 21.6	89.9	6.2
1,4-Dichlorobenzene	19.8	18.3	16.4 – 20.6	92.5	8.2
2-Ethyl-1,3-dimethylbenzene	24.1	22.3	20.8 – 24.7	92.4	5.8
4-Chloro-o-xylene	25.7	21.8	20.2 – 26.0	84.8	9.2
1,1-Dimethylindan	21.7	23.8	21.6 – 26.2	109.5	7.1
p-Methylphenol	26.6	ND‡			
Tetrahydronaphthalene	23.4	23.1	20.2 – 26.3	98.6	10.4
1,2,4-Trichlorobenzene	23.2	19.2	17.6 – 20.9	82.9	7.3
Hexachloro-1,3-butadiene	26.9	30.9	27.4 – 34.3	113.8	9.2
2-Methylbiphenyl	24.4	24.8	22.0 – 27.1	101.4	7.7
1,6-Dimethylnaphthalene	24.2	9.5	7.6 – 11.2	39.3	11.6
2-Isopropylnaphthalene	22.7	20.6	18.8 – 22.3	90.8	6.3
Pentachlorobenzene	26.1	12.4	10.7 – 14.3	47.7	11.3
Hexachlorobenzene	20.1	6.2	5.0 – 7.4	31.1	12.9
2,2',4,4',6,6'-Hexachlorobiphenyl	27.1	28.2	26.1 – 32.3	104.3	7.1
2,2',4,5,5'-Pentachlorobiphenyl	28.2	23.0	17.8 – 27.6	81.5	14.3

*Stripping at 40°C, unsalted method.
†Based on six purging analyses using single ion quantification.
‡ND = not detected.
Reprinted with permission from *Environmental Science and Technology,* 17:571.[4]
Copyright 1983. American Chemical Society.

6040 C.   Purge and Trap Technique

For applications of this technique to analyses for volatile organics, volatile aromatic organics, and volatile halocarbons, see Sections 6210, 6220, and 6230, respectively.

6210   VOLATILE ORGANICS

6210 A.   Introduction

1. Sources and Significance

Many organic compounds have been detected in ground and surface waters. While most groundwater contamination episodes are traceable to leaking underground fuel or solvent storage vessels, landfills, agriculture practices, and wastewater disposal, the most probable cause for contamination of some aquifers and surface waters has never been firmly established. Contamination may be due to past practices of on-site (leachfield) disposal of domestic and industrial wastes or to illegal discharges. Organohalides, particularly the trihalomethanes, are present in most chlorinated water systems, especially those using surface waters as a source of supply. Toxicological studies on animal models have shown that some of these organics have the potential for teratogenesis or carcinogenesis in human beings. To minimize these health risks, sensitive detection and accurate and reproducible quantitation of organics is of paramount importance.

2. Selection of Method

Three gas chromatographic/mass spectrometric (GC/MS) methods for purgeable organics are presented. While all three methods are similar in compound recovery, column materials, and chromatographic conditions, the methods are not interchangeable from a regulatory point of view.[1,2] See each method for list of specific compounds covered.

Method B is a packed-column GC/MS method useful for the determination of purgeable organics in municipal and industrial wastes. Method C, also a packed-column GC/MS method, is applicable to the determination of various volatile organic compounds in finished drinking water, raw source water, or drinking water in any stage of treatment. Method D is identical to Method C, except that a capillary column is used. Methods C and D are intended primarily for the detection of large numbers of contaminants at very low concentrations not detectable with the materials and conditions used in Method B. Quality assurance requirements and instrument and method performance criteria are more stringent for the methods applicable to drinking water.

All three methods are highly sophisticated microanalytical procedures that should be used only by analysts experienced in operation of purge and trap units and GC/MS systems and in evaluation and interpretation of mass spectra.

3. References

1. U.S. ENVIRONMENTAL PROTECTION AGENCY. 1987. National primary drinking water regulations—synthetic organic chemicals; monitoring for unregulated contaminants; final rule. 40 CFR Parts 141 & 142; *Federal Register* 52, No. 130.
2. U.S. ENVIRONMENTAL PROTECTION AGENCY. 1986. Guidelines establishing test procedures for the analysis of pollutants under the Clean Water Act. 40 CFR Part 136; *Federal Register* 51, No. 125.

6210 B.   Purge and Trap Packed-Column
Gas Chromatographic/Mass Spectrometric Method I*

This method[1] is applicable to the determination of purgeable organics† in municipal and industrial discharges. The method may be extended to screen samples for acrolein and acrylonitrile; however, another method[2] for these compounds is preferred.

1. General Discussion

a. Principle: Volatile organic compounds are transferred efficiently from the aqueous to the vapor phase by bubbling an inert gas through a water sample contained in a specially designed purging chamber at ambient temperature. The vapor is swept through a sorbent trap to collect the organics. After purging is completed, the trap is heated and backflushed with the same inert gas to desorb the compounds onto a gas chromatographic column. The gas chromatograph is temperature-programmed to separate the compounds. The detector is a mass spectrometer.[3,4] See Section 6010C for discussion of gas chromatographic and mass spectrometric principles.

b. Interferences: See Section 6010C. Impurities in the purge gas and organic compounds outgassing from the plumbing ahead of the trap account for most contamination problems. Demonstrate that the system is free from contamination under operational conditions by analyzing

method blanks daily (see ¶ 7a). (NOTE: Use blanks for monitoring only; corrections for blank values are unacceptable.) Avoid using non-TFE plastic tubing, non-TFE thread sealants, or flow controllers with rubber components in the purge and trap system.

Samples can be contaminated by diffusion of volatile organics (particularly fluorocarbons and methylene chloride) through the septum seal during shipment and storage. Use a field reagent blank prepared from reagent water and carried through the sampling and handling procedure as a check on such contamination.

Contamination by carryover can occur whenever high-level and low-level concentration samples are analyzed sequentially. To reduce carryover, rinse purging device and sample syringe with reagent water between samples. Follow analysis of an unusually concentrated sample with an analysis of reagent water to check for cross-contamination. For samples containing large amounts of water-soluble materials, suspended solids, high boiling compounds, or high levels of the subject group of compounds, wash purging device with a detergent solution, rinse it with distilled water, and then dry it in an oven at 105°C between analyses. The trap and other parts of the system also are subject to contamination; therefore, frequently bake and purge the entire system. Loss of volatile constituents is also an important source of error.

c. Detection limits: The method detection limit (MDL) is the minimum concentration of a substance that can be measured and reported with 99% confidence that the value is above zero.[5] The MDL concentrations listed in Table 6210:I were obtained with reagent water.[6] The MDL actually obtained in a given analysis will vary, depending on instrument sensitivity and matrix effects.

---

* Approved by Standard Methods Committee, 1988. Accepted by U.S. Environmental Protection Agency as equivalent to EPA Method 624.
† Benzene, bromodichloromethane, bromoform, bromomethane, carbon tetrachloride, chlorobenzene, chloroethane, 2-chloroethylvinyl ether, chloroform, chloromethane, dibromochloromethane, 1,2-dichlorobenzene, 1,3-dichlorobenzene, 1,4-dichlorobenzene, 1,1-dichloroethane, 1,2-dichloroethane, 1,1-dichloroethene, *trans*-1,2-dichloroethene, 1,2-dichloropropane, *cis*-1,3-dichloropropene, *trans*-1,3-dichloropropene, ethylbenzene, methylene chloride, 1,1,2,2-tetrachloroethane, tetrachloroethene, toluene, 1,1,1-trichloroethane, 1,1,2-trichloroethane, trichloroethene, trichlorofluoromethane, and vinyl chloride.

TABLE 6210:I. CHROMATOGRAPHIC CONDITIONS AND METHOD DETECTION LIMITS (MDL)

Compound	Retention Time min	Method Detection Limit* μg/L
Chloromethane	2.3	nd
Bromomethane	3.1	nd
Vinyl chloride	3.8	nd
Chloroethane	4.6	nd
Methylene chloride	6.4	2.8
Trichlorofluoromethane	8.3	nd
1,1-Dichloroethene	9.0	2.8
1,1-Dichloroethane	10.1	4.7
trans-1,2-Dichloroethene	10.8	1.6
Chloroform	11.4	1.6
1,2-Dichloroethane	12.1	2.8
1,1,1-Trichloroethane	13.4	3.8
Carbon tetrachloride	13.7	2.8
Bromodichloromethane	14.3	2.2
1,2-Dichloropropane	15.7	6.0
cis-1,3-Dichloropropene	15.9	5.0
Trichloroethene	16.5	1.9
Benzene	17.0	4.4
Dibromochloromethane	17.1	3.1
1,1,2-Trichloroethane	17.2	5.0
trans-1,3-Dichloropropene	17.2	nd
2-Chloroethylvinyl ether	18.6	nd
Bromoform	19.8	4.7
1,1,2,2-Tetrachloroethane	22.1	6.9
Tetrachloroethene	22.2	4.1
Toluene	23.5	6.0
Chlorobenzene	24.6	6.0
Ethyl benzene	26.4	7.2
1,3-Dichlorobenzene	33.9	nd
1,2-Dichlorobenzene	35.0	nd
1,4-Dichlorobenzene	35.4	nd

Column conditions: Carbopack B (60/80 mesh) coated with 1% SP-1000 packed in a 1.8 m by 3 mm ID glass column with helium carrier gas at 30 mL/min flow rate. Column temperature held at 45°C for 3 min, then programmed at 8°C/min to 220°C and held for 15 min.
* nd = not determined.

d. *Safety:* The toxicity or carcinogenicity of each reagent has not been defined precisely. Benzene, carbon tetrachloride, chloroform, 1,4-dichlorobenzene, and vinyl chloride have been classified tentatively as known or suspected, human or mammalian carcinogens. Prepare primary standards of these compounds in a hood and wear a NIOSH/MESA-approved toxic gas respirator when handling high concentrations.

2. Sampling and Storage

See Section 6010B.1.

3. Apparatus

a. *Purge and trap system:* The purge and trap system consists of three separate pieces

of equipment: purging device, trap, and desorber. Several complete systems now are
available commercially.

1) *Purging device*, designed to accept 5-
mL samples with a water column at least
3 cm deep. Keep the gaseous head space
between the water column and the trap to
a total volume of less than 15 mL. Pass
the purge gas through the water column
as finely divided bubbles with a diameter
of less than 3 mm at the origin. Introduce
the purge gas no more than 5 mm from
the base of the water column. The purging
device illustrated in Figure 6210:1 meets
these design criteria.

2) *Trap*, at least 25 cm long and with an
inside diameter of at least 3 mm, packed
to contain the following minimum lengths
of adsorbents: 1.0 cm of methyl silicone-
coated packing, ¶ 4*b*2), 15 cm of 2,6-di-
phenylene oxide polymer, ¶ 4*b*1), and 8 cm

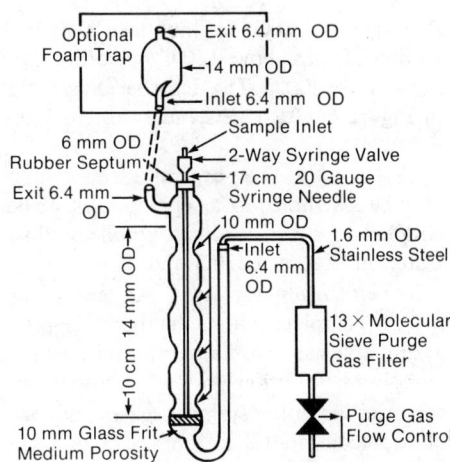

Figure 6210:1. Purging device.

of silica gel, ¶ 4*b*3). Figure 6210:2 illustrates the minimum specifications.

3) *Desorber*, capable of rapidly heating

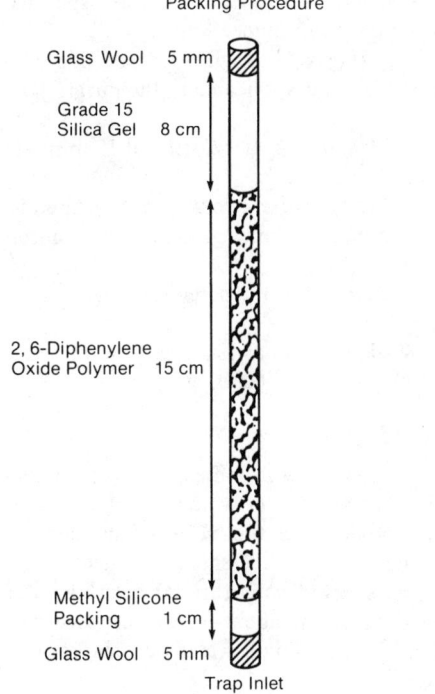

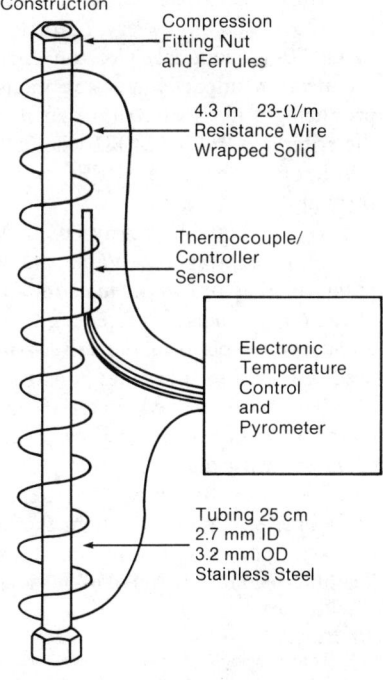

Figure 6210:2. Trap packings and construction to include desorb capability.

the trap to 180°C. Do not heat polymer section of trap above 180°C nor other sections above 200°C. The desorber illustrated in Figure 6210:2 meets these design criteria.

4) *Assembly*: The purge and trap system may be assembled as a separate unit or be coupled to a gas chromatograph as illustrated in Figures 6210:3 and 4.

b. Gas chromatograph:‡ An analytical system complete with a temperature-programmable gas chromatograph suitable for on-column injection and all required accessories including syringes, gases, and analytical column 1.8 m long × 3 mm ID stainless steel or glass, packed with 1% SP-1000 on Carbopack B (60/80 mesh) or equivalent. This column was used to develop the detection limit and precision and bias data presented herein. Guidelines for the use of alternate column packings are provided in ¶ 5a.

c. Mass spectrometer, capable of scanning from 20 to 260 amu every 7 s or less, utilizing 70-V (nominal) electron energy in the electron impact ionization mode, and producing a mass spectrum that meets all the criteria in Table 6210:II when 50 ng of 4-bromofluorobenzene (BFB) is injected through the GC inlet.

d. GC/MS interface: Any GC to MS interface that gives acceptable calibration points at 50 ng or less per injection for each of the compounds of interest and achieves all acceptable performance criteria may be used. GC to MS interfaces constructed of all glass or glass-lined materials are recommended. Glass can be deactivated by silanizing with dichlorodimethylsilane.

e. Data system: Interface to the mass spectrometer computer system that allows the continuous acquisition and storage on machine-readable media of all mass spectra

‡ Gas chromatographic methods are extremely sensitive to the materials used. Mention of trade names by *Standard Methods* does not preclude the use of other existing or as-yet-undeveloped products that give *demonstrably* equivalent results.

TABLE 6210:II. BFB KEY m/z ABUNDANCE CRITERIA

Mass	m/z Abundance Criteria
50	15 to 40% of mass 95
75	30 to 60% of mass 95
95	Base peak, 100% relative abundance
96	5 to 9% of mass 95
173	<2% of mass 174
174	>50% of mass 95
175	5 to 9% of mass 174
176	>95% but <101% of mass 174
177	5 to 9% of mass 176

obtained throughout the chromatographic program. Use computer software that allows searching any GC/MS data file for specific m/z (masses) and plotting such m/z abundances versus time or scan number. This type of plot is defined as an Extracted Ion Current Profile (EICP). Software also should allow integrating the abundance in any EICP between specified time or scan number limits.

f. Syringes, 5-mL glass hypodermic with luerlok tip, if applicable to the purging device.

g. Microsyringes, 25-μL, 0.15-mm ID needle.

h. Syringe valve, two-way, with luer ends.

i. Syringe, 5-mL, gastight with shutoff valve.

j. Bottle, 15-mL, screw-cap, with TFE cap liner.

k. Balance, analytical, capable of accurately weighing 0.0001 g.

4. Reagents

a. Reagent water: Reagent water is defined as a water in which an interferent is not observed at the MDL of the constituents of interest. Generate reagent water by passing tap water through a carbon filter bed containing about 0.5 kg activated carbon§ or by using a water purification system.‖

§ Filtrasorb-300, Calgon Corp., or equivalent.
‖ Millipore Super-Q or equivalent.

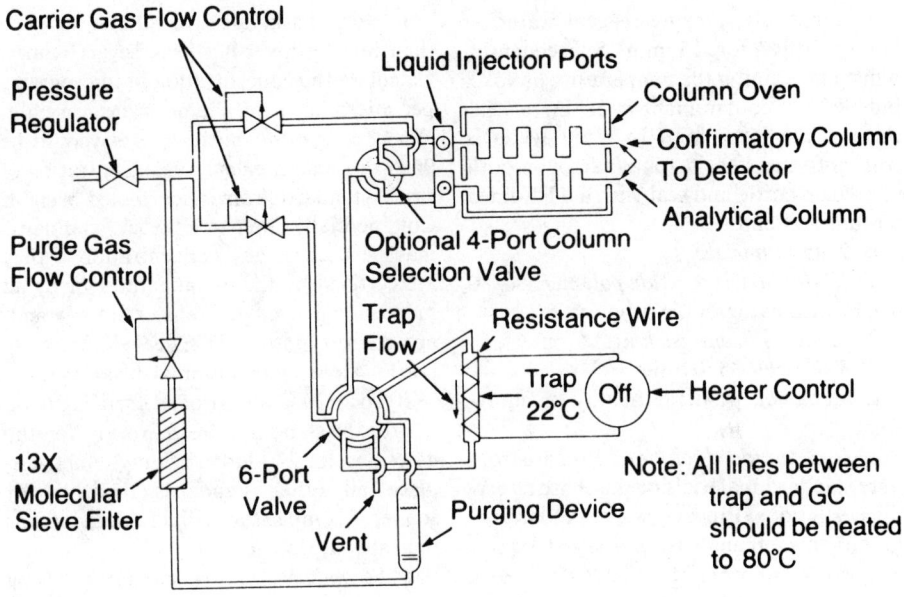

Figure 6210:3. Purge and trap system (purge mode).

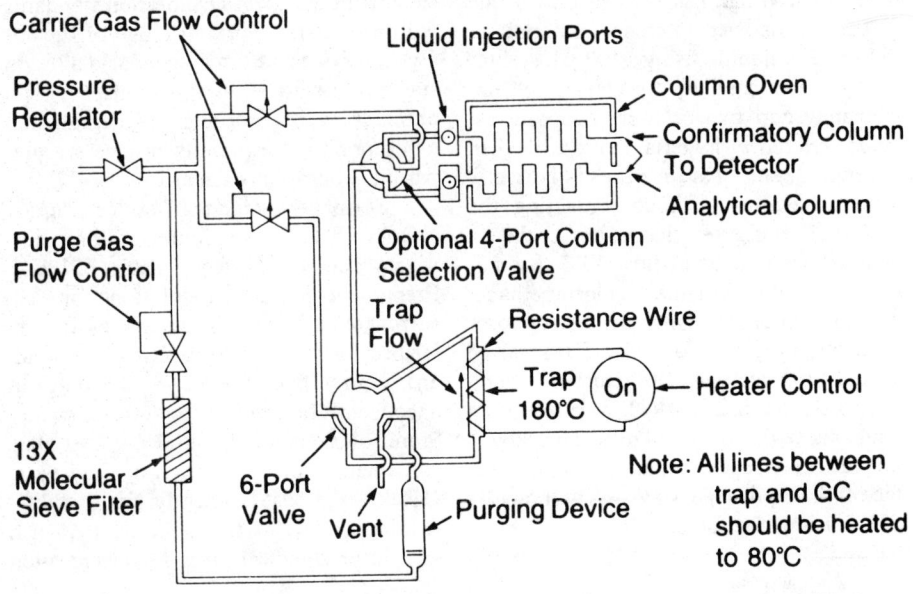

Figure 6210:4. Purge and trap system (desorb mode).

Alternatively, prepare reagent water by boiling water for 15 min. Subsequently, while maintaining the temperature at 90°C, bubble a contaminant-free inert gas through the water for 1 h. While water is still hot, transfer it to a narrow-mouth screw-cap bottle and seal with a TFE-lined septum and cap.

b. Trap materials:

1) *2,6-Diphenylene oxide polymer,* 60/80 mesh, chromatographic grade.#

2) *Methyl silicone packing.***

3) *Silica gel,* 35/60 mesh.††

c. Methanol, pesticide quality or equivalent.

d. Stock standard solutions: Prepare from pure standard materials or purchase as certified solutions. Prepare stock standard solutions in methanol using assayed liquids or gases as appropriate. CAUTION: *Toxic substances; see ¶ 1d.*

Place about 9.8 mL methanol in a 10-mL ground-glass-stoppered volumetric flask. Let stand, unstoppered, for about 10 min or until all alcohol-wetted surfaces have dried. Weigh flask to nearest 0.1 mg.

Add the assayed reference materials as follows: For liquids, using a 100-µL syringe or disposable capillary tip glass pipet, immediately add two or more drops of assayed reference material to flask, then reweigh. Be sure that the drops fall directly into the alcohol without contacting the flask neck. For gases, that is, any of the six halocarbons that boil below 30°C (bromomethane, chloroethane, chloromethane, dichlorodifluoromethane, trichlorofluoromethane, vinyl chloride), fill a 5-mL valved gastight syringe with the reference standard to the 5.0-mL mark. Lower needle to 5 mm above the methanol meniscus. Slowly introduce reference standard above the surface of the liquid (the heavy gas will rapidly dissolve into the methanol).

Reweigh flask, dilute to volume, stopper, then mix by inverting flask several times. Calculate the concentration in micrograms per microliter from the net gain in weight. When compound purity is assayed to be 96% or greater, calculate concentration of stock standard from uncorrected weight. Commercially prepared stock standards may be used at any concentration if they are certified by the manufacturer or by an independent source. Transfer stock standard solution into a TFE-sealed screw-cap bottle. Store, with minimal headspace, at −10 to −20°C and protect from light.

Prepare standards fresh weekly for the gases and for 2-chloroethylvinyl ether. Replace all other standards monthly, or sooner if comparison with check standards indicates a problem.

e. Secondary dilution standards: Using stock standard solutions, prepare in methanol secondary dilution standards that contain the compounds of interest, either singly or mixed together. Prepare secondary dilution standards at concentrations such that the aqueous calibration standards will bracket the working range of the analytical system. Store secondary dilution standards with minimal headspace and check frequently for signs of degradation or evaporation, especially just before preparing calibration standards.

f. Surrogate standard known-addition solution: Select a minimum of three surrogate compounds from Table 6210:III. Prepare stock standard solutions for each surrogate standard in methanol as described above. Prepare a surrogate standard solution from these stock standards at a concentration of 15 µg/mL in water. Store solutions at 4°C in TFE-sealed glass containers with a minimum of headspace. Check solutions frequently for stability. Adding 10 µL of this solution to 5 mL sample or standard gives a concentration equivalent to 30 µg/L of each surrogate standard.

g. Calibration standards: Using a 25-µL

Tenax or equivalent.
** 3% OV-1 on Chromosorb W (60/80 mesh) or equivalent.
†† Davison grade 15 or equivalent.

syringe, prepare at least three concentration levels for each compound by carefully adding 20.0 μL of one or more secondary dilution standards to 50, 250, or 500 mL reagent water. Prepare one standard at a concentration near, but above, the MDL and the others to correspond to the expected range of sample concentrations or to define the working range of the detector. Aqueous calibration standards can be stored up to 24 h, if held in sealed vials with zero headspace. Otherwise, discard after 1 h.

Prepare a known-addition solution containing each of the internal standards as directed in ¶s 4d and e above. Preferably prepare secondary dilution standard at a concentration of 15 μg/mL of each internal standard compound. Adding 10 μL of this standard to 5.0 mL sample or calibration standard gives a concentration equivalent to 30 μg/L.

h. BFB standard: Prepare a 25-μg/mL solution of BFB in methanol.

i. Quality control (QC) check sample: Obtain a check sample concentrate‡‡ con-

taining each compound at a concentration of 10 μg/mL in methanol. If such a sample is not obtainable from an external source, prepare using stock standards prepared independently from those used for calibration.

5. Procedure

a. Operating conditions: Table 6210:I summarizes recommended operating conditions for the gas chromatograph and gives estimated retention times and MDLs that can be achieved under these conditions. An example of the separations obtained with this column is shown in Figure 6210:5. Other packed columns or chromatographic conditions may be used if quality control requirements (¶ 7 below) are met.

b. GC/MS performance tests: At the beginning of each day on which analyses are to be performed, check GC/MS system to see that acceptable performance criteria are achieved for BFB.[7] The performance test must be passed before any samples, blanks, or standards are analyzed, unless the instrument has met DFTPP test[8] earlier.

These performance tests require the following instrumental parameters:

‡‡ For U.S. federal permit-related analyses, use samples obtainable from U.S. EPA Environmental Monitoring and Support Laboratory, Cincinnati, Ohio.

TABLE 6210:III. SUGGESTED SURROGATE AND INTERNAL STANDARDS

Compound	Retention Time* min	Primary m/z	Secondary Masses
Benzene-d₆	17.0	84	
4-Bromofluorobenzene	28.3	95	174, 176
1,2-Dichloroethane-d₄	12.1	102	
1,4-Difluorobenzene	19.6	114	63, 88
Ethylbenzene-d₅	26.4	111	
Ethylbenzene-d₁₀	26.4	98	
Fluorobenzene	18.4	96	70
Pentafluorobenzene	23.5	168	
Bromochloromethane	9.3	128	49, 130, 51
2-Bromo-1-chloropropane	19.2	77	79, 156
1,4-Dichlorobutane	25.8	55	90, 92

* For chromatographic conditions, see Table 6210:I.

Electron energy: 70 V (nominal)

Mass range: 20 to 260 amu

Scan time: To give at least 5 scans per peak but not to exceed 7 s per scan.

At the beginning of each day, inject 2 μL of BFB solution directly on the column. Alternatively, add 2 μL of BFB solution to 5.0 mL of reagent water or standard solution and analyze the solution according to ¶ *d* below. Obtain a background-corrected mass spectrum of BFB and confirm that all the key *m/z* criteria in Table 6210:II are achieved. If all criteria are not achieved, retune mass spectrometer and repeat test until all criteria are achieved.

c. Calibration: Calibrate system daily as follows:

1) System setup—Assemble the purge and trap system. Condition trap overnight at 180°C by backflushing with an inert gas flow of at least 20 mL/min. Condition trap for 10 min once daily before use.

Connect purge and trap system to a gas chromatograph. Operate gas chromatograph using the specified temperature and flow rate conditions. Calibrate purge and trap gas chromatographic system using the internal standard technique, ¶ 2).

2) Internal standard calibration procedure—Select three or more internal standards that are similar in analytical behavior to the compounds of interest, chosen from the compounds listed in Table 6210:III, and demonstrate that the measurement of the internal standard is not affected by method or matrix interferences. Because of such limitations, no internal standard is

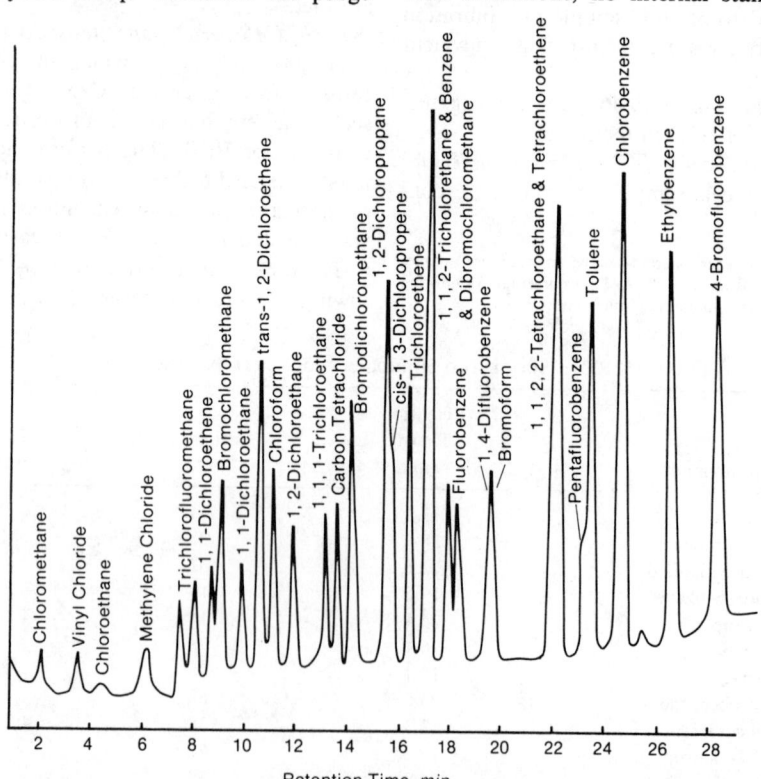

Figure 6210:5. Gas chromatogram of volatile organics. Column: 1% SP-1000 on Carbopack-B; program: 45°C for 3 min, 8°C/min to 220°C; detector: mass spectrometer.

applicable to all samples. The compounds recommended for use as surrogate known additions for quality control have been used successfully as internal standards because of their generally unique retention times. Prepare calibration standards at a minimum of three concentration levels for each compound as described in ¶ 4g. Prepare a known-addition solution containing each of the internal standards using the procedures described in ¶s 4d and e. Preferably prepare the secondary dilution standard at a concentration of 15 µg/mL of each internal standard compound. Adding 10 µL of this standard to 5.0 mL sample or calibration standard is equivalent to adding 30 µg/L. Analyze each calibration standard according to the procedure for samples, adding 10 µL of internal standard known-addition solution directly to the syringe. Tabulate peak height or area responses against concentration for each compound and internal standard, and calculate response factors (RF) for each compound as follows:

$$RF = \frac{(A_s)\,(C_{is})}{(A_{is})\,(C_s)}$$

where:

A_s = response for compound to be measured,
A_{is} = response for internal standard,
C_{is} = concentration of internal standard, and
C_s = concentration of compound to be measured.

Use average RF if RSD is less than 35%. Check calibration by procedure of ¶ 5b3) below, using calibration acceptance criteria found in Table 6210:IV.

3) Calibration check—Verify working calibration curve, calibration factor, or RF on each working day by analyzing a QC check sample [¶ 4i)] according to sample analysis procedure. For each compound, compare the response (Q) with the corresponding calibration acceptance criteria (Table 6210:IV). If the responses for all compounds of interest fall within the des-

ignated ranges, sample analysis may begin. NOTE: Because of the large number of compounds, there is a substantial probability that one or more will not meet the calibration acceptance criteria when all compounds are analyzed. If any individual Q falls outside the range, repeat test only for those compounds that failed to meet the calibration acceptance criteria. If the response does not fall within the range in this second test, prepare a new calibration curve, calibration factor, or RF for that compound.

d. Sample analysis: Adjust purge gas (nitrogen or helium) flow rate to 40 mL/min. Attach trap inlet to purging device and set purge and trap system to purge (Figure 6210:3). Open syringe valve on purging device sample introduction needle.

Let sample come to ambient temperature before introducing it into the syringe. Remove plunger from a 5-mL syringe and attach a closed syringe valve. Open sample bottle (or standard) and carefully pour sample into syringe barrel to just short of overflowing. Replace syringe plunger and compress sample. Open syringe valve and vent any residual air while adjusting sample volume to 5.0 mL. Because this process creates a headspace in the sample bottle and may destroy the validity of the sample if it is stored, fill and hold a second syringe at this time. Add 10.0 µL of the surrogate addition solution (¶ 7f), and 10.0 µL of the internal standard addition solution, ¶ c2), above through the valve bore, then close valve.

Attach syringe-syringe valve assembly to syringe valve on purging device. Open syringe valves and inject sample into purging chamber. Close both valves and purge sample for 11.0 ± 0.1 min at ambient temperature.

After purge time, attach trap to chromatograph, adjust purge and trap system to desorb mode (Figure 6210:4), and begin to temperature program the gas chromatograph. Introduce trapped materials to

TABLE 6210:IV. CALIBRATION AND QC ACCEPTANCE CRITERIA*

Compound	Range for Q $\mu g/L$	Limit for s $\mu g/L$	Range for $\overline{X}$ g/L	Range for P, P_s %
Benzene	12.8–27.2	6.9	15.2–26.0	37–151
Bromodichloromethane	13.1–26.9	6.4	10.1–28.0	35–155
Bromoform	14.2–25.8	5.4	11.4–31.1	45–169
Bromomethane	2.8–37.2	17.9	D–41.2	D–242
Carbon tetrachloride	14.6–25.4	5.2	17.2–23.5	70–140
Chlorobenzene	13.2–26.8	6.3	16.4–27.4	37–160
Chloroethane	7.6–32.4	11.4	8.4–40.4	14–230
2-Chloroethylvinyl ether	D–44.8	25.9	D–50.4	D–305
Chloroform	13.5–26.5	6.1	13.7–24.2	51–138
Chloromethane	D–40.8	19.8	D–45.9	D–273
Dibromochloromethane	13.5–26.5	6.1	13.8–26.6	53–149
1,2-Dichlorobenzene	12.6–27.4	7.1	11.8–34.7	18–190
1,3-Dichlorobenzene	14.6–25.4	5.5	17.0–28.8	59–156
1,4-Dichlorobenzene	12.6–27.4	7.1	11.8–34.7	18–190
1,1-Dichloroethane	14.5–25.5	5.1	14.2–28.5	59–155
1,2-Dichloroethane	13.6–26.4	6.0	14.3–27.4	49–155
1,1-Dichloroethene	10.1–29.9	9.1	3.7–42.3	D–234
trans-1,2-Dichloroethene	13.9–26.1	5.7	13.6–28.5	54–156
1,2-Dichloropropane	6.8–33.2	13.8	3.8–36.2	D–210
cis-1,3-Dichloropropene	4.8–35.2	15.8	1.0–39.0	D–227
trans-1,3-Dichloropropene	10.0–30.0	10.4	7.6–32.4	17–183
Ethyl benzene	11.8–28.2	7.5	17.4–26.7	37–162
Methylene chloride	12.1–27.9	7.4	D–41.0	D–221
1,1,2,2-Tetrachloroethane	12.1–27.9	7.4	13.5–27.2	46–157
Tetrachloroethene	14.7–25.3	5.0	17.0–26.6	64–148
Toluene	14.9–25.1	4.8	16.6–26.7	47–150
1,1,1-Trichloroethane	15.0–25.0	4.6	13.7–30.1	52–162
1,1,2-Trichloroethane	14.2–25.8	5.5	14.3–27.1	52–150
Trichloroethene	13.3–26.7	6.6	18.6–27.6	71–157
Trichlorofluoromethane	9.6–30.4	10.0	8.9–31.5	17–181
Vinyl chloride	0.8–39.2	20.0	D–43.5	D–251

* Q　= concentration measured in QC check sample,
s　= standard deviation of four recovery measurements,
$\overline{X}$　= average recovery of four recovery measurements,
P, P_s = percent recovery measured,
D　= detected; result must be greater than zero.
Criteria calculated assuming a QC check sample concentration of 20 $\mu g/L$.
NOTE—These criteria are based directly on the method performance data in Table 6210:VI. Where necessary, the limits for recovery were broadened to assure applicability of the limits to concentrations below those used to develop Table 6210:VI.

GC column by rapidly heating trap to 180°C while backflushing trap with an inert gas at 20 to 60 mL/min for 4 min. If rapid heating of the trap cannot be achieved, use GC column as a secondary trap by cooling it to 30°C (subambient temperature, if poor peak geometry and random retention time problems persist) instead of initial program temperature.

While trap is being desorbed into gas

chromatograph, empty purging chamber using sample introduction syringe. Wash chamber with two 5-mL flushes of reagent water. Be sure all areas wetted during purging are also wetted during the rinse to maximize flushing.

After desorbing sample for 4 min, recondition trap by returning purge and trap system to purge mode. Wait 15 s, then close syringe valve on purging device to begin gas flow through trap. Maintain trap temperature at 180°C. After approximately 7 min, turn off trap heater and open syringe valve to stop gas flow through trap. The cool trap is ready for the next sample.

If response for any m/z exceeds the working range of the system, dilute sample from the portion in the second syringe and reanalyze with reagent water.

Obtain EICPs for the primary m/z (Table 6210:V) and at least two secondary masses for each compound of interest. Meet the following criteria to make a qualitative identification:

• The characteristic masses of each compound are a maximum in the same or within one scan of each other.
• The retention time is within 30 s of the retention time of the authentic compound.
• The relative peak heights of the three characteristic masses in the EICPs fall within 20% of the relative intensities of these masses in a reference mass spectrum. The reference mass spectrum can be obtained from a standard analyzed in the GC/MS system or from a reference library.

Structural isomers that have very similar mass spectra and less than 30 s difference in retention time can be identified explicitly only if the resolution between authentic isomers in a standard mix is acceptable. Acceptable resolution is achieved if the baseline to valley height between isomers is less than 25% of the sum of the two peak heights. Otherwise, structural isomers are identified as isomeric pairs.

6. Calculation

When a compound has been identified, base quantitation on the integrated abundance from the EICP of the primary characteristic m/z given in Table 6210:V. If the sample produces an interference for the primary m/z, use a secondary characteristic m/z to quantitate.

Calculate concentration in sample using the response factor (RF) and the following equation:

$$\text{Concentration, } \mu g/L = \frac{(A_s)(C_{is})}{(A_{is})(RF)}$$

where:

A_s = response for compound to be measured,
A_{is} = response for internal standard, and
C_{is} = concentration of internal standards.

Report results in micrograms per liter without correction for recovery. Report all quality control data with sample results.

7. Quality Control

a. Quality-control program: The analyst should be experienced in operation of a purge and trap system and a gas chromatographic/mass spectrometer and in interpretation of mass spectra. A minimum quality-control program consists of an initial demonstration of laboratory capability and an ongoing analysis of samples with known additions. Maintain records to document data quality. Compare ongoing data quality checks with established performance criteria to determine if results are acceptable. When results of known additions indicate atypical performance, analyze a quality control check standard to confirm that the operation is in control.

Make an initial, one-time, demonstration of acceptable bias and precision as described in ¶ *b* below.

Because advances are occurring in chromatography, certain options (¶ 5*a*) are permitted to improve separation or reduce

TABLE 6210:V. CHARACTERISTIC MASSES FOR PURGEABLE ORGANICS

Compound	Primary	Secondary
Chloromethane	50	52
Bromomethane	94	96
Vinyl chloride	62	64
Chloroethane	64	66
Methylene chloride	84	49, 51, 86
Trichlorofluoromethane	101	103
1,1-Dichloroethene	96	61, 98
1,1-Dichloroethane	63	65, 83, 85, 98, 100
trans-1,2-Dichloroethene	96	61, 98
Chloroform	83	85
1,2-Dichloroethane	98	62, 64, 100
1,1,1-Trichloroethane	97	99, 117, 119
Carbon tetrachloride	117	119, 121
Bromodichloromethane	127	83, 85, 129
1,2-Dichloropropane	112	63, 65, 114
trans-1,3-Dichloropropene	75	77
Trichloroethene	130	95, 97, 132
Benzene	78	
Dibromochloromethane	127	129, 208, 206
1,1,2-Trichloroethane	97	83, 85, 99, 132, 134
cis-1,3-Dichloropropene	75	77
2-Chloroethylvinyl ether	106	63, 65
Bromoform	173	171, 175, 250, 252, 254, 256
1,1,2,2-Tetrachloroethane	168	83, 85, 131, 133, 166
Tetrachloroethene	164	129, 131, 166
Toluene	92	91
Chlorobenzene	112	114
Ethyl benzene	106	91
1,3-Dichlorobenzene	146	148, 113
1,2-Dichlorobenzene	146	148, 113
1,4-Dichlorobenzene	146	148, 113

cost. For each modification, repeat the procedure in ¶ b below.

Each day, analyze a reagent water blank to demonstrate that interferences from the analytical system are under control.

On an ongoing basis, make known additions to, and analyze, a minimum of 10% of all samples to monitor and evaluate laboratory data quality according to ¶ c below.

Also demonstrate through the analyses of quality control check standards (¶ d below) that the measurement system is in control. The frequency of the check standard analyses should be 10% of all samples

analyzed but it may be reduced if known-addition recoveries meet all specified quality-control criteria.

Maintain performance records (¶ e below).

Also make known additions of surrogate standards to all samples to monitor continuing laboratory performance as described in ¶ f below.

b. Analyst proficiency check: To establish the ability to generate data with acceptable precision and bias, perform the following operations:

Prepare a QC check sample containing

20 μg/L of each compound by adding 200 μL of QC check sample concentrate (¶4*i*) to 100 mL reagent water. Analyze four 5-mL portions according to ¶ 5. Calculate average recovery ($\overline{X}$) in μg/L, and the standard deviation of the recovery (s) in μg/L, for each compound. Compare s and $\overline{X}$ with the corresponding acceptance criteria for precision and bias, respectively. If s and $\overline{X}$ for all compounds of interest meet the acceptance criteria (Table 6210:IV), system performance is acceptable and sample analysis may begin. If any individual s exceeds the precision limit or any individual X falls outside the range for bias, the system performance is unacceptable.

When results for one or more of the compounds fail at least one of the acceptance criteria, proceed as follows: Either

1) Locate and correct the source of the problem and repeat test for all compounds of interest beginning with analyses of four 5-mL portions, as above, or

2) Repeat test only for those compounds that failed to meet the criteria. Repeated failure, however, will confirm a general problem with the measurement system. If this occurs, locate and correct the source of the problem and retest for all compounds of interest beginning with the analytical step indicated.

c. *Analyses of samples with known additions:* On an ongoing basis, make known additions to at least 5% of the samples from each sample site being monitored. For laboratories analyzing one to 20 samples per month, analyze at least one such sample with a known addition per month.

Determine the concentration of the known addition as follows: If the concentration of a specific compound in the sample is being checked against a regulatory concentration limit, make an addition at that limit or one to five times higher than the background concentration determined below, whichever is larger. If the concentration of a specific compound is not being checked against a regulatory limit, add 20 μg/L or one to five times higher than the background concentration as determined below, whichever concentration is larger.

To determine background concentration (B), analyze one 5-mL sample portion. If necessary, prepare a new QC check sample concentrate (¶ 4*i*) appropriate for the background concentration in the sample. Add to a second 5-mL sample portion 10 μL of the QC check sample concentrate and analyze it to determine the concentration after addition (A) of each compound. Calculate each percent recovery (P) as $100(A - B)\%/T$, where T is the true value of the addition.

Compare percent recovery (P) for each compound with the corresponding QC acceptance criteria (Table 6210:IV). These acceptance criteria include an allowance for error in measurement of both background and addition concentrations, assuming an addition:background ratio of 5:1. This error will be accounted for to the extent that the analyst's addition:background ratio approaches 5:1.[9] If the known addition was at a concentration lower than 20 μg/L, use either the tabulated QC acceptance criteria or optional QC acceptance criteria calculated for the specific addition concentration.

To calculate optional acceptance criteria use the equation in Table 6210:VI to calculate accuracy (X') and substitute the addition concentration (T) for C. Calculate overall precision (S') using the equation in Table 6210:VI, substituting X' for $\overline{X}$. Calculate the range for recovery at the addition concentration as $(100\ X'/T) \pm 2.44\ (100\ S'/T)\%$.[9]§§

If any individual P falls outside the designated range for recovery, analysis for that constituent has failed the acceptance criteria. Analyze a check standard containing each compound that failed the criteria as described in ¶ *d* below.

d. *Quality-control check standard anal-*

---

§§ The value 2.44 is two times the value 1.22 derived in the reference.[9]

TABLE 6210:VI. METHOD BIAS AND PRECISION AS FUNCTIONS OF CONCENTRATION*

Compound	Bias, as Recovery, X' $\mu g/L$	Single-Analyst Precision, S_r $\mu g/L$	Overall Precision, S' $\mu g/L$
Benzene	$0.93C + 2.00$	$0.26\overline{X} - 1.74$	$0.25\overline{X} - 1.33$
Bromodichloromethane	$1.03C - 1.58$	$0.15\overline{X} + 0.59$	$0.20\overline{X} + 1.13$
Bromoform	$1.18C - 2.35$	$0.12\overline{X} + 0.34$	$0.17\overline{X} + 1.38$
Bromomethane†	$1.00C$	$0.43\overline{X}$	$0.58\overline{X}$
Carbon tetrachloride	$1.10C - 1.68$	$0.12\overline{X} + 0.25$	$0.11\overline{X} + 0.37$
Chlorobenzene	$0.98C + 2.28$	$0.16\overline{X} - 0.09$	$0.26\overline{X} - 1.92$
Chloroethane	$1.18C + 0.81$	$0.14\overline{X} + 2.78$	$0.29\overline{X} + 1.75$
2-Chloroethylvinyl ether†	$1.00C$	$0.62\overline{X}$	$0.84\overline{X}$
Chloroform	$0.93C + 0.33$	$0.16\overline{X} + 0.22$	$0.18\overline{X} + 0.16$
Chloromethane	$1.03C - 1.81$	$0.37\overline{X} + 2.14$	$0.58\overline{X} + 0.43$
Dibromochloromethane	$1.01C - 0.03$	$0.17\overline{X} - 0.18$	$0.17\overline{X} + 0.49$
1,2-Dichlorobenzene‡	$0.94C + 4.47$	$0.22\overline{X} - 1.45$	$0.30\overline{X} - 1.20$
1,3-Dichlorobenzene	$1.06C + 1.68$	$0.14\overline{X} - 0.48$	$0.18\overline{X} - 0.82$
1,4-Dichlorobenzene‡	$0.94C + 4.47$	$0.22\overline{X} - 1.45$	$0.30\overline{X} - 1.20$
1,1-Dichloroethane	$1.05C + 0.36$	$0.13\overline{X} - 0.05$	$0.16\overline{X} + 0.47$
1,2-Dichloroethane	$1.02C + 0.45$	$0.17\overline{X} - 0.32$	$0.21\overline{X} - 0.38$
1,1-Dichloroethene	$1.12C + 0.61$	$0.17\overline{X} + 1.06$	$0.43\overline{X} - 0.22$
trans-1,2-Dichloroethene	$1.05C + 0.03$	$0.14\overline{X} + 0.09$	$0.19\overline{X} + 0.17$
1,2-Dichloropropane†	$1.00C$	$0.33\overline{X}$	$0.45\overline{X}$
cis-1,3-Dichloropropene†	$1.00C$	$0.38\overline{X}$	$0.52\overline{X}$
trans-1,3-Dichloropropene†	$1.00C$	$0.25\overline{X}$	$0.34\overline{X}$
Ethyl benzene	$0.98C + 2.48$	$0.14\overline{X} + 1.00$	$0.26\overline{X} - 1.72$
Methylene chloride	$0.87C + 1.88$	$0.15\overline{X} + 1.07$	$0.32\overline{X} + 4.00$
1,1,2,2-Tetrachloroethane	$0.93C + 1.76$	$0.16\overline{X} + 0.69$	$0.20\overline{X} + 0.41$
Tetrachloroethene	$1.06C + 0.60$	$0.13\overline{X} - 0.18$	$0.16\overline{X} - 0.45$
Toluene	$0.98C + 2.03$	$0.15\overline{X} - 0.71$	$0.22\overline{X} - 1.71$
1,1,1-Trichloroethane	$1.06C + 0.73$	$0.12\overline{X} - 0.15$	$0.21\overline{X} - 0.39$
1,1,2-Trichloroethane	$0.95C + 1.71$	$0.14\overline{X} + 0.02$	$0.18\overline{X} + 0.00$
Trichloroethene	$1.04C + 2.27$	$0.13\overline{X} + 0.36$	$0.12\overline{X} + 0.59$
Trichlorofluoromethane	$0.99C + 0.39$	$0.33\overline{X} - 1.48$	$0.34\overline{X} - 0.39$
Vinyl chloride	$1.00C$	$0.48\overline{X}$	$0.65\overline{X}$

* X' = expected recovery for one or more measurements of a sample containing a concentration of C,
  S_r = expected single analyst standard deviation of measurements at an average concentration found of $\overline{X}$,
  S' = expected interlaboratory standard deviation of measurements at an average concentration found of $\overline{X}$,
  C = true value for the concentration,
  $\overline{X}$ = average recovery found for measurements of samples containing a concentration of C.
† Estimates based on the performance in a single laboratory.[11]
‡ Due to chromatographic resolution problems, performance statements for these isomers are based on the sums of their concentrations.

ysis: If analysis of any compound fails the acceptance criteria for recovery, prepare and analyze a QC check standard containing each compound that failed as follows:

Prepare the QC check standard by adding 10 µL of QC check sample concentrate (¶ 4i) to 5 mL reagent water. The QC check standard needs only to contain the compounds that failed criteria in the test in ¶ c above. Analyze the QC check standard to determine the concentration, A, of each compound. Calculate each percent recovery (P_s) as 100 $(A/T)\%$, where T is the true value of the standard concentration. Compare percent recovery (P_s) for each compound that failed the test in ¶ c with the corresponding tabulated QC acceptance criteria. If the recovery of any compound falls outside the designated range, laboratory performance is out of control: identify and correct the problem immediately. The analytical result for that compound in the sample without a known addition is suspect.

e. Bias assessment and records: Assess method bias and maintain records. For example, after the analysis of five wastewater samples as in ¶ c above, calculate the average percent recovery $(\overline{P})$ and the standard deviation of the percent recovery (s_p). Express bias assessment as a percent recovery interval from $\overline{P} - 2s_p$ to $\overline{P} + 2s_p$. If $\overline{P} = 90\%$ and $s_p = 10\%$, the recovery interval is expressed as 70–110%. Update bias assessment for each compound regularly, (e.g., after each five to ten new accuracy measurements).

f. Use of surrogate compounds: As a quality control check, make known additions to all samples of surrogate standard addition solutions as described in ¶ *5d*, and calculate the percent recovery of each surrogate compound.

g. Additional quality-assurance practices: Other desirable practices depend on the needs of the laboratory and the nature of the samples. Analyze field duplicates to assess the precision of the environmental measurements. Whenever possible, analyze standard reference materials and participate in relevant performance evaluation studies.

8. Precision and Bias

This method was tested by 15 laboratories using reagent water, drinking water, surface water, and industrial wastewaters with additions at six concentrations over the range 5 to 600 μg/L.[10] Single-operator precision, overall precision, and method bias were found to be related directly to

the compound concentration and essentially independent of the sample matrix. Linear equations describing these relationships are presented in Table 6210:VI.

9. References

1. U.S. ENVIRONMENTAL PROTECTION AGENCY. 1984. Method 624—Purgeables. 40 CFR Part 136, 43373; *Federal Register* 49, No. 209.
2. U.S. ENVIRONMENTAL PROTECTION AGENCY. 1984. Method 603—Acrolein and acrylonitrile. 40 CFR Part 136, 43281; *Federal Register* 49, No. 209.
3. BELLAR, T.A. & J.J. LICHTENBERG. 1974. Determining volatile organics at microgram-per-litre levels by gas chromatography. *J. Amer. Water Works Assoc.* 66:739.
4. BELLAR, T.A. & J.J. LICHTENBERG. 1978. Semi-automated headspace analysis of drinking waters and industrial waters for purgeable volatile organic compounds. *In* C.E. Van Hall, ed. Measurement of Organic Pollutants in Water and Wastewater. STP 686, American Soc. Testing & Materials, Philadelphia, Pa.
5. U.S. ENVIRONMENTAL PROTECTION AGENCY. 1984. Definition and procedure for the determination of the method detection limit. 40 CFR Part 136, Appendix B. *Federal Register* 49, No. 209.
6. OLYNYK, P., W.L. BUDDE & J.W. EICHELBERGER. 1980. Method Detection Limit for Methods 624 and 625. Unpublished report.
7. BUDDE, W.L. & J.W. EICHELBERGER. 1980. Performance Tests for the Evaluation of Computerized Gas Chromatography/Mass Spectrometry Equipment and Laboratories. EPA-600/4-80-025, U.S. Environmental Protection Agency, Environmental Monitoring and Support Lab., Cincinnati, Ohio.
8. EICHELBERGER, J.W., L.E. HARRIS & W.L. BUDDE. 1975. Reference compound to calibrate ion abundance measurement in gas chromatography — mass spectrometry systems. *Anal. Chem.* 47:995.
9. PROVOST, L.P. & R.S. ELDER. 1983. Interpretation of percent recovery data. *Amer. Lab.* 15:58.
10. U.S. ENVIRONMENTAL PROTECTION AGENCY. 1984. EPA Method Study 29,

Method 624—Purgeables. EPA-600/2-84-054, National Technical Information Serv., PB84-209915, Springfield, Va.

11. SLATER, R. & T. PRESSLEY. 1984. Method

Performance Data for Method 624 (memorandum). U.S. Environmental Protection Agency, Environmental Monitoring and Support Lab., Cincinnati, Ohio.

6210 C. Purge and Trap Packed-Column Gas Chromatographic/Mass Spectrometric Method II*

This method[1] is applicable to the determination of purgeable organic compounds† in finished drinking water, raw source water, or drinking water in any treatment stage. The method varies only slightly from Section 6210B, which is intended for use with wastewater samples.

1. General Discussion

a. *Principle:* See Section 6210B.1a.

b. *Interferences:* See Section 6210B.1b. If possible, use this method with a purge and trap system devoted to analysis of low-level samples.

Take special precautions to analyze for methylene chloride. Isolate the analytical and sample storage area from all atmospheric sources of methylene chloride: otherwise random background levels will

result. Construct all gas chromatography carrier gas lines and purge-gas plumbing from stainless steel or copper tubing because methylene chloride will permeate through TFE tubing. Wear clean laboratory clothing not previously exposed to methylene chloride.

c. *Detection limits:* Method detection limits (MDLs)[2] are compound-dependent and vary with purging efficiency and concentration. In a single laboratory using reagent water and known additions of 1.5 μg/L, observed MDLs were in the range of 0.1 to 0.5 μg/L. The applicable concentration range of this method is compound- and instrument-dependent, but is approximately 0.2 to 200 μg/L. Compounds that are inefficiently purged from water will not be detected when present at low concentrations, but they can be measured with acceptable bias and precision when present in sufficient amounts. Determination of some geometrical isomers (i.e., xylenes) may be hampered by coelution.

d. *Safety:* The toxicity or carcinogenicity of each reagent has not been defined precisely. Benzene, carbon tetrachloride, bis-1-chloroisopropyl ether, 1,4-dichlorobenzene, 1,2-dichloroethane, hexachlorobutadiene, 1,1,2,2-tetrachloroethane, 1,1,2-trichloroethane, chloroform, 1,2-dibromoethane, tetrachloroethene, trichloroethene, and vinyl chloride have been classified tentatively as known or suspected human or mammalian carcinogens. Handle pure standard materials and stock standard so-

* Approved by Standard Methods Committee 1988. Accepted by U.S. Environmental Protection Agency as equivalent to EPA Method 524.
† Benzene, bromobenzene, bromochloromethane, bromodichloromethane, bromoform, bromomethane, *sec*-butylbenzene, *tert*-butylbenzene, carbon tetrachloride, chlorobenzene, chloroethane, chloroform, chloromethane, 2-chlorotoluene, 4-chlorotoluene, dibromochloromethane, 1,2-dibromo-3-chloropropane, 1,2-dibromoethane, dibromomethane, 1,2-dichlorobenzene, 1,3-dichlorobenzene, 1,4-dichlorobenzene, dichlorodifluoromethane, 1,1-dichloroethane, 1,2-dichloroethane, 1,1-dichloroethene, *cis*-1,2-dichloroethene, *trans*-1,2-dichloroethene, 1,2-dichloropropane, 1,3-dichloropropane, 2,2-dichloropropane, 1,1-dichloropropane, ethylbenzene, hexachlorobutadiene, isopropylbenzene, methylene chloride, *n*-propylbenzene, styrene, 1,1,1,2-tetrachloroethane, 1,1,2,2-tetrachloroethane, tetrachloroethene, toluene, 1,1,1-trichloroethane, 1,1,2-trichloroethane, trichloroethene, trichlorofluoromethane, 1,2,3-trichloropropane, vinyl chloride, *o*-xylene, *m*-xylene, and *p*-xylene.

lutions of these compounds in a hood and wear a NIOSH/MESA- approved toxic gas respirator when handling high concentrations.

2. Sampling and Storage

See Section 6010B.1.

3. Apparatus

a. Purge and trap system: The purge and trap system consists of three separate pieces of equipment: purging device, trap, and desorber. Several complete systems now are available commercially.

1) *Purging device,* designed to accept 25-mL samples with a water column at least 5 cm deep. For other requirements, see Section 6210B.3a1).

Needle spargers may be used instead of the glass frit shown in Figure 6210:1; however, in either case, introduce purge gas at a point ≤ 5 mm from base of water column.

2) *Trap,* at least 25 cm long and with an inside diameter of at least 3 mm, packed with the following minimum lengths of adsorbents: 1.0 cm of methyl silicone-coated packing, Section 6210B.4b2), 7.7 cm of 2,6-diphenylene oxide polymer, Section 6210B.4b1), 7.7 cm of silica gel, Section 6210B.4b3), and 7.7 cm of coconut charcoal, ¶ 4b below). If analysis is not to be made for dichlorodifluoromethane, the charcoal can be eliminated and the polymer section lengthened to 15 cm. The minimum specifications for the trap are illustrated in Figure 6210:6.

The use of the methyl-silicone-coated packing is recommended, but not mandatory. The packing serves a dual purpose of protecting the diphenylene oxide polymer adsorbant from aerosols and of insuring that it is fully enclosed within the heated

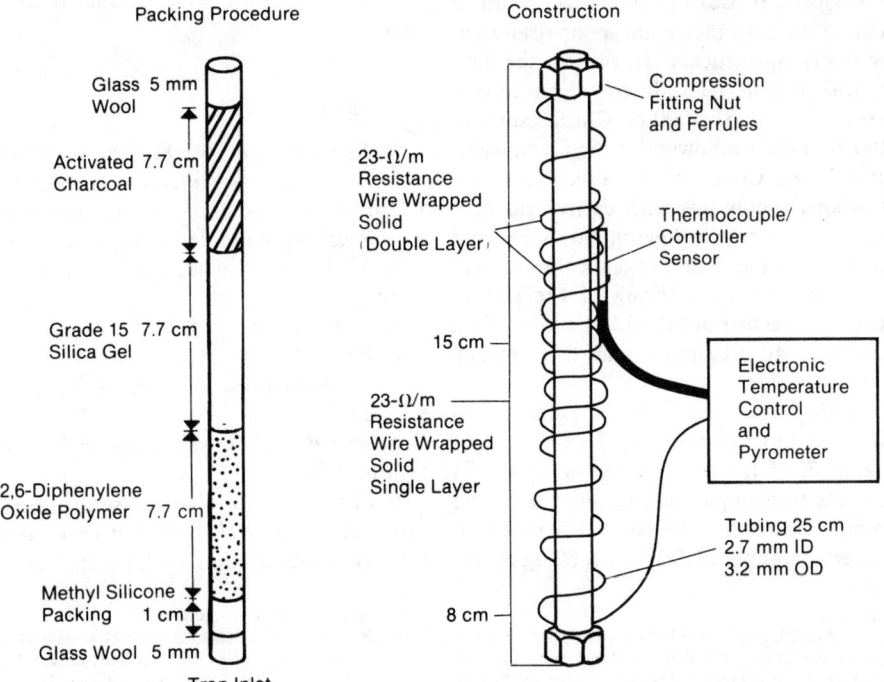

Figure 6210:6. Trap packings and construction to include desorb capability.

zone of the trap, thus eliminating potential cold spots. Alternatively, silanized glass wool may be used as a spacer at the trap inlet.

Before initial use, condition trap overnight at 180°C by backflushing with an inert gas flow of at least 20 mL/min. Vent trap effluent to the room, not to the analytical column. Before daily use, condition trap for 10 min at 180°C with backflushing. The trap may be vented to the analytical column during daily conditioning; however, run the column through the temperature program before sample analysis.

3) *Desorber:* See Section 6210B.3*a*3). Trap failure is characterized by a pressure drop above 21 kPa across the trap during purging or by poor bromoform sensitivities.

b. Gas chromatograph:‡ See Section 6210B.3*b*.

During handling, packing, and programming, active sites can be exposed on the Carbopack-B packing, which can result in tailing peak geometry and poor resolution of many constituents. To protect the analytical column, pack the first 5 cm of column with 3% SP-1000 on Chromosorb-W (60/80 mesh) followed by the Carbopack-B packing. Condition the precolumn and Carbopack columns with carrier gas flow at 220°C overnight. Pneumatic shocks and rough treatment of packed columns will cause excessive fracturing of the Carbopack. If pressure above 415 kPa is required to obtain 40 mL/min carrier flow, repack column.

c. Mass spectrometer, capable of scanning from 35 to 450 amu every 7 s or less, utilizing 70-V (nominal) electron energy in the electron impact ionization mode, and producing a mass spectrum that meets all criteria in Table 6210:II when 50 ng or less

of 4-bromofluorobenzene is injected through the GC inlet. To ensure sufficient precision, the desired scan rate permits acquisition of at least five spectra while a sample component elutes from the gas chromatograph.

d. GC/MS interface: Use an all-glass enrichment device and an all-glass transfer line. Another enrichment device or transfer line can be substituted if the performance specifications described in ¶ 7 below can be achieved.

e. Data system: See Section 6210B.3*e*.

f. Syringes, 25-mL glass hypodermic with luerlok tip.

g. Syringe valves, two-way, with luer ends.

h. Microsyringes, 25-µL with a 5 cm × 0.15 mm ID, 22° bevel needle;§ also with 10- and 100-L capacity.

i. Syringes, 0.5-, 1.0-, and 5-mL, gastight with shutoff valve.

j. Bottles, 15-mL, with TFE-lined screw caps.

4. Reagents

a. Reagent water: See Section 6210B.4*a*.

b. Trap packing materials: See Section 6210B.4*b*. In addition, coconut charcoal, 6/20 mesh sieved to 26 mesh‖, is required.

c. Methanol, pesticide quality or equivalent.

d. Hydrochloric acid: HCl, 1 + 1.

e. Vinyl chloride, 99.9% pure.#

f. Standard stock solutions: See Section 6210B.4*d*.

g. Secondary dilution standards: See Section 6210B.4*e*.

h. Calibration standards: Use directions given in Section 6210B.4*g*, but prepare at least five concentration levels for each com-

---

‡ Gas chromatographic methods are extremely sensitive to the materials used. Mention of trade names by *Standard Methods* does not preclude the use of other existing or as-yet-undeveloped products that give *demonstrably* equivalent results.

§ Hamilton # 702N or equivalent.
‖ Barnebey Cheney, CA-580-26 lot M-2649, or equivalent.
Available from Ideal Gas Products, Inc., Edison, N.J. and from Matheson, East Rutherford, N.J. Certified mixtures of vinyl chloride in nitrogen at 1.0 and 10.0 ppm are available from several sources.

pound. More than one set of standards may be required.

i. *Internal standard known-addition solution:* Prepare a solution containing fluorobenzene and 1,2-dichlorobenzene-d$_4$ in methanol using the procedures described in Section 6210B.4d and e. Prepare the secondary dilution standard at a concentration of 25 μg/mL of each internal standard compound. Adding 10 μL of such a standard to 25.0 mL sample or calibration standard would yield a concentration equivalent to 10 μg/L.

j. *BFB standard:* Prepare a 25 μg/mL solution of bromofluorobenzene in methanol.

k. *Laboratory control standard concentrate:* Using standard stock solutions, prepare a solution containing each compound of interest at a concentration 500 times the MCL in methanol.

5. Procedure

a. *Operating conditions:* Table 6210:VII summarizes recommended operating conditions for the gas chromatograph. An example of the separation achieved is shown in Figure 6210:5.

b. *GC/MS performance tests:* Follow the procedure in Section 6210B.5b using 25 mL reagent water. Use a mass range of 35 to 300 amu, not 20 to 260 amu.

c. *Calibration:* Calibrate system daily after performance tests using the procedure of Section 6210B.5c.

The internal standard to be used is optional but aromatics usually are compared to 1,2-dichlorobenzene-d$_4$ and all other compounds are compared to the internal standard having the closest relative retention time.

To calibrate for vinyl chloride, use a certified gaseous mixture of vinyl chloride in nitrogen. Fill purging device with 25.0 mL reagent water or aqueous calibration standard. Start purging aqueous mixture. Inject a known volume (between 100 and 2000

TABLE 6210:VII. CHROMATOGRAPHIC CONDITIONS FOR VOLATILE ORGANIC COMPOUNDS ON PACKED COLUMNS

Compound	Retention Time min
Vinyl chloride	3.8
Dichlorodifluoromethane	3.8
Methylene chloride	6.4
Trichlorofluoromethane	8.3
1,1-Dichloroethene	9.0
Bromochloromethane	9.3
1,1-Dichloroethane	10.1
trans-1,2-Dichloroethene	10.8
Chloroform	11.4
Dibromomethane	12.1
1,2-Dichloroethane	12.1
2,2-Dichloropropane	12.7
1,1,1-Trichloroethane	13.4
Carbon tetrachloride	13.7
Bromodichloromethane	14.3
1,2-Dichloropropane	15.7
1,1-Dichloropropene	16.0
Trichloroethene	16.5
Benzene	17.0
Dibromochloromethane	17.1
1,2-Dibromoethane	17.9
1,3-Dichloropropane	18.4
Bromoform	19.8
1,1,2,2-Tetrachloroethane	22.1
Tetrachloroethene	22.2
Toluene	23.5
Chlorobenzene	24.6
1,2-Dibromo-3-chloropropane	25.8
Bromobenzene	26.7
Isopropylbenzene	28.5
m-Xylene	29.5
Styrene	29.7
n-Propylbenzene	30.7
o-Xylene	30.9
p-Xylene	30.9
t-Butylbenzene	31.5
2-Chlorotoluene	31.5
Hexachlorobutadiene	32.0
4-Chlorotoluene	32.5
sec-Butylbenzene	32.5
1,2-Dichlorobenzene	35.0
1,4-Dichlorobenzene	35.3

Column conditions: 2 m × 2 mm ID glass column packed with Carbopack B (60–80 mesh) coated with 1% SP-1000. Carrier gas: helium at flow of 30 mL/min. Column temperature held at 45°C for 3 min, then programmed at 8°C/min to 220°C and held until all compounds are eluted.

μL) of calibration gas (at room temperature) directly into purging device with a gastight syringe. Slowly inject gaseous sample through a septum seal at the top of purging device at 2000 μL/min. Do not inject standard through aqueous sample inlet needle. Inject gaseous standard before 5 min of the 11-min purge time have elapsed. Determine the aqueous equivalent concentration of vinyl chloride standard injected from the equation:

$$S = 0.51 \ (C)(V)$$

where:

S = aqueous equivalent concentration of vinyl chloride standard, μg/L,
C = concentration of gaseous standard, ppm (v/v), and
V = volume of standard injected, mL.

d. Sample analysis: Using two 25-mL syringes follow the procedure of Section 6210B.5*d.* Use purge chamber washes of 25 mL. Desorb compounds by backflushing the trap with an inert gas flow rate of 15 mL/min, not 20 to 60 mL/min.

Hold column temperature at 45°C for 3 min, then program at 8°C/min to 220°C and hold until all compounds elute.

When all sample components have been eluted, terminate mass spectrometer data acquisition and store data files. Use appropriate data output software to display full range mass spectra and appropriate extracted ion current profiles (EICPs). See Section 6210B.5*d.* If any ion abundance exceeds the system working range, dilute sample in second syringe with reagent water and analyze.

6. Calculation

When compound has been identified, base quantitation on integrated abundance from the EICP of the primary characteristic m/z given in Table 6210:VIII. If the sample produces an interference for the primary m/z, use a secondary characteristic m/z to quantitate.

Report results in micrograms per liter. Report all quality control data with sample results.

7. Quality Control

a. Quality control program: The analyst should be experienced in measuring purgeable organics at the low microgram-per-liter level. A minimum quality-control program consists of an initial demonstration of laboratory capability and an ongoing analysis of samples with known additions. Maintain records to document data quality. Compare ongoing data quality checks with established performance criteria to determine if results are acceptable. Analyze a quality control check standard to confirm that the operation is in control.

Make an initial demonstration of acceptable accuracy and precision as described in ¶ *b* below. If procedural modifications are made, repeat this initial demonstration.

Each day, analyze a reagent water blank to demonstrate that interferences from the analytical system are under control.

Demonstrate through the analyses of quality control check standards (¶ *c* below) that the measurement system is in control. Analyze check standards equivalent to 10% of all samples analyzed, with a minimum of two determinations per month.

Weekly, demonstrate the ability to analyze low-level samples as described in ¶ *d* below.

The laboratory reagent blank should represent a response of less than 0.1 μg/L.

b. Analyst proficiency check: Prepare a QC check sample concentrate containing each compound to be reported, at a concentration of 500 times the maximum contaminant level (MCL) or 5 μg/mL, whichever is smaller, in methanol. Prepare this check sample using stock standards prepared independently from those used for calibration. Analyze seven 25-mL check samples at 0.2 MCL or 2 μg/L. Produce

TABLE 6210:VIII. CHARACTERISTIC MASSES (*m/z*) FOR PURGEABLE ORGANICS

Compound	Primary Ion	Secondary Ions	Compound	Primary Ion	Secondary Ions
Benzene	78	—	1,2-Dichloropropane	63	112
Bromobenzene	156	77, 158	1,3-Dichloropropane	76	78
Bromochloromethane	128	49, 130	2,2-Dichloropropane	77	97
Bromodichloro-methane	83	85, 127	1,1-Dichloropropene	75	110, 77
			Ethylbenzene	91	106
Bromoform	173	175, 254	Hexachlorobutadiene	225	223, 227
Bromomethane	94	96	Isopropylbenzene	105	120
n-Butylbenzene	91	92, 134	*p*-Isopropyltoluene	119	134, 91
sec-Butylbenzene	105	134	Methylene chloride	84	86, 49
tert-Butylbenzene	119	91, 134	Naphthalene	128	—
Carbon tetrachloride	117	119	*n*-Propylbenzene	91	120
Chlorobenzene	112	77, 114	Styrene	104	78
Chloroethane	64	66	1,1,1,2-Tetrachloro-ethane	131	133, 119
Chloroform	83	85			
Chloromethane	50	52	1,1,2,2-Tetrachloro-ethane	83	131, 85
2-Chlorotoluene	91	126	Tetrachloroethene	166	168, 129
4-Chlorotoluene	91	126	Toluene	92	91
Dibromochloro-methane	129	127	1,1,1-Trichloroethane	97	99, 61
			1,1,2-Trichloroethane	83	97, 85
1,2-Dibromo-3-chloro-propane	75	155, 157	Trichloroethene	95	130, 132
1,2-Dibromoethane	107	109, 188	Trichlorofluoro-methane	101	103
Dibromomethane	93	95, 174	1,2,3-Trichloropropane	75	77
1,2-Dichlorobenzene	146	111, 148	Vinyl chloride	62	64
1,3-Dichlorobenzene	146	111, 148	*m*-Xylene	106	91
1,4-Dichlorobenzene	146	111, 148	*o*-Xylene	106	91
Dichlorodifluoro-methane	85	87	*p*-Xylene	106	91
1,1-Dichloroethane	63	65, 83			
1,2-Dichloroethane	62	98	Internal standards/surrogates:		
1,1-Dichloroethene	96	61, 63			
cis-1,2-Dichloroethene	96	61, 98	Fluorobenzene	96	70
trans-1,2-Dichloro-ethene	96	61, 98	1,2-Dichlorobenzene-d₄	150	115, 152
			p-Bromofluorobenzene	95	174, 176

each sample by injecting 10 μL check sample concentrate into 25 mL reagent water in a glass syringe through the syringe valve.

Calculate average recovery ($\overline{X}$), μg/L, and standard deviation of recovery (s), μg/L, for each compound. Calculate the MDL for each compound.[2] The calculated MDL should be less than the addition level. For each compound, ($\overline{X}$) should be between 90% and 110% of the true value. Additionally, s should be $\leq 35\%$ of $\overline{X}$. If s and $\overline{X}$ for all compounds meet the criteria, system performance is acceptable and sample analysis may begin. If any s exceeds the precision limit or if any $\overline{X}$ falls outside the range for recovery, system performance is unacceptable for that compound.

Because of the large number of compounds, a substantial probability exists that one or more will fail at least one of the acceptance criteria when all compounds are analyzed. When results for one or more of the compounds fail at least one of the acceptance criteria, repeat analysis of seven 25-mL check samples for failed compounds.

c. *Quality-control check standard analysis:* Regularly demonstrate that the measurement system is in control by analyzing a quality control sample for all compounds of interest at the MCL or 10 μg/L, whichever is smaller. Prepare a check standard by adding 50 μL of QC check sample concentrate to 25 mL reagent water in a glass syringe. Analyze and calculate recovery for each compound. Acceptable recovery is between 60% and 140% of the expected value. If recovery for any compound falls outside the designated range, the compound has failed the acceptance criteria and reanalysis is required.

d. *Low-level analysis:* Weekly demonstrate the ability to analyze low-level samples. Prepare a low-level check sample by adding 10 μL QC check sample concentrate to 25 mL reagent water. Analyze and calculate recovery, which should be between 60% and 140% of the expected

value. If a compound fails the test, reanalyze for that compound. Repeated failure confirms a general problem with the measurement system. If this occurs, locate and correct the problem and repeat the test for all compounds of interest.

e. *Additional quality-assurance practices:* Other desirable practices depend on the needs of the laboratory and the nature of the samples. Field duplicates may be analyzed to assess precision of environmental measurements. Whenever possible, analyze standard reference materials and participate in relevant performance evaluation studies.

8. Precision and Bias

This method was tested in a single laboratory using reagent water with additions at concentrations between 1 and 5 μg/L.[1] Single-operator precision and bias data are presented for some selected compounds in Table 6210:IX. Some method detection limits have been calculated from data collected in three laboratories.[1,3]

9. References

1. U.S. ENVIRONMENTAL PROTECTION AGENCY. 1985. Methods for the Determination of Organic Compounds in Finished Drinking Water and Raw Source Water (rev. Sept. 1986). Modified from A. Alford-Stevens, J.W. Eichelberger & W.L. Budde. 1983. Purgeable organic compounds in water by gas chromatography/mass spectrometry, Method 524. Environmental Monitoring and Support Lab., Cincinnati, Ohio.

2. GLASER, J.A., D.L. FOERST, G.D. MCKEE, S.A. QUAVE & W.L. BUDDE. 1981. Trace analyses for wastewaters. *Environ. Sci. Technol.* 15:1426.

3. SLATER, R.W. 1985. Method Detection Limits for Drinking Water Volatiles. Unpublished report, U.S. Environmental Protection Agency, Environmental Monitoring and Support Lab., Cincinnati, Ohio.

TABLE 6210:IX. SINGLE-LABORATORY BIAS AND PRECISION DATA FOR VOLATILE ORGANIC
COMPOUNDS IN REAGENT WATER (PACKED COLUMN)

Compound	Conc. Tested μg/L	Number of Samples	Average Conc. Measured μg/L	Standard Deviation μg/L	Rel. Std. Dev. %
Benzene	1.0	8	0.97	0.036	3.6
Bromobenzene	1.0	8	0.92	0.042	4.6
Bromodichloromethane	1.5	8	1.43	0.096	6.7
Bromoform	2.5	8	2.36	0.23	9.7
Carbon tetrachloride	1.0	8	0.88	0.098	11.1
Chlorobenzene	1.0	8	1.02	0.047	4.6
Chloroform	1.0	8	1.03	0.086	8.4
Dibromochloromethane	1.5	8	1.49	0.10	7.0
1,2-Dibromo-3-chloropropane	3.0	8	3.4	0.63	18.2
1,2-Dibromoethane	1.0	8	0.93	0.13	13.6
Dibromomethane	1.0	8	0.94	0.11	11.4
1,2-Dichlorobenzene	5.0	8	4.95	0.35	7.1
1,4-Dichlorobenzene	5.0	8	5.27	0.72	13.6
Dichlorodifluoromethane	1.0	8	0.96	0.11	11.9
1,1-Dichloroethane	1.0	8	1.05	0.060	5.9
1,2-Dichloroethane	1.0	8	0.97	0.077	7.9
1,1-Dichloroethene	1.0	8	1.09	0.066	6.1
trans-1,2-Dichloroethene	1.0	8	0.98	0.066	6.8
1,2-Dichloropropane	1.0	8	1.01	0.060	5.9
1,3-Dichloropropane	1.0	8	1.00	0.033	3.4
Methylene chloride	1.0	7	0.99	0.045	4.5
Styrene	1.0	8	1.06	0.066	6.2
1,1,2,2-Tetrachloroethane	1.0	8	1.11	0.14	12.8
Tetrachloroethene	1.0	8	0.93	0.10	10.9
Toluene	1.0	8	1.05	0.043	4.1
1,1,1-Trichloroethane	1.0	8	1.05	0.093	8.8
Trichloroethene	1.0	8	0.90	0.12	13.6
Trichlorofluoromethane	1.0	7	1.09	0.072	6.6
Vinyl chloride	1.0	8	0.98	0.11	10.8
o-Xylene	1.0	8	1.02	0.068	6.7
p-Xylene	1.0	8	1.11	0.047	4.2

6210 D. Purge and Trap Capillary-Column Gas Chromatographic/Mass Spectrometric Method*

This procedure[1] is applicable to determining those compounds listed for Method 6210C and the following: n-butylbenzene, naphthalene, p-isopropyltoluene, 1,2,3-trichlorobenzene, 1,2,4-trichlorobenzene, 1,2,4-trimethylbenzene, and 1,3,5-trimethylbenzene. It differs primarily in the type of gas chromatographic column used, requiring either a wide-bore or a narrow-bore capillary column for analysis.

1. General Discussion

See Section 6210C.1 and Tables 6210:X and XI. Detection limits in this method were observed to be in the range of 0.02 to 0.2 μg/L for known additions of 0.1 to 0.5 μg/L.

2. Sampling and Storage

See Section 6010B.1.

3. Apparatus

Use the equipment specified in Section 6210C.3 with one of the columns described below. The capillary interface is required only with a narrow-bore column (Column 3).

a. Gas chromatographic column:

1) Column 1: 60-m-long × 0.75-mm-ID VOCOL† wide-bore capillary column with 1.5-μm film thickness.

2) Column 2: 30-m-long × 0.53-mm-ID DB-624‡ mega-bore capillary column with 3-μm film thickness.

3) Column 3: 30-m-long × 0.32-mm-ID

fused silica capillary column coated with Durabond DB-5‡ with 1-μm film thickness.

b. Mass spectrometer, capable of scanning from 35 to 300 amu every 2 s or less.

c. Capillary interface, to connect the purge and trap concentrator to the capillary column of the gas chromatograph. This interface condenses desorbed materials on an uncoated fused silica capillary pre-column and when flash heated, transfers compounds to the analytical capillary column. In a stream of liquid nitrogen the fused silica is held at −150°C for cryofocusing. After desorption, the interface is heated to 250°C in 15 s or less.

4. Reagents

See Section 6210C.4.

5. Procedure

Retention times for wide-bore and narrow-bore columns are given in Tables 6210:X and XI, respectively. Follow the procedure specified in Section 6210C.5 with the following exceptions:

a. Desorption for wide-bore capillary column: Interface column through all-glass jet separator and desorb as specified in Section 6210C.5d. Hold column temperature at 10°C for 5 min, then program at 6°C/min to 160°C and hold until all compounds elute. See Figures 6210:7 and 8.

b. Desorption for narrow-bore capillary column: A jet separator usually is not required to interface column. Purge as specified in Section 6210B.5d for 11 min. Attach trap to cryogenically cooled interface at −150°C and adjust purge and trap system to desorb mode. Introduce trapped

---

* Approved by Standard Methods Committee 1988. Accepted by U.S. Environmental Protection Agency as equivalent to EPA Method 524.

† Supelco, Inc. or equivalent.

‡ J&W Scientific, Inc. or equivalent.

TABLE 6210:X. CHROMATOGRAPHIC CONDITIONS FOR VOLATILE ORGANIC COMPOUNDS ON WIDE-BORE CAPILLARY COLUMNS

	Retention Time min	
Compound	Column 1*	Column 2†
Dichlorodifluoromethane	1.55	0.70
Chloromethane	1.63	0.73
Vinyl chloride	1.71	0.79
Bromomethane	2.01	0.96
Chloroethane	2.09	1.02
Trichlorofluoromethane	2.27	1.19
1,1-Dichloroethene	2.89	1.57
Methylene chloride	3.60	2.06
trans-1,2-Dichloroethene	3.98	2.36
1,1-Dichloroethane	4.85	2.93
2,2-Dichloropropane	6.01	3.80
cis-1,2-Dichloroethene	6.19	3.90
Chloroform	6.40	4.80
Bromochloromethane	6.74	4.38
1,1,1-Trichloroethane	7.27	4.84
Carbon tetrachloride	7.61	5.26
1,1-Dichloropropene	7.68	5.29
Benzene	8.23	5.67
1,2-Dichloroethane	8.40	5.83
Trichloroethene	9.59	7.27
1,2-Dichloropropane	10.09	7.66
Bromodichloromethane	10.59	8.49
Dibromomethane	10.65	7.93
Toluene	12.43	10.00
1,1,2-Trichloroethane	13.41	11.05
Tetrachloroethene	13.74	11.15
1,3-Dichloropropane	14.04	11.31
Dibromochloromethane	14.39	11.85
1,2-Dibromoethane	14.73	11.83
1-Chlorohexane	15.46	13.29
Chlorobenzene	15.76	13.01
1,1,1,2-Tetrachloroethane	15.94	13.33
Ethylbenzene	15.99	13.39
p-Xylene	16.12	13.69
m-Xylene	16.17	13.68
o-Xylene	17.11	14.52
Styrene	17.31	14.60
Bromoform	17.93	14.88
Isopropylbenzene	18.06	15.46
1,1,2,2-Tetrachloroethane	18.72	16.35
Bromobenzene	18.95	15.86
1,2,3-Trichloropropane	19.02	16.23
n-Propylbenzene	19.06	16.41
2-Chlorotoluene	19.34	16.42
1,3,5-Trimethylbenzene	19.47	16.90
4-Chlorotoluene	19.50	16.72

TABLE 6210:X, CONT.

Compound	Retention Time min	
	Column 1*	Column 2†
tert-Butylbenzene	20.28	17.57
1,2,4-Trimethylbenzene	20.34	17.70
sec-Butylbenzene	20.79	18.09
p-Isopropyltoluene	21.20	18.52
1,3-Dichlorobenzene	21.22	18.14
1,4-Dichlorobenzene	21.55	18.39
n-Butylbenzene	22.22	19.49
1,2-Dichlorobenzene	22.52	19.17
1,2-Dibromo-3-chloropropane	24.53	21.08
1,2,4-Trichlorobenzene	26.55	23.08
Hexachlorobutadiene	26.99	23.68
Naphthalene	27.17	23.52
1,2,3-Trichlorobenzene	27.78	24.18
Internal standards/surrogates:		
Fluorobenzene	8.81	6.45
p-Bromofluorobenzene	18.63	15.71
1,2-Dichlorobenzene-d$_4$	22.26	19.14

* Column 1 — 60 m × 0.75 mm ID VOCOL capillary column. Hold at 10°C for 5 min, then program to 160°C at 6°/min.
† Column 2 — 30 m × 0.53 mm ID DB-624 mega-bore capillary column. Hold at 10°C for 5 min, then program to 160°C at 6°/min.

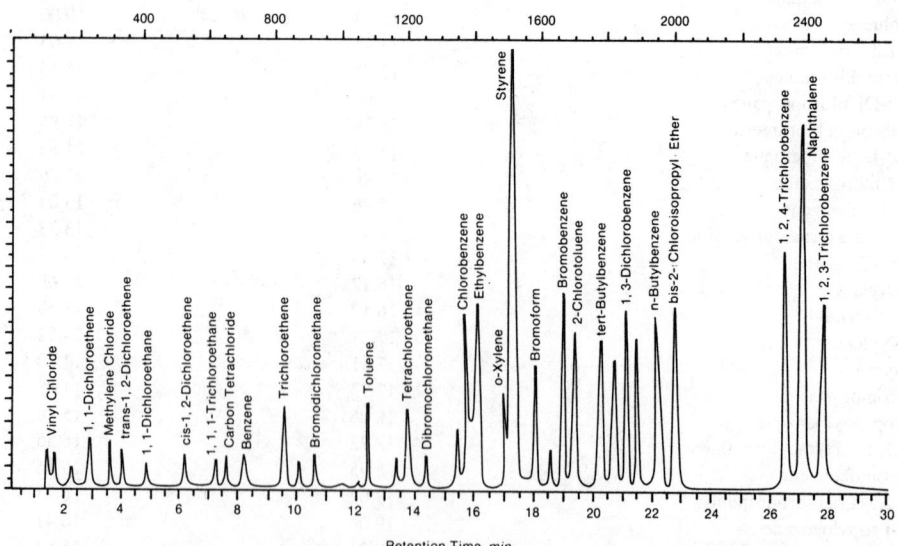

Figure 6210:7. Gas chromatogram of volatile organics (wide-bore capillary column). Column: 60 m × 0.75 mm ID VOCOL capillary; program: 10°C for 5 min, 6°C/min to 160°C.

TABLE 6210:XI. CHROMATOGRAPHIC CONDITIONS FOR VOLATILE ORGANIC COMPOUNDS ON NARROW-BORE CAPILLARY COLUMN

Compound	Retention Time min Column 3*
Dichlorodifluoromethane	0.88
Chloromethane	0.97
Vinyl chloride	1.04
Bromomethane	1.29
Chloroethane	1.45
Trichlorofluoromethane	1.77
1,1-Dichloroethene	2.33
Methylene chloride	2.66
trans-1,2-Dichloroethene	3.54
1,1-Dichloroethane	4.03
cis-1,2-Dichloroethene	5.07
2,2-Dichloropropane	5.31
Chloroform	5.55
Bromochloromethane	5.63
1,1,1-Trichloroethane	6.76
1,2-Dichloroethane	7.00
1,1-Dichloropropene	7.16
Carbon tetrachloride	7.41
Benzene	7.41
1,2-Dichloropropane	8.94
Trichloroethene	9.02
Dibromoethane	9.09
Bromodichloromethane	9.34
Toluene	11.51
1,1,2-Trichloroethane	11.99
1,3-Dichloropropane	12.48
Dibromochloromethane	12.80
Tetrachloroethene	13.20
1,2-Dibromoethane	13.60
Chlorobenzene	14.33
1,1,1,2-Tetrachloroethane	14.73
Ethylbenzene	14.73
p-Xylene	15.30
m-Xylene	15.30
Bromoform	15.70
o-Xylene	15.78
Styrene	15.78
1,1,2,2-Tetrachloroethane	15.78
1,2,3-Trichloropropane	16.26
Isopropylbenzene	16.42
Bromobenzene	16.42
2-Chlorotoluene	16.74
n-Propylbenzene	16.82
4-Chlorotoluene	16.82
1,3,5-Trimethylbenzene	16.99
tert-Butylbenzene	17.31

TABLE 6210:XI, CONT.

Compound	Retention Time min Column 3*
1,2,4-Trimethylbenzene	17.31
sec-Butylbenzene	17.47
1,3-Dichlorobenzene	17.47
p-Isopropyltoluene	17.63
1,4-Dichlorobenzene	17.63
1,2-Dichlorobenzene	17.79
n-Butylbenzene	17.95
1,2-Dibromo-3-chloropropane	18.03
1,2,4-Trichlorobenzene	18.84
Naphthalene	19.07
Hexachlorobutadiene	19.24
1,2,3-Trichlorobenzene	19.24
Internal standard:	
Fluorobenzene	8.81

* Column 3 — 30 m × 0.32 mm ID DB-5 capillary column. Hold at 10°C for 5 min, program at 6°C/min for 10 min, then program to 145°C at 15°C/min.

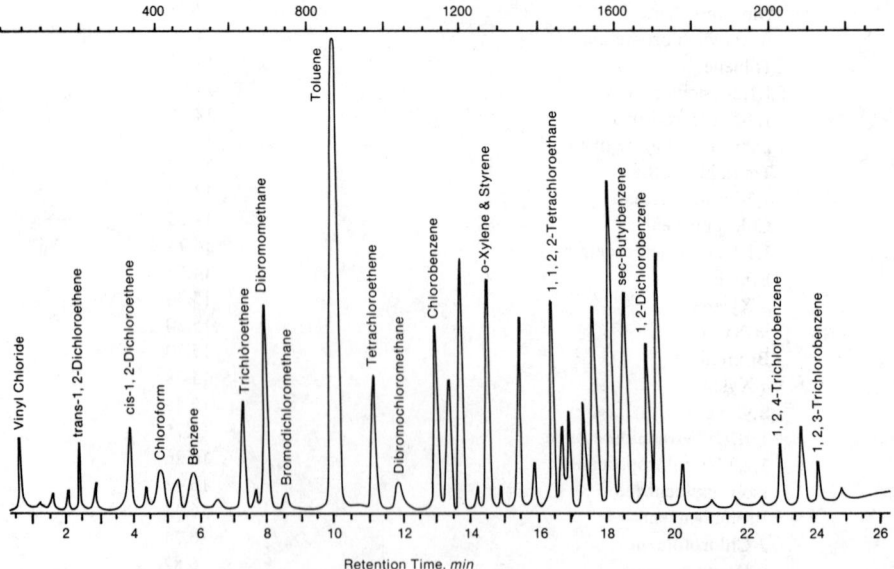

Figure 6210:8. Gas chromatogram of volatile organics (mega-bore capillary column). Column 2: 30 m long × 0.53 mm ID DB-624 mega-bore column; program: 10°C for 5 min, 6°C/min to 160°C.

materials to interface by rapidly heating trap to 180°C while backflushing trap with an inert gas at 4 mL/min for 5.0 ± 0.1 min while introducing sample into interface, empty purging device using sample syringe, rinse twice with 25 mL reagent water; leave syringe valve open to vent purge gas. After desorbing for 5 min, flash heat interface to 250°C to transfer sample to chromatographic column. Hold column temperature at 10°C for 5 min, program at 6°C/min to 70°C, then at 15°C/min to 145°C.

After desorbing sample for 5 min, recondition trap by returning system to purge mode. After about 15 min of heating at 180°C, cool for next sample. See Figure 6210:9.

6. Calculation

See Section 6210C.6.

7. Quality Control

See Section 6210C.7.

8. Precision and Bias

This method was tested in a single laboratory using a wide-bore capillary column and reagent water with additions at concentrations between 0.5 and 10 μg/L.[2] Single-operator precision and bias data are presented in Table 6210:XII.

The method was similarly tested using a cryofocused narrow-bore column with additions between 0.1 and 0.5 μg/L.[1] Bias and precision data are presented in Table 6210:XIII.

9. References

1. U.S. ENVIRONMENTAL PROTECTION AGENCY. 1985. Methods for the Determination of Organic Compounds in Finished Drinking Water

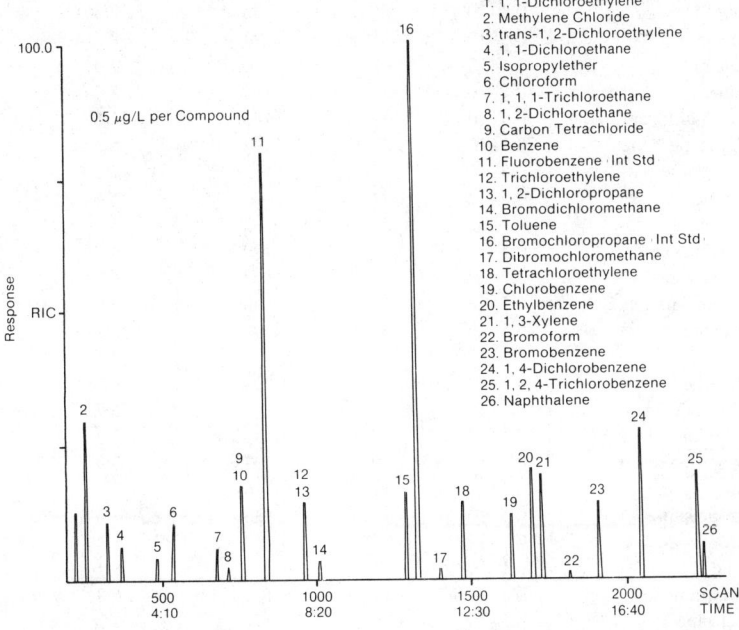

Figure 6210:9. Chromatogram of test mixture (narrow-bore capillary column).

and Raw Source Water (rev. Sept. 1986). Modified from A. Alford-Stevens, J.W. Eichelberger & W.L. Budde. 1983. Purgeable organic compounds in water by gas chromatography/mass spectrometry, Method 524. Environmental Monitoring and Support Lab., Cincinnati, Ohio.

2. SLATER, R.W. 1986. Method Detection Limits for Drinking Water Volatiles. Unpublished report, U.S. Environmental Protection Agency, Environmental Monitoring and Support Lab., Cincinnati, Ohio.

TABLE 6210:XII. SINGLE-LABORATORY BIAS AND PRECISION DATA FOR VOLATILE ORGANIC COMPOUNDS IN REAGENT WATER (WIDE-BORE CAPILLARY COLUMN)

Compound	Conc. Range $\mu g/L$	Number of Samples	Recovery* %	Standard Deviation of Recovery† %	Relative Std. Dev. %
Benzene	0.1–10	31	97	5.5	5.7
Bromobenzene	0.1–10	30	100	5.5	5.5
Bromochloromethane	0.5–10	24	90	5.7	6.4
Bromodichloromethane	0.1–10	30	95	5.7	6.1
Bromoform	0.5–10	18	101	6.4	6.3
Bromomethane	0.5–10	18	95	7.8	8.2
n-Butylbenzene	0.5–10	18	100	7.6	7.6
sec-Butylbenzene	0.5–10	16	100	7.6	7.6
tert-Butylbenzene	0.5–10	18	102	7.4	7.3
Carbon tetrachloride	0.5–10	24	84	7.4	8.8
Chlorobenzene	0.1–10	31	98	5.8	5.9
Chloroethane	0.5–10	24	89	8.0	9.0
Chloroform	0.5–10	24	90	5.5	6.1
Chloromethane	0.5–10	23	93	8.3	8.9
2-Chlorotoluene	0.1–10	31	90	5.6	6.2
4-Chlorotoluene	0.1–10	31	99	8.2	8.3
1,2-Dibromo-3-chloropropane	0.5–10	24	83	16.6	19.9
Dibromochloromethane	0.1–10	31	92	6.5	7.0
1,2-Dibromoethane	0.5–10	24	102	4.0	3.9
Dibromoethane	0.5–10	24	100	5.6	5.6
1,2-Dichlorobenzene	0.1–10	31	93	5.8	6.2
1,3-Dichlorobenzene	0.5–10	24	99	6.8	6.9
1,4-Dichlorobenzene	0.2–20	31	103	6.6	6.4
Dichlorodifluoromethane	0.5–10	18	90	6.9	7.7
1,1-Dichloroethane	0.5–10	24	96	5.1	5.3
1,2-Dichloroethane	0.1–10	31	95	5.1	5.4
1,1-Dichloroethene	0.1–10	34	94	6.3	6.7
cis-1,2-Dichloroethene	0.5–10	18	101	6.7	6.7
trans-1,2-Dichloroethene	0.1–10	30	93	5.2	5.6
1,2-Dichloropropane	0.1–10	30	97	5.9	6.1
1,3-Dichloropropane	0.1–10	31	96	5.7	6.0
2,2-Dichloropropane	0.5–10	12	86	14.6	16.9
1,1-Dichloropropene	0.5–10	18	98	8.7	8.9
Ethylbenzene	0.1–10	31	99	8.4	8.6
Hexachlorobutadiene	0.5–10	18	100	6.8	6.8
Isopropylbenzene	0.5–10	16	101	7.7	7.6
p-Isopropyltoluene	0.1–10	23	99	6.7	6.7

TABLE 6210:XII, CONT.

Compound	Conc. Range μg/L	Number of Samples	Recovery* %	Standard Deviation of Recovery† %	Relative Std. Dev. %
Methylene chloride	0.1–10	30	95	5.0	5.3
Naphthalene	0.1–100	31	104	8.6	8.2
n-Propylbenzene	0.1–10	31	100	5.8	5.8
Styrene	0.1–100	39	102	7.3	7.2
1,1,1,2-Tetrachloroethane	0.5–10	24	90	6.1	6.8
1,1,2,2-Tetrachloroethane	0.1–10	30	91	5.7	6.3
Tetrachloroethene	0.5–10	24	89	6.0	6.8
Toluene	0.5–10	18	102	8.1	8.0
1,2,3-Trichlorobenzene	0.5–10	18	109	9.4	8.6
1,2,4-Trichlorobenzene	0.5–10	18	108	9.0	8.3
1,1,1-Trichloroethane	0.5–10	18	98	7.9	8.1
1,1,2-Trichloroethane	0.5–10	18	104	7.6	7.3
Trichloroethene	0.5–10	24	90	6.5	7.3
Trichlorofluoromethane	0.5–10	24	89	7.2	8.1
1,2,3-Trichloropropane	0.5–10	16	108	15.6	14.4
1,2,4-Trimethylbenzene	0.5–10	18	99	8.0	8.1
1,3,5-Trimethylbenzene	0.5–10	23	92	6.8	7.4
Vinyl chloride	0.5–10	18	98	6.5	6.7
o-Xylene	0.1–31	18	103	7.4	7.2
m-Xylene	0.1–10	31	97	6.3	6.5
p-Xylene	0.5–10	18	104	8.0	7.7

* Recovery was calculated using internal standard of fluorobenzene.
† Standard deviation was calculated by pooling data from three concentration levels.

TABLE 6210:XIII. SINGLE-LABORATORY BIAS AND PRECISION DATA FOR VOLATILE ORGANIC
COMPOUNDS IN REAGENT WATER
(NARROW-BORE CAPILLARY COLUMN)

Compound	Conc. $\mu g/L$	Number of Samples	Recovery* %	Standard Deviation $\mu g/L$	Rel. Std. Dev. %
Benzene	0.1	7	99	6.2	6.3
Bromobenzene	0.5	7	97	7.4	7.6
Bromochloromethane	0.5	7	97	5.8	6.0
Bromodichloromethane	0.1	7	100	4.6	4.6
Bromoform	0.5	7	101	5.4	5.3
Bromomethane	0.5	7	99	7.1	7.2
n-Butylbenzene	0.5	7	94	6.0	6.4
sec-Butylbenzene	0.5	7	110	7.1	6.5
tert-Butylbenzene	0.5	7	110	2.5	2.3
Carbon tetrachloride	0.1	7	108	6.8	6.3
Chlorobenzene	0.1	7	91	5.8	6.4
Chloroethane	0.1	7	100	5.8	5.8
Chloroform	0.1	7	105	3.2	3.0
Chloromethane	0.5	7	101	4.7	4.7
2-Chlorotoluene	0.5	7	99	4.6	4.6
4-Chlorotoluene	0.5	7	96	7.0	7.3
1,2-Dibromo-3-chloropropane	0.5	7	92	10.0	10.9
Dibromochloromethane	0.1	7	99	5.6	5.7
1,2-Dibromoethane	0.5	7	97	5.6	5.8
Dibromoethane	0.5	7	93	5.6	6.0
1,2-Dichlorobenzene	0.1	7	97	3.5	3.6
1,3-Dichlorobenzene	0.1	7	101	6.0	5.9
1,4-Dichlorobenzene	0.1	7	106	6.5	6.1
Dichlorodifluoromethane	0.1	7	99	8.8	8.9
1,1-Dichloroethane	0.5	7	98	6.2	6.3
1,2-Dichloroethane	0.1	7	100	6.3	6.3
1,1-Dichloroethene	0.1	7	95	9.0	9.5
cis-1,2 Dichloroethene	0.1	7	100	3.7	3.7
trans-1,2-Dichloroethene	0.1	7	98	7.2	7.3
1,2-Dichloropropane	0.5	7	96	6.0	6.3
1,3-Dichloropropane	0.5	7	99	5.8	5.9
2,2-Dichloropropane	0.5	7	99	4.9	4.9
1,1-Dichloropropene	0.5	7	102	7.4	7.3
Ethylbenzene	0.5	7	99	5.2	5.3
Hexachlorobutadiene	0.5	7	100	6.7	6.7
Isopropylbenzene	0.5	7	102	6.4	6.3
p-Isopropyltoluene	0.5	7	113	13.0	11.5
Methylene chloride	0.5	7	97	13.0	13.4
Naphthalene	0.5	7	98	7.2	7.3
n-Propylbenzene	0.5	7	99	6.6	6.7
Styrene	0.5	7	96	19.0	19.8
1,1,1,2-Tetrachloroethane	0.5	7	100	4.7	4.7
1,1,2,2-Tetrachloroethane	0.5	7	100	12.0	12.0
Tetrachloroethene	0.1	7	96	5.0	5.2
Toluene	0.5	7	100	5.9	5.9

TABLE 6210:XIII, CONT.

Compound	Conc. μg/L	Number of Samples	Recovery* %	Standard Deviation μg/L	Rel. Std. Dev. %
1,2,3-Trichlorobenzene	0.5	7	102	8.9	8.7
1,2,4-Trichlorobenzene	0.5	7	91	16.0	17.6
1,1,1-Trichloroethane	0.5	7	100	4.0	4.0
1,1,2-Trichloroethane	0.5	7	102	4.9	4.8
Trichloroethene	0.1	7	104	2.0	1.9
Trichlorofluoromethane	0.1	7	97	4.6	4.7
1,2,3-Trichloropropane	0.5	7	96	6.5	6.8
1,2,4-Trimethylbenzene	0.5	7	96	6.5	6.8
1,3,5-Trimethylbenzene	0.5	7	101	4.2	4.2
Vinyl chloride	0.1	7	104	0.2	0.2
o-Xylene	0.5	7	106	7.5	7.1
m-Xylene	0.5	7	106	4.6	4.3
p-Xylene	0.5	7	97	6.1	6.3

* Recovery was calculated using internal standard of fluorobenzene.

6211 METHANE*

6211 A. Introduction

1. Occurrence and Significance

Methane (CH_4) is a colorless, odorless, tasteless combustible gas occasionally found in groundwaters. Escape of this gas from water may cause an explosive atmosphere not only in a utility's tanks, pumphouses, and other facilities, but also on the consumer's property, particularly where water is sprayed through poorly ventilated spaces such as public showers.

The explosive limits of CH_4 in air are 5 to 15% by volume. At sea level, a 5% CH_4 concentration in air theoretically could be reached in a poorly ventilated space sprayed with hot (68°C) water having a CH_4 concentration of only 0.7 mg/L. At higher water temperatures, the vapor pressure of water is so great that no explosive mixture can form. At lower barometric pressures, the theoretical hazardous concentration of methane in water will be reduced proportionally. In an atmosphere of N_2 or other inert gas, at least 12.8% O_2 must be present for there to be an explosion hazard.

Methane also is produced from wastewater and may be present in sewers and wastewater treatment plants (see Section 2720).

2. Selection of Method

The combustible-gas indicator method (B) offers the advantages of simplicity, speed, and great sensitivity. The volumetric method (C) can be made more accurate for concentrations of 4 to 5 mg/L and higher, but will not be satisfactory for very low concentrations. The volumetric method also can be applied to differentiate between CH_4 and other gases, as when a water supply is contaminated by liquid petroleum gas or other volatile combustible materials.

Methane also may be determined with the gas chromatograph as described in Sludge Digester Gas, Section 2720. This method permits differentiation between H_2 and CH_4, and/or its higher homologs.

---

*Approved by Standard Methods Committee, 1988.

6211 B. Combustible-Gas Indicator Method

1. General Discussion

a. Principle: An equilibrium according to Henry's law is established between CH_4 in solution and the partial pressure of CH_4 in the gas phase above the solution. The partial pressure of CH_4 can be determined with a combustible-gas indicator. The operation of the instrument is based on the catalytic oxidation of a combustible gas on a heated platinum filament that is made a part of a Wheatstone bridge. The heat generated by the oxidation of the gas increases the electrical resistance of the filament. The resulting imbalance of the electrical circuit causes deflection of a milliammeter that may be calibrated in terms of percentage of CH_4 or percentage of the lower explosive limit of the gas sampled.

b. Interference: Small amounts of ethane usually are associated with CH_4 in natural gas and presumably would be present in water that contains methane. Hydrogen gas has been observed in well waters and would behave similarly to CH_4 in this procedure. Hydrogen sulfide may interfere if the pH of the water is low enough for an appreciable fraction of the total sulfide to exist in the un-ionized form. The vapors of combustible oils also may interfere. In general, these interferences are of no practical importance because primary interest is in calculating the explosion hazard to which all combustible gases and vapors contribute.

Interference due to H_2S can be reduced by the addition of solid NaOH to the container before sampling.

c. Minimum detectable concentration: The limit of sensitivity of the test is approximately 0.2 mg/L.

d. Sampling: If the water is supersaturated with CH_4, a representative sample cannot be obtained unless the water is under sufficient pressure to keep all of the gas dissolved. Operate wells long enough to insure sampling water coming directly

from the aquifer. Representative samples can be expected only when the well is equipped with a pump operating at sufficient submergence to assure that no gas escapes from the water.

2. Apparatus

*a. Combustible-gas indicator:** Connect a three-way stopcock to the inlet to zero instrument on atmospheric air immediately before obtaining sample reading. For laboratory use, replace the suction bulb with a filter pump throttled to draw gas through the instrument at a rate of approximately 600 mL/min. See Figure 6211:1.

b. Laboratory filter pump.

c. Glass bottle, 4-L, fitted with a two-hole rubber stopper. Extend inlet tube to within 1 cm of bottom. End outlet tube approximately 1 cm below stopper. Use metal or glass tubes, each fitted with stopcocks or with short (approximately 5-cm) lengths of rubber tubing and pinchcocks. The entire assembly should be capable of holding a low vacuum for several hours. Determine volume of assembly by filling with water and measuring volume, or weight, of water contained.

3. Reagent

Sodium hydroxide, NaOH, pellets.

4. Procedure

a. Rough estimation of CH_4 concentration: Fill bottle about half full of water, using a rubber tube connecting sampling tap and inlet tube, with outlet tube open. With both inlet and outlet tubes closed,

---

*Marketed under the following trade names: "Explosimeter," "Methane Gas Detector," and "Methane Tester," all manufactured by Mine Safety Appliance Co., Pittsburgh, Pa. 15235, and "J-W Combustible Gas Indicator," manufactured by Bacharach Instrument Co., Mountain View, Calif. 94043, or equivalent.

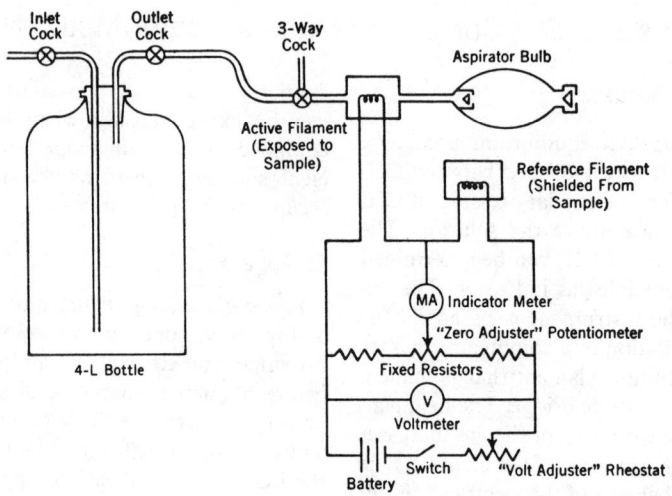

Figure 6211:1. Combustible gas indicator circuit and flow diagram.

shake bottle vigorously for approximately 15 s and let stand for approximately 1 min. Sample gas phase by withdrawing gas from the outlet, leaving inlet open to admit air. If the needle swings rapidly to a high level on the meter and then drops to zero, the CH₄-air mixture is too rich to burn; take a smaller sample for the final test. If needle deflection is too small to be read accurately, take a larger volume of water.

b. Accurate determination: If the water contains H_2S, add approximately 0.5 g NaOH pellets to empty bottle to suppress interference. Evacuate bottle, using filter pump. Fill bottle not more than three-quarters full by connecting inlet tube to sampling cock, with outlet tube closed. After collecting desired volume of water, let bottle fill with air through inlet tube. Close inlet cock, shake bottle vigorously for 60 s, and let stand for at least 2 h. Sample gas phase through outlet tube with inlet cock open. Take reading as rapidly as possible before the entering air has diluted sample appreciably. Measure volume of water sampled.

5. Calculation

The weight of CH_4 (w), in mg, in the sample is given by the equation:

$$w = P \left(\frac{1.928\ V_g}{T + 273} + \frac{890\ V_1}{H} \right)$$

where:
  P = partial pressure of CH_4, kPa,
  T = temperature, °C,
  V_g = volume of gas phase, mL,
  V_1 = volume of liquid phase, mL, and
  H = Henry's law constant, kPa/mole CH_4/ mole of water.

Values for Henry's constant are as follows:

Temperature °C	Henry's Constant H^*	Temperature °C	Henry's Constant H^*
0	2.265	40	5.261
5	2.625	45	5.577
10	3.010	50	5.846
15	3.413	60	6.342
20	3.804	70	6.749
25	4.181	80	6.911
30	4.544	90	7.013
35	4.926	100	7.106

*Multiply given values by 10^6.

For most determinations, it may be assumed that atmospheric pressure is 100 kPa, and that the temperature is 20°C. The concentration of CH_4 in the sample is then given by:

$$mg\ CH_4/L = Rf\left(6.7\ \frac{V_0 - V_1}{V_1} + 0.24\right)$$

where:

 R = scale reading,
 V_0 = total volume of sample bottle, mL,
 V_1 = volume of water sampled, mL, and
 f = factor depending on instrument used.

If the instrument reads directly in percentage of methane, $f = 1.00$. If the instrument reads in percentage of the lower explosive limit of CH_4, $f = 0.05$. For instruments that require additional factors, consult the manufacturer. For example, one commercial instrument with a scale that reads in percentage of the lower explosive limit of combustible gases requires an additional factor of 0.77 for CH_4. Hence, the value of f in the above equation would be 0.77×0.05, or 0.0385.

For more accurate work, or in locations where normal barometric pressure is significantly lower than 100 kPa, use the equation:

$$mg\ CH_4/L = RBf\left(19.277\ \frac{V_0 - V_1}{TV_1} + \frac{8900}{H}\right)$$

where:

 B = barometric pressure, kPa,

and other symbols are as above.

6. Accuracy

The accuracy of the determination is limited by the accuracy of the instrument used. Errors of approximately 10% may be expected. Calibration of instrument on known CH_4-air mixtures will improve accuracy.

7. Bibliography

ROSSUM, J.R., P.A. VILLARRUZ & J.A. WADE. 1950. A new method for determining methane in water. *J. Amer. Water Works Assoc.* 42:413.

6211 C.   Volumetric Method

1. General Discussion

a. Principle: If CH_4 is slowly mixed with an excess of O_2 in the presence of a platinum coil heated to yellow incandescence, most of the CH_4 will be converted to CO_2 and H_2O in a smooth reaction. Several passes of the mixed gases may be needed to burn substantially all the CH_4. An excess of O_2 is mixed with the sample before passage through the assembly. By differential absorption and volumetric changes the product CO_2 is measured.

b. Interference: Low-boiling hydrocarbons other than ethane and vapors from combustible oils interfere. These substances, however, are not likely to be present in water in sufficiently high concentration to affect the results significantly.

c. Minimum detectable concentration: This method is not satisfactory for determining CH_4 in water where the concentration is less than 2 mg/L.

d. Sampling: Collect sample as directed in Method B and observe the same precautions to obtain representative samples (Section 6211B.1*d*). Omit NaOH pellets and fill sample bottle with water up to 90% of capacity.

2. Methane Determination

See Section 2720B for a description of apparatus, reagents, procedure, calculation, and precision and bias

Use percentage of CH_4 found by this method with Henry's law to obtain the CH_4 concentration in original sample. Substitute CH_4 percentage for R (scale reading)

and $f = 1$ in the calculation given under Section 6211B.5 preceding.

3. Bibliography

DENNIS, L.M. & M.L. NICHOLS. 1929. Gas Analysis. Macmillan Co., New York, N.Y.

HALDANE, J.S. & J.I. GRAHAM. 1935. Methods of Air Analysis. Charles Griffin & Co., London.

BUSWELL, A.M. & T.E. LARSON. 1937. Methane in ground waters. *J. Amer. Water Works Assoc.* 29:1978.

BERGER, L.B. & H.H. SCHRENK. 1938. Bureau of Mines Haldane gas analysis apparatus. U.S. Bur. Mines Information Circ. No. 7017.

LARSON, T.E. 1938. Properties and determination of methane in ground waters. *J. Amer. Water Works Assoc.* 30:1828.

6220   VOLATILE AROMATIC ORGANICS

6220 A.   Introduction

1. Sources and Significance

A number of aromatic and certain aliphatic hydrocarbons have been detected in ground and surface waters. The origins of these compounds in water supplies are frequently unknown and often controversial, but in the case of surface waters the sources usually are fuel and/or oil spills. Hydrocarbon contamination of groundwaters has been attributed to leaking underground fuel storage tanks, industrial wastes, landfills, underground (leachfield) disposal of domestic and trade wastes, and in some cases, illegal discharges. Toxicological studies have linked at least some of these compounds, for example, benzene, to adverse human health effects.

2. Selection of Method

Two   gas   chromatographic   (GC) methods   for   purgeable   aromatic com-

pounds are included: Method B is intended for wastewater samples while Method C, which also covers certain unsaturated aliphatics, is for drinking water. Both use a photoionization detector and two chromatography columns, the first for the initial   analysis   and   the   second   for confirmation of compound identity. From a regulatory point of view, the methods are not interchangeable.[1,2] Method D refers to purge and trap gas chromatographic/mass spectrometric (GC/MS) methods suitable for determination of these compounds in water and wastewater. For information on selection of GC/MS methods, see Section 6210A.2. Method E refers to a GC/MS method for numerous compounds, including certain volatile aromatics, that are partitioned into an organic solvent before analysis; this method also is suitable for water and wastewater samples. See each method for specific compounds covered.

9. References

1. U.S. ENVIRONMENTAL PROTECTION AGENCY. 1987. National primary drinking water regulations—synthetic organic chemicals; monitoring for unregulated contaminants; final rule. 40 CFR Parts 141 & 142; *Federal Register* 52, No. 130.

2. U.S. ENVIRONMENTAL PROTECTION AGENCY. 1986. Guidelines establishing test procedures for the analysis of pollutants under the Clean Water Act. 40 CFR Part 136; *Federal Register* 51, No. 125.

6220 B. Purge and Trap Gas Chromatographic Method I*

This purge and trap gas chromatographic method[1] is applicable to the determination of purgeable aromatics, namely, benzene, chlorobenzene, 1,2-dichlorobenzene, 1,3-dichlorobenzene, 1,4-dichlorobenzene, ethylbenzene, and toluene, in municipal and industrial discharges. When analyzing unfamiliar samples for any or all of these compounds, support the identifications by at least one additional qualitative technique. This method includes analytical conditions for a second, confirmatory, gas chromatographic column. Alternatively, analyze by Method 6210B or other gas chromatographic/mass spectrometric (GC/MS) method. Samples containing petroleum byproducts may result in misidentification of alkylbenzene as chlorobenzenes on both columns. Perform confirmation by GC/MS technique in this case.

1. General Discussion

a. Principle: The principle of this method for aromatics is identical with that of Section 6210B. The detector is a photoionization detector. See Section 6010C for discussion of gas chromatographic principles.

b. Interferences: See Section 6210B.1*b* and 6010C.

---

* Approved by Standard Methods Committee, 1988. Accepted by U.S. Environmental Protection Agency as equivalent to EPA Method 602.

c. Detection limits: The method detection limit (MDL) is the minimum concentration of a substance that can be measured and reported with 99% confidence that the value is above zero.[2] The MDL concentrations listed in Table 6220:I were obtained with reagent water.[3] The MDL actually obtained in a given analysis will vary, depending on instrument sensitivity and matrix effects.

This method is applicable in the concentration range from the MDL to $1000 \times$ MDL.[3] Use direct aqueous injection techniques to measure concentration levels above $1000 \times$ MDL.

d. Safety: The toxicity or carcinogenicity of each reagent has not been defined precisely. Benzene and 1,4-dichlorobenzene have been classified tentatively as known or suspected human or mammalian carcinogens. Prepare primary standards of these compounds in a hood and wear a NIOSH/MESA-approved toxic gas respirator when handling high concentrations.

2. Sampling and Storage

See Section 6010B.1.

3. Apparatus

a. Purge and trap system: The purge and trap system consists of three separate pieces of equipment: purging device, trap, and desorber. Several complete systems now are available commercially.

TABLE 6220:I. CHROMATOGRAPHIC CONDITIONS AND METHOD DETECTION LIMITS

	Retention Time min		Method Detection Limit μg/L
Compound	Column 1	Column 2	
Benzene	3.33	2.75	0.2
Toluene	5.75	4.25	0.2
Ethylbenzene	8.25	6.25	0.2
Chlorobenzene	9.17	8.02	0.2
1,4-Dichlorobenzene	16.8	16.2	0.3
1,3-Dichlorobenzene	18.2	15.0	0.4
1,2-Dichlorobenzene	25.9	19.4	0.4

Column 1 conditions: Supelcoport (100/120 mesh) coated with 5% SP-1200/1.75% Bentone-34 packed in a 1.8 m × 2.2 mm ID stainless steel column with helium carrier gas at 36 mL/min flow rate. Column temperature held at 50°C for 2 min, then programmed at 6°C/min to 90°C for a final hold.

Column 2 conditions: Chromosorb W-AW (60/80 mesh) coated with 5% 1,2,3-tris(2-cyanoethoxy)propane packed in a 1.8 m × 2.2 mm ID stainless steel column with helium carrier gas at 30 mL/min flow rate. Column temperature held at 40°C for 2 min, then programmed at 2°C/min to 100°C for a final hold.

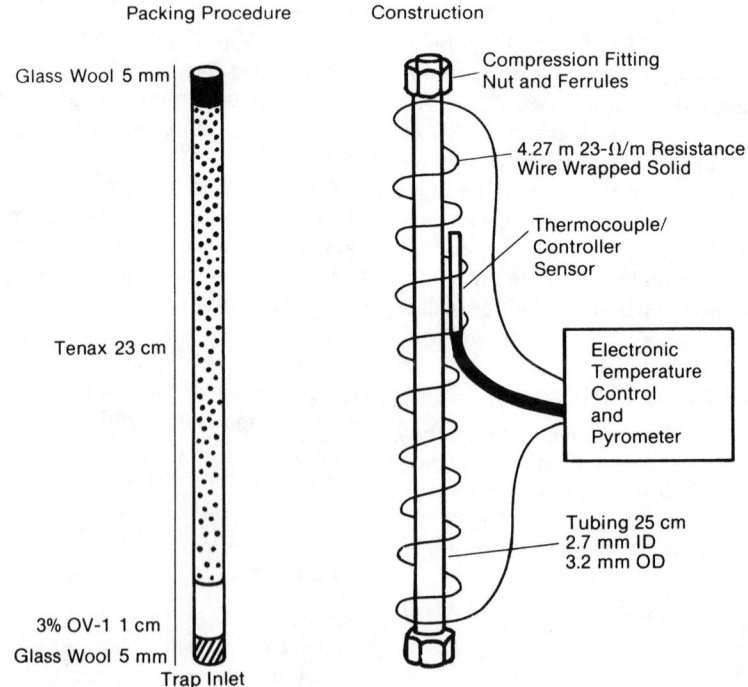

Figure 6220:1. Trap packings and construction to include desorb capability.

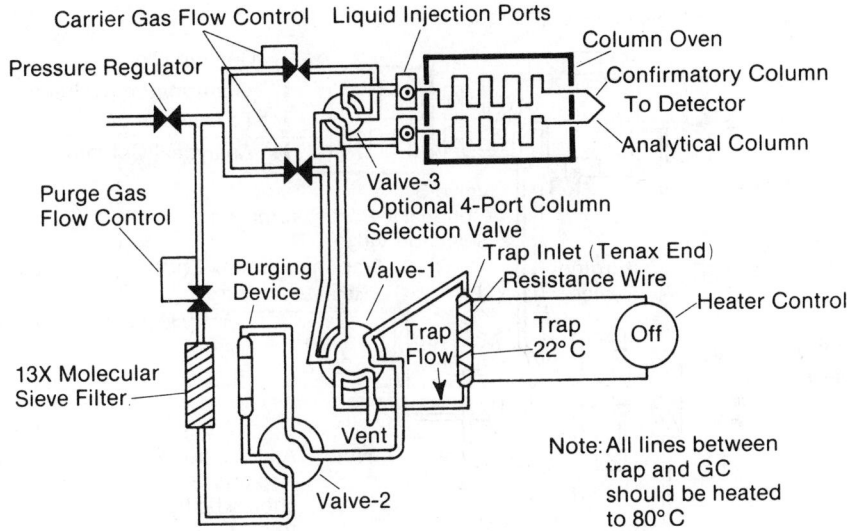

Figure 6220:2. Purge and trap system (purge mode).

1) *Purging device*: See Section 6210B.3a1).

2) *Trap*, at least 25 cm long and with an inside diameter of at least 3 mm, packed with 1 cm of methyl silicone-coated packing and 23 cm of 2,6-diphenylene oxide polymer (¶ 4*d*) as shown in Figure 6220:1. This trap was used to develop the detection limit and precision and bias data for this method.

Alternatively, use either of the two traps described in Section 6210C, although water vapor will preclude the measurement of low concentrations of benzene.

3) *Desorber*: See Section 6210B.3a3). The desorber illustrated in Figure 6220:1 also meets these criteria.

4) *Assembly*: The purge and trap system may be assembled as a separate unit or be coupled to a gas chromatograph as illustrated in Figures 6220:2, 3, and 4.

b. Gas chromatograph:† An analytical

system complete with a temperature-programmable gas chromatograph suitable for on-column injection and all required accessories including syringes, analytical columns, gases, detector, and strip-chart recorder. A data system is recommended for measuring peak areas.

1) *Column 1*, 1.8 m long × 2.2 mm ID stainless steel or glass, packed with 5% SP-1200 and 1.75% Bentone-34 on Supelcoport (100/120 mesh) or equivalent. The detection limit and precision and bias data presented herein were developed with this column. For guidelines for the use of alternate column packings see ¶ 5*a*.

2) *Column 2*, 2.4 m long × 2.2 mm ID stainless steel or glass, packed with 5% 1,2,3-tris(2-cyanoethoxy)propane on Chromosorb W-AW (60/80 mesh) or equivalent.

3) *Photoionization detector:*‡ This type of detector was used to develop detection limit and precision and bias data presented herein. For use of alternate detectors see ¶ 5*a*.

---

† Gas chromatographic methods are extremely sensitive to the materials used. Mention of trade names by *Standard Methods* does not preclude the use of other existing or as-yet-undeveloped products that give *demonstrably* equivalent results.

‡ H-NU Systems, Inc., Model PI-51-02 or equivalent.

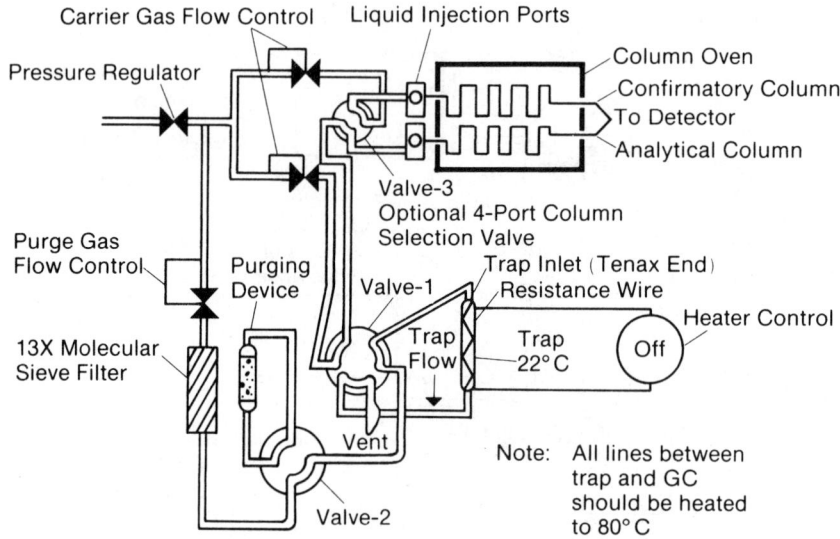

Figure 6220:3. Purge and trap system (dry mode).

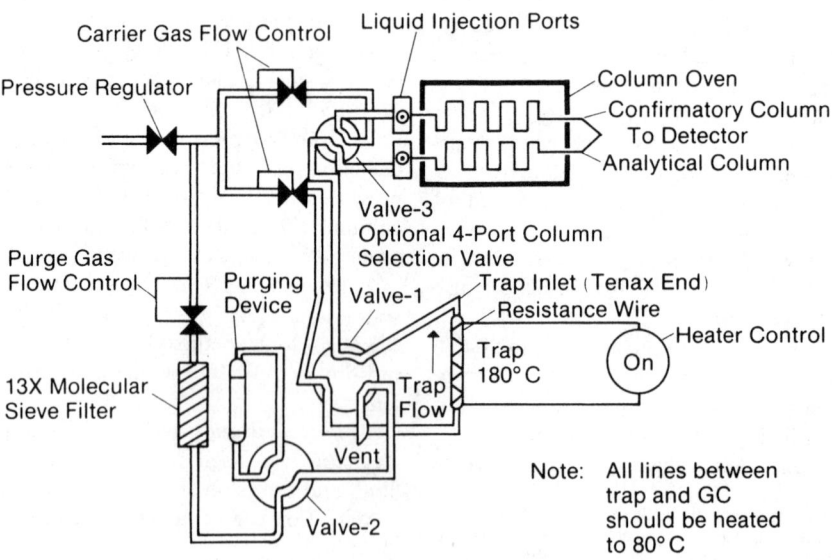

Figure 6220:4. Purge and trap system (desorb mode).

c. *Syringes*, 5-mL glass hypodermic with luerlok tip, if applicable to the purging device.

d. *Microsyringes*, 25-µL, 0.15-mm-ID needle.

e. *Syringe valve*, two-way, with luer ends.

f. *Bottle*, 15-mL, screw-cap, with TFE cap liner.

g. *Balance*, analytical, capable of accurately weighing 0.0001 g.

4. Reagents

a. *Reagent water:* See Section 6210B.4a.

b. *Hydrochloric acid*, 1 + 1: Add 50 mL conc HCl to 50 mL reagent water.

c. *Trap materials:*

1) *2,6-Diphenylene oxide polymer*, 60/80 mesh, chromatographic grade.§

2) *Methyl silicone packing.*‖

d. *Methanol*, pesticide quality or equivalent.

e. *Stock standard solutions:* See Section 6210B.4d. Prepare solutions as directed for liquid reference materials. Store in the dark at 4°C for no longer than 1 month.

f. *Secondary dilution standards:* See Section 6210B.4e.

g. *Calibration standards:* See Section 6210B.4g.

h. *Quality control (QC) check sample concentrate:* See Section 6210B.4i.

5. Procedure

a. *Operating conditions:* Table 6220:I summarizes the recommended operating conditions for the gas chromatograph and gives estimated retention times and MDLs that can be achieved under these conditions. An example of the separations obtained by Column 1 is shown in Figure 6220:5. Other packed columns, chromatographic conditions, or detectors may be used if the requirements for quality control check sample analysis (6210B.7b) are met.

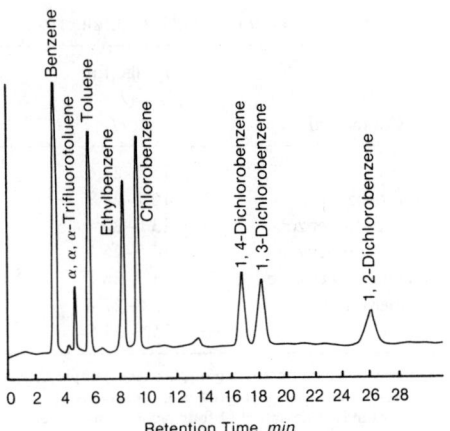

Figure 6220:5. Gas chromatogram of purgeable aromatics. Column: 5% SP-1200/1.75% Bentone-34 on Supelcoport; program: 50°C for 2 min, 6°C/min to 90°C; detector: photoionization, 10.2V.

b. *Calibration:* Calibrate system daily as directed in Section 6210B.5c. Calibration acceptance criteria are given in Table 6220:II.

c. *Sample analysis:* Follow purging procedure given in Section 6210B.5d, but use purge configuration shown in Figure 6220:2 and sample purging time of 12.0 ± 0.1 min.

After purging, disconnect purging device from trap. Dry trap by maintaining a flow of 40 mL/min of dry purge gas through it for 6 min (Figure 6220:3). If purging device has no provision for bypassing purger for this step, insert a dry purger into the device to minimize moisture in the gas.

Desorb, recondition trap, and identify compounds using the procedures of Section 6210B.5d, beginning with "attach trap to chromatograph," and the desorb configuration of Figure 6220:4.

6. Calculation

See Section 6230B.6.

---

§ Tenax or equivalent.
‖ 3% OV-1 on Chromosorb W (60/80 mesh) or equivalent.

TABLE 6220:II. CALIBRATION AND QC ACCEPTANCE CRITERIA*†

Compound	Range for Q $\mu g/L$	Limit for s $\mu g/L$	Range for $\overline{X}$ $\mu g/L$	Range for P, P_s %
Benzene	15.4–24.6	4.1	10.0–27.9	39–150
Chlorobenzene	16.1–23.9	3.5	12.7–25.4	55–135
1,2-Dichlorobenzene	13.6–26.4	5.8	10.6–27.6	37–154
1,3-Dichlorobenzene	14.5–25.5	5.0	12.8–25.5	50–141
1,4-Dichlorobenzene	13.9–26.1	5.5	11.6–25.5	42–143
Ethylbenzene	12.6–27.4	6.7	10.0–28.2	32–160
Toluene	15.5–24.5	4.0	11.2–27.7	46–148

* Criteria calculated assuming a QC check sample concentration of 20 $\mu g/L$.
† Q = concentration measured in QC check sample,
  s = standard deviation of four recovery measurements,
  $\overline{X}$ = average recovery for four recovery measurements,
P_s, P = recovery measured.

NOTE: These criteria are based directly on the method performance data in Table 6220:III. Where necessary, the limits for recovery were broadened to assure applicability of the limits for recovery were broadened to assure applicability of the limits to concentrations below those used to develop Table 6220:III.

7. Quality Control

Follow instructions given in Section 6230B.7*a* through *g*, with the following exceptions: Calculate optional acceptance criteria (¶ 7*c*) using the equations in Table 6220:III for X' and S'; for surrogate compound (¶ 7*f*), use α,α,α- trifluorotoluene.

8. Precision and Bias

This method was tested by 20 laboratories using reagent water, drinking water, surface water, and three industrial wastewaters with additions at six concentrations over the range 2.1 to 550 $\mu g/L$.[3] Single-operator precision, overall precision, and

TABLE 6220:III. METHOD BIAS AND PRECISION AS FUNCTIONS OF CONCENTRATION*

Compound	Bias, as Recovery, X' $\mu g/L$	Single-Analyst Precision, s' $\mu g/L$	Overall Precision, S' $\mu g/L$
Benzene	$0.92C + 0.57$	$0.09\overline{X} + 0.59$	$0.21\overline{X} + 0.56$
Chlorobenzene	$0.95C + 0.02$	$0.09\overline{X} + 0.23$	$0.17\overline{X} + 0.10$
1,2-Dichlorobenzene	$0.93C + 0.52$	$0.17\overline{X} - 0.04$	$0.22\overline{X} + 0.53$
1,3-Dichlorobenzene	$0.96C - 0.04$	$0.15\overline{X} - 0.10$	$0.19\overline{X} + 0.09$
1,4-Dichlorobenzene	$0.93C + 0.09$	$0.15\overline{X} + 0.28$	$0.20\overline{X} + 0.41$
Ethylbenzene	$0.94C + 0.31$	$0.17\overline{X} + 0.46$	$0.26\overline{X} + 0.23$
Toluene	$0.94C + 0.65$	$0.09\overline{X} + 0.48$	$0.18\overline{X} + 0.71$

* X' = expected recovery for one or more measurements of a sample containing a concentration of C,
  s' = expected single-analyst standard deviation of measurements at an average concentration found of $\overline{X}$,
  S' = expected interlaboratory standard deviation of measurements at an average concentration found of $\overline{X}$,
  C = true value for the concentration, $\mu g/L$,
  $\overline{X}$ = average recovery found for measurements of samples containing a concentration of C, $\mu g/L$.

method bias were found to be related directly to compound concentration and essentially independent of the sample matrix. Linear equations to describe these relationships are presented in Table 6220:III.

9. References

1. U.S. ENVIRONMENTAL PROTECTION AGENCY. 1984. Method 602—Purgeable aromatics. 40 CFR Part 136, 43272; *Federal Register* 49, No. 209.

2. U.S. ENVIRONMENTAL PROTECTION AGENCY. 1984. Definition and procedure for the determination of the method detection limit. 40 CFR Part 136, Appendix B. *Federal Register* 49, No. 209.

3. U.S. ENVIRONMENTAL PROTECTION AGENCY. 1984. EPA Method Study 25, Method 602—Purgeable Aromatics. EPA 600/4-84-042, National Technical Information Serv., PB84-196682, Springfield, Va.

6220 C. Purge and Trap Gas Chromatographic Method II*

This method[1] is applicable to the determination of purgeable aromatic and unsaturated compounds in finished drinking water, raw source water, or drinking water in any treatment stage.† It is not applicable to the determination of styrene in chlorinated drinking waters because of its rapid oxidation rate. The method varies only slightly from that of Section 6220B, which is intended for use with wastewater samples. See Section 6230D for determination of halocarbons and aromatics in series.

1. General Discussion

a. Principle: The principle of this purge and trap method is identical to that given in Section 6210B. The detector used is a photoionization detector. See Section 6010C for discussion of gas chromatographic principles.

b. Interferences: See Section 6210B.1*b* and 6010C. For samples containing large amounts of water-soluble materials, suspended solids, high boiling compounds, or high levels of compounds being determined, wash purging device with a soap solution, rinse with distilled water, and then dry in an oven at 105°C between analyses. Excess water will cause a negative baseline deflection in the chromatogram. The method provides for a dry purge period to prevent this problem.

c. Detection limits: In a single laboratory using reagent water and known additions of 0.2 µg/L, observed method detection limit (MDLs)[2] for these compounds were in the range of 0.01 to 0.05 µg/L, depending on the compound. Some laboratories may not be able to achieve these detection limits because results depend on instrument sensitivity and matrix effects. Individual aromatic compounds can be measured at concentrations up to 1500 µg/L. Analysis of complex mixtures containing partially resolved compounds may be hampered by concentration differences larger than a factor of 10.

d. Safety: The toxicity or carcinogenicity

---

* Approved by Standard Methods Committee, 1987. Accepted by U.S. Environmental Protection Agency as equivalent to EPA Method 503.1.

† Benzene, bromobenzene, *n*-butylbenzene, *sec*-butylbenzene, *tert*-butylbenzene, chlorobenzene, 2-chlorotoluene, 4-chlorotoluene, 1,2-dichlorobenzene, 1,3-dichlorobenzene, 1,4-dichlorobenzene, ethylbenzene, hexachlorobutadiene, isopropylbenzene, 4-isopropyltoluene, naphthalene, *n*-propylbenzene, styrene, tetrachloroethene, toluene, 1,2,3-trichlorobenzene, 1,2,4-trichlorobenzene, trichloroethene, 1,2,4-trimethylbenzene, 1,3,5-trimethylbenzene, *o*-xylene, *m*-xylene, and *p*-xylene.

of each reagent has not been defined precisely. Benzene, 1,4-dichlorobenzene, hexachlorobutadiene, tetrachloroethene, and trichloroethene have been classified tentatively as known or suspected human or mammalian carcinogens. Prepare primary standards of these compounds in a hood and wear a NIOSH/MESA-approved toxic gas respirator when handling high concentrations.

2. Sampling and Storage

See Section 6010B.1.

3. Apparatus

a. Purge and trap system: The purge and trap system consists of three separate pieces of equipment: purging device, trap, and desorber. Several complete systems now are available commercially.

1) *Purging device:* See Section 6210C.3*a*1).

2) *Trap:* For trap specifications, see Section 6220B.3*a*2). For conditioning procedure, see Section 6210C.3*a*2).

3) *Desorber:* See Sections 6210B.3*a*3) and 6220B.3*a*3).

4) *Assembly:* See Section 6220B.3*a*4).

b. Gas chromatograph:‡ See Section 6220.3*b.*

1) *Column 1,* 1.5 to 2.5 m × 2.2 mm ID stainless steel or glass, packed with 5% SP-1200 and 1.75% Bentone-34 on Supelcoport (100/120 mesh) or equivalent. Use a flow rate of helium carrier gas of 30 mL/min. Program column temperature to hold

‡ Gas chromatographic methods are extremely sensitive to the materials used. Mention of trade names by *Standard Methods* does not preclude the use of other existing or as-yet-undeveloped products that give *demonstrably* equivalent results.

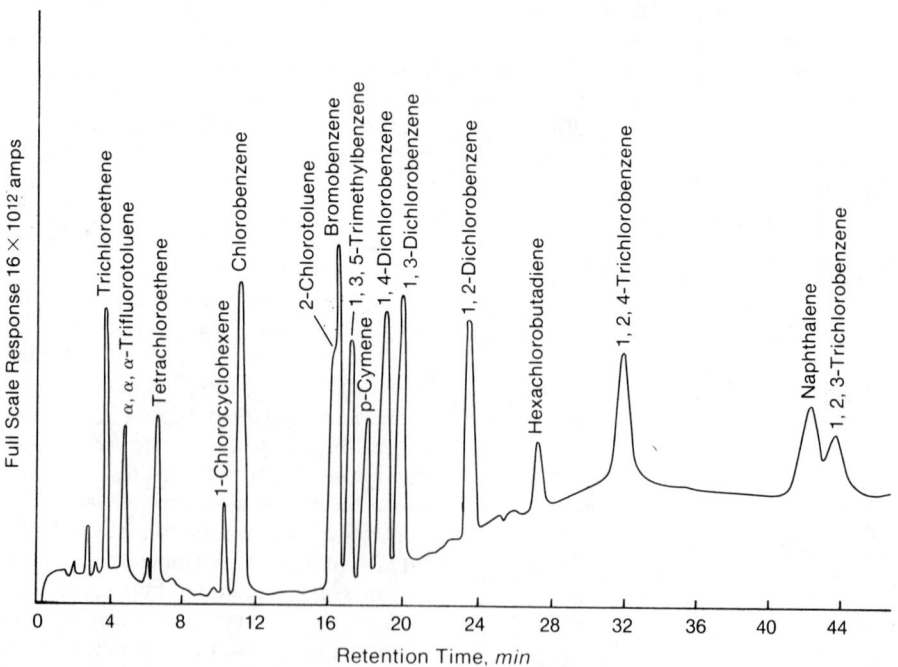

Figure 6220:6. Chromatogram of test mixture. Column: 5% SP-1200/1.75% Bentone-34; program: 50°C for 2 min, 3°C/min to 110°C; detector: photoionization; sample: 0.5 μg/L standard mixture.

at 50°C for 2 min, increase to 110°C at 3°C/ min, and hold at 110°C until all expected compounds have eluted. When column is not in use, hold it at 110°C. Condition new SP-1200/Bentone columns with carrier gas flow at 120°C for several days before connecting to detector. A sample chromatogram obtained with Column 1 is presented in Figure 6220:6.

2) *Column 2,* 1.5 to 2.5 m long × 2.2 mm ID stainless steel or glass, packed with 5% 1,2,3-tris(2-cyanoethoxy) propane on Chromosorb W-AW (60/80 mesh) or equivalent. Use a flow rate of helium carrier gas of 30 mL/min. Program column temperature to hold at 40°C for 2 min, increase to 100°C at 2°C/min, and hold at 100°C until all expected compounds have eluted. A sample chromatogram obtained with Column 2 is presented in Figure 6220:7.

3) *Photoionization detector:* See Section 6220B.3*b*3).

c. Syringes, 5-mL glass hypodermic with luerlok tip.

d. Syringe valves, two-way, with luer ends.

e. Microsyringes, 25-μL with a 5-cm × 0.15-mm-ID, 22° bevel needle,§ also with 10- and 100-μL capacity.

f. Bottles, 15-mL, with TFE-lined screw caps.

4. Reagents

a. Reagent water: See Section 6210B.4*a*.

b. Trap packing materials: See Section 6210B.4*b*1) and 2).

c. Methanol, pesticide quality or equivalent.

d. Hydrochloric acid, HCl, 1+1.

e. Standard stock solutions: See Section 6210B.4*d*.

f. Secondary dilution standards: See Section 6210B.4*e*.

g. Calibration standards: Use directions given in Section 6210B.4*g*, but prepare at least five concentration levels for each com-

_____
§ Hamilton #702N or equivalent.

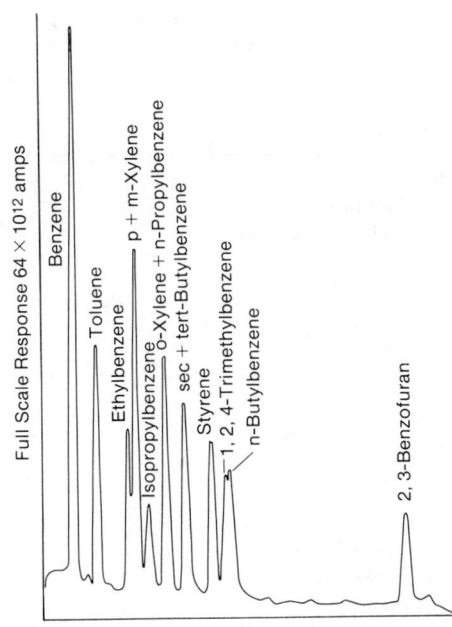

Figure 6220:7. **Chromatogram of test mixture.** Column: 5% 1,2,3-tris (2-cyanoethoxy) propane on Chromosorb-W; program: 40°C for 2 min, 2°C/min to 100°C; detector: photoionization; sample: 2.0 μg/L standard mixture.

pound. More than one set of standards may be required.

5. Procedure

a. Operating conditions: Table 6220:IV summarizes recommended operating conditions for the gas chromatograph and gives estimated retention times.

b. Calibration:

1) External standard calibration procedure—Prepare standards as directed in ¶ 4*g*. Starting with the least concentrated standard, analyze each calibration standard according to ¶ *d* below and tabulate peak height or area responses versus the concentration in the standard. Prepare a calibration curve for each compound. Alternatively, if the ratio of response to

TABLE 6220:IV. RETENTION TIMES FOR AROMATIC COMPOUNDS

Compound	Retention Time s Column 1	Column 2
Benzene	199	165
Trichloroethene	231	142
α, α, α-Trifluorotoluene*	296	168
Toluene	384	255
Tetrachloroethene	406	168
Ethylbenzene	606	375
1-Chlorocyclohexene	637	345
p-Xylene	653	403
Chlorobenzene	689	481
m-Xylene	689	403
o-Xylene	738	518
Isopropylbenzene	768	455
Styrene	834	690
1,4-Bromofluorobenzene	852	740
n-Propylbenzene	879	518
tert-Butylbenzene	975	595
2-Chlorotoluene	985	681
4-Chlorotoluene	990	—
Bromobenzene	999	807
sec-Butylbenzene	1027	595
1,3,5-Trimethylbenzene	1043	612
4-Isopropyltoluene	1090	681
1,2,4-Trimethylbenzene	1090	750
1,4-Dichlorobenzene	1152	975
1,3-Dichlorobenzene	1211	901
Cyclopropylbenzene	1211	—
n-Butylbenzene	1211	765
2,3-Benzofuran	1320	1460
1,2-Dichlorobenzene	1425	1161
Hexachlorobutadiene	1650	1011
1,2,4-Trichlorobenzene	1928	1535
Naphthalene	2545	2298
1,2,3-Trichlorobenzene	2631	1820

* Recommended internal standard.

Column 1 conditions: Supelcoport (100/120 mesh) coated with 5% SP-1200/1.75% Bentone 34 packed in a 1.8 m × 2.2 mm ID stainless steel or glass column with helium carrier at 30 mL/min flow rate. Column temperature held at 50°C for 2 min, then programmed at 3°C/min to 110°C for a final hold.

Column 2 conditions: Chromosorb W-AW(60/80 mesh) coated with 5% 1,2,3-tris(2-cyanoethoxy) propane packed in a 1.8 m × 2.2 mm ID stainless steel or glass column with helium carrier gas at 30 mL/min flow rate. Column temperature held at 40°C for 2 min, then programmed at 2°C/min to 100°C for a final hold.

concentration (calibration factor) is a constant over the working range (< 10% relative standard deviation, RSD), assume linearity through the origin and use the average ratio or calibration factor in place of a calibration curve.

Verify the working calibration curve or calibration factor daily by measuring one or more calibration standards. If the response for any compound varies from the predicted by more than ±20%, repeat the test using a fresh calibration standard. If the results still do not agree, generate a new calibration curve or use a single-point calibration standard. Single-point calibration is a viable alternative to a calibration curve. Prepare single-point standards from the secondary dilution standards in methanol. Prepare at a concentration that produces a response close (±20%) to that of the sample. Do not use less than 20 μL of the secondary dilution standard to produce a single-point calibration standard in reagent water.

2) *Internal standard calibration procedure*—Alternatively, use internal standard calibration techniques. The recommended compound is α,α,α-trifluorotoluene. Add internal standard to sample just before purging. Check validity of internal standard calibration factors daily by analyzing a calibration standard.

c. *Instrument performance:* Check the performance of entire analytical system daily using analyses of reagent blanks, standards, duplicate samples, and the laboratory control standard (see below). All peaks in standard chromatograms shall be sharp and symmetrical. Correct any peak tailing significantly in excess of that shown in the method chromatograms. If only the compounds eluting before ethylbenzene give random responses or unusually wide peak widths, are poorly resolved, or are missing, the problem usually is traceable to the trap/desorber. If negative peaks appear early in the chromatogram, increase dry purge time to 5 min.

Check precision between laboratory replicates. A properly operating system shows an average relative standard deviation of less than 10%. Poor precision generally is traceable to pneumatic leaks, especially around sample purger, or to an improperly adjusted lamp intensity power. Monitor retention times for each compound using calibration standards and the laboratory control standard. If individual retention times vary by more than 10% over an 8-h period or do not fall within 10% of an established norm, locate and correct source of retention data variance.

d. *Sample analysis:* Follow purging procedure given in Section 6210B.5d, but use purge configuration given in Figure 6220:2 and purge time of 12.0 ± 0.1 min.

After purging, adjust purge and trap system to dry purge position (Figure 6220:3) for 4 min. Empty purging device using sample syringe and wash chamber with two 5-mL flushes of reagent water. Desorb, recondition trap, and identify compounds using procedures of Section 6210B.5d, beginning with ". . . attach trap to chromatograph, . . ." and the desorb configuration of Figure 6220:4.

6. Calculation

Determine concentration of individual compounds. If the external standard calibration procedure is used, calculate concentration of compound being measured from the peak response using the calibration curve or calibration factor previously determined. If the internal standard calibration procedure is used, calculate concentration using the response factor (*RF*) determined in Section 6210B.5c2) and the following equation:

$$\text{Concentration, } \mu\text{g/L} = \frac{(A_s)(C_{is})}{(A_{is})(RF)}$$

where:

A_s = response for compound to be measured,
A_{is} = response for internal standard, and
C_{is} = concentration of internal standard.

Report results in micrograms per liter without correction for recovery. Report quality control data with the sample results.

7. Quality Control

Using a check sample concentrate of 100 times the MCL or 1 μg/mL rather than 500 times the MCL or 5 μg/mL and 5-mL rather than 25-mL portions, follow the procedure of Section 6210C.7.

8. Precision and Bias

Single-laboratory precision and bias for most of the compounds in Ohio River water and chlorinated drinking water are presented in Table 6220:V.[3] This method was tested by 20 laboratories using drinking water with added aromatic compounds at six concentrations between 2.2 and 600 μg/L. Single-operator precision, overall precision, and method bias were found to be related directly to compound concentration. Linear equations describing these relationships are presented in Table 6220:VI.[4]

Bias and precision data from multilaboratory studies for several purgeable aromatics in reagent water are presented in Table 6220:VII.[5]

TABLE 6220:V. SINGLE-LABORATORY BIAS AND PRECISION FOR AROMATIC AND UNSATURATED COMPOUNDS IN CHLORINATED DRINKING WATER AND RAW SOURCE WATER

Compound	Matrix Type*	Addition $\mu g/L$	Number of Samples	Average Recovery %	Relative Standard Deviation %
Benzene	A,B	0.40	13	100	2.8
Bromobenzene	A,B	0.50	19	93	6.2
n-Butylbenzene	A	0.40	7	78	15.7
sec-Butylbenzene	A	0.40	7	80	11.0
tert-Butylbenzene	A	0.40	7	88	8.7
Chlorobenzene	A,B	0.50	19	96	5.8
1-Chlorocyclohexene	A,B	0.50	19	89	7.1
4-Chlorotoluene	A,B	0.50	17	91	5.0
1,2-Dichlorobenzene	A,B	0.50	18	92	7.1
1,3-Dichlorobenzene	A,B	0.50	19	91	8.5
1,4-Dichlorobenzene	A,B	0.50	19	95	6.4
Ethylbenzene	A	0.40	7	93	8.5
Hexachlorobutadiene	A	0.50	10	74	16.8
Isopropylbenzene	A	0.40	7	88	8.7
Naphthalene	A,B	0.50	16	92	14.8
n-Propylbenzene	A	0.40	7	83	9.3
Tetrachloroethene	A,B	0.50	19	97	7.8
Toluene	A,B	0.40	13	94	6.6
1,2,3-Trichlorobenzene	A,B	0.50	18	85	10.4
1,2,4-Trichlorobenzene	A,B	0.50	18	86	10.1
Trichloroethene	A,B	0.50	19	97	6.8
α,α,α-Trifluorotoluene†	A,B	0.50	18	88	9.7
1,2,4-Trimethylbenzene	A	0.40	7	75	8.7
1,3,5-Trimethylbenzene	A	0.50	10	92	8.7
m-Xylene	A	0.40	7	90	7.7
o-Xylene	A	0.40	7	90	7.2
p-Xylene	A	0.40	7	85	8.7

* Matrix A is drinking water; matrix B is raw source water.
† Recommended internal standard.

TABLE 6220:VI. SINGLE-ANALYST PRECISION, OVERALL PRECISION, AND BIAS FOR PURGEABLE AROMATICS IN DRINKING WATER*

Compound	Single-Analyst Precision $\mu g/L$	Overall Precision $\mu g/L$	Bias as Mean Recovery ($\overline{X}$) $\mu g/L$
Benzene	$0.11\overline{X} - 0.06$	$0.22\overline{X} + 1.11$	$0.97C + 0.85$
Chlorobenzene	$0.10\overline{X} + 0.12$	$0.16\overline{X} + 0.36$	$0.94C + 0.12$
1,2-Dichlorobenzene	$0.10\overline{X} + 0.42$	$0.18\overline{X} + 0.28$	$0.91C + 0.44$
1,3-Dichlorobenzene	$0.08\overline{X} + 0.33$	$0.15\overline{X} + 0.33$	$0.93C + 0.21$
1,4-Dichlorobenzene	$0.09\overline{X} + 0.39$	$0.15\overline{X} + 0.39$	$0.91C + 0.26$
Ethylbenzene	$0.10\overline{X} + 0.18$	$0.20\overline{X} + 0.68$	$0.97C + 0.41$
Toluene	$0.10\overline{X} + 0.18$	$0.21\overline{X} + 0.16$	$0.94C + 0.17$

* $\overline{X}$ = mean recovery, $\mu g/L$.
   C = true value for concentration, $\mu g/L$.

9. References

1. U.S. ENVIRONMENTAL PROTECTION AGENCY. 1985. Methods for the Determination of Organic Compounds in Finished Drinking Water and Raw Source Water (rev. Sept. 1986). Modified from U.S. Environmental Protection Agency. 1981. The analysis of aromatic chemicals in water by the purge and trap method, Method 503.1. Environmental Monitoring and Support Lab., Cincinnati, Ohio.

2. GLASER, J.A., D.L. FOERST, G.D. MCKEE, S.A. QUAVE & W.L. BUDDE. 1981. Trace analyses for wastewaters. Environ. Sci. Technol. 15:1426.

3. BELLAR, T.A. & J.J. LICHTENBERG. 1982. The determination of volatile aromatic compounds

TABLE 6220:VII. BIAS AND PRECISION DATA FOR PURGEABLE AROMATICS FROM MULTILABORATORY PERFORMANCE EVALUATION STUDIES

Compound	Addition $\mu g/L$	Number of Laboratories	Average Measured Concentration $\mu g/L$	Relative Standard Deviation %	Average Recovery %
Benzene	94.1	9	91.9	18.6	98
	47.0	10	47.0	11.8	100
	18.8	8	18.7	16.4	100
	8.10	11	6.22	40.8	88
Chlorobenzene	41.4	5	39.8	6.20	96
	27.6	7	27.1	12.1	98
	13.8	6	14.3	6.73	104
	5.52	8	5.65	25.3	102
1,2-Dichlorobenzene	96.9	5	72.9	31.6	75
	19.4	4	16.5	18.8	85
1,4-Dichlorobenzene	68.6	5	62.5	22.8	91
	13.7	5	14.6	29.1	107
1,2,4-Trichlorobenzene	80.8	6	77.6	14.3	96
	6.7	6	8.46	30.7	126

in drinking water and raw source water. Unpublished report, U.S. Environmental Protection Agency, Environmental Monitoring and Support Lab., Cincinnati, Ohio.

4. U.S. ENVIRONMENTAL PROTECTION AGENCY. 1984. EPA Method Study 25, Method 602—Purgeable Aromatics. EPA 600/4-84-042, National Technical Information Serv., PB84-19 6682, Springfield, Va.

5. U.S. ENVIRONMENTAL PROTECTION AGENCY. 1984. Analytical methods and monitoring issues associated with volatile organics in drinking water. Office of Drinking Water, Washington, D.C.

6220 D.   Purge and Trap
Gas Chromatographic/Mass Spectrometric Method

See Sections 6210B and C for packed-column methods and Section 6210D for capillary-column method.

6220 E.   Liquid-Liquid Extraction
Gas Chromatographic/Mass Spectrometric Method

See Section 6410B.

6230   VOLATILE HALOCARBONS

6230 A.   Introduction

1. Sources and Significance

Organohalides, particularly trihalomethanes, have been found in most chlorinated water supplies in the United States; typically they are produced in the treatment process. Also detected in many raw source waters and in corresponding finished drinking waters have been common organohalide solvents that are traceable to industrial effluents, on-site waste disposal, sanitary landfills, or illegal discharges. Toxicological studies suggest that chloroform ($CHCl_3$) and other organohalides have detrimental effects on human health. Organohalides in water supplies should be monitored closely so that measures may be taken to minimize (by blending or using alternate forms of disinfection) or eliminate them.

2. Selection of Method

Three gas chromatographic (GC) methods for purgeable halocarbons are presented, as well as gas chromatographic/mass spectrometric (GC/MS) methods. See each method for list of specific compounds covered.

Method B is a packed-column GC method useful for the detection of volatile halocarbons in municipal and industrial wastewater discharges. Method C, also a packed-column GC method, is applicable to the determination of various volatile halocarbons in finished drinking water, raw source water, or drinking water at any stage of treatment. Method D is a capillary-column GC method capable of attaining low microgram-per-liter detection limits for most organohalides. While the methods are similar, they are not interchangeable from a regulatory point of view.[1,2] Packed column methods, in general, cannot attain some of the low detection limits required by certain regulations, and thus should be used only for routine monitoring of drinking water and analyses of wastewaters.

For information on selection of GC/MS methods, see Section 6210A.2.

All of these methods are highly sophisticated microanalytical procedures that should be used only by analysts experienced in the operation of such systems and in evaluation and interpretation of the data they produce.

3. References

1. U.S. ENVIRONMENTAL PROTECTION AGENCY. 1987. National primary drinking water regulations—synthetic organic chemicals; monitoring for unregulated contaminants; final rule. 40 CFR Parts 141 & 142; *Federal Register* 52, No. 130.
2. U.S. ENVIRONMENTAL PROTECTION AGENCY. 1986. Guidelines establishing test procedures for the analysis of pollutants under the Clean Water Act. 40 CFR Part 136; *Federal Register* 51, No. 125.

6230 B. Purge and Trap Packed-Column Gas Chromatographic Method I*

This method[1] is applicable to the determination of purgeable halocarbons† in municipal and industrial discharges. When analyzing unfamiliar samples for any or all of these compounds, support the identifications by at least one additional qualitative technique. This method includes analytical conditions for a second, confirmatory, gas chromatographic column. Alternatively, analyze according to Section 6210B or other gas chromatograph/mass spectrometer (GC/MS) method.[2] The method should be used only by analysts familiar with gas chromatographic techniques.

1. General Discussion

a. Principle: See Section 6210B.*a.* The

---

* Approved by Standard Methods Committee, 1987. Accepted by U.S. Environmental Protection Agency as equivalent to EPA Method 601.

† Bromodichloromethane, bromoform, bromomethane, carbon tetrachloride, chlorobenzene, chloroethane, 2-chloroethylvinyl ether, chloroform, chloromethane, dibromochloromethane, 1,2-dichlorobenzene, 1,3-dichlorobenzene, 1,4-dichlorobenzene, dichlorodifluoromethane, 1,1-dichloroethane, 1,2-dichloroethane, 1,1-dichloroethene, *trans*-1,2-dichloroethene, 1,2-dichloropropane, *cis*-1,3-dichloropropene, *trans*-1,3-dichloropropene, methylene chloride, 1,1,2,2-tetrachloroethane, tetrachloroethene, 1,1,1-trichloroethane, 1,1,2-trichloroethane, trichloroethene, trichlorofluoromethane, and vinyl chloride.

detector is a halide-specific electrolytic conductivity or microcoulometric detector.[3,4]

b. *Interference:* See Section 6210B.1*b.*

c. *Detection limits:* The method detection limit (MDL) is the minimum concentration of a substance that can be measured and reported with 99% confidence that the value is above zero.[5] The MDL concentrations listed in Table 6230:I were obtained by using reagent water.[6] The MDL actually obtained in a given analysis will vary, depending on instrument sensitivity and matrix effects and therefore should be determined by each laboratory performing this analysis.

This method is recommended for use in the concentration range from the MDL to 1000 × MDL. Use direct aqueous injection

TABLE 6230:I. CHROMATOGRAPHIC CONDITIONS AND METHOD DETECTION LIMITS (MDL)

Compound	Retention Time min		Method Detection Limit* μg/L
	Column 1	Column 2*	
Chloromethane	1.50	5.28	0.08
Bromomethane	2.17	7.05	1.18
Dichlorodifluoromethane	2.62	nd	1.81
Vinyl chloride	2.67	5.28	0.18
Chloroethane	3.33	8.68	0.52
Methylene chloride	5.25	10.1	0.25
Trichlorofluoromethane	7.18	nd	nd
1,1-Dichloroethene	7.93	7.72	0.13
1,1-Dichloroethane	9.30	12.6	0.07
trans-1,2-Dichloroethene	10.1	9.38	0.10
Chloroform	10.7	12.1	0.05
1,2-Dichloroethane	11.4	15.4	0.03
1,1,1-Trichloroethane	12.6	13.1	0.03
Carbon tetrachloride	13.0	14.4	0.12
Bromodichloromethane	13.7	14.6	0.10
1,2-Dichloropropane	14.9	16.6	0.04
cis-1,3-Dichloropropene	15.2	16.6	0.34
Trichloroethene	15.8	13.1	0.12
Dibromochloromethane	16.5	16.6	0.09
1,1,2-Trichloroethane	16.5	18.1	0.02
trans-1,3-Dichloropropene	16.5	18.0	0.20
2-Chloroethylvinyl ether	18.0	nd	0.13
Bromoform	19.2	19.2	0.20
1,1,2,2-Tetrachloroethane	21.6	nd	0.03
Tetrachloroethene	21.7	15.0	0.03
Chlorobenzene	24.2	18.8	0.25
1,3-Dichlorobenzene	34.0	22.4	0.32
1,2-Dichlorobenzene	34.9	23.5	0.15
1,4-Dichlorobenzene	35.4	22.3	0.24

Column 1 conditions: Carbopack B (60/80 mesh) coated with 1% SP-1000 packed in a 2.4 m × 3 mm ID stainless steel or glass column with helium carrier gas at 40 mL/min flow rate. Column temperature held at 45°C for 3 min, then programmed at 8°C/min to 220°C and held for 15 min.

Column 2 conditions: Porisil-C (100/120 mesh) coated with *n*-octane packed in a 1.8 m × 3 mm ID stainless steel or glass column with helium carrier gas at 40 mL/min flow rate. Column temperature held at 50°C for 3 min, then programmed at 6°C/min to 170°C and held for 4 min.

* nd—not determined.

techniques to measure concentration levels above 1000 × MDL.

d. *Safety:* The toxicity or carcinogenicity of each reagent has not been defined precisely. Carbon tetrachloride, chloroform, 1,4-dichlorobenzene, and vinyl chloride have been classified tentatively as known or suspected human or mammalian carcinogens. Prepare primary standards of these compounds in a hood and wear NIOSH/MESA-approved toxic gas respirator when handling high concentrations.

2. Sampling and Storage

See Section 6010B.1.

3. Apparatus

a. *Purge and trap system:* The purge and trap system consists of three separate pieces of equipment: purging device, trap, and desorber. Several complete systems now are available commercially.

1) *Purging device:* See Section 6210B.3a1).

2) *Trap:* See Section 6210C.3a2).

3) *Desorber:* See Section 6210B.3a3).

4) *Assembly:* See Section 6210B.3a4).

b. *Gas chromatograph:*‡ An analytical system complete with a temperature-programmable gas chromatograph suitable for on-column injection and all required accessories including syringes, analytical columns, gases, detector, and strip-chart recorder. A data system is recommended for measuring peak areas.

1) *Column 1,* 2.4 m long × 3 mm ID stainless steel or glass, packed with 1% SP-1000 on Carbopack B (60/80 mesh) or equivalent. The detection limit and precision and bias data presented herein were

developed with this column. For guidelines for the use of alternate column packings, see ¶ 5a.

2) *Column 2,* 1.8 m long × 3 mm ID stainless steel or glass, packed with chemically bonded *n*-octane on Porasil-C (100/120 mesh) or equivalent.

3) *Electrolytic conductivity or microcoulometric detector:* The electrolytic conductivity detector was used to develop detection limit and precision and bias data presented herein. For use of alternate detectors, see ¶ 5a.

c. *Syringes,* 5-mL glass hypodermic with luerlok tip, if applicable to the purging device.

d. *Microsyringes,* 25-μL, 0.15-mm ID needle.

e. *Syringe valve,* two-way, with luer ends.

f. *Syringe,* 5-mL, gastight with shutoff valve.

g. *Bottle,* 15-mL, screw-cap, with TFE cap liner.

h. *Balance,* analytical, capable of accurately weighing 0.0001 g.

i. *Vial,* 25-mL or larger, equipped with a screw cap with a hole in the center.§ To prepare for sampling use, wash with detergent, rinse with tap and distilled water, and dry at 105°C. Similarly treat the TFE-faced silicone septum.‖

4. Reagents

a. *Reagent water:* See Section 6210B.4a.

b. *Trap materials:* See Section 6210B.4b. In addition, coconut charcoal, 6/10 mesh sieved to 26 mesh#, is required.

c. *Methanol,* pesticide quality or equivalent.

d. *Stock standard solutions:* See Section 6210B.4d.

---

‡ Gas chromatographic methods are extremely sensitive to the materials used. Mention of trade names by *Standard Methods* does not preclude the use of other existing or as-yet-undeveloped products that give *demonstrably* equivalent results.

§ Pierce #13075 or equivalent.
‖Pierce #12722 or equivalent.
Barnebey Cheney, CA-580-26 lot M-2649, or equivalent.

e. Secondary dilution standards: See Section 6210B.4*e*.

f. Calibration standards: Prepare as in Section 6210B.4*g*, but use 100, 500, or 1000 mL reagent water.

g. Quality control (QC) check sample: See Section 6210B.4*i*.

5. Procedure

a. Operating conditions: Table 6230:I summarizes recommended operating conditions for the gas chromatograph and gives estimated retention times and MDLs that can be achieved under these conditions. An example of the separations obtained with Column 1 is shown in Figure 6230:1. Other packed columns, chromatographic conditions, or detectors may be used if the requirements for quality control check sample analysis (Section 6210B.7*b*) are met.

b. Calibration: Calibrate system daily as follows:

1) System setup—See Section 6210B.5*c*1)

2) Calibration procedure—Use either the internal standard calibration procedure given in Section 6210B.5*c*2), or the external standard calibration procedure as follows:

External standard calibration procedure—Prepare standards as directed in ¶ 4*f*. Analyze each calibration standard according to procedure for samples, and tabulate peak height or area responses versus concentration in standard. Prepare a calibration curve for each compound. Alternatively, if the ratio of response to concentration (calibration factor) is a constant over the working range (< 10% relative standard deviation, RSD), assume linearity through the origin and use the average ratio or calibration factor in place of a calibration curve.

When internal standard calibration procedure is used, select one or more internal standards similar in analytical behavior to

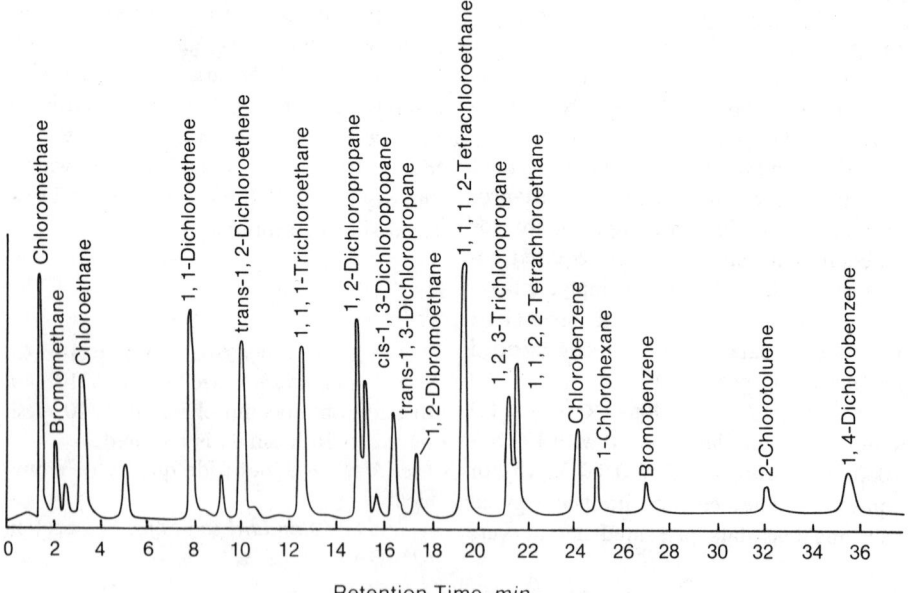

Figure 6230:1. Gas chromatogram of purgeable halocarbons. Column: 1% SP-1000 on Carbopack-B; program: 45°C for 3 min, 8°C/min to 220°C; detector: Hall 700-A electrolytic conductivity.

the compounds of interest and proceed according to Section 6210B.5c2). If the RF value over the working range is a constant (< 10% RSD), the RF can be assumed to be invariant; use average RF for calculations. Alternatively, use the results to plot a calibration curve of response ratios, A_s/A_{is} vs. RF.

3) Calibration check—Use procedure of Section 6210B.5c3) with calibration acceptance criteria of Table 6230:II.

c. Sample analysis: Follow the procedures for purging, desorbing, and trap reconditioning detailed in Section 6210B.5d.

Identify compounds in sample by comparing peak retention times with peaks of

TABLE 6230:II. CALIBRATION AND QC ACCEPTANCE CRITERIA*†

Compound	Range for Q μg/L	Limit for s μg/L	Range for $\overline{X}$ μg/L	Range P, P_s %
Bromodichloromethane	15.2–24.8	4.3	10.7–32.0	42–172
Bromoform	14.7–25.3	4.7	5.0–29.3	13–159
Bromomethane	11.7–28.3	7.6	3.4–24.5	D–144
Carbon tetrachloride	13.7–26.3	5.6	11.8–25.3	43–143
Chlorobenzene	14.4–25.6	5.0	10.2–27.4	38–150
Chloroethane	15.4–24.6	4.4	11.3–25.2	46–137
2-Chloroethylvinyl ether	12.0–28.0	8.3	4.5–35.5	14–186
Chloroform	15.0–25.0	4.5	12.4–24.0	49–133
Chloromethane	11.9–28.1	7.4	D–34.9	D–193
Dibromochloromethane	13.1–26.9	6.3	7.9–35.1	24–191
1,2-Dichlorobenzene	14.0–26.0	5.5	1.7–38.9	D–208
1,3-Dichlorobenzene	9.9–30.1	9.1	6.2–32.6	7–187
1,4-Dichlorobenzene	13.9–26.1	5.5	11.5–25.5	42–143
1,1-Dichloroethane	16.8–23.2	3.2	11.2–24.6	47–132
1,2-Dichloroethane	14.3–25.7	5.2	13.0–26.5	51–147
1,1-Dichloroethene	12.6–27.4	6.6	10.2–27.3	28–167
trans-1,2-Dichloroethene	12.8–27.2	6.4	11.4–27.1	38–155
1,2-Dichloropropane	14.8–25.2	5.2	10.1–29.9	44–156
cis-1,3-Dichloropropene	12.8–27.2	7.3	6.2–33.8	22–178
trans-1,3-Dichloropropene	12.8–27.2	7.3	6.2–33.8	22–178
Methylene chloride	15.5–24.5	4.0	7.0–27.6	25–162
1,1,2,2-Tetrachloroethane	9.8–30.2	9.2	6.6–31.8	8–184
Tetrachloroethene	14.0–26.0	5.4	8.1–29.6	26–162
1,1,1-Trichloroethane	14.2–25.8	4.9	10.8–24.8	41–138
1,1,2-Trichloroethane	15.7–24.3	3.9	9.6–25.4	39–136
Trichloroethene	15.4–24.6	4.2	9.2–26.6	35–146
Trichlorofluoromethane	13.3–26.7	6.0	7.4–28.1	21–156
Vinyl chloride	13.7–26.3	5.7	8.2–29.9	28–163

* Criteria calculated assuming a QC check sample concentration of 20 μg/L.
† Q = concentration measured in QC check sample,
  s = standard deviation of four recovery measurements,
  $\overline{X}$ = average recovery for four recovery measurements,
  P, P_s = recovery measured,
  D = detected; result must be greater than zero.

NOTE: These criteria are based directly on the method performance data in Table 6230:III. Where necessary, the limits for recovery were broadened to assure applicability of the limits to concentrations below those used to develop Table 6230:III.

standard chromatograms. Base the width of retention time window used to make identifications on measurements of actual retention time variations of standards over the course of a day. To calculate a suggested window size, use three times the standard deviation of a retention time for a compound. Analyst's experience is important in interpreting chromatograms.

If response for a peak exceeds the working range of the system, dilute sample from the portion in the second syringe with reagent water and reanalyze. If sample concentrations are above 1000 × MDL the analyst may wish to use direct aqueous injection.

6. Calculation

See Section 6220C.6.

7. Quality Control

a. Quality control program: The analyst should be experienced in operation of purge and trap and gas chromatographic systems and in interpretation of gas chromatograms.

For requirements of a quality-control program, see Section 6210B.7a.

b. Analyst proficiency check: Proceed according to Section 6210B.7b. Use acceptance criteria given in Table 6230:II.

c. Analyses of samples with known additions: On an ongoing basis, make known additions to at least 10% of the samples from each sample site being monitored. For laboratories analyzing one to ten samples per month, analyze at least one such sample with a known addition per month.

Proceed as in Section 6210B.7c, using acceptance criteria of Table 6230:II. To calculate optional acceptance criteria, use the equation in Table 6230:III to calculate bias (X') and substitute the addition concentration (T) for C. Calculate overall precision (S') using the equation in Table 6230:III, substituting X' for $\overline{X}'$. Calculate

the range for recovery at the addition concentration as $(100\ X'/T\ \pm\ 2.44\ (100\ S'/T)\%$.[7]**

If any individual P falls outside the designated range for recovery, analysis for that constituent has failed the acceptance criteria. Analyze a check standard containing each compound that failed the criteria as described in ¶ *d* below.

d. Quality-control check standard analysis: If analysis for any compound fails the acceptance criteria for recovery, prepare and analyze a QC check standard containing each compound that failed according to the procedure given in Section 6210B.7d. NOTE: The frequency for the required analysis of a QC check standard will depend on the number of compounds being tested for simultaneously, the complexity of the sample matrix, and the performance of the laboratory. If the entire list of compounds in Table 6230:II is measured in the sample in ¶ *c* above, the probability that the analysis of a QC check standard will be required is high; therefore, routinely analyze the QC check standard with the known-addition sample.

e. Bias assessment and records: See Section 6210B.7e.

f. Use of surrogate compounds: Monitor both the performance of the analytical system and the effectiveness of the method in dealing with each sample matrix by adding surrogate compounds to each sample, standard, and reagent water blank. A combination of bromochloromethane, 2-bromo-1-chloropropane, and 1,4-dichlorobutane covers the range of the temperature program used in this method. From stock standard solutions prepared as in 6210B.4d, add a volume to give 750 μg of each surrogate to 45 mL reagent water contained in a 50-mL volumetric flask, mix, and dilute to volume for a concentration of 15 ng/μL. Add 10 μL of this surrogate

---

** The value 2.44 is two times the value 1.22 derived in the reference.[7]

TABLE 6230:III. METHOD BIAS AND PRECISION AS FUNCTIONS OF CONCENTRATION*

Compound	Bias, as Recovery, X' $\mu g/L$	Single-Analyst Precision, s' $\mu g/L$	Overall Precision, S' $\mu g/L$
Bromodichloromethane	$1.12C - 1.02$	$0.11\overline{X} + 0.04$	$0.20\overline{X} + 1.00$
Bromoform	$0.96C - 2.05$	$0.12\overline{X} + 0.58$	$0.21\overline{X} + 2.41$
Bromomethane	$0.76C - 1.27$	$0.28\overline{X} + 0.27$	$0.36\overline{X} + 0.94$
Carbon tetrachloride	$0.98C - 1.04$	$0.15\overline{X} + 0.38$	$0.20\overline{X} + 0.39$
Chlorobenzene	$1.00C - 1.23$	$0.15\overline{X} - 0.02$	$0.18\overline{X} + 1.21$
Chloroethane	$0.99C - 1.53$	$0.14\overline{X} - 0.13$	$0.17\overline{X} + 0.63$
2-Chloroethylvinyl ether†	$1.00C$	$0.20\overline{X}$	$0.35\overline{X}$
Chloroform	$0.93C - 0.39$	$0.13\overline{X} + 0.15$	$0.19\overline{X} - 0.02$
Chloromethane	$0.77C + 0.18$	$0.28\overline{X} - 0.31$	$0.52\overline{X} + 1.31$
Dibromochloromethane	$0.94C + 2.72$	$0.11\overline{X} + 1.10$	$0.24\overline{X} + 1.68$
1,2-Dichlorobenzene	$0.93C + 1.70$	$0.20\overline{X} + 0.97$	$0.13\overline{X} + 6.13$
1,3-Dichlorobenzene	$0.95C + 0.43$	$0.14\overline{X} + 2.33$	$0.26\overline{X} + 2.34$
1,4-Dichlorobenzene	$0.93C - 0.09$	$0.15\overline{X} + 0.29$	$0.20\overline{X} + 0.41$
1,1-Dichloroethane	$0.95C - 1.08$	$0.08\overline{X} + 0.17$	$0.14\overline{X} + 0.94$
1,2-Dichloroethane	$1.04C - 1.06$	$0.11\overline{X} + 0.70$	$0.15\overline{X} + 0.94$
1,1-Dichloroethene	$0.98C - 0.87$	$0.21\overline{X} - 0.23$	$0.29\overline{X} - 0.40$
trans-1,2-Dichloroethene	$0.97C - 0.16$	$0.11\overline{X} + 1.46$	$0.17\overline{X} + 1.46$
1,2-Dichloropropane†	$1.00C$	$0.13\overline{X}$	$0.23\overline{X}$
cis-1,3-Dichloropropene†	$1.00C$	$0.18\overline{X}$	$0.32\overline{X}$
trans-1,3-Dichloropropene†	$1.00C$	$0.18\overline{X}$	$0.32\overline{X}$
Methylene chloride	$0.91C - 0.93$	$0.11\overline{X} + 0.33$	$0.21\overline{X} + 1.43$
1,1,2,2-Tetrachloroethene	$0.95C + 0.19$	$0.14\overline{X} + 2.41$	$0.23\overline{X} + 2.79$
Tetrachloroethene	$0.94C + 0.06$	$0.14\overline{X} + 0.38$	$0.18\overline{X} + 2.21$
1,1,1-Trichloroethane	$0.90C - 0.16$	$0.15\overline{X} + 0.04$	$0.20\overline{X} + 0.37$
1,1,2-Trichloroethane	$0.86C + 0.30$	$0.13\overline{X} - 0.14$	$0.19\overline{X} + 0.67$
Trichloroethene	$0.87C + 0.48$	$0.13\overline{X} - 0.03$	$0.23\overline{X} + 0.30$
Trichlorofluoromethane	$0.89C - 0.07$	$0.15\overline{X} + 0.67$	$0.26\overline{X} + 0.91$
Vinyl chloride	$0.97C - 0.36$	$0.13\overline{X} + 0.65$	$0.27\overline{X} + 0.40$

* X' = expected recovery for one or more measurements of a sample containing a concentration of C,
   s' = expected single-analyst standard deviation of measurements at an average concentration found of $\overline{X}$,
   S' = expected interlaboratory standard deviation of measurements at an average concentration found of $\overline{X}$,
   C = true value for the concentration, $\mu g/L$,
   $\overline{X}$ = average recovery found for measurements of samples containing a concentration of C, $\mu g/L$.
† Estimates based on performance in a single laboratory.[8]

addition solution directly into the 5-mL syringe with every sample and reference standard analyzed. Prepare a fresh surrogate solution weekly. If the internal standard calibration procedure is being used, the surrogate compounds may be added directly to the internal standard addition solution, Section 6210B.5c2).

g. *Additional quality-assurance practices:* Other desirable practices depend on the needs of the laboratory and the nature of the samples. MDLs should be verified periodically to prevent an unrecognized rise in MDL due to declining detector response. Field duplicates may be analyzed to assess the precision of the environmental measurements. When doubt exists over the identification of a chromatogram peak, use confirmatory techniques such as gas chromatography with a dissimilar column, specific element detector, or mass spectrometer. Whenever possible, analyze standard reference materials and participate in relevant performance evaluation studies.

8. Precision and Bias

This method was tested by 20 laboratories using reagent water, drinking water, surface water, and three industrial wastewaters with additions at six concentrations over the range 8.0 to 500 μg/L.[6] Single-operator precision, overall precision, and method bias were found to be related directly to the compound concentration and essentially independent of the sample matrix. Linear equations describing these relationships are presented in Table 6230:III.

9. References

1. U.S. ENVIRONMENTAL PROTECTION AGENCY. 1984. Method 601—Purgeable halocarbons. 40 CFR Part 136, 43261; *Federal Register* 49, No. 209.
2. U.S. ENVIRONMENTAL PROTECTION AGENCY. 1984. Method 624—Purgeables. 40 CFR Part 136, 43373; *Federal Register* 49, No. 209.
3. BELLAR, T.A. & J.J. LICHTENBERG. 1974. Determining volatile organics at microgram-per-litre-levels by gas chromatography. *J. Amer. Water Works Assoc.* 66: 739.
4. BELLAR, T.A. & J.J. LICHTENBERG. 1978. Semi-automated headspace analysis of drinking waters and industrial waters for purgeable volatile organic compounds. *In* C. E. Van Hall, ed. Measurement of Organic Pollutants in Water and Wastewater. STP 686, American Soc. Testing & Materials, Philadelphia, Pa.
5. U.S. ENVIRONMENTAL PROTECTION AGENCY. 1984. Definition and procedure for the determination of the method detection limit. 40 CFR Part 136, Appendix B. *Federal Register* 49, No. 209.
6. U.S. ENVIRONMENTAL PROTECTION AGENCY. 1984. EPA Method Study 24, Method 601—Purgeable Halocarbons by the Purge and Trap Method. EPA-600/4-84-064, National Technical Information Serv., PB84-212448, Springfield, Va.
7. PROVOST, L.P. & R.S. ELDER. 1983. Interpretation of percent recovery data. *Amer. Lab.* 15:58.
8. POTTER, B. 1983. Method validation data for EPA Method 601 (memorandum). U.S. Environmental Protection Agency, Environmental Monitoring and Support Lab., Cincinnati, Ohio.

6230 C. Purge and Trap Packed-Column Gas Chromatographic Method II*

This method[1] is applicable to the determination of purgeable halocarbons† in finished drinking water, raw source water, or drinking water in any treatment stage. The method varies only slightly from that of Section 6230B, which is intended for use with wastewater samples.

1. General Discussion

a. Principle: See Sections 6210B.1*a* and 6230B.1*a*.

b. Interferences: See Section 6210C.1*b*.

c. Detection limits: In a single laboratory using reagent water and known additions of 0.2 μg/L, observed method detection limits (MDLs)[2] for these compounds were in the range of 0.01 to 0.05 μg/L, depending on the compound. Some laboratories may not be able to achieve these detection limits because results depend on instrument sensitivity and matrix effects. Analysis of complex mixtures containing partially resolved compounds may be hampered by concentration differences larger than a factor of 10. This problem commonly occurs when analyzing finished drinking waters because of the relatively high trihalomethane content. When such samples are analyzed, chloroform will af-

fect the method detection limit for 1,2-dichloroethane.

d. Safety: The toxicity or carcinogenicity of each reagent has not been defined precisely. Carbon tetrachloride, 1,2-dichloroethane, 1,1,2,2-tetrachloroethane, 1,1,2-trichloroethane, chloroform, 1,2-dibromoethane, tetrachloroethene, trichloroethene, and vinyl chloride have been classified tentatively as known or suspected human or mammalian carcinogens. Prepare primary standards of these compounds in a hood and wear a NIOSH/MESA-approved toxic gas respirator when handling high concentrations.

2. Sampling and Storage

See Section 6010B.1.

3. Apparatus

a. Purge and trap system: The purge and trap system consists of three separate pieces of equipment: purging device, trap, and desorber. Several complete systems now are available commercially.

1) *Purging device:* See Section 6210C.3*a*1).

2) *Trap:* See Section 6210C.3*a*2). If only compounds boiling above 35°C are to be analyzed, both the silica gel and charcoal can be eliminated and the polymer increased to fill the entire trap.

3) *Desorber:* See Section 6210B.3*a*3). Trap failure is characterized by a pressure drop above 21 kPa across the trap during purging or by poor bromoform sensitivities.

4) *Assembly:* See Figures 6210:3 and 4.

b. Gas chromatograph:‡ See Section 6230B.3*b*.

---

* Approved by Standard Methods Committee, 1987. Accepted by U.S. Environmental Protection Agency as equivalent to EPA Method 502.1.
† Bromobenzene, bromochloromethane, bromodichloromethane, bromoform, bromomethane, carbon tetrachloride, chlorobenzene, chloroethane, chloroform, chloromethane, 2-chlorotoluene, 4-chlorotoluene, dibromochloromethane, 1,2-dibromoethane, dibromomethane, 1,2-dichlorobenzene, 1,3-dichlorobenzene, 1,4-dichlorobenzene, dichlorodifluoromethane, 1,1-dichloroethane, 1,2-dichloroethane, 1,1-dichloroethene, *cis*-1,2-dichloroethene, *trans*-1,2-dichloroethene, 1,2-dichloropropane, 1,3-dichloropropane, 2,2-dichloropropane, 1,1-dichloropropene, methylene chloride, 1,1,1,2-tetrachloroethane, 1,1,2,2-tetrachloroethane, tetrachloroethene, 1,1,1-trichloroethane, 1,1,2-trichloroethane, trichloroethene, trichlorofluoromethane, 1,2,3-trichloropropane, and vinyl chloride.

‡ Gas chromatographic methods are extremely sensitive to the materials used. Mention of trade names by *Standard Methods* does not preclude the use of other existing or as-yet-undeveloped products that give *demonstrably* equivalent results.

1) *Column 1:* See Section 6230B.3*b*1) and use conditions given beneath Table 6230:I. For measures to protect the column, see Section 6210C.3*b*.

A sample chromatogram obtained with Column 1 is presented in Figure 6230:1.

2) *Column 2:* See Section 6230B.3*b*2) and use conditions given beneath Table 6230:I. A sample chromatogram obtained with Column 2 is presented in Figure 6230:2.

3) *Electrolytic conductivity or microcoulometric detector:* Halogen-specific systems eliminate misidentifications due to non-organohalides that may be coextracted during the purge step.

4) *Photoionization detector:* A photoionization detector may be inserted between the analytical column and the halide detector to analyze simultaneously for aromatic or unsaturated volatile organic compounds. Most compounds detectable by the method of Section 6220C can be measured this way.

c. Syringes, 5-mL glass hypodermic with luerlok tip.

d. Other equipment: See Section 6210C.3*g–j*.

4. Reagents

See Section 6210C.4*a–h*.

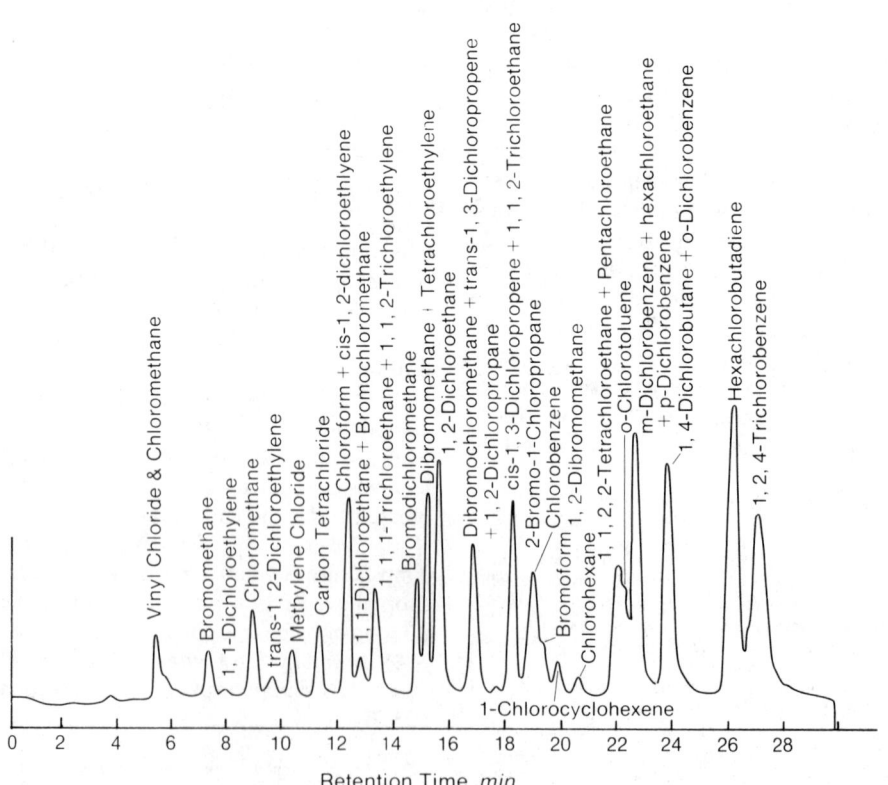

Figure 6230:2. Gas chromatogram of purgeable halocarbons. Column: n-Octane on Porasil-C; program: 50°C for 3 min, 6°C/min to 170°C, detector: electrolytic conductivity.

5. Procedure

a. Operating conditions: Table 6230:I summarizes recommended operating conditions for the gas chromatograph and Table 6230:IV gives estimated retention times.

TABLE 6230:IV. RETENTION TIMES FOR ORGANOHALIDES

Compound	Retention Time s	
	Column 1	Column 2
Chloromethane	90	317
Bromomethane	130	423
Dichlorodifluoromethane	157	nd
Vinyl chloride	160	317
Chloroethane	200	521
Methylene chloride	315	607
Trichlorofluoromethane	431	nd
1,1-Dichloroethene	476	463
Bromochloromethane	509	760
1,1-Dichloroethane	558	754
trans-1,2-Dichloroethene	605	563
cis-1,2-Dichloroethene	605	726
Chloroform	641	725
1,2-Dichloroethane	684	921
Dibromomethane	698	895
1,1,1-Trichloroethane	756	786
Carbon tetrachloride	781	664
Bromodichloromethane	819	877
Dichloroacetonitrile	884	nd
1,2-Dichloropropane	895	997
1,1-Dichloropropene	904	nd
Trichloroethene	948	787
1,3-Dichloropropane	973	nd
Dibromochloromethane	989	997
1,1,2-Trichloroethane	991	1084
1,2-Dibromoethane	1046	1131
2-Chloroethylethyl ether	1056	nd
2-Chloroethylvinyl ether	1080	nd
Bromoform	1154	1150
1,1,1,2-Tetrachloroethane	1163	1302
1,2,3-Trichloropropane	1279	nd
Chlorocyclohexane	1283	nd
1,1,2,2-Tetrachloroethane	1297	nd
Tetrachloroethene	1300	898
1-Chlorocyclohexene	1345	1186
Chlorobenzene	1451	1130
1,2-Dibromo-3-chloropropane	1560	nd
Bromobenzene	1626	nd
2-Chlorotoluene	1927	1320
1,3-Dichlorobenzene	2042	1346
1,2-Dichlorobenzene	2094	1411
1,4-Dichlorobenzene	2127	1340

nd = not determined.

b. Calibration:

1) External standard calibration procedure — Prepare standards as directed in Section 6210C.4*h*. Starting with the least concentrated standard, analyze each calibration standard according to Section 6230B.5*c* and proceed as directed in Section 6220C.5*b*1).

2) Internal standard calibration procedure — Alternatively, use internal standard calibration techniques. Either 2-bromo-1-chloropropane or 1,4-dichlorobutane is recommended. Add internal standard to sample just before purging. Check validity of internal standard calibration factors daily by analyzing a calibration standard. To calibrate for vinyl chloride, use procedure given in Section 6210C.5*c*, but use 5 mL in purging device.

c. Instrument performance: Check performance of entire analytical system daily using analyses of reagent blanks, standards, duplicate samples, and the laboratory control standard (see below). All peaks in standard chromatograms shall be sharp and symmetrical. Correct any peak tailing significantly in excess of that shown in the method chromatograms. Tailing problems generally are traceable to active sites on the GC column or the detector operation. If only the compounds eluting before chloroform give random responses or unusually wide peak widths, are poorly resolved, or are missing, the problem usually is traceable to the trap/desorber. If only brominated compounds show poor peak geometry or do not respond properly at low concentrations, repack the trap. Excessive detector reactor temperatures also can cause low bromoform response. If negative peaks appear in the chromatogram, replace both the ion exchange column and electrolyte in the detector.

Check precision between replicate analyses. A properly operating system shows an average relative standard deviation of less than 10%. Poor precision generally is traceable to pneumatic leaks, especially around sample purger and detector reactor inlet and exit, electronic problems, or sampling and storage problems. Monitor retention times for each organohalide using calibration standards and the laboratory control standard. If individual retention times vary by more than 10% over an 8-h period or do not fall within 10% of an established norm, locate and correct source of retention data variance.

d. Sample analysis: Follow procedure detailed in Section 6210B.5*d*.

6. Calculation

Identify each organohalide in the sample chromatogram by comparing retention time of suspect peak to retention times generated by the calibration standards and the laboratory control standard. See Section 6230B.6.

7. Quality Control

Using a check sample concentrate of 100 times the MCL or 1 μg/mL rather than 500 times the MCL or 5 μg/mL and 5 mL rather than 25-mL portions, follow the procedure of Section 6210C.7.

8. Precision and Bias

Single-laboratory precision and bias for organohalides in Ohio River water and carbon-filtered tap water are presented in Table 6230:V.[1] This method was tested by 20 laboratories using drinking water with added organohalides at six concentrations between 8 and 505 μg/L. Single-operator precision, overall precision, and method bias were found to be related directly to the compound concentration. Linear equations describing these relationships are presented in Table 6230:VI.[3]

9. References

1. U.S. ENVIRONMENTAL PROTECTION AGENCY. 1985. Methods for the Determination of Organic Compounds in Finished Drinking Water

TABLE 6230:V. SINGLE-LABORATORY BIAS AND PRECISION FOR ORGANOHALIDES IN OHIO RIVER WATER AND DRINKING WATER

Compound	Addition $\mu g/L$	Bias as Average Recovery %	Number of Samples	Standard Deviation $\mu g/L$	Relative Standard Deviation %
Bromobenzene	0.40	93	20	0.047	12
Bromochloromethane	0.40	90	19	0.038	9.5
Bromodichloromethane	0.20	100	17	0.013	6.5
Bromoform	0.20	95	17	0.030	15.0
Carbon tetrachloride	0.20	90	17	0.014	7.0
Chlorobenzene	0.40	88	18	0.037	9.3
Chlorocyclohexane	0.40	93	21	0.033	8.3
1-Chlorocyclohexene	0.40	93	21	0.051	12.8
Chloroethane	0.40	93	20	0.071	18
2-Chloroethylethyl ether	0.40	95	18	0.030	7.5
Chloromethane	0.40	93	16	0.034	8.5
2-Chlorotoluene	0.40	85	20	0.037	9.3
Dibromochloromethane	0.20	95	17	0.014	7.0
1,2-Dibromoethane	0.40	93	18	0.050	12.5
Dibromomethane	0.40	100	5	0.032	8.0
1,2-Dichlorobenzene	0.40	95	21	0.053	13
1,3-Dichlorobenzene	0.40	95	21	0.033	8.3
1,4-Dichlorobenzene	0.40	90	20	0.051	13
Dichlorodifluoromethane	0.40	103	12	0.081	20
1,1-Dichloroethane	0.20	95	17	0.012	6.0
1,2-Dichloroethane	0.20	110	17	0.014	7.0
1,1-Dichloroethene	0.40	88	18	0.027	9.3
1,2-Dichloroethene*	0.40	88	20	0.028	7.0
1,2-Dichloropropane	0.40	95	20	0.014	3.5
1,3-Dichloropropane	0.40	98	21	0.026	6.5
1,1-Dichloropropene	0.40	88	18	0.037	9.3
Methylene chloride	0.20	85	17	0.024	12.0
1,1,1,2-Tetrachloroethane	0.40	93	20	0.032	8.0
1,1,2,2-Tetrachloroethane	0.40	95	18	0.036	9.0
Tetrachloroethene	0.20	90	17	0.019	9.5
1,1,1-Trichloroethane	0.40	93	20	0.032	8.0
1,1,2-Trichloroethane	0.40	95	15	0.024	6.0
Trichloroethene	0.20	94	17	0.012	6.0
Trichlorofluoromethane	0.40	90	21	0.037	9.3
1,2,3-Trichloropropane	0.40	100	20	0.038	9.5
Vinyl chloride	0.20	110	12	0.029	15

* = includes *cis-* and *trans-* isomers.
Data obtained using a Tracor Hall Model 700-A detector, with following operating conditions:

Reactor tube:	Nickel 1.6 mm OD
Reactor temperature:	810°C
Reactor base temperature:	250°C
Electrolyte:	100% *n*-propyl alcohol
Electrolyte flow rate:	0.8 mL/min
Reaction gas:	Hydrogen at 40 mL/min
Carrier gas:	Helium at 40 mL/min

TABLE 6230:VI. BIAS, SINGLE-ANALYST PRECISION, AND OVERALL PRECISION FOR ORGANOHALIDES IN DRINKING WATER*

Compound	Bias as Mean Recovery $(\overline{X})$ $\mu g/L$	Single-Analyst Precision $\mu g/L$	Overall Precision $\mu g/L$
Bromodichloromethane	$1.00C + 0.96$	$0.13\overline{X} - 1.41$	$0.18\overline{X} + 3.06$
Bromoform	$1.02C - 1.81$	$0.10\overline{X} + 0.20$	$0.24\overline{X} + 1.25$
Carbon tetrachloride	$1.00C - 2.20$	$0.10\overline{X} + 1.57$	$0.20\overline{X} + 1.09$
Chlorobenzene	$1.00C - 1.39$	$0.07\overline{X} + 1.71$	$0.16\overline{X} + 1.43$
Chloroethane	$1.08C - 1.97$	$0.07\overline{X} + 0.65$	$0.19\overline{X} + 0.39$
Chloroform	$0.90C + 3.44$	$0.05\overline{X} + 5.58$	$0.09\overline{X} + 6.21$
Chloromethane	$0.91C - 0.99$	$0.28\overline{X} + 0.27$	$0.49\overline{X} + 1.51$
Dibromochloromethane	$0.98C + 2.89$	$0.10\overline{X} + 1.55$	$0.23\overline{X} + 0.91$
1,2-Dichlorobenzene	$0.91C + 1.12$	$0.12\overline{X} + 2.02$	$0.17\overline{X} + 2.26$
1,3-Dichlorobenzene	$0.91C - 0.13$	$0.15\overline{X} + 0.64$	$0.24\overline{X} + 1.48$
1,4-Dichlorobenzene	$0.91C + 0.26$	$0.09\overline{X} + 0.39$	$0.15\overline{X} + 0.39$
1,1-Dichloroethane	$0.93C - 2.04$	$0.09\overline{X} + 0.47$	$0.18\overline{X} + 1.13$
1,2-Dichloroethane	$1.03C - 0.41$	$0.06\overline{X} + 1.69$	$0.18\overline{X} + 1.21$
1,1-Dichloroethene	$1.03C - 1.16$	$0.12\overline{X} + 0.13$	$0.31\overline{X} - 0.71$
trans-1,2-Dichloroethene	$0.98C - 1.02$	$0.16\overline{X} + 0.29$	$0.24\overline{X} + 0.95$
1,2-Dichloropropane	$0.98C + 1.19$	$0.19\overline{X} - 0.61$	$0.27\overline{X} - 0.10$
Methylene chloride	$0.97C - 1.50$	$0.08\overline{X} + 1.04$	$0.17\overline{X} + 2.43$
1,1,2,2-Tetrachloroethane	$0.92C - 0.82$	$0.09\overline{X} - 1.42$	$0.20\overline{X} + 1.65$
Tetrachloroethene	$0.96C + 0.35$	$0.17\overline{X} + 0.96$	$0.25\overline{X} + 0.58$
1,1,1-Trichloroethane	$0.92C + 0.02$	$0.14\overline{X} - 0.33$	$0.27\overline{X} - 0.76$
1,1,2-Trichloroethane	$0.84C + 0.83$	$0.06\overline{X} + 0.99$	$0.19\overline{X} + 0.69$
Trichloroethene	$0.92C - 0.10$	$0.13\overline{X} + 0.23$	$0.32\overline{X} - 0.57$
Trichlorofluoromethane	$0.92C + 1.21$	$0.22\overline{X} + 0.03$	$0.30\overline{X} + 0.64$
Vinyl chloride	$1.06C - 1.86$	$0.14\overline{X} - 0.17$	$0.32\overline{X} + 0.07$

*$\overline{X}$ = mean recovery, $\mu g/L$,
　C = true value for the concentration, $\mu g/L$.
For operating conditions, see footnote to Table 6230:V.

and Raw Source Water (rev. September 1986). Modified from: U.S. Environmental Protection Agency. 1981. The determination of halogenated chemicals in water by the purge and trap method, Method 502.1. Environmental Monitoring and Support Lab., Cincinnati, Ohio.

2. GLASER, J.A., D.L. FOERST, G.D. MCKEE, S.A. QUAVE & W.L. BUDDE. 1981. Trace analyses for wastewaters. *Environ. Sci. Technol.* 15:1426.

3. U.S. ENVIRONMENTAL PROTECTION AGENCY. 1984. EPA Method Study 24, Method 601— Purgeable Halocarbons by the Purge and Trap Method. EPA-600/4-84-064, National Technical Information Serv., PB84-212448, Springfield, Va.

6230 D. Purge and Trap Capillary-Column Gas Chromatographic Method*

This procedure[1] is applicable to determining those compounds listed for Method 6230C as well as several others.† It differs primarily in the type of gas chromatographic column used, requiring a wide-bore capillary column and the detector, requiring a high-temperature photoionization detector in series with either an electrolytic conductivity or microcoulometric detector.

1. General Discussion

See Section 6230C.1.

2. Sampling and Storage

See Section 6010B.1.

3. Apparatus

Use the equipment specified in Section 6230C.3 with the following exceptions:

a. *Gas chromatographic column:* See Section 6210D.3a1) and 2).

b. *Photoionization detector:* High-temperature photoionization detector with 10.2-eV lamp.‡

4. Reagents

See Section 6210C.4a–h.

5. Procedure

Retention times are given in Table 6230:VII. Follow the procedure specified

---

* Approved by Standard Methods Committee, 1987. Accepted by U.S. Environmental Protection Agency as equivalent to EPA Method 502.2.
† Benzene, n-butylbenzene, sec-butylbenzene, tert-butylbenzene, 1,2-dibromo-3-chloropropane, ethylbenzene, hexachlorobutadiene, isopropylbenzene, p-isopropyltoluene, naphthalene, n-propylbenzene, styrene, toluene, 1,2,3-trichlorobenzene, 1,2,4-trichlorobenzene, 1,2,4-trimethylbenzene, 1,3,5-trimethylbenzene, o-xylene, m-xylene, and p-xylene.
‡ Tracor Model 703 or equivalent.

TABLE 6230:VII. RETENTION TIMES FOR VOLATILE ORGANIC COMPOUNDS ON PHOTOIONIZATION DETECTOR (PID) AND ELECTROLYTIC CONDUCTIVITY DETECTOR (ECD)

Compound	Retention Time* *min*	
	PID	ECD
Dichlorodifluoromethane	—	8.47
Chloromethane	—	9.47
Vinyl chloride	9.88	9.93
Bromomethane	—	11.95
Chloroethane	—	12.37
Trichlorofluoromethane	—	13.49
1,1-Dichloroethene	16.14	16.18
Methylene chloride	—	18.39
trans-1,2-Dichloroethene	19.30	19.33
1,1-Dichloroethane	—	20.99
2,2-Dichloropropane	—	22.88
cis-1,2-Dichloroethane	23.11	23.14
Chloroform	—	23.64
Bromochloromethane	—	24.16

TABLE 6230:VII, CONT.

Compound	Retention Time* min	
	PID	ECD
1,1,1-Trichloroethane	—	24.77
1,1-Dichloropropene	25.21	25.24
Carbon tetrachloride	—	25.47
Benzene	26.10	—
1,2-Dichloroethane	—	26.27
Trichloroethene	27.99	28.02
1,2-Dichloropropane	—	28.66
Bromodichloromethane	—	29.43
Dibromomethane	—	29.59
Toluene	31.95	—
1,1,2-Trichloroethane	—	33.21
Tetrachloroethene	33.88	33.90
1,3-Dichloropropane	—	34.00
Dibromochloromethane	—	34.73
1,2-Dibromoethane	—	35.34
Chlorobenzene	36.56	36.59
Ethylbenzene	36.72	—
1,1,1,2-Tetrachloroethane	—	36.80
m-Xylene	36.98	—
p-Xylene	36.98	—
o-Xylene	38.39	—
Styrene	38.57	—
Isopropylbenzene	39.58	—
Bromoform	—	39.75
1,1,2,2-Tetrachloroethane	—	40.35
1,2,3-Trichloropropane	—	40.81
n-Propylbenzene	40.87	—
Bromobenzene	40.99	41.03
1,3,5-Trimethylbenzene	41.41	—
2-Chlorotoluene	41.41	41.45
4-Chlorotoluene	41.60	41.63
tert-Butylbenzene	42.92	—
1,2,4-Trimethylbenzene	42.71	—
sec-Butylbenzene	43.31	—
p-Isopropyltoluene	43.81	—
1,3-Dichlorobenzene	44.08	44.11
1,4-Dichlorobenzene	44.43	44.47
n-Butylbenzene	45.20	—
1,2-Dichlorobenzene	45.71	45.74
1,2-Dibromo-3-chloropropane	—	48.57
1,2,4-Trichlorobenzene	51.43	51.46
Hexachlorobutadiene	51.92	51.96
Naphthalene	52.38	—
1,2,3-Trichlorobenzene	53.34	53.37
Internal standards:		
Fluorobenzene	26.84	—
2-Bromo-1-chloropropane	—	33.08

* Dash indicates detector does not respond.

in Section 6230C.5 with the following exceptions:

Cool column oven to less than 10°C and augment carrier gas flow with 24 mL helium/min before entering photoionization detector. Use helium carrier gas at flow rate of 6 mL/min through capillary column.

Hold column temperature at 10°C for 8 min, then program at 4°C/min to 180°C and hold until all compounds elute.

Connect the photoionization and electroconductivity detectors in series with a short piece of uncoated capillary tubing, 0.32 to 0.5 mm ID. See Figure 6230:3.

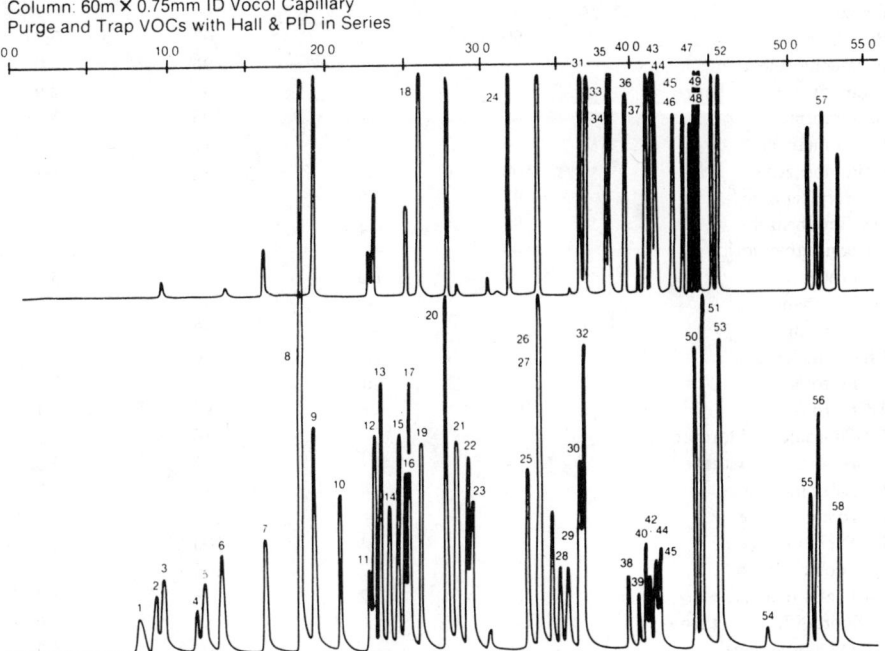

Column: 60m × 0.75mm ID Vocol Capillary
Purge and Trap VOCs with Hall & PID in Series

Figure 6230:3. Dual chromatogram of volatile organics (capillary column) with photoionization (above) and electrolytic conductivity (below) detectors. Column: 60 m × 0.75 mm ID Vocol capillary; purge and trap VOCs with Hall & PID in series.

KEY:
1—dichlorodifluoromethane, 2—chloromethane, 3—vinyl chloride, 4—bromomethane, 5—chloroethane, 6—trichlorofluoromethane, 7—1,1-dichloroethene, 8—methylene chloride, 9—trans-1,2-dichloroethene, 10—1,1-dichloroethane, 11—2,2-dichloropropane, 12—cis-1,2-dichloroethene, 13—chloroform, 14—bromochloromethane, 15—1,1,1-trichloroethane, 16—1,1-dichloropropene, 17—carbon tetrachloride, 18—benzene, 19—1,2-dichloroethane, 20—trichloroethene, 21—1,2-dichloropropane, 22—bromodichloromethane, 23—dibromomethane, 24—toluene, 25—1,1,2-trichloroethane, 26—tetrachloroethene, 27—1,3-dichloropropane, 28—dibromochloromethane, 29—1,2-dibromoethane, 30—chlorobenzene, 31—ethylbenzene, 32—1,1,1,2-tetrachloroethane, 33—m-xylene, 34—p-xylene, 35—o-xylene, 36—styrene, 37—isopropylbenzene, 38—bromoform, 39—1,1,2,2-tetrachloroethane, 40—1,2,3-trichloropropane, 41—n-propylbenzene, 42—bromobenzene, 43—1,3,5-trimethylbenzene, 44—2-chlorotoluene, 45—4-chlorotoluene, 46—tert-butylbenzene, 47—1,2,4-trimethylbenzene, 48—sec-butylbenzene, 49—p-isopropyltoluene, 50—1,3-dichlorobenzene, 51—1,4-dichlorobenzene, 52—n-butylbenzene, 53—1,2-dichlorobenzene, 54—1,2-dibromo-3-chloropropane, 55—1,2,4-trichlorobenzene, 56—hexachlorobutadiene, 57—naphthalene, 58—1,2,3-trichlorobenzene.

TABLE 6230:VIII. SINGLE-LABORATORY BIAS AND PRECISION FOR VOLATILE ORGANIC COMPOUNDS
IN REAGENT WATER*

Compound	Photoionization Detector		Electrolytic Conductivity Detector	
	Recovery %	Standard Deviation of Recovery $\mu g/L$	Recovery* %	Standard Deviation of Recovery $\mu g/L$
Benzene	99	1.2	—	—
Bromobenzene	99	1.7	97	2.7
Bromochloromethane	—	—	96	3.0
Bromodichloromethane	—	—	97	2.9
Bromoform	—	—	106	5.5
Bromomethane	—	—	97	3.7
n-Butylbenzene	100	4.4	—	—
sec-Butylbenzene	97	2.6	—	—
tert-Butylbenzene	98	2.3	—	—
Carbon tetrachloride	—	—	92	3.3
Chlorobenzene	100	1.0	103	3.7
Chloroethane	—	—	96	3.8
Chloroform	—	—	98	2.5
Chloromethane	—	—	96	8.9
2-Chlorotoluene	nd	nd	97	2.6
4-Chlorotoluene	101	1.0	97	3.1
1,2-Dibromo-3-chloropropane	—	—	86	9.9
Dibromochloromethane	—	—	102	3.3
1,2-Dibromoethane	—	—	97	2.7
Dibromomethane	—	—	109	7.4
1,2-Dichlorobenzene	102	2.1	100	1.5
1,3-Dichlorobenzene	104	1.7	106	4.3
1,4-Dichlorobenzene	103	2.2	98	2.3
Dichlorodifluoromethane	—	—	89	5.9
1,1-Dichloroethane	—	—	100	5.7
1,2-Dichloroethane	—	—	100	3.8
1,1-Dichloroethene	100	2.4	103	2.9
cis-1,2 Dichloroethene	nd	nd	105	3.5
trans-1,2-Dichloroethene	93	3.7	99	3.7
1,2-Dichloropropane	—	—	103	3.8
1,3-Dichloropropane	—	—	100	3.4
2,2-Dichloropropane	—	—	105	3.6
1,1-Dichloropropene	103	3.6	103	3.4
Ethylbenzene	101	1.4	—	—
Hexachlorobutadiene	99	9.5	98	8.3
Isopropylbenzene	98	0.9	—	—
p-Isopropyltoluene	98	2.4	—	—
Methylene chloride	—	—	97	2.8
Naphthalene	102	6.3	—	—
n-Propylbenzene	103	2.0	—	—
Styrene	104	1.4	—	—
1,1,1,2-Tetrachloroethane	—	—	99	2.3

TABLE 6230:VIII, CONT.*

Compound	Photoionization Detector		Electrolytic Conductivity Detector	
	Recovery %	Standard Deviation of Recovery $\mu g/L$	Recovery* %	Standard Deviation of Recovery $\mu g/L$
1,1,2,2-Tetrachloroethane	—	—	99	6.8
Tetrachloroethene	101	1.8	97	2.4
Toluene	99	0.8	—	—
1,2,3-Trichlorobenzene	106	1.9	98	3.1
1,2,4-Trichlorobenzene	104	2.2	102	2.1
1,1,1-Trichloroethane	—	—	104	3.4
1,1,2-Trichloroethane	—	—	109	6.2
Trichloroethene	100	0.78	96	3.5
Trichlorofluoromethane	—	—	96	3.4
1,2,3-Trichloropropane	—	—	99	2.3
1,2,4-Trimethylbenzene	99	1.2	—	—
1,3,5-Trimethylbenzene	101	1.4	—	—
Vinyl chloride	109	5.4	95	5.6
o-Xylene	99	0.8	—	—
m-Xylene	100	1.4	—	—
p-Xylene	99	0.9	—	—

* Recoveries and standard deviations were determined from seven samples with 10 $\mu g/L$ of each compound added. Recoveries were determined by internal standard method. Internal standards were fluorobenzene for PID and 2-bromo-1-chloropropane for HECD.
Dash indicates detector does not respond.
nd = not determined.

6. Calculation

See Section 6230C.6. If both detectors have responded to a compound, use the data from the detector showing the greater response unless greater specificity or precision is given by the alternate detector.

7. Quality Control

See Section 6230C.7.

8. Precision and Bias

This method was tested in a single laboratory using reagent water with known additions at 10 $\mu g/L$.[2] Single-laboratory precision and bias data are presented in Table 6230:VIII.

9. References

1. U.S. ENVIRONMENTAL PROTECTION AGENCY. 1986. Volatile organic compounds in water by purge and trap capillary column gas chromatography with photoionization and electrolytic conductivity detectors in series. Environmental Monitoring and Support Lab., Cincinnati, Ohio.

2. HO, J.S. 1986. Method performance data for Method 502.2. Unpublished report.

6230 E. Purge and Trap
Gas Chromatographic/Mass Spectrometric Method

See Sections 6210B and C for packed-column methods and Section 6210D for capillary-column method.

6231 1,2-DIBROMOETHANE (EDB) AND 1,2-DIBROMO-3-CHLOROPROPANE (DBCP)

6231 A. Introduction

1. Sources and Significance

Dibromoethane and dibromochloropropane have been found in groundwater supplies in many areas of the United States; typically they are found in agricultural areas where these compounds have been applied in the past as fumigants. Toxicological studies suggest that they may have detrimental effects on human health, and therefore many states have established maximum contaminant levels for them.

2. Selection of Method

The liquid-liquid extraction gas chromatographic (GC) method (6231B) uses a microextraction and capillary columns and is the preferred method. In addition, these compounds can be detected by the purge and trap gas chromatographic/mass spectrometric (GC/MS) and GC methods (6231C and D), and dibromoethane by closed-loop stripping analysis (see Section 6040). For additional information on applicability, sensitivity, precision, and bias, see specific methods.

6231 B. Liquid-Liquid Extraction Gas Chromatographic Method*

This method[1-3] is applicable to the determination of 1,2-dibromoethane (EDB) and 1,2-dibromo-3-chloropropane (DBCP) in drinking water and untreated groundwater.

1. General Discussion

a. Principle: The sample is extracted with hexane and injected into a gas chromatograph equipped with a linearized electron capture detector for separation and analysis. Identification is confirmed by analyzing the sample with a dissimilar column. See Section 6010C for discussion of gas chromatographic principles.

b. Interferences: Impurities in the extracting solvent usually account for most analytical problems. Analyze solvent blanks on each new bottle of solvent before use. Obtain indirect daily checks on the extracting solvent by monitoring sample blanks; whenever an interference is noted, reanalyze the extracting solvent. If neces-

---

* Approved by Standard Methods Committee, 1987. Accepted by U.S. Environmental Protection Agency as equivalent to EPA Method 504.

sary, remove interference by distillation or column chromatography[3] or, more simply, obtain a new source solvent. Interference-free solvent contains less than 0.1 μg/L individual compound interference. Store solvents in an area free of organochlorine solvents.

Accidental sample contamination can occur through diffusion of volatile organics through the septum seal into the sample bottle during shipment and storage. Sample blanks monitor this.

EDB at low concentrations may be masked by very high levels of dibromochloromethane (DBCM) when the confirmation column is used.

For further information on interferences in gas chromatographic methods, see Section 6010C.

c. Detection limits: The method detection limits (MDL)[4] for EDB and DBCP are 0.01 μg/L. The method is useful over a concentration range from approximately 0.03 to 200 μg/L. Actual detection limits are highly dependent on the characteristics of the gas chromatographic system used.

d. Safety: The toxicity or carcinogenicity of each reagent has not been defined precisely. EDB and DBCP have been classified tentatively as known or suspected human or mammalian carcinogens. Handle pure standard materials and stock standard solutions in a hood or glovebox and wear a NIOSH/MESA-approved toxic gas respirator when handling high concentrations.

2. Sampling and Storage

Collect all samples in duplicate and prepare replicate field blanks with each sample set. A sample set is all of the samples collected from the same general sampling site at approximately the same time. Prepare the field reagent blanks in the laboratory by filling a minimum of two sample bottles with reagent water, sealing, and shipping

to the sampling site along with sample bottles.

Fill sample bottle to overflowing without air bubbles. When sampling from a water tap, open tap and flush until water temperature has stabilized (usually about 10 min). Adjust flow rate to about 500 mL/min and collect duplicate samples from the flowing stream. When sampling from a well, fill a wide-mouth bottle or beaker with sample, and carefully fill duplicate 40-mL sample bottles.

Keep samples at 4°C in an atmosphere free of organic solvent vapors, from day of collection until analysis. Do not add sodium thiosulfate as a dechlorinating agent nor acidify.

Analyze all samples within 28 d of collection.

3. Apparatus

a. Sample containers, 40-mL screw-cap vials† each with a TFE-faced silicone septum.‡ Wash vials and septa with detergent and rinse with tap and distilled water before using. Let vials and septa air dry at room temperature, place in a 105°C oven for 1 h, then remove and let cool in an area free of organics.

b. Vials, auto sampler, screw cap with septa, 1.8 mL.§

c. Microsyringes, 10- and 100-μL.

d. Microsyringe, 25-μL with a 51- by 0.15-mm needle.‖

e. Pipets, 2.0- and 5.0-mL transfer.

f. Volumetric flasks, 10- and 100-mL, glass stoppered.

g. Standard solution storage containers, 15-mL bottles with TFE-lined screw caps.

h. Gas chromatograph: # See Section

---

† Pierce #13075 or equivalent.
‡ Pierce #12722 or equivalent.
§ Varian #96-000099-00 or equivalent.
‖ Hamilton 702N or equivalent.
Gas chromatographic methods are extremely sensitive to the materials used. Mention of trade names by *Standard Methods* does not preclude the use of other existing or as-yet-undeveloped products that give *demonstrably* equivalent results.

6230B.3*b*, except that the system is equipped with a linearized electron capture detector and a capillary column splitless injector.

Two gas chromatography columns are recommended. Column 1 is a highly efficient column that provides separations for EDB and DBCP without interferences from trihalomethanes. Use Column 1 as the primary analytical column unless routinely occurring compounds are not adequately resolved. Use Column 2 as a confirmatory column when GC/MS confirmation is not available.

1) *Column 1,* 30 m long × 0.32-mm ID fused silica capillary with dimethyl silicone mixed phase.** See Table 6231:I. Injector temperature: 200°C; detector temperature: 290°C. See Figure 6231:1 for a sample chromatogram.

2) *Column 2* (confirmation column), 30 m long × 0.32-mm ID fused silica capillary with methyl polysiloxane phase.†† See Table 6231:I. Injector temperature: 200°C; detector temperature: 290°C.

4. Reagents

a. Reagent water: See Section 6210B.4*a*.

b. Hexane extraction solvent, UV grade.‡‡

c. Methanol, pesticide quality or equivalent.

d. Sodium chloride, NaCl: Before using, pulverize and place in a muffle furnace at room temperature. Increase temperature to 400°C for 30 min. Store in capped bottle.

e. 1,2-Dibromoethane, 99%.§§

f. 1,2-Dibromo-3-chloropropane, 99.4%.‖‖

** Durawax-DX3, 0.25-µm film, or equivalent.
†† DB-1, 0.25-µm film, or equivalent.
‡‡ Burdick and Jackson #216 or equivalent.
§§ Such as that available from Aldrich Chemical Company.
‖‖ Such as that available from AMVAC Chemical Corporation, Los Angeles, Calif.

TABLE 6231:I. CHROMATOGRAPHIC CONDITIONS FOR 1,2-DIBROMOETHANE (EDB) AND 1,2-DIBROMO-3-CHLOROPROPANE (DPCP)

Compound	Retention Time *min*	
	Column 1	Column 2
EDB	9.5	8.9
DBCP	17.3	15.0

Column 1 conditions: Durawax-DX 3 (0.25-µm film thickness) in a 30 m long × 0.32-mm ID fused silica capillary column with helium carrier gas at linear velocity of 25 cm/s. Column temperature held isothermal at 40°C for 4 min, then programmed at 8°C/min to 180°C for final hold.

Column 2 conditions: DB-1 (0.25 µm film thickness) in a 30 m long × 0.32-mm ID fused silica capillary column with helium carrier gas at linear velocity of 25 cm/s. Column temperature held isothermal at 40°C for 4 min, then programmed at 10°C/min to 270°C for final hold.

g. Standard stock solutions: See Section 6210B.4*d*. Store in 15-mL bottles with TFE-lined screw caps. Methanol solutions prepared from liquid standard materials are stable for at least 4 weeks when stored at 4°C.

h. Secondary dilution standards: See Section 6210B.4*e*. Dilution standards are as stable as stock solutions.

5. Procedure

a. Operating conditions: Table 6231:I summarizes recommended operating conditions for the gas chromatograph and estimated retention times.

b. Calibration: Prepare calibration standards as directed in Section 6210B.4*g* and analyze according to ¶ 5*d* below. Follow rest of calibration procedure in Section 6210C.5*b*1), but limit variations from predicted response to ± 15% rather than ± 20%.

c. Instrument performance: See Section 6230C.5*c*.

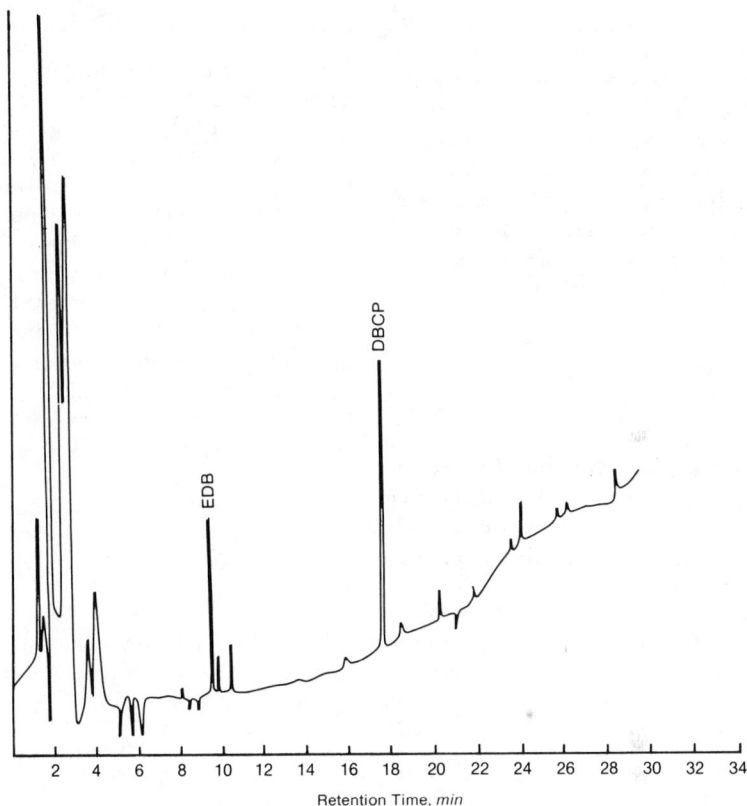

Figure 6231:1. Extract of reagent water with 0.114 μg/L added EDB and DBCP. Column: fused silica capillary; liquid phase: Durawax-DX3; film thickness: 0.25μm; column dimensions: 30 m × 0.317 mm ID.

d. Sample analysis: Let samples and standards come to room temperature. For samples and field blanks, open bottle, discard 5 mL using a 5-mL transfer pipet, and replace container cap. Weigh to nearest 0.1 g; record weight for subsequent volume determination. For calibration standards, QC check standards, and reagent blank, measure 35 mL using a 50-mL graduated cylinder and transfer to a 40-mL sample container.

Remove container cap and add 7 g NaCl. Add 2.0 mL hexane with a transfer pipet. Recap and shake vigorously by hand for 1 min. Let water and hexane phases separate. (If sample is stored at this stage, keep container upside down). Carefully transfer 0.5 mL of hexane layer into an autosampler vial using a disposable glass pipet. Transfer remaining hexane phase, but not any of the water phase, into a second autosampler vial. Hold second vial at 4°C for reanalysis if necessary. Transfer first sample vial to an autosampler set up to inject 2.0-μL portions into the gas chromatograph. Alternatively, manually inject 2-μL portions.

To determine sample volume for samples and field blanks, remove cap and discard remaining sample/hexane mixture. Shake off remaining drops using short, brisk wrist movements. Reweigh empty container with original cap and calculate net weight of sample by difference to the nearest 0.1 g. This net weight is equivalent to the volume of water (in mL) extracted. Alternatively, weigh vial before collection and reweigh full vial. Sample volume then equals gross weight (g) − [tare weight (g) + 5 g].

6. Calculation

Identify EDB and DBCP in sample chromatogram by comparing retention time of suspect peak to retention times generated by calibration and laboratory control standards. Retention times of samples should be within ± 0.1 min of standard for positive identification.

Use calibration curve or calibration factor to calculate uncorrected concentration (C_i) of each compound (e.g., calibration factor × response). Calculate sample volume (V_s) as equal to the net sample weight:

$$V_s = \text{gross weight} - \text{bottle tare}$$

The corrected sample concentration is:

$$\text{Concentration, } \mu g/L = C_i \times \frac{35}{V_s}$$

Round off results to the nearest 0.1 µg/ L or two significant figures.

7. Quality Control

Follow procedures given in Section 6210C.7. Prepare seven check samples containing 0.05 µg/L by adding 35 µg/L of check sample concentrate to 35-mL portions of reagent water. Analyze as in 6210C.7b. Calculate the minimum detection level at the 99% confidence level for seven replicates[4] as 3.143 s. For each compound, $\overline{X}$ should be between 80% and 120% of the true value. To check measurement system control (Section 6210C.7c) use a quality control sample containing 0.25 µg/L. To check low-level analysis (Section 6210C.7d) use a check sample containing 0.05 µg/L.

8. Precision and Bias

Single-laboratory precision and bias at several concentrations in tap water are presented in Table 6231:II[5].

9. References

1. GLAZE, W.H. & C.C. LIN. 1984. Optimization of Liquid-Liquid Extraction Methods for Analysis of Organics in Water. EPA-600/S4-83-052, U.S. Environmental Protection Agency.

TABLE 6231:II. SINGLE-LABORATORY PRECISION AND BIAS FOR EDB AND DBCP IN TAP WATER

Compound	Number of Samples	Addition µg/L	Average Bias %	Relative Standard Deviation %
1,2-Dibromoethane	7	0.03	114	9.5
	7	0.24	98	11.8
	7	50.0	95	4.7
1,2-Dibromo-3-chloropropane	7	0.03	90	11.4
	7	0.24	102	8.3
	7	50.0	94	4.8

2. HENDERSON, J.E., G.R. PEYTON & W.H. GLAZE. 1976. A convenient liquid-liquid extraction method for the determination of halomethanes in water at the parts-per-billion level. *In* L. H. Keith, ed. Identification and Analysis of Organic Pollutants in Water. Ann Arbor Science Publ., Ann Arbor, Mich.

3. RICHARD, J.J. & G.A. JUNK. 1977. Liquid extraction for rapid determination of halomethanes in water. *J. Amer. Water Works Assoc.* 69:62.

4. GLASER, J.A., D. L. FOERST, G. D. MCKEE, S. A. QUAVE & W. L. BUDDE. 1981. Trace analyses for wastewater. *Environ. Sci. Technol.* 15:1426.

5. WINFIELD, T., et al. Analysis of organohalide pesticides in drinking water by microextraction and gas chromatography. In preparation.

6231 C.   Purge and Trap Gas Chromatographic/Mass Spectrometric Method

See Sections 6210C and D for packed-column method and capillary-column method, respectively.

6231 D.   Purge and Trap Gas Chromatographic Method

See Section 6230C for packed-column method (EDB only) and 6230D for capillary-column method.

6232   TRIHALOMETHANES

6232 A.   Introduction

1. Sources and Significance

The trihalomethane (THM) compounds have been found in most chlorinated water supplies in the United States; typically they are produced in the treatment process as a result of chlorination. The formation of these compounds is a function of precursor concentration, contact time, chlorine dose, and pH. Toxicological studies suggest that chloroform is a potential human carcinogen. Consequently, total trihalomethanes are being regulated in potable waters.

2. Selection of Method

Several methods are available for measurement of the trihalomethanes. Some of these are specific for these compounds and others have a much broader spectrum. Method 6232B is a simple liquid-liquid extraction gas chromatographic (GC) method that is highly sensitive and very precise for these compounds. Method C refers to purge and trap gas chromatographic/mass spectrometric (GC/MS) methods that can detect not only THMs but also a wide variety of other compounds. Method D refers to purge and trap GC methods with similar target compounds. All of these methods have approximately the same sensitivity for the trihalomethanes; method choice depends on availability of equipment, operator choice, and the list of desired target compounds. In addition, closed-loop stripping analysis can be used for several of these compounds (see Section 6040).

6232 B.   Liquid-Liquid Extraction
Gas Chromatographic Method*

This method[1-3] is applicable only to the determination of four trihalomethanes (THMs), i.e., chloroform, bromodichloromethane, dibromochloromethane, and bromoform in finished drinking water, drinking water during intermediate stages of treatment, and the raw source water. For other compounds or sample sources, collect precision and bias data on actual samples[4] and provide qualitative confirmation of results by gas chromatography/mass spectrometry (GC/MS) to demonstrate the usefulness of the method.

Make qualitative analyses using GC/MS[5] or the purge and trap method[6] to char-

---

* Approved by Standard Methods Committee, 1987. Accepted by U.S. Environmental Protection Agency as equivalent to EPA Method 501.2

acterize each raw source water if peaks appear as interferences in the analysis.

1. General Discussion

a. Principle: Sample is extracted once with solvent and the extract is injected into a gas chromatograph equipped with a linearized electron capture detector for separation and analysis. Extraction and analysis time is 10 to 50 min per sample, depending on analytical conditions.

Confirmatory evidence is obtained by using dissimilar columns and temperature programming. When component concentrations are sufficiently high (> 50 $\mu g/L$), halogen-specific detectors may be used for improved specificity. Unequivocal confirmatory analyses at high levels (> 50 $\mu g/L$) can be made by using a mass spectrometer instead of the electron capture detector. At levels below 50 $\mu g/L$, only the purge and trap technique using GC/MS can provide unequivocal confirmation.[5,6]

Standards added to organic-free water and samples are extracted and analyzed identically to compensate for possible extraction losses.

The concentrations of the trihalomethanes are summed and reported as total trihalomethanes in micrograms per liter.

See Section 6010 for discussion of gas chromatographic principles.

b. Interferences: Impurities contained in the extracting solvent usually account for most analytical problems. Analyze solvent blanks before using a new bottle of solvent to extract samples. Make indirect daily checks on the extracting solvent by monitoring the sample blanks. Whenever an interference is noted in the sample blank, reanalyze the extracting solvent. Discard extraction solvent if a high level (> 10 $\mu g/L$) of interfering compounds is traced to it. Low-level interferences usually can be removed by distillation or column chromatography;[7] however, it usually is more economical to obtain a new solvent or se-

lect an approved alternative solvent. Interference-free solvent is defined as a solvent containing less than 0.4 $\mu g/L$ individual trihalomethane interference. Protect interference-free solvents by storing in a nonlaboratory area known to be free of organochlorine solvents. *Do not subtract blank values as a correction.*

Accidental sample contamination has been attributed to diffusion of volatile organics through the septum seal on a sample bottle during shipment and storage. Use the sample blank to monitor for this problem.

This liquid-liquid extraction technique efficiently extracts a wide boiling range of nonpolar organic compounds and also extracts the polar organic components of the sample with varying efficiencies. To analyze rapidly for trihalomethanes with sensitivities in the low microgram-per-liter range, use the semi-specific electron capture detector and chromatographic columns that have relatively poor resolving power. Because of these concessions, the probability of encountering chromatographic interferences is high. Trihalomethanes are primarily products of the chlorination process and seldom appear in raw source water. The absence of peaks with retention times similar to the trihalomethanes in raw source water analysis is evidence of an interference-free finished drinking water analysis. Because of these possible interferences, analyze a representative raw source water when analyzing finished drinking water. When potential interferences are noted in the raw source water, use the alternate chromatographic columns to reanalyze the sample set. If interferences still are noted, make confirmatory qualitative identifications as directed in ¶ 1a. If the peaks are confirmed to be other than trihalomethanes and they add significantly to the total trihalomethane value in the finished drinking water, analyze sample set by the purge and trap method.

For further information on interferences in gas chromatographic methods, see Section 6010C.

c. *Detection limits*: The method is useful for trihalomethane concentrations from approximately 0.5 to 200 μg/L. Actual detection limits are highly dependent on the characteristics of the gas chromatographic system used.

2. Sampling and Storage

See Section 6010B.1.

Do not add any preservatives at the time of sample collection if maximum trihalomethane formation is to be measured. If chemical stabilization is not used at time of sampling, add the reducing agent just before extracting the sample.[8,9]

The raw source water sample history should resemble that of the finished drinking water. Take into account the average retention time of the finished drinking water within the water plant when sampling the raw source water.

For septum-seal screw-cap bottles, open bottle and fill to overflowing, place on a level surface, position TFE side of septum seal on convex sample meniscus, and screw cap on tightly. Invert sample and lightly tap cap on a solid surface. The absence of entrapped air indicates a successful seal. If bubbles are present, open bottle, add a few more drops of sample and reseal as above.

Store blanks and samples, collected at a given site (sample set), together in a protected area known to be free from contamination. A sample set is defined as all samples collected at a given site. At a water treatment plant, duplicate raw source water, duplicate finished water, and duplicate sample blanks comprise the sample set. When samples are collected and stored under these conditions, no measurable loss of trihalomethanes has been detected over extended periods of time.[8] If possible, analyze samples within 14 d of collection.

For samples collected soon after chlo-

rination, quenching with reducing agent may not be sufficient to prevent formation of THMs completely because of hydrolysis of intermediates. In that case, acidification is necessary and consistent with the recommended preservation techniques.

3. Apparatus

a. *Extraction vessel:*

1) For sample that does not form an emulsion, use 14-mL (total volume) screw-cap vial† with a TFE-faced silicone septum.‡

2) For sample that forms emulsions (turbid source water), use 15-mL screw-cap centrifuge tube§ with a TFE cap liner.

Prepare extraction flasks according to directions for sample bottles, ¶ 2.

b. *Microsyringes*, 10-, 100-μL.

c. *Microsyringe*, 25-μL with 5.08-cm × 0.02-cm needle.‖

d. *Syringes*, 10-mL glass hypodermic with luerlok top (2 each).#

e. *Syringe valve*, two-way with luer ends (2 each).#

f. *Pipet*, 2.0-mL transfer.

g. *Volumetric flasks, glass-stoppered*, 10 and 100 mL.

h. *Gas chromatograph*, preferably temperature-programmable, with linearized electron capture detector. See ¶ *i*3), below.

i. *Chromatographic columns:***

1) 4-mm ID × 2 m long glass packed with 3% SP-1000 on Supelcoport (100/120 mesh) operated at 50°C with 60 mL/min flow. See Figure 6232:1 for a sample chromatogram and Table 6232:I for retention data.

2) 2-mm ID × 2 m long glass packed

---

† Pierce # 13310 or equivalent.
‡ Pierce # 12718 or equivalent.
§ Corning 8062-15 or equivalent.
‖ Hamilton 702N or equivalent.
Hamilton # 86570 or equivalent.
** Chromatographic methods are extremely sensitive to the materials used. Mention of trade names by *Standard Methods* does not preclude the use of other existing or as-yet-undeveloped products that give *demonstrably* equivalent results.

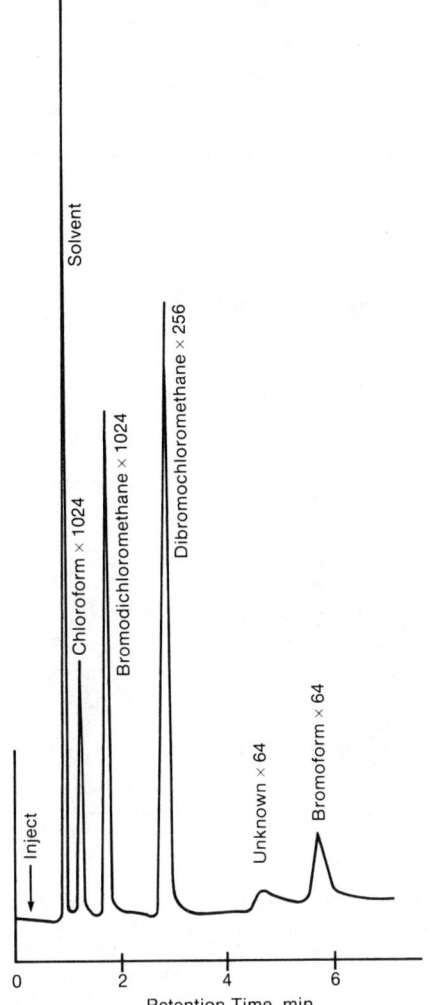

Figure 6232:1. Chromatogram of finished water extract. Column packing: 3% SP-1000; carrier gas: 5% CH$_4$ in argon; carrier flow: 60 mL/min; column temperature: 50°C; detector: electron capture.

with 10% squalane on Chromosorb WAW (80/100 mesh) operated at 67° C with 25 mL/min flow. This column is recommended as the primary analytical column. Trichloroethylene, a common raw source water contaminant, coelutes with bromo-

dichloromethane. See Figure 6232:2 for a sample chromatogram and Table 6232:I for retention data.

3) 2-mm ID × 3 m long glass packed with 6% OV-11/4% SP-2100 on Supelcoport (100/120 mesh); temperature program 45°C for 12 min, then program at 1°C/min to 70°C with a 25 mL/min flow. Hold until expected compounds have eluted. See Figure 6232:3 for a sample chromatogram and Table 6232:I for retention data.

j. Standard storage containers: 15-mL amber screw-cap septum bottles†† with TFE-faced silicone septa.‡‡

4. Reagents

a. Extraction solvent: See ¶ 1*b*. Recommended solvent is pentane.§§ Alternatively, use hexane, methylcyclohexane, or 2,2,4-trimethylpentane.

b. Methyl alcohol.

c. Activated carbon.‖‖

d. Reference standards.##

1) *Bromoform, 96%.*##

2) *Bromodichloromethane, 97%.****

3) *Dibromochloromethane.****†††

4) *Chloroform, 99%.****

e. THM-free water: Generate THM-free water, which is water free of interference

---

†† Pierce #19830 or equivalent.

‡‡ Pierce # 12716 or equivalent.

§§ Pentane has been selected as the best solvent for this analysis because it elutes, on all of the columns, well before any of the trihalomethanes. High altitudes or laboratory temperatures in excess of 24°C may make the use of this solvent impractical. Alternative solvents are acceptable; however, the analyst may experience baseline variations in the elution areas of the trihalomethanes due to coelution of these solvents. The degree of difficulty appears to depend on the design and condition of the electron capture detector. Such problems should be insignificant when concentrations of the coeluting trihalomethane exceed 5 µg/L.

‖‖ Filtrasorb 200, Calgon Corp., Pittsburgh, Pa., or equivalent.

As a precautionary measure, check all standards for purity by boiling point determination or GC/MS assays.

*** Aldrich Chemical Company or equivalent.

††† Columbia Organic Chemicals Company, Inc., Columbia, S.C., or equivalent.

TABLE 6232:I. RETENTION TIMES FOR TRIHALOMETHANES

| | Retention Time _min_ | | |
Trihalomethane	Column 1	Column 2	Column 3
Chloroform	1.0	1.3	4.9
Bromodichloromethane	1.5	2.5*	11.0
Dibromochloromethane	2.6	5.6	23.1
Bromoform	5.5	10.9	39.4

* On this column, trichloroethylene, a common raw source water contaminant, coelutes with bromodichloromethane.

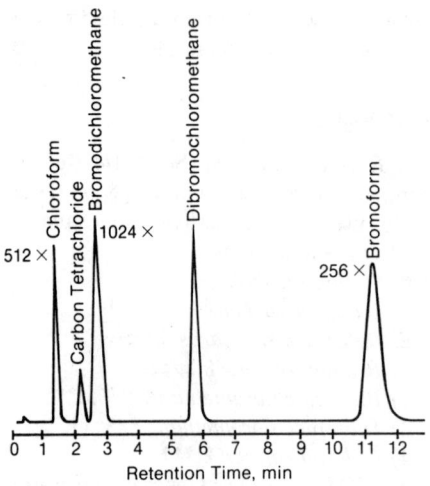

Figure 6232:2. Chromatogram of extract of standard. Column packing: 10% squalane; carrier flow: 25 mL/min; column temperature: 67°C.

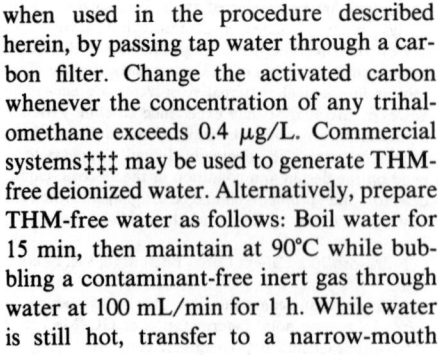

Figure 6232:3. Chromatogram of extract of standard. Column packing: 6% OV-11+4% SP-2100; carrier flow: 25 mL/min; temperature program: 45°C for 12 min, 1°/min to 70°C.

when used in the procedure described herein, by passing tap water through a carbon filter. Change the activated carbon whenever the concentration of any trihalomethane exceeds 0.4 μg/L. Commercial systems‡‡‡ may be used to generate THM-free deionized water. Alternatively, prepare THM-free water as follows: Boil water for 15 min, then maintain at 90°C while bubbling a contaminant-free inert gas through water at 100 mL/min for 1 h. While water is still hot, transfer to a narrow-mouth

‡‡‡ Millipore SuperQ Water System or equivalent.

screw-cap bottle with a TFE seal. Test THM-free water daily by analyzing for trihalomethanes.

f. Standard stock solutions: Place 9.8 mL methyl alcohol in a 10.0-mL ground-glass-stoppered volumetric flask. Let flask stand unstoppered about 10 min or until all alcohol-wetted surfaces dry. Weigh to nearest 0.1 mg. Using a 100-μL syringe, immediately add 2 to 3 drops of reference standard to flask, and reweigh. Be sure that reference standard falls directly into the alcohol without contacting flask neck. Dilute to volume, stopper, then mix by inverting flask several times. Transfer standard solution to a dated and labeled 15-mL screw-cap bottle with a TFE cap liner.

Calculate concentration in micrograms per microliter from the net gain in weight. Store solution at 4°C. Such standards are stable up to 4 weeks when stored under these conditions. If more time elapses, discard solutions.

CAUTION: *Trihalomethanes are toxic; prepare primary dilutions in a hood. Preferably also use a NIOSH/MESA-approved toxic gas respirator when handling high concentrations of such materials.*

g. *Calibration standards:* Either:

1) From standard stock solutions, prepare a multicomponent secondary dilution mixture in methyl alcohol so that a 20-μL injection into 100 mL THM-free water will generate a calibration standard that produces a response close ($\pm$ 25%) to that of the sample; or

2) Construct a calibration curve for each trihalomethane, using a minimum of three different concentrations. Two of the concentrations must bracket each sample.

In either case, prepare standards by injecting not more than 20 μL of alcoholic standards into 100 mL of THM-free water with a 25-μL microsyringe.§§§ Rapidly inject alcoholic standard into expanded area of filled volumetric flask, then remove needle immediately. Mix aqueous standards by inverting flask three times only. Discard contents in neck of flask. Fill sample syringe from standard solution in expanded area of flask as directed in ¶ 5. Never use pipets to dilute or transfer samples and aqueous standards. Aqueous standards, when stored with a headspace, are not stable; discard after 1 h. Store

aqueous standards according to ¶s 2 and 5.

Extract and analyze aqueous calibration standards identically with samples.

Calibration procedures[8] that require delivery of less than 20 μL of methanolic standards to 10.0-mL volumes of water contained in the sample syringe are acceptable only if the methanolic standard is delivered by the solvent flush technique.[10]

h. *Quality control check standard mixture:* From standard stock solutions, prepare a secondary dilution mixture in methyl alcohol containing 10.0 ng of each compound/μL.‖‖‖ Daily, inject 20.0 μL of this mixture into 100 mL of organic-free water and analyze (¶ 7a).

5. Procedure

Remove plungers from two 10-mL syringes and attach a closed syringe valve to each. Open sample bottle (or standard); if chemical reducing agent has not been added to the sample, add just before analysis at the ratio of 2.5 to 3 mg/40 mL or add 1 mg directly to sample in extraction flask. Carefully pour sample into one of the syringe barrels until it overflows. Replace plunger and compress sample. Open syringe valve and vent any residual air while adjusting sample volume to 10.0 mL. Close valve.

Fill second syringe in an identical manner from same sample bottle and reserve for replicate analysis (see ¶ 7 below).

Pipet 2.0 mL extraction solvent into a clean extraction flask, carefully inject contents of syringe into extraction flask, seal with a TFE-faced septum, and shake vigorously for 1 min. Let stand until phases separate (60 s). If phases do not separate

---

§§§ Hamilton 702N or equivalent. Variations in needle geometry will adversely affect the ability to deliver reproducible volumes of methanolic standards into water.

‖‖‖ For U.S. federal permit-related analyses, use samples obtainable from U.S. EPA Environmental Monitoring and Support Laboratory, Cincinnati, Ohio.

on standing, centrifuge to facilitate separation.

Analyze sample by injecting 3.0 µL (solvent flush technique)[10] of the upper (organic) phase into the gas chromatograph.

6. Calculation

Locate each trihalomethane in the sample chromatogram by comparing retention time of suspect peak to average retention time and variance computed in ¶7g. The retention time of the suspect peak must fall within the limits established by those statistics for a single column identification.

Calculate concentration of each trihalomethane by comparing peak heights or peak areas of samples to those of standards as follows, using only two significant figures:

$$\text{Trihalomethane, µg/L} = \frac{S}{P} \times C$$

where:

S = sample peak height or area,
P = standard peak height or area that gives a detector response within ± 25% of sample response, and
C = concentration of standard giving response, P, µg/L.

Calculate total trihalomethane concentration as:

$$\text{TTHM} = \text{CHCl}_3 + \text{CHBrCl}_2 + \text{CHBr}_2\text{Cl} + \text{CHBr}_3$$

where:

TTHM = total trihalomethane concentration, µg/L, and $CHCl_3$, $CHBrCl_2$, $CHBr_2Cl$, $CHBr_3$ = individual THM concentrations, µg/L.

Calculate limit of detection (LOD) for each trihalomethane not detected using the following criterion:

$$\text{LOD, µg/L} = \frac{A \times F \times (2 \text{ µg/L})}{(B \times F)}$$

where:

A = five times the noise level in millimeters at the exact retention time of the trihalomethane or the baseline displacement in millimeters from theoretical zero at exact retention time for the trihalomethane,
B = peak height in millimeters of 2-µg/L quality check standard, and
F = attenuation factor.

Report results obtained from lower limit of detection estimates along with data for samples.

7. Quality Control

a. *Quality check standards analysis*: Daily, before any samples are analyzed, extract and analyze a 2-µg/L quality check standard (¶ 4h). Calculate instrument status checks and lower limit of detection estimates on basis of response factor calculations at five times the noise level. The data also can be used to estimate concentration in samples. From this information, determine appropriate standard dilutions to be made.

b. *Interference monitoring:* Analyze sample blank and raw source water to monitor for potential interferences as described in ¶ 1b.

c. *Samples with known additions:* In a laboratory analyzing more than 10 samples daily, analyze a laboratory-generated known addition that closely duplicates the average finished drinking water in trihalomethane composition and concentration as each 10th sample (See ¶ 4g).

In a laboratory analyzing fewer than 10 samples daily, each time the analysis is made, analyze at least one laboratory-generated known-addition sample that closely duplicates the average finished drinking water in trihalomethane composition and concentration (See ¶ 4g).

d. *Duplicate analysis:* Randomly select and analyze in duplicate 10% of all samples. Maintain an up-to-date log on bias and precision data collected on known-addition samples and duplicate samples. If results are significantly different from those cited in ¶ 8 below, check entire analytical scheme to determine why the laboratory's precision and bias limits are excessive.

e. External quality-control samples: Quarterly, add an external reference trihalomethane quality control sample to organic-free water and analyze. The results from this sample should agree within 20% of the true value for each trihalomethane. If not, check each step in the standard generation procedure to solve the problem.

f. Electron capture system calibration: Be aware of the linear response characteristics of the electron capture system. Generate and recheck calibration curves quarterly for each trihalomethane over the concentration range encountered in samples to confirm linear response range of the system. Quantitative data can not be calculated from nonlinear responses. Whenever nonlinear responses occur, dilute sample and reanalyze.

g. Retention times: Maintain a record of retention times for each trihalomethane using data gathered from known-addition samples and standards. Daily calculate average retention time for each trihalomethane and the variance for the analyses. If individual trihalomethane retention time varies by more than 10% over an 8-h period or does not fall within 10% of an established norm, the system is "out of control." Seek and correct the source of retention data variation.

8. Precision and Bias

The single-laboratory precision and bias data in Table 6232:II were generated by adding known amounts of trihalomethanes to organic-free water as described in ¶ 4g. The mixtures were analyzed as true unknowns.

9. References

1. U.S. ENVIRONMENTAL PROTECTION AGENCY. 1980. Analysis of trihalomethanes in drinking water by liquid/liquid extraction. 40 CFR Part 141, Appendix C, Part II.

2. MIEURE, J.P. 1977. A rapid and sensitive method for determining volatile organohalides in water. *J. Amer. Water Works Assoc.* 69:60.

3. REDING, R., ET AL. 1978. THM's in drinking water: Analysis by LLE and comparison to purge and trap. *In* C. E. Van Hall, ed. Measurement of Organic Pollutants in Water and Wastewater. STP 686, American Soc. Testing & Materials, Philadelphia, Pa.

4. Handbook for Analytical Quality Control in Water and Waste Water Laboratories. 1972. Analytical Quality Control Lab., National Environmental Research Center, Cincinnati, Ohio.

5. BUDDE, W. L., & J. W. EICHELBERGER. 1979. Organic Analysis Using Gas Chromatography-Mass Spectrometry. Ann Arbor Science Publ., Ann Arbor, Mich.

6. U.S. ENVIRONMENTAL PROTECTION AGENCY. 1979. The Analysis of Trihalomethanes in Finished Water by the Purge and Trap Method. Environmental Monitoring & Support Lab., Environmental Research Center, Cincinnati, Ohio.

7. RICHARD, J. J. & G. A. JUNK. 1977. Liquid extraction for rapid determination of halomethanes in water. *J. Amer. Water Works Assoc.* 69:62.

8. BRASS, H. J., et al. 1977. *In* R. B. Pojasek, ed. National Organic Monitoring Survey: Sampling and Purgeable Organic Compounds, Drinking Water Quality Through Source Protection, p. 398. Ann Arbor Science Publ., Ann Arbor, Mich.

9. KOPFLER, F. C., et al. 1976. GC/MS determination of volatiles for the National Organics Reconnaissance Survey (NORS) of Drinking Water. *In* L. H. Keith, ed. Identification and Analysis of Organic Pollutants in Water. Ann Arbor Science Publ., Ann Arbor, Mich.

10. WHITE, L. D., et al. 1970. Convenient optimized method for the analysis of selected solvent vapors in industrial atmosphere. *Amer. Ind. Hyg. Assoc. J.* 31:225.

TABLE 6232:II. SINGLE-LABORATORY PRECISION AND BIAS

Compound	Dose Levels μg/L	Number of Samples	Mean μg/L	Precision (Relative Standard Deviation) %	Bias (Recovery) %
CHCl₃	9.1	5	10	11	110
CHCl₃	69	3	73	5.3	106
CHBrCl₂	1.2	5	1.3	9.8	108
CHBrCl₂	12	2	15	1.4	125
CHBr₂Cl	2.7	5	2.0	17	74
CHBr₂Cl	17	3	16	9.9	94
CHBr₃	2.9	5	2.2	10	76
CHBr₃	14	3	16	12	114

6232 C.   Purge and Trap Gas Chromatographic/Mass Spectrometric Method

See Sections 6210B and C for packed-column methods and Section 6210D for capillary-column method.

6232 D.   Purge and Trap Gas Chromatographic Method

See Sections 6230B and C for packed-column methods and Section 6230D for capillary column method.

6410   EXTRACTABLE BASE/NEUTRALS AND ACIDS

6410 A.   Introduction

1. Sources and Significance

The semivolatile compounds covered by this section include many classes of compounds, each characterized by different sources. The compounds include polynuclear aromatic hydrocarbons, often as by-products of petroleum processing or combustion; phthalates, used as plasticizers; phenolics, found most often in wood preservatives; and organochlorine pesticides, found most often in agricultural runoff or in wastewaters draining such areas. Many of the listed compounds are toxic or carcinogenic. However, they generally are relatively insoluble in water so they do not occur frequently in potable waters or most wastewaters.

2. Selection of Method

Method 6410B is a broad-spectrum gas chromatographic/mass spectrometric (GC/MS) packed- or capillary-column method for detection of these compounds following liquid-liquid extraction. Although this method can be used to determine all the listed compounds, it is not the most sensitive method for individual classes of compounds, which are detected at lower concentrations by GC methods such as those listed in 6420C (phenols),

6440B and C (polynuclear aromatic hydrocarbons), and 6630C and D (organochlorine pesticides and PCBs). In some cases, notably the pesticides, the GC method is substantially more sensitive than the GC/MS method. In other cases, such as the phenols, there is less difference between the methods.

6410 B.   Liquid-Liquid Extraction Gas Chromatographic/Mass Spectrometric Method*

This method[1] is applicable to the determination of organic compounds that are partitioned into an organic solvent and are amenable to gas chromatography,† in municipal and industrial discharges.

---

* Approved by Standard Methods Committee, 1987. Accepted by U.S. Environmental Protection Agency as equivalent to EPA Method 625.

† *Base/neutral extractables*: acenaphthene, acenaphthylene, anthracene, aldrin, benzo(a)anthracene, benzo(b)fluoranthene, benzo(k)fluoranthene, benzo-(a)pyrene, benzo(ghi)perylene, benzyl butyl phthalate, β-BHC, δ-BHC, bis(2-chloroethyl) ether, bis(2-chloroethoxy) methane, bis(2-ethylhexyl) phthalate, bis(2-chloroisopropyl) ether more correctly known as 2,2-oxybis (1-chloropropane), 4-bromophenyl phenyl ether, chlordane, 2-chloronaphthalene, 4-chlorophenyl phenyl ether, chrysene, 4,4'-DDD, 4,4'-DDE, 4,4'-DDT, dibenzo(a,h)anthracene, di-*n*-butylphthalate, 1,3-dichlorobenzene, 1,2-dichlorobenzene, 1,4-dichlorobenzene, 3,3'-dichlorobenzidine, dieldrin, diethyl phthalate, dimethyl phthalate, 2,4-dinitrotoluene, 2,6-dinitrotoluene, di-*n*-octylphthalate, endosulfan sulfate, endrin aldehyde, fluoranthene, fluorene, heptachlor, heptchlor epoxide, hexachlorobenzene, hexachlorobutadiene, hexachloroethane, indeno(1,2,3-cd)pyrene, isophorone, naphthalene, nitrobenzene, *N*-nitrosodi-*n*-propylamine, PCB-1016, PCB-1221, PCB-1232, PCB-1242, PCB-1248, PCB-1254, PCB-1260, phenanthrene, pyrene, toxaphene, 1,2,4-trichlorobenzene.

Acid extractables: 4-chloro-3-methylphenol, 2-chlorophenol, 2,4-dichlorophenol, 2,4-dimethylphenol, 2,4-dinitrophenol, 2-methyl-4,6-dinitrophenol, 2-nitrophenol, 4-nitrophenol, pentachlorophenol, phenol, 2,4,6-trichlorophenol.

The method may be extended to include the following compounds: benzidine, α-BHC, γ-BHC, endosulfan I, endosulfan II, endrin, hexachlorocyclopentadiene, *N*-nitrosodimethylamine, *N*-nitrosodiphenylamine.

1. General Discussion

a. Principle: A measured volume of sample is extracted serially with methylene chloride at a pH above 11 and again at a pH below 2. The extract is dried, concentrated, and analyzed by GC/MS.[2,3] Qualitative compound identification is based on retention time and relative abundance of three characteristic masses (m/z). Quantitative analysis uses internal-standard techniques with a single characteristic m/z.

b. Interferences:

1) General precautions—See Section 6010C. Method interferences may be caused by contaminants in solvents, reagents, glassware, and other sample-processing hardware that lead to discrete artifacts and/or elevated baselines in detector output. Routinely demonstrate that all materials are free from interferences under the conditions of the analysis by running laboratory reagent blanks as described in Section 6210B.7a.

Clean all glassware thoroughly[4] as soon as possible after use by rinsing with the last solvent used in it, followed by detergent washing with hot water and rinsing with tap water and distilled water. Drain glassware dry and heat in a muffle furnace at

400°C for 15 to 30 min. Some thermally stable materials, such as PCBs, may not be eliminated by this treatment. Solvent rinses with acetone and pesticide-quality hexane may be substituted for the baking. Thorough rinsing with such solvents usually eliminates PCB interference. Do not heat volumetric ware in a muffle furnace. After drying and cooling, seal and store glassware in a clean environment to prevent accumulation of dust or other contaminants. Store inverted or capped with aluminum foil.

Use high-purity reagents and solvents to minimize interference. Purification of solvents by distillation in all-glass systems may be required.

Matrix interferences may be caused by coextracted contaminants. The extent of matrix interferences will vary considerably depending on the sample.

2) Special precautions—Benzidine can be lost by oxidation during solvent concentration. Under the alkaline conditions of the extraction step, α-BHC, γ-BHC, endosulfan I and II, and endrin are subject to decomposition. Hexachlorocyclopentadiene is subject to thermal decomposition in the inlet of the gas chromatograph, chemical reaction in acetone solution, and photochemical decomposition. N-nitrosodimethylamine is difficult to separate from the solvent under the chromatographic conditions described. N-nitrosodiphenylamine decomposes in the gas chromatographic inlet and cannot be separated from diphenylamine. Other methods may be preferred for these compounds.[1]

The base-neutral extraction may cause significantly reduced recovery of phenol, 2-methylphenol, and 2,4-dimethylphenol. Results obtained under these conditions are minimum concentrations.

The packed gas chromatographic columns recommended for the basic fraction may not be able to resolve certain isomeric pairs including the following: anthracene and phenanthrene; chrysene and benzo-(a)anthracene; and benzo(b)fluoranthene and benzo(k)fluoranthene because retention time and mass spectra for these pairs are not sufficiently different to make unambiguous identification possible. Use alternative techniques, such as the method for polynuclear aromatic hydrocarbons (Section 6440B), to identify and quantify these compounds.

In samples containing many interferences, use chemical ionization (CI) mass spectrometry to make identification easier. Tables 6410:I and II give characteristic CI ions for most compounds covered by this method. Use of CI mass spectrometry to support electron ionization (EI) mass spectrometry is encouraged but not required.

c. Detection limits: The method detection limit (MDL) is the minimum concentration of a substance that can be measured and reported with 99% confidence that the value is above zero.[5] The MDL concentrations listed in Tables 6410:I and II were obtained with reagent water.[6] The MDL actually obtained in a given analysis will vary, depending on instrument sensitivity and matrix effects.

d. Safety: The toxicity or carcinogenicity of each reagent has not been defined precisely. Benzo(a)anthracene, benzidine, 3,3'-dichlorobenzidine, benzo(a)pyrene, α-BHC, β-BHC, δ-BHC, γ-BHC, dibenzo(a,h)anthracene, n-nitrosodimethylamine, 4,4'-DDT, and polychlorinated biphenyls (PCBs) have been tentatively classified as known or suspected, human or mammalian carcinogens. Prepare primary standards of these compounds in a hood and wear a NIOSH/MESA-approved toxic gas respirator when handling high concentrations.

2. Sampling and Storage

Collect grab samples in 1-L amber glass bottles fitted with a screw cap lined with TFE. Foil may be substituted for TFE if the sample is not corrosive. If amber bottles

TABLE 6410:I. CHROMATOGRAPHIC CONDITIONS, METHOD DETECTION LIMITS, AND CHARACTERISTIC MASSES FOR BASE/NEUTRAL EXTRACTABLES

Compound	Retention Time min	Method Detection Limit µg/L	Characteristic Masses					
			Electron Impact			Chemical Ionization		
			Primary	Secondary	Secondary	Methane	Methane	Methane
1,3-Dichlorobenzene	7.4	1.9	146	148	113	146	148	150
1,4-Dichlorobenzene	7.8	4.4	146	148	113	146	148	150
Hexachloroethane	8.4*	1.6	117	201	199	199	201	203
bis(2-Chloroethyl) ether	8.4	5.7	93	63	95	63	107	109
1,2-Dichlorobenzene	8.4	1.9	146	148	113	146	148	150
bis(2-Chloroisopropyl) ether*	9.3	5.7	45	77	79	77	135	137
N-Nitrosodi-n-propylamine			130	42	101			
Nitrobenzene	11.1	1.9	77	123	65	124	152	164
Hexachlorobutadiene	11.4	0.9	225	223	227	223	225	227
1,2,4-Trichlorobenzene	11.6	1.9	180	182	145	181	183	209
Isophorone	11.9	2.2	82	95	138	139	167	178
Naphthalene	12.1	1.6	128	129	127	129	157	169
bis(2-Chloroethoxy) methane	12.2	5.3	93	95	123	65	107	137
Hexachlorocyclopentadiene†	13.9		237	235	272	235	237	239
2-Chloronaphthalene	15.9	1.9	162	164	127	163	191	203
Acenaphthylene	17.4	3.5	152	151	153	152	153	181
Acenaphthene	17.8	1.9	154	153	152	154	155	183
Dimethyl phthalate	18.3	1.6	163	194	164	151	163	164
2,6-Dinitrotoluene	18.7	1.9	165	89	121	183	211	223
Fluorene	19.5	1.9	166	165	167	166	167	195
4-Chlorophenyl phenyl ether	19.5	4.2	204	206	141			
2,4-Dinitrotoluene	19.8	5.7	165	63	182	177	223	251
Diethyl phthalate	20.1	1.9	149	177	150	169	223	251
N-Nitrosodiphenylamine†	20.5	1.9	169	168	167	169	170	198
Hexachlorobenzene	21.0	1.9	284	142	249	284	286	288
α-BHC†	21.1		183	181	109			
4-Bromophenyl phenyl ether	21.2	1.9	248	250	141	249	251	277
γ-BHC†	22.4		183	181	109			

TABLE 6410:I, CONT.

Compound	Retention Time (min)	Method Detection Limit (μg/L)	Characteristic Masses					
			Electron Impact			Chemical Ionization		
			Primary	Secondary	Secondary	Methane	Methane	Methane
Phenanthrene	22.8	5.4	178	179	176	178	179	207
Anthracene	22.8	1.9	178	179	176	178	179	207
β-BHC	23.4	4.2	181	183	109			
Heptachlor	23.4	1.9	100	272	274			
δ-BHC	23.7	3.1	183	109	181			
Aldrin	24.0	1.9	66	263	220			
Dibutyl phthalate	24.7	2.5	149	150	104	149	205	279
Heptachlor epoxide	25.6	2.2	353	355	351			
Endosulfan I†	26.4		237	338	341			
Fluoranthene	26.5	2.2	202	101	100	203	231	243
Dieldrin	27.2	2.5	79	263	279			
4,4′-DDE	27.2	5.6	246	248	176			
Pyrene	27.3	1.9	202	101	100	203	231	243
Endrin†	27.9		81	263	82			
Endosulfan II†	28.6		237	339	341			
4,4′-DDD	28.6	2.8	235	237	165			
Benzidine†	28.8	44	184	92	185	185	213	225
4,4′-DDT	29.3	4.7	235	237	165			
Endosulfan sulfate	29.8	5.6	272	387	422			
Endrin aldehyde			67	345	250			

Compound								
Butyl benzyl phthalate	29.9	2.5	149	91	206	149	299	327
bis(2-Ethylhexyl) phthalate	30.6	2.5	149	167	279	149	229	257
Chrysene	31.5	2.5	228	226	229	228	229	257
Benzo(a)anthracene	31.5	7.8	228	229	226	228		
3,3'-Dichlorobenzidine	32.2	16.5	252	254	126			
Di-n-octyl phthalate	32.5	2.5	149	149				
Benzo(b)fluoranthene	34.9	4.8	252	253	125	252	253	281
Benzo(k)fluoranthene	34.9	2.5	252	253	125	252	253	281
Benzo(a)pyrene	36.4	2.5	252	253	125	252	253	281
Indeno(1,2,3-cd)pyrene	42.7	3.7	276	138	277	276	277	305
Dibenzo(a,h)anthracene	43.2	2.5	278	139	279	278	279	307
Benzo(ghi)perylene	45.1	4.1	276	138	277	276	277	305
N-Nitrosodimethylamine†			42	74	44			
Chlordane‡	19–30		373	375	377			
Toxaphene‡	25–34		159	231	233			
PCB 1016‡	18–30		224	260	294			
PCB 1221‡	15–30	30	190	224	260			
PCB 1232‡	15–32		190	224	260			
PCB 1242‡	15–32		224	260	294			
PCB 1248‡	12–34		294	330	262			
PCB 1254‡	22–34	36	294	330	362			
PCB 1260‡	23–32		330	362	394			

* The proper chemical name is 2,2'-oxybis(1-chloropropane).

† See introductory section of text.

‡ These compounds are mixtures of various isomers. (See Figures 6410:2 through 12.)

Column conditions: Supelcoport (100/120 mesh) coated with 3% SP-2250 packed in a 1.8 m long × 2 mm ID glass column with helium carrier gas at 30 mL/min flow rate. Column temperature held isothermal at 50°C for 4 min, then programmed at 8°C/min to 270°C and held for 30 min.

TABLE 6410:II. CHROMATOGRAPHIC CONDITIONS, METHOD DETECTION LIMITS, AND CHARACTERISTIC MASSES FOR ACID EXTRACTABLES

Compound	Retention Time min	Method Detection Limit µg/L	Characteristic Masses					
			Electron Impact			Chemical Ionization		
			Primary	Secondary	Secondary	Methane	Methane	Methane
2-Chlorophenol	5.9	3.3	128	64	130	129	131	157
2-Nitrophenol	6.5	3.6	139	65	109	140	168	122
Phenol	8.0	1.5	94	65	66	95	123	135
2,4-Dimethylphenol	9.4	2.7	122	107	121	123	151	163
2,4-Dichlorophenol	9.8	2.7	162	164	98	163	165	167
2,4,6-Trichlorophenol	11.8	2.7	196	198	200	197	199	201
4-Chloro-3-methylphenol	13.2	3.0	142	107	144	143	171	183
2,4-Dinitrophenol	15.9	42	184	63	154	185	213	225
2-Methyl-4,6-dinitrophenol	16.2	24	198	182	77	199	227	239
Pentachlorophenol	17.5	3.6	266	264	268	267	265	269
4-Nitrophenol	20.3	2.4	65	139	109	140	168	122

Column conditions: Supelcoport (100/120 mesh) coated with 1% SP-1240DA packed in a 1.8 m long × 2 mm ID glass column with helium carrier gas at 30 mL/min flow rate. Column temperature held isothermal at 70°C for 2 min then programmed at 8°C/min to 200°C.

are not available, protect samples from light. Wash and rinse bottle and cap liner with acetone or methylene chloride, and dry before use. Follow conventional sampling practices[7] but do not rinse bottle with sample. Collect composite samples in refrigerated glass containers. Optionally, use automatic sampling equipment as free as possible of plastic tubing and other potential sources of contamination; incorporate glass sample containers for collecting a minimum of 250 mL. Refrigerate sample containers at 4°C and protect from light during compositing. If the sampler includes a peristaltic pump, use a minimum length of compressible silicone rubber tubing, but before use, thoroughly rinse it with methanol and rinse repeatedly with distilled water to minimize contamination. Use an integrating flow meter to collect flow-proportional composites.

Fill sample bottles and, if residual chlorine is present, add 80 mg sodium thiosulfate per liter of sample and mix well. Ice all samples or refrigerate at 4°C from time of collection until extraction.

Extract samples within 7 d of collection and analyze completely within 40 d of extraction.

3. Apparatus

a. Separatory funnel, 2-L, with TFE stopcock.

b. Drying column, chromatographic, 400 mm long × 19 mm ID, with coarse frit filter disk.

c. Concentrator tube, Kuderna-Danish, 10-mL, graduated.‡ Check calibration at volumes used. Use ground-glass stopper to prevent evaporation.

d. Evaporative flask, Kuderna-Danish, 500-mL.§ Attach to concentrator tube with springs.

e. Snyder column, Kuderna-Danish, three-ball macro.‖

f. Snyder column, Kuderna-Danish, two-ball micro.#

g. Vials, 10- to 15-mL, amber glass, with TFE-lined screw cap.

h. Continuous liquid-liquid extractor, equipped with TFE or glass connecting joints and stopcocks requiring no lubrication.**

i. Boiling chips, approximately 10/40 mesh. Heat to 400°C for 30 min or extract in a Soxhlet extractor with methylene chloride.

j. Water bath, heated, with concentric ring cover and temperature control to ± 2°C. Use bath in a hood.

k. Balance, analytical, capable of accurately weighing 0.0001 g.

l. Gas chromatograph:†† An analytical system complete with a temperature-programmable gas chromatograph and all required accessories including syringes, analytical columns, and gases. Use chromatograph with the injection port designed for on-column injection when packed columns are used and for splitless injection when capillary columns are used.

1) *Column for base/neutrals*, 1.8 m long × 2 mm ID glass, packed with 3% SP-2250 on Supelcoport (100/200 mesh) or equivalent. This column was used to develop the detection limit and precision and bias data presented herein. Guidelines for the use of alternate columns (e.g. DB-5 fused silica capillary) are provided in ¶ 5*b*.

2) *Column for acids*, 1.8 m long × 2 mm ID glass, packed with 1% SP-1240DA on Supelcoport (100/120 mesh) or equivalent. The detection limit and precision and

---

‡ Kontes K-570050-1025 or equivalent.
§ Kontes K-570001-0500 or equivalent.

‖ Kontes K-503000-0121 or equivalent.
Kontes K-569001-0219 or equivalent.
** Hershberg-Wolf Extractor, Ace Glass Co., Vineland, N.J., P/N 6841-10, or equivalent.
†† Gas chromatographic methods are extremely sensitive to the materials used. Mention of trade names by *Standard Methods* does not preclude the use of other existing or as-yet-undeveloped products that give *demonstrably* equivalent results.

bias data presented herein were developed with this column. For guidelines for the use of alternate columns (e.g. DB-5 fused silica capillary) see ¶ 5b.

m. Mass spectrometer, capable of scanning from 35 to 450 amu every 7 s or less, utilizing 70-V (nominal) electron energy in the electron impact ionization mode, and producing a mass spectrum that meets all the criteria in Table 6410:III when 50 ng of decafluorotriphenyl phosphine [DFTPP; bis(perfluorophenyl) phenyl phosphine] is injected through the GC inlet.

n. GC/MS interface: See Section 6210B.3d.

o. Data system: See Section 6210B.3e.

4. Reagents

a. Reagent water: See Section 6210B.4a.

b. Sodium hydroxide solution, NaOH, 10N: Dissolve 40 g NaOH in reagent water and dilute to 100 mL.

c. Sodium sulfate, Na_2SO_4, granular, anhydrous. Purify by heating at 400°C for 4 h in a shallow tray.

d. Sodium thiosulfate, $Na_2S_2O_3 \cdot 5H_2O$, granular.

e. Sulfuric acid, H_2SO_4, 1 + 1: Slowly add 50 mL conc H_2SO_4 to 50 mL reagent water.

TABLE 6410:III. DFTPP KEY MASSES AND ABUNDANCE CRITERIA

Mass	m/z Abundance Criteria
51	30–60% of mass 198
68	Less than 2% of mass 69
70	Less than 2% of mass 69
127	40–60% of mass 198
197	Less than 1% of mass 198
198	Base peak, 100% relative abundance
199	5–9% of mass 198
275	10–30% of mass 198
365	Greater than 1% of mass 198
441	Present but less than mass 443
442	Greater than 40% of mass 198
443	17–23% of mass 442

f. Acetone, methanol, methylene chloride, pesticide quality or equivalent.

g. Stock standard solutions: Prepare from pure standard materials or purchase as certified solutions. Prepare by accurately weighing about 0.0100 g of pure material, dissolve in pesticide-quality acetone or other suitable solvent, and dilute to volume in a 10-mL volumetric flask; 1 μL = 1.00 μg compound. When compound purity is assayed to be 96% or greater, use the weight without correction to calculate concentration of the stock standard. Use commercially prepared stock standards at any concentration if certified by the manufacturer or by an independent source.

Transfer stock standard solutions into TFE-sealed screw-cap bottles. Store at 4°C and protect from light. Check stock standard solutions frequently for signs of degradation or evaporation, especially just before preparing calibration standards. Replace stock standard solutions after 6 months, or sooner if comparison with check standards indicates a problem.

h. Surrogate standard known-addition solution: Select a minimum of three surrogate compounds from Table 6410:IV. Prepare a surrogate standard solution containing each selected surrogate compound at a concentration of 100 μg/mL in acetone. Adding 1.00 mL to 1000 mL sample is equivalent to a concentration of 100 μg/L of each surrogate standard. Store at 4°C in TFE-sealed glass container. Check solution frequently for stability. Replace solution after 6 months, or sooner if comparison with quality-control check standards indicates a problem.

i. DFTPP standard: Prepare a 25-μg/mL solution of DFTPP in acetone.

j. Calibration standards: Prepare calibration standards at a minimum of three concentration levels for each compound by adding appropriate volumes of one or more stock standards to a volumetric flask. To each calibration standard or standard mixture, add a known constant amount of one

TABLE 6410:IV. SUGGESTED INTERNAL AND SURROGATE STANDARDS

Base/Neutral Fraction	Acid Fraction
Aniline-d_5	2-Fluorophenol
Anthracene-d_{10}	Pentafluorophenol
Benzo(a)anthracene-d_{12}	Phenol-d_5
4,4'-Dibromobiphenyl	2-Perfluoromethyl phenol
4,4'-Dibromooctafluorobiphenyl	
Decafluorobiphenyl	
2,2'-Difluorobiphenyl	
4-Fluoroaniline	
1-Fluoronaphthylene	
2-Fluoronaphthylene	
Naphthalene-d_8	
Nitrobenzene-d_5	
2,3,4,5,6-Pentafluorobiphenyl	
Phenanthrene-d_{10}	
Pyridine-d_5	

or more internal standards (such as those listed in Table 6410:IV), and dilute to volume with acetone. Prepare one calibration standard at a concentration near, but above, the MDL and others corresponding to the expected range of sample concentrations or defining the working range of the GC/MS system.

k. Quality control (QC) check sample concentrate: Obtain a check sample concentrate‡‡ containing each compound at a concentration of 100 µg/mL in acetone. Multiple solutions may be required. PCBs and multicomponent pesticides may be omitted. If such a sample is not available from an external source, prepare using stock standards prepared independently from those used for calibration.

5. Procedure

a. Extraction: Extraction by means of a separatory funnel, ¶ 1), is most common, but if emulsions will prevent acceptable solvent recovery, use continuous extraction, ¶ 2).

‡‡ For U.S. federal permit-related analysis, use samples obtainable from U.S. EPA Environmental Monitoring and Support Laboratory, Cincinnati, Ohio.

1) Separatory funnel extraction—Normally use a sample volume of 1 L. For sample volumes of 2 L, use 250-, 100-, and 100-mL volumes of methylene chloride for the serial extraction of the base/neutrals and 200-, 100-, and 100-mL volumes of methylene chloride for the acids.

Mark water meniscus on side of sample bottle for later determination of sample volume. Pour entire sample into a 2-L separatory funnel. Pipet 1.00 mL surrogate standard solution into separatory funnel and mix well. Check pH with wide-range pH paper and adjust to pH > 11 with NaOH solution.

Add 60 mL methylene chloride to sample bottle, seal, and shake for 30 s to rinse inner surface. Transfer solvent to separatory funnel and extract sample by shaking for 2 min with periodic venting to release excess pressure. Let organic layer separate from water phase for a minimum of 10 min. If emulsion interface between layers is more than one-third the volume of the solvent layer, use mechanical techniques to complete phase separation. The optimum technique depends on the sample, but may include stirring, filtering emulsion through glass wool, centrifuging, or other physical methods. Collect methylene chloride extract in a 250-mL erlenmeyer flask.

If the emulsion cannot be broken (recovery of less than 80% of the methylene chloride, corrected for water solubility of methylene chloride) in the first extraction, transfer sample, solvent, and emulsion into extraction chamber of a continuous extractor and proceed as described in ¶ 2) below.

Add a second 60-mL volume of methylene chloride to sample bottle and repeat extraction procedure, combining extracts in the erlenmeyer flask. Perform a third extraction in the same manner.

After the third extraction, adjust pH of aqueous phase to <2 using H_2SO_4. Serially extract acidified aqueous phase three times with 60-mL portions of methylene chloride. Collect and combine extracts in a 250-mL erlenmeyer flask and label combined extracts as the acid fraction.

For each fraction, assemble a Kuderna-Danish (K-D) concentrator by attaching a 10-mL concentrator tube to a 500-mL evaporative flask. Other concentration devices or techniques may be used if the requirements of ¶ 7 are met.

Pour combined extract through a solvent-rinsed drying column containing at least 10 cm anhydrous Na_2SO_4 or more and collect extract in concentrator. Rinse erlenmeyer flask and column with 20 to 30 mL methylene chloride to complete transfer.

Add one or two clean boiling chips to the evaporative flask and attach a three-ball Snyder column. Prewet Snyder column by adding about 1 mL methylene chloride to the top. Place K-D apparatus on a hot water bath (60 to 65°C) in a hood so that concentrator tube is partially immersed in the hot water and entire lower rounded surface of flask is bathed with hot vapor. Adjust vertical position of apparatus and water temperature as required to complete concentration in 15 to 20 min. At proper rate of distillation the column balls actively

chatter but the chambers are not flooded with condensed solvent. When the apparent volume of liquid reaches 1 mL, remove K-D apparatus and let drain and cool for at least 10 min.

Remove Snyder column and rinse flask and its lower joint into the concentrator tube with 1 to 2 mL methylene chloride, preferably using a 5-mL syringe.

Add another one or two clean boiling chips to concentrator tube for each fraction and attach a two-ball micro-Snyder column. Prewet Snyder column by adding about 0.5 mL of methylene chloride to the top. Place K-D apparatus on a hot-water bath (60 to 65°C) so that concentrator tube is partially immersed in hot water and continue concentrating as directed above without further solvent addition until apparent volume of liquid reaches about 0.5 mL. After cooling, remove Snyder column and rinse flask and its lower joint into the concentrator tube with approximately 0.2 mL acetone or methylene chloride. Adjust final volume to 1.0 mL with solvent. Stopper concentrator tube and store refrigerated if further processing will not be done immediately. If extract is to be stored longer than 2 d, transfer to a TFE-sealed screw-cap vial and label base/neutral or acid fraction as appropriate.

Determine original sample volume by refilling sample bottle to mark and transferring liquid to a 1000-mL graduated cylinder. Record sample volume to nearest 5 mL.

2) Continuous extraction—Mark water meniscus on side of sample bottle, and determine sample volume later as described in ¶ 1). Check pH with wide-range pH paper and adjust to pH >11 with NaOH solution. Transfer sample to continuous extractor and, using a pipet, add 1.00 mL surrogate standard solution and mix well. Add 60 mL methylene chloride to sample bottle, seal, and shake for 30 s to rinse inner surface. Transfer solvent to extractor. Repeat rinse with an additional 50- to 100-

mL portion methylene chloride and add rinse to extractor.

Add 200 to 500 mL methylene chloride to distilling flask, add sufficient reagent water to ensure proper operation, and extract for 24 h. Let cool and detach distilling flask. Dry, concentrate, and seal extract as in ¶ 1) above.

Charge a clean distilling flask with 500 mL methylene chloride and attach it to continuous extractor. Carefully, while stirring, adjust pH of aqueous phase to less than 2 with H_2SO_4. Extract for 24 h. Dry, concentrate, and seal extract as in ¶ 1) above.

b. *GC/MS operating conditions:* Table 6410:I summarizes the recommended gas

chromatographic operating conditions for the base/neutral fraction and Table 6410:II for the acid fraction. Included in these tables are retention times and MDLs that can be achieved under these conditions. Examples of the separations obtained with these columns are shown in Figures 6410:1 through 12. Other packed or capillary (open-tubular) columns or chromatographic conditions may be used if the requirements of ¶ 7 are met.

c. *GC/MS performance tests:* At the beginning of each day on which analyses are to be performed, check GC/MS system to see if acceptable performance criteria are achieved for DFTPP.[8] Each day that benzidine is to be determined, the tailing factor

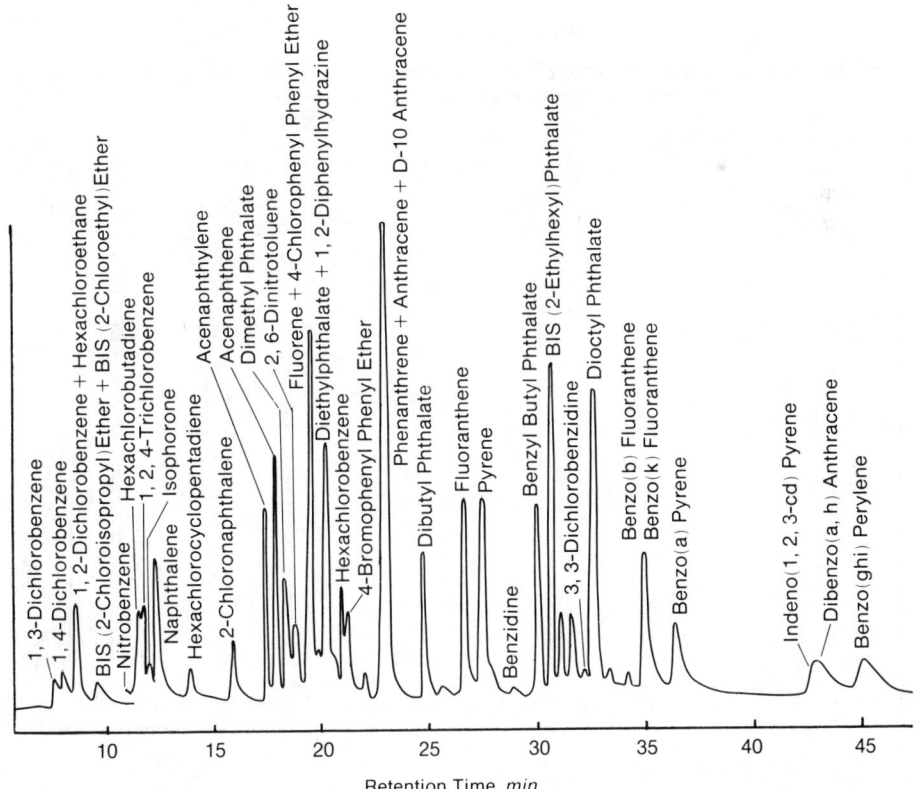

Figure 6410:1. Gas chromatogram of base/neutral fraction. Column: 3% SP-2250 on Supelcoport; program: 50°C for 4 min, 8°C/min to 270°C; detector: mass spectrometer.

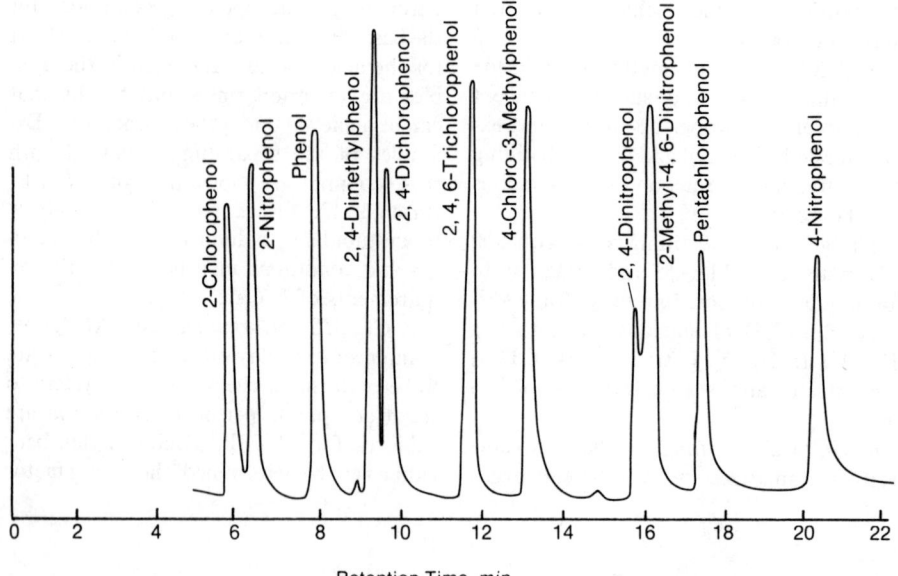

Figure 6410:2. Gas chromatogram of acid fraction. Column: 1% SP-1240DA on Supelcoport; program: 70°C for 2 min, 8°C/min to 200°C; detector: mass spectrometer.

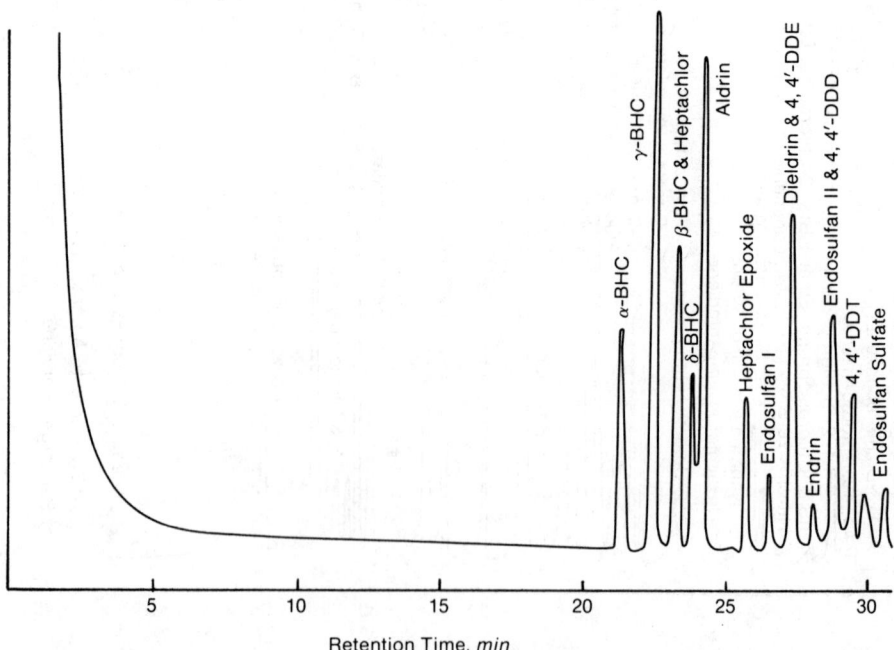

Figure 6410:3. Gas chromatogram of pesticide fraction. Column: 3% SP-2250 on Supelcoport; program: 50°C for 4 min, 8°C/min to 270°C; detector: mass spectrometer.

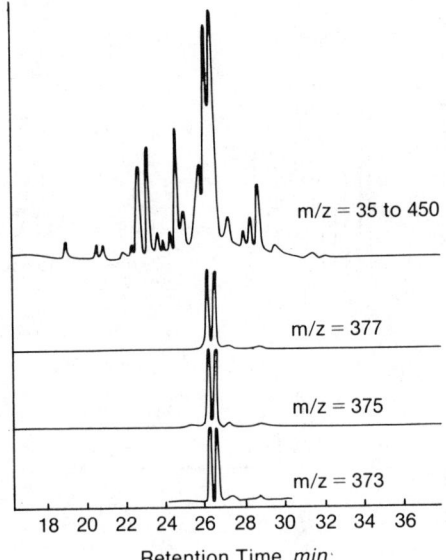

Figure 6410:4. Gas chromatogram of chlordane. Column: 3% SP-2250 on Supelcoport; program: 50°C for 4 min, 8°C/min to 270°C; detector: mass spectrometer.

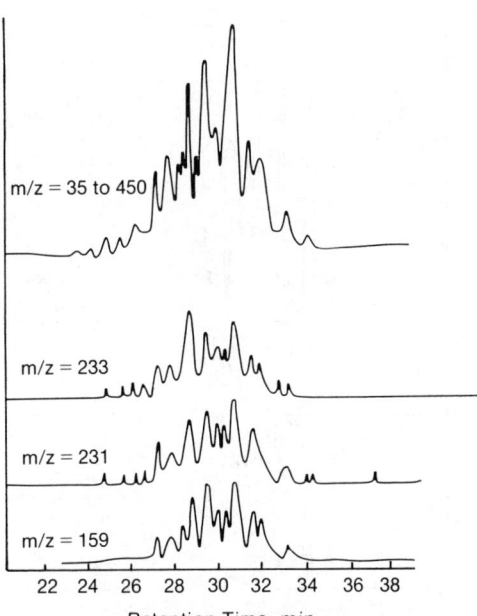

Figure 6410:5. Gas chromatogram of toxaphene. Column: 3% SP-2250 on Supelcoport; program: 50°C for 4 min, 8°C/min to 270°C; detector: mass spectrometer.

criterion described in ¶ 2) must be achieved. Each day that the acids are to be determined, the tailing factor criterion described in ¶ 3) must be achieved.

These performance tests have the requirements given in Section 6210B.5*b*, but use following conditions:

Electron energy: 70 V (nominal)
Mass range: 35 to 450 amu
Scan time: To give at least 5 scans per peak but not to exceed 7 s per scan.

1) *DFTPP performance test*—At beginning of each day, inject 2 μL (50 ng) DFTPP standard solution. Obtain a background-corrected mass spectrum of DFTPP and confirm that all the key *m/z* criteria in Table 6410:III are achieved. If not, retune mass spectrometer and repeat test until all criteria are achieved. Meet

performance criteria before any samples, blanks, or standards are analyzed. The tailing factor tests in ¶s 2) and 3) may be performed simultaneously with the DFTPP test.

2) *Column performance test for base/neutrals* — At beginning of each day that base/neutral fraction is to be analyzed for benzidine, calculate benzidine tailing factor. Inject 100 ng benzidine either separately or as a part of a standard mixture that may contain DFTPP, and calculate tailing factor, which must be less than 3.0. Calculation of the tailing factor is illustrated in Figure 6410:13.[9] Replace column packing if tailing factor criterion cannot be met.

3) *Column performance test for acids*— At beginning of each day that acids are to

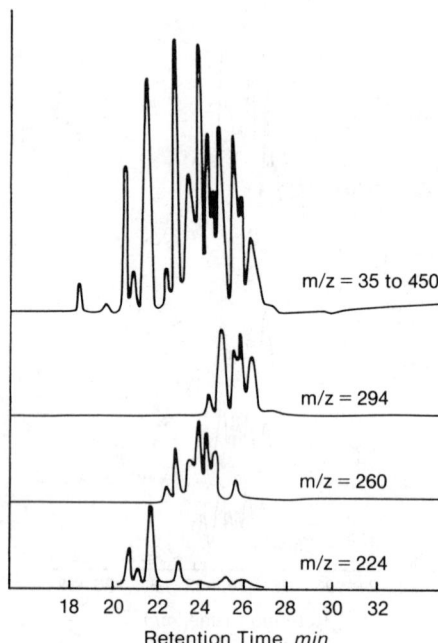

Figure 6410:6. Gas chromatogram of PCB-1016. Column: 3% SP-2250 on Supelcoport; program: 50°C for 4 min, 8°C/min to 270°C; detector: mass spectrometer.

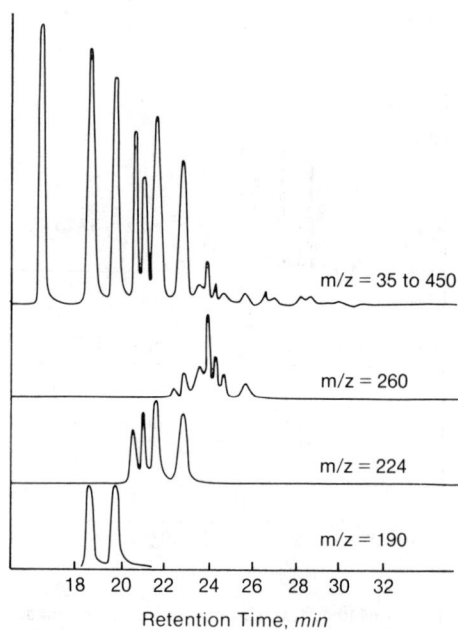

Figure 6410:7. Gas chromatogram of PCB-1221. Column: 3% SP-2250 on Supelcoport; program: 50°C for 4 min, 8°C/min to 270°C; detector: mass spectrometer.

be determined, inject 50 ng pentachlorophenol either separately or as a part of a standard mix that may contain DFTPP. The tailing factor for pentachlorophenol must be less than 5. Calculation of the tailing factor is illustrated in Figure 6410:13.[9] Replace column packing if tailing factor criterion cannot be met.

d. Calibration of GC/MS system: Calibrate system daily after performance tests.

Select three or more internal standards similar in analytical behavior to the compounds of interest. Demonstrate that the measurement of the internal standards is not affected by method or matrix interferences. Some recommended internal standards are listed in Table 6410:IV. Use base peak *m/z* as the primary *m/z* for quantification. If interferences are noted, use one of the next two most intense *m/z* quantities for quantification. Using injections of 2 to 5 μL, analyze each calibration standard according to ¶ *e* below and tabulate area of primary characteristic *m/z* (Tables 6410:I and II) against concentration for each compound and internal standard. Calculate response factors (RF) for each compound by the equation given in Section 6210B.5c2). If the RF value over the working range is a constant (< 35% RSD), it can be assumed to be invariant; use the average RF for calculations. Alternatively, use the results to plot a calibration curve of response ratios, A_s/A_{is} vs. *RF*.

Verify working calibration curve or RF on each working day by measuring one or

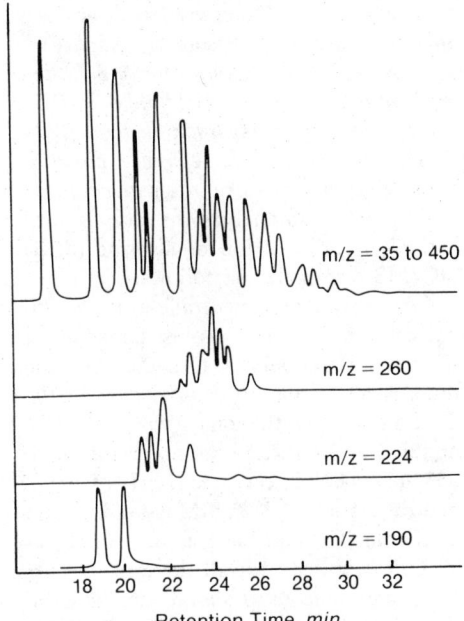

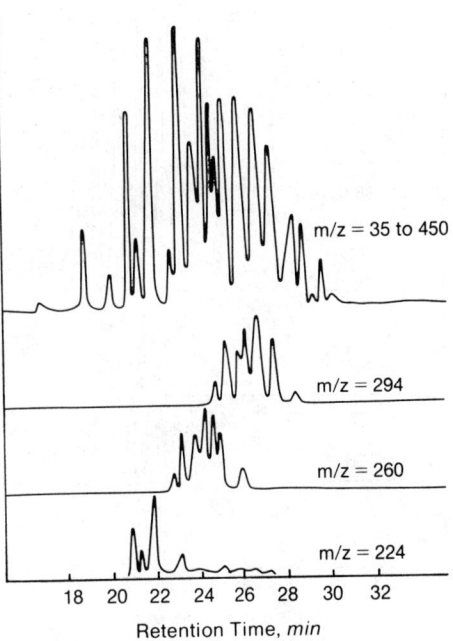

Figure 6410:8. Gas chromatogram of PCB-1232. Column: 3% SP-2250 on Supelcoport; program: 50°C for 4 min, 8°C/min to 270°C; detector: mass spectrometer.

Figure 6410:9. Gas chromatogram of PCB-1242. Column: 3% SP-2250 on Supelcoport; program: 50°C for 4 min, 8°C/min to 270°C; detector: mass spectrometer.

more calibration standards. If the response for any compound varies from the predicted response by more than 20%, repeat test using a fresh calibration standard. Alternatively, prepare a new calibration curve for that compound.

e. Sample analysis: Add internal standard to sample extract, mix thoroughly, and immediately inject 2 to 5 µL of sample extract or standard into GC/MS system using solvent-flush technique[10] to minimize losses due to adsorption, chemical reaction, or evaporation. Smaller (1.0-µL) volumes may be injected if automatic devices are used. Record volume injected to nearest 0.05 µL. If response for any m/z exceeds

the working range of the GC/MS system, dilute extract and reanalyze. Make all qualitative and quantitative measurements as described below and in ¶ 6. When extract is not being used, store at 4°C, protected from light, in screw-cap vial equipped with unpierced TFE-lined septum.

Obtain EICPs for the primary m/z and the two other masses listed in Tables 6410:I and II. See ¶ *d* for masses to be used with internal and surrogate standards. Use the following criteria to make a qualitative identification:

• The characteristic masses of each compound maximize in the same or within one scan of each other.

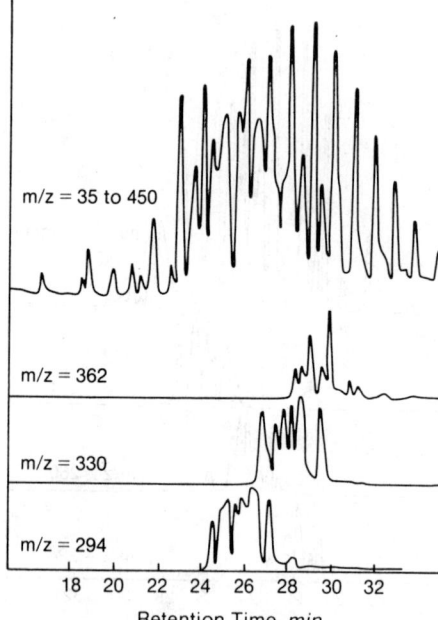

m/z = 35 to 450

m/z = 362

m/z = 330

m/z = 294

18 20 22 24 26 28 30 32
Retention Time, *min*

Figure 6410:10. Gas chromatogram of PCB-1248.
Column: 3% SP-2250 on Supel-
coport; program: 50°C for 4 min,
8°C/min to 270°C; detector: mass
spectrometer.

• The retention time falls within ± 30 s
of the retention time of the authentic compound.

• The relative peak heights of the three
characteristic masses in the EICPs fall
within ± 20% of the relative intensities of
these masses in a reference mass spectrum
obtained from a standard analyzed in the
GC/MS system or from a reference library.

Structural isomers that have very similar
mass spectra and less than 30 s difference
in retention time can be identified explicitly
only if the resolution between authentic isomers in a standard mix is acceptable. Acceptable resolution is achieved if the
baseline to valley height between the isomers is less than 25% of the sum of the

two peak heights. Otherwise, structural isomers are identified as isomeric pairs.

*f. Screening procedure for 2,3,7,8-tet-
rachlorodibenzo-p-dioxin (2,3,7,8-TCDD):*
CAUTION: *In screening a sample for 2,3,7,8-
TCDD, do not handle reference material
without taking extensive safety precautions.*
It is sufficient to analyze the base/neutral
extract by selected ion monitoring (SIM)
GC/MS techniques, as follows:

Concentrate base/neutral extract to a final volume of 0.2 mL. Adjust temperature
of base/neutral column to 220°C. Operate
mass spectrometer to acquire data in the
SIM mode using the ions at m/z 257, 320,
and 322 and a dwell time no greater than
333 ms/mass. Inject 5 to 7 μL of base/
neutral extract. Collect SIM data for a total
of 10 min. The possible presence of 2,3,7,8-
TCDD is indicated if all three masses exhibit simultaneous peaks at any point in
the selected ion current profiles. For each
occurrence where the possible presence of
2,3,7,8-TCDD is indicated, calculate and
retain the relative abundances of each of
the three masses. False positives may be
caused by the presence of single or coeluting combinations of compounds whose
mass spectra contain all of these masses.
Conclusive results of the presence and concentration level of 2,3,7,8-TCDD can be
obtained only from a properly equipped
laboratory using a specialized test
method.[11]

6. Calculation

When a compound has been identified,
base quantitation on the integrated abundance from the EICP of the primary characteristic m/z given in Tables 6410:I and
II. Use base peak m/z for internal and
surrogate standards. If sample produces an
interference for the primary m/z, use a secondary characteristic m/z to quantitate.

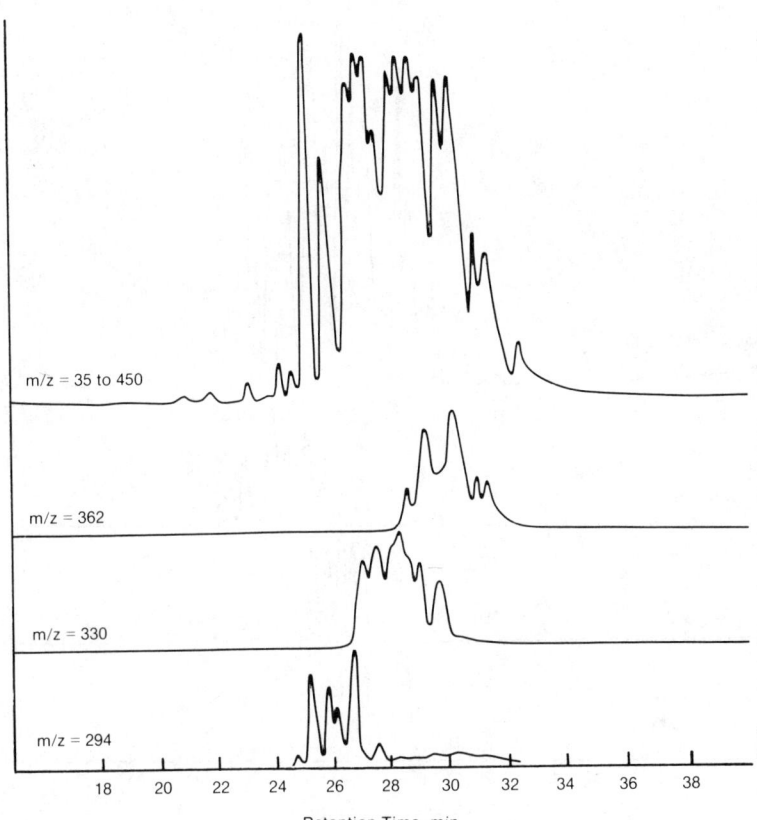

Figure 6410:11. Gas chromatogram of PCB-1254. Column: 3% SP-2250 on Supelcoport; program: 50°C for 4 min, 8°C/min to 270°C; detector: mass spectrometer.

Calculate sample concentration using the response factor (RF) determined in ¶ 5d and the equation:

$$\text{Concentration, } \mu\text{g/L} = \frac{(A_s)\,(I_s)}{(A_{is})\,(RF)\,(V_e)}$$

where:

A_s = area of characteristic m/z for compound or surrogate standard to be measured,

A_{is} = area of characteristic m/z for internal standard,

I_s = amount of internal standard added to each extract, μg, and

V_e = volume of water extracted, L.

Report results in μg/L without correction for recovery. Report all QC data with sample results.

7. Quality Control

a. Quality control program: See Section 6210B.7*a*.

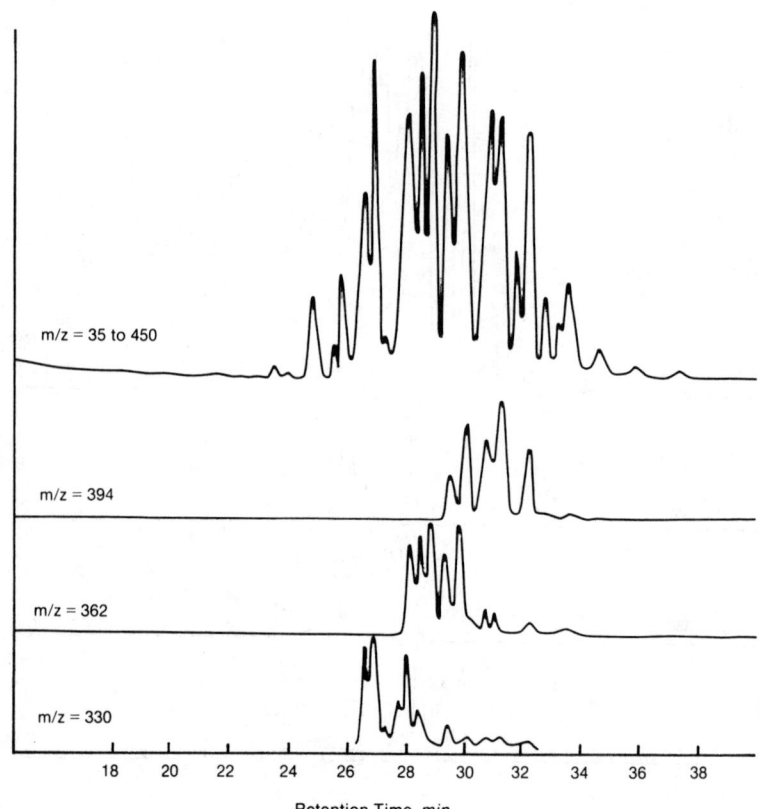

Figure 6410:12. Gas chromatogram of PCB-1260. Column: 3% SP-2250 on Supelcoport; program: 50°C for 4 min, 8°C/min to 270°C; detector: mass spectrometer.

b. Analyst proficiency check: Proceed according to Section 6210B.7*b*. Use Table 6410:V for acceptance criteria.

c. Analyses of samples with known additions: Make known additions to at least 5% of the samples from each sample site being monitored. Procedure is as directed in Section 6210B.7*c*, but use 100 µg/L instead of 20 µg/L. In determining background concentration of each compound, add to the second sample portion 1.0 mL

of QC check sample concentrate. Use quality acceptance criteria given in Table 6410:V.

d. Quality-control check standard analysis: Proceed as in Section 6210B.7*d*, but prepare QC check standard with 1.0 mL QC check standard concentrate and 1 L reagent water.

e. Accuracy assessment and records: See Section 6210B.7*e*.

f. Use of surrogate compounds: As a qual-

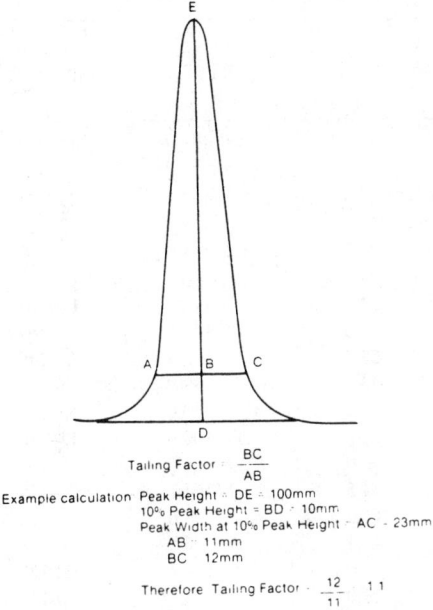

$$\text{Tailing Factor} = \frac{BC}{AB}$$

Example calculation: Peak Height = DE = 100mm
10% Peak Height = BD = 10mm
Peak Width at 10% Peak Height = AC = 23mm
AB = 11mm
BC = 12mm

$$\text{Therefore: Tailing Factor} = \frac{12}{11} = 1.1$$

Figure 6410:13. Tailing factor calculation.

ity control check, make known additions to all samples of surrogate standard solution as described in ¶ 5a1), and calculate percent recovery of each surrogate compound.

g. Additional quality-assurance practices: Other desirable practices depend on the needs of the laboratory and the nature of the samples. Analyze field duplicates to assess precision of environmental measurements. Whenever possible, analyze standard reference materials and participate in relevant performance evaluation studies. Certain compounds, such as phthalates, are common laboratory contaminants. When these are measured above the detection limits in sample blanks, locate their source and repeat the analysis after taking corrective action.

8. Precision and Bias

This method was tested by 15 laboratories using reagent water, drinking water, surface water, and industrial wastewaters with additions at six concentrations over the range 5 to 1300 µg/L.[3] Single-operator precision, overall precision, and method bias were found to be related directly to the compound concentration and essentially independent of the sample matrix. Linear equations describing these relationships are presented in Table 6410:VI.

9. References

1. U.S. ENVIRONMENTAL PROTECTION AGENCY. 1984. Method 625—Base/neutrals and acids. 40 CFR Part 136, 43385; *Federal Register* 49, No. 209.
2. U.S. ENVIRONMENTAL PROTECTION AGENCY. 1977. Sampling and Analysis Procedures for Screening of Industrial Effluents for Priority Pollutants. Environmental Monitoring and Support Lab., Cincinnati, Ohio.
3. U.S. ENVIRONMENTAL PROTECTION AGENCY. 1984. EPA Method Study 30, Method 625—Base/Neutrals, Acids, and Pesticides. EPA-600/4-84-053, National Technical Information Serv., PB84-206572, Springfield, Va.
4. AMERICAN SOCIETY FOR TESTING AND MATERIALS. 1978. Standard practices for preparation of sample containers and for preservation of organic constituents. ASTM Annual Book of Standards, Part 31, D3694-78. Philadelphia, Pa.
5. U.S. ENVIRONMENTAL PROTECTION AGENCY. 1984. Definition and procedure for the determination of the method detection limit. 40 CFR Part 136, Appendix B. *Federal Register* 49, No. 209.

TABLE 6410:V. QC ACCEPTANCE CRITERIA*

Compound	Test Concentration $\mu g/L$	Limits for s $\mu g/L$	Range for $\overline{X}$ $\mu g/L$	Range for P, P_s %
Acenaphthene	100	27.6	60.1–132.3	47–145
Acenaphthylene	100	40.2	53.5–126.0	33–145
Aldrin	100	39.0	7.2–152.2	D–166
Anthracene	100	32.0	43.4–118.0	27–133
Benzo(a)anthracene	100	27.6	41.8–133.0	33–143
Benzo(b)fluoranthene	100	38.8	42.0–140.4	24–159
Benzo(k)fluoranthene	100	32.3	25.2–145.7	11–162
Benzo(a)pyrene	100	39.0	31.7–148.0	17–163
Benzo(ghi)perylene	100	58.9	D–195.0	D–219
Benzyl butyl phthalate	100	23.4	D–139.9	D–152
δ-BHC	100	31.5	41.5–130.6	24–149
β-BHC	100	21.6	D–100.0	D–110
bis(2-Chloroethyl) ether	100	55.0	42.9–126.0	12–158
bis(2-Chloroethoxy) methane	100	34.5	49.2–164.7	33–184
bis(2-Chloroisopropyl) ether†	100	46.3	62.8–138.6	36–166
bis(2-Ethylhexyl) phthalate	100	41.1	28.9–136.8	8–158
4-Bromophenyl phenyl ether	100	23.0	64.9–114.4	53–127
2-Chloronaphthalene	100	13.0	64.5–113.5	60–118
4-Chlorophenyl phenyl ether	100	33.4	38.4–144.7	25–158
Chrysene	100	48.3	44.1–139.9	17–168
4,4'-DDD	100	31.0	D–134.5	D–145
4,4'-DDE	100	32.0	19.2–119.7	4–136
4,4'-DDT	100	61.6	D–170.6	D–203
Dibenzo(a,h)anthracene	100	70.0	D–199.7	D–227
Di-n-butyl phthalate	100	16.7	8.4–111.0	1–118
1,2-Dichlorobenzene	100	30.9	48.6–112.0	32–129
1,3-Dichlorobenzene	100	41.7	16.7–153.9	D–172
1,4-Dichlorobenzene	100	32.1	37.3–105.7	20–124
3,3'-Dichlorobenzidine	100	71.4	8.2–212.5	D–262
Dieldrin	100	30.7	44.3–119.3	29–136
Diethyl phthalate	100	26.5	D–100.0	D–114
Dimethyl phthalate	100	23.2	D–100.0	D–112
2,4-Dinitrotoluene	100	21.8	47.5–126.9	39–139
2,6-Dinitrotoluene	100	29.6	68.1–136.7	50–158
Di-n-octylphthalate	100	31.4	18.6–131.8	4–146
Endosulfan sulfate	100	16.7	D–103.5	D–107
Endrin aldehyde	100	32.5	D–188.8	D–209
Fluoranthene	100	32.8	42.9–121.3	26–137
Fluorene	100	20.7	71.6–108.4	59–121
Heptachlor	100	37.2	D–172.2	D–192
Heptachlor epoxide	100	54.7	70.9–109.4	26–155
Hexachlorobenzene	100	24.9	7.8–141.5	D–152
Hexachlorobutadiene	100	26.3	37.8–102.2	24–116
Hexachloroethane	100	24.5	55.2–100.0	40–113
Indeno(1,2,3-cd)pyrene	100	44.6	D–150.9	D–171
Isophorone	100	63.3	46.6–180.2	21–196
Naphthalene	100	30.1	35.6–119.6	21–133

TABLE 6410:V, CONT.

Compound	Test Concentration $\mu g/L$	Limits for s $\mu g/L$	Range for $\overline{X}$ $\mu g/L$	Range for P, P_s %
Nitrobenzene	100	39.3	54.3–157.6	35–180
N-Nitrosodi-n-propylamine	100	55.4	13.6–197.9	D–230
PCB-1260	100	54.2	19.3–121.0	D–164
Phenanthrene	100	20.6	65.2–108.7	54–120
Pyrene	100	25.2	69.6–100.0	52–115
1,2,4-Trichlorobenzene	100	28.1	57.3–129.2	44–142
4-Chloro-3-methylphenol	100	37.2	40.8–127.9	22–147
2-Chlorophenol	100	28.7	36.2–120.4	23–134
2,4-Dichlorophenol	100	26.4	52.5–121.7	39–135
2,4-Dimethylphenol	100	26.1	41.8–109.0	32–119
2,4-Dinitrophenol	100	49.8	D–172.9	D–191
2-Methyl-4,6-dinitrophenol	100	93.2	53.0–100.0	D–181
2-Nitrophenol	100	35.2	45.0–166.7	29–182
4-Nitrophenol	100	47.2	13.0–106.5	D–132
Pentachlorophenol	100	48.9	38.1–151.8	14–176
Phenol	100	22.6	16.6–100.0	5–112
2,4,6-Trichlorophenol	100	31.7	52.4–129.2	37–144

* s = standard deviation for four recovery measurements,
$\overline{X}$ = average recovery for four recovery measurements,
P, P_s = percent recovery measured, and
D = detected; results must be greater than zero.
† The proper chemical name is 2,2′-oxybis(1-chloropropane).
NOTE: These criteria are based directly upon the method performance data in Table 6410:VI. Where necessary, the limits for recovery were broadened to assure applicability of the limits to concentrations below those used to develop Table 6410:VI.

TABLE 6410:VI. METHOD BIAS AND PRECISION AS FUNCTIONS OF CONCENTRATION*

Compound	Bias as Recovery, X' $\mu g/L$		Single-Analyst Precision, $s,'$ $\mu g/L$		Overall Precision, S' $\mu g/L$	
Acenaphthene	$0.96C$ +	0.19	$0.15\overline{X}$ −	0.12	$0.21\overline{X}$ −	0.67
Acenaphthylene	$0.89C$ +	0.74	$0.24\overline{X}$ −	1.06	$0.26\overline{X}$ −	0.54
Aldrin	$0.78C$ +	1.66	$0.27\overline{X}$ −	1.28	$0.43\overline{X}$ +	1.13
Anthracene	$0.80C$ +	0.68	$0.21\overline{X}$ −	0.32	$0.27\overline{X}$ −	0.64
Benzo(a)anthracene	$0.88C$ −	0.60	$0.15\overline{X}$ +	0.93	$0.26\overline{X}$ −	0.28
Benzo(b)fluoranthene	$0.93C$ −	1.80	$0.22\overline{X}$ +	0.43	$0.29\overline{X}$ +	0.96
Benzo(k)fluoranthene	$0.87C$ −	1.56	$0.19\overline{X}$ +	1.03	$0.35\overline{X}$ +	0.40
Benzo(a)pyrene	$0.90C$ −	0.13	$0.22\overline{X}$ +	0.48	$0.32\overline{X}$ +	1.35
Benzo(ghi)perylene	$0.98C$ −	0.86	$0.29\overline{X}$ +	2.40	$0.51\overline{X}$ −	0.44
Benzyl butyl phthalate	$0.66C$ −	1.68	$0.18\overline{X}$ +	0.94	$0.53\overline{X}$ +	0.92
β-BHC	$0.87C$ −	0.94	$0.20\overline{X}$ −	0.58	$0.30\overline{X}$ −	1.94
δ-BHC	$0.29C$ −	1.09	$0.34\overline{X}$ +	0.86	$0.93\overline{X}$ −	0.17
bis(2-Chloroethyl) ether	$0.86C$ −	1.54	$0.35\overline{X}$ −	0.99	$0.35\overline{X}$ +	0.10
bis(2-Chloroethoxy) methane	$1.12C$ −	5.04	$0.16\overline{X}$ +	1.34	$0.26\overline{X}$ +	2.01
bis(2-Chloroisopropyl) ether†	$1.03C$ −	2.31	$0.24\overline{X}$ +	0.28	$0.25\overline{X}$ +	1.04
bis(2-Ethylhexyl) phthalate	$0.84C$ −	1.18	$0.26\overline{X}$ +	0.73	$0.36\overline{X}$ +	0.67
4-Bromophenyl phenyl ether	$0.91C$ −	1.34	$0.13\overline{X}$ +	0.66	$0.16\overline{X}$ +	0.66
2-Chloronaphthalene	$0.89C$ +	0.01	$0.07\overline{X}$ +	0.52	$0.13\overline{X}$ +	0.34
4-Chlorophenyl phenyl ether	$0.91C$ +	0.53	$0.20\overline{X}$ −	0.94	$0.30\overline{X}$ −	0.46
Chrysene	$0.93C$ −	1.00	$0.28X$ +	0.13	$0.33\overline{X}$ −	0.09
4,4'-DDD	$0.56C$ −	0.40	$0.29\overline{X}$ −	0.32	$0.66\overline{X}$ −	0.96
4,4'-DDE	$0.70C$ −	0.54	$0.26\overline{X}$ −	1.17	$0.39\overline{X}$ −	1.04
4,4'-DDT	$0.79C$ −	3.28	$0.42\overline{X}$ +	0.19	$0.65\overline{X}$ −	0.58
Dibenzo(a,h)anthracene	$0.88C$ +	4.72	$0.30\overline{X}$ +	8.51	$0.59\overline{X}$ +	0.25
Di-n-butyl phthalate	$0.59C$ +	0.71	$0.13\overline{X}$ +	1.16	$0.39\overline{X}$ +	0.60
1,2-Dichlorobenzene	$0.80C$ +	0.28	$0.20\overline{X}$ +	0.47	$0.24\overline{X}$ +	0.39
1,3-Dichlorobenzene	$0.86C$ −	0.70	$0.25\overline{X}$ +	0.68	$0.41\overline{X}$ +	0.11
1,4-Dichlorobenzene	$0.73C$ −	1.47	$0.24\overline{X}$ +	0.23	$0.29\overline{X}$ +	0.36
3,3'-Dichlorobenzidine	$1.23C$ −	12.65	$0.28\overline{X}$ +	7.33	$0.47\overline{X}$ +	3.45
Dieldrin	$0.82C$ −	0.16	$0.20\overline{X}$ −	0.16	$0.26\overline{X}$ −	0.07
Diethyl phthalate	$0.43C$ +	1.00	$0.28\overline{X}$ +	1.44	$0.52\overline{X}$ +	0.22
Dimethyl phthalate	$0.20C$ +	1.03	$0.54\overline{X}$ +	0.19	$1.05\overline{X}$ −	0.92
2,4-Dinitrotoluene	$0.92C$ −	4.81	$0.12\overline{X}$ +	1.06	$0.21\overline{X}$ +	1.50
2,6-Dinitrotoluene	$1.06C$ −	3.60	$0.14X$ +	1.26	$0.19X$ +	0.35
Di-n-octylphthalate	$0.76C$ −	0.79	$0.21\overline{X}$ +	1.19	$0.37\overline{X}$ +	1.19
Endosulfan sulfate	$0.39C$ +	0.41	$0.12\overline{X}$ +	2.47	$0.63\overline{X}$ −	1.03
Endrin aldehyde	$0.76C$ −	3.86	$0.18\overline{X}$ +	3.91	$0.73\overline{X}$ −	0.62
Fluoranthene	$0.81C$ +	1.10	$0.22\overline{X}$ −	0.73	$0.28\overline{X}$ −	0.60
Fluorene	$0.90C$ −	0.00	$0.12\overline{X}$ +	0.26	$0.13\overline{X}$ +	0.61
Heptachlor	$0.87C$ −	2.97	$0.24\overline{X}$ −	0.56	$0.50\overline{X}$ −	0.23
Heptachlor epoxide	$0.92C$ −	1.87	$0.33\overline{X}$ −	0.46	$0.28\overline{X}$ +	0.64
Hexachlorobenzene	$0.74C$ +	0.66	$0.18\overline{X}$ −	0.10	$0.43\overline{X}$ −	0.52
Hexachlorobutadiene	$0.71C$ −	1.01	$0.19\overline{X}$ +	0.92	$0.26\overline{X}$ +	0.49
Hexachloroethane	$0.73C$ −	0.83	$0.17\overline{X}$ +	0.67	$0.17\overline{X}$ +	0.80
Indeno(1,2,3-cd)pyrene	$0.78C$ −	3.10	$0.29\overline{X}$ +	1.46	$0.50\overline{X}$ +	0.44
Isophorone	$1.12C$ +	1.41	$0.27\overline{X}$ +	0.77	$0.33\overline{X}$ +	0.26
Naphthalene	$0.76C$ +	1.58	$0.21\overline{X}$ −	0.41	$0.30\overline{X}$ −	0.68

TABLE 6410:VI, CONT.

Compound	Bias as Recovery, X' $\mu g/L$	Single-Analyst Precision, s_r' $\mu g/L$	Overall Precision, S' $\mu g/L$
Nitrobenzene	$1.09C - 3.05$	$0.19\overline{X} + 0.92$	$0.27\overline{X} + 0.21$
N-Nitrosodi-n-propylamine	$1.12C - 6.22$	$0.27\overline{X} + 0.68$	$0.44\overline{X} + 0.47$
PCB-1260	$0.81C - 10.86$	$0.35\overline{X} + 3.61$	$0.43\overline{X} + 1.82$
Phenanthrene	$0.87C - 0.06$	$0.12\overline{X} + 0.57$	$0.15\overline{X} + 0.25$
Pyrene	$0.84C - 0.16$	$0.16\overline{X} + 0.06$	$0.15\overline{X} + 0.31$
1,2,4-Trichlorobenzene	$0.94C - 0.79$	$0.15\overline{X} + 0.85$	$0.21\overline{X} + 0.39$
4-Chloro-3-methylphenol	$0.84C + 0.35$	$0.23\overline{X} + 0.75$	$0.29\overline{X} + 1.31$
2-Chlorophenol	$0.78C + 0.29$	$0.18\overline{X} + 1.46$	$0.28\overline{X} + 0.97$
2,4-Dichlorophenol	$0.87C + 0.13$	$0.15\overline{X} + 1.25$	$0.21\overline{X} + 1.28$
2,4-Dimethylphenol	$0.71C + 4.41$	$0.16\overline{X} + 1.21$	$0.22\overline{X} + 1.31$
2,4-Dinitrophenol	$0.81C - 18.04$	$0.38\overline{X} + 2.36$	$0.42\overline{X} + 26.29$
2-Methyl-4,6-dinitrophenol	$1.04C - 28.04$	$0.10\overline{X} + 42.29$	$0.26\overline{X} + 23.10$
2-Nitrophenol	$1.07C - 1.15$	$0.16\overline{X} + 1.94$	$0.27\overline{X} + 2.60$
4-Nitrophenol	$0.61C - 1.22$	$0.38\overline{X} + 2.57$	$0.44\overline{X} + 3.24$
Pentachlorophenol	$0.93C + 1.99$	$0.24\overline{X} + 3.03$	$0.30\overline{X} + 4.33$
Phenol	$0.43C + 1.26$	$0.26\overline{X} + 0.73$	$0.35\overline{X} + 0.58$
2,4,6-Trichlorophenol	$0.91C - 0.18$	$0.16\overline{X} + 2.22$	$0.22\overline{X} + 1.81$

* X' = expected recovery for one or more measurements of a sample containing a concentration of C,

s_r' = expected single-analyst standard deviation of measurements at an average concentration found of $\overline{X}$,

S' = expected interlaboratory standard deviation of measurements at an average concentration found of $\overline{X}$,

C = true value for the concentration, and

$\overline{X}$ = average recovery found for measurements of samples containing a concentration of C.

† The proper chemical name is 2,2'-oxybis(1-chloropropane).

6. OLYNYK, P., W.L. BUDDE & J.W. EICHEL-BERGER. 1980. Method Detection Limit for Methods 624 and 625. Unpublished report.

7. AMERICAN SOCIETY FOR TESTING AND MATERIALS. 1976. Standard practices for sampling water. ASTM Annual Book of Standards, Part 31, D3370-76. Philadelphia, Pa.

8. EICHELBERGER, J.W., L.E. HARRIS & W.L. BUDDE. 1975. Reference compound to calibrate ion abundance measurement in gas chromatography-mass spectrometry. *Anal. Chem.* 47:995.

9. McNAIR, N.M. & E.J. BONELLI. 1969. Basic Chromatography. Consolidated Printing, Berkeley, Calif.

10. BURKE, J.A. 1965. Gas chromatography for pesticide residue analysis; some practical aspects. *J. Assoc. Offic. Anal. Chem.* 48:1037.

11. U.S. ENVIRONMENTAL PROTECTION AGENCY. 1984. Method 613—2,3,7,8-Tetrachlorodibenzo-*p*-dioxin. 40 CFR Part 136, 43368; *Federal Register* 49, No. 209.

6420 PHENOLS

6420 A. Introduction

1. Sources and Significance

Phenols are found in many wastewaters and some raw source waters in the United States. They generally are traceable to industrial effluents or landfills. These compounds have a low taste threshold in potable waters and also may have a detrimental effect on human health at higher levels.

2. Selection of Method

For methods of determining total phenols in water and wastewater, see Section 5530.

The methods presented in this section are intended for the determination of individual phenolic compounds. For specific compounds covered, see each method. Method 6420B is a gas chromatographic (GC) method using liquid-liquid extraction and either flame ionization detection (FID) or derivatization and electron capture detection (ECD) to determine a wide variety of phenols at relatively low concentrations. In addition, Method 6420C, a liquid-liquid extraction gas chromatographic/mass spectrometric (GC/MS) method, can be used to determine the phenols at slightly higher concentrations.

6420 B. Liquid-Liquid Extraction Gas Chromatographic Method*

This method[1] is applicable to the determination of phenol and certain substituted phenols† in municipal and industrial discharges. When analyzing unfamiliar samples for any or all of these compounds, support the identifications by at least one additional qualitative technique. Alternatively, use the derivatization, cleanup, and electron capture detector gas chromatography (ECD/GC) procedure to confirm measurements made by the flame ionization detector gas chromatographic (FID/GC) procedure. The method for base/neutrals and acids (Section 6410B) provides gas chromatograph/mass spectrometer (GC/MS) conditions appropriate for qualitative and quantitative confirmation of results using the extract produced.

1. General Discussion

a. Principle: See Section 6010C for discussion of gas chromatographic principles.

---

* Approved by Standard Methods Committee, 1987. Accepted by U.S. Environmental Protection Agency as equivalent to EPA Method 604.
† 4-Chloro-3-methylphenol, 2-chlorophenol, 2,4-dichlorophenol, 2,4-dimethylphenol, 2,4-dinitrophenol, 2-methyl-4,6-dinitrophenol, 2-nitrophenol, 4-nitrophenol, pentachlorophenol, phenol, 2,4,6-trichlorophenol.

A measured volume of sample is acidified and extracted with methylene chloride. The extract is dried and exchanged to 2-propanol during concentration. The extract is separated by gas chromatography and phenols are measured with a flame ionization detector.[2]

The method provides for a derivatization and column chromatography cleanup procedure to aid in the elimination of interferences.[2,3] Derivatives are analyzed by an electron capture detector.

b. Interferences:

1) General precautions—See Section 6410B.1b.

2) Other countermeasures—The cleanup procedure in ¶ 5c can be used to overcome many of these inteferences, but unique samples may require additional cleanup to achieve the method detection limits.

The basic sample wash (¶ 5a) may cause low recovery of phenol and 2,4-dimethylphenol. Results obtained under these conditions are minimum concentrations.

c. Detection limits: The method detection limit (MDL) is the minimum concentration of a substance that can be measured and reported with 99% confidence that the value is above zero.[4] The MDL concentrations listed in Tables 6420:I and II were obtained by using reagent water.[5] Similar results were achieved with representative wastewaters. The MDL actually obtained in a given analysis will vary, depending on instrument sensitivity and matrix effects.

d. Safety: The toxicity or carcinogenicity of each reagent used in this method has not been defined precisely. Take special care in handling pentafluorobenzyl bromide, which is a lachrymator, and 18-crown-6-ether, which is highly toxic.

2. Sampling and Storage

See Section 6410B.2.

3. Apparatus

Use all the apparatus specified in Section 6410B.3a–g and i–k, and in addition:

a. Chromatographic column, 100 mm long × 10 mm ID, with TFE stopcock.

b. Reaction flask, 15- to 25-mL round-bottom, with standard tapered joint, fitted with a water-cooled condenser and U-

TABLE 6420:I. CHROMATOGRAPHIC CONDITIONS AND METHOD DETECTION LIMITS

Compound	Retention Time min	Method Detection Limit µg/L
2-Chlorophenol	1.70	0.31
2-Nitrophenol	2.00	0.45
Phenol	3.01	0.14
2,4-Dimethylphenol	4.03	0.32
2,4-Dichlorophenol	4.30	0.39
2,4,6-Trichlorophenol	6.05	0.64
4-Chloro-3-methylphenol	7.50	0.36
2,4-Dinitrophenol	10.00	13.0
2-Methyl-4,6-dinitrophenol	10.24	16.0
Pentachlorophenol	12.42	7.4
4-Nitrophenol	24.25	2.8

Column conditions: Supelcoport (80/100 mesh) coated with 1% SP-1240DA packed in a 1.8 m long × 2 mm ID glass column with nitrogen carrier gas at 30 mL/min flow rate. Column temperature was 80°C at injection, programmed immediately at 8°C/min to 150°C final temperature. MDLs determined with an FID.

TABLE 6420:II. SILICA GEL FRACTIONATION AND ELECTRON CAPTURE GAS CHROMATOGRAPHY OF PFBB DERIVATIVES

Parent Compound	Percent Recovery By Fraction*				Retention Time min	Method Detection Limit μg/L
	1	2	3	4		
2-Chlorophenol	—	90	1	—	3.3	0.58
2-Nitrophenol	—	—	9	90	9.1	0.77
Phenol	—	90	10	—	1.8	2.2
2,4-Dimethylphenol	—	95	7	—	2.9	0.63
2,4-Dichlorophenol	—	95	1	—	5.8	0.68
2,4,6-Trichlorophenol	50	50	—	—	7.0	0.58
4-Chloro-3-methylphenol	—	84	14	—	4.8	1.8
Pentachlorophenol	75	20	—	—	28.8	0.59
4-Nitrophenol	—	—	1	90	14.0	0.70

Column conditions: Chromosorb W-AW-DMCS (80/100 mesh) coated with 5% OV-17 packed in a 1.8 m long × 2.0 mm ID glass column with 5% methane/95% argon carrier gas at 30 mL/min flow rate. Column temperature held isothermal at 200°C. MDLs determined with an ECD.

* Eluant composition:
    Fraction 1 - 15% toluene in hexane.
    Fraction 2 - 40% toluene in hexane.
    Fraction 3 - 75% toluene in hexane.
    Fraction 4 - 15% 2-propanol in toluene.

shaped drying tube containing granular calcium chloride.

c. Gas chromatograph:‡ An analytical system complete with a temperature-programmable gas chromatograph suitable for on-column injection and all required accessories including syringes, analytical columns, gases, detector, and strip-chart recorder. Preferably use a data system for measuring peak areas.

1) Column for underivatized phenols, 1.8 m long × 2 mm ID glass, packed with 1% SP1240DA on Supelcoport (80/100 mesh) or equivalent. The detection limit and precision and bias data presented herein were developed with this column. For guidelines for the use of alternate columns (e.g., capillary or megabore) see ¶ 5b1).

2) Column for derivatized phenols, 1.8 m long × 2 mm ID, glass, packed with 5% OV-17 on Chromosorb W-AW-DMCS (80/100 mesh) or equivalent. This column was used to develop the detection limit and precision and bias data presented herein. For guidelines for the use of alternate columns (e.g., capillary or megabore) see ¶ 5b1).

3) Detectors, flame ionization (FID) and electron capture (ECD). Use the FID to determine parent phenols. Use the ECD when determining derivatized phenols. For guidelines for use of alternative detectors see ¶ 5b1).

4. Reagents

Use reagents listed in Section 6410B.4a–f, and in addition:

a. Sodium hydroxide solution, NaOH, 1N: Dissolve 4 g NaOH in reagent water and dilute to 100 mL.

‡ Gas chromatographic methods are extremely sensitive to the materials used. Mention of trade names by Standard Methods does not preclude the use of other existing or as-yet-undeveloped products that give demonstrably equivalent results.

b. Sulfuric acid, H_2SO_4, *1N:* Slowly add 58 mL conc H_2SO_4 to 500 mL reagent water and dilute to 1 L.

c. Potassium carbonate, K_2CO_3, powdered.

d. Pentafluorobenzyl bromide (α-bromopentafluorotoluene), 97% minimum purity. (CAUTION: *This chemical is a lachrymator.*)

e. 18-Crown-6-ether (1,4,7,10,13,16-hexaoxacyclooctadecane), 98% minimum purity. (CAUTION: *This chemical is highly toxic.*)

f. Derivatization reagent: Add 1 mL pentafluorobenzyl bromide and 1 g 18-crown-6-ether to a 50-mL volumetric flask and dilute to volume with 2-propanol. Prepare fresh weekly. Prepare in a hood. Store at 4°C and protect from light.

g. Acetone, hexane, methanol, methylene chloride, 2-propanol, toluene, pesticide quality or equivalent.

h. Silica gel, 100/200 mesh.§ Activate at 130°C overnight and store in a desiccator.

i. Stock standard solutions: Prepare from pure standard materials or purchase as certified solutions. Prepare as directed in Section 6410B.4*g*, but dissolve material in 2-propanol.

j. Calibration standards: Prepare standards appropriate to chosen means of calibration.

1) *External standards:* Prepare at a minimum of three concentration levels for each compound by adding volumes of one or more stock standards to a volumetric flask and diluting to volume with 2-propanol. Prepare one standard at a concentration near, but above, the MDL (see Table 6420:I or II) and the others to correspond to the expected range of sample concentrations or to define the working range of the detector.

2) *Internal standards:* Prepare at a minimum of three concentration levels for each compound by adding volumes of one or more stock standards to a volumetric flask.

To each calibration standard, add a known constant amount of one or more internal standards, and dilute to volume with 2-propanol. Prepare one standard at a concentration near, but above, the MDL and the others to correspond to the expected range of sample concentrations or to define the working range of the detector.

k. Quality control (QC) check sample concentrate: Obtain a check sample concentrate ‖ containing each compound at a concentration of 100 µg/mL in 2-propanol. If such a sample is not available from an external source, prepare using stock standards prepared independently from those used for calibration.

5. Procedure

a. Extraction: Mark water meniscus on side of sample bottle for later determination of volume. Pour entire sample into a 2-L separatory funnel. For samples high in organic content, solvent wash sample at basic pH as prescribed in next paragraph, to remove potential interferences. During wash, avoid prolonged or exhaustive contact with solvent, which may result in low recovery of some phenols, notably phenol and 2,4-dimethylphenol. For relatively clean samples, omit wash and extract directly.

To wash, adjust pH to 12.0 or greater with NaOH solution. Add 60 mL methylene chloride and shake the funnel for 1 min with periodic venting to release excess pressure. Discard solvent layer. Repeat wash up to two additional times if significant color is being removed.

Before extraction, adjust to pH of 1 to 2 with H_2SO_4. Extract three times with methylene chloride as directed in Section 6410B.5*a*1). Assemble Kuderna-Danish apparatus, concentrate extract to 1 mL,

---

§ Davison grade 923 or equivalent.

‖ For U.S. federal permit-related analyses, use samples obtainable from U.S. EPA Environmental Monitoring and Support Laboratory, Cincinnati, Ohio.

and remove, drain, and cool K-D apparatus as directed in Section 6410B.5a1).

Increase temperature of hot water bath to 100°C. Remove Snyder column and rinse flask and its lower joint into concentrator tube with 1 to 2 mL 2-propanol. Preferably use a 5-mL syringe for this operation. Attach a two-ball micro-Snyder column to concentrator tube and prewet column by adding about 0.5 mL 2-propanol to the top. Place micro-K-D apparatus on water bath so that concentrator tube is partially immersed in hot water. Adjust vertical position of apparatus and water temperature so as to complete concentration in 5 to 10 min. (CAUTION: If temperature is raised too quickly the sample may be blown out of the K-D apparatus). At proper rate of distillation the column balls actively chatter but the chambers are not flooded. When the apparent volume of liquid reaches 2.5 mL, remove K-D apparatus and let drain and cool for at least 10 min. Add 2 mL 2-propanol through top of micro-Snyder column and resume concentrating as before. When the apparent volume of liquid reaches 0.5 mL, remove K-D apparatus and let drain and cool for at least 10 min.

Remove micro-Snyder column and rinse lower joint into concentrator tube with a minimum amount of 2-propanol. Adjust extract volume to 1.0 mL. Stopper concentrator tube and store at 4°C if further processing will not be done immediately. If extract is to be stored longer than 2 d, transfer to a TFE-sealed screw-cap vial. If sample extract requires no further cleanup, proceed with chromatographic analysis (¶ b). If sample requires further cleanup, proceed to ¶ c.

Determine original sample volume by refilling sample bottle to mark and transferring liquid to a 1000-mL graduated cylinder. Record sample volume to nearest 5 mL.

b. *Flame ionization detector gas chromatography (FID/GC):*

1) Operating conditions—Table 6420:I summarizes the recommended operating conditions for the gas chromatograph and gives retention times and MDLs that can be achieved under these conditions. An example of the separations obtained with this column is shown in Figure 6420:1. Other packed or capillary (open-tubular) columns, chromatographic conditions, or detectors may be used if the requirements of ¶ 7 are met.

2) Calibration—To calibrate the system for underivatized phenols, establish gas chromatographic operating conditions equivalent to those given in Table 6420:I. Calibrate using the external or the internal standard technique as follows:

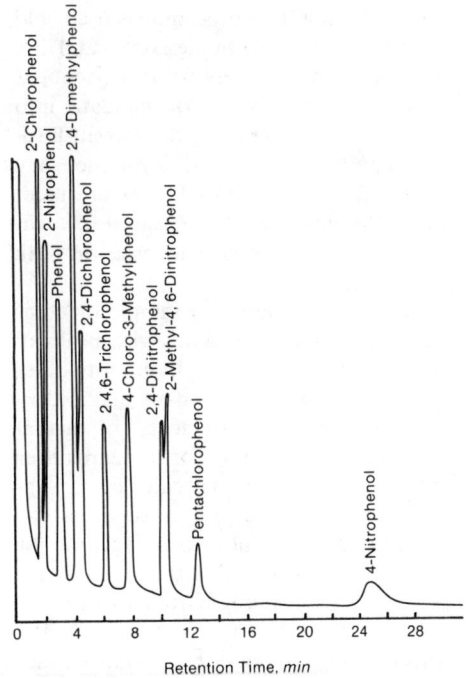

Figure 6420:1. Gas chromatogram of phenols. Column: 1% SP-1240DA on Supelcoport; program: 80°C at injection, immediate 8°C/min to 150°C; detector: flame ionization.

a) External standard calibration procedure—Prepare standards as directed in ¶ 4*j*1) and follow the procedure of ¶ 3) below. Tabulate data and obtain calibration curve or calibration factor as directed in Section 6230B.5*b*2).

b) Internal standard calibration procedure — Prepare samples as directed in ¶ 4*j*2) and follow the procedure of ¶ 3) below. Tabulate data and calculate response factors as directed in Section 6210B.5*c*2).

Verify working calibration curve, calibration factor, or RF on each working day by measuring one or more calibration standards. If the response for any compound varies from the predicted response by more than ±15%, prepare a new calibration curve for that compound.

3) Sample analysis—If the internal standard calibration procedure is used, add internal standard to sample extract and mix thoroughly immediately before injecting 2 to 5 μL sample extract or standard into gas chromatograph using the solvent-flush technique.[6] Smaller (1.0-μL) volumes may be injected if automatic devices are used. Record volume injected to nearest 0.05 μL and resulting peak size in area or peak height units.

Identify compounds in sample by comparing peak retention times with peaks of standard chromatograms. Base width of retention time window used to make identifications on measurements of actual retention time variations of standards over the course of a day. To calculate a suggested window size, use three times the standard deviation of a retention time for a compound. Analyst's experience is important in interpreting chromatograms.

If the response for a peak exceeds the working range of the system, dilute extract and reanalyze.

If peak response can not be measured because of interferences, use the alternative gas chromatographic procedure (¶ *c* below).

c. Derivatization and electron capture detector gas chromatography (ECD/GC):

1) Derivatization—Pipet 1.0 mL of the 2-propanol solution of standard or sample extract into a glass reaction vial. Add 1.0 mL derivatizing reagent (¶ 4*f*); this is sufficient to derivatize a solution having a total phenolic content not exceeding 0.3 mg/mL. Add about 3 mg K_2CO_3 and shake gently. Cap mixture and heat for 4 h at 80°C in a hot water bath. Remove from hot water bath and let cool. Add 10 mL hexane and shake vigorously for 1 min. Add 3.0 mL distilled, deionized water and shake for 2 min. Decant a portion of the organic layer into a concentrator tube and cap with a glass stopper.

2) Cleanup—Place 4.0 g silica gel in a chromatographic column. Tap column to settle silica gel and add about 2 g anhydrous Na_2SO_4 to the top. Pre-elute column with 6 mL hexane. Discard eluate and just before exposing Na_2SO_4 layer to air, pipet onto the column 2.0 mL hexane solution, ¶ 1) above, that contains the derivatized sample or standard. Elute column with 10.0 mL hexane and discard eluate. Elute column, in order, with 10.0 mL 15% toluene in hexane (Fraction 1), 10.0 mL 40% toluene in hexane (Fraction 2), 10.0 mL 75% toluene in hexane (Fraction 3), and 10.0 mL 15% 2-propanol in toluene (Fraction 4). Prepare all elution mixtures on a volume:volume basis. Elution patterns for the phenolic derivatives are shown in Table 6420:II. Fractions may be combined as desired, depending on the specific phenols of interest or level of interferences.

3) Operating conditions—Table 6420:II summarizes the recommended operating conditions for the gas chromatograph and gives retention times and MDLs that can be achieved under these conditions. An example of the separations obtained with this column is shown in Figure 6420:2.

4) Calibration—Calibrate system daily by preparing a minimum of three 1-mL portions of calibration standards, ¶ 4*j*1),

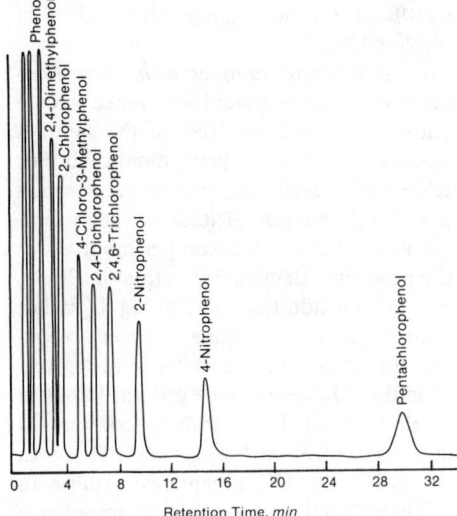

Figure 6420:2. Gas chromatogram of PFB derivatives of phenols. Column: 5% OV-17 on Chromosorb W-AW-DMCS; temperature: 200°C; detector: electron capture.

containing each of the phenols of interest and derivatized as above. Analyze 2 to 5 μL of each column eluate collected as in ¶ 5) below and tabulate peak height or area responses against calculated equivalent mass of underivatized phenol injected. Prepare a calibration curve for each compound.

Before using any cleanup procedure, process a series of calibration standards through the procedure to validate elution patterns and to assure absence of interferences from the reagents.

5) *Sample analysis*—Inject 2 to 5 μL column fractions into the gas chromatograph using the solvent-flush technique. Smaller (1.0-μL) volumes can be injected if automatic devices are used. Record volume injected to nearest 0.05 μL and resulting peak size in area or peak height units. If peak response exceeds linear range of system, dilute extract and reanalyze.

6. Calculation

a. FID/GC analysis: Determine concentration of individual compounds. If the external standard calibration procedure is used, calculate amount of material injected from peak response using calibration curve or calibration factor determined previously. Calculate sample concentration from the equation:

$$\text{Concentration, } \mu g/L = \frac{(A)\,(V_t)}{(V_i)\,(V_s)}$$

where:

A = amount of material injected, ng,
V_i = volume of extract injected, μL,
V_t = volume of total extract, μL, and
V_s = volume of water extracted, mL.

If the internal standard calibration procedure is used, calculate concentration in sample using the response factor (RF) determined above and the equation:

$$\text{Concentration, } \mu g/L = \frac{(A_s)\,(I_s)}{(A_{is})\,(RF)\,(V_o)}$$

where:

A_s = response for compound to be measured,
A_{is} = response for internal standard,
I_s = amount of internal standard added to each extract, μg, and
V_o = volume of water extracted, L.

b. Derivatization and ECD/GC analysis: To determine concentration of individual compounds in the sample, use the equation:

$$\text{Concentration, } \mu g/L = \frac{(A)\,(V_t)\,(B)\,(D)}{(V_i)\,(V_s)\,(C)\,(E)}$$

where:

A = mass of underivatized phenol represented by area of peak in sample chromatogram, determined from calibration curve in ¶ 5c4), ng,
V_i = volume of eluate injected, μL,
V_t = total volume of column eluate or combined fractions from which V_i was taken, μL,

V_s = volume of water extracted in ¶ 5a, mL,
B = total volume of hexane added in ¶ 5c1), mL,
C = volume of hexane sample solution added to cleanup column in ¶ 5c2), mL,
D = total volume of 2-propanol extract before derivatization, mL, and
E = volume of 2-propanol extract carried through derivatization in ¶ 5c1), mL.

Report results in μg/L without correction for recovery. Report QC data with sample results.

7. Quality Control

a. Quality control program: See Section 6210B.7a.

b. Analyst proficiency check: To establish the ability to generate data with acceptable bias and precision, perform the following operations:

Using a pipet, prepare QC check samples at a concentration of 100 μg/L by adding 1.00 mL of 100 μg/mL QC check sample concentrate to each of four 1-L portions reagent water. Analyze check samples according to the method of ¶ 5 and proceed with the check described in Section

6210B.7b. Use acceptance criteria given in Table 6420:III.

c. Analyses of samples with known additions: On an ongoing basis, make known additions to at least 10% of the samples from each sample site being monitored. For laboratories analyzing one to ten samples per month, analyze at least one such sample with a known addition per month. Use the procedure detailed in Section 6210B.7c, but use an addition of 100 μg/L rather than 20 μg/L and compare percent recovery for each compound with the corresponding QC acceptance criteria found in Table 6420:III. If the known addition was at a concentration lower than 100 μg/L, use either the QC acceptance criteria in Table 6420:III or optional QC acceptance criteria calculated for the specific addition concentration based on the equations in Table 6420:IV.

d. Quality-control check standard analysis: If analysis of any compound fails to meet the acceptance criteria for recovery, prepare and analyze a QC check standard containing each compound that failed.

TABLE 6420:III. QC ACCEPTANCE CRITERIA*

Compound	Test Conc. μg/L	Limit for s μg/L	Range for $\overline{X}$ μg/L	Range for P, P_s %
4-Chloro-3-methylphenol	100	16.6	56.7–113.4	49–122
2-Chlorophenol	100	27.0	54.1–110.2	38–126
2,4-Dichlorophenol	100	25.1	59.7–103.3	44–119
2,4-Dimethylphenol	100	33.3	50.4–100.0	24–118
4,6-Dinitro-2-methylphenol	100	25.0	42.4–123.6	30–136
2,4-Dinitrophenol	100	36.0	31.7–125.1	12–145
2-Nitrophenol	100	22.5	56.6–103.8	43–117
4-Nitrophenol	100	19.0	22.7–100.0	13–110
Pentachlorophenol	100	32.4	56.7–113.5	36–134
Phenol	100	14.1	32.4–100.0	23–108
2,4,6-Trichlorophenol	100	16.6	60.8–110.4	53–119

* s = standard deviation of four recovery measurements,
$\overline{X}$ = average recovery for four recovery measurements, and
P, P_s = percent recovery measured.
NOTE: These criteria are based directly upon the method performance data in Table 6420:IV. Where necessary, the limits for recovery were broadened to assure applicability of the limits to concentrations below those used to develop Table 6420:IV.

TABLE 6420:IV. METHOD BIAS AND PRECISION AS FUNCTIONS OF CONCENTRATION*

Compound	Bias, as Recovery, X' $\mu g/L$	Single-Analyst Precision, s'_s $\mu g/L$	Overall Precision, S' $\mu g/L$
4-Chloro-3-methylphenol	$0.87C - 1.97$	$0.11\overline{X} - 0.21$	$0.16\overline{X} + 1.41$
2-Chlorophenol	$0.83C - 0.84$	$0.18\overline{X} + 0.20$	$0.21\overline{X} + 0.75$
2,4-Dichlorophenol	$0.81C + 0.48$	$0.17\overline{X} - 0.02$	$0.18\overline{X} + 0.62$
2,4-Dimethylphenol	$0.62C - 1.64$	$0.30\overline{X} - 0.89$	$0.25\overline{X} + 0.48$
4,6-Dinitro-2-methylphenol	$0.84C - 1.01$	$0.15\overline{X} + 1.25$	$0.19\overline{X} + 5.85$
2,4-Dinitrophenol	$0.80C - 1.58$	$0.27\overline{X} - 1.15$	$0.29\overline{X} + 4.51$
2-Nitrophenol	$0.81C - 0.76$	$0.15\overline{X} + 0.44$	$0.14\overline{X} + 3.84$
4-Nitrophenol	$0.46C + 0.18$	$0.17\overline{X} + 2.43$	$0.19\overline{X} + 4.79$
Pentachlorophenol	$0.83C + 2.07$	$0.22\overline{X} - 0.58$	$0.23\overline{X} + 0.57$
Phenol	$0.43C + 0.11$	$0.20\overline{X} - 0.88$	$0.17\overline{X} + 0.77$
2,4,6-Trichlorophenol	$0.86C - 0.40$	$0.10\overline{X} + 0.53$	$0.13\overline{X} + 2.40$

* X' = expected recovery for one or more measurements of a sample containing a concentration of C,
s'_s = expected single-analyst standard deviation of measurements at an average concentration found of $\overline{X}$,
S' = expected interlaboratory standard deviation of measurements at an average concentration found of $\overline{X}$,
C = true value for the concentration, and,
$\overline{X}$ = average recovery found for measurements of samples containing a concentration of C.

NOTE: The frequency for the required analysis of a QC check standard will depend on the number of compounds being tested for simultaneously, the complexity of the sample matrix, and the performance of the laboratory.

Prepare the QC check standard by adding 1.0 mL of QC check sample concentrate to 1 L reagent water and proceed as in Section 6210B.7d using Table 6420:III.

e. *Bias assessment and records:* Assess method bias and maintain records as directed in Section 6210B.7e.

f. *Additional quality-assurance practices:* See Section 6210B.7g.

8. Precision and Bias

This method was tested by 20 laboratories using reagent water, drinking water, surface water, and three industrial wastewaters with known additions at six concentrations over the range 12 to 450 $\mu g/$ L.[7] Single-operator precision, overall precision, and method bias were found to be related directly to compound concentration and essentially independent of sample matrix. Linear equations describing these relationships for a flame ionization detector are presented in Table 6420:IV.

9. References

1. U.S. ENVIRONMENTAL PROTECTION AGENCY. 1984. Method 604—Phenols. 40 CFR Part 136, 43290; *Federal Register* 49, No. 209.

2. U.S. ENVIRONMENTAL PROTECTION AGENCY. 1984. Determination of phenols in industrial and municipal wastewaters. Final rep. EPA Contract 68-03-2625, Environmental Monitoring and Support Lab., Cincinnati, Ohio.

3. KAWAHARA, F.K. 1968. Microdetermination of derivatives of phenols and mercaptans by means of electron capture gas chromatography. *Anal. Chem.* 40:100.

4. U.S. ENVIRONMENTAL PROTECTION AGENCY. 1984. Definition and procedure for the determination of the method detection limit. 40 CFR Part 136, Appendix B. *Federal Register* 49, No. 209.

5. U.S. ENVIRONMENTAL PROTECTION AGENCY. Development of detection limits, EPA Method 604, Phenols. Special letter report for EPA

Contract 68-03-2625, Environmental Monitoring and Support Lab., Cincinnati, Ohio.

6. BURKE, J.A. 1965. Gas chromatography for pesticide residue analysis; some practical aspects. *J. Assoc. Offic. Anal. Chem.* 48:1037.

7. U.S. ENVIRONMENTAL PROTECTION AGENCY. 1984. EPA Method Study 14, Method 604—Phenols. EPA-600/4-84-044, National Technical Information Serv., PB84-196211, Springfield, Va.

6420 C.   Liquid-Liquid Extraction
Gas Chromatographic/Mass Spectrometric Method

See Section 6410B.

6431   POLYCHLORINATED BIPHENYLS (PCBs)

6431 A.   Introduction

1. Sources and Significance

The polychlorinated biphenyls (PCBs) are found principally in water supplies contaminated by transformer oils in which PCBs were originally used as a heat-exchange medium. Although the use of these compounds has been banned, there are still numerous transformers in existence that contain PCBs, which results in their occasional discharge into potable water or wastewater. These compounds are toxic, bioaccumulative, and extremely stable, and thus there is a need to monitor them in wastewaters.

2. Selection of Method

The liquid-liquid extraction (LLE) gas chromatographic (GC) method is used to monitor both the PCBs and the organochlorine pesticides simultaneously. This method has excellent sensitivity. The LLE gas chromatographic/mass spectrometric (GC/MS) method also can be used to detect PCBs, but with substantially less sensitivity.

6431 B. Liquid-Liquid Extraction
Gas Chromatographic Method

See Sections 6630B and C.

6431 C. Liquid-Liquid Extraction
Gas Chromatographic/Mass Spectrometric Method

See Section 6410B.

6440 POLYNUCLEAR AROMATIC HYDROCARBONS

6440 A. Introduction

1. Sources and Significance

The polynuclear aromatic hydrocarbons (PAHs) often are by-products of petroleum processing or combustion. Many of these compounds are highly carcinogenic at relatively low levels. Although they are relatively insoluble in water, their highly hazardous nature merits their monitoring in potable waters and wastewaters.

2. Selection of Method

Method 6440B encompasses both a high-performance liquid chromatographic (HPLC) method with UV and fluorescence detection and a gas chromatographic (GC) method using flame ionization detection. Method 6440C is a gas chromatographic/

mass spectrometric (GC/MS) method that also can detect these compounds at somewhat higher concentrations. Certain of

these compounds may also be measured by closed-loop stripping analysis, Section 6040.

6440 B.   Liquid-Liquid Extraction Chromatographic Method*

This method[1] is applicable to the determination of certain polynuclear aromatic hydrocarbons (PAH)† in municipal and industrial discharges. When analyzing unfamiliar samples for any or all of these compounds, support the identifications by at least one additional qualitative technique. The method for base/neutrals and acids (Section 6410B) provides gas chromatograph/mass spectrometer (GC/MS) conditions appropriate for qualitative and quantitative confirmation of results using the extract produced.

1. General Discussion

a. Principle: A measured volume of sample is extracted with methylene chloride. The extract is dried, concentrated, and separated by the high-performance liquid chromatographic (HPLC) or gas chromatographic (GC) method. If other analyses having essentially the same extraction and concentration steps are to be performed, extraction of a single sample will be sufficient for all the determinations. Ultraviolet (UV) and fluorescence detectors are used with HPLC to identify and measure the

PAHs. A flame ionization detector is used with GC.[2]

The method provides a silica gel column cleanup to aid in eliminating interferences. When cleanup is required, sample concentration levels must be high enough to permit separate treatment of subsamples before the solvent-exchange steps.

Chromatographic conditions (¶ 5d) appropriate for the simultaneous measurement of combinations of these compounds may be selected but they do not adequately resolve the following four pairs of compounds: anthracene and phenanthrene; chrysene and benzo(a)anthracene; benzo-(b)fluoranthene and benzo(k)fluoranthene; and dibenzo(a,h)anthracene and indeno(1,2,3-cd)pyrene. Unless reporting the sum of an unresolved pair is acceptable, use the liquid chromatographic method, which does resolve all 16 listed PAHs.

b. Interferences: See Section 6410B.1b for precautions concerning glassware, reagent purity, and matrix interferences. Interferences in liquid chromatographic techniques have not been assessed fully. Although HPLC conditions described allow for unique resolution of specific PAHs, other PAH compounds may interfere.

c. Detection limits: The method detection limit (MDL) is the minimum concentration of a substance that can be measured and reported with 99% confidence that the value is above zero.[3] The MDL concentrations listed in Table 6440:I were obtained with reagent water.[4] Similar results were

* Approved by Standard Methods Committee, 1987. Accepted by U.S. Environmental Protection Agency as equivalent to EPA Method 610.
† Acenaphthene, acenaphthylene, anthracene, benzo-(a)anthracene, benzo(a)pyrene, benzo(b)fluoranthene, benzo(ghi)perylene, benzo(k)fluoranthene, chrysene, dibenzo(a,h)anthracene, fluoranthene, fluorene, indeno(1,2,3-cd)pyrene, naphthalene, phenanthrene, and pyrene.

TABLE 6440:I. HIGH-PERFORMANCE LIQUID CHROMATOGRAPHY CONDITIONS AND METHOD DETECTION LIMITS

Compound	Retention Time min	Column Capacity Factor k'	Method Detection Limit µg/L*
Naphthalene	16.6	12.2	1.8
Acenaphthylene	18.5	13.7	2.3
Acenaphthene	20.5	15.2	1.8
Fluorene	21.2	15.8	0.21
Phenanthrene	22.1	16.6	0.64
Anthracene	23.4	17.6	0.66
Fluoranthene	24.5	18.5	0.21
Pyrene	25.4	19.1	0.27
Benzo(a)anthracene	28.5	21.6	0.013
Chrysene	29.3	22.2	0.15
Benzo(b)fluoranthene	31.6	24.0	0.018
Benzo(k)fluoranthene	32.9	25.1	0.017
Benzo(a)pyrene	33.9	25.9	0.023
Dibenzo(a,h)anthracene	35.7	27.4	0.030
Benzo(ghi)perylene	36.3	27.8	0.076
Indeno(1,2,3-cd)pyrene	37.4	28.7	0.043

HPLC column conditions: Reverse phase HC-ODS Sil-X, 5 µm particle size, in a 25 cm × 2.6 mm ID stainless steel column. Isocratic elution for 5 min using acetonitrile/water (4 + 6), then linear gradient to 100% acetonitrile over 25 min at 0.5 mL/min flow rate. If columns having other internal diameters are used, adjust flow rate to maintain a linear velocity of 2 mm/s.
* The MDL for naphthalene, acenaphthylene, acenaphthene, and fluorene were determined using a UV detector. All others were determined using a fluorescence detector.

achieved with representative wastewaters. MDLs for the GC method were not determined. The MDL actually obtained in a given analysis will vary, depending on instrument sensitivity and matrix effects. This method has been tested for linearity of known-addition recovery from reagent water and has been demonstrated to be applicable over the concentration range from 8 × MDL to 800 × MDL,[4] with the following exception: benzo(ghi)perylene recovery at 80 × and 800 × MDL were low (35% and 45%, respectively).

d. Safety: The toxicity or carcinogenicity of each reagent has not been defined precisely. The following compounds have been classified tentatively as known or suspected, human or mammalian carcinogens: benzo(a)anthracene, benzo(a)pyrene, and dibenzo(a,h)anthracene. Prepare primary standards of these compounds in a hood and wear NIOSH/MESA-approved toxic gas respirator when handling high concentrations.

2. Sampling and Storage

For collection and general storage requirements, see Section 6410B.2. Because PAHs are light-sensitive, store samples, extracts, and standards in amber or foil-wrapped bottles to minimize photolytic decomposition.

3. Apparatus

Use all the apparatus specified in Section 6410B.3a–g and i–k, and in addition:

a. Chromatographic column, 250 mm long × 10 mm ID with coarse frit filter disk at bottom and TFE stopcock.

b. High-performance liquid chromatograph (HPLC): An analytical system complete with column supplies, high-pressure syringes, detectors, and compatible strip-chart recorder. Preferably use a data system for measuring peak areas and retention times.

1) *Gradient pumping system,* constant flow.

2) *Reverse phase column,* HC-ODS Sil-X, 5-μm particle diam, in a 25-cm × 2.6-mm ID stainless steel column.‡ This column was used to develop MDL and precision and bias data presented herein. For guidelines for the use of alternate column packings see ¶ 5*d*1).

3) *Detectors,* fluorescence and/or UV. Use the fluorescence detector for excitation at 280 mm and emission greater than 389 nm cutoff.§ Use fluorometers with dispersive optics for excitation utilizing either filter or dispersive optics at the emission detector. Operate the UV detector at 254 nm and couple it to the fluorescence detector. These detectors were used to develop MDL and precision and bias data presented herein. For guidelines for the use of alternate detectors see ¶ 5*d*1).

c. Gas chromatograph:‖ An analytical system complete with temperature-programmable gas chromatograph suitable for on-column or splitless injection and all required accessories including syringes, analytical columns, gases, detector, and strip-chart recorder. Preferably use a data system for measuring peak areas.

1) *Column,* 1.8 m long × 2 mm ID glass, packed with 3% OV-17 on Chromosorb W-AW-DCMS (100/120 mesh) or equivalent. This column was used to develop the retention time data in Table 6440:II. For

guidelines for the use of alternate columns (e.g. capillary or megabore) see ¶ 5*d*2).

2) *Detector,* flame ionization. This detector is effective except for resolving the four pairs of compounds listed in ¶ 1*a*. With the use of capillary columns, these pairs may be resolved with GC. For guidelines for the use of alternate detectors see ¶ 5*d*2).

4. Reagents

a. Reagent water: See Section 6210B.4*a*.

b. Sodium thiosulfate, $Na_2S_2O_3 \cdot 5H_2O$, granular.

c. Cyclohexane, methanol, acetone, methylene chloride, pentane, pesticide quality or equivalent.

d. Acetonitrile, HPLC quality, distilled in glass.

e. Sodium sulfate, Na_2SO_4, granular, anhydrous. Purify by heating at 400°C for 4 h in a shallow tray.

f. Silica gel, 100/200 mesh, desiccant.# Before use, activate for at least 16 h at 130°C in a shallow glass tray, loosely covered with foil.

g. Stock standard solutions: Prepare as directed in Section 6410B.4*g*, using acetonitrile as the solvent.

h. Calibration standards: Prepare standards appropriate to chosen means of calibration following directions in Section 6420B.4*j*, except that acetonitrile is the diluent instead of 2-propanol. See Table 6440:I for MDLs.

i. Quality control (QC) check sample concentrate: Obtain a check sample concentrate** containing each compound at the following concentrations in acetonitrile: 100 μg/mL of any of the six early-eluting PAHs (naphthalene, acenaphthylene, acenaphthene, fluorene, phenanthrene, and anthracene); 5 μg/mL of benzo(k)fluoran-

---

‡ Perkin Elmer No. 089-0716 or equivalent.
§ Corning 3-75 or equivalent.
‖ Gas chromatographic methods are extremely sensitive to the materials used. Mention of trade names by *Standard Methods* does not preclude the use of other existing or as-yet-undeveloped products that give *demonstrably* equivalent results.

#Davison, grade 923 or equivalent.
** For U.S. federal permit-related analyses, use samples obtainable from U.S. EPA Environmental Monitoring and Support Laboratory, Cincinnati, Ohio.

TABLE 6440:II. GAS CHROMATOGRAPHIC CONDITIONS AND RETENTION TIMES

Compound	Retention Time min
Naphthalene	4.5
Acenaphthylene	10.4
Acenaphthane	10.8
Fluorene	12.6
Phenanthrene	15.9
Anthracene	15.9
Fluoranthene	19.8
Pyrene	20.6
Benzo(a)anthracene	24.7
Chrysene	24.7
Benzo(b)fluoranthene	28.0
Benzo(k)fluoranthene	28.0
Benzo(a)pyrene	29.4
Dibenzo(a,h)anthracene	36.2
Indeno(1,2,3-cd)pyrene	36.2
Benzo(ghi)perylene	38.6

GC column conditions: Chromosorb W-AW-DCMS (100/120 mesh) coated with 3% OV-17 packed in a 1.8 × 2 mm ID glass column with nitrogen carrier gas at 40 mL/min flow rate. Column temperature held at 100°C for 4 min, then programmed at 8°C/min to a final hold at 280°C.

thene; and 10 μg/mL of any other PAH. If such a sample is not available from an external source, prepare using stock standards prepared independently from those used for calibration.

5. Procedure

a. Extraction: Mark water meniscus on side of sample bottle for later determination of volume. Pour entire sample into a 2-L separatory funnel and extract as directed in Section 6410B.5*a*1) without any pH adjustment.

After extraction, concentrate by adding one or two clean boiling chips to the evaporative flask and attach a three-ball Snyder column. Prewet Snyder column by adding about 1 mL methylene chloride to the top. Place K-D apparatus on a hot water bath (60 to 65°C) in a hood so that the concentrator tube is partially immersed in the hot water, and the entire lower rounded surface of flask is bathed with hot vapor. Adjust vertical position of apparatus and water temperature as required to complete the concentration in 15 to 20 min. At proper rate of distillation the column balls actively chatter but the chambers are not flooded with condensed solvent. When the apparent volume of liquid reaches 1 mL, remove K-D apparatus and let drain and cool for at least 10 min.

Remove Snyder column and rinse flask and its lower joint into concentrator tube with 1 to 2 mL methylene chloride. Preferably use a 5-mL syringe for this operation. Stopper concentrator tube and store refrigerated if further processing will not be done immediately. If extract is to be stored longer than 2 d, transfer to a TFE-sealed screw-cap vial and protect from light. If sample extract requires no further cleanup, proceed with gas or liquid chromatographic analysis (¶s *c* through *f* below). If sample requires further cleanup, first follow procedure of ¶ *b* before chromatographic analysis.

Determine original sample volume by refilling sample bottle to mark and transferring liquid to a 1000-mL graduated cylinder. Record sample volume to nearest 5 mL.

b. *Cleanup and separation:* Use procedure below or any other appropriate procedure; however, first demonstrate that the requirements of ¶ 7 can be met.

Before using silica-gel cleanup technique, exchange extract solvent to cyclohexane. Add 1 to 10 mL sample extract (in methylene chloride) and a boiling chip to a clean K-D concentrator tube. Add 4 mL cyclohexane and attach a two-ball micro-Snyder column. Prewet column by adding 0.5 mL methylene chloride to the top. Place micro-K-D apparatus on a boiling (100°C) water bath so that concentrator tube is partially immersed in hot water. Adjust vertical position of apparatus and water temperature so as to complete concentration in 5 to 10 min. At proper rate of distillation the column balls actively chatter but the chambers are not flooded. When apparent volume of liquid reaches 0.5 mL, remove K-D apparatus and let drain and cool for at least 10 min. Remove micro-Snyder column and rinse its lower joint into concentrator tube with a minimum amount of cyclohexane. Adjust extract volume to about 2 mL.

To perform silica-gel column cleanup, make a slurry of 10 g activated silica gel in methylene chloride and place in a 10-mm-ID chromatographic column. Tap column to settle silica gel and elute with methylene chloride. Add 1 to 2 cm anhydrous Na_2SO_4 to top of silica gel. Pre-elute with 40 mL pentane. Elute at rate of about 2 mL/min. Discard eluate and just before exposure of Na_2SO_4 layer to the air, transfer all the cyclohexane sample extract onto column using an additional 2 mL cyclohexane. Just before exposure of Na_2SO_4 layer to air, add 25 mL pentane and continue elution. Discard this pentane eluate. Next, elute column with 25 mL methylene

chloride/pentane (4 + 6) (v/v) into a 500-mL K-D flask equipped with a 10-mL concentrator tube. Concentrate collected fraction to less than 10 mL as in ¶ 5a. After cooling, remove Snyder column and rinse flask and its lower joint with pentane.

c. *Reconcentration:* Concentrate further as follows:

1) For high-performance liquid chromatography—To extract in a concentrator tube, add 4 mL acetonitrile and a new boiling chip. Attach a two-ball micro-Snyder column and concentrate solvent as in ¶ 5a (but set water bath at 95 to 100°C.) After cooling, remove micro-Snyder column and rinse its lower joint into the concentrator tube with about 0.2 mL acetonitrile. Adjust extract volume to 1.0 mL.

2) For gas chromatography—To achieve maximum sensitivity with this method, concentrate extract to 1.0 mL. Add a clean boiling chip to methylene chloride extract in concentrator tube. Attach a two-ball micro-Snyder column. Prewet column by adding about 0.5 mL methylene chloride to the top. Place micro-K-D apparatus on a hot water bath (60 to 65°C) and continue concentration as in ¶ 5b. Remove micro-Snyder column and rinse its lower joint into concentrator tube with a minimum amount of methylene chloride. Adjust final volume to 1.0 mL and stopper concentrator tube.

d. *Operating conditions:*

1) High-performance liquid chromatography—Table 6440:I summarizes the recommended operating conditions for HPLC and gives retention times, capacity factors, and MDLs that can be achieved under these conditions. Preferably use the UV detector for determining naphthalene, acenaphthylene, acenapthene, and fluorene and the fluorescence detector for the remaining PAHs. Examples of separations obtained with this HPLC column are shown in Figures 6440:1 and 2. Other HPLC columns, chromatographic conditions, or detectors may be used if the requirements of ¶ 7 are met.

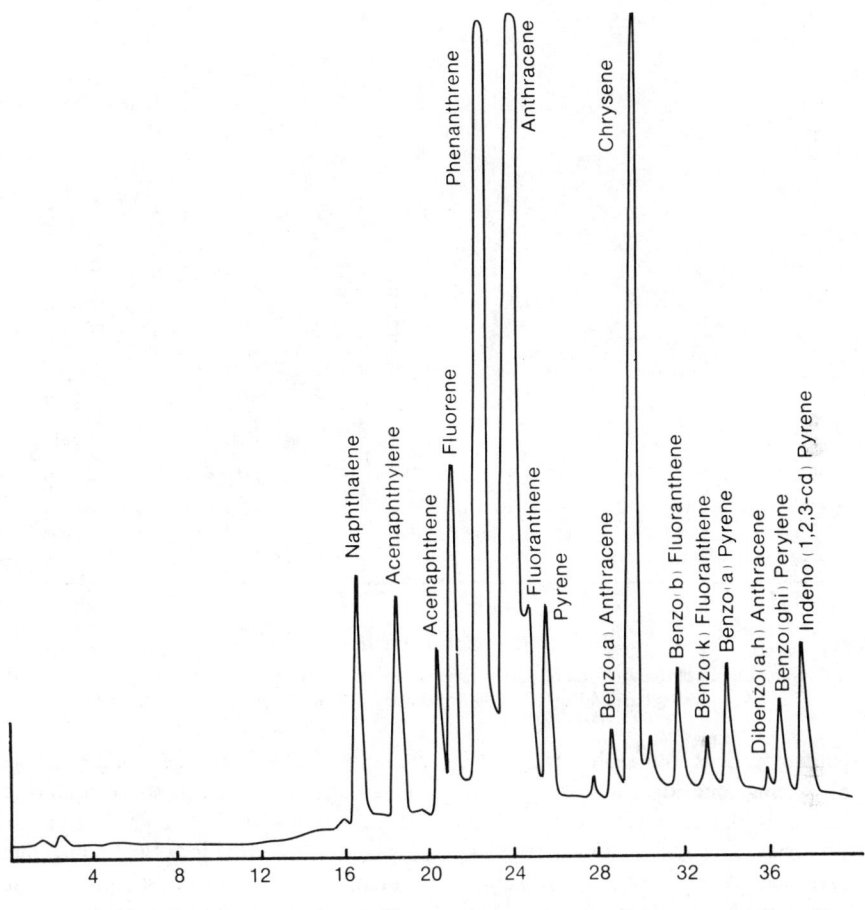

Figure 6440:1. Liquid chromatogram of polynuclear aromatic hydrocarbons. Column: HC—ODS SIL-X; mobile phase: 40% to 100% acetonitrile in water; detector: ultraviolet at 254 nm.

2) Gas chromatography—Table 6440:II summarizes the recommended operating conditions for the gas chromatograph and gives retention times that were obtained under these conditions. An example of the separations is shown in Figure 6440:3. Other packed or capillary (open-tubular) columns, chromatographic conditions, or detectors may be used if the requirements of ¶ 7 are met.

e. Calibration: Calibrate system daily using either external or internal standard procedure.

1) External standard calibration procedure—Prepare standards as directed in ¶ 4*h* and follow either procedure of ¶*f* below. Tabulate data and obtain calibration curve or calibration factor as directed in Section 6230B.5*b*2).

2) Internal standard calibration procedure—Prepare standards as directed in ¶ 4*h* and follow either procedure of ¶*f* below. Tabulate data and calculate response factors as directed in Section 6210B.5*c*2).

Verify working calibration curve, cali-

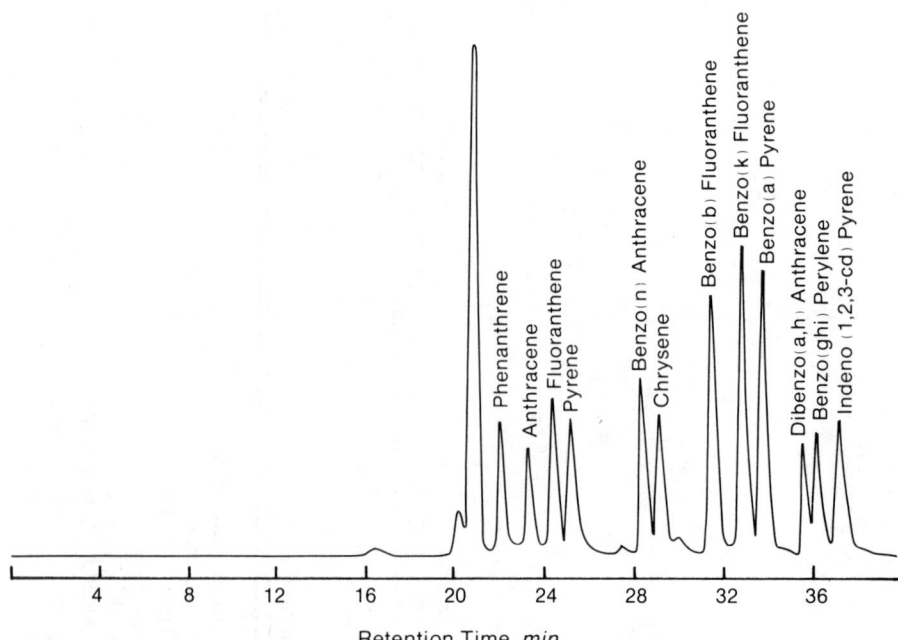

Figure 6440:2. Liquid chromatogram of polynuclear aromatic hydrocarbons. Column: HC—ODS SIL-
X; mobile phase: 40% to 100% acetonitrile in water; detector: fluorescence.

bration factor, or RF on each working day
by measuring one or more calibration
standards. If the response for any compound varies from the predicted response
by more than ± 15%, repeat test using a
fresh calibration standard. Alternatively,
prepare a new calibration curve for that
compound.

Before using any cleanup procedure,
process a series of calibration standards
through the procedure to validate elution
patterns and the absence of interferences
from the reagents.

   f. Sample analysis:

1) High-performance liquid chromatography—If the internal standard calibration
procedure is being used, add internal standard to sample extract and mix thoroughly.
Immediately inject 5 to 25 µL sample extract or standard into HPLC using a high-
pressure syringe or a constant-volume sample injection loop. Record volume injected

to nearest 0.1 µL and resulting peak size
in area or peak height units. Re-equilibrate
HPLC column at initial gradient conditions for at least 10 min between injections.

Identify compounds in sample by comparing peak retention times with peaks of
standard chromatograms. Base width of retention time window used to make identifications on measurements of actual
retention time variations of standards over
the course of a day. To calculate a suggested window size use three times the
standard deviation of a retention time for
a compound. Analyst's experience is important in interpreting chromatograms.

If the response for a peak exceeds the
working range of the system, dilute extract
with acetonitrile and reanalyze.

If peak response cannot be measured because of interferences, further cleanup is
required.

2) Gas chromatography—See Section

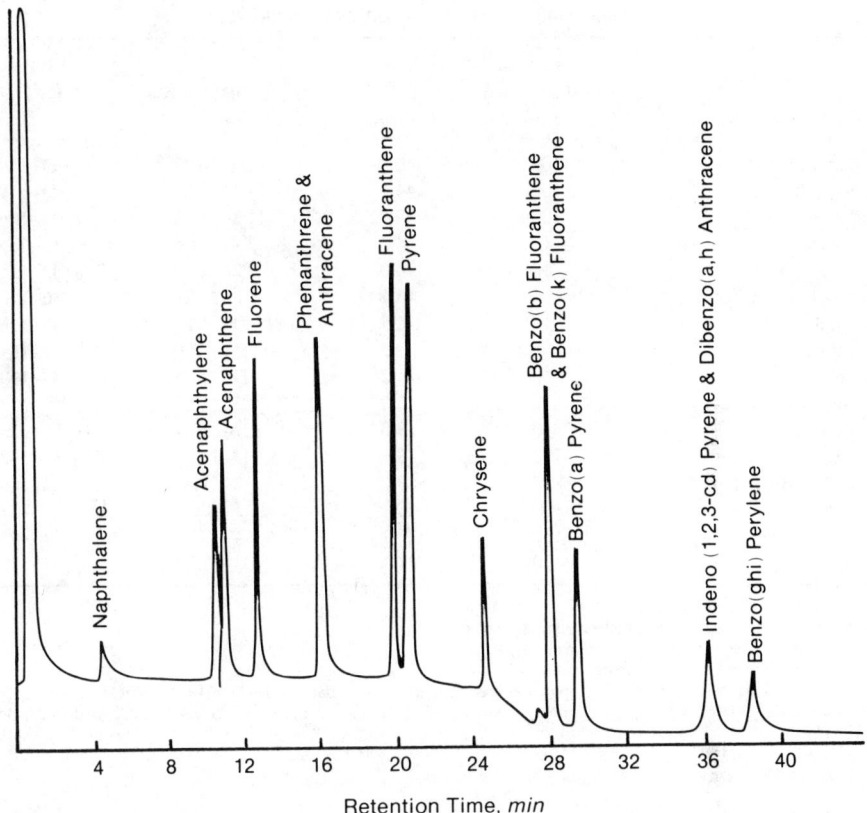

Figure 6440:3. Gas chromatogram of polynuclear aromatic hydrocarbons. Column: 3% OV-17 on Chromosorb W-AW-DCMS; program: 100°C for 4 min, 8°C/min to 280°C; detector: flame ionization.

6420B.5*b*3). If peak response cannot be measured because of interferences, further cleanup is required.

6. Calculation

Determine concentration of individual compounds using the procedures given in Section 6420B.6*a*. Report results in µg/L without correction for recovery. Report QC data with sample results.

7. Quality Control

a. Quality-control program: See Section 6210B.7*a*.

b. Analyst proficiency check: To establish the ability to generate data with acceptable precision and bias, proceed as follows: Using a pipet, prepare QC check samples at test concentrations shown in Table 6440:III by adding 1.00 mL of QC check sample concentrate (¶ 4*i*) to each of four 1-L portions of reagent water. Analyze QC check samples according to the method of ¶ 5. Calculate average recovery and standard deviation of the recovery, compare with acceptance criteria, and evaluate and correct system performance as directed in Section 6210B.7*b*, using acceptance criteria given in Table 6440:III.

TABLE 6440:III. QC ACCEPTANCE CRITERIA*

Compound	Test Conc. $\mu g/L$	Limit for s $\mu g/L$	Range for $\overline{X}$ $\mu g/L$	Range for P, P_s %
Acenaphthene	100	40.3	D–105.7	D–124
Acenaphthylene	100	45.1	22.1–112.1	D–139
Anthracene	100	28.7	11.2–112.3	D–126
Benzo(a)anthracene	10	4.0	3.1– 11.6	12–135
Benzo(a)pyrene	10	4.0	0.2– 11.0	D–128
Benzo(b)fluoranthene	10	3.1	1.8– 13.8	6–150
Benzo(ghi)perylene	10	2.3	D– 10.7	D–116
Benzo(k)fluoranthene	5	2.5	D– 7.0	D–159
Chrysene	10	4.2	D– 17.5	D–199
Dibenzo(a,h)anthracene	10	2.0	0.3– 10.0	D–110
Fluoranthene	10	3.0	2.7– 11.1	14–123
Fluorene	100	43.0	D–119	D–142
Indeno(1,2,3-cd) pyrene	10	3.0	1.2– 10.0	D–116
Naphthalene	100	40.7	21.5–100.0	D–122
Phenanthrene	100	37.7	8.4–133.7	D–155
Pyrene	10	3.4	1.4– 12.1	D–140

* s = standard deviation of four recovery measurements,
$\overline{X}$ = average recovery for four recovery measurements,
P, P_s = percent recovery measured, and
D = detected; result must be greater than zero.
NOTE: These criteria are based directly upon the method performance data in Table 6440:IV. Where necessary, the limits for recovery were broadened to assure applicability of the limits to concentrations below those used to develop Table 6440:IV.

c. *Analyses of samples with known additions:* See Section 6420B.7c. Prepare QC check sample concentrate according to ¶ 4i and use Tables 6440:III and IV. On an ongoing basis, make known additions to at least 10% of the samples from each sample site being monitored. For laboratories analyzing one to ten samples per month, analyze at least one such sample with a known addition per month. Use the procedure described in Section 6210B.7c.

d. *Quality-control check standard analysis:* See Section 6420B.7d. Prepare QC check standard according to ¶ 4i and use Table 6440:III. If all compounds in Table 6440:III are to be measured in the sample in ¶ c above, it is probable that the analysis of a QC check standard will be required; therefore, routinely analyze the QC check standard with the known-addition sample.

e. *Accuracy assessment and records:* See Section 6210B.7e.

f. *Additional quality-assurance practices:* See Section 6210B.7g.

8. Precision and Bias

This method was tested by 16 laboratories using reagent water, drinking water, surface water, and three industrial wastewaters with known additions at six concentrations over the range 0.1 to 425 μg/L.[5] Single-operator precision, overall precision, and method bias were found to be related directly to compound concentration and essentially independent of sample matrix. Linear equations describing these relationships are presented in Table 6440:IV.

TABLE 6440:IV. METHOD BIAS AND PRECISION AS FUNCTIONS OF CONCENTRATION*

Compound	Bias as Recovery, X' $\mu g/L$	Single-Analyst Precision, s_r $\mu g/L$	Overall Precision, S' $\mu g/L$
Acenaphthene	$0.52C + 0.54$	$0.39\overline{X} + 0.76$	$0.53\overline{X} + 1.32$
Acenaphthylene	$0.69C - 1.89$	$0.36\overline{X} + 0.29$	$0.42\overline{X} + 0.52$
Anthracene	$0.63C - 1.26$	$0.23\overline{X} + 1.16$	$0.41\overline{X} + 0.45$
Benzo(a)anthracene	$0.73C + 0.05$	$0.28\overline{X} + 0.04$	$0.34\overline{X} + 0.02$
Benzo(a)pyrene	$0.56C + 0.01$	$0.38\overline{X} + 0.01$	$0.53\overline{X} - 0.01$
Benzo(b)fluoranthene	$0.78C + 0.01$	$0.21\overline{X} + 0.01$	$0.38\overline{X} - 0.00$
Benzo(ghi)perylene	$0.44C + 0.30$	$0.25\overline{X} + 0.04$	$0.58\overline{X} + 0.10$
Benzo(k)fluoranthene	$0.59C + 0.00$	$0.44\overline{X} - 0.00$	$0.69\overline{X} + 0.01$
Chrysene	$0.77C - 0.18$	$0.32\overline{X} - 0.18$	$0.66\overline{X} - 0.22$
Dibenzo(a,h)anthracene	$0.41C + 0.11$	$0.24\overline{X} + 0.02$	$0.45\overline{X} + 0.03$
Fluoranthene	$0.68C + 0.07$	$0.22\overline{X} + 0.06$	$0.32\overline{X} + 0.03$
Fluorene	$0.56C - 0.52$	$0.44\overline{X} - 1.12$	$0.63\overline{X} - 0.65$
Indeno(1,2,3-cd)pyrene	$0.54C + 0.06$	$0.29\overline{X} + 0.02$	$0.42\overline{X} + 0.01$
Naphthalene	$0.57C - 0.70$	$0.39\overline{X} - 0.18$	$0.41\overline{X} + 0.74$
Phenanthrene	$0.72C - 0.95$	$0.29\overline{X} + 0.05$	$0.47\overline{X} - 0.25$
Pyrene	$0.69C - 0.12$	$0.25\overline{X} + 0.14$	$0.42\overline{X} - 0.00$

* X' = expected recovery for one or more measurements of a sample containing a concentration of C,
s_r = expected single-analyst standard deviation of measurements at an average concentration found of $\overline{X}$,
S' = expected interlaboratory standard deviation of measurements at an average concentration found of $\overline{X}$,
C = true value for concentration, and
$\overline{X}$ = average recovery found for measurements of samples containing a concentration of C.

9. References

1. U.S. ENVIRONMENTAL PROTECTION AGENCY. 1984. Method 610—Polynuclear aromatic hydrocarbons. 40 CFR Part 136, 43344; *Federal Register* 49, No. 209.
2. U.S. ENVIRONMENTAL PROTECTION AGENCY. 1982. Determination of polynuclear aromatic hydrocarbons in industrial and municipal wastewaters. EPA-600/4-82-025, National Technical Information Serv., PB82-258799, Springfield, Va.
3. U.S. ENVIRONMENTAL PROTECTION AGENCY. 1984. Definition and procedure for the determination of the method detection limit. 40 CFR Part 136, Appendix B. *Federal Register* 49, No. 209.
4. COLE, T., R. RIGGIN & J. GLASER. 1980. Evaluation of method detection limits and analytical curve for EPA Method 610 PNA's. International Symp. Polynuclear Aromatic Hydrocarbons, 5th, Battelle's Columbus Lab., Columbus, Ohio.
5. U.S. ENVIRONMENTAL PROTECTION AGENCY. 1984. EPA Method Study 20, Method 610—PNA's. EPA-600/4-84-063, National Technical Information Serv., PB84-211614, Springfield, Va.

6440 C.   Liquid-Liquid Extraction
Gas Chromatographic/Mass Spectrometric Method

See Section 6410B.

6630   ORGANOCHLORINE PESTICIDES

6630 A.   Introduction

1. Sources and Significance

The organochlorine pesticides commonly occur in waters that have been affected by agricultural discharges. Some of the listed compounds are degradation products of other pesticides detected by this method. Several of the pesticides are bioaccumulative and relatively stable, as well as toxic or carcinogenic; thus they require close monitoring.

2. Selection of Method

Methods 6630B and C consist of gas chromatographic (GC) procedures following liquid-liquid extraction of water samples. They are relatively sensitive methods that can be used to detect numerous pesticides. Differences between the methods are minimal after extraction. Method 6630D is a gas chromatographic/mass spectrometric (GC/MS) method that can detect all of the target compounds, but at much higher concentrations. All these methods also are useful for determination of polychlorinated biphenyls (PCBs).

6630 B.   Liquid-Liquid Extraction
Gas Chromatographic Method I*

1. General Discussion

a. Application: This gas chromatographic procedure is suitable for quantitative determination of the following specific compounds: BHC, lindane, heptachlor, aldrin, heptachlor epoxide, dieldrin, endrin, captan, DDE, DDD, DDT, methoxychlor, endosulfan, dichloran, mirex, and pentachloronitrobenzene. Under favorable circumstances, strobane, toxaphene, chlordane (tech.), and others also may be determined when relatively high concentrations of these complex mixtures are pres-

ent and the chromatographic fingerprint is recognizable in packed or capillary column analysis. Trifluralin and certain organophosphorus pesticides, such as parathion, methylparathion, and malathion, which respond to the electron-capture detector, also may be measured. However, the usefulness of the method for organophosphorus or other specific pesticides must be demonstrated before it is applied to sample analysis.

b. Principle: In this procedure the pesticides are extracted with a mixed solvent, diethyl ether/hexane or methylene chloride/hexane. The extract is concentrated

---

*Approved by Standard Methods Committee, 1985.

by evaporation and, if necessary, is cleaned up by column adsorption-chromatography. The individual pesticides then are determined by gas chromatography. Although procedures detailed below refer primarily to packed columns, capillary column chromatography may be used provided that equivalent results can be demonstrated. See Section 6010C.2a1) for discussion of gas chromatographic principles and 6010C.2b2) for discussion of electron-capture detector.

As each component passes through the detector a quantitatively proportional change in electrical signal is measured on a strip-chart recorder. Each component is observed as a peak on the recorder chart. The retention time is indicative of the particular pesticide and peak height/peak area is proportional to its concentration.

Variables may be manipulated to obtain important confirmatory data. For example, the detector system may be selected on the basis of the specificity and sensitivity needed. The detector used in this method is an electron-capture detector that is very sensitive to chlorinated compounds. Additional confirmatory identification can be made from retention data on two or more columns where the stationary phases are of different polarities. A two-column procedure that has been found particularly useful is specified. If sufficient pesticide is available for detection and measurement, confirmation by a more definitive technique, such as mass spectrometry, is desirable.

c. *Interference:* See Sections 6010C.2a2) and 6010C.2b2). Some compounds other than chlorinated compounds respond to the electron-capture detector. Among these are oxygenated and unsaturated compounds. Sometimes plant or animal extractives obscure pesticide peaks. These interfering substances often can be removed by auxiliary cleanup techniques. A magnesia-silica gel column cleanup and separation procedure is used for this pur-

pose. Such cleanup usually is not required for potable waters.

1) Polychlorinated biphenyls (PCBs)—Industrial plasticizers, hydraulic fluids, and old transformer fluids that contain PCBs are a potential source of interference in pesticide analysis. The presence of PCBs is suggested by a large number of partially resolved or unresolved peaks that may occur throughout the entire chromatogram. Particularly severe PCB interference will require special separation procedures.

2) Phthalate esters—These compounds, widely used as plasticizers, cause electron-capture detector response and are a source of interferences. Water leaches these esters from plastics, such as polyethylene bottles and plastic tubing. Phthalate esters can be separated from many important pesticides by the magnesia-silica gel column cleanup. They do not cause response to halogen-specific detectors such as microcoulometric or electrolytic conductivity detectors.

d. *Detection limits:* The ultimate detection limit of a substance is affected by many factors, for example, detector sensitivity, extraction and cleanup efficiency, concentrations, and detector signal-to-noise level. Lindane usually can be determined at 10 ng/L in a sample of relatively unpolluted water; the DDT detection limit is somewhat higher, 20 to 25 ng/L. Increased sensitivity is likely to increase interference with all pesticides.

e. *Sample preservation:* Some pesticides are unstable. Transport under iced conditions, store at 4°C until extraction, and do not hold more than 7 d. When possible, extract upon receipt in the laboratory and store extracts at 4°C until analyzed. Analyze extracts within 40 d.

2. Apparatus

Clean thoroughly all glassware used in sample collection and pesticide residue analyses. Clean glassware as soon as possible after use. Rinse with water or the solvent that was last used in it, wash with soapy water, rinse with tap water, distilled water, redistilled acetone, and finally with pesticide-quality hexane. As a precaution, glassware may be rinsed with the extracting solvent just before use. Heat heavily contaminated glassware in a muffle furnace at 400°C for 15 to 30 min. High-boiling-point materials, such as PCBs, may require overnight heating at 500°C, but no borosilicate glassware can exceed this temperature without risk. Do not heat volumetric ware. Clean volumetric glassware with special reagents.† Rinse with water and pesticide-quality hexane. After drying, store glassware to prevent accumulation of dust or other contaminants. Store inverted or cover mouth with aluminum foil.

a. *Sample bottles:* 1-L capacity, glass, with TFE-lined screw cap. Bottle may be calibrated to minimize transfers and potential for contamination.

b. *Evaporative concentrator,* Kuderna-Danish, 500-mL flask and 10-mL graduated lower tube fitted with a 3-ball Snyder column, or equivalent.

c. *Separatory funnels,* 2-L capacity, with TFE stopcock.

d. *Graduated cylinders,* 1-L capacity.

e. *Funnels,* 125-mL.

f. *Glass wool,* filter grade.

g. *Chromatographic column,* 20 mm in diam and 400 mm long, with coarse fritted disk at bottom.

h. *Microsyringes,* 10- and 25-μL capacity.

i. *Hot water bath.*

j. *Gas chromatograph,* equipped with:

1) *Glass-lined injection port.*

2) *Electron-capture detector.*

3) *Recorder:* Potentiometric strip chart, 25-cm, compatible with detector and associated electronics.

4) *Borosilicate glass column,* 1.8 m × 4-mm ID or 2-mm ID.

Variations in available gas chromatographic instrumentation necessitate different operating procedures for each. Therefore, refer to the manufacturer's operating manual as well as gas chromatography catalogs and other references (see Bibliography). In general, use equipment with the following features:

• Carrier-gas line with a molecular sieve drying cartridge and a trap for removal of oxygen from the carrier gas. A special purifier‡ may be used. Use only dry carrier gas and insure that there are no gas leaks.

• Oven temperature stable to ±0.5°C or better at desired setting.

• Chromatographic columns—A well-prepared column is essential to an acceptable gas chromatographic analysis. Obtain column packings and pre-packed columns from commercial sources or prepare column packing in the laboratory.

It is inappropriate to give rigid specifications on size or composition to be used because some instruments perform better with certain columns than do others. Columns with 4-mm ID are used most commonly. The carrier-gas flow is approximately 60 mL/min. When 2-mm-ID columns are used, reduce carrier-gas flow to about 25 mL/min. Adequate separations have been obtained by using 5% OV-210 on 100/120 mesh dimethyl-dichlorosilane-treated diatomaceous earth§ in a 2-m column. The 1.5% OV-17 and 1.95% QF-1 column is recommended for confirmatory analysis. Two additional column options are included: 3% OV-1 and mixed-phase 6% QF-1 + 4% SE-30, each on dimethyl-dichlorosilane-treated diatomaceous earth, 100-120 mesh. OV-210, which is a refined form of QF-1, may be substituted for QF-1. A column is suitable when it effects adequate and reproducible resolution.

---

† No Chromix, Godax, 6 Varick Place, New York, N.Y., or equivalent.

‡ Hydrox, Matheson Gas Products, P. O. Box E, Lyndhurst, N.J., or equivalent.

§ Gas Chrom Q, Applied Science Labs., Inc., P. O. Box 440, State College, Pa., or equivalent.

Sample chromatograms are shown in Figures 6630:1 through 6630:4.

Alternately, use fused silica capillary ‖ columns, 30 m long with a 0.32-mm ID and 0.25-μm film thickness, or equivalent. See Figure 6630:5. To confirm identification use a column of different polarity.#

3. Reagents**

Use solvents, reagents, and other materials for pesticide analysis that are free from

interferences under the condition of the analysis. Specific selection of reagents and distillation of solvents in an all-glass system may be required. "Pesticide quality" solvents usually do not require redistillation; however, always determine a blank before use.

a. Hexane.

b. Petroleum ether, boiling range 30 to 60°C.

c. Diethyl ether: CAUTION: *Explosive peroxides tend to form.* Test for presence of peroxides†† and, if present, reflux over granulated sodium-lead alloy for 8 h, distill in a glass apparatus, and add 2% methanol.

‖ J & W Scientific, DB-5, DB-1701, or equivalent.
J & W Scientific, DB-1, or equivalent.
**Gas chromatographic methods are extremely sensitive to the materials used. Mention of trade names by *Standard Methods* does not preclude the use of other existing or as-yet-undeveloped products that give *demonstrably* equal results.

††Use E. M. Quant™, MCB Manufacturing Chemists, Inc., 2909 Highland Ave., Cincinnati, Ohio, or equivalent.

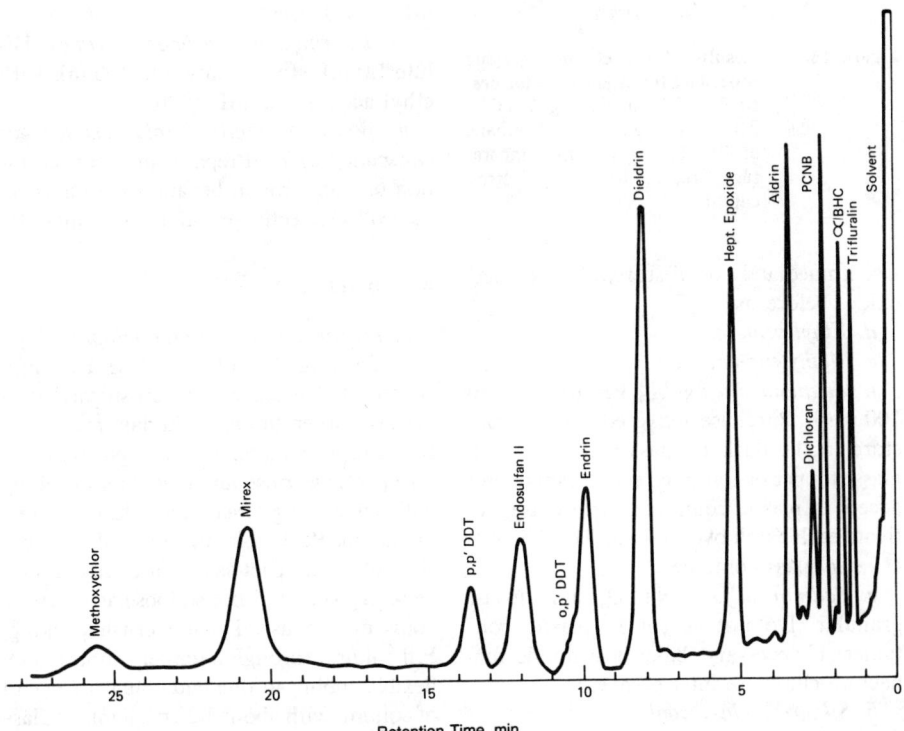

Figure 6630:1. Results of gas chromatographic procedure for organochlorine pesticides. Column packing: 1.5% OV-17 + 1.95% QF-1; carrier gas: argon/methane at 60 mL/min; column temperature: 200°C; detector: electron capture in pulse mode.

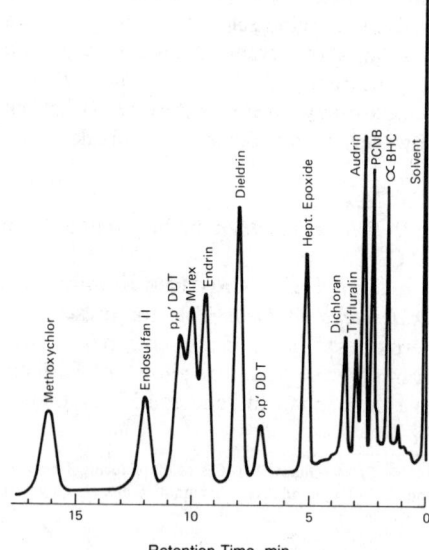

Retention Time, min

Figure 6630:2. Results of gas chromatographic procedure for organochlorine pesticides. Column packing: 5% OV-210; carrier gas: argon/methane at 70 mL/min; column temperature: 180°C; detector: electron capture.

Use immediately or, if stored, test for peroxides before use.

   d. Ethyl acetate.

   e. Methylene chloride.

   f. Magnesia-silica gel,‡‡ PR grade, 60 to 100 mesh. Purchase activated at 676°C and store in the dark in glass container with glass stopper or foil-lined screw cap; do not accept in plastic container. Before use, activate each batch overnight at 130°C in foil-covered glass container.

   g. Sodium sulfate, Na_2SO_4, anhydrous, granular: Do not accept in plastic container. If necessary, bake in a muffle furnace to eliminate interferences.

   h. Silanized glass wool.

   i. Column packing:

   1) Solid support—Dimethyl dichlorosi-

lane-treated diatomaceous earth,§§ 100 to 120 mesh.

   2) Liquid phases—OV-1, OV-210, 1.5% OV-17 (SP 2250) + 1.95% QF-1 (SP 2401), and 6% QF-1 + 4% SE-30, or equivalent.

   j. Carrier gas: One of the following is required:

   1) *Nitrogen gas,* purified grade, moisture- and oxygen-free.

   2) *Argon-methane* (95 + 5%) for use in pulse mode.

   k. Pesticide reference standards: Obtain purest standards available (95 to 98%) from gas chromatographic and chemical supply houses.

   l. Stock pesticide solutions: Dissolve 100 mg of each pesticide in ethyl acetate and dilute to 100 mL in a volumetric flask; 1.00 mL = 1.00 mg.

   m. Intermediate pesticide solutions: Dilute 1.0 mL stock solution to 100 mL with ethyl acetate; 1.0 mL = 10 μg.

   n. Working standard solutions for gas chromatography: Prepare final concentration of standards in hexane solution as required by detector sensitivity and linearity.

4. Procedure

   a. Preparation of chromatograph:

   1) Packing the column—Use a column constructed of silanized borosilicate glass because other tubing materials may catalyze sample component decomposition. Before packing, rinse and dry column tubing with solvent, e.g., methylene chloride, then methanol. Pack column to a uniform density not so compact as to cause unnecessary back pressure and not so loose as to create voids during use. Do not crush packing. Fill column through a funnel connected by flexible tubing to one end. Plug other end of column with about 1.3 cm silanized glass wool and fill with aid of *gentle* vibration or tapping but do not use an electric vi-

---

‡‡Florisil™ or equivalent.

§§Gas-Chrom Q™, Supelcoport, or equivalent.

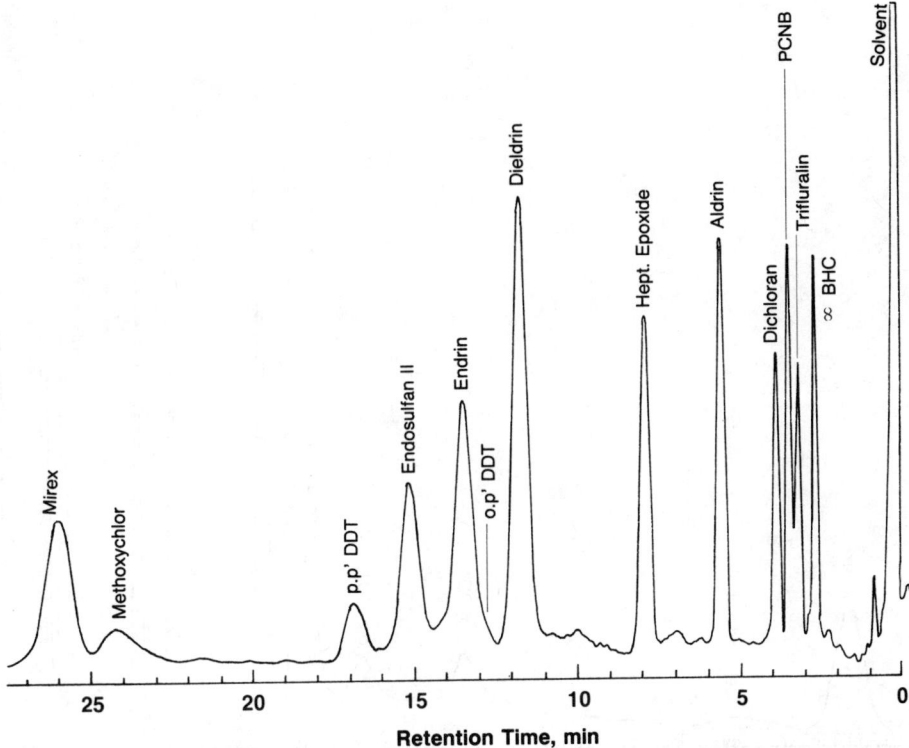

Figure 6630:3. Chromatogram of pesticide mixture. Column packing: 6% QF-1 + 4% SE-30; carrier gas; argon/methane at 60 mL/min; column temperature: 200°C; detector: electron capture.

brator because it tends to fracture packing. Optionally, apply a vacuum to plugged end. Plug open end with silanized glass wool.

2) Conditioning—Proper thermal and pesticide conditioning are essential to eliminate column bleed and to provide acceptable gas chromatographic analysis. The following procedure provides excellent results: Connect packed column to the injection port. *Do not* connect column to detector; however, maintain gas flow through detector by using the purge-gas line, or in dual-column ovens, by connecting an unpacked column to the detector. Adjust carrier-gas flow to about 50 mL/min and slowly (over a 1-h period) raise oven temperature to 230°C. After 24 to 48

h at this temperature the column is ready for pesticide conditioning.

Adjust oven temperature and carrier-gas flow rate to approximate operating levels. Make six consecutive 10-μL injections of a concentrated pesticide mixture through column at about 15-min intervals. Prepare this injection mixture from lindane, heptachlor, aldrin, heptachlor epoxide, dieldrin, endrin, and *p,p'*-DDT, each compound at a concentration of 200 ng/μL. After pesticide conditioning, connect column to detector and let equilibrate for at least 1 h, preferably overnight. Column is then ready for use.

3) Injection technique

a) Develop an injection technique with constant rhythm and timing. The "solvent

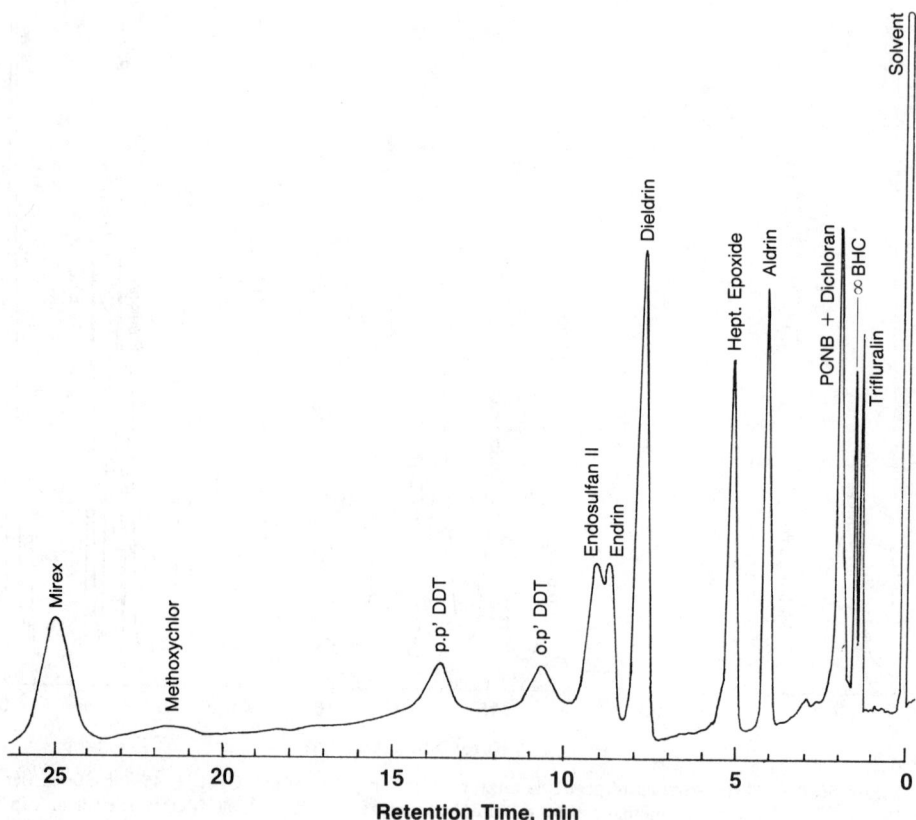

Figure 6630:4. Chromatogram of pesticide mixture. Column packing: 3% OV-1; carrier gas: argon/methane at 70 mL/min; column temperature: 180°C; detector: electron capture.

flush" technique described below has been used successfully and is recommended to prevent sample blowback or distillation within the syringe needle. Flush syringe with solvent, then draw a small volume of clean solvent into syringe barrel (e.g., 1 μL in a 10-μL syringe). Remove needle from solvent and draw 1 μL of air into barrel. Draw 3 to 4 μL of sample extract into barrel. Remove needle from sample extract and draw approximately 1 μL air into barrel. Record volume of sample extract between air pockets. Rapidly insert needle through inlet septum, depress plunger, withdraw syringe. After each injection

thoroughly clean syringe by rinsing several times with solvent.

b) Inject standard solutions of such concentration that the injection volume and peak height of the standard are approximately the same as those of the sample.

b. Treatment of samples:

1) Sample collection—Fill sample bottle to neck. Collect samples in duplicate.

2) Extraction of samples—Shake sample well and accurately measure all the sample in a 1-L graduated cylinder in two measuring operations if necessary (or use a precalibrated sample bottle to avoid transfer operation). Pour sample into a 2-L separatory funnel. Rinse sample bottle and cyl-

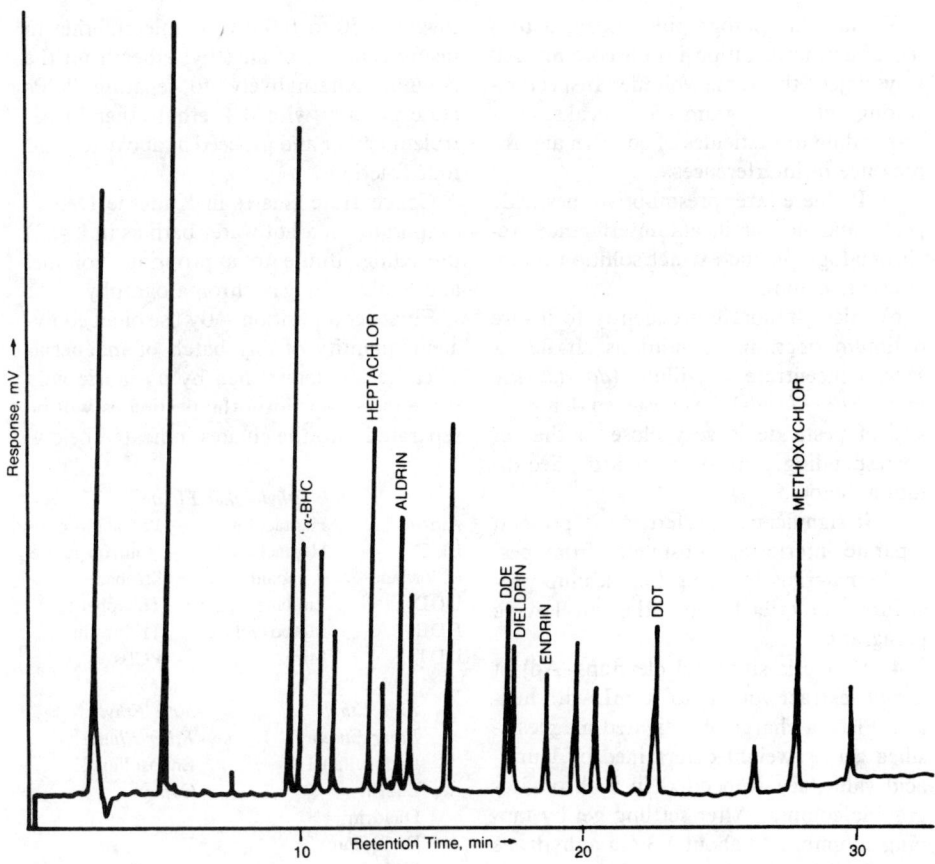

Figure 6630:5. Chromatogram of pesticide mixture. Column DB-5, 30 m long, multilevel program temperature, electron-capture detector.

inder with 60 mL 15% diethyl ether or methylene chloride in hexane, pour this solvent into separatory funnel, and shake vigorously for 2 min. Let phases separate for at least 10 min.

Drain water phase from separatory funnel into sample bottle and carefully pour organic phase through a 2-cm-OD column containing 8 to 10 cm of Na_2SO_4 into a Kuderna-Danish apparatus fitted with a 10-mL concentrator tube. Pour sample back into separatory funnel.

Rinse sample bottle with 60 mL mixed solvent, use solvent to repeat sample extraction, and pass organic phase through Na_2SO_4. Complete a third extraction with 60 mL of mixed solvent that was used to rinse sample bottle again, and pass organic phase through Na_2SO_4. Wash Na_2SO_4 with several portions of hexane and drain well. Fit Kuderna-Danish apparatus with a three-ball Snyder column and reduce volume to about 7 mL in a hot water bath (90 to 95°C). At this point all methylene chloride present in the initial extracting solvent has been distilled off. Cool, remove concentrator tube from Kuderna-Danish apparatus, rinse ground-glass joint, and dilute to 10 mL with hexane. Make initial gas chromatographic analysis at this dilution.

3) Gas chromatography—Inject 3 to 4 µL of extract solution into a column. Always inject the same volume. Inspect resulting chromatogram for peaks corresponding to pesticides of concern and for presence of interferences.

a) If there are presumptive pesticide peaks and no significant interference, rechromatograph the extract solution on an alternate column.

b) Inject standards frequently to insure optimum operating conditions. If necessary, concentrate or dilute (*do not use methylene chloride*) the extract so that peak size of pesticide is very close to that of corresponding peaks in standard. (See dilution factor, ¶ 5*a*).

c) If significant interference is present, separate interfering substances from pesticide materials by using the cleanup procedure described in the following paragraph.

4) Magnesia-silica gel cleanup—Adjust sample extract volume to 10 mL with hexane. Place a charge of activated magnesia-silica gel ‖ ‖ (weight determined by lauric-acid value, see Appendix) in a chromatographic column. After settling gel by tapping column, add about 1.3 cm anhydrous granular Na_2SO_4 to the top. Pre-elute column, after cooling, with 50 to 60 mL petroleum ether. Discard eluate and just before exposing sulfate layer to air, quantitatively transfer sample extract into column by careful decantation and with subsequent petroleum ether washings (5 mL maximum). Adjust elution rate to about 5 mL/min and, separately, collect the eluates in 500-mL Kuderna-Danish flasks equipped with 10-mL receivers.

Make first elution with 200 mL 6% ethyl ether in petroleum ether, and the second with 200 mL 15% ethyl ether in petroleum ether. Make third elution with 200 mL 50% ethyl ether-petroleum ether and the fourth with 200 mL 100% ethyl ether. Fol-

low with 50 to 100 mL petroleum ether to insure removal of all ethyl ether from the column. Alternatively, to separate PCBs elute initially with 0% ethyl ether in petroleum ether and proceed as above to yield four fractions.

Concentrate eluates in Kuderna-Danish evaporator in a hot water bath as in ¶ 4*b*2) preceding, dilute to appropriate volume, and analyze by gas chromatography.

Eluate composition—By use of an equivalent quantity of any batch of magnesia-silica gel as determined by its lauric acid value (see Appendix) the pesticides will be separated into the eluates indicated below:

6% Ethyl Ether Eluate		
Aldrin	Heptachlor	Pentachloro-
BHC	Heptachlor	nitrobenzene
Chlordane	epoxide	Strobane
DDD	Lindane	Toxaphene
DDE	Methoxychlor	Trifluralin
DDT	Mirex	PCBs

15% Ethyl Ether Eluate	50% Ethyl Ether Eluate
Endosulfan I	Endosulfan II
Endrin	Captan
Dieldrin	
Dichloran	
Phthalate esters	

If present, certain thiophosphate pesticides will occur in each of the above fractions as well as in the 100% ether fraction. For additional information regarding eluate composition and the procedure for determining the lauric acid value, refer to the FDA Pesticide Analytical Manual (see Bibliography). For elution pattern test procedure see Appendix, Section 4.

5) Determination of extraction efficiency—Add known amounts of pesticides in ethyl acetate solution to 1 L water sample and carry through the same procedure as for samples. Dilute an equal amount of intermediate pesticide solution (¶ 3*m* above) to the same final volume. Call peak height from standard "*a*" and peak height

---

‖ ‖ Florisil™ or equivalent.

from sample to which pesticide was added "*b*," whereupon the extraction efficiency equals *b/a*. Periodically determine extraction efficiency and a control blank to test the procedure. Also analyze one set of duplicates with each series of samples as a quality-control check.

5. Calculation

a. Dilution factor: If a portion of the extract solution was concentrated, the dilution factor, *D*, is a decimal; if it was diluted, the dilution factor exceeds 1.

b. Determine pesticide concentrations by direct comparison to a single standard when the injection volume and response are within 10% of those of the sample pesticide of interest (Table 6630:I). Calculate concentration of pesticide:

$$\mu g/L = \frac{A \times B \times C \times D}{E \times F \times G}$$

where:

A = ng standard pesticide,
B = peak height of sample, mm, or area count,
C = extract volume, μL,
D = dilution factor,
E = peak height of standard, mm, or area count,
F = volume of extract injected, μL, and
G = volume of sample extracted, mL.

TABLE 6630:I. RETENTION RATIOS OF VARIOUS ORGANOCHLORINE PESTICIDES RELATIVE TO ALDRIN

Liquid phase*	1.5% OV-17 + 1.95% QF-1	5% OV-210	3% OV-1	6% QF-1 + 4% SE-30
Column Temperature	200°C	180°C	180°C	200°C
Argon/methane carrier flow	60 mL/min	70 mL/min	70 mL/min	60 mL/min
Pesticide	RR	RR	RR	RR
∝-BHC	0.54	0.64	0.35	0.49
PCNB	0.68	0.85	0.49	0.63
Lindane	0.69	0.81	0.44	0.60
Dichloran	0.77	1.29	0.49	0.70
Heptachlor	0.82	0.87	0.78	0.83
Aldrin	1.00	1.00	1.00	1.00
Heptachlor epoxide	1.54	1.93	1.28	1.43
Endosulfan I	1.95	2.48	1.62	1.79
p,p'-DDE	2.23	2.10	2.00	1.82
Dieldrin	2.40	3.00	1.93	2.12
Captan	2.59	4.09	1.22	1.94
Endrin	2.93	3.56	2.18	2.42
o,p'-DDT	3.16	2.70	2.69	2.39
p,p'-DDD	3.48	3.75	2.61	2.55
Endosulfan II	3.59	4.59	2.25	2.72
p,p'-DDT	4.18	4.07	3.50	3.12
Mirex	6.1	3.78	6.6	4.79
Methoxychlor	7.6	6.5	5.7	4.60
Aldrin (Min absolute)	3.5	2.6	4.0	5.6

* All columns glass, 180 cm × 4 mm ID, solid support Gas-Chrom Q (100/200 mesh).

Typical chromatograms of representative pesticide mixtures are shown in Figures 6630:1 through 6630:5.

Report results in micrograms per liter without correction for efficiency.

6. Precision and Bias

Ten laboratories in an interlaboratory study selected their own water samples and added four representative pesticides to replicate samples, at two concentrations in acetone. The added pesticides came from a single source. Samples were analyzed with and without magnesia-silica gel cleanup. Precision and recovery data are given in Table 6630:II.

7. Bibliography

MILL, P.A. 1968. Variation of Florisil activity: simple method for measuring absorbent capacity and its use in standardizing Florisil columns. *J. Assoc. Off. Anal. Chem.* 51:29.

FOOD AND DRUG ADMINISTRATION. 1968 (revised 1978). Pesticide Analytical Manual, 2nd ed. U.S. Dep. Health, Education & Welfare, Washington, D.C.

MONSANTO CHEMICAL COMPANY. 1970. Monsanto Methodology for Arochlors—Analysis of Environmental Materials for Biphenyls, Analytical Chemistry Method 71-35. St. Louis, Mo.

U.S. ENVIRONMENTAL PROTECTION AGENCY. 1971. Method for Organic Pesticides in Water and Wastewater. National Environmental Research Center, Cincinnati, Ohio.

STEERE, N.V., ed. 1971. Handbook of Laboratory Safety. Chemical Rubber Company, Cleveland, Ohio.

GOERLITZ, D.F. & E. BROWN. 1972. Methods for analysis of organic substances in water. *In* Techniques of Water Resources Investigations of the United States Geological Survey, Book 5, Chapter A3, p. 24. U.S. Dep. Interior, Geological Survey, Washington, D.C.

U.S. ENVIRONMENTAL PROTECTION AGENCY. 1978. Method for Organochlorine Pesticides in Industrial Effluents. National Environmental Research Center, Cincinnati, Ohio.

TABLE 6630:II. PRECISION AND BIAS DATA FOR SELECTED ORGANOCHLORINE PESTICIDES

Pesticide	Level Added ng/L	Pre-treatment	Mean Recovery ng/L	Recovery %	Precision* ng/L S_T	S_o
Aldrin	15	No cleanup	10.42	69	4.86	2.59
	110		79.00	72	32.06	20.19
	25	Cleanup†	17.00	68	9.13	3.48‡
	100		64.54	65	27.16	8.02‡
Lindane	10	No cleanup	9.67	97	5.28	3.47
	100		72.91	73	26.23	11.49‡
	15	Cleanup†	14.04	94	8.73	5.20
	85		59.08	70	27.49	7.75‡
Dieldrin	20	No cleanup	21.54	108	18.16	17.92
	125		105.83	85	30.41	21.84
	25	Cleanup	17.52	70	10.44	5.10‡
	130		84.29	65	34.45	16.79‡
DDT	40	No cleanup	40.30	101	15.96	13.42
	200		154.87	77	38.80	24.02
	30	Cleanup†	35.54	118	22.62	22.50
	185		132.08	71	49.83	25.31

* S_T = overall precision and S_o = single-operator precision.
† Use of magnesia-silica gel column cleanup before analysis.
‡ $S_o < S_T/2$.

U.S. ENVIRONMENTAL PROTECTION AGENCY. 1978. Method for Polychlorinated Biphenyls in Industrial Effluents. National Environmental Research Center, Cincinnati, Ohio.

U.S. ENVIRONMENTAL PROTECTION AGENCY. 1979. Handbook for Analytical Quality Control in Water and Wastewater Laboratories. National Environmental Research Center, Analytical Quality Control Laboratory, Cincinnati, Ohio.

U.S. ENVIRONMENTAL PROTECTION AGENCY. 1980. Analysis of Pesticide Residues in Human and Environmental Samples. Environmental Toxicology Div., Health Effects Laboratory, Research Triangle Park, N.C.

Appendix—Standardization of Magnesia-Silica Gel* Column by Weight Adjustment Based on Adsorption of Lauric Acid

A rapid method for determining adsorptive capacity of magnesia-silica gel is based on adsorption of lauric acid from hexane solution. An excess of lauric acid is used and the amount not adsorbed is measured by alkali titration. The weight of lauric acid adsorbed is used to calculate, by simple proportion, equivalent quantities of gel for batches having different adsorptive capacities.

1. Reagents

a. *Ethyl alcohol,* USP or absolute, neutralized to phenolphthalein.

b. *Hexane,* distilled from all-glass apparatus.

c. *Lauric acid solution:* Transfer 10.000 g lauric acid to a 500-mL volumetric flask, dissolve in hexane, and dilute to 500 mL; 1.00 mL = 20 mg.

d. *Phenolphthalein indicator:* Dissolve 1 g in alcohol and dilute to 100 mL.

e. *Sodium hydroxide,* 0.05N: Dilute 25 mL 1N NaOH to 500 mL with distilled water. Standardize as follows: Weigh 100 to 200 mg lauric acid into 125-mL erlenmeyer flask; add 50 mL neutralized ethyl alcohol and 3 drops phenolphthalein indicator; titrate to permanent end point; and calculate milligrams lauric acid per milliliter NaOH (about 10 mg/mL).

2. Procedure

Transfer 2.000 g magnesia-silica gel to a 25-mL glass-stoppered erlenmeyer flask. Cover loosely with aluminum foil and heat overnight at 130°C. Stopper, cool to room temperature, add 20.0 mL lauric acid solution (400 mg), stopper, and shake occasionally during 15 min. Let adsorbent settle and pipet 10.0 mL supernatant into a 125-mL erlenmeyer flask. Avoid including any gel. Add 50 mL neutral alcohol and 3 drops phenolphthalein indicator solution; titrate with 0.05N NaOH to a permanent end point.

3. Calculation of Lauric Acid Value and Adjustment of Column Weight

Calculate amount of lauric acid adsorbed on gel as follows:

Lauric acid value = mg lauric acid/g gel = 200 − (mL required for titration × mg lauric acid/mL 0.05N NaOH).

To obtain an equivalent quantity of any batch of gel, divide 110 by lauric acid value for that batch and multiply by 20 g. Verify proper elution of pesticides by the procedure given below.

---

*Florisil™ or equivalent.

4. Test for Proper Elution Pattern and Recovery of Pesticides

Prepare a test mixture containing aldrin, heptachlor epoxide, *p,p'*-DDE, dieldrin, parathion, and malathion. Dieldrin and parathion should elute in the 15% eluate; all but a trace of malathion in the 50% eluate, and the others in the 6% eluate.

6630 C.   Liquid-Liquid Extraction Gas Chromatographic Method II*

This method[1] is applicable to the determination of organochlorine pesticides and PCBs†‡ in municipal and industrial discharges. When analyzing unfamiliar samples for any or all of these compounds, support the identifications by at least one additional qualitative technique. This method includes analytical conditions for a second, confirmatory gas chromatographic column. Alternatively, analyze by a gas chromatographic/mass spectrometric (GC/MS) method for base/neutrals and acids using the extract produced by this method.

1. General Discussion

a. Principle: A measured volume of sample is extracted with methylene chloride. The extract is dried and exchanged to hexane during concentration. If other determinations having essentially the same extraction and concentration steps are to be performed, a single sample extraction is sufficient. The extract is separated by gas chromatography and the compounds are measured with an electron capture detector.[2] See Section 6010C for discussion of gas chromatographic principles.

The method provides procedures for magnesia-silica gel column cleanup and elemental sulfur removal to aid in the elimination of interferences. When cleanup is required, sample concentration levels must be high enough to permit separate treatment of subsamples. Chromatographic conditions appropriate for the simultaneous measurement of combinations of compounds may be selected.

b. Interferences: See Section 6410B.1*b*1) for precautions concerning glassware, reagent purity, and matrix interferences.

Phthalate esters may interfere in pesticide analysis with an electron capture detector. These compounds generally appear in the chromatogram as large, late-eluting peaks, especially in the 15 and 50% fractions from magnesia-silica gel. Common flexible plastics contain phthalates that are easily extracted during laboratory operations. Cross-contamination of clean glassware routinely occurs when plastics are handled during extraction steps, especially when solvent-wetted surfaces are handled. Minimize interferences from phthalates by avoiding use of plastics. Exhaustive cleanup of reagents and glassware may be required to eliminate phthalate contamination.[3,4] Phthalate ester interference can be avoided by using a microcoulometric or electrolytic conductivity detector.

c. Detection limits: The MDL is the minimum concentration of a substance that can be measured and reported with 99% confidence that the value is above zero.[5] The

---

* Approved by Standard Methods Committee, 1987. Accepted by U.S. Environmental Protection Agency as equivalent to EPA Method 608.
† Aldrin, α-BHC, β-BHC, δ-BHC, γ-BHC, chlordane, 4,4'-DDD, 4,4'-DDE, 4,4'-DDT, dieldrin, endosulfan I, endosulfan II, endosulfan sulfate, endrin, endrin aldehyde, heptachlor, heptachlor epoxide, toxaphene, PCB-1016, PCB-1221, PCB-1232, PCB-1242, PCB-1248, PCB-1254, PCB-1260.
‡ The PCBs constitute a class of 209 compounds. This procedure is designed to determine nine commercial formulations known as the Aroclors, each of which is a mixture of PCBs.

MDL concentrations listed in Table 6630:III were obtained by using reagent water.[6] Similar results were achieved with representative wastewaters. The MDL actually obtained in a given analysis will vary, depending on instrument sensitivity and matrix effects. This method has been tested for linearity of known-addition recovery from reagent water and is applicable over the concentration range from $4 \times$ MDL to $1000 \times$ MDL with the following exceptions: Chlordane recovery at $4 \times$ MDL was low (60%); toxaphene recovery was linear over the range of $10 \times$ MDL to $1000 \times$ MDL.[6] It is difficult to determine MDLs for mixtures such as these. To calculate the MDLs given, a few of the GC peaks in each mixture were used. Depending on the particular peaks selected, these results may or may not be reproducible in other laboratories.

d. Safety: The toxicity or carcinogenicity of each reagent has not been defined precisely. The following compounds have been

TABLE 6630:III. CHROMATOGRAPHIC CONDITIONS AND METHOD DETECTION LIMITS*

	Retention Time *min*		Method Detection Limit
Compound	Column 1	Column 2	$\mu g/L$
α-BHC	1.35	1.82	0.003
γ-BHC	1.70	2.13	nd
β-BHC	1.90	1.97	nd
Heptachlor	2.00	3.35	0.003
δ-BHC	2.15	2.20	0.009
Aldrin	2.40	4.10	0.004
Heptachlor epoxide	3.50	5.00	0.083
Endosulfan I	4.50	6.20	0.014
4,4'-DDE	5.13	7.15	0.004
Dieldrin	5.45	7.23	0.002
Endrin	6.55	8.10	0.006
4,4'-DDD	7.83	9.08	0.011
Endosulfan II	8.00	8.28	0.004
4,4'-DDT	9.40	11.75	0.012
Endrin aldehyde	11.82	9.30	0.023
Endosulfan sulfate	14.22	10.70	0.066
Chlordane	mr	mr	0.014
Toxaphene	mr	mr	0.24
PCB-1016	mr	mr	nd
PCB-1221	mr	mr	nd
PCB-1232	mr	mr	nd
PCB-1242	mr	mr	0.065
PCB-1248	mr	mr	nd
PCB-1254	mr	mr	nd
PCB-1260	mr	mr	nd

Column 1 conditions: Supelcoport (100/120 mesh) coated with 1.5% SP-2250/1.95% SP-2401 packed in a 1.8 m long × 4 mm ID glass column with 5% methane/95% argon carrier gas at 60 mL/min flow rate. Column temperature held isothermal at 200°C, except for PCB-1016 through PCB-1248 at 160°C.

Column 2 conditions: Supelcoport (100/120 mesh) coated with 3% OV-1 packed in a 1.8 m long × 4 mm ID glass column with 5% methane/95% argon carrier gas at 60 mL/min flow rate. Column temperature held isothermal at 200°C for the pesticides; at 140°C for PCB-1221 and 1232; and at 170°C for PCB-1016 and 1242 to 1268.

*mr = multiple peak response. See Figures 6630:2 through 10.

nd = not determined.

classified tentatively as known or suspected, human or mammalian carcinogens: 4,4'-DDT, 4,4'-DDD, the BHCs, and the PCBs. Prepare primary standards of these compounds in a hood and wear a NIOSH/MESA-approved toxic gas respirator when handling high concentrations. Treat and dispose of Hg used for sulfur removal as a hazardous waste.

2. Sampling and Storage

For collection and storage requirements, see Section 6410B.2. If samples will not be extracted within 72 h of collection, adjust pH to the range 5.0 to 9.0 with NaOH or H_2SO_4. Record volume of acid or base used. If aldrin is to be determined, add sodium thiosulfate when residual chlorine is present.

3. Apparatus

Use apparatus specified in Section 6410B.3a–e, g, and i–k.

In addition:

a. Chromatographic column, 400 mm long × 22 mm ID, with TFE stopcock and coarse frit filter disk.§

b. Gas chromatograph:‖ An analytical system complete with gas chromatograph suitable for on-column injection and all required accessories including syringes, analytical columns, gases, and strip-chart recorder. Preferably use a data system for measuring peak areas.

1) Column 1, 1.8 m long × 4 mm ID, glass, packed with 1.5% SP-2250/1.95% SP-2401 on Supelcoport (100/120 mesh) or equivalent. This column was used to develop the detection limit and precision and bias data presented herein. For guidelines for the use of alternate column packings see ¶ 5c.

2) Column 2, 1.8 m long × 4 mm ID, glass, packed with 3% OV-1 on Supelcoport (100/120 mesh) or equivalent.

3) Detector, electron-capture. This detector was used to develop the detection limit and precision and bias data presented herein. For use of alternate detectors see ¶ 5c.

4. Reagents

This method requires reagents described in Section 6410B.4a–e, and in addition:

a. Acetone, hexane, isooctane, methylene chloride, pesticide quality or equivalent.

b. Ethyl ether, nanograde redistilled in glass if necessary. Demonstrate before use freedom from peroxides by means of test strips.# Remove peroxides by procedures provided with the test strips. After cleanup, add 20 mL ethyl alcohol preservative per liter of ether.

c. Magnesia-silica gel,** 60/100 mesh. Purchase activated at 1250°F and store in the dark in glass containers with ground-glass stoppers or foil-lined screw caps. Before use, activate each batch for at least 16 h at 130°C in a foil-covered glass container; let cool.

d. Mercury, triple-distilled.

e. Copper powder, activated.

f. Stock standard solutions: Prepare as directed in Section 6410B.4g, using isooctane as the solvent.

g. Calibration standards: See Section 6420B.4j. Dilute with isooctane and use MDL values from Table 6630:III.

h. Quality control (QC) check sample concentrate: Obtain a check sample concentrate†† containing each compound at the following concentrations in acetone: 4,4'-DDD, 10 μg/mL; 4,4'-DDT, 10 μg/mL; endosulfan II, 10 μg/mL; endosulfan sulfate, 10 μg/mL; endrin, 10 μg/mL; any

---

§ Kontes K-42054 or equivalent.

‖ Gas chromatographic methods are extremely sensitive to the materials used. Mention of trade names by *Standard Methods* does not preclude the use of other existing or as-yet-undeveloped products that give *demonstrably* equivalent results.

E. Merck, EM Science Quant or equivalent.

** Florisil or equivalent.

†† For U.S. federal permit-related analyses, use samples obtainable from U.S. EPA Environmental Monitoring and Support Laboratory, Cincinnati, Ohio.

other single-component pesticide, 2 μg/ mL. If this method will be used only to analyze for PCBs, chlordane, or toxaphene, the QC check sample concentrate should contain the most representative multicomponent compound at a concentration of 50 μg/mL in acetone. If such a sample is not available from an external source, prepare using stock standards prepared independently from those used for calibration.

5. Procedure

a. Extraction: Mark water meniscus on side of sample bottle for later determination of volume. Pour entire sample into a 2-L separatory funnel and extract with methylene chloride as directed in Section 6410B.5*a*1), without any pH adjustment or solvent wash.

After extracting and concentrating with a three-ball Snyder column, increase temperature of hot water bath to about 80°C. Momentarily remove Snyder column, add 50 mL hexane and a new boiling chip, and reattach Snyder column. Concentrate extract as before but use hexane to prewet column. Complete concentration in 5 to 10 min.

Remove Snyder column and rinse flask and its lower joint into the concentrator tube with 1 to 2 mL hexane. Preferably use a 5-mL syringe for this operation. Stopper concentrator tube and store refrigerated if further processing will not be done immediately. If extract is to be stored longer than 2 d, transfer to a TFE-sealed screwcap vial. If extract requires no further cleanup, proceed with gas chromatographic analysis. If further cleanup is required, follow procedure of ¶ *b* before chromatographic analysis.

Determine original sample volume by refilling sample bottle to mark and transferring liquid to a 1000-mL graduated cylinder. Record sample volume to nearest 5 mL.

b. Cleanup and separation: Use either procedure below or any other appropriate procedure; however, first demonstrate that the requirements of ¶ 7 can be met. The magnesia-silica gel column allows for a select fractionation of compounds and eliminates polar interferences. Elemental sulfur, which interferes with the electron-capture gas chromatography of certain pesticides, can be removed by the technique described below.

1) *Magnesia-silica gel column cleanup*— Place a weight of magnesia-silica gel (nominally 20 g) predetermined by calibration, ¶ *d*3), into a chromatographic column. Tap column to settle gel and add 1 to 2 cm anhydrous Na_2SO_4 to the top. Add 60 mL hexane to wet and rinse. Just before exposure of the Na_2SO_4 layer to air, stop elution of hexane by closing stopcock on column. Discard eluate. Adjust sample extract volume to 10 mL with hexane and transfer it from K-D concentrator tube onto column. Rinse tube twice with 1 to 2 mL hexane, adding each rinse to the column. Place a 500-mL K-D flask and clean concentrator tube under chromatographic column. Drain column into flask until Na_2SO_4 layer is nearly exposed. Elute column with 200 mL 6% ethyl ether in hexane (v/v) (Fraction 1) at a rate of about 5 mL/ min. Remove K-D flask and set aside. Elute column again, using 200 mL 15% ethyl ether in hexane (v/v) (Fraction 2), into a second K-D flask. Elute a third time using 200 mL 50% ethyl ether in hexane (v/v) (Fraction 3). The elution patterns for the pesticides and PCBs are shown in Table 6630:IV. Concentrate fractions for 15 to 20 min as in ¶ *a*, using hexane to prewet the column, and set water bath temperature at about 85°C. After cooling, remove Snyder column and rinse flask and its lower joint into concentrator tube with hexane. Adjust volume of each fraction to 10 mL with hexane and analyze by gas chromatography, ¶s *c* through *e* below.

2) *Sulfur interference removal*—Elemental sulfur usually will elute entirely in

TABLE 6630:IV. DISTRIBUTION OF CHLORINATED PESTICIDES AND PCBs INTO MAGNESIA-SILICA GEL COLUMN FRACTIONS[5]

Compound	Recovery by Fraction* %		
	1	2	3
Aldrin	100	—	—
α-BHC	100	—	—
β-BHC	97	—	—
δ-BHC	98	—	—
γ-BHC	100	—	—
Chlordane	100	—	—
4,4'-DDD	99	—	—
4,4'-DDE	98	—	—
4,4'-DDT	100	—	—
Dieldrin	0	100	—
Endosulfan I	37	64	—
Endosulfan II	0	7	91
Endosulfan sulfate	0	0	106
Endrin	4	96	—
Endrin aldehyde	0	68	26
Heptachlor	100	—	—
Heptachlor epoxide	100	—	—
Toxaphene	96	—	—
PCB-1016	97	—	—
PCB-1221	97	—	—
PCB-1232	95	4	—
PCB-1242	97	—	—
PCB-1248	103	—	—
PCB-1254	90	—	—
PCB-1260	95	—	—

* Eluant composition:
　Fraction 1–6% ethyl ether in hexane
　Fraction 2–15% ethyl ether in hexane
　Fraction 3–50% ethyl ether in hexane

Fraction 1 of the magnesia-silica gel column cleanup. To remove sulfur interference from this fraction or the original extract, pipet 1.00 mL concentrated extract into a clean concentrator tube or TFE-sealed vial. Add 1 to 3 drops of mercury and seal.[7] Mix for 15 to 30 s. If prolonged shaking (2 h) is required, use a reciprocal shaker. Alternatively, use activated copper powder for sulfur removal.[8] Analyze by gas chromatography.

c. *Gas chromatography operating conditions:* Table 6630:III summarizes the recommended operating conditions for the gas chromatograph and gives retention times and MDLs that can be achieved under these conditions. Examples of separations obtained with Column 1 are shown in Figures 6630:6 to 15. Other packed or capillary (open-tubular) columns,[9] chromatographic conditions, or detectors may be used if the requirements of ¶ 7 are met.

d. *Calibration:* Calibrate system daily by either external or internal procedure. NOTE: For quantification and identification of mixtures such as PCBs, chlordane, and toxaphene, take extra precautions.[9–11]

1) External standard calibration proce-

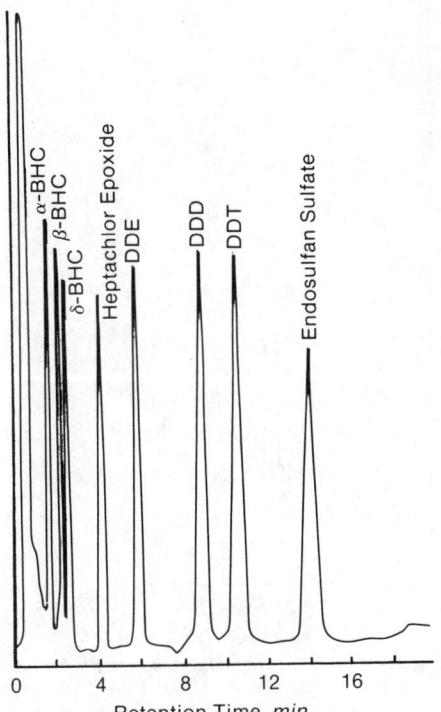

Figure 6630:6. Gas chromatogram of pesticides.
Column: 1.5% SP-2250/1.95% SP-2401 on Supelcoport; temperature: 200°C; detector: electron capture.

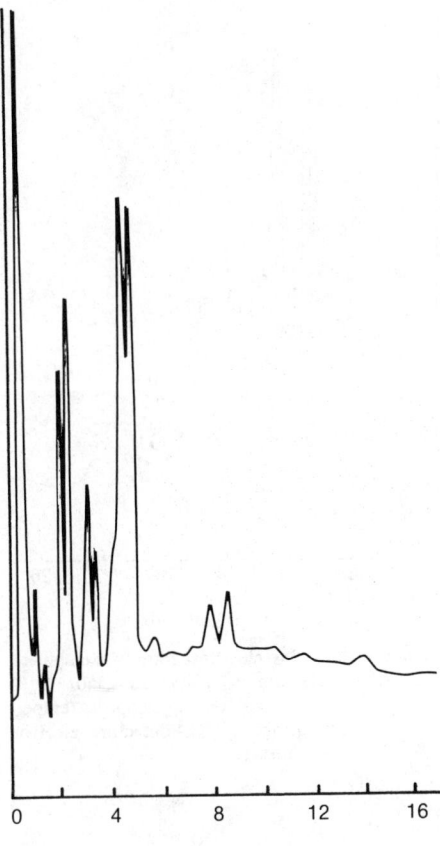

Figure 6630:7. Gas chromatogram of chlordane.
Column: 1.5% SP-2250/1.95% SP-2401 on Supelcoport; temperature: 200°C; detector: electron capture.

dure—Prepare standards as directed in ¶ 4g and follow procedure of Section 6420B.5b3). Tabulate data and obtain calibration curve or calibration factor as directed in Section 6230B.5b2).

2) Internal standard calibration procedure—Prepare standards as directed in ¶ 4g and follow procedure of Section 6420B.5b3). Tabulate data and calculate response factors as directed in Section 6210B.5c2).

Verify working calibration curve, calibration factor, or RF on each working day by measuring one or more calibration standards. If the response for any compound varies from the predicted response by more than ±15%, repeat test using a fresh calibration standard. Alternatively, prepare a new calibration curve for that compound.

3) Magnesia-silica gel standardization—Gel from different batches or sources may vary in adsorptive capacity. To standardize the amount used, use the lauric acid value[12], which measures the adsorption from a hexane solution of lauric acid (mg/g gel). Determine the amount to be used for each column by dividing 110 by this ratio and multiplying the quotient by 20 g.

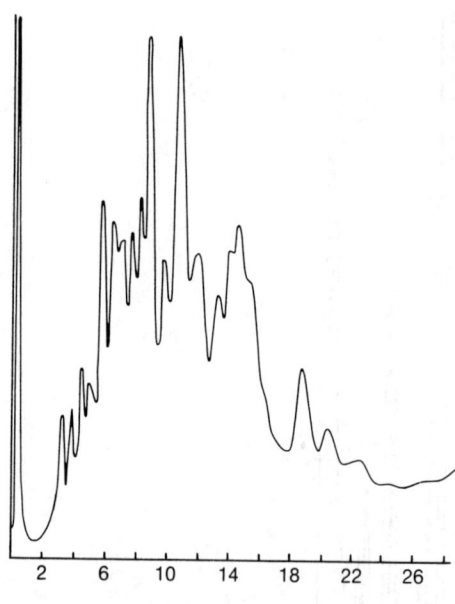

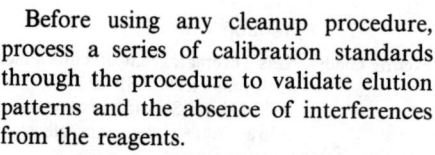

Retention Time, *min*

Figure 6630:8. Gas chromatogram of toxaphene. Column: 1.5% SP-2250/1.95% SP-2401 on Supelcoport; temperature: 200°C; detector: electron capture.

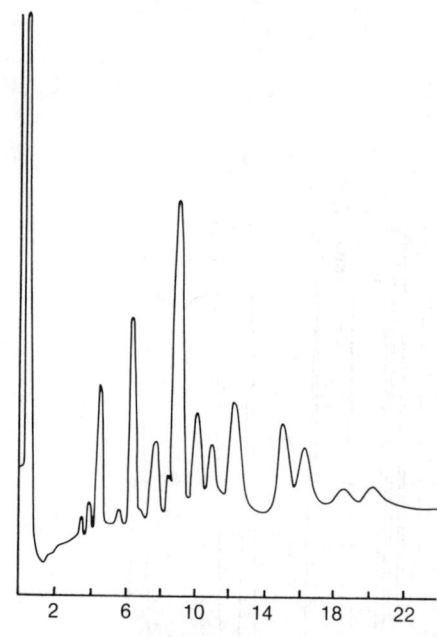

Retention Time, *min*

Figure 6630:9. Gas chromatogram of PCB-1016. Column: 1.5% SP-2250/1.95% SP-2401 on Supelcoport; temperature: 160°C; detector: electron capture.

Before using any cleanup procedure, process a series of calibration standards through the procedure to validate elution patterns and the absence of interferences from the reagents.

e. Sample analysis: See Section 6420B.5*b*3). If peak response cannot be measured because of interferences, further cleanup is required.

6. Calculation

Determine concentration of individual compounds using procedures given in Section 6420B.6*a*.

If it is apparent that two or more PCB (Aroclor) mixtures are present, the Webb and McCall procedure[13] may be used to identify and quantify the Aroclors, de-

pending on the Aroclors present. Other techniques also are available.

For multicomponent mixtures (chlordane, toxaphene, and PCBs) match peak retention times in standards with peaks in sample. Quantitate every identifiable peak unless interference with individual peaks persists after cleanup. Add peak height or peak area of each identified peak in chromatogram. Calculate as total response in sample versus total response in standard. Environmental degradation of these compounds may make identification difficult. This method is suitable only for intact mixtures such as the original Aroclor or pesticide formulation and it is not suitable for the other altered mixtures that are sometimes found in the environment. In these instances the GC peak pattern would not match the standard.

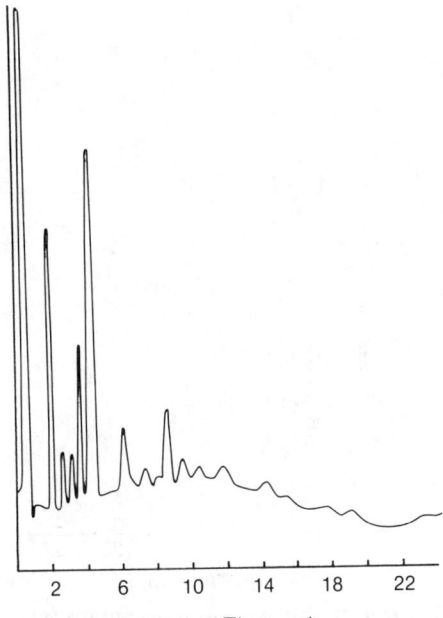

Retention Time, *min*

Figure 6630:10. Gas chromatogram of PCB-1221.
Column: 1.5% SP-2250/1.95%
SP-2401 on Supelcoport; temperature: 160°C; detector: electron
capture.

Report results in micrograms per liter
without correction for recovery data. Report QC data with the sample results.

7. Quality Control

a. Quality control program: See Section
6210B.7*a*.

b. Analyst proficiency check: To establish
the ability to generate data with acceptable
precision and bias, proceed as follows: Using a pipet, prepare QC check samples at
test concentrations shown in Table 6630:V
by adding 1.00 mL of QC check sample
concentrate (¶ 4*h*) to each of four 1-L portions of reagent water. Analyze QC check
samples according to the method beginning
in ¶ 5*a*. Calculate average recovery and
standard deviation of the recovery, compare with acceptance criteria and evaluate

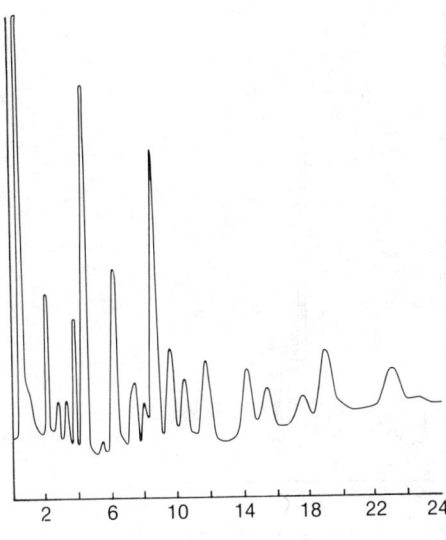

Retention Time, *min*

Figure 6630:11. Gas chromatogram of PCB-1232.
Column: 1.5% SP-2250/1.95%
SP-2401 on Supelcoport; temperature: 160°C; detector: electron
capture.

and correct system performance as directed
in Section 6210B.7*b*.

*c. Analyses of samples with known ad-
ditions:* See Section 6420B.7*c*. Prepare QC
check sample concentrates according to
¶ 4*h* and use Tables 6630:III and IV.

*d. Quality-control check standard anal-
ysis:* See Section 6420B.7*d*. Prepare QC
check standard according to ¶ 4*h* and use
Table 6630:V. If all compounds in Table
6630:V are to be measured in the sample
in ¶ *c* above, it is probable that the analysis
of a QC check will be required; therefore,
routinely analyze the QC check standard
with the known-addition sample.

e. Bias assessment and records: See Section 6210B.7*e*.

f. Additional quality-assurance practices:
See Section 6230B.7*g*.

8. Precision and Bias

This method was tested by 20 laboratories using reagent water, drinking water,

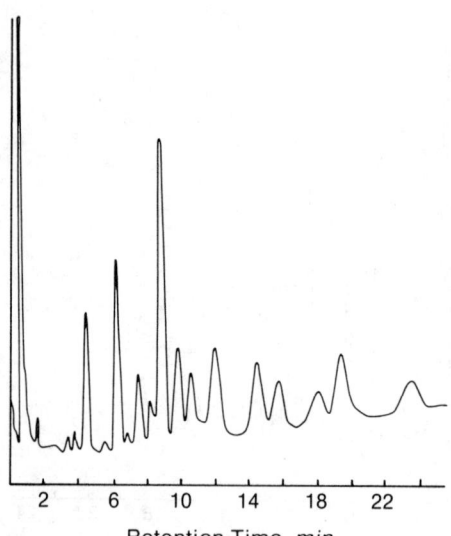

Figure 6630:12. Gas chromatogram of PCB-1242.
Column: 1.5% SP-2250/1.95%
SP-2401 on Supelcoport; temperature: 160°C; detector: electron
capture.

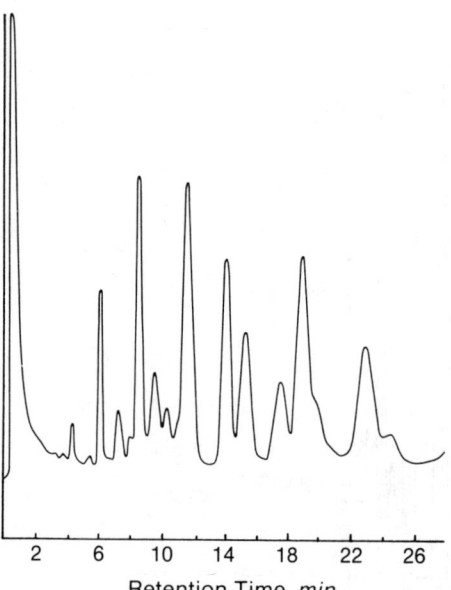

Figure 6630:13. Gas chromatogram of PCB-1248.
Column: 1.5% SP-2250/1.95%
SP-2401 on Supelcoport; temperature: 160°C; detector: electron
capture.

surface water, and industrial wastewaters
with known additions at six concentrations
over the range 0.5 to 30 μg/L for single-
component pesticides and 8.5 to 400 μg/
L for multicomponent samples.[14] Single-
operator precision, overall precision, and
method bias were found to be related directly to the compound concentration and
essentially independent of sample matrix.
Linear equations describing these relationships are presented in Table 6630:VI.

9. References

1. U.S. ENVIRONMENTAL PROTECTION
   AGENCY. 1984. Method 608 — Organochlorine pesticides and PCBs. 40 CFR Part 136,
   43321; *Federal Register* 49, No. 209.
2. U.S. ENVIRONMENTAL PROTECTION
   AGENCY. 1982. Determination of pesticides
   and PCBs in industrial and municipal wastewaters. EPA-600/4-82-023, National Technical Information Serv., PB82-214222,
   Springfield, Va.
3. GIAM, C.S., H.S. CHAN & G.S. NEF. 1975.
   Sensitive method for determination of phthalate ester plasticizers in open-ocean biota samples. *Anal. Chem.* 47:2225.
4. GIAM, C.S. & H.S. CHAN. 1976. Control of
   Blanks in the Analysis of Phthalates in Air
   and Ocean Biota Samples. U.S. National Bur.
   Standards, Spec. Publ. 442.
5. U.S. ENVIRONMENTAL PROTECTION
   AGENCY. 1984. Definition and procedure for
   the determination of the method detection
   limit. 40 CFR Part 136, Appendix B. *Federal
   Register* 49, No. 209.
6. U.S. ENVIRONMENTAL PROTECTION
   AGENCY. 1980. Method Detection Limit and
   Analytical Curve Studies, EPA Methods 606,
   607, and 608. Special letter rep. for EPA Contract 68-03-2606, Environmental Monitoring
   and Support Lab., Cincinnati, Ohio.
7. GOERLITZ, D.F. & L.M. LAW. 1971. Note on
   removal of sulfur interferences from sediment
   extracts for pesticide analysis. *Bull. Environ.
   Contam. Toxicol.* 6:9.
8. U.S. ENVIRONMENTAL PROTECTION
   AGENCY. 1980. Manual of Analytical

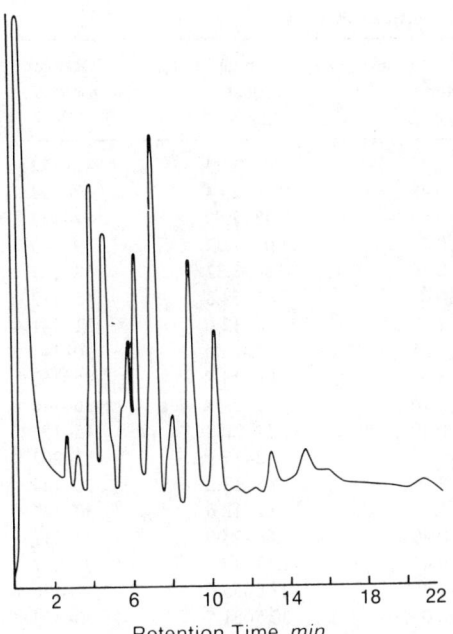

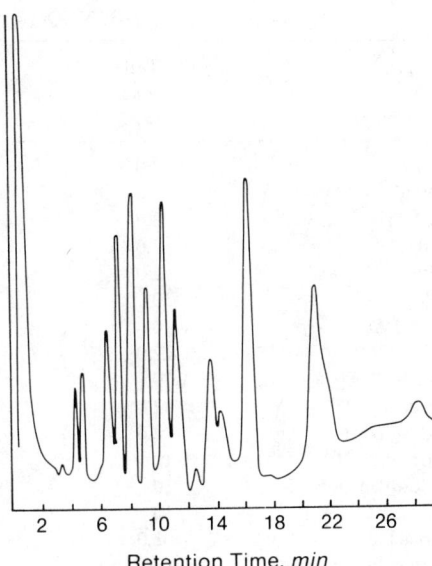

Retention Time, *min*

Figure 6630:14. Gas chromatogram of PCB-1254. Column: 1.5% SP-2550/1.95% SP-2401 on Supelcoport; temperature: 200°C; detector: electron capture.

Retention Time, *min*

Figure 6630:15. Gas chromatogram of PCB-1260. Column: 1.5% SP-2250/1.95% SP-2401 on Supelcoport; temperature: 200°C; detector: electron capture.

Methods for the Analysis of Pesticides in Human and Environmental Samples. EPA-600/8-80-038, Health Effects Research Lab., Research Triangle Park, N.C.

9. ALFORD-STEVENS, A. et al. 1986. Characterization of commercial Aroclors by automated mass spectrometric determination of polychlorinated biphenyls by level of chlorination. *Anal. Chem.* 58:2014.

10. ALFORD-STEVENS, A. 1986. Analyzing PCB's. *Environ. Sci. Technol.* 20:1194.

11. ALFORD-STEVENS, A. 1987. Mixture analytes. *Environ. Sci. Technol.* 21, 137.

12. MILLS, P.A. 1968. Variation of florisil activity: Simple method for measuring absorbent capacity and its use in standardizing florisil columns. *J. Assoc. Offic. Anal. Chem.* 51:29.

13. WEBB, R.G. & A.C. McCALL. 1973. Quantitative PCB standards for election capture gas chromatography. *J. Chromatog. Sci.* 11:366.

14. U.S. ENVIRONMENTAL PROTECTION AGENCY. 1984. EPA Method Study 18, Method 608—Organochlorine Pesticides and PCBs. EPA-600/4-84-061, National Technical Information Serv., PB84-211358, Springfield, Va.

TABLE 6630:V. QC ACCEPTANCE CRITERIA*

Compound	Test Conc. $\mu g/L$	Limit for s $\mu g/L$	Range for $\overline{X}$ $\mu g/L$	Range for $\overline{P}$, P_s %
Aldrin	2.0	0.42	1.08–2.24	42–122
α-BHC	2.0	0.48	0.98–2.44	37–134
β-BHC	2.0	0.64	0.78–2.60	17–147
δ-BHC	2.0	0.72	1.01–2.37	19–140
γ-BHC	2.0	0.46	0.86–2.32	32–127
Chlordane	50	10.0	27.6–54.3	45–119
4,4′-DDD	10	2.8	4.8–12.6	31–141
4,4′-DDE	2.0	0.55	1.08–2.60	30–145
4,4′-DDT	10	3.6	4.6–13.7	25–160
Dieldrin	2.0	0.76	1.15–2.49	36–146
Endosulfan I	2.0	0.49	1.14–2.82	45–153
Endosulfan II	10	6.1	2.2–17.1	D–202
Endosulfan sulfate	10	2.7	3.8–13.2	26–144
Endrin	10	3.7	5.1–12.6	30–147
Heptachlor	2.0	0.40	0.86–2.00	34–111
Heptachlor epoxide	2.0	0.41	1.13–2.63	37–142
Toxaphene	50	12.7	27.8–55.6	41–126
PCB-1016	50	10.0	30.5–51.5	50–114
PCB-1221	50	24.4	22.1–75.2	15–178
PCB-1232	50	17.9	14.0–98.5	10–215
PCB-1242	50	12.2	24.8–69.6	39–150
PCB-1248	50	15.9	29.0–70.2	38–158
PCB-1254	50	13.8	22.2–57.9	29–131
PCB-1260	50	10.4	18.7–54.9	8–127

* s = standard deviation of four recovery measurements,
     $\overline{X}$ = average recovery for four recovery measurements,
P, P_s = percent recovery measured, and
     D = detected; result must be greater than zero.
NOTE: These criteria are based directly on the method performance data in Table 6630:VI. Where necessary, the limits for recovery were broadened to assure applicability of the limits to concentrations below those used to develop Table 6630:VI.

TABLE 6630:VI. METHOD PRECISION AND BIAS AS FUNCTIONS OF CONCENTRATION*

Compound	Bias, as Recovery, X' $\mu g/L$	Single-Analyst Precision, s_r $\mu g/L$	Overall Precision, S' $\mu g/L$
Aldrin	$0.81C + 0.04$	$0.16\overline{X} - 0.04$	$0.20\overline{X} - 0.01$
α-BHC	$0.84C + 0.03$	$0.13\overline{X} + 0.04$	$0.23\overline{X} - 0.00$
β-BHC	$0.81C + 0.07$	$0.22\overline{X} - 0.02$	$0.33\overline{X} - 0.05$
δ-BHC	$0.81C + 0.07$	$0.18\overline{X} + 0.09$	$0.25\overline{X} + 0.03$
γ-BHC	$0.82C - 0.05$	$0.12\overline{X} + 0.06$	$0.22\overline{X} + 0.04$
Chlordane	$0.82C - 0.04$	$0.13\overline{X} + 0.13$	$0.18\overline{X} + 0.18$
4,4'-DDD	$0.84C + 0.30$	$0.20\overline{X} - 0.18$	$0.27\overline{X} - 0.14$
4,4'-DDE	$0.85C + 0.14$	$0.13\overline{X} + 0.06$	$0.28\overline{X} - 0.09$
4,4'-DDT	$0.93C - 0.13$	$0.17\overline{X} + 0.39$	$0.31\overline{X} - 0.09$
Dieldrin	$0.90C + 0.02$	$0.12\overline{X} + 0.19$	$0.16\overline{X} + 0.16$
Endosulfan I	$0.97C + 0.04$	$0.10\overline{X} + 0.07$	$0.18\overline{X} + 0.08$
Endosulfan II	$0.93C + 0.34$	$0.41\overline{X} - 0.65$	$0.47\overline{X} - 0.20$
Endosulfan sulfate	$0.89C - 0.37$	$0.13\overline{X} + 0.33$	$0.24\overline{X} + 0.35$
Endrin	$0.89C - 0.04$	$0.20\overline{X} + 0.25$	$0.24\overline{X} + 0.25$
Heptachlor	$0.69C + 0.04$	$0.06\overline{X} + 0.13$	$0.16\overline{X} + 0.08$
Heptachlor epoxide	$0.89C + 0.10$	$0.18\overline{X} - 0.11$	$0.25\overline{X} - 0.08$
Toxaphene	$0.80C + 1.74$	$0.09\overline{X} + 3.20$	$0.20\overline{X} + 0.22$
PCB-1016	$0.81C + 0.50$	$0.13\overline{X} + 0.15$	$0.15\overline{X} + 0.45$
PCB-1221	$0.96C + 0.65$	$0.29\overline{X} - 0.76$	$0.35\overline{X} - 0.62$
PCB-1232	$0.91C + 10.79$	$0.21\overline{X} - 1.93$	$0.31\overline{X} + 3.50$
PCB-1242	$0.93C + 0.70$	$0.11\overline{X} + 1.40$	$0.21\overline{X} + 1.52$
PCB-1248	$0.97C + 1.06$	$0.17\overline{X} + 0.41$	$0.25\overline{X} - 0.37$
PCB-1254	$0.76C + 2.07$	$0.15\overline{X} + 1.66$	$0.17\overline{X} + 3.62$
PCB-1260	$0.66C + 3.76$	$0.22\overline{X} - 2.37$	$0.37\overline{X} - 4.86$

* X' = expected recovery for one or more measurements of a sample containing a concentration of C,
   s_r = expected single-analyst standard deviation of measurements at an average concentration found of $\overline{X}$,
   S' = expected interlaboratory standard deviation of measurements at an average concentration found of $\overline{X}$,
   C = true value for the concentration, and
   $\overline{X}$ = average recovery found for measurements of samples containing a concentration of C.

6630 D.   Liquid-Liquid Extraction
Gas Chromatographic/Mass Spectrometric Method

See Section 6410B.

6640 CHLORINATED PHENOXY ACID HERBICIDES

6640 A.   Introduction

1. Sources and Significance

The chlorinated phenoxy herbicides occur in raw source waters and in corresponding finished drinking waters that have been subjected to industrial discharges or agricultural effluents affected by these herbicides. These compounds are used extensively for weed control and are very potent herbicides even at low concentrations. Esters and salts of 2,4-D and silvex have been used as aquatic herbicides in lakes, streams, and irrigation canals. Toxicological studies suggest that these compounds have detrimental effects on human health; therefore they should be closely monitored in water supplies.

2. Selection of Method

The gas chromatographic method presented here (6640B) utilizes derivatization and GC with electron-capture detector (ECD) to detect these compounds.

6640 B.   Liquid-Liquid Extraction
Gas Chromatographic Method*

1. General Discussion

a. *Principle:* Chlorinated phenoxy acid herbicides such as 2,4-D [2,4-dichlorophenoxyacetic acid], silvex [2-(2,4,5-trichlorophenoxy) propionic acid], 2,4,5-T [2,4,5-trichlorophenoxyacetic acid], and similar chemicals may be determined by a gas chromatographic procedure.

Because these compounds may occur in water in various forms (e.g., acid, salt, ester) a hydrolysis step is included to permit determination of the active part of the herbicide.

Chlorinated phenoxy acids and their esters are extracted from the acidified water sample with ethyl ether. The extracts are hydrolyzed and extraneous material is removed by a solvent wash. The acids are converted to methyl esters and are further cleaned up on a microadsorption column. The methyl esters are determined by gas chromatography.

b. *Interference:* See Section 6630B.1c. Organic acids, especially chlorinated acids, cause the most direct interference. Phenols, including chlorophenols, also may inter-

---

*Approved by Standard Methods Committee, 1985.

fere. Alkaline hydrolysis and subsequent extraction eliminate many of the predominant chlorinated insecticides. Because the herbicides react readily with alkaline substances, loss may occur if there is alkaline contact at any time except in the controlled alkaline hydrolysis step. Acid-rinse glassware and glass wool and acidify sodium sulfate (Na_2SO_4) to avoid this possibility.

c. Detection limits: The practical lower limits for measurement of phenoxy acid herbicides depend primarily on sample size and instrumentation used. If the extract from a 1-L sample is concentrated to 2.00 mL and 5.0 µL of concentrate is injected into the electron-capture gas chromatograph, reliable measurement of 50 ng 2,4-D/L, 10 ng silvex/L, and 10 ng 2,4,5-T/L is feasible. Concentrating extract to 0.50 mL permits detection of approximately 10 ng 2,4-D/L, 2 ng silvex/L, and 2 ng 2,4,5-T/L. The sensitivity of the electron-capture detector often is affected adversely by extraneous material in sample or reagents. Concentrating the extract progressively amplifies this complication. Thus, the practical lower limits of measurement are difficult to define.

2. Apparatus

Clean glassware with detergent in the usual manner, rinse in dilute HCl, and finally rinse in distilled water. To assure removal of organic matter, follow the procedure given in Section 6630B.2.

a. Sample bottles: 1-L capacity, glass, with TFE-lined screw caps. Bottles may be calibrated to minimize transfers and potential for contamination.

b. Evaporative concentrator, Kuderna-Danish, 250-mL flask and 5-mL volumetric receiver.†

c. Snyder columns, three-ball macro, one-ball micro.

d. Separatory funnels, 2-L and 60-mL

sizes with TFE stopcocks and taper ground-glass stoppers.†

e. Pipets, Pasteur, disposable, 140 mm long and 5 mm ID, glass.

f. Microsyringes, 10-µL.

g. Sand bath, fluidized,‡ or water bath.

h. Erlenmeyer flask, 250-mL with ground-glass mouth to fit Snyder columns.

i. Gas chromatographic system: See Section 6630B.2j. Operating parameters that produce satisfactory chromatograms for herbicide analyses are: injector temperature, 215°C; oven temperature, 185°C; and carrier-gas flow, 70 mL/min in a 4-mm-ID column.

3. Reagents§

Check all reagents for purity by the gas chromatographic procedure. Save time and effort by selecting high-quality reagents that do not require further preparation. Some purification of reagents may be necessary as outlined below. If more rigorous treatment is indicated, obtain reagent from an alternate source.

a. Diethyl ether, reagent grade. See 6630B.3c.

b. Toluene, pesticide quality, distilled in glass, or equivalent.

c. Sodium sulfate, Na_2SO_4, anhydrous, granular. Store at 130°C.

d. Sodium sulfate solution: Dissolve 50 mg anhydrous Na_2SO_4 in distilled water and dilute to 1 L.

e. Sodium sulfate, acidified: Add 0.1 mL conc H_2SO_4 to 100 g Na_2SO_4 slurried with enough ethyl ether to just cover the solid. Remove diethyl ether by vacuum drying. Mix 1 g of resulting solid with 5 mL distilled water and confirm that mixture pH is below 4. Store at 130°C.

f. Sulfuric acid, H_2SO_4, conc.

---

†Kontes or equivalent.

‡TeCam or equivalent.
§Chromatographic methods are extremely sensitive to the materials used. Mention of trade names by *Standard Methods* does not preclude the use of other existing or as-yet-undeveloped products that give *demonstrably* equal results.

g. Sulfuric acid, H_2SO_4, 1 + 3. Store in refrigerator.

h. Potassium hydroxide solution: Dissolve 37 g KOH pellets in distilled water and dilute to 100 mL.

i. Boron trifluoride-methanol, 14% boron trifluoride by weight.

j. Magnesia-silica gel,|| PR grade, 60 to 100 mesh. Purchase activated at 676°C and store at 130°C.

k. Glass wool, filtering grade, acid-washed.

l. Herbicide standards, acids, and methyl esters, analytical reference grade or highest purity available.

m. Stock herbicide solutions: Dissolve 100 mg herbicide or methyl ester in 60 mL diethyl ether; dilute to 100 mL in a volumetric flask with hexane; 1.00 mL = 1.00 mg.

n. Intermediate herbicide solution: Dilute 1.0 mL stock solution to 100 mL in a volumetric flask with a mixture of equal volumes of diethyl ether and toluene; 1.00 mL = 10.0 μg.

o. Standard solution for chromatography: Prepare final concentration of methyl ester standards in toluene solution according to the detector sensitivity and linearity.

4. Procedure

a. Sample extraction: Accurately measure sample (850 to 1000 mL) in a 1-L graduated cylinder (or use a precalibrated sample bottle to avoid transfer operations). Acidify to pH 2 with conc H_2SO_4 and pour into a 2-L separatory funnel. Rinse sample bottle and cylinder with 150 mL ethyl ether, add ether to separatory funnel, and shake vigorously for 1 min. Let phases separate for at least 10 min. Occasionally, emulsions prevent adequate separation. If emulsion forms, drain off separated aqueous layer, invert separatory funnel,

––––––––––
||Florisil™ or equivalent.

and shake rapidly. CAUTION: *Vent funnel frequently to prevent excessive pressure buildup.* Collect extract in a 250-mL ground-glass-stoppered erlenmeyer flask containing 2 mL KOH solution. Extract sample twice more, using 50 mL diethyl ether each time, and combine extracts in erlenmeyer flask.

b. Hydrolysis: Add 15 mL distilled water and a small boiling stone and fit flask with a three-ball Snyder column. Remove ether on a steam bath and continue heating for a total of 60 min. Transfer concentrate to a 60-mL separatory funnel. Extract twice, with 20 mL diethyl ether each time, and discard ether layers. The herbicides remain in the aqueous phase.

Acidify by adding 2 mL cold (4°C) 1 + 3 H_2SO_4. Extract once with 20 mL and twice with 10 mL diethyl ether each. Collect extracts in a 125-mL erlenmeyer flask containing about 0.5 g acidified anhydrous Na_2SO_4. Let extract remain in contact with Na_2SO_4 for at least 2 h.

c. Esterification: Fit a Kuderna-Danish apparatus with a 5-mL volumetric receiver. Transfer diethyl ether extract to Kuderna-Danish apparatus through a funnel plugged with glass wool. Use liberal washing of ether. Crush any hardened Na_2SO_4 with a glass rod. Before concentrating, add 0.5 mL toluene. Reduce volume to less than 1 mL on a sand or hot water bath heated to 60 to 70°C. Attach a Snyder micro-column to Kuderna-Danish receiver and concentrate to less than 0.5 mL.

Alternatively, if quantitative recovery is demonstrated, concentrate extract by placing concentrator ampule in a water bath at 70°C. Reduce volume to less than 1 mL using a gentle stream of clean, dry nitrogen (filtered through activated carbon). CAUTION: *Do not use new plastic tubing between the carbon trap and the sample as interferences may be introduced.* Rinse internal wall of ampule with hexane during concentration, never let extract go to dryness, and keep ampule solvent level below water level in the bath. Adjust final volume to 1 mL with hexane.

Cool and add 0.5 mL boron trifluoride-methanol reagent. Use the small one-ball Snyder column as an air-cooled condenser and hold contents of receiver at 50°C for 30 min in the sand bath. Cool and add enough Na₂SO₄ solution (¶ 3d above) so that the toluene-water interface is in the neck of the Kuderna-Danish volumetric receiver flask (about 4.5 mL). Stopper flask with a ground-glass stopper and shake vigorously for about 1 min. Let stand for 3 min for phase separation.

Pipet solvent layer from receiver to top of a small column prepared by plugging a disposable Pasteur pipet with glass wool and packing with 2.0 cm Na₂SO₄ over 1.5 cm magnesia-silica gel adsorbent. Collect eluate in a 2.5-mL graduated centrifuge tube. Complete transfer by repeatedly rinsing volumetric receiver with small quantities of toluene until a final eluate volume of 2.0 mL is obtained. Check calibration of centrifuge tubes to insure that graduations are correct.

d. Gas chromatography: Analyze a suitable portion, 5 to 10 μL, by gas chromatography, using at least two columns for identification and quantification. Inject standard herbicide methyl esters frequently to insure optimum operating conditions. Always inject the same volume. Adjust sample volume extract with toluene, if necessary, so that the sizes of the peaks obtained are close to those of the standards (see ¶ 5a below). For sample chromatograms, see Figures 6640:1 and 2.

e. Determination of recovery efficiency: Add known amounts of herbicides to 1-L water sample, carry through the same procedure as the samples, and determine recovery efficiency. Periodically determine recovery efficiency and a control blank to test the procedure. Analyze one set of duplicates with each series of samples as a quality-control check.

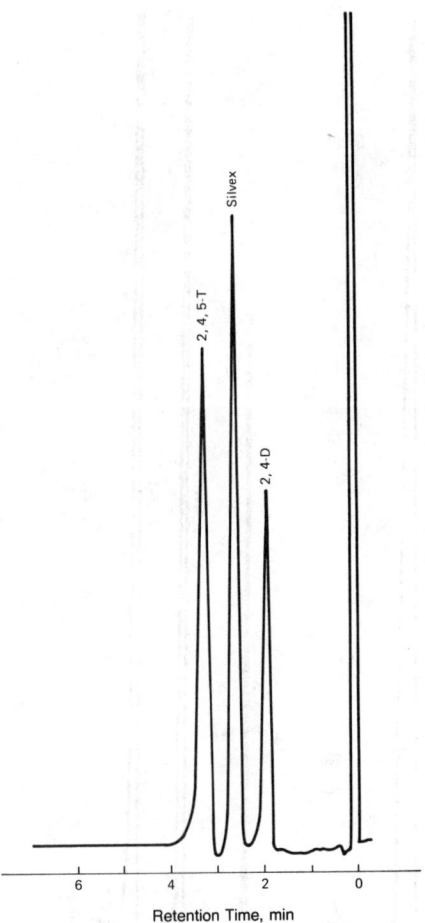

Figure 6640:1. Results of gas chromatographic procedure for chlorinated phenoxy acid herbicides. Column: 1.5% OV-17 + 1.95% QF-1; carrier gas: argon (5%)/methane at 70 mL/min; column temperature: 185°C; detector: electron capture.

5. Calculation

a. Dilution factor: If a portion of the extract solution was concentrated, the dilution factor, *D,* is less than 1; if it was diluted, the dilution factor exceeds 1.

Compare peak height or area of a standard to peak height or area of sample to

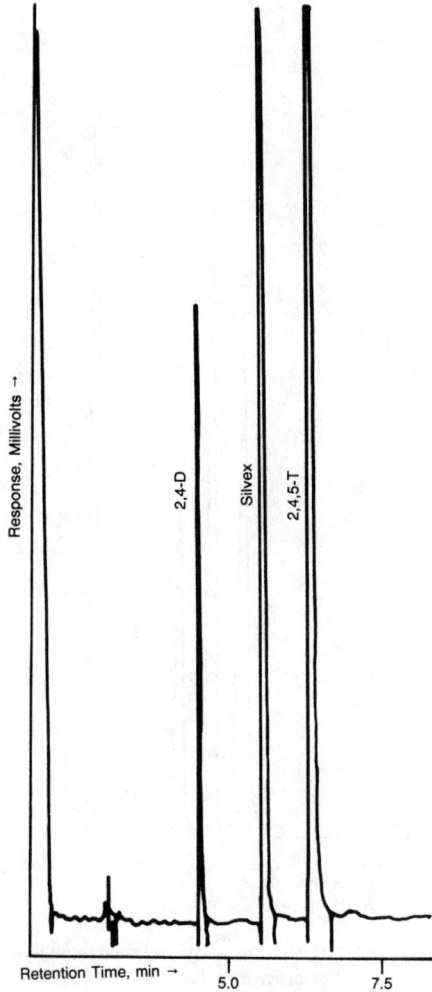

Figure 6640:2. Chromatogram of herbicide mixture. DB-5 column, 30 m long, electron-capture detector.

TABLE 6640:I. RETENTION TIMES FOR METHYL ESTERS OF SOME CHLORINATED PHENOXY ACID HERBICIDES RELATIVE TO 2,4-D METHYL ESTER

	Relative Retention Time for Given Liquid Phase*	
Herbicide	1.5% OV-17 + 1.95% QF-1	5% OV-210
2,4-D	1.00	1.00
Silvex	1.34	1.22
2,4,5-T	1.72	1.51
2,4-D (min absolute)	2.00	1.62

* All columns glass, 180 cm × 4 mm ID, solid support Gas Chrom Q (100/120 mesh); column temperature 185°C; argon/methane carrier flow, operated in pulse mode, 70 mL/min.

B = peak height or area of sample, mm or mm²,
C = extract volume, μL,
D = dilution factor,
E = peak height or area of standard, mm or mm²,
F = volume injected, μL, and
G = volume of sample extracted, mL.

Report results as the methyl ester in micrograms per liter without correction for recovery efficiency.

6. Precision and Bias

Single-laboratory precision and recovery data are presented in Tables 6640:II and III. These data were obtained by analyzing surface water samples from six sources with three added herbicides.

7. Bibliography

METCALF, L.D. & A.A. SCHMITZ. 1961. The rapid preparation of fatty acid esters for gas chromatographic analysis. *Anal. Chem.* 33:363.
GOERLITZ, D.F. & W.L. LAMAR. 1967. Determination of phenoxy acid herbicides in water by electron-capture and microcoulometric gas chromatography. U.S. Geol. Surv. Water-Supply Paper 1817-C.

determine amount of herbicide injected (see Table 6640:I).

Calculate concentration of herbicide:

$$\mu\text{g/L} = \frac{A \times B \times C \times D}{E \times F \times G}$$

where:

A = weight of herbicide standard injected, ng,

TABLE 6640:II. PRECISION OF PHENOXY ACID HERBICIDES FROM DOSED SURFACE WATER

Material	Concentration Range ng/L	Number of Samples	Recovery Average %	Single-Operator Precision $S_o\%$
2,4-D	300–515	11	93	5.0
Silvex	70–290	12	94	6.5
2,4,5-T	90–290	12	100	8.0

TABLE 6640:III. RECOVERY OF PHENOXY ACID HERBICIDES FROM DOSED SURFACE WATER

Sample	Herbicide	Amount Added ng	Recovery %
1	2,4-D	308	94
	Silvex	70	87
	2,4,5-T	92	86
2	2,4-D	308	82
	Silvex	70	99
	2,4,5-T	92	110
3	2,4-D	308	90
	Silvex	70	81
	2,4,5-T	92	104
4	2,4-D	470	91
	Silvex	126	91
	2,4,5-T	140	86
5	2,4-D	470	97
	Silvex	126	86
	2,4,5-T	140	96
6	2,4-D	515	99
	Silvex	222	98
	2,4,5-T	221	104

PART 7000

EXAMINATION OF WATER

AND WASTEWATER FOR

RADIOACTIVITY

7010 INTRODUCTION*

7010 A. General Discussion

1. Occurrence and Monitoring

The radioactivity in water and wastewater originates from both natural sources and human activities. The latter include operations concerned with the nuclear fuel cycle, from mining to reprocessing; medical uses of radioisotopes; industrial uses of radioisotopes; worldwide fallout from atmospheric testing of nuclear devices; and enhancement of the concentration of naturally occurring radionuclides. Monitoring programs for water and wastewater should be designed to assess realistically the degree of environmental radioactive contamination. In some cases, for example, compliance monitoring for drinking water, the conditions are spelled out.[1] In others, it may be necessary to examine the individual situation[2] for consideration of the critical radionuclide(s), the critical pathway by which the critical radionuclide moves through the environment, and a critical population group that is exposed to the particular radionuclide(s) moving along this particular pathway. Use of the critical nuclide-pathway-population approach narrows the list of possible radionuclides to monitor. A list of the most hazardous radionuclides can be selected by examining the radioactivity concentration standards given by the International Committee on Radiation Protection (ICRP)[3], the Federal Radiation Council (FRC)[4], the National Committee on Radiation Protection and Measurement (NCRP)[2], the U.S. Environ-

mental Protection Agency[1], and also agencies in other countries. Individual states within the United States may have their own radioactivity concentration standards if they are Nuclear Regulatory Commission (NRC) agreement states. With few exceptions, these numerical values for radioactivity concentrations in air and water are comparable if certain qualifying assumptions are applied.

Monitoring programs should provide adequate warning of unsafe environmental conditions so that proper precautions can be taken, and of course, to assure that conditions are safe when they are indeed safe. In either circumstance, it is necessary to establish base lines for the kinds and quantities of radionuclides present naturally and to measure additions to this natural background. In this way, measurements may be made to provide information for sound judgments regarding the hazardous or non-hazardous nature of increased concentrations.

2. Types of Measurement

Meaningful measurements require careful application of good scientific techniques. The types of measurements to be made are determined by the objectives of the testing. Gross alpha and gross beta measurements are relatively inexpensive, can be completed quickly, and are useful for screening to determine whether further analysis for specific radionuclides is merited. However, gross measurements give no information about the isotopic composition

---

*Approved by Standard Methods Committee, 1988.

of the sample, cannot be used to estimate radiation dose, and have poor sensitivity if the concentration of dissolved solids is high. Accurate gross beta and especially gross alpha measurements require careful preparation of standards to determine self-absorption and the ability to prepare samples in a similar manner.

Specific radionuclide measurements are required if dose estimates are to be made, results of gross analyses exceed a certain level, or long-term trends are being monitored. Specific analyses usually are more expensive and time-consuming than a gross analysis. Specific measurements identify radionuclides by the energy of emitted radiation, chemical techniques, half-life, or a combination of these characteristics. Gamma-emitting radionuclides can be measured rapidly and with a minimum of sample preparation by using gamma spectrometry. Measurements requiring chemical separations make it possible to increase the sensitivity by increasing the sample quantity measured.

Knowledge of the chemical and radiochemical characteristics of the radionuclide being measured are critical for satisfactory results. Gross alpha and gross beta results will not provide accurate information about radionuclides having energies significantly different from the energy of the calibration standard. During concentration of water samples by evaporation, radionuclides present in elemental form (e.g., radioiodine, polonium) or as compounds (e.g., tritium, carbon-14) may be lost by volatilization. If the sample is ignited, the chance of volatilization loss is even greater. Groundwater generally contains nuclides of the uranium and thorium series. Use special care in sampling and analyzing such samples because members of these series often are not in secular equilibrium.

3. References

1. U.S. ENVIRONMENTAL PROTECTION AGENCY. 1986. Water Pollution Control; Radionuclides; Advance Notice of Proposed Rulemaking, 40 CFR, Part 141, 34836; *Federal Register* 51, No. 189.
2. NATIONAL COMMITTEE ON RADIATION PROTECTION AND MEASUREMENTS. 1959. Maximum Permissible Body Burdens and Maximum Permissible Concentrations of Radionuclides in Air and Water for Occupational Exposure. NBS Handbook No. 69, pp. 1, 17, 37, 38, & 93.
3. INTERNATIONAL COMMISSION ON RADIATION PROTECTION. 1979. Limits for Intakes of Radionuclides by Workers. ICRP Publ. 30, Pergamon Press, New York, N.Y.
4. FEDERAL RADIATION COUNCIL. 1961. Background Material for the Development of Radiation Protection Standards. Rep. No. 2 (Sept.), U.S. Government Printing Off., Washington, D.C.

4. Bibliography

CORYELL, C.D. & N. SUGARMAN, eds. 1951. Radiochemical Studies: The Fission Products. McGraw-Hill Book Co., New York, N.Y.

COMAR, C.I. 1955. Radioisotopes in Biology and Agriculture. McGraw-Hill Book Co., New York, N.Y.

FRIEDLANDER, G., J.W. KENNEDY & J.M. MILLER. 1964. Nuclear and Radiochemistry, 2nd ed. John Wiley & Sons, New York, N.Y.

LEDERER, C.M., J.M. HOLLANDER & I. PERLMANN. 1967. Table of Isotopes, 6th ed. John Wiley & Sons, New York, N.Y.

INTERNATIONAL ATOMIC ENERGY AGENCY. 1971. Disposal of Radioactive Wastes into Rivers, Lakes, and Estuaries. IAEA Safety Ser. No. 36, St. 1/PUB 283.

NATIONAL COUNCIL ON RADIATION PROTECTION AND MEASUREMENTS. 1976. Environmental Radiation Measurements. NCRP Rep. No. 50, Washington, D.C.

7010 B.   Sample Collection and Preservation

1. Collection

The principles of representative sampling of water and wastewater apply to sampling for radioactivity examinations (see Section 1060).

Because a radioactive element often is present in submicrogram quantities, a significant fraction may be lost by adsorption on the surface of containers used in the examination. Similarly, a radionuclide may be largely or wholly adsorbed on the surface of suspended particles.

Sample containers vary in size from 0.5 L to 18 L, depending on required analyses. Use containers of plastic (polyethylene or equivalent) or glass, except for tritium samples (use glass only). When radioactive industrial wastes or comparable materials are sampled, consider the possibility of deposition of radioactivity on surfaces of glassware, plastic containers, and equipment that may cause a loss of radioactivity and possible contamination of subsequent samples collected in inadequately cleansed containers.

2. Preservation

For general information on sample preservation see Section 1060. Add preservative at time of collection unless sample will be separated into suspended and dissolved fractions, but do not delay acid addition beyond 5 d. Use conc hydrochloric (HCl) or nitric (HNO_3) acid to obtain a pH <2, except for radiocesium (use only conc HCl) and radioiodine and tritium (use no preservative). Hold acidified sample at least 16 h before analysis. For further details see references.[1-3]

Test preservatives and reagents for radioactive content.

3. Wastewater Samples

Wastewater often contains larger amounts of nonradioactive suspended and dissolved solids than does water and often most of the radioactivity is in the solid phase. Generally, the use of carriers in the analysis is ineffective without prior conversion of the solid phase to the soluble phase; even then high fixed solids may interfere with radioanalytical procedures. Table 7010:I shows the usual solubility characteristics of common radioelements in wastewater.

The radioelements may exhibit unusual chemical characteristics because of the presence of complexing agents or the method of waste production. For example, tritium may be combined in an organic compound when used in the manufacture of luminous articles; radioiodine from hospitals may occur as complex organic compounds, compared to elemental and iodide forms found in fission products from the processing of spent nuclear fuels; uranium and thorium daughter products often exist as inorganic complexes other than oxides after processing in uranium mills; the strontium 90 titanate waste from a radiois-

TABLE 7010:I. USUAL DISTRIBUTION OF COMMON RADIOELEMENTS BETWEEN THE SOLID AND LIQUID PHASES OF WASTEWATER

In Solution	In Suspension
HCO_3	Ce
Co	Cs
Cr	Mn
Cs	Nb
H	P
I	Pm
K	Pu
Ra	Ra
Rn	Sc
Ru	Th
Sb	U
Sr	Y
	Zn
	Zr

otope heat source would be quite insoluble compared to most other strontium wastes. Valuable information on the chemical composition of wastes, the behavior of radioelements, and the quantity of radioisotopes in use appears in the literature.[4,5]

4. References

1. U.S. GEOLOGICAL SURVEY. 1977. Methods for Determination of Radioactive Substances in Water and Fluvial Sediments. U.S. Government Printing Off., Washington, D.C.

2. HARLEY, J.H., ed. 1972. Environmental Measurements Procedures Manual. HASL-300, U.S. Dep. Energy, New York, N.Y.

3. U.S. ENVIRONMENTAL PROTECTION AGENCY. 1978. Manual for the Interim Certification of Laboratories Involved in Analyzing Public Drinking Water Supplies. EPA 600/8-78-008, Washington, D.C.

4. INTERNATIONAL ATOMIC ENERGY AGENCY. 1960. Disposal of Radioactive Wastes. International Atomic Energy Agency, Vienna, Austria.

5. NEMEROW, N.L. 1963. Industrial Waste Treatment Addison-Wesley, Reading, Mass.

7010 C. Counting Room

Design and construction of a counting room may vary widely. Provide a room free of dust and fumes that may affect electrical stability of instruments. Stabilize and reduce background radiation by making the walls, floor, and ceiling out of at least 5 cm of concrete but avoid using shales, granites, and sands containing sufficient natural activity to affect instrument background.

A modern chemical laboratory can be used to process routine environmental samples, but preferably segregate monitoring work from other laboratory operations.

Provide air-conditioning and humidity control depending on the number of instruments to be used and the prevailing climatic conditions. Generally, electronic instruments perform best when the temperature remains constant within 3°C and does not exceed 30°C. Keep the temperature inside the instrument chassis below the maximum specified by the manufacturer.

Humidity affects instrument performance even more than extremes of temperature because of moisture buildup on critical components. This causes leakage and arcing, and shortens the life of these components. A humidity between 30 and 80% usually is satisfactory.

Most counting instruments are supplied with constant-voltage regulators suitable for controlling minor fluctuations in line voltage. For unusual fluctuations use an auxiliary voltage regulator transformer. Use a manually reset voltage-sensitive device in series with a voltage regulator placed in the main power line to instruments to protect them in case of power failure or fluctuating line voltage.

Store samples containing appreciable activity at a distance so as not to affect instrument background counting rate.

Cover floors and desk tops with a material that can be cleaned easily or replaced as necessary.

7010 D. Counting Instruments

1. General Principles

Geiger-Mueller and proportional counters operate on the principle that the expenditure of energy by a radiation event causes ionization of counter gas and electron collection at the anode of the counting chamber. Through gas or electronic amplification, or both, the ion-collection event triggers an electronic scaler recorder.

The principle of scintillation counters is similar in that quanta of light produced by the interaction of a radiation event and the detection phosphor are seen by a photomultiplier tube. The tube converts the light pulse into an amplified electrical pulse that is recorded by an electronic scaler. Thallium-activated sodium iodide crystals and silver-activated zinc sulfide screens form useful scintillation detectors for counting gamma and alpha radioactivity, respectively. Semiconductor detectors are relatively new and are undergoing rapid changes. Alpha and gamma spectrometers using, respectively, silicon surface-barrier and lithium-drifted or intrinsic germanium detectors are available.

Characteristic of most counters is a background or instrument counting rate usually due to cosmic radiation, radioactive contaminants in instrument parts and counting room construction material, and the nearness of radioactive sources. The background is roughly proportional to the size or mass of the counting chamber or detector, but it can be reduced by metal shielding.

Instrument "noise," or the false recording of radiation events, may be caused by faulty circuitry, too sensitive a gain setting, high humidity, and variable line voltage or transients. Control these problems by using constant-voltage transformers with transient filters, properly adjusting gain setting as specified by the manufacturer, and air-conditioning the counting room.

2. Types of Counters

The internal proportional counter accepts counting pans within the counting chamber and at the beta operating voltage, records all alpha, all beta, and some gamma radiation emitted into the counting gas. Theoretically, half the radiation is emitted in the direction of the counting pan. Some of the beta radiation, but only 1 to 2% of the alpha radiation, is back-scattered into the counting gas by sample solids, the counting pan, or the walls of the counting chamber. For substantially weightless samples, considerably more than 60% of the beta radiation and about 50% of the alpha radiation are counted. However, take considerable care in sample preparation to prevent the sample or counting pan from distorting the electrical field of the counter and depressing the counting rate. Avoid nonconducting surfaces, airborne dusts, and vapor from moisture or solvents.

End-window Geiger-Mueller counting tubes are rugged and stable counting detectors. Samples usually are mounted 5 to 15 mm from the window. Under these conditions, most alpha and weak beta radiations are stopped completely by the air gap and mica window and are not counted. Counting efficiencies for mixed fission products frequently are less than 10% for substantially weightless samples having an area less than that of the window. Because most Geiger-Mueller tubes have diameters of about 2.5 cm, the pan size—and, as a consequence, the sample volume—is restricted. Under these conditions, detectability is low and uncertain, particularly for unknown sources of radiation. On the other hand, Geiger-Mueller tubes are excellent for counting samples of tracers or purified radionuclides. Prepare standard sample mounts that yield reproducible counting efficiencies for which counting is not affected by the electrical conductance of sample pans.

Thin-window (polyester plastic film* less than 250 µg/cm^2) tubes approximately 5 cm in diameter provide counting efficiencies intermediate between those of conventional Geiger-Mueller tubes and internal counters. Counting alpha activity in these thin-window counters is satisfactory.

Thin-window counters with chamber diameter greater than 60 mm and sample mount diameter greater than 50 mm are preferable to Geiger-Mueller tubes. These counting chambers have good operational stability and less interference from nonconducting surfaces and moisture vapors than internal proportional counters.

3. Internal Proportional Counters

a. Uses: Internal proportional counters are suitable for determining alpha activity at the alpha operating plateau and alpha-plus-beta activity at the beta operating plateau. The alpha or beta activity, or both, can refer to a single or to several radionuclides.

Use instruments consisting of a counting chamber, a preamplifier, and a scaler with high-voltage power supply, timer, and register. Use the specified counting gas and accessories, make adjustments for sensitivity, and use in accordance with manufacturer's operating instructions.

b. Plateau (alpha or beta): Find the operating voltage where the counting rate is constant, i.e., varies less than 5% over a 150-V change in anode voltage:

1) With the instrument in operating order, place the alpha or beta standard (see Section 7010E.3) in the chamber, close, and flush with counter gas for 2 to 5 min.

2) Use the manufacturer's recommended operating voltage and count for a convenient time giving an acceptable coefficient of variation, preferably 2%. Repeat at voltages higher and lower than the suggested operating voltage in increments of 50 V. (CAUTION: *Instrument damage will result from prolonged continuous discharge at too high a voltage.*)

3) Plot relative counting rate (ordinate) against anode voltage (abscissa). A plateau at least 150 V long with a slope of 5% or less should result (see Figure 7010:1). Select an anode voltage near the center of this plateau as the operating voltage.

c. *Counter stability:* Check instrument stability at the operating voltage by counting the plateau source daily (see Section 7010E.3). If the source count is within two standard deviations of the previously determined mean count rate, proceed as in d following. If the source count is not so reproduced, repeat the test. If stability is not attained, service the instrument.

d. *Background:* Determine the background (with an empty counting pan in the counting chamber). Use a background counting time as long as the longest sample-counting time. Make control charts as an aid in stability testing.

e. *Sample counting:* Place dry sample on a counting pan in the counting chamber and ground pan to chamber piston. Flush with counter gas and count for a preset time, or preset count, to give the desired counting precision (see Section 7010G).

4. End-Window Counters

End-window counters may be used for beta-gamma and absorption examinations. Most alpha and soft beta radiations are stopped by the air gap and window. Use a sample pan with a diameter less than that of the window and, for maximum efficiency, place it as close to the window as possible. House detector inside a 5-cm-thick lead shield to improve counting sensitivity by decreasing background by about

---

*Mylar or equivalent.

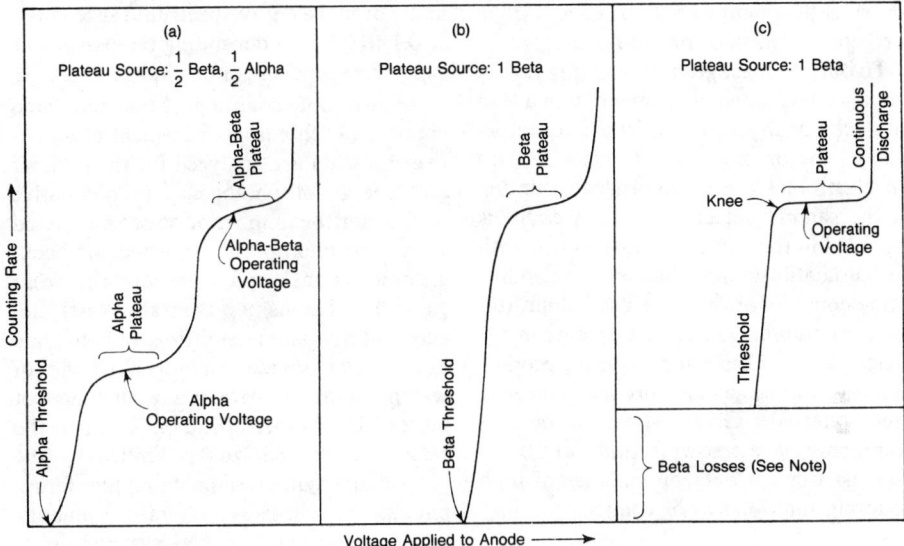

Figure 7010:1. Shape of counting rate—anode voltage curves. Key: (a) and (b) are for internal proportional counter with P-10 gas; (c) is for end-window Geiger-Mueller counter with Geiger gas. (Note: Beta losses are dependent on energy of radiation and thickness of window and air path.)

50%. Use associated equipment consisting of a scaler having a timer, a register, and a high-voltage power supply.

See ¶s 3a-e above for operation and calibration.

Coincidence correction: Geiger-Mueller counters commonly have resolving times of 100 to 400 μs; therefore, correct data on samples of high counting rate for loss in counts.

5. Thin-Window Proportional Counter

The thin-window proportional counter, without heavy shielding and/or anti-coincidence guard circuitry, has application for counting moderate to high levels and for counting residues that adhere poorly to the counting pan. The counters detect alpha and low-energy beta emitters. They are about one-half as sensitive as internal proportional counters because the geometry of counting is not as favorable and absorption losses (air path and window) are greater.

Because the sample is outside, this counter is less affected than the internal proportional counter by contamination from loose residues, losses due to residue moisture, and poor electrical conductance. See ¶s 3a-e above for operation and calibration.

6. Low-Background Beta Counter

The low-background counter is useful for measurements as low as 0.1 and as high as 50 pCi/sample. Higher activity levels, to about 1000 pCi, can be counted if other beta detectors are not available. The counters are designed primarily for beta emitters having a maximum beta energy above 0.3 MeV but are suitable also for determining alpha activity when operated in the proportional region.

The detector window thickness usually is less than 1.0 mg/cm^2 and attenuation of high-energy beta rays is relatively minor. Use sample pans with a diameter less than that of the window. Counting efficiencies for weightless samples vary from 30 to 55% for beta radiation of moderate energy.

To obtain a background counting rate of 1 cpm or less use an instrument with a lead or steel shield and an anticoincidence device of one or more guard detectors with the electronics needed to prevent counting in the sample detectors when a count is recorded in the guard. An instrument with an automatic sample changer is desirable. Some counters of this type use helium-isobutane or similar gases that operate in the Geiger region. Instruments using proportional counting gas also are available and are preferred. Geiger-Mueller counters commonly have resolving times of 100 to 400 μs. Correct data on samples of high counting rate for loss of counts.

7. Gamma Spectrometer

Gamma spectrum analysis may be made with a minimum of sample preparation. Unless a very complex spectrum with overlapping photopeaks is obtained, chemical separation followed by gamma analysis for quantitative measurements on each sample fraction is unnecessary. This method normally excludes nongamma-emitting radionuclides, and those having photon emission energies less than 0.01 MeV usually are measured with considerable uncertainty. Two types of gamma spectrometers are in use: the sodium iodide, thallium-activated [NaI(Tl)] crystal system using a scintillation phenomenon and germanium diodes, some with lithium activation [Ge(Li)]. For details on operation and calibration of gamma spectrum analyzer see manufacturer's instructions.

a. Principle: Gamma photons from a sample enter the sensitive volume of the detector and interact with detector atoms. The interactions are converted to an electrical voltage pulse proportional to the photon energy. Pulses are stored in sequence in finite energy-equivalent increments (such as 0.02 MeV/channel for NaI systems and 0.001 MeV/channel for Ge sys-

tems) over the entire spectrum range (such as 0.1 to 2 MeV) depending on instrument capabilities and operator choice.

After sample counting, the accumulated counts in each energy increment of an entire spectrum are analyzed for the number and energy of photopeaks (a qualitative test) or for the number of pulses associated with each photopeak, corrected for background count and interference from gamma emissions (a quantitative test). Because each gamma-emitting radionuclide usually has several photopeaks, one of which yields the greatest abundance of pulses, the number of radionuclides in the sample to be analyzed is limited by the probability that overlapping or interfering photopeaks will cause errors in a quantitative estimation. For NaI systems, analysis of four to eight components is practical. With more complex mixtures, use chemical separation followed by gamma spectrum analysis of each fraction, or use a Ge system.

1) In NaI crystal systems, the interaction of gamma photons with the detector gives rise to pulses of light that are proportional in intensity to the gamma photon energy. Light pulses enter a photomultiplier tube (PMT) and are converted to electrical voltage pulses proportional to the light (scintillation) intensity. Because of the components involved, the resolving time is about 10^{-9} s.

2) In Ge diode detector systems, the interaction of gamma photons with the detector ionizes detector atoms. A bias voltage applied to the detector allows collection of electrons proportional to the deposited photon energy with a resolving time of 10^{-9} to 10^{-13} s. Such a system has exceptionally high resolution.

b. Components: A gamma spectrometer consists of a detector, pulse-height analyzer system, data readout capability, and a shielded enclosure. Connect the detector to a preamplifier and a high-voltage power supply. Place detector and sample in a metal shield (10 to 20 cm steel or equivalent) to reduce external radiation background level. The pulse-height analyzer system consists of a linear amplifier, a biased linear amplifier, an analog-to-digital converter (ADC), memory storage, and a logic control mechanism. The logic control capability permits data storage in various modes and display or recall of data. The data readout system contains one or more of the following: an oscilloscope for visual display, a readout indicator, an electric typewriter or digital printer, a magnetic-tape recorder, a strip-chart recorder, an x-y recorder or plotter, and a computer terminal with associated capabilities. The oscilloscope is useful in aligning the instrument with standards such as ^{60}Co, ^{137}Cs, and ^{207}Bi. Computer capability is essential in data reduction and in complex spectrum stripping procedures.

A common scintillation (NaI) detector is a crystal 10 cm in diameter by 10 cm thick hermetically sealed in a container that is optically coupled to a photomultiplier tube.

A common high-resolution germanium detector consists of a diode of over 30 cm^3 sensitive volume encased in a 7.6-cm-diam vacuum-sealed cylinder with a dip-stick immersed in liquid nitrogen in a large cryostat, a preamplifier, and a detector bias voltage supply. The detector is cooled with liquid nitrogen to protect the diode and to reduce electronic noise generation. Both lithium-drifted and intrinsic germanium detectors are operated at the temperature of liquid nitrogen. The former detector must be maintained at the liquid nitrogen temperature at all times while the latter can be shipped or stored at ambient temperature. Use a linear amplifier that will maintain the pulse resolution provided by the detector.

A single-channel gamma spectrum analyzer is similar to a multichannel analyzer but is limited to the examination of a single energy range at a time. The instrument is best used for continuous monitoring of a waste having fixed radionuclide composition, making gross gamma measurements, or measuring a single gamma-emitting radionuclide. It is similar to the multichannel analyzer except that the design is comparatively inexpensive.

8. Alpha Spectrometer

Semiconductor particle detectors, that is, silicon surface-barrier detectors, are used for alpha spectrometry. Detector performance is affected primarily by resolution, active area, and depletion depth. Chemical separation of the sample followed by monomolecular electrodeposition on counting planchets is required. Count under a high vacuum. For details on operation and calibration of alpha spectrum analysis see manufacturer's instructions.

a. Principle: Alpha emissions from a sample enter a sensitive detector volume and interact with detector atoms. Ionization of the detector and collection of the electrons by use of a bias voltage applied to the detector allows the generation of an electrical voltage pulse proportional to the deposited alpha particle energy.

b. Components: An alpha spectrometer consists of a detector, a vacuum chamber for the detector and sample, a preamplifier, a detector bias voltage supply, a linear amplifier, a biased amplifier, a mechanical vacuum pump, a multichannel analyzer (with ADC and memory storage) or single-channel analyzer, and data readout capability similar to that discussed under gamma spectrometry.

9. Alpha Scintillation Counter

When an alpha particle bombards an impure crystal of zinc sulfide, a portion of the kinetic energy is transformed into visible light. The sulfide scintillates more efficiently when it contains silver impurities and when the duration of the light pulse is shortened by the presence of nickel ions.

The alpha scintillation counter consists

of a phosphor detector coupled to a photomultiplier, a high-voltage supply, an amplifier-discriminator, and a scaler. Generally, the photomultiplier tube has a window diameter greater than that of the sample unless the phosphor is coupled to a light-focusing optical system.

Place the silver-activated and nickel-quenched zinc sulfide phosphor near, or in contact with, the alpha-emitting sample and arrange it so that a photomultiplier tube observes the light pulses, which are amplified and recorded on the scaler.

Mount solid samples in a thin layer (less than 3 mg/cm^2) on a planchet. Place the phosphor between the sample and the photomultiplier tube. Enclose sample and detector in a light-tight chamber 3 to 5 mm from the phototube window. Under these conditions counting efficiency is from 35 to 40%.

Gaseous samples contained in a dome-shaped cell coated with zinc sulfide "paint" are observed more efficiently than solid samples. See Section 7500-Ra for a description of such a system.

10. Liquid Beta Scintillation Counter

When a sample containing radionuclides is mixed with an organic liquid scintillator, light is produced. The flashes of light are detected and amplified by one or more photomultiplier tubes.

Liquid scintillation counters are well suited for counting low-energy beta emitters such as tritium or carbon 14 because self-absorption losses are eliminated. Counting efficiencies approach 100% for high-energy betas, but for tritium the efficiency is much lower because the beta pulses, lowest in energy, are at the level of the "dark current" pulses from the photomultiplier tube and are discriminated against to reduce background. Most liquid scintillation instruments use two photomultipliers in coincidence to reduce background from "dark current." Most liquid scintillation counter systems incorporate at least a two-channel analyzer, which permits more than one beta emitter to be counted at the same time if the respective maximum beta energies differ by a factor of at least three.

Dissolve or suspend the sample in a liquid scintillation solution (see Section 7500-^{3}H.B.3a). Place sample in a transparent or translucent bottle to enable the light flashes to be transmitted to the phototube. Use a calibration standard containing the same radionuclide prepared in the same medium. Determine background by counting a bottle containing both solvent and scintillator. Measure background at least once daily.

11. Bibliography

NADER, J.S., G.R. HAGEE & L.R. SETTER. 1954. Evaluating the performance of the internal counter. *Nucleonics* 12(6):29.

CROUTHAMEL, C.E., ed. 1960. Applied Gamma-Ray Spectrometry. Pergamon Press, New York, N.Y., Vol. II.

TAYLOR, J.M. 1963. Semiconductor Particle Detectors. Butterworths, Washington, D.C.

GOULDING, F.S. 1964. A survey of the applications and limitations of various types of detectors in radiation energy measurement. *IEEE Trans. Nucl. Sci.* NS-11:177.

GOULDING, F.S. 1964. Semiconductor detectors—their properties and applications. *Nucleonics* 22(5):54.

HEATH, R.L. 1964. Scintillation Spectrometry, Gamma Ray Spectrum. IDO-16880, Technical Information Div., U.S. Atomic Energy Comm., Washington, D.C., Vols. 1 & 2.

SIEGBAHN, K., ed. 1965. Alpha, Beta, and Gamma Ray Spectroscopy. Vol. 1. North Holland Publishing Co., Amsterdam.

DEARNALEY, G. & D.C. NORTHROP. 1966. Semiductor Counters for Nuclear Radiations, 2nd ed. John Wiley & Sons, New York, N.Y.

DEARNALEY, G. 1966. Nuclear detection by solid state devices. *J. Sci. Instrum.* 43:869.

LOS ALAMOS SCIENTIFIC LABORATORY, RADIOCHEMISTRY GROUP J-11. 1967. Collected Radiochemical Procedures, 3rd ed. Rep. No. LA-1721, U.S. Atomic Energy Comm., Washington, D.C.

BOLOGNA, J.A. & S.B. HELMICH. 1969. An Atlas of Gamma Ray Spectra, LA 4312.

FRENCH, W.R., JR., R.L. LaSHURE & J.L. CURRAN. 1969. Lithium drifted germanium detectors. *Amer. J. Phys.* 37:11.

FRENCH, W.R., JR., W.M. WEHRBEIN & S.E. MOORE. 1969. Measurement of photoelectric, Compton, and pair-production cross sections in germanium. *Amer. J. Phys.* 37:391.

MCKENZIE, J.M. 1969. Index to the Literature of Semiconductor Detectors. U.S. Government Printing Off., Washington, D.C.

GOULDING, F.S. & Y. STONE. 1970. Semiconductor radiation detectors. *Science* 170:281.

HEATH, R.L. 1974. Gamma Ray Spectrum Catalogue, Ge(Li) and Si(Li) Spectrometry. ANCR-1000-2, National Technical Information Serv., Springfield, Va.

7010 E. Laboratory Reagents and Apparatus

See Section 1070 for basic standards applying to laboratory reagents and apparatus. The following special instructions are pertinent:

1. Reagents and Distilled Water

Periodically check the background radioactivity of all solutions and reagents used. Discard those having a radioactivity level that significantly interferes with the test.

2. Apparatus

Before reuse, thoroughly decontaminate apparatus and glassware with detergents and complexing agents, followed, if necessary, by acid and distilled-water rinses. Segregate equipment and glassware for storage and reuse on samples of comparable activity, i.e., keep apparatus for background and low-level counting separate from that for higher levels by using distinctive markings and different storage cabinets or laboratories. Preferably use single-use counting pans, planchets, and auxiliary supplies. Glassware that is slightly radiocontaminated may be entirely satisfactory for use in chemical tests but is unsatisfactory for radioanalysis.

3. Radioactivity Sources

a. Solutions: Use standard solutions having calibrations traceable to sources of radioactivity certified by the National Institute of Standards and Technology (NIST), formerly National Bureau of Standards (NBS).

b. Plateau or check sources:

1) Alpha—Uranium oxide (U_3O_8) plated, not less than 45 mm in diameter, having an alpha activity of about 10 000 cpm, or plutonium or americium plated as a weightless alpha standard source.

2) Beta—Cover uranium oxide (U_3O_8), plated as described above, with 8 to 10 mg/cm² aluminum foil, or use cesium 137, strontium 90-yttrium 90 in equilibrium, or any other long-lived beta emitter.

7010 F. Expression of Results

Preferably report results of radioactivity analyses in terms of picocuries per liter (pCi/L) at 20°C or, for samples of specific gravity significantly different from 1.00, picocuries per gram where 1 pCi = 10^{-12} Ci = 2.22 dpm. For samples normally containing 1000 to 1 000 000 pCi/unit volume or weight, use the nanocurie (nCi) unit (1 nCi = 10^{-9} Ci = 1000 pCi). For values higher than 1000 nCi, use the microcurie (μCi) unit (1 μCi = 10^{-6} Ci = 10^6 pCi).

Report results in such a way that they do not imply greater or lesser precision than can be obtained by the method used. See Section 1030.

"Gross alpha" implies unknown alpha

sources in which natural uranium, ^{239}Pu, or ^{241}Am has been used to determine self-absorption and efficiency factors.

"Gross beta" implies unknown sources of beta, including some gamma radiation, and calibration with ^{137}Cs or ^{90}Sr-^{90}Y in equilibrium.

7010 G. Statistics

Section 1030 discusses the statistics of analytical problems as applied to chemical constituents. These remarks also are generally applicable to radioactivity examinations.

1. Standard Deviation and Counting Error

The variability of any measurement is described by the standard deviation, which can be obtained from replicate determinations. There is an inherent variability in radioactivity measurements because disintegrations occur in a random manner described by the Poisson distribution. This distribution is characterized by the standard deviation of a large number of events, N, that equals its square root, or:

$$\sigma(N) = N^{1/2}$$

For ease in mathematical application, the normal (Gaussian) approximation to the Poisson distribution ordinarily is used. This approximation, which generally is valid at $N \geq 20$, is the particular normal distribution with a mean of N and standard deviation of $N^{1/2}$.

More often, the concern is not with the standard deviation of the number of counts but rather with the deviation in rate (number of counts per unit time):

$$R' = \frac{N}{t}$$

where:

   t = time of observation.

The standard deviation in the counting rate, $\sigma(R')$, is calculated by usual methods for propagation of error:

$$\sigma(R') = \frac{N^{1/2}}{t} = \frac{(R't)^{1/2}}{t} = \left(\frac{R'}{t}\right)^{1/2}$$

In practice, all counting instruments have a background counting rate, B, when no sample is present. With a sample, the counting rate increases to R_o. The counting rate R due to the sample is:

$$R = R_o - B$$

By propagation-of-error methods, calculate the standard deviation of R as follows:

$$\sigma(R) = \left(\frac{R_o}{t_1} + \frac{B}{t_2}\right)^{1/2}$$

where:

   t_1, t_2 = times at which the gross sample and background counting rates were measured, respectively.

Counting times depend on the limit of detection required (see ¶ 2 below). It is desirable to divide the counting time into equal periods to check constancy of the observed counting rate. For low-level counting, use $t_1 = t_2$. The error thus calculated includes only the error caused by inherent variability of the radioactive disintegration process. Report it as the "counting error."

Use a confidence level of 95%, or 1.96 standard deviations, as the counting error.

2. Limit of Detection

Different conventions with differing terminology and mathematics have been used to estimate the lower limit of detection (LLD) or the minimum detectable activity (MDA).[1-3] To eliminate confusion and the production of noncomparable data, it is proposed that the Environmental Measurements Laboratory procedure[1] be used exclusively. The basis of this procedure is hypothesis testing. LLD is defined as the smallest quantity of sample radioactivity that will yield a net count for which there is a predetermined level of confidence that radioactivity is present. Two errors may occur: Type I, in which a false conclusion is reached that radioactivity is present, and Type II, with a false conclusion that radioactivity is absent.

The LLD may be approximated as

$$LLD \simeq (K_\alpha + K_\beta) S_O$$

where:

K_α = value for the upper percentile of the standardized normal variate corresponding to the preselected risk of concluding falsely that activity is present (α),

K_β = the corresponding value for the predetermined degree of confidence for detecting presence of activity $(1 - \beta)$, and

S_O = estimated standard error of the net sample counting rate.

For sample and background counting rates that are similar (as is expected at or near the LLD) and for α and β equal to 0.05, the smallest amount of radioactivity that has a 95% probability of being detected is,

$$LLD_{95} = 4.66\ S_b$$

where:

S_b = standard deviation of the instrument background counting rate.

The LLD thus calculated is in units of counts per minute; to convert to concentration use the appropriate factors of sample volume, counting efficiency, etc. Note that the approximation $LLD = 4.66\ S_b$ can be used only for determinations where S_b is known so that $S_O = \sqrt{2}\ S_b$ and there are no counting interferences. Examples of appropriate determinations are tritium, gross alpha or beta, or any single nuclide determination.

Where tracers are added to determine yield or more than one radionuclide is counted in a sample, use the general form of the equation above, for which the 95% confidence level would be $LLD \simeq 3.29\ S_O$.

3. References

1. HARLEY, J.H., ed. 1972. Environmental Measurements Procedures Manual. HASL-300, U.S. Dep. Energy, New York, N.Y.
2. NATIONAL COUNCIL ON RADIATION PROTECTION AND MEASUREMENTS. 1978. A Handbook of Radioactivity Measurement Procedures. NCRP Rep. No. 58, Washington, D.C.
3. AMERICAN NATIONAL STANDARDS INSTITUTE. 1974. American National Standard Specifications and Performance of On-Site Instrumentation for Continuously Monitoring Radioactivity in Effluents. ANSI N13 10-1974, Inst. Electrical Electronics Engineers, Inc., New York, N.Y.

4. Bibliography

JARRETT. A.A. 1946. Statistical Methods Used in the Measurement of Radioactivity (Some Useful Graphs). Document No. AECU-262 (June 17). U.S. Atomic Energy Comm., Washington, D.C.

7020 QUALITY ASSURANCE*

7020 A. Basic Quality Assurance Program

1. Program Elements

The continuous application of quality assurance principles to radiological measurements results in consistent data, but does not guarantee accuracy. Accurate data are both precise and unbiased. To obtain them requires day-to-day control over instrumentation, chemical processing, and associated factors. Use of three allied but independent procedures is most useful: (a) "blind" replicate analysis of real samples to evaluate within laboratory precision or reproducibility; (b) cross-check analysis of natural samples among several laboratories to determine agreement with other laboratories; and (c) analysis of standard samples to determine accuracy, i.e., the agreement of results with a known value. Process real samples as part of the normal laboratory workload but submit replicates to analysis without the knowledge of the analyst. Obtain natural samples for inter-laboratory checking by collecting larger-than-normal samples and subdividing them for distribution. Standard samples may be natural samples or carefully prepared simulated environmental samples to which have been added known or accurately determined concentrations of appropriate radionuclides.

To conduct cross-check and standard analysis evaluations most effectively, use an independent referee laboratory. Standard samples may be used within a single laboratory, but exercise extreme care to prepare the samples by an accurate means divorced from the analytical method being studied. Statistically analyze all quality-assurance data to permit making firm conclusions about validity. Use control charts to simplify measuring the amounts by which the mean value of several individual determinations is biased from the true value. If a mean value fails to fall within the limits established by the mean chart, the analytical results are not from the expected normal distribution.

The overriding factor in a successful quality-control program is proper selection of criteria for acceptable and attainable accuracy. Through experimentation and experience, select criteria that reflect both the capabilities of the laboratory and the requirements of the users of the laboratory output. Secondly, select a sufficient number of samples to test properly whether the analytical data meet the established criteria.

2. Standard Sample Sources

At present standard samples of environmental materials and standardized radio-isotope solutions are available from the International Atomic Energy Agency (IAEA),† the U.S. National Institute of Standards and Technology,‡ and the U.S. Environmental Protection Agency.§ Participation is possible in intercomparison programs sponsored by IAEA, EPA, or the World Health Organization.||

---

*Approved by Standard Methods Committee, 1988.

†International Atomic Energy Agency, Div. of Research and Laboratories, Kaerntner Ring, 1010 Vienna, Austria.
‡National Institute of Standards and Technology (formerly National Bureau of Standards), Center for Radiation Research, Radiochemistry Section, Washington, D.C. 20234.
§Environmental Monitoring Systems Laboratory, Nuclear Radiation Division, Radioanalysis Branch, P.O. Box 93478, Las Vegas, Nev. 89193-3478.
||World Health Organization, Geneva, Switzerland.

3. Bibliography

INHORN, S.L., ed. 1978. Quality Assurance Practices for Health Laboratories American Public Health Assoc., Washington, D.C.

WATSON, J.E. 1980. Upgrading Environmental Radiation Data. Health Physics Soc. Rep. HPSR-1, EPA 520/1-80-012.

7020 B. Quality Assurance for Wastewater Samples

Generally it is not feasible to perform collaborative (interlaboratory) analyses of wastewater samples because of the variable composition of elements and solids from one facility to the next. The methods included herein have been evaluated by use of homogeneous samples and are useful for nonhomogeneous samples after sample preparation (wet or dry oxidation and/or fusion and solution) resulting in homogeneity. A potential problem or characteristic of reference samples used for collaborative testing is that they may be deficient in radioelements exhibiting interferences because of decay during shipment of short-half-life radionuclides. Generally, however, analytical steps have been incorporated into the methods to eliminate these interferences, even though they may not be necessary for the reference samples under study.

7110 GROSS ALPHA AND GROSS BETA RADIOACTIVITY (TOTAL, SUSPENDED, AND DISSOLVED)*

7110 A. Introduction

1. Occurrence

a. *Natural radioactivity:* Uranium, thorium, and radium are naturally occurring radioactive elements that have a long series of radioactive daughters that emit alpha or beta and gamma radiations until a stable end-element is produced. These naturally occurring elements, through their radioactive daughter gases, radon and thoron, cause an appreciable airborne particulate activity and contribute to the radioactivity of rain and groundwaters. Additional naturally radioactive elements include potassium 40, rubidium 87, samarium 147, lutetium 176, and rhenium 187.

b. *Artificial radioactivity:* With the development and operation of nuclear reactors and other atom-smashing machines, large quantities of radioactive elements are being produced. These include almost all the elements in the periodic table.

2. Significance

The Environmental Protection Agency has established maximum contaminant levels for radium 226, radium 228, and gross alpha as follows: combined radium 226 and 228, 5 pCi/L; gross alpha (including rad-

*Approved by Standard Methods Committee, 1985.

ium 226 but excluding radon and uranium), 15 pCi/L. For beta particles and photon radioactivity from artificial radionuclides in community water systems, the maximum contaminant levels are intended to produce an annual dose equivalent to the total body or any internal organ of less than 4 millirem/year. Specifically, if the average annual concentration of gross beta activity is less than 50 pCi/L and if the average annual concentrations of tritium and strontium 90 are less than 20 000 pCi/L and 8 pCi/L, respectively, no further analyses are required. If the gross beta activity exceeds 50 pCi/L, the major radioactive contaminants must be identified and organ and whole body doses calculated; the combined doses shall not exceed 4 millirem/year.

With the simpler techniques for routine measurement of gross beta activity, the presence of contamination may be determined in a matter of hours, whereas days may be required to make the radiochemical analyses necessary to identify radionuclides present.

Regular measurements of gross alpha and gross beta activity in water may be invaluable for early detection of radioactive contamination and indicate the need for supplemental data on concentrations of more hazardous radionuclides.

3. Bibliography

U.S ENVIRONMENTAL PROTECTION AGENCY. 1976. Drinking Water Regulations. Radionuclides. *Fed. Reg.* 41:28402.

7110 B.   Counting Method

1. General Discussion

a. Selection of counting instrument: The thin-window, heavily shielded, gas-flow, anticoincidence-circuitry proportional counter is the recommended instrument for counting gross alpha and beta radioactivity because of its superior operating characteristics. These include a very low background and a high sensitivity to detect and count an alpha and beta radiation range that is reasonable but not so wide as that of internal proportional counters. Calibrate the instrument by adding standard nuclide portions to media comparable to the samples and preparing, mounting, and counting the standards exactly as the samples.

An internal proportional or Geiger counter also may be used; however, the internal proportional counter has a higher background for beta counting than the thin-window counter and alpha activity cannot be determined separately with a Geiger counter. Alpha activity can be measured with either a thin-window or internal proportional counter; counting efficiency is higher for the internal counter.

b. Calibration standard: When gross beta activity is assayed in samples containing mixtures of naturally radioactive elements and fission products, the choice of a calibration standard may influence the beta results significantly because self-absorption factors and counting chamber characteristics are beta-energy-dependent.

A standard solution of cesium 137, certified by the National Institute of Standards and Technology (NIST, formerly National Bureau of Standards, NBS) or traceable to a certified source, is recommended for calibration of counter efficiency and self-absorption for gross beta determinations. The half-life of cesium 137 is about 30 years. The daughter products after beta decay of cesium 137 are stable barium 137 and metastable barium 137, which in turn disintegrates by gamma emission. For this reason, the standardization of cesium 137 solutions may be stated in terms of the gamma emission rate per milliliter or per gram. To convert gamma rate to equivalent beta disintegration rate, multiply the calibrated gamma emission rate by 1.29.

Strontium 90 in equilibrium with its daughter yttrium 90 also is a suitable gross beta standard; its use is recommended by EPA.[1]

Note that gross beta results are meaningless unless the calibration standard also is reported.

c. *Radiation lost by self-absorption:* The radiation from alpha emitters having an energy of 8 MeV and from beta emitters having an energy of 60 KeV will not escape from the sample if the emitters are covered by a sample thickness of 5.5 mg/cm². The radiation from a weak alpha emitter will be stopped if covered by only 4 mg/cm² of sample solids. Consequently, for low-level counting it is imperative to evaporate all moisture and preferable to destroy organic matter before depositing a thin film of sample solids from which radiation may enter the counter. In counting water samples for gross beta radioactivity, a solids thickness of 10 mg/cm² or less on the bottom area of the counting pan is recommended. For the most accurate results, determine the self-absorption factor as outlined below.

d. *Calibration of overall counter efficiency:* Correct observed counting rate for geometry, back-scatter, and self-absorption (sample absorption).

Although it is useful to know the variation in these individual factors, determine overall efficiency by preparing standard sample sources and unknowns.

1) For measuring mixed fission products or beta radioactivity of unknown composition, use a standard solution of cesium 137 or strontium 90 in equilibrium with its daughter yttrium 90.

Prepare a standard (known disintegration rate) in an aqueous solution of sample solids similar in composition to that present in samples. Dispense increments of solution in tared pans and evaporate. Make a series of samples having a solids thickness of 1 to 10 mg/cm² of bottom area in the counting pan. Evaporate carefully to obtain uniform solids deposition. Dry (103 to 105°C), weigh, and count. Calculate the ratio of counts per minute to disintegrations per minute (efficiency) for different weights of sample solids. Plot efficiency as a function of sample thickness and use the resulting calibration curve to convert counts per minute (cpm) to disintegrations per minute (dpm).

2) If other radionuclides are to be tested, repeat the above procedure, using certified solutions of each radionuclide. Avoid unequal distribution of sample solids, particularly in the 0- to 3-mg/cm² range, in both calibration and sample preparation.

3) For alpha calibration, proceed as above, using a standard solution of natural uranium salt (not depleted uranium), plutonium 239, or americium 241. Recount alpha standard at the beta operating voltage and determine alpha amplification factor (¶5b below). Report calibration standard used with results.

2. Apparatus

a. *Counting pans,* of metal resistant to corrosion from sample solids or reagents, about 50 mm diam, 6 to 10 mm in height, and thick enough to be serviceable for one-time use. Stainless steel planchets are recommended for acidified samples.

b. *Thin end-window proportional counter,* capable of accommodating a counting pan.

c. *Alternate counters:* Other beta counters are internal proportional and Geiger counters.

d. *Membrane filter,** 0.45-μm pore diam.

e. *Gooch crucibles.*

f. *Counting gas,* as recommended by the instrument manufacturer.

---

*Type HA, Millipore Filter Corp., Bedford, Mass., or equivalent.

3. Reagents

a. *Methyl orange indicator solution.*

b. *Nitric acid,* HNO_3, *1N.*

c. *Clear acrylic solution:* Dissolve 50 mg clear acrylic† in 100 mL acetone.

d. *Ethyl alcohol, 95%.*

e. *Conducting fluid:*‡ Prepare according to manufacturer's directions (for internal counters).

f. *Standard certified cesium 137 or strontium 90-yttrium 90 solution.*§

g. *Standard certified americium 241,*§ *plutonium 239,*§ *or natural uranium solution.* § For natural uranium, use material in secular equilibrium.

h. *Reagents for wet-combustion procedure:*

1) *Nitric acid,* HNO_3, *6N.*

2) *Hydrogen peroxide solution:* Dilute 30% H_2O_2 with an equal volume of water.

4. Procedure

a. *Total sample activity:*

1) For each 20 cm² of counting pan area, take a volume of sample containing not more than 200 mg residue for beta examination and not more than 100 mg residue for alpha examination. The specific conductance test helps to select the appropriate sample volume.

2) Evaporate by either of the following techniques:

a) Add sample directly to a tared counting pan in small increments, with evaporation at just below boiling temperature. This procedure is not recommended for large samples.

b) Place sample in a borosilicate glass beaker or evaporating dish, add a few drops of methyl orange indicator solution, add 1N HNO_3 dropwise to pH 4 to 6, and evap-

---

†Lucite or equivalent.
‡Anstac 2M, Chemical Development Corporation, Danvers, Mass., or equivalent.
§Pesticide and Radiation Quality Assurance Branch, Environmental Monitoring Systems Laboratory, P. O. Box 15027, Las Vegas, Nev. 89114.

---

orate on a hot plate or steam bath to near dryness. Avoid baking solids on evaporation vessel. Transfer to a tared counting pan with the aid of a rubber policeman and distilled water from a wash bottle. Using a rubber policeman, thoroughly wet walls of evaporating vessel with a few drops of acid and transfer washings to counting pan. (Excess alkalinity or mineral acidity is corrosive to aluminum counting pans.)

3) Complete drying in an oven at 103 to 105°C, cool in a desiccator, weigh, and keep dry until counted.

4) Treat sample residues having particles that tend to be airborne with a few drops of clear acrylic solution, then air- and oven-dry and weigh.

5) With a thin end-window counter count alpha and/or beta activity.

6) Store sample in a desiccator and count for decay if necessary. Avoid heat treatment because it will increase the escape rate of gaseous daughter products.

b. *Activity of dissolved matter:* Proceed as in ¶ 4a1) above, using a sample filtered through a 0.45-μm membrane filter.

c. *Activity of suspended matter:*

1) For each 10 cm² of membrane filter area, take a volume of sample not to exceed 50 mg suspended matter for alpha assay and not to exceed 100 mg for beta assay.

2) Filter sample through membrane filter with suction; then wash sides of filter funnel with a few milliliters of distilled water.

3) Transfer filter to a tared counting pan and oven-dry.

4) If sample is to be counted in an internal counter, saturate membrane with alcohol and ignite. (When beta or alpha activity is counted with another type of counter, ignition is not necessary provided that the sample is dry and flat.) When burning has stopped, direct flame of a Meker burner down on the partially ignited sample to fix sample to pan.

5) Cool, weigh, and count alpha and beta activities.

6) If sample particles tend to be airborne,

treat sample with a few drops of clear acrylic solution, air-dry, and count.

7) Alternatively, prepare membrane filters for counting in internal counters by wetting filters with conducting fluid, drying, weighing, and counting. (Include weight of membrane filter in the tare.)

d. Activity of suspended matter (alternate): If it is impossible to filter sewage, highly polluted waters, or industrial wastes through membrane filters in a reasonable time, proceed as follows:

1) Determine total and dissolved activity by the procedures given in ¶s 4*a* and 4*b* and estimate suspended activity by difference.

2) Filter sample through an ashless mat or filter paper of stated porosity. Dry, ignite, and weigh suspended fixed residue. Transfer and fix a thin uniform layer of sample residue to a tared counting pan with a few drops of clear acrylic solution. Dry, weigh, and count in a thin end-window counter for alpha and beta activity.

e. Activity of nonfatty semisolid samples: Use the following procedure for samples of sludge, vegetation, soil, etc.:

1) Determine total and fixed solids of representative samples according to Section 2540.

2) Reduce fixed solids of a granular nature to a fine powder with pestle and mortar.

3) Transfer a maximum of 100 mg fixed solids for alpha assay and 200 mg fixed solids for beta assay for each 20 cm^2 of counting pan area (see NOTE below).

4) Distribute solids to uniform thickness in a tared counting pan by (*a*) spreading a thick aqueous cream of solids that is weighed after oven-drying, or (*b*) dispensing dry solids of known weight and spreading with acetone and a few drops of clear acrylic solution.

5) Oven-dry at 103 to 105°C, weigh, and count.

NOTE: The fixed residue of vegetation and similar samples usually is corrosive to aluminum counting pans. To avoid difficulty, use stainless steel pans or treat a weighed amount of fixed residue with HCl or HNO$_3$ in the presence of methyl orange indicator to pH 4 to 6, transfer to an aluminum counting pan, dry at 103 to 105°C, reweigh, and count.

f. Alternate wet-combustion procedure for biological samples: Some samples, such as fatty animal tissues, are difficult to process according to ¶ 4*e* above. An alternate procedure consists of acid digestion. Because a highly acid and oxidizing state is created, volatile radionuclides may be lost under these conditions.

1) To a 2- to 10-g sample in a tared silica dish or equivalent, add 20 to 50 mL 6*N* HNO$_3$ and 1 mL 15% H$_2$O$_2$ and digest at room temperature for a few hours or overnight. Heat gently and, when frothing subsides, heat more vigorously but without spattering, until nearly dry. Add two more 6*N* HNO$_3$ portions of 10 to 20 mL each, heat to near boiling, and continue gentle treatment until dry.

2) Ignite in a muffle furnace for 30 min at 600°C, cool in a desiccator, and weigh.

3) Continue the test as described in ¶s 4*e*3)-5) above.

5. Calculation and Reporting

a. Alpha activity: Calculate alpha activity, in picocuries per liter, by the equation

$$\text{Alpha} = \frac{\text{net cpm} \times 1000}{2.22\,e\,v}$$

where:

e = calibrated overall counter efficiency (see ¶1*d*, above), and

v = volume of sample counted, mL.

Express the counting error in picocuries per sample size by dividing picocuries per sample by sample size expressed in appropriate units. Similarly, calculate and report alpha activity in picocuries or nanocuries per kilogram of moist biological material or per kilogram of moist and per kilogram of dry silt.

b. *Beta activity:* Calculate and report gross beta activity and counting error in picocuries or nanocuries per liter of fluid, per kilogram of moist (live weight) biological material, or per kilogram of moist and per kilogram of dry silt, according to ¶s *a*, above, and *c*, below.

To calculate picocuries of beta activity per liter, determine the value of *e* in the above equation as described in ¶1*d*, above.

When counting beta activity in the presence of alpha activity by gas-flow proportional counting systems (at the beta plateau) alpha particles also are counted. Because alpha particles are more readily absorbed by increased sample thickness than beta particles, alpha/beta count ratios vary. Therefore, prepare a calibration curve by counting standards (americium 241 or plutonium 239) with increasing solids thickness, first on the alpha plateau, then on the beta plateau. Plot the ratios of the two counts against mg/cm^2 thickness, determine the alpha amplification factor (*M*), and correct the amplified alpha count on the beta plateau for the sample.

$$M = \frac{\text{net cpm on beta plateau}}{\text{net cpm on alpha plateau}}$$

If significant alpha activity is indicated by the sample alpha plateau count, determine beta activity by counting the sample at the beta plateau and calculating:

$$\text{Beta, pCi/L} = \frac{B - AM}{2.22 \times D \times V}$$

where:
  B = net beta counts at the beta plateau,
  A = net alpha counts at the alpha plateau,
  M = alpha amplification factor (from ratio plot),
  2.22 = cpm/pCi,
  D = beta counting efficiency, and
  V = sample volume, liters.

Some gas-flow proportional counters have electronic discrimination to eliminate alpha counts at the beta operating voltage. For these instruments the alpha amplification factor will be less than 1.

Where greater precision is desired, for example, when the count of alpha activity at the beta plateau is a substantial fraction of the net counts per minute of gross beta activity, the beta counting error equals $(E_a^2 + E_b^2)^{1/2}$, where E_a is the alpha counting error and E_b the gross beta counting error.

c. *Counting error:* Determine the counting error, E (in picocuries per sample), at the 95% confidence level from:

$$E = \frac{1.96 \, \sigma(R)}{2.22 \, e}$$

where $\sigma(R)$ is calculated as shown in Section 7010G, using $t_1 = t_2$ (in minutes); and e, the counter efficiency, is defined and calculated as in ¶1*d*.

d. *Miscellaneous information to be reported:* In reporting radioactivity data, identify adequately the sample, sampling station, date of collection, volume of sample, type of test, type of activity, type of counting equipment, standard calibration solutions used (particularly when counting standards other than those recommended in ¶1*d* are used), time of counting (particularly if short-lived isotopes are involved), weight of sample solids, and kind and amount of radioactivity. So far as possible, tabulate the data for ease of interpretation and incorporate repetitious items in the table heading or in footnotes. Unless especially inconvenient, do not change quantity units within a given table. For low-level assays, optimally report the counting error to assist in the interpretation of results.

6. Precision and Bias

In a collaborative study of two sets of paired water samples containing known additions of radionuclides, 15 laboratories determined the gross alpha activity and 16 analyzed gross beta activity. The samples contained simulated water minerals of approximately 350 mg fixed solids/L. The alpha results of one laboratory were rejected as outliers.

The average recoveries of added gross alpha activity were 86, 87, 84, and 82%. The precision (random error) at the 95% confidence level was 20 and 24% for the two sets of paired samples. The method was biased low, but not seriously.

The average recoveries of added gross beta activity were 99, 100, 100, and 100%. The precision (random error) at the 95% confidence level was 12 and 18% for the two sets of paired samples. The method showed no bias.

7. Reference

1. U.S. ENVIRONMENTAL PROTECTION AGENCY. 1980. Prescribed Procedures of Measurement of Radioactivity in Drinking Water. EPA-600/4-80-032.

8. Bibliography

BURTT, B.P. 1949. Absolute beta counting. *Nucleonics* 5:8, 28.

GOLDIN, A.S., J.S. NADER & L.R. SETTER. 1953. The detectability of low-level radioactivity in water. *J. Amer. Water Works Assoc.* 45:73.

SETTER, L.R., A.S. GOLDIN & J.S. NADER. 1954. Radioactivity assay of water and industrial wastes with internal proportional counter. *Anal. Chem.* 26:1304.

SETTER, L.R. 1964. Reliability of measurements of gross beta radioactivity in water. *J. Amer. Water Works Assoc.* 56:228.

THATCHER, L.L., V.J. JANZER & K.W. EDWARDS. 1977. Techniques of water resources investigations of the US Geological Survey. Chap. A5 *in* Methods for Determination of Radioactive Substances in Water and Fluvial Sediments. Stock No. 024-001-02928-6, U.S. Government Printing Off., Washington, D.C.

JOHNS, F.B., ET AL. 1979. Radiochemical Analytical Procedures for Analysis of Environmental Samples. EMLS-LV-0539-17, Environmental Monitoring Systems Lab., Off. Research & Development, U.S. Environmental Protection Agency, Las Vegas, Nev.

7500-Cs RADIOACTIVE CESIUM*

7500-Cs A. Introduction

Radioactive cesium has been considered one of the more hazardous radioactive nuclides produced in nuclear fission. Upon ingestion, like potassium, cesium distributes itself throughout the soft tissue and has a relatively short residence time in the body. Half-lives of ^{134}Cs and ^{137}Cs are 2 and 30 years, respectively, both being beta- and gamma-emitters. The EPA Interim Primary Drinking Water Regulations' limit of 4 mrem/year is equivalent to 80 and 200 pCi/L, respectively; the recommended detection limit for ^{134}Cs is 10 pCi/L.

---

*Approved by Standard Methods Committee, 1988.

7500-Cs B. Precipitation Method

1. General Discussion

Principle: If the activity of cesium is high, radioactive cesium can be determined directly by gamma-counting a large liquid sample (4 L) or the sample can be evaporated to dryness and counted. For lower-level environmental samples, add cesium carrier to an acidified sample and collect the cesium as phosphomolybdate. This is purified and precipitated as Cs_2PtCl_6 for counting. If total radiocesium determined by beta-counting exceeds 30 pCi/L, determine ^{134}Cs and ^{137}Cs by gamma spectrometry.

2. Apparatus

a. Magnetic stirrer with TFE-coated magnet bar.

b. Centrifuge, bench-size clinical, and centrifuge tubes.

c. Filter papers* and glass fiber filter, 2.4 cm diam.

d. pH paper, wide range, 1 to 11 pH.

e. Filtering apparatus: See Section 7500-Sr.B.2c.

f. Counting instruments: Use either a low-background beta counter (see Section 7010D.4) or a gamma spectrometer (see Section 7010D.5).

3. Reagents

a. Ammonium phosphomolybdate reagent, $H_{12}Mo_{12}N_3O_{40}P$: Dissolve 100 g molybdic acid (85% MoO_3) in a mixture of 240 mL distilled water and 140 mL conc ammonium hydroxide (NH_4OH). When solution is complete, filter and add 60 mL conc nitric acid (HNO_3). Separately mix 400 mL conc HNO_3 and 960 mL distilled water. After both solutions cool to room temperature, add, with constant stirring,

the $(NH_4)_6Mo_7O_{24}$ solution to the HNO_3 solution. Let stand for 24 h. Filter† and discard insoluble material.

Collect filtrate in a 3-L beaker and heat to 50 to 55°C (never above 55°C). Remove from heating unit. Add 25 g sodium dihydrogen phosphate (NaH_2PO_4) dissolved in 100 mL distilled water, stir occasionally for 15 min, and let settle (approximately 30 min). Filter and wash precipitate with 1% potassium nitrate (KNO_3) and finally with distilled water. Dry precipitate and paper at 100°C for 3 to 4 h. Transfer solid $(NH_4)_3PMo_{12}O_{40}$ to a weighing bottle and store in a desiccator.

b. Chloroplatinic acid, 0.1*M:* Dissolve 51.8 g $H_2PtCl_6 \cdot 6H_2O$ in distilled water and dilute to 1000 mL.

c. Cesium carrier: Dissolve 1.267 g cesium chloride (CsCl) in distilled water and dilute to 100 mL; 1 mL = 10 mg Cs.

d. Calcium chloride, 3*M:* Dissolve 330 g $CaCl_2$ in distilled water and dilute to 1000 mL.

e. Ethanol, 95%.

f. Hydrochloric acid, HCl, conc, 6*N*, 1*N*.

g. Sodium hydroxide, NaOH, 6*N*.

4. Procedure

a. To a 1-L sample, add 1.0 mL cesium carrier and enough conc HCl to make the solution about 0.1*N* HCl (about 8.6 mL). Slowly add 1 g $(NH_4)_3PMo_{12}O_{40}$ and stir for 30 min using a magnetic stirrer at 800 rpm. Let precipitate settle for at least 4 h and discard supernatant by decanting or using suction (provided by an inverted glass funnel connected to a vacuum source). Using a stream of 1*N* HCl, quantitatively transfer precipitate to a centrifuge tube. Centrifuge and discard supernatant.

---

*Whatman No. 41, 9 cm diam; Whatman No. 42, 2.4 cm diam; or equivalent.

†Whatman No. 42 filter paper or equivalent.

Wash precipitate with 20 mL 1*N* HCl and discard wash solution.

b. Dissolve precipitate by dropwise addition of 3 to 5 mL 6*N* NaOH. Heat over a flame for several minutes to remove ammonium ions. (Moist pH paper turns green as long as NH_3 vapors are evolved.) Dilute to 20 mL with distilled water. Add 10 mL 3*M* $CaCl_2$ and adjust to pH 7 with 6*N* HCl to precipitate $CaMoO_4$. Stir, centrifuge, and filter‡ supernatant into a 50-mL centrifuge tube. Wash precipitate remaining in the original centrifuge tube with 10 mL distilled water, filter through the same filter paper, and combine the wash with filtrate. Discard precipitate and filter paper.

c. Add 2 mL 0.1*M* H_2PtCl_6 and 5 mL ethanol. Cool and stir in ice bath for 10 min. Using distilled water transfer to a tared glass-fiber filter. Wash with successive portions of distilled water, 1*N* HCl, and ethanol.

d. Dry at 110°C for 30 min, cool, weigh, mount on a nylon disk and ring with polyester plastic§ cover and beta-count or gamma-scan for ^{134}Cs and ^{137}Cs.

‡Whatman No. 41 filter paper or equivalent.
§Mylar or equivalent.

5. Calculation

Calculate the concentration of radiocesium as follows:

$$Cs, \ pCi/L = \frac{C}{2.22 \times E\,V\,R}$$

where:

C = net count rate, cpm,
E = counter efficiency,
V = volume of sample, L, and
R = fractional chemical yield
$= \dfrac{\text{recovered } Cs_2PtCl_6, \ mg \times 0.3945}{\text{added Cs carrier, mg}}$

6. Bibliography

FINSTON, H.L. & M.T. KINSLEY. 1961. The Radiochemistry of Cesium. Rep. NAS-NS-3035, U.S. Atomic Energy Comm.

KRIEGER, H.L. 1976. Interim Radiochemical Methodology for Drinking Water. EPA-600/4-75-008 (revised), U.S. Environmental Protection Agency, Environmental Monitoring and Support Lab., Cincinnati, Ohio.

7500-I RADIOACTIVE IODINE*

7500-I A Introduction

1. Occurrence and Significance

Radioiodine that results from testing nuclear devices or is released during use and processing of reactor fuels is a major concern in radioactivity monitoring. Fission products may contain iodine 129 through

*Approved by Standard Methods Committee, 1988.

iodine 135. Iodine 129 has a half-life of 1.6 $\times 10^7$ years but a relatively low specific activity (1.73×10^{-4} Ci/g for ^{129}I as compared to 1.24×10^5 Ci/g for ^{131}I). The half-life of ^{131}I is 8 d while for the other isotopes it is shorter (35 min to 21 h). At present, only ^{131}I is likely to be found in water. When ingested or inhaled, it concentrates in the thyroid gland and may cause thyroid cancer.

The EPA drinking water maximum contaminant level for ^{131}I is 3 pCi/L.

2. Selection of Method

Of the three methods, the precipitation method (B) is preferred because it is simple and involves the least time. Method C, in which iodide is concentrated by absorption on an anion resin, purified, and counted in a beta-gamma coincidence system, is sensitive and accurate. Method D uses distillation. With each method it is possible to

reach the EPA recommended detection limit of 1 pCi ^{131}I/L.

3. Bibliography

KLEINBERG, J. & G.A. COWAN. 1960. The Radiochemistry of Fluorine, Chlorine, Bromine and Iodine. Rep. NAS-NS-3005, U.S. Atomic Energy Comm.

BRAUER, F.P., J.H. KAYE & R.E. CONNALY. 1970. X-ray and β-γ. Coincidence Spectrometry Applied to Radiochemical Analysis of Environmental Samples. Advances in Chemistry Ser., No. 93, Radionuclides in the Environment, pp. 231-253. American Chemical Soc.

7500-I B.   Precipitation Method

1. General Discussion

Principle: Iodate carrier is added to an acidified sample and, after reduction with Na_2SO_3 to iodide, the ^{131}I is precipitated with $AgNO_3$. The precipitate is dissolved and purified with zinc powder and H_2SO_4 and the solution is reprecipitated as PdI_2 for counting.

2. Apparatus

a. Counting instrument: Low-background beta counter (see Section 7010D) or beta-gamma coincidence counter.

b. Fine-fritted glass funnel.

c. Filter apparatus: Two-piece filter funnel with filtering equipment.*

d. Filter materials: Filter paper;† glass-fiber filter, 2.4 cm diam; or 0.8-μm pore-diam membrane filter, 4.7 cm diam.

3. Reagents

a. Ammonium hydroxide, NH_4OH, 6N.
b. Ethanol, 95%.
c. Hydrochloric acid, HCl, 6N.

d. Iodate carrier: Dissolve 1.685 g KIO_3 in distilled water and dilute to 100 mL. Store in dark flask; 1 mL = 10 mg I.

e. Nitric acid, HNO_3, conc.

f. Palladium chloride, $PdCl_2$: Dissolve 3.3 g $PdCl_2$ in 100 mL 6N HCl; 1 mL = 20 mg Pd.

g. Silver nitrate, $AgNO_3$, 0.1M: Dissolve 17 g $AgNO_3$ in distilled water and dilute to 1000 mL. Store in dark flask.

h. Sodium sulfite, Na_2SO_3, 1M (freshly prepared): Dissolve 6.3 g Na_2SO_3 in distilled water and dilute to 50 mL.

i. Sulfuric acid, H_2SO_4, 2N.

j. Zinc, powder, reagent grade.

4. Procedure

a. To a 2000-mL sample, add 15 mL conc HNO_3 and 1.0 mL iodate carrier. Mix well. Add 4 mL freshly prepared 1M Na_2SO_3 and stir for 30 min. Add 20 mL 0.1M $AgNO_3$, stir for 1 h, and let settle for 1 h. Decant and discard as much of the supernatant as possible. Filter remainder through a glass-fiber filter and discard filtrate.

b. Transfer filter to a centrifuge tube and

---

*Fisher Filtrator or equivalent.
†Whatman No. 42 or equivalent.

slurry with 10 mL distilled water. Add 1 g zinc powder and 2 mL 2N H_2SO_4 and stir frequently for at least 30 min. Filter, with vacuum, through a fine-fritted glass funnel and collect filtrate in an erlenmeyer flask. Wash both residue and filter with a minimum quantity of distilled water and add wash water to filtrate. Discard residue.

c. Add 2 mL 6N HCl and heat in water bath at 80°C for 10 min. Add 1 mL 0.2M PdCl$_2$ and digest for at least 5 min. Centrifuge and discard supernatant.

d. Dissolve precipitate in 5 mL 6N NH$_4$OH and heat in boiling water bath for 5 min. Filter through a glass-fiber filter and collect filtrate in a centrifuge tube. Discard filter and residue.

e. Neutralize filtrate with 6N HCl, add 2 mL in excess, and heat in a water bath. Add 1 mL 0.2M PdCl$_2$ to reprecipitate PdI$_2$ and digest for 10 min. Cool slightly and transfer to a tared filter with distilled water. Wash successively with 5-mL portions of distilled water and 95% ethanol. Dry in a vacuum oven at 60°C for 1 h, weigh precipitate, mount, and beta-count.

f. If final PdI$_2$ precipitate on a glass-fiber filter is counted in a low-background beta counter, the background counting rate is relatively high (about 1.3 cpm). If precipitate is collected on a 0.8-μm membrane filter and dried for 30 min at 70°C it may

be counted in a beta-gamma coincidence scintillation system with a background rate of less than 0.1 cpm.

If a low-background counter is used, confirm identity of ^{131}I by recounting precipitate after about 1 week to check the half-life.

5. Calculation

Calculate concentration of radioiodine as follows:

$$^{131}\text{I, pCi/L} = \frac{C}{2.22 \times EVR \times A}$$

where:

C = net count rate, cpm,

E = counting efficiency of ^{131}I as function of mass of PdI$_2$ precipitate,

V = volume of sample, L,

R = fractional chemical yield

$= \dfrac{\text{recovered PdI}_2 \times 0.0704}{\text{added iodine carrier}}$, and

A = ^{131}I decay factor for the time interval between sample collection and measurement.

6. Bibliography

U.S. ENVIRONMENTAL PROTECTION AGENCY. 1980. Prescribed Procedures for Measurement of Radioactivity in Drinking Water. EPA-600/4-80-032, Environmental Monitoring and Support Lab., Cincinnati, Ohio.

7500-I C. Ion-Exchange Method

1. General Discussion

Principle: A known amount of inactive iodine in the form of KI is added as a carrier and the sample is taken through an oxidation-reduction step using hydroxylamine and sodium bisulfite to convert all iodine to iodide. Iodine, as the iodide, is concentrated by absorption on an anion-exchange column. Following an NaCl wash, iodine is eluted with sodium hypochlorite. Iodine in the iodate form is reduced to I$_2$, extracted into CCl$_4$, and back-extracted as iodide into water. The iodine finally is precipitated as PdI$_2$.

2. Apparatus

a. Counting instrument: Low-background beta counter (Section 7010D.4) or beta-gamma coincidence counter.

b. Chromatographic column, 2 cm × 15 cm.

c. Vacuum filter holder, 2.5 cm² filter area.

d. Filter paper, 2.4 cm diam.

e. Vacuum oven.

3. Reagents

a. Iodine carrier: Weigh approximately 13 g dried KI to the nearest 0.1 mg. Dissolve in a 1-L volumetric flask containing 100 mL distilled water. Add 10 mL 1*M* NaHSO₃ and dilute to mark with distilled water. Concentration of carrier I, mg/L = g KI × 0.7644.

b. Ethanol, absolute.

c. Hydroxylamine hydrochloride, 1*M:* Dissolve 6.95 g NH₂OH·HCl in distilled water and dilute to 100 mL.

d. Nitric acid, HNO₃, conc, 8*N*, 1.6*N*.

e. Sodium bisulfite, 1*M:* Dissolve 1.04 g NaHSO₃ in distilled water and dilute to 10 mL.

f. Sodium hydroxide, 12*N:* Dissolve 480 g NaOH in distilled water and dilute to 1 L.

g. Sodium hypochlorite, NaOCl, 5%: Use available household bleach.

h. Anion-exchange resin.†

i. Carbon tetrachloride, CCl₄, reagent grade.

j. Hydrochloric acid, HCl, 3*N*, 1*N*.

k. Palladium chloride: Dissolve 3.3 g PdCl₂ in 100 mL 6*N* HCl; 1 mL = 20 mg Pd.

l. Sodium chloride, NaCl, 2*M:* Dissolve 117 g NaCl in distilled water and dilute to 1 L.

m. Hydroxylamine hydrochloride wash solution: Add 20 mL conc HNO₃ and 20 mL 1*M* NH₂OH·HCl to 100 mL distilled water.

4. Procedure

a. To 1 L sample in a beaker add, while stirring, 2.0 mL iodine carrier and 5 mL 5% NaOCl, and heat for 2 to 3 min to complete oxidation. After the interchange reaction (2 to 3 min), slowly add 5 mL conc HNO₃. Add 25 mL 1*M* NH₂OH·HCl and stir. Let reaction go on for a few seconds, add 10 mL 1*M* NaHSO₃, and adjust pH to 6.5 with 12*N* NaOH or 1.6*N* HNO₃. Stir thoroughly for a few minutes. (Stir samples containing a large amount of organic material, such as muddy water, for 45 min.) Filter through a glass-fiber filter to remove suspended matter. Discard residue.

b. Pour 20 mL anion-exchange resin into a column and wash sides down with distilled water. Pass sample through ion-exchange column at a flow rate of 20 mL/min. Discard effluent. Wash column with 200 mL distilled water and then with 100 mL 2*M* NaCl at a flow rate of 4 mL/min. Discard wash solutions.

c. Add 50 mL 5% NaOCl in 10- to 20-mL increments, stirring the resin as needed to eliminate gas bubbles, and maintain a flow rate of 2 mL/min. To the eluted volume of 50 to 60 mL, collected in a beaker, carefully add 10 mL conc HNO₃ to make sample 2 to 3*N* in HNO₃ and transfer to a separatory funnel. (Add acid slowly with stirring until vigorous reaction subsides.)

d. Add 50 mL CCl₄ and 10 mL 1*M* NH₂OH·HCl. Extract iodine into organic phase by shaking for about 2 min. Let phases separate and transfer organic phase to another separatory funnel. Add 25 mL CCl₄ and 5 mL 1*M* NH₂OH·HCl to the first separatory funnel and shake for 2 min. Combine organic phase with the one obtained from the first extraction. Discard aqueous phase. Add 20 mL NH₂OH·HCl wash solution to the organic phase and shake for 2 min. Let phases separate and

---

*Whatman No. 42 or equivalent.

†Dowex 1 × 8, 50-100 mesh, chloride form, or equivalent.

transfer organic phase to a clean separatory funnel. Discard wash solution.

e. Add 25 mL distilled water and 10 drops 1*M* NaHSO$_3$ to organic phase. Shake for 2 min, let phases separate, and discard organic phase. Transfer aqueous phase to a beaker. Add 10 mL 3*N* HCl. Using a stirrer-hot plate, boil and stir the sample until it evaporates to 10 to 15 mL or begins to turn yellow.

f. Add 1.0 mL PdCl$_2$ solution dropwise. Rinse sides of beaker with 1*N* HCl and add sufficient 1*N* HCl to make a volume of 30 mL. Continue stirring until cool. Place beaker in a stainless steel tray and store at about 4°C overnight.

g. Filter through a tared filter mounted in a filter holder. Wash residue with 1*N* HCl and then with absolute alcohol. Dry in a vacuum oven at 60°C for 1 h. Cool in a desiccator, weigh precipitate, then seal it between polyester tape and polyester plastic film,‡ with the film over the precipitate. Count with a beta-gamma coincidence system.

5. Calculation

Calculate ^{131}I, pCi/L, as in B.5.

6. Bibliography

AMERICAN SOCIETY FOR TESTING AND MATE-RIALS. 1972 Book of ASTM Standards. Part 23. D 2334-68, Philadelphia, Pa.

GABAY, J.J., C.J. PAPERIELLO, S. GOODYEAR, J.C. DALY & J.M. MATUSZEK. 1974. A method of determining ^{129}I in milk and water. *Health Phys.* 26:89.

---

‡Mylar or equivalent.

7500-I D.   Distillation Method

1. General Discussion

Principle: Iodine carrier is added to an acidified sample and iodine is distilled into a caustic solution. The distillate is acidified and the iodine is extracted into CCl$_4$. After back-extraction as iodide, the iodine is purified as PdI$_2$ for counting.

2. Apparatus

a. Distillation apparatus and 3-L round-bottom flask.

b. Separatory funnel, 60 mL.

c. Filter apparatus: Two-piece filter funnel with filtering equipment.*

d. Filter paper: See B.2*d.*

3. Reagents

a. Ammonium hydroxide, NH$_4$OH, conc.

b. Carbon tetrachloride, CCl$_4$.

c. Ethanol, 95%.

d. Hydrochloric acid, HCl, 6*N*, 1*N*.

e. Iodide carrier: Dissolve 2.616 g KI in distilled water, add 2 drops NaHSO$_3$, and dilute to 100 mL. Store in dark flask. 1 mL = 20 mg I.

f. Nitric acid, HNO$_3$, conc.

g. Palladium chloride: Dissolve 3.3 g PdCl$_2$ in 100 mL 6*N* HCl; 1 mL = 20 mg Pd.

h. Sodium bisulfite, NaHSO$_3$, 1*M:* Dissolve 5.2 g NaHSO$_3$ in distilled water and dilute to 50 mL. Prepare only in small quantities.

i. Sodium hydroxide, NaOH, 0.5*N.*

j. Sodium nitrite, NaNO$_2$, 1*M:* Dissolve 69 g NaNO$_2$ in distilled water and dilute to 1 L.

k. Sulfuric acid, H$_2$SO$_4$, 12*N.*

l. Tartaric acid, C$_4$H$_6$O$_6$, 50%: Dissolve 50 g C$_4$H$_6$O$_6$ in distilled water and dilute to 100 mL.

---

*Fisher Filtrator or equivalent.

4. Procedure

a. To a 2000-mL sample in a 3-L round-bottom flask, add 15 mL 50% $C_4H_6O_6$ and 1.0 mL iodide carrier. Mix well, cautiously add 25 mL cold conc HNO_3, and close distillation apparatus (Figure 7500-I:1).

b. Connect an air line to still inlet, adjust flow rate to about 2 bubbles/s, and distill for at least 15 min into 15 mL 0.5N NaOH. Cool and transfer NaOH solution to a 60-mL separatory funnel. Discard still residue.

c. Adjust distillate to slightly acid with 1 mL 12N H_2SO_4 and oxidize with 1 mL 1M $NaNO_2$. Add 10 mL CCl_4 and shake for 1 to 2 min. Transfer organic layer to a clean 60-mL separatory funnel containing 2 mL 1M $NaHSO_3$.

d. Add 5 mL CCl_4 and 1 mL 1M $NaNO_2$ to original separatory funnel containing the aqueous layer and shake for 2 min. Combine organic fractions. Repeat and discard aqueous layer.

e. Shake separatory funnel thoroughly until CCl_4 layer is decolorized; let phases separate and transfer aqueous layer to a centrifuge tube. Add 2 mL 1M $NaHSO_3$ to the separatory funnel containing CCl_4 and shake for several minutes. When phases separate, add aqueous layer to centrifuge tube. Add 1 mL distilled water to separatory funnel and shake for several minutes. When the phases separate, add aqueous layer to centrifuge tube. Discard organic layer.

f. To combined aqueous fractions, add 2 mL 6N HCl and heat in water bath at 80°C for 10 min. Add 1.0 mL $PdCl_2$ solution dropwise, with stirring, and digest for 15 min.

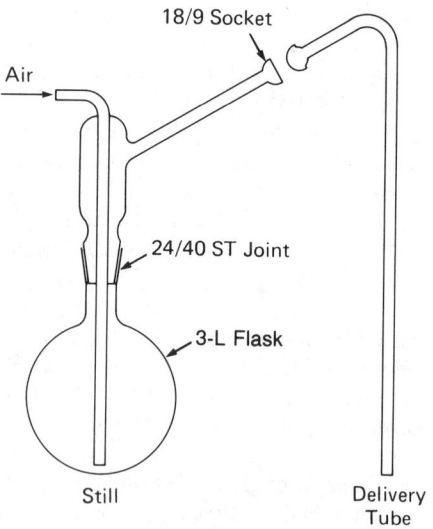

Figure 7500-I:1. Distillation apparatus for iodine analysis (not to scale).

g. Cool, stir precipitate, and transfer to a tared filter mounted in a two-piece funnel. Let precipitate settle by gravity for uniform deposition, then apply suction. Wash residue with 10 mL 1N HCl, 10 mL distilled water, and then with 10 mL 95% ethanol. Dry in a vacuum oven at 60°C for 1 h. Cool in desiccator, weigh, mount, and make beta count.

5. Calculation

Calculate the concentration of radioiodine as given in B.5.

6. Bibliography

AMERICAN SOCIETY FOR TESTING AND MATERIALS. 1972 Book of ASTM Standards. Part 23. D 2334-68, Philadelphia, Pa.

7500-Ra RADIUM*

7500-Ra A. Introduction

1. Isotopes

Radium has four naturally occurring isotopes—11.6-d radium 223, 3.6-d radium 224, 1600-year radium 226, and 5.75-year radium 228. Radium 223 is a member of the uranium 235 series, radium 224 and radium 228 are members of the thorium series, and radium 226 is a member of the uranium 238 series. The contribution of radium 228 (a beta emitter) to the total radium alpha activity is negligible because of the 1.9-year half-life of its first alpha-emitting daughter product, thorium 228. The other three radium isotopes are alpha emitters; each gives rise to a series of relatively short-lived daughter products, including three more alpha emitters.

2. Regulatory Limits

The 1976 Environmental Protection Agency Interim Primary Drinking Water regulations established a maximum contaminant level for combined radium 226 and 228 of 5 pCi/L and for radium 226 of 3 pCi/L.

3. Selection of Method

The principles of the two common methods for measuring radium are (*a*) alpha-counting a barium-radium sulfate precipitate that has been purified, and (*b*) measurement of radon 222 produced from the radium 226 in a sample or in a soluble concentrate isolated from the sample.

The determination of radium by precipitation (Method B) includes all alpha-emitting radium isotopes; it is a screening technique particularly applicable to drinking water. As long as the concentration of radium is less than the ^{226}Ra plus ^{228}Ra drinking water standard, examination by a more specific method is seldom needed. This method also is applicable to sewage and industrial wastes, provided that steps are taken to destroy organic matter and eliminate other interfering ions (see Gross Alpha and Gross Beta Radioactivity, Section 7110). However, avoid igniting sample ash or a fusion will be necessary.

The emanation technique (Method C), based on measurement of radon 222, is nearly, but not absolutely, specific for radium 226. Procedures for soluble, suspended, and total radium 226 are given.

The sequential precipitation method (Method D) can be used to measure either radium 228 alone or radium 228 and radium 226. The EPA-recommended detection limits are satisfied by this method.

---

*Approved by Standard Methods Committee, 1988.

7500-Ra B. Precipitation Method

1. General Discussion

a. Application: This method is suitable for determination of the alpha-emitting isotopes of radium.

b. Principle: Because of the difference in half-lives of the nuclides in the series including the alpha-emitting Ra isotopes, these isotopes can be identified by the rate of ingrowth and decay of their daughters in a barium sulfate precipitate.[1-3] The ingrowth of alpha activity from radium 226 increases at a rate governed primarily by the 3.8-d half-life radon 222. The ingrowth of alpha activity in radium 223 is complete by the time a radium-barium precipitate can be prepared for counting. The ingrowth of the first two alpha-emitting daughters of radium 224 is complete within a few minutes and the third alpha daughter activity increases at a rate governed by the 10.6-h half-life of lead 212. The activity of the radium 224 itself, with a 3.6-d half-life, also is decreasing, leading to a rather complicated ingrowth and decay curve.

Lead and barium carriers are added to the sample containing alkaline citrate, then sulfuric acid (H_2SO_4) is added to precipitate radium, barium, and lead as sulfates. The precipitate is purified by washing with nitric acid (HNO_3), dissolving in alkaline EDTA, and reprecipitating as radium-barium sulfate after pH adjustment to 4.5. This slightly acidic EDTA keeps other naturally occurring alpha emitters and the lead carrier in solution.

2. Apparatus

a. Counting instruments: One of the following is required:

1) *Internal proportional counter,* gas-flow, with scaler and register.

2) *Alpha scintillation counter,* silver-activated zinc sulfide phosphor deposited on thin polyester plastic, with photomultiplier tube, scaler, timer, and register; or

3) *Proportional counter,* thin end-window, gas-flow, with scaler and register.

b. Membrane filter holder, or stainless steel or TFE filter funnels, with vacuum source.*

c. Membrane filters† or *glass fiber filters.‡*

3. Reagents

a. Citric acid, 1*M:* Dissolve 210 g $H_3C_6H_5O_7 \cdot H_2O$ in distilled water and dilute to 1 L.

b. Ammonium hydroxide, conc and 5*N:* Verify strength of old 5*N* NH_4OH solution before use.

c. Lead nitrate carrier: Dissolve 160 g $Pb(NO_3)_2$ in distilled water and dilute to 1 L; 1 mL = 100 mg Pb.

d. Stock barium chloride solution: Dissolve 17.79 g $BaCl_2 \cdot 2H_2O$ in distilled water and dilute to 1 L in a volumetric flask; 1 mL = 10 mg Ba.

e. Barium chloride carrier: To a 100-mL volumetric flask, add 20.00 mL stock $BaCl_2$ solution using a transfer pipet, dilute to 100 mL with distilled water, and mix; 1 mL = 2.00 mg Ba.

f. Methyl orange indicator solution.

g. Phenolphthalein indicator solution.

h. Bromcresol green indicator solution: Dissolve 0.1 g bromcresol green sodium salt in 100 mL distilled water.

i. Sulfuric acid, H_2SO_4, 18*N.*

j. Nitric acid, HNO_3, conc.

k. EDTA reagent, 0.25*M:* Add 93 g disodium ethylenediaminetetraacetate dihydrate to distilled water, dilute to 1 L, and mix.

---

*Fisher Filtrator or equivalent.
†Millipore Type HAWP or equivalent.
‡No. 934-AH, diameter 2.4 cm, H. Reeve Angel and Co., or equivalent.

l. Acetic acid, conc.

m. Ethyl alcohol, 95%.

n. Acetone.

o. Clear acrylic solution:§ Dissolve 50 mg clear acrylic in 100 mL acetone.

p. Standard radium 226 solution: Prepare as directed in Method C, ¶s 3*d-f*, except that in ¶ *f* (standard radium 226 solution), add 0.50 mL BaCl$_2$ stock solution (¶ 3*a*) before adding the ^{226}Ra solution; 1 mL final standard radium solution so prepared contains 2.00 mg Ba/mL and approximately 3 pCi ^{226}Ra/mL after the necessary correcting factors are applied.

4. Procedure for Radium in Drinking Water and for Dissolved Radium

a. To 1 L sample in a 1500-mL beaker, add 5 mL 1*M* citric acid, 2.5 mL conc NH$_4$OH, 2 mL Pb(NO$_3$)$_2$ carrier, and 3.00 mL BaCl$_2$ carrier. In each batch of samples include a distilled water blank.

b. Heat to boiling and add 10 drops methyl orange indicator.

c. While stirring, slowly add 18*N* H$_2$SO$_4$ to obtain a permanent pink color; then add 0.25 mL acid in excess.

d. Boil gently 5 to 10 min.

e. Set beaker aside and let stand until precipitate has settled (3 to 5 h or more).‖

f. Decant and discard clear supernate. Transfer precipitate to a 40-mL or larger centrifuge tube, centrifuge, decant, and discard supernate.

g. Rinse wall of centrifuge tube with a 10-mL portion of conc HNO$_3$, stir precipitate with a glass rod, centrifuge, and

discard supernate. Repeat rinsing and washing two more times.

h. To precipitate, add 10 mL distilled water and 1 to 2 drops phenolphthalein indicator solution. Stir and loosen precipitate from bottom of tube (using a glass rod if necessary) and add 5*N* NH$_4$OH, dropwise, until solution is definitely alkaline (red). Add 10 mL EDTA reagent and 3 mL 5*N* NH$_4$OH. Stir occasionally for 2 min. Most of the precipitate should dissolve, but a slight turbidity may remain.

i. Warm in a steam bath to clear solution (about 10 min), but do not heat for an unnecessarily long period.# Add conc acetic acid dropwise until red color disappears; add 2 or 3 drops bromcresol green indicator solution and continue to add conc acetic acid dropwise, while stirring with a glass rod, until indicator turns green (aqua).** BaSO$_4$ will precipitate. Note date and time of precipitation as zero time for ingrowth of alpha activity. Digest in a steam bath for 5 to 10 min, cool, and centrifuge. Discard supernate. The final pH should be about 4.5, which is sufficiently low to destroy the Ba-EDTA complex, but not Pb-EDTA. A pH much below 4.5 will precipitate PbSO$_4$.

j. Wash Ba-Ra sulfate precipitate with distilled water and mount in a manner suitable for counting as given in ¶s *k, l,* or *m* following.

k. Transfer Ba-Ra sulfate precipitate to a tared stainless steel planchet with a minimum of 95% ethyl alcohol and evaporate under an infrared lamp. Add 2 mL acetone and 2 drops clear acrylic solution, disperse precipitate evenly, and evaporate under an infrared lamp. Dry in oven at 110°C, weigh, and determine alpha activity, preferably with an internal proportional counter. Cal-

---

§Lucite or equivalent.

‖If original concentrations of isotopes of radium other than ^{226}Ra are of interest, note date and time of this original precipitation as the separation of the isotopes from their parents; use a minimal settling time and complete procedure through ¶ *j* without delay. Assuming the presence of and separation of parents, decay of ^{223}Ra and ^{224}Ra begins at the time of the first precipitation, but ingrowth of decay products is timed from the second precipitation (¶ *i*). The time of the first precipitation is not needed if the objective is to check the final precipitate for its ^{226}Ra content only.

#If solution does not clear in 10 min, cool, add another mL 5*N* NH$_4$OH, let stand 2 min, and heat for another 10-min period.

**The end point is most easily determined by comparison with a solution of similar composition that has been adjusted to pH 4.5 using a pH meter.

culate net counts per minute and weight of precipitate.

l. Weigh a membrane filter, a counting dish, and a weight (glass ring) as a unit. Transfer precipitate to tared membrane filter in a holder and wash with 15 to 25 mL distilled water. Place membrane filter in dish, add glass ring, and dry at 110°C. Weigh and count in one of the counters mentioned under ¶ 2a above. Calculate net counts per minute and weight of precipitate.

m. Add 20 mL distilled water to the Ba-Ra sulfate precipitate, let settle in a steam bath, cool, and filter through a special funnel with a tared glass fiber filter. Dry precipitate at 110°C to constant weight, cool, and weigh. Mount precipitate on a nylon disk and ring with an alpha phosphor on polyester plastic film,[4] and count in an alpha scintillation counter. Calculate net counts per minute and weight of precipitate.

n. If the isotopic composition of the precipitate is to be estimated, perform additional counting as mentioned in the calculation below.

o. Determination of combined efficiency and self-absorption factor: Prepare standards from 1 L distilled water and the standard radium 226 solution (¶ 3p preceding). Include at least one blank. The barium content will impose an upper limit of 3.0 mL on the volume of the standard radium 226 solution that can be used. If x is volume of standard radium 226 solution added, then add $(3.00 - x)$ mL $BaCl_2$ carrier (¶ 3e above). Analyze standards as samples, beginning with ¶ 4a, but omit 3.00-mL $BaCl_2$ carrier.

From the observed net count rate, calculate the combined factor, *bc,* from the formula:

$$bc = \frac{net\ cpm}{ad \times 2.22 \times pCi\ radium\ 226}\ ^{\dagger\dagger}$$

---
††See calculation that follows.

where:

> ad = ingrowth factor (see below) multiplied by chemical yield.

If all chemical yields on samples and standards are not essentially equal, the factor *bc* will not be a constant. In this event, construct a curve relating the factor *bc* to varying weights of recovered $BaSO_4$.

5. Calculation

$$Radium,\ pCi/L = \frac{net\ cpm}{a\,b\,c\,d\,e \times 2.22}$$

where:

> a = ingrowth factor (as shown in the following tabulation):

Ingrowth h	Alpha Activity from ^{226}Ra
0	1.000
1	1.016
2	1.036
3	1.058
4	1.080
5	1.102
6	1.124
24	1.489
48	1.905
72	2.253

> b = efficiency factor for alpha counting,
> c = self-absorption factor,
> d = chemical yield, and
> e = sample volume, L.

The calculations are based on the assumption that the radium is radium 226. If the observed concentration approaches 3 pCi/L, it may be desirable to follow the rate of ingrowth and estimate the isotopic content[2,3] or, preferably, to determine radium 226 by radon 222.

The optimum ingrowth periods can be selected only if the ratios and identities of the radium isotopes are known. The number of observed count rates at different ages must be equal to or greater than the num-

TABLE 7500-Ra:I. CHEMICAL AND RADIOCHEMICAL COMPOSITION OF SAMPLES USED TO
DETERMINE BIAS AND PRECISION OF RADIUM 226 METHOD

Radionuclide Composition	Samples			
	Pair 1		Pair 2	
	A	B	C	D
Radium 226,* pCi/L	12.12	8.96	25.53	18.84
Thorium 228,* pCi/L	none	none	25.90	19.12
Uranium, natural, pCi/L	105	77.9	27.7	20.5
Lead 210,* pCi/L	11.5	8.5	23.7	17.5
Strontium 90,* pCi/L	49.1	36.3	13.9	10.2
Cesium 137, pCi/L	50.3	37.2	12.7	9.5
NaCl, mg/L	60	60	300	300
CaSO$_4$, mg/L	30	30	150	150
MgCl$_2$·6H$_2$O, mg/L	30	30	150	150
KCl, mg/L	5	5	10	10

* Daughter products were in substantial secular equilibrium.

ber of radium isotopes present in a mixture. In the general case, suitable ages for counting are 3 to 18 h for the first count; for isotopic analysis, additional counting at 7, 14, or 28 d is suggested, depending on the number of isotopes in mixture. The amounts of the various radium isotopes can be determined by solving a set of simultaneous equations.[4] This approach is most satisfactory when radium 226 is the predominant isotope; in other situations, the approach suffers from statistical counting errors.

6. Precision and Bias

In a collaborative study, 20 laboratories analyzed four water samples for total (dissolved) radium. The radionuclide composition of these reference samples is shown in Table 7500-Ra:I. Note that Samples C and D had a ^{224}Ra concentration equal to that of ^{226}Ra.

The four results from each of two laboratories and two results from a third laboratory were rejected as outliers. The average recoveries of radium 226 from the remaining A, B, C, and D samples were 97.5, 98.7, 94.9, and 99.4%, respectively.

At the 95% confidence level, the precision (random error) was 28% and 30% for the two sets of paired samples. The method is biased low for radium 226, but not seriously. The method appears satisfactory for radium 226 alone or in the presence of an equal activity of radium 224 when correction for radium 224 interference is made from a second count.

For the determination of ^{224}Ra in Samples C and D, the results of two laboratories were excluded. Hence the average recoveries were 51 and 45% for Samples C and D, respectively. At the 95% confidence level, the precision was 46% for this pair of samples. The results indicated that the method for ^{224}Ra is seriously biased low. When the recoveries for radium 224 did not agree with those for radium 226, this may have been due, in part, to incomplete instructions given in the method to account for the transitory nature of ^{224}Ra activity. The method as given here contains footnotes calling attention to the importance of the time of counting. Still uncertain is the degree of separation of radium 224 from its parent, thorium 228, in ¶s 4a through g above.

Radium 223 and radium 224 analysis by this method may be satisfactory, but special refinements and further investigations are required.

7. References

1. KIRBY, H.W. 1954. Decay and growth tables for naturally occurring radioactive series. *Anal. Chem.* 26:1063.

2. SILL, C. 1960. Determination of radium-226, thorium-230, and thorium-232. Rep. No. TID 7616 (Oct.), U.S. Atomic Energy Comm., Washington, D.C.

3. GOLDIN, A.S. 1961. Determination of dissolved radium. *Anal. Chem.* 33:406.

4. HALLDEN, N.A. & J.H. HARLEY. 1960. An improved alpha-counting technique. *Anal. Chem.* 32:1961.

7500-Ra C.   Emanation Method

1. General Discussion

a. Application: This method is suitable for the determination of soluble, suspended, and total radium 226 in water. In this method, *total* radium 226 means the sum of suspended and dissolved radium 226. Radon means radon 222 unless otherwise specified.

b. Principle: Radium in water is concentrated and separated from sample solids by coprecipitation with a relatively large amount of barium as the sulfate. The precipitate is treated to remove silicates, if present, and to decompose insoluble radium compounds, fumed with phosphoric acid to remove sulfite (SO_3^{2-}), and dissolved in hydrochloric acid (HCl). The completely dissolved radium is placed in a bubbler, which is then closed and stored for a period of several days to 4 weeks for ingrowth of radon. The bubbler is connected to an evacuated system and the radon gas is removed from the liquid by aeration, dried with a desiccant, and collected in a counting chamber. The counting chamber consists of a dome-topped scintillation cell coated inside with silver-activated zinc sulfide phosphor; a transparent window forms the bottom (Figure 7500-Ra:1). The chamber rests on a photomultiplier tube during counting. About 4 h after radon collection, the alpha-counting rate of radon and decay products is at equilibrium, and a count is obtained and related to radium 226 standards similarly treated.

The counting gas used to purge radon from the liquid to the counting chamber may be helium, nitrogen, or aged air.

Some radon (emanation) techniques employ a minimum of chemistry but require high dilution of the sample and large chambers for counting the radon 222.[1] Others

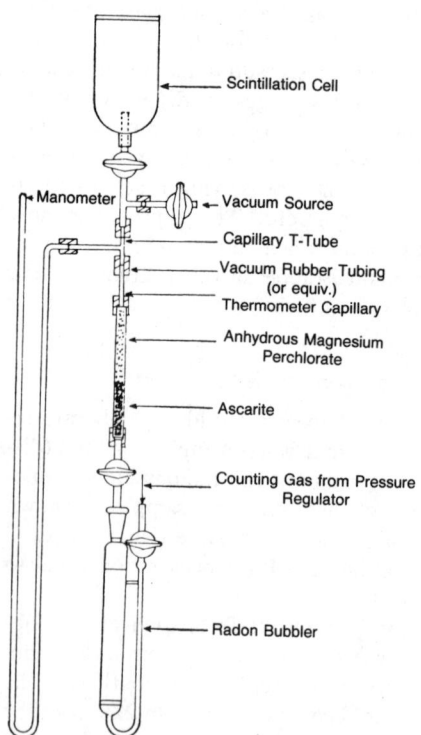

Figure 7500-Ra:1. De-emanation assembly.

involve more chemical separation, concentration, and purification of radium 226 before de-emanation into counting cells of either the ionization or alpha scintillation types. The method[2] given here requires a moderate amount of chemistry coupled with a sensitive alpha scintillation count of radon 222 plus daughter products in a small chamber.[3]

c. *Concentration techniques:* The chemical properties of barium and radium are similar; therefore, because barium does not interfere with de-emanation, as much as 100 mg may be used to aid in coprecipitating radium from a sample to be placed in a single radon bubbler. However, because some radium 226 is present in barium salts, reagent tests are necessary to account for radium 226 introduced in this way.

d. *Interferences:* Only the gaseous alpha-emitting radionuclides, radon 219 (actinon) and radon 220 (thoron), can interfere. Interference from these radionuclides would be expected to be very rare in water not contaminated by such industrial wastes as uranium mill elements.[2] The half-lives of these nuclides are only 3.92 and 54.5 s, respectively, so only their alpha-emitting decay products interfere.

Interference from stable chemicals is limited. Small amounts of lead, calcium, and strontium, collected by the barium sulfate, do not interfere. However, lead may cause deterioration of platinum ware. Calcium at a concentration of 300 mg/L and other dissolved solids (in brines) at 269 000 mg/L cause no difficulty.[4]

The formation of precipitates in excess of a few milligrams during the radon 222 ingrowth period is a warning that modifications[2] may be necessary because radon 222 recovery may be impaired.

e. *Minimum detectable concentration:* The minimum detectable concentration depends on counter characteristics, background-counting rate of scintillation cell, length of counting period, and contamination of apparatus and environment by radium 226. Without reagent purification, the overall reagent blank (excluding background) should be between 0.03 and 0.05 pCi radium 226, which may be considered the minimum detectable amount under routine conditions.

2. Apparatus

The de-emanation assembly is shown in Figure 7500-Ra:1, and its major components are described in ¶s *b-e*, below.

a. *Scintillation counter assembly* with a photomultiplier (PM) tube 5 cm or more in diameter, normally mounted, face up, in a light-tight housing. The photomultiplier tube, preamplifier, high-voltage supply, and scaler may be contained in one chassis; or the PM tube and preamplifier may be used as an accessory with a proportional counter or a separate scaler. A high-voltage safety switch should open automatically when the light cover is removed, to avoid damage to the photomultiplier tube.

Use a preamplifier with a variable gain adjustment. Equip counter with a flexible ground wire attached to the chassis and to the neck of the scintillation cell by an alligator clip or similar device. Ascertain operating voltage by determining a plateau using ^{222}Rn in the scintillation cell as the alpha source; the slope should not exceed 2%/100 V. Calibrate and use counter and scintillation cell as a unit when more than one counter is available. The background-counting rate for the counter assembly without the scintillation cell in place should be 0.00 to 0.03 cpm.

b. *Scintillation cells,*[2,3] Lucas-type, preferably having a volume of 95 to 140 mL, made in the laboratory, or commercially available.*

c. *Radon bubblers,* capacity 18 to 25 mL.† Use gas-tight glass stopcocks and a

---

*William H. Johnston Laboratories, 3617 Woodland Ave., Baltimore, Md. 21215.
†Available from Corning Glass Works, Special Sales Section, Corning, N.Y. 14830.

fritted glass disk of medium porosity.‡ Use one bubbler for a standard ^{226}Ra solution and one for each sample and blank in a batch.[2]

d. Manometer, open-end capillary tube or vacuum gauge having volume that is small compared to volume of scintillation cell, 0 to 760 mm Hg.

e. Gas purification tube, 7 to 8 mm OD standard-wall glass tubing, 100 to 120 mm long, constricted at lower end to hold glass wool plug; thermometer capillary tubing.

f. Sample bottles, polyethylene, 2- to 4-L capacity.

g. Membrane filters.§

h. Gas supply: Helium, nitrogen, or air aged in high-pressure cylinder with two-stage pressure regulator and needle valve. Helium is preferred.

i. Silicone grease, high-vacuum.

j. Sealing wax, low-melting.‖

k. Laboratory glassware: Excepting bubblers, decontaminate all glassware before and between uses by heating for 1 h in EDTA decontaminating solution at 90 to 100°C, then rinse in water, 1N HCl, and again in distilled water to dissolve barium (radium) sulfate, Ba(Ra)SO$_4$.

Removal of previous samples from bubblers and rinsing is described in ¶ 5*a*17). More extensive cleaning of bubblers requires removal of wax from joints, silicone grease from stopcocks, and the last traces of barium-radium compounds.

l. Platinum ware: Crucibles (20 to 30 mL) or dishes (50 to 75 mL), large dish (for flux preparation), and platinum-tipped tongs (preferably Blair type). Clean platinum ware by immersion and rotation in a molten bath of potassium pyrosulfate, remove, cool, rinse in hot tap water, digest in hot 6N HCl, rinse in distilled water, and finally flame over a burner.

‡Corning or equivalent.
§Type HAWP, Millipore Filter Corp., Bedford, Mass., or equivalent.
‖Pyseal, Fisher Scientific Co., Pittsburgh, Pa., or equivalent.

3. Reagents

a. Stock barium chloride solution: Dissolve 17.79 g BaCl$_2 \cdot$2H$_2$O in distilled water and dilute to 1 L; 1 mL = 10 mg Ba.

b. Dilute barium chloride solution: Dilute 200.0 mL stock barium chloride solution to 1000 mL as needed; 1 mL = 2.00 mg Ba. Let stand 24 h and filter through a membrane filter.

Optionally, add approximately 40 000 dpm of ^{133}Ba to this solution before dilution. Take account of the stable barium carrier added with the ^{133}Ba and with the diluting solution, so that the final barium concentration is near 2 mg/L. The use of ^{133}Ba provides a convenient means of checking on the recovery of ^{226}Ra from the sample; see 7, below. Use the BaCl$_2$ solution containing ^{133}Ba in steps described in ¶s 5*a*3), *b*8), and *c*3). Do *not* use in *d* below; instead, use a separate dilution of stock BaCl$_2$ solution for preparing ^{226}Ra standard solutions.

c. Acid barium chloride solution: To 20 mL conc HCl in a 1-L volumetric flask, add dilute BaCl$_2$ solution to the mark and mix.

d. Stock radium 226 solution: Take every precaution to avoid unnecessary contamination of working area, equipment, and glassware, preferably by preparing ^{226}Ra standards in a separate area or room reserved for this purpose. Obtain a National Institute of Standards and Technology (NIST, formerly National Bureau of Standards, NBS) gamma ray standard containing 0.1 μg ^{226}Ra as of date of standardization. Using a heavy glass rod, cautiously break neck of ampule, which is submerged in 300 mL acid BaCl$_2$ solution in a 600-mL beaker. Chip ampule unit until it is thoroughly broken or until hole is large enough to give complete mixing. Transfer solution to a 1-L volumetric flask, rinse beaker with acid BaCl$_2$ solution, dilute to mark with same solution, and mix; 1 mL = approximately 100 pg ^{226}Ra.

Determine the time in years, t, since the NIST standardization of the original ^{226}Ra solution. Calculate pCi ^{226}Ra/mL as:

$$\text{pCi } ^{226}\text{Ra} = [1 - (4.3 \times 10^{-4})(t)] [100] [0.990]$$

e. Intermediate radium 226 solution: Dilute 100 mL stock radium 226 solution to 1000 mL with acid BaCl$_2$ solution; 1 mL = approximately 10 pCi ^{226}Ra.

f. Standard radium 226 solution: Add 30.0 mL intermediate radium 226 solution to a 100-mL volumetric flask and dilute to mark with acid BaCl$_2$ solution; 1 mL = approximately 3 pCi ^{226}Ra and contains about 2 mg Ba. See ¶ *d* et seq. above for correction factors.

g. Hydrochloric acid, HCl, conc, 6N, 1N, and 0.1N.

h. Sulfuric acid, H$_2$SO$_4$, conc and 0.1N.

i. Hydrofluoric acid, HF, 48%, in a plastic dropping bottle. (CAUTION.)

j. Ammonium sulfate solution: Dissolve 10 g (NH$_4$)$_2$SO$_4$ in distilled water and dilute to 100 mL in a graduated cylinder.

k. Phosphoric acid, H$_3$PO$_4$, 85%.

l. Ascarite, 8 to 20 mesh.

m. Magnesium perchlorate, anhydrous desiccant.

n. EDTA decontaminating solution: Dissolve 10 g disodium ethylenediaminetetraacetate dihydrate and 10 g Na$_2$CO$_3$ in distilled water and dilute to 1 L in a graduated cylinder.

o. Special reagents for total and suspended radium:

1) *Flux:* Add 30 mg BaSO$_4$, 65.8 g K$_2$CO$_3$, 50.5 g Na$_2$CO$_3$, and 33.7 g Na$_2$B$_4$O$_7 \cdot$ 10H$_2$O, to a 500-mL platinum dish. Mix thoroughly and heat cautiously to expel water, then fuse and mix thoroughly by swirling. Cool flux, grind in a porcelain mortar to pass a 10- to 12-mesh (or finer) screen, and store in an airtight bottle.

2) *Dilute hydrogen peroxide solution:* Dilute 10 mL 30% H$_2$O$_2$ to 100 mL in a graduated cylinder. Prepare daily.

4. Calibration of Scintillation Counter Assembly

a. Test bubblers by adding about 10 mL distilled water and passing air through them at the rate of 3 to 5 mL (free volume)/ min. Air should form many fine bubbles rather than a few large ones; the latter condition indicates nonuniform pores. Do not use bubblers requiring excessive pressure to initiate bubbling. Fritted-glass disks of medium porosity (¶ *2c*) usually are satisfactory.

b. Apply silicone grease to stopcocks of a bubbler and, with gas inlet stopcock closed, add 1 mL stock BaCl$_2$ solution and 10 mL (30 pCi) standard radium 226 solution, and fill bubbler two-thirds to three-fourths full with additional acid BaCl$_2$ solution.

c. With bubbler in a clamp or rack, dry joint with lint-free paper or cloth, warm separate parts of the joint, apply sealing wax sparingly to the male part, and make the connection with a twisting motion to spread the wax uniformly in the ground joint. Let cool. Establish zero ingrowth time by purging liquid with counting gas for 15 to 20 min according to ¶ *4j* below and adjust inlet pressure to produce a froth a few millimeters thick. Close stopcocks, record date and time, and store bubbler, preferably for 3 weeks or more (with most samples) before collecting and counting ^{222}Rn. A much shorter ingrowth period of 16 to 24 h is convenient for a standard bubbler. Obtain an estimate of ^{222}Rn present at any time from the B columns in Table 7500-Ra:II.

d. Attach scintillation cell as shown in Figure 7500-Ra:1; # substitute a glass tube with a stopcock for bubbler so that the

---

#The system as described and shown in Figure 7500-Ra:1 is considered minimal. In routine work, use manifold systems and additional, more precise needle valves. An occasional drop of solution will escape from the bubbler; provide enough free space beyond the outlet stopcock to accommodate this liquid, preventing its entrance into the gas-purifying train.

TABLE 7500-Ra:II. FACTORS FOR DECAY OF RADON 222, GROWTH OF RADON 222 FROM RADIUM 226, AND CORRECTION OF RADON 222 ACTIVITY FOR DECAY DURING COUNTING

Time	Factor for Decay of Radon 222 $A = e^{-\lambda t}$		Factor for Growth of Radon 222 from Radium 226 $B = 1 - e^{-\lambda t}$		Factor for Correction of Radon 222 Activity for Decay during Counting $C = \lambda t / (1 - e^{-\lambda t})$
	Hours	Days	Hours	Days	Hours
0.0	1.0000		0.000 00		1.000
0.2	0.9985		0.001 51		1.001
0.4	0.9970		0.003 01		1.001
0.6	0.9955		0.004 52		1.002
0.8	0.9940		0.006 02		1.003
1	0.9925	0.8343	0.007 52	0.1657	1.004
2	0.9850	0.6960	0.014 99	0.3040	1.008
3	0.9776	0.5807	0.022 40	0.4193	1.011
4	0.9703	0.4844	0.029 75	0.5156	1.015
5	0.9630	0.4041	0.037 05	0.5959	1.019
6	0.9557	0.3372	0.044 29	0.6628	1.023
7	0.9485	0.2813	0.051 48	0.7187	1.027
8	0.9414	0.2347	0.058 61	0.7653	1.031
9	0.9343	0.1958	0.065 69	0.8042	1.034
10	0.9273	0.1633	0.072 72	0.8367	1.038
11	0.9203	0.1363	0.079 69	0.8637	1.042
12	0.9134	0.1137	0.086 62	0.8863	1.046
13	0.9065	0.0948	0.093 49	0.9052	1.050
14	0.8997	0.0791	0.100 31	0.9209	1.054
15	0.8929	0.0660	0.107 07	0.9340	1.058
16	0.8862	0.0551	0.1138	0.9449	1.062
17	0.8795	0.0459	0.1205	0.9541	1.066
18	0.8729	0.0383	0.1271	0.9617	1.069
19	0.8664	0.0320	0.1336	0.9680	1.073
20	0.8598	0.0267	0.1402	0.9733	1.077
21	0.8534	0.0223	0.1466	0.9777	1.081
22	0.8470	0.0186	0.1530	0.9814	1.085
23	0.8406	0.0155	0.1594	0.9845	1.089
24	0.8343	0.0129	0.1657	0.9871	1.093
25	0.8280	0.0108	0.1720	0.9892	1.097
26	0.8218	0.0090	0.1782	0.9910	1.101
27	0.8156	0.0075	0.1844	0.9925	1.105
28	0.8095	0.0063	0.1905	0.9937	1.109
29	0.8034	0.0052	0.1966	0.9948	1.113
30	0.7973	0.0044	0.2027	0.9956	1.118

TABLE 7500-RA:II, CONT.

Time	Factor for Decay of Radon 222 $A = e^{-\lambda t}$		Factor for Growth of Radon 222 from Radium 226 $B = 1 - e^{-\lambda t}$		Factor for Correction of Radon 222 Activity for Decay during Counting $C = \lambda t /(1 - e^{-\lambda t})$
	Hours	Days	Hours	Days	Hours
31	0.7913	0.0036	0.2087	0.9964	1.122
32	0.7854	0.0030	0.2146	0.9970	1.126
33	0.7795	0.0025	0.2205	0.9975	1.130
34	0.7736	0.0021	0.2264	0.9979	1.134
35	0.7678	0.0018	0.2322	0.9982	1.138
36	0.7620	0.0015	0.2380	0.9985	1.142
37	0.7563	0.0012	0.2437	0.9988	1.146
38	0.7506	0.0010	0.2494	0.9990	1.150
39	0.7449	0.0009	0.2551	0.9991	1.154
40	0.7393	0.0007	0.2607	0.9993	1.159
41	0.7338	0.0006	0.2662	0.9994	1.163
42	0.7283	0.0005	0.2717	0.9995	1.167
43	0.7228	0.0004	0.2772	0.9996	1.171
44	0.7173	0.0003	0.2827	0.9997	1.175
45	0.7120	0.0003	0.2880	0.9997	1.179
46	0.7066	0.0002	0.2934	0.9998	1.184
47	0.7013	0.0002	0.2987	0.9998	1.188
48	0.6960	0.0002	0.3040	0.9998	1.192
49	0.6908	0.0001	0.3092	0.9999	1.196
50	0.6856	0.0001	0.3144	0.9999	1.201
51	0.6804	0.0001	0.3196	0.9999	1.205
52	0.6753	0.0001	0.3247	0.9999	1.209
53	0.6702	0.0001	0.3298	0.9999	1.213
54	0.6652	0.0001	0.3348	0.9999	1.218
55	0.6602	0.0000	0.3398	1.0000	1.222
56	0.6552	0.0000	0.3448	1.0000	1.226
57	0.6503	0.0000	0.3497	1.0000	1.231
58	0.6454	0.0000	0.3546	1.0000	1.235
59	0.6405	0.0000	0.3595	1.0000	1.239
60	0.6357	0.0000	0.3643	1.0000	1.244

compressed gas can be turned on or off conveniently. Open stopcock on scintillation cell, close stopcock to gas, and gradually open stopcock to vacuum source to evacuate cell. Close stopcock to vacuum source and check manometer reading for 2 min to test system, especially the scintillation cell, for leaks.

e. Open stopcock to counting gas and cautiously admit gas to scintillation cell until atmospheric pressure is reached.

f. Center scintillation cell on photomultiplier tube, cover with light-tight hood and, after 10 min, obtain a background counting rate (preferably over a 100- to 1000-min period, depending on concentration of ^{226}Ra in samples). *Do not expose phototube to external light with the high voltage applied.*

g. Repeat Steps d through f above for each scintillation cell.

h. If the leakage test and background are satisfactory, continue calibration.

i. With scintillation cell and standard bubbler (¶ 4c) on vacuum train, open stopcock on scintillation cell and evacuate scintillation cell and purification system (Figure 7500-Ra:1) by opening stopcock to vacuum source. Close stopcock to vacuum source. Check system for leaks as in Step d above.

j. Adjust gas regulator (diaphragm) valve so that a very slow stream of gas will flow with the needle valve open. Attach gas supply to inlet of bubbler.

k. Note time as beginning of an approximately 20-min de-emanation period. Very cautiously open bubbler outlet stopcock to equalize pressure and transfer all or most of the fluid in the inlet side arm to bubbler chamber.

l. Close outlet stopcock and very cautiously open inlet stopcock to flush remaining fluid from side arm and fritted disk. Close inlet stopcock.

m. Repeat Steps h and l above, four or five times, to obtain more nearly equal pressures on the two sides of bubbler.

n. With outlet stopcock fully open, cautiously open inlet stopcock so that gas flow produces a froth a few millimeters thick at surface of bubbler solution. Maintain flow rate by gradually increasing pressure with regulator valve and continue de-emanation until pressure in cell reaches atmospheric pressure. Total elapsed time for the de-emanation should be 15 to 25 min.

o. Close stopcocks to scintillation cell, close bubbler inlet and outlet, shut off and disconnect gas supply, and record date and time as the ends of the ^{222}Rn ingrowth and de-emanation periods and as the beginnings of decay of ^{222}Rn and ingrowth of decay products.

p. Store bubbler for another ^{222}Rn ingrowth in the event a subsequent de-emanation is desired (Table 7500-Ra:II). The standard bubbler may be kept indefinitely.

q. Four hours after de-emanation, when daughter products are in virtual transient equilibrium with ^{222}Rn, place scintillation cell on photomultiplier tube, cover with light-tight hood, let stand for at least 10 min, then begin counting. Record date and time counting was started and finished.

r. Correct net counting rate for ^{222}Rn decay (Table 7500-Ra:II) and relate it to picocuries ^{226}Ra in standard bubbler (see ¶ 6a). Unless the scintillation cell is physically damaged, the calibration will remain essentially unchanged for years. Occasional calibration is recommended.

s. Repeat Steps h through r above on each scintillation cell.

t. To remove ^{222}Rn and prepare scintillation cell for reuse, evacuate and cautiously refill with counting gas. Routinely, repeat evacuation and refilling twice, and repeat process more times if the cells have contained a high ^{222}Rn activity. (Decay products with a half-life of approximately 30 min will remain in the cell. Do not check background on cells until activity of decay products has had time to decay to insignificance.)

5. Procedure

a. Soluble radium 226:

1) Using a membrane filter, filter at least 1 L sample or a volume containing up to 30 pCi ^{226}Ra and transfer to a polyethylene bottle as soon after sampling as possible. Save the suspended matter for determination by the procedure described in 5b, below. Record sample volume filtered if suspended solids are to be analyzed as in the procedure for ^{226}Ra in suspended matter.

2) Add 20 mL conc HCl/L of filtrate and continue analysis when convenient.

3) Add 50 mL dilute BaCl$_2$ solution, with vigorous stirring, to 1020 mL acidified filtrate [¶ 2) preceding] in a 1.5-L beaker. In each batch of samples include a reagent blank consisting of distilled water plus 20 mL conc HCl.

4) Cautiously, with vigorous stirring, add 20 mL conc H$_2$SO$_4$. Cover beaker and let precipitate overnight.

5) Filter supernate through a membrane filter, using 0.1N H$_2$SO$_4$ to transfer Ba-Ra precipitate to filter, and wash precipitate twice with 0.1N H$_2$SO$_4$.

6) Place filter in a platinum crucible or dish, add 0.5 mL HF and 3 drops (0.15 mL) (NH$_4$)$_2$SO$_4$ solution, and evaporate to dryness.

7) Carefully ignite over a small flame until carbon is burned off; cool. (After filter is charred a Meker burner may be used.)

8) Add 1 mL H$_3$PO$_4$ with a calibrated dropper and heat on hot plate at about 200°C. Gradually raise temperature and maintain at about 300 to 400°C for 30 min.

9) Swirl vessel over a low Bunsen flame, adjusted to avoid spattering, while covering the walls with hot H$_3$PO$_4$. Continue to heat for a minute after precipitate fuses into a clear melt (just below redness) to insure complete removal of SO$_3$.

10) Fill cooled vessel one-half full with 6N HCl, heat on steam bath, then gradually add distilled water to within 2 mm of top of vessel.

11) Evaporate on boiling steam bath until there are no more vapors of HCl.

12) Add 6 mL 1N HCl, swirl, and warm to dissolve BaCl$_2$ crystals.

13) Close gas inlet stopcock, add a drop of water to the fritted disk of the fully greased and tested radon bubbler, and transfer sample from platinum vessel to bubbler with a medicine dropper. Use dropper to rinse vessel with at least three 2-mL portions of distilled water. Add distilled water until bubbler is two-thirds to three-fourths full.

14) Dry, wax if necessary, and seal joint. Establish zero ingrowth time as instructed in ¶ 4c preceding.

15) Close stopcocks, record date and time, and store bubbler for ^{222}Rn ingrowth, preferably for 3 weeks for low concentrations of radium 226.

16) De-emanate and count ^{222}Rn as instructed for calibrations in ¶s 4i through r, with sample replacing standard bubbler.

17) The sample in the bubbler may be stored for a second ingrowth or it may be discarded and the bubbler cleaned for reuse. (A bubbler is readily cleaned while in an inverted position by attaching a tube from a beaker containing 100 mL 0.1N HCl to the inlet and attaching another tube from outlet to a suction flask. Alternately open and close outlet and inlet stopcocks to pass the acid rinse water sequentially through the fritted disk, accumulate in the bubbler, and flush into the suction flask. Drain bubbler with the aid of vacuum, heat ground joint gently to melt wax, and separate joint. More extensive cleaning, as indicated in ¶ 2k above, may be necessary if the bubbler contained more than 10 pCi ^{226}Ra.)

b. Radium 226 in suspended matter:

1) Suspended matter in water usually contains siliceous materials that require fusion with an alkaline flux to insure recovery of radium. Dry suspended matter (up to 1000 mg inorganic material) retained on the membrane filter specified in ¶ 5a1) above in a tared platinum crucible and ignite as in ¶ 5a7).

2) Weigh crucible to estimate residue.

3) Add 8 g flux/g residue, but not less than 2 g flux, and mix with a glass rod.

4) Heat over a Meker burner until melting begins, being careful to prevent spattering. Continue heating for 20 min after bubbling stops, with an occasional swirl of the crucible to mix contents and achieve a uniform melt. A clear melt usually is obtained only when the suspended solids are present in small amount or have a high silica content.

5) Remove crucible from burner and rotate as melt cools to distribute it in a thin layer on crucible wall.

6) When cool, place crucible in a covered beaker containing 120 mL distilled water, 20 mL conc H_2SO_4, and 5 mL dilute H_2O_2 solution for each 8 g flux. (Reduce acid and H_2O_2 in proportion to flux used.) Rotate crucible to dissolve melt if necessary.

7) When melt is dissolved, remove and rinse crucible into beaker. Save crucible for Step 10) below.

8) Heat solution and slowly add 50 mL dilute $BaCl_2$ solution with vigorous stirring. Cover beaker and let stand overnight for precipitation. (Precipitation with cool sample solution also is satisfactory.)

9) Add about 1 mL dilute H_2O_2 and, if yellow color (from titanium) deepens, add more H_2O_2 until there is no further color change.

10) Continue analysis according to ¶s 5a5) through 16).

11) Calculate result as directed in ¶s 6a and b, taking into account that the suspended solids possibly were contained in a sample volume other than 1 L [see ¶ 5a1)].

c. Total radium 226:

1) Total ^{226}Ra in water is the sum of soluble and suspended ^{226}Ra as determined in 5a and b preceding, or it may be determined directly by examining the original water sample that has been acidified with 20 mL conc HCl/L sample and stored in a polyethylene bottle.

2) Thoroughly mix acidified sample and take 1020 mL or a measured volume containing not more than 1000 mg inorganic suspended solids.

3) Add 50 mL dilute $BaCl_2$ solution and slowly, with vigorous stirring, add 20 mL conc H_2SO_4/L sample. Cover and let precipitate overnight.

4) Filter supernate through membrane filter and transfer solids to filter as in ¶ 5a5) preceding.

5) Place filter and precipitate in tared platinum crucible and proceed as in ¶s 5b2) through 10) above but with the following changes in the procedure given in ¶ 5b8): Omit adding dilute $BaCl_2$ solution, digest for 1 h on a steam bath, and filter immediately after digestion without stirring up $BaSO_4$. (If these changes are not made, filtration will be very slow.)

6) Calculate total radium 226 concentration as directed in ¶s 6a and b.

6. Calculations

a. Calculate the ^{226}Ra in a bubbler, including reagent blank, as follows:

$$^{226}Ra \text{ in pCi} = \frac{R_s - R_b}{R_c} \times \frac{1}{1 - e^{-\lambda t_1}}$$

$$\times \frac{1}{e^{-\lambda t_2}} \times \frac{\lambda t_3}{1 - e^{-\lambda t_3}}$$

where:

λ = decay constant for ^{222}Rn, 0.007 55/h,

t_1 = time interval allowed for ingrowth of ^{222}Rn, h,

t_2 = time interval between de-emanation and counting, h,

t_3 = time interval of counting, h,

R_s = observed counting rate of sample in scintillation cell, cph,

R_b = (previously) observed background counting rate of scintillation cell with counting gas, cph,

R_c = calibration constant for scintillation cell [i.e., observed net counts per hour, corrected by use of ingrowth and decay factors (C/AB from below) per picocurie of Ra in standard], or:

$$^{226}\text{Ra in pCi} = \frac{(R_s - R_b)}{R_c} \times \frac{C}{AB}$$

where:

A = factor for decay of ^{222}Rn (see Table 7500-Ra:II),

B = factor for growth of ^{222}Rn from ^{226}Ra (see Table 7500-Ra:II), and

C = factor for correction of ^{222}Rn activity for decay during counting (see Table 7500-Ra:II).

For nontabulated times, obtain decay factors for ^{222}Rn by multiplying together the appropriate tabulated "day" and "hour" decay factors, interpolating for less than 0.2 h if indicated by the precision desired. Obtain radon 222 growth factors for nontabulated times most accurately, especially for short periods (e.g., in calibrations), by calculation from ^{222}Rn decay factors given in Column A and using formula given in heading for Column B (of Table 7500-Ra:II). Linear interpolations are satisfactory for routine samples. Obtain the decay-during-counting factors by linear interpolation for all nontabulated times.

In calculating cell calibration constants, use the same equation, but picocuries of ^{226}Ra is known and R_c is unknown.

b. Convert the activity into picocuries per liter of soluble, suspended, or total ^{226}Ra by the following equation:

$$^{226}\text{Ra, pCi/L} = \frac{(D - E) \times 1000}{\text{mL sample}}$$

where:

D = pCi ^{226}Ra found in sample, and

E = pCi ^{226}Ra found in reagent blank.

7. Recovery of Barium (Radium 226) (Optional)

If ^{133}Ba was added in reagent *b*, check recovery of Ba by removing sample from bubbler, adjusting its volume appropriately, gamma-counting it under standard-ized conditions, and comparing the result with the count obtained from a 50-mL portion (evaporated if necessary to reduce volume) of dilute barium solution also counted under standardized conditions; add 1 mL H_3PO_4 to the latter portion before counting. The assumption that the Ba and ^{226}Ra are recovered to the same extent is valid in the method described.

Note that ^{226}Ra and its decay products interfere slightly even if a gamma spectrometer is used. The technique works best when the ratio of ^{133}Ba to ^{226}Ra is high.

Determinations of recovery are particularly helpful with irreplaceable samples, both in gaining experience with the method and in applying the general method to unfamiliar media.

8. Precision and Bias

In a collaborative study, seven laboratories analyzed four water samples for dissolved radium 226 by this method. No result was rejected as an outlier. The average recoveries of added radium 226 from Samples A, B, C, and D (below) were 97.1, 97.3, 97.6, and 98.0%, respectively. At the 95% confidence level, the precision (random error) was 6% and 8% for the two sets of paired samples. Because of the small number of participating laboratories and the low values for random and total errors, there was no evidence of laboratory systematic errors. Neither radium 224 at an activity equal to that of the radium 226 nor dissolved solids up to 610 mg/L produced a detectable error in the results.

Test samples consisted of two pairs of simulated moderately hard and hard water samples containing known amounts of added radium 226 and other radionuclides. The composition of the samples with respect to nonradioactive substances was the same for a pair of samples but varied for the two pairs. The radiochemical composition of the samples is given in Table 7500-Ra:I.

9. References

1. HURSH, J.B. 1954. Radium-226 in water supplies of the U.S. *J. Amer. Water Works Assoc.* 46:43.
2. RUSHING, D.E., W.J. GARCIA & D.A. CLARK. 1964. The analysis of effluents and environmental samples from uranium mills and of biological samples for radium, polonium, and uranium. *In* Radiological Health and Safety in Mining and Milling of Nuclear Materials. International Atomic Energy Agency, Vienna, Austria, Vol. 11, p. 187.
3. LUCAS, H.F. 1957. Improved low-level alpha scintillation counter for radon. *Rev. Sci. Instrum.* 28:680.
4. RUSHING, D.E. 1967. Determination of dissolved radium-226 in water. *J. Amer. Water Works Assoc.* 59:593.

7500-Ra D.   Sequential Precipitation Method (PROPOSED)

1. General Discussion

a. Application: This method can be used to determine soluble radium 228 alone or soluble radium 228 plus radium 226.

b. Principle: Radium 228 and radium 226 in water are concentrated and separated by coprecipitation with barium and lead as sulfates and purified by EDTA chelation. After 36-h ingrowth of actinium 228 from radium 228, actinium 228 is carried on yttrium oxalate, purified, and beta-counted. Radium 226 in the supernatant is precipitated as the sulfate, purified, and alpha-counted (Method B) or it is transferred to a radon bubbler and determined by the emanation procedure (Method C), which is the preferred method.

If analysis of radium 226 is not required, the procedure for radium 228 may be terminated by beta-counting the yttrium oxalate precipitate with a follow-up precipitation of barium sulfate for yield determination. If it is determined that radium 228 is absent, the radium 226 fraction may be alpha-counted directly. If radium 228 is present, radium 226 must be determined by radon emanation.

c. Sampling and storage: To drinking water or a filtered sample of turbid water, add 2 mL conc nitric acid (HNO$_3$)/L sample at the time of collection or immediately after filtration.

2. Apparatus

a. Counting instruments: One of the following is required:

1) *Internal proportional counter,* gas flow, with scaler, timer, and register; or a thin end-window (polyester plastic)* proportional counting chamber with scaler, timer, register amplifier, and preferably having an anticoincident system (low background).

2) *Scintillation counter assembly:* See ¶ C.2a. This equipment is necessary only if radium 226 is determined sequentially with radium 228 and is analyzed by emanation of radon.

b. Centrifuge, bench-size clinical, with polypropylene tubes.

c. Filter funnels, for 2.4-cm filter paper.

d. Stainless steel pans, 5.1 cm.

e. Infrared drying lamp assembly.

f. Magnetic stirrer hot plate.

g. Membrane filters, 47-mm diam, 0.45-μm pore diam.†

3. Reagents

a. Acetic acid, conc.

b. Acetone, anhydrous.

c. Ammonium hydroxide, NH$_4$OH, conc.

d. Ammonium oxalate solution: Dissolve

---

*Mylar or equivalent.
†Gelman Ga-6 or equivalent.

25 g $(NH_4)_2C_2O_4$ in distilled water and dilute to 500 mL.

e. *Ammonium sulfate solution:* Dissolve 20 g $(NH_4)_2SO_4$ in a minimum of distilled water and dilute to 100 mL.

f. *Ammonium sulfide solution:* Dilute 10 mL $(NH_4)_2S$ (20 to 24%) to 100 mL with distilled water.

g. *Barium carrier standardized:* Dissolve 2.846 g $BaCl_2 \cdot 2H_2O$ in distilled water, add 0.5 mL conc HNO_3, and dilute to 100 mL; 1 mL = 16 mg Ba.

h. *Citric acid, 1M:* See Section B.3a.

i. *EDTA reagent, 0.25 M:* See Section B.3k.

j. *Ethanol, 95%.*

k. *Lead carrier: Solution A:* Dissolve 2.397 g $Pb(NO_3)_2$ in distilled water, add 0.5 mL conc HNO_3, and dilute to 100 mL; 1 mL = 15 mg Pb. *Solution B:* Dilute 10 mL Solution A to 100 mL with distilled water; 1 mL = 1.5 mg Pb.

l. *Methyl orange indicator solution:* Dissolve 0.1 g methyl orange powder in 100 mL distilled water.

m. *Nitric acid, HNO_3,* conc, 6N, and 1N.

n. *Sodium hydroxide, 18N:* Dissolve 720 g NaOH in 500 mL distilled water and dilute to 1 L.

o. *Sodium hydroxide, 10N:* Dissolve 400 g NaOH in 500 mL distilled water and dilute to 1 L.

p. *Sodium hydroxide, NaOH, 1N.*

q. *Strontium-yttrium mixed carrier: Solution A:* Dilute 10.0 mL yttrium carrier to 100 mL. *Solution B:* Dissolve 0.4348 g $Sr(NO_3)_2$ in distilled water and dilute to 100 mL. Combine equal volumes of Solutions A and B; 1 mL = 0.9 mg Sr and 0.9 mg Y.

r. *Sulfuric acid, H_2SO_4, 18N.*

s. *Yttrium carrier:* Add 12.7 g Y_2O_3 (Section 7500-Sr.B.3d) to an erlenmeyer flask containing 20 mL distilled water. Heat to boiling and, while stirring with a magnetic stirring hot plate, add small portions of conc HNO_3. (About 30 mL is necessary to dissolve the Y_2O_3. Small additions of dis-

tilled water also may be needed to replace water lost by evaporation.) After total dissolution, add 70 mL conc HNO_3 and dilute to 1 L with distilled water; 1 mL = 10 mg Y.

4. Procedure

a. *Radium 228:*

1) For 1 L sample add 5 mL 1M citric acid and a few drops methyl orange indicator. The solution should be red. Add 10 mL lead carrier (Solution A), 2.0 mL barium carrier, and 2 mL yttrium carrier; stir well. Heat to incipient boiling and maintain at this temperature for 30 min.

2) Add conc NH_4OH until a definite yellow color is obtained; add a few drops excess. Precipitate lead and barium sulfates by adding 18N H_2SO_4 until the red color reappears; add 0.25 mL excess. Add 5 mL $(NH_4)_2SO_4$ solution/L sample. Stir frequently and hold at about 90°C for 30 min.

3) Cool and filter with suction through a membrane filter. Quantitatively transfer precipitate to filter. Carefully place filter in a 250-mL beaker. Add about 10 mL conc HNO_3 and heat gently until the filter dissolves completely. Using conc HNO_3 transfer precipitate to a centrifuge tube. Centrifuge and discard supernatant.

4) Wash precipitate with 15 mL conc HNO_3, centrifuge, and discard supernatant. Repeat wash and centrifuge again. Add 25 mL EDTA reagent, heat in a hot water bath, and stir well. Add a few drops 10N NaOH if the precipitate does not dissolve readily.

5) Add 1 mL strontium-yttrium mixed carrier and stir thoroughly. Add a few drops 10N NaOH if any precipitate forms. Add 1 mL $(NH_4)_2SO_4$ solution and stir thoroughly. Add conc acetic acid until $BaSO_4$ precipitates; add 2 mL excess. The pH should be about 4.5. Digest in a hot water bath (80°C) until precipitate settles. Centrifuge and discard supernatant.

6) Add 20 mL EDTA reagent, heat in a hot water bath, and stir until precipitate

dissolves. Repeat Step 5. Note time of last $BaSO_4$ precipitation as zero time for ingrowth of ^{228}Ac. Dissolve precipitate in 20 mL EDTA reagent, add 0.5 mL yttrium carrier and 1 mL lead carrier (Solution B). If any precipitate forms, dissolve by adding a few drops 10N NaOH. Mix well, cap tube, and age at least 36 h.

7) Add 0.3 mL $(NH_4)_2S$ solution and mix well. Add 10N NaOH dropwise with vigorous stirring until PbS precipitates; add 10 drops excess. Stir intermittently for about 10 min. Centrifuge and decant supernatant into a clean tube.

8) Add 1 mL lead carrier (Solution B), 0.1 mL $(NH_4)_2S$ solution, and a few drops 10N NaOH. Repeat precipitation of PbS. Centrifuge and filter supernatant through filter paper‡ into a clean tube. Wash filter with a few milliliters of distilled water. Discard residue.

9) Add 5 mL 18N NaOH (make at least 2N in OH$^-$). Because of the short half-life of ^{228}Ac (6.13 h) complete the following procedure without delay. Mix well and digest in a hot water bath until $Y(OH)_3$ coagulates. Centrifuge and decant supernatant into a beaker. Cover beaker and save supernatant for ^{226}Ra analysis, ¶s b or c below. Note time of $Y(OH)_3$ precipitation; this is the end of ^{228}Ac ingrowth and beginning of ^{228}Ac decay. (t_3 = time in minutes between last $BaSO_4$ and first $Y(OH)_3$ precipitations.) Dissolve precipitate in 2 mL 6N HNO_3. Heat and stir in a hot water bath about 5 min. Add 5 mL distilled water and reprecipitate $Y(OH)_3$ with 3 mL 10N NaOH. Heat and stir in a hot water bath until precipitate coagulates. Centrifuge and discard supernatant.

10) Dissolve precipitate with 1 mL 1N HNO_3 and heat in hot water bath for several minutes. Dilute to 5 mL with distilled water and add 2 mL ammonium oxalate solution. Heat to coagulate, centrifuge, and discard supernatant. Add 10 mL distilled

water, 6 drops 1N HNO_3, and 6 drops ammonium oxalate solution. Heat and stir in a hot water bath for several minutes. Centrifuge and discard supernatant. Transfer quantitatively to a tared stainless-steel planchet using a minimum quantity of distilled water. Dry under an infrared lamp to constant weight and count in a low-background beta counter. (t_1 = time in minutes between first $Y(OH)_3$ precipitation and counting.)

If analysis of radium 226 is not required, complete Steps b1) and 3) below to obtain the fractional barium yield to be used in calculating ^{228}Ra activity.

b. Radium by precipitation:

1) To the supernatant saved in ¶ a9) above add 4 mL conc HNO_3 and 2 mL $(NH_4)_2SO_4$ solution, mixing well after each addition. Add conc acetic acid until $BaSO_4$ precipitates; add 2 mL excess. Digest on a hot plate until precipitate settles. Centrifuge and discard supernatant.

2) Add 20 mL EDTA reagent, heat in a hot water bath, and stir until precipitate dissolves. Add a few drops 10N NaOH if precipitate does not dissolve readily. Add 1 mL strontium-yttrium mixed carrier and 1 mL lead carrier (Solution B), and stir thoroughly. Add a few drops 10N NaOH if any precipitate forms. Add 1 mL $(NH_4)_2SO_4$ solution and stir thoroughly. Add conc acetic acid until $BaSO_4$ precipitates; add 2 mL excess. Digest in a hot water bath until precipitate settles. Centrifuge, discard supernatant, and note time.

3) Wash precipitate with 10 mL distilled water. Centrifuge and discard supernatant. Transfer quantitatively to a tared stainless-steel planchet using a minimum quantity distilled water. Dry under an infrared lamp to constant weight. If after sufficient beta decay of the actinium fraction ^{228}Ra is found to be absent, make a direct alpha count for ^{226}Ra. If ^{228}Ra is present, determine ^{226}Ra by radon emanation, ¶ c below.

4) Count immediately in an alpha proportional counter.

c. Radium 226 by radon: Transfer the final precipitate obtained in *b* above to a small beaker using a rubber policeman and 14 mL EDTA reagent. Add a few drops 10*N* NaOH and heat on a hot plate to dissolve. Cool and transfer to a radon bubbler (Figure 7500-Ra:1) rinsing beaker with 1 mL EDTA reagent. Proceed as in Method C beginning with 5*a*14).

5. Calculation

a. Calculation of ^{228}Ra concentration:

$$^{228}Ra, pCi/L = \frac{C}{2.22 \times EVR}$$

$$\times \frac{\lambda t_2}{(1 - e^{-\lambda t_2})} \times \frac{1}{(1 - e^{-\lambda t_3})} \times \frac{1}{e^{-\lambda t_1}}$$

where:

C = average net count rate, cpm,
E = counter efficiency, for ^{228}Ac,
V = sample volume, L,
R = fractional chemical yield of yttrium carrier, ¶ 4*a*10), multiplied by fractional chemical yield of barium carrier, ¶ *b*3),

λ = decay constant of ^{228}Ac, 0.001 884/ min,
t_1 = time between first Y(OH)$_3$ precipitation and start of counting, min.
t_2 = counting time, min, and
t_3 = ingrowth time of ^{228}Ac between last BaSO$_4$ precipitation and first Y(OH)$_3$ precipitation, min.

The factor $\lambda t_2/(1 - e^{-\lambda t_2})$ corrects average count rate to count rate at beginning of counting time.

b. Calculation of ^{226}Ra (plus any ^{224}Ra and ^{223}Ra) concentration: See Section B.5.

c. Calculation of ^{226}Ra (emanation) concentration: See Section C.6.

6. Bibliography

JOHNSON, J.O. 1971. Determination of radium 228 in natural waters. Radiochemical Analysis of Water. U.S. Geol. Surv. Water Supply Paper 1696-G, U.S. Government Printing Off., Washington, D.C.

KRIEGER, H.L. 1976. Interim Radiochemical Methodology for Drinking Water. EPA-600/ 4-75-008 (revised), U.S. Environmental Protection Agency, Environmental Monitoring and Support Lab., Cincinnati, Ohio.

7500-Sr   TOTAL RADIOACTIVE STRONTIUM AND STRONTIUM 90*

7500-Sr A.   Introduction

The important radioactive nuclides of strontium produced in nuclear fission are ^{89}Sr and ^{90}Sr. Strontium 90 is one of the most hazardous of all fission products. It decays slowly, with a half-life of 28 years. Upon ingestion, strontium is concentrated in the bone. The total occupational permissible exposure to ^{90}Sr allows a concentration in water of 100 pCi/L, as compared to 10 000 pCi/L for ^{89}Sr, which has a half-

life of only 50.5 d. The 1976 Environmental Protection Agency Interim Primary Drinking Water regulations limit the concentration of ^{90}Sr in water to 8 pCi/L when other sources of intake are not considered.

The method presented in this section is designed to measure total radioactive strontium (^{89}Sr and ^{90}Sr) or ^{90}Sr alone in drinking water or in filtered raw water. It is applicable to sewage and industrial wastes provided that steps are taken to destroy organic matter and eliminate other interfering ions.

*Approved by Standard Methods Committee, 1988.

7500-Sr B. Precipitation Method

1. General Discussion

a. Principle: A known amount of inactive strontium ions, in the form of strontium nitrate, $Sr(NO_3)_2$, is added as a "carrier." The carrier, alkaline earths, and rare earths are precipitated as the carbonate to concentrate the radiostrontium. The carrier, along with the radionuclides of strontium, is separated from other radioactive elements and inactive sample solids by precipitation as $Sr(NO_3)_2$ from fuming nitric acid solution. The strontium carrier, together with the radionuclides of strontium, finally is precipitated as strontium carbonate, $SrCO_3$, which is dried, weighed to determine recovery of carrier, and measured for radioactivity. The activity in the final precipitate is due to radioactive strontium only, because all other radioactive elements have been removed. A correction is applied to compensate for losses of carrier and activity during the various purification steps. A delay in the count will give an increased counting rate due to the ingrowth of ^{90}Y.

b. Concentration techniques: Because of the very low amount of radioactivity, a large sample must be taken and the activity concentrated by precipitation. $Sr(NO_3)_2$ and barium nitrate, $Ba(NO_3)_2$, carriers are added to the sample. Sodium carbonate is then added to concentrate radiostrontium by precipitation of alkaline earth carbonates along with other radioactive elements. The supernate is discarded. The precipitate is dissolved and reprecipitated to remove interfering radionuclides.

c. Interference: Radioactive barium (^{140}Ba, ^{140}La) interferes in the determination of radioactive strontium inasmuch as it precipitates with the radioactive strontium. Eliminate this interference by adding inactive $Ba(NO_3)_2$ carrier and separating this from the strontium by precipitating barium chromate in acetate buffer solution. Ra-

dium isotopes also are eliminated by this treatment.

In hard water, some calcium nitrate may be coprecipitated with $Sr(NO_3)_2$ and can cause errors in recovery of the final precipitate and in measuring its activity. Eliminate this interference by repeated precipitations of strontium as the nitrate followed by leaching the $Sr(NO_3)_2$ with acetone (CAUTION).

For total radiostrontium, count the precipitate within 3 to 4 h after the final separation and before ingrowth of ^{90}Y.

d. Determination of ^{90}Sr: Because it is impossible to separate the isotopes ^{89}Sr and ^{90}Sr by any chemical procedure, the amount of ^{90}Sr is determined by separating and measuring the activity of ^{90}Y, its daughter. After equilibrium is reached, the activity of ^{90}Y is exactly equal to the activity of ^{90}Sr. Two alternate procedures are given for the separation of ^{90}Y. In the first method, ^{90}Y is separated by extraction into tributyl phosphate from concentrated nitric acid (HNO_3) solution. It is back-extracted into dilute HNO_3 and evaporated to dryness for beta counting. The second method consists of adding yttrium carrier, separating by precipitation as yttrium hydroxide, $Y(OH)_3$, and finally precipitating yttrium oxalate for counting.

2. Apparatus

a. Counting instruments: Use either an internal proportional counter, gas-flow, with scaler, timer, and register; or a thin end-window (polyester plastic film*) proportional or G-M counting chamber with scaler, timer, register amplifier, and preferably having an anticoincident system (low background).

---

*Mylar, E.I. du Pont de Nemours, Wilmington, Del., or equivalent.

b. Filter paper,† 2.4 cm diam; or glass fiber filters, 2.4 cm diam.

c. Two-piece filtering apparatus for 2.4-cm filters such as TFE filter holder,‡ stainless steel filter holder, or equivalent.

d. Stainless steel pans, about 50 mm diam and 7 mm deep, for counting solids deposited on pan bottom. For counting precipitates on 2.4-cm filters, use nylon disk with ring§ on which the filter samples are mounted and covered by 0.25 mil film.

3. Reagents

a. Strontium carrier, 10 mg Sr^{2+}/mL, standardized: Carefully add 24.16 g $Sr(NO_3)_2$ to a 1-L volumetric flask and dilute with distilled water to the mark. For standardization, pipet three 10.0-mL portions of strontium carrier solution into 40-mL centrifuge tubes and add 15 mL $2N$ Na_2CO_3 solution. Stir, heat in a boiling water bath for 15 min, and cool. Filter $SrCO_3$ precipitate through a tared fine-porosity sintered-glass crucible of 15-mL size. Wash precipitate with three 5-mL portions of water and then with three 5-mL portions of absolute ethanol (or acetone). Wipe crucible with absorbent tissue and dry to constant weight in an oven at 110°C (20 min). Cool in a desiccator and weigh.

$$Sr, \text{ mg/mL} = \frac{(\text{mg } SrCO_3) \ (0.5935)}{10}$$

b. Barium carrier, 10 mg Ba^{2+}/mL: Dissolve 19.0 g $Ba(NO_3)_2$ in distilled water and dilute to 1 L.

c. Rare earth carrier, mixed: Dissolve 12.8 g cerous nitrate hexahydrate, $Ce(NO_3)_3 \cdot 6H_2O$, 14 g zirconyl chloride octahydrate, $ZrOCl_2 \cdot 8H_2O$, and 25 g ferric

chloride hexahydrate, $FeCl_3 \cdot 6H_2O$, in 600 mL distilled water containing 10 mL conc HCl, and dilute to 1 L.

d. Yttrium carrier: Dissolve 12.7 g yttrium oxide,‖ Y_2O_3, in 30 mL conc HNO_3 by stirring and warming. Add an additional 20 mL conc HNO_3 and dilute to 1 L with distilled water; 1 mL is equivalent to 10 mg Y, or approximately 34 mg $Y_2(C_2O_4)_3 \cdot 9H_2O$. Determine exact equivalence by precipitating yttrium carrier in acid solution according to ¶s 4*c*2)-8), below or by extracting yttrium carrier in acid solution according to ¶s 4*b*3)-11), below.

e. Acetate buffer solution: Dissolve 154 g $NH_4C_2H_3O_2$ in 700 mL distilled water, add 57 mL conc acetic acid, adjust pH to 5.5 by dropwise addition of conc acetic acid or $6N$ NH_4OH as necessary, and dilute to 1 L.

f. Acetic acid, $6N$.

g. Acetone, anhydrous.

h. Ammonium hydroxide, NH_4OH, $6N$.

i. Hydrochloric acid, HCl, $6N$.

j. Methyl red indicator, 0.1%: Dissolve 0.1 g methyl red in 100 mL distilled water.

k. Nitric acid, HNO_3, fuming (90%), conc, $14N$, $6N$, and $0.1N$.

l. Oxalic acid, saturated solution: Dissolve approximately 11 g $H_2C_2O_4 \cdot 2H_2O$ in 100 mL distilled water.

m. Sodium carbonate solution, 1 *M:* Dissolve 124 g $Na_2CO_3 \cdot H_2O$ in distilled water and dilute to 1 L.

n. Sodium chromate solution, 0.5*M:* Dissolve 117 g $Na_2CrO_4 \cdot 4H_2O$ in distilled water and dilute to 1 L.

o. Sodium hydroxide, $6N$: Dissolve 240 g NaOH in distilled water and dilute to 1 L.

p. Tributyl phosphate, reagent grade: Shake with an equal volume of $14N \ HNO_3$

---

†Whatman No. 42 or equivalent.
‡Flurolon Laboratory, Box 305, Caldwell, N.J.
§Control Molding Corp., Staten Island, N.Y., or equivalent.

‖ Yttrium oxide, Code 1118, American Potash and Chemical Corp., West Chicago, Ill., or equivalent. Yttrium oxide of purity less than Code 1118 may require purification because of radioactivity contamination.

to equilibrate. Separate and discard the HNO_3 washings.

4. Procedure

a. Total radiostrontium:

1) To 1 L of drinking water, or a filtered sample of raw water in a beaker, add 2.0 mL conc HNO_3 and mix. Add 2.0 mL each of strontium and barium carriers and mix well. (A precipitate of $BaSO_4$ may form if the water is high in sulfate ion, but this will cause no difficulties.) A smaller sample may be used if it contains at least 25 pCi strontium. The suspended matter that has been filtered off may be digested [see Gross Alpha and Gross Beta Radioactivity, 7110B.4f1)], diluted, and analyzed separately.

2) Heat to boiling, then add 20 mL 6N NaOH and 20 mL 1M Na_2CO_3. Stir and let simmer at 90 to 95°C for about 1 h.

3) Set beaker aside until precipitate has settled (about 1 to 3 h).

4) Decant and discard clear supernate. Transfer precipitate to a 40-mL centrifuge tube and centrifuge. Discard supernate.

5) Add, dropwise (CAUTION—effervescence), 4 mL conc HNO_3. Heat to boiling, stir, then cool under running water.

6) Add 20 mL fuming HNO_3, cool 5 to 10 min in ice bath, stir, and centrifuge. Discard supernate.

7) Add 4 mL distilled water, stir, and heat to boiling to dissolve the strontium. Centrifuge while hot to remove remaining insolubles and decant supernate to a clean centrifuge tube. Add 2 mL 6N HNO_3, heat to boiling, centrifuge while hot, and combine supernate with aqueous supernate. Discard insoluble residue of SiO_2, $BaSO_4$, etc.

8) Cool combined supernates, then add 20 mL fuming HNO_3, cool 5 to 10 min in ice bath, stir, centrifuge, and discard supernate.

9) Add 4 mL distilled water and dissolve by heating. Repeat Step 8) preceding.

10) Repeat Step 9) preceding if more than 200 mg Ca were present in the sample.

11) After last HNO_3 precipitation, invert tube in a beaker for about 10 min to drain off most excess HNO_3. Add 20 mL anhydrous acetone, stir thoroughly, cool, and centrifuge. Discard supernate (CAUTION).

12) Dissolve precipitate of $Sr(NO_3)_2$ + $Ba(NO_3)_2$ in 10 mL distilled water and boil for 30 s to remove any remaining acetone.

13) Add 0.25 mL (5 drops) mixed rare earth carrier and precipitate rare earth hydroxides by making solution basic with 6N NH_4OH. Digest in a boiling water bath for 10 min. Cool, centrifuge, and decant supernate to a clean tube. Discard precipitate.

14) Repeat Step 13) preceding.

Note the time of rare earth precipitation, which marks the beginning of the ^{90}Y ingrowth period. Do not delay procedure more than a few hours after the separation; otherwise, false results will be obtained because of ingrowth of ^{90}Y.

15) Add 2 drops methyl red indicator and then add 6N acetic acid dropwise with stirring until indicator changes from yellow to red.

16) Add 5 mL acetate buffer solution, heat to boiling, and add dropwise, with stirring, 2 mL Na_2CrO_4 solution. Digest in a boiling water bath for 5 min. Cool, centrifuge, and decant supernate to a clean tube. Discard residue.

17) Add 2 mL 6N NaOH, add 5 mL 1M Na_2CO_3 solution, and heat to boiling. Cool in an ice bath (about 5 min) and centrifuge. Discard supernate.

18) Add 15 mL distilled water, stir, centrifuge, and discard wash water.

19) Repeat Step 18), and proceed either as in Step 20)a) or 20)b), below. *Save this precipitate if a determination of ^{90}Sr is required.*

20) Either

a) Slurry precipitate with a small volume of distilled water and transfer to a tared stainless steel pan; dry under an infrared

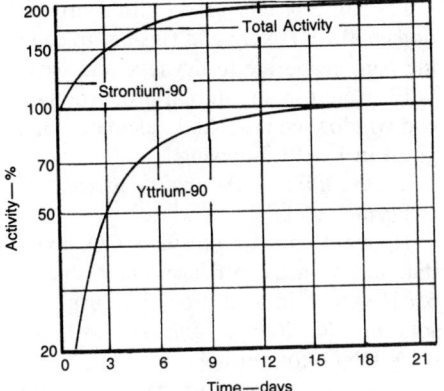

Figure 7500-Sr:1. Yttrium 90 vs. strontium 90 activity as a function of time.

lamp, cool, weigh, and count# the precipitate of $SrCO_3$,** or

b) Transfer precipitate to a tared paper or glass filter mounted in a two-piece funnel. Allow gravity settling for uniform deposition and then apply suction. Wash precipitate with three 5-mL portions of water, three 5-mL portions of 95% alcohol, and three 5-mL portions of ethyl ether or acetone. Dry in an oven at 110 to 125°C for 15 to 30 min, cool, weigh,** mount on a nylon disk and ring with polyester plastic film cover, and count.

21) Calculation

$$\text{Total Sr activity in pCi/L} = \frac{b}{adf \times 2.22}$$

where:

a = beta counter efficiency [see Step 22) below],

d = $\dfrac{\text{mg final } SrCO_3 \text{ precipitate}}{\text{mg } SrCO_3 \text{ in 2 mL of carrier}}$

  = correction for carrier recovery [see Step 23) below],

f = sample volume, L,

b = beta activity, net cpm = $(i/t) - k$,

i = total counts accumulated,

t = time of counting, min, and

k = background, cpm.

22) Counting efficiency— As a first estimate, when mounting sample according to Step 20)a), convert counts per minute to disintegrations per minute, based on the beta activity of cesium 137 standard solutions having a sample thickness equivalent to that of the $SrCO_3$ precipitate. More precise measurements may follow a second count after substantial ingrowth of ^{90}Y from ^{90}Sr, but this precision is not warranted for the usual total radiostrontium determination. When mounting samples according to Step 20)b), determine self-absorption curves by separately precipitating standard solutions of ^{89}Sr and ^{90}Sr as the carbonate (see gross beta in Section 7110).

23) Correction for carrier recovery— 20 mg Sr are equivalent to 33.7 mg $SrCO_3$. Should more than traces of stable strontium be present in the sample, it would act as carrier; hence its determination by flame photometric or atomic absorption spectrometric method would be required.

b. Strontium 90 by extraction of yttrium 90.†† Store $SrCO_3$ precipitate, as in ¶ 4a20), for at least 2 weeks to allow ingrowth of ^{90}Y and then proceed as directed here or in an alternate procedure in Section ¶4c following.

1) Transfer of precipitate to separatory funnel—Either

a) Place a small funnel upright into mouth of a 60-mL separatory funnel; then

---

#Strontium 90 in thick samples is counted with low efficiency; hence, a first count within hours favors ^{89}Sr counting, and a recount after 3 to 6 d that exceeds the first count provides a rough estimate of the ^{90}Y ingrowth—see Figure 7500-Sr:1 and R.J. Velten (1966) below.

**When a determination of total strontium is not required, weigh precipitate [Step 20)a) or 20)b)] for carrier recovery but do not count. Then proceed with ^{90}Sr determination according to ¶4b following.

---

††See footnote to Step 20a) when a determination for only ^{90}Sr is required.

place pan with precipitate, as in Step 20)a), in funnel and add, dropwise, 1 mL 6N HNO$_3$ (CAUTION—*effervescence*); tilt pan to empty into funnel and rinse pan twice with 2-mL portions of 6N HNO$_3$; or

b) Uncover precipitate from filter, as in Step 20)b), and transfer filter with forceps to upright funnel in mouth of 60-mL separatory funnel as in ¶ a) above. Dislodge bulk of precipitate into funnel stem. Dropwise, add with caution 1 mL 6N HNO$_3$ to filter, removing residual precipitate and dissolving bulk precipitate. Rinse filter and funnel twice with 2-mL portions 6N HNO$_3$.

2) Remove filter or pan and add 10 mL fuming HNO$_3$ to separatory funnel through upright funnel.

3) Remove upright funnel and add 1 mL yttrium carrier in a separatory funnel.

4) Add 5.0 mL tributyl phosphate reagent, shake thoroughly for 3 to 5 min, allow phases to separate, and transfer aqueous layer to a second 60-mL separatory funnel.

5) Add 5.0 mL tributyl phosphate reagent, shake 5 min, allow phases to separate, and transfer aqueous layer to a third 60-mL separatory funnel.

6) Combine organic extractants in the first and second funnels into one funnel and wash organic phase twice with 5-mL portions 14N HNO$_3$. Record time as the beginning of ^{90}Y decay (combine acid washings with aqueous phase in third funnel if a second ingrowth of ^{90}Y is desired).

7) Back-extract ^{90}Y from combined organic phases with 10 mL 0.1N HNO$_3$ for 5 min.

8) Continue as in ¶s 4c6)-8) below or transfer aqueous phase from Step 7) immediately above into a 50-mL beaker and evaporate on a hot plate to 5 to 10 mL.

9) Repeat Step 7) above and transfer aqueous phase to beaker in Step 8) preceding; evaporate to 5 to 10 mL.

10) Transfer residual solution in beaker to a tared stainless steel counting pan and evaporate.

11) Rinse beaker twice with 2-mL portions of 0.1N HNO$_3$; add rinsings to counting pan, evaporate to dryness, and weigh.

12) Count in an internal proportional or end-window counter and calculate ^{90}Sr as given in ¶ 4c9) following.

c. Strontium 90 by oxalate precipitation of yttrium 90:‡‡

1) Quantitatively transfer SrCO$_3$ precipitate to a 40-mL centrifuge tube with 2 mL 6N HNO$_3$. Add acid dropwise during dissolution (CAUTION—*effervescence*). Use 0.1N HNO$_3$ for rinsing.

2) Add 1 mL yttrium carrier, 2 drops methyl red indicator and, *dropwise,* add conc NH$_4$OH to the methyl red end point.

3) Add 5 mL more conc NH$_4$OH and *record the time,* which is the end of ^{90}Y ingrowth and the beginning of decay; centrifuge and decant supernate to a beaker (save supernate and washings for a second ingrowth if desired).

4) Wash precipitate twice with 20-mL portions hot distilled water.

5) Add 5 to 10 drops of 6N HNO$_3$, stir to dissolve precipitate, add 25 mL distilled water, and heat in a water bath at 90°C.

6) Gradually add 15 to 20 drops saturated oxalic acid reagent with stirring and adjust to pH 1.5 to 2.0 (pH meter or indicator paper) by adding conc NH$_4$OH dropwise. Digest precipitate for 5 min and cool in an ice bath with occasional stirring.

7) Transfer precipitate to a tared glass fiber filter in a two-piece funnel. Let precipitate settle by gravity (for uniform deposition) and apply suction. Wash precipitate in sequence with 10 to 15 mL hot distilled water and then three times with 95% ethyl alcohol and three times with diethyl ether.

8) Air-dry precipitate with suction for 2 min, weigh, mount on a nylon disk and ring with polyester plastic film cover,

---

‡‡See footnote to Step 20a) when a determination for only ^{90}Sr is required.

count, and calculate ^{90}Sr as follows.
9) Calculation

$$^{90}\text{Sr pCi/L} = \frac{\text{net cpm}}{a\,b\,c\,d\,f\,g \times 2.22}$$

where:

a = counting efficiency for ^{90}Y,
b = chemical yield of extracting or precipitating ^{90}Y,
c = ingrowth correction factor if not in secular equilibrium,
d = chemical yield of strontium determined
gravimetrically or by flame photometry,
f = volume of original sample, L,
g = ^{90}Y decay factor, $e^{-\lambda t}$, and
e = base of natural logarithms,
λ = 0.693/$T_{1/2}$, where $T_{1/2}$ for ^{90}Y is 64.2
h, and
t = time between separation and counting, h.

5. Precision and Bias

In a collaborative study of two sets of
paired, moderately hard water samples
containing known additions of radio-

nuclides, 12 laboratories determined the total radiostrontium and 10 laboratories
determined ^{90}Sr. The results of one sample
from one laboratory were rejected as outliers.

The average recoveries of added total
radiostrontium from the four samples were
99, 99, 96, and 93%. The precision (random error) at the 95% confidence level was
10 and 12% for the two sets of paired
samples. The method was slightly biased
on the low side.

6. Bibliography

HAHN, R.B. & C.P. STRAUB. 1955. Determination
of radioactive strontium and barium in water.
J. Amer. Water Works Assoc. 47:335.
GOLDIN, A.S., R.J. VELTEN & G.W. FRISHKORN.
1959. Determination of radioactive strontium. *Anal. Chem.* 31:1490.
GOLDIN, A.S. & R.J. VELTEN. 1961. Application
of tributyl phosphate extraction to the determination of strontium 90. *Anal. Chem.*
33:149.
VELTEN, R.J. 1966. Resolution of Sr-89 and Sr-
90 in environmental media by an instrumental
technique. *Nucl. Instrum. Methods* 42:169.

7500-³H   TRITIUM*

7500-³H A.   Introduction

Tritium exists fairly uniformly in the environment as a result of natural production
by cosmic radiation and residual fallout
from nuclear weapons tests. This background level gradually is being increased
by the use of nuclear reactors to generate
electricity, although tritium from this

source is only a small proportion of environmental tritium. Nuclear reactors and
fuel-processing plants are localized sources
of tritium because of discharges during normal operation. This industry is expected to
become the major source of environmental
tritium contamination in the future. Tritium is produced in light-water nuclear reactors by ternary fission, neutron capture

---

*Approved by Standard Methods Committee, 1988.

in coolant additives, control rods and plates, and activation of deuterium. About 1% of the tritium in the primary coolant is released in gaseous form to the atmosphere; the remainder eventually is released in liquid waste discharges. Most tritium produced in reactors remains in the fuel and is released when fuel is reprocessed.

Naturally occurring tritium is most abundant in precipitation and lowest in aged water because of its physical decay by beta emission to helium. The maximum beta energy of tritium is 0.018 MeV and its half-life is 12.26 years.

The Environmental Protection Agency Interim Primary Drinking Water Regulations set a maximum contaminant level of 20 000 pCi/L.

7500-^{3}H B.   Liquid Scintillation Spectrometric Method

1. General Discussion

a. Principle: A sample is treated by alkaline permanganate distillation to hold back most quenching materials, as well as radioiodine and radiocarbon. Complete transfer of tritiated water is assured by distillation to near dryness. A subsample of distillate is mixed with scintillation solution and the beta activity is counted on a coincidence-type liquid scintillation spectrometer. The scintillation solution consists of 1,4-dioxane, naphthalene, POPOP, and PPO.* The spectrometer is calibrated with standard solutions of tritiated water; then background and unknown samples are prepared and counted alternately, thus nullifying errors that could result from instrument drift or from aging of the scintillation solution.

b. Interferences: Sample distillation effectively removes nonvolatile radioactivity and the usual quenching materials. For waters containing volatile organic or radioactive materials, use wet oxidation (Section 4500-N$_{org}$) to remove interference from quenching due to volatile organic material. Distillation at about pH 8.5 holds back volatile radionuclides such as iodides and bicarbonates. Double distillation with an appropriate delay (10 half-lives) between

distillations may be required to eliminate interference from volatile daughters of radium isotopes. Some clear-water samples collected near nuclear facilities may be monitored satisfactorily without distillation, especially when the monitoring instrument is capable of discriminating against beta radiation energies higher than those in the tritium range.

2. Apparatus

a. Liquid scintillation spectrometer, coincidence-type.

b. Liquid scintillation vial: 20-mL; polyethylene, low-K glass, or equivalent bottles.

c. Distillation apparatus: 250-mL roundbottom distillation flask, connecting sidearm adapter, condenser, and heating mantle.

3. Reagents

a. Scintillation solution: Thoroughly mix 4 g PPO, 0.05 g POPOP, and 120 g solid naphthalene in 1 L spectroquality 1,4-dioxane. Store in dark bottle. Solution is stable for 2 months. Alternatively, use a commercially prepared scintillation solution available from suppliers of liquid scintillation materials.

b. Low-background water: Use water with no detectable tritium activity (most deep well waters are low in tritium).

c. Standard tritium solution: Dilute avail-

---

*POPOP = 1,4-di-2-(5-phenyloxazolyl) benzene; PPO = (2,5- diphenyloxazole).

able tritium standard solution to approximately 1000 dpm/mL with low-background water.

    d. Sodium hydroxide, NaOH, pellets.

    e. Potassium permanganate, KMnO₄.

4. Procedure

Add three pellets NaOH and 0.1 g KMnO₄ to 100 mL sample in 250-mL distillation flask. Distill at 100 to 105°C, discard first 10 mL distillate, and collect next 50 mL. Thoroughly mix 4 mL distillate with 16 mL scintillation solution in tightly capped vial.

Prepare low-background water and standard tritium solution in same manner as samples.

Hold samples, background, and standards in the dark for 3 h. Count samples containing less than 200 pCi/mL for 100 min and samples containing more than 200 pCi/mL for 50 min.

5. Calculations and Reporting

a. Calculate and report tritium, ³H, in picocuries per milliliter (pCi/mL) or its equivalent, nanocuries per liter (nCi/L) as follows:

$$^3H = \frac{(C - B)}{(E \times 4 \times 2.22)}$$

where:

    C = gross counting rate for sample, cpm,

    B = background counting rate, cpm,

    E = counting efficiency, $(S - B)/D$,

    S = gross counting rate for standard solution, cpm, and

    D = tritium activity in standard sample, dpm, corrected for decay to time of counting.

b. Calculate the counting error at the 95% confidence level based on the equation for $\sigma(R)$ given in Section 7010G. A total count of 40 000 within 1 h for a background count rate of about 50 cpm gives a counting error slightly in excess of 1% at the 95% confidence level.

6. Precision and Bias

Samples with tritium activity above 200 pCi/mL can be analyzed with precision of less than ±6% at the 95% confidence level and those with 1 pCi/mL can be analyzed with a precision of less than ±10%.

7. Bibliography

LIBBY, W.F. 1946. Atmospheric helium-3 and radiocarbon from cosmic radiation. *Phys. Rev.* 69:671.

NATIONAL COUNCIL ON RADIATION PROTECTION, SUBCOMMITTEE ON PERMISSIBLE INTERNAL DOSE. 1959. Maximum Permissible Body Burdens and Maximum Permissible Concentrations of Radionuclides in Air and in Water for Occupational Exposure. NBS Handbook 69 (June), National Bur. Standards, Washington, D.C.

INTERNATIONAL COMMISSION ON RADIATION PROTECTION. 1960. Report of Committee II on permissible dose for internal radiation, 1959. *Health Phys.* 3:41.

BUTLER, F.E. 1961. Determination of tritium in water and urine. *Anal. Chem.* 33:409.

FOOD AND AGRICULTURE ORGANIZATION, INTERNATIONAL ATOMIC ENERGY AGENCY & WORLD HEALTH ORGANIZATION. 1966. Methods of Radiochemical Analysis. World Health Org., Geneva.

SMITH, J.M. 1967. The Significance of Tritium in Water Reactors. General Electric Co., San Jose, Calif.

YOUDEN, W.J. 1967. Statistical Techniques for Collaborative Tests. Assoc. Official Analytical Chemists, Washington, D.C.

PETERSON, H.T.J., J.E. MARTIN, C.L. WEAVER & E.D. HARWARD. 1969. Environmental tritium contamination from increasing utilization of nuclear energy sources. Seminar on Agricultural and Public Health Aspects of Environmental Contamination by Radioactive Materials, International Atomic Energy Assoc., Vienna, pp. 35-60.

SODD, V.J. & K.L. SCHOLZ. 1969. Analysis of tritium in water; a collaborative study. *J. Assoc. Offic. Anal. Chem.* 52:1.

WEAVER, C.L., E.D. HARWARD & H.T. PETERSON. 1969. Tritium in the environment from nuclear power plants. *Pub. Health Rep.* 84, 363.

U.S. ENVIRONMENTAL PROTECTION AGENCY. 1975. Tentative Reference Method for Measurement of Tritium in Environmental Waters. EPA 600/4-75-013, Environmental Monitoring and Support Lab., U.S. Environmental Protection Agency, Las Vegas, Nev.

7500-U   URANIUM*

7500-U A.   Introduction

1. Occurrence

Uranium, the heaviest naturally occurring element, is a mixture of three radioactive isotopes: uranium 238 (99.275%), uranium 235 (0.72%), and uranium 234 (0.005%). Most drinking-water sources, especially ground waters, contain soluble carbonates and bicarbonates that complex and keep uranium in solution.

Although the National Interim Primary Drinking Water Regulations normally do not require measurement of uranium in drinking water, samples that have a gross alpha activity greater than 15 pCi/L may require uranium analysis. The regulations allow the exclusion of alpha radioactivity due to uranium to determine whether a water meets the maximum contaminant level (MCL) of 15 pCi/L of alpha activity. An MCL for uranium of 10 pCi/L is under consideration.

2. Selection of Method

Method B, a radiochemical procedure, determines total uranium alpha activity without making an isotopic uranium analysis. Method C, a fluorometric method, measures the mass of uranium based on its fluorescence when fused in a mixture of

sodium fluoride-sodium carbonate-potassium carbonate. Analytical results are expressed as micrograms per liter, μg/L. In natural uranium, when uranium 234 is in equilibrium with uranium 238, 1 μg = 0.68 pCi (= 1.50 dpm) of uranium 234 plus uranium 238.

3. Bibliography

GRIMALDI, F. S. et al. 1954. Collected Papers on Methods of Analysis for Uranium and Thorium. Bull. 1006, U.S. Geological Survey.

BLANCHARD, R. 1963. Uranium Decay Series Disequilibrium in Age Determination of Marine Calcium Carbonates. Ph.D. Thesis, Washington Univ., St. Louis, Mo.

BARKER, F. B. et al. 1965. Determination of uranium in natural waters. U.S. Geological Survey, Water Supply Paper 1696-C, U.S. Government Printing Off., Washington, D.C.

EDWARD, K. W. 1968. Isotopic analysis of uranium in natural waters by alpha spectroscopy. U.S. Geological Survey, Water Supply Paper 1696-F, U.S. Government Printing Off., Washington, D.C.

THATCHER, L. L., V. J. JANZER & K. W. EDWARDS. 1977. Methods for Determination of Radioactive Substances in Water and Fluvial Sediments. Book 5, Chapter A5. Techniques of Water-Resources Investigations of the United States Geological Survey. U.S. Government Printing Off., Washington, D.C.

KRIEGER, H. L. & E. L. WHITTAKER. 1980. Prescribed procedures for measurement of radioactivity in drinking water. EPA-600/4-80-032, U.S. Environmental Protection Agency.

---

*Approved by Standard Methods Committee, 1988.

7500-U B. Radiochemical Method (PROPOSED)

1. General Discussion

a. *Principle:* The sample is acidified with hydrochloric acid and boiled to eliminate carbonate and bicarbonate ions. Uranium is coprecipitated with ferric hydroxide and subsequently separated. The ferric hydroxide is dissolved, passed through an anion-exchange column, and washed with acid, and the uranium is eluted with dilute hydrochloric acid. The acid eluate is evaporated to near dryness, the residual salt is converted to nitrate, and the alpha activity is counted.

b. *Interference:* The only alpha-emitting radionuclide that may be carried through this procedure is protactinium 231. However, this isotope, which is a decay product of uranium 235, causes very little interference. Check reagents for uranium contamination by analyzing a complete reagent blank.

c. *Sampling*: Preserve sample by adjusting its pH to 2 with HCl at time of collection.

2. Apparatus

a. *Counting instrument,* gas-flow proportional or alpha scintillation counting system.

b. *Ion-exchange column,* approximately 13 mm ID $\times$ 150 mm long with 100-mL reservoir.

c. *Membrane filter apparatus,* 47-mm diam.

3. Reagents

a. *Ammonium hydroxide,* NH_4OH, 5N, 1%.

b. *Anion-exchange resin.**

c. *Ferric chloride carrier:* Dissolve 9.6 g $FeCl_3 \cdot 6H_2O$ in 100 mL 0.5N HCl; 1 mL = 20 mg Fe^{3+}.

d. *Hydriodic acid,* HI, 47%.

e. *Hydrochloric acid,* HCl, conc, 8N, 6N, 0.1N.

f. *Iodic acid,* 1 mg/mL: Dissolve 100 mg HIO_3 in 100 mL 4N HNO_3.

g. *Nitric acid,* HNO_3, conc, 4N.

h. *Sodium hydrogen sulfite,* 1%: Dissolve 1 g $NaHSO_3$ in 100 mL 6N HCl.

i. *Uranium standard solution:*† Dissolve 177.3 mg natural undepleted uranyl acetate, $UO_2(C_2H_3O_2)_2 \cdot 2H_2O$, in 1000 mL 0.2N HNO_3; 1 mL = 100 μg U = 150 dpm U = 67.6 pCi U.

4. Calibration

Determine counting efficiency, E, for a known amount of uranium standard solution (about 750 dpm) evaporated from 6 to 8 mL of 1 mg/mL HIO_3 solution in a 50-mm-diam stainless steel planchet. After flaming planchet, count for at least 50 min. Run a reagent blank with the standard portions and count.

$$\text{Counting efficiency, } E = \frac{C - B}{D}$$

where:

C = gross alpha count rate of standard, cpm,
B = alpha background count rate, cpm, and
D = disintegration rate of uranium standard, dpm.

Determine uranium recovery factor by adding a measured amount of uranium standard to the same volume of sample and taking it through the entire procedure. Alpha count the separated, evaporated, and flamed uranium planchet.

$$\text{Recovery factor, } R = \frac{C' - B'}{DE}$$

*Dowex 1×4, 100-200 mesh, chloride form, or equivalent.

†Standard radioactive solutions with uranium isotopes in equilibrium are available from the U.S. Environmental Protection Agency, Environmental Monitoring Systems Laboratory, P. O. Box 15027, Las Vegas, Nev.

where:

C' = gross count rate of sample with added uranium, cpm,

B' = count of reagent blank, cpm,

D = disintegration rate of uranium standard, dpm, and

E = counting efficiency.

5. Procedure

a. To 1 L unacidified sample in a 1500-L beaker, add 5 mL conc HCl and 1 mL FeCl$_3$ carrier. In each batch of samples include a distilled-water blank. Cover with watch glass and heat to boiling for 20 min. If pH is greater than 1, add conc HCl dropwise to bring pH to 1. While sample is boiling, gently add 5N NH$_4$OH from a polyethylene squeeze bottle with the delivery tube inserted between the watch glass and the beaker lip. Add 5N NH$_4$OH until turbidity persists while boiling continues; then add 10 mL more. Continue boiling for 10 min more, then set aside for 30 min to cool and settle. After sufficient settling, decant and filter supernate through a 47-mm, 0.45-μm membrane filter using a large filtering apparatus. Slurry the remaining precipitate, transfer to the filtering apparatus, and filter with suction. Complete transfer using 1% solution of NH$_4$OH delivered from a polyethylene squeeze bottle. Place filtering apparatus over a clean 250-mL filtering flask, add 25 mL 8N HCl to dissolve precipitate, and filter. Wash precipitate and filter with an additional 25 mL 8N HCl.

b. Prepare an ion-exchange column by slurrying the anion-exchange resin with 8N HCl and pouring it into a 13-mm-ID column to give a resin bed height of about 80 mm. Transfer solution to the 100-mL reservoir of the ion-exchange column. Rinse side-arm filtering flask twice with 25-mL portions of 8N HCl. Combine in the ion-exchange reservoir. Pass sample solution through the anion-exchange column at a flow rate of not more than 5 mL/min. After sample has passed through column, elute the iron (and plutonium if present) with six column volumes of freshly prepared 8N HCl containing 1 mL 47% HI /9 mL 8N HCl. Wash column with two additional column volumes of 8N HCl. Discard all washes. Elute uranium into a 100-mL beaker with six column volumes of 0.1N HCl. Evaporate acid eluate to near dryness and convert residue to the nitrate form by three successive treatments with 5-mL portions of conc HNO$_3$, evaporating to near dryness each time. *Do not bake.* Dissolve residue (of which there may be very little visible) in 2 mL 4N HNO$_3$. Using a transfer pipet, transfer to a marked planchet. Complete transfer by rinsing beaker three times with 2-mL portions of 4N HNO$_3$. Evaporate planchet contents to dryness under a heat lamp, flame to remove traces of HIO$_3$, cool, and count for alpha activity.

c. To regenerate anion-exchange resin column, pass three column volumes of 1% NaHSO$_3$ in 6N HCl through the column, follow with six column volumes of 6N HCl, and then three column volumes of distilled water. Do not let resin become dry. When ready for the next set of samples, equilibrate by passing six column volumes of 8N HCl through the column.

6. Calculations

$$\text{Uranium alpha activity, pCi/L} = \frac{C'' - B'}{2.22 \times ERV}$$

where:

C'' = gross count rate of sample, cpm,

V = volume of sample, L, and other factors are as defined above.

7500-U C. Fluorometric Method (PROPOSED)

1. General Discussion

a. Principle: For samples containing more than 20 μg U/L, uranium is determined directly. A dried water residue is fused in a sodium fluoride-sodium carbonate-potassium carbonate flux. The material is solidified in a disk and fluorescence due to uranium is measured under ultraviolet light in a reflection-type fluorometer.

For samples containing less than 20 μg U/L, uranium is first separated from quenching elements and excessive salt concentrations by coprecipitation as uranyl phosphate on aluminum phosphate. The aluminum phosphate is dissolved in dilute nitric acid containing magnesium nitrate as a salting agent and the coprecipitated uranium is extracted into ethyl acetate. The extract is dried and fused in a sodium fluoride-sodium carbonate-potassium carbonate flux and the fluorescence is measured.

b. Interferences: The fluorescence of uranium in a fluoride matrix can be quenched or enhanced by either cations or anions. When uranium is present in low concentrations (less than 20 μg/L) these interferences can be eliminated by the coprecipitation of uranium on aluminum phosphate and subsequent uranium extraction into ethyl acetate. Carbonate ions form soluble uranium complexes that prevent coprecipitation on aluminum phosphate. Carbonates are removed by acidification and expelled from solution as carbon dioxide.

c. Sampling: Preserve sample by acidifying to pH 2 with HNO_3 at time of collection.

2. Apparatus

a. Fluorometer: A reflection-type instrument of high sensitivity.

b. Fusion dishes, gold or platinum, 0.4 mm thick × 19 mm diam.

c. Muffle furnace, with controlled temperature.

d. Microlite pipet, 100 μL.

3. Reagents

a. Aluminum nitrate, 0.24N: Dissolve 15 g $Al(NO_3)_3 \cdot 9H_2O$ in 500 mL water.

b. Ammonium hydroxide, NH_4OH, conc.

c. Diammonium hydrogen phosphate, 0.33N: Dissolve 7.26 g $(NH_4)_2HPO_4$ in 500 mL water.

d. Ethyl acetate, $CH_3COOC_2H_5$, reagent grade.

e. Flux: Mix together 9 parts sodium fluoride, NaF, 45.5 parts sodium carbonate, Na_2CO_3, and 45.5 parts potassium carbonate, K_2CO_3, by weight, in a ball mill.

f. Magnesium nitrate, 7N: Dissolve 449 g $Mg(NO_3)_2 \cdot 6H_2O$ in 350 mL water containing 32 mL conc HNO_3. Warm if necessary to dissolve. Cool and dilute to 500 mL.

g. Nitric acid, HNO_3, conc and 0.1N.

h. Phenolphthalein indicator solution, alcoholic.

i. Sodium thiosulfate, $Na_2S_2O_3$, crystals.

j. Uranium standard stock solution, 1000 μg/mL: Weigh 0.1179 g U_3O_8 into a 100-mL beaker and dissolve in 10 mL 1N HNO_3, warming on a hot plate as required. Transfer to a 100-mL volumetric flask and dilute to volume.

k. Uranium standard solution, 10 μg/mL: Dilute 5.0 mL uranium standard stock solution to 500 mL in a volumetric flask with 0.1N HNO_3.

l. Uranium standard solution, 1 μg/mL: Dilute 10.0 mL uranium standard solution, 10 μg/mL, to 100 mL in a volumetric flask with 0.1N HNO_3.

m. Uranium standard solution, 0.1 μg/mL: Dilute 10.0 mL uranium standard solution, 1.0 μg/mL, to 100 mL in a volumetric flask with 0.1N HNO_3.

4. Procedure

a. Samples containing more than 20 μg U/L: Transfer two 100-μL portions filtered sample to each of two gold dishes and evaporate to dryness under a heat lamp. To one gold dish add 100 μL uranium standard solution (0.1 μg/mL for samples with 20 to 400 μg/L or 1.0 μg/mL for samples containing more than 400 μg U/L) and evaporate to dryness under a heat lamp. Using a balance sensitive to at least 1 mg, weigh 400 ± 4 mg flux into each of the two gold dishes. Prepare a blank flux sample by weighing 400 ± 4 mg flux into a clean gold dish. Set the three gold dishes in a stainless steel support and place in a preheated muffle furnace at 625°C. After 15 min remove from furnace and cool in a desiccator for 30 min.

Read in a fluorometer as directed below.

b. Samples containing less than 20 μg U/ L: To a 1-L sample in a 1500-mL beaker, add 2 mL conc HNO_3 (if sample had not been acidified) and 5 mL each of $Al(NO_3)_3$ and $(NH_4)_2HPO_4$ solutions, and mix. If sample was chlorinated and if residual chlorine is present, add one crystal of $Na_2S_2O_3$. Heat to near boiling to expel CO_2. Add 5 drops phenolphthalein indicator solution and enough conc NH_4OH to make solution pink. Reduce heat and digest for 30 min. Remove from heat, cool, and settle for 1 h. Decant and filter clarified supernatant through a 47-mm glass fiber filter, transferring settled precipitate at the very end. Wash beaker and filter with small portions of water. Fold filter into thirds (similar to the folding of a letter) and transfer to a 50-mL screw-cap centrifuge tube. (If any precipitate remains on the inside edges of the filtering apparatus, gently wipe with the folded filter before transferring.) Add 15 mL $7N$ $Mg(NO_3)_2 \cdot 6H_2O$ to dissolve the aluminum phosphate. Add 10 mL ethyl acetate, securely cap the tube, and mix thoroughly for 1 min using a vortex mixer. Centrifuge at 2000 rpm for 5 min. Using a Pasteur transfer pipet, transfer about 9 mL of the top ethyl acetate layer to a 30-mL beaker. Repeat ethyl acetate extraction twice more. Slowly evaporate combined ethyl acetate fractions to dryness, add 1 mL conc HNO_3, and dissolve residue. Using the same Pasteur pipet, transfer the acid to a 5-mL volumetric flask. Using the pipet, wash beaker with 1 mL water and transfer to the volumetric flask. Repeat washing twice more. Gently mix, cool, dilute to volume with water, and shake thoroughly.

Treat prepared sample as indicated above for U concentrations greater than 20 μg/L.

5. Fluorometric Determination

Place gold dish containing sample plus added uranium in fluorometer. Following the manufacturer's suggested technique, adjust voltage to maximize the reading. Remove dish, insert background sample, and adjust null voltage to zero. Repeat voltage adjustment with the two gold dishes until balance is obtained between full-scale deflection and zero.

Insert gold dish containing sample and record output.

6. Calculations

Express results in micrograms per liter and calculate as follows:

$$\text{Uranium, μg U/L} = \frac{5\left[\dfrac{R_s - R_b}{R_{ss} - R_s}\right] \times a}{b \times V}$$

where:

R_s = % fluorescence of sample,
R_b = % fluorescence of blank,
R_{ss} = % fluorescence of sample with added uranium,
a = weight of added uranium, μg,
b = volume of concentrate, mL, and
V = volume of volumetric flask, mL.

If the sample has not been concentrated, delete the factors "5" and "b" from the above equation.

7. Precision and Bias

Single-laboratory precision was evaluated by replicate analyses of a sample with added uranium at a concentration of 10 μg U/L. The coefficient of variation was ±15%.

Single-laboratory accuracy was evaluated over the range of 1 to 10 μg U/L. Percent accuracy was calculated from the following equation:

$$\% \text{ Accuracy} = \frac{100 \ (x_j - x_t)}{x_t}$$

where:

x_j = measured uranium concentration in individual sample and

x_t = known uranium concentration of the sample.

Average percent accuracy, A, is calculated from the equation:

$$\% \text{ Accuracy} = \frac{\Sigma \ \% \text{ Accuracy}}{N}$$

where:

$\Sigma \ \%$ Accuracy = sum of individual accuracy determinations and

N = number of determinations.

The single-laboratory average percent accuracy was ±104%.

PART 8000

TOXICITY TEST METHODS

FOR AQUATIC ORGANISMS

8010 INTRODUCTION*

8010 A. General Discussion

1. Uses of Toxicity Tests

Toxicity tests are necessary in water pollution evaluations because chemical and physical tests alone are not sufficient to assess potential effects on aquatic biota.[1,2] For example, the interaction of chemical factors and the toxic effects of complex matrices cannot be determined. Different kinds of aquatic organisms are not equally susceptible to the same toxic substances nor are organisms equally susceptible throughout the life cycle. Even previous exposure to toxicants can alter susceptibility.

Toxicity tests are useful for a variety of purposes that include determining: (a) suitability of environmental conditions for aquatic life, (b) favorable and unfavorable environmental factors, such as DO, pH, temperature, salinity, or turbidity, (c) effect of environmental factors on waste toxicity, (d) toxicity of wastes to a test species, (e) relative sensitivity of aquatic organisms to an effluent or toxicant, (f) amount of waste treatment needed to meet water pollution control requirements, (g) effectiveness of waste treatment methods, (h) permissible effluent discharge rates, and (i) compliance with water quality standards, effluent requirements, and discharge permits.

2. Test Procedures

There is a need to use correct and uniform tests and terminology (see Section 8010B, Terminology), to apply relevant tests for meeting legal, effluent testing, monitoring, and research requirements, and to meet the uniqueness of a given environment.[3–14]

The procedures given below allow measurement of biological responses to known and unknown concentrations of materials in both fresh and saline waters. These toxicity tests are applicable to routine monitoring requirements as well as research needs. Refer to Part 9000 for microbiological methods and Part 10000 for field and other types of biological laboratory methods for water quality evaluations. Refer to Section 10900 for identification aids for aquatic organisms.

Reasonable uniformity of procedures and of data presentation is essential. The use of standardized methods described below will ensure adequate uniformity, reproducibility, and general usefulness of results without interfering unduly with the adaptability of the tests to local circumstances.

Quality assurance practices for toxicity test methods include all aspects of the test that affect the quality of the data. These include sampling and handling, source and condition of test organisms, and the test procedures themselves. While these practices are generally less defined than those for chemical procedures, guidelines are available for single compound testing and general laboratory practices[15] and for effluent evaluations in technical guidance

*Approved by Standard Methods Committee, 1989.

manuals for conducting acute and short-term chronic toxicity tests with effluents.[16-18]

3. References

1. TARZWELL, C.M. 1958. The use of bioassays in the safe disposal of electroplating wastes. *Amer. Electroplaters Soc. 44th Annu. Tech. Proc.*:60.

2. TARZWELL, C.M. 1971. Bioassays to determine allowable waste concentrations in the aquatic environment. I. Measurement of pollution effects on living organisms. *Proc. Royal Soc. London B.* 177:279.

3. BROWN, L.M. 1975. Concepts and outlook in testing the toxicity of substances to fish. *In* G.E. Glass, Bioassay Techniques and Environmental Chemistry. Ann Arbor Science Publ., Ann Arbor, Mich.

4. GREEN, J.C., W.E. MILLER, T. SHIROYAMA & T.E. MALONEY. 1975. Utilization of algal assays to assess the effects of municipal, industrial, and agricultural wastewater effluents upon phytoplankton production in the Snake River System. *Water, Air, Soil Pollut.* 4:415.

5. FRY, F.E.J. 1947. Effects of the environment on animal activity. Univ. Toronto Stud. Biol. Ser. 55, *Publ. Ont. Fish Res. Lab.* 68:1.

6. DOUDOROFF, P. et al. 1951. Bioassay methods for the evaluation of acute toxicity of industrial wastes to fish. *Sewage Ind. Wastes* 23:1380.

7. LLOYD, R. & D.H.M. JORDAN. 1963. Predicted and observed toxicities of several sewage effluents to rainbow trout. *J. Proc. Inst. Sewage Purif.* Pt.2:167.

8. BALL, I.R. 1967. The relative susceptibilities of some species of freshwater fish to poisons—I. Ammonia. *Water Res.* 1:767.

9. SPRAGUE, J.B. 1969. Measurement of pollutant toxicity to fish. I. Bioassay methods for acute toxicity. *Water Res.* 3:793.

10. Water Quality Criteria. Report of the National Technical Advisory Committee to the Secretary of the Interior. 1968. Federal Water Pollution Control Admin., U.S. Dep. Interior, U.S. Government Printing Off. I-X, Washington, D.C.

11. NATIONAL ACADEMY OF SCIENCE AND NATIONAL ACADEMY OF ENGINEERS. 1972. Water Quality Criteria. U.S. Government Printing Off., Washington, D.C.

12. COMMITTEE ON METHODS FOR TOXICITY TESTS WITH AQUATIC ORGANISMS. 1975. Methods for Acute Toxicity Tests with Fish, Macroinvertebrates, and Amphibians. Ecol. Res. Ser., EPA-660/3-75-009, U.S. Environmental Protection Agency.

13. KOPPERDAHL, F.R. 1976. Guidelines for Performing Static Acute Toxicity Fish Bioassays in Municipal and Industrial Waste Waters. Rep. to California State Water Resources Control Board, Sacramento.

14. MOUNT, D.I. & C. STEPHAN. 1967. A method for establishing acceptable toxicant limits for fish—Malathion and the butoxyethanol ester of 2,4-D. *Trans. Amer. Fish. Soc.* 96:185.

15. U.S. ENVIRONMENTAL PROTECTION AGENCY. 1987. Federal Insecticide, Fungicide and Rodenticide Act (FIFRA); Good Laboratory Practice Standards. Proposed Rule. 40 CFR Part 160; *Federal Register* 52:48920.

16. PELTIER, W.H. & C.I. WEBER, eds. 1985. Methods for Measuring the Acute Toxicity of Effluents to Freshwater and Marine Organisms. EPA-600/4-85-013, Environmental Monitoring and Support Lab., U.S. Environmental Protection Agency, Cincinnati, Ohio.

17. HORNING, W.B. & C.I. WEBER, eds. 1985. Short-Term Methods for Estimating the Chronic Toxicity of Effluents and Receiving Waters to Freshwater Organisms. EPA-600/4-85-014, Environmental Monitoring and Support Lab., U.S. Environmental Protection Agency, Cincinnati, Ohio.

18. WEBER, C.I., W.B. HORNING, D.J. KLEMM, T.W. NEIHEISEL, P.A. LEWIS, E.L. ROBINSON, J. MENKEDICK & F. KESSLER, eds. 1988. Short-Term Methods for Estimating the Chronic Toxicity of Effluents and Receiving Waters to Marine and Estuarine Organisms. EPA-600/4-87-028, Environmental Monitoring and Support Lab., U.S. Environmental Protection Agency, Cincinnati, Ohio.

4. Bibliography

ROBERTS, M.H., JR., J.E. WARINNER, C. TSAI, D. WRIGHT & L.E. CRONIN. 1982. Comparison of estuarine species sensitivities to three toxicants. *Arch. Environ. Contam. Toxicol.* 11:681.

8010 B. Terminology

An aquatic toxicity test is a procedure in which the responses of aquatic organisms are used to detect or measure the presence or effect of one or more substances, wastes, or environmental factors, alone or in combination.

1. General Terms

Acclimate—to accustom test organisms to different environmental conditions, such as temperature, light, and water quality.[1]

Response—the measured biological effect of the material tested. In acute toxicity tests this usually is death. In biostimulation tests it is biomass increase.

Control—test organisms exposed to dilution water alone and/or the natural water to which they are normally exposed.

Range assay—a preliminary assay designed to establish approximate toxicity of a solution. Test design incorporates multiple, widely spaced, concentrations with single replicates; exposure is 8 to 24 h.

Screening assay—an assay to determine if an impact is likely to be observed; test design incorporates one concentration, multiple replicates, exposure 24 to 96 h.

Definitive assay—an assay designed to establish concentration at which a particular end point occurs. Exposures for these tests are longer than for screen or range assays, incorporating multiple concentrations at closer intervals and multiple replicates.

2. Toxicity Terms

Dose—amount of toxicant that enters the organism.[2] Dose and concentration are not interchangeable.

Toxicity—adverse effect to a test organism caused by "pollutants," generally a poison or mixture of poisons. Toxicity is a resultant of concentration and time, modified by variables such as temperature, chemical form, and availability.

Exposure time—time of exposure of test organism to test solution.

Acute toxicity—a relatively short-term lethal or other effect, usually defined as occurring within 4 d for fish and macroinvertebrates and shorter times (2 d) for smaller organisms.

Chronic toxicity—long-term effects that may be related to changes in appetite, growth, metabolism, reproduction, and even death or mutations.[1]

Lethal concentration (LC)—toxicant concentration producing death of test organism. Usually defined as median (50%) lethal concentration, LC50, i.e., concentration killing 50% of exposed organisms at a specific time of observation, for example, 96-h LC50.[2]

Effective concentration (EC)—toxicant concentration affecting a specific response, such as respiration rate, loss of equilibrium, in a given time; for example, 96-h EC50.[2]

Asymptotic LC50—toxicant concentration at which LC50 approaches a constant for a prolonged exposure time.

Median tolerance limit (TLm)—test material concentration at which 50% of test organisms survive for a specified exposure time. This term has been superseded by median lethal concentration (LC50).[3]

No-observed-effect concentration (NOEC)—in a full- or partial- life-cycle test, the highest toxicant concentration in which the values for the measured parameters are not statistically significantly different from the control.

Lowest-observed-effect concentration (LOEC)—in a full- or partial- life-cycle test, the lowest toxicant concentration in which the values for the measured parameters are statistically significantly different from the control.

3. Biostimulation Terms

Limiting nutrient—the nutrient most needed for growth in relation to the quantities of other nutrients.

Nutrient—a specific substance required for organism growth.

Specific growth rate—unit rate of mass change for a population of organisms.

Maximum standing crop—maximum weight of organisms during a test, specified as wet or dry weight.

4. Flow Terms

Static test—test in which solutions and test organisms are placed in test chambers and kept there for the duration of the test.[2]

Recirculation test—static test with circulation of test solution through test chamber. Test solution may be treated by aeration, filtration, sterilization, etc., to maintain water quality.[2]

Renewal test—static test with periodic exposure (usually at 24-h intervals) of test organisms to fresh test solution of the same composition. This is accomplished by transferring test organisms or replacing test solution.[2]

Flow-through test—test in which solution is replaced continuously in test chambers for the test duration.[2]

Mixing—agitation performed by mechanical stirrers, pumps (air, water), or inflow currents. Note that aeration or vigorous mixing may increase losses of volatile substances.

5. Evaluation of Results

Maximum allowable toxicant concentration (MATC)—toxicant concentration that may be present in a receiving water without causing significant harm to productivity or other uses. MATC is determined by long-term tests of either partial life cycle with sensitive life stages or a full life cycle of the test organism.

Chronic value (ChV)—geometric mean of the NOEC and LOEC from partial- and full-life-cycle tests.

Application factor—a factor applied to acute toxicity tests to estimate toxicant concentration that is safe for chronic or lifetime exposure of test organism.[3,4]

$$\text{Application factor (AF)} = \frac{\text{MATC}}{\text{96-h LC50}}$$

Acute-chronic ratio—another way to relate acute and chronic toxicity. The acute-chronic ratio (ACR) is the inverse of the AF and is usually greater than 1; its use should prevent confusion between "application factors" and "safety factors."

6. References

1. KOPPERDAHL, F.R. 1976. Guidelines for Performing Static Acute Toxicity Fish Bioassays in Municipal and Industrial Waste Waters. Rep. to California State Water Resources Control Board, Sacramento.

2. COMMITTEE ON METHODS FOR TOXICITY TESTS WITH AQUATIC ORGANISMS. 1975. Methods for Acute Toxicity Tests with Fish, Macroinvertebrates, and Amphibians. Ecol. Res. Ser., EPA-660/3-75-009, U.S. Environmental Protection Agency.

3. NATIONAL ACADEMY OF SCIENCE AND NATIONAL ACADEMY OF ENGINEERS. 1972. Water Quality Criteria. U.S. Government Printing Off., Washington, D.C.

4. MOUNT, D.I. 1977. An Assessment of Application Factors in Aquatic Toxicology. U.S. EPA-600/3-77-085, U.S. Environmental Protection Agency, Corvallis, Ore.

8010 C.   Basic Requirements for Toxicity Tests

1. General Requirements

The basic requirements and desirable conditions for toxicity tests are: (*a*) an abundant supply of water of desired quality (see 8010E.4*b*), (*b*) an adequate and effective flowing water system constructed of nonpolluting or absorbing materials, (*c*) adequate space and well-planned holding, culturing, and testing equipment and facilities, and (*d*) an adequate source of healthy experimental organisms. Much valuable information and advice are available for planning and constructing water supply systems.[1-5]

2. Requirements for Specific Test Purposes

The facilities, equipment, and water supplies needed for effective tests depend on the type of tests and their objectives. For effluent and monitoring compliance tests, use dilution water from outside the zone of influence of the waste. Use a water supply free from pollution and choose facilities and equipment as indicated above for the following tests: (*a*) determination of most sensitive species and life stage, (*b*) effect of different toxicants, (*c*) effects of water quality and environmental factors alone and in combination with toxicants, and (*d*) maximum concentration of waste that does not taint the flesh of edible organisms.

3. References

1. U.S. DEPARTMENT OF COMMERCE. 1970. Aquarium Design Criteria, special ed. National Fisheries Center Aquarium.
2. CLARK, J.R. & R.L. CLARK, eds. 1964. Sea Water Systems for Experimental Aquariums. U.S. Fish & Wildl. Serv. Bur. Sports Fish & Wildl. Res. Rep. 63, U.S. Government Printing Off., Washington, D.C.
3. SPOTTE, S. 1973. Marine Aquarium Keeping—The Science, the Animals, the Art. Wiley Interscience Publ., New York, N.Y.
4. LASKER, R. & L.L. VLYMER. 1969. Experimental Seawater Aquarium. U.S. Fish & Wildl. Serv. Bur. Commercial Fish. Circ. 334, U.S. Government Printing Off., Washington, D.C.
5. TARZWELL, C.M. 1962. Development of water quality criteria for aquatic life. *J. Water Pollut. Control Fed.* 34:1178.

8010 D.   Conducting Toxicity Tests

1. Types of Toxicity Tests: Their Uses, Advantages, and Disadvantages

Toxicity tests are classified according to (*a*) duration—short-term, intermediate, and/or long-term, (*b*) method of adding test solutions—static, recirculation, renewal, or flow-through, and (*c*) purpose—effluent quality monitoring, single compound testing, relative toxicity, relative sensitivity, taste or odor, or growth rate, etc.

Use short-term tests for routine monitoring suitable for effluent discharge permit requirements and for exploratory tests. Short-term definitive tests determine LC50 or EC50. These tests also indicate toxicant concentrations to be used in intermediate and long-term tests. Short-term tests are valuable for quickly supplying an estimate of toxicity, for assessing relative toxicity of different toxicants or wastes to a selected test organism, or relative sensitivity of different organisms to different conditions of such variables as temperature and pH. They also indicate the maximum allowable concentrations for very short exposures, such as those that might occur to organisms passing through a thermal electric power plant or a zone of heated water.

Use intermediate tests when LC50 determination requires additional time, for studies of life stages of long-life-cycle organisms, and to indicate toxicant concentrations for life-cycle tests.

Use long-term tests for estimating the chronic or MATC values. Long-term testing may include full-life-cycle or partial-life-cycle testing. Exposures may be 7 d for rapid chronic tests involving specific portions of an organism's life cycle, 21 to 28 d for traditional partial-life-cycle assays, and several months and longer for full-life-cycle tests.

Exercise caution when static tests are used for evaluation of solutions with high BOD levels. These tests can be conducted successfully with incorporation of rigorous dissolved oxygen monitoring and acceptable aeration.

Volatile or unstable toxicants decrease in concentration during the test, so the exposure of test organisms becomes progressively less. Metabolic products may build up and undesirably high concentrations of CO_2 or NH_3 may occur. Toxicant concentration may be reduced by sorption on sediments and test chamber walls or by combination with the mucus or metabolic products of the test organisms and in their bodies.

Flow-through tests are desirable for high-BOD samples and for those that contain unstable or volatile substances. Organisms with high metabolic rates are difficult to maintain in standing water, whereas flow-through tests provide well-oxygenated test solutions, nonfluctuating toxicant concentrations, and continuous removal of metabolic wastes. Use flow-through tests whenever there is evidence or expectation of rapid toxicity changes of the test solution. Such a change is indicated when the survival time of test animals in a fresh solution is significantly shorter than the survival time in a corresponding 2-d-old solution, provided that adequate DO is present throughout both tests. Flow-through tests of industrial effluents and chemicals that are removed appreciably from solution by precipitation, by test organisms, or by other means are preferable to static tests and their modifications.

The LC50 values may be useful measures of acute toxicity but they *do not* represent concentrations that are safe or harmless in aquatic habitats subject to pollution. Concentrations of wastes that are not demonstrably toxic in 96 h may be very toxic under conditions of continuous exposure in a receiving water. Thus the 96-h LC50 may represent only a small fraction of long-term toxicity. When estimating safe discharge rates or dilution ratios for effluents or other pollutants on the basis of acute toxicity evaluations, use AFs determined by life-cycle tests. Even the provision of an apparently ample margin of safety can fail to accomplish its purpose when there is cumulative toxicity that cannot be predicted from acute toxicity results.

No single, simple AF is valid for all wastes or toxicants. The constituents of a complex waste responsible for acute toxicity may be, but are not necessarily, the constituents responsible for chronic or cumulative toxicity demonstrable in diluted waste that is no longer acutely toxic. The chronic toxicity may be lethal after a long time or it may cause only nonlethal impairment of function. Knowledge of acute toxicity of a waste often can be very helpful in predicting and preventing acute damage to aquatic life in receiving waters as well as in regulating toxic waste discharges.

2. Short-Term Toxicity Tests

a. Range-finding tests: For effluents or materials of unknown toxicity conduct short-term (usually 24-h or 48-h), small-scale range-finding or exploratory tests to determine approximate concentration range to be covered in full-scale, short-term tests. For effluents with low or slow-acting toxicity, 48- or 96-h tests may be necessary. Expose test organisms to a wide range of concentrations, usually in a logarithmic ratio, such as 0.01, 0.1, 1, 10, and 100%. Attempt to include concentrations that kill all organisms and others that kill very few or no organisms. For short-term, definitive tests, select a geometrically spaced series of concentrations between the highest concentration that killed no, or only a few, test organisms and the lowest concentration that killed most or all test organisms.

Prepare test concentrations as described in Section 8010F.2*b*.

b. Short-term definitive tests: Because death is an important, easily detected adverse effect, the most commonly used tests are for acute lethality. These tests are most appropriate for routine monitoring and checking conformity with NPDES requirements.[1]

Short-term tests may be static, renewal, recirculation, or flow-through. Exposure periods for these tests usually are 48 h or 96 h. Static or static renewal tests often are used when the test organisms are phyto- or zooplankton because these organisms are easily washed out in flow-through tests. Test solutions may be renewed daily if required because of oxygen demand, if the toxicant is unstable or volatile, or in the case of whole effluents, daily variation in the composition of the effluent. Renewals may also be less frequent. If the test material has high BOD, is volatile, or is relatively unstable, use the renewal or flow-through technique.

Test duration is determined by the toxicant and the test objectives and usually is the same for different groups of organisms. For short-life-cycle organisms such as phytoplankton, the usual exposure time can cover many generations. Determine test duration in part by the length of the life cycle. Generally, expose fish and large invertebrates in static tests for 4 d and in flow-through tests for an equal period unless composition of the toxicant is variable. In this case longer exposure may be useful to assess impacts of toxicant variability. Expose *Daphnia* for 48 h. Short-term tests have been limited arbitrarily to 96 h, but longer tests sometimes are desirable because death does not occur always within the 96-h period. When some test animals, though still alive, are dying or evidently affected after 96-h exposure, prolong the test or express the results of the test as a 96-h EC50, defining the observed effect. If tests are continued for longer periods, feed the test organisms.

For feeding requirements see Sections 8210 through 8910. Record feeding and ensure that it is equivalent in each container.

Special tests may be conducted on altered or treated samples of effluent to obtain additional toxicity information. For example, effluent dilution water mixtures may be aged 24 to 48 h before adding the test organisms, to determine changes in toxicity. When special tests are conducted, describe methods in detail.

3. Intermediate-Term Toxicity Tests

No sharp time separation exists between short- and intermediate- or between intermediate- and long-term tests. Usually tests lasting 10 d or less are considered short-term while intermediate tests last from 11 to 90 d. However, the length of the test organism's life cycle helps to determine what is short-term, intermediate, or long-term.

Intermediate-length tests may be static, renewal, or flow-through, but flow-through tests are recommended for most situations. For conduct of tests see Section 8010F.3a.

4. Long-Term, Partial- or Complete-Life-Cycle Toxicity Tests

With few exceptions, use flow-through tests with exposure extending over as much of the life cycle as possible. Continue tests from egg to egg or beyond, or for several life cycles for smaller forms. The EPA defines chronic tests with invertebrates, used in effluent evaluations, as tests in which a mean of three broods are produced.[2] Determine the maximum concentrations of toxicant not producing harmful effects with continuous exposure. The overall objective of this type of test is to determine MATCs or chronic value (ChV) of effluents, toxicants, or wastes. Use life-cycle tests to determine AF and the effects on growth, reproduction, development of sex products, maturation, spawning, success of spawning and hatching, survival of larvae or fry, growth and survival of different life stages, deformities, behavior, and bioaccumulation, although bioaccumulation (or bioconcentration) also is determined with more mature animals in specially designed tests.[3]

In life-cycle or partial-life-cycle tests, ensure that water quality factors such as temperature, pH, salinity, and DO follow the natural seasonal cycle unless the test objective is to study one of these factors. It may be essential that the natural annual cycle be duplicated if the development of sex products, spawning, and development of eggs and larvae are to be normal. Do not let toxicant concentrations vary by more than ±15% from the selected concentration because of uptake by test organisms, absorption, precipitation, or other factors.

In these tests, select five or more concentrations on the basis of short- or intermediate-term tests and set up at least in duplicate. Vary exposure chambers, spawning chambers, and other equipment to meet the needs of the different organisms. (See Sections 8111 through 8910.) Other apparatus, water supplies, and analytical determinations are listed in Section 8010E.

5. Short-Term Tests for Estimating Chronic Toxicity

Tests are available to estimate long-term effects of a toxicant or effluent after a relatively short (7-d) exposure. End points for the tests, called chronic estimator or rapid bioassessment tests, include lethality, reproductive potential, and growth. Methods for fresh- and saltwater invertebrates and fishes are being used widely.[4,5]

6. Special-Purpose Toxicity Tests

a. Relative sensitivity to a toxicant: To rank the sensitivity of different species to a toxicant, use a standard water and standard exposure conditions. Select exposure conditions (e.g., temperature, DO, pH, CO_2, and salinity) in a favorable range for the test species and keep conditions constant throughout the test.

b. Relative sensitivity of various toxicants to selected species: These tests resemble sensitivity tests because the selected test conditions, dilution waters, and test species are kept constant and standard. Prevent any change in sensitivity of test organisms during the tests. If possible, select species from several different groups: an alga, microcrustacean, macrocrustacean, insect, mollusk, or fish.

c. Flesh tainting tests: CAUTION: *Perform such tests only when there is assurance that the intake of potentially tainting substances through consumption of the organism is safe. In cases where sufficient information on the substances is not available, replace consumption with smell.* Use these tests to determine the maximum concentrations of wastes and materials that do not taint the flesh of edible aquatic organisms. Expose organisms that are large enough to supply portions for a taste panel. Set up exposure tanks as for other flow-through tests. Perform range-finding tests over a wide concentration range to determine the concentrations for a more definitive series of tests.

After exposure, prepare test organisms for taste testing. Clean, prepare for cooking, wrap in aluminum foil, and bake in an oven. When organisms are cooked, divide them into portions, wrap in aluminum foil, assign a code number, and distribute to a taste panel while still warm, along with samples of unexposed organisms similarly cooked, wrapped, and coded. Record the observations of the panel on a prepared form and determine the highest concentration of test material not causing detectable tainting based on either taste or smell. Several tests may be necessary.[6-9]

d. Growth-rate determinations: Growth rate is an important response of both algae and fish to toxicants and environmental factors. This section discusses the topic with respect to fish.[10] For a discussion related to algae see Section 8111G.3c. Always report details of the method of feeding fish in growth studies. Three techniques are available:

Unrestricted food supply—Provide attractive and palatable food (usually live food such as *Daphnia* or tubificid worms) uninterruptedly in greater quantities than fish can consume. Make a mass balance on food consumed by weighing food introduced and uneaten food removed.

Intermittent satiated food supply—Provide all the attractive food that fish can consume at time of feeding once or twice daily. After fish cease to feed, remove all uneaten food.

Uniformly restricted food supply—Once per day provide all fish with an amount of food that they will consume completely and without exception. Ideally, hold fish separately in individual aquariums or compartments. For fish held together feed so that all fish have an equal opportunity to consume food. Uniformity of temperature and DO helps to ensure equal feeding of a group of fish.

While growth studies usually have been conducted with unrestricted and intermittent satiated feeding techniques, it is recommended that each study include at least one test series using uniformly restricted food supply. Only this technique can reveal whether growth rate differences are not the result of the effect of the toxicant on appetite or food consumption rate. The presence of an abundant food supply can obscure toxic effects. For example, fish exposed to toxicants such as cyanide or pentachlorophenol increase food consumption rate to compensate partially for loss of efficiency of food utilization caused by the toxicant. This may not be possible in natural conditions where food supply may be limited.

Ideally, include a series of tests with different, uniformly restricted food rations with the lowest ration near that which results in no growth (or loss of weight) in the control. This is the maintenance level. Determine the effect of the variable under study at any level of food availability and consumption by relating observed growth rates to, for example, toxicant concentration, at each feeding level.[11]

Juvenile fish may gain enough weight in 2 to 3 weeks to determine growth rate satisfactorily. Longer exposures with weighings at intervals of approximately 10 d are needed to determine long-term effects such as acclimation or accumulative toxicity.

Report results as growth rates computed as follows:

$$\text{Growth rate} = \frac{\text{weight gain, g}}{\text{time interval, d}} \times \frac{1}{\text{mean weight, g}}$$

where:

mean weight
= [weight at start of time interval (g)
+ weight at end of time interval (g)]
÷ 2

Determine dry weight, wet weight, and fat (lipid) content of fish at the beginning and end of a test. Weight gain due to increased fat content is not universally considered true growth; some investigators consider that true growth occurs only when there is an increase of protein. However, fat storage is important ecologically and bioenergetically because fat can be used as an energy source during periods of malnutrition and weight loss.[11] Fat content also is important in understanding the dynamics of toxicant uptake, storage, and depuration.

7. References

1. KOPPERDAHL, F.R. 1976. Guidelines for Performing Static Acute Toxicity Fish Bioassays in Municipal and Industrial Waste Waters. Rep. to California State Water Resources Control Board, Sacramento.
2. PELTIER, W.H. & C.I. WEBER, eds. 1985. Methods for Measuring the Acute Toxicity of Effluents to Freshwater and Marine Organisms. EPA-600/4-85-013, Environmental Monitoring and Support Lab., U.S. Environmental Protection Agency, Cincinnati, Ohio.
3. AMERICAN SOCIETY FOR TESTING AND MATERIALS. 1987. Standard practice for conducting bioconcentration tests with fishes and saltwater bivalve molluscs. E-1022-84, Annual Book of ASTM Standards, Vol. 11.04. American Soc. for Testing & Materials, Philadelphia, Pa.
4. HORNING, W.B. & C.I. WEBER, eds. 1985. Short-Term Methods for Estimating the Chronic Toxicity of Effluents and Receiving Waters to Freshwater Organisms. EPA-600/4-85-014, Environmental Monitoring and Support Lab., U.S. Environmental Protection Agency, Cincinnati, Ohio.
5. WEBER, C.I., W.B. HORNING, D.J. KLEMM, T.W. NEIHEISEL, P.A. LEWIS, E.L. ROBINSON, J. MENKEDICK & F. KESSLER, eds. 1988. Short-Term Methods for Estimating the Chronic Toxicity of Effluents and Receiving Waters to Marine and Estuarine Organisms. EPA-600/4-87-028, Environmental Monitoring and Support Lab., U.S. Environmental Protection Agency, Cincinnati, Ohio.
6. SURBER, E.W., J.N. ENGLISH & G.N. McDERMOTT. 1962. Tainting of fish by outboard motor exhaust wastes as related to gas and oil consumption. PHS Publ. No. 999-WP-25, Environmental Health Ser.:170, U.S. Public Health Serv., Washington, D.C.
7. AMAN, C.W. 1955. The relation of taste and odor to flavor. *Taste Odor Control J.* 21(10):1.
8. BALDWIN, R.E., D.H. STRONG & J.H. TORRIE. 1961. Flavor and aroma of fish taken from four fresh-water sources. *Trans. Amer. Fish. Soc.* 90:176.
9. DAWSON, E.H. & B.L. HARRIS. 1951. Sensory Methods for Measuring Differences in Food Quality. Agr. Inform. Bull. 34, U.S. Dep. Agriculture, Washington, D.C.
10. BROWN, M.E., ed. 1957. The Physiology of Fishes. Vol. 1, Metabolism. Academic Press Inc., New York, N.Y.
11. WARREN, C.E. & P. DOUDOROFF. 1971. Biology and Water Pollution Control. W.B. Saunders Co., Philadelphia, Pa.

8010 E. Preparing Organisms for Toxicity Tests

1. Selecting Test Organisms

The prime considerations in selecting test organisms are: their sensitivity to the factors under consideration; their geographical distribution, abundance, and availability within a practical size range throughout the year; their recreational, economic, and ecological importance; the availability of culture methods for rearing them in the laboratory and a knowledge of their requirements; and their general physical condition and freedom from parasites and disease. To select a best species consider available information on sensitivity or determine sensitivity with short-term tests. Select the test species based on the considerations listed above as well as organism size and life-cycle length. Generally, organisms not over 5 to 8 cm long and having a short life cycle are most desirable, but some tests require larger organisms with long life cycles.

For studies to determine effluent effects, select species representative of fauna in the area impacted. In most cases, use of laboratory-cultured species is preferable to use of those collected from the field. Laboratory-cultured organisms are of known age and quality. This allows for use of the most sensitive life stages throughout the year. Their use is also more cost-effective and allows for better quality assurance and quality control. When using field-collected organisms make comparative tests to relate sensitivity of the selected test species to the most sensitive locally important species. For each series of tests, collect organisms at one time from a single source. Use organisms that are nearly uniform in size with the largest individual not more than 50% longer than the shortest. Use organisms of the same age group or life stage. Report time, place, source and culture history for cultured organisms and method of collection, transportation, handling and acclimation.

In designing a test, consider any unusual past conditions to which the organisms have been exposed (pesticides, effluents from industries, waste treatment plants, return flows, etc.). Interactive effects of a new toxicant mixed with those presently being discharged to the receiving water may be important. Do not collect test organisms from polluted areas where they are in poor condition, diseased, parasitized, or deformed, or where they have unusually high body burdens of chemicals.

Knowledge of the environmental requirements and food habits is important in selecting test organisms. Because methods for laboratory holding and culturing throughout the life cycle are limited, it may be necessary to collect certain life stages of selected organisms from the field for testing.

2. Collecting Test Organisms

Many smaller invertebrates and fish can be collected along the shore in dip nets, in coarse plankton nets, or by hand. Catch larger species that occur near shore in seines. Traps and fyke nets are good for freshwater areas but are selective for some species. Use various types of trawls to sample in the sea. Otter trawls are effective for collecting benthic species and midwater trawls for pelagic species. Various dredges are available to collect benthic species from different types of bottoms or to collect different sizes of organisms. Catch commercially important species such as lobster, blue crab, and dungeness crab in traps or take by deep-water trawls. Harvest species that colonize surfaces, such as barnacles, from plates of wood, plastic, or glass suspended in the water. Insure that organisms are not damaged during collection, transfer, and transport. When seining or using trawls, make short hauls. Do not collect significant amounts of plant materials, debris, mud, sand, or gravel in net or in bag of seine because these will injure the animals. Always leave seine bag in the water at end of haul, stretch out wings of seine, open bag entrance, dip out organisms with a bucket, and transfer directly to prepared holding tanks. Do not expose delicate, easily damaged species to air. Take out larger, more hardy species with soft mesh dip nets. Do not collect too many animals at one time. After bringing a trawl up to the boat, very quickly bring it over the side without letting the catch hit the boat. Immerse that portion of net containing collection in a tank of water. Open trawl and remove desired animals by dipping with a bucket or a hand net with small soft mesh. Have adequate quantities of clean water available in tanks before beginning a haul. Transfer organisms to tanks as rapidly and carefully as possible. If organisms are to be transported any distance by boat, hold in aerated live boxes. If they are transported by truck, put them in large baffled and insulated tanks filled with water from area in which they were collected. Aerate the water and maintain at temperature of collection. Determine water temperature, salinity, DO, and pH at the collecting site. Do not handle organisms more than necessary. Make transfers with suitable containers or hand nets, or for small organisms, by large-bore pipets. Use hand nets made of soft material with several layers around the net rim and free from sharp points or projections. Clean and sterilize all equipment before use. Do not crowd organisms during transport. Oxygenation, water exchange, and cooling may reduce distress.

Observe collected animals for possible injury resulting from transport to the laboratory. Examine smaller forms under a dissecting microscope. Criteria for assessing injury depend on the species and are more difficult for sluggish ones. Useful criteria include loss of appendages, inability to maintain a normal body posture (e.g., dorsal side uppermost), abnormal locomotion, refusal to feed, discoloration, or uncoordinated movements of the mouth or other body parts.

For additional information on collecting aquatic organisms, see Part 10000.

3. Handling, Holding, and Conditioning Test Organisms

During transport to the laboratory, organisms sometimes are crowded, bruised, and otherwise stressed, thereby increasing their susceptibility to disease. To avoid introducing disease into stock tanks, treat organisms during transit or on arrival in accordance with procedures in ¶ 5 below and as suggested for each of the different groups (Sections 8210 through 8910). Hold field-collected fish in quarantine for at least 7 d to observe for parasites and disease,

and to recover from collection and transport stress; observe invertebrates for at least 2 d. If more than 10% of the collected animals die after the second day or if they are parasitized or diseased beyond control, do not use them. Clean and sterilize all contacted containers and equipment and collect another supply from a different area if possible.

Because it is not always possible to collect from unpolluted areas and the collector cannot always be sure that a particular organism has not been exposed to a toxicant, sample collected individuals to determine if they have accumulated pesticides, heavy metals, and toxic materials to be studied. Check animals or materials collected as food for test organisms for disease and content of pesticides, heavy metals, and toxic materials to be studied. Feed test organisms daily during quarantine.

After quarantine period, transfer disease-free animals to regular stock tanks. Discard organisms that touch dry surfaces, are dropped, or are injured during handling. To avoid unnecessary stress, do not subject organisms to rapid temperature or water-quality changes. In general, change water temperature less than 3°C in any 24-h period. For stenothermal, deep-water species use an even smaller rate of temperature change. Preferably keep DO concentrations at or near saturation but never at less than 60% of saturation. After transfer to stock holding tanks, begin a slow acclimation to laboratory conditions such as temperature, salinity, and hardness. The period of acclimation will be governed by type of organism and extent of changes in water quality. For forms with a life cycle of several months or more, use an acclimation period of at least 2 to 3 weeks. Inspect organisms closely and frequently to determine stress, unusual behavior, parasites or disease, changes in color, or failure to eat. Avoid crowding. Provide adequate flow-through water so that characteristics such as DO, pH, CO_2, salinity, hardness, and NH_3 are favorable. Check temperature and DO frequently. Do not let metabolic products accumulate. Generally, use a flow-through rate of 6 to 10 tank volumes/ d. Usually, greater amounts of flow-through water are required for smaller organisms on a weight-volume basis. For small organisms, use a water flow of at least 3 L/d/g. When brood stock are being held, periodic or continuous treatment for parasite and disease control may be required.[1,2] Clean tanks and equipment thoroughly and often, removing or flushing out all growths and wastes, preferably daily but at least twice per week. Remove all uneaten food within 24 h. Use different sets of nets and other equipment for different groups of organisms and clean and sterilize them between uses. When handling is necessary, clean hands and nets before touching organisms. Cover tanks and containers to prevent organisms from jumping out. Shield tanks with curtains or by some other means to protect organisms from nearby movements and noise. Provide photoperiods and light intensities favorable to the organisms (see Section 8010F.3f). Begin acclimation to test conditions at a suitable interval in advance of testing.

It is of utmost importance that animals be kept in excellent condition before the tests. Make no abrupt changes in environmental conditions; preferably follow natural seasonal variations in environmental conditions such as temperature and daylight patterns. Do not supersaturate with gases, especially in winter when very cold water is brought into the laboratory and warmed. If there is danger of gas bubble disease, keep incoming water in an open system and let it cascade over baffles or otherwise aerate it to bring dissolved gases into equilibrium with the air.[3]

Acclimate freshwater arthropods by rearing them in the dilution water at the test temperatures, unless temperature is one of the factors being studied. Acclimate other organisms to the dilution water and test temperatures by gradually changing the water from 100% holding water to 100% dilution water over a period of several days. Keep all organisms in 100% dilution water for at least 2 d before use. Do not use a group of organisms if more than 5% die during the 48 h immediately before the beginning of the test.[4] If a group fails to meet these criteria discard or re-treat, hold, and reacclimate if necessary.

Make necessary provisions for organisms that require a special substrate, cover, or materials to use for clinging, support, the building of cases, or hiding.

Hold cold-water, freshwater organisms between 5 and 15°C. Hold warm-water organisms between 10 and 25°C, depending on season. Hold aquatic invertebrates within the temperature range of the water from which they were obtained unless they are being acclimated for special temperature tests.

4. Culturing Test Organisms

The advantage of cultured test organisms over field-collected animals is that the age, life history, and existing conditions are documented and thus these organisms are more likely to behave predictably. For organisms used extensively in effluent biomonitoring programs, EPA has developed a series of methods[5-7] that are adaptable to most laboratories. Culturing of test organisms requires strict adherence to standard protocol, 7-d/week monitoring, and an adequate facility.

a. Facilities, construction materials, and equipment: Do not use construction materials in contact with dilution water that contain leachable substances or adsorb significant amounts of substances from the water. Stainless steel probably is the best construction material for freshwater systems. Glass significantly adsorbs some trace organics. Do not use rubber or plastics containing fillers, additives, stabilizers, plasticizers, etc. Fluorocarbon plastic, nylon, and their equivalents usually are acceptable. Glass-fiber-reinforced polyester, polyester resin, and epoxy resins also may be used. Test the toxicity of all materials before purchasing large quanitities. Clean and flush all new tanks, troughs, and similar equipment with dilution water for several days before use. Use a glass or titanium interface between the water and heating elements for marine waters and glass or stainless steel for fresh waters. If concrete tanks are used, leach with frequent water changes over a period of weeks before use.

Provide adequate space for test organisms, holding facilities, water storage reservoirs, and water supply systems. Provide distribution of hot and cold water and mixing facilities to obtain any desired temperature. Aerate or vigorously mix to prevent gas supersaturation caused by heating dilution water. Use air compressors with water seals and filters to prevent oil from entering air lines and contaminating tanks. When large volumes of air are needed, use low-pressure blowers. Do not locate air intakes in shops or furnace rooms or near outlets from hoods, chemical laboratories, or vehicle exhausts. Provide acclimation and culturing tanks with temperature control and aeration. Design holding facilities for ease of cleaning and prevention of bacterial growths. For holding and culturing fish and many macroinvertebrates, preferably use round tanks of at least 1 to 3 m diam (Figure 8010:1). Provide a standpipe drain in the center, threaded below the tank floor so that, when the standpipe is removed, the opening is flush with the tank bottom. Slope tank bottom gently to center. Use tanks with smooth surfaces to facilitate cleaning, to prevent injuries to organisms, and to insure that no material will collect in corners, cracks, and crevices. Introduce water into a circular tank as a jet along the edge and above the surface to create a circular movement of water around the central standpipe. Fit another pipe, with half-moon cutouts at its base, over the standpipe

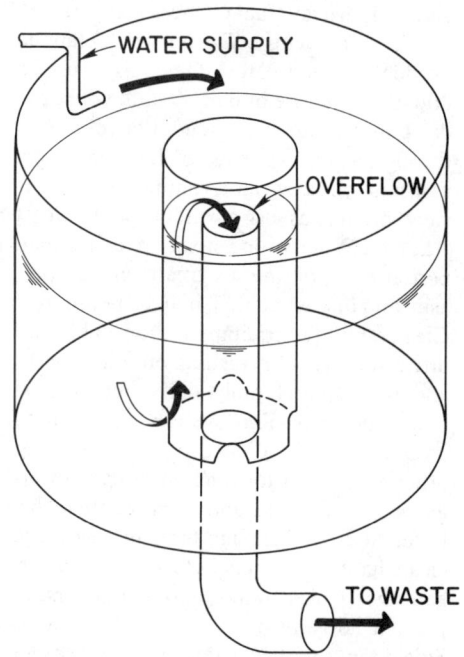

Figure 8010:1. Holding tank design for fish and macroinvertebrates.

and screen, so that the outflowing water passes up through the outside pipe, then down the standpipe. This results in a circular current and a certain amount of self-cleaning. Square or rectangular tanks may be used for special purposes or when space is scarce. Provide standpipes at one end for draining, with threads for securing the pipe on the underside. Ensure that tank corners are rounded and that surfaces are smooth.

b. Water supply: Provide a flowing water system for holding, spawning, and rearing a variety of aquatic organisms. Reconstituted fresh water and artificial seawater are not cost-effective for large-scale rearing or for flow-through tests. Use natural, unpolluted water supplies that have low turbidity, high DO, low BOD, and an annual temperature cycle that approximates that of the test organisms.

1) Freshwater supplies—A good fresh-

water supply is constant in quality and does not contain more than the designated amounts of the following: suspended solids, 20 mg/L; total organic carbon or chemical oxygen demand (TOC or COD), 10 mg/L; un-ionized NH_3, 20 μg/L; total residual chlorine, 0.01 mg/L; total organophosphorus pesticides, 50 ng/L;* total organochlorine pesticides plus PCBs, 50 ng/L.* Consider water to be of constant quality if the monthly ranges of hardness, alkalinity, conductivity, TOC or COD, and salinity are less than 10% of the average values and the pH range is less than 0.4 units.

Check municipal water supplies to determine their acceptability from the standpoint of, for example, copper, lead, zinc, fluoride, and free or combined chlorine concentrations. If a satisfactory freshwater supply is not available or if a standard water is required for comparative toxicity tests, relative sensitivity tests, or tests to determine the effects of hardness, pH, or total alkalinity on the toxicity of various materials, use a reconstituted standard water.

Prepare standard fresh water (Tables 8010:I and 8010:II) by adding reagent-grade chemicals to glass-distilled and/or deionized water. For special studies, determine that the distilled and/or deionized water contains less than the indicated constituents:

Conductivity	1 μmho/cm
Total organic carbon (TOC)	1 mg/L
or chemical	
oxygen demand (COD)	2 mg/L
Boron, fluoride	100 μg/L each
Un-ionized ammonia	20 μg/L
Aluminum, arsenic, chromium,	
cobalt, copper, iron, lead,	
nickel, zinc	1 μg/L each
Total residual chlorine	3 μg/L

Cadmium, mercury, silver	100 μg/L each
Total organophosphorus pesticides	50 ng/L*
Total organochlorine pesticides plus polychlorinated biphenyls (PCBs)	50 ng/L*

Carbon-filtered deionized water usually is acceptable. Determine conductivity of distilled and/or deionized water for each batch of reconstituted water. Check other constituents periodically. If the water is prepared from a dechlorinated water (use sodium bisulfite), test the reconstituted water to determine that first instar daphnids survive for 48 h (Section 8711).[4]

The pH, alkalinity, and hardness of a receiving water influence toxicity of some materials, especially metals. Therefore, it is desirable to have a supply of both hard and soft waters.

It is advantageous to have water with temperatures between 3 and 7°C during the winter and between 20 and 25°C at peak summer temperatures. For general use, the pH should be in the range of 7 to 8.2 and dissolved CO_2 should be 1 mg/L or less.

2) Marine water supplies—Use unpolluted marine water with low turbidity and settleable solids and a pH and salinity favorable for the test organism. Ensure that annual salinity variations are not so wide as to be harmful to the organisms. In general, it is preferable to have a source of higher-salinity water (e.g., ocean water) from which brackish water can be prepared by dilution.

If a suitable marine water supply is not available, use artificial seawater for limited culturing and toxicity testing. Prepare artificial seawater by adding the compounds listed in Table 8010:III to 800 mL glass-distilled or deionized water, in the order listed. Make sure that each salt is dissolved before adding the next. Make up to 1 L with distilled or deionized water. The salinity should be 34 $\pm$ 0.5 g/kg and pH 8.0 $\pm$ 0.2. Obtain desired salinity at time of

*No individual pesticide should exceed the allowable concentration limit set in the National Water Quality Guidelines, EPA, as set in accordance with the Federal Water Pollution Control Act 92-500 as amended 1972.

TABLE 8010:I. RECOMMENDED COMPOSITION FOR RECONSTITUTED FRESH WATER

Water Type	Salts Required mg/L				Water Quality		
	NaHCO$_3$	CaSO$_4$· 2H$_2$O	MgSO$_4$	KCl	pH*	Hardness mg CaCO$_3$/L	Alkalinity mg CaCO$_3$/L
Very soft	12	7.5	7.5	0.5	6.4–6.8	10–13	10–13
Soft	48	30	30	2.0	7.2–7.6	40–48	30–35
Moderately hard	96	60	60	4.0	7.4–7.8	80–100	60–70
Hard	192	120	120	8.0	7.6–8.0	160–180	110–120
Very hard	384	240	240	16.0	8.0–8.4	280–320	225–245

* Approximate equilibrium pH after aeration with fish in water.

use by dilution with deionized water. Alternatively, prepare marine water from commercially available salt mixes.

To increase salinity of a natural water, use a strong natural brine prepared by freezing and then partially thawing seawater. Natural brine also may be prepared by evaporation of natural seawater with aeration and low heat (maximum of 35°C). This is satisfactory if only limited amounts of water are needed; for larger volumes, use commercial sea salts or a stronger solution of artificial seawater.

In the preparation of artificial seawater, be sure that an undesirable concentration of metals does not occur. Even reagent chemicals contain traces of several metals and their extensive use can result in a buildup of metals. If large volumes of artificial seawater are not required, remove

TABLE 8010:II. QUANTITIES OF REAGENT-GRADE CHEMICALS TO BE ADDED TO AERATED SOFT RECONSTITUTED FRESH WATER FOR BUFFERING pH[4,8]

Desired pH*	Quantity of Chemical to Be Added mL/L water		
	1.0N NaOH	1.0M KH$_2$PO$_4$	0.5M H$_3$BO$_3$
6.0	1.3	80.0	—
6.5	5.0	30.0	—
7.0	19.0	30.0	—
7.5	—	—	—
8.0	19.0	20.0	—
8.5	6.5	—	40.0
9.0	8.8	—	30.0
9.5	11.0	—	20.0
10.0	16.0	—	18.0

* Approximate equilibrium pH with fish in water. Do not aerate after adding these chemicals.

TABLE 8010:III. PROCEDURE FOR PREPARING
RECONSTITUTED SEAWATER*[9,10]

Compound in Order of Addition	Final Concentration mg/L
NaF	3
$SrCl_2 \cdot 6H_2O$	20
H_3BO_3	30
KBr	100
KCl	700
$CaCl_2 \cdot 2H_2O$	1 470
Na_2SO_4	4 000
$MgCl_2 \cdot 6H_2O$	10 780
NaCl	23 500
$Na_2SiO_3 \cdot 9H_2O$	20
Na_4EDTA*	1
$NaHCO_3$	200

* Tetrasodium ethylenediaminetetraacetate. Omit when toxicity tests are conducted with metals. Omit when tests are conducted with plankton or larvae. Strip the medium of trace metals.[11,12]

the metals by passing the seawater through a column containing a cation-exchange resin in the sodium form.

The suitability of any artificial saltwater is enhanced by aging, with aeration, and by the introduction of nitrogen-fixing bacteria from water in which healthy aquatic organisms have been maintained.

c. Food and feeding:

1) Culture of microorganisms—Phytoplankton and zooplankton may be cultured as test organisms for studying biostimulation, toxicity, etc. They may be cultured also as food for other organisms such as copepods, *Daphnia* and other microcrustaceans, the larvae and adults of mollusks, and young and adult fish.

a) Culture medium for freshwater algae—Prepare reconstituted fresh water by adding reagent-grade macro- and micronutrients to glass-distilled and/or deionized water in the concentrations given in Tables 8010:IV.A and B.

Prepare a separate stock solution of each macronutrient salt in 1000 times the specified final concentration in glass-distilled or deionized water. Combine trace metals and

EDTA in a single micronutrient stock solution in glass-distilled or deionized water at 1000 times the final concentration of each. Note that in toxicity testing of metals the effect of EDTA must be established.

To prepare algal culture medium, add 1 mL of each macronutrient stock solution to 900 mL glass-distilled or deionized water, then add 1 mL trace metal EDTA mixture and make up to 1 L with glass-distilled or deionized water.

Whenever axenic algal cultures are used, or if bacterial growth interferes with the test, prepare media aseptically: To 800 mL glass-distilled or deionized water add 1 mL of each macronutrient stock solution in the order listed, mixing after each addition. Filter-sterilize by passing through a sterile 0.2-μm-porosity membrane filter (pre-rinsed with 100 mL double-distilled water) into an autoclave-sterilized container. Add 1 mL filter-sterilized micronutrient solution, and make up to 1 L with sterile distilled or deionized water.

Store uninoculated sterile reference medium in the dark to avoid photochemical changes.

When sterility is desired in algal tests, check sterility periodically by adding 1 mL inoculated test culture to tubes of sterile nutrient test medium and incubate in the dark at the test temperature for 2 weeks. The appearance of opalescence in the test medium indicates contamination.

Prepare sterile bacterial nutrient test medium by adding the following quantities of chemicals to 1 L glass-distilled water:

Sodium glutamate	250 mg
Sodium acetate	250 mg
Glycine	250 mg
Sucrose	250 mg
Sodium lactate	250 mg
DL alanine	250 mg
Nutrient agar	50 mg

Bring medium to a boil, dispense to test tubes, and sterilize by autoclaving.

TABLE 8010:IV.A. MACRONUTRIENT STOCK SOLUTION

Compound	Concentration mg/L	Element	Resulting Concentration mg/L
NaNO₃	25.5	N	4.20
		Na	11.0
NaHCO₃	15.0	C	2.14
K₂HPO₄	1.04	K	0.469
		P	0.186
MgSO₄·7H₂O	14.7	S	1.91
		Mg	2.90
MgCl₂	5.70		
CaCl₂·2H₂O	4.41	Ca	1.20

b) Culture medium for marine algae—To artificial seawater (Table 8010:III) add nutrients listed in Table 8010:V to give the indicated concentrations in the algal culture medium.

When an unpolluted seawater is used, prepare the medium by enriching filter-sterilized seawater with micronutrients at one-half the indicated concentrations. If sterile techniques are required, follow procedures for fresh water. When sterilization is performed by autoclaving, add vitamins after autoclaving. When filter-sterilization is used, use a positive pressure of 72 kPa.

2) Mass production of algae as food for other organisms—Rearing zooplankton, various filter-feeders, the larvae of crustaceans, and fish requires large quantities of phytoplankters. These needs must be met with an apparatus capable of producing continuous amounts of desired organisms at high densities. Such an apparatus (Figure 8010:2) permits easy assembly, cleaning, and sterilization and efficient utilization of light energy, and is constructed for continuous use.†

---

†For details of construction and use, contact Dr. Richard Steele, U.S. EPA Environmental Research Laboratory, Narragansett, R.I. 02882.

TABLE 8010:IV.B. MICRONUTRIENT STOCK SOLUTION

Compound	Concentration μg/L	Element	Resulting Concentration μg/L
H₃BO₃	186	B	32.5
MnCl₂	264	Mn	115
ZnCl₂	3.27	Zn	1.57
CoCl₂	0.780	Co	0.354
CuCl₂	0.009	Cu	0.004
Na₂MoO₄·2H₂O	7.26	Mo	2.88
FeCl₃	96.0	Fe	33.0
Na₂EDTA·2H₂O	300	—	—

TABLE 8010:V. NUTRIENTS FOR ALGAL
CULTURE MEDIUM IN SEAWATER

Compound	Concentration		Concentration of Nutrient
NaNO$_3$	25.0	mg/L	4.2 mg N/L
K$_2$HPO$_4$	1.05	mg/L	0.19 mg P/L
FeCl$_3$	72.6	μg/L	
MnCl$_2$	2.30	μg/L	
ZnCl$_2$	2.10	μg/L	
Na$_2$MoO$_4$·2H$_2$O	2.50	μg/L	
CuCl$_2$	0.20	μg/L	
Na$_2$EDTA	300	μg/L	
Vitamins:			
Thiamine	0.100	mg/L	
Biotin	0.50	μg/L	
B$_{12}$	0.50	μg/L	

The main body of the unit is a 60-cm
section of 15-cm-diam borosilicate glass
drainage pipe. The top section is a 15-cm
to 5-cm concentric reducer and the bottom
section is a 15-cm to 5-cm ell. Each section
accommodates a No. 12 silicone rubber
stopper held in place by a carboy or similar
type clamp. Hold the three sections together by aluminum ring clamps and seal
adjoining surfaces by silicone "O" rings
made from small-diameter tubing. Use material that is autoclavable and nontoxic. Set
up assembly, light source, culture medium
supply bottle, air filter, pumps, and stands
as shown in Figure 8010:3.

Use this or similar device to supply cells
on a periodic or continuous basis. As the
cells are withdrawn, add more medium.
The following species have been grown at
the indicated concentrations: *Skeletonema
costatum,* 4.3 × 10^6 cells/mL; *Dunaliella
tertiolecta,* 4.4 × 10^6 cells/mL; *Isochrysis
galbana,* 7.0 × 10^6 cells/mL; *Monochrysis
lutheri,* 5.0 × 10^6 cells/mL.

3) Food for macroinvertebrates and
fish—A suitable food is essential for rearing
various macroinvertebrate and fish life
stages. Distinguish carnivores from herbivores to supply the correct type of food.
Organisms taken as food differ for different

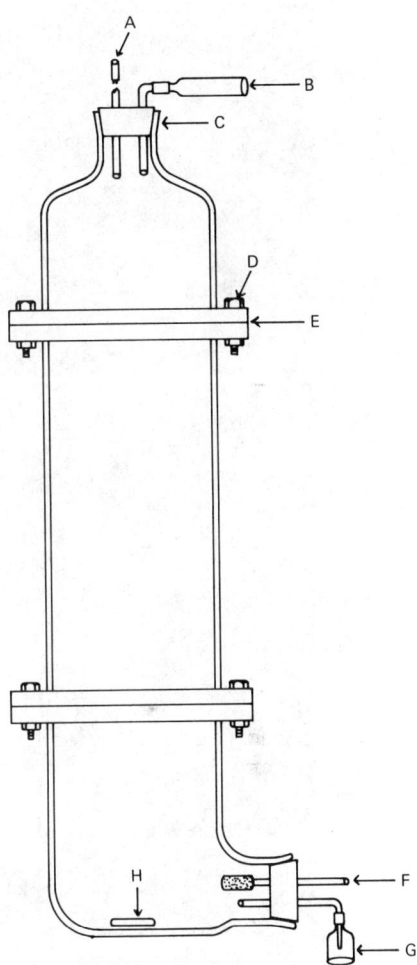

Figure 8010:2. Algal culture units. A—medium
inlet tube; B—air exhaust tube;
C—No. 12 silicone stopper with
two holes (may be held in place
with carboy clamp); D—aluminum flange clamp ring; E—silicone tube, 0.24 cm OD, in a
groove between adjoining glass
surfaces; F—air inlet tube; G—
aseptic filling bell for withdrawing
cells from culture; H—magnetic
stirring bar.

life stages of a species. As organisms grow
they require progressively larger food organisms. Many feed on pelagic organisms

Figure 8010:3. Method of lighting, free-standing frames, and placement of medium source, air pumps, and other apparatus for mass algal culture devices.

whose movements should be sufficient to attract the predator but slow enough so they can be caught readily. Use food organisms that are nutritious, easily digested, uncontaminated, and readily obtainable. Distribute zooplankton food in rearing tanks to match distribution of organisms using it. Provide an adequate amount of food with a ratio of number of prey to predator varying from 50:1 to 200:1. If a small number of organisms is reared in a large tank, provide more food organisms

to insure that enough are captured. Some algae and diatoms used for food have a tendency to settle to the bottom. Circular movement of the water in rearing tanks, provided as described in Section 8010E.4a, keeps food materials in suspension.

When using cultured microorganisms as food, be aware of possible environmental changes they may cause. In addition to the possible presence of toxic metabolites, algal blooms may produce excess oxygen and result in supersaturation and gas bubble disease.[3] Use live food whenever possible. Analyze food for toxicants, especially pesticides and heavy metals. Supplement natural foods with commercially available dried and pelleted foods.‡ (See 8910 for fish feeding.) These foods should be attractive to the organisms, supply necessary nutrients and trace elements, and contain binders to insure pellet stability.[13] Nutritionally deficient diets may cause significant differences in sensitivity of test organisms to toxicants.

Methods for rearing freshwater organisms have been described by Needham et al.[14] Methods for rearing larvae of marine animals with special reference to their food organisms have been summarized by Hirano and Oshima.[15] May[16] has reviewed the literature on laboratory feeding larvae of marine fish. Braughn and Schoettger[17] have reviewed needs of freshwater fish. Piper et al.[18] discuss standard fish hatchery culture facilities and operations.

d. Cleaning containers and equipment:

1) Cleaning holding, acclimation, testing, and dilution water tanks—Clean test containers and toxicant delivery systems before use. Wash new containers with laboratory detergent and rinse with 100% acetone, water, acid (such as 5% HNO_3), and twice with tap water. After each test wash the system appropriately, e.g., acid to remove metals and bases; detergent, sodium hypochlorite § (NaOCl) solution (200 mg/ L), organic solvent, or activated carbon to remove organic compounds, etc. Immediately before testing, rinse again with dilution water.[4] Labor-saving devices, useful when many tests are performed, are available.

‡Some foods that have been widely used include Glenco Trout Food, Glenco, Minn. 53336; Biorell and Tetraminc, available from local pet shops; Oregon Moist, Warrenton, Ore. 97146; and Cerophyll, Cerophyll Laboratories, Inc., Kansas City, Mo. 64111. The latter has been used for the small forms and as a food for organisms providing food for the higher species.

§Do not use acid and hypochlorite together.

2) Removal of unused food and wastes—Do not let unused food or fecal material accumulate. Whenever possible, build holding and testing containers with sloping bottoms so food and feces can be removed easily with a siphon. The amount and frequency of cleaning depends on the organism, ratio of dilution water to weight and volume of organisms, and feeding schedule. Clean holding containers at least once every other day. If growths occur on sides of containers, dislodge with a rubber spatula and let settle for removal.

5. Parasites and Disease

a. Stress in relation to parasites and disease: Unexpected and often unexplained mortalities in experimental and control animals interfere with acute or chronic test results. While many factors may be responsible for the death of an animal, diseases due to specific pathogens are among the most significant. In general obtain fish and other animals from pathogen-free stocks (specific hatcheries, etc.) rather than stressing populations by parasite and disease controls. Also optimize laboratory conditions for the individual species to prevent the fostering of disease conditions.

When large numbers of organisms are retained in a relatively small space, undesirable growths, diseases, and parasites become a problem. If the water is unpolluted and poor in nutrients, problems often can be controlled by strict sanitation. However, if the water is somewhat enriched by organic materials and potential toxicants are present, problems increase greatly. Pathogens and parasites that might be very rare in natural waters become potential and ever-present dangers in intensive culture. Bacteria grow on uneaten food, fecal material, and other wastes and compete for DO. They provide a potential source of unwanted growths, disease, and toxic products. Even with good flow-through, each corner, crevice, and dead circulation area of a tank may become a trouble area.

Filtration and/or sterilization of water, regular cleaning of holding vessels, strict sanitation practices, and sterilization of equipment are essential for healthy animals. Uniform food distribution, limiting the amount of unused food, and removal of unused food and waste materials are also important.

Organisms exposed to toxicants become stressed, weakened, and much more susceptible to parasites and disease. Because other environmental factors contribute to reduced resistance, pay careful attention to nutrition, oxygen supply, and water quality.

b. Control methods: UV light and ozonation have been used successfully to control disease and parasites. Antibiotics used in holding tanks reduce bacterial populations. To reduce mortality and to avoid introduction of disease into stock tanks, treat with a wide-spectrum antibiotic immediately after collection, during transport, or on arrival at the laboratory. Holding in a tetracycline-based antibiotic,|| 15 mg/L for 24 to 48 h, can be very helpful. Other chemotherapeutic agents are available, but use care in their application because some are toxic at low concentrations.[20] Do not use treated organisms for tests for at least 10 d after treatment. If contamination is suspected, disinfect tanks and containers with 200 mg NaOCl/L for 1 h.

For larval tests, use strict sanitary measures including sterilization of utensils and containers, filtration and UV sterilization of water, and removal of metabolic products. If disease signs appear in larval cultures, discard the entire culture.

For tests using adult fish and shellfish, early diagnosis and prompt treatment, when available, can prevent losses.

||Terramycin or equivalent.

6. References

1. MOUNT, D.I. & W.A. BRUNGS. 1967. A device for continuous treatment of fish in holding chambers. *Trans. Amer. Fish. Soc.* 96:55.

2. CLINE, T.F. & G. POST. 1972. Therapy for trout eggs infected with *Saprolegnia. Progr. Fish-Cult.* 34:148.

3. RUCKER, R.R. & K. HODGEBOOM. 1953. Observations on gas-bubble disease of fish. *Progr. Fish-Cult.* 15:24.

4. COMMITTEE ON METHODS FOR TOXICITY TESTS WITH AQUATIC ORGANISMS. 1975. Methods for Acute Toxicity Tests with Fish, Macroinvertebrates, and Amphibians. Ecol. Res. Ser., EPA-660/3-75-009, U.S. Environmental Protection Agency.

5. PELTIER, W.H. & C.I. WEBER, eds. 1985. Methods for Measuring the Acute Toxicity of Effluents to Freshwater and Marine Organisms. EPA-600/4-85-013, Environmental Monitoring and Support Lab., U.S. Environmental Protection Agency, Cincinnati, Ohio.

6. HORNING, W.B. & C.I. WEBER, eds. 1985. Short-Term Methods for Estimating the Chronic Toxicity of Effluents and Receiving Waters to Freshwater Organisms. EPA-600/4-85-014, Environmental Monitoring and Support Lab., U.S. Environmental Protection Agency, Cincinnati, Ohio.

7. WEBER, C.I., W.B. HORNING, D.J. KLEMM, T.W. NEIHEISEL, P.A. LEWIS, E.L. ROBINSON, J. MENKEDICK & F. KESSLER, eds. 1988. Short-Term Methods for Estimating the Chronic Toxicity of Effluents and Receiving Waters to Marine and Estuarine Organisms. EPA-600/4-87-028, Environmental Monitoring and Support Lab., U.S. Environmental Protection Agency, Cincinnati, Ohio.

8. MARKING, L.L. & V.K. DAWSON. 1973. Toxicity of Quinaldine Sulfate to Fish. Invest. Fish Control 48, U.S. Bur. Sport Fish & Wildlife, Washington, D.C.

9. KESTER, E., I. DREDALL, D. CONNERS & R. PYTOWICZ. 1967. Preparation of artificial seawater. *Limnol. Oceanogr.* 12:176.

10. ZAROOGIAN, G.E., G. PESCH & G. MORRISON. 1969. Formulation of an artificial sea water media suitable for oyster larvae development. *Amer. Zool.* 9:1141.

11. ZILLIOUS, E.J., H.R. FOUCK, J.C. PRAGER & J.A. CARDIN. 1973. Using *Artemia* to assay

oil dispersement toxicities. *J. Water Pollut. Control Fed.* 45:2389.

12. DAVEY, E.W., J.H. GENTILE, S.J. ERICKSON & P. BETZER. 1970. Removal of trace metals from marine culture medium. *Limnol. Oceanogr.* 15:486.

13. MEYERS, S.P. & Z.P. ZEIN-ELDIN. 1972. Binders and pellet stability in development of crustacean diets. Proc. 3rd Annu. Workshop World Maricult. Soc., p. 351.

14. NEEDHAM, J.G., P.S. GALTSOFF, F.E. LUTZ & P.S. WELSH. 1937. Culture Methods for Invertebrate Animals. Comstock Publ. Co., Inc., Ithaca, N.Y. XXXII.

15. HIRANO, R. & Y. OSHIMA. 1963. Rearing of larvae of marine animals with special reference to their food organisms. *Bull. Jap. Soc. Sci. Fish.* 29:282.

16. MAY, R.C. 1970. Feeding larval marine fishes in the laboratory, A review. Calif. Mar. Res. Comm., CalCOFI Rep. 14:76.

17. BRAUGHN, J.L. & R.A. SCHOETTGER. 1975. Acquisition and Culture of Research Fish: Rainbow Trout, Fathead Minnows, Channel Catfish, and Bluegills. Ecol. Res. Ser. EPA-660/3-75-011, U.S. Environmental Protection Agency.

18. PIPER, R.G. et al. Fish Hatchery Management. U.S. Dep. Interior, U.S. Fish & Wildlife Serv., Washington, D.C.

19. HESSELBERG, R.J. & R.M. BURRESS. 1967. Labor Saving Devices for Bioassay Laboratories. Invest. Fish Control 21, U.S. Bur. Sport Fish & Wildlife, Washington, D.C.

20. WILLFORD, W.A. 1967. Toxicity of 22 Therapeutic Compounds to Six Fishes. Invest. Fish. Control 18, U.S. Bur. Sport Fish & Wildlife, Washington, D.C.

8010 F. Toxicity Test Systems, Materials, and Procedures

1. Water Supply Systems and Testing Equipment

a. *Composition of materials used:* Construct all components of a test system, including water heating and cooling units, constant-level troughs and head boxes, valves and fittings, diluters, pumps, mixing equipment, tanks, and exposure chambers, from inert materials. Acceptable materials include glass, perfluorocarbon plastics, and No. 316 stainless steel. Unplasticized plastics may be used in the dilution portion including holding and acclimation tanks. Avoid contact with brass, copper, lead, or rubber.

b. *Temperature regulation:* Obtain dilution water of the desired temperature by mixing hot and cold water of constant temperatures in the correct proportions, by heat exchangers, or by heaters or coolers in constant-level troughs and head boxes. A heated room or high-low incubator with thermostatic controls usually is suitable for static tests on warm-water organisms. Hold dilution water in tanks until it reaches ambient temperature for conducting static tests. For cold-water species use a specially insulated constant-temperature room or large water bath equipped with temperature controls and adequate circulating water. A satisfactory design for a small laboratory to conduct short-term static tests has been described.[1,2] Special facilities required for different groups of organisms are described in Sections 8111 through 8910.

c. *Toxicant delivery system:* In flow-through tests use metering pumps or other devices for accurate delivery of toxicant or test material into the dilution water. Most toxicant delivery systems have been designed for fresh water and may not be applicable to all wastes. If necessary, stir test chamber contents to maintain suspended solids and nonhomogeneous wastes in flow-through and static tests. Deliver dilution water from constant-head troughs or head boxes by siphons, constricted tubing, noz-

zles, or pumps. Deliver toxicants by siphons from constant-head reservoirs,
pumps, calibrated glass nozzles, or Mariotte bottles.[3] Mix dilution water and toxicant in tanks with baffles or stirrers or in
mixing troughs.[4] Since the introduction of
the serial diluter,[5] various methods and
types of diluters have been described.[6-25]

The choice of a toxicant-delivery system
for flow-through toxicity tests depends on
such factors as dilution, flow rate, quantity
of toxicant available, and the presence of
suspended solids. The proportional diluter[16] was designed to handle a dilution
factor (i.e., the factor by which a concentration is multiplied to calculate the next
lower concentration) between 0.50 and
0.90. The serial diluter has been modified
to provide a narrow range of concentrations.[17] Other diluters operate well with a
dilution factor of 0.50.[22,24] Flow rates
through test chambers may vary from 400
mL/min for the proportional diluter[5] to
6000 mL/min.[22] Some stock solutions[19,21-23]
and other systems have been modified to
handle larger volumes of toxicant and high
suspended solids concentrations.[20,24] These
diluters are most suitable for effluent toxicity tests. Several diluters have been designed for flow-through toxicity tests with
embryo-larval stages of aquatic organisms.[21,23] One of these systems was designed
to eliminate the air/water interface for tests
using volatile compounds and compounds
with very low solubility.[21]

The basic components of a flow-through
system are shown in Figure 8010:4. The
diluent water reservoir is large enough to
provide water for at least 5 d. If dilution
water is added to this reservoir continuously, a smaller capacity is preferred. Dilution water flows at a constant rate by
gravity from this reservoir to a constant-
head diluent-supply head box through a
nonmetallic float-controlled valve or other
device and then to the diluter. Provide head
box with heating or cooling equipment and
a thermostat to maintain constant temper-

ature. Equip test containers with an overflow system designed to prevent organisms
from entering outlets. Clean test containers
as described in 8010E.4d.[25]

The constant-head toxicant supply is a
constant-level tank, a Mariotte bottle, or
other device. If toxicant is added at greater
than a drip rate, adjust toxicant temperature to that of the diluter via a heat exchange; otherwise, the toxicant heat
exchanger in Figure 8010:4 is not necessary. If the toxicant stock solution is unstable, renew it before it degrades. If
metering pumps are used, the toxicant supply system need not maintain a constant
head.

A simple valve control system for regulation of flow rates of dilution water and
toxicant solution has been described.[26] For
more toxic materials, less toxicant is required and a Mariotte bottle or syringe
pump that delivers a very slow but constant
flow is useful.[27,28]

A diluter meters dilution water from the
constant-head box and toxicant from the
constant-head tank or other containers and
mixes them in the proper proportions for
each of the test chambers. After proper
calibration of the diluter, make toxicant
stock solutions to the proper concentration.
Shield the toxicant supply reservoir from
light.

Provide a mixing chamber between diluter and test container for each concentration. If duplicate test containers are
used, run separate delivery tubes from the
mixing chamber to each duplicate. Use flow
rates through test containers of at least 6
tank water volumes/24 h. Do not let rates
through test containers vary temporally or
between containers by more than ±10%.
Calibrate the toxicant and dilution water
volumes used in each portion of the toxicant delivery system and the flow rate
through each test container. Check operation of toxicant delivery system daily during test.

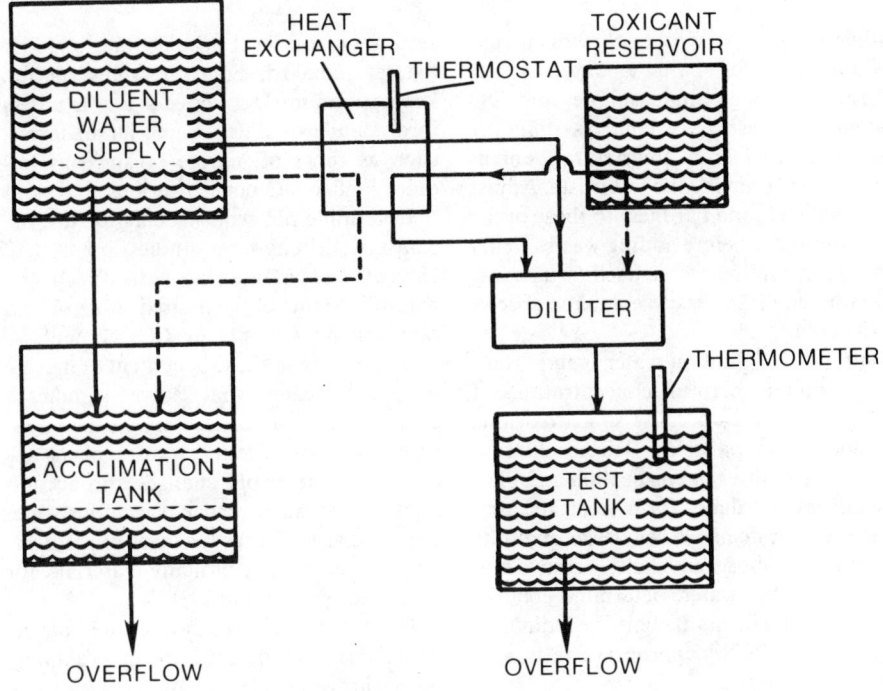

Figure 8010:4. Basic components of flow-through system.

2. Preparing Test Materials

a. Dilution water: Whenever possible, test toxicity of effluents on site where ample supplies of toxicant and dilution water are available. On-site testing permits temperature, DO, pH, hardness, salinity, turbidity, and other qualities of the dilution water to vary with those of the receiving water. Convey the effluent sample to testing chambers with as little modification as possible. Do not unnecessarily aerate, heat, cool, or agitate. In cases where the testing facility is remote from the effluent discharge site, artificial or reconstituted water may be used as the diluent. If the diluted effluent is low in DO, adjust flow-through and loading in the test chambers so that DO is not reduced significantly; aerate as a last resort. Hold temperature at or near that of the receiving water.

If the receiving water is deficient in DO and has temperatures above the locally applicable water quality standards, bring these into compliance so that allowable levels of a specific waste can be assessed meaningfully. Determine toxicity of test waste in conjunction with other contaminants present in receiving water by taking dilution water from receiving water just outside the area of effect of test waste. This is especially necessary when effluents contain metal salts, cyanide complexes, ammonium compounds, or other materials, the toxicity of which is greatly influenced by changes in pH, hardness, etc. If there are wide variations in receiving water quality characteristics, determine waste toxicity at the upper and lower limits of the range.

Evaluate receiving water stress on aquatic biota by using two controls, one with the receiving water and another (either natural or synthetic) with an un-polluted water of similar quality. Do not let calcium, magnesium, sulfate, and DO content for freshwater controls differ by more than 10% from the natural content of the water receiving the test waste. Adjust pH, alkalinity, and hardness to those of the receiving water before adding wastes to determine if dilution water itself is unfavorable for the more sensitive aquatic species in the area.

Turbidity of dilution water is important in determining harmful concentrations of potential toxicants because some toxicants are sorbed on particles. Turbid dilution water may limit visual inspection and photosynthesis of algae, form deposits, and clog water systems. When large amounts of settleable solids significantly remove toxicants from the water, determine concentrations of toxicants in bottom sediments and their toxicity by appropriate tests with benthic organisms.

When the purpose of the test is other than to determine toxicity of an effluent, use for dilution water only a nonpolluted natural or synthetic water of constant and reproducible quality that is favorable for aquatic life. Warm or cool dilution water to test temperature and bring to equilibrium with atmospheric gases before use.

Use standard water conditions and organisms for comparative toxicity and sensitivity tests. Use reconstituted fresh or marine water [Sections 8010E.4b1) and 2)] if natural supply is not suitable. Because of the effects of water quality on toxicity, use both hard and soft water for tests on freshwater organisms.

Many marine organisms spend a portion of their life cycles in estuaries. In life-cycle tests, change dilution water in accordance with their requirements at different life stages. If the effects of temperature are not being studied, keep it within a favorable range.

For warm-water species keep DO at the highest of the following: 4 mg/L, 60% of saturation, or level specified by state or federal standard. For cold-water species, keep minimum DO above 5 mg/L unless local standards differ. Some larval forms, such as those of marine crustaceans, require higher DO concentrations.

Determine pH requirements for test organisms. In long-term studies, other than effluent and monitoring tests, keep pH within 0.4 units of the desired value. Avoid rapid changes in pH or CO_2 content. A rapid increase in the CO_2 content of marine waters indicates that some significant change has occurred that should be investigated at once. Freshwater organisms are more tolerant of pH changes and accommodate to much wider variations than strictly marine forms. Changes in pH drastically alter toxicity of many materials, for example, cyanide and NH_3.

In working with estuarine and marine organisms and different life stages that may be marine or estuarine, salinity is of prime importance. Use the natural salinity for each test species and its different developmental stages.[29]

Keep acidity, total alkalinity, and hardness of dilution water constant. Alkalinity and hardness influence toxicity of some metals and total alkalinity is an important factor in photosynthesis and algal growth.

b. Toxicant solution: Prepare toxicant solution in advance and add immediately to the dilution water for static tests. If a toxicant is unstable, determine its stability and replace as necessary. If possible, measure toxicant concentrations during the test. Prepare all solutions for each series of tests from the same source sample. Disperse undissolved material uniformly by shaking.

If solvents are necessary, use acetone, dimethylformamide (DMF), ethanol, methanol, isopropanol, acetonitrile, dimethylacetamide, ethylene glycol, or triethylene glycol to prepare stock solutions. Certain surfactants[24] may be useful. Use only the minimal amount of solvent necessary to disperse the toxicant. Do not exceed 0.5 mg/L in static and 0.1 mg/L in flow-through test solutions.

If a solvent is used, make two sets of controls, one containing no solvent and the other containing the highest concentration of solvent used.

Some effluents, especially oily wastes, are difficult to distribute evenly. Agitate test tank contents with a magnetic stirring device or other mixer to maintain solids in suspension. The nature of the test material governs the preparation of test concentrations and frequency of test medium replacement. Common problems include insolubility, adsorption on exposed surfaces, decomposition, photolysis, loss of volatiles, high BOD, and bacterial growth. These can change the apparent test concentration and lead to erroneous results.

Effluent samples that vary in composition with time may require a series of tests to develop toxicity data.

Store effluent samples, in containers from which all the air has been expelled, at $4 \pm 1°C$. Do not store samples longer than absolutely necessary. (Choose sample containers to minimize changes in concentration of constituents in sample.)

Thoroughly mix test materials before use. Use material directly as a stock solution of toxicant or prepare a stock solution using filtered dilution water. Make stock solutions with dilution water on a volume-to-volume basis. If the effluent is liquid, designate the percentage waste in each test concentration. If the waste is a solid, dilute on a weight-to-volume basis, e.g., milligrams per liter.

If the waste contains both solids and liquids, mix thoroughly to disperse before using as a stock toxicant, and provide agitation in the stock reservoir and test containers. If larger organisms are tested, a propeller placed under a screen or perforated false bottom may be used to maintain consistency of the solution. If the solids settle out rapidly and do not contact pelagic organisms, test only the liquid portion. After thorough mixing, let settle and decant or drain off the liquid for use as test toxicant. If the solid waste portion is toxic, set up test chambers having a certain weight-to-volume ratio of bottom material and expose benthic and burrowing organisms of the receiving water area. Mix wastes and let settle before adding organisms. If waste contains sparingly water-soluble materials check solubility and if below or at very low toxic concentrations, solvents, emulsifying agents, or water-miscible solvents may be used to disperse.

c. Test organisms: Select test organisms as described in Section 8010E.1 and handle as indicated in Sections 8010E.3 and 4. For long-term tests, use only test organisms in excellent condition. At the end of the test, control organisms still should be in good condition.

3. Test Procedures

a. Experimental design: Expose test organisms in at least duplicate containers. By using more organisms and replicate test containers for each toxicant concentration, variability can be evaluated. Use only true replicates with no water connection between test containers. Each test consists of a minimum of five test concentrations and a control, with an additional control if a solvent is used. Arrange test containers at random in the testing area. If replicates are used, randomize each series of test containers separately. Distribute organisms randomly to test containers either adding one at a time to each container if there are to be less than 11 organisms per container or two at a time if there are to be more. In short-term static tests, add organisms to intermediate containers and then add them to test chambers containing the toxicant all at the same time.

Most short-term, acute tests with fish and invertebrates require control survival of 90% or more to be considered acceptable. If this is not achieved, repeat the test.

This level of acceptance also may be used with other end points such as spawning, normal development, or general apparent health. Lower levels of survival may be acceptable in longer exposure assays or in tests that use life stages where survival, even under ideal conditions, is limited. Examples of acceptable reduced survival include short-term tests for estimating the chronic toxicity of effluent to freshwater and marine organisms[30,31]. In these assays acceptable survival for *Mysidopsis bahia, Cyprinodon variegatus* is 80% and for *Ceriodaphnia* and *Pimephales promelas,* 85%.

In short-term static or renewal tests with fish, use 10 or more test organisms in each toxicant concentration. Use larger numbers of organisms per test concentration for smaller organisms. The number of organisms exposed in each test concentration is governed by size of the organism; expected normal mortality; extent of cannibalism; availability of dilution water, toxicant, and test organisms; and desired test precision. Test precision depends on variability of organism response, number of organisms exposed to each concentration, number of replications, and toxicant concentration and its variability.

For a given test increasing the number of test organisms increases precision. It is recommended that the 95% confidence interval be less than ± 30% of the mean. With test organisms for which culture methods are not available, this precision may be difficult or impossible to attain. As a general rule, use at least 10 test organisms at each concentration; preferably use 20 or more, especially with small organisms and in tests with larvae where cannibalism and natural mortality are high.

Determine length and weight of representative organisms before the test to establish loading rates and acceptability with respect to size variation. After acclimation has begun, do not handle test organisms. Increases in weight or growth may be determined by adding more animals than required initially so that some may be removed to make necessary measurements. At the end of the assay weight and length may be measured to determine sublethal impacts for computation of an EC50 or for confirmation of loading rates and sizes, if required.

b. Selecting test concentrations: Express liquid waste concentrations as percent on a volume-to-volume basis. Express concentrations of nonaqueous wastes and of individual chemicals as milligrams or micrograms per liter. Clearly indicate what the weight represents as inclusion of water of hydration as part of the weight of the solute (e.g., $CuSO_4 \cdot 5H_2O$). When an impure chemical is tested, especially in a formulation containing added inert ingredients, indicate the chemical composition by weight and whether the LC50 value is based on concentration of total material or active ingredient.

Although an LC50 may be determined by using any appropriate series of test concentrations, the geometric series of concentration values is simplest to use. Multiply the highest and succeeding concentrations by a constant factor (0.5 to 0.6) to obtain concentrations that are evenly spaced on a logarithmic scale.

The magnitude of concentration intervals to establish an LC50 or EC50 by interpolation depends on the required degree of precision and on the experimental data.

c. Loading: For static tests do not exceed an organism loading of 0.8 g/L in the test container. In tests with small organisms and tropical forms, decrease loading to as low as 0.1 g/L and accommodate large test organisms by using larger or duplicate test containers. Limit the number of test organisms per volume of test solution so that during the test (*a*) DO remains greater than 60% of saturation for cold-water species and greater than 40% of saturation for warm-water species; (*b*) toxicant concentration is not lowered significantly; (*c*) concentrations of metabolic products (e.g.,

NH_3, CO_2) do not become too high; and (*d*) organisms are not stressed by crowding. Do not let the concentration of un-ionized ammonia exceed 20 μg NH_3-N/L (Table 8010:VI).

For flow-through studies, use a flow rate of at least 6 tank volumes/24 h to maintain desirable temperature and DO and safe concentrations of metabolites. If the DO concentration drops below the desired level, increase the turnover rate within the diluter. If this is inadequate, aerate the dilution water before mixing with effluent.[24]

d. Physical and chemical determinations:

1) Dilution water analysis—For fresh water, measure hardness, alkalinity, pH, TOC, COD, and suspended solids once every 30 d and at the beginning and end of the test. If water quality is variable, test more frequently. Tap water can be dechlorinated by active aeration (using air stones) for 24 h, filtration through activated carbon, or use of sodium thiosulfate. If a treated tap water is used, measure residual chlorine by one of the methods given in Section 4500-Cl.[33]

Analyze weekly for pH, alkalinity, and hardness to define test-water variability. If characteristics are affected by the toxicant, test samples from each toxic concentration at least once every other week. For brackish or marine dilution water, measure salinity, pH, DO, and temperature two or three times daily; and suspended solids and

TOC or COD at least once every 30 d and at the beginning and end of each test.

2) Toxicant analysis—For flow-through life-cycle tests it may not be necessary to make routine detailed analyses, but make periodic tests to insure that the correct ratio of effluent to dilution water is maintained in exposure tanks.

For studies to determine water quality criteria measure concentration of toxicants in each container at the beginning and at least once during the test, in at least one container at the next-to-lowest toxicant concentration at least weekly, and in at least one container whenever a malfunction is detected in any part of the toxicant delivery system. For replicate test containers use a ratio of the highest measured concentration to the lowest measured concentration of less than 1.15; if this is exceeded, check the toxicant delivery system and analyze additional samples from test containers to determine if the sampling or analytical method is sufficiently precise. Do not accept measured toxicant concentrations differing by more than $\pm$ 15% from the calculated concentration.

Record temperature at least hourly throughout the test (24 h/d) in at least one test container and make additional measurements on dilution water and other test solutions. Measure DO, pH, and salinity at the beginning of the test and daily thereafter in the control, high, medium, and low

TABLE 8010:VI. PERCENTAGE OF AMMONIA UN-IONIZED IN DISTILLED WATER*

Temperature °C	Percentage Un-ionized at Given pH								
	6.0	6.5	7.0	7.5	8.0	8.5	9.0	9.5	10.0
5	0.01	0.04	0.11	0.40	1.1	3.6	10	27	54
10	0.02	0.06	0.18	0.57	1.8	5.4	15	36	64
15	0.03	0.08	0.26	0.83	2.6	7.7	21	45	72
20	0.04	0.12	0.37	1.2	3.7	11	28	55	80
25	0.05	0.17	0.51	1.7	5.1	14	35	63	84
30	0.07	0.23	0.70	2.3	7.0	19	43	70	88

* Prepared from data given in Sillen and Martell.[32]

toxicant concentrations. Generally, variation of $\pm 0.5°C$ is allowable, but do not exceed $\pm 1.0°C$.

Take water samples for chemical analysis at the center of the exposure tank; do not include surface scum or material from tank bottom or sides. If analytical results are not affected by storage, collect daily, equal-volume grab samples and composite for a week. Analyze sufficient samples throughout the test to determine whether the concentration of toxicant is reasonably constant. If it is not, analyze enough samples weekly to show the variability of toxicant concentration. If methods are available, determine in the next-to-lowest concentration the loss of toxicant. If the loss is more than 10%, attempt to alleviate by using either a faster flow rate or a lower loading.

When possible and necessary, analyze mature and immature test organisms for toxicant residues. For larger organisms analyze muscle and liver and possibly gills, blood, brain, bone, kidney, GI tract, gonads, and skin. For large organisms, analysis of whole specimens does not replace analysis of individual tissues, especially muscle.

e. Biological data and observations: In short-term tests with macroinvertebrates and fish, count the number of dead or affected organisms in each container at least daily throughout the test. With certain fast-acting biocides it may be useful to count the number of dead or affected organisms in each container at 1.5, 3, 6, 12, and 24 h after the beginning of the test. Remove dead organisms as soon as observed. It is more important to obtain data that will define the shape of the toxicity curve than to obtain data at prespecified times.

Death is the adverse effect most often used to reflect acute toxicity. The usual criterion for death is no movement, especially no gill movement in fish, and no reaction to gentle prodding. Death is not easily determined for some invertebrates. Cessation of movement of antennae, mouth

parts, or other organs may be used. When death cannot be determined, use EC50 rather than LC50. The effect usually used for determining EC50 with daphnids, midge larvae, copepods, and other organisms is immobilization, defined as inability to move, except for minor activity of appendages. Other effects can be used to determine EC50, but always report the effect and its definition. Also report such effects as erratic swimming, loss of reflex, discoloration, changes in behavior, excessive mucus production, hyperventilation, opaque eyes, curved spine, hemorrhaging, molting, and cannibalism.

In short-term tests, organism reactions during the first few hours may indicate the nature of the toxicant and serve as a guide for further tests.

In long-term partial- or full-life-cycle tests use a photographic method for counting and measuring small test organisms.[34] This is rapid and accurate, and does not entail handling the organism. With this method, use exposure tanks with glass bottoms and drains that allow the water level to be drawn down. To count and measure test organisms, draw water down to a depth of 2 to 3 cm and transfer tank to a light box having fluorescent lights under a square millimeter grid of adequate size. Photograph aquarium bottom; this shows the organisms over the grid. On an enlargement of the picture, count and measure organisms.

f. Photoperiod and artificial light: In long-term studies to determine water quality requirements for those species requiring annual-light-cycle photoperiods, simulate the natural seasonal daylight and darkness periods at the locality or some central location.[35] (See Section 8910 for more information on light cycles for fish.)

Use wide-spectrum fluorescent tubes as a light source similar to daylight.* Some

*For example, OPTIMA 50, Duro-Test Corp., North Bergen, N.J. 07047, or equivalent.

organisms require subdued light, others need a place to hide, and some, such as lake trout eggs, require darkness during certain life stages. Base exposure to light on what is normal to, and required by, the species. Measure light intensity at the water surface. In short-term tests a standard photoperiod of 16 h light, 8 h dark is suggested but often usual laboratory lighting is adequate.

g. Exposure chambers: For organisms weighing more than 0.5 g, use a test solution between 15 and 30 cm deep. In short-term tests, these organisms often are exposed in about 15 L solution in 20-L wide-mouth, soft-glass bottles. Fabricate test containers of other sizes by welding (not soldering) stainless steel, by gluing double-strength or stronger window glass with clear silicone adhesive, or by modifying glass bottles, battery jars, or beakers to provide screened overflow holes or V-notches. Because silicone adhesives absorb some organochlorine and organophosphorous pesticides, expose as little of the adhesive as possible to the water. Place extra beads of adhesive for added strength only on the outside of containers. Expose smaller organisms in 4-L wide-mouth, soft-glass bottles or battery jars that contain 2 to 3 L solution. Expose daphnids, midge larvae, copepods, and other small organisms in loosely covered beakers or other containers. Disposable plastic containers may be used for tests involving compounds that do not react with the plastic. Disposable plastic containers are recommended for the 7-d *Ceriodaphnia* survival and reproduction assay[36] and *Champia parvula* sexual reproduction assay[30].

Keep surface areas small in relation to volume to limit sorption on vessel walls. With flow-through tests keep liquid surface area/volume ratio small to reduce loss of volatiles. For various exposure chamber designs see Sections 8210 through 8910.

4. References

1. HENDERSON, D. & C.M. TARZWELL. 1957. Bioassays for the control of industrial effluents. *Sewage Ind. Waste* 29:1002.

2. LENNON, R.E. & C.R. WALKER. 1964. Investigations in Fish Control I. Laboratories and Methods for Screening Fish-Control Chemicals. U.S. Bur. Sport Fish Wildl. Circ. 185, U.S. Government Printing Off., Washington, D.C.

3. MCALLISTER, W.A., JR., W.L. MAUCH & F.L. MAYER, JR. 1972. A simplified device for metering chemicals in intermittent-flow bioassays. *Trans. Amer. Fish. Soc.* 101:555.

4. LOWE, J.I. 1964. Chronic exposure of spot, *Leiostomus xanthurus*, to sublethal concentrations of toxaphene in seawater. *Trans. Amer. Fish. Soc.* 93:396.

5. MOUNT, D.I. & R.E. WARNER. 1965. A Serial Dilution Apparatus for Continuous Delivery of Various Concentrations of Material in Water. PHS Publ. No. 999-WP-23, Environ. Health Ser., U.S. Dep. Health, Education & Welfare, Washington, D.C.

6. MOUNT, D.I. & C. STEPHAN. 1967. A method for establishing acceptable toxicant limits for fish—Malathion and the butoxy-ethanol ester of 2,4-D. *Trans. Amer. Fish. Soc.* 96:185.

7. CLINE, T.F. & G. POST. 1972. Therapy for trout eggs infected with *Saprolegnia. Progr. Fish-Cult.* 34:148.

8. CHANDLER, J.H., H.O. SANDERS & D.F. WALSH. 1974. An improved chemical delivery apparatus for use in intermittent-flow bioassays. *Bull. Environ. Contam. Toxicol.* 12:123.

9. SCHIMMEL, S.C., D.J. HANSEN & J. FORESTER. 1974. Effects of aroclor 1254 on laboratory-reared embryos and fry of sheepshead minnows (*Cyprinodon variegatus*). *Trans. Amer. Fish Soc.* 103:582.

10. FREEMAN, R.A. 1971. A constant flow delivery device for chronic bioassay. *Trans. Amer. Fish. Soc.* 100:135.

11. BENGTSSON, B.E. 1972. A simple principle for dosing apparatus in aquatic systems. *Arch. Hydrobiol.* 70:413.

12. GRANMO, A. & S.C. KOLLBERG. 1972. A new simple water flow system for accurate continuous flow tests. *Water Res.* 6:1597.

13. BENOIT, D.A. & F.A. PUGLISI. 1973. A sim-

plified flow-splitting chamber and siphon for proportional diluters. *Water Res.* 7:1915.

14. LICHATOWICH, J.A., P.W. O'KEEFE, J.A. STRAND & W.L. TEMPLETON. 1973. Development of methodology and apparatus for the bioassay of oil. *In* Proc. Joint Conf. Prevention and Control of Oil Spills, p. 659. American Petroleum Inst., U.S. Environmental Protection Agency & U.S. Coast Guard, Washington, D.C.

15. ABRAM, F.S.H. 1973. Apparatus for control of poison concentration in toxicity studies with fish. *Water Res.* 7:1875.

16. MOUNT, D.I. & W.A. BRUNGS. 1967. A simplified dosing apparatus for fish toxicology studies. *Water Res.* 1:21.

17. THATCHER, T.O. & J.F. SANTNER. 1966. Acute toxicity of LAS to various fish species. *Proc. 21st Ind. Waste Conf.,* Purdue Univ., Eng. Ext. Bull. No. 121:996.

18. SHUMWAY, D.L. & J.R. PALENSKY. 1973. Impairment of the flavor of fish by water pollutants. Ecological Research Series No. EPA-R3-73-101. U.S. Environmental Protection Agency, Washington, D.C.

19. DEFOE, D.L. 1975. Multichannel toxicant injection system for flow-through bioassays. *J. Fish. Res. Board Can.* 32:544.

20. RILEY, C.W. 1975. Proportional diluter for effluent bioassays. *J. Water Pollut. Control Fed.* 47:2620.

21. BIRGE, W.J., J.A. BLACK, J.E. HUDSON & D.M. BRUSER. 1979. Embryo-larval toxicity tests with organic compounds. *In* L.L. Marking & R.A. Kimerle, eds. Aquatic Toxicology. ASTM STP 667, American Soc. Testing & Materials, Philadelphia, Pa., p. 313.

22. GARTON, R.R. 1980. A simple continuous-flow toxicant delivery system. *Water Res.* 14:227.

23. BENOIT, D.A., V.R. MATTSON & D.L. OLSON. 1982. A continuous-flow mini-diluter system for toxicity testing. *Water Res.* 16:457.

24. PELTIER, W. 1978. Methods for measuring the acute toxicity of effluents to aquatic organisms. EPA-600/4-78-012, U.S. Environmental Protection Agency, Washington, D.C.

25. LEMKE, A.E. 1964. A new device for constant-flow test chambers. *Progr. Fish-Cult.* 26:136.

26. JACKSON, H.W. & W.A. BRUNGS. 1966. Biomonitoring of industrial effluents. *Proc. 21st Ind. Waste Conf.,* Purdue Univ., Eng. Ext. Bull. 121:117.

27. SURBER, E.W. & T.O. THATCHER. 1963. Laboratory studies of the effects of alkyl benzene sulfonate (ABS) on aquatic invertebrates. *Trans. Amer. Fish. Soc.* 92:152.

28. BURROWS, R.E. 1949. Prophylactic treatment for control of fungus, *Saprolegnia parasitica. Progr. Fish-Cult.* 11:97.

29. DAVEY, E.W., J.H. GENTILE, S.J. ERICKSON & P. BETZER. 1970. Removal of trace metals from marine culture medium. *Limnol. Oceanogr.* 15:486.

30. WEBER, C.I., W.B. HORNING, D.J. KLEMM, T.W. NEIHEISEL, P.A. LEWIS, E.L. ROBINSON, J. MENKEDICK & F. KESSLER, eds. 1988. Short-Term Methods for Estimating the Chronic Toxicity of Effluents and Receiving Waters to Marine and Estuarine Organisms. EPA-600/4-87-028, Environmental Monitoring and Support Lab., U.S. Environmental Protection Agency, Cincinnati, Ohio.

31. HORNING, W.B. & C.I. WEBER, eds. 1989. Short-Term Methods for Estimating the Chronic Toxicity of Effluents and Receiving Waters to Freshwater Organisms, 2nd ed. EPA-600/4-89-001, Environmental Monitoring and Support Lab., U.S. Environmental Protection Agency, Cincinnati, Ohio.

32. SILLEN, L.C. & A.E. MARTELL. 1964. Stability constants of metal ion complexes. Spec. Publ. 17, Chemical Soc., London, England.

33. ANDREW, R.W. & G.E. GLASS. 1974. Amperometric methods for determining residual chlorine, ozone and sulfite. U.S. Environmental Protection Agency, National Water Quality Lab., Duluth, Minn.

34. MCKIM, J.M. & D.A. BENOIT. 1971. Effect of long-term exposures to copper on survival, reproduction and growth of brook trout *Salvelinus fontinalis* (Mitchill). *J. Fish. Res. Board Can.* 28:655.

35. DRUMMOND, R.A. & W.F. DAWSON. 1970. An inexpensive method for simulating a diel pattern of lighting in the laboratory. *Trans. Amer. Fish. Soc.* 99:434.

36. HORNING, W.B. & C.I. WEBER, eds. 1985. Short-Term Methods for Estimating the Chronic Toxicity of Effluents and Receiving

Waters to Freshwater Organisms. EPA-600/4-85-014, Environmental Monitoring and Support Lab., U.S. Environmental Protection Agency, Cincinnati, Ohio.

5. Bibliography

COMMITTEE ON METHODS FOR TOXICITY TESTS WITH AQUATIC ORGANISMS. 1975. Methods for Acute Toxicity Tests with Fish, Macroinvertebrates, and Amphibians. EPA-660/3-75-009, U.S. Environmental Protection Agency, Corvallis, Ore.

SPRAGUE, J.B. & A. FOGELS. 1977. Watch the Y in bioassay. In Proc. 3rd Aquatic Toxicity Workshop. Canadian Environmental Protection Service Tech. Rep. No. EPA-5-AR-77-1.

LEE, D.R. 1980. Reference toxicants in quality control of aquatic bioassays. In A.L. Bui-kema, Jr. & J. Cairns, Jr., eds. Aquatic Invertebrate Bioassays. American Soc. Testing & Materials, Philadelphia, Pa.

AMERICAN SOCIETY FOR TESTING AND MATERIALS. 1984. Standard practice for conducting acute toxicity tests with fishes, macroinvertebrates, and amphibians. E 729-80. Annual Book of ASTM Standards. American Soc. Testing & Materials, Philadelphia, Pa.

U.S. ENVIRONMENTAL PROTECTION AGENCY. 1985. Methods for Measuring the Acute Toxicity of Effluents to Freshwater and Marine Organisms, 3rd ed. EPA-600/4-85-013, U.S. Environmental Protection Agency.

AMERICAN SOCIETY FOR TESTING AND MATERIALS. 1986. New Standard Guide for Conducting Acute Toxicity Tests on Aqueous Effluents with Fishes, Macroinvertebrates, and Amphibians, Draft 8. American Soc. Testing & Materials, Philadelphia, Pa.

8010 G. Calculating, Analyzing, and Reporting Results of Toxicity Tests

The precision of a biological test is limited by a number of factors including the normal biological variation among individuals of a species. Studies with a randomly selected species do not give accurate information on toxicity of a compound to other species and life stages or to an entire biota. A test with one species yields an accurate estimate of the toxicity only to others of that species of similar size, age, and physiological condition, in water with the same or similar characteristics, and under similar test conditions.

1. Analyzing Results of Quantal Toxicity Tests[1]

Responses are of two kinds; quantal and quantitative. In a *quantal* test, an organism either shows the response under study or does not show it, for example, it dies or does not die. Thus, at any concentration greater than that tolerated without effect, a certain percentage of test organisms will show the response within some stated time. In a *quantitative* or graded test, each organism responds to a variable degree, e.g., amount of growth.

Quantal tests are designed to estimate the concentration of a test material that affects 50% of the test organisms. Determine either death or a sublethal effect, such as immobilization, fatigue in a swimming test,[2] avoidance reaction,[3] or a significant effect on such factors as growth, fertility, or tissue structure. For all responses except death, substitute "EC50" for "LC50" in the text below and substitute the word or phrase describing the sublethal response for "mortality." Always report 95% confidence limits for LC50 and EC50 values. The most widely used methods for calculating an LC50 and confidence limits are: probit,[4] logit,[5] moving average,[6] and Litchfield-Wilcoxon.[7] These methods, however, are not interchangeable. Therefore, choose a specific method for analyses and use it throughout a study. Computer programs for computing LC50 values are available on diskette.*

a. Estimating LC50 and EC50 by probit method using graphical analysis: Tabulate observations of mortality as in Table 8010:VII for at least one selected exposure time; the time ordinarily selected usually is the longest one used in the test, often 96 h. Only one successive 0% and one successive 100% mortality value, plus those nearest the center of the range of concentrations, are used.

To construct the graph, plot percentage mortality as the ordinate against concentration as the abscissa on probit paper. Death is plotted on a probit or probability scale and concentration on a logarithmic scale (Figure 8010:5). Because the probit scale never reaches 0 or 100%, plot any such points with an arrow indicating their true position.

Next, fit a line to the points by eye. Give most consideration to points between 16

_____
* Statistical Support Staff, Biological Methods Branch, Environmental Monitoring and Support Laboratory, U.S. Environmental Protection Agency, Cincinnati, Ohio.

and 84% mortality and try to minimize total vertical deviations of the line from the points. If there is doubt about placing the line, draw it as horizontally as possible because this acknowledges more variability in the data.

Read the concentration causing 50% mortality from the fitted line; this is the estimated LC50 for the selected exposure time. Report this as the result of the test. In the example, in Figure 8010:5, the estimated 96-h LC50 is approximately 4.4%, that is, the estimated concentration that would kill the average or typical test organism in 96 h. Graphical estimates of LC50 (as shown in Table 8010:VII) or EC50 usually are very close to those obtained by formal probit analysis with a computer.[8,9] However, confidence limits are not obtained by graphical interpolation.

b. Confidence limits of the LC50: Confidence limits of the LC50 also should be calculated and can be obtained from the probit analyses used to determine the LC50. Use simplified nomographic methods for field work or when a computer is not available.[7] For more precise calcu-

TABLE 8010.VII. EXPERIMENTAL DATA FROM HYPOTHETICAL TOXICITY TEST SUBJECTED TO PROBIT ANALYSIS

Concentration of Waste % by volume	No. of Test Organisms	Number of Test Organisms Dead at							
		2 h	4 h	6 h	8 h	24 h	48 h	72 h	96 h
10	10	1	4	7	9	10	10	10	10
7.5	10	0	1	2	6	9	9	10	10
5.6	10	0	0	0	2	7	7	8	9
4.2	10	0	0	0	0	1	4	4	4
3.2	10	0	0	0	0	0	1	1	1
0	10	0	0	0	0	0	0	0	0
LC50, %, estimated from graph		10	10	9	7.1	5.2	4.7	4.5	4.4
LC50, estimated by probit analysis		—	—	8.96	7.02	5.27	4.70	4.46	4.34
95% confidence limits		—	—	7.60	5.82	4.53	3.95	3.87	3.49
		—	—	10.5	8.42	6.12	5.59	5.14	5.40
Slope of probit line		—	—	10.9	8.42	10.1	7.03	9.54	11.3

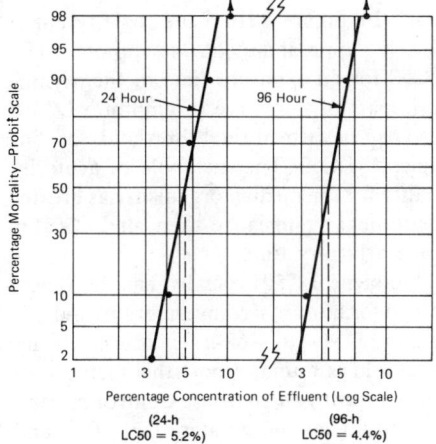

Percentage Mortality—Probit Scale

98
95
90

24 Hour   96 Hour

70
50
30
10
5
2

1   3   5   10   3   5   10

Percentage Concentration of Effluent (Log Scale)

(24-h   (96-h
LC50 = 5.2%)   LC50 = 4.4%)

Figure 8010:5. Examples of median lethal concentration determinations at two representative times by probit analysis and line of best fit.

lations computer programs are available.[8] Even when a formal probit analysis is carried out with a computer, make a graph such as Figure 8010:5 to check the reasonableness of the computed LC50.

Test for significant difference between two LC50s obtained from duplicate tests by examining confidence limits for overlap. If there is no overlap the LC50s are significantly different. However, the LC50s still may be different if the confidence limits overlap. Test for significant differences more exactly by the formula:

$$f_{1,2} = \frac{1.96\ \text{SE}_{\text{diff}}}{\text{LC50}_{1,2}}$$
$$= \text{antilog}\ \sqrt{(\log f_1)^2 + (\log f_2)^2}$$

where:

  f = factor for 95% confidence limits of LC50, i.e., the confidence limits are LC50 $\times$ f and LC50 $\div$ f (f = antilog of two standard deviations of the log LC50).

This formula has been adapted to simple nomogram use.[7] If the ratio (greater LC50)/(smaller LC50) exceeds the value

for $f_{1,2}$, the LC50s are significantly different. If LC50s calculated from duplicate tests are significantly different, test the populations to determine if they differ in length, weight, age, or sex. Use parametric statistical tests such as Student's t test and ANOVAs to determine if such differences in populations are significant and, therefore, could influence the LC50s. If assumptions of analysis of variance are not met, then nonparametric tests for significant differences may be used. The Mann-Whitney U test or the Wilcoxon two-sample test are nonparametric tests that are analogous to the t test and ANOVA with two classes.

The confidence limits about the LC50 do *not* describe variability of the LC50 under conditions other than those tested. The limits indicate the accuracy of the estimate of replicate tests at the same time under the same conditions. A precision of about 10% is sometimes attainable, but better than that is not to be expected unless more than 10 organisms are exposed at each concentration.

  c. Other methods of analyzing results: The graph for estimating the LC50 (Figure 8010:5) can be constructed with an arithmetic scale for percentage mortality. However, the probit scale is preferred because it usually gives a straight line.

Logits have been used instead of probits with equivalent results.[5] Reciprocal transformations and angle (arcsine) transformations with estimation of LC50 by a moving average also have been used.[6] These methods have limitations but the last one is recommended for estimating LC50 and confidence limits when fewer than two partial lethalities have been obtained. Median effective time (ET50) for mortality at each concentration is estimated by plotting percentage mortality on a probit scale against time on a logarithmic scale and then using probit analysis techniques similar to those given above.[4,10–12] Procedures are the same,

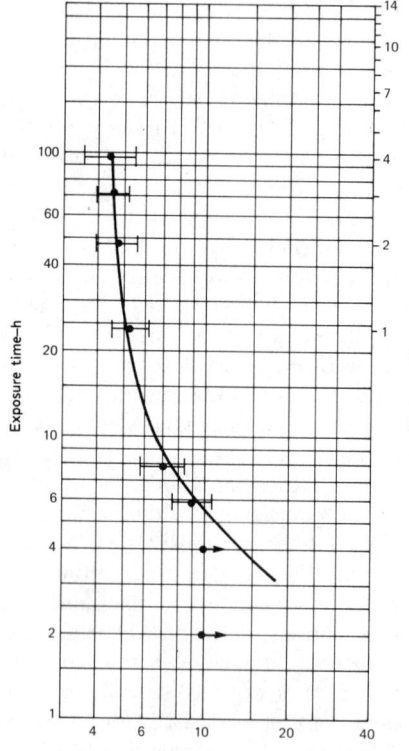

Median Concentration of Effluent—per cent by volume

Figure 8010:6. Toxicity curve, drawn from LC50s determined in Figure 8010:5. Curve almost asymptotic. The 95% confidence limits are shown for each LC50.

although more frequent observations of mortality may be required.

   d. Plotting toxicity curves: Most tests provide information on mortality at times before the final selected time. Use such information to plot a toxicity curve. Estimate LC50 from a graph plotted as in Figure 8010:5 for each observation time. Use the series of LC50s to construct a toxicity curve as the experiment proceeds (Figure 8010:6).

   A toxicity curve gives an overall picture of test progress and indicates when acute lethality has ceased. This is indicated by the curve becoming asymptotic to the time

axis.[1] In Figure 8010:6 the toxicity curve closely approaches an asymptote. The LC50 for an exposure time in the asymptotic part of the curve (asymptotic LC50) also can be termed the "threshold" or "incipient" LC50. The threshold of acute lethality for the median organism has greater theoretical significance than an LC50 for some arbitrary time.

   Incipient LC50s usually can be determined for most macroinvertebrates and fish within a 96- to 168-h exposure.[1] If no threshold is found, report this fact.

   e. Mortality in controls: Control mortality should not be greater than 10% and preferably not more than 5%. More than this is unsatisfactory and requires repetition of the test. Sometimes with long-term tests or with some invertebrates that have considerable mortality under the best possible conditions, it is necessary to use Abbott's formula:[13]

$$P = \frac{P^* - C}{1 - C}$$

where:

P and P^* = corrected and observed proportions responding to the experimental stimulus and

C = proportion responding in the control.

This approach does not solve the problem of probable interaction effects of the toxicant with whatever is causing mortality in the control.

2. Analyzing Results of Graded or Quantitative Toxicity Tests

   In quantitative or graded tests, each organism gives a response that is measurable on a continuous graded scale. For example, each organism might show a measurable percentage increase in body weight. Because there usually are many test organisms, a series of several graded measurements is generated for each test concentration. A one-way analysis of variance may be used initially to assess significance of differences.

Any of several techniques such as the Student-Newman-Keuls test,[14] Duncan's new multiple-range test,[15,16] or Dunnett's test[17] may be used to determine whether responses at a given concentration are significantly different from responses of the control group. Computer programs for computing these statistics are available.†

These techniques are applicable to most graded responses described in Sections 8111 through 8750.

3. Reporting Results

Report the LC50 with specified exposure time and 95% confidence limits. Graphically show the toxicity curve or a list of the LC50s for different exposure times. State mortality in controls.

Provide descriptions of: (a) test organisms, their species, number, source, weight, condition, acclimation to test conditions, treatment for disease and parasites before use, and observations on behavior during the test; (b) tested material, its source, storage, and physical and chemical properties; (c) dilution water, its source, physical and chemical properties and their variations during the test, pretreatment, additives, unusual constituents, or known contaminants; (d) test solution, its physical and chemical properties, especially the concentration of toxic component, if measurable, and the test temperature; (e) test method, a brief mention if standard, or its description if different, plus the specific experimental design; (f) test conditions, including type of container with volume and depth of solution, number of organisms and loading rate, toxicant delivery system, flow rate, or frequency of renewal; and (g) the criterion of response.

† Statistical Support Staff, Biological Methods Branch, Environmental Monitoring and Support Laboratory, U.S. Environmental Protection Agency, Cincinnati, Ohio.

4. References

1. SPRAGUE, J.B. 1970. Review paper, Measurement of pollutant toxicity to fish—II. Utilizing and applying bioassay results. Water Res. 4:3.
2. BRETT, J.R. 1967. Swimming performance of sockeye salmon (Oncorhynchus nerka) in relation to fatigue time and temperature. J. Fish. Res. Board Can. 24:1731.
3. SPRAGUE, J.B. 1968. Avoidance reactions of rainbow trout to zinc sulfate solutions. Water Res. 2:367.
4. FINNEY, D.J. 1971. Probit Analysis, 3rd ed. Cambridge Univ. Press, London & New York.
5. BERKSON, J. 1953. A statistically precise and relatively simple method of estimating the bioassay with quantal response based on the logistic function. J. Amer. Statist. Assoc. 48:565.
6. PICKERING, O.H. & W.N. VIGOR. 1965. The acute toxicity of zinc to eggs and fry of the fathead minnow. Progr. Fish-Cult. 27:153.
7. LITCHFIELD, J.T. & F. WILCOXON. 1949. A simple method of evaluating dose-effect experiments. Pharmacol. Exp. Ther. 96:99.
8. DIXON, W.J., ed. 1970. BMD biomedical computer programs. In Automatic Computation Ser. No. 2, 2nd ed., Univ. California Press, Los Angeles.
9. WALSH, G.E., C.H. DEANS, & L.L. MCLAUGHLIN. 1987. Comparison of the EC50s of algal toxicity tests calculated by four methods. Environ. Toxicol. Chem. 6:767.
10. LITCHFIELD, J.T. 1949. A method for rapid graphic solution of time-percent effect curves. Pharmocol. Exp. Ther. 97:399.
11. SHEPARD, M.P. 1955. Resistance and tolerance of young speckled trout (Salvelinus fontinalis) to oxygen lack, with special reference to low oxygen acclimation. J. Fish. Res. Board Can. 12:387.
12. SPRAGUE, J.B. 1973. The ABC's of pollutant bioassay using fish. In J. Cairns & K.L. Dickson, eds. Biological Methods for the Assessment of Water Quality. ASTM STP 528, p. 6. American Soc. Testing & Materials, Philadelphia, Pa.
13. TATTERSFIELD, F. & H.M. MORRIS. 1924. An apparatus for testing the toxic values of contact insecticides under controlled conditions. Bull. Entomol. Res. 14:223.
14. KEULS, M. 1952. The use of the "studentized

range" in connection with an analysis of variance. *Euphytica* 1:112.

15. DUNCAN, D.B. 1955. Multiple range and multiple F tests. *Biometrics* 11:1.

16. MIDDLEBROOKS, E.J., D.B. PORCELLA, E.A. PEARSON, P.H. MCGAUHEY & G.A. ROHLICH. 1971. Biostimulation and algal growth kinetics of wastewater. *J. Water Pollut. Control Fed.* 43:454.

17. DUNNETT, C.W. 1955. A multiple comparison procedure for comparing several treatments with a control. *J. Amer. Statist. Assoc.* 50:1096.

5. Bibliography

ABBOTT, W.S. 1925. A method of computing the effectiveness of an insecticide. *J. Econ. Entomol.* 18:265.

LITCHFIELD, J.T., JR. & F. WILCOXON. 1949. A simplified method of evaluating dose-effect experiments. *J. Pharm. Exp. Therapeut.* 96:99.

JENSEN, A.L. 1972. Standard error of LC50 and sample size of fish bioassays. *Water Res.* 6:85.

KENDALL, M.G. & A. STUART. 1973. The Advanced Theory of Statistics, Vol. 3, 3rd ed. Hafner Publishing Co., New York, N.Y.

ZAR, J.H. 1974. Biostatistical Analysis. Prentice-Hall, Englewood Cliffs, N.J.

STEPHAN, C.E. 1977. Methods for calculating an LC 50. *In* F.L. Mayer & J.L. Hamelink, eds. Aquatic Toxicology and Hazard Evaluation. ASTM STP 634, American Soc. Testing & Materials, Philadelphia, Pa.

FINNEY, D.J. 1978. Statistical Methods in Biological Assay, 3rd ed. Griffin Press, London, England.

8010 H.　Interpreting and Applying Results of Toxicity Tests

1. Interpretation of Results

The 48- and 96-h LC50 values produced from standard acute toxicity tests are useful measures of relative acute lethal toxicity to test organisms under specified conditions. These values do not necessarily have any direct meaning in terms of "safe" or "hazardous" conditions in natural water. Long-term exposure to much lower concentrations may be lethal to fish and other organisms and/or may cause nonlethal impairment of their function. Similarly, short-term exposure to these or higher values of total contaminants may cause no discernible effect.

2. Influence of Test Conditions

The results obtained in a toxicity test, in large part, depend on the conditions and nature of exposure; they are the product of operationally defined procedures. Therefore, it is important to select the testing procedures carefully to provide appropriate conditions and ensure that the results are applicable to the water quality problem at hand. At the outset define the problem

carefully and succinctly and establish how the results of the toxicity test will assist in the problem solution. Selection of type and species of test organism, duration of test, and physical and chemical conditions of test are key factors in obtaining useful, interpretable toxicity test results.

Although not always possible, it is best to prescribe toxicity test conditions that are as close as possible to the natural environmental conditions. For some variables such as pH, temperature, and contaminants, even small differences between the laboratory test conditions and those in the natural environment can reduce the utility of the results.[1] Some situations dictate that "standard" toxicity tests (i.e., incorporating a standard set of test conditions such as 96-h exposure, standard chemical and physical conditions, and standard test organisms) should be run in addition to those incorporating conditions more similar to those of the environment of concern. This practice provides results that could be compared with those reported in the literature as well as a potential route for adapting or interpreting results of other standard toxicity tests.

In some instances, in-stream toxicity tests with caged fish or other organisms are a more reliable means by which to evaluate toxicity properly. Such procedures can test for the impact of a complex variety of contaminants. They can aid in identifying the more and less toxic components of a wastewater through comparison of results (mortality rate, concentrations, and durations of exposure) with the results of tests on known components of the discharge. Also they can integrate the multitude of environmental factors affecting aquatic organism impact.

Factors such as expected rate of dilution, concentration/time relationships in the water, contaminant availability, and duration and pattern of organism exposure in the environment of concern compared to the test conditions, as well as functioning and sensitivity of the test organisms compared to those of concern, must be considered. A hazard assessment approach of the type described in recent literature[2-4] can be applied to the interpretation of toxicity test results. This is the coordinated evaluation of the physics, aquatic chemistry, and biology of the system, and the toxicology of the contaminant(s) to determine the impact on aquatic organisms or other specified beneficial uses of the water.

3. Statistical Interpretation

Another important aspect is statistics. It is crucial in conducting or interpreting the results of a toxicity test to understand clearly that statistically significant differences between control and test organisms, upstream and downstream populations, etc., are not necessarily changes or differences that have ecological or beneficial use impact significance. Furthermore, trends or other changes that appear to have biological/beneficial use significance may not be statistically demonstrable because of sample size or other limitations.

While the application and interpretation of toxicity tests in water quality management programs may appear to be more difficult than the more frequently used chemical test, they actually may prove to be similar. Toxicity tests generally provide a sounder basis upon which technically valid and cost-effective contaminant control programs can be based. If properly conducted, the tests evaluate the availability of contaminant forms that are present and incorporate appropriate durations of exposure of aquatic organisms. Additional references should be reviewed.[5-28]

4. References

1. LEE, G.F. 1973. Chemical aspects of bioassay techniques for establishing water quality criteria. *Water Res.* 7:1525.
2. LEE, G.F. & R.A. JONES. 1983. The role of water chemistry in measuring toxicity. *In* Aquatic Toxicology and Hazard Assessment: 6th Symposium. (Oct. 1981). American Soc. Testing & Materials, Philadelphia, Pa.
3. CAIRNS, J., JR., K.L. DICKSON & A.W. MAKI, eds. 1978. Estimating the Hazard of Chemical Substances to Aquatic Life. ASTM STP 657, American Soc. Testing & Materials, Philadelphia, Pa.
4. DICKSON, K.L., A.W. MAKI & J. CAIRNS, JR., eds. 1979. Analyzing the Hazard Evaluation Process. Water Quality Section, American Fisheries Soc., Bethesda, Md.
5. AMERICAN FISHERIES SOCIETY. 1979. A Review of the EPA Red Book: Quality Criteria for Water. American Fisheries Soc., Bethesda, Md.
6. BATES, J.M. & C.I. WEBER. 1981. Ecological Assessments of Effluent Impacts on Communities of Indigenous Aquatic Organisms. ASTM STP 730, American Soc. Testing & Materials, Philadelphia, Pa.
7. BERGMAN, H., ed. In press. Hazard Assessment for Complex Effluents. Proc. Cody Workshop on Environmental Assessment of Effluents. Soc. Environmental Toxicology & Chemistry, Cody, Wyo.
8. BRANSON, D.R., D.N. ARMENTROUT, W.M.

PARKER & C. VAN HALL. 1981. Effluent monitoring step by step. *Environ. Sci. Technol.* 15:513.

9. BUIKEMA, A.L., JR. & J. CAIRNS, JR., eds. 1980. Aquatic Invertebrate Bioassays. ASTM STP 715, American Soc. Testing & Materials, Philadelphia, Pa.

10. CAIRNS, J. & K.L. DICKSON. 1978. Field and laboratory protocols for evaluating the effects of chemical substances on aquatic life. *J. Testing Evaluation* 6:81.

11. CAIRNS, J., K.L. DICKSON & G.F. WESTLAKE, eds. 1976. Biological Monitoring of Water and Effluent Quality. ASTM STP 607, American Soc. Testing & Materials, Philadelphia, Pa.

12. CONWAY, R.A., ed. 1982. Environmental Risk Analysis for Chemicals. Reinhold Publishing Corp., New York, N.Y.

13. CRAIG, G.R. 1982. Toxicity of liquid industrial wastes to aquatic organisms. *In* Waste Treatment and Utilization—Theory and Practice of Waste Management, Vol. 2. Pergamon Press, New York, N.Y.

14. KENAGA, E.E. 1979. Aquatic test organisms and methods useful for assessment of chronic toxicity of chemicals. *In* Analyzing the Hazard Evaluation Process. Water Quality Section, American Fisheries Soc., Washington, D.C.

15. LAUGHLIN, T.A., C.R. CRIPE & R.J. LIVINGSTON. 1978. Field and laboratory avoidance reactions by blue crab to stormwater runoff. *Trans. Amer. Fish. Soc.* 107:78.

16. LEE, G.F. & R.A. JONES. In press. Water quality hazard assessment for domestic wastewaters. *In* H. Bergman, ed. Hazard Assessment for Complex Effluents. Proc. Cody Workshop on Environmental Assessment of Effluents. Soc. Environmental Toxicology & Chemistry, Cody, Wyo.

17. LEE, G.F. & R.A. JONES. 1983. Active vs. passive water quality monitoring programs for domestic and industrial wastewater discharges. *J. Water Pollut. Control Fed.* 55:405.

18. LEE, G.F., R.A. JONES & B.W. NEWBRY.

1982. Alternative approach to assessing water quality impact of wastewater effluents. *J. Water Pollut. Control Fed.* 54:165.

19. LEE, G.F., R.A. JONES, B.W. NEWBRY & T.J. HEINEMANN. 1982. Use of hazard assessment approach for evaluating impact of chlorine and ammonia in Pueblo, CO, domestic wastewaters on water quality in the Arkansas River. *In* Aquatic Toxicology and Hazard Assessment. Proc. ASTM 5th Symp. on Aquatic Toxicology, STP 766, American Soc. Testing & Materials, Philadelphia, Pa.

20. MARKING, L.L. & R.A. KIMERLE, eds. 1979. Aquatic Toxicology. ASTM STP 667, American Soc. Testing & Materials, Philadelphia, Pa.

21. MAYER, F.L. & J.L. HAMELINK, eds. 1976. Aquatic Toxicology and Hazard Evaluation. ASTM STP 634, American Soc. Testing & Materials, Philadelphia, Pa.

22. MAYER, F.L. & J.L. HAMELINK, eds. 1976. Quality Criteria for Water. U.S. Government Printing Off., Washington, D.C.

23. MOUNT, D.I. 1968. Chronic toxicity of copper to fathead minnows, *Pimephales promelas* Rafinesque. *Water Res.* 2:215.

24. NEWBRY, B.W., G.F. LEE, R.A. JONES & T.J. HEINEMANN. 1983. Studies on the water quality hazard of domestic wastewater treatment plant chlorine. *In* Water Chlorination: Environmental Impact and Health Effects. Ann Arbor Science Publ., Ann Arbor, Mich.

25. SPRAGUE, J.B. 1971. Review paper, Measurement of pollutant toxicity to fish—III. Sublethal effects and "safe" concentrations. *Water Res.* 5:245.

26. U.S. ENVIRONMENTAL PROTECTION AGENCY. 1976. Quality Criteria for Water. U.S. Government Printing Off., Washington, D.C.

27. U.S. ENVIRONMENTAL PROTECTION AGENCY. 1980. Notice of water quality criteria documents; availability. *Fed. Register* 45:231 (Nov. 28).

28. WARNER, R.E. 1967. Bioassays for microchemical environment contaminants. *Bull. World Health Org.* 36:181.

8030 MUTAGENICITY

8030 A. Introduction

A general method for detecting the presence of mutagens with a sensitive microorganism has been devised and is used widely for testing environmental materials.

8030 B. *Salmonella* Test

1. Principle

The *Salmonella* test,[1,2] also known as the Ames test, utilizes a set of histidine-requiring strains containing different types of mutations in the histidine operon. The standard test strains also contain other mutations that increase their sensitivity to chemical mutagens by various means. In this method tester strain cultures are grown in a nutrient broth and placed on agar plates containing nutrients, a microsomal preparation (S-9) obtained from rat liver, and the potential mutagen at various concentrations. For most mutagens, there is a linear dose-response curve for a certain concentration range of the mutagen. In the toxic range, there is a decrease in the number of revertants, i.e., cells that are able to grow in the absence of histidine. The number of revertants is a measure of the ability of the mutagen to produce change in DNA and therefore a measure of mutagenicity.

2. Test Procedure and Interpretation

The procedure is becoming standardized; in general, for initial screening of a chemical, use a three log concentration range in the presence and in the absence of the S-9 preparation. Interpret test results cautiously with regard to extrapolation to carcinogenicity, although the correlation is highly significant. Report mutagenicity test results as revertants per microgram of test compound taken from the linear portion of the dose-response curve.

3. References

1. AMES, B.N., J. McCANN & E. YAMASKI. 1975. Methods for detecting carcinogens and mutagens with the *Salmonella*/mammalian mutagenicity test. *Mutation Res.* 31:347.
2. MARON, D.M. & B.N. AMES. 1983. Revised methods for the *Salmonella* mutagenicity test. *Mutation Res.* 113:173.

8110　TEST PROCEDURES FOR ALGAE

Test procedures using algae are valuable for determining the primary productivity of a water as well as for testing toxicity of a water. The procedures that follow measure the response of certain algae to the nutritional status of a water (biostimulation, Section 8111) or measure the response of algae to materials that interfere with their normal metabolism (toxicity, Section 8112). Taken individually or together, the tests allow the assessment of the quality of a water to determine effects of point and nonpoint discharges on algal growth.

8111　BIOSTIMULATION (ALGAL PRODUCTIVITY)*

8111 A.　General Principles

The algal assays consist of three steps: (*a*) selection and measurement of appropriate factors or conditions during the assay (for example, biomass indicators such as total cell carbon); (*b*) presentation and statistical evaluation of measurements; and (*c*) interpretation of results.

Interpretation of results involves assessment of receiving water to determine its nutritional status and its sensitivity to change, potential effects of materials on algal growth in receiving waters, effects of changes in waste treatment processes on algal growth in receiving waters, impact of nutrients in tributary waters on algal growth in lakes and receiving waters, and effects of measures such as those used in lake restoration.

The maximum specific growth rate and the maximum standing crop are responses that can be estimated from growth measurements. The maximum specific growth rate is related to concentration of rate-limiting nutrient present. The maximum standing crop is proportional to the initial amount of limiting nutrient available.

The algal test procedure for determining primary productivity of a water sample is based on Liebig's "Law of the Minimum," which states that growth is limited by the substance that is present in minimal quantity in respect to the need of the organism. Biostimulants are substances that increase algal growth or the potential for algal growth.

Algal species used in biostimulation tests are selected to allow for a standardized test of growth response using a well-characterized organism under standard laboratory conditions. See Sections 10010, 10200, and 10300 for methods appropriate to field studies.

---

*Approved by Standard Methods Committee, 1988.

Effects of various substances on maximum specific growth rate and maximum crop of selected algal species cultured under specified conditions are measured. Results are assessed by comparing growth in the presence of selected substances to growth in controls. Experimental designs must incorporate sufficient replication to permit statistical evaluation of results.

8111 B. Planning and Evaluating Algal Assays

1. Sampling

Because water quality may vary greatly with time and point of collection, establish sampling programs to obtain representative and comparable data. It may be valuable to sample both epilimnion and hypolimnion in stratified water bodies. Use transection lines to secure adequate samples and locate sampling stations. In rivers and streams, take samples upstream and downstream from suspected nutrient sources and from tributary streams. When general conditions are evaluated, include samples from a number of natural waters having a range of representative water qualities.

Because the nutrient contents of natural waters and wastewaters often vary daily and seasonally, use composite or frequent grab sampling.

2. Test Variables

Deficiency of any essential nutrient may limit algal growth, but tests usually are made for those few nutrients most likely to be growth-limiting (nitrogen, phosphorus, trace elements).

To evaluate the potential effect of a substance on receiving waters, consider the following factors: amount and distribution, chemical and/or physical nature, fate and persistence, pathways by which it will reach the receiving water, dilution by the receiving body, and selection of appropriate test water.[1,2]

When the algal assay is used to measure stimulation of growth by a given effluent, include the following in the overall evaluation: effluent conditions, growth measurements and test organisms, concentration of growth-limiting nutrient, and potential nutrient concentration and changes in availability.

3. References

1. NATIONAL EUTROPHICATION RESEARCH PROGRAM. 1971. Algal Assay Procedure: Bottle Test. U.S. Environmental Protection Agency, Pacific Northwest Environmental Research Lab., Corvallis, Ore.
2. EUTROPHICATION AND LAKE RESTORATION BRANCH. 1974. Marine Algal Assay Procedure: Bottle Test. U.S. Environmental Protection Agency, Pacific Northwest Environmental Research Lab., Corvallis, Ore.

4. Bibliography

McGAUHEY, P.H., D.B. PORCELLA & G.L. DUGAN. 1970. Eutrophication of surface waters—Indian Creek reservoir. First Progress Rep., FWQA Grant No. 16010 DNY. U.S. Environmental Protection Agency, Pacific Northwest Environmental Research Lab., Corvallis, Ore.
MALONEY, T.E., W.E. MILLER & T. SHIROYAMA. 1971. Algal responses to nutrient additions in natural waters. Spec. Symp., American Soc. Limnology & Oceanography. Special Symposium on Nutrients and Eutrophication: Limiting-Nutrient Controversy 1:134.
MILLER, W.E. & T.E. MALONEY. 1971. Effects of secondary and tertiary wastewater effluents on algal growth in a lake-river system. J. Water Pollut. Control Fed. 43:2361.
MALONEY, T.E., W.E. MILLER & N.L. BLIND. 1972. Use of algal assays in studying eutrophication problems. Proc. Int. Conf. Water Pollut. Res. 6th. p. 205. Pergamon Press, Oxford, England & New York, N.Y.
SCHERFIG, J., P.S. DIXON, R. APPLEMAN & C.A.

JUSTICE. 1973. Effect of Phosphorus Removal on Algal Growth. Ecol. Res. Ser. 660/3-75-015, U.S. Environmental Protection Agency.

MILLER, W.E., T.E. MALONEY & J.C. GREENE. 1974. Algal productivity in 49 lakes as determined by algal assays. *Water Res.* 8:667.

SPECHT, D.T. 1975. Seasonal variation of algal biomass production potential and nutrient limitation in Yaquina Bay, Oregon. *In* E.J. Middlebrooks, D.H. Falkenborg, and T.E. Maloney, eds. Proceedings Workshop on Biostimulation and Nutrient Assessment, Utah State Univ., Logan, Sept. 10-12, 1975. PRWG 168-1; also published as Biostimulation and Nutrient Assessment. Ann Arbor Science Publ., Ann Arbor, Mich.

8111 C.   Apparatus

1. Sampling and Sample Preparation

a. Sampler, nonmetallic.

b. Sample bottles, borosilicate glass, linear polyethylene, polycarbonate, or polypropylene, capable of being autoclaved.

c. Membrane filter apparatus, for use with 47-mm petri prefilter pads and 0.45-μm-porosity filters.

d. Autoclave or pressure cooker, capable of producing 108 kPa at 121°C.

2. Culturing and Incubation

a. Culture vessels: Use permanently numbered erlenmeyer flasks of good-quality borosilicate glass. Use the same brand of glass throughout the laboratory. When trace nutrients are being studied, use special glassware such as high-silica glass, polycarbonate, or coated (e.g., silicone) glassware. While flask size is not critical, the surface-to-volume ratios of the growth medium are, because of CO_2 limitation. Use the following:

> 40 mL sample in 125-mL flask
> 60 mL sample in 250-mL flask
> 100 mL sample in 500-mL flask

Eliminate those flasks having anomalous growth from future tests.

b. Culture closures: Use demonstrably nontoxic foam plugs,* loose-fitting alumi-num foil, or inverted beakers to permit some gas exchange and prevent contamination. Determine for each batch of closures whether that batch has any significant effect on maximum specific growth rate and/or maximum standing crop.

c. Constant-temperature room: Provide constant-temperature room, or equivalent incubator, capable of maintaining temperature of 18 ± 2°C (marine) to 24 ± 2°C (freshwater).

d. Illumination: Use "cool-white" fluorescent lighting to provide 4304 lux ± 10% or 2152 lux ± 10% measured adjacent to the flask at the liquid level with closure in place.

An illumination meter can be used in place of a spectroradiometer to obtain the required light energy level. For example, the energy level output of a bank of six 144-cm "cool white" fluorescent lamps (GE 40-W, @ 60 Hz) was approximately 1300 μW/cm² (range, 380 to 760 nm) at a distance of 67 cm, as measured with an ISCO Model SRC spectroradiometer. With the same measurement geometry, a Weston Model 756 Illumination Meter reading was 4304 lux. All reflecting surfaces were matte white. Therefore, utilizing a calibrated illumination meter, one may, by adjusting the height of the lights, achieve a known energy level output of 1300 μW/cm². For further discussion of problems of differences in absorption of light by photosyn-

---

*Gaymar white, polyurethane foam plugs, VWR Scientific or Gaymar Industries, Inc., 701 Seneca St., Buffalo, N.Y. 14210, or *demonstrably nontoxic* equivalent.

thesizing organisms and by the human eye and their measurement, see Tyler.[1]

e. *Light meter:* Calibrate meter against a standard light source or light meter.

3. General Equipment

a. *Analytical balance* capable of weighing 100 g with a precision of ± 0.1 mg.

b. *Microscope and illuminator,* good quality, general purpose.

c. *Hemocytometer or plankton counting slide.*

d. *pH meter* to measure to ± 0.1 pH unit.

e. *Dry-heat oven* capable of operating at up to 120°C.

f. *Centrifuge* capable of a relative centrifugal force of at least 1000 × g.

g. *Spectrophotometer or colorimeter* for use at 600 to 750 nm.

4. Optional Equipment

a. *Electronic particle (cell) counter.*

b. *Fluorometer,* suitable for chlorophyll a.

c. *Shaker table,* capable of 100 oscillations/min.

5. Reference

1. TYLER, J.E. 1973. Applied radiometry. *Oceanogr. Mar. Biol. Annu. Rev.* 11:11.

8111 D. Sample Handling

1. Sampling Procedure

Use a nonmetallic water sampler and autoclavable storage container. Do not reuse containers when toxic or nutrient contamination is suspected. Leave a minimum of air space in the transport container and keep it in the dark at 0 to 4°C.

2. Removal of Indigenous Algae

To use unialgal test species, "remove" indigenous algae before assay by either separation or destruction. Filter, autoclave, or autoclave and filter, according to information sought.[1] In either case, prepare sample as soon as possible (within 24 h) after collection.

Use membrane filtration to remove indigenous algae before determining deficiencies in soluble nutrients that have not been taken up by filtrable organisms or to predict effect of added nutrients to a test water at a specific time. Pretreat 0.45-μm-porosity membrane filters by filtering at least 50 mL distilled or deionized water. Discard filtrate and use another collecting vessel. Filter the quantity of sample as needed under reduced pressure of 51 kPa or more. If the sample contains a large amount of suspended material, filter through an appropriate prefilter (for example, glass fiber), also washed with water, before filtration through the 0.45-μm-porosity filter.

Use autoclaving to determine amount of algal biomass that can be grown from all nutrients in the water, including those contained in filtrable organisms. Autoclave freshwater samples at 108 kPa and 121°C for 30 min or 10 min/L of sample, whichever is longer. Pasteurize marine or estuarine samples for 4 h at 60°C. After autoclaving, cool and let sample equilibrate either in an air or CO_2-rich atmosphere to restore lost CO_2. If an electronic particle counter is to be used for cell counting, pass the CO_2-equilibrated sample through a 0.45-μm membrane filter.

3. Storage

Changes occur in water samples during storage regardless of storage conditions. The extent and nature of these changes is not well known; therefore, keep storage duration to a minimum after sample preparation. Store samples in full containers with no air space. Before sample preparation, store samples in the dark at 0 to 4°C. If prolonged storage is anticipated, prepare sample first and then store in the dark at 0 to 4°C.

4. Reference

1. JOINT INDUSTRY-GOVERNMENT TASK FORCE ON EUTROPHICATION. 1969. Provisional Algal Assay Procedure. U.S. Environmental Protection Agency, Pacific Northwest Environmental Research Lab., Corvallis, Ore.

8111 E.   Synthetic Algal Culture Medium

See Section 8010E.4c1).

8111 F.   Inoculum

1. Recommended Test Algae

a. Freshwater algae:

1) *Selenastrum capricornutum* Printz.

2) *Microcystis aeruginosa* Kutz (*Anacystis cyanae* Drouet and Daily).

3) *Anabaena flos-aquae* (Lyngb.) de Brebisson.

4) *Cyclotella* sp.

5) *Nitzschia* sp.

6) *Synedra* sp.

For any of the latter three algae, add 20 mg Si/L (101.214 mg $Na_2SiO_3 \cdot 9H_2O/L$) to the culture medium as noted in Section 8010E.4c.

b. Marine algae:

1) *Dunaliella tertiolecta* Butcher (DUN Clone) Woods Hole Oceanographic Institution.

2) *Thalassiosira pseudonana* (Hasle and Heimdal) (CN Clone) (old *Cyclotella nana*) Univ. Rhode Island. Do not shake.

3) *Skeletonema costatum* (Greville) Clevel.

2. Sources of Test Algae

Obtain algal cultures from: Dr. Richard C. Starr, Culture Collection of Algae, Department of Botany, University of Texas, Austin, Tex. 78712 or Special Studies Branch, Corvallis Environmental Research Laboratory, Environmental Protection Agency, 200 SW 35th Street, Corvallis, Ore. 97330.

Other sources of cultures are: Virginia Institute of Marine Science, Gloucester Point, Va. 23062; Chesapeake Biological Laboratory, Box 38, Solomons, Md. 20688; Dr. Robert Guillard, Woods Hole Oceanographic Institution, Woods Hole, Mass. 02543; or Graduate School of Oceanography, University of Rhode Island, Narragansett, R.I. 02881.

After receipt of cultures, check authenticity and purity.

3. Maintaining Stock Cultures

a. Medium: See Section 8010E.4c1).

b. Incubation conditions:

1) Freshwater species—Temperature 24 ± 2°C under continuous cool-white fluorescent lighting at 4304 lux ± 10% for *S. capricornutum* and the diatoms; 2152 lux ± 10% for *M. aeruginosa* and *A. flos-aquae*; shake at 110 oscillations/min. If

other species are used, always relate growth of those species to *S. capricornutum* to insure comparability.

2) Marine species—Temperature 18 ± 2°C under continuous cool-white fluorescent lighting at 4304 lux ± 10% for *D. tertiolecta* (shake at 110 oscillations/min) and for *T. pseudonana* (do *not* shake but swirl daily). Higher temperatures (up to 24°C) may be justified for appropriate test species used in the Gulf of Mexico and other warm-water marine systems. If other species are used, always relate growth of those species to *D. tertiolecta* or *S. costatum* to insure comparability.

c. First stock transfer: Upon receipt of inoculum species, transfer a portion to the algal culture medium. (Example: 1 mL of inoculum in 100 mL in a 500-mL erlenmeyer flask).

d. Subsequent stock transfers: Make a new stock transfer, using aseptic technique, as the first operation on opening a stock culture. The volume transfered is not critical so long as enough cells are included to overcome significant growth lag. Make weekly stock transfers to provide a continuing supply of "healthy" cells. Check algal cultures microscopically to insure that the stock cultures remain unialgal.

e. Age of inoculum: Use cultures 1 to 3 weeks old as a source of inoculum. For *Selenastrum, Dunaliella,* and the diatoms, a 1-week incubation often is sufficient to provide enough cells. The blue-green species require a longer time to achieve maximum crop and 2 to 3 weeks may be required to provide test inocula.

4. Preparing Inoculum

Centrifuge stock culture and discard supernatant. Resuspend sediment cells in an appropriate volume of glass-distilled water containing 15 mg $NaHCO_3$/L for freshwater species and artificial seawater minus nutrients for marine species (Section 8010E.4*c*1); Table 8010:II) diluted to appropriate salinity, and again centrifuge. Resuspend sedimented algae in the proper solution and use as the inoculum.

5. Amount of Inoculum

Count cells suspended in the prepared inoculum and pipet into the test water to give a starting cell concentration as follows:

S. capricornutum	10^3 cells/mL
M. aeruginosa	50×10^3 cells/mL
A. flos-aquae	50×10^3 cells/mL
Diatoms	10^3 cells/mL
D. tertiolecta	10^2 cells/mL

Calculate volume of transfer to result in the above concentrations in the test flasks (e.g., for *S. capricornutum*, if there are 5 $\times 10^5$ cells/mL in stock culture, transfer 0.2 mL/100 mL test water).

8111 G.   Test Conditions and Procedures

1. Temperature

Keep temperature at 18 ± 2°C for marine species and 24 ± 2°C for freshwater species.

2. Illumination

See Section 8111F.3*b*. Measure light intensity adjacent to the flask at the liquid level. A convenient and acceptable way to achieve these two light intensities is to set up the illumination for 4304 lux and then place cheesecloth over the blue-green algal culture flask to reduce the light intensity at the liquid level to 2052 lux.

3. Procedure

a. Preparation of glassware: Wash all glassware with detergent (nonphosphate or sodium carbonate) and rinse thoroughly with tap water. Never use chromic acid or similar cleaning solutions. Then rinse with a warm 10% (v/v) solution of reagent-grade HCl. Fill vials and centrifuge tubes with 10% HCl solution for a few minutes. Fill all large containers to about one-tenth capacity with HCl solution and swirl to bathe entire inner surface. After HCl rinse, rinse glassware five times with tap water, then five times with deionized water.

Place pipets in 10% HCl solution for 12 h or longer and rinse at least 10 times with tap water in an automatic pipet washer, then three times with deionized water. Preferably use disposable pipets to eliminate the need for pipet washing and to minimize possibility of contamination.

Dry cleaned glassware at 105°C in an oven and store either in closed cabinets or on open shelves with tops covered with aluminum foil.

Before use, autoclave culture flasks covered with aluminum foil at 108 kPa for 15 min. After autoclaving, prerinse flasks with culture medium and invert on absorbent paper for 20 to 30 min to drain.

b. pH control: To insure the availability of CO_2, keep the pH below 8.5 by using optimum surface-to-volume ratios (Section 8111C.2), continuously shaking the flask (approximately 100 oscillations/min), and/or ventilating with air or air/CO_2 mixture.

c. Growth measurements: Describe the growth of a test alga in the bottle test[1] by maximum specific growth rate and maximum standing crop.[2] Generally, these measurements are made at different times, the former early in the test and the latter near the end.

1) Maximum specific growth rate—The maximum specific growth rate (μ_{max}) for an individual flask is the largest specific growth rate (μ) occurring at any time during incubation. Determine μ_{max} for a set of replicate flasks by averaging μ_{max} of the individual flasks.

The specific growth rate, μ, is defined by:

$$\mu, \mathrm{d}^{-1} = \frac{\ln (X_2/X_1)}{t_2 - t_1}$$

where:

X_2 = biomass concentration at end of selected time interval,

X_1 = biomass concentration at beginning of selected time interval, and

$t_2 - t_1$ = elapsed time between selected intervals, d.

If biomass (dry weight) is determined indirectly, e.g., by cell counts, compute specific growth rate directly without converting to biomass if the factor relating the direct determination to biomass remains constant for the time period. Cell counts, chlorophyll content, and other measures of biomass do not necessarily remain constant in relation to one another or to dry weight.

a) Laboratory measurements—The maximum specific growth rate occurs during the logarithmic phase of growth, usually between Day 0 and Day 5; therefore, measure biomass at least daily during the first 5 d of incubation. Indirect measurements of biomass, such as cell counts, normally will be required because of the difficulty in making accurate gravimetric measurements at low cell densities.

b) Computation of maximum specific growth rate—Calculate maximum specific growth rate (μ_{max}) by using the equation in ¶ 1) to determine the daily specific growth rate (μ) for each replicate flask. Average all replicates, using the largest value for each flask. Alternatively, prepare a semilog plot of biomass concentration versus time for each flask. Ideally, the exponential growth phase is drawn on the plot. If it appears that the data describe two straight lines, use the line of steepest slope. A linear regression analysis of the data also may be used to determine the best-fit straight line. Select two data points that most closely fit the line and determine the specific growth rate (μ). Average the largest specific growth rates for the replicate flasks to obtain μ_{max}.

2) Maximum standing crop—Maximum standing crop in any flask is defined as the maximum algal biomass achieved during incubation. For practical purposes, it may be assumed that the maximum standing crop has been achieved when the increase in biomass is less than 5%/d.

After attaining the maximum standing crop, determine the dry weight of algal biomass gravimetrically. If biomass is determined indirectly, convert results to an equivalent dry weight using appropriate conversion factors.

4. Biomass Monitoring

Several methods may be used, but relate each to dry weight.

a. Dry weight: Use either the aluminum dish or membrane filter method. To use the first, centrifuge a suitable portion of algal suspension, wash sedimented cells three times in distilled water containing 15 mg $NaHCO_3$/L, transfer to tared crucibles or aluminum cups, dry overnight in a hot-air oven at 105°C, and weigh. This method is more sensitive than the second method but cells may be lost during washing.

For the second method, filter a measured portion of algal suspension through a tared 0.45-μm-pore-diam membrane filter. Dry filters for several hours at 60°C in an oven. Place filters in folded sheets of paper or aluminum weighing dishes on which the weights or codes are written. Cool in a desiccator and weigh. Filter a suitable portion of culture under a vacuum of 51 kPa (50 mL or less as the volume or cell density dictates). Rinse filter funnel with 50 mL distilled water containing appropriate salts (8111F.4) using a wash bottle and let the rinsings pass through the filter. Dry at 60°C, cool in desiccator, and weigh. To correct for loss of weight of filters during washing, wash two blank filters with 50 mL distilled water, pouring it through slowly under reduced vacuum. Dry and weigh filters and record weight loss. This correction is not large, but is essential for meaningful results on dilute cultures.

b. Cell counting:

1) Electronic particle counting—Suspend *S. capricornutum* cells in a 1% NaCl electrolyte solution in a ratio of 1.0 mL cell suspension to 9 mL of 0.22-μm-filtered saline (10:1 dilution). Pass the resulting suspension through a 100-μm-diam aperture. Each cell that passes through the aperture causes a voltage drop proportional to its displaced electrolyte volume, which is recorded as a count. A knowledge of both the number of particles (cells) counted per unit volume of sample (usually 0.5 mL) and the mean particle (cell) volume displaced allows changes in cell biomass (in microliters per liter) to be calculated. Equations have been developed that accurately relate volume to dry weight.

2) Direct microscopic counting—Use a hemocytometer or plankton counting cell (Section 10200F.2). For filamentous algae break up the algal filaments by using a syringe, an ultrasonic bath, a high-speed blender, or vigorous stirring with glass beads. Each of these techniques has drawbacks, but expelling the sample forcefully through a syringe against the inside of the flask is most satisfactory. Other methods of biomass measurement such as dry weight, absorbance, or chlorophyll fluorescence are more precise than cell counts for growth assessment of filamentous algae.

c. Absorbance: Measure absorbance with a spectrophotometer or colorimeter at a wavelength of 750 nm. Report instrument make and model, geometry and path length of the cuvette, wavelength used, and the equivalence to biomass (absorbance units per milligram dry weight per liter).

Limit photometric measurement of absorbance to a range of $0.05 < D < 1.0$, where D represents optical density.

d. Chlorophyll: All algae contain chlorophyll and measuring this pigment can yield some insight into the relative amount of algal biomass present.[2,3] Measure chlorophyll either in vivo or in vitro by fluorescence.

1) In vivo procedure—Swirl flasks to homogenize. Pipet a portion of cell suspension (5 mL minimum) into a small beaker or vial. Zero fluorometer with a distilled water blank before each sample reading. Pour well-mixed sample into cuvette and read fluorescence.

2) In vitro procedure—See Section 10200G.

e. *Total cell carbon:* Determine by carbon analyzer. Report equivalence between total cell carbon and dry weight in milligrams per liter.

5. References

1. NATIONAL EUTROPHICATION RESEARCH PROGRAM. 1971. Algal Assay Procedure: Bottle Test. U.S. Environmental Protection Agency, Pacific Northwest Environmental Research Lab., Corvallis, Ore.

2. JOINT INDUSTRY-GOVERNMENT TASK FORCE ON EUTROPHICATION. 1969. Provisional Algal Assay Procedure. U.S. Environmental Protection Agency, Pacific Northwest Environmental Research Lab., Corvallis, Ore.

3. STRICKLAND, J.D.H. & T.R. PARSONS. 1965. A Manual of Sea Water Analysis, 2nd rev. ed. Fisheries Research Board Canada Bull. 125.

6. Bibliography

WEISS, C.M. & R.W. HELMS. 1971. Interlaboratory Precision Test—An Eight Laboratory Evaluation of the Provisional Algal Assay Procedure: Bottle Test. Dep. Environmental Science & Engineering, School Public Health, Univ. North Carolina, Chapel Hill.

8111 H.   Effect of Additions

The quantity of cells produced in a given medium is limited by the concentration of nutrient present in the lowest relative quantity with respect to the needs of the organism. If a quantity of the limiting nutrient is added to the medium, cell production increases until this additional supply is depleted or until some other nutrient becomes limiting. Additions of substances other than that which is limiting would yield no increase in cell production. Nutrient additions may be made singly or in combination and the growth response compared to that of untreated controls to identify those substances that limit growth rate or cell production. The selection of additives, e.g., nitrogen, phosphorus, iron, wastewater effluents, will depend on the requirements of the test.

In all cases, keep the volume of added nutrient solution as small as possible, but make it large enough to yield a potentially measurable response. Relate the concentrations of additions to nutrient levels in the sample. To assess the effect of nutrient additions, compare treated sample to an untreated control. For highly productive controls, flask-to-flask variations may be high and might mask the effect of small additions of the limiting nutrient.

It is sometimes necessary to check the test water for the presence of toxic material. To do this, treat the sample with an appropriate dilution of the complete synthetic medium. If no growth, or less than expected growth, occurs, toxic materials are suspected. In some situations, sample dilution or addition of a chelating agent will eliminate toxic effects.

8111 I. Data Analysis and Interpretation

I. Reporting Requirements

The fundamental measure used in the algal assay to determine biostimulation is the amount of suspended solids (dry weight) produced and determined gravimetrically. Other biomass indicators may be used, but all results must include experimentally determined conversion factors and the dry weight of suspended solids. Use several biomass indicators whenever possible, because biomass indicators respond differently to any given nutrient-limiting condition.

Report results of addition assays with results from two types of reference samples: the assay reference medium and untreated water samples. Present entire growth curves for each of the two types of reference sample. Report results of individual assays as maximum specific growth rate (with time of occurrence) and maximum standing crop (with time at which it was reached).

To determine the nutrients that limit growth rate by single-nutrient additions, treat a number of replicate flasks with single nutrients, determine the maximum specific growth rate for each flask, and compare the averages by Student's t test or other appropriate statistical tests.

To identify growth-rate-limiting nutrients by multiple-nutrient tests, make analysis-of-variance calculations. Account for possible interaction between different nutrients by using factorial analysis.[1]

The methods for finding growth-rate-limiting nutrients are used also to determine the nutrient that limits maximum standing crop. Determine the available concentration of growth-limiting nutrient by comparing maximum standing crop in an untreated sample with a maximum standing crop in reference medium.

Report both maximum specific growth rate and maximum standing crop with their confidence intervals. Base the calculation of confidence interval for the average values on at least five samples. Consequently, make a minimum of five replications when an unfamiliar source water is first analyzed. Use these results to calculate the standard deviation. For subsequent samples from the same source use only three replicates and report with the confidence interval established for that source water.

In algal assays, occasionally one flask among replicates shows a considerable growth difference from the others. Eliminate such outliers from the results (and flasks from apparatus) if they fall outside the 95% confidence limits.

The overall evaluation of assay results consists of first determining whether a result is significant when considered as a laboratory measurement. Several methods are available, such as the Student's t test and analysis-of-variance techniques.

The second part of the evaluation is the correlation of laboratory assay results to effects observed or predicted in the field. Specific guidelines are not yet available, but note the general considerations in Section 8111B.

2. References

1. EUTROPHICATION AND LAKE RESTORATION BRANCH. 1974. Marine Algal Assay Procedure: Bottle Test. U.S. Environmental Protection Agency, Pacific Northwest Environmental Research Lab., Corvallis, Ore.

8112   TOXICITY TESTING WITH PHYTOPLANKTON*

8112 A.   Introduction

The phytoplankton are primary producers in the aquatic community and, as such, are at the base of aquatic food chains. Because of this, they must be tested in bioassays that predict and determine the

___
*Approved by Standard Methods Committee, 1988.

potential effects of a substance on the aquatic environment. The same general principles and techniques used in determining biostimulation (Section 8111) are used to determine toxicity to phytoplankton. The procedure applies to freshwater, estuarine, and marine phytoplankton.

8112 B.   Inoculum

In addition to the marine or estuarine algae listed in 8111F.1b, *Monochrysis lutheri* Droop may be used. Maintain the test species in full-strength media [Section 8010E.4c1) and Tables 8010:III and 8010:IV.A and B]. Test species must be in the logarithmic growth phase; therefore, transfer them to fresh culture medium every 4 to 5 d.

8112 C.   Test Conditions and Procedures

1. Maximum Specific Growth Rate

Add test material to test vessels to give desired concentrations. Prepare triplicate vessels for each concentration. Use dilutions of culture medium to simulate chemical conditions of specific receiving waters. For optimum surface-to-volume ratios, see Section 8111C.2a.

The maximum specific growth rate (μ_{max}) occurs during the logarithmic phase of growth, usually between Day 0 and Day 5. Therefore, measure biomass at least daily during the first 5 d of incubation. Indirect measurements of biomass, such as chlorophyll *a* or cell numbers, usually will be required because accurate gravimetric measurements at low cell densities are dif-

ficult. See Section 8111G.3c and 8111G.4 for methods.

Test a geometric series of concentrations initially (see Section 8010F.3b). After this preliminary test, progressively bisect intervals on a logarithmic scale. Narrow the range of test concentrations to determine the concentration that reduces the maximum specific growth rate (μ_{max}) to 50% that of the control. This requires that two of the concentrations tested fall on each side of the concentration that inhibited (μ_{max}) to 50% (see Section 8010G).

Compare the maximum specific growth rate (μ_{max}) to that obtained in the synthetic freshwater or artificial seawater culture medium. Regional and seasonal variations in quality make natural waters unsuitable as standard test media for comparative toxicity tests. Therefore, use a synthetic freshwater medium and/or artificial seawater. Add various concentrations of toxicants to the culture medium in triplicate and inoculate with test species.

2. Other Tests

For other types of tests, such as those to determine effluent requirements or compliance with water quality standards, take dilution water from the receiving body near the outfall but outside its influence. Remove undesirable organisms before making growth rate tests with selected sensitive species (Section 8111D). Determine maximum specific growth rates (μ_{max}) in test vessels and compare with controls and EC50s based on percent of growth reduction. An alternative approach that provides a number that may relate to natural conditions should be reviewed.[1]

3. Reference

1. MILLER, W.E., J.C. GREENE & T. SHIROYAMA. 1978. The *Selenastrum capricornutum* Printz Algal Assay Bottle Test. U.S. Environmental Protection Agency Rep. EPA-600/9-78-018, National Technical Information Serv., U.S. Dep. Commerce, Springfield, Va.

4. Bibliography

ERICKSON, S.J., N. LACKIE & T.E. MALONEY. 1970. A screening technique for estimating copper toxicity to estuarine phytoplankton. *J. Water Pollut. Control Fed.* 42:R270.

WALSH, G.E. 1972. Effects of herbicides on photosynthesis and growth of marine unicellular algae. *Hyacinth Control J.* 10:45.

GREEN, J.C., W.E. MILLER, T. SHIROYAMA & E. MERWIN. 1975. Toxicity of Zinc to the Green Alga *Selenastrum capricornutum* as a Function of Phosphorus or Ionic Strength. U.S. Environmental Protection Agency Rep. EPA-660/3-75-034, National Technical Information Serv., U.S. Dep. Commerce, Springfield, Va.

GREEN, J.C., W.E. MILLER, T. SHIROYAMA, R.A. SOLTERO & K. PUTNAM. 1976. Use of algal assays to assess the effects of municipal and smelter wastes upon phytoplankton production. *In* Terrestrial and Aquatic Ecological Studies of the Northwest. Eastern Washington State College Press, Cheney.

8310 TOXICITY TEST PROCEDURES FOR CILIATED PROTOZOA*

8310 A. Introduction

1. Ecological Importance

Protozoa, algae, and bacteria form the broad base of aquatic food chains. Ciliated protozoa are the most numerous animals of the estuarine benthos[1,2] and may be more important than bacteria as regenerators of nutrients, particularly nitrogen and phosphorus.[3,4] Further, some ciliates can concentrate certain persistent pesticides and related chemicals[5-7] and thereby aid in their translocation to higher trophic levels. Thus it is possible that the effects of such toxicants could be exerted at higher trophic levels either through disruption of nutrient cycles or through biological concentration of the toxicants higher in the food chain.[8,9]

---

*Approved by Standard Methods Committee, 1988.

The procedures described herein have been used successfully to test toxicity and bioaccumulation of toxic materials.[5,6] Responses measured are the effects on population growth rate and maximum population density and degree of accumulation and concentration of toxicants.

2. References

1. BORROR, A.C. 1963. Morphology and ecology of the benthic ciliated protozoa of Alligator Harbor, Florida. *Arch. Protistenk.* 106:465.

2. FENCHEL, T. 1967. The ecology of marine microbenthos. I. The quantitative importance of ciliates as compared with metazoans in various types of sediments. *Ophelia* 4:121.

3. JOHANNES, R.E. 1965. Influence of marine protozoa on nutrient regeneration. *Limnol. Oceanogr.* 10:434.

4. JOHANNES, R.E. 1968. Nutrient regeneration in lakes and oceans. *In* M.R. Droop & E.J. Ferguson Wood, eds. Advances in Microbiology of the Sea 1:203. Academic Press, New York, N.Y.

5. COOLEY, N.R., J.M. KELTNER, JR. & J. FORESTER. 1972. Mirex and Aroclor® 1254: Effect on and accumulation by *Tetrahymena pyriformis* strain W. *J. Protozool.* 19:636.

6. COOLEY, N.R., J.M. KELTNER, JR. & J. FORESTER. 1973. The polychlorinated biphenyls, Aroclors® 1248 and 1260: Effect on and accumulation by *Tetrahymena pyriformis. J. Protozool.* 20:443.

7. GREGORY, W.W., JR., J.K. REED & L.E. PRIESTER, JR. 1969. Accumulation of parathion and DDT by some algae and protozoa. *J. Protozool.* 16:69.

8. BURDICK, G.E., E.J. HARRIS, H.J. DEAN, T.M. WALKER, J. SKEA & D. COLBY. 1964. The accumulation of DDT in lake trout and the effect on reproduction. *Trans. Amer. Fish. Soc.* 93:127.

9. BUTLER, P.A. 1969. Monitoring pesticide pollution. *Bio-Science* 19:889.

8310 B.   Selecting and Preparing Test Organisms

1. Obtaining Test Organisms

Tetrahymena pyriformis is the recommended test organism because it occurs in freshwater and salt marshes, has worldwide distribution, and is grown readily in axenic culture. Its physiology has been studied extensively.[1,2] Strain W is used in this test; however, several other strains can be used. These may be obtained from the American Type Culture Collection (ATCC), 12301 Parklawn Drive, Rockville, Md. 20852. Instructions for cultivation will be furnished when requested. If other strains are used, note their source.

2. Holding and Culturing Test Organisms

a. Culture of test organisms: Use standard bacteriological techniques to prepare and autoclave culture media and to transfer axenic cultures of *T. pyriformis*. Maintain stock cultures at 26 ± 0.5°C. Slant culture tubes containing 10 mL of medium at 60° from the vertical to enhance aeration. Keep duplicate stock cultures at room temperature and hold them upright to retard growth by decreasing aeration. Grow cultures axenically in 18- × 150-mm bacteriological culture tubes capped with polypropylene or stainless steel closures and supported in vinyl-coated 40-tube racks.

b. Culture medium: Use the same medium for stock and test cultures: 2% (w/v) proteose peptone, 0.2% (w/v) yeast extract, 0.5% glucose, and 90μM Fe:EDTA chelate/L. Adding Fe:EDTA chelate eliminates population growth variation due to differences in iron content of different lots of proteose peptone. Prepare chelate by the method of Conner and Cline.[3] Omit Fe:EDTA chelate when testing metal toxicity.

Prepare medium in distilled water or artificial seawater [Section 8010E.4c1)b)] diluted to a lower salinity, dispense among culture tubes, and autoclave for 15 min at 108 kPa. Before inoculation let stand for a time long enough to restore DO. Best growth occurs in the distilled water formulation, although there is no significant difference between the growth rates of populations grown in the distilled water medium and the 5 g/kg salinity medium.

2. References

1. CORLISS, J.O. 1970. The comparative systematics of species comprising the hymenostome ciliate genus *Tetrahymena. J. Protozool.* 17:198.
2. ELLIOTT, A.M., ed. 1973. Biology of *Tetrahymena.* Dowden, Hutchinson & Ross, Inc., Stroudsburg, Pa.
3. CONNER, R.L. & S.G. CLINE. 1964. Iron deficiency and the metabolism of *Tetrahymena pyriformis. J. Protozool.* 11:486.

8310 C.   Toxicity Test Procedures

1. Preparation of Toxicant Stock Solutions

Prepare a sterile stock solution for each concentration as described in Section 8010F.3b. For water-insoluble toxicants, use acetone and polyethylene glycol 200 as solvents at a final concentration of 0.1%. During range-finding tests always test for interaction of toxicant and solvent at the solvent concentrations used.

2. Conducting Tests

a. *Test procedures:* Use test procedures described in Section 8010F. Select and prepare test concentrations of toxicants as described in Sections 8010F.2b and 3b. Prepare 11 or 12 optically matched culture tubes for each concentration, 10 tubes for test solutions and a control and special controls for solvent and emulsifier if used. Inoculate each tube with 1 mL of a 65-h culture diluted by adding 3 mL culture to 7 mL sterile medium. Adjust volume so that each tube contains 10 mL of mixed toxicant and inoculum. Run tests for 96 h in an incubator in which temperature is either kept constant or varied daily and seasonally in accordance with the natural temperature regime of the area. Concurrently use five concentrations of toxicant (see Section 8010F.3b). Because large variations usually are observed among replicates, use six replicates to insure statistical validity at $p = 0.05$. Analyze effects of toxicant on growth rate and population density by analysis of variance (see Section 8010G).

Run tests to determine bioaccumulation for 7 d in replicate using 10 cultures per concentration plus controls.

b. *Analytical procedures:* Chemically analyze toxicant solutions at beginning and end of test (see Section 8010F.3d).

At end of test, pool all cultures from each concentration and concentrate cells by centrifugation at 1000 g. Wash cells twice in toxicant-free medium and store, if necessary, before analysis, in sealed tubes at $-20°C$. Typically 100 mL of a 96-h control culture yields about 1 g packed cells. Analyze medium and cells separately.[1,2] Calculate residues on a wet-weight basis. To obtain dry weight values prepare replicate cultures and dry cells at 60°C for 65 h.

3. References

1. SLATER, J.V. & A.M. ELLIOTT. 1951. Volume change in *Tetrahymena* in relation to age of the culture. *Proc. Amer. Soc. Protozool.* 2:20.
2. HATCH, W.R. & W.L. OTT. 1968. Determination of sub-microgram quantities of mercury by atomic absorption spectrophotometry. *Anal. Chem.* 40:2085.

8310 D. Evaluating and Reporting Results

1. Evaluating Data

Measure population density in a spectrophotometer* as absorbance at 540 nm. Use blanks of sterile proteose-peptone medium with the appropriate concentration of toxicant and carrier. Absorbance values cannot be converted to absolute numbers of ciliates;[1] therefore, use mean absorbance to estimate differences in population density between cultures. Estimate effects of toxicants on growth rate and population density from absorbance measurement of control and test cultures at 0, 4, 8, 16, 24, 36, 48, 60, 72, 84, and 96 h. Plot these data to determine the period of exponential growth for each population. Because growth in this period is logarithmic, estimate growth rate by the quantity b of the least squares estimate of $y = a + bx$. The calculated line gives a close fit to the data, generally $r \geq 0.9$. Determine effect of toxicant on growth rate and population density as described in Sections 8010G.1 and 2.

---

*The Bausch and Lomb Spectronic 20 has been used successfully in this type of study because it accepts large culture tubes. Other available instruments that can be adapted to large tubes and should be satisfactory are Coleman Junior and Junior II and Turner Model 330 Spectrophotometers.

8410 TOXICITY TEST PROCEDURES USING SCLERACTINIAN CORAL*

8410 A. Introduction

1. Ecological Importance

The scleractinian corals[1-7] comprise one of the most conspicuous and important components of many tropical reef ecosystems. They are especially sensitive to environmental perturbations and are valuable indicator organisms of water quality in shallow tropical marine environments.

The reef-forming corals flourish only within a narrow range of chemical and physical conditions including clear water,

---

*Approved by Standard Methods Committee, 1988.

low inorganic and organic nutrients, low sedimentation rate, and tropical open-ocean temperature and salinity.

Corals are valued for their beauty. They also form protecting reefs about islands and lagoons.[8] These massive structures provide an essential environment for many organisms and are the basis for an entire ecosystem. Reef corals are slow to reestablish themselves; the process often requires decades. Once the corals are killed, the imbalance may lead to erosion and substantial environmental modification such that the area becomes permanently unsuited for coral recovery.

2. Sensitivity to Environmental Conditions

Corals lack specialized circulatory and excretory systems and are highly sensitive to changes in their physical-chemical environment. Unlike most reef animals, corals cannot migrate or burrow to protect themselves from localized short-term environmental stress, nor can they isolate themselves temporarily from harsh conditions by withdrawing into a shell or tube.

3. References

1. WELLS, J.W. 1954. Recent corals of the Marshall Island, Bikini and nearby atolls. Part 2, Oceanogr. (Biologic). *Geol. Survey Paper* 260:385. Pl. 94-185. Washington, D.C.

2. WELLS, J.W. 1956. Scleractinia. *In* R.C. Moore, ed. Treatise on Invertebrate Paleontology. F328-F444, Geological Soc. America.

3. VAUGHAN, T.W. & J.W. WELLS. 1943. Revision of the suborders, families and genera of the Scleractinia. *Geol. Soc. Amer. Spec. Papers* 44:1.

4. CROSSLAND, C. 1952. Madreporaria, Hydrocorallinae, Heliopora and Tubipora. Great Barrier Reef Expedition. VI:86. British Museum (Natural History).

5. HICKSON, S.J. 1924. An Introduction to the Study of Corals. Manchester Univ. Press, London & New York.

6. VAUGHAN, T.W. 1907. Recent Madreporaria of the Hawaiian Islands and Laysan. Bull. LIX, Smithsonian Inst., U.S. National Museum, Washington, D.C.

7. SMITH, F.G. 1971. Atlantic Reef Corals. Univ. Miami Press, Miami, Fla.

8. YONGE, C.M. 1963. The biology of coral reefs. *Advan. Mar. Biol.* I(4):209.

8410 B. Selecting and Preparing Test Organisms

1. Suggested Test Species

a. *Tropical, Indo-Pacific area* (species listed in order of their importance):

1) Branching *Acropora*
Acropora formosa (Dana)
Acropora is not present in Hawaii. Substitute the branching form of *Montipora verrucosa* Lamarck.

2) Finely branched *Pocillopora*
Pocillopora damicornis (Linnaeus)
Synonyms: *Pocillopora caespitosa* Dana, *Pocillopora bulbosa*
Alternate: *Pocillopora brevicornis* Lamarck.

3) Branching *Porites*
Porites compressa Dana or small *Porites lobata* Dana
Alternate: *Porites andrewsi* Vaughan.

4) Representative of solitary hermatypic corals: *Fungia scutaria* Lamarck.

5) Representative of the ahermatypic (lacking zooxanthellae) corals: *Tubastrea aurea* (Quoy and Giamard).

b. *Tropical Atlantic area* (test groups listed in order of their importance):

1) Branching *Acropora*
Acropora cervicornis (Lamarck).

2) Branching *Porites*
Porites porites (Pallas)
Alternate: *Porites furcata* (Lamarck).
Pocillopora is not present in the Atlantic.

3) Other widely used corals:
Meandrina meandrites (Linnaeus) forma *meandrites*
Montastrea annularis (Ellis and Solander)
Montastrea cavernosa (Linnaeus) forma *areolata*.

4) Representative of solitary hermatypic corals: *Scolymia lacera* (Pallas) formerly *Mussa lacera* (Pallas).

5) Representative of ahermatypic (lack-

ing zooxanthellae) corals: *Tubastrea aurea* (Quoy and Giamard).

This list of recommended species is not intended to restrict selection of test species, but rather to act as a guide in choosing test organisms.

2. Selecting Test Organisms

Because basic differences exist between Caribbean and Indo-Pacific coral fauna, it can be difficult to find comparable species from the two areas. Wells[1] estimated that 36 species occur in the Caribbean as opposed to at least 500 for the Indo-Pacific area. Few coral species are common to both oceans. Genera studied intensively in the Indo-Pacific include *Acropora, Pocillopora, Porites*, and *Fungia. Meandrina, Montastrea*, and *Manicina* often are used as laboratory subjects in the Caribbean.

Except for *Acropora cervicornis*, all suggested species have been used successfully. *A. cervicornis* is listed as the primary test species for the Atlantic because it is widely distributed, very sensitive, and similar to the important Indo-Pacific species *A. formosa.* Although *A. cervicornis* could eventually prove to be too delicate to maintain in some experimental systems, it has been used in short-term laboratory studies[2] and in field transplants.[3] With proper care it should be possible to culture this species in the laboratory.

The ahermatypic corals do not contain symbiotic algae and long-term maintenance requires that *Tubastrea* be kept in flowing seawater with a source of plankton for food. Although this organism is not an important reef-former, it has been listed because it is common to both oceans and lacks symbiotic algae. It has been used as a comparison species in tests designed to assess the role of zooxanthellae in hermatypic corals.[4]

Whenever possible, use the suggested species because of their previous use as laboratory subjects, widespread occurrence,

and relative importance as reef-builders. Preferably use small-polyp forms.[5]

In determining adverse effects of toxicants, the high species diversity of coral reefs dictates the use of at least several genera. Include members of other locally dominant species as well as species that may be the most sensitive. Ideally, make comparative tests to determine tolerance of all locally common species so that the most sensitive species can be used in long-term studies.

3. Collecting, Handling, and Holding Test Organisms

a. Collecting corals: Guidelines for the operation of collecting boats in tropical reef areas are given by Domm.[6] The field procurement program should produce large numbers of small, unattached specimens that are similar in size, morphology, genetic makeup, and environmental history. Take extreme care in selecting specimens because corals display a wide variety of growth forms within a given species.[7–10]

For a branching species, use a single large colony as the source of all specimens for a test. Clip branches of approximately 10 g wet weight from the colony with pruning shears. Selection of specimens of solitary and massive corals will depend on local availability. Unattached species, such as *Fungia*, are collected easily, whereas the massive and encrusting forms generally grow firmly attached to the reef and must be pried or broken away with a pry bar or chisel. Because corals vary widely in size, morphology, and availability, it is difficult to give firm rules on collection for all species.

b. Field preparation: If broken fragments are left in a protected reef area where there is no heavy wave action to roll them about, coral tissue will cover the surface within 1 month and a small colony completely covered with living tissue will be produced. Use a substrate in the holding area consisting of gravel-size reef rubble and not subject to abnormally high temperatures or land runoff. On unprotected coasts use a deep offshore area for holding. Corals may be allowed to heal over in a laboratory seawater system.

Larger corals are tagged easily by tying commercially available vinyl "spaghetti" tags to the specimen. Use monofilament tags on small specimens. Small coral colonies readily overgrow a monofilament or vinyl strand and produce a solidly tagged colony.

c. Handling and holding corals: Keep animals submerged in seawater at all times and handle lightly and only when necessary. When transferring corals, submerge transfer container, place coral in container, and transport submerged coral to its new location; then again submerge container and remove coral. If coral is to be transported over a distance and must remain in a transfer container for more than 5 min, insure that the DO remains within ± 10% saturation and that the temperature does not change by more than 1°C. A boat supplied with a live well is desirable for transportation. If a large volume of water is used to contain a small volume of coral, there is less danger of these limits being reached. Preferably renew water in containers at a rate sufficient to keep DO and temperature within safe levels, rather than aerating. It can be assumed that when DO has been altered adversely by coral metabolic activity, other biologically important characteristics also have been affected adversely. Aeration will not correct these other changes.

d. Long-term holding: For long-term holding in the laboratory, use a substrate of natural reef rubble with its associated organisms covering the container bottom to a depth of 5 cm. In addition to invertebrate herbivores included with coral rubble, maintain small herbivorous fish (such as *Acanthurus triostegus*) in the tanks to control algal growth. Use holding tanks of several hundred liters capacity and maintain by pumping a continuous flow of seawater at a rate sufficient to flush tanks at least hourly. If DO concentrations or temperatures between inlet and outlet change by more than 10% or 1°C, respectively, increase flow rate.

4. Culturing Test Organisms

a. Water supply: Take seawater from an area that supports good, natural coral growths. For static studies, an artificial seawater can be used [Section 8010E.4b2)].

Successful laboratory maintenance of corals requires continuous flow of large amounts of uncontaminated open-ocean water. Use the seawater system, pumps, pipe, cleanouts, and delivery systems described in Sections 8010E and F.

b. Food and feeding: Specific nutritional requirements of hermatypic corals are more complex than those of most other organisms because of the symbiotic relationship between the coral and the zooxanthellar algae it contains. It is due also to the rapid calcification process that produces the massive coral skeleton.

Symbiotic algae utilize sunlight to produce food, a portion of which is transferred to the coral.[11] In addition, corals capture plankton[12] and are capable of digesting bacteria[13,14] and absorbing dissolved organics from the water.[14,15] Coral also takes up various inorganic nutrients (e.g., HCO_3^-) for the calcification process and certain inorganic plant nutrients.[16-19] Corals continue to grow when kept in full sunlight in filtered, continuous-flow seawater systems but cease to grow and eventually die when deprived of light even if supplied with unfiltered seawater.[4,20]

Observe feeding behavior during tests. The carnivorous feeding habits of 15 Pacific species of corals have been described by Abe.[21] In the Atlantic, *Manicina areolata*, *Montastrea cavernosa*, and *Porites porites* have been used in feeding studies[22,23] and readily will ingest freshly hatched nauplii of the brine shrimp, *Artemia.* Zooplankton captured with a fine mesh net also can be used.

5. Parasites and Predators

For a general discussion, see Section 8010E.5. The known coral parasites and predators have been reviewed.[24] Many of these can enter laboratory tanks in larval form and grow rapidly to maturity while feeding on coral tissue. White patches appear, especially on the undersides of the coral, when predators are present and feeding. Generally, predators will be attached to the coral near damaged areas. Corals of the genus *Montipora*, and possibly other Acroporids, are attacked by the Polyclad flatworm *Prosthiostomum.*[25] *Porites* is eaten by the acolid nudibranch *Phestilla sibogae,* while *Tubastrea* is eaten by a similar species, *Phestilla melanobranchia.*[26] *Fungia* is attacked by the wentletrap *Epitomium ulu.*[27] If predators of these types appear, remove them by drawing them up into an ordinary household basting pipet. Damaged specimens of larval corals, as well as adult corals, may be destroyed rarely by holotrich ciliate protozoa.[28]

6. References

1. WELLS, J.W. 1956. Scleractinia. *In* R.C. Moore, ed. Treatise on Invertebrate Paleontology. F328-F444, Geological Soc. America.
2. PEARSE, V.B. & L. MUSCATINE. 1971. Role of symbiotic algae (zooxanthellae) in coral calcification. *Biol. Bull.* 141:350.
3. SHINN, E.A. 1966. Coral growth rate, an environmental indicator. *J. Paleontol.* 40:233.
4. FRANZISKET, L. 1969. Riffkorallen konnen autotroph leben. *Naturwissenschaften* 56:144.
5. FISHELSON, L. 1973. Ecological and biological phenomena influencing coral-species composition on the reef tables at Eilat (Gulf of Aqaba, Red Sea). *Marine Biol.* 19:183.
6. DOMM, S.B. 1971. The safe use of open boats in the coral reef environment. *Atoll Res. Bull.* 143:1.
7. VAUGHAN, T.W. & J.W. WELLS. 1943. Revision of the suborders, families and genera of the Scleractinia. *Geol. Soc. Amer. Spec. Papers* 44:1.
8. WOOD-JONES, F. 1907. On the growth forms and supposed species in corals. *Zool. Soc. London, Proc.,* 518.
9. WOOD-JONES, F. 1910. Corals and Atolls. L. Reeve, London.
10. MARAGOS, J.E. 1972. A Study of the Ecology of Hawaiian Reef Corals. Ph.D. thesis, Univ. Hawaii, Honolulu.
11. MUSCATINE, L. & E. CERNICHIARI. 1969. Assimilation of photosynthetic products of zooxanthellae by a reef coral. *Biol. Bull.* 137:506.
12. YONGE, C.M. 1973. The nature of reef-building (hermatypic) corals. *Bull. Mar. Sci.* 23:1.
13. DISALVO, L.H. 1971. Ingestion and assimilation of bacteria by two scleractinian coral species. *In* H.M. Lenhoff, L. Muscatine & L.V. Davis, eds. Experimental Coelenterate Biology. p. 129. Univ. Hawaii Press, Honolulu.
14. SOROKIN, Y.I. 1973. On the feeding of some scleractinian corals with bacteria and dissolved organic matter. *J. Amer. Soc. Limnol. Oceanogr.* 18:380.
15. STEPHENS, G.C. 1962. Uptake of organic material by aquatic invertebrates. I. Uptake of glucose by the solitary coral *Fungia scutaria. Biol. Bull.* 123:648.
16. FRANZISKET, L. 1973. Uptake and accumulation of nitrate and nitrite by reef corals. *Naturwissenschaften* 60:552.
17. FRANZISKET, L. 1974. Nitrate uptake by reef corals. *Int. Rev. Gesamten Hydrobiol.* 59:1.
18. GOREAU, T.F., N.I. GOREAU & C.M. YONGE. 1971. Reef corals; autotrophs or heterotrophs? *Biol. Bull.* 141:247.
19. KAWAGUTI, S. 1953. Ammonia metabolism of the reef corals. *Biol. J. Okayama Univ.* 1:171.
20. YONGE, C.M. & A.G. NICHOLLS. 1931. Studies on the physiology of corals. IV. The structure, distribution and physiology of the zooxanthellae. *Sci. Rep. Great Barrier Reef Exped.* 1:135.
21. ABE, N. 1938. Feeding behavior and the nematocyst of *Fungia* and 15 other species of corals. *Palau Trop. Biol. Sta. Stud.* 1:469.
22. COLES, S.L. 1969. Quantitative estimates of feeding and respiration for three scleractinian corals. *J. Amer. Soc. Limnol. Oceanogr.* 14:949.

23. LEHMAN, J.T. & J.E. PORTER. 1973. Chemical activation and feeding in the Caribbean reef-building coral *Montastrea cavernosa*. *Biol. Bull.* 145:140.

24. ROBERTSON, R. 1970. Review of the predators and parasites of stony corals with special reference to symbiotic prosobranch gastropods. *Pac. Sci.* 24:43.

25. JOKIEL, P.L. & S.J. TOWNSLEY. 1974. Biology of the polyclad *Prosthiostomum* sp., a new coral parasite from Hawaii. *Pac. Sci.* 28:361.

26. HARRIS, L.G. 1971. Nudibranch associations as symbioses. *In* T.C. Cheng, ed. Aspects of the Biology of Symbiosis. p. 77. University Park Press, Baltimore, Md.

27. BOSCH, H.F. 1965. A gastropod parasite of solitary corals in Hawaii. *Pac. Sci.* 19:267.

28. SISSON, R.F. 1973. Life cycle of a coral captured in color. *Nat. Geogr. Mag.* 143:780.

8410 C. Toxicity Test Procedures

1. General Considerations

Many coral reef communities are "biologically accommodated," rather than "physically controlled."[1] Because corals rapidly modify water chemistry of closed systems, give special attention to loading (Section 8010F.3c). Closed recirculating systems using carbonate and charcoal filters have been used to hold corals for several months.[2-4] Lighting systems needed to simulate reef environments tend to raise water temperatures. Because of these factors, use only continuous-flow systems for long-term tests. For facilities, equipment, and construction materials, see Sections 8010E.3 and 4 and 8010F.1.

2. Preparing Test Materials

a. Dilution water: Use tropical open-ocean surface water as the dilution medium. Take water from an offshore area, away from strong terrestrial influence (see Section 8010C).

b. Toxicant solutions: See Section 8010F.2b.

c. Test organisms: See Sections 8010E and F for general procedures.

3. Test Procedures and Conditions

For general procedures, see Section 8010F. Use at least 20 coral colonies in each test concentration and in controls. The number and size of exposure chambers will be governed by size of colonies used. Load as indicated in Section 8010F.3c. Select toxic concentrations as described in Section 8010F.3b. Replace tank volume hourly at a flow rate that varies no more than ±5%. Use toxicant delivery systems described in Section 8010F.1. Keep test chambers clean. Maintain salinity at 33 to 35.0 g/kg or at that of the area from which the coral were collected. Hold temperatures at 27.0 ± 1°C or at those of the natural habitat. Keep DO within ±10% of saturation. The pH may vary from 8.1 to 8.4. Use light source described in Section 8010F.3f with a midday intensity at the water surface. Cover top of coral colonies with at least 2 cm water. Illuminate for tropical conditions, 12 h light and 12 h dark with appropriate twilight periods (Section 8010F.3f). Supply nutrients in long-term studies. The need for animal food varies among species and probably with environmental conditions within a single species. When animal food is needed, feed as described in Section 8410B.4b.

4. Conducting Tests

Conduct flow-through tests of effluent or toxicants under local conditions of light, temperature, and water quality. Carry out tests in large outdoor aquariums exposed to full sunlight. Supply tanks with a continuous flow of seawater pumped from the receiving water.

Measure and report frequently physical, chemical, and biological characteristics of the water. The light intensity (especially ultraviolet) in shallow tanks can be excessive. Simulate light levels at the intended depth by screening part of the natural light with a filter such as glass-fiber window screening.[5,6]

Treat coral colonies as described in Sections 8410B.3*b* and *c*. Acclimate coral fragments that have developed polyps on all sides in holding facilities for 30 d before use. If more than 10% of colonies exhibit damage, discard and use a new group. Because it will not be known if corals used in a test are functionally autotrophic or heterotrophic[7] under stress conditions, run duplicate series for each concentration with and without a source of zooplankton. Use field-collected zooplankton or rear in the laboratory as described in Section 8712C.

Use uniform-size colonies weighing approximately 10 g wet weight or more, depending on available space.

The various life stages of corals have been photographed[8] and described.[9,10] Coral planulae have been used successfully to test relative wastewater toxicity.[11] The advantage of using larvae is their small size, permitting use of small containers and large numbers of organisms. However, free planulae are quite unlike adult colonies: they are noncalcifying, nonfeeding, solitary, planktonic organisms with some mobility. Planulae may not reflect accurately the effects of environmental alterations on adult colonies. In some species, planulae may be the most resistant life stage.[10]

Planulae are not always available.[10,12–17] They usually can be obtained from *Pocillopora damicornis* in the Pacific and from *Agaricia agaricites* in the Atlantic. Gather planulae by collecting mature coral heads and placing them in an aquarium with a continuous flow of seawater. Pass the outlet water from the aquarium into a submerged upright cylinder closed at the ends with plankton netting. This retains planulae expelled by the coral polyps. Hold them for up to several days until needed.[18] Release of planulae has been stimulated by warming the water to 35°C for a few minutes;[9,10] however, this is not recommended because it tends to produce damaged and immature specimens.

Conduct static tests with planulae in 150-mL borosilicate glass beakers containing 100 mL test solution held in a water bath or air-conditioned room. Use flow-through systems for medium- and long-term studies. Conduct these tests in 100-mL or larger glass containers, tightly covered with plankton mesh and containing 20 organisms. Replace container volume hourly.

Because some planulae will settle in the containers, also measure setting success. The extremely high mortality involved in setting and early colony growth[18,19] indicates that this life stage is most sensitive to environmental disruption.

Coral larvae can be settled on glass slides[20,21] to produce new colonies for testing. Newly settled corals can be substituted for planulae as described previously.

5. References

1. SANDERS, H.L. 1968. Marine benthic diversity: A comparative study. *Amer. Natur.* 102:243.
2. ABE, N. 1937. Post-larval development of the coral *Fungia actiniformis* var. *palawensis* Doderlein. *Palao Trop. Biol. Sta. Stud.* 1:73.
3. YONGE, C.M., M.J. YONGE & A.G. NICHOLLS. 1932. Studies on the physiology of corals. VI. The relationship between respiration and the production of oxygen by their zooxanthellae. *Sci. Rep. Great Barrier Reef Exped.* 2:213.
4. CATALA, R. 1964. Carnival Under the Sea. R. Sicard, Paris, France.
5. COLES, S.L. 1973. Some Effects of Temperature and Related Physical Factors on Hawaiian Reef Corals. Ph.D. dissertation, Dep.

Zoology, Univ. Hawaii, Honolulu (unpublished).

6. JONES, R.S. & R.H. RANDALL. 1973. A study of biological impact caused by natural and man-induced changes on a tropical reef. Univ. Guam Tech. Rep. No. 7:1.

7. GOREAU, T.F., N.I. GOREAU & C.M. YONGE. 1971. Reef corals; autotrophs or heterotrophs? Biol. Bull. 141:247.

8. SISSON, R.F. 1973. Life cycle of a coral captured in color. Nat. Geogr. Mag. 143:780.

9. EDMONDSON, C.H. 1929. Growth of Hawaiian Corals. Bull. No. 58:1, Bernice Pauahi Bishop Museum, Honolulu, Hawaii.

10. EDMONDSON, C.H. 1946. Behavior of coral planulae under altered saline and thermal conditions. Occas. Pap. Bernice Pauahi Bishop Mus. 18:283.

11. ENGINEERING SCIENCE, INC. 1971. Water Quality Program for Oahu with Special Emphasis on Waste Disposal. Final Report of Work Area 5. Chapter III, Toxicity Bioassays. Dep. Public Works, City and County of Honolulu, Hawaii.

12. WILSON, H.V. 1888. On the development of Manicina areolata. J. Morphol. 2:191.

13. DUERDEN, J.E. 1904. The coral Siderastrea radians and its postlarval development. Carnegie Inst. Wash. Publ. 20:1.

14. YABE, H. & EGUCHI. 1939. Ecological studies on Rhizopsammia minuta var. mutsuensis. Jubilee Publ. Prof. H. Yabe 60th Birthday Vol. 1:175. (Japanese with English summary).

15. YONGE, C.M. 1935. Studies on the biology of Tortugas corals. I. Observations on Meandrina areolata Linn. Carnegie Inst. Wash. Publ. 452:185.

16. MARSHALL, S.M. & T.A. STEPHENSON. 1933. The breeding of reef animals. Part I. The corals. Sci. Rep. Great Barrier Reef Exped. 3:219.

17. KAWAGUTI, S. 1941. On the physiology of reef corals. V. Tropisms of coral planulae, considered as a factor of distribution of the reefs. Palao Trop. Biol. Sta. Stud. 2:319.

18. HARRIGAN, J.F. 1972. The Planulae Larvae of Pocillopora damicornis: Lunar Periodicity of Swarming and Substratum Selection Behavior. Ph.D. dissertation, Dep. Zoology, Univ. Hawaii, Honolulu.

19. FISHELSON, L. 1973. Ecology of coral reefs in the Gulf of Aqaba (Red Sea) influenced by pollution. Oecologia (Berlin) 12:55.

20. CROSSLAND, C. 1952. Madreporaria, Hydrocorallinae, Heliopora and Tubipora. Great Barrier Reef Expedition. VI:86. British Museum (Natural History).

21. ABE, N. 1938. Feeding behavior and the nematocyst of Fungia and 15 other species of corals. Palau Trop. Biol. Sta. Stud. 1:469.

8410 D.   Evaluating and Reporting Results

Follow general procedures described in Section 8010G for data analysis and reporting results.

1. Lethal Response

Most corals are colonial; therefore, they present the problem of what should be counted as one individual. Although coral fragments differ in total number of polyps, a lethal concentration generally will kill all polyps and thus allow treatment of the colony as an individual. Interdependence between polyps has been suggested.[1,2] Exceptions result from strong gradients affecting only one portion of the colony. For example, accumulation of sediments around the base of the colony may kill only those polyps contacted. High light intensity may kill only the exposed polyps. These problems usually can be eliminated by proper test design.

Death of the colony becomes obvious when polyps become insensitive to stimulation and opaque in appearance. Within a few hours after death, corals begin to disintegrate. At the end of a test, return corals to optimal conditions and observe for recovery.

2. Visible Sublethal Effects

Healthy corals generally contain dense concentrations of zooxanthellae, containing plant pigments that give the polyps a characteristic color. In response to sublethal toxicant concentrations, the zooxanthellae are extruded over a period of hours or days, leaving the polyps transparent. This may be a response to high temperature, low light intensity, and starvation,[3] as well as to runoff of fresh water and silt[4] and estuarine effluents.[5]

Under some conditions, the first sign of damage is destruction of the thin tissue covering the septae or cenosarc; the skeleton then becomes visible. Some corals will extrude mesenterial filaments when subjected to stress.

Irritated or damaged corals may produce large amounts of mucus. Often distressed corals will contract their normally expanded polyps. Edmondson[6] and Mayer[7] reported that corals cease feeding long before lethal temperatures are reached. Samoan corals exposed to pesticides show abnormal feeding behavior long before death occurs.

Abe[8] described a method of using carmine granules to observe ciliary currents on the surfaces of corals. This may be a useful technique for observing signs of distress.

3. Other Tests

In-situ field measurement of response can be used in support of laboratory tests. Transplanting corals into a different environment frequently has been used to evaluate coral response to various environmental differences.[9-14]

Locate transplant stations to give a span of extreme gradients while being similar in all other respects (depth, wave action, salinity, substrate type, etc.). Use as a platform for attachment of transplanted coral a section of vinyl-coated steel wire fencing material firmly anchored to the reef with iron stakes or concrete blocks. Tie the corals to the platform with plastic-coated electrical wire. In areas of fine sediment,

support the frame off the bottom, unless sedimentation is the factor being investigated.

Select, handle, tag, and field-prepare specimens in the same manner as for laboratory tests. Use larger heads weighing up to several hundred grams because space is not limited.

Continue such studies for a minimum of 1 year and check corals for mortality, growth, or other responses monthly. Monitor the environment for significant water quality factors such as salinity, temperature, DO, light, sediment load, dissolved nutrients, particulate matter, and dissolved organic carbon.

In-situ measurements of coral calcification and oxygen metabolism have been carried out in submerged enclosures.[12,15] This approach can be used in connection with field transplants.

4. References

1. HARRIGAN, J.F. 1972. The Planulae Larvae of *Pocillopora damicornis:* Lunar Periodicity of Swarming and Substratum Selection Behavior. Ph.D. dissertation, Dep. Zoology, Univ. Hawaii, Honolulu.
2. GOREAU, T.F. 1963. Calcium carbonate deposition by coralline algae and corals in relation to their roles as reef builders. *Ann. N.Y. Acad. Sci.* 109:127.
3. YONGE, C.M. & A.G. NICHOLLS, 1931. Studies on the physiology of corals. IV. The structure, distribution and physiology of the zooxanthellae. *Sci. Rep. Great Barrier Reef Exped.* 1:135.
4. GOREAU, T.F. 1964. Mass expulsion of zooxanthellae from Jamaican reef communities after Hurricane Flora. *Science* 145:383.
5. WELLS, J.M., A.H. WELLS & J.G. VANDERWALKER. 1973. *In situ* studies in benthic reef communities. *Helgolander wiss. Meeresunters* 24:78.
6. EDMONDSON, G.H. 1928. The ecology of a Hawaiian coral reef. Bull. No. 45:1, Bernice Pauahi Bishop Museum, Honolulu, Hawaii.
7. MAYER, J.W. 1915. On the development of the coral *Agaricia fragilis* Dana. *Amer. Acad. Arts Sci. Proc.* 51·483.

8. ABE, N. 1938. Feeding behavior and the nematocyst of *Fungia* and 15 other species of corals. *Palau Trop. Biol. Sta. Stud.* 1:469.
9. SHINN, E.A. 1966. Coral growth rate, an environmental indicator. *J. Paleontol.* 40:233.
10. WOOD-JONES, F. 1907. On the growth forms and supposed species in corals. *Zool. Soc. London, Proc.*, 518.
11. MARAGOS, J.E. 1972. A Study of the Ecology of Hawaiian Reef Corals. Ph.D. thesis, Univ. Hawaii, Honolulu.
12. GLYNN, P.W. & R.H. STEWART. 1973. Distribution of coral reefs in the Pearl Islands (Gulf of Panama) in relation to thermal conditions. *Limnol. Oceanogr.* 18:367.
13. MARSHALL, S.M. & A.P. ORR. 1931. Sedimentation on Low Isles Reef and its relation to coral growth. *Sci. Rep. Great Barrier Reef Exped.* 1:93.
14. JOKIEL, P.L. & S.L. COLES. 1974. Effects of heated effluent on hermatypic corals at Kahe Point, Oahu. *Pac. Sci.* 28:1.
15. BARNES, D.J. & D.L. TAYLOR. 1973. *In situ* studies of calcification in the coral *Montastrea annularis. Helgolander wiss. Meeresunters* 24:284.

8510   TOXICITY TEST PROCEDURES FOR ANNELIDS*

8510 A.   Introduction

The phylum Annelida includes three classes: Polychaeta, Oligochaeta, and Hirudinea. Polychaetes are an important, often predominant, component of marine and estuarine biota. In subtidal benthic environments, they comprise about 30 to 75% of the macroinvertebrate species and individuals. They include a variety of feeding types with the majority being either filter or detritus feeders. Deposit-feeding polychaetes affect surface sediments by their burrowing and irrigating habits. They are important food for snails, large crustaceans, fish, and birds. Many species have short life cycles.

Oligochaetes are among the most common benthic invertebrates in all types of aquatic environments. Particular species assemblages are recognized indicators of environmental quality. In grossly polluted freshwater habitats, oligochaetes dominate the benthic fauna, whereas in estuarine areas they and polychaete worms are often the most common benthic organisms. They feed mainly on bacteria, although other feeding types occur. They affect surface sediments as do the polychaetes. They are an important primary or alternate food for leeches, crustaceans, fish, and birds.

Hirudinea are leeches, either free-living or parasitic; they have not been used as test organisms for toxicity tests.

The following procedures are intended to serve as guidelines for the use of polychaetes and oligochaetes in various toxicity tests. These procedures also can be adapted to testing sediments.

---

*Approved by Standard Methods Committee, 1985.

8510 B. Selecting and Preparing Test Organisms

1. Selecting Test Organisms

In accord with the criteria listed in Section 8010E.1, the recommended test species include (but are not restricted to) the following:

a. *Marine polychaetes:*
1) Family Nereidae
   Neanthes arenaceodentata (*Neanthes acuminata* and *Neanthes caudata* of some authors) (New England, Florida, California coasts, Europe).
   Neanthes succinea (all of U.S. coasts).
   Neanthes virens (east coast of U.S.).
2) Family Capitellidae
   Capitella capitata (cosmopolitan).
3) Family Ctenodrilidae
   Ctenodrilus serratus (cosmopolitan).
4) Other species used in toxicity tests but not discussed include: *Nereis diversicolor, Arenicola cristata*, and *Abarenicola pacifica.*

b. *Freshwater oligochaetes:*
1) Family Tubificidae
   Limnodrilus hoffmeisteri (cosmopolitan).
   Tubifex tubifex (cosmopolitan).
   Branchiura sowerbyi (cosmopolitan).
2) Family Lumbriculidae
   Stylodrilus heringianus (holarctic).

c. *Marine oligochaetes:*
1) Family Tubificidae
   Monopylephorus cuticulatus (N.E. Pacific).
   Tubificoides "fraseri" (all North American coasts).
   Limnodriloides verrucosus (all North American coasts).

d. *Other freshwater and marine oligochaetes* used in toxicity tests but not discussed include: *Lumbriculus variegatus, Quistadrilus multisetosus, Spirosperma ferox, Spirosperma nikolskyi, Rhyacodrilus montana, Vamchaetadrilus pacifica, Ilyodrilus frantzi, Nais communis, Paranais frici*, and *Paranais litoralis.*

2. Collecting Test Organisms

a. *Marine polychaetes:*
1) *Neanthes arenaceodentata, Neanthes succinea*, and *Capitella capitata* inhabit intertidal mud flats in estuarine areas and the fouling communities on pilings, boat floats, or submerged objects. To obtain worms, bring substrate or fouling material into the laboratory, place in white enameled pans, and cover with seawater. After a period of time, the worms come to the surface; remove them with a fine brush and transfer to petri dishes containing seawater. Examine each specimen under a dissecting microscope and discard all injured worms. Transfer uninjured specimens to 4-L aquariums or shallow trays.

2) *Ctenodrilus* occurs on fouling organisms attached to floats and pilings. Because this species is minute, use a dissecting microscope to look for it. Because only a small number can be collected at one time, establish a laboratory colony [Section 8510B.3a3)].

b. *Freshwater and marine oligochaetes:*
Limnodrilus hoffmeisteri and *Tubifex tubifex* inhabit muddy sediments and are particularly common in areas of gross organic pollution. Examination of preserved specimens usually is necessary for positive identification. In live culture these species can

be separated on the basis of the presence (*T. tubifex*) or absence (*L. hoffmeisteri*) of hair setae. *Branchiura sowerbyi* is a larger worm with gills on posterior segments; it is common in muddy, warm-water areas. *Stylodrilus heringianus* is common in areas of clean, fine sand and is identified by the presence of extra rings and nonretractable penes.

The marine species *Monopylephorus cuticulatus*, *Tubificoides fraseri*, and *Tectidrilus verrucosus*, are found in muddy sediments and are separable based on size (*M. cuticulatus* > *T. fraseri* > *L. verrucosus*), setae (*M. cuticulatus* has occasional irregular hair setae), and color (*L. verrucosus* has a papillate skin and greenish tinge, the others are dark red). Positive identification requires examination of preserved specimens.

To obtain worms, sieve the sediments through a 0.5-mm sieve and sort specimens under a dissecting microscope. Discard badly damaged worms. Transfer uninjured specimens to aerated aquariums or shallow trays for holding and feeding.

3. Culturing

a. Marine polychaetes:

1) Condition of animals—Discard animals injured during collection. Some species such as *Neanthes arenaceodentata* can regenerate a tail, thus it is not always necessary to discard worms missing tails when establishing a culture. Save worms with gametes in the coelom for starting cultures, but normally do not use them for toxicity tests.

2) Food and feeding—Cultures of the polychaete species mentioned here can be maintained without sediment; therefore, the worms must be fed. Also feed worms in long-term experiments. Cultures of the larger species (i.e., *N. arenaceodentata*) have better survival, growth, and tube pro-

duction if fed a mixed diet that consists of a macroalga for tube construction and a commercially prepared invertebrate or fish diet for nutrition. The green alga, *Enteromorpha* sp., is convenient because it grows abundantly in most estuarine areas of North America. Collect in quantity, wash with seawater, dry, and store indefinitely. Before use, soak the alga in seawater and knead to separate individual filaments. Other macroalgae, for example, cultured brown *Ectocarpus siliculosus*, produce excellent results in polychaete cultures but are not as convenient to use. Place sufficient macroalga in culture containers to allow worms to construct tubes. Add commercially prepared diet* to the worm cultures three times weekly. Vigorously mix flakes with a small amount of seawater to moisten and break them up before adding them to the cultures. To minimize overfeeding, examine each culture container before adding the commercial diet. If most of the diet material is uneaten, do not add more and add less food at subsequent feedings.

A powdered diet is suitable for small species (*Ctenodrilus serratus* and *Capitella capitata*) and the larvae of *N. arenaceodentata*. Prepare a fine powder from dried *Enteromorpha* sp. or one of the commercial diets by grinding the dry material in a blender and sieving it to smaller than 0.061 mm. *C. serratus* can be fed at a rate of 0.1 mL/worm/week of a mixture of 1.0 g powder/100 mL seawater.

Feed living *Dunaliella* sp. to larval *N. succinea* until the larvae settle. For culturing instructions see Section 8010E.4c1)*b*). Feed *Dunaliella* sp. at the rate of 10 mL culture (minimum cell count of 20 000 cells/mL)/L of worm culture or at a rate great enough to maintain a green color in the seawater. After the larvae settle, feed *Enteromorpha* sp. until the swimming reproductive epitoke stage is reached.

3) Producing test organisms

---

*Prawn Flakes, Plankton Flakes, TetraMin, or equivalent.

a) *Capitella capitata*—Laboratory-cultured specimens begin to mature in about 15 to 25 d after hatching. A mature female develops white masses of eggs in the coelom from about segment 10 posteriorly and a mature male develops specialized setae on the dorsal surface of segments 8 and 9. The female lays fertilized eggs along the inside lining of her tube where larval development continues until the trochophore larvae emerge 4 to 6 d later. To obtain free-swimming trochophore larvae, examine tubes under a dissecting microscope to detect those containing eggs or larvae. Recently fertilized eggs appear white, but as they mature they become grey-green and can be seen moving about. Place tubes containing larvae in a petri dish. Under a dissecting microscope open the tubes to free the trochophores. One female provides 200 to 300 trochophores. Remove and discard the female and the tube containing any larvae that did not swim free. Use the free-swimming larvae in tests or let them develop for later use.

b) *Neanthes succinea*—Take nearly mature epitokes from the field or laboratory colony and hold until they complete sexual metamorphosis. Mature epitokes swim to the water surface and release gametes. If fertilization is successful, separate the zygotes into several 4-L jars containing aerated water and let them develop to the three-setiger stage (about 1 week). These larvae are ready for use in tests. One fertilization provides more than 2000 larvae.

c) *Neanthes arenaceodentata*—Before spawning, either the male or female enters the tube or burrow of another worm. If the worms are of different sex, they remain together and spawn within the tube. The female dies within 1 d after spawning and the male incubates the eggs for about 3 weeks, at which time they have 18 to 21 setigerous segments. At that time the young worms leave the tube, begin feeding, and construct their own tubes. Feed them *Enteromorpha* as indicated in Section 8510B.3*a*2). Under laboratory conditions (20°C) sexual maturity is reached in 3 to 4 months. It is impossible to distinguish the sex of immature forms morphologically. Distinguish by observing whether or not they fight when placed together. Use a female with maturing eggs in her coelom as a known individual to identify the sex of immature worms. The most convenient time to obtain young juveniles is shortly after they have left the parent's tube and have begun to feed.

d) *Ctenodrilus serratus*—This species reproduces asexually about every 14 d at 20°C by transverse division. Each individual produces about five to eight new specimens. Large colonies can be maintained with minimum care.

b. Freshwater and marine oligochaetes:

1) Condition of animals—Oligochaetes show great regenerative abilities, hence it is not always necessary to discard injured specimens. Mature individuals with well-developed clitellar regions are particularly important for culture establishment. Keep cultures in the dark or under natural light/dark regimes.

2) Food and feeding—Oligochaetes feed mainly on bacteria in sediments; therefore, in experiments with natural sediments additional feeding is unnecessary. Short-term experiments do not require feeding; for long-term experiments (> 10 d) provide sediment.

Condition sterile sediments by preparing an inoculum of *Enteromorpha* (for marine worms) or lettuce (for freshwater worms) consisting of the aqueous material remaining after decay of the plant fibers in diluent water. Add inoculum directly to culture containers in a volume not to exceed 10% of the total. Preferably use sediments of fine sand with some silt content, rather than more muddy sediments in which the worms are difficult to find. Check cultures periodically for spoilage; if this occurs, clean cultures and restart. Oligochaetes have no larval stage. No separate feeding regime is required for juveniles.

3) Producing test organisms—Gravid worms in culture lay eggs that hatch in 3 to 14 d, depending on species and temperature. Newly hatched worms lack the full component of adult setae but rapidly develop these.

Freshwater species generally grow better in mixed culture. The following combinations are recommended: *L. hoffmeisteri* and *T. tubifex*, *L. hoffmeisteri* and *B. sowerbyi*, *T. tubifex* and *B. sowerbyi*, *S. heringianus* and *L. hoffmeisteri*. *T. fraseri* is a parthenogenic species and is particularly amenable to culturing; *L. verrucosus* and *M. cuticulatus* can be cultured as pure species.

4. Parasites and Diseases

Microbial growth can result from overfeeding, improper conditioning of food, or insufficient DO. Prevent fungal growths by proper sanitation and periodic care. To minimize overfeeding, examine each aquarium before feeding. If most of the diet is uneaten, do not add more and add less food at subsequent feedings. Generally there is adequate DO in 4-L aquariums; however, aeration can be increased to correct for any deficiency. The internal protozoan parasitic gregarines have been observed to reduce the vitality of some species of laboratory populations of polychaetes. Gregarines are widespread in polychaetes and oligochaetes but it is not known if they cause similar problems in these species.

5. Bibliography

KORSCHELT, E. 1931. Art und Daver der ungeschlechtlichen Fortpflanzungen bei *Ctenodrilus*. *Zool. Anz.* 93:227.

THORSON, G. 1946. Reproduction and larval development of Danish marine bottom invertebrates with special reference to the planktonic larvae in the sound (Oresund). *Komm. Dan. Fisk.-Havundergelser. Ser. Plankton. Meddel.* 4:1.

REISH, D.J. 1957. The life history of the polychaetous annelid *Neanthes caudata* (delle Chiaje), including a summary of development in the family Nereidae. *Pac. Sci.* 11:216.

RICHARDS, T.L. 1969. Physiological Ecology of Selected Polychaetous Annelids Exposed to Different Temperature, Salinity, and Dissolved Oxygen Combinations. Ph.D. dissertation, Univ. Maine.

AKESSON, B. 1970. *Orphryotrocha labronica* as a test animal for the study of marine pollution. *Helgolander wiss. Meeresunters* 20:293.

REISH, D.J. 1970. The effects of varying concentrations of nutrients, chlorinity, and dissolved oxygen on polychaetous annelids. *Water Res.* 4:721.

REISH, D.J. 1974. The establishment of laboratory colonies of polychaetous annelids. *Thallassia Jugoslav.* 10:181.

BAILY, H.C. & D.H.W. LIU. 1980. *Lumbriculus variegatus*, a benthic oligochaete, as a bioassay organism. *In* J.C. Eaton, P.R. Parrish & A.C. Hendricks, eds. Aquatic Toxicology. ASTM STP 707, American Soc. Testing & Materials, Philadelphia, Pa.

BRINKHURST, R.O., P.M. CHAPMAN & M.A. FARRELL. 1983. A comparative study of respiration rates of some aquatic oligochaetes in relation to sublethal stress. *Int. Rev. Ges. Hydrobiol.* 68:685.

REISH, D.J. 1985. The use of *Neanthes arenaceodentata* as a laboratory experimental animal. *Tethys* 11:335.

8510 C. Toxicity Test Procedures

1. General Procedures

Use exploratory tests (see Section 8010D) to determine toxicant concentrations for short-term tests. Prepare dilution water and toxicant solutions and introduce them into test containers as described in Section 8010F.

2. Water Supply

a. Artificial seawater: See Section 8010E.4*b*2). Use a salinity of approximately 35.5 g/kg and a pH of about 7.8 for marine populations; use lower salinity for estuarine worms.

b. Natural seawater: Determine and report quality routinely. Maintain dilution water salinity at or near selected or normal concentration. During a test, do not allow salinity to vary by more than ± 3 g/kg. In all except effluent tests, filter seawater through a 0.45-μm membrane filter.

c. Distilled or tap water: Determine quality (hardness, alkalinity, chemical constituents) and report routinely. Use a neutral (pH 7.0) water.

3. Exposure Chambers

a. Marine polychaetes: Use 4-L aquariums or glass jars for short-term and intermediate static and renewal tests and for long-term tests where flow-through facilities are not available. Cover aquariums to prevent entrance of foreign materials. Do not add more than 2.5 L test solution to each 4-L aquarium. Use 500-mL erlenmeyer flasks, containing 100 mL seawater, for either short-term or long-term experiments when only one organism is placed in each flask. Close the flask with a No. 7 TFE stopper fitted with a glass tube for aeration. Use small stender dishes (30 mL) for larval tests. For flow-through tests, use exposure chambers described in Section 8720B.3b3). In the case of cannibalistic species such as *Neanthes arenaceodentata*, isolate individuals during testing. Container size depends on biomass; maintain loading densities below 0.5 g/L for static conditions and below 0.5 g/L/d for flow-through tests at 20°C. For tests with sediment, use glass crystalling dishes of the appropriate size for species tested and number of individuals per dish. Fill dishes with sediment 1 to 4 cm deep. Let clean seawater flow over top of sediment.

b. Freshwater and marine oligochaetes: Conduct toxicity tests similarly to polychaete larval tests. Use shallow disposable polyethylene petri dishes with covers for static or replacement tests. Container size depends on biomass; maintain loading densities below 0.5 g/L, and preferably be-

low 0.2 g/L. Place 10 worms in each container per test concentration plus controls. Run duplicate tests. Worms can be tested individually, with 20 individuals per test concentration. For flow-through tests, use exposure chambers described in Section 8720B.3b3).

4. Conducting the Toxicity Tests

a. Setting up the test chambers: For static and renewal tests, set up as described in Section 8010D. In short-term tests, do not clean exposure containers. In long-term tests in which the organisms are fed, remove unused food and other materials as described in Section 8010E.4d. It is unnecessary to provide a bottom substrate for any but long-term oligochaete toxicity tests. Photoperiod and light intensity do not appear to be a factor in polychaete tests; however, test oligochaetes either in the dark or with a natural light/dark simulation. Keep temperatures within ± 2°C of the natural habitat unless the effect of temperature is being tested.

1) Marine polychaetes—Use a minimum of 20 worms for each test concentration. For cannibalistic species, such as *N. arenaceodentata,* use one worm per container. For other species, place 2, 5, or 10 worms in each container (depending on biomass and container size, see Section 8010F.3c). For tests with sediments, use a minimum of three replicate dishes per sediment concentration and 10 to 15 worms per dish. Individuals of cannibalistic species do not have to be separated with sediment in the dishes.

2) Freshwater and marine oligochaetes—Use a minimum of 20 worms for each test concentration, preferably in two replicates of 10 worms each. Although the worms will intertwine when healthy, toxified individuals remain separate and toxic effects (e.g., the progressive disintegration of posterior segments) will be manifest.

b. Duration and type of test:

1) Short-term tests—The length of short-term or acute tests depends on the length of the organisms's life cycle (see Section 8010F.3*a*). Short-term tests may be static if they last for 4 d or less or they may be renewal tests of up to 10 d.

2) Intermediate-length tests—Use tests of intermediate length for determining adult polychaete survival. For most species conduct these renewal or flow-through tests for 28 d. Use 10-d tests to determine survival of juvenile or adult polychaetes in contaminated sediments.

3) Long-term tests—Long-term tests involve only polychaetes and are either partial life-cycle tests beginning with the trochophore larval stage and continuing through sexual maturity or life-cycle tests beginning with the newly settled larval stage and continuing through reproduction and subsequent larval settlement of the offspring. Use renewal or flow-through tests.

Select and prepare test concentrations as described in Section 8010F.2*b*. Measure and mix dilution water and stock toxicant solutions by proportional diluters and deliver to exposure chambers as described in Section 8010F.1. Make tests in flow-through exposure chambers similar to those used for the dungeness crab [Sections 8720B.3*b*3) and 8720C.4*c*2)]. Renewal tests using 4-L exposure chambers may be necessary if flowing seawater is unavailable.

The duration of long-term tests depends on length of life-cycle of the organism and varies from about 1 month with *C. capitata* to 3 or more months with *N. succinea, N. virens,* and *N. arenaceodentata.*

c. Test organisms: See Section 8510B.

d. Performing tests:

1) Short-term tests—Set up and conduct renewal tests as described in Section 8010D.2. Determine survival of adults by checking exposure chambers at 1, 2, 4, 8, 18, and 24 h, then once or twice daily thereafter. Dead specimens generally are pale and swollen and lie on the bottom; live specimens usually respond to physical stimulation. If the tests are more than 4 d long, renew solutions, preferably daily but at least every fourth day. In short-term tests with polychaete larvae, determine survival after 96 h by microscopic examination. The absence of larvae generally indicates death because decomposition of small larvae is rapid.

2) Intermediate-length tests—Set up test chambers described in ¶ 4*a* above to determine adult lethality (LC50 or incipient LC50). Examine test containers daily to determine survival. Report results as 28-d LC50. If no organisms are killed after a certain length of exposure, report the period beyond which there is no further kill and the percentage killed in each test concentration. Calculate asymptotic LC50. For contaminated sediment tests, after 10 d sieve contents of each replicate dish and count number of survivors. If a graded series of mixed sediments has been used, calculate LC50 based on percentage of contaminated sediment.

3) Polychaete life-cycle tests beginning with the trochophore larval stages—Set up as described previously in this section. Conduct test through sexual maturity with test periods varying from 3 to 4 weeks with *C. capitata,* and from 2 to 3 months or longer for *N. arenaceodentata, N. succinea,* and *N. virens.* Feed larvae as described in Section 8510B.3*a*2). Determine survival at least twice weekly for *C. capitata* and once a week for *Neanthes.* During the early part of the study, count organisms on bottom of exposure chambers. If a renewal test is being conducted, decant supernatant fluid, examine under a dissecting microscope, and replace fluid with fresh test solution. For flow-through tests, remove chambers from exposure box, count organisms, and replace chamber. If no organisms are observed by the third examination, terminate that test chamber.

When *C. capitata* is the test organism, remove test chambers after about 15 to 16 d and every 2 d thereafter to check with a dissecting microscope for presence of eggs in the coelom and later for the presence of zygotes along the sides of the tube. Remove females when developing eggs are in the trochophore stage and count the larvae. Discard females and larvae after counting larvae and recording other data such as length and number dead or deformed larvae. Continue to examine each exposure chamber every 2 d for detection of females incubating larvae until all females have been removed and the total number of larvae recorded.

For *N. succinea,* set up exposure chambers as described in ¶ *4a* above with 25 larvae in each 1-L exposure chamber or 10 larvae in each flow-through exposure chamber. Use 10 chambers per concentration tested. Because these worms fight and are cannibalistic when crowded, prepare additional exposure chambers or reduce numbers in each test chamber to five individuals after the first month. Continue tests until animals reach the epitoke stage; then determine individual lengths and total weights and compare with those in the control.

4) Polychaete life-cycle tests beginning with the newly settled larval stage—These tests will vary in duration from about 1 month for *C. capitata* to 3 months or more for *N. arenaceodentata.* Set up tests as described previously with newly settled larvae. Use a minimum of two specimens per flask and 10 flasks per concentration. As tests progress, count organisms as above. Examine for survival once or twice per week as in ¶ *d3*).

For *N. arenaceodentata,* use recently emerged larvae having approximately 18 to 21 setigerous segments. If a renewal test is conducted, place four worms in each of five 4-L exposure chambers with 2.5 L test solution. Set up five containers for each test concentration and control. For flow-through tests, place two larvae in each of 10 exposure chambers for each test concentration and the controls. At 25 d, ex-

amine worms by viewing from outside the container for the presence of eggs in the coelom. If necessary, move mature worms among the replicates of a given treatment to pair males with females. Mature eggs reach 450 μm diam and are yellowish-orange. Examine at 5-d intervals until eggs are noted and then at 2- to 3-d intervals to determine whether eggs are being laid. The females die within 1 d after laying eggs and the males incubate them for about 3 weeks. The life cycle is complete when the worm leaves the male's tube. Remove males and count larvae.

5. Bibliography

BELLAN, G., D.J. REISH & J.P. FORET. 1972. The sublethal effects of a detergent on the reproduction, development and settlement in the polychaetous annelid *Capitella capitata. Mar. Biol.* 14:183.

REISH, D.J., F.M. PILTZ, J.M. MARTIN & J.Q. WORD. 1974. The induction of abnormal polychaete larvae by heavy metals. *Mar. Pollut. Bull.* 5:125.

OSHIDA, P.S., A.J. MEARNS, D.J. REISH & C.S. WORD. 1976. The Effects of Hexavalent and Trivalent Chromium on *Neanthes arenaceodentata* (Polychaeta: Annelida). *Southern California Coastal Water Research Project,* TM 225.

REISH, D.J., C.E. PESCH, J.H. GENTILE, G. BELLAN & D. BELLAN-SANTINI. 1978. Interlaboratory calibration experiments using the polychaetous annelid *Capitella capitata. Mar. Environ. Res.* 1:109.

REISH, D.J. 1980. The effect of different pollutants on ecologically important polychaete worms. EPA 600/3-80-053, U.S. Environmental Protection Agency, U.S. Government Printing Off., Washington, D.C.

CHAPMAN, P.M., M.A. FARRELL & R.O. BRINKHURST. 1982. Relative tolerances of selected aquatic oligochaetes to individual pollutants and environmental factors. *Aquat. Toxicol.* 2:47.

CHAPMAN, P.M., M.A. FARRELL & R.O. BRINKHURST. 1982. Effects of species interactions on the survival and respiration of *Limnodrilus hoffmeisteri* and *Tubifex tubifex* (Oligochaeta, Tubificidae) exposed to various pollutants and environmental factors. *Water Res.* 16:1405.

PESCH, C.E. & G.L. HOFFMAN. 1983. Interlaboratory comparison of a 28-day toxicity test with the polychaete *Neanthes arenaceoden-*

tata. In W.E. Bishop, R.D. Cardwell & B.B. Heidolph, eds. Aquatic Toxicology and Hazard Assessment, 6th Symp., American Soc. Testing & Materials, Philadelphia, Pa.

PESCH, C.E., P.S. SCHAVER & M.A. BALBONI. 1986. Effect of diet on copper toxicity to *Neanthes arenaceodentata* (Annelida: Polychaeta). *In* T.M. Poston & R. Purdy, eds. Aquatic Toxicology and Environmental Fate, 9th Symp., American Soc. Testing & Materials, Philadelphia, Pa.

CHAPMAN, P.M. & D.G. MITCHELL. 1986. Acute tolerance tests with the oligochaete *Nais commonis* (Naididae) and *Ilyodrilus frontzi* (Tubificidae). *Hydrobiologia* 137:61.

CHAPMAN, P.M. In press. Oligochaete respiration as a measure of sediment toxicity in Puget Sound, Washington. *Hydrobiologia.*

8510 D.   Data Evaluation

1. Short-Term and Intermediate Adult Survival Studies

Determine the LC50 values for each exposure period as described in Section 8010G. A useful alternative is to determine LT50 (time to 50% mortality) in comparative exposures to single toxicant, effluent, water, or sediment concentrations. LT50s also provide useful ancillary information for LC50 studies.

2. Polychaete Life-Cycle Studies Beginning with the Trochophore and Settled Larval Stages

The number of females forming and laying eggs and the number of offspring produced are inversely related to sublethal toxicant concentrations at levels below the LC50. They provide a more subtle measure of effects than the LC50. Record life-cycle data for each concentration of toxicant as follows: number of females forming eggs, number of females laying eggs, and number of eggs and live offspring produced. Compare these data, expressed on a percentage basis, for all test concentrations with those obtained from the controls. Use statistical and reporting techniques described in Sections 8010G and H.

8610   TOXICITY TEST PROCEDURES USING MOLLUSKS*

8610 A.   Introduction

1. Suitability for Toxicity Tests

Oysters, clams, scallops, and mussels are distributed widely and are of great value as food for man. These mollusks and others are suitable for toxicity evaluation in short- and long-term tests. Oyster and clam embryos have been used to measure effects of chemical and environmental variables in estuarine and marine environments.

Methods for using freshwater mollusks in standard tests are being developed; methods for laboratory culture and maintenance of some hydrobiid snails are available.[1]

*Approved by Standard Methods Committee, 1988.

Toxicants affect bivalves by interfering with fertilization, normal embryonic development, growth (shell deposition), byssal thread secretion, reproduction, and normal tissue histology. These toxic effects are the bases for short- and long-term toxicity tests. Adult bivalves generally are not suitable for determination of acute lethal concentrations of toxicants because of their ability to close their shells and protect themselves from toxicant.

2. Reference

1. VAN DER SCHALIE, H. & G.M. DAVIS. 1968. Culturing *Oncomelania* snails (Prosobranchia: *Hydrobiidae*) for studies of oriental Schistosomiasis. *Malacologia* 6:321.

8610 B. Selecting and Preparing Test Organisms

1. Selecting Test Organisms

For oil and heavy metals, the bay scallop is the most sensitive species. Some adult mollusks are resistant to many materials and accumulate them to high concentrations. In comparison to fish and to other invertebrates, oyster larvae may be more or less sensitive, especially to pesticides. Species recommended for testing include:

Crassostrea gigas	Pacific oyster
Crassostrea virginica	Eastern oyster
Ostrea lurida	Olympia oyster
Argopecten irradians irradians	Bay scallop
Mytilus edulis	Mussel
Mercenaria mercenaria	Quahog
Spisula solidissima	Surf clam
Mulinia lateralis	Coot clam
Macoma balthica	
Rangia cuneata	

2. Water Supply and Water System

Tests require a marine laboratory with a supply of clean, unfiltered estuarine or open-ocean water of the desired temperature and salinity range. (Section 8010F.4b and 8010F.1 and 2a). If necessary, use artificial seawater.

For filter-feeding mollusks use a natural seawater containing planktonic organisms.

In long-term growth tests, supply adult oysters with a minimum of 5 L of unfiltered seawater/h/oyster. Supply clams and scallops with comparable amounts of natural seawater rich in plankton. In flow-through tests, distribute water from a constant-head tank (Section 8010F.1).

3. Collecting, Conditioning, and Culturing Test Organisms

Collect test organisms from the field, purchase from commercial dealers, or rear them.

Use natural seawater rich in plankton for growing adults or spawners. Clean intake pipes and entire water system frequently to insure that growth in pipes does not remove plankton. If there is not sufficient food, or if a continuous flow-through of unfiltered seawater results in problems of competitors, parasites, and disease, produce planktonic food in rearing chambers [Section 8010E.4c2)].

a. Oysters: For tests with adults, or for producing embryos, use adults of the Pacific or Eastern oyster, 7.5 to 15 cm in height. Cull oysters to singles and condition at 2- to 4-week intervals, depending on need. In the laboratory, hold each lot of oysters as a separate population in conditioning trays. Oysters collected from December to April need longer conditioning than those collected between April and

July. After August, oysters begin resorbing their unspawned gametes and are unsatisfactory spawners. To have spawning oysters after August, collect during spring months and keep in year-round cool water, or in refrigerated, flowing seawater in laboratory at less than 12°C.[1]

Clean oysters of fouling organisms and other extraneous materials. Use 15 oysters per conditioning tray (58 × 46 × 8 cm). Provide each tray with a minimum of 7 L/h of flowing seawater at 20 ± 1°C. Oysters require 2 to 6 weeks of thermal conditioning before they are ready to spawn and can be held 2 to 4 weeks for use as spawners before discarding. Conditioned mature mollusks will spawn in either natural or artificial seawater. When they are induced to spawn in natural seawater, transfer immediately to receiving water or artificial medium for collection of gametes.

Check oysters daily and remove moribund individuals. If any die, empty tray and clean with detergent and warm water. Scrub remaining oysters with clean seawater, rinse several times, and replace in tray.

Clean accumulated feces and silt from trays at least once and preferably twice a week. Should an unplanned spawning occur, discard all oysters in that tray. When fertilized eggs are required, rinse conditioned oysters with clean seawater and place individually in spawning dishes. Raise water temperature 5 to 10°C to induce spawning and add a sperm suspension as a further inducement. Prepare sperm suspension by opening a male oyster, rupturing gonad, and gently washing sperm free from gonadal tissue into a 1-L beaker with a fine jet of 20°C seawater. Take care not to rupture other body organs during the process and use enough water so that DO remains near saturation. After a spawning attempt, return oysters that have not spawned to conditioning trays. Discard females once they have spawned. Place any surplus males that spawn in a separate tray and use for making sperm suspensions.

b. Clams: Collect clams from the field or secure from commercial fishermen. Keep adults in live boxes or in containers with adequate flow-through water during transit; then place in suitable holding trays. Use specimens in long-term tests or for production of larvae for short-term tests. If adults are not taken during normal spawning season, thermally condition for spawning by placing in trays with flow-through seawater at 18 to 22°C, depending on species, for 3 to 4 weeks. Stimulate adult clams with ripe gonads to spawn naturally by increasing water temperature to 24 to 28°C, depending on species, and add a sperm suspension as described for oysters. Produce larvae only from naturally liberated eggs. A stock of 30 to 40 thermally conditioned adults is adequate to assure spawning at any time. Place eggs in test solutions as soon as possible, and not more than 2 h after fertilization.

To use larvae in longer-term tests in the absence of natural food, use a mixture of *Isochrysis galbana* and *Monochrysis lutheri* or some other cultured algae[2] grown in standard enriched seawater described in Section 8010E.4c1)b).

c. Mussels: Collect *Mytilus edulis* from pilings, rocks, floating boat docks, jetties, and other suitable habitats. Those from floating docks or platforms are preferred because they are easiest to collect and clean. Because larger specimens may contain maturing gametes that may be freed during tests, use only small specimens in intermediate-term tests with adults. Use specimens of uniform size, 15 to 20 mm wide.[3] Separate this size class by passing mussels through a 20-mm-diam hole drilled in a piece of sheet metal; mussels of the correct size class will pass through this hole but not through one 15 mm in diameter. Discard all other specimens. Clean surface and trim byssal threads, but do not remove byssal thread stalk.

Acclimate cleaned specimens in aquariums for 1 week at not more than 4°C above field water temperature but below 26°C and at a density of about 10 specimens/4 L seawater. Provide aeration and filtration in static aquariums. Observe holding aquariums daily for deaths. Discard all specimens if more than 10% die during acclimation. Occasionally some 15- to 20-mm animals have mature gametes and will spawn. Change water to prevent fouling and remove the mature specimen if it can be identified. If spawning is widespread discard all specimens.

d. Scallops: Collect bay scallops from the field or purchase from fishermen taking special precautions to insure proper handling. Use adults as described for the oyster. Culture methods and conditioning techniques are available.[1,4–7] Condition at 20 to 22°C for 3 to 8 weeks in same type of tray used for oysters. Check gonads periodically to determine development. Place a finger in a gaping scallop to hold shells open so the gonads can be seen. The bay scallop is a functional hermaphrodite. The testis comprises the anterior border of the gonad and the ovary the posterior portion. When ripe, the ovarian portion is reddish orange and the testis cream-colored. After gonads have ripened, induce spawning by raising water temperature to 27 to 30°C. Procedures for spawning, handling ova, fertilization, and rearing larvae have been described.[5] Feed larvae with marine phytoplankton cultured as described in Section 8010E.4c1)b).

4. Parasites and Diseases

Cyclopoid copepods may be present in the mantle cavity or digestive tract of mollusks.[8] Species inhabiting the mantle cavity do not cause known pathological damage. However, cyclopoid copepods, such as *Mytilicola intestinalis,* inhabiting the digestive tract, may damage cellular linings and cause a higher incidence of mortality. Ex-amine digestive tracts of a randomly selected sample of 20 mollusks for cyclopoid copepods before using a group taken in one collection, especially in areas where *M. intestinalis* is known to occur. Do not use mollusks if incidence of infestation exceeds 10%.

5. References

1. LOOSANOFF, V.L. & H.C. DAVIS. 1963. Rearing of bivalve mollusks. *Advan. Mar. Biol.* 1:1.
2. CHANLEY, P. & M. CASTAGNA. 1966. Larval development of the pelecypod *Lyonsia hyalina. Nautilus* 79(4):123.
3. REISH, D.J. & J.C. AYRES, JR. 1968. Studies on the *Mytilus edulis* community in Alamitos Bay, California. III. The effects of reduced dissolved oxygen and chlorinity concentrations on survival and byssal thread production. *Veliger* 10:384.
4. BELDING, D.L. 1910. A Report upon the Scallop Fishery of Massachusetts, Including the Habits, Life History of *Pecten irradians,* Its Rate of Growth and Other Factors of Economic Value. Spec. Rep., Comm. Fish & Game, Boston, Mass.
5. CASTAGNA, M. & W. DUGGAN. 1971. Rearing the bay scallop, *Argopecten irradians. Proc. Nat. Shellfish. Assoc.* 61:80.
6. TURNER, H. & J.E. HANKS. 1960. Experimental stimulation of gametogenesis in *Hydroides dianthus* and *Pecten irradians* during the winter. *Biol. Bull.* 119:145.
7. SASTRY, A.N. 1966. Temperature effects in reproduction of the bay scallop, *Argopecten irradians* Lamarck. *Biol. Bull.* 130:118.
8. CHENG, T.C. 1967. Marine mollusks as hosts for symbiosis with a critical review of known parasites of commercially important species. *Advan. Mar. Biol.* 5.

6. Bibliography

GUTSELL, J.S. 1930. Natural history of the bay scallop. *Bull. U.S. Bur. Fish.* 46:569.
LOOSANOFF, V.L. & H.C. DAVIS. 1951. Delayed spawning of lamellibranchs by low temperature. *J. Mar. Res.* 10:197.
DAVIS, H.C. 1953. On food and feeding of larvae of the American oyster, *C. virginica. Biol. Bull.* 104:334.
TUBIASH, H.S. & P.E. CHANLEY. 1963. Bacterial

necrosis of bivalve larvae. *Bacteriol. Proc.* 1963.

DAVIS, H.C. & A. CALABRESE. 1964. Combined effects of temperature and salinity on development of eggs and growth of larvae of *M. mercenaria* and *C. virginica. U.S. Bur. Commer. Fish. Bull.* 63:643.

GALTSOFF, P.S. 1964. The American oyster, *Crassostrea virginica. U.S. Bur. Commer. Fish. Bull.* 64:1.

MATTHIESSEN, G.C. & R.C. TONER. 1966. Possible Methods of Improving the Shellfish Industry of Martha's Vineyard, Duke's County, Massachusetts. Marine Research Foundation Inc., Edgartown, Mass.

BROWN, B.E. 1972. The effect of copper and zinc on the metabolism of the mussel *Mytilus edulis. Mar. Biol.* 16:108.

FAVRETTO, L. & F. TUNIS. 1974. Typical level of lead in *Mytilus galloprovincialis* Lmk from the Gulf of Trieste. *Rev. Int. Oceanogr. Med.* 33:67.

8610 C. Conducting the Toxicity Tests

1. Short-Term Tests

a. Oyster embryo tests: Produce oyster eggs in the laboratory and fertilize to initiate development. They usually become shelled, straight-hinge larvae in 48 h. Use normality of development in receiving water samples to determine its quality. When embryos are cultured in artificial seawater use normality of development as a criterion of relative toxicity of added toxicants.

Use artificial seawater[1] [Section 8010E.4*b*2), Table 8010:II] for spawning adults and culturing embryos.[2] Artificial seawater has been proposed as a standard testing medium.[3,4] Methodology for oyster embryo culture was developed in studies of the eastern oyster, *Crassostrea virginica,*[5] and adapted for the Pacific oyster, *C. gigas,*[6] and the blue mussel, *Mytilus edulis.*[7] Use this method for other bivalves that can spawn under controlled laboratory conditions.

Use the following general steps for tests with fertilized oyster eggs on a year-round basis:

1) Two hours before spawning is desired, place 15 thermally conditioned ripe female oysters in an equal number of borosilicate glass baking dishes about 22 × 12 × 8 cm filled with filtered, UV-light-treated seawater or artificial seawater at 10°C.

2) Raise water temperature by placing dishes in a water bath at 28 to 30°C.

3) About 30 min before spawning is desired, add to each dish 20 mL sperm suspension, prepared as in Section 8610B.3*a*. The combination of increased temperature and sperm usually induces accelerated pumping by the oysters and one or more ripe females to spawn. The sperm fertilizes the eggs as they are discharged.

4) If spawning is not achieved in about 1 h or oysters stop vigorous pumping activity, replace water in spawning dishes with fresh 20°C water and repeat the process. Pipet additional sperm suspension into water being drawn in by the oysters to initiate spawning.

5) About 30 to 45 min after spawning, pour eggs from a single female (usually 6 to 40 × 10⁶) into a 2-L beaker. Determine egg density from two counts, in a Sedgwick-Rafter cell, of number of eggs in 1-mL samples of a 1:99 dilution of homogeneous egg suspension.

6) Bring temperature of control and test water to 20°C ± 0.5°C, (25°C ± 0.5°C in southern areas), before inoculation with oyster embryos. Add enough embryo suspension to each test container to give a population density of 20 000 to 30 000/L. Use at least 10% of the cultures as controls and at least two replicates of each experimental condition.

7) Incubate cultures in a 20 ± 1°C water bath for 48 h (25 ± 1°C in southern areas) and then pour through a 37-μm sieve to retain and concentrate larvae.

8) Wash larvae into a 100-mL graduated cylinder. Take a 2-mL sample containing 150 to 250 larvae with an automatic pipet and preserve in vials with 3% neutral formalin for microscopic examination.

9) Count preserved larvae in a Sedgwick-Rafter cell and record number of normal and abnormal larvae. Normal larvae are fully shelled, even though they may be misshapen or undersized. This criterion avoids the need to make an excessive number of value judgments to classify a larva as normal or abnormal. The percentage of normal larvae is the basic measure of biological response. Further details, test variability, and computer methods for data processing and analysis are available.[6]

b. Scallop, clam, and mussel embryo tests: Make tests with bay scallop, clam, and mussel embryos using procedures described for oyster embryos.

c. Oyster shell deposition test:

1) General considerations—This 96-h test demonstrates the comparative toxicity of pollutants to young oysters. Conduct tests in flowing, unfiltered seawater at a temperature between 15 and 30°C. Actively feeding oysters extend their mantle edges to the periphery of the shell or valves. However, the body can contract to occupy a much smaller area. If peripheral valve edges are mechanically ground away, oysters respond by depositing new shell to replace this loss.[8]

New shell growth is primarily linear during the first week and the deposition rate is an index of the animals' reaction to ambient water quality. With acceptable water conditions, 25 mm and larger oysters deposit as much as 1 mm/d of peripheral new shell. Small oysters are more suitable than large ones because typically they form new shell deposits within a broader temperature range than mature oysters. Interpretation

of test data is independent of minor fluctuations in temperature and salinity during the 96-h exposure because the simultaneous shell deposition in control oysters is considered to be the norm or 100%.

2) Procurement and preparation of oysters—Cull oysters about 25 to 50 mm in height (i.e., the long axis) with reasonably flat, rounded shape to singles, brush clean, and maintain in trays in natural environment. To test, reclean oysters and remove 3 to 5 mm of shell periphery by hand-holding oyster against an electric disk grinder. Insure uniform removal from shell rim to produce a smoothly-rounded blunt profile. Discard oysters damaged by removing too wide a rim of shell.

Fabricate test aquariums of glass, clear acrylic, or wood treated with fiberglass (64 × 38 × 10 cm deep) to provide adequate space for 20 oysters. Deliver unfiltered seawater from a constant-head trough or head box through a diluter system (Section 8010F.1) or by calibrated siphons through a mixing trough (Figure 8610:1 and Lowe[9]) into which toxicant is metered.* Prepare toxicant stock solutions so that a delivery of 1 or 2 mL/min will produce desired concentration. When pumps are used (Figure 8610:1), install baffles in trough to ensure adequate mixing and aeration before water enters aquariums. The aquariums contain about 18 L at 75% capacity. A flow rate of 100 L/h will provide 5 L/h/oyster.

3) Test procedure—Use a preliminary exposure series to determine a suitable range of toxicant concentrations (see Section 8010D.2a). Expose five oysters for 48 h to concentrations of 100, 10, 1.0, 0.1, and 0.01 mg/L to bracket range of toxicant concentrations required to determine 96-h EC50. Lower concentrations may need to be checked.

---

*Sage syringe pump, Sage Instruments, Inc., Cambridge, Mass. 02139, MilRoyal® controlled pump, Milton Roy Co., St. Petersburg, Fla. 33733, or equivalent.

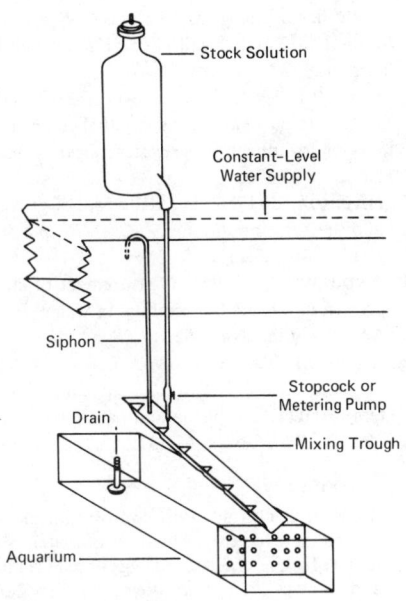

- Stock Solution
- Constant-Level Water Supply
- Siphon
- Stopcock or Metering Pump
- Drain
- Mixing Trough
- Aquarium

Figure 8610:1. Diagram of constant-flow apparatus.[9]

Prepare and distribute oysters randomly so that each control and test aquarium contains 20 individuals. Place oysters with left, cupped valve downward and anterior hinged ends all oriented in one direction. Establish one control aquarium, another receiving only toxicant solvent, and one aquarium for each toxicant concentration. After 96 h, remove all oysters and measure shell increments. Because shell deposition is not uniform on periphery, record length of longest "finger" of new shell, measured to nearest 0.5 mm.

4) Calculation—Calculate ratio of mean growth of a group of test oysters to mean growth of control oysters to provide a percentage index of response. For 96-h exposure, calculate concentrations allowing 50% relative shell growth and 5% relative shell growth.

2. Long-Term Tests

a. Oyster growth test:

1) General considerations—Oysters will grow from setting size to sexual maturity in about 3 to 4 months under optimum conditions. Use 50 to 100 individuals to determine chronic effects. Toxic effects may be manifested by accumulation of chemical residues, changes in resistance to disease, or interference with reproduction.

Evaluate oyster growth by recording weight increases. If oysters are weighed under water, day-to-day changes in shell deposition can be detected.[10,11]

For meaningful weight data, carefully clean oysters and remove all fouling organisms and debris. Do not destroy shell integrity before each weighing. Insure that oyster's valves are closed when it is exposed to air and that shell is free from air bubbles when weighed. Weighings can be replicated to 0.05 g and, under satisfactory conditions, a 15-g oyster (underwater weight) will gain 1 g or more per week.

Sublethal effects may become apparent only slowly; extended exposure may be required.[12]

2) Test procedure—Place small (2- to 3-cm) single oysters, randomly selected from a population of approximately known age, in test aquariums large enough to accommodate their anticipated growth. Supply water and toxicant as described for 96-h shell deposition test. Use a minimum of 5 L seawater/h/oyster. Select toxicant concentration on basis of 96-h tolerance data (Section 8010D.2a).

Position groups of 50 oysters on shallow compartmented racks to facilitate handling and identification of individual animals.[12] At weekly intervals remove oysters, clean, and keep immersed in water until weighed. Clean containers at this time.

Weigh individual immersed oysters with a top-loading balance with suspension attachment. Estimate weights to nearest 0.01 g. Record changes in shell length at each weighing. Expose a suitable number of extra oysters in test and control groups to serve as periodic subsamples to determine residue accumulations and histological changes.

b. Scallop, clam, and mussel growth tests: Bay scallops, clams, and mussels can be used in growth tests similar to those described for oysters if provisions are made to meet their special requirements.

3. References

1. ZAROOGIAN, G.E., G. PESCH & G. MORRISON. 1969. Formulation of an artificial sea water medium suitable for oyster larvae development. *Amer. Zool.* 9:1144.
2. CALABRESE, A., R.S. COLLIER, D.A. NELSON & J.R. MACINNES. 1973. The toxicity of heavy metals to embryos of the American oyster *Crassostrea virginica. Mar. Biol.* 18:162.
3. TARZWELL, C.M. 1969. Standard methods for determination of relative toxicity of oil dispersants and mixtures of dispersants and various oils to aquatic life. *In* Proc. Joint Conf. on Prevention and Control of Oil Spills. p. 179. American Petroleum Inst.
4. LaROCHE, G., R. EISLER & C.M. TARZWELL. 1970. Bioassay procedures for oil and oil dispersant toxicity evaluation. *J. Water Pollut. Control Fed.* 42:1982.
5. LOOSANOFF, V.L. & H.C. DAVIS. 1963. Rearing of bivalve mollusks. *Advan. Mar. Biol.* 1:1.
6. WOELKE, C.E. 1972. Development of a Receiving Water Quality Bioassay Criterion Based on the 48-h Pacific Oyster (*Crassostrea gigas*) Embryo. Tech. Rep. 9, Washington Dep. Fisheries, Olympia.
7. DIMICK, R.E. & W.P. BREESE. 1965. Bay mussel embryo bioassay. *In* Proc. 12th Pacific Northwest Industrial Waste Conf., Univ. Washington, Seattle. p. 165.
8. BUTLER, P.A. 1965. Reaction of some estuarine mollusks to environmental factors. *In* C.M.

Tarzwell, ed. Biological Problems in Water Pollution. U.S. Public Health Serv. Publ. 999-WP-25, p. 92.
9. LOWE, J.I. 1964. Chronic exposure of spot, *Leiostomus xanthurus,* to sub-lethal concentrations of toxaphene in seawater. *Trans. Amer. Fish. Soc.* 93:396.
10. HAVINGA, B. 1928. The daily rate of growth of oysters during summer. *J. Cons. Perma. Int. Explor. Mer.* 3:231.
11. ANDREWS, J.D. 1961. Measurement of shell growth in oysters by weighing in water. *Proc. Nat. Shellfish. Assoc.* 52:1.
12. LOWE, J.I., P.D. WILSON, A.J. RICK & A.J. WILSON, JR. 1971. Chronic exposure of oysters to DDT, toxaphene and parathion. *Proc. Nat. Shellfish. Assoc.* 61:231.

4. Bibliography

OKUBO, K. & T. OKUBO. 1962. Study of the bioassay method for the evaluation of water pollution. II. Use of fertilized eggs of sea urchins and bivalves. *Bull. Tokai Reg. Fish. Res. Lab.* No. 32.

WOELKE, C.E. 1965. Bioassays of pulp mill wastes with oysters. *In* C.M. Tarzwell, ed. Biological Problems in Water Pollution. U.S. Public Health Serv. Publ. 999-WP-25, p. 67.

WOELKE, C.E. 1965. Development of a Bioassay Method Using the Marine Alga, *Monochrysis lutheri.* Washington Dep. Fisheries Shellfish Progress Rep., Olympia.

WOELKE, C.E. 1967. Measurement of water quality with the Pacific oyster embryo bioassay. *Spec. Tech. Pub.* 416:112. American Soc. Testing & Materials, Philadelphia, Pa.

WOELKE, C.E. 1968. Application of shellfish bioassay results to the Puget Sound pulp mill problem. *Northwest Sci.* 42(4):125.

WOELKE, C.E., T.D. SCHINK & E.W. SANBORN. 1970. Development of an *in situ* marine bioassay with clams. Annu. Rep. Oct. 1, 1969-Sept. 30, 1970. July 6, 1970. Washington Dep. Fisheries, Olympia.

8610 D.   Reporting and Analyzing Results

Except for special studies, analyze data, calculate results, and report results as described in Section 8010G.

8710 MICROCRUSTACEANS

Microcrustaceans are arthropods that have a very important role in aquatic ecosystems. These planktonic animals harvest smaller organisms and serve as food for many larger animals. The extreme diversity of microcrustaceans requires a complex classification scheme.

Many microcrustaceans exhibit an interesting type of circadian rhythm, which is the vertical migrations normally regulated by light. The movement is typically up into or toward surface waters in the evening hours and back down into the deeper waters at other times, except for some upward movement around dawn.

The carbon content of a wide variety of zooplankton ranges from approximately 30 to 40% of the dry weight, and the nitrogen and phosphorus values generally lie in the ranges of 5 to 10% and 0.5 to 1%, respectively.[1]

The freshwater cladoceran, *Daphnia,* commonly known as the water flea, is used as a freshwater toxicity test organism (see Section 8711). It belongs to the subclass Branchiopoda, of the class Crustacea. The marine copepod, *Acartia tonsa,* which belongs to the subclass Ostracoda, is used for testing the effects of toxicants in the marine environment (see Section 8712). There are newer short-term methods for estimating chronic toxicity that have not yet been incorporated into this text.

Reference

1. PARSONS, T.R., M. TAKAHASHI & B. HARGRAVE. 1984. Biological Oceanographic Processes. Pergamon Press, Oxford, England.

8711 TOXICITY TEST PROCEDURES FOR *DAPHNIA* *

8711 A. Introduction

Because balloting for the 17th edition could not resolve substantive differences, the 16th edition text, with format changes only, is included herein. Substantial changes are planned.

1. General Considerations

Daphnia have been used in tolerance studies for over a century. Results obtained by different investigators using *Daphnia* are comparable and in general, *Daphnia* are less tolerant of toxic substances than are fish.[1]

Daphnia magna is the largest of the *Daphnia,* reaching a maximum size of over 5 mm. Large numbers can be reared in a relatively small space. Neonates (first-instar young) are 0.8 and 1.0 mm long and can be observed without optical aids. This stage is the one most commonly used for tolerance studies. Species smaller than *D. magna* and *D. pulex* may require special handling.

Individual female *D. magna* have been known to live for as long as 4 months at 20°C. They have been cultured in natural waters and in dechlorinated tap water. Foods include bacteria, algae, and yeast, together with soil extracts and organic materials such as cotton-seed meal, herring meal, powdered dried grass, and enriched trout fry granules. *Daphnia* have been reared individually in small vessels and in mass culture in large aquariums.[2-6]

---

* Approved by Standard Methods Committee, 1981.

Reproduction can be restricted to the production of females by diploid parthenogenesis when suitable culture conditions are maintained, thereby insuring a supply of experimental animals whose genetic variability is limited to the heterozygosity of the parent.[7] Genetic uniformity is expected within parthenogenetic clones derived from a single female.[8] When culture conditions are optimum *D. magna* females release their first broods of young within 10 d at 20°C and 7 d at 25°C. Thereafter young are released every 3 to 4 d at 20°C and 2 to 3 d at 25°C. Twenty or more young per brood are produced as long as culture conditions remain adequate. A single female may release over 400 young during her lifetime.

2. References

1. KEMP, H.T., J.P. ABRAMS & R.C. OVERBECK. 1971. Water Quality Criteria Data Book Vol. 3. Effects of Chemicals on Aquatic Life, Selected Data from the Literature through 1968. Environmental Protection Agency, Water Pollution Center Res. Ser. 18050GWV05/71, U.S. Government Printing Off., Washington, D.C.

2. NEEDHAM, J.G., P.S. GALTSOFF, F.E. LUTZ & P.S. WELCH. 1937. Culture Methods for Invertebrate Animals. Comstock Publ. Co. Reprinted by Dover Publ., Inc., New York, N.Y.

3. NAUMANN, E. 1933. *Daphnia magna* Straus als Versuchstier. *Kgl. Fysiogr. Sallsk. Lund Forhandl.* 3:15.

4. ANDERSON, B.G. 1944. The toxicity thresholds of various substances found in industrial wastes as determined by the use of *Daphnia magna. Sewage Works J.* 16:1156.

5. PARKER, B.L. & J.E. DEWEY. 1969. Further improvements on the mass rearing of *Daphnia magna. J. Econ. Entomol.* 62:725.

6. BIESINGER, K.E. & G.M. CHRISTENSEN. 1972. Effects of various metals on survival, growth, reproduction and metabolism of *Daphnia magna. J. Fish. Res. Board Can.* 29:1691.

7. BACCI, G., G. COGNETTI & A.M. VACCARI. 1961. Endomeiosis and sex determination in *Daphnia pulex. Experientia* 17:505.

8. HEBERT, P.P. & R.D. WARD. 1972. Inheritance during parthenogenesis in *Daphnia magna. Genetics* 71:639.

8711 B. Selecting and Preparing Test Organisms

1. Obtaining Test Organisms

Obtain *Daphnia* from established cultures or by field collection. While *Daphnia* of any age can be available at any time, it is more convenient to use neonates than older animals for testing. When reared at 20°C, they undergo their first ecdysis about 30 h after release from the brood chamber and at 25°C after about 20 h. Neonates can be segregated from stock animals at 24-h intervals when reared at 20°C and at 12-h intervals at 25°C. Neonates are less tolerant of many substances than older animals. *Daphnia* are more susceptible to most substances at ecdysis than between molts.

2. Culturing

One of the simplest media is the manure-soil medium[1,2] supplemented periodically with yeast. Make this medium by mixing 5 g dried sheep manure, 25 g garden soil or sandy muck, and 1 L pond, spring, or tap water. Let stand for 2 d at room temperature, then strain through bolting cloth with mesh openings approximately 0.15 mm. During straining, work some of the finer soil particles through the cloth. Set filtrate aside for 1 week or more and discard sediment. To make final medium, mix 1 part filtrate with 6 to 8 parts pond, spring, or dechlorinated tap water. The original filtrate may be kept indefinitely before final medium is made.

Use final medium for individual or mass culturing. For individual rearing, dispense 100 mL medium into 125-mL wide-mouth flint-glass bottles or equivalent vessels and inoculate with one daphnid per bottle. Beginning 1 d after inoculation, add 1 mL of a suspension containing 1 mg active dry yeast in water to each bottle on alternate days. For mass rearing use 3.8-L wide-mouth glass jars filled with 3 L medium to which a suspension of 30 mg active dry yeast is added on alternate days. Once the cultures are initiated, do not change the culture medium, but occasionally add water to replace that lost by evaporation and in removal of young. To retard evaporation, use a perforated cover that permits air diffusion. Aeration is unnecessary because the critical DO for *Daphnia* is less than 15% of saturation at 20°C.

With a stock of 100 individually reared stock females, over 300 neonates can be available daily. When stock females begin reproducing, periodically remove young, preferably every 24 h at 20°C or 12 h at 25°C. When stock animals reach old age and their reproductive rate drops, replace with young females in fresh medium.

Other methods of culturing, such as those using algae,[3] have been used successfully. Biesinger and Christensen[4] developed a medium consisting of a suspension of 0.5 g powdered dried grass* and 10 g enriched trout fry granules in 250 mL unpolluted lake, well, or river water mixed vigorously in a blender for 5 min. Strain suspension through a stainless steel screen with about 0.066-cm opening and add an additional 50 mL water to rinse blender. Refrigerate suspension. Before a portion is withdrawn, mix thoroughly. Feed it at the rate of 1 mL/week/L.

2. References

1. ANDERSON, B.G. 1944. The toxicity thresholds of various substances found in industrial wastes as determined by the use of *Daphnia magna. Sewage Works J.* 16:1156.

2. ANDERSON, B.G. 1950. The apparent thresholds of toxicity to *Daphnia magna* for chlorides of various metals when added to Lake Erie water. *Trans. Amer. Fish. Soc.* 78:96.

3. PARKER, B.L. & J.E. DEWEY. 1969. Further improvements on the mass rearing of *Daphnia magna. J. Econ. Entomol.* 62:725.

4. BIESINGER, K.E. & G.M. CHRISTENSEN. 1972. Effects of various metals on survival, growth, reproduction and metabolism of *Daphnia magna. J. Fish. Res. Board Can.* 29:1691.

---

*Cerophyll, Cerophyll Laboratories, Inc., 4722 Broadway, Kansas City, Mo. 64111, or equivalent.

8711 C.   Toxicity Test Procedures

1. Static Tests

a. Setting up the tests: Prepare test materials, dilution water, and toxicant solutions as described in Sections 8010F.2 and 3. Select test concentrations as described in Section 8010F.3*b*. Make up test solutions and controls in 100-mL quantities in 125-mL wide-mouth flint-glass bottles or equivalent vessels.

b. Performing the tests: After preparing test solutions, segregate neonate *Daphnia* that have been released from the mothers' brood chambers during the preceding 24 h at 20°C or 12 h at 25°C and collect in one vessel. Wash in three changes of diluent, allowing 5 min in each wash solution to reduce carryover of materials. Introduce 10 neonates into each test vessel and the control. Use a piece of 8-mm glass tubing 20 cm long, with one end drawn to a diameter of 2 mm and the other end fire-polished and fitted with a 2-mL-capacity rubber bulb, for collecting and transferring neonates. Use identical tubing with both ends fire-polished, and fitted with a rubber bulb, for handling adults.

After introducing neonates to test solutions, observe them regularly, usually after 1, 2, 4, 8, and 16 h and daily thereafter. Record number of motile animals in each test vessel. Consider an animal nonmotile if it shows no independent movement even after a test vessel is rotated. Nonmotile animals are not necessarily dead. At threshold concentrations of such substances as ethanol, acetone, and chlorobutanol, animals may show no movement and the heart may have ceased to beat but on transfer to dilution water they will recover. However, if such animals are maintained in the test medium they will die. Continue observations for 5 d or as long as most of the animals in some of the test solutions remain motile. Run tests in triplicate.

Do not feed animals during tests. *D. magna* will live for as long as 1 week without food in well-balanced salt solutions.[1] Different species and longer-term tests will require modifications of standard conditions.

2. Long-Term Tests

a. Determination of reproductive impairment: Reproduction may be impaired at much lower toxicant concentrations than those causing acute toxicity, sometimes as low as 0.000 03 of that producing acute toxicity. Precede these tests by acute tolerance tests to establish the maximum concentration to be used.

b. Preparation of test medium: Prepare test medium in the same way as regular culture medium, except using water representative of that receiving the effluent discharge. Prepare a series of 6 to 10 1-L quantities of medium to which graded amounts of toxicant have been added. Use as the highest concentration of toxicant the equivalent to the LC50 or EC50 at 96 h. Reduce each successive concentration in a geometric progression by a factor of three or more in preliminary experiments. Use undiluted culture medium as control. Dispense each liter of test medium in 100-mL quantities to each of 10 bottles.

c. Performing tests: Segregate and collect neonate *Daphnia* as for static tests. Do not wash animals. Introduce one neonate into each bottle. On the following day and on alternate days thereafter, add a 1-mL suspension containing 1 mg active dry yeast to each bottle. Make daily observations and note dead or immobilized animals. As the animals grow and reproduce, remove young and record their number. Replace lost water. Continue observations until control animals have released at least six broods of young, which will take about 21 d at 25°C and 30 d at 20°C. Handle animals as individual cultures of stock animals.

At end of observation period analyze results and test for significance the differences in the number of young produced. Note time of appearance of first broods and number of broods.

If the number of young produced in the lowest concentration of toxicant differs significantly from that in the controls, repeat the test with reduced concentrations of toxicant until no significant differences are obtained.

After reaching toxicant concentrations where no differences are observed between tests and controls in number of young and broods and time of appearance of first broods, make a final set of tests with 30 or more animals in each of four concentrations of toxicant bracketed about the lowest concentration for which significant differences in production of young were found. Use an equal number of animals for the control.

3. Reference

1. STAMPER, W.R. 1969. The determination of the optimal combination of concentrations of sodium, potassium, calcium and magnesium chlorides for the survival of *Daphnia pulex*. Ph.D. thesis, Pennsylvania State Univ.

8711 D. Evaluating and Reporting Results

Assemble, analyze, evaluate, and report data as described in Section 8010G.

8712 TOXICITY TEST PROCEDURES FOR THE CALANOID COPEPOD, ACARTIA TONSA (DANA)*

8712 A. Introduction

This method provides data on the effects of toxicants on a marine copepod, *Acartia*

---

*Approved by Standard Methods Committee, 1988.

tonsa. Use general procedures described in Sections 8010D, E, and F. *A. tonsa* is sensitive; therefore, handle gently and only when necessary.

8712 B. Selecting and Preparing Test Organisms

1. Collecting and Holding Test Organisms

a. Collection: Collect *A. tonsa* by towing ($\geq$ 4 km/h) a plankton net (aperture 150 to 250 μm) at a depth of 1 to 3 m. Transfer animals carefully by pouring gently into insulated containers three-fourths filled with ambient seawater. Remove predators immediately, especially jellyfish and ctenophores. Do not exceed a copepod density of 250/L to assure that the DO concentration remains adequate. Measure and record temperature, salinity, DO, and pH at time of collection and maintain these conditions during initial holding stages.

b. Holding: Immediately on return to the laboratory, transfer samples to 2.3 L (190 $\times$ 100 mm) borosilicate crystallizing dishes. Adjust volume of water in each dish to 2 L with filtered seawater at ambient temperature and salinity and add the algal diet for adult copepods, described in ¶ 2b below, in the quantity necessary to attain the density of organisms listed in Table

8712:I. Incubate cultures at collection temperature and illuminate as described in Section 8010F.3f. After 24 h, begin to acclimate cultures to 20°C and 20 g/kg salinity. Change salinity and temperature in increments of 3 g/kg and 3°C/d. If organisms are held in the original vessel during acclimation, alternately siphon off and add seawater of a different salinity. Transfer organisms either by gently pouring, pipetting, or siphoning (use 1-cm-ID tubing) to new vessels. During acclimation, maintain a daily feeding schedule to provide the density of algae listed in Table 8712:I.

c. Sorting and identification: For basic information on taxonomy and biology of *Acartia* and other coastal calanoids consult other sources.[2-7] Separate *A. tonsa* from other organisms collected in the tow. To facilitate copepod capture, reduce culture volume from 2.0 L to 0.5 L by slowly siphoning the seawater, with a 150-μm plankton netting screen over the siphon intake. Carefully draw individual adults up into a wide-bore ($\geq$2 mm) transfer pipet and place in depression slides. Identify and transfer to food-enriched filtered seawater to 20 g/kg salinity at 20°C [see ¶ 2a1) and Section 8010E.4c1)b)]. Exclude all nauplii and juvenile forms.

d. Water supply:

1) Natural water—Collect sufficient seawater from the study area to perform tests. This seawater, when adjusted to 20 g/kg salinity and 20°C, must support survival of adult *A. tonsa* for at least 96 h, the complete life cycle of the copepod, and with proper enrichment, the growth of food algae. Use Niskin or Van Dorn samplers and collect seawater from 3 to 10 m depth. Record salinity, DO, and pH. Transport collected seawater to laboratory in glass or polyethylene carboys aged in seawater. Filter through a 1.0-μm acid-washed filter (glass fiber, cellulose-acetate, nylon, or polycarbonate). Rinse containers with deionized water, fill with filtered seawater, and store at 4°C.

2) Artificial seawater—If no suitable natural seawater is available use the synthetic seawater formulation described in Section 8010E.4b2). Heinle[7] found a commercial seawater* suitable for the culture of both *A. tonsa* and *Eurytemora affinis*. Algal assays indicate that this product is unsuitable for tests involving heavy metals; use it only for culture until more extensive comparative data are available.

2. Culturing Test Organisms

a. Production of food for test organisms:

1) Culture medium for algae—See Section 8010E.4c1)b).

2) Growth chambers—Grow algal cultures in standard screw-capped test tubes, flasks, or in the fill-and-draw semicontinuous system described below.

3) Culturing and harvesting algae for food—Grow algae for feeding copepods and other zooplankton in filtered natural

or synthetic seawater at 20 g/kg salinity and 20°C with 2500 to 5000 lux continuous illumination for 14 h daylight and 10 h darkness (14L:10D). Use nutrient enrichments of Guillard and Ryther.[8]

Dispense enriched seawater [see Section 8010E.4c1)b)] into screw-capped test tubes (50 mL) or erlenmeyer flasks. After autoclaving for 15 min at 108 kPa and 121°C, cool and equilibrate with atmospheric gases for 48 h. Check sterility as described in Section 8010E.4c1)a).

The recommended algal culture system (Figure 8712:1) is a fill-and-draw type in which cultures are maintained near their maximum log-phase cell density and growth rate. Draw off medium and organisms and replace with fresh medium so that culture will have reached same cell density within 24 h. When longer than 24-h intervals occur between harvests, draw off and replace proportionally greater amounts of culture. Scale this system up or down depending on food needs. Simultaneously keep a series of tube cultures of each of the four algal foods in case of contamination of the large cultures.

Determine algal cell densities by direct microscopic counts using a hemacytometer, Palmer-Maloney chamber, or Utermohl chamber (inverted microscope).[9,10] An electronic particle counter is an accurate and rapid method for determining unialgal densities. If visual counts are necessary, relate these to specific absorbance at 750 nm with a spectrophotometer (see Section 8111G.4). A curve relating cells per milliliter to absorbancy can be used to replace the cell count.

b. Mass culture of test organisms: This system provides large quantities of *A. tonsa* of standard age for tests.

1) Growth chambers—Use mass culture units of Mullin and Brooks[11] and Frost.[12] The culture vessel in Figure 8712:2 is a borosilicate glass aspirator bottle with a capacity of 13 to 45 L, depending on the number of copepods needed. Gently mix

*Instant Ocean®, available from Aquarium Systems, Inc., 1450 East 298th St., Wickliffe, Ohio 44092.

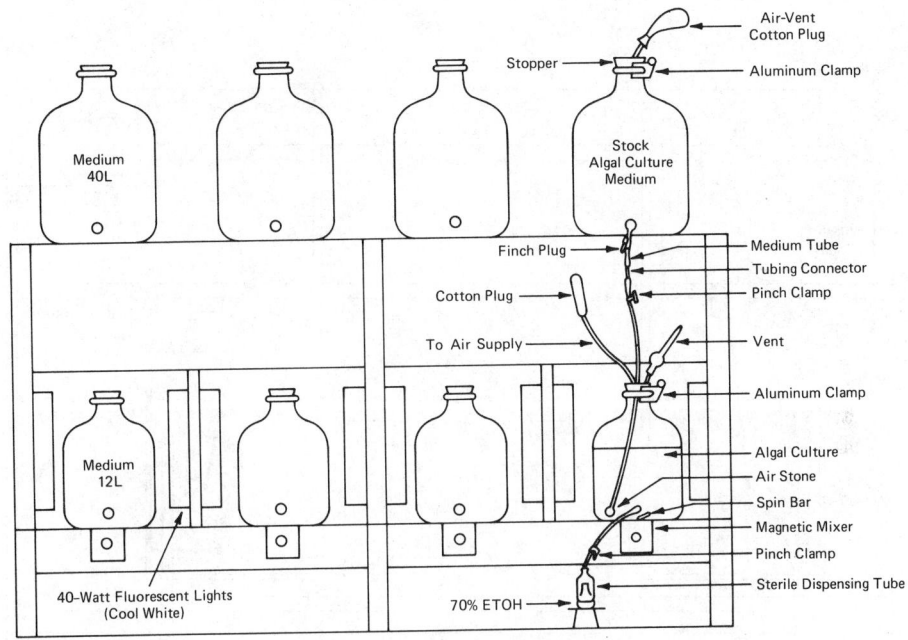

Figure 8712:1. Algal culture system.

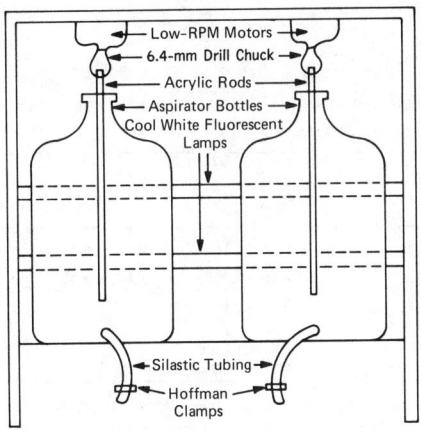

Figure 8712:2. Apparatus for mass copepod culture (static).

the contents with a low-speed motor (≤ 25 rpm) with a stirring rod mounted above the culture vessel. Slow mixing maintains algal food in suspension where plank-tonic copepods normally feed. Keep water movement gentle and free of vortices such as those produced by magnetic stirrers. Use cool white fluorescent lights to provide 1000 lux illumination incident to the culture surface on a 14L:10D cycle.

Alternatively use a continuous-flow system (Figure 8712:3). Fit a cylindrical vessel (15 to 40 L) with a standpipe and drain. Collar the standpipe with 200-μm plastic† screen to prevent loss of eggs and nauplii. Mount a low-speed motor (≤ 25 rpm) with stirrer above the culture vessel. Add filtered seawater by a peristaltic pump or siphon from a constant-head tank at a rate of 1 to 3 tank volumes/24 h. Feed by mixing the four algal species (Table 8712:I) in a single container and introduce them into the culture chambers with a pump (Figure 8712:3) at a rate necessary to maintain a food cell density of 2 to 5 × 10^7 cells/L. Place the

---

†Nitex or equivalent.

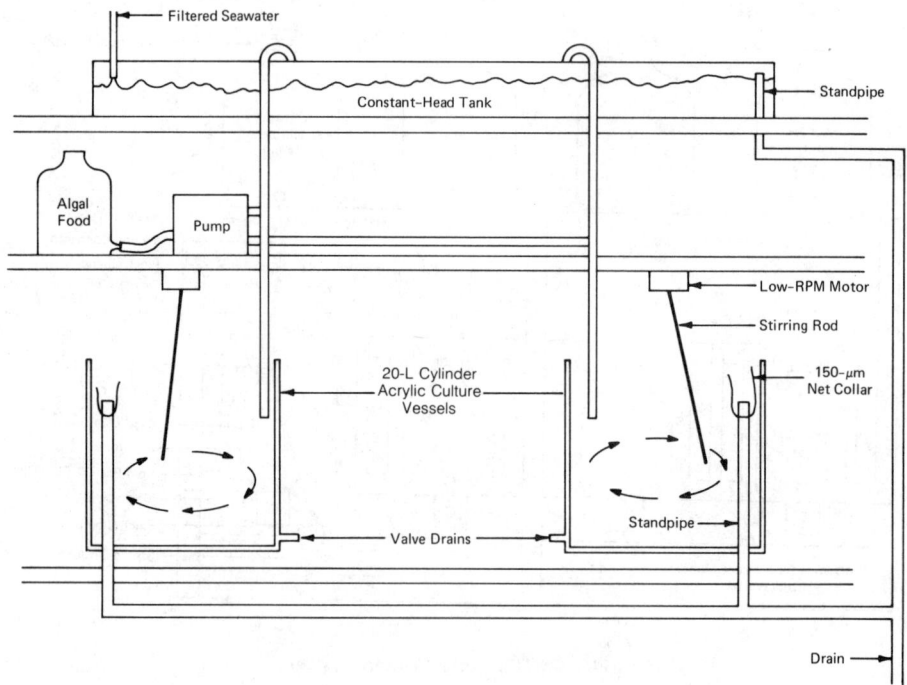

Figure 8712:3. Apparatus for mass copepod culture (flowing).

TABLE 8712:I. COMPOSITION OF ALGAL DIET AND RECOMMENDED CONCENTRATIONS FOR ADULT AND NAUPLIAR FEEDING AND EGG-LAYING[1]

Species	Adult & Copepodite	Naupliar	Egg-Laying
Skeletonema costatum	5.0×10^6	5.0×10^5	1.5×10^7
Thalassiosira pseudonana	7.0×10^6	7.0×10^5	2.1×10^7
Isochrysis galbana	5.0×10^6	5.0×10^5	1.5×10^7
Rhodomonas baltica	3.0×10^6	3.0×10^5	9.0×10^6
Total cells/L of copepod culture	2.0×10^7	2.0×10^6	6.0×10^7

stirring rod in the culture vessel so that the lower 25% of the vessel receives only gentle mixing. This permits a quiet area for mating and egg laying and keeps sediment from being stirred up and clogging the screen. Drain and clean the system every 3 to 4 weeks. Brush the screen daily with a fine brush to avoid clogging.

2) Algal diet for copepods—Although various algal diets have been used for copepod cultures[10,13-15] use the algal diet given in Table 8712:I.[1] *Skeletonema costatum* has been added because it is naturally occurring food for *A. tonsa.*

3) Setting up cultures of test organisms—*A. tonsa* females are each capable of producing more than 30 eggs/d when fed the recommended ration. If 250 or more

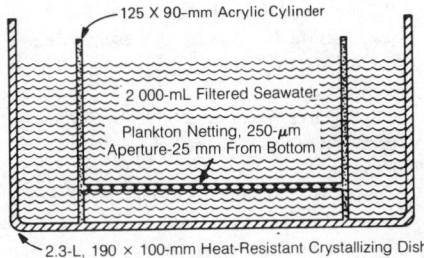

Figure 8712:4 Generation cage (after Heinle).

mental stage in the life cycle of *A. tonsa* at 20°C and 20 g/kg is as follows:

Stage	Length d
Egg (newly oviposited)	1
Nauplius (6 instars)	7
Copepodite (6 instars)	6
Adult (until gravid)	3
Total life cycle	17

gravid females are kept as breeders, theoretically over 4000 eggs will be produced within 24 h. For this potential number of adults, use a 40-L culture vessel. Generally, the relationship between culture volume and organism density is 10:1.

For production and collection of copepod eggs, use generation cages (Figure 8712:4), consisting of a clear acrylic plastic cylinder 125 mm in diam and 90 mm high, with one end covered by plankton netting having an aperture size of 250 µm. Place this cylinder in a 2.3-L (190 × 100 mm) crystallizing dish with covered end 25 mm from bottom. Add 2 L filtered seawater or artificial seawater and place 50 to 100 gravid females in each screen-bottom cylinder. Feed three times the adult algal diet (Table 8712:I). The plankton netting allows eggs to pass through and hatch in crystallizing dish where they are protected from adult predation. After 24 h remove adults by gently lifting each generation cage from crystallizing dish and quickly immersing it in another dish with three times the usual food density. Carefully siphon remaining seawater from all dishes containing eggs and nauplii into a glass aspirator bottle containing filtered seawater. Adjust final volume and feed nauplioid culture as indicated in Table 8712:I. If a second mass culture is desired, repeat after 24 h, using required number of generation cages to produce needed organisms.

The average length of each developmental stage in the life cycle of *A. tonsa* at 20°C and 20 g/kg is as follows:

During the first 6 d of mass culture only naupliar stages are present. Feed daily with 2×10^6 cells/L. On the third and seventh days, slowly siphon off 50% of culture medium and replace with clean medium. Cover intake end of siphon with 60-µm netting to prevent loss of nauplii.

After the seventh day, copepodites should be present. From this time on, feed 2×10^7 cells/L/d, replacing 50% of the culture volume with filtered seawater every third day. Within 16 to 17 d the population will have reached maturity and can be used for tests or to start new cultures. Average adult life span at 20°C is approximately 30 d.

Maintain a non-age-standardized mass culture in reserve. Use gravid females from original generation cages to start a 12-L system. Feed adult food ration and replace 50% of culture water every third day. In addition, harvest approximately one-third of culture, including organisms, periodically (10 to 14 d) to keep population at about 50 adults and copepodites.

4) Harvesting test organisms—Harvest mass cultures of copepods that have reached the adult stage as follows: Reduce culture volume 75% using a slow siphon with its intake covered with 60-µm plankton netting. Transfer carefully the remaining 25%, including organisms, to 2.3-L borosilicate glass crystallizing dishes to provide a total volume of 2 L/dish. Because of organism fragility do not constrict discharge tube to reduce flow; instead, control flow through ventral tubulation on aspi-

rator bottle (Figure 8712:2) by minimizing head between culture vessel and crystallizing dish. A slow flow minimizes turbulence and opportunity for organisms to collide with vessel walls.

Further concentrate harvested animals in crystallizing dishes by siphoning off culture medium. Facilitate capture by using the positive phototactic response of the animals.

5) Other organisms—The culture system designed for *A. tonsa* has worked well for *Eurytemora affinis* and *Pseudodiaptomus coronatus*. However, the generation cages are suitable only for *A. tonsa* because it releases eggs individually. Both *E. affinis* and *P. coronatus* produce egg sacs.

3. References

1. WILSON, D.F. & K.K. PARRISH. 1971. Remating in a planktonic marine calanoid copepod. *Mar. Biol.* 9:202.
2. CONOVER, R.J. 1956. Oceanography of Long Island Sound, 1952-1954. VI. Biology of *Acartia clausi* and *A. tonsa*. *Bull. Bingham Oceanogr. Coll.* 15:156.
3. HEINLE, D.R. 1966. Production of a calanoid copepod *Acartia tonsa*, in the Patuxent River Estuary. *Chesapeake Sci.* 7:59.
4. HEINLE, D.R. 1969. Effects of Temperature on the Population Dynamics of Estuarine Copepods. Ph.D. thesis, Univ. Maryland, College Park.
5. WILSON, C.B. 1932. The Copepods of the Woods Hole Region. Massachusetts. Smithsonian Inst., U.S. National Museum Bull.
6. ROSE, M. 1933. Faune de France. No. 26. Copepodes Pelagiques. Librarie de la Faculté des Sciences, Reprinted 1970 by Kraus Reprint, Nendeln, Lichtenstein.
7. HEINLE, D.R. 1969. Culture of calanoid copepods in synthetic seawater. *J. Fish. Res. Board Can.* 26:150.
8. GUILLARD, R.R. & J.H. RYTHER. 1962. Studies of marine planktonic diatoms. I. *Cyclotella nana* Hustedt and *Detonula confervacia* (Cleve) Grant. *Can. J. Microbiol.* 8:229.
9. PALMER, C.M. & T.E. MALONEY. 1954. A new counting slide for nannoplankton. *J. Amer. Soc. Limnol. Oceanogr.*, Spec. Publ. 21:1.
10. SCHWOERBEL, J. 1970. Methods of Hydrobiology. Pergamon Press, New York, N.Y.
11. MULLIN, M.M. & E.R. BROOKS. 1967. Laboratory culture, growth rate and feeding behavior of a planktonic marine copepod. *J. Amer. Soc. Limnol. Oceanogr.* 12:657.
12. FROST, B.W. 1972. Effects of size and concentration of food particles on the feeding behavior of the marine planktonic copepod *Calanus pacificus*. *J. Amer. Soc. Limnol. Oceanogr.* 17:805.
13. ZILLIOUX, E.J. & D.F. WILSON. 1966. Culture of a planktonic calanoid copepod through multiple generations. *Science* 151:996.
14. KATONA, S.K. 1970. Growth characteristics of the copepods *Eurytemora affinis* and *E. herdmani* in laboratory cultures. *Helgolander wiss. Meeresunters* 20:373.
15. NASSOGNE, A. 1970. Influence of food organisms on the development and culture of pelagic copepods. *Helgolander wiss. Meeresunters* 20:333.

8712 C.   Toxicity Test Procedures

With adult *A. tonsa* and the previously described culture method, make short-term tests as described below.

1. Range-Finding Tests

Use 10 adult *Acartia* per replicate with three replicates per test concentration and control. For a test container, use a suitable flat-bottom borosilicate glass dish containing 100 mL seawater ≥ 2.0 cm deep. Generally use a broad range of concentrations (see Sections 8010F.3b and 8010D.2a). Prepare toxicant solutions as described in Section 8010F.2b. Capture 10 adult *Acartia* for each test chamber from stock cultures with a wide-bore transfer pipet and transfer to a 20-mL beaker containing 5 mL filtered seawater. Adjust final volume to 15 mL. Add animals and 15 mL medium to 85 mL of toxicant-dosed medium in test chamber by immersing beaker and rinsing gently. Expose for 96 h. Observe and record number of dead, moribund, and living copepods after 1.5, 3, 6, 12, 24, 48, 72, and 96 h exposure. To ascertain if a motionless animal is dead, touch it gently with a sealed glass capillary probe. If control mortalities exceed 15%, reject results.

2. Definitive Short-Term Tests

Follow general culture conditions and handling as described in Sections 8010E and F. Specifically test 15 adults in each of four replicate test vessels per toxicant concentration and control. Select test concentrations of toxicants as described in Section 8010F.3b. Expose animals and collect data as described in Sections 8010D and F.

8712 D. Evaluating and Reporting Results

Make calculations, present data, and analyze and express results as described in Section 8010G.

8720 TOXICITY TESTING PROCEDURES FOR MACROCRUSTACEANS*

8720 A. Introduction

Crustaceans are a large, mostly marine group of organisms. The class, Crustacea, contains more than 25 000 species grouped into the two subclasses, Entomostraca and Malacostraca. Many of the smaller Entomostraca, the copepods and cladocerans, commonly are referred to as microcrustaceans and are important macroplanktonic organisms in both marine and fresh waters. Toxicity testing methods for the microcrustaceans are given in Sections 8710 through 8712. The subclass Malacostraca contains the larger and economically important crustaceans. Although most are marine there are some important freshwater forms. In the family Mysidae, *Mysis relicta* of the Great Lakes is an important food of salmonids. The order Amphipoda, containing over 3000 species, is almost exclusively marine but has important freshwater species in the genera *Hyalella*, *Gammarus*, *Crangonyx*, and *Pontoporeia*. The order Isopoda also is almost exclusively marine but has a few freshwater forms. Of greater economic importance is the order Decapoda, containing the lobsters, spiny lobsters, crabs, shrimp, prawns, and crayfish. Crustaceans are relevant es-

*Approved by Standard Methods Committee, 1988.

pecially for determining the toxicity of pesticides in the aquatic environment because of their phylogenic relationship to the insects for whose control many pesticides have been developed.

8720 B.   Selecting and Preparing Test Species

1. Selecting Test Organisms

The general principles governing the selection of test organisms are described in 8010E.1. The following are suggested as test organisms:

Freshwater species:
Gammarus lacustris
Gammarus pseudolimnaeus
Gammarus fasciatus
Hyalella azteca
Pontoporeia affinis
Mysis relicta
Palaemonetes cummingi
Palaemonetes kadiakensis
Crayfish — *Cambarus*
    Potamobius
    Orconectes rusticus
Marine and brackish water species:
Palaemonetes pugio — grass shrimp
Palaemonetes vulgaris
Palaemonetes intermedius
Crangon septemspinosa — sand shrimp
Penaeus duorarum — pink shrimp
Penaeus aztecus — brown shrimp
Penaeus setiferus — white shrimp
Homarus americanus — American
    lobster
Callinectes sapidus — blue crab
Cancer irroratus — rock crab
Cancer borealis — jonah crab
Cancer magister — dungeness crab
Panopeus herbstii — mud crab
Rhithropanopeus harrisii — mud crab
Menippe mercenaria — stone crab

Accurately identify test animals before use. Most inshore macrocrustaceans in the United States have been classified and regional keys are available.[1-5]

2. Collecting and Handling Test Organisms

Collect smaller forms in dip nets or coarse plankton nets; collect larger near-shore forms in small-mesh seines. Trawl for most marine forms. For collecting and transporting methods see Sections 10500B and 8010E.2.

3. Holding, Acclimating, and Culturing Test Organisms

a. Water supply: See Section 8010E.4*b*.

b. Acclimating, holding, and maintaining stock cultures: See Section 8010E.3 and 4. Risks in handling most adult crustaceans usually are not great because of their rigid exoskeleton and general durability. Both larval and adult forms of many species are cannibalistic and readily attack their fellows in the soft-shell stage. Hold juveniles and adults in individual compartments in long troughs or divided tanks. Form the compartments with perforated separators that slide into slots on the sides. Use stainless steel for freshwater forms and glass, acrylic, plastic, or plywood covered with fiberglass for marine forms. Provide rigid, transparent covers to prevent loss of the highly motile specimens. Use perforated separators to ensure a flow of water through each compartment to remove metabolic products and provide DO. The crustacean growth process, which involves a periodic ecdysis or sloughing of the rigid exoskeleton, imposes a lack of uniformity in test animals that is not readily detectable in advance. In the pre-ecdysis stage and during ecdysis animals are heavily stressed and more sensitive to unsatisfactory environmental conditions and toxicants.[6]

1) Amphipods—Collect freshwater amphipods from their natural habitat or rear in the laboratory. A few species have been reared through three or more generations. Maintain several stock cultures and supplement by collections from the natural habitat. In the laboratory, a new generation can be produced in about 8 to 20 weeks, depending on species and water temperature.[7-9] For holding and for tests, use a known favorable temperature and a light intensity of 540 to 1600 lux at the water surface.

Keep amphipods collected from the field in flow-through systems with water temperatures held initially near the temperature of the water from which they were collected. Gradually change temperature to that at which tests are to be conducted. Handle test animals carefully and as little as possible. Use small dip nets having soft flexible netting and no projections.

Feed young and adults on aspen, maple, and birch leaves that have been soaked for several weeks in flowing water. Test leaves to insure that they are free from pesticides or other toxicants. Supply leaves in sufficient quantities to support a stable population but not to the extent that they cause DO depletion or excessive fungal growth. Supply some leaves at all times to provide cover. Use commercial fish food to supplement the leaf diet.

2) Crayfish—Collect specimens from their natural habitat by trapping, seining, or by hand (Section 8010E.2). General procedures for holding and acclimating are as described in Sections 8010E.3 and 4.

Because crayfish are cannibalistic, hold all but the young stages in separate compartments. Suitable holding, acclimating, and culturing chambers are stainless steel, glass, fiberglass-covered wood, or plastic troughs, 180 cm long, 30 cm wide, and 20 cm deep, with a divider down the center to make two long troughs. Make shallow channels on the sides and central divider every 15 cm into which separators can be slipped to make 12 compartments on each side, each approximately 15 × 15 cm square and 20 cm deep. This size is suitable for crayfish. The number and size of compartments depends on the size and number of organisms to be tested. To hold a large number of small crayfish, remove the separators to make a tank of the desired length. Provide separators with a large number of perforations so they operate as screens. Control water depth in test chambers by a standpipe in the last compartment of the trough.[9] When cleaning the separators, temporarily raise them a short distance from the bottom to allow excess food and wastes to be washed out and remove the standpipe in the last compartment to insure strong flows. Clean routinely with a siphon and a brush to loosen materials from compartments, screens, walls, and bottoms. Supply water adjusted to the desired temperature and DO to the two head compartments by a siphon from a constant-head box. Use a minimum flow of 10 trough volumes/d. Adjust volume to maintain favorable water quality in each compartment. Water depth required depends on size of organisms but 15 cm is preferred. Provide each set of troughs with a transparent lid.

For life-cycle studies beginning with eggs or newly hatched young, collect ovigerous females and place in flow-through troughs under natural water conditions. Begin acclimation to different conditions after 2 d. Hold animals in troughs until young hatch. Remove compartment dividers to provide freedom of movement of young. Clean as described in Section 8010E.4d.

Use macerated fish food for juveniles and adults. Alternatively use prepared dry fish food. Use very finely divided pieces of fish and commercial fish food pellets as food for the newly hatched.

3) Crabs—Successful static culture of brachyuran crab larvae has been made for several species of Atlantic coast crabs.[10-13] Long-term static or renewal bioassays with these species have been performed.[14-16] Culture of dungeness crab, *Cancer magister*, larvae has been reported.[17-19] Culturing crab larvae requires a favorable water supply and control of competitors, predators, and disease. Filter water and disinfect by UV light treatment. For unpolluted open ocean water little or no treatment is required. If the supply is from an estuary receiving organic wastes, purify before use. Filter seawater for the flow-through system by gravity flow through a coarse, quartz sand filter and adjust to the desired salinity, approximately 25 to 30 g/kg, by adding fresh water. To remove other organisms, refilter under pressure through sequential layers of 40/60-mesh garnet, 20/30-mesh silica sand, and 0.3 cm hard coal.* Follow by filtration through a polishing filter† and treat with UV light. Use constant-level head boxes equipped with heating, cooling, and stirring devices, to deliver constant measured flows by siphons, selected nozzles, or constant and accurate delivery pumps.

Collect ovigerous females or purchase from fishermen and place in holding tanks or in flow-through troughs similar to, but larger than, those described for crayfish. Acclimate and condition as described in Section 8010E.3. When eggs are ready to hatch, transfer females to static tanks provided with aerated and UV-disinfected water at 30 g/kg and 13°C. As eggs hatch, dip out swimming first-stage larvae with beakers and transfer to culture beakers with large-bore pipets.

Dungeness crab larvae have long delicate spines that make their culture in flowing

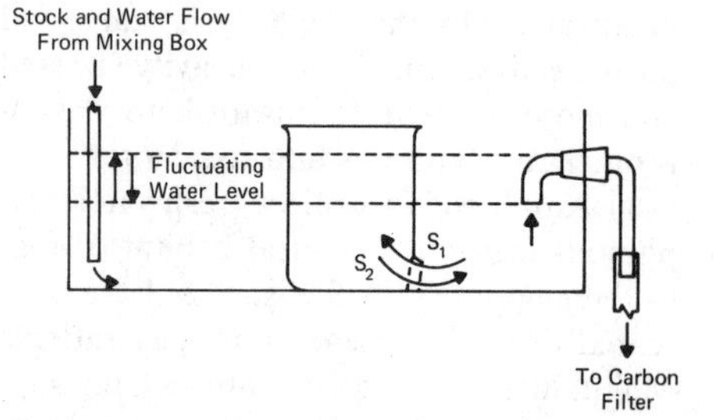

Figure 8720:1. Rearing and exposure beaker and automatic siphon for dungeness crab larvae.[19]

systems difficult. Culture larvae to the fourth and fifth stage in 250-mL beakers that have a hole 15 mm in diameter blown through their sides near the bottom. Using silicone cement, fasten a plastic‡ screen having 360-µm openings over this hole on the inside of the beaker and plastic screen with 210-µm openings over the hole on the outside.[19] Because of the lip created by blowing the glass, the two screens are 3 to 4 mm apart. The larger-mesh screen on the inside is less likely to catch and damage spines of larval crabs while the smaller-mesh screen on the outside does not come in contact with larvae but does retain food organisms, brine shrimp nauplii.

Set the 250-mL beakers in glass trays or aquariums large enough to accomodate 10 beakers and provide a depth of at least 10 cm.[19] Supply trays with a constant flow of water by a tube that discharges near the tray bottom. Provide an automatic siphon at the outlet so there is continual filling and drawdown (Figure 8720:1). Construct the automatic siphon so that when the water reaches the high point and the siphon is activated the beakers contain approximately 200 mL and when the siphon is broken the beakers contain about 150 mL.

The automatic siphon consists of a silicone rubber stopper drilled to receive an 8-mm-ID, right-angle glass tube on one end

---

*Filter design patented by Microfloc Corp., Corvallis, Ore. 97330.
†The Commercial Filter Corp., Lebanon, Ind. 46052, Fulflo, Model F15-10, or equivalent.

‡Nitex or equivalent.

and a 5-mm-ID, right-angle glass tube on the other end. In a 1.3-cm-diam hole blown through side of tray, insert stopper with 8-mm hole on inside of tray. Insert tubes into stopper as shown in Figure 8720:1. Placement of hole inside of tray controls water level in beakers at 200 mL. The distance between the top of the inside hole in the stopper and the bottom of the inside siphon leg is equal to the difference in depth between 200 mL and 150 mL. Make the siphon intake perfectly flat and smooth to prevent air from being drawn[19] into siphon. Adjust tube diameters to give a 15-min cycle — 10 min filling and 5 min drawdown.

When culture chambers are set up and functioning, place 10 first-stage larvae in each beaker with a smooth large-bore pipet. The larvae can be fed nonliving food but preferably feed first-stage brine shrimp nauplii at the rate of 70 for each crab larva 3 times/week through the third stage, then 100 brine shrimp for each crab larva. Keep density of crab larvae low and that of food organisms high, to minimize crab larvae contacts that may result in cannibalism. Before feeding, transfer larvae to clean chlorine-disinfected and rinsed beakers. Use a temperature of 12 to 13°C, pH 8, and a salinity of 25 to 30 g/kg. Adjust the photoperiod to correspond with natural conditions or, if the cycle is off-season, to correspond to the normal annual cycle of light and dark. Exclude natural light and use fluorescent light (Section 8010F.3f). Under these conditions survival of 80 to 90% through the fourth zoeal stage has been attained. Larvae usually begin molting into the fifth zoeal stage by the 45th day. Mortalities then increase.

Juvenile and adult dungeness crabs are much less susceptible to disease than larvae. With strict sanitation and unpolluted open-sea water, sand filtration alone provides sufficient water quality control. Hold juvenile and adult crabs in trough compartments similar to, but larger than, those used for crayfish. To allow sufficient space for each juvenile crab use compartments 15 × 15 cm and 15 to 20 cm deep. For adult crabs use 30- × 30-cm or 40- × 40-cm compartments with a depth of about 30 cm. For large specimens use deeper water. For ease of supplying water, arrange troughs on stands having three shelves with space on each for two troughs. Feed cut-up or macerated fresh fish, clams, or mussels, or commercial dried fish foods to juveniles and adults. Remove unused food within 24 h to reduce fouling.

Routinely clean sides and bottoms of compartments and remove wastes with vacuum or siphon cleaners. Raise screen separators a few millimeters and flush as suggested for crayfish troughs.

4) American lobster, *Homarus americanus* — Obtain adult lobsters by trapping or purchase from lobster fishermen. Ovigerous females can be obtained most readily in the early spring from lobster fishermen who have permits. Select females with brownish eggs because these eggs will hatch within a few weeks to a few months, depending in part on water temperature.[20] Immediately place ovigerous females in holding tanks 300 × 100 × 30 cm. Use 25 to 30 lobsters per tank at a ratio of 2 females to 1 male. Fasten the claws of these lobsters with elastic bands but not wooden pegs. Pass uncontaminated seawater continuously through the tank at a rate that maintains the DO at or above 80% of saturation. Maintain salinity between 30 g/kg and that normal to seawater. Maintain temperature at or above 15°C.[20] Feed clams, quahogs, bay scallop viscera, fish (such as alewives), crabs, abalone scraps, squid, or commercial dry pelleted foods.

To provide an egg-hatching tank, place a partition across the lower end of the holding tank 30 cm from the end. Locate standpipe to maintain desired water levels at one side of this 100- × 30-cm area. Remove a piece 5 cm deep and 30 cm long from the top central portion of the partition for the outflow from the hatching tank. Fit screen

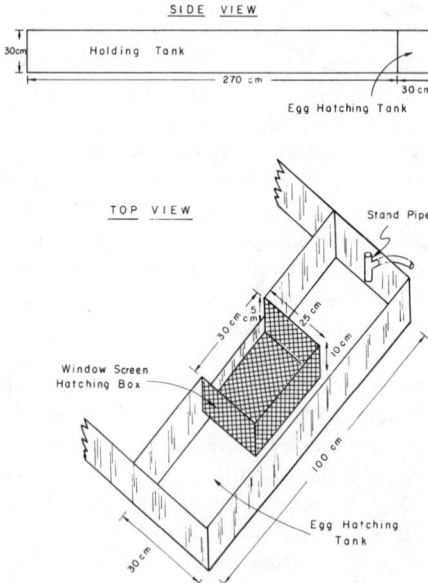

Figure 8720:2. Egg-hatching tank for lobsters.

box, 25 cm wide, 15 cm deep, and 30 cm long with a notch 5 × 30 cm in the top of one side of the frame, against the notch in the partition. Use plastic window screen with 2-mm-square[4] openings for screen box. Place female lobsters with eggs about to hatch in this hatching box (Figure 8720:2). Supply with flowing seawater at a rate sufficient to maintain DO above 80% of saturation. Maintain temperature at 19 to 20°C.

The larval development stages of *H. americanus* have been fully described.[21,22] The larval period extends from the time of hatching to the fourth molt or attainment of the fifth stage. Duration of the larval period depends somewhat on water temperature.[23] During the first three stages larvae are free-swimming, move toward light, and remain near the water surface. After the fourth stage they become bottom-crawlers and seek dark places, lead a nocturnal life, and acquire the defensive instincts of the adult lobster.

Construct from molded fiberglass a special 40-L culture tank for rearing larval stages[24] with a combined water circulator and overflow device at the center (Figure 8720:3).[24] For tests use a 5-L tank. Maintain water flow for this modified tank between 0.5 and 1 L/min through the standpipe at the top. Discharge into the tank at the bottom through small slits in the manifold at the bottom of the overflow circulator.

Maintain temperature variations to within ± 1°C and regulate flow rate using ball valves.§ If water supply contains silt or suspended matter, pass through a sand filter. If supply is unpolluted open-ocean water, use without treatment. If water is contaminated, use fine filtration and disinfection by ozone or UV light. Use a 10-μm wound polypropylene filter and a 40-W UV unit‖ for fine filtration.[24] For more information on filtration refer to Spotte.[25] For flows of >40 L/min use a diatomaceous earth filter. In some instances recirculating systems may be advantageous, especially when treatment with streptomycin is required to prevent growth of filamentous bacteria or when large amounts of water are required. The maximum concentration of lobster larvae is about 45/L. At higher concentrations there are more lobster-to-lobster contacts and cannibalism.

As eggs begin to hatch, wash first-stage larvae over the 5- × 30-cm notch in the partition into the screen box. Dip larvae from this box with a small beaker and place in modified Hughes larval rearing chamber by submerging the beaker and gently removing it. Stock with <225 larvae.

Feed newly hatched larvae foods such as clams, mussels, liver, or other meat, chopped in a blender at a ratio of 1 part meat to 2 parts seawater. Continually drip

---

§Chemcock ball valve, Celanese Piping Systems, Aquafin Corp., Burbank, Calif., or equivalent.
‖PVCL-1, Aquafine Corp., Burbank, Calif., or equivalent.

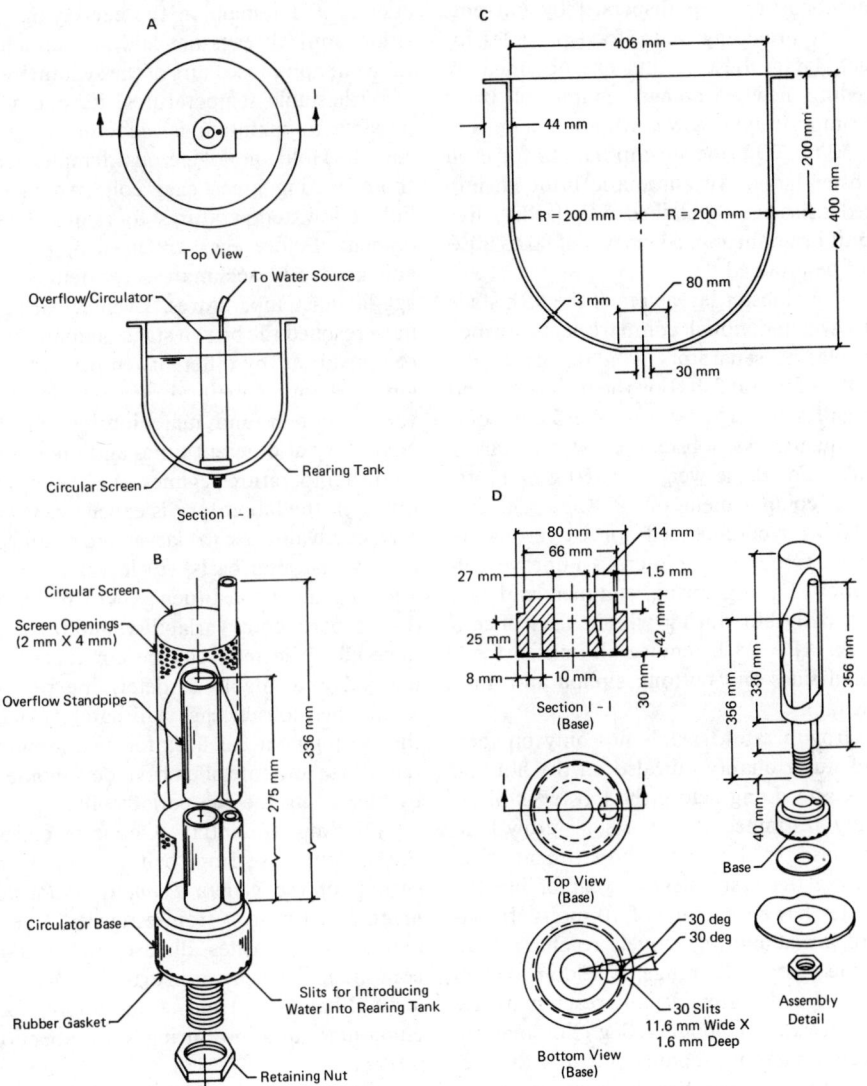

Figure 8720:3. Hughes lobster-rearing tank.[24] A—general views; B—views of overflow/circulator; C—details of rearing tank construction; D—construction and assembly details for rearing tank and overflow/circulator. This is a 40-L tank. For bioassays, scale to 5-L volume.

this mixture into the rearing tanks from a stock bottle, the contents of which are mixed continuously by a magnetic stirrer. To insure that food is well dispersed and held in suspension for as long as possible in the rearing chamber, introduce water under pressure through openings in overflow circulator. Because lobster larvae are cannibalistic, keep dispersed by currents and by providing many food particles for each larva. Best results are obtained by feeding newly hatched nauplii of brine shrimp, *Artemia salina*. Maintain a density of 50 to 200 brine shrimp nauplii for each lobster larva. An automatic brine shrimp feeder has been described.[26] By feeding live adult brine shrimp, a survival of 80 to 90% can be attained.[24]

When lobster larvae reach the fifth stage place in individual compartments formed by placing separators in a trough as described for crayfish. For the fifth stage and juveniles use 15- × 15- × 15-cm compartments. As lobsters grow use larger tanks. For those weighing 460 g or more, use a compartment 60 × 45 × 30 cm. Feed as recommended for ovigerous females. Ground, whole crabs improve coloration. During spring, summer, and fall, feed daily but during winter, feed once a week. After 24 h remove all unused food. Clean sides and bottom; siphon and flush out tanks.

Growth rate depends not only on food and water quality but also on the holding tank size. Long before the lobster is physically restrained it reduces its growth in response to holding compartment size. During the first calendar year of life the lobster has an average of 10 molts. In nature, larval molting actively reaches a peak in the 15 to 20°C range; it seldom occurs below 5°C. Lobsters usually reach maturity at a weight of about 460 g. For mating, place a male in a compartment with a female immediately after she has molted and is in the soft stage. Success of mating decreases with time after molting. When temperatures of 22 to 24°C are maintained year round, lobsters reach maturity in 2 years.[27] The rate of egg development depends in part on temperature. For extrusion of eggs, place females in a deeper tank because they need at least 45 cm of water over them. Provide the egg-laying tank with a rough or nonslip bottom that allows female to assume and remain in the egg-laying position until all eggs are laid and attached to the nonplumose hairs of the swimmerets.

With stable temperatures, it should be possible to maintain larval cultures year round. Hold nonovigerous females and those bearing green eggs collected in the fall at low temperatures to retard development. Before eggs are needed, remove and gradually acclimate some females to egg-laying temperatures. Even when eggs have reached the brown stage, hatching can be spread out by different temperature regimes. Another method of producing larvae is to rear and mate lobsters in the laboratory at different times and under different temperature regimes. Although culturing in the laboratory is expensive it has certain advantages: (*a*) larvae are produced on a year-round basis; (*b*) larvae are of a known genetic constituency, which can reduce experimental variability; and (*c*) complete-life-cycle tests can be conducted. A method is available to determine beforehand when lobster eggs will hatch.[28] Once the eye pigment has been formed, monitor the course and rate of embryo development by measuring the eye periodically.

5) *Shrimp (Natantia)*—Obtain by collection or purchase from bait dealers. Seine shrimp of the genera *Penaeus*, *Palaemonetes*, and *Crangon* from estuaries. Check animals for parasites, disease, and general condition. For general instructions on collecting, handling, transfering, holding, acclimating, and culturing, see Section 8010E.

a) *Palaemonetes*—Three marine species and three freshwater species of the genus *Palaemonetes* have been reared through metamorphosis. They are suitable for life-cycle studies and can be brought from the field for direct use or for laboratory rearing. Place field-collected adult shrimp in suitable flow-through aquarium water. Feed freshwater species macerated parts of local fishes;[29] feed marine forms macerated mollusks or fish.[30] Examine shrimp periodically to detect ovigerous females. When eggs are nearly ready to hatch, remove desired number of females from tank and put into individual containers. Keep females in these containers, preferably with flow-through water, until eggs begin to hatch. During this period feed macerated fish or other suitable food. After eggs hatch, remove female and feed prelarvae or prezoeae 1-d-old *Artemia salina* nauplii. The rearing procedure for larvae is similar for all six species.[31] Use equipment and procedures similar to those for the dungeness crab, but with rearing chambers with a capacity of 1 L set in a deeper tray. Place 10 larvae in each beaker and feed with newly hatched brine shrimp nauplii and ideally maintain at 25°C. Filter and disinfect water. During the larval period, provide 14 h light and 10 h dark cycle[32] (see Section 8010F.3*f*). Inspect larvae and feed daily. If sediments or wastes tend to collect, remove them daily with a siphon. At 25°C the larval period lasts 16 to 24 d. The average length of larval life is between 19 and 20 d. To rear through entire life cycle, immediately place females that have laid and hatched their eggs in an aquarium with males. Mating takes place, producing a second batch of fertilized eggs. The egg incubation period depends on temperature; usually 24 to 28 d are required. The number of eggs laid varies. There are six larval stages, the first being the protozoea. The seventh stage is a post-larval or juvenile shrimp, which marks the end of metamorphosis.

Keep larvae of marine species in seawater adjusted to 25 g/kg salinity. Feed with newly hatched *Artemia salina* nauplii. Rear at between 23 and 27°C using procedures similar to those used for freshwater species. Larval development has been described.[33-35] Remove chelipeds of ovigerous females with fine surgical scissors to prevent removal of eggs. When rearing larvae to a particular age, maintain a 10 to 15% surplus to compensate for mortality and to provide for other uses. The larvae

are relatively hardy to temperature and salinity and can be reared at 25°C and a salinity between 15 and 25 g/kg.

b) *Penaeus*—Hold shrimp in glass tanks of at least 30 L capacity. Provide each tank with flow-through water, 2 to 3 cm of sand over the bottom, and a screen over the top to prevent the shrimp from jumping out. Avoid overloading; keep no more than 22 to 24 animals in a 30-L tank. For *Penaeus* spp. use a minimum flow of 7.5 L/g/d; flows up to 22 L/g/d may be desirable to insure DO above 60% of saturation and the removal of metabolic products. Acclimate to laboratory test conditions for about 2 weeks. For short-term or medium-length tests with adults and juveniles, shrimp can be field-collected. Cut-up fish is a satisfactory food. Cut a fillet from mullet, grouper, or other abundant species into 1-cm² pieces; feed one piece for each shrimp once each 2 to 3 d, depending on the size of the shrimp. Remove uneaten food daily. For larval tests or life-cycle studies collect gravid females offshore, let them spawn, and rear larvae at least to the postlarval stage. Penaeid shrimp can be reared from egg to postlarvae in the laboratory.[36-38] Mock and Murphy[39] have described improvements on procedures for spawning females, rearing larvae, and culturing the diatom *Skeletonema* as food for protozoeal stages. If diatoms are used as food, use airlift pumps to prevent accumulation of the diatoms.[40] Feed freshly hatched brine shrimp, *Artemia*, to the mysis and postlarval stages. See Mock[41] for additional data on culture of algae and their feeding to shrimp. Equipment and procedures for continuous mass culture of algae as food are described in Section 8010E.4*c*2).

The protozoeal stages, 1 through 3, of the penaeid shrimp require algae as food. Because larval shrimp are pelagic and unable to search for food during the early part of their life cycle, maintain the required density of the phytoplankton *Skeletonema costatum* and *Tetraselmis* sp. Add these to larval culture chambers according to stage of development, number of larvae present, and volume of water:

Protozoeal I	*Skeletonema*	50 000 cells/mL
Protozoeal II	*Skeletonema*	150 000 cells/mL
Protozoeal III	*Tetraselmis*	20 000 cells/mL
Mysis I	*Artemia* nauplii	3/mL
Mysis II	*Artemia* nauplii	3/mL
Mysis III	*Artemia* nauplii	3/mL
Postlarvae I-IV	*Artemia* nauplii	3/mL

Maintain phytoplankton in continuous culture or harvest and freeze to use later. Algae culture production units shown in Figures 8010:2 and 8010:3, Section 8010E.4*c*2), will produce daily 7.5 L of culture containing 4.3×10^6 *Skeletonema costatum*/mL or 7.0×10^6 *Isochrysis galbana*/mL and several other species of algae at similar concentrations.

Add algae as food for the larval shrimp as either a fresh or frozen concentrate. Concentrate algae by centrifuging and discard the growth medium.

Use a temperature of 28 to 30°C and a salinity of 27 to 35 g/kg. Omit antibiotics from the larval culture medium if the EDTA (disodium salt) is substituted at a concentration of 10 mg/L seawater.[42] Feed juvenile shrimp with fresh pieces of fish, clams, or mussels[42] or with prepared dried foods.[43] Knowledge of environmental requirements is essential for correct feeding and rearing of shrimp.[44,45]

4. Parasites, Diseases, and Harmful Growths

Much remains to be learned concerning harmful growths, parasites, and diseases of crustaceans.[46] For general problems and control procedures see Section 8010E.5.

Adult shellfish in recirculated or flow-through systems are susceptible to biotoxins and pathogens. Remove metabolites and dead individuals from recirculating systems.

Juvenile and adult lobsters, crabs, and shrimps are subject to bacterial and fungal infections. *Gaffkya*, a bacterial pathogen, is particularly prevalent in tank-held lobsters, while *Vibrio* disease occurs in tank-held postlarval adult shrimp. Most captive crustaceans are subject to "shell disease," produced by chitin-destroying bacteria. A systemic fungal disease has been described in European prawns and several fungal infections occur in wild shrimp populations.

The larval stages of the lobster and several other crustaceans are prone to infections of the ubiquitous marine bacterium *Leucothrix mucor*, which has produced mortalities of over 90% in larval cultures.[47] The exuvia and the new exoskeleton after molting become entangled in the long dense filaments of the bacteria and the larvae are unable to swim or feed adequately. This organism also can produce high mortalities by causing pelagic eggs to sink and by interfering with the filtering apparatus of larval forms and the functioning of gills.[48]

In some instances it may be necessary to culture larvae in artificial seawater to avoid *L. mucor* infection. Place ovigerous females in a bath of malachite green (5 mg/L) for 1 min *only* or rinse them several times in artificial seawater of the correct salinity that contains streptomycin, 2 mL/L, from a stock solution containing 2 g/L of antibiotic. Maintaining a 1-mg/L concentration of antibiotic throughout the larval culture period prevents infections. Twice daily cleaning also is a good preventive method. Seawater, filtered and exposed to UV radiations, should be nearly bacteria-free.

A disease of lobster larvae tentatively has been associated with the phycomycete *Haliphtorus*. It appears as a scab on the first segment of the thoracic appendages up to and surrounding the first row of gills. Thorough cleaning and UV treatment of the water supply is the only known treatment. In most cases these scabs adhere to both old and new carapaces and thus cause a mechanical impediment to molting. Mortality appears to be restricted to larvae and young juveniles. No deaths of specimens with a carapace length over 27 mm have been observed.

The fungus *Lagenidium* sp. causes serious problems in rearing larval shrimp.[49] The disease first becomes apparent in the second protozoeal stage and disappears as the shrimp reach the first mysis stage. Shrimp become immobilized by replacement of muscle tissue by fungal mycelium.

Parasites have been found in many species of crustaceans and their presence can influence results. In *Uca*, an ectoparasitic isopod is found on the gills, nematodes in the gut, and metacercaria in the green glands. Species of *Lagenidium* similar to the one that occurs in shrimp occur in other marine crustaceans. *L. callinectes* occurs in eggs and larvae of the blue crab.[50,51] The blue crab has a barnacle *(Octolasmus lowei)* living in association with its gills and gill chamber, metacercariae in various organs, and the sacculinid *Loxothylacus taxanas* beneath its abdomen.

Saprolegia parasitica attacks larvae of the shrimp *Palaemonetes kadiakensis*.[52] Amphipods sometimes are parasitized by larval stages of acanthocephalan worms.[53]

5. References

1. SMITH, R.I., ed. 1964. Keys to Marine Invertebrates of the Woods Hole Region, 1st ed. Systematics-Ecology Program, Marine Biology Lab., Woods Hole, Mass.

2. WILLIAM, A.B. 1965. Marine decapod crustaceans of the Carolinas. *Fish Bull.* 65:10298.

3. WASS, M.L. 1955. The decapod crustaceans of Alligator Harbor and adjacent inshore areas of northwestern Florida. *Quart. J. Fla. Acad. Sci.* 18:129.

4. LIGHT, S.F., R.I. SMITH, F.A. PIDELKA, E.P. ABBOTT & F.M. WEESNER. 1957. Intertidal Invertebrates of the Central California Coast, 2nd ed. Univ. California Press, Berkeley.

5. HALSINGER, J.R. 1972. The Freshwater Amphipod Crustaceans (Gammaridae) of North America. Biota of Freshwater Ecosystems,

Identification Manual No. 5, Water Pollut. Res. Ser. No. 5, PB-222 926/80.

6. HUBSCHMAN, J.H. 1967. Effects of copper on the crayfish *Orconectes rusticus* (Girard). I. Acute toxicity. *Crustaceana* 12:33.

7. CLEMENS, H.P. 1950. Life Cycle and Ecology of *Gammarus fasciatus* Say. Franz T. Stone Inst. Hydrobiol. Contrib. 12, Ohio State Univ., Columbus.

8. COOPER, W.E. 1965. Dynamics and productivity of a natural population of a freshwater amphipod *Hyalella azteca. Ecol. Monogr.* 35:377.

9. SMITH, W.E. 1973. Thermal tolerances of two species of *Gammarus. Trans. Amer. Fish. Soc.* 102:431.

10. COSTLOW, J.D. & C.G. BOOKHOUT. 1960. A method of developing brachyuran crab eggs in vitro. *Limnol. Oceanogr.* 5:212.

11. COSTLOW, J.D., C.G. BOOKHOUT & R. MONROE. 1962. Salinity-temperature effects on the larval development of the crab *Panopeus herbstii* Milne-Edwards reared in the laboratory. *Physiol. Zool.* 35:78.

12. COSTLOW, J.D. & C.G. BOOKHOUT. 1962. The larval development of *Sesarma recticulatum* Say reared in the laboratory. *Crustaceana* 4:281.

13. COSTLOW, J.D. & C.G. BOOKHOUT. 1971. The effect of cyclic temperature on larval development in the mud-crab *Rhithropanopeus harrisii.* D.J. Crisp. 4th European Mar. Biol. Symp., p. 211. Cambridge Press, London.

14. BOOKHOUT, C.G., A.J. WILSON, JR., T.W. DUKE & J.I. LOWE. 1972. Effects of mirex on the larval development of two crabs. *Water, Air Soil Pollut.* 1:165.

15. EPIFANIO, C.E. 1971. Effects of dieldrin in seawater on the development of two species of crab larvae, *Leptodius floridanus* and *Panopeus herbstii. Mar. Biol.* 11:356.

16. LOWE, J.I. 1965. Chronic exposure of blue crabs, *Callinectes sapidus* to sublethal concentrations of DDT. *Ecology* 46:899.

17. REED, P.H. 1969. Culture methods and effects of temperature and salinity on survival and growth of dungeness crab, *Cancer magister* larvae in the laboratory. *J. Fish. Res. Board Can.* 26:389.

18. BUCHANAN, D.V., R.E. MILLEMANN & N.E. STEWART. 1970. Effects of the insecticide sevin on various stages of the dungeness crab,

Cancer magister. J. Fish. Res. Board Can. 27:93.

19. BUCHANAN, D.V., M.J. MYERS & R.S. CALDWELL. 1975. An improved flowing water apparatus for culture of brachyuran crab larvae. (unpublished).

20. HUGHES, J.T. & G.C. MATTHIESSEN. 1962. Observations on the biology of the American lobster, *Homarus americanus. Limnol. Oceanogr.* 7:414.

21. HERRICK, F.H. 1896. The American lobster: A study of its habits and development. *Bull. U.S. Fish Comm.* 15:1.

22. HERRICK, F.H. 1911. Natural history of the American lobster. *Bull. U.S. Bur. Fish.* 29:147.

23. TEMPLEMAN, W. 1948. Growth per molt in the American lobster. *Bull. Newfoundland Govern. Lab.* 18:26.

24. HUGHES, J.T., R.A. SHLESER & G. TCHOBANOGLOUS. 1974. A rearing tank for lobster larvae and other aquatic species. *Progr. Fish-Cult.* 36:129.

25. SPOTTE, S.H. 1970. Fish and Invertebrate Culture, Water Management in Closed Systems. John Wiley & Sons, Inc., New York, N.Y.

26. SMITH, R.A., J.A. HOLMAN & R.H. KRAMER. 1974. Automatic brine shrimp feeder. *Progr. Fish-Cult.* 36:133.

27. HUGHES, J.T., J.J. SULLIVAN & R. SHLESER. 1972. Enhancement of lobster growth. *Science* 177:1110.

28. PERKINS, H.C. 1972. Developmental rates at various temperatures of embryos of the northern lobster (*Homarus americanus* Milne-Edwards). *Fish. Bull.* 70:96.

29. BROAD, A.C. & J.H. HUBSCHMAN. 1963. The larval development of *Palaemonetes kadiakensis,* M.J. Rathbun, in the laboratory. *Trans. Amer. Microsc. Soc.* 82:185.

30. HUBSCHMAN, J.H. & A.C. BROAD. 1974. The larval development of *Palaemonetes intermedius* Holthuis 1949 (Decapoda, Palaemonidae) reared in the laboratory. *Crustaceana* 26:89.

31. DOBKIN, S. 1963. The larval development of *Palaemonetes paludosus* (Gibbes 1850) (Decapoda Palaemonidae) reared in the laboratory. *Crustaceana* 6:41.

32. HUBSCHMAN, J.H. & J.A. ROSE. 1969. *Palaemonetes kadiakensis* Rathbun: Post embryonic growth in the laboratory (Decapoda, Palaemonidae). *Crustaceana* 16:81.

33. FAXON, W. 1879. On the development of *Palaemonetes vulgaris. Bull. Mus. Comp. Zool.* (Harvard) 5:303.

34. BROAD, A.C. 1957. Larval development of *Palaemonetes pugio* Holthuis. *Biol. Bull.* 112:144.

35. BROAD, A.C. 1957. The relationship between diet and larval development of *Palaemonetes. Biol. Bull.* 112:162.

36. COOK, H.L. & M.A. MURPHY. 1966. Rearing penaeid shrimp from eggs to postlarvae. *Proc. 19th Annu. Conf. S.E. Assoc. Game Fish. Comm.* 19:283.

37. COOK, H.L. & M.A. MURPHY. 1969. The culture of larval penaeid shrimp. *Trans. Amer. Fish. Soc.* 98:751.

38. COOK, H.L. 1967. A method of rearing penaeid shrimp larva for experimental studies. *FAO (Food Agr. Organ. U.N.) Fish. Rep.* 3:709.

39. MOCK, C.R. & M.A. MURPHY. 1970. Techniques for raising penaeid shrimp from egg to postlarvae. *Proc. 1st Annu. Workshop World Maricult. Soc.* 1:143.

40. SALSER, B.R. & C.R. MOCK. 1973. An airlift circulator for algal culture tanks. *Proc. 4th Annu. Workshop World Maricult. Soc.* 4:295.

41. MOCK, C.R. 1974. Larval Culture of Penaeid Shrimp at the Galveston Biological Laboratory. NOAA (Nat. Ocean. Atmos. Admin.) Tech. Rep. NMFS (Nat. Mar. Fish. Serv.) Circ. 388:33.

42. MOCK, C.R. 1973. Shrimp culture in Japan. *Mar. Fish. Rev.* 35(3-4):71.

43. MEYERS, S.P. & Z.P. ZEIN-ELDIN. 1972. Binders and pellet stability in development of crustacean diets. *Proc. 3rd Annu. Workshop World Maricult. Soc.* 3:351.

44. MOCK, C.R., R.A. NEAL & B.R. SALSER. 1973. A closed raceway for the culture of shrimp. *Proc. 4th Annu. Workshop World Maricult. Soc.* 4:247.

45. ZEIN-ELDIN, Z.P. & G.W. GRIFFITH. 1969. An appraisal of the effects of salinity and temperature on growth and survival of postlarval penaeids. *FAO (Food Agr. Organ. U.N.) Fish. Rep.* 3:1015.

46. ANDERSON, J.I.W. & D.A. CONROY. 1968. The significance of disease in preliminary attempts to raise crustacea in sea water. *Bull. Off. Inform. Epizoot.* 69:1239.

47. BROCK, T.D. 1966. The habitat of *Leucothrix mucor,* a widespread marine organism. *Limnol. Oceanogr.* 11:303.

48. JOHNSON, P.W., J.M. SIEBURTH, A. SASTRY,

C.R. ARNOLD & M.S. DOTY. 1971. *Leucothrix mucor* infestation of benthic crustacea, fish eggs and tropical algae. *Limnol. Oceanogr.* 16:962.

49. LIGHTNER, D.V. & C.T. FONTAIN. 1973. A new fungus disease of the white shrimp *Penaeus setiferus*. *J. Invertebr. Pathol.* 22:94.

50. COUCH, J.H. 1942. A new fungus on crab eggs. *J. Elisha Mitchell Sci. Soc.* 58:158.

51. ROGERS-TALBERT, R. 1948. The fungus *Lagenidium callinectes* Couch on eggs of the blue crab in Chesapeake Bay. *Biol. Bull.* 95:214.

52. HUBSCHMAN, J.H. & J.A. SCHMITT. 1969. Primary mycosis in shrimp larvae. *J. Invertebr. Pathol.* 13:351.

53. SPENCER, L.T. 1974. Parasitism of *Gammarus lacustris* (Crustacea; Amphipoda) by *Polymorphus minutus* (Acanthocephala) in Colorado. *Amer. Midland Natur.* 91:505.

6. Bibliography

HADLEY, P.B. 1906. Regarding the rate of growth of the American lobster, *Homarus americanus*. *36th Annu. Rep., Comm. Inland Fish. R.I.*, p.153.

TEMPLEMAN, W. 1934. Mating in the American lobster. *Contrib. Can. Biol. Fish.* 8:423.

TEMPLEMAN, W. 1936. Further contributions to mating in the American lobster. *J. Biol. Board Can.* 2:223.

MACKAY, D.C.G. 1943. Temperature and world distribution of the genus *Cancer*. *Ecology* 24:113.

BOUSFIELD, E.L. 1958. Fresh water amphipod crustaceans of glaciated North America. *Can. Field Natur.* 72:55.

VERNBERG, F.J. & R.E. TASHIAN. 1959. Studies on the physiological variation between tropical and temperate zone fiddler crabs of the genus *Uca*. I. Thermal death limits. *Ecology* 40:589.

COSTLOW, J., C.G. BOOKHOUT & R. MONROE. 1960. The effect of salinity and temperature on larval development of *Sesarma cinerium* (Boxc) reared in the laboratory. *Biol. Bull.* 118:183.

KINNE, O. 1963. The effect of temperature and salinity on marine and brackish water animals. *Oceanogr. Mar. Biol. Annu. Rev.* 1:301.

COSTLOW, J. & C.G. BOOKHOUT. 1964. An approach to the ecology of marine invertebrate larvae. *Symp. Exp. Mar. Ecol., Occas. Publ.* 2:69. Graduate School Oceanography, Univ. Rhode Island.

VERNBERG, F.J. & J.D. COSTLOW. 1966. Studies on the physiological variation between tropical and temperate zone fiddler crabs of the genus *Uca*. IV. Oxygen consumption of larvae and young crabs reared in the laboratory. *Physiol. Zool.* 39:36.

HUGHES, J.T. 1968. Grow your own lobsters commercially. *Ocean Ind.* 3(12):46.

SAILA, S., J. FLOWERS & J.T. HUGHES. 1968. Fecundity of the American lobster *Homarus americanus*. *Trans. Amer. Fish. Soc.* 98:537.

BALLARD, B.S. & R.E. TASHIAN. 1969. Osmotic accomodation in *Callinectes supidus* Rathbun. *Comp. Biochem. Physiol.* 29:671.

VERNBERG, F.J. 1969. Acclimation of intertidal crabs. *Amer. Zool.* 9:333.

HARGRAVE, G.T. 1970. The utilization of benthic microflora by *Hyalella azteca* (Amphipoda). *J. Anim. Ecol.* 39:427.

NIMMO, D.R., A.J. WILSON, JR. & R.R. BLACKMAN. 1970. Localization of DDT in the body organs of pink and white shrimp. *Bull. Environ. Contam. Toxicol.* 5:333.

RICE, A.L. & D.I. WILLIAMSON. 1970. Methods for rearing larval decapod crustacea. *Helgolander wiss. Meeresunters* 20.

SASTRY, A.N. 1970. Culture of brachyuran crab larvae using a recirculating sea water system in the laboratory. *Helgolander wiss. Meeresunters* 20:406.

SASTRY, A.N. 1971. Culture of brachyuran crab larvae under controlled conditions. *In* M. Uda, ed. The Ocean World. p. 475. Joint Oceanographic Assembly. Japan Soc. Promotion Science, Tokyo.

HYNES, H.B.N. & F. HARPER. 1972. The life histories of *Gammarus lacustris* and *Gammarus pseudolimnaeus* in southern Ontario. *Crustaceana*, Suppl. 3:329.

JENIO, F., JR. 1972. The *Gammarus* of Elm Spring, Union County, Illinois (Amphipoda: Gammaridae). Ph.D. dissertation, Southern Illinois Univ., Carbondale.

REES, C.P. 1972. The distribution of the amphipod *Gammarus pseudolimnaeus* Bousfield as influenced by oxygen concentration, substratum and current velocity. *Trans. Amer. Microsc. Soc.* 19:514.

STRONG, D.R., JR. 1972. Life history variation among populations of an amphipod (*Hyalella azteca*). *Ecology* 53:1103.

BARLOCHER, F. & B. KENDRICK. 1973. Fungi and food preferences of *Gammarus pseudolimnaeus*. *Arch. Hydrobiol.* 72:501.

LOCKWOOD, A.P.M. & C.B.E. INMAN. 1973. Changes in the apparent permeability to water at moult in the amphipod *Gammarus duebeni* and the isopod *Idotea linearis*. *Comp. Biochem. Physiol.* 44A:943.

VERNBERG, W.B., P. DeCOURSEY & W.J. PAD-

GETT. 1973. Synergistic effects of environmental variables on larvae of *Uca pugilator*. *Mar. Biol.* 22:307.

HYNES, H.B.N., N.K. KAUSHIK, M.A. LOCK, D.L. LUSH, Z.S.J. STOCKER, R.R. WALLACE & D.P. WILLIAMS. 1974. Benthos and allochthonous organic matter in streams. *J. Fish. Res. Board Can.* 31:545.

NILSSON, L.M. 1974. Energy budget of a laboratory population of *Gammarus pulex* (Amphipoda). *Oikos* 25:35.

8720 C. Conducting the Toxicity Tests

1. General Considerations

Procedures for crustaceans in general follow the methods outlined in Sections 8010D and F. However, many adult crustaceans and some larvae must be segregated into individual compartments because of their aggressive and cannibalistic tendencies. As with most other groups, larval and juvenile crustaceans usually are significantly more sensitive than adults and therefore are preferred for short-term tests.

Determining toxic effects on decapods is complicated by three factors: initial paralysis, delayed response, and much greater sensitivity at molting periods. A true sublethal effect may be shown either by increased irritability or by inactivity. Penaeid shrimp may lie motionless for days without dying and become covered by silt, or they may be so irritable that they damage themselves by hitting aquarium walls when lights are turned on or when someone walks by. These effects may be reversible, disappearing when clean water is restored. A more definitive indication of toxicity is partial or complete paralysis of adults or lack of swimming by larvae. Crabs may lie paralyzed for days before dying. In general, paralysis is not a reversible effect. In nature, a paralyzed animal would be easy prey for predators and would not survive for long. The second factor that may complicate interpretation of crustacean tests is the "delayed response." After short exposure to toxic materials, test animals appear unharmed. If they are held in clean water, however, they may begin to die, the mortality eventually reaching 100%, even though no deaths occurred in the first few days after the exposure period.

Because of delayed mortality, place all surviving individuals from each test exposure in compartmented tanks receiving clean dilution water and hold under favorable environmental conditions for 2 weeks to detect delayed effects.

For tests to assess effects of periodic spills, discharges, or dumping, base results on several series of short-term tests at the indicated exposures made over a period of weeks to determine the cumulative effects of intermittent exposure. This is essential because at least some decapods show cumulative effects and three 8-h exposures at a given concentration, even though they are as much as 3 weeks apart, are equivalent to a 24-h exposure to the same concentration.[1]

Tests with embryonic and larval stages may last from a few hours to as much as 30 d, depending on the species and larval stages used. Toxicity criteria include egg hatchability, rate and success of molting, swimming ability, tendency to lose appendages, and metamorphosis of the larvae as well as death. Tests with selected larvae, juveniles, and adults may measure acute toxicity with paralysis or death as the end point. Long-term and life-cycle tests usually measure effects on respiration, behavior, osmo-regulation, growth, molting rates, reproduction, and general well-being.

2. Water Supply—Dilution Water and Water Distribution System

See Section 8010F.1.

3. Equipment and Materials

See Section 8010F.

4. Test Procedures

a. Amphipods:

1) Short-term tests—Make exploratory or range-finding tests as described in Section 8010D.2 under either flow-through or static conditions. For static exposure chambers, use 5- to 6-L-capacity wide-mouth glass containers containing 4 L test solution. For long-term tests use flow-through test chambers. Prepare toxicant solutions and place in exposure chambers as described in Section 8010D. In exploratory and short-term definitive tests keep water temperature constant using water baths or constant-temperature rooms. Preferably use a test temperature of 16 to 18°C. For effluent tests match test temperature to that of receiving water as described in Section 8010F.2. Use juveniles for these tests. Use five organisms in each of the widely spaced concentrations for exploratory tests. Use five or more concentrations of toxicant in definitive tests with duplicate test chambers for each concentration and the controls. Place 10 organisms in each test container, thus obtaining 20 test organisms for each control and concentration tested. Select organisms randomly and immediately place in exposure chambers as described in Section 8010F.3a. Do not feed. Use a photoperiod of 16 h light and 8 h dark with changes from L to D and D to L made during a 0.5-h twilight period. Supply light as described in Section 8010F.3f and at an intensity of 540 to 1600 lux at the water surface. Maintain DO as described in Section 8010F.2a. Limit static tests, in which the organisms are not fed, to 4 d.

2) Partial- or complete-life-cycle tests—Use long-term tests to determine waste toxicity to various species of amphipods.[2-7] Use equipment for the water system described in the literature[2-7] and in Section 8010F.1. Prepare toxic solutions as described in Section 8010F.2b and select test concentrations on the basis of short-term test results (Section 8010F.3b). When methods of analysis are available, sample test concentrations immediately after tests are begun and at monthly or more frequent intervals thereafter. Analyze as described in Section 8010F.3d to determine actual exposure concentrations.

Use glass aquariums with volumes of 8 to 20 L. To enhance mixing, deliver toxicant near one end of exposure chamber and position screened standpipe at other end. Maintain a minimum solution depth of 15 cm. Hold test solution flow rate to each chamber sufficient to maintain DO and remove metabolic products. Stabilize test solution flow to chambers before introducing test organisms. For effluent tests use DO and temperature of receiving water. For other tests, maintain DO above 60% saturation but do not supersaturate.

Use naturally occurring photoperiods during the stages of development under investigation (Section 8010F.3f).

Start partial- or full-life-cycle tests with newly hatched young, juveniles, or adults. To test newly hatched young, collect ovigerous females about to shed young from the holding tanks of field-collected specimens or from culture tanks. Place each one in a clean 400- to 600-mL beaker containing dilution water to which it has been acclimated to test temperatures, feed as described in Section 8720B.3b1), and hold until young are shed. Record number produced by each female.

When sufficient newly hatched young are available, begin test by selecting at random 30 young. Place them in each of duplicate growth aquariums for each test concentration and the controls. Feed test organisms as described in Section 8720B.3b1) and clean aquariums as described in Section 8010E.4*d*. Observe test chamber daily; remove dead organisms and preserve for future study. At the end of 60 d siphon the contents of each test chamber into pans and record all living young. Randomly select 15 from each exposure chamber and return them to their respective exposure chambers. Use the methods described in Clemens[8] for measuring the remaining organisms to determine growth. Preserve all organisms for toxicant accumulation and histopathological analyses. Observe test chambers daily and record behavior and number dead. As the amphipods approach maturity and mate, look for gravid females about to shed young. Place one in each beaker and handle as for the F_1 generation. Record number of young produced by each female and randomly place 30 in each duplicate exposure chamber for each test concentration and control. Terminate adult exposure, record number and sex of each animal, measure animals, and preserve for additional analyses. Continue exposure of young and record survival in each concentration after 1 month. Continue exposure to 60 d and then terminate. Count, measure length and weight, record condition of all survivors, and preserve for other studies.

b. Crayfish:

1) Short-term tests—Follow general procedures in Sections 8010D and F for range-finding and definitive tests. Make special modifications to apparatus for crayfish. Preferably, make short-term tests with juveniles and young adults. For a test series use animals of similar size, with the largest individual being no more than 50% longer than the shortest, and in approximately the same intermolt or molt stage. Expose all organisms in individual compartments. Duration of exposure depends on study objectives and may be 96 h, 1 week, or 2 weeks. Make exposures in the compartmented troughs described in 8720B.3*b*2). Tests may be static, renewal, or flow-

through. If the test lasts more than 96 h, use flow-through procedures with some feeding and thorough cleaning of compartments.

2) Partial-life-cycle tests—Make exposures in the compartmented troughs described in Section 8720B.3*b*2) with one trough used for each concentration tested and each control. Run tests in duplicate. Prepare dilution water and stock solutions of toxicants and arrange test troughs as described for other crustaceans. Use modified proportional diluters (Section 8010F.1) to deliver toxicant to exposure chambers. Place 100 newly hatched young in each of the large chambers formed by removing the screens and converting 10 compartments into one large chamber. Adjust water quality to follow the normal seasonal cycle in local favorable waters. Regulate water flow, feed, and clean as described in Section 8720B.3*b*2) and in the general procedures. Before the young become cannibalistic, remove them from the common exposure chamber and count; randomly select 20 individuals and place one in each of the 20 compartments of their respective exposure troughs. Count, measure, weigh, and preserve. Make daily observations on mortality and condition and record results. Continue exposure until animals become adults. Record numbers, weight, condition, and sex. Preserve organisms for additional studies.

c. Brachyura—Crabs: Numerous investigators have used short- and long-term tests.[9–16] Basic procedures are those for fish and other organisms (see Sections 8010D and F). Methods in current use for the dungeness crab are presented as a guide for conducting tests with this and other species of the group.[17,18]

1) Short-term tests—Use zoeal stages because juveniles and adults are more resistant to toxicants. Conduct short-term studies to determine relative sensitivity, relative toxicity, or test concentrations for long-term tests under conditions of tem-

perature, pH, salinity, etc., described for rearing the dungeness crab, Section 8720B.3*b*3). Condition and prepare dilution water as described for crab rearing.

For short-term tests of egg hatchability and early larval development to the first zoeal stage, treat ovigerous females with eggs hatching to control bacterial growths, parasites, and diseases (Section 8720B.4). Then place in tanks with flowing seawater at 12 to 13°C and a salinity of 25 to 30 g/kg. When initial hatching occurs, carefully remove unhatched eggs from the egg mass and place 30 directly into each of the 250-mL test chambers described in Section 8720B.3*b*3), which have been set up 10 in each tray for each test concentration and the controls. Measure, then mix, dilution water and toxicant as described in Section 8010F.1. Operate the automatic siphon on a 15-min or longer cycle [Section 8720B.3*b*3)]. Development from egg to pre-zoeal stage to the first zoeal stages usually requires less than 24 h. Determine hatching success, molting from prezoeal to first zoeal stage, and percentage of motile first-stage zoeae. To make short-term tests with first or later zoeal stages, treat female crabs as in Section 8720B.3*b*3), place directly into filtered, UV-disinfected water at 13°C and 30 g/kg salinity, and hold until hatching occurs. When large numbers of first-stage zoeae come near the surface or collect in corners, dip them out with a beaker and place in beaker exposure chambers, 10 larvae in each beaker, and 10 beakers for each concentration and for controls. Expose for 96 to 168 h. Use larvae of a comparable stage of development.

To conduct tests with later zoeal stages, rear the larvae in seawater to the selected stage as described. Before beginning the tests, collect megalops stages for testing from the field and hold for short periods until they metamorphose into the early juvenile stage so that this stage can be used in the test. Determine DO and pH during the test. When possible, make chemical analyses to determine concentrations to which organisms actually are exposed. Use photoperiods normal for the season during which the larval stages develop. Because of the cannibalistic tendencies of crabs beyond the larval stages, expose each individual in a separate exposure chamber. The compartmented troughs described in Section 8720B.3*b*3) provide the necessary exposure chambers and form a compact unit for testing of each effluent or toxicant concentration. With other than effluent toxicity tests that are carried out as described in Section 8010F, recommended test conditions for these stages are 12 to 13°C, 25 to 33 g/kg salinity, and DO above 60% saturation. Feed test organisms macerated fish or mollusks.

2) *Partial-life-cycle tests*—Conduct flow-through tests in the beaker test chambers described for rearing. Supply dilution water and toxicants by pumps or siphons from constant-head tanks and stock solution containers. Use procedures described in Section 8010F.1. Obtain eggs or first-zoeal-stage larvae from females, as described for short-term tests. To begin the long-term study with eggs, place 30 eggs in each of 10 beaker test chambers in the tray, making 300 eggs for each concentration. After hatching, determine number of dead, deformed, and active first-zoeal-stage larvae and randomly reduce number of active larvae to 20 in each test beaker. Continue exposure through the desired number of zoeal stages. Transfer test organisms to clean beakers three times per week. At that time, feed 70 first-stage brine shrimp nauplii to each crab larva. After the third zoeal stage, increase feeding to 100 nauplii per crab larva. Examine for bacterial growths or disease. If needed, treat larvae continuously with streptomycin as described in the rearing procedures. If necessary, pass effluents through a filter to remove coarse materials to prevent clogging of proportional diluter and screens. In partial-life-cycle tests continue test from the egg stage

through the fourth zoeal stage. Usually, zoeae molt into the fourth zoeal stage by the 45th day. If mortality in controls is not excessive, carry on until the early juvenile stage. In other effluent tests, condition dilution water and keep water quality the same as that used to rear crab larvae.

When crabs reach the juvenile stage, transfer to compartmented trough chambers, one to each compartment; continue exposure. Enumerate, measure, and weigh all crabs to be discarded; preserve for additional tests.

Feed juvenile crabs pieces of fish or some other marine animal every 2 or 3 d and remove excess food after 24 h. Clean exposure chambers by scraping down sides to remove algae; brush crabs to remove growth twice per week. In tests with small juvenile crabs, provide a sand substrate if it is found that they have difficulty in shedding the old exoskeleton during ecdysis. The criteria for death of zoeae are cessation of heartbeat and swimming, failure to recover after transfer to pure water, and opaqueness. The criterion of death for juveniles is absence of movement after stimulation. Terminate test at 4 months and record data.

d. American lobster, Homarus americanus: By laboratory rearing it is possible to produce any live stage for short-term tests at most times of year.

1) Short-term tests—Conduct short-term static, renewal, or flow-through tests with juvenile or adult lobsters in compartmented rearing troughs [Section 8720B.3b4)]. Use dilution water as described in Section 8010F.2a. Use procedures given in Sections 8010D and F for dungeness crab and crayfish. Preferably make flow-through tests. For tests longer than 96 h, feed lobsters, clean, and carry out other functions described for rearing [Section 8720B.3b4)]. Use the adaptation of the Hughes rearing chamber for larval stages (Figure 8720:3). Filter, disinfect, and treat for parasites and disease as for dunge-

ness crab larvae. Obtain eggs and larvae, handle, and transfer as described in Section 8720B.3b4). When conducting larval tests with effluents, filter receiving water and other effluent to remove other organisms and clogging materials. Because water movement is necessary to keep larvae separated and off the bottom, use flow-through tests with flow rates of 0.5 to 1 L/min. Meter stock toxic solutions and dilution water by a proportional diluter; discharge into a mixing chamber and then pump to larval test chamber to provide necessary currents. Feed with newly hatched nauplii of brine shrimp at a rate of at least 50 per animal [Section 8720B.3b4)]. Use five concentrations and a control. Stock each 10-L exposure chamber with a minimum of 100 first-stage larvae. Continue tests for 96 to 168 h. To stock each duplicate exposure chamber and the control, dip larvae from screen box of hatching tank with a small beaker and place in a glass tray for counting by the photographic method (Section 8010F.3e). At termination, count survivors photographically or visually. Use the care, feeding, and cleaning procedures given for rearing. Controls give an indication of cannibalism; see Abbott's formula (Section 8010G.1e).

2) Partial-life-cycle and life-cycle tests—Determine test concentrations from short-term test results. Handle test animals as in rearing [Section 8720B.3b4)]. To speed up the tests and complete various life stages within a shorter time, expose at 22 to 24°C. To obtain test concentrations, use proportional diluters (Section 8010F). Use duplicate larval rearing chambers for each test concentration and control and place in operation before introducing test organisms.

Dip larvae from screen box of hatching tank with a small beaker and place in a glass tray to count by the photographic method, Section 8010F.3e. Place 150 first-stage larvae in each chamber. Immediately after photographing, dip larvae from tray and gently float them from beaker into larval exposure chamber. Concentrate larvae that cannot be dipped in one corner of tray by tipping and immerse that corner in exposure chamber to transfer all larvae. Maintain food density (brine shrimp) above 750/L. Check density periodically by dipping out a measured amount of test solution and counting. Observe all exposure chambers daily. At completion of larval stages and attainment of fifth stage, shut off water flow, reduce volume by a screened siphon, dip out lobsters, and place in a glass tray. To insure recovery of all lobsters from larval rearing chamber, disconnect it, remove overflow circulator, and wash those not dipped out into a tray. Randomly select 20 lobsters and place one in each of the 20 compartments of compartmented troughs and continue exposure. Anesthetize remaining lobsters in the tray, photograph to measure length, remove, count, and preserve.

Set up a compartmented trough for each larval exposure chamber, i.e., 40 lobsters for each exposure concentration and the controls. Feed and care for lobsters and clean troughs as described previously. Transfer lobsters to larger compartmented troughs as they grow, randomly reducing the number to 20 for each test concentration when they reach about 250 g. Weigh, measure and preserve those discarded. Continue exposure in each test concentration and control until test objectives are obtained or until maturity. When animals are mature and ready to mate, randomly select two pairs, one from each side of the compartmented trough, and place in mating chambers immediately after female has molted and is in the soft stage. After mating, remove males. Measure, weigh, and preserve all lobsters from each exposure concentration and controls. Place two females in compartmented troughs for each concentration. Remove separators so each has the entire length of one side of the trough. Continue exposure until they are nearly ready to extrude eggs. Place them in deep egg-laying chambers with special rough bottom until eggs are extruded and fastened. Then place them in egg-laying tanks, one for each toxicant concentration and controls. Continue exposure through egg hatching. Collect and place larvae in larval exposure chambers, two for each test concentration and controls, and expose as for F_1 generation. Count and record all live larvae produced, all dead first-stage larvae, and all unhatched eggs. Measure, weigh, and preserve females. Continue F_2 larvae only through fourth stage. Count and measure all larvae. Record number of active larvae, number deformed, and number dead. Measure, weigh, and preserve. Also record and report results of other tests such as histology, disease, parasites, and toxicant accumulation.

e. Shrimp:

1) *Palaemonetes*—By manipulating temperature and photoperiods it has been possible to induce spawning in the laboratory[19] and so make various life stages available throughout the year.

a) Short-term tests—For adults and the six larval stages use glass jars or aquariums as described in Section 8010F.3. Use static, renewal, or flow-through tests.

Rear adults and larvae as described in Section 8720B.3*b*5)a). A large number of ovigerous females is needed to produce larvae required for tests. Feed adults macerated fish or mollusks and feed the larvae brine shrimp daily before and during tests. Rear a 10 to 15% surplus of larvae to insure an adequate number for the test. Ideally, hatch and rear larvae for a series of tests at one time and in mass culture. Conduct short-term tests with larvae of specified ages. Most first-stage larvae metamorphose to postlarvae in 18 to 21 d. Therefore, set up two or more series of tests to cover the larval and postlarval stages. Because of variability of each age group of larvae and increased sensitivity before and during molting, run two to three replicates simultaneously for each test concentration in each series of tests.

In static tests, monitor DO in each test container and control to indicate if and when to renew test solutions. Do not let DO concentrations fall more than 0.5 mg/ L below the minimum levels occurring at that period in the receiving water. For other types of tests, use the same DO and temperature at which test organisms are reared, about 25°C and 60% of saturation or above. For marine species use salinities between 15 and 25 g/kg but do not vary salinity more than ± 1 g/kg in a test series.

b) *Life-cycle tests*—If feasible, begin tests with ovigerous females. Place 25 females with well-developed eggs in a compartmented trough from which the separators have been removed. Stock six or seven of these double-compartmented troughs for five test concentrations and controls. Feed with macerated fish, mollusks, or other suitable organisms. Observe daily to detect females with eggs nearly ready to hatch. Remove these females, place in separate glass jars supplied with appropriate test concentration, and hold until eggs hatch. Renew test solutions as required. Supply brine shrimp nauplii as food for those larvae hatching first. When hatching is complete, remove females and transfer larvae by pipet or gentle pouring into test chambers receiving toxicant. These chambers are the same as the rearing chambers for dungeness crabs (Figure 8720:1). Alternatively use rearing chambers like those for lobster larvae. For dungeness crab chambers, use 1-L capacity and set up in duplicate 20 exposure chambers for each concentration. Stock with five larvae each. If the exposure chamber for lobster larvae is used (Figure 8720:3), stock with 50 larvae each in duplicate. After transferring active larvae, check hatching chamber to determine percentage of eggs that did not hatch, first-stage larvae that died, and total number of live larvae. Count, measure, and preserve. Continue

exposure with daily feeding of brine shrimp until the postlarval stage, which should occur in 18 to 21 d. Randomly select 25 postlarvae from each duplicate group for each exposure concentration. Place groups of 25 postlarvae on each side of a compartmented trough similar to the troughs in which the ovigerous females were exposed. There will be two compartments in the trough, one for each of the two groups of 25 postlarvae for each concentration and the controls. Preserve those not needed for further exposure to determine lengths, weights, histopathological effects, and toxicant accumulation. Feed with macerated fish and/or mollusks and continue exposure until eggs develop and are about to hatch. Select at random and transfer five ovigerous females from each group to hatching chambers; continue exposure. Determine hatching success and number and percentage survival of active first-stage larvae. Count, measure, weigh, and preserve those not taken for further exposure.

2) *Penaeus*

a) *Short-term tests*—Use flow-through short-term tests for penaeid shrimp. Hold, acclimate, and culture shrimp as described in Section 8720B.3b5). For delivery of toxicants see Section 8010F.1. For flow-through tests use aquariums 60 × 30 × 30 cm deep. Place 2 to 3 cm clean sand in the bottom of aquariums and cover tops with screens. After exposure chambers are operating under equilibrium conditions place 20 to 22 juvenile shrimp having a total weight of not more than 50 g in each aquarium. Use duplicate aquariums for each concentration and control. Use uniform-sized shrimp. Conduct tests at 25°C and salinity between 25 and 34 g/kg, depending on the water supply. Maintain salinity to within ± 1 g/kg.[20] Use a flow rate of not less than 7 L/d/g of organism.

If tests are for effluents, use procedures described in Section 8010D.

Use short-term test with larval stages to determine toxicant concentrations for long-term studies. Obtain larval stages of shrimp from gravid females brought in from offshore. While this is a time-consuming and expensive process, once secured, the first protozoeal shrimp are obtained readily. Handle and feed as described in Section 8720B.3b5)b). Set up desired test toxicants in duplicate for five test concentrations and controls with 100 mL each in 250-mL beakers, each covered with a watch glass. For range-finding tests use concentrations of 0.01, 0.1, 1, 10, and 100%. Set up definitive tests at the concentrations indicated by the range-finding tests and as described in Section 8010D. Using a smooth-bore pipet, introduce 10 larvae into each test beaker. Place beakers in a temperature-controlled shaker bath. Add food (algae for protozoeal stages and brine shrimp for mysis stages) as indicated in Section 8720B.3b5)b). Test duration is governed by length of stage tested. At conclusion of test, examine beaker contents in petri dishes under a microscope and determine number of deformed, dead, and live larvae.

b) Partial-life-cycle tests—Begin with egg and continue to postlarval or juvenile stage (Section 8010D). Maintain temperature at 28 to 28.5°C and salinity at 28 to 30 g/kg. As soon as possible acclimate ovigerous females brought in from offshore to 28 to 28.5°C. While on the collecting boat, control temperature in insulated chests. Females usually spawn the first night after collection. As soon as spawning is complete, remove them from the spawning chamber and take samples to estimate the number of eggs in the chamber.[21] Set up a series of exposure beakers as described for the dungeness crab [Section 8720B.3b3)]. Use screens that will retain shrimp larvae and brine shrimp but allow algae to pass through. Set up 10 test beakers in a glass tray for each test concentration and controls. Larvae usually pass through five nauplii stages to the protozoea stage in about 35 h. On the basis of egg count, take portions from the mixed spawning tank containing about 10 nauplii and put in each exposure beaker; this will give 100 larvae for each test concentration and control. Just before the nauplii metamorphose to the protozoeal stage add algal food to each tank. Maintain toxicant flow at 50 mL/min. Add sufficient algae to each exposure tray to supply the quantities indicated in Section 8720B.3b5)b). Add only enough algal concentrate to maintain the desired concentration. Culture algae in units shown in Section 8010E.4c2). Monitor algal cell concentrations in the exposure trays daily.

Feed algae throughout the protozoeal stages. At transformation to the mysis stage, phase out algal feed and add newly hatched brine shrimp. Because algae tend to remove metabolites, increase the flow-through in the mysis stage. Feed brine shrimp nauplii to excess. Do not let their concentration drop below 3/mL. Sample by pipet daily and count to insure adequate concentrations.[21] Feed brine shrimp to larvae through the fourth day of the postlarval stages. At that time remove them from beakers, place in glass dishes, and count. Count active, deformed, and dead larvae. Record percentage living and dead. Transfer living post-larvae to compartmented troughs, one trough for each concentration. Change diet to macerated fish, mollusks, or dry food.[22,23] Remove or flush out sediment and organic material as described previously. Examine for presence of disease or parasites. For bacterial or fungal growth, treat as in Section 8010E.5 and 8720B.4. If test chambers become too crowded, set up additional duplicate chambers. If desired, continue into the juvenile stage. At the end of the study, collect all organisms from each tank, weigh, measure, and make histological and toxicant accumulation studies as appropriate.

5. References

1. SMITH, W.E. 1973. Thermal tolerances of two species of *Gammarus. Trans. Amer. Fish. Soc.* 102:431.

2. ARTHUR, J.W. & E.N. LEONARD. 1970. Effects of copper on *Gammarus pseudolimnaeus, Physa integra* and *Campeloma decisum* in soft water. *J. Fish. Res. Board Can.* 27:1277.

3. ARTHUR, J.W. & J.G. EATON. 1971. Chloramine toxicity to the amphipod, *Gammarus pseudolimnaeus* and the fathead minnow, *Pimephales promelas. J. Fish. Res. Board Can.* 28:1841.

4. NEBEKER, A.V. & F.A. PUGLISI. 1974. Effect of polychlorinated biphenyls (PCB's) in survival and reproduction of *Daphnia, Gammarus* and *Tanytarsus. Trans. Amer. Fish. Soc.* 103:722.

5. OSEID, D.M. & L.L. SMITH. 1974. Chronic toxicity of hydrogen sulfide to *Gammarus pseudolimnaeus. Trans. Amer. Fish. Soc.* 103:819.

6. ARTHUR, J.W., R.W. ANDREW, V.R. MATTSON, D.T. OLSON, G.E. GLASS, B.J. HALLIGAN & C.T. WALBRIDGE. 1975. Comparative toxicity of sewage-effluent disinfection to freshwater aquatic life. Ecol. Res. Ser., EPA-600/3-75-012, U.S. Environmental Protection Agency.

7. ARTHUR, J.W., A.E. LEMKE, V.R. MATTSON & J.B. HALLIGAN. 1974. Toxicity of sodium nitrilotriacetate (NTA) to the fathead minnow and an amphipod in soft water. *Water Res.* 8:187.

8. CLEMENS, H.P. 1950. Life Cycle and Ecology of *Gammarus fasciatus* Say. Franz T. Stone Inst. Hydrobiol. Contrib. 12, Ohio State Univ., Columbus.

9. COSTLOW, J.D., C.G. BOOKHOUT & R. MONROE. 1962. Salinity-temperature effects on the larval development of the crab *Panopeus herbstii* Milne-Edwards reared in the laboratory. *Physiol. Zool.* 35:78.

10. BOOKHOUT, C.B., A.J. WILSON, JR., T.W. DUKE & J.I. LOWE. 1972. Effects of mirex on the larval development of two crabs. *Water, Air Soil Pollut.* 1:165.

11. EPIFANIO, C.E. 1971. Effects of dieldrin in seawater on the development of two species of crab larvae, *Leptodius floridanus* and *Panopeus herbstii. Mar. Biol.* 11:356.

12. LOWE, J.I. 1965. Chronic exposure of blue crabs, *Callinectes sapidus* to sublethal concentrations of DDT. *Ecology* 46:899.

13. BUCHANAN, D.V., R.E. MILLEMANN & N.E. STEWART. 1970. Effects of the insecticide sevin on various stages of the dungeness crab, *Cancer magister. J. Fish. Res. Board Can.* 27:93.

14. COLLIER, R.S., J.E. MILLER, M.A. DAWSON & F.P. THURBERG. 1973. Physiological response of the mud crab *Eurypanopeus depressus* to cadmium. *Bull. Environ. Contam. Toxicol.* 10:378.

15. THURBERG, F.P., M.A. DAWSON & R.S. COLLIER. 1973. Effects of copper and cadmium on osmoregulation and oxygen consumption in two species of estuarine crabs. *Mar. Biol.* 23:171.

16. VERNBERG, W.B. & J. VERNBERG. 1972. The synergistic effects of temperature, salinity and mercury on survival and metabolism of the adult fiddler crab, *Uca pugilator. Fish. Bull.* 70:415.

17. ARMSTRONG, D.A., D.V. BUCHANAN, M.H. MALLON, R.S. CALDWELL & R.E. MILLIMAN. 1975. Toxicity of the insecticide methoxychlor to the dungeness crab, *Cancer magister* Dana. Oregon State Univ. Marine Science Center, Newport (unpublished).

18. CALDWELL, R.S., D.V. BUCHANAN, D.A. ARMSTRONG, M.H. MALLON & R.E. MILLIMAN. 1975. Toxicity of pesticides to the dungeness crab *Cancer magister* Dana. I. The fungicide captan. Oregon State Univ. Marine Science Center, Newport (unpublished).

19. LITTLE, G. 1968. Induced winter breeding and larval development in the shrimp, *Palaemonetes pugio* Holthuis (Caridea Palaemonidae). *In* Studies on Decapod Larval Development. *Crustaceana,* Suppl. 2:19.

20. BAHNER, L.H., C.D. CRAFT & D.R. NIMMO. 1975. A salt water flow-through bioassay method with controlled temperature and salinity. *Progr. Fish-Cult.* 37:126.

21. MOCK, C.R. & M.A. MURPHY. 1970. Techniques for raising penaeid shrimp from egg to postlarvae. *Proc. 1st Annu. Workshop World Maricult. Soc.* 1:143.

22. MOCK, C.R. 1974. Larval Culture of Penaeid Shrimp at the Galveston Biological Laboratory. NOAA (Nat. Ocean. Atmos. Admin.) Tech. Rep. NMFS (Nat. Mar. Fish. Serv.) Circ. 388:33.

23. MOCK, C.R. 1973. Shrimp culture in Japan. *Mar. Fish. Rev.* 35(3-4):71.

6. Bibliography

GRUNBAUM, B.W., B.V. SIEGEL, A.R. SCHULZ & P.L. KIRK. 1955. Determination of oxygen uptake by tissue growth in all glass differential microrespirometer. *Mikrochim. Acta* 1955: 1069.

ROBERTS, J.L. 1957. Thermal acclimation of metabolism of *Pachygrapsus crassipes* Randall. II. Mechanisms and the influence of season and latitude. *Physiol. Zool.* 30:242.

EISLER, R. 1969. Acute toxicities of insecticides to marine decapod crustaceans. *Crustaceana* 16:302.

PORTMAN, J.E. 1972. Results of acute toxicity tests with marine organisms, using a standard method. *In* M. Ruivo, ed. Marine Pollution and Sea Life. Fishing News (Brooks) Ltd., London.

SANDIFER, P.A. 1973. Effects of temperature and salinity on larval development of grass shrimp, *Palaemonetes vulgaris* (Decapoda Caridea). *Fish. Bull.* 71:115.

8720 D. Reporting Results

Analyze and evaluate data and report results as recommended in Section 8010G and H.

8750 TOXICITY TEST PROCEDURES FOR AQUATIC INSECTS*

8750 A. Introduction

1. Ecological Importance

Aquatic insects are important components of lake and stream biota. In trout streams, they comprise 50 to 90% of the macroinvertebrate species. Such groups as mayflies, stoneflies, caddisflies, and midges are major food items for many species of fish.[1] Many aquatic insects are more sensitive to pollutants than are fish.

2. Suitability for Toxicity Tests

The wide variety of aquatic insects, their abundance in unpolluted streams, their

sensitivity to low concentrations of pollutants, and the ease of maintenance of many species under laboratory conditions make them useful test animals. Procedures using aquatic insects have been developed for determining acceptable environmental conditions or concentrations of toxicants.[2] Most studies have been short-term tests, but the procedures can be used for long-term tests.

Toxicants may interfere with survival, growth, reproduction, emergence, and metabolism of aquatic insects. Because effects of long-term exposure to sublethal concentrations of toxicants may be more impor-

*Approved by Standard Methods Committee, 1988.

tant than effects of infrequent short-term exposure to higher concentrations, flow-through, long-term tests are recommended.

3. References

1. HYNES, H.B. 1970. Ecology of Running Waters. Univ. Toronto Press, Buffalo, N.Y. & Toronto, Ont., Canada.
2. SURBER, E.W. & T.O. THATCHER. 1963. Laboratory studies of the effects of alkyl benzene sulfonate on aquatic invertebrates. *Trans. Amer. Fish. Soc.* 92:152.

4. Bibliography

EDMUNDSON, W.T. 1959. Freshwater Biology, 2nd ed. John Wiley & Sons, Wiley Interscience, New York, N.Y.
MACAN, T.T. 1963. Freshwater Ecology. John Wiley & Sons, Wiley Interscience, New York, N.Y.
HYNES, H.B. 1970. Biology of Polluted Waters. Univ. Toronto Press, Buffalo, N.Y. & Toronto, Ont., Canada.
GAUFIN, A.R. 1972. Water Quality Requirements of Aquatic Insects. Final Rep. Contract 14-12-438, Water Quality Off., U.S. Environmental Protection Agency.

8750 B.   Selecting and Preparing Test Organisms

1. Species Selection

Use insects that are important food for fishes, readily available and abundant, relatively easy to keep and culture in the laboratory, and most sensitive to the materials under investigation.

a. *Suggested test organisms:*
1) Stoneflies
   Pteronarcys dorsata
   Pteronarcys californica
   Hesperoperla lycorias
   Hesperoperla pacifica
2) Mayflies
   Hexagenia bilineata
   Hexagenia limbata
   Hexagenia rigida
   Ephemerella subvaria
3) Caddisflies
   Brachycentrus americanus
   Brachycentrus occidentalis
   Clistoronia magnifica
b. *Other species that have been used:*
1) Stoneflies
   Isogenus frontalis
   Perlesta placida
   Paragnetina media
   Phasganophora capitata
   Acroneuria californica
2) Mayflies
   Ephemerella cornuta

   Ephemerella grandis
   Ephemerella doddsi
   Ephemerella needhami
   Ephemerella tuberculata
   Stenonema ithaca
3) Caddisflies
   Hydropsyche betteni
   Macronemum zebratum
   Arctopsyche grandis
   Hydropsyche bifida
4) Diptera
   Chironomus plumosus
   Chironomus attenuatus
   Chironomus tentans
   Chironomus californicus
   Glyptochironomus labiferus
   Goeldichironomus holoprasinus
   Tanypus grodhausi
   Tanytarsus (paratanytarsus) dissimilis

For each test, use early instar larvae or nymphs when possible, especially for growth studies. Many of the listed species complete a generation in one summer. Use late instars for adult emergence tests. For all tests, insects that are cultured are preferable because their source and history are known.

2. Collecting Test Animals

When cultured animals are not available, collect all test specimens from clean, natural waters rich in aquatic insects (see Section 10500, Benthic Macroinvertebrates). Collect larger stream species from riffle areas of clean, well-aerated gravel rubble streams with hand screens or bottom samplers such as the Surber sampler. Stir bottom and let current carry dislodged insects downstream into net.

Immediately after collection, gently place net contents in a 15- to 20-L insulated container partly filled with stream water. Transport to laboratory. Remove and discard larger rocks after it has been determined that they are free of insects. If transportation time exceeds 30 min, provide for aeration and temperature control. In laboratory, swirl water in containers and dip it out. Pour through a screen-bottom container (of a mesh that will retain insects required), held partly submerged in a tank of water. Wash screenings into a holding tank. If it is desired to separate insects, wash into a large white enamel pan containing 3 to 5 cm of water. Remove desired species with a large-bore pipet or small spoon-shaped screen and place in holding tanks. For riffle insects use oval or round flow-through tanks[1] provided with rocks for cover and paddle wheels to provide a current in dilution water.[1] Alternatively collect insects by gently picking up rocks, rubble, or gravel, and carefully washing or picking, then placing desired insects in insulated containers for transport to laboratory.

To obtain benthic insects, sample bottom materials with Eckman, Petersen, or Ponar dredges. Empty dredge into a large pail, add water, and swirl by hand. Partly submerge an appropriate mesh washing screen, pour a portion of swirling sample into it, and wash by moving up and down in the water. Place washed insects in an insulated container and continue until enough insects have been collected.

Chironomids probably will be the dominant insect species in silt bottom material. However, other important immature insects such as dragonflies, damselflies, several species of Diptera, beetles, and mayflies may be found in and on silt bottoms. The mayfly, *Hexagenia limbata*, is a large species often occurring in great abundance in soft, unpolluted muds rich in organic matter that occur in deep pools, ponds, lakes, and reservoirs. Obtain these by collecting top 8 cm of mud and washing as described previously.

3. Holding, Acclimating, and Culturing

a. General considerations: As soon after collection as possible, examine insects for injury. Place all uninjured specimens in holding chambers, supply them with food, and hold for at least 1 week for observation and acclimation to desired temperature. Acclimate stream species in flowing water. Keep in oval troughs that have a current of water or in stainless steel wire cages in running water.[1] In these troughs include flat stones covered with attached algae as cover and food for herbivorous species. Supply insects with materials to build larval and pupal cases. For caddisflies, use sand grains, small pieces of wood, and plant materials retained by a 16-mesh screen. Permit insects that construct tubes or cases to do so. Hold benthic species in aquariums provided with a 3- to 5-cm layer of unsterilized mud from the site where they were collected. *Hexagenia* require a substrate in which to burrow.[2] For chironomids use the highly organic ooze that overlies the bottom where they were collected.

Provide water, DO, and other conditions as described in Section 8010E and F. Maintain final holding temperature within 3°C of temperature at which organisms were collected. For long holding periods, maintain natural seasonal temperatures. When aquatic insects are collected in winter at water temperatures of 1°C or lower, acclimate them to higher temperatures if they are to be used in short-term tests (Section 8010E.3).

Different species require different light intensities. Stoneflies require stones under which they can hide from direct light. Fix light cycle at a certain day length, or vary it seasonally to correspond with natural annual photoperiod. For *Chironomus plumosus*, use a 16-h photoperiod. Lamps and fixtures are described in Section 8010F.3*f*.

b. Food and feeding: *Acroneuria, Brachycentrus, Isogenus,* and *Paragnetina* are predators requiring live food. Feed to excess with small midges, blackfly larvae, mosquitoes, or small caddisfly larvae from an unpolluted environment.[2] Feed *Pteronarcys* and *Ephemerella* to excess with coarse, chopped maple, birch, or aspen leaves that have fallen naturally and have been dried and then soaked in test water for at least 2 weeks before feeding. Feed *Hexagenia, Hydropsyche,* and *Arctopsyche* finely ground leaves and fish-food pellets. If the substrate is rich in organic matter, additional food may not be required for *Hexagenia*. Avoid overfeeding with fish food because it causes DO depletion. The larvae of some Hydropsychidae are highly carnivorous and cannibalistic; keep them well-fed with plankton, microcrustacea, blackfly larvae, and other organisms, collected from fish hatcheries, ponds, lakes, and streams with a net of No. 20 bolting silk.

Feed chironomids twice per week. Keep in jars supplied with algal culture medium [Section 8010E.4*c*1)a)] inoculated with algae including diatoms. Alternatively use a mixture of 5 g fish food plus 1 g powdered dried grass* blended in 1 L of water. Add about 100 mL of this suspension to each culture per feeding. If there is no flow-through, remove 100 mL of test solution before feeding. Use 10-L culture jars containing 8 L or less of medium with a screen cover to retain adults.[4,5] Keep in a constant-temperature room at 21 to 24°C. For long-term studies follow natural temperature cycle of water from which chironomids were taken. Because the jars have a mud substrate, do not clean them or overfeed the organisms. Collect emerging adults for breeding in wire screen cylinders placed over the culture jars.[6,7]

4. References

1. SURBER, E.W. & T.O. THATCHER. 1963. Laboratory studies of the effects of alkyl benzene sulfonate on aquatic invertebrates. *Trans. Amer. Fish. Soc.* 92:152.
2. FREMLING, C.R. & G.L. SCHOENING. 1973. Artificial substrates for *Hexagenia* may-fly nymphs. *In* Proc. 1st Int. Conf. Ephemeroptera, p. 209.
3. NEBEKER, A.V. & F.A. PUGLISI. 1974. Effect of polychlorinated biphenyls on survival and reproduction of *Daphnia, Gammarus* and the midge *Tantarus. Trans. Amer. Fish Soc.* 103:722.
4. NEBEKER, A.V. 1972. Effect of high winter water temperatures on adult emergence of aquatic insects. *Water Res.* 5:777.
5. BAY, E.C. 1967. An inexpensive filter-aquarium for rearing and experimenting with aquatic invertebrates. *Turtox News* 45:146.
6. BREVER, K.D. 1965. A rearing technique for the colonization of chironomid midges. *Ann. Entomol. Soc. Amer.* 58:135.
7. NEBEKER, A.V. & A.E. LEMKE. 1968. Preliminary studies on the tolerance of aquatic insects to heated waters. *J. Kans. Entomol. Soc.* 41:413.

5. Bibliography

USINGER, R.L., ed. 1956. Aquatic Insects of California—With Keys to North American Genera and California Species. Univ. California Press, Berkeley.
ROBACK, S.S. 1957. The Immature Tendipedids of the Philadelphia Area. Monogr. Acad. Natural Sci. Philadelphia No. 9, Philadelphia, Pa.
FREMLING, C.R. 1967. Methods for mass-rearing

---

*Cerophyl, Agri-Tech, Inc., Kansas City, Mo. 64112, or equivalent.

Hexagenia (Ephemeroptera: Ephemeridae). *Trans. Amer. Fish. Soc.* 96:407.

NEBEKER, A.V. 1972. Effect of low oxygen concentration on survival and emergence of aquatic insects. *Trans. Amer. Fish. Soc.* 101:675.

GAUFIN, A.R., R. CLUBB & R. NEWELL. 1974. Studies on the tolerance of aquatic insects to low oxygen concentrations. *Great Basin Natur.* 31:45.

ANDERSON, N.H. 1977. Continuous rearing of the Limnephilid caddisfly, *Clistoronia magnifica* (Banks). Proc. 2nd Symp. on Trichoptera, 1977. Junk, The Hague, Netherlands.

MERRILL, R.W. & K.W. CUMMINS. 1978. An Introduction to the Aquatic Insects of North America. Kendell/Hunt Publishing Co., Dubuque, Iowa.

PENNAK, R. 1978. Fresh Water Invertebrates of the United States, 2nd ed. John Wiley & Sons Inc., New York, N.Y.

8750 C. Toxicity Test Procedures

1. General Procedures

Conduct tests as described in Section 8010D. If possible, use a minimum of 20 specimens for each toxicant concentration with an additional 40 animals for growth studies. Two species may be tested in the same tank if precautions are taken to avoid predation.

Do not use static testing with stream insects unless air stones or water movement can simulate natural water conditions. Use static tests with certain lake or reservoir species if required DO levels are maintained. For long-term tests, see Section 8010D.

a. Test tanks: Use glass and stainless steel aquariums of either 8-L or 20-L size for quiet-water species. For stream species, use round or oval, stainless steel or epoxy-painted troughs [1,2,3] (90 cm long, 15 cm wide, and 15 cm deep) in which natural stream flow is simulated. Set tanks side by side so paddle wheels on one long shaft can be used to circulate water in them all.[1] Jetted incoming water from the diluter also can maintain adequate flow.

b. Flow rate: Use flows to each tank of no less than 6 to 10 tank volumes/24 h. In aquariums without water-circulating devices use much higher flows for stream species to simulate stream flow. In oval test tanks use velocities near 0.5 cm/s. For quiet-water forms, such as *Hexagenia* and *Chironomus*, do not disturb mud substrate with water flow.

c. Aeration: Aeration is unnecessary; however, use if desired with nonvolatile toxicants to increase or control water movement, especially for tank tests with lake and reservoir species or if DO levels drop.

d. Cleaning: See Section 8010E.4*d*. Siphon out detritus on tank bottom weekly during long-term testing. If a mud substrate is used, no cleaning is necessary. Avoid overfeeding.

e. Substrate: For all stream riffle species use fine-mesh stainless steel screens formed into cylinders or cubes, which provide 10 to 15 cm^2/insect. Place cages in oval troughs or in glass cylinders.[1] For 30- to 90-d adult emergence tests, obtain clean rocks, 5 to 10 cm in diameter (one for every three insects) from collection site for a substrate. Provide fine screen or sticks that protrude above water surface for adult emergence tests.

f. Light and photoperiod: See Section 8010F.3*f*. Use natural photoperiod at time of testing for locality in which test is conducted. Increase day length during adult emergence tests by 0.5 h every 2 weeks.

g. Temperature: See Section 8010F.1*b*. Use 10°C as a winter temperature. For trout stream insects, use summer temperatures near 15°C. Increase temperature during adult emergence tests by 1°C each week up to a maximum of 5°C above initial temperature. When using warm-water stream or lake insects, follow natural temperature cycle.

h. Time of year: Under natural conditions, most species emerge as adults in spring. Therefore start adult emergence tests no later than March 1st. *Hexagenia limbata* and most midge species are exceptions, emerging throughout summer in most localities.

2. Toxicant Preparation

See Sections 8010F.1 and 2*b*.

3. Test Procedures for *Hexagenia*

Use *Hexagenia* for short-term survival (96 to 168 h), survival for 5 to 60 d, adult emergence, or full-life-cycle tests (90 to 120 d). Use a minimum of 20 organisms per aquarium of not less than 8 L capacity. Use a water depth of 8 to 20 cm. Provide a fine organic ooze substrate 4 to 5 cm deep and as similar as possible to that where naiads occur naturally. When using newly hatched *Hexagenia* to start a test, use 50/tank. When *Hexagenia* eggs are used as a source of larvae, pipet them into petri dishes (about 200/dish) with 200 mL test water at about 20°C, and let hatch.

When substrate is mud, determine survival by counting number of dead animals that have left their burrows and/or by counting number of new burrows formed after disturbing mud surface sufficiently to destroy entrances to old burrows. If counts do not agree, use the latter. For acute toxicity tests, alternatively use an artificial substrate of epoxy resin to facilitate observation and monitoring of test animals.[4] For growth or emergence tests, set up an additional set of containers so that naiads can be removed periodically for measurement. Remove 10 naiads from their burrows after 20 to 60 d to determine growth. Do not remove more than 50% of surviving animals before conclusion of these tests. Keep a record of total number removed. Use these animals to provide additional data on growth and emergence. Record body length, head capsule width, and live weight.

In acute toxicity tests, determine survival after 1.5, 3, 6, and 12 h and twice daily thereafter. As a sign of death, use failure of specimens to respond by movement to gentle probing or flashlight illumination. In longer-term studies, check tanks daily to remove and record dead animals and cast naiad skins, which indicate successful molting.

For growth studies, determine initial range and mean of total length, head capsule width, and weight from specimens in holding tank. Kill all animals in warm water (40 to 50°C) before measuring. Take measurements twice during testing, using animals that are to be discarded. Obtain final measurements for all survivors. Make two counts: number of adults and cast skins; if different, use cast skins because some adults may have escaped.

Determine and record percentage of adults that emerge, sex, incidence of incomplete emergence (i.e., half-out of nymphal skin, wings unsuccessfully unfolded, etc.), adult length, weight, and head capsule width, and number of mature eggs.

4. Test Procedures for *Chironomus*

Follow procedures described in Section 8010F. For each concentration, use duplicate 20-L aquariums with mud or Cerophyl® substrate and screen covers. Maintain flow to each test container at about 2 L/h. Use a mud substrate similar to that for *Hexagenia*. Use lighting and photoperiod as described in Section 8010F.3*f*. Do not feed animals during short-term tests. Feed during 30-d and emergence tests as in Section 8750B.3*b*. If prepared food is used, add Cerophyl® or about 100 mL food suspension to each container twice per week.

For long-term tests, place 50 first-instar larvae (about 1.5 mm long and less than 24 h old) in each test aquarium. Transfer larvae with an eyedropper. Determine number of emerging adult males and females. Count both adults and pupal cases. If counts differ, use pupal case count. At 25 ± 1°C emergence takes about 1 month. To determine success of fertilization of eggs, take 50 eggs and determine the percent hatchability. If it is impossible to separate and count eggs, hatch fertilized egg masses in beakers with same test water from which adults emerged. Count 60 larvae into a petri dish and examine for injured larvae. Transfer often will injure early instar larvae; if a correction for this is not made in the count, errors in percent survival result. After examination, count 50 larvae and return to test chamber and rear them to adult stage. End points for taking and analyzing data are emergence of adults, egg production, and hatching of young. Repeat complete test at least once.

5. References

1. SURBER, E.W. & T.O. THATCHER. 1963. Laboratory studies of the effects of alkyl benzene sulfonate on aquatic invertebrates. *Trans. Amer. Fish. Soc.* 92:152.
2. NEBEKER, A.V. 1972. Effect of high winter water temperatures on adult emergence of aquatic insects. *Water Res.* 5:777.
3. NEBEKER, A.V. & A.E. LEMKE. 1968. Preliminary studies on the tolerance of aquatic insects to heated waters. *J. Kans. Entomol. Soc.* 41:413.
4. FREMLING, C.R. & G.L. SCHOENING. 1973. Artificial substrates for *Hexagenia* may-fly nymphs. *In* Proc. 1st Int. Conf. Ephemeroptera, p. 209.

6. Bibliography

SANDERS, H.O. & O.B. COPE. 1968. The relative toxicities of several pesticides to naiads of three species of stone-flies. *Limnol. Oceanogr.* 13:112.
GAUFIN, A.R. & S. HERN. 1971. Laboratory studies on tolerance of aquatic insects to heated waters. *J. Kans. Entomol. Soc.* 44:240.
ROUSSEL, J.S. 1972. Standard methods for detection of insecticide resistance of *Diabrotica* and *Hypera* beetles. *Bull. Entomol. Soc. Amer.* 18:179.
FRIESEN, M.K. 1979. Use of eggs of the burrowing mayfly *Hexagenia rigida* in toxicity testing. *In* E. Scherer, ed. Toxicity Tests for Freshwater Organisms. *Can. Spec. Publ. Fish. Aquat. Sci.* 44:27.
AMERICAN SOCIETY FOR TESTING AND MATERIALS. 1980. Standard Practice for Conducting Acute Toxicity Tests with Fishes, Macroinvertebrates, and Amphibians. ASTM Standard E 729-80, Philadelphia, Pa.
FROMLING, C.R. & W.L. MAUCK. 1980. Methods for using nymphs of burrowing mayflies (Ephemeroptera, *Hexagenia*) as toxicity test organisms. *In* A.L. Birikema & J. Carrs, eds. Aquatic Invertebrate Bioassays. ASTM STP 715, American Soc. Testing & Materials, Philadelphia, Pa.

8750 D.   Data Evaluation

Analyze, evaluate, and report data from various tests as described in Section 8010G.

8910   TOXICITY TEST PROCEDURES FOR FISH*

8910 A.   Introduction

1. Suitability for Toxicity Tests

Fish have been used more widely for toxicity tests than any other group of aquatic organisms. The number of species extensively used, however, has been limited to approximately 15. Only a few freshwater and only a single marine species have been

used routinely in life-cycle tests.[1] Until recently, few marine fish were cultured or reared in the laboratory.

2. Reference

1. COMMITTEE ON METHODS FOR TOXICITY TESTS WITH AQUATIC ORGANISMS. 1975. Methods for Acute Toxicity Tests with Fish, Macroinvertebrates, and Amphibians. Ecol. Res. Ser. EPA-660/3-75-009, U.S. Environmental Protection Agency, Duluth, Minn.

*Approved by Standard Methods Committee, 1985.

8910 B.   Fish Selection and Preparation

1. Selection of Test Species

General guidelines for selecting test organisms are outlined in Section 8010E.1. A prime consideration in the selection of a fish species is its sensitivity to the effluent, material, or environmental factors under consideration. In addition, the relative sensitivity of the fish compared to that of other organisms in the ecosystem should be considered. Because a number of freshwater fish have been reared through more than one generation in hatcheries, it is possible to conduct life-cycle studies with some fish. If it is desirable to use a fish that is not reared in a hatchery, collection of various life stages from the field for partial life-cycle tests is necessary. In a test, use fish obtained from groups of the same source.

Adult fish may live indefinitely at levels of environmental factors and toxicant concentrations that may be unsuitable for the survival of early life stages. Therefore, preferably use the most sensitive known life stage for tests.

Select test species according to the following criteria: (a) the species should occur or be closely related to a species that occurs in the receiving water for the waste being tested; (b) the species should be available

in numbers sufficient for the tests; (c) the species should be capable of being held in the laboratory in healthy conditions (i.e., active, feeding, proper orientation, absence of lesions, etc.) for at least 1 month; and (d) the species should represent an important trophic link or economic resource in the ecosystem of the receiving water. Consider relative sensitivities of the different species and life stages if the data are available.

The following is a partial list of freshwater, estuarine, and marine fish that have been used for toxicity tests:[1,2]

a. Freshwater fish:

Clupeidae:
  Alosa
    pseudoharengus   Alewife
  Dorosoma petenense   Threadfin shad
Salmonidae:
  Coregonus artedii   Lake herring
  Coregonus
    clupeaformis   Lake whitefish
  Prosopium
    williamsoni   Mountain whitefish
  Oncorhynchus
    gorbuscha   Pink salmon
  Oncorhynchus keta   Chum salmon

Oncorhynchus	
kisutch	Coho salmon
Oncorhynchus nerka	Sockeye salmon
Oncorhynchus	
tschawytscha	Chinook salmon
Salmo clarki	Cutthroat trout
Salmo gairdneri	Rainbow trout
Salmo salar	Atlantic salmon
Salmo trutta	Brown trout
Salvelinus fontinalis	Brook trout
Salvelinus	
namaycush	Lake trout
Osmeridae:	
Osmerus mordax	Rainbow smelt
Esocidae:	
Esox lucius	Northern pike
Cyprinidae:	
Carassius auratus	Goldfish
Cyprinus carpio	Carp
Notropis atherinoides	Emerald shiner
Notemigonus	
crysoleucas	Golden shiner
Pimephales notatus	Bluntnose minnow
Pimephales promelas	Fathead minnow
Catostomidae:	
Catostomus	
commersoni	White sucker
Ictaluridae:	
Ictalurus melas	Black bullhead
Ictalurus natalis	Yellow bullhead
Ictalurus nebulosus	Brown bullhead
Ictalurus punctatus	Channel catfish
Cyprinodontidae:	
Jordanella floridae	Flagfish
Poeciliidae:	
Gambusia affinis	Mosquitofish
Poecilia reticulata	Guppy
Percichthyidae:	
Morone chrysops	White bass
Centrarchidae:	
Lepomis macrochirus	Bluegill
Micropterus	
dolomieui	Smallmouth bass
Micropterus	
salmoides	Largemouth bass
Pomoxis annularis	White crappie

Pomoxis	
nigromaculatus	Black crappie
Percidae:	
Perca flavescens	Yellow perch
Stizostedion	
canadense	Saugar
Stizostedion v	
canadense	Walleye pike

b. *Marine and estuarine fish:*

Clupeidae:	
Brevoortia patronus	Gulf menhaden
Brevoortia tyrannus	Atlantic menhaden
Clupea harengus	Atlantic herring
Harengula	
pensacolae	Scaled sardine
Sardinops sagax	Pacific sardine
Engraulidae:	
Anchoa mitchilli	Bay anchovy
Cyprinodontidae:	
Cyprinodon	
variegatus	Sheepshead minnow
Fundulus	
heteroclitus	Mummichog
Fundulus similis	Longnose killifish
Fundulus parvipinnis	California killifish
Atherinidae:	
Menidia beryllina	Tidewater silverside
Menidia menidia	Atlantic silverside
Gasterosteidae:	
Gasterosteus	
aculeatus	Threespine stickleback
Percichthyidae:	
Morone saxatilis	Striped bass
Serranidae:	
Centropristis striata	Black sea bass
Sparidae:	
Lagodon rhomboides	Pinfish
Sciaenidae:	
Leiostomus	
xanthurus	Spot
Micropogon	
undulatus	Atlantic croaker
Mugilidae:	
Mugil cephalus	Striped mullet
Mugil curema	White mullet
Pleuronectidae:	

| *Pseudopleuronectes* | |
| *americanus* | Winter flounder |

2. Collecting and Handling Test Fish

Collecting equipment and methods are described in Sections 8010E.2 and 10600. Handling and holding are discussed in Section 8010E.3. To maintain healthy fish, avoid unnecessary stress. Do not subject fish to rapid changes in temperature or water quality.

a. Freshwater fish: Whenever possible, obtain species routinely raised in hatcheries. Some minnows can be obtained from bait dealers; large species can be obtained from fishermen. Salmonid fish usually are available from private, state, and federal hatcheries. Obtain trout certified free from infectious pancreatic necrosis, furunculosis, kidney disease, and whirling disease. When fish cannot be obtained from hatcheries, appropriate field collection is acceptable. Collecting permits usually are required by state agencies.[2]

b. Marine and estuarine fish: Collect various life stages of marine fish in the field for laboratory tests. Vertical movement of early larval stages may necessitate nighttime collection. Many marine fish and most marine fish larvae are extremely fragile; handle carefully during collection, sorting, and transfer. For sorting and transferring larvae during and after collection, use a large-bore 3- to 5-mm-ID pipet. Whenever possible, transfer by dipping or gently pouring. Fine-mesh dip nets also are suitable if transfers are made quickly.

3. Holding and Acclimating

See Sections 8010E and F for additional discussion.

Keep fish stocks in tanks, small ponds, live boxes, or screen pens, depending on fish size and number. Provide bottom-dwelling fish with a silica-sand substrate. Use good-quality dilution water, as discussed in Section 8010E.4b, for acclima-

tion. Feed fish natural or commercially available prepared foods daily during acclimation. Detailed information on handling, holding, care, and feeding of fish is available.[1-4] Food requirements may vary with the species and size of fish; use care in selecting the diet. Provide fish obtained from a hatchery with food to which they are accustomed. Many fish can be maintained for long periods on dried food but live food supplements are desirable. Do not overfeed. Analyze food and reject if contaminated with PCBs, pesticides, or heavy metals.

To maintain fish in good condition during holding and acclimation, watch carefully for signs of disease, stress, physical damage, and mortality. Remove dead and abnormal individuals immediately. If mortality rate during acclimation exceeds 10%, discard the population.

Handle fish carefully and as quickly as possible.[1,2] For extensive handling such as weighing, measuring, or taking other data, anesthetize fish.[3] Before using fish from any source, determine body burdens of pesticides, PCBs, heavy metals, and other toxicants.

For short-term tests, use fish of the same size and if possible, the same year class. The length of the largest fish should not be more than 1.5 times the length of the shortest fish. Acclimate fish to laboratory conditions for 14 d before the test.

4. Parasites and Disease

a. Stress in relation to parasites and disease: Unexpected and often unexplained mortalities in experimental and control animals interfere with acute or chronic tests. While many factors may be responsible for the death of an animal, diseases due to specific pathogens are among the most significant. Optimize laboratory conditions for each particular species to prevent the development of disease.

When large numbers of organisms are

retained in a relatively small space, undesirable growths, diseases, and parasites may become a problem. If the water is unpolluted and poor in nutrients, problems often can be controlled by strict sanitation. However, if the water is somewhat enriched by organic materials and potential toxicants are present, problems increase greatly. Pathogens and parasites that might be very rare in natural waters become ever-present dangers in intensive culture. Uneaten food, fecal material, and other wastes provide a potential source of unwanted growths, bacteria, disease, and toxic products. Even with adequate flow-through, insufficient velocities may create dead areas in a tank that may become trouble areas.

Filtration and/or sterilization of water, regular cleaning of holding vessels, sterilization of equipment, and securing disease-free fish with adequate feeding are the first lines of defense.

Organisms exposed to toxicants are stressed, weakened, and much more susceptible to parasites and disease. Because various environmental factors may contribute to reduced resistance, pay careful attention to nutrition, oxygen supply, and water quality. Do not feed fish during the 2 d before initiating short-term tests.

b. Control methods:

1) General—Ultraviolet light and ozonation have been used successfully to control disease and parasites. Antibiotics used in holding tanks reduce bacterial populations. To reduce mortality and to avoid introduction of disease into stock tanks, treat fish with a broad-spectrum antibiotic immediately after collection, during transport, or on arrival at the laboratory. Holding fish in an antibiotic solution (such as 15 mg tetracycline/L) for 24 to 48 h generally is effective. Other chemotherapeutic agents are available, but use care in their application because some are toxic at low concentrations.[5] Do not use treated organisms for tests until at least 10 d after treatment. Clean and disinfect tanks and containers with 200 mg sodium hypochlorite (NaOCl)/L for 1 h after removal of diseased fish. Dechlorinate with sodium thiosulfate and rinse before reusing tanks.

For tests with fish larvae use strict sanitary measures including sterilization of utensils and containers, filtration and UV sterilization of water, and removal of feces and waste food. If signs of disease appear in a larval culture, discard the entire culture.

For tests using adult fish, secure healthy, disease-free fish. Treat in emergencies.

2) Control methods for freshwater fish—Treat freshwater fish to cure or prevent disease by the methods in ¶ *b*1) above and Table 8910:I. These methods have been found dependable, but their efficacy may be altered by temperature or water quality. If fish are severely diseased, destroy the entire lot. A number of good reviews of fish diseases and parasites and methods for their control have been published.[5-15]

3) Control methods for marine and estuarine fish—Species such as the clupeids are particularly susceptible to bacterial (usually *Vibrio*) infections. Tetracycline-based antibiotics in the food provide limited protection.

Ectoparasitic protozoa (particularly the ciliate *Cryptocaryon* and the dinoflagellate *Oodinium*) may multiply quickly to epizootic proportions in marine aquariums. Treat with formalin and cupric acetate.[15]

Treat monogenetic trematodes on gills and body surfaces by brine and sodium pyrophosphate dips.

Treatments for other microbial and parasitic diseases (such as lymphocystis, fungus, and parasitic copepods) are available.[16]

Published information on related topics includes a summary of advances and unsolved problems in marine fish larval culture,[17] a description of a successful larval culture system,[18] and a discussion of disinfection of water supplies.[19]

TABLE 8910:I. RECOMMENDED PROPHYLACTIC AND THERAPEUTIC TREATMENTS FOR FRESHWATER FISH TO BE USED FOR EXPERIMENTAL PURPOSES

Disease	Chemical	Concentration mg/L	Application
External bacteria	Hyamine 1622® or 3500®*	1–2 AI†	30–60 min in flow-through system‡
	Nitrofurazone (water mix)	3–5 AI	30–60 min in flow-through system‡
	Neomycin sulfate	25	30–60 min in flow-through system‡
	Oxytetracycline hydrochloride (water-soluble)	25 AI	30–60 min in flow-through system‡
Monogenetic trematodes, fungi, and external protozoa§	Formalin *plus* zinc-free malachite green oxalate	25 ± 0.1	1–2 h in static system, 30–60 min in flow-through system‡
	Formalin	150–250	
	KMnO₄	2–6	1–2 h in static system, 30–60 min in flow-through system‡
	NaCl	14 000–30 000 2000–4000	5–10 min dip 24 h minimum, but may be continued indefinitely
	Para-dimethyl aminobenzenediazo sodium sulfonate (35% AI)‖	20	30–60 min in flow-through system‡
Parasitic copepods	Trichlorfon #	0.25 AI	Weekly for up to 4 weeks if necessary in static or flow-through systems. Do not use at > 27°C.

* Benzalkonium chloride.
† AI = active ingredient.
‡ Add concentrated stock solution to the inflowing water by a drip system or by the technique of Brungs and Mount.[8]
§ One treatment usually is sufficient except for *Ichthyophthirius*, which must be treated daily or every other day until no sign of the protozoans remains. This may take 4 to 5 weeks at 10°C and 11 to 13 d at 15 to 21°C. A temperature of 32°C is lethal to *Ichthyophthirius* in 1 week.
‖ Dexon® or equivalent.
Masoten® or equivalent.

5. Culturing Test Fish

a. Freshwater fish: More than 30 species of freshwater fish have been reared for stocking fresh waters. The cultural methods can be adapted to laboratory scale to produce different life stages of fish.[20-34]

Methods are given below for four freshwater species commonly used in toxicity experiments: the brook trout, *Salvelinus fontinalis*; the fathead minnow, *Pimephales promelas*; the bluegill sunfish, *Lepomis macrochirus*; and the channel catfish, *Ictalurus punctatus*. These species are representative of test organisms that can be found in various freshwater habitats, ranging from free-flowing streams and rivers to stagnant ponds and lakes.

1) Brook trout, *Salvelinus fontinalis*—Collect sexually mature individuals by seining or from stock cultures. Breeding generally occurs in the fall (October-January), although spawning time may be regulated by artificial control of the photoperiod, or, less commonly, by injection of pituitary hormones.[35-38] Separate mature animals by sex into holding tanks such as those described in Section 8010E.4a. Maintain flow of water through the tanks at least at 6 to 10 tank volumes/d, depending on the amount needed to keep the oxygen concentration above 60% saturation and un-ionized ammonia below 20 μg NH_3-N/L. To synchronize reproductive activity, design temperature and photoperiod regimes to simulate the conditions presented in Table 8910:II. Dawn and dusk light intensity variations may be simulated if desired.[39] During the pre-spawning period, feed the fish to a slight excess with a diet consisting of any of the various commercially available pelleted trout foods. However, during the brooding period do not use a diet containing cottonseed meal, which often is found in commercial foods.[40] Siphon unused food and wastes from the tanks daily and brush clean the interior surfaces at least weekly.

Spawning may be accomplished by photoperiod-induced breeding or by artificial fertilization. For laboratory spawning use holding tanks constructed of No. 316 stainless steel or other appropriate material, with dimensions of 90 $\times$ 30 $\times$ 40 cm and a water depth of 30 cm. Construct spawning nest of double-strength glass or stainless steel consisting of an outer box 28 $\times$ 33 $\times$ 7.5 cm which has three equally spaced 2.5-cm-diam holes drilled midway up each end. Cover the holes with 10-mesh stainless steel wire to allow water in the box to drain down to 2.5 cm when box is

removed from spawning tank. Place a bottomless egg retainer insert within the spawning box. The egg retainer consists of a 27-$\times$ 32-cm grid of 2.5-cm-square compartments constructed from 1.3-cm-wide strips of 7-mesh stainless steel screen. Cover egg retainer grid by a 2-mm-mesh stainless steel screen to which 1.3- to 2.5-cm smooth gravel is attached with silicone adhesive. The dimensions of the spawning nest have been selected for females weighing less than 150 g. For larger fish, increase size of box.

After acclimation to the appropriate photoperiod, introduce two males and four females (sexually mature) into each spawning tank. Place one cleaned and sterilized nest box in each spawning tank and remove embryos from the box at a fixed time each day (preferably after 1300 h Evansville time, so fish are not disturbed during the morning).

Fertilized eggs also may be obtained by a process called stripping. Detailed techniques are described elsewhere,[40-43] but the process basically involves stripping eggs and milt into a container with gentle mixing by hand or with a feather so that the eggs are exposed to the milt but are not broken. Two or more pairs of trout often are stripped into the same container to decrease the possibility that eggs will remain unfertilized if one of the male fish is sterile.[40] After mixing, let eggs stand for 5 to 10 min and rinse to remove the milt. Set the fertilized eggs aside for 20 to 30 min to harden before transferring them to incubation chambers.

Place the fertilized eggs (embryos), whether obtained from spawning tanks or by stripping, in incubation chambers until hatching. For small-scale cultures, such as those that may be manipulated experimentally during a partial-life-cycle toxicity test (Section 8910C.2), incubation cups may be constructed from 8-cm sections of 5-cm-OD plastic or glass tubing with nylon or stainless steel screening cemented over one

TABLE 8910:II. EVANSVILLE, INDIANA, PHOTOPERIOD

Dawn to Dusk Time h	Date	Day Length h:min
0600–1815	Mar. 1	12:15
0600–1900	15	13:00
0600–1930	Apr. 1	13:30
0600–2015	15	14:15
0600–2045	May 1	14:45
0600–2115	15	15:15
0600–2130	June 1	15:30
0600–2145	15	15:45
0600–2145	July 1	15:45
0600–2130	15	15:30
0600–2100	Aug. 1	15:00
0600–2030	15	14:30
0600–2000	Sept. 1	14:00
0600–1930	15	13:30
0600–1845	Oct. 1	12:45
0600–1815	15	12:15
0600–1770	Nov. 1	11:30
0600–1700	15	11:00
0600–1645	Dec. 1	10:45
0600–1630	15	10:30
0600–1630	Jan. 1	10:30
0600–1645	15	10:45
0600–1715	Feb. 1	11:15
0600–1745	15	11:45

end to retain the embryos. Oscillate incubation cups (containing less than 250 embryos) in holding tank (or test) water by a rocker arm driven by a 2-rpm electric motor.[44] Incubation chambers for mass culture of trout generally consist of wire-screen trays with rectangular or oblong mesh (15- × 3.5-mm openings) stacked vertically in deep flow-through troughs[40,43] or spaced along horizontal troughs.[45,46] The embryos are treated twice weekly to control fungus by adding 3.75 g malachite green dissolved in 3 L water into the water flow (8 L/min) at a rate of 30 mL/min.[40] Remove dead embryos that turn white by "shocking" (agitation of the embryos) every 2 d. Maintain incubation chambers in the dark because exposure to light may result in premature hatching or death.[40] The rate of hatching is determined by the temperature regime; the optimum range, 7 to 10°C, produces hatching in 44 to 68 d.

On hatching, the alevins escape through the rectangular mesh of the trays into the troughs, where they can be maintained at least until the yolk sac is absorbed. Up to 25 000 alevins are maintained routinely in a hatchery trough of standard size (of approximately 250 L).[41]

After development to the free-swimming fry stage, provide at least one complete exchange of water per hour (1.2 L/min/100 fry is preferred). Maintain water pH between 6.5 and 8.5, dissolved oxygen above 5 mg/L, and dissolved solids above 50 mg/L, and insure that the water is free of pollutants.[40] Feed fry dry commercial food, slightly to excess, as often as 10 times/d.[40,43,45] Daily rations generally average 7 to 9% body weight.

When trout reach the fingerling stage, transfer to large raceways or reduce the numbers to a loading of approximately 1 g/L/d.[40] Detailed feeding regimes are given elsewhere,[40] but daily rations for juvenile trout are gauged by appetites, generally 4 to 5% of body weight.[40,43,45] As fish grow, grade and sort them into separate tanks according to size to prevent cannibalism. Do grading monthly.[41]

2) Fathead minnow, *Pimephales promelas*—Adult fathead minnows can be obtained from the field, from laboratory cultures, or from bait dealers. This species spawns naturally from the end of April until August, but in the laboratory it spawns continuously at temperatures between 18 and 29°C and the photoperiod in Table 8910:II. It reproduces naturally at less than 1 year of age; the life span is short, frequently not more than 15 months.

Holding tanks and spawning chambers for fathead minnows are similar to those described for brook trout [Sections 8010E.4*a* and 8910B.5*a*1)]; the species is not very habitat-sensitive so various other systems of ponds, tanks, and raceways may be used.[43,46,47] Because fathead minnows deposit eggs on the underside of submerged objects, substitute spawning substrates for trout nests. Spawning substrates may be constructed from inverted halves of tile 7.5 cm ID and 7 to 10 cm long, or equivalent. Flow rate, DO requirements, aeration, cleaning, and operation are as described for brook trout.

Photoperiods used throughout the life cycle simulate dawn-to-dusk times at Evansville, Ind. (Table 8910:II). Maintain temperature at approximately 25°C.

For laboratory culture or life-cycle toxicity tests, place eight males and eight females in each spawning chamber along with four spawning substrates. Feed fish once or twice a day with frozen adult brine shrimp as the main diet and supplement by pelleted trout food. Once spawning begins, remove four males, retaining those males guarding well-established territories under the tiles. Introduce a fifth substrate tile as cover for females. Fry hatch after 4 to 6 d at 25°C, so remove spawning substrates with eggs for hatching in a separate incubator unit at about 3-d intervals.

If mass cultures are desired, use large tanks, raceways, or ponds for rearing.

3) Bluegill sunfish, *Lepomis macrochirus*—Breeding and cultivation of bluegills may be carried out in a variety of ponds or tanks by controlled natural reproduction.[40,43] Adult bluegills average 100 to 150 g in weight (12 to 18 cm in length) and generally spawn when 1 year old. The breeding period is long (May to August) and spawning takes place more than once during the season.[40,43]

Stock spawning ponds with a 1:1 or, preferably, a 2:3 ratio of male to female adult bluegills. Although bluegills do not require the highly controlled environment required by brook trout, maintain adequate DO and water quality conditions (see Section 8010E.4*b*). Bluegills adapt readily to a wide variety of commercial feeds; feed to satiation.

Provide spawning ponds with small piles of gravel in the shallow water (0.5 to 1.0 m) around the edges.[43] Male bluegills will use this material to build nesting areas. Space gravel piles at least 1 m apart to reduce fighting between males guarding territories.[46]

If dense spawning for mass fry-production is desired, place spawning stalls side by side around the perimeter of the ponds.[43] Make stalls 1 m long and enclose on three sides by wood or concrete, place gravel on the bottom, and orient the open side toward the pond center. Hatching takes place less than 5 d after spawning and the male protects the young fry. By slipping a screen over the open end, fry can be captured after hatching but before they have dispersed. Stock fry in growing ponds at densities of 40 fry/m².

Fry may be reared in spawning ponds

with adults, but reduce stocking densities for brooding. For fairly high densities of fry (40/m^2) maintain only two or three pairs of spawning adults per 4000 m^2. If large fry are desired for harvest, use a lower stock density: one breeding pair per 4000 m^2 to produce 10 fry/m^2. This low stocking density will produce fry averaging 1.5 g (2.5 to 5 cm) at the autumn harvest following a summer spawning.[40] Sort juvenile bluegills periodically according to size to facilitate grading of specimens for toxicity tests. Rearing from laboratory-fertilized eggs also can be attempted.

4) Channel catfish, *Ictalurus punctatus*—To establish breeding stock, collect adult catfish 3 or more years old, preferred size 1 to 4.5 kg. Segregate adults by sex in holding ponds until spawning is desired. Use stocking density of 100 g/m^2 when additional growth is desired or 200 g/m^2 for no additional growth.[48] Optimal temperature is between 21 and 26°C; maintain pH between 6.3 and 7.5. Feed adult catfish commercially available fish food pellets supplemented with fresh or frozen cut fish and live fathead minnows. Daily rations should equal approximately 3% of the weight of the stock.[40,43,48]

The spawning season in nature begins in the spring (April or May); increase the daily ration then to 4% of body weight. To condition adults for spawning, keep sexually ripe individuals in strong currents for 10 d to 2 weeks. Methods have been developed for spawning in ponds and pens,[40,47] but aquarium spawning is most efficient in terms of space and rates of successful spawning.[40] In this method, stop feeding at the onset of spawning. Inject females intraperitoneally with hormones: three doses of 2.2 to 22 mg acetone-dried fish pituitary material at varying intervals (from 6 h in warm water to 24 h in cold climates) or a single dose of 2200 IU/kg body weight of human chorionic gonadotropin.[40] Most injected fish will spawn within 16 to 24 h after the last injection.

Pair the catfish according to size in troughs (23 to 240 L) provided with flowing water and tarpaper mats as spawning substrate.[40,47] Fertilized eggs adhere to each other in oval glutinous masses. New eggs are golden, later turning pink as the embryos develop. After spawning, remove spawners and eggs. Use the troughs for additional spawning.

Incubate embryos either in hatching troughs[40,43] or in open-mouth hatchery jars.[41,43] Hatching troughs may be of any convenient size but at least 25 cm deep and supplied with running water.[40] Retain the egg mass in a wire-mesh basket suspended in the hatching trough and place a paddle-like agitator driven by an electric motor alongside each basket to insure mixing of the eggs.[40,43] If hatchery jars are used, place each egg mass in a separate 6- to 8-L jar. Introduce a gentle flow of water just above the mass by a rubber tube to simulate the agitation provided in nature by the fanning activities of the male.

Catfish embryos hatch in 8 to 10 d at 24°C. The fry have light-colored bodies with pink yolk sacs. Remove fry from hatching troughs or hatchery jars when the yolk sac disappears, approximately 3 to 5 d after hatching. Transfer fry by siphoning through a large-bore glass tube into rearing troughs.

Catfish fry can be reared according to the methods established for trout except that warmer temperatures (24 to 28°C)[41] are necessary. For the first 4 or 5 d in the rearing troughs, feed fry sparingly with finely ground fish food 10 times/d. Siphon off uneaten food after 2 h. Increase daily rations until they equal approximately 4 to 5% of total body weight. Use a 1:1 mixture of fish scraps and fish food pellets when the fry begin to come to the surface.[40,43]

Juvenile catfish reach fingerling size 6 to 9 months after hatching. Reduce the ration then to 3% of body weight. The stocking rate in ponds during the second year is up to 1 fish/2 m^2. During the third year, stocking densities are 1 fish/4 m^2 with a daily ration of 2% of body weight until the spawning season.

b. *Marine and estuarine fish:* Culture methods for marine and estuarine fish species are less developed than those for freshwater species. General information about marine fish culture methods can be found in several sources.[18,19,49-51] Specific techniques have been described for rearing herring, bay anchovy, scaled sardine, and Pacific sardine from fertilized eggs.[50,52-56] The sheepshead minnow and common mummichog have been reared routinely through parts of the life cycle.[57-63] Mullet, croaker, black sea bass, spot, striped bass embryos, and anchovy have been cultured to or beyond yolk-sac absorption.[64-70] Difficulty in rearing many marine fish species through certain life stages, e.g., yolk-sac absorption, may limit test duration. Use of hormones to induce egg production artificially may produce over-ripe eggs with resultant poor fertilization and emergence of larvae.[33,65] Handle gametes, developing embryos, and larvae with care; zygotes of some species are especially susceptible to mechanical damage[34] and even gentle handling can result in significant reduction in percent hatch of embryos. For any species, maintain adult brood stock, eggs, and larvae under conditions approximating those in the natural environment.

A method is described below for culture of the sheepshead minnow, *Cyprinodon variegatus,* which at present is the only estuarine species routinely cultured and used in egg-to-embryo or embryo-larval toxicity tests.

The sheepshead minnow thrives over a wide range of salinities and temperatures. Acclimate adult fish $\geq$ 27 mm standard length to laboratory conditions for at least 2 weeks at a salinity of at least 15 g/kg and a temperature of 30°C. Hold photoperiod at 12 h light, 12 h dark. During this period feed liberally on fresh or frozen adult brine shrimp. Eggs from natural spawning may be obtained by placing a pair of adult fish in a spawning chamber about 12 $\times$ 18 $\times$ 10 cm high. Place spawning trays (2 cm deep, formed by 0.5-mm nylon screen attached to a frame and covered with 2-mm nylon screen) in the spawning chambers. Larger spawning tanks with several spawning pairs may be used, provided that each male has space to establish a territory. As embryos are deposited they fall through the screen into the trays, thus preventing predation by adult fish and allowing easy removal. Each pair may spawn a maximum of 10 to 30 embryos/d but average production is about 8/pair/d.

Alternatively, sheepshead minnows may be induced to produce eggs by hormone injection.[60,61] Inject each female intraperitoneally with 50 IU human chorionic gonadotropic hormone. Repeat after 2 d. On the third day most females can be readily stripped to obtain ripe eggs. Strip or dissect eggs into filtered seawater in a beaker and add macerated testes.[63] The number of eggs produced per female by this method is 100 to 200 depending on fish size. This method has the advantage of producing eggs at specified times; however, the fish typically are sacrificed to obtain eggs and sperm, thus reducing brood stock.

Fertilized eggs may be hatched in flowing or static water systems. For a flowing water system, place embryos in a hatching chamber formed by gluing a 9-cm-high collar of 0.5-mm mesh nylon screen around a petri dish. Suspend hatching chambers in flow-through seawater aquariums with self-starting siphons. As the water level in the aquarium changes, water in the hatching baskets is exchanged gently. Alternatively, place embryos in separatory funnels and aerate gently.[71,72] Sheepshead minnow fry hatch after 5 d at 30°C and a salinity between 15 and 20 g/kg. As embryos hatch, transfer to a rearing aquarium and immediately feed newly hatched brine shrimp (live or frozen) or a dry food. Supplement a dry-food diet occasionally with live organisms. Survival from fertile eggs to 28 d is approximately 85%.[62,63] Juveniles become sexually distinguishable when about 24 mm long and females produce eggs within 3 months after hatching.

6. References

1. COMMITTEE ON METHODS FOR TOXICITY TESTS WITH AQUATIC ORGANISMS. 1975. Methods for Acute Toxicity Tests with Fish, Macroinvertebrates, and Amphibians. Ecol. Res. Ser. EPA-660/3-75-009, U.S. Environmental Protection Agency, Duluth, Minn.

2. AMERICAN SOCIETY FOR TESTING & MATERIALS COMMITTEE E-35. 1980. Standard Practices for Conducting Acute Toxicity Tests with Fishes, Macroinvertebrates and Amphibians. ASTM Des. No. E-729, American Soc. Testing & Materials, Philadelphia, Pa.

3. HUNN, J.B., R.A. SCHOETTGER & E.W. WHEALDON. 1968. Observations on the handling and maintenance of bioassay fish. Progr. Fish-Cult. 30:164.

4. LEWIS, W.M. 1962. Maintaining Fishes for Experimental and Instructional Purposes. Southern Illinois Univ. Press, Carbondale.

5. WILLFORD, W.A. 1967. Toxicity of 22 Therapeutic Compounds to Six Fishes. Invest. Fish Control 18, U.S. Bur. Sport Fish. & Wildlife, Washington, D.C.

6. RUCKER, R.R. & K. HODGEBOOM. 1953. Observations on gas-bubble disease of fish. Progr. Fish-Cult. 15:24.

7. MARKING, L.L. & V.K. DAWSON. 1973. Toxicity of Quinaldine sulfate to fish. Invest. Fish Control 48, U.S. Bur. Sport Fish. & Wildlife, Washington, D.C.

8. BRUNGS, W.A. & D.I. MOUNT. 1967. A device for continuous treatment of fish in holding chambers. Trans. Amer. Fish. Soc. 96:55.

9. SNIESZKO, S.F. 1970. A Symposium on Diseases of Fishes and Shellfishes. Spec. Publ. 5, American Fisheries Soc., Washington, D.C.

10. HOFFMAN, G.L. 1967. Parasites of North American Freshwater Fishes. Univ. California Press, Berkeley & Los Angeles.

11. VAN DUIJN, D., JR. 1973. Diseases of Fishes, 3rd ed. Charles C. Thomas Co., Springfield, Ill.

12. REICHENBACK-KLINKE, H. & E. ELKAN. 1965. The Principal Diseases of Lower Vertebrates. Academic Press, London & New York.

13. DAVIS, H.S. 1953. Culture and Diseases of Game Fishes. Univ. California Press, Berkeley & Los Angeles.

14. HOFFMAN, G.L. & F.P. MEYER. 1974. Parasites of Freshwater Fishes: A Review of Their Control and Treatment. T.F.H. Publications Inc., Ltd., Neptune City, N.J.

15. NIGRELLI, R.F. & G.D. RUGGIERI. 1966. Enzootics in the New York aquarium caused by Cryptocaryon irritans Brown, 1951 (=Ichthyophthirius marinus 1961) a histophagous ciliate in the skin, eyes and gills of marine fishes. Zoologica 51:97.

16. SINDERMANN, C.J. 1970. Principal Diseases of Marine Fish and Shellfish. Academic Press, New York, N.Y.

17. HOUDE, E.D. 1973. Some recent advances and unresolved problems in the culture of marine fish larvae. Proc. World Maricult. Soc. 3:83.

18. HOUDE, E.D. & A.J. RAMSEY. 1971. A culture system for marine fish larvae. Progr. Fish-Cult. 33:156.

19. HOFFMAN, G.L. 1974. Disinfection of contaminated water by ultraviolet irradiation with emphasis on whirling disease (Myxosoma cerebralis), and its effects on fish. Trans. Amer. Fish. Soc. 103:541.

20. NATIONAL ACADEMY OF SCIENCES. 1973. Nutrient requirements of trout, salmon and catfish. Publ. Off. Nat. Acad. Sci. Washington, D.C. 11:1.

21. STALNAKER, C.B. & R.E. GRESSWELL. 1974. Early life history and feeding of young mountain white fish. EPA-660/3-73-019, Off. Research & Development, U.S. Environmental Protection Agency. U.S. Government Printing Off., Washington, D.C.

22. CARLSON, A.R. & J.G. HALE. 1972. Successful spawning of largemouth bass Micropterus salmoides (Lacepede) under laboratory conditions. Trans. Amer. Fish. Soc. 101:539.

23. HOKANSON, K.E.F., J.H. MCCORMICK, B.R. JONES & J.H. TUCKER. 1973. Thermal requirements for maturation, spawning, and embryo survival of the brook trout Salvelinus fontinalis. J. Fish. Res. Board Can. 30:975.

24. MCCORMICK, J.H., K.E.F. HOKANSON & B.R. JONES. 1972. Effects of temperature on growth and survival of young brook trout,

Salvelinus fontinalis. J. Fish. Res. Board Can. 29:1107.

25. HOKANSON, K.E.F., J.H. MCCORMICK & B.R. JONES. 1973. Temperature requirements for embryos and larvae of the northern pike, *Esox lucius* (Linnaeus). *Trans. Amer. Fish. Soc.* 102:89.

26. SIEFERT, R.E. 1972. First food of larval yellow perch, white sucker, bluegill, emerald shiner and rainbow smelt. *Trans. Amer. Fish Soc.* 101:219.

27. PICKERING, Q.H. 1974. Chronic toxicity of nickel to the fathead minnow. *J. Water Pollut. Control Fed.* 46:760.

28. MOUNT, D.I. & C.E. STEPHAN. 1969. Chronic toxicity of copper to the fathead minnow (*Pimephales promelas,* Rafinesque) in soft water. *J. Fish. Res. Board Can.* 26:2449.

29. EATON, J.G. 1973. Chronic toxicity of a copper, cadmium and zinc mixure to the fathead minnow (*Pimephales promelas,* Rafinesque). *Water Res.* 7:1723.

30. EATON, J.G. 1974. Chronic cadmium toxicity to the bluegill *(Lepomis macrochirus* Rafinesque). *Trans. Amer. Fish. Soc.* 103:729.

31. SMITH, W.E. 1973. A cyprinodontid fish, *Jordanella floridae,* as a laboratory animal for rapid chronic bioassays. *J. Fish. Res. Board Can.* 30:329.

32. MCKIM, J.M. & D.A. BENOIT. 1974. Duration of toxicity tests for establishing "no effect" concentrations for copper with brook trout (*Salvelinus fontinalis*). *J. Fish. Res. Board Can.* 31:449.

33. STEVENS, R.E. 1966. Hormone-induced spawning of striped bass for reservoir stocking. *Progr. Fish-Cult.* 28:19.

34. BLAXTER, J.H.S. 1969. Development: Eggs and larvae. *In* Fish Physiology, 3, Reproduction and Growth, Bioluminescence, Pigments and Poisons. Academic Press, New York, N.Y.

35. HENDERSON, N.E. 1963. Influence of light and temperature on the reproductive cycle of the eastern brook trout *Salvelinus fontinalis* (Mitchell). *J. Fish. Res. Board. Can.* 20:859.

36. PYLE, E.A. 1969. The effect of constant light or constant darkness on the growth and sexual maturity of brook trout. Fish. Res. Bull. No. 31. The nutrition of trout, Cortland Hatchery Rep. No. 36:13.

37. HOOVER, E.E. & H.E. HUBBARD. 1937. Modification of the sexual cycle in trout by control of light. *Copeia* 1937:206.

38. ATZ, J.W. & G.E. PICKFORD. 1959. The use of pituitary hormones in fish culture. *Endeavor* 18:125.

39. DRUMMOND, R.A. & W.F. DAWSON. 1970. An impressive method for simulating diel patterns of lighting in the laboratory. *Trans. Amer. Fish. Soc.* 99:434.

40. BARDACH, J.E., J.R. RYTHER & W.O. MCLARNEY. 1972. Aquaculture: The Farming and Husbandry of Freshwater and Marine Organisms. Wiley Interscience, John Wiley & Sons, Inc., New York, N.Y.

41. DAVIS, H.S. 1967. Culture and Diseases of Game Fishes. Univ. California Press, Berkeley & Los Angeles.

42. LEITRITZ, E. 1959. Trout and Salmon Culture (Hatchery Methods). Fishery Bull. No. 107, California Dep. Fish & Game, Sacramento.

43. HUET, M. 1970. Textbook of Fish Culture: Breeding and Cultivation of Fish. Thanet Press, Margate, England.

44. MOUNT, D.I. 1968. Chronic toxicity of copper to fathead minnows, *Pimephales promelas* Rafinesque. *Water Res.* 2:215.

45. HASKELL, D.C. 1959. Trout growth in hatcheries. *N.Y. Fish Game J.* 6:204.

46. BREDER, C.M. & D.E. ROSEN. 1960. Modes of Reproduction in Fishes. Natural History Press, Garden City, N.Y.

47. BROWN, E.E. 1977. World Fish Farming: Cultivation and Economics. Avi Publishing Company, Inc., Westport, Conn.

48. GRIZZELL, R.A., E.G. SULLIVAN & D.W. DILLON. 1968. Catfish farming—an agricultural enterprise. Rep. 4-26433, U.S. Dep. Agriculture.

49. SHELBOURNE, J.E. 1964. The artificial propagation of marine fish. *Adv. Mar. Biol.* 2:1.

50. MAY, R.C. 1970. Feeding larval marine fishes in the laboratory: A review. Calif. Mar. Res. Comm., CalCOFI Rep. 14:76.

51. MAY, R.C. 1971. An Annotated Bibliography of Attempts to Rear the Larvae of Marine Fishes in the Laboratory. NOAA (Nat. Ocean. Atmos. Admin.) Tech. Rep. NMFS-SSRF 632.

52. SAKSENA, V.P. & E.D. HOUDE. 1972. Effect of food level on the growth and survival of laboratory reared larvae of the bay anchovy, *Anchoa mitchilli,* Valenciennes and scaled sar-

dine, *Harengula pensacolae*, Goode and Bean. *J. Exp. Mar. Biol. Ecol.* 8:249.

53. BLAXTER, J.H.S. 1968. Rearing herring larvae to metamorphosis and beyond. *J. Mar. Biol. Assoc. U.K.* 48:17.

54. HOUDE, E.D. & B.J. PALKO. 1970. Laboratory rearing of the clupeid fish, *Harengula pensacolae*, from fertilized eggs. *Mar. Biol.* 5:354.

55. SAKSENA, V.P., C. STEINMETZ & E.D. HOUDE. 1972. Effect of temperature on growth and survival of laboratory reared larvae of the scaled sardine *Harengula pensacolae*, Goode and Bean. *Trans. Amer. Fish. Soc.* 101:691.

56. LASKER, R. 1964. An experimental study of the effect of temperature on the incubation time, development and growth of Pacific sardine embryos and larvae. *Copeia* 1964:399.

57. BOYD, J.F. & R.C. SIMMONS. 1974. Continuous laboratory production of fertile *Fundulus heteroclitus*, Walbaum eggs lacking chorionic fibrils. *J. Fish. Biol.* 6:389.

58. MIDDAUGH, D.P. & J.M. DEAN. 1977. Comparative sensitivity of eggs, larvae and adults of the estuarine teleosts, *Fundulus heteroclitus* and *Menidia menidia* to cadmium. *Bull. Environ. Contam. Toxicol.* 17:645.

59. MIDDAUGH, D.P. & P.W. LEMPESIS. 1976. Laboratory spawning and rearing of a marine fish, the silverside *Menidia menidia. Mar. Biol.* 35:295.

60. HANSEN, D.J. & P.R. PARRISH. 1975. Suitability of sheepshead minnows (*Cyprinodon variegatus*) for life cycle toxicity tests. *In* F.L. Mayer & J.L. Hamelink, Aquatic Toxicity and Hazard Identification. ASTM STP 634, American Soc. Testing & Materials, Philadelphia, Pa.

61. BAYLIFF, W.H. 1950. The life history of the silverside *(Menidia menidia* Linnaeus). *Chesapeake Biol. Lab. Publ.* 50:1.

62. SCHIMMEL, S.C., D.J. HANSEN & J. FORESTER. 1974. Effects of Aroclor 1254 on laboratory-reared embryos and fry of sheepshead minnows (*Cyprinodon variegatus*). *Trans. Amer. Fish. Soc.* 103:582.

63. HANSEN, D.J., S.C. SCHIMMEL & J. FORESTER. 1973. Aroclor 1254 in eggs of sheeps-

head minnows: Effects on fertilization success and survival of embryos and fry. *Proc. 27th Annu. Conf. S.E. Assoc. Game Fish Comm.*: 420.

64. KUO, C., Z. H. SHEHADEH & K.K. MILISEN. 1973. A preliminary report on the development, growth and survival of laboratory reared larvae of the grey mullet, *Mugil cephalus* L. *J. Fish. Biol.* 5:459.

65. MIDDAUGH, D.P. & R.L. YOAKUM. 1974. The use of chorionic gonadotropin to induce laboratory spawning of the Atlantic Croaker, *Micropogon undulatus*, with notes on subsequent embryonic development. *Cheasapeake Sci.* 15:110.

66. HOFF, F.H. 1972. Artificial spawning of black seabass, *Centropristis striata*, aided by chorionic gonadotropin hormones. Florida Dep. Natural Resources Marine Research Lab. (mimeograph).

67. DAWSON, C.E. 1959. A study of the biology and life history of the spot *Leiostomus xanthurus* Lacépède with special reference to South Carolina. *Bears Bluff Lab. Contrib.* 28:1.

68. RANEY, E.C. 1952. The life history of the striped bass, *Roccus saxatilis*, Walbaum. *Bull. Bingham Oceanogr. Coll.* 14:5.

69. KRAMER, D. & J.R. ZWEIEL. 1970. Growth of anchovy larvae, *Engraulis mordax*, Girard in the laboratory as influenced by temperature. *Rep. Calif. Coop. Oceanic Fish. Invest.* 14:84.

70. O'CONNELL, C.P. & L.P. RAYMOND. 1970. The effect of food density on survival and growth of early post yolk sac larvae of the northern anchovy, *Engraulis mordax*, Girard, in the laboratory. *J. Exp. Mar. Biol. Ecol.* 5:187.

71. HANSEN, D.J. 1978. Laboratory culture of sheepshead minnows (*Cyprinodon variegatus*). *In* Bioassay Procedures for the Ocean Disposal Permit Program. EPA-600/9-78-010, U.S. Environmental Protection Agency.

72. HANSEN, D.J., P.R. PARRISH, S.C. SCHIMMEL & L.R. GOODMAN. 1978. Life-cycle toxicity test using sheepshead minnows (*Cyprinodon variegatus*). *In* Bioassay Procedures for the Ocean Disposal Permit Program. EPA-600/9-78-010, U.S. Environmental Protection Agency.

8910 C.   Test Procedures

1. Short-Term Tests

a. General test procedures: Short-term testing can be used to determine relative toxicity of substances, for definitive testing of LC50 or EC50, and to estimate toxicant concentrations for intermediate- and long-term tests.

Short-term tests may be static, static with renewal, flow-through, or recirculating, depending on the objective of the test, the life stage being tested, and the character of the toxicant or effluent (see 8010D.1 and 2).

Although any life stage may be used, short-term tests are performed most commonly with small species (less than 5 g body weight) or juvenile forms of large fish species (see Section 8910B.1). The life stage selected depends on test purpose, availability, and laboratory facilities (see Section 8910B.1). If culturing is necessary to obtain a particular life stage, see Section 8910B.5.

Typically expose 20 fish to each toxicant concentration with 10 fish/tank. Although larger sample sizes are desirable, do not use a larger sample if it compromises some other consideration of good testing. Select fish of uniform size, with the longest no more than 1.5 times the length of the shortest. For juvenile and adult fish, terminate feeding 48 h before initiating tests. For all tests, limit fish weight/L test solution. This practice minimizes oxygen depletion, metabolic waste accumulation, and crowding-induced stress. In flow-through tests, use a maximum test chamber loading of 5 g/L at 20°C or less or 2.5 g/L above 20°C. For static testing, do not load above 0.8 g/L at 20°C or less and 0.4 g/L above 20°C.

b. Specific test procedures:

1) Freshwater fish

a) Equipment and physical conditions—Use test equipment made of glass, No. 316 stainless steel, or perfluorocarbon plastics.* Use unplasticized plastics such as polyethylene, polypropylene, and polyvinylchloride† in the water delivery system. In static and flow-through systems with low rates of exchange, avoid certain types of TFE stoppers; pretest stoppers to insure absence of toxicity.

Select temperature appropriate for the species being tested, do not vary it by more than ± 2°C during a 96-h test, and do not vary it by more than ±1°C during any 48 h. Simulate the photoperiod at the test site or use the photoperiod at Evansville, Ind., which is especially suitable for brook trout (Table 8910:II). This table is arranged so that photoperiod adjustments are made only at dusk. The listed dawn and dusk times need not correspond to the actual time during the test being conducted. For example, a test started on March 1 would require the photoperiod for the Evansville test date of March 1. The lights may be turned on at any time on that day, so long as they remain on for 12 h and 15 min. Fifteen days later, the photoperiod would be changed to 13 h. Gradual changes in light intensity at dawn and dusk may be simulated within the photoperiod shown, but they should not last more than 0.5 h from "full on" to "full off" and vice versa.[1]

Supply dilution water from an unpolluted surface source, well, or spring. If the source potentially is contaminated with pathogens, disinfect with UV irradiation before using. Analyze water in control and assay chambers daily for pH, alkalinity, dissolved oxygen, and temperature. Maintain temperature at ±2°C of the acclimation system temperature and the DO concentration ≥ 5.0 mg/L (≥ 60% saturation). Avoid aerating because aeration may alter the results of toxicity tests.

The length of a short-term test varies, generally from 24 to 96 h. Longer test periods may be used, depending upon the characteristics and variability of the waste. Dilution factors in the receiving stream

---

*Teflon® or equivalent.

†Tygon® or equivalent.

provide guidance in determining effluent concentrations for range-finding tests.

For information pertaining to species selection, collection, holding, acclimation, disease control, and culturing see Sections 8010E and 8910B. Calculating, analyzing, and reporting procedures are found in Section 8010G.

b) Test procedure

Range-finding test: If the approximate toxicity of test material is unknown, conduct an abbreviated range-finding test to determine the concentrations that should be used in the definitive tests. Do this in static tests using five fish per chamber exposed to three to five widely spaced toxicant concentrations plus a control. This test usually can be completed within 24 h.

Definitive test: To determine LC50 or EC50 use a 96-h test with a minimum of five toxicant concentrations arranged exponentially, and a control according to the results of the range-finding test. If more than 10% of the fish in a control system die, repeat the test.

2) Marine fish—Conduct toxicity tests requiring saline dilution water with relatively contaminant-free natural seawater of known salinity. If this is not possible, use artificial seawater (see Section 8010E.4b) or commercially available seawater.‡ Because flow-through tests require large volumes of dilution water, a substantial expense may be involved if a commercial product is used. By contrast, the cost of providing commercially formulated saline dilution water for static testing is minimal.

The test procedure for marine species is essentially identical to that for freshwater fish. Additionally, determine salinity daily.

If effluent testing is to be done, preferably use euryhaline species such as *Fundulus heteroclitus* (mummichog), *Cyprinodon variegatus* (sheepshead minnow), and *Menidia menidia* (Atlantic silverside) rather than stenohaline fish. Effluent toxicity testing often requires exposing fish to 50 to 100% effluent. This exposure may create a salinity-induced stress on stenohaline organisms that could invalidate test results unless salinity is controlled by adding appropriate amounts of sea salt.

2. Partial-Life-Cycle Tests

a. General test procedures: Partial-life-cycle tests may be initiated with any life stage: newly spawned eggs, newly hatched larvae, juveniles, or sexually immature adults. The life stage selected depends on the species, relative sensitivity of various life stages, laboratory space and facilities, availability of different stages, and test purpose.

Procedures for conducting these tests are the same as those described below (Section 8910C.3) for complete-life-cycle tests and differ only in test duration, which depends on species, life stage used, laboratory facilities, and test purpose. For many fish, 1 to 4 weeks (30 d) is typical, but for long-lived species, 6 months might be appropriate.

b. Specific test procedures: Follow the procedures outlined in 8910C.3b for culture procedures. Species other than those listed in Section 8910C.3b may be used if culture technology is adequate.

3. Complete-Life-Cycle Tests

a. General test procedures: Use newly spawned eggs, newly hatched larvae, juveniles, or sexually immature fish to initiate a test. The life stage selected depends on species, laboratory space and facilities, availability of the life stages, and test purpose.

Expose enough fish to each concentration of toxicant to insure adequate numbers of each sex at maturity but low enough to prevent mortality due to crowding. Additional fish may be introduced to tanks with similar treatment to provide specimens for histological examination, residue analysis,

---

‡Rila marine mix, Instant Ocean,® or equivalent.

or selected physiological measures of condition. The exact number of fish required depends on life stage at start of test and test species (see Section 8910C.3*b*).

At start of test, measure total length and weight of all fish. Repeat measurement for any fish that die during the test. To prevent injury, anesthetize§ large fish before handling. Larvae and small fish may be measured by a photographic method.[2] At the end of a test, record sex, gonadal condition, length, and weight of each fish.

For viability and hatchability tests, incubate eggs from each spawning at an optimum temperature. If effluent tests are being made, set the temperature to that of receiving water. Count live and dead eggs and remove dead eggs daily. Evaluate egg viability for all spawnings by incubating eggs until development clearly is observed (some defined stage of embryogenesis is reached). Determine hatchability for all spawnings in all exposure chambers or from a predetermined number of spawnings when the species tested is one that spawns continuously (many times per female). Count number of dead, deformed, and normal larvae hatched daily, using a dissecting microscope if necessary.

To evaluate larval growth and survival at each toxicant concentration, collect a uniform number of normal larvae (usually 20 to 50) at random from two or more successful hatches and place in chambers for that toxicant concentration. If there is a prolonged hatching period, use the median hatching date as the test starting date. Determine length and number of larvae upon transfer to growth chambers, preferably by the photographic method. Determine total length of larvae at selected intervals and at end of test. Count, measure, and remove dead larvae daily.

For methods of toxicant mixing and delivery see Section 8010F.1*c*. Install an automatically triggered emergency alarm system to indicate failure of diluter, temperature controller, or water supply.

Locate spawning tanks, exposure tanks, and growth chambers appropriate for the test species. Design each growth chamber so that test solutions can be drained down to 2.5 to 3 cm and the chamber transferred to a fluorescent light box provided with a millimeter grid for photographing fish.[2]

Construct incubation cups or chambers from 8-cm sections of 5-cm-OD glass tubing by cementing nylon screen to retain embryos and larvae over one end of tube. Oscillate incubation cups in test water by a rocker arm driven by a 2-rpm electric motor[3] or provide necessary aeration and movement in some other suitable way.

Monitor fish and embryos maintained for physiological, biochemical, and histological tests carefully. As a minimum, report all pertinent data for each test container at the beginning, about a third of the way through, and at end of test. Include number and weight of individuals, number of spawnings, number of embryos, and total lengths of normal, deformed, and injured mature and immature males and females. Count and record all survivors and mortalities. Calculate mean incubation time for median spawning and hatch dates if known. The hatchability, fry survival, growth, and percent deformities also may be determined.

If possible, measure toxicant concentration in all tanks at each concentration weekly. Use composites of equal-volume daily grab samples for 1 week if it has been shown that analytical results for the test compound are unaffected by storage. Analyze enough samples throughout the test to determine whether toxicant concentration is constant. If this is not possible, analyze enough samples weekly to establish variability of toxicant concentration (see Section 8010F.3*d*).

Record temperature continuously. Measure oxygen levels daily in each tank. Analyze water from the control and one

---

§MS-222 or equivalent.

exposure tank at least weekly for pH, hardness, alkalinity, and conductivity in freshwater systems and pH and salinity in marine systems. If any characteristic is affected by the toxicant, analyze that characteristic at least 5 d/week, rotating among tanks so that each is analyzed once every other week.

When possible, analyze mature fish and/or eggs, larvae, and juveniles for toxicant residues. Analyze the whole organism or muscles on each dead fish and a composite of all fish surviving the entire test period at each exposure concentration.

b. Specific test procedures: Procedures used for life-cycle tests are described below for two freshwater and one marine species. Other species are used more rarely.

1) Brook trout, *Salvelinus fontinalis* (freshwater)—This test procedure is not a complete life-cycle test, but rather it follows the life cycle from the yearling stage through spawning, egg hatching, and development for 90 d because of the longevity of brook trout.

a) Equipment and physical conditions—For description of suitable diluter systems see Section 8010F. Set up duplicate tanks for each test concentration and control. Premix each concentration before delivery to duplicate spawning tanks and growth chambers.

Construct alevin-to-juvenile growth chambers with dimensions of 18 × 15 × 18 cm of glass or stainless steel with a glass bottom. Maintain water depth at about 13 cm. Design each chamber so that the water can be drained down to a depth of 2 to 3 cm to allow the chamber to be placed over a millimeter grid on a fluorescent light box for photographing fish.

Construct spawning tanks of No. 316 stainless steel, with dimensions of 80 × 30 × 40 cm. Use a 30-cm water depth. Place a spawning substrate or nest[4] in spawning tanks at the appropriate time. Use spawning nest, 28 × 33 × 7.5 cm, made of double-strength glass or stainless steel.

Large fish may require a larger nest. Drill three 2.5-cm holes in each end, 2.5 cm from the bottom, and cover with 10-mesh stainless steel wire to allow water in the box to drain to a depth of 2.5 cm when the box is removed from the spawning chamber. Place a bottomless screen egg-retainer (27 × 32 × 1.3 cm with 2.5-cm square compartments, constructed from 1.3-cm-wide strips of 7-mesh stainless steel screen) in the spawning box. Place 2-mesh stainless steel screen, 27 × 32 cm, to which 1.3- to 2.5-cm gravel is attached with silicone adhesive, on top of the screen egg retainer. Use smooth gravel to prevent injury to active, spawning fish. This spawning box is readily removed from the spawning tanks to collect eggs for transfer to incubation cups. For spawning stocks, select yearling fish that will not grow too large in the confinement of the spawning box. Fish weighing not more than 50 to 70 g at time of selection and 150 g at spawning are appropriate. If fish weigh more than 150 g, use a larger spawning box.

Simulate the photoperiod and dawn-to-dusk times at Evansville, Ind., or at local areas and provide conditions described above in Section 8910C.1*b*.

Provide water flow of 6 to 10 tank volumes/d to maintain oxygen levels above 60% saturation. Remove unused food and wastes from growth chambers daily. Brush interior surfaces to remove attached growths at least weekly.

b) Exposure procedures—To begin the test, collect juveniles from the field no later than March 1 and acclimate for at least 1 month or use cultured stock of equivalent age. Judge suitability of fish for testing on the basis of acceptance of food, apparent lack of disease, and occurrence of less than 2% mortality during acclimation and no mortality during the 2 weeks before the test.

Begin exposure by placing at least 12 acclimated yearling brook trout in each duplicate tank at each test concentration and suitable controls using a stratified random assignment (see 8010F.3a). This allows about a 4-month exposure to toxicant before the onset of secondary or rapid-growth phase of the gonads. Extra test animals may be included at the beginning so that fish can be removed periodically for special examination or for chemical analysis.

Use a particulate or pelleted trout food. Feed fish the largest particle or pellet they will take, at least twice daily. Base amount on a reliable hatchery feeding schedule. Analyze each batch of food for pesticides.

Record mortalities daily and measure total length and weight of fish directly at initiation of tests and every 3 months thereafter. Do not feed fish for 24 h before weighing. Lightly anaesthetize them to facilitate measuring.

When secondary sexual characteristics are well-developed (approximately 2 weeks before spawning), separate males, females, and undeveloped fish in each tank and randomly reduce number of sexually mature fish to two males and four females per tank. Record number of mature, immature, deformed, and injured males and females in each tank and number from each category to be discarded. Thoroughly clean, sterilize, and rinse the spawning substrates and place one in each spawning tank. As soon as spawning begins, set up incubation cups (as described in 8910B.5a) or a suitable alternate system to receive embryos for hatching. Remove embryos from the substrate at a fixed time each day (preferably after 1300 h, Evansville time, so fish are not disturbed during the morning).

Randomly select 50 embryos from the first eight spawnings of 50 embryos or more in each duplicate spawning chamber and place in an egg incubator cup. Count remaining embryos from the first eight spawnings and all embryos from subsequent spawnings and place them in separate incubator cups to determine viability as evidenced by development to a specific stage, e.g., formation of neural keel after

11 to 12 d at 9°C. Remove and record number of dead embryos from each spawn. Never place more than 250 embryos in one incubator cup. Incubate all embryos to determine viability and discard after 12 d. Discarded embryos may be analyzed chemically or used for other measurements.

Obtain additional information on hatchability and alevin survival by transferring embryos from control tanks immediately after spawning (a) to tanks having test concentrations where spawning is reduced or absent and (b) to tanks where an effect is seen on survival of embryos or alevins, and by transferring embryos from those test concentrations to control tanks. Always reserve two growth chambers in each duplicate spawning tank for embryos produced in that tank.

Remove dead embryos daily from incubator cups. When hatching begins, record number of alevins hatching daily in each cup. On completion of hatching in any cup, transfer fish to a culture dish and randomly sample 25 alevins. Count dead or deformed alevins. Transfer 25 selected alevins to a growth chamber and place it over the light box to measure by the photographic method. After photographing, return alevins to incubator cup. Never net alevins, but transfer by gentle pouring or by large-bore pipets. Transport in growth chambers containing a 2.5-cm depth of test solution. Preserve unused alevins in formalin for subsequent histological examination. Record length and weight of discarded alevins separately from the data for fish kept for subsequent exposure.

For 90-d growth and survival exposures randomly select 25 alevins from each duplicate incubator cup for each test concentration and control. Because embryos from one spawn may hatch over a 3- to 6-d period, use the median hatch date to establish the start time of the 90-d growth and survival period. For growth tests select two groups of 25 alevins that are less than 3 weeks apart in age. Use any remaining groups only for hatchability testing. After photographing to determine length, preserve for weight determination. To equalize effects of incubator cups on growth, keep all groups selected for 90-d exposure in the incubator cups for 3 weeks after the median hatch date, then release into growth chambers. Begin feeding immediately. Keep separate the two groups selected from the duplicate exposure chambers for each test concentration. Record mortalities daily, total lengths at 30 to 60 d after hatching (by the photographic method), and total length and weight at 90 d after hatching. At the end of the test cease feeding juveniles for 24 h and then weigh. Terminate survival and growth studies after 3 months, at which time fish may be used for chemical analysis of tissue and physiological measurements of toxicant-related effects.

End exposure of all parental fish after 3 weeks in which no spawning occurs in any tank. Record mortality and weight, measure total length of parental fish, and check sex and condition of gonads (e.g., reabsorption, degree of maturation, spent ovaries).

Report, for each tank of a partial-life-cycle test, number and individual weights and total lengths of immature males and females at initiation of test, after 3 months, at reduction in numbers, and at end of test. Report individual weights and total lengths of normal, deformed, and injured fish, number maturing, number dying during test, number of spawnings and eggs, hatchability and fry survival, growth, and deformities. Calculate a mean incubation time based on date of spawning and median hatch dates.

For additional information on the life cycle of brook trout consult other sources.[4-13]

2) *Fathead minnow, Pimephales promelas* (freshwater)

a) *Equipment and physical system*—See Sections 8010F and 8910C.3. The physical systems are similar to those for brook trout [8910C.3*b*1)a)].

Use one of two arrangements of test tanks (glass or stainless steel with glass ends): duplicate spawning tanks for each of the five or more test concentrations and controls, measuring 30 × 30 × 90 cm with a 30-cm-square portion at one end, screened off and divided in half to form two larval chambers for the progeny; deliver test water separately to larval and spawning chambers of each tank, with about one-third of the water volume going to each larval chamber. Alternatively, use duplicate spawning tanks measuring 30 × 30 × 60 cm plus duplicate progeny tanks for each spawning tank. Use a larval tank with minimum dimensions of 30 × 30 × 30 cm, divided to form two separate larval chambers with separate standpipes, or separate 30 × 15 × 30-cm tanks. Supply test solutions and water for controls as in Section 8010F.1. Maintain a water depth of 15 cm in all tanks.

Flow rate, oxygen requirements, aeration, cleaning, and operation are as described for the brook trout.

Fathead minnows deposit eggs on the underside of submerged objects. For spawning substrates use inverted halves of tile 7.5 cm ID and 7 to 10 cm long, or equivalent. Place tiles parallel to the long axis of the spawning tank so that each end is readily accessible to the fish.

Fasten incubation cups, such as those described for the brook trout, to a rocker arm with a vertical travel distance of 3 to 5 cm.[3] Lamps and illumination are described in Section 8010F.3*f*.

Photoperiods simulate dawn-to-dusk times at Evansville, Ind. (Table 8910:II). Make adjustments in day length on the 1st and 15th day of every month. Regardless of actual starting date, adjust the photoperiod so that the mean or estimated hatching date of the fish used to start the experiment corresponds to the Evansville day length for December 1. This gives a consistent prespawning exposure. The dawn and dusk times listed in Table 8910:II need not correspond to the actual times at the testing location.

Maintain temperature at 25°C ± 2°C and record continuously.

b) Exposure procedures—To start tests, use a mixture of approximately equal numbers of embryos or larvae from at least three different females. Set aside enough embryos or larvae at the start of the test to supply an adequate number of fish for acute mortality tests used in determining application factors. To determine concentrations to be used in life-cycle tests measure acute mortality in advance with juvenile or young adult fish.

Begin life-cycle test by distributing 50 embryos or 1- to 5-d-old larvae in each duplicate spawning tank for each test concentration by a stratified random assignment. If 1- to 5-d-old larvae are not available, use fish up to 30 d old. Extra fish may be added at the beginning so that some can be removed periodically for special examinations.

During the test, feed fish once or twice a day with live brine shrimp nauplii for 30 to 60 d and thereafter frozen adult brine shrimp supplemented by pelleted trout food, *Daphnia*, and chopped earthworms. Feed larvae once daily with live young zooplankton from mixed cultures of small copepods, rotifers, and protozoans. Check each batch of commercial feed for pesticides.

When test fish are 60 ± 2 d old, discard injured or crippled individuals and randomly reduce the number in each tank to 15. Record number, length, and weight of discarded and deformed fish. If necessary to obtain 15 fish per tank, select one or two fish for transfer from one duplicate to the other. Continue routine feeding and cleaning until fish mature and are almost ready to spawn. Place five spawning tiles in each duplicate spawning tank, separated fairly widely to reduce fighting between the guarding male fish. Place tiles so undersides and guard males can be seen from the tank end. During spawning, reduce number of sexually maturing males to no more than four per tank. Reserve the fifth tile as cover for females. Do not remove males having well-established territories under tiles where a recent spawn has occurred.

Daily, remove embryos from spawning tiles starting at 1200 h Evansville time (Table 8910:II). Loosen embryos from spawning tiles and at the same time separate them from one another by lightly placing a finger on the egg mass and moving it in a circular pattern with increasing pressure until the embryos begin to roll. Wash groups of embryos into separate containers and return tiles to spawning aquariums. Count embryos, select those needed for incubation, and discard remainder after counting. Check all embryos for different stages of development. If more than one distinct developmental stage is present, consider each stage as one spawning and handle separately as described below.

Randomly select 50 unbroken embryos from a single spawn and place in an incubator cup to determine viability and hatchability. Count and discard remaining embryos. Determine viability and hatchability on each spawn of more than 49 embryos until the number of spawns (> 49 embryos) in each tank equals the number of females in that tank. Subsequently, test for hatchability only embryos from every third spawning of more than 49 embryos. Remove weekend spawns from tiles and count embryos but discard.

If no spawning occurs for a week, cease parental fish testing. Record total length and weight, sex, and gonadal condition of parental fish. Do not freeze fish before determining sex and gonadal conditions.

Each day, record live and dead embryos in incubator cups, remove dead embryos, and clean cup screens. The total number of embryos accounted for always should add up to within two of 50; if not, discard the entire batch. After 4 to 6 d, when larvae begin to hatch, cease handling or removing them from cups until all have hatched. At that time, if enough larvae are still alive, select 40 at random and transfer immediately to a larval growth chamber to determine survival and growth of second generation. Count and discard entire cup groups not used for survival and growth studies.

Select larvae for 30 and 60 d growth and survival exposures from early spawned embryos in each duplicate tank. Plan their distribution for hatchability tests so that a new group of larvae is ready to be tested as soon as possible after the previously tested group is removed from the larval chambers. Record mortality and larval length at 30 and 60 d after hatching. Weigh them when the larval test is ended. Do not feed fish (larvae, juveniles, or adults) for 24 h before weighing.

Transfer 50 of the 60-d post-hatch fish from each growth chamber to the corresponding spawning chamber and adjust the photoperiod to December 1. Follow procedures used for the F_1 generation to determine survival of embryos, larvae, and juveniles of the F_2 generation. Cease testing adult fish on completion of spawning. Continue post-hatch study to 60 d.

Fish and embryos obtained from the test also may be used for physiological, biochemical, histological, and other tests for toxicant-produced effects.

Record the following data for each test tank and the controls: total number and length of normal and deformed individuals at the end of 30 and 60 d for each generation; total length, weight, and number of each sex, both normal and deformed, at the end of the tests; mortality during tests; number of spawns and embryos produced in each and total embryo production by each generation; percentage of embryos hatching; number and percentage of larvae surviving and growth of fry as well as deformities produced. For additional infor-

mation on the life cycle of the fathead minnow and flow-through life-cycle tests, consult other sources.[14-23]

3) Life-cycle tests with other freshwater species—Partial-life-cycle toxicity tests have been performed with the bluegill, *Lepomis macrochirus*, and the flagfish, *Jordanella floridae*.[24-30] When culture techniques are available, other native freshwater species may be used as appropriate.

4) Sheepshead minnow, *Cyprinodon variegatus* (marine)

a) Equipment and physical system—Use test apparatus similar to that used for freshwater fish with spawning aquariums at least 30 × 18 × 20 cm. Place a spawning chamber in each aquarium as described in 8910B.5b.

b) Exposure procedures—Begin with adult fish or preferably with embryos. Secure embryos either by natural spawning or by hormone-induced spawning (as described in 8910B.5b). Keep water temperature above 22°C, preferably at 30°C, with salinities above 15 g/kg. When starting with adults, set up five or six spawning aquariums for each test concentration and controls. Use breeding fish, all from the same stock, that have been kept in holding tanks for at least 2 weeks, during which less than 2% mortality occurred. Feed fish a combination of frozen adult brine shrimp and dry trout food. Maintain water flow through spawning aquariums at 6 to 10 tank volumes/d. Use natural seawater filtered to remove planktonic larvae 15 μm and larger.

When embryos are produced remove them from spawning chambers and place in each of eight hatching chambers for each concentration being tested and the controls. Start toxicant dosing in exposure chambers before the hatching chambers are placed inside them. Construct hatching chambers by cementing a 9-cm-wide strip of 500-μm nylon screen around a petri dish. Place the hatching chambers in 90- × 30- × 30-cm exposure chambers in 7 cm of water with flow-through of the toxicant. As embryos hatch, feed fry with newly hatched brine shrimp nauplii. Clean screens on incubation cups and chambers daily. Check and record daily survival of embryos and fry, which constitute the first filial (F_1) generation.

On the first day after hatching remove each chamber and count and measure fry photographically. During the first 2 weeks feed with newly hatched brine shrimp nauplii. During the following 2 weeks supplement this diet with dry trout pellets or dry mollie flakes. After 4 weeks count and measure fish by the photographic method and reduce the number to 50 for each test concentration and controls. Record length, weight, condition, and number of living, deformed, and dead fish remaining. Determine total number dying in each test concentration and controls. Preserve specimens for future tests or discard. Place the 50 selected fish, 25 each, in growth chambers having a glass bottom and provisions for drawing the water level down to 1 to 2 cm. Feed a mixed diet of brine shrimp and dry trout food twice daily and examine daily for dead specimens. At 8 weeks measure again by the photographic method. Twice daily, feed dry food supplemented with frozen adult brine shrimp until maturity. Check each batch of food for pesticides, PCBs, and other toxicants. Clean all exposure aquariums and spawning and hatching chambers two to three times per week. Siphon out all wastes.

As fish approach sexual maturity, place separate pairs in spawning chambers, five from each duplicate exposure chamber; i.e., 10 pair for each test concentration and controls, and continue exposure. Count, measure, and weigh all unused fish from each duplicate exposure chamber. Record number deformed and dying in each test concentration and controls, condition of fish, and other pertinent data. Preserve some fish for whole-body analysis. As fertilized eggs are produced, remove at a specified time daily, count, and place 25 F_2's in a hatching chamber as for the F_1 generation. Record the total number of embryos produced in each chamber, time required to hatch, hatching success, and survival of embryos. Test those not placed in hatching chambers for fertility and record percent fertile.

If no spawning occurs in high toxicant concentrations, transfer embryos from controls and incubate in the high concentrations. Further, if spawning does occur at the high test concentrations, place embryos from these concentrations in control aquariums.

Keep pairs in each spawning chamber until all needed embryos have been obtained. At termination, measure and weigh spawning pairs and record all other pertinent data. Preserve for toxicant analyses.

Expose embryos of F_2 generation in hatching chambers in their respective duplicate exposure chambers for each test concentration and controls as before. Count and measure by photographic method as for the F_1 generation. Feed fish and record results. At the end of 4 weeks terminate the test. Weigh and measure all fish; record number of deformed fish and determine number that died. Preserve for histological examination and tissue analyses. Determine effects of each test concentration and calculate safe levels. Analyze, handle, and report data as in Section 8010G. During tests, periodically record temperature, oxygen concentration, pH, and salinity. If possible, chemically analyze test water for toxicant at the beginning, at various times during exposure, and on completion of tests. Analyze lots of 10 fish from highest and lowest exposure concentrations and from controls for toxicant accumulation. Analyze dilution water for toxicant at beginning and end of test.

For additional information about the life cycle of *Cyprinodon variegatus*, and testing procedures, consult other sources.[31-34]

5) *Life-cycle tests with other marine*

fishes—Life-cycle tests may be performed with other marine fishes such as *Fundulus heteroclitus* and *Menidia menidia*, although standard procedures presently are not available. For information on life cycle and culture of these species, consult other sources.[31,35-38]

4. References

1. DRUMMOND, R.A. & W.F. DAWSON. 1970. An impressive method for simulating diel patterns of lighting in the laboratory. *Trans. Amer. Fish. Soc.* 99:434.
2. MARTIN, 1967. A method of measuring lengths of juvenile salmon from photographs. *Progr. Fish-Cult.* 29:238.
3. MOUNT, D.I. 1968. Chronic toxicity of copper to fathead minnows, *Pimephales promelas* Rafinesque. *Water Res.* 2:215.
4. BENOIT, D.A. 1974. Artificial laboratory spawning substrate for brook trout (*Salvelinus fontinalis* Mitchell). *Trans. Amer. Fish. Soc.* 103:144.
5. HOOVER, E.E. & H.E. HUBBARD. 1937. Modification of the sexual cycle in trout by control of light. *Copeia* 1937:206.
6. ATZ, J.W. & G.E. PICKFORD. 1959. The use of pituitary hormones in fish culture. *Endeavor* 18:125.
7. MCKIM, J.M. & D.A. BENOIT. 1971. Effect of long-term exposures to copper on survival, reproduction and growth of brook trout, *Salvelinus fontinalis* (Mitchell). *J. Fish. Res. Board Can.* 28:655.
8. ALLISON, L.N. 1951. Delay of spawning in eastern brook trout by means of artificially prolonged light intervals. *Progr. Fish-Cult.* 13:111.
9. CARSON, B.W. 1955. Four years progress in the use of artificially controlled light to induce early spawning of brook trout. *Progr. Fish-Cult.* 17:99.
10. HALE, J.G. 1968. Observations on brook trout, *Salvelinus fontinalis* spawning in 10-gallon aquaria. *Trans. Amer. Fish. Soc.* 97:299.
11. HENDERSON, N.E. 1962. The annual cycle in the testes of the eastern brook trout. *Salvelinus fontinalis* (Mitchell). *Can. J. Zool.* 40:631.
12. HENDERSON, N.E. 1963. Influence of light and temperature on the reproductive cycle of the eastern brook trout *Salvelinus fontinalis* (Mitchell). *J. Fish Res. Board Can.* 20:859.
13. WYDOSKI, R.S. & E.L. COOPER. 1966. Maturation and fecundity of brook trout from infertile streams. *J. Fish. Res. Board Can.* 23:623.
14. PICKERING, Q.H. 1974. Chronic toxicity of nickel to the fathead minnow. *J. Water Pollut. Control Fed.* 46:760.
15. MOUNT, D.I. & C.E. STEPHAN. 1969. Chronic toxicity of copper to the fathead minnow (*Pimephales promelas*, Rafinesque) in soft water. *J. Fish. Res. Board Can.* 26:2449.
16. EATON, J.G. 1973. Chronic toxicity of a copper, cadmium and zinc mixure to the fathead minnow (*Pimephales promelas*, Rafinesque). *Water Res.* 7:1723.
17. BRUNGS, W.A. 1969. Chronic toxicity of zinc in the fathead minnow, *Pimephales promelas* Rafinesque. *Trans. Amer. Fish. Soc.* 98:272.
18. BRUNGS, W.A. 1971. Chronic effects of low dissolved oxygen concentrations on the fathead minnow (*Pimephales promelas*, Rafinesque). *J. Fish. Res. Board Can.* 28:1119.
19. CARLSON, D.R. 1967. Fathead minnow, *Pimephales promelas*, Rafinesque, in the Des Moines River, Boone County, Iowa and the Skunk River drainage, Hamilton and Story Counties, Iowa. *Iowa State J. Sci.* 41:363.
20. MARKUS, H.C. 1934. Life history of the fathead minnow (*Pimephales promelas*). *Copeia* 1934:116.
21. MOUNT, D.I. & C.E. STEPHAN. 1967. A method for establishing acceptable toxicant limits for fish—malathion and the butoxyethanol ester of 2,4-D. *Trans. Amer. Fish. Soc.* 96:185.
22. PICKERING, Q.H. & T.O. THATCHER. 1970. The chronic toxicity of linear alkylate sulfonate (LAS) to *Pimephales promelas*, Rafinesque. *J. Water Pollut. Control Fed.* 42:243.
23. PICKERING, Q.H. & W.N. VIGOR. 1965. The acute toxicity of zinc to eggs and fry of the fathead minnow. *Progr. Fish-Cult.* 27:153.
24. EATON, J.G. 1974. Chronic cadmium toxicity to the bluegill (*Lepomis macrochirus* Rafinesque). *Trans. Amer. Fish. Soc.* 103:729.
25. SMITH, W.E. 1973. A cyprinodontid fish, *Jordanella floridae*, as a laboratory animal for rapid chronic bioassays. *J. Fish. Res. Board Can.* 30:329.
26. BREDER, C.M. 1936. The reproduction be-

havior of North American sunfish. *Zoologica* 21:1.

27. EATON, J.G. 1970. Chronic malathion toxicity to the bluegill, *Lepomis macrochirus*. *Water Res.* 4:673.

28. MCCOMISH, T.S. 1968. Sexual differentiation of bluegills by the urogenital opening. *Progr. Fish-Cult.* 30:28.

29. FOSTER, N.R., J. CAIRNS, JR. & R.L. KAESLER. 1969. The flagfish *Jordanella floridae*, as a laboratory animal for behavioral bioassay studies. *Proc. Acad. Natur. Sci. Philadelphia* 121:129.

30. BENOIT, D.A. 1975. Chronic effects of copper on survival, growth and reproduction of the bluegill (*Lepomis macrochirus*). *Trans. Amer. Fish. Soc.* 104:353.

31. HANSEN, D.J. & P.R. PARRISH. 1975. Suitability of sheepshead minnows (*Cyprinodon variegatus*) for life cycle toxicity tests. *In* F.L. Mayer & J.L. Hamelink, Aquatic Toxicity and Hazard Identification. ASTM STP 634, American Soc. Testing & Materials, Philadelphia, Pa.

32. SCHIMMEL, S.C., D.J. HANSEN & J. FORESTER. 1974. Effects of Aroclor 1254 on laboratory-reared embryos and fry of sheepshead minnows (*Cyprinodon variegatus*). *Trans. Amer. Fish. Soc.* 103:582.

33. HANSEN, D.J., S.C. SCHIMMEL & J. FORESTER. 1973. Aroclor 1254 in eggs of sheepshead minnows: Effects on fertilization success and survival of embryos and fry. *Proc. 27th Annu. Conf. S.E. Assoc. Game Fish Comm.*: 420.

34. HANSEN, D.J. 1978. Laboratory culture of sheepshead minnows (*Cyprinodon variegatus*). *In* Bioassay Procedures for the Ocean Disposal Permit Program. EPA-600/9-78-010, U.S. Environmental Protection Agency.

35. BOYD, J.F. & R.C. SIMMONS. 1974. Continuous laboratory production of fertile *Fundulus heteroclitus*, Walbaum eggs lacking chorionic fibrils. *J. Fish. Biol.* 6:389.

36. HILDEBRAND, S.F. 1923. Notes on habits and development of eggs and larvae of the silversides *Menidia menidia* and *Menidia beryllina*. *U.S. Bur. Fish. Bull.* 38:113.

37. RUBINOFF, I. 1958. Raising the atherinid fish *Menidia menidia* in the laboratory. *Copeia* 1958:146.

38. RUBINOFF, I. & E. SHAW. 1960. Hybridization in two sympatric species of atherinid fishes, *Menidia menidia* Linnaeus and *Menidia beryllina* Cope. *Amer. Mus. Natur. Hist.* No. 1999:1.

PART 9000

MICROBIOLOGICAL

EXAMINATION OF

WATER

9010 INTRODUCTION

9010 A. General Discussion

The following sections describe procedures for making microbiological examinations of water samples to determine sanitary quality. The methods are intended to indicate the degree of contamination with wastes. They are the best techniques currently available; however, their limitations must be understood thoroughly.

Tests for detection and enumeration of indicator organisms, rather than of pathogens, are used. The coliform group of bacteria, as herein defined, is the principal indicator of suitability of a water for domestic, industrial, or other uses. The cultural reactions and characteristics of this group of bacteria have been studied extensively.

Experience has established the significance of coliform group density as a criterion of the degree of pollution and thus of sanitary quality. The significance of the tests and the interpretation of results are well authenticated and have been used as a basis for standards of bacteriological quality of water supplies.

The membrane filter technique, which involves a direct plating for detection and estimation of coliform densities, is as effective as the multiple-tube fermentation test for detecting bacteria of the coliform group. Modification of procedural details, particularly of the culture medium, has made the results comparable with those given by the multiple-tube fermentation procedure. Although there are limitations

in the application of the membrane filter technique, it is equivalent when used with strict adherence to these limitations and to the specified technical details. Thus, two standard methods are presented for the detection of bacteria of the coliform group.

It is customary to report results of the coliform test by the multiple-tube fermentation procedure as a Most Probable Number (MPN) index. This is an index of the number of coliform bacteria that, more probably than any other number, would give the results shown by the laboratory examination; it is not an actual enumeration. By contrast, direct plating methods such as the membrane filter procedure permit a direct count of coliform colonies. In both procedures coliform density is reported conventionally as the MPN or membrane filter count per 100 mL. Use of either procedure permits appraising the sanitary quality of water and the effectiveness of treatment processes. Because it is not necessary to provide a quantitative assessment of coliform bacteria for all samples, a qualitative, presence-absence test is included.

Fecal streptococci also are indicators of fecal pollution and methods for their detection and enumeration are given. A multiple-tube dilution and a membrane filter procedure are included.

Methods for the differentiation of the coliform group are included. Such differentiation generally is considered of limited value in assessing drinking water quality because the presence of any coliform bacteria renders the water potentially unsatisfactory and unsafe. Speciation may provide information on colonization of a distribution system and further confirm the validity of coliform results.

Coliform group bacteria present in the gut and feces of warm-blooded animals generally include organisms capable of producing gas from lactose in a suitable culture medium at 44.5 ± 0.2°C. Inasmuch as coliform organisms from other sources often cannot produce gas under these conditions, this criterion is used to define the fecal component of the coliform group. Both the multiple-tube dilution technique and the membrane filter procedure have been modified to incorporate incubation in confirmatory tests at 44.5°C to provide estimates of the density of fecal organisms, as defined. Procedures for fecal coliforms also include a 24-h multiple-tube test using A-1 medium and a 7-h rapid method. This differentiation yields valuable information concerning the possible source of pollution in water, and especially its remoteness, because the *nonfecal* members of the coliform group may be expected to survive longer than the *fecal* members in the unfavorable environment provided by the water.

The heterotrophic plate count may be determined by pour plate, spread plate, or membrane filter method. It provides an approximate enumeration of total numbers of viable bacteria that may yield useful information about water quality and may provide supporting data on the significance of coliform test results. The heterotrophic plate count is useful in judging the efficiency of various treatment processes and may have significant application as an in-plant control test. It also is valuable for checking quality of finished water in a distribution system as an indicator of microbial regrowth and sediment buildup in slow-flow sections and dead ends.

Experience in the shipment of un-iced samples by mail indicates that noticeable changes may occur in type or numbers of bacteria during such shipment for even limited periods of time. Therefore, refrigeration during transportation is recommended to minimize changes, particularly when ambient air temperature exceeds 13°C.

Procedures for the isolation of certain pathogenic bacteria and protozoa are presented. These procedures are tedious and complicated and are not recommended for routine use. Likewise, tentative procedures for enteric viruses are included but their routine use is not advocated.

Examination of routine bacteriological samples cannot be regarded as providing complete information concerning water quality. Always consider bacteriological results in the light of information available concerning the sanitary conditions surrounding the sample source. For a water supply, precise evaluation of quality can be made only when the results of laboratory examinations are interpreted in the light of sanitary survey data. Consider inadequate the results of the examination of a single sample from a given source. When possible, base evaluation of water quality on the examination of a series of samples collected over a known and protracted period of time.

Pollution problems of tidal estuaries and other bodies of saline water have focused attention on necessary modification of existing bacteriological techniques so that they may be used effectively. In the following sections, applications of specific techniques to saline water are not discussed because the methods used for fresh waters also can be used satisfactorily with saline waters.

Methods for examination of the waters of swimming pools and other bathing places are included. The standard procedures for the plate count, fecal coliforms, and fecal streptococci are identical with those used for other waters. Procedures for *Staphylococcus* and *Pseudomonas aeruginosa,* organisms commonly associated with the upper respiratory tract or the skin, are included.

Procedures for aquatic fungi, actinomycetes, and nematodes are included.

Sections on rapid methods for coliform testing and on the recovery of stressed organisms are included. Because of increased interest and concern with analytical quality control, this section continues to be expanded.

The bacteriological methods in Part 9000, developed primarily to permit prompt and rapid examination of water samples, have been considered frequently to apply only to routine examinations. However, these same methods are basic to, and equally valuable in, research investigations in sanitary bacteriology and water treatment. Similarly, all techniques should be the subject of investigations to establish their specificity, improve their procedural details, and expand their application to the measurement of the sanitary quality of water supplies or polluted waters.

9010 B.   U.S. EPA Regulations for Drinking Water Quality

In the United States the quality of public water supplies is judged in terms of the 1975 U.S. Environmental Protection Agency (EPA) Interim Primary Drinking Water Regulations.[1] These regulations provide for a minimum number of samples to be examined per month and establish the maximum number of coliform organisms (Maximum Contaminant Level, MCL) allowable per 100 mL of finished water. They also require that analyses be made in a certified laboratory.[2] EPA is in the process of adopting permanent regulations.

1. Sampling

Make bacteriological examinations on samples collected at representative points throughout the distribution system. Select the frequency of sampling and the location of sampling points to insure accurate determination of the bacteriological quality of the treated water supply, which may be controlled in part by the known quality of the untreated water and thus by the need for treatment. Base the minimum number of samples to be collected and examined each month on the population served by the supply. It is important to examine repetitive samples from a designated point, as well as samples from a number of widely distributed sampling points. Take samples at reasonably evenly spaced time intervals.

2. Application

For the multiple-tube fermentation technique the maximum number of allowable coliform organisms is prescribed in terms of standard portion volume (10 mL or 100 mL) and the number of portions examined. The absence of gas in all tubes, when five 10-mL portions are examined by the fermentation tube method (equivalent to an MPN of less than 2.2 coliforms/100 mL), generally is interpreted to indicate that the single sample meets the standards. A positive confirmed phase for coliform organisms in three or more tubes (10-mL portions) or five portions (100-mL portions) indicates the need for immediate remedial action and additional examinations. Analyze repeat samples of finished water from the same location that consistently show three or more positive 10-mL portions by the completed test. Similarly, for the membrane filter technique, the standard portion volume is 100 mL, the quality limit is 1 coliform colony/100 mL, and the action limit is more than 4 coliform colonies/100 mL. Analyze repeat samples of finished water from the same location that consistently give positive results by the verification procedure.

Although the heterotrophic plate count is not required in the EPA Interim Primary Drinking Water Regulations, its use may be required in conjunction with modification of the turbidity limit.

These standards also specify limiting concentrations of chemical and physical constituents of water as related to its safety and potability.

The World Health Organization has established International Standards of Drinking Water Quality.[3] These are similar to the U.S. drinking water standards, but have been modified and liberalized to apply to water supply conditions in all parts of the world.

3. References

1. U.S. ENVIRONMENTAL PROTECTION AGENCY. 1976. National Interim Primary Drinking Water Regulations. EPA-570/9-76-003, Washington, D.C.

2. U.S. ENVIRONMENTAL PROTECTION AGENCY. 1978. Manual for the Interim Certification of Laboratories Involved in Analyzing Public Drinking Water Supplies. EPA-600/8-78-008, Washington, D.C.

3. WORLD HEALTH ORGANIZATION. 1983. Guidelines for Drinking-Water Quality. World Health Org., Geneva.

9020   QUALITY ASSURANCE*

9020 A.   Introduction

1. General Discussion

The growing emphasis on water quality standards, enforcement, and monitoring requires the establishment and effective operation of a quality assurance program to substantiate the validity of analytical data.

A laboratory quality assurance program is the orderly application of the practices necessary to remove or reduce errors that may occur in laboratory operations, as caused by personnel, equipment, supplies, sampling procedures, and analytical methodology.

The program must be practical and integrated. It should require only a reasonable amount of time or it will be bypassed. When properly administered, a balanced, conscientiously applied quality assurance program will yield uniformly high-quality data without interfering with the primary analytical functions of the laboratory. Detailed descriptions of quality control practices are available.[1,2] Generally, 15% of total analyst time should be spent on different aspects of a quality assurance program. Intralaboratory and interlaboratory quality control practices should be documented and the records should be available for inspection. *The quality assurance guidelines discussed below are recommended as a minimal program for a microbiology laboratory.*

2. References

1. INHORN, S.L., ed. 1977. Quality Assurance Practices for Health Laboratories. American Public Health Assoc., Washington, D.C.

---

*Approved by Standard Methods Committee, 1988.

2. BORDNER, R.H., J.A. WINTER & P.V. SCAR-PINO, eds. 1978. Microbiological Methods for Monitoring the Environment, Water and Wastes. EPA 600/8-78-017, Environmental Monitoring & Support Lab., U.S. Environmental Protection Agency, Cincinnati, Ohio.

9020 B. Intralaboratory Quality Control Guidelines

All laboratories have some intralaboratory quality control practices that have evolved from common sense and the principles of controlled experimentation. Special problems exist in microbiology because analytical standards, known additions, and reference samples usually are not available. Personal judgment is required more frequently. An effective program must control all factors, from sample collection through data reporting, that can influence the results. The factors include sampling techniques, sample storage and holding, facilities, personnel, equipment, supplies, media, and analytical test procedures. It is especially important that laboratories performing only a limited amount of microbiological testing exercise strict quality control. These guidelines will assist laboratories in establishing and improving quality control programs.

1. Facilities and Personnel

a. Ventilation: Plan well-ventilated laboratories that can be maintained free of dust, drafts, and extreme temperature changes. Central air conditioning is recommended to reduce contamination, permit more stable operation of incubators, and decrease moisture problems with media and analytical balances.

b. Space utilization: Design and operate the laboratory to minimize through traffic and visitors. Provide a separate area for preparing and sterilizing media, glassware, and equipment. Use a special work area such as a vented laminar-flow hood for dispensing and preparing sterile media, transferring microbial cultures, or working with pathogenic materials. In smaller laboratories it may be necessary, although undesirable, to carry out these activities in the same room.

c. Laboratory bench areas: Provide a minimum of 2 m linear bench space per analyst and additional areas for preparation and support activities. For stand-up work, typical bench dimensions are 90 to 97 cm high and 70 to 76 cm deep. For sit-down activities such as microscopy and plate counting, benches are 75 to 80 cm high. Specify bench tops of stainless steel, epoxy plastic, or other smooth, impervious surface that is inert and corrosion-resistant and has a minimum number of seams. Install even, glare-free lighting with about 1000 lux intensity at the working surface.

d. Walls and floors: Assure that walls are covered with a smooth finish that is easily cleaned and disinfected. Specify floors of sealed, smooth concrete, vinyl, asphalt tile, or other impervious, washable surface.

e. Air monitoring: Maintain high standards of cleanliness in work areas. Monitor air, at least monthly, with air density plates and bench tops with RODAC plates or the swab method.[1] The number of colonies on the air density plate test should not exceed $160/m^2/15$ min exposure (15 colonies/plate/15 min).

f. Laboratory cleanliness: Regularly clean laboratory rooms and wash benches, shelves, floors, and windows. Wet-mop floors and treat with a disinfectant solution; do not sweep or dry-mop. Wipe bench tops and treat with a disinfectant before and after use. Do not permit laboratory to become cluttered.

g. *Personnel:* Ideally, bacteriological testing should be done by a professional microbiologist. If that is not possible, have a professional microbiologist or trained analyst available for guidance and assistance.

Clearly define work assignments. Train and evaluate the analyst in basic laboratory procedures. The supervisor periodically should review procedures of sample collecting and handling, media and glassware preparation, sterilization, routine testing procedures, counting, data handling, and quality control techniques to identify and eliminate problems. Management should encourage laboratory personnel to take additional training to advance skills and knowledge.

2. Laboratory Equipment and Instrumentation

Verify that each item of equipment meets the user's needs for precision and minimization of bias. Perform equipment maintenance on a regular basis as recommended by the manufacturer or obtain preventive maintenance contracts on autoclave, balances, microscopes, and other equipment, whenever economically feasible. Directly record all quality control checks in a permanent log book.

Use the following equipment control procedures:

a. *Thermometer/temperature-recording instruments:* Check accuracy of thermometers or recording instruments semiannually against a certified National Institute of Standards and Technology (NIST, formerly National Bureau of Standards, NBS) thermometer or one traceable to NIST and conforming to NIST specifications. For general purposes use thermometers graduated in increments of 0.5° or less. For a 44.5°C water bath, use a submersible thermometer graduated to 0.2°C or less. Record temperature check data in a quality control log. Mark NIST calibration corrections on each thermometer used with an incubator, refrigerator, or freezer. When possible, equip incubators and water baths with temperature-recording instruments providing a continuous record of operating temperature.

b. *Balance:* Wipe balance before and after each use with a soft brush made of such material as camel's hair. Clean balance pans after each use and wipe spills up immediately with a damp towel. Inspect weights with each use and discard if corroded. Check weights monthly against certified weights. For weighing 2 g or less, use an analytical balance with a sensitivity less than 1 mg at a 10-g load. For larger quantities, use a pan balance with a sensitivity of 0.1 g at a 150-g load.

c. *pH meter:* Standardize pH meter with at least two standard buffers (pH 4.0, 7.0, or 10.0) and compensate for temperature before each series of tests. Date buffer solutions when opened and check monthly against another pH meter if available. See Section 4500-H$^+$.

d. *Water deionization unit:* Proper mixed-bed deionization resin columns produce satisfactory reagent-grade water. Commercial systems are available that combine prefiltration, mixed-bed resins, activated carbon, and final filtration cartridges to produce a reagent-grade water. The life of such cartridges can be extended greatly if the intake water is distilled or treated by reverse osmosis. To avoid contamination do not store such water. Deionization systems tend to produce the same quality water until resins or activated carbon are near exhaustion and quality abruptly becomes unacceptable.

Monitor deionized water continuously or daily with a conductivity meter and analyze at least annually for trace metals. Replace cartridges at intervals recommended by the manufacturer or as indicated by analytical results. Filter product water through a 0.22-μm-pore-diam membrane filter to remove bacterial contamination.

e. *Water still:* Stills produce water of a

good grade that characteristically deteriorates slowly over time as corrosion, leaching, and fouling occur. These conditions can be controlled with proper maintenance and cleaning. Stills efficiently remove dissolved substances but not dissolved gases or volatile organic chemicals. Freshly distilled water may contain chlorine and ammonia (NH_3). On storage, additional NH_3 and CO_2 are absorbed from the air. Use softened water as the source water to reduce frequency of cleaning the still. Drain and clean still and reservoir according to manufacturer's instructions and usage.

f. Reverse osmosis units: Commercial reverse osmosis (RO) units consisting of prefilters and reverse osmosis cartridges are sold as water purification systems. These RO units remove only about 90% of impurities; consequently, do not use water from an RO unit for microbiological analyses. The major use of RO units should be initial cleanup of water before deionization or distillation. Commercial units combining prefiltration, reverse osmosis, and mixed-bed ion-exchange resins are recommended.

g. Media dispensing apparatus: Check accuracy of volumes dispensed with a graduated cylinder at start of each volume change and periodically throughout extended runs. If the unit is used more than once per day, pump a large volume of hot distilled water through the dispenser to rinse. Correct leaks, loose connections, or malfunctions immediately. At the end of the work day, break apparatus down into parts, wash, rinse with reagent water, and dry. Lubricate parts according to manufacturer's instructions or at least once per month.

h. Hot-air oven: Test performance quarterly with commercially available spore strips or spore suspensions. Monitor temperature with a thermometer accurate in the 160 to 180°C range and record results. Use heat-indicating tape to identify sup-

plies and materials that have been exposed to sterilization temperatures.

i. Autoclave: Record items sterilized, temperature, pressure, and time for each run. Optimally use a recording thermometer. Check operating temperature weekly with a minimum/maximum thermometer. Test performance with spore strips or suspensions monthly. Use heat-indicating tape to identify supplies and materials that have been sterilized.

j. Refrigerator: Check and record temperature daily and clean monthly. Identify and date materials stored. Defrost as required and discard outdated materials quarterly.

k. Freezer: Check and record temperature daily. A recording thermometer and alarm system are highly desirable. Identify and date materials stored. Defrost and clean semiannually; discard outdated materials.

l. Membrane filter equipment: Before use, assemble filtration units and check for leaks. Coat units with silicone to improve drainage. Discard units if inside surfaces are scratched. Wash and rinse filtration assemblies thoroughly after use, wrap in nontoxic paper or foil or place in noncorrosive container, and autoclave. See Section 9222B.1.

m. Ultraviolet sterilization lamps: Disconnect unit monthly and clean lamps by wiping with a soft cloth moistened with ethanol. Test lamps quarterly with UV light meter* and replace if they emit less than 70% of initial output or if plate count agar spread plates containing 200 to 250 microorganisms, exposed to the light for 2 min, do not show a count reduction of 99%.

n. Safety cabinet (hood): Check filters monthly for plugging or dirt accumulation and clean or replace as needed. Once per month expose plate count agar plates to air

---

*Shortwave UV meter, UV Products, Inc., San Gabriel, Calif., or equivalent.

flow for 1 h. Incubate plates at 35°C for 24 h and examine for contamination. A properly operating safety cabinet should produce no growth on the plates. Disconnect UV lamps and clean monthly by wiping with a soft cloth moistened with ethanol. Check lamps' efficiency as specified above. Inspect cabinet for leaks and rate of air flow quarterly. Use a pressure monitoring device to measure efficiency of hood performance. Maintain as directed by the manufacturer.

o. Water bath: Keep an appropriate thermometer (¶ 2a, above) immersed in the water bath; monitor and record temperature daily unless a recording thermometer and alarm system are used. Use only stainless steel, plastic-coated, or other corrosion-proof racks. Clean bath as needed.

p. Incubator (air or water jacketed): Check and record temperature twice daily (morning and afternoon) on the shelf areas in use. If a glass thermometer is used, submerge bulb and stem in water or glycerine to the stem mark. For best results use a recording thermometer and alarm system. Place incubator in an area where temperature is maintained at 16 to 27°C.

q. Microscopes: Use lens paper to clean optics and stage after each use. Cover microscope when not in use.

Permit only trained technicians to use fluorescence microscope and light source. Monitor fluorescence lamp with a light meter and replace when a significant loss in fluorescence is observed. Log lamp operation time, efficiency, and alignment. Periodically check lamp alignment, particularly when the bulb has been changed; realign if necessary. Use known positive 4+ fluorescence slides as controls.

3. Laboratory Supplies

a. Glassware: Before each use, examine glassware and discard items with chipped edges or etched inner surfaces. Particularly examine screw-capped dilution bottles and flasks for chipped edges that could leak and contaminate the work area or create aerosols. Inspect glassware after washing for excessive water beads and rewash. Make the following tests for clean glassware as necessary:

1) pH check—Because some cleaning solutions are difficult to remove completely, spot check batches of clean glassware for pH reaction, especially if soaked in alkali or acid. To test clean glassware for an alkaline or acid residue add a few drops of 0.04% bromthymol blue (BTB) or other pH indicator and observe the color reaction. BTB should be blue-green (in the neutral range).

To prepare 0.04% bromthymol blue indicator solution, add 16 mL 0.01N NaOH to 0.1 g BTB and dilute to 250 mL with distilled water.

2) Test for inhibitory residues on glassware and plasticware—Certain wetting agents or detergents used in washing glassware may contain bacteriostatic or inhibiting substances requiring 6 to 12 rinsings to remove all traces and insure freedom from residual bacteriostatic action. Perform this test annually and before using a new supply of detergent. If prewashed, presterilized plasticware is used, test it for inhibitory residues.

a) Procedure—Wash and rinse six petri dishes according to usual laboratory practice and designate as Group A.

Wash six petri dishes as above, rinse 12 times with successive portions of reagent water, and designate as Group B.

Rinse six petri dishes with detergent wash water (in use concentration), dry without further rinsing, and designate as Group C.

Sterilize dishes in Groups A, B, and C by the usual procedure.

To test presterilized plasticware, set up six sterile plastic petri dishes, designate as Group D, and proceed.

Add not more than 1 mL of a culture of *E. aerogenes* known to contain 50 to 150 colony-forming units to dishes in Groups A through D, and proceed according to the procedure described for the heterotrophic plate count (Section 9215). Preliminary testing may be necessary to obtain the specified count range. To help assure counts within the specified count range, inoculate three plates of each group with 0.1 mL and the other three plates of each group with 1 mL.

b) Interpretation of results—Difference in averaged counts on plates in Groups A through D should be less than 15% if there are no toxic or inhibitory effects.

Differences in averaged counts of less than 15% between Groups A and B and greater than 15% between Groups A and C indicates that the cleaning detergent has inhibitory properties that are eliminated during routine washing.

b. *Utensils and containers for media preparation:* Use utensils and containers of borosilicate glass, stainless steel, aluminum, or other corrosion-resistant material (see Section 9030). Do not use copper utensils.

c. *Reagent-grade water quality:* The quality of water obtainable from a water purification system differs with the system used and its maintenance. See 2*d, e,* and *f* above. Acceptable limits of water quality are given in Table 9020:I. If these limits are not met, investigate and correct. Although pH measurement of purified water is characterized by drift, extreme readings are indicative of chemical contamination.

1) Test for bacteriological quality of reagent water—The test is based on the growth of *Enterobacter aerogenes* in a chemically defined minimal growth medium. The presence of a toxic agent or a growth-promoting substance will alter the 24-h population by an increase or decrease of 20% or more when compared to a control. Perform the test at least annually, when the source of reagent water is changed, and when an analytical problem occurs.

The test is complex, requires skill and experience, and is not easily done on an infrequent basis. It requires work over 4 d, an ultrapure water from an independent source as a control, high-purity reagents, and extreme cleanliness of culture flasks, petri dishes, test tubes, pipets, etc. The test is very sensitive to toxicants and results cannot be related directly to routine analyses.

a) Apparatus and material—Use borosilicate glassware and rinse in water freshly redistilled from a glass still before sterilizing it with dry heat; steam sterilization will recontaminate these specially cleaned items. Test sensitivity and reproducibility depend in part on cleanliness of sample containers, flasks, tubes, and pipets. It often is convenient to set aside new glassware for exclusive use in this test. Use any strain of coliform IMViC type $-$ $-$ $+$ $+$ *(E. aerogenes)* obtained from an ambient water or wastewater sample.

b) Reagents—Use only reagents and chemicals of ACS grade. Test sensitivity is controlled in part by the reagent purity. Prepare reagents in water freshly redistilled from a glass still.

Sodium citrate solution: Dissolve 0.29 g sodium citrate, $Na_3C_6H_6O_7 \cdot 2H_2O$, in 500 mL water.

Ammonium sulfate solution: Dissolve 0.26 g $(NH_4)_2SO_4$ in 500 mL water.

Salt-mixture solution: Dissolve 0.26 g magnesium sulfate, $MgSO_4 \cdot 7H_2O$; 0.17 g calcium chloride, $CaCl_2 \cdot 2H_2O$; 0.23 g ferrous sulfate, $FeSO_4 \cdot 7H_2O$; and 2.50 g sodium chloride, NaCl, in 500 mL water.

Phosphate buffer solution: Stock phosphate buffer solution (9050C.1*a*) diluted 1:25 in water.

Boil all reagent solutions 1 to 2 min to kill vegetative cells. Store solutions in sterilized glass-stoppered bottles in the dark at 5°C for up to several months provided that they are tested for sterility before each use. Because the salt-mixture solution will develop a slight turbidity within 3 to 5 d as the ferrous salt converts to the ferric state,

TABLE 9020:I. QUALITY OF PURIFIED WATER USED IN MICROBIOLOGY TESTING

Test	Monitoring Frequency	Limit
Chemical tests:		
Conductivity	Continuously or with each use	> 0.5 megohms resistance or < 2 μmhos/cm at 25°C
pH	With each use	5.5–7.5
Total organic carbon	Monthly	< 1.0 mg/L
Heavy metals, single (Cd, Cr, Cu, Ni, Pb, and Zn)	Annually*	< 0.05 mg/L
Heavy metals, total	Annually*	≤ 0.1 mg/L
Ammonia/organic nitrogen	Monthly	< 0.1 mg/L
Total chlorine residual	Monthly or with each use	< detection limit
Bacteriological tests:		
Heterotrophic plate count (See Section 9215)	Monthly	< 1000 cfu/mL
Water quality test (See 3c1)	Annually and for a new source	0.8–3.0 ratio
Use test (see 3d)	Annually and for a new source	Student's t ≤ 2.78

* Or more frequently if there is a problem.

prepare the salt-mixture solution without FeSO₄ for long-term storage. To use the mixture, add an appropriate amount of freshly prepared and freshly boiled iron salt. Discard solutions with a heavy turbidity and prepare a new solution. Discard phosphate buffer solution if it becomes turbid.

c) Samples—To prepare test samples collect 150 to 200 mL laboratory reagent water and control (redistilled) water in sterile borosilicate glass flasks and boil for 1 to 2 min. Avoid longer boiling to prevent chemical changes.

d) Procedure—Label five flasks or tubes, A, B, C, D, and E. Add water samples, media reagents, and redistilled water to each flask as indicated in Table 9020:II.

Add a suspension of *E. aerogenes* (IMViC type − − + +) of such density that each flask will contain 30 to 80 cells/mL, prepared as directed below. Cell den-

sities below this range result in inconsistent ratios while densities above 100 cells/mL result in decreased sensitivity to nutrients in the test water. Make an initial bacterial count by plating triplicate 1-mL portions from each culture flask in plate count agar. Incubate Flasks A through E at 35°C for 24 ± 2 h. Prepare final plate counts from each flask, using dilutions of 1, 0.1, 0.01, 0.001, and 0.0001 mL.

e) Preparation of bacterial suspension—On the day before making the distilled-water suitability test, inoculate a strain of *E. aerogenes* onto a nutrient agar slant with a slope approximately 6.3 cm long contained in a 125- × 16-mm screw-cap tube. Streak entire agar surface to develop a continuous-growth film and incubate 18 to 24 h at 35°C.

f) Harvesting of viable cells—Pipet 1 to 2 mL sterile dilution water from a 99-mL water blank onto the 18- to 24-h culture.

TABLE 9020:II. REAGENT ADDITIONS FOR WATER QUALITY TEST

Media Reagents	Control Test mL		Optional Tests mL		
	Control A	Test Water B	Carbon/Nitrogen Available C	Nitrogen Source D	Carbon Source E
Sodium citrate solution	2.5	2.5	—	2.5	—
Ammonium sulfate solution	2.5	2.5	—	—	2.5
Salt-mixture solution	2.5	2.5	2.5	2.5	2.5
Phosphate buffer (7.3 ± 0.1)	1.5	1.5	1.5	1.5	1.5
Unknown water	—	21.0	21.0	21.0	21.0
Redistilled water	21.0	—	5.0	2.5	2.5
Total volume	30.0	30.0	30.0	30.0	30.0

Emulsify growth on slant by gently rubbing bacterial film with pipet, being careful not to tear agar; pipet suspension back into original 99-mL water blank.

g) Dilution of bacterial suspension—Make a 1:100 dilution of original bottle into a second water blank, a further 1:100 dilution of second bottle into a third water blank, a 1:10 dilution of third bottle into a fourth water blank, shaking vigorously after each transfer. Pipet 1.0 mL of the fourth dilution $(1:10^5)$ into each of Flasks A, B, C, D, and E. This procedure should result in a final dilution of the organisms to a range of 30 to 80 viable cells per milliliter of test solution.

h) Verification of bacterial density—Variations among strains of the same organism, different organisms, media, and surface area of agar slopes possibly will necessitate adjustment of the dilution procedure to arrive at a specific density range between 30 to 80 viable cells. To establish the growth range numerically for a specific organism and medium, make a series of plate counts from the third dilution to determine bacterial density. Choose proper volume from this third dilution, which, when diluted by the 30 mL in Flasks A, B, C, D, and E, will contain 30 to 80 viable cells/mL. If the procedures are standardized as to slant surface area and laboratory technique, it is possible to reproduce results on repeated experiments with the same strain of microorganism.

i) Procedural difficulties—Problems in this method may be due to: storage of test water sample in soft-glass containers or in glass containers without liners for metal caps; use of chemicals in reagent preparation not of analytical-reagent grade or not of recent manufacture; contamination of reagent by distilled water with a bacterial background (to avoid this, make a heterotrophic plate count on all media reagents before starting the suitability test, as a check on stock solution contamination); failure to obtain desired initial bacterial concentration or incorrect choice of dilution used to obtain 24-h plate count; delay in pouring plates; and prolongation of incubation time beyond 26-h limit, resulting in desensitized growth response.

j) Calculation—For growth-inhibiting substances:

$$\text{Ratio} = \frac{\text{colony count/mL, Flask B}}{\text{colony count/mL, Flask A}}$$

A ratio of 0.8 to 1.2 (inclusive) shows no toxic substances; a ratio of less than 0.8 shows growth-inhibiting substances in the water sample. For nitrogen and carbon sources that promote growth:

$$\text{Ratio} = \frac{\text{colony count/mL, Flask C}}{\text{colony count/mL, Flask A}}$$

For nitrogen sources that promote growth:

$$\text{Ratio} = \frac{\text{colony count/mL, Flask D}}{\text{colony count/mL, Flask A}}$$

For carbon sources that promote bacterial growth:

$$\text{Ratio} = \frac{\text{colony count/mL, Flask E}}{\text{colony count/mL, Flask A}}$$

Do not calculate the last three ratios when the first ratio indicates a toxic reaction. For these ratios a value above 1.2 indicates an available source for bacterial growth.

k) Interpretation of results—The colony count from Flask A after 20 to 24 h at 35°C will depend on number of organisms initially planted in Flask A and strain of *E. aerogenes* used. For this reason, run the control, Flask A, for each individual series of tests. However, for a given strain of *E. aerogenes* under identical environmental conditions, the terminal count should be reasonably constant when the initial plant is the same. The difference in initial plant of 30 to 80 will be about threefold larger for the 80 organisms initially inoculated in Flask A, provided that the growth rate remains constant. Thus, it is essential that initial colony counts on Flasks A and B be approximately equal.

When the ratio exceeds 1.2, assume that growth-stimulating substances are present. However, this procedure is extremely sensitive and ratios up to 3.0 have little significance in actual practice. Therefore, if the ratio is between 1.2 and 3.0, do not make Tests C, D, and E, except in special circumstances.

Usually Flask C will be very low and Flasks D and E will have a ratio of less than 1.2 when the ratio of Flask B to Flask A is between 0.8 and 1.2. Limiting factors of growth in Flask A are nitrogen and organic carbon. An extremely large amount of ammonia nitrogen with no organic carbon could increase the ratio in Flask D above 1.2, or absence of nitrogen with high carbon concentration could give ratios above 1.2 in Flask E, with a B:A ratio between 0.8 and 1.2.

A ratio below 0.8 indicates that the water contains toxic substances, and this ratio includes all allowable tolerances. As indicated in the preceding paragraph, the ratio could go as high as 3.0 from 1.2 without any undesirable consequences.

Specific corrective measures cannot be recommended in specific instances of defective distillation apparatus. However, make a careful inspection of the distillation equipment and review production and handling of distilled water to help locate and correct the cause of difficulty.

Feedwater to a still often is passed through a deionizing column and a carbon filter. If these columns are well maintained, most inorganic and organic contaminants will be removed. If maintenance is poor, input water may be degraded to a quality lower than that of raw tap water.

The best distillation system is made of quartz or high-silica-content borosilicate glass with special thermal endurance. Tin-lined stills are not recommended. For connecting plumbing, use stainless steel, borosilicate glass, or special plastic pipes made of polyvinyl chloride (PVC). Use stainless steel storage reservoirs and protect them from dust.

1) Test sensitivity—Taking copper as one relative measurement of distilled water toxicity, maximum test sensitivity is 0.05 mg Cu/L in a distilled water sample.

d. Use test for evaluation of reagent water, media, and membranes: When a new lot of culture medium, membrane filters, or a new source of reagent-grade water is to be used, or at annual testing of water, make comparison tests of the current lot in use (reference lot) against the new lot (test lot) as follows:

1) Procedure—Use a single batch of control (redistilled or distilled water polished by deionization) water, glassware, membrane filters, or other needed materials as specified to control all other variables except the one under study. Make parallel pour or spread plate or membrane filter plate tests on reference lot and test lot, according to procedures in Sections 9215 and 9222. As a minimum, make single analyses on five positive water samples. Replicate analyses and additional samples can be tested to increase the sensitivity of detecting differences between reference and test lots.

When comparing sources of water, perform the tests in parallel using control water and test water separately for all water used in the tests (dilution, rinse, media preparation, etc.).

2) Counting and calculations—After incubation, compare bacterial colonies from the two lots for size and appearance. If colonies on the test lot plates are atypical or noticeably smaller than colonies on the reference lot plates, record the evidence of inhibition or other problem, regardless of count differences. Count plates and calculate the individual count per 1 mL or per 100 mL. Transform the count to logarithms and enter the log-transformed results for the two lots in parallel columns. Calculate the difference, d, between the two transformed results for each sample, in-

cluding the $+$ or $-$ sign, the mean, $\bar{d}$, and the standard deviation s_d of these differences (see Section 1010B).

Calculate Student's t statistic, using the number of samples as n:

$$t = \frac{\bar{d}}{s_d/\sqrt{n}}$$

3) Interpretation—Use the critical t value, from a Student's t table[2] for comparison against the calculated value. At the 0.05 significance level this value is 2.78 for five samples (four degrees of freedom). If the calculated t value does not exceed 2.78, the lots do not produce significantly different results and the test lot is acceptable. If the calculated t value exceeds 2.78, the lots produce significantly different results. If the test lot t exceeds the reference lot result, the test lot is more stimulatory. If the test lot result is less than that of the reference lot, the test lot is less stimulatory.

If the colonies are atypical or noticeably smaller on the test lot and the Student's t exceeds 2.78, review test conditions, repeat the test, and/or reject the test lot and obtain another one.

e. Reagents: Because reagents are an integral part of microbiological analyses, their quality must be assured. Use only chemicals of ACS or equivalent grade because impurities can inhibit bacterial growth, provide nutrients, or fail to produce the desired reaction. Date chemicals and reagents when received and when first opened for use. Make reagents to volume in volumetric flasks and transfer for storage to good-quality inert plastic or borosilicate glass bottles with borosilicate, polyethylene, or other plastic stoppers or caps. Label prepared reagents with name and concentration, date prepared, and initials of preparer. Include positive and negative control cultures with each series of cultural or biochemical tests.

f. Dyes and stains: In microbiological

analyses, organic chemicals are used as selective agents (e.g., brilliant green), as indicators (e.g., phenol red lactose), and as microbiological stains (e.g., Gram stain). Dyes from commercial suppliers vary from lot to lot in percent dye, dye complex, insolubles, and inert materials. Because dyes for microbiology must be of proper strength and stability to produce correct reactions, use only dyes certified by the Biological Stain Commission. Check bacteriological stains before use with at least one positive and one negative control culture and record results.

g. *Membrane filters and pads:* The quality and performance of membrane filters vary with the manufacturer, type, brand, and lot. These variations result from differences in manufacturing methods, materials, quality control, and storage conditions.

1) Membrane filters and pads for water analyses should meet the following specifications:

a) Filter diam 47 mm, mean pore diam 0.45 μm. Alternate filter and pore sizes may be used if the manufacturer provides data verifying performance equal to or better than that of 47-mm diam, 0.45-μm pore size filter. At least 70% of filter area must be pores.

b) When filters are floated on reagent water, the water diffuses uniformly through the filters in 15 s with no dry spots on the filters.

c) Flow rates are at least 55 mL/min/ cm^2 at 25°C and a differential pressure of 93 kPa.

d) Filters are nontoxic, free of bacterial-growth-inhibiting or stimulating substances, and free of materials that directly or indirectly interfere with bacterial indicator systems in the medium; ink grid is nontoxic. The arithmetic mean of five counts on filters must be at least 90% of the arithmetic mean of the counts on five agar spread plates using the same sample volumes and agar media.

e) Filters retain the organisms from a 100-mL suspension of *Serratia marcescens* containing 1×10^3 cells.

f) Water-extractables in filter do not exceed 2.5% after the membrane is boiled in 100 mL reagent water for 20 min, dried, cooled, and brought to constant weight.

g) Absorbent pad has diam 47 mm, thickness 0.8 mm, and is capable of absorbing 2.0 $\pm$ 0.2 mL Endo broth.

h) Pads release less than 1 mg total acidity calculated as $CaCO_3$ when titrated to the phenolphthalein end point with 0.02N NaOH.

i) If filter and absorbent pad are not sterile, they should not be degraded by sterilization at 121°C for 10 min. Confirm sterility by absence of growth when a membrane filter is placed on a pad saturated with tryptone glucose extract broth or tryptone glucose extract agar and incubated at 35°C for 24 h.

2) Standardized tests:

Standardized tests are available for evaluating retention, recovery, extractables, and flow rate characteristics of membrane filters.[2]

Some manufacturers provide information beyond that required by specifications and certify that their membranes are satisfactory for water analysis. They report retention, pore size, flow rate, sterility, pH, percent recovery, and limits for specific inorganic and organic chemical extractables.

To maintain quality control inspect each lot of membranes before use and during testing to insure they are round and pliable, with undistorted gridlines after autoclaving. After incubation, colonies should be well-developed with well-defined color and shape as defined by the test procedure. The gridline ink should not channel growth along the ink line nor restrict colony development. Colonies should be distributed evenly across the membrane surface.

h. *Culture media:* Because cultural methods depend on properly prepared media, use the best available materials and techniques in media preparation, storage, and application. For control of quality, use commercially prepared media whenever available but note that such media may vary in quality among manufacturers and even from lot to lot from the same manufacturer. A reference standard plate count agar is available from APHA for use by manufacturers who, after appropriate testing, may certify that their medium meets the specifications and standards of APHA. Because such standardization is not available for other media listed in Part 9000, test as described in 3d above.

Order media in quantities to last no longer than 1 year. Use media on a first-in, first-out basis. When practical, order media in quarter pound (114 g) multiples rather than one pound (454 g) bottles, to keep the supply sealed as long as possible. Record kind, amount, and appearance of media received, lot number, expiration date, and dates received and opened. Check inventory quarterly for reordering. Discard media that are caked, discolored, or show other deterioration.

Because temperature, light, and moisture conditions differ among laboratories, it is impossible to establish absolute shelf-life limits for unopened bottles of media. A conservative limit for unopened bottles is 2 years at room temperature. If bottles of media are older than 1 year, compare recovery of recent pure culture isolates and natural samples using the old medium and another proven lot.

Use open bottles of media within 6 months after opening. Once bottles are opened, store them in a desiccator immediately after use.

1) Preparation of media—Prepare media in containers that are at least twice the volume of the medium being prepared. Stir media, particularly agars, while heating. Avoid scorching or boil-over by using a boiling water bath for small batches of media and by continually attending to larger volumes heated on a hot plate or gas burner. Preferably use hot plate-magnetic stirrer combinations. Identify and date prepared media. Prepare all media in deionized or distilled water of proven quality. Measure water volumes and media with graduates or pipets conforming to NIST and APHA standards, respectively. For potentially polluted samples, do not use blow-out pipets. After preparation and storage, melt agar media in boiling water or flowing steam.

Check pH of a portion of each medium after sterilization and cooling. Check pH of solid medium with a surface probe. Record results. Make minor adjustments in pH (<0.5 pH units) with NaOH or HCl solution to the pH specified in formulation. If the pH difference is larger than 0.5 units, discard the batch and resolve the problem. Incorrect pH values may indicate a problem with reagent water quality, medium deterioration, or improper preparation. Review instructions for preparation and check water pH. If water pH is unsatisfactory, prepare a new batch of medium using water from a new source (see 9020B.2d and e). If water is satisfactory, remake medium and check; if pH is again incorrect, prepare medium from another bottle.

Record pH problems in the media record book and report to the manufacturer if the medium is indicated as the source of error. Examine prepared media for unusual color, darkening, or precipitation and record observations. Consider variations of sterilization time and temperature as possible causes for problems. If any of the above occur, discard the medium.

2) Sterilization—Expose media to sterilization temperatures for the minimal time specified. A double-walled autoclave permits maintenance of full pressure and temperature in the jacket between loads and reduces chance for heat damage. Follow manufacturer's directions for sterilization of specific media. The required exposure time varies with form and type of material, type of medium, presence of carbohydrates,

and volume. Table 9020:III gives guidelines for typical items.

Remove sterilized media from autoclave as soon as chamber pressure reaches zero. Never reautoclave media.

Check effectiveness of sterilization with each run by using *Bacillus stearothermophilus* spore suspensions or strips (commercially available) inside glassware. Sterilization at 121°C for 15 min kills the spores; if growth of the autoclaved spores occurs after incubation at 55°C, sterilization was inadequate.

Sterilize nonautoclavable solutions or media by filtration through a 0.22-μm-pore-diam filter in a sterile filtration and receiving apparatus. Filter and dispense medium in a safety cabinet or biohazard hood if available. Sterilize glassware (pipets, petri dishes, sample bottles) in an autoclave or an oven at 170°C for 2 h. Sterilize equipment, supplies, and other solid or dry materials that are heat-sensitive, by exposing to ethylene oxide in a gas sterilizer. Use commercially available spore strips or suspensions to check dry heat and ethylene oxide sterilization.

3) Use of agars and broths—Temper melted agars in a water bath at 44 to 46°C

until used but do not hold longer than 3 h. To monitor agar temperature, expose a bottle of water or medium to the same heating and cooling conditions as the agar. Insert a thermometer in the monitoring bottle to determine when the temperature is 44 to 46°C and suitable for use in pour plates. After pouring agar plates for streaking, dry agar surfaces by keeping dish slightly open for at least 15 min in a bacteriological hood to avoid contamination.

Handle sterile fermentation tubes carefully to avoid entrapping air in inner tubes thereby producing false positive reactions. Examine freshly prepared tubes to determine that gas bubbles are absent.

4) Storage of media—Prepare sterile media in amounts that will be used within holding time limits given in Table 9020:IV. If fermentation tube media are refrigerated, incubate overnight before use and check for false positive gas bubbles. Prepare media that are to be stored for more than 1 week in screw-capped or tightly-capped tubes and flasks to prevent loss of moisture. Seal prepoured agar plates in plastic bags and refrigerate to retain moisture.

To check for loss of moisture in broth tubes, mark the original liquid level in sev-

TABLE 9020:III. TIME AND TEMPERATURE FOR AUTOCLAVE STERILIZATION

Material	Time at 121°C
Membrane filters and pads	10 min
Carbohydrate-containing media (lauryl tryptose, BGB broth, etc.)	12–15 min
Contaminated materials and discarded cultures	30 min
Membrane filter assemblies (wrapped), sample collection bottles (empty)	15 min
Dilution water, 99 mL in screw-cap bottles	15 min
Rinse water volumes of 0.5 to 1 L	30 min
Rinse water in excess of 1 L	Adjust for volume; check for sterility

TABLE 9020:IV. HOLDING TIMES FOR
PREPARED MEDIA

Medium	Holding Time
Membrane filter (MF) broth in screw-cap flasks at 4°C	96 h
MF agar in plates with tight-fitting covers at 4°C	2 weeks
Agar or broth in loose-cap tubes at 4°C	1 week
Agar or broth in tightly closed screw-cap tubes at 4°C	3 months
Poured agar plates with loose-fitting covers in sealed plastic bags at 4°C	2 weeks
Large volume of agar in tightly closed screw-cap flask or bottle at 4°C	3 months

eral tubes of each batch and monitor loss of moisture. If estimated loss exceeds 10%, discard tubes. Protect media containing dyes from light; if color changes are observed, discard medium.

Prepared sterile broths and agars available from commercial sources may offer advantages when analyses are done intermittently, when staff is not available for preparation work, or when cost can be balanced against other factors of laboratory operation. Check performance of these media as described in ¶ 5 below. Time limits for holding prepared media are given in Table 9020:IV.

5) Quality control of prepared media—Maintain in a bound book a complete record of each batch of medium prepared with name of preparer and date, name and lot number of medium, amount of medium weighed, volume of medium prepared, sterilization time and temperature, pH measurements and adjustments, and preparations of labile components. Include ste-

rility and positive and negative control culture checks on all media as described below.

4. Analytical Quality Control Procedures

a. General quality control procedures:

1) For membrane filter tests, check sterility of media, membrane filters, dilution and rinse water, and glassware and equipment, as a minimum at the end of each series of samples, using sterile water as the sample.

2) For multiple-tube procedures, check sterility of media, dilution water, and glassware. To test sterility of media, subject a representative portion of each batch to incubation at an appropriate temperature for 24 to 48 h and observe for growth. Check each batch of dilution water for sterility by adding 20 mL water to 100 mL of a nonselective broth. Alternatively, aseptically pass 100 mL or more dilution water through a membrane filter and place filter on growth medium suitable for heterotrophic bacteria. Incubate at 35 ± 0.5°C for 24 h and observe for growth. If any contamination is indicated, reject analytical data from samples tested with these materials and request immediate resampling.

3) For each lot of medium check analytical procedures by testing with known positive and negative control cultures for the organism(s) under test. See Table 9020:V for examples of test cultures.

4) Perform duplicate analyses on 5% of samples and on at least one sample per test run.

5) In laboratories with more than one analyst, have each make parallel analyses on at least one positive sample monthly.

b. Measurement of method precision: Calculate precision of duplicate analyses for each different type of sample examined, for example, drinking water, ambient water, wastewater, etc. according to the following procedure:

TABLE 9020:V. CONTROL CULTURES FOR MICROBIOLOGICAL TESTS

Group	Control Culture	
	Positive	Negative
Total coliforms	Escherichia coli	Staphylococcus aureus
	Enterobacter aerogenes	Pseudomonas sp.
Fecal coliforms	E. coli	E. aerogenes
		Streptococcus faecalis
Fecal streptococci	Streptococcus faecalis	Staphylococcus aureus
		E. coli

1) Make duplicate analyses on the first 15 positive samples of a specific type. Have each set of duplicates analyzed by the same analyst, but include all analysts within the laboratory. Record duplicate analyses as D_1 and D_2.

2) Calculate the logarithm of each result. If either of a set of duplicate results is zero, add 1 to both values before calculating the logarithms.

3) Calculate the range (R) for each pair of transformed duplicates as the mean ($\overline{R}$) of these ranges.

See sample calculation in Table 9020:VI.

4) Thereafter, analyze 10% of routine samples in duplicate. Transform the duplicates as in ¶ 2) and calculate their range. If the range is greater than 3.27 $\overline{R}$, analyst variability is excessive. Determine if increased imprecision is acceptable; if not, discard all analytical results since the last precision check (see Table 9020:VII). Identify and resolve the analytical problem before making further analyses.

5) Update the criterion used in ¶ 4) by periodically repeating the procedures of ¶s 1) through 3) using the most recent sets of 15 duplicate results.

TABLE 9020:VI. CALCULATION OF PRECISION CRITERION

Sample No.	Duplicate Analyses		Logarithms of Counts		Range of Logarithms (R_{log}) ($L_1 - L_2$)
	D_1	D_2	L_1	L_2	
1	89	71	1.9494	1.8513	0.0981
2	38	34	1.5798	1.5315	0.0483
3	58	67	1.7634	1.8261	0.0627
.	.	.	.	.	.
.	.	.	.	.	.
.	.	.	.	.	.
14	7	6	0.8451	0.7782	0.0669
15	110	121	2.0414	2.0828	0.0414

Calculations:

1) Σ of R_{log} = 0.0981 + 0.0483 + 0.0627 + ... + 0.0669 + 0.0414

        = 0.718 89

2) $\overline{R} = \dfrac{\Sigma R_{log}}{n} = \dfrac{0.718\ 89}{15} = 0.0479$

3) Precision criterion = 3.27 $\overline{R}$ = 3.27 (0.0479) = 0.1566

TABLE 9020:VII. DAILY CHECKS ON PRECISION OF DUPLICATE COUNTS*

Date of Analysis	Duplicate Analyses		Logarithms of Counts		Range of Logarithms	Acceptance of Range†
	D_1	D_2	L_1	L_2		
8/29	71	65	1.8513	1.8129	0.0383	A
8/30	110	121	2.0414	2.0828	0.0414	A
8/31	73	50	1.8633	1.6990	0.1643	U

* Precision criterion = $(3.27\overline{R})$ = 0.1566.
† A = acceptable; U = unacceptable.

c. *Quality control on multiple-tube dilution tests:*

1) Completed test—For routine analyses, do the completed test (9221B.3) on 10% of positive samples. If no positives result from potable water samples, complete at least one positive source water quarterly.

2) Check for suppression—For public water supply samples with a history of heavy growth without gas in presumptive-phase tubes, submit such tubes to the confirmed phase (9221B.2) to check for coliform bacteria.

d. *Quality control on membrane filter procedures:*

1) Colony verification—For each type of test conducted, verify colonies monthly from a known positive sample. If laboratory has two or more analysts, require each to count typical colonies on the same membrane from one positive sample per month. Verify colonies on the membrane and compare the analysts' counts to the verified count.

2) Total coliform analyses—Verify by picking at least 10 colonies and testing for lactose fermentation or alternatively by rapid biochemical tests or multi-test systems:

For drinking water samples[3] verify all sheen colonies counted as coliforms, regardless of the amount of sheen, when this number is 5 to, and including, 10/100 mL. When the number exceeds 10/100 mL, randomly pick and verify at least 10 colonies

representative of all sheen types. If no positives result from testing drinking water samples, analyze at least one known positive source water quarterly.

a) Lactose fermentation—Transfer growth from at least 10 sheen colonies counted as coliforms to lauryl tryptose and brilliant green lactose bile broth tubes. Incubate at 35 ± 0.5°C and examine for gas production at 24 and 48 h. If only the lauryl tryptose tube from a particular colony produces gas, transfer growth from the positive tube to another brilliant green lactose bile broth tube and retest. Gas production in brilliant green lactose bile broth within 48 h verifies total coliform organisms and excludes false positive results. Ideally pick at least 10 non-sheen-producing colonies to determine that false negatives do not occur.

b) Alternative biochemical tests—These tests require using isolated (more than 2 mm apart) and pure colonies. If a mixed colony is suspected, streak growth on M-Endo medium and pick an isolated colony before proceeding to ¶ c) or d) or use the fermentation tube method described in ¶ a) above. Proceed to ¶ c) or d).

c) Rapid test—Perform the cytochrome oxidase (CO) test for indophenol and *o*-nitrophenyl-β-D-galactopyranoside (ONPG) test for β-D-galactosidase. A kit for these tests is available commercially. A negative CO and positive ONPG test verify the colony to be a coliform.

Cytochrome oxidase test—Use the procedure described in Section 9225E.7.

ONPG test—Prepare monosodium phosphate solution, 1.0M, by dissolving 6.9 g $NaH_2PO_4 \cdot H_2O$ in 45 mL reagent water, adding 3 mL 30% NaOH and adjusting to pH 7.0. Dilute to 50 mL and store in refrigerator. Prepare ONPG solution by dissolving 80 mg o-nitrophenyl-β-D-galactopyranoside (ONPG) in 15 mL water at 37°C and adding 5 mL 1M NaH_2PO_4 (solution should be colorless). Store in a refrigerator. Before using warm a portion sufficient for the number of tests to be done to 37°C.

Emulsify a large loopful of growth from each 18 to 24 h slant culture in a 10- × 75-mm fermentation tube or spot plate. Add one drop toluene to each tube or well and shake well. Let tubes stand for 5 min in 35°C water bath. Add 0.25 mL buffered ONPG solution to each tube and reincubate in 35°C water bath. Read tubes or spot plates at 0.5, 1, and 24 h. A positive result is development of yellow color.

d) Multi-test systems—Inoculate from the colony into a commercially available multi-test system that includes lactose fermentation and/or CO and ONPG tests.

3) Fecal coliform analyses—Verify by picking at least 10 isolated colonies from membranes on M-FC medium containing typical blue colonies and transferring to lauryl tryptose broth. Incubate at 35°C for 24 and 48 h and examine for gas production. Transfer growth from positive tubes to EC broth and incubate at 44.5°C for 24 h. Gas production in EC broth verifies presence of fecal coliform organisms.

4) Analyses for fecal streptococci—Verify by picking at least 10 isolated esculin positive colonies from m-E agar or red colonies from m-enterococcus agar. Transfer to brain heart infusion (BHI) agar and broth. Perform catalase test on 24-h-old cultures. Transfer catalase-negative cultures (possible fecal streptococci) to 40% bile BHI broth and incubate at 35°C. Transfer also to BHI broth and incubate at 45°C. Growth in 40% bile and at 45°C verify fecal streptococci.

5) *Escherichia coli* analyses—Verify by picking at least 10 well-isolated yellow or yellow-brown colonies from the urease test to nutrient agar plates or slants and to trypticase soy broth. Incubate the agar and broth cultures for 24 h at 35°C. With a platinum loop, remove an ample amount of growth from the nutrient agar and deposit on a filter paper previously saturated with cytochrome oxidase (CO) reagent freshly prepared that day. Development within 15 s of a deep purple color where the bacteria were deposited is a positive test.

Transfer growth from the trypticase soy broth to Simmons' citrate agar, tryptone broth, and EC broth in a fermentation tube. Incubate the Simmons' citrate agar for 24 h and tryptone broth for 48 h at 35°C. Incubate EC broth in a water bath at 44.5°C for 24 h. Make certain the water level is above the level of the EC broth in the tube. Add 0.5 mL Kovac's indole reagent to the 48-h tryptone broth culture and shake gently. Development of a deep red color in the alcohol layer indicates a positive test. *E. coli* is EC gas positive, indole positive, oxidase negative, and does not grow on citrate medium.

Alternately, use commercially available multi-test identification systems to verify colonies. Select a multi-test system for Enterobacteriaceae that includes lactose fermentation and/or ONPG and CO tests.

6) Enterococci analyses—Verify by picking at least 10 well-isolated pink to red colonies with black or reddish-brown precipitate from EIA agar into brain heart infusion (BHI) broth and onto a BHI agar slant. Incubate broth tubes for 24 h and slants for 48 h at 35°C. After 24 h incubation, transfer a loopful of material from BHI broth to Bile Esculin Agar (BEA) and BHI broth with 6.5% NaCl and incubate for 48 h at 35°C. Transfer another loopful of material from the 24 h BHI broth to

another tube of BHI broth and incubate for 48 h at 45°C. Observe for growth. Gram strain growth from the 48 h BHI agar slant. Gram-positive cocci that grow in BEA and hydrolyze esculin, and that grow in BHI broth at 10 and 45°C and in BHI broth + 6.5% NaCl are considered to be enterococci.

5. Records and Data Reporting

Keep records of microbiological analyses for at least 5 years. Actual laboratory reports may be kept, or data may be transferred to tabular summaries, provided that the following information is included: date, place, and time of sampling, name of sample collector; identification of sample; date of receipt of sample and analysis; person(s) responsible for performing analysis; analytical method used; the raw data and the calculated results of analysis. Verify that each result was entered correctly from the bench sheet and initialled by the analyst. If an information storage and retrieval system is used, double check data on the printouts.

6. Data Handling

a. Distribution of bacterial populations: In most chemical analyses the distribution of analytical results follows the Gaussian curve, which has symmetrical distribution of values about the mean (see Section 1010B). Microbial distributions are not necessarily symmetrical. Bacterial counts often are characterized as having a skewed distribution because of many low values and a few high ones. These characteristics lead to an arithmetic mean that is considerably larger than the median. The frequency curve of this distribution has a long right tail, such as that shown in Figure 9020:1, and is said to display positive skewness.

Application of the most rigorous statistical techniques requires the assumption of symmetrical distributions such as the nor-

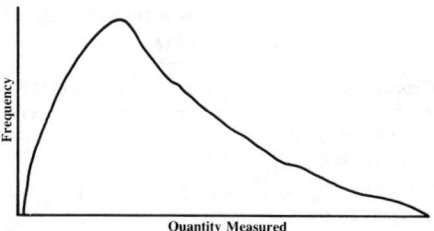

Figure 9020:1. Frequency curve (positively skewed distribution).

mal curve. Therefore it usually is necessary to convert skewed data so that a symmetrical distribution resembling the normal distribution results. An approximately normal distribution can be obtained from positively skewed data by converting numbers to their logarithms, as shown in Table 9020:VIII. Comparison of the frequency tables for the original data (Table 9020:IX) and their logarithms (Table 9020:X) shows that the logarithms approximate a symmetrical distribution.

b. Central tendency measures of skewed

TABLE 9020:VIII. COLIFORM COUNTS AND THEIR LOGARITHMS

MPN Coliform Count No./100 mL	log MPN
11	1.041
27	1.431
36	1.556
48	1.681
80	1.903
85	1.929
120	2.079
130	2.114
136	2.134
161	2.207
317	2.501
601	2.779
760	2.881
1020	3.009
3100	3.491
$\bar{x} = 442$	$\bar{x}_g$ = antilog 2.1825 = 152

TABLE 9020:IX. COMPARISON OF FREQUENCY
OF MPN DATA

Class Interval	Frequency (MPN)
0 to 400	11
400 to 800	2
800 to 1200	1
1200 to 1600	0
1600 to 2000	0
2000 to 2400	0
2400 to 2800	0
2800 to 3200	0

$$\log \bar{x}_g = \frac{\Sigma (\log x_i)}{n} = \frac{32.737}{15} = 2.1825$$

geometric mean

$$\bar{x}_g = \text{antilog } (2.1825) = 152$$

and arithmetic mean

$$\bar{x} = \frac{\Sigma x_i}{n} = \frac{6632}{15} = 442$$

Therefore, although regulations or tradition may require or cause microbiological data to be reported as the arithmetic mean or median, the preferred statistic for summarizing microbiological data is the geometric mean.

distribution: If the logarithms of numbers from a positively skewed distribution are approximately normally distributed, the original data have a log-normal distribution. The best estimate of central tendency of log-normal data is the geometric mean, defined as:

$$\bar{x}_g = n \sqrt{(x_1)(x_2) \cdots (x_n)}$$

and

$$\log \bar{x}_g = \frac{\Sigma (\log x_i)}{n}$$

that is, the geometric mean is equal to the antilog of the arithmetic mean of the logarithms. For example, the following means calculated from the data in Table 9020:VIII are drastically different.

TABLE 9020:X. COMPARISON OF FREQUENCY
OF LOG MPN DATA

Class Interval	Frequency (log MPN)
1.000 to 1.300	1
1.300 to 1.600	2
1.600 to 1.900	1
1.900 to 2.200	5
2.200 to 2.500	1
2.500 to 2.800	2
2.800 to 3.100	2
3.100 to 3.400	0
3.400 to 3.700	1

7. References

1. BORDNER, R.H., J.A. WINTER & P.V. SCARPINO, eds. 1978. Microbiological Methods for Monitoring the Environment, Water and Wastes. EPA-600/8-78-017, Environmental Monitoring & Support Lab., U.S. Environmental Protection Agency, Cincinnati, Ohio.
2. AMERICAN SOCIETY FOR TESTING AND MATERIALS. 1977. Annual Book of ASTM Standards, Part 31, Water. American Soc. Testing & Materials, Philadelphia, Pa.
3. U.S. ENVIRONMENTAL PROTECTION AGENCY. 1982. Manual for the Certification of Laboratories Analyzing Drinking Water, Criteria and Procedures. EPA-570/9-82-002, U.S. Environmental Protection Agency, Washington, D.C.

8. Bibliography

GREENBERG, A.E., J.S. THOMAS, T.W. LEE & W.R. GAFFEY. 1967. Interlaboratory comparisons in water bacteriology. *J. Amer. Water Works Assoc.* 59:237.
GELDREICH, E.E. 1971. Application of bacteriological data in potable water surveillance. *J. Amer. Water Works Assoc.* 63:225.
GELDREICH, E.E. 1975. Handbook for Evaluating Water Laboratories, 2nd ed. EPA-670/9-75-006, Municipal Environmental Research Lab., U.S. Environmental Protection Agency, Cincinnati, Ohio.
ELMUND, G.K., K.L. MARTIN, D.C. ERICKSON

& S.M. MORRISON. 1983. Interlaboratory total coliform testing for quality assurance. *In* Water Quality and Treatment: Advances in Laboratory Technology. Proc. 11th AWWA Water Quality Technology Conf., American Water Works Assoc., Denver, Colo.

BORDNER, R.H. 1985. Quality assurance for microbiological analyses of water. *In* J.K. Taylor & T.W. Stanley, eds. Quality Assurance for Environmental Measurements. ASTM STP 867, American Soc. Testing & Materials, Philadelphia, Pa.

9020 C. Interlaboratory Quality Control

1. Discussion

An interlaboratory quality control program is a system of agreed-upon requirements and laboratory practices necessary to maintain minimal quality standards among a group of participant laboratories. To establish such a program the participants first adopt uniform sampling procedures and standardized analytical methodology. Minimal standards are set for laboratory operations (personnel, facilities, equipment, supplies, data handling, and quality control).

After methods and laboratory operational standards are established, an independent agency inspects laboratory facilities and conducts methods validation and performance evaluation studies. Both method validation and performance evaluation studies specify the analytical methodology but the former is done to establish precision and bias (recovery) of selected methods while the latter evaluates laboratory performance and establishes acceptance limits. A critical part of interlaboratory quality control is to identify problems and follow up with technical assistance.

An example of interlaboratory quality assurance is the federal/state program for the certification of water supply laboratories, developed under the Safe Drinking Water Act and its amendments.[1]

In the certification program, public water supply laboratories must be certified according to minimal criteria and procedures and quality assurance described in the EPA manual on certification:[2] criteria are established for laboratory operations and methodology; on-site inspections are required by the certifying state agency or its surrogate to verify minimal standards; annually, laboratories are required to perform acceptably on unknown samples in formal studies, as samples are available; the responsible authority follows up on problems identified in the on-site inspection or performance evaluation and requires corrections within a set period of time. Individual state programs may exceed the federal criteria.

On-site inspections of laboratories in the present certification program show that primary causes for discrepancies in drinking water laboratories have been inadequate equipment, improperly prepared media, incorrect analytical procedures, and insufficiently trained personnel. The EPA manual of microbiological methods[3] provides added guidance on sampling, analytical methodology, and quality-assurance practices.

2. References

1. Safe Drinking Water Act, Public Law 93-523. Dec. 16, 1974. 88 Stat. 1660, 42 U.S. Code (USC) 300 f.
2. U.S. ENVIRONMENTAL PROTECTION AGENCY. 1982. Manual for the Certification of Laboratories Analyzing Drinking Water, Criteria and Procedures. EPA-570/9-82-002, U.S. Environmental Protection Agency, Washington, D.C.

3. BORDNER, R.H., J.A. WINTER & P.V. SCAR-
PINO, eds. 1978. Microbiological Methods for
Monitoring the Environment, Water and
Wastes. EPA 600/8-78-017, Environmental
Monitoring & Support Lab., U.S. Environmen-
tal Protection Agency, Cincinnati, Ohio.

9030   LABORATORY APPARATUS*

9030 A.   Introduction

This section contains specifications for microbiological laboratory equipment. For testing and maintenance procedures related to quality control, see Section 9020.

9030 B.   Equipment Specifications

1. Incubators

Incubators must maintain a uniform and constant temperature at all times in all areas, that is, they must not vary more than ±0.5°C in the areas used. Obtain such accuracy by using a water-jacketed or anhydric-type incubator with thermostatically controlled low-temperature electric heating units properly insulated and located in or adjacent to the walls or floor of the chamber and preferably equipped with mechanical means of circulating air.

Incubators equipped with high-temperature heating units are unsatisfactory, because such sources of heat, when improperly placed, frequently cause localized overheating and excessive drying of media, with consequent inhibition of bacterial growth. Incubators so heated may be operated satisfactorily by replacing high-temperature units with suitable wiring arranged to operate at a lower temperature and by installing mechanical air-circulation devices. It is desirable, where ordinary room temperatures vary excessively, to keep laboratory incubators in special rooms maintained at a few degrees below the recommended incubator temperature.

Alternatively, use special incubating rooms well insulated and equipped with properly distributed heating units, forced air circulation, and air exchange ports, provided that they conform to desired temperature limits. When such rooms are used, record the daily temperature range in areas where plates or tubes are incubated. Provide incubators with open metal wire or perforated sheet shelves so spaced as to assure temperature uniformity throughout the chamber. Leave a 2.5-cm space between walls and stacks of dishes or baskets of tubes.

Maintain an accurate thermometer, traceable to the National Institute of Standards and Technology (NIST, formerly National Bureau of Standards, NBS), with the bulb immersed in liquid (glycerine, water,

*Approved by Standard Methods Committee, 1988.

or mineral oil) on each shelf in use within the incubator and record daily temperature readings (preferably morning and afternoon). It is desirable, in addition, to maintain a maximum and minimum registering thermometer within the incubator on the middle shelf to record the gross temperature range over a 24-h period. At intervals, determine temperature variations within the incubator when filled to maximum capacity. Install a recording thermometer whenever possible, to maintain a continuous and permanent record of temperature.

Ordinarily, a water bath with a gabled cover to reduce water and heat loss, or a solid heat sink incubator, is required to maintain a temperature of 44.5 ± 0.2°C. If satisfactory temperature control is not achieved, provide water recirculation. Keep water depth in the incubator sufficient to immerse tubes to upper level of media.

2. Hot-Air Sterilizing Ovens

Use hot-air sterilizing ovens of sufficient size to prevent internal crowding; constructed to give uniform and adequate sterilizing temperatures of 170 ± 10°C; and equipped with suitable thermometers. Optionally use a temperature-recording instrument.

3. Autoclaves

Use autoclaves of sufficient size to prevent internal crowding; constructed to provide uniform temperatures within the chambers (up to and including the sterilizing temperature of 121°C); equipped with an accurate thermometer the bulb of which is located properly on the exhaust line so as to register minimum temperature within the sterilizing chambers (temperature-recording instrument is optional); equipped with pressure gauge and properly adjusted safety valves connected directly with saturated-steam supply lines equipped with appropriate filters to remove particulates and oil droplets or directly to a suitable special steam generator (do not use steam from a boiler treated with amines for corrosion control); and capable of reaching the desired temperature within 30 min. Confirm, by chemical or toxicity tests, that the steam supply has not been treated with amines or other corrosion-control chemicals that will impart toxicity.

Use of a vertical autoclave or pressure cooker is not recommended because of difficulty in adjusting and maintaining sterilization temperature and the potential hazard. If a pressure cooker is used in emergency or special circumstances, equip it with an efficient pressure gauge and a thermometer the bulb of which is 2.5 cm above the water level.

4. Gas Sterilizers

Use a sterilizer equipped with automatic controls capable of carrying out a complete sterilization cycle. As a sterilizing gas use ethylene oxide (CAUTION: *Ethylene oxide is toxic—avoid inhalation, ingestion, and contact with the skin. Also, ethylene oxide forms an explosive mixture with air at 3-80% proportion.*) diluted to 10 to 12% with an inert gas. Provide an automatic control cycle to evacuate sterilizing chamber to at least 0.06 kPa, to hold the vacuum for 30 min, to adjust humidity and temperature, to charge with the ethylene oxide mixture to a pressure dependent on mixture used, to hold such pressure for at least 4 h, to vent gas, to evacuate to 0.06 kPa, and finally, to bring to atmospheric pressure with sterile air. The humidity, temperature, pressure, and time of sterilizing cycle depend on the gas mixture used.

Store overnight sample bottles with loosened caps that were sterilized by gas, to allow last traces of gas mixture to dissipate. Incubate overnight media sterilized by gas, to insure dissipation of gas.

In general, mixtures of ethylene oxide

with chlorinated hydrocarbons such as freon are harmful to plastics, although at temperatures below 55°C, gas pressure not over 35 kPa, and time of sterilization less than 6 h, the effect is minimal. If carbon dioxide is used as a diluent of ethylene oxide, increase exposure time and pressure, depending on temperature and humidity that can be used.

Determine proper cycle and gas mixture for objects to be sterilized and confirm by sterility tests.

5. Optical Counting Equipment

a. Pour and spread plates: Use Quebec-type colony counter, dark-field model preferred, or one providing equivalent magnification (1.5 diameters) and satisfactory visibility.

b. Membrane filters: Use a binocular microscope with magnification of 10 to 15×. Provide daylight fluorescent light source at angle of 60 to 80° above the colonies; use low-angle lighting for nonpigmented colonies.

c. Mechanical tally.

6. pH Equipment

Use electrometric pH meters, accurate to at least 0.1 pH units, for determining pH values of media.

7. Balances

Use balances providing a sensitivity of at least 0.1 g at a load of 150 g, with appropriate weights. Use an analytical balance having a sensitivity of 1 mg under a load of 10 g for weighing small quantities (less than 2 g) of materials. Single-pan rapid-weigh balances are most convenient.

8. Media Preparation Utensils

Use borosilicate glass or other suitable noncorrosive equipment such as stainless steel. Use glassware that is clean and free of residues, dried agar, or other foreign materials that may contaminate media.

9. Pipets and Graduated Cylinders

Use pipets of any convenient size, provided that they deliver the required volume accurately and quickly. The error of calibration for a given manufacturer's lot must not exceed 2.5%. Use pipets having graduations distinctly marked and with unbroken tips. Bacteriological transfer pipets or pipets conforming to APHA standards may be used. *Do not pipet by mouth; use a pipet aid.*

Use graduated cylinders meeting ASTM Standards (D-86 and D-216) and with accuracy limits established by NIST where appropriate.

10. Pipet Containers

Use boxes of aluminum or stainless steel, end measurement 5 to 7.5 cm, cylindrical or rectangular, and length about 40 cm. When these are not available, paper wrappings for individual pipets may be substituted. To avoid excessive charring during sterilization, use best-quality sulfate pulp (kraft) paper. *Do not use copper or copper alloy cans or boxes as pipet containers.*

11. Refrigerator

Use a refrigerator maintaining a temperature of 1 to 4.4°C to store samples, media, reagents, etc. Do not store volatile solvents, food, or beverages in a refrigerator with media. Frost-free refrigerators may cause excessive media dehydration on storage longer than 1 week.

12. Temperature-Monitoring Devices

Use glass or metal thermometers graduated to 0.5°C to monitor most incubators and refrigerators. Use thermometers graduated to 0.1°C for incubators operated above 40°C. Use continuous recording devices that are equally sensitive. Verify accuracy by comparison with a NIST-certified thermometer, or equivalent.

13. Dilution Bottles or Tubes

Use bottles or tubes of resistant glass, preferably borosilicate glass, closed with glass stoppers or screw caps equipped with liners that do not produce toxic or bacteriostatic compounds on sterilization. Do not use cotton plugs as closures. Mark graduation levels indelibly on side of dilution bottle or tube. Plastic bottles of nontoxic material and acceptable size may be substituted for glass provided that they can be sterilized properly.

14. Petri Dishes

For the plate count, use glass or plastic petri dishes about 100 × 15 mm. Use dishes the bottoms of which are free from bubbles and scratches and flat so that the medium will be of uniform thickness throughout the plate. For the membrane filter technique use loose-lid glass or plastic dishes, 60 × 15 mm, or tight-lid dishes, 50 × 12 mm. Sterilize petri dishes and store in metal cans (aluminum or stainless steel, but not copper), or wrap in paper—preferably best-quality sulfate pulp (kraft)—before sterilizing.

15. Membrane Filtration Equipment

Use filter funnel and membrane holder made of seamless stainless steel, glass, or autoclavable plastic that does not leak and is not subject to corrosion. Field laboratory kits are acceptable but standard laboratory filtration equipment and procedures are required.

16. Fermentation Tubes and Vials

Use fermentation tubes of any type, if their design permits conforming to medium and volume requirements for concentration of nutritive ingredients as described subsequently. Where tubes are used for a test of gas production, enclose a shell vial, inverted. Use tube and vial of such size that the vial will be filled completely with medium, at least partly submerged in the tube, and large enough to make gas bubbles easily visible.

17. Inoculating Equipment

Use wire loops made of 22- or 24-gauge nickel alloy* or platinum-iridium for flame sterilization. Reusable transfer loops of aluminum or stainless steel are satisfactory. Use loops at least 3 mm in diameter. Sterilize by dry heat or steam. Single-service hardwood applicators also may be used. Make these 0.2 to 0.3 cm in diameter and at least 2.5 cm longer than the fermentation tube; sterilize by dry heat and store in glass or other nontoxic containers.

18. Sample Bottles

For bacteriological samples, use sterilizable bottles of glass or plastic of any suitable size and shape. Use bottles capable of holding a sufficient volume of sample for all required tests and an adequate air space, permitting proper washing, and maintaining samples uncontaminated until examinations are completed. Ground-glass-stoppered bottles, preferably wide-mouthed and of resistant glass, are recommended. Plastic bottles of suitable size, wide-mouthed, and made of nontoxic materials such as polypropylene that can be sterilized repeatedly are satisfactory as sample containers. Presterilized plastic bags, with or without dechlorinating agent, are available commercially and may be used. Plastic containers eliminate the possibility of breakage during shipment and reduce shipping weight.

Metal or plastic screw-cap closures with liners may be used on sample bottles provided that no toxic compounds are produced on sterilization.

Before sterilization, cover tops and necks of sample bottles having glass closures with aluminium foil or heavy kraft paper.

---

*Chromel, nichrome, or equivalent.

19. Bibliography

COLLINS, W.D. & H.B. RIFFENBURG. 1923. Contamination of water samples with material dissolved from glass containers. *Ind. Eng. Chem.* 15:48.

CLARK, W.M. 1928. The Determination of Hydrogen Ion Concentration, 3rd ed. Williams & Wilkins, Baltimore, Md.

ARCHAMBAULT, J., J. CUROT & M.H. McCRADY. 1937. The need of uniformity of conditions for counting plates (with suggestions for a standard colony counter). *Amer. J. Pub. Health* 27:809.

BARKWORTH, H. & J.O. IRWIN. 1941. The effect of the shape of the container and size of gas tube in the presumptive coliform test. *J. Hyg.* 41:180.

RICHARDS, O.W. & P.C. HEIJN. 1945. An improved dark-field Quebec colony counter. *J. Milk Technol.* 8:253.

COHEN, B. 1957. The measurement of pH, titratable acidity, and oxidation-reduction potentials. *In* Manual of Microbiological Methods. Society of American Bacteriologists. McGraw-Hill Book Co., New York, N.Y.

MORTON, H.E. 1957. Stainless-steel closures for replacement of cotton plugs in culture tubes. *Science.* 126:1248.

McGUIRE, O.E. 1964. Wood applicators for the confirmatory test in the bacteriological analysis of water. *Pub. Health Rep.* 79:812.

BORDNER, R.H., J.A. WINTER & P.V. SCARPINO, eds. 1978. Microbiological Methods for Monitoring the Environment, Water and Wastes. EPA-600/8-78-017, Environmental Monitoring & Support Lab., U.S. Environmental Protection Agency, Cincinnati, Ohio.

AMERICAN PUBLIC HEALTH ASSOCIATION. 1985. Standard Methods for the Examination of Dairy Products, 15th ed. American Public Health Assoc., Washington, D.C.

9040　WASHING AND STERILIZATION*

Cleanse all glassware thoroughly with a suitable detergent and hot water, rinse with hot water to remove all traces of residual washing compound, and finally rinse with laboratory-pure water. If mechanical glassware washers are used, equip them with influent plumbing of stainless steel or other nontoxic material. Do not use copper piping to distribute water. Use stainless steel or other nontoxic material for the rinse water system.

Sterilize glassware, except when in metal containers, for not less than 60 min at a temperature of 170°C, unless it is known from recording thermometers that oven temperatures are uniform, under which exceptional condition use 160°C. Heat glassware in metal containers to 170°C for not less than 2 h.

Sterilize sample bottles not made of plastic as above or in an autoclave at 121°C for 15 min.

For plastic bottles loosen caps before autoclaving to prevent distortion.

*Approved by Standard Methods Committee, 1988.

9050 PREPARATION OF CULTURE MEDIA*

9050 A. General Procedures

1. Storage of Culture Media

Store dehydrated media (powders) in tightly closed bottles in the dark at less than 30°C in an atmosphere of low humidity. Do not use them if they discolor or become caked and lose the character of a free-flowing powder. Purchase dehydrated media in small quantities that will be used within 6 months after opening. Additionally, use stocks of dehydrated media containing selective agents such as sodium azide, bile salts or derivatives, antibiotics, sulfur-containing amino acids, etc., of relatively current lot number (within a year of purchase) so as to maintain optimum selectivity. See also Section 9020.

Prepare culture media in batches that will be used in less than 1 week. However, if the media are contained in screw-capped tubes they may be stored for up to 3 months. See Table 9020:IV for specific details. Store media out of direct sun and avoid contamination and excessive evaporation.

Liquid media in fermentation tubes, if stored at refrigeration or even moderately low temperatures, may dissolve sufficient air to produce, upon incubation at 35°C, a bubble of air in the tube. Incubate fermentation tubes that have been stored at a low temperature overnight before use and discard tubes containing air.

Fermentation tubes may be stored at approximately 25°C; but because evaporation may proceed rapidly under these conditions—resulting in marked changes in concentration of the ingredients—do not store at this temperature for more than 1 week.

Discard tubes with an evaporation loss exceeding 1 mL.

2. Adjustment of Reaction

State reaction of culture media in terms of hydrogen ion concentration, expressed as pH.

The increase in hydrogen ion concentration (decrease in pH) during sterilization will vary slightly with the individual sterilizer in use, and the initial reaction required to obtain the correct final reaction will have to be determined. The decrease in pH usually will be 0.1 to 0.2 but occasionally may be as great as 0.3 in double-strength media. When buffering salts such as phosphates are present in the media, the decrease in pH value will be negligible.

Make tests to control adjustment to required hydrogen ion concentration with a pH meter. Measure pH of prepared medium as directed in Section 4500-H$^+$. Titrate a known volume of medium with a solution of NaOH to the desired pH. Calculate amount of NaOH solution that must be added to the bulk medium to reach this reaction. After adding and mixing thoroughly, check reaction and adjust if necessary. The required final pH is given in the directions for preparing each medium. If a specific pH is not prescribed, adjustment is unnecessary.

The pH of reconstituted dehydrated media seldom will require adjustment if made according to directions. Such factors as errors in weighing dehydrated medium or overheating reconstituted medium may produce an unacceptable final pH. Measure pH, especially of rehydrated selective media, regularly to insure quality control and media specifications.

*Approved by Standard Methods Committee, 1988.

3. Sterilization

After rehydrating a medium, dispense promptly to culture vessels and sterilize within 2 h. Do not store nonsterile media.

Sterilize all media, except sugar broths or broths with other specifications, in an autoclave at 121°C for 15 min after the temperature has reached 121°C. When the pressure reaches zero, remove medium from autoclave and cool quickly to avoid decomposition of sugars by prolonged exposure to heat. To permit uniform heating and rapid cooling, pack materials loosely and in small containers. Sterilize sugar broths at 121°C for 12 to 15 min. The maximum elapsed time for exposure of sugar broths to any heat (from time of closing loaded autoclave to unloading) is 45 min. Preferably use a double-walled autoclave to permit preheating before loading to reduce total needed heating time to within the 45-min limit.

4. Bibliography

BUNKER, G.C. & H. SCHUBER. 1922. The reaction of culture media. *J. Amer. Water Works Assoc.* 9:63.

AMERICAN PUBLIC HEALTH ASSOCIATION. 1972. Standard Methods for the Examination of Dairy Products, 13th ed. American Public Health Assoc., New York, N.Y.

LENNETTE, E.H., E.H. SPAULDING & J.P. TRUANT, eds. 1974. Manual of Clinical Microbiology. American Soc. Microbiology, Washington, D.C.

9050 B. Water

1. Specifications

To prepare culture media and reagents, use only distilled or demineralized water that has been tested and found free from traces of dissolved metals and bactericidal or inhibitory compounds. Toxicity in distilled water may be derived from fluoridated water high in silica. Other sources of toxicity are silver, lead, and various unidentified organic complexes. Where condensate return is used as feed for a still, toxic amines or other boiler compounds may be present in distilled water. Residual chlorine or chloramines also may be found in distilled water prepared from chlorinated water supplies. If chlorine compounds are found in distilled water, neutralize them by adding an equivalent amount of sodium thiosulfate or sodium sulfite.

Distilled water also should be free of contaminating nutrients. Such contamination may be derived from flashover of organics during distillation, continued use of exhausted carbon filter beds, deionizing columns in need of recharging, solder flux residues in new piping, dust and chemical fumes, and storage of water in unclean bottles. Store distilled water out of direct sunlight to prevent growth of algae. Good housekeeping practices usually will eliminate nutrient contamination.

See Section 9020 for distilled water suitability test.

2. Bibliography

STRAKA, R.P. & J.L. STOKES. 1957. Rapid destruction of bacteria in commonly used diluents and its elimination. *Appl. Microbiol.* 5:21.

GELDREICH, E.E. & H.F. CLARK. 1965. Distilled water suitability for microbiological applications. *J. Milk Food Technol.* 28:351.

MacLEOD, R.A., S.C. KUO & R. GELINAS. 1967. Metabolic injury to bacteria. II. Metabolic injury induced by distilled water or Cu^{++} in the plating diluent. *J. Bacteriol.* 93:961.

9050 C. Media Specifications

The need for uniformity dictates the use of dehydrated media. Never prepare media from basic ingredients when suitable dehydrated media are available. Follow manufacturer's directions for rehydration and sterilization. Commercially prepared media in liquid form (sterile ampule or other) also may be used if known to give equivalent results. See Section 9020 for quality-control specifications.

The terms used for protein source in most media, for example, peptone, tryptone, tryptose, were coined by the developers of the media and may reflect commercial products rather than clearly defined entities. It is not intended to preclude the use of alternative materials provided that they produce equivalent results.

NOTE—The term "percent solution" as used in these directions is to be understood to mean "grams of solute per 100 mL solution."

1. Dilution Water

a. Buffered water: To prepare stock phosphate buffer solution, dissolve 34.0 g potassium dihydrogen phosphate (KH_2PO_4), in 500 mL distilled water, adjust to pH 7.2 ± 0.5 with $1N$ sodium hydroxide (NaOH), and dilute to 1 L with distilled water.

Add 1.25 mL stock phosphate buffer solution and 5.0 mL magnesium chloride solution (81.1 g $MgCl_2 \cdot 6H_2O/L$ distilled water) to 1 L distilled water. Dispense in amounts that will provide 99 ± 2.0 mL or 9 ± 0.2 mL after autoclaving for 15 min.

b. Peptone water: Prepare a 10% solution of peptone in distilled water. Dilute a measured volume to provide a final 0.1% solution. Final pH should be 6.8.

Dispense in amounts to provide 99 ± 2.0 mL or 9 ± 0.2 mL after autoclaving for 15 min.

Do not suspend bacteria in any dilution water for more than 30 min at room temperature because death or multiplication may occur.

2. Culture Media

Specifications for individual media are included in subsequent sections. Details are provided where use of a medium first is described.

9060 SAMPLES*

9060 A. Collection

1. Containers

Collect samples for microbiological examination in bottles that have been cleansed and rinsed carefully, given a final rinse with distilled water, and sterilized as directed in Sections 9030 and 9040. For some applications samples may be collected in presterilized plastic bags.

2. Dechlorination

Add a reducing agent to containers intended for the collection of water having residual chlorine or other halogen unless they contain broth for direct planting of sample. Sodium thiosulfate ($Na_2S_2O_3$) is a satisfactory dechlorinating agent that neutralizes any residual halogen and prevents continuation of bactericidal action during

*Approved by Standard Methods Committee, 1988.

sample transit. The examination then will indicate more accurately the true microbial content of the water at the time of sampling.

For sampling chlorinated wastewater effluents add sufficient $Na_2S_2O_3$ to a clean sample bottle to give a concentration of 100 mg/L in the sample. In a 120-mL bottle 0.1 mL of a 10% solution of $Na_2S_2O_3$ will neutralize a sample containing about 15 mg/L residual chlorine. For drinking water samples, the concentration of dechlorinating agent may be reduced: 0.1 mL of a 3% solution of $Na_2S_2O_3$ in a 120-mL bottle will give a final concentration of 18 mg/L in the sample and will neutralize up to 5 mg/L residual chlorine. In emergency disinfection with higher concentrations of chlorine add sufficient dechlorination agent to give a concentration of 100 mg/L in the sample.

Cap bottle and sterilize either by dry or moist heat, as directed (Section 9040). Presterilized plastic bags containing $Na_2S_2O_3$ are available commercially.

Collect water samples high in copper or zinc and wastewater samples high in heavy metals in sample bottles containing a chelating agent that will reduce metal toxicity. This is particularly significant when such samples are in transit for 4 h or more. Use 372 mg/L of the disodium salt of ethylenediaminetetraacetic acid (EDTA). Adjust EDTA solution to pH 6.5 before use. Add EDTA separately to sample bottle before bottle sterilization (0.3 mL 15% solution in a 120-mL bottle) or combine it with the $Na_2S_2O_3$ solution before addition.

3. Sampling Procedures

When the sample is collected, leave ample air space in the bottle (at least 2.5 cm) to facilitate mixing by shaking, before examination. Collect samples that are representative of the water being tested, flush or disinfect sample ports, and use aseptic techniques to avoid sample contamination.

Keep sampling bottle closed until it is to be filled. Remove stopper and cap as a unit; do not contaminate inner surface of stopper or cap and neck of bottle. Fill container without rinsing, replace stopper or cap immediately, and if used, secure hood around neck of bottle.

a. Potable water: If the water sample is to be taken from a distribution-system tap without attachments, select a tap that is supplying water from a service pipe directly connected with the main, and is not, for example, served from a cistern or storage tank. Open tap fully and let water run to waste for 2 or 3 min, or for a time sufficient to permit clearing the service line. Reduce water flow to permit filling bottle without splashing. If tap cleanliness is questionable, apply a solution of sodium hypochlorite (100 mg NaOCl/L) to faucet before sampling; let water run for additional 2 to 3 min after treatment. Do not sample from leaking taps that allow water to flow over the outside of the tap. In sampling from a mixing faucet remove faucet attachments such as screen or splash guard, run hot water for 2 min, then cold water for 2 to 3 min, and collect sample as indicated above.

If the sample is to be taken from a well fitted with a hand pump, pump water to waste for about 5 min before collecting sample. If the well is equipped with a mechanical pump, collect sample from a tap on the discharge. If there is no pumping machinery, collect a sample directly from the well by means of a sterilized bottle fitted with a weight at the base; take care to avoid contaminating samples by any surface scum.

In drinking water evaluation, collect samples of finished water and from distribution sites selected to assure systematic coverage during each month. Carefully choose distribution system sample locations to include dead-end sections to demonstrate bacteriological quality throughout the network and to ensure that localized contamination does not occur through cross-connections, breaks in the distribution lines, or reduction in positive pressure. Sample locations may be public sites (police and fire stations, government office buildings, schools, bus and train stations, airports, community parks), commercial establishments (restaurants, gas stations, office buildings, industrial plants), private residences (single residences, apartment buildings, and townhouse complexes), and special sampling stations built into the distribution network. Establish sampling program in consultation with state and local health authorities.

b. Raw water supply: In collecting samples directly from a river, stream, lake, reservoir, spring, or shallow well, obtain samples representative of the water that is the source of supply to consumers. It is undesirable to take samples too near the bank or too far from the point of drawoff, or at a depth above or below the point of drawoff.

c. Surface waters: Stream studies may be short-term, high-intensity efforts. Select bacteriological sampling locations to include a baseline location upstream from the study area, industrial and municipal waste outfalls into the main stream study area, tributaries except those with a flow less than 10% of the main stream, intake points for municipal or industrial water facilities, downstream samples based on stream flow time, and downstream recreational areas. Dispersion of wastewaters into the receiving stream may necessitate preliminary cross-section studies to determine completeness of mixing. Where a tributary stream is involved, select the sampling point near the confluence with the main stream. Samples may be collected from a boat or from bridges near critical study points. Choose sampling frequency to be reflective of stream or water body conditions. For example, to evaluate waste dis-

charges, sample every 4 to 6 h and advance the time over a 7- to 10-d period.

To monitor stream and lake water quality establish sampling locations at critical sites. Sampling frequency may be seasonal for recreational waters, daily for water supply intakes, hourly where waste treatment control is erratic and effluents are discharged into shellfish harvesting areas, or even continuous.

d. Bathing beaches: Sampling locations for recreational areas should reflect water quality within the entire recreational zone. Include sites from upstream peripheral areas and locations adjacent to drains or natural contours that would discharge stormwater collections or septic wastes. Collect samples in the swimming area from a uniform depth of approximately 1 m. Consider sediment sampling of the water-beach (soil) interface because of exposure of young children at the water's edge.

To obtain baseline data on marine and estuarine bathing water quality include sampling at low, high, and ebb tides.

Relate sampling frequency directly to the peak bathing period, which generally occurs in the afternoon. Preferably, collect daily samples during the recognized bathing season; minimum sampling includes Friday, Saturday, Sunday, and holidays. When limiting sampling to days of peak recreational use, preferably collect a sample in the morning and the afternoon. Correlate bacteriological data with turbidity levels or rainfall over the watershed to make rapid assessment of water quality changes.

e. Sediments and sludges: The bacteriology of bottom sediments is important in water supply reservoirs, in lakes, rivers, and coastal waters used for recreational purposes, and in shellfish-growing waters. Sediments may provide a stable index of the general quality of the overlying water, particularly where there is great variability in its bacteriogical quality.

Sampling frequency in reservoirs and

lakes may be related more to seasonal changes in water temperatures and storm water runoff. Bottom sediment changes in river and estuarine waters may be more erratic, being influenced by stormwater runoff, increased flow velocities, and sudden changes in the quality of effluent discharges.

Bacteriological examination of sludges from water and wastewater treatment processes is desirable to determine the impact of their disposal into receiving waters, ocean dumping, or burial in landfill operations. Sludge monitoring also may indicate the effectiveness of wastewater treatment processes.

f. Manual sampling: Take samples from a river, stream, lake, or reservoir by holding the bottle near its base in the hand and plunging it, neck downward, below the surface. Turn bottle until neck points slightly upward and mouth is directed toward the current. If there is no current, as in the case of a reservoir, create a current artificially by pushing bottle forward horizontally in a direction away from the hand. When sampling from a boat, obtain samples from upstream side of boat. If it is not possible to collect samples from these situations in this way, attach a weight to base of bottle and lower it into the water. In any case, take care to avoid contact with bank or stream bed; otherwise, water fouling may occur.

g. Sampling apparatus: Special apparatus that permits mechanical removal of bottle stopper below water surface is required to collect samples from depths of a lake or reservoir. Various types of deep sampling devices are available. The most common is the ZoBell J-Z sampler, which uses a sterile 350-mL bottle and a rubber stopper through which a piece of glass tubing has been passed. This tubing is connected to another piece of glass tubing by a rubber connecting hose. The unit is mounted on a metal frame containing a cable and a messenger. When the messenger is released, it strikes the glass tubing at a point that has been slightly weakened by a file mark. The glass tube is broken by the messenger and the tension set up by the rubber connecting hose is released and the tubing swings to the side. Water is sucked into the bottle as a consequence of the partial vacuum created by sealing the unit at time of autoclaving. Commercial adaptations of this sampler and others are available.

Bottom sediment sampling also requires special apparatus. The sampler described by Van Donsel and Geldreich has been found effective for a variety of bottom materials for remote (deep water) or hand (shallow water) sampling. Construct this sampler preferably of stainless steel and fit with a sterile plastic bag. A nylon cord closes the bag after the sampler penetrates the sediment. A slide bar keeps the bag closed during descent and is opened, thereby opening the bag, during sediment sampling.

For sampling wastewaters or effluents the techniques described above generally are adequate; in addition see Section 1060.

4. Size of Sample

The volume of sample should be sufficient to carry out all tests required, preferably not less than 100 mL.

5. Identifying Data

Accompany samples by complete and accurate identifying and descriptive data. Do not accept for examination inadequately identified samples.

6. Bibliography

ZOBELL, C.E. 1941. Apparatus for collecting water samples from different depths for bacteriological analysis. *J. Mar. Res.* 4:173.

PUBLIC HEALTH LABORATORY SERVICE WATER SUB-COMMITTEE. 1953. The effect of sodium thiosulphate on the coliform and *Bacterium coli* counts of non-chlorinated water samples. *J. Hyg.* 51:572.

SHIPE, E.L. & A. FIELDS. 1956. Chelation as a

method for maintaining the coliform index in water samples. *Pub. Health Rep.* 71:974.

HOATHER, R.C. 1961. The bacteriological examination of water. *J. Inst. Water Eng.* 61:426.

COLES, H.G. 1964. Ethylenediamine tetra-acetic acid and sodium thiosulphate as protective agents for coliform organisms in water samples stored for one day at atmospheric temperature. *Proc. Soc. Water Treat. Exam.* 13:350.

VAN DONSEL, D.J. & E.E. GELDREICH. 1971. Relationships of Salmonellae to fecal coliforms in bottom sediments. *Water Res.* 5:1079.

DAHLING, D.R. & B.A. WRIGHT. 1984. Processing and transport of environmental virus samples. *Appl. Environ. Microbiol.* 47:1272.

9060 B.   Preservation and Storage

1. Holding Time and Temperature

Start microbiological examination of a water sample promptly after collection to avoid unpredictable changes. If samples cannot be processed within 1 h after collection, use an iced cooler for storage during transport to the laboratory. If it is known that the results will be used in legal action, employ a special messenger to deliver samples to the laboratory within 6 h and maintain chain of custody.

Hold temperature of all stream pollution, drinking, and wastewater samples below 10°C during a maximum transport time of 6 h. Refrigerate these samples upon receipt in the laboratory and process within 2 h. When local conditions necessitate delays in delivery of samples longer than 6 h, consider either making field examinations using field laboratory facilities located at the site of collection or using delayed-incubation procedures.

Unfortunately, these requirements seldom are realistic in the case of individual potable water samples shipped directly to the laboratory by mail, bus, etc., but the time elapsing between collection and examination should not exceed 24 h. Where refrigeration of individual water samples sent by mail is not possible, a thermos-type insulated sample bottle (or equivalent) that can be sterilized may be used. Record time and temperature of storage of all samples and consider this information in the interpretation of data.

2. Bibliography

CALDWELL, E.L. & L.W. PARR. 1933. Present status of handling water samples—Comparison of bacteriological analyses under varying temperatures and holding conditions, with special reference to the direct method. *Amer. J. Pub. Health* 23:467.

COX, K.E. & F.B. CLAIBORNE. 1949. Effect of age and storage temperature on bacteriological water samples. *J. Amer. Water Works Assoc.* 41:948.

PUBLIC HEALTH LABORATORY SERVICE WATER SUB-COMMITTEE. 1952. The effect of storage on the coliform and *Bacterium coli* counts of water samples. Overnight storage at room and refrigerator temperatures. *J. Hyg.* 50:107.

PUBLIC HEALTH LABORATORY SERVICE WATER SUB-COMMITTEE. 1953. The effect of storage on the coliform and *Bacterium coli* counts of water samples. Storage for six hours at room and refrigerator temperatures. *J. Hyg.* 51:559.

MCCARTHY, J.A. 1957. Storage of water sample for bacteriological examinations. *Amer. J. Pub. Health* 47:971.

LONSANE, B.K., N.M. PARHAD & N.U. RAO. 1967. Effect of storage temperature and time on the coliform in water samples. *Water Res.* 1:309.

LUCKING, H.E. 1967. Death rate of coliform bacteria in stored Montana water samples. *J. Environ. Health* 29:576.

MCDANIELS, A.E. & R.H. BORDNER. 1983. Effect of holding time and temperature on coliform numbers in drinking water. *J. Amer. Water Works Assoc.* 75:458.

MCDANIELS, A.E. et al. 1985. Holding effects on coliform enumeration in drinking water samples. *Appl. Environ. Microbiol.* 50:755.

9211 RAPID DETECTION METHODS*

9211 A. Introduction

There is a generally recognized need for methods that permit rapid estimation of the bacteriological quality of water. Applications of rapid methods may range from analysis of wastewater to potable water quality assessment. In the latter case, during emergencies involving water treatment plant failure, line breaks in a distribution network, or other disruptions to water supply caused by disasters, there is urgent need for rapid assessment of the sanitary quality of water.

Ideally, rapid procedures would be reliable and have sensitivity levels equal to those of the standard tests routinely used. However, sensitivity of a rapid test may be compromised because the bacterial limit sought may be below the minimum bacterial concentration essential to rapid detection. Rapid tests fall into two categories, those involving modified conventional procedures and those requiring special instrumentation and materials.

---

*Approved by Standard Methods Committee, 1988.

9211 B. Seven-House Fecal Coliform Test (SPECIALIZED)

This method[1,2] is similar to the fecal coliform membrane filter procedure (see Section 9222D) but uses a different medium and incubation temperature to yield results in 7 h that generally are comparable to those obtained by the standard fecal coliform method.

1. Medium

M-7 h FC agar: This medium may not be available in dehydrated form and may require preparation from the basic ingredients.

Proteose peptone No. 3		
or polypeptone	5.0	g
Yeast extract	3.0	g
Lactose	10.0	g
d-Mannitol	5.0	g
Sodium chloride, NaCl	7.5	g
Sodium lauryl sulfate	0.2	g
Sodium desoxycholate	0.1	g
Bromcresol purple	0.35	g
Phenol red	0.3	g

Agar	15.0	g
Distilled water	1	L

Heat in boiling water bath. After ingredients are dissolved heat additional 5 min. Cool to 55 to 60°C and adjust pH to 7.3 ± 0.1 with $0.1N$ NaOH (0.35 mL/L usually required). Cool to about 45°C and dispense in 4- to 5-mL quantities to petri plates with tight-fitting covers. Store at 2 to 10°C. Discard after 30 d.

2. Procedure

Filter an appropriate sample volume through a membrane filter, place filter on the surface of a plate containing M-7 h FC agar medium, and incubate at 41.5°C for 7 h. Fecal coliform colonies are yellow (indicative of lactose fermentation).

3. References

1. VAN DONSEL, D.J., R.M. TWEDT & E.E. GELDREICH. 1969. Optimum temperature for

quantitation of fecal coliforms in seven hours on the membrane filter. *Bacteriol. Proc. Abs.* No. G46, p. 25.

2. REASONER, D.J., J.C. BLANNON & E.E. GELDREICH. 1979. Rapid seven hour fecal coliform test. *Appl. Environ. Microbiol.* 38:229.

9211 C. Special Techniques (SPECIALIZED)

Special rapid techniques are summarized in Table 9211:I. Most are not sensitive enough for potable water quality measurement or are not specific. They may be useful in monitoring wastewater effluents and natural waters but require reagents not generally available, are tedious, or require special handling or incubation schemes incompatible with most water laboratory schedules. Except for the colorimetric test, none are suitable for routine use but they may be used as research tools. The user should refer to the literature citations for the technique listed in the table for procedural details, conditions for use, and method limitations. Only the adenosine triphosphate (ATP) procedure (the firefly bioluminescence system), the colorimetric test to estimate total microbial density, and a radiometric fecal coliform procedure that uses a ^{14}C-labeled substrate can be recommended.

Correlate initial concentration of bacteria with ATP concentration by extracting ATP from serial dilutions of a bacterial suspension, or for the ^{14}C radiometric method, standardize by determining the $^{14}CO_2$ released by known concentrations of fecal coliform organisms in natural samples, not pure cultures. In using any rapid procedure, determine the initial bacterial density by using an appropriate procedure such as heterotrophic plate count (Section 9215) or total (9221) or fecal (9222) coliforms, and correlate with results from the special rapid technique.

1. Bioluminescence Test (Total Viable Microbial Measurement)

The firefly luciferase test for ATP in living cells is based on the reaction between the luciferase enzyme, luciferin (enzyme substrate), magnesium ions, and ATP. Light is emitted during the reaction and can be measured quantitatively and correlated with the quantity of ATP extracted from known numbers of bacteria. When all reactants except ATP are in excess, ATP is the limiting factor. Addition of ATP drives the reactions, producing a pulse of light that is proportional to the ATP concentration.

The assay is completed in less than 1 h.[1-3] For monitoring microbial populations in water, the ATP assay is limited primarily by the need to concentrate bacteria from the sample to achieve the minimum ATP sensitivity level, which is 10^5 cells/mL. When combined with membrane filtration of a 1-L sample, ATP assay can provide the sensitivity level needed.

2. Radiometric Detection (Fecal Coliforms)

In this test, $^{14}CO_2$ is released from a ^{14}C-labeled substrate.[14] The technique permits presumptive detection of as few as 2 to 20 fecal coliform bacteria in 4.5 h. The test uses M-FC broth, uniformly labeled ^{14}C-mannitol, and two-temperature incubation; 2 h at 35°C followed by 2.5 h at 44.5°C for fecal coliform specificity. Add labeled substrate at start of 44.5°C incubation. Use membrane filtration to concentrate organisms from sample and place membrane filter in M-FC broth in a sealable container. The $^{14}CO_2$ released is trapped by exposure to Ba(OH)$_2$-saturated filter paper disk. ^{14}C activity is assayed by liquid scintillation spectrometry. Except for the use of the ^{14}C-

TABLE 9211:I. SPECIAL RAPID TECHNIQUES

Microbial Group	Rapid Method	Test Time h	Sensitivity cells/mL	Reference
Nonspecific microflora	Bioluminescence	1	100 000	1–3
	Chemiluminescence	1	500 000	3–5
	Impedance	3–12	100 000	6–9
	Colorimetric	0.02	10 000	10
	Epifluorescence/ fluorometric	< 1–several	—	11–13
Fecal coliforms	Radiometric	4–5	2–20	14
	Glutamate decarboxylase	10–13	0.01–500 000	15–17
	Electrochemical	1–7	1 000 000	18–20
	Impedance	6–12	200–100 000	6–9
	Gas chromatographic assay	9–12	> 50	21
	Colorimetric	8–20	5–130 000	22
	Potentiometric	3.5–15	0.1–> 10 000 000	23
Gram-negative bacteria	Limulus assay	2	500–3000	24–27
	Fluorescent antibody	2–3	—	28–30

mannitol substrate and liquid scintillation spectrometry to count the activity of the $^{14}CO_2$ released by the fecal coliforms, this procedure is similar to those given in Section 9222.

3. References

1. CHAPPELLE, E.W. & G.L. PICCIOLO. 1975. Laboratory Procedures Manual for the Firefly Luciferase Assay for Adenosine Triphosphate (ATP). NASA GSFC Doc. X-726-75-1, National Aeronautics & Space Admin., Washington, D.C.

2. PICCIOLO, G.L., E.W. CHAPPELLE, J.W. DEMING, R.R. THOMAS, D.A. NIBLE & H. OKREND. 1981. Firefly luciferase ATP assay development for monitoring bacterial concentration in water supplies. EPA-600/S2:81-014, U.S. Environmental Protection Agency, Cincinnati, Ohio; NTIS No. PB 88-103809/AS, National Technical Information Serv., Springfield, Va.

3. NELSON, W.H., ed. 1985. Instrumental Methods for Rapid Microbiological Analysis. VCH Publishers, Inc., Deerfield Beach, Fla.

4. SEITZ, W.R. & M.P. NEARY. 1974. Chemi-luminescence and bioluminescence. Anal. Chem. 46:188A.

5. OLENIAZ, W.S., M.A. PISANO, M.H. ROSENFELD & R.L. ELGART. 1968. Chemiluminescent method for detecting microorganisms in water. Environ. Sci. Technol. 2:1030.

6. WHEELER, T.G. & M.C. GOLDSCHMIDT. 1975. Determination of bacterial cell concentrations by electrical measurements. J. Clin. Microbiol. 1:25.

7. SILVERMAN, M.P. & E.F. MUNOZ. 1979. Automated electrical impedance technique for rapid enumeration of fecal coliforms in effluents from sewage treatment plants. Appl. Environ. Microbiol. 37:521.

8. MUNOZ, E.F. & M.P. SILVERMAN. 1979. Rapid, single-step most-probable-number method for enumerating fecal coliforms in effluents from sewage treatment plants. Appl. Environ. Microbiol. 37:527.

9. FIRSTENBERG-EDEN, R. & G. EDEN. 1984. Impedance Microbiology. John Wiley & Sons, Inc., New York, N.Y.

10. WALLIS, C. & J.L. MELNICK. 1985. An instrument for the immediate quantification of bacteria in potable waters. Appl. Environ. Microbiol. 49.1251.

11. BITTON, G., R.J. DUTTON & J.A. FORAN. 1984. A new rapid technique for counting microorganisms directly on membrane filters. *Stain Technol.* 58:343.

12. SIERACKI, M.E., P.W. JOHNSON & J.M. SIEBURTH. 1985. Detection, enumeration, and sizing of planktonic bacteria by image-analyzed epifluoresence microscopy. *Appl. Environ. Microbiol.* 49:799.

13. McCoy, W.F. & B.H. OLSON. 1985. Fluorometric determination of the DNA concentration in municipal drinking water. *Appl. Environ. Microbiol.* 49:811.

14. REASONER, D.J. & E.E. GELDREICH. 1978. Rapid detection of water-borne fecal coliforms by $^{14}CO_2$ release. *In* A.N. Sharpe & D.S. Clark, eds. Mechanizing Microbiology. Charles C. Thomas, Publisher, Springfield, Ill.

15. MORAN, J.W. & L.D. WITTER. 1976. An automated rapid test for *Escherichia coli* in milk. *J. Food Sci.* 41:165.

16. MORAN, J.W. & L.D. WITTER. 1976. An automated rapid method for measuring fecal pollution. *Water Sewage Works* 123:66.

17. TRINEL, P.A., N. HANOUNE & H. LeCLERC. 1980. Automation of water bacteriological analysis: running test of an experimental prototype. *Appl. Environ. Microbiol.* 39:976.

18. WILKINS, J.R., G.E. STONER & E.H. BOYKIN. 1974. Microbial detection method based on sensing molecular hydrogen. *Appl. Microbiol.* 27:947.

19. WILKINS, J.R. & E.H. BOYKIN. 1976. Analytical notes—electrochemical method for early detection of monitoring of coliforms. *J. Amer. Water Works Assoc.* 68:257.

20. GRANA, D.C. & J.R. WILKINS. 1979. Description and field test results of an *in situ* coliform monitoring system. NASA Tech. Paper 1334, National Aeronautics & Space Admin., Washington, D.C.

21. NEWMAN, J.S. & R.T. O'BRIEN. 1975. Gas chromatographic presumptive test for coliform bacteria in water. *Appl. Environ. Microbiol.* 30:584.

22. WARREN, L.S., R.E. BENOIT & J.A. JESSEE. 1978. Rapid enumeration of faecal coliforms in water by a colorimetric β-galactosidase assay. *Appl. Environ. Microbiol.* 35:136.

23. JOUENNE, T., G.-A. JUNTER & G. CARRIERE. 1985. Selective detection and enumeration of fecal coliforms in water by potentiometric measurement of lipoic acid reduction. *Appl. Environ. Microbiol.* 50:1208.

24. TENCATE, J.W., H.R. BULER, A. STURK & J. LEVIN. 1985. Bacterial Endotoxins. Structure, Biomedical Significance, and Detection with the *Limulus* Amebocyte Lysate Test. Alan R. Liss, Inc., New York, N.Y.

25. JORGENSEN, J.H., J.C. LEE, G.A. ALEXANDER & H.W. WOLF. 1979. Comparison of *Limulus* assay, standard plate count, and total coliform count for microbiological assessment of renovated wastewater. *Appl. Environ. Microbiol.* 37:928.

26. JORGENSEN, J.H. & G.A. ALEXANDER. 1981. Automation of the *Limulus* amebocyte lysate test by using the Abbott MS-2 microbiology system. *Appl. Environ. Microbiol.* 41:1316.

27. TSUGI, K., P.A. MARTIN & D.M. BUSSEY. 1984. Automation of chromogenic substrate *Limulus* amebocyte lysate assay method for endotoxin by robotic system. *Appl. Environ. Microbiol.* 48:550.

28. ABSHIRE, R.L. 1976. Detection of enteropathogenic *Escherichia coli* strains in wastewater by fluorescent antibody. *Can. J. Microbiol.* 22:365.

29. ABSHIRE, R.L. & R.K. GUTHRIE. 1973. Fluorescent antibody techniques as a method for the detection of fecal pollution. *Can. J. Microbiol.* 19:201.

30. THOMASON, B.M. 1981. Current status of immunofluorescent methodology. *J. Food Protect.* 44:381.

9211 D. Coliphage Detection (PROPOSED)

Coliphages are bacteriophages that infect and replicate in coliform bacteria and appear to be present wherever total and fecal coliforms are found. Correlations between coliphages and coliform bacteria in fresh water generally show that coliphages may be used to indicate the sanitary quality of water.[1-5] Because coliphages are more resistant to chlorine disinfection than total or fecal coliforms, they may be a better indicator of disinfection efficiency than coliform bacteria.[4] The quantitative relationship between coliphages and coliform bacteria in disinfected waters is different from that in natural fresh waters because of differences in their survival rates.

1. Materials and Culture Media

a. Host culture: Escherichia coli C, ATCC No. 13706.

b. Media:

1) Tryptic(ase) soy agar (TSA), to maintain E. coli C host stock cultures:

Tryptone (pancreatic digest of casein) or equivalent	15.0 g
Soytone (soybean peptone) or equivalent	5.0 g
Sodium chloride, NaCl	5.0 g
Agar	15.0 g
Distilled water.	1 L

pH should be 7.3 ± 0.1 at 25°C; if necessary, adjust pH with 0.1 or 1.0N NaOH or HCl. Heat to boiling to dissolve, then autoclave for 15 min at 121°C. For agar slants, dispense 5 to 8 mL in 16- × 125-mm screw-capped tubes before sterilizing; for plates, dispense 20 to 25 mL per petri dish after autoclaving and cooling to about 45°C.

2) Tryptic(ase) soy broth (TSB):

Tryptone (pancreatic digest of casein), or equivalent. . . .	17.0 g
Soytone (soybean peptone), or equivalent	3.0 g
Dextrose	2.5 g
Sodium chloride, NaCl	5.0 g
Dipotassium hydrogen phosphate, K_2HPO_4.	2.5 g
Distilled water.	1 L

pH should be 7.3 ± 0.1 at 25°C; adjust with 0.1 or 1.0N NaOH or HCl, if necessary. Warm and agitate to dissolve completely. Dispense in appropriate volumes

as needed; sterilize in autoclave for 15 min at 121°C.

3) Modified tryptic(ase) soy agar (MTSA): To the ingredients of TSB, add ammonium nitrate, NH_4NO_3, 1.60 g; strontium nitrate, $Sr(NO_3)_2$, 0.21 g; and agar, 15 g. pH should be 7.3 ± 0.1 at 25°C; if necessary, adjust pH with 0.1 or 1.0N NaOH or HCl. Heat to boiling to dissolve, dispense 5.5 mL in 16- × 25-mm screw-capped tubes, and autoclave for 15 min at 121°C.

4) Glycerine: Add 10% (w/v) to tryptic(ase) soy broth before autoclave sterilization.

5) 2,3,5-triphenyl tetrazolium chloride (TPTZ), 1% (w/v) in ethanol. Add to MTSA tempered at 45 to 46°C to enhance plaque visibility. Prepare fresh weekly.

2. Procedure

a. Frozen host preparation: Inoculate E. coli C from a stock agar slant (on TSA) into a tube(s) containing 10 mL TSB and 10% glycerine (w/v) and incubate overnight at 35°C. Then inoculate each tube into a flask containing 25 mL TSB plus 10% glycerine and incubate at 35 ± 0.5°C until an optical density of 0.5 at 520 nm is obtained (equivalent to about 1×10^9 E. coli C cells/mL). Measure optical density using a spectrometer. Zero spectrometer with sterile TSB plus 10% glycerine.

Aseptically dispense 4.5-mL portions of cell suspension in sterile plastic test tubes, cap, chill to 9°C, and freeze at −20°C. Store for no longer than 6 weeks in non-frost-free freezer to reduce loss of frozen host culture viability.

b. Assay procedure: The procedure is directly applicable to samples containing more than 5 coliphage/100 mL; if sample contains more than 1000 coliphage/100 mL, dilute sample 1:5 or 1:10 with sterile distilled water before proceeding.

Thaw tube(s) of frozen host E. coli C in 44.5°C water bath. Use one tube of host culture per sample. Add 1.0 mL of host *E. coli* C culture, 5 mL sample or dilution, and 0.08 mL TPTZ to each of four tubes of MTSA (melted and held at about 45°C).

Mix thoroughly and pour into separate 100- × 15-mm labeled petri dishes, cover, and let agar gel. Incubate inverted plates at 35°C. Count plaques after incubating for 4 to 6 h.

3. Interpreting and Reporting Results

Bacteriophage infect and multiply in sensitive bacteria. This results in lysis of the bacterial cells and a release of phage particles to infect adjacent cells. As the infected coliform bacteria are lysed, visible clear areas known as plaques develop in the lawn of confluent bacterial growth. Count plaques on each plate and record. Obtain the number of plaques/100 mL of sample by summing the plaques on the four plates and multiplying by 5. If a diluted sample has been used, additionally multiply by the reciprocal of the dilution factor.

Based on coliphage counts, estimate total and fecal coliform numbers as shown below.[4] Independently verify equations for specific types of samples and locations.

Total coliforms:

$$\log y = 0.627 \ (\log x) + 1.864$$

where:

y = total coliforms/100 mL and
x = coliphages/100 mL.

Fecal coliforms:

$$\log y = 0.805 \ (\log x) + 0.895$$

where:

y = fecal coliforms/100 mL and
x = coliphages/100 mL.

4. References

1. WENTZEL, R.S., P.E. O'NEILL & J.F. KITCHENS. 1982. Evaluation of coliphage detection as a rapid indicator of water quality. *Appl. Environ. Microbiol.* 43:430.
2. ISBISTER, J.D. & J.L. ALM. 1982. Rapid coliphage procedure for water treatment processes. *In* Proc. Amer. Water Works Assoc. Water Quality Technol. Conf., Seattle, Wash., Dec. 6-9, 1981.
3. ISBISTER, J.D., J.A. SIMMONS, W.M. SCOTT & J.F. KITCHENS. 1983. A simplified method for coliphage detection in natural waters. *Acta Microbiol. Polonica* 32:197.
4. KOTT, Y., N. ROZE, S. SPERBER & N. BETZER. 1974. Bacteriophages as viral pollution indicators. *Water Res.* 8:165.
5. KENNEDY, J.D., JR., G. BITTON & J.L. OBLINGER. 1985. Comparison of selective media for assay of coliphages in sewage effluent and lake water. *Appl. Environ. Microbiol.* 49:33.

9212 STRESSED ORGANISMS*

9212 A. Introduction

1. General Discussion

Indicator bacteria, including coliforms and fecal streptococci, may become stressed or injured in waters and wastewaters. These injured bacteria are incapa-

---

* Approved by Standard Methods Committee 1988.

ble of growth and colony formation under standard conditions because of structural or metabolic damage. As a result, a substantial portion of the indicator bacteria present, i.e., 10 to greater than 90%, may not be detected.[1] These false negative bacteriological findings could result in an inaccurate definition of water quality, or even worse, lead to the erroneous acceptance of a potentially hazardous condition resulting from the presence of resistant pathogens.[2]

Stressed organisms are present under ordinary circumstances in chlorinated drinking water and wastewater effluents, saline waters, polluted natural waters, and relatively clean surface waters. High numbers of injured indicator bacteria may be associated with partial or inadequate disinfection and the presence of metal ions or other toxic substances. These and other factors, including extremes of temperature and pH and solar radiation, may lead collectively to significant underestimations of the number of viable indicator bacteria.

Recent publications support the health significance of injured coliform bacteria.[3-5] These reports show that enteropathogenic bacteria are less susceptible than coliforms to injury under conditions similar to those in drinking water and that injured pathogens retain the potential for virulence. Hence, methods allowing for the enumeration of injured coliform bacteria yield more sensitive determinations of potential health risks. This conclusion is further supported by the observation that viruses and waterborne pathogens that form cysts also are more resistant to environmental stressors than indicator bacteria.

with concentrated medium, and exposure to untempered liquefied agar media. Excessive numbers of nonindicator bacteria also interfere with detection of indicators by causing injury.[7]

3. References

1. McFeters, G.A., J.S. Kippin & M.W. LeChevallier. 1986. Injured coliforms in drinking water. *Appl. Environ. Microbiol.* 51:1.
2. LeChevallier, M.W. & G.A. McFeters. 1985. Enumerating injured coliforms in drinking water. *J. Amer. Water Works Assoc.* 77:81.
3. LeChevallier, M.W., A. Singh, D.A. Schiemann & G.A. McFeters. 1985. Changes in virulence of waterborne enteropathogens with chlorine injury. *Appl. Environ. Microbiol.* 50:412.
4. Singh, A. & G.A. McFeters. 1986. Repair, growth and production of heat-stable enterotoxin by *E. coli* following copper injury. *Appl. Environ. Microbiol.* 51:738.
5. Singh, A., R. Yeager & G.A. McFeters. 1986. Assessment of in vivo revival, growth, and pathogenicity of *Escherichia coli* strains after copper- and chlorine-induced injury. *Appl. Environ. Microbiol.* 52:832.
6. McFeters, G.A., S.C. Cameron & M.W. LeChevallier. 1982. Influence of diluents, media and membrane filters on the detection of injured waterborne coliform bacteria. *Appl. Environ. Microbiol.* 43:97.
7. LeChevallier, M.W. & G.A. McFeters. 1985. Interactions between heterotrophic plate count bacteria and coliform organisms. *Appl. Environ. Microbiol.* 49:1338.

2. Sample Handling and Collection

Certain laboratory manipulations following sample collection also may produce injury or act as a secondary stress to the organisms.[6] These include excessive sample storage time, prolonged holding time (more than 30 min) of diluted samples before inoculation into growth media and of inoculated samples before incubation at the proper temperature, incorrect media formulations, incomplete mixing of sample

4. Bibliography

Clark, H.F., E.E. Geldreich, H.L. Jeter & P.W. Kabler. 1951. The membrane filter in sanitary bacteriology. *Pub. Health Rep.* 66:951.
McKee, J.E., R.T. McLaughlin & P. Lesgourgues. 1958. Application of molecular filter techniques to the bacterial assay of sewage. III. Effects of physical and chemical disinfection. *Sewage Ind. Wastes* 30:245.
Rose, R.E. & W. Litsky. 1965. Enrichment procedure for use with the membrane filter for the isolation and enumeration of fecal streptococci from water. *Appl. Microbiol.* 13:106.

MAXCY, R.B. 1970. Non-lethal injury and limitations of recovery of coliform organisms on selective media. *J. Milk Food Technol.* 33:445.

LIN, S.D. 1973. Evaluation of coliform tests for chlorinated secondary effluents. *J. Water Pollut. Control Fed.* 45:498.

BRASWELL, J.R. & A.W. HOADLEY. 1974. Recovery of *Escherichia coli* from chlorinated secondary sewage. *Appl. Microbiol.* 28:328.

STEVENS, A.P., R.J. GRASSO & J.E. DELANEY. 1974. Measurements of fecal coliform in estuarine water. *In* D.D. Wilt, ed., Proceedings of the 8th National Shellfish Sanitation Workshop, U.S. Dep. Health, Education, & Welfare, Washington, D.C.

BISSONNETTE, G.K., J.J. JEZESKI, G.A. MCFETERS & D.S. STUART. 1975. Influence of environmental stress on enumeration of indicator bacteria from natural waters. *Appl. Microbiol.* 29:186.

BISSONNETTE, G.K., J.J. JEZESKI, G.A. MCFETERS & D.S. STUART. 1977. Evaluation of recovery methods to detect coliforms in water. *Appl. Environ. Microbiol.* 33:590.

9212 B.   Recovery Enhancement

This section describes some general procedures and considerations regarding recovery of stressed indicator organisms.

For chlorinated samples, insure that sufficient dechlorinating agent is present in the sample bottle (see Section 9060A.2).[1] Collect water samples with elevated concentrations of heavy-metal ions in a sample bottle containing a chelating agent[2] (see Section 9060A.2) and minimize sample storage time (see Section 9060B). Use buffered peptone dilution water rather than buffered water (see Section 9050C.1) when preparing dilutions of samples containing heavy-metal ions. After making dilutions, inoculate test media within 30 min.

Resuscitation of stressed or injured organisms is enhanced by inoculating samples and initially culturing organisms in an enriched, noninhibitory medium at a moderate temperature.

Although no simple test is available to establish the presence of injured bacteria in a given sample, bacteria in water known to contain stressors such as disinfectants or heavy metals frequently will be injured.[1,3] When multiple-tube fermentation test results consistently are higher than those obtained from parallel membrane filter tests, or there is other indication of suboptimal recovery, consider injury probable and use one or more of the following procedures.

1. Recovery of Injured Total Coliform Bacteria Using Membrane Filtration

Use m-T7 agar[4] in the procedure described for the membrane filter test (see Section 9222A).

Proteose peptone No. 3	5.0 g
Yeast extract	3.0 g
Lactose	20.0 g
Tergitol 7	0.4 mL
Polyoxyethylene ether W1	5.0 g
Bromthymol blue	0.1 g
Bromcresol purple	0.1 g
Agar	15.0 g
Distilled water	1 L

Adjust to pH 7.4 with $0.1N$ NaOH after sterilization at 121°C for 15 min. Aseptically add 1.0 µg penicillin G/mL when medium has cooled to about 45°C.

After filtering sample place filter on m-T7 agar and incubate at 35°C for 22 to 24 h. Coliform colonies are yellow. Verify not less than 10% of coliform colonies by the procedure in Section 9222A. 5*f.* With some drinking water samples with many noncoliform bacteria, confluent growth occurs. To obtain reliable results, carefully distinguish target yellow colonies from background growth.

2. Recovery of Injured Fecal Coliform Bacteria Using Membrane Filtration

a. Enrichment-temperature acclimation: Use two-layer agar (M-FC agar with a nonselective overlay medium) with a 2-h incubation at 35°C followed by 22 h at 44.5°C.[5] Prepare the M-FC agar plate in advance but do not add the overlay agar more than 1 h before use.

Alternatively, use a preenrichment in phenol red lactose broth incubated at 35°C for 4 h followed by M-FC agar at 44.5°C for 22 h.[6]

As a third option, prepare enrichment two-layer medium containing specific additives and incubate for 1.5 h at room temperature (22 to 26°C) followed by 35°C for 4.5 h and 44.5°C for 18 h.[7]

b. Temperature acclimation:[8] Modify elevated temperature procedure by preincubation of M-FC cultures for 5 h at 35°C, followed by 18 ± 1 h at 44.5°C. Use a commercially available temperature-programmed incubator to make the change from 35 to 44.5°C after the 5 h preincubation period to eliminate inconvenience and provide a practical method of analysis.

c. Deletion of suppressive agent:[9] Eliminate rosolic acid from M-FC medium and incubate cultures at 44.5°C ± 0.2°C for 24 h. Fecal coliform colonies are intense blue on the modified medium and are distinguished from the cream, gray, and pale-green colonies typically produced by nonfecal coliforms.

d. Alternative medium-temperature acclimation: Use m-T7 medium with an 8 h incubation at 37°C followed by 12 h at 44.5°C.[10]

e. Verification of stressed fecal coliform bacteria: Modifications of media and procedures may decrease selectivity and differentiation of fecal coliform colonies. Therefore, if any procedural modifications are used, verify not less than 10% of the blue colonies from a variety of samples. Use lauryl tryptose broth (Section 9221A)

(35°C for 48 h) with transfer of gas-producing cultures to EC broth (Section 9221C) (44.5°C for 24 h). Gas production at 44.5°C confirms the presence of fecal coliforms.

3. Recovery of Stressed Fecal Streptococci Using Membrane Filtration

Using bile broth medium yields fecal streptococcus recoveries comparable with multiple-tube fermentation tests.[11] Preincubate membrane filters on an enrichment medium for 2 h at 35°C and follow by plating on m-Enterococcus agar (Section 9230) for 48 ± 2 h at 35°C.

Verification of stressed fecal streptococci—Verify not less than 10% of the colonies from a variety of samples using the confirmed test procedure given in Section 9230B.5.

4. References

1. McFeters, G.A. & A.K. Camper. 1983. Enumeration of coliform bacteria exposed to chlorine. *In* A.I. Laskin, Advances in Applied Microbiology. 29:177.
2. Domek, M.J., M.W. LeChevallier, S.C. Cameron & G.A. McFeters. 1984. Evidence for the role of metals in the injury process of coliforms in drinking water. *Appl. Environ. Microbiol.* 48:289.
3. LeChevallier, M.W. & G.A. McFeters. 1985. Interactions between heterotrophic plate count bacteria and coliform organisms. *Appl. Environ. Microbiol.* 49:1338.
4. LeChevallier, M.W., S.C. Cameron & G.A. McFeters. 1983. New medium for the improved recovery of coliform bacteria from drinking water. *Appl. Environ. Microbiol.* 45:484.
5. Rose, R.E., E.E. Geldreich & W. Litsky. 1975. Improved membrane filter method for fecal coliform analysis. *Appl. Microbiol.* 29:532.
6. Lin, S.D. 1976. Membrane filter method for recovery of fecal coliforms in chlorinated sewage effluents. *Appl. Environ. Microbiol.* 32:547.
7. Stuart, D.S, G.A. McFeters & J.E. Schillinger. 1977. Membrane filter technique for quantification of stressed fecal coliforms in the

aquatic environment. *Appl. Environ. Microbiol.* 34:42.

8. GREEN, B.L., E.M. CLAUSEN & W. LITSKY. 1977. Two-temperature membrane filter method for enumerating fecal coliform bacteria from chlorinated effluents. *Appl. Environ. Microbiol.* 33:1259.

9. PRESSWOOD, W.G. & D. STRONG. 1977. Modification of M-FC medium by eliminating rosolic acid. Amer. Soc. Microbiol. Abs. Annu. Meeting. ISSN-0067-2777:272.

10. LECHEVALLIER, M.W., P.E. JAKANOSKI, A.K. CAMPER & G.A. MCFETERS. 1984. Evaluation of m-T7 agar as a fecal coliform medium. *Appl. Environ. Microbiol.* 48:371.

11. LIN, S.D. 1974. Evaluation of fecal streptococci tests for chlorinated secondary effluents. *J. Environ. Eng. Div., Proc. Amer. Soc. Civil Engr.* 100:253.

9213 MICROBIOLOGICAL EXAMINATION OF RECREATIONAL WATERS*

9213 A. Introduction

1. Significance of Microorganisms

Recreational waters can be categorized as freshwater swimming pools, whirlpools, and naturally occurring fresh and marine surface waters. Historically, these waters have been examined for coliform and/or heterotrophic bacteria. Although detection of coliform bacteria in water indicates that it may be unsafe to drink, other bacteria have been isolated from recreational waters that may suggest public health risks through body contact, ingestion, or inhalation.

Bacteria suggested as indicators of recreational water quality include the coliform group, species of *Pseudomonas, Streptococcus, Staphylococcus*, and in rare cases, *Legionella.*

Viral diseases associated with untreated recreational water include those caused by Coxsackie A and B, adenovirus types 3 and 4, hepatitis A, and a variety of gastroenteritis viruses. The last group may be responsible for a substantial proportion of undefined gastroenteritis cases.

Such organisms as *Mycobacterium, Candida albicans*, and species of *Naegleria* and *Acanthamoeba* also may be important, especially because they may produce spores and cysts that are more resistant than indicator bacteria.

Routine examination for pathogenic microorganisms is not recommended except for special studies or for investigations of water-related illness, at which time microbiological assays should be focused on the known or suspected pathogen.

Described below are acceptable, available methods for microbial indicators of recreational water quality. Consider the type of water to be examined in selecting the microbiological method(s) or indicator(s) to be used. No single procedure is adequate to isolate all microorganisms from water and the presence of one microorganism does not signify presence or absence of any other.

2. Bibliography

COWAN, S.T. & K.J. STEEL. 1965. Manual for the Identification of Medical Bacteria. Cambridge Univ. Press, Cambridge, England.

*Approved by Standard Methods Committee, 1988.

9213 B. Swimming Pools

1. General Discussion

a. *Characteristics:* A swimming pool is a body of water of limited size contained in a holding structure.[1] The water generally is chlorinated potable water but it also may be derived from thermal springs or salt water. The modern pool has a recirculating system so that the water can be filtered and disinfected. Pools should be characterized as disinfected or untreated.

b. *Monitoring requirements:*

1) General—Monitor water quality in pools for changes in chemical and physical characteristics that may result in irritation to the bather's skin, eyes, and mucosal barriers or may adversely affect disinfection. Microorganisms of concern typically are those from the bather's body and its orifices and include those causing infections of the ear, upper respiratory tract, skin, and intestinal or genitourinary tracts. Water quality depends on the efficacy of disinfection, the number of bathers in the pool at any one time, and the total number of bathers per day.

2) Disinfected indoor pools—Periodically determine the residual levels of disinfectant and turbidity during periods of peak bather load, or when the turbidity is above 1 NTU and the heterotrophic plate count (HPC) exceeds 500 colony-forming units/mL. The heterotrophic plate count is the primary indicator of disinfection efficiency. Other supporting indicators may include normal skin flora that are likely to be shed, including species of *Streptococcus*, *Staphylococcus*, and *Pseudomonas*. These organisms account for a large percentage of swimming-pool-associated illnesses and may be relatively resistant to the effect of chlorine.[2] In special circumstances *Myco-* *bacterium*, *Legionella*, or *Candida albicans* may be significant.

3) Disinfected outdoor pools—Fecal coliform bacteria are the primary indicators of contamination from animal pets, rodents, stormwater runoff, and human sources. Supporting indicators include the heterotrophic plate count and species of *Streptococcus, Staphylococcus, and Pseudomonas.*[3,4]

4) Untreated pools—The primary indicator may be fecal coliform bacteria. Supporting indicators are those described for disinfected pools.

2. Samples

a. *Containers:* Collect samples for bacteriological examination of swimming pool water as directed in Section 9060A. Use containers with capacities of 120 to 480 mL, depending on analyses to be made. Add sodium thiosulfate, $Na_2S_2O_3$, as a disinfectant neutralizer in an amount sufficient to provide an approximate concentration of 100 mg/L in the sample. Do this by adding 0.1 mL of 10% solution of $Na_2S_2O_3$ to a 120-mL bottle or 0.4 mL to a 480-mL bottle. After adding $Na_2S_2O_3$, stopper or cap the bottle and sterilize.

b. *Sampling procedure:* Collect samples in the area of, and during the time of, maximum bather density. Information on bathing load also will be helpful in subsequent interpretation of laboratory results.

For pools equipped with a filter, samples may be collected conveniently from sampling cocks provided in the return and discharge lines from the filter. Alternatively, carefully remove cap of a sterile sample bottle and hold the bottle near its base at an angle of 45°. Plunge bottle vertically into the water approximately 20 cm to fill, while making sure that the dechlorinating agent (necessary only for disinfected pools) is not washed out. Collect samples where water depth is approximately 1 m. To prevent sample contamination during collection, attach bottle to a device such as a sampling pole or wear sterile rubber or plastic gloves.

Because most bacteria shed by bathers are in body oils, saliva, and mucus discharges that layer near the surface, collect additional samples of the surface microlayer in 1-m-deep water. Collect microlayer samples by plunging a sterile glass plate (approximately 20 cm by 20 cm) vertically through the water surface and withdrawing it upward at a rate of approximately 6 cm/s. Remove surface film and water layer adhering to both sides of plate with a sterile silicone rubber scraper and collect in a sterile glass bottle. Repeat until desired volume is obtained. To minimize microbial contamination, wrap glass plate and scraper in metal foil and sterilize by autoclaving before use. Wear sterile rubber or plastic gloves during sampling or hold glass plate with forceps, clips, or tongs.

Determine residual chlorine or other disinfectant at poolside at the time of sample collection.

c. Sample storage: See Section 9060B.

d. Sample volume: See Section 9222.

e. Sample dilution: If sample dilutions are required, use 0.1% peptone water as diluent to optimize recovery of stressed organisms (see Section 9222 for suggested sample volume). Because peptone water has a tendency to foam, avoid including air bubbles when pipetting to assure accurate measure. Process sample promptly to prevent bacterial growth.

3. Heterotrophic Plate Count

Determine the plate count as directed in Section 9215. Use at least two plates per dilution.

4. Tests for Fecal Coliforms

Test for fecal coliforms according to the multiple-tube fermentation technique (Section 9221), the membrane filter technique (Section 9222), or rapid methods (Section 9211).

5. Test for Staphylococci (PROPOSED)

a. Mannitol salt agar:

Beef extract.............	1.0	g
Proteose peptone.........	10.0	g
Sodium chloride, NaCl....	75.0	g
d-Mannitol..............	10.0	g
Agar....................	15.0	g
Phenol red..............	0.025	g
Distilled water...........	1	L

Sterilize by autoclaving; pH should be 7.4.

b. Procedure: Use the membrane filter technique (see Section 9222) to prepare sample. Place filter on mannitol salt agar and incubate at 37°C for 36 h. *Staphylococcus* colonies are yellow or surrounded by yellow zone.

6. Test for *Staphylococcus aureus*

Use a modified multiple-tube procedure.

a. Media:

1) *M-staphylococcus broth:*

Tryptone................	10.0	g
Yeast extract............	2.5	g
Lactose.................	2.0	g
Mannitol................	10.0	g
Dipotassium hydrogen phosphate, K₂HPO₄.....	5.0	g
Sodium chloride, NaCl....	75.0	g
Sodium azide, NaN₃......	0.049	g
Distilled water...........	1	L

Sterilize by boiling for 4 min; pH should be 7.0 ± 0.2. For 10-mL inocula prepare and use double-strength medium.

2) *Lipovitellin-salt-mannitol agar:* This medium may not be available in dehydrated form and may require preparation

from the basic ingredients or by addition of egg yolk to a dehydrated base.

Beef extract..............	1.0	g
Polypeptone	10.0	g
Sodium chloride, NaCl	75.0	g
d-Mannitol	10.0	g
Agar	15.0	g
Phenol red	0.025	g
Egg yolk	20.0	g
Distilled water	1	L

Sterilize by autoclaving; pH should be 7.4 ± 0.2.

b. *Procedure:* Inoculate tubes of M-staphylococcus broth as directed in Section 9221. Incubate at 35 ± 1°C for 24 h. Hold original enrichment sample but streak from positive (turbid) tubes on plates of lipovitellin-salt-mannitol agar and incubate at 35 ± 1°C for 48 h. Opaque (24 h), yellow (48 h) zones around the colonies are positive evidence of lipovitellin-lipase activity (opaque) and mannitol fermentation (yellow).

Restreak negative plates from the original enrichment tube cultures before discarding tube. Lipovitellin-lipase activity has a 95% positive correlation with coagulase production. If necessary, confirm positive isolates as catalase-negative, coagulase-positive, fermenting mannitol, fermenting glucose anaerobically, yielding typical microscopic morphology, and gram-positive.

7. Tests for *Pseudomonas aeruginosa*

Tests for *P. aeruginosa* are presented in Sections 9213E and F and include a membrane filter procedure and a multiple-tube technique.

8. Test for *Streptococci*

Determine fecal streptococci as described in Section 9230, and if necessary, make additional biochemical tests to identify species.

9. References

1. CENTERS FOR DISEASE CONTROL. 1983. Swimming Pools—Safety and Disease Control through Proper Design and Operation. DHHS—CDC No. 83-8319, Centers for Disease Control, Atlanta, Ga.
2. SEYFRIED, P.L., R.S. TOBIN, N.E. BROWN & P.F. NESS. 1985. A prospective study of swimming related illness. II. Morbidity and the microbiological quality of water. *Amer. J. Pub. Health* 75:1071.
3. FAVERO, M.S. & C.H. DRAKE. 1964. Comparative study of microbial flora of iodated and chlorinated pools. *Pub. Health Rep.* 79:251.
4. KEIRN, M.A. & H.D. PUTNAM. 1968. Resistance of staphylococci to halogens as related to a swimming pool environment. *Health Lab. Sci.* 3:180.

10. Bibliography

WORKING PARTY OF THE PUBLIC HEALTH LABORATORY SERVICE. 1965. A bacteriological survey of swimming baths in primary schools. *Monthly Bull. Min. Health & Pub. Health Lab. Serv.* 24:116.

GRUN, L.R. & H. KLEYBRINK. 1972. Staphylokokken-Mikrokokken im Badewasser. *Zentralbl. Bakteriol. Parasitenk. Infektionskr. Hyg. Abt. Orig. B.* 155:384.

ADAMS, J.C. 1972. Unusual organism which gives a positive elevated temperature test for fecal coliforms. *Appl. Microbiol.* 23:172.

GUNN, B.A., W.E. DUNKELBERG, JR. & J.R. CRUTZ. 1972. Clinical evaluation of 2% LSM medium for primary isolation and identification of staphylococci. *Amer. J. Clin. Pathol.* 57:236.

HATCHER, R.F. & B.C. PARKER. 1974. Investigations of Freshwater Surface Microlayers. VPI-SRRC-BULL 64. Virginia Polytechnic Inst. and State Univ., Blacksburg.

CLARK, N.A. & W.F. HILL. 1977. Disinfection of drinking water, swimming-pool-water and treated sewage effluent. *In* Disinfection, Sterilization and Preservation. Lea & Febiger, Philadelphia, Pa.

9213 C.   Whirlpools

1. General Discussion

a. Characteristics: A whirlpool is a shallow pool with a maximum water depth of 1.2 m; it has a closed-cycle water system, a heated water supply, and usually a hydrojet recirculation system. It may be constructed of plastic, fiberglass, redwood, or epoxy-lined surfaces. Whirlpools are designed for recreational as well as therapeutic use and may accommodate one or more bathers. They are located in homes, apartments, hotels, athletic facilities, rehabilitation centers, and hospitals. The bather population is usually transient; even whirlpools in homes may not be used exclusively by family members.

b. Monitoring requirements: Whirlpool-associated infections are common. To ensure whirlpool safety, test water for proper germicide concentration, pH, and temperature. Other scheduled maintenance and testing operations have been described.[1-4]

To make microbiological assays in the event of an outbreak of whirlpool-associated infection, collect samples at the time of, or as close to the time of, the outbreak as possible.[5] Measure pH and temperature of the water, as well as the free and combined chlorine (or other halogen). Refrigerate sample immediately and preferably process microbiologically within 6 to 12 h.

Microbiological analyses should include direct assay for *P. aeruginosa* by means of selective-differential or enrichment technique(s) as described in 9213E and 9213F. Although *P. aeruginosa* is the most frequently incriminated pathogen, also measure staphylococci and streptococci. The standard indices of pollution, i.e., coliform bacteria, are insufficient to judge the microbiological quality of whirlpool water.

Assays for coliform bacteria might indicate the efficacy of the disinfectant used, but in all likelihood, they will not represent the etiologic agent of whirlpool-associated infections.

2. References

1. CENTERS FOR DISEASE CONTROL. 1981. Suggested Health and Safety Guidelines for Public Spas and Hot Tubs. DHHS-CDC #99-960. United States Government Printing Off., Washington, D.C.
2. SOLOMON, S.L. 1985. Host factors in whirlpool-associated *Pseudomonas aeruginosa* skin disease. *Infect. Control* 6:402.
3. HIGHSMITH, A.K., P.N. LEE, R.F. KHABBAZ & V.P. MUNN. 1985. Characteristics of *Pseudomonas aeruginosa* isolated from whirlpools and bathers. *Infect. Control* 6:407.
4. GROOTHUIS, D.G., A.H. HAVELAAR & H.R. VEENENDAAL. 1985. A note on legionellas in whirlpools. *J. Appl. Bacteriol.* 58:479.
5. HIGHSMITH, A.K. & M.S. FAVERO. 1985. Microbiological aspects of public whirlpools. *Clin. Microbiol. Newsletter* 7:9.

3. Bibliography

GELDREICH, E.E., A.K. HIGHSMITH & W.J. MARTONE. 1985. Public whirlpools—the epidemiology and microbiology of disease. *Infect. Control* 6:392.

9213 D.   Natural Bathing Beaches

1. General Discussion

A natural bathing beach is any shoreline area of a stream, lake, ocean, impoundment, or hot spring that is used for recreation. A wide variety of pathogenic microorganisms can be transmitted to humans through use of natural fresh and marine recreational waters that may be contaminated by wastewater.[1,2] These include enteropathogenic agents, such as *Salmonella, Shigella,* enteroviruses, protozoa, and multicellular parasites; human pathogens or "opportunists," such as *P. aeruginosa, Klebsiella, Vibrio parahemolyticus, V. vulnificus, Aeromonas hydrophila,* and *C. albicans,* which may multiply in recreational waters in the presence of sufficient nutrients; organisms carried into the water from the skin and upper orifices of the recreationists, such as *Staphylococcus aureus;* and other organisms, e.g., pathogenic mycobacteria and leptospira, *Francisella tularensis,* and pathogenic *Naegleria* species (amoebic meningoencephalitis).[3–9]

Methods suitable for the routine examination of recreational waters are not available for all of the above organisms. Even with the methods described herein, and particularly with reference to the marine environment, there may be local conditions that compromise the accuracy or selective and differential characteristics of these methods.

Measure fecal contamination of naturally occurring recreational waters with the fecal coliform test. Because of the possibility of false positives in thermal waters, verify suspected fecal coliform positives by commercial multi-test systems [Section 9222B.5*f* 2)b)].

An application of the fecal streptococcus test in natural waters is in the development of fecal coliform:fecal streptococcus ratios. Fecal coliform:fecal streptococcus ratios of 4.0 or higher typically indicate domestic waste while ratios of 0.6 or lower are common to discharges from farm animals or stormwater runoff. Optimally, determine this ratio near the point of waste discharge because the ratio changes with distance or time. Because several biotypes of *Streptococcus faecalis* are ubiquitous in the environment, it may be difficult to interpret the sanitary significance of densities of fecal streptococci below 100 organisms/100 mL.

To determine the ratio, use M-FC and KF agar or PSE agar in the membrane filter technique or the multiple-tube procedures described for fecal coliforms and fecal streptococci (see Sections 9221, 9222, and 9230).

The test for *E. coli* can be applied to fresh, estuarine, and marine waters. Methods are available for *P. aeruginosa, Salmonella,* and *Klebsiella.* The enumeration of *P. aeruginosa, Aeromonas hydrophila,* and *Klebsiella* species in recreational waters can be of considerable value with reference to the discharge of highly nutritive wastes, e.g., pulp mill wastes, effluents from textile finishing plants, into receiving waters.

2. Samples

a. Containers: Collect samples as directed in Section 9060A. The size of the container varies with the number and variety of tests to be performed. Adding $Na_2S_2O_3$ to the bottle is unnecessary.

b. Sampling procedure: Collect samples 0.3 m below the water surface in the areas of greatest bather density. Take samples over the range of environmental and climatic conditions, especially during times when maximal pollution can be expected, i.e., periods of tidal, current, and wind influences, stormwater runoff, sewage bypassing. See Section 9213B.2*b* for methods of sample collection and Section 9222 for suggested sample volumes.

c. Sample storage: See Section 9060B.

3. Tests for *Escherichia coli*

1) mTEC agar:*

Proteose peptone	5.0	g
Yeast extract	3.0	g
Lactose	10.0	g
Sodium chloride, NaCl	7.5	g

*Difco or equivalent.

Dipotassium phosphate,		
K_2HPO_4	3.3	g
Monopotassium phosphate		
KH_2PO_4	1.0	g
Sodium lauryl sulfate	0.2	g
Sodium desoxycholate	0.1	g
Bromcresol purple	0.08	g
Bromphenol red	0.08	g
Agar	15.0	g
Distilled water	1	L

Sterilize by autoclaving; pH should be 7.3 ± 0.2. Pour 4 to 5 mL liquefied agar into culture dishes (50 × 10 mm). Store in refrigerator.

2) Urea substrate:*

Urea	2.0	g
Phenol red	10	mg
Distilled water	100	mL

Sterilization is unnecessary. Adjust pH to 5.0 ± 0.2. Store at 2 to 8°C. Use within 1 week.

b. Procedure: Filter sample through a membrane filter (see Section 9222), place membrane on mTEC agar, incubate at 30°C for 2 h to rejuvenate injured or stressed bacteria, and then incubate at 44.5°C for 22 h. Transfer filter to a filter pad saturated with urea substrate. After 15 min, count yellow or yellow-brown colonies with the aid of a fluorescent lamp and a magnifying lens. *E. coli* produces yellow or yellow-brown colonies. Verify a portion of these differentiated colonies on a commercial multi-test system.

4. Tests for Fecal Streptococci

Perform tests for fecal streptococci by the multiple-tube technique (Section 9230B) or membrane filter technique (Section 9230C).

5. Tests for *Pseudomonas aeruginosa*

Perform tests for *P. aeruginosa* as directed in Sections 9213E and F. Use the multiple-tube test with turbid samples but note that the procedures may not be applicable to marine samples.

6. Tests for *Salmonella / Shigella*

See Section 9260.

7. References

1. CABELLI, V.J. 1980. Health Effects Criteria for Marine Recreational Waters. EPA-600/1-80-031, U.S. Environmental Protection Agency, Research Triangle Park, N.C.
2. DUFOUR, A.P. 1984. Health Effects Criteria for Fresh Recreational Waters. EPA-600/1-84-004, U.S. Environmental Protection Agency, Research Triangle Park, N.C.
3. KESWICK, B.H., C.P. GERBA & S.M. GOYAL. 1981. Occurrence of enteroviruses in community swimming pools. *Amer. J. Pub. Health* 71:1026.
4. DUTKA, B.J. & K.K. KWAN. 1978. Health indicator bacteria in water surface microlayers. *Can. J. Microbiol.* 24:187.
5. CABELLI, V.J., H. KENNEDY & M.A. LEVIN. 1976. *Pseudomonas aeruginosa* and fresh recreational waters. *J. Water Pollut. Control Fed.* 48:367.
6. SHERRY, J.P., S.R. KUCHMA & B.J. DUTKA. 1979. The occurrence of *Candida albicans* in Lake Ontario bathing beaches. *Can. J. Microbiol.* 25:1036.
7. STEVENS, A.R., R.L. TYNDALL, C.C. COUTANT & E. WILLAERT. 1977. Isolation of the etiological agent of primary amoebic meningoencephalitis from artificially heated waters. *Appl. Environ. Microbiol.* 34:701.
8. WELLINGS, F.M., P.T. AMUSO, S.L. CHANGE & A.L. LEWIS. 1977. Isolation and identification of pathogenic *Naegleria* from Florida lakes. *Appl. Environ. Microbiol.* 34:661.
9. N'DIAYE, A., P. GEORGES, A. N' GO & B. FESTY. 1985. Soil amoebas as biological markers to estimate the quality of swimming pool waters. *Appl. Environ. Microbiol.* 49:1072.

8. Bibliography

OLIVIERI, V.P., C.W. DRUSE & K. KAWATA. 1977. Microorganisms in Urban Stormwater. EPA-600/2-77-087, U.S. Environmental Protection Agency, Cincinnati, Ohio.

9213 E.   Membrane Filter Technique for *Pseudomonas aeruginosa*

1. Laboratory Apparatus

See Section 9222B.1.

2. Culture Media

a. M-PA agar: This agar may not be available in dehydrated form and may require preparation from the basic ingredients.

L-lysine HCl	5.0	g
Sodium chloride, NaCl	5.0	g
Yeast extract	2.0	g
Xylose	2.5	g
Sucrose	1.25	g
Lactose	1.25	g
Phenol red	0.08	g
Ferric ammonium citrate	0.8	g
Sodium thiosulfate, Na$_2$S$_2$O$_3$	6.8	g
Agar	15.0	g
Distilled water	1	L

Adjust to pH 6.5 ± 0.1 and sterilize by autoclaving. Cool to 55 to 60°C; readjust to pH 7.1 ± 0. 2 and add the following dry antibiotics per liter of agar base: sulfapyridine,* 176 mg; kanamycin,† 8.5 mg; nalidixic acid,‡ 37.0 mg; and cycloheximide,§ 150 mg. After mixing dispense in 3-mL quantities in 50- × 12-mm petri plates. Store poured plates at 2 to 10°C. Discard unused medium after 1 month.

b. Modified M-PA agar: To M-PA agar add 1.5 g MgSO$_4$.7H$_2$O/L and reduce concentrations of Na$_2$S$_2$O$_3$ to 5 g/L and xylose to 1.25 g/L (adding 0.1 g sodium desoxycholate is optional).

c. Milk agar (Brown and Scott Foster Modification):
Mixture A:

Instant nonfat milk ‖	100	g
Distilled water	500	mL

Mixture B:

Nutrient broth	12.5	g
Sodium chloride, NaCl	2.5	g
Agar	15.0	g
Distilled water	500	mL

Separately sterilize Mixtures A and B; cool rapidly to 55°C; aseptically combine mixtures and pour into 100- × 15-mm petri plates, about 20 mL/plate.

3. Procedure

a. Presumptive tests: Filter 200 mL or smaller portions of natural waters or up to 500 mL of swimming pool waters through sterile membrane filters. Place each membrane on a poured plate of modified M-PA agar so that there is no air space between the membrane and the agar surface. Invert plates and incubate at 41.5 ± 0.5°C for 72 h.

Typically, *P. aeruginosa* colonies are 0.8 to 2.2 mm in diameter and flat in appearance with light outer rims and brownish to greenish-black centers. Count typical colonies, preferably from filters containing 20 to 80 colonies. Use a 10-to 15-power magnifier as an aid in colony counting.

b. Confirmation tests: Use milk agar to confirm a number of typical and atypical colonies. Make a single streak (2 to 4 cm long) from an isolated colony on a milk agar plate and incubate at 35 ± 1.0 °C for 24 h. *P. aeruginosa* hydrolyzes casein and produces a yellowish to green diffusible pigment.

---

*Nutritional Biochemicals, Cleveland, Ohio.
†Bristol-Myers, Syracuse, N.Y.
‡Calbiochem, La Jolla, Calif.
§Actidione, Upjohn Company, Kalamazoo, Mich., or equivalent.

‖ Carnation or equivalent.

4. Interpretation and Calculation of Density

Except as noted above, confirmation is not required routinely. In the absence of confirmation, report results as "presumptive." Calculate and record as the number of *P. aeruginosa*/100 mL.

5. Bibliography

DRAKE, C.H. 1966. Evaluation of culture media for the isolation and enumeration of *Pseudomonas aeruginosa*. Health Lab. Sci. 3:10.

BROWN, M.R.W. & J.H. SCOTT FOSTER. 1970. A simple diagnostic milk medium for *Pseudomonas aeruginosa*. J. Clin. Pathol. 23:172.

LEVIN, M.A. & V.J. CABELLI. 1972. Membrane filter technique for enumeration of *Pseudomonas aeruginosa*. Appl. Microbiol. 24:864.

DUTKA, B.J. & K.K. KWAN. 1977. Confirmation of the single-step membrane filter procedure for estimating *Pseudomonas aeruginosa* densities in water. Appl. Environ. Microbiol. 33:240.

BRODSKY, M.H. & B.W. CIEBIN. 1978. Improved medium for recovery and enumeration of *Pseudomonas aeruginosa* from water using membrane filters. Appl. Environ. Microbiol. 36:26.

9213 F. Multiple-Tube Technique for *Pseudomonas aeruginosa*

1. Laboratory Apparatus

See Section 9221.

2. Culture Media

a. Asparagine broth: This medium may not be available in dehydrated form and may require preparation from the basic ingredients.

Asparagine, DL	3.0	g
Anhydrous dipotassium hydrogen phosphate, K_2HPO_4	1.0	g
Magnesium sulfate, $MgSO_4 \cdot 7H_2O$	0.5	g
Distilled water	1	L

Adjust pH to 6.9 to 7.2 before sterilization.

b. Acetamide broth: This medium may not be available in dehydrated form and may require preparation from the basic ingredients.

Acetamide	10.0	g
Sodium chloride, NaCl	5.0	g
Anhydrous dipotassium hydrogen phosphate, K_2HPO_4	1.39	g
Anhydrous potassium dihydrogen phosphate, KH_2PO_4	0.73	g
Magnesium sulfate, $MgSO_4 \cdot 7H_2O$	0.5	g
Phenol red	0.012	g
Distilled water	1	L

Adjust pH to 6.9 to 7.2 before sterilization.

Prepare acetamide agar slants as above except add 15 g agar, boil to dissolve agar and dispense in 8-mL quantities in 16-mm tubes. After autoclaving, incline tubes while cooling to provide a large slant surface.

3. Procedure

a. Presumptive test: Inoculate five 10-mL, five 1-mL, and five 0.1-mL samples into asparagine broth. Use 10 mL single-strength broth for inocula of 1 mL or less and 10 mL double-strength broth for 10-mL inocula. For swimming pools, higher dilutions may be necessary. Incubate inoculated tubes at 35 to 37°C. After 24 h and again after 48 h of incubation, examine tubes under long-wave ultraviolet light (black light) in a darkened room. Production of a greenish fluorescent pigment constitutes a positive presumptive test.

b. Confirmed test: Confirm positive tubes by inoculating 0.1 mL of culture into acetamide broth or onto the surface of acet-amide agar slants. A positive confirmed reaction is the development of a high pH as indicated by a purple color within 24 to 36 h of incubation at 35 to 37°C.

c. Computing and recording MPN: Refer to Table 9221:V and to Section 9221D.

9215   HETEROTROPHIC PLATE COUNT*

9215 A.   Introduction

1. Applications

The heterotrophic plate count (HPC), formerly known as the standard plate count, is a procedure for estimating the number of live heterotrophic bacteria in water and measuring changes during water treatment and distribution or in swimming pools. Colonies may arise from pairs, chains, clusters, or single cells, all of which are included in the term "colony-forming units" (CFU). The final count also depends on interaction among the developing colonies; choose that combination of procedure and medium that produces the greatest number of colonies within the designated incubation time. Three different methods and four different media are described.

2. Selection of Method

The pour plate method (9215B) is simple to perform and can accommodate volumes of sample or diluted sample ranging from 0.1 to 2.0 mL. The colonies produced are relatively small and compact, showing less tendency to encroach on each other than those produced by surface growth. On the other hand, submerged colonies often are slower growing, are difficult to transfer, and are not described in published studies. A thermostatically controlled water bath is essential for tempering the agar, but even so, significant heat shock from the transient exposure of the sample to 45 to 46°C agar may occur.

The spread plate method (9215C) causes no heat shock and all colonies are on the agar surface where they can be distinguished readily from particles and bubbles. Colonies can be transferred quickly and compared easily to published descriptions. However, this method is limited by the small volume of sample or diluted sample that can be absorbed by the agar: 0.1 to 0.5 mL, depending on the degree to which the prepoured plates have been dried. To use this procedure, maintain a supply of suitable predried, absorbent agar plates.

The membrane filter method (9215D) permits testing large volumes of low-turbidity water and is the method of choice for low-count waters (< 1 to 10 CFU/mL). This method produces no heat shock but adds the expense of the membrane filter. Further disadvantages include the smaller display area, the need to detect colonies by reflected light against a white

---

*Approved by Standard Methods Committee, 1988.

background if colored filters or contrast stains are not used, and possible variations in membrane filter quality (see Section 9020B.3g).

3. Work Area

Provide a level table or bench top with ample area in a clean, draft-free, well-lighted room or within a horizontal-flow laminar hood. Use table and bench tops having nonporous surfaces and disinfect before any analysis is made.

4. Samples

Collect water as directed in Section 9060A. Initiate analysis as soon as possible after collection to minimize changes in bacterial population. The recommended maximum elapsed time between collection and examination of samples is 8 h (maximum transit time 6 h, maximum processing time 2 h). When analysis cannot begin within 8 h, maintain sample at a temperature below 4°C but do not freeze. Do not let maximum elapsed time between collection and analysis exceed 24 h.

Hold or transport bottled water samples obtained from retail outlets unrefrigerated under the same ambient conditions as those found in the retail store until tested. Examine freshly bottled samples (less than 48 h old) within 6 h of collection if unrefrigerated and within 24 h if refrigerated.

5. Sample Preparation

Mark each plate with sample number, dilution, date, and any other necessary information before examination. Prepare at least duplicate plates for each volume of sample or dilution examined. For the pour or spread plate methods use glass (65 cm²) or disposable plastic (57 cm²) petri dishes.

Thoroughly mix all samples or dilutions by rapidly making about 25 complete up-and-down (or back-and-forth) movements. Optionally, use a mechanical shaker to shake samples or dilutions for 15 s.

6. Media

Compare new lots of media with current lot in use according to Section 9020B.3h.

a. Plate count agar (tryptone glucose yeast agar): Use for pour and spread plate methods. This high-nutrient agar, widely used in the past, gives lower counts than R2A or NWRI agar. It is included for laboratories wishing to make media comparisons or to extend the continuity of old data.

Tryptone	5.0	g
Yeast extract	2.5	g
Glucose	1.0	g
Agar	15.0	g
Distilled water	1	L

pH should be 7.0 ± 0.2 after autoclaving at 121°C for 15 min.

b. m-HPC agar:† Use this high-nutrient medium only for the membrane filter method.

Peptone	20.0	g
Gelatin	25.0	g
Glycerol	10.0	mL
Agar	15.0	g
Distilled water	1	L

Mix all ingredients except glycerol. Adjust pH to 7.1, if necessary, with $1N$ NaOH, heat to dissolve, add glycerol, and autoclave at 121°C for 5 min.

c. R2A agar: Use for pour, spread plate, and membrane filter methods. This low-nutrient agar gives higher counts than high-nutrient formulations.

Yeast extract	0.5	g
Proteose peptone No. 3 or polypeptone	0.5	g
Casamino acids	0.5	g
Glucose	0.5	g
Soluble starch	0.5	g
Dipotassium hydrogen phosphate, K_2HPO_4	0.3	g

---

†Formerly called m-SPC agar.

Magnesium sulfate hepta-
hydrate, MgSO₄·7H₂O ..   0.05 g
Sodium pyruvate ....   0.3 g
Agar ....   15.0 g
Distilled water ....   1 L

Adjust pH to 7.2 with solid K_2HPO_4 or KH_2PO_4 before adding agar. Heat to dissolve agar and sterilize at 121°C for 15 min.

d. NWRI agar (HPCA): Use for pour, spread plate, and membrane filter methods. This medium is likely to produce higher colony counts than the three media described above. It may not be available in dehydrated form and may require preparation from the basic ingredients.

Peptone ....   3.0 g
Soluble casein ....   0.5 g
K_2HPO_4 ....   0.2 g
$MgSO_4$ ....   0.05 g
$FeCl_3$ ....   0.001 g
Agar ....   15.0 g
Distilled water ....   1 L

Adjust pH to 7.2 before autoclaving for 15 min at 121°C.

7. Incubation

U.S. EPA revised regulations will describe the pour plate method. In testing to meet these regulations, incubate at 35°C for 48 h. Otherwise, select from among recommended times and temperatures for monitoring changes in water quality. The highest counts typically will be obtained from 5- to 7-d incubation at a temperature of 20 to 28°C. If unable to provide incubators controlled at 20 to 28°C, use a dust-free, room-temperature cabinet. Include time and temperature in reporting results.

During incubation maintain humidity within the incubator so that plates will have no moisture weight loss greater than 15%. This is especially important if prolonged incubation is used. A pan of water placed at the bottom of the incubator may be sufficient but note that to prevent rusting or oxidation the inside walls and shelving

should be of high-grade stainless steel or anodized aluminum. For long incubation in nonhumidified incubators, seal plates in plastic bags.

8. Counting and Recording

a. Pour and spread plates: Count all colonies on selected plates promptly after incubation. If counting must be delayed temporarily, store plates at 5 to 10°C for no more than 24 h, but avoid this as routine practice. Record results of sterility controls on the report for each lot of samples.

Use an approved counting aid, such as the Quebec colony counter, for manual counting. If such equipment is not available, count with any other counter provided that it gives equivalent magnification and illumination. Automatic plate counting instruments are available. These generally use a television scanner coupled to a magnifying lens and an electronics package. Their use is acceptable if evaluation in parallel with manual counting gives comparable results.

In preparing plates, plant sample volumes that will give from 30 to 300 colonies/plate. The aim is to have at least one dilution giving colony counts between these limits, except as provided below.

Ordinarily, do not plant more than 2.0 mL of sample; therefore, when the total number of colonies developing from 2.0 mL is less than 30, disregard the rule above and record result observed. With this exception, consider only plates having 30 to 300 colonies in determining the plate count. Compute bacterial count per milliliter by multiplying average number of colonies per plate by the reciprocal of the dilution used. Report counts as CFU per milliliter.

If there is no plate with 30 to 300 colonies, and one or more plates have more than 300 colonies, use the plate(s) having a count nearest 300 colonies. Compute the count by multiplying average count per plate by the reciprocal of the dilution used and report as estimated CFU per milliliter.

If plates from all dilutions of any sample have no colonies, report the count as less than one ($<$ 1) times the reciprocal of the corresponding lowest dilution. For example, if no colonies develop on the 1:100 dilution, report the count as less than 100 ($<$ 100) estimated CFU/mL.

If the number of colonies per plate far exceeds 300, do not report result as "too numerous to count" (TNTC). If there are fewer than 10 colonies/cm^2, count colonies in 13 squares (of the colony counter) having representative colony distribution. If possible, select seven consecutive squares horizontally across the plate and six consecutive squares vertically, being careful not to count a square more than once. Multiply sum of the number of colonies in 13 representative square centimeters by 5 to compute estimated colonies per plate when the plate area is 65 cm^2. When there are more than 10 colonies/cm^2, count four representative squares, take average count per square centimeter, and multiply by the appropriate factor to estimate colonies per plate. The factor is 57 for disposable plastic plates and 65 for glass plates. When bacterial counts on crowded plates are greater than 100 colonies/cm^2, report result as greater than ($>$) 6500 times the reciprocal of the highest dilution plated for glass plates or greater than ($>$) 5700 times the reciprocal for plastic plates. Report as estimated colony-forming units per milliliter.

If spreading colonies (spreaders) are encountered on the plate(s) selected, count colonies on representative portions only when colonies are well distributed in spreader-free areas and the area covered by the spreader(s) does not exceed one-half the plate area.

When spreading colonies must be counted, count each of the following types as one: a chain of colonies that appears to be caused by disintegration of a bacterial clump as agar and sample were mixed; a spreader that develops as a film of growth between the agar and bottom of petri dish; and a colony that forms in a film of water at the edge or over the agar surface. The last two types largely develop because of an accumulation of moisture at the point from which the spreader originates. They frequently cover more than half the plate and interfere with obtaining a reliable plate count.

Count as individual colonies similar-appearing colonies growing in close proximity but not touching, provided that the distance between them is at least equal to the diameter of the smallest colony. Count impinging colonies that differ in appearance, such as morphology or color, as individual colonies.

If plates have excessive spreader growth, report as "spreaders" (Spr). When plates are uncountable because of missed dilution, accidental dropping, and contamination, or the control plates indicate that the medium or other material or labware was contaminated, report as "laboratory accident" (LA).

b. Membrane filter method: Count colonies on membrane filters using a stereoscopic microscope at 10 to 15 $\times$ magnification. Preferably place petri dish on microscope stage slanted at 45° and adjust light source vertical to the colonies. Optimal colony density per filter is 20 to 200. If colonies are small and there is no crowding, a higher limit is acceptable.

Count all colonies on the membrane when there are 1 to 2, or fewer, colonies per square. For 3 to 10 colonies per square count 10 squares and obtain average count per square. For 10 to 20 colonies per square count 5 squares and obtain average count per square. Multiply average count per square by 100 times the reciprocal of the dilution to give colonies per milliliter. If there are more than 20 colonies per square, record count as $>$ 2000 times the reciprocal of the dilution. Report averaged counts as estimated colony-forming units.

Make estimated counts only when there are discrete, separated colonies without spreaders.

9. Computing and Reporting Counts

The term "colony-forming units" (CFU) is descriptive of the methods used; therefore, report all counts as colony-forming units. Include in the report the method used, the incubation temperature and time, and the medium. For example: CFU/mL, pour plate method, 35°C/48 h, plate count agar or 28°/5 d, R2A agar; or CFU/mL, spread plate method, 20°/7 d, NWRI agar.

To compute the heterotrophic plate count, CFU/mL, multiply total number of colonies or average number (if duplicate plates of the same dilution) per plate by the reciprocal of the dilution used. Record dilutions used and number of colonies on each plate counted or estimated.

When colonies on duplicate plates and/or consecutive dilutions are counted and results are averaged before being recorded, round off counts to two significant figures only when converting to colony-forming units.

Avoid creating fictitious precision and accuracy when computing colony-forming units by recording only the first two left-hand digits. Raise the second digit to the next higher number when the third digit from the left is 5, 6, 7, 8, or 9; use zeros for each successive digit toward the right from the second digit. For example, report a count of 142 as 140 and a count of 155 as 160, but report a count of 35 as 35.

10. Personal Errors

Avoid inaccuracies in counting due to carelessness, damaged or dirty optics that impair vision, or failure to recognize colonies. Laboratory workers who cannot duplicate their own counts on the same plate within 5% and the counts of other analysts within 10% should discover the cause and correct such disagreements.

9215 B.   Pour Plate Method

1. Samples and Sample Preparation

See 9215A.4 and 9215A.5.

2. Sample Dilution

Prepare water used for dilution blanks as directed in Section 9050C.

a. Selecting dilutions: Select the dilution(s) so that the total number of colonies on a plate will be between 30 and 300 (Figure 9215:1). For example, where a heterotrophic plate count as high as 3000 is suspected, prepare plates with 10^{-2} dilution.

For most potable water samples, plates suitable for counting will be obtained by plating 1 mL and 0.1 mL undiluted sample and 1 mL of the 10^{-2} dilution.

b. Measuring sample portions: Use a sterile pipet for initial and subsequent transfers from each container. If pipet becomes contaminated before transfers are completed, replace with a sterile pipet. Use a separate sterile pipet for transfers from each different dilution. Do not prepare dilutions and pour plates in direct sunlight. Use caution when removing sterile pipets from the container; to avoid contamination, do not drag pipet tip across exposed ends of pipets in the pipet container or across lips and necks of dilution bottles. When removing sample, do not insert pipets more than 2.5 cm below the surface of sample or dilution.

c. Measuring dilutions: When discharging sample portions, hold pipet at an angle of about 45° with tip touching bottom of petri dish or inside neck of dilution bottle.

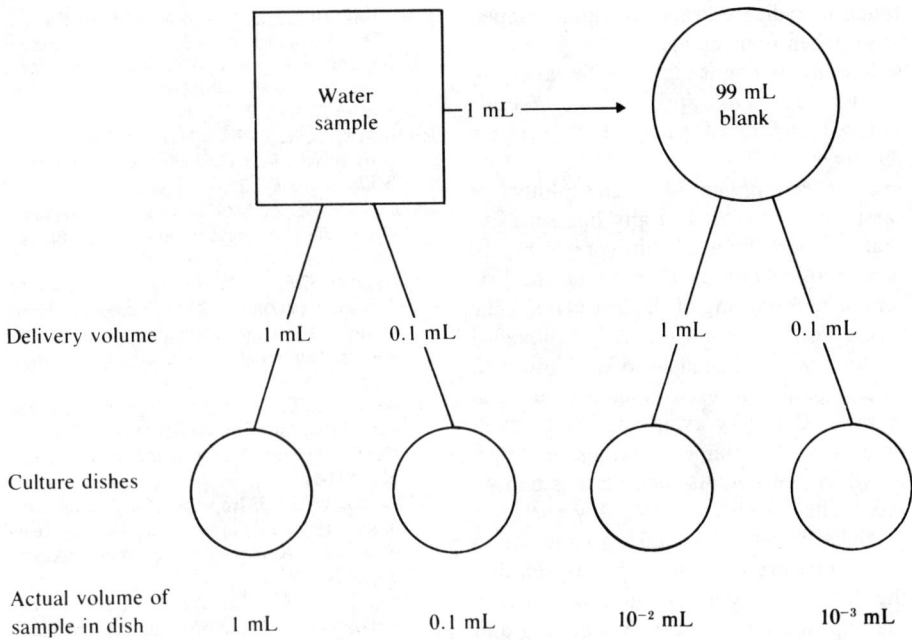

Figure 9215:1. Preparation of dilutions.

Lift cover of petri dish just high enough to insert pipet. Allow 2 to 4 s for liquid to drain from 1-mL graduation mark to tip of pipet. If pipet is not a blow-out type, touch tip of pipet *once* against a dry spot on petri dish bottom. Less preferably, use a cotton-plugged blow-out-type pipet and gently blow out remaining volume of sample dilution. When 0.1-mL quantities are measured, let diluted sample drain from chosen reference graduation until 0.1 mL has been delivered. Remove pipet without retouching it to dish. Pipet 1 mL, 0.1 mL, or other suitable volume into sterile petri dish before adding melted culture medium. Use decimal dilutions in preparing sample volumes of less than 0.1 mL; in examining sewage or turbid water, do not measure a 0.1-mL inoculum of original sample, but prepare an appropriate dilution. Prepare at least two replicate plates for each sample dilution used. After depositing test portions for each series of plates, pour culture me-dium and mix carefully. Do not let more than 20 min elapse between starting pipet-ting and pouring plates.

3. Plating

a. Melting medium: Melt sterile solid agar medium in boiling water or by exposure to flowing steam in a partially closed container, but avoid prolonged exposure to unnecessarily high temperatures during and after melting. Do not resterilize plating medium. If the medium is melted in two or more batches, use all of each batch in order of melting, provided that the contents remain fully melted. Discard melted agar that contains precipitate.

Maintain melted medium in a water bath between 44 and 46°C until used. In a separate container place a thermometer in water or medium that has been exposed to the same heating and cooling as the plating medium. Do not depend on the sense of touch to indicate proper medium temperature when pouring agar.

Use plate count agar, R2A agar, or NWRI agar as specified in Section 9215A.6. Before using a new lot of medium test its suitability.

b. Pouring plates: Limit the number of samples to be plated in any one series so that no more than 20 min (preferably 10 min) elapse between dilution of the first sample and pouring of the last plate in the series. Pour at least 10 to 12 mL liquefied medium maintained at 44 to 46°C into each dish by gently lifting cover just high enough to pour. Carefully avoid spilling medium on outside of container or on inside of dish lid when pouring. As each plate is poured mix melted medium thoroughly with test portions in petri dish, taking care not to splash mixture over the edge, by rotating the dish first in one direction and then in the opposite direction, or by rotating and tilting. Let plates solidify (within 10 min) on a level surface. After medium solidifies, invert plates and place in incubator.

c. Sterility controls: Check sterility of medium and dilution water blanks by pouring control plates for each series of samples. Prepare additional controls to determine contamination of plates, pipets, and room air.

4. Incubation

See Section 9215A.7.

5. Counting, Recording, Computing, and Reporting

See Sections 9215A.8 and 9215A.9.

6. Bibliography

BREED, R.S. & W.D. DOTTERER. 1916. The number of colonies allowable on satisfactory agar plates. Tech. Bull. 53, New York Agricultural Experiment Sta.

BUTTERFIELD, C.T. 1933. The selection of a dilution water for bacteriological examinations. *J. Bacteriol.* 23:355; *Pub. Health Rep.* 48:681.

ARCHAMBAULT, J., J. CUROT & M.H. MCCRADY. 1937. The need of uniformity of conditions for counting plates (with suggestions for a standard colony counter). *Amer. J. Pub. Health* 27:809.

RICHARDS, O.W. & P.C. HEIJN. 1945. An improved dark-field Quebec colony counter. *J. Milk Technol.* 8:253.

BERRY, J.M., D.A. MCNEILL & L.D. WITTER. 1969. Effect of delays in pour plating on bacterial counts. *J. Dairy Sci.* 52:1456.

GELDREICH, E.E., H.D. NASH, D.J. REASONER & R.H. TAYLOR. 1972. The necessity of controlling bacterial populations in potable waters: Community water supply. *J. Amer. Water Works Assoc.* 64:596.

GELDREICH, E.E. 1973. Is the total count necessary? Proc. 1st Annu. Water Quality Technol. Conf., American Water Works Assoc., Paper No. VII-1.

GINSBURG, W. 1973. Improved total count techniques. Proc. 1st Annu. Water Quality Technol. Conf., American Water Works Assoc., Paper No. VIII.

DUTKA, B.J., A.S.Y. CHAU & J. COBURN. 1974. Relationship of heterotrophic bacterial indicators of water pollution and fecal sterols. *Water Res.* 8:1047.

KLEIN, D.A. & S. WU. 1974. Stress: a factor to be considered in heterotrophic microorganism enumeration from aquatic environments. *Appl. Microbiol.* 37:429.

GELDREICH, E.E., H.D. NASH, D.J. REASONER & R.H. TAYLOR. 1975. The necessity for controlling bacterial populations in potable waters: Bottled water and emergency water supplies. *J. Amer. Water Works Assoc.* 67:117.

BELL, C.R., M.A. HOLDER-FRANKLIN & M. FRANKLIN. 1980. Heterotrophic bacteria in two Canadian rivers.—I. Seasonal variation in the predominant bacterial populations. *Water Res.* 14:449.

REASONER, D.J. & E.E. GELDREICH. 1981. Influence of medium, methods and incubation time and temperature on the bacterial count of potable water. 81st Annu. Meeting, American Soc. Microbiology, Dallas, Tex., Paper No. N27.

MEANS, E.G., L. HANAMI, G.F. RIDGWAY & B.H. OLSON. 1981. Evaluating mediums and plating techniques for enumerating bacteria in water distribution systems. *J. Amer. Water Works Assoc.* 73:585.

AMERICAN PUBLIC HEALTH ASSOCIATION. 1985. Standard Methods for the Examination of Dairy Products, 15th ed. American Public Health Assoc., Washington, D.C.

REASONER, D.J. & E.E. GELDREICH. 1985. A new medium for the enumeration and subculture of bacteria from potable water. *Appl. Environ. Microbiol.* 49:1.

9215 C. Spread Plate Method

1. Laboratory Apparatus

a. Glass rods: Bend 4-mm-diam fire-polished glass rods, 200 mm in length, 45° about 40 mm from one end. Sterilize before using.

b. Pipet, glass, 1.1 mL, with tempered, rounded tip. Do not use disposable plastic pipets.

c. Turntable (optional).*

d. Incubator or drying oven, set at 42°C, or laminar-flow hood.

2. Media

See 9215A.6*a, c,* and *d.* If R2A agar is used best results are obtained at 28°C with 7 d incubation; if NWRI is used, incubate at 20°C for 7 d.

3. Preparation of Plates

Pour 15 mL of the desired medium into sterile 100 × 15 or 90 × 15 petri dishes; let agar solidify. Predry plates inverted so that there is a 2- to 3-g water loss overnight with lids on. See Figure 9215:2, Table 9215:I, or Figure 9215:3. Use predried plates immediately after drying. For predrying and using plates the same day, pour 25 mL agar into petri dish and dry in a laminar-flow hood at room temperature (24 to 26°C) with the lid off to obtain the desired 2- to 3-g weight loss. See Figure 9215:3.

4. Procedure

Prepare sample dilutions as directed in 9215B.2.

a. Glass rod: Pipet 0.1 or 0.5 mL sample onto surface of predried agar plate. Using a sterile bent glass rod, distribute inoculum over surface of the medium by rotating the dish by hand or on a turntable. Let inoculum be absorbed completely into the medium before incubating.

b. Pipet: Pipet desired sample volume (0.1, 0.5 mL) onto the surface of the predried agar plate while dish is being rotated on a turntable. Slowly release sample from pipet while making one to-and-fro motion, starting at center of the plate and stopping 0.5 cm from the plate edge before returning to the center. Lightly touch the pipet to the plate surface. Let inoculum be absorbed completely by the medium before incubating.

5. Incubation

See 9215A.7.

6. Counting, Recording, Computing, and Reporting

See 9215A.8 and 9215A.9.

7. Bibliography

BUCK, J.D. & R.C. CLEVENDOR. 1960. The spread plate as a method for the enumeration of marine bacteria. *Limnol. Oceanogr.* 5:78.

CLARK, D.S. 1967. Comparison of pour and surface plate methods for determination of bacterial counts. *Can. J. Microbiol.* 13:1409.

VAN SOESTBERGAN, A.A. & C.H. LEE. 1969. Pour plates or streak plates. *Appl. Microbiol.* 18:1092.

CLARK, D.S. 1971. Studies on the surface plate method of counting bacteria. *Can. J. Microbiol.* 17:943.

GILCHRIST, J.E., J.E. CAMPBELL, C.B. DONNELLY, J.T. PEELER & J.M. DELANEY. 1973. Spiral plate method for bacterial determination. *Appl. Microbiol.* 25:244.

PTAK, D.M. & W. GINSBURG. 1976. Pour plate vs. streak plate method. Proc. 4th Annu.

---

*Fisher Scientific, hand operated, No. 08-758 or Lab-Line motor driven, No. 1580, or equivalent.

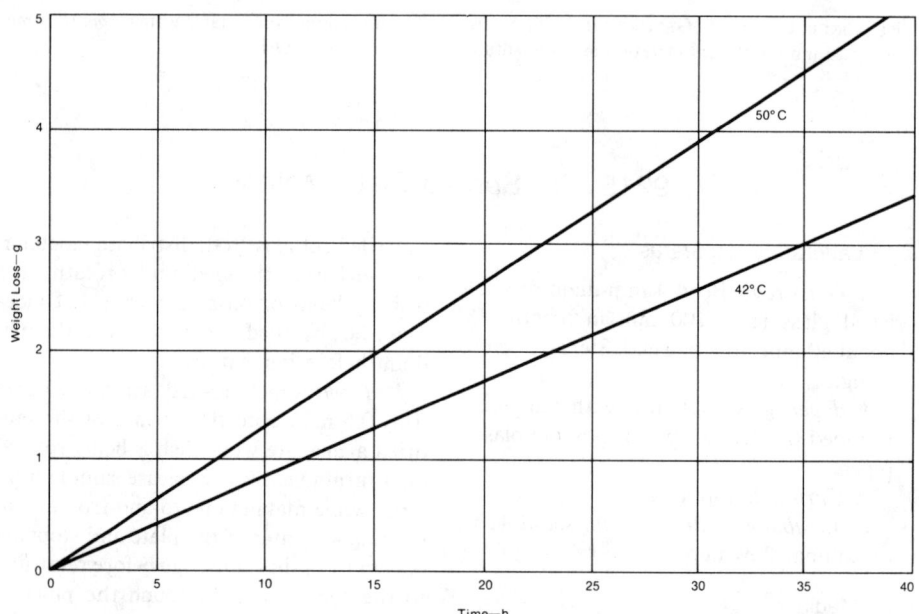

Figure 9215:2. Drying weight loss of 15-mL agar plates stored separately, inverted with lids on. Source: Unpublished data. Water Purification Lab., Chicago Dep. Water.

TABLE 9215:I. EFFECT OF TEMPERATURE OF DRYING ON WEIGHT LOSS OF 15-ML AGAR PLATES
STORED SEPARATELY*

	Time for Plates to Lose 1 to 4 g of Water (Avg. for 5 Plates) h							
Temp. °C	Plates Inverted with Lids On				Plates Inverted with Lids Removed			
	1 g	2 g	3 g	4 g	1 g	2 g	3 g	4 g
24	32	64	95	125	3.7	7.0	10.5	14.0
37	17	35	51	67	1.7	3.5	5.3	7.0
50	6	12	18	24	0.7	1.3	1.9	2.7
60	4	8	12	16	—	—	—	—

* Referenced in Canada Centre for Inland Waters Manual, Burlington, Ont.

Water Quality Technol. Conf., American Water Works Assoc., Paper No. 2B-5.

DUTKA, B.J., ed. 1978. Methods for Microbiological Analysis of Waters, Wastewaters and Sediments. Inland Waters Directorate, Scientific Operation Div., Canada Centre for Inland Waters, Burlington, Ont.

KAPER, J.B., A.L. MILLS & R.R. COLWELL. 1978. Evaluation of the accuracy and precision of enumerating aerobic heterotrophs in water samples by the spread method. *Appl. Environ. Microbiol.* 35:756.

YOUNG, M. 1979. A modified spread plate technique for the determination of concentrations of viable heterotrophic bacteria. STP 673:41-51, American Soc. Testing & Materials, Philadelphia, Pa.

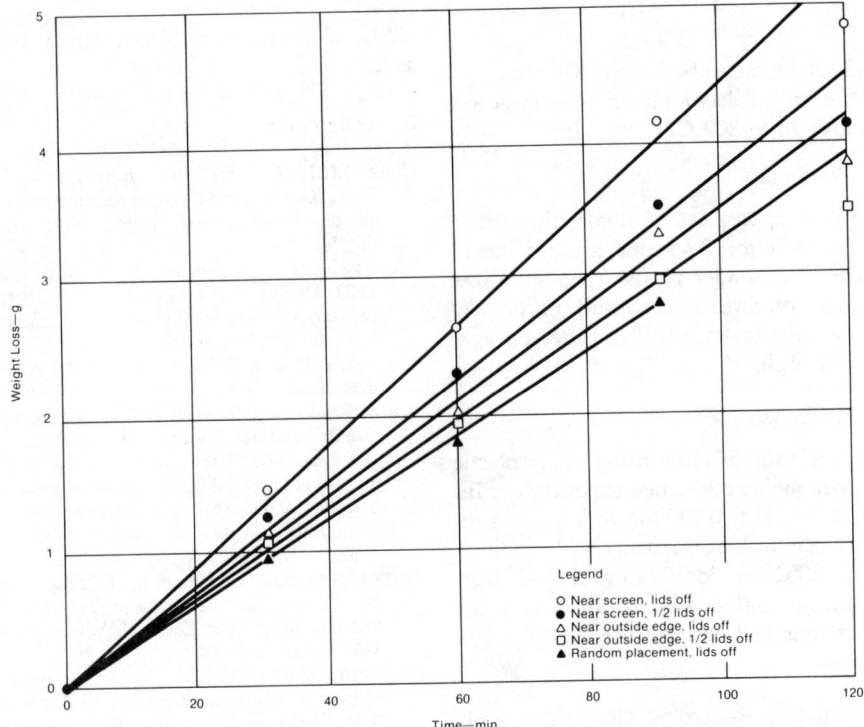

Figure 9215:3. Weight loss of 25-mL agar plates (100 × 15 mm) dried separately in a laminar-flow hood at room temperature (24 to 26°C), relative humidity (30 to 33%), and air velocity 0.6 m/s. Source: Unpublished data. Alberta Environmental Centre, Vengreville, Alta.

GELDREICH, E.E. 1981. Current status of microbiological water quality criteria. *ASM News* 47:23.

TAYLOR, R.H., M.J. ALLEN & E.E. GELDREICH.

1981. Standard plate count: A comparison of pour plate and spread plate methods. Proc. 9th Annu. Water Quality Technol. Conf., American Water Works Assoc.

9215 D.   Membrane Filter Method

1. Laboratory Apparatus

See Section 9222B.1.

2. Media

See 9215A.6. Use m-HPC agar, or alternatively R2A or NWRI agar.

3. Preparation of Plates

Dispense 5-mL portions of sterile medium into 50- × 9-mm petri dishes. Let solidify at room temperature. Prepared plates may be stored inverted in a plastic bag or tight container in a refrigerator, preferably for no longer than 1 week.

4. Sample Size

The volume to be filtered will vary with the sample. Select a maximum sample size to give 20 to 200 CFU per filter.

5. Procedure

Filter appropriate volume through a sterile 47-mm, 0.45-μm, gridded membrane filter, under partial vacuum. Rinse funnel with three 20- to 30-mL portions of sterile dilution water. Place filter on agar in petri dish.

6. Incubation

Place dishes in close fitting box or plastic bag containing moistened paper towels. Incubate at 35 ± 0.5°C for 48 h if using m-HPC agar, or longer if using R2A medium, or at 20°C for 7d if using NWRI agar. Duplicate plates may be incubated for other time and temperature conditions as desired.

7. Counting, Recording, Computing, and Reporting

See 9215A.8 and 9215A.9. Report as CFU/mL, membrane filter method, time, medium.

8. Bibliography

CLARK, H.F., E.E. GELDREICH, H.L. JETER & P.W. KABLER. 1951. The membrane filter in sanitary bacteriology. *Pub. Health Rep.* 66:951.

STOPERT, E.M., W.T. SOKOSKI & J.T. NORTHAM. 1962. The factor of temperature in the better recovery of bacteria from water by filtration. *Can. J. Microbiol.* 8:809.

TAYLOR, R.H. & E.E. GELDREICH. 1979. A new membrane filter procedure for bacterial counts in potable water and swimming pool samples. *J. Amer. Water Works Assoc.* 71:402.

CLARK, J.A. 1980. The influence of increasing numbers of non-indicator organisms upon the detection of indicator organisms by the membrane filter and presence-absence tests. *Can. J. Microbiol.* 20:827.

DUTKA, B.J., ed. 1981. Membrane Filtration, Applications, Techniques, and Problems. Marcel Dekker, Inc., New York, N.Y. and Basel, Switzerland.

HOADLEY, A.W. 1981. Effect of injury on the recovery of bacteria on membrane filters. *In* B. J. Dutka, ed. Membrane Filtration, Applications, Techniques, and Problems. pp. 413-450. Marcel Dekker, Inc., New York, N.Y. and Basel, Switzerland.

9216   DIRECT TOTAL MICROBIAL COUNT (PROPOSED)*

9216 A.   Introduction

Direct total cell counts of bacteria in water or wastewater usually exceed counts obtained from heterotrophic plate counts and most probable number methods because, unlike those procedures, direct counts preclude errors caused by viability-related phenomena such as selectivity of growth media, cell clumping, and slow growth rates.

---

*Approved by Standard Methods Committee, 1988.

9216 B. Epifluorescence Microscopic Method

1. General Discussion

The epifluorescence microscopic method produces direct total cell counts with relative speed (20 to 30 min from time of sampling) and sensitivity. It does not permit differentiation of bacterial cells on the basis of taxonomy, metabolic activity, or viability, and it cannot be used to estimate the microbial biomass because of considerable variation in the volume of individual cells. The method requires an experienced technician who can distinguish microbial cells from debris on the basis of morphology.

The method consists of sample fixation for storage, staining with a chemical fluorochrome, vacuum filtration onto a nonfluorescing polycarbonate membrane, and enumeration by counting with an epifluorescence microscope.

2. Apparatus

a. Microscope, vertical UV illuminator for epifluorescence with flat field 100× oil immersion objective lens, to give total magnification of at least 1000×.

b. Counting graticule, ocular lens micrometer* calibrated with stage micrometer.*

c. Filters,† including excitation filters (KP 490 and LP 455), beam splitter (LP 510), and barrier filter (LP 520 using mercury lamp, HBO 50).

d. Blender or vortex mixer.

e. Filtration unit, suitable for use with 25-mm-diam membrane filters.

f. Membrane filters, polycarbonate,‡ 25-mm-diam, 0.2-μm pore size (purchase non-fluorescent or prepare by soaking membrane in Irgalan black [2 g/L in 2% acetic acid] for 24 h, then rinse in water and air dry); cellulosic§ 25-mm-diam, 5-μm pore size.

g. Syringes, 3-mL, disposable, with disposable syringe filters, 0.2-μm pore size.

h. Test tubes, glass, screw-capped, 13- × 125-mm.

3. Reagents

a. Phosphate buffer: Dissolve 13.6 g KH_2PO_4 in water and dilute to 1 L. Adjust to pH 7.2 if necessary; filter through 0.2-μm membrane filter.

b. Fixative, 5.0% (w/v) glutaraldehyde in phosphate buffer. Prepare fresh daily.

c. Fluorochrome, 0.1% (w/v) acridine orange‖ in phosphate buffer.

d. Immersion oil, low fluorescing.#

4. Procedure

Collect water samples as directed in Section 9060. Add 9.0 mL sample to test tube containing 1.0 mL fixative. Fixed samples can be stored at 4°C for up to 3 weeks without significant decrease in cell numbers.

Disperse and dilute samples from mesotrophic or eutrophic sources to obtain reproducible results. Mix sample using blender or vortex mixer, then make tenfold dilutions in phosphate buffer as necessary. Clean water samples may not require dilution but larger sample volumes (> 100 mL) may be required to obtain reliable counts.

Place 1 mL sample or dilution on a non-

---

fluorescent polycarbonate filter supported by a cellulosic membrane filter in filter holder. Using disposable sterile syringe filters, add 1 mL fluorochrome and wait 2 min, then add about 3 mL filtered phosphate buffer to promote more even cell distribution. Alternatively, combine fluorochrome with sample in a small clean vial, let react, and add mixture to filter holder. Filter with vacuum (about 13 kPa). Wash with 2 mL phosphate buffer and filter. Remove polycarbonate filter with forceps and air dry for 1 to 2 min. The filter can be cut into quarter sections and saved if needed. Place dried filter on a drop of immersion oil on a clean glass microscope slide. Add a small drop of immersion oil to filter surface. Gently cover filter with a clean glass cover slip. Samples can be stored in the dark for several months without significant loss of fluorescence.

Examine at least 10 randomly selected fields on the filter using the 100× oil immersion lens to establish that distribution of microbial cells is uniform and that individual cells can be enumerated (if not, dilute sample and repeat). Preferably count 10 to 50 cells per field. Count number of cells in at least 20 squares using the calibrated counting graticule.

5. Calculations

Calculate the average number of cells per filter. Obtain effective filter area from specifications of filtration unit. Extrapolate to determine number of cells per milliliter of sample:

$$\text{Total cells/mL} = (\text{avg cells/square}) \\ \times (\text{squares/filter}) \\ \times (\text{dilution factor}) / \text{sample volume, mL.}$$

6. Bibliography

HOBBIE, J. E., R. J. DALEY & S. JASPER. 1977. Use of nuclepore filters for counting bacteria by fluorescence microscopy. *Appl. Environ. Microbiol.* 33:1225.

AMERICAN SOCIETY FOR TESTING AND MATERIALS. 1987. Standard test method for enumeration of aquatic bacteria by epifluorescence microscopy counting procedure. ASTM D4455-85, Annual Book of ASTM Standards, Vol. 11.02, Water. American Soc. Testing & Materials, Philadelphia, Pa.

SIERACK, M. E., P. W. JOHNSON & J. McH. SIEBURTH. 1985. Detection, enumeration, and sizing of planktonic bacteria by image-analyzed epifluorescence microscopy. *Appl. Environ. Microbiol.* 49:799.

9221　MULTIPLE-TUBE FERMENTATION TECHNIQUE FOR MEMBERS OF THE COLIFORM GROUP*

9221 A.　Introduction

The coliform group comprises all aerobic and facultative anaerobic, gram-negative, nonspore-forming, rod-shaped bacteria that ferment lactose with gas and acid formation within 48 h at 35°C.

The standard test for the coliform group may be carried out either by the multiple-tube fermentation technique (through the presumptive-confirmed phases or completed test) described herein or by the membrane filter (MF) technique (Section 9222). Each technique is applicable within

---

*Approved by Standards Methods Committee, 1988.

the limitations specified and with due consideration of the purpose of the examination.

For the multiple-tube fermentation technique, results of the examination of replicate tubes and dilutions are reported in terms of the Most Probable Number (MPN) of organisms present. This number, based on certain probability formulas, is an estimate of the mean density of coliforms in the sample.

The precision of each test depends on the number of tubes used. The most satisfactory information will be obtained when the largest sample inoculum examined shows gas in some or all of the tubes and the smallest sample inoculum shows no gas in all or a majority of the tubes. Bacterial density (9221D.2) can be estimated by the formula given or from the table using the number of positive tubes in the multiple dilutions. The number of sample portions selected will be governed by the desired precision of the result. MPN tables are based on the assumption of a Poisson distribution (random dispersion). However, if the sample is not adequately shaken before the portions are removed or if clumping of bacterial cells occurs, the MPN value will be an underestimate of the actual bacterial density.

Interest in the multiple-tube technique, the numerous investigations into the precision of the MPN, and the expression of test results as MPNs, should not lead to regarding this method as a statistical exercise rather than a means of estimating the coliform density of a water and its sanitary quality. The best assessment for water treatment effectiveness or the sanitary quality of an untreated water depends on the interpretation of bacteriological results and of all other information obtained by engineering or sanitary surveys.

1. Water of Drinking Water Quality

When drinking water is analyzed to determine if the quality meets the standards of the U.S. Environmental Protection Agency (EPA), use five 10-mL portions of sample. Use five culture bottles if 100-mL sample portions are inoculated. To obtain a more precise estimate of bacterial numbers in treated drinking water, which should contain no coliforms per 100 mL, use 10 replicate tubes, each containing 10-mL or 100-mL sample portions. Use the confirmed phase (9221B.2) when examining all drinking waters, natural waters, and effluents. The completed test (9221B.3) is the reference standard to be applied to selected samples on a seasonal basis and at least quarterly for quality control.

For the routine examination of most public water supplies, particularly those that are disinfected, the object of the coliform test is to determine compliance with EPA standards as a measure of the efficiency of treatment plant operation or water/effluent quality. A high proportion of coliform occurrences in a distribution system may be attributed not to treatment failure at the plant or the well source, but to bacterial regrowth in the mains. Because it is difficult to distinguish between coliform regrowth and new contamination, assume all coliform occurrences to be new contamination unless otherwise demonstrated. It is expected that more than 95% of all samples examined will yield negative results. An occasional positive result, unless repeated from the same sampling point (resampling), usually is of limited significance but should not be ignored. An increase in the number of positive samples over a period of time or an abrupt increase in a short period of time indicates a change in water quality, the significance of which should be determined by a sanitary survey. Promptly make necessary corrections, based on the sanitary survey, and verify for effectiveness by additional bacteriological testing.

2. Water of Other than Drinking Water Quality

In the examination of nonpotable waters inoculate a series of tubes with appropriate decimal dilutions of the water (multiples and submultiples of 10 mL), based on the probable coliform density. Use the pre-sumptive-confirmed phase of the multiple-tube procedure. Use the more labor-inten-sive completed test (9221B.3) as a quality control measure on at least 10% of coli-form-positive nonpotable water samples on a seasonal basis. The object of the exami-nation of nonpotable water generally is to estimate the density of bacterial contami-nation, determine a source of pollution, en-force water quality standards, or trace the survival of microorganisms. Each objective requires a numerical value for reporting results. The multiple-tube fermentation technique may be used to obtain statisti-cally valid MPN estimates of coliform den-sity. Examine a sufficient number of samples to yield representative results for the sampling station. Generally, the geo-metric mean or median value of the results of a number of samples will yield a value in which the effect of sample-to-sample var-iation is minimized. The direct-count, membrane filter technique may prove the better procedure to accomplish this objec-tive.

3. Other Samples

The multiple-tube fermentation tech-nique is applicable to the analysis of salt or brackish waters as well as muds, sedi-ments, and sludges. Follow the precautions given above on portion sizes and numbers of tubes per dilution.

To prepare solid or semisolid samples weigh the sample and add diluent to make a 10^{-1} dilution. For example, place 50 g sample in sterile blender jar, add 450 mL sterile phosphate buffer or 0.1% peptone dilution water, and blend for 1 to 2 min at low speed (8000 rpm). Prepare the appro-priate decimal dilutions of the homoge-nized slurry as quickly as possible to minimize settling.

9221 B. Standard Total Coliform Multiple-Tube (MPN) Fermentation Techniques

1. Presumptive Phase

Use lauryl tryptose broth in the pre-sumptive portion of the multiple-tube test. Lactose broth may be used as an alternative medium provided that it has been dem-onstrated not to increase the frequency of false positives nor mask coliforms present in drinking water samples. If the medium has been refrigerated after sterilization, in-cubate overnight at 35°C before use. Dis-card tubes showing growth and/or bubbles.

a. Reagents and culture media:

1) *Lauryl tryptose broth:*

Tryptose	20.0	g
Lactose	5.0	g
Dipotassium hydrogen phosphate, K_2HPO_4. . . .	2.75	g
Potassium dihydrogen phosphate, KH_2PO_4. . . .	2.75	g
Sodium chloride, NaCl . . .	5.0	g
Sodium lauryl sulfate	0.1	g
Distilled water	1	L

Add dehydrated ingredients to distilled water, mix thoroughly, and heat to dis-solve. pH should be 6.8 ± 0.2 after steri-lization. Before sterilization, dispense sufficient medium, in fermentation tubes with an inverted vial, to cover inverted vial at least partially after sterilization. Alternatively, omit inverted vial and add 0.01 g/L bromcresol purple to presumptive medium to determine acid production, the indicator of a positive result in this part of the coliform test. Close tubes with metal or heat-resistant plastic caps.

Make lauryl tryptose broth of such strength that adding 100-mL or 10-mL portions of sample to medium will not reduce ingredient concentrations below those of the standard medium. Prepare in accordance with Table 9221:I.

Close tubes with metal or heat-resistant plastic caps.

2) *Lactose broth:*

Beef extract.	3.0	g
Peptone	5.0	g
Lactose	5.0	g
Distilled water	1	L

Add dehydrated ingredients to water, mix thoroughly, and heat to dissolve. pH should be 6.9 ± 0.2 after sterilization. Before sterilization, dispense in fermentation tubes of such dimensions that the liquid in the inoculated tube will cover the inverted vial at least partially after sterilization. Alternatively, omit inverted vial and add 0.01 g/L bromcresol purple to presumptive medium to determine acid production, the indicator of a positive result in this part of the coliform test. Close tubes with metal or heat-resistant plastic caps.

Make lactose broth of such strength that adding 100-mL or 10-mL portions of sample to medium will not reduce ingredient concentrations below those of the standard medium. Prepare in accordance with Table 9221:II.

b. Procedure:

1) Arrange fermentation tubes in rows of five tubes each in a test tube rack. The number of five tube rows and the sample volumes selected depend upon the quality and character of the water to be examined. For potable water use five 10-mL portions or ten 10-mL portions; for nonpotable water use five tubes per dilution (of 10, 1, 0.1 mL, etc.).

In making dilutions and measuring diluted sample volumes, follow the precautions given in Section 9215B.2. Use Figure 9215:1 as a guide to preparing dilutions. Shake sample and dilutions vigorously about 25 times. Inoculate each tube of the set of five with replicate sample volumes (in increasing decimal dilutions, if decimal quantities of the sample are used). Mix test portions in the medium by gentle agitation.

2) Incubate inoculated tubes at 35 ± 0.5°C. After 24 ± 2 h shake each tube gently and examine it for gas or acidic growth (distinctive yellow color) and, if no gas or acidic growth has formed, reincubate and reexamine at the end of 48 ± 3 h. Record presence or absence of gas or acid production. If the inner vial is omitted,

TABLE 9221:I. PREPARATION OF LAURYL TRYPTOSE BROTH

Inoculum mL	Amount of Medium in Tube mL	Volume of Medium + Inoculum mL	Dehydrated Lauryl Tryptose Broth Required g/L
1	10 or more	11 or more	35.6
10	10	20	71.2
10	20	30	53.4
100	50	150	106.8
100	35	135	137.1
100	20	120	213.6

TABLE 9221:II. PREPARATION OF LACTOSE BROTH

Inoculum mL	Amount of Medium in Tube mL	Volume of Medium + Inoculum mL	Dehydrated Lactose Broth Required g/L
1	10 or more	11 or more	13.0
10	10	20	26.0
10	20	30	19.5
100	50	150	39.0
100	35	135	50.1
100	20	120	78.0

growth with acidity signifies a positive presumptive reaction.

c. *Interpretation:* Production of gas or acidic growth in the tubes within 48 ± 3 h constitutes a positive presumptive reaction. Submit tubes with a positive presumptive reaction to the confirmed phase (9221B.2).

The absence of acidic growth or gas formation at the end of 48 ± 3 h of incubation constitutes a negative test. An arbitrary 48-h limit for observation doubtless excludes occasional members of the coliform group that grow very slowly (see Section 9212).

2. Confirmed Phase

The confirmed phase is outlined in Figure 9221:1.

a. *Reagents and culture media:* Use brilliant green lactose bile broth fermentation tubes for the confirmed phase.

Brilliant green lactose bile broth:

Peptone	10.0	g
Lactose	10.0	g
Oxgall	20.0	g
Brilliant green	0.0133	g
Distilled water	1	L

Add dehydrated ingredients to water, mix thoroughly, and heat to dissolve. pH should be 7.2 ± 0.2 after sterilization. Before sterilization, dispense, in fermentation tubes with an inverted vial, sufficient medium to cover inverted vial at least partially after sterilization. Close tubes with metal or heat-resistant plastic caps.

b. *Procedure:* Submit all primary tubes showing any amount of gas or acidic growth within 24 h of incubation to the confirmed phase. If active fermentation or acidic growth appears in the primary tube earlier than 24 h, transfer to the confirmatory medium, preferably without waiting for the full 24-h period to elapse. If additional primary tubes show acidic growth at the end of a 48-h incubation period, submit these to the confirmed phase.

Gently shake or rotate primary tube showing gas or acidic growth to resuspend the organisms. With a sterile metal loop 3 mm in diameter, transfer one loopful of culture to a fermentation tube containing brilliant green lactose bile broth or insert a sterile wooden applicator at least 2.5 cm into the culture, promptly remove, and plunge applicator to bottom of fermentation tube containing brilliant green lactose bile broth. Remove and discard applicator. Repeat for all other positive presumptive tubes.

Incubate the inoculated brilliant green lactose bile broth tube for 48 ± 3 h at 35 ± 0.5°C.

Formation of gas in any amount in the inverted vial of the brilliant green lactose bile broth fermentation tube at any time

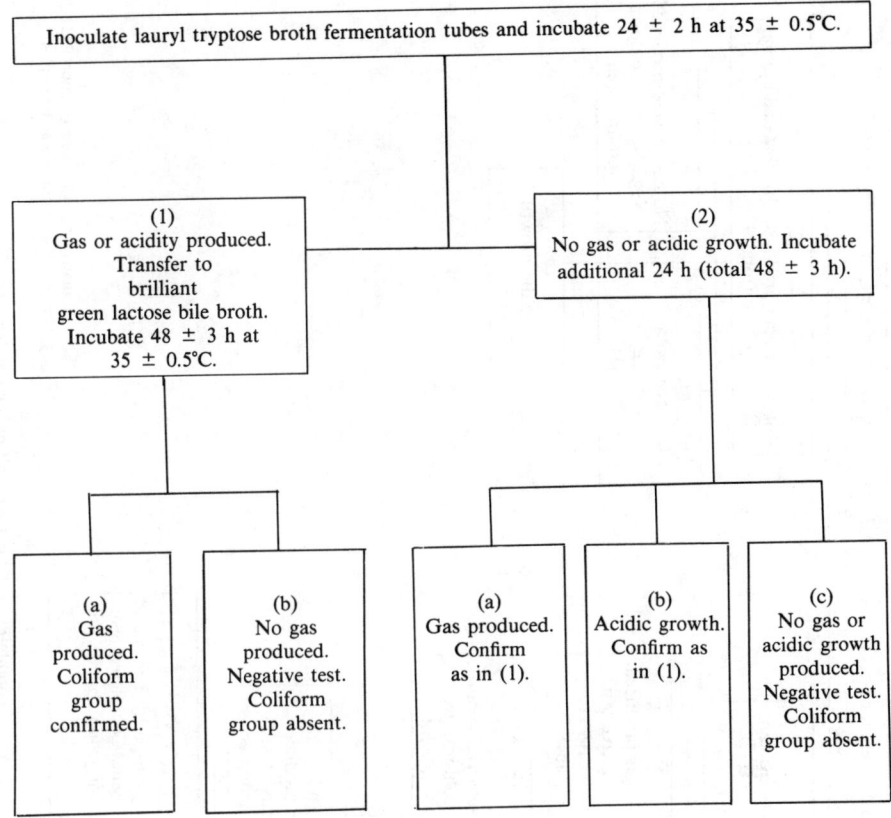

Figure 9221:1. Schematic outline of confirmed phase.

within 48 ± 3 h constitutes a positive confirmed phase. Calculate the MPN value from the number of positive brilliant green lactose bile tubes as described in Section 9221D.

c. Alternative procedure: Use this alternative only for polluted water or wastewater known to produce positive results consistently.

If all presumptive tubes are positive in two or more consecutive dilutions within 24 h, submit to the confirmed phase only the tubes of the highest dilution (smallest sample inoculum) in which all tubes are positive and any positive tubes in still higher dilutions. Submit to the confirmed phase all tubes in which gas or acidic growth is produced only after 48 h.

3. Completed Test

To establish definitively the presence of coliform bacteria and to provide quality control data, use the completed test on all positive confirmed tubes (see Figure 9221:2). Double confirmation into brilliant green lactose bile broth for total coliforms and EC broth for fecal coliforms (see Sec-

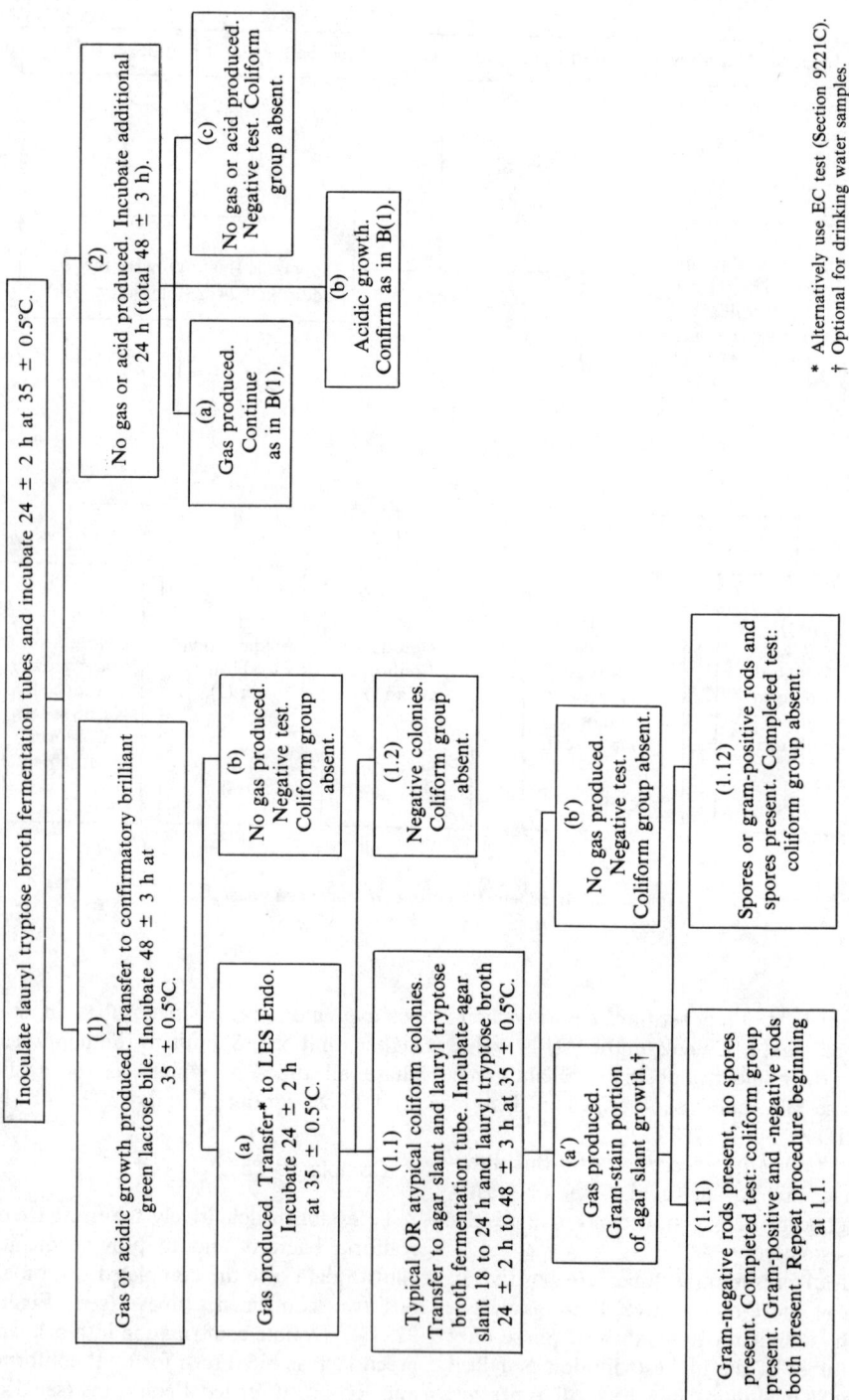

Figure 9221:2. Schematic outline of completed test for total coliform detection.

* Alternatively use EC test (Section 9221C).
† Optional for drinking water samples.

tion 9221C below) may be used. Consider positive EC broth elevated temperature (44.5°C) results as a positive completed test response. Parallel positive brilliant green lactose bile broth cultures with negative EC broth cultures indicate the presence of nonfecal coliforms and must be submitted to the completed test procedure to obtain an MPN test value.

a. Culture medium and reagents:

1) *Nutrient agar:*

Peptone	5.0	g
Beef extract	3.0	g
Agar	15.0	g
Distilled water	1	L

Add ingredients to water, mix thoroughly, and heat to dissolve. pH should be 6.8 ± 0.2 after sterilization. Before sterilization, dispense in screw-capped tubes. After sterilization, immediately place tubes in an inclined position so that the agar will solidify with a sloped surface. Tighten screw caps after cooling and store in a protected, cool storage area.

2) *Gram-stain reagents:*

a) *Ammonium oxalate-crystal violet (Hucker's):* Dissolve 2 g crystal violet (90% dye content) in 20 mL 95% ethyl alcohol; dissolve 0.8 g $(NH_4)_2C_2O_4 \cdot H_2O$ in 80 mL distilled water; mix the two solutions and age for 24 h before use; filter through paper into a staining bottle.

b) *Lugol's solution, Gram's modification:* Grind 1 g iodine crystals and 2 g KI in a mortar. Add distilled water, a few milliliters at a time, and grind thoroughly after each addition until solution is complete. Rinse solution into an amber glass bottle with the remaining water (using a total of 300 mL).

c) *Counterstain:* Dissolve 2.5 g safranin dye in 100 mL 95% ethyl alcohol. Add 10 mL to 100 mL distilled water.

d) *Acetone alcohol:* Mix equal volumes of ethyl alcohol (95%) with acetone.

b. Procedure:

1) Using aseptic technique, streak one LES Endo agar (Section 9222B.2) plate from each tube of brilliant green lactose bile broth showing gas, as soon as possible after the observation of gas. Streak plates in a manner to insure presence of some discrete colonies separated by at least 0.5 cm. Observe the following precautions when streaking plates to obtain a high proportion of successful isolations if coliform organisms are present: (*a*) Use a sterile 3-mm-diam loop or an inoculating needle slightly curved at the tip; (*b*) tap and incline the fermentation tube to avoid picking up any membrane or scum on the needle; (*c*) insert end of loop or needle into the liquid in the tube to a depth of approximately 0.5 cm; and (*d*) streak plate for isolation with curved section of the needle in contact with the agar to avoid a scratched or torn surface. Flame loop between second and third quadrants to improve colony isolation.

Incubate plates (inverted) at 35 ± 0.5°C for 24 ± 2 h.

2) The colonies developing on LES Endo agar are defined as *typical* (pink to dark red with a green metallic surface sheen); *atypical* (pink, red, white, or colorless colonies without sheen) after 24 h incubation; or *negative* (all others). From each plate pick one or more typical, well-isolated coliform colonies or, if no typical colonies are present, pick two or more colonies considered most likely to consist of organisms of the coliform group, and transfer growth from each isolate to a single-strength lauryl tryptose broth fermentation tube and onto a nutrient agar slant. (The latter is unnecessary for drinking water samples.)

If needed, use a colony magnifying device to provide optimum magnification when colonies are picked from the LES Endo agar plates. When transferring colonies, choose well-isolated colonies and barely touch the surface of the colony with a flame-sterilized, air-cooled transfer needle to minimize the danger of transferring a mixed culture.

Incubate secondary broth tubes (lauryl tryptose broth with inverted fermentation vials inserted) at 35 ± 0.5°C for 24 ± 2 h; if gas is not produced within 24 ± 2 h reincubate and examine again at 48 ± 3 h. Microscopically examine Gram-stained preparations from those 24-h nutrient agar slant cultures corresponding to the secondary tubes that show gas.

3) Gram-stain technique—The Gram stain may be omitted from the completed test for potable water samples only because the occurrences of gram-positive bacteria and spore-forming organisms surviving this selective screening procedure are infrequent in drinking water.

Various modifications of the Gram stain technique exist. Use the following modification by Hucker for staining smears of pure culture; include a gram-positive and a gram-negative culture as controls.

Prepare separate light emulsions of the test bacterial growth and positive and negative control cultures on the same slide using drops of distilled water on the slide. Air-dry and fix by passing slide through a flame and stain for 1 min with ammonium oxalate-crystal violet solution. Rinse slide in tap water and drain off excess; apply Lugol's solution for 1 min.

Rinse stained slide in tap water. Decolorize for approximately 15 to 30 s with acetone alcohol by holding slide between the fingers and letting acetone alcohol flow across the stained smear until the solvent flows colorlessly from the slide. Do not over-decolorize. Counterstain with safranin for 15 s, rinse with tap water, blot dry with absorbent paper or air dry, and examine microscopically. Gram-positive organisms are blue; gram-negative organisms are red. Results are acceptable only when controls have given proper reactions.

c. *Interpretation:* Formation of gas in the secondary tube of lauryl tryptose broth within 48 ± 3 h and demonstration of gram-negative, nonspore-forming, rod-shaped bacteria from the agar culture constitute a positive result for the completed test, demonstrating the presence of a member of the coliform group. If no gas is produced in the secondary tube of lauryl tryptose broth within 48 ± 3 h, adjust original MPN result calculated from the confirmed test accordingly (Section 9221D).

4. Bibliography

MEYER, E.M. 1918. An aerobic spore-forming bacillus giving gas in lactose broth isolated in routine water examination. *J. Bacteriol.* 3:9.

HUCKER, G.J. & H.J. CONN. 1923. Methods of Gram Staining. N.Y. State Agr. Exp. Sta. Tech. Bull. No. 93.

NORTON, J.F. & J.J. WEIGHT. 1924. Aerobic spore-forming lactose fermenting organisms and their significance in water analysis. *Amer. J. Pub. Health* 14:1019.

HUCKER, G.J. & H.J. CONN. 1927. Further Studies on the Methods of Gram Staining. N.Y. State Agr. Exp. Sta. Tech. Bull. No. 128.

PORTER, R., C.S. MCCLESKEY & M. LEVINE. 1937. The facultative sporulating bacteria producing gas from lactose. *J. Bacteriol.* 33:163.

COWLES, P.B. 1939. A modified fermentation tube. *J. Bacteriol.* 38:677.

SHERMAN, V.B.D. 1967. A Guide to the Identification of the Genera of Bacteria. Williams & Wilkins, Baltimore, Md.

BUCHANAN, R.E. & N.E. GIBBONS, eds. 1974. Bergey's Manual of Determinative Bacteriology, 8th ed. Williams & Wilkins, Baltimore, Md.

GELDREICH, E.E. 1975. Handbook for Evaluating Water Bacteriological Laboratories, 2nd ed. EPA-670/9-75-006, U.S. Environmental Protection Agency, Cincinnati, Ohio.

EVANS, T.M., C.E. WARVICK, R.J. SEIDLER & M.W. LeCHEVALLIER. 1981. Failure of the most-probable number techniques to detect coliforms in drinking water and raw water supplies. *Appl. Environ. Microbiol.* 41:130.

SEIDLER, R.J., T.M. EVANS, J.R. KAUFMAN, C.E. WARVICK & M.W. LeCHEVALLIER. 1981. Limitations of standard coliform enumeration techniques. *J. Amer. Water Works Assoc.* 73:538.

GERHARDS, P., ed. 1981. Manual of Methods for General Bacteriology. American Soc. Microbiology, Washington, D.C.

GREENBERG, A.E. & D.A. HUNT, eds. 1985. Laboratory Procedures for the Examination of Seawater and Shellfish, 5th ed. American Public Health Assoc., Washington, D.C.

9221 C. Fecal Coliform MPN Procedure

Elevated-temperature tests for the separation of organisms of the coliform group into those of fecal origin and those derived from nonfecal sources are available. Modifications in technical procedures, standardization of methods, and detailed studies of members of the coliform group found in the feces of various warm-blooded animals compared with those from other environmental sources have established the value of a fecal coliform determination. The test can be performed by one of the multiple-tube procedures described here or by membrane filter methods as described in Section 9222. The procedure using EC medium yields adequate information about the source of the coliform group (fecal or nonfecal) when used in confirmation. Do not use it for direct isolation of coliforms from water because prior enrichment is required in a presumptive medium for optimum recovery of fecal coliforms. The procedure using A-1 broth is a single-step method not requiring confirmation.

The fecal coliform test (using EC medium) is applicable to investigations of stream pollution, raw water sources, wastewater treatment systems, bathing waters, seawaters, and general water-quality monitoring. The procedure is not recommended as a substitute for the total coliform test in the examination of potable waters, because no coliform bacteria of any kind should be tolerated in a treated water. The test using A-1 medium is applicable to seawater and treated wastewater.

1. Fecal Coliform Test (EC Medium)

The fecal coliform test differentiates between coliforms of fecal origin (intestines of warm-blooded animals) and coliforms from other sources. Use EC medium or, for a more rapid test of the quality of shellfish waters and treated wastewaters, use A-1 medium in a direct test.

a. EC medium:

Tryptose or trypticase....	20.0	g
Lactose	5.0	g
Bile salts mixture or bile salts No. 3	1.5	g
Dipotassium hydrogen phosphate, K_2HPO_4....	4.0	g
Potassium dihydrogen phosphate, KH_2PO_4....	1.5	g
Sodium chloride, NaCl...	5.0	g
Distilled water	1	L

Add dehydrated ingredients to water, mix thoroughly, and heat to dissolve. pH should be 6.9 ± 0.2 after sterilization. Before sterilization, dispense in fermentation tubes, each with an inverted vial, sufficient medium to cover the inverted vial at least partially after sterilization. Close tubes with metal or heat-resistant plastic caps.

b. Procedure: Submit all presumptive fermentation tubes showing any amount of gas or heavy growth within 48 h of incubation to the confirmed test.

1) Gently shake or rotate presumptive fermentation tubes showing gas or heavy growth. With a sterile 3-mm-diam metal loop or sterile wooden applicator stick, transfer growth from each presumptive fermentation tube to EC broth.

2) Incubate inoculated EC broth tubes in a water bath at 44.5 ± 0.2°C for 24 ± 2 h.

Place all EC tubes in water bath within 30 min after inoculation. Maintain a sufficient water depth in water bath incubator to immerse tubes to upper level of the medium.

c. Interpretation: Gas production in an EC broth culture within 24 h or less is considered a positive fecal coliform reaction. Failure to produce gas (growth sometimes occurs) constitutes a negative reaction indicating a source other than the intestinal tract of warm-blooded animals. Calculate MPN from the number of positive EC broth tubes as described in Section 9221D.

2. Fecal Coliform Direct Test (A-1 Medium)

a. A-1 broth: This medium may not be available in dehydrated form and may require preparation from the basic ingredients.

Lactose	5.0	g
Tryptone	20.0	g
Sodium chloride, NaCl ...	5.0	g
Salicin	0.5	g
Polyethylene glycol		
p-isooctylphenyl ether*	1.0	mL
Distilled water	1	L

Heat to dissolve solid ingredients, add polyethylene glycol *p*-isooctylphenyl ether, and adjust to pH 6.9 ± 0.1. Before sterilization dispense in fermentation tubes with an inverted vial sufficient medium to cover the inverted vial at least partially after sterilization. Close with metal or heat-resistant plastic caps. Sterilize by autoclaving at 121°C for 10 min. Store in dark at room temperature for not longer than 7 d. Ignore formation of precipitate.

Make A-1 broth of such strength that

---

*Triton X-100, Rohm and Haas Co., or equivalent.

adding 10-mL sample portions to medium will not reduce ingredient concentrations below those of the standard medium. For 10-mL samples prepare double-strength medium.

b. Procedure: Inoculate tubes of A-1 broth as directed in Section 9221B.1*b*1). Incubate for 3 h at 35 ± 0.5°C. Transfer tubes to a water bath at 44.5 ± 0.2°C and incubate for an additional 21 ± 2 h.

c. Interpretation: Gas production in any A-1 broth culture within 24 h or less is a positive reaction indicating coliforms of fecal origin. Calculate MPN from the number of positive A-1 broth tubes as described in Section 9221D.

3. Bibliography

PERRY, C.A. & A.A. HAJNA. 1933. A modified Eijkman medium. *J. Bacteriol.* 25:419.

PERRY, C.A. & A.A. HAJNA. 1944. Further evaluation of EC medium for the isolation of coliform bacteria and *Escherichia coli. Amer. J. Pub. Health* 34:735.

GELDREICH, E.E., H.F. CLARK, P.W. KABLER, C.B. HUFF & R.H. BORDNER. 1958. The coliform group. II. Reactions in EC medium at 45°C. *Appl. Microbiol.* 6:347.

GELDREICH, E.E., R.H. BORDNER, C.B. HUFF, H.F. CLARK & P.W. KABLER. 1962. Type distribution of coliform bacteria in the feces of warm-blooded animals. *J. Water Pollut. Control Fed.* 34:295.

GELDREICH, E.E. 1966. Sanitary significance of fecal coliforms in the environment. FWPCA Publ. WP-20-3 (Nov.). U.S. Dep. Interior, Washington, D.C.

ANDREWS, W.H. & M.W. PRESNELL. 1972. Rapid recovery of *Escherichia coli* from estuarine water. *Appl. Microbiol.* 23:521.

OLSON, B.H. 1978. Enhanced accuracy of coliform testing in seawater by a modification of the most-probable-number method. *Appl. Microbiol.* 36:438.

STRANDRIDGE, J.H. & J.J. DELFINO. 1981. A-1 Medium: Alternative technique for fecal coliform organism enumeration in chlorinated wastewaters. *Appl. Environ. Microbiol.* 42:918.

9221 D. Estimation of Bacterial Density

1. Precision of Fermentation Tube Test

Unless a large number of sample portions is examined, the precision of the fermentation tube test is rather low. For example, even when the sample contains 1 coliform organism/mL, about 37% of 1-mL tubes may be expected to yield negative results because of random distribution of the bacteria in the sample. When five tubes, each with 1 mL sample, are used under these conditions, a completely negative result may be expected less than 1% of the time.

Even when five fermentation tubes are used, the precision of the results obtained is not of a high order. Consequently, exercise great caution when interpreting the sanitary significance of coliform results obtained from the use of a few tubes with each sample dilution, especially when the number of samples from a given sampling point is limited.

2. Computing and Recording of MPN

Record the number of positive findings of coliform group organisms (either confirmed or completed) and compute in terms of the Most Probable Number (MPN). The MPN values, for a variety of planting series and results, are given in Tables 9221:III, IV, and V. Included in these tables are the 95% confidence limits for each MPN value determined. If the sample volumes used are those found in the tables, report the value corresponding to the number of positive and negative results in the series as the MPN/100 mL.

The sample volumes indicated in Tables 9221:III and IV relate more specifically to finished waters. Table 9221:V illustrates MPN values for combinations of positive and negative results when five 10-mL, five 1.0-mL, and five 0.1-mL volumes of samples are tested. When the series of decimal dilutions is different from that in the table, select the MPN value from Table 9221:V

TABLE 9221:III. MPN INDEX AND 95% CONFIDENCE LIMITS FOR VARIOUS COMBINATIONS OF POSITIVE AND NEGATIVE RESULTS WHEN FIVE 10-ML PORTIONS ARE USED

No. of Tubes Giving Positive Reaction Out of 5 of 10 mL Each	MPN Index/ 100 mL	95% Confidence Limits (Approximate)	
		Lower	Upper
0	< 2.2	0	6.0
1	2.2	0.1	12.6
2	5.1	0.5	19.2
3	9.2	1.6	29.4
4	16.0	3.3	52.9
5	> 16.0	8.0	Infinite

TABLE 9221:IV. MPN INDEX AND 95% CONFIDENCE LIMITS FOR VARIOUS COMBINATIONS OF POSITIVE AND NEGATIVE RESULTS WHEN TEN 10-ML PORTIONS ARE USED

No. of Tubes Giving Positive Reaction Out of 10 of 10 mL Each	MPN Index/ 100 mL	95% Confidence Limits (Approximate)	
		Lower	Upper
0	< 1.1	0	3.0
1	1.1	0.03	5.9
2	2.2	0.26	8.1
3	3.6	0.69	10.6
4	5.1	1.3	13.4
5	6.9	2.1	16.8
6	9.2	3.1	21.1
7	12.0	4.3	27.1
8	16.1	5.9	36.8
9	23.0	8.1	59.5
10	> 23.0	13.5	Infinite

for the combination of positive tubes and calculate according to the following formula:

MPN value (from table)

$$\times \frac{10}{\text{largest volume tested}} = \text{MPN}/100 \text{ mL}$$

TABLE 9221:V. MPN INDEX AND 95% CONFIDENCE LIMITS FOR VARIOUS COMBINATIONS OF POSITIVE RESULTS WHEN FIVE TUBES ARE USED PER DILUTION (10 mL, 1.0 mL, 0.1 mL)

Combination of Positives	MPN Index/ 100 mL	95% Confidence Limits		Combination of Positives	MPN Index/ 100 mL	95% Confidence Limits	
		Lower	Upper			Lower	Upper
0-0-0	< 2	—	—	4-2-0	22	9.0	56
0-0-1	2	1.0	10	4-2-1	26	12	65
0-1-0	2	1.0	10	4-3-0	27	12	67
0-2-0	4	1.0	13	4-3-1	33	15	77
				4-4-0	34	16	80
1-0-0	2	1.0	11	5-0-0	23	9.0	86
1-0-1	4	1.0	15	5-0-1	30	10	110
1-1-0	4	1.0	15	5-0-2	40	20	140
1-1-1	6	2.0	18	5-1-0	30	10	120
1-2-0	6	2.0	18	5-1-1	50	20	150
				5-1-2	60	30	180
2-0-0	4	1.0	17	5-2-0	50	20	170
2-0-1	7	2.0	20	5-2-1	70	30	210
2-1-0	7	2.0	21	5-2-2	90	40	250
2-1-1	9	3.0	24	5-3-0	80	30	250
2-2-0	9	3.0	25	5-3-1	110	40	300
2-3-0	12	5.0	29	5-3-2	140	60	360
3-0-0	8	3.0	24	5-3-3	170	80	410
3-0-1	11	4.0	29	5-4-0	130	50	390
3-1-0	11	4.0	29	5-4-1	170	70	480
3-1-1	14	6.0	35	5-4-2	220	100	580
3-2-0	14	6.0	35	5-4-3	280	120	690
3-2-1	17	7.0	40	5-4-4	350	160	820
4-0-0	13	5.0	38	5-5-0	240	100	940
4-0-1	17	7.0	45	5-5-1	300	100	1300
4-1-0	17	7.0	46	5-5-2	500	200	2000
4-1-1	21	9.0	55	5-5-3	900	300	2900
4-1-2	26	12	63	5-5-4	1600	600	5300
				5-5-5	≥ 1600	—	—

When more than three dilutions are used in a decimal series of dilutions, use the results from only three of these in computing the MPN. To select the three dilutions to be used in determining the MPN index, choose the highest dilution that gives positive results in all five portions tested (no lower dilution giving any negative results) and the two next succeeding higher dilutions. Use the results at these three volumes in computing the MPN index. In the examples given below, the significant dilution results are shown in boldface. The number in the numerator represents positive tubes; that in the denominator, the total tubes planted; the combination of positives simply represents the total number of positive tubes per dilution:

Example	1 mL	0.1 mL	0.01 mL	0.001 mL	Combination of positives	MPN Index /100 mL
a	5/5	5/5	2/5	0/5	5-2-0	5000
b	5/5	4/5	2/5	0/5	5-4-2	2200
c	0/5	1/5	0/5	0/5	0-1-0	20

In c, select the first three dilutions so as to include the positive result in the middle dilution.

When a case such as that shown below in line d arises, where a positive occurs in a dilution higher than the three chosen according to the rule, incorporate it in the result for the highest chosen dilution, as in e:

Example	1 mL	0.1 mL	0.01 mL	0.001 mL	Combination of positives	MPN Index /100mL
d	5/5	3/5	1/5	1/5	5-3-2	1400
e	5/5	3/5	2/5	0/5	5-3-2	1400

When it is desired to summarize with a single MPN value the results from a series of samples, use the geometric mean or the median.

Table 9221:V shows the most likely positive tube combinations. If unlikely combinations occur with a frequency greater than 1% it is an indication that the technique is faulty or that the statistical assumptions underlying the MPN estimate are not being fulfilled. The MPN for combinations not appearing in the table, or for other combinations of tubes or dilutions, may be estimated by Thomas' simple formula:

MPN/100 mL

$$= \frac{\text{no. of positive tubes} \times 100}{\sqrt{\left(\begin{array}{c}\text{mL sample in} \\ \text{negative tubes}\end{array} \times \begin{array}{c}\text{mL sample in} \\ \text{all tubes}\end{array}\right)}}$$

While the MPN tables and calculations are described for use in the coliform test, they are equally applicable to determining the MPN of any other organisms provided suitable test media are available.

3. Bibliography

McCRADY, M.H. 1915. The numerical interpretation of fermentation tube results. *J. Infect. Dis.* 12:183.

McCRADY, M.H. 1918. Tables for rapid interpretation of fermentation tube results. *Can. J. Pub. Health* 9:201.

HOSKINS, J.K. 1933. The most probable number of *B. coli* in water analysis. *J. Amer. Water Works Assoc.* 25:867.

HOSKINS, J.K. 1934. Most Probable Numbers for evaluation of *Coli-Aerogenes* tests by fermentation tube method. *Pub. Health Rep.* 49:393 (Reprint 1621).

HOSKINS, J.K. & C.T. BUTTERFIELD. 1935. Determining the bacteriological quality of drinking water. *J. Amer. Water Works Assoc.* 27:1101.

HALVORSON, H.O. & N.R. ZIEGLER. 1933-35. Application of statistics to problems in bacteriology. *J. Bacteriol.* 25:101; 26:331,559; 29:609.

SWAROOP, S. 1938. Numerical estimation of *B. coli* by dilution method. *Indian J. Med. Res.* 26:353.

DALLA VALLE, J.M. 1941. Notes on the most probable number index as used in bacteriology. *Pub. Health Rep.* 56:229.

THOMAS, H.A., JR. 1942. Bacterial densities from fermentation tube tests. *J. Amer. Water Works Assoc.* 34:572.

WOODWARD, R.L. 1957. How probable is the Most Probable Number? *J. Amer. Water Works Assoc.* 49:1060.

McCarthy, J.A., H.A. Thomas & J.E. Delaney. 1958. Evaluation of reliability of coliform density tests. *Amer. J. Pub. Health* 48:12.

U.S. Environmental Protection Agency. 1975. Interim primary drinking water standards. *Fed. Reg.* 40(51):11990 (Mar. 14, 1975).

de Man, J.C. 1977. MPN tables for more than one test. *European J. Appl. Microbiol.* 4:307.

9221 E.　Presence-Absence (P-A) Coliform Test

The presence-absence (P-A) test for the coliform group is a simple modification of the multiple-tube procedure. Simplification, by use of one large test portion (100 mL) in a single culture bottle to obtain qualitative information on the presence or absence of coliforms, is justified on the theory that no coliforms should be present in 100 mL of a drinking water sample. The P-A test also provides the optional opportunity for further screening of the culture to isolate other indicators (fecal coliform, *Aeromonas, Staphylococcus, Pseudomonas,* fecal streptococcus, and *Clostridium*) on the same qualitative basis. Additional advantages include the possibility of examining a larger number of samples per unit of time and comparative studies with the membrane filter procedure indicate that the P-A test may maximize coliform detection in samples containing many organisms that could overgrow coliform colonies and cause problems in detection.

The P-A test is intended for use on routine sample submissions collected from distribution systems or water treatment plants. Initially, examine approximately 100 samples by either membrane filter or multiple-tube methods as well as the P-A test. After it is established that quantitative methods usually give negative results for coliforms, the P-A test alone may be used. When sample locations produce a positive P-A result for coliforms, examine subsequent repeat samples by a quantitative procedure until negative results are obtained from two consecutive samples. Analyze infrequently collected samples such as those from private wells by multiple-tube or membrane filter methods.

1. P-A Coliform Test

a. P-A broth: This medium may not be available in dehydrated form and may require preparation from the combination of other commercially available dehydrated media.

Lactose broth (dehydrated)	13.0	g
Lauryl tryptose broth (dehydrated)	17.5	g
Bromcresol purple	0.0085	g
Distilled water	1	L

Make this formulation triple strength when examining 100-mL samples. Dissolve the lactose broth and lauryl tryptose broth ingredients sequentially in water without heating, using a stirring device. Dissolve the bromcresol purple in 10 mL 0.1N NaOH and add to the broth solution. Dispense 50 mL prepared medium into a screw-cap 250-mL milk dilution bottle. A fermentation tube insert is not necessary. Autoclave for 12 min at 121°C with the total time in the autoclave limited to 30 min or less. pH should be 6.8 ± 0.2 after sterilization.

b. Procedure: Shake sample approximately 25 times and inoculate 100 mL into a P-A culture bottle. Mix thoroughly by inverting bottle four or five times to achieve even distribution of the triple-strength medium throughout the sample. Incubate at 35 ± 0.5°C and inspect after 24 and 48 h for acid reactions. A distinct yellow color forms in the medium when acid conditions exist following lactose fermentation. If gas also is being produced, gently shaking the bottle will result in a foaming reaction. Any amount of gas and/or acid constitutes a

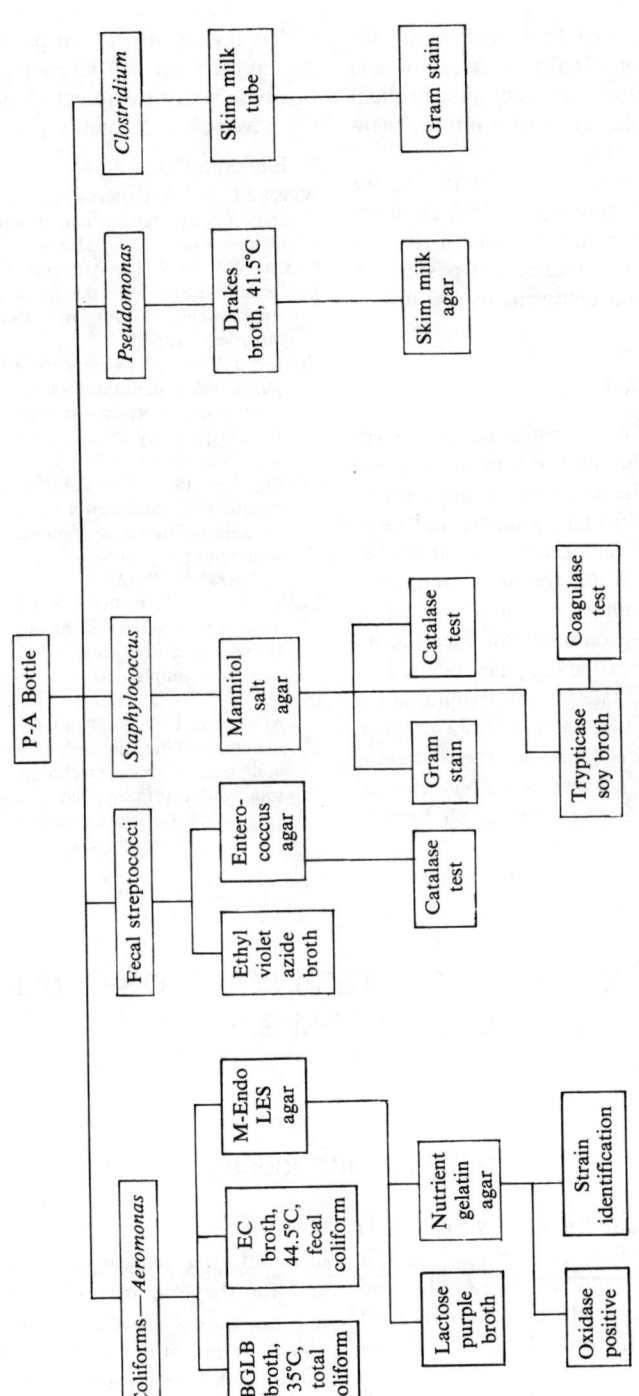

Figure 9221:3. Schematic outline of optional P-A confirmations for various indicator organisms.

positive presumptive test requiring confirmation. Transfer all cultures that show acid reaction or acid and gas reaction to brilliant green lactose bile (BGLB) broth for incubation at 35 ± 0.5°C.

c. Interpretation: Gas production in the BGLB broth culture within 48 h confirms the presence of coliform bacteria. Report result as presence-absence test positive or negative for total coliforms in 100 mL of sample.

2. Optional P-A Tests

Detection of fecal coliforms, fecal streptococci, and other indicator bacteria by the P-A test may be achieved by appropriate selection of confirmatory media and incubation time and temperature as outlined in Figure 9221:3. Water treatment and other adverse environmental conditions often place great stress on indicator bacteria, resulting in an extended lag phase before logarithmic growth takes place. Extending the P-A test incubation period to 72 or 96 h often will allow isolation of other indicator organisms. Do not use isolations of indicator bacteria after the usual 48 h incubation for regulatory purposes; however, the information will allow the laboratory to advise water treatment plants of potential adverse water quality problems.

3. Bibliography

WEISS, J.E. & C.A. HUNTER. 1939. Simplified bacteriological examination of water. *J. Amer. Water Works Assoc.* 31:689.

CLARK, J.A. 1969. The detection of various bacteria indicative of water pollution by a presence-absence (P-A) procedure. *Can. J. Microbiol.* 15:771.

CLARK, J.A. & L.T. VLASSOFF. 1973. Relationships among pollution indicator bacteria isolated from raw water and distribution systems by the presence-absence (P-A) test. *Health Lab. Sci.* 10:163.

CLARK, J.A. 1980. The influence of increasing numbers of nonindicator organisms upon the detection of indicator organisms by the membrane filter and presence-absence tests. *Can. J. Microbiol.* 26:827.

CLARK, J.A., C.A. BURGER & L.E. SABATINOS. 1982. Characterization of indicator bacteria in municipal raw water, drinking water and new main samples. *Can. J. Microbiol.* 28:1002.

JACOBS, N.J., W.L. ZEIGLER, F.C. REED, T.A. STUKEL & E.W. RICE. 1986. Comparison of membrane filter, multiple-fermentation-tube, and presence-absence techniques for detecting total coliforms in small community water systems. *Appl. Environ. Microbiol.* 51:1007.

9222   MEMBRANE FILTER TECHNIQUE FOR MEMBERS OF THE COLIFORM GROUP*

9222 A.   Introduction

The membrane filter (MF) technique is highly reproducible, can be used to test relatively large volumes of sample, and yields numerical results more rapidly than the multiple-tube procedure. The membrane filter technique is extremely useful in monitoring drinking water emergencies and for the examination of a variety of natural waters. However, the MF technique has limitations, particularly in testing waters with high turbidity and noncoliform (background) bacteria. For such waters

---

*Approved by Standard Methods Committee, 1988.

and also when the membrane filter technique has not been used previously, it is desirable to conduct parallel tests with the multiple-tube fermentation technique (Section 9221) to demonstrate applicability and comparability.

1. Definition

As applied to the membrane filter technique, the coliform group may be defined as comprising all aerobic and facultative anaerobic, gram-negative, nonspore-forming, rod-shaped bacteria that develop a red colony with a metallic sheen within 24 h at 35°C on an Endo-type medium containing lactose. When purified cultures of coliform bacteria are tested they produce a negative cytochrome oxidase (CO) and positive β-galactosidase (ONPG) reaction. Generally, all red, pink, blue, white, or colorless colonies lacking sheen are considered noncoliforms by this technique.

2. Applications

Turbidity caused by the presence of algae or other interfering material may not permit testing of a sample volume sufficient to yield significant results. Low coliform estimates may be caused by the presence of high numbers of noncoliforms or of toxic substances. The MF technique is applicable to the examination of saline waters, but not wastewaters that have received only primary treatment followed by chlorination because of turbidity in high volume samples or wastewaters containing toxic metals or toxic organics such as phenols. For the detection of stressed total coliforms in treated drinking water and chlorinated secondary or tertiary wastewater effluents use a method designed for stressed organism recovery (see Section 9212B.1). For fecal coliforms use the multiple-tube fermentation technique (Section 9221C) if the results are to be used in enforcement actions. Alternatively, a modified membrane filter technique for fecal coliforms (Section 9212)

in chlorinated wastewater may be used if parallel testing with the multiple-tube fermentation technique shows comparability for each site-specific type of sample.

The standard volume to be filtered for drinking water samples is 100 mL (see Section 9010B). This may be distributed among multiple membranes if necessary. Because treated drinking water should contain no coliforms per 100 mL, water plant laboratories should consider testing 1-L samples of finished water, provided that particulates are not present to interfere with filtration or development of discrete colonies. In such situations, divide the sample into four portions of 250 mL for analysis and total any coliforms on each membrane into a single report on 1 L examined. Smaller or larger samples may be used for other waters or special analyses.

Statistical comparisons of results obtained by the multiple-tube method and the membrane filter technique show that the membrane filter is more precise (compare Tables 9221:I, II, and III with Table 9222:II). Although data from each test yield approximately the same water quality information, numerical results are not identical (see Section 9010B for drinking water). For raw water sources it would be expected that 80% of the membrane filter test results would be within the 95% confidence limits of the multiple-tube completed test results. Results from the multiple-tube test would be expected to be higher than membrane filter results because of a built-in positive statistical bias.

3. Bibliography

CLARK, H.F., E.E. GELDREICH, H.L. JETER & P.W. KABLER. 1951. The membrane filter in sanitary bacteriology. *Pub. Health Rep.* 66:951.

KABLER, P.W. 1954. Water examinations by membrane filter and MPN procedures. *Amer. J. Pub. Health* 44:379.

THOMAS, H.A. & R.L. WOODWARD. 1956. Use

of molecular filter membranes for water potability control. *J. Amer. Water Works Assoc.*
48:1391.

McCARTHY, J.A., J.E. DELANEY & R.J. GRASSO.
1961. Measuring coliforms in water. *Water
Sewage Works* 108:238.

LIN, S. 1973. Evaluation of coliform test for chlorinated secondary effluents. *J. Water Pollut.
Control Fed.* 45:498.

MANDEL, J. & L.F. NANNI. 1978. Measurement
evaluation. *In* S.L. Inhorn, ed. Quality Assurance Practices for Health Laboratories, p.
209. American Public Health Assoc., Washington, D.C.

9222 B.   Standard Total Coliform Membrane Filter Procedure

1. Laboratory Apparatus

For membrane filter analyses use glassware and other apparatus composed of material free from agents that may have
detrimental effects on bacterial growth.
Carefully note any deviations from the recommendations presented below and make
quantitative tests to demonstrate that such
deviations have not introduced agents or
factors resulting in conditions less favorable for bacterial growth.

Sterilize glassware as described in Washing and Sterilization, Section 9040.

a. Sample bottles: See Section 9030B.18.

b. Dilution bottles: See Section
9030B.13.

c. Pipets and graduated cylinders: See
Section 9030B.9. Before sterilization, cover
opening of graduated cylinders with metal
foil or a suitable paper substitute.

d. Containers for culture medium: Use
clean borosilicate glass flasks presterilized
to reduce bacterial contamination. Any size
or shape of flask may be used, but erlenmeyer flasks with metal caps, metal foil
covers, or screw caps provide for adequate
mixing of the medium contained and are
convenient for storage.

e. Culture dishes: Use sterile borosilicate
glass or disposable plastic petri-dish type,
60×15 mm, 50×12 mm, or other appropriate size. Wrap clean, glass culture
dishes before sterilization, singly or in convenient numbers, in metal foil if sterilized
by dry heat, or suitable paper substitute

when autoclaved. Because glass and some
disposable plastic culture dishes have loose-
fitting covers, take precautions during incubation to prevent possible loss of moisture by evaporation with resultant drying
of medium and to maintain a humid environment for optimum colony development.

Disposable plastic dishes that are tight-
fitting and meet the specifications noted
above also may be used. Suitable sterile
plastic dishes are available commercially.

f. Filtration units: The filter-holding assembly (constructed of glass, autoclavable
plastic, porcelain, or stainless steel) consists
of a seamless funnel fastened to a base by
a locking device or held in place by magnetic force. The design should permit the
membrane filter to be held securely on the
porous plate of the receptacle without mechanical damage and allow all fluid to pass
through the membrane during filtration.

Separately wrap the two parts of the assembly in heavy wrapping paper, sterilize
by autoclaving, and store until use. Alternatively treat unwrapped parts by ultraviolet radiation before reusing units during
a filtration series. Field units may be san-
itized by igniting methyl alcohol or immersing in boiling water for 5 min. Do not
ignite plastic parts. Sterile, disposable field
units may be used.

For filtration, mount receptacle of filter-
holding assembly in a 1-L filtering flask
with a side tube or other suitable device

(manifold to hold three to six filter assemblies) such that a pressure differential (34 to 51 kPa) can be exerted on the filter membrane. Connect flask to an electric vacuum pump, a filter pump operating on water pressure, a hand aspirator, or other means of securing a pressure differential (138 to 207 kPa). Connect a flask of approximately the same capacity between filtering flask and vacuum source to trap carry-over water.

g. *Membrane filter:* Use membrane filters (for additional specifications, see Section 9020) with a rated pore diameter such that there is complete retention of coliform bacteria. Use only those filter membranes that have been found, through adequate quality control testing and *certification by the manufacturer,* to exhibit: full retention of the organisms to be cultivated, stability in use, freedom from chemical extractables that may inhibit bacterial growth and development, a satisfactory speed of filtration (within 5 min), no significant influence on medium pH (beyond $\pm$ 0.2 units), and no increase in number of confluent colonies or spreaders compared to control membrane filters. Use membranes grid-marked in such a manner that bacterial growth is neither inhibited nor stimulated along the grid lines when the membranes and entrapped bacteria are incubated on a suitable medium. Preferably use fresh stocks of membrane filters and if necessary store them in an environment without extremes of temperature and humidity. Obtain no more than a year's supply at any one time.

Preferably use presterilized membrane filters for which the manufacturer has certified that the sterilization technique has neither induced toxicity nor altered the chemical or physical properties of the membrane. If membranes are sterilized in the laboratory, autoclave for 10 min at 121°C. At the end of the sterilization period, let the steam escape rapidly to minimize accumulation of water of condensation on filters.

h. *Absorbent pads* consist of disks of filter paper or other material certified for each lot by the manufacturer to be of high quality and free of sulfites or other substances that could inhibit bacterial growth. Use pads approximately 48 mm in diameter and of sufficient thickness to absorb 1.8 to 2.2 mL of medium. Presterilized absorbent pads or pads subsequently sterilized in the laboratory should release less than 1 mg total acidity (calculated as $CaCO_3$) when titrated to the phenolphthalein end point, pH 8.3, using $0.02N$ NaOH and produce pH levels of 7 $\pm$ 0.2. Sterilize pads simultaneously with membrane filters available in resealable kraft envelopes, or separately in other suitable containers. Dry pads so they are free of visible moisture before use. See sterilization procedure described for membrane filters above and Section 9020 for additional specifications on absorbent pads.

i. *Forceps:* Smooth-tipped, without corrugations on the inner sides of the tips. Sterilize before use by dipping in 95% ethyl or absolute methyl alcohol and flaming.

j. *Incubators:* Use incubators to provide a temperature of 35 $\pm$ 0.5°C and to maintain a high level of humidity (approximately 90% relative humidity).

k. *Microscope and light source:* Count colonies on membrane filters using a magnification of 10 to 15 diameters and a cool white fluorescent light source adjusted to give maximum sheen discernment. Optimally use a binocular wide-field dissecting microscope. Do not use a microscope illuminator with optical system for light concentration from an incandescent light source for discerning coliform colonies on Endo-type media.

2. Materials and Culture Media

The need for uniformity dictates the use of dehydrated media. Never prepare media from basic ingredients when suitable dehydrated media are available. Follow manufacturer's directions for rehydration and sterilization. Commercially prepared media in liquid form (sterile ampule or other) also may be used if known to give equivalent results. See Section 9020 for media quality control specifications.

Test each new medium lot for satisfactory productivity by preparing dilutions of a culture of *Enterobacter aerogenes* (Section 9020) and filtering appropriate volumes to give 20 to 80 colonies per filter. With each new lot of Endo-type medium, verify a minimum 10% of coliform colonies, obtained from natural samples, to establish the differential accuracy of the medium lot.

a. LES Endo agar:

Yeast extract	1.2	g
Casitone or trypticase	3.7	g
Thiopeptone or thiotone...	3.7	g
Tryptose	7.5	g
Lactose	9.4	g
Dipotassium hydrogen phosphate, K_2HPO_4.....	3.3	g
Potassium dihydrogen phosphate, KH_2PO_4.....	1.0	g
Sodium chloride, NaCl	3.7	g
Sodium desoxycholate.....	0.1	g
Sodium lauryl sulfate	0.05	g
Sodium sulfite, Na_2SO_3....	1.6	g
Basic fuchsin	0.8	g
Agar	15.0	g
Distilled water	1	L

Rehydrate in 1 L distilled water containing 20 mL 95% ethanol (which controls background growth and coliform colony size). Bring to a near boil to dissolve agar, then promptly remove from heat and cool to 45 to 50°C. Dispense in 5- to 7-mL quantities into lower section of 60-mm glass or plastic petri dishes. If dishes of any other size are used, adjust quantity to give an equivalent depth. Do not expose plates to direct sunlight; store in the dark at 2 to 10°C and discard unused medium after 2 weeks.

b. M-Endo medium*:

Tryptose or polypeptone...	10.0	g
Thiopeptone or thiotone ...	5.0	g
Casitone or trypticase	5.0	g
Yeast extract	1.5	g
Lactose	12.5	g
Sodium chloride, NaCl	5.0	g
Dipotassium hydrogen phosphate, K_2HPO_4.....	4.375	g
Potassium dihydrogen phosphate, KH_2PO_4.....	1.375	g
Sodium lauryl sulfate	0.050	g
Sodium desoxycholate.....	0.10	g
Sodium sulfite, Na_2SO_3....	2.10	g
Basic fuchsin	1.05	g
Agar	15.0	g
Distilled water	1	L

Rehydrate in 1 L distilled water containing 20 mL 95% ethanol. Heat to near boiling to dissolve agar, then promptly remove from heat and cool to below 50°C. Do not sterilize by autoclaving. Final pH should be between 7.1 and 7.3.

Store finished medium in the dark at 2 to 10°C and discard any unused medium after 96 h.

Liquid medium (without agar) and absorbent pads may be used if pads are certified free of sulfite.

3. Samples

Collect samples as directed in Sections 9060A and B.

4. Coliform Definition

All organisms that produce a red colony with a metallic sheen within 24 h incubation at 35°C on an Endo-type medium are considered members of the coliform group. The sheen may cover the entire colony or may appear only in a central area or on the periphery. The coliform group thus defined is based on the production of alde-

---

*Dehydrated Difco M-Endo Broth MF (No. 0749), dehydrated BBL m-Coliform Broth (No. 11119), or equivalent may be used provided 1.5% agar is added in preparation.

hydes from fermentation of lactose. While this biochemical characteristic is part of the metabolic pathway of gas production in the multiple-tube test, some variations in metallic sheen development may be observed among coliform strains. However, this slight difference in indicator definition is not considered to change its public health significance, particularly if suitable studies have been conducted to establish the relationship between results obtained by the membrane filter and those obtained by the standard tube dilution procedure.

Verify lactose fermentation or reaction to CO and ONPG by sheen colonies to avoid false positive results, especially for drinking water samples. Preferably verify all colonies; alternatively, verify at least 10% of the colonies when there are more than 50 coliform-type colonies present from a wastewater sample. For coliform counts of more than 5/100 mL from drinking water, verify a minimum of five colonies included in the direct count. Coliforms occasionally may produce atypical colonies. Experience in recognizing such colonies may be gained by verifying all types of sheen and non-sheen colonies. See Section 9222B.5*f* for specific verification procedure.

5. Procedures

Generally, an enrichment procedure may improve the assessment of drinking water quality. However, this step may be eliminated in the routine examination of drinking water where repeated determinations have shown that adequate results are obtained by a single-step MF technique. Enrichment usually is not necessary in the examination of nonpotable water or wastewater. Verify all samples of drinking water giving positive results as described above.

a. Selection of sample size: Size of sample will be governed by expected bacterial density, which in drinking water samples will be limited only by the degree of turbidity or by the noncoliform growth on the medium (Table 9222:I).

An ideal sample volume will yield about 50 coliform colonies and not more than 200 colonies of all types. Analyze drinking waters by filtering 100 to 500 mL or more, or by filtering replicate smaller sample volumes such as duplicate 50-mL or four replicates of 25-mL portions. Analyze other waters by filtering three different volumes (diluted or undiluted), depending on the expected bacterial density. See Section 9215B.2 for preparation of dilutions. When less than 20 mL of sample (diluted or undiluted) is to be filtered, add approximately 10 mL sterile dilution water to the funnel before filtration. This increase in water volume aids in uniform dispersion of the bacterial suspension over the entire effective filtering surface.

b. Filtration of sample: Using sterile forceps, place a sterile membrane filter (grid side up) over porous plate of receptacle. Carefully place matched funnel unit over receptacle and lock it in place. Filter sample under partial vacuum. With filter still in place, rinse funnel by filtering three 20- to 30-mL portions of sterile dilution water. Upon completion of final rinse and the filtration process disengage vacuum, unlock and remove funnel, immediately remove membrane filter with sterile forceps, and place it on selected medium with a rolling motion to avoid entrapment of air. Insert a sterile rinse water sample (100 mL) after filtration of a series of 10 samples to check for possible cross-contamination or contaminated rinse water. Incubate the control membrane culture under the same conditions as the sample.

Use sterile filtration units at the beginning of each filtration series as a minimum precaution to avoid accidental contamination. A filtration series is considered to be interrupted when an interval of 30 min or longer elapses between sample filtrations. After such interruption, treat any further sample filtration as a new filtration

TABLE 9222:I. SUGGESTED SAMPLE VOLUMES FOR MEMBRANE FILTER TOTAL COLIFORM TEST

Water Source	Volume (X) To Be Filtered mL							
	100	50	10	1	0.1	0.01	0.001	0.0001
Drinking water	X							
Swimming pools	X							
Wells, springs	X	X	X					
Lakes, reservoirs	X	X	X					
Water supply intake			X	X	X			
Bathing beaches			X	X	X			
River water				X	X	X	X	
Chlorinated sewage				X	X	X		
Raw sewage				X	X	X	X

series and sterilize all membrane filter holders in use. Decontaminate this equipment between successive filtrations by using an ultraviolet (UV) sterilizer, flowing steam, or boiling water. In the UV sterilization procedure, a 2-min exposure to UV radiation is sufficient. Do not expose membrane-filter culture preparations to random UV radiation leaks that might emanate from the sterilization cabinet. Eye protection is recommended; either safety glasses or prescription-ground glasses afford adequate eye protection against stray radiation from a UV sterilization cabinet that is not light-tight during the exposure interval. Clean UV tube regularly and check it periodically for effectiveness to insure that it will produce a 99.9% bacterial kill in a 2-min exposure. See also Section 9020.

c. *Enrichment technique:* Place a sterile absorbent pad in the upper half of a sterile culture dish and pipet enough enrichment medium (1.8 to 2.0 mL lauryl tryptose broth) to saturate pad. Carefully remove any excess liquid from absorbent pad. Aseptically place filter through which the sample has been passed on pad. Incubate filter, without inverting dish, for 1.5 to 2 h at 35 ± 0.5°C in an atmosphere of at least 90% relative humidity.

If the agar-based medium is used, remove enrichment culture from incubator, lift filter from enrichment pad, and roll it onto the agar surface. Incorrect filter placement is at once obvious, because patches of unstained membrane indicate entrapment of air. Where such patches occur, carefully reseat filter on agar surface. If the liquid medium is used, prepare final culture by removing enrichment culture from incubator and separating the dish halves. Place a fresh sterile pad in bottom half of dish and saturate it with 1.8 to 2.0 mL of final M-Endo medium. Transfer filter, with same precautions as above, to new pad. Discard used enrichment pad.

With either the agar or the liquid medium, invert dish and incubate for 20 to 22 h at 35 ± 0.5°C. Proceed to ¶ e below.

d. *Alternative single-step direct technique:* If the agar-based medium is used, place prepared filter directly on agar as described in preceding section and incubate for 22 to 24 h at 35 ± 0.5°C.

If liquid medium is used, place a pad in the culture dish and saturate with 1.8 to 2.0 mL M-Endo medium. Place prepared filter directly on pad, invert dish, and incubate for 22 to 24 h at 35 ± 0.5°C.

e. *Counting:* The typical coliform colony has a pink to dark-red color with a metallic surface sheen. The sheen area may vary in size from a small pinhead to complete coverage of the colony surface. Count sheen colonies with the aid of a low-power (10 to 15 magnifications) binocular wide-field dissecting microscope or other optical device, with a cool white fluorescent light source directed from above, and as nearly perpendicular as possible to the plane of the filter. Colonies that lack sheen may be pink, red, white, or colorless and are considered to be noncoliforms. The total count of colonies (coliform and noncoliform) on Endo-type medium has no relation to the total number of bacteria present in the original sample. However, a high count of noncoliform colonies may interfere with the maximum development of coliforms. Anaerobic incubation at 35°C for 24 h for some groundwater samples may suppress noncoliform colonies but must be carefully evaluated to insure no loss of coliform recovery.

Samples of disinfected water or wastewater effluent may include stressed organisms that grow relatively slowly and produce maximum sheen in 22 to 24 h. Organisms from undisinfected sources may produce sheen at 16 to 18 h, and the sheen subsequently may fade after 24 to 30 h.

f. Coliform verification: Typical sheen colonies may be produced occasionally by noncoliform organisms. Verify by a test for lactose fermentation or by using alternative procedures involving either a rapid (4 h) test of two key biochemical reactions or a multi-test system for speciation. See Section 9020B.4*d*.

1) Lactose fermentation—Verify all sheen colonies included in the direct count or a minimum of five such colonies from drinking water samples by transferring growth from each colony to lauryl tryptose broth; incubate at 35 ± 0.5°C for 48 h. Gas formed in lauryl tryptose broth within 48 h verifies the colony as a coliform.

2) Alternative coliform verifications— Apply this alternative coliform verification procedure to isolated colonies on the membrane filter culture. If a mixed culture is suspected or if colony separation is less

than 2 mm, streak the growth to M-Endo medium to assure culture purity or submit the mixed growth to the fermentation tube method.

a) Rapid test—A rapid verification of colonies utilizes test reactions for cytochrome oxidase (CO) and β-galactosidase (ONPG).† Coliform reactions are CO negative and ONPG positive within 4 h incubation of tube culture or micro (spot) test procedure.

b) Commercial multi-test systems— Verify the colony by inoculating in a multitest identification system for Enterobacteriaceae that includes lactose fermentation and/or ONPG and CO test reactions.

6. Calculation of Coliform Density

Report coliform density as (total) coliforms/100 mL. Compute the count, using membrane filters with 20 to 80 coliform colonies and not more than 200 colonies of all types per membrane, by the following equation:

(Total) coliform colonies/100 mL

$$= \frac{\text{coliform colonies counted} \times 100}{\text{mL sample filtered}}$$

For verified coliform counts, adjust the initial count based upon the positive verification percentage and report as "verified coliform count per 100 mL."

Percentage verified coliforms

$$= \frac{\text{number of verified colonies}}{\text{total number of sheen colonies}} \times 100$$

a. Water of drinking water quality: With water of good quality, the occurrence of coliforms generally will be minimal. Therefore, count all coliform colonies (disregarding the lower limit of 20 cited above)

---

†ONPG (orthonitrophenol galactoside) is the substrate for the β-galactosidase test.

and use the formula given above to obtain coliform density.

If confluent growth occurs, that is, growth covering either the entire filtration area of the membrane or a portion thereof, and colonies are not discrete, report results as "confluent growth with (or without) coliforms" and request a new sample from the same location. On reexamination, divide the 100-mL sample into duplicate 50-mL test portions to reduce interference from overcrowding of the MF surface. If the total number of bacterial colonies, coliforms plus noncoliforms, exceeds 200 per membrane, or if the colonies are not distinct enough for accurate counting, report results as "too numerous to count" (TNTC). The presence of coliforms in such cultures showing no sheen may be indicated by placing the entire membrane filter culture into a sterile tube of brilliant green lactose bile broth. As an alternative, brush the entire filter surface with a sterile loop or applicator stick and inoculate this growth to the tube of brilliant green lactose bile broth. If gas is produced from this culture within 48 h at $35 \pm 0.5°C$, conclude that coliforms are present. In any case, request a new sample and select more appropriate volumes to be filtered per membrane, remembering that the standard drinking water portion is 100 mL. Thus, instead of filtering 100 mL per membrane, 50-mL portions may be filtered through each of two membranes, 25-mL portions may be filtered through each of four membranes, etc. Total the coliform counts observed on all membranes and report as number per 100 mL.

b. *Water of other than drinking water quality:* As with potable water samples, if no filter has a coliform count falling in the ideal range, total the coliform counts on all filters and report as number per 100 mL. For example, if duplicate 50-mL portions were examined and the two membranes had five and three coliform colonies,

respectively, report the count as eight coliform colonies per 100 mL, i.e.,

$$\frac{[(5 + 3) \times 100]}{(50 + 50)}$$

Similarly, if 50-, 25-, and 10-mL portions were examined and the counts were 15, 6, and < 1 coliform colonies, respectively, report the count as 25/100 mL, i.e.,

$$\frac{[(15 + 6 + 0) \times 100]}{(50 + 25 + 10)}$$

On the other hand, if 10-, 1.0-, and 0.1-mL portions were examined with counts of 40, 9, and < 1 coliform colonies, respectively, select the 10-mL portion only for calculating the coliform density because this filter had a coliform count falling in the ideal range. The result is 400/100 mL, i.e.,

$$\frac{(40 \times 100)}{10}$$

In this last example, if the membrane with 40 coliform colonies also had a total bacterial colony count greater than 200, report the coliform count as $\geq 400/100$ mL.

Report confluent growth or membranes with colonies too numerous to count as described in *a* above. Request a new sample and select more appropriate volumes for filtration.

c. *Statistical reliability of membrane filter results:* Although the statistical reliability of the membrane filter technique is greater than that of the MPN procedure, membrane counts really are not absolute numbers. Table 9222:II illustrates some 95% confidence limits. These values assume that bacteria are distributed randomly and follow a Poisson distribution. For results with counts, c, greater than 20 organisms, calculate the approximate 95% confidence limits using the following normal distribution equations:

TABLE 9222:II. 95% CONFIDENCE LIMITS FOR MEMBRANE FILTER COLIFORM RESULTS USING 100-ML SAMPLE

Number of Coliform Colonies Counted	95% Confidence Limits	
	Lower	Upper
0	0.0	3.7
1	0.1	5.6
2	0.2	7.2
3	0.6	8.8
4	1.0	10.2
5	1.6	11.7
6	2.2	13.1
7	2.8	14.4
8	3.4	15.8
9	4.0	17.1
10	4.7	18.4
11	5.4	19.7
12	6.2	21.0
13	6.9	22.3
14	7.7	23.5
15	8.4	24.8
16	9.4	26.0
17	9.9	27.2
18	10.7	28.4
19	11.5	29.6
20	12.2	30.8

Upper limit $= c + 2 \sqrt{c}$
Lower limit $= c - 2 \sqrt{c}$

7. Bibliography

FIFIELD, C.W. & C.P. SCHAUFUS. 1958. Improved membrane filter medium for the detection of coliform organisms. *J. Amer. Water Works Assoc.* 50:193.

MCCARTHY, J.A. & J.E. DELANEY. 1958. Membrane filter media studies. *Water Sewage Works* 105:292.

RHINES, C.E. & W.P. CHEEVERS. 1965. Decontamination of membrane filter holders by ultraviolet light. *J. Amer. Water Works Assoc.* 57:500.

GELDREICH, E.E., H.L. JETER & J.A. WINTER. 1967. Technical considerations in applying the membrane filter procedure. *Health Lab. Sci.* 4:113.

WATLING, H.R. & R.J. WATLING. 1975. Note on the trace metal content of membrane filters. *Water SA* 1:28.

LIN, S.D. 1976. Evaluation of Millipore HA and HC membrane filters for the enumeration of indicator bacteria. *Appl. Environ. Microbiol.* 32:300.

STANDRIDGE, J.H. 1976. Comparison of surface pore morphology of two brands of membrane filters. *Appl. Environ. Microbiol.* 31:316.

GELDREICH, E.E. 1976. Performance variability of membrane filter procedure. *Pub. Health Lab.* 34:100.

GRABOW, W.O. & M. DU PREEZ. 1979. Comparison of m-Endo LES, MacConkey and Teepol media for membrane filtration counting of total coliform bacteria in water. *Appl. Environ. Microbiol.* 38:351.

DUTKA, B.D., ed. 1981. Membrane Filtration Applications, Techniques and Problems. Marcel Dekker, Inc., New York, N.Y.

EVANS, T.M., R.G. SEIDLER & M.W. LECHE-VALLIER. 1981. Impact of verification media and resuscitation on accuracy of the membrane filter total coliform enumeration technique. *Appl. Environ. Microbiol.* 41:1144.

FRANZBLAU, S.G., B.J. HINNEBUSCH, T.M. KELLEY & N.A. SINCLAIR. 1984. Effect of non-coliforms on coliform detection in potable groundwater: improved recovery with an anaerobic membrane filter technique. *Appl. Environ. Microbiol.* 48:142.

MCFETERS, G.A., J.S. KIPPIN & M.W. LECHE-VALLIER. 1986. Injured coliforms in drinking water. *Appl. Environ. Microbiol.* 51:1.

9222 C. Delayed-Incubation Total Coliform Procedure

Modification of the standard membrane filter technique permits membrane shipment or transport after filtration to a distant laboratory for incubation and completion of the test. This delayed-incubation test may be used where it is impractical to apply conventional procedures. It also may be used: (*a*) where it is not possible to maintain the desired sample temperature during transport; (*b*) when the elapsed time between sample collection and analysis would exceed the approved time limit; (c) where the sampling location is remote from laboratory services; (d) when it is necessary to monitor streams for water quality or pollution control activities by a standardized procedure; or (e) for other reasons that prevent analysis of the sample at or near the sample site.

Data secured by the delayed-incubation test have yielded results consistent with those from the immediate standard test in independent studies of samples from both fresh and salt waters. Determine the applicability of the delayed-incubation test for a specific water source by comparing with results of test procedures using conventional methods.

To conduct the delayed-incubation test, filter sample in the field immediately after collection, place filter on the transport medium, and ship to the laboratory. Complete the coliform determination in the laboratory by transferring the membrane to standard M-Endo or LES Endo medium, incubating at 35 ± 0.5°C for 20 to 22 h, and counting typical coliform colonies that develop. Transport media are designed to keep coliform organisms viable and generally do not permit visible growth during transit time. Bacteriostatic agents in holding/preservative media suppress growth of microorganisms en route but allow normal coliform growth after transfer to a fresh medium.

The delayed-incubation test follows the methods outlined for the total coliform membrane filter procedure, except as indicated below. Two alternative methods are given, one using the M-Endo preservative medium and the other the LES MF holding medium.

1. Apparatus

a. *Culture dishes:* Use disposable, sterile, moisture-tight plastic petri dishes (50 × 12 mm). Such containers are light in weight and are less likely to break in transit. In an emergency or when plastic dishes are unavailable, use sterile glass petri dishes wrapped in plastic film or similar material. See Section 9222B.1e for specifications.

b. *Field filtration units:* See Section 9222B.1f for specifications. Disinfect by adding methyl alcohol to the filtering chamber, igniting the alcohol, and covering unit to produce formaldehyde. Ultraviolet light disinfection also may be used in the field if an appropriate power source is available (115 V, 60 Hz). Glass or metal filtration units may be sterilized by immersing in boiling water for 5 min. Use a hand aspirator to obtain necessary vacuum.

2. Materials and Transport Media

a. *M-Endo methods:*

1) *M-Endo preservative medium:* Prepare as described in Section 9222B.2b. After cooling to below 45°C, aseptically add 3.84 g sodium benzoate (USP grade)/ L or 3.2 mL 12% sodium benzoate solution/100 mL of medium. Mix ingredients and dispense in 5- to 7-mL quantities to 50- × 12-mm petri plates. Store poured plates at 2 to 10°C. Discard unused medium after 96 h.

2) *Sodium benzoate solution:* Dissolve 12 g $NaC_7H_5O_2$ in sufficient distilled water to make 100 mL. Sterilize by autoclaving or by filtering through a 0.22-μm pore size membrane filter. Discard after 6 months.

3) *Cycloheximide:** Optionally add cycloheximide to M-Endo preservative medium. It may be used for samples that previously have shown overgrowth by fungi, including yeasts. Prepare modification by aseptically adding 50 mg/100 mL M-Endo preservative medium. Store cycloheximide solution at 5 to 10°C and discard after 6 months. Cycloheximide is a powerful skin irritant; handle with caution according to the manufacturer's directions.

---

*Actidione®, manufactured by the Upjohn Company, Kalamazoo, Mich., or equivalent.

b. LES method:
LES MF holding medium:

Tryptone	3.0	g
M-Endo broth MF	3.0	g
Dipotassium hydrogen phosphate, K_2HPO_4	3.0	g
Paraaminobenzoic acid	1.2	g
Agar	15.0	g
Distilled water	1	L

Rehydrate in distilled water. Heat to near boiling to dissolve agar, promptly remove from heat, and cool to below 50°C. Aseptically add 1.0 g sodium benzoate, 1.0 g sulfanilamide, and 0.5 g cycloheximide. After mixing, dispense in 5- to 7-mL quantities to 50- × 12-mm petri dishes. Store poured plates at 2 to 10°C. Discard unused medium after 2 weeks. Final pH should be 7.1 ± 0.1.

3. Procedure

a. Sample preservation and shipment: Place an absorbent pad in the bottom of a sterile petri dish and saturate with selected coliform holding medium (see Section 9222C.2 above). Remove membrane filter from filtration unit with sterile forceps and roll it, grid side up, onto the surface of the selected coliform holding medium. Protect membrane from moisture loss by tightly closing plastic petri dish. Prevent membrane dehydration during transit. Place culture dish containing membrane in an appropriate shipping container and send to the laboratory for test completion. The sample can be held without visible growth for a maximum of 72 h on the holding/preservative medium. This usually allows use of the mail or a common carrier. Visible growth occasionally begins on the transport medium when high temperatures are encountered during transit.

b. Transfer: At the laboratory, transfer membrane from plastic dish in which it was shipped to a second sterile petri dish containing M-Endo or LES Endo medium.

c. Incubation:

1) M-Endo method—Transfer membrane from M-Endo preservative medium to a pad and petri dish containing M-Endo medium without the growth-suppressing reagents and incubate at 35 ± 0.5°C for 20 to 22 h.

2) LES method—Transfer membrane from LES MF holding medium to LES Endo agar (see Section 9222B.2) and incubate at 35 ± 0.5°C for 20 to 22 h. If distinct colonies are observable without magnification at time of transfer, store petri dish containing transferred membrane at 5 to 10°C until it can be incubated at 35 ± 0.5°C for 16 to 18 h. This manipulation of incubation time permits control over the problems of overgrowth and sheen dissipation that interfere with the coliform colony count.

4. Estimation of Coliform Density

Proceed as in Section 9222B.6 above. Record times of collection, filtration, and laboratory examination, and calculate the elapsed time. Report elapsed time with coliform results.

5. Bibliography

GELDREICH, E.E., P.W. KABLER, H.L. JETER & H.F. CLARK. 1955. A delayed incubation membrane filter test for coliform bacteria in water. *Amer. J. Pub. Health* 45:1462.

PANEZAI, A.K., T.J. MACKLIN & H.G. COLES. 1965. *Coli-aerogenes* and *Escherichia coli* counts on water samples by means of transported membranes. *Proc. Soc. Water Treat. Exam.* 14:179.

McCARTHY, J.A. & J.E. DELANEY. 1965. Methods for measuring the coliform content of water. Sec. III. Delayed holding procedure for coliform bacteria. PHS Res. Grant WP 00202 NIH Rep.

BREZENSKI, F.T. & J.A. WINTER. 1969. Use of the delayed incubation membrane filter test for determining coliform bacteria in sea water. *Water Res.* 3:583.

9222 D.   Fecal Coliform Membrane Filter Procedure

Fecal coliform bacterial densities may be determined either by the multiple-tube procedure or by a membrane filter (MF) technique. If the MF procedure is used for chlorinated effluents, demonstrate that it gives comparable information to that obtainable by the multiple-tube test before accepting it as an alternative. The MF procedure uses an enriched lactose medium and incubation temperature of 44.5 ± 0.2°C for selectivity and gives 93% accuracy in differentiating between coliforms found in the feces of warm-blooded animals and those from other environmental sources. Because incubation temperature is critical, submerge waterproofed (plastic bag enclosures) MF cultures in a water bath for incubation at the elevated temperature or use an appropriate, accurate solid heat sink incubator. Areas of application for this method are stated in the introduction to the multiple-tube fecal coliform procedures, Section 9221C.

1. Materials and Culture Medium

a. M-FC medium: The need for uniformity dictates the use of dehydrated media. Never prepare media from basic ingredients when suitable dehydrated media are available. Follow manufacturer's directions for rehydration. Commercially prepared media in liquid form (sterile ampule or other) also may be used if known to give equivalent results. See Section 9020 for quality control specifications.

M-FC medium:

Tryptose or biosate	10.0	g
Proteose peptone No. 3 or		
polypeptone	5.0	g
Yeast extract	3.0	g
Sodium chloride, NaCl	5.0	g
Lactose	12.5	g
Bile salts No. 3 or bile salts		
mixture	1.5	g
Aniline blue	0.1	g
Distilled water	1	L

Rehydrate in distilled water containing 10 mL 1% rosolic acid in 0.2N NaOH.* Heat to near boiling, promptly remove from heat, and cool to below 50°C. Do not sterilize by autoclaving. Dispense 5- to 7-mL quantities to 50- × 12-mm petri plates and let solidify if agar is used. Final pH should be 7.4. Store finished medium at 2 to 10°C and discard unused medium after 2 weeks.

Test each medium lot for satisfactory productivity by preparing dilutions of a culture of *Escherichia coli* (Section 9020) and filtering appropriate volumes to give 20 to 80 colonies per filter. With each new lot of medium verify 10 or more colonies obtained from several natural samples, to establish the absence of false positives. For most samples M-FC medium may be used without the 1% rosolic acid addition, provided there is no interference with background growth. Such interference may be expected in stormwater samples collected during the first runoff (initial flushing) after a long dry period.

b. Culture dishes: Use tight-fitting plastic dishes because the MF cultures are submerged in a water bath during incubation. Enclose groups of fecal coliform cultures in plastic bags or seal individual dishes with waterproof (freezer) tape to prevent leakage during submersion. Specifications for plastic culture dishes are given in Section 9222B.1e above.

c. Incubator: The specificity of the fecal coliform test is related directly to the incubation temperature. Static air incubation is undesirable because of potential heat layering within the chamber and the slow recovery of temperature each time the incubator is opened during daily operations. To meet the need for greater tem-

---

*Rosolic acid reagent will decompose if sterilized by autoclaving. Store stock solution in the dark at 2 to 10°C and discard after 2 weeks or sooner if its color changes from dark red to muddy brown.

perature control use a water bath or a heat-sink incubator. A temperature tolerance of 44.5 ± 0.2°C can be obtained with most types of water baths that also are equipped with a gable top for the reduction of water and heat losses. A circulating water bath is excellent but may not be essential to this test if the maximum permissible variation of 0.2°C in temperature can be maintained with other equipment.

2. Procedure

a. Selection of sample size: Select volume of water sample to be examined in accordance with the information in Table 9222:III. Use sample volumes that will yield counts between 20 and 60 fecal coliform colonies per membrane.

When the bacterial density of the sample is unknown, filter several decimal volumes to establish fecal coliform density. Estimate volume expected to yield a countable membrane and select two additional quantities representing one-tenth and ten times this volume, respectively.

b. Filtration of sample: Follow the same procedure and precautions as prescribed under Section 9222B.5*b* above.

c. Preparation of culture dish: Place a sterile absorbent pad in each culture dish and pipet approximately 2 mL M-FC me-dium, prepared as directed above, to saturate pad. Carefully remove any excess liquid from culture dish. Place prepared filter on medium-impregnated pad as described in Section 9222B above.

As a substrate substitution for the nutrient-saturated absorbent pad, add 1.5% agar to M-FC broth.

d. Incubation: Place prepared cultures in waterproof plastic bags or seal petri dishes, submerge in water bath, and incubate for 24 ± 2 h at 44.5 ± 0.2°C. Anchor dishes below water surface to maintain critical temperature requirements. Place all prepared cultures in the water bath within 30 min after filtration. Alternatively, use an appropriate, accurate solid heat sink incubator.

e. Counting: Colonies produced by fecal coliform bacteria on M-FC medium are various shades of blue. Pale yellow colonies may be atypical *E. coli*; verify for gas production in mannitol at 44.5°C. Nonfecal coliform colonies are gray to cream-colored. Normally, few nonfecal coliform colonies will be observed on M-FC medium because of selective action of the elevated temperature and addition of rosolic acid salt reagent. Count colonies with a low-power (10 to 15 magnifications) binocular wide-field dissecting microscope or other optical device.

TABLE 9222:III. SUGGESTED SAMPLE VOLUMES FOR MEMBRANE FILTER FECAL COLIFORM TEST

Water Source	Volume (X) To Be Filtered mL						
	100	50	10	1	0.1	0.01	0.001
Lakes, reservoirs	X	X					
Wells, spring	X	X					
Water supply intake		X	X	X			
Natural bathing waters		X	X	X			
Sewage treatment plant, secondary effluent			X	X	X		
Farm ponds, rivers				X	X	X	
Stormwater runoff				X	X	X	
Raw municipal sewage					X	X	X
Feedlot runoff					X	X	X

3. Calculation of Fecal Coliform Density

Compute the density from the sample quantities that produced MF counts within the desired range of 20 to 60 fecal coliform colonies. This colony density range is more restrictive than the 20 to 80 total coliform range because of larger colony size on M-FC medium. Calculate fecal coliform density as directed in Section 9222B.6 above. Record densities as fecal coliforms per 100 mL.

4. Bibliography

GELDREICH, E.E., H.F. CLARK, C.B. HUFF & L.C. BEST. 1965. Fecal-coliform-organism medium for the membrane filter technique. *J. Amer. Water Works Assoc.* 57:208.

ROSE, R.D., E.E. GELDREICH & W. LITSKY. 1975. Improved membrane filter method for fecal coliform analysis. *Appl. Microbiol.* 29:532.

LIN, S.D. 1976. Membrane filter method for re-covery of fecal coliforms in chlorinated sewage effluents. *Appl. Environ. Microbiol.* 32:547.

PRESSWOOD, W.G. & D.K. STRONG. 1978. Modification of M-FC medium by eliminating rosolic acid. *Appl. Environ. Microbiol.* 36:90.

GREEN, B.L., W. LITSKY & K.J. SLADEK. 1980. Evaluation of membrane filter methods for enumeration of faecal coliforms from marine waters. *Mar. Environ. Res.* 67:267.

SARTORY, D.P. 1980. Membrane filtration faecal coliform determinations with unmodified and modified M-FC medium. *Water SA* 6:113.

GRABOW, W.O.K., C.A. HILNER, C.A. & P. COUBROUGH. 1981. Evaluation of standard and modified M-FC, MacConkey, and Teepol media for membrane filter counting of fecal coliform in water. *Appl. Environ. Microbiol.* 42:192.

RYCHERT, R.C. & G.R. STEPHENSON. 1981. Atypical *Escherichia coli* in streams. *Appl. Environ. Microbiol.* 41:1276.

PAGEL, J.E., A.A. QURESHI, D.M. YOUNG & L.T. VLASSOFF. 1982. Comparison of four membrane filter methods for fecal coliform enumeration. *Appl. Environ. Microbiol.* 43:787.

9222 E. Delayed-Incubation Fecal Coliform Procedure

This delayed-incubation procedure is comparable to the delayed-incubation total coliform procedure (Section 9222C). It eliminates the need for a field water bath incubator and frees the field investigator from the time-consuming task of counting colonies. Examination at a central laboratory, rather than in the field, permits colony confirmation and complete biochemical identification of the organisms, as necessary.

Results obtained by this delayed method have been consistent with results from the immediate standard test under various laboratory and field use conditions. However, determine test applicability for a specific water source by comparison with the standard membrane filter test, especially for saline waters, chlorinated wastewaters, and waters containing toxic substances. Use the delayed incubation test only when the standard immediate fecal coliform test cannot be performed.

To conduct the delayed-incubation test filter sample in the field immediately after collection, place filter on M-VFC holding medium (see ¶ 2a below), and ship to the laboratory. Complete fecal coliform test by transferring filter to M-FC medium, incubating at 44.5°C for 24 ± 2 h, and counting fecal coliform colonies.

The M-VFC medium keeps fecal coliform organisms viable but prevents visible growth during transit. Membrane filters can be held for up to 3 d on M-VFC holding medium with little effect on the fecal coliform counts.

1. Apparatus

a. Culture dishes: See Section 9222C.1a for specifications.

b. Field filtration units: See Section 9222C.1*b.*

2. Materials and Transport Medium

a. M-VFC holding medium:
This medium may not be available in dehydrated form and may require preparation from the basic ingredients.

Casitone, vitamin-free	0.2	g
Sodium benzoate	4.0	g
Sulfanilamide	0.5	g
Ethanol (95%)	10.0	mL
Distilled water	1	L

Heat to dissolve medium and sterilize by filtration through a membrane filter (pore diam, 0.22 μm). If only 100 mL are to be prepared, preferably add 2 mL of a 1:100 aqueous solution of casitone rather than 0.02 g dry reagent. Final pH should be 6.7.

Store finished medium at 2 to 10°C and discard unused medium after 2 weeks.

b. M-FC medium: Prepare as described in Section 9222D.1*a.*

3. Procedure

a. Membrane filter transport: Place an absorbent pad in a tight-lid plastic petri dish and saturate with M-VFC holding medium. After filtering sample remove membrane filter from filtration unit and place it on medium-saturated pad. Use only tight-lid dishes to prevent moisture loss; however, avoid having excess liquid in the dish. Place culture dish containing membrane in an appropriate shipping container and send to laboratory. Membranes can be held on the transport medium at ambient temperature for a maximum of 72 h with little effect on fecal coliform counts.

b. Transfer: At the laboratory remove membrane from holding medium and place it in another dish containing M-FC medium.

c. Incubation: After transfer of filter to M-FC medium, place tight-lid dishes in waterproof plastic bags and submerge in a water bath at 44.5°C ± 0.2°C for 24 ± 2 h or use a solid heat sink incubator.

d. Counting: Colonies produced by fecal coliform bacteria are various shades of blue. Nonfecal coliform colonies are gray to cream-colored. Count colonies with a binocular wide-field dissecting microscope at 10 to 15 magnifications.

4. Estimation of Fecal Coliform Density

Count as directed in Section 9222D.2*e* above and compute fecal coliform density as described in Section 9222D.3. Record time of collection, filtration, and laboratory examination, and calculate and report elapsed time.

5. Bibliography

TAYLOR, R.H., R.H. BORDNER & P.V. SCARPINO. 1973. Delayed incubation membrane-filter test for fecal coliforms. *Appl. Microbiol.* 25:363.

9222 F. *Klebsiella* Membrane Filter Procedure

Klebsiella, a genus included in the coliform group as defined herein, may be associated with coliform regrowth in water supply distribution systems and is often a major component of the coliform population in paper mill, textile, and other industrial wastes. The normal coliform population in human and other warm-blooded animal feces may contain 30 to 40% *Klebsiella* strains. Approximately 4% of bacterial pneumonia cases and 18% of urinary tract infections are caused by pathogenic strains of *Klebsiella.* Of these clinical strains, 85% are fecal coliforms. Environmental sources, such as vegetation, paper mills, textile production, sugar cane, and farm produce, contribute 71 to 88% of all *Klebsiella* in the total coliform population. *Klebsiella* occasionally may become established in the sediments of water supply distribution networks as a result of inadequate source water protection, unsatisfactory treatment protocols, or changes in the integrity of the pipe environment.

Rapid quantitation may be achieved in a membrane filter procedure by modifying M-FC agar base through substitution of inositol for lactose and adding carbenicillin or by using M-Kleb agar. These methods reduce the necessity for biochemical testing of pure strains. Preliminary verification of differentiated colonies is recommended.

1. Apparatus

a. Culture dishes: See Section 9222B.1e for specifications.

b. Filtration units: See Section 9222B.1f.

2. Materials and Culture Medium

a. Modified M-FC agar (M-FCIC agar): This medium may not be available in dehydrated form and may require preparation from the basic ingredients:

Tryptose or biosate	10.0	g
Proteose peptone No. 3 or		
polypeptone	5.0	g
Yeast extract	3.0	g
Sodium chloride, NaCl. . . .	5.0	g
Inositol	10.0	g
Bile salts No. 3 or bile salts		
mixture	1.5	g
Aniline blue	0.1	g
Agar	15.0	g
Distilled water	1	L

Heat medium to boiling and add 10 mL 1% rosolic acid dissolved in 0.2N NaOH. Cool to below 45°C and add 50 mg carbenicillin.* Do not sterilize by autoclaving. Final pH should be 7.4 ± 0.1.

b. M-Kleb agar:

---

*Available from Geopen, Roerig-Pfizer, Inc. New York, N.Y.

Phenol red agar	31.0	g
Adonitol	5.0	g
Aniline blue	0.1	g
Sodium lauryl sulfate	0.1	g
Distilled water	1	L

Sterilize by autoclaving for 15 min at 121°C. After autoclaving, cool to 50°C in a water bath; add 20 mL 95% ethyl alcohol (not denatured) and 0.05 g carbenicillin/ L. Shake thoroughly and dispense aseptically into 50- × 9-mm plastic culture plates. The final pH should be 7.4 ± 0.2. Prepared medium can be held for 20 d at 4°C.

3. Procedure

a. See Section 9222B.5 for selection of sample size and filtration procedure. Place membrane filter on agar surface; incubate for 24 ± 2 h at 35 ± 0.5°C. *Klebsiella* colonies on M-FCIC agar are blue or bluish-gray. Most atypical colonies are brown or brownish. Occasional false positive occurrences are caused by *Enterobacter* species. *Klebsiella* colonies on M-Kleb agar are deep blue to blue gray, whereas other colonies most often are pink or occasionally pale yellow. Count colonies with a low-power (10 to 15 magnifications) binocular wide field dissecting microscope or other optical device.

b. Verification: Verify *Klebsiella* colonies from the first set of samples from ambient waters and effluents and when *Klebsiella* is suspect in water supply distribution systems. Verify a minimum of five typical colonies by transferring growth from a colony or pure culture to a commercial multi-test system for gram-negative speciation. Key tests for *Klebsiella* are citrate (positive), indole (negative), motility (negative), lysine decarboxylase (positive), ornithine decarboxylase (negative), and urease (positive). *Klebsiella* of nonfecal origin can be identified by correlation of indole production and liquefaction of pectin with the ability to grow at 10°C and a negative fecal coliform response.

4. Bibliography

DUNCAN, D.W. & W.E. RAZELL. 1972. *Klebsiella* biotypes among coliforms isolated from forest environments and farm produce. *Appl. Microbiol.* 24:933.

STRAMER, S.L. 1976. Presumptive identification of *Klebsiella pneumoniae* on M-FC medium. *Can. J. Microbiol.* 22:1774.

BAGLEY, S.T. & R.J. SEIDLER. 1977. Significance of fecal coliform-positive *Klebsiella. Appl. Environ. Microbiol.* 33:1141.

KNITTEL, M.D., R.J. SEIDLER, C. ABY & L.M. CABE. 1977. Colonization of the botanical environment by *Klebsiella* isolates of pathogenic origin. *Appl. Environ. Microbiol.* 34:557.

EDMONSON, A.S., E.M. COOK, A.P.D. WILCOCK & R. SHINEBAUM. 1980. A comparison of the properties of *Klebsiella* isolated from different sources. *J. Med. Microbiol.* (U.K.) 13:541.

SMITH, R.B. 1981. A Critical Evaluation of Media for the Selective Identification and Enumeration of *Klebsiella.* M.S. thesis, Dep. Civil & Environmental Engineering, Univ. Cincinnati, Ohio.

NIEMELA, S. I. & P. VAATANEN. 1982. Survival in lake water of *Klebsiella pneumoniae* discharged by a paper mill. *Appl. Environ. Microbiol.* 44:264.

GELDREICH, E.E. & E.W. RICE. 1987. Occurrence, significance, and detection of *Klebsiella* in water systems. *J. Amer. Water Works Assoc.*, 79:74.

9225 DIFFERENTIATION OF THE COLIFORM BACTERIA*

9225 A. Introduction

Identification of bacteria that constitute the coliform group sometimes is necessary to determine the nature of pollution. Differential tests for identification must be used with the knowledge that all strains taxonomically assigned to the coliform group do not conform necessarily to the coliform definition stated in this manual because they may not ferment lactose, or if they do, they may not produce gas. Furthermore, gram-negative bacteria other than coliforms ferment lactose and produce sheen (e.g., *Aeromonas* spp.) and not all strains of a species will react uniformly in media. Mutable and injured organisms may not give classical responses. The traditional "IMViC" tests (i.e., indole, methyl red, Voges-Proskauer, and citrate utilization) are useful for coliform differentiation, but do not provide complete identification. Additional biochemical tests are necessary. Commercially available kits for identification are useful and economical alternatives to traditional differential media.

*Approved by Standard Methods Committee, 1988.

9225 B.   Culture Purification

1. Procedure

A pure culture is essential for accurate identification. Obtain a pure culture by picking carefully from the center of a well-isolated colony that gives typical responses on an appropriate solid medium or membrane filter, and streaking on a tryptic soy or nutrient agar plate. Better distribution of colonies in the subculture is obtained if a portion of the picked colony is emulsified in peptone broth or distilled water and then streaked. When picking a colony from a primary culture on a selective medium, be aware that viable cells, which have not formed colonies themselves, may surround the picked colony. Incubate the subculture at 35 ± 0.5°C for 24 h and test a single well-isolated colony by the Gram stain to confirm the sole presence of gram-negative, nonspore-forming rods. Also determine that the culture is oxidase-negative. Oxidase-positive, gram-negative, nonspore-forming rods are not coliform bacteria, but may be *Aeromonas,* which is not regarded as an indicator of fecal pollution.

Variation in organisms of the coliform group occurs occasionally and mixed reactions in differential media may indicate a pure culture undergoing variation. Persistent plus-minus reactions in differential media indicate a mixed culture caused by inadequate purification.

2. Bibliography

PTAK, D.J., W. GINSBURG & B.F. WILLEY. 1974. Aeromonas, the great masquerader. Proc. American Water Works Assoc. Water Quality Technology Conf., Dallas, Tex., Dec. 2, 1974.

9225 C.   Differentiation

1. Definition

Coliforms are defined here as gram-negative nonspore-forming rods that ferment lactose with gas formation within 48 h at 35°C or, as applied to the membrane filter method, produce a dark red colony with a metallic sheen within 24 h on an Endo-type medium containing lactose. However, anaerogenic (non-gas-producing) lactose-fermenting strains of *Escherichia coli* and coliforms that do not produce metallic sheen on Endo medium may be encountered. These organisms, as well as typical coliforms, can be considered indicator organisms, but they are excluded from the current definition of coliforms; testing beyond that given here is required to distinguish them.

2. Characteristics and Tests

Family nomenclature is given in Table 9225:I. Lactose-aerogenic species are noted. Table 9225:II presents characteristics of indicator organisms on several substrates that provide a basis for further differentiation. Not all species yield characteristic reactions all of the time. IMViC tests may be sufficient for some purposes, but generally will provide only partial identification. Decarboxylase tests are useful; the lysine decarboxylase test helps distinguish *Citrobacter* from *Arizona* and ornithine decarboxylase, urea slant agar[8], and

TABLE 9225:I NOMENCLATURE OF ENTEROBACTERIACEAE[1-3]

Tribe	Genus	Species
Escherichieae	*Escherichia*	*E. coli**
	Shigella	*S. dysenteriae, S. flexneri, S. boydii,* *S. sonnei*
Edwardsielleae	*Edwardsiella*	*E. tarda*
Salmonelleae	*Salmonella*	*S. cholerae-suis, S. typhi,* *S. enteritidis*
	Arizona	*A. hinshawii**
	Citrobacter	*C. freundii,* C. diversus**
Klebsielleae	*Klebsiella*	*K. pneumoniae,* K. ozaenae,** *K. rhinoscleromatis,* K. oxytoca**
	Enterobacter	*E. cloacae,* E. aerogenes,** *E. agglomerans†*
	Serratia	*S. marcescens, S. liquefaciens,* *S. rubidaea*
	Hafnia	*H. alvei**
Proteeae	*Proteus*	*P. vulgaris, P. mirabilis*
	Providencia	*P. alcalifaciens, P. stuartii,* *P. rettgeri*
Erwinieae	*Morganella*	*M. morganii*
	Erwinia	(Amylovara, Herbicola, Caratova groups)
	Pectobacterium	
Yersinieae	*Yersinia*	*Y. pestis, Y. enterocolitica,* *Y. pseudotuberculosis, Y. ruckeri,* *Y. intermedia, Y. fredericksenii*

* Species that may ferment lactose with gas.
† Previous designations: *Erwinia lathyri* and *Erwinia herbicola.*

motility tests are useful in distinguishing *Klebsiella* from *Enterobacter.*

Preparing differential media and reagents may not be as economical for many laboratories as using commercially prepared and prepackaged multiple-test kits, which reduce quality-control work. Prepackaged kits are simple to store and use, and give reproducible and accurate results; however, test reactions periodically with known stock cultures of bacteria to assure accuracy and reproducibility of results. Make further tests if the kit provides equivocal results. Several commercial kits that

TABLE 9225:II DIFFERENTIATION OF POTENTIAL INDICATOR ORGANISMS OF THE ENTEROBACTERIACEAE*[3-7]

Species	Lactose†	Indole	Methyl Red	Voges-Proskauer	Simmons Citrate†	Motility	Ornithine	Lysine†
E. coli	91.6	98	99	0	0.2	62.2	57.1	80.6
A. hinshawii	68.9 (d15.6)	5.1	100	0	96.8 (d2.5)	100	100	99.4
C. freundii	39.4 (d50.8)	6.7	99.5	0	90.4	95.7	17.2	0
C. diversus	32.7 (d51.4)	100	100	0	99.1	92.9	100	0
K. pneumoniae	98.7	0	11.3	93.7	96.8	0	0	97.2
K. oxytoca	‡	100	‡	‡	‡	‡	‡	‡
K. rhinoscleromatis	6 (d70)	0	100	0	0	0	0	0
K. ozaenae	72.6 (d6.3)	0	99.7	0	28.1 (d32.4)	0	4	35.8 (d6.8)
E. cloacae	76.3 (d21.8)	0	3.3	100	98.9	92.4	93.7	0
E. aerogenes	92.5	0.8	1.6	100	92.6	91.7	95.9	97.5
E. agglomerans	40.5 (d11)	18.7	44.8	67.9	66.6 (d19.2)	89.4	0	0
H. alvei	2.8 (d11.9)	0	35	83.6	5.6 (d63)	94.1	98.6	99.6

* Numbers represent percent positive for each substrate and species.
† Percent positive 1 to 2 days; d = percent positive delayed reactions 3 days or more.
‡ Biochemically K. oxytoca is very similar to K. pneumoniae except for indole production.

have been evaluated extensively are described below.*

3. References

1. EDWARDS, P.R. & W.H. EWING. 1972. Identification of Enterobacteriaceae, 3rd ed. Burgess Publ. Co., Minneapolis, Minn.
2. KRIEG, N.R., ed. 1984. Bergey's Manual of Systematic Bacteriology, Vol. I. Williams & Wilkins Co., Baltimore, Md.
3. BRENNER, D.J., J.J. FARMER, F.W. HICKMAN, M.A. ASBURY & A.G. STEIGERWALT. 1977. Taxonomic and Nomenclature Changes in Enterobacteriaceae. U.S. Dep. Health, Education & Welfare Publ. No. (CDC) 78-8356, Center for Disease Control, Atlanta, Ga.
4. EWING, W.H. & M.A. FIFE. 1972. Biochemical Characterization of Enterobacter agglomerans. U.S. Dep. Health, Education & Welfare Publ. No. (CDC) 73-8173, Center for Disease Control, Atlanta, Ga.

5. EWING, W.H. 1973. Differentiation of Enterobacteriaceae by Biochemical Reactions. U.S. Dep. Health Education & Welfare Publ. No. (CDC) 74-8270, Center for Disease Control, Atlanta, Ga.
6. EWING, W.H., B.R. DAVIS & W.J. MARTIN. 1972. Biochemical Characterization of Escherichia coli. U.S. Dep. Health Education & Welfare Publ. No. (HSM) 72-8109, Center for Disease Control, Atlanta, Ga.
7. EWING, W.H. 1971. Biochemical Characterization of Citrobacter freundii and Citrobacter diversus. Center for Disease Control, Atlanta, Ga.
8. LENNETTE, E.H., A. BALOWS, W.J. HAUSLER, JR. & H.J. SHADOMY, eds. 1985. Manual of Clinical Microbiology, 4th ed. American Soc. for Microbiology, Washington, D.C.
9. WASHINGTON, J.A., II. 1977. Kits for identification of microorganisms. Lab. Medicine 2:1.

4. Bibliography

BORMAN, E.K., C.A. STUART & K.M. WHEELER. 1944. Taxonomy of the family Enterobacteriaceae. J. Bacteriol. 48:351.
EWING. W.H. 1966. Enterobacteriaceae: Taxonomy and Nomenclature. U.S. Dep. Health Education & Welfare, National Center for Disease Control, Atlanta, Ga.
WOLFE, M.W. & D. AMSTERDAM. 1968. New diagnostic system for the identification of lactose-fermenting gram-negative rods. Appl. Microbiol. 16:1528.
ELLER, C. & F.F. EDWARDS. 1968. Nitrogen deficient medium in the differentiation and isolation of Klebsiella and Enterobacter from feces. Appl. Microbiol. 16:896.
PTAK, D.J., W. GINSBURG & B.F. WILLEY. 1973. Identification and incidence of Klebsiella in chlorinated water supplies. J. Amer. Water Works Assoc. 65:604.

---

*System Name	Description
API	Twenty microcapsules on a card can provide 21 test results.
Enterotube	Plastic cylinder containing 12 media-filled compartments. A single needle thrust sequentially inoculates all compartments and provides 15 test results.
Inolex Enteric 1 & 2	A micromethod with 10 capillary units on a card. With two different cards provides 20 test results.
Microlab ID	A micromethod giving 15 test results in four steps.
Minitek	Reagent-impregnated disks are placed mechanically in small wells. Up to 34 disks provide 36 test results.

Further information is available.[9]

9225 D.   Significance of Coliform Types

1. Discussion

The significance of various coliform organisms in water has been and is a subject of considerable study. Collectively, the coliforms are referred to as indicator organisms because they indicate the presence of human or animal feces. All types of coliform organisms may occur in feces. Although E. coli nearly always will be found in fresh pollution from warm-blooded animals, other coliform organisms may be found in fresh pollution in the absence of E. coli. The genera Enterobacter, Klebsiella, Citrobacter, and Escherichia usually represent the majority of isolations made from raw and treated municipal water supplies, with *Enterobacter* being isolated most frequently, but all coliforms do not necessarily originate from sewage. *E. coli* is more readily affected by conventional water treatment than other coliforms. Coliform organisms, some of which are free-living saprophytes, can multiply on leather washers, wood, swimming pool ropes, and jute packing and may produce slimes inside pipes. Differentiation of coliform types is valuable in determining the source of increased coliform densities resulting from the multiplication of organisms on or in organic materials. The presence of a large number of coliform organisms of the same type in water from a well or spring, or from a single tap on a distribution system, suggests that such multiplication has occurred.

Industrial wastes containing high concentrations of bacterial nutrients are capable of promoting large aftergrowths of coliforms in effluents and receiving waters. Aftergrowths also may appear in treated water distribution systems. *Aeromonas* and other oxidase-positive, gram-negative bacteria, as well as *Erwinia,* can be expected in raw and treated waters. *Aeromonas* and *Erwinia* spp. are not indicator bacteria as defined herein.

2. Bibliography

Bordner, R.H. & B.J. Carroll, eds. 1972. Proc. Seminar on the Significance of Fecal Coliforms in Industrial Wastes. U.S. Environmental Protection Agency, Off. Enforcement, National Field Investigations Center, Denver, Colo.

Clark, J.A. & J.E. Pagel. 1977. Pollution indicator bacteria associated with raw and drinking water supplies. *Can. J. Microbiol.* 4:465.

Gorden, R.W. & C.B. Fliermans. 1978. Survival and viability of *Escherichia coli* in a thermally altered reservoir. *Water Res.* 12:343.

9225 E.   Media, Reagents, and Procedures

Commercially available media and reagents can reduce work and cost; however, include negative and positive controls with known stock cultures to assure accuracy and reliability. Detailed methods are available.[1] Expected test results are shown in Table 9225:II.

1. Indole Test

Indole is a product of the metabolism of tryptophane.

a. Reagents:

1) *Medium:* Use tryptophane broth. Dissolve 10.0 g tryptone or trypticase/L distilled water. Dispense in 5-mL portions in test tubes and sterilize.

2) *Test reagent:* Dissolve 5 g paradimethylaminobenzaldehyde in 75 mL isoamyl (or normal amyl) alcohol, ACS grade, and add 25 mL conc HCl. The reagent should be yellow. Some brands of paradimethylaminobenzaldehyde are not satisfactory and some good brands become unsatisfactory on aging.

The amyl alcohol solution should have a pH value of less than 6.0. Purchase both amyl alcohol and benzaldehyde compound in as small amounts as will be consistent with the volume of work to be done.

b. Procedure: Inoculate 5-mL portions of medium from a pure culture and incubate at 35 0.5°C for 24 ± 2 h. Add 0.2 to 0.3 mL test reagent and shake. Let stand for about 10 min and observe results.

A dark red color in the amyl alcohol surface layer constitutes a positive indole test; the original color of the reagent, a negative test. An orange color probably indicates the presence of skatole, a breakdown product of indole.

2. Methyl Red Test

The methyl red test measures terminal pH; i.e., it tests for acid production from glucose by bacteria. The metabolic reactions demonstrated by gas production and the methyl red and Voges-Proskauer tests are related and are useful in differentiating members of the coliform group.

a. Reagents:

1) *Medium:* Use buffered glucose broth. Dissolve 5.0 g proteose peptone or equivalent peptone, 5.0 g glucose, and 5.0 g dipotassium hydrogen phosphate (K_2HPO_4) in 1 L distilled water. Dispense in 5-mL portions in test tubes and sterilize by autoclaving at 121°C for 12 to 15 min, making sure that total time of exposure to heat is not longer than 30 min.

2) *Indicator solution:* Dissolve 0.1 g methyl red in 300 mL 95% ethyl alcohol and dilute to 500 mL with distilled water.

b. Procedure: Inoculate 10-mL portions of medium from a pure culture. Incubate at 35°C for 5 d. To 5 mL of the culture add 5 drops methyl red indicator solution.

Incubation for 48 h is adequate for most cultures, but do not incubate for less than 48 h. If test results are equivocal at 48 h repeat with cultures incubated for 4 or 5 d. In such cases incubate duplicate cultures at 22 to 25°C. Testing of culture portions at 2, 3, 4, and 5 d may provide positive results sooner.

Record a distinct red color as methyl red-positive and a distinct yellow color as methyl red-negative. Record a mixed shade as questionable and possibly indicative of incomplete culture purification.

3. Voges-Proskauer Test

This tests for acetoin, produced metabolically by certain coliform bacteria in peptone glucose medium.

a. Reagents:

1) *Media:* Use the medium described for the methyl red differential test or, alternatively, the salt peptone glucose medium. Dissolve 10.0 g polypeptone or proteose peptone, 5.0 g sodium chloride (NaCl), and 10.0 g glucose in 1 L distilled water; pH should be 7.0 to 7.2. Dispense in 5-mL portions in test tubes and sterilize by autoclaving at 121°C for 12 to 15 min, making sure that total time of exposure to heat is not longer than 30 min.

2) *Naphthol solution:* Dissolve 5 g purified α-naphthol (melting point 92.5°C or higher) in 100 mL absolute ethyl alcohol. When stored at 5 to 10°C, this solution is stable for 2 weeks.

3) *Potassium hydroxide, 7N:* Dissolve 40 g KOH in 100 mL distilled water.

b. Procedure: Inoculate 5 mL of either medium and incubate for 48 h at 35 ± 0.5°C. To 1 mL of culture add 0.6 mL naphthol solution and 0.2 mL KOH solution. Development of a pink to crimson color within 5 min constitutes a positive test. Do not read after 5 min. Discard tubes developing a copper color.

4. Sodium Citrate Test

This procedure tests the ability of bacteria to utilize citrate as the sole source of carbon.

a. Alternate media: Use either Koser's citrate broth or Simmons' citrate agar. To make Koser's citrate broth dissolve 1.5 g sodium ammonium hydrogen phosphate ($NaNH_4HPO_4 \cdot 4H_2O$), 1.0 g dipotassium hydrogen phosphate (K_2HPO_4), 0.2 g magnesium sulfate heptahydrate ($MgSO_4 \cdot 7H_2O$), and 3.0 g sodium citrate dihydrate crystals in 1 L distilled water. Dispense in 5-mL portions in test tubes and sterilize. To make Simmons' citrate agar, add 0.2 g $MgSO_4 \cdot 7H_2O$, 1.0 g ammonium dihydrogen phosphate ($NH_4H_2PO_4$), 1.0 g K_2HPO_4, 2.0 g sodium citrate dihydrate, 5.0 g NaCl, 15.0 g agar, and 0.08 g bromthymol blue to 1 L distilled water. Tube for long slants.

b. Procedure:

1) Lightly inoculate liquid medium with a straight needle, never with a pipet. Incubate at 35 ± 0.5°C for 72 to 96 h. Record

visible growth as positive, no growth as negative.

2) Inoculate agar medium with straight needle, using both a stab and a streak. Incubate 48 h at 35 ± 0.5°C. Record growth on the medium with (usually) a blue color as a positive reaction; record absence of growth as negative.

5. Motility Test

a. Medium: Use motility test medium made by adding 3.0 g beef extract, 10.0 g peptone, 5.0 g NaCl, and 4.0 g agar to 1 L distilled water. Adjust pH to 7.4, dispense in 8-mL portions in test tubes, and sterilize.

b. Procedure: Inoculate by stabbing into the top of the medium column to a depth of 5 mm. Incubate for 1 to 2 d at 35°C. If negative, incubate an additional 5 d at 22 to 25°C.

Diffuse growth through the medium from the point of inoculation is positive. Alternatively, prepare the medium without agar and examine a young culture using the hanging drop slide technique for motile organisms.

6. Decarboxylase Test

This procedure tests the ability of bacteria to metabolize the amino acids lysine and ornithine.

a. Reagents:

1) *Media:* Use a basal medium made according to the Moeller or Falkow methods.[1] For the Moeller method, dissolve 5.0 g peptone (Orthana special, thiotone, or equivalent), 5.0 g beef extract, 0.625 mL bromcresol purple (1.6%), 2.5 mL cresol red (0.2%), 0.5 g glucose, and 5.0 mg pyridoxal in 1 L distilled water and adjust to pH 6.0 to 6.5. For the Falkow method, dissolve 5.0 g peptone, 3.0 g yeast extract, 1.0 g glucose, and 1.0 mL bromcresol purple (1.6%) in 1 L distilled water and adjust to pH 6.7 to 6.8. For either basal medium divide into three portions: make no addition to the first portion, add enough L-lysine dihydrochloride to the second portion to make a 1% solution, and add L-ornithine dihydrochloride to the third to make 1% (for the Falkow method, add only 0.5% of the L-amino acid). After adding ornithine readjust pH of the medium. Dispense in 3- to 4-mL portions in screw-capped test tubes and sterilize by autoclaving at 121°C for 10 min. A floccular precipitate in the ornithine medium does not interfere with its use.

2) *Mineral oil:* Use mineral oil sterilized by autoclaving at 121°C for 30 to 60 min depending on the size of the container.

b. Procedure: Lightly inoculate each of the three media, add a layer of about 10 mm thickness of mineral oil, and incubate at 37°C for up to 4 d. Examine tubes daily. A color change from yellow to violet or reddish-violet constitutes a positive decarboxylase test; a change to bluish gray indicates a weak positive; no color change represents a negative test. See Table 9225:II.

7. Oxidase Test

This procedure tests for the enzyme cytochrome oxidase, as a means of differentiating between groups of bacteria. Coliform bacteria are oxidase negative.

a. Reagents:

1) *Media:* Use either nutrient agar or tryptic soy agar plates to streak cultures and produce isolated colonies. From these obtain the inoculum for oxidase testing on impregnated filter paper. Do not use any medium that includes a carbohydrate in its formulation. Use only tryptic soy agar if reagent is dropped on colonies.

Tryptic soy agar:

Tryptone	15.0	g
Soytone	5.0	g
Sodium chloride, NaCl	5.0	g
Agar	15.0	g
Distilled water	1.0	L

pH should be 7.3 ± 0.2 after sterilization.

2) *Tetramethyl-paraphenylenediamine dihydrochloride,* 1% aqueous solution, freshly prepared or refrigerated for no longer than 1 week. Impregnate a filter paper strip* with this solution. Alternatively, prepare a 1% solution of dimethylparaphenylenediamine hydrochloride. Single-use reagent ampules, commercially available, are convenient and economical but use them with caution. When the reagent is to be dropped directly on colonies, use tryptic soy agar plates because nutrient agar plates give inconsistent results; when smearing a portion of a picked colony on reagent-impregnated filter paper, do not transfer any medium with the culture material.

b. *Procedure:* Remove some of a colony from agar plate with a nichrome or platinum wire, a wooden or plastic applicator stick, or a glass rod and smear on the test strip. Do not use iron or other reactive wire because it will cause false positive reactions. A dark purple color that develops within 10 s is a positive oxidase test. Test positive and negative cultures concurrently. If the liquid reagent is used, drop it on colonies on the culture plate. Oxidase-positive colonies develop a pink color that successively becomes maroon, dark red, and finally, black.

---

*Whatman No. 1 or equivalent.

8. Reference

1. LENNETTE, E.H., A. BALOWS, W.J. HAUSLER, JR. & H.J. SHADOMY, eds. 1985. Manual of Clinical Microbiology, 4th ed. American Soc. for Microbiology, Washington, D.C.

9. Bibliography

CLARK, W.M. 1915. The final hydrogen ion concentrations of cultures of *Bacillus coli. Science* 42:71.

CLARK, W.M. & W.A. LUBS. 1915. The differentiation of bacteria of the colon-aerogenes family by the use of indicators. *J. Infect. Dis.* 17:160.

LEVINE, M. 1916. On the significance of the Voges-Proskauer reaction. *J. Bacteriol.* 1:153.

LEVINE, M. 1921. Notes on *Bact. coli* and *Bact. aerogenes. Amer. J. Pub. Health* 11:21.

KOSER, S.A. 1924. Correlation of citrate utilization by members of the colon-aerogenes group with other differential characteristics and with habitat. *J. Bacteriol.* 9:59.

SIMMONS, J.S. 1926. A culture medium for differentiating organisms of typhoid-colon-aerogenes groups and for isolation of certain fungi. *J. Infect. Dis.* 39:309.

KOVACS, N. 1928. A simplified method for detecting indol formation by bacteria. *Z. Immunitatsforsch.* 56:311; *Chem. Abstr.* 22:3425.

RUCHHOFT, C.C., J.G. KALLAS, B. CHINN & E.W. COULTER. 1930 & 1931. Coli-aerogenes differentiation in water analysis. *J. Bacteriol.* 21:407; 22:125.

EPSTEIN, S.S. & R.H. VAUGHN. 1934. Differential reactions in the coli group of bacteria. *Amer. J. Pub. Health* 24:505.

BARRITT, M.W. 1936. The intensification of the Voges-Proskauer reaction by the addition of alpha-naphthol. *J. Pathol. Bacteriol.* 42:441.

VAUGHN, R., N.B. MITCHELL & M. LEVINE. 1939. The Voges-Proskauer and methyl red reactions in the coli-aerogenes group. *J. Amer. Water Works Assoc.* 31:993.

BARRY, A.L. & K.L. BERSOHN. 1969. Methods for storing oxidase test reagents. *Appl. Microbiol.* 17:933.

9230 FECAL STREPTOCOCCUS AND ENTEROCOCCUS GROUPS*

9230 A.   Introduction

1. Fecal Streptococcus Group

The fecal streptococcus group consists of a number of species of the genus *Streptococcus*, such as *S. faecalis, S. faecium, S. avium, S. bovis, S. equinus,* and *S. gallinarum*. They all give a positive reaction with Lancefield's Group D antisera[1] and have been isolated from the feces of warm-blooded animals. In addition, *S. avium* sometimes reacts with Lancefield's Group Q antisera. *S. faecalis* subsp. *liquefaciens* and *S. faecalis* subsp. *zymogenes* are differentiated based on the ability of these strains to liquefy gelatin and hemolyze red cells. However, the validity of these subspecies is questionable.[2,3]

The normal habitat of fecal streptococci is the gastrointestinal tract of warm-blooded animals. *S. faecalis* and *S. faecium* once were thought to be more human-specific than other *Streptococcus* species. Other species have been observed in human feces but less frequently.[4] Similarly, *S. bovis, S. equinus,* and *S. avium* are not exclusive to animals, although they usually occur at higher densities in animal feces.[5] Certain streptococcal species predominate in some animal species and not in others, but it is not possible to differentiate the source of fecal contamination based on speciation of fecal streptococci.

The fecal streptococci have been used with fecal coliforms to differentiate human fecal contamination from that of other warm-blooded animals. Previous editions of *Standard Methods* suggested that the ratio of fecal coliforms (FC) to fecal streptococci (FS) could provide information about the source of contamination. A ratio greater than four was considered indicative

of human fecal contamination, whereas a ratio of less than 0.7 was suggestive of contamination by nonhuman sources. The value of this ratio has been questioned because of variable survival rates of fecal streptococcus group species. *S. bovis* and *S. equinus* die off rapidly, once exposed to aquatic environments, whereas *S. faecalis* and *S. faecium* tend to survive longer.[6] Furthermore, disinfection of wastewaters appears to have a significant effect on the ratio of these indicators, which may result in misleading conclusions regarding the source of contaminants.[7] The ratio is affected also by the methods for enumerating fecal streptococci. The KF membrane filter procedure has a false-positive rate ranging from 10 to 90% in marine and fresh waters.[8-10] For these reasons, the FC/FS ratio cannot be recommended, and should not be used as a means of differentiating human and animal sources of pollution.

2. Enterococcus Group

The enterococcus group is a subgroup of the fecal streptococci that includes *S. faecalis, S. faecium, S. gallinarum,* and *S. avium.* The enterococci are differentiated from other streptococci by their ability to grow in 6.5% sodium chloride, at pH 9.6, and at 10°C and 45°C.

The enterococci portion of the fecal streptococcus group is a valuable bacterial indicator for determining the extent of fecal contamination of recreational surface waters. Studies at marine and fresh water bathing beaches indicated that swimming-associated gastroenteritis is related directly to the quality of the bathing water and that enterococci are the most efficient bacterial indicator of water quality.[11,12] Water qual-

---

*Approved by Standard Methods Committee, 1988.

ity guidelines based on enterococcal density have been proposed for recreational waters.[13] For recreational fresh waters the guideline is 33 enterococci/100 mL while for marine waters it is 35/100 mL. Each guideline is based on the geometric mean of at least five samples per 30-d period during the swimming season.

3. Selection of Method

The multiple-tube technique is applicable primarily to raw and chlorinated wastewater and sediments, and can be used for fresh and marine waters. The membrane filter technique also may be used for fresh and saline water samples, but is unsuitable for highly turbid waters.

4. References

1. SNEATH, P.H.A., N.S. MAIR, M.E. SHARPE & J.G. HOLT. eds. 1986. Bergey's Manual of Systematic Bacteriology. Vol. 2, Williams & Wilkins, Baltimore, Md.

2. JACOB, A.E., G.J. DOUGLAS & S.J. HOBBS. 1973. Self-transferable plasmids determining the haemolysin and bacteriocin of *Streptococcus faecalis* var. *zymogenes*. *J. Bacteriol.* 121:863.

3. OLIVER, D.R., B.L. BROWN & D.B. CLEWELL. 1977. Characterization of plasmids determining haemolysin and bacteriocin production in *Streptococcus faecalis* 5952. *J. Bacteriol.* 130:948.

4. WATANABE, T., H. SHIMOHASHI, Y. KAWAI, & M. MUTAI. 1981. Studies on streptococci. I. Distribution of fecal streptococci in man. *Microbiol. Immunol.* 25:257.

5. THOMAS, C.D. & M. A. LEVIN. 1978. Quantitative analysis of group D streptococci. Abs. Annual Meeting, American Soc. Microbiology, p. 210.

6. FEACHAM, R. 1975. An improved role for faecal coliform to faecal streptococci ratios in the differentiation between human and non-human pollution sources. *Water Res.* 9:689.

7. ROSSER, P.A.E. & D.P. SARTORY. 1982. A note on the effect of chlorination of sewage effluents on faecal coliform to faecal streptococci ratios in the differentiation of faecal pollution sources. *Water S.A.* 8:66.

8. FUJIOKA, R.S., A.A. UENO & O.T. NARIKAWA. 1984. Recovery of False Positive Fecal Streptococcus on KF Agar from Marine Recreational Waters. Tech. Rep. No. 168, Water Resources Research Center, Univ. Hawaii at Manoa, Honolulu.

9. OLIVIERI, V.P., C.W. KRUSE, K. KAWATA & J.E. SMITH. 1977. Microorganisms in Urban Stormwater. EPA-600/2-77-087, U.S. Environmental Protection Agency, Edison, N.J.

10. ERICKSEN, T.H., C. THOMAS & A. DUFOUR. 1983. Comparison of two selective membrane filter methods for enumerating fecal streptococci in freshwater samples. Abs. Annual Meeting, American Soc. Microbiology, p. 279.

11. CABELLI, V.J. 1983. Health Effects Criteria for Marine Waters. EPA-600/1-80-031, U.S. Environmental Protection Agency, Cincinnati, Ohio.

12. DUFOUR, A.P. 1984. Health Effects Criteria for Fresh Recreational Waters. EPA-600/1-84-004, U.S. Environmental Protection Agency, Cincinnati, Ohio.

13. U.S. ENVIRONMENTAL PROTECTION AGENCY. 1986. Ambient Water Quality Criteria for Bacteria—1986. EPA-440/5-84-002, U.S. Environmental Protection Agency, Washington, D.C.

9230 B.   Multiple-Tube Technique

1. Materials and Culture Media

a. Azide dextrose broth:[1]

Beef extract	4.5	g
Tryptone or polypeptone	15.0	g
Glucose	7.5	g
Sodium chloride, NaCl	7.5	g
Sodium azide, NaN₃	0.2	g
Distilled water	1	L

pH should be 7.2 ± 0.2 at 25°C after sterilization.

b. Pfizer selective enterococcus (PSE) agar:[2]

Peptone C	17.0	g
Peptone B	3.0	g
Yeast extract	5.0	g
Bacteriological bile	10.0	g

Sodium chloride, NaCl	5.0	g
Sodium citrate	1.0	g
Esculin	1.0	g
Ferric ammonium citrate ..	0.5	g
Sodium azide, NaN₃	0.25	g
Agar	15.0	g
Distilled water	1	L

pH should be 7.1 ± 0.2 after sterilization. Hold medium for not more than 4 h at 45 to 50°C before plates are poured.

2. Presumptive Test Procedure

Inoculate a series of tubes of azide dextrose broth with appropriate graduated quantities of sample. Use sample of 10 mL portions or less. Use double-strength broth for 10-mL inocula. The portions used will vary in size and number with the sample character. Use only decimal multiples of 1 mL (see Section 9221 for suggested sample sizes).

Incubate inoculated tubes at 35 ± 0.5°C. Examine each tube for turbidity at the end of 24 ± 2 h. If no definite turbidity is present, reincubate, and read again at the end of 48 ± 3 h.

3. Confirmed Test Procedure

Subject all azide dextrose broth tubes showing turbidity after 24- or 48-h incubation to the confirmed test.

Streak a portion of growth from each positive azide dextrose broth tube on PSE agar. Incubate the inverted dish at 35 ± 0.5°C for 24 ± 2 h. Brownish-black colonies with brown halos confirm the presence of fecal streptococci.

Brownish-black colonies with brown halos may be transferred to a tube of brain-heart infusion broth containing 6.5% NaCl. Growth in 6.5% NaCl broth and at 45°C indicates that the colony belongs to the enterococcus group.

4. Computing and Recording of MPN

Estimate fecal streptococci densities from the number of tubes in each dilution series that are positive on PSE agar. Similarly, estimate enterococci densities from the number of tubes in each dilution series containing streptococci that can grow in 6.5% NaCl broth. Compute the combination of positives and record as the most probable number (MPN). Refer to Section 9221D.

5. References

1. MALLMAN, W.L. & E.B. SELIGMANN. 1950. A comparative study of media for the detection of streptococci in water and sewage. *Amer. J. Pub. Health* 40:286.

2. ISENBERG, H.D., D. GOLDBERT & J. SAMPSON. 1972. Laboratory studies with a selective enterococcus medium. *Health Lab. Sci.* 9:289.

9230 C.   Membrane Filter Techniques

1. Laboratory Apparatus

See Section 9222B.1.

2. Materials and Culture Media

a. mE agar for enterococci:[1]

Peptone................	10.0	g
Sodium chloride, NaCl	15.0	g
Yeast extract	30.0	g
Esculin	1.0	g
Actidione (cycloheximide) ...	0.05	g
Sodium azide, NaN₃	0.15	g
Agar	15.0	g
Distilled water	1	L

Heat to dissolve ingredients, sterilize, and cool in a water bath at 44 to 46°C. Mix 0.25 g nalidixic acid in 5 mL reagent-grade water, add a few drops of 0.1N NaOH to dissolve the antibiotic, and add to the basal medium. Add 0.15 g 2,3,5-triphenyl tetrazolium chloride and mix well to dissolve. Pour the agar into 9- × 50-mm petri dishes to a depth of 4 to 5 mm (approximately 4 to 6 mL), and let solidify. The final pH should be 7.1 ± 0.2. Store poured plates in the dark at 2 to 10°C. Discard after 30 d. (NOTE: This medium is recommended for culturing enterococci in fresh and marine recreational waters.)

b. EIA substrate:[1]

Esculin	1.0	g
Ferric citrate	0.5	g
Agar	15.0	g
Distilled water	1	L

The pH should be 7.1 ± 0.2 before autoclaving. Heat to dissolve ingredients, sterilize, and cool in a water bath at 44 to 46°. Pour medium into 50-mm petri dishes to a depth of 4 to 5 mm (approximately 4 to 6 mL) and let solidify. Store poured plates in the dark at 2 to 10°C. Discard after 30 d.

c. m Enterococcus agar for fecal streptococci:[2]

Tryptose	20.0	g
Yeast extract	5.0	g
Glucose	2.0	g
Dipotassium phosphate, K$_2$HPO$_4$	4.0	g
Sodium azide, NaN$_3$	0.4	g
2,3,5-Triphenyl tetrazolium chloride	0.1	g
Agar	10.0	g
Distilled water	1	L

Heat to dissolve ingredients. Do not autoclave. Dispense into 9- × 50-mm petri plates to a depth of 4 to 5 mm (approximately 4 to 6 mL), and let solidify. Prepare fresh medium for each set of samples. (NOTE: This medium is recommended for Group D streptococci in fresh and marine waters.)

d. Brain-heart infusion broth:

Infusion of calf brain	200	g

Infusion of beef heart	250	g
Proteose peptone	10.0	g
Glucose	2.0	g
Sodium chloride, NaCl	5.0	g
Disodium hydrogen phosphate, Na$_2$HPO$_4$	2.5	g
Distilled water	1	L

The pH should be 7.4 after sterilization.

e. Brain-heart infusion agar: Add 15.0 g agar to the ingredients for brain-heart infusion broth. The pH should be 7.4 after sterilization. Tube for slants.

f. Bile esculin agar:[3]

Beef extract	3.0	g
Peptone	5.0	g
Oxgall	40.0	g
Esculin	1.0	g
Ferric citrate	0.5	g
Agar	15.0	g
Distilled water	1	L

Heat to dissolve ingredients. Dispense 8 to 10 mL into tubes for slants or an appropriate volume into a flask for subsequent pouring into plates. Autoclave at 121°C for 15 min. Do not overheat because this may cause darkening of the medium. Cool to 44 to 46°C and slant the tubes or dispense 15 mL into 15- × 100-mm petri dishes. The final pH should be 6.6 ± 0.2 after sterilization. Store at 4 to 10°C.

3. Procedures

a. mE Method:[1]

1) Selection of sample size and filtration—Filter appropriate sample volumes through a 0.45-μm, gridded, sterile membrane to give 20 to 60 colonies on the membrane surface. Transfer filter to agar medium in petri dish, avoiding air bubbles beneath the membrane.

2) Incubation—Invert culture plates and incubate at 41°C ± 0.5°C for 48 h.

3) Substrate test—After 48 h incubation, carefully transfer filter to EIA medium. Incubate at 41°C ± 0.5°C for 20 min.

4) Counting—Pink to red enterococci

colonies develop a black or reddish-brown precipitate on the underside of the filter. Count colonies using a fluorescent lamp and a magnifying lens.

b. m Enterococcus method:[2]

1) Selection of sample size and filtration—See ¶ 3a.

2) Incubation—Let plates stand for 30 min, then invert and incubate at 35 ± 0.5°C for 48 h.

3) Counting—Count all light and dark red colonies as enterococci. Count colonies using a fluorescent lamp and a magnifying lens.

4. Calculation of Fecal Streptococci or Enterococci Density

Compute density from sample quantities producing membrane filter counts within the desired 20- to 60-fecal streptococcus or enterococcus colony range. Calculate as in Section 9222B.6. Record densities as fecal streptococci or enterococci per 100 mL.

5. Verification Tests

Pick selected typical colonies from a membrane and streak for isolation onto the surface of a brain-heart infusion agar plate. Incubate at 35°C ± 0.5°C for 24 to 48 h.

Transfer a loopful of growth from a well-isolated colony on brain-heart infusion agar into a brain-heart infusion broth tube and to each of two clean glass slides. Incubate the brain-heart infusion broth at 35 ± 0.5°C for 24 h. Add a few drops of freshly prepared 3% hydrogen peroxide to the smear on a slide. The appearance of bubbles constitutes a positive catalase test and indicates that the colony is not a member of the fecal streptococcus group. If the catalase test is negative, i.e., no bubbles, make a Gram stain of the second slide. Fecal streptococci and enterococci are gram-positive, ovoid cells, 0.5 to 1.0 μm in diameter, mostly in pairs or short chains.

Transfer a loopful of growth from the brain-heart infusion broth to each of the following media: bile esculin agar (incubate at 35 ± 0.5°C for 48 h); brain-heart infusion broth (incubate at 45 ± 0.5°C for 48 h); brain-heart infusion broth with 6.5% NaCl (incubate at 35 ± 0.5°C for 48 h).

Growth of catalase-negative, gram-positive cocci on bile esculin agar and at 45°C in brain-heart infusion broth verifies that the colony is of the fecal streptococcus group. Growth at 45°C and in 6.5% NaCl broth indicates that the colony belongs to the enterococcus group.

6. Serological Verification of Group D Fecal Streptococci

An alternate verification test for Group D streptococci can be performed using the precipitin method of Lancefield.[4] This test is highly specific for *S. faecalis, S. faecium, S. avium, S. gallinarum, S. bovis,* and *S. equinus.*

a. Antigen preparation: Pick typical single colonies from the membrane filter and streak for isolation on brain-heart infusion agar or blood agar plates. Pick a well-isolated colony and inoculate into 30 to 50 mL of Todd-Hewitt broth.[5] Incubate at 35°C for 24 h under aerobic conditions. Concentrate bacterial suspension by centrifuging (3000 × g for 5 min). Draw supernatant off and resuspend cells in 0.5 mL saline solution. Autoclave resuspended cells for 15 min at 121°C. Centrifuge the bacteria and decant clear supernatant fluid containing the group antigen.

b. Capillary precipitin test: Antisera for this test may be obtained from commercial sources.

Dip a 1.2- to 1.5-mm-OD capillary tube into antiserum and draw up about 1 cm of serum. Place a finger over upper end of tube so that no air will be drawn up and carefully wipe off excess antiserum. Dip tube into streptococcal antigen extract solution and draw up an equal volume of antigen. Carefully wipe off excess extract. Place a finger over upper end of tube and force lower end into plasticine to plug lower opening. Invert tube and place it in plasticine groove of a capillary holding rack.

A positive test for Group D antigen is characterized by a white precipitate that appears at the antigen-antiserum interface within 15 min and usually by 5 min. If no reaction has occurred by 30 min, the test is negative. Examination of the tubes is more effective if they are read in a bright light against a dark background.

Serological verification of Group D streptococci also can be done using commercially available agglutination tests.* The slide agglutination tests are simple and appear to be reliable. Group D streptococci are verified directly from isolated colonies on membrane filter plates or from broth culture tubes. To verify presumptive enterococci cultures, test also for salt tolerance (growth in 6.5% NaCl broth).

7. Identification of Individual Species within Fecal Streptococcus and Enterococcus Groups

Table 9230:I shows some of the key biochemical reactions for identifying fecal streptococci, enterococci, and species within these two groups.

---

*Phadebact Strep D test, Pharmacia Diagnostics, Piscataway, N.J. and Streptex Test, Burroughs-Wellcome Co., Research Triangle Park, N.C.

8. References

1. LEVIN, M.A., J.R. FISCHER & V.J. CABELLI. 1975. Membrane filter technique for enumeration of enterococci in marine waters. *Appl. Microbiol* 30:66.

2. SLANETZ, L.W. & C.H. BARTLEY. 1957. Numbers of enterococci in water, sewage and feces determined by the membrane filter technique with an improved medium. *J. Bacteriol.* 74:591.

3. PHILLIPS, E. & P. NASH. 1985. Culture media. *In* E.H. Lennette, A. Ballows, W.J. Hausler, Jr. & H.J. Shadomy, eds. Manual of Clinical Microbiology, 5th ed. American Soc. Microbiology, Washington, D.C.

4. LANCEFIELD, R.C. 1933. A serological differentiation of human and other groups of hemolytic streptococci. *J. Exp. Med.* 57:571.

5. Difco Laboratories. 1984. Difco Manual, 10th ed.

6. FACKLAM, R.R. & R.B. CAREY. 1985. Streptococci and aerococci. *In* E.H. Lennette, A. Ballows, W.J.Hausler, Jr. & H.J. Shadomy, eds. Manual of Clinical Microbiology, 5th ed. American Soc. Microbiology, Washington, D.C.

7. GROSS, J.C., M.P. HOUGHTON & L.B. SENTERFIT. 1975. Presumptive speciation of *Streptococcus bovis* and other Group D streptococci from human sources by using arginine and pyruvate tests. *J. Clin. Microbiol.* 1:54.

8. COWAN, S.T. & K.J. STEEL. 1965. Manual for the Identification of Medical Bacteria. Cambridge Univ. Press, Cambridge, England.

9. BRIDGE, P.D. & P.H.A. SNEATH. 1983. Numerical taxonomy of streptococcus. *J. Gen. Microbiol.* 129:565.

TABLE 9230:I. SELECTED KEY BIOCHEMICAL CHARACTERISTICS OF THE STREPTOCOCCUS SPECIES WITHIN THE FECAL STREPTOCOCCUS AND ENTEROCOCCUS GROUPS*

| Test | Fecal Streptococcus Group | | | | | |
| | Enterococcus Group | | | | | |
	S. faecalis	S. faecium	S. avium	S. gallinarium	S. bovis	S. equinus
Catalase	−	−	−	−	−	−
40% Bile	+	+	+	+	+	+
Esculin[6]	+	+	+	+	+	+
Growth at 45°C	+	+	+	+	+	+
Growth in 6.5% NaCl	+	+	+	+	−	−
Growth at 10°C	+	+	+	+	−	−
Pyruvate utilization[7]	+	−	−	−	−	−
Phosphatase activity[8]	+	−	+	+	−	−
Arginine hydrolysis[8]	+	+	−d	−	−	−
L-Sorbose fermentation[9]	−	−	+	−	−	−
Lactose fermentation[8]	+	+	+	+	+	+
n-acetyl-β-glucoseaminidase activity[9]	−d	−	−	+	−	−
Starch[6]	−	−	−	−	+	−
Arabinose[6]	−	+	+	−	−	−

* + = 90% or more of strains are positive
− = 90% or more of strains are negative
d = reactions variable

9240 IDENTIFICATION OF IRON AND SULFUR BACTERIA*

9240 A. Introduction

The group of nuisance organisms collectively designated "iron and sulfur bacteria" is neither morphologically nor physiologically homogeneous, yet it may be characterized by the ability to transform or deposit significant amounts of iron or sulfur, usually in the form of objectionable slimes. However, iron and sulfur bacteria are not the sole producers of bacterial slimes.

The organisms in this group may be filamentous or single-celled, autotrophic or heterotrophic, aerobic or anaerobic. According to conventional bacterial classifi-

*Approved by Standard Methods Committee, 1988.

cation, these organisms are assigned to a variety of orders, families, and genera. They are studied as "iron and sulfur bacteria," because these elements and their transformations may be important in water treatment and distribution systems and may be especially bothersome in waters for industrial use, such as cooling and boiler waters. Iron bacteria may cause fouling and plugging of wells and distribution systems and sulfate-reducing bacteria may cause rusty water and tuberculation of pipes. These organisms also may cause odor, taste, frothing, color, and increases in turbidity in waters.

The nutrient supply of iron and sulfur bacteria may be wholly or partly inorganic and they may extract it, if attached in a gelatinous substrate, from a low concentration in flowing water. This seems quite important in the case of certain sulfur bacteria utilizing small amounts of hydrogen sulfide or in the case of organisms such as *Gallionella,* which obtain their energy from the oxidation of ferrous iron. *Thiobacillus ferrooxidans* and *Ferrobacillus ferrooxidans,* which contribute to the problem of acid mine drainage, can be identified by tests for transformation of ferrous to ferric iron or oxidation of reduced sulfur compounds. Temperature, light, pH, and oxygen supply also affect the growth of these organisms. Under different environmental conditions some bacteria may appear either as iron or as sulfur bacteria.

9240 B.   Iron Bacteria

1. General Characteristics

"Iron bacteria" are considered to be capable of metabolizing reduced iron present in their aqueous habitat and of depositing it in the form of hydrated ferric oxide on or in their mucilaginous secretions. A somewhat similar mechanism is used by bacteria that utilize manganese. The large amount of brown slime so produced will impart a reddish tinge and an unpleasant odor to drinking water and may render the supply unsuitable for domestic or industrial purposes. Bacteria of this type obtain energy by the oxidation of iron from the ferrous to the ferric state; the ferric form is precipitated as ferric hydroxide ($FeOH_3$). Iron may be obtained from the pipe itself or from the water within it. The amount of ferric hydroxide deposited is very large in comparison with the enclosed cells.

Some bacteria that do not oxidize ferrous iron nevertheless may cause it to be dissolved or deposited indirectly. In their growth, either they liberate iron by utilizing organic radicals to which the iron is attached or they alter environmental conditions to permit the solution or deposition of iron. In consequence, less ferric hydroxide may be produced, but taste, odor, and fouling may be engendered.

2. Collection of Samples and Identification

Identification of nuisance iron bacteria usually has been made on the basis of microscopic examination of the suspected material. Directly examine bulked activated sludge, masses of microbial growth in lakes, rivers, and streams, and slime growths in cooling-tower waters. Suspected development of iron bacteria in water wells or in distribution systems may require special efforts to secure samples useful for identification. Continued heavy deposition of iron caused by the oxidation of ferrous iron by air or by other environmental changes often hides the sheaths or stalks of iron bacteria. The cells within the filaments often die and disintegrate and the filaments tend to be fragmented or crushed by the mass of the iron precipitate.

Settle or centrifuge samples of water drawn directly from wells and examine sediment microscopically. Place a portion of sediment on a microscope slide, cover with a cover slip, and examine under a low-power microscope for filaments and iron-encrusted filaments. The material trapped by filters placed in front of back-surge valves often has yielded excellent specimens of iron bacteria. Water pumped from wells may be passed through a 0.45-μm membrane filter and the filter examined microscopically after drying and clearing with immersion oil applied directly to the membrane. Phase-contrast microscopes have made possible the examination of unstained culture material. Use india ink or lactophenol blue for staining when conventional light microscopy is used. To dissolve iron deposits place several drops of 1N HCl at one edge of cover slip and draw it under cover slip by applying filter or blotting paper to opposite edge. Reducing compounds such as sodium ascorbate also may be used to dissolve deposits and permit observation of cellular structure. To verify that the material is iron, add a solution of potassium ferrocyanide to a sample on a slide, cover, and draw 1N HCl under cover slip. A blue precipitate of Prussian blue will form as iron around cells or filaments is dissolved.

Identify organisms by comparing with available drawings or photographs of iron bacteria.[1-14] Some examples are given in Figures 9240:1 through 9240:5.

3. References

1. LUESCHOW, L.A. & K.M. MACKENTHUN. 1962. Detection and enumeration of iron bacteria in municipal water supplies. *J. Amer. Water Works Assoc.* 54:751.
2. STARKEY, R.L. 1945. Transformations of iron by bacteria in water. *J. Amer. Water Works Assoc.* 37:963.

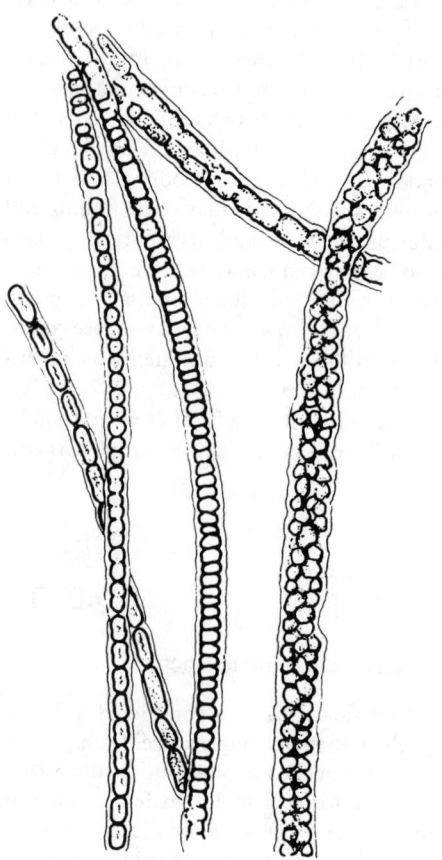

Figure 9240:1. Filaments of *Crenothrix polyspora* showing variation of size and shape of cells within the sheath. Note especially the multiple small round cells, or "conidia", found in one of the filaments. This distinctive feature is the reason for the name *polyspora*. Young growing colonies usually are not encrusted with iron or manganese. Older colonies often exhibit empty sheaths that are heavily encrusted. Cells may vary considerably in size: Rod-shaped cells average 1.2 to 2.0 μm in width by 2.4 to 5.6 μm in length; coccoid cells of "conidia" average 0.6 μm in diameter.

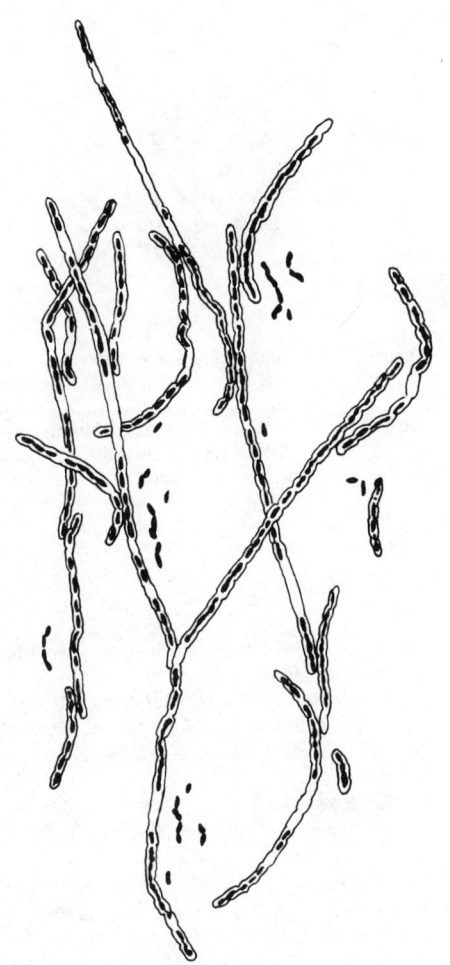

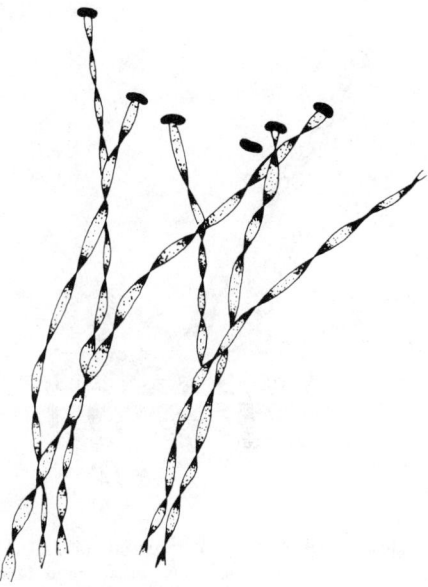

Figure 9240:3. Laboratory culture of *Gallionella ferruginea*, showing cells, stalks excreted by cells, and branching of stalks where cells have divided. A precipitate of inorganic iron on and around the stalks often blurs the outlines. Cells at tip of stalk average 0.4 to 0.6 μm in width by 0.7 to 1.1 μm in length.

Figure 9240:2. Filaments of *Sphaerotilus natans*, showing cells within the filaments and some free "swarmer" cells. Filaments show false branching and areas devoid of cells. Individual cells within the sheath may vary in size, averaging 0.6 to 2.4 μm in width by 1.0 to 12.0 μm in length; most strains are 1.1 to 1.6 μm wide by 2.0 to 4.0 μm long.

3. STOKES, J.L. 1954. Studies on the filamentous sheathed iron bacterium *Sphaerotilus natans*. *J. Bacteriol.* 67:278.

4. KUCERA, S. & R.S. WOLFE. 1957. A selective enrichment method for *Gallionella ferruginea*. *J. Bacteriol.* 74:344.

5. WAITZ, S. & J.B. LACKEY. 1958. Morphological and biochemical studies on the organism *Sphaerotilus natans*. *Quart. J. Fla. Acad. Sci.* 21:335.

6. WOLFE, R.S. 1958. Cultivation, morphology, and classification of the iron bacteria. *J. Amer. Water Works Assoc.* 50:1241.

7. WOLFE, R.S. 1960. Observations and studies of *Crenothrix polyspora. J. Amer. Water Works Assoc.* 52:915.

8. WOLFE, R.S. 1960. Microbial concentration of iron and manganese in water with low concentrations of these elements. *J. Amer. Water Works Assoc.* 52:1335.

9. DONDERO, N.C., R.A. PHILIPS & H. HEUKELEKIAN. 1961. Isolation and preservation of cultures of *Sphaerotilus. Appl. Microbiol.* 9:219.

Figure 9240:5. Single-celled iron bacterium *Siderocapsa treubii*. Cells are surrounded by a deposit of ferric hydrate. Individual cells average 0.4 to 1.5 μm in width by 0.8 to 2.5 μm in length.

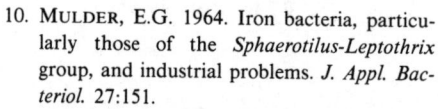

Figure 9240:4. Mixture of fragments of stalks of *Gallionella ferruginea* and inorganic iron-manganese precipitate found in natural samples from wells. Fragmented stalks appear golden yellow to orange when examined under the microscope.

treubii in certain waters of the Niederrhein. *Gewasser Abwasser* 39/40:41.

10. MULDER, E.G. 1964. Iron bacteria, particularly those of the *Sphaerotilus-Leptothrix* group, and industrial problems. *J. Appl. Bacteriol.* 27:151.

11. DRAKE, C.H. 1965. Occurrence of *Siderocapsa*

12. BUCHANAN, R.E. & N.E. GIBBONS, eds. 1974. Bergey's Manual of Determinative Bacteriology, 8th ed. Williams & Wilkins, Baltimore, Md.

13. EDMONDSON, W.T., ed. 1959. Ward & Whipple's Fresh Water Biology, 2nd ed. John Wiley & Sons, New York, N.Y.

14. SKERMAN, V.B.D. 1967. A Guide to the Identification of the Genera of Bacteria, 2nd ed. Williams & Wilkins, Baltimore, Md.

9240 C.   Sulfur Bacteria

1. General Characteristics

The bacteria that oxidize or reduce significant amounts of organic sulfur compounds exhibit a wide diversity of morphological and biochemical characteristics. One group, the sulfate-reducing bacteria, consists of single-celled forms that grow anaerobically and reduce sulfate, SO_4^{2-}, to hydrogen sulfide, H_2S. A second group, the photosynthetic green and purple sulfur bacteria, grows anaerobically in the light and uses H_2S as a hydrogen donor for photosynthesis. The sulfide is oxidized to sulfur or sulfate. A third group, the aerobic sulfur-oxidizers, oxidizes reduced sulfur compounds aerobically to obtain energy for chemoautotrophic growth.

The sulfur bacteria of most importance in the water and wastewater field are the sulfate-reducing bacteria, which include *Desulfovibrio*, and the single-celled aerobic sulfur-oxidizers of the genus *Thiobacillus*. The sulfate-reducing bacteria contribute greatly to tuberculations and galvanic corrosion of water mains and to taste and odor problems in water. *Thiobacillus*, by its production of sulfuric acid, has contributed to the destruction of concrete sewers and the acid corrosion of metals.

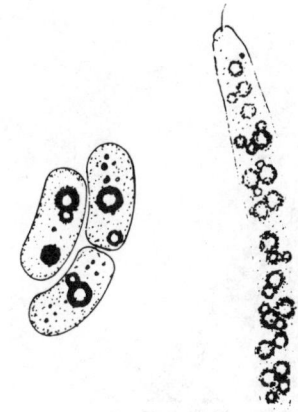

Figure 9240:6. Photosynthetic purple sulfur bacteria: Large masses of cells have brown-orange to purple color—may appear chalky if there is a large amount of sulfur within the cells. Left: cells of *Chromatium okenii* (5.0 to 6.5 μm wide by 8 to 15 μm long) containing sulfur granules. Right: *Thiospirillum jenense* (3.5 to 4.5 μm wide by 30 to 40 μm long); cell contains sulfur granules and polar flagellum is visible.

2. Collection of Samples and Identification

Identification of nuisance sulfur bacteria usually has been made on the basis of microscopic examination of the suspected material. Examine directly samples of slimes suspended in waters, scrapings from exposed surfaces, or sediments.

Three groups of sulfur bacteria may be recognized microscopically; green and purple sulfur bacteria; large, colorless filamentous sulfur bacteria; and large, colorless, nonfilamentous sulfur bacteria. The fourth group, consisting of sulfate-reducing bacteria and sulfur-oxidizing bacteria of the genus *Thiobacillus*, cannot be identified by appearance alone.

a. Green and purple sulfur bacteria:

1) Green sulfur bacteria most frequently occur in waters containing H_2S. They are small, ovoid to rod-shaped, nonmotile organisms, generally less than 1 μm in diameter, and with a yellowish-green color

in masses. Sulfur globules are seldom if ever deposited within the cells.

2) Purple sulfur bacteria (Figure 9240:6) occur in waters containing H_2S. They are large, generally stuffed with sulfur globules, and often so intensely pigmented as to make individual cells appear red. Large, dense, highly colored masses are detected easily by the naked eye.

b. Colorless filamentous sulfur bacteria: Colorless filamentous sulfur bacteria (Figures 9240:7 and 9240:8) occur in waters where both oxygen and H_2S are present. They may form mats with a slightly yellowish-white appearance due to deposition of internal sulfur globules. They generally are large and may be motile with a characteristic gliding movement. Identify by comparing organisms with available photographs.[1-3]

c. Colorless nonfilamentous sulfur bac-

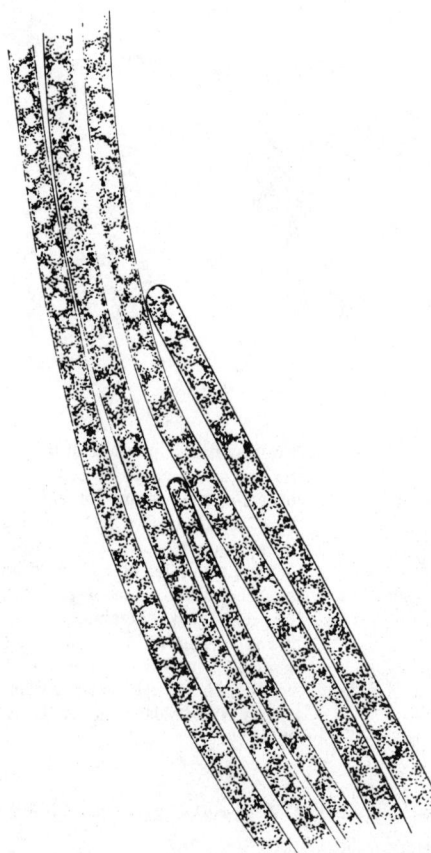

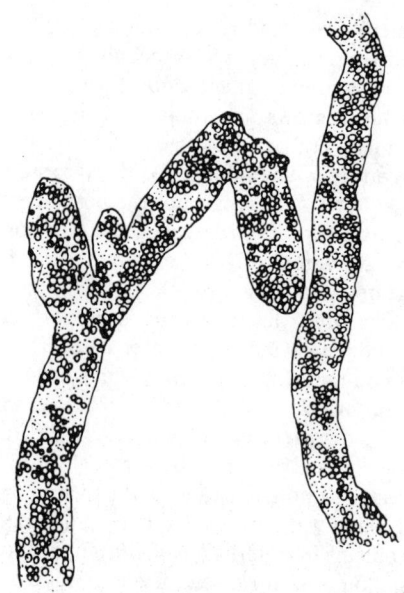

Figure 9240:8. Colorless filamentous sulfur bacteria: portion of a colony of *Thiodendron mucosum*, showing branching of the mucoid filament. Individual cells (1.0 to 2.5 μm wide by 3 to 9 μm long) have been found within the jelly-like material of the filaments. The long axis of the cells runs parallel to the long axis of the filaments.

Figure 9240:7. Colorless filamentous sulfur bacteria: *Beggiatoa alba* trichomes, containing granules of sulfur. Filaments are composed of a linear series of individual rod-shaped cells that may be visible when not obscured by light reflecting from sulfur granules. Trichomes are 2 to 15 μm in diameter and may be up to 1500 μm long; individual cells, if visible, are 4.0 to 16.0 μm long.

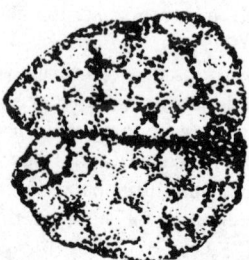

Figure 9240:9. Colorless nonfilamentous sulfur bacteria: dividing cell of *Thiovolum majus*, containing sulfur granules. Cells may measure 9 to 17 μm in width by 11 to 18 μm in length and are generally found in nature in a marine littoral zone rich in organic matter and hydrogen sulfide.

teria: Colorless, nonfilamentous sulfur bacteria (see Figure 9240:9, for example) usually are associated with decaying algae. They are extremely motile, ovoid to rod-shaped with sulfur globules and possible calcium carbonate deposits. They generally are very large.

d. Colorless small sulfur bacteria and sulfate-reducing bacteria: The small single-celled bacteria, *Thiobacillus* spp., and the sulfate-reducing bacteria, such as *Desulfovibrio,* cannot be identified by direct microscopic examination. *Thiobacillus* types are small, colorless, motile, and rod-shaped and are found in an environment containing H₂S. Sulfur globules are absent. Identify *Thiobacillus* types, *Desulfovibrio,* or other sulfate-reducing bacteria physiologically.'

3. References

1. LACKEY, J.B. & E.W. LACKEY. 1961. The habitat and description of a new genus of sulfur bacterium. *J. Gen. Microbiol.* 26:28.
2. FAUST, L. & R.S. WOLFE. 1961. Enrichment and cultivation of *Beggiatoa alba. J. Bacteriol.* 81:99.
3. MORGAN, G.B. & J.B. LACKEY. 1965. Ecology of a sulfuretum in a semitropical environment. *Z. Allg. Mikrobiol.* 5:237.

9240 D. Enumeration, Enrichment, and Isolation of Iron and Sulfur Bacteria (PROPOSED)

There are no good means of enumerating iron and sulfur bacteria other than the sulfate-reducing bacteria and the thiobacilli. Laboratory cultivation and isolation of pure cultures is difficult and successful isolation is uncertain. This is especially true of attempts to isolate filamentous bacteria from activated sludge or other sources where many different bacterial types are present.

1. Media

a. Casitone-glycerol-yeast autolysate broth (CGY): This medium may not be available in dehydrated form and may require preparation from the basic ingredients. It may be solidified by adding 1.5% agar.

Casitone	5.0 g
Glycerol	10.0 g
Yeast autolysate	1.0 g
Distilled water	1 L

b. Isolation medium (iron bacteria): This medium may not be available in dehydrated form and may require preparation from the basic ingredients.

Glucose.	0.15	g
Ammonium sulfate, (NH₄)₂SO₄.	0.5	g
Calcium nitrate, Ca(NO₃)₂ . .	0.01	g
Dipotassium hydrogen phosphate, K₂HPO₄	0.05	g
Magnesium sulfate, MgSO₄ · 7H₂O	0.05	g
Potassium chloride, KCl . . .	0.05	g
Calcium carbonate, CaCO₃ . .	0.1	g
Agar	10.0	g
Vitamin B₁₂	0.01	mg
Thiamine	0.4	mg
Distilled water	1	L

c. Maintenance (SCY) medium (iron bacteria): This medium may not be available in dehydrated form and may require preparation from the basic ingredients.

Sucrose	1.0	g
Casitone	0.75	g
Yeast extract.	0.25	g
Trypticase soy broth without dextrose	0.25	g
Agar	10.0	g
Vitamin B₁₂	0.01	mg
Thiamine	0.4	mg
Distilled water	1	L

d. Mn agar: This medium may not be

available in dehydrated form and may require preparation from the basic ingredients.

Manganous carbonate,		
MnCO$_3$.	2.0	g
Beef extract	1.0	g
Ferrous ammonium sulfate,		
Fe(NH$_4$)$_2$(SO$_4$)$_2$.	150	mg
Sodium citrate	150	mg
Yeast extract.	75	mg
Cyanocobalamin	0.005	mg
Agar	10.0	g
Distilled water	1	L

Prepare and sterilize the medium without cyanocobalamin. Separately sterilize cobalamin by filtration and add aseptically just before medium solidifies.

e. Iron oxidizing medium (Thiobacillus ferrooxidans): This medium may not be available in dehydrated form and may require preparation from the basic ingredients.

Basal salts:
Ammonium sulfate,		
(NH$_4$)$_2$SO$_4$	3.0	g
Potassium chloride, KCl . . .	0.10	g
Dipotassium hydrogen phosphate,		
K$_2$HPO$_4$.	0.50	g
Magnesium sulfate,		
MgSO$_4$ · 7H$_2$O	0.50	g
Calcium nitrate,		
Ca(NO$_3$)$_2$	0.01	g
H$_2$SO$_4$, 10N.	1.0	mL
Distilled water	700	mL

Energy source:
Ferrous sulfate, FeSO$_4$ · 7H$_2$O,		
14.74% solution (w/v). . .	300	mL

Separately sterilize basal salts and energy source and combine when cool. Store in the refrigerator and discard after 2 weeks. A precipitate will form and the medium will be opalescent and green. The pH should be 3.0 to 3.6.

f. Ferrous sulfide agar (Gallionella ferruginea): This medium may not be avail-

able in dehydrated form and may require preparation from the basic ingredients.

Agar layer:
Ferrous sulfide, FeS		
(washed precipitate and liquid)	500	mL
Sodium sulfide, Na$_2$S	15.6	g
Ferrous ammonium sulfate,		
Fe(NH$_4$)$_2$(SO$_4$)$_2$ · 6H$_2$O . .	78.4	g
Boiling distilled water.	1	L
Agar (liquid) (30 g/L)	500	mL

Liquid overlay:
Ammonium chloride, NH$_4$Cl	1.0	g
Dipotassium phosphate,		
K$_2$HPO$_4$	0.5	g
Magnesium sulfate,		
MgSO$_4$ · 7H$_2$O	0.2	g
Calcium chloride, CaCl$_2$ . . .	0.1	g
Distilled water	1	L

Prepare FeS by reacting equal molar quantities of Na$_2$S and Fe(NH$_4$)$_2$(SO$_4$)$_2$ in boiling distilled water. Let precipitate settle from the hot solution in a completely filled and stoppered bottle. Wash precipitate four times by decanting supernatant and replacing with boiling water. Store FeS in a glass stoppered bottle completely filled with additional boiling distilled water.

Add equal volumes of FeS and 3% agar at 45°C. Prepare slants in screw-capped tubes. Prepare liquid overlay, bubble CO$_2$ through it for 10 to 15 s, and add several milliliters to agar slant.

A variation of the basic medium requires adding 0.5 mL formalin (40% formaldehyde solution) to a screw-capped dilution bottle containing 10 mL FeS agar and 100 mL liquid overlay. Add 0.001% bromthymol blue and 0.004% bromcresol purple to liquid overlay.

g. Sulfate-reducing medium: This medium may not be available in dehydrated form and may require preparation from the basic ingredients.

Sodium lactate	3.5	g
Beef extract	1.0	g
Peptone	2.0	g

Magnesium sulfate,
$MgSO_4 \cdot 7H_2O$ . . . . . . 2.0   g
Sodium sulfate, Na_2SO_4. . . . 1.5   g
Dipotassium phosphate,
K_2HPO_4 . . . . 0.5   g
Ferrous ammonium sulfate,
$Fe(NH_4)_2(SO_4)_2 \cdot 6H_2O$ . . 0.392 g
Calcium chloride, $CaCl_2$ . . . 0.10   g
Sodium ascorbate . . . . . . 0.10   g
Distilled water 1   L

pH should be 7.5 ± 0.3 after sterilization. Prepare medium excluding ferrous ammonium sulfate and sodium ascorbate, dispense in screw-capped test tubes, and sterilize. For use, the tubes must be completely filled; therefore, in a flask sterilize extra medium to be added to tubes for filling. On day of use, prepare separate solutions of ferrous ammonium sulfate (3.92 g/100 mL) and sodium ascorbate (1.00 g/ 100 mL), sterilize by filtration through a 0.45-µm membrane filter, and aseptically add 0.1 mL each solution/10 mL basal medium.

h. Sulfate-reducing medium (Thiobacillus thioparus): This medium may not be available in dehydrated form and may require preparation from the basic ingredients.

Sodium thiosulfate,
$Na_2S_2O_3 \cdot 5H_2O$. . . . . . 10.0   g
Dipotassium hydrogen phosphate,
K_2HPO_4 . . . . 2.0   g
Magnesium sulfate,
$MgSO_4 \cdot 7H_2O$ . . . . 0.1   g
Calcium chloride,
$CaCl_2 \cdot 2H_2O$ . . . . 0.1   g
Ammonium sulfate,
$(NH_4)_2SO_4$. . . . . 0.1   g
Ferric chloride, $FeCl_3 \cdot 6H_2O$ 0.02   g
Distilled water 1   L

pH should be 7.8 after sterilization. Separately sterilize $Na_2S_2O_3$ and $(NH_4)_2SO_4$ and add before use.

i. Sulfur medium (Thiobacillus thiooxidans): This medium may not be available

in dehydrated form and may require preparation from the basic ingredients.

Sulfur, elemental. . . . . . . 10.0   g
Potassium dihydrogen phosphate,
KH_2PO_4 . . . . 3.0   g
Magnesium sulfate,
$MgSO_4 \cdot 7H_2O$ . . . . 0.5   g
Ammonium sulfate,
$(NH_4)_2SO_4$. . . . . 0.3   g
Calcium chloride,
$CaCl_2 \cdot 2H_2O$ . . . . 0.25   g
Ferric chloride, $FeCl_3 \cdot 6H_2O$ 0.02   g
Distilled water 1   L

pH should be 4.8 after sterilization. Weigh sulfur into 250-mL flasks using 1 g/flask. Add 100 mL medium to each flask and sterilize with intermittent steam (30 min for each of 3 consecutive d).

2. Iron Bacteria

a. Sphaerotilus-Leptothrix:

1) Iron bacteria, especially those belonging to the *Sphaerotilus-Leptothrix* group, thrive in media too dilute to support the proliferation of more rapidly growing organisms. One medium[1] is partially selective for *Sphaerotilus* (BOD dilution water supplemented with 100 mg/L sodium lactate). Dispense 50 mL of this medium in French square bottles and autoclave at 69 kPa for 15 min. To inoculate sample add 25-mL portions of stream water or 1-, 5-, and 10-mL portions of settled wastewater or process liquor to duplicate bottles of medium. Incubate at 22 to 25°C for 5 d and observe for filamentous growth. Isolate pure cultures by picking a filament from the BOD-lactate broth and streaking on 0.05% meat extract agar. After incubating for 24 h at 25°C, pick typical curling filaments with the aid of a dissecting microscope and transfer to casitone-glycerol-yeast autolysate (CGY) broth. If a pellicle with no underlying turbidity develops in 2 to 3 d, transfer a filament to a CGY agar slant, incubate at 25°C until growth is visible, and store in a refrigerator.

A detailed key for identifying filamentous microorganisms in complex mixtures such as wastewater and activated sludge is available.[2]

2) Isolation and maintenance media have proven quite successful for identifying various groups of filamentous organisms, including iron bacteria.[3] Prepare agar slants of these media and aseptically pipet 3 mL sterile tap water onto surface of slants. Inoculate tubes and incubate at room temperature until turbid growth has developed in liquid layer. The cells will remain viable for 3 months in the refrigerator.

3) Another good maintenance medium for cultivating the *Sphaerotilus* group is CGY.[4]

4) *Leptothrix (Sphaerotilus discophorous)* can be distinguished from *Sphaerotilus natans* by its ability to oxidize manganous ion. Use Mn-agar as the differential medium.[5]

b. *Thiobacillus ferrooxidans:* Although this organism also is a sulfur-oxidizing bacterium,[6,7] its main importance has been in acid mine drainage. A medium suitable for enumeration of the MPN is available.[8] Some oxidation of iron occurs during sterilization but the loss of ferrous iron is not appreciable. The medium has a precipitate (probably ferrous and ferric phosphates), is opalescent and green, has a pH of 3.0 to 3.6, and contains 9000 mg/L ferrous iron. Growth of the organism is manifested by a decrease in pH and an increase in concentration of oxidized iron. With practice and use of uninoculated controls, an increase of deep orange-brown color can be seen in positive enrichment tubes or flasks as compared to negative ones. Shake test-tube dilutions daily because these organisms are highly aerobic.

c. *Gallionella ferruginea:* For cultivation of this organism use ferrous sulfide agar.[9,10] Inoculate tubes with a drop of suspension of a suspected *Gallionella* deposit. Growth at room temperature usually occurs in 18 to 36 h and appears as a white deposit on sides of test tube. The ring of colonies occurring at a certain level reflects a balance between upward diffusion of ferrous ions and downward diffusion of oxygen molecules. Supplement ferrous sulfide agar with formalin[11] to use it for isolation of pure cultures.

d. *Other iron bacteria:* An acid-tolerant (pH 3.5 to 5.0) filamentous iron-oxidizing *Metallogenium* has been isolated with a medium[12] containing $(NH_4)_2SO_4$, 0.1%; $CaCO_3$, 0.01%; $MgSO_4$, 0.02%; K_2HPO_4, 0.001%; potassium acid phthalate, 0.4%; and 250 mg/L ferrous iron from an acidified $FeSO_4 \cdot 7H_2O$, solution. Add 0.4% formalin to 100 mL of the isolating medium in a 250-mL erlenmeyer flask.

For heterotrophic iron-precipitating bacteria[13] use a ferric ammonium citrate medium consisting of: $(NH_4)_2SO_4$, 0.5 g/L; $NaNO_3$, 0.5 g/L; K_2HPO_4, 0.5 g/L; $MgSO_4 \cdot 7H_2O$, 0.5 g/L; and ferric ammonium citrate, 10.0 g/L. Adjust pH to 6.6 to 6.8 and sterilize. To make the medium solid add 15 g/L agar.

Alternatively grow iron bacteria by combining 20 mL liquid broth medium, 10 mL raw water or inoculum, and 3 g iron oxide. Incubate 48 to 72 h at 25°C on a wrist-action shaker to produce good, visible growth.

3. Sulfur Bacteria

a. *Sulfate-reducing bacteria:*
1) To enumerate sulfate-reducing bacteria such as *Desulfovibrio,* use a sulfate-reducing medium.[14] Inoculate tubes and fill completely with sterile medium to create anaerobic conditions. For comparative purposes, incubate one or two uninoculated controls with each set of inoculated tubes. To sample volumes greater than 10 mL, pass sample through a 0.45-μm membrane filter and transfer filter to screw-cap test tube with medium. If sulfate-reducing bacteria are present, tubes will show black-

ening within 4 to 21 d of incubation at 20 to 30°C.

2) An agar medium suitable for growth and enumeration of sulfate-reducing bacteria also is available.[15] The medium consists of trypticase soy agar, 4.0%, fortified with additional agar, 0.5%, to which is added 60% sodium lactate (0.4% v/v), hydrated magnesium sulfate, 0.2%, and ferrous ammonium sulfate, 0.2%. Adjust pH to 7.2 to 7.4 and sterilize. Medium should be clear and free from precipitate. Inoculate all plates within 1 or at most 4 h after agar hardens to prevent saturation with oxygen. To prevent moisture condensation on petri dish covers, replace covers with sterile absorbent tops until 10 to 15 min after agar hardens. Place uninverted plates in desiccator or Brewer jars and replace atmosphere with tank hydrogen or nitrogen by successive evacuation and gas replacement. Incubate at room temperature (21 to 24°C) or at 28 to 30°C, the optimum temperature for these organisms. Growth and blackening around the colonies is typical of sulfate-reducing bacteria and may occur at any time between 2 and 21 d, although the usual time is 2 to 7 d.

3) Media suitable for enumeration of various species of sulfate-reducing bacteria have been evaluated.[16,17]

b. *Photosynthetic purple and green sulfur bacteria:* Because these organisms are so specialized and rarely cause problems in water and wastewater treatment processes, methods for their isolation and enumeration are not included here. An excellent review is available.[18]

c. *Thiobacillus spp.*: The growth and physiology of different species of the single-celled sulfur-oxidizing bacteria of the genus *Thiobacillus* have been evaluated carefully.[19,20] Media[21] suitable for enumeration of *Thiobacillus thioparus* and *Thiobacillus thiooxidans* by an MPN technique are listed in 9240D.1. Inoculate medium and incubate for 4 to 5 d at 25 to 30°C. Growth of thiobacilli produces elemental sulfur,

which sinks to the bottom with a coincident decrease in pH and turbidity of the medium. Chemical tests for formation of sulfate are necessary to confirm presence of *Thiobacillus.*

d. *Beggiatoa:* Methods for enrichment and isolation of *Beggiatoa* depend on use of a hay extract.[22] Extract dried hay at 100°C in large volumes of water, changing water three times during extraction. The final wash water has an amber color. After draining extracted hay, dry it on trays at 37°C. Prepare enrichment medium by suspending 8 g extracted and dried hay in 1 L tap water; dispense 70 mL per 125-mL erlenmeyer flask and sterilize by autoclaving. Inoculate 5-mL portions of mud containing decaying plant material from small ponds, lakes, and streams and incubate at room temperature. In successful enrichments a strong odor of H_2S is noticeable in the flasks and *Beggiatoa* growth appears within 10 d as a white film on surface of medium and submerged upper walls of flask. To isolate organism, wash portions of white surface film several times in sterile tap water, place on surface of agar plates (1% agar and 0.2% beef extract), and incubate at 28°C. From those plates in which filaments have migrated to the periphery of the agar surface and away from contaminants, cut out agar blocks containing single, isolated filaments of *Beggiatoa* and place, filament side down, on fresh plates of same medium. Incubate again at 28°C. After growth has progressed to the extent that isolated single filaments are present, repeat isolation procedure.

An inorganic medium for *Beggiatoa* also has been described.[23]

4. References

1. ARMBRUSTER, E.H. 1969. Improved technique for isolation and identification of *Sphaerotilus. Appl. Microbiol.* 17:320.

2. FARQUHAR, G.J. & W.C. BOYLE. 1971. Identification of filamentous microorganisms in ac-

tivated sludge. *J. Water Pollut. Control Fed.* 43:604.

3. VANVEEN, W.L. 1973. Bacteriology of activated sludge, in particular the filamentous bacteria. *Antonie van Leeuwenhoek* (Holland) 39:189.

4. DONDERO, N.C., R.A. PHILIPS & H. HEU-KELEKIAN. 1961. Isolation and preservation of cultures of *Sphaerotilus. Appl. Microbiol.* 9:219.

5. MULDER, E.G. & W.L. VANVEEN. 1963. Investigations on the *Sphaerotilus-Leptothrix* group. *Antonie van Leeuwenhoek* (Holland) 29:121.

6. UNZ, R.F. & D.G. LUNDGREN. 1961. A comparative nutritional study of three chemoautotrophic bacteria: *Ferrobacillus ferrooxidans, Thiobacillus ferrooxidans,* and *Thiobacillus thiooxidans. Soil Sci.* 92:302.

7. MCGORAN, C.J.M., D.W. DUNCAN & C.C. WALDEN. 1969. Growth of *Thiobacillus ferrooxidans* on various substrates. *Can. J. Microbiol.* 15:135.

8. SILVERMAN, M.P. & D.C. LUNDGREN. 1959. Studies on the chemoautotrophic iron bacterium *Ferrobacillus ferrooxidans. J. Bacteriol.* 77:642.

9. KUCERA, S. & R.S. WOLFE. 1957. A selective enrichment method for *Gallionella ferruginea. J. Bacteriol.* 74:344.

10. WOLFE, R.S. 1958. Cultivation, morphology, and classification of the iron bacteria. *J. Amer. Water Works Assoc.* 50: 1241.

11. NUNLEY, J.W. & N.R. KRIEG. 1968. Isolation of *Gallionella ferruginea* by use of formalin. *Can. J. Microbiol.* 14:385.

12. WALSH, F. & R. MITCHELL. 1972. A pH dependent succession of iron bacteria. *Environ. Sci. Technol.* 6:809.

13. CLARK, F.M., R.M. SCOTT & E. BONE. 1967. Heterotrophic, iron-precipitating bacteria. *J. Amer. Water Works Assoc.* 59:1036.

14. LEWIS, R.F. 1965. Control of sulfate-reducing bacteria. *J. Amer. Water Works Assoc.* 57:1011.

15. IVERSON, W.P. 1966. Growth of *Desulfovibrio* on the surface of agar media. *Appl. Microbiol.* 14:529.

16. LECHEVALIER, H.A. & D. PRAMER. 1970. The Microbes, 1st ed. J.B. Lippincott Co., Philadelphia, Pa.

17. MARA, D.D. & D.J.A. WILLIAMS. 1970. The evaluation of media used to enumerate sulphate reducing bacteria. *J. Appl. Bacteriol.* 33:543.

18. PFENNIG, N. 1967. Photosynthetic bacteria. *Annu. Rev. Microbiol.* 21:285.

19. HUTCHINSON, M., K.I. JOHNSTONE & D. WHITE. 1965. The taxonomy of certain thiobacilli. *J. Gen. Microbiol.* 41:357.

20. HUTCHINSON, M., K.I. JOHNSTONE & D. WHITE. 1966. Taxonomy of the acidophilic thiobacilli. *J. Gen. Microbiol.* 44:373.

21. STARKEY, R.L. 1937. Formation of sulfide by some sulfur bacteria. *J. Bacteriol.* 33:545.

22. SCOTTEN, H.L. & J.L. STOKES. 1962. Isolation and properties of *Beggiatoa. Arch. Mikrobiol.* 42:353.

23. KOWALLIK, U. & E.G. PRINGSHEIM. 1966. The oxidation of hydrogen sulfide by beggiatoa. *Amer. J. Bot.* 53:801.

9250 DETECTION OF ACTINOMYCETES*

9250 A. Introduction

1. General Discussion

Earthy-musty odors affect the quality and public acceptance of municipal water

supplies in many parts of the world. They are among the naturally occurring odors that plant operators find most difficult to remove by conventional treatment. As

*Approved by Standard Methods Committee, 1988.

early as 1929, it was assumed that these odors could be attributed to volatile metabolites formed during normal actinomycete development.[1] Two such compounds, geosmin and 2-methylisoborneol, have been isolated[2-8] and identified as the agents responsible for earthy-musty odor problems in surface water.[8-10] Both, however, are produced also by some filamentous blue-green algae.[11-15] Geosmin and 2-methylisoborneol have threshold odor concentrations well below the microgram-per-liter level. Thus, traces of these products are sufficient to impart a disagreeable odor to water or a muddy flavor to fish. In areas periodically plagued by this problem, it is prudent to enumerate actinomycetes. Identification of their relative abundance in a drinking water source can provide yet another means to assess water quality. The methods described are well-established techniques that have been used with success in the isolation and enumeration of actinomycetes related to public water supplies.[16,17] Actinomycetes also have been recognized as a cause of disruptions in wastewater treatment. Massive growths are capable of producing thick foam in the activated sludge process.[18,19]

Of the general properties of actinomycetes, the most striking is their fungal-type morphology. Although actinomycetes were looked upon initially as fungi, later research revealed that they were filamentous, branching bacteria.[20] The actinomycetes are represented most commonly by saprophytic forms that have an extensive impact on the environment by decomposing and transforming a wide variety of complex organic residues. Widely distributed in nature, actinomycetes constitute a considerable proportion of the population of soil and lake and river muds. Most actinomycetes from which geosmin and 2-methylisoborneol have been identified are members of the genus *Streptomyces*, which is considered the most likely to be significant in water supply problems.

2. Samples

a. Collection: Collect samples as directed in Section 9060A.

b. Storage: Analyze samples as promptly after collection as possible. Store water samples below 10°C if they cannot be processed promptly.

3. References

1. ADAMS, B.A. 1929. *Cladothrix dichotoma* and allied organisms as a cause of an "indeterminate" taste in chlorinated water. *Water & Water Eng.* 31:327.
2. GERBER, N.N. & H.A. LECHEVALIER. 1965. Geosmin, an earthy-smelling substance isolated from actinomycetes. *Appl. Microbiol.* 13:935.
3. GERBER. N.N. 1968. Geosmin, from microorganisms, is trans-1,10-dimethyl-trans-9-decalol. *Tetrahedron Lett.* 25:2971.
4. MARSHALL, J.A. & A.R. HOCHSTETLER. 1968. The synthesis of (±)-geosmin and the other 1,10-dimethyl-9-decalol isomers. *J. Org. Chem.* 33:2593.
5. ROSEN, A.A., R.S. SAFFERMAN, C.I. MASHNI & A.H. ROMANO. 1968. Identity of odorous substances produced by *Streptomyces griseoluteus. Appl. Microbiol.* 16:178.
6. MEDSKER, L.L., D. JENKINS & J.F. THOMAS. 1969. Odorous compounds in natural waters: 2-exo-hydroxy-2-methylbornane, the major odorous compound produced by several actinomycetes. *Environ. Sci. Technol.* 3:476.
7. GERBER, N.N. 1969. A volatile metabolite of actinomycetes, 2-methylisoborneol. *J. Antibiot.* 22:508.
8. ROSEN, A.A., C.I. MASHNI & R.S. SAFFERMAN. 1970. Recent developments in the chemistry of odour in water: The cause of earthy/musty odour. *Water Treat. Exam.* 19:106.
9. PIET, G.J., B.C.J. ZOETEMAN & A.J.A. KRAAYEVELD. 1972. Earthy-smelling substances in surface waters of the Netherlands. *Water Treat. Exam.* 21:281.
10. YURKOWSKI, M. & J.A.L. TABACHEK. 1974. Identification, analysis, and removal of geosmin from muddy-flavored trout. *J. Fish. Res. Board Can.* 31:1851.
11. SAFFERMAN, R.S., A.A. ROSEN, C.I. MASHNI & M.E. MORRIS. 1967. Earthy-smelling sub-

stances from a blue-green alga. *Environ. Sci. Technol.* 1:429.

12. MEDSKER, L.L., D. JENKINS & J.F. THOMAS. 1968. Odorous compounds in natural waters. An earthy-smelling compound associated with blue-green algae and actinomycetes. *Environ. Sci. Technol.* 2:461.

13. KIKUCHI, T., T. MIMURA, K. HARIMAYA, H. YANO, M. ARIMOTO, Y. MASADA & T. IN-OUE. 1973. Odorous metabolites of blue-green alga *Schizothrix muelleri* Nageli collected in the southern basin of Lake Biwa. Identification of geosmin. *Chem. Pharm. Bull.* 21:2342.

14. TABACHEK, J.L. & M. YURKOWSKI. 1976. Isolation and identification of blue-green algae producing muddy odor metabolites, geosmin and 2-methylisoborneol, in saline lakes in Manitoba. *J. Fish. Res. Board Can.* 33:25.

15. IZAGUIRRE, G., C.J. HWANG, S.W. KRASNER & M.J. MCGUIRE. 1982. Geosmin and 2-methylisoborneol from cyanobacteria in three

16. SAFFERMAN, R.S. & M.E. MORRIS. 1962. A method for the isolation and enumeration of actinomycetes related to water supplies. Robert A. Taft Sanitary Engineering Center Tech. Rep. W62-10, U.S. Public Health Serv., Cincinnati, Ohio.

17. KUSTER, E. & S.T. WILLIAMS. 1964. Selection of media for isolation of *Streptomyces*. *Nature* 202:928.

18. LECHEVALIER, H.A. 1975. Actinomycetes of sewage-treatment plants. Environ. Protection Technol. Ser., EPA-600/2-75-031, U.S. Environmental Protection Agency, Cincinnati, Ohio.

19. LECHEVALIER, M.P. & H.A. LECHEVALIER. 1974. *Nocardia amarae*, sp. nov., an actinomycete common in foaming activated sludge. *Int. J. Syst. Bacteriol.* 24:278.

20. LECHEVALIER, H.A. & M.P. LECHEVALIER. 1967. Biology of actinomycetes. *Annu. Rev. Microbiol.* 21:71.

water supply systems. *Appl. Environ. Microbiol.* 43:708.

9250 B.   Actinomycete Plate Count

1. General Discussion

A plating method using a double-layer agar technique has been adapted for determining actinomycete density. Because only the thin top layer of the medium is inoculated with sample, surface colonies predominate and identification and counting of colonies is facilitated.

2. Preparation and Dilution

Prepare and dilute samples as directed in Section 9215 or 9610. Dilutions up to 1:1000 (10^{-3}) usually are suitable for raw water, while treated waters may be examined directly. For soil samples, use dilutions from 1:1000 (10^{-3}) to 1:1 000 000 (10^{-6}).

3. Medium

Starch-casein agar:

Soluble starch	10.0	g
Casein	0.3	g
Potassium nitrate, KNO_3	2.0	g
Sodium chloride, NaCl	2.0	g
Dipotassium hydrogen phosphate, K_2HPO_4	2.0	g
Magnesium sulfate, hydrate, $MgSO_4 \cdot 7H_2O$	0.05	g
Calcium carbonate, $CaCO_3$	0.02	g
Ferrous sulfate, hydrate, $FeSO_4 \cdot 7H_2O$	0.01	g
Agar	15.0	g
Distilled water	1	L

No pH adjustment is required. Medium is used to prepare double-layer plates. Store medium for bottom layer in bulk or in tubes in 15-mL amounts. Store medium for surface layer in tubes in 17.0-mL amounts.

4. Procedure

a. Plating: Prepare three plates for each dilution to be examined. Aseptically transfer 15 mL of sterile starch-casein agar to a petri dish and let agar solidify, thus forming the bottom layer. To a test tube containing 17.0 mL liquefied starch-casein agar at 45 to 48°C, add 2 mL of appropriately diluted sample and 1 mL of the antifungal antibiotic, cycloheximide,* prepared in distilled water (1 mg/mL) and sterilized by autoclaving for 15 min at 121°C. Pipet 5 mL of inoculated agar over the hardened bottom layer with gentle swirling to obtain even distribution of the surface layer.

b. Incubation: Invert and incubate at 28°C until no new colonies appear. Usually this requires 6 to 7 d.

c. Counting: Plates suitable for counting contain 30 to 300 colonies. Identify actinomycetes by gross colony appearance. If necessary, verify by microscopic examination at a magnification of 50 to 100×,

---

*Actidione®, Upjohn and Company, Kalamazoo, Mich., or equivalent.

as shown in Figure 9250:1. Actinomycete colonies, because of filamentous growth, typically have a fuzzy colonial border. Table 9250:I lists the distinguishing characteristics commonly used to differentiate actinomycete from other bacterial colonies. Cycloheximide generally suppresses fungal growth; however, fungal colonies, if present, can be recognized by their wooly appearance. Microscopically, fungi reveal a considerably larger cell diameter than actinomycetes.

5. Calculation

Report actinomycetes per milliliter of water or gram (dry weight) of soil. If three plates are used per sample, the average number of colonies on all plates (total number of colonies/3), times 2, times the reciprocal of the dilution (10/1, 100/1, 1000/1, etc.) equals the actinomycete colony count per milliliter of original sample. For solid or semisolid samples, correct for water content and report actinomycete colonies per gram, dry weight, of sample.

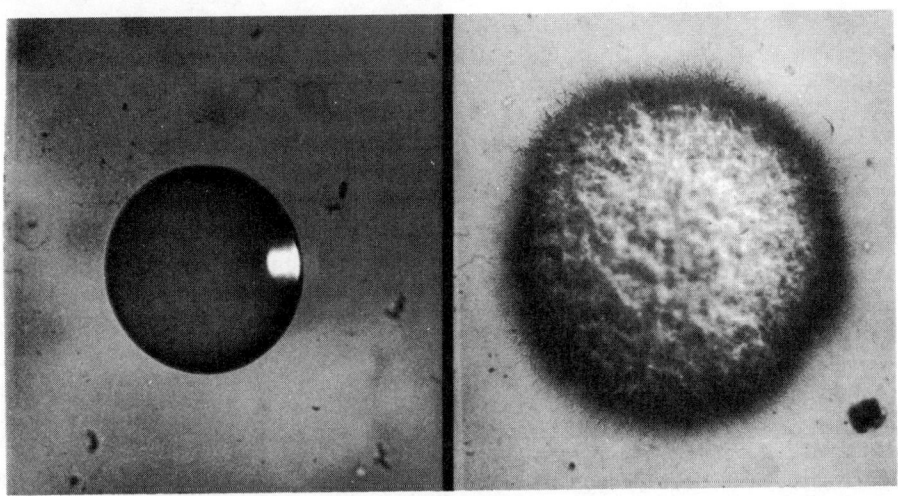

Figure 9250:1. Bacterial colonies—typical colony type vs. actinomycete colony type, 50×. Left: A typical bacterial colony characterized by a smooth mucoid appearance and a relatively distinct smooth border. Right: An actinomycete colony characterized by the mass of branching filaments that result in the fuzzy appearance of its border and by the dull powdery appearance of the spore-laden, aerial hyphae.

TABLE 9250:I. GENERAL MACROSCOPIC PROPERTIES OF BACTERIAL COLONIES ON SOLID MEDIUM

Characteristic	Typical Colony Type	Actinomycete Colony Type*
Appearance	Shiny or opalescent	When young it is composed of hyphae, but in some species these may later fragment. Substrate and surface hyphae have no distinctive color. As the colony matures, fluffy aerial hyphae that carry spores form and give to colonies of different species various colors and sometimes a chalky appearance. Soluble pigments, either melanin or brightly colored type, that diffuse into the medium, also are common.
Texture	Soft	Strong and leathery
Degree of adherence to solid medium	Weak	Strong
Edge of colony	Regular, continuous, and not different from colony as a whole	Irregular, intermittent, slightly less dense than colony as a whole, and of hyphal appearance.

* Actinomycetes are authentic bacteria by all modern criteria, except for their hyphal character and mode of spore formation.

9260   DETECTION OF PATHOGENIC BACTERIA*

9260 A.   Introduction

1. General Discussion

A wide variety of enteric pathogenic bacteria may occur in water supplies and in wastewater. With increasing demands on water resources, the potential for contamination of surface and groundwater by enteric microorganisms could be expected to increase. Waterborne outbreaks of disease continue to occur. Pathogenic bacteria that have been transmitted by water or wastewater include *Salmonella, Shigella, Campylobacter,* enteropathogenic *Escherichia coli, Vibrio cholerae, Leptospira,* and *Yersinia.* While not enteric, a recently recognized bacterial family, the Legionellaceae, is widely distributed in the aquatic environment and pneumonia outbreaks associated with tap water and an aerosol transmission route have been reported.

Routine examination of water and wastewater for pathogenic bacteria is not recommended. No single procedure is available that can be used to isolate and identify all pathogens. Therefore, negative findings for specific pathogens are provisional because state-of-the-art methodology may not be sufficiently sensitive to detect low levels of pathogens. For example, salmonellae are extremely common in the environment but isolation techniques even for these ubiquitous organisms involve relatively complicated procedures that ex-

---

* Approved by Standard Methods Committee, 1988.

ceed the capabilities of many water laboratories. Routine examination of water and wastewater for pathogens is limited by factors such as lack of facilities, untrained personnnel, insufficient laboratory time, high costs, and inadequate methods. In view of the foregoing, it is apparent that there is a strong need for intensive research in this area and that such research should be encouraged at every opportunity.

The coliform test may not always be a suitable indicator of the microbiological safety of water.[1-4] Under some circumstances pathogenic microorganisms can be isolated from waters containing few if any coliform bacteria. The Legionellaceae, because they are not enteric, would not be expected to demonstrate a relationship to fecal bacteria. Nevertheless, the coliform test has been, and continues to be, a useful tool for assessing the quality of water. The pathogen isolation procedures that follow are offered *as a starting point* for the investigator who may wish to initiate a research study, for example, to obtain background data on the numbers, types, and frequency of occurrence of pathogens in water. Because methods for many of the pathogens included in Sections 9260, 9711 (pathogenic protozoa), and 9510 (enteric viruses) may be changing rapidly, refer to current literature or consult with investigators who are working with the pathogen of interest.

2. References

1. AHMED, J., I.A. POSNI & M.A. SIDIQUI. 1964. Bacteriological examination of drinking water of Karachi and isolation of enteric pathogens. *Pakistan J. Sci. Ind. Res.* 7:103.
2. SELIGMANN, R. & R. REITLER. 1965. Enteropathogens in water with low *Esch. coli* titers. *J. Amer. Water Works Assoc.* 57:1572.
3. GREENBERG, A.E. & H.J. ONGERTH. 1966. Salmonellosis in Riverside, California. *J. Amer. Water Works Assoc.* 58:1145.
4. HENDRICKS, C.W. 1978. Exceptions to the coliform and fecal coliform test. *In* G. Berg, ed. Indicators of Viruses in Water and Food. Ann Arbor Science, Ann Arbor, Mich.

3. Bibliography

KABLER, P. 1959. Removal of pathogenic microorganisms by sewage treatment processes. *Sewage Ind. Wastes* 31:1373.

9260 B. General Qualitative Isolation and Identification Procedures for *Salmonella*

The methods presented below for the isolation of *Salmonella* from water or wastewater are not standardized. The procedures may need modification to fit a particular set of circumstances and methods comparisons are encouraged.

Rather than a specific protocol for *Salmonella* detection in water, a brief summary of methods suitable for recovery of these organisms is given. Methods currently available have been used in numerous field investigations to demonstrate *Salmonella* in both fresh and marine water environments. The occurrence of *Salmonella* in water is highly variable; there are limitations and variations in both the sensitivity and selectivity of accepted *Salmonella* isolation procedures for the detection of the more than 1700 *Salmonella* serotypes currently recognized. Thus, a negative result by any of these methods does not imply the absence of salmonellae, nor does it imply the absence of other pathogens.

1. Concentration Techniques

Generally, it is necessary to examine a relatively large sample to isolate pathogenic organisms. These organisms usually are present in small numbers compared to coliforms, because their sporadic occurrence is related to the incidence of disease or infection at a given period.

a. *Swab technique:* Prepare swabs from cheesecloth 23 cm wide, folded five times at 36-cm lengths and cut lengthwise to within 10 cm from the head into strips approximately 4.5 cm wide. Securely wrap the uncut or folded end of each swab with 16-gauge wire for use in suspending the swab in water. Place the swabs in kraft-type bags and sterilize at 121°C for 15 min. Place swab just below the surface of the sampling location for 1 to 3 d. (Longer swab exposure will not increase entrapment of pathogens.) Gauze pads of similar thickness—for example, maternity pads—may be substituted. During sampling, particulate matter and microorganisms are concentrated from the water passing through or over the swab. After exposure, retrieve the swab, place it in a sterile plastic bag, ice, and send to the laboratory. Maximum storage-transit time allowable is 6 h. Do not transport swabs in enrichment media; ambient transport temperature may cause sufficient proliferation of competitive organisms to mask salmonellae. In the laboratory, place pad or portions of it in enrichment media. When flasks of enrichment medium containing iced swabs are to be incubated at 40 to 41°C, place flasks in a 44.5°C water bath for 5 min before incubation in an air incubator.

b. *Diatomaceous earth technique:* Place an absorbent pad (not a membrane filter) on a membrane filter funnel receptacle, assemble funnel, and add 2.5 g sterile diatomaceous earth* to pack the funnel neck loosely. Apply vacuum and filter 2 L of sample. After filtration, disassemble funnel, divide resulting "plug" of diatomaceous earth and absorbent pad in half aseptically with a knife-edged, sterile spatula, and add

to suitable enrichment media. Alternatively, place entire plug in enrichment medium.

c. *Large-volume sampler:* Use a filter composed of borosilicate glass microfibers bonded with epoxy resin to examine several liters or more of sample, provided that sample turbidity does not limit filtration.[1] The filter apparatus consists of a 2.5- × 6.4-cm cartridge filter and a filter holder.† Sterilize by autoclaving at 121°C for 15 min. Place sterile filter apparatus (connected in series with tubing to a 20-L water bottle reservoir and vacuum pump) in the 20-L sample container appropriately calibrated to measure volume of sample filtered. Apply vacuum and filter an appropriate volume. When filtration is complete, remove filter and place in a selective enrichment medium.

d. *Membrane filter technique:* To examine low-turbidity water, filter several liters through a sterile 142-mm-diam membrane of 0.45-μm pore size.[2] For turbid waters, precoat the filter: make 1 L of sterile diatomaceous earth suspension (5 g/L distilled water) and filter about 500 mL. Without interrupting filtration, quickly add sample (1 L or more) to remaining suspension and filter. After filtration place membrane in a sterile blender jar containing 100 mL sterile 0.1% peptone water and homogenize at high speed for 1 min. Add entire homogenate to 100 mL double-strength selective enrichment medium. Alternatively, use multiple 47-mm-diam membrane filters to filter the sample. Immerse each membrane aseptically in 50 mL single-strength selective enrichment medium and incubate.

Qualitative detection of *Salmonella* in suspect potable water also may be achieved successfully by further analysis of selected M-Endo MF cultures (from 100 mL sample volume) that contain significant background growth and total coliforms.[3] After

---

* Celite, Mansville Corp., Denver, Colo., or equivalent.

† Balston Type AA filter with Type 90 holder, or equivalent.

completing routine coliform count, place entire filter with mixed growth into 10 mL tetrathionate broth (containing 1:50 000 brilliant green dye) for *Salmonella* enrichment before differential colony isolation on brilliant green agar. This unique approach requires no special large sample collections and can be an extension of the routine total coliform analysis.

2. Enrichment

Selectively enrich the concentrated sample in a growth medium that suppresses growth of coliform bacteria. Sample enrichment is essential, because the pathogens usually are present in low numbers and solid selective media for colony isolation are somewhat toxic, even to pathogens. No single enrichment medium can be recommended that allows optimum growth of all *Salmonella* serotypes. Use two or more selective enrichment media in parallel for optimum detection.

a. Dulcitol selenite broth inhibits many nonpathogenic enterobacteria during the early hours of incubation following inoculation and allows *Salmonella* strains to multiply rather rapidly. Optimum incubation time for maximum recovery of *Salmonella* is 24 h. However, recovery of relatively slow-growing organisms such as variants of *S. enteritidis* (Montevideo, Enteritidis, and Worthington) require longer incubation periods. Make repeated streakings from the same inoculated medium after each 24-h period. Streak from broth cultures that develop turbidity and any orange-red color due to selenite reduction onto suitable selective solid media.

b. Tetrathionate broth may yield more salmonellae than selenite broth. However, extend incubation beyond 24 h, with repeat streaking from the same tube several times during the first day and daily up to 5 d to increase potential recovery of all serotypes that may be present. Transfer 1 mL tetrathionate broth culture to a fresh tube of the same medium for continued incubation to enrich further for *Salmonella* growth and enhance recovery on streak plates. Improve suppression of nonpathogenic organisms by adding 1:50 000 brilliant green dye and, in highly turbid marine waters, add sodium lauryl sulfate. Improve sensitivity by adding 3 mg *l*-cystine/L tetrathionate broth.

c. Selenite cystine broth may yield *Salmonella* other than those isolated by other enrichment broths with both freshwater and marine samples.

3. Selective Growth

Further separation of pathogens from the remaining nonpathogenic bacterial population is facilitated by proper choice of incubation temperature for primary enrichment followed by secondary differentiation on selective solid media. These two factors, incubation temperature and choice of media, are interrelated. More *Salmonella* may be recovered at either 35 or 37°C with bismuth sulfite agar than at higher temperatures. Great skill at screening for these pathogens is necessary because of the competing growth of various nonpathogens. Use of an incubation temperature of 41.5°C for both primary enrichment broth and differential media enhances the isolation of many salmonellae while reducing the number of interfering organisms. Some *Salmonella* serotypes, among them *S. typhi,* will not grow at this elevated temperature.

Solid media commonly used for enteric pathogen detection may be classed into three groups: (*a*) differential media with little or no inhibition toward nonpathogenic bacteria, such as EMB (containing sucrose); (*b*) selective media containing bile salts or sodium desoxycholate as inhibitors, such as MacConkey's agar, desoxycholate agar, or xylose lysine desoxycholate (XLD) agar; and (*c*) selective media containing brilliant green dye, such as brilliant green agar or bismuth sulfite agar. Any medium selected must provide optimum suppression of coliforms while permitting good recovery of the pathogenic group. Streaking duplicate plates, one heavily and one lightly, often aids in recognition of enteric pathogens in the presence of large numbers of interfering organisms.

a. *Brilliant green agar:* Typical well-isolated *Salmonella* colonies grown on this medium are pinkish white with a red background. *S. typhi* and a few other species of *Salmonella* grow poorly because of the brilliant green dye content. Lactose-fermenters not subject to growth suppression will form greenish colonies or may produce other colorations. Occasionally, slow lactose-fermenters (*Proteus, Citrobacter,* and *Pseudomonas*) will produce colonies resembling those of a pathogen. Suppress spreading effect of pseudomonads by increasing agar concentration to 2%. In some instances, *Proteus* has been observed to "swarm"; reduce this tendency by using agar plates dried to remove surface moisture. If suspect *Salmonella* colonies are not observed after 24 h incubation, reincubate for an additional 24 h to permit slow-growing or partially inhibited organisms to develop visible colonies. If typical colonies are not observed or if the streak plate is crowded, isolate in pure culture a few colonies for biochemical characterization tests. Nonlactose-fermenting colonies in close proximity to lactose-fermenting colonies may be masked.

b. *Bismuth sulfite agar (Wilson and Blair medium[4]):* Luxuriant growth of many *Salmonella* species (including *S. typhi*) can be expected on this medium. Examine bismuth sulfite plates after 24 h incubation for suspect colonies; reincubate for 24 h to detect slow-growing strains. Typical colonies usually develop a black color, with or without a metallic sheen, and frequently this blackening extends beyond the colony to give a "halo" effect. A few species of *Salmonella* develop a green coloration; therefore, isolate some of these colony

types when typical colonies are absent. As with brilliant green agar, typical colony coloration may be masked by numerous bordering colonies after 48 h incubation. A black color also is developed by other H_2S-producing colonies, for example, *Proteus* and certain coliforms.

c. *Xylose lysine desoxycholate agar:* Compared to brilliant green dye, sodium desoxycholate is only slightly toxic to fastidious *Salmonella*. *Salmonella* and *Arizona* organisms produce black-centered red colonies. Coliform bacteria, *Proteus,* and many *Enterobacter* produce yellow colonies. Optimum incubation time is 24 h. If plates are incubated longer, an alkaline reversion and subsequent blackening occurs with H_2S-positive nonpathogens (*Citrobacter, Proteus vulgaris,* and *P. mirabilis*).

d. *Xylose lysine brilliant green agar:* This medium is especially good for *Salmonella* from marine samples. The brilliant green inhibits many *Proteus, Enterobacter,* and *Citrobacter* species.

4. Biochemical Reactions

Many enteric organisms of little or no pathogenicity share certain major biochemical characteristics with *Salmonella*. The identification of pathogens by colony characteristics on selective solid media has limitations inherent in the biological variations of certain organisms and cannot be relied on for even tentative identification. Suspected colonies grown on selective solid media must be purified and further characterized by biochemical reactions; final verification is based on serological identification. Usually a large number of cultures will be obtained from the screening procedure.

Commercially available differential media kits (see Section 9225) may be used as an alternative to Phases 1, 2, and 3 described below, before serological confirmation. These kits give 95 to 98% agreement with conventional tests, although more significant tests will be necessary to achieve further differentiation among strains of *Enterobacteriaceae*.

When such kits are not used, follow a sequential pattern of biochemical testing that will result in a greater saving of media and time for laboratory personnel.[5]

Phase 1—Preliminary screening, phenylalanine deaminase activity: Discard phenylalanine deaminase-positive cultures immediately as indicative of the *Proteus* group. In this test, spot isolates on phenylalanine agar and incubate for 24 h at either 35 or 37°C. Phenylalanine deaminase activity is indicated by a green zone that develops around the colony after flooding the plate with a 0.5 M FeCl$_3$ solution. Subject phenylalanine deaminase-negative cultures to the biochemical tests of Phase 2.

Simultaneously, test for inability to ferment lactose on a selective agar, such as MacConkey's agar.

Phase 2—Biochemical tests: The tests used are:

Medium	Purpose of Test
TSI	Fermentation pattern: H$_2$S production; *o*-nitrophenyl β-D-galactoside (ONPG) for β-D-galactosidase
SIM	Production of indole and H$_2$S, motility

Conformance to the typical biochemical patterns of the *Salmonella* determines whether to process cultures further (Phase 3). Aberrant cultures may be encountered that do not conform to all the classical reactions attributed to each pathogenic group. In all cases, therefore, review reactions as a whole and do not discard cultures on the basis of a small number of apparent anomalies.

Phase 3—Fermentation reactions: Test fermentation reactions in dextrose, mannitol, maltose, dulcitol, xylose, rhamnose, and inositol broths to characterize further the biochemical capabilities of the isolates. This additional sorting reduces the possible number of positive cultures to be processed for serological confirmation. If the testing laboratory is equipped for serological identification, this series of biochemical tests may be eliminated.

5. Genus Identification by Serological Techniques

Upon completion of the recommended biochemical tests, inoculate the suspected *Salmonella* pure culture onto a brain-heart infusion agar slant and incubate for 18 to 24 h at 35°C. With wax pencil (china marker) divide an alcohol-cleaned glass slide into four sections. Prepare a dense suspension of test organism by suspending growth from an 18- to 24-h agar slant in 0.5 mL 0.85% NaCl solution. Place a drop of *Salmonella* "O" polyvalent antiserum in the first section and antiserum plus 0.85% NaCl in the second section. Using a clean inoculating loop, transfer a loopful of bacterial suspension to the third section containing 0.85% NaCl solution and to the fourth section containing 0.85% NaCl solution plus antiserum. Gently rock slide back and forth. If agglutination is not apparent in the fourth section at the end of 1 min, the test is negative. All other sections should remain clear.

If the biochemical reactions are characteristic of *S. typhi* and the culture reacts with "O" polyvalent antiserum, check other colonies from the same plate for Vi antigen reaction. If there is no agglutination with *Salmonella* Vi antiserum, the culture is not *S. typhi*. Identification of *Salmonella* serotypes requires determination of H antigens and phase of the organism as described by Edwards and Ewing.[5] Isolates yielding biochemical reactions consistent for *Salmonella* and positive with polyvalent "O" antiserum may be identified as "*Salmonella* sp., serotype or bioserotype undetermined."

6. References

1. LEVIN, M.A., J.R. FISCHER & V.J. CABELLI. 1974. Quantitative large-volume sampling technique. *Appl. Microbiol.* 28:515.
2. PRESNELL, M.W. & W.H. ANDREWS. 1976. Use of the membrane filter and a filter aid for concentrating and enumerating indicator bacteria and *Salmonella* from estuarine waters. *Water Res.* 10:549.
3. CANLAS, L. 1975. Personal communication. Guam Environmental Protection Agency, Agana, Guam.
4. WILSON, W.J. & E.M. McV. BLAIR. 1926. Combination of bismuth and sodium sulfite affording enrichment and selective medium for typhoid and paratyphoid groups of bacteria. *J. Pathol. Bacteriol.* 29:310.
5. EDWARDS, P.R. & W.H. EWING. 1972. Identification of Enterobacteriaeceae, 3rd ed. Burgess Publ. Co., Minneapolis, Minn.

7. Bibliography

LEIFSON, E. 1935. New culture media based on sodium desoxycholate for the isolation of intestinal pathogens and for enumeration of colon bacilli in milk and water. *J. Pathol. Bacteriol.* 40:581.
MÜLLER, G. 1947. Der Nachweis von Keimer der Typhus-Paratyphusgruppe im Wasser. H.H. Nolke Verlag, Hamburg, Germany.
GREENBERG, A.E., R.W. WICKENDEN & T.W. LEE. 1957. Tracing typhoid carriers by means of sewage. *Sewage Ind. Wastes* 29:1237.
McCOY, J.H. 1964. *Salmonella* in crude sewage, sewage effluent, and sewage polluted natural waters. *In* Int. Conf. Water Pollut. Res., 1st, London, 1962. Vol. 1:205. Macmillan, New York, N.Y.
BREZENSKI, F.T., R. RUSSOMANNO & P. DE-FALCO, JR. 1965. The occurrence of *Salmonella* and *Shigella* in post-chlorinated and nonchlorinated sewage effluents and receiving waters. *Health Lab. Sci.* 2:40.
RAJ, H. 1966. Enrichment medium for selection of *Salmonella* from fish homogenate. *Appl. Microbiol.* 14:12.
SPINO, D.E. 1966. Elevated temperature technique for the isolation of *Salmonella* from streams. *Appl. Microbiol.* 14:591.
GALTON, M.M., G.K. MORRIS & W.T. MARTIN. 1968. *Salmonella* in foods and feeds. Review of isolation methods and recommended procedures. Public Health Serv. Bur. Disease Prevention & Environmental Control, National Center for Disease Control, Atlanta, Ga.
BREZENSKI, F.T. & R. RUSSOMANNO. 1969. The detection and use of *Salmonella* in studying polluted tidal estuaries. *J. Water Pollut. Control Fed.* 41:725.
CHERRY, W.B., J.B. HANKS, B.M. THOMASON, A.M. MURLIN, J.W. BIDDLE & J.M. GROOM. 1972. Salmonellae as an index of pollution of surface waters. *Appl. Microbiol.* 24:334.

9260 C.   Immunofluorescence Identification Procedure for *Salmonella*

The direct fluorescent antibody (FA) technique is a rapid and effective means of detecting salmonellae in fresh- and sea-water samples. It may be used as a screening technique to provide rapid results for large numbers of samples, such as those from recreational or shellfish-harvesting waters. Sample volumes used depend on the degree of contamination. Where gross pollution is present, use smaller samples. When background information is absent, analyze a 2-L sample, using the diatomaceous earth concentration technique.

1. Apparatus for Fluorescence Microscopy

Standard fluorescent antibody microscopy equipment may be obtained separately or in a package containing the essential instrumentation (*a-f*):

a. Light microscope with microscope stand.

b. Light source, providing energy in the short-wavelength region of the spectrum. A high-pressure mercury 200-W arc enclosed in a quartz envelope, a 75- to 150-

W xenon high-pressure lamp, or a low-voltage 100-W quartz halogen lamp may satisfy this requirement. A significant portion of the energy should be emitted in the ultraviolet and blue region of the spectrum.

c. Power pack to provide constant voltage and wattage output for the selected lamp.

d. Basic filters including heat-absorbing filter (KG-1 or KG-2, or equivalent): red-absorbing filter (BG-38 or equivalent); exciter filter (BG-12 or equivalent, BG-12 being also a blue filter); and barrier filter (OG-1 or blue-absorbing filter). New interference excitation filters having very high transmission in the blue portion of the spectrum (490 nm) are available. Barrier filters used with these have a sharp cutoff at 500 to 510 nm.

e. Optics: The fluorescence microscope must have high-quality optics. A 100× objective with an iris diaphragm to reduce the numerical aperture (N.A.) for dark-field work is essential. Because the N.A. is similar for all 100× objectives (1.25 to 1.30), base selection on desire for a flat-field (plano) lens.

f. Cardioid dark-field condenser for illuminating specimen. A 95× oil immersion objective with built-in iris diaphragm is desirable. True dark-field illumination can be achieved only if the objective N.A. is smaller than the condenser N.A., i.e., of the illuminating cone of light. (Difference in N.A. between objective and condenser should be at least 0.05). Reduce N.A. of an oil immersion objective by using the built-in diaphragm or by putting a funnel stop onto the objective.

g. FA pre-cleaned micro slides, 7.6- × 2.5-cm, 0.8- to 1.0-mm thickness.

h. Cover glass for FA slides, No. 1½, 0.16- to 0.19-mm thickness.

i. Staining assembly consisting of dish, cover, and slide rack with handle. Five dishes are required; for Kirkpatrick's fixative, 95% ethanol, first PBS rinse, second PBS rinse, and reagent water.

j. Moist chamber used to incubate slides containing smears with added conjugate. A simple chamber consists of water-saturated toweling with a culture dish bottom (150 by 20 mm) placed over the wet toweling.

2. Reagents

a. Nondrying immersion oil, Type A (low fluorescence).*

b. FA Kirkpatrick fixative, consisting of 60 mL absolute ethanol, 30 mL chloroform, and 10 mL formaldehyde.†

c. Phosphate-buffered saline (PBS): Add 10 g buffer‡ to 1000 mL freshly prepared distilled water. Stir until the powder dissolves completely. Adjust with 0.2 N NaOH to pH 8.0.

d. FA mounting fluid: Use standardized reagent-grade glycerine adjusted to pH 9.0 with 0.2 N NaOH and intended for mounting slides to be viewed with the FA microscope.

e. Reagent (laboratory pure) water: Use double-distilled water from an all-glass still or other high-quality analytical-grade laboratory water.

f. FA Salmonella panvalent conjugate is a fluorescein-conjugated anti-*Salmonella* globulin.§ To rehydrate, add 5 mL reagent water to a vial of conjugate. Determine working dilution (see ¶ 5e). Store unused rehydrated conjugate in a freezer, preferably at −60°C. Avoid repeated freezing and thawing.

3. Concentration Technique

Place an absorbent pad on a membrane filter funnel and add sufficient sterile diatomaceous earth‖ to pack funnel neck loosely. Filter 2 L of sample. Rinse funnel with 50 to 100 mL sterile phosphate-buffered dilution water or 0.1% peptone water.

---

* R. P. Cargille Laboratories, Inc. Cedar Grove, N.J., or equivalent.
† Difco No. 3188 or equivalent.
‡ Difco Bacto-FA Buffer, dried, or equivalent.
§ Difco or equivalent.
‖ Celite, Mansville Corp., Denver, Colo., or equivalent.

Disassemble funnel and remove resulting "plug" of diatomaceous earth and the absorbent pad. Repeat with a second 2-L sample.

4. Enrichment

Immerse one plug and absorbent pad in a flask containing 300 mL selenite cystine broth. Immerse second plug and absorbent pad in a flask containing 300 mL tetrathionate broth supplemented with 3 mL 1:1000 aqueous solution of brilliant green dye and 3 mg *l*-cystine. Incubate at either 35 or 37°C for 24 h.

5. Fluorescent Antibody Reaction and Analysis

a. Prepare spot plates of brilliant green agar (BGA) and xylose lysine brilliant green (XLBG) agar by placing 1 drop (about 0.01 mL, delivered with a wire loop) of the enrichment medium (selenite cystine or tetrathionate broth) at each of four separate points on the agar surface.[1] Space drops on agar plate so that FA microscope slide will cover two inoculation points. This is essential because glass slide impression smears of the inoculated points will be made after incubation of plates.

b. Incubate BGA and XLBG plates at either 35 or 37°C for 3 h. After incubation make impression smears by taking a *clean* FA microscope glass slide and placing it over two inoculated points on the medium. Press down lightly, being careful not to move glass slide horizontally. Do not apply too much pressure, because it will cause movement of the slide and collection of additional agar. Repeat this process for the other two inoculation points and for inoculation points on second agar medium. Prepare a total of four FA slides in this manner.

c. Air-dry smears and fix for 2 min in Kirkpatrick's fixative. Rinse slides briefly in 95% ethanol and let air dry. *Do not blot.*

d. Cover fixed smears with 1 drop of

Salmonella panvalent conjugate. Before use, dilute commercial conjugate and determine appropriate working dilution. Most batches are effective at a 1:4 dilution but this will vary with the type of fluorescence equipment used, light source, alignment, magnification, cultures, etc. Determine working dilution (titer) of each lot of conjugate.

e. To determine conjugate titer use a known 18- to 24-h *Salmonella* culture grown in veal infusion broth and make smears on FA glass slide. Dilute conjugate and treat as outlined in *c* and *d* above. For example, if the following results are obtained:

Dilution of Conjugate	Fluorescence
1:2	4+
1:4	4+
1:6	4+
1:8	2+
1:10	1+

use the second highest dilution giving 4+ fluorescence. In the above example use a 1:4 dilution of conjugate. Diluting conjugate insures minimum cross-reactivity. Prepare fresh diluted conjugate daily.

f. After covering each smear with 1 drop of appropriate dilution of conjugate, place slides in a moist chamber to prevent evaporation of staining reagent. After 30 min wash away excess reagent by dipping slides into phosphate-buffered saline (pH 8.0). Place slides in second bath of buffered saline for 10 min. Remove, rinse in distilled water, and drain dry. *Do not blot.*

g. Place a small drop of mounting fluid on the smear and cover with a No. 1½ cover slip. Seal edges of cover slip with clear fingernail polish. Examine sealed slides within a few hours while fluorescence is of optimum intensity. Examine under a fluorescence microscope unit fitted with appropriate filters.

h. Include a positive control slide with

each set of samples. This checks conjugate reactivity and FA equipment generally.

6. Recording and Interpreting Results

The intensity of organisms fluorescing in any given field is important in assessing positive *Salmonella* smears. If the majority of cells present fluoresce (4+ or 3+) the smear is positive. Carefully scrutinize smears showing only a few scattered fluorescing cells. Critical examination of cellular morphology may distinguish between these cells and salmonellae. The degree of fluorescence is the criterion on which positivity is based. Consider weakly fluorescing cells (2+ and 1+) negative. Confirm all positive FA results by cultural techniques (see Section 9260B).

Reaction	Description	Fluorescence Intensity
Positive	Brilliant yellow-green fluorescence, cells sharply outlined.	4+
Positive	Bright yellow-green fluorescence, cells sharply outlined with dark center.	3+
Negative	Dull yellow-green fluorescence, cells not sharply outlined.	2+
Negative	Faint green fluorescence discernible in dense areas, cells not outlined.	1+
Negative	No fluorescence.	0

7. Quantitative Immunofluorescence Microspectrofluorometric Microscopy

To make such analyses use a system consisting of analyzing and illumination sections. The analyzing section includes an eyepiece monochromator assembly and a photomultiplier-photometer. The eyepiece uses a beam splitter that reflects to the monochromator and the observer's eye allowing for simultaneous visual observation and quantitative analysis of the yellow-green fluorescence intensity. The photometer package provides meter readout in milliamperes so that visual observation of fluorescence can be correlated with objective reading. Microspectrofluorometry can be done with a conventional fluorescence microscope.

8. Reference

1. KATZ, I.J. & F.T. BREZINSKI. 1973. Detection of *Salmonella* by fluorescent antibody. U.S. Environmental Protection Agency, Edison, N.J.

9. Bibliography

SCHULTE, S.J., J.S. WITZEMAN & W.M. HALL. 1968. Immunofluorescent screening for *Salmonella* in foods: comparison with culture methods. *J. Amer. Org. Agr. Chem.* 51:1334.

THOMASON, B.M. & J.G. WALLS. 1971. Preparation and testing of polyvalent conjugates for F.A. detection of Salmonellae. *Appl. Microbiol.* 22:876.

THOMASON, B.M. 1971. Rapid detection of *Salmonella* microcolonies by fluorescent antibody. *Appl. Microbiol.* 22:1064.

9260 D.   Quantitative *Salmonella* Procedures

1. Multiple-Tube Enrichment Technique

Because of the high ratio of coliform bacteria to pathogens, large samples (1 L or more) are required. Any concentration method in Section 9260B.1 may be used but preferably concentrate the sample by the membrane filter technique (Section 9260B.1*d*). After blending the membrane with 100 mL sterile 0.1% peptone water, use a quantitative MPN procedure by proportioning homogenate into a five-tube, three-dilution multiple-tube procedure using either dulcitol selenite or tetrathionate broth as the selective enrichment medium. Incubate for 24 h at 35°C and streak from each tube to plates of brilliant green and xylose lysine desoxycholate agars. Incubate for 24 h at 35°C. Select from each plate at least one colony suspected of being *Salmonella,* inoculate a slant each of triple sugar iron (TSI) and lysine iron agars, and incubate for 24 h at 35°C. Test cultures giving a positive reaction for *Salmonella* by serological techniques (see Section 9260B.5). From the combination of *Salmonella* negative and positive tubes, calculate the MPN/10 L of original sample (see Section 9221E).

2. Membrane Filter Procedure for *S. typhi*

A quantitative procedure for *S. typhi* is available. The method utilizes M-bismuth sulfite broth and the membrane filter procedure for bacterial concentration. It is applicable only with samples low in organic and particulate materials, because quantities of 100 mL or more generally are filtered. After filtration (see Section 9222B.5), incubate filter on a pad saturated with M-bismuth sulfite broth for 18 to 20 h at 35° C. Transfer to a fresh pad saturated with M-bismuth sulfite broth and continue incubation to give a total of 30 h. Transfer suspect colonies (smooth, glistening colonies with jet-black centers surrounded by a thin clear white border) to triple sugar iron (TSI) agar and incubate at 35° C for 18 h. Proceed with additional biochemical and serological procedures as described under qualitative methods (Section 9260B.3 and 4).

3. Bibliography

CLARK, H.F., E.E. GELDREICH, H.L. JETER & P.W. KABLER. 1951. The membrane filter in sanitary bacteriology—Culture of *Salmonella typhosa* from water samples on a membrane filter. *Pub. Health Rep.* 66:951.

9260 E.   *Shigella*

While most shigellosis epidemics are spread by contaminated food or by person-to-person contact, they also may be caused by contaminated drinking water. Outbreaks of waterborne shigellosis frequently result from accidental interruption of water treatment, flood-carried excreta contami-nation of well water supplies, cross-connections between contaminated water pipes and potable water supply lines, untreated water supplies, or wastewater seepage into water supply lines.

Shigellae have been found in various polluted waters, but methodology is qualitative and low in sensitivity. Instability of some biochemical characteristics can occur in *Shigella* strains introduced into the water environment. Coliform bacteria and most strains of *Proteus vulgaris* are antagonistic to *Shigella*.

1. Concentration Techniques

See Section 9260B.1. Use a sample of 1 to 10 L.

2. Enrichment

Choose a selective enrichment medium to minimize accumulation of volatile acid by-products derived from growth of coliforms and other antagonistic organisms. Use nutrient broth adjusted to pH 8.0 (a less favorable growth pH for coliforms) and incubate for 6 to 18 h at 35°C. Streak cultures at 6 and 18 h to selective differential agar plates to optimize *Shigella* recovery. Alternatively, use an autocytotoxic enrichment medium prepared by adding 4-chloro-2-cyclopentylphenyl β-D-galactopyranoside (CPPG) to a final concentration of 0.001M in lactose broth buffered to pH 6.5 with citrate buffer (at a final concentration of 0.05M). Solubilize in lactose broth by use of a magnetic stirrer and careful heating at 45°C. Sterilize lactose-CPPG broth by membrane filtration. Add sample concentrates to lactose-CPPG broth and incubate for 24 h at 35°C.

3. Selective Growth

Use xylose lysine desoxycholate (XLD) agar for primary isolation of *Shigella* strains. Suspected *Shigella* colonies are red. Incubate for 24 h at 35°C. The major interfering organisms on this differential agar include strains of *Proteus, Providencia,* and *Pseudomonas.*

4. Biochemical Reactions and Serological Identification

Follow procedures described in Section 9260B.4. For serological identification use the slide agglutination technique (see Section 9260B.5) with *Shigella* antisera (polyvalent and type-specific sera).

5. Bibliography

TAYLOR, W.I. & B. HARRIS. 1965. Isolation of Shigellae, II. Comparison of plating media and enrichment broths. *Amer. J. Clin. Pathol.* 44:476.
TAYLOR, W.I. & D. SCHELHART. 1968. Isolation of *Shigella*, V. Comparison of enrichment broths and stools. *Appl. Microbiol.* 16:1383.
TAYLOR, W.I. & D. SCHELHART. 1968. Isolation of *Shigella*, VI. Performance of media with stool specimens. *Appl. Microbiol.* 16:1387.
HENTGES, D.J. 1969. Inhibitions of *Shigella flexneri* by the normal intestinal flora. II. Mechanism of inhibition by coliform organisms. *J. Bacteriol.* 97:513.
PARK, C.E., M.K. RAYMAN, R. SZABO & Z.K. STANKSEWICZ. 1976. Selective enrichment of *Shigella* in the presence of *Escherichia coli* by use of 4-chloro-2-cyclopentylphenyl β-D-galactopyranoside. *Can. J. Microbiol.* 22:654.

9260 F.   Pathogenic *Escherichia coli*

Escherichia coli is a normal inhabitant of the digestive tract. However disease-causing *E. coli* have been isolated from tap water,[1] drinking water sources,[2] and mountain streams.[3] The distribution of these organisms is worldwide.[4] It is unlikely that *E. coli* could initiate disease by transmission through a properly treated potable water. Historically, at least in the United States, these organisms reportedly cause disease almost exclusively in infants, but because infants normally are given boiled or sterilized water, such waterborne infections probably are infrequent. Diarrhea of travelers may be caused by pathogenic *E. coli.*[5]

At present, three basic types of enteropathogy are associated with strains of *E. coli:* enteropathogenic (EPEC), enterotoxigenic (ETEC), and enterorinvasive (EIEC).[6] At least six mechanisms for expressing virulence have been reported, ranging from the production of cholera-like toxins to enteroadherence. The various pathogenic *E. coli* types may be associated with one or more of these mechanisms.

1. Examination Procedures

Routine examination of potable water supplies for pathogenic *E. coli* is impractical because of the difficulties inherent in determining the pathogenic nature of isolated *E. coli* strains. Currently, the most definitive means of determining the pathogenic nature of isolates are the immunological detection of the cholera-like (LT) toxin present in certain ETEC strains, the determination of cytopathic effects of whole bacteria cells or extracts on selected cell culture types, and animal bioassay using such methods as the rabbit ileal loop inoculation.[7] The pathogenic *E. coli* are commonly associated with certain serovars of which more than 60 have been identified. Unfortunately, these serological groupings are not limited to pathogenic strains; thus, use serotyping as a means of determining the pathogenicity of *E. coli* isolates with caution.

Isolate pathogenic *E. coli* from a potable water supply by use of the membrane filter technique (Section 9222B), preferably with M-FC broth.[8] Pick characteristic blue colonies, purify, and determine IMViC reactions (Section 9225E). Test isolates that are IMViC reaction $+ + - -$ and that produce gas from lactose by serological techniques[9] to select those that belong to serotypes common to pathogenic *E. coli*. There is, however, no biochemical marker that will separate pathogenic from nonpathogenic strains and the relationship between serotype and pathogenicity is questionable.[10] Serological, *in vivo* and *in vitro* tests will be required to determine if an isolate is pathogenic.

2. References

1. Ewing, W.H. 1962. Sources of *Escherichia coli* cultures that belong to O-antigen groups associated with infantile diarrheal disease. *J. Infect. Dis.* 110:114.

2. Seigneurin, R., R. Magnin & M.L. Achard. 1951. Types d'*Escherichia coli* isolés des eaux d'alimentation. *Ann. Inst. Pasteur* 89:473.

3. Petersen, N. & N.J. Boring. 1960. A study of coliform densities and *Escherichia coli* serotypes in two mountain streams. *Amer. J. Hyg.* 71:134.

4. Wachsmith, K. 1984. Laboratory detection of enterotoxins. *In* P. Ellner, ed. Infectious Diarrheal Diseases. Marcel Dekker, Inc., New York, N.Y.

5. Merson, M.H., G.K. Morris, D.A. Sack, J.G. Wells, J.C. Feeley, B. Sack, W.B. Creech, A.Z. Kapikian & E.J. Gangarosa. 1976. Travellers diarrhea in Mexico: A prospective study of physicians and family members attending a congress. *N. England J. Med.* 294:1299.

6. Baldini, M.M. 1983. Summary work shop on enteropathogenic *Escherichia coli. J. Infect. Dis.* 147:1108.

7. Spira, W.M., R.B. Sack & J.L. Froelich. 1983. Simple adult rabbit model for *Vibrio cholerae* and enterotoxigenic *Escherichia coli. Infect. Immun.* 32:739.

8. Glantz, P.J. & T.M. Jacks. 1968. An evaluation of the use of *Escherichia coli* serogroups as a means of tracing microbial pollution of water. *Water Resour. Res.* 4:625.

9. Edwards, P.R. & W.H. Ewing. 1972. Identification of Enterobacteriaeceae, 3rd ed. Burgess Publishing Co., Minneapolis, Minn.

10. Sack, R.B. 1975. Human diarrheal disease caused by enterotoxigenic *Escherichia coli. Ann. Rev. Microbiol.* 29:333.

9260 G. *Campylobacter jejuni*

Campylobacter jejuni has been isolated from inadequately treated or untreated river water[1] and mountain streams, and both endemic and epidemic waterborne disease has been reported.[2] *Campylobacter* is sensitive to chlorine and thus is unlikely to be transmitted through, or remain viable in, properly treated potable water. *C. coli* also is a human pathogen. It has biochemical characteristics nearly identical with those of *C. jejuni* and is considered closely related.

The occurrence of campylobacters in natural water is extremely variable and it is not yet known which of the organisms isolated are pathogenic to humans. Introduced organisms persist in cold water up to several weeks[3] but survival in surface waters at 25°C is for 2 d or less. Because *C. jejuni* is carried in the intestinal flora of a wide variety of wild and domestic mammals and birds, it is frequently isolated from waters with high levels of fecal coliforms. Isolation of *C. jejuni* or *C. coli* is difficult because of the competitive growth of other organisms. Other members of the genus occasionally may be found in surface waters but they can be distinguished from *C. jejuni* and *C. coli* by their growth and biochemical characteristics.

1. Concentration Techniques

Methods for isolating campylobacters are not standardized and must be considered research procedures subject to modification. Examine relatively large samples to isolate these organisms.

For low-turbidity water, filter several liters through a sterile 47-mm-diam membrane filter of 0.45-μm pore size. Place filter in a liquid selective enrichment medium or place right (face) side up on a selective plate medium. Incubate at 42°C overnight in a microaerobic atmosphere (approximately 5 to 10% oxygen and 3 to 10% CO_2),[4] remove filter, and incubate the plate for another 48 h at 42°C. Discard filter or continue the isolation attempt by placing filter in selected enrichment broth and incubating for 48 h at 42°C. For water likely to be heavily contaminated, prefilter through a 0.6-μm filter.

For turbid water pressure filtration is necessary. Use a stainless steel filtration device with a 1.5-L reservoir.* Assemble with the following filter sequence: Place a 142-mm, 3.0-μm filter on screen inside reservoir with a 124-mm prefilter on top of it. In the bottom tubing adapter place a 47-mm, 1.2-μm filter. Then place Swinnex filter holders in parallel with a 47-mm, 0.60-μm filter in the upstream filter holder and a 47-mm, 0.45-μm filter in the downstream holder. Add 1 L sample to the reservoir, seal, and apply pressure of about 350 kPa. After filtration, remove the 0.45-μm-pore-diam filter and culture on selective plate medium as indicated above.

2. Enrichment

Selectively enrich the concentrated sample in a growth medium that suppresses growth of coliform bacteria. Oosterom's selective medium is an antibiotic-containing thioglycollate medium enriched with 1.5% ox bile.[5] This medium, when incubated at 42°C in a microaerobic atmosphere, permits *C. jejuni* to multiply while inhibiting competing microflora. Alternatively, use Campy-thio medium[6] with 1.5% ox bile, which may be slightly more selective. After incubation for 48 h streak cultures on a selective plate medium.

3. Selective Growth

Use either Campy-BAP[6] or Butzler's medium[7] for primary isolation of *C. jejuni* or *C. coli*. Incubate for 48 h at 42°C. Distin-

---

* Millipore No. 316 or equivalent.

guish suspected *C. jejuni* or *C. coli* colonies by their spreading mucoid appearance, although they may appear also as regular, convex, entire colonies. The predominant interfering organisms are *Pseudomonas aeruginosa* and those Enterobacteriaceae strains resistant to the antibiotics.

4. Biochemical Reactions

Check suspect colonies for purity before making biochemical tests.

Phase 1—Preliminary screening: Perform spot catalase test. Make smear and Gram stain those colonies showing a positive reaction. If carbol fuchsin (0.8%) is used as the counterstain campylobacters appear as curved or "S" shaped gram-negative rods. In older cultures coccoid forms may be seen. Observe motility using a dark-field or phase-contrast microscope. Darting motility is characteristic. The oxidase test

usually is positive although it may be delayed. Organisms identified by these presumptive tests are almost always campylobacters, and usually are *C. jejuni* or *C. coli.*

Phase 2: The tests are minimal for determining whether suspect strains belong to the genus *Campylobacter.*

Test	Results for Campylo-bacters
Oxidative and fermentative glucose utilization	−
Nitrate reduction	+
Growth in 3.5% NaCl	−
H₂S production in TSI	−

To distinguish *C. jejuni* or *C. coli* from other *Campylobacter* species make the following tests:

Species	Catalase	Growth at 25°C	Growth at 42°C	Growth in Hippurate	Nalidixic Acid Sensitivity
C. jejuni	+	−	+	+	+
C. coli	+	−	+	−	+
C. fetus	+	+	±	−	−
C. sputorum	−	+	−	−	−

5. Serological Identification

Serological identification involves the indirect hemagglutination technique or the slide agglutination test. These specialized tests must be done in reference laboratories.

6. References

1. KNILL, M., W.G. SUCKLING & A.D. PEARSON. 1978. Environmental isolation of heat-tolerant *Campylobacter* in the Southampton area. *Lancet* 2:1002.

2. MENTZING, L.-O. 1981. Waterborne outbreaks of *Campylobacter* enteritis in Central Sweden. *Lancet* No. 8242 (Vol. II for 1981):352.

3. BLASER, M.J., H.L. HARDESTY, B. POWERS & W.-L.L. WANG. 1980. Survival of *Campylo-*

bacter fetus subsp *jejuni* in biological milieux. *J. Clin. Microbiol.* 11:309.

4. SMIBERT, R.M. 1978. The genus *Campylobacter. Ann. Rev. Microbiol.* 32:673.

5. OOSTEROM, J., M.J. VEREIJKEN & G.B. ENGELS. 1981. *Campylobacter* isolation. *Vet. Quart.* 3:104.

6. BLASER, M.J., I.D. BERKOWITZ, F.M. LAFORCE, J. CRAVENS, L.B. RELLER & W.-L.L. WANG. 1979. *Campylobacter* enteritis: clinical and epidemiologic features. *Ann. Intern. Med.* 91:179.

7. BUTZLER, J.P. & M.B. SKIRROW. 1979. *Campylobacter* enteritis. *Clin. Gastroenterol.* 8:737.

7. Bibliography

SMIBERT, R.M. 1978. The genus *Campylobacter. Annu. Rev. Microbiol.* 32:673.

9260 H. *Vibrio cholerae*

Vibrio cholerae is the etiological agent of cholera, a mild to severe form of diarrheal disease.[1] The O1 serogroup of *V. cholerae* commonly is associated with major epidemics in developing countries although sporadic outbreaks have been reported in the United States and elsewhere.[1,2] Non-O1 *V. cholerae* may cause invasive disease and mild to severe forms of diarrhea.[1] *V. cholerae* is autochthonous flora to brackish and estuarine waters possibly contributing to waterborne transmission of cholera and its persistence in the environment.[3]

1. Concentration Techniques

See Section 9260B.1. Choose a sample volume that is appropriate for the sample type and level of sensitivity desired.

2. Enrichment Procedures

Enrich samples in alkaline peptone water (1% peptone, 1% NaCl, pH 8.4), using appropriate concentration of broth relative to sample volume (see Tables 9221:I and II). Streak cultures on thiosulfate-citrate-bile salts-sucrose agar (TCBS) after 6 and 15 to 18 h of incubation at 35°C. Some brands of TCBS have been reported to be superior to other brands for isolation of *V. cholerae*.[4]

3. Selective Growth

Incubate TCBS plates for 24 h at 35°C. Suspected *V. cholerae* colonies will appear yellow, indicating sucrose fermentation. Predominant interfering bacteria may include *V. alginolyticus, V. anguillarum, V. fluvialis, V. metschnikovii,* and *Aeromonas* spp.

4. Presumptive Tests to Differentiate *V. cholerae*

Streak sucrose-positive colonies on a nonselective agar medium, such as tryptic soy agar with 1% w/v NaCl.

The following tests describe *V. cholerae:*

Test	Reaction
Gram-negative rod	+
Motility	+
Cytochrome oxidase	+
Indole	+
Glucose fermentation	+
Sensitivity to 150 μg 0/129/mL[5]	+

Discard isolates that do not react as indicated above.

5. Classification of Isolates as *V. cholerae*

The following tests are recommended to identify isolates as *V. cholerae*. When only the detection of the O1 serotype is required, substitute agglutination assays for some of the following tests:

Test[5]	Reaction*
ONPG	+
Nitrate reduction	+
Indole	+
0/129 sensitivity:	
10 μg	+
150 μg	+
Swarming	−
Luminescence	v
Thornley's arginine dihydrolase	−
Lysine decarboxylase	+
Ornithine decarboxylase	+
Growth at 42°C	+

*v differs for strains within the species.

Test[5]	Reaction*
Growth at % NaCl:	
0%	+
3%	+
6%	v
8%	−
10%	−
Voges-Proskauer reaction	+
Gas from glucose fermentation	−
Fermentation to acid:	
L-arabinose	−
m-inositol	−
D-mannose	v
Sucrose	+
Enzyme production:	
Alginase	−
Amylase	+
Chitinase	+
Gelatinase	+
Lipase	+
Utilization as sole source of carbon:	
γ-aminobutyrate	−
Cellobiose	−
L-citrulline	−
Ethanol	−
D-gluconate	+
D-glucuronate	−
L-leucine	−
Putrescine	−
Sucrose	+
D-xylose	−

*v differs for strains within the species.

6. Serological Identification

The slide agglutination technique can be used for serological identification of *V. cholerae* (see Section 9260B.5). The O1 serogroup is divided into Ogawa, Inaba, and Hikojima serotypes. Non-O1 *V. cholerae* also can be serotyped; however, only specialized laboratories have this capability.

7. Biotypes of Serogroup O1 *V. cholerae*

Use the following tests to classify *V. cholerae* O1 as El Tor or classical biotypes:

Test[5]	Biovar	
	Classical	El Tor
Hemolysis of sheep erythrocytes	−	v*
Voges-Proskauer reaction	−	+
Chicken erythrocyte agglutination	−	+
Sensitivity:		
Polymyxin B (50 IU)	+	−
Mukerjee classical phage IV	+	−
Mukerjee El Tor phage 5	−	+

*v different reaction within the serovar.

8. Other Procedures

Environmental samples also may be examined by fluorescent-antibody techniques, provided that highly specific antisera are available.[6]

9. References

1. FARMER, J.J., F.W. HICKMAN-BRENNER & M.T. KELLY. 1985. Vibrio. *In* E.H. Lennette, A. Balows, W.J. Hausler & H.J. Shadomy, eds. Manual of Clinical Microbiology, 4th ed. American Soc. Microbiology, Washington, D.C.

2. BLAKE, P.A., D.T. ALLEGRA, J.D. SNYDER, T.J. BARRETT, L. McFARLAND, C.T. CARAWAY, J.C. FEELEY, J.P. CRAIG, J.V. LEE, N.D. PUHR & R.A. FELDMAN. 1980. Cholera—a possible endemic focus in the United States. *N. England J. Med.* 302:305.

3. COLWELL, R.R., P.A. WEST, D. MANEVAL, E.F. REMMERS, E.L. ELLIOT & N.E. CARLSON. 1984. Ecology of pathogenic vibrios in Chesapeake Bay. *In* R.R. Colwell, ed. Vibrios in the Environment. John Wiley & Sons, New York, N.Y.

4. WEST, P.A., E. RUSSEK, P.R. BRAYTON & R.R. COLWELL. 1982. Statistical evaluation of a quality control method for isolation of pathogenic *Vibrio* species on selected thiosulfate-citrate-bile salts sucrose agars. *J. Clin. Microbiol.* 16:1110.

5. WEST, P.A. & R.R. COLWELL. 1984. Identification and classification of Vibrionaceae—an overview. *In* R.R. Colwell, ed. Vibrios in the

Environment. John Wiley & Sons, New York, N.Y.

6. Xu, H.S., N.C. Roberts, L.B. Adams, P.A. West, R.J. Sieveling, A. Huq, M.I. Huq, R. Rahman & R.R. Colwell. 1984. An indirect fluorescent antibody staining procedure for detection of *Vibrio cholerae* serovar O1 cells in aquatic environmental samples. *Microbiol. Methods* 2:2221.

10. Bibliography

U.S. Department of Health, Education and Welfare. 1965. Proceedings Cholera Research Symposium. Public Health Serv. Publ. No. 1328. Washington, D.C.

Gorback, S.L., J.B. Banwell, N.P. Pierce, B.O. Chatterjee & R.C. Mitra. 1970. Intestinal microflora in a chronic carrier of *Vibrio cholerae. J. Infect. Dis.* 121:383.

9260 I.   Pathogenic Leptospires

The occurrence of pathogenic leptospires in natural waters is extremely variable. Many factors make interpretation of results difficult, for example, intermittent leptospire discharge from infected wildlife or farm animals and the effects of stormwater runoff and flooding of contaminated land.[1] Leptospire persistence in warm, slow-moving waters of pH 6.0 to 8.0[2-5] and moderate levels of bacterial nutrients[6] also complicate interpretation. Even when pathogenic leptospires are present, their detection is difficult because of the competitive growth of other organisms[7] and the need to differentiate between pathogenic and saprophytic strains.[3,7-10] Failure to isolate pathogenic leptospires from natural waters does not necessarily indicate their absence.

These factors explain why qualitative methodology evolved to detect leptospires in polluted water. Long-term incubation on various media is necessary because of the relatively slow growth of the organisms. During incubation, check inoculated media weekly for the appearance of leptospires and for culture contamination using dark-field microscopy. Upon detection, characterize the leptospire isolates further by various biochemical and serological tests to separate pathogenic and saprophytic strains. Use animal tests for pathogenic leptospires but only on primary pure-culture isolates because pathogenic strains may become avirulent through subsequent culture passages.

1. Preliminary Concentration Technique

Pathogenic leptospires tend to concentrate in near-shore bottom sediments of streams and farm ponds. Gently agitate bottom sediment before sampling to insure collection of bacteria-laden material from the sediment-water interface. Use the bacteriological bottom sampler or standard sample bottles (see Section 9060A) to collect this finely suspended material. Upon return to the laboratory, or preferably at a field site, shake sample vigorously to release entrapped bacteria from the sediment and prefilter immediately through either filter paper* or a membrane filter absorbent pad. Pass prefiltered sample through a Swinney hypodermic adapter containing a fiberglass prefilter and a membrane filter of 0.45-μm pore size to separate leptospires (which can pass through the pores into the filtrate) from other organisms. Rinse with an equal volume of sterile dilution water.

2. Enrichment

Inoculate portions of sample filtrate (1 mL and 0.1 mL) into Fletcher's semisolid medium containing 10% rabbit serum.[10] Incubate inoculated medium at 30°C for 6 weeks. Examine each tube at least weekly

---

*Whatman No. 1 or equivalent.

for leptospiral growth and culture contamination: use dark-field illumination and $250\times$ magnification.[11] Strains of *Vibrio, Spirillum,* or *Paraspirillum* are the most common contaminants observed, particularly when filtrate volumes greater than 0.1 mL are examined.[11]

Leptospires are helicoidal, usually 6 to 20 μm long with each coil about 0.2 to 0.3 μm in diameter. The coils of leptospires are more compact than those of other spirochaetes.[12] If leptospires are not observed microscopically within a 6-week incubation period, the test is negative.

As an alternate enrichment procedure, inoculate spread plates of SM agar[13] or bovine albumin polysorbate 80 medium[14,15] with 0.1 to 1.0 mL sample filtrate. Incubate at 30°C for 7 to 9 d. With bovine albumin polysorbate 80 medium use an agar overlay of 0.7% distilled water agar. Regardless of agar medium used, prepare it 1 to 2 d before inoculation to condition the agar and promote even spreading of the inoculum over the surface. Identify all colonies morphologically by dark-field microscopy before making biochemical and serological tests or animal inoculations.

3. Differentiation of Leptospires

Detection of pathogenic leptospires in lakes and streams indicates leptospirosis in domestic or wild animals that frequent these waters and signals a health risk to bathers. It is critically important to differentiate pathogenic from saprophytic leptospire strains.

a. Culture reactions: Saprophytic leptospires grow well in Stuart's medium containing 10% rabbit serum supplemented with 10 μg copper sulfate $(CuSO_4)/mL$[4,5] or 100 μg 8-azaguanine/mL.[2,5] Only saprophytic leptospires grow in a 10% rabbit serum medium at 13°C.[2] Saprophytic strains demonstrate higher oxidase response[16] and higher egg yolk decomposition activity[17] than pathogenic leptospires.

Optimum incubation temperature for pathogenic leptospires is 30°C. Incubate all tests for 5 d. Use no single test to differentiate saprophytic from pathogenic leptospires.[18]

b. Verification of pathogenicity: Use commercial antisera permitting tentative identification of pathogenic leptospires. Make final verification for pathogenicity by intraperitoneal injection of the suspect strain into guinea pigs. After 4 weeks, sacrifice animals, test blood for serum antibody titers above 100, and demonstrate presence of pathogenic leptospires by cultivation of aseptically removed kidney tissue in Stuart's medium.

4. References

1. CRAWFORD, R.P., J.M. HEINEMANN, W.F. MCCULLOCH & S.L. DIESCH. 1971. Human infections associated with waterborne leptospires, and survival studies on serotype pomona. *J. Amer. Vet. Med. Assoc.* 159:1477.
2. GALTON, M.M., R.W. MENGES & J.H. STEELE. 1958. Epidemiological patterns of leptospirosis. *Ann. N.Y. Acad. Sci.* 70:427.
3. JOHNSON, R.C. & V.G. HARRIS. 1967. Differentiation of pathogenic and saprophytic leptospires. I. Growth at low temperatures. *J. Bacteriol.* 94:27.
4. OKAZAKI, W. & L.M. RINGEN. 1957. Some effects of various environmental conditions on the survival of *Leptospira pomona. Amer. J. Vet. Res.* 18:219.
5. RYU, E. & C.K. LIU. 1966. The viability of leptospires in the summer paddy water. *Jap. J. Microbiol.* 10:51.
6. DIESCH, S.L., W.F. MCCULLOCH, J.L. BRAUN & R.P. CRAWFORD, JR. 1969. Environmental studies on the survival of leptospires in a farm creek following a human leptospirosis outbreak in Iowa. Proc. Annu. Conf., *Bull. Wildlife Dis. Assoc.* 5:166.
7. CHANG, S.L., M. BUCKINGHAM & M.P. TAYLOR. 1948. Studies of *Leptospira icterohemorrhagiae.* IV. Survival in water and sewage: Destruction in water by halogen compounds, synthetic detergents and heat. *J. Infect. Dis.* 82:256.
8. JOHNSON, R.C. & P. ROGERS. 1964. Differentiation of pathogenic and saprophytic lep-

tospires with 8-azaguanine. *J. Bacteriol.* 88:1618.

9. FUZI, M. & R. CSOKA. 1960. Differentiation of pathogenic and saprophytic leptospire by means of a copper sulfate test. *Zentralbl. Bakteriol. Parasitenk. Infektionskr. Hyg. Abt. Orig.,* I. 179:231.

10. CRAWFORD, R.P., J.L. BRAUN, W.F. MCCULLOCH & S.L. DIESCH. 1969. Characterization of leptospires isolated from surface waters in Iowa. *Bull. Wildlife Dis. Assoc.* 5:157.

11. BRAUN, J.L., S.L. DIESCH & W.F. MCCULLOCH. 1968. A method for isolating leptospires from natural surface waters. *Can. J. Microbiol.* 14:1011.

12. TURNER, L.H. 1970. Leptospirosis III. Maintenance, isolation and demonstration of leptospires. *Trans. Roy. Soc. Trop. Med. Hyg.* 64:623.

13. BASEMAN, J.B., R.C. HENNEBERRY & C.D. COX. 1966. Isolation and growth of leptospira on artificial media. *J. Bacteriol.* 91:1374.

14. ELLINGHAUSEN, H.C., JR. & W.G. MCCULLOUGH. 1965. Nutrition of *Leptospira pomona* and growth of 13 other serotypes. Fractionation of oleic albumin complex and a medium of bovine albumin and polysorbate 80. *Amer. J. Vet. Res.* 26:45.

15. TRIPATHY, D.N. & L.E. HANSON. 1971. Agar overlay method for growth of leptospires in solid medium. *Amer. J. Vet. Res.* 32:1125.

16. FUZI, M. & R. CSOKA. 1961. Rapid method for differentiation of parasitic and saprophytic leptospirae. *J. Bacteriol.* 81:1008.

17. FUZI, M. & R. CSOKA. 1961. An egg-yolk reaction test for the differentiation of leptospira. *J. Pathol. Bacteriol.* 82:208.

18. KMETY, E., I. OKESJI, P. BAKASS & B. CHORVATH. 1966. Evaluation of methods for differentiating pathogenic and saprophytic leptospira strains. *Ann. Soc. Belge. Med. Trop.* 46:111.

9260 J. Legionellaceae

The Legionellaceae have been implicated in outbreaks of disease occurring since 1947.[1] Two forms of disease are recognized: a pneumonic form called Legionnaires' Disease and a nonpneumonic form called Pontiac fever. The first species was isolated following the historic outbreak associated with the Legionnaires' Convention in Philadelphia, Pa., in 1976. Epidemiological findings and animal studies have shown that the organism is transmitted via the airborne route[2] and is ubiquitous in moist environments. The reservoirs for most outbreaks have been either contaminated air conditioning cooling tower water or contaminated potable water distribution systems.[3,4] *Legionella* species also have been isolated in non-disease-related circumstances from a wide variety of aquatic environments such as lakes, streams, reservoirs, and sewage.[5,6] The organisms are able to survive for prolonged periods in laboratory distilled and tap water.[7]

The Legionellaceae are composed of a single genus, *Legionella,* and more than 22 different species.[8] The organisms are gram-negative, aerobic, non-spore-forming bacteria. They are 0.5 to 0.7 μm wide and 2 to 20 μm long. They possess polar, subpolar, and/or lateral flagella. All require cysteine and iron salts for growth.

Although *Legionella* originally were isolated in guinea pigs and embryonated hen's eggs, it has been shown that plating directly on artificial media is more sensitive than animal inoculation for *L. pneumophila.*[9] The most widely used medium is an ACES (*N*-2-acetamideo-2-aminoethanesulfonic acid) buffered (pH 6.9) charcoal yeast extract (BCYE) agar supplemented with cysteine, ferric pyrophosphate, and optimally, alpha-ketoglutarate (BCYE-alpha).[10]

No one medium will be optimal for the

recovery of *Legionella* from every environmental site; thus different selective media with various antibiotic combinations in a BCYE base may be necessary.[10–12] Also, pretreating samples with hydrochloric acid-potassium chloride, pH 2.2, is useful for eliminating non-*Legionella* organisms.[13] The two most commonly used selective media are GPVA medium (BCYE-alpha supplemented with glycine anisomycin, vancomycin, and polymyxin B) and CCVC medium (BCYE-alpha supplemented with polymyxin B, cephalothin, vancomycin and cycloheximide). The GPVA medium is less inhibitory to some *Legionella* species. Use CCVC medium in combination with a less selective medium.

1. Sample Collection

Collect water samples from the littoral zone or from cooling towers, condenser coils, storage tanks, showers, water taps, etc. In most instances, a 1-L water sample is sufficient. Larger volumes of water (1 to 10 L)[6] may be needed in water having low bacterial counts. In addition to collecting water samples, it may be useful to swab various fixtures (e.g., shower heads) and plate directly on selective media. Transport samples to the laboratory in insulated containers. Refrigerate samples that cannot be processed immediately. Treat chlorinated water with sodium thiosulfate (see Section 9060A.2).

2. Immunofluorescence Procedure

If laboratory facilities for bacterial isolation are unavailable or if rapid presumptive identification is required, examine samples by direct fluorescent antibody staining (DFA). Centrifuge 100 mL at 3500 × g for 30 min at room temperature and reconstitute the sedimented material in 6 to 10 mL filter-sterilized (0.2 μm filter) water from sample. Prepare smears for DFA by filling two 1.5-cm circles on a microscope slide with the concentrate. Air-dry sample smears, gently heat-fix, treat with 10% formalin for 10 min, rinse with phosphate-buffered saline (pH 7.6), and react with specific fluorescent antibodies.[6,14] The DFA procedure is less than absolutely specific[15] and it lacks viability determination. Many environmental bacteria (i.e., *Pseudomonas* spp. and *Xanthomonas-Flavobacterium* group) cross-react with the *Legionella* DFA reagents. To determine whether organisms are viable, use secondary staining with a tetrazolium dye.[16] Confirm *Legionella* using direct isolation procedures.

3. Media and Reagents

a. Buffered charcoal yeast extract alpha base:[14]

Norit SG charcoal	2.0	g
Yeast extract	10.0	g
ACES buffer	10.0	g
Ferric pyrophosphate, soluble	0.25	g
L-cysteine, HCl·H₂O	0.4	g
Agar	17.0	g
Potassium alpha-ketoglutarate	1.0	g
Distilled water	1.0	L

Dissolve yeast extract, agar, charcoal, glycine, and alpha-ketoglutarate in approximately 850 mL water; boil. Dissolve 10 g ACES buffer in 100 mL warm water, adjust pH to 6.9 with 1N KOH and add. Autoclave 15 min at 121°C. Cool to 50°C. Dissolve 0.4 g cysteine and 0.25 g ferric pyrophosphate in 10 mL of water each and filter sterilize separately (0.22 μm). After base has cooled, add cysteine, ferric pyrophosphate, and dyes in that order. Adjust pH to 6.9 with sterile 1N KOH and dispense.

b. GPVA medium:[11,12]*

Glycine	0.3%	
Polymyxin B	100	units/mL
Vancomycin	5	μg/mL
Anisomycin	80	μg/mL

---

*Available commercially.

To cooled BCYE-alpha base with glycine, add filter-sterilized antibiotics and mix. Adjust pH to 6.9 with sterile $1N$ KOH and dispense.

c. *CCVC medium*[11]†

Cephalothin	4	µg/mL
Colistin	16	µg/mL
Vancomycin	0.5	µg/mL
Cycloheximide	80	µg/mL

To cooled BCYE-alpha base add filter-sterilized antibiotics and mix. Adjust to pH 6.9 with sterile $1N$ KOH and dispense.

d. *Acid treatment reagent*,[11] pH 2.0 ($0.2M$ KCl/HCl):

Solution A—$0.2M$ KCl (14.9 g/L in distilled water).

Solution B—$0.2M$ HCl (16.7 mL/L $10N$ HCl in distilled water).

Mix 18 parts of Solution A with 1 part of Solution B. Check pH against a pH 2.0 standard buffer. Dispense into screw-cap tubes in 1.0-mL volumes and sterilize by autoclaving.

e. *Alkaline neutralizer reagent*[11] ($0.1N$ KOH):

Stock solution—$0.1N$ KOH (6.46 g/L in deionized water). Dilute 10.7 mL of stock solution with deionized water to 100 mL. Dispense into screw-cap tubes in convenient volumes and sterilize by autoclaving. pH of *d* and *e* combined in equal volumes should be 6.9.

4. Sample Preparation

a. *Low-bacterial-count water:* Concentrate water that has a low total bacterial count either by filtration[11] or continuous-flow centrifugation.[17] Filter samples through sterile 47-mm filter funnel assemblies containing a 0.2-µm porosity polycarbonate filter.‡ After filtration, immediately remove the filter aseptically and place it *soiled side down* in a 50-mL centrifuge tube or similar-size vessel containing 10 mL sterile tap water or phosphate buffer. If more than one filter is required to concentrate a sample combine them.

b. *High-bacterial-count water:* Process water that has a high total bacterial count directly. Place 10 mL sample in a 50-mL centrifuge tube or similar-size vessel containing 10 mL of sterile tap water or phosphate buffer.

c. *Sample dispersion:* Disperse organisms from filter or aggregates by mixing with a vortex mixer (3×30 s) or in a sonic bath for 10 min.

d. *Plating:* Plate acid-treated and non-acid-treated samples on two types of BCYE: plain and selective with antibiotics.

1) No acid treatment—Inoculate three plates each of BCYE-alpha and selective BCYE-alpha (GPVA or CCVC) with 0.1 mL of suspension. Spread with a sterile smooth glass rod. Save remainder of specimen for acid treatment and store at 4°C.

2) Acid treatment—Place 1.0 mL of suspension in a sterile 13×100-mm screw-capped tube containing 1.0 mL acid treatment reagent and mix. Final pH of mixture should be approximately 2.2. Let stand for 15 min at room temperature, neutralize by adding 1.0 mL alkaline neutralizer reagent, and mix. Inoculate 0.1 mL onto three plates each of BCYE-alpha and selective BCYE-alpha (GPVA or CCVC) and spread with a sterile smooth glass rod.

3) Incubation—Incubate all plates at 35°C in a humidified atmosphere ($> 50\%$) for up to 10 d. A candle jar or humidified CO_2 incubator (2 to 5% CO_2) is acceptable.

e. *Total bacterial count examination:* Determine the adequacy of processing for each high-bacterial-count water. Some samples may require dilution, concentration, or animal inoculation. If the total count of the acid-treated sample exceeds 300 colonies on BCYE selective medium,

---

† This medium may not be available in dehydrated form and may require preparation from the basic ingredients.
‡ Nuclepore Corp., 7035 Commerce Circle, Pleasanton, Calif., or equivalent.

make a further 10-fold dilution of the sample stored at 4°C. Repeat acid-treatment and plating.

If the total count of the non-acid-treated sample is less than 30 colonies on BCYE agar, concentrate and treat the collected water as previously described for low-bacterial-count water.

5. Examination of Cultures of Legionellae

With the aid of a dissecting microscope, examine all cultures daily after 48 h incubation for the presence of opaque bacterial colonies that have a "ground-glass" appearance. Place plates with *Legionella*-like colonies in a biological safety cabinet equipped with a burner, a bacteriological needle, and a loop. Aseptically pick each suspect colony onto BCYE-alpha agar and a BCYE agar plate prepared without L-cysteine. Streak the inoculated portion of each plate with a sterile loop to provide areas of heavy growth and incubate for 24 h.

Reincubate plates without growth an additional 24 h. Plates demonstrating growth on only BCYE-alpha agar are presumptive *Legionella*. Confirm *Legionella* by slide agglutination or direct immunofluorescence. If these confirmatory techniques are not available, send subcultures of the presumptive legionellae to a reference laboratory for further identification. Because there are many types in some species, especially *L. pneumophila* serogroup 1, investigation of environmental sites as possible reservoirs of epidemic-causing strains may be useful.[18] Effective investigatory techniques include monoclonal antibody subtyping, electrophoretic isoenzyme analysis, restriction endonuclease tests, and plasmid analysis.

6. References

1. McDade, J.E., C.C. Shepard, D.W. Frasier, T.R. Tsai, M.A. Redus, W.T. Dowdle & The Laboratory Investigation Team. 1977. Legionnaires's Disease: isolation of a bacterium and demonstration of its role in other respiratory disease. *N. England J. Med.* 297:1197.

2. Berendt, R.F. *et al.* 1980. Dose-response of guinea pigs experimentally infected with aerosols of *Legionella pneumophila. J. Infect. Dis.* 141:186.

3. Fliermans, C.B., W.B. Cherry, L.H. Orrison, S.J. Smith, D.L. Tison & D.H. Pope. 1981. Ecological distribution of *Legionella pneumophila. Appl. Environ. Microbiol.* 41:9.

4. Tobin, J.O.H., R.A. Swan & C.L.R. Bartlett. 1981. Isolation of *Legionella pneumophila* from water systems: methods and preliminary results. *Brit. Med. J.* 282:515.

5. Cherry, W.B., G.W. Gorman, L.H. Orrison, C.W. Moss, A.G. Steigerwalt, H.W. Wilkinson, S.E. Johnson, R.M. McKinney & D.J. Brenner. 1982. *Legionella jordanis:* a new species of *Legionella* isolated from water and sewage. *J. Clin. Microbiol.* 15:290.

6. Fliermans, C.B., W.B. Cherry, L.H. Orrison & L. Thacker. 1979. Isolation of *Legionella pneumophila* from nonepidemic related aquatic habitats. *Appl. Environ. Microbiol.* 37:1239.

7. Skaliy, P. & H.V. McEachern. 1979. Survival of the Legionnaires's Disease bacterium in water. *Ann. Intern. Med.* 90:662.

8. Brenner, D.J., A.G. Steigerwalt, G.W. Gorman, H.W. Wilkinson, W.F. Bibb, M. Hackel, R.L. Tyndall, J. Campbell, J.C. Feeley, W.L. Thacker, P. Skaliy, W.T. Martin, B.J. Brake, B.S. Fields, H.W. McEachern & L. K. Corcoran. 1985. Ten new species of *Legionella. Int. J. System. Bacteriol.* 35:50.

9. Feeley, J.C., R.J. Gibson, G.W. Gorman, N.C. Langford, J. K. Rasheed, D.C. Macel & W.B. Baine. 1979. Charcoal-yeast extract agar: primary isolation medium for *Legionella pneumophila. J. Clin. Microbiol.* 10:437.

10. Edelstein, P.H. 1982. Comparative studies of selective media for isolation of *Legionella pneumophila* from potable water. *J. Clin. Microbiol.* 16:697.

11. Gorman, G.W., J.M. Barbaree & J.C. Feeley. 1983. Procedures for the Recovery of

Legionella from Water. Developmental Manual, Centers for Disease Control, Atlanta, Ga.

12. WADOWSKY, R.M. & R.B. YEE. 1981. Glycine-containing selective medium for isolation of *Legionellaceae* from environmental specimens. *Appl. Environ. Microbiol.* 42:768.

13. BOPP, C.A., J.W. SUMNER, G. K. MORRIS & J.G. WELLS. 1981. Isolation of *Legionella* spp. from environmental water samples by low-pH treatment and use of selective medium. *J. Clin. Microbiol.* 13:714.

14. JONES, G.L. & G.A. HEBERT. 1979. Legionnaires—the disease, the bacterium and methodology. U.S. Dep. Health, Education, & Welfare, Centers for Disease Control, Atlanta, Ga.

15. EDELSTEIN, P.H., R.M. MCKINNEY, R.D. MEYER, M.A.C. EDELSTEIN, C.J. KRAUSE & S.M. FINEGOLD. 1980. Immunologic diagnosis of Legionnaires' Disease: cross reactions with anaerobic and microaerophilic organisms and infections caused by them. *J. Infect. Dis.* 141:652.

16. FLIERMANS, C.B., R.J. SORACCO & D.H. POPE. 1981. Measure of *Legionella pneumophila* activity *in situ. Curr. Microbiol.* 6:89.

17. VOSS, L., K.S. BUTTON, M.S. RHEINS & O.H. TUOVINEN. 1984. Sampling methodology for enumeration of Legionella spp. in water distribution systems. *In* C. Thornsberry, A. Balows, J.C. Feeley & W. Jakubowski, eds. *Legionella,* Proceedings of the 2nd International Symposium. American Soc. Microbiology, Washington, D.C.

18. BARBAREE, J.M., G.W. GORMAN, W.T. MARTIN, B.S. FIELDS & W.E. MORRILL. 1987. Protocol for sampling environmental sites for Legionellae. *Appl. Environ. Microbiol.* 53:1454.

7. Bibliography

CENTERS FOR DISEASE CONTROL, NATIONAL INSTITUTE OF ALLERGY AND INFECTIOUS DISEASES & WORLD HEALTH ORGANIZATION. 1979. International Symposium on Legionnaire's Disease. *Ann. Intern. Med.* 90:489.

BLACKMAN, J.A., F.W. CHANDLER, W.B. CHERRY, A.C. ENGLAND, J.C. FEELEY, M.D. HICKLIN, R.M. MCKINNEY & H.W. WILKINSON. 1981. Legionellosis. *Amer. J. Pathol.* 103:427.

DUFOUR, A. & W. JAKUBOWSKI. 1982. Drinking water and Legionnaire's Disease. *J. Amer. Water Works Assoc.* 74:631.

THORNSBERRY, C., A. BALOWS, J.C. FEELEY & W. JAKUBOWSKI, eds. 1984. *Legionella,* Proceedings of the 2nd International Symposium. American Soc. Microbiology, Washington, D.C.

9260 K.   Yersinia enterocolitica

Yersinia enterocolitica is a gram-negative bacterium that can cause acute gastroenteritis and can be found in water in cold or temperate areas of the United States. Many wild, domestic, and farm animals are reservoirs of this organism, including wild animals associated with water habitats (beavers, minks, muskrats, nutrias, otters, and racoons).[1,2] The organism can grow at temperatures as low as 4°C with a generation time of 3.5 to 4.5 h if at least trace amounts of organic nitrogen are present.[3] In the U.S. during the period 1971 to 1978 there were two reported incidents of waterborne gastroenteritis possibly caused by *Yersinia,*[3-5] although the documentation is poor.

Yersinia has been isolated from untreated surface and ground waters in the Pacific Northwest, New York, and other regions of North America, with highest isolations during the colder months.[6-8] Concentrations have ranged from 3 to 7900/ 100 mL. *Yersinia* isolations do not correlate with levels of total and fecal coliforms or total plate count bacteria.[7] There is little information on *Yersinia* survival in natural waters and water treatment processes.

In studies of *Yersinia* in chlorinated-dechlorinated secondary effluent and receiving (river) water, the organism was isolated in 27% of the effluent samples, 9% of the upstream samples, and 36% of the downstream samples.[9] Mean total and fecal coliform reductions in effluent chlorination were 99.93 and 99.95%, respectively. In a survey of untreated and treated (chlorination or filtration plus chlorination) drinking water supplies, *Yersinia* was found in 14.0 and 5.7% of the samples, respectively.[7] Water samples with less than 2.2 coliforms/100 mL were 6.9% *Yersinia*-positive and those with more than 2.2 coliforms/100 mL were 15.9% *Yersinia*-positive. *Yersinia* isolation did not correlate with total or fecal coliforms in this study.

Because of the existence of animal reservoirs, the widespread occurrence and persistence of *Yersinia* in natural and treated water in at least some geographic areas, the evidence for possible waterborne outbreaks, and the lack of definitive information on its reduction by treatment processes, this pathogen is of potential importance in drinking water.

1. Concentration and Cultivation

A membrane filter method for enumerating and isolating *Yersinia enterocolitica* is available. The method may be used for examining large volumes of low-turbidity water and for presumptively identifying the organism without transfering colonies to multiple confirmatory media.

Filter sample through a membrane filter (see Section 9260B.1*d*). Place membrane filter on a cellulose pad saturated with m-YE recovery broth. Incubate for 48 h at 25°C. Aseptically transfer the membrane to a lysine-arginine agar substrate and incubate anaerobically at 35°C. After 1 h, puncture a hole in the membrane next to each yellow to yellow-orange colony with a needle, transfer the membrane to a urease-saturated absorbent pad, and incubate at 25°C for 5 to 10 min. Immediately count all distinctly green or deep bluish-purple colonies. The green or bluish colonies are sorbitol-positive, lysine- and arginine-negative, and urease-positive. They may be presumptively identified as *Y. enterocolitica*.

2. References

1. WETZLER, T.F. & J. ALLARD. 1977. *Yersinia enterocolitica* from trapped animals in Washington State. Paper presented at International Conf. Disease in Nature Communicable to Man. Yellow Bay, Mont.
2. WETZLER, T.F., J.T. REA, G. YUEN & W. TURNBERG. 1978. *Yersinia enterocolitica* in waters and wastewaters. Paper presented at 106th Annual Meeting, American Public Health Assoc., Los Angeles, Calif.
3. HIGHSMITH, A.K., J.C. FEELEY, P. SKALIY, J.G. WELLS & B.T. WOOD. 1977. The isolation and enumeration of *Yersinia enterocolitica* from well water and growth in distilled water. *Appl. Environ. Microbiol.* 34:745.
4. EDEN, K.V., M.L. ROSENBERG, M. STOOPLER, B.T. WOOD, A.K. HIGHSMITH, P. SKALIY, J.G. WELLS & J.C. FEELEY. 1977. Waterborne gastrointestinal illness at a ski-resort—isolation of *Yersinia enterocolitica* from drinking water. *Pub. Health Rep.* 92:245.
5. KEET, E. 1974. *Yersinia enterocolitica* septicemia. *N. Y. State J. Med.* 74:2226.
6. HARVEY, S., J.R. GREENWOOD, M.J. PICKETT & R.A. MAH. 1976. Recovery of *Yersinia enterocolitica* from streams and lakes of California. *Appl. Environ. Microbiol.* 32:352.
7. WETZLER, T.F., J.R. REA, G.J. MA & M. GLASS. 1979. Non-association of *Yersinia* with traditional coliform indicators. *In* Proc. Annu. Meeting American Water Works Assoc., American Water Works Assoc., Denver, Colo.
8. SHAYEGANI, M., I. DEFORGE, D.M. MC-GLYNN & T. ROOT. 1981. Characteristics of *Yersinia enterocolitica* and related species isolated from human, animal, and environmental sources. *J. Clin. Microbiol.* 14:304.
9. TURNBERG, W.L. 1980. Impact of Renton Treatment Plant effluent upon the Green-Duwamish River. Masters Thesis, Univ. Washington, Seattle.
10. BARTLEY, T.D., T.J. QUAN, M.T. COLLINS & S.M. MORRISON. 1982. Membrane filter technique for the isolation of *Yersinia enterocolitica*. *Appl. Environ. Microbiol.* 43:829.

3. Bibliography

HIGHSMITH, A.K., J.C. FEELEY & G.K. MORRIS. 1977. *Yersinia enterocolitica:* a review of the bacterium and recommended laboratory methodology. *Health Lab. Sci.* 14:253.

BOTTONE, E. J. 1977. *Yersinia enterocolitica:* a panoramic view of a charismatic microorganism. *CRC Critical Rev. Microbiol.* 5:211.

9510   DETECTION OF ENTERIC VIRUSES*

9510 A.   Introduction

1. Occurrence

Viruses excreted with feces or urine from any species of animal may pollute water. Especially numerous, and of particular importance to health, are the viruses that infect the gastrointestinal tract of man and are excreted with the feces of infected individuals. These viruses are transmitted most frequently from person to person by the fecal-oral route. However, they also are present in domestic sewage which, after various degrees of treatment, is discharged to either surface waters or the land. Consequently, enteric viruses may be present in sewage-contaminated surface and ground waters that are used as sources of drinking water. The viruses known to be excreted in relatively large numbers with feces include polioviruses, coxsackieviruses, echoviruses, and other enteroviruses, adenoviruses, reoviruses, rotaviruses, the hepatitis A (infectious hepatitis) virus(es), and the Norwalk-type agents that can cause acute infectious nonbacterial gastroenteritis. With the possible exception of hepatitis A, each group or subgroup consists of a number of different serological types; thus more than 100 different human enteric viruses are recognized. Other viruses may be present in domestic sewage, but not usually in large numbers.[1−5]

In temperate climates enteroviruses occur at peak levels in sewage during the late summer and early fall. However, hepatitis A virus (HAV), Norwalk-type viruses, and rotaviruses may be important exceptions because the incidence of the diseases due to these viruses increases in the colder months. Quantitative information on seasonal patterns of occurrence in water and wastewater of these latter viruses is lacking because they can not be assayed readily with conventional cell culture techniques. The Norwalk-type viruses have not been cultivated in any cell cultures, although immunochemical assay methods have been developed to detect them as antigens.[6,7] Human rotaviruses and HAV have been cultivated recently in cell cultures, but the techniques are difficult and require concomitant use of immunoassays such as immunofluorescence to detect virus growth.[8−12]

Viruses are not normal flora in the intestinal tract; they are excreted only by infected individuals, mostly infants and young children. Infection rates vary considerably from area to area, depending on sanitary and socioeconomic conditions. Viruses usually are excreted in numbers several orders of magnitude lower than those

---

*Approved by Standard Methods Committee, 1985.

of coliform bacteria. Because enteric viruses multiply only within living, susceptible cells, their numbers cannot increase in sewage. Sewage treatment, dilution, natural inactivation, and water treatment further reduce viral numbers. Thus, although large outbreaks of waterborne viral disease may occur when massive sewage contamination of a water supply takes place,[13,14] waterborne transmission of viral infection and disease in technologically advanced nations depends on whether minimal quantities of viruses are capable of producing infections. It has been demonstrated that infection can be produced experimentally by a very few virus units,[15] although the risk of infection increases with increasing ingested doses.[16] The risk of infection incurred by an individual in a community with a water supply containing a very few virus units has not been determined.[17]

It has been hypothesized that transmission of small numbers of viruses through water supplies may produce inapparent infections. However, the subsequent transmission of viruses from these cases of inapparent infections to susceptible contacts probably involves large quantities of viruses. This may result in a considerable amount of disease transmission in a community, epidemiologically consistent with contact and not with transmission from a common source such as water.

Most recognized waterborne virus disease outbreaks in the U.S. have been caused by obvious sewage contamination of untreated or inadequately treated private and semipublic water supplies. Virus disease outbreaks in community water supply systems usually are caused by contamination through the distribution system.[18,19]

2. Testing for Viruses

The routine examination of water and wastewater for enteric viruses is not recommended now. However, in special circumstances such as wastewater reclamation, disease outbreaks, or special research studies, it may be prudent or essential to conduct virus testing. Such testing should be done only by competent and specially trained water virologists having adequate facilities.

Laboratories planning to concentrate viruses from water and wastewater should do so with the clear understanding that the available methodology has important limitations.[20] Even the most current methods for concentrating viruses from water still are being researched and continue to be modified and improved. The efficiency of a virus concentration method may vary widely depending on water quality. Furthermore, none of the available virus detection methods have been tested adequately with representatives from all of the virus groups of public health importance. Most virus concentration methods have achieved adequate virus recoveries with water or wastewater samples that have been contaminated experimentally with known quantities of a few specific enteric viruses. Although method effectiveness in field trials is difficult to evaluate, some virus concentration methods have been used successfully to recover naturally occurring enteric viruses. Some of these methods require expensive equipment and materials for sample processing and all virus assay and identification procedures require expensive cell culture and related virology laboratory facilities.

Detecting viruses in water through recovery of infectious virus requires three general steps: (a) collecting a representative sample, (b) concentrating the viruses in the sample, and (c) identifying and estimating quantities of the concentrated viruses. Particular problems associated with the detection of viruses of public health interest in the aquatic environment are: (a) the small size of virus particles (about 20 to 100 nm in diameter), (b) the low virus concentrations in water and the variability in amounts and types that may be present, (c) the inherent instability of viruses as biological entities, (d) the various dissolved and suspended materials in water and wastewater that interfere with virus detection procedures, and (e) the present limitations of virus estimation and identification methods.

3. Selection of Concentration Method

The densities of enteric viruses in water and wastewater usually are so low that virus concentration is necessary, except possibly for raw sewage in certain areas or seasons.[21,22] Numerous methods for concentrating waterborne enteric viruses have been proposed, tested under laboratory conditions with experimentally contaminated samples, and in some cases used to detect viruses under field conditions.[23,24]

Virus concentration methods often are capable of processing only limited volumes of water of a given quality. In selecting a virus concentration method consider the probable virus density, the volume limitations of the concentration method for that type of water, and the presence of interfering constituents. A sample volume less than 1 L and possibly as small as a few milliliters may suffice for recovery of viruses from raw or primary treated sewage. For drinking water and other relatively nonpolluted waters, the virus levels are likely to be so low that hundreds or perhaps thousands of liters must be sampled to increase the probability of virus detection.

Three different techniques used to concentrate viruses from water are described herein: absorption to and elution from microporous filters (Methods B and C); aluminum hydroxide adsorption-precipitation (Method D); and polyethylene glycol (PEG) hydroextraction-dialysis (Method E).[23,24] A separate technique (Method F) for recovering viruses from solids in small volumes of water also is described. Virus concentration by adsorption to and elution from microporous filters can be used for both small volumes of wastewater and large volumes of natural and finished waters. The aluminum hydroxide adsorption-precipitation and PEG hydroextraction-dialysis methods are impractical for processing large fluid volumes. However, they are suitable for concentrating viruses from wastewater or other waters having relatively high virus densities and for second-step concentration (reconcentration) of viruses in primary eluates obtained by processing large sample volumes through microporous filters.

4. Recovery Efficiencies

In examining a particular water include a preliminary evaluation of virus recovery efficiency. To do this add a known quantity of one or more test virus types to the required volume of sample, process the sample by the concentration method, and assay the concentrate for test viruses to determine virus recovery efficiency. Ideally, such seeded samples should be used whenever field samples are processed. If seeded samples are used concurrently with field samples, take appropriate steps, including disinfection and sterilization and the use of aseptic technique, to prevent accidental contamination of samples.

5. References

1. BITTON, G. 1980. Introduction to Environmental Virology. John Wiley & Sons, New York, N.Y.

2. WORLD HEALTH ORGANIZATION SCIENTIFIC GROUP. 1979. Human Viruses in Water, Wastewater and Soil. Tech. Rep. Ser. 639, World Health Org., Geneva, Switzerland.

3. SOBSEY, M.D. 1975. Enteric viruses and drinking water supplies. *J. Amer. Water Works Assoc.* 67:414.

4. FEACHEM, R., H. GARELICK & J. SLADE. 1981. Enteroviruses in the environment. *Trop. Dis. Bull.* 78:185.

5. ENVIRONMENTAL RESEARCH CENTER. 1978. Human Viruses in the Aquatic Environment: A Status Report with Emphasis on the EPA Research Program. EPA-507/9-78/006, U.S.

Environmental Protection Agency, Cincinnati, Ohio.

6. BLACKLOW, N.R. & G. CUKOR. 1980. Viral gastroenteritis agents, Chap. 90 *in* E.H. Lennette, A. Balows, W.J. Hausler, Jr. & J.P. Truant, eds. Manual of Clinical Microbiology, 3rd ed. American Soc. Microbiology, Washington, D.C.

7. KAPIKIAN, A.Z., R.H. YOLKEN, H.B. GREENBERG, R.G. WYATT, A.R. KALICA, R.M. CHANOCK & H.W. KIM. 1979. Gastroenteritis viruses. *In* E.H. Lennette & N.J. Schmidt, eds. Diagnostic Procedures for Viral, Rickettsial and Chlamydial Infections. American Public Health Assoc., Washington, D.C.

8. PROVOST, P.J., W.J. MCALEER & M.R. HILLEMAN. 1982. *In vitro* cultivation of hepatitis A virus. *In* W. Szmuness, H.J. Alter & J.E. Maynard, eds. Viral Hepatitis. 1981. International Symposium, The Franklin Institute Press, Philadelphia, Pa.

9. WYATT, R.G., W.D. JAMES, E.H. BOHL, K.W. THEIL, L.J. SAIF, A.R. KALICA, H.B. GREENBERG, A.Z. KAPIKIAN & R.M. CHANOCK. 1980. Human rotavirus type 2: Cultivation *in vitro*. *Science* 207:189.

10. SATO, K., Y. INABA, T. SHINOZAKI, R. FUJII & M. MATUMOTO. 1981. Isolation of human rotavirus in cell cultures. *Arch. Virol.* 69:155.

11. LOCARNINI, S.A., A.G. COULEPIS, H. ZHUANG, E.G. WESTAWAY & I.D. GUST. 1982. Characterization of the replication cycle of hepatitis A virus *in vitro*. *In* W. Szmuness, H.J. Alter & J.E. Maynard, eds. Viral Hepatitis. 1981. International Symposium, The Franklin Institute Press, Philadelphia, Pa.

12. FEINSTONE, S.M., L.F. BARKER & R.H. PURCELL. 1979. Hepatitis A and B. Chap. 29 *in* E.H. Lennette & N.J. Schmidt, eds. Diagnostic Procedures for Viral, Rickettsial and Chlamydial Infections. American Public Health Assoc., Washington, D.C.

13. VISNAWATHAN, R. 1957. Epidemiology. *Indian J. Med. Res.* 45:1 (supplementary number).

14. MELNICK, J.L. 1957. A water-borne urban epidemic of hepatitis. *In* Hepatitis Frontiers. Little, Brown & Co., Boston, Mass.

15. PLOTKIN, S.A. & M. KATZ. 1967. Minimal infective doses of viruses for man by the oral route. *In* G. Berg, ed. Transmission of Viruses by the Water Route. Interscience Publ., New York, N.Y.

16. AKIN, E. 1981. A review of infective dose data for enteroviruses and other enteric microorganisms in human subjects. *In* Microbial Health Considerations of Soil Disposal of Domestic Wastewaters. EPA-600/9-83-017, U.S. Environmental Protection Agency, Washington, D.C.

17. PAYMENT, P. 1981. Isolation of viruses from drinking water at the Pont-Viau water treatment plant. *Can. J. Microbiol.* 27:417.

18. CRAUN, G.F. & J.L. MCCABE. 1973. Review of the causes of waterborne disease outbreaks. *J. Amer. Water Works Assoc.* 65:74.

19. CRAUN, G.F. 1981. Outbreaks of waterborne disease in the United States: 1971-1978. *J. Amer. Water Works Assoc.* 73:360.

20. SOBSEY, M.D. 1982. Quality of currently available methodology for monitoring viruses in the environment. *Environ. Internat.* 7:39.

21. BURAS, N. 1974. Recovery of viruses from waste-water and effluent by the direct inoculation method. *Water Res.* 8:19.

22. BURAS, N. 1976. Concentration of enteric viruses in wastewater and effluent: A two year survey. *Water Res.* 10:295.

23. SOBSEY, M.D. 1976. Methods for detecting enteric viruses in water and wastewater. *In* G. Berg, H.L. Bodily, E.H. Lennette, J.L. Melnick & T.G. Metcalf, eds. Viruses in Water. American Public Health Assoc., Washington, D.C.

24. GERBA, C.P. & S.M. GOYAL. 1982. Methods in Environmental Virology. Marcel Dekker, New York.

9510 B. Virus Concentration from Small Sample Volumes by Adsorption to and Elution from Microporous Filters (PROPOSED)

1. General Discussion

Viruses can be concentrated from aqueous samples by reversibly adsorbing them to microporous filters and then eluting them from the filters in a small liquid volume.[1] The virus-containing sample is pressure-filtered through microporous filters having large surface areas to which viruses adsorb, presumably by both electrostatic and hydrophobic interactions.[2] Two general types of adsorbent filters are available: electronegative (negative surface charge) and electropositive (positive surface charge). The former filters are composed of either cellulose esters or fiberglass with organic resin binders. They adsorb viruses most efficiently in the presence of multivalent cations such as Al^{3+} and Mg^{2+} and/or at low pH, usually pH 3.5. The latter filters are composed of either fiberglass or cellulose and a positively charged organic, polymeric resin. They adsorb viruses efficiently over a wide pH range without added polyvalent salts. If the sample is neutral or acidic, it can be processed with these filters without chemical conditioning.

Electropositive filters have given virus recoveries comparable to those with electronegative filters[3-5] and have been used in field studies,[6,7] but there is still insufficient documentation from different laboratories using a variety of virus types and water samples to include a specific method for them now.

Adsorbed viruses usually are eluted from the surfaces of microporous filters by pressure-filtering a small volume of eluent fluid through the filters in situ. The eluent is either a slightly alkaline proteinaceous fluid such as beef extract or a more alkaline buffer such as glycine-NaOH, pH 10.5 to 11.5. If glycine-NaOH is used as eluent, preferably use pH 10.5 because of the greater likelihood of virus inactivation at the higher pH.[8,9]

Microporous filter methods suffer from three main limitations. Sample suspended matter tends to clog the adsorbent filter, thereby limiting the volume that can be processed and possibly interfering with the elution process.[10] Dissolved and colloidal organic matter in some waters can interfere with virus adsorption to filters, presumably by competing with viruses for adsorption sites,[11-13] and they also can interfere with virus elution. Finally, viruses adsorbed to suspended matter may be removed in any clarification procedure applied before virus adsorption. These solids-associated viruses are lost from the sample unless special efforts are made to recover the solids and process them for viruses.[10] A method for recovering solids-associated viruses from small volumes of water and wastewater is given in Section 9510F. Despite these limitations, virus concentration by adsorption to and elution from microporous filters is a most promising technique for detecting viruses.

2. Equipment and Apparatus

a. Adsorbent filter holder, 47-, 90-, or 142-mm diam, equipped with pressure relief valve.

b. Pressure vessel, 12- or 20-L capacity.

c. Positive pressure source up to about 400 kPa with regulator: laboratory air line, air pump, or cylinder of compressed air or nitrogen gas.

d. Autoclavable vinyl plastic tubing with plastic or metal connectors (quick-disconnect type), for connecting positive pressure source, pressure vessel, and filter holder in series.

e. pH meter.

f. Beakers, 50- to 500-mL.

g. Laboratory balance.

h. *Graduated cylinders,* 25- to 100-mL.

i. *Pipets,* 1-, 5-, and 10-mL.

3. Materials

a. *Virus adsorbent filter:* Use either:

1) *Cellulose nitrate filter,* 0.45-μm porosity.*

2) *Fiberglass-acrylic resin filter,* 0.45-μm porosity.† Filter media available commercially only as flat sheets can be cut to the desired disk diameter with scissors.

b. *Prefilter:* Use one or more cellulose nitrate or fiberglass-acrylic resin filters or equivalent, with porosities greater than 0.45 μm to prevent clogging of the virus adsorbent filter by suspended matter. Place prefilters on top of the 0.45-μm-porosity virus adsorbent filter in the same filter holder.

4. Reagents

a. *Hydrochloric acid,* HCl, 0.1, 1.0, and 10N.

b. *Sodium hydroxide,* NaOH, 0.1, 1.0, and 10N.

c. *Aluminum chloride,* $AlCl_3 \cdot 6H_2O$, 0.15N, or magnesium chloride, $MgCl_2 \cdot 6H_2O$, 5N.

d. *Sodium thiosulfate,* $Na_2S_2O_3 \cdot 5H_2O$, 0.5% (w/v).

e. *Sodium chloride,* 0.14N, pH 3.5: Dissolve 8.18 g in 1 L distilled water and adjust to pH 3.5 with HCl.

f. *Virus eluent:* Use either:

1) *Glycine-NaOH,* pH 10.5 or 11.5: Prepare 0.05M glycine solution, autoclave, and adjust to pH 10.5 or 11.5 with 1 to 10N NaOH. Add phenol red, 0.0005%, as a pH indicator.

2) *Beef extract,* 3%, pH 9.0: Dissolve 30 g beef extract paste or 24 g beef extract powder in 1000 mL distilled water, adjust to pH 9.0 with 1 to 10N NaOH, and sterilize by autoclaving.

*Type HA, Millipore Corp., Bedford, Mass., or equivalent.

†No. 8025-035, Filterite Corp., Timonium, Md., or equivalent.

g. *Glycine-HCl,* pH 1.5: Prepare 0.05M glycine solution, autoclave, and adjust to pH 1.5 with 1 to 10N HCl. Add phenol red, 0.0005%, as a pH indicator.

h. *Nutrient broth,* 10×, pH 7.5: Dissolve 8.0 g nutrient broth in 90 mL distilled water, adjust to pH 7.5, dilute to 100 mL with distilled water, and sterilize by autoclaving.

i. *Antibiotics:* Use either:

1) *Penicillin-streptomycin,* 10×: Contains 5000 IU penicillin/mL and 5000 μg streptomycin/mL. Use commercially available form or prepare by dissolving powdered sodium or potassium penicillin-G and streptomycin sulfate in distilled water and sterilizing by filtration. Store frozen.

2) *Gentamycin-kanamycin,* 100×: Contains 5000 μg/mL each of gentamycin (base) and kanamycin (base). Prepare by combining aseptically equal volumes of commercially available sterile gentamycin and kanamycin solutions, 10 000 μg/mL, respectively, or by dissolving powdered gentamycin sulfate and kanamycin sulfate in distilled water and sterilizing by filtration. Store refrigerated or frozen.

j. *Hanks balanced salt solution,* 10×: Use commercially available form or prepare following a standard protocol.[14,15]

k. *Sodium hypochlorite,* 5.25% available chlorine (household bleach).

5. Procedure

a. *Sterilization of apparatus, materials, and reagents:* Most reagents, virus adsorbent filters, filter holders, tubing, and labware can be sterilized by autoclaving or made virus-free by streaming steam. To sterilize filters load into their holders; if several filters are to be placed in one holder, place filter with smallest porosity on the bottom with progressively larger filters on top. Do not use an automatic drying cycle when autoclaving virus adsorbent filters. Sterilize apparatus and material that cannot be autoclaved or treated with streaming steam by treating with 10-mg/L free chlorine solution, pH 7.0, for 30 min and rinse or flush with 50-mg/L sterile $Na_2S_2O_3$ solution. Do not treat adsorbent filters with chlorine. Use aseptic technique during all virus concentration operations to prevent extraneous microbial contamination.

b. Sample size and choice of filter size: Sample size and, hence, filter diameter depend partly on water quality and the probable virus concentration. Single-stage microporous filter adsorption-elution methods have been used to recover viruses from 100 mL raw sewage on 47-mm-diam filters[16] and from 3.8 to 4.6 L secondary and tertiary sewage effluent on 90- or 142-mm-diam filters.[12,16,17] Based on the diameter and solids-holding characteristics of the filters, the scale and volume capacity of the apparatus and materials, and the quality of the samples, the practical limits for sample size are 20, 8, and 2 L for 142-, 90-, and 47-mm-diam filters, respectively.

c. Sample collection and storage: Collect samples aseptically in sterile containers. If they contain residual chlorine, immediately add $Na_2S_2O_3$ solution to give a final concentration of 50 mg/L. Process samples as soon as possible after collection; do not hold samples for more than 2 h at up to 25°C or 48 h at 2 to 10°C. Do not freeze samples unless they cannot be processed within 48 h; then freeze and store at −70°C or less.

d. Sample processing: Adjust sample to pH 3.5 and $0.0015N$ $AlCl_3$ or to between pH 6.0 and 3.5 and $0.1N$ $MgCl_2$. Make sample adjustments either in a pressure vessel or in another appropriate container. Mix sample vigorously during addition of 1.0 or $0.1N$ HCl and $AlCl_3$ solution (1 part solution to 100 parts sample) or $MgCl_2$ solution (1 part solution to 50 parts sample). Because $AlCl_3$ is an acid salt, it may decrease sample pH slightly. Do not let sample pH fall below 3.0.

Place sample in a pressure vessel connected to a source of positive pressure and connect pressure vessel outlet to inlet of virus adsorbent filter holder. With pressure relief valve on filter holder opened, apply a slight positive pressure to purge air from filter holder. When sample just begins to flow from pressure relief valve, quickly close valve and continue filtration at a rate not exceeding 28 mL/min/cm² of filter area (about 130, 250, and 4000 mL/min for 47-, 90-, and 142-mm-diam filters, respectively). After filtering entire sample let positive pressure source purge excess fluid from filter holder.

Wash filters with $0.14N$ NaCl to remove excess Al^{3+} or Mg^{2+} from virus adsorbent filter. Use about 1.5 mL NaCl solution/cm² filter area (25, 100, and 240 mL for 47-, 90-, and 142-mm-diam filters, respectively). Place wash solution in a pressure vessel connected to filter holder inlet, use positive pressure to filter solution through virus adsorbent filter, discard filtrate, and let positive pressure purge virus adsorbent filter of excess wash solution.

Elute viruses from filters with a recommended eluent. Use about 0.45 mL eluent/cm² filter surface area (about 7.5, 28, and 71 mL for 47-, 90-, and 142-mm-diam filters, respectively). With pressure relief valve on filter holder open, add eluent to filter holder so that it completely covers filter surface. When eluent begins to discharge from pressure relief valve, quickly close valve. If pH 11.5 glycine-NaOH is the eluent, place a sterile beaker under filter outlet and apply positive pressure so that filtrate flows slowly from filter holder outlet. Collect filtrate in sterile beaker and, when filtrate no longer flows, slowly increase pressure to force retained fluid from filters. Quickly check eluate (filtrate) pH. If it is less than 11.0, elute with additional pH 11.5 glycine-NaOH until an eluate with a pH ≥ 11.0 is obtained. Immediately after checking pH, adjust eluate to a pH between 9.5 and 7.5 with pH 1.5 glycine-HCl or $0.1N$ HCl while mixing vigorously. Complete elution and eluate pH adjustment to 7.5 to 9.5 in 5 min or less to avoid the possibility of appreciable virus inactivation.

If pH 10.5 glycine-NaOH is the eluent, proceed as with pH 11.5 glycine-NaOH, but pass the eluate through the filters a total of five times. For each elution, collect the filtrate, readjust to pH 10.5 with 1.0 or $0.1N$ NaOH, and then pass through the filter. After the fifth elution, adjust filtrate to pH 7.4 with glycine-HCl, pH 1.5, or $0.1N$ HCl.

If 3% beef extract, pH 9.0, is the eluent, place a sterile beaker under filter outlet, apply a slight positive pressure to eluent-containing filter holder so that filtrate flows slowly from the outlet, and collect filtrate. Slowly increase pressure to force additional retained fluid from filters.

Measure eluate volume and add 1/10 of the measured volume each of penicillin-streptomycin or gentamycin-kanamycin, Hanks balanced salt solution, and 10X nutrient broth (add last item to glycine eluates only). Adjust sample to pH 7.4 with glycine-HCl or $0.1N$ HCl while mixing vigorously. Store at either 4 or $-70°C$, depending on the time until virus assay. Maximum storage at 4°C is 48 h.

6. References

1. FARRAH, S.R., C.P. GERBA, C. WALLIS & J.L. MELNICK. 1976. Concentration of viruses from large volumes of tapwater using pleated membrane filters. *Appl. Environ. Microbiol.* 31:221.

2. FARRAH, S.R., D.O. SHAH & L.O. INGRAM. 1981. Effects of chaotropic and antichaotropic agents on the elution of poliovirus adsorbed to membrane filters, *Proc. Nat. Acad. Sci. U.S.* 18:1229.

3. SOBSEY, M.D. & B.L. JONES. 1979. Concentration of poliovirus from tap water using positively charged microporous filters. *Appl. Environ. Microbiol.* 37:588.

4. SOBSEY, M.D. & J.S. GLASS. 1980. Poliovirus concentration from tap water with electropositive adsorbent filters. *Appl. Environ. Microbiol.* 40:201.

5. SOBSEY, M.D., R.S. MOORE & J.S. GLASS. 1981. Evaluating adsorbent filter performance for enteric virus concentrations in tap water. *J. Amer. Water Works Assoc.* 73:542.

6. CHANG, L.T., S.R. FARRAH & G. BITTON. 1981. Positively charged filters for virus recovery from wastewater treatment plant effluents. *Appl. Environ. Microbiol.* 42:921.

7. HEJKAL, T.W., B. KESWICK, R.L. LABELLE, C.P. GERBA, Y. SANCHEZ, G. DREESMAN, B. HAFKIN & J.L. MELNICK. 1982. Viruses in a community water supply associated with an outbreak of gastroenteritis and infectious hepatitis. *J. Amer. Water Works Assoc.* 74:318.

8. SOBSEY, M.D., J.S. GLASS, R.J. CARRICK, R.R. JACOBS & W.A. RUTALA. 1980. Evaluation of the tentative standard method for enteric virus concentration from large volumes of tap water. *J. Amer. Water Works Assoc.* 72:292.

9. SOBSEY, M.S., J.S. GLASS, R.R. JACOBS & W.A. RUTALA. 1980. Modification of the tentative standard method for improved virus recovery efficiency. *J. Amer. Water Works Assoc.* 72:350.

10. WELLINGS, F.M., A.L. LEWIS & C.W. MOUNTAIN. 1976. Demonstration of solids-associated virus in wastewater and sludge. *Appl. Environ. Microbiol.* 31:354.

11. FARRAH, S.R., S.M. GOYAL, C.P. GERBA, C. WALLIS & P.T.B. SHAFFER. 1976. Characteristics of humic acid and organic compounds concentrated from tapwater using the aquella virus concentrator. *Water Res.* 10:897.

12. WALLIS, C. & J.L. MELNICK. 1967. Concentration of viruses from sewage by adsorption on Millipore membranes. *Bull. World Health Org.* 36:219.

13. SOBSEY, M.D., C. WALLIS, M. HENDERSON & J.L. MELNICK. 1973. Concentration of enteroviruses from large volumes of water. *Appl. Microbiol.* 26:529.

14. SCHMIDT, N.J. 1979. Tissue culture technics for diagnostic virology. *In* E.H. Lennette & N.J. Schmidt, eds. Diagnostic Procedures for Viral and Rickettsial Infections, 5th ed. American Public Health Assoc., Washington, D.C.

15. ROVOZZO, G.C. & C.N. BURKE. 1973. A Manual of Basic Virological Techniques. Prentice-Hall, Englewood Cliffs, N.J.

16. RAO, V.C., U. CHANDORKAR, N.U. RAO, P. KUMARAN & S.B. LAKHE. 1972. A simple method for concentrating and detecting viruses in wastewater. *Water Res.* 6:1565.

17. GERBA, C.P., S.R. FARRAH, S.M. GOYAL, C. WALLIS & J.L MELNICK. 1978. Concentration of enteroviruses from large volumes of tap water, treated sewage, and seawater. *Appl. Environ. Microbiol.* 35:540.

9510 C. Virus Concentration from Large Sample Volumes by Adsorption to and Elution from Microporous Filters (PROPOSED)

1. General Discussion

This section describes a two-stage process for concentrating viruses from large sample volumes. Viruses in eluate volumes too large to be conveniently and economically assayed directly in cell cultures, such as those obtained from processing large volumes of water through cartridge or large disk filters, can be concentrated further (reconcentrated) by several alternative methods. Viruses in proteinaceous eluates can be reconcentrated by either "organic flocculation,"[1,2] aluminum hydroxide adsorption-precipitation (Section 9510D), or polyethylene glycol hydroextraction-dialysis (Section 9510E). These reconcentration techniques can be used for both proteinaceous and organic buffer eluates from all types of water. Organic flocculation, now used widely, involves precipitating viruses by acidifying eluates to pH 3.5, recovering the precipitate by centrifugation, and then resuspending it in a small volume of alkaline buffer.[1]

Additionally, viruses in nonproteinaceous eluates such as glycine-NaOH can be reconcentrated by adsorption to and elution from small microporous filters. The eluate is adjusted to pH and ionic conditions for optimum virus adsorption, filtered through a secondary adsorbent, and adsorbed viruses are eluted with a small volume of eluent. This procedure can be used only for reconcentrating primary eluates obtained from processing drinking water and other highly finished waters because of potential interfering substances likely to be present in primary eluates from natural and less finished waters.

Figure 9510:1 shows the alternative microporous filter adsorption-elution and reconcentration methods.

For general information on microporous filter techniques, see Section 9510B.1.

2. Equipment and Apparatus

a. *Apparatus for first-stage concentration* (Figure 9510:2):

1) *First-stage virus adsorbent filter holder.*

2) *Chemical additive system.* Use either:

a) *Fluid proportioner* with four feed pumps (quadraplex) and a mixing chamber.*

b) *Venturi-type proportioning injector*† with plastic or metal connectors (quick-disconnect type) and a length of vinyl tubing for the chemical feed line.[3] To feed two separate additives, attach a "Y" or "T" connector and two lengths of vinyl tubing to the chemical feed port, or alternatively, use two separate proportioning injectors. It may be necessary to use a bypass system

---

*Johanson and Son Machine Corp., Clifton, N.J., or equivalent.
†Models 202-P, 203-P or 204-P, Dema Engineering Co., St. Louis, Mo., or equivalent.

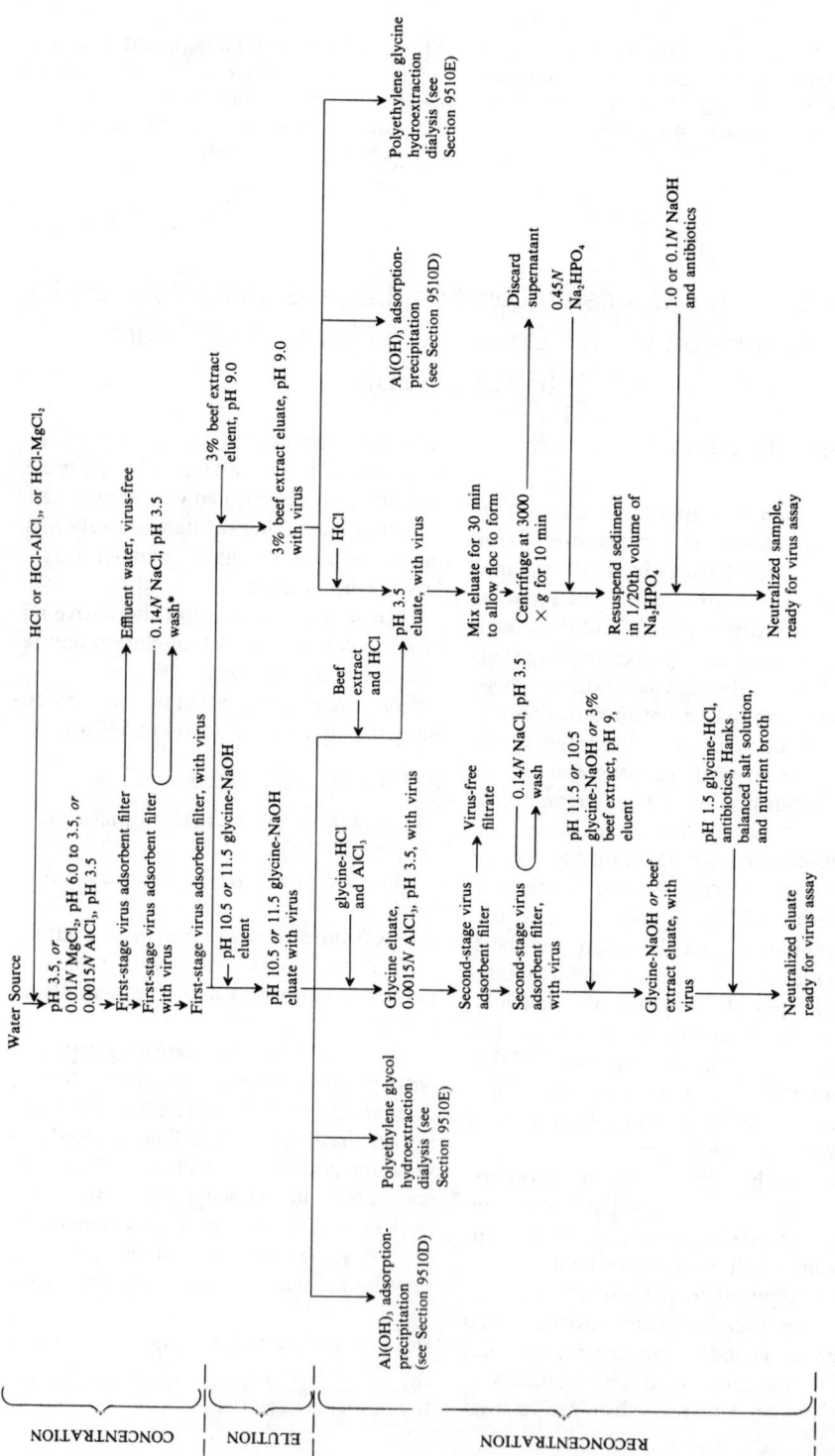

Figure 9510:1. Two-stage microporous filter adsorption-elution method for concentrating viruses from large volumes of water.

* When using MgCl₂ or AlCl₃ to enhance virus adsorption.

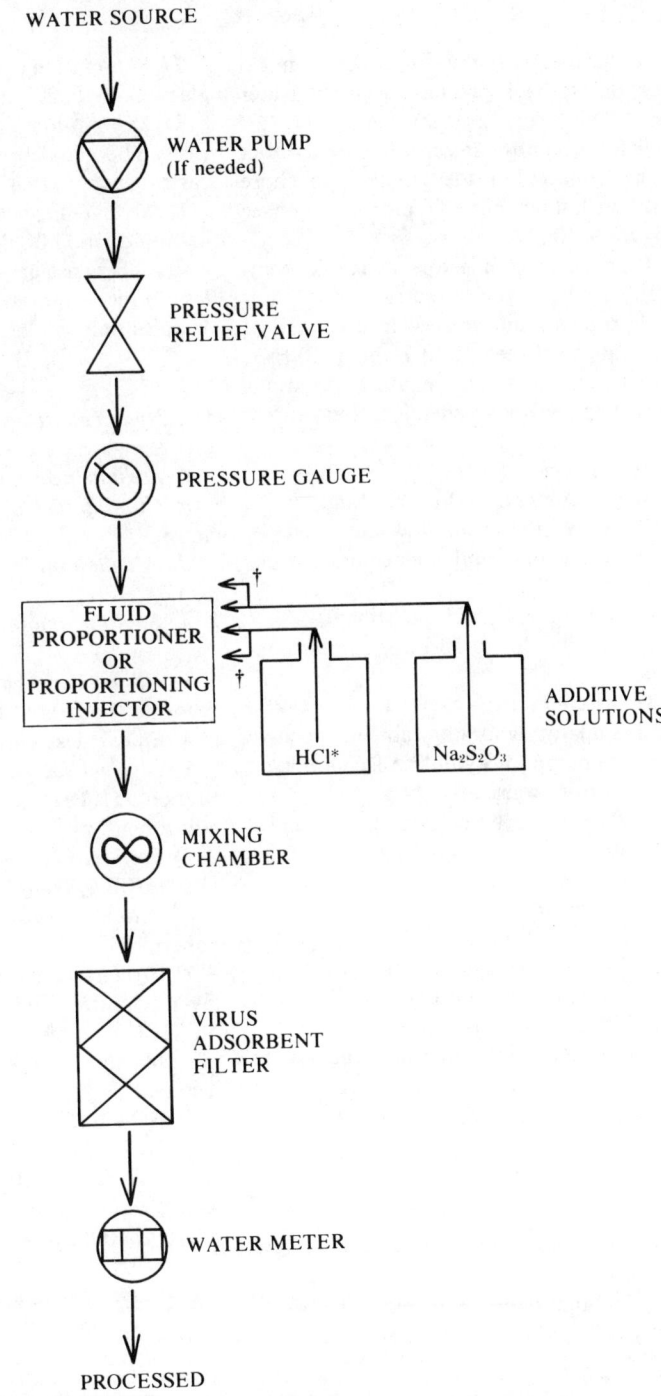

Figure 9510:2. Schematic of apparatus for first-stage concentration.

* Plus AlCl₃ or MgCl₂, if used.
† Required for fluid proportioner but not proportioning injector.

with the injector to prevent loss of chemical feed due to back pressure from the water line.[4] This bypass system consists of "T" pipe fittings on the injector inlet and outlet ports connected by a length of flexible hose with an in-line shut-off/control valve (see Figure 9510:2).

Proportioning injectors available commercially will process water at flow rates of 3 to 33 L/min with water-to-chemical feed ratios between 10 to 1 and 1110 to 1. Select equipment and operating conditions providing a water-to-chemical feed ratio of 100 to 1.

3) *Water flow meter.*

4) *Pressure gauge,* 0 to 400 kPa.

5) *Vinyl plastic tubing,* autoclavable, with plastic or metal connectors (quick-disconnect type).

6) *Pressure relief valve* (optional).

7) *Carboys,* 20- to 50-L, or similar containers.

8) *Positive pressure source* up to 400 kPa with regulator: laboratory air line, positive pressure pump, or cylinder of compressed air or nitrogen gas.

9) *Pump* (if source water is not under pressure).

b. pH meter.

c. Laboratory balance.

d. Beakers, 2- or 4-L.

e. Pressure vessel, 4-L.

f. Graduated cylinders, 1- and 2-L.

g. Pipets, 1-, 5-, and 10-mL.

h. Centrifuge with rotor and buckets for 250- to 500-mL-capacity bottles.‡

i. Centrifuge bottles, 250- to 500-mL.

3. Materials

a. First-stage virus adsorbent filters: Use one of the following:

1) 293-mm-diam, 8.0- and 1.2-μm-porosity cellulose nitrate filter series.§

2) 17.8-cm-long, 8.0-μm-porosity fiberglass-epoxy filter tube.‖

3) 25.4-cm-long, 0.25- or 0.45-μm-porosity fiberglass-acrylic resin pleated filter cartridge.#

b. Second-stage virus adsorbent filters: 47-mm-diam, 3.0-, 0.45-, and 0.25-μm-porosity fiberglass-acrylic resin filter series. Use to reconcentrate highly finished water samples only.#

4. Reagents

a. Hydrochloric acid, HCl, 0.06, 1,** and 6N.

b. Sodium hydroxide, NaOH, 10N.

c. Aluminum chloride, $AlCl_3 \cdot 6H_2O$, 0.15 and 6N.**

d. Magnesium chloride, $MgCl_2 \cdot 6H_2O$, 10N.**

e. Sodium thiosulfate, $Na_2S_2O_3 \cdot 5H_2O$, 0.5% (w/v).

f. Sodium hypochlorite, 5.25% available chlorine (household bleach).

g. Eluent: Use either:

1) *Glycine-NaOH,* pH 10.5 or 11.5: See Section 9510B.4*f*1). Use within 2 h of pH adjustment.

2) *Beef extract,* 3%, pH 9.0.†† See Section 9510B.4*f*2).

h. Eluate neutralizing solution: Use either:

1) *Glycine-HCl,* pH 1.5: Prepare 0.05M glycine solution and adjust to pH 1.5 with 6N HCl. Add phenol red, 0.0005%, as a pH indicator. Use within 2 h of pH adjustment.

2) *HCl,* 1.0N.

i. Nutrient broth, 10×, pH 7.5: Dissolve 8.0 g nutrient broth in 90 mL distilled water, adjust to pH 7.5 with 10N NaOH, dilute to 100 mL with distilled water, and sterilize by autoclaving.

j. Disodium phosphate, 0.45N: Dissolve

---

‡Required for alternative reconcentration procedure using 3% beef extract.
§Millipore Corp., Bedford, Mass., or equivalent.

‖Balston, Inc., Lexington, Mass., or equivalent.
#Filterite Corp., Timonium, Md., or equivalent.
**Recommended for first-stage virus adsorption.
§§For alternative reconcentration procedure: organic flocculation.

40.2 g $Na_2HPO_4 \cdot 7H_2O$ in 1 L distilled water and sterilize by autoclaving.

k. *Antibiotics:* See Section 9510B.4*i.*

l. *Sodium chloride, 0.14N:* Dissolve 8.18 g NaCl in 1 L distilled water.

m. *Hanks balanced salt solution,* 10×: See Section 9510B.4*j.*

5. Procedure

a. *Sterilization of apparatus, materials, and reagents:* See Section 9510B.5*a.*

b. *Sample size:* For drinking water use a minimum sample of 400 L, although 2000 L or more may have to be processed to detect viruses at a concentration of 1 to 2 infectious units/400 L.

c. *Preparation of feed solutions:* Use an HCl additive solution to adjust sample pH to 3.5 for virus adsorption to filters. If acidification to pH 3.5 is inadequate for obtaining maximum virus adsorption, add either $AlCl_3$ or $MgCl_2$ solution.

When only HCl is used, prepare additive solution as follows: Determine concentration of HCl additive solution by titrating a 1-L sample of dechlorinated water to pH 3.5 with 0.06N HCl and noting volume required. The volume, in milliliters, of titrant required is equal to the volume of 6N HCl needed/L distilled water for making the additive solution. Make at least 5 L additive solution for 400 L of sample.

When $AlCl_3$ is used to enhance virus adsorption use pH 3.5 and a final concentration of added $AlCl_3$ of 0.0015N. Because $AlCl_3$ is an acid salt, titrate a 1-L sample to about pH 4.0 with 0.06N HCl, add $AlCl_3$ to a concentration of 0.0015N and continue titration to pH 3.5, noting volume of titrant used. Prepare additive solution by adding titrant volume (mL) of 6.0N HCl/L of 0.15N $AlCl_3$.

When $MgCl_2$ is used to enhance virus adsorption, use a pH between 3.5 and 6.0 and a final concentration of added $MgCl_2$ of 0.1N. To prepare additive solution titrate

a 1-L sample to desired pH with 0.06N HCl as previously described and note volume of titrant used. Add the titrant volume of 6.0N HCl/L 10N $MgCl_2$ to make the additive solution.

d. *Preparation of chemical additive system:*

1) When using a fluid proportioner, operate at a pressure of 100 to 700 kPa and a water flow rate of 4 to 40 L/min. Adjust each of the four chemical additive pumps of the proportioner for a ratio of 1 to 200 (1 part chemical additive to 200 parts water). Use two pumps, operating reciprocally, for each additive so that the overall dilution for each additive is 1 to 100. One additive is either HCl, $HCl-AlCl_3$ or $HCl-MgCl_2$; the other additive, 0.5% $Na_2S_2O_3$, is needed only when processing samples containing chlorine. Place lines from the two pumps of each additive solution into the additive containers and manually operate the pump metering rods to fill feed lines and purge them of air. Connect fluid proportioner to source water and operate briefly without a virus adsorbent in place. Sample conditioned water from proportioner outlet and check pH. The pH should be 3.5 ± 0.3.

2) When using a Venturi-type proportioning injector, connect injector assembly to water source and to adsorbent filter inlet, and place additive feed line(s) into additive container(s). Position valve on injector outlet to drain line position (away from adsorbent filter). Begin flow of sample. Adjust screw-operated control valve on chemical feed of proportioning injector until water collected from drain line is at the desired pH as measured with a pH meter. If $Na_2S_2O_3$ is used to neutralize chlorine, check to insure that chlorine is absent. Connect virus concentrator assembly to source water by attaching concentrator inlet hose to valved outlet of a pressurized water source or to outlet of a water pump, the inlet of which has been placed in the source water. Operate for several minutes

without a virus adsorbent in place to purge the unit of chlorine solution. Collect a sample from outlet of meter to insure absence of chlorine.

e. *First-stage concentration:* After preparing concentration apparatus and additive solutions and checking conditioned water for proper pH and absence of chlorine, attach a virus adsorbent filter to outlet of chemical additive system. Attach water meter and effluent hose to virus adsorbent outlet. Record initial meter reading and add to this value the desired volume to be processed plus an additional 1 or 2% (to account for volume of either 1 or 2 additive solutions, respectively). This gives meter reading at which sampling is to be stopped. Turn on water and start a timer (or record starting time). Shortly after filtration begins collect a sample from filter outlet and check for absence of chlorine and for appropriate pH value. Also check flow rate. Do not use a flow rate above 40 L/min. Recheck pH and chlorine residual several times during sample processing. When desired volume has been processed, turn water off. Purge filter holder of excess water with positive pressure from an air or nitrogen gas source.

f. *Washing and virus elution:* If AlCl$_3$ or MgCl$_2$ has been used, wash excess Al^{3+} or Mg^{2+} from filter with 4 L 0.14N NaCl. Omit washing if only HCl was used. Place wash solution in a 4-L pressure vessel and pass through filter with positive pressure. Purge filter of excess wash solution with positive pressure and discard entire filtrate.

Using aseptic technique, elute virus from filter as soon as possible in the field or after returning to the laboratory. If filter holders with adsorbed viruses must be returned to the laboratory, seal filter holder openings, place filter holder in a sterile plastic bag, and chill.

Use pH 10.5 or 11.5 glycine-NaOH or 3% beef extract, pH 9.0, to elute viruses from first-stage adsorbent filters. Because some viruses are inactivated when pH 11.5

glycine-NaOH is used, alternatively elute with pH 10.5 glycine-NaOH or 3% beef extract, pH 9.0.[1,3,4]

To elute, place eluent in a pressure vessel. Use minimum eluent volumes of 1 L and 300 mL for cartridge and 293-mm-diam disk filters, respectively. To elute with pH 11.5 glycine-NaOH, connect pressure vessel to inlet of filter holder and with pressure relief valve on filter holder open, apply a small positive pressure to the system so that eluent fills void volume of filter holder. When eluent begins to discharge from pressure relief valve, quickly close it. Filter remaining eluent slowly through filter within 1 to 2 min and collect filtrate (eluate) in a sterile 2- or 4-L beaker. When filtrate no longer appears, slowly increase pressure to force additional fluid from filter. If using pH 11.5 glycine-NaOH eluent, immediately check filtrate pH and if it is less than 11.0, elute with 1 L more of pH 11.5 glycine-NaOH. Immediately after checking pH, adjust filtrate to a pH between 7.5 and 9.5 with pH 1.5 glycine-HCl while mixing vigorously. Complete elution and pH adjustment to 7.5 to 9.5 in 5 min or less to avoid possibility of appreciable virus inactivation.

To elute with pH 10.5 glycine-NaOH, use either batch or continuous-flow eluent recirculation. For the batch method, begin elution as with pH 11.5 glycine-NaOH. Collect the filtrate, measure pH, and readjust to pH 10.5 with 1.0 or 0.1N NaOH while mixing vigorously. Then, using this eluate, elute filters four more times, readjusting filtrate to pH 10.5 before each elution. After the fifth elution, adjust filtrate to pH 7.4 with pH 1.5 glycine-HCl or 1.0N HCl while mixing vigorously.

Alternatively, elute with pH 10.5 glycine-NaOH by continuous recirculation. Place eluent in a sterile beaker. Attach short lengths of sterile vinyl or rubber tubing to inlet and outlet openings of filter holder and place free ends of tubing in eluent beaker; slip midsection of filter inlet tubing into a peristaltic or roller pump. Open pressure relief valve on filter holder and operate pump at slow speed so that eluent fills void volume of filter holder. When eluent begins to discharge from pressure relief valve, quickly close it. Increase pump speed so that eluent recirculates through filter assembly and beaker at a minimum flow rate of 100 mL/min. After 5 min recirculation, remove filter inlet tube from beaker and pump remaining fluid from filter assembly. Connect filter inlet to positive pressure source to force additional eluent from filter. Adjust eluate to pH 7.4 with pH 1.5 glycine-HCl or 1.0N HCl while mixing vigorously.

To elute with 3% beef extract, pH 9.0, follow the procedure described above for pH 11.5 glycine-NaOH. Adjust collected filtrate to pH 7.4 with pH 1.5 glycine-HCl or 1N HCl while mixing vigorously. The 5 min time limit to complete elution with pH 11.5 glycine-NaOH is not necessary when beef extract is used.

g. Reconcentration of primary eluates: Further concentrate (reconcentrate) viruses in primary eluates either by organic flocculation, Al(OH)$_3$ adsorption-precipitation (Section 9510D), polyethylene glycol hydroextraction-dialysis (Section 9510E), or adsorption to and elution from microporous filters. The latter technique can be used only for glycine or other organic buffer eluates.

To concentrate (reconcentrate) viruses in glycine eluates by adsorption to and elution from filters, adjust to pH 3.5 with pH 1.5 glycine-HCl and add AlCl$_3$ to a final concentration of 0.0015N while mixing vigorously. Transfer sample to a 4-L pressure vessel. Filter through a 47-mm-diam 3.0-, 0.45-, and 0.25-μm-porosity fiberglass-acrylic resin filter series at a flow rate of no more than 130 mL/min and discard filtrate. Rinse filters with 25 mL 0.14N NaCl to remove excess Al^{3+}. Pipet NaCl solution directly into filter inlet or place in a small pressure vessel connected to the

inlet. Use positive pressure to pass NaCl solution through filter and discard filtrate. Elute adsorbed viruses from filter with 7-mL portions of either pH 10.5 or 11.5 glycine-NaOH or 3% beef extract, pH 9.0. Pipet 7 mL eluent directly into filter holder inlet or into a small pressure vessel connected to filter inlet and connect to a positive pressure source. Carefully apply positive pressure so that eluate flows slowly from filter outlet into a sterile container. When filtrate no longer flows from outlet, increase pressure to force retained fluid from filters. If using pH 11.5 glycine-NaOH, measure eluate pH and immediately adjust to pH between 7.5 and 9.5 with pH 1.5 glycine-HCl. Repeat this elution procedure with another 7-mL portion of pH 11.5 glycine-NaOH. Complete reconcentration within 5 min. If neither eluate portion had an initial pH of 11.0 or more, repeat elution procedure with additional 7-mL portions of pH 11.5 glycine-NaOH until an eluate portion has a pH of at least 11.0. Combine all eluates.

If using pH 10.5 glycine-NaOH, elute five successive times with 7-mL volumes of eluent. After each elution, readjust eluate to pH 10.5 with 0.1N NaOH while mixing vigorously. After the fifth elution, adjust eluate to pH 7.4 with pH 1.5 glycine-HCl or 0.1N HCl while mixing vigorously.

If using 3% beef extract, pH 9.0, elute with two 7-mL volumes, combine filtrates, and adjust to pH 7.4.

Measure total eluate volume. For glycine eluates, add 1/10th the measured sample volume of 10× Hanks balanced salt solution and 10× nutrient broth. To all eluates add appropriate volumes of antibiotics (1/10th volume penicillin-streptomycin or 1/100th volume gentamycin-kanamycin, or both). Store at 4 or −70°C, depending on time until virus assay.

Further concentrate viruses in beef extract eluates by precipitation at pH 3.5 (organic flocculation). Viruses in glycine eluates also can be reconcentrated by this technique by first supplementing them with beef extract to a final concentration of 1 to 3%. Use sterile beef extract paste (about 80% beef extract) or sterile 20% beef extract solution made from powder to bring glycine eluates to the desired beef extract concentration. While mixing vigorously, adjust eluate to pH 3.5 by adding $1N$ HCl dropwise. Continue to mix at slow speed for 30 min and centrifuge at $3000 \times g$ for 10 min. Decant and discard supernatant. With vigorous mixing, resuspend sediment in 1/20 the initial sample volume of $0.45N$ Na_2HPO_4. Add antibiotics (1/10 final sample volume penicillin-streptomycin, 1/100 final sample volume gentamycin-kanamycin, or both) and while mixing vigorously adjust to pH 7.4 with 1.0 or $0.1N$ NaOH. Check electrical conductivity of sample. If conductivity is $>13\ 000\ \mu$mhos, dialyze sample against Hanks balanced salt solution before assay. Store at 4 or $-70°$C, depending on time until virus assay.

6. References

1. KATZENELSON, E., B. FATTAL & T. HOSTO-VESKY. 1976. Organic flocculation: an efficient second-step concentration method for the detection of viruses in tapwater. *Appl. Environ. Microbiol.* 32:638.
2. BITTON, G., B.N. FELDBERG & S.R. FARRAH. 1979. Concentration of enteroviruses from seawater and tap water by organic flocculation using non-fat dry milk and casein water. *Air Soil Pollut.* 10:187.
3. PAYMENT, P. & M. TRUDEL. 1980. A simple low cost apparatus for conditioning large volumes of water for virological analysis. *Can. J. Microbiol.* 26:548.
4. PAYMENT, P. & M. TRUDEL. 1981. Improved method for the use of proportioning injectors to condition large volumes of water for virological analysis. *Can. J. Microbiol.* 27:455.

9510 D. Virus Concentration by Aluminum Hydroxide Adsorption-Precipitation (PROPOSED)

1. General Discussion

Viruses can be concentrated from small volumes of water, wastewater, and adsorbent filter eluates by precipitation with aluminum hydroxide.[1-4] This process probably involves both electrostatic interactions between the negatively charged virus surface and the positively charged aluminum hydroxide [$Al(OH)_3$] surfaces and coordination of the virus surface by hydroxo-aluminum complexes.[5] Viruses are adsorbed to an $Al(OH)_3$ precipitate that is either added to the sample or formed in the sample from a soluble aluminum salt and a base such as sodium carbonate (Na_2CO_3) or sodium hydroxide (NaOH). Viruses are allowed to adsorb to the $Al(OH)_3$ precipitate and the virus-containing precipitate is collected by filtration or centrifugation. The recovered precipitate may be inoculated directly into laboratory hosts for virus assay or the viruses are eluted from the precipitate with an alkaline buffer or a proteinaceous solution before virus assay.

The major limitations of this method are that sample size is limited to perhaps a few liters, soluble organic matter can interfere with virus adsorption, and virus recovery from the precipitate may be incomplete. Virus adsorption may be improved by forming the $Al(OH)_3$ precipitate in the sample instead of adding it preformed. Although virus adsorption can be maximized by using large amounts of $Al(OH)_3$, the adsorbed viruses become more difficult to elute. Therefore, some intermediate amount of $Al(OH)_3$ is used to achieve maximum virus recovery. Also, $Al(OH)_3$ is a relatively nonspecific adsorbent so that other substances may be concentrated with viruses. The presence of such impurities may cause the concentrated sample to be toxic for the cell cultures normally used for virus assay.

Several modifications of the Al(OH)$_3$ adsorption-precipitation procedure have been used to concentrate viruses from water, wastewater, and eluates from adsorbent filters. Initially, preformed Al(OH)$_3$ precipitates were made by adding Na$_2$CO$_3$ to AlCl$_3$ solutions and the Al(OH)$_3$ precipitate was resuspended in 0.15N NaCl. This was added to the wastewater and the mixture was stirred gently for 1 h or more to allow viruses to adsorb to the precipitate. The precipitate was recovered by filtration, resuspended in cell culture media, and inoculated into cell cultures.[3,4] More recent procedural modifications include: (a) Al(OH)$_3$ precipitate formation within the sample,[1,2,6–8] (b) recovery of the Al(OH)$_3$ precipitate by centrifugation followed by elution of viruses from the precipitate with alkaline eluents,[1,2,6,7] and (c) a large-volume method in which the precipitate is formed in the sample and collected on a cartridge filter, and viruses are eluted from the precipitate on the filter with alkaline eluent.[9] The method described here is for relatively small sample volumes and uses Al(OH)$_3$ that is either preformed or generated within the sample. The latter modification is preferable because some viruses are not adsorbed efficiently by preformed precipitates.[10]

2. Equipment and Apparatus

a. *Centrifuge*, with rotor and buckets, capable of operating at about 1900 $\times$ g.

b. *Centrifuge bottles and tubes.*

c. *Beaker*, 100-mL or larger.

d. *pH meter.*

e. *Magnetic stirrer* and stirring bars or alternative mixing device.

f. *Graduated cylinders*, 100-mL or larger.

g. *Pipets*, 1-, 5-, and 10-mL.

h. *Laboratory balance.*

i. *Vacuum-type filter holder or Buchner filter funnel*,* 47-mm diam or larger.

j. *Filter flask.**

k. *Spatula*,* flat blade, metal or autoclavable plastic.

l. *Vacuum source*,* vacuum pump or laboratory vacuum line.

3. Materials

Filter:* Fiberglass-acrylic resin filter† or microporous filter, 0.45-μm porosity,‡ 47-mm diam or larger. To prevent virus adsorption, filter 0.1% polyoxyethylene sorbitan monooleate solution (¶ 4h) through the filters, using about 1 mL solution/cm^2 of filter surface area. Rinse filter with distilled water, using about 10 mL/cm^2 of filter surface area. Sterilize treated filters by autoclaving.

4. Reagents

a. *Hydrochloric acid*, HCl, 0.1 and 1.0N.

b. *Sodium hydroxide*, NaOH, 0.1 and 1.0N.

c. *Sodium carbonate*, Na$_2$CO$_3$, 4N.§

d. *Aluminum chloride*, AlCl$_3$, 0.075N§ or 0.9N.

e. *Sodium chloride*, NaCl, 0.14N.

f. *Beef extract*, 3%, pH 7.4: Dissolve 3 g beef extract paste or 2.4 g beef extract powder in 90 mL distilled water, adjust to pH 7.4 with 1.0 or 0.1N NaOH, dilute to 100 mL with distilled water, and sterilize by autoclaving.

g. *Antibiotics:* Use either:

1) *Penicillin-streptomycin*, 10$\times$, containing 5000 IU penicillin/mL and 5000 μg streptomycin/mL. Available commercially or prepare by dissolving powdered sodium or potassium penicillin-G and streptomy-

---

*Required for optional method for collecting Al(OH)$_3$ precipitate from sample.
†Millipore AP20 or equivalent.
‡Millipore HA or equivalent.
§For alternative procedure using preformed Al(OH)$_3$ precipitate.

cin sulfate in distilled water and sterilizing by filtration.

2) *Gentamycin-kanamycin,* 100×, containing 5000 μg/mL each of gentamycin base and kanamycin base (Section 9510B.4*i*).

h. Polyoxyethylene sorbitan monooleate,‖ 0.1% (v/v) in distilled water.

5. Procedure

a. Sterilization of apparatus, materials, and reagents: See Section 9510B.5*a*.

b. Preparation of preformed Al(OH)₃ precipitate: While mixing 100 mL 0.075*N* AlCl₃ at room temperature, slowly add 4*N* Na₂CO₃ solution to form precipitate and adjust to pH 7.2. Continue mixing for 15 min and, if necessary, add more Na₂CO₃ to maintain pH 7.2. Centrifuge at 1100 × g for 15 min and discard supernatant. Resuspend sediment in 0.14*N* NaCl and recentrifuge. Discard supernatant, resuspend sediment in 0.14*N* NaCl, and sterilize by autoclaving. Cool, centrifuge again, decant supernatant, and resuspend Al(OH)₃ sediment in 50 mL sterile 0.14*N* NaCl. Store at 4°C.

c. Sample size, collection, and storage: Process samples of no more than several liters because the method is too cumbersome and time-consuming for larger volumes. See Section 9510B.5*c* for sample collection and storage procedures.

d. Sample processing: Do not prefilter sample[11,12] because substantial virus losses can occur. Adjust sample to pH 6.0 with 1.0 or 0.1*N* HCl while mixing vigorously. Form Al(OH)₃ precipitate in sample by adding 1 part 0.9*N* AlCl₃ solution to 100 parts sample to give a final 0.009*N* Al³⁺ concentration. Check sample pH and readjust to 6.0 with 1.0 or 0.1*N* NaOH or HCl, if necessary. Mix slowly for 15 min at room temperature.

Alternatively, use preformed Al(OH)₃ precipitate by adding 1 part stock Al(OH)₃ suspension/100 parts sample and mix slowly for 2 h at 4 to 10°C to allow for virus adsorption.

Collect virus-containing Al(OH)₃ precipitate by centrifugation or filtration. To collect precipitate by centrifugation, centrifuge at 1700 × g for 15 to 20 min, discard supernatant, and resuspend sediment in 1/1000 to 1/20 original sample volume of 3% beef extract, pH 7.4.

To collect precipitate by filtration, vacuum filter sample through a treated filter (¶ 3 above) held in a vacuum-type filter holder or Buchner funnel, using additional filters if filter clogs before entire sample is filtered. Carefully scrape precipitate from filter(s) with a sterile spatula and resuspend in 1/1000 to 1/20 original sample volume of 3% beef extract, pH 7.4.

Regardless of collection method, vigorously mix the Al(OH)₃ beef extract suspension and, if necessary, adjust to pH 7.4 with 0.1*N* HCl or NaOH. Continue mixing for a total of 10 min. Centrifuge at 1900 × g for 30 min. Decant supernatant, add 1/10 the volume of the concentrate of penicillin-streptomycin solution or 1/100 volume of gentamycin-kanamycin and store at 4 or −70°C.

6. References

1. PAYMENT, P., C.P. GERBA, C. WALLIS & J.L. MELNICK. 1976. Methods for concentrating viruses from large volumes of estuarine water on pleated membranes. *Water Res.* 10:893.

2. FARRAH, S.R., S.M. GOYAL, C.P. GERBA, C. WALLIS & J.L. MELNICK. 1977. Concentration of enteroviruses from estuarine water. *Appl. Environ. Microbiol.* 33:1192.

3. WALLIS, C. & J.L. MELNICK. 1967. Concentration of viruses on aluminum hydroxide precipitates. *In* G. Berg, ed. Transmission of Viruses by the Water Route. Interscience Publ., New York, N.Y.

4. WALLIS, C. & J.L. MELNICK. 1967. Virus concentration on aluminum and calcium salts. *Amer. J. Epidemiol.* 85:459.

‖Tween 80®, ICI United States, Inc., Wilmington, Del., or equivalent. Required for optional method for collecting Al(OH)₃ precipitate from sample.

5. COOKSON, J.T., JR. 1974. The chemistry of virus concentration by chemical methods. *Develop. Ind. Microbiol.* 15:160.
6. LYDHOLM, B. & A.L. NIELSEN. 1979. Methods for detection of virus in wastewater applied to samples from small scale treatment systems. *Water Res.* 14:169.
7. SELNA, M.W. & R.P. MIELE. 1977. Virus sampling in wastewater-field experiences. *J. Environ. Eng. Div., Proc. Amer. Soc. Civil Eng.* 103:693.
8. DOBBERKAU, H.J., R. WALTER & S. RUDIGER. 1981. Methods for virus concentration from water. *In* M. Goddard & M. Butler, eds. Viruses and Wastewater Treatment. Pergamon Press, New York, N.Y.
9. FARRAH, S.R., C.P. GERBA, C. WALLIS & J.L. MELNICK. 1978. Concentration of poliovirus from tapwater onto membrane filters with aluminum chloride at ambient pH levels. *Appl. Environ. Microbiol.* 35:624.
10. FARRAH, S.R., G.M. GOYAL, C.P. GERBA, R.H. CONKLIN & E.M. SMITH. 1978. Comparison between adsorption of poliovirus and rotavirus by aluminum hydroxide and activated sludge flocs. *Appl. Environ. Microbiol.* 35:360.
11. SOBSEY, M.D., C.P. GERBA, C. WALLIS & J.L. MELNICK. 1977. Concentration of enteroviruses from large volumes of turbid estuary water. *Can. J. Microbiol.* 23:770.
12. HOMMA, A., M.D. SOBSEY, C. WALLIS & J.L. MELNICK. 1973. Virus concentration from sewage. *Water Res.* 7:945.

9510 E. Hydroextraction-Dialysis with Polyethylene Glycol (PROPOSED)

1. General Discussion

Polyethylene glycol (PEG) hydroextraction is an ultrafiltration process in which the sample is placed in a cellulose dialysis bag and exposed to PEG, a hygroscopic material. Water and microsolutes leave the sample by passing across the semipermeable dialysis membrane into the hygroscopic PEG.[1] Viruses and other macrosolutes, including PEG, cannot cross the dialysis membrane. The sample volume in the dialysis bag is reduced by water loss to the PEG, thereby concentrating viruses and other macrosolutes. The viruses retained in the dialysis bag are recovered by opening the bag, collecting the remaining sample, and eluting any viruses possibly adsorbed to the inner walls of the bag with a small volume of slightly alkaline proteinaceous solution such as 3% beef extract, pH 9.0. The collected concentrate and eluate are combined and assayed for viruses.

The main limitations of this method are that only small samples (less than 1 L) can be processed conveniently, virus elution from the walls of the dialysis bag may be incomplete unless the elution is done painstakingly, and other macrosolutes in the sample that are concentrated with viruses may interfere with virus assays by being cytotoxic.

Initial investigations of this method reported low and highly variable virus recoveries from wastewater.[2,3] The type of dialysis tubing and eluent solution as well as the thoroughness of the elution step have been found to influence virus recovery efficiency. More recently, with modified procedures, efficient and consistent virus recoveries have been obtained from wastewater and from adsorbent filter eluates.[4,5]

2. Equipment and Apparatus

a. Beakers, 100-mL or larger.

b. Graduated cylinders, 100-mL or larger.

c. *Dialysis tubing clamps.* *

d. *Pan,* approximately $30 \times 30 \times 12$ cm, autoclavable.

e. *Magnetic stirrer* and stirring bars or alternative mixing device.

f. *Centrifuge,* with rotor and buckets, capable of operating at about $1900 \times g$.

g. *pH meter.*

h. *Pipets,* 1-, 5-, and 10-mL.

i. *Tape roller*† or similar device to aid in washing the inside walls of dialysis bags with eluting fluid.

j. *Ultrasonic   disruptor-emulsifier,*‡ probe type, capable of generating 100 W of acoustical output.

3. Materials

a. *Dialysis tubing,* seamless, regenerated cellulose, 4.8-nm average pore diameter.§

b. *Polyethylene glycol (PEG),* ‖ dry flakes.

4. Reagents

See Section 9510D.3.

5. Procedure

a. *Sterilization of apparatus, materials, and reagents:* See Section 9510B.5a. Do not sterilize PEG.

b. *Sample size, collection, and storage:* Process samples of no more than a few hundred milliliters. See Section 9510B.5c for sample collection and storage procedures.

c. *Preparation of dialysis tubing:* Cut a length of dialysis tubing long enough to accommodate entire sample. Close one end with a clamp. Do not tie knots to close dialysis tubing. Fill tubing bag with distilled water, sterilize by autoclaving, and let cool.

d. *Sample processing:* Aseptically remove

dialysis bag from distilled water and drain. Fill bag with sample and close open end with a second clamp. Place bag in a pan containing a 5-cm layer of PEG, making sure that bag does not touch pan walls. Cover tubing with an additional 5 cm PEG and store at 4°C (for about 18 h) until sample volume has been reduced to no more than a few milliliters. (If PEG 6000 is used the process time is reduced to 4 to 6 h.) Although sample may be allowed to dewater completely, do not let it remain in this state.

Remove dialysis bag from PEG and quickly wash PEG from outside of bag with sterile distilled water. Remove clamp from one end of bag and carefully collect sample concentrate. Add about 1/200 to 1/20 the original sample volume of 3% beef extract, pH 9.0, and clamp closed. Thoroughly wash inside walls of bag with beef extract by rubbing fluid from one end to the other several times using either fingers or a roller device. Remove clamp from one end of bag and collect fluid, kneading or squeezing to recover the last traces. Add recovered fluid to previously collected sample concentrate.

Adjust to pH 7.5 with 1.0 or 0.1N HCl while mixing vigorously. To disperse solids-associated viruses in sample, stir overnight (about 18 h) in the cold (about 4°C) or treat with ultrasonics at 100 W for 1 to 2 min. Prevent sample temperature from rising above 37°C during ultrasonic treatment by chilling in an ice bath. Centrifuge at $1900 \times g$ for 30 min. Decant supernatant, add 1/10 the volume of the concentrate of penicillin-streptomycin solution or 1/100 volume of gentamycin-kanamycin, and store at 4 or −70°C.

6. References

1. SOBSEY, M.D. 1976. Methods for detecting enteric viruses in water and wastewater. *In* G. Berg, H.L. Bodily, E.H. Lennette, J.L. Melnick & T.G. Metcalf, eds. Viruses in Water. American Public Health Assoc., Washington, D.C.

2. CLIVER, D.O. 1967. Detection of enteric viruses

---

*Fisher Scientific No. 8-670-11A, or equivalent.

†Optional. Fisher Scientific No. 14-245-21 or equivalent.

‡Optional.

§Made by Union Carbide Corp. and available from many scientific supply companies.

‖Carbowax® 20 000 or 6000 or equivalent.

by concentration with polyethylene glycol. *In* G. Berg., ed. Transmission of Viruses by the Water Route. Interscience Publ., New York, N.Y.

3. SHUVAL, H.I., S. CYMBALISTA, B. FATTAL & N. GOLDBLUM. 1967. Concentration of enteric viruses in water by hydro-extraction and two-phase separation. *In* G. Berg, ed. Transmission of Viruses by the Water Route. Interscience Publ., New York, N.Y.

4. WELLINGS, F.M., A.L. LEWIS, C.W. MOUNTAIN & L. V. PIERCE. 1975. Demonstration of virus in groundwater after effluent discharge onto soil. *Appl. Microbiol.* 29:751.

5. RAMIA, S. & S.A. SATTAR. 1979. Second-step concentration of viruses in drinking and surface waters using polyethylene glycol hydroextraction. *Can. J. Microbiol.* 25:587.

9510 F. Recovery of Viruses from Suspended Solids in Water and Wastewater (PROPOSED)

1. General Discussion

Viruses in the aquatic environment often are associated with solids or particulate matter, either adsorbed to particulate surfaces or embedded within the solid.[1-3] Both freely suspended and solids-associated viruses are concentrated from water by the methods described above. There is evidence that solids-associated viruses are not eluted efficiently from adsorbent filters or from Al(OH)$_3$ precipitates and organic flocs. Recovery of solids-associated viruses by microporous filter methods employing in-situ elution is inconsistent.[2] Solids-associated viruses on adsorbent filters are eluted more efficiently by disrupting filters in elution fluid than by in-situ elution,[2] but this is cumbersome and time-consuming, especially for large-diameter disk filters and cartridge filters.

For small volumes of water and wastewater, solids-associated viruses can be recovered expediently by separating the solids by centrifuging, decanting the supernatant, and eluting viruses from the solids by resuspending in a small volume of eluent.[4] Viruses in the supernatant can be concentrated by one of the procedures described in Sections 9510B, C, D, or E. Viruses eluted from the resuspended solids are separated from the solids by centrifuging and are assayed directly or concentrated further by organic flocculation.[5,6] Major limitations of these methods are incomplete virus elution and poor virus recoveries due to interferences from sample constituents.

2. Equipment and Apparatus

a. Centrifuge, with rotor and buckets for 250- to 1000-mL-capacity bottles, capable of operating at about 1250 × g.

b. Centrifuge bottles, 250- to 1000-mL.

c. pH meter.

d. Laboratory balance.

e. Graduated cylinder, 250-mL or larger.

f. Beaker, 250-mL or larger.

g. Sample bottles, 250-mL or larger.

h. Magnetic stirrer and stirring bars, or alternative mixing device.

i. Pipets, 1-, 5-, and 10-mL.

3. Reagents

a. Hydrochloric acid, HCl, 0.1 and 1.0*N*.

b. Sodium hydroxide, NaOH, 0.1 and 1.0*N*.

c. Eluent: Dissolve 10 g beef extract, 1.34 g disodium phosphate heptahydrate, Na$_2$HPO$_4$·7H$_2$O, and 0.12 g citric acid in 90 mL distilled water, adjust to pH 7.0 with 1*N* HCl or NaOH, dilute to 100 mL with distilled water, and sterilize by autoclaving.

d. Antibiotics: See Section 9510B.4*i*.

4. Procedure

a. *Sterilization of apparatus, materials, and reagents:* See Section 9510B.5*a*.

b. *Sample size, collection, and storage:* Collect and process samples of no more than 10 L, depending on capacity of centrifuge. See Section 9510B.5*c* for sample collection and storage procedures.

c. *Sample processing:* Aseptically transfer 250- to 1000-mL sample volumes to centrifuge bottles and centrifuge at 1250 × *g* for 20 min. Decant and pool supernatants for subsequent processing for viruses by one of the methods for water or wastewater described previously.

Elute viruses from the sedimented solids by resuspending in eluent. Use 40 mL eluent per quantity of sediment from 250 mL of original sample. Pool resuspended sediments from multiple centrifuge bottles in a sterile beaker. Alternatively, keep resuspended sediments from small numbers of centrifuge bottles in the bottles and process them individually. While vigorously mixing with a magnetic stirrer, adjust to pH 7.0 by slowly adding 1*N* NaOH or HCl, if necessary. Reduce mixing speed and continue mixing for 30 min. During this period, check sample pH and readjust to pH 7.0 as necessary. As an alternative to mixing for 30 min, sonicate samples at 100 W for 15 min in a rosette cooling cell maintained at 4°C. Return sample to centrifuge bottles. Centrifuge at 1250 × *g* and 4°C for 15 min, collect supernatant for subsequent assay or further concentration, and discard the sediment.

If desired, further concentrate viruses from this supernatant by organic flocculation (see Section 9510E). For supernatants that will be assayed directly for viruses with no further concentration, adjust to pH 7.4, add 1/10 the volume of sample of penicillin-streptomycin or 1/100 volume of gentamycin-kanamycin and store at 4 or −70°C.

5. References

1. SCHAUB, S.A. & B.P. SAGIK. 1975. Association of enteroviruses with natural and artificially introduced colloidal solids in water and infectivity of solids-associated virions. *Appl. Microbiol.* 30:212.

2. WELLINGS, F.M., A.L. LEWIS & C.W. MOUNTAIN. 1976. Viral concentration techniques for field sample analysis. *In* L.B. Baldwin, J.M. Davidson & J.F. Gerber, eds. Virus Aspects of Applying Municipal Waste to Land. Univ. Florida, Gainesville.

3. WELLINGS, F.M., A.L. LEWIS & C.W. MOUNTAIN. 1974. Virus survival following wastewater spray irrigation of sandy soils. *In* J.F. Malina, Jr. & B.P. Sagik, eds. Virus Survival in Water and Wastewater Systems. Univ. Texas, Austin.

4. BERG, G. & D.R. DAHLING. 1980. Method for recovering viruses from river water solids. *Appl. Environ. Microbiol.* 39:850.

5. GERBA, C.P. 1982. Detection of viruses in soil and aquatic sediments. *In* C.P. Gerba & S.M. Goyal, eds. Methods in Environmental Virology. Marcel Dekker, Inc., New York, N.Y.

6. FARRAH, S.R. 1982. Isolation of viruses associated with sludge particles. *In* C.P. Gerba & S.M. Goyal, eds. Methods in Environmental Virology. Marcel Dekker, Inc., New York, N.Y.

9510 G. Assay and Identification of Viruses in Sample Concentrates (PROPOSED)

1. Storage of Sample Concentrates

Because it often is impossible to assay sample concentrates immediately, store them at room temperature (about 25°C) for up to 2 h or at refrigerator temperatures (4 to 10°C) for up to 48 h to minimize virus losses. Freeze samples requiring storage longer than 48 h at −70°C or less. Do not freeze samples at −10 to −20°C because extensive inactivation of some enteric viruses may occur. Store sample concentrates from finished waters in separate freezers or physically separated from other virus-containing material in common freezers.

2. Decontamination of Sample Concentrates

Sample concentrates, especially those from wastewater, are likely to be contaminated with bacteria and fungi that can overgrow cell cultures and interfere with virus detection and assay. Do not decontaminate by centrifugation or filtration because virus losses are likely to occur. For many samples, especially those from finished waters, contamination is controlled adequately by antibiotics such as penicillin-streptomycin or gentamycin-kanamycin that are added immediately after the sample is obtained. To provide additional protection against fungal contamination, add amphotericin B or nystatin at concentrations of 2.5 and 50 μg/mL, respectively.[1] If penicillin-streptomycin or gentamycin-kanamycin are inadequate, use one or more additional antibiotics such as aureomycin, neomycin, or polymyxin B. To maximize the antibiotic effects, incubate samples for 1 to 3 h at 25 to 37°C after adding the antibiotics. Bacterial destruction is further enhanced by freezing at −70°C after incubation with antibiotics. Keep samples frozen until assayed for viruses. To determine if antibiotic treatment has been effective, plate a small subsample on a general-purpose medium such as plate count agar by the spread plate technique and incubate at 37°C for 24 to 48 h.

If extensive bacterial contamination persists after antibiotic treatment, treat with chloroform. Add 1/10 volume of sample of chloroform ($CHCl_3$) and mix vigorously for 30 min at room temperature or homogenize 1 to 2 min at 4 to 10°C. For phase separation, centrifuge at $\geq 1000 \times g$ or store overnight in a refrigerator. Separate sample (upper layer) from $CHCl_3$ (bottom layer) by aspirating with a pipet and bubble with filter-sterilized air for about 15 min to remove dissolved $CHCl_3$. It may be necessary to place sample in a sterile, shallow container and expose it to the atmosphere in a sterile air environment (laminar air flow clean bench or biological safety cabinet) for up to several hours to remove remaining traces of $CHCl_3$. Do not use ether to decontaminate samples because of the hazard of explosion or fire.

3. Laboratory Facilities and Host Systems for Virus Assay

Because viruses are obligate, intracellular parasites, they grow (multiply) only in living host cells. This ability to multiply in, and thereby destroy, their host cells is the basis for virus detection and assay. The two major host cell systems for human enteric viruses are whole animals (usually mice) and mammalian cell cultures of primate origin.

A complete description of facilities, equipment, materials, and methods for conducting virus assays is beyond the scope of this book; see standard handbooks on virology and cell culture.[1-4] Virus assay is beyond the capability of most water and wastewater microbiology laboratories. It should be done only by a trained virologist working in specially equipped virology laboratory facilities. Take particular care to prevent samples or inoculated hosts from becoming contaminated with viruses from other sources and to prevent virus cross-contamination arising from sample concentrates or inoculated hosts. Process and handle samples in a Class II Type I biological safety cabinet[5] or in a "sterile" room or cubicle. The use of such cabinets or facilities is mandatory for testing drinking water or other finished water samples.

There is no single, universal host system

for all enteric viruses. Some enteric viruses, notably hepatitis A virus, human rotaviruses, and Norwalk-type gastroenteritis viruses, cannot be assayed routinely in any convenient laboratory host systems. However, most of the known enteric viruses can be detected by using two or more cell culture systems and perhaps suckling mice. The latter previously were considered essential for the detection of group A coxsackie-viruses, but recent studies indicate that the RD cell line may be nearly as sensitive as suckling mice for the isolation of these viruses as well as other enteroviruses.[6,7] In general, the more different host systems used, the greater the enteric virus recovery rate. However, the number of different host systems used is limited by practical and economic considerations.

There have been numerous comparative studies on relative sensitivities of various cell culture systems for enteric virus detection,[6-26] but no systematic, comprehensive study has been reported for enteric virus recoveries from water and wastewater. Primary or secondary human embryonic kidney (HEK) cell cultures appear to be the single most sensitive host system for enteric virus isolations, but they are becoming increasingly more difficult to obtain regularly and, when available from commercial sources, they are expensive. Primary or secondary African green, cynomolgus, or rhesus monkey or baboon kidney cells are sensitive hosts for many enteroviruses and reoviruses, but are not particularly suitable for recovering adenoviruses or group A coxsackieviruses. BGM, a continuous line derived from African green monkey kidney cells, may be comparable in sensitivity to primary monkey kidney cells for enteric virus recovery.[7,18,21,25,26] A number of other continuous cell lines as well as human fetal diploid cell strains have been evaluated for enteric virus recoveries. Some human fetal diploid cell strains give virus isolation rates comparable to primary monkey kidney cells, but

plentiful supplies of specific human fetal diploid cell strains are not readily available and many are difficult to maintain. Furthermore, each different cell strain must be characterized for virus susceptibility. Most continuous cell lines generally are less effective than primary cells, but comparable isolation rates for some enteric virus groups have been obtained with Hep-2[11] and HeLa[17,25] cells.

Assay the entire sample concentrate for enteric viruses, using at least two different host systems and dividing entire sample equally among the hosts. Preferably use primary (or secondary) HEK cells with either primary (or secondary) monkey kidney or BGM cells for the recovery of most enteroviruses, adenoviruses, and reoviruses. Additional use of either suckling mice or RD cells provides for enhanced recovery of group A coxsackieviruses. Different host systems may be substituted for these if it is demonstrated that they have equivalent sensitivity.

4. Virus Quantitation Procedures for Sample Concentrates

a. Advantages and disadvantages of different quantitation procedures: Virus assays in suckling mice or other animals are quantal assays and in cell cultures they can be done either by quantal (most probable number or 50% endpoint) or enumerative (plaque) methods. Selection between cell culture assay methods depends on the sample and the choice between achieving either maximum virus sensitivity or maximum precision and accuracy in estimating virus concentration. The plaque technique generally is more precise and accurate than the quantal assay because relatively large numbers of individual infectious units can be counted directly as discrete, localized areas of infection (plaques). Quantal assays are more sensitive than monolayer plaque assays, but are less sensitive than an agar cell suspension plaque assay.[26]

Because virus plaques are discrete areas of infection arising from a single infectious virus unit, it is relatively easy to recover viruses from individual plaques and then to inoculate them into additional cell cultures to obtain a pure virus culture for identification. However, large proportions of so-called "false-positive" plaques that do not confirm as virus-positive when material from these plaques is further passaged in cell cultures have been reported.[27,28] Whether this problem is due to nonviral, plaque-like areas of cytotoxicity from the sample or to technical inability to passage viruses successfully from the initial plaques remains uncertain.[27-29]

The use of specific plaque assay conditions for optimizing the recovery of certain enteric virus groups may preclude efficient recovery of other enteric groups requiring different plaque assay conditions. Furthermore, some viruses, such as adenoviruses, do not form plaques efficiently under any conditions. Cytotoxicity due to water or wastewater constituents in sample concentrates is difficult to control in plaque assay systems because the agar overlay medium is difficult to remove and replace.

A potential limitation of quantal assays is the possibility that two or more different virus types will be inoculated into the same cell culture and thus produce a simple positive culture. This not only results in an underestimation of virus concentration but also requires separation of the individual virus types by further passage in cell culture. Such mixed cultures may go undetected unless virus isolates are identified serologically. Recent results indicate that mixed positive cultures are encountered rarely when samples are divided into small portions for inoculation into a series of replicate cell cultures.[7,25]

Cytotoxicity due to constituents of sample concentrates usually can be controlled in quantal assay cell cultures by replacing the culture medium before the cells die.

   b. *Cell culture procedures for virus iso-* *lation and assay:* To assay sample concentrates in cell cultures by quantal or plaque methods, drain the medium from newly confluent cultures and inoculate with unit volumes of sample. Use no more than 0.06 mL sample/cm^2 of cell layer surface, e.g., maximum volumes of 1.5, 4.5, and 7.4 mL in cell culture flasks with areas of 25, 74, and 150 cm^2, respectively. If samples are expected to contain such large quantities of viruses that it would be difficult to make reliable estimates of concentration, inoculate cell cultures with small sample volumes or dilutions of concentrates. Allow viruses to adsorb to cells for 2 h at 37 ± 0.5°C. Redistribute inoculum over the cell layer manually every 15 min or keep cultures on a mechanical rocker during the adsorption period. Add liquid maintenance medium to cultures for quantal assays or agar-containing medium for plaque assays. Incubate at 37°C and invert plaque assay cultures so that cell (agar) side of culture faces up.

Microscopically examine quantal assay cultures for the appearance of cytopathic effects (CPE) daily during the first 3 d and then periodically for a total of at least 14 d. Do not change cell culture medium unless cytotoxicity or cell deterioration occurs. Freeze cultures developing CPE at −70°C when more than 75% of the cells become involved. After 14 or more days, freeze at −70°C all remaining cultures, including those remaining negative for CPE as well as controls. Thaw cultures and clarify culture fluid-cell lysate by slow-speed centrifugation or filtration through sterile 0.22- or 0.45-μm porosity filters. Inoculate clarified material from each initial (first-passage) culture into a second (second-passage) culture by transferring 20% of the total initial culture into newly confluent cell cultures of the same type. Microscopically examine second-passage cultures for development of CPE periodically over a period of 14 or more days. Consider second-passage cultures developing CPE as confirmed virus-positive. Freeze and store at $-70°C$ for virus identification. Discard as negative any virus cultures negative for CPE after this second incubation period of 14 or more days.

Periodically examine plaque assay cultures for appearance of plaques over a 14-d period. Mark and tally plaques as they appear. Transfer viruses from each plaque directly to at least two newly confluent, liquid-medium cell cultures of the same type[27] before plaques become too large and grow together or before the entire cell layer deteriorates. Do not store material obtained from plaques before transfer to new cell cultures, as this may result in loss of virus titer and unsuccessful transfers. Microscopically examine these second-passage cultures periodically over 14 d for development of CPE. Freeze cultures developing CPE at $-70°C$ for virus identification.

c. *Virus isolation and assay in mice:* To detect group A and B coxsackieviruses in mice, inoculate samples into animals no older than 24 h using standard procedures.[2,3,8] Use either the intracerebral or intraperitoneal route, inoculating 0.02 and 0.05 mL, respectively. Observe mice daily over a 14-d period for development of weakness, tremors, and either flaccid (due to group A coxsackieviruses) or spastic (due to group B coxsackieviruses) paralysis. Sacrifice animals developing symptoms, and using sterile technique, prepare 20% tissue suspensions in Hanks balanced salt solution of the entire skinned, eviscerated torso or just the brain and legs. Store suspensions at $-70°C$ until used for further passage and identification. For second passage in mice, follow general procedures used for the initial inoculations. However, making a second passage in cell cultures is preferable to making a second passage in mice because it is easier to do subsequent virus identification by neutralization tests.

d. *Estimating virus concentration:* Determining the amount of virus in a sample concentrate depends on the assay used. If a sample concentrate is assayed in cell cultures by the plaque technique, count all plaques and calculate the virus concentration, expressed as plaque-forming units (PFU).

If a sample concentrate is assayed by the quantal method, estimate the virus concentration by the most probable number (MPN) method and express as most probable number of infectious units (MPNIU), or by a 50% end point method and express as 50% infectious or lethal dose (ID_{50} or LD_{50}).[2,4,30–32] If the undiluted sample concentrate or a single sample dilution was inoculated into a series of replicate cell cultures (or mice), calculate the MPNIU from the number of confirmed CPE-negative cultures (or mice), q, per total number of cultures (or mice) inoculated, n, according to the formula

$$MPN = -\ln(q/n)$$

If more than one sample dilution was inoculated into cell cultures (or mice), calculate the MPNIU from the formula developed by Thomas:[33]

$$MPN/mL = \frac{P}{\sqrt{NQ}}$$

where:
  P = total number of positive cultures (or mice) from all dilutions,
  N = total mL sample inoculated for all dilutions, and
  Q = total mL sample in all negative cultures (or mice).

In using this formula, exclude from the computation all dilutions containing only positive cultures (or mice).

For MPN values obtained from a single sample dilution, the 95% confidence interval is based on the standard error of the binomial distribution when more than 30 cultures (or mice) are inoculated or from the confidence coefficient table of Crow[32,34] when 30 or fewer cultures (or mice) are inoculated.

Make 50% end-point estimates arithmetically by either the Reed-Muench or Karber method.[2,4] These methods require results from several equally spaced sample dilutions, preferably with about the same number of dilutions above and below the 50% end point, and may not be useful for sample concentrates containing relatively low virus levels.

e. Identification of virus isolates: Identify enteric viruses isolated from sample concentrates by standard serological techniques, although preliminary identification of genus (enterovirus, reovirus, or adenovirus) sometimes can be made on the basis of information obtained from the isolation procedure. Enteric viruses recovered in suckling mice are likely to be either group A or B coxsackieviruses. For enteric viruses isolated in cell cultures, preliminary identification of genus often can be made from the characteristic appearance of cytopathic effects (CPE) in infected cell cultures.

Confirm preliminary identification of suspected adenovirus and reovirus isolates by detecting their respective group specific antigens by complement fixation tests using clarified, second-passage cell-culture lysate as the antigen. Identify specific reovirus serotypes by hemagglutination-inhibition (HI) or neutralization (Nt) tests. Adenovirus serotypes can be separated into four groups on the basis of their ability (or inability) to hemagglutinate rhesus monkey or rat erythrocytes.[2,3,8] Except for type 18, the first 28 numbered adenoviruses can be identified as to specific serotype by HI. Alternatively, identify all adenovirus serotypes by Nt tests using either individual type-specific antisera or intersecting antisera pools. Also identify specific enterovirus serotypes by neutralization tests in cell cultures using intersecting pools of hyperimmune sera.[2,3,8,35] Use mice for Nt tests for group A and B coxsackieviruses only if the virus isolates fail to propagate in cell cultures.[36] Because polioviruses often are the most prevalent enteroviruses in water and wastewater, test enterovirus isolates for neutralization by an antisera pool against the three types of poliovirus before making neutralization tests with intersecting antisera pools.

5. References

1. PAUL, J. 1975. Cell and Tissue Culture, 5th ed. Churchill Livingstone, New York, N.Y.

2. LENNETTE, E.H. & N.J. SCHMIDT, eds. 1969. Diagnostic Procedures for Viral and Rickettsial Infections, 4th ed. American Public Health Assoc., Washington, D.C.

3. LENNETTE, E.H., A. BALOWS, W.J. HAUSLER & J.P. TRUANT, eds. 1980. Manual of Clinical Microbiology, 3rd ed. American Soc. Microbiology, Washington, D.C.

4. ROVOZZO, G.C. & C.N. BURKE. 1973. A Manual of Basic Virological Techniques. Prentice-Hall, Englewood Cliffs, N.J.

5. U.S. PUBLIC HEALTH SERVICE. 1976. Guidelines for Research Involving Recombinant DNA Molecules. Appendix D-I, Biological Safety Cabinets. National Inst. Health, Bethesda, Md.

6. SCHMIDT, N.J., H.H. HO & E.H. LENNETTE. 1975. Propagation and isolation of group A coxsackieviruses in RD cells. *J. Clin. Microbiol.* 2:183.

7. SCHMIDT, N.J., H.H. HO, J.L. RIGGS & E.H. LENNETTE. 1978. Comparative sensitivity of various cell culture systems for isolation of viruses from wastewater and fecal samples. *Appl. Environ. Microbiol.* 36:480.

8. HSIUNG, G.D. 1973. Diagnostic Virology, revised ed. Yale Univ. Press, New Haven, Conn.

9. KELLY, S., J. WINSSER & W. WINKELSTEIN. 1957. Poliomyelitis and other enteric viruses in sewage. *Amer. J. Pub. Health* 47:72.

10. KELLY, S. & W.W. SANDERSON. 1962. Comparison of various tissue cultures for the isolation of enteroviruses. *Amer. J. Pub. Health* 52:455.

11. PAL, S.R., J. MCQUILLIN & P.S. GARDNER. 1963. A comparative study of susceptibility of primary monkey kidney cells, Hep 2 cells and

HeLa cells to a variety of faecal viruses. *J. Hyg.*, Camb. 61:493.

12. LEE, L.H., C.A. PHILLIPS, M.A. SOUTH, J.L. MELNICK & M.D. YOW. 1965. Enteric virus isolations in different cell cultures. *Bull. World Health Org.* 32:657.

13. SCHMIDT, N.J., H.H. HO & E.H. LENNETTE. 1965. Comparative sensitivity of human fetal diploid kidney cell strains and monkey kidney cell cultures for isolation of certain human viruses. *Amer. J. Clin. Pathol.* 43:297.

14. BERQUIST, K.R. & G.J. LOVE. 1966. Relative efficiency of three tissue culture systems for the primary isolation of viruses from feces. *Health Lab. Sci.* 3:195.

15. HERRMANN, E.C. 1967. The usefulness of human fibroblast cell lines for the isolation of viruses. *Amer. J. Epidemiol.* 85:200.

16. FAULKNER, R.S. & C.E. VAN ROOYEN. 1969. Studies on surveillance and survival of viruses in sewage in Nova Scotia. *Can. J. Pub. Health* 60:345.

17. LUND, E. & C.E. HEDSTROM. 1969. A study on sampling and isolation methods for the detection of virus in sewage. *Water Res.* 3:823.

18. SHUVAL, H., B. FATTAL, S. CYMBALISTA & N. GOLDBLUM. 1969. The phase-separation method for the concentration and detection of viruses in water. *Water Res.* 3:225.

19. SCHMIDT, N.J. 1972. Tissue culture in the laboratory diagnosis of virus infections. *Amer. J. Clin. Pathol.* 57:820.

20. COONEY, M.K. 1973. Relative efficiency of cell cultures for detection of viruses. *Health Lab. Sci.* 4:295.

21. DAHLING, D.R., G. BERG & D. BERMAN. 1974. BGM: A continuous cell line more sensitive than primary rhesus and African green kidney cells for the recovery of viruses from water. *Health Lab. Sci.* 11:275.

22. SCHMIDT, N.J., H.H. HO & E.H. LENNETTE. 1976. Comparative sensitivity of the BGM cell line for the isolation of enteric viruses. *Health Lab. Sci.* 13:115.

23. HATCH, M.H. & G.E. MARCHETTI. 1971. Isolation of echoviruses with human embryonic lung fibroblast cells. *Appl. Microbiol.* 22:736.

24. RUTALA, W.A., D.F. SHELTON & D. ARBITER. 1977. Comparative sensitivities of viruses to cell cultures and transport media. *Amer. J. Clin. Pathol.* 67:397.

25. IRVING, L.G. & F.A. SMITH. 1981. One-year

survey of enteroviruses, adenoviruses and reoviruses isolated from effluent at an activated-sludge purification plant. *Appl. Environ. Microbiol.* 41:51.

26. MORRIS, R. & W.M. WAITE. 1980. Evaluation of procedures for recovery of viruses from water—II detection systems. *Water Res.* 14:795.

27. FANNIN, K.F., S.H. ABID, J.J. BERTUCCI, J.M. REED, S.C. VANA & C. LUE-HING. 1978. Significance of reporting infectious viral or plaque forming unit viral concentrations from environmental samples. *Abs. Annu. Meeting, American Soc. Microbiology, Washington, D.C.*

28. LEONG, L.Y.C., S.J. BARRETT & R.R. TRUSSEL. 1978. False-positives in testing of secondary sewage for enteric viruses. *Abs. Annu. Meeting, American Soc. Microbiology, Washington, D.C.*

29. KEDMI, S. & B. FATTAL. 1981. Evaluation of the false-positive enteroviral plaque phenomenon occurring in sewage samples. *Water Res.* 15:73.

30. CHANG, S.L., G. BERG, K.A. BUSCH, R.E. STEVENSON, N.A. CLARKE & P.W. KABLER. 1958. Application of the Most Probable Number method for estimating concentrations of animal viruses by tissue culture technique. *Virology* 6:27.

31. CHANG, S.L. 1965. Statistics of the infective units of animal viruses. *In* G. Berg, ed. Transmission of Viruses by the Water Route. Interscience Publ., New York, N.Y.

32. SOBSEY, M.D. 1976. Field monitoring techniques and data analysis. *In* L.B. Baldwin, J.M. Davidson & J.F. Gerber, eds. Virus Aspects of Applying Municipal Waste to Land. Univ. Florida, Gainsville.

33. THOMAS, H.A., JR. 1942. Bacterial densities from fermentation tube tests. *J. Amer. Water Works Assoc.* 34:572.

34. CROW, E.L. 1956. Confidence intervals for a proportion. *Biometrika* 43:423.

35. MELNICK, J.L., V. RENNICK, B. HAMPIL, N.J. SCHMIDT & H.H. HO. 1973. Lyophilized combination pools of enterovirus equine antisera: Preparation and test procedures for the identification of field strains of 42 enteroviruses. *Bull. World Health Org.* 48:263.

36. MELNICK, J.L., N.J. SCHMIDT, B. HAMPIL & H.H. HO. 1977. Lyophilized combination

pools of enterovirus equine antisera: Preparation and test procedures for the identification of field strains of 19 group A coxsackievirus serotypes. *Intervirology* 8:1720.

9610 DETECTION OF FUNGI*

9610 A. Introduction

1. Significance

Fungi, including yeasts and filamentous species or molds, are ubiquitously distributed, achlorophyllous, heterotrophic organisms with organized nuclei and usually with rigid walls. They may be found wherever nonliving organic matter occurs, although some species are pathogenic and others are parasitic. In spring water near the source, the number of fungus spores usually is minimal. Unpolluted stream water has relatively large numbers of species representing the true aquatic fungi (species possessing flagellated zoospores and gametes), aquatic Hyphomycetes, and soil fungi. Moderately polluted water may carry cells or spores of the three types; however, it has fewer true aquatic fungi and aquatic Hyphomycetes, and soil fungi are more numerous. Heavily polluted water has large numbers of soil fungi. The group designated as soil fungi includes yeast-like fungi, many species of which have been isolated from polluted waters.

The association between fungal densities and organic loading suggests that fungi may be useful indicators of pollution. Unfortunately, no single species or group of fungi has been identified as important in this role. There may be some exceptional special cases; for example, the principal distinction between the yeasts *Candida lam-*

bica and *C. krusei* is the ability to use pentose sugars. Because the former species grows well on pentoses, it could be used as an indicator of pulp and paper mill wastes, which are high in such sugars. Certain species of yeasts and filamentous fungi are characteristic of warmer waters and may be useful indicators of thermal pollution.

Because fungi possess broad enzymatic capabilities, they can degrade actively most complex natural substances and certain synthetic compounds, including some pesticides. Most fungi are aerobic or microaerophilic, although a few species show limited anaerobic metabolism and a very few are capable of totally anaerobic growth.[1] Some species do not require light.

2. Occurrence and Survival

Fungi are present in, and have been recovered from, diverse, remote, and extreme aquatic habitats including lakes, ponds, rivers, streams, estuaries, marine environments, wastewaters, sludge, rural and urban stormwater runoff, well waters, acid mine drainage, asphalt refineries, jet fuel systems, and aquatic sediments.

a. Fungi in potable water: Fungi have been found in potable water[2-7] and on the inner surface of distribution system pipes.[8] Either they survive water treatment or they enter the system after treatment and remain viable. Tuberculate macroconidia of *Histoplasma capsulatum*[9] can pass through

*Approved by Standard Methods Committee, 1988.

a 0.75-m rapid sand filter. Plain sedimentation or alum flocculation and settling removed 80 to 99% of the spores. If these relatively large (8- to 14-μm) globose, tuberculate macroconidia pass through treatment, it is not surprising that other fungi, typically with smaller spores, are found in treated water.

Having survived treatment or having been introduced after treatment, fungal spores can remain viable for extended periods of time. Pathogenic fungi have been stored effectively in sterile distilled water for relatively long periods.[10] Spores of *H. capsulatum,* stored in raw Ohio River water and sterile tap water remained highly infective for mice after 400 d.[11]

Tastes and odors in potable water are associated with the presence of procaryotic organisms such as bacteria, actinomycetes, and cyanobacteria. However, fungi may be involved.[6,7]

Propagules from 19 genera of filamentous fungi have been isolated from a chlorinated surface water system and a nonchlorinated groundwater distribution system.[2] The mean number of colony-forming units (CFU) was 18/100 mL in the groundwater system and 34/100 mL in the surface system.

In Finland,[3] fungi were isolated from rivers, lakes, and ponds supplying nine communities with sand-filtered water, three with artificially recharged groundwater, of which two used chemical coagulation, and three with chemically coagulated and disinfected water. Mesophilic fungi were common in all raw water samples; however, thermotolerant fungi were more abundant in river than in lake water. Chemical coagulation and disinfection proved far more efficient in removing fungi than sand filtration and disinfection. *Aspergillus fumigatus* was the most common fungus.

Five chlorinated groundwater systems[4] in the U.S. yielded an average count per positive sample of about 5.5 CFU/100 mL. In France[5] yeasts were recovered from 50% of 38 samples and filamentous fungi from 81%.

Except for *Aspergillus fumigatus*[3,4], *A. flavus,*[4,5] and *A. niger,*[4] the fungi isolated from potable water usually are not considered to be medically important. Fungus infections may be significant for individuals with compromised immune systems. Most of the fungi are common soil saprobes.

b. Fungi in recreation waters: Some fungi pathogenic to humans may be expected in recreational waters such as pools and beaches and in accompanying washing facilities such as shower stalls.

Trichophyton mentagrophytes, the cause of tinea pedis or athlete's foot, has been isolated[12] from the wooden flooring of a shower stall. Seven species of pathogenic and potentially pathogenic fungi were isolated from 361 samples of beach sand in Hawaii.[13] Beach sand on the German Baltic coast, in Portugal, and on the Adriatic coast yielded *Epidermophyton.*[14]

c. Survival on chlorination: An unidentified yeast was isolated from other organisms that survived chlorination of wastewater effluents.[15] This yeast survived 1 mg free chlorine/L for 20 min in contrast with *E. coli,* which failed to survive 5 min contact with 0.03 mg free chlorine/L. The amount of chlorine for fungus control is known at least for *C. albicans.* It has been shown[16] that cells of *C. albicans* were inactivated effectively with 4 mg chlorine/L in 30 min when the initial cell count was 10^5 cells/mL. In an Illinois study with *C. parapsilosis,*[17,18] a commonly isolated yeast known to cause health problems in the tropics, greater amounts of chlorine were required to inactivate the organism than coliform bacteria. Mechanisms of inactivation by chlorine on assimilative stages of yeasts and other microorganisms have been suggested.[19] Fungal cells, especially conidia, can survive much higher doses of chlorine than coliform bacteria,[20] including a 10-min exposure to 10 mg chlorine/L when the initial spore count was approximately 10^6/mL.

3. Growth Patterns and Identification

In water there are two basic patterns of fungal growth. True aquatic fungi produce zoospores or gametes that are motile by means of flagella, either of the whiplash or tinsel type. Some fungi, particularly the Trichomycetes (fungi that inhabit the hind gut of certain worms, mosquito larvae, etc.), have amoeboid stages. Aquatic fungi typically are collected by exposing suitable baits (solid foodstuffs) in the habitat being examined or in a sample within the laboratory. Relatively little work on these fungi in polluted water has been done in the U.S. They have been studied more extensively in polluted waters in England, Germany, and Japan.

The second fungal growth form is non-motile in all stages of the life cycle. Growth and reproduction usually are asexual (anamorphic). Three growth processes have been recognized: (*a*) filamentous growth with blastic spores or spores produced in special structures; (*b*) filamentous growth with the filaments breaking up in an arthric (fragmenting) manner to form separate spores called arthroconidia as in *Geotrichum* and related genera; and (*c*) single-celled growth produced on each parent cell, called budding, typical of the yeasts.

Identification of fungi, which are considerably larger than bacteria, is dependent on colonial morphology on a solid medium, growth and reproduction morphology, and, for yeasts, physiological activity in laboratory cultures. Increasing numbers of fungi usually indicate increasing organic loadings in water or soil. Large numbers of similar fungi suggest excessive organic load while a highly diversified mycobiota indicates populations adjusted to the environmental organics. Despite their wide occurrence, little attention has been given to the presence and ecological significance of fungi in aquatic habitats. The relevance of fungi and their activities in water is emphasized by increasing knowledge of their pathogenicity for humans, animals, and plants; their role as food or energy sources; their activity in natural purification processes; and their function in sediment formation.

A survey of the literature of fungi occurring in water, wastewater, and related organically polluted substrata listed 984 species[21]: 133 species were assigned to the Mastigomycotina (fungi with flagellated zoospores); 79 to the Zygomycotina, mostly mucoraceous fungi; 161 to the Ascomycotina, including perfect (teleomorphic) states of some of those assigned to the Fungi Imperfecti; 18 to the Basidiomycotina, including perfect states of several yeasts; and 593 to the Deuteromycotina or Fungi Imperfecti. Of the total, 133 species were zoosporic, 131 species were yeast-like, and 718 species were filamentous. Most zoosporic species were recovered from mildly polluted or unpolluted waters; of the remaining species fewer than half were recovered in numbers large enough to indicate membership in a population for even a brief period of time. The significance of fungi in both aquatic and terrestrial environments has been discussed in detail.[22–38]

Quantitative enumeration of fungi is not equivalent to that of unicellular bacteria because a fungal colony may develop from a single cell (spore), an aggregate of cells (a cluster of spores or a single multi-celled spore), or from a mycelial or pseudomycelial fragment (containing more than one viable cell). It is assumed that each fungal colony developing in laboratory culture originates from a single colony-forming unit (CFU), which may or may not be a single cell.

4. References

1. TABAK, H. & W.B. COOKE. 1968. Growth and metabolism of fungi in an atmosphere of nitrogen. *Mycologia* 60:115.
2. NAGY, L.A. & B.H. OLSON. 1982. The occurrence of filamentous fungi in water distribution systems. *Can. J. Microbiol.* 28:667.
3. NIEMI, R.M., S. KUNTH & K. LUNDSTROM. 1982. Actinomycetes and fungi in surface waters and potable water. *Appl. Environ. Microbiol.* 43:378.
4. ROSENZWEIG, W.D., H. MINNIGH & W.O. PIPES. 1986. Fungi in potable water distribution systems. *J. Amer. Water Works Assoc.* 78:53.
5. HINZELIN F. & J.C. BLOCK. 1985. Yeast and filamentous fungi in drinking water. *Environ. Tech. Letters* 6:101.
6. BURMAN, N.P. 1965. Symposium on consumer complaints. 4. Taste and odour due to stagnation and local warming in long lengths of piping. *Proc. Soc. Water Treat. Exam.* 14:125.
7. BAYS, L.R., N.P. BURMAN & W.M. LEWIS. 1970. Taste and odour in water supplied in Great Britain. A survey of the present position and problems for the future. *Proc. Soc. Water Treat. Exam.* 19:136.
8. NAGY, L.A. & B.H. OLSON. 1985. Occurrence and significance of bacteria, fungi, and yeasts associated with distribution pipe surfaces. Proc. American Water Works Assoc., Water Quality Technology Conf., p. 213.
9. METZLER, D.F., C. RITTER & R.L. CULP. 1956. Combined effect of water purification processes on the removal of *Histoplasma capsulatum* from water. *Amer. J. Pub. Health* 46:1571.
10. CASTELLANI, A. 1963. The cultivation of pathogenic fungi in sterile distilled water. *Commentarii* 1(10):1.
11. COOKE, W.B. & P.W. KABLER. 1953. The survival of *Histoplasma capsulatum* in water. *Lloydia* 16:252.
12. AJELLO, L. & M.E. GETZ. 1954. Recovery of dermatophytes from shoes and shower stalls. *J. Invest. Derm.* 22:17.
13. KISHIMOTO, R.A. & G.E. BAKER. 1969. Pathogenic and potentially pathogenic fungi isolated from beach sands and selected soils of Oahu, Hawaii. *Mycologia* 61:539.
14. MULLER, G. 1973. Occurrence of dermatophytes in the soils of European beaches. *Sci. Total Environ.* 2:116.
15. ENGELBRECHT, R.S., D.H. FOSTER, E.O. GREENING & S.H. LEE. 1974. New microbial indicators of waste water efficiency. Environ. Protect. Technol. Ser. No. 670/2-73-082.
16. JONES, J. & J.A. SCHMITT. 1978. The effect of chlorination on the survival of cells of *Candida albicans*. *Mycologia* 70:684.
17. ENGELBRECHT, R.S. & C.N. HAAS. 1977. Acid-fast bacteria and yeasts as disinfection indicators: Enumeration methodology. Proc. American Water Works Assoc. Water Quality Technology Conf. 1977, p. 1.
18. HAAS, C.N. & R.S. ENGELBRECHT. 1980. Chlorine dynamics during inactivation of coliforms, acid-fast bacteria, and yeasts. *Water Res.* 14:1749.
19. HAAS, C.N. & R.S. ENGELBRECHT. 1980. Physiological alterations of vegetative microorganisms resulting from chlorination. *J. Water Pollut. Control Fed.* 52:1976.
20. ROSENZWEIG, D.W., H.A. MINNIGH & W.O. PIPES. 1983. Chlorine demand and inactivation of fungal propagules. *Appl. Environ. Microbiol.* 45:182.
21. COOKE, W.B. 1986. The Fungi of "Our Mouldy Earth". *Beihefte zur Nova Hedwigia* 85:1.
22. BARLOCHER, F. & B. KENDRICK. 1976. Hyphomycetes as intermediates of energy flow in streams. *In* E. B. Jones, ed. Recent Advances in Aquatic Mycology. Elek Science, London, England.
23. COOKE, W.B. 1961. Pollution effects on the fungus population of a stream. *Ecology* 42:1.
24. COOKE, W.B. 1965. The enumeration of yeast populations in a sewage treatment plant. *Mycologia* 57:696.
25. COOKE, W.B. 1970. Our Mouldy Earth. FWPCA Res. Contract Ser. Publ. No. CWR, Cincinnati, Ohio.
26. COOKE, W.B. 1971. The role of fungi in waste treatment. *CRC Critical Rev. Environ. Control.* 1:581.
27. COOKE, W.B. 1976. Fungi in sewage. *In* E.B.G. Jones, ed. Recent Advances in Aquatic Mycology. Elek Science, London, England.
28. COOKE, W.B. 1979. The Ecology of Fungi. CRC Press, Boca Raton, Fla.
29. DICK, M.W. 1971. The ecology of Saprolegniales in the lentic and littoral muds with a

general theory of fungi in the lake ecosystem. *J. Gen. Microbiol.* 65:325.

30. HARLEY, J.L. 1971. Fungi in ecosystems. *J. Appl. Ecol.* 8:627.

31. MEYERS, S.P., D.G. AHEARN & W.L. COOK. 1970. Mycological studies of Lake Champlain. *Mycologia* 62:504.

32. NOELL, J. 1973. Slime-inhabiting geofungi in a polluted stream (winter/spring). *Mycologia* 65:57.

33. PARK, D. 1972. Methods of detecting fungi in organic detritus in water. *Trans. Brit. Mycol. Soc.* 58:281.

34. QURESHI, A.A. & B.J. DUTKA. 1974. A preliminary study on the occurrence and distribution of geofungi in Lake Ontario near the Niagara River. Proc. 17th Conf. Great Lakes Research, International Soc. Great Lakes Research.

35. SHERRY, J.P. & A.A. QURESHI. 1986. Isolation and enumeration of fungi using membrane filtration. *In* B.J. Dutka, ed. Membrane Filtration Applications, Techniques, and Problems. Marcel Dekker, New York, N.Y.

36. SIMARD, R.E. 1971. Yeasts as an indicator of pollution. *Marine Poll. Bull.* 12:123.

37. SPARROW, F.K. 1968. Ecology of fresh water fungi. *In* G. C. Ainsworth & A.S. Sussman, eds. The Fungi, An Advanced Treatise. Vol. 3. The Fungal Population. Academic Press, New York, N.Y.

38. TOMLINSON, T.G. & I.L. WILLIAMS. 1975. Fungi, *In* C.R. Curds & H.A. Hawkes, eds. Ecological Aspects of Used-Water Treatment. Vol. 1. The Organisms and their Ecology. Academic Press, New York, N.Y.

5. Bibliography

EMERSON, R. 1958. Mycological organization. *Mycologia* 50:589.

COOKE, W.B. 1958. Continuous sampling of trickling filter populations. I. Procedures. *Sewage Ind. Wastes* 30:21.

COOKE, W.B. & A. HIRSCH. 1958. Continuous sampling of trickling filter populations. II. Populations. *Sewage Ind. Wastes* 30:139.

COOKE, W.B. 1959. Trickling filter ecology. *Ecology* 40:273.

SPARROW, F.K. 1959. Fungi. (Ascomycetes, Phycomycetes); including W.W. Scott, Key to genera, Fungi Imperfecti (Aquatic Hyphomycetes only). In W.T. Edmondson, ed. Ward & Whipple's Fresh Water Biology, 2nd ed. John Wiley & Sons, New York, N.Y.

COOKE, W.B. 1963. A Laboratory Guide to Fungi in Polluted Waters, Sewage, and Sewage Treatment Systems, Their Identification and Culture. USPHS Publ. 999-WP-1, Cincinnati, Ohio.

FULLER, M.S. & R.O. PAYTON. 1964. A new technique for the isolation of aquatic fungi. *BioScience* 14:45.

WILLOUGHBY, L.C. & V.G. COLLINS. 1966. A study of fungal spores and bacteria in Blelham Tarn and its associated streams. *Nova Hedwigia* 12:150.

COOKE, W.B. & G.S. MATSUURA. 1969. Distribution of fungi in a waste stabilization pond system. *Ecology* 50:689.

BROCK, T.D. 1970. Biology of Microorganisms. Prentice-Hall, Englewood Cliffs, N.J.

COOKE, W.B. 1970. Fungi in the Lebanon sewage treatment plant and in Turtle Creek, Warren Co., Ohio. *Mycopathol. Mycol. Appl.* 42:89.

PATERSON, R.A. 1971. Lacustrine fungal communities. *In* J. Cairns, ed. Structure and Function of Microbial Communities. Symposium, American Microscopical Soc., Burlington, Vt. 1969. Virginia Polytechnic Inst. & State Univ. Res. Div. Mono. 3:209.

JONES, E.B.G. 1971. Aquatic Fungi. *In* C. Booth, ed. Methods in Microbiology 4:335. Academic Press, New York, N.Y.

FARR, D.F. & R.A. PATERSON. 1974. Aquatic fungi in rivers: Their distribution and response to pollutants. VPI-WRRC Bull. 68, Virginia Water Resources Research Center, Virginia Polytechnic Inst. & State Univ., Blacksburg.

GARETH JONES, E.B., ed. 1976. Recent Advances in Aquatic Mycology. Elek Science, London.

FULLER, M.S., ed. 1978. Lower Fungi in the Laboratory. *Palfrey Contrib. in Botany* 1:1-212. Athens, Ga.

ALEXOPOULOS, C.J. & C.W. MIMS. 1979. Introductory Mycology, 3rd ed. John Wiley & Sons, New York, N.Y.

9610 B.   Pour Plate Technique

1. Samples

a. Containers: Collect samples as directed in Section 9060A. Alternatively, use cylindrical plastic vials with snap-on caps. These vials usually are sterile as received. Transport them in an upright position to minimize the chance of leakage and discard after use.

b. Storage: Hold samples for not more than 24 h. If analysis is not begun promptly after sample collection, refrigerate.

2. Media

For counting, neopeptone-glucose-rose bengal-aureomycin® agar is the usual medium of choice, although experience may indicate that Czapek agar (for *Aspergillus, Penicillium,* and related fungi) and yeast extract-malt extract-glucose agar or Diamalt agar (for yeasts) may be preferable. For inventory, use neopeptone-glucose agar.

a. Neopeptone-glucose-rose bengal aureomycin® agar: Add 5.0 g neopeptone, 10.0 g glucose, 3.5 mL rose bengal solution (1 g/100 mL distilled water), and 20.0 g agar to 1 L distilled water. Because this medium is used for making pour plates, prepare and store basal agar either in bulk, or more conveniently, in tubes in 10-mL amounts. Sterilize by autoclaving; the final pH should be about 6.5.

Separately prepare a solution of chlortetracycline or tetracycline (1.0 g water-soluble antibiotic/150 mL distilled water) and refrigerate. Before use, sterilize by filtration. To complete the medium, add 0.05 mL sterile solution to 10 mL melted basal agar at about 45°C.

This medium may not be available in dehydrated form and may require preparation from the basic ingredients. Dehydrated Cooke's rose bengal agar may be used in place of the agar base.

This medium is useful for isolating a broad spectrum of fungal species.

b. Czapek (or Czapek-Dox) agar: Dissolve 30.0 g sucrose, 3.0 g sodium nitrate (NaNO$_3$), 1.0 g dipotassium hydrogen phosphate (K$_2$HPO$_4$), 0.5 g magnesium sulfate (MgSO$_4$), 0.5 g potassium chloride (KCl), 0.01 g ferrous sulfate (FeSO$_4$), and 15.0 g agar in 1 L distilled water. The pH should be 7.3 after sterilization.

This medium is useful for isolating species of *Aspergillus, Penicillium, Paecilomyces,* and some other fungi with similar physiological requirements.

c. Yeast extract-malt extract-glucose agar: Dissolve 3.0 g yeast extract, 3.0 g malt extract, 5.0 g neopeptone (or equivalent), 10.0 g glucose, and 20.0 g agar in 1 L distilled water. No adjustment of pH is required.

This medium is useful for isolating yeasts.

d. Diamalt agar: Dissolve 150 g diamalt and 20.0 g agar in 1 L distilled water. No adjustment of pH is required. The medium will be turbid but filtration is unnecessary.

This medium is useful in the purification of yeast isolates and for study of yeast species in various specified tests.

e. Neopeptone-glucose agar: Dissolve 5.0 g neopeptone (or equivalent), 10.0 g glucose, and 20.0 g agar in 1 L distilled water. The pH should be about 6.5 after sterilization. (This medium is known also as Emmons' Sabouraud Agar or Emmons' Sabouraud Dextrose Agar.)

This medium is useful for maintaining stock cultures. It is comparable to neopeptone-glucose-rose bengal aureomycin® agar but contains neither rose bengal nor an antibiotic.

3. Procedure

As many as 40 samples can be analyzed simultaneously by a single analyst by the following procedure; however, 20 samples represents the optimum number.

a. Preparation and dilution: To a sterile 250-mL erlenmeyer flask add 135 mL sterile distilled water and 15 mL sample to obtain a 1:10 sample dilution. Use a sterile measuring device for each sample, or, less preferably, rinse the measure with sterile distilled water between samples. Mix sample well before withdrawing the 15-mL portion. Shake flask on a rotary shaker at about 120 to 150 oscillations/min for about 30 min or transfer flask contents to a blender jar, cover and blend at low speed for 1 min or high speed for 30 s. Preferably use a sterile blender jar and appurtenances for each sample or wash jar thoroughly between samples and rinse with sterile distilled water. Further dilutions may be made by adding 45 mL sterile distilled water to 5 mL of a 1:10 diluted suspension.

For stream water samples a dilution of 1:10 usually is adequate. Dilute samples with large amounts of organic material, such as sediments, to 1:100 or 1:1000. Dilute stream bank or soil samples to 1:1000 or 1:10 000.

b. Plating: Prepare five plates for each dilution to be examined. To use neopeptone-glucose-rose bengal-aureomycin® agar, aseptically transfer 10 mL of medium at 45°C to a 9-cm petri dish. Add 1 mL of appropriate sample dilution and mix thoroughly by tilting and rotating dish (see plating procedure under heterotrophic plate count, Section 9215). Alternatively add to petri dish 1 mL sample, 0.05 mL antibiotic solution, and 10 mL liquefied agar medium at 43 to 45°C. Solidify agar

as rapidly as possible. (In arid areas use more medium to prevent dehydration during incubation.)

c. Incubation: Stack plates but do not invert. Incubate at room conditions of temperature (20 to 24°C) and lighting but avoid direct sunlight. Examine and count plates after 3, 5, and 7 d.

d. Counting and inventory: The fungus plate count will provide the basis for rough quantitative comparisons among samples; the inventory will give relative importance of at least the more readily identifiable species or genera.

In preparing plates, use sample portions that will give about 50 to 60 colonies on a plate. Determine this volume by trial and error. When first examining a new habitat plate at least two sample dilutions. Estimates of up to 300 colonies may be made, but discard more crowded plates. The medium containing rose bengal tends to produce discrete colonies and permits slow-growing organisms to develop.

The inventory includes the direct identification of fungi based on colonial morphology and the counting of colonies assignable to various species or genera. When discrete colonies cannot be identified and identification is important, use a nichrome wire with its tip bent in an L-shape to pick from each selected colony and streak on a slant of neopeptone-glucose agar (¶ 2e). If five plates are used per sample, the average number of colonies on all plates (total number of colonies counted/ 5) times the reciprocal of the dilution (10/ 1, 100/1, 1000/1, etc.) equals the fungus colony count per milliliter of original sample. For solid or semisolid samples, use a correction for the water content to report fungus colonies per gram dry weight. Determine water content by drying paired 15-mL portions of original sample at 100°C overnight; the difference between wet and dry weights is the amount of water lost from the sample.

9610 C.   Spread Plate Technique

The spread plate technique is an alternative procedure for obtaining quantitative data on colony-forming units.

1. Samples

See Section 9610B.1.

2. Media

Use any of the following media. Aureomycin®-rose bengal-glucose-peptone agar (ARGPA) (e) and streptomycin-terramycin®-malt extract agar (STMEA) (f) are useful in analyzing sewage and polluted waters.[1]

a. Neopeptone-glucose-rose bengal aureomycin® agar: See Section 9610B.2a.

b. Czapek (or Czapek-Dox) agar: See Section 9610B.2b.

c. Yeast extract-malt extract-glucose agar: See Section 9610B.2c.

d. Diamalt agar: See Section 9610B.2d.

e. Aureomycin®-rose bengal-glucose-peptone agar: Dissolve 10.0 g glucose, 5.0 g peptone, 1.0 g potassium dihydrogen phosphate (KH_2PO_4), 0.5 g magnesium sulfate ($MgSO_4 \cdot 7H_2O$), 0.035 g rose bengal, and 20.0 g agar in 800 mL distilled water and sterilize. Dissolve 70.0 mg aureomycin® hydrochloride in 200 mL distilled water, sterilize by filtration, and add to the cooled (42 to 45°C) agar base. The pH should be 5.4. Pour 25-mL portions into sterile petri dishes (100 × 15 mm) and let agar harden. Poured plates may be held up to 4 weeks at 4°C.

f. Streptomycin-terramycin®-malt extract agar: Dissolve 30.0 g malt extract, 5.0 g peptone, and 15.0 g agar in 800 mL distilled water and sterilize. Dissolve 70.0 mg each of streptomycin and terramycin® in separate 100-mL portions distilled water, sterilize by filtration, and add to the cooled (42 to 45°C) agar base. The pH should be 5.4. Pour about 20-mL portions into sterile petri dishes (60 × 15 mm) and let agar harden. Poured plates may be held up to 4 weeks at 4°C.

3. Procedure

a. Preparation and dilution: See Sections 9215A.5 and 9610B.3a. Make dilutions with buffered water (Section 9050C.1) and select dilutions that yield 20 to 150 colonies per plate.

b. Plating: Pre-dry plates separately with lids off in a laminar-flow hood at room temperature and about 30% relative humidity for 1 to 1.5 h. Prepare at least three plates, or five plates if data are to be analyzed statistically, per sample or dilution. Using a sterile pipet, transfer 1.0 mL of sample or dilution onto surface of a pre-dried agar plate. Spread sample over entire agar surface using a sterile L-shaped glass rod or use a mechanical device to rotate plate and ensure proper sample distribution.

c. Incubation: With dish covers on, let plates dry at room temperature, invert plates, and incubate at 15°C for 7 d in an atmosphere of high humidity (90 to 95%). Alternatively incubate at 20°C for 5 to 7 d. Slow-growing fungi may not produce noticeable colonies until 6 or 7 d.

d. Counting and recording: Using a Quebec colony counter, count all colonies on each selected plate. If counting must be delayed temporarily, hold plates at 4°C for not longer than 24 h. Depending on colony size, plates with as many as 150 colonies can be counted but the optimal maximum number is 100 colonies.

Record results as colony-forming units (CFU)/100 mL original sample. For solid or semisolid samples report CFU/g wet or dry, preferably dry. If three or more plates are used per sample, use average number of colonies times the reciprocal of the dilution (see 9610B) to give colony count. If no plates have colonies, record count as < 1 for the highest dilution. If there are more than 150/plate, record as Too Numerous To Count (TNTC) but indicate a count of > 150 for the appropriate dilution. If colonies are crowded and overlapping with spreaders, record as "obscured" (OBSC) and repeat analysis with higher dilution or earlier observations.

4. Reference

1. EL-SHAARAWI, A., A.A. QURESHI & B.J. DUTKA. 1977. Study of microbiological and physical parameters in Lake Ontario adjacent to the Niagara River. *J. Great Lakes Res.* 3:196.

9610 D. Membrane Filter Technique

For general information on the membrane filter technique and apparatus needed see Section 9222. However, except for comparisons of different manufacturers' membranes, no critical tests have been reported for membrane filters for fungal isolation efficiency. Media components, pH levels, and antibiotics have been used in routine plating procedures. It appears that the reported procedures are satisfactory.

1. Samples

See Sections 9610B.1 and 9060A.

2. Media

Use unmodified or modified aureomycin®-rose bengal-glucose-peptone agar (MARGPA) or modified streptomycin-terramycin®-malt extract agar (MSTMEA).[1] These media are prepared identically to the unmodified media described in 9610C.2*e* and *f* except that the concentration of each antibiotic is increased from 70 mg/L to 200 mg/L. Dispense media in portions of 5 to 7 mL in glass or plastic petri dishes (60 × 15 mm); plastic dishes with tight-fitting lids are preferred.

3. Procedure

a. Preparation and dilution: See Sections 9215A.5 and 9610B.3*a.* Select dilutions to yield 20 to 100 colonies per membrane.

b. Filtration: Filter appropriate volumes of well-shaken sample or dilution, in triplicate, through membrane filters with pore diameter of 0.45 or 0.8 μm. See Section 9222.

c. Incubation: Transfer filters to dishes, invert dishes or not, and incubate at 15°C for 5 d in a humid atmosphere. Alternatively incubate at 20°C for 3 d, or longer depending on fungi present.

d. Counting and recording: Using a binocular dissecting microscope at a magnification of 10×, count all colonies on each selected plate. If counting must be delayed temporarily, hold plates at 4°C for not longer than 24 h. Ideal plates have 20 to 80 colonies per filter. See Section 9610C.

4. Reference

1. QURESHI, A.A. & B.J. DUTKA. 1978. Comparison of various brands of membrane filter for their ability to recover fungi from water. *Appl. Environ. Microbiol.* 32:445.

9610 E. Technique for Yeasts

Of the total number of fungal colonies obtained from polluted waters, as many as 50% may be yeast colonies. Solid media such as those described above do not permit growth of all yeasts; thus, a quantitative enrichment technique may be useful in addition to the plate count (see also Fungi Pathogenic to Humans, Section 9610H).

1. Media

For enrichment, use yeast nitrogen base-glucose broth; for isolation, use yeast extract-malt extract-glucose agar or diamalt agar.

a. Yeast nitrogen base-glucose broth: Dissolve 13.4 g yeast nitrogen base in 1 L distilled water; sterilize by filtration. Prepare 500 mL each of 2% and 40% aqueous glucose solutions and sterilize separately by filtration. To make final medium, aseptically add to a sterile 250-mL erlenmeyer flask 25 mL yeast nitrogen base solution and 25 mL of either 2% or 40% glucose solutions to make 1% or 20% final glucose concentrations. Stopper flask with a gauze-wrapped cotton stopper and store until used.

b. Yeast extract-malt extract-glucose agar: See Section 9610B.2c.

c. Diamalt agar: See Section 9610B.2d.

2. Procedure

a. Sample preparation and dilution: Prepare as directed in Section 9610B.

b. Enrichment: In 250-mL erlenmeyer flasks prepare one flask each of yeast nitrogen base medium containing 1% and 20% glucose. Inoculate with 1 mL of appropriate sample dilution and incubate at room temperature on a rotary shaker operating at 120 to 150 oscillations/min for at least 64 h. Shaken cultures are necessary to prevent overgrowth by filamentous fungi.

c. Isolation: Remove flasks from shaker and let settle 4 to 5 h. Yeast cells, if present, will settle to the bottom, bacteria and filamentous fungi will remain in suspension, and filamentous fungi will float on the surface or will be attached to the glass surface at or above the meniscus. With a nichrome wire loop remove a loopful of sediment at the sediment-supernatant interface from a tilted flask and smear-streak on yeast extract-malt extract-glucose agar. Use three plates per flask. Incubate at room temperature but out of direct sunlight for 2 to 3 d. It is not necessary to invert dishes. To obtain pure cultures, pick from reasonably isolated colonies and restreak on the same medium or on diamalt agar plates. Obtain pure cultures of as many different colonies as can be recognized.

d. Counting: It is impossible to obtain a meaningful plate count after this type of enrichment isolation. If it is assumed that one cell in the original sample will produce one or more colonies on the plates after enrichment, it can be stated that yeasts, or specific types of yeasts, occur at a minimal number dependent on the highest positive dilution. The reciprocal of this dilution is the indicated number of yeasts in the sample.

3. Bibliography

LODDER, J., ed. 1970. The Yeasts, A Taxonomic Study, 2nd ed. North Holland Publ. Co., Amsterdam.

BUCK, J.D. 1975. Distribution of aquatic yeasts—effect of inoculation temperature and chloramphenicol concentration on isolation. *Mycopathologia* 56:73.

9610 F.   Zoosporic Fungi

1. Occurrence and Significance

Most fungi found in lacustrine (lake) and lotic (river) habitats that reproduce asexually by motile, uniflagellate spores and have determinate growth of the fungal body belong to the class Chytridiomycetes. Fungi with indeterminate growth, asexual reproduction by motile, biflagellate spores, and sexual reproduction involving oogonia and antheridia, are members of the class Oomycetes. A reduction in numbers of species of both classes tends to occur in polluted areas of rivers but more species of Oomycetes than of Chytridiomycetes can be found in polluted situations. Species of the Oomycete genera *Saprolegnia* (notably *S. ferax*) and *Leptomitus* appear to be more tolerant than other forms. Bioassay studies indicate that Oomycetes are more tolerant to zinc, cyanide, and mannitol than are Chytridiomycetes. The latter appear to be more tolerant to treatment with surfactants than do the Oomycetes.

Some Chytridiomycetes may parasitize planktonic and other algae. In the case of epidemic fungal infections of phytoplankton species, the activities of fungi may affect the composition of phytoplankton communities by delaying the time of algal maxima and by reducing the population of certain algae so that other phytoplankters will replace the infected algal populations. In the case of nonepidemic infections, fungi may not influence algal populations; instead, they may infect only phytoplankters during periods of decline and thus only hasten decomposition of the algae.

Filamentous Oomycetes, particularly members of the Saprolegniaceae and Pythiaceae, are found in virtually all types of freshwater habitats and damp-to-wet soils. Most of the nearly 250 species involved occur as saprobes on dead and decaying organic matter such as insect exuviae, algae, and submerged vascular plant remains. A few occur as parasites of algae, aquatic invertebrates, fish, and vascular plants; none are associated with human disease.

Rarely do any of these fungi develop in sufficient numbers to be observed or collected directly. Consequently, various techniques have been devised for their collection and isolation.

2. Sampling and Baiting

Collect samples in sterile 35-mL plastic vials, refrigerate, and start analysis within 6 to 8 h. Place each sample in a sterile plate (20 × 100 mm) and dilute with 10 to 15 mL sterile distilled or deionized water. Add three to four split hemp seed halves (*Cannabis sativa*), or whole seeds of mustard *(Brassica)* or sesame *(Sesamum)* as bait to each culture. Incubate at 18 to 23°C and examine daily for fungal growth on the bait. As growth becomes evident, usually within 72 h, remove the infected bait, wash it thoroughly with water from a wash bottle, and transfer to a fresh plate of water containing two to three halves of hemp or other seed. Genera may be identified from spore arrangement within the sporangium and the manner in which spores are released. Specific determination requires microscopic examination of the sexual reproductive structures.

To collect the few naturally occurring parasites or pathogens, place the host organisms in a plate containing sterile water and hemp seed.

3. Isolation

Although most filamentous Oomycetes can be cultivated on plain cornmeal agar, selective media for isolating *Saprolegnia* from fresh water have been developed.[1]

Obtain axenic cultures by drawing spores into a micropipet as they emerge from the sporangium. Less preferably, use hyphal tips, but note that several different genera and species frequently occur on a single piece of bait. Transfer the spore suspension or hyphal tip to a plate of cornmeal agar. When growth on the agar has occurred, remove bacteria-free hyphal tips aseptically by cutting out a small block of agar. Transfer to fresh medium or water. If growth is not free from contamination after one transfer, make additional transfers to insure pure cultures. Other methods have been outlined.[2]

4. Dilution Plating

Make serial dilutions with sterile distilled water (1:100 000 to 1:700 000) and spread 1 mL over surface of a *freshly* prepared cornmeal agar plate. Remove each developing colony and transfer to water for identification. This method permits numerical estimation and determination of composition of the Oomycete community but requires at least 10 plates.

5. References

1. Ho, H.H. 1975. Selective media for the isolation of *Saprolegnia* spp. from fresh water. *Can. J. Microbiol.* 21:1126.

2. SEYMOUR, R.L. 1970. The Genus *Saprolegnia*. *Nova Hedwigia* Beihefte 19:1.

6. Bibliography

WILLOUGHBY, L.G. 1962. The occurrence and distribution of reproductive spores of Saprolegniales in fresh water. *J. Ecol.* 50:733.

9610 G.   Aquatic Hyphomycetes

1. Occurrence and Significance

Freshwater Hyphomycetes are a very specialized group of conidial fungi that usually occur on partially decayed, submerged leaves and occasionally wood of angiosperms. The mycelium, which is branched and septate, ramifies through the leaf tissue, especially in petioles and veins. The conidiophores project into the water and the conidia that usually develop are liberated under water. Mature conidia also can be found in the surface foam of most rivers, streams, and lakes. The conidia of the majority of these fungi are hyaline, thin-walled, and either tetraradiately branched, that is, with four divergent arms, or sigmoid (S-shaped) with the curvature in more than one plane. A special feature of the conidia is that while suspended in water, even for long periods, they do not germinate. However, if they come to rest on a solid surface, germ tubes are produced within a few hours. The size and morphology of these spores make them potentially more prominent in plankton analysis work than the spores of other fungi. Ecological investigations of freshwater Hyphomycetes have been limited to substrate, habitat, dispersal, and their role in the enhancement of leaf substrates as food for aquatic invertebrates. The most common substrates of these organisms are submerged, decaying leaves of angiosperms such as alder (*Alnus*), oak (*Quercus*), hazelnut (*Corylus*), elm (*Ulmus*), maple (*Acer*), chestnut (*Castanea*), blackberry (*Rubus*), ash (*Fraxinus*), and willow (*Salix*). Submerged gymnosperm leaves usually are free of aquatic Hyphomycetes. The usual habitat of these fungi is well-oxygenated water, such as alpine brooks, mountain streams, and fast-flowing rivers. However, they also have been found in slow-running, often contaminated, rivers, stagnant or temporary pools, melting snow, and soil. There is often an increase in the numbers of species and individuals of aquatic Hyphomycetes from autumn until spring, with a decline between April and June.

2. Sample Collection and Storage

For most freshwater environments, collect foam or partially decayed, submerged, angiosperm leaves in sterile bottles. Refrigerate sample until it is examined.

3. Sample Treatment and Analysis

Wash the leaf samples in sterile distilled water and place one to three leaves in a sterile petri dish about 1 cm deep containing sterile pond, river, or lake water. Incubate at room temperature. Within 1 to

2 d, the mycelium and conidia develop. Conidiophores and conidia can be observed with a dissecting microscope on any portion of a leaf surface, but most frequently are seen on petioles and veins. When released, the conidia either remain suspended in the water or settle to the dish bottom. Using a dissecting microscope, pick up single conidia with a micropipet. Transfer each conidium to a microscope slide in a drop of water for identification. The conidium may be transferred with a sterile needle to a plate of 2% malt extract agar for colony production. Search for conidia in foam samples with a dissecting microscope and isolate single conidia as described above. Submerge mycelial plugs from stock culture isolates of aquatic Hyphomycetes in autoclaved pond water in deep petri dishes; conidiogenesis usually occurs within 2 to 10 d.

Conidia in all stages of development can be preserved on slides with lactophenol mounting medium in which either acid fuchsin or cotton blue is dissolved and sealed with clear fingernail polish. To permit good adherence of the nail polish, avoid excessive amounts of mounting medium.

9610 H.   Fungi Pathogenic to Humans

1. Occurrence and Significance

Routine isolations of fungi from polluted streams and wastewater treatment plants usually have yielded relatively few species pathogenic to human and other higher animals. *Geotrichum candidum*, an arthroconidium-producing fungus, for which there is, in the United States, presumptive evidence of an association with disease, is isolated almost universally. When its teleomorphic stage (an ascus) develops, it is known as *Endomyces candidus. Rhinocladiella mansonii*, now called *Exophiala mansonii*, a causal agent of one form of chromomycosis, usually in the tropics, is equally widespread. It has been listed also as *Phialophora jeanselmei* and *Trichosporium heteromorphum. Aspergillus fumigatus*, a causal agent of pulmonary aspergillosis, is commonly isolated. *Pseudallescheria (Petriellidium, Allescheria) boydii* is a causal agent of eumycotic mycetomas and other eumycotic conditions grouped[1] under the heading "Pseudallescheriasis." Infection may follow from a puncture wound with contaminated materials or breathing such contaminated materials as sprays at wastewater treatment plants or contaminated air. It usually is recovered in its anamorphic state, *Scedosporium (Monosporium) apiospermum*. The presence of these fungi in stream water probably represents soil runoff because virtually all zoopathogenic fungi exist saprobically in soil as their natural reservoir. Other zoopathogenic fungi occasionally are recovered in low frequencies from streams, polluted or not. Another fungus, the yeast *Candida albicans*, can be recovered in varying numbers from wastewater treatment plant effluents, streams receiving such effluents, and recreational waters. In humans this fungus is usually a commensal organism, like *Geotrichum candidum*, coexisting in harmony with its host organism; up to 80% of normal, healthy adults have detectable levels of *C. albicans* in their feces, while about 35% harbor it in their oral cavities in the absence of any overt disease. A very large proportion of the female population has vaginal candidiasis in varying degrees of severity. The presence of *C. albicans* in raw wastewater, wastewater treatment plant effluent, or contaminated water is not surprising. *C. albicans* has been isolated from these habitats on routine media heavily supplemented with antibacterial drugs, but not on media or with techniques described in Sections 9610B or E. It also has been isolated from estuarine and marine habitats on a maltose-yeast nitrogen base-chloramphenicol-cycloheximide medium.

2. Identification of *C. albicans*

C. albicans can be identified among the white and pink yeasts growing on an 0.8-μm black membrane filter on maltose-yeast nitrogen base-chloramphenicol-cyclohex-imide medium. From each colony, inoculate a 0.5-mL portion of calf or human blood serum, incubate at 37°C for 2 to 3 h, transfer a drop or two to a slide, and examine microscopically for the production of germ tubes from a majority of the cells. Of the white yeasts, only *C. albicans* produces these short hyphae from the parent cell within 2 to 3 h incubation.[2]

3. References

1. RIPPON, J. 1982. Medical Mycology. W. B. Saunders Co., Philadelphia, Pa.
2. BUCK, J.D. & B.M. BUBACIS. 1978. Membrane filter procedure for enumeration of *Candida albicans* in natural waters. *Appl. Environ. Microbiol.* 35:237.

4. Bibliography

PAGAN, E.F. 1970. Isolation of human pathogenic fungi from river water. Ph.D. dissertation, Botany Dep., Ohio State Univ., Columbus.

9711   PATHOGENIC PROTOZOA*

9711 A.   Introduction

1. Significance

The pathogenic protozoa of primary concern in drinking water and wastewater are *Giardia lamblia, Entamoeba histolytica,* and *Cryptosporidium.* These organisms cause diarrhea or gastroenteritis of varying severity, and waterborne outbreaks have been attributed to each agent. *Giardia* is the most frequently identified etiologic agent in waterborne outbreaks. Between 1965 and 1984, 90 waterborne outbreaks and more than 23 000 cases of giardiasis were reported.[1] *E. histolytica* causes amebiasis or amebic dysentery. While it has not been a frequent cause of waterborne outbreaks in recent times, it is still endemic in the United States, causing an average of 28 deaths per year.[2] *Cryptosporidium* is a protozoan of increasing concern as a human pathogen.[3] It can cause a cholera-like diarrhea that is self-limiting in immunocompetent individuals but it may be prolonged and life-threatening in immune-deficient persons.[4] *Cryptosporidium* has been associated with traveler's diarrhea[5] and drinking water has been implicated as the vehicle of transmission in at least one outbreak.[6] Reports on detection methodology for *Cryptosporidium* in water and wastewater are available.[7,8]

Methods for protozoa are not well standardized and should be considered research procedures subject to modification.

2. References

1. CRAUN, G. F. & W. JAKUBOWSKI. 1987. Status of waterborne giardiasis outbreaks and moni-

* Approved by Standard Methods Committee, 1988.

toring methods. *In* International Symposium on Water Related Health Issues Proceedings. TPS 87-3, American Water Resources Assoc., Bethesda, Md.

2. CENTERS FOR DISEASE CONTROL. 1984. Morbidity and mortality weekly report-annual summary, 1983. U. S. Dep. Health & Human Services, Public Health Serv., Atlanta, Ga.

3. FAYER, R. & B. L. P. UNGAR. 1986. *Cryptosporidium* spp and cryptosporidosis. *Microbiol. Rev.* 50:458.

4. CURRENT, W. L. 1984. *Cryptosporidium* and cryptosporidiosis. *In* Acquired Immune Deficiency Syndrome. A. R. Liss, Inc., New York, N.Y.

5. SOAVE, R. & P. MA. 1985. Cryptosporidiosis traveler's diarrhea in two families. *Arch. Intern. Med.* 145:70.

6. D'ANTONIO, R. G., R. E. WINN, J. P. TAYLOR, T. L. GUSTAFSON, W. L. CURRENT, M. M. RHODES, G. W. GARY, JR. & R. A. ZAJAC. 1985. A waterborne outbreak of cryptosporidiosis in normal hosts. *Ann. Intern. Med.* 103:886.

7. ROSE, J. B., C. E. MUSIAL, M. J. ARROWOOD, C. R. STERLING & C. P. GERBA. 1985. Development of a method for the detection of *Cryptosporidium* in drinking water. *In* Proceedings, 13th Annual Water Quality Technology Conference. American Water Works Assoc., Denver, Colo.

8. MUSIAL, C. E., M. J. ARROWOOD, C. R. STERLING & C. P. GERBA. 1987. Detection of *Cryptosporidium* in water by using polypropylene cartridge filters. *Appl. Environ. Microbiol.* 53:687.

9711 B.   *Giardia lamblia*

1. General Discussion

The causative agent of giardiasis is a flagellated protozoan shed in the feces of man and animals, most often in the cyst stage, although with cases of severe watery diarrhea, the fragile trophozoite reproductive stage may be shed. As few as 10 cysts ingested in capsules have been found to be infective for humans[1] and it is possible that fewer viable cysts ingested in drinking water may be sufficient to initiate infection. An infected individual may shed more than 10^6 cysts/g of stool.[2] Cysts may survive for up to 2 months in drinking water at 8°C.[3] Animal or human fecal contamination of water supplies could lead to waterborne transmission of giardiasis.[4]

Most documented waterborne outbreaks[5] associated with municipal supplies and recreational areas have occurred as the result of drinking surface water where the only treatment was chlorination. *Giardia* cysts are more resistant to disinfection than coliform bacteria.[6-8] Contrary to earlier belief, they can be inactivated by chlorine under appropriate conditions of temperature, pH, contact time, and dose.[9-11] *Giardia* cysts, when present in water supplies, probably would occur at low concentrations.[12-15]

Analyzing water for *Giardia* cysts requires concentration, purification, detection, and identification. Cysts have been concentrated from water using sand,[14] membrane,[16-18] or microporous depth filters.[19,20] Purification of cysts from extraneous debris generally is accomplished by using flotation techniques with solutions of zinc sulfate,[19] sucrose,[17,21] potassium citrate,[20,22] or Percoll-sucrose.[23] Detection and identification are based on microscopic examination of purified concentrates using chromogenic[24,25] or fluorescent antibody[20,23] staining techniques and morphological criteria.

Although methods for detecting cysts in water have been available since 1975,[14] no comparative studies of method efficiency, precision, or sensitivity have been reported with a variety of waters under different conditions.

A modification of the method described below has been used to detect cysts in raw and treated supplies.[26,27] The method was developed by a consensus panel[19] and should be regarded as experimental, with limited recovery efficiency data. Modifications in the procedures may be appropriate, depending on local water quality conditions, availability of equipment, and the experience of the analyst. A negative result may reflect intermittent shedding patterns, poor recovery efficiency with a given water, and insufficient sample volume or sampling frequency. The significance of a positive finding is difficult to assess unless suitable infectivity determinations and/or epidemiological studies are performed. Figure 9711:1 shows an analytical flow chart.

2. Sampling

Collect samples of the raw water source, at the treatment plant before disinfection, and from the distribution system. The total number of samples will depend on the study objectives and available resources. Collect a sample using the apparatus shown in Figures 9711:2 and 3, consisting of an inlet hose, plastic filter holder with 25-cm-long yarn-wound filter (1-μm porosity)* outlet hose, water meter, and a limiting orifice flow-control device with a flow rate of 6.3 $\times$ 10^{-5} m^3/s.† The yarn may be orlon, polypropylene, or other suitable material that does not release fibers during subsequent processing. Components of the apparatus need not be sterile, but thoroughly drain and rinse the equipment between samples. Use aseptic technique during sample collection to protect the sample collector and to prevent sample cross-contamination. A line pressure of 100 to 130 kPa is satisfactory; if water under

pressure is unavailable, use a pump. If a pump is used, install it downstream (on the effluent end) of the filter. This will eliminate the potential for cross-contamination of samples by cysts that might remain in the pump from a previous sampling.

The volume of water to be sampled depends on the intent of the investigation; a minimum sample size of 380 L is suggested.

To collect a sample, connect inlet hose to an appropriate sampling tap. The direction of water flow is from the outside to the inside of the yarn-wound filter cartridge. Record time and meter reading and open sampling tap. Collect sample, turn off sampling tap, record time and meter reading, and disconnect sampling apparatus. Take care to maintain opening on the inlet hose above the level of opening on outlet hose to prevent filter backwashing loss of particulate matter. Drain residual water as completely as possible from the sampling apparatus. When unit is completely drained, open filter holder, aseptically remove filter cartridge, place it in a labelled plastic bag, and seal bag. Place labelled bag within a second plastic bag and seal. If additional samples are to be collected with the same filter device, thoroughly rinse the influent hose and filter holder with the water to be sampled before installing another filter cartridge. Refrigerate samples or place on wet ice as soon as possible after collection. Transport to the laboratory on wet ice and process as soon as practical but within 48 h. Do not freeze. Minimize shipping and storage times.

3. Sample Processing

a. Extraction: Handle filter samples aseptically. Using a razor knife, separate the filter fibers from the support core by making a longitudinal cut down the entire length of the filter cartridge. Cut the fibers into two or more approximately equal portions consisting of an outer and an inner layer. There often will be a fairly distinct

---

* Commercial Filters Division, Carborundum Co., Lebanon, Ind., or equivalent.

† Eaton Corp., Carol Stream, Ill., or equivalent.

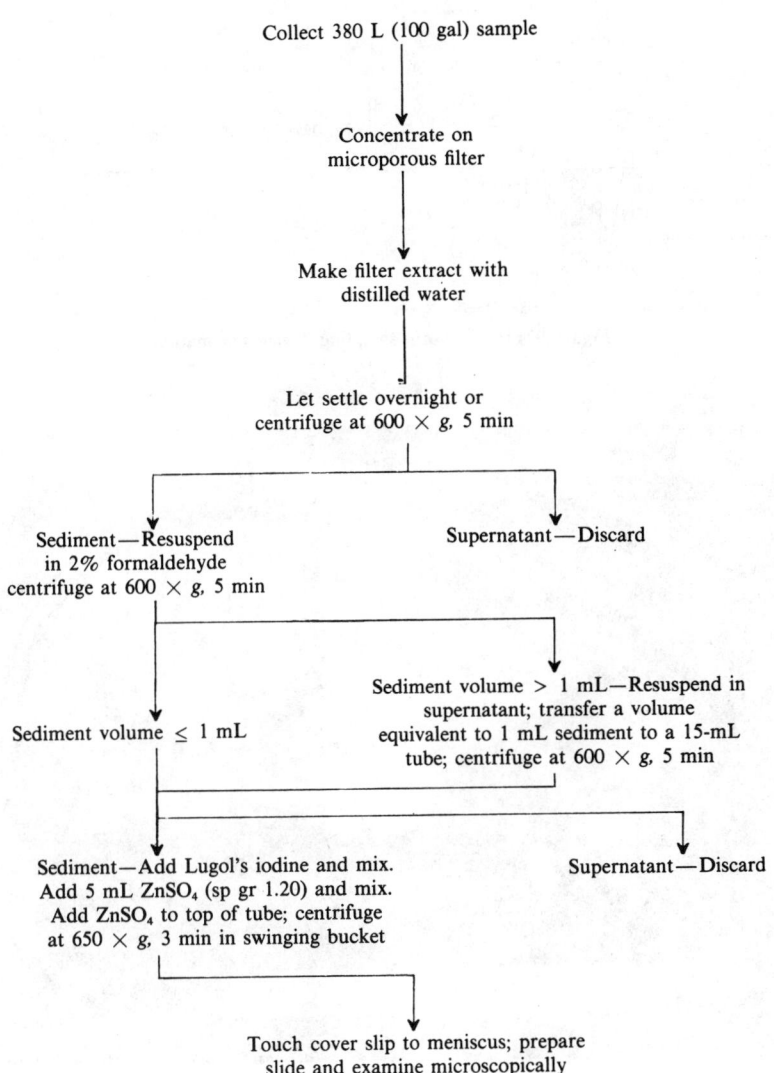

Figure 9711:1. **Giardia method flow chart.**

line of demarcation based on the depth to which sample particulate matter has penetrated the filter cartridge. Alternatively, locate the end of yarn on the outside of the cartridge, unwind the fiber, and divide into equal portions. Wash each portion separately with 1 L distilled water by kneading or by shaking for approximately 10 min. Express fluid from the fiber portions and combine all fluid. If the fiber mat retains

significant amounts of particulate matter, repeat the extraction process on the portions until the fiber appears clean. If further processing is not possible at this point, preserve the extract by adding a sufficient volume of 37% (v/v) formaldehyde to produce a final concentration of 2% (v/v) formaldehyde. Refrigerate the preserved extract until concentration.

b. Extract concentration: Concentrate

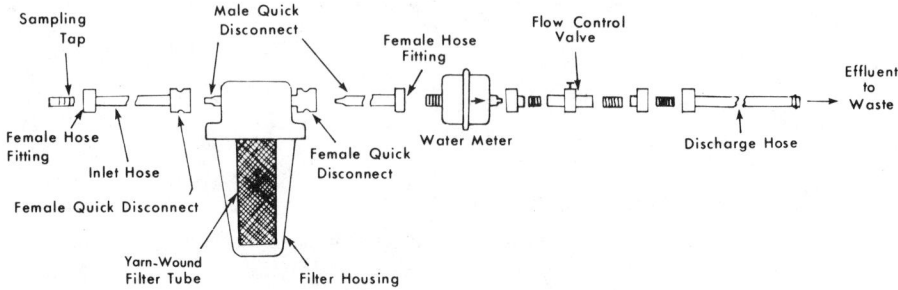

Figure 9711:2. *Giardia* **sampling device schematic.**

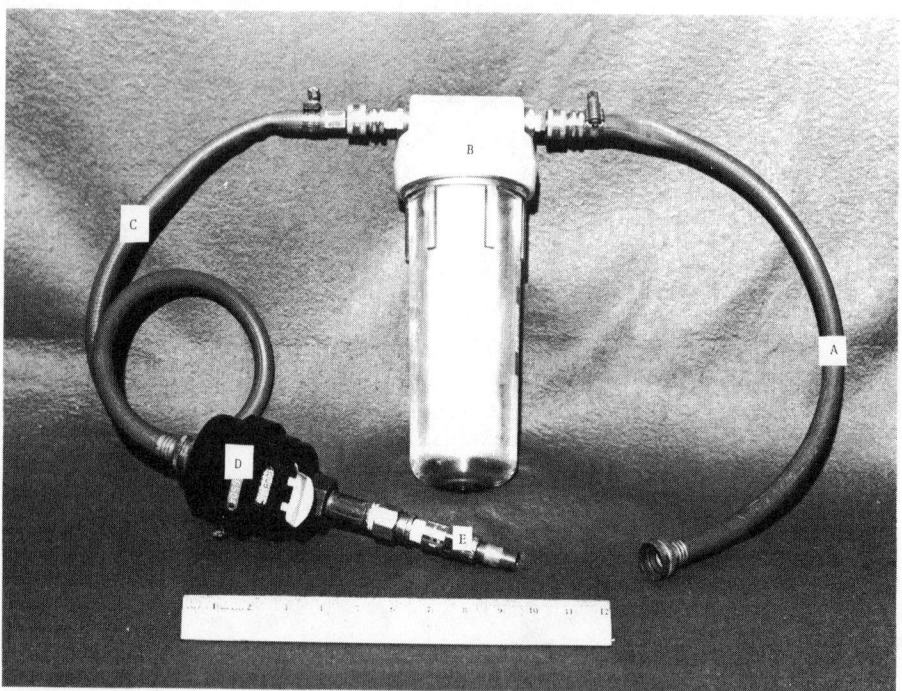

Figure 9711:3. *Giardia* **sampling device. A—inlet hose; B—filter housing; C—outlet hose; D—water meter; E—limiting orifice flow controller.**

filter extract by letting extract settle in a refrigerator overnight or by centrifuging at $600 \times g$ for 5 min. Decant or aspirate supernatant fluids and resuspend pellets in a volume of 2% (v/v) formaldehyde equal to the total volume of the combined sediments. If little sediment is present, suspend the sediments in 10 mL of 2% (v/v) formaldehyde. Recentrifuge combined sediments for 5 min at $600 \times g$. If the sediment volume is ≤ 1 mL, continue processing by decanting or aspirating the entire supernatant. Discard supernatant. If sediment volume is > 1 mL, resuspend it in the supernatant fluid, transfer a volume of suspension equivalent to 1 mL of sediment to a 15-mL tube, and again centrifuge for 5 min at $600 \times g$. Decant and discard supernatant.

Add 2 to 3 drops of Lugol's iodine (dilute

stock solution 1:5 with sp gr 1.20 zinc sulfate solution) to the pelleted material and mix with an applicator stick. Add 5 mL zinc sulfate (ZnSO$_4$) solution (sp gr 1.20) and mix on a vortex mixer. Bring volume up to the top of the tube with additional ZnSO$_4$ solution, being careful to avoid spillover. Using a swinging bucket rotor, centrifuge for 3 min at 650 $\times$ *g*. Touch a clean, grease-free, No. 1 22- $\times$ 22-mm cover slip to the meniscus and let tube and contents remain static for 2 to 3 min after placement. Be careful to handle the cover slip by the edges. Carefully lift cover slip straight up, place on a slide, and seal cover slip with vaspar.

c. Microscopic examination: Scan entire cover-slip area with a 100 $\times$ magnification on a bright-field microscope using reduced light. Alternatively, use fluorescent antibody techniques.[20,23] Identify suspect organisms at a magnification of 450 to 1000 $\times$ using an ocular micrometer. Identification must be made by an individual with a demonstrated proficiency for recognizing and differentiating intestinal protozoa. Take care not to confuse *Giardia* with yeasts, diatoms, *Coccidia,* and other organisms. Positive identification of *Giardia* requires observing structures of the right size and shape and having at least two internal morphological characteristics (nuclei, median bodies, axonemes). Examine sediment from the entire sample in the above manner before declaring a sample to be negative.

4. Reporting Results

Report dimensions of *Giardia* cyst-like structures larger than 8 μm and less than 19 μm. In addition, report any internal morphological features observed. Photomicrographs of suspect cysts may be useful in documenting or corroborating identification. Record and report non-*Giardia* microorganisms detected.

5. Cultivation

No in-vitro method for cultivating *Giardia* from the cyst stage is available currently. Trophozoites may be cultured;[28] however, it is less likely that trophozoites would be present in water supplies in detectable numbers or that they would survive long.

6. Viability

Practical methods for determining the viability of cysts in water samples are not available. The gerbil animal model[29] may be used for infectivity assays of unpreserved water sample concentrates. However, infection in this animal may be erratic.[30] A sample containing viable cysts may or may not produce infection. Consequently, a negative result does not necessarily mean that viable cysts are absent.

7. References

1. RENDTORFF, R. C. 1954. The experimental transmission of human intestinal protozoan parasites. II. *Giardia lamblia* cysts given in capsules. *Amer. J. Hyg.* 59:209.

2. DANCIGER, M. & M. LOPEZ. 1975. Numbers of *Giardia* in the feces of infected children. *Amer. J. Trop. Med. Hyg.* 24:237.

3. BINGHAM, A. K., E. L. JARROLL, E. A. MEYER & S. RADULESCU. 1979. *Giardia* sp.: physical factors of excystation *in vitro* and excystation vs. eosin exclusion as determinants of viability. *Exp. Parasitol.* 47:284.

4. KIRNER, J. C., J. D. LITTLER & L. A. ANGELO. 1978. A waterborne outbreak of giardiasis in Camas, Washington. *J. Amer. Water Works Assoc.* 70:35.

5. CRAUN, G. F. & W. JAKUBOWSKI. 1987. Status of waterborne giardiasis outbreaks and monitoring methods. *In* International Symposium on Water Related Health Issues Proceedings. TPS 87-3, American Water Resources Assoc., Bethesda, Md.

6. HOFF, J. C. 1979. Disinfection resistance of *Giardia* cysts: Origins of current concepts and research in progress. *In* W. Jakubowski & J. C. Hoff, eds. EPA Symposium on Waterborne Transmission of Giardiasis. EPA-600/9-79-

001, U.S. Environmental Protection Agency, Cincinnati, Ohio.

7. RICE, E. W. & J. C. HOFF. 1981. Inactivation of *Giardia lamblia* cysts by ultraviolet irradiation. *Appl. Environ. Microbiol.* 42:546.

8. WICKRAMANAYAKE, G. G., A. J. RUBIN & O. J. SPROUL. 1985. Effects of ozone and storage temperature on *Giardia* cysts. *J. Amer. Water Works Assoc.* 77:74.

9. JARROLL, E. L., A. K. BINGHAM & E. A. MEYER. 1981. Effect of chlorine on *Giardia lamblia* cyst viability. *Appl. Environ. Microbiol.* 41:483.

10. RICE, E. W., J. C. HOFF & F. W. SCHAEFER, III. 1982. Inactivation of *Giardia* cysts by chlorine. *Appl. Environ. Microbiol.* 43:250.

11. HOFF, J. C. 1986. Inactivation of microbial agents by chemical disinfectants. Publ. No. EPA-600/2-86/067, Water Engineering Research Lab., U.S. Environmental Protection Agency, Cincinnati, Ohio.

12. MOORE, G. T., W. M. CROSS, D. MCGUIRE, C. S. MOLLOHAN, N. N. GLEASON, G. P. HEALY & L. H. NEWTON. 1969. Epidemic giardiasis at a ski resort. *N. England J. Med.* 281:402.

13. CHANG, S. L. & P. W. KABLER. 1956. Detection of cysts of *Entamoeba histolytica* in tap water by the use of membrane filter. *Amer. J. Hyg.* 64:170.

14. SHAW, P. K., R. E. BRODSKY, D. O. LYMAN, B. T. WOOD, C. P. HIBLER, G. R. HEALY, K. I. E. MACLEOD, W. STAHL & M. G. SCHULTZ. 1977. A communitywide outbreak of giardiasis with evidence of transmission by a municipal water supply. *Ann. Intern. Med.* 87:426.

15. AKIN, E. W. & W. JAKUBOWSKI. 1986. Drinking water transmission of giardiasis in the United States. *Water Sci. Technol.* 18:219.

16. HAUSLER, W. J., JR., W. E. DAVIS & N. P. MOYER. 1984. Development and testing of a filter system for isolation of *Giardia lamblia* cysts from water. *Appl. Environ. Microbiol.* 47:1346.

17. WALLIS, P. M. & J. M. BUCHANAN-MAPPIN. 1985. Detection of *Giardia* cysts at low concentrations in water using Nuclepore membranes. *Water Res.* 19:331.

18. ISAAC-RENTON, J. L., C. P. J. FUNG & A. LOCHAN. 1986. Evaluation of a tangential-flow multiple-filter technique for detection of *Giardia lamblia* cysts in water. *Appl. Environ. Microbiol.* 52:400.

19. JAKUBOWSKI, W. 1984. Detection of *Giardia* cysts in drinking water: state-of-the-art. *In* S. L. Erlandsen & E. A. Meyer, eds. *Giardia* and Giardiasis. Plenum Press, New York, N.Y.

20. RIGGS, J. L., K. NAKAMURA & J. CROOK. 1984. Identifying *Giardia lamblia* by immunofluorescence. *In* M. Pirbazari & J. S. Devinny, eds. Proceedings of the 1984 Specialty Conference on Environmental Engineering. American Soc. Civil Engineers, New York, N.Y.

21. SCHAEFER, F. W., III, & E. W. RICE. 1981. *Giardia* methodology for water supply analysis. *In* Proceedings, 9th Annual Water Quality Technology Conference. American Water Works Assoc., Denver, Colo.

22. SORENSON, S. K., J. L. RIGGS, P. D. DILEANIS & T. J. SUK. 1986. Isolation and detection of *Giardia* cysts from water using direct immunofluorescence. *Water Resour. Bull.* 22:843.

23. SAUCH, J. F. 1985. Use of immunofluorescence and phase-contrast microscopy for detection and identification of *Giardia* cysts in water samples. *Appl. Environ. Microbiol.* 50:1434.

24. JAKUBOWSKI, W. 1984. *Giardia* methods workshop. *In* Proceedings, 12th Annual Water Quality Technology Conference. American Water Works Assoc., Denver, Colo.

25. SPAULDING, J. J., R. E. PACHA & G. W. CLARK. 1983. Quantitation of *Giardia* cysts by membrane filtration. *J. Clin. Microbiol.* 18:713.

26. BELOSEVIC, M., G. M. FAUBERT, J. D. MACLEAN, C. LAW & N. A. CROLL. 1983. *Giardia lamblia* infections in Mongolian gerbils: an animal model. *J. Infect. Dis.* 147:222.

27. WALLIS, P. M. & H. M. WALLIS. 1986. Excystation and culturing of human and animal *Giardia.* spp. by using gerbils and TYI-S-33 medium. *Appl. Environ. Microbiol.* 51:647.

8. Bibliography

RENDTORFF, R. C. & C. J. HOLT. 1954. The experimental transmission of human intestinal parasites. IV. Attempts to transmit *Endamoeba coli* and *Giardia lamblia* cysts by water. *Amer. J. Hyg.* 60:327.

MEYER, E. A. & J. A. CHADD. 1967. Preservation of *Giardia* trophozoites by freezing. *J. Parasitol.* 53:1108.

CHRISTIE, D. W., R. S. ANDERSON, E. T. BELL & G. L. GALLAGHER. 1971. Ulceration of the ileum and giardiasis in a beagle. *Vet. Rec.* 88:214.

FRANKOWSKI, I. 1975. Detectability of parasite eggs and cysts of lamblia in the feces as a function of the time of viewing the preparation under the microscope. *Wiad. Lek.* 28:1841.

SHEFFIELD, H. G. & B. BJORVATN. 1977. Ultrastructure of the cyst *Giardia lamblia. Amer. J. Trop. Med. Hyg.* 26:23.

ERLANDSEN, S. L. & E. A. MEYER, eds. 1984. *Giardia* and Giardiasis. Plenum Press, New York, N.Y.

WALLIS, P. M. & B. R. HAMMOND. 1988. Advances in *Giardia* Research. University of Calgary Press, Calgary, Alberta, Canada.

9711 C. *Entamoeba histolytica*

1. Occurrence

Fecal contamination of drinking water is a major source of amebiasis, although the oral-fecal route and consumption of uncooked vegetables are important. Plumbing defects involving cross-connections between sewer and water lines, back-siphonage from toilets, drainage from defective sewer lines over an open water cooler, and leaking low-pressure water lines submerged in wastewater have caused disease outbreaks. *E. histolytica* occurs in wastewater in low densities and 1 to 5 cysts/L have been found in the effluent of a wastewater treatment plant. Cysts may remain viable for many days but their initial densities are low and these numbers are reduced drastically in the receiving stream through dilution, water temperature changes, and settling.

2. Concentration Techniques

Use a sample of 4 L or more and membrane filters of 7 to 10 μm pore size if turbidity is not limiting. Filter sample but avoid drying the filter; discontinue suction, transfer filter to side wall of a 100-mL beaker, and repeatedly flush filter surface with several milliters sterile distilled water. Alternatively use the large-volume concentration and processing technique given above for *Giardia*.

3. Direct Microscopic Examination

Place washings in a Sedgwick-Rafter counting cell and examine under low-power magnification for cysts and/or trophozoites.

4. Cultivation

Inoculate sample concentrate into modified liver infusion medium,[1] incubate at 37°C for at least 3 but not more than 6 d, and examine microscopically for trophozoites. By concentrating and culturing replicate sample portions an analysis for the most probable number of *E. histolytica* can be made.

5. References

1. CHANG, S. L. & P. W. KABLER. 1956. Detection of cysts of *Entamoeba histolytica* in tap water by the use of membrane filters. *Amer. J. Hyg.* 64:170.

6. Bibliography

CHANG, S. L. & G. M. FAIR. 1941. Viability and destruction of the cysts of *Entamoeba histolytica. J. Amer. Water Works Assoc.* 33:1705.

RUDOLFS, W., L. L. FALK & R. A. RAGOTZKIE. 1951. Contamination of vegetables grown in polluted soil. II. Field and laboratory studies on entamoeba cysts. *Sewage Ind. Wastes* 23:478.

9810 NEMATOLOGICAL EXAMINATION*

9810 A. Introduction

1. Occurrence

Free-living nematodes usually are benthal or wet-soil dwellers, thriving in aerobic habitats plentifully supplied with bacteria and other microbial food. Hence, they propagate in slow sand filters, flourish in aerobic biological wastewater treatment plants, and appear in large numbers in secondary effluents. Surface waters receiving such effluents may carry much larger numbers of nematodes than those free of waste discharges.[1-3] Because of their active motility and resistance to chlorination, nematodes are not susceptible to conventional water supply treatment and may enter the distribution system in large numbers.

Nematodes of wastewater-treatment origin may carry ingested human enteric pathogens. While information on free-living nematodes may supplement bacteriological data on pollution history,[3] these organisms bear only a very remote relation to infection transmission potential.

Plant-parasitic nematodes include aquatic forms that feed on algae and submerged aquatic plants and nonaquatic forms that parasitize fungi and higher plants. The aquatic forms generally are included with the free-living nematodes, while the nonaquatic groups are dealt with as plant parasites. In plant-nematode-infested areas, land runoff and effluents from crop-processing factories may carry eggs, females, and/or nematode-infested plant tissue. Surface waters receiving such runoff and effluents, when used for irrigation, may spread the infestation.

2. Description of Organisms

Nematodes, commonly known as round-, eel- or threadworms, are invertebrates without appendages. Nematodes parasitizing animals usually are macroscopic; the smallest, such as pinworms and hookworms, are about 1 cm long but the microfilariae (juveniles) of blood or tissue threadworms are microscopic (about 0.25 mm long). Freshwater, soil, and plant-parasitic nematodes usually are microscopic, although some large or swollen forms occasionally are visible with a 6 to 10× hand lens. A microscope is essential for accurate identification. Most soil and freshwater nematodes are about 0.5 to 1 mm long and 0.02 to 0.05 mm wide, with both ends tapered. Sexes are separate in most genera, but some are hermaphroditic (laying fertilized eggs) or parthenogenetic (laying diploid eggs).

A nematode has a head, a body, and a tail, which are indistinctly separated. The head usually consists of a six-lipped mouth and a stoma leading to the esophagus. In many forms feeding on zoomicrobes the stoma is enlarged into a buccal cavity with or without teeth. Nematodes adapted to feed on bacteria, yeast, and minute algae usually have a tubular stoma while in those forms adapted to suck cell fluids, the stoma has become modified into a hollow stylet. The tail is posterior to the anus, usually tapered, and may end abruptly or be elongated.

Between the head and tail is the lengthy body containing the alimentary canal—esophagus, esophageal bulb, and intestine. The body cavity is the space between body wall and the alimentary canal; it contains parts of the reproductive, nervous, and excretory systems. The latter two are hard to discern without staining and high magnification, although the excretory canal and pore usually are easy to see. Many nematodes have caudal glands in the wide part of the tail while others have papilla-like phasmids on the ventro-lateral sides of the tail.

---

*Approved by Standard Methods Committee, 1985.

The female reproductive system consists of one or two ovaries, oviducts, uteri, and vagina, with the latter open to the outside through the genital pore or vulva. In most free-living nematodes, the genital pore is located posterior to the midpoint of the body; in some plant-parasitic genera it is located in the anterior one-fourth to accommodate the much-expanded gonads. The counterparts in males are the testes, seminal vesicles, vas deferens, and ejaculatory duct. A pair of conspicuous spicules (the male copulatory organs) are located near the opening of the ejaculatory duct.

The life cycle of free-living nematodes consists of the egg, four larval stages, and one adult stage. Most eggs are hard to identify in raw water or wastewater effluent. In finished water free from most microfaunal forms, eggs from live or dead females in the distribution system can be recognized readily. Juveniles resemble adults, but lack a reproductive system. Newly hatched juveniles are about one-fifth adult size. These developmental stages are illustrated in Figure 9810:1.

Plant-parasitic nematodes are divided into parasites of the aerial parts and parasites of the underground roots and stems. Some infected plant roots harbor swollen females and/or egg masses. Potatoes, beets, and carrots are common host-plants. Control of nematodes in infested soils is difficult. Cleaning infested plant parts before marketing can release nematodes, galls, and eggs into wastewater.

3. References

1. CHANG, S.L., R.L. WOODWARD & P.W. KABLER. 1969. Survey of free-living nematodes and amoebas in municipal supplies. *J. Amer. Water Works Assoc.* 52:613.
2. CHAUDHURI, N., R. SIDDIQI & R.S. ENGELBRECHT. 1964. Source and persistence of nematodes in surface waters. *J. Amer. Water Works Assoc.* 56:73.

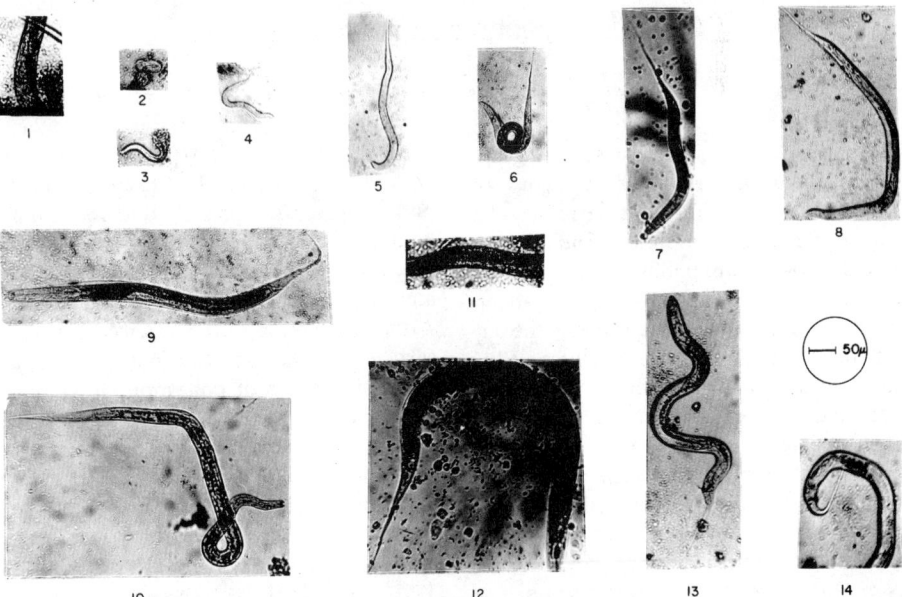

Figure 9810:1. Life cycle of nematodes. 1. Section of female worm showing 3 ova. 2. Ovum freed from a female worm with a well-formed larva. 3. Larva hatching out from an ovum. 4. 1st-stage larva with tail. 5. 2nd-stage larva. 6–7. 3rd-stage larvae. 8. 4th-stage larva. 9–10. Female worms. 11. Section of mature female worm showing striated cuticle. 12. Mature female worm with fully developed uterus and one ovum. 13. Male worm with spicules and gubernaculum. 14. Posterior section of male worm showing spicules on side view.

3. CHANG, S.L. 1972. Zoomicrobial indicators of water pollution. Presented at 72nd Annual Meeting of American Soc. Microbiology, Philadelphia, Pa., Apr. 23-28, 1972.

4. Bibliography

PETERS, B.G. 1930. A biological investigation of sewage. *J. Helminth.* 8:133.

COBB, N.A. 1935. Contributions to a Science of Nematology. Collected Papers between 1914 and 1935. Williams & Wilkins, Baltimore, Md.

CHITWOOD, B.G. & M.W. ALLEN. 1959. Nemata. Chap. 15 *in* W.T. Edmondson, ed. Freshwater Biology, 2nd ed. John Wiley & Sons, New York, N.Y.

EDMONDSON, W.T., ed. 1959. Ward & Whipple's Fresh Water Biology, 2nd ed. John Wiley & Sons, New York, N.Y.

CHANG, S.L. 1960. Survival and protection against chlorination of human enteric pathogens in free-living nematodes isolated from water supplies. *Amer. J. Trop. Med. Hyg.* 9:136.

SASSER, J.N. & W.R. JENKINS, eds. 1960. Nematology: Fundamental and Recent Advances with Emphasis on Plant Parasitic and Soil Forms. Univ. North Carolina, Chapel Hill.

CHANG, S.L. & P.W. KABLER. 1962. Free-living nematodes in aerobic treatment effluent. *J. Water Pollut. Control Fed.* 34:1356.

CALAWAY, W.T. 1963. Nematodes in wastewater treatment. *J. Water Pollut. Control Fed.* 35:1006.

ENGELBRECHT, R.S. & J.H. AUSTIN. 1964. Nematodes and their detection in public water supplies. Presented to Kentucky-Tennessee Sect., American Water Works Assoc., Sept. 15, 1964.

WALTERS, J.V. & R.R. HOLCOMB. 1967. Isolation of an enteric pathogen from sewage-borne nematodes. (abstr.) *Nematologica* 13:155.

CHANG, S.L. 1970. Interactions between animal viruses and higher forms of microbes. *J. San. Eng. Div., Proc. Amer. Soc. Civil Eng.* 96:151.

FERRIS, V.R. & J.M. FERRIS. 1979. Thread Worms. *In* C.W. Hart, Jr. & S.L.H. Fuller, eds. Pollution Ecology of Estuarine Invertebrates. Academic Press, New York, N.Y.

9810 B.   Technique for Nematodes

1. Samples

Collect all samples from sampling stations where bacteriological samples are taken so that findings can be interpreted together. If examination is intended to obtain data on stream pollution, take samples above and below an outfall, with an appropriate number of stations below the outfall to ascertain nematode settling. When samples are taken for plankton analysis, nematodes may be included but report them separately.

In plant-nematode-infested areas, especially in waters polluted with effluents from cleaning processes, also collect samples of effluent.

a. Sample collection: Collect samples as for bacteriological examination (Section 9060) with the following modifications: Use sample not less than 2 L, preferably 4 L. Square plastic containers are convenient for collection and shipment of samples. Containers need not be sterilized. Wash them thoroughly with tap water and rinse with distilled water. For examination of potable water, samples of 20 to 100 L may be used.

b. Sample storage: It is preferable to examine material containing live nematodes. Although sample refrigeration is unnecessary, analyze as soon as possible and at least within 2 d of collection. If samples cannot be examined within 2 d, preserve them. For taxonomic purposes, kill the nematodes by gentle heat, because live nematodes placed directly in fixative become distorted. Kill nematodes by placing a vial containing them in a water bath at 57°C for 15 min. If a water bath is not available, heat-relax nematodes by adding equal quantities of boiling water and sample (at room temperature). After heat-relaxing treatment, place nematodes immediately in 4% formalin (4 mL commercial 37 to 40% formaldehyde solution plus 96 mL water). For fixing and storing mass collections, add an 8% formalin solution to an equal quantity of sample.

2. Nematode Count Procedure

a. Sample concentration: If convenient, concentrate sample at site of collection. Filter sample as directed under Section 9222B.5*b* but replace the membrane filter with a woven nylon strainer (pore size 25 to 30 μm). It usually will be possible to filter 4 L of moderately polluted water through a single strainer in a reasonable time. If strainer clogging occurs, use two or more strainers for the entire sample. As last amount of water disappears from the surface, disconnect holder top. Remove strainer with a pair of forceps and place it on wall of a clean 100-mL beaker containing 2 to 4 mL phosphate-buffered dilution water or distilled water. Using a capillary pipet, flush surface of strainer 8 to 10 times with water in beaker to dislodge nematodes. If more than one strainer is used to concentrate a single sample, pool washings.

After filtering sample, wash strainer-holder assembly in running tap water and rinse with distilled water before reuse. Sterilization is unnecessary.

b. Counting: Transfer 1 mL thoroughly mixed wash water from the beaker to a Sedgwick-Rafter counting chamber. Using 100× magnification, scan entire chamber and count all nematodes, juveniles as well as adults. If the number of nematodes counted per milliliter of concentrate exceeds 10, multiply the number by milliliters concentrate, divide by liters of sample, and report nematodes per liter. If the number of nematodes is between 5 and 10/mL concentrate, count nematodes in a second milliliter of concentrate; if the number is less than 5/mL, count entire concentrate. It may be useful to separate live nematode count from dead; this can be done only in unpreserved samples.

In examining samples taken from waters suspected of carrying plant-parasitic nematodes, look for egg masses and females, and count them.

3. Identification of Common Aquatic Nematodes

Pour scanned portions of sample concentrate into a 10-mL conical centrifuge tube and centrifuge at 1000 rpm for 5 min. Without disturbing sediment, remove supernate with a capillary pipet until 4 to 5 drops remain, mix, and transfer a drop of the remainder to a microscope slide. Place a ring of vaseline around the drop and cover with a coverslip. Examine for essential anatomical features according to the illustrated key in Section 9810C.

4. Interpretation

Because free-living nematodes in open waters of lakes and rivers are chiefly, if not entirely, attributable to pollution by secondary effluents from aerobic wastewater treatment plants, their number provides a rough, but reliable, indication of level of pollution. Adult nematodes, especially large ones, settle to the bottom at rates depending on water quiescence or flow rate. Hence, their presence in samples taken at specified distances from the point of discharge provides useful information on the pollution history. The rapidity with which results are obtained gives this analysis an obvious advantage.

When nematological results are examined in conjunction with bacteriological findings, inferences can be made about pollution history, presence of toxic substances, anaerobiosis, and related matters. For instance, a combination of high fecal bacterial count and very low nematode count suggests that the water may be polluted by a primary effluent, by a secondary effluent from an anaerobic biological treatment plant, or by effluent from a stabilization pond or lagoon. A very low fecal bacterial count and a high nematode count strongly indicate that the water has been used as a carrier of disinfected secondary effluent because nematodes are much more resistant to chlorine and other common water disinfectants than most bacteria. A moderately high fecal bacteria count and a low nematode count, with mostly small or juvenile forms, indicates that the polluted water is stagnant or very slow-flowing, thus permitting settling of large nematodes. If examination reveals a large percentage of dead or dying nematodes, long-time anaerobiosis or the presence of a nematicide is likely.

If nematode enumeration is included in examining a raw water source, and if treatment processes include prechlorination, sedimentation, and filtration, a much lower nematode count in the finished water indicates effective chlorination that immobilized nematodes and facilitated their removal by flocculation. In such a treatment practice, use the following as a guide to treatment efficiency:

Nematode Removal %	Efficiency of Treatment
90 or more	Very good
75–90	Good
50–74	Fair
50 or less	Poor

When repeated examination reveals nematode counts equal to, or greater than, 20/L of raw water, consider reducing the nematode load in the finished water by special treatment. When plant-parasitic nematodes are found in an irrigation water, its use on cropland without special treatment may be undesirable.

9810 C. Illustrated Key to Freshwater Nematodes

1. General Discussion

The following key was devised so that persons trained in biology, but not necessarily in nematology, could use it. The illustrations include original drawings, photocopies of published drawings, or photocopies on which figures were redrawn. The two most important references were Goodey[1] and Chitwood & Chitwood.[2] Other publications used as references and for illustrative material are listed in the bibliography.

Published literature indicates that several genera in this key contain species predominantly associated with terrestrial habitats. Presence of such nematodes suggests runoff from banks or higher ground in which various plant species (often food sources for these nematodes) are growing. These genera are indicated by an asterisk (*).

2. Key

Refer to
couplet No.

1.   Cephalic setae indistinct or absent 2

     Cephalic setae absent but setae-like head appendages present 64

     Cephalic setae present 69

2.(1)   Stylet present 3

     Stylet absent 38

3.(2)   Base of stylet knobbed or flanged 4

     Stylet knobs or flanges absent 29

4.(3)   Valvate median esophageal bulb present. 5

     Valvate median esophageal bulb absent 22

5.(4)   Females eel-like 6

     Females swollen 21

6.(5)   Vulva at mid-body 7

          Vulva on lower third of body 14

7.(6)   Esophagus not overlapping intestine 8

          Esophagus overlapping intestine 11

8.(7)   Stylet length less than 50 μm 9

          Stylet length
          greater than 80 μm . . . . *Dolichodorus*

9.(8)   Tail terminus pointed . . . . *Tetylenchus*

          Tail terminus not pointed 10

10.(9)   Tail terminus knobbed . . . . *Psilenchus**

          Tail terminus never knobbed or pointed . . . . *Tylenchorhynchus**

11.(7)   Labium offset 12

          Labium flattened, amalgamated or nearly so 13

12.(11)   Stylet massive, 40-50 μm long . . . . . . *Hoplolaimus**

          Stylet long and thin, longer than 90 μm . . . . *Belonolaimus**

Refer to couplet No.

13.(11) Body 0.5-1.0 mm long, tail tip not mucronate . . . .*Radopholus**

Body 2-3 mm long, tail tip usually mucronate . . . . *Hirschmanniella*

14.(6) Cuticle heavily annulated, stylet elongate 15

Cuticle not heavily annulated, stylet short 17

15.(14) Cuticular sheath absent 16

Cuticular sheath present . . . . *Hemicycliophora*

16.(15) Annules with cuticular spines or scales . . . .*Criconema**

Annules plain without spines or scales . . . . *Criconemoides**

17.(14) Body death position straight 18

Body death position spiral . . . . *Helicotylenchus**

18.(17) Median esophageal bulb distinct but not pronounced 19

Median esophageal bulb well developed . . . . *Aphelenchoides*

*Refer to
couplet No.*

19.(18)   Esophagus overlapping intestine 20

Esophagus not overlapping intestine . . . . *Tetylenchus**

20.(19)   Median bulb and valves small,
stylet usually weak . . . . *Ditylenchus**

Median bulb valves and stylet
well developed, labium flattened . . . . . . *Pratylenchus**

21.(5)   Female body soft, white, with
few or no internal eggs . . . . *Meloidogyne**

Female body a rigid brown cyst
usually with many internal eggs . . . . *Heterodera**

22.(4)   Stylet short, less than 100 μm 23

Stylet long, greater than 100 μm . . . . *Xiphinema**

23.(22)   Stylet complex 24

Stylet simple 25

24.(23)   Stylet with anterior arch-like portion. . . . . . . *Diphtherophora**

Stylet with dorsal thickening piece. . . . . *Tylencholaimellus**

25.(23)   Stylet knobs elongate, flange-like. 26

Stylet knobs round 27

*Refer to
couplet No.*

26.(25)   Filiform tail . . . . *Aulolaimoides*

Round tail . . . . *Enchodelus*

27.(25)   Tail rounded 28

Tail pointed . . . . *Nothotylenchus*

28.(27)   Esophagus base elongate . . . . *Tylencholoaimus**

Esophagus base oval . . . . *Doryllium*

29.(3)   Valvate median esophageal bulb absent 30

Valvate median esophageal bulb present 37

30.(29)   Stomal walls not cuticularized 31

Stomal walls cuticularized *(Actinolaimus,
Metactinolaimus, Paractinolaimus)* . . . . *Actinolaiminae*

31.(30)   Esophagus with basal expansions 32

Esophagus
expanding uniformly . . . . *Oionchus*

Refer to couplet No.

32.(31) Terminal fifth or sixth
         of esophagus an ovoid bulb 33

         Posterior third of esophagus swollen 36

33.(32) Stylet axial, positioned centrally 34

         Stylet not axial, originating from tooth in stoma wall . . . . *Campydora**

34.(33) Gonads paired; vulva usually near mid-body 35

         Gonad single, posterior to vulva; vulva anterior to mid-body . . . . *Tyleptus**

35.(34) Stylet slender . . . . *Leptonchus**

         Stylet not slender . . . . *Dorylaimoides**

36.(32) Stylet axial, positioned centrally (*Dorylaimus,
         Eudorylaimus, Labronema, Mesodory-
         laimus, Thornia, Laimydorus, Prodorylaimus*). . . . . Dorylaiminae

         Stylet not axial, originating from tooth in stoma wall. . . . . *Nygolaimus*

37.(29) Tail pointed . . . . *Seinura**

         Tail rounded . . . . *Aphelenchus**

38.(2)   Teeth present, prominent 39

          Teeth absent, minute, or indistinct 50

39.(38)   Esophagus without mid-region expansion 40

          Esophagus expanded at mid-region 49

40.(39)   Tail pointed or tapering 41

          Tail rounded 47

41.(40)   Male tail without setae 42

          Male tail with setae . . . . *Oncholaimus*

42.(41)   Stoma with denticles 43

          Stoma without denticles 45

43.(42)   Denticles scattered or in longitudinal rows 44

          Denticles in transverse rows . . . . *Mylonchulus*

44.(43)   Denticles situated on longitudinal rib of stoma . . . . *Prionchulus*

            Denticles scattered on stoma wall . . . . *Sporonchulus*

45.(42)   Tooth anteriorly directed 46

            Tooth retrorse . . . . *Anatonchus*

46.(45)   Tooth in basal part of stoma . . . . *Iotonchus*

            Tooth in anterior part of stoma . . . . *Mononchus*

47.(40)   Stoma with prominent medial or apical tooth 48

            Stoma with small basal tooth . . . . *Bathyodontus*

48.(47)   Stoma with 3 teeth, without small
           basal tooth, caudal glands opening terminally . . . . *Enoplocheilus*

            Stoma with large anterior and small
            basal tooth, caudal glands opening ventrally . . . . *Mononchulus*

49.(39)   Lip region with rib-like armature . . . . *Mononchoides*

            Lip region without rib-like armature . . . . *Diplogaster*

*Refer to
couplet No.*

50.(38) Esophagus with
basal expansions. 51

Esophagus uniformly
cylindrical 60

51.(50) Esophagus without
mid-region expansion 52

Esophagus expanded at mid-region. 55

52.(51) Amphids distinct 53

Amphids indistinct 54

53.(52) Stoma walls anteriorly inflated with minute tooth . . . . *Microlaimus*

Stoma walls without tooth and with straight, tapering sides . . . . *Leptolaimus*

54.(52) Stoma with 3 rod-like thickenings . . . . *Rhabdolaimus*

Stoma without rod-like thickenings . . . . *Monochromadora*

Refer to
couplet No.

55.(51)   Gonads paired 56

Gonads single 58

56.(55)   Stomal walls straight, amalgamated 57

Stomal walls separated, not straight . . . . *Alloionema*

57.(56)   Moderately swollen metacorpus, stoma not excessively elongate . . . . *Rhabditis*

Elongate, cylindrical metacorpus, stoma elongate . . . . *Cylindrocorpus*

58.(55)   Tail with sharp terminus 59

Tail bluntly conical . . . . *Cephalobus*

59.(58)   Anterior part of stoma a broad, open chamber . . . . *Panagrolaimus*

Stoma narrow, collapsed . . . . *Eucephalobus*

60.(50)   Stoma absent or indistinct 61

Stoma distinct 63

61.(60)   Lip region narrow, tooth absent 62

Lip region broad, small denticle apparent in stomal area . . . . . . *Tripyla*

62.(61)   Amphid aperture appearing as large slit . . . . *Amphidelus*

Amphid aperture appearing
as minute pores . . . . *Alaimus*

63.(60)   Stoma narrow and long . . . . *Cryptonchus*

Stoma wide and shallow . . . . *Bathyonchus*

64.(1)   Body symmetrical 65

Body asymmetrical, bearing
series of protuberances on side . . . . *Bunonema**

65.(64)   Lip appendages not elaborate 66

Lip appendages elaborate 68

Refer to
couplet No.

66.(65) Lateral lip appendages thorn-like, directed laterally . . . . *Diploscapter*

Lateral lip appendages not thorn-like or directed laterally 67

67.(66) Papillae or setae horn-like. *Macrolaimus*

Lips flap-like and pointed anteriorly . . . . *Teratocephalus*

68.(65) Lip appendages forked and elaborately fringed. . . . . *Acrobeles**

Lip appendages membranous and wing-like. . . . . *Wilsonema**

69.(1) Post-cephalic setae absent 70

Post-cephalic setae present (may be very faint ex. *Tobrilus*) 92

70.(69) Stylet absent 71

Stylet present 91

*Refer to
couplet No.*

71.(70)   Teeth absent, minute or indistinct 72

Teeth usually present, prominent 85

72.(71)   Esophagus with basal expansions. 73

Esophagus uniformly cylindrical 82

73.(72)   Amphids oval, spiral, or stirrup-shaped 74

Amphids circular 80

74.(73)   Amphids spiral 75

Amphids not spiral 79

75.(74)   Cuticular punctations absent 76

Cuticular punctations present. 78

76.(75)   Esophageal bulb without valves 77

Esophageal bulb valvate . . . . *Plectus & Anaplectus*

Refer to
couplet No.

77.(76)   Esophageal-intestinal valve elongate . . . . *Paraplectonema*

Esophageal-intestinal valve shortened . . . . *Paraphanolaimus*

78.(75)   Labial region characteristically flap-like . . . . *Euteratocephalus*

Labial region not flap-like, lips bluntly rounded. . . . . *Ethmolaimus*

79.(74)   Amphids oval . . . . *Greenenema*

Amphids stirrup-shaped . . . . *Chronogaster*

80.(73)   Esophageal-intestinal valve shortened 81

Esophageal-intestinal valve elongate . . . . *Desmolaimus*

81.(80)   Excretory pore and large
excretory gland present. . . . . *Domorganus*

Excretory pore and gland indistinct or absent . . . . *Monhystera*

Refer to
couplet No.

82.(72)   Stoma wide and shallow, conspicuous, tail filiform . . . . *Prismatolaimus*

Stoma narrow, elongate, collapsed or inconspicuous. 83

83.(82)   Gonad single . . . . *Cylindrolaimus*

Gonads paired 84

84.(83)   Amphids inconspicuous . . . . *Tripyla*

Amphids conspicuous . . . . *Aphanolaimus*

85.(71)   Terminal fifth or sixth of
esophagus an ovoid bulb 86

Esophagus uniformly cylindrical,
stoma with massive teeth . . . . *Ironus*

86.(85)   Cuticular punctations present. 87

Cuticular punctations absent 89

Refer to
couplet No.

87.(86)   Amphids not spiral 88

Amphids spiral . . . . *Achromadora*

88.(87)   Four longitudinal rows of cuticular markings present . . . . *Chromadora*

No longitudinal rows of cuticular markings present . . . . . . *Prochromadorella*

89.(86)   Amphids distinct 90

Amphids indistinct. . . . . *Butlerius*

90.(89)   Female gonad double, amphid hook-shaped . . . . *Anonchus*

Female gonad single, amphid circular. . . . . *Monhystrella*

91.(70)   Lip region annulated, not set off. . . . . *Atylenchus*

Lip region smooth, set off . . . . *Eutylenchus*

92.(69) Esophagus with basal expansion 93

Esophagus uniformly cylindrical 98

93.(92) Cuticular punctation present, amphids not circular. 94

Cuticular punctation absent, amphids circular 97

94.(93) Ocelli (eye spots) present 95

Ocelli absent 96

95.(94) Stoma with three equal-sized teeth . . . . *Chromadorina*

Stoma with at least one large tooth . . . . *Punctodora*

96.(94) Cuticle with lateral
longitudinal rows of punctation . . . . *Hypodontolaimus*

Cuticle without lateral differentiations . . . . *Chromadorita*

Refer to
couplet No.

97.(93)   Esophageal bulb valvate. . . . . *Prodesmodora*

Esophageal bulb without valves . . . . *Odontolaimus*

98.(92)   Amphid anterior on body 99

Amphid posteriorly located . . . . *Bastiania*

99.(98)   Amphid spiral . . . . *Paracyatholaimus*

Amphid cup-shaped or obscure 100

100.(99) Stomal teeth massive . . . . *Oncholaimus*

Stomal teeth small . . . . *Tobrilus*

3. References

1. GOODEY, T. 1963. Soil and Freshwater Nematodes, 2nd ed. (Revised by J.B. Goodey). The Methuen Co., London, & John Wiley & Sons, New York, N.Y.

2. CHITWOOD, B.G. & M.B. CHITWOOD. 1937. An Introduction to Nematology, Section I: Anatomy (rev. ed., 1950). Monumental Printing Co., Baltimore, Md.

4. Bibliography

THORNE, G. 1939. A monograph of the nematodes of the superfamily Dorylaimoidea. *Capita. Zool.* 8:1.

GERLACH, S.A. 1954. Brasilianische Meeres-Nematoden 1. *Bol. Inst. Oceanog.* 5:3.

CHITWOOD, B.G. & A.C. TARJAN. 1957. A redescription of *Atylenchus decalineatus* Cobb, 1913 (Nematoda: Tylenchinae). *Proc. Helminth. Soc. Wash.* 24:48.

ANDRÁSSAY, I. 1959. Nematoden aus dem Psammon des Adige-Flusses, I. *Mem. Mus. Civ. Stor. Nat., Verona* 7:163.

HOPPER, B.E. & E.J. CAIRNS. 1959. Taxonomic Keys to Plant, Soil and Aquatic Nematodes. Alabama Polytechnic Inst., Auburn.

CHITWOOD, B.G. 1960. A preliminary contribution on the marine nemas (Adenophorea) of Northern California. *Trans. Amer. Microsc. Soc.* 79:347.

LUC, M. 1960. *Dolichodorus profundus* n. sp. (Nematoda-Tylenchida). *Nematologica* 5:1.

LOOF, P.A.A. 1961. The nematode collection of Dr. J.G. de Man. *Meded. Lab. Fytopath.* 190:169.

THORNE, G. 1964. Nematodes of Puerto Rico: Belondiroidea new superfamily, Leptonchidae, Thorne, 1935, and Belonenchidae new family (Nemata, Adenophorea, Dorylaimida). Univ. Puerto Rico Agr. Exp. Sta. Tech. Paper 39.

EDWARD, J.C. & S.L. MISRA. 1966. *Criconema vishwanathum* n. sp. and four other hitherto described Criconematinae. *Nematologica* 11:566.

HOPPER, B.E. & S.P. MEYERS. 1967. Folliculous marine nematodes on turtle grass, *Thalassia testudinum* König, in Biscayne Bay, Florida. *Bull. Mar. Sci.* 17:471.

MULVEY, R.H. & H.J. JENSEN. 1967. The Mononchidae of Nigeria. *Can. J. Zool.* 45:667.

ALLEN, M.W. & E.M. NOFFSINGER. 1968. Revision of the genus *Anaplectus* (Nematoda: Plectidae). *Proc. Helminth. Soc. Wash.* 35:77.

ANDRÁSSAY, I. 1968. Fauna Paraguayensis 2. Nematoden aus den Galeriewaldern des Acaray-Flusses. *Opusc. Zool. Boest.* 8:167.

DEGRISSE, A. 1968. Bijdrage tot de morfologie en de systematiek van Criconematidae (Taylor, 1936) Thorne, 1949 (Nematoda). Plantenatlas Sleutel, Gent.

ANDRÁSSAY, I. 1973. Nematoden aus strand- und höhoenbiotopen von Kuba. *Acta Zool. Hung.* 19(3-4):233.

FERRIS, V.R., J.M. FERRIS, & J.P. TJEPKEMA. 1973. Genera of Freshwater Nematodes (Nematoda) of Eastern North America. Biota of Freshwater Ecosystems. Identification Manual No. 10, U.S. Environmental Protection Agency.

TARJAN, A.C. & B.E. HOOPER. 1974. Nomenclatorial Compilation of Plant and Soil Nematodes. Society of Nematologists. O.E. Painter Printing Co., DeLeon Springs, Fla.

PART 10000

BIOLOGICAL EXAMINATION

OF WATER

10010 INTRODUCTION

Water quality affects the abundance, species composition, stability, productivity, and physiological condition of indigenous populations of aquatic organisms. Therefore, the nature and health of the aquatic communities is an expression of the quality of the water. Biological methods used for assessing water quality include the collection, counting, and identification of aquatic organisms; biomass measurements; measurements of metabolic activity rates; measurements of the toxicity, bioconcentration, and bioaccumulation of pollutants; and processing and interpretation of biological data.

Information from these types of measurements may serve one or more of the following purposes:

1. To explain the cause of color and turbidity and the presence of objectionable odors, tastes, and visible particulates in water;

2. To aid in the interpretation of chemical analyses, for example, in relating the presence or absence of certain biological forms to oxygen deficiency or supersaturation in natural waters;

3. To identify the source of a water that is mixing with another water;

4. To explain the clogging of pipes, screens, or filters, and to aid in the design and operation of water and wastewater treatment plants;

5. To determine optimum times for treatment of surface water with algicides and to monitor treatment effectiveness;

6. To determine the effectiveness of drinking water treatment stages and to aid in determining effective chlorine dosage within a water treatment plant as such dosage is related to organic materials in water;

7. To identify the nature, extent, and biological effects of pollution;

8. To indicate the progress of self-purification in bodies of water;

9. To aid in explaining the mechanism of biological wastewater treatment methods and to serve as an index of the effectiveness of treatment;

10. To aid in determining the condition and effectiveness of unit processes in a wastewater treatment plant;

11. To document short- and long-term variability in water quality caused by natural phenomena and/or human activities;

12. To provide data on the status of an aquatic system on a regular basis.

The specific nature of a problem and the reasons for collecting samples will dictate which communities of aquatic organisms will be examined and which sampling and analytical techniques will be used.

The following communities of aquatic organisms are considered in specific sections that follow:

1. PLANKTON: A community of plants (phytoplankton) and animals (zooplankton), usually swimming or suspended in water, nonmotile or insufficiently motile to overcome transport by currents. In fresh water they generally are small or microscopic in size; in the marine or estuarine environment, larger forms are observed more frequently.

2. PERIPHYTON (AUFWUCHS): A community of microscopic plants and animals associated with the surfaces of submersed objects. Some are attached, some move about. Many of the protozoa and other minute invertebrates and algae found in the plankton also occur in the periphyton.

3. MACROPHYTON: The larger plants of

all types. They are sometimes attached to the bottom (benthic), sometimes free-floating, sometimes totally submersed, and sometimes partly emergent. Complex types usually have true roots, stems, and leaves; the macroalgae are simpler but may have stem- and leaf-like structures.

4. MACROINVERTEBRATES: The invertebrates defined here are those retained by the US Standard No. 30 sieve. They are generally bottom-dwelling organisms (benthos).

5. FISH: As defined in this publication, they are only the finned fish.

6. AMPHIBIANS, AQUATIC REPTILES, BIRDS, AND MAMMALS: These vertebrates also may be affected directly or indirectly by spills or other discharges of pollutants and may be useful in monitoring the presence of toxic substances or long-term changes in water quality. Discussion of these organisms is not included.

Large numbers of bacteria and fungi are present in the plankton and periphyton and constitute an essential element of the total aquatic ecosystem. Although their interactions with living and dead organic matter profoundly affect the larger aquatic organisms, techniques for their investigation are not included herein (see Part 9000).

Field observations are indispensable for meaningful biological interpretations, but many biological factors cannot be evaluated directly in the field. These must be analyzed as field data or field samples within the laboratory. Because the significance of the analytical result depends on the representativeness of the sample, attention is given to field methods as well as to associated laboratory procedures.

Before sampling begins, clearly define study objectives. For example, the frequency of a repetitive sampling program may vary from hourly, for a detailed study of diel variability, to every third month

(quarterly) for a general assessment of seasonal conditions, depending on objectives. The scope of the study must be adjusted to limitations in personnel, time, and budget. Before the development of a study plan, examine historic data for the study area and conduct a literature search of work by previous investigators.

Whenever practicable, biologists should collect their own samples. Much of the value of an experienced biologist lies in personal observations of conditions in the field and in the ability to recognize signs of environmental changes as reflected in the various aquatic communities.

The primary orientation of Part 10000 is toward field collection and associated laboratory analyses to aid in determining the status of aquatic communities under field conditions and to aid in interpreting the influence of past and present environmental conditions. Many other types of studies may be, and are being, conducted that are oriented more toward laboratory research. Such laboratory studies will develop further basic knowledge of community and/or organism responses under controlled conditions and will aid in predicting effects of future changes in environmental conditions on the aquatic communities. However, such studies are not within the scope of this presentation.

The complex interrelationships existing in an aquatic environment are reflected in the organization of the following sections. Field and laboratory procedures relating to one section also may be appropriate for other sections; consequently, frequent cross-references between sections have been made.

The methods selected are necessary for the appraisal of water quality. Principal emphasis is on methods and equipment, rather than on interpretation or application of results.

10200 PLANKTON*

10200 A. Introduction

The term "plankton" refers to those microscopic aquatic forms having little or no resistance to currents and living free-floating and suspended in open or pelagic waters. Planktonic plants, "phytoplankton," and planktonic animals, "zooplankton," are covered in this section. The phytoplankton (microscopic algae) occur as unicellular, colonial, or filamentous forms. Many are photosynthetic and are grazed upon by zooplankton and other aquatic organisms. Other organisms occurring in the same environment are dealt with elsewhere: zoosporic fungi in Section 9610F; aquatic hyphomycetes in Section 9610G; and bacteria in Part 9000. The zooplankton in fresh water comprise principally protozoans, rotifers, cladocerans, and copepods; a greater variety of organisms occurs in marine waters.

1. Significance

Plankton, particularly phytoplankton, long have been used as indicators of water quality.[1-4] Some species flourish in highly eutrophic waters while others are very sensitive to organic and/or chemical wastes. Some species develop noxious blooms, sometimes creating offensive tastes and odors[5] or anoxic or toxic conditions resulting in animal deaths or human illness.[6] The species assemblage of phytoplankton and zooplankton also may be useful in assessing water quality.[7]

Because of their short life cycles, plankters respond quickly to environmental changes, and hence their standing crop and species composition are more likely to indicate the quality of the water mass in which they are found. They strongly influ-

ence certain nonbiological aspects of water quality (such as pH, color, taste, and odor), and in a very practical sense, they are a part of water quality. Certain taxa often are useful in determining the origin or recent history of a given water mass. Because of their transient nature, and often patchy distribution, however, the utility of plankters as water quality indicators may be limited. Information on plankton as indicators is interpreted best in conjunction with concurrently collected, physicochemical and other biological data.

Planktonic organisms predominate in ponds, lakes, and oceans. Potamoplankton develop in large rivers with slow-moving waters that approach lentic conditions. Because their origin can be uncertain and the duration of their exposure to pollutants unknown, plankters generally are less valuable as water quality indicators in lotic than in lentic environments.

2. References

1. PALMER, C.M. 1969. A composite rating of algae tolerating organic pollution. *J. Phycol.* 5:78.
2. PALMER, C.M. 1963. The effect of pollution on river algae. *Bull. N.Y. Acad. Sci.* 108:389.
3. RAWSON, D.S. 1956. Algal indicators of trophic lake types. *Limnol. Oceanogr.* 1:18.
4. STOERMER, E.F. & J.J. YANG. 1969. Plankton Diatom Assemblages in Lake Michigan. Spec. Rep. No. 47, Great Lakes Research Div., Univ. Michigan, Ann Arbor.
5. PRESCOTT, G.W. 1968. The Algae: A Review. Houghton Mifflin Co., Boston, Mass.
6. CARMICHAEL, W., ed. 1981. The Water Environment, Algal Toxins and Health. Plenum Press, New York, N.Y.
7. GANNON, J.E. & R.S. STEMBERGER. 1978. Zooplankton (especially crustaceans and rotifers) as indicators of water quality. *Trans. Amer. Microsc. Soc.* 97:16.

*Approved by Standard Methods Committee, 1989.

10200 B.   Sample Collection

1. General Considerations

The frequency and location of sampling is dictated by the purpose of the study.[1] Locate sampling stations as near as possible to those selected for chemical and bacteriological sampling to insure maximum correlation of findings. Establish a sufficient number of stations in as many locations as necessary to define adequately the kinds and quantities of plankton in the waters studied. The physical nature of the water (standing, flowing, or tidal) will influence greatly the selection of sampling stations. The use of sampling sites selected by previous investigators usually will assure the availability of historical data that will lead to a better understanding of current results and provide continuity in the study of an area.

In stream and river work, locate stations upstream and downstream from suspected pollution sources and major tributary streams and at appropriate intervals throughout the reach under investigation. If possible, locate stations on both sides of the river because lateral mixing of river water may not occur for great distances downstream. In a similar manner, investigate tributary streams suspected of being polluted but take care in the interpretation of data from a small stream because much of the plankton may be periphytic in origin, arising from scouring of natural substrates by the flowing water. Plankton contributions from adjacent lakes, reservoirs, and backwater areas, as well as soil organisms carried into the stream by runoff, also can influence data interpretation. The depth from which water is discharged from upstream stratified reservoirs also can affect the nature of the plankton.

Because water of rivers and streams usually is well mixed vertically, subsurface sampling, i.e., the upper meter or a composite of two or more strata, often is adequate for collection of a representative sample. There may be problems caused by stratification due to thermal discharges or mixing of warmer or colder waters from tributaries and reservoirs. Always sample in the main channel of a river and avoid sloughs, inlets, or backwater areas that reflect local habitats rather than river conditions. In rivers that are mixed vertically and horizontally, measure plankton populations by examining periodic samples collected at midstream 0.5 to 1 m below the surface.

If it can be determined or correctly assumed that the plankton distribution is uniform and normal, use a scheme of random sampling to accomodate statistical testing. Include both random selection of sampling sites and transects as well as the random collection of samples at each selected site. On the other hand, if it is known or assumed that plankton distribution is variable or patchy, include additional sampling sites, collect composite samples, and increase sample replication. Use appropriate statistical tests to determine population variability.

In sampling a lake or reservoir use a grid network or transect lines in combination with random procedures. Take a sufficient number of samples to make the data meaningful. Sample a circular lake basin at strategic points along a minimum of two perpendicular transects extending from shore to shore; include the deepest point in the basin. Sample a long, narrow basin at several points along a minimum of three regularly spaced parallel transects that are perpendicular to the long axis of the basin, with the first near the inlet and the last near the outlet. Sample a large bay along several parallel transects originating near shore and extending to the lake proper. Because many samples are required to appraise completely the plankton assemblage, it may be necessary to restrict sampling to strategic points, such as the vicinity of water intakes and discharges, constrictions within the water body, and major bays that may influence the main basin.

In lakes, reservoirs, and estuaries where plankton populations can vary with depth, collect samples from all major depth zones or water masses. The sampling depths will be determined by the water depth at the station, the depth of the thermocline or an isohaline, or other factors. In shallow areas of 2 to 3 m depth, subsurface samples collected at 0.5 to 1 m may be adequate. In deeper areas, collect samples at regular depth intervals. In estuaries sample above and below the pyncocline. Depth intervals for sampling vary for estuaries of different sizes and depths, but use depths representative of the vertical range. Composite sampling above and below the pyncocline often is used. In marine sampling, the intent and scope of the study will determine the collection extent.

Over the continental shelf, take samples at stations approximately equidistant from the shore seaward. Take a vertical series from surface to near bottom at each station, gradually adding more stations across the shelf. It is important to sample the entire vertical range over a continental shelf. Benthic grab samples may be taken to collect dormant resting cells or cysts. Beyond the shelf in pelagic waters, sample in the photic zone from the surface to the thermocline for phytoplankton and to deeper depths for zooplankton. Sampling depths vary, but often are at 10- to 25-m intervals above the thermocline, then at 100- to 200-m intervals below the thermocline to 1000 m, and thereafter at 500- to 1000-m intervals.

Samples usually are referred to as "surface" or "depth" (subsurface) samples. The latter are samples taken from some stated depth, whereas surface samples may be interpreted as samples collected as near the water surface as possible. A "skimmed" sample of the surface film plankton (neuston)[2] can be revealing; however, ordinarily do not include a disproportionate quantity of surface film in a surface sample because a neustonic flora[3] as well as plankton often are trapped on top or at the surface film together with pollen, dust, and other detritus. Various methods have been used for sampling surface organisms.

Sampling frequency depends on the intent of the study as well as the range of seasonal fluctuations, the immediate meteorological conditions, adequacy of equipment, and availability of personnel. Select a sampling frequency at some interval shorter than community turnover time. This requires consideration of life-cycle length, competition, predation, flushing, and current displacement. Frequent plankton sampling is desirable because of normal temporal variability and migratory character of the plankton community. Daily vertical migrations occur in response to sunlight, and random horizontal migrations or drifts are produced by winds, shifting currents, and tides. Ideally, collect daily samples and, when possible, sample at different times during the day and at different depths. When this is not possible, weekly, biweekly, monthly, or even quarterly sampling still may be useful for determining major population changes.

In river, stream, and estuarine regions subject to tidal influence, expect fluctuations in plankton composition over a tidal cycle. A typical sampling pattern at a station within an estuary includes a vertical series of samples taken from the surface, across the pyncocline, to near bottom, collected at 3-h intervals, over at least two complete tidal cycles. Once a characteristic pattern is recognized the sampling routine may be modified.

A useful series of monographs on oceanographic methodology has been published.[4-7] Representative taxonomic references for estuarine and marine phytoplankton include diatoms,[8-11] dinoflagellates,[12-14] coccolithophores,[15] and cyanophyceae[16] (cyanobacteria).

2. Sampling Procedures

Once sampling locations, depths, and frequency have been determined, prepare for field sampling. Label sample containers with sufficient information to avoid confusion or error. On the label indicate date, cruise number, sampling station, study area (river, lake, reservoir), type of sample, and depth. Use waterproof labels. When possible, enclose collection vessels in a protective container to avoid breakage. If samples are to be preserved immediately after collection, add preservative to container before sampling. Sample size depends on type and number of determinations to be made; the number of replicates depends on statistical design of the study and statistical analyses selected for data interpretation. Always design a study around an objective with a statistical approach rather than fit statistical analyses to data already collected.

In a field record book note sample location, depth, type, time, meteorological conditions, turbidity, water temperature, salinity, and other significant observations. Engineer's field notebooks with waterproof paper are very suitable. Field data are invaluable when analytical results are interpreted and often help to explain unusual changes caused by the variable character of the aquatic environment. Collect coincident samples for chemical analyses to help define environmental variations having a potential effect on plankton.

a. Phytoplankton: In oligotrophic waters or where phytoplankton densities are expected to be low collect a sample of up to 6 L. For richer, eutrophic waters collect a sample of 0.5 to 1 L.

For qualitative and quantitative evaluations collect whole (unfiltered and unstrained) water samples with a water collection bottle consisting of a cylindrical tube with stoppers at each end and a closing device. Lower the open sampler to the desired depth and close by dropping a weight, called a messenger, which slides down the supporting wire or cord and trips the closing mechanism. If possible, obtain composite samples from several depths or pool samples from one depth from several casts. The most commonly used samplers that operate on this principle are the Kemmerer,[17] Van Dorn[18] (Figure 10200:1), Niskin, and Nansen samplers.

Because these samplers collect whole water samples, both net and nannoplankton are collected. Different size categories of phytoplankton can be separated in the field or laboratory by filtering through netting of the appropriate mesh size. Typically, for net plankton use a No. 20 net with mesh openings of 76 to 80 μm (silk bolting cloth or nylon monofilament screen cloth). Because of their small size, most pico- and nannoplankton cells will pass through collection nets. Select appropriate mesh sizes for concentrating the various size categories of phytoplankton typical of the aquatic system under study.[19,20]

The Van Dorn usually is the preferred sampler for standing crop, primary productivity, and other quantitative determinations because its design offers no inhibition to free flow of water through the cylinder. In deep-water situations, the Niskin bottle is preferred. It has the same design as the Van Dorn sampler except that the Niskin sampler can be cast in a series on a single line for simultaneous sampling at multiple depths with the use of auxiliary messengers. Because the triggering devices of these samplers are very sensitive, avoid

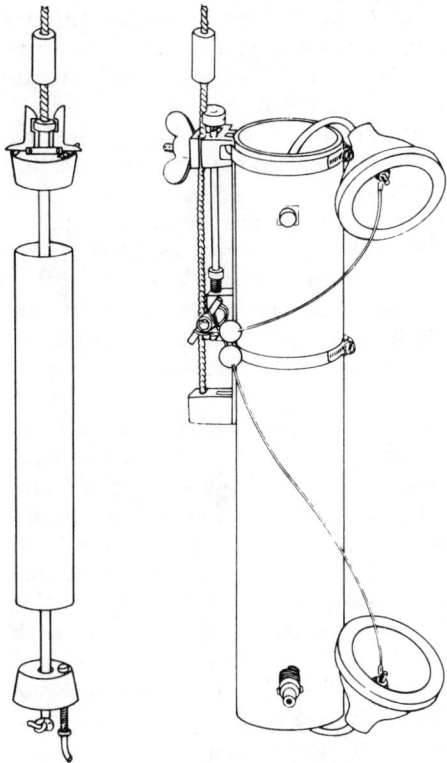

Figure 10200:1. Structural features of common water samplers, Kemmerer (left) and Van Dorn (right).

rough handling. Always lower the sampler into the water; do not drop. Kemmerer and Van Dorn samplers have capacities of 0.5 L or more. Polyethylene or polyvinyl chloride sampling devices are preferred to metal samplers because the latter liberate metallic ions that may contaminate the sample. Use polyethylene or glass sample storage bottles. Metallic ion contamination can lead to significant errors when algal assays or productivity measurements are made.

For shallow waters use the Jenkins surface mud sampler,[21] one of the bottle samplers modified so that it is held horizontally (Figure 10200:1),[22] or an appropriate bacteriological sampler.[23]

For greater speed of collection and to obtain large, accurately measured quan-

tities of organisms, use a pump. Diaphragm and peristaltic pumps are less damaging to organisms than centrifugal pumps.[24] Centrifugal pump impellers can damage organisms as can passage through the hose.[25] Lower a weighted hose, attached to a suction pump, to the desired depth, and pump water to the surface. The pump is advantageous because it supplies a homogeneous sample from a given depth or an integrated sample from the surface to a particular depth. If a centrifugal pump is used, draw samples from the line before they reach the impeller. For samples to be analyzed for organochlorine compounds use TFE tubing.

To examine live samples fill containers partially and store in a refrigerator or ice chest in the dark, or preferably, hold at ambient temperature. Examine specimens promptly after collection.

If it is impossible to examine living material or if phytoplankton are to be counted later, preserve the sample. For a sample that will be preserved, fill the container completely. The most suitable phytoplankton preservative is Lugol's solution, which can be used for most forms including the naked flagellates. Unfortunately, acidic Lugol's solution (or formalin) dissolves the coccoliths of Coccolithophores, which are common in estuarine and marine waters.

Lugol's solution: To preserve samples with Lugol's solution add 0.3 mL Lugol's solution to 100 mL sample and store in the dark. For long-term storage add 0.7 mL Lugol's solution per 100 mL sample and buffered formaldehyde to a minimum of 2.5% final concentration after 1 h. Prepare Lugol's solution by dissolving 20 g potassium iodide (KI) and 10 g iodine crystals in 200 mL distilled water containing 20 mL glacial acetic acid.[26] Replacing glacial acetic acid with sodium acetate makes the solution neutral or slightly alkaline. This allows preservation of Coccolithophores, but would be less effective for other flagellates.

Other acceptable preservatives are:

Formalin: To preserve samples with formalin, add 40 mL buffered formalin (20 g sodium borate + 1 L 37% formaldehyde) to 1 L of sample immediately after collection. In estuarine and marine collections, adjust pH to at least 7.5 with sodium borate for samples containing Coccolithophores.

Merthiolate: To preserve samples with merthiolate add 36 mL merthiolate solution to 1 L of sample and store in the dark. Prepare merthiolate solution by dissolving 1.0 g merthiolate, 1.5 g sodium borate ($Na_2B_2O_4$), and 1.0 mL Lugol's solution in 1 L distilled water. Merthiolate-preserved samples are not sterile, but can be kept effectively for 1 year, after which time formalin must be added.[27]

"M³" fixative: Prepare by dissolving 5 g KI, 10 g iodine, 50 mL glacial acetic acid, and 250 mL formalin in 1 L distilled water (dissolve the iodide in a small quantity of water to aid in solution of the iodine). Add 20 mL fixative to 1 L sample and store in the dark.

Glutaraldehyde: Preserve samples by adding neutralized glutaraldehyde to yield a final concentration of 1 to 2%.

Other commonly used preservatives include 95% alcohol, and 6-3-1 preservative, (6 parts water, 3 parts 95% alcohol, and 1 part formalin). Use equal volumes of preservative and sample.

To retain color in preserved plankton, store samples in the dark or add 1 mL saturated copper sulfate ($CuSO_4$) solution/ L.

Most preservatives distort and disrupt certain cells,[28,29] especially those of delicate forms such as *Euglena, Cryptomonas, Synura, Chromulina,* and *Mallamonas.* Lugol's iodine solution usually is least damaging for these phytoflagellates. To become familiar with live specimens and preservation-caused distortions, use reference collection from biological supply houses or consult experienced co-workers.

b. Zooplankton: The choice of sampler depends on the type of zooplankton, the kind of study (distribution, productivity, etc.) and the body of water being investigated. Zooplankton populations invariably are distributed in a patchy way, making both sampling and data interpretation difficult.

For collecting microzooplankton (20 to 200 μm) such as protozoa, rotifers, and immature microcrustacea, use the bottle samplers described for phytoplankton. The small zooplankters usually are sufficiently abundant to yield adequate samples in 5- to 10-L bottles; however, composite samples over depth and time are recommended. Water bottle samplers are suitable especially for discrete-depth samples. If depth-integrated samples are desired, use pumps or nets. The larger and more robust microzooplankters (e.g., loricate forms and crustacea) may be concentrated by passing the whole water through a 20-μm mesh net. If quantitative estimates of other non-loricate, delicate forms are required, do not screen. Fix 0.5 to 5 L of whole water for enumeration of these forms.

Bottle samplers usually are unsuitable for collecting larger zooplankton, such as mature microcrustacea, that, unlike the smaller forms, are much less numerous and are sufficiently agile to avoid capture. Although comparatively large water volumes, and consequently adequate numbers of microcrustacea, can be sampled with a pump, avoidance by larger, more agile zooplankters at the pump head can cause sampling error. Consequently, larger trap samplers or nets are the preferred collection methods.

The Juday trap[30] operates on the same principle as the water bottle samplers but is generally larger (10 L). The larger size makes the Juday trap more suitable for collecting zooplankters, especially larger copepods. However, it is awkward to use and its 10-L capacity is inadequate for oligotrophic lakes or other water bodies with few zooplankters. Because it is constructed of metal it is unsuited if heavy metals analyses are required.

The Schindler-Patalas trap[31] (Figure 10200:2) usually is preferred to the Juday trap because it is constructed of clear acrylic plastic and is transparent. It can be lowered into the water with minimal disturbance and is suitable for collecting larger zooplankters. Models of 10- to 12-L capacity are available but the 30-L size is preferred. It has no mechanical closing mechanism and thus is convenient for cold-weather sampling when mechanical devices tend to malfunction. Like the Juday trap, it can be fitted with nets of various mesh sizes, but the No. 20 mesh net is used most often.

Plankton nets are preferred to bottles and traps for sampling where plankters are few or where only qualitative data or a large biomass is needed for analysis. Because they were designed originally for qualitative sampling, modifications are required for quantitative work.

The mesh size, type of material, orifice size, length, hauling method, type of tow, and volume sampled will depend on the particular needs of the study.[32,33] Type of netting and mesh size determine filtration efficiency, clogging tendencies, velocity, drag, and the condition of the sample after collection. Silk, formerly the common mesh material in plankton nets, is not recommended because of shrinkage of mesh openings and rotting with age. Nylon monofilament mesh is preferred because of its mesh size accuracy and durability. Nylon nets of different mesh sizes still are labelled by the silk rating system: characteristics of commonly used nylon plankton nets are listed in Table 10200:I. Finer mesh sizes clog more readily than coarser mesh; a compromise must be made between mesh size small enough to retain desired organisms effectively and a size large enough to preclude a serious clogging problem. If clogging occurs, reduce its effects by decreasing the length of tow.

The maximum volume, V_M, of water that can be filtered through a net during a vertical tow can be estimated with the formula,

$$V_M = \pi r^2 d$$

where:
  r = radius of net orifice and
  d = depth to which net is lowered.

This volume is a maximum because clogging of the net's meshes by phytoplankton and other particles and, for fine netting, even the netting itself can cause some water to be diverted from the net's path.[34,35] Keep net towing distance as short as practical to alleviate clogging. If the net has a pro-

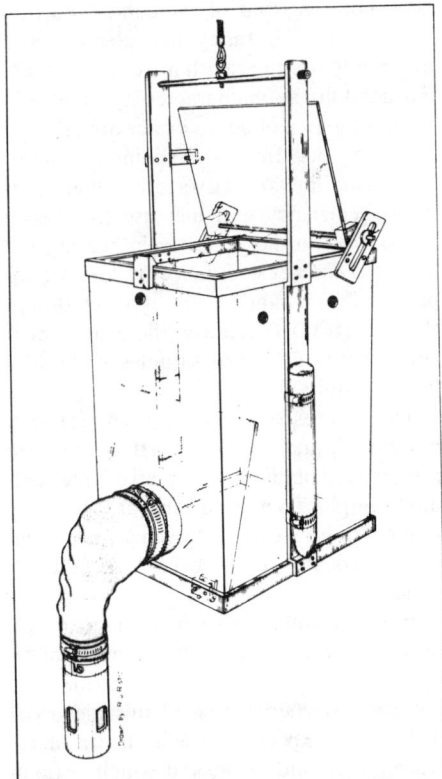

Figure 10200:2. The Schindler-Patalas plankton trap.

TABLE 10200:I. CHARACTERISTICS OF COMMONLY USED PLANKTON NETS

Silk No.	Size of Aperture μm	Approx. Open Area %	Classification
000	1024	58	Largest zooplankton and ichthyoplankton
00	752	54	Larger zooplankton and ichthyoplankton
0	569	50	Large zooplankton and ichthyoplankton
2	366	46	Large microcrustacea
6	239	44	Microcrustacea
10	158	45	Microcrustacea and most rotifers
20	76	45	Net phyto- and zooplankton
25	64	33	Nannoplankton

nounced green or brown color after towing, clogging probably has occurred.

To estimate sampling volume, V_A, mount a calibrated flow meter midway between the net rims and mouth center (the meter is mounted off-center to avoid flow reduction associated with the towing bridle).[36] Equip meter with lock mechanisms to prevent it turning in reverse or while in air. Record flow-meter readings before and after collecting sample. Calculate filtration efficiency, E, from:

$$E = V_A / V_M$$

If E is less than about 0.8, substantial clogging has occurred. Take steps to increase efficiency. Clogging not only decreases the volume filtered, but also leads to biased samples because filtration efficiency is nonuniform during the tow.[33]

Various types of plankton nets are shown in Figure 10200:3. Simple conical nets have been used for many years with little modification in design or improvement in accuracy. Their major source of error is that the filtration characteristics of conical nets usually are unknown. Filtration efficiency in No. 20 mesh cone nets ranges from 40 to 77%. To improve efficiency, place a po-

rous cylinder collar or nonporous truncated cone in front of the conical portion of the net. The Juday net exemplifies a commonly used net with a truncated cone. For good filtration characteristics the ratio of filtering area of net to orifice area should be at least 3:1. Bridles attaching the net to the towing line also adversely influence filtration efficiency and increase turbulence in front of the net, thereby increasing the potential for net avoidance by larger zooplanters. The tandem, Bongo net design (Figure 10200:3) reduces these influences and permits duplicate samples to be collected simultaneously.

Three types of tows are used: vertical, horizontal, and oblique. Vertical tows are preferred to obtain an integrated water column sample. To make a vertical tow, lower the weighted net to a given depth, then raise vertically at an even speed of 0.5 m/s.

In small water bodies haul the net hand over hand with a steady, unhurried motion approximating the speed of 0.5 m/s. In large bodies where long net hauls and vessel drifting is expected, use a davit, meter wheel, angle indicator, and winch. Attach a 3- to 5-kg weight to hold the net down. Determine depth of the net by multiplying

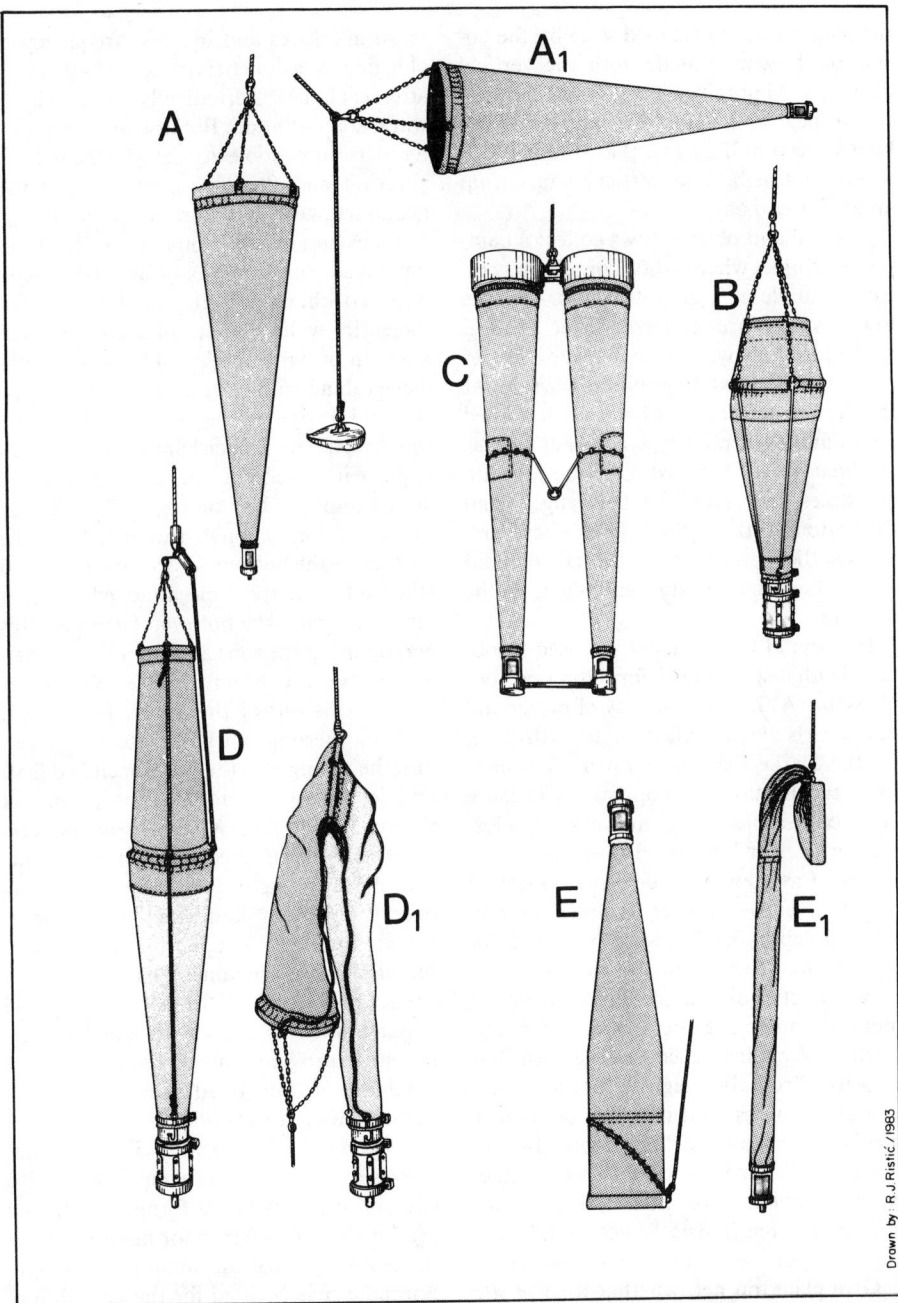

Drawn by R.J.Ristić /1983

Figure 10200:3. Examples of commonly used plankton sampling nets. (A) Simple conical tow-net; A—rigged for vertical tows; A$_1$—for oblique or horizontal tows; (B) Wisconsin (Birge) tow-net with truncated cone to improve filtration efficiency; (C) Bongo net, can be fitted with flow meters and opening/closing mechanisms; (D) Wisconsin net fitted with messenger-activated closing mechanism, D—open, D$_1$—closed; (E) Free-fall net, E—open, E$_1$—closed.

the length of the extended wire by the co-sine of the wire's angle with the vertical direction. Maintain wire angle as close to the vertical as possible by controlling the boat's speed null against the wind drift, or wherever feasible, do vertical hauls from an anchored boat.

Vertical and oblique tows collect a com-posite sample, whereas horizontal tows col-lect a sample at a discrete depth. Oblique tows usually are preferred over vertical tows in shallow water or wherever a longer net tow is required. For oblique tows, lower the net or sampler to some predetermined depth and then raise at a constant rate as the boat moves forward. Oblique tows do not necessarily sample a true angle from the bottom to the surface. Under best con-ditions the pattern is somewhat sigmoid due to boat acceleration and slack in the tow line.

Horizontal tows usually are used to ob-tain depth distribution information on zoo-plankton. Although a variety of horizontal samplers is available (see Figure 10200:4), use the Clarke-Bumpus sampler[37] for quan-titative collection of zooplankton because of its built-in flowmeter and opening-clos-ing device. For horizontal tows use a boat equipped as above and determine sampler depth as above. Lower sampler to pre-selected depth, open, tow at that depth for 5 to 10 min, then close and raise it.

A variety of zooplankton sampling methods can be used in flowing water. The method of choice depends largely on flow velocity. Properly weighted bottles, traps and pump hoses, and nets can be used in medium- to slow-flowing waters. In tur-bulent, well-mixed waters, collect surface water by bucket and filter it through the appropriate mesh size. Select sample size based on concentration of zooplankters.

Give plankton net, whether used *in situ* or as a concentration device in field or lab-oratory, proper care and maintenance. Do not let particulate matter dry on the net because it can significantly reduce size of mesh apertures and increase frequency of clogging. Wash net thoroughly with water after each use. Periodically clean with a warm soap solution. Because nylon net ma-terial is susceptible to deterioration from abrasion and sunlight, guard against un-necessary wear and store in the dark.

Traps and nets do not work well in shal-low areas with growths of aquatic vegeta-tion. To obtain an integrated sample for the entire water column in such areas, use a length of light-weight rubber or polyeth-ylene tubing with netting attached over one end and a rope on the other.[38] Attach net-ting by tape or rubber bands that will stay in place in water, but can be removed easily after sampling. Use tubing of 5- to 10-cm diam and long enough to reach from the surface to the bottom. Lower the open end (the end with the rope attached) until it almost touches the bottom. Then pull this end up using the rope and keep the covered end above the water surface. When the open end is out of the water, let the end with the netting fall back into the water, pull the tubing into the boat, open end first, and let the water in the tube drain out through the netting. When the zooplankton has been concentrated in a small volume, just above the netting, remove the netting over a container and catch the concen-trated sample. Wash netting and end of tubing into the container to assure that all the zooplankton is collected. This method is not limited to areas with aquatic vege-tation. It provides an excellent method of obtaining an integrated sample from any shallow area. In standing waters, collect tow samples by filtering 1 to 5 m^3 of water.

Preserve zooplankton samples with 70% ethanol or 5% buffered formalin. Ethanol preservative is preferred for materials to be stained in permanent mounts or stored. Formalin may be used for the first 48 h of preservation with subsequent transfer to 70% ethanol. Formalin preservative may cause distortion of pleomorphic forms such as protozoans and rotifers. Make formalin

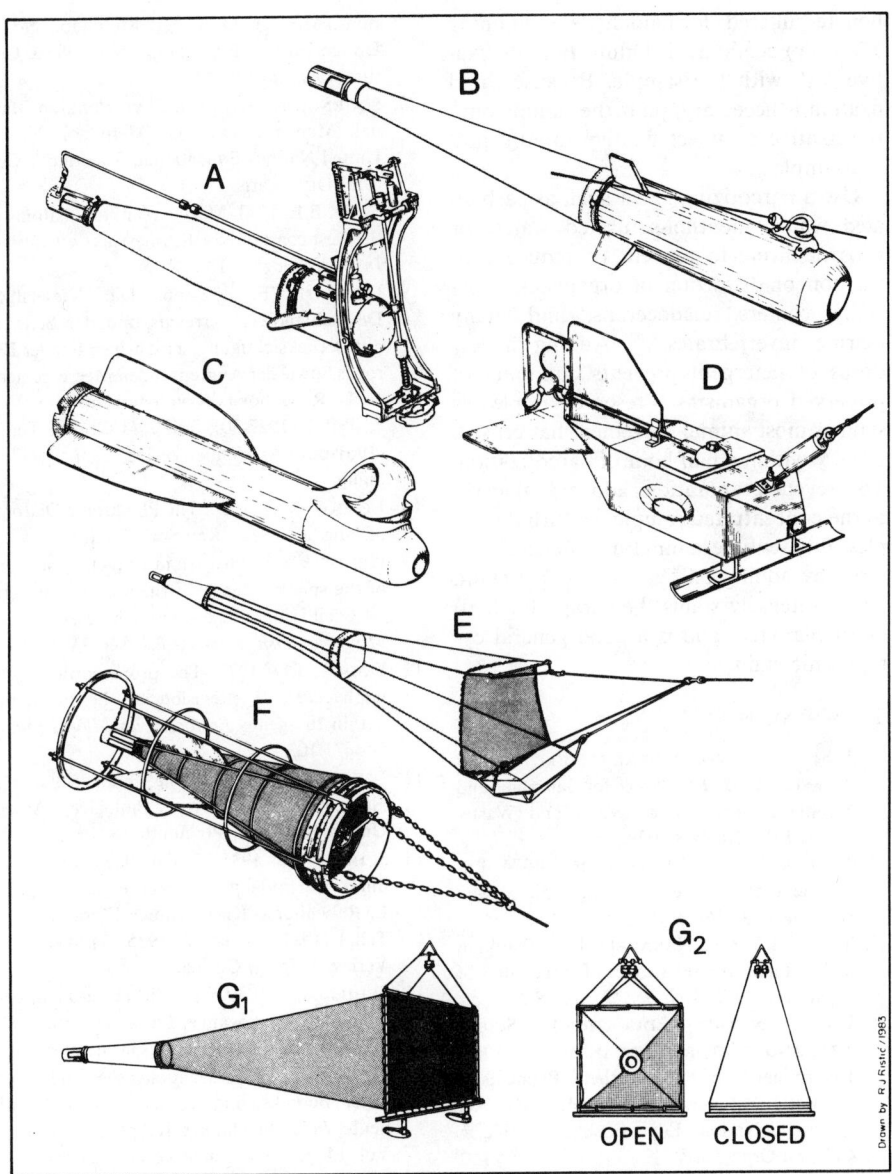

Figure 10200:4. Examples of commonly used high-speed zooplankton samplers. (A) Clarke-Bumpus sampler; (B) Miller sampler; (C) Hardy plankton indicator; (D) Hardy continuous plankton recorder; (E) Issacs-Kidd mid-water trawl; (F) Gulf V sampler; (G) Tucker trawl, G_1-sideview, G_2-front view open and closed.

in sucrose-saturated water to minimize carapace distortion and loss of eggs in crustaceans, especially cladocerans.[39] Bouin's fixative produces reasonable results for soft-bodied microzooplankton.[40] This fixative is picric acid saturated in calcium carbonate-buffered formaldehyde containing 5% (v/v) acetic acid. Dilute Bouin's fixative 1:19 with the sample. Because rapid fixation is necessary, pour the sample onto the fixative or inject fixative rapidly into the sample.

Use a narcotizing agent such as carbonated water, menthol-saturated water, or neosynephrine to prevent or reduce contraction or distortion of organisms, especially rotifers, cladocerans, and many marine invertebrates.[41,42] Adding a few drops of detergent prevents clumping of preserved organisms. Preserve samples as soon as most animal movement has ceased, usually within a half hour of narcotization. To prevent evaporation, add 5% glycerin to the concentrated sample. In turbid samples, differentiate animal and detrital material by adding 0.04% rose bengal stain, which intensely stains the carapace (shell) of zooplankters and is a good general cytoplasmic stain.

5. References

1. U.S. ENVIRONMENTAL PROTECTION AGENCY. 1982. Handbook for Sampling and Sample Preservation of Water and Wastewater. EPA-600/4-82-029.

2. PARKER, B.C. & R.F. HATCHER. 1974. Enrichment of surface freshwater microlayers with algae. *J. Phycol.* 10:185.

3. TAGUCHI, S. & K. NAKAJIMA. 1971. Plankton and seston in the sea surface of three inlets of Japan. *Bull. Plankton Soc. Japan* 18:20.

4. UNITED NATIONS EDUCATIONAL, SCIENTIFIC AND CULTURAL ORGANIZATION. 1966. Determination of Photosynthetic Pigments in Sea-water. Monogr. Oceanogr. Methodol. No. 1. United Nations Educational, Scientific & Cultural Org., Paris.

5. UNITED NATIONS EDUCATIONAL, SCIENTIFIC AND CULTURAL ORGANIZATION. 1968. Zooplankton Sampling. Monogr. Oceanogr. Methodol. No. 2. United Nations Educational, Scientific & Cultural Org., Paris.

6. UNITED NATIONS EDUCATIONAL, SCIENTIFIC AND CULTURAL ORGANIZATION. 1973. A Guide to the Measurement of Marine Primary Production under Some Special Conditions. Monogr. Oceanogr. Methodol. No. 3. United Nations Educational, Scientific & Cultural Org., Paris.

7. SOURNIA, A., ed. 1978. Phytoplankton Manual. Monogr. Oceanogr. Methodol. No. 6. United Nations Educational, Scientific & Cultural Org., Paris.

8. CUPP, E.E. 1943. Marine plankton diatoms of the west coast of North America. *Bull. Scripps Inst. Oceanogr.* 5:1.

9. HUSTEDT, F. 1927-66. Die Kieselalgen Deutschlands, Österreichs und der Schweiz mit Berucksichtigung der Übrigen Lander Europas Sowie der Angrenzenden Meeresgebiete. *In* L. Rabenhorst, Kryptogamen-Flora. Vol. 7: Teil 1 (1927-30); Teil 2 (1931-59); Teil 3 (1961-66). Akademie Verlag, Leipzig, Germany.

10. LEBOUR, M.V. 1930. The Planktonic Diatoms of Northern Seas. Ray Soc., London.

11. HENDEY, N.I. 1964. An introductory account of the smaller algae of British coastal waters, V. Bacillariophyceae (Diatoms). *Fish. Invest. Min. Agr. Fish. Food (G.B.)*, Ser. IV:1.

12. DODGE, J.D. 1975. The prorocentrales (Dinophyceae), II. Revision of the taxonomy within the genus *Prorocentrum. Bot. Limnol. Soc.* 71:103.

13. LEBOUR, M.V. 1925. The Dinoflagellates of Northern Seas. Marine Biological Assoc. United Kingdom, Plymouth.

14. SCHILLER, J. 1931-37. Dinoflagellatae (Peridineae) in monographischer Behandlung. *In* L. Rabenhorst, Kryptogamen-Flora. Vol. 10; Teil 1 (1931-33); Teil 2 (1935-37). Akademie Verlag, Leipzig, Germany.

15. SCHILLER, J. 1930. Coccolithineae. *In* L. Rabenhorst, Kryptogamen-Flora. Vol. 10, p. 89. Akademie Verlag, Leipzig, Germany.

16. GEITLER, L. 1932. Cyanophyceae von Europa unter Berucksichtigung der anderen Kontinente. *In* L. Rabenhorst, Kryptogamen-Flora. Vol. 14, p. 1. Akademie Verlag, Leipzig, Germany.

17. WELCH, P.S. 1948. Limnological Methods. Blakiston Co., Philadelphia, Pa.

18. STRICKLAND, J.D.H. & T.R. PARSONS. 1968. A Practical Manual of Sea Water Analysis. Fish. Res. Board Can. Bull. No. 167. Queen's Printer, Ottawa, Ont.

19. DUSSART, B.M. 1965. Les différentes catégories de plancton. *Hydrobiologia* 26:72.

20. SIEBURTH, J.McN., V. SMETACEK & J. LENZ. 1978. Pelagic ecosystem structure: Heterotrophic compartments of plankton and their relationship to plankton size fractions. *Limnol. Oceanogr.* 23:1256.

21. MORTIMER, C.H. 1942. The exchange of dissolved substances between mud and water in lakes. *J. Ecol.* 30:147.

22. VOLLENWEIDER, R.A. 1969. A Manual on Methods for Measuring Primary Production in Aquatic Environments. IBP Handbook No. 12. Blackwell Scientific Publ., Oxford, England.

23. GELDREICH, E.E., H.D. NASH, D.F. SPINO & D.J. REASONER. 1980. Bacterial dynamics in a water supply reservoir: a case study. *J. Amer. Water Works Assoc.* 72:31.

24. BEERS, J.R. 1978. Pump sampling. *In* A. Sournia, ed. Phytoplankton Manual. United Nations Educational, Scientific and Cultural Org., Paris.

25. EXTON, R.J., W.M. HOUGHTON, W. ESAIAS, L.W. HAAS & D. HAYWARD. 1983. Spectral differences and temporal stability of phycoerythrin fluorescence in estuaries and coastal waters due to the domination of labile cryptophytes and stable cyanobacteria. *Limnol. Oceanogr.* 28:1225.

26. WETZEL, R.G. & G.E. LIKENS. 1979. Limnological Analyses. W.B. Saunders Co., Philadelphia, Pa.

27. WEBER, C.I. 1968. The preservation of phytoplankton grab samples. *Trans. Amer. Microsc. Soc.* 87:70.

28. PAERL, H.W. 1984. An evaluation of freeze fixation as a phytoplankton preservation method for microautoradiography. *Limnol. Oceanogr.* 29:417.

29. SILVER, M.W. & P.J. DAVOLL. 1978. Loss of ^{14}C activity after chemical fixation of phytoplankton: Error source for autoradiography and other productivity measurements. *Limnol. Oceanogr.* 23:362.

30. JUDAY, C. 1916. Limnological apparatus. *Trans. Wis. Acad. Sci.* 18:566.

31. SCHINDLER, D.W. 1969. Two useful devices for vertical plankton and water sampling. *J. Fish. Res. Board Can.* 26: 1948.

32. SCHWOERBEL, J. 1970. Methods of Hydrobiology. Pergamon Press, Toronto, Ont.

33. TRANTER, D.J., ed. 1968. Reviews on Zooplankton Sampling Methods. United Nations Educational, Scientific & Cultural Org., Switzerland.

34. GANNON, J.E. 1980. Towards improving the use of zooplankton in water quality surveillance of the St. Lawrence Great Lakes. Proc. 1st Biol. Surveillance Symp., 22nd Conf. Great Lakes Res. Can. Tech. Rep. Fish. Aquat. Sci. 976, pp. 87-109.

35. ROBERTSON, A. 1968. Abundance, distribution, and biology of plankton in Lake Michigan with the addition of a Research Ships of Opportunity project. Spec. Rep. No. 35, Great Lakes Research Div., Univ. Michigan, Ann Arbor.

36. EVANS, M.S. & D.W. SELL. 1985. Mesh size and collection characteristics of 50-cm diameter conical plankton nets. *Hydrobiologia* 122:97.

37. CLARKE, G.L. & D.F. BUMPUS. 1940. The Plankton Sampler: An Instrument for Quantitative Plankton Investigations. Spec. Publ. No. 5, Limnological Soc. America.

38. PENNAK, R.W. 1962. Quantitative zooplankton sampling in littoral vegetation areas. *Limnol. Oceanogr.* 7:487.

39. HANEY, J.F. & D.J. HALL. 1973. Sugar-coated *Daphnia*; A preservation technique for Cladocera. *Limnol. Oceanogr.* 18:331.

40. COATS, D.W. & J.F. HEINBOKEL. 1982. A study of reproduction and other life cycle phenomena in plankton protists using an acridine orange fluorescence technique. *Mar. Biol.* 67:71.

41. GANNON, J.E. & S.A. GANNON. 1975. Observations on the narcotization of crustacean zooplankton. *Crustaceana* 28(2):220.

42. STEEDMAN, H.F. 1976. Narcotizing agents and methods. *In* H.F. Steedman, ed. Zooplankton Fixation and Preservation. Monogr. Oceanogr. Methodol. No. 4. United Nations Educational, Scientific & Cultural Org., Paris.

10200 C. Concentration Techniques

The organisms contained in water samples sometimes must be concentrated in the laboratory before analysis. Three techniques for concentrating phytoplankton, namely, sedimentation, membrane filtration, and centrifugation, are described below. A special technique for zooplankton also is given.

1. Sedimentation

Sedimentation is the preferred method of concentration because it is nonselective (unlike filtration) and nondestructive (unlike filtration or centrifugation), although many of the picoplankton, the smaller nannoplankton, and actively swimming flagellates (in unpreserved samples) may not settle completely. The volume concentrated varies inversely with the abundance of organisms and is related to sample turbidity. It may be as small as 1 mL for use with an inverted microscope or as large as 1 L for general phytoplankton and zooplankton enumeration.

Allow 1 h settling/mm of column depth. For a treated sample (10 mL detergent/L) allow about 0.5 h settling/mm depth.[1] The sample may be concentrated in a series of steps by quantitatively transferring the sediment from the initial container to sequentially smaller ones. Use cylindrical settling chambers with thin, clear glass bottoms. Fill settling chambers without forming a vortex, keep them vibration-free, and move them carefully to avoid nonrandom distribution of settled matter. Carefully siphon or decant the supernatants to obtain the desired final volume (5 mL for diatom mounts). Store the concentrated sample in a closed, labeled glass vial.

2. Membrane Filtration

The filtration method permits use of high magnification for enumerating small plankters including flagellates and cyanobacteria. However, delicate forms such as "naked" flagellates are distorted by even gentle filtration. When populations are dense and the content of detritus is high, the filter clogs quickly and silt may crush the organisms or obscure them from view.

Pour a measured volume of well-mixed sample into a funnel equipped with a membrane filter having a pore diameter of 0.45 μm. Apply a vacuum of less than 50 kPa to the filter until about 0.5 cm of sample remains on filter. Break vacuum, then apply low vacuum (about 12 kPa) to remove remaining water but not to dry the filter.

For samples with a low phytoplankton and silt content the method does not require counting of individual plankters to assemble enumeration data and it increases the probability of observing less abundant forms.[2] Samples also may be concentrated on a filter, inverted onto a microscope slide, and quick-frozen, permitting the removal of the filter and transfer of plankton to the slide.[3,4]

3. Centrifugation

Plankton can be concentrated by batch or continuous centrifugation. Centrifuge batch samples at 1000 g for 20 min. The Foerst continuous centrifuge is no longer recommended as a quantitative device but it may be desirable to continue its use in existing programs to assure continuity with previously collected data. Although centrifugation accelerates sedimentation, it may damage fragile organisms.

4. Zooplankton Concentration

Zooplankton samples often need to be concentrated in the field, especially when large water bottles or pump methods of sampling are used. Moreover, samples obtained by nets or other methods sometimes need to be concentrated further for storage or preparation for examination. When only small volume reductions are needed, pour sample back into the bucket of traps or nets. In processing large volumes of water as with pump sampling, use larger plankton buckets or funnels with greater water volume retention and filtration surface area. Construct a filter funnel similar to that shown in Figure 10200:5 of clear acrylic plastic or other suitable material.[5] The volume of the apparatus and the mesh size depend on volume of water to be filtered and size of organisms to be retained. The mesh size of the filter funnel normally is the same as that of the net or other field sampling device.

5. References

1. FURET, J.E. & K. BENSON-EVANS. 1982. An evaluation of the time required to obtain sedimentation of fixed algal particles prior to enumeration. *Brit. Phycol. J.* 17:253.
2. MCNABB, C.D. 1960. Enumeration of freshwater phytoplankton concentrated on the membrane filter. *Limnol. Oceanogr.* 5:57.
3. HEWES, C.D. & O. HOLM-HANSEN. 1983. A method for recovering nanoplankton from filters for identification with the microscope: The filter-transfer-freeze (FTF) technique. *Limnol. Oceanogr.* 28:389.
4. HEWES, C.D., F.M.H. REID & O. HOLM-HANSEN. 1984. The quantitative analysis of nano-

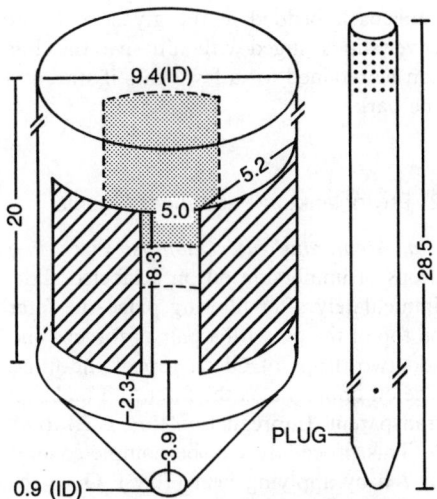

Figure 10200:5. Filter funnel for concentrating zooplankton. This device, originally designed for rotifers, can be modified for other zooplankters by changing the dimensions and mesh size. Dimensions in centimeters. (After Likens and Gilbert.[5])

plankton: A study of methods. *J. Plankton Res.* 6:601.
5. LIKENS, G.E. & J.J. GILBERT. 1970. Notes on quantitative sampling of natural populations of planktonic rotifers, *Limnol. Oceanogr.* 15:816.

10200 D.   Preparing Slide Mounts

1. Phytoplankton Semi-Permanent Wet Mounts

Agitate the settled sample concentrate and withdraw a subsample with an accurately calibrated automatic pipet. Clean and calibrate the pipet regularly. To prepare wet mounts transfer 0.1 mL to a glass slide, place a cover slip over the sample, and ring the cover slip with an adhesive such as clear nail polish to prevent evaporation. For semipermanent slides, mix glycerin with the sample; as the sample ages the water evaporates, leaving the organisms imbedded in the glycerin. If the cover slip is ringed with adhesive, the slide can be retained for a few years if stored in the dark.

2. Phytoplankton Permanent Mounts

a. Membrane filter mounts: Place two drops of immersion oil on a labeled slide. Immediately after filtering place the filter on top of the oil with a pair of forceps and add two drops of oil on top of the filter. The oil impregnates the filter and makes it transparent. Impregnation time is 24 to 48 h. This procedure can be completed in 1 to 2 h by applying heat (70°C). Once the filter has cleared, place a few additional drops of oil on it and cover with a cover slip. The mounted filter is now ready for microscopic examination. Alternatively, mount membrane filters in mounting medium.* Immerse filters in 1-propanol to displace residual water and transfer to xylol for several minutes to clear filters. Place a section of filter or entire filter on a microscope slide with the mounting medium, cover with a cover glass, and dry at low temperature.[1]

b. Sedimented slide mounts: Two techniques are available for making permanent, resin mounts of natural phytoplankton that has been deposited by sedimentation on a microscope slide or cover glass and dehydrated by ethanol vapor substitution.[2,3]

3. Diatom Mounts

Samples concentrated for diatom analysis by settling or centrifugation may contain dissolved materials, such as marine salts, formalin, and detergents, that will leave interfering residues. Wash well with distilled water before slide preparation.

Transfer several drops of washed concentrate by means of a large-bore disposable pipet or large-bore dropper to a cover glass on a hot plate warmed enough to increase the evaporation rate but not enough to cause boiling (use a large-bore pipet or dropper to prevent possible selective filtration, thus exclusion, of larger forms or those forming colonies or chains). If the cleaned material is very concentrated, improve distribution of diatoms by adding the drops to a cover glass already flooded with distilled water. Evaporate to dryness. Repeat addition and evaporation until a sufficient quantity of sample has been transferred to the cover glass, but avoid producing a residue so dense that organisms cannot be recognized. If in doubt about the density, examine under a compound microscope. After evaporation, incinerate the residue on the cover glass on a hot plate at 300 to 500°C; alternatively, use a muffle furnace. This usually requires 20 to 45 min. Mount as described below.

Treat samples concentrated for diatom analysis by membrane filtration as described by Patrick and Reimer.[4] Mix equal volumes of conc nitric acid (HNO_3) and sample. CAUTION: *When working with conc HNO_3 wear safety goggles and an acid-resistant apron and gloves, and work under a hood.* Add a few grains of potassium dichromate ($K_2Cr_2O_7$)[5] to facilitate digestion of the filter and cellular organic matter. Add more dichromate if solution color changes from yellow to green. Place sample on a hot plate and boil down to approximately one-third the original volume. Alternatively, let treated sample stand overnight. This cleaning process destroys organic matter and leaves only diatom shells (frustules). Cool, wash with distilled water, and mount as described above. Transfer cleaned frustules to a cover glass and dry as described above.

Place a drop of mounting medium in the center of a labeled slide. Use 25- by 75-mm slides with frosted ends. Using a suitable

---

*Permount, Fisher Scientific Co., or equivalent.

microscopic mounting medium† assures permanent, easily handled mounts for examination under oil immersion. Heat the slide to near 90°C for 1 to 2 min before applying the heated cover slip with its sample residue to hasten evaporation of solvent in the mounting medium. Remove the slide to a cool surface and, during cooling (5 to 10 s), apply firm but gentle pressure to the cover glass with a broad, flat instrument.

4. Zooplankton Mounts

For zooplankton analyses, withdraw a 5-mL subsample from the concentrate and dilute or concentrate further as necessary. Transfer sample to a counting cell or chamber (see below) for analysis as a wet mount. Use polyvinyl lactyl phenol‡ for preparing semipermanent zooplankton mounts. The mounts are good for about a year, after which time the clearing agent causes deterioration of organisms. For long-term storage ring cover slip with clear lacquer (fingernail polish) to retard mountant crystallization. For permanent mounting, other mountants are available.§

---

†Hyrax, Custom Research and Development, Inc., 8500 Mt. Vernon Road, Auburn, Calif., or equivalent.
‡Biomedical Specialists, Box 1687, Santa Monica, Calif.
§CMC-10, Master's Chemical Co., P.O. Box 2382, Des Plaines, Ill.; Hydramount, Biomedical Specialists, Box 1687, Santa Monica, Calif.; or equivalent.

For the protozoan portion of the microzooplankton, a protargol staining procedure[6] not only provides a permanent mount but also reveals the cytological details often necessary for identification. This procedure is qualitative and is especially important in taxonomic studies of the ciliated protozoa.

5. References

1. MILLIPORE FILTER CORPORATION. 1966. Biological examination of water, sludge and bottom materials. Millipore Techniques, Water Microbiology, p. 25.
2. SANFORD, G.R., A. SANDS & C.R. GOLDMAN. 1962. A settle-freeze method for concentrating phytoplankton in quantitative studies. Limnol. Oceanogr. 14:790.
3. CRUMPTON, W.G. & R.G. WETZEL. 1981. A method for preparing permanent mounts of phytoplankton for critical microscopy and cell counting. Limnol. Oceanogr. 26:976.
4. PATRICK, R. & C.W. REIMER. 1967. The Diatoms of the United States. Vol. 1. Monogr. 13, Philadelphia Acad. Natur. Sci.
5. HOHN, M.H. & J. HELLERMAN. 1963. The taxonomy and structure of diatom populations for three eastern North American rivers using three sampling methods. Trans. Amer. Microsc. Soc. 62:250.
6. SMALL, E.B. & D.H. LYNN. 1985. Phylum Ciliophora Doflein, 1901. In J.J. Lee, S.H. Hunter & E.C. Bovee, eds. An Illustrated Guide to the Protozoa. Soc. Protozoology, Lawrence, Kansas.

10200 E. Microscopes and Calibrations

1. Compound Microscope

Use either a standard or an inverted compound microscope for algal identification and enumeration. Equip either type with a mechanical stage capable of moving all parts of a counting cell past the objective lens. Standard equipment is a set of $10\times$ or $12.5\times$ oculars and $10\times$, $20\times$, $40\times$, and $100\times$ objectives. Use objectives to provide adequate working distance for the counting chamber. Magnification requirements vary with the plankton fraction being investigated, the type of microscope, counting chamber used, and optics. The Sedgwick-Rafter chamber limits magnification to approximately 200×. The Palmer-Maloney cell permits magnification up to 500×. Inverted microscopes are limited in resolution by their optics. The useful upper limit of magnification for any objective is 1000 times the numerical aperture (NA). Above this magnification, no greater detail can be resolved. Use combinations of oculars, intermediate magnifiers, and objectives to obtain the greatest magnification without exceeding the useful limit of magnification. When the limit is exceeded, empty magnification results. Empty magnification occurs where the image is larger but no greater resolution is achieved. Always maximize resolution. Optics providing contrast enhancement such as phase contrast or differential interference contrast are useful.

2. Stereoscopic Microscope

The stereoscopic microscope is essentially two complete microscopes assembled into a binocular instrument to give a stereoscopic view and an erect rather than an inverted image. Use this microscope for the study and counting of large plankters such as mature microcrustacea. Include 10× to 15× paired oculars in combination with 1× to 8× objectives. This combination of optics bridges the gap between the hand lens and the compound microscope and provides magnification ranging from 10× to 120×. Alternatively, use a good-quality zoom-type instrument with comparable magnification.

3. Inverted Compound Microscope

The inverted compound microscope often is used routinely for plankton counting in many laboratories.[1-3] This instrument is unique in that the objectives are below a movable stage and the illumination comes from above, thus permitting viewing of organisms that have settled to the bottom of a chamber. Place samples in a cylindrical settling chamber having a thin, clear glass bottom. Chambers of various capacities are available; the appropriate size depends on the density of organisms. After a suitable period of settling (see Section 10200C.1), count organisms in the settling chamber.

The major advantage of the inverted microscope is that by a simple rotation of the nosepiece a specimen can be examined (or counted) directly in the settling chamber at any desired magnification. Although not recommended, oil immersion objectives have some useful applications. No preparation or manipulation other than settling is required. Generally, examine a preserved sample. Techniques are available for samples with an abundance of organisms that tend to float.[4]

4. Epifluorescence Microscope

An epifluorescence microscope may be either standard or inverted. It uses incident light to excite electrons in intracellular compounds, such as pigments or absorbed stains, with the energy emitted during electron return to the ground state being measured as fluorescent light. The technique has been applied to the microscopic identification of chlorophyll-containing cells (autotrophs) and nonpigmented heterotrophic plankton; fluorescent stains such as primulin or proflavin also have been used to differentiate nannoplanktonic primary and secondary producers.[5-7] Excitation and emission wavelengths are unique for each pigment and stain and require distinct light filter combinations and light sources. Select the filter combinations for the particular application. Epifluorescence microscopy is particularly useful for the enumeration of picoplankton and heterotrophic flagellate populations common to most aquatic systems. Concentrate samples by membrane filtration. Use epifluorescence microscopy as a complementary procedure to standard light microscope counting techniques.

5. Microscope Calibration

Microscope calibration is essential. The usual equipment for calibration is a Whipple grid (ocular micrometer, reticle, or reticule) placed in an eyepiece of the microscope and a stage micrometer that has a standardized, accurately ruled scale on a glass slide. The Whipple disk (Figure 10200:6) has an accurately ruled grid subdivided into 100 squares. One square near the center is subdivided further into 25 smaller squares. The outer dimensions of the grid are such that with a 10× objective and a 10× ocular, it delimits an area of approximately 1 mm² on the microscope stage. Because this area may differ from one microscope to another, carefully calibrate the Whipple grid for each microscope.

With the ocular and stage micrometers parallel and in part superimposed, match the line at the left edge of the Whipple grid with the zero mark on the stage micrometer scale (Figure 10200:7). Determine the width of the Whipple grid image to the nearest 0.01 mm from the stage micrometer scale. Should the width of the image of the Whipple grid be exactly 1 mm (1000 μm), the larger squares will be 1/10 mm (100 μm) on a side and each of the smaller squares 1/50 mm (20 μm).

When the microscope is calibrated at higher magnifications, the entire scale on the stage micrometer will not be seen; make measurements to the nearest 0.001 mm. Additional details for calibration are available.[8]

5. References

1. WETZEL, R.G. & G.E. LIKENS. 1979. Limnological Analyses. W. B. Saunders Co., Philadelphia, Pa.

2. LUND, J.W.G., C. KIPLING & E.D. LeCREN. 1958. The inverted microscope method of estimating algal numbers and the statistical basis of estimations by counting. *Hydrobiologia* 11:143.

3. SICKO-GOAD, L. & E.F. STOERMER. 1984. The

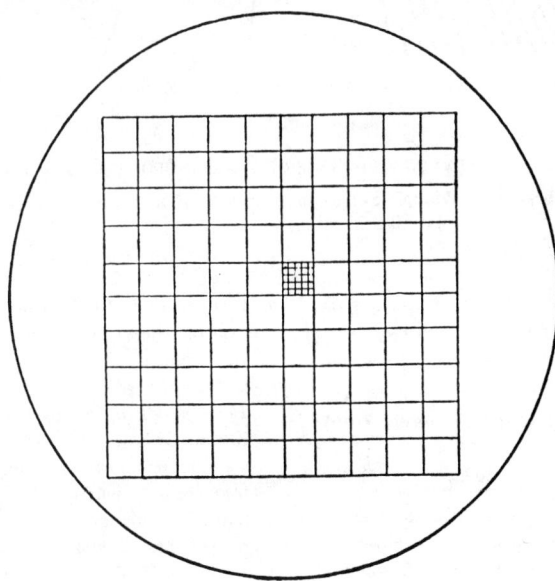

Figure 10200:6. Ocular micrometer ruling. A Whipple micrometer reticule is illustrated.

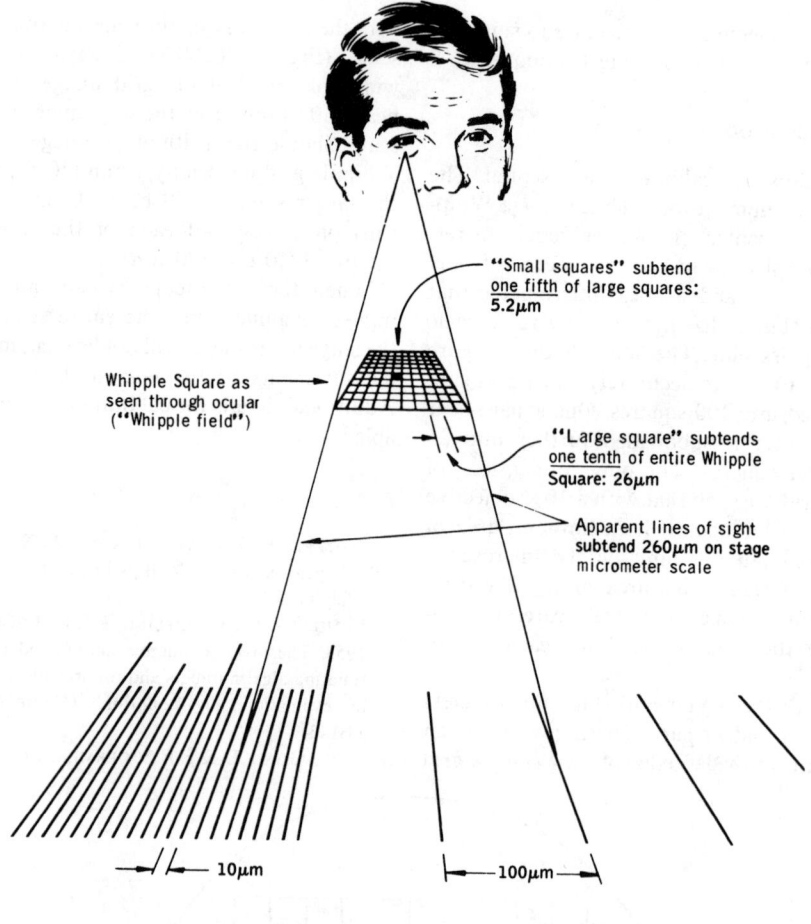

"Small squares" subtend
<u>one fifth</u> of large squares:
5.2µm

Whipple Square as
seen through ocular
("Whipple field")

"Large square" subtends
<u>one tenth</u> of entire Whipple
Square: 26µm

Apparent lines of sight
subtend 260µm on stage
micrometer scale

— 10µm

—100µm—

← —————— PORTION OF MAGNIFIED IMAGE OF STAGE MICROMETER SCALE —————— →

Figure 10200:7. Calibration of Whipple Square, as seen with 10× ocular and 43× objective (approximately 430× total magnification).

need for uniform terminology concerning phytoplankton cell size fractions and examples of picoplankton from the Laurentian Great Lakes. *J. Great Lakes Res.* 10:90.

4. REYNOLDS, C.S. & G.H.M. JAWORSKI. 1978. Enumeration of natural *Microcystis* populations. *Brit. Phycol. J.* 13:269.

5. DAVIS, P.G. & J. McN. SIEBURTH. 1982. Differentiation of phototrophic and heterotrophic nanoplankton populations in marine waters by epifluorescence microscopy. *Ann. Inst. Oceanogr.* 58:249.

6. CARON, D.A. 1983. Techniques for enumeration of heterotrophic and phototrophic nanoplankton, using epifluorescence microscopy, and comparison with other procedures. *Appl. Environ. Microbiol.* 46:491.

7. SHERR, E.B. & B.F. SHERR. 1983. Double-staining epifluorescence techniques to assess frequency of dividing cells and bacteriovory in natural populations of heterotrophic microprotozoa. *Appl. Environ. Microbiol.* 46:1388.

8. JACKSON, H.W. & L.G. WILLIAMS. 1962. Calibration and use of certain plankton counting equipment. *Trans. Amer. Microsc. Soc.* 81:96.

10200 F. Phytoplankton Counting Techniques

1. Counting Units

Some phytoplankton are unicellular while others are multicellular (colonial). The variety of configurations poses a problem in enumeration. For example, should a four-celled colony of *Scenedesmus* (Plate 3, A) be reported as one colony or four individual cells? Listed below are suggestions for reporting:

Enumeration Method	Counting Unit	Reporting Unit
Total cell count	One cell	Cells/mL
Natural unit count[1] (clump count)	One organism (any unicellular organism or natural colony)	Units/mL
Areal standard unit count*	400 μm²	Units/mL

Making a total cell count is time-consuming and tedious, especially when colonies consist of thousands of individual cells. The natural unit or clump is the most easily used system; however, it is not necessarily accurate because sample handling and preserving may dislodge cells from the colony. The unit method also may not be quantitatively accurate nor reflect abundance of biomass or biovolume. Whatever method is chosen, identify it in reporting results.

If the distribution of organisms is random and the population fits a Poisson distribution, the counting error may be estimated.[2] For example, the approximate 95% confidence limits, as a percentage of the number of units counted (N), equals:

$$\frac{2}{\sqrt{N}}\,(100\%)$$

Thus, if 100 units are counted, the 95% confidence limits approximate ± 20%. For a count of 400 units, the limits are about 10%.

2. Counting Procedures

To enumerate plankton use a counting cell or chamber that limits the volume and area for ready calculation of population densities.

When counting with a Whipple grid, establish a convention for tallying organisms lying on an outer boundary line. For example, in counting a "field" (entire Whipple square), designate the top and left boundaries as "no-count" sides, and the bottom and right boundaries as "count" sides. Thus, tally every plankter touching a "count" side from the inside or outside but ignore any touching a "no-count" side. If significant numbers of filamentous or other large forms cross two or more boundaries of the grid, count them separately at a lower magnification and include their number in the total count.

To identify organisms use standard bench references (see Section 10900G).

Do not count dead cells or broken diatom frustules. Tally empty centric and pennate diatoms separately as "dead centric diatoms" or "dead pennate diatoms" for use in converting the diatom species proportional count to a count per milliliter.

Magnification is important in phytoplankton identification and enumeration. Although magnifications of 100× to 200× are useful for counting large organisms or colonies, much higher magnifications often

*Areal standard unit equals area of four small squares in Whipple grid at a magnification of 200.

are required. It is useful to categorize techniques for phytoplankton counting according to the magnifications provided.

a. Low-magnification (up to 200×) methods: The Sedgwick-Rafter (S-R) cell is a device commonly used for plankton counting because it is easily manipulated and provides reasonably reproducible data when used with a calibrated microscope equipped with an eyepiece measuring device such as the Whipple grid.

The greatest disadvantage associated with the cell is that objectives providing high magnification cannot be used. As a result, the S-R cell is not appropriate for examining nannoplankton. The S-R cell is approximately 50 mm long by 20 mm wide by 1 mm deep. The total area of the bottom is approximately 1000 mm² and the total volume is approximately 1000 mm³ or 1 mL. Carefully check the exact length and depth of the cell with a micrometer and calipers before use.

1) Filling the cell—Before filling the S-R cell with sample, place the cover glass diagonally across the cell and transfer sample with a large-bore pipet (Figure 10200:8). Placing cover slip in this manner will help prevent formation of air bubbles in cell corners. The cover slip often will rotate slowly and cover the inner portion of the S-R cell during filling. Do not overfill because this would yield a depth greater than 1 mm and produce an invalid count. Do not permit large air spaces caused by evaporation to develop in the chamber during a lengthy examination. To prevent formation of air spaces, occasionally place a small drop of distilled water on edge of cover glass.

Before counting let the S-R cell stand for at least 15 min to settle plankton. Count plankton on the bottom of the S-R cell. Some phytoplankton, notably some blue-green algae or motile flagellates in unpreserved samples, may not settle but rise to the underside of the cover slip. When this occurs, count these organisms and add to total of those counted on the cell bottom to derive total number of organisms. Count algae in strips or fields.

2) Strip counting—A "strip" the length of the cell constitutes a volume approximately 50 mm long, 1 mm deep, and the width of the total Whipple grid.

The number of strips to be counted is a function of the precision desired and the number of units (cells, colonies, or filaments) per strip. Derive number of plankton in the S-R cell from the following:

$$\text{No./mL} = \frac{C \times 1000 \text{ mm}^3}{L \times D \times W \times S}$$

where:

C = number of organisms counted,
L = length of each strip (S-R cell length), mm,
D = depth of a strip (S-R cell depth), mm,
W = width of a strip (Whipple grid image width), mm, and
S = number of strips counted.

Multiply or divide number of cells per milliliter by a correction factor to adjust for sample dilution or concentration.

3) Field counting—On samples containing many plankton (10 or more plankters per field), make field counts rather than strip counts. Count plankters in random fields each consisting of one Whipple grid. The number of fields counted will depend on plankton density and statistical accuracy desired (see 10200F.1). Calculate the number of plankton per milliliter as follows:

$$\text{No./mL} = \frac{C \times 1000 \text{ mm}^3}{A \times D \times F}$$

where:

C = number of organisms counted,
A = area of a field (Whipple grid image area), mm²,
D = depth of a field (S-R cell depth), mm, and
F = number of fields counted.

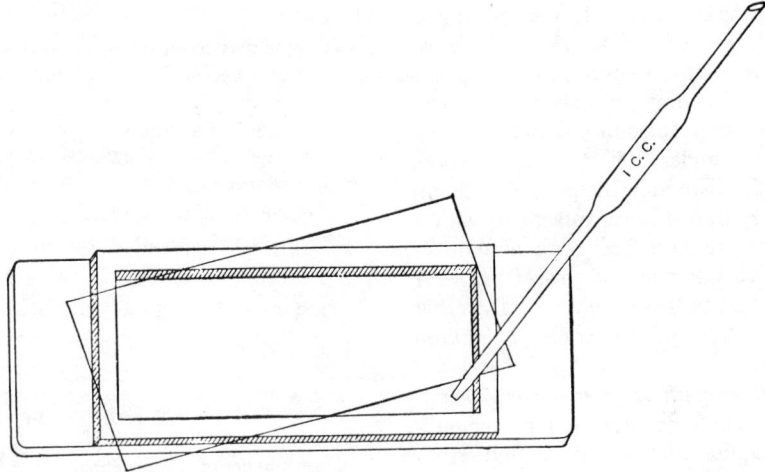

Figure 10200:8. Counting cell (Sedgwick-Rafter), showing method of filling. Source: WHIPPLE, G.C., G.M. FAIR, and M.C. WHIPPLE. 1927. The Microscopy of Drinking Water. John Wiley & Sons, New York, N.Y.

Multiply or divide the number of cells per milliliter by a correction factor to adjust for sample dilution or concentration.

b. Intermediate magnification (low to 500×) methods: The Palmer-Maloney (P-M) nannoplankton cell[3] is designed specifically for nannoplankton enumeration. It has a circular chamber with a 17.9-mm diam, 0.4-mm depth, and 0.1-mL volume. The shallow depth permits use of 40 to 45× objectives with sufficient working distance. The principal disadvantage of the P-M cell is that these magnifications (400 to 450×) often are insufficient for nannoplankton identification and enumeration.

Because a relatively small sample portion is examined in the P-M cell do not use it unless the sample contains a dense population (10 or more plankters per field). Such a small sample portion from a less dense population causes serious underestimation of density.

Introduce sample with a pipet into one of the 2- by 5-mm channels on the side of the chamber with the cover slip in place. After a 10-min settling period count the plankters in random fields, with the number of fields depending on density and variety of plankton and the statistical accuracy desired. Strips may be counted in this or any other circular cell by measuring the effective diameter and counting two perpendicular strips that cross at the center. Calculate the number per milliliter as follows:

$$\text{No./mL} = \frac{C \times 1000 \text{ mm}^3}{A \times D \times F}$$

where:
  C = number of organisms counted,
  A = area of a field (Whipple grid image), mm^2,
  D = depth of a field (P-M cell depth), mm, and
  F = number of fields counted.

Multiply or divide the number of cells per milliliter by a correction factor to adjust for sample dilution or concentration.

Another readily available chamber is the standard medical hemacytometer used for enumerating blood cells. It has a ruled grid machined into a counting plate and is fitted with a ground-glass cover slip. The grid is divided into 1-mm^2 divisions; the chamber is 0.1 mm deep. Introduce sample by pipet and view under 450× magnification. Count all cells within the grid. The chamber comes from the manufacturer with a detailed instruction sheet containing directions on calculations and proper usage. A disadvantage to these counting cells is that the sample must have a very high plankton density to yield statistically reliable data.

c. *High-magnification methods:* Examination of phytoplankton at high magnification requires the use of oil immersion objectives. Suitable procedures include using inverted microscope chambers, membrane filter mounts, sedimented slide mounts, the Lackey drop method, and diatom mounts.

1) Inverted microscope counts—Prepare a sample for examination by filling the settling chamber. After the desired settling time (see Section 10200C.1), transfer the chamber to the microscope stage. Count perpendicular strips across the center of the bottom cover glass. Strip counts may be made by using a Whipple grid or special counting oculars that have a pair of adjustable parallel hairs and a single cross hair. Determine the width of the strip with a stage micrometer and tally organisms as they pass the single cross hair that functions as a reference point. Hold strip width constant for any series of samples. Alternatively examine random nonoverlapping fields until at least 100 units of the dominant species are counted. For highest accuracy, particularly because algae distribution may be nonuniform, count the entire chamber floor. Alternatively, make a random field-minimum count to attain a precision level of at least 85%.[4]

$$\text{Strip count (No./mL)} = \frac{C \times A_t}{L \times W \times S \times V}$$

where:

C = number of organisms counted,
A_t = total area of bottom of settling chamber, mm^2,
L = length of a strip, mm,
W = width of a strip (Whipple grid image width), mm,
S = number of strips counted, and
V = volume of sample settled, mL.

$$\text{Field count (No./mL)} = \frac{C \times A_t}{A_f \times F \times V}$$

where:

A_f = area of a field (Whipple grid image area), mm^2,
F = number of fields counted,

and other terms are as defined above.

2) Membrane filter mounts—Concentrate sample as directed in Section 10200C.2 and prepare membrane filter as directed in Section 10200D.2*a*.

Examine samples, concentrated on unlined membrane filters and mounted in oil as described above. Count enough random fields to ensure desired level of statistical accuracy (see 10200F.1). Select magnification level and size of microscope field (quadrat) such that the most abundant species appear in at least 70% but not more than 90% of microscopic fields examined (80% is optimum). Adjust microscope field size by using part or all of the Whipple grid. Examine 30 random microscope fields and record number of fields in which each species occurred. Report results as organisms per milliliter, calculated as follows:

$$\text{No./mL} = \frac{N \times Q}{V \times D}$$

where:

N = density (organisms/field) from Table 10200:II,
Q = number of fields per filter,
V = milliliters filtered, and
D = dilution factor (0.96 for 4% formalin preservative).

TABLE 10200:II. CONVERSION TABLE FOR MEMBRANE FILTER TECHNIQUE (Based on 30 Scored Fields)

Total Occurrence	F^* %	$N\dagger$
1	3.3	0.03
2	6.7	0.07
3	10.0	0.10
4	13.3	0.14
5	16.7	0.18
6	20.0	0.22
7	23.3	0.26
8	26.7	0.31
9	30.0	0.35
10	33.3	0.40
11	36.7	0.45
12	40.0	0.51
13	43.3	0.57
14	46.7	0.63
15	50.0	0.69
16	53.3	0.76
17	56.7	0.83
18	60.0	0.91
19	63.3	1.00
20	66.7	1.10
21	70.0	1.20
22	73.3	1.32
23	76.7	1.47
24	80.0	1.61
25	83.3	1.79
26	86.7	2.02
27	90.0	2.30
28	93.3	2.71
29	96.7	3.42
30	100.0	?

$$* F = \frac{\text{total number of species occurrences} \times 100}{\text{total number of fields examined}}$$

$\dagger N$ = number of organisms per field.

3) Sedimented slide mounts—Examine mounts prepared as directed in Section 10200D.2*b*.

4) Lackey drop method—The Lackey drop (microtransect) method[5] is a simple method of obtaining counts of considerable accuracy with samples containing a dense plankton population. It is similar to the S-R strip count.

Prepare slides as directed in Section 10200D.1. Oil immersion objectives can be used with the semipermanent slides. Count organisms in enough strips to ensure desired level of statistical accuracy (see 10200F.1). Calculate number of organisms per milliliter as follows:

$$No./mL = \frac{C \times A_t}{A_s \times S \times V}$$

where:
  C = number of organisms counted,
  A_t = area of cover slip, mm^2,
  A_s = area of one strip, mm^2,
  S = number of strips counted, and
  V = volume of sample under the cover slip, mL.

5) Diatom mounts—Prepare samples as directed in Section 10200D.3.

For diatom species proportional count, examine diatom samples under oil immersion at a magnification of at least 900×. Scan lateral strips the width of the Whipple grid until at least 250 cells are counted. Available time and accuracy required dictate the number of cells to be counted. Determine percentage abundance of each species from tallied counts and calculate counts per milliliter of each species by multiplying percent abundance by total live and dead diatom count obtained from the plankton counting chamber. For greater accuracy distinguish between living and dead diatoms at the species level.

6) Phytoplankton staining technique—Staining algae permits differentiation between "live" and "dead" diatoms.[6] This permits enumerating total phytoplankton in a single sample without sacrificing detailed diatom taxonomy. It also results in permanent reference slides. The procedure is most useful when diatoms are major components of phytoplankton and it is important to distinguish between living and dead diatoms.

Preferably preserve samples in Lugol's solution or alternatively in formalin (see

10200B.3). For analysis thoroughly mix the sample and filter a portion through a 47-mm-diam membrane filter (pore diam 0.45 or 0.65 μm). Use a vacuum of 16 to 20 kPa and never let sample dry. Add 2 to 5 mL aqueous acid fuchsin solution (dissolve 1 g acid fuchsin in 100 mL distilled water to which 2 mL glacial acetic acid has been added; filter) to the filter and let stand for 20 min. After staining, filter sample, wash briefly with distilled water, and filter again. Administer successive rinses of 50%, 90%, and 100% propanol to the sample while filtering. Soak for 2 min in a second 100% propanol wash, filter, and add xylene. At least two washes are required; let the final one soak 10 min before filtering. Trim the xylene-soaked filter and place on a microscope slide on which there are several drops of mounting medium.† Apply several more drops of medium to top of filter and install a cover glass. Carefully squeeze out excess mounting medium. Make the final mount permanent by lacquering the edges of the cover glass.

Count organisms using the most appropriate magnification. "Live" diatoms typically are red while "dead" ones are unstained. Oil immersion is necessary for species identifications of diatoms and many other algae. Count either strips or random

---

†Permount, Fisher Scientific Co., or equivalent.

fields and calculate plankton densities per milliliter:

$$\text{No./mL} = \frac{C \times A_t}{A_c \times V}$$

where:

C = number of organisms counted,
A_t = total area of effective filter before trimming and mounting,
A_c = area counted (strips or fields), and
V = volume of sample filtered, mL.

3. References

1. INGRAM, W.M. & C.M. PALMER. 1952. Simplified procedures for collecting, examining, and recording plankton in water. *J. Amer. Water Works Assoc.* 44:617.

2. STRICKLAND, J.D.H. & T.R. PARSONS. 1968. A Practical Manual of Sea Water Analysis. Fish. Res. Board Can. Bull. No. 167. Queen's Printer, Ottawa, Ont.

3. PALMER, C.M. & T.E. MALONEY. 1954. A New Counting Slide for Nannoplankton. Spec. Publ. No. 21, American Soc. Limnology & Oceanography.

4. SOURNIA, A., ed. 1978. Phytoplankton Manual. Monogr. Oceanogr. Methodol. No. 6. United Nations Educational, Scientific & Cultural Org., Paris.

5. LACKEY, J.B. 1938. The manipulation and counting of river plankton and changes in some organisms due to formalin preservation. *Pub. Health Rep.* 53:2080.

6. OWEN, B.B., JR., M. AFZAL & W.R. CODY. 1978. Staining preparations for phytoplankton and periphyton. *Brit. Phycol. J.* 13:155.

10200 G.   Zooplankton Counting Techniques

1. Subsampling

Count entire samples with <200 zooplankters without subsampling. Most zooplankton samples will contain more organisms than can be enumerated practically, therefore, use a subsampling procedure. Before subsampling, remove and enumerate all large uncommon organisms such as fish larvae in fresh water or coe-

lenterates, decapods, fish larvae, etc., in salt water. Subsample by the pipet or splitting method.

In the pipet method, adjust sample to a convenient volume in a graduated cylinder or Imhoff cone. Concentrating the plankton by using a rubber bulb and clear acrylic plastic tube with fine mesh netting fitted on the end is convenient and accurate (Figure 10200:9). For picoplankton and the smaller microzooplankton, use sedimentation techniques described for concentrating phytoplankton. Transfer sample to a beaker or other wide-mouth vessel for subsampling with a Hensen-Stempel or similar wide-bore pipet. Gently stir sample completely and randomly with the pipet and quickly withdraw 1 to 5 mL. Transfer to a suitable counting chamber.

Alternatively, subsample by splitting with any of a number of devices of which the Folsom plankton splitter[1] is best known (Figure 10200:10). Level splitter before using. Place sample in the splitter and divide into subsplits. Rinse splitter into the subsamples. Repeat until a workable number (200 to 500 individuals) is obtained in a subsample. Exercise care to provide unbiased splits. Even when using the Folsom splitter unbiased subsamples cannot be unquestioningly assumed;[2] therefore, count animals in several subsamples from the same sample to verify that the splitter is unbiased and to determine the sampling error introduced by using it.

Another method permits abundance estimates of more equivalent levels of precision among taxa than obtained with either the Hensen-Stempel pipet or the Folsom splitter.[3] Normal counting procedures tally organisms on the basis of their abundance in a sample. Therefore, in a sample with a dominant organism making up 50% of total numbers, the tally of the dominant taxon will be large and have a small error. However, error about the subdominants will increase as the tally of each taxon decreases. By accepting one level of precision, the technique[3] has been developed to obtain the same error about dominants and subdominants, permitting quantitative comparisons between taxa over successive times or between stations.

2. Enumeration

Using a compound microscope and a

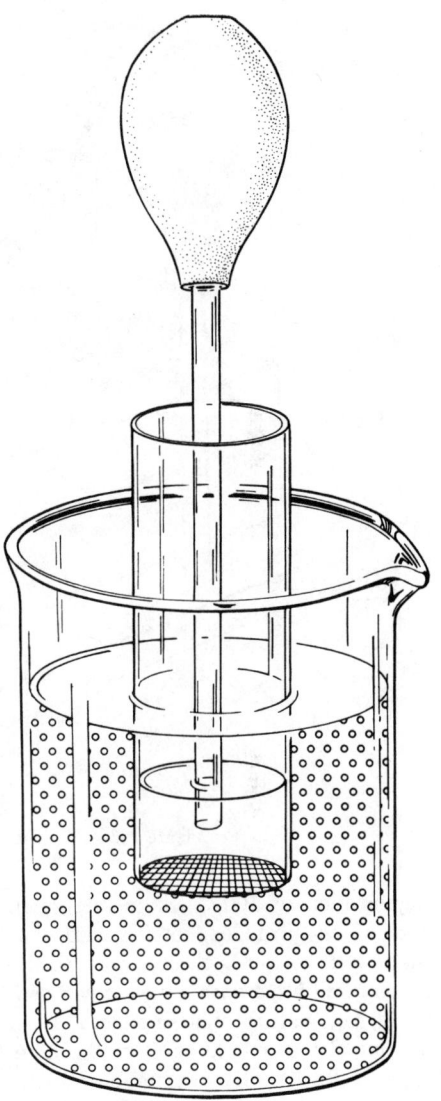

Figure 10200:9. A simple, efficient device for concentrating plankton. The tube is lowered into the beaker containing the sample. Water filtering into the tube is removed with the rubber bulb. The filter is nylon monofilament screen cloth that is glued to the bottom of the tube. The mesh size should be sufficiently small to prevent zooplankters from entering the filtrate (after Dodson and Thomas[5]).

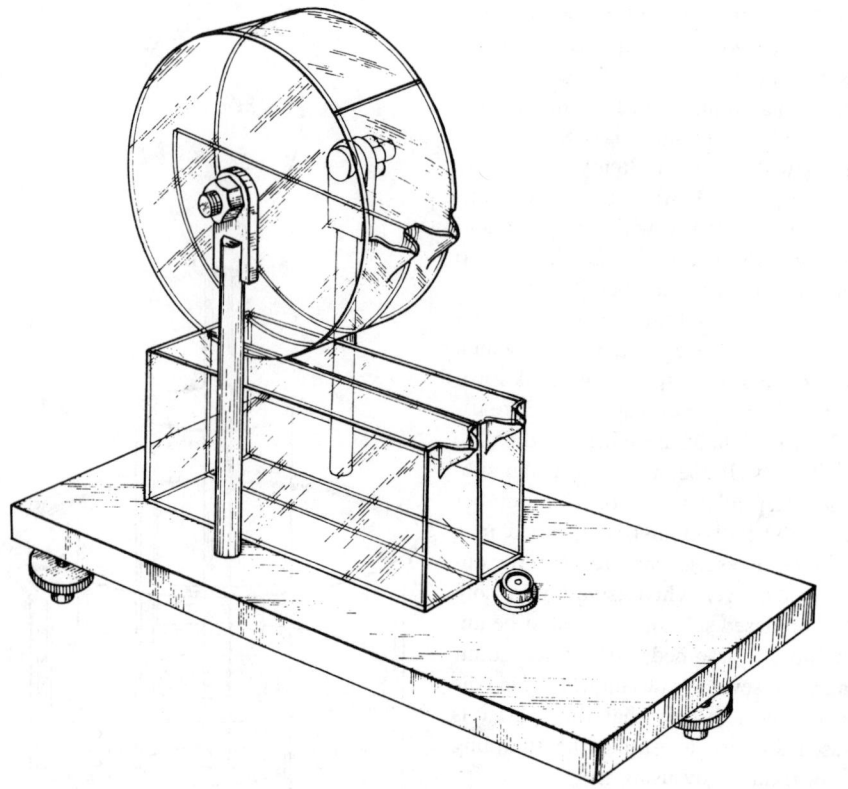

Drawn by: R.J.Ristić/1983

Figure 10200:10. The Folsom plankton splitter.

magnification of 100×, enumerate small zooplankton (protozoa, rotifers, and nauplii) in a 1- to 5-mL clear acrylic plastic counting cell fitted with a glass cover slip. For larger, mature microcrustacea use a counting chamber holding 5 to 10 mL. A Sedgwick-Rafter cell is not suitable because of size. An open counting chamber 80 by 50 mm and 2 mm deep is desirable; however, an open chamber is difficult to move without jarring and disrupting the count. A mild detergent solution placed on the chamber before counting reduces organism movements or special counting trays with parallel or circular grooves or partitions[4,5] can be used. Count microcrustacea with a binocular dissecting microscope at 20× to 40× magnification. If identification is questionable, remove organisms with a mi-crobiological transfer loop and examine at a higher magnification under a compound microscope.

Report smaller zooplankton as number per liter and larger forms as number per cubic meter:

$$\text{No./m}^3 = \frac{C \times V'}{V'' \times V'''}$$

where:

C = number of organisms counted,
V' = volume of the concentrated sample, mL,
V'' = volume counted, mL, and
V''' = volume of the grab sample, m³.

To obtain organisms per liter divide by 1000.

3. References

1. LONGHURST, A.R. & D.L.R. SEIBERT. 1967. Skill in the use of Folsom's plankton sample splitter. *Limnol. Oceanogr.* 12:334.

2. MCEWEN, G.F., M.W. JOHNSON & T.R. FOLSOM. 1954. A statistical analysis of the Folsom sample splitter based upon test observations. *Arch. Meteorol. Geophys. Bioklimatol.*, Ser. A, 6:502.

3. ALDEN, R.W., III, R.C. DAHIYA & R.J. YOUNG, JR. 1982. A method for the enumeration of zooplankton samples. *J. Exp. Mar. Biol. Ecol.* 59:185.

4. GANNON, J.E. 1971. Two counting cells for the enumeration of zooplankton micro-crustacea. *Trans. Amer. Microsc. Soc.* 90:486.

5. DODSON, A.N. & W.H. THOMAS. 1964. Concentrating plankton in gentle fashion. *Limnol. Oceanogr.* 9:455.

10200 H. Chlorophyll

The concentration of photosynthetic pigments is used extensively to estimate phytoplankton biomass.[1,2] All green plants contain chlorophyll *a*, which constitutes approximately 1 to 2% of the dry weight of planktonic algae. Other pigments that occur in phytoplankton include chlorophylls *b* and *c*, xanthophylls, phycobilins, and carotenes. The important chlorophyll degradation products found in the aquatic environment are the chlorophyllides, pheophorbides, and pheophytins. The presence or absence of the various photosynthetic pigments is used, among other features, to separate the major algal groups.

The three methods for determining chlorophyll *a* in phytoplankton are the spectrophotometric,[3-5] the fluorometric,[6-8] and the high-performance liquid chromatographic (HPLC) techniques.[9] Fluorometry is more sensitive than spectrophotometry, requires less sample, and can be used for in-vivo measurements.[10] These optical methods can significantly under- or overestimate chlorophyll *a* concentrations,[11-18] in part because of the overlap of the absorption and fluorescence bands of co-occurring accessory pigments and chlorophyll degradation products.

Pheophorbide *a* and pheophytin *a*, two common degradation products of chlorophyll *a*, can interfere with the determination of chlorophyll *a* because they absorb light and fluoresce in the same region of the spectrum as does chlorophyll *a*. If these pheopigments are present, significant errors in chlorophyll *a* values will result. Pheopigments can be measured either by spectrophotometry or fluorometry, but in marine and freshwater environments the fluorometric method is unreliable when chlorophyll *b* co-occurs. Upon acidification of chlorophyll *b*, the resulting fluorescence emission of pheophytin *b* is coincident with that of pheophytin *a*, thus producing underestimation and overestimation of chlorophyll *a* and pheopigments, respectively.

HPLC is a useful method for quantifying photosynthetic pigments[9,13,15,16,19-21] including chlorophyll *a*, accessory pigments (e.g., chlorophylls *b* and *c*), and chlorophyll degradation products (chlorophyllides, pheophorbides, and pheophytins). Pigment distribution is useful for quantitative assessment of phytoplankton community composition and zooplankton grazing activity.[22]

1. Pigment Extraction

Conduct work with chlorophyll extracts in subdued light to avoid degradation. Use opaque containers or wrap with aluminum foil. The pigments are extracted from the plankton concentrate with aqueous acetone and the optical density (absorbance) of the

extract is determined with a spectrophotometer. The ease with which the chlorophylls are removed from the cells varies
considerably with different algae. To
achieve consistently the complete extraction of the pigments, disrupt the cells mechanically with a tissue grinder.

Glass fiber filters are preferred for removing algae from water. The glass fibers
assist in breaking the cells during grinding,
larger volumes of water can be filtered, and
no precipitate forms after acidification. Inert membrane filters such as polyester filters may be used where these factors are
irrelevant.

a. Equipment and reagents:

1) *Tissue grinder:** Successfully macer-
ating glass fiber filters in tissue grinders
with grinding tube and pestle of conical
design may be difficult. Preferably use
round-bottom grinding tubes with a matching pestle having grooves in the TFE tip.

2) *Clinical centrifuge.*

3) *Centrifuge tubes,* 15-mL graduated,
screw-cap.

4) *Filtration equipment,* filters, glass fi-
ber† or membrane (0.45-μm porosity, 47-
mm diam); vacuum pump; solvent-resistant disposable filter assembly, 1.0-μm pore
size;‡ 10-mL solvent-resistant syringe.

5) *Saturated magnesium carbonate solution:* Add 1.0 g finely powdered $MgCO_3$
to 100 mL distilled water.

6) *Aqueous acetone solution:* Mix 90 parts
acetone (reagent grade BP 56°C) with 10
parts saturated magnesium carbonate solution.

b. Extraction procedure:

1) Concentrate sample by centrifuging
or filtering as soon as possible after collection. If processing must be delayed, hold
samples on ice or at 4°C and protect from

exposure to light. Use opaque bottles because even brief exposure to light during
storage will alter chlorophyll values. Samples on filters taken from water having pH
7 or higher may be placed in airtight plastic
bags and stored frozen for 3 weeks. Samples from acidic water must be processed
promptly to prevent chlorophyll degradation. Use glassware and cuvettes that are
clean and acid-free.

2) Place sample in a tissue grinder, cover
with 2 to 3 mL 90% aqueous acetone solution, and macerate at 500 rpm for 1 min.
Use TFE/glass grinder for a glass-fiber filter and glass/glass grinder for a membrane
filter.

3) Transfer sample to a screw-cap centrifuge tube, rinse grinder with a few milliliters 90% aqueous acetone, and add the
rinse to the extraction slurry. Adjust total
volume to a constant level, 5 to 10 mL,
with 90% aqueous acetone. Use solvent
sparingly and avoid excessive dilution of
pigments. Steep samples at least 2 h at 4°C
in the dark.

4) Clarify by filtering through a solvent-
resistant disposable filter (to minimize retention of extract in filter and filter holder,
force 1 to 2 mL air through the filter after
the extract), or by centrifuging in closed
tubes for 20 min at 500 g. Decant clarified
extract into a clean, calibrated, 15-mL,
screw-cap centrifuge tube and measure total volume. Proceed as in 2, 3, or 4 below.

2. Spectrophotometric Determination of Chlorophyll

a. Equipment and reagents:

1) *Spectrophotometer,* with a narrow
band (pass) width (0.5 to 2.0 nm) because
the chlorophyll absorption peak is relatively narrow. At a spectral band width of
20 nm the chlorophyll *a* concentration may
be underestimated by as much as 40%.

2) *Cuvettes,* with 1-, 4-, and 10-cm path
lengths.

3) *Pipets,* 0.1- and 5.0-mL.

---

*Kontes Glass Co., Vineland, N.J. 08360: Glass/glass
grinder, Model No. 8855: Glass/TFE grinder, Model
886000; or equivalent.
†Whatman GF/F (0.7 μm), GFB (1.0 μm), Gelman AE
(1 μm),[23] or equivalent.
‡Gelman Acrodisc or equivalent.

4) *Hydrochloric acid,* HCl, $0.1 N$.

b. *Determination of chlorophyll* a *in the presence of pheophytin* a: Chlorophyll *a* may be overestimated by including pheopigments that absorb near the same wavelength as chlorophyll *a*. Addition of acid to chlorophyll *a* results in loss of the magnesium atom, converting it to pheophytin *a*. Acidify carefully to a final molarity of not more than $3 \times 10^{-3} M$ to prevent certain accessory pigments from changing to absorb at the same wavelength as pheophytin *a*.[13] When a solution of pure chlorophyll *a* is converted to pheophytin *a* by acidification, the absorption-peak-ratio (OD664/OD665) of 1.70 is used in correcting the apparent chlorophyll *a* concentration for pheophytin *a*.

Samples with an OD664 before/OD665 after acidification ratio ($664_b/665_a$) of 1.70 are considered to contain no pheophytin *a* and to be in excellent physiological condition. Solutions of pure pheophytin show no reduction in OD665 upon acidification and have a $664_b/665_a$ ratio of 1.0. Thus, mixtures of chlorophyll *a* and pheophytin *a* have absorption peak ratios ranging between 1.0 and 1.7. These ratios are based on the use of 90% acetone as solvent. Using 100% acetone as solvent results in a chlorophyll *a* before-to-after acidification ratio of about 2.0.[3]

Spectrophotometric procedure—Transfer 3 mL clarified extract to a 1-cm cuvette and read optical density (OD) at 750 and 664 nm. Acidify extract in the cuvette with 0.1 mL $0.1 N$ HCl. Gently agitate the acidified extract and read OD at 750 and at 665 nm, 90 s after acidification. The volumes of extract and acid and the time after acidification are critical for accurate, consistent results.

The OD664 before acidification should be between 0.1 and 1.0. For very dilute extracts use cuvettes having a longer path length. If a larger cell is used, add a proportionately larger volume of acid. Correct OD obtained with larger cuvettes to 1 cm before making calculations.

Subtract the 750-nm OD value from the readings before (OD 664 nm) and after acidification (OD 665 nm).

Using the corrected values calculate chlorophyll *a* and pheophytin *a* per cubic meter as follows:

Chlorophyll *a*, mg/m^3

$$= \frac{26.7 \, (664_b - 665_a) \times V_1}{V_2 \times L}$$

Pheophytin *a*, mg/m^3

$$= \frac{26.7 \, [1.7 \, (665_a) - 664_b] \times V_1}{V_2 \times L}$$

where:

V_1 = volume of extract, L,
V_2 = volume of sample, m^3,
L = light path length or width of cuvette, cm, and
664_b, 665_a = optical densities of 90% acetone extract before and after acidification, respectively.

The value 26.7 is the absorbance correction and equals $A \times K$

where:

A = absorbance coefficient for chlorophyll *a* at 664 nm = 11.0, and
K = ratio expressing correction for acidification,

$$= \frac{\left(\dfrac{664_b}{665_a}\right)_{\text{pure chlorophyll } a}}{\left(\dfrac{664_b}{665_a}\right)_{\text{pure chlorophyll } a} - \left(\dfrac{664_b}{665_a}\right)_{\text{pure pheophytin } a}}$$

$$= \frac{1.7}{1.7 - 1.0} = 2.43$$

c. *Determination of chlorophyll* a, b, *and* c *(trichromatic method):*

Spectrophotometric procedure—Transfer extract to a 1-cm cuvette and measure optical density (OD) at 750, 664, 647, and 630 nm. Choose a cell path length or dilution to give an OD664 between 0.1 and 1.0.

Use the optical density readings at 664, 647, and 630 nm to determine chlorophyll a, b, and c, respectively. The OD reading at 750 nm is a correction for turbidity. Subtract this reading from each of the pigment OD values of the other wavelengths before using them in the equations below. Because the OD of the extract at 750 nm is very sensitive to changes in the acetone-to-water proportions, adhere closely to the 90 parts acetone:10 parts water (v/v) formula for pigment extraction. Turbidity can be removed easily by filtration through a disposable, solvent-resistant filter attached to a syringe or by centrifuging for 20 min at 500 g.

Calculate the concentrations of chlorophyll a, b, and c in the extract by inserting the corrected optical densities in following equations:[5]

a) $C_a = 11.85(OD664) - 1.54(OD647) - 0.08(OD630)$

b) $C_b = 21.03(OD647) - 5.43(OD664) - 2.66(OD630)$

c) $C_c = 24.52(OD630) - 7.60(OD647) - 1.67(OD664)$

where:

C_a, C_b, and C_c = concentrations of chlorophyll a, b, and c, respectively, mg/L, and

OD664, OD647, and OD630 = corrected optical densities (with a 1-cm light path) at the respective wavelengths.

After determining the concentration of pigment in the extract, calculate the amount of pigment per unit volume as follows:

Chlorophyll a, mg/m^3

$$= \frac{C_a \times \text{extract volume, L}}{\text{Volume of sample, m}^3}$$

3. Fluorometric Determination of Chlorophyll a

The fluorometric method for chlorophyll a is more sensitive than the spectrophotometric method and thus smaller samples can be used. Calibrate the fluorometer spectrophotometrically with a sample from the same source to achieve acceptable results. Optimum sensitivity for chlorophyll a extract measurements is obtained at an excitation wavelength of 430 nm and an emission wavelength of 663 nm. A method for continuous measurement of chlorophyll a in vivo is available, but is reported to be less efficient than the in-vitro method given here, yielding about one-tenth as much fluorescence per unit weight as the same amount in solution. Pheophytin a also can be determined fluorometrically.[24]

a. Equipment and reagents: In addition to those listed under 1a and 2a above:

Fluorometer,§ equipped with a high-intensity F4T.5 blue lamp, photomultiplier tube R-446 (red-sensitive), sliding window orifices $1\times$, $3\times$, $10\times$, and $30\times$, and filters for light emission (CS-2-64) and excitation (CS-5-60). A high-sensitivity door is preferable.

b. Extraction procedure: Prepare sample as directed in 1b above.

1) Calibrate fluorometer with a chlorophyll solution of known concentration as follows: Prepare chlorophyll extract and analyze spectrophotometrically. Prepare serial dilutions of the extract to provide concentrations of approximately 2, 6, 20, and 60 μg chlorophyll a/L. Make fluorometric readings for each solution at each sensitivity setting (sliding window orifice): $1\times$, $3\times$, $10\times$, and $30\times$. Using the values obtained, derive calibration factors to convert fluorometric readings in each sensitivity level to concentrations of chlorophyll a, as follows:

§Model 111, Sequoia-Turner Corp., 755 Ravendale Dr., Mountain View, Calif. or equivalent.

$$F_s = \frac{C_a'}{R_s}$$

where:

F_s = calibration factor for sensitivity setting S,

R_s = fluorometer reading for sensitivity setting S, and,

C_a' = concentration of chlorophyll a determined spectrophotometrically, $\mu g/L$.

2) Measure sample fluorescence at sensitivity settings that will provide a mid-scale reading. (Avoid using the $1\times$ window because of quenching effects.) Convert fluorescence readings to concentrations of chlorophyll a by multiplying the readings by the appropriate calibration factor.

c. *Determination of chlorophyll* a *in the presence of pheophytin* a: This method normally is not applicable to freshwater samples. See discussion under 10200G and 1c above.

1) Equipment and reagents—In addition to those listed under 1a and 2a above, pure chlorophyll $a\|$ (or a plankton chlorophyll extract with a spectrophotometric before-and-after acidification ratio of 1.70 containing no chlorophyll b).

2) Fluorometric procedure—Calibrate fluorometer as directed in ¶ 3b1). Determine extract fluorescence at each sensitivity setting before and after acidification. Calculate calibration factors (F_s) and before-and-after acidification fluorescence ratio by dividing fluorescence reading obtained before acidification by the reading obtained after acidification. Avoid readings on the $1\times$ scale and those outside the range of 20 to 80 fluorometric units.

3) Calculations—Determine the "corrected" chlorophyll a and pheophytin a in sample extracts with the following equations:[8,24]

Chlorophyll a, mg/m³ $= F_s \dfrac{r}{r-1}(R_b - R_a)$

Pheophytin a, mg/m³ $= F_s \dfrac{r}{r-1}(rR_a - R_b)$

where:

F_s = conversion factor for sensitivity setting S (see ¶ 2b, above),

R_b = fluorescence of extract before acidification,

R_a = fluorescence of extract after acidification, and

$r = R_b/R_a$, as determined with pure chlorophyll a for the instrument. Redetermine r if filters or light source are changed.

d. *Extraction of whole water, nonfiltered samples:* Alternatively, to prevent cell lysis during filtration, extract whole water sample.

1) Equipment and reagents—Fluorometer equipped with a high-sensitivity R928 phototube# with output impedance of 36 ma/W at 675 nm and a high-sensitivity door. Place neutral density filter (40-60N) in the rear light path**, selected to permit reagent blanking on the highest sensitivity scale.

2) Extraction procedure—Decant 1.5 mL sample into screw-cap test tube and add 8.5 mL 100% acetone. Mix with vortex mixer and hold in the dark for 6 h at room temperature. Filter through glass fiber filter†† or centrifuge. Measure fluorescence as described in Section 10200H.3 and estimate concentrations as in ¶ 3c. Because humic substances interfere, if they are present filter a sample portion (see 10200H. 1b) and process filtrate with sample. Subtract filtrate (blank) fluorescence from that of sample.

|| Purified chlorophyll a, Sigma Chemical Company, St. Louis, Mo., or equivalent.

Hammamatsu Corp., Middlesex, N.J., or equivalent.
** If using Model 10-005, Sequoia-Turner Corp., or equivalent.
†† Whatman GF/F or equivalent.

4. High-Performance Liquid Chromatographic Determination of Algal Chlorophylls and Their Degradation Products
(PROPOSED)

a. *Equipment and reagents:* In addition
to those listed for pigment extraction, ¶ 1a
above:

1) *High-pressure liquid chromatograph*
capable of a flow rate of 2.0 mL/min.

2) *High-pressure injector valve* equipped
with a 100-μL sample loop.

3) *Guard column* (4.0 $\times$ 0.5 cm, C_{18}
packing material, 3-μm particle size, or
equivalent protection system) for extending
life of primary column.

4) *Reverse-phase HPLC column.‡‡*

5) *Fluorescence detector* capable of excitation at 430 $\pm$ 30 nm and measuring
emission at wavelengths greater than 600
nm.

6) *Data recorder device:* Strip chart recorder or, preferably, an electronic integrator.

7) *Syringe,* glass, 250-μL.

8) *HPLC eluants:* System A (80:15:5;
methanol: Type I reagent water: ion-pairing solution) and System B (80:20; meth-
anol:acetone). Use HPLC-grade solvents;
measure volumes before mixing. Filter
eluants through a solvent-resistant 0.4-μm
filter before use and degas with helium.
Prepare the ion-pairing (IP) solution from
15 g tetrabutylammonium acetate§§ and 77
g ammonium acetate ‖‖ made up to 1 L
with Type I reagent water.[15]

9) *Calibration standards:* Individually
dissolve 1 mg each pure chlorophyll a and
b ‖‖ in 100 mL 90% acetone. Determine
the exact concentrations spectrophotometrically (ϵ_{664} for chlorophyll a in 90%
acetone = 87.67 L g^{-1} cm^{-1}; ϵ_{647} for chlorophyll b in 90% acetone = 51.36 L g^{-1}

cm^{-1})[5]. Prepare pheophytin $a + a'$ and b
$+ b'$ standards from the primary chlorophyll a and b standards by acidification
with hydrochloric acid; correct respective
concentrations for Mg^{2+} loss. Extract chlorophyll c with 90% acetone from diatoms,
purify by thin-layer chromatography
(TLC)[25] and calibrate spectrophotometrically (ϵ_{631} for a mixture containing equal
amounts of chlorophylls c_1 and c_2 in 90%
acetone containing 1% pyridine = 42.6 L
g^{-1} cm^{-1}; the absence of this small amount
of pyridine is presumed to cause only small
differences in the absorption properties of
chlorophyll c.[26] Alternatively, determine
the chlorophyll c content of a 90% acetone
extract made from diatoms, spectrophotometrically (chlorophyll $c_1 + c_2$, μg/mL
= $24.36E_{630} - 3.73E_{664}$)[5] and use as standard. Prepare chlorophyllide a from diatoms,[27] purify by TLC[25] and calibrate
spectrophotometrically in 90% acetone
(ϵ_{664} for chlorophyllide a = 128 L g^{-1}
cm^{-1}).[28] Prepare pheophorbide a by acidification of chlorophyllide a, purify by
TLC,[25] and calibrate spectrophotometrically in 90% acetone (ϵ_{665} for pheophorbide
a = 69.8 L g^{-1} cm^{-1}).[28] Standards stored
under nitrogen in the dark at $-20°C$ are
stable for about one month.

b. *Procedure:*

1) Set up and equilibrate the HPLC system with solvent System A at a flow rate
of 2 mL/min. Adjust fluorometer sensitivity to provide full-scale reading with the
most concentrated chlorophyll a standard.

2) Calibrate HPLC system by preparing
working standards from the primary standards (on day of use). Once retention times
of the standards are determined for a particular system, simplify standardization by
preparing serial dilutions from mixed
standards. Prepare separately mixed standards for the chlorophylls and chlorophyllide a and for the pheophytins and
pheophorbide a. Mix 1-mL portions of
standards with 300 μL ion-pairing solution
and equilibrate for 5 min before injection

---

‡‡Microsorb C_{18} column, 10 cm long, 3-μm particle size,
Rainin Co., or equivalent.
§§Fluka Chemical Corp., 980 South Second Street, Ron-
konkoma, N.Y., or equivalent.
‖ ‖Sigma Chemical Company, or equivalent.

(use of ion-pairing agents greatly enhances separation of dephytolated pigments, chlorophyllide *a*, chlorophyll *c*, and pheophorbide *a*). Prepare blanks by mixing 1 mL 90% acetone with 300 μL IP solution. Rinse syringe twice with 150 μL standard and draw about 250 μL standard into syringe for injection. Place syringe in injector valve, overfilling the 100-μL sample loop. Construct calibration curves by plotting fluorescence peak areas (or heights) against standard pigment concentrations.

3) Prepare samples for injection by mixing a 1-mL portion of the 90% acetone pigment extract with 300 μL IP solution.

4) Use a two-step solvent program to optimize separation of the chlorophylls from their degradation products.[15] After injection, change from solvent System A to System B over 5 min and follow with System B for 15 min at a flow rate of 2 mL/min. Re-equilibrate the column with System A for 5 min before the next injection for a total analysis time of approximately 25 min. Degas the solvent systems with helium during analysis. Increase lifetime of HPLC column by storing it in 100% methanol between runs. Periodically flush the HPLC system with reagent water (Type I) to avoid buildup of ion-pairing agents.

5) Calculate individual pigment concentrations using the following formula:

$$C_i = \frac{A_s \, F_i \, V_E}{V_I \, V_S}$$

where:

C_i = individual pigment concentration, ng/L,
A_s = area of individual pigment peak from sample injection,
F_i = standard response factor (ng pigment / 0.1 mL standard divided by corresponding peak area),
V_I = injection volume (0.1 mL),
V_E = extraction volume, mL, and
V_S = sample volume, L.

6) This method is designed only for quantification of chlorophylls and their degradation products. Detect carotenoid pigments, which also are present in the 90% acetone extracts but do not fluoresce, by absorbance spectroscopy (at about 440 nm).[21]

7) The elution order and approximate retention times for the major chlorophyll pigments and their degradation products are shown in Figure 10200:11. The detection limits ($s/n = 2$) vary with fluorometer configuration and flow rate; however, they range from 10 to 100 pg per injection for most chlorophylls and their degradation products.[15,21,29] The accuracy of the HPLC method depends primarily on purity of pigment standards. Preferably measure absorption spectra (350 to 750 nm) of the standards and compare with published

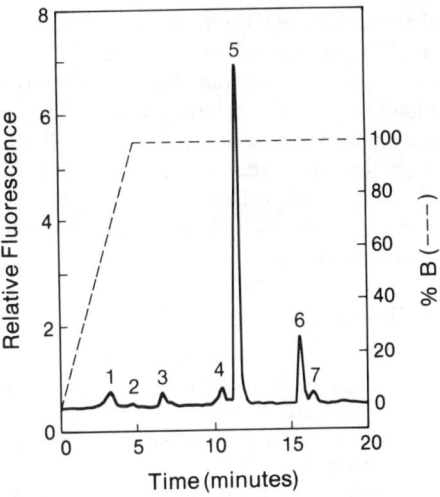

Figure 10200:11. Reverse-phase HPLC chromatogram for a fivefold dilution of EPA sample. Injection volume 100 μL; peaks detected by fluorescence spectroscopy (λ_{ex}: 400–460 nm; λ_{em}: >600 nm). Peak identities are: 1—chlorophyllide *a*; 2—chlorophyll *c*; 3—pheophorbide *a*; 4—chlorophyll *b*; 5—chlorophyll *a*; 6—pheophytin *a*; and 7—pheophytin *a'*. The chlorophyll *b* degradation products, pheophytin *b* and pheophytin *b'*, were below detection limits. Peak identities confirmed by on-line diode array spectroscopy (350–550 nm).

data. Pigment purity also can be assessed by HPLC analysis, providing there are no co-eluting contaminants whose absorption and fluorescence bands overlap with the standards. HPLC and spectrophotometrically derived pigment concentrations for available EPA standards agree reasonably well ($\pm 20\%$) if spectrophotometric results are corrected for the presence of pheopigments and the HPLC results are expressed as pigment equivalents (e.g., chlorophyll a equivalents = chlorophyllide a + chlorophyll a + chlorophyll a', providing the proper molecular weight corrections are applied).[30] Thus, if significant amounts of chlorophyll derivatives are present, pigment concentrations determined spectrophotometrically will be overestimated. The agreement between HPLC and fluorometrically derived results depends on the presence of accessory chlorophylls b, c, and their derivatives. Triplicate injections of a fivefold dilution of an EPA sample gave coefficients of variation of 7.5% (chlorophyllide a), 9.1% (chlorophyll c), 13.4% (pheophorbide a), 9.6% (chlorophyll b), 0.5% (chlorophyll a), 6.2% (pheophytin a), and 22.9% (pheophytin a'), with an average value of 10% for the seven pigments analyzed.

5. References

1. ROTT, E. 1980. Spectrophotometric and chromatographic chlorophyll analysis: comparison of results and discussion of the trichromatic method. *Ergebn. Limnol.* (Suppl. to *Arch. Hydrobiol.*) 14:37.

2. MARKER, A.F.H., E. A. NUSCH, H. RAI & B. RIEMANN. 1980. The measurement of photosynthetic pigments in freshwaters and standardization of methods: Conclusions and recommendations. *Ergebn. Limnol.* (Suppl. to *Arch. Hydrobiol.*) 14:91.

3. LORENZEN, C.J. 1967. Determination of chlorophyll and pheo-pigments: spectrophotometric equations. *Limnol. Oceanogr.* 12:343.

4. FITZGERALD, G.P. & S.L. FAUST. 1967. A spectrophotometric method for the estimation of percentage degradation of chlorophylls to pheo-pigments in extracts of algae. *Limnol. Oceanogr.* 12:335.

5. JEFFREY, S.W. & G.F. HUMPHREY. 1975. New spectrophotometric equations for determining chlorophylls a, b, and c in higher plants, algae and natural phytoplankton. *Biochem. Physiol. Pflanzen* 167:191.

6. YENTSCH, C.S. & D.W. MENZEL. 1963. A method for the determination of phytoplankton chlorophyll and phaeophytin by fluorescence. *Deep Sea Res.* 10:221.

7. LOFTUS, M.E. & J.H. CARPENTER. 1971. A fluorometric method for determining chlorophylls *a*, *b*, and *c*. *J. Mar. Res.* 29:319.

8. HOLM-HANSEN, O., C.J. LORENZEN, R.W. HOLMES & J.D.H. STRICKLAND. 1965. Fluorometric determination of chlorophyll. *J. Cons. Cons. Perma. Int. Explor. Mer* 30:3.

9. ABAYCHI, J.K. & J.P. RILEY. 1979. The determination of phytoplankton pigments by high-performance liquid chromatography. *Anal. Chim. Acta* 107:1.

10. LORENZEN, C.J. 1966. A method for the continuous measurement of *in vivo* chlorophyll concentration. *Deep Sea Res.* 13:223.

11. JACOBSEN, T.R. 1978. A quantitative method for the separation of chlorophylls *a* and *b* from phytoplankton pigments by high-pressure liquid chromatography. *Mar. Sci. Comm.* 4:33.

12. BROWN, L.M., B.T. HARGRAVE & M.D. MACKINNON. 1981. Analysis of chlorophyll *a* in sediments by high-performance liquid chromatography. *Can. J. Fish. Aquat. Sci.* 38:205.

13. GIESKES, W.W. & G.W. KRAAY. 1983. Unknown chlorophyll *a* derivatives in the North Sea and the tropical Atlantic Ocean revealed by HPLC analysis. *Limnol. Oceanogr.* 28:757.

14. GOWEN, R.J., P. TETT & B.J.B. WOOD. 1983. Changes in the major dihydroporphyrin plankton pigments during the spring bloom of phytoplankton in two Scoottish sea-lochs. *J. Mar. Biol Assoc. U.K.* 63:27.

15. MANTOURA, R.F.C. & C.A. LLWEWLLYN. 1983. The rapid determination of algal chlorophyll and carotenoid pigments and their breakdown products in natural waters by reverse-phase high-performance liquid chromatography. *Anal. Chim. Acta* 151:297.

16. GIESKES, W.W.C. & G.W. KRAAY. 1984.

Phytoplankton, its pigments, and primary production at a central North Sea station in May, July and September 1981. *Neth. J. Sea Res.* 18:71.

17. HALLEGRAEFF, G.M. & S.E. JEFFREY. 1985. Description of new chlorophyll *a* alteration products in marine phytoplankton. *Deep Sea Res.* 32:697.

18. TREES, C.C., M.C. KENNICUTT II & J.M. BROOKS. 1985. Errors associated with the standard fluorometric determination of chlorophylls and phaeopigments. *Mar. Chem.* 17:1.

19. ESKINS, K., C.R. SCHOFIELD & H.J. DUTTON. 1977. High-performance liquid chromatography of plant pigments. *J. Chromatogr.* 135:217.

20. WRIGHT, S.W. & J.D. SHEARER. 1984. Rapid extraction and high performance liquid chromatography of chlorophylls and carotenoids from marine phytoplankton. *J. Chromatogr.* 294:281.

21. BIDIGARE, R.R., M.C. KENNICUTT II & J.M. BROOKS. 1985. Rapid determination of chlorophylls and their degradation products by high-performance liquid chromatography. *Limnol Oceanogr.* 30:432.

22. JEFFREY, S.W. 1974. Profiles of photosynthetic pigments in the ocean using thin-layer chromatography. *Mar. Biol.* 26:101.

23. PHINNEY, D.A. & C.S. YENTSCH. 1985. A novel phytoplankton chlorophyll technique: toward automated analysis. *J. Plankton Res.* 7:633.

24. STRICKLAND, J.D.H. & T.R. PARSONS. 1968. A Practical Manual of Sea Water Analysis. Fish. Res. Board Can. Bull. No. 167. Queen's Printer, Ottawa, Ont.

25. JEFFREY, S.W. 1981. An improved thin-layer chromatographic technique for marine phytoplankton pigments. *Limnol. Oceanogr.* 26:191.

26. JEFFREY, S.W. 1972. Preparation and some properties of crystalline chlorophyll c_1 and c_2 from marine algae. *Biochim. Biophys. Acta* 279:15.

27. BARRETT, J. & S.W. JEFFREY. 1971. A note on the occurrence of chlorophyllase in marine algae. *J. Exp. Mar. Biol. Ecol.* 7:255.

28. LORENZEN, C.J. & J. NEWTON DOWNS. 1986. Specific absorption coefficients of chlorophyllide *a* and pheophorbide *a* in 90 percent acetone, and comments on the fluorometric determination of chlorophyll and pheopigments. *Limnol. Oceanogr.* 31:449.

29. SARTORY, D.P. 1985. The determination of algal chlorophyllous pigments by high performance liquid chromatography and spectrophotometry. *Water Res.* 19:605.

30. MURRAY, A.P., C.F. GIBBS & A.R. LONGMORE. 1986. Determination of chlorophyll in marine waters: Intercomparison of a rapid HPLC method with full HPLC, spectrophotometric and fluorometric methods. *Mar. Chem.* 19:211.

10200 I.   Determination of Biomass (Standing Crop)

Biomass is a quantitative estimate of the total mass of living organisms within a given area or volume. It may include the mass of a population (species biomass) or of a community (community biomass) but gives no information on community structure or function. The most accurate methods for estimation of biomass are dry weight, ash-free dry weight, and volume of living organisms. Indirect methods include estimates of total carbon, caloric content, nitrogen, lipids, carbohydrates, silica (diatoms), and chlorophyll (algae). Adenosine triphosphate[1] (ATP) and deoxyribonucleic acid[2,3] (DNA) also have been used as indirect estimates. All estimates of biomass can be affected by the presence of organic and inorganic detritus; ATP and DNA analyses include contributions from the bacterial flora.[4]

1. Chlorophyll *a*

Chlorophyll *a* is used as an algal biomass indicator.[5] Assuming that chlorophyll *a* constitutes, on the average, 1.5% of the dry weight of organic matter (ash-free weight) of algae, estimate the algal biomass by multiplying the chlorophyll *a* content by a factor of 67.

2. Biovolume (Cell Volume)

Plankton data derived on a volume-per-volume basis often are more useful than numbers per milliliter.[6] Determine cell volume by using the simplest geometric configuration that best fits the shape of the cell being measured (such as sphere, cone, cylinder).[7] Cell sizes of an organism can differ substantially in different waters and from the same waters at different times during the year; therefore, average measurements from 20 individuals of each species for each sampling period. Calculate the total biovolume of any species by multiplying the average cell volume in cubic micrometers by the number per milliliter.

Compute total wet algal volume as:

$$V_t = \sum_{i=1}^{n} (N_i \times V_i)$$

where:

V_t = total plankton cell volume, mm³/L,
N_i = number of organisms of the ith species/ L, and
V_i = average volume of cells of ith species, μm³.

3. Cell Surface Area

An estimation of cell surface area is valuable in analyzing interactions between the cell and surrounding waters. Compute average surface area in square micrometers and multiply by the number per milliliter of the species being considered.

4. Displacement Volume

This method[8] measures an equivalent volume of liquid that is displaced by the sample. Displacement volume may be determined by several methods; for simple, direct measurement proceed as follows: Place sample in sieve of mesh size equal to

or smaller than net used in capture; let sample drain and transfer to a measured volume of water in a graduated cylinder; measure the new volume containing sample plus known volume. The displacement volume equals the new volume minus original measured volume of water.

5. Gravimetric Methods

The biomass of the plankton community can be estimated from gravimetric determinations, although silt and organic detritus interfere. Determine dry weight by placing 100 mg wet concentrated sample in a clean, ignited, and tared porcelain crucible and dry at 105°C for 24 h. Alternatively, filter a known volume of sample through 0.45-μm-pore-diam membrane or a prerinsed, dried, and preweighed glass-fiber filter. (Note that the small sample used in direct filtration may lead to error if not handled properly.) Cool sample in a desiccator and weigh. Obtain ash-free weight by igniting the dried sample at 500°C for 1 h. Cool, rewet ash with distilled water, and bring to constant weight at 105°C. The ash is rewetted to restore water of hydration of clays and other minerals; this may amount to as much as 10% of weight lost during incineration.[9] The ash-free weight is preferred to dry weight to compare mixed assemblages. The ash content may constitute 50% or more of the dry weight in phytoplankton having inorganic structures, such as the diatoms. In other forms the ash content is only about 5% of dry weight.

6. Adenosine Triphosphate (ATP)

Methods of measuring adenosine triphosphate (ATP) in plankton provide the only means of determining the total viable plankton biomass. ATP occurs in all plants and animals, but only in living cells; it is not associated with nonliving particulate material. The ratio of ATP to biomass varies from species to species, but appears to be constant enough to permit reliable estimates of biomass from ATP measurements.[10] The method is simple and relatively inexpensive and the instrumentation is stable and reliable. The method also has many potential applications in entrainment and bioassay work, especially plankton mortality studies.

   a. Equipment and reagents:

   1) *Glassware:* clean, sterile, dry borosilicate glass flasks, beakers, and pipets.

   2) *Filters:* 47-mm-diam, 0.45-μm-porosity membrane filters.

   3) *Filtration equipment.*

   4) *Freezer* ($-20°C$).

   5) *Boiling water bath.*

   6) *Detection instruments* designed specifically for measuring ATP.*

   7) *Microsyringes:* 10-, 25-, 50-, 100-, and 250-μL.

   8) *Reaction cuvettes and vials.*

   9) *Tris buffer* (0.02M, pH 7.75): Dissolve 7.5 g trishydroxymethylaminomethane in 3 L distilled water and adjust to pH 7.75 with 20% HCl. Autoclave 150-mL portions at 115°C for 15 min.

   10) *Luciferin-luciferase enzyme preparation:†* Rehydrate frozen ($-20°C$) lyophilized extracts of firefly lanterns with Tris buffer as directed by the supplier; let stand at room temperature 2 to 3 h, then centrifuge at 300 g for 1 min and decant the supernatant into a clean, dry test tube; let stand at room temperature for 1 h.

   11) *Purified ATP standard:* Dissolve 12.3 mg disodium ATP in 1 L distilled water and dilute 1.0 mL to 100 mL with Tris buffer; 0.2 mL = 20 ng ATP.

   b. Procedure:

   1) Calibration—To determine the calibration factor (F), prepare a series of dilutions of purified ATP standard and record the light emission from several portions of each concentration of standard. Correct mean area of standards by sub-

tracting peak reading or mean area of several blanks using 0.2 mL Tris buffer. Calculate calibration factor F_s as:

$$F_s = \frac{C}{A_s}$$

where:

   F_s = calibration factor at sensitivity S,
   A_s = peak reading or mean area under standard ATP curve corrected for blank, and
   C = concentration of ATP in standard solution, ng/mL.

   2) Sample analysis—Collect a 1- to 2-L sample in a clean, sterile sampler. Pass through a 250-μm net to remove large zooplankton[10] and filter through a 47-mm 0.45-μm-porosity filter by applying a vacuum of about 30 kPa. (IMPORTANT: Break the suction before the last film of water is pulled through the filter.) Quickly place filter in a small beaker. Immediately cover filter with 3 mL boiling Tris buffer, using an automatic pipet. Place beaker in boiling water bath for 5 min and, with a Pasteur pipet, transfer extract to a clean, dry, calibrated test tube. Rinse filter and beaker with 2 mL boiling Tris buffer; combine extracts, record volume, bring volume up to 5 mL with Tris buffer, cover tubes with parafilm and, if samples cannot be analyzed immediately, freeze at $-25°C$. Extracts may be stored for many months in a freezer. Prepare at least triplicate extracts of each sample.

   The analytical procedure depends on detection equipment used. If a scintillation counter is used, pipet 0.2 mL enzyme preparation into a glass vial. Measure the light emission of the enzyme preparation (blank) for 2 to 3 min at sensitivity settings near that anticipated for the sample. Add 0.2 mL sample extract to the vial, record the time, and swirl. Start recording light output 10 s after combining ATP extract and enzyme preparation; record output for 2 to 3 min, using the same time period for all

---

*Beckman, JRB, Turner Designs, or equivalent.
†Dupont, Sigma Chemical, or equivalent.

samples. Determine the mean of areas under the curves obtained and correct by subtracting mean of areas under the curves obtained from blanks prepared as directed in Strickland and Parsons.[11]

c. Calculations: Calculate concentration of ATP as:

$$ATP, \text{ng/L} = \frac{A_c \times V_e \times F_s}{V_s}$$

where:

A_c = mean corrected area under extract curves,
V_e = extract volume, mL,
V_s = volume of sample, L, and
F_s = calibration factor.

If an ATP content of 2.4 μg ATP/mg dry weight organic matter is assumed,[12] total living plankton biomass (*B*), as dry weight organic matter, is given as:

$$B, \text{mg/L} = \frac{ATP}{(2.4)(1000)}$$

7. References

1. HOLM-HANSEN, O. & C.R. BOOTH. 1966. The measurement of adenosine triphosphate in the ocean and its ecological significance. *Limnol. Oceanogr.* 11:510.
2. HOLM-HANSEN, O., W.H. SUTCLIFFE, JR. & J. SHARP. 1968. Measurement of deoxyribonucleic acid in the ocean and its ecological significance. *Limnol. Oceanogr.* 13:507.
3. HOLM-HANSEN, O. 1969. Determination of

microbial biomass in ocean profiles. *Limnol. Oceanogr.* 14:740.
4. PAERL, H.W., M.M. TILZER & C.R. GOLDMAN. 1976. Chlorophyll *a* vs. ATP as algal biomass indicators in lakes. *J. Phycol.* 12:242.
5. CREITZ, G.I. & F.A. RICHARDS. 1955. The estimation and characterization of plankton populations by pigment analysis. *J. Mar. Res.* 14:211.
6. KUTKUHN, J.H. 1958. Notes on the precision of numerical and volumetric plankton estimates from small sample concentrations. *Limnol. Oceanogr.* 3:69.
7. VOLLENWEIDER, R.A. 1969. A Manual on Methods for Measuring Primary Production in Aquatic Environments. IBP Handbook No. 12. Blackwell Scientific Publ., Oxford, England.
8. JACOBS, F. & G.C. GRANT. 1978. Guidelines for zooplankton sampling in quantitative baseline and monitoring programs. EPA-600/3-78-026, U.S. Environmental Protection Agency.
9. NELSON, D.J. & D.C. SCOTT. 1962. Role of detritus in the productivity of a rock-outcrop community in a Piedmont stream. *Limnol. Oceanogr.* 7:396.
10. RUDD, J.W.M. & R.D. HAMILTON. 1973. Measurement of adenosine triphosphate (ATP) in two precambrian shield lakes of northwestern Ontario. *J. Fish. Res. Board Can.* 30:1537.
11. STRICKLAND, J.D.H. & T. R. PARSONS. 1968. A Practical Manual of Sea Water Analysts. Fish. Res. Board Can. Bull. No. 167. Queen's Printer, Ottawa, Ont.
12. WEBER, C.I. 1973. Recent developments in the measurement of the response of plankton and periphyton to changes in their environment. *In* G. Glass, ed. Bioassay Techniques and Environmental Chemistry. Ann Arbor Science Publ., Ann Arbor, Mich.

10200 J.   Metabolic Rate Measurements

The physiological condition and the spectrum of biological interactions of the aquatic community must be considered for evaluation of the state of natural waters. Earlier, numbers, species composition, and biomass were the prime considerations. Recognition of the limitations of this approach led to the measurement of rates of metabolic processes such as photosynthesis (productivity), nitrogen fixation, respiration, and electron transport. These provide a better understanding of the complex nature of the aquatic ecosystem. An indication of photosynthetic efficiency can be determined by the productivity index (mg C fixed/unit chlorophyll a).[1]

1. Nitrogen Fixation

The ability of an organism to fix nitrogen is a great competitive advantage and plays a major role in population dynamics. Two reliable methods for estimating nitrogen fixation rates in the laboratory are the ^{15}N isotope tracer method[2,3] and the acetylene reduction method.[4] Because the rate of nitrogen fixation varies greatly with different organisms and with the concentration of combined nitrogen, nitrogen fixation rates cannot be used to estimate biomass of nitrogen-fixing organisms. However, the acetylene reduction method is useful in measuring nitrogen budgets and in algal assay work.[5]

2. Productivity, Oxygen Method

Productivity is defined as the rate at which inorganic carbon is converted to an organic form. Chlorophyll-bearing organisms (phytoplankton, periphyton, macrophytes) serve as primary producers in the aquatic food chain. Photosynthesis ultimately results in the formation of a wide range of organic compounds, release of oxygen, and reduction of carbon dioxide (CO_2) in the surrounding waters. Primary productivity[6] can be determined by measuring the changes in oxygen and CO_2 concentrations.[7] In poorly buffered waters, pH can be a sensitive property for detecting variations in the system. As CO_2 is removed during photosynthesis, the pH rises. This shift can be used to estimate both photosynthesis and respiration.[8] The sea and many fresh waters are too highly buffered to make this useful, but it has been applied successfully to productivity studies in some lake waters.

Two methods of measuring the rate of carbon uptake and net photosynthesis in situ are the oxygen method[9] and the carbon 14 method.[10] In both methods, clear (light) and darkened (dark) bottles are filled with water samples and suspended at regular depth intervals for an incubation period of several hours or samples are incubated under controlled conditions in environmental growth chambers in the laboratory.

The basic reactions in algal photosynthesis involve uptake of inorganic carbon and release of oxygen, summarized by the relationship:

$$CO_2 + H_2O \rightarrow (CH_2O)_x + O_2$$

The chief advantages of the oxygen method are that it provides estimates of gross and net productivity and respiration and that analyses can be performed with inexpensive laboratory equipment and common reagents. The dissolved oxygen (DO) concentration is determined at the beginning and end of the incubation period. Productivity is calculated on the assumption that one atom of carbon is assimilated for each molecule of oxygen released.

a. Equipment:

1) *BOD bottles,* numbered, 300-mL, clear and opaque borosilicate glass, with ground-glass stopper and flared mouth, for sample incubation. Acid-clean the bottles, rinse thoroughly with distilled water, and just before use, rinse with water being tested. Do not use phosphorus-containing detergents.

If suitable opaque bottles are not available, make clear BOD bottles opaque by painting them black and wrapping with black waterproof tape. As a further precaution, wrap entire bottle in aluminum foil or place in a light-excluding container during incubation.

2) *Supporting line or rack* that does not shade suspended bottles.

3) *Nonmetallic opaque acrylic Van Dorn sampler* or equivalent, of 3- to 5-L capacity.

4) *Equipment and reagents for dissolved oxygen determinations:* See Section 4500-O.

5) *Pyrheliometer.*

6) *Submarine photometer.*

7) *Thermometer.*

b. Procedure:

1) Obtain a profile of the input of solar radiation for the photoperiod with a pyrheliometer.

2) Determine depth of euphotic zone (the region that receives 1% or more of surface illumination) with a submarine photometer. Select depth intervals for bottle placement. The photosynthesis-depth curve will be approximated closely by placing samples at intervals equal to one-tenth the depth of the euphotic zone. Estimate productivity in relatively shallow water with fewer depth intervals.

3) Measure oxygen concentration with probe or by titration and temperature and salinity to determine whether water is supersaturated with respect to oxygen (see Table 4500-O:I). If water is supersaturated, bubble nitrogen gas through sample to lower initial oxygen concentration to less than 80% saturation.

4) Keep samples out of direct sunlight during handling. Introduce samples taken from each preselected depth into duplicate clear, darkened, and initial-analysis bottles. Insert delivery tube of sampler to bottom of sample bottle and fill so that three volumes of water are allowed to overflow. Remove tube slowly and close bottle. Use water from the same grab sample to fill a "set" (one light, one dark, and one initial bottle).

5) Immediately treat (fix) samples taken for the chemical determination of initial dissolved oxygen (see Section 4500-O) with manganous sulfate ($MnSO_4$), alkaline iodide, and sulfuric acid (H_2SO_4) or check with an oxygen probe. Analyses may be delayed several hours if necessary, if samples are fixed or iced and stored in the dark.

6) Suspend duplicate paired clear and darkened bottles at the depth from which the samples were taken and incubate for at least 2 h, but never longer than it takes for oxygen-gas bubbles to form in the clear bottles or DO to be depleted in the dark bottles.

7) At the end of the exposure period, immediately determine DO as described above.

c. Calculations: The increase in oxygen concentration in the light bottle during incubation is a measure of net production which, because of the concurrent use of oxygen in respiration, is somewhat less than the total (or gross) production. The loss of oxygen in the dark bottle is used as an estimate of total plankton respiration. Thus:

Net photosynthesis = light bottle DO − initial DO

Respiration = initial DO − dark bottle DO

Gross photosynthesis = light bottle DO − dark bottle DO

Average results from duplicates.

1) Calculate the gross or net production for each incubation depth and plot:

mg carbon fixed/m^3
= mg oxygen released/L $\times$ 12/32 $\times$ 1000 $\times$ K

where K is the photosynthetic quotient (PQ), ranging from 1 to 2, depending on the nitrogen supply.[11,12]

Use the factor 12/32 to convert oxygen to carbon; under ideal conditions 1 mole of O_2 (32 g) is released for each mole of carbon (12 g) fixed.

2) Productivity is defined as the rate of production and generally is reported in grams carbon fixed per square meter per day. Determine the productivity of a vertical column of water 1 m square by plotting productivity for each exposure depth and graphically integrating the area under the curve.

3) Using the solar radiation profile and

photosynthesis rate during incubation adjust the data to represent phytoplankton productivity for the entire photoperiod. Because photosynthetic rates vary widely during the daily cycle,[13,14] do not attempt to convert data to other test circumstances.

3. Productivity, Carbon 14 Method

A solution of radioactive carbonate ($^{14}CO_3^{2-}$) is added to light and dark bottles that have been filled with sample as described for the oxygen method. After incubation in situ, collect the plankton on a membrane filter, treat with hydrochloric acid (HCl) fumes to remove inorganic carbon 14, and assay for radioactivity. The quantity of carbon fixed is proportional to the fraction of radioactive carbon assimilated.

This procedure differs from the oxygen method in that it affords a direct measurement of carbon uptake and measures only net photosynthesis.[15] It is basically more sensitive than the oxygen method, but fails to account for organic materials that leach from cells[16,17] during incubation.

a. Equipment and reagents:

1) *Pyrheliometer.*

2) *Submarine photometer.*

3) *BOD bottles and supporting apparatus:* See ¶s 2a1) and 2), above.

4) *Membrane-filtering device and 25-mm filters* with pore diameters of 0.22, 0.30, 0.45, 0.80, and 1.2 μm.

5) *Counting equipment for measuring radioactivity:* Scaler with end-window tube, gas flow meter, or liquid scintillation counter (see Section 7010D). The thin-window tube is the least expensive detector and, when used with a small scaler, provides acceptable data at modest cost.

6) *Fuming chamber:* Use a glass desiccator with a depth of about 1.4 cm conc HCl in desiccant chamber. The fuming chamber is recommended for filter decontamination.[18,19]

7) *Syringe or pipet,* nonmetallic.

8) *Chemical reagents:* See Sections 4500-CO_2 (Carbon Dioxide) and 2320 (Alkalinity).

9) *Radioactive carbonate solutions:*

a) *Sodium chloride dilution solution, 5%* NaCl (w/v): Add 0.3 g sodium carbonate (Na_2CO_3) and one pellet sodium hydroxide (NaOH) per liter. Use for marine studies only.

b) *Carrier-free radioactive carbonate solution,* commercially available in sealed vials having approximately 5 μCi ^{14}C/mL. Confirm absence of suspended and dissolved toxic metals[20] or filter and pass through an ion-exchange column.*

c) *Working solutions* with activities of 1, 5, and 25 μCi ^{14}C/2 mL. For studies of fresh water use carrier-free radioactive carbonate and for studies of marine water prepare by diluting carrier-free radioactive carbonate solution with NaCl dilution solution.

d) *Stock ampules:* Prepare ampules containing 2 mL of required working solution. Fill ampules and autoclave sealed ampules at 121°C for 20 min.[21]

b. Procedure:

1) Obtain a record of incident solar radiation for the photoperiod with a pyrheliometer.

2) Determine depth intervals for sampling and incubation as described above.

3) Use duplicate light and dark bottles at each depth. Also use dark bottles or bottles harvested at time zero. Fill bottles with sample, add 2 mL radioactive carbonate solution (using a nonmetallic pipet) to the bottom of each bottle, and mix thoroughly by repeated inversion. The concentration of carbon 14 should be approximately 10 μCi/L in relatively productive waters, to 100 μCi/L, or higher, in oligotrophic (open ocean) waters. To obtain statistical significance, have at least 1000 cpm in the filtered sample. Take duplicate samples at each depth to determine

---

*Chelex 100 or equivalent.

initial concentration of inorganic carbon (CO_2, HCO_3^-, and CO_3^{2-}) available for photosynthesis (see Section 4500-CO_2). For estuarine and marine samples, estimate total inorganic carbon concentrations with a simple titration procedure[22] and make initial temperature, salinity, and pH measurements.

4) Incubate samples for up to 4 h. If measurements are required for the entire photoperiod, overlap 4-h periods from dawn until dusk. A 4-h incubation period may be sufficient provided energy input is used as the basis for integrating incubation period to entire photoperiod. For incubation procedure, see ¶ 2b6) above.

5) After incubating remove sample bottles and immediately place in the dark. Filter unpreserved samples without delay. Avoid sample preservation to avoid lysing cells or determine extracellular products.

6) Filter two portions of each sample through a membrane filter, taking care that the largest pore size is consistent with quantitative retention of plankton. Although the 0.45-μm pore filter usually is adequate, determine the efficiency of sample retention immediately before analysis, with a wide range of pore sizes.[23,24] Apply approximately 30 kPa of vacuum during filtration. Excess vacuum may cause extensive cell rupture and loss of radioactivity through the membrane.[25] Use maximum sample volume consistent with rapid filtration (1 to 2 min), but do not clog filter.

7) Place membranes in HCl fumes for 20 min. Count filters as soon as possible, although extended storage in a desiccator is acceptable.

8) Determine radioactivity by counting with an end-window tube, windowless gas flow detector, or liquid scintillation counter.

9) Determine counting geometry of thin-window and windowless gas flow detectors.[26] Using three ampules of carbon 14, prepare a series of barium carbonate ($BaCO_3$) precipitates on tared 0.45-μm

membrane filters as directed below. The precipitates will contain the same amount of carbon 14 activity but will have different thicknesses ranging from 0.5 to 6.0 mg/cm^2. Dilute each ampule to 500 mL with a solution of 1.36 g Na_2CO_3/L CO_2-free distilled water. Pipet 0.5-mL portions into each of seven conical flasks containing 0, 0.5, 1.5, 2.5, 3.5, 4.5, and 5.5 mL, respectively, of a solution of 1.36 g Na_2CO_3/L CO_2-free distilled water. Add, respectively, 0.3, 0.6, 1.2, 1.8, 2.4, 3.0, and 3.6 mL 1.04% barium chloride ($BaCl_2$) solution. Let $BaCO_3$ precipitate stand 2 h with gentle swirling every half hour. Collect each precipitate on a filter (using an apparatus with a filtration area comparable to that of the samples). With suction, dry filters without washing; place in a desiccator for 24 h, weigh, and count. The counting rate increases exponentially with decreasing precipitate thickness. Extrapolate graphically (or mathematically) to zero precipitate thickness and multiply the zero-thickness counting rate by 1000 to correct for ampule dilution. This represents the amount of activity added to each sample bottle used to determine fraction of carbon 14 taken up in light and dark bottles.

c. Calculations:

1) Subtract the mean dark-bottle or time-zero sample count from the mean light-bottle counts for each replicate pair.

2) Determine the total dissolved inorganic carbon available for photosynthesis (carbonate, bicarbonate, and free CO_2) from pH and alkalinity measurements; make direct measurement of total CO_2 according to Section 4500-CO_2 or the methods described in the literature.[27-30]

3) Determine quantity of carbon fixed by using the following relationship:

mg carbon fixed/L =

$$\frac{\text{counting rate of filtered sample}}{\text{total activity added to sample}}$$

$$\times \frac{300}{\text{volume filtered}}$$

$\times$ mg/L initial inorganic carbon $\times$ 1.064†

4) Integrate productivity for the entire depth of euphotic zone and express as grams carbon fixed per square meter per day [see ¶ 2c2) above].

5) Using the solar radiation records and photosynthesis rates during incubation, adjust data to represent phytoplankton productivity for the entire photoperiod. If samples were incubated for less than the full photoperiod, apply a correction factor.

4. References

1. GUNDERSEN, K. 1973. *In-situ* determination of primary production by means of the new incubator, ISIS. *Helgolander wiss. Meeresunters.* 24:465.

2. BURRIS, R.H., F.J. EPPLING, H.B. WAHLIN & P.W. WILSON. 1942. Studies of biological nitrogen fixation with isotopic nitrogen. *Proc. Soil Sci. Soc. Amer.* 7:258.

3. NEESS, J.C., R.C. DUGDALE, V.A. DUGDALE & J.J. GOERING. 1962. Nitrogen metabolism in lakes. I. Measurement of nitrogen fixation with N^{15}. *Limnol. Oceanogr.* 7:163.

4. STEWART, W.D.P., G.P. FITZGERALD & R.H. BURRIS. 1967. *In situ* studies on N_2 fixation using the acetylene reduction technique. *Proc. Nat. Acad. Sci.* 58:2071.

5. STEWART, W.D.P., G.P. FITZGERALD & R.H. BURRIS. 1970. Acetylene reduction assay for determination of phosphorus availability in Wisconsin lakes. *Proc. Nat. Acad. Sci.* 66:1104.

6. GOLDMAN, C.R. 1968. Aquatic primary production. *Amer. Zoologist* 8:31.

7. ODUM, H.T. 1957. Primary production measurements in eleven Florida springs and a marine turtle-grass community. *Limnol. Oceanogr.* 2:85.

8. BEYERS, R.J. & H.T. ODUM. 1959. The use of carbon dioxide to construct pH curves for the measurement of productivity. *Limnol. Oceanogr.* 4:499.

9. GAARDER, T. & H.H. GRAN. 1927. Investigations of the production of plankton in Oslo Fjord. *Rapp. Proces-Verbaux. Reunions Cons. Perma. Int. Explor. Mer* 42:1.

10. STEEMAN-NIELSEN, E. 1952. The use of radioactive carbon (C-14) for measuring organic production in the sea. *J. Cons. Perma. Int. Explor. Mer* 18:117.

11. WILLIAMS, P.J. LEB., R.C.T. RAINE & J.R. BRYAN. 1979. Agreement between the ^{14}C and oxygen methods of measuring phytoplankton production: Reassessment of the photosynthetic quotient. *Oceanol. Acta* 2:411.

12. DAVIES, J.M. & P.J. LEB. WILLIAMS. 1984. Verification of ^{14}C and O_2 derived primary organic production using an enclosed system. *J. Plankton Res.* 6:457.

13. RYTHER, J.H. 1956. Photosynthesis in the ocean as a function of light intensity. *Limnol. Oceanogr.* 1:61.

14. FEE, E.J. 1969. A numerical model for the estimation of photosynthetic production, integrated over time and depth, in natural waters. *Limnol. Oceanogr.* 14:906.

15. STEEMAN-NIELSEN, E. 1964. Recent advances in measuring and understanding marine primary production. *J. Ecol.* 52(Suppl.):119.

16. ALLEN, M.B. 1956. Excretion of organic compounds by *Chlamydomonas*. *Arch. Mikrobiol.* 24:163.

17. FOGG, G.E. & W.D. WATT. 1965. The kinetics of release of extracellular products of photosynthesis by phytoplankton. *In* C. R. Goldman, ed. Primary Productivity in Aquatic Environments. Suppl. 18, Univ. California Press, Berkeley.

18. WETZEL, R.G. 1965. Necessity for decontamination of filters in C^{14} measured rates of photosynthesis in fresh waters. *Ecology* 46:540.

19. MCALLISTER, C.D. 1961. Decontamination of filters in the C^{14} method of measuring marine photosynthesis. *Limnol. Oceanogr.* 6:447.

20. CARPENTER, E.J. & J.S. LIVELY. 1980. Review of estimates of algal growth using ^{14}C tracer techniques. *In* P.G. Falkowski, ed. Primary Productivity in the Sea. Brookhaven Symp. Biol. No. 31. Plenum Press, New York, N.Y.

21. STRICKLAND, J.D.H. & T.R. PARSONS. 1968. A Practical Manual of Sea Water Analysis. Fish. Res. Board Can. Bull. No. 167. Queen's Printer, Ottawa, Ont.

†Correction for isotope effect.

22. PARSONS, T.R., Y. MAITA & C.M. LALLI. 1984. A Manual of Chemical and Biological Methods for Seawater Analysis. Pergamon Press, New York, N.Y.

23. LASKER, R. & R.W. HOLMES. 1957. Variability in retention of marine phytoplankton by membrane filters. *Nature* 180:1295.

24. HOLMES, R.W. & C.G. ANDERSON. 1963. Size fractionation of C^{14}-labelled natural phytoplankton communities. *In* C.H. Oppenheimer, ed. Symposium on Marine Microbiology. Charles C. Thomas, Springfield, Ill.

25. ARTHUR, C.R. & F.H. RIGLER. 1967. A possible source of error in the C^{14} method of measuring primary productivity. *Limnol. Oceanogr.* 12:121.

26. JITTS, H.R. & B.D. SCOTT. 1961. The deter-mination of zero-thickness activity in Geiger counting of C^{14} solutions used in marine productivity studies. *Limnol. Oceanogr.* 6:116.

27. SAUNDERS, G.W., F.B. TRAMA & R.W. BACHMANN. 1962. Publ. No. 8, Great Lakes Research Div., Univ. Michigan, Ann Arbor.

28. DYE, J.F. 1944. The calculation of alkalinities and free carbon dioxide in water by use of nomographs. *J. Amer. Water Works Assoc.* 36:859.

29. MOORE, E.W. 1939. Graphic determination of carbon dioxide and the three forms of alkalinity. *J. Amer. Water Works Assoc.* 31:51.

30. PARK, K., D.W. HOOD & H.T. ODUM. 1958. Diurnal pH variation in Texas bays and its application to primary production estimations. *Publ. Inst. Mar. Sci. Univ. Tex.* 5:47.

10300   PERIPHYTON*

10300 A.   Introduction

1. Definition and Significance

Microorganisms growing on stones, sticks, aquatic macrophytes, and other submerged surfaces are useful in assessing the effects of pollutants on lakes, streams, and estuaries. Included in this group of organisms, here designated periphyton,[1,2] are the zoogleal and filamentous bacteria, attached protozoa, rotifers, and algae, and the free-living microorganisms that swim, creep, or lodge among the attached forms.

Unlike the plankton, which often do not respond fully to the influence of pollution in rivers for a considerable distance downstream, the periphyton show dramatic responses immediately below pollution sources. Examples are the beds of *Sphaerotilus* and other "slime organisms" commonly observed in streams below discharges of organic wastes. Because the abundance and composition of the periphyton at a given location are governed by the water quality at that point, observations of their condition generally are useful in evaluating conditions in bodies of water.

The use of periphyton in assessing water quality often is hindered by the lack of suitable natural substrates at the desired sampling station. Furthermore, it often is difficult to collect quantitative samples from these surfaces. To circumvent these problems artificial substrates have been used to provide a uniform surface type, area, and orientation.[3]

2. References

1. ROLL, H. 1939. Zur terminologie des periphytons. *Arch. Hydrobiol.* 35:39.

*Approved by Standard Methods Committee, 1988.

2. YOUNG, O.W. 1945. A limnological investigation of periphyton in Douglas Lake, Michigan. *Trans. Amer. Microsc. Soc.* 64:1.

3. SLADECKOVA, A. 1962. Limnological investigation methods for the periphyton community. *Bot. Rev.* 28:286.

10300 B. Sample Collection

1. Station Selection

In rivers, locate stations a short distance upstream and at one or more points downstream from the suspected pollution source or intended study area. In large rivers, sample both sides of the stream. Because the effects of a pollutant depend on the assimilative capacity of the stream and on the nature of the pollutant, progressive changes in water quality downstream from the pollution source may be caused entirely by dilution and cooling—as in the case of nutrients, toxic industrial wastes, and thermal pollution—or by gradual mineralization of degradable organics. Cursory examination of shoreline and bottom periphyton growths downstream from an outfall may disclose conspicuous zones of biological response to water quality that will be useful in determining appropriate sites for sampling stations. When an intensive sampling program is not feasible, a minimum of two sampling stations, one in a reference area upstream from a pollution source and the other in the community downstream from the source, where complete mixing with the receiving water has occurred, will provide data on the periphyton community.

In lakes, reservoirs, lentic waters, and other standing-water bodies where zones of pollution may be arranged concentrically, locate stations in areas adjacent to a waste outfall and in unaffected areas.

2. Sample Collection

a. Natural substrates: Collect qualitative samples by scraping submerged stones, sticks, pilings, and other available substrates. Many devices have been developed to collect quantitative samples from irregular surfaces, but success rarely is achieved.

b. Artificial substrates: The most widely used artificial substrate is the standard, plain, 25- by 75-mm glass microscope slide, but other materials such as clear vinyl plastic also are suitable. Do not change substrate type during a study because colonization varies with substrate. In small, shallow streams and in the littoral regions of lakes where light is transmitted to the bottom, place slides or other substrates in frames anchored to the bottom. In large, deep streams or standing-water bodies where turbidity varies widely, place slides vertically with the slide face at right angles to the prevailing current. A floating rack, as shown in Figure 10300:1, is suitable. Expose several slides for each type of analysis to assure collecting sufficient material and to determine variability in results caused by normal differences in colonization of individual slides. In addition to effects of pollutants, length of substrate exposure and seasonal changes in temperature and other natural environmental conditions may have a profound effect on sample composition. No community on an artificial substrate is completely representative of the natural community.

Place, expose, and handle all artificial substrate samplers in conditions as nearly identical as possible, whether they are replicate samplers at a particular sampling location or samplers at different locations. Sampler type and/or construction cause changes in surrounding physical conditions that in turn affect periphyton growth. Variations of 10 to 25% between sample rep-

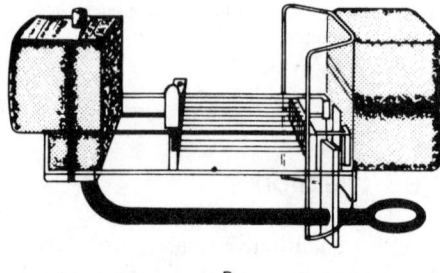

B

Figure 10300:1. Periphyton sampler. Floating slide rack constructed of clear vinyl plastic and styrofoam, used in streams and lakes. Source: PATRICK, R., M.H. HOHN & J.H. WALLACE. 1954. A new method of determining the pattern of the diatom flora. *Bull. Philadelphia Acad. Natur. Sci.* 259:1.

licates are not uncommon. Therefore, to reduce sampling error and increase interpretive power, reduce the magnitude of all possible test variables and use sufficient replication.

c. Exposure period: Colonization on clean slides proceeds at an exponential rate for the first 1 or 2 weeks and then slows. Because exposures of less than 2 weeks may result in very sparse collections, and exposures of more than 2 weeks may result in loss of material due to sloughing, 2 weeks usually constitutes the optimum sampling interval during the summer. This exposure period precludes collecting sexually mature thalli of larger, slow-growing filamentous algae such as *Cladophora* and *Stigeoclonium*. To obtain optimum growth during the winter, use a longer exposure period. For the most exacting work, determine the optimum exposure period by pretesting colonization over a period of about 6 weeks.

Secondary problems associated with macroinvertebrate infestation and grazing may occur, often within 7 to 14 d. To reduce the confounding influence of grazing, increase substrate sampling area and expose for 7 to 10 d.

3. Sample Preservation

Preserve samples that are taken for counting and identification in 5% neutralized formalin or merthiolate (see Section 10200B.4).

Preserve slides intact in bottles of suitable size or scrape into containers in the field. Air-dry slides for dry and ash-free dry weight in the field and store in a 3.0- × 7.7-cm glass bottle. Place slides for chlorophyll analyses in acetone in the field or collect and freeze with trichlorotrifluoroethane* or CO_2 and hold on dry ice until returned to the laboratory. Store all samples in the dark.

4. Bibliography

COOKE, W.B. 1956. Colonization of artificial bare areas by microorganisms. *Bot. Rev.* 22:613.

HOHN, M.H. 1966. Artificial substrate for benthic diatoms—collection, analysis, and interpretation. *In* K.W. Cummings, C.A. Tryon, Jr., & R.T. Hartman, eds. Organism-Substrate Relationships in Streams. Spec. Publ. No. 4. p. 87. Pymatuning Lab. Ecology. Univ. Pittsburgh, Pittsburgh, Pa.

KEVERN, N.R., J.L. WILHM & G.M. VAN DYNE. 1966. Use of artificial substrata to estimate the productivity of periphyton communities. *Limnol. Oceanogr.* 11:499.

ARTHUR, J.W. & W.B. HORNING. 1969. The use of artificial substrates in pollution surveys. *Amer. Midland Natur.* 82:83.

TIPPETT, R. 1970. Artificial surfaces as a method of studying populations of benthic micro-algae in fresh water. *Brit. Phycol. J.* 5:187.

ERTL, M. 1971. A quantitative method of sampling periphyton from rough substrates. *Limnol. Oceanogr.* 16:576.

ANDERSON, M.A. & S.L. PAULSON. 1972. A simple and inexpensive woodfloat periphyton sampler. *Progr. Fish-Cult.* 34:225.

NORTH AMERICAN BENTHOLOGICAL SOCIETY. 1974–1985. (Annual) Current and Select Bibliographies on Benthic Biology. Springfield, Ill.

NEROZZI, A. & P. SILVER. 1983. Periphytic community analysis in a small oligotropic lake. *Proc. Penn. Acad. Sci.* 57:138.

WETZEL, R., ed. 1983. Periphyton of Freshwater Ecosystems. Developments in Hydrobiology

*Freon or equivalent.

17. Dr. W. Jung BV Publishers, The Hague, The Netherlands.

HAMILTON, P.B. & H.C. DUTHIE. 1984. Periphyton colonization of rock surfaces in a boreal forest stream studied by scanning electron microscopy and track autoradiography. *J. Phycol.* 20:525.

NIELSEN, T.S., W.H. FUNK, H.L. GIBBONS & R.M. DUFFNER. 1984. A comparison of periphyton growth on artificial and natural substrates in the Upper Spokane River, Washington, USA. *Northwest Sci.* 58:243.

PIP, E. & G.G.C. ROBINSON. 1984. A comparison of algal periphyton composition on 11 species of submerged macrophytes. *Hydrobiol. Bull.* 18:109.

POULIN, M., L. BERARD-THERRIAULT & A. CARDINAL. 1984. Benthic diatoms from hard substrates of marine and brackish waters of Quebec Canada 3. Fragilarioideae, Fragilariales, Fragilariaceae. *Nat. Can.* (Que). 111:349.

STEVENSON, R.J. 1984. How currents on different sides of substrates in streams affect mechanisms of benthic algal accumulation. *Int. Rev. Ges. Hydrobiol.* 69:241.

VYMAZAL, J. 1984. Short-term uptake of heavy metals by periphytic algae. *Hydrobiologia* 119:171.

AUSTIN, A. & J. DENISEGER. 1985. Periphyton community changes along a heavy metals gradient in a long narrow lake. *Environ. Exper. Bot.* 25:41.

FLOWER, R.J. 1985. An improved epilithon sampler and its evaluation in two acid lakes. *Brit. Phycol. J.* 20:109.

LAMBERTI, G.A. & V.H. RESH. 1985. Comparability of introduced tiles and natural substrates for sampling lotic bacteria, algae, and macroinvertebrates. *Freshwater Biol.* 15:21.

PIEKARCZYK, R. & E. MCARDLE. 1985. Pioneer colonization and interaction of photosynthetic and heterotrophic microorganisms on an artifical substrate of polyurethane foam in E.J. Beck Lake, Illinois, USA. *Trans. Ill. State Acad. Sci.* 78:81.

10300 C.   Sample Analysis

1. Sedgwick-Rafter Counts

Remove periphyton from slides with a razor blade and rubber policeman. Do not include periphyton on the edges of the slide. Disperse scrapings in 100 mL or other suitable volume of preservative with vigorous shaking, or use a blender. Transfer a 1-mL portion to a Sedgwick-Rafter cell, and make a strip count as described in Section 10200F.2*a*. If material in the Sedgwick-Rafter cell is too dense to count directly, discard and replace with a diluted sample.

Express the counts as cells or filaments per square millimeter of substrate area, calculated as follows:

1) Cells/mL suspended scrapings

$$= \frac{\text{actual count/strip}}{\text{volume of 1 strip, mL}}$$

2) Cells/mm^2 slide surface

$$= \text{cells/mL suspended scrapings}$$
$$\times \frac{\text{total volume of scrapings}}{\text{area of slide or slides, mm}^2}$$

2. Inverted Microscope Method Counts

Using an inverted microscope for periphyton counts permits magnifications higher than those possible with the Sedgwick-Rafter cell. Prepare scrapings as in 10300C.1 and transfer a measured portion, after serial dilution if necessary, into a standardized plankton sedimentation chamber. After a suitable period of settling (see Section 10200C.1), count organisms in the settling chamber by counting all organisms within a known number of strips or random fields. Calculate algal density per unit area of substrate as follows:

$$\text{Organisms/mm}^2 = \frac{N \times A_t \times V_t}{A_c \times V_s \times A_s}$$

where:

N = number of organisms counted,

A_t = total area of chamber bottom, mm^2,

V_t = total volume of original sample suspension, mL,

A_c = area counted (strips or fields), mm^2,

V_s = sample volume used in chamber, mL, and

A_s = surface area of slide or substrate, mm^2.

Separation of periphyton from silt and detritus may be enhanced by adding a drop or less of a saturated iodine solution to the counting chamber just before counting. This is especially useful when Chlorophyta are the predominant organisms because iodine stains starch food reserves blue. Iodine can be added even to preserved samples.

3. Diatom Species Proportional Counts

Preparation of permanent diatom mounts from periphyton samples differs from preparation of mounts from plankton samples because of the need to remove extracellular organic matter (such as gelatinous materials). If this is not removed it will produce a thick brown or black carbonaceous deposit on the cover glass when the sample is incinerated. Decompose organic substances by oxidation with ammonium persulfate or with HNO$_3$ or 30% H$_2$O$_2$ and K$_2$Cr$_2$O$_7$ (see Section 10200D.3) before mounting sample. To oxidize with persulfate place approximately 5 mL sample in a disposable 10-mL vial. Let stand 24 h, withdraw supernatant liquid by aspiration, replace with a 5% solution of (NH$_4$)$_2$S$_2$O$_8$, and mix thoroughly. Do not exceed a total volume of 8 mL. Heat vial to approximately 90°C for 30 min. Let stand 24 h, withdraw supernatant liquid, and replace with distilled water. After three changes of distilled water, with a disposable pipet transfer a drop of the diatom suspension to a cover glass, evaporate to dryness, and prepare and count a mount as described for plankton (Section 10200).

Count as least 500 frustules and express results as organisms per square millimeter.

4. Stained Sample Preparation and Counting

Staining periphyton samples permits distinguishing algae from detritus and "live" from "dead" diatoms. This distinction is especially important because periphyton often acts as a graveyard for dead diatoms of planktonic as well as periphytic origin. In this method all algal components of periphyton may be studied in one preparation, without sacrificing detailed diatom taxonomy.[1] It yields permanent slides for reference collections.

Thoroughly mix preserved samples in the preservative solution by using a blender for a few seconds. Prepare acid fuchsin stain by dissolving 1 g acid fuchsin in 100 mL distilled water, adding 2 mL glacial acetic acid, and filtering. Place a measured sample in a centrifuge tube with 10 to 15 mL acid fuchsin stain. Mix sample and stain several times during a 20-min staining period; centrifuge at 1000 g for 20 min.

Decant stain, being careful not to disturb sediment, or siphon off supernatant. Add 10 to 15 mL 90% propanol, mix, centrifuge for 20 min, and decant supernatant. Repeat using two washes of 100% propanol and one wash of xylene. Centrifuge, decant xylene, and add fresh xylene. At this stage, store sample in well-sealed vials or prepare slides.

Slides for quantitative periphyton examinations require random dispersion of a known amount of xylene suspension. Use a microstirrer to break up clumps of algae before removing sample portion from xylene suspension. Count a number of drops of suspended sample into a thin ring of mounting medium* on a slide. Mix the xylene suspension and medium with a spatula until the xylene has evaporated. Warm the

---

*Hyrax, Custom Research and Development, Inc., 8500 Mt. Vernon Rd., Auburn, Calif., or equivalent.

slide on a hot plate at 45°C and cover sample with a cover slip.

Count organisms on the prepared slides using the magnification most appropriate to the desired level of taxonomic identification. Count strips or random fields. Calculate algal density per unit area of substrate:

$$\text{Organisms/area sampled} = \frac{N \times A_t \times V_t}{A_c \times V_s \times A_s}$$

where the terms are as defined in 10300C.2.

5. Dry and Ash-Free Weight

Collect at least three replicate slides for weight determinations.[2] Slides air-dried in the field can be stored indefinitely if protected from abrasion, moisture, and dust. While weights can be obtained from the material used for chlorophyll determinations, preferably use slides expressly designated for dry and ash-free weight analysis.

a. Equipment:

1) *Analytical balance,* with a sensitivity of 0.1 mg.

2) *Drying oven,* double-wall, thermostatically controlled to within ± 1°C.

3) *Electric muffle furnace* with automatic temperature control.

4) *Crucibles,* porcelain, 30-mL capacity.

5) *Single-edge razor blades or rubber policeman.*

b. Procedure:

1) If dry and ash-free weights are to be obtained from the material used for chlorophyll determinations, combine the particulate matter and the acetone extract from each slide, evaporate acetone in a hood on a steam bath or in an explosion-proof oven, dry to constant weight at 105°C, and ignite for 1 h at 500°C. If weights are to be obtained from field-dried material, re-wet dried material with distilled water and remove from slides with a razor blade or rubber policeman. Place scrapings from each slide in a separate pre-washed, prefired, tared crucible; dry to constant weight at 105°C; cool in a desiccator and weigh; and ignite for 1 h at 500°C.

2) Re-wet ash with distilled water and dry to constant weight at 105°C. This reintroduces water of hydration of clay and other minerals, which is not driven off at 105°C but is lost during ashing. If not corrected for, this water loss will be recorded as volatile organic matter.[3]

c. Calculations: Calculate the mean weight from the slides and report as dry weight and ash-free weight per square meter of exposed surface. If 25- by 75-mm slides are used, then

$$\text{g/m}^2 = \frac{\text{g/slide (average)}}{0.003\ 75}$$

6. Chlorophyll and Pheophytin

The chlorophyll content of attached communities is a useful index of the phytoperiphyton biomass. Because quantitative chlorophyll determinations require the collection of periphyton from a known surface area, use artificial substrates. Extract the pigments with aqueous acetone (see Section 10200G) and use a spectrophotometer or fluorometer for analysis. If immediate pigment extraction is not possible, samples may be stored frozen for as long as 30 d if kept in the dark.[4] The ease with which chlorophylls are removed from cells varies considerably with different algae; to achieve complete pigment extraction disrupt the cells mechanically with a grinder, blender, or sonic disintegrator, or freeze them. Grinding is the most rigorous and effective of these methods.

The Autotrophic Index (AI) is a means of determining the trophic nature of the periphyton community (see Section 10200G). It is calculated as follows:

$$\text{AI} = \frac{\text{Biomass (ash-free weight of organic matter), mg/m}^2}{\text{Chlorophyll } a,\ \text{mg/m}^2}$$

Normal AI values range from 50 to 200; larger values indicate heterotrophic associations or poor water quality. Nonviable organic material affects this index. Depending on the community, its location and growth habit, and method of sample collection, there may be large amounts of nonliving organic material that may inflate the numerator and produce disproportionately high AI values. Nonetheless, the AI is a useful means of describing changes in periphyton communities between sampling locations.

a. Equipment and reagents: See Section 10200G.

b. Procedure: In the field, place individual glass microscope slides used as substrates directly into 100 mL of a mixture of 90% aqueous acetone and 10% saturated $MgCO_3$ solution. Immediately store on dry ice in the dark. (NOTE: Vinyl plastic is soluble in acetone. If vinyl plastic is used as the substrate, scrape periphyton from it before solvent extraction.) If extraction cannot be carried out immediately, freeze samples in the field and keep frozen until processed.

Rupture cells by grinding in a tissue homogenizer and steep in acetone for 24 h in the dark at or near 4°C.

To determine pigment concentration, follow the procedures given in Section 10200G.

c. Calculation: After determining pigment concentration in the extract, calculate amount of pigment per unit surface area of sample as follows:

mg chlorophyll a/m^2

$$= \frac{C_a \times \text{volume of extract, L}}{\text{area of substrate, m}^2}$$

where:
   C_a is as defined in Section 10200G.

7. References

1. OWEN, B.B., JR. 1977. The effect of increased temperatures on algal communities of artificial stream channels. Ph.D. dissertation, Univ. Alberta, Edmonton.
2. NEWCOMBE, C.L. 1950. A quantitative study of attachment materials in Sodon Lake, Michigan. *Ecology* 31:204.
3. NELSON, D.J. & D.C. SCOTT. 1962. Role of detritus in the productivity of a rock outcrop community in a piedmont stream. *Limnol. Oceanogr.* 7:396.
4. GRZENDA, A.R. & M.L. BREHMER. 1960. A quantitative method for the collection and measurement of stream periphyton. *Limnol. Oceanogr.* 5:190.

8. Bibliography

EATON, J.W. & B. MOSS. 1966. The estimation of numbers and pigment content in epipelic algal populations. *Limnol. Oceanogr.* 11:584.
MOSS, B. 1968. The chlorophyll *a* content of some benthic algal communities. *Arch. Hydrobiol.* 65:51.
WETZEL, R. ed. 1983. Periphyton of Freshwater Ecosystems. Developments in Hydrobiology 17. Dr. W. Jung BV Publishers, The Hague, The Netherlands.
DELBECQUE, E.J.P. 1985. Periphyton on Nymphaeids: An evaluation of methods and separation techniques. *Hydrobiologia* 124:85.
TREES, C.C., M.C. KENNICUTT & J. M. BROOKS. 1985. Errors associated with the standard fluorometric determination of chlorophylls and phaeopigments. *Mar. Chem.* 17:1.

10300 D.   Productivity

The productivity of periphyton communities is a function of water quality, substrate, and seasonal patterns in temperature and solar illumination. It may be estimated from temporal changes in standing crop (biomass) or from the rate of oxygen evolution or carbon uptake.[1]

1. Biomass Accumulation

a. Ash-free dry weight: The accumulation

rate of organic matter on artificial substrates by attachment, growth, and reproduction of colonizing organisms has been used widely to estimate the productivity of streams and reservoirs.[2,3] To use this method, expose several replicate clean substrates for a predetermined period, scrape the accumulated material from the slides, and ash as described previously.

$$P = \frac{\text{mg ash-free weight/slide}}{tA}$$

where:

P = net productivity, mg ash-free weight/ m^2/d,

t = exposure time, d, and

A = area of a slide, m^2.

Obtain estimates of seasonal changes in standing crop of established communities by placing many replicate substrates at a sampling point and then retrieving a few at a time at regular intervals. The recommended collection interval ranges from 2 to 4 weeks for a year or longer.[2] Gain in ash-free weight per unit area from one collection period to the next is a measure of net production.

b. *ATP estimates:* Measurement of adenosine triphosphate (ATP) has been used in recent years to estimate microbial biomass in water. This technique is applicable to periphyton.[4] It provides an additional tool for assessing the magnitude and rate of biomass accumulation on substrates in natural waters. At present, the procedure should be limited to communities colonizing artificial substrates.

1) Equipment and reagents—See Section 10200H.5a.

2) Procedure—Either scrape periphyton from an exposed artificial substrate or, if standard glass microscope slides are used, place them in polyethylene slide mailers containing preheated (99°C) Tris buffer. Immerse in a boiling water bath for 10 min to extract ATP. If samples are not assayed immediately, freeze at −25°C; they may be stored in a freezer for up to several months. Complete analysis as directed in Section 10200H.5b. Slides exposed in waters containing high turbidity may collect substantial amounts of particulates including clays. ATP sorbs to these materials; the sorption results in a quenching effect.

3) Calculations—See Section 10200H.5c.

2. Standing Water Productivity Measured by Oxygen Method

Hourly and daily rates of oxygen evolution and carbon uptake by periphyton growing in standing water can be studied by confining this community briefly in bottles, bell jars, or other chambers. In contrast, the metabolism of organisms in flowing water is highly dependent on current velocity and cannot be determined with precision under static conditions. Productivity estimates for flowing waters and those for standing waters present different problems; therefore, separate procedures are given.

Productivity and respiration of epilithic and epipelic periphyton in littoral regions of lakes and ponds can be determined by inserting transparent and opaque bell jars or open-ended plastic chambers into substrata along transects perpendicular to the shoreline.[5,6] Chambers are left in place for one-half the photoperiod. The DO concentration in a chamber is determined at the beginning and end of the exposure period. Gross productivity is the sum of the net gain in DO in the transparent chamber and the oxygen used in respiration. Values obtained are doubled to determine productivity for the entire photoperiod.

Failure to account for changes in DO in chambers caused by plankton photosynthesis and respiration may cause serious errors in the estimates of periphyton metabolism. It is essential that these values be obtained at the time the periphyton is studied by using the light- and dark-bottle method (see Section 10200I).

a. Equipment and reagents:

1) *Clear and darkened glass or plastic* chambers,* approximately 20 cm in diameter and 30 cm high, with a median lateral port, sealed with a serum bottle stopper for removal of small water samples for DO analyses or for the insertion of an oxygen probe. Fit the chamber with a small, manually operated, propeller-shaped stirring paddle.

2) *Dissolved oxygen probe, or equipment and reagents required for Winkler dissolved oxygen determinations:* See Section 4500-O.

b. Procedure: At each station place both a transparent and an opaque chamber over the substrate at sunrise or noon and leave in place for one-half the photoperiod. In extremely productive environments or to define the hourly primary productivity changes throughout the day, use incubation periods shorter than one-half the photoperiod. The minimum incubation period giving reliable results is 2 h. Determine DO concentration at the beginning of the incubation period.

Include a set of Gaarder-Gran light- and dark-bottle productivity and respiration measurements with each set of chambers to obtain a correction for phytoplankton metabolism. Incubate for the same time period as the chambers. See Section 10200I.

At end of exposure period, carefully mix the water in the chambers and determine DO concentration.

c. Calculations: When the exposure period is one-half of the photoperiod, calculate gross primary productivity of the periphyton community as:

$$P_G = \frac{2\,[V_c(C'_{fc} - C'_{ic}) + V_o(C'_{io} - C'_{fo})]}{A}$$

where:

P_G = gross production, mg $O_2/m^2/d_{12h}$,
V_c = volume of clear chamber, L,
C'_{fc} and
$\quad C'_{ic}$ = final and initial concentrations, respectively, of DO in the clear chamber, mg/L, corrected for phytoplankton metabolism,
V_o = volume of opaque chamber, L,
C'_{io} and
$\quad C'_{fo}$ = initial and final concentrations, respectively, of DO in the opaque chamber, mg/L, corrected for phytoplankton metabolism, and
A = substrate area, m².

Correct for the effects of phytoplankton metabolism in the overall oxygen change in the clear chamber by the following equations:

$$C'_{fc} = C_{fc} - C_{flb}$$
$$C'_{ic} = C_{ic} - C_{ilb}$$
$$C'_{fo} = C_{fo} - C_{fdb}$$
$$C'_{io} = C_{io} - C_{idb}$$

where:

C_{fc} = final DO concentration in clear chamber, mg/L,
C_{flb} = final DO concentration in light bottle, mg/L,
C_{ic} = initial DO concentration in clear chamber, mg/L,
C_{ilb} = initial DO concentration in light bottle, mg/L,
C_{fo} = final DO concentration in opaque chamber, mg/L,
C_{fdb} = final DO concentration in dark bottle, mg/L,
C_{io} = initial DO concentration in opaque chamber, mg/L, and
C_{idb} = initial DO concentration in dark bottle, mg/L.

Calculate periphyton community respiration by:

$$R = \frac{24\,V_0(C'_{io} - C'_{fo})}{tA}$$

---

*Plexiglas or equivalent.

where:

R = community respiration, mg $O_2/m^2/d_{24h}$, and

t = length of exposure, h.

Determine the net periphyton community (P_N) as the difference:

$$P_N = P_G - R$$

If the incubation time is different from one-half the photoperiod, modify the daily gross production calculation as follows:

$$P_G = \frac{t_p[V_c(C'_{fc} - C'_{ic}) + V_o(C'_{io} - C'_{fo})]}{tA}$$

where:

t_p = length of the daily photoperiod, h.

Community respiration and net production calculations for incubation periods other than one-half the photoperiod are not changed.

3. Standing Water Productivity Measured by Carbon 14 Method

The approach is similar to that described above for the oxygen method. Transparent and opaque chambers are placed over the substrate, carbon 14-labeled Na_2CO_3 is injected into the chamber by syringe, mixed well, and allowed to incubate with the periphyton for one-half the photoperiod. The concentration of dissolved inorganic carbon available for photosynthesis is determined by titration. At the end of the incubation period, the periphyton is removed from the substrate and assayed for carbon 14.[5,7]

a. Equipment and reagents:

1) *Incubation chamber:* See Section 10300D.2a.

2) *Special equipment and reagents:* See Section 10200I.

3) *Carbon 14-labeled solution of sodium carbonate,* having a specific activity of approximately 10 μCi/mL.

4) *Other equipment and reagents:* See Section 4500-CO_2.

b. Procedure: At each station place a transparent and opaque chamber over the substrate and add approximately 10 μCi carbon 14/L of chamber volume. Mix water in the chambers well, taking care to avoid disturbing the periphyton. Determine concentration of dissolved inorganic carbon as described in Section 2320. At end of exposure period, remove surface centimeter of periphyton enclosed in the chamber, freeze, and store frozen in a vacuum desiccator.

Immediately before analysis, expose sample to fumes of HCl for 10 to 15 min to drive off all inorganic carbon 14 retained in the periphyton. Combust sample (or portion) by the Van Slyke method[7] and assay radioactivity by one of the following methods: (*a*) flush CO_2 produced by combustion into a gas-flow counter or electrometer; (*b*) take it up in a 0.1N solution of Na_2CO_3, precipitate as $BaCO_3$ on a membrane filter, and count with an end-window tube; or (*c*) assay as the Na_2CO_3 solution by the liquid scintillation technique.

c. Calculations:

$$P_N = \frac{\text{activity in sample}}{\text{activity added}}$$
$$\times \frac{\text{dissolved inorganic carbon}}{\text{area of substrate}} \times 1.064$$

where:

P_N = net productivity for exposure period, and

1.064 = correction for isotope effect.

4. Flowing Water Productivity Measured by Oxygen Method

Primary productivity of the periphyton community in a stream or river ecosystem can be related to demand changes in DO. These changes are the integrated effects of photosynthesis, affected by light levels and turbidity, that is carried out during the photoperiod by stream plankton, periphyton, and the submerged portions of macrophytes. Respiration results from metabolism of plant communities, aquatic animals, and attached and free-floating microbial heterotrophs. Water depth, turbulence, and water temperature all influence the process of reaeration. Oxygen also can enter by accrual of groundwater and surface change. Daily fluctuations in photosynthetic production of oxygen are imposed on the relatively steady demand of respiratory activity. However, this latter process may fluctuate greatly in streams receiving a significant load of organic wastes, particularly under intermittent loads such as oxygen demand from urban stormwater runoff. Respiration rates also may vary diurnally under certain conditions, but the factors involved are not well understood.

The rate of change in stream DO (q) in grams per cubic meter per hour is represented by the following function of the photosynthetic rate (p), respiration (r), reaeration (d), and accrual from groundwater inflow and surface runoff (a):[8]

$$q = p - r + d + a$$

If the equation is multiplied through by depth in meters (z), the resulting values are in terms of grams oxygen per square meter per hour. Figure 10300:2 illustrates this conceptual relationship between q, primary productivity, and respiration of the stream plant community.

The procedure measures the time-variable oxygen concentrations in a stream over a 24-h period. Compensations are made for oxygen changes due to physical factors (accrual and reaeration) and the rate of oxygen change due to biological activity that is separated into components due to respiration and primary production. The metabolic rates are the sum of the activity of the entire stream community. Planktonic productivity and respiration

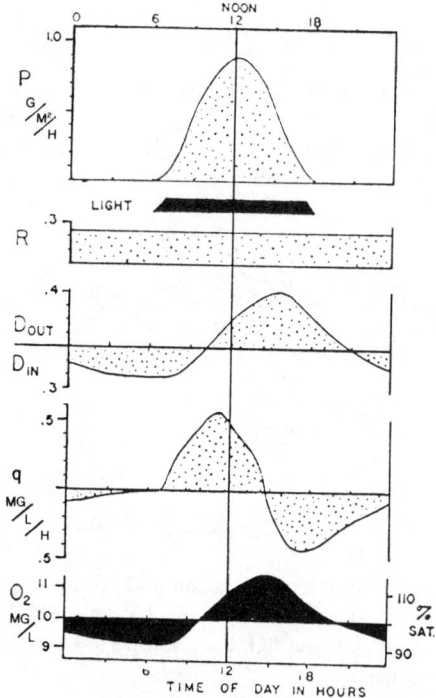

Figure 10300:2. Component processes in the oxygen metabolism of a section of a hypothetical stream during the course of a cloudless day. Production (P), respiration (R), and diffusion (D) are given on an areal basis. The combined effect of these rate processes for a stream 1 m deep is given in mg/L/h (q). The actual oxygen values that would result in a stream with a long homogeneous community are given in the lower-most curve. Source: ODUM, H.T. 1956. Primary production in flowing waters. *Limnol. Oceanogr.* 1:102.

can be separated easily from overall community activity by the use of the light- and dark-bottle oxygen technique (see Section 10200I). However, in most small streams planktonic production is insignificant. The component of production and respiration due to macrophytes is very difficult to separate from periphytic metabolic activity in systems where vascular plants are common.

Because periphyton attach to plant surfaces as well as nonliving substrates, radiotracer techniques are required to separate the component of production due to macrophytes from that due to attached algae.[9] When vascular plants are present use techniques discussed in Section 10400 to estimate their contribution to net primary productivity.

Respiration by fish and benthic fauna also is difficult to quantitate directly and usually is not separated from periphyton respiration. If compartmentalized animal metabolism is required, calculate this contribution from laboratory respiration rates extrapolated to the field situation based on animal population sizes.[10,11]

Estimate primary productivity in flowing water by either the free water demand method or the Thomas-O'Connell[12] chamber method. The first does not introduce artificiality to the system; however, it is difficult to separate the components of metabolic activity except for the contribution due to plankton. The chamber method measures periphyton activity alone.[13–16]

Depending on the hydrologic characteristics of the stream system, accrual and reaeration may be significant. Accrual can be accounted for by simple mixing equations if estimates of the accrued flow and its oxygen concentration are known. In practice, select for study reaches that do not incur significant accrual. Measure reaeration rates either directly[13–16] or by estimation from physical and hydrodynamic features of the stream itself.[15,16]

a. Equipment:

1) *BOD bottles,* for light- and dark-bottle measurements. See Section 10200I.

2) *DO meter and probe* for measurement of DO.

3) *Bottom chamber,* 60 × 20 × 10 cm, with 32-cm lengthwise dividing baffle, rheostat-controlled submersible pump, temperature thermistor, and DO probe.[12] Use clear and opaque plastic sleeves for covering chamber and petri dishes or other means of placing periphyton within chambers.

4) *Current meter,* capable of detecting water current velocities ranging from 0.03 to 3 m/s in water depths as shallow as 0.3 m.

5) *Tape measure (30 m) and depth staff,* or similar equipment, as required to measure stream cross sections.

6) *Fluorometer,* capable of detecting fluorescent dye concentration at 0.5 to 100 μg/L (required only if direct measurement of reaeration is made).

7) *Liquid scintillation counter,* capable of sensitive detection of ^{85}Kr and ^{3}H (required only if direct measurement of reaeration is made).

b. Procedure:

1) Light- and dark-chamber method— Grow samples of typical periphyton communities on artificial substrate or collect natural material. Transfer identical portions to both clear and opaque chambers, taking care to use sufficient periphyton to make the ratio of chamber volume to periphyton area equivalent to the ratio of stream volume to periphyton substrate area. Measure current in the stream and match the circulation rate in the clear and opaque chambers to the current. Measure DO concentrations initially in both clear and opaque chambers and after 1 to 3 h to estimate the rate of oxygen increase or decrease. Make concurrent measurements of phytoplankton activity using light- and dark-bottle techniques as described in Section 10200I.2. Incubate light and dark bottles for the same time interval as the chambers.

Make several measurements during the photoperiod to define daily primary productivity. In addition, collect sufficient natural substrate samples of the study reach to estimate periphyton biomass (see Section 10300B). At end of incubation period harvest enclosed periphyton and determine ash-free biomass (see Section 10300B).

2) *Free-water diurnal curve methods*— Measure, hourly or continuously, DO concentration and water temperature for a 24-h period at one or two stations, depending on stream conditions, precision desired, and availability of equipment. If similar conditions exist for some distance upstream from the reach being studied, diurnal measurements of DO at a single station are sufficient to determine productivity. Where upstream conditions are significantly different from those in the reach being studied, make measurements at the upstream and downstream limits of the reach.

If the single-station method is used, measure depth at several points along the study reach to define average depth. Map and/or make physical surveys to estimate magnitude of possible sources of accrual via effluents or tributary streams and springs. If the two-station method is used, measure the wetted cross-sectional stream area as well as current velocity at several points to define flow (in cubic meters per second) and average cross-sectional area. Correct for phytoplankton activity by light- and dark-bottle measurements (see Section 10200I.2).

3) *Direct measurement of reaeration*[14]— Under special circumstances it may be desirable to estimate reaeration directly although the results may not be more accurate than those of the empirical formulations usually used. The tracer gas technique is satisfactory, but is difficult and requires sophisticated equipment not routinely available. Use this method with care and with full recognition of its restrictions. Depending on stream flow, release 10 to 250 μCi ^{85}Kr with 5 to 125 μCi ^{3}H at the upstream end of the reach together with sufficient fluorescent dye to produce a concentration of 10 μg/L when completely mixed across the river cross section. Make fluorometric measurements at the downstream end of the reach until the dye peak

appears, then collect water samples to measure the ^{85}Kr/^{3}H ratio by liquid scintillation techniques. Record time of travel for the dye peak from the injection point.

c. Calculations:

1) *Chamber method*—Calculation is analogous to that used for the bell jar technique discussed in Section 10300D.2.

$$P_n = \frac{V_c(C'_{fc} - C'_{ic})\, B}{t W_c}$$

where:

P_n = hourly rate of net primary production, mg O_2/m^2/h,

V_c = volume of clear chamber, L,

B = average periphyton biomass estimated for the study reach, mg/m^2,

t = incubation period, h,

W_c = total biomass of periphyton contained in clear chamber, mg,

C'_{fc} = final oxygen concentration in clear chamber, corrected for phytoplankton metabolism, mg/L:

   C'_{fc} = $C_{fc} - C_{flb}$

   C_{fc} = final DO in clear chamber,

   C_{flb} = final DO in light bottle, and

C'_{ic} = initial oxygen concentration in clear chamber corrected for light-bottle measurement, mg/L:

   C'_{ic} = $C_{ic} - C_{ilb}$

   C_{ic} = initial DO in clear chamber, and

   C_{ilb} = initial DO in light bottle.

$$r = \frac{V_o(C'_{io} - C'_{fo})\, B}{t W_o}$$

where:

r = hourly periphyton respiration rate, mg O_2/m^2/h,

V_o = volume of opaque chamber, L,

B = average periphyton biomass for the study reach, mg/m^2,

W_o = total biomass of periphyton contained in opaque chamber, mg,

C'_{io} = initial oxygen concentration in opaque chamber, corrected for phytoplankton respiration, mg/L:

   C'_{io} = $C_{io} - C_{idb}$

C_{io} = initial DO in opaque chamber, mg/L,

C_{idb} = initial DO in dark bottle, mg/L, and

C'_{fo} = final oxygen concentration in opaque chamber, mg/L:

$C'_{fo} = C_{fo} - C_{fdb}$

C_{fo} = final DO in opaque chamber, mg/L, and

C_{fdb} = final DO in dark bottle, mg/L.

For each pair of chamber measurements,

$$P_g = P_n + r$$

where:

P_g = hourly gross periphytic primary production, mg $O_2/m^2/h$.

P_G is the area under the curve of primary production per hour through the photoperiod, mg $O_2/m^2/d$ (see Figure 10300:3). Also,

$$R = \left(\frac{\sum_{1}^{n} r_n}{n}\right) \times 24$$

where:

R = total periphyton community respiration, mg $O_2/m^2/d$, and

n = number of observations.

Thus,

$$P_N' = P_G - R$$

where:

P_N = net periphytic production, mg $O_2/m^2/d$.

2) Free water methods

a) Calculation of reaeration or diffusion for both the single and upstream-downstream methods—Calculate k_2 from radiotracer data as follows:

$$K_{Kr} = \frac{-1}{t} \ln \frac{(C_{Kr}/C_H)_d}{(C_{Kr}/C_H)_u}$$

and

$$k_2 = \frac{K_{Kr}}{0.83}$$

where:

k_2 = reaeration coefficient (base e), d^{-1},

K_{Kr} = base e transfer coefficient for ^{85}Kr, d^{-1},

t = time of travel, d,

$(C_{Kr}/C_H)_u$ = ratio of released radioactivities ($\mu Ci/mL$) ^{85}Kr to ^3H at the upstream station, and

$(C_{Kr}/C_H)_d$ = ratio of radioactivities ($\mu Ci/mL$) ^{85}Kr to ^3H at the downstream station.

The reaeration coefficient also can be calculated from an equation relating the rate of energy dissipation in a stream to k_2.[14,15]

$$k_2 = K \frac{\Delta h}{t}$$

where:

K = escape coefficient,

Δh = change in water surface elevation in a stream reach, and

t = time of flow through a stream reach.

This can be expressed in terms of hydrodynamic and physical data:

$$k_{2_{20}} = K' \frac{\Delta H}{\Delta X} \times V$$

where:

K' = 28.3 $\times$ 10^3 s/m $\cdot$ d for stream flows between 0.028 and 0.28 m^3/s; 21.3 $\times$ 10^3 s/m $\cdot$ d for stream flows between 0.28 and 0.56 m^3/s; and 15.3 $\times$ 10^3 s/m $\cdot$ d for stream flows above 0.56 m^3/s,

$k_{2_{20}}$ = reaeration coefficient, d^{-1}, at 20°C,

$\frac{\Delta H}{\Delta X}$ = slope, m/km, and

V = velocity, m/s.

Convert $k_{2_{20}}$ to the temperature of the stream by the following equation:

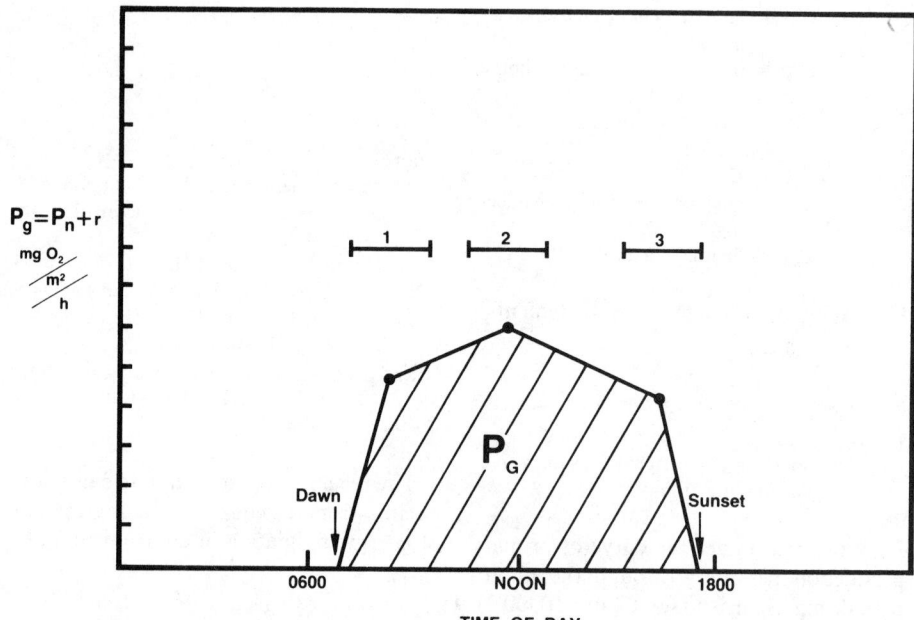

Figure 10300:3. Gross periphytic primary production (P_G) determined by the O'Connell-Thomas Chamber. P_G is the area under the curve obtained by graphical integration planimetry. Each point is the run $P_g = P_n + r$ for incubation periods 1, 2, and 3, which are denoted by the indicated lines.

$$k_{2_T} = k_{2_{20}} (1.024)^{(T-20)}$$

where:

k_{2_T} = k_2 at ambient water temperature, d^{-1}, and

T = ambient water temperature, °C.

Convert to D in mg/L/h:

$$D = \frac{k_{2_T} C_s}{24}$$

where:

C_s = oxygen concentration at saturation at ambient stream temperatures, mg/L.

b) **Single-station method—Calculation of primary productivity and respiration** from diurnal oxygen and temperature measurements at a single station is sum-

marized in Figure 10300:4 and Table 10300:I.

Tabulate hourly DO measurements and temperatures. Determine C_s (DO of air-saturated H_2O at each temperature from Table 4500-O:I) and compute uncorrected DO consumption, milligrams per liter per hour, for each period:

$$\Delta DO_{\substack{\text{hours} \\ \text{1 to 2}}} = DO_{\text{hour 2}} - DO_{\text{hour 1}}$$

Plot on the half hour, as shown in Figure 10300:4b.

Calculate the net primary production and respiration of phytoplankton as shown in Section 10200I. Determine the 24-h average hourly plankton respiration, $(\sum_{1}^{n} r_p)/n$, in milligrams per liter per hour every half hour. Calculate the hourly net phytoplank-

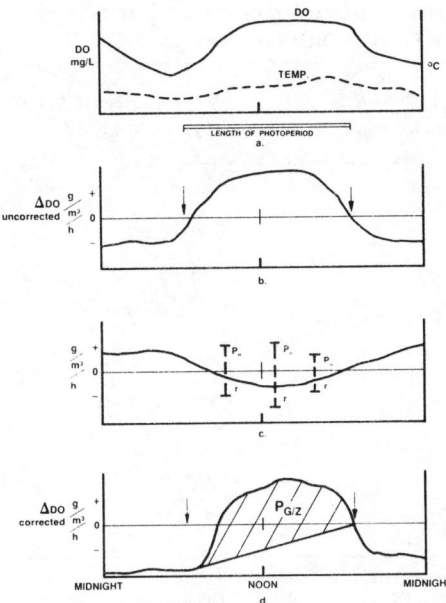

Figure 10300:4. Calculation of gross primary production at a single station. P_g, g $O_2/m^2/h$ = area of corrected rate of change curve integrated for the length of the photoperiod multiplied by average water depth, z, for the reach in meters.

ton production and tabulate for the approximate hours during the photoperiod. Plot as shown on Figure 10300:4c.

Calculate and tabulate k_{2T} and substitute D for each C_s, as outlined in ¶ a), above. Plot as shown in Figure 10300:4c.

Correct each ΔDO for diffusion and phytoplankton metabolism:

$ΔDO_{corrected}$, mg/L/h
$$= ΔDO_{uncorrected} - D - P_p - R_p$$

Plot each point as shown in Figure 10300:4d.

Gross primary productivity of the benthic and attached populations are computed as the area under the curve in Figure 10300:4d from sunrise to sunset. This is primary production in grams per cubic meter per day. Multiply by the average depth

for a reach, z meters, to obtain P_G in grams per square meter per day. Calculate community respiration:

$$R = 24 z F$$

where:
R = community respiration, g/m²/d,
z = depth, m, and
F = average hourly ΔDO for the dark period (without regard to sign), mg/L/h.

Calculate net primary productivity P_N as:

$$P_N = P_G - R$$

c) Upstream-downstream method—Calculation of primary productivity and respiration for a stream reach from upstream and downstream pairs of diurnal curves of oxygen and water temperature is summarized in Figure 10300:5 and Table 10300:II.

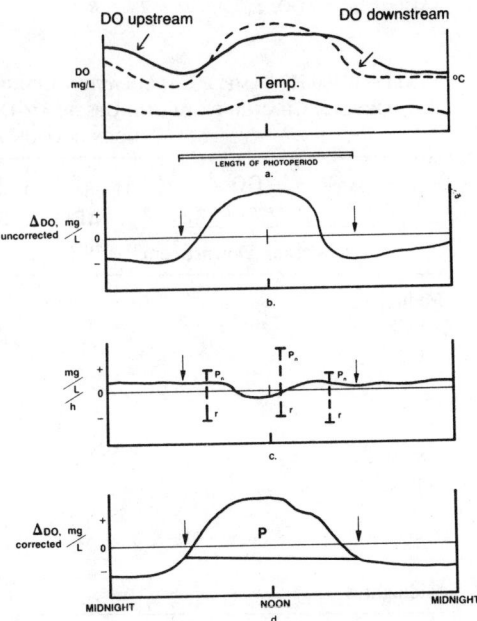

Figure 10300:5. Calculation of gross periphytic primary productivity from upstream-downstream diurnal curves. P is the area under the corrected rate of change graph.

TABLE 10300:I. SAMPLE CALCULATION LEDGER FOR COMPUTATION OF CORRECTED RATE OF OXYGEN CHANGE FROM A SINGLE-STATION DIURNAL CURVE

Time h	DO mg/L	Water Temp. $°C$	C_s* mg/L	Uncorrected ΔDO† $mg/L/h$	P_p‡ $mg/L/h$	R_p§ $mg/L/h$	k_2 d^{-1}	D $mg/L/h$	Corrected ΔDO‖ $mg/L/h$
Midnight									
0030									
0100									
0230									
.									
.									
.									
Noon									
1230									
1300									
.									
.									
.									
Midnight									

* DO concentration at 100% saturation for a given water temperature, from Table 4500-O:I.
† Hourly rate of change of DO. For example, for noon to 1300, $DO_{1200-1300} = DO_{1300} - DO_{1200}$; plot at 1230.
‡ Phytoplankton net production.
§ Phytoplankton respiration rate.
‖ $\Delta DO_{corrected} = \Delta DO_{uncorrected} - D - P_p - R_p$

TABLE 10300:II. SAMPLE CALCULATION LEDGER FOR COMPUTATION OF CORRECTED RATES OF OXYGEN CHANGE FROM THE UPSTREAM-DOWNSTREAM DIURNAL CURVES OF OXYGEN CONCENTRATION AND TEMPERATURE

Time h	DO mg/L		Uncorrected ΔDO mg/L	Water Temp. $°C$	C_s* mg/L	P_p† mg/L	R_p‡ mg/L	k_2 d^{-1}	Corrected ΔDO‖ mg/L
	Upstream	Downstream							
Midnight									
0100									
0200									
.									
.									
.									
Noon									
1300									
.									
.									
Midnight									

* DO concentration at 100% saturation for a given water temperature, from Table 4500-O:I.
† Change in oxygen concentration in the light bottle per hour multiplied by travel time between the upstream and downstream station.
‡ Change in oxygen concentration in the dark bottle multiplied by travel time between the upstream and downstream station.
‖ $\Delta DO_{corrected} = \Delta DO_{uncorrected} - D - P_p - R_p$

Alternatively, calculate as below, with oxygen change expressed as the difference between stations rather than as change per hour. The calculations are analogous. Multiply the area under a curve of oxygen change between two stations, corrected for diffusion and plankton metabolism and expressed in milligrams per liter, by the discharge in cubic meters per hour, and divide by the water surface area between the two stations. This, multiplied by 24, yields gross primary productivity in grams per square meter per day.

To compute gross primary productivity by this method, tabulate upstream and downstream DO and average water temperature for the reach at each hour. Calculate ΔDO between upstream and downstream stations for each hour as

$$\Delta DO = DO_{downstream} - DO_{upstream}$$

Tabulate C_s and determine the planktonic activity. Correct for planktonic respiration by relating average hourly dark bottle DO change to the time of travel in the stream reach; correct for planktonic production by the hourly change in DO in the light bottle times the time of travel (see Table 10300:II).

Calculate or tabulate k_2 and convert it to the total oxygen diffusion for the reach. Because diffusion, D, is expressed as milligrams per liter per hour, multiply it by the travel time to obtain the diffusion correction.

Correct each hourly upstream-downstream ΔDO as shown in Table 10300:II. Integrate the area under this ΔDO curve from sunrise to sunset to give P as in Figure 10300:5d.

$$P_G, \text{g/m}^2/\text{d} = \frac{Q}{A} P$$

where:

Q = flow, m³/h, and
A = reach area, m² (average reach width × reach length).

Respiration, R, g $O_2/\text{m}^2/\text{d}$

$$=\frac{\Delta DO_{dark} \times Q \times 24}{A}$$

and

Net production, $P_N = P_G - R$

5. References

1. VOLLENWEIDER, R.A., ed. 1969. A Manual on Methods for Measuring Primary Production in Aquatic Environments. IBP Handbook No. 12. F.A. Davis Co., Philadelphia, Pa.
2. SLADECEK, V. & A. SLADECKOVA. 1964. Determination of periphyton production by means of the glass slide method. *Hydrobiologia* 23:125.
3. KING, D.L. & R.C. BALL. 1966. A qualitative and quantitative measure of *aufwuchs* production. *Trans. Amer. Microsc. Soc.* 82:232.
4. CLARK, J.R., D.I. MESSENGER, K.L. DICKSON & J. CAIRNS, JR. 1978. Extraction of ATP from *aufwuchs* communities. *Limnol. Oceanogr.* 23:1055.
5. WETZEL, R.G. 1963. Primary productivity of periphyton. *Nature* 197:1026.
6. WETZEL, R.G. 1964. A comparative study of the primary production of higher aquatic plants, periphyton, and phytoplankton in a large shallow lake. *Int. Rev. Ges. Hydrobiol.* 49:1.
7. ARONOFF, S. 1956. Techniques in Radiochemistry. Iowa State College Press, Ames.
8. ODUM, H.T. 1956. Primary production in flowing waters. *Limnol. Oceanogr.* 1:102.
9. ALLEN, H.L. 1971. Primary productivity, chemo-organotrophy, and nutritional interactions of epiphytic algae and bacteria or macrophytes in the littoral of a lake. *Ecol. Monogr.* 41:97.
10. HALL, C.A.S. 1972. Migration and metabolism in a temperate stream ecosystem. *Ecology* 53:585.
11. NIXON, S.W. & C.A. OVIATT. 1974. Ecology of a New England salt marsh. *Ecol. Monogr.* 43:463.
12. THOMAS, N.A. & R.L. O'CONNELL. 1966. A method for measuring primary production by stream benthos. *Limnol. Oceanogr.* 11:386.
13. COPELAND, B.J. & W.R. DUFFER. 1964. Use of a clear plastic dome to measure gaseous diffusion rates in natural waters. *Limnol. Oceanogr.* 9:494.

14. TSIVOGLOU, E.C. & L.A. NEAL. 1976. Tracer measurement of reaeration. III. Predicting the capacity of inland streams. *J. Water Pollut. Control Fed.* 48:2669.

15. GRANT, R.S. 1976. Reaeration-coefficient measurements of 10 small streams in Wisconsin. U.S. Geol. Surv. Water Resources Publ. 76-96.

16. ODUM, H.T. & C.M. HOSKIN. 1958. Comparative studies of the metabolism of marine water. *Publ. Inst. Mar. Sci. Univ. Tex.* 4:115.

6. Bibliography

POMEROY, L.R. 1959. Algal productivity in salt marshes. *Limnol. Oceanogr.* 4:386.

CASTENHOLZ, R.W. 1961. An evaluation of a submerged glass method of estimating production of attached algae. *Verh. Int. Ver. Limnol.* 14:155.

WHITFORD, L.A. & G.J. SCHUMACHER. 1964. Effect of a current on respiration and mineral uptake in *Spirogyra* and *Oedogonium. Ecology* 45:168.

DUFFER, W.R. & T.C. DORRIS. 1966. Primary productivity in a southern Great Plains stream. *Limnol. Oceanogr.* 11:143.

MCINTIRE, C.D. 1966. Some factors affecting respiration of periphyton communities in lotic environments. *Ecology* 47:918.

CUSHING, C.E. 1967. Periphyton productivity and radionuclide accumulation in the Columbia River, Washington, USA. *Hydrobiologia* 29:125.

HANSMANN, E.W., C.B. LANE & J.D. HALL. 1971. A direct method of measuring benthic primary production in streams. *Limnol. Oceanogr.* 16:822.

SCHINDLER, D.W., V.E. FROST & R.V. SCHMIDT. 1973. Production of epilithiphyton in two lakes of the experimental lakes area, northwestern Ontario. *J. Fish. Res. Board Can.* 30:1511.

NORTH AMERICAN BENTHOLOGICAL SOCIETY. 1974–1985 (annual). Current and Select Bibliographics on Benthic Biology. Springfield, Ill.

10300 E.   Interpreting and Reporting Results

Although several systems have been developed to organize and interpret periphyton data, no single method is universally accepted. The methods may be qualitative or quantitative. Qualitative methods deal with the taxonomic composition of the communities in zones of pollution, whereas quantitative methods deal with community structure using diversity indices, similarity indices, and numerical indices of saprobity.

1. Qualitative Methods (Indicator Species and Communities)

The saprobity system developed by Kolkwitz and Marsson is a widely used method of interpreting periphyton data. This scheme divides polluted stream reaches into polysaprobic, α and β mesosaprobic, and oligosaprobic zones, and lists the characteristics of each. The system has been refined[1,2] and enlarged by Fjerdingstad[3,4] and Sladecek.[5,6]

2. Quantitative Methods

These methods use cell counts per unit area of substrate and numerical indices of pollution or water quality. Considerable data on cell densities and species composition of periphyton collected on glass slides in polluted rivers in England are available.[7]

Other indices include the Shannon-Weaver,[8] Simpson's,[9] and Pinkham-Pearson.[10] The saprobity system[11] also may be used where code numbers assigned for the saprobial value and the abundance of individual species are used to calculate a Mean Saprobial Index. Results also may be expressed by the truncated-log normal distribution of diatom species[12,13] as well as the Autotrophic Index (AI).[14]

3. References

1. KOLKWITZ, R. 1950. Oekologie der saprobien. *Ver Wasser-, Boden, Lufthyg. Schriftenreihe* (Berlin) 4:1.

2. LIEBMANN, H. 1951. Handbuch der Frischwasser und Abwasserbiologie. Bd. I. Oldenbourg, Munich, Germany.

3. FJERDINGSTAD, E. 1964. Pollution of streams estimated by benthal phytomicroorganisms. I. A saprobic system based on communities of organisms and ecological factors. *Int. Rev. Ges. Hydrobiol.* 49:63.

4. FJERDINGSTAD, E. 1965. Taxonomy and saprobic valency of benthic phytomicroorganisms. *Int. Rev. Ges. Hydrobiol.* 50:475.

5. SLADECEK, V. 1966. Water quality system. *Verh. Int. Ver. Limnol.* 16:809.

6. SLADECEK, V. 1973. System of water quality from the biological point of view. *Arch. Hydrobiol.* 7:1.

7. BUTCHER, R.W. 1946. Studies in the ecology of rivers. VI. The algal growth in certain highly calcareous streams. *J. Ecol.* 33:268.

8. SHANNON, C.E. 1948. The Mathematical Theory of Communication. Univ. Illinois Press, Urbana.

9. SIMPSON, E.H. 1949. Measurement of diversity. *Nature* 163:688.

10. PINKHAM, C.F.A. & J.G. PEARSON. 1976. Applications of a new coefficient of similarity to pollution surveys. *J. Water Pollut. Control Fed.* 48:717.

11. PANTLE, R. & H. BUCK. 1955. Die biologische überwachung der Gewasser und der Darstellung der Ergebnisse. *Gas-Wasserfach* 96:604.

12. PATRICK, R., M.H. HOHN & J.H. WALLACE. 1954. A new method for determining the pattern of the diatom flora. *Bull. Philadelphia Acad. Natur. Sci.* 259:1.

13. PATRICK, R. 1973. Use of algae, especially diatoms, in the assessment of water quality. *In* J. Cairns, Jr., ed. Biological Methods for the Assessment of Water Quality. ASTM STP 528, American Soc. Testing & Materials, Philadelphia, Pa.

14. WEBER, C. 1973. Recent developments in the measurement of the response of plankton and periphyton to changes in their environment. *In* G. Glass, ed. Bioassay Techniques and Environmental Chemistry. Ann Arbor Science Publ., Ann Arbor, Mich.

4. Bibliography

FJERDINGSTAD, F. 1950. The microflora of the River Mølleaa, with special reference to the relation of the benthal algae to pollution. *Folia Limnol. Scand.* 5:1.

BLUM, J.L. 1956. The ecology of river algae. *Bot. Rev.* 22:291.

YOUNT, J.L. 1956. Factors that control species numbers in Silver Springs, Florida. *Limnol. Oceanogr.* 1:286.

BUTCHER, R.W. 1959. Biological assessment of river pollution. *Proc. Linnean Soc. London* 170:159.

HOHN, M.H. 1959. The use of diatom populations as a measure of water quality in selected areas of Galveston and Chocolate Bay, Texas. *Publ. Inst. Mar. Sci. Univ. Tex.* 5:206.

HOHN, M.H. 1961. Determining the pattern of the diatom flora. *J. Water Pollut. Control Fed.* 33:48.

PATRICK, R. 1963. The structure of diatom communities under varying ecological conditions. *Ann. N.Y. Acad. Sci.* 108:359.

SCHLICHTING, H.E., JR. & R.A. GEARHEART. 1966. Some effects of sewage effluent upon phyco-periphyton in Lake Murray, Oklahoma. *Proc. Okla. Acad. Sci.* 46:19.

SLADECKOVA, A. & V. SLADECEK. 1966. Periphyton as indicator of reservoir water quality. *Technol. Water* (Czech) 7:507.

TAYLOR, M.P. 1967. Thermal effects on the periphyton community in the Green River. Tennessee Valley Authority, Div. Health & Safety, Water Qual. Br., Biol. Sect., Chattanooga, Tenn.

PATRICK, R. 1968. The structure of diatom communities in similar ecological conditions. *Amer. Natur.* 102:173.

DICKMAN, M. 1969. A quantitative method for assessing the toxic effects of some water soluble substances, based on changes in periphyton community structure. *Water Res.* 3:963.

BESCH, W.K., M. RICARD & R. CANTIN. 1970. Use of benthic diatoms as indicators of mining pollution in the N.W. Miramichi River. *Tech. Rep. Fish. Res. Board Can.* 202:1.

NUSCH, E.A. 1970. Ecological and systematic studies of the Peritricha (Protozoa, Ciliata) in the periphyton community of reservoirs and dammed rivers with different degrees of saprobity. *Arch. Hydrobiol.* (Suppl.)37:243.

ROSE, F.L. & C.D. MCINTIRE. 1970. Accumulation of dieldrin by benthic algae in laboratory streams. *Hydrobiologia* 35:481.

WHITTON, B.A. 1970. Toxicity of zinc, copper and lead to Chlorophyta from flowing waters. *Arch. Mikrobiol.* 72:353.

BURROWS, E.M. 1971. Assessment of pollution effects by the use of algae. *Proc. Roy. Soc. Lond. Ser. B.* 177:295.

PATRICK, R. 1971. The effects of increasing light and temperature on the structure of diatom communities. *Limnol. Oceanogr.* 16:405.

ARCHIBALD, R.E.M. 1972. Diversity of some South African diatom associations and its relation to water quality. *Water Res.* 6:1229.

CAIRNS, J., JR., B.R. LANZA & B.C. PARKER. 1972. Pollution-related structural and functional changes in aquatic communities with emphasis on freshwater algae and protozoa. *Proc. Acad. Natur. Sci. Philadelphia* 124:79.

OLSON, T.A. & T.O. ODLAUG. 1972. Lake Superior Periphyton in Relation to Water Quality. Water Pollut. Control Res. Ser., 18080 DEM 02/72. Univ. Minnesota School Public Health, Minneapolis.

HANSMANN, E.W. 1973. Effects of logging on periphyton in coastal streams of Oregon. *Ecology* 54:194.

RUTHVEN, J.A. & J. CAIRNS, JR. 1973. Response of fresh-water protozoan artificial communities to metals. *J. Protozool.* 20:127.

MIDWEST BENTHOLOGICAL SOCIETY. 1964-1973 (annual). Current and Select Bibliographies on Benthic Biology. Springfield, Ill.

NORTH AMERICAN BENTHOLOGICAL SOCIETY. 1974-1985 (annual). Current and Select Bibliographies on Benthic Biology. Springfield, Ill.

BAXTER, R.M. 1977. Environmental effects of dams and impoundments. *Annu. Rev. Ecol. Systematics* 8:255.

WEITZEL, R.L., ed. 1979. Methods of Measurement of Periphyton Communities: A Review. ASTM Spec. Tech. Publ. 690, American Soc. Testing & Materials, Philadelphia, Pa.

STEVENSON, R.J. 1984. Epilithic and epipelic diatoms in the Sandusky River USA with emphasis on species diversity and water pollution. *Hydrobiologia* 114:161.

KOSINSKI, R.J. 1984. The effect of terrestrial herbicides on the community structure of stream periphyton. *Environ. Pollut. Ser. A, Ecol. Biol.* 36:165.

LINDSTROM, E.A. & T.S. TRASAN. 1984. Influence of current velocity on periphyton distribution and succession in a Norwegian soft water river. *Verh. Int. Ver. Limnol.* 22:1965.

MCGUIRE, M.J., R.M. JONES, E.G. MEANS, G. IZAGUIRRE & A.E. PRESTON. 1984. Controlling attached blue-green algae with copper sulfate. *J. Amer. Water Works Assoc.* 76:60.

PARKER, B.C., G.J. SCHUMACHER & L.A. WHITFORD. 1984. Some rarely reported algae of the Appalachian Mountains, Eastern North America: Why so rare? *Va. J. Sci.* 35:197.

10400   MACROPHYTON*

10400 A.   Introduction

1. General Discussion

The macrophyton consists principally of aquatic vascular flowering plants, but it also includes the aquatic mosses, liverworts, ferns, and the larger marine algae. Like other primary producers, these plants respond to the quality of the water in which they grow; thus their analysis is important for water quality evaluation. They often constitute a dominant factor in the habitat of other aquatic organisms.

Freshwater forms range in size from the tiny watermeal (*Wolffia* spp.), about the size of a pinhead, to plants such as the cattail (*Typha* spp.), up to 4 m high, and finally to cypress trees (*Taxodium* spp.),

---

* Approved by Standard Methods Committee, 1988.

up to 50 m high. Higher aquatic plants often are clustered in large numbers, many in pure stands, covering extensive areas of shallow lakes, reservoirs, marshes, and canals. A few of the larger freshwater algae (*Chara* spp., *Nitella* spp., and *Cladophora* spp.) resembling higher plants in size, form, and habit sometimes are included in the macrophyton. In marine water, the intertidal rockweeds (*Fucus* spp. and *Ascophyllum* spp.) and offshore kelps (*Fucus* spp. and *Macrocystis* spp.) are conspicuous. Vascular marine or estuarine plants, such as the eelgrass (*Zostera* spp.) and the marshgrass (*Spartina* spp.), are essential to the aquatic ecosystem.

Three growth forms of macrophyton generally are recognized: floating, submersed, and emersed. Floating plants are not rooted; their principal foliage or crown floats on the water surface. All or most of the foliage of submersed plants grows beneath the water surface: the plants may or may not have roots. Growing tips of submersed plants may emerge to flower. Emersed plants may be subdivided into erect-leaved plants and floating-leaved plants. Their principal foliage is in the air at or above the water surface; they are attached by roots to the bottom mud. Floating-leaved emergents lack the supportive tissue found in the erect-leaved emersed plants. In some cases the same species may grow as a floating or emersed type, or as a submersed or emersed type. Submersed and emersed vascular plants typically are rooted to the bottom but they may be found detached and floating.

The distribution and abundance of higher plants is subject to considerable spatial and temporal variation. Among the many factors that determine their presence and density are sediment type, water turbidity, nutrient concentrations, water depth, shoreline disturbance, herbivore grazing, and human activities. Zonation in the littoral zone of lakes and shallow, slow-moving streams is common. Emergent macrophytes generally are found in the most shallow portion of the littoral zone. During periods of low water level they may occupy the terrestrial as well as the aquatic habitat. The depth of inhabitation seldom exceeds 1 m. Floating-leaved plants commonly are found in the lower littoral areas at depths between 1 and 3 m. Submersed plants may occur from the edge of the shore to the interface of the littoral and profundal zones, but rarely extend beyond a depth of 10 m because of limitation of underwater light.

2. Bibliography

BUTCHER, R.W. 1933. Studies on the ecology of rivers. I. On the distribution of macrophytic vegetation in the rivers of Britain. *J. Ecol.* 21:58.

WELCH, P.S. 1948. Limnological Methods. Blakiston Co., Philadelphia, Pa.

MILLIGAN, H.F. 1969. Management of aquatic vascular plants and algae. *In* Eutrophication: Causes, Consequences, Correctives. National Acad. Science, Washington, D.C.

BOYD, C.E. 1971. The limnological role of aquatic macrophytes and their relationship to reservoir management. *In* G.E. Hall, ed. Reservoir Fisheries and Limnology. Spec. Publ. Amer. Fish. Soc. No. 8.

HUTCHINSON, G.E. 1975. A Treatise on Limnology. John Wiley & Sons, New York, N.Y.

WESTLAKE, D.F. 1975. Macrophytes. *In* B.A. Whitton, ed. River Ecology. Univ. California Press, Berkeley & Los Angeles.

WETZEL, R.G. 1975. Limnology. W.B. Saunders Co., Philadelphia, Pa.

WILE, I. 1975. Lake restoration through mechanical harvesting of aquatic vegetation. *Verh. Int. Ver. Limnol.* 19:660.

WOOD, R.D. 1975. Hydrobotanical Methods. University Park Press, Baltimore, Md.

RASCHKE, R.L. 1978. Macrophyton. *In* W. T. Mason, Jr., ed. Methods for the Assessment and Prediction of Mineral Mining Impacts on Aquatic Communities: A Review and Analysis. U.S. Dep. Interior, Harpers Ferry, W. Va.

DENNIS, W.M. & B.G. ISOM, eds. 1984. Ecological Assessment of Macrophyton: Collection, Use, and Meaning of Data. ASTM STP 843, American Soc. Testing & Materials, Philadelphia, Pa.

10400 B. Preliminary Survey

1. General Considerations

A macrophyte survey includes species identification, location, and quantity. More detailed studies may involve functioning of aquatic plants in nutrient and heavy metal uptake and turnover, use of plants as indicator organisms, and effects of plants on water quality conditions.

Several sampling protocols are required to meet diverse survey needs. The usefulness of a given study and the appropriate types of statistical analyses are determined and fixed initially.

To develop a good sample design, determine what information is desired, prevailing environmental conditions, the life and growth form of the species being sampled, and the methods for obtaining reproducible data that are comparable to other or future studies. In defining reporting requirements, consider such as the use of scientific names; the selection of appropriate descriptors such as frequency, density, biomass, cover, diversity, productivity, and outer limit of vegetation growth; and the use of proper statistical techniques.

2. Pre-Field Investigations

During pre-field investigations assemble maps, charts, aerial photographs, taxonomic keys, and past reports. Maps, charts, and aerial photographs help determine access routes, project size, plant community distribution, habitat characteristics that may influence plant distribution, and sampling obstacles or hazards. These items also provide a base for doing field work and reporting results. These may be available locally from municipal engineering departments, zoning boards, planning commissions, drainage districts, and land conservation commissions. At the state level information may come from natural resource agencies, natural history survey organizations, and transportation departments. At the federal level, the Geological Survey, Soil Conservation Service, Bureau of Land Management, Forest Service, Park Service, Fish and Wildlife Service, Tennessee Valley Authority, Bureau of Indian Affairs, Army Corps of Engineers, Environmental Protection Agency, and the National Oceanographic and Atmospheric Administration have many maps, charts, and aerial photographs available. A final source is private map companies that publish hydrographic maps for fishermen and recreational boaters.

Past reports provide historical records useful for planning sampling logistics and interpreting results. Comparable studies taken at different times provide a dynamic look at the vegetation. An often-overlooked resource is a herbarium storing pressed and mounted plant specimens. These generally are located in universities and natural history museums.

3. Field Reconnaissance

Sampling efficiency is improved and a sampling scheme can be refined during a field reconnaissance. It provides an opportunity to learn the species of plants present and to sketch their distribution. The species-area curve technique frequently is used to determine the likelihood of finding more species during a preliminary survey. A field reconnaissance allows the investigator to answer logistical questions that are the bane of all sampling efforts.

4. Bibliography

ONDOK, J.P. & J. KVET. 1978. Selection of sampling areas in assessment of production. *In* D. Dykyjova & J. Kvet, eds. Pond Littoral Ecosystems: Structure and Functioning. Ecological Studies 28. Springer-Verlag, Berlin.

MACEINA, M.J., J.V. SHIREMAN, K.A. LANGELAND & D.E. CANFIELD, JR. 1984. Prediction of submersed plant biomass by use of a recording fathometer. *J. Aquat. Plant Manage.* 22:35.

Also see Section 10400A.2.

10400 C.   Vegetation Mapping Methods

Mapping vegetation stands may be necessary. This should be done during the preliminary survey.[1]

1. Baseline Method

Vegetation maps constructed using the baseline method or the basepoint-stadia rod-alidade method generally are limited to pure stands of floating or emersed littoral macrophyton in small bodies of water. In clear water, the outline of pure stands of submersed vegetation can be determined by using a viewing box (usually a wooden box with a watertight glass lens) from the surface or by underwater observation by a diver using a snorkel or SCUBA (self-contained underwater breathing apparatus). The baseline method and the basepoint-stadia rod-alidade method provide accurate maps of vegetation in areas up to 1×10^5 m^2 where most of the vegetation outline is visible. The baseline method uses intercepting lines from each end of a predetermined base line to closely spaced markers (i.e., chaining pins) around the stand. By presetting the map scale, the ratio between the length of the base line and its reduction on the map (drawn on a plane table) can be determined. The basepoint-stadia rod-alidade method is a modification of the baseline method in that the distance between the vegetational outline and the base point is determined with an alidade, range finder, or portable Loran-C unit. One unit on the stadia rod, as viewed between the cross-hairs of the alidade, is equivalent to a distance of 3.05 m between the stadia rod and the alidade. Chaining pins are not required when this method is used. In practice, more readings closely spaced along the vegetational outline usually are taken using the basepoint-stadia rod-alidade method.

It is not necessary to measure all distances and angles; some can be determined trigonometrically. After all angles and distances are calculated, fill in irregularities through inspection and use of other maps and photographs.

2. Line Intercept Method

The line intercept method is preferable for mapping mixed stands and/or large areas. Select sampling points at equal intervals along a base line. Choose the length of the intervals by the degree of accuracy desired: the closer the sampling points, the more accurate the map. Run transects perpendicularly from the base line to the boundary of the plant stand. Use an intercept line (transect line) of plastic-coated wire rope to prevent stretching. If line flotation is a problem, use lead weights applied at regular intervals to sink the line and act as interval markers on the rope to designate sampling units. Use 0.5-m segments (in dense vegetation) or 5-m sampling units (in sparse vegetation) for determining plant species that vertically intercept the line at each segment. In sparse vegetation, place a ring of brass or PVC pipe, 0.25 m^2, over the sample point on the intercept line and record plant species within the ring. During underwater surveys, record data with a wax or soft lead pencil on a writing board constructed of plastic overlays. Construct the vegetation map by placing points where plant species are found within an outline map (or aerial photograph) of the sample area. After encircling the points on the map by single species, determine the enclosed area with planimetry.

Determine relative frequency at sample points along an intercept line with a set sampler consisting of a 2-cm steel tube, 2 m long, to which five 0.75-cm by 25-cm steel rods are attached on 40-cm centers. Record vegetation touching each of the five points within 2.54 cm of the distal tip. If more than one plant species is touching, record the plant nearest the tip. If no plant is touched, record bare ground. Use field data to derive frequency of plant species at sample points, as well as by contour elevation and intercept line segment.

Determine percent cover of vegetation from vegetation maps by planimetry: digital compensating polar planimeters are the most accurate.

3. Remote Sensing

a. General considerations: Remote sensing techniques have been used to detect, assess, and monitor aquatic macrophytes. These techniques include analog aircraft and satellite serial photography, digital aircraft and satellite multispectral scanning in the visible, infrared, and thermal bands; microwave techniques, primarily side-looking airborne radar (SLAR); and shuttle imaging radar (SIR). They provide a synoptic view of large areas, allowing quick surveys to further delineate areas of interest and revisit capacity, both at relatively low cost.

In selecting remote sensing system(s) consider expected results, time for project completion, and available resources. The larger the area, the greater the advantage of remote sensing. Remote sensing also lends itself to studies over time, because each image is a historical record.

For determining general associations of a widespread macrophytic growth, satellite imagery firms recommend ground resolutions of between 30 and 80 m. Widespread multitemporal coverage is available at scales ranging from 1:100 000 to 1:1 000 000 at a reasonable cost in the form of paper prints, transparencies, and digital formats. Landsat Multispectral Scanner (MSS) and Thematic Mapper (TM) have been used to identify areas with emergent aquatic vegetation or topped-out submergent vegetation. Surface roughness is a requisite for an imaging return with SLAR and SIR. Landsat provides limited capability for species discrimination, but availability, cost, and repeat cycles make it useful in determining presence of large populations over time. For a detailed vegetation survey, including discrete species identification, use much larger-scale imagery, (1:10 000 or greater). High-altitude aerial coverage available through the National High Altitude Mapping Program (NHAP) and other sources at original scales of approximately 1:60 000 to 1:120 000 provides both good initial areal coverage and the capacity of enlargement up to five times.[2]

After determining scale, select film/filter or sensor combinations. These include black and white imagery, color infrared, and black and white-infrared imagery; color infrared film used with a yellow filter is most generally useful. Other combinations have been described.[3-7] For detailed flight planning consider growth stage of plants, water depth and clarity, tidal conditions, cloud cover, and sun angle.[2,8,9]

Available resources ultimately determine remote sensing activities.

b. Aerial photography:

1) Equipment—For small-format aerial photography, use a good-quality 35-mm single-lens reflex (SLR) camera with manual or through-the-lens metering or any good 70-mm camera system, preferably with motor drive. Intervalometers providing for exact exposure intervals for stereo photography are available for both systems. A 28-mm-focal-length lens with a 35-mm SLR camera gives wide coverage at low altitudes; 40-mm and 80-mm lenses have been successful with 70-mm camera systems. When photographing with either black-and-white or color film, use a skylight or haze filter. When photographing with color infrared film below 1700 m, use a Wratten 12 filter; at altitudes above 1700 m use a Wratten 15 filter. Time of year and condition of vegetation also may influence the choice of yellow/orange filter. In turbid waters, color film* provides better water penetration and is more useful in

---

* Kodak Ektachrome Professional film type 5036 or equivalent.

mapping submerged vegetation than color infrared film.[3] Color infrared† yields more detail, is more useful in mapping emergent vegetation or wetlands, and may provide more detail in clear, nonturbid waters.

Preferably mount camera in the belly of a high-wing, single-engine aircraft for low-altitude, small-format photography. Belly mounts require special aircraft modification, but provide a stable platform, protection for the camera, and good access. Alternatively, camera may be mounted on the aircraft door.[6]

2) Procedures—Because sun angle is critical in obtaining high-quality aerial photography, schedule flight for a time when solar altitude is between 40 and 68 degrees.[7]

Set camera at designated ASA reading for the film (assume 100 to 125 for color infrared film) and shoot in the automatic exposure mode. At a typical airspeed of about 190 km/h (approximately 120 mph) a shutter speed of 1/250 or 1/500 is adequate. Determine proper f-stop from an aerial exposure computer; in general, f 5.6 to 11 (f 8 is optimum) gives acceptable color exposure. Proper exposure of color infrared film depends on such factors as time of day and year, altitude, humidity, and type of landscape.

Process film through the manufacturer, a photo lab, or aerial photography service. Development of color infrared film is available solely through the manufacturer.

c. Fathometry: Recording fathometers are best applied in water more than 1 m deep where the instruments can determine accurately the height and distribution of subsurface macrophytes.

A recording fathometer can be mounted on most boats and can accurately determine one-dimensional (percent cover) and two-dimensional (percent vertical area) profiles of submersed vegetation. Make lin-ear and planimetric measurements on chart tracings that provide permanent records for ready comparison over time. Calculate percent cover by dividing the linear measurement for a macrophyte species or community by the total chart paper length for any given transect. Dividing the area of the tracing with macrophytes by the total water area gives percent vertical area. Use a fathometer accurate to the nearest 0.1 m.

Mount the transducer for the recording fathometer with brackets on the boat's transom. Keep speed and recorder speed constant to produce tracings of similar length and resolution. Only a few minutes are required for replication of transects several kilometers long. Unless gross morphological differences exist, species discrimination on the chart tracings is difficult or impossible. Mark boundaries of monotypic colonies and community types with a fix line on the chart tracings. Dense vegetation mats that reach the surface may impede boat movement, prevent the transducer signal from reaching the hydrosoil, and merge tracings of macrophytes with the transducer line. Tracing patterns from water less than 1 m deep may be difficult to interpret.[10]

4. References

1. RASCHKE, R.L. 1984. Mapping—Surface or ground surveys. In W.M. Dennis & B.G. Isom, eds. Ecological Assessment of Macrophyton: Collection, Use, and Meaning of Data. Spec. Tech. Publ. 843, American Soc. Testing & Materials, Philadelphia, Pa.

2. AVERY, E.T. & G.L. BERLINE. 1985. Interpretation of Aerial Photographs, 4th ed. Burgess Publishing Co., Minneapolis, Minn.

3. ANDREWS, D.S., D.H. WEBB & A.L. BATES. 1984. The use of aerial remote sensing in quantifying submersed aquatic macrophytes. In W.M. Dennis & B.G. Isom, eds. Ecological Assessment of Macrophyton: Collection, Use, and Meaning of Data. Spec. Tech. Publ. 843, American Soc. Testing & Materials, Philadelphia, Pa.

† Kodak Aerochrome Infrared film type 2443 or equivalent.

4. Long, K.S. 1979. Remote Sensing of Aquatic Plants. Tech. Rep., Waterways Experiment Sta., U.S. Army Corps of Engineers, Vicksburg, Miss., Vol. A-79-2.

5. Breedlove, B.W. & W.M. Dennis. 1984. The use of small-format aerial photography in aquatic macrophyton sampling. *In* W.M. Dennis & B.G. Isom, eds. Ecological Assessment of Macrophyton: Collection, Use, and Meaning of Data. Spec. Tech. Publ. 843, American Soc. Testing & Materials, Philadelphia, Pa.

6. Meyer, M.P. & P.D. Grumstrup. 1978. Remote sensing application in agriculture and forestry. Res. Rep. 78-1, IAFHE RSL.

7. Benton, A.R., Jr. 1976. Monitoring aquatic plants in Texas. Tech. Rep. RSC-76, Texas A & M Remote Sensing Center, College Station, Tex.

8. Colwell, R.H., ed. 1983. Manual of Remote Sensing, 2nd ed. American Soc. Photogrammetry & Remote Sensing, Falls Church, Va.

9. Slama, C., ed. 1980. Manual of Photogrammetry, 4th ed. American Soc. Photogrammetry & Remote Sensing, Falls Church, Va.

10. Maceina, M.J. & J.V. Shireman. 1980. The use of a recording fathometer for determination of distribution and biomass of hydrilla. *J. Aquat. Plant Manage.* 18:34.

5. Bibliography

Schmid, W.D. 1965. Distribution of aquatic vegetation as measured by line intercept with SCUBA. *Ecology* 48:816.

Osborne, J.A. 1977. Ground Truth Measurements of *Hydrilla verticillata* Royle and Those Factors Influencing Underwater Light Penetration to Coincide with Remote Sensing and Photographic Analysis. Res. Rep. No. 2, Bur. Aquatic Plant Research & Control, Florida Dep. Natural Resources, Tallahassee.

10400 D. Population Estimates

1. Sampling Design

The design of a sampling program depends on study aims, collection methods, variation and distribution of vegetation, personnel and funds available, and accuracy expected. Variation in space usually is not random; distribution is determined by water depth, shoreline activity, sediment type, or other factors. The parametric statistic for estimating the true population mean assumes that the population being sampled has a normal distribution and that all sample units have the same probability of being selected. Avoid fixed sampling stations in sampling programs to determine population means, unless they are chosen at random at the beginning of the study. Because normally distributed plant populations may not be a characteristic of contiguous plant communities, use parametric statistics with caution.

The simple random sampling design is best applied to homogeneous, noncontiguous plant communities. The number of stations required to obtain an estimate of the true population mean with a predetermined level of confidence and permissible error can be determined by applying the data from a pilot study to the following equation:

$$N = \left(\frac{t \times S}{d \times \bar{x}} \right)^2$$

where:

N = number of sampling stations,

t = Student's t at a given probability level; because N is unknown, set $t = 2.0$; t is approximately equal to 2.0 for $N > 30$,

S = standard deviation,

$\bar{x}$ = estimator of true population mean usually determined by conducting a pilot study; and

d = permissible error of the final mean; $d = 0.1$ is recommended for vegetation studies ($\pm 10\%$).

An estimate of sampling program cost may be obtained as the sum of initial fixed cost (such as cost of equipment purchase)

and variable cost (cost per sample multiplied by number of samples).

Apply stratified random sampling to populations having many homogeneous stands. This design is best applied to populations with obvious gradients and, in practice, to gain precision by the minimized variance within strata. Determine placement of strata by a pilot study. To maximize precision, place stratum boundaries around homogeneous areas; generally, the fewer strata, the greater precision. Allocate sampling in stratified random sampling design according to:

$$\frac{N_i}{N} = \frac{W_i S_i}{\sum(W_i S_i)}$$

where:

N_i = number of samples in stratum i,
N = total number of samples,
W_i = a weight reflecting the size (number of quadrats, for example) of stratum i, and
S_i = standard deviation of sampled characteristic within stratum i.

Means for population measurement taken along randomly placed stations on a transect line do not represent large areas or lake populations unless the transect line is placed randomly. Arbitrarily placed transect lines within a sampling area may or may not reflect the true variation of the vegetation within.

2. Collection Methods

a. Field inventory/reconnaissance:

1) Manual collection—If water depth, clarity, temperature, flow, and other circumstances permit, collect specimens by hand. Under ideal conditions, manual collection by wading, snorkeling, or with SCUBA in deeper water habitats permits a detailed and comprehensive evaluation of the macrophyte community.

2) Drag chains—Construct drag chains by welding sharpened U-shaped hooks to a short length (0.6 to 1.0 m) of medium-weight chain. Attach chain to a rope and pull it through the water. Attach a float to the end of the rope to prevent its loss if the chain is snagged and/or dropped. The drag chain can be used readily from a slow-moving or stationary boat and is most efficient in collection of submersed macrophyte species with tall growth forms.

3) Rakes and tongs—Rakes with various handle lengths and oyster tongs may be useful in collecting macrophytes. A rope may be attached to the rake handle for sampling in deep water or to facilitate sampling over a wider radius.

4) Grab samplers—Devices developed for sampling benthic organisms, such as the Ekman, Ponar, and similar grab samplers (see Section 10500B.3), may be used to collect macrophytes. The light weight of the Ekman grab makes it preferable for the rapid and numerous samplings often required for survey inventories.

5) Recording fathometers—Use to determine height and distribution of subsurface macrophytes. Species with similar morphology usually cannot be distinguished from chart tracing; use supplemental methods to identify species.

b. Quantitative sampling: Numerical data collected to describe vegetation commonly include such measures of abundance as density, frequency, cover, and biomass/standing crop.[1,2] Collect these data from plots or quadrats or, less frequently, by plotless sampling techniques. The choice of analytical method depends on vegetation density and types, water depth, flow, height of vegetation in the water column, and nature of the sediment.

1) Line intercepts—This plotless sampling technique entails use of a weighted nylon or lead core line laid along the bottom between two known points or oriented by a compass reading. For dense floating mat vegetation, a floating line may be laid on top of the mat. A surveyor measures the linear distance occupied by various species that underlie the transect line. Express

these as a percentage of the total line length for individual species as well as for all species combined. If frequency data are desired, mark the line in increments (e.g., 1 m) and treat species presence/absence in a manner similar to data from quadrat sampling. The line intercept has been used to characterize and map aquatic communities[3] and to correlate distribution of macrophytes with selected environmental factors[4]. In aquatic environments, the line-intercept method is time-consuming and may require a diver equipped with SCUBA. Problems arise in determining whether a plant underlies the transect line.[5]

2) Belt transects—This technique is similar to the line transect. Data are collected along a fixed line, but from a two-dimensional plot or belt. The belt can be treated as a series of contiguous quadrats or quadrat location may be selected on the basis of a fixed interval or water depth. Use floating or sinking frames.

3) Quadrats—Quadrats can be used for such population and community estimates as frequency, cover, density, and biomass/standing crop. The sampling area of quadrat samplers varies from about 0.023 m^2 for the standard Ekman grab to 0.39 m^2 for the WES sampler described in ¶ b) below.

With the exception of frames, most sampling devices described have been used to obtain estimates of standing crop (aboveground biomass). Standing crop generally is used because of the difficulty in collecting underground plant parts, such as rhizome and roots.

a) Manual samplers—These are relatively simple devices delineating sampling areas from which the macrophytes are removed manually or with cutting shears. Although they can be used in deep water and manipulated by a diver, they work best in shallow water. They are relatively inexpensive and can be constructed easily or purchased from commercial sources.

Frames are suitable for sampling in shallow water. For sampling short, erect plants, use a square sinking frame constructed of metal. For dense or tangled vegetation, a square assembly frame with pins or wing nuts at the corners or a fixed-corner three-sided frame may be useful. Decide whether to include only macrophytes rooted within the frame or also overlapping plants. In deep water, difficulty and bias may occur in sampling tall submersed vegetation. For macrophytes forming a dense floating mat, use a floating frame constructed of wood or PVC pipe.

Box samplers are useful for sampling where water is shallow and the bottom consists of unconsolidated sediments. The sampler consists of an open-ended box with a metal cutting flange at the bottom and lateral handles; a sampler constructed of ¼-in. plexiglass with dimensions 0.5 m × 0.5 m × 0.6 m and aluminum cutting flange and corner reinforcements weighs about 12 kg. Unlike frame samplers, box samplers require no decision about what plants to include. With modifications, a box sampler can be used in deep water.[6]

Benthic dome (BeD) samplers[7] may be used for sampling in deep flowing waters. The sampler consists of a plastic dome with a stainless steel circular collar that can be pushed into the substrate. It weighs approximately 11 kg and has a sampling area of 0.25 m^2.

Various samplers [8,9] developed for macroinvertebrate sampling also may be used to collect macrophytes. These include the Surber sampler, suitable for shallow rivers with moderate current [see Section 10500B.3b1)]; the stovepipe (cylindrical) sampler, suitable for wadable waters with unconsolidated sediment bottoms [see Section 10500B.3c3)]; and the Ekman grab sampler, best suited for soft sediment bottoms with short, erect vegetation [see Section 10500B.3a6)].

b) Mechanically operated samplers—Mechanical sampling decreases sample collection time, increases accuracy of standing crop and biomass estimates, and is subject to less bias than many manual methods. However, mechanical sampling devices are costly and complex, and require a floating platform with winches, cables, and booms. The samplers described below are useful in deep water.

CAUTION: *Use extreme caution for safe operation.*

The Louisiana box sampler is an open-ended 35-cm-high box made of sheet metal or similar material that samples a 61- × 61-cm quadrat (sampling area = 0.37 m²).[10] It can be used from a V-hulled or pontoon boat and is hoisted above the water with a cable and boom. A quick-release mechanism lets the sampler fall free through the water column. Aquatic vegetation is trapped against the bottom and severed by cutting edges along the base of sampler. A nylon net sack over the top retains severed plant fragments. A diver inserts a horizontal cutting blade in a slot at the level of the substrate before the sampler is hoisted to the surface. Manual insertion of the cutting plate by a diver makes use of the Louisiana box sampler comparatively efficient. In soft sediments, the sampler may penetrate too deeply and require lifting before the cutting plate can be inserted. Rocks, stumps, roots, and other debris may prevent complete closure of the cutting door.

The Osborne sampler[11] is a stainless steel box having outside dimensions of 50 cm × 50 cm × 60 cm high and a sampling area of 0.25 m². The sampler weighs 110 kg and is operated by winch and cable from a pontoon boat. After hoisting and suspending the sampler alongside the pontoon boat, a quick-release mechanism allows free-fall through the water column. Tempered steel blades along the bottom edge of the sampler cut vegetation during the descent. A wire mesh screen fastened to the top prevents loss of plant fragments. A hinged slotted door is closed with a lift cable and the sampler is winched to the pontoon boat

platform for removal of macrophytes and sediments. Because the sampler penetrates and collects sediments, the sample includes roots and rhizomes and can be used to estimate biomass as well as standing crop. Efficient operation and accurate biomass estimates require an unconsolidated substrate free of rocks and other debris.

The Waterways Experiment Station (WES)[12] sampler is made of perforated stainless steel and operated from a pontoon boat with an overhead beam that allows it to be hydraulically raised and lowered through a circular opening in the pontoon's platform. Two types are available: one is cylindrical with a sampling area of 0.28 m² and the other is square with sampling area of 0.39 m². Rotating cutting blades at the base of each sampler sever vegetation as the samplers are lowered. The bottom cutting plate of each is closed hydraulically. A major advantage of the WES sampler is its capability to obtain plant samples from any depth. The Louisiana Box and Osborne samplers, once released, free-fall to the substrate, whereas the hydraulic operation of the WES sampler controls its descent. The size and weight of the trailer and pontoon boat for the WES sampler restrict its use in certain water bodies and requires an improved ramp for launching. Although the WES square sampling head is reported to provide a more accurate estimate of standing crop than the circular one, a substantial underestimate of actual standing crop is reported.[12]

3. Sample Preparation and Analysis

a. Biomass:

1) Fresh weight (wet weight)—Wash samples free of silt and debris, place in a nylon bag (mesh size 0.75 cm) and spin in a garment washer at 560 rpm for 6 to 7 min to remove excess moisture. Weigh sample to nearest 0.01 g.

2) Dry weight—Dry subsample (not less than 10%) in a forced-air oven at 105°C

for 48 h or until a constant weight is achieved. The coefficient of variation for a series of subsamples *should not exceed 10%*. Calculate dry weight by dividing dry weight of subsample by fresh weight of subsample times fresh weight of sample.

3) Ash-free dry weight—Transfer dried subsample to a covered and preweighed crucible. Ignite at 550°C for 6 h. Calculate ash-free dry weight by determining the ratio between ash and dry weight times dry weight of sample.

b. Chlorophyll content: Extract fresh plant material with acetone made basic with $MgCO_3$. Grind the plant material and centrifuge at 2500 rpm for 10 to 15 min. Wash residue with acetone and add filtered washings to extract. Dry overnight over anhydrous Na_2SO_4 and dilute to 90% acetone with water. Determine chlorophyll (see Section 10200G).

c. Carbon content: Most entire plants contain 46 to 48% carbon on a dry-weight basis. This factor can be used to calculate carbon content and make comparisons.

d. Caloric content: Determine energy content by bomb calorimetry.

e. Species identification:

1) Sample preparation—Use fresh specimens for identification wherever possible. Avoid immature plants or plants lacking flowers. Because aquatic plants contain from 80 to 95% water and have less supportive tissue than terrestrial forms, a different procedure is required for drying, preserving, and mounting them. Collect plants during peak growing season at the time of flowering and/or fruit development if practical. Include the entire plant as the specimen (stems, rhizomes, leaves, roots, flowers, and fruit).

After collection of plants, either press them in the field[13,14] or wrap specimen in several layers of paper and submerge in water. Label wrapped specimens with date and location of collection on an index card and place sample and card in a plastic bag. Preferably use an ice chest containing crushed ice for storage in the field but do not store for longer than 4 to 6 h. Plants can be kept for several days under refrigeration at 4°C.

Prepare a dry mount by centering the plant on 100% rag herbarium paper. Clean the plant of all silt and residue. Place emergent plants immediately on paper because they take on a natural posture. Place a limp plant in a shallow pan of water and slide the herbarium paper under it; with a slow motion, raise the paper at a 30° angle while keeping the plant centered. The leaves and stems should lie flat on the paper. Drain off excess water, cover with wax paper to prevent plant from sticking to blotters, and place in a plant press between paper and blotters. Store at room temperature for 3 to 5 d, changing the blotters at least every other day until the plant is sufficiently dry for permanent mounting.

To prepare a wet mount place specimen in an airtight glass vessel filled with 1 part 10% formalin, 3 parts water, and a trace of powdered copper. Plants will remain lifelike and retain their color for many years in this condition.

2) Identification—A stereomicroscope is needed to identify many plants, especially aquatic grasses and sedges. Observe vegetative and floral structures by dissecting them, under magnification, with forceps and fine needle probes.

Preferably identify to species. Numerous references are available to assist in identifying aquatic macrophytes (see Section 10900G).

4. Data Presentation

Express fresh weight (wet weight), dry weight, and ash-free weight as grams or kilograms per square meter. Determine significant digits for dry weight and ash-free dry weight from the accuracy of the scale used to obtain fresh weight: do not use more significant digits than those used for expressing fresh weight. Report pigment as grams chlorophyll per gram dry plant matter and caloric value as gram calories per gram dry plant matter.

5. References

1. COX, G.W. 1967. Laboratory Manual of General Ecology. William C. Brown Co., Dubuque, Iowa.

2. KERSHAW, K.A. 1971. Quantitative and Dynamic Ecology. Edward Arnold Co., London, England.

3. LIND, C.T. & G. COTTAM. 1969. The submerged aquatics of University Bay: A study in eutrophication. *Amer. Midland Natur.* 81:353.

4. SCHMID, W.D. 1965. Distribution of aquatic vegetation as measured by line intercept with scuba. *Ecology* 46:816.

5. RASCHKE, R.L. & P.C. RUSANOWSKI. 1984. Aquatic macrophyton field collection methods and laboratory analyses. *In* W.M. Dennis & B.G. Isom, eds. Ecological Assessment of Macrophyton: Collection, Use, and Meaning of Data. Spec. Tech. Publ. 843, American Soc. Testing & Materials, Philadelphia, Pa.

6. PUCKERSON, L.L. & G.E. DAVID. 1975. An *in situ* quantitative epibenthic sampler. U.S. Dep. Interior National Park Serv. Rep., Everglades National Park, Homestead, Fla.

7. RASCHKE, R.L. & P.J. FREY. 1981. Benthic dome (BeD) sampler. *Progr. Fish-Cult* 43:56.

8. EDMONDSON, W.T. & G.G. WINBERG. 1971. A Manual on Methods for the Assessment of Secondary Productivity in Fresh Waters. IBP Handbook No. 17. Blackwell Scientific Publ., Oxford and Edinburgh, U.K.

9. ISOM, B.G. 1978. Benthic macroinvertebrates. *In* W.T. Mason, Jr., ed. Methods for the Assessment and Prediction of Mineral Mining Impacts on Aquatic Communities: A Review and Analysis. U.S. Dep. Interior, Harpers Ferry, W.Va.

10. MANNING, J.H. & R.E. JOHNSON. 1975. Water level fluctuation and herbicide application: An integrated control method for *Hydrilla* in a Louisiana reservoir. *J. Aquat. Plant Manage.* 13:11.

11. OSBORNE, J.A. 1984. The Osborne submersed aquatic plant sampler for obtaining biomass measurements. *In* W.M. Dennis & B.G. Isom, eds. Ecological Assessment of Macrophyton: Collection, Use, and Meaning of Data. Spec. Tech. Publ. 843, American Soc. Testing & Materials, Philadelphia, Pa.

12. SABOL, B.M. 1984. Development and use of the Waterways Experiment Station's hydraulically operated submersed aquatic plant sampler. *In* W.M. Dennis & B.G. Isom, eds. Ecological Assessment of Macrophyton: Collection, Use, and Meaning of Data. Spec. Tech. Publ. 843, American Soc. Testing & Materials, Philadelphia, Pa.

13. HAYNES, R.R. 1984. Techniques for collecting aquatic and marsh plants. *Ann. Missouri Bot. Gard.* 71:229.

14. TSUDA, R.T. & I.A. ABBOTT. 1985. Collection, handling, preservation, and logistics. *In* M.M. Littler & D.S. Littler, eds. Handbook of Phycological Methods. Cambridge University Press, Cambridge, England.

6. Bibliography

FORSBERG, C. 1959. Quantitative sampling of subaquatic vegetation. *Oikos* 10:233.

STEEL, R.G.D. & J.H. TORRIE. 1960. Principles and Procedures of Statistics with Special Reference to the Biological Sciences. McGraw-Hill Book Co., New York, N.Y.

FAGER, E.W., A.O. PLECHSIG, R.F. FORD, R.I. CLUTTER & R.J. GHELARDI. 1966. Equipment for use in ecological studies using SCUBA. *Limnol. Oceanogr.* 11:503.

LIVERMORE, D.F. & W.E. WUNDERLICH. 1969. Mechanical removal of organic production from waterways. *In* Eutrophication: Causes, Consequences, Correctives. National Acad. Sciences, Washington, D.C.

Also see Sections 10400A.2 and 10400C.5.

10400 E. Productivity

1. General Discussion

The complexity and heterogeneity of form, function, phenology, and distribution of aquatic macrophytes have resulted in diverse ways of determining their productivity. These methods can be grouped broadly as biomass methods, based on standing crop or on changes in crop, and metabolic methods, based on estimates of inorganic carbon or oxygen exchange resulting from photosynthesis. The biomass methods generally are simpler than the metabolic methods and require little specialized equipment or expertise and fewer assumptions. Biomass methods integrate responses to environmental conditions and provide estimates of above-ground production only. They are best used for long-term comparisons (several months to a year) because they are easily confounded by seasonal changes. Biomass methods are insensitive to losses due to fragmentation, herbivory, and secretion or leaching of organics. In contrast, metabolic methods provide instantaneous measures of photosynthesis and thus reflect responses of plants to different environmental conditions. Metabolic methods for estimating plant productivity also can provide insight into factors controlling distribution and success. The principal drawbacks of the metabolic methods are that they require specialized equipment and assumptions that may be tenuous and are based on photosynthetic rates (typically measured over a period of minutes to hours) and require extrapolation to net assimilation over longer periods.

In choosing a method consider why macrophyte production is of interest and the use for the data, the habit (growth form and phenology) and habitat of the population, and the cost and effort required to obtain the desired information. The common methods are listed in Table 10400:I and are described below.

2. Biomass Methods

a. Biomass harvest methods: Measurement of standing crop varies from a simple, one-time sample of maximum biomass to complicated evaluations of seasonal biomass dynamics by methods originally intended for grassland plants. [1-3] These methods are applicable mainly to emergent and submersed macrophytes. Preferably evaluate productivity of floating plants by using relative growth rate, permanent quadrats, and random samples[4] or by the turnover and metabolic methods discussed in ¶s 2*b* and 3, below.

1) Standing crop measures—Peak standing crop is measured by the above-ground biomass (usually as dry weight per unit area) at the time of apparent maximum standing crop (usually time of flowering). This single measurement does not account for biomass carried over from the previous season, losses of material before the peak, or growth after the peak, and therefore generally underestimates net above-ground annual production (NAAP) to an extent depending on the relationship of annual turnover to maximum biomass.[5] The method provides a reasonable estimate of NAAP for many submersed species.

The seasonal biomass accumulation method[6] is a modification of the peak standing crop method that considers only the positive changes in live material. Live above-ground biomass is determined at the beginning of the growing season and at the time of maximum standing crop; the NAAP is calculated as the difference. This method accounts for yearly carryover of living material, but in some cases this will result in a further underestimate of NAAP relative to the peak standing crop method.[2]

When recruitment is continuous during

TABLE 10400:I. METHODS USED TO DETERMINE MACROPHYTE PRODUCTION*

Method	Emergent		Floating		Submersed	
	Deciduous	Evergreen	Deciduous	Evergreen	Deciduous	Evergreen
Biomass harvest:						
Standing crop	+	−	+	−	+	−
Biomass dynamics	+	−	−	−	−	−
Biomass tagging:						
Turnover, growth increment, and summed shoot maximum	+	+	+	+	+	+
Cohort	+	−	+	−	+	−
Below-ground biomass	+	+	+	+	+	+
Oxygen measurement:						
Light and dark bottle	−	−	−	−	+	+
Open system	−	−	−	−	+	+
Radiocarbon incorporation	−	−	−	−	+	+
Inorganic carbon exchange:						
Continuous CO_2 exchange	+	+	+	+	+	+
Discrete inorganic C measurement	−	−	−	−	+	+
Potentiometric C flux	−	−	−	−	+	+

*+ designates applicable method; − designates method not commonly applied. Evergreen implies retention of substantial above-ground biomass year round; deciduous implies that 10% or less of seasonal maximum biomass is present year round.

the growing season, the biomass peak is less well-defined and greater losses may occur before the seasonal maximum. Under such conditions, standing crop methods yield poor estimates of the NAAP.[3]

2) Biomass dynamics methods—These methods are applicable to emergent plants when dead material remains near the site of decomposition.

The Smalley method[7] estimates net production on the basis of samplings of live and dead material (per unit area) at regular 3- to 6-week intervals. Net production equals the increase in material between samplings: a decrease in live and dead material indicates no net production; an increase in live material and decrease in dead material indicates production equal to the increase in live material; and a decrease in live material and increase in dead biomass with a negative sum indicates no production, while a positive sum indicates production. The method underestimates production if dead material from other areas is present or if new growth is undetected when mortality is high. It is sensitive to sampling frequency and requires a homogeneous area large enough to accommodate replicate sampling. A modification of this method can be used where standing crop varies little from year to year; in this case net above-ground production is assumed to equal the summed losses of dead material.[8]

More complex procedures based on harvests from a series of paired plots have been proposed.[9,10]

b. Biomass tagging methods:

1) Biomass turnover and growth increment methods—Leaf turnover and biomass marking studies involve marking individual plant leaves to follow production, growth, and loss over the year or growing season. For plants with large, long-lived leaves and basal growth, e.g., *Vallisneria, Zostera,* macroalgae, determine short-term growth of individual leaves. Use these methods for studies of populations (species) rather than communities. Marking methods are particularly useful for evergreen plants, where there may be little seasonal change in biomass, and for plants where the ratio of production to biomass is either very much greater than, or very much less than, one. Turnover measures estimate production and biomass loss. They are considered better than harvest techniques, both with respect to accuracy and for the additional phenological information. Tagging methods require major efforts in regular censusing and SCUBA diving for sampling deep, submersed populations.

Where leaves or plant parts are about the same size seasonally, the method is relatively simple, requiring only periodic censusing of plant parts. These methods are not appropriate for species with much branching or different rates of leaf production and loss per branch, and are confounded by intense grazing or by sloughing of newly produced parts.

a) Biomass turnover—To determine turnover time of leaves and shoots, where individuals are easy to distinguish, or where vegetative spread is insignificant or easy to account for and leaf size is fairly constant, choose 10 to 50 individual plants and mark each leaf of each plant. If all new leaves are initiated to the inside of older leaves tag only the newest leaves initially. Mark new leaves or stems with anything that does not interfere with normal growth and is not easily lost, e.g., staples, hole punches, plastic bird rings, fishing line, indelible markers. At regular intervals, revisit plant, tag new leaves, and record the number of new leaves and total number of leaves present. Because new leaves may not be fully expanded at a sampling visit, use a convention concerning the developmental stage of leaves.[11] The sampling interval (weeks to months) depends on research needs and the likelihood of losing the youngest tagged leaf. Compute the annual leaf turnover for each individual plant as the number of new leaves produced annually divided by the maximum number of leaves present at any time in the season. Because the turnover rate is the ratio of production to biomass, calculate NAAP by multiplying the turnover ratio by the maximum above-ground biomass. If vegetative spread is common, modify this method so that all stems or leaves of a species within plots of a given area are tagged.[12,13] Revisit plots at regular intervals, tag new leaves, and record total number of leaves present. Calculate production as above. This method will not account for changes in plant size (increase in weight between years), for mortality, or for recruitment into the population, because only plants initially present that survive the season are included.

b) Growth increment measurements— If leaf size changes throughout the growing season or if several types of above-ground parts are present, use methods that account for such differences. Such methods frequently are used for seagrasses.[14-16] Mark each leaf of every plant within a quadrat at a set level above the bottom, relative to a stationary stake or frame. If plants are not buried deeply, use a set distance above the base (i.e., rhizome or root-shoot interface). After a predetermined time, remove all leaves or shoots at the level of the stake or at the set distance. Weigh unmarked leaves produced during the interval. Remove and weigh growth on older leaves (the portion below the marking, but above the ground or base). The combined weights are the net leaf production during the interval. These data also can be used to calculate relative growth rates as [(ln Δ weight)/Δ time]. Make detailed measurements of leaf growth rates by marking plants at the reference level more frequently, e.g., every other day.[17]

A modification of this method[18] permits computation of the growth rate of individual leaves and production of different plant parts.

c) Summed shoot maximum method— Where nondestructive measures are required determine shoot size and number and estimate production.[3] Choose permanent quadrats of a size allowing easy enumeration of plant parts. Label every stem in each quadrat. On regular sampling visits label and count new stems. Develop length:weight regressions using plants collected from areas near but outside the quadrats. This is required only once or twice during the year, depending on the characteristics of study species. Estimate net above-ground (shoot) primary productivity using the number of new parts produced and their weights (based on length:weight regressions). Estimate production per unit area as the mean leaf turnover, multiplied by maximum shoot mass.

For certain species, more complex variations of this method have been used.[19,20]

2) Cohort methods—Cohort methods often are used to determine net production of aquatic plants subject to substantial biomass loss before the seasonal maximum biomass is attained. They are useful for species in which groups of individuals, or subunits, initiated at the same time (cohorts), can be identified, that is, for plants where shoots emerge only during one time, or several discrete times, during an annual cycle.

The Allen curve method has been adapted for aquatic macrophytes. It provides an estimate of net production from tables of the numbers and weights of all individuals. It is particularly appropriate for populations for which shoot death and initiation occur throughout the growing season. The method can account for periods of negative production during the year.

Tag all members of a cohort (all leaves or shoots) shortly after initiation or emergence. This is usually the maximum number of individuals present at any time in the cohort, because mortality will decrease the number thereafter.[21] Some new individuals (but still members of the same cohort) may be present, however, by the second sampling visit.[3] Record the number of individuals and their mean weight (dry weight per leaf or shoot). Determine weight from size:weight regressions constructed from data for plants outside the study plot, but using members of the same cohort. Alternatively, harvest adjacent plots of cohort members to estimate the average weight per individual. Repeat for several replicate plots. Revisit plots regularly and record number and mean weight of individuals. Visit frequently enough to minimize potential for loss of young stems or leaves before they have been counted. Plot values (see Figure 10400:1) and determine total area beneath the curve by planimetery or gravimetrically, to estimate net annual above-ground production. Repeat if more than one cohort emerges per year; net above-ground annual production then is the sum of the areas beneath several curves.[3,21] If negative production occurs during the year add this loss back to yield net production. Losses usually can be

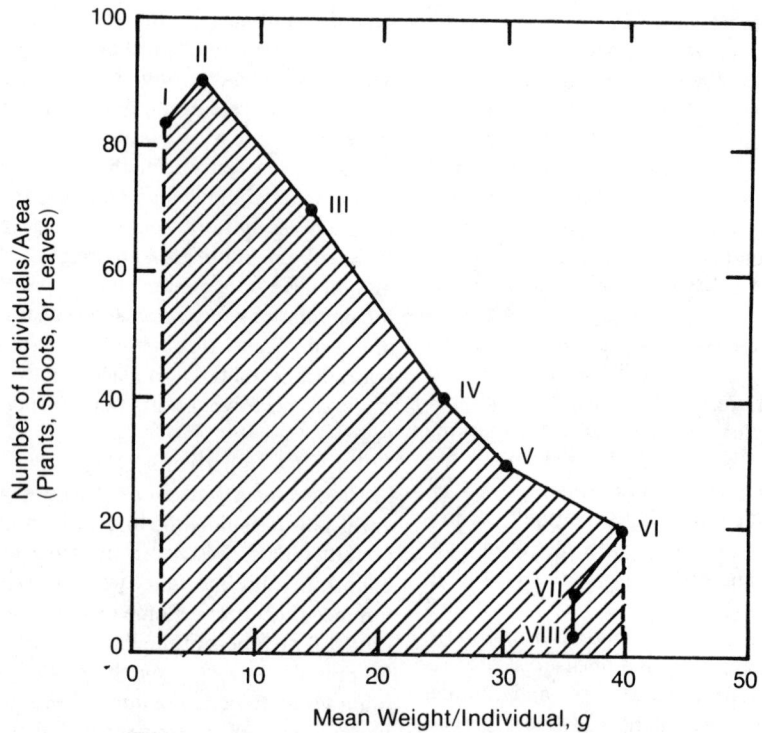

Figure 10400:1. Allen curve for a cohort of a population of aquatic macrophytes. Sample data are indicated by Roman numerals. Net annual above-ground production is proportional to the shaded area.

avoided by initiating studies after winter senescence.[21]

An adaptation of the cohort production method[22] recognizes the hierarchical structure of many aquatic macrophytes. This method may be used when differential turnover rates for plant subunits would confound the simple cohort method.

c. Estimates of below-ground production: The below-ground portions of aquatic macrophytes often comprise a substantial portion of the plant's biomass and net production, and may have important ecosystem implications because of metabolic activity and decay.[5] The amount of below-ground biomass varies dramatically between species, within a species, and seasonally for populations of the same species growing in different habitats.

1) Peak biomass—The maximum below-ground biomass is commonly taken as the net annual below-ground production. This may substantially over- or underestimate actual production, depending on biomass turnover. For many submersed nonevergreens, it provides a reasonable estimate of production, especially where shoot turnover approaches 1 and little below-ground biomass overwinters. For most floating-leaved and emergent macrophytes, it will overestimate biomass.

2) Seasonal biomass accumulation—A more appropriate, but more difficult, method involves repeated sampling and determination of live below-ground biomass at regular intervals during the year. For marsh plants, use sediment cores (25 cm deep, 6- to 10-cm diam) taken at regular

intervals (2 to 4 weeks).[23,24] Sieve the cores to remove below-ground biomass and determine weight of live portion. Live material usually can be distinguished by its light color, texture, and turgor. Stains also may be used; chlorazol black stains dead material[25] and tetrazolium stains live material. Use the seasonal change in live biomass as an estimate of production. As for shoots, the loss of plant material before the maximum biomass is attained and any production after the maximum biomass is attained are not accounted for.

3) Root turnover measures—For many applications, the most reasonable way to determine below-ground production is by extrapolation from shoot production or turnover. For many plants, production and turnover of below-ground biomass is directly related to growth and turnover of above-ground plant parts. However, the seasonal translocation of carbohydrate reserves between the below-ground and above-ground portions by most aquatic macrophytes requires that these turnover-based estimates be used only for long-term, annual, production estimates, unless morphological observations indicate otherwise. The establishment of a constant relationship between shoot and below-ground turnover[18,26] justifies such extrapolations.

3. Metabolic Methods

Metabolic methods estimate macrophyte productivity on the basis of short-term measures of inorganic carbon and oxygen exchange resulting from macrophyte photosynthesis and respiration. These short-term (usually < 3 h) measures must be integrated over time and space and converted to an estimate of organic matter production. Metabolic methods give short-term assessment of processes related to productivity that reflect a metabolic response to existing environmental conditions and they provide information concerning factors governing plant productivity.

The photosynthetic inorganic carbon budgets of aquatic macrophytes are more complex that those of terrestrial plants. When complex carbon or oxygen exchange patterns occur, a combination of metabolic methods may be used,[27,28] or a biomass method may be preferable.

a. Oxygen measurement: Determining changes in dissolved oxygen (DO) resulting from photosynthesis and respiration is the most common metabolic method for estimating macrophyte productivity. Changes in DO can be determined for plants in chambers (in the field or laboratory) or for the whole system when macrophyte metabolism dominates the oxygen dynamics. Measure DO concentrations with polarographic oxygen electrodes or by titration (Section 4500-O). For the most part, the methodology is similar to that described for phytoplankton and periphyton; however, several additional considerations warrant attention.

1) Light and dark bottle (chamber) method—This method is similar to those described for other primary producers, where portions of plants or the above-ground portions of several plants are enclosed in either clear or opaque containers. The change in oxygen per unit plant per unit time is an expression of plant photosynthesis (or respiration in the dark bottle) that is used to calculate productivity during the incubation period. Daily or longer-term production is estimated, using additional information on environmental conditions during the incubation period and during the period over which a production estimate is needed, and the relationship between the response of the plant segment incubated and the response of the entire plant (or population) to those environmental conditions.[29–33] See Sections 10200I.2 and 10300D.2 for procedural details. Mixing to simulate natural water movement is desirable.[34–41] Equipment modifications have been discussed,[42–48] as have specific analytical requirements.[29,49–52]

Determining productivity for periods longer than the test hours requires including respirations during the dark (24-h net production = daytime net production minus nighttime respiration or daytime gross production minus 24-h respiration) and extrapolating from the incubation period to these longer periods. Estimate net photosynthesis during the daylight period from productivity measurements in 3- to 4-h intervals from dawn to dusk daily, or by multiplying the production measured during a 3- to 4-h midday incubation by the fraction of total daily light received during the incubation period. For monthly or yearly periods there are several methods for extrapolating.[53,54] Most estimates of macrophyte production by these methods rely on a single midday incubation conducted at weekly-to-monthly intervals. There are three ways to estimate production for longer periods. First, measure productivity at 3- to 4-h intervals from dawn to dusk, on a clear day, several times during the growing season, to establish a relationship between midday photosynthesis and total daily photosynthesis. Estimate daily production on the basis of midday incubation at regular intervals (weekly to monthly), using the relationship between light during the incubation period to daily light and photosynthesis during the incubation period to daily photosynthesis as determined above. Estimate production on intervening days, when photosynthesis is not measured, on the basis of the available light on those days. Second, by using experimental evidence (P-I curves from plants incubated at different depths) or literature values, establish a relationship between light and productivity and a midday incubation (weekly to monthly) as a scaling factor. Carefully estimate production on the basis of available light for any day.[55,56] Third, use more involved modeling that accounts for growth characteristics and environmental conditions to provide estimates of macrophyte production.[57,58]

The epiphyton associated with a macrophyte may influence the determination of photosynthesis and respiration,[59–63] but the respective contributions to photosynthesis cannot be resolved by the oxygen method.[61] Alternatively, use the ^{14}C method for estimating macrophyte photosynthesis.

Short-term measures of macrophyte production using the light-dark bottle method do not account for the loss of organic substances that may be as much as 10% of the recently fixed carbon.[64] Other losses of fixed carbon, as well as sloughing, grazing, and fragmentation, also are unaccounted for. Further, the light-dark bottle oxygen method cannot predict the allocation of photosynthate to below-ground growth or reproduction.[5,65] The inability to account for such factors is a shortcoming of all metabolic methods, but oxygen methods provide reasonable estimates of short-term biomass net accumulation for some plants.[33,52]

2) Open-system oxygen method—In flowing-water systems where macrophytes dominate primary production, analysis of diurnal oxygen curves can be used to estimate macrophyte production.[66,67] The change in the oxygen content of a parcel of water is the result of both community metabolism and oxygen diffusion across the water surface (see Section 10300D.4). Entry of groundwater or surface runoff is assumed negligible during measurements. Criteria[68] for determining the suitability of a system for open-system monitoring include extent of uniform area, high biological activity, water depth and residence time (influencing observable changes in water chemistry), turbulence (influencing spatial variation for monitoring and gaseous exchange with the atmosphere), and uniformity of the channel (allowing the calculation of production per unit area and providing uniform residence time). The accuracy and sensitivity of several methods have been compared.[69]

The single-station method (Section 10300D.4) is used most commonly. In addition to the radiotracer method presented, the reaeration coefficient can be determined by direct methods or calculations based on physical parameters.[69–71] To use a single station assume stream homogeneity above the region of measurement.

To use an automated two-station system, calculate a continuous function (Fourier series) to determine an exact solution to the oxygen mass balance equation, rather than the approximate finite difference solution.[72] This system models net productivity so that the oxygen concentrations at the downstream station can be predicted accurately from upstream values. Additionally, it provides detailed information, such as hourly variations in net production and seasonal changes in community photosynthetic characteristics.[73]

For analytical details, see Section 10300D.4a and b.

Determine production from the change in DO between the two stations:

$$dc/dt = K(c_s - c) + \beta(t)$$

where:

dc/dt = change in DO concentration between stations,

K = reaeration coefficient,

c_s = saturation DO concentration, mg/L,

c = DO concentration, mg/L,

$\beta(t)$ = net productivity, and

t = flow time between stations.

For short time intervals, write t as a Fourier cosine series:

$$\beta(t) = A_o/2 + \sum_{n=1}^{\infty} A_n \cos nwt$$

where:

A_n = Fourier coefficients and

$w = 2\pi/48$.

Determine enough Fourier coefficients to solve for $\beta(t)$.[72] After solving the equation

plot estimated net photosynthetic rates as a function of incoming solar radiation to evaluate the solution for net production. An anomalous relationship between light and net production, or an unreasonable variation in net photosynthesis over time suggests solution errors. The estimate of K is the weak point of the method; an iterative approach to estimation of K, based on the morning and afternoon relationship between net production and light,[69] has been proposed.

b. Radiocarbon incorporation method: The radiocarbon (^{14}C) tracer method[74] has been used extensively for estimating the productivity of virtually all aquatic primary producers. It is based on the measurement of radiocarbon incorporation when a small but known quantity is made available to a plant. The proportion of tracer added to that incorporated indicates the fraction of stable carbon that is incorporated. This method is more sensitive than oxygen methods and thus can be used where photosynthetic rates are very low or where carbon consumption is excessive.

Although this method directly measures the incorporation of external inorganic carbon, the relationship of this incorporation to net or gross photosynthesis is not without controversy.[75] The general consensus is that the ^{14}C method for macrophytes provides a rate of carbon incorporation between net and gross photosynthesis, but closer to net photosynthesis.[30,33,42,47,52,76] The primary drawback of the ^{14}C method is that it provides no measure of dark-period respiration. If the photoassimilation of ^{14}C is assumed to estimate net photosynthesis for macrophytes (gross photosynthesis minus respiration), then respiration during the dark period must be measured separately to account for carbon lost at night in the calculation of 24-h net production (24-h net primary production = gross photosynthesis − 24-h respiration, or here, 24-h NPP = daytime ^{14}C uptake − nighttime respiration).[52] As noted for the oxygen method, respiration usually is assumed to be identical during the day and night for macrophytes.[33,52]

For procedural details see above for the light and dark bottle oxygen method and Section 10300D.3. Analysis of organic fractions into which carbon 14 may be incorporated has been described [59,77,78] as have procedures for carbon 14.[79-83]

c. Inorganic carbon exchange methods: Changes in the inorganic carbon concentration in air or water surrounding aquatic macrophytes as a result of photosynthesis and respiration can be determined by several methods. These methods provide a highly sensitive, direct measure of photosynthetic carbon uptake, thus requiring few assumptions, and can provide additional information on plant physiology. They require considerable expertise and expensive equipment. Carbon-exchange methods that use closed chambers have the same potential problems as the light-dark chamber oxygen method; however, the formation of bubbles during macrophyte photosynthesis usually does not interfere.

1) *Continuous infrared gas analysis (IRGA)*—The continuous IRGA method is based on measuring the change in CO_2, resulting from photosynthesis and respiration, in an air stream that has passed over the leaf or leaves of a plant. The change in CO_2 concentration is determined from the difference in concentration between a reference and a sample air flow. For submersed plants, enclose portions of the plant in a cuvette filled with a bathing solution (e.g., water from the site of collection), and bubble a gas stream through the solution. The CO_2 in the air stream equilibrates with the CO_2 (all components of the carbonate system) in the water. The change in CO_2 concentration of the air stream reflects plant metabolism and can be measured as the air stream passes through the IR analyzer. Details of equipment and procedures have been discussed.[76,84-90]

2) *Discrete inorganic carbon exchange*—Metabolic inorganic carbon exchange with the air or water also can be measured by determining the change in inorganic carbon in sealed containers after a discrete incubation period, using IRGA,[63,87,91-93] gas chromatography,[94] or a total organic carbon analyzer operating in the inorganic carbon mode (see Section 5310). The change in CO_2 in the air, for emergent and floating plants, or the change in dissolved inorganic carbon (DIC) in the water, for submersed plants, is determined for subsamples of the incubation medium. This method is analogous to the light and dark bottle oxygen method, except that changes in the CO_2 or DIC concentrations are determined, rather than the change in DO concentration.[33] This method requires more expensive equipment than the oxygen method does; however, in low-DIC waters it is much easier to measure the change in DIC resulting from plant metabolism than the corresponding change in DO (due to the relative abundance of DO and DOC and the high sensitivity of the DIC method). See references cited above for procedural details.

The change in CO_2 in the light is net photosynthesis. A darkened chamber can be used to determine respiration. Extrapolation to estimate production and additional methodological considerations are covered in the discussion for the light and dark bottle oxygen method. As for other carbon exchange methods, the exchange of CO_2 with the atmosphere, as well as marl formation, can result in errors in the determination of carbon exchange.

3) *Potentiometric measurement of inorganic carbon flux*—The physical-chemical relationships of the carbonate buffer systems in natural waters dictate a definable relationship between DIC (or alkalinity) and pH. On the basis of such relationships, changes in the pH of the water resulting from plant-mediated changes in DIC concentrations (due to photosynthesis and respiration) can be used

to determine the inorganic carbon exchange for submersed macrophytes. The method presented here is based on the estimation of the DIC or total carbon (C_T) determined by a Gran titration.[95] This approach requires only a good pH meter capable of readings to 0.1 pH unit, an electrode, and careful laboratory procedure. However, it is not suitable for waters of very high or low pH, and has limitations similar to those of the oxygen method. While it has been applied primarily to laboratory studies,[96-98] it is adaptable for field use.

To measure inorganic carbon flux,[99] incubate and collect sample as described for the light and dark bottle method. Use 125-mL gastight bottle or polypropylene syringe for sampling. Determine temperature of a 2.0- to 100-mL water sample and add sample to a titration vessel containing a magnetic stir bar, or add stir bar to collection bottle for direct titration. Assure that titration vessel is partially sealed around the outlet to provide protection from the atmosphere. Record initial pH. Stir sample and use a 0.5- to 5-mL-capacity piston syringe-buret to titrate stepwise with HCl or other appropriate acid of known normality. Use acid of such normality that the total solution volume is not changed by more that 10% by the end of titration. Record pH and volume of titrant added at three points between pH 7.6 and 6.7 (or lower if necessary) and another three points between pH 4.4 and 3.7.

Calculate net photosynthesis by the change in carbon in the light bottle or container, and respiration by the change in carbon in the dark container as follows: Calculate F_2 from pH readings in the lower pH range as

$$F_2 = [\text{antilog} \ (a - \text{pH})] \times [(V_s + v)/V_s]$$

where:

F_2 = antilogarithmic Gran functions for pH change with titrant additions,

a = any convenient number above the pH range, e.g., 5,

V_s = sample volume, mL, and

v = titrant volume, mL.

Plot F_2 against v and fit with a straight line, locating the intersection of the line on the v axis (v_2). Calculate F_1 from pH readings in the higher pH range as

$$F_1 = [\text{antilog} \ (b - \text{pH})] \\ \times (V_s + v) \times (v_2 - v)$$

where:

F_1 = antilogarithmic Gran functions for pH change with titrant additions and

b = any convenient number above the pH range, e.g., 8.

Plot F_1 against v, locating the intersection of the best-fitting straight line with the v axis (v_1). Then,

$$V_1 = v_1 \times \frac{1000}{V_s} \times n$$

$$V_2 = v_2 \times \frac{1000}{V_s} \times n$$

$$C_T = V_2 - V_1 = \frac{1000}{V_s} \times n \times (v_2 - v_1)$$

where:

V_1 = acidity, meq/L,

V_2 = total alkalinity, meq/L,

C_T = total DIC, mmol/L, and

n = normality of the acid titrant.

Use of antilog paper simplifies this procedure. The method can be used for water samples with an initial pH of less than 7 if an equal amount of NaOH is added to all samples.[96,99]

A formulation of a relationship of pH to C_T for a water of constant alkalinity and changing pH (e.g., as a result of photosynthesis or respiration) eliminates the need for titrations to determine the change in C_T in an incubation vessel resulting from macrophyte photosynthesis, as only the initial and final pH would be required.[99]

4. Data Presentation

Express seasonal and annual rates of macrophyte production in units of carbon or dry weight per unit area of colonization or littoral region. Occasionally, energy units (kcal) may be used. For most applications, the net annual above-ground production, expressed as grams C per square meter per year or grams dry weight (or ash-free dry weight) per square meter per year is suitable. Express the results of short-term (minutes, hours, days) measurements of production or photosynthesis as carbon fixed per unit shoot dry weight or per unit chlorophyll *a*. For emergent plants, report as carbon uptake per unit leaf or shoot surface area. The value of data collected depends on clear statements of how the production values are calculated and expressed (e.g., dried at 70 or 105°C, dry weight or ash-free dry weight, per unit area colonized or per area lake bottom, etc.) and on the provision of ancillary information to allow the data to be re-expressed and compared with values from other studies (chlorophyll content, ash content, above:below ground weight, etc.).

5. References

1. LINTHURST, R.A. & R.J. REIMOLD. 1978. An evaluation of methods for estimating the net aerial primary productivity of estuarine angiosperms. *J. Appl. Ecol.* 15:919.

2. SHEW, D.M., R.A. LINTHURST & E.D. SENECA. 1981. Comparisons of production computation methods in a southeastern North Carolina salt marsh. *Estuaries* 4:97.

3. DICKERMAN, J.A., A.J. STEWART & R.G. WETZEL. 1986. Estimates of net above ground production: Sensitivity to sampling frequency. *Ecology* 67:650.

4. TUCKER, C.S. & T.A. DEBUSK. 1981. Productivity and nutritive value of *Pistia stratioles* and *Eichhornia crassipes. J. Aquat. Plant Manage.* 19:61.

5. WESTLAKE, D.F. 1982. The primary productivity of water plants. *In* J.J. Symoens & P. Compere, eds. Studies on Aquatic Vascular Plants. Royal Botanical Soc. Belgium, Brussels.

6. MILNER, C. & R.E. HUGHES. 1968. Methods for the Measurement of Primary Production of Grasslands. IBP Handbook No. 6. Blackwell Scientific Publ., Oxford, England.

7. KIRBY, C.J. & J.G. GOSSELINK. 1976. Primary production in a Louisiana Gulf coast *Spartina alterniflora* marsh. *Ecology* 58:1052.

8. VALIELA, I., J.M. TEAL & W.J. SASS. 1975. Production and dynamics of salt marsh vegetation and the effects of experimental treatment with sewage sludge. *J. Appl. Ecol.* 12:973.

9. WIEGERT, R.G. & F.C. EVANS. 1964. Primary production and the disappearance of dead vegetation in an old field. *Ecology* 45:49.

10. LOMNICKI, R.A., E. BANDOLA & K. JANKOWSKA. 1968. Modification of the Wiegert-Evans method for the estimation of net primary production. *Ecology* 49:147.

11. MOELLER, R.E. 1978. Seasonal changes in biomass, tissue chemistry, and net production of the evergreen hydrophyte, *Lobelia dortmanna. Can. J. Bot.* 56:1425.

12. SAND-JENSEN, K. 1975. Biomass, net production and growth dynamics in an eelgrass (*Zostera marina* L.) population in Vellerup Vig, Denmark. *Ophelia* 14:185.

13. SAND-JENSEN, K. & M. SONDERGAARD. 1978. Growth and production of isoetids in oligotrophic Lake Kalguard, Denmark. *Verh. Int. Ver. Limnol.* 20:659.

14. ZIEMAN, J.C. & R.G. WETZEL. 1980. Productivity in seagrasses: Methods and rates. *In* R.C. Phillips & C.P. McRoy, eds. Handbook of Seagrass Biology. Garland STPM Press, New York, NY.

15. ZIEMAN, J.C. 1974. Methods for the study of the growth and production of turtle grass, *Thalassia testudinum* Konig. *Aquaculture* 4:139.

16. ZIEMAN, J.C. 1975. Quantitative and dynamic aspects of the ecology of turtle grass *Thalassia testudinum. Estuarine Res.* 1:541.

17. BROUNS, J.J.W.M. & F.M.L. HEIJS. 1986. Production and biomass of the seagrass *Enhalus acoroides* (L.f) Royle and its epiphytes. *Aquat. Bot.* 25:21.

18. JACOBS, R.P.W.M. 1979. Distribution and aspects of the production and biomass of eelgrass, *Zostera marina* L., at Roscoff, France. *Aquat. Bot.* 7:151.

19. JACKSON, D., S.P. LONG & C.F. MASON. 1986. Net primary production, decomposition and export of *Spartina anglica* on a Suffolk salt-marsh. *J. Ecol.* 74:647.

20. DAWSON, F.H. 1976. The annual production of the aquatic macrophyte *Ranunculus penicillatus* var. *calcareus* (R.W. Butcher) C.D.K. Cook. *Aquat. Bot.* 2:51.

21. MATHEWS, C.P. & D.F. WESTLAKE. 1969. Estimation of production by populations of higher plants subject to high mortality. *Oikos* 20:156.

22. CARPENTER, S.R. 1980. Estimating net shoot production by a hierarchical cohort method of herbaceous plants subject to high mortality. *Amer. Midland Natur.* 140:163.

23. VALIELA, I., J.M. TEAL & N.Y. PERSSON. 1976. Production dynamics of experimentally enriched salt march vegetation: Belowground biomass. *Limnol. Oceanogr.* 21:245.

24. GALLAGHER, J.L. & F.G. PLUMLEY. 1979. Underground biomass profiles and productivity in Atlantic coastal marshes. *Amer. J. Bot.* 66:156.

25. WILLIAMS, D.D. & N.E. WILLIAMS. 1974. A counterstaining technique for use in sorting benthic samples. *Limnol. Oceanogr.* 19:152.

26. WIUM-ANDERSON, S. & J. BORUM. 1984. Biomass variation and autotrophic production of an epiphyte-macrophyte community in coastal Danish area: I. Eelgrass (*Zostera marina* L.) biomass and net production. *Ophelia* 23:33.

27. KEELEY, J.E. & G. BUSCH. 1984. Carbon assimilation characteristics of the aquatic CAM plant *Isoetes howellii. Plant Physiol.* 76:525.

28. BOSTON, H.L. & M.S. ADAMS. 1986. The contribution of crassulacean acid metabolism to the annual productivity of two aquatic vascular plants. *Oecologia* 68:615.

29. HILL, B.H., J.R. WEBSTER & A.E. LINKINS. 1984. Problems in the use of closed chambers for measuring photosynthesis by a lotic macrophyte. *In* W.M. Dennis & B.G. Isom, eds. Ecological Assessment of Macrophyton: Collection, Use, and Meaning of Data. Spec. Tech. Publ. 843, American Soc. Testing & Materials, Philadelphia, Pa.

30. WETZEL, R.G. 1965. Techniques and problems of primary productivity measurements in higher aquatic plants and periphyton. *In* C.R. Goldman, ed. Primary Production in Aquatic

Environments. Mem. Ist. Ital. Idrobiol., 18 Suppl., Univ. California Press, Berkeley.

31. WESTLAKE, D.F. 1978. Rapid exchange between plant and water. *Verh. Int. Ver. Limnol.* 20:2363.

32. MOESLUND, B., M.G. KELLY & N. THYSSEN. 1981. Storage of carbon and transport of oxygen in river macrophytes: Mass-balance, and the measurement of primary productivity in rivers. *Arch. Hydrobiol.* 93:45.

33. KEMP, M.W., M.R. LEWIS & T.W. JONES. 1986. Comparison of methods for measuring production by the submersed macrophyte, *Potamogeton perfoliatus* L. *Limnol. Oceanogr.* 31:1322.

34. RAVEN, J.A. 1970. Exogenous inorganic carbon sources in plant photosynthesis. *Biol. Rev.* 45:167.

35. SMITH, F.A. & N.A. WALKER. 1980. Photosynthesis by aquatic plants: Effects of unstirred layers in relation to assimilation of CO_2 and HCO_3 and carbon isotope discrimination. *New Phytol.* 86:245.

36. BLACK, M.A., S.C. MABERLY & D.H.N. SPENCE. 1981. Resistance to carbon dioxide fixation in four submerged freshwater macrophytes. *New Phytol.* 69:54.

37. WESTLAKE, D.F. 1967. Some effects of low-velocity current on the metabolism of aquatic macrophytes. *J. Exp. Bot.* 18:187.

38. MADSEN, T.V. & M. SONDERGAARD. 1983. The effects of current velocity on the photosynthesis of *Callitrishe stagnalia* Scop. *Aquat. Bot.* 15:187.

39. JENKINS, J.T. & M.C.F. PROCTER. 1985. Water velocity, growth-form and diffusion resistances to photosynthesis CO_2 uptake in aquatic bryophytes. *Plant Cell. Environ.* 8:317.

40. RODGERS, J.H., JR., K.L. DICKSON & J. CAIRNS, JR. 1978. A chamber for *in situ* evaluation of periphyton productivity in lotic systems. *Arch. Hydrobiol.* 84:389.

41. DAWSON, F.H., D.F. WESTLAKE & G.I. WILLIAMS. 1981. An automatic system to study the responses of respiration and photosynthesis by submerged macrophytes to environmental variables. *Hydrobiol.* 77:277.

42. WETZEL, R.G. 1964. A comparative study of the primary productivity of higher aquatic plants, periphyton, and phytoplankton in a large shallow lake. *Int. Rev. Ges. Hydrobiol.* 49:1.

43. LOVE, R.J.R. & G.G.C. ROBINSON. 1977. The primary productivity of submerged macrophytes in West Blue Lake Manitoba. *Can. J. Bot.* 55:118.

44. LITTLER, M.M. & K.E. ARNOLD. 1985. Electrodes and chemicals. *In* M.M. Littler and D.S. Littler, eds. Handbook of Phycological Methods, Ecological Field Methods: Macroalgae. Cambridge University Press, Cambridge, England.

45. HATCHER, B.G. 1977. An apparatus for measuring photosynthesis and respiration of intact large marine algae and comparison of results with those from experiments with tissue segments. *Mar. Biol.* 43:381.

46. LOBBAN, C.S. 1978. Translocation of ^{14}C in *Macrocystis pyrifera* (giant kelp). *Plant Physiol.* 61:585.

47. BITTAKER, H.F. & R.L. IVERSON. 1976. *Thalassia testudinum* productivity: A field comparison of measurement methods. *Mar. Biol.* 37:39.

48. VOOREN, C.M. 1981. Photosynthetic rates of benthic algae from the deep coral reef of Curacao. *Aquat. Bot.* 10:143.

49. ADAMS, M.S., J. TITUS & M. MCCRAKEN. 1974. Depth distribution of photosynthetic activity in a *Myriophyllum spicatum* community in Lake Wingra. *Limnol. Oceanogr.* 19:377.

50. GUILIZZONI, P. 1977. Photosynthesis of the submergent macrophyte *Ceratophyllum demersum* in Lake Wingra. *Trans. Wis. Acad. Sci. Arts Lett.* 65:152.

51. PRIDDLE, J. 1980. The production ecology of benthic plants in some antarctic lakes II. Laboratory physiological studies. *J. Ecol.* 68:155.

52. YIPKIN, Y., S. BEER, E.P.H. BEST, T. KAIRESALO & K. SALONEN. 1986. Primary production of macrophytes: terminology, approaches and a comparison of methods. *Aquat. Bot.* 26:129.

53. WESTLAKE, D.F. 1974. Methods for measuring production rates—Calculation of day rates per unit of lake surface (b) periphyton and macrophytes. *In* R.A. Vollenweider, ed. A Manual on Methods for Measuring Primary Production in Aquatic Environments, 2nd ed. IBP Handbook No. 12. Blackwell Scientific Publ., Oxford, England.

54. WETZEL, R.G. & G.E. LIKENS. 1979. Limnological Analyses. W.B. Saunders Co., Philadelphia, Pa.

55. GOULDER, R. 1970. Day-time variations in the rates of production by two natural communities of submerged freshwater macrophytes. *J. Ecol.* 58:521.

56. MCCRACKEN, M.D., M.S. ADAMS, J. TITUS & W. STONE. 1975. Diurnal course of photosynthesis in *Myriophyllum spicatum* and *Oedogonium*. *Oikos* 26:355.

57. TITUS, J.E., R.A. GOLDSTEIN, M.S. ADAMS, J.B. MANKIN, R.V. O'NEILL, P.R. WEILER, JR., H.H. SHUGART & R.S. BOOTH. 1975. A production model for *Myriophyllum spicatum* L. *Ecology* 56:1129.

58. BEST, E.P.H. 1981. A preliminary model for the growth of *Ceratophyllum demersum* L. *Verh. Int. Ver. Limnol.* 21:1484.

59. ALLEN, H.A. 1971. Primary productivity, chemo-organotrophy, and nutritional interactions of epiphytic algae and bacteria on macrophytes in the littoral of a lake. *Ecol. Monogr.* 41:97.

60. HUTCHINSON, G.E. 1975. A Treatise on Limnology. Vol. 3. Limnological Botany. John Wiley & Sons, New York, N.Y.

61. SAND-JENSEN, K. 1977. Effects of epiphytes on eelgrass photosynthesis. *Aquat. Bot.* 3:55.

62. CATTANEO, A. & J. KALFF. 1980. The relative contribution of aquatic macrophytes and their epiphytes to the production of macrophyte beds. *Limnol. Oceanogr.* 25:280.

63. KAIRESALO, T. 1983. Photosynthesis and respiration with an *Equisetum fluviatile* L. stand in Lake Paajarvi, southern Finland. *Arch. Hydrobiol.* 96:317.

64. HOUGH, R.A. & R.G. WETZEL. 1972. A ^{14}C-assay for photorespiration in aquatic plants. *Plant Physiol.* 49:987.

65. WESTLAKE, D.R. 1975. Primary production of freshwater macrophytes. *In* J.P. Cooper, ed. Photosynthesis and Productivity in Different Environments. IBP Handbook No. 3. Cambridge University Press, Cambridge, England.

66. ODUM, H.T. 1956. Primary production in flowing waters. *Limnol. Oceanogr.* 1:103.

67. EDWARDS, R.W. & M. OWENS. 1962. The effects of plants on river conditions. IV. The oxygen balance of a chalk stream. *J. Ecol.* 50:207.

68. KINSEY, D.W. 1985. Open-flow systems. *In* M.M. Littler & D.S. Littler, eds. Handbook of Phycological Methods, Ecological Field

Methods: Macroalgae. Cambridge University Press, Cambridge, England.

69. KOSINSKI, R.J. 1984. A comparison of the accuracy and precision of several open-water oxygen productivity techniques. *Hydrobiol.* 119:139.

70. OWENS, M. 1974. Methods of measuring production rates in running water. *In* R.A. Vollenweider, ed. A Manual on Methods for Measuring Primary Production in Aquatic Environments. IBP Handbook No. 12. Blackwell Scientific Publ., Oxford, England.

71. WILCOCK, R.J. 1982. Simple predictive equations for calculating stream reaeration coefficients. *N. Zealand Sci.* 25:53.

72. KELLY, M.G., G.M. HORNBERGER & B.J. COSBY. 1974. Continuous automated measurements of rates of photosynthesis and respiration in an undisturbed river community. *Limnol. Oceanogr.* 19:305.

73. KELLY, M.G., N. THYSSEN & B. MOESLAND. 1983. Light and the annual variation of oxygen- and carbon-based measurements of productivity in a macrophyte-dominated river. *Limnol. Oceanogr.* 28:503.

74. STEEMANN NIELSEN, E. 1952. The use of radio-active carbon (C^{14}) for measuring organic production in the sea. *J. Cons. Perm. Int. Explor. Mer* 18:117.

75. PETERSON, B.J. 1980. Aquatic primary productivity and the ^{14}C-CO_2 method: A history of the productivity problem. *Annu. Rev. Ecol. System.* 11:359.

76. TITUS, J.E., M.S. ADAMS, T.D. GUSTAFSON, W.H. STONE & D.F. WESTLAKE. 1979. Evaluation of differential infrared gas analysis for measuring gas exchange by submersed aquatic plants. *Photosynthetica* 13:294.

77. HOUGH, R.A. 1979. Photosynthesis, respiration, and organic carbon release in *Elodea canadensis* Michx. *Aquat. Bot.* 7:1.

78. ARNOLD, K.E. & M.M. LITTLER. 1985. The carbon-14 method for measuring primary productivity. *In* M.M. Littler & D.S. Littler, eds. Handbook of Phycological Methods, Ecological Field Methods: Macroalgae. Cambridge University Press, Cambridge, England.

79. BURNISON, B.K. & K.T. PEREZ. 1974. A simple method for the dry combustion of ^{14}C-labeled materials. *Ecology* 55:899.

80. BEER, S., A.J. STEWART & R.G. WETZEL. 1982. Measuring chlorophyll *a* and ^{14}C-labeled photosynthate in aquatic angiosperms by use of a tissue solubilizer. *Plant Physiol.* 69:57.

81. FRANKO, D.A. 1986. Measurement of algal chlorophyll-a and carbon assimilation by a tissue solubilizer method: A critical analysis. *Arch. Hydrobiol.* 106:327.

82. HISCOX, J.D. & G.F. ISRAELSTAM. 1979. A method for the extraction of chlorophyll from leaf tissue without maceration. *Can. J. Bot.* 57:1332.

83. FILBIN, G.J. & R.A. HOUGH. 1984. Extraction of ^{14}C-labeled photosynthate from aquatic plants with dimethyl sulfoxide (DMSO). *Limnol. Oceanogr.* 29:426.

84. BROWSE, J.A. 1985. Measurement of photosynthesis by infrared gas analysis. *In* M.M. Littler & D.S. Littler, eds. Handbook of Phycological Methods, Ecological Field Methods: Macroalgae. Cambridge University Press, Cambridge, England.

85. VAN, T.K., W.T. HALLER & G. BOWES. 1976. Comparison of the photosynthetic characteristics of three submersed aquatic plants. *Plant Physiol.* 58:761.

86. CEULEMANS, R., F. KOCKELBERGH & I. IMPENS. 1986. A fast, low cost and low power requiring device for improving closed loop CO_2 measuring systems. *J. Exp. Bot.* 37:1234.

87. BARKO, J.W., P.G. MURPHY & R.G. WETZEL. 1977. An investigation of primary production and ecosystem metabolism in a Lake Michigan dune pond. *Arch. Hydrobiol.* 81:155.

88. SALE, P.J.M., P.T. ORR, G.S. SHELL & D.J.C. ERSKINE. 1985. Photosynthesis and growth rates in *Salivinia molesta* and *Eichhornia crassipes. J. Appl. Ecol.* 22:125.

89. BROWSE, J.A. 1979. An open-circuit infrared gas analysis system for measuring aquatic plant photosynthesis at physiological pH. *Aust. J. Plant Physiol.* 6:493.

90. VINCENT, W.F. & C. HOWARD-WILLIAMS. 1986. Antarctic stream ecosystems: physiological ecology of a blue-green algae epilithon. *Freshwater Biol.* 16:219.

91. TITUS, J.E. & W.H. STONE. 1982. Photosynthetic response of two submersed macrophytes to dissolved inorganic carbon concentrations and pH. *Limnol. Oceanogr.* 27:151.

92. MADSEN, T.V. 1985. A community of submerged aquatic cam plants in Lake Kalgaard, Denmark. *Aquat. Bot.* 23:97.

93. SALONEN, K. 1981. Rapid and precise determination of total inorganic carbon and some gases in aqueous solutions. *Water Res.* 15:403.

94. STAINTON, M.P. 1973. A syringe gas-stripping procedure for gas-chromatographic determination of dissolved inorganic carbon in fresh water and carbonates in sediments. *J. Fish. Res. Board Can.* 30:1441.

95. GRAN, G. 1952. Determination of the equivalence point in potentiometric tirations. Part II. *Analyst* 77:661.

96. DENNY, P., P.T. ORR & D.J. ERSKINE. 1983. Potentiometric measurements of carbon dioxide flux of submerged aquatic macrophytes in pH-stated natural waters. *Freshwater Biol.* 13:507.

97. ALLEN, E.D. & D.H.N. SPENCE. 1981. The differential ability of aquatic plants to utilize the inorganic carbon supply in freshwaters. *New Phytol.* 87:269.

98. MABERLY, S.C. & D.H.N. SPENCE. 1983. Photosynthetic inorganic carbon use by freshwater plants. *J. Ecol.* 71:705.

99. TALLING, J.F. 1973. The application of some electrochemical methods to the measurement of photosynthesis and respiration in fresh waters. *Freshwater Biol.* 3:335.

6. Bibliography

EDWARDS, R.W. & M. OWENS. 1960. The effects of plants on river conditions. I. Summer crops and estimates of net productivity of macrophytes in a chalk stream. *J. Ecol.* 48:151.

WESTLAKE, D.F. 1963. Comparisons of plant productivity. *Biol. Rev.* 38:385.

WESTLAKE, D.F. 1965. Some basic data for investigations of the productivity of aquatic macrophytes. *In* C.R. Goldman, ed. Primary Production in Aquatic Environments. Mem. Ist. Ital. Idrobiol., 18 Suppl., Univ. California Press, Berkeley.

WESTLAKE, D.F. 1967. Some effects of low-velocity currents on the metabolism of aquatic macrophytes. *J. Exp. Bot.* 18:187.

WETZEL, R.G. 1969. Excretion of dissolved organic compounds of aquatic macrophytes. *Bioscience* 19:539.

VOLLENWEIDER, R.A., ed. 1974. A Manual on Methods for Measuring Primary Production in Aquatic Environments, 2nd ed. IBP Handbook 12. Blackwell Scientific Publ., Oxford, England.

COOPER, J.P., ed. 1975. Photosynthesis and Productivity in Different Environments. IBP Handbook 3. Cambridge University Press, Cambridge, England.

HUTCHINSON, G.E. 1975. A Treatise on Limnology. John Wiley & Sons, New York, N.Y.

WETZEL, R.G. 1975. Limnology. W.B. Saunders Co., Philadelphia, Pa.

BOTT, T.L., J.T. BROCK, C.E. CUSHINGS, S.V. GREGORY, D. KING & R.C. PETERSEN. 1978. A comparison of methods for measuring primary productivity and community respiration in streams. *Hydrobiologia* 60:3.

CATTANEO, A. & J. KALFF. 1978. Seasonal changes in the epiphyte community of natural and artificial macrophytes in Lake Memphremagog. (Que.–Vt.). *Hydrobiologia* 60:135.

ONDOK, J.P. & J. KVET. 1978. Selection of sampling areas in assessment of production. *In* D. Dykyjova & J. Kvet, eds. Pond Littoral Ecosystems: Structure and Functioning. Ecological Studies 28. Springer-Verlag, Berlin.

GILLESPIE, D.M. & A.C. BENKE. 1979. Methods for calculating cohort production from field data—some relationships. *Limnol. Oceanogr.* 24:171.

LEWIS, M.R., W.M. KEMP, J.J. CUNNINGHAM & J.C. STEVENSON. 1982. A rapid technique for preparation of aquatic macrophyte samples for measuring ^{14}C incorporation. *Aquat. Bot.* 13:203.

SCHUBAUER, J.P. & C.S. HOPKINSON. 1984. Above- and belowground emergent macrophyte production and turnover in a coastal marsh ecosystem, Georgia. *Limnol. Oceanogr.* 29:1052.

10500 BENTHIC MACROINVERTEBRATES*

10500 A.   Introduction

1. Definition

Benthic macroinvertebrates are animals inhabiting the substratum of lakes, streams, estuaries, and marine waters. They may construct attached cases, tubes, or nets that they live on or in, or roam freely over rocks, organic debris, and other substrates during all or part of their life cycle. Although very young specimens of many forms are small, macroinvertebrates are considered by definition to be visible to the unaided eye and are retained on a U.S. Standard No. 30 sieve (0.595-mm openings). The standard sieve opening for marine benthic fauna also is 0.595 mm (U.S. Standard No. 30 sieve); however, to accommodate some historical data bases a 1.0 mm, U.S. Standard No. 18 sieve may be used. Programs already using a No. 30 sieve should continue to do so and all new programs should use a No. 30 sieve to collect organisms from marine sediments. Assessment of species composition, richness, diversity, evenness, and major taxonomic patterns may be enhanced significantly by use of a No. 30 sieve. The major disadvantages of this sieve size are the time and cost to process samples. Included among the macroinvertebrates are sponges, coelenterates, flatworms, nematodes, roundworms, annelids, mollusks, echinoderms, macrocrustaceans, insects, and other invertebrates.

2. Response to Environment

The composition and density (number of individuals per unit area) of macroinvertebrate communities in streams, lakes, estuaries, and marine waters are reasonably stable from year to year in unperturbed environments. However, seasonal fluctuations associated with life-cycle dynamics of individual species may result in extreme variation at specific sites within any calendar year.

Most aquatic habitats, particularly free-flowing streams and waters with acceptable water quality and substrate conditions, support diverse macroinvertebrate communities in which there is a reasonably balanced distribution of species among the total number of individuals present. Such communities respond to changing habitat quality by adjustments in community structure. However, many estuaries are dominated by a few species. Small changes in their relative numbers may not be indicative of changes in water quality.

Macroinvertebrate community responses to environmental perturbations are useful in assessing the impact of municipal, industrial, oil, and agricultural wastes, and impacts from other land uses on surface water bodies. Three situations for which patterns of macroinvertebrate community structure change have been documented are organic loading, substrate alteration, and toxic chemical pollution. Severe organic pollution usually results in a restriction in the variety of macroinvertebrates to only the most tolerant ones and a corresponding increase in density of those tolerating the polluted conditions, usually low dissolved oxygen concentration. On the other hand, siltation and toxic chemical pollution may not only reduce but eliminate the entire macroinvertebrate community from an affected area. Not all cases conform to those described because conditions may be mediated by other environmental and biological factors.

Assessing the impact of a pollution source generally involves comparison of macroinvertebrate communities and their

---

* Approved by Standards Methods Committee, 1988.

habitats at sites influenced by pollution with those collected from adjacent unaffected sites. The procedure includes sampling and analyzing both communities and subsequently determining whether the presumed pollution-affected community differs from the nonaffected one. The basic information required for most community structure analyses is a count of individuals per species. From the count data, the communities can be characterized and compared according to community structure, density, biomass, diversity, or other analyses. Equally desirable is a characterization of the dissolved oxygen concentration, substrate, water depth, type of sediment, grain size, and total organic carbon (TOC).

3. Bibliography

NEEDHAM, J.G. & P.R. NEEDHAM. 1941. A Guide to the Study of Freshwater Biology. Comstock Publishing Co., Ithaca, N.Y.

HUTCHINSON, G.E. 1957. A Treatise on Limnology. John Wiley & Sons, New York, N.Y.

HYNES, H.B.N. 1963. The Biology of Polluted Waters. Liverpool Univ. Press, England.

MACAN, T.T. 1963. Freshwater Ecology. John Wiley & Sons, New York, N.Y.

RUTTNER, F. 1966. Fundamentals of Limnology. Univ. Toronto Press, Toronto, Ont.

MACKENTHUN, K.M. & W.M. INGRAM. 1967. Biological Associated Problems in Freshwater Environments. Federal Water Pollution Control Admin., Washington, D.C.

HYNES, H.B.N. 1970. The Ecology of Running Waters. Univ. Toronto Press, Toronto, Ont.

ODUM, E.P. 1971. Fundamentals of Ecology, 3rd ed. Saunders Publishing Co., Philadelphia, Pa.

McNULTY, J.K. 1971. Effects of sewage pollution in Biscayne Bay, Florida: sediments and the distribution of benthic and fouling macroorganisms. Bull. Mar. Sci. Gulf Carib. 11:394.

KINNER, P., D. MAURER & V. LEATHEM. 1974. Benthic invertebrates in Delaware Bay: Animal-sediment associations of the dominant species. Int. Rev. Ges. Hydrobiol. 59:685.

COULL, B.C., ed., 1977. Ecology of Marine Benthos. University of South Carolina Press, Columbia.

NICHOLS, J.A. 1977. Benthic community structure near the Woods Hole sewage outfall. Int. Rev. Ges. Hydrobiol. 52:235.

PEARSON, T.H. & R. ROSENBERG. 1978. Macrobenthic succession in relation to organic enrichment and pollution of the marine environment. Oceanogr. Mar. Biol. Annu. Rev. 16:229.

FAUCHALD, K. & P.A. JUMARS. 1979. The diet of worms: A study of polychaete feeding guilds. Oceanogr. Mar. Biol. Annu. Rev. 17:193.

EWING, R.M. & D.M. DAUER. 1982. Macrobenthic community structure of the lower Chesapeake Bay. I. Old Plantation Creek, Kings Creek, Cherrystone Inlet and the adjacent offshore area. Int. Rev. Ges. Hydrobiol. 67:777.

TOURTELLOTE, G.H. & D.M. DAUER. 1983. Macrobenthic communities of the lower Chesapeake Bay. II. Lynnhaven Roads, Lynnhaven Bay, Broad Bay, and Linkhorn Bay. Int. Rev. Ges. Hydrobiol. 68:59.

HAWTHORNE, S.D. & D.M. DAUER. 1983. Macrobenthic communities of the lower Chesapeake Bay. III. Southern Branch of the Elizabeth River. Int. Rev. Ges. Hydrobiol. 68:193.

DAUER, D.M., T.L. STOKES, JR., H.R. BARKER, JR., R.M. EWING & J.W. SOURBEER. 1984. Macrobenthic communities of the lower Chesapeake Bay. IV. Baywide transects and inner continental shelf. Int. Rev. Ges. Hydrobiol. 69:1.

RESH, V.H. & D.M. ROSENBERG. 1984. The Ecology of Aquatic Insects. Praeger Publ., New York, N.Y.

10500 B.   Sample Collection

1. General Considerations

Before conducting a survey, determine specific objectives and define clearly the information sought. Discussion with water chemists, hydrologists, and individuals from other disciplines will be helpful. Ultimate selection of a methodology will depend on whether a stream, lake, reservoir, or marine area is to be studied. For example, to determine whether the macroinvertebrate community downstream from a discharge is damaged, only a few sampling stations upstream and downstream from the discharge, usually two or more, are needed. However, if the objective is to delimit the extent of damage from a discharge or series of discharges, it is necessary to have reference stations upstream from all discharges, to bracket each discharge with stations, and to establish stations downstream. In marine waters, it may be necessary to sample a nearby estuary or, for open ocean waters, to sample some distance from the discharge point.

Characterize the physicochemical properties of faunal sampling station substrate and overlying water. Measure such properties as particle size characteristics (mean diameter, sorting coefficient, skewness, and kurtosis); size class distribution of sand, silt, and clay; organic content and toxic pollutant concentration; temperature, salinity, hardness, alkalinity, dissolved oxygen, and nutrient (total and dissolved) concentrations; and biochemical oxygen demand and water depth.

After gaining a thorough understanding of the factors involved with a particular body of water, select specific areas to be sampled. There is no set number of sampling stations that will be sufficient to monitor all possible waste discharges. No water quality survey is routine, nor can one be conducted totally on a "cookbook" basis. However, if some basic rules such as the following are adhered to carefully, a sound survey can be designed:

1. Always establish a reference station(s) upstream or at a point remote from all wastewater discharges of concern. Because most surveys are made to determine the damage that pollution causes to aquatic life, this will be the basis for comparison of the biota in polluted and unpolluted areas. Preferably have at least two reference stations, one well away from, or upstream from, the discharge and the other directly above, or in the immediate vicinity of, the effluent discharge, but not subject to its influence. Whenever it is feasible, use reference stations having physicochemical characteristics similar to those of the substrate and overlying water of the receiving area.

2. Locate a station immediately downstream or in the affected area, in the immediate vicinity of each discharge.

3. If the discharge does not mix completely on entering the body of water, but channels along one side or disperses in a specific direction, subdivide stations into left-bank, midchannel, and right-bank sections of the stream, into concentric arcs in lakes and oceanic waters, or any other configuration that will meet study objectives.

4. Establish stations at various distances downstream from the last discharge of concern to determine the linear extent of damage. In the marine environment, an estuary nearby may be sampled or in open ocean waters samples may be taken in an area comparable with respect to currents, depth, sediment characteristics, and salinity.

5. To permit comparison of macroinvertebrate communities, be sure that all sampling stations are ecologically similar. For example, select stations that are similar with respect to bottom substrate (sand, gravel, rock, mud, organic content, or toxic chemicals concentration), depth, presence of riffles and pools, stream width, gradient, flow velocity, bank or shore cover, salinity, nutrient and dissolved oxygen concentrations, hardness, and wave exposure.

6. Collect samples for physical and chemical analyses close to biological sampling stations to assure correlation of findings; take such samples at the same time and from the same grab. Collect substrate samples for physicochemical analyses from the upper few centimeters where most organisms live.

7. Locate sampling stations for macroinvertebrates in an area not influenced by atypical conditions.

8. Discharges in areas near a coast may be subject to salt water intrusion (salt water wedge). In such areas, macroinvertebrate

populations change drastically; document this effect.

9. When sampling in flowing water optimally initiate sampling at the most downstream station and then proceed up-current to minimize disruptions induced by the sampling itself.

For a long-term biological monitoring program collect macroinvertebrates at each station at least once during each of the annual seasons. More frequent sampling may be necessary if the characteristics of the effluents change or if spills occur. Make allowance for collections at night where "drift" organisms are of special concern. In general, the most critical period for macroinvertebrates in streams is during periods of high temperature and low flow, whereas in estuarine and marine environments it is the period of maximum stratification and poor vertical mixing. If available time and funds limit sampling frequency, make at least one survey during the critical time.

2. Sampling Design

A sample usually is defined as a portion taken from some larger aggregate about which inferences are to be made. The problem in collecting a representative sample arises from the variation usually encountered in successive samples. Without knowledge of sample variation, the degree to which the data truly represent the population cannot be known. Therefore, take replicate samples of a population or population aggregate if definitive statistical inferences about the population are to be made.[1]

Standardize sampling design to consider the following requirements:

1. Define the set of all samples that can be selected (i.e., separate the population into sampling units). For example, if the area known to contain the population (the sampling universe) to be sampled is 1000 m² and 1 m² is to be sampled, there are 1000 samples in the set of all available sampling sites.

2. Assign to each possible sample a known probability of selection (randomize to give every sample equal probability of being selected). Using the situation cited above, divide the area to be sampled into 1000 discrete units.

3. Select each sample by a process that assigns the correct probability of selection to all samples. Use a table of random numbers to select sites for sampling or, alternatively, base on proportional allocation of sampling effort according to bottom or substrate type. In this case, sampling effort for a given substrate is based on ratio of habitat type to total substrate; however, take samples randomly from each stratum.

4. Choose a method for computation of estimates from the sample that provides a unique value for each sample, i.e., mean or standard deviation.

Standardize acquisition and recording of data when practical. Record and report in metric units.

One of the most difficult decisions to make when planning collection of aquatic invertebrates is how many replicate samples to take at each station or substation to obtain reliable information. Most taxa are not distributed uniformly over the bottom of a river, lake, estuary, or ocean. Different habitats (sand, mud, gravel, or organic material) support different densities and species of organisms. Even on a relatively homogeneous bottom, animals tend to aggregate. Therefore, take replicate samples to evaluate this variability.

Use at least three replicate samples per station to describe the macroinvertebrate community. More replicates may be necessary to achieve the desired level of precision. If a station must be divided into substations, sample each substation as described above.[2] Ideally, conduct a baseline survey to determine station substrate characteristics and the number of samples necessary to achieve the desired level of

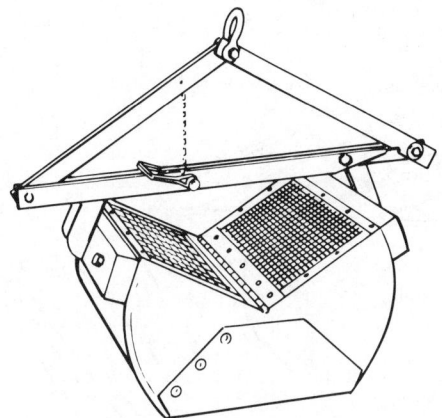

Figure 10500:1. Ponar grab.

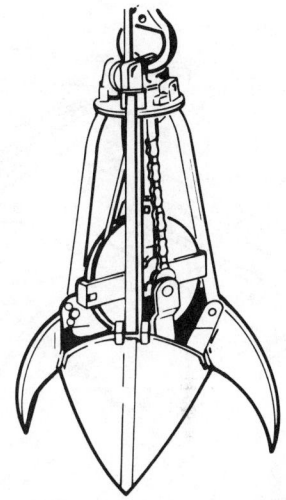

Figure 10500:2. Orange-peel sampler.

accuracy. Take a large number of replicate faunal grabs at a representative station of all sedimentary types present. After determining number of species and densities, predict number of grabs necessary for a reliable estimate of the density of individuals from:

$$N = \left(\frac{t \times s}{D \times x}\right)^2$$

where:

s = standard deviation of samples from preliminary survey,

t = tabulated t value at 0.05 level with degrees of freedom of preliminary survey,

x = mean density of preliminary survey, and

D = required level of precision expressed as decimal (0.30 to 0.35 usually yields a statistically reliable estimate).

3. Sampling Devices, Quantitative

Quantitative and qualitative samplers have been designed to collect organisms from stream and lake bottoms. The most common quantitative sampling devices are the Ponar, Petersen, and Ekman grabs and the Surber or square-foot stream bottom sampler, all described below.

a. Grab samplers:

1) *The Ponar grab* (Figure 10500:1) is used increasingly in medium to deep rivers, lakes, and reservoirs.[3] It is similar to the Petersen grab in size, weight, lever system, and sample compartment, but has side plates and a screen on top of the sample compartment to prevent sample loss during closure. With one set of weights, the standard 23- by 23-cm sampler weighs 20 kg. A 15- by 15-cm petite Ponar may be used. The large surface disturbance associated with a Ponar grab can be reduced by installing hinged, rather than fixed, screen tops, thereby reducing the pressure wave associated with the sampler's descent. This sampler is best used for sand, gravel, or small rocks with mud but it can be used in all substrates except bedrock.

2) *The orange-peel grab* (Figure 10500:2) is a multi-jawed round grab with a canvas closure at the top serving as a portion of the sample compartment. The 1600-cm^3 size generally is used, although larger sizes are available. The area sampled and volume of material collected depend on depth of penetration.[4] This grab is suited for use in marine environments and deep lakes with sandy substrates.

3) *The Petersen grab* (Figure 10500:3) is

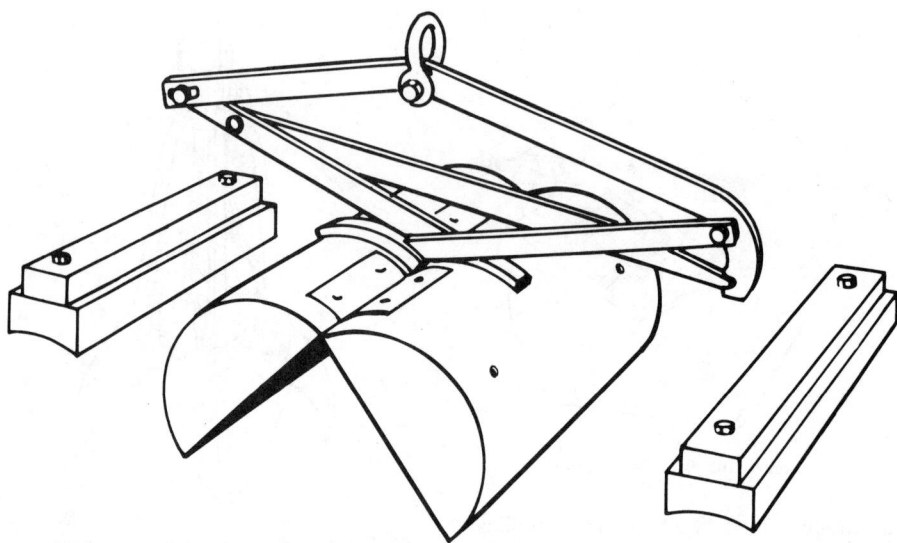

Figure 10500:3. Petersen grab.

used widely for sampling hard bottoms such as sand, gravel, marl, and clay in swift currents and deep water. It is an iron, clam-type grab manufactured in various sizes that will sample an area of from 0.06 to 0.09 m². It weighs approximately 13.7 kg, but may weigh as much as 31.8 kg when auxiliary weights are bolted to its sides. The primary advantage of the extra weights is to make the grab stable in swift currents and to give additional cutting force in fibrous or firm bottom materials. Modify the sampler by adding end plates, by cutting large strips out at the top of each side, and by adding a hinged 30-mesh screen as in the Ponar grab.[5]

To use the Petersen grab, set the hinged jaws and lower to the bottom slowly to avoid disturbing lighter bottom materials. Ease rope tension to release the catch. As the grab is raised the lever system closes the jaws.

4) *The Van Veen grab* (Figure 10500:4) is useful in sampling in the open sea and in large lakes. The long arms tend to stabilize the sampler without disturbing the

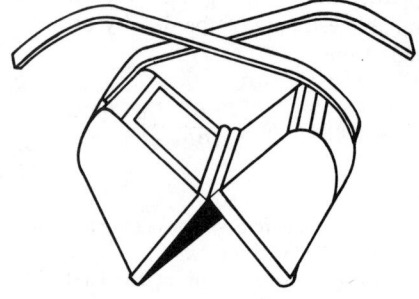

Figure 10500:4. Van Veen grab.

water at the water-substrate interface. It is basically an improved version of the Petersen grab and is useful on substrates of mud, gravel, pebbles, and sand. The sampler is heavy; lower it from a boat or ship platform with mechanical or hydraulic lifts.

5) *The Smith-McIntyre grab* (Figure 10500:5) has the heavy steel construction of the Petersen, but its jaws are closed by strong coil springs.[6] Chief advantages are its stability and easier control in rough

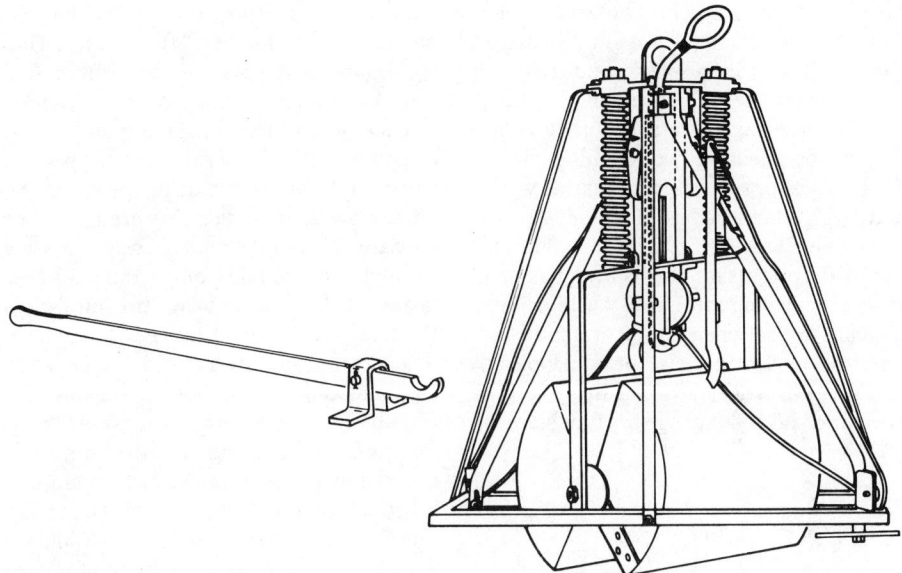

Figure 10500:5. Smith-McIntyre grab.

water. Its bulk and heavy weight require operation from a large boat equipped with a winch. The 45.4-kg grab can sample an area of 0.2 m,[2,7,8] but smaller models (0.1 m² or 0.05 to 0.06 m²) are available.

6) *The Shipek grab* (Figure 10500:6) is designed to take a sample 0.04 m² in surface area and approximately 10 cm deep at the center. The sample compartment is composed of two concentric half cylinders.

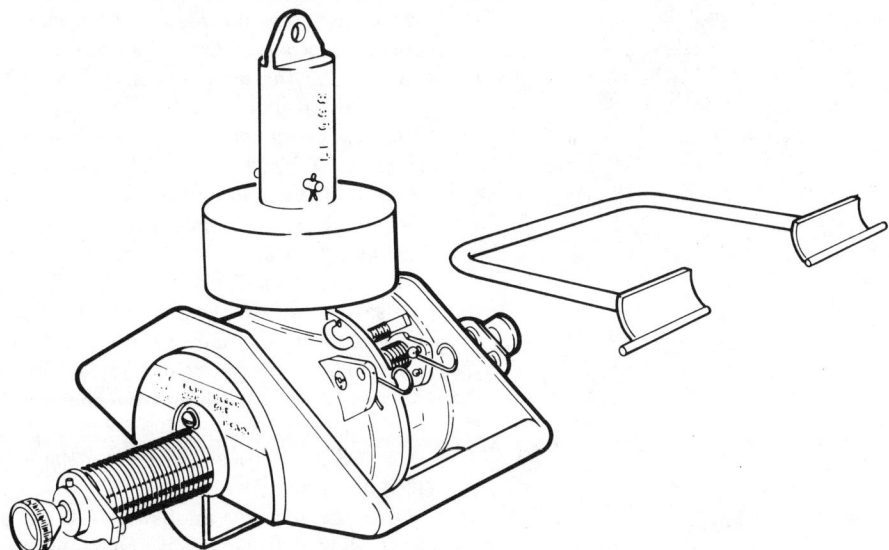

Figure 10500:6. Shipek grab.

When the grab touches bottom, inertia from a self-contained weight releases a catch and helical springs rotate the inner half cylinder by 180°. The sample bucket may be disengaged from the upper semi-cylinder by releasing two retaining latches. This grab is for special use in marine waters and large inland bodies of water.

7) *The Ekman grab* (Figure 10500:7) is useful only for sampling silt, muck, and sludge in water with little current. It is difficult to use when rocky or sandy bottoms are present because small pebbles or grit prevent proper jaw closure. The grab is made of 12- to 20-gauge brass or stainless

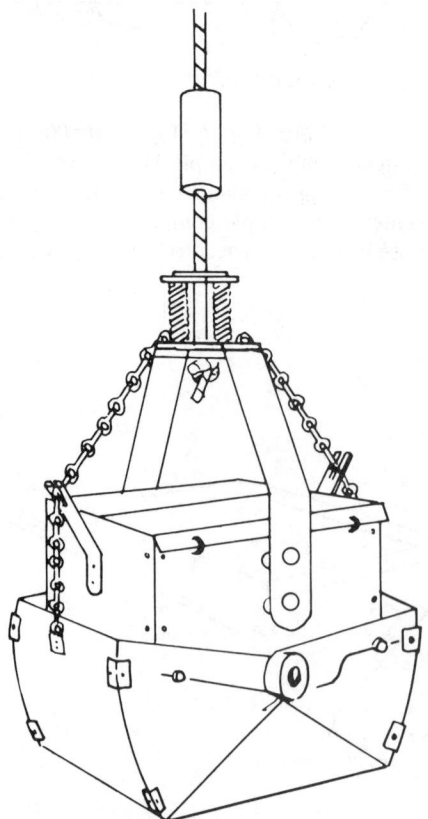

Figure 10500:7. Ekman grab.

steel and weighs approximately 3.2 kg. The box-like part holding the sample has spring-operated jaws on the bottom that must be cocked manually (exercise caution in cocking and handling the grab because of possible injuries if jaws are tripped accidentally). At the top of the grab are two hinged overlapping lids that are held open partially during descent by water passing through the sample compartment. These lids are held shut by water pressure when the sampler is being retrieved. The grab is made in three sizes: 15 × 15 cm, 23 × 23 cm, and 30 × 30 cm, but the smallest size usually is adequate and is most desirable for replicate sampling. A taller model of this sampler, either 23 cm or 30.5 cm tall, is better than the 15-cm version. To prevent sample overflow and loss, place a Standard U.S. No. 30 sieve insert in the top for deep sediments.

b. Riffle samplers:

1) *The Surber or square-foot riffle sampler* (Figure 10500:8)[9] consists of two brass frames, each 30.5 cm (1 foot) square, hinged together along one edge. When in use, the two frames are locked at right angles, one frame marking off the area of substrate to be sampled, and the other supporting a net to collect organisms washed into it from the sample area.

The net usually is 69 cm long with the first few centimeters and the wings constructed of heavier material (canvas, taffeta) to increase durability. Standard mesh size is 9 threads/cm. While a smaller mesh size might increase the number of smaller invertebrates and young instars collected, it also will clog more easily and exert more resistance to the current than a larger mesh. This could result in a loss of organisms due to backwashing from the sample net. This sampler is specific for macrobenthos; many microcomponents of the benthos are not collected.

Use this sampler in shallow (30 cm or less), flowing water. When it is used in deeper water some organisms may be carried over the top of the sampler. Position sampler securely on the stream bottom parallel to water flow with the net portion downstream. Take care not to disturb the substrate upstream from sampler. Leave no gaps under the edges of the frame that would allow water to wash under the net. Fill gaps that may occur along the back edge of the sampler by carefully shifting rocks and gravel along the outside edge. When the sampler is in place (it may be necessary to hold it in place with one hand in a strong current), carefully turn over and rub lightly all rocks and large stones with the hands to dislodge organisms clinging to them. Examine each stone for organisms, larval or pupal cases, etc., that may be clinging to it before discarding. Scrape attached algae, insect cases, etc., from the stones into the sampler net. Stir remaining gravel and sand with the hands or a stick to a depth of 5 to 10 cm, depending on the substrate, to dislodge bottom-dwelling organisms. It may be necessary to hand-pick some mussels and snails that are not carried into the net by the current.

Remove sample by inverting net into sample container. Carefully examine net for small organisms clinging to it. Remove these, preferably with forceps to avoid damage, and include in sample. Rinse sampler net after each use.

A common problem in using the Surber sampler is that organisms wash under the bottom edge of the sampler. The following modifications have been suggested for different substrates:

For loose gravel—Extend bottom edge of Surber frame to 5 or more cm allowing for insertion of frame into substrate to a greater depth. This method works well in soft substrates such as sand and gravel where the current causes substrate shifting.

For coarse gravel and rock—Add serrated extension to back edge of frame to secure it and reduce washing from under this edge. This method is helpful in hard gravel and rock substrates where sinking the entire frame is impossible.

For gravel and bedrock—Add a 5-cm band of flexible material to bottom edge of sampler to create a seal in rocky, uneven substrates. Make band of foam rubber or fine-textured synthetic sponge. Remove organisms that stick to foam and include in sample.

c. Core or cylindrical samplers: Use core or cylindrical samplers to sample sediments in depth. They are better than a Surber sampler when used in combination with sieving of fine sediments contained within the small sample area, 13 to 26 cm^2. Effi-

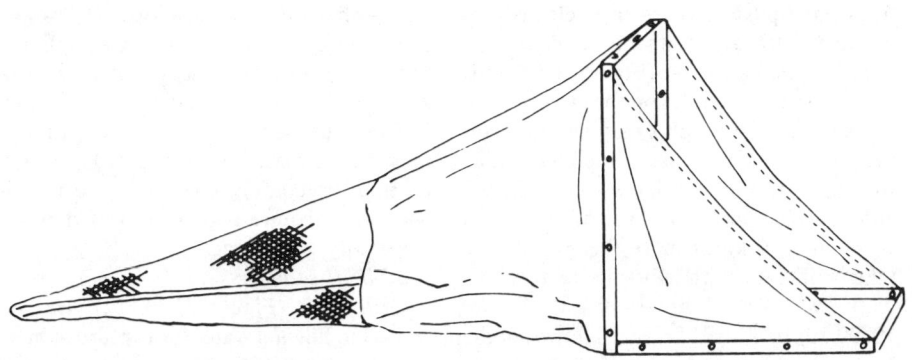

Figure 10500:8. Surber or square-foot sampler.

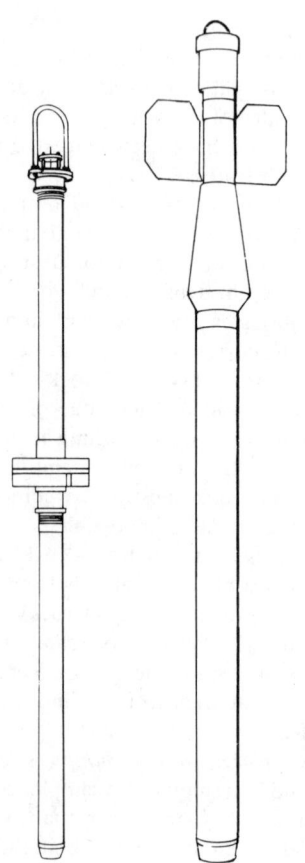

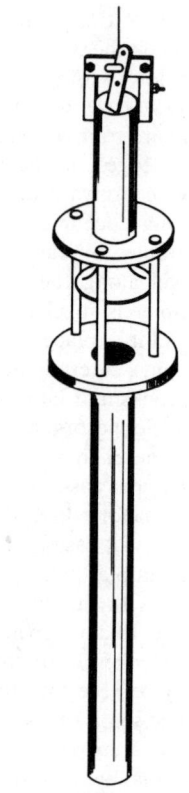

Figure 10500:10. KB corer.

Figure 10500:9. Phleger core sampler.

cient use as surface samplers requires dense animal populations. Core samplers vary from hand-pushed tubes to explosive-driven and automatic-surfacing models.[10]

1) *The Phleger corer* (Figure 10500:9) is widely used. It operates on the gravity principle. Styles and weights vary among manufacturers; some use interchangeable weights that allow variations between 7.7 and 35.0 kg, while others use fixed weights weighing 41.0 kg or more. Length of core taken will vary with substrate texture.

2) *The KB core sampler* (Figure 10500:10) or a modification known as the Kajak-Brinkhurst corer, may be useful in obtaining estimates of the standing stock

of benthic macroinvertebrates inhabiting soft sediments.[11]

3) *The Wilding or stovepipe sampler* (Figure 10500:11) is made in various sizes and with many modifications. It is especially useful for quantitatively sampling a bottom with dense, vascular plant growth by hand. It may be used to sample vegetation, mud-water interface sediment, or most shallow stream substrates. Large volumes of vegetation, when sampled in this way, may require a great deal of time for laboratory processing.

d. Drift samplers:

Drift nets (Figure 10500:12) are anchored in flowing water for capture of macroinvertebrates that have migrated or have been dislodged from the bottom substrates

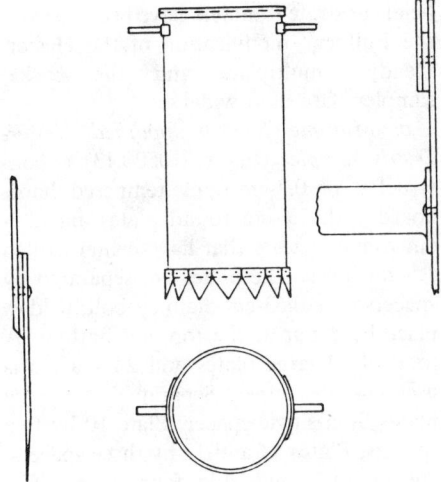

Figure 10500:11. Wilding or stovepipe sampler.

into the current. Drift organisms are important to the stream ecosystem because they are prey for stream fish and thus should be considered in the study of fish populations. Drift organisms respond to pollutional stresses by increased drift from an affected area; therefore, drift is important in water-quality investigations, especially of spills of toxic materials. Drift also is a factor in recolonizing denuded areas and it contributes to recovery of disturbed streams.

Use nets having a 929-cm^2 upstream opening and mesh equivalent to U.S. Standard No. 30 screen (0.595-mm pore size). After placing the net in the water, frequently remove organisms and debris to prevent clogging and subsequent diversion of water at the net opening. Use replicate

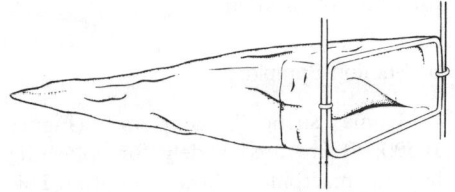

Figure 10500:12. Drift net sampler.

samples as appropriate to meet study objectives. Set drift-net samples for any specified time (usually 3 h) but use the same time for each station. Sampling between dusk and 1 AM is optimum.

The total quantity (numbers or biomass) of organisms drifting past a given station per 24 h divided by the total stream discharge is the best measure of drift intensity. Report data in terms of (number/24 h)/(m^3/24 h) or (biomass/24 h)/(m^3/24 h).[12–14]

4. Sampling Devices, Qualitative

When sampling qualitatively, search for as many different organisms as possible. Collect samples by any method that will capture representative species.

a. Bottom nets are the most versatile collecting devices for shallow, flowing water, and are useful also for shoreline collecting in lakes. When combined with a standardized kicking technique,[15] bottom nets are appropriate for quantitatively sampling macroinvertebrates.[16]

b. Tow nets or trawls range from simple sled-mounted nets to complicated devices incorporating teeth that dig into the bottom. Some models feature special apparatus to hold the net open during towing and to close it during descent and retrieval. Available styles have been discussed.[10,17,18]

5. Sampling Devices, Artificial Substrate Samplers

Artificial substrate samplers are devices of standard composition and configuration placed in the water for a predetermined exposure period for colonization by macroinvertebrate communities. Because many physical variables encountered in bottom sampling are minimized, e.g., depth, light penetration, temperature differences, and species substrate preferences, artificial substrate sampling complements other types of sampling. Like natural submerged substrates such as logs and pilings, artificial substrates are colonized primarily by larval aquatic insects, crustaceans, coelenterates, bryozoans, and to some extent worms and mollusks. The organisms that colonize artificial substrates are primarily drift organisms, such as immature larvae and eggs, carried by water currents. Colonization rates should be similar, so the numbers and kinds of organisms reflect capacity to support aquatic life.

Position artificial substrates in the euphotic zone (0.3 m) for maximum abundance and diversity of macroinvertebrates.[19] Optimum time for substrate colonization is 6 weeks for most waters in the U.S. For uniformity of depth, suspend sampler from floats on a 3.2-mm steel cable. If vandalism is a problem, use subsurface floats or place sampler on the bottom. Regardless of installation technique, use uniform procedures.

At shallow water stations (less than 1.2 m deep), install samplers so that the exposure occurs midway in the water column at low flow. For samplers installed in July when the water depth is about 1.2 m and the August average low flow is 0.6 m, install 0.3 m above the bottom. Take care not to let samplers touch bottom or they may become covered with silt, thereby increasing the sampling error. In shallow streams with sheet rock bottoms, secure artificial substrates to 0.95-cm steel rods that are driven into the substrate or secure to rods that are mounted on low, flat, rectangular blocks.

Before removing samples from the water, enclose them in an oversized plastic bag (double wrapping) that is tightly sealed to prevent possible loss of organisms or use a large dip net (openings equivalent to a U.S. Standard No. 30 sieve) when the sample is removed. Disassemble sampler and brush in a pan of water in the field or add preservative to the bag containing the intact sampler, and disassemble and brush later in the laboratory.

Although many different styles of artificial substrate samplers have been tested,[20] the Fullner[21] modification of the Hester-Dendy[22] multiplate and the basket sampler[19] are used widely.

a. *Multiple-plate or modified Hester-Dendy sampler* (Figure 10500:13) is constructed of 0.3-cm-thick tempered hardboard with 7.6-cm round plates and 2.5-cm round spacers that have center-drilled 1.6-cm holes. The plates are separated by spacers on a 0.63-cm-diam eyebolt, held in place by a nut at the top and bottom. A total of 14 large plates and 24 spacers is used in each sampler. Separate the top nine plates by a single spacer, Plate 10 by two spacers, Plates 11 and 12 by three spacers, and Plates 13 and 14 by four spacers. The sampler is approximately 14 cm long and 7.6 cm in diameter, has an exposed surface area of approximately 1160 cm², and weighs about 0.45 kg. Do not reuse samplers exposed to oils and chemicals that may inhibit colonization. Because it is cylindrical, the sampler fits a wide-mouth container for shipping and storage. The sampler is inexpensive, compact, and lightweight.[19,21,22]

b. *The basket sampler*[19] (Figure 10500:14) is a cylindrical "barbecue" basket 28 cm long and 17.8 cm in diameter, filled with approximately 30 5.1-cm-diam rocks or rocklike material weighing 7.7 kg. A hinged side door allows access to the contents. The sampler provides an estimated 0.24 m² of surface area for colonization. The factors governing proper installation and collection are the same as those described for the multiplate sampler. Some investigators prefer using the basket because natural substrate materials are used for colonization.

6. Suction Samplers

"Dome suction" samplers[23] (Figure 10500:15) are used widely for collecting benthic macroinvertebrate samples. This sampler can be placed directly on specific

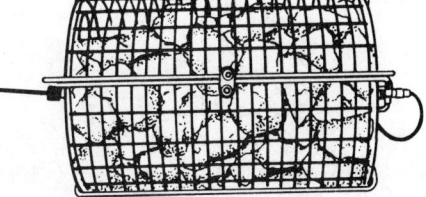

Figure 10500:14. Basket sampler.

sampling sites but a scuba diver is required to collect samples.[24] Improved accuracy of locating sampling sites and ability to collect a large number of replicate samples outweigh the disadvantage of using a diver. Suction samplers have been used widely in sampling marine environments but they have obvious depth limitations.

7. References

1. SNEDECOR, G.W. & W.G. COCHRAN. 1967. Statistical Methods. Iowa State Univ. Press, Ames.

2. ELLIOTT, J.M. 1971. Some Methods for the Statistical Analysis of Samples of Benthic Invertebrates. Freshwater Biological Assoc., Sci. Publ. No. 25.

3. POWERS, C.F. &. A. ROBERTSON. 1967. Design and Evaluation of an All-Purpose Benthos Sampler. Spec. Rep. No. 30, Great Lakes Research Div., Univ. Michigan, Ann Arbor.

4. MERNA, J.W. 1962. Quantitative sampling with the orange peel dredge. *Limnol. Oceanogr.* 7:432.

5. WEBER, C.I., ed. 1973. Biological Field and Laboratory Methods for Measuring the Quality of Surface Waters and Effluents. EPA-670/4-73-001, U.S. Environmental Protection Agency.

6. SMITH, W. & A.D. MCINTYRE. 1954. A spring-loaded bottom sampler. *J. Mar. Biol. Assoc. U.K.* 33:257.

7. MCINTYRE, A.D. 1971. Efficiency of Marine Bottom Samplers. *In* N.A. Holme & A.D. McIntyre, eds. Methods for the Study of Marine Benthos. IBP Handbook No. 16, p. 140. Blackwell Scientific Publications, Oxford, England.

8. WIGLEY, R.L. 1967. Comparative efficiency of Van Veen and Smith-McIntyre grab sam-

Figure 10500:13. Hester-Dendy artificial substrate unit.

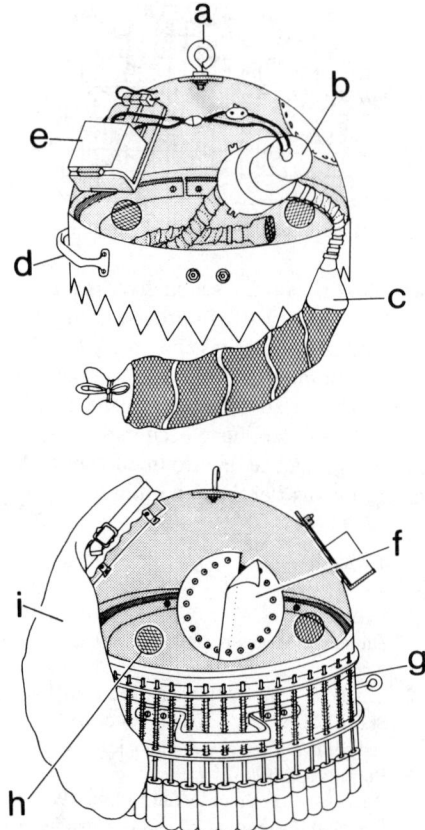

Figure 10500:15. Dome sampler with serrated band (rear view) and polyurethane cylinder band (side view). a. eye bolt. b. bilge pump. c. net bag. d. handle. e. battery. f. arm-hole cover. g. self-adjusting contour rod. h. screened port. i. rock bag. Source: GALE, W.F. & J.D. THOMPSON. 1975. A suction sample for quantitatively sampling benthos on rocky substrates in rivers. *Trans. Amer. Fish. Soc.* 104: 398.

plers as recorded by motion pictures. *Ecology* 48:168.

9. SURBER, E. 1937. Rainbow trout and bottom fauna production in one mile of stream. *Trans. Amer. Fish. Soc.* 66:193.

10. BARNES, H. 1959. Oceanographic and Marine Biology. George Allen and Unwin, Ltd., London, England.

11. BRINKHURST, R.O., K.E. CHUA & E. BATOOSINGH. 1969. Modifications in sampling procedures as applied to studies on the bacteria and tubificid oligochaetes inhabiting aquatic sediments. *J. Fish Res. Board Can.* 26:2581.

12. WATERS, T.F. 1961. Standing crop and drift of stream bottom organisms. *Ecology* 42:532.

13. DIMOND, J.B. 1967. Pesticides and Stream Insects. Bull. No. 2, Maine Forest Serv., Augusta, & Conservation Foundation, Washington, D.C.

14. WATERS, T.F. 1972. The drift of stream insects. *Annu. Rev. Entomol.* 17:253.

15. FROST, S., A. HUN & W. KERSHAW. 1971. Evaluation of a kicking technique for sampling stream bottom fauna. *Can. J. Zool.* 49:167.

16. CROSSMAN, J.S., J. CAIRNS, JR. & R.L. KAESLER. 1973. Aquatic Invertebrate Recovery in the Clinch River Following Hazardous Spills and Floods. Water Resour. Res. Center Bull. 63, Virginia Polytechnic Inst. & State Univ., Blacksburg.

17. WELCH, P.S. 1948. Limnological Methods. Blakiston Co., Philadelphia, Pa.

18. USINGER, R.L. 1956. Aquatic Insects of California, with Keys to North American Genera and California Species. Univ. California Press, Berkeley.

19. MASON, W.T., JR., C.I. WEBER, P.A. LEWIS & E.C. JULIAN. 1973. Factors affecting the performance of basket and multiplate macroinvertebrate samplers. *Freshwater Biol.* 3:409.

20. BEAK, T.W., T.C. GRIFFING & G. APPLEBY. 1974. Use of artificial substrates to assess water pollution. *In* Proceedings Biological Methods for the Assessment of Water Quality. American Soc. Testing & Materials, Philadelphia, Pa.

21. FULLNER, R.W. 1971. A comparison of macroinvertebrates collected by basket and modified multiple-plate samplers. *J. Water Pollut. Control Fed.* 43:494.

22. HESTER, F.E. & J.B. DENDY. 1962. A multiple-plate sampler for aquatic macroinvertebrates. *Trans. Amer. Fish. Soc.* 91:420.

23. GALE, W. & J. THOMPSON. 1975. A suction sampler for quantitatively sampling benthos

on rocky substrates in rivers. *Trans. Amer. Fish. Soc.* 104:398.

24. SIMMONS, G.M., JR. 1977. The Use of Underwater Equipment in Freshwater Research. VPI-SG-77-03, Virginia Polytechnic Inst. & State Univ., Blacksburg.

8. Bibliography

MACAN, T.T. 1958. Methods of sampling bottom fauna in stony streams. *Mitt. Int. Ver. Limnol.* 8:1.

DICKSON, K.L., J. CAIRNS, JR. & J.C. ARNOLD. 1971. An evaluation of the use of a basket type artificial substrate for sampling macroinvertebrate organisms. *Trans. Amer. Fish. Soc.* 100:553.

AMERICAN SOCIETY FOR TESTING AND MATERIALS. 1986. 1986 Annual Book of Standards. Vol. 11.04, Standards D-4342-84, D-4343-84, D-4344-84, D-4345-84, D-4347-84, D-4387-84, D-4401-84, D-4407-84, D-4556-85, D-4557-85, D-4558-85. American Soc. Testing & Materials, Philadelphia, Pa.

10500 C. Sample Processing and Analysis

1. Sample Processing

After collecting a bottom grab sample, transfer it to either specially designed sieve tables (or hoppers) or a container. If container (such as small trash can) is used, dilute with ambient water, and swirl. Pour slurry gradually into a sieve bucket. Gently wash slurry over screen to prevent damaging or losing specimens. Slurries that clog the screen require removal of screened material. A series of one or two coarser screens (e.g., 1-cm and 0.5-cm mesh) will hold back larger materials such as leaves, sticks, shells, and gravel while permitting organisms and smaller materials to pass through to the bottom sieve. Carefully check rocks, sticks, shells, and other objects for attached or burrowed organisms before discarding.

Wash residual on the screen into a container. A cheesecloth bag is very useful because it does not restrict the quantity of wash water. Label containers with a collection code but do not affix labels to lids. Similar labels can be written with pencil or indelible ink on high-rag-content paper and placed in the container. Record label code on a field sheet that describes location, date, type of sample, collector's name, and other pertinent information.

Use laboratory elutriation devices[1,2] to reduce time required to sort benthic organisms from samples containing large amounts of silt, mud, or clay. Wash screened material into a container and fix the contents in a solution of 10% buffered formalin or 70% ethanol.[3-6] If ethanol is used, do not fill more than one-half the container with screened material. Preserve animals with calcareous shells or exoskeletons, i.e., mussels, snails, crayfish, and ostracods, in ethanol.[6,7]

Some macroinvertebrates (worm-like species) are identified more easily if they are relaxed to prevent constriction during preservation. Common relaxants include aqueous solutions of 10% isopropyl alcohol, 0.15% propylene phenoxytol, or 5 to 7% magnesium chloride. Narcotize organisms before fixing them. Fix annelid specimens in 5 to 10% formalin before preserving them in 70 to 80% ethanol (note that alcohol is not a satisfactory tissue fixative). Fixation stabilizes tissue proteins to retain characteristics of the living form.[8]

For qualitative samples place rocks, sticks, and other objects in a white pan partially filled with water. Many animals will float free from these objects and can be removed with forceps.

Assign identification numbers either in the field or at the laboratory and transcribe information from the labels to a permanent ledger. The ledger provides a convenient reference in identifying number of samples collected at various places, time of sampling, and water characteristics.

Filter organisms taken from artificial substrates with a U.S. Standard No. 30 sieve, fix with 10% buffered formalin, and preserve in 10% buffered formalin or 70% ethanol.

2. Sorting and Identification

Whether organisms are sorted in the field or the laboratory, follow consistent procedures. Before processing a sample, transfer information from the label to a data sheet that provides space for scientific names and number of individuals. Place sample directly in a shallow white tray with water for sorting. To facilitate sorting organisms from detritus, stain organisms with rose bengal (200 mg/L) in the formalin or ethanol preservative for at least 24 h.[9] Staining may be inappropriate for such organisms as oligochaetes that require clearing for identification. Examine entire sample and separate organisms unless they occur in very large numbers. If a subsample is sorted, take care that rare forms are not excluded. As organisms are picked from the sample sort under a scanning lens or stereoscopic microscope and separate them into different taxonomic categories (e.g., Odonata, Coleoptera, and Ephemeroptera) and identify on the data sheet. Place animals in separate vials according to category and fill vials with 5% formalin or 70% ethanol. Label vials with sample number, date, sampling location, names of organisms, etc.

Identify animals in each vial using a stereoscopic and compound microscope, according to need, and available experience and resources. Identify organisms to species level if possible. Additional sources of information on laboratory techniques and identification of macroinvertebrates are available (see Bibliography).

7. References

1. WORSWICK, J.M. & M.T. BARBOUR. 1974. An elutriation apparatus for macroinvertebrates. Limnol. Oceanogr. 19:538.

2. LAUFF, G.H., K.W. CUMMINS, C.H. ERIKSON & M. PARKER. 1961. A method for sorting bottom fauna samples by elutriation. Limnol. Oceanogr. 6:462.

3. EDMONDSON, W.T., ed. 1959. Ward and Whipple's Freshwater Biology, 2nd ed. John Wiley & Sons, New York, N.Y.

4. COOK, D.G. & R.O. BRINKHURST. 1973. Marine Flora and Fauna of the Northeastern United States, Annelida: Oligochaeta. NOAA Tech. Rep. NMFS CIRC-374, U.S. Dep. Commerce, National Oceanic Atmospheric Admin., National Marine Fisheries Serv.

5. KLEMM, D.J. 1982. Leeches (Annelida Hirudinea:) of North America. EPA-600/3-82-025, Environmental Monitoring & Support Lab., U.S. Environmental Protection Agency, Cincinnati, Ohio.

6. PENNAK, R.W. 1978. Freshwater Invertebrates of the United States, 2nd ed. John Wiley & Sons, New York, N.Y.

7. BURCH, J.B. 1972. Freshwater Sphaeriacean Clams (Mollusca: Pelecypoda) of North America. U.S. Environmental Protection Agency.

8. KLEMM, D.J., ed. 1985. A Guide to the Freshwater Annelida (Polychaeta, Naidid and Tubificid Oligochaeta, and Hirudinea) of North America. Kendall/Hunt Publ. Co., Dubuque, Iowa.

9. MASON, W.T., JR. & P.P. YEVICH. 1967. The use of phloxine B and rose bengal stains to facilitate sorting benthic samples. Trans. Amer. Microsc. Soc. 86:221.

4. Bibliography

HARTMAN, O. 1941. Polychaetous annelids. Part IV. Pectinariidae, with a review of all species from the western hemisphere. Allan Hancock Pacific Exp. 7:325.

HARTMAN, O. 1944. Polychaetous annelids. Part IV. Paraonidae, Magelonidae, Longosomidae, Ctenodrillidae, and Sabellariidae. Allan Hancock Pacific Exp. 10:311.

HARTMAN, O. 1945. The Marine Annelids of North Carolina. Duke University Press, Durham, N.C.

HARTMAN, O. 1947. Polychaetous annelids. Part VII. Capitellidae. Allan Hancock Pacific Exp.

PETTIBONE, M.H. 1963. Marine polychaete worms of the New England region. I. Families Aphroditidae through Trochochaetidae. *U.S. Nat. Mus. Bull.* 227:1.

SMITH, R.I., ed. 1964. Keys to marine invertebrates of the Woods Hole Region. Contrib. No. 11, Systematics-Ecology Program, Marine Biological Lab., Woods Hole, Mass.

McDAIN, J.C. 1968. The Caprellidae (Crustacea: Amphipoda) of the Western North Atlantic. Smithsonian Institute Bull. 278, Washington, D.C.

HOLME, N.A. & A.D. McINTYRE. 1971. Methods for the Study of Marine Benthos. IBP Handbook No. 16. Blackwell Scientific Publications, Oxford, England.

FOSTER, N.M. 1971. Spionidae (Polychaete) of the Gulf of Mexico and the Caribbean Sea. Stud. Fauna Curacao other Caribbean Islands 36.

GOSNER, K.L. 1971. Guide to Identification of Marine and Estuarine Invertebrates. Cape Hatteras to the Bay of Fundy. Wiley-Interscience, New York, N.Y.

LEWIS, P.A. 1972. References for the Identification of Freshwater Macroinvertebrates. EPA-R4-F2-006, U.S. Environmental Protection Agency.

WASS, M.L. et al. 1972. A checklist of the biota of the lower Chesapeake Bay. Special Scientific Rep. No. 65, Virginia Inst. Marine Science, Gloucester Point.

BOUSFIELD, E.L. 1973. Shallow-Water Gammaridean Amphipoda of New England. Cornell University Press, Ithaca, N.Y.

DAY, J.H. 1973. New Polychaeta from Beaufort, with a key to all species recorded from North Carolina. U.S. Circ. No. 375, National Oceanic Atmospheric Admin., National Marine Fisheries Serv., Washington, D.C.

WATLING, L. & D. MAURER. 1973. Guide to the Macroscopic Estuarine and Marine Invertebrates of the Delaware Bay Region. Delaware Bay Rep. Ser. Vol. 5, p. 178. Univ. Delaware, Newark.

WILLIAMS, A.B. 1974. Marine flora and fauna of the northeastern United States. Crustacean: Decapoda. U.S. Circ. No. 389, National Oceanic Atmospheric Admin., National Marine Fisheries Serv., Washington, D.C.

FOX, R.S. & K.H. BYNUM. 1975. The amphipod crustaceans of North Carolina estuarine waters. *Chesapeake Sci.* 16:223.

GARDINER, S.L. 1975. Errant polychaete annelids from North Carolina. *J. Elisha Mitchell Sci. Soc.* 91:77.

MORRIS, P.A. 1975. A Field Guide to Shells of the Atlantic and Gulf Coasts and the West Indies. Houghton Mifflin Co., Boston, Mass.

PENNAK, R.W. 1978. Freshwater Invertebrates of the United States, 2nd ed. Wiley-Interscience, New York, N.Y.

EWING, R.M. & D.M. DAUBER. 1981. A new species of *Amastigos* (Polychaeta: Capitellidae) from the Chesapeake Bay and Atlantic coast of the United States with notes on the Capitellidae of the Chesapeake Bay. *Proc. Biol. Soc. Wash.* 94:163.

HEARD, R.W. 1982. Guide to common tidal marsh invertebrates of the Northeastern Gulf of Mexico. Alabama Sea Grant Consortium, MASGP-79-004.

MERRITT, R.W. & K.W. CUMMINS. 1984. An Introduction to the Aquatic Insects of North America, 2nd ed. Kendall/Hunt Publishing Co., Dubuque, Iowa.

BRINKHURST, R.O. 1986. Guide to the Freshwater Aquatic Microdrile Oligochaetes of North America. Canadian Spec. Publ. Fisheries & Aquatic Science 84, Dep. Fisheries & Oceans, Ottawa, Ont.

10500 D.   Data Evaluation and Presentation

There are two basic approaches used in evaluating effects of pollutants on aquatic life. The first is to make a qualitative analysis of fauna and flora "above and below" or "before and after," thereby determining species present or absent. Then, through an understanding of the responses of various species to specific pollutants, determine the significance of damage or change.

The second approach is to make a quantitative inventory of the number of specimens, species, and structure of the aquatic community affected by the pollutant and to compare with reference information. In most pollution surveys these approaches are integrated because each provides valuable interpretive information.

1. Qualitative Data Evaluation

No two aquatic organisms react identically to a pollutant because of complex interrelationships between genetic factors and environmental conditions. However, certain groups are intolerant of pollution. For example, operculate snails, immature stages of certain mayflies, stoneflies, caddisflies, riffle beetles, hellgrammites, and echinoderms are sensitive to most pollutants. Pollution-tolerant macroinvertebrates such as certain sludgeworms, midge larvae (bloodworms), leeches, pulmonate snails, and some polychaetes usually increase in number under organically enriched conditions. Facultative organisms, those that tolerate moderate pollution, include most snails, sowbugs, scuds, and blackfly larvae. Tolerant organisms may be found in either clean or polluted situations so that their presence is not definitive. However, a population of tolerant organisms combined with an absence of intolerant ones is a good indication of the presence of pollution. The same species found in different geographical areas may well react differently or be present in different numbers throughout the year.

2. Quantitative Data Evaluation

Statistical methods of data evaluation and mathematical expressions of community structure are invaluable in data analysis. Statistical analyses of biological data commonly include determining the mean and confidence interval and use such tests as chi-square, Student's t, regression, correlation, one- and two-way analysis of variance, robust analyses, and numerous nonparametric tests. The use of mathematical expressions of community structure to derive numerical indices of diversity of aquatic communities is based on the general, though not invariably true, assumption that the greater the diversity of aquatic life, the greater the structural and functional stability of the system.

Diversity indices, although limited, condense considerable biological data into a single numerical value. Diversity indices in current use include d (diversity per individual), which follows concepts of information theory, the SCI (Sequential Comparison Index), and cluster analyses, among others.[1-6] More sophisticated multivariate statistical analyses, such as cluster and ordination, principal component, MANOVA, and discriminant analyses generally are more appropriate and less subject to criticism.[7]

To evaluate statistically the data collected in a pollution survey, identify the sources of variability commonly found. Variability in macroinvertebrate data comes from the methods of sampling and the distribution of organisms. Perhaps the major source is sampling error. Organisms generally are clustered in relation to habitat distribution; therefore, random samples often show high variability among replicates. In statistical analyses of quantitative data, large numbers of samples often are required to detect statistically significant differences. Exercise care in using parametric statistical methods because the basic assumption of normal distribution is not always true. Data often have to be transformed before being tested. Do not assume that a statistically significant difference is ecologically significant.

3. Data Presentation

Data presentation may take many forms. The basic techniques include tables, bar graphs (horizontal and vertical), pie diagrams, pictorial charts (ideographs), line graphs, frequency distribution tables and graphs, histograms, frequency polygons, and cumulative frequency polygons. These may be superimposed on maps. Several reports that may be useful in analyzing macroinvertebrate data have been included in the bibliography.

4. References

1. WILHM, J.L. 1967. Comparison of some diversity indices applied to populations of benthic macroinvertebrates in a stream receiving organic wastes. *J. Water Pollut. Control Fed.* 39:1673.
2. WILHM, J.L. & T.C. DORIS. 1968. Biological parameters for water quality criteria. *Bioscience* 18:477.
3. WILHM, J.L. 1970. Range of diversity index in benthic macroinvertebrate populations. *J. Water Pollut. Control Fed.* 42:R221.
4. WILHM, J.L. 1972. Graphic and mathematical analyses of biotic communities in polluted streams. *Annu. Rev. Entomol.* 17:223.
5. CAIRNS, J., JR., D. W. ALBAUGH, F. BUSEY & M.D. CHANAY. 1968. The sequential comparison index—a simplified method for non-virologists to estimate relative differences in biological diversity in stream pollution studies. *J. Water Pollut. Control Fed.* 40:1607.
6. BOESCH, D.F. 1977. Application of Numerical Classification in Ecological Investigations of Water Pollution. Ecol. Res. Ser., EPA-600/3-77-033, U.S. Environmental Protection Agency.
7. SMITH, W., V.R. GIBSON, L.S. BROWN-LEGER & J.F. GRASSLE. 1979. Diversity as an indicator of pollution. Cautionary results from microcosm experiments. *In* J.P. Grassle, G.P. Patil, W. Smith & C. Taille, eds. Ecological Diversity in Theory and Practice. International Publ. House, Fairland, Md.

5. Bibliography

BECK, W.M. 1955. Suggested method for reporting biotic data. *Sewage Ind. Wastes* 27:1193.

INGRAM, W.M. 1960. Effective methods for collecting and recording data from water pollution surveys. *In* C.M. Tarzwell, compiler. Biological Problems in Water Pollution, p. 260. U.S. Dep. Health, Education & Welfare, Cincinnati, Ohio.

INGRAM, W.M. & A.F. BARTSCH. 1960. Graphic expression of biological data in water pollution reports. *J. Water Pollut. Control Fed.* 32:297.

PIELOU, E.C. 1966. The measurement of diversity in different types of biological collections. *J. Theor. Biol.* 13:131.

LLOYD, M., J.H. ZAR & J.R. KARR. 1968. On the calculation of information—Theoretical measures of diversity. *Amer. Midland Natur.* 79:257.

CAIRNS, J., JR., K.L. DICKSON, R.E. SPARKS & W.T. WALLER. 1970. A preliminary report on rapid biological information systems for water pollution control. *J. Water Pollut. Control Fed.* 45:685.

CAIRNS, J., JR. 1971. A simple method for the biological assessment of the effects of waste discharges on aquatic bottom-dwelling organisms. *J. Water Pollut. Control Fed.* 43:755.

ERMAN, D.C. & W.T. HELM. 1971. Comparison of some species importance values and ordination techniques used to analyze benthic invertebrate communities. *Oikos* 22:240.

CLIFFORD, H.T. & W. STEPHENSON. 1975. An Introduction to Numerical Classification. Academic Press, New York, N.Y.

BOESCH, D.F. 1977. Application of Numerical Classification in Ecological Investigations of Water Pollution. EPA-600/3-77-033, U.S. Environmental Protection Agency Research Lab., Corvallis, Ore.

10600   FISH*

10600 A.   Introduction

1. Ecological Importance

Fish are a major component of most aquatic habitats with over 775 species in North American streams, rivers, and lakes. They are the focus of economically important sport and commercial fisheries; licensing fees for both private and commercial sectors provide funds for state and federal agencies. They are an impor-

---

*Approved by Standard Methods Committee, 1989.

tant source of food for humans and are a key unit in many natural food webs. Because they are a prominent component of aquatic ecosystems they have an impact on the physicochemical properties of the system, plankton, macrophytes, and other aquatic organisms. They also serve as environmental indicators. Changes in the composition of a fish assemblage often indicate a variation in pH, salinity, temperature regime, solutes, flow clarity, dissolved oxygen, substrate composition, or pollution level. The gain or loss of certain species is a common consequence of eutrophication. Because fish are conspicuous they often are the primary indicators of the toxification of streams and lakes. In extreme cases the presence of dead or moribund fish may adversely affect potability and recreational use of waters, create foul odors, and obstruct shorelines.

Fish share many physiological properties with mammals and are used in both the laboratory and the field by the environmental manager and health specialist as assay tools.

2. Definitions

A *population* is a group of individuals of any one kind of organism occupying a particular space. Its study includes definition of taxonomic position, habitat and mobility, diet, numbers of individuals by age, size, weight, sex, fecundity, and sources of mortality.

An *assemblage* is a group of several populations sharing a common geographical area. The study of their coordinated activity is key to the understanding of the environmental system.

3. Scope of Analysis

An analysis of a target unit commences with review of existing data followed by sampling and preservation, identification, demographic estimation, and pathological examination.

The guidelines provided here are directed to the general practitioner who may need specialists such as the commercial and sport fisherman, fishery biologist, taxonomist, histopathologist, population statistician, systems ecologist, and toxicologist. Adapting to the particular situation is the key element in the study of fish in their natural habitat.[1-5]

4. References

1. EVERHART, W.H. & W.D. YOUNGS. 1981. Principles of Fishery Science. Cornell Univ. Press, Ithaca, N.Y.
2. NIELSEN, L. & D.L. JOHNSON, eds. 1983. Fisheries Techniques. American Fisheries Soc., Bethesda, Md.
3. RICKER, W.E., ed. 1975. Methods for the Assessment of Fish Production in Fresh Waters, 3rd ed. IBP Handbook No. 3. Blackwell Scientific Publ., Oxford, England.
4. ROYCE, W.F. 1984. Introduction to the Practice of Fishery Science. Academic Press, Orlando, Fla.
5. TEMPLETON, R.G. 1984. Freshwater Fisheries Management. Fishing News Books Ltd., Farnham, Surrey, England.

10600 B.   Data Acquisition

1. Planning and Organization

a. Objectives/variables: Before collecting new data define study objectives. Understand the relative importance of such variables as time of day and year, weather, flow/flood and tidal conditions, method and its selectivity, and competence and experience of the investigators.

b. Regulations: A detailed understanding of licensing and permit requirements for the collection of specimens is essential.

Most states have strictly enforced regulations on both the collection and disposition of fish specimens. Effective public relations usually involves the guidance of local residents and waterfront associations on the activities planned.

  c. Units: Carefully choose units of measure, giving attention to the conventions and expectations of those sponsoring and using the data. Units may influence methods used or be fixed by the intent to duplicate earlier efforts.

  d. Site inspection: Make an early and thorough visual examination of the study site. Use glasses with polarized lenses to facilitate examination of bottom features and detection of fish. Binoculars allow field identification of fish and help in the logging of needed behavioral information. In clear water use face plate and snorkel or SCUBA (self contained underwater breathing apparatus) to define habitat, identify fish, and observe behavior. Use wet or dry suits during colder periods and even under ice but note that special safety training is needed.

  e. Data forms: Print data forms on good-quality bond paper. Waterproof plastic is available and is usable for both graphite and ballpoint pen. Include the following information:

- Date and time of collection or observation;
- Exact location using the Universal Mercator System (UTM) or a local variant; township, range and section numbers; county and state; physical features such as a stream confluence, islands, bays, etc.; and station identification number or code;
- Site conditions as required, e.g., presence of ice, flood state, tidal stage, meteorological events such as air temperature, occurrence of storms and rainfall in last 48 h, water temperature, discharge water temperature, vegetational cover on nearby shores, etc.;
- Purpose of activity or project;

- Description of collections or observations made including preservatives, photograph numbers, and gear type;
- Personnel and their functions;
- Name of person recording the data; and
- Chain of custody signatures and dates.

  f. Description: Collected materials may define the species represented, describe the population of a particular species, describe a species assemblage, or characterize impacts of some event such as a chemical spill. The detailed analysis may include:

- Preliminary species assignment for each specimen;
- Number of individuals of each species;
- Standard, fork, or total length for each specimen;
- Sex, if discernible;
- Maturity as indicated by gonadal condition and coloration;
- Weight of each specimen (displacement volume is used in some cases);
- Description of unusual features such as tags, deformities, lesions, tumors, or parasites; and
- Materials taken for aging, such as scales, spines, or opercular bones.

  g. Conduct of field workers: Provide adequate advance notice of activities to conservation officers, wardens, local law-enforcement agencies, and officers of lake or watershed associations. Inform them in detail of the actions planned. Understand trespass law for the study area and request access.

Be sensitive to the use of waterfront areas. Avoid, where possible, damage to amenity plantings and capture of favored specimens, e.g., a pair of large smallmouth bass holding a territory next to a dock. Deal pleasantly and in a well-informed manner with the questions of onlookers. Display the name and address of the study group through such means as name tags, arm patches, or equipment decals.

Demonstrate regard for safety. Wear life preservers when appropriate. Avoid wearing waders on board vessels when in deep water. Follow local restrictions on boat speed. Handle gear proficiently. Use failsafe switches on all electrical gear — especially electroshocking equipment.

Identify all gear with name, address, telephone number, and permit or license numbers of the using agency.

Position gear inconspicuously to minimize tampering and vandalism. Avoid navigational channels and other heavily used sites. Whenever appropriate, submerge indicator buoys and mark their location by paired range points such as navigational aids or landmarks. When transferring gear from one locality to another be careful not to translocate organisms. Maintain gear in a professional manner. Avoid use of tattered nets and casually repaired and dirty equipment.

Dispose of processed specimens by assignment to a museum or academic institution or burial in a certified landfill or isolated area. Pay attention to local legislation regarding humane care of vertebrates including fishes. Do not overharvest. A 2-h evening set of a gill net may yield a sufficient number of specimens for a particular study while an overnight set would be wasteful. Avoid sampling nontarget species.

Be prepared to sample on short notice (such as after a storm or flood) or at awkward times (such as at night or in the winter).

2. Existing Data

Published and unpublished data already exist for most larger lakes and river systems. Natural history museums and academic institutions are primary sources and often can provide preserved materials as well as the names of local specialists. Members of the American Fisheries Society, the American Society of Ichthyologists and Herpetologists, and the Ecological Society of America also can provide information.

Private engineering and environmental consulting firms often maintain detailed regional files. State agencies such as departments of health, environment, conservation, fisheries, wildlife management, and planning, and federal agencies such as the U.S. Fish and Wildlife Service, U.S. Environmental Protection Agency, Bureau of Reclamation, and U.S. Geological Survey are good resources. Large data sets may be available from power utilities, refineries, beverage producers, and chemical companies having riparian facilities. Commercial fishermen and master sport fishermen of an area are other important sources. Local libraries, newspapers, and local residents may provide useful material, including photographs, dates, and even specimens.

3. Collection and Observation Methods

a. Angling: The use of hook and line is an ancient means and involves relatively simple gear[1] but its effective use is a matter of skill. The services of a competent fisherman are often valuable. The techniques depend on the resourcefulness and skill of the angler, may be time-consuming and, when highly refined gear is used, may be expensive.

b. Set line: A set line is a heavy line anchored at each end bearing regularly spaced leaders having hooks. It is widely used for commercial and private fishing. It usually is fished overnight on the bottom and can be used to great depths.

c. Trolling: Trolling is towing a hook and line behind a vessel. It is effective for larger fish of open and deep waters. Use metal line equipped with metal weights or wing depressors to achieve desired depths. Specialized lures that reflect sonar often are used with acoustical electrical gear to determine depth and to locate target fish. Trolling may be the most economical means of capturing the target species.

d. Spear and arrow: The use of barbed and/or hooked (some automated) spears is of limited utility. It often is prohibited. Sport bow-and-arrow fishing may yield large numbers of carp, gars, or other larger fish in shallow water. Spearing through ice often is effective, especially for sturgeon and larger esocids and percids.

e. Nets — General remarks: Netting is used in static gear such as traps and weirs and in active gear such as seines and trawls. Netting may be made of cotton, plastic, or metal. Nets of natural fiber are subject to microbial decomposition and have been supplanted by other materials but are still of value where theft is likely. Plastic netting is exceedingly durable but is weakened by ultraviolet irradiation; avoid prolonged exposure to sunlight. Netting is available in colors that may hold some sampling advantage. Mesh sizes are measured in terms of "bar," i.e., along the edge of the frame, "diagonal," i.e., from opposite angles of the frame, and "stretched," i.e., from opposite angles when the net is under tension. Knotting varies; some knots are abrasive to captured fish.

f. Hoop, fyke, and trap nets: Elongated, tapered nets supported on hoops and variously divided into chambers with secondary tapered net sections and anchored to the bottom are common. Usually they are used at depths less than 3 m. They may be kept in place for a protracted period but usually are visited daily to remove the catch from the inner chamber or cod. In slowly moving water orient the cod end into the current. The basic hoop or ring net may be converted into a fyke net by adding panels of netting at the open end. Those added to the sides are called wings; the single panel placed at the center of the mouth may be quite long and is called the leader. Wings and leaders usually are equipped with floats and weights and are placed to deflect and trap or confuse the normal movements of fishes along shore into the net. The main body of a typical hoop or fyke net may range from 5 to 10 m in length and up to 2.5 m in diameter.

A net hung on rectangular framing is called a trap net (Figure 10600:1). On commercial fishing grounds researchers usually contract for the catches of larger trap nets rather than use their own. Another net type of this general design is the pound net.

Hoop, fyke, and trap nets are especially effective for larger fishes and in live capture such as for spawning stock and fish to be used in mark-and-recapture population assessment.

Nets of this type are available through net supply houses and commonly are built to specification. They require a boat to set and inspect. They often are conspicuous and thus attractive to the public. They require surveillance and the clear posting of their ownership and purpose.

g. Traps: The term "fish trap" usually is reserved for smaller, portable units commonly made of galvanized wire. They are fitted with one or more conical inserts and an opening to remove the catch.

Smaller devices usually called "minnow traps" may be useful. Many types are available, including highly durable plastic units that can be easily stacked, stored and transported. Minnow traps may be placed in a well, a cave pool, near hydroelectric facilities, or in other awkward localities with good results. They may be made more effective through baiting.

h. Weirs: Weirs are stationary traps usually installed along the course of a stream or river. They are of complex design and may be incorporated into a fish ladder and dam. They guide fish into a sampling or capture sector called the "pot."

i. Gill nets: Nets constructed of thin line with mesh large enough for the target species to penetrate partially are called "gill nets." The fish become entrapped while attempting to swim through the net and are harvested. Gill nets are composed of panels of netting of the same or diverse mesh size suspended between a stronger "float line" equipped with flotation devices and attached along the upper working edge and

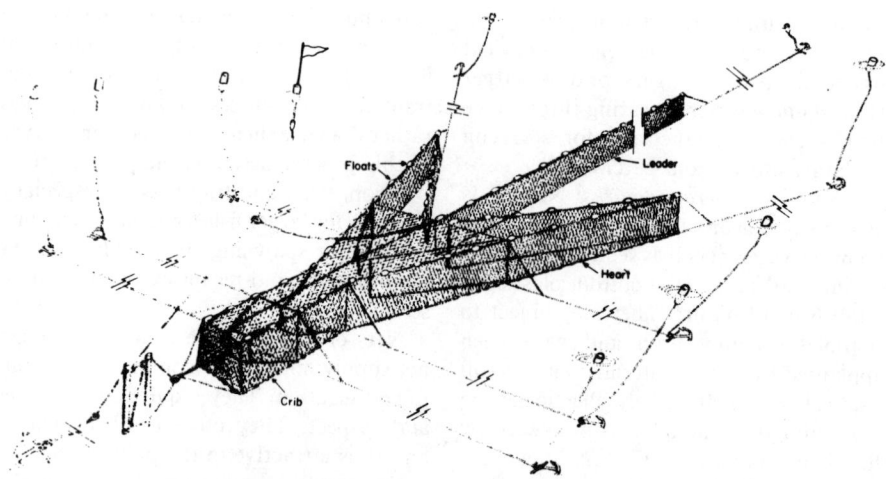

Figure 10600:1. Diagram of a sunken trap net. Source: RICKER, W.E., ed. 1971. Methods for Assessment of Fish Production in Fresh Waters. IBP Handbook No. 3. Blackwell Scientific Publ., Oxford, England.

another heavier "lead line" attached along the lower working edge. Adjust the weights and floats to position the net on the bottom, surface, or at an intermediate depth. The ends of the gill net are equipped with anchors, tether lines, and buoys.

The gill net may be set under ice through appropriate holes. Gill nets may be used in standing water at depths 2 to 3 m greater than the depth of the net; they are less successful in flowing water.

The setting and recovery of gill nets require special attention. The sampling site must be free of ensnaring objects such as submerged trees. The boat used to set the net must be free of projections that can catch the net during the payout and the net boat must proceed at a speed in harmony with the workers discharging the net. The net must be well pleated and free for setting; the anchors, tethers, and buoys must be ready for quick release as well. At the conclusion of the set the tether usually is pulled to extend the net to its full working length. The tether line must be long enough so that the buoy is not pulled out of sight. The marker buoy can be set below

the surface to reduce vandalism and paired range points may be used to relocate the site. If the tether line and buoys are lost, attempt retrieval by grappling hook or SCUBA. Failure to recover lost gill netting can be environmentally disastrous because it will continue to catch and kill fish. Report lost gill netting to local fisheries authorities.

Overnight setting is common; however, a test set during dawn or dusk or a short night set may be advisable to avoid the capture of an excessive number of specimens. Lifting the gill net is best done during calm weather and early in the morning. The captured fish, if relatively few, are removed from the net as it is lifted or retained in the net and placed in a sturdy box for subsequent "picking" on shore. If the net is to be reset immediately, carefully repleat it. Recording position (including direction of entry) and mesh size of gilled fish may provide valuable information. On completion of the "picking" remove twigs, leaves, and other matter from the net and clean and dry it.

j. Trammel flag or tangle nets: Trammel

nets are composed of three panels of netting hung together; they catch fish by entanglement. The central panel is of smaller mesh and the fish is ensnared by passing through the coarse panel to form a bag in the central net. Trammel nets have several commercial applications but are used infrequently in fisheries biology.

k. Trawls: A trawl is a towed net. The mouth of the net is maintained by either a frame, as in the beam trawl, or with hydraulic planes called "otter boards" or "doors" working together with weights and floats as in the "otter trawl." Trawls are specialized to work at the surface, in midwater, or on the bottom. In the surface trawl, buoying devices predominate; in the midwater trawl they are balanced against the weight; and in the bottom trawl the weights predominate. The bottom trawl usually has abrasion skirting on the lower surfaces, rollers that facilitate movement over obstructions, and special chains (ticklers) that run along the lower leading lip of the mouth. The tickler stimulates fish to rise up off the bottom and into the net. Some trawls have one or more conical inserts before the cod or terminal part. The cod is held closed with a cinch line that can be pulled to release the catch onto the sorting deck or tables. The trawl is pulled by a bridle and warp worked from a hydraulic winch. Smaller trawls may be worked strenuously by hand. The length of warp required depends on speed and depth of sampling; however, a 30° angle of warp to the water surface is typical. The speed of towing relates to the gear but is around 2 or 3 knots. Depth and the character of the bottom may be defined by echo sounder. The duration of a tow ranges from a few minutes to several hours. Night trawling may yield larger catches but is more difficult in many inland waters because of navigational aids, anchorage buoys, and other obstructions.

l. Ichthyoplankton sampling: Ichthyoplankton consists of the eggs and very young stages of fish (larvae).[2,3] Sampling may be either by plankton nets or bulk water sampling. Nets having a mouth diameter smaller than the main body of the net may be towed faster than the usual 2 to 3 knots. The towing bridle of the net affects sample collection and a number of designs such as the double net or "bongo net" have been devised to reduce this influence. Measure water volume sampled by a calibrated cylinder with an internal propeller. Other devices may be lowered to a given depth triggered to open and then triggered to close again, providing a sample from a known depth. The Clarke-Bumpus is a quantitative plankton sampler.[4] In vertical sampling the net is lowered to the bottom or some prescribed depth and then pulled upwards, sampling the water column. Mesh size for commercially available plankton netting ranges from 0.158 to 0.790 mm but standard meshes have yet to be specified.

Bulk water sampling of plankton consists of collecting a known volume of water and separating the plankton by filtration and/or centrifugation.

m. Seines: A seine is a simple panel of netting pulled by a bridle at each end (Figure 10600:2). In many smaller seines the bridle is attached to pulling poles or "brails." The upper line of the seine is equipped with floats and the lower with weights. Some seines are fitted with a central bag of smaller mesh that traps the fish. Seines range in size from 1 to about 100 m long and from 1 to 3 m deep. Mesh size depends on the target species. The seine is an effective device for sampling smaller fishes.

Seines may be worked over shorelines relatively free of obstructions. Pulling may be either parallel, angled, or perpendicular to shore. Two samplers, wearing waders, form the net into a gentle "U" while pulling. After a suitable distance the seine, with the lead line on the bottom, is pulled to and up on the shore. A series of shorter

Figure 10600:2. **Bag seine in operation in a small stream.** Source: RICKER, W.E., ed. 1971. Methods for Assessment of Fish Production in Fresh Waters. IBP Handbook No. 3. Blackwell Scientific Publ., Oxford, England.

passes may be more productive than one long one.

Small seines may be used over cobbled stream beds by placing the net poles firmly in the bottom and then rolling the cobbles upcurrent of the net; this action dislodges fish to drift into the net. Benthic species are especially prone to capture in this manner.

Usually work large seines perpendicular to shore. Fix a long warp on shore and play it out by a boat working offshore to a set distance. Swing the boat parallel to shore and release the net. Return the boat to shore with a second warp. Draw the seine to shore by hand or by some kind of draft engine using the two warps.

A block seine may be used to block the mouth of an embayment. This sampling is effective when water elevations change. A self-contained block net is the closing seine consisting of a circular panel of netting strung on a weighted bottom ring and a floating upper ring. Drop the device onto the bottom enclosing the fish and remove them with a dip net. The technique is to a degree quantitative but emplacement of the gear causes fish to leave the study area.

The commercial purse seine is a larger version of the closing net and is used in deeper water for the capture of schooling fish. The lower edge is equipped with a pursing warp or line that allows the net to be drawn together under the surrounded school. After closure the fish are dip netted or hydraulically removed.

n. Lift, dip, and throw nets: These nets may be operated by a single worker. The lift net consists of a square panel of net held open by a pair of diagonal braces and is lifted at the crossing point by a cord. Such nets usually are baited while resting on the bottom and lifted when a sufficient number of fish have been gathered.

Dip or pole nets are conical nets attached to a ring frame which in turn is attached to a pole. They are effective in the capture of salmonids, smelt, and various clupeids (such as blueback herring and alewife).

Throw or cast nets are circular nets often having a pursable and weighted margin operated by a draw string. The net is thrown over a school of fish or a baited sector of bottom, the margin pursed and the closed net removed from the bottom.

o. Goin dredge: The Goin dredge is a wooden box with a net bottom and one wall missing. Handles or hand holes are placed on the three remaining sides and are used to work the device through vegetation.

p. Electrofishing: Electrical devices also may be used for collecting.[5-7] They are particularly useful in areas where uneven bottoms, fast-flowing water, or obstructions make conventional collecting techniques difficult or impossible. They are nonselective in terms of size and type of fish caught. An electrical field in the water is produced by passing a current between two submersed electrodes or between one electrode and the ground. Depending upon design, electrical devices produce either alternating current (AC) or direct current (DC). AC stuns fish in its field, allowing them to be dipped from the water, whereas DC induces galvanotaxis so that the fish move toward one of the poles, from where they are recovered. DC devices are particularly effective in turbid water or in waters with numerous obstructions or heavy vegetation. AC devices are more likely to kill fish. Effectiveness of electrofishing is affected by such environmental factors as water hardness and availability of electrolytes. Addition of salt to raise conductivity may be necessary.

Electrofishing gear ranges from large, gasoline-motor rigs (Figure 10600:3) mounted in a boat to small, portable backpack units powered by a gasoline motor or a battery. The electric seine consists of a conventional seine in which electrodes of an AC source run along the lead and float lines, thus greatly increasing collecting efficiency.

q. Ichthyocides: Ichthyocides are the least selective method of collecting but are the most efficient in terms of percentage of total population sampled.[8] Chemicals used may act as a metabolic poison, as a vasoconstrictor (suffocant), or as an anesthetic. Most have some serious fault in terms of their value as a fish-collecting tool, such as slow reaction time, lack of sufficient killing power except in high doses, deleterious side effects to either the environment or user, or expense. Rigorous time-consuming permitting procedures usually are required. Among the chemicals or chemical preparations used are cresol, sodium cyanide, sodium hypochlorite, antimycin A and B, quinaldine, tricaine methanosulfonate (MS-222), urethane, and rotenone; of these, only antimycin and rotenone now are approved for use by the U.S. Environmental Protection Agency.

Rotenone has been most widely used as

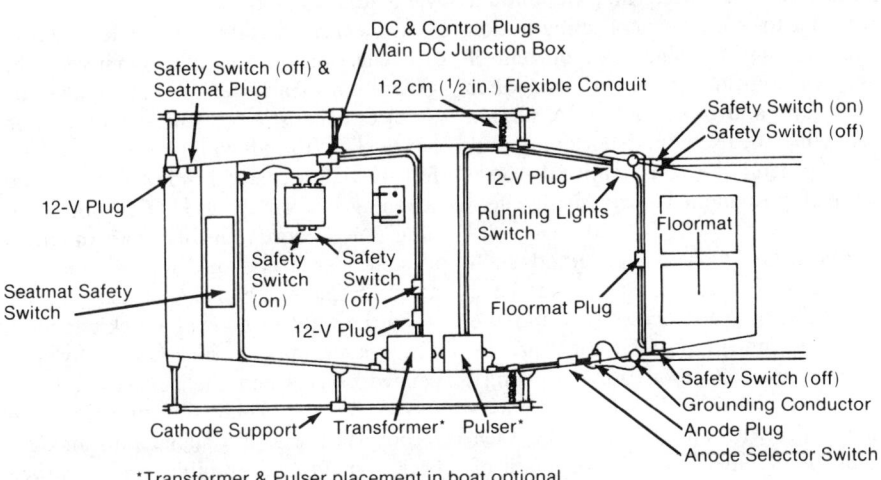

Figure 10600:3. Diagram of well-equipped electrofishing boat. Source: NOVOTNY, D.W. & G.R. PRIEGEL. 1974. Electrofishing Boats, Improved Designs and Operational Guidelines to Increase the Effectiveness of Boom Shockers. Tech. Bull. No. 73, Dep. Natural Resources, Madison, Wis.

a general collecting tool. It is the commercial name for a crystalline ketone ($C_{23}H_{22}O_6$) found in six genera of leguminous plants, particularly in the genus *Derris*. Rotenone functions as a vasoconstrictor and fish affected by it literally suffocate. There is no damage to the body tissues; fish so killed are safe to eat. It is rapidly detoxified by potassium permanganate, is adversely affected by light and high temperatures, and when subjected to such conditions, breaks down rapidly.

Rotenone is available as powder, resin, crystals, or liquid emulsions. A 5% preparation is the usual strength employed in collecting. The powdered preparations are least stable and the resins and crystalline products the most stable. A ready-to-use liquid emulsion is most convenient, but it is relatively expensive. If weight and space problems exist, use crystals or resins. Immediately before use mix resins and crystals with solvents. For the resins, mix 100 g fragmented rotenone (broken up before mixing) with 1 L commercial-grade acetone and 100 mL emulsifier. For the crystals, mix 20 g with 3.8 L acetone. Crystals are the recommended form. The amount of rotenone to be used per collecting station varies according to volume and current of water, reaction time of the toxicant, water temperature, and other factors. A sample bioassay may be useful in determining dosage. In general, be conservative because good-quality rotenone is surprisingly effective.

r. Concussion: Concussive methods such as explosive devices and substances have been used; however, most states and the federal government prohibit their use. A simple concussive method called "tunking" or "stoning" consists of striking an emergent boulder with another rock and then turning the boulder over for collection of the fishes underneath it. Sculpins and minnows may be captured in this manner.

s. Creel census: The systematic collection of data from sport fishermen is a primary means of analysis. Field examination of fish caught, coupled with a standard interview of the fisherman, is common. Allocation of fishing diaries among fishermen followed by their systematic retrieval is another method. Statistically rigorous sampling methods and well-designed interview protocols are essential for the generation of reliable results.

t. Slurp gun: The slurp gun consists of a valved cylinder fitted with a plunger. It is used with SCUBA or while snorkeling and is especially effective for selective sampling of fish at nesting sites, or within otherwise inaccessible interstices.

u. Stomach and gut examination: The contents of the digestive tracts of fish often include examples of additional fish species not otherwise sampled.

v. Serendipity: Useful specimens can be collected after various kinds of fish mortalities. The release of toxic substances, lethal changes in water temperature, anoxic water, construction of cofferdams, dewatering of power-plant flumes, drying up of natural water bodies, and the stranding of fish after floods exemplify events that can yield study materials.

w. Acoustic methods: Acoustic means of fish detection and quantification involve the generation of a series of sound pulses by an electroacoustic transducer mounted on a vessel, a towed body, or a fixed land feature.[9] Vertical and side or horizontal scanning are widely used. For side scanning, one or more transducers are mounted at some depth and oriented horizontally, as on the piers of a bridge crossing a river or on the walls or floor of a lock chamber. The sound impulse is reflected by suspended objects and the bottom. It is received by the same transducer, which reconverts the sound signal into an electrical signal that is amplified and displayed and/or recorded on a paper tape, VCR tape, liquid display, photographic film, etc. The usual paper tape provides a profile of the water showing the bottom and the objects suspended above it. Objects having high "acoustic impedance," such as fish with air bladders, are especially vivid targets and produce strong signals. Gas bubbles, plankton, particulate matter, and density differences associated with a thermocline, floating leaves, and other objects may produce acoustic traces. In the more typical recording type of sounding apparatus a single, relatively stable fish will produce a chevron-shaped "echogram", with the point of the chevron pointing upwards. Several fish usually can be resolved into a corresponding number of such figures, but a school of fish will produce a cloud-like mark, complicating analysis. Under these circumstances estimation of biomass usually is undertaken.

Verify species being recorded by capture or direct observation (e.g., SCUBA). Weighing captured specimens permits extrapolation to biomass estimation.

Apart from quantification, sonar methods provide high-quality data on the vertical and horizontal location of fish stocks.

x. Tags and tagging: The marking of a fish or group of fish followed by their release and recapture provides information on movement, rates of growth, and population characteristics. The methods include: dyes and stains (including fluorescent materials), finclips, attachment of tags (Figure 10600:4), emplacement of encoded wires, sonic (transmitter) tags, the "PIT" tag (see below), branding, radioisotopes, and marker chemicals that can be detected easily or influence ageable features such as the scales or spines. Beware the influence the mark has on the fish, for example, jaw tags can suppress growth rate through interference with feeding; brightly colored tags may increase the role of predation; large tags may impair swimming or cause entanglement with plants or other objects; infection may be induced. Marking is an important tool because it permits identification of genetic strains of hatchery stocks, observation of the behavior of individual fish submitted to particular chemicals or other stressors, observation of dominance and social rank, etc.

The passive integrated transponder (PIT)* is an innovation having considerable utility.[10] Small (10-mm × 2.1-mm) glass-imbedded electrical units called "PIT tags" may be injected into the abdominal cavity of a fish using a syringe-like device (see Figure 10600:5). Fish 5 cm and larger may be marked in this fashion. The tag is activated by a hand-held excitatory unit that elicits a distinctive signal permitting recognition of the individual fish and immediate data processing by computer. The specimen may be processed with little or no handling by using a water vessel or a flow-through cylinder (as is done with migratory salmon moving through a weir) during the tag excitation. The tags last for several years. Salmon smolt can be tagged and individually recognized years later as the adult fish return for spawning.

y. Snorkeling and SCUBA: Snorkeling is the process of using a face plate to facilitate underwater vision and a "J"-shaped breathing tube, to allow sustained observation of underwater conditions. SCUBA (self-contained underwater breathing apparatus) consists of a compressed-air tank, a gas-flow regulator, and a hose and mouth piece. Ancillary but crucial gear includes a weight belt for adjusting buoyancy, fins, an inflatable life vest, and either a wetsuit or a dry suit.

Depths less than 15 m are accessible to those with good training. Greater depths require considerable training and experience.

Underwater study permits precise gear emplacement, direct observation of gear function, rapid definition of niche allocation of the species represented, assessment of impact of chemical spills in deeper water

---

*Available from BioSonics, Inc., 3670 Stone Way North, Seattle, Wash.

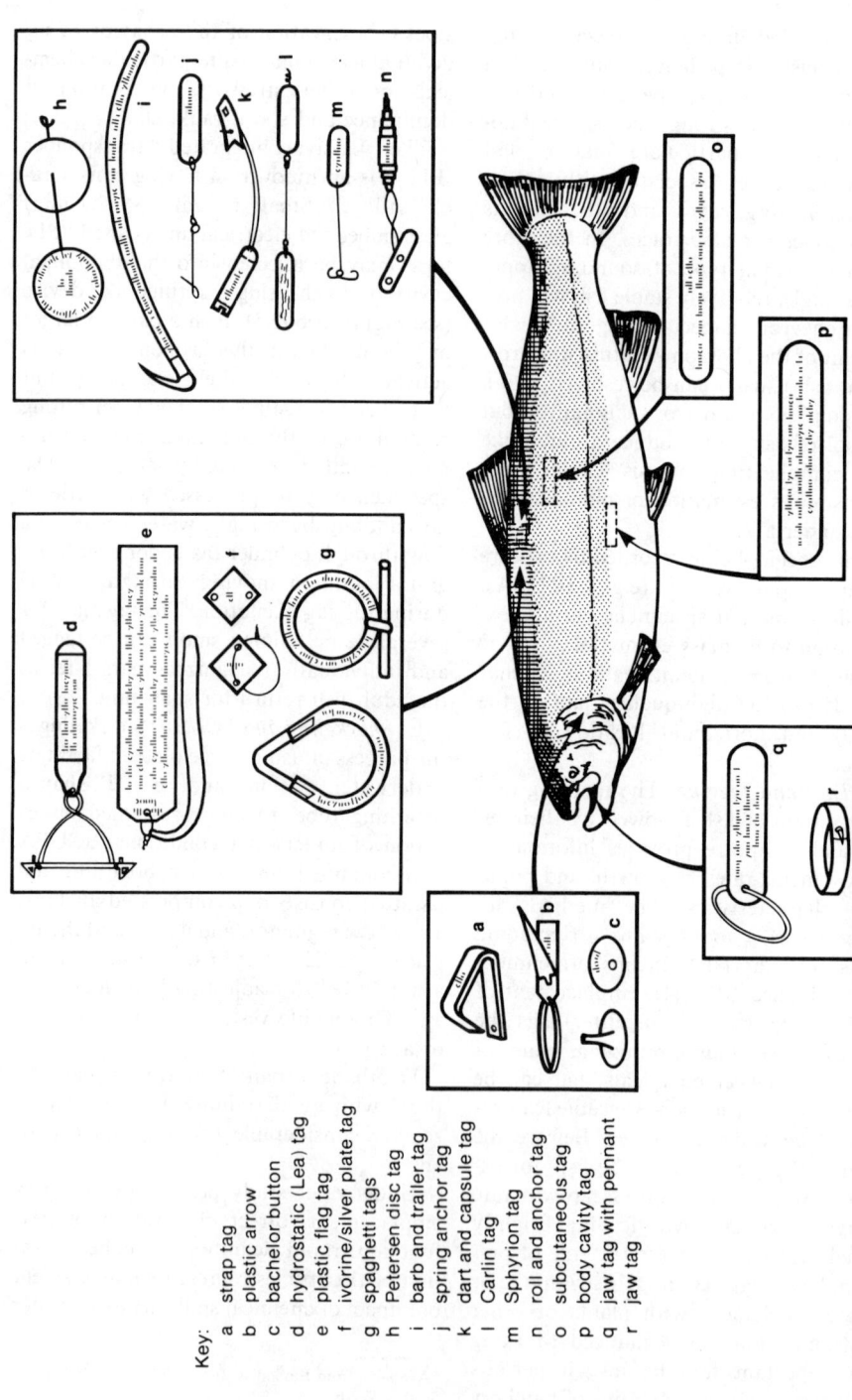

Figure 10600:4. Types of tags commonly used. Source: STOTT, B. 1971. Marking and tagging. *In* W.E. Ricker, ed. Methods for Assessment of Fish Production in Fresh Waters. IBP Handbook No. 3. Blackwell Scientific Publ., Oxford, England.

Key:

a strap tag
b plastic arrow
c bachelor button
d hydrostatic (Lea) tag
e plastic flag tag
f ivorine/silver plate tag
g spaghetti tags
h Petersen disc tag
i barb and trailer tag
j spring anchor tag
k dart and capsule tag
l Carlin tag
m Sphyrion tag
n roll and anchor tag
o subcutaneous tag
p body cavity tag
q jaw tag with pennant
r jaw tag

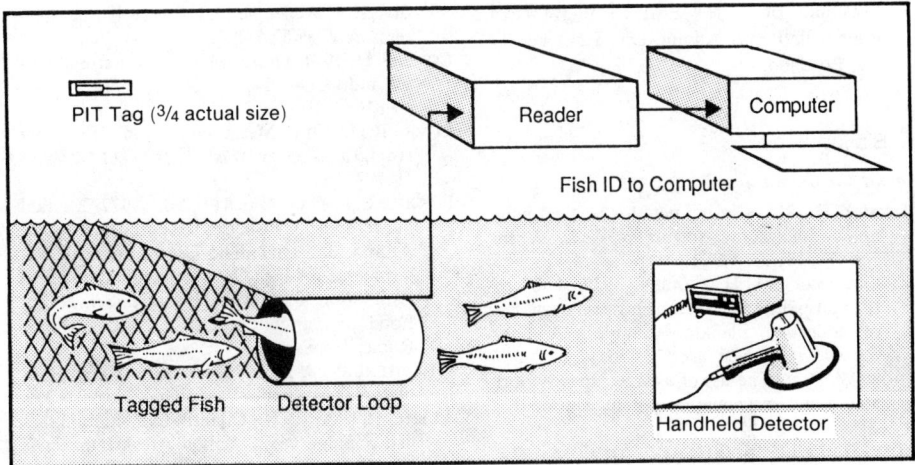

Figure 10600:5. Passive integrated transponder (PIT) tagging system. Source: BioSonics, Inc., Seattle, Wash.

(when dead and moribund specimens may be resting on the bottom), the study of behavior, and estimation of population size. A team of divers can swim side-by-side along a defined course noting fish present and making simultaneous records of observation using underwater recording materials. An underwater pad consisting of sanded vinyl plastic can be marked with the usual graphite pencil and erased with a conventional rubber eraser. Aluminum foil taped onto a plastic sheet, again marked with a pencil, is effective and has the added advantage of providing a permanent record.

4. References

1. MCCLANE, A.J., ed. 1974. McClane's New Standard Fishing Encyclopedia and International Angling Guide, 2nd ed. Holt, Rinehart & Winston, New York, N.Y.

2. SNYDER, D.E. 1976. Terminology for intervals of larval fish development. *In* J. Boreman, ed. Great Lakes Fish Egg and Larvae Identification: Proceedings of a Workshop, p. 41. U.S. Fish Serv., National Powerplant Team, Ann Arbor, Mich.

3. BOWLES, R.R. & J.V. MERRINER. 1978. Evaluation of ichthyoplankton sampling gear used in power plant studies. *In* L.D. Jensen, ed. Proc. 4th Annual Workshop on Entrainment and Impingement, 5 Dec. 1978, Chicago, Ill.

4. CLARKE, G.L. & D.F. BUMPUS. 1939. Brief account of a plankton sampler. *Int. Rev. Ges. Hydrobiol. Hydrogr.* 39:190.

5. NOVOTNY, D. & G.R. PRIEGEL. 1974. Electrofishing Boats, Improved Designs and Operational Guidelines to Increase the Effectiveness of Boom Shockers. Tech. Bull. No. 73, Dep. Natural Resources, Madison, Wis.

6. VIBERT, R., ed. 1967. Fishing with Electricity, Its Application to Biology and Management, Contributions to a Symposium. European Inland Fisheries Advisory Comm., FAO, Fishing News Books Ltd., London, U.K.

7. WEBSTER, D.A., J.L. FORNEY, R.H. GIBBS, JR., J.H. SEVERS & W.F. VAN WOERT. 1955. A comparison of alternating and direct currents in fishery work. *N.Y. Fish Game J.* 2:106.

8. SCHNICK, R.A., F.P. MEYER & D.L. GRAY. 1986. A Guide to Approved Chemicals in Fish Production and Fishery Resource Management. National Fish Research Lab., La Crosse, Wis.

9. MITSON, R.B. 1983. Fisheries Sonar. Fishing News Books Ltd., Farnham, Surrey, England.

10. PRENTICE, E.F., C.W. SIMS & D. L. PARK. 1985. A Study to Determine the Biological

Feasibility of a New Fish Tagging System. Bonneville Power Admin. Div. Fish & Wildlife, Portland, Ore.

5. Bibliography

Larval and Immature Fishes

NANSEN, F. 1915. Closing nets for vertical hauls and for horizontal towing. *J. Cons. Perma. Int. Explor. Mer* 67:1.

GIBBONS, S.G. & J.H. FRASER. 1937. The centrifugal pump and suction hose as a method of collecting plankton samples. *J. Cons. Perma Int. Explor. Mer* 12:155.

ARON, W. 1958. The use of a large capacity portable pump for plankton patchiness. *J. Mar. Res.* 16:158.

BARY, B.M., J.G. DESTEFANO, M. FORSYTH & J. VON DEN KERKHOF. 1958. A closing high-speed plankton catcher for use in vertical and horizontal towing. *Pacific Sci.* 12:46.

CLARKE, W.D. 1964. The jet net, a new high-speed plankton sampler. *J. Mar. Res.* 22:284.

ARON, W., E.H. AHLSTROM, B.M. BARY, A.W.H. BE & W.D. CLARKE. 1965. Towing characteristics of plankton sampling gear. *Limnol. Oceanogr.* 10:333.

BEERS, J.R., G.L. STEWART & J.D.H. STRICKLAND. 1967. A pumping system for sampling small plankton. *J. Fish. Res. Board Can.* 24:1.

FRASER, J.H. 1968. Zooplankton sampling. *Nature* 211:915.

CROCE, N.D. & A. CHIARABINI. 1971. A suction pipe for sampling mid-water and bottom organisms in the sea. *Deep Sea Res.* 18:851.

SHERMAN, K. & K.A. HONEY. 1971. Size Selectivity of the Gulf III and Bongo Zooplankton Samplers. *Int. Comm. Northwest Atl. Fish. Res. Bull.* 8:45.

Mature Fishes

HASKELL, D.C., J. MACDOUGAL & D. GEDULDIG. 1954. Reactions and motion of fish in a direct current electric field. *N.Y. Fish & Game J.* 1:47.

LAGLER, K. F. 1956. Freshwater Fishery Biology, 2nd ed. Wm. C. Brown Co., Dubuque, Iowa.

HURLBERT, S.H. 1971. The nonconcept of species diversity: a critique and alternative parameters. *Ecology* 52:577.

LOVE, R.H. 1971. Dorsal-aspect target strength of an individual fish. *J. Acoust. Soc. Amer.* 49:816.

LOVE, R.H. 1971. Measurements of fish target strength: a review. *Fish Bull. NOAA/NMFS* 69:703.

FORBES, S.T. & O. NAKKEN, eds. 1972. Manual of Methods for Fisheries Resource Survey and Appraisal. Part 2: The Use of Acoustic Instruments for Fish Detection and Abundance Estimation. FAO Man. Fisheries Sci. No. 5, Food & Agriculture Org. United Nations, Rome, Italy.

WEATHERLEY, A.H. 1972. Growth of Fish Populations. Academic Press, New York, N.Y.

KUSHLAN, J.A. 1974. Quantitative sampling of fish populations in shallow, freshwater environments. *Trans. Amer. Fish Soc.* 103:348.

URICK, R.J. 1975. Principles of Underwater Sound. McGraw-Hill, New York, N.Y.

GULLAND, J.A., ed. 1977. Fish Population Dynamics, John Wiley & Sons, New York, N.Y.

GRINSTED, B.G., R.M. GENNINGS, G.R. HOOPER, C.A. SCHULTZ & D.A. HORTON. 1978. Estimation of standing crop of fishes in the predator-stocking-evaluation reservoirs. *Proc. Annu. Conf. S.E. Assoc. Game Fish Comm.* 30:120.

BURCZYNSKI, J. 1979. Introduction to the Use of Sonar Systems for Estimating Fish Biomass. FAO Fish. Tech. Pap. No. 191, Food & Agriculture Org. United Nations, Rome, Italy.

SHORROCKS, B. 1979. The Genesis of Diversity. University Park Press, Baltimore, Md.

HOCUTT, C.H. & J.R. STAUFFER, JR., eds. 1980. Biological Monitoring of Fish. Lexington Books, D.C. Heath and Co., Lexington, Mass.

WELCH, H.E. & K.H. MILLS. 1981. Marking fish by scarring soft fin rays. *Can. J. Fish Aquat. Sci.* 38:1168.

NAKKEN, O. & S.C. VENEMA, eds. 1983. Symposium on Fishery Acoustics. Fish. Rep. 300, Food & Agriculture Org. United Nations, Unipub, Ann Arbor, Mich.

BROWNIE, C., D.R. ANDERSON, K.P. BURNHAM & D.S. ROBSON. 1985. Statistical Inference from Band Recovery Data. Colorado Fish & Wildlife Res. Unit, Fort Collins.

10600 C. Sample Preservation

The decision to preserve specimens depends on study objectives. Preserved material may be necessary to confirm identity of a species, to evaluate certain demographic characteristics of the population, or to estimate incidence of parasitic infection or disease. It also may be essential evidence in legal proceedings.

Do not preserve specimens unless there is clear need for them. Fixatives are toxic and can be dangerous if used improperly. Preserved specimens require expensive and time-consuming curation.

Fix specimens in 10% formalin (a 9:1 ambient water dilution of 100% formalin). Formaldehyde gas reaches saturation in water at about 37% by weight and by convention this saturated solution is called 100% formalin. Fix fish less than 10 cm in total length without opening the visceral cavity. Larger specimens require injection of preservative into the visceral cavity or slitting of right ventral body wall for about 25% of body length. Specimens larger than 25 cm total length (and especially oily species) usually require injection of concentrated formalin into the dorsal muscle mass.

The placement of fish in the sampler container, i.e., head up or down, depends on intended use. A good ratio of specimen mass to preservative is 1:1 with the level of the preservative submerging the specimen by an inch or more.

Preserve ichthyoplankton in 10% formalin. When crustaceans or organic detritus is present use 20% formalin. To facilitate sorting, 1 g rose bengal stain/L fixative may be added to stain eggs and larvae selectively.

Ideally use wide-mouthed containers with a durable, screw-type plastic cover. If using metal caps, add about 1 g sodium borate/L preserved material.

Formaldehyde is highly allergenic; minimize direct contact. Formaldehyde is best transported in tightly sealed plastic containers.

After several days to two or more weeks in the fixative (depending on the size of the fish), transfer specimens to 70% ethyl alcohol for long-term preservation.

Small amounts of 10% formalin may be released into wastewater collection systems. However, bulk release may be harmful and/or illegal. Plan carefully for safe disposal of fixatives and preservatives.

Isopropyl alcohol is a less expensive and less flammable substitute for ethyl alcohol, but weigh these advantages against the fact that isopropyl is not a good fixative and may damage the specimens for histology. Both alcohols are highly flammable when stored in bulk.

For pathology fix whole fish or organs for at least 24 h in 10 times their volume of neutral buffered formalin before further processing:

37% Formaldehyde		
(100% formalin)	100	mL
Distilled water	900	mL
Sodium phosphate monobasic,		
NaH$_2$PO$_4 \cdot$H$_2$O	4	g
Sodium phosphate dibasic,		
Na$_2$HPO$_4$	6.5	g

Fix fish less than 8 cm long but open the ventral body cavity from the anus to below the head. Fixation is most effective on live fish or as soon after death as possible. Delay results in autolysis. In larger fish that cannot be preserved whole, preserve representative sections of organs about 5 × 5 × 2 mm in size. Obtain such tissue blocks from organ areas having irregular color, size, or consistency, using a scalpel or razor blade and fix immediately. Except in large fish, fix the entire brain. See Figure 10600:6 for morphological and anatomical guidance.

Rigorously document preserved materials. Place labels composed of highly resistant bond paper, bearing the key facts of lot number, collection date, locality, name(s) of collector(s), and other particulars written in graphite or waterproof india ink, inside each container or attach to larger specimens held in plastic bags.

Ship fixed specimens packed in absorbent paper or cloth (cheesecloth or paper towels) lightly moistened with the fixative or preservative and sealed in a plastic bag in turn sealed in a second or third bag. Pad package and place in a box or canister for

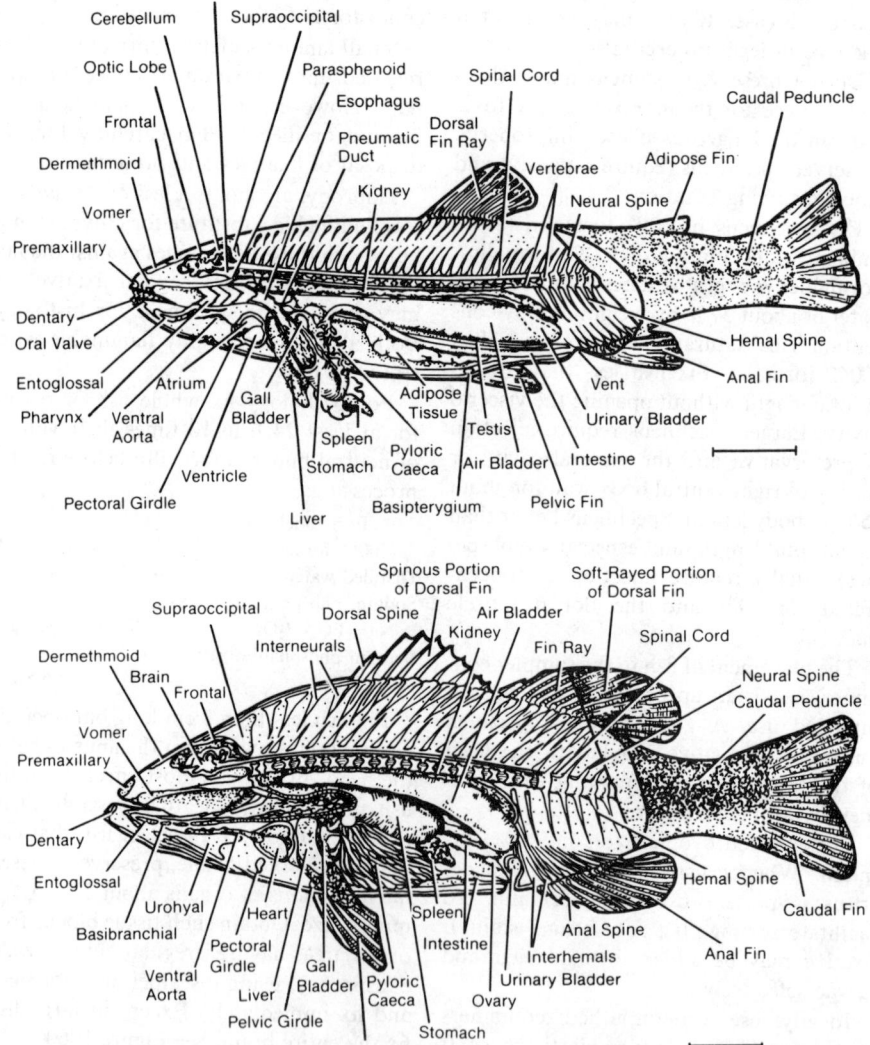

Figure 10600:6. Key organs and external body parts of a soft-rayed (upper) and spiny-rayed (lower) fish. Source: LAGLER, K.F. 1956. Freshwater Fishery Biology, 2nd ed. Wm. C. Brown Co., Dubuque, Iowa.

transport with a copy of the chain-of-custody documents. Shipment in bulk fixative or preservative via the United States mails is prohibited.

If preserved specimens are no longer needed after completion of a study, offer them to a regional or national museum. Planning for eventual disposal of specimens is both economical and provides for the best long-term use of the materials collected.

10600 D. Analysis of Collections

1. Identification

a. General remarks: Identification of fish is based on diagnostic characters such as body form, color and size, shape and position of fins, meristic features such as the number of rays in a fin or the number of scales in a specific series, the presence of distinctive organs such as barbels, or the lateral line and various proportions such as the ratio of the length of the head to the total length of the body (see Figure 10600:6). Diagnostic features may vary with age, sex, reproductive condition, social status, time of year, and habitat of the fish. Diagnostic keys and other descriptive materials are available for all regions of North America and a list of selected works appears in the bibliography by regional category:

General
Pacific Slope and Northwest
  Alaska, British Columbia, Washington, Oregon (W), California
Southwest and Intermountain
  Oregon (E), Wyoming (W), Colorado (W), Idaho, New Mexico, Arizona, Utah, Nevada
Great Plains
  Alberta, Saskatchewan, Manitoba, Montana, North Dakota, South Dakota, Wyoming (E), Colorado (E), Nebraska, Kansas, Oklahoma, Texas

Piedmont and Florida
  Delaware, Maryland, New Jersey, Virginia, Pennsylvania (E), North Carolina, South Carolina, Georgia, Florida, Alabama, New York
Mississippi-Ohio Valley
  Iowa, Illinois, Indiana, Ohio, Pennsylvania (W), Missouri, Kentucky, West Virginia, Arkansas, Tennessee, Louisiana, Mississippi, New York
Great Lakes
  Minnesota, Wisconsin, Michigan, Ontario
Northeastern
  Quebec, Newfoundland, New Brunswick, Prince Edward Island, Nova Scotia, Maine, New Hampshire, Vermont, Massachusetts, Connecticut, Rhode Island, New York

Common names of North American fishes are listed in a special publication of The American Fisheries Society.[1]

Identification may be performed on both fresh and preserved specimens. Fresh materials are essential for color. Fixed specimens are suitable for determining meristic or mensural characteristics. Use a dissecting binocular microscope with illumination and dissecting tools to examine specimens less than 10 cm long.

b. Ichthyoplankton: The identification of fish eggs and larval fish is a special discipline and selected works are cited in the bibliography. Enumeration often involves knowledge of adult populations such as sexual maturity, spawning migration, spawning groups, and presence/absence of the species in the watershed. Intensive studies may involve captive spawning, rearing, and documentation of larval development of selected species in the watershed. Obtain specialized assistance for identification of ichthyoplankton and the more difficult taxa such as the Cyprinidae (minnows), Clupeidae (herrings), Catostomidae (suckers), Poeciliidae (livebearers), Cyprinodontidae (killifishes), Percidae (perches), and others.

c. Rare and endangered forms: Pay special attention to rare and endangered forms that are protected by law. If a rare or endangered form is present a special permit or memorandum of understanding may be required. Do not intentionally collect rare or endangered forms; if they are accidentally taken and fixed contact the responsible agency and transfer specimens to a designated museum.

2. Diet

Examination of the contents of the digestive tract provides information on the amount and kind of foods eaten. Stomach contents may be extracted from some living fish by thrusting a smooth and moistened glass or metal cylinder down the esophagus into the stomach. Most analyses, however, involve the sacrifice of the fish. After capture, preserve either the entire fish or the viscera.

To characterize diet, use a binocular dissecting microscope with the organisms suspended in water in a shallow transparent container. Define the contents to the lowest practical taxon and express as frequency of occurrence. Enumeration of each dietary item by weight, volume, or caloric content may be done in special studies.

3. Structure of Populations and Assemblages

a. General remarks: The actual properties of a population such as number, average size, and weight are estimated statistics.[2] For ecological and management purposes evaluate the numbers of individuals and the biomass (total weight) along with the factors that regulate these. Four key variables define number: natality, i.e., number of individuals being added through reproduction, mortality, immigration, and emigration.

Biomass is the net result of dietary income and its conversion efficiency, respiration, defecation, and loss to other compartments of the ecosystem such as predators, parasites, and pathogens.

The ideal static characterization of a population consists of graphs that show by sex:

Age frequency and maturation
Length frequency
Weight frequency
Age versus length
Age versus weight
Length versus weight

b. Population size: One of the simplest and most practical population estimation methods is the Petersen ratio, using marked fish. Collect a sample of fish, mark, and release. Collect a second or recapture sample including both marked and unmarked fish. The method is based on the general assumption of random distribution, that is, the number of marked fish recovered is to the total catch in the second sample as the total number of marked fish is to the total population. The larger the number of marked fish, the more reliable the estimate. It is particularly important to mark large numbers of the smaller species, if possible, because individuals of such species are less easily recovered. Estimate population size using the mark-recapture method by the formula

$$\hat{N} = MC/R$$

where:

$\hat{N}$ = estimate of population size,
M = number of fish marked and released,
C = recapture sample size, including both marked and unmarked fish, and
R = number of marked fish recaptured.

The estimate is of the population present at the time of the first (marking and release) sample, not at the time of recapture.

c. Age and growth: Age and growth rates are useful for determining the effects of water quality on fish populations. The three methods used for determining age of fish are comparison of length-frequency distribution, recovery of marked fish of known age, and interpretation of layers laid down on hard parts of the fish. Determination of age distribution by length-frequency studies often is adequate for the first 2 to 4 years, but usually fails to separate older age groups reliably because of increasing overlap in length distribution. It also becomes less reliable as one approaches the equator because breeding seasons are more protracted in warmer areas, causing yearly age groups to be less well defined. Another frequently used method of determining age is interpreting and counting growth zones or growth checks that appear in the hard parts of fish. Those considered to be formed once a year are called year marks, annual marks, annual rings, or annuli (Figure 10600:7). They are formed during alternate periods of faster and slower growth (or no growth at all) and reflect various environmental or internal influences. In a temperate region, the period of little or no growth usually occurs only once annually, beginning in winter and extending into spring or early summer. Generally, the more the seasons differ with respect to temperature, the sharper the annual marks will be. The most distinct annual rings are developed in temperate climates of the northern and southern hemispheres. Scales and bony

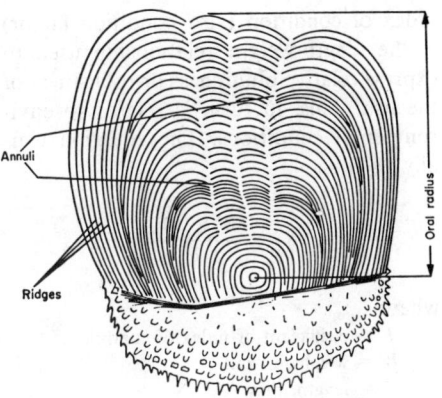

Figure 10600:7 Fish scale. Source: RICKER, W.E., ed. 1971. Methods for Assessment of Fish Production in Fresh Waters. IBP Handbook No. 3. Blackwell Scientific Publ., Oxford, England.

structures also have been of value in the study of seasonal growth. Still another method of age and growth determination is by marking or tagging fish. Most tagging methods are not applicable to small fish, or if they are, they may cause mortality, so that recaptures are few. Instead of using tags, small fish may be marked by fin clipping or scarring or by injecting dye.

Scales are the structures most often used in age determinations of fish because they are easily removed without injury to the specimen. For scaleless fish, removal of other structures (otoliths, vertebrae, fin spines) may necessitate sacrifice of the individual. Take scales usually from the upper mid-side of the body where they are large and symmetrical. Wherever scales are taken from, remove them from the same part of the body in all individuals from the same sample. Several scales may be needed for analysis, because an annulus that appears doubtful on one scale may be clear in another. In addition, some species may have regenerated, i.e., replaced, scales that may not contain all annuli.

d. Index of condition: The coefficient or

index of condition (also condition factor) is the length-weight relationship used to express relative plumpness or robustness of the fish. This, in turn, is related to environmental conditions. The equation usually used is:

$$K = \frac{(W \times 10^5)}{L^3}$$

where:

K = coefficient or index of condition,
W = weight, g, and
L = length, mm.

The gonadosomal coefficient (weight of gonads divided by the remaining weight of the body) and hepatosomal coefficient (weight of the liver divided by the remaining weight of the body) also are used as indicators of physiological condition.

e. *Diversity:* Diversity indices are used to quantify the structure of the species assemblage in a particular habitat over time.[3] Usually, the number of species represented and the relative numbers of each species represented are incorporated into a single number. Margalef's index is illustrative.

$$d = \frac{(S - 1)}{\ln N}$$

where:

d = Margalef Index,
S = number of species represented in sample, and
N = total number of individuals in the sample.

4. References

1. ROBINS, C. R., R.M. BAILEY, C.E. BOND, J.R. BOOKER, E.A. LACHNER, R.N. LEA & W.B. SCOTT, eds. 1980. A List of Common and Scientific Names of Fishes from the United States and Canada, 4th ed. Spec. Publ. No. 12, American Fisheries Soc., Bethesda, Md.

2. CUSHING, D.H. 1975. Marine Ecology and Fisheries. Cambridge Univ. Press, Cambridge, U.K.

3. PIELOU, E.C. 1975. Ecological Diversity. Wiley-Interscience Publ., John Wiley & Sons, New York, N.Y.

5. Bibliography

Larval and Immature Fishes

HUBBS, C.L. 1943. Terminology of early stages of fishes. *Copeia* 1943(4):160.

MANSUETI, A.J. & J.D. HARDY, JR. 1967. Development of Fishes of the Chesapeake Bay Region: An Atlas of Egg, Larval and Juvenile Stages. Natural Resources Inst., Univ. Maryland, College Park.

LIPPSON, A.J. & R.L. MORAN. 1974. Manual for Identification of Early Developmental Stages of Fishes of the Potomac River Estuary. Martin Marietta Corp. Environ. Tech. Center, Baltimore, Md.

JONES, P.W., F.D. MARTIN & J.D. HARDY, JR. 1978. Development of Fishes of the Mid-Atlantic Bight—An Atlas of Egg, Larval and Juvenile Stages. Vol. I. Acipenseridae through Ictaluridae. U.S. Fish & Wildlife Serv. Biol. Serv. Program FWS/OBS - 78/11.

HARDY, J.D., JR. 1978. Development of Fishes of the Mid-Atlantic Bight—An Atlas of Egg, Larval, and Juvenile Stages. Vol. II. Anguillidae through Syngnathidae. U.S. Fish & Wildlife Serv. Biol. Serv. Program FWS/OBS -78/12.

HARDY, J.D., JR. 1978. Development of Fishes of the Mid-Atlantic Bight—An Atlas of Egg, Larval and Juvenile Stages. Vol. III. Aphredoderidae through Rachycentridae. U.S. Fish & Wildlife Serv. Biol. Serv. Program FWS/OBS -78/12.

JOHNSON, G.D. 1978. Development of Fishes of the Mid-Atlantic Bight—An Atlas of Egg, Larval and Juvenile Stages. Vol. IV. Carangidae through Ephippidae. U.S. Fish & Wildlife Serv. Biol. Serv. Program FWS/OBS - 78/12.

FRITZSCHE, R.A. 1978. Development of Fishes of the Mid-Atlantic Bight—An Atlas of Egg, Larval and Juvenile Stages. Vol. V. Chaetodontidae through Ophidiidae. U.S. Fish & Wildlife Serv. Biol. Serv. Program FWS/OBS - 78/12.

MARTIN, F.D. & G.E. DREWRY. 1978. Development of Fishes of the Mid-Atlantic Bight—An Atlas of Egg, Larval and Juvenile Stages. Vol. VI. Stramateidae through Ogcocephalidae. U.S. Fish & Wildlife Serv. Biol. Serv. Program FWS/OBS -78/12.

WANG, J.C.S. & R.J. KERNEHAN. 1979. Fishes of the Delaware Estuaries: A Guide to the Early Life Histories. EA Communications, Ecological Analysts, Inc., Towson, Md.

SNYDER, D.E. 1981. Contributions to a Guide to the Cypriniform Fish Larvae of the Upper Colorado River System in Colorado. Biol. Sci. Ser. No. 3, Bur. Land Management, Colorado.

WANG, J.C.S. 1981. Taxonomy of the Early Life Stages of Fishes—Fishes of the Sacramento - San Joaquin Estuary and Moss Landing Harbor, Elkhorn Slough, Calif. Ecological Analysts, Inc., Concord, Calif.

ANER, N.A., ed. 1982. Identification of Larval Fishes of the Great Lakes Basin with Emphasis on the Lake Michigan Drainage. Great Lakes Fisheries Research Center, Ann Arbor, Mich.

Mature Fishes

General

PERLMUTTER, A. 1961. Guide to Marine Fishes. New York Univ. Press, New York, N.Y.

LEE, D.S., C.R. GILBERT, C.H. HOCUTT, R.E. JENKINS, D.E. MCALLISTER & J.R. STAUFFER, JR. 1980 et seq. Atlas of North American Freshwater Fishes. North Carolina State Mus. Natural History, Raleigh.

CASTRO, J.I. 1983. Sharks of the North American Waters. Texas A & M Univ. Press, College Station.

KUEHNE, R.A. & R.W. BARBOUR. 1983. The American Darters. Univ. Press of Kentucky, Lexington.

PAGE, L.M. 1983. Handbook of Darters. TFH Publications, Inc., Neptune, N.J.

Pacific Slope and Northwest

MCGINNIS, S.M. 1958. Fishes of California. Univ. California Press, Berkeley.

MCPHAIL, J.D. & C.C. LINDSEY. 1970. Freshwater Fishes of Northwestern Canada and Alaska. Bull. 173, Fisheries Research Board Canada, Ottawa, Ont.

MILLER, D.J. & R.N. LEA. 1972. Guide to the Coastal Marine Fishes of California. Fish. Bull. 157, California Dep. Fish & Game, Sacramento.

HART, J.L. 1973. Pacific Fishes of Canada. Bull. 180, Fisheries Research Board Canada, Ottawa, Ont.

MOYLE, P.B. 1976. Inland Fishes of California. Univ. California Press, Berkeley.

WYDOSKI, R.S. & R.R. WHITNEY. 1979. Inland Fishes of Washington. Univ. Washington Press, Seattle.

MORROW, J.E. 1980. The Freshwater Fishes of Alaska. Alaska Northwest Publishing Co., Anchorage.

ESCHMEYER, W.N., E.S. HERALD & H. HAM-MAN. 1983. A Field Guide to Pacific Coast Fishes. The Peterson Field Guide Series, Houghton Mifflin Co., Boston, Mass.

Southwest and Intermountain

KOSTER, W.J. 1957. Guide to the Fishes of New Mexico. Univ. New Mexico Press, Albuquerque.

LARIVERS, I. 1962. Fish and Fisheries of Nevada. Nevada State Fish & Game Comm., Carson City.

EVERHART, W.H. & W.R. SEAMAN. 1971. Fishes of Colorado. Colorado Game and Parks Div., Denver.

MINCKLEY, W.L. 1973. Fishes of Arizona. Arizona Game & Fish Dep., Phoenix.

SIMPSON, J.C. & R.L. WALLACE. 1978. Fishes of Idaho. Univ. Idaho Press, Moscow.

Great Plains

KNAPP, F.T. 1953. Fishes Found in the Fresh Waters of Texas. Ragland Studio and Lithograph Printing Co., Brunswick, Ga.

BAILEY, R.M. & M.O. ALLUM. 1962. Fishes of South Dakota. Misc. Publ. Mus. Zool. Univ. Mich. 119, Ann Arbor.

SIGLER, W.F. & R.R. MILLER. 1963. Fishes of Utah. Utah Game & Fish Dep., Salt Lake City.

BAXTER, G.T. & J.R. SIMON. 1970. Wyoming Fishes. Bull. 4, Wyoming Game and Fish Comm., Cheyenne.

EVERHART, W.H. & W.R. SEAMAN. 1971. Fishes of Colorado. Colorado Game & Parks Div., Denver.

MORRIS, J., L. MORRIS & L. WITT. 1972. The Fishes of Nebraska. Nebraska Game and Parks Comm., Lincoln.

MILLER, R.J. & H.W. ROBINSON. 1973. The Fishes of Oklahoma. Oklahoma State Univ. Press, Stillwater.

SCOTT, W.B. & E.J. CROSSMAN. 1973. Freshwater Fishes of Canada. Bull. 184, Fisheries Research Board Canada, Ottawa, Ont.

HOESE, H.D. & R.H. MOORE. 1977. Fishes of the Gulf of Mexico: Texas, Louisiana, and Adjacent Waters. Texas A & M Univ. Press, College Station.

Piedmont and Florida

HILDEBRAND, S.F. & W.C. SCHROEDER. 1928. Fishes of Chesapeake Bay. Fishery Bull. 43, U.S. Bur. Fisheries.

SMITH-VANIZ, W.F. 1968. Freshwater Fishes of Alabama. Agricultural Experiment Sta., Auburn Univ. Press, Auburn, Ala.

DAHLBERG, M.D. & D.C. SCOTT. 1971. The

Freshwater Fishes of Georgia. Bull. Georgia Acad. Sci. 29.

DAHLBERG, M.D. 1975. Guide to Coastal Fishes of Georgia and Nearby States. Univ. Georgia Press, Athens.

JENKINS, R.E. & N.M. BURKHEAD. In press. Freshwater Fishes of Virginia. Univ. Virginia Press, Charlottesville.

Mississippi-Ohio Valley

COOK, F.A. 1959. Freshwater Fishes of Mississippi. Mississippi Game & Fish Comm., Jackson.

DOUGLAS, N.H. 1974. Freshwater Fishes of Louisiana. Claitors Publ. Div., Baton Rouge, La.

CLAY, W.M. 1975. The Fishes of Kentucky. Kentucky Dep. Fish & Wildlife Resources, Frankfort.

PFLIEGER, W.L. 1975. The Fishes of Missouri. Missouri Dep. Conservation, Columbia.

HOESE, H.D. & R.H. MOORE. 1977. Fishes of the Gulf of Mexico: Texas, Louisiana and Adjacent Waters. Texas A & M Univ. Press, College Station.

SMITH, P.W. 1979. The Fishes of Illinois. Univ. Illinois Press, Urbana.

TRAUTMAN, M.B. 1981. The Fishes of Ohio, 2nd ed. Ohio State Univ. Press, Columbus.

Great Lakes

HUBBS, C.L. & K.F. LAGLER. 1958. Fishes of the Great Lakes Region. Univ. Michigan Press, Ann Arbor.

SCOTT, W.B. & E.J. CROSSMAN. 1973. Freshwater Fishes of Canada, Bull. 184. Fisheries Research Board Canada, Ottawa, Ont.

TRAUTMAN, M.B. 1981. The Fishes of Ohio, 2nd ed. Ohio State Univ. Press, Columbus.

BECKER, G.C. 1983. Fishes of Wisconsin. Univ. Wisconsin Press, Madison.

Northeastern Region

TEE-VAN, J., ed. 1948 et seq. Fishes of the Western North Atlantic. Mem. Sears Foundat. Mar. Res., New Haven, Conn.

BIGELOW, H.B. & W.C. SCHROEDER. 1953. Fishes of the Gulf of Maine. Fish Bull. 74, U.S. Fish & Wildlife Serv.

LEIM, A.H. & W.B. SCOTT. 1966. Fishes of the Atlantic Coast of Canada. Bull. 155, Fisheries Research Board Canada, Ottawa, Ont.

SCOTT, W.B. & E.J. CROSSMAN. 1973. Freshwater Fishes of Canada. Bull. 184, Fisheries Research Board Canada, Ottawa, Ont.

SMITH, C.L. 1985. The Inland Fishes of New York State. New York State Dep. Environmental Conservation, Albany, N.Y.

10600 E.   Investigation of Fish Kills

Fish kills vary, from the individual fish that dies of old age to the catastrophic kill, from partial to complete, and from natural occurrence to the results of human activity. No single investigative procedure can be appropriate for all situations. The following brief description may serve as an aid in investigating kills. Getting to the scene promptly before the evidence has decomposed or drifted away is vital. If surveillance of a particular body of water or area is involved, have available preset plans and equipment on a standby basis.[1,2]

1. Causes of Fish Kills

Fish kills may be caused by such natural events as acute temperature change, storms, ice and snow cover, decomposition of natural materials, salinity change, spawning mortalities, parasites, and bacterial and viral epidemics. Human-caused fish kills may be attributed to municipal or industrial wastes, agricultural activities, and water control activities.

2. Classification of Kills

One dead fish in a stream may be called a fish kill; however, in a practical sense, adopt some minimal range in number of dead fish observed, plus additional qualifications, in reporting and classifying fish kills. Any fish kill is significant if it affects fish of sport or commercial value, results from a suspected negligent discharge or malfunctioning waste treatment facility, or causes widespread environmental damage.

The following definitions, based on a stream about 60 m wide and 2 m deep, are suggested as guidelines. For other size streams, make proportional adjustments.

MINOR KILL: 1 to 100 dead or dying fish confined to a small area or stream stretch. If recurrent, it could be significant; investigate.

MODERATE KILL: 100 to 1000 dead or dying fish of various species in 1 to 2 km of stream or equivalent area of a lake or estuary.

MAJOR KILL: 1000 or more dead or dying fish of many species in a reach of stream up to 16 km or greater, or equivalent area of a lake or estuary.

3. Investigation Techniques

In preparing for a field investigation, study area maps and determine the zone of fish mortality and the access to it. Identify waste dischargers. Contact participating laboratories to discuss number and size of samples that will be submitted, types of analyses required, dates of sample receipt, method of sample shipment, date by which results are needed, and to whom results are to be reported. Use two information record forms for fish kill investigations, an initial contact form and a field investigation form.

On all fish kill investigations take a thermometer, dissolved oxygen test kit, conductivity and pH meters; or a general chemical kit, biological sampling gear, sample bottles, and other specimen containers. Include in the investigating team at least one person who is experienced in investigating fish kills.

The field investigation consists of visual observations, sampling of fish, water, and other biota, and physical measurements of the environment. The first local observer of the kill is a useful guide to the area, which should be reconnoitered initially to establish that a fish kill actually has occurred.

If a fish kill has taken place, immediately start fish sampling because collection of dying or recently dead fish is critical. For purposes of comparison, if possible, collect healthy fish from an unaffected area.

Examine moribund or recently dead animals immediately for external and internal abnormalities. Record all changes in color, size, location and consistency of organs. Record the location of lesions by organ (Figure 10600:6). Colored photographs, with indication of scale, taken at the site are an excellent way of documenting observations and greatly help the consulting pathologist. If photography is not possible describe exactly what was seen, recording the number and size of abnormalities. Provide healthy fish of the same species when possible.

Do not freeze samples for pathology. If fixatives are unavailable place samples in plastic bags on ice and rush to the pathologist.

If virology or bacteriology testing is indicated freeze additional specimens of key organs such as the liver, kidney, spleen, heart, and brain and other parts showing abnormality or lesions, label, and forward for analysis.

Bleed dying or dead fish at collection time to obtain at least 1 g blood. Collect a blood sample in a chemically clean, solvent-washed glass bottle with a TFE-lined screw cap.

Identify and count dead fish. In a large river count dead fish from a fixed station such as a bridge during a fixed period of time. Extrapolate to the total time involved. Alternatively, in a large river or lake, make a shore count and project to entire area of kill. In smaller water bodies traverse entire area to enumerate dead fish.

Collect water samples representative of unpolluted and polluted areas in accordance with the instruction given in Section 10200. As a minimum, measure temperature, pH, dissolved oxygen, and conductivity. Make additional tests depending on suspected causes of the fish kill. Take samples for examination of plankton, periphyton, macrophyton, and macro-invertebrates.

Record observations on water appearance, streamflow, and weather conditions. Color photographs are valuable in recording conditions.

4. References

1. BURDICK, G.E. 1965. Some problems in the determination of the cause of fish kills. *In* Biological Problems in Water Pollution. Publ. No. 999-WP-25, U.S. Public Health Serv., Washington, D.C.
2. SMITH, L.L., JR., et al. 1956. Procedures for Investigation of Fish Kills. A Guide for Field Reconnaissance and Data Collection. Ohio River Sanitary Comm., Cincinnati, Ohio.

5. Bibliography

FEDERAL WATER POLLUTION CONTROL ADMINISTRATION. 1967 & 1968. Pollution Caused Fish Kills. Publ. No. CWA-7, U.S. Dep. Interior.

FEDERAL WATER POLLUTION CONTROL COMMISSION. 1970. Investigating Fish Mortalities. Publ. No. CWT-5, U.S. Dep. Interior (also available as No. 0-380-257, U.S. Government Printing Off., Washington, D.C.).

ANDERSON, B.G. & D.L. MITCHUM. 1974. Atlas of Trout Histology. Wyoming Game and Fish Dep., Cheyenne.

GRIZZLE, J.M. & W.A. ROGERS. 1976. Anatomy and Histology of the Channel Catfish. Agricultural Experiment Sta., Auburn Univ., Auburn, Ala.

ROBERTS, R.J., ed. 1978. Fish Pathology. Bailliere Trindall, Cassell Ltd., London, England.

GORMAN, D.B. 1982. Histology of the Striped Bass. AFS Monogr. No. 3, American Fisheries Society, Bethesda, Md.

ROBERTS, R.J., ed. 1982. Microbial Diseases of Fish. Academic Press, New York, N.Y.

YASUTAKE, W.T. & J.H. WALES. 1983. Microscopic Anatomy of Salmonids: An Atlas. U.S. Fish & Wildlife Serv. Resource Publ. 150, U.S. Dep. Interior, Washington, D.C.

CAIRNS, V.W., P.V. HODSON & J.O. NRIAGU, eds. 1984. Contaminant Effects on Fisheries. John Wiley & Sons, New York, N.Y.

ELLIS, A.E. 1985. Fish and Shellfish Pathology. Academic Press, New York, N.Y.

10900   IDENTIFICATION OF AQUATIC ORGANISMS*

Experienced aquatic biologists will be familiar with most organisms illustrated in Plates 1 through 38, and seldom will need the assistance of keys to identify organisms to the level illustrated. Because these plates are not intended for critical identification, specific (species) names are not cited. Types most likely to be observed are illustrated. For the convenience of those less familiar with the organisms referred to in preceding

sections, a series of short keys is presented to enable them to identify most organisms to the level illustrated by the plates.

In conformity with preceding sections, organisms are arbitrarily divided into microscopic and macroscopic, depending on whether or not they pass through a U.S. Standard No. 30 sieve. For the study of microscopic forms, use a compound microscope. For examination of the smaller macroscopic organisms and to resolve the finer structures of larger forms, use a wide-field stereoscopic microscope.

---

*Approved by Standard Methods Committee, 1988.

10900 A.   Procedure in Identification

Critical identification of a specimen often is time-consuming, even for an experienced biologist. Before looking at any key or other aid to identification, carefully study the specimen for one to several minutes. If necessary, find other examples and compare them with the unknown.

It is important to know where or under what conditions the organism lived before attempting to identify it. For example, did it come from fresh water—a lake or a stream? Is it marine—from the open ocean, shoreline, or estuary? Was it a free swimmer or floater in the water? Was it a bottom organism, attached, crawling, or burrowing? Finally, turn to the following key to major groups.

Only the more common types of aquatic organisms are illustrated here, with special attention to those most frequently used in water quality evaluation. When specimens do not fit obviously into one of the types listed, consult a professional biologist, a microbiologist for the bacteria and fungi, or some of the references provided. Descriptions of color and movement refer to freshly collected or living specimens, or, in the case of microscopic forms, to those preserved as described in Section 10200.

Sizes of the organisms illustrated in Plates 1 through 38 are shown in parentheses in the legend. These are intended to represent *common* sizes, not absolute maxima or minima. Exceptional individuals and even whole localized populations may be encountered that are considerably larger or smaller than the sizes cited.

10900 B.   Key to Major Groups of Aquatic Organisms (Plates 1–38)

Beginning with couplet 1a and 1b of the Keys, compare the descriptions given with the subject specimen. A choice must be made between statement "a" and statement "b." Proceed to the couplet number indicated at the right and repeat the process. Continue until the name of an organism or a plate number is cited instead of another couplet number. Additional information is provided in many of the plate legends.

*Refer to
Couplet
No.*

1a.   Macroscopic: The organism, mass, or colony is visible to the naked eye   13
1b.   Microscopic: Not readily visible to the naked eye   2

1. Key to Microscopic Organisms

2a.   Specimen a single living cell or a mass or colony of relatively independent cells (shapeless, rounded, or threadlike).   3
2b.   Specimen a many-celled, highly organized plant or animal   7
3a.   Cells contain one or more pigments, including chlorophyll *a* (overall color may range through various shades of green, blue, red, brown, or yellow). ALGAE (for details, see Section 10900D following, "Key for Identification of Freshwater Algae")   4

Refer to
Couplet
No.

3b. Cells typically colorless, lacking chlorophyll *a*   12

    4a. Nuclei present; pigment confined to chloroplasts   5

    4b. Nuclei, plastids, or vacuoles absent (pseudovacuoles may be present in certain fila-
        mentous forms). Pigment generally diffused throughout cytoplasm. BLUE-GREEN AL-
        GAE, Plates 1 and 2.

5a. Cell wall permanently rigid, composed of SiO_2, geometrical in appearance, and with regular
    patterns of fine markings; composed of two essentially similar halves, one placed over the
    other as a cover. Golden brown to greenish in color. DIATOMS, Plates 5 and 6.

5b. Cell wall, if present, capable of sagging or bending, rigidity depending on internal pressure
    of cell contents. Cell walls usually of one piece   6

    6a. Cells or colonies nonmotile. Usually some shade of green. NONMOTILE GREEN ALGAE,
        Plates 3 and 4.

    6b. Cells or colony move by means of relatively long whiplike flagella. PIGMENTED
        FLAGELLATES, Plates 11 and 12.

7a. Body with cilia (hairlike structures used for locomotion)   8

7b. Body without cilia.   9

    8a. Body generally covered with cilia, usually somewhat elongate or wormlike, bilaterally
        symmetrical. Minute FLATWORMS (Platyhelminthes), relatives of *Planaria*, Plate 19.

    8b. Cilia confined to one or two crowns at anterior end, which often present the illusion
        of rotating wheels. Internal jaws present. ROTIFERS (Rotifera), Plate 17.

9a. Long slender unsegmented worms that move by sinuous crawling or thrashing motion.
    ROUNDWORMS (Nemathelminthes), Plate 18.

9b. Possess external skeleton and jointed appendages   10

    10a. Crawl about or swim by means of jointed appendages thrust out from between two
        clamlike shells. All appendages can be withdrawn entirely within shells when disturbed.
        OSTRACODS (Ostracoda), Plate 21.

    10b. Swim rapidly by means of a pair of enlarged jointed appendages (antennae) that cannot
        be withdrawn inside carapace or shell.   11

11a. Locomotor appendages (antennae) branched. Microcrustacea, CLADOCERA (Cladocera),
    Plate 20.

11b. Locomotor appendages (antennae) unbranched; body tapers toward rear. Microcrustacea,
    COPEPODS (Copepoda), Plate 21.

    12a. Ingest and digest food internally (ingested food of various colors may be visible through
        body wall). Single-celled or colonial, attached or free-living. PROTOZOANS (Protozoa),
        Plates 13, 14, and 15.

    12b. Digest food externally and absorb products through cell wall. Often secrete masses
        of slime. BACTERIA and FUNGI, Plate 38.

2. Key to Macroscopic Organisms

13a. Specimen a mass of filaments or a glob of gelatinous or semisolid material containing many
    tiny units, requiring microscopic examination to determine details of structure   2

13b. Specimen a well-organized unit or colony   14

    14a. Organism plantlike; flowerlike structures, if present, do not respond when touched,
        generally are colored some shade of green, brown, or red   16

    14b. Organism animal-like; usually responds rapidly when touched, whether attached or
        free-living.   15

15a. Internal backbone present (vertebrates)   17

Refer to
Couplet
No.

15b. No internal backbone present (macroinvertebrates)*. 18

    16a. Plant structure relatively simple. Attachment structures may be present, but no true roots or fibrous tissue. Larger ALGAE, Plate 7 and Color Plates A (*Nitella*) and F (*Chara and Batrachospermum*).

    16b. Plant structure usually includes true roots, stems, and leaves. Fibers or vascular tissue usually present; flowers or seeds may be observed. (One atypical group, "watermeal," consists only of tiny roundish masses, 0.5 to 1 mm in diameter, often misidentified as algae.) HIGHER PLANTS, Plates 8, 9, and 10.

17a. Side appendages, if present, are flat fins. FISHES, Plate 36.

17b. Side appendages, if present, are footlike, with separate digits. AMPHIBIANS, Plate 37.

3. Key to Macroinvertebrates

    18a. Body bilaterally symmetrical (with right and left sides, but may be superficially coiled into a spiral); animal usually not attached but may live inside an attached cocoon or case, or crawl about; usually solitary 23

    18b. Symmetry not bilateral 19

19a. Body typically radially symmetrical 21

19b. Body or colony nonsymmetrical 20

    20a. Body mass generally porous; not a colony, sometimes finger- or antler-like. Freshwater representatives generally are fragile, colored green or brown; marine forms tougher, various colors. SPONGES (Porifera), Plate 16.

    20b. Body mass otherwise 22

21a. Animals with soft smooth bodies and tentacles around a mouth; no anus. Solitary or colonial. Larger colonies usually have rigid limy skeleton of massive, branched, or fan-shaped form. HYDRAS, SEA ANEMONES, JELLYFISHES, CORALS, etc. (Coelenterata), Plate 35A, B.

21b. Body covering usually spiny, soft or rigid, flattened or elongate, typically having five radii, with or without spines or arms; anus present. Solitary. Marine only. STARFISHES and relatives (Echinodermata), Plate 34.

    22a. Colony a jellylike mass, a network of branching tubes, a plant-like tuft, or a lacy limy crust or mass. MOSS ANIMALS (Bryozoa), Plate 16.

    22b. Exclusively marine. Surface of body or colony relatively smooth but tough. Solitary forms, sac-like, with two external openings. Exhibit all degrees of colonialism. Compound forms range from thin slimy masses, with organisms arranged in tiny radial patterns to huge, shapeless masses resembling tough frozen gelatin. SEA SQUIRTS, SEA PORK (Ascidiacea, Urochorda, Chordata), not illustrated.

23a. Animal living within a hard limy shell, soft body (Mollusca) 29

23b. Animal without a limy shell 24

    24a. Jointed legs present (may not be functional). Body may be hard or soft 30

    24b. Jointed legs absent, body covering mostly soft, animal pliable (a hardened head capsule may be present) 25

25a. Body girded by annulations or creases at regular intervals, dividing it into many small segments much wider than long 26

25b. Segments present or absent; if present, not much wider than they are long. 27

    26a. Body with suction disk at one or both ends, in length usually less than 10 times its width. LEECHES (Annelida, Hirudinea), Plate 19.

---

* Invertebrates retained on a U.S. Standard No. 30 sieve.

Refer to
Couplet
No.

26b. Body without suction disks, in length usually more than 10 times its width; hairs or bristles often evident. SEGMENTED WORMS (Annelida), Plate 19.

27a. Body unsegmented, long and slender, appearing smooth, evenly tapered to a fine point at one end. ROUNDWORMS (Nematoda), Plate 18.

27b. Body otherwise   28

    28a. Body flat, elongate, or oblong; unsegmented head is spade-shaped. Pigmented spots on top of head often give the animal a cross-eyed appearance. FLATWORMS (Turbellaria), Plate 19.

    28b. Body segmented, cylindrical, oblong, or capsule-like; may or may not have a head capsule and thick fleshy knobs on underside. Larvae of TWO-WINGED FLIES (Diptera), Plate 29.   30

29a. Shell consisting of two hinged halves. BIVALVES (Pelecypoda), Plate 33.

29b. Shell entire, usually spiral but may be "coolie hat"-shaped. SNAILS (Gastropoda), Plate 32.

    30a. Body with functional legs.   31

    30b. Body without functional legs, mummy- or capsule-like, living in a cocoon. PUPAE (Insecta), Plate 22.   38

31a. Body with three pairs of legs. Larvae, nymphs, and some adults (Insecta)   42

31b. Body with more than three pairs of legs   32

    32a. Body compact, spider-like, with four conspicuous pairs of legs (two other pairs of appendages present). WATER MITES (Acari), Plate 35.

    32b. Body with at least five conspicuous pairs of legs. CRUSTACEANS (Crustacea)   33

4. Key to Crustaceans

33a. Sides of body compressed.   34

33b. Body flattened horizontally.   36

    34a. Eyes on stalks   35

    34b. Eyes, if present, only seen as spots on sides of head. SCUDS (Amphipoda), Plate 21.

35a. Pincers on first pair of legs strong and large; other legs stout, cylindrical, and used for walking. CRAYFISH, also marine lobster (Decapoda), Plate 21.

35b. Pincers on first pair of legs weak and small; other legs, thin and flattened, are used for swimming. SHRIMPS (Mysidea and others), Plate 21.

    36a. Eyes on stalks, shells generally broad, various shapes (marine and brackish water). CRABS (Decapoda), not illustrated.

    36b. Eyes not on stalks.   37

37a. Body covering hard; divided into broad head, truncate body, and sharp tail sections (marine). HORSESHOE CRABS (Arthropoda), Plate 35.

37b. Body with three or more joints. SOWBUGS (Isopoda), Plate 21.

5. Key to Insect Pupae

    38a. Back of pupa with small, paired, hook-bearing plates. CADDISFLIES (Trichoptera), Plate 27.

    38b. Back without paired hook-bearing plates but may have knobs or bristles   39

39a. Developing wings (pads) held free from body. BEETLES (Coleoptera), Plate 30.

39b. Wing pads closely appressed to body, mummy-like, or appendages not evident   40

Refer to
Couplet
No.

40a. With one closely appressed pair of wing pads, but not fused to body; or capsule-like, appendages not evident. TWO-WINGED FLIES (Diptera), Plate 28.

40b. Two pairs of wing pads. 41

41a. First two or three abdominal segments with spiracles (holes for breathing) on each side; body without numerous projections. AQUATIC MOTHS (Lepidoptera), not illustrated.

41b. Body differing from above, may have numerous knobs or other projections on back. HELLGRAMMITES (Neuroptera and Megaloptera), Plate 26.

6. Key to Insect Larvae, Nymphs, and Some Adults

42a. Animal flea-like, with a bifid projecting appendage on the underside. SPRINGTAILS (Collembola), Plate 35.

42b. Animal otherwise 43

43a. Body ending in long segmented filaments. 44

43b. Long filaments absent or, if present, not segmented 45

44a. Two tail filaments, legs ending in two claws. STONEFLIES (Plecoptera), Plate 23.

44b. Three tail filaments (with few exceptions); middle filament may be slightly smaller than laterals, legs ending in one claw. MAYFLIES (Ephemeroptera), Plate 24.

45a. Back of body covered with two hard wing covers, a pair of membranous wings underneath the covers. ADULT BEETLES (Coleoptera), Plate 30.

45b. Back without hard wing covers 46

46a. Body with exposed membranous wings or wing pads on back 47

46b. Body without membranous wings or wing pads (larvae) 49

47a. Membranous wings present; held flat and in a V-shape on back. Mouth parts formed into a long, sharply pointed beak folded underneath body. TRUE BUGS (Hemiptera), Plate 31.

47b. Membranous wings absent, wing pads present. Mouth parts formed into an extendable, scoop-like mask that covers face. (Odonata) 48

48a. Body ending in three oblong, fan-like plates. DAMSELFLIES (Zygoptera), Plate 25.

48b. Fan-like plates absent. DRAGONFLIES (Anisoptera), Plate 25.

49a. Mouth parts formed into slender curved rods nearly half as long as body (less than 10 mm). SPONGILLA FLIES (Neuroptera), not illustrated.

49b. Mouth parts adapted for biting or chewing. 50

50a. Body with five paired knobs on underside of abdominal segments, legs on first three segments short and stubby. Often found on lily pads. AQUATIC MOTHS (Lepidoptera), not illustrated.

50b. Body without paired knobs on underside of abdomen 51

51a. Sides of each abdominal segment with a slender, tapering process 52

51b. Sides of each abdominal segment without a tapering process, but may have hair-like or tubular processes 53

52a. Body ending in a pair of hook-bearing fleshy legs or in a single tapering filament. HELLGRAMMITES and relatives (Megaloptera), Plate 26.

52b. Body otherwise. BEETLES (Coleoptera), Plate 30.

53a. Body covering mostly hard; knobs, hairlike processes, or other special ornamentation may be present on back, or else body is entirely soft except for a hardened head capsule. BEETLES (Coleoptera), Plate 30.

Refer to
Couplet
No.

53b. Most of body soft except for a hardened head capsule and with one to three hard plates
on the back of first body segments; tubular processes may be present on sides of the body
in various arrangements. Body may end in a pair of hook-bearing legs. Most larvae living
in portable cases made of bits of sticks, leaves, or sand or in attached fibrous cases.
CADDISFLIES (Trichoptera), Plate 27.

10900 C.   List of Common Types of Aquatic Organisms (Plates 1–38), by Trophic Level

	Plate No.		*Plate No.*
1. Producers (plants)		Sponges (Porifera) and bryozoa	16
		Rotifers (Rotatoria)	17
Blue-green algae (Cyanophyta):		Roundworms (Nematoda)	18
coccoid.	1	Flatworms (Platyhelminthes) and segmented worms (Annelida)	19
Blue-green algae (Cyanophyta):			
filamentous	2	Crustaceans: cladoceran types	20
Nonmotile green algae (Chlorophyta):		Crustaceans: common types	21
coccoid.	3	Insects: types of pupae	22
Nonmotile green algae (Chlorophyta):		Insects: stoneflies (Plecoptera)	23
filamentous	4	Insects: mayflies (Ephemeroptera)	24
Diatoms (Bacillariophyceae): pennate . .	5	Insects: damsel- and dragonflies	
Diatoms (Bacillariophyceae): centric . .	6	(Odonata)	25
Types of larger marine algae	7	Insects: hellgrammites and relatives	
Higher plants: floating (Spermatophyta		(Megaloptera)	26
and Pteridophyta).	8	Insects: caddisflies (Trichoptera).	27
Higher plants: submersed (Spermatophyta)	9	Insects: two-winged flies (Diptera). . . .	28
		Insects: two-winged flies (Diptera). . . .	29
Higher plants: emersed (Spermatophyta)	10	Insects: beetles (Coleoptera)	30
		Insects: true bugs (Hemiptera)	31
Pigmented flagellates: single cells (various phyla).	11	Mollusks: snails (Gastropoda)	32
		Mollusks: bivalves (Pelecypoda)	33
Pigmented flagellates: colonial types (various phyla).	12	Echinoderm types	34
		Miscellaneous invertebrates.	35
		Fish types	36
2. Consumers (animals)		Amphibian types	37
Nonpigmented flagellates (Protozoa) . .	13	**3. Reducers (bacteria and fungi)**	
Amoebas (Protozoa).	14		
Ciliates (Protozoa).	15	Bacteria and fungi.	38

ACKNOWLEDGMENTS

Plates 1 through 38, which follow on succeeding pages, present over 200 aquatic organisms commonly found in natural, polluted, and treated waters. These plates were drawn especially for this work by Eugene Schunk of the Cincinnati Art Service, Inc. In a number of instances, it would have been impossible to illustrate a certain organism for the purposes of this manual were it not for the courtesy of other publishers, who permitted illustrations from their publications to be incorporated herein. The following organisms were so reproduced:

Plate

5: B—*Diatoma,*

F—*Achnanthes,*

G—*Gomphonema,*

H—*Cymbella,* and

K—*Surirella,* courtesy of Veb Gustav Fischer Verlag, Jena. Source: Die Susswasser—Flora Mitteleuropas, Heft 10, by F. Hustedt. 1930.

6: C—*Coscinodiscus,* and

D—*Melosira,* courtesy of E. Schweizerbart'sche Verlagsbuchhandlung, Stuttgart. Source: Das Phytoplankton des Susswassers, Die Binnengewasser, Band XVI, Teil II, Halfte II, by G. Huber-Pestalozzi and F. Hustedt, 1942. Plates CVIII-CXVI and CXXIII.

F—*Skeletonema,* courtesy of Academische Verlagsgesellschaft, Leipzig. Source: Die Kieselalgen, by F. Hustedt. In: L. Rabenhorst, Kryptogamen-Flora von Deutschland, Osterreich und der Schweiz, Band VII, 1930.

16: E—*Membranipora monostachys,* reprinted by permission of G.P. Putnam's Sons, Inc., New York. Source: Field Book of Seashore Life, by R.W. Miner. Copyright 1950 by the author. Plate 236, page 817.

17: I—*Notholca* Robert W. Pennak, Fresh-Water Invertebrates of the United States, Copyright © 1953, The Ronald Press Company, New York. Figure 116N, page 190, adapted for Figure 171, courtesy of The Ronald Press.

21: A—*Asellus* (sowbug),

C—*Mysis* (shrimp),

D—*Diaptomus* (copepod),

E—*Cypridopsis* (ostracod),

F—*Cyclops* (copepod), and

G—*Cambarus* (crayfish, crawdad), courtesy of Holden-Day, Inc., San Francisco, California. Source: Needham & Needham's Guide to the Study of Freshwater Biology, 1951. Figures 1 and 10, Plate 14, page 37; Figures 16, 18 and 20, Plate 24, page 61; and Figure 9, Plate 14, page 37.

22: Dr. Harold Walters

32: A—*Pomacea* (apple snail),

B—*Marisa,*

E—*Tarebia,*

I—*Lymnaea* (pond snail),

J—*Helisoma* (orb snail), and

M—*Lanx* (limpet) courtesy of John Wiley & Sons, Inc., New York. Source: Ward & Whipple, Fresh Water Biology (2nd ed.), W.T. Edmondson, Editor, 1959. Figures 43.31A(A), 43.31B(B), 43.62B(E), 43.13(I), 43.20(J) and 43.14(M).

C—*Campeloma,*

D—*Bithynia* (faucet snail),

F—*Pleurocera* (river snail),

Plate

G—*Valvata,*

H—*Littorina* (periwinkle),

K—*Nassa* (mud snail),

L—*Ferrissia* (limpet), and

N—*Physa,* courtesy of R.M. Sinclair, Advisor for Biological Sampling and Analysis (American Public Health Association) 13th ed.

34: Connecticut State Geological and Natural History Survey: Echinoderms of Connecticut, by Wesley Roswell Coe, 1912.

35: C—*Limulus* (horseshoe crab) courtesy of Western Publishing Company, Inc., Golden Press Division, Racine, Wisconsin. Source: Seashores, a Golden Nature Guide, 1955. Page 79.

37: C—*Ambystoma* (terrestrial adult), courtesy of Dover Publications, Inc., New York. Source: Biology of the Amphibia, by G.K. Noble, 1931. Figure 147C, page 471.

D—*Ambystoma* (aquatic larva), courtesy of the New York State Museum and Science Service, Albany, New York. Source: The Salamanders of New York, by Sherman C. Bishop, 1941. Figure 33b, page 166. [Bulletin 324, New York State Museum, Albany.]

E—*Necturus,* courtesy of Dover Publications, Inc., New York. Source: Biology of the Amphibia, by G.K. Noble, 1931. Figure 35B, page 99.

G—*Siren intermedia* (siren), reprinted from Sherman Bishop: Handbook of Salamanders. Copyright 1943 by Comstock Publishing Company, Inc. Used by permission of Cornell University Press.

38: A—(a) micrococcus, (b) streptococcus, (c) sarcina, (d) bacillus, (e) vibrio, (f) spirillum, courtesy of John Wiley & Sons, Inc., New York. Source: Ward & Whipple, Fresh Water Biology (2nd ed.), W.T. Edmondson, Editor, 1959. Figure 3.1.

(k) actinomycete growth form, Selman A. Waksman, The Actinomycetes. Copyright © 1957, The Ronald Press Company, New York. Figure 2–6, page 18, adapted for Figure 37A(k), courtesy The Ronald Press.

B—*Tetracladium* and (e) (f), *Achlya,* courtesy of John Wiley & Sons, Inc., New York. Source: Ward & Whipple, Fresh Water Biology (2nd ed.), W.T. Edmondson, Editor, 1959. Figures 4.119 and 4.79.

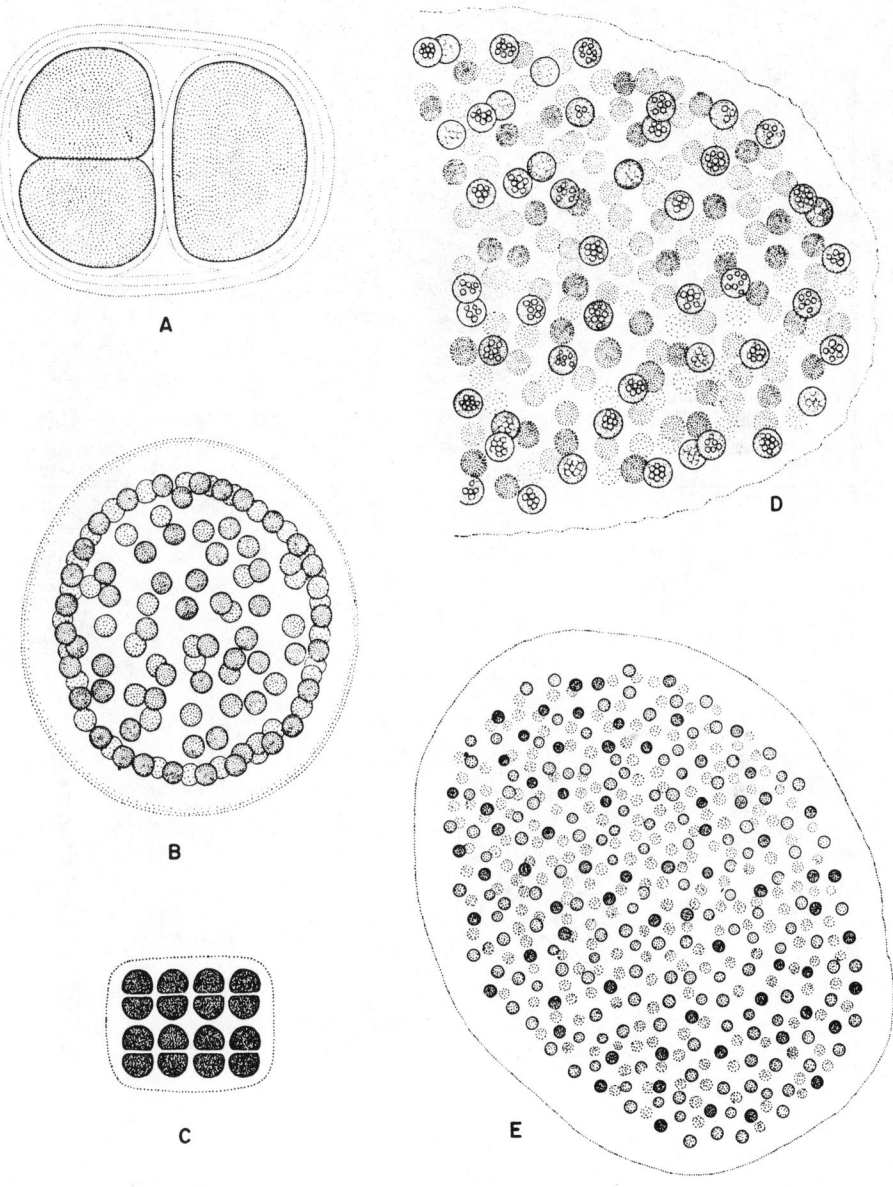

Plate 1. Blue-green algae: Coccoid (Phylum Cyanophyta). Dimensions refer to individual cells.

A—*Anacystis*	(4–20 μm)	D—*Anacystis* sp.	(4–6 μm)
B—*Gomphosphaeria*	(3–6 μm)	E—*Anacystis* sp.	(3–4 μm)
C—*Agmenellum*	(2–6 μm)		

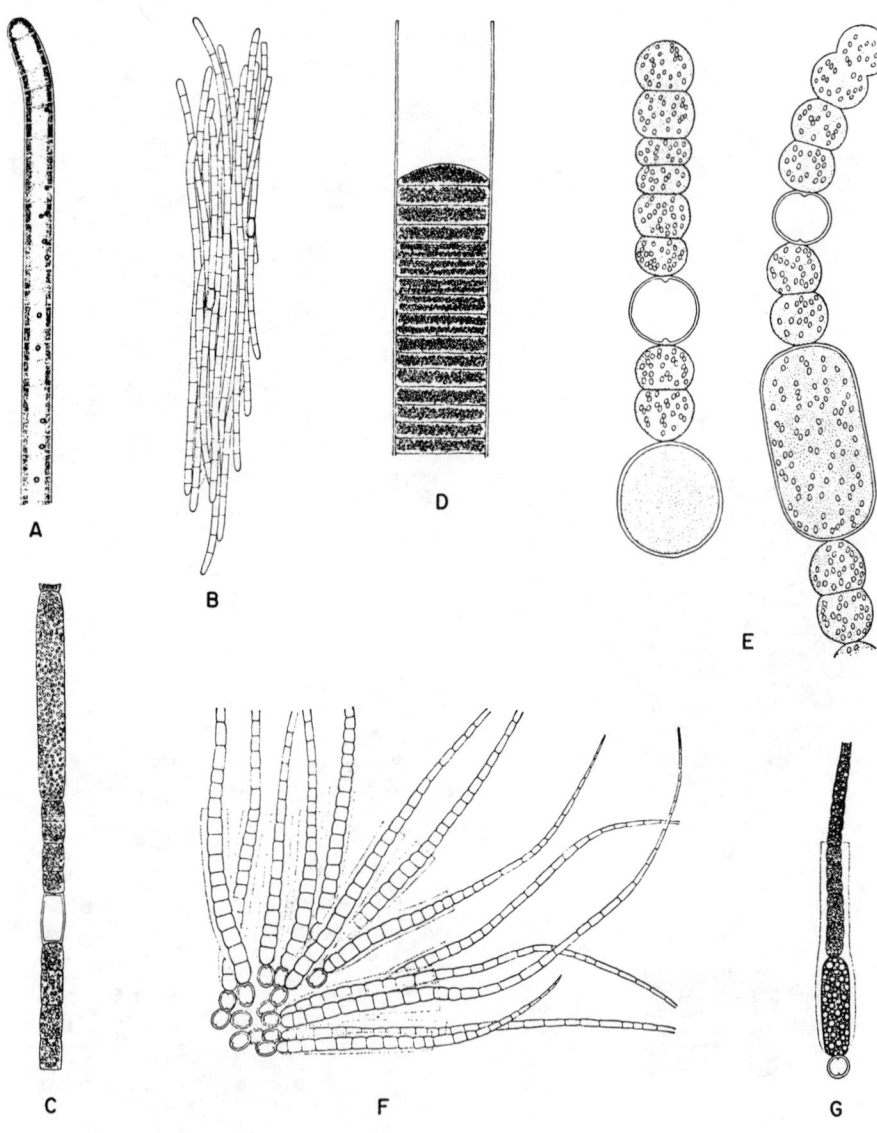

Plate 2. Blue-green algae: Filamentous (Phylum Cyanophyta). Most dimensions refer to diameter of individual filaments.

A—*Oscillatoria*   (4–20 μm)   E—*Anabaena*   (5–12 μm)
B—*Aphanizomenon,*   (5–6 μm)   F—*Gleotrichia,*   (Cells 7–9
   aggregate of filaments   portion of colony   μm diameter
C—*Aphanizomenon,* detail   near akinete)
D—*Lyngbya*   (4–20 μm)   G—*Gleotrichia,* detail

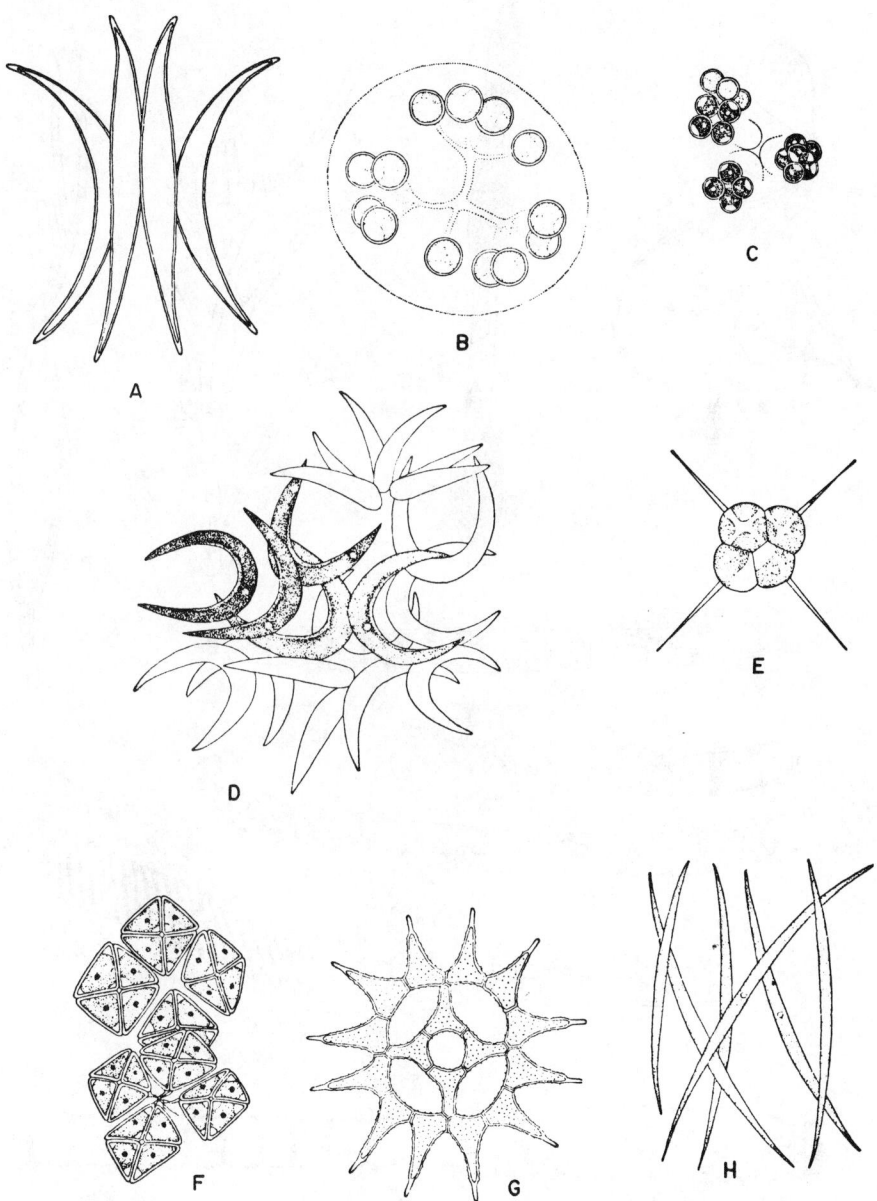

Plate 3. Nonmotile green algae: Coccoid (Phylum Chlorophyta). Dimensions refer to individual cells.

A—*Scenedesmus*	(4–6 μm diameter)	E—*Tetrastrum*	(5–9 μm)
B—*Dictyosphaerium*	(8–14 μm)	F—*Crucigenia*	(5–8 μm)
C—*Westella*	(5–7 μm)	G—*Pediastrum*	(10–20 μm)
D—*Selenastrum*	(6–7 μm)	H—*Ankistrodesmus*	(2–3 μm)

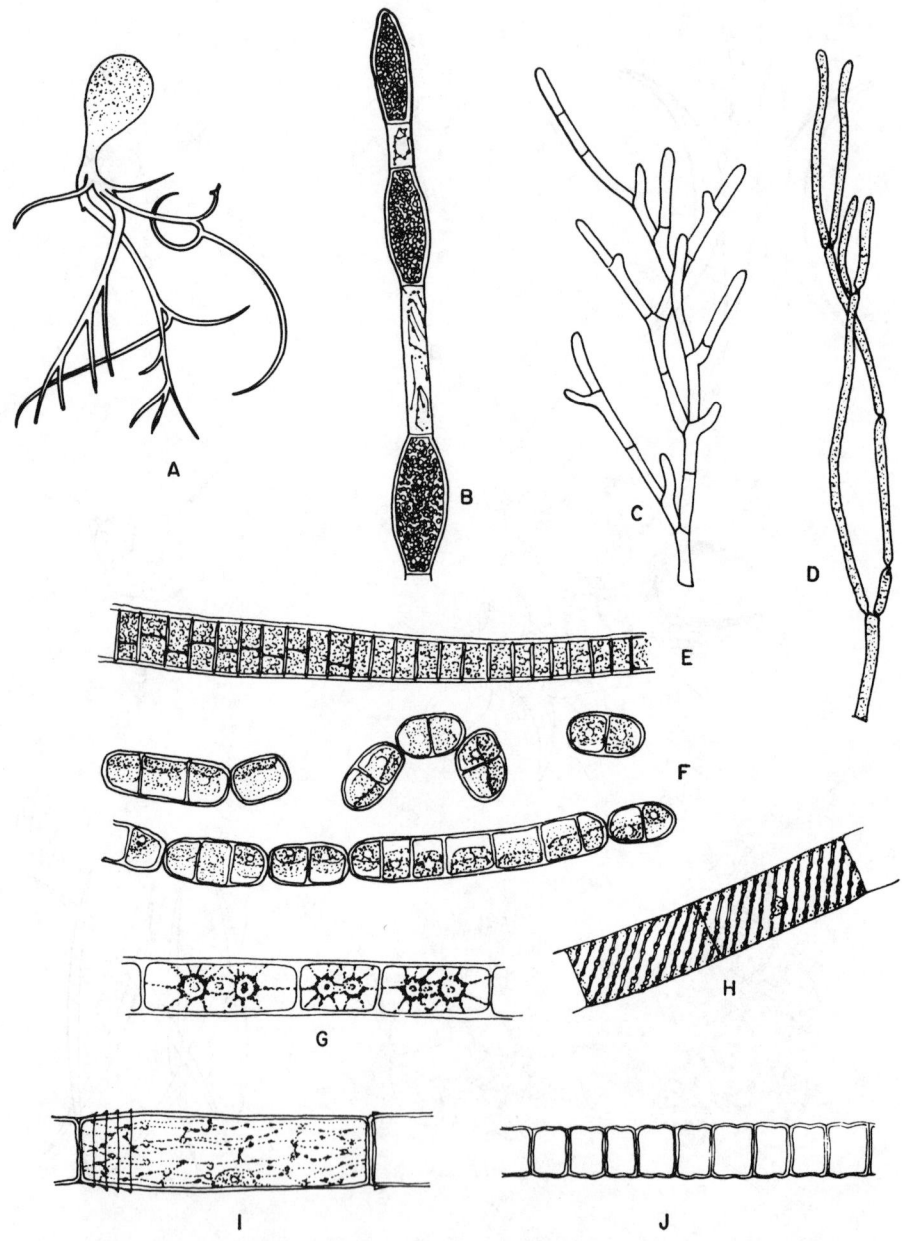

Plate 4. Nonmotile green algae: Filamentous (Phylum Chlorophyta). Dimensions refer to diameters of filaments or to mass.

A—*Botrydium*	(1000–2000 μm)	F—*Stichococcus*	(3 μm)
B—*Pithophora*	(50–100 μm)	G—*Zygnema*	(20–35 μm)
C—*Microthamnion*	(2–4 μm)	H—*Spirogyra*	(15–100 μm)
D—*Dichotomosiphon*	(50–100 μm)	I—*Oedogonium*	(6–40 μm)
E—*Schizomeris*	(12–18 μm)	J—*Hyalotheca*	(12–30 μm)

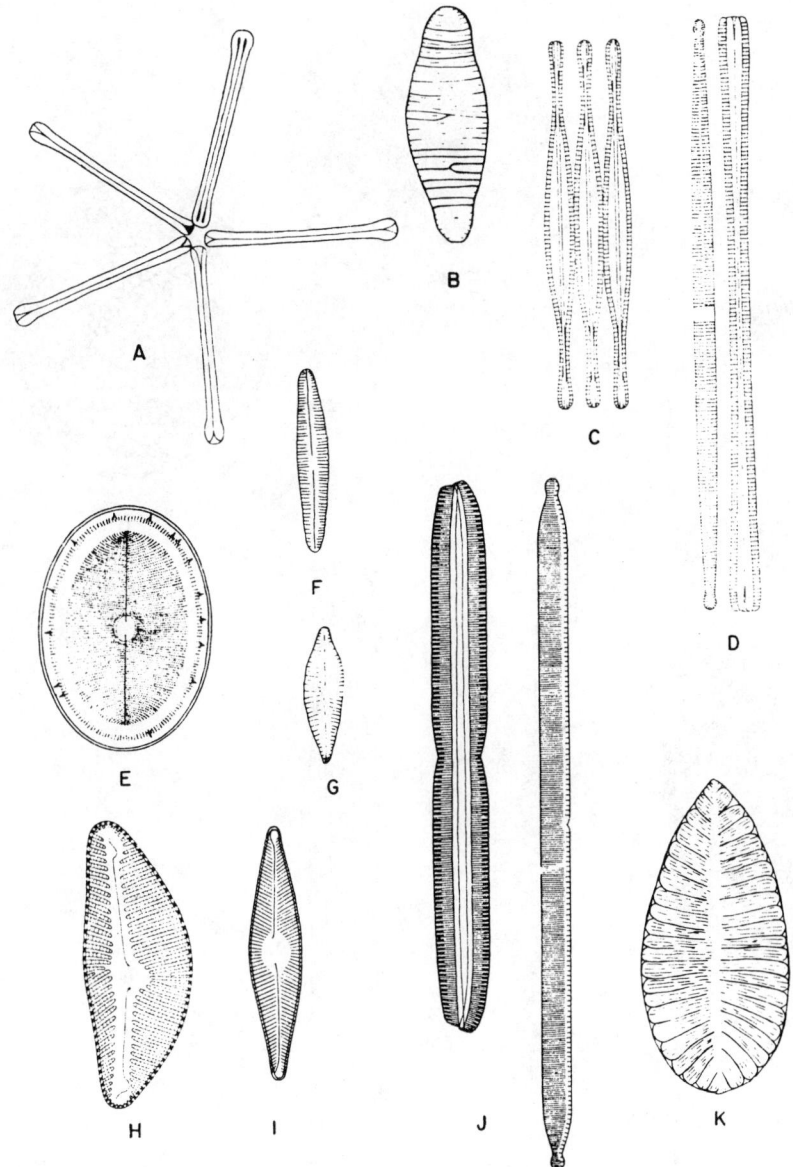

Plate 5. Diatoms: Pennate (Phylum Chrysophyta, Class Bacillariophyceae). Dimensions refer to length of cells unless otherwise specified.

A—*Asterionella*	(300 μm, entire colony)	G—*Gomphonema*	(20 μm)
B—*Diatoma*	(20 μm)	H—*Cymbella*	(15 μm)
C—*Fragilaria*	(100 μm)	I—*Navicula*	(30 μm)
D—*Synedra*	(200 μm)	J—*Nitzschia*	(100 μm)
E—*Cocconeis*	(10 μm)	K—*Surirella*	(20 μm)
F—*Achnanthes*	(10 μm)		

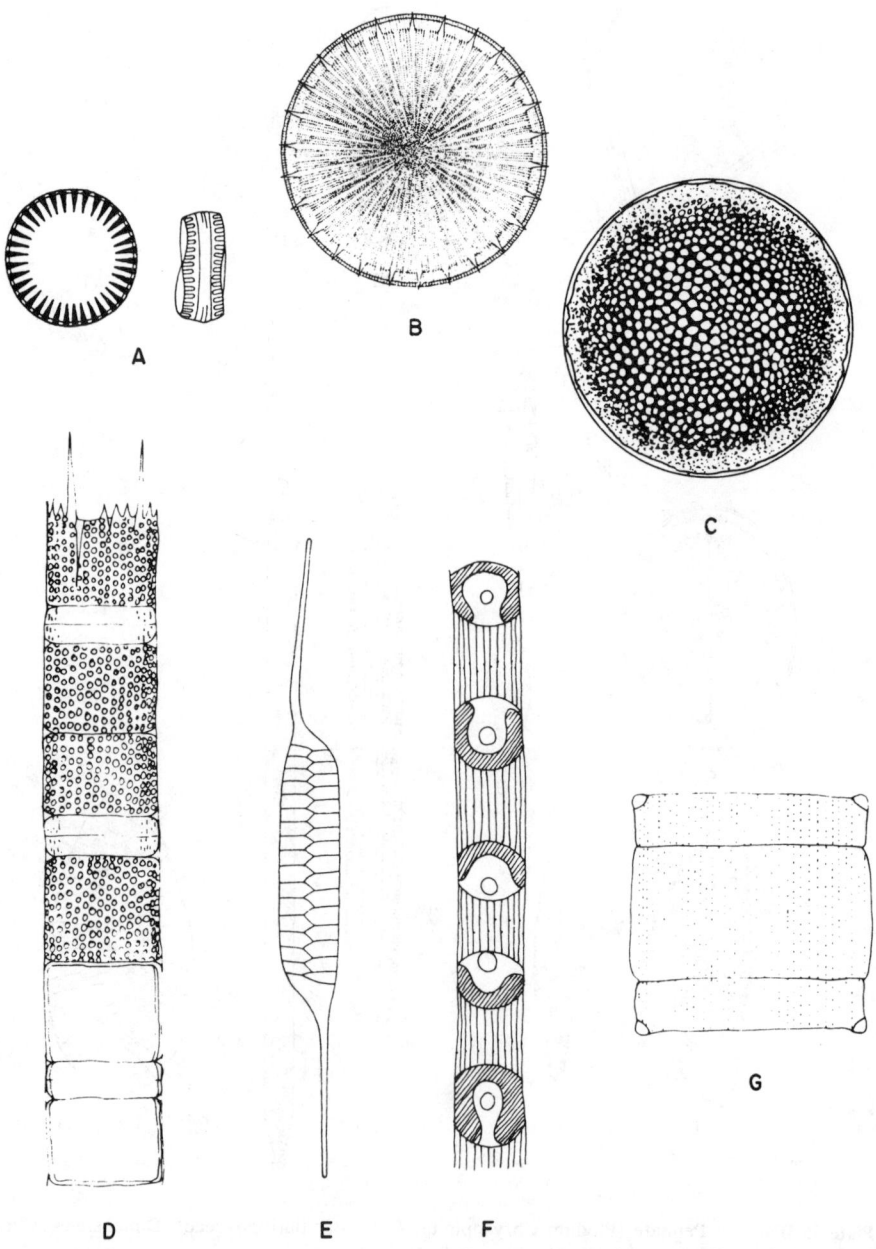

Plate 6. Diatoms: Centric (Phylum Chrysophyta, Class Bacillariophyceae). Dimensions refer to diameter.

A—*Cyclotella*   (10 μm)   E—*Rhizosolenia*   (5–15 μm)
B—*Stephanodiscus*   (30 μm)   F—*Skeletonema*   (3–18 μm)
C—*Coscinodiscus*   (20 μm)   G—*Biddulphia*   (100 μm)
D—*Melosira*   (3–12 μm)

Plate 7. Types of larger marine algae (green, brown, and red).

Green algae (Phylum Chlorophyta):
A—*Enteromorpha*   (40 cm)
B—Sea lettuce, *Ulva*   (20 cm)
Brown algae (Phylum Phaeophyta):
C—Rockweed, *Fucus*   (75 cm)
D—Giant kelp, *Nereocystis*   (20 m)

Red algae (Phylum Rhodophyta):
E—*Gracilaria*   (50 cm)
F—*Corallina*   (4 cm)

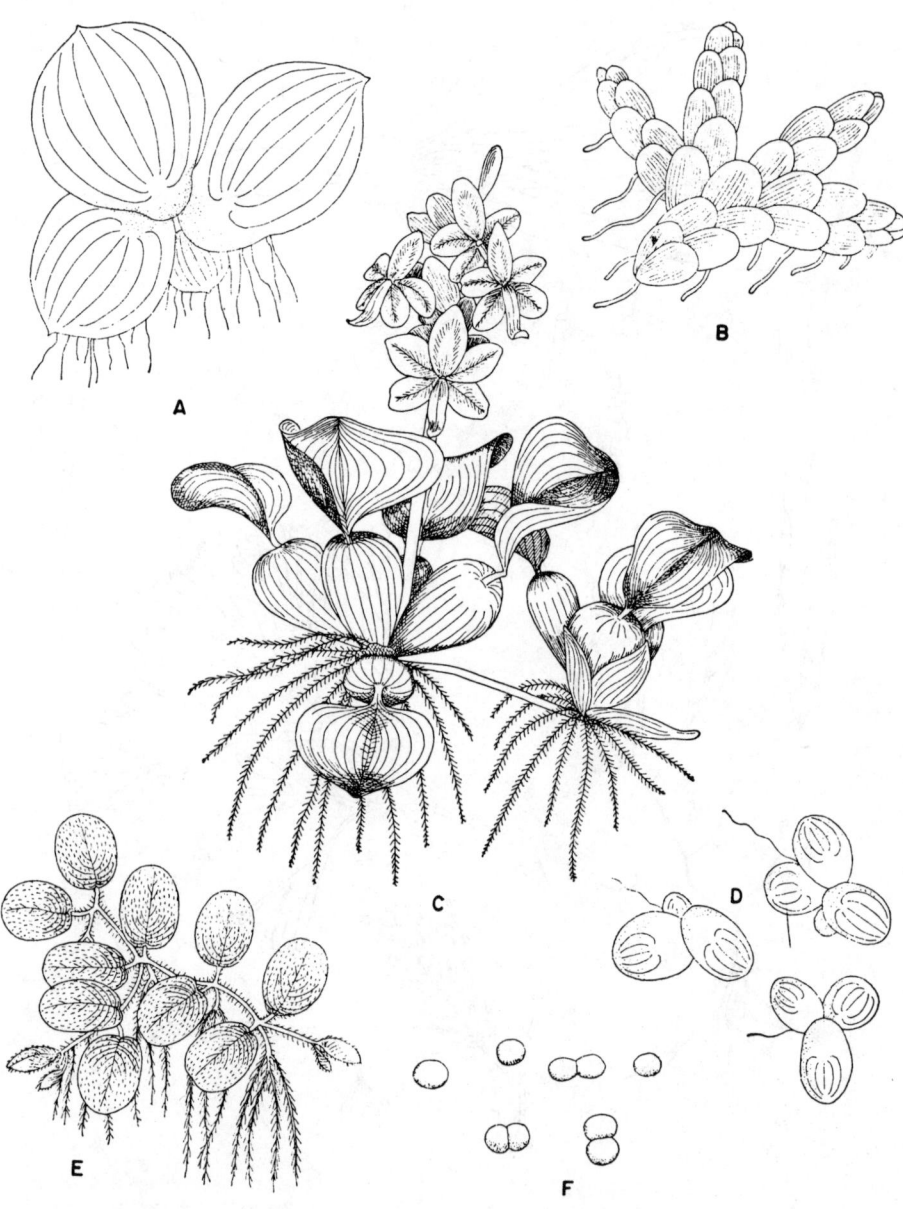

Plate 8. Higher plants: Floating plants.

A—Great duckweed, *Spirodela*
   (Phylum Spermatophyta, 8 mm)
B—Water velvet, *Azolla*
   (Phylum Pteridophyta, 1 cm)
C—Water hyacinth, *Eichhornia*
   (Phylum Spermatophyta, 22 cm)

D—Lesser duckweed, *Lemna*
   (Phylum Spermatophyta, 5 mm)
E—Water fern, *Salvinia*
   (Phylum Pteridophyta, 4 cm)
F—Watermeal, *Wolffia*
   (Phylum Spermatophyta, 1–1.5 mm)

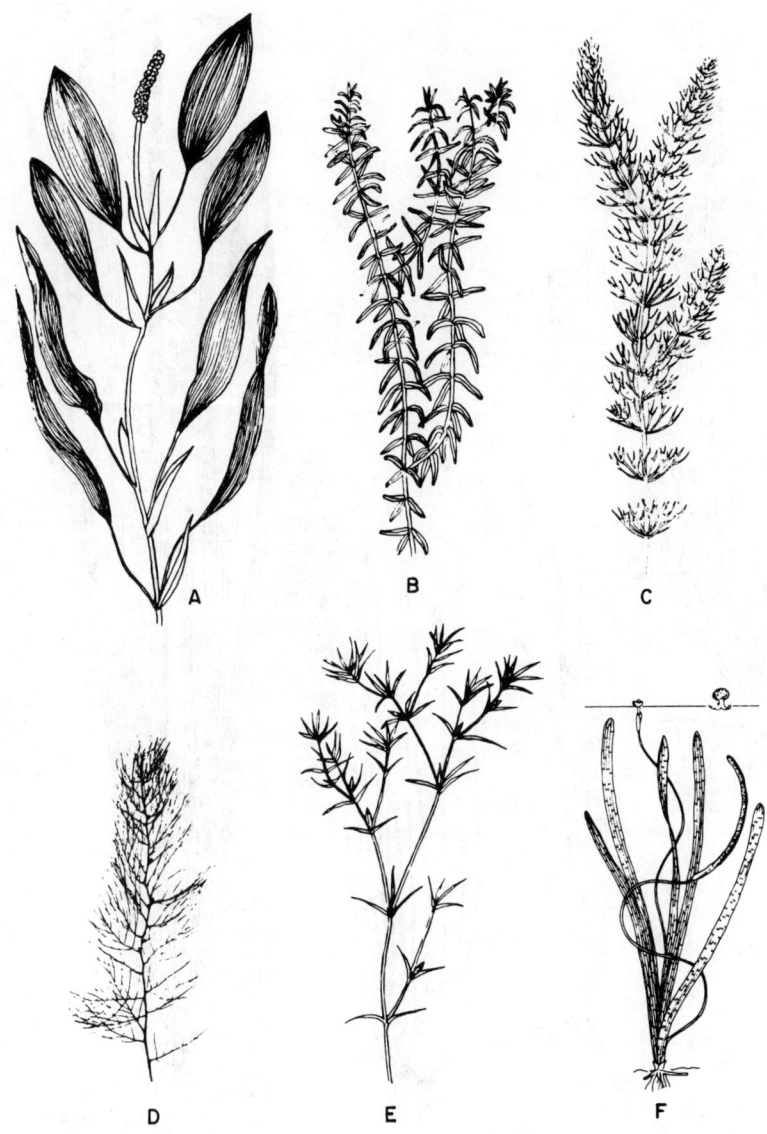

Plate 9. Higher plants: Submersed (all forms illustrated are Spermatophytes).

A—Pondweed, *Potamogeton*   (30–60 cm)   E—Naiad, *Najas*   (60 cm)
B—Waterweed, *Elodea*   (15 cm)   F—Eelgrass, *Vallisneria*   (45 cm)
C—Coontail, *Ceratophyllum*   (30 cm)
D—Water milfoil,
    Myriophyllum   (30 cm)

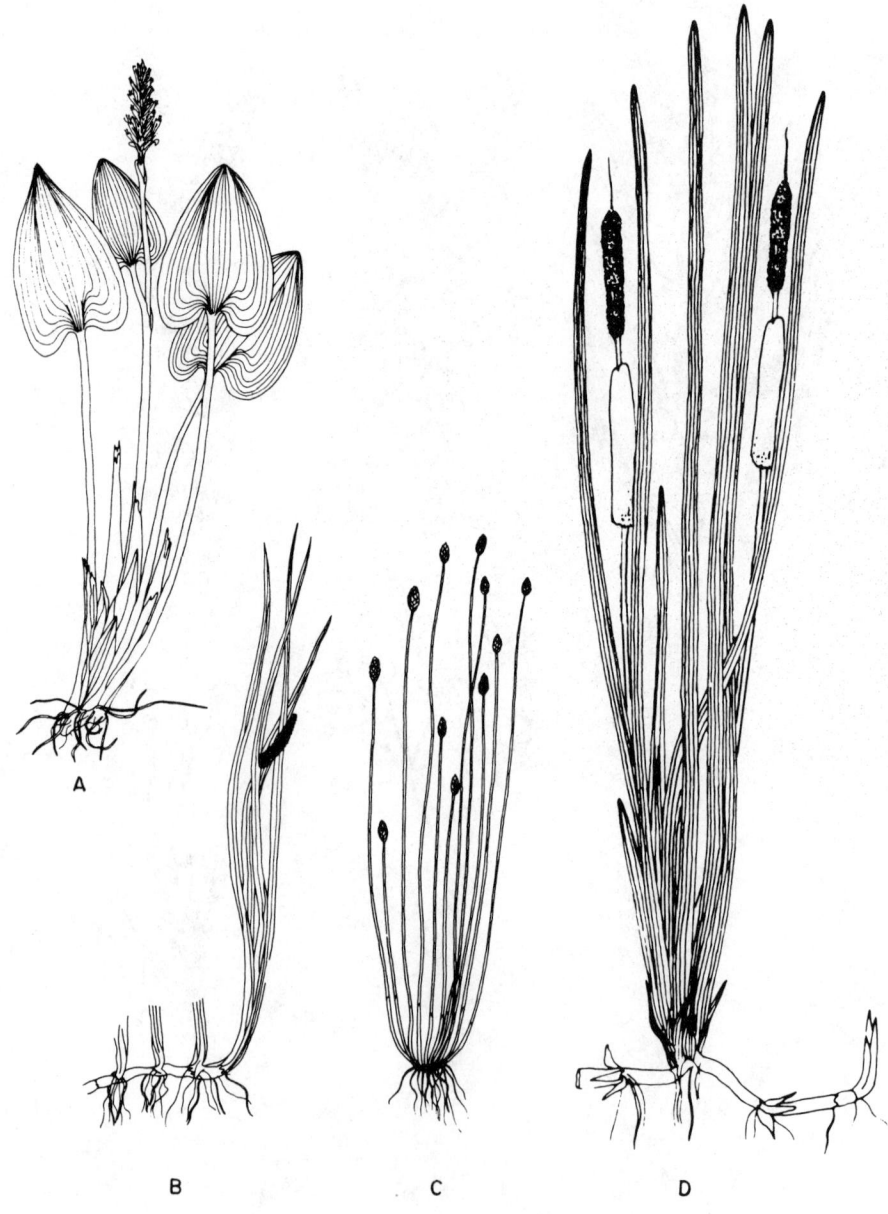

Plate 10. Higher plants: Emersed (all forms illustrated are Spermatophytes).

A—Pickerelweed, *Pontederia*   (60 cm) | C—Spike rush, *Eleocharis*   (30 cm)
B—Sweetflag, *Acorus*   (30 cm) | D—Cattail, *Typha*   (1–2 m)

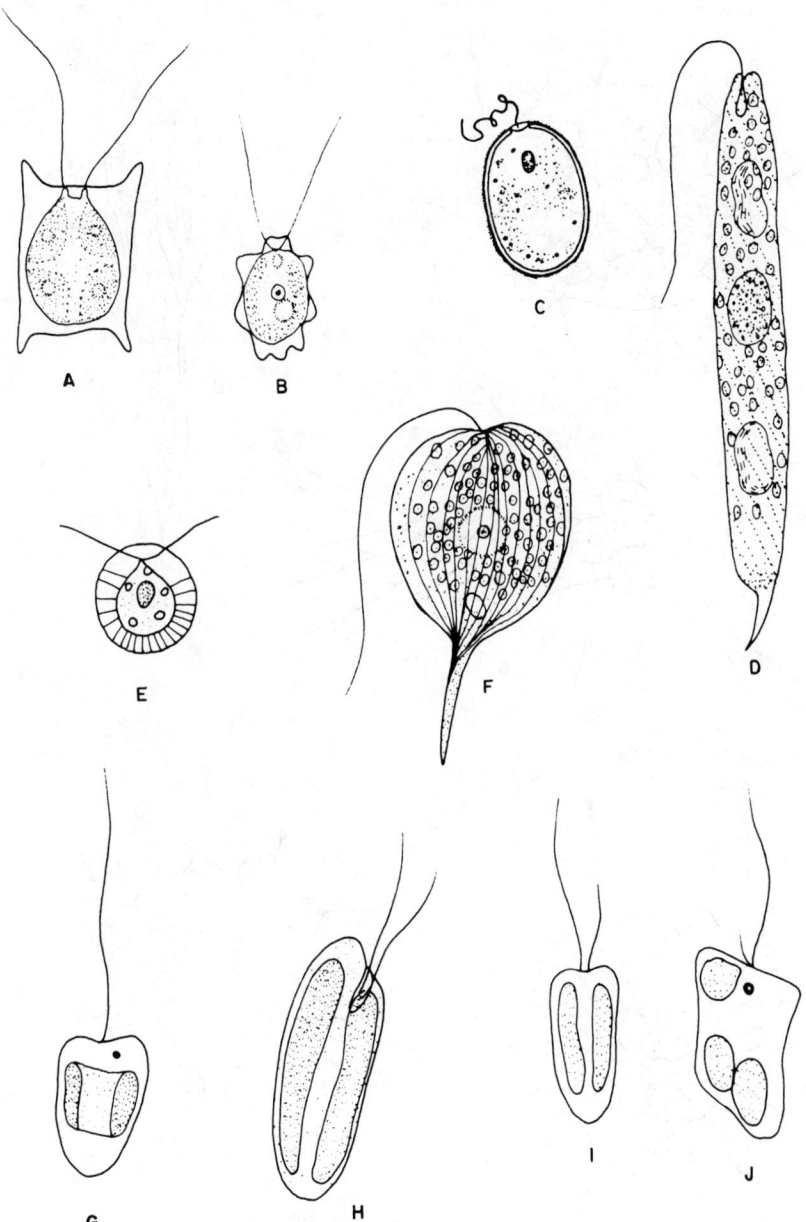

Plate 11. Pigmented flagellates: Single-celled (various phyla).

A—*Pteromonas*	(9–18 μm)	F—*Phacus*	(20–50 μm)
B—*Lobomonas*	(5–14 μm)	G—*Chromulina*	(4–10 μm)
C—*Trachelomonas*	(15–30 μm)	H—*Cryptomonas*	(6–12 μm)
D—*Euglena*	(10–25 μm)	I—*Ochromonas*	(7–14 μm)
E—*Haematococcus*	(40–45 μm)	J—*Chloramoeba*	(10–15 μm)

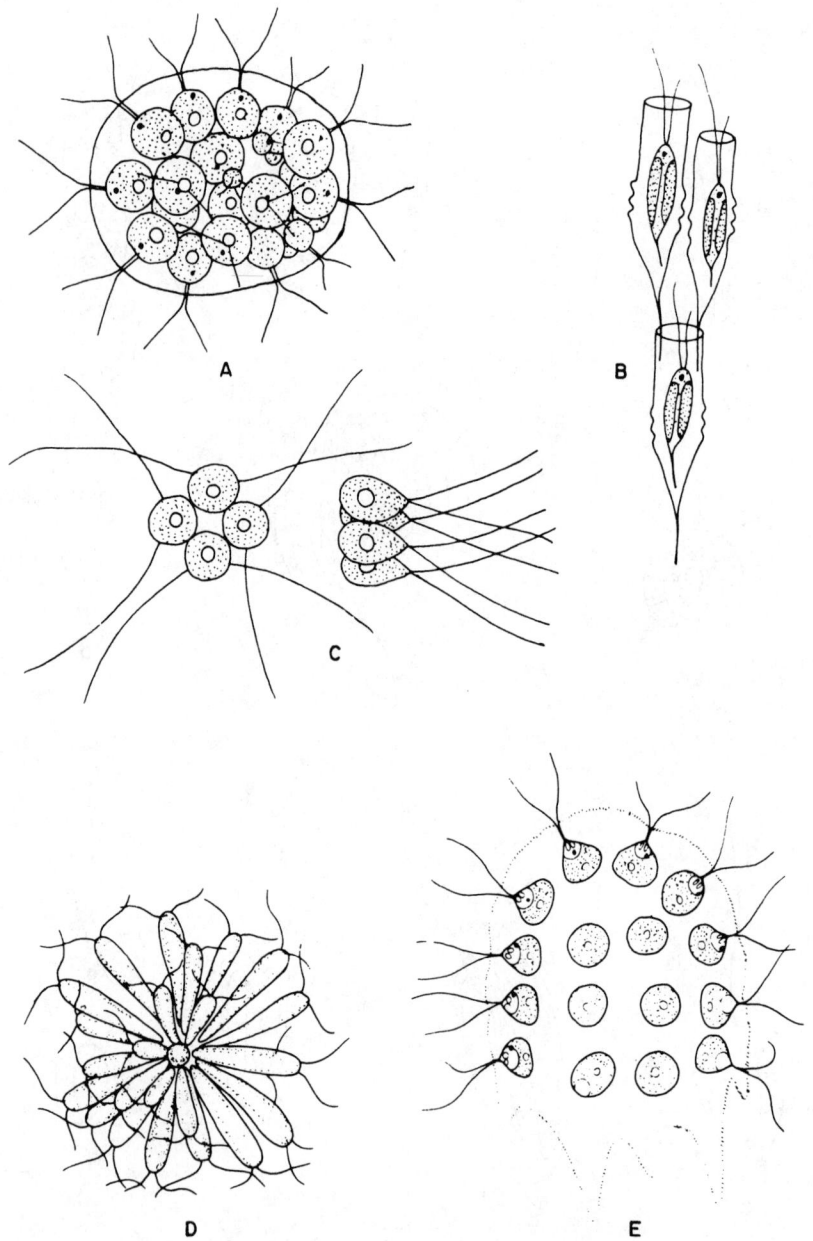

Plate 12. Pigmented flagellates: Colonial types (various phyla). Dimensions refer to individual cells unless otherwise specified.

A—*Pleodorina*	(8-10 μm)	D—*Synura*	(10-15 μm)
B—*Dinobryon*	(7-12 μm)	E—*Platydorina*	(66-70 μm
C—*Gonium*	(7-12 μm)		colony)

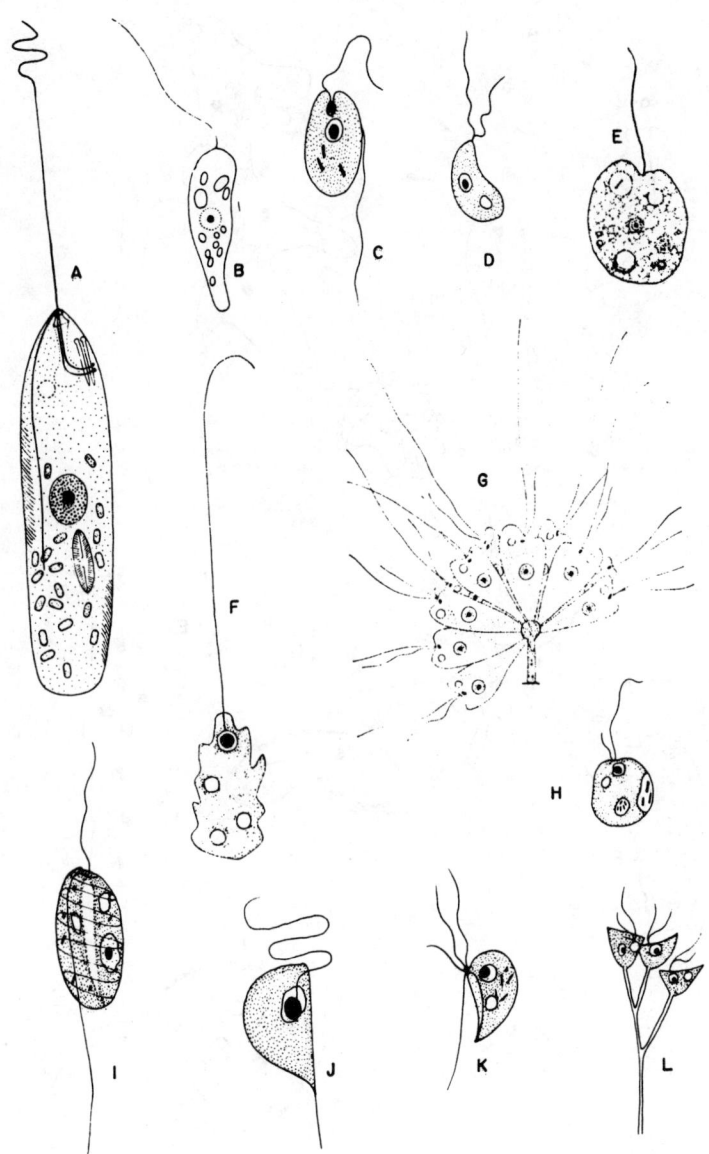

Plate 13. Nonpigmented flagellates (Phylum Protozoa).

A—*Peranema*	(40–70 μm)	G—*Anthophysa*	(5–6 μm)
B—*Astasia*	(40–50 μm)	H—*Monas*	(5–16 μm)
C—*Bodo*	(11–22 μm)	I—*Anisonema*	(14–60 μm)
D—*Dinomonas*	(15–16 μm)	J—*Cercomonas*	(10–36 μm)
E—*Oikomonas*	(5–20 μm)	K—*Tetramitus*	(11–30 μm)
F—*Mastigamoeba*	(28–200 μm)	L—*Dendromonas*	(8 μm)

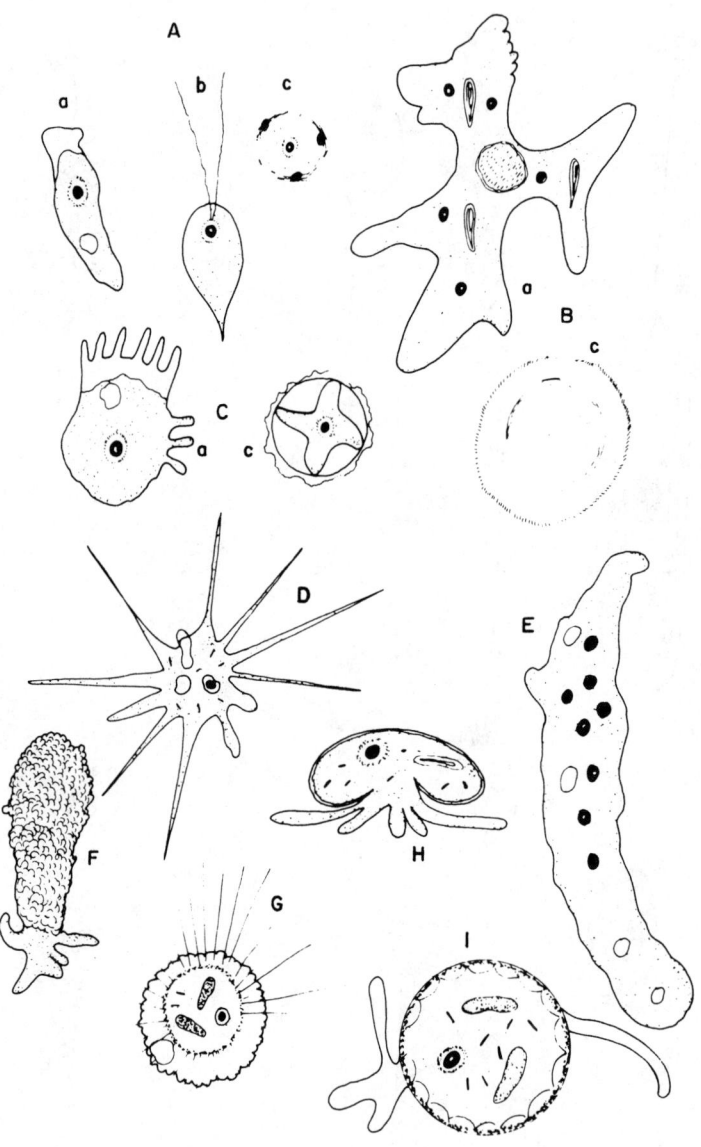

Plate 14. Amoebas (Phylum Protozoa). (a) Amoeboid stages, (b) flagellated stages, (c) cyst stages.

A—*Naegleria*	(10–36 μm)	E—*Pelomyxa*	(0.25–3 mm)
B—*Amoeba* sp.	(30–600 μm)	F—*Difflugia*	(40 μm)
C—*Acanthamoeba*		G—*Actinophrys*	(25–50 μm)
(*Hartmannella*)	(15–25 μm)	H—*Arcella* (side view)	(30–260 μm)
D—*Amoeba radiosa*	(30–120 μm)	I—*Arcella* (top view)	(30–260 μm)

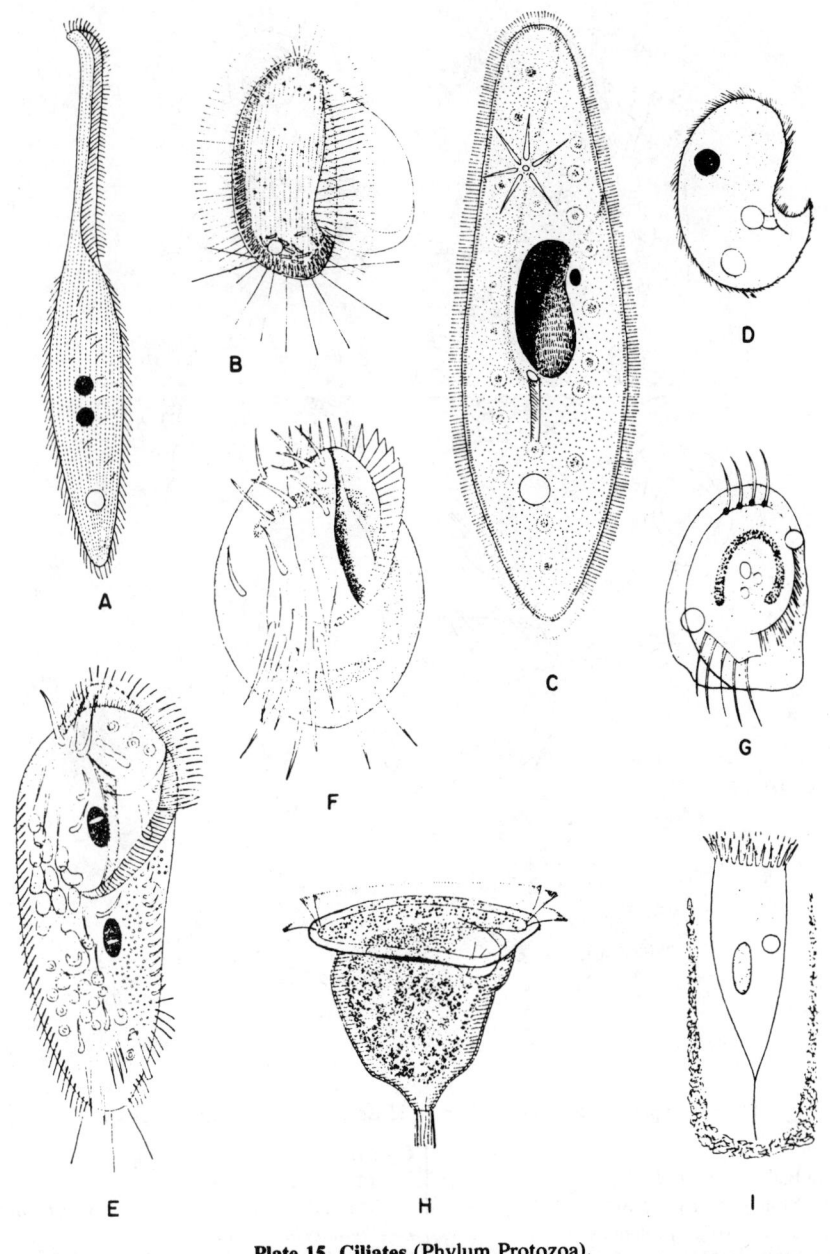

Plate 15. Ciliates (Phylum Protozoa).

A—*Lionotus*	(100 μm)	F—*Euplotes*	(70–195 μm)
B—*Pleuronema*	(38–120 μm)	G—*Aspidisca*	(30–50 μm)
C—*Paramoecium*	(50–330 μm)	H—*Vorticella*	(40–175 μm)
D—*Colpoda*	(12–110 μm)	I—*Tintinnidium*	(40–200 μm)
E—*Stylonychia*	(100–300 μm)		

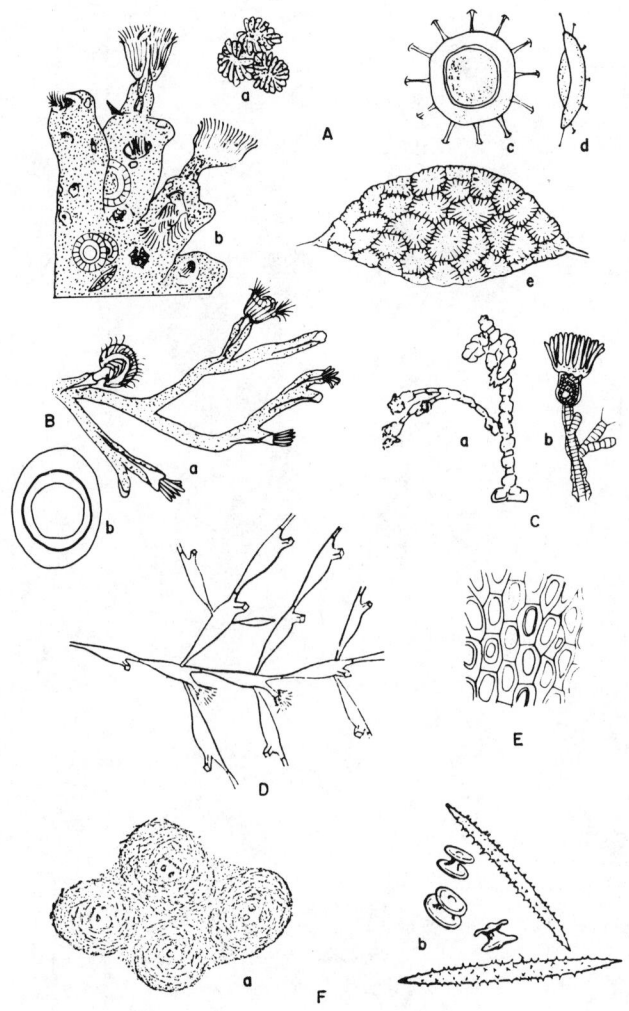

Plate 16. Sponges (Phylum Porifera) and Bryozoans (Phylum Bryozoa).

Bryozoa:
A—Jellyball, *Pectinatella*
  (a) Young colony (15 mm)
  (b) Section (highly magnified)
  (c) Statoblast (1 mm)
  (d) Statoblast (1 mm)
  (e) Colony on a plant stem (10 cm)
B—*Plumatella*
  (a) Colony (4 cm)
  (b) Statoblast (0.5 mm)

C—*Urnatella* (5 mm)
  (a) Colony (7 mm)
  (b) Individual zooid at tip of stalk (0.5 mm)
D—*Paludicella* (6 mm)
E—*Membranipora*, an encrusting marine form
    (individuals 1 mm, colonies unlimited)
Porifera:
F—*Trochospongilla*
  (a) Gemmules in a colony (1 mm)
  (b) Spicules (0.2 mm)

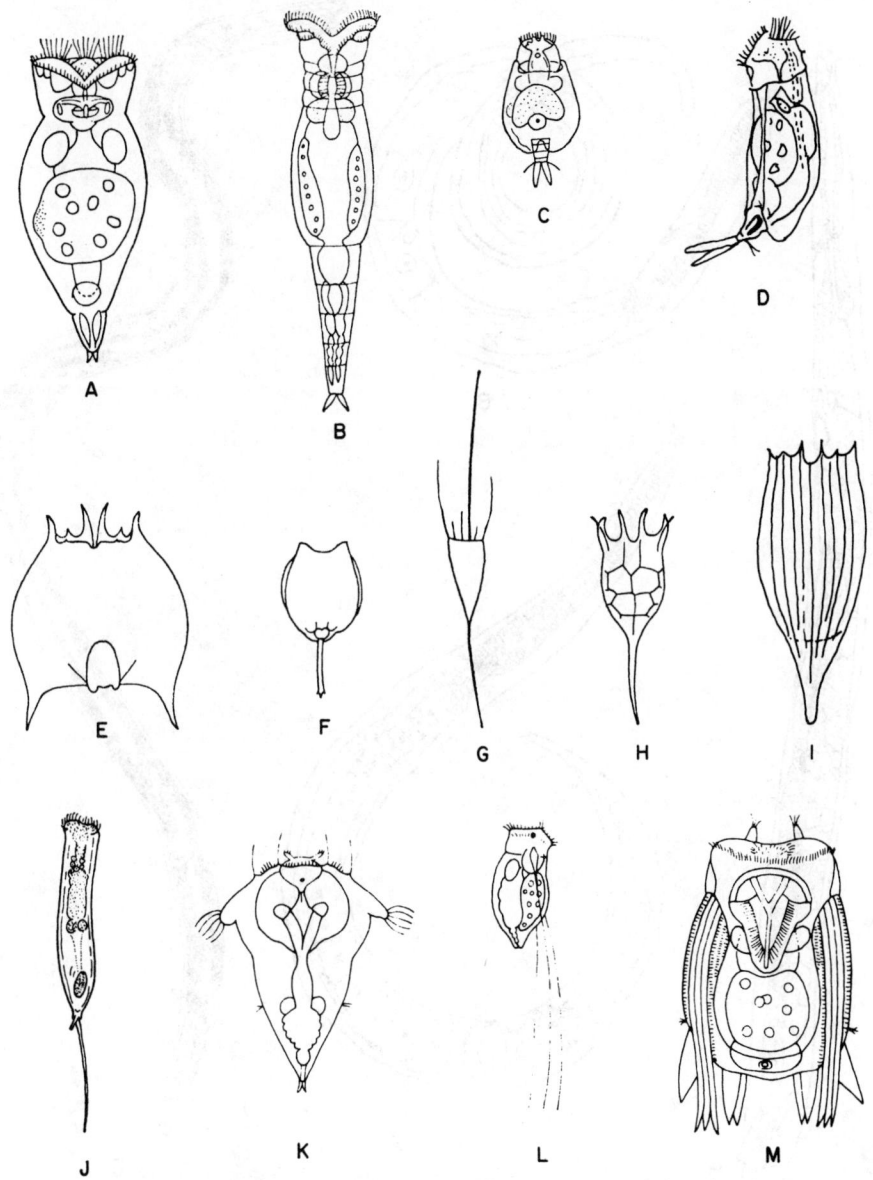

Plate 17. Rotifers (Phylum Rotatoria). Dimensions include spines.

A—*Epiphanes*	(600 μm)	H—*Keratella*	(200 μm)
B—*Philodina*	(400 μm)	I—*Notholca*	(200 μm)
C—*Euchlanis*	(250 μm)	J—*Trichocerca*	(600 μm)
D—*Proales*	(450 μm)	K—*Synchaeta*	(260 μm)
E—*Brachionus*	(200 μm)	L—*Filinia*	(150 μm)
F—*Monostyla*	(150 μm)	M—*Polyarthra*	(175 μm)
G—*Kellicottia*	(1 mm)		

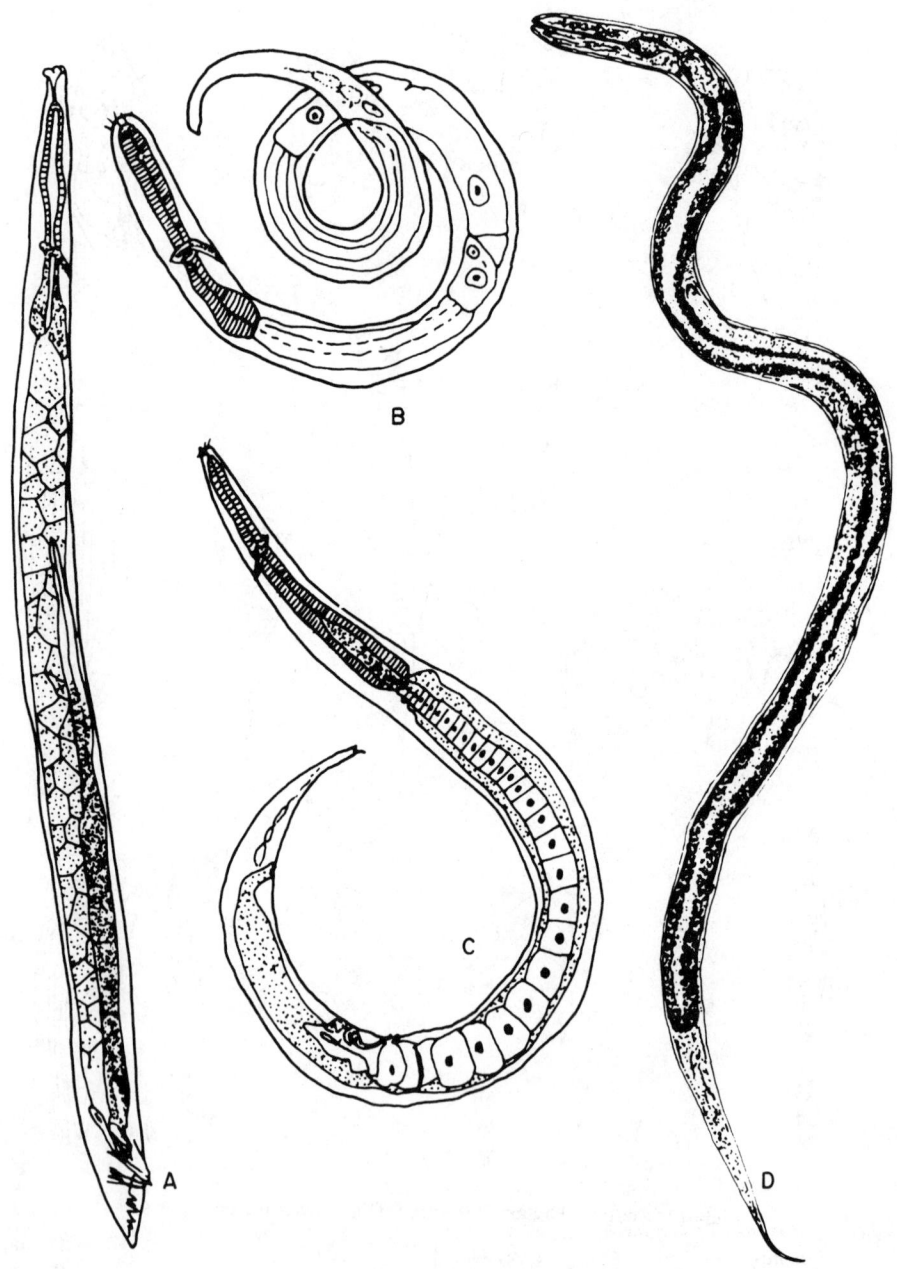

Plate 18. Roundworms (Phylum Nemathelminthes).

A—*Rhabditis* (male)   (1.6–1.9 mm)   D—*Diplogasteroides*
B—*Achromadora* (female)   (0.3–0.7 mm)   (female)   (1.5–1.85 mm)
C—*Monhystera* (female)   (0.8–1.0 mm)

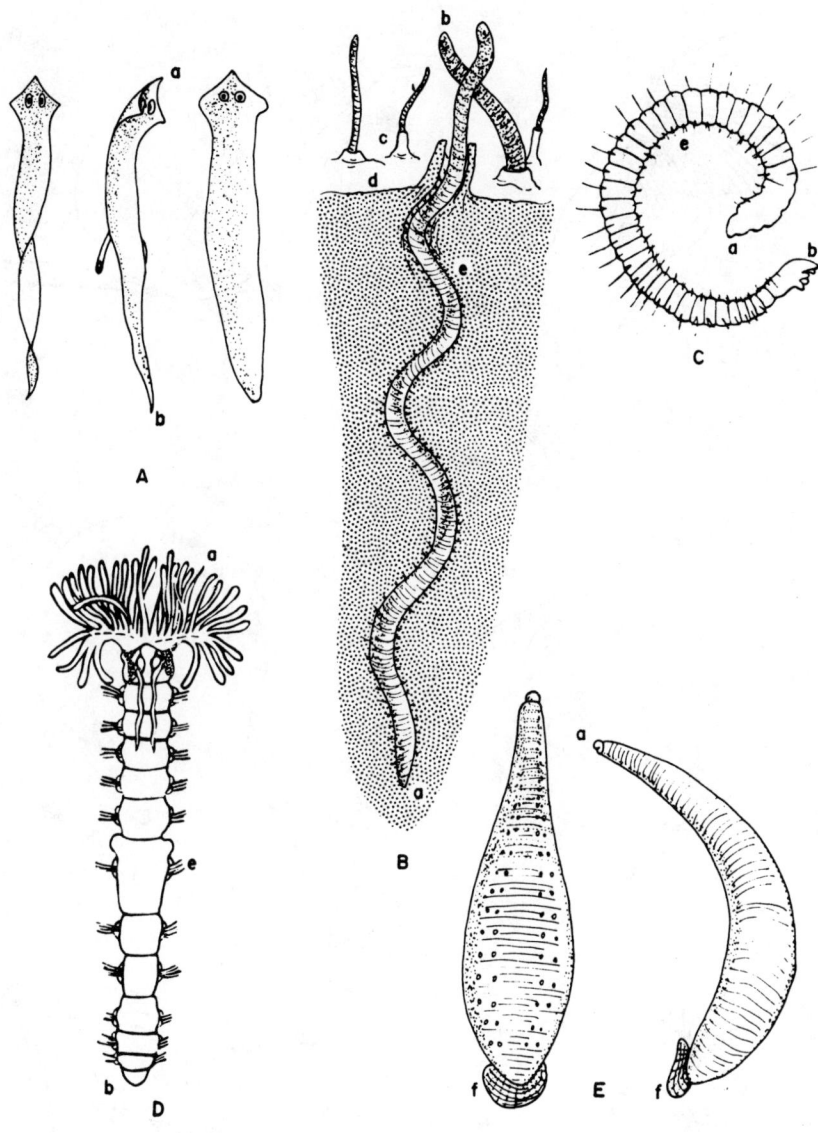

Plate 19. Flatworms (Phylum Platyhelminthes) **and segmented worms** (Phylum Annelida) (a) Anterior end, (b) posterior end. (c) tubes, (d) mud surface, (e) setae, (f) sucker disk.

Platyhelminthes:
A—*Planaria,* a free-living flatworm (5–13 mm)
Annelida:
B—*Tubifex,* a sludgeworm (25–50 mm)
C—*Dero,* a bristle worm (3–7 mm)

D—*Manayunkia,* a freshwater tube-building polychaet worm similar to certain common marine forms (5 mm)
E—Leech (50 mm)

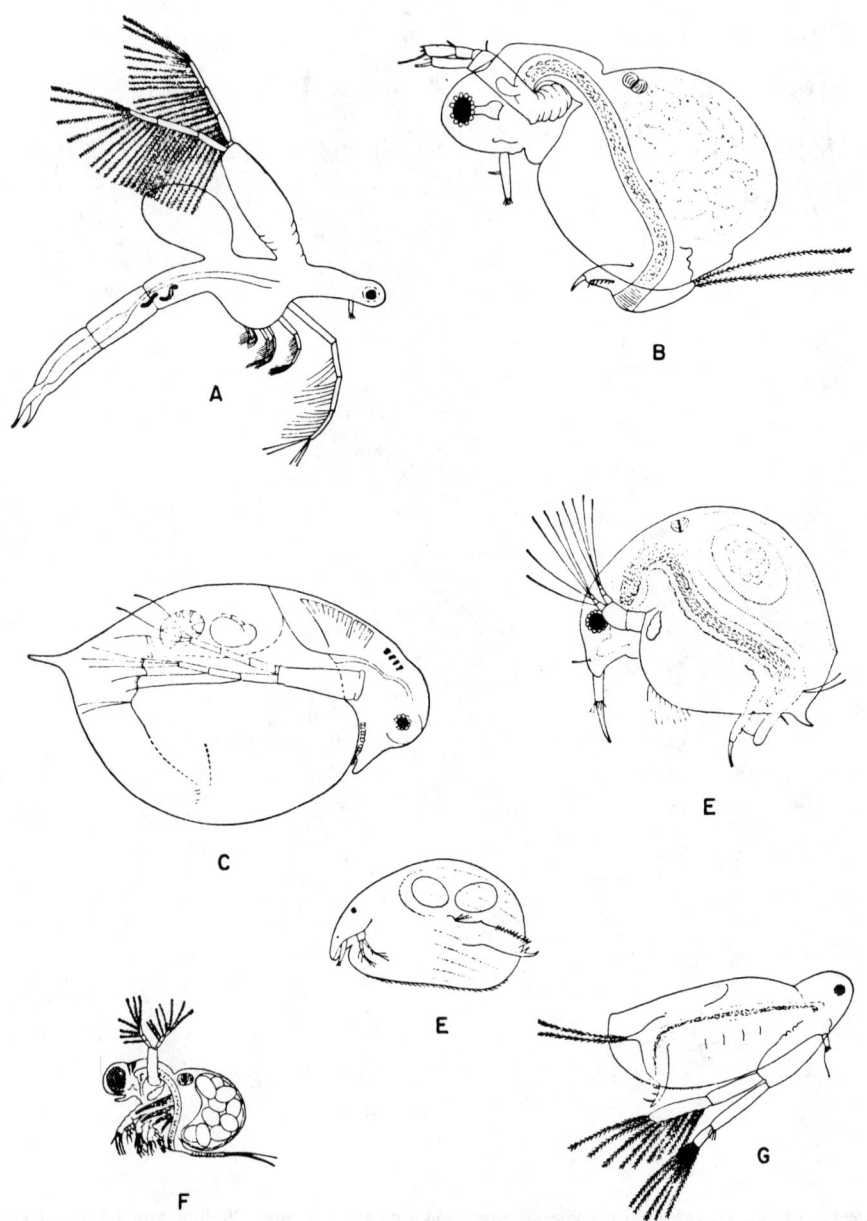

Plate 20. Crustaceans (Phylum Arthropoda, Class Crustacea): Types of cladocerans (Order Clado-
    cera).

A—*Leptodora*	(9 mm)	E—*Bosmina*	(0.4 mm)
B—*Moina*	(1.5 mm)	F—*Polyphemus*	(1.5 mm)
C—*Daphnia*	(2 mm)	G—*Diaphanosoma*	(1.5 mm)
D—*Alona*	(0.4 mm)		

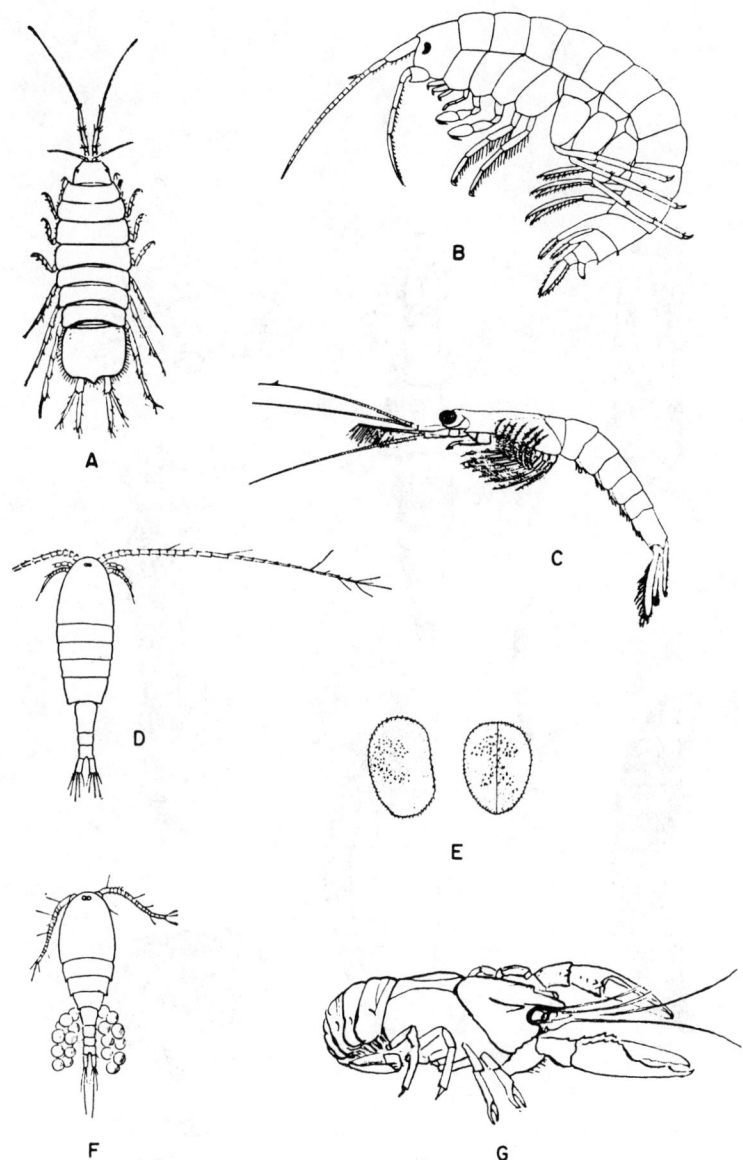

Plate 21. Crustaceans (Phylum Arthropoda, Class Crustacea): Selected common types.

A—Sowbug, *Asellus*, Order Isopoda (20 mm)
B—Scud, *Gammarus*, Order Amphipoda (15 mm)
C—Shrimp, *Mysis*, Order Decapoda (20 mm)
D—Copepod, *Diaptomus*, Order Copepoda (2 mm)

E—Ostracod, *Cypridopsis*, Order Ostracoda (1 mm)
F—Copepod, *Cyclops*, Order Copepoda (1 mm)
G—Crayfish, crawdad, *Cambarus*, Order Decapoda (150 cm)

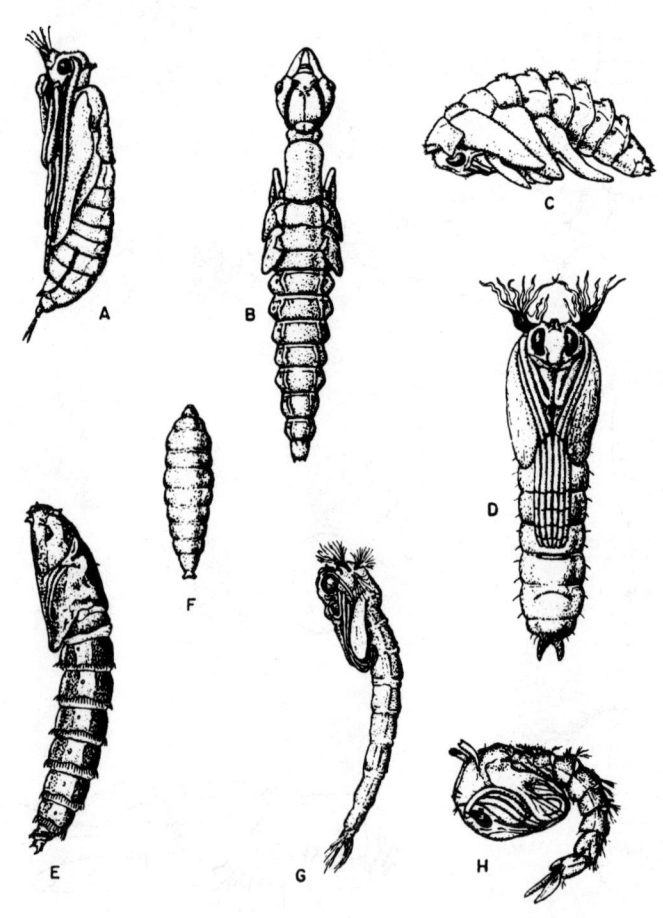

Plate 22. Types of insect pupae.

A—Caddisfly, *Goera*, Order Trichoptera
B—Hellgrammite, *Corydalis*, Order Megaloptera
C—Beetle, *Cybister*, Order Coleoptera
D—Cranefly, *Antocha*, Tipulidae

E—Blowfly, *Tabanus*, Tabinidae
F—Sewage fly, *Limnophorus*, Anthomyidae
G—Midge, *Chironomus*, Chironomidae
H—Mosquito, *Culex*, Culicidae

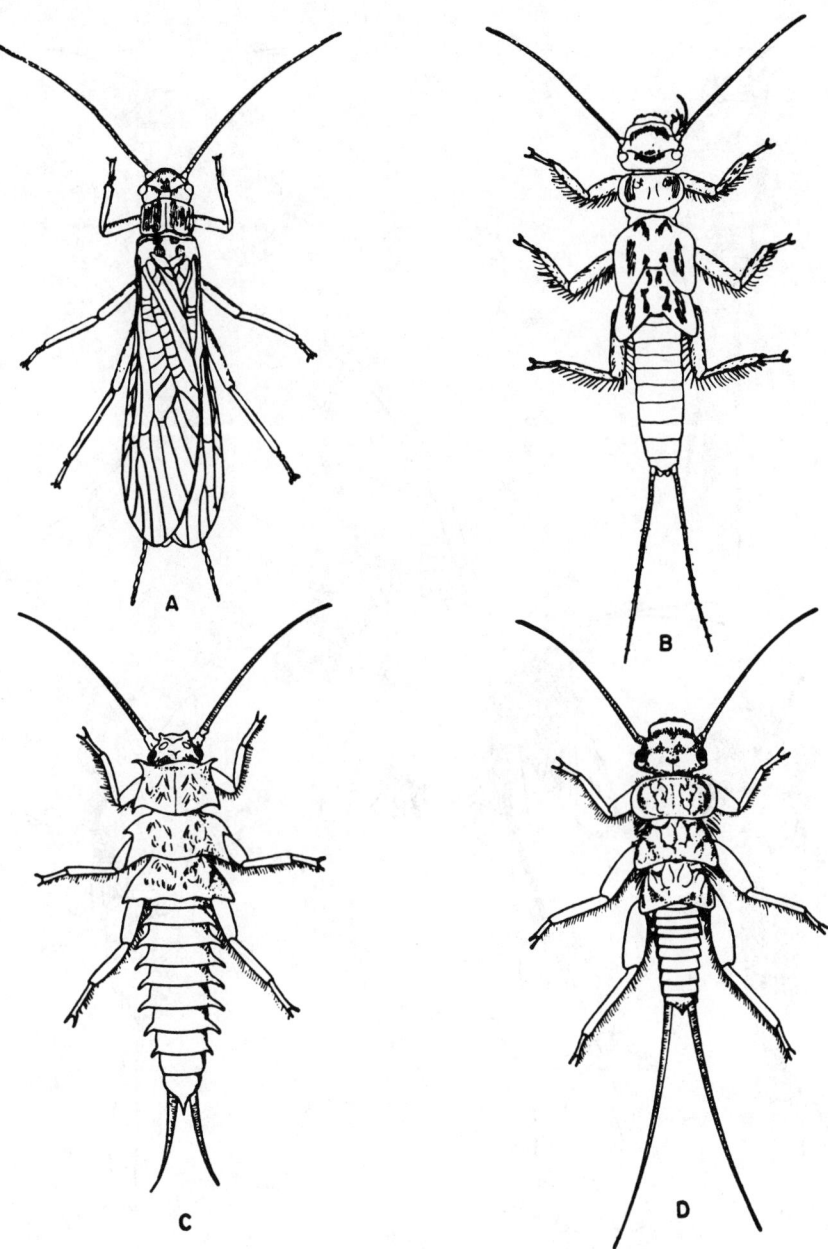

Plate 23. Stoneflies (Order Plecoptera).

A—Adult *Isoperla*, Isoperlidae (14–23 mm)
B—Nymph *Isoperla*,
    Isoperlidae (10–14 mm)

C—Nymph *Pteronarcys*,
    Pteronarcidae (10–40 mm)
D—Nymph *Acroneuria*, Perlidae (20–30 mm)

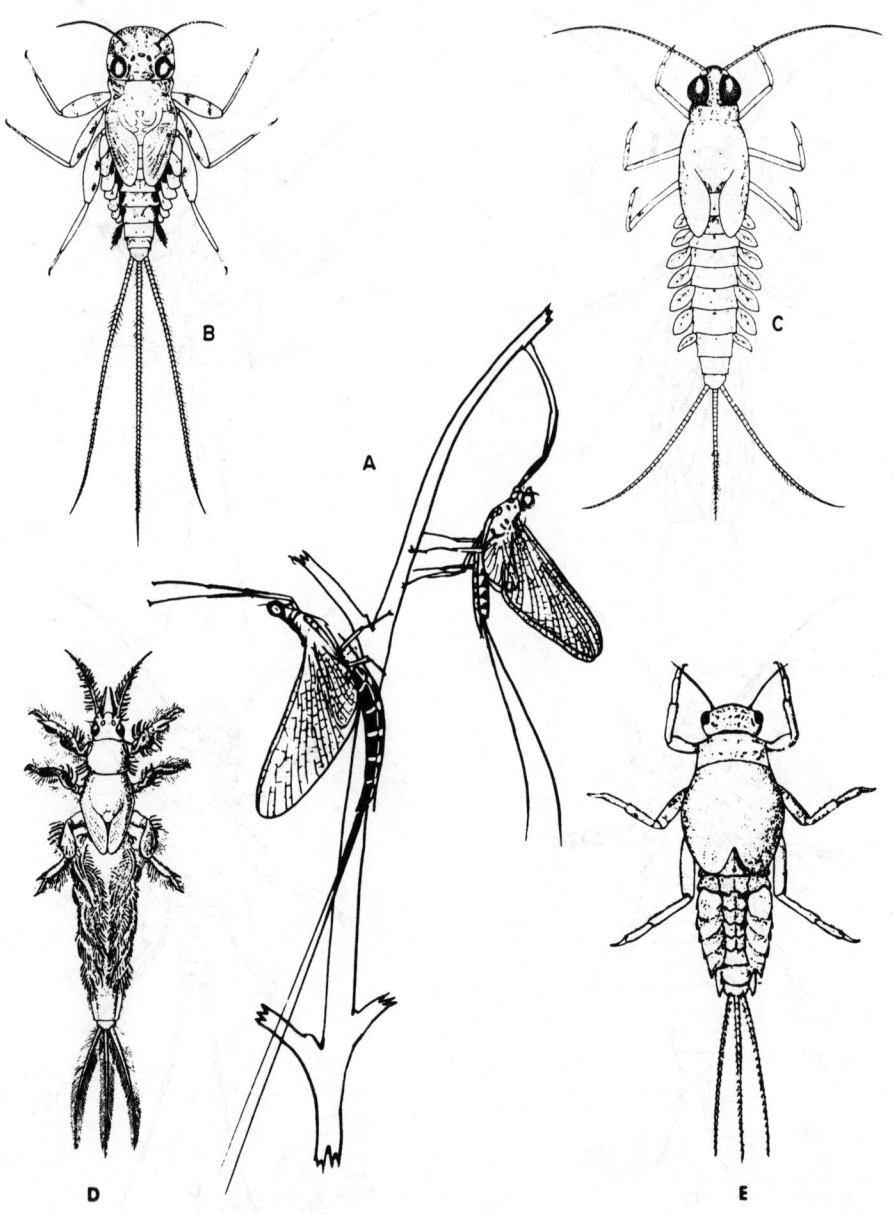

Plate 24. Mayflies (Order Ephemeroptera).

A—Adult mayfly, Heptageniidae (12–18 mm)
B—Nymph *Stenonema*, Heptageniidae (10–14 mm)
C—Nymph *Baetis*, Baetidae (7–14 mm)
D—Nymph *Hexagenia*, Ephemeridae (20–30 mm)
E—Nymph *Ephemerella*, Ephemerellidae (8–15 mm)

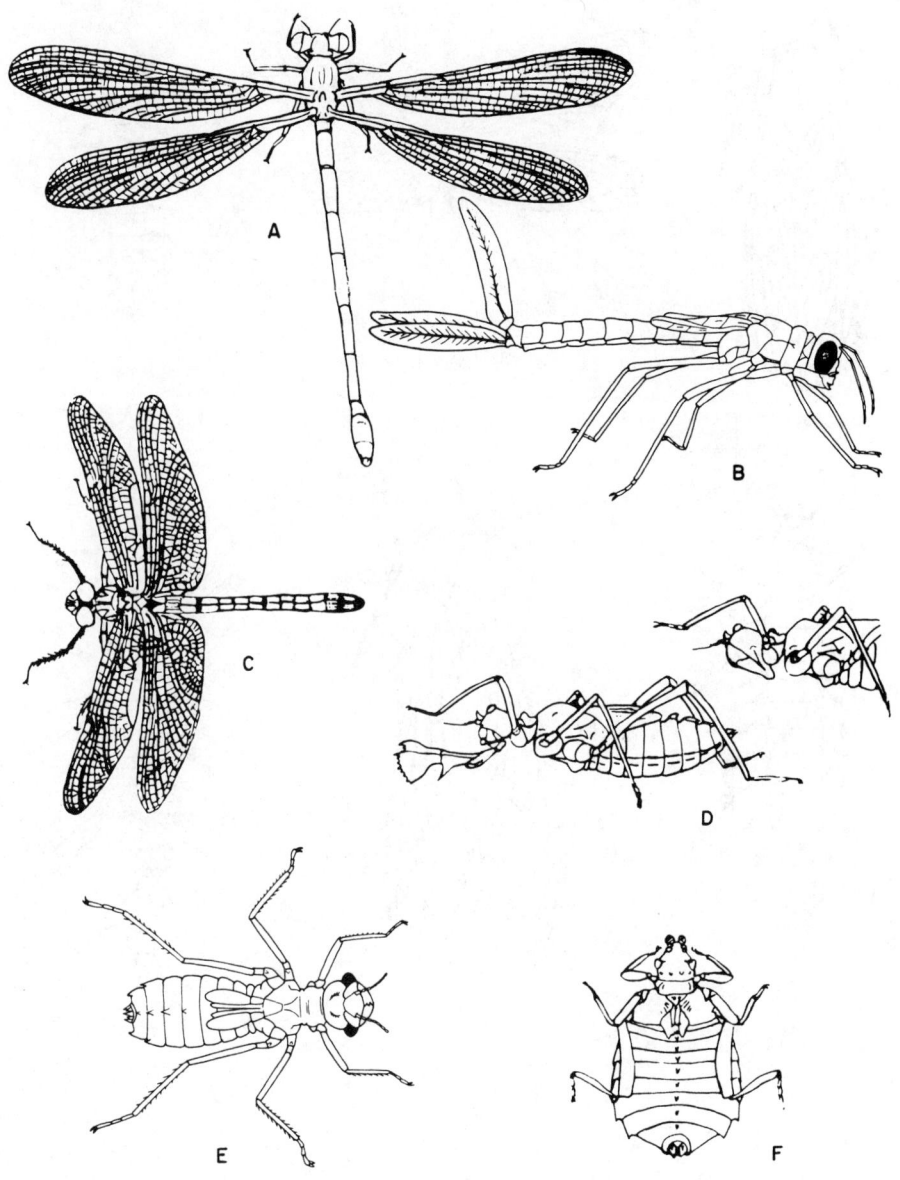

Plate 25. Damselflies, dragonflies (Order Odonata)

A—Adult damselfly (35-55 mm)
B—Damselfly nymph *Lestes*, Coenagrionidae (20-30 mm)
C—Adult dragonfly *Macromia*, Libellulidae (50-70 mm)
D—Dragonfly nymph *Macromia*, showing "mask" both extended and contracted, Libellulidae (15-45 mm)
E—Dragonfly nymph *Helocordulia*, Libellulidae (15-45 mm)
F—Dragonfly nymph *Hagenius*, Gomphidae (15-20 mm)

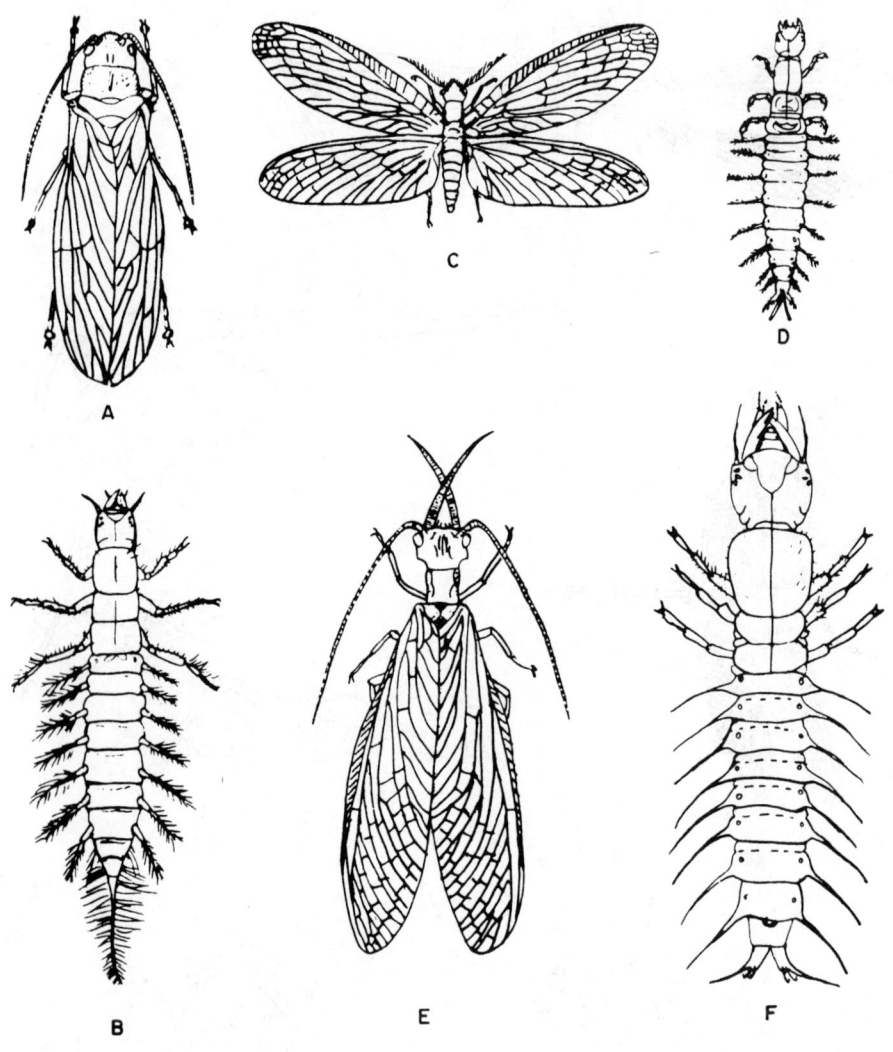

Plate 26. Hellgrammite and relatives.

A—Adult alderfly *Sialis*, Sialidae (9-15 mm)
B—Alderfly larva *Sialis*, Sialidae (15-30 mm)
C—Adult fishfly *Chauliodes*, Corydalidae (15-30 mm)
D—Fishfly larva *Chauliodes*, Corydalidae (20-40 mm)

E—Adult dobsonfly, *Corydalus* (25-70 mm)
F—Dobsonfly larva or hellgrammite, *Corydalus* (25-90 mm)

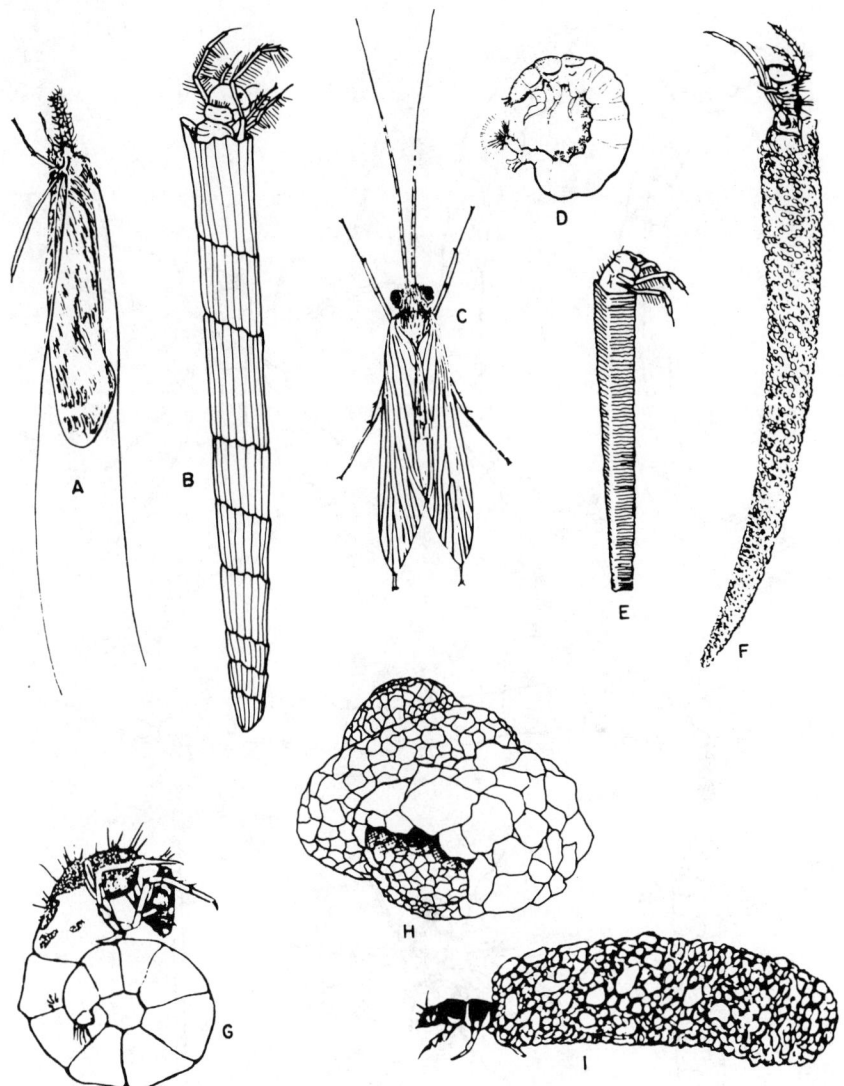

Plate 27. Caddisflies (Order Trichoptera).

A—Adult *Triaenodes*, Leptoceridae (10–20 mm)

B—Larva and case, *Triaenodes*, Leptoceridae (10–14 mm)

C—Adult *Hydropsyche*, Hydropsychidae (20–30 mm)

D—*Hydropsyche* larva, Hydropsychidae (20–30 mm)

E—Larva and case, *Brachycentrus*, Brachycentridae (12–16 mm)

F—Larva and case, *Leptocella*, Leptoceridae (14–18) mm

G—*Helicopsyche* larva, Helicopsychidae (6–10 mm)

H—*Helicopsyche* case, Helicopsychidae (4–6 mm)

I—Larva and case, *Ochrotricha*, Hydroptilidae (4–6 mm)

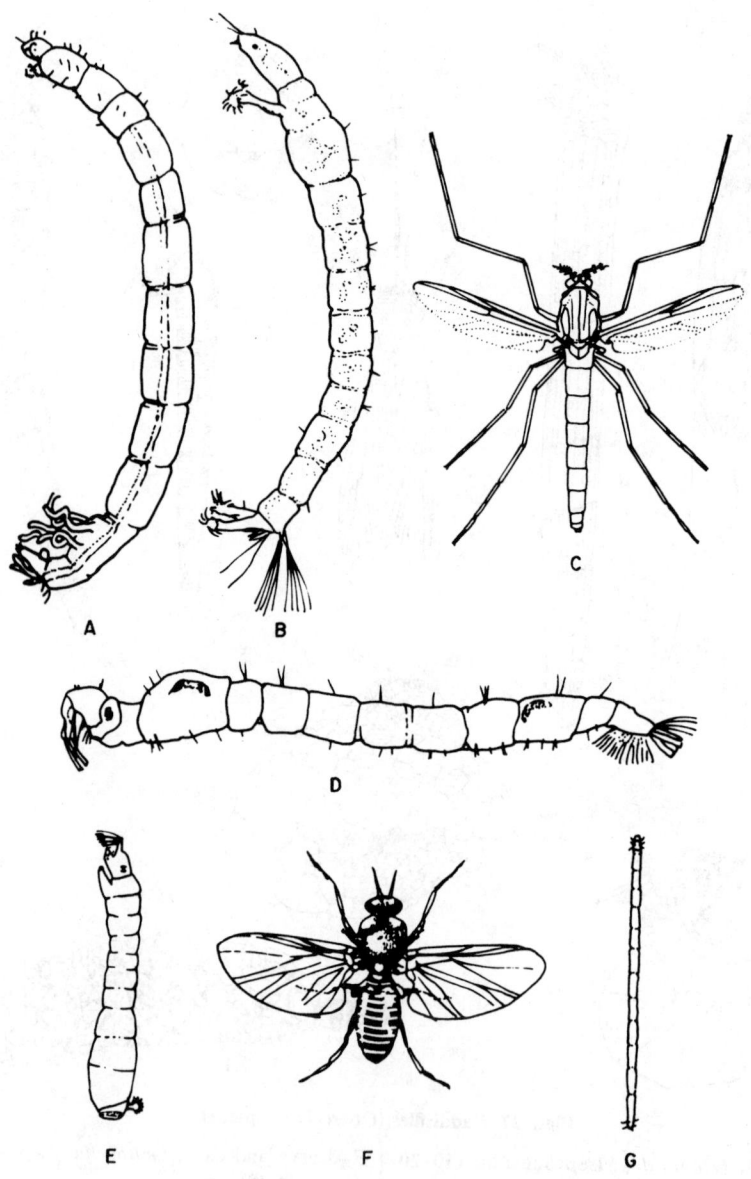

Plate 28. Two-winged flies (Order Diptera).

A—Larva midge *Chironomus*, Chironomidae (5-30 mm)

B—Larva midge *Ablabesmyia*, Chironomidae (5-10 mm)

C—Adult midge, Chironomidae (4-12 mm)

D—Larva phantom midge *Chaoborus*, Culicidae (8-12 mm)

E—Larva black fly *Simulium*, Simuliidae (3-8 mm)

F—Adult black fly *Simulium*, Simuliidae (2-6 mm)

G—Larva biting midge, Ceratopogonidae (3-12 mm)

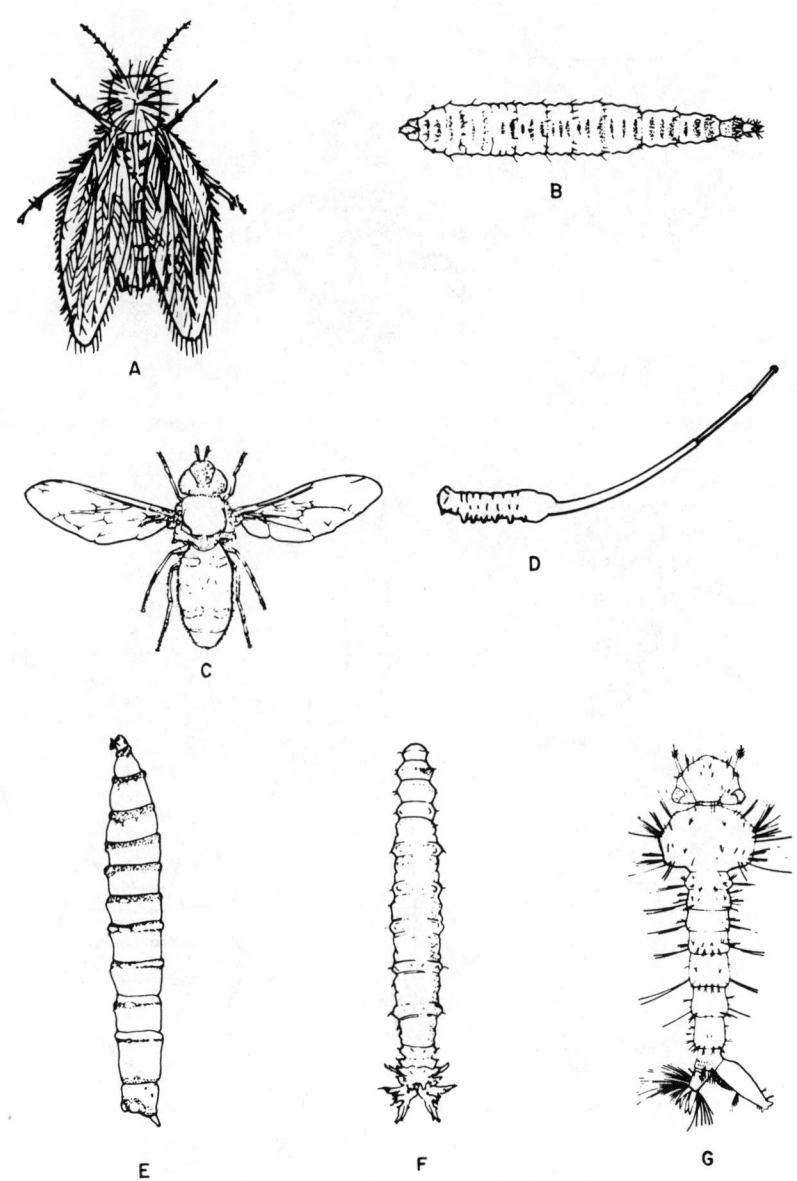

Plate 29. Two-winged flies (Order Diptera)

A—Adult sewage fly *Psychoda*, Psychodidae (2-5 mm)

B—Larva sewage fly *Psychoda*, Psychodidae (4-6 mm)

C—Adult drone fly, Syrphidae (10-15 mm)

D—Rat-tailed maggot *Eristalis*, Syrphidae (15-30 mm)

E—*Tabanus* larva, Tabanidae (30-40 mm)

F—Larva cranefly *Tipula*, Tipulidae (30-40 mm)

G—Larva mosquito *Aedes*, Culicidae (10-15 mm)

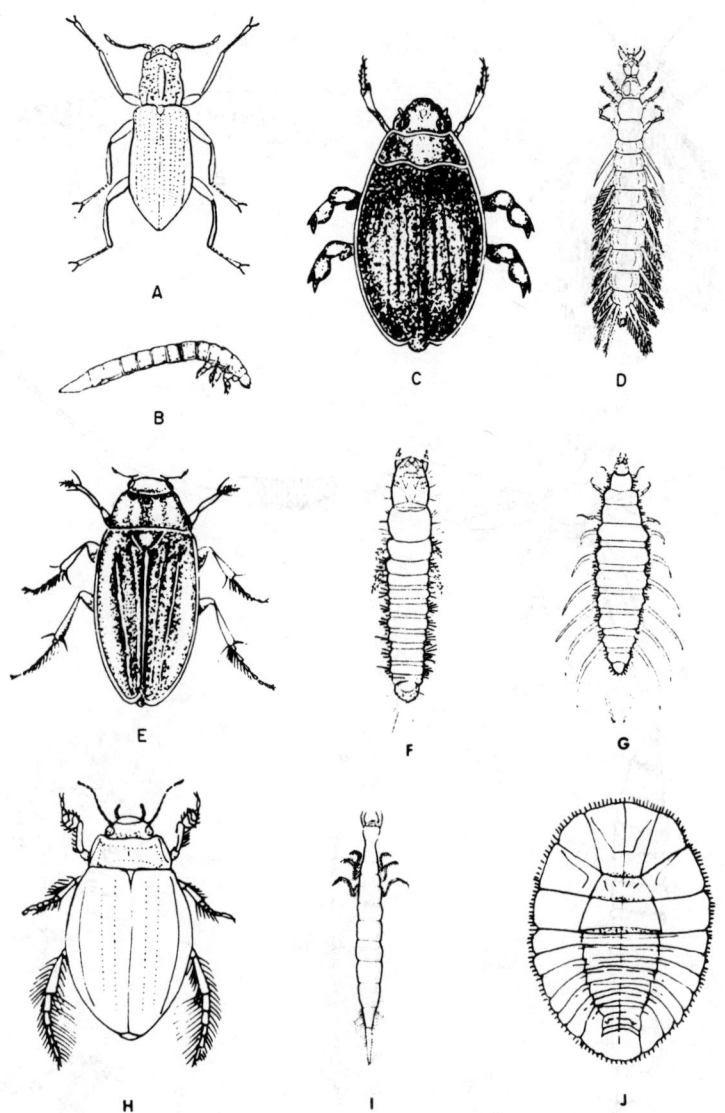

Plate 30. Beetles (Order Coleoptera).

A—Adult riffle beetle *Stenelmis*, Elmidae (2-5 mm)

B—Larva *Narpus*, Elmidae (4-10 mm)

C—Adult whirligig beetle *Dineutus*, Gyrinidae (7-15 mm)

D—Larva *Dineutus*, Gyrinidae (10-30 mm)

E—Adult water scavenger beetle *Hydrophilus*, Hydrophilidae (2-40 mm)

F—Larva *Berosus*, Hydrophilidae (5-20 mm)

G—Larva *Enochrus*, Hydrophilidae (10-25 mm)

H—Adult predacious diving beetle *Dytiscus*, Dytiscidae (2-40 mm)

I—Larva *Cybister*, Dytiscidae (10-25 mm)

J—Larva water penny *Psephenus*, Psephenidae (3-10 mm)

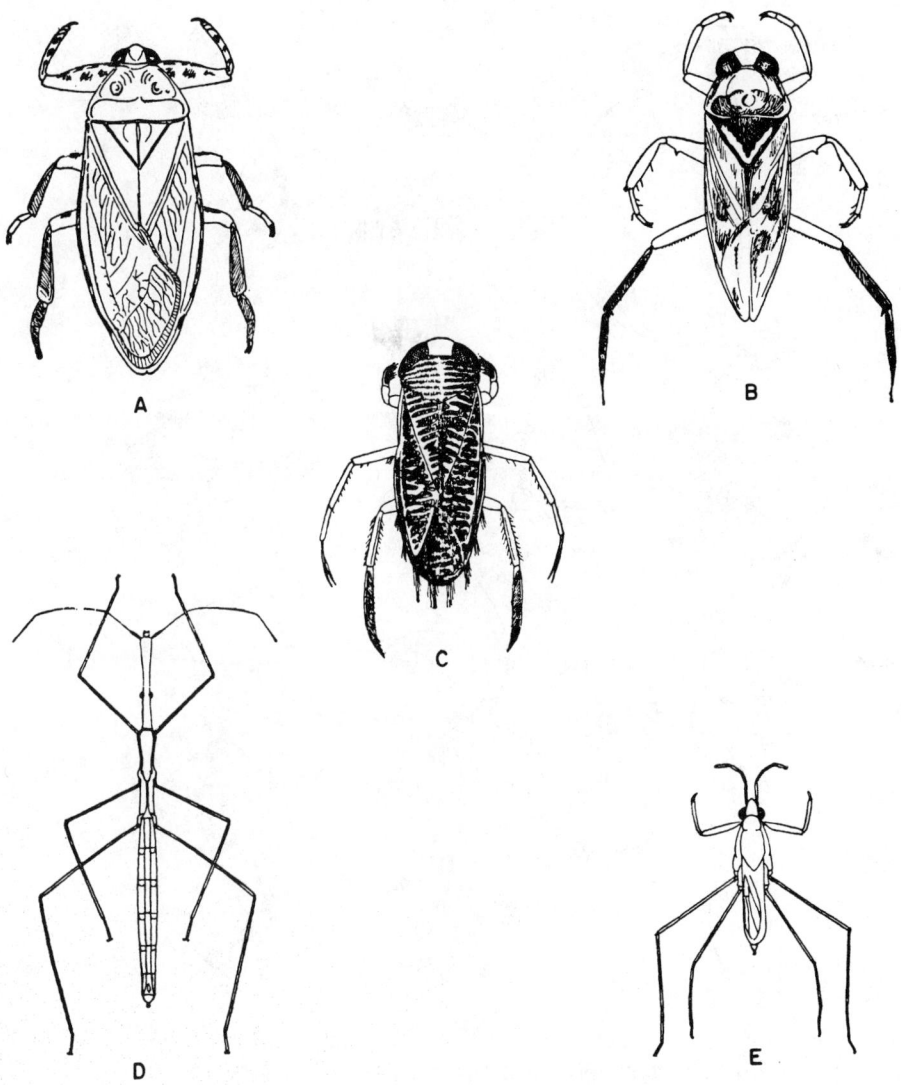

Plate 31. True bugs (Order Hemiptera, all adults).

A—Electric light bug, *Lethocerus*, Belostomidae (20–70 mm)

B—Backswimmer, *Notonecta*, Notonectidae (5–17 mm)

C—Water boatman *Sigara*, Corixidae (3–12 mm)

D—Marsh treader *Hydrometra*, Hydrometridae (8–11 mm)

E—Water strider *Gerris*, Gerridae (2–15 mm)

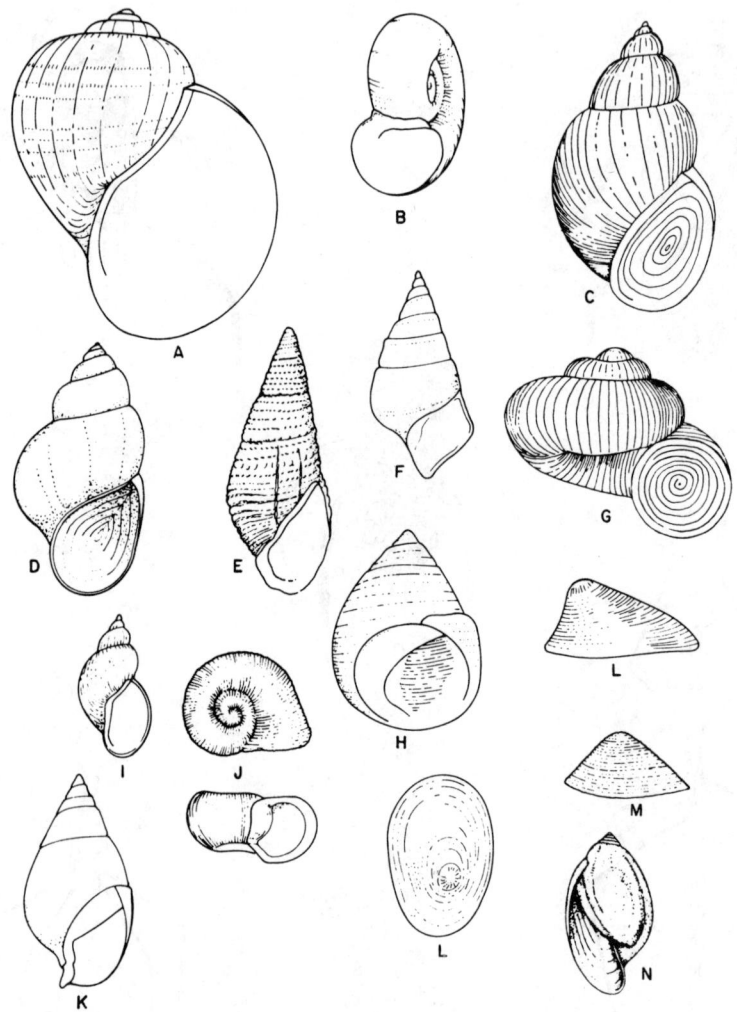

Plate 32. Mollusks (Phylum Mollusca): Snails (Class Gastropoda).

Gill-breathing families:

A—Apple snail *Pomacea*, Pilidae (5 cm)
B—*Marisa*, Pilidae (15 mm)
C—*Campeloma*, Viviparidae (4 cm)
D—Faucet snail *Bithynia*, Amnicolidae (2 cm)
E—*Tarebia*, Thiaridae (15 mm)
F—River snail *Pleurocera*, Pleuroceridae (3 cm)
G—*Valvata*, Valvatidae (1 cm)
H—Periwinkle *Littorina*, Littorinidae (marine, 2 cm)

Lung breathers:

I—Pond snail *Lymnaea*, Lymnaeidae (15 mm)
J—Orb snail *Helisoma*, Planorbidae (1 cm)
K—Mud snail *Nassa*, Nassidae (marine, 2 cm)
L—Limpet *Ferrissia*, Ancylidae (2 mm)
M—Limpet *Lanx*, Lancidae (10 mm)
N—Pouch snail *Physa*, Physidae (5 mm)

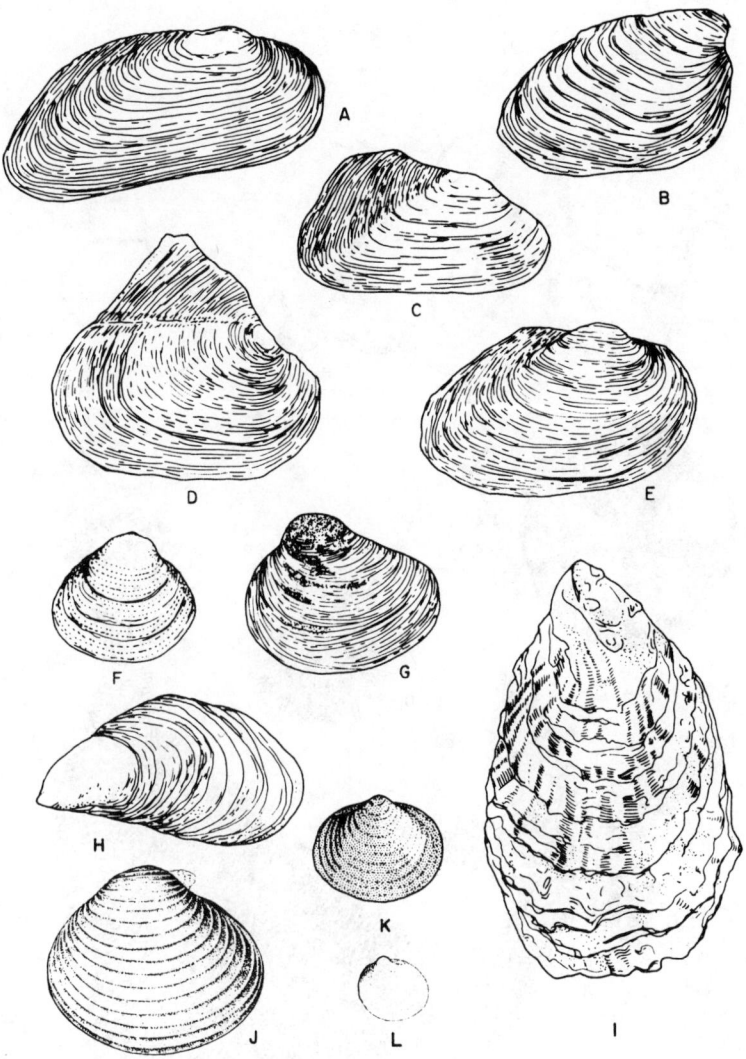

Plate 33. Mollusks (Phylum Mollusca): Bivalves (Class Pelecypoda)

A—Spectacle case *Margaritifera*, Margaritiferidae (10 cm)

B—Pearly mussel *Pleurobema*, Unionidae (10 cm)

C—Pearly mussel *Gonidea*, Unionidae (10 cm)

D—Winged lampshell *Proptera*, Lampsilinae (13 cm)

E—Papershell *Anodonta*, Anodontidae (14 cm)

F—Marsh clam *Polymesoda*, Corbiculidae (marine, 4 cm)

G—Rangia clam *Rangia*, Mactridae (marine, 5 cm)

H—Edible mussel *Mytilus*, Mytilidae (marine, 6 cm)

I—Oyster *Crassostrea*, Ostreidae (marine, 9 cm)

J—Asiatic clam *Corbicula*, Corbiculidae (4 cm)

K—Fingernail clam *Sphaerium*, Sphaeriidae (1 cm)

L—Peashell clam *Pisidium*, Sphaeriidae (5 mm)

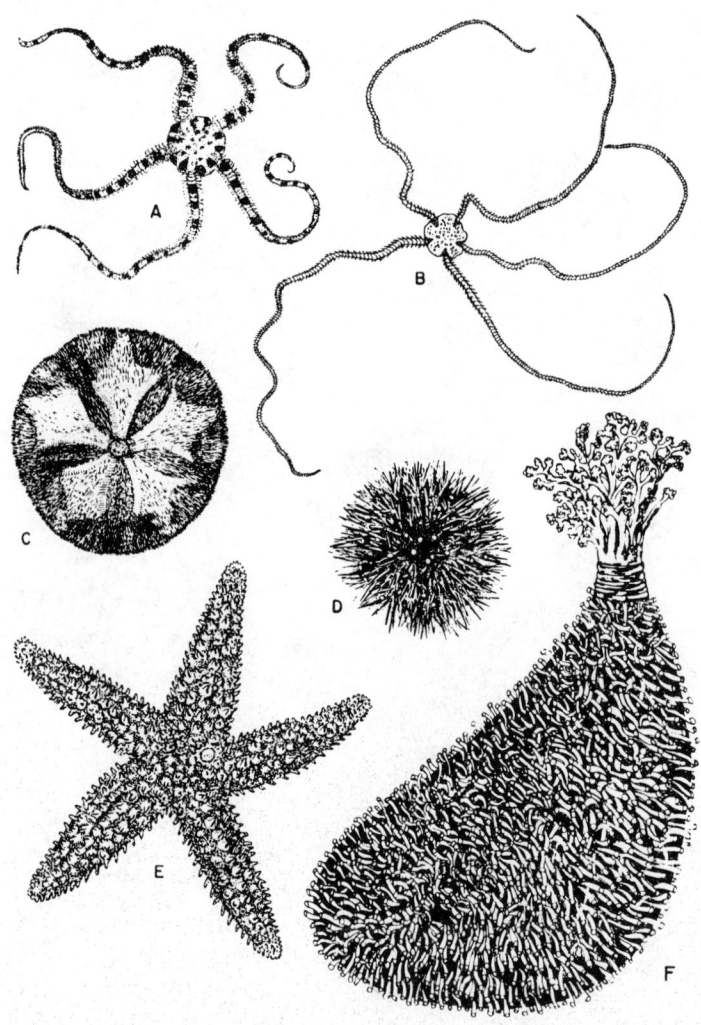

Plate 34. Echinoderm types (Phylum Echinodermata, all marine).

A—Brittle star, class Ophiuroidea: *Ophiopholis* (disc 15 mm)

B—Brittle star, class Ophiuroidea: *Amphioplus* (disc 5 mm)

C—Sand dollar, class Echinoidea: *Echina-rachnius* (7 cm)

D—Sea urchin, class Echinoidea: *Strong-ylocentrotus* (6 cm)

E—Starfish, class Asteroidea: *Asterias* (15 cm)

F—Sea cucumber, class Holothuroidea: *Thyone* (10 cm)

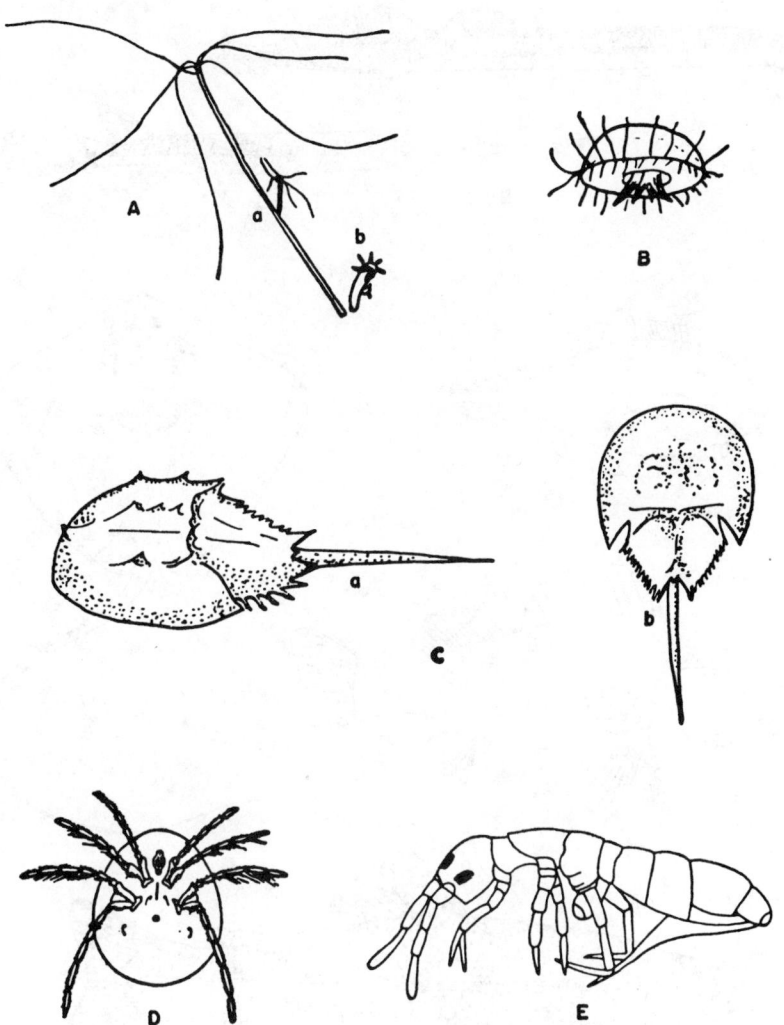

Plate 35. Miscellaneous invertebrates.

Freshwater coelenterates (Phylum Coelenterata):

A—*Hydra*, at (a) extended (2 cm) with bud, and at (b) contracted
B—Jellyfish (Medusa) stage of *Craspedacusta* (2 cm)

Arthropods (Phylum Arthropoda):

C—Horseshoe crab (Class Arachnoidea), marine: *Limulus* (30 cm); (a) shows side view and (b) top view.
D—Water mite (Class Arachnoidea): *Limnochares* (3 mm)
E—Springtail (Class Insecta, Order Collembola): *Orchesella* (2 mm)

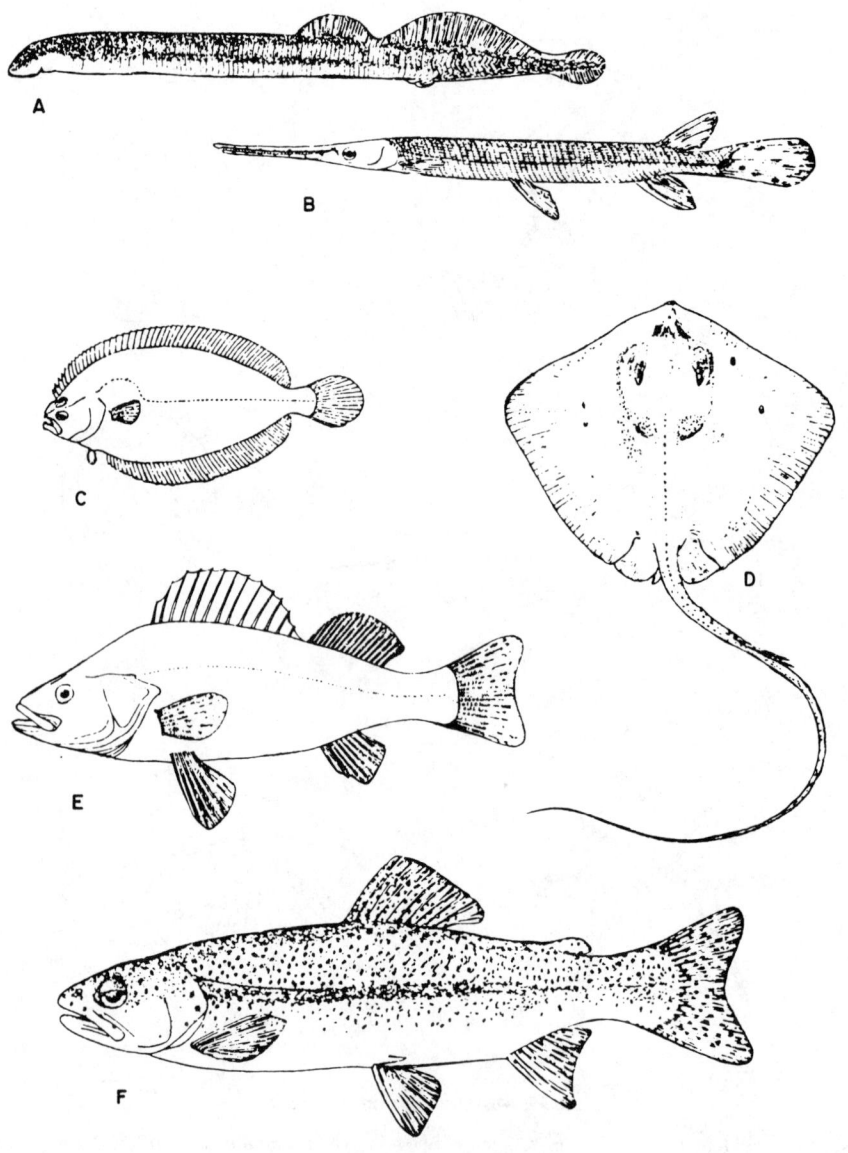

Plate 36. Some types of fishes (Phylum Chordata).

A—Jawless fish (Class Agnatha): lamprey, *Petromyzon* (30–45 cm)

B—Ganoid fish (Class Osteichthys, or Pisces): long-nosed gar, *Lepisosteus* (2.4 m)

C—Flatfish (Class Osteichthys): flounder, *Paralichthys* (30–60 cm)

D—Cartilage fish (Class Chondrichthys): stingray, *Dasyatis* (2 m)

E—Spiny-rayed fish (Class Osteichthys): perch, *Perca* (30 cm)

F—Soft-rayed fish (Class Osteichthys): rainbow trout, *Salmo* (30 cm)

Plate 37. Types of amphibians (Phylum Chordata, Class Amphibia)

Frogs and toads (Order Salientia):

A—The "tadpole" larva (note the developing leg protruding from the body of the tadpole)

B—An adult frog, *Rana* (20 cm). Salientia with dry warty skins are usually called toads.

Salamanders (Order Caudata):

C—*Ambystoma* (20 cm). Adult is typically terrestrial.

D—*Ambystoma* larva is aquatic. Salamander larvae typically have gills.

E—Water dog or mud puppy, *Necturus* (to 30 cm). Larval gills are retained by the adult.

F—An adult aquatic salamander with a flat tail, *Diemictylus* (9 cm).

G—An aquatic salamander. The hind legs have been lost; *Siren* (1 m).

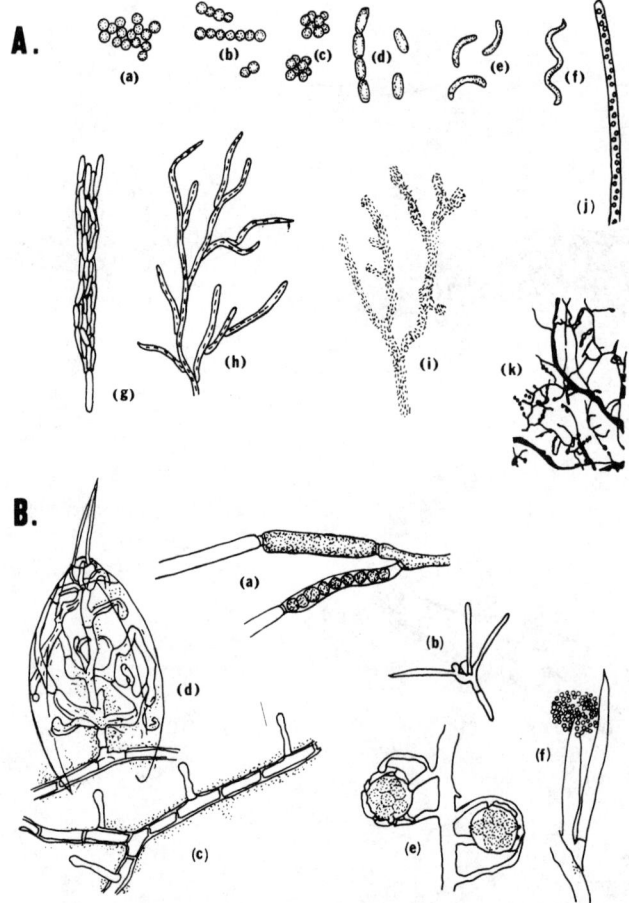

Plate 38. Bacteria and Fungi. (Diameter of most bacterial cells is less than 2 μm, though *Beggiatoa* may range up to 16 μm in diameter, and be of indefinite length.)

A—Bacteria:

(Cellular forms and arrangements)

(a) micrococcus
(b) streptococcus
(c) sarcina
(d) bacillus
(e) vibrio
(f) spirillum

(Sewage organisms)

(g) *Sphaerotilus* ("sewage fungus") cells
(h) A *Sphaerotilus* growth form
(i) A growth form of *Zoogloea*
(j) *Beggiatoa* (sulfur bacterium)
(k) An actinomycete growth form from compost

B—Fungi:

(a) *Leptomitus*, showing zoospores and cellulin plugs (diameter 8.5–16 μm)
(b) *Tetracladium* (diameter 2.5–3.5 μm)
(c) *Zoophagus*, showing mycelial pegs
(d) *Zoophagus* with rotifer impaled on mycelial peg (diameter 3 μm)
(e) *Achlya*, showing oospores
(f) *Achlya*, showing extruded encysted zoospores (Oogonia 50–60 μm, oospores 18.5–22 μm, encysted zoospores 3–5 μm)

10900 D.　Key for Identification of Freshwater Algae Common in Water Supplies and Polluted Waters (Color Plates A-F)

By C. Mervin Palmer

Beginning with 1a and 1b, choose one of the two contrasting statements and follow this procedure with the "a" and "b" statements of the number given at the end of the chosen statement. Continue until the name of the alga is given instead of another key number. (Where recent changes in names of algae have been made, the new name is given followed by the old name in parentheses.)

Refer to
Couplet
No.

1a.　Plastid (separate color body) absent; complete protoplast pigmented; generally blue-green; iodine starch test * negative (blue-green algae) 4

1b.　Plastid or plastids present; parts of protoplast free of some or all pigments; generally green, brown, red, etc., but not blue-green; iodine starch test * positive or negative 2

　　2a.　Cell wall permanently rigid (never showing evidence of collapse), and with regular pattern of fine markings (striations, etc.); plastids brown to green; iodine starch test * negative; flagella absent; wall of two essentially similar halves, one placed over the other as a cover (diatoms) 29

　　2b.　Cell wall, if present, capable of sagging, wrinkling, bulging, or rigidity, depending on existing turgor pressure of cell protoplast; regular pattern of fine markings on wall generally absent; plastids green, red, brown, etc.; iodine starch test * positive or negative; flagella present or absent; cell wall continuous and generally not of two parts 3

3a.　Cell or colony motile; flagella present (often not readily visible); anterior and posterior ends of cell different from one another in contents and often in shape (flagellate algae) . 51

3b.　Nonmotile; true flagella absent; ends of cells often not differentiated (green algae and associated forms) 77

1. Blue-Green Algae

　　4a.　Cells in filaments (or much elongated to form a thread) 5

　　4b.　Cells not in (or as) filaments 23

5a.　Heterocysts present 6

5b.　Heterocysts absent 14

　　6a.　Heterocyst located at one end of filament 7

　　6b.　Heterocysts at various locations in filament 9

7a.　Filaments radially arranged in a gelatinous bead . . . . *Rivularia*

7b.　Filaments isolated or irregularly grouped 8

　　8a.　Filament gradually narrowed to one end . . . . *Calothrix*

　　8b.　Filament not gradually narrowed to one end . . . . *Cylindrospermum*

9a.　Filament unbranched 10

9b.　Filament with occasional (false) branches 13

---

* Add 1 drop of Lugol's (iodine) solution, diluted 1:1 with distilled water. In about 1 min, if positive, starch is stained blue and later black. Other structures (such as nucleus, plastids, cell wall) may also stain, but turn brown to yellow.

10a. Crosswalls in filament much closer together than width of filament . . . *Nodularia*

10b. Crosswalls in filament at least as far apart as width of filament.   11

11a. Filaments normally in tight parallel clusters; heterocysts and spores cylindric to long oval in shape . . . . *Aphanizomenon*

11b. Filaments not in tight parallel clusters; heterocysts and spores often round to oval   12

12a. Filaments in a common gelatinous mass . . . . *Nostoc*

12b. Filaments not in a common gelatinous mass. . . . . *Anabaena*

13a. False branches in pairs. . . . . *Scytonema*

13b. False branches, single . . . . *Tolypothrix*

14a. Filament or elongated cell attached at one end, with one or more round cells (spores) at the other . . . . *Entophysalis (Chamaesiphon)*

14b. Filament generally not attached at one end; no terminal spores present   15

15a. Filament with regular spiral from throughout   16

15b. Filament not spiral, or with spiral form limited to a portion of filament   17

16a. Filament septate . . . . *Arthrospira*

16b. Filament nonseptate . . . . *Spirulina*

17a. Filament very narrow, only 0.5 to 2.0 μm wide. . . . . *Schizothrix*

17b. Filament 3 to 95 μm wide.   18

18a. Filaments loosely aggregated or not in clusters   19

18b. Filaments tightly aggregated and surrounded by a common gelatinous secretion that may be invisible.   22

19a. Filament surrounded by wall-like sheath that frequently extends beyond the ends of the filament of cells; filament generally without movement.   20

19b. Filament not surrounded by a wall-like sheath; filament may show movement   21

20a. Cells separated from one another by a space . . . . *Johannesbaptistia*

20b. Cells in contact with adjacent cells . . . . *Lyngbya*

21a. All filaments short, with less than 20 cells; one or both ends of filament sharply pointed . . . . *Raphidiopsis*

21b. Filaments long, with more than 20 cells; filaments commonly without sharp-pointed ends . . . . *Oscillatoria*

22a. Filaments arranged in a tight, essentially parallel bundle. . . . . *Microcoleus*

22b. Filaments arranged in irregular fashion, often forming a mat . . . . . *Phormidium*

23a. Cells in a regular pattern of parallel rows, forming a plate . . . . *Agmenellum (Merismopedia)*

23b. Cells not regularly arranged to form a plate.   24

24a. Cells regularly arranged near surface of a spherical gelatinous bead   25

24b. Gelatinous bead, if present, not spherical   26

25a. Cells ovate to heart-shaped, connected to center of bead by colorless stalks . . . . *Gomphosphaeria*

25b. Cells round, without gelatinous stalks . . . . . . *Gomphosphaeria (Coelosphaerium* type*)*

26a. Cells cylindric-oval . . . . *Coccochloris (Aphanothece)*

26b. Cells spherical.   27

27a. Two or more distinct layers of gelatinous sheath around each cell or cell cluster . . . . *Anacystis (Gloeocapsa)*

27b. Gelatinous sheath around cells not distinctly layered.   28

28a. Cells isolated or in colonies of 2 to 32 cells . . . . *Anacystis (Chroococcus)*

28b. Cells in colonies composed of many cells . . . . . . *Anacystis (Microcystis, Polycystis)*

Refer to
Couplet
No.

2. Diatoms

29a. Front (valve) view circular in outline; markings radial in arrangement; cells may form a
   filament (centric diatoms) 30
29b. Front (valve) view elongate, not circular; transverse markings in one or two longitudinal
   rows; cells, if grouped, not forming a filament (pennate diatoms) 32
   30a. Cells in persistent filaments with valve faces in contact; therefore, cells commonly
      seen in side (girdle) view . . . .*Melosira*
   30b. Cells isolated or in fragile filaments, often seen in front (valve) view 31
31a. Radial markings (striations), in valve view, extending from center to margin; short spines
   often present around margin (valve view) . . . . *Stephanodiscus*
31b. Area of prominent radial markings, in valve view, limited to approximately outer half of
   circle, marginal spines generally absent. . . . . *Cyclotella*
   32a. Cell longitudinally symmetrical in valve view 33
   32b. Cell longitudinally unsymmetrical (two sides unequal in shape), at least in valve
      view 49
33a. Raphe at or near the edge of the valve. 34
33b. Raphe or pseudoraphe median or submedian 35
   34a. Marginal, keeled raphe areas lie opposite one another on the two valves
      . . . . *Hantzschia*
   34b. Marginal, keeled raphe areas lie diagonal to one another on the two valves
      . . . . *Nitzschia*
35a. Cell transversely symmetrical in valve view 36
35b. Cell transversely unsymmetrical (two ends unequal in shape or size), at least in valve
   view 44
   36a. Cell round-oval in valve view, not more than twice as long as it is wide. *Cocconeis*
   36b. Cell elongate, more than twice as long as it is wide 37
37a. Cell flat (girdle face wide, valve face narrow) . . . .*Tabellaria*
37b. Girdle and valve faces about equal in width. 38
   38a. Cell with several markings (septa) extending without interruption across the valve
      face; no marginal line of pores present . . . .*Diatoma*
   38b. Cross-markings (striations or costae) on valve surface, interrupted by either longi-
      tudinal space (pseudoraphe), or line (raphe), or line of pores (carinal dots) 39
39a. Cells attached side by side to form a ribbon of several to many cells. . . . . . *Fragilaria*
39b. Cells isolated or in pairs. 40
   40a. Cell narrow, linear, often narrowed to both ends; true raphe absent . . . . *Synedra*
   40b. Cell commonly "boat-shape" in valve view; true raphe present 41
41a. Cell longitudinally unsymmetrical in girdle view; sometimes with attachment stalk
   . . . . *Achnanthes*
41b. Cell symmetrical in girdle as well as valve view; generally not attached 42
   42a. Area without striations extending as a transverse belt around middle of cell
      . . . .*Stauroneis*
   42b. No continuous clear belt around middle of cell 43
43a. Cell with coarse transverse markings (costae), which appear as solid lines even under high
   magnification . . . . *Pinnularia*
43b. Cell with fine transverse markings (striae), which appear as lines of dots under high
   magnification . . . . *Navicula*
   44a. Cells attached together at one end only to form radiating colony . . . . *Asterionella*
   44b. Cell not forming a loose radiating colony 45

Refer to
Couplet
No.

45a. Cells in fan-shaped colonies . . . . *Meridion*
45b. Cells isolated or in pairs   46
    46a. Prominent wall markings in addition to striations present just below lateral margins
        on valve surface of cell. . . . . *Surirella*
    46b. Wall markings along sides of valve limited to striations   47
47a. Cell elongate, sides almost parallel except for terminal knobs . . . . *Asterionella*
47b. Sides of cell converging toward one end   48
    48a. Cells bent in girdle view . . . . *Rhoicosphenia*
    48b. Cells straight in girdle view . . . . *Gomphonema*
49a. Valves with transverse septa or costae . . . . *Epithemia*
49b. Valves with no transverse septa or costae   50
    50a. Raphe located almost through center of valve. . . . . *Cymbella*
    50b. Raphe excentric, near concave edge of valve. . . . . *Amphora*

3. Flagellate Algae

51a. Cell in a loose, rigid conical sac (lorica); isolated or in a branching colony . . *Dinobryon*
51b. Case or sac, if present, not conical; colony, if present, not branching.   52
    52a. Cells isolated or in pairs   53
    52b. Cells in a colony of four or more cells   71
53a. Prominent transverse groove encircles cell   54
53b. Cell without transverse groove   56
    54a. Cell with prominent rigid projections, one forward and two or three on posterior
        end . . . . *Ceratium*
    54b. Cell without several rigid polar projections   55
55a. Portions above and below transverse groove about equal. . . . . *Peridinium*
55b. Front portion distinctly larger than posterior portion. . . . . *Massartia*
    56a. Cell with long bristles extending from surface plates . . . . *Mallomonas*
    56b. Cell without bristles and surface plates.   57
57a. Cell protoplast enclosed in loose rigid covering (lorica)   58
57b. Cell with tight membrane or wall but no loose rigid covering   60
    58a. Lorica flattened; cell with two flagella . . . . *Phacotus*
    58b. Lorica not flattened; cell with one flagellum.   59
59a. Lorica often opaque, generally dark brown to red; plastid green . . . . . *Trachelomonas*
59b. Lorica often transparent, colorless to light brown; plastid light brown . . . *Chrysococcus*
    60a. Plastids brown to red to olive or blue-green   61
    60b. Plastids grass green.   64
61a. Plastids blue-green to blue. . . . . *Chroomonas*
61b. Plastids brown to red to olive green   62
    62a. Plastid brown; one or two flagella   63
    62b. Plastids red, red-brown, or olive green; two flagella . . . . *Rhodomonas*
63a. Anterior end of cell oblique; two flagella. . . . . *Cryptomonas*
63b. Anterior end of cell rounded or pointed; one flagellum. . . . . *Chromulina*
    64a. Cell with colorless rectangular wing . . . . *Pteromonas*
    64b. No wing extending from cell   65
65a. Cells flattened; margin rigid. . . . . *Phacus*
65b. Cell not flattened; margin rigid or flexible.   66
    66a. Pyrenoid present in the single plastid; no paramylon; margin not flexible; two or
        more flagella per cell   67

Refer to
Couplet
No.

66b. Pyrenoid absent; paramylon present; several plastids per cell; margin flexible or rigid;
one flagellum per cell 70
67a. Cells fusiform (tapering at each end) . . . . *Chlorogonium*
67b. Cells not fusiform, generally almost spherical 68
68a. Plastids numerous . . . . *Vacuolaria*
68b. Plastids few, commonly one 69
69a. Two flagella per cell . . . . *Chlamydomonas*
69b. Four flagella per cell . . . . *Carteria*
70a. Cell flexible in form; paramylon a capsule or disk; cell elongate . . . . *Euglena*
70b. Cell rigid in form; paramylon ring-shaped; cell almost spherical . . . . *Lepocinclis*
71a. Plastids brown 72
71b. Plastids green 73
72a. Cells in contact with one another . . . . *Synura*
72b. Cells separated from one another by space . . . . *Uroglenopsis*
73a. Colony flat, one cell thick . . . . *Gonium*
73b. Colony rounded, more than one cell thick 74
74a. Cells in contact with one another 75
74b. Cells separated from one another by space 76
75a. Cells radially arranged . . . . *Pandorina*
75b. Cells all facing one direction . . . . *Pyrobotrys (Chlamydobotrys)*
76a. Cells more than 400 per colony . . . . *Volvox*
76b. Cells less than 75 per colony . . . . *Eudorina*

4. Green Algae and Associated Forms

77a. Cells jointed together to form a net . . . . *Hydrodictyon*
77b. Cells not forming a net 78
78a. Cells attached side by side to form a plate or ribbon one cell wide and thick; number
of cells commonly two, four, or eight . . . . *Scenedesmus*
78b. Cells not attached side by side 79
79a. Cells isolated or in nonfilamentous or nontubular thalli 80
79b. Cells in filaments or other tubular or threadlike thalli 111
80a. Cells isolated and narrowest at the center because of incomplete fissure (desmids) . 81
80b. Cells isolated or in clusters but without central fissure 84
81a. Each half of cell with three spinelike or pointed knobular extensions . . . . *Staurastrum*
81b. Cell margin with no such extensions 82
82a. Semicells with a median incision or depression 83
82b. Semicells with no median incision or depression . . . . *Cosmarium*
83a. Margin with rounded lobes . . . . *Euastrum*
83b. Margin with sharp-pointed teeth . . . . *Micrasterias*
84a. Cells elongate 85
84b. Cells round to oval or angular 92
85a. Cell radiating from a central point . . . . *Actinastrum*
85b. Cells isolated or in irregular clusters 86
86a. Cells with terminal spines . . . . *Schroederia*
86b. Cells without terminal spines 87
87a. Cells with colorless attachment area at one end . . . . *Characium*
87b. No attachment area at one end of cell 88
88a. Plastids two per cell; unpigmented area across center of cell . . . . . . *Closterium*
88b. Cell with plastid that continues across the center 89

Refer to
Couplet
No.

89a. Cell 5 to 10 times as long as it is broad   90
89b. Cell 2 to 4 times as long as it is broad   91
  90a. Pyrenoid absent, or one per cell . . . . *Ankistrodesmus*
  90b. Pyrenoids several per cell . . . . *Closteriopsis*
91a. Cells semicircular; cell ends pointed but with no terminal spines . . . . . . *Selenastrum*
91b. Cells arcuate but less than semicircular; cell ends pointed and each with a short spine
  . . . . *Closteridium*
  92a. Cells regularly arranged in a tight, flat colony . . . . *Pediastrum*
  92b. Cells not in a tight, flat regular colony   93
93a. Cells angular   94
93b. Cells round to oval   95
  94a. Two or more spines at each angle . . . . *Polyedriopsis*
  94b. Spines none or less than two at each angle . . . . *Tetraedron*
95a. Cells with long sharp spines   96
95b. Long sharp spines absent   99
  96a. Cells round   97
  96b. Cells oval   98
97a. Cells isolated . . . . *Golenkinia*
97b. Cells in colonies . . . . *Micractinium*
  98a. Each cell end has one spine . . . . *Diacanthos*
  98b. Each cell end has more than one spine . . . . *Chodatella*
99a. Colony of definite regular form, round to oval   100
99b. Colony, if present, not a definite oval or sphere; or cells may be isolated   104
  100a. Colony a tight sphere of cells   101
  100b. Colony a loose sphere of cells enclosed by a common membrane   102
101a. Sphere solid, slightly irregular, no connecting processes between cells *Planktosphaeria*
101b. Sphere hollow, regular; short connecting processes between cells . . . . *Coelastrum*
  102a. Cells round   103
  102b. Cells oval . . . . *Oocystis*
103a. Cells connected to center of colony by branching stalk . . . . *Dictyosphaerium*
103b. No stalk connecting cells . . . . *Sphaerocystis*
  104a. Oval cells, enclosed in a somewhat spherical, often orange-colored matrix
  . . . . *Botryococcus*
  104b. Cells rounds, isolated or in colorless matrix   105
105a. Adjoining cells with straight flat walls between their protoplasts   106
105b. Adjoining cells with rounded walls between their protoplasts   107
  106a. Cells embedded in a common gelatinous matrix . . . . *Palmella*
  106b. No matrix or sheath outside of cell walls . . . . *Phytoconis (Protococcus)*
107a. Cells loosely arranged in a large gelatinous matrix . . . . *Tetraspora*
107b. Cells isolated or tightly grouped in a small colony   108
  108a. Cells located inside of protozoa . . . . *Zoochlorella*
  108b. Cells not inside of protozoa   109
109a. Plastid filling 2/3 or less of cell   110
109b. Plastid filling 3/4 or more of cell . . . . *Chlorococcum*
  110a. Cell diameter 2 μm or less; reproduction by cell division . . . . *Nannochloris*
  110b. Cell diameter 2.5 μm or more; reproduction by internal spores . . . . . . *Chlorella*
111a. Cells attached end to end in an unbranched filament   112
111b. Thallus branched, or more than one cell wide   119
  112a. Plastids in form of one or more marginal spiral ribbons . . . . *Spirogyra*
  112b. Plastids not in form of spiral ribbons   113

Refer to
Couplet
No.

113a. Filaments, when breaking, separating through middle of cells   114
113b. Filaments, when breaking, separating irregularly or at ends of cells   115
    114a. Starch test positive; cell margin straight; one plastid, granular . . . . . *Microspora*
    114b. Starch test negative; cell margin slightly bulging; several plastids . . . . *Tribonema*
115a. Marginal indentations between cells . . . . *Desmidium*
115b. No marginal indentations between cells   116
    116a. Plastids, two per cell . . . . *Zygnema*
    116b. Plastid, one per cell (sometimes appearing numerous)   117
117a. Some cells with walls having transverse wrinkles near one end; plastid an irregular net
  . . . . *Oedogonium*
117b. No apical wrinkles in wall; plastid not porous   118
    118a. Plastid a flat or twisted axial ribbon . . . . *Mougeotia*
    118b. Plastid an arcuate marginal band . . . . *Ulothrix*
119a. Thallus a flat plate of cells . . . . *Hildenbrandia*
119b. Thallus otherwise   120
    120a. Thallus a long tube without crosswalls . . . . *Vaucheria*
    120b. Thallus otherwise   121
121a. Thallus a leathery strand with regularly spaced swellings and a continuous surface membrane of cells . . . . *Lemanea*
121b. Thallus otherwise   122
    122a. Filament unbranched . . . . *Schizomeris*
    122b. Filament branched   123
123a. Branches in whorls (clusters)   124
123b. Branches single or in pairs   126
    124a. Thallus embedded in gelatinous matrix . . . . *Batrachospermum*
    124b. Thallus not embedded in gelatinous matrix   125
125a. Main filament one cell thick . . . . *Nitella*
125b. Main filament three cells thick . . . . *Chara*
    126a. Most of filament surrounded by a layer of cells . . . . *Compsopogon*
    126b. Filament not surrounded by a layer of cells   127
127a. End cell of branches with a rounded or blunt-pointed tip   128
127b. End cell of branches with a sharp-pointed tip   130
    128a. Plastids green; starch test positive   129
    128b. Plastids red; starch test negative . . . . *Audouinella*
129a. Some cells dense, swollen, dark green (spores); other cells light green, cylindric
  . . . . *Pithophora*
129b. All cells essentially alike, light to medium green, cylindric . . . . *Cladophora*
    130a. Filaments embedded in gelatinous matrix   131
    130b. Filaments not embedded in gelatinous matrix   132
131a. Cells of main filament much wider than even the basal cells of branches
  . . . . *Draparnaldia*
131b. No abrupt change in width of cells from main filament to branches . . . . *Chaetophora*
    132a. Branches very short, with no cross-walls . . . . *Rhizoclonium*
    132b. Branches long, with cross-walls   133
133a. Branches ending in an abrupt spine having a bulbous base . . . . *Bulbochaete*
133b. Branches gradually reduced in width, ending in a long pointed cell, with or without
  color . . . . *Stigeoclonium*

10900 E.   Recent Changes in Names of Algae

Old Name	New Name
Aphanocapsa	*Anacystis*
Aphanothece	*Coccochloris*
Chamaesiphon	*Entophysalis*
Chantransia	*Audouinella*
Chlamydobotrys	*Pyrobotrys*
Chroococcus	*Anacystis*
Clathrocystis	*Anacystis*
Coelosphaerium	*Gomphosphaeria*
Encyonema	*Cymbella*
Gloeocapsa	*Anacystis*
Gloeothece	*Coccochloris*
Merismopedia	*Agmenellum*
Microcystis	*Anacystis*
Odontidium	*Diatoma*
Polycystis	*Anacystis*
Protococcus	*Phytoconis*
Sphaerella	*Haematococcus*
Synechococcus	*Coccochloris*

10900 F.   Index to Illustrations

NOTE: Arabic numerals refer to black and white plate numbers, capital letters to color plates. See also "Recent Changes in Names of Algae," Section 10900E, preceding. Family names are not generally included.

Organism	Plate	Organism	Plate	Organism	Plate
Ablabesmyia	28	*Azolla*	8	Chlorophyta	7
Acanthamoeba	14	Bacillariophyceae	5, 6	Chondrichthys	36
Achlya	38	Bacillus	38	Chordata	36, 37
Achnanthes	F, 5	Backswimmer	31	*Chromulina*	D, 11
Achromadora	18	Bacteria	38	*Chroococcus:* see *Anacystis*	
Acorus	10	*Baetis*	24	*Chrysococcus*	D
Acroneuria	23	*Batrachospermum*	F	Chrysophyta	5, 6
Actinastrum	E	Beetles	22, 30	Ciliates (Protozoa)	15
Actinomycetes	38	*Beggiatoa*	38	Cladocerans	20
Actinophrys	14	*Berosus*	30	*Cladophora*	D, F
Aedes	29	*Biddulphia*	6	Clams	33
Agmenellum	C, D, 1	*Bithynia*	32	*Closterium*	B
Agnatha	36	Bivalves	33	*Coccochloris*	D
Alderflies	26	*Bodo*	13	*Cocconeis*	D, 5
Algae, blue-green	1, 2	*Bosmina*	20	*Coelastrum*	E
Algae, brown	7	*Botrydium*	4	Coelenterates	35
Algae, marine	7	*Botryococcus*	E	*Coelosphaerium:*	
Algae, nonmotile green	3, 4	*Brachionus*	17	see *Gomphosphaeria*	
Algae, red	7	*Brachycentrus*	27	Coleoptera	22, 30
Alona	20	Brittle stars	34	*Colpoda*	15
Ambystoma	37	Bryozoa	16	*Compsopogon*	F
Amoebas (Protozoa)	14	Bugs, true	31	Coontail	9
Amoeba sp.	14	*Bulbochaete*	F	Copepod	21
Amphibians	37	Caddisflies	22, 27	*Corallina*	7
Amphioplus	34	*Calothrix*	D	*Corbicula*	33
Amphipod	21	*Cambarus*	21	*Corydalus*	22, 26
Anabaena	A, B, C, 2	*Campeloma*	32	*Coscinodiscus*	6
Anacystis	A, B, C, 1	*Carteria*	C	*Craspedacusta*	35
Anisonema	13	Cattail	10	*Crassostrea*	33
Ankistrodesmus	D, 3	Caudata	37	Crayfish	21
Annelid	19	*Ceratium*	A	*Crucigenia*	3
Anodonta	33	*Ceratophyllum*	9	Crustaceans, cladocerans	20
Anthophysa	13	Ceratopogonidae	28	Crustaceans, common	
Antocha	22	*Cercomonas*	13	types	21
Aphanizomenon	A, 2	*Chaetophora*	F	*Cyptomonas*	11
Arachnoidea	35	*Chaoborus*	28	*Culex*	22
Arcella	14	*Chara*	F	Cyanophyta	1, 2
Arthropods	35	*Chauliodes*	26	*Cybister*	22, 30
Arthrospira	C	Chironomidae	28	*Cyclops*	21
Asellus	21	*Chironomus*	22, 28	*Cyclotella*	B, D, 6
Aspidisca	15	*Chlamydomonas*	C	*Cylindrospermum*	E
Astasia	13	*Chloramoeba*	11	*Cymbella*	B, F, 5
Asterias	34	*Chlorella*	B, C	*Cypridopsis*	21
Asterionella	A, B, 5	*Chlorococcum*	C	Damselfly	25
Audouinella	F	*Chlorogonium*	C	*Daphnia*	20

Organism	Plate	Organism	Plate	Organism	Plate
Dasyatis	36	Flatworm	19	*Leptocella*	27
Decapoda	21	Flies, two-winged	22, 28, 29	*Leptodora*	20
Dendromonas	13	Flounder	36	*Leptomitus*	38
Dero	19	*Fragilaria*	B, E, 5	*Lestes*	25
Desmidium	E	Frogs	37	*Lethocerus*	31
Diaphanosoma	20	*Fucus*	7	*Limnochares*	35
Diaptomus	21	Fungi	38	*Limnophorus*	22
Diatoma	B, 5	*Gammarus*	21	Limpet	32
Diatoms, centric	6	Gar, long-nosed	36	*Limulus*	35
Diatoms, pennate	5	Gastropoda	32	*Lionotus*	15
Dichotomosiphon	4	Gemmule	16	*Littorina*	32
Dictyosphaerium	3	*Gerris*	31	*Lobomonas*	11
Diemictylus	37	*Gleotrichia*	2	*Lymnaea*	32
Difflugia	14	*Goera*	22	*Lyngbya*	C, F, 2
Dineutus	30	*Gomphonema*	C, F, 5	*Macromia*	25
Dinobryon	A, B, 12	*Gomphosphaeria*	A, E, 1	*Mallomonas*	A
Dinomonas	13	*Gonidea*	33	*Manayunkia*	19
Diplogasteroides	18	*Gonium*	E, 12	*Margaritifera*	33
Diptera	22, 28, 29	*Gracilaria*	7	*Marisa*	32
Dobsonflies	22, 26	*Haematococcus*	11	Marsh treader	31
Dragonflies	25	*Hagenius*	25	*Mastigamoeba*	13
Draparnaldia	F	*Hartmannella*	14	Mayflies	24
Duckweed	8	*Helicopsyche*	27	Megaloptera	22
Dytiscus	30	*Helisoma*	32	*Melosira*	B, 6
Echinarachnius	34	Hellgrammite	26	*Membranipora*	16
Echinoderms	34	*Helocordulia*	25	*Meridion*	D
Eelgrass	9	Hemiptera	31	*Merismopedia:*	
Eichhornia	8	Heptageniidae	24	see *Agmenellum*	
Eleocharis	10	*Hexagenia*	24	*Micractinium*	E
Elodea	9	*Hildenbrandia*	D	*Micrasterias*	D
Enochrus	30	Holothuroidea	34	Micrococcus	38
Enteromorpha	7	Horseshoe crab	35	*Microcoleus*	D
Entophysalis	D	*Hyalotheca*	4	*Microcystis:* see *Anacystis*	
Ephemerella	24	*Hydra*	35	*Microspora*	F
Ephemeroptera	24	*Hydrodictyon*	A	*Microthamnion*	4
Epiphanes	17	*Hydrometra*	31	Midges	28
Eristalis	29	*Hydrophilus*	30	*Moina*	20
Euastrum	E	*Hydropsyche*	27	Mollusks, bivalves	33
Euchlanis	17	*Isoperla*	23	Mollusks, snails	32
Eudorina	E	Isopoda	21	*Monas*	13
Euglena	C, E, 11	Jellyball	16	*Monhystera*	18
Euplotes	15	Jellyfish	35	*Monostyla*	17
Ferrissia	32	*Kellicottia*	17	Mosquito (larva)	29
Filina	17	Kelp	7	*Mougeotia*	E
Fish flies	26	*Keratella*	17	Mud puppy	37
Fishes	36	Lamprey	36	Mussels	33
Flagellates, nonpigmented		*Lanx*	32	*Myriophyllum*	9
(protozoa)	13	Leech	19	*Mysis*	21
Flagellates, pigmented,		*Lemanea*	D	*Mytilus*	33
colonial types	12	*Lemna*	8	Myxophyceae: see	
Flagellates, pigmented,		*Lepisosteus*	36	Cyanophyta	
single-celled	11	*Lepocinclis*	C	*Naegleria*	14

Organism	Plate	Organism	Plate	Organism	Plate
Naiad	9	Planaria	19	Sea lettuce	7
Najas	9	Plants, higher, emersed	10	Sea urchins	34
Narpus	30	Plants, higher, floating	8	Segmented worms	19
Nassa	32	Plants, higher, submersed	9	Selenastrum	3
Navicula	B, D, 5	Platydorina	12	Sewage fly	22, 29
Necturus	37	Platyhelminthes	19	Shrimp	21
Nemathelminthes	18	Plecoptera	23	Sialis	26
Nereocystis	7	Pleodorina	12	Sigara	31
Nitella	A	Pleurobema	33	Simulium	28
Nitzschia	C, 5	Pleurocera	32	Siren	37
Nodularia	E	Pleuronema	15	Skeletonema	6
Notholca	17	Plumatella	16	Snails	32
Notonecta	31	Polyarthra	17	Sowbug	21
Ochromonas	11	Polymesoda	33	Spectacle case	33
Ochrotricha	27	Polyphemus	20	Spermatophytes	8, 9, 10
Odonata	25	Pomacea	32	Sphaerium	33
Oedogonium	F, 4	Pondweed	9	Sphaerocystis	E
Oikomonas	13	Pontederia	10	Sphaerotilus	38
Oocystis	E	Potamogeton	9	Spicule	16
Ophioplus	34	Proales	17	Spike rush	10
Orchesella	35	Proptera	33	Spirillum	38
Oscillatoria	B, C, 2	Protococcus: see Phytoconis		Spirodela	8
Osteichthys	36	Protozoa	13, 14, 15	Spirogyra	B, C, 4
Ostracod	21	Psephenus	30	Sponges	16
Oyster	33	Psychoda	29	Springtail	35
Palmella	B	Pteridophytes	9	Starfishes	34
Paludicella	16	Pteromonas	11	Statoblast	16
Pandorina	A	Pteronarcys	23	Staurastrum	A, D
Paralichthys	36	Pyrobotrys	C	Stauroneis	E
Paramoecium	15	Rana	37	Stenelmis	30
Pectinatella	16	Rangia	33	Stenonema	24
Pediastrum	E, 3	Rat-tailed maggot	29	Stephanodiscus	E, 6
Pelecypoda	33	Ray	36	Stichococcus	4
Pelomyxa	14	Rhabditis	18	Stigeoclonium	C, F
Peranema	13	Rhizoclonium	D	Stingray	36
Perca	36	Rhizosolenia	6	Stoneflies	23
Perch	36	Rhodomonas	D	Streptococcus	38
Peridinium	A	Rhodophyta	7	Strongylocentrotus	34
Periwinkle	32	Rivularia	B	Stylonychia	15
Petromyzon	36	Rockweed	7	Surirella	D, 5
Phacotus	D	Rotifers	17	Sweet flag	10
Phacus	C, E, 11	Roundworms	18	Synchaeta	17
Phaeophyta	7	Salamanders	37	Synedra	A, B, 5
Philodina	17	Salientia	37	Synura	A, 12
Phormidium	C, F	Salmo	36	Syrphidae	29
Physa	32	Salvinia	8	Tabanus	29
Phytoconis	F	Sand dollars	34	Tabellaria	A, B
Pickerelweed	10	Sarcina	38	Tadpole	37
Pinnularia	D	Scenedesmus	E, 3	Tarebia	32
Pisces	36	Schizomeris	4	Tetracladium	38
Pisidium	33	Scud	21	Tetraedron	C
Pithophora	4	Sea cucumbers	34	Tetramitus	13

Organism	Plate	Organism	Plate	Organism	Plate
Tetraspora	F	*Typha*	10	Watermeal	8
Tetrastrum	3	*Ulothrix*	D, F	Water milfoil	9
Thyone	34	*Ulva*	7	Water mite	35
Tintinnidium	15	*Urnatella*	16	Water penny	30
Tipula	29	*Uroglenopsis*	A	Water strider	31
Toads	37	*Vallisneria*	9	Water velvet	8
Tolypothrix	F	*Valvata*	32	Waterweed	9
Trachelomonas	B, 11	*Vaucheria*	F	*Westella*	3
Triaenodes	22, 27	Vibrio	38	*Wolffia*	8
Tribonema	B	*Volvox*	A	Worms	18, 19
Trichocerca	17	*Vorticella*	15	*Zoogloea*	38
Trichoptera	27	Water boatman	31	*Zoophagus*	38
Trochospongilla	16	Water dog	37	*Zygnema*	E, 4
Trout	36	Water fern	8		
Tubifex	19	Water hyacinth	8		

10900 G.   Selected Taxonomic References

The most useful references for the nonspecialist are listed below. These references are primarily regional and will aid the biologist in the identification of both freshwater and marine plants and animals. Each reference is listed once under the broadest classification, i.e., a general reference to the identification of invertebrates is listed under "Invertebrates, General" and is not repeated under each individual phylum. As a rule, more academic and specialized reports on a single genus or family are not listed; however, these often are listed in the bibliography of a cited reference.

1. General, Introductory

MINER, R.W. 1950. Field Book of Seashore Life. G.P. Putnam's Sons, New York, N.Y.

DAVIS, C.C. 1955. The Marine and Fresh-Water Plankton. Michigan State Univ. Press, East Lansing.

EDMONDSON, W.T., ed. 1959. Ward and Whipple's Fresh Water Biology, 2nd ed. John Wiley & Sons, New York, N.Y.

NEEDHAM, J.G. & P.R. NEEDHAM. 1962. A Guide to the Study of Fresh-Water Biology, 5th ed. Holden-Day Inc., San Francisco, Calif.

NEWELL, G.E. & R.C. NEWELL. 1963. Marine Plankton, A Practical Guide. Hutchinson Educational Ltd., London, England.

REID, G.K. 1967. Pond Life. A Guide to Common Plants and Animals of North American Ponds and Lakes. Golden Press, New York, N.Y.

RICKETTS, E.F. & J. CALVIN. 1968. Between Pacific Tides, 4th ed. Revised by J.W. Hedgpeth. Stanford Univ. Press, Stanford, Calif.

REISH, D.J. 1983. Marine Life of Southern California. Kendall/Hunt Publ. Co., Dubuque, Iowa.

2. Algae

General

SMITH, G.M. 1950. The Fresh-Water Algae of the United States. McGraw-Hill, New York, N.Y.

TAYLOR, W.R. 1957. Marine Algae of the Northeastern Coast of North America, 2nd ed. Univ. Michigan Press, Ann Arbor.

PALMER, C.M. 1959. Algae in Water Supplies. U.S. Pub. Health Serv. Publ. No. 657, Washington, D.C.

PRESCOTT, G.W. 1968. The Algae: A Review. Houghton Mifflin Co., Boston, Mass.

WHITFORD, L.A. & G.J. SCHUMACHER. 1973. A Manual of Fresh-Water Algae. Sparks Press, Raleigh, N.C.

DAVES, C.J. 1974. Marine Algae of the West Coast

of Florida. Univ. Miami Press, Coral Gables, Fla.

ABBOTT, I.A. & G.J. HOLLENBERG. 1976. Marine Algae of California. Stanford Univ. Press, Stanford, Calif.

EDWARDS, P. 1976. Illustrated Guide to the Seaweeds and Sea Grasses in the Vicinity of Port Arkansas, Texas. Univ. Texas Press, Austin.

GEORGE, A. 1976. A guide to algal keys (excluding seaweeds). *Brit. Phycol. J.* 11:49.

LEE, T. 1977. The Seaweed Handbook: An Illustrated Guide to Seaweeds from North Carolina to the Arctic. Marine Press, Boston, Mass.

ABBOTT, I.A. & E.Y. DAWSON. 1978. How to Know the Seaweeds, 2nd ed. Wm. C. Brown Co., Dubuque, Iowa.

PRESCOTT, G.W. 1978. How to Know the Fresh Water Algae, 3rd ed. Wm. C. Brown Co., Dubuque, Iowa.

HUMM, H.J. 1979. The Marine Algae of Virginia. Univ. Virginia Press, Charlottesville.

LOBBAN, C.S. & M.J. WYNNE, eds. 1981. The Biology of Seaweeds. Univ. California Press, Berkeley.

ROUND, F.E. 1981. The Ecology of Algae. Cambridge Univ. Press, New York, N.Y.

Blue-Green Algae

HUMM, H.J. 1962. Key to the Genera of Marine Bluegreen Algae of Southeastern North America. Va. Fish. Lab. Spec. Sci. Rep. No. 28.

WELCH, H. 1964. An introduction to the bluegreen algae, with a dichotomous key to all the genera. *Limnol. Soc. S. Afr. News Letter* 1:25.

FOGG, G.E., W.D.P. STEWART, P. FAY & A.E. WASLBY. 1973. The Blue-Green Algae. Academic Press, London, England.

HUMM, H.J. & S.B. HICKS. 1980. Introduction and Guide to the Marine Blue-Green Algae. Wiley-Interscience, Somerset, England.

VANLANDINGHAM, S.L. 1982. Guide to the Identification, Environmental Requirements and Pollution Tolerance of Blue-Green Algae (Cyanophyta). EPA-600/3-82-073, U.S. Environmental Protection Agency.

Green Algae

TRANSEAU, E.N. 1951. The Zygnemataceae. Ohio State Univ. Press, Columbus.

Red Algae

WRAY, J.L. 1977. Calcareous Algae. Elsevier Science Publishing Co., Amsterdam, Netherlands.

KAPRAUN, D.F. 1980. An Illustrated Guide to the Benthic Marine Algae of Coastal North Carolina. I. Rhodophyta. Univ. North Carolina Press, Chapel Hill.

Phytoplankton and Diatoms

CUPP, E.E. 1943. Marine Plankton Diatoms of the West Coast of North America. Bull. Scripps Inst. Oceanogr. 5:1.

HUSTEDT, F. 1955. Marine littoral diatoms, Beaufort, North Carolina. Duke Univ. Mar. Sta. Bull. 6:5.

PATRICK, R. & C.W. REIMER. 1966. The Diatoms of the United States. Vol. 1. Philadelphia Acad. Natur. Sci. Monogr. No. 13, Philadelphia, Pa.

WEBER, C.I. 1971. A Guide to the Common Diatoms at Water Pollution Surveillance System Stations. U.S. Environmental Protection Agency, National Environmental Research Center, Cincinnati, Ohio.

YAMAJI, I. 1973. Illustrations of the Marine Plankton of Japan. Hoikusha Publ. Co. Ltd., Osaka.

BONEY, A.D. 1975. Phytoplankton. Studies in Biology, No. 52. Edward Arnold, London, England.

PATRICK, R. & C.W. REIMER. 1975. The Diatoms of the United States Exclusive of Alaska and Hawaii. Vol. 2, Fragilariaceae, Eunotiaceae, Acanthaceae, Naviculaceae. Philadelphia Acad. Natur. Sci. Monogr. No. 13, Philadelphia, Pa.

SOURNIA, A. 1978. Phytoplankton Manual. Monogr. on Oceanographic Methodology No. 6, United Nations Educational, Scientific & Cultural Org., Paris.

BELCHER, H. & E. SWALE. 1979. An Illustrated Guide to River Phytoplankton. Her Majesty's Stationery Off., London, England.

VINYARD, W.C. 1980. Diatoms of North America. Mad River Press, Eureka, Calif.

MARSHALL, H.G. 1986. Identification Manual for Phytoplankton of the United States Atlantic Coast. EPA-600/4-86-003, U.S. Environmental Protection Agency, Cincinnati, Ohio.

3. Fungi

General

COOKE, W.B. 1986. The Fungi of "Our Mouldy Earth." *Beiheft Nova Hedwigia* Berlin, Stuttgart 85:1.

FULLER, M.S. & A. JAWORSKI, eds. 1987. Zoosporic Fungi in Teaching and Research. Southeastern Publ. Corp., Athens, Ga.

Phycomycetes

SPARROW, F.K. 1960. Aquatic Phycomycetes, 2nd ed. Univ. Michigan Press, Ann Arbor.

DICK, M.W. 1969. Morphology and taxonomy of the Oomycetes, with special reference to Saprolegniaceae, Leptomytaceae and Pythiaceae. 1. Sexual reproduction. *New Phytol.* 68:751.

DICK, M.W. 1973. Saprolegniales. *In* G.C. Ainsworth, F.K. Sparrow & A.S. Sussman, eds. The Fungi. Vol. IV B. Academic Press, New York, N.Y.

SPARROW, F.K. 1973. Chytridiomycetes, Hyphochytridiomycetes. *In* G.C. Ainsworth, F.K. Sparrow & A.S. Sussman, eds. The Fungi. Vol. IV B. Academic Press, New York, N.Y.

KARLING, J.S. 1977. Chytridiomycetarum Iconographia. J. Cramer, Germany.

BARR, D.J.S. 1978. Taxonomy and phylogeny of Chytrids. *BioSystems* 10:153.

KARLING, J.S. 1980. The Simple Biflagellate Holocarpic Phycomycetes, 2nd ed. J. Cramer, Germany.

Ascomycetes

INGOLD, C.T. 1954. Aquatic Ascomycetes: Discomycetes from lakes. *Trans. Brit. Mycol. Soc.* 37:1.

INGOLD, C.T. 1955. Aquatic Ascomycetes: Further species from the English Lake District. *Trans. Brit. Mycol. Soc.* 38:157.

INGOLD, C.T. 1976. The morphology and biology of freshwater fungi excluding the Phycomycetes. *In* E.B. Gareth Jones, ed. Recent Advances in Aquatic Mycology. John Wiley & Sons, New York, N.Y.

Fungi Imperfecti

RANZONI, F.V. 1953. The aquatic Hyphomycetes of California. *Farlowia* 4:353.

CRANE, J.L. 1968. Freshwater Hyphomycetes of the northern Appalachian highlands including New England and three coastal plain states. *Amer. J. Bot.* 55:996.

KENDRICK, W.B., ed. 1971. Taxonomy of Fungi Imperfecti. Univ. Toronto Press, Toronto, Ont.

KENDRICK, W.B. & J.W. CARMICHAEL. 1973. Hyphomycetes. *In* G.C. Ainsworth, F.K. Sparrow & A.S. Sussman, eds. The Fungi, An Advanced Treatise. Vol. IV A. Academic Press, New York, N.Y.

INGOLD, C.T. 1975. An Illustrated Guide to Aquatic and Waterborne Hyphomycetes (Fungi Imperfecti) with Notes on their Biology. Freshwater Biol. Assoc. Sci. Publ. 30.

INGOLD, C.T. 1976. The morphology and biology of freshwater fungi excluding the Phycomycetes. *In* E.B. Gareth Jones, ed. Recent Advances in Aquatic Mycology. John Wiley & Sons, New York, N.Y.

CARMICHAEL, J.W., W.B. KENDRICK, I.L. CONNORS & L. SIGLER. 1980. Genera of Hyphomycetes. Univ. Alberta Press, Edmonton.

4. Higher Plants

MUENSCHER, W.C. 1944. Aquatic Plants of the United States. Comstock Publishing Co., Ithaca, N.Y.

FERNALD, M.L. 1950. Gray's Manual of Botany, 8th ed. D. Van Nostrand Co., New York, N.Y.

GLEASON, H.A. 1952. The New Brittain and Brown Illustrated Flora of the Northeastern United States and Canada. Lancaster Press, Inc., Lancaster, Pa.

FASSETT, N.C. 1960. A Manual of Aquatic Plants (with a revision appendix by E.C. Ogden). Univ. Wisconsin Press, Madison.

EYLES, D.E. & J.L. ROBERTSON. 1963. Guide and Key to the Aquatic Plants of the Southeastern United States. U.S. Fish Wildl. Serv. Circ. 158.

STEWARD, A.N., L.R. DENNIS & H.M. GILKEY. 1963. Aquatic Plants of the Pacific Northwest, with Vegetative Keys, 2nd ed. Oregon State Univ. Press, Corvallis.

RADFORD, A.E., H.E. AHLES & C.R. BELL. 1968. Manual of Vascular Flora of the Carolinas, Univ. North Carolina Press, Chapel Hill.

MASON, H.L. 1969. Flora of the Marshes of California. Univ. California Press, Berkeley.

WILSON, L.W. 1969. Common Aquatic Weeds. Agr. Handbook No. 352, U.S. Dep. Agriculture.

CORRELL, D.S. & H.B. CORRELL. 1972. Aquatic and Wetland Plants of Southwestern United States. 16030 DNL, U.S. Environmental Protection Agency.

MOUL, E.T. 1973. Higher Plants of the Marine Fringe. Marine Flora and Fauna of the Northeastern U.S. Circ. 384, National Oceanic Atmospheric Admin., National Marine Fisheries Serv., U.S. Government Printing Off., Washington, D.C.

HOTCHKISS, N. 1978. Common Marsh, Underwater and Floating-Leaved Plants of the United States and Canada. Dover Publications, New York, N.Y.

GODFREY, R.K. & J.W. WOOTEN. 1979. Aquatic and Wetland Plants of Southeastern United States, Monocotyledons. Univ. Georgia Press, Athens.

GODFREY, R.K. & J.W. WOOTEN. 1981. Aquatic and Wetland Species of the Southeastern United States, Dicotyledons. Univ. Georgia Press, Athens.

LEWIS, R.R., III. 1982. Creation and Restoration

of Coastal Plant Communities. CRC Press, Boca Raton, Fla.

5. Invertebrates, General

HYMAN, L.H. 1940-67. The Invertebrates. Vols. 1-6. McGraw-Hill, New York, N.Y.

GOSNER, K.L. 1971. Guide to Identification of Marine and Estuarine Invertebrates; Cape Hatteras to the Bay of Fundy. Wiley-Interscience, New York, N.Y.

WATLING, L. & D. MAURER. 1973. Guide to the Macroscopic Estuarine and Marine Invertebrates of the Delaware Bay Region. Vol. 5. College Marine Studies, Univ. Delaware, Newark.

KOZLOFF, E.N. 1974. Keys to the Marine Invertebrates of Puget Sound, the San Juan Archipelago, and Adjacent Regions. Univ. Washington Press, Seattle.

FOTHERINGHAM, N. & S. BRUNENMEISTER. 1975. Common Marine Invertebrates of the Northwestern Gulf Coast. Gulf Publishing Co., Houston, Tex.

SMITH, R.I. & J.T. CARLTON, eds. 1975. Light's Manual: Intertidal Invertebrates of the Central California Coast, 3rd ed. Univ. California Press, Berkeley.

MILNE, L. & M. MILNE. 1976. Invertebrates of North America. Doubleday & Co., New York, N.Y.

PENNAK, R.W. 1978. Freshwater Invertebrates of the United States, 2nd ed. John Wiley & Sons, New York, N.Y.

MORRIS, R.H., D.P. ABBOTT & E.C. HADERLIE. 1980. Intertidal Invertebrates of California. Stanford Univ. Press, Stanford, Calif.

HEARD, R.W. 1982. Guide to Common Tidal Marsh Invertebrates of the Northeastern Gulf of Mexico. Mississippi Alabama Sea Grant Consortium, MASGP-79.

BARNES, R.D. 1987. Invertebrate Zoology, 5th ed. Saunders College, Philadelphia, Pa.

6. Protozoa

ELLIS, B.F. & A.R. MESSINA. 1940 to date. Catalogue of Foraminifera. Spec. Publ., American Mus. Natural History, New York, N.Y.

CUSHMAN, J.A. 1948. Foraminifera, 4th ed. Harvard Univ. Press, Cambridge, Mass.

JAHN, T.L. & F.F. JAHN. 1949. How to Know the Protozoa. Wm. C. Brown Co., Dubuque, Iowa.

LOEBLICH, A.R., JR. & H. TAPPAN. 1964. Sarcodina, chiefly "Thecamoebians" and Foraminiferida. *In* R.C. Moore, ed. Treatise on Invertebrate Paleonotology, Part C, Protista 2. 2 vols. Geological Soc., New York, N.Y.

KUDO, R.R. 1966. Protozoology, 5th ed. C.C. Thomas, Springfield, Ill.

LEWIS, K.B. 1970. A key to the recent genera of the foraminiferida. New Zealand Oceanogr. Inst. Mem. No. 45.

BICK, H. 1972. Ciliated Protozoa. An Illustrated Guide to the Species Used as Bacteriological Indicators in Fresh Water Biology. World Health Org., Geneva.

WESTPHAL, A. 1976. Protozoa. Blackie, Glasgow.

CORLISS, J.O. 1979. The Cilated Protozoa: Characterization, Classification, and Guide to the Literature, 2nd ed. Pergamon Press, New York, N.Y.

7. Porifera

DeLAUBENFELS, M.W. 1932. The marine and fresh water sponges of California. *Proc. U.S. Nat. Mus.* 81, Art. 4.

PENNEY, J.T. & A.A. RACEK. 1968. Comprehensive Revision of a Worldwide Collection of Freshwater Sponges (Porifera: Spongillidae). U.S. Nat. Mus. Bull. 272.

FRY, W.C., ed. 1970. The Biology of Porifera. Symp. Zool. Soc. London No. 25. Academic Press, London.

BERGGUIST, P.R. 1978. Sponges. Hutchinson & Co., London.

8. Cnidaria

FRASER, C.M. 1937. Hydroids of the Pacific Coast of Canada and the United States. Univ. Toronto Press, Toronto, Ont.

FISHER, W.K. 1938. Hydrocorals of the North Pacific Ocean. *Proc. U.S. Nat. Mus.* 84:493.

CARLGREN, O. 1952. Actiniaria from North America. *Arkiv. Zool.* (ser. 2) 3:373.

DURHAM, J.W. & J.L. BARNARD. 1952. Stony Corals of the Eastern Pacific Collected by The Velero III and Velero IV. Allan Hancock Pacific Exped., Univ. So. Calif. 16:1.

FRASER, C.M. 1954. Hydroids of the Atlantic Coast of North America. Univ. Toronto Press, Toronto, Ont.

HAND, C. 1954-1955. The Sea Anemones of Central California. Parts 1-3. *Wasmann J. Biol.* 12:345; 13:37; 13:189.

KRAMP, P.L. 1961. Synopsis of the medusae of the world. *J. Mar. Biol. Assoc. U.K.* 40:1.

CALDER, D.R. 1971. Hydroids and Hydromedusae of Southern Chesapeake Bay. Va. Inst. Mar. Sci., Special Papers in Marine Science No. 1.

LARSON, R.S. 1976. Cnidaria: Scyphozoa. Marine Flora and Fauna of the Northeastern U.S. Circ. 397, National Oceanic Atmospheric Admin., National Marine Fisheries Serv., U.S. Government Printing Off., Washington, D.C.

CAIRNS, S.D. 1986. A Revision of the Northwest Atlantic Stylasteridae (Coelenterata: Hydrozoa). Smithsonian Contrib. Zool. No. 418.

9. Rotifera

DONNER, J. 1966. Rotifers. Warne, London & New York.

RUTTNER-KOLISKO, A. 1974. Plankton Rotifers—Biology and Taxonomy. *Die Binnengewasser* 26: Suppl. 1.

KOSTE, W. 1978. Rotaria—Die Radertiere Mitteleuropas. Borntraeger, Berlin.

STEMBERGER, R.S. 1979. A Guide to the Rotifers of the Laurentian Great Lakes. EPA-600/4-79-021, U.S. Environmental Protection Agency, Cincinnati, Ohio.

10. Platyhelminthes

HYMAN, L.H. 1953. The polyclad flatworms of the Pacific Coast of North America. *Bull. Amer. Mus. Nat. Hist.* 100:269.

RISER, N.W. & M.P. Morse, eds. 1974. Biology of the Turbellaria. McGraw-Hill Co., New York, N.Y.

11. Nemathelminthes

HOPE, W.D. & D.G. MURPHY. 1972. A taxonomic hierarchy and checklist of the genera and higher taxa of marine nematodes. Smithsonian Contrib. Zool. No. 137.

TARJAN, A.C., R.P. ESSER & S.L. CHANG. 1977. An illustrated key to the nematodes found in fresh water. *J. Water Pollut. Control Fed.* 49:2318.

12. Nemertea

CORREA, D.D. 1964. Nemerteans from California and Oregon. *Proc. Calif. Acad. Sci.* 31:19.

13. Annelida

General

KLEMM, D.J., ed. 1985. A Guide to the Freshwater Annelida (Polychaeta, Naidid and Tubificid Oligochaeta, and Hirudinea of North America). Kendall/Hunt Publ. Co., Dubuque, Iowa.

Polychaeta

HARTMAN, O. 1968. Atlas of the Errantiate Polychaetous Annelids from California. Allan Hancock Foundation, Univ. Southern California, Los Angeles.

HARTMAN, O. 1969. Atlas of the Sedentariate Polychaetous Annelids from California. Allan

Hancock Foundation, Univ. Southern California, Los Angeles.

DAY, J.H. 1973. New Polychaeta from North Carolina, with a Key to All Species Recorded from North Carolina. Circ. 375, National Oceanic Atmospheric Admin., National Marine Fisheries Serv., Seattle, Wash.

BANSE, K. & K.D. HOBSON. 1974. Benthic Errantiate Polychaetes of British Columbia and Washington. Fish. Res. Bd. Can. Bull. No. 185.

GARDINER, S.L. 1975. Errant polychaete annelids from North Carolina. *J. Elisha Mitchell Sci. Soc.* 91:77.

FAUCHALD, K. 1977. The Polychaete Worms; Definitions and Keys to the Order, Families, and Genera. Los Angeles Co. Mus. Natur. Hist. Sci. Ser. 28, Los Angeles, Calif.

HOBSON, K.D. & K. BANSE. 1981. Sedentariate and Archiannelid Polychaetes of British Columbia and Washington. Can. Fish. Aquatic Sci. Bull. No. 209.

PETTIBONE, M.H. Marine Polychaete Worms of the New England Region. 1. Aphroditae through Trochochaetidae. U.S. Nat. Mus. Bull. 227.

UEBELACKER, J.M. & B.G. JOHNSON. 1984. Taxonomic Guide to the Polychaetous Annelids of the Northern Gulf of Mexico. Barry A. Vittor & Assoc., Mobile, Ala.

KLEMM, D.J. 1985. Freshwater Polychaeta (Annelida). *In* D.J. Klemm, ed. A Guide to the Freshwater Annelida (Polychaeta, Naidid and Tubificid Oligochaeta, and Hirudinea of North America). Kendall/Hunt Publ. Co., Dubuque, Iowa.

Oligochaeta and Hirudinea

MANN, K.H. 1961. Leeches (Hirudinea), Their Structure, Physiology, Ecology and Embryology. Pergamon Press, Oxford.

BRINKHURST, R.O. & B.G.M. JAMIESON. 1971. Aquatic Oligochaeta of the World. Oliver & Boyd, Edinburgh.

SAWYER, R.T. 1972. North American Freshwater Leeches, Exclusive of the Piscicolidae, with Key to All Species. Ill. Natur. Hist. Monogr. No. 46.

EDWARDS, C.A. & J.K. LOFTY. 1977. Biology of Earthworms, 2nd ed. Chapman and Hall, London.

HILTUNEN, J.K. & D.J. KLEMM. 1980. A Guide to the Naididae (Annelida: Clitellata: Oligochaeta) of North America. EPA-600/4-80-031, U.S. Environmental Protection Agency, Cincinnati, Ohio.

BRINKHURST, R.O. 1982. British and other Marine and Estuarine Oligochaetes. Linn. Soc. Synopsis Brit. Fauna, New Ser. No. 21.

KLEMM, D. J. 1982. Leeches (Annelida: Hirudinea) of North America. EPA-600/3-82-025, U.S. Environmental Protection Agency, Cincinnati, Ohio.

STIMPSON, K.S., D.J. KLEMM & J.K. HILTUNEN. 1982. A Guide to the Freshwater Tubificidae (Annelida: Clitellata: Oligochaeta) of North America. EPA-600/3-82-033, U.S. Environmental Protection Agency, Cincinnati, Ohio.

BRINKHURST, R.O. 1986. Guide to the Freshwater Aquatic Microdrile Oligochaeta of North America. Can. Fish. Aquatic Sci. Spec. Bull. No. 87.

SAWYER, R.T. 1986. Leech Biology and Behavior. II. Feeding Biology, Ecology, and Systematics, Oxford Univ. Press, New York, N.Y.

14. Sipuncula and Echiura

STEPHEN A.C. & S.J. EDMONDS. 1982. The Phyla Sipuncula and Echiura. British Mus. Natural History, London.

15. Crustaceans

General

KAESTNER, A. 1970. Invertebrate Zoology. Vol. 3. Crustacea. Wiley-Interscience, New York, N.Y.

Branchiopoda

LINDER, F. 1952. Contributions to the morphology and taxonomy of the Branchiopoda Notostraca, with special reference to the North American species. Proc. U.S. Nat. Mus. 102:1.

LONGHURST, A.R. 1955. A review of the Notostraca. Bull. Brit. Mus. (Natur. Hist.) Zool. 3:1.

BROOKS, J.L. 1957. The systematics of North American Daphnia. Mem. Conn. Acad. Sci. 13.

GOULDEN, C.E. 1968. The Systematics and Evolution of the Moinidae. Trans. Amer. Phil. Soc. N.S. 58:6.

Ostracoda

KORNIKER, L.S. 1981. Revision, Distribution, Ecology, and Ontogeny of the Ostracoda Subfamily Cyclasteropinae (Myodocopina: Cylindroleberidae). Smithsonian Contrib. Zool. No. 319.

KORNIKER, L.S. 1986. Sarsiellidae of the Western Atlantic and Northern Gulf of Mexico, and Revision of the Sarsiellinae (Ostracoda: My-

odocopina). Smithsonian Contrib. Zool. No. 415.

KORNIKER, L.S. 1986. Cylindroleberididae of the Western North Atlantic, Northern Gulf of Mexico, and Zoogeography of the Myodocopina (Ostracoda). Smithsonian Contrib. Zool. No. 425.

Copepoda

COULL, B.C. 1977. Copepoda: Harpactioidea. Marine Flora and Fauna of the Northeastern U.S. Circ. 399, National Oceanic Atmospheric Admin., National Marine Fisheries Serv., U.S. Government Printing Off., Washington, D.C.

DAWSON, J.K. & G. KNATZ, 1980. Illustrated key to the planktonic copepods of San Pedro Bay, California. Tech. Rep. Allan Hancock Foundation 2:1.

Cirripedia

CORNWALL, I.E. 1955. The Barnacles of British Columbia. B.C. Prov. Mus. Handbook No. 7.

NEWMAN, W.A. 1976. Revision of the Balanomorph barnacles; including a catalog of the species. San Diego Soc. Natur. Hist. Mem. 9:1.

Cumacea

LIE, U.M. 1969. Cumacea from Puget Sound and off the NW coast of Washington, with descriptions of two new species. Crustaceana 17:19.

JONES, N.S. 1976. British Cumaceans. Synopsis of the British Fauna (New Series) No. 7, Academic Press. London.

WATLING, L. 1979. Crustacea: Cumacea. Marine Flora and Fauna of the Northeastern U.S. Circ. 423, National Oceanic and Atmospheric Admin., National Marine Fisheries Serv., U.S. Government Printing Off., Washington, D.C.

Mysidacea

BANNER, A.H. 1948-1950. A taxonomic study of the Mysidacea and Euphausicea (Crustacea) of the Northeastern Pacific. Parts 1-3. Trans. Roy. Can. Inst. 26:345; 27:65; 28:1.

TATTERSALL, W.M. 1951. A Review of the Mysidacea of the U.S. National Museum. U.S. Nat. Mus. Bull. 201.

STUCK, K.C., H.M. PERRY & R.W. HEARD. 1979. An annotated key to the Mysidacea of the north central Gulf of Mexico. Gulf Res. Rep. 6:225.

KATHMAN, R.D., W.C. AUSTIN, J.C. SALTMAN & J.D. FULTON. 1986. Identification Manual to the Mysidacea and Euphausiacea of the

Northeast Pacific. Can. Fish Aquatic Sci. Special Publ. No. 93.

Tanaidacea and Isopoda

VAN NAME, W.C. 1936. The American land and freshwater isopod crustaceans. *Bull. Amer. Mus.* 71:1.

SCHULTZ, G.A. 1969. How to Know the Marine Isopod Crustaceans. W.C. Brown, Dubuque, Iowa.

WILLIAMS, W.D. 1972. Freshwater Isopods (Asellidae) of North America. Biota of Freshwater Ecosystems. Ident. Manual No. 7, U.S. Environmental Protection Agency, U.S. Government Printing Off., Washington, D.C.

Amphipoda

BOUSFIELD, E.L. 1958. Freshwater amphipod crustaceans of glaciated North America. *Can. Field Natur.* 72:55.

MCCAIN, J.C. 1968. The Caprellidae (Crustacea: Amphipoda) of the Western North America. U.S. Nat. Mus. Bull. 278.

BARNARD, J.L. 1969. The Families and Genera of Marine Gammaridean Amphipods. U.S. Nat. Mus. Bull. 271.

HOLSINGER, J.R. 1972. The Fresh Water Amphipod Crustaceans (Gammaridae) of North America. Biota of Fresh Water Ecosystems. Ident. Manual No. 5, U.S. Environmental Protection Agency, U.S. Government Printing Off., Washington, D.C.

LAUBITZ, D.R. 1972. The Caprellida (Crustacea, Amphipoda) of Atlantic and Arctic Canada. Publ. Biol. Oceanogr. Nat. Mus., Canada, No. 4.

BOUSFIELD, E.L. 1973. Shallow-Water Gammaridean Amhipoda of New England. Cornell Univ. Press, Ithaca, N.Y.

FOX, R.S. & K.H. BYNUM. 1975. The amphipod crustaceans of North Carolina estuarine waters. *Chesapeake Sci.* 16:223.

LINCOLN, R.J. 1979. British Marine Amphipoda: Gammaridea. British Museum (Natural History), London.

BRUSCA, G.J. 1981. Annotated keys to the Hyperiidea (Crustacea: Amphipoda) of North American coastal waters. *Tech. Rep. Allan Hancock Foundation* 5:1.

BARNARD, J.L. & C.M. BARNARD. 1983. Freshwater Amphipoda of the World. Hayfield Assoc., Mt. Vernon, Va.

Decapoda

GARTH, J.S. 1958. Brachyura of the Pacific coast of America. Oxyrhyncha. *Allan Hancock Pac. Exped.* 21:1.

WILLIAMS, A.B. 1965. Marine decapod crustaceans of the Carolinas. *Fish. Bull.* 65:1.

GARTH, J.S. & W. STEPHENSON. 1966. Brachyura of the Pacific Coast of America. Brachyrhyncha: Portunidae. Allan Hancock Monogr. Mar. Biol. 1.

MANNING, R.B. 1969. Stomatopod crustacea of the western Atlantic. *Stud. Trop. Oceanogr.* 8:1.

PEREZ, F.E. 1969. Western Atlantic shrimps of the genus *Penaeus. Fish. Bull.* 67:1.

HOBBS, H.H., JR. 1972. Crayfish (Astacidae) of North and Middle America. Biota of Freshwater Ecosystems. Ident. Manual No. 9, U.S. Environmental Protection Agency, U.S. Government Printing Off., Washington, D.C.

MANNING, R.B. 1972. Stomatopod crustacea. Eastern Pacific expeditions of the New York Zoological Society. *Zoologica* 56:95.

HOBBS, H.H., JR. 1974. A Check-List of the North and Middle American Crayfishes (Decapoda:Astacidae and Cambaridea). Smithsonian Contrib. Zool. No. 166.

WILLIAMS, A.B. 1974. Crustacea: Decapoda. Marine Flora and Fauna of Northeastern U.S. Circ. 389, National Oceanic Atmospheric Admin., National Marine Fisheries Serv., U.S. Government Printing Off., Washington, D.C.

BUTLER, T.H. 1980. Shrimps of the Pacific Coast of Canada. Can. Fish Aquatic Sci. Bull. No. 202.

HOBBS, H.H., JR. 1981. The crayfishes of Florida. Smithsonian Contrib. Zool. No. 318.

HART, J.F.L. 1982. Crabs and their Relatives of British Columbia. British Columbia Prov. Mus. Handbook No. 40, Victoria.

HART, C.W., JR. & J. CLARK. 1987. An Interdisciplinary Bibliography of Freshwater Crayfishes (Astacoidea and Parastacoidea) from Aristotle through 1985. Smithsonian Contrib. Zool. No. 455.

16. Insects

General and Introductory

USINGER, R.L., ed. 1956. Aquatic Insects of California. With Keys to North American Genera and California Species. Univ. California Press, Berkeley.

BORROR, D.J. & R.E. WHITE. 1970. A Field Guide to the Insects of America North of Mexico. Peterson Field Guide Ser., Houghton Mifflin Co., Boston, Mass.

BORROR, D.J., D.M. DeLONG & C.A. TRIPLEHORN. 1976. An Introduction to the Study of Insects, 4th ed. Holt, Rinehart & Winston, New York, N.Y.

BLAND, R.G. & H.E. JAQUES. 1978. How to

ALGAE COLOR PLATES

A through F

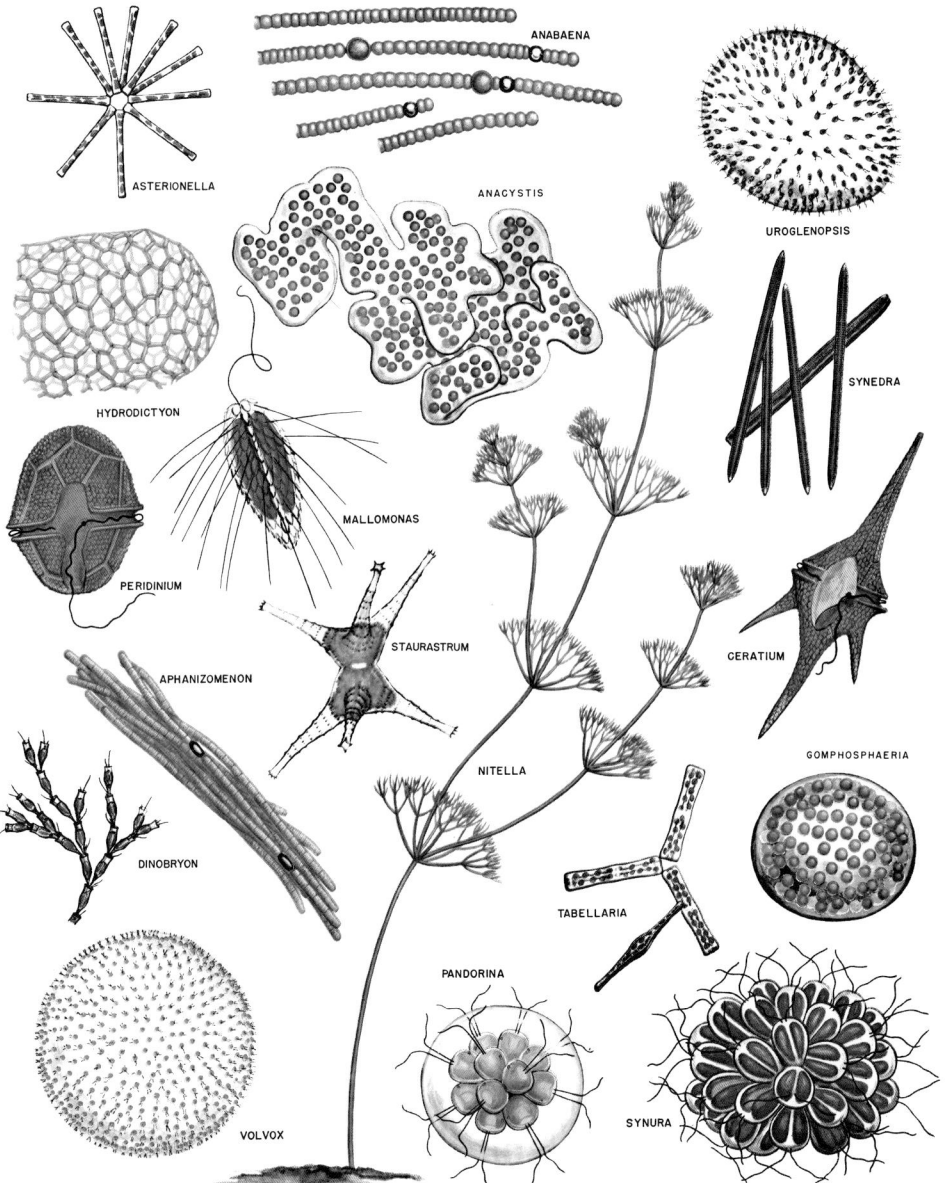

Plate A. Taste and odor algae

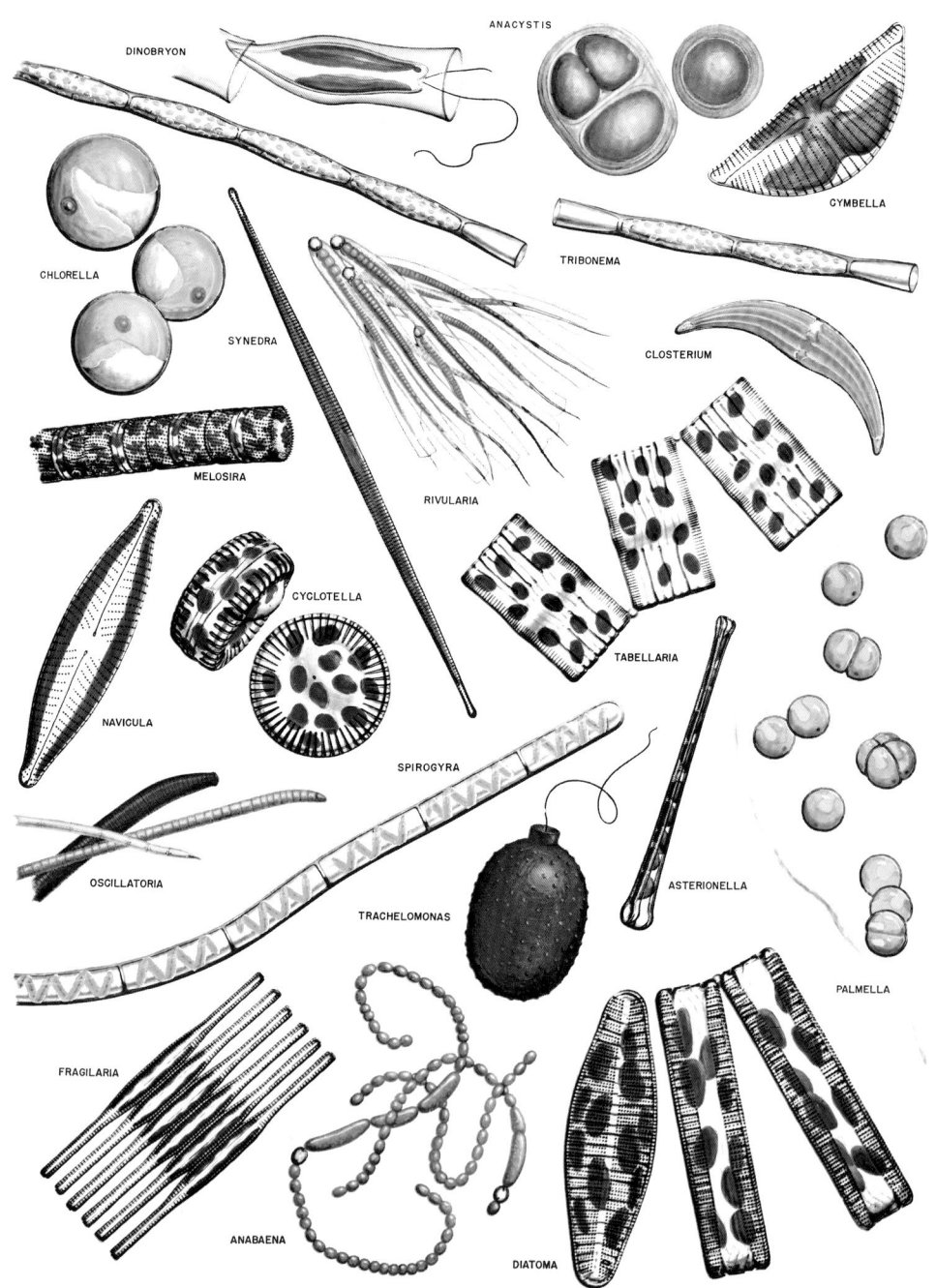

Plate B. Filter clogging algae

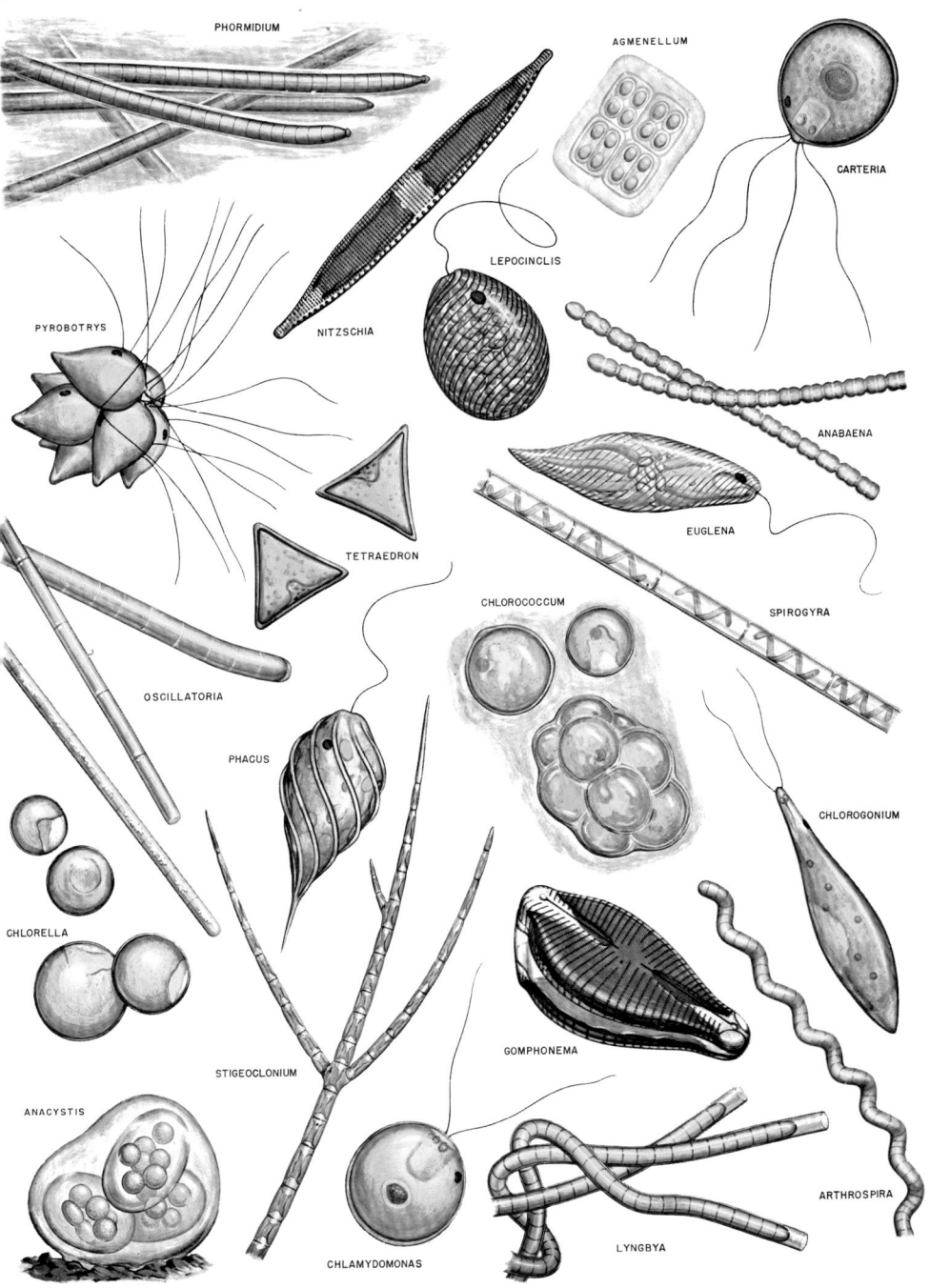

Plate C. Polluted water algae

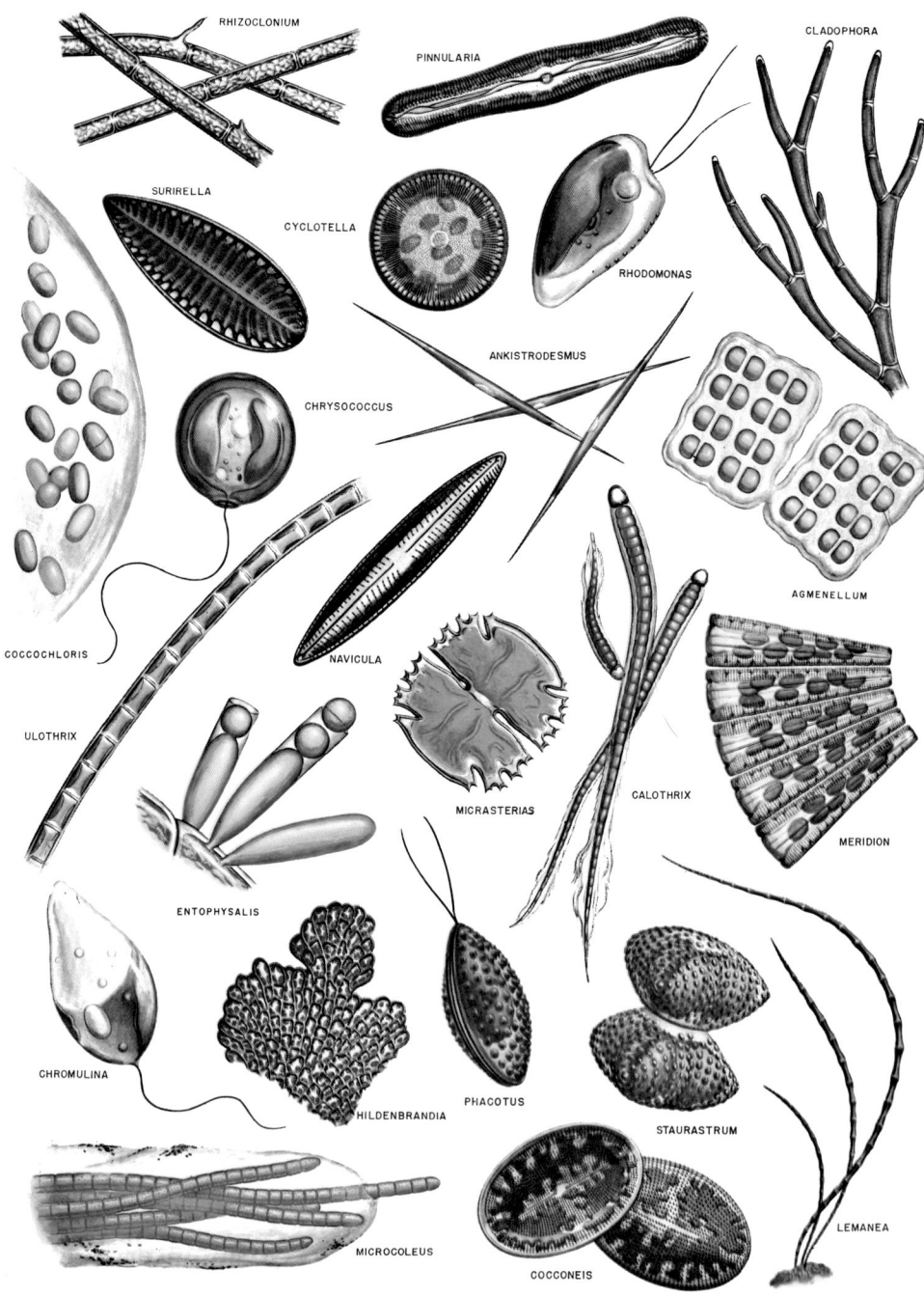

Plate D. Clean water algae

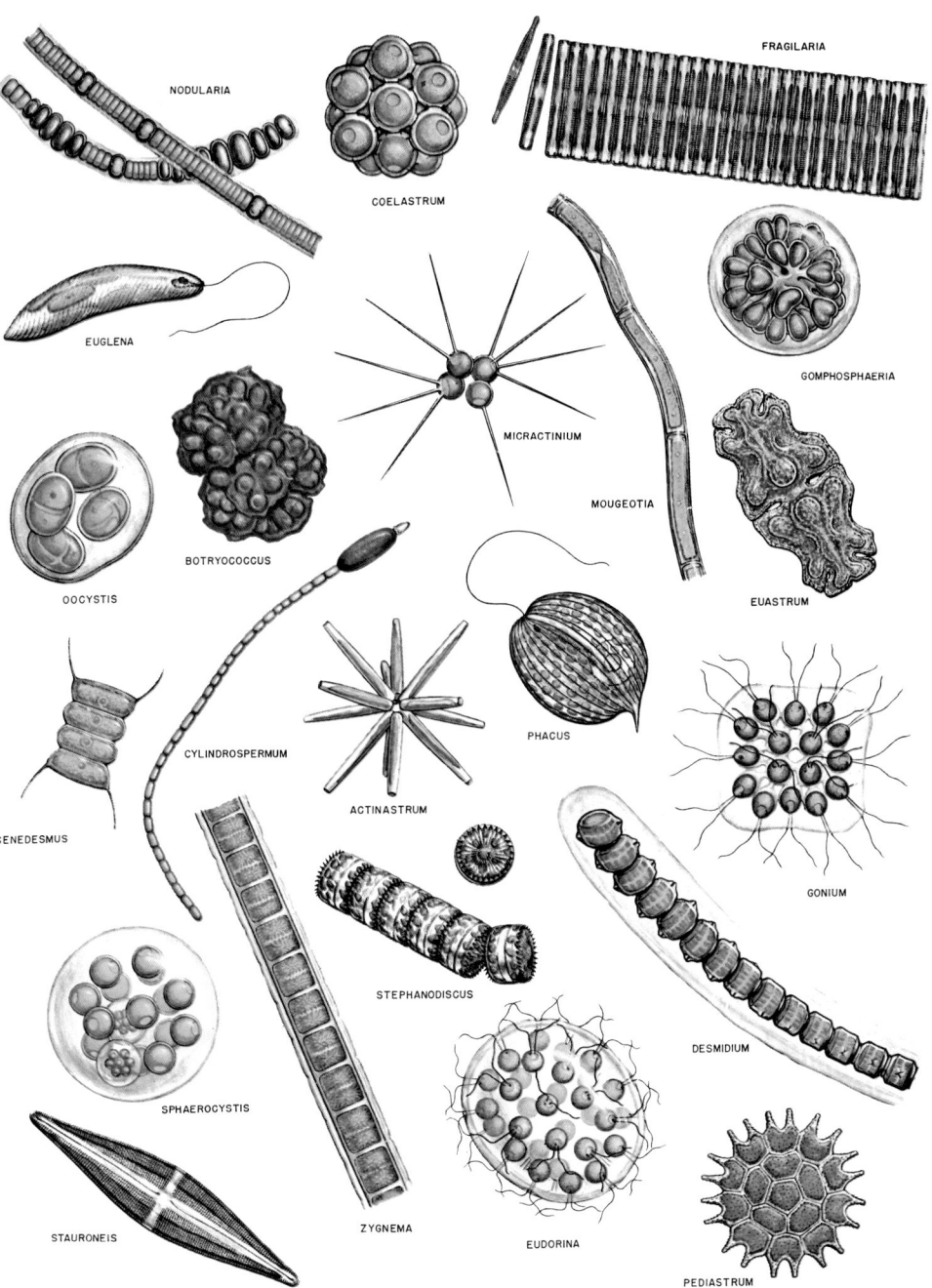

Plate E. Plankton and other surface water algae

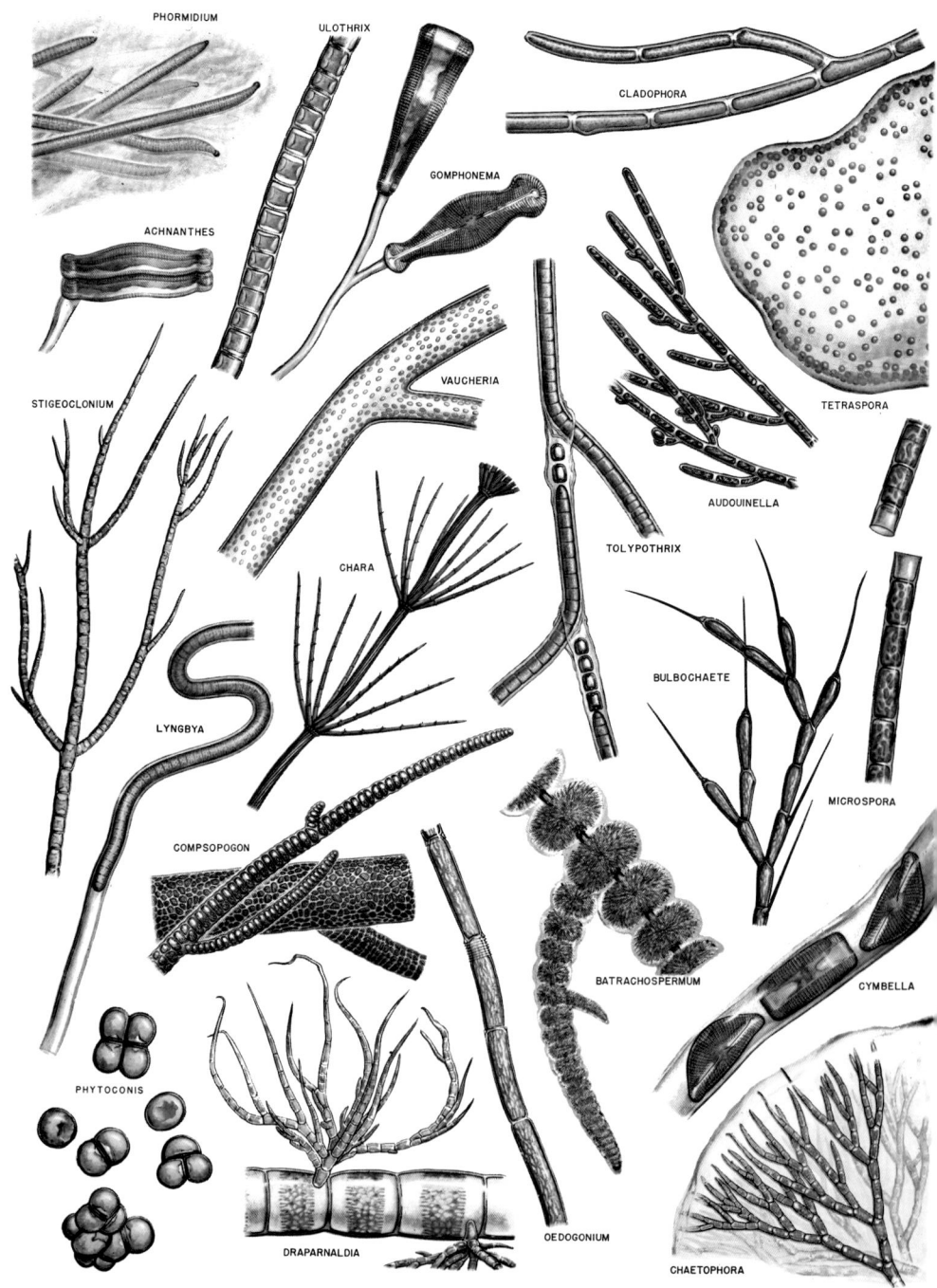

Plate F. Algae growing on reservoir walls

Know the Insects, 3rd ed. Wm. C. Brown Co., Dubuque, Iowa.

CHU, H.F. 1979. How to Know the Immature Insects. Wm. C. Brown Co., Dubuque, Iowa.

LEHMKUHL, D.M. 1979. How to Know the Aquatic Insects. Wm. C. Brown Co., Dubuque, Iowa.

MCCAFFERTY, W.P. 1981. Aquatic Entomology. The Fishermen's and Ecologists' Illustrated Guide to Insects and Their Relatives. Science Books International, Boston, Mass.

BRIGHAM, A.R., W.U. BRIGHAM & A. GNILKA, eds. 1982. Aquatic Insects and Oligochaetes of North and South Carolina. Midwest Aquatic Enterprises, Mahomet, Ill.

MERRITT, R.W. & K.W. CUMMINS, eds. 1983. An Introduction to the Aquatic Insects of North America, 2nd ed. Kendall/Hunt Publ. Co., Dubuque, Iowa.

Mayflies (Ephemeroptera)

BURKS, B.D. 1953. The mayflies, or Ephemeroptera, of Illinois. *Bull. Ill. Natur. Hist. Surv.* 26:1.

LEONARD, J.W. & F.A. LEONARD. 1962. Mayflies of Michigan Trout Streams. Cranbrook Inst. Science, Mich.

EDMUNDS, G.F., R.K. ALLEN & W.L. PETERS. 1963. An annotated key to the nymphs of the families and subfamilies of mayflies (Ephemeroptera). *Univ. Utah Biol. Ser.* 13:1.

EDMUNDS, G.F., JR., S.L. JENSEN & L. BERNER. 1976. The Mayflies of North and Central America. Univ. Minnesota Press, Minneapolis.

BEDNARIK, A.F. & W.P. MCCAFFERTY. 1979. Biosystemic Revision of the genus *Stenomenia* (Ephemeroptera: Heptageniidae). Can. Fish Aquatic Sci. Bull. No. 201.

Dragonflies and Damselflies (Odonata)

NEEDHAM, J.G. & M.J. WESTFALL, JR. 1955. A Manual of the Dragonflies of North America, Including the Greater Antilles, and the Provinces of the Mexican Border. Univ. California Press, Berkeley.

WALKER, E.M. & P.S. CORBET. 1975. The Odonata of Canada and Alaska, Vol. III. Univ. Toronto Press, Toronto, Ont.

WALKER, E.M. 1983. The Odonata of Canada and Alaska, Vols. I and II. Univ. Toronto Press, Toronto, Ont.

Stoneflies (Plecoptera)

FRISON, T.H. 1935. The stoneflies, or Plecoptera, of Illinois. *Bull. Ill. Natur. Hist. Surv.* 20:281.

FRISON, T.H. 1942. Studies of North American Plecoptera, with special reference to the fauna of Illinois. *Bull. Ill. Natur. Hist. Surv.* 22:235.

BERNER, L. 1950. The Mayflies of Florida. Univ. Florida Press, Gainesville.

JEWETT, S.G. 1960. The stoneflies (Plecoptera) of California. *Bull. Calif. Insect Surv.* 6:125.

HITCHCOCK, S.W. 1974. Guide to the Insects of Connecticut. Part VII. The Plecoptera or Stoneflies of Connecticut. State Geol. Natur. Hist. Surv. Conn. Bull. 107.

Megaloptera and Neuroptera

ROSS, H.H. 1937. Studies of nearctic aquatic insects. I. Nearctic alder flies of the genus *Sialis. Bull. Ill. Natur. Hist. Surv.* 21:57.

PARFIN, S.I. & A.B. GURNEY. 1956. The Spongilla-flies, with special reference to those of the Western Hemisphere (Sisyridae, Neuroptera). *Proc. U.S. Nat. Mus.* 105:421.

Caddisflies (Trichoptera)

ROSS, H.H. 1944. The caddisflies, or Trichoptera of Illinois. *Bull. Ill. Natur. Hist. Surv.* 23:1.

WIGGINS, G.B. 1977. Larvae of the North American Caddisfly Genera (Trichoptera). Univ. Toronto Press, Toronto, Ont.

SCHUSTER, G.A. & D.A. ETNIER. 1978. A Manual for the Identification of the Larvae of the Caddisfly genera *Hydropsyche* Pictet and *Symphitopsyche* Ulmer in Eastern and Central North America (Trichoptera: Hydropsychidae). EPA-600/4-78-060, U.S. Environmental Protection Agency, Cincinnati, Ohio.

Diptera

JOHANNSEN, O.A. 1934, 1935, 1937. Aquatic Diptera. Parts I-IV (Pt. V by L.C. Thomsen). Memoirs Cornell Univ. Agr. Exp. Sta. Reproduced in 1969 by Entomological Reprint Specialists, Los Angeles, Calif.

MASON, W.T., JR. 1968. An Introduction to the Identification of Chironomid Larvae. Div. Pollution Surveillance, Federal Water Pollution Control Admin., Cincinnati, Ohio.

BRYCE, D. & A. HOBART. 1972. The biology and identification of the larvae of the Chironomidae (Diptera). *Entomol. Gaz.* 23:175.

SIMPSON, K.W. & R.W. BODE. 1980. Common Larvae of Chironomidae (Diptera) from New York State Streams and Rivers with Particular Reference to the Fauna of Artificial Substrates. Bull. No. 439, New York State Mus., Albany.

Manual of Nearctic Diptera, Volume 1. 1981. Monogr. No. 27, Biosystems Research Inst., Ottawa, Ont.

OLIVER, D.R. 1981. Chironomidae. *In* Manual of

Nearctic Diptera, Vol. 1. Monogr. No. 27, Biosystems Research Inst., Ottawa, Ont.

PETERSON, B.V. 1981. Simuliidae. *In* Manual of Nearctic Diptera, Vol. 1. Monogr. No. 27, Biosystems Research Inst., Ottawa, Ont.

TESKEY, H.J. 1981. Key to Families—Larvae. *In* Manual of Nearctic Diptera, Vol. 1. Monogr. No. 27, Biosystems Research Inst., Ottawa, Ont.

CRANSTON, P.S. 1982. A Key to the Larvae of the British Orthocladiinae (Chironomidae). Freshwater Biol. Assoc. Sci. Publ. No. 45.

BODE, R.W. 1983. Larvae of North American *Eukiefferiella* and *Tvetenia* (Diptera: Chironomidae). Bull. N.Y. State Mus. No. 452, Albany.

Beetles (Coleoptera)

JAQUES, H.E. 1951. How to Know the Beetles. Wm. C. Brown Co., Dubuque, Iowa.

DILLON, E.S. & L.S. DILLON. 1961. Manual of Common Beetles of Eastern North America. Harper and Row, New York, N.Y.

BROWN, H.D. 1972. Aquatic Dryopoid Beetles (Coleoptera) of the United States. Biota of Freshwater Systems. Ident. Manual No. 6, U.S. Environmental Protection Agency, U.S. Government Printing Off., Washington, D.C.

ARNETT, R.H. 1973. The Beetles of the United States. American Entomological Inst., Ann Arbor, Mich.

Hemiptera

HUNGERFORD, H.B. 1919. The biology and ecology of aquatic and semi-aquatic Hemiptera. *Univ. Kans. Sci. Bull.* 9:1.

HUNGERFORD, H.B. 1948. The Corixidae of the Western Hemisphere (Hemiptera). *Univ. Kans. Sci. Bull.* 32:1.

17. Mollusks

General

MCLEAN, J.H. 1969. Marine Shells of Southern California. Los Angeles Co. Mus. Natur. Hist., Sci. Ser. 24, Zool. No. 11.

ANDREWS, J. 1971. Sea Shells of the Texas Coast. Univ. Texas Press, Austin.

ABBOTT, R.T. 1974. American Seashells, 2nd ed. Van Nostrand Reinhold Co., New York, N.Y.

KEEN, A.M. & E. COAN. 1974. Marine Mollusca Genera of Western North America: With Illustrated Key. Stanford Univ. Press, Stanford, Calif.

EMERSON, W.K. & M.K. JACOBSON. 1976. The American Museum of Natural History Guide to Shells—Land, Freshwater, and Marine

from Nova Scotia to Florida. Alfred A. Knopf, New York, N.Y.

CLARKE, A.H. 1981. The Freshwater Molluscs of Canada. National Mus. Natural Science/National Mus. Canada.

Snails and Related Gastropods

WALTER, H.J. & J.B. BURCH. 1957. Key to the Genera of Fresh Water Gastropods (Snails and Limpets) Occurring in Michigan. Mus. Zool. Univ. Mich. Circ. No. 3, Ann Arbor.

RUSSELL, H.D. 1971. Index Nudibranchia. A Catalogue of the Literature 1554-1965. Delaware Mus. Natur. Hist., Dover.

BURCH, J.B. 1962. How to Know Eastern Land Snails. Wm. C. Brown Co., Dubuque, Iowa.

BURCH, J.B. 1982. Freshwater Snails (Mollusca: Gastropoda) of North America. EPA-600/3-82-026, U.S. Environmental Protection Agency, Cincinnati, Ohio.

Bivalves (Pelecypoda)

CLARKE, A.H., JR. & C.O. BERG. 1959. The freshwater mussels of central New York with an illustrated key to the species of north-eastern North America. *Cornell Univ. Agr. Exp. Sta. Mem.* 267:1.

HEARD, W.H. & J. BURCH. 1966. Keys to the Genera of Freshwater Pelecypods of Michigan. Mus. Zool. Univ. Mich. Circ. No. 4, Ann Arbor.

PARMALEE, P.W. 1967. The Fresh-Water Mussels of Illinois. Popular Sci. Ser. Vol. VIII, Illinois State Museum, Springfield.

BURCH, J.B. 1975. Freshwater Sphaeriacean Clams (Mollusca: Pelecypoda) of North America. Malacological Publications, Hamburg, Mich.

BURCH, J.B. 1975. Freshwater Unionacean Clams (Mollusca: Pelecypoda) of North America. Malacological Publications, Hamburg, Mich.

BUCHANAN, A.C. 1980. Mussels (Naiades) of the Meramec River Basin, Missouri. Aquatic Ser. No. 17, Missouri Dep. Conservation, Columbia.

MACKIE, G.L., D.S. WHITE & T.W. ZDEBA. 1980. A Guide to Freshwater Mollusks of the Laurentian Great Lakes with special emphasis on the Genus *Pisidium*. EPA-600/3-80-068, U.S. Environmental Protection Agency Research Lab., Duluth, Minn.

VOKES, H.E. 1981. Genera of the Bivalvia: A Systematic and Bibliographic Catalogue. Paleontological Research Inst., Ithaca, N.Y.

18. Bryozoa

OSBORN, R.C. 1950-1953. Bryozoa of the Pacific Coast of North America. Parts 1-3. *Allan Hancock Foundation Pac. Exped.* 14:1.

MATURO, F.J.S. 1957. A study of the Bryozoa of Beaufort, North Carolina and vicinity. *J. Elisha Mitchell Sci. Soc.* 73:11.

RYLAND, J.S. 1960. Bryozoans. Hutchinson Univ. Library, London.

SOULE, J.D. & D.F. SOULE. 1969. Systematics and biogeography of burrowing bryozoans. *Inst. Zool.* 9:791.

WOOLLACOTT, R.M. & W.J. NORTH. 1971. Bryozoans of California and northern Mexico kelp beds. *Nova Hedwigia,* 32:455.

19. Echinoderms

FISHER, W.K. 1911-1930. Asteroidea of the North Pacific and adjacent waters. U.S. Nat. Mus. Bull. 76.

MORTENSEN, T. 1928-1951. Monograph of the Echinoidea. 5 vols. Reitzel, Copenhagen.

CLARK, A.M. 1962. Starfishes and Their Relations. British Mus. (Natural History), London.

PAWSON, D.L. & H.B. FELL. 1965. A revised classification of the dendrochirate holothurians. *Breviora* 214:1.

GRAY, I.E., M.E. DOWNEY & M.J. CERAME-VIVAS. 1968. Seastars of North Carolina. *Fish. Bull.* 67:127.

KYTE, M.A. 1969. A synopsis and key to the recent Ophiuroidea of Washington state and southern British Columbia. *J. Fish Res. Board Can.* 26:1727.

NICHOLS, D. 1969. Echinoderms, 4th ed. Hutchinson Univ., London.

AUSTIN, W.C. & M.P. HAYLOCK. 1973. British Columbia Marine Faunistic Survey Report: Ophiuroidea from the Northeast Pacific. Fish. Res. Board Can. Tech. Rep. 426.

DOWNEY, M.E. 1973. Starfishes from the Caribbean and Gulf of Mexico. Smithsonian Contrib. Zool. No. 126.

PAWSON, D.L. 1977. Echinodermata: Holothuroidea. Marine Flora and Fauna of the Northeastern U.S. Circ. No. 405, National Oceanic Atmospheric Admin., National Marine Fisheries Serv., Washington, D.C.

LAMBERT P. 1981. The Sea Stars of British Columbia. British Columbia Prov. Mus. Handbook No. 39, Victoria.

20. Urochordata (Tunicates)

VAN NAME, W.G. 1945. The north and south American ascidians. *Bull. Amer. Mus. Natur. Hist.* 84:1.

BERRILL, N.J. 1950. The Tunicata. Royal Soc., London.

MILLAR, R.H. 1970. British Ascidians. Academic Press, London.

MILLAR, R.H. 1971. The biology of ascidians. *Advan. Mar. Biol.* 9:1.

21. Fishes

See Section 10600D for references to the identification of fishes.

22. Amphibians

STEBBINS, R.C. 1966. A Field Guide to Western Reptiles and Amphibians. Houghton Mifflin Co., Boston, Mass.

COCHRAN, D.M. & C.J. GOIN. 1970. The New Field Book of Reptiles and Amphibians. Putnam Nature Field Book. G.P. Putnam's Sons, New York, N.Y.

CONANT, R. 1975. Field Guide to the Reptiles and Amphibians of Eastern and Central North America, 2nd ed. Houghton Mifflin Co., Boston, Mass.

INDEX

Page numbers followed by "i" and "t" denote references to illustrations and tables, respectively.

A

Abbreviations, *inside back cover*
Absorbent pads
  for coliforms 9–85
Absorbent solutions
  for nitrogen (ammonia) 4–116
Absorber assembly
  for arsenic 3–75i
Absorption cell
  for mercury 3–29
Absorption system
  for chlorine dioxide 4–76, 76–i
Absorption tube
  for arsenic 3–75
Absorptivity
  calculation 1–48
Abundance criteria
  BFB 6–30t
Acartia tonsa, *see* Copepods
Acceptance limits
  duplicate and known-additions analyses
    1–8t
Accuracy 1–14i
  AA methods 3–18t
  definition 1–4
  ion chromatographic method 4–6t
Acenaphthene
  liquid-liquid extraction chromatographic
    method 6–148
  liquid-liquid extraction GC/MS method
    6–113
Acenaphthylene
  liquid-liquid extraction chromatographic
    method 6–148
  liquid-liquid extraction GC/MS method
    6–113
Acetamine broth
  for *Pseudomonas aeruginosa* 9–53
Acetate buffer solution
  for aluminum 3–65
  for bromide 4–12
  for chlorine 4–52
  for cyanides 4–38
  for fluoride 4–92

Acetate *(cont.)*
  for iron 3–103
  for metals 3–27
  for strontium (radioactive) 7–49
  for sulfate 4–207
  for zinc 3–159
Acetic acid
  for cyanide 4–38
  for organic halogen 5–29
  for volatile acids 5–72
Acetone alcohol
  for coliforms 9–73
Acetone solution
  for chlorophyll 10–32
Acetylene
  for metals 3–20
Acid extractables
  characteristic masses 6–118t
  chromatographic conditions 6–118t
  detection limits 6–118t
  gas chromatogram 6–124i
  liquid-liquid extraction GC/MS method
    6–113
    bias and precision 6–134t
  methods 6–112
  quality control acceptance criteria 6–132t
  standards 6–120, 6–121t
Acid fuchsin stain
  for periphyton 10–52
  for phytoplankton 10–28
Acid hydrolysis
  for phosphorus 4–170
Acidification
  metals samples 3–2, 3–6
Acidity
  definition 2–30
  methods 2–30
  titration method 2–30
Acids
  organic
    chromatographic separation method
      5–70
    methods 5–69
  preparation *inside front cover*, 1–44
  safety precautions 1–63

Acids *(cont.)*
  volatile
    distillation method 5–72
    methods 5–69
  volatile fatty 5–69
Acid treatment reagent
  for Legionellaceae 9–151
Acid zirconyl SPADNS reagent
  for fluoride 4–90
Acoustic methods
  for fish sampling 10–122
Acrylic solution
  for radioactivity, gross alpha and beta
    7–18
  for radium 7–31
Actinomycetes
  colony type 9–129i
  methods 9–126
  plate count 9–128
  samples 9–127
Action tendency scale
  for flavor rating assessment 2–28
Activated carbon
  for odor determinations 2–19
  for organic halogen 5–30, 5–32
  for phosphorus 4–174
Activated sludge, *see* Sludge
Activity coefficients
  for calcium carbonate saturation 2–43t,
    2–44t
Acute toxicity, *see* Toxicity
Additions
  in biostimulation studies 8–50
Adenosine triphosphate, *see* ATP
Adjustment of reaction
  culture media 9–29
ADMI tristimulus filter method
  for color 2–9
Adsorption
  reagent water preparation 1–57
Adsorption assembly
  for organic halogen 5–29
Adsorption cells and holders
  for color 2–9
Adsorption column
  for organic halogen 5–29
  for thiocyanate 4–43
Adsorption-elution method
  for virus concentration 9–159, 9–163,
    9–164i
    large water volumes 9–163
    small water volumes 9–159
Adsorption precipitation method
  for virus concentration 9–170

Aerial photography
  for macrophyton mapping 10–72
Agars
  bile esculin 9–111
  bismuth sulfite 9–134
  brain-heart infusion 9–111
  brilliant green 9–134
  buffered charcoal yeast extract (BCYE)
    9–150
  Czapek 9–188
  diamalt 9–188
  ferrous sulfide 9–122
  for microbiological examinations 9–16
  for sulfate-reducing bacteria 9–125
  LES Endo 9–86
  lipovitellin-salt-mannitol 9–47
  mannitol salt 9–47
  mE 9–110
  mE 9–110
  m-Enterococcus 9–111
  M-FC, modified (M-FCIC) 9–98
  m-HPC 9–55
  M-Kleb 9–98
  milk 9–52
  Mn 9–121
  M-PA 9–52
    modified 9–52
  M-7h FC 9–36
  m-T7 9–43
  m-TEC 9–50
  neopeptone-glucose 9–188
  neopeptone-glucose-rose bengal-aureomy-
    cin 9–188
  nutrient 9–73
  NWRI (HPCA) 9–56
  Pfizer selective enterococcus (PSE) 9–109
  plate count 9–55
  R2A 9–55
  Simmons' citrate 9–105
  starch-cassein 9–128
  tryptic(ase) soy 9–40
    modified 9–40
  tryptic soy 9–106
  xylose lysine brilliant green 9–134
  xylose lysine desoxycholate 9–134
  yeast extract-malt extract-glucose 9–188
Age
  fish 10–131
Agitator
  for chlorine 4–55
Air
  for metals by AA 3–20
Air monitoring
  microbiological laboratories 9–5

Air scrubber solution
for nitrogen (ammonia) 4–127
Alcohol
for fish sample preservation 10–127
Aldrin
liquid-liquid extraction GC method 6–170
liquid-liquid extraction GC/MS method 6–113
Algae
biomass determination 10–39
biostimulation studies 8–46
blue-green 10–145i, 146i
key 10–183
clean-water
color plates following 10–200
cultures for biostimulation studies 8–17
culture units 8–18, 8–19i, 8–20i, 8–86, 8–87i
diatoms 10–149i, 10–150i
key 10–185
filter-clogging *color plates following* 10–200
flagellate
key 10–186
food
for copepods 8–86, 8–88t
for shrimp 8–99
freshwater
culture medium 8–17, 8–18t
key 10–183
green 10–147i, 10–148i
key 10–187
index to illustrations 10–191
marine
culture medium 8–18, 8–19t
large 10–151i
name changes 10–190
planktonic *color plates following* 10–200
polluted-water
color plates following 10–200
productivity, *see* Biostimulation
reservoir walls
color plates following 10–200
surface-water
color plates following 10–200
taste-and odor-causing
color plates following 10–200
test procedures 8–42
Alizarin fluorine blue solution
for fluoride 4–92
Alkalies
preparation *inside front cover*, 1–44
safety precautions 1–63

Alkali iodide azide reagent
for DO 4–152
Alkaline neutralizer reagent
for Legionellaceae 9–151
Alkaline pyrogallol reagent
for sludge digester gas 2–89
Alkalinity
bicarbonate
nomograph 4–16i
calculation method 4–19
carbonate
nomograph 4–17i
definition 2–35
methods 2–35
nomographic determination 4–14
relationships, calculation 2–38t
titration method 2–35
Alkalinity difference technique
calcium carbonate precipiation potential 2–48
Allen curve
macrophyton productivity 10–83, 10–84i
Alpha radioactivity, gross
method 7–15
regulatory limits 7–15
Alum flocculation modification
for DO 4–158
Aluminon buffer reagent
for beryllium 3–80
Aluminon method
for beryllium 3–80
Aluminum
AA method 3–25
low concentrations 3–27
micro quantities 3–36
Eriochrome cyanine R method 3–64
fluoride interference correction curves 3–67i
inductively coupled plasma method 3–54
methods 3–63
pyrocatechol violet method, automated 3–68
precision and bias 3–72t, 3–73t
Aluminum chloride solution
for sulfides 4–194
Aluminum hydroxide adsorption-precipitation method
for virus concentration 9–170
Aluminum hydroxide suspension
for chloride 4–68
Aluminum nitrate solution
for metals 3–25
for uranium 7–59

Aluminum solution
  for aluminum 3–25, 3–64, 3–68
Aluminum sulfate-silver sulfate-boric acid
    buffer
  for nitrogen (nitrate) 4–134
Alum solution
  for DO 4–158
Ames test, see Salmonella test
Amine-sulfuric acid solution
  for sulfide 4–196
Aminoantipyrine solution
  for phenols 5–52
Aminonaphtholsulfonic acid reagent
  for silica 4–188
Aminonaphtholsulfonic acid-sodium
    bisulfite-sodium sulfite
  for silica 4–190
Ammonia
  removal from distilled water 1–56
  un-ionized 8–29, 8–29t
  also see Nitrogen (ammonia)
Ammonia electrode filling solution
  for nitrogen (nitrate) 4–140
Ammonia manifold 4–127i
Ammonia nitrogen, see Nitrogen (ammonia)
Ammonia selective electrode method
  for nitrogen (ammonia) 2–122
    using known additions 4–124
    precision and bias 4–114t
Ammonia solutions
  for nitrogen (ammonia) 4–128
Ammonium acetate buffer solution
  for aluminum and beryllium 3–27
  for iron 3–103
Ammonium acetate solution
  for metals digestion 3–10
Ammonium bifluoride solution
  for sulfite 4–202
Ammonium chloride buffer solution
  for sulfate 4–210
Ammonium chloride-EDTA solution
  for nitrogen (nitrate) 4–136
Ammonium chloride solution
  for BOD 5–5
  for calcium 3–88
  for cyanates 4–41
  for nitrogen (nitrate) 4–138
  for organic halogen 5–29
Ammonium hydroxide solution
  for calcium 3–88
  for copper 3–96

Ammonium (cont.)
  for iodide 4–107
  for lead 3–108
  for phenols 5–52
  for radium 7–30
  for selenium 3–135, 3–140
  for zinc 3–161
  preparation inside front cover
Ammonium molybdate solution
  for phosphorus 4–175, 4–177, 4–179
  for silica 4–185, 4–190
Ammonium nitrate solution
  for metals 3–36
Ammonium oxalate-crystal violet
    (Hucker's) solution
  for coliforms 9–73
Ammonium oxalate solution
  for calcium 3–88
  for radium 228 7–44
Ammonium persulfate-phosphoric acid
    reagent
  for vanadium 3–156
Ammonium phosphate
  for metals 3–36
Ammonium phosphomolybdate reagent
  for cesium (radioactive) 7–22
Ammonium purpurate indicator solution
  for calcium 3–86
Ammonium pyrrolidine dithiocarbamate
    (APDC) solution
  for metals 3–23
Ammonium solution
  for nitrogen (ammonia) 4–118, 4–120
Ammonium sulfate solution
  for nitrogen (nitrate) 4–134
  for radium 226 7–37
  for radium 228 7–45
  for reagent water bacteriological quality
    9–9
Ammonium sulfide solution
  for radium 228 7–45
Ammonium thiocyanate reagent
  for silver 3–143
Amoebas 10–158i
Amperometric method
  for chlorine dioxide 4–80
Amperometric titration method
  for chlorine 4–54
    low-level 4–57
  for chlorine dioxide 4–78
  for iodine 4–105

Amphibians 10–181i
Amphipods
  holding, acclimating, and culturing 8–93
  life-cycle toxicity tests 8-105
  short-term toxicity tests 8-105
  also see Crustaceans
Ampules
  for organic carbon 5–25
  for selenium 3–140
Ampule sealer
  for COD 5–14
Amyl alcohol-benzaldehyde solution
  for coliform differentiation 9–104
Analytical methods
  collaborative testing 1–24
  correctness 1–20
  development and evaluation 1–22
Analyzer assembly
  for organic halogen 5–29, 5–30i
Anatomy
  fish 10–128i
Angling
  for fish sampling 10–116
Anion-cation balance
  for checking analyses 1–20
Anion column separations 4–4i, 4–5i
Anions
  ion chromatography method 4–2
Anion solutions
  for ion chromatography 4–4
Annelids
  collection 8–66
  culturing 8–67
  exposure chambers 8–70
  food 8–67
  parasites and diseases 8–69
  test organism selection 8–66
  toxicity test data evaluation 8–73
  toxicity test methods 8–65
  water supply 8–69
Anthracene
  liquid-liquid extraction chromatographic
    method 6–148
  liquid-liquid extraction GC/MS method
    6–113
Antibiotics
  for virus concentration 9–160, 9–171
  use in toxicity tests 8–22
    for crustaceans 8–100
Antifoaming agents
  for selenium 3–52

Antigen
  for Group D streptococci 9–112
Antimony
  AA methods
    electrothermal 3–36
    flame 3–20
  inductively coupled plasma method 3–54
  methods 3–73
Antimony solution
  for antimony 3–20
Antimony test
  for sulfide 4–192
A-1 broth
  for fecal coliforms 9–76
A-1 medium test
  for fecal coliforms 9–76
Apparatus
  laboratory 1–41
  microbiological tests 9–24
  radioactivity analyses 7–11
Appearance
  method 2–1
Application factor
  definition 8–4
Aqua regia
  for metals 3–20
Aquatic humic substances
  definition 5–37
  diethylaminoethyl method 5–38
  methods 5–37
  XAD method 5–40
Aquatic hyphomycetes, *see* Hyphomycetes
Aquatic insects, *see* Insects
Aquatic organisms
  identification 10–136
  key 10–137
  methods 10–1
  taxonomic references 10–194
  trophic levels 10–142
Argentometric method
  for chloride 4–68
Argon
  for metals 3–58
Aromatics, *see* Aromatics, volatile; Polynu-
    clear aromatic hydrocarbons
Aromatics solutions
  for volatile aromatics 6–69
Aromatics, volatile
  chromatographic conditions 6–66t
  detection limits 6–66t
  gas chromatogram 6–69i, 6–72i, 6–73i

Aromatics *(cont.)*
  methods 6–64
    purge and trap GC method 6–65, 6–71
      bias and precision 6–70t, 6–76t, 6–77t
      calibration criteria 6–70t
      quality control acceptance criteria
        6–70t
    retention times 6–74t
Arrows
  for fish sampling 10–116
Arsenic
  AA method
    electrothermal 3–36
  hydride generation/AA method
    manual 3–43
  inductively coupled plasma method 3–54
  mercuric bromide stain method 3–76
  methods 3–74
  safety precautions 1–63
  silver diethyldithiocarbamate method
    3–74
Arsenic solutions
  for arsenic 3–46, 3–75
Arsenious acid
  for iodide 4–109
Arsenite solution
  for chlorine 4–52
Arsine generator
  for arsenic 3–75, 3–75i, 3–76, 3–77i
Arthropods 10–179i
Ascarite
  for radium 226 7–37
Ascorbic acid method
  for phosphorus 4–177
    automated 4–178
      precision and bias 4–178t
Ascorbic acid solution
  for aluminum 3–65
  for phosphorus 4–177
Ash-free dry weight
  periphyton 10–54
Ashing, dry
  for metals samples 3–11
Asparagine broth
  for *Pseudomonas aeruginosa* 9–53
Assays, virus 9–176
Assemblage
  fish
    definition 10–114
Atomic absorption spectrometric equipment
  calibration 3–22
  for metals 3–17, 3–36

Atomic absorption spectrometric method
    1–50
  bias and precision 3–18t
  cold-vapor
    for metals 3–28
  concentration ranges 3–16t
  electrothermal 3–32
    matrix modifiers 3–33, 3–33t
    precision 3–40t, 3–41t
    relative error 3–42t
    sensitivity and concentration ranges
      3–34, 3–35t
  flame 3–13
  for aluminum 3–25
    low concentrations 3–27
    micro quantities 3–36
  for antimony 3–20
    micro quantities 3–36
  for arsenic
    micro quantities 3–36
  for barium 3–25
    micro quantities 3–36
  for beryllium 3–25
    low concentrations 3–27
    micro quantities 3–36
  for bismuth 3–20
  for cadmium 3–20
    low concentrations 3–23
    micro quantities 3–36
  for calcium 3–20
  for cesium 3–20
  for chromium 3–20
    low concentrations 3–23
    micro quantities 3–36
  for cobalt 3–20
    low concentrations 3–23
    micro quantities 3–36
  for copper 3–20
    low concentrations 3–23
    micro quantities 3–36
  for gold 3–20
  for iridium 3–20
  for iron 3–20
    low concentrations 3–23
    micro quantities 3–36
  for lead 3–20
    low concentrations 3–23
    micro quantities 3–36
  for lithium 3–20
  for magnesium 3–20
  for manganese 3–20
    low concentrations 3–23

Atomic absorption spectrometric method
(cont.)
    micro quantities 3–36
    for mercury 3–29
    for metals
        control ranges 3–19t
        single-operator precision 3–19t
    for molybdenum 3–25
        micro quantities 3–36
    for nickel 3–20
        low concentrations 3–23
        micro quantities 3–36
    for osmium 3–25
    for palladium 3–20
    for platinum 3–20
    for potassium 3–20
    for rhenium 3–25
    for rhodium 3–20
    for ruthenium 3–20
    for selenium
        micro quantities 3–36
    for silicon 3–25
    for silver 3–20
        low concentrations 3–23
        micro quantities 3–36
    for sodium 3–20
    for strontium 3–20
    for thallium 3–20
    for thorium 3–25
    for tin 3–20
        micro quantities 3–36
    for titanium 3–25
    for vanadium 3–25
    for zinc 3–20
        low concentrations 3–23
    hydride generation
        for arsenic 3–43
        for metals 3–43
        for selenium 3–43
        for selenium 3–50
Atomizer
  for arsenic 3–46
  for selenium 3–46
ATP
  periphyton 10–55
  plankton 10–40
  standard 10–41
ATP procedure
  for total viable microorganisms 9–37
Aufwuchs, see Periphyton
Aureomycin-neopeptone-glucose-rose bengal
  agar
  for fungi 9–188

Aureomycin-rose bengal-glucose-peptone
  agar
  for fungi 9–190
    modified 9–191
Autoclaves
  for microbiological laboratories 9–7
  specifications 9–25
  sterilization times and temperatures 9–16t
Automated analyses 1–51
  for aluminum (PCV) 3–68
  for chloride (ferricyanide) 4–73
  for fluoride (complexone) 4–92
  for nitrogen (ammonia) (phenate) 4–126
  for nitrogen (nitrate) (cadmium reduction) 4–137
  for nitrogen (nitrate) (hydrazine reduction) 4–141
  for phosphorus (ascorbic acid reduction) 4–178
  for silica (heteropoly blue) 4–189
  for sulfate (methylthymol blue) 4–208
Automated analytical equipment
  for aluminum 3–68, 3–70i, 3–71i
  for chloride 4–73, 4–74i,
  for fluoride 4–92, 4–93i
  for nitrogen (ammonia) 4–126, 4–127i
  for nitrogen (nitrate) 4–138, 4–138i,
    4–141, 4–142i
  for phosphorus 4–179, 4–180i
  for silica 4–189, 4–190i
  for sulfate 4–209, 4–209i
Autotrophic Index
  for periphyton 10–53
Azide alkali iodide reagent
  for DO 4–152
Azide dextrose broth
  for fecal streptococcus 9–109
Azide modification
  for DO 4–152
Azides
  safety precautions 1–63

B

Background correction
  Smith-Hieftje
    for metals by AA 3–15
  Zeeman
    for metals by AA 3–14
Bacteria 10–182i
  actinomycetes
    methods 9–126

Bacteria *(cont.)*
colonies
description 9–130t
typical vs. actinomycete type 9–129i
direct total count 9–64
iron 9–116i, 9–117i, 9–118i
characteristics 9–115
enumeration, enrichment, and isolation
9–121
identification 9–114, 9–115
isolation medium 9–121
maintenance medium 9–121
methods 9–115
samples 9–115
pathogenic 9–130
rapid detection methods 9–36, 9–38t
sulfate-reducing
enumeration, enrichment, and isolation
9–124
sulfur 9–119i, 9–120i
characteristics 9–118
enumeration, enrichment, and isolation
9–121
identification 9–114, 9–119
methods 9–118
photosynthetic 9–125
samples 9–119
also see Coliform group bacteria and specific genera and species
Bacterial density
coliforms 9–77
Bacteriologial examination
methods 9–1
recreational waters 9–45
samples 9–31
also see Microbiological examination
Bacteriological quality
reagent-grade water 9–9
Baiting
fungi, zoosporic 9–193
Balances
for base/neutral and acid extractables
6–119
for microbiological examinations 9–6,
9–26
for solids 2–73
Barbituric acid-pyridine reagent
for cyanide 4–31
Barium
AA method 3–25
micro quantities 3–36
inductively coupled plasma method
3–54

Barium *(cont.)*
methods 3–78
recovery in radium 226 method 7–43
Barium carrier
for radium 228 7–45
for strontium (radioactive) 7–49
Barium chloride crystals
for sulfate 4-207
Barium chloride solution
for radium 7–30
for radium 226 7–36
for sulfate 4–205, 4–209
Barium diphenylaminesulfonate
for chlorine 4–60
Barium solution
for barium 3–25
Baseline method
for macrophyton mapping 10–71
Base/neutral extractables
characteristic masses 6–115t
chromatographic conditions 6–115t
detection limits 6–115t
gas chromatogram 6–123i
liquid-liquid extraction GC/MS method
6–113
bias and precision 6–134t
methods 6–112
quality control acceptance criteria 6–132t
standards 6–120, 6–121t
Base/neutral extractable solutions
for base/neutral and acid extractables
6–120
Basepoint-stadia rod-alidade method
for macrophyton mapping 10–71
Basic trihalomethane formation potential
definition 5–73
method 5–79
precision and bias 5–80t
Basket sampler
for macroinvertebrates 10–106, 10–107i
Batch adsorption procedure
for organic halogen 5–32
Batch method
for cation removal by ion exchange 1–46
Bathing waters
microbiological examination 9–49
microbiological samples 9–33, 9–50
also see Swimming pools
Bathocuproine method
for copper 3–97
Beef extract
for virus concentration 9–160, 9–166,
9–171

Beetles 10–174i
Beggiatoa 9–120i
  enumeration, enrichment, and isolation
    9–125
Below-ground production
  macrophyton productivity 10–84
Belt transects
  for macrophyton sampling 10–76
Benthic dome samplers
  for macrophyton sampling 10–76
Benthic macroinvertebrates, *see* Macroinver-
    tebrates, benthic
Benzaldehyde-amyl alcohol solution
  for coliform differentiation 9–104
Benzene
  purge and trap GC method 6–65, 6–71
    capillary column 6–93
  purge and trap GC/MS method 6–27,
    6–42
    capillary column 6–50
Benzene-isobutanol solvent
  for phosphorus 4–175
Benzidine
  liquid-liquid extraction GC/MS method
    6–113
Benzo(a)anthracene
  liquid-liquid extraction chromatographic
    method 6–148
  liquid-liquid extraction GC/MS method
    6–113
Benzo(a)pyrene
  liquid-liquid extraction chromatographic
    method 6–148
  liquid-liquid extraction GC/MS method
    6–113
Benzo(b)fluoranthene
  liquid-liquid extraction chromatographic
    method 6–148
  liquid-liquid extraction GC/MS method
    6–113
Benzo(ghi)perylene
  liquid-liquid extraction chromatographic
    method 6–148
  liquid-liquid extraction GC/MS method
    6–113
Benzo(k)fluoranthene
  liquid-liquid extraction chromatographic
    method 6–148
  liquid-liquid extraction GC/MS method
    6–113
Beryllium
  AA method 3–25
    low concentration 3–27

Beryllium *(cont.)*
  AA method *(cont.)*
    micro quantities 3–36
    aluminon method 3–80
    inductively coupled plasma method 3–54
    methods 3–79
    safety precautions 1–63
Beryllium compounds
  safety precautions 1–63
Beryllium solution
  for beryllium 3–26, 3–80
Beta radioactivity, gross
  method 7–15
  regulatory limits 7–16
BFB
  abundance criteria 6–30t
BFB standard
  for volatile organics 6–33, 6–45
BHC
  liquid-liquid extraction GC method 6–170
  liquid-liquid extraction GC/MS method
    6–113
Bias 1–14i
  BTFP analysis 5–80t
  closed loop stripping GC/MS analysis
    6–23t
  definition 1–4, 1–14
  for method validation 1–22
  nitrogen (ammonia) methods 4–112t
  organic halogen method 5–36t
  pesticide method 6–168t
  phosphorus methods 4–169t
  purge and trap GC/MS method
    for volatile organics 6–49t
  TFP analysis 5–78t
Bile esculin agar 9–111
Bioassays, *see* Toxicity tests
Biochemical oxygen demand, *see* BOD
Biochemical reactions
  Campylobacter jejuni 9–144
  coliform verification 9–19
  Salmonella 9–134
  Shigella 9–141
  Vibrio cholerae 9–145
Biological examination
  introduction 10–1
Biological hazards
  monitors 1–65
  safety precautions 1–65
Biological samples
  alpha and beta radioactivity 7–19
Biological wastes
  disposal 1–70

Bioluminescence test
  for total viable microorganisms 9–37
Biomass
  algae 8–49
  fish 10–130
  macrophyton 10–77
  periphyton 10–54
  plankton 10–39
Biomass dynamics methods
  macrophyton productivity 10–81
Biomass methods
  macrophyton productivity 10–80
Biomass tagging methods
  macrophyton productivity 10–82
Biomass turnover
  macrophyton productivity 10–82
Biostimulation
  additions 8–50
  culture medium 8–17
  data analysis and interpretation 8–51
  inoculum 8–46
  methods 8–42
  samples 8–45
  sampling equipment 8–44
  terminology 8–4
  test conditions and procedures 8–47
Biovolume
  plankton 10–40
Birge tow-net
  for plankton 10–11i
Bismuth
  AA method 3–20
  methods 3–81
Bismuth solution
  for bismuth 3–20
Bismuth sulfite agar
  for *Salmonella* 9–134
Bis(2-hydroxyethyl) dithiocarbamate
    solution
  for zinc 3–161
Bivalves 10–177i
Block heater
  for COD 5–14
Bluegill sunfish, *see* Sunfish, bluegill
Boats
  for electrofishing 10–121i
BOD
  carbonaceous vs. nitrogenous 5–2
  dilution requirements 5–3
  5-day BOD test 5–4
  methods 5–2
  sampler assembly 4–151i

Boiling chips
  for base/neutral and acid extractables
    6–119
Bongo net
  for plankton 10–11i
Borate buffer solution
  for basic trihalomethane formation poten-
    tial 5–79
  for nitrogen (ammonia) 4–115
  for zinc 3–163
Borax solution
  for silica 4–186
Boric acid-silver sulfate-aluminum sulfate
    buffer
  for nitrogen (nitrate) 4–134
Boric acid solution
  for nitrogen (ammonia) 4–121
Borohydride reagent
  for selenium 3–51
Boron
  carmine method 4–10
  curcumin method 4–7
  inductively coupled plasma method 3–54
  methods 4–7
Boron solution
  for boron 3–57, 4–8
Boron trifluoride-methanol
  for herbicides 6–184
Bottle, stripping
  for organic constituents (individual) 6–13i
Bottles
  for methane apparatus 6–61
  for organic constituents (individual)
    6–11
  for plankton productivity 10–43
  for sulfides 4–194
  for trihalomethane formation potential
    5–75
  gas washing
    for surfactants 5–57
    for selenium 3–138
  also see Glassware
Bottom nets
  for macroinvertebrates 10–105
Box samplers
  for macrophyton sampling 10–76
Brachyura
  see Crabs
Brain-heart infusion
  for fecal streptococcus 9–111
Brain-heart infusion agar
  for fecal streptococcus 9–111

Brilliant green agar
for *Salmonella* 9–134
Brilliant green lactose bile broth
for coliforms 9–70
Brines
metals analysis by AA 3–14
Bromate-bromide solution
for phenols 5–52
Bromcresol green indicator solution
for alkalinity 2–37
for radium 7–30
Bromide
ion chromatography method 4–2
methods 4–11
Bromide-bromate solution
for phenols 5–52
Bromide solution
for bromide 4–12
for ion chromatography 4–4
Bromobenzene
purge and trap GC method 6–71, 6–87
capillary column 6–93
purge and trap GC/MS method 6–42
capillary column 6–50
Bromochloromethane
purge and trap GC method 6–87
capillary column 6–93
purge and trap GC/MS method 6–42
capillary column 6–50
Bromodichloromethane
liquid-liquid extraction GC method 6–104
purge and trap GC method 6–79, 6–87
capillary column 6–93
purge and trap GC/MS method 6–27,
6–42
capillary column 6–50
Bromoform
liquid-liquid extraction GC method 6–104
purge and trap GC method 6–79, 6–87
capillary column 6–93
purge and trap GC/MS method 6–27,
6–42
capillary column 6–50
Bromomethane
purge and trap GC method 6–79, 6–87
capillary column 6–93
purge and trap GC/MS method 6–27,
6–42
capillary column 6–50
Bromophenyl phenyl ether
liquid-liquid extraction GC/MS method
6–113

Bromphenol blue indicator
for acidity 2–33
Bromphenol blue-diphenylcarbazone reagent
indicator
for chloride 4–69
Brook trout
see Trout, brook
Broths
acetamide 9–53
A-1 9–76
asparagine 9–53
azide dextrose 9–109
brilliant green lactose bile 9–70
casitone-glycerol-yeast autolysate (CGY)
9–121
dulcitol selenite 9–133
for microbiological examinations 9–16
glucose 9–105
Koser's citrate 9–105
lactose 9–69, 9–70t
lauryl tryptose 9–68, 9–69t
M-staphylococcus 9–47
P-A 9–80
salt peptone glucose 9–105
selenite cystine 9–133
tetrathionate 9–133
tryptic(ase) soy 9–40
tryptophane 9–104
yeast nitrogen base-glucose 9–192
Bryozoans 10–160i
Buffered charcoal yeast extract (BCYE)
agar
for Legionellaceae 9–150
Buffered glucose broth
for coliform differentiation 9–105
Buffer solutions
acetate
for aluminum 3–65
for bromide 4–12
for chlorine 4–52
for cyanide 4–38
for fluoride 4–92
for strontium (radioactive) 7–49
for sulfate 4–207
for zinc 3–159
aluminon
for beryllium 3–80
ammonium
for sulfate 4–210
ammonium acetate
for aluminum and beryllium 3–27
for iron 3–103

Buffer solutions *(cont.)*
  ammonium acetate *(cont.)*
  borate
    for basic trihalomethane formation potential 5–79
    for nitrogen (ammonia) 4–115
    for zinc 3–163
  CDTA
    for fluoride 4–88
  citric acid-ammonium hydroxide
    for iodide 4–107
  EDTA
    for hardness 2–54
  for pH 4–97
  hexamethylaminetetraamine
    for aluminum 3–69
  phosphate
    for BOD 5–4
    for chlorine 4–56, 4–60, 4–65
    for distilled water bacteriological quality 9–9
    for microbial counts 9–65
    for phenols 5–52
    for trihalomethane formation potential 5–76
  phosphate-carbonate
    for mercury 3–119
  silver sulfate-boric acid
    for nitrogen (nitrate) 4–134
Bugs, true 10–175i
Bunsen coefficient
  for oxygen in water 2–98t
Burner
  for AA method 3–17
Burner head
  nitrous oxide
    for metals by AA 3–25
Butanol-chloroform reagent
  for organic acids 5–70
Butylbenzene
  purge and trap GC method 6–71
    capillary column 6–93
  purge and trap GC/MS method 6–42
    capillary column 6–50
Butyl benzyl phthalate
  liquid-liquid extraction GC/MS method 6–113

C

Caddisflies 10–171i
Cadmium
  AA method 3–20
    low concentrations 3–23
    micro quantities 3–36

Cadmium *(cont.)*
  AA method *(cont.)*
    dithizone method 3–83
    inductively coupled plasma method 3–54
    methods 3–82
Cadmium copper granules
  for nitrogen (nitrate) 4–135
Cadmium reduction method
  for nitrogen (nitrate) 4–135
    automated 4–137
Cadmium solution
  for cadmium 3–20, 3–83
Calcium
  AA method 3–20
  EDTA titrimetric method 3–85
  inductively coupled plasma method 3–54
  methods 3–85
  permanganate titrimetric method 3–87
Calcium carbonate
  indices of precipitation/dissolution tendency 2–42
  saturation index 2–42
Calcium carbonate deposition test
  calcium carbonate precipitation potential 2–48
Calcium carbonate precipitation potential
  calculation method 2–48
  experimental methods 2–48
  methods 2–47
Calcium carbonate saturation
  activity coefficients 2–43t, 2–44t
  effects 2–41
  equilibrium constants 2–43t, 2–44t
  indices 2–40
  methods 2–40
  precipitation/dissolution indices 2–42, 2–47
Calcium carbonate saturation indices
  alkalinity difference technique 2–46
  calculation method 2–42
  computer codes 2–49, 2–50t
  saturometry method 2–46
Calcium chloride solution
  for BOD 5–5
  for cesium (radioactive) 7–22
Calcium hydroxide solution
  for pH 4–100
Calcium hypochlorite solution
  for cyanides amenable to chlorination 4–35
Calcium nitrate
  for metals 3–36
Calcium solution
  for calcium 3–21

Calcium solution *(cont.)*
  for hardness 2–56
  for metals 3–20
Calculation method
  for carbon dioxide and alkalinity 4–19
  for hardness 2–53
  for hydrogen sulfide 4–198
  for magnesium 3–114
Caldwell-Lawrence diagrams
  for calcium carbonate saturation 2–46
Calibration
  AA equipment 3–22
  for analytical quality control 1–8
Calibration check standard
  definition 1–4
Calibration concentrations
  for metals by ICP method 3–55t
Calibration criteria
  for volatile aromatics
    purge and trap gas chromatographic
      method 6–70t
  for volatile halocarbons 6–83t
  for volatile organics
    purge and trap GC/MS method
      6–36t
Calibration curves
  metals analyses 3–3
  verification 1–48
Calibration gases
  for sludge digester gas 2–92
Calibration standards
  for phenols 6–140
Calmagite indicator
  for hardness 2–55
Caloric content
  macrophyton 10–78
Campylobacter jejuni
  biochemical reactions 9–144
  concentration techniques 9–143
  enrichment 9–143
  identification 9–144
  methods 9–143
  selective growth 9–143
Candida albicans
  identification 9–196
Capillary injector
  for organic constituents (individual) 6–15
Capillary interface
  for volatile organics 6–50
Capillary precipitin test
  for Group D streptococci 9–112
Capitella capitata
  culturing 8–68
Carbon, activated, *see* Activated carbon

Carbonate-phosphate buffer solution
  for mercury 3–119
Carbonate solution
  for plankton productivity 10–45
Carbonate-tartrate reagent
  for tannin and lignin 5–68
Carbon content
  macrophyton 10–78
Carbon dioxide
  calculation method 4–19
  methods 4–13
  nomographic method 4–14, 4–18i
  removal from distilled water 1–56
  titrimetric method 4–19
Carbon disulfide
  for organic constituents (individual) 6–15
Carbon exchange methods
  macrophyton productivity 10–88
Carbon 14-labeled sodium carbonate solution
  for periphyton productivity 10–57
Carbon 14 method
  for periphyton productivity 10–57
  for plankton productivity 10–45
Carbon solution
  for organic carbon 5–20
Carbon tetrachloride
  for silver 3–143
  purge and trap GC 6–79, 6–87
    capillary column 6–93
  purge and trap GC/MS method 6–27,
    6–42
    capillary column 6–50
Carbon, total organic (TOC), *see* Organic
  carbon (total)
Carcinogenic compounds
  safety precautions 1–64
Carmine method
  for boron 4–10
Carmine reagent
  for boron 4–10
Carrier-free radioactive carbonate solution
  for plankton productivity 10–45
Carrier gas
  for organic carbon 5–20
  for organic constituents (individual)
    6–16
  for pesticides (organochlorine) 6–162
  for sludge digester gas 2–92
Casitone-glycerol-yeast autolysate (CGY)
  broth
  for iron bacteria 9–121
Catalysts
  for nitrogen (organic) 4–144

Catalytic reduction method
  for iodide 4-109
Catfish, channel
  culturing 8-128
Cation-exchange column
  for aluminum 3-68
Cations
  removal by ion exchange 1-46
CDTA buffer
  for fluoride 4-88
Cell culture procedure
  virus isolation and assay 9-179
Cell surface area
  plankton 10-40
Cell volume
  plankton 10-40
Centrifugation
  for plankton concentration 10-16
Centrifuge
  for acids, volatile 5-72
  for radium 228 7-44
  for selenium 3-135
  for virus concentration 9-166, 9-171,
    9174, 9-175
Centrifuge bottles
  for selenium 3-135
Centrifuge tubes
  for chlorophyll 10-32
Ceric ammonium sulfate solution
  for iodide 4-109
Certification
  water laboratories 9-23
Cesium
  AA method 3-20
  methods 3-90
  radioactive
    method 7-21
    precipitation method 7-22
    regulatory limits 7-21
Cesium carrier
  for cesium (radioactive) 7-22
Cesium solution
  for cesium 3-21
Chain-of-custody procedures
  samples 1-34
Chambers
  for periphyton productivity 10-56, 10-59
Channel catfish, see Catfish, channel
Characteristic masses
  acid extractables 6-118t
  base/neutral extractables 6-115t
  volatile organics 6-38t, 6-47t

Chelating resin
  for metals 3-36
Chelation
  metals in microbiological samples 9-32
Chemical additive system
  for virus concentration 9-163
Chemical hazards
  in laboratories 1-62
Chemical monitors
  for laboratory safety 1-69
Chemical oxygen demand (COD), see COD
Chemicals
  for fish collection 10-121
  safety data sheets 1-59, 1-62, 1-70
  safety precautions 1-62
  waste disposal 1-70
  also see Reagents; Ichthyocides
Chemical spill kits 1-61
Chironomus
  toxicity test procedures 8-118
Chloramines
  procedure 4-57, 4-61, 4-63
Chloramine-T solution
  for bromide 4-12
  for cyanide 4-31
Chlordane
  gas chromatogram 6-125i, 6-175i
  liquid-liquid extraction GC method 6-170
  liquid-liquid extraction GC/MS method
    6-113
Chloride
  argentometric method 4-68
  automated equipment setup 4-74i
  ferricyanide method, automated 4-73
  ion chromatography method 4-2
  mercuric nitrate method 4-69
  methods 4-67
  potentiometric method 4-71
  titration curve 4-72i
Chloride solution
  for ion chromatography 4-4
  for chloride 4-73
Chlorinated phenoxy acid herbicides, see
    Herbicides
Chlorination
  fungi survival 9-184
  nitrogen effects on 4-111
Chlorine dioxide
  amperometric method 4-78, 4-80
  DPD method 4-79
  equivalent weights 4-83t
  generation-absorption system 4-76, 4-6i

Chlorine *(cont.)*
  iodometric method 4–76
  methods 4–75
Chlorine dioxide solution
  for chlorine dioxide 4–76
Chlorine (residual)
  amperometric titration method 4–54
    low level 4–57
  combined 4–56
  DPD colorimetric method 4–62
  DPD ferrous titrimetric method 4–58
  free 4–56, 4–61, 4–63
  iodometric electrode method 4–65
  iodometric method 4½8, 4–51
  methods 4–45
  syringaldazine (FACTS) method 4–64
Chlorine solution
  for chlorine 4–62
  for trihalomethane formation potential
    5–75
Chlorobenzene
  purge and trap GC method 6–65, 6–71,
    6–79, 6–87
    capillary column 6–93
  purge and trap GC/MS method 6–27,
    6–42
    capillary column 6–50
Chlorodibromomethane
  liquid-liquid extraction gas chromato-
    graphic method 6–104
Chloroethane
  purge and trap GC method 6–79, 6–87
    capillary column 6–93
  purge and trap GC/MS method 6–27,
    6–42
    capillary column 6–50
Chloroethoxy methane bis-(2-)
  liquid-liquid extraction GC/MS method
    6–113
Chloroethyl ether bis(2-)
  liquid-liquid extraction GC/MS method
    6–113
Chloroethylvinyl ether
  purge and trap GC method 6–79
  purge and trap GC/MS method 6–27
Chloroform
  for cadmium 3–83
  liquid-liquid extraction GC method 6–104
  purge and trap GC method 6–79, 6–87
    capillary column 6–93
  purge and trap GC/MS method 6–27, 6–42
    capillary column 6–50

Chloroform-butanol reagent
  for organic acids 5–70
Chloroform extraction method
  for phenols 5–51
Chloroform solution
  for organic halogen 5–30
Chloroisopropyl ether bis-(2-)
  liquid-liquid extraction GC/MS method
    6–113
Chloromethane
  purge and trap GC method 6–79, 6–87
    capillary column 6–93
  purge and trap GC/MS method 6–27,
    6–42
    capillary column 6–50
Chloromethylphenol
  liquid-liquid extraction GC method 6–137
  liquid-liquid extraction GC/MS method
    6–113
Chloronaphthalene
  liquid-liquid extraction GC/MS method
    6–113
Chlorophenol
  liquid-liquid extraction GC method 6–137
  liquid-liquid extraction GC/MS method
    6–113
Chlorophenyl phenyl ether
  liquid-liquid extraction GC/MS method
    6–113
Chlorophyll
  chromatogram 10–37i
  determination 8–49, 10–31
  extraction 10–31
  high-performance liquid chromatographic
    method 10–36
  in macrophyton 10–78
  in periphyton 10–53
  spectrophotometric method 10–32
Chlorophyll *a*
  algal biomass indicator 10–39
  fluorometric method 10–34
Chlorophyll solutions
  for chlorophyll 10–36
Chloroplatinic acid
  for cesium (radioactive) 7–22
Chloro-6-(trichloro methyl) pyridine
  for BOD 5–5
Chlorotoluene
  purge and trap GC method 6–71, 6–87
    capillary column 6–93
  purge and trap GC/MS method 6–42
    capillary column 6–50

Chromaticity diagrams
  for color 2–7i
Chromatium okenii 9–119i
Chromatogram, *see* Gas chromatogram; Liquid chromatogram
Chromatograph
  for chlorophyll 10–36
  for sludge digester gas 2–91
  ion
    for anions 4–3
  also see Gas chromatograph; High-performance liquid chromatograph
Chromatographic column
  for aquatic humic substances 5–38, 5–40
  for base/neutral and acid extractables 6–119
  for chlorophyll 10–36
  for pesticides (organochlorine) 6–160
  for phenols 6–138, 6–139
  for polynuclear aromatic hydrocarbons 6–149, 6–150
  for selenium 3–131, 3–139
  for volatile halocarbons 6–88
Chromatographic conditions
  for acid extractables 6–118t
  for base/neutral extractables 6–115t
  for dibromoethane and dibromochloropropane 6–100t
  for organochlorine pesticides and PCBs 6–171t
  for PCBs 6–171t
  for phenols 6–138t
  for volatile aromatics 6–66t
  for volatile halocarbons 6–80t, 6–81
  for volatile organics 6–28t, 6–45t
    narrow-bore capillary column 6–53t
    wide-bore capillary column 6–51t
Chromatographic method
  for chlorophyll 10–36
  for herbicides 6–182
  for pesticides (organochlorine) 6–158
  for polynuclear aromatic hydrocarbons 6–148
    conditions 6–151t
    method detection limits 6–149t
    retention times 6–151t
  for sludge digester gas 2–91
  also see Gas chromatographic-mass spectrometric method; Gas chromatographic method
Chromatographic separation method
  for organic acids 5–70

Chromium
  AA method 3–20
    low concentrations 3–23
    micro quantities 3–36
  colorimetric method 3–91
  inductively coupled plasma method 3–54
  methods 3–90
Chromium solution
  for chromium 3–21, 3–92
Chronic toxicity, *see* Toxicity
Chrysene
  liquid-liquid extraction chromatographic method 6–148
  liquid-liquid extraction GC/MS method 6–113
Ciliates 10–159i
Citrate-cyanide reducing solution
  for lead 3–108
Citric acid
  for iodide 4–107
  for radium 7–30
Citric buffer solution
  for iodide 4–107
Clams
  conditioning 8–75
  embryo toxicity tests 8–78
  growth tests 8–80
  toxicity tests 8–74
  also see Mollusks
Clarke-Bumpus sampler
  for zooplankton 10–12, 10–13i
Cleaning
  microbiological laboratories 9–5
Cleansers
  for laboratory glassware 1–41
Cleanup procedure
  for phenols 5–50
Closed-loop stripping apparatus
  for organic constituents (individual) 6–11, 6–13i
Closed-loop stripping GC/MS analysis
  for organic constituents
    bias 6–23t, 6–25t
    precision 6–24t, 6–25t
  for organic constituents (individual) 6–11
    detection limits 6–12t
    operating conditions 6–20t
Closed reflux colorimetric method
  for COD 5–15
Closed reflux titrimetric method
  for COD 5–14
Clostridium
  presence-absence test 9–82i

Closures
  for laboratory glassware 1–41
Cobalt
  AA method 3–20
    low concentrations 3–23
    micro quantities 3–36
  inductively coupled plasma method 3–54
  methods 3–94
Cobalt solution
  for cobalt 3–21
Cobalt thiocyanate active substances
  definition 5–64
  method 5–64
Cobaltothiocyanate reagent
  for nonionic surfactants 5–65
Cobaltous chloride solution
  for nitrogen (ammonia) 4–118
COD
  closed reflux colorimetric method 5–15
  closed reflux titrimetric method 5–14
  definition 5–10
  digestion reagent quantities 5–15t
  methods 5–10
  open reflux method 5–12
Coelenterates 10–179i
Cohort methods
  macrophyton productivity 10–83
Cold-vapor AA method
  for mercury 3–29
  for metals 3–28
Coliform group bacteria
  coliphage rapid detection method 9–39
  counts 9–21t
  culture purification 9–100
  decarboxylase (ornithine and lysine) test
    9–102t, 9–106
  definition 9–66, 9–87
  density estimation 9–77, 9–89
  differentiation 9–99
  drinking water tests 9–67
  E. coli, pathogenic 9–141
  fecal
    A-1 medium test 9–76
    delayed incubation procedure 9–96
    EC medium test 9–75
    in bathing waters 9–50
    in swimming pools 9–47
    membrane filter procedure 9–94, 9–95t
    MPN procedure 9–75
    radiometric detection 9–37
    recovery enhancement 9–44
    seven-hour test 9–36
    verification 9–20

Coliform; fecal (cont.)
    IMViC tests 9–100, 9–102t
    in bathing waters 9–50
    indole test 9–104
    lactose test 9–102t
    lysine test 9–102t
    membrane filter technique 9–83
      confidence limits 9–91t
      sample volumes 9–88t
    methyl red test 9–104
    motility test 9–102t, 9–106
    MPN 9–77t, 9–78t
    multiple-tube fermentation technique
      9–66
      USEPA regulations 9–3
    nonpotable water tests 9–68
    ornithine test 9–102t
    oxidase test 9–106
    presence-absence test 9–80
    sample size
      membrane filter technique 9–87, 9–88t
    significance 9–103, 9–131
    sodium citrate test 9–105
    total
      completed test 9–71, 9–72i
      confirmed phase 9–70, 9–71i
      counting 9–89
      delayed incubation membrane filter
        technique 9–91
      direct technique 9–88
      enrichment technique 9–88
      Gram stain technique 9–74
      membrane filter procedure 9–84
      MPN test 9–68
      presumptive phase 9–68
      recovery enhancement 9–43
    verification
      lactose fermentation 9–89
      multi-test systems 9–89
      rapid test 9–89
    Voges-Proskauer test 9–102t
    also see Bacteria; Enterobacteriaceae
Coliphage detection method
  for coliforms 9–39
Collaborative testing
  analytical methods 1–24
Color
  chromaticity diagrams 2–7i
  definition 2–2
  field method 2–3
  filtration system 2–5i
  hues for dominant wavelengths 2–6t
  methods 2–2

Color *(cont.)*
   nonstandard laboratory methods 2–3
   platinum-cobalt method 2–2
   spectrophotometric method 2–4
   standards 2–3
   tristimulus filter method 2–8
   visual comparison method 2–2
Color development solution
   for nitrogen (nitrate) 4–141
Color solutions
   for chloride 4–73
   for fluoride 4–92
   for nitrogen (ammonia) 4–118
   for nitrogen (nitrate) 4–138
   for silica 4–185
Color standards
   for silica 4–187t
Colorimeter
   for aluminum 3–68
Colorimetric equipment 1–47
   also see Filter photometer; Spectrophoto-
      meter
Colorimetric methods 1–47
   for aluminum 3–64
   for arsenic 3–74
   for beryllium 3–80
   for boron 4–7, 4–10
   for bromide 4–11
   for cadmium 3–83
   for chlorine 4–62, 4–64
   for chromium 3–91
   for COD 5–15
   for copper 3–95, 3–97
   for cyanide 4–31
   for fluoride 4–89
   for iodide 4–107, 4–109
   for iodine 4–103
   for nitrogen (ammonia) 4–117, 4–120
   for nitrogen (nitrate) 4–135, 4–141
   for nitrogen (nitrite) 4–129
   for nitrogen (organic) 4–144, 4–147
   for ozone 4–162
   for phenols 5–51, 5–54
   for phosphorus 4–173, 4–175, 4–177
   for selenium 3–135
   for silica 4–184, 4–188, 4–189
   for silver 3–142
   for sulfide 4–195
   for sulfite 4–201
   for tannin and lignin 5–68
   for thiocyanate 4–42
   for vanadium 3–155
   for zinc 3–158, 3–160, 3–162

Column
   ion exchange 1–46i
Column method
   for cation removal 1–46
Column packing
   for pesticides (organochlorine) 6–162
   for sludge digester gas 2–92
Combined reagent
   for phosphorus 4–177
Combustible-gas indicator
   for methane 6–61
      circuit and flow diagram 6–62i
Combustible-gas indicator method
   for methane 6–61
Combustion infrared method
   for organic carbon 5–18
Communities
   periphyton 10–66
Compensation
   in photometric measurements 1–49
Completed test
   for coliforms (total) 9–71, 9–72i
Complexing agents
   for hardness 2–55
Complexone method
   for fluoride 4–92
Computer codes
   for calcium carbonate saturation indices
      2–49, 2–50t
Concentration apparatus
   for viruses 9–163
   for zooplankton 10–29i
Concentration ranges
   for AA methods 3–15, 3–16t
Concentrations
   uniform reagent 1–45
   units 1–27, 1–43
Concentration techniques
   for *Campylobacter jejuni* 9–143
   for *Entamoeba histolytica* 9–203
   for *Giardia lamblia* 9–199
   for leptospires 9–147
   for plankton 10–16
   for *Salmonella* 9–131, 9–137
   for sulfide 4–194
   for viruses 9–156, 9–157
   for *Yersinia enterocolitica* 9–154
   for zooplankton 10–16
Concentrator tube
   for base/neutral and acid extractables
      6–119
Concussion
   for fish sampling 10–122

Condenser, cardioid dark-field
  for *Salmonella* 9–137
Condition index
  for fish 10–131i
Conductivity
  computation 1–20
  definition 1–4, 2–57
  for checking analyses 1–21
  methods 2–57
  potassium chloride solutions 2–60t
Conductivity cell
  for conductivity 2–59
Conductivity instruments
  for conductivity 2–59
Conductivity method
  for salinity 2–62
Conductivity water
  for conductivity 2–60
Confidence coefficient
  definition 1–4
Confidence interval
  definition 1–4
Confidence limit
  definition 1–4
Confirmed test
  for coliforms (total) 9–70, 9–71i
  for fecal streptococcus 9–110
  for *Pseudomonas aeruginosa* 9–54
Congo red paper
  for copper 3–96
Constant-flow apparatus
  for oyster shell deposition test 8–79i
Constant-temperature bath
  for odor determinations 2–19
Constant-temperature room
  for biostimulation 8–44
Consumers
  aquatic animals 10–142
Containers 1–41
  for samples 1–35
  also see Bottles; Glassware
Contaminants
  reagent water 1–57
Continuous-flow analyses 1–51
Continuum source corrector
  for AA methods 3–14
Control charts
  analysis 1–10
  duplicate analyses 1–11i
  for analytical quality control 1–9
  means chart 1–9, 1–10i, 1–12i
  range chart 1–9, 1–10t, 1–11i

Control cultures
  for microbiological tests 9–18t
Control ranges
  for metals by AA 3–19t
Conversion factors
  metric to English units *inside back cover*
  milligrams to milliequivalents 1–27t
Copepods
  algae as food 8–86, 8–88t
  collection 8–85
  culture apparatus 8–87i, 8–88i
  culturing 8–86
  food 8–86
  generation cages 8–89, 8–89i
  growth chambers 8–86
  holding 8–85
  identification 8–85
  toxicity test method 8–90
  water supply 8–86
Copper
  AA method 3–20
    low concentrations 3–23
    micro quantities 3–36
  bathocuproine method 3–97
  for nitrogen (organic) 4–144
  inductively coupled plasma method 3–54
  methods 3–94
  neocuproine method 3–95
Copper solution
  for copper 3–21, 3–96, 3–98
Copper sulfate solution
  for nitrogen (nitrate) 4–136, 4–138, 4–141
  for selenium 3–140
Copper sulfate-sulfamic acid flocculation
      modification
  for DO 4–158
Copper sulfate-sulfamic acid inhibitor solu-
      tion
  for DO 4–158
Copper-cadmium granules
  for nitrogen (nitrate) 4–135
Corals
  collecting 8–58
  culturing 8–59
  field preparation 8–58
  food supply 8–59
  handling 8–59
  holding 8–59
  lethal response 8–63
  parasites and predators 8–60
  planulae toxicity tests 8–62
  sublethal effects 8–63
  test conditions 8–61

Corals *(cont.)*
    test species 8–57
    toxicity tests 8–56
    transplanting 8–64
    water supply 8–59
Core samplers
    for macroinvertebrates 10–104, 10–104i
Correction curves
    aluminum in presence of fluoride 3–67i
Correction factors
    radon 222 activity 7–38t
Correctness
    analyses 1–20
Cotton, grease-free
    for oil and grease 5–47
Counters (biological)
    for biostimulation studies 8–49
    for direct microbial count 9–65
    for microorganisms 9–26
Counters (radioactivity) 7–5
    alpha scintillation 7–9
    alpha spectrometer 7–9
    beta 7–7
    counting rate-anode voltage curves 7–7i
    end-window 7–6
    for alpha and beta radiation 7–16
    for cesium 7–22
    for iodine (radioactive) 7–24, 7–26
    for periphyton productivity 10–45, 10–59
    for radium 7–30
    for radium 226 7–35
    for radium 228 7–44
    for strontium 7–48
    for uranium 7–57
    gamma spectrometer 7–8
    Geiger-Mueller 7–5,6,7
    internal proportional 7–6
    liquid beta scintillation 7–10
    low-background beta 7–7
    proportional 7–5
        thin-window 7–7
Counterstain
    for coliforms 9–73
Counting cell
    for phytoplankton 10–25i
    for plankton 10–24
Counting methods
    for algal biomass 8–49
    for bacteria, heterotrophic 9–56
    for coliforms 9–89
    for organisms in toxicity tests 8–30
    for periphyton 10–51

Counting methods *(cont.)*
    for phytoplankton 10–23
    for yeasts 9–192
Counting pans
    for alpha and beta radioactivity 7–17
    for strontium 7–49
Counting rate-anode voltage curves
    for radioactivity counters 7–7i
Counting room
    radioactivity analyses 7–4
Cover glass
    for *Salmonella* slides 9–137
Crabs
    culture chambers 8–94, 8–94i, 8–95
    food 8–95
    holding, acclimating, and culturing 8–94
    toxicity tests 8–106
        short-term 8–106
    water supply 8–94
    also see Crustaceans
Crayfish
    culture chambers 8–93
    food 8–93
    holding, acclimating, and culturing 8–93
    toxicity tests 8–106
    also see Crustaceans
Creel census
    for fish sampling 10–122
Crenothrix polyspora 9–116i
Crucibles
    for acids (organic) 5–70
    for calcium 3–88
Crustaceans 10-164i, 10–165i
    acclimation 8–92
    collection 8–92
    culturing 8–92
    diseases 8–100
    growths, harmful 8–100
    handling 8–92
    holding 8–92
    key 10–140
    parasites 8–100
    sublethal effects 8–104
    test species 8–92
    toxic effects 8–104
    toxicity tests 8–91, 8–104
    also see Amphipods, Crabs, Crayfish,
        Lobsters, Shrimp
        Daphnia, Copepods
Ctenodrilus serratus 8–68
Culture chambers
    for biostimulation 8–44

Culture chambers *(cont.)*
  for crabs 8–94, 8–94i, 8–95
  for crayfish 8–93
  for fish 8–14, 8–14i, 8–125, 8–126, 8–129
  for lobsters 8–96, 8–97i, 8–98
  for macroinvertebrates 8–13
  for macroinvertebrates 8–14i
  for shrimp 8–99
Culture closures
  for biostimulation 8–44
Culture dishes
  for fecal coliforms 9–94
  for total coliforms 9–84, 9–92
Culture medium
  basal
    for coliform differentiation 9–106
  CCVC 9–150
  EC 9–75
  ferric ammonium citrate 9–124
  for algae 8–17
  for biostimulation 8–17
  for coliforms 9–86
  for *Daphnia* 8–82
  for heterotrophic plate count 9–55, 9–63
  for *Metallogenium* 9–124
  for microbiological examinations
    holding times 9–17t
    pH adjustment 9–29
    preparation 9–15, 9–29
    preparation 9–29
    quality control 9-13, 9–14, 9–17
    specifications 9–31
    sterilization 9–15, 9–30
    storage 9–16, 9–29
    utensils for preparing 9–26
    water for preparing 9–30
  for protozoa 8–54
  for *Pseudomonas aeruginosa* 9–52, 9–53
  GPVA 9–150
  iron-oxidizing 9–122
  isolation, iron bacteria 9–121
  LES MF 9–93
  maintenance (SCY), iron bacteria 9–121
  M-Endo 9–86
  M-VFC 9–97
  sulfate-reducing 9–122
    for *Thiobacillus thioparus* 9–122
  sulfur
    for *Thiobacillus thiooxidans* 9–123
  also see Agars, Broths
Culture medium containers
  for total coliforms 9–84

Culture medium dispensing apparatus
  maintenance 9–7
Culture tubes
  for selenium 3–134
Cultures
  algae
    for biostimulation studies 8–46
  coliforms 9–100
Cupferron solution
  for chromium 3–92
Cupferron solution
  for nickel 3–122
Curcumin method
  for boron 4–7
Curcumin reagent
  for boron 4–8
Current meter
  for periphyton productivity 10–59
Cuvettes
  for chlorophyll 10–32
CVCC medium
  for Legionellaceae 9–150
Cyanates
  method 4–41
Cyanide
  amenable to chlorination after distillation
    method 4–34
  amenable to chlorination without distil-
    lation
    method 4–36
  colorimetric method 4–31
  dissociable
    method 4–38
  distillation apparatus 4–28, 4–29i
  methods 4–20
  preliminary treatment 4–25
  safety precautions 1–64
  selective electrode method 4–33
  solid wastes 4–22
  spot test for sample screening 4–40
  titrimetric method 4–30
  total after distillation 4–28
  weak
    method 4–38
Cyanide citrate reducing solution
  for lead 3–108
Cyanide selective electrode method
  for cyanide 4–33
Cyanide solution
  for cyanide 4–31, 4–33
Cyanogen chloride
  method 4–39

Cycloheximide
for coliforms 9–92
Cylinder, graduated
for microbiological tests 9–26
Cyprinodon variegatus, see Minnow, sheepshead
Cytochrome oxidase test
for coliform verification 9–19
Czapek agar
for fungi 9–188

D

Damselflies 10–169i
Daphnia
culturing 8–82
long-term toxicity tests 8–84
reproductive impairment test 8–84
static toxicity tests 8–83
toxicity tests 8–81
Data
observations 1–4t
quality 1–14
rejection criteria 1–3
Data quality
glossary of terms 1–4
Data system
for chlorophyll 10–36
for organic constituents (individual) 6–15
for volatile organics 6–30
DDD
liquid-liquid extraction GC method 6–170
liquid-liquid extraction GC/MS method
6–113
DDE
liquid-liquid extraction GC method 6–170
liquid-liquid extraction GC/MS method
6–113
DDT
liquid-liquid extraction GC method 6–170
liquid-liquid extraction GC/MS method
6–113
Decafluorotriphenyl phosphine (DFTPP)
solution
for base/neutral and acid extractables
6–120
masses and abundance criteria 6–120t
Decarboxylase test
for coliform differentiation 9–102t, 9–106
Decay factors
radon 222 7–38t
Dechlorinating agent
for nitrogen (ammonia) 4–115

Dechlorination
microbiological samples 9–31
odor samples 2–18
Decontamination
virus sample concentrate 9–177
De-emanation assembly
for radium 226 7–34i
Deionization unit
maintenance 9–6
Delayed-incubation procedure
for fecal coliforms 9–96
for total coliforms 9–91
Densitometer
for salinity 2–64
Density method
for salinity 2–64
Derivatives
phenols
recovery 6–139t
Derivatization reagent
for phenols 6–140
Desiccator
for solids 2–73
Desorber
for volatile halocarbons 6–87
for volatile organics 6–29
Detection limits 1–18
AA methods 3–15
definition 1–4
determination 1–18
for acid extractables 6–118t
for base/neutral extractables 6–115t
for closed loop stripping GC/MS method
6–12t
for metals by ICP method 3–55t
for organochlorine pesticides and PCBs
6–171t
for PCBs 6–171t
for phenols 6–138t
for volatile aromatics 6–66t
for volatile halocarbons 6–80t
for volatile organics 6–28t
relationships 1–19, 1–19i
Detectors
flame ionization
for polynuclear aromatic hydrocarbons
6–150
fluorescence
for polynuclear aromatic hydrocarbons
6–150
for chlorophyll 10–36
for gas chromatographic methods 6–7

Detectors *(cont.)*
  for high-performance liquid chromato-
      graphic methods 6–9
  for phenols 6–139
  for volatile aromatics 6–67
  for volatile halocarbons 6–81, 6–88, 6–93
  ultraviolet
      for polynuclear aromatic hydrocarbons
          6–150
Dialysis tubing
  for virus concentration 9–174
Diamalt agar
  for fungi 9–188
Diaminonaphthalene solution
  for selenium 3–135
Diammonium hydrogen phosphate solution
  for magnesium 3–113
  for sulfide 4–196
  for uranium 7–59
Diatomaceous earth technique
  for *Salmonella* concentration 9–132
Diatomaceous-silica filter aid suspension
  for oil and grease 5–46
Diatom mounts
  counting 10–27
Diatoms 10–149i, 10–150i
  key 10–185
  mounts 10–18
  species proportional count 10–27,
      10–52
Dibenzo(a,h)anthracene
  liquid-liquid extraction chromatographic
      method 6–148
  liquid-liquid extraction GC/MS method
      6–113
Dibromochloromethane
  purge and trap GC method 6–79, 6–87
      capillary column 6–93
  purge and trap GC/MS method 6–27,
      6–42
      capillary column 6–50
Dibromochloropropane (DBCP)
  chromatographic conditions 6–100t
  gas chromatogram 6–101i
  liquid-liquid extraction GC method 6–98
      bias and precision 6–102t
  methods 6–98
  purge and trap GC method
      capillary column 6–93
  purge and trap GC/MS method 6–42
      capillary column 6–50
  retention time 6–100t

Dibromochloropropane solution
  for dibromochloropropane 6–100
Dibromoethane (EDB)
  chromatographic conditions 6–100t
  gas chromatogram 6–101i
  liquid-liquid extraction GC method
      6–98
      bias and precision 6–102t
  methods 6–98
  purge and trap GC method 6–87
      capillary column 6–93
  purge and trap GC/MS method 6–42
      capillary column 6–50
  retention time 6–100t
Dibromoethane solution
  for dibromoethane 6–100
Dibromomethane
  purge and trap GC method 6–87
      capillary column 6–93
  purge and trap GC/MS method 6–42
      capillary column 6–50
Dibutyl phthalate
  liquid-liquid extraction GC/MS method
      6–113
Dichloramine
  procedure 4–57, 4–61, 4–63
Dichlorobenzene
  liquid-liquid extraction GC/MS method
      6–113
  purge and trap GC method 6–65, 6–71,
      6–79, 6–87
      capillary column 6–93
  purge and trap GC/MS method 6–27,
      6–42
      capillary column 6–50
Dichlorobenzidine
  liquid-liquid extraction GC/MS method
      6–113
Dichlorodifluoromethane
  purge and trap GC method 6–79, 6–87
      capillary column 6–93
  purge and trap GC/MS method 6–42
      capillary column 6–50
Dichloroethane
  purge and trap GC method 6–79, 6–87
      capillary column 6–93
  purge and trap GC/MS method 6–27,
      6–42
      capillary column 6–50
Dichloroethene
  purge and trap GC method 6–79, 6–87
      capillary column 6–93

Dichloroethene *(cont.)*
  purge and trap GC/MS method 6–27,
    6–42
    capillary column 6–50
Dichlorophenol
  liquid-liquid extraction GC method 6–137
  liquid-liquid extraction GC/MS method
    6–113
Dichloropropane
  purge and trap GC method 6–79, 6–87
    capillary column 6–93
  purge and trap GC/MS method 6–27,
    6–42
    capillary column 6–50
Dichloropropene
  purge and trap GC method 6–79, 6–87
    capillary column 6–93
  purge and trap GC/MS method 6–27,
    6–42
    capillary column 6–50
Dichromate solution
  for chlorine 4–49
Dieldrin
  liquid-liquid extraction GC method 6–170
  liquid-liquid extraction GC/MS method
    6–113
Diethylaminoethyl cellulose
  for aquatic humic substances 5–38
Diethylaminoethyl method
  for aquatic humic substances 5–38
Diethyl ether
  for herbicides 6–183
  for pesticides (organochlorine) 6–161
Diethylthalate
  liquid-liquid extraction GC/MS method
    6–113
Differentiation
  coliforms 9–99
  Enterobacteriaceae 9–102t
    kits 9–103
    Klebsiella 9–101
  fecal streptococci and enterococci 9–112,
    9–114t
  leptospires 9–148
Digester gas, *see* Sludge digester gas
Digestion
  for metals 3–6
    nitric acid 3–8
    nitric acid-perchloric acid 3–10
    nitric acid-perchloric acid-hydrofluoric
      acid 3–11

Digestion *(cont.)*
  for metals *(cont.)*
    nitric acid-sulfuric acid 3–9
    sludge samples 3–7
    solid samples 3–7
  for phosphorus 4–171, 4–172
    perchloric acid 4–171
    persulfate 4–172
    sulfuric acid-nitric acid 4–172
  hydrochloric acid
    for selenium 3–134
  hydrogen peroxide
    for selenium 3–132
  nitric acid-hydrochloric acid
    for metals 3–8
  permanganate
    for selenium 3–133
  persulfate
    for selenium 3–132
Digestion apparatus
  for nitrogen (organic) 4–145, 4–147
Digestion rack
  for phosphorus 4–172
Digestion reagent
  for nitrogen (organic) 4–145
Digestion vessels
  for COD 5–14
Digestion vials
  for selenium 3–133
Dihydroxybenzoic acid solution
  for trihalomethane formation potential
    5–76
Dilution
  for BOD 5–3, 5–7
  for heterotrophic plate count 9–58, 9–59i
  for odor intensities 2–22t
  reagents 1–44
  swimming pool samples 9–47
Dilution bottles
  for microbiological tests 9–27
Dilution water, *see* Water, dilution
Dimethylglyoxime method
  for nickel 3–123
Dimethylphenol
  liquid-liquid extraction GC method 6–137
  liquid-liquid extraction GC/MS method
    6–113
Dinitrophenol
  liquid-liquid extraction GC method 6–137
  liquid-liquid extraction GC/MS method
    6–113

Dinitrotoluene
  liquid-liquid extraction GC/MS method
    6-113
Di-n-octyl phthalate
  liquid-liquid extraction GC/MS method
    6-113
Diphenylcarbazide solution
  for chromium 3-92
Diphenylcarbazone-bromphenol blue indi-
    cator reagent
  for chloride 4-70
Diphenylcarbazone-xylene cyanol nitric acid
    reagent
  for chloride 4-69
Dip nets
  for fish sampling 10-120
Direct membrane filter technique
  for coliforms 9-88
Direct-sensing membrane diffusion method
  for dissolved gas supersaturation 2-94
Direct total microbial count 9-64
  epifluorescence microscopic method 9-65
Discordancy
  data 1-4t
Disease
  annelids 8-69
  crustaceans 8-100
  fish 8-122, 8-124t
  mollusks 8-76
  toxicity tests 8-21
Disodium bathocuproine disulfonate solu-
    tion
  for copper 3-98
Disodium phosphate
  for virus concentration 9-166
Displacement volume
  plankton 10-40
Disruptor-emulsifier, ultrasonic
  for virus concentration 9-174
Dissolved oxygen (DO), see Oxygen (dis-
    solved)
Distillation
  for ammonia-free water 4-115
  for fluoride 4-85
  for nitrogen (ammonia) 4-115
  reagent-water 1-56
Distillation apparatus
  for cyanide 4-28, 4-29i
  for fluoride 4-86, 4-86i
  for iodine (radioactive) 7-27, 7-28i
  for nitrogen (ammonia) 4-115

Distillation apparatus (cont.)
  for nitrogen (organic) 4-147, 4-148i
  for phenols 5-50
  for tritium 7-54
Distillation method
  for acids (volatile) 5-72
  for iodine (radioactive) 7-27
Distillation reagent
  for fluoride 4-92
Distilled water, see Water, distilled
Dithizone method
  for cadmium 3-83
  for lead 3-107
  for mercury 3-119
  for silver 3-142
  for zinc 3-158, 3-160
Dithizone solution
  for cadmium 3-83
  for lead 3-108
  for mercury 3-119
  for silver 3-143
  for zinc 3-159, 3-161
  preparation 1-49
Diversity indices
  for macroinvertebrates 10-112
DPD colorimetric method
  for chlorine 4-62
  for chlorine dioxide 4-79
DPD ferrous titrimetric method
  for chlorine 4-58
DPD indicator solution
  for chlorine 4-60
Drag chains
  for macrophyton sampling 10-75
Dragonflies 10-169i
Dredge, Goin
  for fish 10-120
Drift net samplers
  for macroinvertebrates 10-104, 10-105i
Drinking water, see Water, drinking
Dropper
  for arsenic and selenium 3-46
Dry mounts
  macrophyton 10-78
Drying column, chromatographic
  for base/neutral and acid extractables
    6-119
Drying tube
  for mercury 3-29
Dulcitol selenite broth
  for Salmonella enrichment 9-133

Duplicates
  acceptance limits 1–8t
  control charts 1–11i
  for analytical quality control 1–8
Dye
  for aluminum 3–65
  for microbiological examinations 9–13
  specifications 1–43

                            E

Earthy-musty-smelling compounds
  closed-loop stripping GC/MS analysis
    bias 6–23t
    data 6–21t
    detection limits 6–12t
    precision 6–24t
EC medium
  for fecal coliforms 9–75
EC medium test
  for fecal coliforms 9–75
Echinoderms 10–178i
EDTA-hydroxylamine solution
  for selenium 3–135
EDTA reagent
  for aluminum 3–65
  for beryllium 3–80
  for calcium 3–86
  for chlorine dioxide 4–79
  for cyanide 4–37
  for hardness 2–54, 2–56
  for nitrogen (ammonia) 4-118, 4–127
  for radium 7–30
  for radium 226 7–37
  for sulfate 4–210
  for sulfite 4–200
EDTA titrimetric method
  for calcium 3–85
  for hardness 2–53
Eelworms, see Nematodes
Effective concentration
  definition 8–3
  estimation in quantal toxicity tests 8–34
Egg-hatching tank
  for lobsters 8–95, 8–96i
Egg-laying tank
  for lobsters 8–95
EIA substrate 9–111
Ekman grab
  for macroinvertebrates 10–102, 10–102i
  for macrophyton 10–75, 10–76

Electrical conductivity method
  for salinity 2–62
Electrical hazards 1–66
Electrode
  for chloride 4–71
  for chlorine 4–55, 4–66
  for DO 4–161
  for fluoride 4–88
  for nitrogen (ammonia) 4–114t
  for nitrogen (nitrate) 4–134
  for pH 4–96
  membrane
    calibration 4–161
    effect of stirring 4–161i
    for DO 4–159
    salting out effect on 4–160i
    temperature effects on 4–160i
  problems 4–101
  selective ion 1–51
  storage 4–100
Electrode filling solution
  for nitrogen (nitrate) 4–134
Electrode method
  for cyanide 4–33
  for DO 4–158
  for fluoride 4–87
  for nitrogen (ammonia) 4–122
  for nitrogen (nitrate) 4–133, 4–140
Electrofishing
  boat and equipment 10–121i
  devices 10–120
Electrolytic conductivity detector
  for gas chromatographic methods 6–7
Electrometer
  for nitrogen (ammonia) 4–123
Electrometric method
  for pH 4–95
Electron capture detector
  for gas chromatographic methods 6–7
Electrothermal atomic absorption spectro-
    metric method, see Atomic absorption
    spectrometric method, electrothermal
Eluent
  for anions 4–3
  for chlorophyll 10–36
  for virus concentration 9–160, 9–166,
    9–175
Elutriation devices
  for macroinvertebrate analysis 10–109
Emanation method
  for radium 226 7–34
End-point detection apparatus
  for chlorine 4–55

Endosulfan
   liquid-liquid extraction GC method 6–170
   liquid-liquid extraction GC/MS method
      6–113
Endosulfan sulfate
   liquid-liquid extraction GC method 6–170
   liquid-liquid extraction GC/MS method
      6–113
Endrin
   liquid-liquid extraction GC method 6–170
   liquid-liquid extraction GC/MS method
      6–113
Endrin aldehyde
   liquid-liquid extraction GC method 6–170
   liquid-liquid extraction GC/MS method
      6–113
Enrichment
   Campylobacter jejuni 9–143
   coliforms 9–88
   leptospires 9–147
   Salmonella 9–133, 9–138
   Shigella 9–141
   Vibrio cholerae 9–145
   yeasts 9–192
Enslow test
   calcium carbonate precipiation potential
      2–48
Entamoeba histolytica
   concentration technique 9–203
   method 9–203
Enterobacteriaceae
   differentiation 9–100, 9–102t
   nomenclature 9–101t
   also see Coliform group bacteria
Enterococcus
   definition 9–108
   differentiation 9–113
   differentiation 9–114t
   mE method 9–111
   methods 9–108
   verification 9–112
Epifluorescence microscopic method
   for direct total microbial count 9–65
Equilibrium constants
   for calcium carbonate saturation 2–43t,
      2–44t
Equivalency testing
   analytical methods 1–24
Equivalent weights
   for chlorine dioxide calculation 4–83t
Eriochrome Black T indicator
   for hardness 2–55

Eriochrome Blue Black R indicator
   for calcium 3–86
Eriochrome cyanine R method
   for aluminum 3–64
Eriochrome cyanine R solution
   for aluminum 3–65
Eriochrome cyanine solution
   for aluminum 3–65
Erlenmeyer flask
   for herbicides 6–183
   for phosphorus 4–171
Error
   types
      definition 1–5
Escherichia coli
   culture for coliphage detection 9–40
   in bathing waters 9–50
   pathogenic 9–141
Ethanol-hydrochloric acid cleaning solution
   for aluminum 3–68
Ethylbenzene
   purge and trap GC method 6–65, 6–71
   capillary column 6–93
Ethylbenzene
   purge and trap GC/MS method 6–27,
      6–42
   capillary column 6–50
Ethylene diamine
   for cyanide pretreatment 4–26
Ethyl ether
   for organochlorine pesticides and PCBs
      6–172
Ethylhexyl phthalate bis-(2-)
   liquid-liquid extraction GC/MS method
      6–113
Evaporating dishes
   for boron 4–8
   for silica 4–183
   for solids 2–73
Evaporative concentrator
   for herbicides 6–183
   for pesticides (organochlorine) 6–160
Evaporative flask
   for base/neutral and acid extractables
      6–119
Evaporator
   for selenium 3–139
Exposure chambers
   for amphipods 8–105
   for annelids 8–70
   for insects 8–117
   for oyster shell deposition test 8–78
   for toxicity tests 8–31

Exposure period
  for periphyton sample collection 10–50
Extraction
  chlorophyll 10–31
Extraction method
  for *Giardia lamblia* 9–198
  for oil and grease 5–45
    in sludges 5–46
Extraction solvent
  for trihalomethanes 6–107
Extraction vessel
  for trihalomethanes 6–106
Eye washes 1–60

                            F

FACTS method
  for chlorine 4–64
FA Kirkpatrick fixative
  for *Salmonella* 9–137
Falkow basal medium
  for coliform differentiation 9–106
FA Salmonella panvalent conjugate 9–137
Fathead minnow, *see* Minnow, fathead
Fathometers
  for macrophyton sampling 10–75
Fathometry
  for macrophyton mapping 10–73
Fecal coliform test
  A1 medium 9–76
  EC medium 9–75
Fecal coliforms, *see* Coliform group bacteria, fecal
Fecal streptococcus group, *see* Streptococcus, fecal
Fermentation reactions
  Salmonella 9–135
Fermentation tubes
  for microbiological tests 9–27
Ferric ammonium citrate medium
  for heterotrophic iron-precipitating bacteria 9–124
Ferric ammonium sulfate solution
  for sulfite 4–202
Ferric chloride solution
  for BOD 5–5
  for cyanide pretreatment 4–26
  for sulfide 4–196
  for uranium 7–57
Ferric nitrate-mercuric thiocyanate solution
  for chloride 4–73

Ferric nitrate solution
  for chloride 4–73
  for thiocyanate 4–43
Ferricyanide method
  for chloride 4–73
Ferroin indicator solution
  for COD 5–12
Ferrous ammonium sulfate solution
  for chlorine 4–60
Ferrous ammonium sulfate solution
  for COD 5–12, 5–14
  for iodide 4–109
  for nitrogen (nitrite) 4–130
Ferrous sulfide agar
  for iron bacteria (*Gallionella ferruginea*) 9–122
Fiber suppressor
  for anions 4–3
Field counting
  phytoplankton 10–24
Field filtration units
  for delayed-incubation total coliforms 9–92
Field method
  for color 2–3
Filter
  for organic constituents (individual)
    extraction 6–15i
    flow rate 6–15i
    resistance effect on compound recovery 6–16i
  for phosphorus 4–170
  for sulfate 4–205, 4–206
  for virus concentration 9–160, 9–166, 9–171
Filter crucibles
  for calcium 3–88
Filter funnel
  for phenols 5–51
  for zooplankton concentration 10–17i
Filter holder
  for floatables 2–67
  for organic constituents (individual) 6–13
  for virus concentration 9–159
Filter, optical
  for *Salmonella* 9–137
Filter paper
  for phenols 5–51
Filter photometer
  for aluminum 3–64
  for anionic surfactants 5–61
  for arsenic 3–75

Filter photometer *(cont.)*
  for beryllium 3–80
  for boron 4–8, 4–10
  for bromide 4–12
  for cadmium 3–83
  for chlorine 4–62, 4–65
  for chromium 3–92
  for color 2–9
  for copper 3–96, 3–98
  for cyanide 4–31
  for fluoride 4–90
  for iodide 4–107, 4–109
  for iodine 4–104
  for iron 3–102
  for manganese 3–115
  for nickel 3–122
  for nitrogen (ammonia) 4–117, 4–120
  for nitrogen (nitrate) 4–135
  for nitrogen (nitrite) 4–129
  for ozone 4–163
  for phosphorus 4–173, 4–177
  for silica 4–185, 4–188
  for silver 3–143
  for sulfate 4–207
  for sulfide 4–196
  for sulfite 4–202
  for tannin and lignin 5–68
  for thiocyanate 4–42
  for vanadium 3–155
  for zinc 3–158, 3–160, 3–163
Filters
  for organic constituents (individual)
    6–14
Filtration
  metals 3–5
  phosphorus 4–170
Filtration equipment
  for chlorophyll 10–32
  for coliforms 9–84
    delayed-incubation procedure 9–92
  for color 2–5, 2–5i
  for direct microbial count 9–65
  for *Salmonella* 9–132
  for solids 2–74
  for sulfate 4–205
  for surfactants 5–57
Fire extinguishers 1–60
Fish 10–180i
  acclimation 8–122
  age and growth determination 10–131
  analysis of collections 10–129
  anatomical diagrams 10–128i

Fish *(cont.)*
  assemblage
    definition 10–114
  collection 8-122, 10–116
  culturing 8–124, 8–129
  data acquisition 10–114
  definition 10–2
  diet analysis 10–130
  diseases 8–122, 8–124t
  diversity indices 10–132
  ecological importance 10–113
  food 8–19
  freshwater 8–124
  handling 8–122
  holding 8–122
  holding and culturing tanks 8–14,
    8–14i
  identification 10–129
  incubation cups 8–135
  marine and estuarine 8–129
  methods 10–113
  parasites 8–122, 8–124t
  photoperiod 8–126t
  population
    definition 10–114
    estimation 10–130
  preservation 10–127
  scales 10–131, 10–131i
  test species 8–120
  toxicity tests 8–120, 8–133
    complete-life-cycle 8–134
    partial-life-cycle 8–134
    short-term 8–133
Fish kills 10–134
Five-day test
  for BOD 5–4
Fixative
  for direct microbial counts 9–65
  for *Salmonella* 9–137
Fixed matter, *see* Solids, fixed
Flagellates 10–155i, 10–156i, 10–157i
Flame emission photometric method
  for lithium 3–110
  for potassium 3–125
  for sodium 3–146
  for strontium 3–150
Flame ionization detector
  for gas chromatographic methods 6–8
Flame photometer
  for lithium 3–110
  for sodium 3–147
Flame photometry 1–50

Flameless atomic absorption spectrometric method, *see* Atomic absorption spectrometric method, electrothermal
Flasks, *see* Glassware
Flatworms 10–163i
Flavor
  methods 2–23
  relationship to taste 2–23
  threshold test 2–24
Flavor profile analysis
  for taste 2–29
Flavor rating assessment
  action tendency scale 2–28
  for taste 2–27
Flavor threshold number 2–25t
  dilutions 2–26t
Flavor threshold test
  for taste 2–24
Flesh-tainting tests 8–8
Flies 10–167i, 10–168i, 10–169i, 10–170i, 10–171i, 10–172i, 10–173i
Floatable oil tube
  for oil and grease 2–69, 2–69i
Floatables
  methods 2–65
  particulate
    method 2–66
    precision and bias 2–68t
Floatables sampler-mixer 2–66, 2–66i
Flotation funnel 2–67, 2–67i
  mounting 2–67i
Flow meter, gas
  for surfactants 5–57
Fluid proportioner
  for virus concentration 9–163
Fluoranthene
  liquid-liquid extraction chromatographic method 6–148
  liquid-liquid extraction GC/MS method 6–113
Fluorene
  liquid-liquid extraction chromatographic method 6–148
  liquid-liquid extraction GC/MS method 6–113
Fluorescence detector
  for high-performance liquid chromatographic methods 6–10
Fluorescent antibody reaction
  for *Salmonella* 9–138

Fluorescent antibody technique
  for Legionellaceae 9–150
  for *Salmonella* 9–136
Fluoride
  automated equipment manifold 4–93i
  complexone method (automated) 4–92
  distillation 4–85
    apparatus 4–86i
  methods 4–84
    interferences 4–85t
  preliminary treatment 4–84
  selective-electrode method 4–87
  SPADNS method 4–89
Fluoride solution
  for fluoride 4–88, 4–90, 4–92
  for ion chromatography 4–4
Fluorochrome
  for direct microbial counts 9–65
Fluorometer
  for chlorophyll a 10–34
  for periphyton productivity 10–59
  for uranium 7–59
Fluorometric equipment
  for selenium 3–137
Fluorometric method
  for chlorophyll a 10–34
  for selenium 3–137
  for uranium 7–59
Flux
  for radium 226 7–37
  for uranium 7–59
Folin phenol reagent
  for tannin and lignin 5–68
Folsom plankton splitter
  for zooplankton subsamples 10–29, 10–30i
Food
  for brook trout 8–137
  for fish 8–19
  for macroinvertebrates 8–19
  for minnows
    fathead 8–139
    sheepshead 8–140
  for toxicity test organisms 8–17
Forceps
  for coliforms 9–85
Formaldehyde solution
  for cyanides 4–37
  for fish sample preservation 10–127
Formalin
  for fish preservation 10–127
  for phytoplankton preservation 10–8

Free fall net
  for plankton 10–11i
Free fatty acid phase
  for phenols 562
Freezer
  for microbiological laboratory 9–7
Freon, see Trichlorotrifluoroethane
Frequency curve 9–21i
Fume hoods 1–61
Fuming chamber
  for plankton productivity 10–45
Fungi 10–182i
  growth patterns 9–185
  membrane filter technique 9–191
  methods 9–183
  pathogenic
    methods 9–195
  pour plate technique 9–188
  samples 9–188
  spread plate technique 9–190
  survival of chlorination 9–184
  zoosporic 9–192
Furnace, graphite
  for AA method (electrothermal) 3–36
Fusion dishes
  for uranium 7–59
Fyke nets
  for fish 10–117

G

Gallic acid method
  for vanadium 3–155
Gallic acid solution
  for vanadium 3–156
Gallionella ferruginea 9–117i, 9–118i
  enumeration, enrichment, isolation 9–124
  ferrous sulfide agar for 9–122
Gas
  compressed
    safety precautions 1–67
  sludge digester, see Sludge digester gas
  supersaturation
    direct-sensing   membrane   diffusion
      method 2–94
    methods 2–94
Gas analysis apparatus
  for sludge digester gas 2–89
Gas chromatogram
  acid extractables 6–124i
  base/neutral extractables 6–123i

Gas chromatogram (cont.)
  chlordane 6–125i, 6–175i
  dibromochloropropane 6–101i
  dibromoethane 6–101i
  organochlorine pesticides 6–175i
  PCB-1016 6–126i, 6–176i
  PCB-1221 6–126i, 6–177i
  PCB-1232 6–127i, 6–177i
  PCB-1242 6–127i, 6–178i
  PCB-1248 6–128i, 6–178i
  PCB-1254 6–129i, 6–179i
  PCB-1260 6–130i, 6–179i
  pesticides 6–124i
  phenol derivatives 6–143i
  phenols 6–141i
  polynuclear aromatic hydrocarbons
    6–153, 6–154, 6–155
  toxaphene 6–125i, 6–176i
  trihalomethanes 6–107i, 6–108i
  volatile aromatics 6–69i, 6–72i, 6–73i
  volatile halocarbons 6–88i, 6–95i
  volatile organics 6–34i, 6–52i, 6–54i,
    6–55i
Gas chromatograph
  for base/neutral and acid extractables
    6–119
  for dibromoethane and dibromochloropro-
    pane 6–99
  for organic constituents (individual) 6–6,
    6–15
  for organochlorine pesticides 6–168
  for organochlorine pesticides and PCBs
    6–172
  for pesticides 6–183
  for phenols 6–139
  for polynuclear aromatic hydrocarbons
    6–183
  for trihalomethanes 6–106
  for volatile aromatics 6–67, 6–72
  for volatile halocarbons 6–81
  for volatile organics 6–30, 6–44, 6–50
Gas chromatographic/mass spectrometric
    interface
  for volatile organics 6–30, 6–44
Gas chromatographic/mass spectrometric
    method
  closed-loop stripping
    for organic constituents (individual)
      6–11
  for base/neutral and acid extractables
    6–113

Gas chromatographic/mass spectrometric
  method *(cont.)*
  for volatile organics 6–27, 6–42
    capillary column 6–50
Gas chromatographic method 1–51
  detectors 6–7
  for dibromoethane and dibromochloropro-
    pane 6–98
  for herbicides 6–182
  for organic constituents (individual) 6–6
  for pesticides 6–158, 6–170
  for phenols 6–137
  for polynuclear aromatic hydrocarbons 6–
    148
    conditions 6–151t
    retention times 6–151t
  for trihalomethanes 6–104
  for volatile aromatics 6–65, 6–71
  for volatile halocarbons 6–79, 6–87
    capillary column 6–93
Gas collection apparatus
  for sludge digester gas 2–88i
Gas extraction
  for organics concentration 6–10
Gas flow meter
  for sulfite 4–201
Gas purification tube
  for radium 226 7–36
Gas supply
  for radium 226 7–36
Gas washing bottle
  for sulfite 4–201
  for surfactants 5–57
Generation cage
  for copepods 8–89, 8–89i
Generation system
  for chlorine dioxide 4–76, 4–76i
Gentamycin-kanamycin
  for virus concentration 9–160, 9–172
Geosmin
  mass spectrum 6–22i
  also see Earthy-musty-smelling compounds
Giardia lamblia
  analytical flow chart 9–199i
  concentration technique 9–199
  extraction 9–198
  identification 9–201
  method 9–197
  sampler 9–200i
Gill nets
  for fish 10–117
Glass electrodes, *see* Electrodes

Glass rods
  for heterotrophic plate count 9–61
Glasses, safety
  for laboratory personnel 1–61
Glassware 1–41
  for aluminum 3–64
  for biostimulation studies 8–47
  for bromide 4–12
  for cadmium 3–83
  for chlorine 4–56, 4–62
  for chlorine dioxide 4–81
  for copper 3–98
  for flavor threshold test 2–25
  for herbicides 6–183
  for iodide 4–107
  for iodine 4–104
  for iron 3–102
  for lead 3–108
  for mercury 3–119
  for microbiological tests 9–8, 9–28
  for odor determinations 2–18
  for organic constituents (individual) 6–11
  for ozone 4–163
  for pesticides (organochlorine) 6–159
  for phosphorus 4–173
  for radium 226 7–36
  for silver 3–143
  for sodium 3–147
  for zinc 3–158
  inhibitory residue detection 9–8
  pH check 9–8
  protection for shipment 1–42
  reagent storage 1–44
  volumetric 1–42
  also see Bottles
Glossary
  statistical terms 1–4
Gloves
  for laboratory personnel 1–61
Glucose broth, buffered
  for coliform differentiation 9–105
Glucose-glutamic acid check
  for BOD 5–5
Glucose-glutamic acid solution
  for BOD 5–5
Glucose-neopeptone agar
  for fungi 9–188
Glucose-neopeptone-rose bengal aureomycin
  agar
  for fungi 9–188
Glucose-peptone-aureomycin-rose bengal
  agar

Glucose-peptone-aureomycin-rose bengal agar *(cont.)*
  for fungi 9–190
Glucose-peptone-aureomycin-rose bengal agar
  for fungi
    modified 9–191
Glucose-yeast extract-malt extract agar
  for fungi 9–188
Glucose-yeast nitrogen base broth
  for yeasts 9–192
Glutaraldehyde preservative
  for phytoplankton 10–8
Glycerol-casitone-yeast autolysate broth (CGY)
  for iron bacteria 9–121
Glycine-hydrochloric acid
  for virus concentration 9–160, 9–166
Glycine-sodium hydroxide
  for virus concentration 9–160, 9–166
Glycine solution
  for chlorine 4–60
  for chlorine dioxide 4–79
  for ozone 4–163
Gold
  AA method 3–20
  methods 3–99
Gold solution
  for gold 3–21
GPVA medium
  for Legionellaceae 9–150
Grab samplers
  for macrophyton 10–75
Gram stain technique
  for coliform bacteria 9–74
Grass medium
  for *Daphnia* culture 8–83
Gravimetric method
  for magnesium 3–113
  for plankton biomass 10–40
  for silica 4–183
  for sulfate
    with residue drying 4–206
    with residue ignition 4–204
Grease
  trichlorotrifluoroethane-soluble floatable method 2–69
Grease and oil, *see* Oil and grease
Growth chambers
  for copepods 8–86
  for trout 8–136
  also see Culture chambers

Growth factors
  radon 222 7–38t
Growth increment measurements
  macrophyton productivity 10–82
Growth rate
  determination 8–9
  maximum specific 8–r8
  phytoplankton 8–52
  specific 8–4
Growths, harmful
  on crustaceans 8–100
Guard column
  for anions 4–3
Gulf V sampler
  for zooplankton 10–13i
Gut examination
  fish 10–122
Gutzeit generator, *see* Arsine generator

H

Halocarbons
  liquid-liquid extraction GC method 6–104
Halocarbons, volatile
  calibration criteria 6–83t
  chromatogram 6–82i
  chromatographic conditions 6–80t
  detection limits 6–80t
  gas chromatogram 6–88i, 6–95i
  methods 6–78
  purge and trap GC method 6–79, 6–87
    bias and precision 6–85t, 6–91t, 6–92t, 6–96t
    capillary column 6–93
  quality control acceptance criteria 6–83t
  retention times 6–89t, 6–93t
Halogens, organic, *see* Organic halogen (dissolved); Halocarbons, volatile
Hardness
  calculation method 2–53
  definitions 2–52
  EDTA titrimetric method 2–53
  interference with inhibitors 2–54t
  methods 2–52
Hardy continuous plankton recorder 10–13i
Hardy plankton indicator 10–13i
Hatching chambers
  for sheepshead minnow 8–140
Hay extract
  for *Beggiatoa* culture 9–125

Heater
  for organic constituents (individual) 6–11,
    6–13i
Heating block
  for COD 5–14
Helium gas
  for organic constituents (individual) 6–16
Hellgrammites 10–170i
Hemacytometer
  for phytoplankton counting 10–25
Heptachlor
  liquid-liquid extraction GC method 6–170
  liquid-liquid extraction GC/MS method
    6–113
Heptachlor epoxide
  liquid-liquid extraction GC method 6–170
  liquid-liquid extraction GC/MS method
    6–113
Heptoxime method
  for nickel 3–121
Heptoxime reagent
  for nickel 3–122
Herbicides, chlorinated phenoxy acid
  chromatogram 6–186i
  liquid-liquid extraction GC method 6–182
  recovery from surface water 6–187t
  results 6–185i
  retention times 6–186t
  also see Pesticides
Herbicide solutions
  for herbicides 6–184
Hester-Dendy sampler
  for macroinvertebrates 10–106, 10–107i
Heteropoly blue method
  for silica 4–188
    automated 4–189
Heterotrophic plate count, see Plate count,
  heterotrophic
Hexachlorobenzene
  liquid-liquid extraction GC/MS method
    6–113
Hexachlorobutadiene
  liquid-liquid extraction GC/MS method
    6–113
  purge and trap GC method 6–71
    capillary column 6–93
  purge and trap GC/MS method 6–42
    capillary column 6–50
Hexachlorocyclopentadiene
  liquid-liquid extraction GC/MS method
    6–113

Hexachloroethane
  liquid-liquid extraction GC/MS method
    6–113
Hexagenia
  toxicity test procedures 8–118
Hexamethylenetetraamine buffer solution
  for aluminum 3–69
Hexamethylenetetramine solution
  for turbidity 2–14
High-performance liquid chromatographic
    method
  detectors 6–9
  for chlorophyll 10–36
  for organic constituents (individual) 6–9
  for polynuclear aromatic hydrocarbons
    6–148
    conditions 6–149t
    method detection limits 6–149t
Holding tanks
  for lobsters 8–95
Homarus americanus, see Lobsters
Hoop nets
  for fish 10–117
Host systems
  for virus assays 9–177
Hughes lobster-rearing tank 8–97i
Humic substances, aquatic, see Aquatic
  humic substances
Hydrazine reduction method, automated
  for nitrogen (nitrate) 4–141
Hydrazine sulfate solution
  for nitrogen (nitrate) 4–141
  for turbidity 2–14
Hydride generation AA method
  for arsenic and selenium 3–43
  for metals 3–43
  for selenium 3–50
Hydride generator, continuous
  for selenium 3–51, 3–51i
Hydrocarbons
  in oil and grease 5–47
Hydrochloric acid
  alternate to sulfuric 1–44
  for alkalinity 2–37
  for aquatic humic substances 5–38
  for chlorine dioxide 4–82
  for iron 3–103
  for selenium 3–51, 3–132, 3–133
  for volatile aromatics 6–69
  for zinc 3–158
  uniform solutions inside front cover

Hydrochloric acid digestion
  for selenium 3–134
Hydrochloric acid-ethanol cleaning solution
  for aluminum 3–68
Hydrochloric acid-glycine
  for virus concentration 9–160, 9–166
Hydrochloric acid-nitric acid
  for metals 3–20, 3–57
Hydrochloric acid-nitric acid digestion
  for metals 3–8
Hydrochloric acid reagent
  for aluminum 3–69
Hydroextraction-dialysis method
  for virus concentration 9–173
Hydrofluoric acid-nitric acid-perchloric acid
    digestion
  for metals 3–11
Hydrogen peroxide digestion
  for selenium 3–132
Hydrogen peroxide solution
  for alpha and beta radioactivity 7–18
  for radium 226 7–37
  for selenium 3–133
Hydrogen sulfide
  ionization constant 4–198t
  proportion in dissolved sulfide 4–199i
  un-ionized, calculation 4–198
Hydroxide ion
  nomograph 4–15i
Hydroxylamine-EDTA solution
  for selenium 3–135
Hydroxyamine hydrochloride-orthophenan-
    throline solution
  for aluminum 3–68, 3–69
Hydroxylamine hydrochloride solution
  for cadmium 3–83
  for copper 3–96
  for iodine (radioactive) 7–26
  for iron 3–103
  for mercury 3–119
  for metals 3–24
  for nickel 3–122
  for selenium 3–133
  for silver 3–143
Hydroxylamine sulfate-sodium chloride
    solution
  for mercury 3–29
Hydroxyquinoline solution
  for aluminum and beryllium 3–27
Hyphomycetes
  methods 9–194

Hyphomycetes (cont.)
  samples 9–194
Hypochlorite solution
  for trihalomethane formation potential
    5–75
Hypochlorous acid reagent
  for nitrogen (ammonia) 4–120

                          I

Ichthyocides
  for fish sampling 10–121
Ichthyoplankton
  preservation 10–127
  sampling 10–119
ICP, see Inductively coupled plasma (ICP)
    methods
Ictalurus punctatus, see Catfish, channel
Identification
  Campylobacter jejuni 9–144
  Giardia lamblia 9–201
  macrophyton species 10–78
  Salmonella 9–131
  Shigella 9–141
  Vibrio cholerae 9–146
  viruses 9–176
Illumination
  for biostimulation 8–44, 8–47
  for coliforms 9–85
  for Salmonella 9–136
  for toxicity tests 8–30
  for turbidity 2–13
Immersion oil
  for direct microbial counts 9–65
  for Salmonella 9–137
Immunofluorescence procedure
  for Legionallaceae 9–150
  for Salmonella 9–136
IMViC tests
  for coliform differentiation 9–100, 9–102t
  also see Indole test; Methyl red test;
      Sodium citrate test;
      Voges-Proskauer test
Incubation
  for heterotrophic plate count 9–56, 9–64
Incubation bottles
  for BOD 5–4
Incubation cups
  for fish 8–135

Incubator
  for BOD 5–4
  for coliforms 9–85, 9–94
  for heterotrophic plate count 9–61
  for trihalomethane formation potential
      5–75
  maintenance 9–8
  specifications 9–24
Indeno(1,2,3-cd)pyrene
  liquid-liquid extraction chromatographic
      method 6–148
  liquid-liquid extraction GC/MS method
      6–113
Indicator
  bromcresol green
    for alkalinity 2–37
    for radium 7–30
  bromphenol blue
    for acidity 2–33
  Calmagite
    for hardness 2–55
  congo red paper
    for copper 3–96
  desk reagents inside front cover
  dimethylaminobenzalrhodanine
    for cyanide 4–30
  diphenylcarbazone-bromphenol blue
    for chloride 4–69
  DPD
    for chlorine 4–60
  Eriochrome Black T
    for hardness 2–55
  Eriochrome Blue Black R
    for calcium 3–86
  ferroin
    for COD 5–12
  for hardness 2–55
  hydroxynaphthylazonitronaph-
      tholsulfonic acid
    for hardness 2–55
  leuco crystal violet
    for iodide 4–107
  MBTH
    for cyanide pretreatment 4–26
  metacresol purple
    for acidity 2–33
    for alkalinity 2–37
  methyl orange
    desk reagent inside front cover
    for radium 228 7–45
  methyl red
    for calcium 3–88

Indicator (cont.)
  methyl red (cont.)
    for coliform differentiation 9–105
    for magnesium 3–113
    for strontium (radioactive) 7–49
    for sulfate 4–205
    for zinc 3–160
  methyl red-methylene blue
    for nitrogen (ammonia) 4–121
  murexide
    for calcium 3–86
  phenol red
    for bromide 4–12
  phenolphthalein
    desk reagent inside front cover
    for organic acids 5–71
    for pesticide column standardization
        6–169
  potassium chromate
    for chloride 4–68
  starch
    for chlorine 4–49
    for DO 4–153
    for sulfite 4–200
  syringaldazine
    for chlorine 4–65
  thymol blue
    for cadmium 3–84
    for organic acids 5–71
Indicator acidifier reagent
  for chloride 4–69
Indicator circuit
  for methane 6–62i
Indicator organisms
  microbiological 9–102t
  periphyton 10–66
  also see Enterobacteriaceae; Coliform
      group bacteria
Indigo colorimetric method
  for ozone 4–162
Indigo solution
  for ozone 4–163
Indole test
  for coliform differentiation 9–104
Inductively coupled plasma (ICP) methods
    1–50
  for metals 3–53
    precision and bias 3–61t
Inductively coupled plasma source
  for metals 3–56
Infrared gas analysis
  macrophyton productivity 10–88

Inhibitors
for hardness 2–55
interferences 2–54t
Inoculating equipment
for microbiological tests 9–27
Inoculum
for biostimulation studies 8–46
phytoplankton 8–52
Inorganic nonmetals
analytical quality control 4–1
Insecticides, see Pesticides
Insects
acclimation 8–115
adults
key 10–141
collection 8–115
culturing 8–115
exposure chambers 8–117
food 8–116
holding 8–115
larvae
key 10–141
nymphs
key 10–141
pupae 10–166i
key 10–140
test species 8–114
toxicity tests 8–113, 8–117
Instrument detection limit
definition 1–4
determination 1–18
Insulating material
for organic halogen adsorption columns
5–29
Interference
causes 1–52
compensation for 1–49
counteracting 1–52
in fluoride methods 4–85t
in gas chromatographic methods 6–6
in metals analyses 3–3
in photometric methods 1–48
removal method for sulfide 4–194
types 1–52
Interlaboratory quality control 9–23
Internal standard
definition 1–5
Inverted microscope counts
for periphyton 10–51
for phytoplankton 10–26
Iodate solution
for chlorine 4–49

Iodate solution (cont.)
for iodine (radioactive) 7–24
Iodic acid
for uranium 7–57
Iodide
catalytic reduction method 4–109
leuco crystal violet method 4–107
methods 4–106
Iodide azide alkali reagent
for DO 4–152
Iodide solution
for iodide 4–107, 4–109
for iodine (radioactive) 7–27
Iodine
amperometric titration method 4–105
leuco crystal violet method 4–103
methods 4–102
radioactive
distillation method 7–27
ion exchange method 7–25
methods 7–23
precipitation method 7–24
regulatory limits 7–24
Iodine solution
for chlorine 4–52
for iodine 4–104
for iodine (radioactive) 7–26
for lead 3–108
for sulfide 4–197
Iodometric electrode method
for chlorine 4–65
Iodometric method
for chlorine 4–48, 4–51
for chlorine dioxide 4–76
for DO 4–150
alum flocculation modification 4–158
azide modification 4–152
copper sulfate-sulfamic acid flocculation
modification 4–158
permanganate modification 4–157
for sulfide 4–197
for sulfite 4–200
Ion chromatograph
for anions 4–3
Ion chromatography method 1–51
for anions 4–2
precision and bias 4–6t
Ion exchange
reagent water preparation 1–56
Ion-exchange column
for anions 4–3
separations 4–4i, 4–5i

Ion-exchange column *(cont.)*
  for boron 4–8
  for cation removal 1–46i
  for sulfate 4–209
  for surfactants 5–65
  for uranium 7–57
Ion-exchange method
  for ammonia-free water 4–115
  for iodine (radioactive) 7–25
Ion-exchange resins, *see* Resins, ion-exchange
Ionization constant
  hydrogen sulfide 4–198t
Ion-selective electrode method
  for fluoride 4–87
Ion sums
  for checking analyses 1–21
Iridium
  AA method 3–20
  methods 3–100
Iridium solution
  for iridium 3–21
Iron
  AA method 3–20
    low concentrations 3–23
    micro quantities 3–36
  inductively coupled plasma method 3–54
  methods 3–100
  phenanthroline method 3–102
    light path lengths 3–104t
Iron bacteria, *see* Bacteria, iron
Iron bacteria isolation medium 9–121
Iron bacteria maintenance (SCY) medium
  9–121
Iron masking solution
  for aluminum 3–68, 3–69
Iron solution
  for iron 3–21, 3–103
Isobutanol-benzene solvent
  for phosphorus 4–175
Isolation
  fungi
    zoosporic 9–193
  Salmonella 9–131
  yeasts 9–192
  Yersinia enterocolitica 9–154
Isolation medium
  for iron bacteria 9–121
Isophorone
  liquid-liquid extraction GC/MS method
    6–113

Isopropylbenzene
  purge and trap GC method 6–71
    capillary column 6–93
  purge and trap GC/MS method 6–42
    capillary column 6–50
Isopropyltoluene
  purge and trap GC method 6–71
    capillary column 6–93
  purge and trap GC/MS method
    capillary column 6–50
Issacs-Kidd mid-water trawl
  for zooplankton 10–13i

                               J

Jackson candle method
  for turbidity 2–12
Jackson candle turbidimeter 2–12
Jenkins surface mud sampler 10–7
Juday trap
  for zooplankton 10–8

                               K

Kanamycin-gentamycin
  for virus concentration 9–160, 9–172
KB core sampler
  for macroinvertebrates 10–104, 10–104i
Kemmerer sampler
  for phytoplankton 10–6, 10–7i
Keys
  algae
    blue-green 10–183
    diatom 10–185
    flagellate 10–186
    freshwater 10–183
    green 10–187
  aquatic organisms 10–137
  crustaceans 10–140
  insects 10–141
    larvae 10–141
    nymphs 10–141
    pupae 10–140
  macroinvertebrates 10–139
  macroscopic organisms 10–138
  microscopic organisms 10–137
  nematodes 9–208

Kjeldahl method
  for nitrogen (organic)
    bias and precision 4–147t
    semi-micro 4–147
Kjeldahl nitrogen, see Nitrogen, kjeldahl
Klebsiella
  differentiation 9–101
  membrane filter procedure 9–97
  verification 9–98
Known additions
  acceptance limits 1–8t
  recovery 1–7
Known additions method
  for interference detection 3–3
Koser's citrate broth
  for coliform differentiation 9–105

L

Laboratories
  certification 9–23
  for virus assays 9–177
  hazard management 1–69
  hazards 1–62
  management
    safety 1–58
  microbiological 9–5
  safety equipment 1–60
  ventilation 9–5
  waste disposal 1–70
Laboratory apparatus 1–41
  microbiological 9–6
Laboratory control standard
  definition 1–5
Laboratory quality assurance
  microbiological examinations 9–4
Laboratory reagents and techniques 1–41
Lackey drop method
  for phytoplankton counting 10–27
Lactose broth
  for coliforms 9–69, 9–70t
Lactose fermentation test
  for coliform verification 9–19, 9–89
Lactose test
  coliform group bacteria 9–102t
Lamps
  for atomic absorption spectrometry 3–17
Lanthanum nitrate solution
  for fluoride 4–92

Lanthanum solution
  for metals 3–20
Lauric acid solution
  for pesticide column standardization 6–169
Lauryl tryptose broth
  for coliforms 9–68, 9–69t
Lead
  AA method 3–20
    low concentrations 3–23
    micro quantities 3–36
  dithizone method 3–107
  inductively coupled plasma method 3–54
  methods 3–106
Lead acetate paper test
  for sulfide 4–192
Lead acetate solution
  for arsenic 3–75
Lead carrier
  for radium 228 7–45
Lead nitrate carrier
  for radium 7–30
Lead solution
  for lead 3–21, 3–108
Legionellaceae
  immunofluorescence procedure 9–150
  methods 9–149
  plating 9–151
Lepomis macrochirus, see Sunfish, bluegill
Leptospires, pathogenic
  concentration technique 9–147
  differentiation 9–148
  enrichment 9–147
  methods 9–147
LES Endo agar
  for coliforms 9–86
LES MF holding medium
  for coliforms 9–93
Lethal concentration
  confidence limits 8–34
  definition 8–3
  estimation 8–34, 8–35i
Lethal response
  corals 8–63
Leuco crystal violet indicator
  for iodide 4–107
Leuco crystal violet method
  for iodide 4–107
  for iodine 4–103
Life cycle
  nematodes 9–205, 9–205i

Life-cycle toxicity tests, *see* Toxicity tests,
  life-cycle
Lift nets
  for fish sampling 10–120
Ligand-exchange chromatographic resin
  for selenium 3–140
Light and dark bottle method
  macrophyton productivity 10–85
Lighting, *see* Illumination
Lignin
  method 5–67
Lignin solutions
  for lignin 5–68
Limiting nutrient
  definition 8–4
Limit of quantitation
  definition 1–5
Limits, upper
  for metals by ICP method 3–55t
Line intercept method
  for macrophyton mapping 10–71,
  10–75
Linear alkylbenzene sulfonate, *see* Surfactants, anionic
Linear alkylbenzene sulfonate solution
  for anionic surfactants 5–61
Lipovitellin-salt-mannitol agar
  for *Staphylococcus aureus* 9–47
Liquid chromatogram
  for chlorophyll 10–37i
Liquid chromatographic methods
  for organic constituents (individual)
  6–9
Liquid-liquid extraction chromatographic
  method
  for polynuclear aromatic hydrocarbons
  6–148
  bias and precision 6–157t
Liquid-liquid extraction GC method
  for dibromoethane and dibromochloropropane 6–98
  bias and precision 6–102t
  for organochlorine pesticides and PCBs
  6–170
  bias and precision 6–181t
  for phenols 6–137
  bias and precision 6–145t
  for polynuclear aromatic hydrocarbons
  6–148
  bias and precision 6–157t for trihalomethanes 6–104
  bias and precision 6–112t

Liquid-liquid extraction GC/MS method
  for base/neutral and acid extractables
  6–113
  bias and precision 6–134t
  tailing factor calculation 6–136i
Liquid-liquid extraction high-performance
  liquid chromatographic method
  for polynuclear aromatic hydrocarbons
  6–148
Liquid-liquid extractor
  for base/neutral and acid extractables
  6–119
Liquid scintillation spectrometer
  for tritium 7–54
Liquid scintillation spectrometric method
  for tritium 7–54
Liquid scintillation vial
  for tritium 7–54
Lithium
  AA method 3–20
  flame emission photometric method 3–110
  inductively coupled plasma method 3–54
  methods 3–109
Lithium solution
  for lithium 3–21, 3–111
  for sodium 3–147
Lobsters
  culture chambers 8–96
  culturing 8–95
  egg-hatching tank 8–95, 8–96i
  egg-laying tanks 8–95
  food 8–95
  holding 8–95
  holding tanks 8–95
  rearing tank 8–97i
  toxicity tests 8–108
  water supply 8–96
Log-normal distribution 1–3
Louisiana box sampler
  for macrophyton 10–77
Lower limit of detection
  definition 1–4
  determination 1–18
Luciferin-luciferase enzyme preparation
  for plankton ATP 10–41
Lugol's solution
  for phytoplankton preservation 10–7
  Gram's modification
    for coliform bacteria 9–73
Lysine test
  for coliform differentiation 9–106
  for coliform group bacteria 9–102t

M

Macrocrustaceans
  toxicity tests 8–91
  also see Crustaceans
Macroinvertebrates
  benthic
    data evaluation and presentation 10–111
    definition 10–95
    identification techniques 10–110
    methods 10–95
    preservation 10–109
    processing and analysis 10–109
    samplers 10–99i, 10–100i, 10–101i, 10–102i, 10–103i
    samples 10–96
    survey guidelines 10–97
  definition 10–2
  food in toxicity tests 8–19
  holding and culturing tanks 8–14, 8–14i
  key 10–139
  response to environment 10–95
Macrokjeldahl method
  for nitrogen (organic) 4–144
Macronutrient stock solution
  for toxicity tests 8–18t
Macroorganisms, aquatic
  key 10–138
Macrophyton 10–152i, 10–153i, 10–154i
  biomass measurement 10–77
  caloric content 10–78
  carbon content 10–78
  chlorophyll measurement 10–78
  definition 10–1
  dry mounts 10–78
  mapping methods 10–71
  methods 10–68
  population estimates 10–74
  preliminary surveys 10–70
  productivity 10–80
    biomass methods 10–80
    metabolic methods 10–85
    methods 10–81t
  remote sensing 10–72
  samplers 10–71, 10–75
  samples 10–70, 10–74
    preservation 10–78
  species identification 10–78
  wet mounts 10–78
  also see Plants
Magnesia-silica gel
  for herbicides 6–184

Magnesia-silica gel *(cont.)*
  for PCBs 6–172
  for pesticides (organochlorine) 6–162, 6–169
Magnesia silica gel column
  for pesticides (organochlorine)
    standardization 6–169
Magnesium
  AA method 3–20
  calculation method 3–114
  gravimetric method 3–113
  inductively coupled plasma method 3–54
  methods 3–112
Magnesium carbonate solution
  for chlorophyll 10–32
Magnesium CDTA inhibitor
  hardness 2–55
Magnesium chloride reagent
  for cyanide 4–28
Magnesium nitrate
  for uranium 7–59
Magnesium solution
  for magnesium 3–21
Magnesium sulfate monohydrate
  for oil and grease 5–47
Magnesium sulfate solution
  for BOD 5–5
Maintenance (SCY) medium
  for iron bacteria 9–121
Malonic acid reagent
  for ozone 4–163
Malt extract-streptomycin-terramycin agar
  for fungi 9–190
    modified 9–191
Malt extract-yeast extract-glucose agar
  for fungi 9–188
Management
  laboratory hazards 1–69
Manganese
  AA method 3–20
    low concentrations 3–23
    micro quantities 3–36
  inductively coupled plasma method 3–54
  methods 3–114
  persulfate method 3–115
Manganese solution
  for manganese 3–21, 3–116
Manganous sulfate solution
  for DO 4–152
  for nitrogen (ammonia) 4–120
Mannitol salt agar
  for staphylococci 9–47

Manometer
 for radium 226 7–36
 for sludge oxygen consumption rate
　2–81
Manure-soil medium
 for *Daphnia* culture 8–82
Mapping
 macrophyton 10–71
Marble test
 calcium carbonate precipitation potential
　2–48
Mass spectrometer
 for base/neutral and acid extractables
　6–120
 for gas chromatographic methods 6–8
 for volatile organics 6–30, 6–44, 6–50
Material Safety Data Sheets 1–59, 1–62,
　1–70
Maximum allowable toxicant concentration
 definition 8–4
Maximum specific growth rate
 measurement 8–48
 phytoplankton 8–52
Maximum standing crop
 definition 8–4
 measurement 8–49
Maximum trihalomethane potential
 definition 5–74
Mayflies 10–168i
MBTH indicator solution
 for cyanide pretreatment 4–26
Mean Saprobial Index
 periphyton 10–66
Means chart 1–9, 1–10i, 1–12i
Mechanical hazards
 in laboratories 1–67
Medium, culture, *see* Culture medium
Membrane-diffusion instruments
 for dissolved gas supersaturation 2–95
Membrane-diffusion method
 for dissolved gas supersaturation 2–94
　time response 2–96
Membrane electrode method
 for DO 4–158
Membrane filter apparatus
 for AA method (electrothermal) 3–36
 for aquatic humic substances 5–38
 maintenance 9–7
Membrane filter counts
 for plankton 10–26
Membrane filter mounts
 for phytoplankton 10–18

Membrane filter technique
 for coliform group bacteria 9–83
　confidence limits 9–91t
　fecal 9–94, 9–95t, 9–96
　Klebsiella 9–97
　recovery enhancement 9–43
　total 9–84, 9–91
 for fungi 9–191
 for heterotrophic bacteria 9–54, 9–57,
　9–63
 for phytoplankton concentration
　10–27t
 for plankton concentration 10–16
 for *Pseudomonas aeruginosa* 9–52
 for *Salmonella* 9–132, 9–140
 for streptococcus (fecal) 9–110
　recovery enhancement 9–44
 quality control 9–19
Membrane filters
 for coliforms 9–85
 for direct microbial count 9–65
 quality control 9–13
Membrane filtration equipment 9–27
Membrane suppressor
 for anions 4–3
mE method
 for enterococcus 9–111
M-Endo medium
 for coliforms 9–86, 9–92
m Enterococcus agar 9–111
m Enterococcus method 9–112
Mercuric bromide paper
 for arsenic 3–76
Mercuric bromide stain method
 for arsenic 3–76
Mercuric nitrate method
 for chloride 4–69
Mercuric nitrate solution
 for chloride 4–70
 for vanadium 3–156
Mercuric sulfate
 for COD 5–12
Mercuric sulfate-potassium dichromate-sul-
　furic acid solution
 for COD 5–16
Mercuric sulfate-silver nitrate reagent
 for manganese 3–116
Mercuric sulfate solution
 for nitrogen (organic) 4–145
Mercuric thiocyanate-ferric nitrate
　solution
 for chloride 4–73

Mercuric thiocyanate solution
for chloride 4–73
Mercury
AA method 3–29
dithizone method 3–119
for nitrogen (organic) 4–144
methods 3–118
safety precautions 1–64
sample preservation 3–2
Mercury solution
for mercury 3–29, 3–119
Merthiolate
for phytoplankton preservation 10–8
Metabolic methods
macrophyton productivity 10–85
Metabolic rate measurement
plankton 10–42
Metacresol purple indicator
for acidity 2–33
for alkalinity 2–37
Metallogenium
isolation 9–124
Metals
AA method
electrothermal 3–32
flame 3–13
acid-extractable
definition 3–1
pretreatment 3–6
analytical quality control 3–3
cold-vapor AA method 3–28
dissolved 3–1
hydride generation AA method 3–43
ICP method 3–53
wavelengths, limits, and calibration concentrations 3–55t
methods 3–1
microbiological samples 9–32
sample contamination 3–3
samples 1–30, 3–1
digestion 3–6
pretreatment 3–5
sludge and solid samples
digestion 3–7
suspended 3–1
total 3–1
Metal solutions
for metals 3–15, 3–20, 3–25, 3–27, 3–56, 3–57
Methane
combustible gas indicator circuit and flow diagram 6–62i

Methane (cont.)
combustible-gas indicator method 6–61
methods 6–60
volumetric method 6–63
Method detection limit
definition 1–5, 1–18
determination 1–18
for method validation 1–22
Methyl isobutyl ketone (MIBK)
for metals 3–23
Methyl orange indicator solution
desk reagent inside front cover
for radium 228 7–45
Methyl red indicator solution
for calcium 3–88
for coliform differentiation 9–105
for magnesium 3–113
for strontium (radioactive) 7–49
for sulfate 4–205
for zinc 3–160
Methyl red-methylene blue indicator solution
for nitrogen (ammonia) 4–121
Methyl red test
for coliform differentiation 9–104
Methyldinitrophenol
liquid-liquid extraction GC method 6–137
liquid-liquid extraction GC/MS method 6–113
Methylene blue active substances
method 5–59
Methylene blue method
for anionic surfactants 5–59
for sulfide 4–195
Methylene blue-methyl red indicator solution
for nitrogen (ammonia) 4–121
Methylene blue solution
for anionic surfactants 5–61
for sulfide 4–196
Methylene chloride
purge and trap GC method 6–79, 6–87
capillary column 6–93
purge and trap GC/MS method 6–27, 6–42
capillary column 6–50
Methylisoborneol
mass spectra 6–21i
also see Earthy-musty-smelling compounds
Methylthymol blue method, automated
for sulfate 4–208

Methylthymol blue reagent
  for sulfate 4–209
M FC agar, modified (M-FCIC)
  for Klebsiella 9–98
M FC broth
  for fecal coliforms 9–94
m HPC agar
  for heterotrophic plate count 9–55
Mice
  virus assay procedures 9–180
Microbial count, direct total 9–64
Microbiological examination
  control cultures 9–18t
  laboratory apparatus 9–24
  laboratory quality assurance 9–4
  methods 9–1
  precision 9–18t
  recreational waters 9–45
  samples 9–31
  whirlpools 9–49
  also see Bacteriological examination
Microbiological sampling
  sediments 9–33
  sludges 9–33
Microcolumn procedure
  for total organic halogen 5–31
Micrometer ruling
  for plankton counting 10–21i
Micronutrient stock solutions
  for toxicity tests 8–18t
Microorganisms
  aquatic
    key 10–137
  culture for toxicity test organism food
    8–17
  pathogenic
    safety precautions 1–65
  total viable
    bioluminescence test 9–37
  also see specific organisms and groups
Microporous filter method
  for virus concentration 9–159, 9–163
Microscopes
  calibration 10–21
  compound
    for plankton 10–19
  epifluorescence
    for plankton 10–20
  for coliforms 9–85
  for direct microbial count 9–65
  for plankton 10–19
  for *Salmonella* 9–136

Microscopes *(cont.)*
  inverted
    for plankton 10–26
  inverted compound
    for plankton 10–20
  maintenance 9–8
  stereoscopic
    for plankton 10–20
Microspectrofluorometric microscopy
  for *Salmonella* 9–139
Microtransect method
  for plankton counting 10–27
Milk agar
  for *Pseudomonas aeruginosa* 9–52
Miller sampler
  for zooplankton 10–13i
Milligrams-milliequivalents conversion
  1–27t
Minnow
  fathead
    culturing 8–127
    food 8–139
    photoperiods 8–138
    spawning chambers 8–138
    spawning substrates 8–138
    toxicity tests 8–138
  sheepshead
    culturing 8–129
    food 8–140
    hatching chambers 8–140
    spawning chambers 8–140
    toxicity tests 8–140
M Kleb agar
  for Klebsiella 9–98
Mn agar
  for iron bacteria 9–121
Moeller basal medium
  for coliform differentiation 9–106
Moist chamber
  for *Salmonella* 9–137
Molar solution 1–43
Mollusks 10–176i, 10–177i
  collecting 8–74
  conditioning 8–74
  culturing 8–74
  diseases 8–76
  parasites 8–76
  test species 8–74
  toxicity tests 8–73, 8–77
  water supply 8–74
  also see Clams, Mussels, Oysters, Scallops

Molybdate-vanadate reagent
  for phosphorus 4–174
Molybdenum
  AA method 3–25
    micro quantities 3–36
  inductively coupled plasma method 3–54
  methods 3–120
Molybdenum solution
  for molybdenum 3–26
Molybdosilicate method
  for silica 4–184
Monitoring
  laboratory safety 1–69
  radioactivity 7–1
  swimming pools 9–46
  whirlpools 9–49
Monochloramine
  procedures 4–57, 4–61, 4–63
Most probable number, see MPN
Motility test
  for coliform differentiation 9–102t,
    9–106
Motility test medium
  for coliform differentiation 9–106
Mounting fluid
  for Salmonella 9–137
Mounts
  diatoms 10–18
    periphyton 10–52
  macrophyton 10–78
  membrane filter
    phytoplankton 10–18
  plankton 10–17
  sedimented slide
    phytoplankton 10–18
  wet
    macrophyton 10–78
    phytoplankton 10–17
    zooplankton 10–19
M-PA agar
  for Pseudomonas aeruginosa 9–52
  modified 9–52
MPN 9–77t, 9–78t
  computation and recording 9–77
  definition 9–1
  frequency 9–22t
  Thomas' formula 9–79
MPN test, see Multiple-tube fermentation
  technique
M 7-h FC agar
  for coliforms 9–36
M-staphylocuccus broth 9–47

mTEC agar
  for E. coli 9–50
M3 fixative
  for phytoplankton preservation 10–8
m T7 agar
  for total coliform recovery enhancement
    9–43
Muffle furnace
  for total solids 2–73
Multiple-plate sampler
  for macroinvertebrates 10–106, 10–107i
Multiple-tube fermentation technique
  for coliforms 9–66
    fecal 9–75
    total 9–68
    USEPA regulations 9–3
  for Pseudomonas aeruginosa 9–53
  for Salmonella 9–140
  for streptococcus (fecal) 9–109
  quality control 9–19
Multitest systems
  for coliform verification 9–20, 9–89
Murexide indicator solution
  for calcium 3–86
Mussels
  collection 8–75
  conditioning 8–75
  culturing 8–75
  embryo tests 8–78
  growth tests 8–80
  toxicity tests 8–74
  also see Mollusks
Mutagenicity
  Ames test 8–41
  Salmonella test 8–41
  test methods 8–41
MVFC holding medium
  for fecal coliforms 9–97

N

Nannoplankton
  mesh size for 10–6
Nansen sampler
  for phytoplankton 10–6
Naphthalene
  liquid-liquid extraction chromatographic
    method 6–148
  liquid-liquid extraction GC/MS method
    6–113

Naphthalene *(cont.)*
  purge and trap GC method 6–71
    capillary column 6–93
  purge and trap GC/MS method
    capillary column 6–50
Naphthol solution
  for coliform differentiation 9–105
Naphthylethylenediamine dihydrochloride-
    sulfanilamide solution
  for nitrogen (nitrate) 4–141
  for nitrogen (nitrite) 4–129
Natantia, *see* Shrimp
National Bureau of Standards, *see* National
    Institute of Standards and Technology
National Institute of Standards and
    Technology
  pH standards 4–99t
  standard substances 1–43
Neanthes arenaceodentata
  culturing 8–68
Neanthes succinea
  culturing 8–68
Nematodes
  characteristics 9–204
  counting 9–207
  examination techniques 9–206
  identification 9–207
  in water supplies 9–208
  key, illustrated 9–208
  life cycle 9–205, 9–205i
  methods 9–204
  samples 9–206
  water treatment effects 9–208
  also see Roundworms; Worms
Neocuproine method
  for copper 3–95, 3–96
Neopeptone-glucose agar
  for fungi 9–188
Neopeptone-glucose-rose bengal aureomycin
    agar
  for fungi 9–188
Nephelometer
  for turbidity 2–13
Nephelometric method
  for turbidity 2–13
Nesslerization method
  for nitrogen (ammonia) 4–117
Nessler reagent
  for nitrogen (ammonia) 4–118
Nessler tubes 1–42
Net plankton
  mesh size for 10–6

Nets
  for fish 10–117, 10–118i, 10–120
  for plankton 10–9, 10–10t, 10–11i
Neutralization agent
  for nitrogen (ammonia) 4–116
Nickel
  AA method 3–20
    low concentrations 3–23
    micro quantities 3–36
  dimethylglyoxime method 3–123
  heptoxime method 3–121
  inductively coupled plasma method 3–54
  methods 3–121
  safety precautions 1–63
Nickel nitrate
  for AA method (electrothermal) 3–36
Nickel solution
  for nickel 3–21
Nickel sulfate solution
  for nickel 3–122
Niskin sampler
  for phytoplankton 10–6
Nitrate
  ion chromatography method 4–2
Nitrate electrode method
  for nitrogen (nitrate) 4–133
Nitrate-nitrite manifold
  for nitrogen (nitrate) 4–142i
Nitrate nitrogen, *see* Nitrogen (nitrate)
Nitrate solution
  for nitrogen (nitrate) 4–4, 4–132, 4–133,
    4–134, 4–138, 4–139, 4–141
  for organic halogen 5–30
Nitrate-nitrite manifold
  for nitrogen (nitrate) 4–138
Nitric acid
  for lead 3–108
  for metals 3–56
  for metals samples 3–2
  uniform solutions *inside front cover*
Nitric acid digestion
  for metals 3–8
Nitric acid-hydrochloric acid digestion
  for metals 3–8
Nitric acid-hydrochloric acid solution
  for metals 3–20, 3–57
Nitric acid indicator-acidifier reagent
  for chloride 4–69
Nitric acid-perchloric acid digestion
  for metals 3–10
Nitric acid-perchloric acid-hydrofluoric acid
    digestion
  for metals 3–11

Nitric acid-silver nitrate reagent
  for sulfate 4–205
Nitric acid-sulfuric acid digestion
  for metals 3–9
  for phosphorus 4–172
Nitric acid-sulfuric acid solution
  for phosphorus 4–171
Nitrification inhibitor
  for BOD 5–5
Nitrite nitrogen, *see* Nitrogen (nitrite)
Nitrite solution
  for nitrogen (nitrate) 4–136, 4–139
  for nitrogen (nitrite) 4–4, 4–130
Nitrobenzene
  liquid-liquid extraction GC/MS method
    6–113
Nitrogen
  effect on chlorination 4–111
  forms 4–110
  methods 4–110
Nitrogen (ammonia)
  color standards 4–118, 4–119t
  concentration ratios in known-addition
    method 4–125t
  distillation step 4–115
  methods 4–111
    bias and precision 4–112t
  nesslerization method 4–117
  phenate method 4–120
    automated 4–126
  selective electrode method 4–122
    known addition method 4–124
  titrimetric method 4–121
  also see Ammonia
Nitrogen fixation
  plankton 10–43
Nitrogen gas
  for chlorine dioxide 4–82
Nitrogen (kjeldahl)
  definition 4–144
  also see Nitrogen (organic)
Nitrogen (nitrate)
  cadmium reduction method 4–135
    automated 4–137
  electrode method 4–133
  hydrazine reduction method, automated
    4–141
  methods 4–131
  titanous chloride reduction method 4–140
  ultraviolet spectrophotometric screening
    method 4–132
Nitrogen (nitrite)
  colorimetric method 4–129

Nitrogen (nitrite) *(cont.)*
  ion chromatography method 4–2
  methods 4–128
Nitrogen (organic)
  distillation apparatus 4–148i
  macrokjeldahl method 4–144
    bias and precision 4–147t
  methods 4–143
  semimicrokjeldahl method 4–147
Nitrogen trichloride
  procedure 4–61, 4–63
Nitrophenol
  liquid-liquid extraction GC method 6–137
  liquid-liquid extraction GC/MS method
    6–113
Nitrosodimethylamine
  liquid-liquid extraction GC/MS method
    6–113
Nitrosodiphenylamine
  liquid-liquid extraction GC/MS method
    6–113
Nitrosodipropylamine
  liquid-liquid extraction GC/MS method
    6–113
Nitrous oxide
  for metals 3–25
Nomograph
  for bicarbonate alkalinity 4–16i
  for carbon dioxide 4–18i
  for carbonate alkalinity 4–17i
  for hydroxide ion 4–15i
Nomographic method
  for carbon dioxide and alkalinity 4–14
Nonionic surfactant solution
  for nonionic surfactants 5–65
Nonmetals, inorganic
  analytical quality control 4–1
  methods 4–1
Normal distribution 1–2, 1–2i
Normal solution
  definition 1–43
Nutrient agar
  for coliforms 9–73
Nutrient broth
  for virus concentration 9–160, 9–166
Nutrient medium
  for algal toxicity tests 8–17, 8–18t
Nutrients
  effect of additions in biostimulation studies
    8–50
  for algal culture medium 8–19t
NWRI (HPCA) agar
  for heterotrophic plate count 9–56

O

Occupational Safety and Health Act 1–59

O'Connell-Thomas chamber
periphyton production results 10–62i

Odor
algae causing
color plate following 10–200
characterization 2–20
intensities 2–22t
methods 2–16
organic compounds causing
method 6–11
also see Earthy-musty-smelling compounds
panel selection 2–20
threshold numbers 2–21t

Odor-free water
for odor determinations 2–19
generator 2–19, 2–19i

Oil
trichlorotrifluoroethane-soluble floatable
method 2–69

Oil and grease
definition 5–41
extraction method
for sludges 5–46
Soxhlet 5–45
hydrocarbons in
method 5–47
methods 5–41
partition-gravimetric method 5–43
partition-infrared method 5–44

Oil, mineral
for coliform differentiation 9–106

Oil, reference
for oil and grease 5–44

ONPG test
for coliform verification 9–19

Open reflux method
for COD 5–12

Open system oxygen method
macrophyton productivity 10–86

Orange-peel grab
for macroinvertebrates 10–99, 10–99i

Organic acids, *see* Acids, organic

Organic carbon (total)
combustion infrared method 5–18
fractions 5–17
methods 5–17
persulfate-ultraviolet oxidation method
5–22
precision and bias 5–24t
wet-oxidation method 5–24

Organic carbon analyzer
for aquatic humic substances 5–38

Organic compounds
safety in sampling 1–32
samples 1–30
toxic and carcinogenic
safety precautions 1–64

Organic constituents (aggregate)
analytical quality control 5–2
introduction to methods 5–1
samples 5–1

Organic constituents (individual)
closed-loop stripping GC/MS analysis
6–11
precision and bias 6–25t
detection limits 6–12t
concentration by gas extraction 6–10
gas chromatographic methods 6–6
methods 6–1, 6–1t, 6–6
samples 6–4

Organic halogen (dissolved)
adsorption-pyrolysis-titrimetric method
5–27
methods 5–26
precision and bias 5–36t

Organic nitrogen, *see* Nitrogen (organic)

Organics solutions
for organic constituents (individual) 6–17

Organics, aromatic
polynuclear
liquid-liquid extraction chromatographic method 6–148
also see Aromatics, volatile

Organics, volatile
characteristic masses 6–38t, 6–47t
chromatographic conditions 6–28t, 6–45t,
6–51t, 6–53t
detection limits 6–28t
gas chromatogram 6–34i, 6–52i, 6–54i
methods 6–26
preservatives 6–5, 6–5t
purge and trap GC/MS method 6–27,
6–42
bias and precision 6–40t, 6–56t, 6–58t
calibration criteria 6–36t
capillary column 6–50
quality control acceptance criteria
6–36t
samples 6–4

Organochlorine pesticides
chromatographic conditions 6–171t
detection limits 6–171t
retention times 6–171t
also see Pesticides, organochlorine

Organochlorine pesticide solutions
  for organochlorine pesticides 6–172
Organohalides, *see* Halocarbons
Ornithine test
  for coliform differentiation 9–102t, 9–106
Orsat-type apparatus
  for sludge digester gas 2–89
Orthophenanthroline-hydroxyamine hydro-
    chloride solution
  for aluminum 3–68, 3–69
Osborne aquatic plant sampler
  for macrophyton 10–77
Osmium
  AA method 3–25
  methods 3–123
Osmium solution
  for osmium 3–26
Oven
  for COD 5–14
  for heterotrophic plate count 9–61
  for microbiological laboratory 9–7
  for selenium 3–133
  for sterilization 9–25
Oxalic acid solution
  for silica 4–185, 4–190
  for strontium (radioactive) 7–49
Oxidase test
  for coliform differentiation 9–106
Oxygen
  Bunsen coefficient 2–98t
  solubility in water 4–154t
Oxygen (dissolved)
  in streams
    sample calculations 10–64t
  in toxicity tests 8–26
  iodometric methods 4–150
    alum flocculation modification 4–158
    azide modification 4–152
    copper sulfate-sulfamic acid flocculation
      modification 4–158
    permanganate modification 4–157
  membrane electrode method 4–158
  methods 4–149
  samplers 4–151, 4–151i
Oxygen consumption rate test
  for sludge 2–81
Oxygen demand (biochemical), *see* BOD
Oxygen demand (chemical), *see* COD
Oxygen measurement
  macrophyton productivity 10–85
Oxygen metabolism
  stream 10–58i
Oxygen method
  for periphyton productivity

Oxygen method *(cont.)*
  flowing water 10–57
  standing water 10–55
  for plankton productivity 10–43
Oxygen probe
  for oxygen consumption rate of sludge
    2–81
Oysters
  collection 8–74
  conditioning 8–74
  culturing 8–74
  embryo tests 8–77
  growth tests 8–79
  shell deposition tests 8–78
    constant-flow apparatus 8–79i
  toxicity tests 8–74
  also see Mollusks
Ozone (residual)
  indigo colorimetric method 4–162
  method 4–162

P

PA broth
  for coliforms 9–80
Palaemonetes
  culturing 8–98
  food 8–98
  toxicity tests
    life-cycle 8–110
    short-term 8–109
Palladium
  AA method 3–20
  methods 3–124
Palladium chloride
  for iodine (radioactive) 7–24, 7–26, 7–27
Palladium solution
  for palladium 3–21
Palmer-Maloney nannoplankton cell 10–25
Panel selection
  for flavor rating assessment 2–27
  for taste 2–24
Paradimethylaminobenzaldehyde reagent
  for coliform differentiation 9–104
Paradimethylaminobenzalrhodanine
    indicator
  for cyanide 4–30
Parasites
  annelids 8–69
  coral 8–60
  crustaceans 8–100
  fish 8–122, 8–124t
  mollusks 8–76

Parasites *(cont.)*
  toxicity tests 8–21
Partition-gravimetric method
  for oil and grease 5–43
Partition-infrared method
  for oil and grease 5–44
Passive integrated transponder (PIT)
  for fish sampling 10–123, 10–125i
Pathogenic fungi, *see* Fungi, pathogenic
Pathogens
  detection 9–130
  safety precautions 1–65
  *also see Campylobacter jejuni; Entamoeba
    histolytica; Escherichia coli*, entero-
    pathogenic; *Giardia lamblia*; Legion-
    allaceae; Leptospires, pathogenic;
    Protozoa pathogenic; *Salmonella;
    Shigella; Vibrio;* Viruses; *Yersinia en-
    terocolitica*
PCB-1016
  gas chromatogram 6–126i, 6–176i
PCB-1221
  gas chromatogram 6–126i, 6–177i
PCB-1232
  gas chromatogram 6–127i, 6–177i
PCB-1242
  gas chromatogram 6–127i, 6–178i
PCB-1248
  gas chromatogram 6–128i, 6–178i
PCB-1254
  gas chromatogram 6–129i, 6–179i
PCB-1260
  gas chromatogram 6–130i, 6–179i
PCBs
  chromatographic conditions 6–171t
  detection limits 6–171t
  liquid-liquid extraction GC method 6–170
    bias and precision 6–181t
  liquid-liquid extraction GC/MS method
    6–113
  methods 6–146
  recovery by fraction 6–174t
  retention times 6–171t
PCB solutions
  for PCBs 6–172
Peak biomass
  macrophyton productivity 10–84
Penaeus
  culturing 8–99
  food 8–99
  toxicity tests
    life cycle 8–111
    short-term 8–110

Penicillin-streptomycin
  for virus concentration 9–160, 9–171
Pentachlorophenol
  liquid-liquid extraction GC method 6–137
  liquid-liquid extraction GC/MS method
    6–113
Peptone-aureomycin-rose bengal-glucose
    agar
  for fungi 9–190
    modified 9–191
Perchlorate salts
  safety precautions 1–64
Perchloric acid
  safety precautions 1–63
Perchloric acid digestion
  for phosphorus 4–171
Perchloric acid-hydrofluoric acid-nitric acid
    digestion
  for metals 3–11
Perchloric acid-nitric acid digestion
  for metals 3–10
Performance audits
  for analytical quality assessment 1–12,
    1–13t
Periphyton
  chlorophyll and pheophytin 10–53
  counting technique 10–51
  data analysis 10–66
  definition 10–1, 10–48
  dry and ash-free weight 10–53
  methods 10–48
  production 10–63i
  productivity 10–54
    carbon 14 method, standing water
      10–57
    oxygen method, flowing water 10–57
    oxygen method, standing water 10–55
  sampler 10–50i
  samples 10–49
  staining 10–52
Permanganate digestion
  for selenium 3–133
Permanganate modification
  iodometric method for DO 4–157
Permanganate titrimetric method
  for calcium 3–87
Personnel
  certification of analytical competence 1–7
Personnel
  microbiological laboratories 9–6
Persulfate digestion
  for phosphorus 4–172
  for selenium 3–132

Persulfate method
   for manganeses 3–115
Persulfate-ultraviolet oxidation method
   for organic carbon 5–22
      precision and bias 5–24t
Pesticide solutions
   for organochlorine pesticides 6–162,
      6–172
Pesticides
   organochlorine
      bias and precision 6–168t, 6–181t
      chromatograms 6–124i, 6–161i, 6–162i,
         6–163i, 6–164i, 6–165i, 6–175i
      liquid-liquid extraction GC method
         6–158, 6–170
      methods 6–158
      quality control acceptance criteria
         6–180t
      recovery by fraction 6–174t
      retention ratios 6–167t
   also see Herbicides
Petersen grab
   for macroinvertebrates 10–99, 10–100i
Petri dishes
   for heterotrophic plate count 9–55
   for microbiological tests 9–27
Pfizer selective enterococcus (PSE) agar
   for fecal streptococcus 9–109
pH
   culture media
      adjustment 9–29
   electrometric method 4–95
   equipment for microbiological tests 9–26
   metal extraction in AA method 3–24
   method 4–94
   standards 4–99t
      NIST 4–98t
pH meter
   calibration 4–100
   for acidity 2–32
   for microbiological examinations 9–6
   for nitrogen (ammonia) 4–123
   for nitrogen (nitrate) 4–140
   for pH 4–96
pH/millivolt meter
   for chlorine 4–66
Phenanthrene
   liquid-liquid extraction chromatographic
      method 6–148
   liquid-liquid extraction GC/MS method
      6–113
Phenanthroline method
   for iron 3–102

Phenanthroline method *(cont.)*
   for sulfite 4–201
Phenanthroline solution
   for iron 3–103
   for sulfite 4–202
Phenate method
   for nitrogen (ammonia) 4–120
      automated 4–126
Phenate reagent
   for nitrogen (ammonia) 4–120
Phenol
   liquid-liquid extraction GC method 6–137
   liquid-liquid extraction GC/MS method
      6–113
Phenol derivatives
   gas chromatogram 6–143i
Phenol red indicator
   for bromide 4–12
Phenol solution
   for phenols 5–51, 6–140
Phenolphthalein indicator solution
   desk reagent *inside front cover*
   for acids (organic) 5–71
   for pesticide chromatographic column
      standardization 6–169
Phenols
   chloroform extraction method 5–51
   chromatographic conditions 6–138t
   cleanup procedure 5–50
   definition 5–48
   gas chromatogram 6–141i
   GC method acceptance criteria 6–144t
   liquid-liquid extraction GC method 6–137
      bias and precision 6–145t
   method detection limits 6–138t
   methods 5–48, 6–137
   photometric method 5–54
   recovery of derivatives 6–139t
Phenylalanine deaminase activity
   Salmonella 9–135
Phenylarsine oxide solution
   for chlorine 4–51, 4–57
   for iodine 4–106
   for nitrogen (ammonia) 4–116
Pheophytin
   in periphyton 10–53
Phleger corer
   for macroinvertebrates 10–104, 10–104i
Phosphate, *see* Phosphorus
Phosphate buffer
   for direct microbial count 9–65
   for BOD 5–4
   for chlorine 4–56, 4–60, 4–65

Phosphate buffer *(cont.)*
  for phenols 5–52
  for reagent water bacteriological quality
    9–9
  for trihalomethane formation potential
    5–76
Phosphate buffered saline
  for *Salmonella* 9–137
Phosphate-carbonate buffer solution
  for mercury 3–119
Phosphate manifold 4–180i
  quality control acceptance criteria 6–180t
Phosphate solution
  for ion chromatography 4–4
  for phosphorus 4–174, 4–177, 4–179
Phosphoric acid-ammonium persulfate
    reagent
  for vanadium 3–156
Phosphoric acid solution
  for metals by electrothermal AA 3–36
  for organic carbon 5–25
  for phenols 5–50
Phosphoric acid-sulfamic acid solution
  for chlorine 4–52
  analysis steps for various forms 4–168i
  ion chromatography method 4–2
  methods
Phosphoric acid-sulfanilamide-naphthyl-
    ethylenediaminedihydro
  for nitrogen (nitrate) 4–141
Phosphorus
  acid hydrolysis 4–170
  ascorbic acid method 4–177
    automated 4–178
    precision and bias 4–178t
  digestion 4–171, 4–172
  filtration 4–170
  forms 4–166
  methods 4–166
    bias and precision 4–169t
  stannous chloride method 4–175
  vanadomolybdophosphoric acid colori-
    metric method 4–173
Photodiode array detector
  for high-performance liquid chromato-
    graphic methods 6–9
Photographic method
  for toxicity test organism counting 8–30
Photoionization detector
  for gas chromatographic methods 6–8
Photometer
  for sulfate 4–207

Photometer *(cont.)*
  also see Spectrophotometer; Filter pho-
    tometer
Photometric method
  for lithium 3–110
  for phenols 5–54
  general requirements 1–47
  interferences 1–48
Photoperiod
  for freshwater fish culture 8–126t
  for toxicity tests 8–30
Physical examination
  methods 2–1
  quality control 2–1
Physical hazards
  in laboratories 1–66
Phytoplankton
  counting cell 10–25i
  counting technique 10–23
  definition 10–3
  inoculum for toxicity tests 8–52
  maximum specific growth rate 8–52
  membrane filter mounts 10–18
  mounts 10–17
  analysis steps for various forms 10–18
  preservatives 10–7
  samples 10–6
  sedimented slide mounts 10–18
  staining and preparation technique 10–27
  toxicity tests 8–52
  also see Plankton
Pimephales promelas, see Minnow, fathead
Pipet containers
  for microbiological tests 9–26
Pipets
  for heterotrophic plate count 9–61
  for microbiological tests 9–26
  for odor determinations 2–19
Pisicides, *see* Ichthyocides
Planchet
  for suspended solids 2–75
Plankton
  biomass (standing crop) determination
    10–39
  concentration techniques 10–16
  definition 10–1, 10–3
  color plates following 10–200
  metabolic rate measurements 10–42
  methods 10–3
  microscopes 10–19
  mounting 10–17
  net 10–6
  nitrogen fixation measurement 10–43

Plankton *(cont.)*
  productivity
    carbon 14 method 10–45
    oxygen method 10–43
  samples 10–4
  staining technique 10–27
  subsampling device 10–30i
  also see Phytoplankton; Zooplankton
Plants
  aquatic 10–152i, 10–153i, 10–154i
  also see Macrophyton
Plasma emission spectroscopic method
  for metals 3–53
Plate count
  actinomycetes 9–128
  heterotrophic
    computing and reporting 9–56
    counting and recording 9–56
    dilutions 9–59i
    for swimming pools 9–47
    incubation 9–56
    media 9–55
    membrane filter method 9–63
    methods 9–54
    pour plate method 9–58
    samples 9–55
    spread plate method 9–61
    sterility controls 9–60
    swimming pool waters 9–47
    work area 9–55
  standard, *see* Plate count, heterotrophic
Plate count agar
  for heterotrophic plate count 9–55
Plates
  for heterotrophic plate count
    weight loss on drying 9–62i, 9–63i
Plating
  for heterotrophic plate count
    pour plate method 9–59
    spread plate method 9–61
  for Legionellaceae 9–151
  for zoosporic fungi 9–193
Platinizing solution
  for conductivity 2–59
Platinum
  AA method 3–20
  methods 3–124
Platinum-cobalt method
  for color 2–2
Platinum solution
  for platinum 3–21
Platinum ware
  for radium 226 7–36

Polarographic method 1–50
  for chlorine 4–54
Polychaetes, *see* Annelids
Polychlorinated biphenyls, *see* PCBs
Polyethylene glycol method
  for virus concentration 9–173
Polynuclear aromatic hydrocarbons
  chromatographic conditions 6–151t
  chromatographic method
    QC acceptance criteria 6–156t
  detection limits 6–149t
  gas chromatogram 6–155i
  liquid chromatogram 6–153i, 6–154i
  liquid-liquid extraction chromatographic
    method 6–148
    bias and precision 6–157t
  methods 6–147
  retention times 6–151t
Polynuclear aromatic hydrocarbon solutions
  for polynuclear aromatic hydrocarbons
    6–150
Ponar grab
  for macroinvertebrates 10–99, 10–99i
  for macrophyton 10–75
Population
  fish
    definition 10–114
  macrophyton 10–74
Post column reactor
  for high-performance liquid chromato-
    graphic methods 6–9
Potassium
  AA method 3–20
  flame photometric method 3–125
  inductively coupled plasma method 3–54
  methods 3–125
Potassium antimonyl tartrate solution
  for phosphorus 4–177, 4–179
Potassium bi-iodate solution
  for chlorine 4–57
  for DO 4–153
Potassium bromide solution
  for chlorine dioxide 4–82
  for mercury 3–119
Potassium chloride solution
  for aquatic humic substances 5–39
  for conductivity 2–60, 2–60t
  for metals 3–25
Potassium chloroplatinate solution
  for nitrogen (ammonia) 4–118
Potassium chromate indicator solution
  for chloride 4–68

Potassium chromate solution
　for silica 4-186
Potassium cyanide-sodium hydroxide
　solution
　for cadmium 3-83
Potassium cyanide solution
　for zinc 3-161, 3-163
Potassium dichromate solution
　for chlorine 4-49
　for COD 5-12, 5-14
Potassium dichromate-sulfuric
　　acid-mercuric sulfate solution
　for COD 5-16
Potassium ferricyanide solution
　for phenols 5-52
Potassium fluoride solution
　for DO 4-153
　for pH 4-100
Potassium hydrogen phthalate solution
　for acidity 2-32
　for COD 5-12
Potassium hydrogen tartrate solution
　for pH 4-100
Potassium hydroxide solution
　for coliform differentiation 9-105
　for herbicides 6-184
　for sludge digester gas 2-89
Potassium iodate solution
　for chlorine 4-49, 4-66
Potassium iodide-iodate titrant
　for sulfite 4-200
Potassium iodide solution
　for arsenic 3-75
　for chlorine 4-56, 4-66, 4-60
Potassium nitrate solution
　for cyanide 4-33
Potassium oxalate solution
　for DO 4-157
Potassium permanganate solution
　for calcium 3-88
　for chlorine 4-63
　for chromium 3-92
　for DO 4-157
　for mercury 3-29, 3-119
　for selenium 3-133
Potassium peroxymonosulfate solution
　for iodide 4-107
Potassium persulfate
　for organic carbon 5-25
Potassium persulfate solution
　for arsenic and selenium 3-46
　for mercury 3-29, 3-119

Potassium solution
　for potassium 3-21, 3-126
Potassium tetrachloromercurate
　for sulfite 4-202
Potassium thiocyanate solution
　for iodide 4-109
Potentiometer
　problems 4-101
Potentiometric titration method 1-51
　for chloride 4-71
Pour plate method
　for bacteria 9-54
　　counting and recording 9-56
　for fungi 9-188
　for heterotrophic plate count 9-58
Power pack
　for *Salmonella* 9-137
Practical quantitation limit
　determination 1-19
Practical Salinity Scale
　for salinity 2-62, 2-63
Precipitation/dissolution quantity indices
　calcium carbonate saturation 2-47
Precipitation method
　for cesium (radioactive) 7-22
　for iodine (radioactive) 7-24
　for radium 7-30
　for strontium (radioactive) 7-48
Precision 1-14i
　AA methods 3-18t
　BTFP analysis 5-80t
　computation 1-15, 1-15t, 1-16t
　definition 1-5, 1-15
　flame AA methods 3-19t
　for method validation 1-22
　GC method for volatile halocarbons 6-85t
　ion chromatographic method 4-6t
　microbiological examination procedures
　　9-17, 9-18t
　nitrogen (ammonia) methods 4-112t
　organic halogen method 5-36t
　pesticide method 6-168t
　phosphorus methods 4-169t
　TFP analysis 5-78t
Predators
　coral 8-60
Prefilter
　for virus concentration 9-160
Preliminary treatment
　BOD 5-7
　cyanides 4-25
　pesticides 6-164

Preliminary treatment *(cont.)*
  selenium 3–131
  sulfides 4–194
Presence-absence test
  for *Clostridium* 9–82i
  for coliforms 9–80
    fecal 9–82i
  for *Pseudomonas* 9–82i
  for staphylococcus 9–82i
  for streptococci (fecal) 9–82i
Preservation
  benthic macroinvertebrates 10–109
  metals 3–2
  microbiological samples 9–35
  organics samples 11
  phytoplankton 10–7
  radioactivity samples 7–3
  samples 1–39
  volatile organic compounds 6–4, 6–5t
  zooplankton 10–12
Preservatives
  for phytoplankton 10–7
    6-3-1 10–8
  for volatile organics samples 6–5, 6–5t
  for volatile organics samples 6–5t
Pressure source
  for virus concentration 9–159, 9–166
Pressure vessel
  for virus concentration 9–159
Presumptive phase
  for coliforms 9–68
  for *Pseudomonas aeruginosa* 9–53
  for streptococcus (fecal) 9–110
Probit analysis
  toxicity tests 8–34, 8–34t
Producers
  aquatic organisms 10–142
Production
  periphyton 10–62i, 10–63i
Productivity
  algae, *see* Biostimulation
  macrophyton 10–80
    biomass methods 10–80
    metabolic methods 10–85
    methods 10–81t
  periphyton 10–54
    carbon 14 method 10–57
    oxygen method 10–55, 10–57
  plankton
    carbon 14 method 10–45
    oxygen method 10–43
Propanol (2)
  for chlorine 4–65

Proportioning injector, Venturi type
  for virus concentration 9–163
Propylbenzene
  purge and trap GC method 6–71
    capillary column 6–93
  purge and trap GC/MS method 6–42
    capillary column 6–50
Protective clothing
  for laboratory personnel 1–61
Protozoa 10–157i, 10–158i, 10–159i
  ciliated
    culturing 8–54
    holding 8–54
    toxicity tests 8–53
  pathogenic
    methods 9–196
Pseudomonas
  presence-absence test 9–82i
Pseudomonas aeruginosa
  culture media 9–52, 9–53
  in bathing waters 9–51
  in swimming pools 9–48
  membrane filter technique 9–52
  multiple-tube technique 9–53
Pump
  air
    for mercury 3–29
  for aquatic humic substances 5–40
  for organic constituents (individual) 6–13
  for phytoplankton sampling 10–7
Purge and trap gas chromatographic (GC)
    method
  for volatile aromatics 6–65, 6–71
    bias and precision 6–70t, 6–76t, 6–77t
  for volatile halocarbons 6–79, 6–87
    bias and precision 6–85t, 6–91t, 6–92t,
      6–96t
    capillary column 6–93
Purge and trap gas chromatographic/mass
    spectrometric (GC/MS) method
  for volatile organics 6–27, 6–42
    capillary column 6–50
    bias and precision 6–40t, 6–49t, 6–56t,
      6–58t
Purge and trap system
  for volatile aromatics 6–65, 6–67i, 6–68i,
    6–72
  for volatile halocarbons 6–81, 6–87
  for volatile organics 6–28, 6–31i, 6–43
Purge gas
  for chlorine dioxide 4–82
  for organic carbon 5–20

Purging device
  for volatile organics 6–29, 6–29i, 6–43
Pyrene
  liquid-liquid extraction chromatographic
    method 6–148
  liquid-liquid extraction GC/MS method
    6–113
Pyridine-barbituric acid reagent
  for cyanide 4–31
Pyrocatechol violet (PCV) method, auto-
    mated
  for aluminum
    precision and bias 3–72t, 3–73t
Pyrocatechol violet (PCV) solution
  for aluminum 3–69
Pyrocatechol violet method, automated
  for aluminum 3–68
Pyrogallol reagent
  for sludge digester gas 2–89

                           Q

Quadrat frames
  for macrophyton sampling 10–76
Quadrats
  for macrophyton sampling 10–76
Qualitative tests
  for sulfide 4–192
Quality assessment
  definition 1–5
  laboratory analyses 1–12
Quality assurance
  definition 1–5
  laboratory analyses 1–6
  planning 1–6
Quality control
  definition 1–5
  inorganic nonmetals analyses 4–1
  laboratory analyses 1–7
  metals analyses 3–3, 3–17
  microbiological examinations 9–4
    interlaboratory 9–23
    intralaboratory 9–5
    statistics 9–21
  organic halogen analyses 5–32
  organics analyses 5–2
  physical and aggregate property methods
    2–1
  radioactivity analyses 7–14
Quality control acceptance criteria
  for organochlorine pesticides and PCBs
    6–180t

Quality control acceptance criteria (cont.)
  for phenol GC method 6–144t
  for volatile aromatics
    purge and trap gas chromatographic
      method 6–70t
  for volatile halocarbons 6–83t
  for volatile organics
    purge and trap GC/MS method 6–36t
Quality control check sample
  for volatile organics 6–33
  for phenols 6–140

                           R

Radiation
  safety precautions 1–66
Radioactive wastes
  disposal 1–71
Radioactivity
  alpha 7–15
  apparatus 7–11
  beta 7–15
  counters 7–5
  counting room 7–4
  expression of results 7–11
  general considerations 7–1
  methods 7–1
  monitoring 7–1
  quality assurance 7–14
  reagents 7–11
  samples 7–3
  standard sources 7–11, 7–14
  statistics 7–12
  wastewater samples 7–3
Radiocarbon incorporation method
  macrophyton productivity 10–87
Radiochemical method
  for uranium 7–57
Radiochemical monitors
  for laboratory safety 1–69
Radioelements
  phase distribution in wastewater 7–3t
Radiometric detection
  for coliforms (fecal) 9–37
Radium
  isotopes 7–29
  methods 7–29
  precipitation method 7–30
  regulatory limits 7–29
Radium 226
  emanation method 7–34

Radium 226 solution
  for radium 7–31
  for radium 226 7–36, 7–37
Radium 228
  sequential precipitation method 7–44
Radon 222
  decay, growth, and activity correction
    factors 7–38t
Radon bubblers
  for radium 226 7–35
Rakes
  for macrophyton sampling 10–75
Random error
  definition 1–5
Range charts 1–9, 1–10t, 1–11i
Rapid detection methods
  for bacteria 936, 9–38t
Rapid test
  for coliform verification 9–19, 9–89
Rare earth carrier
  for radioactive strontium 7–49
Reaction cell
  for arsenic and selenium 3–46, 3–45i
Reaction flask
  for mercury 3–29
  for phenols 6–138
Reagent blanks
  analysis for analytical quality control 1–7
Reagents
  dilutions 1–44
  drying 1–43
  for microbiological examinations 9–13
  general guidelines 1–42
  safety data sheets 1–59, 1–62, 1–70
  specifications 1–42
  storage 1–44
  threshold limit values 1–65t
  uniform concentrations inside front cover,
    1–45
  also see Chemicals
Rearing tanks, see Culture chambers
Recorder
  for AA method 3–17
  for sludge digester gas 2–91
Recovery
  stressed organisms 9–43
Recreational waters
  microbiological examination 9–45
Reducers
  aquatic bacteria and fungi 10–142
Reducing agent
  for silica 4–188, 4–190
Reduction column
  for nitrogen (nitrate) 4–135, 4–136i

Reference solution
  for fluoride 4–90
Reflux apparatus
  for COD 5–12
Reflux method
  for COD 5–12, 5–14, 5–15
Refrigerator
  for microbiological laboratory 9–7, 9–26
Regenerant solution
  for anions 4–3
Regeneration column
  for ion-exchange resins 1–46i
Regulatory limit
  alpha radioactivity 7–15
  cesium (radioactive) 7–21
  iodine (radioactive) 7–24
  radium 7–29
  strontium (radioactive) 7–47
  tritium 7–54
  uranium 7–56
Relative error
  electrothermal AA method 3–42t
Remote sensing
  for macrophyton surveys 10–72
Replicate
  definition 1–5
Reservoirs
  algae growing on walls
    color plates following 10–190
Residue, see Solids
Resin
  for selenium 3–131, 3–140
  for thiocyanate 4–43
  ion-exchange 1–45
    batch method for cation removal 1–46
    column method for cation removal
      1–46, 1–46i
    for surfactants 5–65
    methods 1–46
    regeneration 1–46
Resin cartridges
  for selenium 3–140
Resin pretreatment
  for selenium 3–131
Respirators 1–62
Respirometric device
  for sludge oxygen consumption rate 2–81
Results
  expression 1–27
Resuscitation
  stressed organisms 9–43
Retention ratios
  pesticides 6–167t

Retention times
  for dibromoethane and dibromochloropropane 6–100t
  for earthy-musty-smelling organics 6–21t
  for herbicides 6–186t
  for organochlorine pesticides and PCBs 6–171t
  for PCBs 6–171t
  for trihalomethanes 6–108i
  for volatile aromatics 6–74t
  for volatile halocarbons 6–80t, 6–89t, 6–93t
Reverse osmosis
  reagent water preparation 1–56
Reverse osmosis units
  for microbiological laboratory 9–7
Reverse phase column
  for polynuclear aromatic hydrocarbons 6–150
Rhenium
  AA method 3–25
  methods 3–127
Rhenium solution
  for rhenium 3–26
Rhodium
  AA method 3–20
  methods 3–127
Rhodium solution
  for rhodium 3–21
Rochelle salt solution
  for nitrogen (ammonia) 4–118
Root turnover measurements
  macrophyton productivity 10–85
Rose bengal-glucose-peptone-aureomycin agar
  for fungi 9–190
    modified 9–191
Rose bengal-neopeptone-glucose-aureomycin agar
  for fungi 9–188
Rotameter
  for selenium 3–138
Rotenone
  for fish sampling 10–21
Rotifers 10–161i
Roundworms 10–162i
  also see Nematodes
R2A agar
  for heterotrophic plate count 9–55
Ruggedness
  analytical methods 1–23
Ruthenium
  AA method 3–20

Ruthenium (cont.)
  methods 3–128
Ruthenium solution
  for ruthenium 3–21

S

Safety 1–58
  chemicals 1–62
  committees and officers 1–58, 59
  government regulation 1–59
  laboratory management 1–58
  reagents 1–63
  sampling operations 1–32
  taste tests 2–24
  wall chart 1–61
Safety cabinet (hood) 9–7
Safety containers 1–60
Safety equipment 1–60
Safranin dye counterstain
  for coliforms 9–73
Salinity
  computation 2–65
  density method 2–64
  electrical conductivity method 2–62
  methods 2–61
Salmonella
  biochemical reactions 9–135
  concentration technique 9–131, 9–137
  enrichment 9–133, 9–138
  identification 9–131
  immunofluoroescence technique 9–136, 9–139
  isolation 9–131
  multiple-tube enrichment technique 9–140
  quantitative procedures 9–140
  selective growth 9–133
  serological technique 9–135
Salmonella test
  for mutagenicity 8–41
Salmonella typhi
  membrane filter procedure 9–140
Salt bridge
  for chlorine 4–55
Salting out
  effect on membrane electrodes 4–160, 4–160i
Salt-mixture solution
  for reagent water bacteriological quality 9–9
Salt peptone glucose broth
  for coliforms 9–105

Salvelinus fontinalis, see Trout, brook
Sample bottles
  cleaning 1–41
  for microbiological tests 9–27, 9–31
  for odor 2–19
  protection for shipment 1–42
  also see Glassware
Sample containers
  for dibromoethane and dibromochloropro-
    pane 6–99
  for metals 3–2
Sample dispensers
  for metals by AA method 3–36
Sample introduction apparatus
  for sludge digester gas 2–92
Samplers
  artificial substrate 10–105
  automatic 1–35
  basket 10–106, 10–107i
  benthic dome 10–76
  bottle 10–7
  bottom 9–34
  bottom net 10–105
  box 10–76
  drift net 10–104, 10–105i
  Ekman grab 10–75, 10–76, 10–102,
    10–102i
  for biostimulation 8–44
  for DO 4–151, 4–151i
  for fish 10–116, 10–118i, 10–120i
  for floatables 2–66
  for *Giardia lamblia* 9–198, 9–200i
  for macroinvertebrates 10–99, 10–99i,
    10–101i, 10–102i, 10–103i
  for macrophyton 10–71, 10–75
  for periphyton 10–50i
  for phytoplankton 10–6, 10–7i
  for plankton 10–6, 10–7i
  for plankton productivity 10–43
  for *Salmonella* 9–132
  for zooplankton 10–8, 10–13i
  frame 10–76
  KB core 10–104, 10–104i
  Louisiana box 10–77
  microbiological 9–34
  multiple-plate (Hester-Dendy) 10–106,
    10–107i
  orange-peel grab 10–99, 10–99i
  Osborne aquatic plant 10–77
  Petersen grab 10–99
  Petersen grab 10–99i, 10–99i
  Phleger corer 10–104, 10–104i
  Ponar grab 10–99, 10–99i

Samplers *(cont.)*
  qualitative 10–105
  riffle 10–102
  Shipek grab 10–101, 10–101i
  Smith-McIntyre grab 10–100, 10–101i
  stovepipe 10–76
  suction 10–106, 1–108i
  Surber 10–76, 10–102, 10–103i
  tow net (trawl) 10–105
  Van Veen 10–100, 10–100i
  Waterways Experiment Station 10–77
  Wilding (stovepipe) 10–104, 10–105i
  ZoBell J-Z 9–34
Samples
  chain-of-custody procedures 1–34
  collection 1–30, 1–34
  composite 1–33
  containers 1–35
  for acidity 2–32
  for actinomycetes 9–127
  for aluminum 3–64
  for bathing waters 9–50
  for benthic macroinvertebrates 10–109
  for beryllium 3–80
  for biostimulation studies 8–45
  for BOD 5–4
  for boron 4–7
  for carbon dioxide 4–19
  for chloride 4–67
  for chlorine (residual) 4–48
  for chlorine dioxide 4–75, 4–81
  for chromium 3–90
  for COD 5–11
  for coliforms
    delayed incubation technique 9–93
    fecal 9–95, 9–95t
    total 9–86, 9–88t
  for copper 3–95
  for cyanide 4–25
  for dibromoethane and dibromochloropro-
    pane 6–99
  for DO 4–150
  for fish 10–116, 10–127
  for fish kill investigations 10–135
  for flavor rating assessment 2–27
  for fluoride 4–84
  for fungi 9–188
    zoosporic 9–193
  for *Giardia lamblia* 9–198
  for heterotrophic plate count 9–55
  for hyphomycetes 9–194
  for iron 3–101
  for iron bacteria 9–115

Samples *(cont.)*
　for lead 3–107
　for Legionellaceae 9–150
　　treatment 9–151
　for macroinvertebrates 10–96
　for macrophyton 10–70, 10–74, 10–78
　for manganese 3–115
　for mercury 3–2
　for metals 3–1, 3–3, 3–5
　for methane 6–61, 6–63
　for microbiological examinations 9–31
　　chelation of metals 9–32
　　containers 9–31
　　dechlorination 9–31
　　from bathing waters 9–33
　　from raw water supplies 9–33
　　from surface waters 9–33
　　from water distribution systems 9–32
　　from wells 9–32
　　identifying data 9–34
　　preservation and storage 9–35
　　procedures 9–32
　　shipment 9–2
　　sites 9–33
　　size 9–34
　　USEPA tests 9–3
　for nematodes 9–206
　for nitrogen (ammonia) 4–113, 4–123
　for nitrogen (nitrate) 4–132
　for nitrogen (nitrite) 4–129
　for nitrogen (organic) 4–143
　for oil and grease 5–42
　for organic carbon 5–19
　for organic constituents 5–1, 6–17
　for organic halogen 5–29
　for organochlorine pesticides and PCBs
　　6–172
　for periphyton 10–49, 10–50
　for pesticides 6–159, 6–164
　for phenols 5–49
　for phosphorus 4–168
　for phytoplankton 10–6
　for plankton 10–4
　for polynuclear aromatic hydrocarbons
　　6–149
　for potassium 3–125
　for radioactivity 7–3
　for silica 4–182
　for silver 3–142
　for sludge digester gas 2–88
　for sodium 3–146
　for solids 2–72
　for strontium 3–150

Samples *(cont.)*
　for sulfate 4–204
　for sulfide 4–192
　for sulfite 4–200
　for sulfur bacteria 9–119
　for swimming pool water 9–46
　for trihalomethane formation 5–74
　for trihalomethanes 6–106
　for turbidity 2–12
　for uranium 7–57, 7–59
　for viruses 9–161, 9–167, 9–172, 9–174,
　　9–176
　for zinc 3–157
　for zooplankton 10–8
　grab 1–32
　identification 1–31
　integrated 1–33
　laboratory intercomparison 1–13
　lakes and reservoirs 1–31
　locations for 1–32
　manual procedures 9–34
　metal-containing 1–30
　number 1–36, 1–36i
　organic-containing 1–30
　performance evaluation 1–12
　preservation 1–39
　　metals 3–1
　　microbiological 9–35
　preservatives 1–40
　quantity 1–36
　refrigeration 1–40
　requirements 1–37t
　safety precautions 1–32
　solid and semisolid for coliforms 9–68
　storage 1–39
　streams and rivers 1–31
　suspended solids for viruses 9–176
　types 1–32
　water distribution systems 1–31
　wells 1–31
Sample screening test
　for cyanide 4–40
Sample tubes
　for turbidity 2–13
Sampling stations
　for macroinvertebrates 10–97
　for periphyton 10–49
　for plankton 10–4
Sand bath
　for herbicides 6–183
Sanitary requirements
　in taste tests 2–24

Saprobity system
  periphyton analysis 10–66
Saturometric determination
  calcium carbonate precipiation potential
    2–48
Scales
  fish 10–131, 10–131i
Scallops
  collection 8–76
  conditioning 8–76
  culturing 8–76
  embryo tests 8–78
  growth tests 8–80
  toxicity tests 8–74
  also see Mollusks
Schindler-Patalas trap
  for zooplankton 10–9, 10–9i
Scintillation cells
  for radium 226 7–35
Scintillation counter assembly
  for radium 226 7–35
Scintillation solution
  for tritium 7–54
Scleractinian corals, *see* Corals
Screening method
  for nitrogen (nitrate) 4–132
SCUBA
  for fish sampling 10–123
Seasonal biomass accumulation
  macrophyton productivity 10–84
Seawater
  for annelids 8–69
  for crab culturing 8–94
  metal-free 3–36
  metals analysis by AA 3–14
  reconstituted 8–17t
Sedgwick-Rafter counting cell
  for phytoplankton 10–24, 10–25i
Sedgwick-Rafter counts
  for periphyton 10–51
Sedimentation
  for plankton concentration 10–16
Sedimented slide mounts
  for phytoplankton 10–18
Sediments
  microbiological sampling 9–33
Seeding
  for BOD 5–6
Segmented worms 10–163i
Seines
  for fish 10–119, 10–120i
Selective growth
  Campylobacter jejuni 9–143

Selective growth *(cont.)*
  Salmonella 9–133
  Shigella 9–141
  Vibrio cholerae 9–145
Selective ion electrode 1–51
Selenite cystine broth
  for *Salmonella* enrichment 9–133
Selenium
  AA method
    micro quantities 3–36
  colorimetric method 3–135
  fluorometric method 3–137
  hydride generation AA method
    continuous 3–50
    manual 3–43
  inductively coupled plasma method 3–54
  methods 3–128
  nonvolatile organic
    method 3–139
  preliminary treatment 3–131
  speciation 3–129, 3–130i
  volatile
    method 3–137
Selenium solution
  for selenium 3–51, 3–47, 3–135
Semi-micro-kjeldahl method
  for nitrogen (organic) 4–147
Semisolids
  coliform tests 9–68
  total, fixed, and volatile solids determi-
    nations 2–78
Sensitivity
  AA methods 3–15
Separation
  sulfides, soluble and insoluble 4–194
Separator column
  for anions 4–3
Separatory funnels
  for base/neutral and acid extractables
    6–119
  for cadmium 3–83
  for chromium 3–92
  for copper 3–96
  for herbicides 6–183
  for iron 3–103
  for lead 3–108
  for mercury 3–119
  for nickel 3–122
  for oil and grease 5–43, 5–44
  for pesticides 6–160
  for phenols 5–51
  for selenium 3–135, 3–137
  for silver 3–143

Separatory *(cont.)*
  for surfactants 5–57, 5–61, 5–65
  for zinc 3–158, 3–160
Sequential precipitation method
  for radium 228 7–44
Serological technique
  Salmonella 9–135
  also see Identification
Set lines
  for fish 10–116
Settleable solids
  method 2–78
  also see Solids
Settled sludge volume
  test method 2–83
Settling vessel
  for settled sludge volume 2–83, 2–83i
  for zone settling rate 2–84, 2–85i
Seven-hour fecal coliform test 9–36
Sheepshead minnow, *see* Minnow, sheepshead
Shellfish, *see* Crustaceans; Mollusks
Shields, safety 1–60
Shigella
  biochemical reactions 9–141
  enrichment 9–141
  identification 9–141
  methods 9–140
  selective growth 9–141
Shipek grab
  for macroinvertebrates 10–101, 10–101i
Shipment
  samples, microbiological 9–2
Shoes, safety
  for laboratory personnel 1–61
Short-cut method
  for cyanides and thiocyanates 4–36
Showers, safety 1–60
Shrimp
  holding acclimating, and culturing 8–98
  toxicity tests 8–109
  also see Crustaceans; *Palaemonetes*;
    Penaeus
Siderocapsa treubii 9–118i
Significant figures 1–28
Silica
  color standards 4–187t
  gravimetric method 4–183
  heteropoly blue method 4–188
    automated 4–189
  inductively coupled plasma method 3–54
  light path lengths 4–185t
  mcthods 4–181

Silica *(cont.)*
  molybdate-reactive
    method 4–189
  molybdosilicate method 4–184
Silica gel
  for hydrocarbons 5–48
  for phenols 6–140
  for polynuclear aromatic hydrocarbons
    6–150
Silica solution
  for silica 4–185
  for silicon 3–26
Silicic acid
  for organic acids 5–70
Silicon
  AA method 3–25
Silver
  AA method 3–20
    low concentrations 3–23
    micro quantities 3–36
  dithizone method 3–142
  inductively coupled plasma method 3–54
  methods 3–141
Silver diethyldithiocarbamate method
  for arsenic 3–74
Silver diethyldithiocarbamate reagent
  for arsenic 3–75
Silver foil test
  for sulfide 4–192
Silver nitrate-mercuric sulfate reagent
  for manganese 3–116
Silver nitrate-nitric acid reagent
  for sulfate 4–205
Silver nitrate solution
  for chloride 4–68
  for cyanide 4–30
  for iodine (radioactive) 7–24
Silver-silver chloride electrode
  for chloride 4–71
Silver-silver sulfide electrode test
  for sulfide 4–192
Silver solution
  for silver 3–21, 3–143
Silver sulfate-aluminum sulfate boric acid
    buffer
  for nitrogen (nitrate) 4–134
Silver sulfate-sulfuric acid reagent
  for COD 5–12
Simmons' citrate agar
  for coliforms 9–105
Simulated distribution system trihalome-
    thane concentration
  definition 5–74

Simulated (cont.)
  method 5–82
Siphon device
  for crab culture apparatus 8–94
Slide rack
  for periphyton sampling 10–49, 10–50i
Slides
  fluorescent antibody micro
    for Salmonella 9–137
Sludge
  digestion for metals 3–7
  microbiological sampling 9–33
  oil and grease extraction 5–46
  oxygen consumption rate test 2–81
  settled volume test 2–83
  specific gravity test 2–86
  test methods 2–81
  zone settling rate test 2–84
Sludge digester gas
  collection apparatus 2–88i
  GC method 2–91
  methods 2–87
  volumetric method 2–88
Sludge volume index
  method 2–84
Slurp gun
  for fish sampling 10–122
Smalley method
  macrophyton productivity 10–81
Smith-Hieftje background correction
  for metals by AA 3–15
Smith-McIntyre grab
  for macroinvertebrates 10–100, 10–101i
Snails 10–176i
Snorkeling
  for fish sampling 10–123
Snyder column
  for base/neutral and acid extractables
    6–119
  for herbicides 6–183
Sodium
  AA method 3–20
  flame emission photometric method 3–146
  inductively coupled plasma method 3–54
  methods 3–145
Sodium acetate buffer solution
  for aluminum 3–65
Sodium acetate solution
  for iron 3–103
  for zinc 3–158
Sodium arsenite solution
  for chlorine 4–60
  for fluoride 4–90

Sodium arsenite solution (cont.)
  for nitrogen (ammonia) 4–116
Sodium azide solution
  for chromium 3–92
Sodium benzoate solution
  for coliforms 9–92
Sodium bicarbonate-sodium carbonate solution
  for anions 4–3
Sodium bicarbonate solution
  for chlorine dioxide 4–79
Sodium bisulfite-sodium sulfite-aminonaph-
    tholsulfonic acid
  for silica 4–190
Sodium bisulfite solution
  for iodine (radioactive) 7–26, 7–27
  for manganses 3–117
  for silica 4–188
  for uranium 7–57
Sodium borohydride
  safety precautions 1–64
Sodium borohydride solution
  for arsenic and selenium 3–46
Sodium carbonate-sodium bicarbonate
    solution
  for anions 4–3
Sodium carbonate-sodium sulfate solution
  for lithium 3–110
Sodium carbonate solution
  for alkalinity 2–36
  for periphyton production 10–57
  for strontium (radioactive) 7–49
Sodium chloride
  for dibromoethane and dibromochloropro-
    pane 6–100
Sodium chloride-hydroxylamine sulfate
  for mercury 3–29
Sodium chloride solution
  for chloride 4–68
  for iodide 4–109
  for iodine (radioactive) 7–26
  for organic halogen 5–29
  for plankton productivity 10–45
  for virus concentration 9–160
Sodium chromate solution
  for strontium (radioactive) 7–49
Sodium citrate solution
  for copper 3–96, 3–98
  for reagent water bacteriological quality
    9–9
  for zinc 3–159, 3–161
Sodium citrate test
  for coliform differentiation 9–105

Sodium cyanide inhibitor
  for hardness 2–55
Sodium dihydrogen phosphate solution
  for cyanide 4–32
Sodium dihydrogen phosphate-sulfuric acid
    solution
  for anionic surfactants 5–61
Sodium hydrogen sulfite solution,
  see Sodium bisulfite solution
Sodium hydroxide-glycine
  for virus concentration 9–160, 9–166
Sodium hydroxide-potassium cyanide solu-
    tion
  for cadmium 3–83
Sodium hydroxide-sodium thiosulfate
    reagent
  for nitrogen (organic) 4–145
Sodium hydroxide solution
  for acidity 2–32
  for acids (organic) 5–71
  for aquatic humic substances 5–38, 5–39
  for cyanates 4–41
  for cyanide 4–28, 4–30, 4–32, 4–33
  for iodine (radioactive) 7–26
  for nitrogen (ammonia) 4–115, 4–123
  for nitrogen (nitrate) 4–140, 4–141
  for pesticide chromatographic column
    standardization 6–169
  for phenols 5–50, 6–139
  for radium 228 7–45
  for strontium (radioactive) 7–49
  for sulfate 4–210
  for thiocyanate 4–43
  uniform inside front cover
Sodium hypochlorite solution
  for iodine (radioactive) 7–26
  for nitrogen (ammonia) 4–127
Sodium iodide prereductant solution
  for arsenic and selenium 3–46
Sodium molybdate-sodium tungstate reagent
  for tannin and lignin 5–68
Sodium nitrite solution
  for iodine (radioactive) 7–27
  for manganese 3–117
Sodium nitroprusside solution
  for nitrogen (ammonia) 4–128
Sodium oxalate
  for calcium 3–88
Sodium oxalate solution
  for nitrogen (nitrite) 4–130
Sodium phenate solution
  for nitrogen (ammonia) 4–127

Sodium phosphate solution
  for chlorine dioxide 4–82
Sodium potassium tartrate solution
  for cadmium 3–83
  for nitrogen (ammonia) 4–127
Sodium selenate solution
  for selenium 3–134
Sodium solution
  for sodium 3–21, 3–147
Sodium sulfate
  for base/neutral and acid extractables
    6–120
  for organic constituents (individual) 6–17
  for pesticides (organochlorine) 6–162
  for polynuclear aromatic hydrocarbons
    6–150
Sodium sulfate-sodium carbonate solution
  for lithium 3–110
Sodium sulfate solution
  for herbicides 6–183
Sodium sulfate-sulfuric acid solution
  for sludge digester gas 2–89
Sodium sulfide inhibitor
  for hardness 2–55
Sodium sulfide solution
  for zinc 3–161
Sodium sulfite-sodium bisulfite-aminonaph-
    tholsulfonic acid
  for silica 4–190
Sodium sulfite solution
  for BOD 5–5
  for iodine (radioactive) 7–24
  for lead 3–108
  for nitrogen (ammonia) 4–116
  for silica 4–188
  for trihalomethane formation potential
    5–76
Sodium tartrate solution
  for nickel 3–122
Sodium thiosulfate solution
  for acidity 2–33
  for bromide 4–12
  for chlorine 4–49, 4–51
  for DO 4–153
  for iodide 4–108
  for nitrogen (ammonia) 4–116
  for zinc 3–159
Sodium tungstate-sodium molybdate reagent
  for tannin and lignin 5–68
Solids
  coliform tests 9–68
  definition 2–71

Solids *(cont.)*
  dissolved
    method 2–74
  fixed
    definition 2–71
    method 2–77
    solid and semisolid samples 2–78
  methods 2–71
  settleable
    definition 2–71
  suspended
    method 2–75
    recovery of viruses 9–175
  total
    definition 2–71
    method 2–72
    solid and semisolid samples 2–78
  total dissolved
    use in checking analyses 1–20, 1–21
  volatile
    method 2–77
    solid and semisolid samples 2–78
Solid samples
  digestion for metals 3–7
  total, fixed, and volatile solids in
    method 2–78
Solid wastes
  cyanide in 4–22
Solochrome cyanine R solution
  for aluminum 3–65
Solubility
  oxygen in water 4–154t
Solutions
  storage 1–44
Solvents
  threshold limit values 1–64t
Soxhlet extraction method
  for oil and grease 5–45
SPADNS method
  for fluoride 4–89
SPADNS solution
  for fluoride 4–90
Spaerotilus-Leptothrix group bacteria
  enumeration, enrichment, isolation 9–123
Spawning chambers
  for brook trout 8–136
  for minnows 8–138
Spawning substrates
  for minnows 8–138
Spears
  for fish sampling 10–116
Specific gravity
  definition 2–86

Specific gravity *(cont.)*
  sludge
    method 2–86
    temperature correction factor 2–87t
Specific growth rate, *see* Growth rate,
    specific
Spectrometer
  AA
    for arsenic and selenium 3–46
    for mercury 3–29, 3–30i
  for metals by ICP method 3–56
  for radioactivity analyses 7–9
  gamma
    for radioactivity analyses 7–8
  also see Atomic absorption spectrometric
      equipment
Spectrophotometer
  for aluminum 3–64
  for arsenic 3–75
  for beryllium 3–80
  for boron 4–8, 4–10
  for bromide 4–12
  for cadmium 3–83
  for chlorine 4–62, 4–65
  for chlorophyll 10–32
  for chromium 3–92
  for COD 5–16
  for color 2–5
  for copper 3–96, 3–98
  for cyanide 4–31
  for fluoride 4–90
  for iodide 4–107, 4–109
  for iodine 4--04
  for iron 3--02
  for lead 222
  for manganese 3–116
  for mercury 3–119
  for nickel 3–122
  for nitrogen (ammonia) 4–117, 4–120
  for nitrogen (nitrate) 4–132, 4–135
  for nitrogen (nitrite) 4–129
  for oil and grease 5–44
  for ozone 4–163
  for phenols 5–51, 5–54
  for phosphorus 4–173, 4–177
  for selenium 3–135
  for silica 4–185, 4–188
  for silver 3–143
  for strontium 3–150
  for sulfate 4–207
  for sulfide 4–195
  for sulfite 4–202
  for surfactants 5–61, 5–65

Spectrophotometer *(cont.)*
  for tannin and lignin 5–68
  for thiocyanate 4–42
  for vanadium 3–155
  for zinc 3–158, 3–160, 3–162
Spectrophotometric method
  for chlorophyll 10–32
  for color 2–4
    ADMI 2–9
    ordinates 2–6
  for metals, *see* Atomic absorption spectro-
    metric method
  also see Colorimetric method
Spectrophotometric screening method
  for nitrogen (nitrate) 4–132
Sphaerotilus natans 9–117i
Sponges 10–160i
Spot test
  for cyanide sample screening 4–40
Spread plate method
  for bacteria 9–54
    counting and recording 9–56
  for fungi 9–190
  for heterotrophic plate count 9–61
Square-foot riffle sampler
  for macroinvertebrates 10–102, 10–103i
Stabilizer reagent
  for nitrogen (ammonia) 4–117
Staining
  periphyton 10–52
  phytoplankton 10–27
Staining assembly
  for *Salmonella* 9–137
Staining technique
  for phytoplankton 10–27
Stains
  for microbiological examinations
    quality control 9–13
Standard deviation 1–2
  significant figures 1–29
Standard plate count, *see* Plate count,
    heterotrophic
Standard substances, NIST 1–43
Standards
  external
    analysis for quality control 1–7
  for method validation
    preparation 1–23
  for quality control
    definitions 1–5
  metals by AA 3–15
  pH 4–98t, 4–99t

Standing crop
  macrophyton productivity 10–80
  maximum
    definition 8–4
    measurement 8–49
  plankton 10–39
Stannous chloride method
  for phosphorus 4–175
Stannous chloride solution
  for arsenic 3–75
  for mercury 3–29
  for phosphorus 4–175
Staphylococcus
  in swimming pools 9–47
  presence-absence test 9–82i
Staphylococcus aureus
  in swimming pools 9–47
Starch-cassein agar
  for actinomycete plate count 9–128
Starch indicator
  for chlorine 4–49
  for DO 4–153
  for sulfite 4–200
Statistics 1–2
  distributions 1–2i
  glossary 1–4
  microbiological examination quality con-
    trol 9–21
Sterilization
  autoclave times and temperatures 9–16t
  culture medium 9–15, 9–30
  glassware for microbiological tests 9–28
Sterilization lamps 9–7
Sterilizers, gas 9–25
Stills
  for microbiological laboratories 9–6
  for water distillation 1–56
Stirrer
  for organic constituents (individual) 6–14
  for pH 4–97
  for sulfate 4–207
Stirring
  effect on electrode response 4–161i
Stomach contents
  fish 10–122, 10–130
Stoneflies 10–167i
Storage
  culture medium 9–16
  in laboratories 1–60
  microbiological samples 9–35
  solutions 1–44

Stovepipe sampler
  for macroinvertebrates 10–104, 10–105i
  for macrophyton sampling 10–76
Streptococci
  Group D
    capillary precipitin test 9–112
    verification 9–112
  in swimming pools 9–48
  fecal
    confirmed test 9–110
    definition 9–108
    differentiation 9–113, 9–114y
    in bathing waters 9–51
    membrane filter technique 9–110
    m Enterococcus method 9–112
    methods 9–108
    multiple-tube technique 9–109
    presence-absence test 9–82i
    presumptive test 9–110
    recovery enhancement 9–44
    verification 9–20, 9–112
Streptomycin-penicillin
  for virus concentration 9–160, 9–171
Streptomycin-terramycin-malt extract agar
  for fungi 9–190
Stress
  effect on disease and parasites in toxicity
    tests 8–21
Stressed organisms
  methods 9–41
Strip counting
  phytoplankton 10–24
Strontium
  AA method 3–20
  flame emission photometric method 3–150
  graphical computation of concentration 3–
    151i
  inductively coupled plasma method 3–54
  methods 3–149
  radioactive
    method 7–47
    90
      method 7–47
      relation to yttrium 90 7–51i
      precipitation method 7–48
      regulatory limits 7–47
Strontium carrier
  for strontium (radioactive) 7–49
Strontium solution
  for strontium 3–21, 3–150
Strontium-yttrium carrier
  for radium 228 7–45

Styrene
  purge and trap GC method 6–71
    capillary column 6–93
  purge and trap GC/MS method 6–42
    capillary column 6–50
Sublation
  for surfactants 5–56
  for surfactants
    recovery 5–59t
Sublator
  for surfactants 5–56, 5–57i
Sublethal effects
  corals 8–63
  crustaceans 8–104
Substrates
  for periphyton collection 10–49
Suction samplers
  for macroinvertebrates 10–106, 10–108i
Sulfamic acid-copper sulfate inhibitor solu-
    tion
  for DO 4–158
Sulfamic acid-phosphoric acid solution
  for chlorine 4–52
Sulfamic acid solution
  for sulfite 4–202
Sulfanilamide-naphthylethylenediamine
    dihydrochloride solution
  for nitrogen (nitrate) 4–141
  for nitrogen (nitrite) 4–129
Sulfanilamide solution
  for selenium 3–52
Sulfate
  gravimetric method
    with residue drying 4–206
    with residue ignition 4–204
  ion chromatography method 4–2
  methods 4–204
  methylthymol blue method, automated
    4–208
  turbidimetric method 4–207
Sulfate manifold 4–209i
Sulfate-reducing medium
  for sulfur bacteria 9–122
Sulfate solution
  for ion chromatography 4–4
  for sulfate 4–207, 4–210
Sulfide
  antimony test 4–192
  concentration and interference removal
    methods 4–194
  forms 4–191
  iodometric method 4–197

Sulfide *(cont.)*
    lead acetate paper test 4–192
    methods 4–191
        analytical flow paths 4–193i
    methylene blue method 4–195
    proportion of hydrogen sulfide in 4–199i
    qualitative tests 4–192
    silver foil test 4–192
    silver-silver sulfide electrode test 4–192
    soluble and insoluble separation 4–194
Sulfite
    iodometric method 4–200
    methods 4–199
    phenanthroline method 4–201
Sulfite solution
    for sulfite 4–202
Sulfur bacteria, *see* Bacteria, sulfur
Sulfur dioxide evolution apparatus
    for sulfite 4-201, 4–202i
Sulfur medium (*Thiobacillus thiooxidans*)
    9–123
Sulfuric acid-amine solution
    for sulfide 4–196
Sulfuric acid-mercuric sulfate-potassium
        dichromate solution
    for COD 5–16
Sulfuric acid-nitric acid digestion
    for metals 3–9
    for phosphorus 4–172
Sulfuric acid-nitric acid solution
    for phosphorus 4–172
Sulfuric acid-silver sulfate reagent
    for COD 5–12
Sulfuric acid-sodium dihydrogen phosphate
        solution
    for anionic surfactants 5–61
Sulfuric acid-sodium sulfate solution
    for sludge digester gas 2–89
Sulfuric acid solution
    alternate to hydrochloric 1–44
    for alkalinity 2–37
    for anions 4–3
    for arsenic and selenium 3–46
    for chlorine dioxide 4–79
    for chromium 3–92
    for DO 4–153
    for mercury 3–119
    for nitrogen (ammonia) 4–116, 4–122,
        4–127
    for phenols 6–140
    for phosphorus 4–172, 4–175, 4–177,
        4–179
    for selenium 3–133

Sulfuric acid solution *(cont.)*
    uniform concentrations *inside front cover*
Summed shoot maximum method
    macrophyton productivity 10–83
Sunfish, bluegill
    culturing 8–127
Supersaturation
    dissolved gases
        direct-sensing membrane-diffusion
            method 2–94
        methods 2–94
Suppressor column
    for anions 4–3
Surber sampler
    for macroinvertebrates 10–102, 10–103i
    for macrophyton 1–76
Surface waters
    microbiological samples 9–33
Surfactants
    anionic
        methylene blue method 5–59
    methods 5–55
    methylene blue method 5–59
    nonionic
        method 5–64
    recovery by sublation 5–59t
    separation by sublation 5–56
Surfactant solution
    for nonionic surfactants 5–65
Surrogate standard
    definition 1–5
Suspended matter, *see* Solids, suspended
Swab technique
    for *Salmonella* concentration 9–132
Swimming pools
    microbiological examination 9–46
    samples 9–46
    also see Bathing waters
Syringaldazine indicator
    for chlorine 4–65
Syringaldazine method (FACTS)
    for chlorine 4–64
Syringes
    for arsenic and selenium 3–46
    for dibromoethane and dibromochloropro-
        pane 6–99
    for direct microbial count 9–65
    for organic carbon 5–19, 5–23
    for organic constituents (individual) 6–14,
        6–15
    for trihalomethanes 6–106
    for volatile aromatics 6–69, 6–73
    for volatile halocarbons 6-81, 6-88

Syringes *(cont.)*
   for volatile organics 6–30, 6–44

T

Tagging
   for fish sampling 10–123, 10–124i
      passive integrated transponder (PIT)
      10–125i
Tailing factor calculation
   liquid-liquid extraction GC/MS method
   6–136i
Tangle nets
   for fish sampling 10–118
Tanks
   for toxicity tests, *see* Culture chambers;
      Exposure chambers
Tannin
   method 5–67
Tannin solution
   for tannin 5–68
Tartaric acid solution
   for cadmium 3–83
   for iodine (radioactive) 7–27
Tartrate-carbonate reagent
   for tannin and lignin 5–68
Taste
   algae causing
      color plate following 10–190
   characterization 2–24
   flavor profile analysis 2–29
   flavor rating assessment 2–27
   flavor threshold test 2–24
   methods 2–23
   organic compounds causing
      method 6–11
      also see Earthy-musty-smelling
         compounds
   relationship to flavor 2–23
   safety precautions in tests 2–24
Taste- and odor-causing organics
   closed-loop stripping GC/MS detection
      limits 6–12t
   also see Earthy-musty-smelling
      compounds
Taxonomic references
   for aquatic organisms 10–194
Temperature
   AA method (electrothermal) 3–33
   biostimulation studies 8–47
   effect on electrodes 4–159, 4–160i
   methods 2–80

Temperature *(cont.)*
   odor samples 2–18
   solids determinations 2–72
   toxicity tests 8–23
Temperature correction factors
   specific gravity 2–87t
Temperature-monitoring devices
   for microbiological examination 9–26
Terramycin-malt extract-streptomycin agar
   for fungi 9–190
      modified 9–191
Test tubes
   for direct microbial count 9–65
   for silver 3–143
   also see Glassware
Tetrachloroethane
   purge and trap GC method 6–79, 6–87
      capillary column 6–93
   purge and trap GC/MS method 6–27,
      6–42
      capillary column 6–50
Tetrachloroethene
   purge and trap GC method 6–71, 6–79,
      6–87
      capillary column 6–93
   purge and trap GC/MS method 6–27,
      6–42
      capillary column 6–50
Tetramethylparaphenyleneamine dihydro-
      chloride solution
   for coliform differentiation 9–107
Tetrathionate broth
   for *Salmonella* enrichment 9–133
Thallium
   AA method 3–20
   inductively coupled plasma method 3–54
   methods 3–152
Thallium solution
   for thallium 3–22
Thermometer
   for conductivity 2–59
   for microbiological examinations 9–6
   for microbiological incubator 9–26
   for odor determinations 2–19
   for temperature 2–80
   reversing 2–80
Thioacetamide solution
   for chlorine 4–60
Thiobacillus
   enumeration. enrichment, and isolation
      9–125
Thiobacillus ferrooxidans
   enumeration, enrichment, isolation 9–124

Thiobacillus ferrooxidans (cont.)
iron-oxidizing medium for 9–122
Thiocyanate
method 4–36, 4–42
Thiocyanate solution
for thiocyanate 4–43
Thiodendron mucosum 9–120i
Thiospirillum jenense 9–119i
Thiovolum majus 9–120i
Thomas' formula
for coliform group density 9–79
Thorium
AA method 3–25
methods 3–153
Thorium solution
for thorium 3–26
Threadworms, *see* Nematodes
Threshold limit values
for reagents 1–65t
for solvents 1–64t
Threshold odor number 2–21t
definition 2–21
Throw nets
for fish sampling 10–120
Thymol blue indicator solution
for cadmium 3–84
for organic acids 5–71
Tin
AA method 3–20
micro quantities 3–36
methods 3–153
Tin solution
for tin 3–22
Tissue grinder
for chlorophyll 10–32
Titanium
AA method 3–25
methods 3–154
Titanium solution
for titanium 3–26
Titanous chloride reduction method
for nitrogen (nitrate) 4–140
Titration
potentiometric 1–51
Titration curve
for chloride 4–72i
inflection point 2–30
Titration method
for acidity 2–30
for alkalinity 2–35
Titration vessel
for acidity 2–32

Titrator
for acidity 2–32
for chlorine dioxide 4–81
Titrimetric method
for carbon dioxide 4–19
for COD 5–14
for cyanide 4–30
for nitrogen (ammonia) 4–121
for sulfite 4–200
Tolerance limit, median
definition 8–3
Toluene
for herbicides 6–183
purge and trap GC method 6–65, 6–71
capillary column 6–93
purge and trap GC/MS method 6–27, 6–42
capillary column 6–50
Tongs
for macrophyton sampling 10–75
Total dissolved solids
for checking analyses 1–20, 1–21
Total organic carbon, *see* Organic carbon (total)
Total organic halogen, *see* Organic halogen (dissolved)
Tow nets
for macroinvertebrates 10–105
Tows
for zooplankton 10–10
TOX, *see* Organic halogen (dissolved)
Toxaphene
gas chromatogram 6–125i, 6–176i
liquid-liquid extraction GC method 6–170
liquid-liquid extraction GC/MS method 6–113
Toxic compounds
safety precautions 1–63
Toxic effects
crustaceans 8–104
Toxicant delivery system
for toxicity tests 8–23
Toxicants
analysis in toxicity tests 8–29
Toxicant solutions
for ciliated protozoa 8–55
for toxicity tests 8–26
Toxicity
definitions 8–3
Toxicity curve 8–36i
plotting methods 8–36

Toxicity tests
  annelids 8–65
  basic requirements 8–5
  Chironomus 8–118
  cleaning of equipment 8–21
  copepods 8–85
  corals 8–56
  counting procedures 8–30
  crustaceans 8–81, 8–91
  culture apparatus 8–13
  Daphnia 8–81
  dilution water 8–29
  disease 8–21
  experimental design 8–27
  exposure chambers 8–31
  fish 8–120, 8–133
  flesh-tainting 8–8
  flow-through 8–4
  flow-through systems 8–25i
  food supply in growth-rate determinations
     8–9
  general information 8–1
  growth-rate determinations 8–9
  Hexagenia 8–118
  holding and culturing tanks 8–13, 8–14i
  illumination 8–30
  insects, aquatic 8–113
  intermediate-term 8–7
     for annelids 8–71
  interpretation 8–38
  life-cycle 8–8
     amphipods 8–105
     annelids 8–71
     crabs 8–107
     crayfish 8–106
     fish 8–134
     lobsters 8–108
     Palaemonetes shrimp 8–109
     Penaeus shrimp 8–111
     shrimp 8–110
     also see Toxicity tests, long-term
  loading 8–28
  long-term 8–8
  long-term
     clam growth 8–80
     Daphnia 741
     mussel growth 8–80
     oyster growth 8–79
     scallop growth 8–80
     also see Toxicity tests, life-cycle
  macrocrustaceans 8–91
  methods 8–1

Toxicity tests *(cont.)*
  mollusks 8–73
  mortality in controls 8–36
  organisms
     collecting 8–11
     culturing 8–13
     food 8–17
     handling, holding, and conditioning
        8–12
     preparation 8–10
     selection 8–10
  parasites 8–21
  photoperiods 8–30
  phytoplankton 8–52
  protozoa, ciliated 8–53
  quantal 8–33
  quantitative 8–36
  range-finding 8–7
  recirculation 8–4
  renewal 8–4
  results 8–33
  short-term 8–7
     amphipods 8–105
     annelids 8–71
     clam embryos 8–78
     crabs 8–106
     crayfish 8–106
     fish 8–133
     lobsters 8–108
     mollusks 8–77
     mussel embryos 8–78
     oyster shell deposition 8–78
     Palaemonetes shrimp 8–109
     Penaeus shrimp 8–110
     scallop embryos 8–78
  special-purpose 8–8
  static 8–4
  stress effects 8–21
  temperature control 8–23
  terminology 8–3
  toxicant delivery systems 8–23
  toxicant solutions 8–26, 8–28
  types 8–5
  water supply 8–14
  water supply systems 8–24
Trammel net
  for fish 10–118
Transplantation
  corals 8–64
Trap
  for volatile aromatics 6–66i, 6–67
  for volatile halocarbons 6–87

Trap *(cont.)*
  for volatile organics 6–29, 6–29i, 6–43, 6–43i
Trap materials
  for volatile halocarbons 6–81
  for volatile organics 6–32
Trap nets
  for fish 10–117, 10–118i
Traps
  for fish 10–117
Trawls
  for fish 10–119
  for macroinvertebrates 10–105
Tributyl phosphate
  for strontium (radioactive) 7–49
Trichlorobenzene
  liquid-liquid extraction GC/MS method 6–113
  purge and trap GC method 6–71
    capillary column 6–93
  purge and trap GC/MS method
    capillary column 6–50
Trichloroethane
  purge and trap GC method 6–79, 6–87
    capillary column 6–93
  purge and trap GC/MS method 6–27, 6–42
    capillary column 6–50
Trichloroethene
  purge and trap GC method 6–71, 6–79, 6–87
    capillary column 6–93
  purge and trap GC/MS method 6–27, 6–42
    capillary column 6–50
Trichlorofluoromethane
  purge and trap GC method 6–79, 6–87
    capillary column 6–93
  purge and trap GC/MS method 6–27, 6–42
    capillary column 6–50
Trichlorophenol
  liquid-liquid extraction GC method 6–137
  liquid-liquid extraction GC/MS method 6–113
Trichlorophenol solution
  for organic halogen 5–30
Trichloropropane
  purge and trap GC method 6–87
    capillary column 6–93
  purge and trap GC/MS method 6–42
    capillary column 6–50

Trichlorotrifluoroethane
  for oil and grease 5–43
Trichlorotrifluoroethane-soluble floatable oil and grease
  method 2–69
Trichromatic method
  for chlorophyll *a, b,* and *c* 10–33
Trihalomethane formation
  methods 5–73
Trihalomethane formation potential
  definition 5–73
  method 5–75
  precision and bias 5–78t
Trihalomethanes
  gas chromatogram 6–107i, 6–108i
  liquid-liquid extraction GC method 6–104
    bias and precision 6–112t
  methods 6–104
  retention time 6–108i
Trihalomethane solutions
  for trihalomethanes 6–108, 6–109
Trimethylbenzene
  purge and trap GC method 6–71
    capillary column 6–93
  purge and trap GC/MS method
    capillary column 6–50
Triphenyl tetrazolium chloride
  for coliphage detection 9–40
Tris buffer
  for plankton ATP 10–41
Tristimulus filter method
  for color 2–8
    ADMI 2–9
Tritium
  liquid scintillation spectrometric method 7–54
  methods 7–53
  regulatory limit 7–54
Tritium solution
  for tritium 7–54
Trolling
  for fish sampling 10–116
Trophic levels
  aquatic organisms 10–142
Trout, brook
  culturing 8–125
  food 8–137
  growth chambers 8–136
  photoperiod 8–136
  spawning tanks 8–136
  toxicity tests 8–136

Tryptic(ase) soy agar
  for coliphage detection 9–40
    modified 9–40
Tryptic(ase) soy broth
  for coliphage detection 9–40
Tryptic soy agar
  for coliform differentiation 9–106
Tryptone glucose yeast agar
  for heterotrophic plate count 9–55
Tryptophane broth
  for coliforms 9–104
Tubing
  for mercury apparatus connections 3–29
  for organic constituents (individual) 6–13,
    6–15
Tucker trawl
  for zooplankton 10–13i
Turbidimeter
  calibration 2–14
  for turbidity 2–12, 2–13
  Jackson candle 2–12
Turbidimetric method
  for sulfate 4–207
Turbidity
  Jackson candle method 2–12
  methods 2–11
  nephelometric method 2–13
  removal as interference 1–48
  removal in color determinations 2–2
Turbidity suspension 2–14
  for turbidity 2–14
Two-winged flies 10–172i, 10–173i

U

Ultimate trihalomethane formation potential
  definition 5–74
  method 5–81
Ultrafiltration
  for virus concentration 9–173
Ultraviolet lamps 9–7
Ultraviolet radiation
  safety precautions 1–66
Ultraviolet spectrophotometric screening
    method
  for nitrogen (nitrate) 4–132
Uncertainty, total 1–17
Units of expression 1–27
  metric-English equivalents ibc

Uranium
  fluorometric method 7–59
  methods 7–56
  radiochemical method 7–57
  regulatory limits 7–56
Uranium solutions
  for uranium 7–57, 7–59
Urea solution
  for silver 3–143
Urea substrate
  mTEC 9–51
USEPA
  drinking water microbiological quality
    regulations 9–3

V

Valve
  injector
    for chlorophyll 10–36
  pressure-reducing
    for AA apparatus 3–17
  T-junction
    for AA apparatus 3–25
Vanadate-molybdate reagent
  for phosphorus 4–174
Vanadium
  AA method 3–25
  gallic acid method 3–155
  inductively coupled plasma method 3–54
  interferences 3–155t
  methods 3–154
Vanadium solution
  for vanadium 3–26, 3–155
Vanadomolybdophosphoric acid colorimet-
    ric method
  for phosphorus 4–173
Van Dorn sampler
  for phytoplankton 10–6, 10–7i
Van Veen grab sampler
  for macroinvertebrates 10–100, 10–100i
Vapor pressure
  water, fresh 2–100t
Vegetation, aquatic, see Macrophyton
Vent
  for AA apparatus 3–17
Ventilation
  microbiological laboratories 9–5
Vials
  for base/neutral and acid extractables
    6–119

Vials *(cont.)*
  for dibromoethane and dibromochloropro-
    pane 6–99
  for organic constituents (individual) 6–14
  for selenium 3–133
  for trihalomethane formation potential
    5–75
  for volatile halocarbons 6–81
  for volatile organics 6–4
Vibrio cholerae
  biochemical reactions 9–145
  enrichment procedures 9–145
  identification 9–146
  methods 9–145
  O-1 serogroup 9–145
    biotypes 9–146
  selective growth 9–145
Vinyl chloride
  purge and trap GC method 6–79, 6–87
    capillary column 6–93
  purge and trap GC/MS method 6–27,
    6–42
    capillary column 6–50
Viruses
  assay 9–176
    cell culture procedure 9–179
    host systems 9–177
    in mice 9–180
    laboratory facilities 9–177
  concentrating 9–156
  concentrating
    adsorption and elution method 9–159,
      9–164i
    aluminum hydroxide adsorption-precip-
      itation method 9–170
    apparatus 9–165i
    hydroextraction-dialysis method 9–173
  identification 9–176
  methods 9–155
  quantitation procedures 9–178
  recovery from suspended solids 9–175
  sample concentrate decontamination
    9–177
  sample concentrate storage 9–176
Visual comparison method
  for color 2–2
Voges-Proskauer test
  for coliform differentiation 9–102t
Volatile acids, *see* Acids, volatile
Volatile matter, *see* Solids, volatile
Volatile organic compounds
  sample preservation 11t

Volatile organic *(cont.)*
  also see Organics, volatile
Volumetric glassware 1-42
Volumetric method
  for methane 6–63
  for sludge digester gas 2–88

                          W

Wash solution
  for nitrogen (nitrate) 4–138
  for surfactants 5–61
Washing
  glassware for microbiological tests 9–28
Waste disposal
  in laboratories 1–70
Wastewater
  radioactivity analyses
    quality assurance 7–15
  radioelement phase distribution 7–3t
Water
  ammonia-free
    for nitrogen (ammonia) 4–115
  buffered
    for culture media 9–31
  cadmium-free
    for cadmium 3–83
  carbon-dioxide-free inside front cover
  carbon-dioxide-free
    for acidity 2–32
  carbon-free
    for aquatic humic substances 5–38
    for organic carbon 5–20
  chlorine-demand free
    for trihalomethane formation potential
      5–76
    for chlorine 4–52
  cooling
    for AA method (electrothermal) 3–36
  copper-free redistilled 3–96
  deionized
    for anions 4–3
    preparation 1–56
  deionized distilled ifc
  dilution
    for BOD 5–5
    for culture media 9–31
    for toxicity tests 8–25, 8–61, 8–69
  distilled
    for anions 4–3
    preparation *inside front cover*, 1–56

Water *(cont.)*
  drinking
    coliform group tests 9–67
    USEPA microbiological quality regu-
      lations 9–3
  for copepod culture 8–86
  for coral culture 8–59
  for crabs 8–94
  for culture media 9–30
  for lobsters 8–95
  for mollusks 8–74
  for organics analysis 1–57
  for toxicity tests 8–14
  for volatile organics 6–30
  fresh
    reconstituted 8–16t
    standard 8–15
    vapor pressure 2–100t
  iodine-demand-free
    for iodide 4–107
  mercury-free 3–119
  metal-free 3–20
  nitrite-free
    for nitrogen (nitrite) 4–129
  nonpotable
    coliform group tests 9–68
  odor-free 2–19
  organic-free
    for organics 6–17
    for trihalomethane formation potential
      5–76
  peptone
    for culture media 9–31
  purified
    bacteriological quality test 9–9
  reagent *inside front cover*
    contaminants 1–57
  reagent-grade 1–54
    contaminant removal 1–55t
    preparation 1–56
    purification processes 1–55t
    specifications 1–55t
    types 1–54, 1–57
  redistilled *inside front cover*
  samples
    for fish kill investigations 10–135
  sea
    reconstituted 8–17t
  THM-free
    for trihalomethanes 6–107
  turbidity-free
    for turbidity 2–14

Water bath
  for base/neutral and acid extractables
    6–119
  for BOD 5–4
  for microbiological examination 9–8
  for organic constituents (individual) 6–14
Water distribution systems
  microbiological samples 9–32
Water pollution
  macroinvertebrate response 10–95
Water quality
  for microbiological examinations
    monitoring 9–9, 9–10t
Water quality test
  reagents 9–11t
Water supplies
  raw
    microbiological samples 9–33
Water supply systems
  for toxicity tests 8–24
Waterways Experiment Station sampler
  for macrophyton 10–77
Wavelengths
  for metals by ICP 3–55t
Weirs
  for fish collection 10–117
Wells
  microbiological samples 9–32
Wet combustion procedure
  alpha and beta radioactivity in biological
    samples 7–19
Wet-oxidation method
  for organic carbon 5–24
Whipple grid
  for plankton counting 10–21, 10–21i,
    10–22i
Whirlpools
  microbiological examination 9–49
Wilding sampler
  for macroinvertebrates 10–104, 10–105i
Wisconsin tow-net
  for plankton 10–11i
Work area
  for heterotrophic plate count 9–55
World Health Organization
  drinking water regulations 9–4
Worms 10–162i, 10–163i
  also see Annelids, Nematodes

X

XAD method
  for aquatic humic substances 5–40

XAD resin
for aquatic humic substances 5–40
Xylene
purge and trap GC method 6–71
capillary column 6–93
purge and trap GC/MS method 6–42
capillary column 6–50
Xylose lysine brilliant green agar
for *Salmonella* 9–134
Xylose lysine desoxycholate agar
for *Salmonella* 9–134

Y

Yeast autolysate-casitone-glycerol broth
(CGY)
for iron bacteria 9–121
Yeast extract-malt extract-glucose agar
for fungi 9–188
Yeast nitrogen base-glucose broth
for yeasts 9–192
Yeasts
methods 9–191
Yersinia enterocolitica
concentration technique 9–154
isolation 9–154
methods 9–153
Yttrium carrier
for radium 228 7–45
for strontium (radioactive) 7–49
Yttrium 90 vs. strontium 90 activity 7–51i
Yttrium-strontium carrier
for radium 228 7–45

Z

Zeeman background correction
for metals by AA 3–14
Zinc
AA method 3–20
low concentrations 3–23
dithizone method 3–158, 3–160
for arsenic 3–75, 3–77
inductively coupled plasma method 3–54
methods 3–157
zincon method 3–162
Zinc acetate solution
for cyanides 4–38
for sulfides 4–195
Zinc solution
for zinc 3–22, 3–158, 3–160, 3–163
Zinc sulfate solution
for nitrogen (ammonia) 4–117
Zincon method
for zinc 3–162
Zincon reagent
for zinc 3–163
Zirconyl-acid reagent
for fluoride 4–90
Zone settling rate
test method 2–84
Zooplankton
concentration device 10–29i
concentration techniques 10–16
counting methods 10–28
definition 10–3
mounts 10–19
preservation 10–12
samplers 10–13i
samples 10–8
subsampling device 10–30i
also see Microcrustaceans; Protozoa
Zoosporic fungi, *see* Fungi, zoosporic

ABBREVIATIONS

The following symbols and abbreviations are used throughout this book:

Abbreviation	Referent
AA	atomic absorption
A or amp	ampere(s)
AC	alternating current
ACS	American Chemical Society
amu	atomic mass units
APHA	American Public Health Association
ASTM	American Society for Testing and Materials
AWWA	American Water Works Association
BOD	biochemical oxygen demand
°C	degree(s) Celsius
c	count(s)
Ci	curie(s)
cm, cm^2, cm^3	centimeter(s), square centimeter(s), cubic centimeter(s)
COD	chemical oxygen demand
conc	concentrated
cpm	counts per minute
cps	counts per second
d	day
DC	direct current
diam	diameter
DO	dissolved oxygen
DOX	Dissolved Organic Halogen
dpm	disintegrations per minute
g	gram(s)
g	gravity, unit acceleration of
GC	gas chromatograph
GC/MS	gas chromatograph/mass spectrometer
h	hour
IC	ion chromatograph
ICP	inductively coupled plasma
ID	inside diameter

Abbreviation	Referent
IU	international unit(s)
KeV	kiloelectron volt(s)
kg	kilogram(s)
kPa	kilopascal
L	liter(s)
M	mole or molar
m, m^2, m^3	meter(s), square meter(s), cubic meter(s)
MCL	maximum contaminant level
me	milliequivalent(s)
meV	megaelectron volt(s)
mg	milligram(s)
min	minute(s)
mL	milliliter(s)
mm, mm^2, mm^3	millimeter(s), square millimeter(s), cubic millimeter(s)
mol wt	molecular weight
MPN	most probable number
MS	mass spectrometer
mV	millivolt(s)
μA	microampere(s)
μCi	microcurie(s)
μg	microgram(s)
μL	microliter(s)
μm	micrometer(s)
N	normal
NBS	see NIST
nCi	nanocurie(s)
ng	nanogram(s)
NIST	National Institute of Standards and Technology (formerly National Bureau of Standards)
No.	number
NTU	nephelometric turbidity unit(s)
OD	outside diameter